EIGHTH EDITION

Environmental Science

Earth as a Living Planet

Daniel B. Botkin

Professor Emeritus
Department of Ecology, Evolution, and Marine Biology
University of California, Santa Barbara

President
The Center for the Study of the Environment
Santa Barbara, California

Edward A. Keller

Professor of Environmental Studies and Earth Science
University of California, Santa Barbara

WILEY

JOHN WILEY & SONS, INC.

VICE PRESIDENT AND EXECUTIVE PUBLISHER Kaye Pace
SENIOR ACQUISITIONS EDITOR Rachel Falk
ASSISTANT EDITOR Jenna Paleski
PRODUCTION SERVICES MANAGER Dorothy Sinclair
SENIOR PRODUCTION EDITOR Janet Foxman
MARKETING MANAGER Kristine Ruff
CREATIVE DIRECTOR Harry Nolan
DESIGNER Wendy Lai
SENIOR PHOTO EDITOR Elle Wagner
SENIOR ILLUSTRATION EDITOR Anna Melhorn
SENIOR MEDIA EDITOR Linda Muriello
PRODUCTION SERVICES Furino Production
COVER IMAGE AlaskaStock/Corbis

This book was set in Adobe Garamond by Prepare and printed and bound by Courier/Kendallville. The cover was printed by Courier/Kendallville.

This book is printed on acid-free paper. ∞

Founded in 1807, John Wiley & Sons, Inc. has been a valued source of knowledge and understanding for more than 200 years, helping people around the world meet their needs and fulfill their aspirations. Our company is built on a foundation of principles that include responsibility to the communities we serve and where we live and work. In 2008, we launched a Corporate Citizenship Initiative, a global effort to address the environmental, social, economic, and ethical challenges we face in our business. Among the issues we are addressing are carbon impact, paper specifications and procurement, ethical conduct within our business and among our vendors, and community and charitable support. For more information, please visit our website: *www.wiley.com/go/citizenship*.

Library of Congress Cataloging-in-Publication Data:
Botkin, Daniel B.
Environmental science : earth as a living planet / Daniel B. Botkin, Edward A.
Keller. -- 8th ed.
p. cm.
Includes index.
ISBN 978-0-470-52033-8 (hardback)
1. Environmental sciences. 2. Human ecology. I. Keller, Edward A., 1942- II. Title.
GE105.B68 2011
363.7--dc22

Main-Book ISBN 978-0-470-52033-8
Binder-Ready Version ISBN 978-0-470-91781-7

Printed in the United States of America

10 9 8 7 6 5 4 3 2 1

DEDICATIONS

For my sister, Dorothy B. Rosenthal

who has been a source of inspiration, support, ideas,
and books to read, and is one of my harshest and best critics.

Dan Botkin

and

For Valery Rivera

who contributed so much to this book and
is a fountain of inspiration in our work and lives.

Ed Keller

About the Authors

Photo by Maguire Neblet

Daniel B. Botkin is President of The Center for the Study of Environment, and Professor Emeritus of Ecology, Evolution, and Marine Biology, University of California, Santa Barbara, where he has been on the faculty since 1978, serving as Chairman of the Environmental Studies Program from 1978 to 1985. For more than four decades, Professor Botkin has been active in the application of ecological science to environmental management. He is the winner of the Mitchell International Prize for Sustainable Development and the Fernow Prize for International Forestry, and he has been elected to the California Environmental Hall of Fame.

Trained in physics and biology, Professor Botkin is a leader in the application of advanced technology to the study of the environment. The originator of widely used forest gap-models, he has conducted research on endangered species, characteristics of natural wilderness areas, the biosphere, and global environmental problems including possible ecological effects of global warming. During his career, Professor Botkin has advised the World Bank about tropical forests, biological diversity, and sustainability; the Rockefeller Foundation about global environmental issues; the government of Taiwan about approaches to solving environmental problems; the state of California on the environmental effects of water diversion on Mono Lake. He served as the primary advisor to the National Geographic Society for its centennial edition map on "The Endangered Earth." He directed a study for the states of Oregon and California concerning salmon and their forested habitats.

He has published many articles and books about environmental issues. His latest books are *Beyond the Stoney Mountains: Nature in the American West from Lewis and Clark to Today* (Oxford University Press), *Strange Encounters: Adventures of a Renegade Naturalist* (Penguin/Tarcher), *The Blue Planet* (Wiley), *Our Natural History: The Lessons of Lewis and Clark* (Oxford University Press), *Discordant Harmonies: A New Ecology for the 21st Century* (Oxford University Press), *and Forest Dynamics: An Ecological Model* (Oxford University Press).

Professor Botkin was on the faculty of the Yale School of Forestry and Environmental Studies (1968–1974) and was a member of the staff of the Ecosystems Center at the Marine Biological Laboratory, Woods Hole, MA (1975–1977). He received a B.A. from the University of Rochester, an M.A. from the University of Wisconsin, and a Ph.D. from Rutgers University.

Edward A. Keller was chair of the Environmental Studies and Hydrologic Sciences Programs from 1993 to 1997 and is Professor of Earth Science at the University of California, Santa Barbara, where he teaches earth surface processes, environmental geology, environmental science, river processes, and engineering geology. Prior to joining the faculty at Santa Barbara, he taught geomorphology, environmental studies, and earth science at the University of North Carolina, Charlotte. He was the 1982–1983 Hartley Visiting Professor at the University of Southampton, a Visiting Fellow in 2000 at Emmanuel College of Cambridge University, England, and receipent of the Easterbrook Distinguished Scientist award from the Geological Society of America in 2004.

Professor Keller has focused his research efforts into three areas: studies of Quaternary stratigraphy and tectonics as they relate to earthquakes, active folding, and mountain building processes; hydrologic process and wildfire in the chaparral environment of Southern California; and physical habitat requirements for the endangered Southern California steelhead trout. He is the recipient of various Water Resources Research Center grants to study fluvial processes and U.S. Geological Survey and Southern California Earthquake Center grants to study earthquake hazards.

Professor Keller has published numerous papers and is the author of the textbooks *Environmental Geology, Introduction to Environmental Geology* and (with Nicholas Pinter) *Active Tectonics* (Prentice-Hall). He holds bachelor's degrees in both geology and mathematics from California State University, Fresno; an M.S. in geology from the University of California; and a Ph.D. in geology from Purdue University.

Preface

What Is Environmental Science?

Environmental science is a group of sciences that attempt to explain how life on the Earth is sustained, what leads to environmental problems, and how these problems can be solved.

Why Is This Study Important?

- We depend on our environment. People can live only in an environment with certain kinds of characteristics and within certain ranges of availability of resources. Because modern science and technology give us the power to affect the environment, we have to understand how the environment works, so that we can live within its constraints.
- People have always been fascinated with nature, which is, in its broadest view, our environment. As long as people have written, they have asked three questions about ourselves and nature:

    What is nature like when it is undisturbed by people?

    What are the effects of people on nature?

    What are the effects of nature on people?

Environmental science is our modern way of seeking answers to these questions.

- We enjoy our environment. To keep it enjoyable, we must understand it from a scientific viewpoint.
- Our environment improves the quality of our lives. A healthy environment can help us live longer and more fulfilling lives.
- It's just fascinating.

What Is the "Science" in Environmental Science?

Many sciences are important to environmental science. These include biology (especially ecology, that part of biology that deals with the relationships among living things and their environment), geology, hydrology, climatology, meteorology, oceanography, and soil science.

How Is Environmental Science Different from other Sciences?

- It involves many sciences.
- It includes sciences, but also involves related nonscientific fields that have to do with how we value the environment, from environmental philosophy to environmental economics.
- It deals with many topics that have great emotional effects on people, and therefore are subject to political debate and to strong feelings that often ignore scientific information.

What Is Your Role as a Student and as a Citizen?

Your role is to understand how to think through environmental issues so that you can arrive at your own decisions.

What Are the Professions That Grow Out of Environmental Science?

Many professions have grown out of the modern concern with environment, or have been extended and augmented by modern environmental sciences. These include park, wildlife, and wilderness management; urban planning and design; landscape planning and design; conservation and sustainable use of our natural resources.

Goals of This Book

Environmental Science: Earth as a Living Planet provides an up-to-date introduction to the study of the environment. Information is presented in an interdisciplinary perspective necessary to deal successfully with environmental problems. The goal is to teach you, the student, how to think through environmental issues.

Critical Thinking

We must do more than simply identify and discuss environmental problems and solutions. To be effective, we must know what science is and is not. Then, we need to develop critical thinking skills. Critical thinking is so important that we have made it the focus of its own chapter, Chapter 2. With this in mind, we have also developed *Environmental Science* to present the material in a factual and unbiased format. Our goal is to help you think through the issues, not tell you what to think. To this purpose, at the end of each chapter, we present "Critical Thinking Issues." Critical thinking is further emphasized throughout the text in analytical discussions of topics, evaluation of perspectives, and integration of important themes, which are described in detail later.

Interdisciplinary Approach

The approach of *Environmental Science* is interdisciplinary in nature. Environmental science integrates many disciplines, including the natural sciences, in addition to fields such as anthropology, economics, history, sociology, and philosophy of the environment. Not only do we need the best ideas and information to deal successfully with our environmental problems, but we also must be aware of the cultural and historical contexts in which we make decisions about the environment. Thus, the field of environmental science also integrates the natural sciences with environmental law, environmental impact, and environmental planning.

Themes

Our book is based on the philosophy that six threads of inquiry are of particular importance to environmental science. These key themes, called threads of inquiry, are woven throughout the book.

These six key themes are discussed in more detail in Chapter 1. They are also revisited at the end of each chapter and are emphasized in the Closer Look boxes, each of which is highlighted by an icon suggesting the major underlying theme of the discussion. In many cases, more than one theme is relevant.

Human Population

Underlying nearly all environmental problems is the rapidly increasing human population. Ultimately, we cannot expect to solve environmental problems unless the total number of people on Earth is an amount the environment can sustain. We believe that education is important to solving the population problem. As people become more educated, and as the rate of literacy increases, population growth tends to decrease.

Sustainability

Sustainability is a term that has gained popularity recently. Speaking generally, it means that a resource is used in such a way that it continues to be available. However, the term is used vaguely, and it is something experts are struggling to clarify. Some would define it as ensuring that future generations have equal opportunities to access the resources that our planet offers. Others would argue that sustainability refers to types of developments that are economically viable, do not harm the environment, and are socially just. We all agree that we must learn how to sustain our environmental resources so that they continue to provide benefits for people and other living things on our planet.

A Global Perspective

Until recently it was common to believe that human activity caused only local, or at most regional, environmental change. We now know that human activities can affect the environment globally. An emerging science known as Earth System Science seeks a basic understanding of how our planet's environment works as a global system. This understanding can then be applied to help solve global environmental problems. The emergence of Earth System Science has opened up a new area of inquiry for faculty and students.

The Urban World

An ever-growing number of people are living in urban areas. Unfortunately, our urban centers have long been neglected, and the quality of the urban environment has suffered. It is here that we experience the worst of air pollution, waste disposal problems, and other stresses on the environment. In the past we have centered our studies of the environment more on wilderness than the urban environment. In the future we must place greater focus on towns and cities as livable environments.

People and Nature

People seem to be always interested—amazed, fascinated, pleased, curious—in our environment. Why is it suitable for us? How can we keep it that way? We know that people and our civilizations are having major effects on the environment, from local ones (the street where you live) to the entire planet (we have created a hole in the Earth's ozone layer) which can affect us and many forms of life.

Science and Values

Finding solutions to environmental problems involves more than simply gathering facts and understanding the scientific issues of a particular problem. It also has much to do with our systems of values and issues of social justice. To solve our environmental problems, we must understand what our values are and which potential solutions are socially just. Then we can apply scientific knowledge about specific problems and find acceptable solutions.

Organization

Our text is divided into four parts. Part I **Introductory Chapters** provides a broad overview of the key themes in Environmental Science, introduces the scientific method and the fundamentals of a scientific approach to the environment: Earth as a system; basic biochemical cycles; population dynamics, focusing on the human population; and environmental economics. Part II **Ecology Chapters** explains the scientific basis of ecosystems, biological diversity, ecological restoration and environmental health. Part III **Resource- Management** is about management of our environmental resources: agriculture and environment; forests, parks, wilderness; wildlife and fisheries;as well as chapters on energy: basic principles of energy, fossil fuels and environment, alternative energy, and nuclear energy. Part IV: **Where People Have A Heavy Hand** discusses water pollution; climate change and air pollution; urban environments, and integrated waste management. The section ends with a capstone chapter, integrating and summarizing the main messages of the book.

Special Features

In writing *Environmental Science* we have designed a text that incorporates a number of special features that we believe will help teachers to teach and students to learn. These include the following:

- A **Case Study** introduces each chapter. The purpose is to interest students in the chapter's subject and to raise important questions on the subject matter For example, in Chapter 11, Agriculture, Aquaculture, and Environment, the opening case study tells about a farmer feeding his pigs trail mix, banana chips, yogurt-covered raisins, dried papaya, and cashews, because growing corn for biofuels is raising the costs of animal feed so much.
- **Learning Objectives** are introduced at the beginning of each chapter to help students focus on what is important in the chapter and what they should achieve after reading and studying **the chapter.**
- **A Closer Look** is the name of special learning modules that present more detailed information concerning a particular concept or issue. For example, **A Closer Look 13.2** discusses the reasons for conserving endangered species.
- Many of these special features contain figures and data to enrich the reader's understanding, and relate back to the book themes.
- Near the end of each chapter, a **Critical Thinking Issue** is presented to encourage critical thinking about the environment and to help students understand how the issue may be studied and evaluated. For example Chapter 22 presents a critical thinking issue about **How Can Urban Sprawl Be Controlled?**
- Following the Summary, a special section, **Reexamining Themes and Issues**, reinforces the six major themes of the textbook.
- **Study Questions** for each chapter provide a study aid, emphasizing critical thinking.
- **Further Readings** are provided with each chapter so that students may expand their knowledge by reading additional sources of information (both print and electronic) on the environment.
- **References** cited in the text are provided at the end of the book as notes for each chapter. These are numbered according to their citation in the text. We believe it's important that introductory textbooks carefully cite sources of information used in the writing. These are provided to help students recognize those scholars whose work we depend on, and so that students may draw upon these references as needed for additional reading and research.

Changes in the Eighth Edition

Environmental science is a rapidly developing set of fields. The scientific understanding of environment changes rapidly. Even the kinds of science, and the kinds of connections between science and our ways of life change. Also, the environment itself is changing rapidly: Populations grow; species become threatened or released from near-extinction; our actions change. To remain contemporary, a textbook in environmental science requires frequent updating and with this edition we have examined the entire text and worked to streamline and update every chapter.

Other changes and special features in the eighth edition include:

- A new capstone chapter, Chapter 24, which features a case study on the Gulf oil spill, and revisits the critical themes of the text.
- An updated Chapter on Global Warming, presenting balanced coverage of this important Environmental Science topic.

Combined Chapters

- The former chapters Air Pollution and Indoor Air Pollution have been folded into Chapter 21 to streamline the coverage of Air Pollution and Ozone Depletion.
- Former chapters on Agricultural Production and Environmental Effects of Agriculture have been combined into one.
- Biodiversity and Biogeography have been combined into one chapter.
- Biological Productivity and Energy Flow has been combined with Ecological Restoration.
- Minerals and the Environment and Waste Management have been integrated into one chapter, Materials Management.

New and updated Case Studies, Closer Look Boxes, and Critical Thinking Issues

Updated videos and resources are available to engage students in the key issues and topics of environmental science and provides resources for instructors, including PowerPoints, test bank, prelecture and post-lecture online quizzes, Lecture Launcher PowerPoints with clicker questions, and a variety of news video clips and animations.

Augmentation of Web Site References

Valid information is becoming increasingly available over the Web, and easy access to these data is of great value. Government data that used to take weeks of library search are available almost instantly over the Web. For this reason, we have greatly augmented the number of Web site references and have gathered them all on the book's companion Web site.

Updated Case Studies

Each chapter begins with a case study that helps the student learn about the chapter's topic through a specific example. A major improvement in the eighth edition is the replacement of some older case studies with new ones that discuss current issues and are more closely integrated into the chapter.

Updated Critical Thinking Issues

Each chapter ends with a discussion of an environmental issue, with critical thinking questions for the students. This is one of the ways that the text is designed to help students learn to think for themselves about the analysis of environmental issues. Answers to the end of chapter questions are available for instructor use on the Book Companion Site.

Supplementary Materials

Environmental Science, Eighth Edition, features a full line of teaching and learning resources developed to help professors create a more dynamic and innovative learning environment. For students, we offer tools to build their ability to think clearly and critically. For the convenience of both the professors and students, we provide teaching and learning tools on the Instructor and Student Companion Sites and, through the Wiley Resource Kit.

For Students

Student Web Site (www.wiley.com/college/botkin)

A content-rich Web site has been created to provide enrichment activities and resources for students. These features include review of Learning Objectives, online quizzing, Virtual Field Trips, interactive Environmental Debates, a map of regional case studies, critical thinking readings, glossary and flashcards, Web links to important data and research in the field of environmental studies, and video and animations covering a wide array of selected topics.

Also Available to Package

Environmental Science: Active Learning Laboratories and Applied Problem Sets, 2e by Travis Wagner and Robert Sanford both of University of Southern Maine, is designed to introduce environmental science students to the broad, interdisciplinary field of environmental science by presenting specific labs that use natural and social science concepts to varying degrees and by encouraging a "hands on" approach to understanding the impacts from the environmental/human interface. The laboratory and homework activities are designed to be low-cost and to reflect a sustainability approach in practice and in theory. *Environmental Science: Active Learning Laboratories and Applied Problem Sets, 2e* is available stand-alone or in a package with *Environmental Science, 8e.* Contact your Wiley representative for more information.

Earth Pulse

Utilizing full-color imagery and National Geographic photographs, ***EarthPulse*** takes you on a journey of discovery covering topics such as *The Human Condition, Our Relationship with Nature, and Our Connected World.* Illustrated by specific examples, each section focuses on trends affecting our world today. Included are extensive full-color world and regional maps for reference. *EarthPulse* is available only in a package with *Environmental Science, 8e.* Contact your Wiley representative for more information or visit www.wiley.com/college/earthpulse.

For Instructors

Instructor's Resource Guide

The Instructor's Resource Guide (IRG), prepared by James Yount of Brevard Community College, is available on the Botkin/Keller Web site (www.wiley.com/college/botkin). The IRG provides useful tools to highlight key concepts from each chapter. Each chapter includes the following topics: Lecture Launchers that incorporate technology and opening thought questions; Discussion of Selected Sections from the text, which highlight specific definitions, equations, and examples; and Critical Thinking Activities to encourage class discussion.

Test Bank

The Test Bank, updated and revised by Anthony Gaudin of Ivy Tech Community College, is available on the Botkin/Keller Web site (www.wiley.com/college/botkin). The Test Bank includes approximately 2,000 questions, in multiple-choice, short-answer, and essay formats. The Test Bank is provided in a word.doc format for your convenience to use and edit for your individual needs. For this edition, the author has created many new questions and has labeled the boxed applications according to the six themes and issues set forth in the text. In addition, the author has created questions for the theme boxes and emphasized the themes in many of the questions throughout the test bank.

Respondus Text Bank Network

The Respondus Test Bank is available in the Wiley Resource Kit and on the Wiley Botkin/Keller Web site (www.wiley.com/college/botkin) and provides tests and quizzes for *Environmental Science* Eighth Edition for easy publication into your LMS course, as well as for printed tests. The Respondus Test Bank includes all of the files from the Test Bank, Practice Quizzes, and Pre and Post Lecture Questions in a dynamic computerized format. *For schools without a campus-wide license to Respondus, Wiley will provide one for no additional cost.

Video Lecture Launchers

A rich collection of videos have been selected to accompany key topics in the text. Accompanying each of the videos is contextualized commentary and questions that can further develop student understanding and can be assigned through the Wiley Resource Kit.

PowerPoint™ Presentations

Prepared by Elizabeth Joy Johnson these presentations are tailored to the text's topical coverage and are designed to convey key concepts, illustrated by embedded text art.

Advanced Placement® Guide for Environmental Science

Prepared by Brian Kaestner of Saint Mary's Hall these are available on the Instructor's Resource Web site (www.wiley.com/college/botkin). The Advanced Placement Guide provides a useful tool for high school instructors who are teaching the AP® Environmental Science course. This guide will help teachers to focus on the key concepts of every chapter to prepare students for the Advanced Placement® Exam. Each chapter includes a Chapter Overview that incorporates critical thinking questions, Key Topics important to the exam, and Web links to Laboratories and Activities that reinforce key topics.

Instructor's Web Site

All instructor resources are available on the instructor section of the Wiley Botkin/Keller Web site (www.wiley.com/college/botkin) and within the Wiley Resource Kit.

Completion of this book was only possible due to the cooperation and work of many people. To all those who so freely offered their advice and encouragement in this endeavor, we offer our most sincere appreciation. We are indebted to our colleagues who made contributions.

We greatly appreciate the work of our editor Rachel Falk at John Wiley & Sons, for support, encouragement, assistance, and professional work. We extend thanks to our production editor Janet Foxman, who did a great job and made important contributions in many areas; to Wendy Lai for a beautiful interior design; Ellinor Wagner for photo research; Anna Melhorn for the illustration program. Thanks go out also for editorial assistance from Alissa Etrheim. The extensive media package was enhanced through the efforts of Linda Muriello and Daniela DiMaggio.

Acknowledgments

Reviewers of This Edition

Kevin Baldwin, Monmouth College
James Bartolome, University of California Berkeley
Leonard K. Bernstein, Temple University
Renée E. Bishop, Penn State Worthington Scranton
Edward Chow, University of Michigan- Flint
Katherine Cushing, San Jose State University
Syma Ebbin, Eastern Connecticut State University
Jodee Hunt, Grand Valley State University
John Kraemer, Southeast Missouri State University
John M. Lendvay, University of San Francisco
Bryan Mark, Ohio State University
Mary O'Sullivan, Elgin Community College
Stephen Overmann, Southeast Missouri State University
Gad Perry, Texas Tech University
Randall Repic, University of Michigan Flint
Jennifer Rubin, Rochester Community and Technical College
Ashley Rust, Metropolitan State College
Anthony J. Sadar, Geneva College
Dork Sahagian, Lehigh University
Santhosh Seelan, University of North Dakota
Rich Stevens, Monroe Community College
Kevin Stychar, Texas A & M University-Corpus Christi
Marleen A. Troy, Wilkes University
Timothy Welling, SUNY Dutchess Community College
Don Williams, Park University
Kim Wither, Texas A & M University-Corpus Christi
Caralyn B. Zehnder, Georgia College & State University

Reviewers of Previous Editions

Marc Abrams, Pennsylvania State University
David Aborn, University of Tennessee, Chattanooga
John All, Western Kentucky University
Diana Anderson, Northern Arizona University
Mark Anderson, University of Maine
Robert J. Andres, University of North Dakota
Walter Arenstein, Ohlone College
Daphne Babcock, Collin County Community College
Marvin Baker, University of Oklahoma
Michele Barker-Bridges, Pembroke State University (NC)
James W. Bartolome, University of California, Berkeley
Colleen Baxter, Georgia Military College
Laura Beaton, York College
Susan Beatty, University of Colorado, Boulder
David Beckett, University of Southern Mississippi
Brian Beeder, Morehead State University
Mark Belk, Brigham Young University
Elizabeth Bell, Mission College
Mary Benbow, University of Manitoba
Kristen Bender, California State University, Long Beach
Anthony Benoit, Three Rivers Technical Community College
Leonard K. Bernstein, Temple University
William B.N. Berry, University of California, Berkeley
Joe Beuchel, Triton College
Renée E. Bishop, Penn State Worthington Scranton
Alan Bjorkman, North Park University
Charles Blalack, Kilgore College
Christopher P. Bloch, Texas Tech University
Grady Blount, Texas A&M University, Corpus Christi
Charles Bomar, University of Wisconsin—Stout
Gary Booth, Brigham Young University
Rene Borgella, Ithaca College
John Bounds, Sam Houston State University
Jason E. Box, Ohio State University
Judy Bramble, DePaul University
Scott Brame, Clemson University
Vincent Breslin, SUNY, Stony Brook
Joanne Brock, Kennesaw State University
Robert Brooks, Pennsylvania State University
Bonnie Brown, Virginia Commonwealth University
Robert I. Bruck, North Carolina State University
Grace Brush, Johns Hopkins University
Kelly D. Cain, University of Wisconsin
John Campbell, Northwest Community College (WY)
Rosanna Cappellato, Emory University
Annina Carter, Adirondack Community College
Elaine Carter, Los Angeles City College
Ann Causey, Prescott College (AZ)
Simon Chung, Northeastern Illinois State
W.B. Clapham, Jr., Cleveland State University
Richard Clements, Chattanooga State Technical Community College
Thomas B. Cobb, Bowling Green State University
Jennifer Cole, Northeastern University
Peter Colverson, Mohawk Valley Community College
Terence H. Cooper, University of Minnesota
Jeff Corkill, Eastern Washington University
Harry Corwin, University of Pittsburgh
Kelley Crews, University of Texas
Ellen Crivella, University of Phoenix
Nate Currit, Pennsylvania State University
Rupali Datta, University of Texas at San Antonio
William Davin, Berry College
Craig Davis, Ohio State University
Craig Davis, University of Colorado
Jerry Delsol, Modesto Junior College
Michael L. Denniston, Georgia Perimeter College
David S. Duncan, University of South Florida
Jim Dunn, University of Northern Iowa
Jean Dupon, Menlo College
David J. Eisenhour, Morehead State University
Brian D. Fath, Towson University
Richard S. Feldman, Marist College
Robert Feller, University of South Carolina
James L. Floyd, Community College of Baltimore County
Deborah Freile, Berry College
Andrew Friedland, Dartmouth College
Carey Gazis, Central Washington University
Nancy Goodyear, Bainbridge College
Douglas Green, Arizona State University
Paul Grogger, University of Colorado

James H. Grosklags, Northern Illinois University
Herbert Grossman, Pennsylvania State University
Gian Gupta, University of Maryland
Lonnie Guralnick, Western Oregon University
Raymond Hames, University of Nebraska
John P. Harley, Eastern Kentucky University
Syed E. Hasan, University of Missouri
Bruce Hayden, University of Virginia
David Hilbert, San Diego State University
Joseph Hobbs, University of Missouri
Kelley Hodges, Gulf Coast Community College
Alan Holyoak, Manchester College
Donald Humphreys, Temple University
Walter Illman, The University of Iowa
Dan F. Ippolito, Anderson University
James Jensen, SUNY, Buffalo
David Johnson, Michigan State University
Marie Johnson, United States Military Academy
Gwyneth Jones, Bellevue Community College
S. B. Joshi, York University
Jerry H. Kavouras, Lewis University
Dawn G. Keller, Hawkeye Community College
Deborah Kennard, Mesa State College
Frances Kennedy, State University of West Georgia
Eric Keys, Arizona State University
Jon Kenning, Creighton University
Julie Kilbride, Hudson Valley Community College
Chip Kilduff, Rensselaer Polytechnic Institute
Rita Mary King, The College of New Jersey
John Kinworthy, Concordia University
Thomas Klee, Hillsborough Community College
Sue Kloss, Lake Tahoe Community College
Mark Knauss, Shorter College
Ned Knight, Linfield College
Peter Kolb, University of Idaho
Steven Kolmes, University of Portland
Allen H. Koop, Grand Valley State University
Janet Kotash, Moraine Valley Community College
John Kraemer, Southeast Missouri State University
Matthew Laposata, Kennesaw State University
Kim Largen, George Mason University
Ernesto Lasso de la Vega, International College
Mariana Leckner, American Military University
Henry Levin, Kansas City Community College
Jeanne Linsdell, San Jose State University
Hugo Lociago, University of California, Santa Barbara
John. F. Looney, Jr., University of Massachusetts, Boston
Don Lotter, Imperial Valley College
Tom Lowe, Ball State University
Stephen Luke, Emmanuel College
Tim Lyon, Ball State University
John S. Mackiewicz, University at Albany, State University of New York
T. Anna Magill, John Carroll University
Stephen Malcolm, Western Michigan University
Mel Manalis, University of California, Santa Barbara
Steven Manis, Mississippi Gulf Coast Community College
Heidi Marcum, Baylor University
Bryan Mark, Ohio State University
Susan Masten, Michigan State University
Eric F. Maurer, University of Cincinnati
Timothy McCay, Colgate University
Michael D. McCorcle, Evangel University
Mark A. McGinley, Monroe Community College
Deborah L. McKean, University of Cincinnati
Kendra McSweeney, Ohio State University
James Melville, Mercy College
Chris Migliaccio, Miami-Dade Community College-Wolfson
Earnie Montgomery, Tulsa Junior College, Metro Campus
Michele Morek, Brescia University
James Morris, University of Southern Carolina
Jason Neff, University of Colorado, Boulder
Zia Nisani, Antelope Valley College
Jill Nissen, Montgomery College
Kathleen A. Nolan, St. Francis College
Walter Oechel, San Diego State University
C. W. O'Rear, East Carolina University
Natalie Osterhoudt, Broward Community College
Nancy Ostiguy, Pennsylvania State University
Stephen Overmann, Southeast Missouri State University
Martin Pasqualetti, Arizona State University
William D. Pearson, University of Louisville
Steven L. Peck, Brigham Young University
Clayton Penniman, Central Connecticut State University
Julie Phillips, De Anza College
John Pichtel, Ball State University
David Pimental, Cornell University
Frank X. Phillips, McNeese State University
Thomas E. Pliske, Florida International University
Rosann Poltrone, Arapahoe Community College
John Pratte, Kennesaw State University
Michelle Pulich Stewart, Mesa Community College
Maren L. Reiner, University of Richmond
Randall, Repic, University of Michigan, Flint
Bradley R. Reynolds, University of Tennessee at Chattanooga
Jennifer M. Rhode, Georgia College and State University
Veronica Riha, Madonna University
Melinda S. Ripper, Butler County Community College
Donald C. Rizzo, Marygrove College
Carlton Rockett, Bowling Green State University
Angel Rodriguez, Broward Community College
Thomas K. Rohrer, Carnegie Mellon University
John Rueter, Portland State University
Julie Sanford, Cornerstone University
Robert M. Sanford, University of Southern Maine
Jill Scheiderman, SUNY, Dutchess Community College
Jeffrey Schneider, SUNY, Oswego
Peter Schwartzman, Knox College
Roger Sedjo, Resources for the Future, Washington, D.C.
Christian Shorey, University of Iowa
Joseph Shostell, Pennsylvania State University, Fayette
Joseph Simon, University of South Florida
Daniel Sivek, University of Wisconsin
Patricia Smith, Valencia Community College
James H. Speer, Indiana State University
Lloyd Stark, Pennsylvania State University
Richard T. Stevens, Monroe Community College
Meg Stewart, Vassar College
Iris Stewart-Frey, Santa Clara University
Richard Stringer, Harrisburg Area Community College
Steven Sumithran, Eastern Kentucky University
Janice Swab, Meredith College

Karen Swanson, William Paterson University
Laura Tamber, Nassau Community College (NY)
Todd Tarrant, Michigan State University
Jeffrey Tepper, Valdosta State University
Tracy Thatcher, Cal Poly, San Luis Obispo
Michael Toscano, Delta College
Richard Vance, UCLA
Thomas Vaughn, Middlesex Community College
Charlie Venuto, Brevard Community College
Richard Waldren, University of Nebraska, Lincoln
Sarah Warren, North Carolina State University
William Winner, Oregon State
Wes Wood, Auburn University
Jeffery S. Wooters, Pensacola Junior College
Bruce Wyman, McNeese State University
Carole L. Ziegler, University of San Diego
Ann Zimmerman, University of Toronto
Richard Zingmark, University of South Carolina

Brief Contents

Chapter 1
Key Themes in Environmental Sciences 1

Chapter 2
Science as a Way of Knowing: Critical Thinking about the Environment 22

Chapter 3
The Big Picture: Systems of Change 41

Chapter 4
The Human Population and the Environment 59

Chapter 5
Ecosystems: Concepts and Fundamentals 80

Chapter 6
The Biogeochemical Cycles 104

Chapter 7
Dollars and Environmental Sense: Economics of Environmental Issues 127

Chapter 8
Biological Diversity and Biological Invasions 143

Chapter 9
Ecological Restoration 169

Chapter 10
Environmental Health, Pollution, and Toxicology 185

Chapter 11
Agriculture, Aquaculture, and the Environment 211

Chapter 12
Landscapes: Forests, Parks and Wilderness 235

Chapter 13
Wildlife, Fisheries, and Endangered Species 257

Chapter 14
Energy: Some Basics 286

Chapter 15
Fossil Fuels and the Environment 303

Chapter 16
Alternative Energy and The Environment 326

Chapter 17
Nuclear Energy and the Environment 345

Chapter 18
Water Supply, Use, and Management 368

Chapter 19
Water Pollution and Treatment 398

Chapter 20
The Atmostphere, Climate, and Global Warming 428

Chapter 21
Air Pollution 461

Chapter 22
Urban Environments 497

Chapter 23
Materials Management 519

Chapter 24
Our Environmental Future 551

Appendix A-1

Glossary G-1

Notes N-1

Photo Credits P-1

Index I-1

Contents

Chapter 1 Key Themes in Environmental Sciences 1

CASE STUDY Amboseli National Reserve: A Story of Change 2

1.1 Major Themes of Environmental Science 4

A CLOSER LOOK 1.1 A Little Environmental History 5

1.2 Human Population Growth 6
- Our Rapid Population Growth 6
- Famine and Food Crisis 6

1.3 Sustainability and Carrying Capacity 8
- Sustainability: The Environmental Objective 8
- Moving toward Sustainability: Some Criteria 9
- The Carrying Capacity of the Earth 10

1.4 A Global Perspective 10

1.5 An Urban World 11

1.6 People and Nature 12

1.7 Science and Values 13
- The Precautionary Principle 14
- Placing a Value on the Environment 15

CRITICAL THINKING ISSUE EASTER ISLAND 17

Summary 19

Reexamining Themes and Issues 19

Key Terms 20

Study Questions 20

Further Reading 21

Chapter 2 Science as a Way of Knowing: Critical Thinking about the Environment 22

CASE STUDY Birds at Mono Lake: Applying Science to Solve an Environmental Problem 23

2.1 Understanding What Science Is - and What It Isn't 24
- Science as a Way of Knowing 25
- Disprovability 25

2.2 Observations, Facts, Inferences, and Hypotheses 26
- Controlling Variables 27
- The Nature of Scientific Proof 27
- Theory in Science and Language 29
- Models and Theory 29

A CLOSER LOOK 2.1 The Case of the Mysterious Crop Circles 30
- Some Alternatives to Direct Experimentation 30
- Uncertainty in Science 31
- Leaps of Imagination and Other Nontraditional Aspects of the Scientific Method 31

2.3 Measurements and Uncertainty 32
- A Word about Numbers in Science 32
- Dealing with Uncertainties 32
- Accuracy and Precision 32

A CLOSER LOOK 2.2 Measurement of Carbon Stored in Vegetation 33

2.4 Misunderstandings about Science and Society 34
- Science and Decision Making 34
- Science and Technology 34
- Science and Objectivity 35
- Science, Pseudoscience, and Frontier Science 35

CRITICAL THINKING ISSUE HOW DO WE DECIDE WHAT TO BELIEVE ABOUT ENVIRONMENTAL ISSUES? 36

2.5 Environmental Questions and the Scientific Method 37

Summary 37

Reexamining Themes and Issues 38

Key Terms 38

Study Questions 39

Further Reading 40

Chapter 3 The Big Picture: Systems of Change 41

CASE STUDY Trying to Control Flooding of the Wild Missouri River 42

3.1 Basic Systems Concepts 44
- Static and Dynamic Systems 44
- Open Systems 44

A CLOSER LOOK 3.1 Simple Systems 45
- The Balance of Nature: Is a Steady State Natural? 46
- Residence Time 46

WORKING IT OUT 3.1 AVERAGE RESIDENCE TIME (ART) 47
- Feedback 48

3.2 System Responses: Some Important Kinds of Flows 50
- Linear and Nonlinear Flows 50
- Lag Time 50
- Selected Examples of System Responses 50

A CLOSER LOOK 3.2 Exponential Growth Defined, and Putting Some Numbers on it 52

WORKING IT OUT 3.2 EXPONENTIAL GROWTH 53

3.3 Overshoot and Collapse 53

3.4 Irreversible Consequences 53

3.5 Environmental Unity 54

3.6 Uniformitarianism 54

3.7 Earth as a System 55

3.8 Types of Change 55

CRITICAL THINKING ISSUE IS THE GAIA HYPOTHESIS SCIENCE? 56
Summary 56
Reexamining Themes and Issues 57
Key Terms 57
Study Questions 58
Further Reading 58

Chapter 4 The Human Population and the Environment 59

CASE STUDY Pandemics and World Population Growth 60
4.1 Basic Concepts of Population Dynamics 61
The Human Population as an Exponential Growth Curve 62
A CLOSER LOOK 4.1 A Brief History of Human Population Growth 64
4.2 Projecting Future Population Growth 65
Exponential Growth and Doubling Time 65
Human Population as a Logistic Growth Curve 65
WORKING IT OUT 4.1 FORECASTING POPULATION CHANGE 66
4.3 Age Structure 67
A CLOSER LOOK 4.2 The Prophecy of Malthus 68
4.4 The Demographic Transition 69
Potential Effects of Medical Advances on the Demographic Transition 71
4.5 Longevity and Its Effect on Population Growth 71
Human Death Rates and the Rise of Industrial Societies 72
4.6 The Human Population's Effects on the Earth 73
4.7 The Human Carrying Capacity of Earth 73
4.8 Can We Achieve Zero Population Growth? 74
Age of First Childbearing 74
Birth Control: Biological and Societal 75
National Programs to Reduce Birth Rates 75
CRITICAL THINKING ISSUE WILL THE DEMOGRAPHIC TRANSITION HOLD IN THE UNITED STATES? 76
Summary 76
Reexamining Themes and Issues 77
Key Terms 78
Study Questions 78
Further Reading 78

Chapter 5 Ecosystems: Concepts and Fundamentals 80

CASE STUDY Sea Otters, Sea Urchins, and Kelp: Indirect Effects of Species on One Another 81
5.1 The Ecosystem: Sustaining Life on Earth 83
Basic Characteristics of Ecosystems 83
5.2 Ecological Communities and Food Chains 84
A Simple Ecosystem 84
An Oceanic Food Chain 85
Food Webs Can Be Complex: The Food Web of the Harp Seal 85
A CLOSER LOOK 5.1 Land and Marine Food Webs 86
5.3 Ecosystems as Systems 88
5.4 Biological Production and Ecosystem Energy Flow 89
The Laws of Thermodynamics and the Ultimate Limit on the Abundance of Life 90
WORKING IT OUT 5.1 SOME CHEMISTRY OF ENERGY FLOW 91
5.5 Biological Production and Biomass 93
Measuring Biomass and Production 93
WORKING IT OUT 5.2 GROSS AND NET PRODUCTION 94
5.6 Energy Efficiency and Transfer Efficiency 94
5.7 Ecological Stability and Succession 95
Patterns in Succession 96
5.8 Chemical Cycling and Succession 99
5.9 How Species Change Succession 99
Facilitation 99
Interference 100
Life History Differences 100
CRITICAL THINKING ISSUE SHOULD PEOPLE EAT LOWER ON THE FOOD CHAIN? 101
Summary 102
Reexamining Themes and Issues 102
Key Terms 103
Study Questions 103
Further Reading 103

Chapter 6 The Biogeochemical Cycles 104

CASE STUDY Methane and Oil Seeps: Santa Barbara Channel 105
6.1 Earth Is a Peculiar Planet 106
Space Travelers and Our Solar System 107
The Fitness of the Environment 108
The Rise of Oxygen 108
Life Responds to an Oxygen Environment 109
6.2 Life and Global Chemical Cycles 111
6.3 General Aspects of Biogeochemical Cycles 112
6.4 The Geologic Cycle 113
The Tectonic Cycle 113
The Hydrologic Cycle 115
The Rock Cycle 116
6.5 Some Major Global Biogeochemical Cycles 117
The Carbon Cycle 117
The Carbon-Silicate Cycle 119
The Nitrogen Cycle 120
The Phosphorus Cycle 121
CRITICAL THINKING ISSUE HOW ARE HUMAN ACTIVITIES LINKED TO THE PHOSPHORUS AND NITROGEN CYCLES? 124

Summary 125
Reexamining Themes and Issues 125
Key Terms 126
Study Questions 126
Further Reading 126

Chapter 7 Dollars and Environmental Sense: Economics of Environmental Issues 127

CASE STUDY Cap, Trade, and Carbon Dioxide 128
7.1 Overview of Environmental Economics 129
7.2 Public-Service Functions of Nature 130
7.3 The Environment as a Commons 130
7.4 Low Growth Rate and Therefore Low Profit as a Factor in Exploitation 132
Scarcity Affects Economic Value 133
7.5 Externalities 133
7.6 Valuing the Beauty of Nature 134
7.7 How Is the Future Valued? 135
7.8 Risk-Benefit Analysis 136
CRITICAL THINKING ISSUE **GEORGES BANK: HOW CAN U.S. FISHERIES BE MADE SUSTAINABLE?** 139
Summary 140
Reexamining Themes and Issues 140
Key Terms 141
Study Questions 141
Further Reading 142

Chapter 8 Biological Diversity and Biological Invasions 143

CASE STUDY Citrus Greening 144
8.1 What Is Biological Diversity? 145
Why Do People Value Biodiversity? 145
8.2 Biological Diversity Basics 146
The Number of Species on Earth 147
8.3 Biological Evolution 149
The Four Key Processes of Biological Evolution 149
■ **A CLOSER LOOK 8.1 Natural Selection: Mosquitoes and the Malaria Parasite** 151
Biological Evolution as a Strange Kind of Game 154
8.4 Competition and Ecological Niches 154
The Competitive Exclusion Principle 154
Niches: How Species Coexist 155
Measuring Niches 156
8.5 Symbiosis 157
A Broader View of Symbiosis 158
A Practical Implication 158
8.6 Predation and Parasitism 158
A Practical Implication 158
8.7 How Geography and Geology Affect Biological Diversity 158
Wallace's Realms: Biotic Provinces 159
Biomes 161
Convergent and Divergent Evolution 162
8.8 Invasions, Invasive Species, and Island Biogeography 164
Biogeography and People 165
CRITICAL THINKING ISSUE **POLAR BEARS AND THE REASONS PEOPLE VALUE BIODIVERSITY** 166
Summary 166
Reexamining Themes and Issues 167
Key Terms 168
Study Questions 168
Further Reading 168

Chapter 9 Ecological Restoration 169

CASE STUDY THE FLORIDA EVERGLADES 170
9.1 What Is Ecological Restoration? 171
9.2 Goal of Restoration: What Is "Natural"? 172
9.3 What Is Usually Restored? 173
Rivers, Streams, and Wetlands Restoration: Some Examples 173
Prairie Restoration 176
■ **A CLOSER LOOK 9.1 Island Fox on Santa Cruz Island** 178
9.4 Applying Ecological Knowledge to Restore Heavily Damaged Lands and Ecosystems 180
9.5 Criteria Used to Judge the Success of Restoration 180
CRITICAL THINKING ISSUE **HOW CAN WE EVALUATE CONSTRUCTED ECOSYSTEMS?** 181
Summary 182
Reexamining Themes and Issues 183
Key Terms 183
Study Questions 184
Further Reading 184

Chapter 10 Environmental Health, Pollution, and Toxicology 185

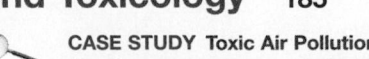

CASE STUDY Toxic Air Pollution and Human Health: Story of a Southeast Houston Neighborhood 186
10.1 Some Basics 187
Terminology 188
Measuring the Amount of Pollution 189
10.2 Categories of Pollutants 189
Infectious Agents 189
■ **A CLOSER LOOK 10.1 Sudbury Smelters: A Point Source** 189

Environmentally Transmitted Infectious Disease 190
Toxic Heavy Metals 191
Toxic Pathways 191
Organic Compounds 193

■ **A CLOSER LOOK 10.2 Mercury and Minamata, Japan** 193

Persistent Organic Pollutants 194

■ **A CLOSER LOOK 10.3 Dioxin: How Dangerous Is It?** 195

Hormonally Active Agents (HAAs) 196

■ **A CLOSER LOOK 10.4 Demasculinization and Feminization of Frogs** 196

Nuclear Radiation 198
Thermal Pollution 198
Particulates 199
Asbestos 199
Electromagnetic Fields 199
Noise Pollution 200
Voluntary Exposure 201

10.3 General Effects of Pollutants 201

Concept of Dose and Response 201
Dose-Response Curve (LD-50, ED-50, and TD-50) 203
Threshold Effects 204
Ecological Gradients 205
Tolerance 205
Acute and Chronic Effects 205

10.4 Risk Assessment 205

CRITICAL THINKING ISSUE **IS LEAD IN THE URBAN ENVIRONMENT CONTRIBUTING TO ANTISOCIAL BEHAVIOR?** 207

Summary 208
Reexamining Themes and Issues 208
Key Terms 209
Study Questions 209
Further Reading 210

Chapter 11 Agriculture, Aquaculture, and the Environment 211

CASE STUDY Biofuels and Banana Chips: Food Crops vs. Fuel Crops 212

11.1 An Ecological Perspective on Agriculture 213

The Plow Puzzle 214

11.2 Can We Feed the World? 214

How We Starve 216

11.3 What We Grow on the Land 218

Crops 218
Livestock: The Agriculture of Animals 218

11.4 Soils 221

Restoring Our Soils 222

■ **A CLOSER LOOK 11.1 The Great American Dust Bowl** 223

11.5 Controlling Pests 224

Pesticides 224

11.6 The Future of Agriculture 227

Increased Production per Acre 227
Increased Farmland Area 227
New Crops and Hybrids 227
Better Irrigation 228
Organic Farming 228
Eating Lower on the Food Chain 228

11.7 Genetically Modified Food: Biotechnology, Farming, and Environment 228

New Hybrids 229
The Terminator Gene 229
Transfer of Genes from One Major Form of Life to Another 229

11.8 Aquaculture 230

Some Negatives 231

CRITICAL THINKING ISSUE **WILL THERE BE ENOUGH WATER TO PRODUCE FOOD FOR A GROWING POPULATION?** 231

Summary 232
Reexamining Themes and Issues 233
Key Terms 234
Study Questions 234
Further Reading 234

Chapter 12 Landscapes: Forests, Parks and Wilderness 235

CASE STUDY Jamaica Bay National Wildlife Refuge: Nature and the Big City 236

12.1 Forests and Forestry 237

How People Have Viewed Forests 237
Forestry 238
Modern Conflicts over Forestland and Forest Resources 238
World Forest Area and Global Production and Consumption of Forest Resources 239
How Forests Affect the Whole Earth 241
The Ecology of Forests 243

■ **A CLOSER LOOK 12.1 The Life of a Tree** 243

Forest Management 244
Can We Achieve Sustainable Forestry? 246
Deforestation 247

12.2 Parks, Nature Preserves, and Wilderness 248

What's the Difference between a Park and a Nature Preserve? 249

■ **A CLOSER LOOK 12.2 A Brief History of Parks Explains Why Parks Have Been Established** 250

Conflicts Relating to Parks 251

12.3 Conserving Wilderness 252

What It Is, and Why It Is of Growing Importance 252
Conflicts in Managing Wilderness 253

CRITICAL THINKING ISSUE **CAN TROPICAL FORESTS SURVIVE IN BITS AND PIECES?** 254

Summary 254
Reexamining Themes and Issues 255
Key Terms 256
Study Questions 256
Further Reading 256

Chapter 13 Wildlife, Fisheries, and Endangered Species 257

CASE STUDY Stories Told by the Grizzly Bear and the Bison 258

13.1 Traditional Single-Species Wildlife Management 260
- Carrying Capacity and Sustainable Yields 260
- An Example of Problems with the Logistic Curve 262

13.2 Improved Approaches to Wildlife Management 263
- Time Series and Historical Range of Variation 263
- Age Structure as Useful Information 264
- Harvests as an Estimate of Numbers 264

13.3 Fisheries 265
- The Decline of Fish Populations 267
- Can Fishing Ever Be Sustainable? 269

■ **A CLOSER LOOK 13.1 King Salmon Fishing Season Canceled: Can We Save Them from Extinction?** 269

13.4 Endangered Species: Current Status 271

■ **A CLOSER LOOK 13.2 Reasons for Conserving Endangered Species—and All Life on Earth** 272

13.5 How a Species Becomes Endangered and Extinct 274
- Causes of Extinction 275

13.6 The Good News: We Have Improved the Status of Some Species 276

■ **A CLOSER LOOK 13.3 Conservation of Whales and Other Marine Mammals** 277

13.7 Can a Species Be Too Abundant? If So, What to Do? 279

13.8 How People Cause Extinctions and Affect Biological Diversity 279

13.9 Ecological Islands and Endangered Species 280

13.10 Using Spatial Relationships to Conserve Endangered Species 281

CRITICAL THINKING ISSUE SHOULD WOLVES BE REESTABLISHED IN THE ADIRONDACK PARK? 281

Summary 282

Reexamining Themes and Issues 283

Key Terms 284

Study Questions 284

Further Reading 285

Chapter 14 Energy: Some Basics 286

CASE STUDY National Energy Policy: From Coast-to-Coast Energy Crisis to Promoting Energy Independence 287

14.1 Outlook for Energy 288
- Energy Crises in Ancient Greece and Rome 288
- Energy Today and Tomorrow 288

14.2 Energy Basics 289

14.3 Energy Efficiency 290

■ **A CLOSER LOOK 14.1 Energy Units** 291

14.4 Energy Sources and Consumption 293
- Fossil Fuels and Alternative Energy Sources 293

14.5 Energy Conservation, Increased Efficiency, and Cogeneration 294
- Building Design 295
- Industrial Energy 296
- Values, Choices, and Energy Conservation 296

14.6 Sustainable-Energy Policy 297
- Energy for Tomorrow 297

■ **A CLOSER LOOK 14.2 Micropower** 299

CRITICAL THINKING ISSUE USE OF ENERGY TODAY AND IN 2030 300

Summary 301

Reexamining Themes and Issues 301

Key Terms 302

Study Questions 302

Further Reading 302

Chapter 15 Fossil Fuels and the Environment 303

CASE STUDY Peak Oil: Are We Ready for It? 304

15.1 Fossil Fuels 305

15.2 Crude Oil and Natural Gas 306
- Petroleum Production 306
- Oil in the 21st Century 308
- Natural Gas 309
- Coal-Bed Methane 309
- Methane Hydrates 310
- The Environmental Effects of Oil and Natural Gas 311

■ **A CLOSER LOOK 15.1 The Arctic National Wildlife Refuge: To Drill or Not to Drill** 312

15.3 Coal 314
- Coal Mining and the Environment 316
- Mountaintop Removal 317

■ **A CLOSER LOOK 15.2 The Trapper Mine** 318
- Underground Mining 319
- Transporting Coal 319
- The Future of Coal 320

15.4 Oil Shale and Tar Sands 321
- Oil Shale 321
- Tar Sands 322

CRITICAL THINKING ISSUE WHAT WILL BE THE CONSEQUENCES OF PEAK OIL? 323

Summary 324

Reexamining Themes and Issues 324

Key Terms 325

Study Questions 325

Further Reading 325

Chapter 16 Alternative Energy and the Environment 326

CASE STUDY Using Wind Power in New Ways for an Old Application 327

16.1 Introduction To Alternative Energy Sources 327

16.2 Solar Energy 328
- Passive Solar Energy 329
- Active Solar Energy 329
- Solar Thermal Generators 331
- Solar Energy and the Environment 332

16.3 Converting Electricity From Renewable Energy Into A Fuel For Vehicles 333

A CLOSER LOOK 16.1 Fuel Cells—An Attractive Alternative 333

16.4 Water Power 334
- Small-Scale Systems 334
- Water Power and the Environment 335

16.5 Ocean Energy 335

16.6 Wind Power 336
- Basics of Wind Power 336
- Wind Power and the Environment 337
- The Future of Wind Power 338

16.7 Biofuels 338
- Biofuels and Human History 338
- Biofuels and the Environment 338

16.8 Geothermal Energy 339
- Geothermal Systems 340
- Geothermal Energy and the Environment 341
- The Future of Geothermal Energy 341

CRITICAL THINKING ISSUE SHOULD WIND TURBINES BE INSTALLED IN NANTUCKET SOUND? 341

Summary 342

Reexamining Themes and Issues 343

Key Terms 344

Study Questions 344

Further Reading 344

Chapter 17 Nuclear Energy and the Environment 345

CASE STUDY Indian Point: Should a Nuclear Power Installation Operate Near One of America's Major Cities? 346

17.1 Current Role of Nuclear Power Plants in World Energy Production 346

17.2 What Is Nuclear Energy? 348
- Conventional Nuclear Reactors 348

A CLOSER LOOK 17.1 Radioactive Decay 350

17.3 Nuclear Energy and the Environment 353
- Problems with the Nuclear Fuel Cycle 353

17.4 Nuclear Radiation in the Environment, and Its Effects on Human Health 354
- Ecosystem Effects of Radioisotopes 354

A CLOSER LOOK 17.2 Radiation Units and Doses 356
- Radiation Doses and Health 358

17.5 Nuclear Power Plant Accidents 358
- Three Mile Island 358
- Chernobyl 359

17.6 Radioactive-Waste Management 360
- Low-Level Radioactive Waste 360
- Transuranic Waste 361
- High-Level Radioactive Waste 361
- What Should the United States Do with Its Nuclear Wastes? 362

17.7 The Future of Nuclear Energy 363
- Possible New Kinds of Nuclear Power Plants 364

CRITICAL THINKING ISSUE SHOULD THE UNITED STATES INCREASE OR DECREASE THE NUMBER OF NUCLEAR POWER PLANTS? 364

Summary 365

Reexamining Themes and Issues 366

Key Terms 367

Study Questions 367

Further Reading 367

Chapter 18 Water Supply, Use, and Management 368

CASE STUDY Palm Beach County, Florida: Water Use, Conservation, and Reuse 369

18.1 Water 370
- A Brief Global Perspective 370
- Groundwater and Streams 372
- Interactions between Surface Water and Groundwater 372

18.2 Water Supply: A U.S. Example 373
- Precipitation and Runoff Patterns 375
- Droughts 375
- Groundwater Use and Problems 375
- Desalination as a Water Source 376

18.3 Water Use 376
- Transport of Water 378
- Some Trends in Water Use 379

18.4 Water Conservation 380
- Agricultural Use 380
- Public Supply and Domestic Use 383
- Industry and Manufacturing Use 384

18.5 Sustainability and Water Management 384
- Sustainable Water Use 384
- Groundwater Sustainability 384
- Water Management 384
- A Master Plan for Water Management 384
- Water Management and the Environment 385
- Virtual Water 386
- Water Footprint 387

18.6 Wetlands 387
- Natural Service Functions of Wetlands 388
- Restoration of Wetlands 389

18.7 Dams and the Environment 390
- Removal of Dams 392

18.8 Global Water Shortage Linked to Food Supply 393

CRITICAL THINKING ISSUE **WHAT IS YOUR WATER FOOTPRINT?** 394

Summary 395

Reexamining Themes and Issues 396

Key Terms 397

Study Questions 397

Further Reading 397

Chapter 19 Water Pollution and Treatment 398

CASE STUDY America's "First River": A Success Story 399

19.1 Water Pollution 400

A CLOSER LOOK 19.1 What Is the Value of Clean Water to New York City? 402

19.2 Biochemical Oxygen Demand (BOD) 403

19.3 Waterborne Disease 404
- Fecal Coliform Bacteria 404

19.4 Nutrients 405
- Eurtrophication 405

A CLOSER LOOK 19.2 Cultural Eutrophication in the Gulf of Mexico 407

19.5 Oil 408

19.6 Sediment 408

19.7 Acid Mine Drainage 409

19.8 Surface-Water Pollution 410
- Reducing Surface-Water Pollution 410

19.9 Groundwater Pollution 412
- Principles of Groundwater Pollution: An Example 412
- Long Island, New York 413

A CLOSER LOOK 19.3 Water for Domestic Use: How Safe Is It? 414

19.10 Wastewater Treatment 414
- Septic-Tank Disposal Systems 415
- Wastewater Treatment Plants 415
- Primary Treatment 416
- Secondary Treatment 416
- Advanced Wastewater Treatment 417
- Chlorine Treatment 417

A CLOSER LOOK 19.4 Boston Harbor: Cleaning Up a National Treasure 417

19.11 Land Application of Wastewater 418
- Wastewater and Wetlands 418
- Louisiana Coastal Wetlands 419
- Phoenix, Arizona: Constructed Wetlands 419

19.12 Water Reuse 420

19.13 Conditions of Stream Ecosystems in the United States 421

19.14 Water Pollution and Environmental Law 421

CRITICAL THINKING ISSUE **IS WATER POLLUTION FROM PIG FARMS UNAVOIDABLE?** 423

Summary 424

Reexamining Themes and Issues 425

Key Terms 426

Study Questions 426

Further Reading 427

Chapter 20 The Atmostphere, Climate, and Global Warming 428

CASE STUDY What Does History Tell Us about Global Warming's Potential Consequences for People? 429

20.1 Fundamental Global Warming Questions 430

20.2 Weather and Climate 431
- The Climate Is Always Changing at a Variety of Time Scales 432

20.3 The Origin of the Global Warming Issue 433

20.4 The Atmosphere 434
- Structure of the Atmosphere 434
- Atmospheric Processes: Temperature, Pressure, and Global Zones of High and Low Pressure 435
- Energy and the Atmosphere: What Makes the Earth Warm 436

20.5 How We Study Climate 438
- The Instrumental Record 438
- The Historical Record 438
- The Paleo-Proxy Record 438
- Proxy Climate Methods 438

20.6 The Greenhouse Effect 441
- How the Greenhouse Effect Works 441

20.7 The Major Greenhouse Gases 443
- Carbon Dioxide 443
- Methane 444
- Chlorofluorocarbons 444
- Nitrous Oxide 444

20.8 Climate Change and Feedback Loops 444
- Possible Negative Feedback Loops for Climate Change 444
- Possible Positive Feedback Loops for Climate Change 445

20.9 Causes of Climate Change 445
- Milankovitch Cycles 445
- Solar Cycles 446
- Atmospheric Transparency Affects Climate and Weather 446
- The Surface of Earth and Albedo (reflectivity) Affects Climate and Weather 447
- Roughness of the Earth's Surface Affects the Atmosphere 447
- The Chemistry of Life Affects the Atmosphere 447
- Climate Forcing 447

20.10 The Oceans and Climate Change 448
- El Niño and Climate 449

20.11 Forecasting Climate Change 450
- Past Observations and Laboratory Research 450
- Computer Simulations 450

20.12 Potential Rates of Global Climate Change 451

20.13 Potential Environmental, Ecological, and Human Effects of Global Warming 451
- Changes in River Flow 451
- Rise in Sea Level 452
- Glaciers and Sea Ice 452

A CLOSER LOOK 20.1 Some Animals and Plants in Great Britain Are Adjusting to Global Warming 454
- Changes in Biological Diversity 455

Agricultural Productivity 455
Human Health Effects 456

20.14 Adjusting to Potential Global Warming 456
International Agreements to Mitigate Global Warming 456

CRITICAL THINKING ISSUE **WHAT IS VALID SCIENCE IN THE GLOBAL WARMING DEBATE?** 457

Summary 458
Reexamining Themes and Issues 459
Key Terms 459
Study Questions 460
Further Reading 460

Chapter 21 Air Pollution 461

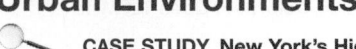

CASE STUDY **Sustainable Skylines: Dallas and Kansas City** 462

21.1 Air Pollution in the Lower Atmosphere 462
A Brief Overview 462
Stationary and Mobile Sources of Air Pollution 463
General Effects of Air Pollution 463
The Major Air Pollutants 464
Criteria Pollutants 465

■ A CLOSER LOOK 21.1 **Acid Rain** 469
Air Toxics 471
Variability of Air Pollution 472
Urban Air Pollution: Chemical and Atmospheric Processes 473
Future Trends for Urban Air Pollution 476
Developing Countries 476

21.2 Controlling Common Pollutants of the Lower Atmosphere 477
Particulates 477
Automobiles 477
Sulfur Dioxide 477
Air Pollution Legislation and Standards 478
The Cost of Controlling Outdoor Air Pollution 479

21.3 High-Altitude (Stratospheric) Ozone Depletion 481
Ultraviolet Radiation and Ozone 481
Measuring Stratospheric Ozone 482
Ozone Depletion and CFCs 483
Simplified Stratospheric Chlorine Chemistry 483
The Antarctic Ozone Hole 484
Polar Stratospheric Clouds 484
Environmental Effects of Ozone Depletion 484
The Future of Ozone Depletion 486

21.4 Indoor Air Pollution 486
Sources of Indoor Air Pollution 487
Pathways, Processes, and Driving Forces 489
Symptoms of Indoor Air Pollution 491

21.5 Controlling Indoor Air Pollution 491
Making Homes and Other Buildings Radon Resistant 493
Design Buildings to Minimize Indoor Air Pollution 493

CRITICAL THINKING ISSUE **SHOULD CARBON DIOXIDE BE REGULATED ALONG WITH OTHER MAJOR AIR POLLUTANTS?** 493

Summary 494
Reexamining Themes and Issues 494
Key Terms 495
Study Questions 496
Further Reading 496

Chapter 22 Urban Environments 497

CASE STUDY **New York's High Line Park in the Sky** 498

22.1 City Life 499
22.2 The City as a System 499
22.3 The Location of Cities: Site and Situation 500
The Importance of Site and Situation 500

■ A CLOSER LOOK 22.1 **Should We Try to Restore New Orleans?** 503
Site Modification 504

22.4 An Environmental History of Cities 505
The Rise of Towns 505
The Urban Center 505
The Industrial Metropolis 505
The Center of Civilization 505

22.5 City Planning and the Environment 506
City Planning for Defense and Beauty 506
The City Park 506

■ A CLOSER LOOK 22.2 **A Brief History of City Planning** 508

22.6 The City as an Environment 508
The Energy Budget of a City 508
The Urban Atmosphere and Climate 509
Solar Energy in Cities 509
Water in the Urban Environment 509
Soils in the City 510
Pollution in the City 510

22.7 Bringing Nature to the City 511
Cities and Their Rivers 511

■ A CLOSER LOOK 22.3 **Design with Nature** 511
Vegetation in Cities 512
Urban "Wilds": The City as Habitat for Wildlife and Endangered Species 513
Animal Pests 514

CRITICAL THINKING ISSUE **HOW CAN URBAN SPRAWL BE CONTROLLED?** 516

Summary 516
Reexamining Themes and Issues 517
Key Terms 518
Study Questions 518
Further Reading 518

CHAPTER 23 Materials Management 519

CASE STUDY **Treasures of the Cell Phone** 520

23.1 The Importance of Resources to Society 521
23.2 Materials Management: What It Is 522
23.3 Mineral Resources 523
How Mineral Deposits Are Formed 523
23.4 Figuring Out How Much Is Left 524
Mineral Resources and Reserves 524

Availability and Use of Our Mineral Resources 524
U.S. Supply of Mineral Resources 525

23.5 Impacts of Mineral Development 526
Environmental Impacts 526
Social Impacts 527
Minimizing the Environmental Impact of Mineral Development 527

■ A CLOSER LOOK 23.1 Golden, Colorado: Open-Pit Mine Becomes a Golf Course 529

23.6 Materials Management and Our Waste 529
History of Waste Disposal 529

23.7 Integrated Waste Management 50
Reduce, Reuse, Recycle 530
Recycling of Human Waste 531

23.8 Municipal Solid-Waste Management 532
Composition of Solid Waste 532
Onsite Disposal 532
Composting 532
Incineration 533
Open Dumps (Poorly Controlled Landills) 533
Sanitary Landfills 533
Reducing the Waste that Ends Up in a Landfill 536

23.9 Hazardous Waste 537

■ A CLOSER LOOK 23.2 "e-waste": A Growing Environmental Problem 538

23.10 Hazardous-Waste Legislation 539
Resource Conservation and Recovery Act 539
Comprehensive Environmental Response, Compensation, and Liability Act 540

23.11 Hazardous-Waste Management: Land Disposal 540

23.12 Alternatives to Land Disposal of Hazardous Waste 542
Source Reduction 542
Recycling and Resource Recovery 542
Treatment 542
Incineration 542

23.13 Ocean Dumping 543

■ A CLOSER LOOK 23.3 Plastics in the Ocean 544

23.14 Pollution Prevention 545

23.15 Sustainable Resource Management 546

CRITICAL THINKING ISSUE **CAN WE MAKE RECYCLING A MORE FINANCIALLY VIABLE INDUSTRY?** 546

Summary 547

Reexamining Themes and Issues 548

Key Terms 549

Study Questions 549

Further Reading 550

Chapter 24 Our Environmental Future 551

CASE STUDY The Oil Spill in the Gulf of Mexico in 2010 552

24.1 Imagine an Ecotopia 555

24.2 The Process of Planning a Future 556

24.3 Environment and Law: A Horse, a Gun, and a Plan 557
The Three Stages in the History of U.S. Environmental Law 557

24.4 Planning to Provide Environmental Goods and Services 558

24.5 Planning for Recreation on Public Lands 559
Who Stands for Nature? Skiing at Mineral King 560
How Big Should Wildlands Be? Planning a Nation's Landscapes 561

24.6 How You Can Be an Actor in the Environmental Law Processes 562
Citizen Actions 562
Mediation 562

24.7 International Environmental Law and Diplomacy 563

24.8 Global Security and Environment 563

24.9 Challenges to Students of the Environment 564

CRITICAL THINKING ISSUE **IS IT POSSIBLE TO DERIVE SOME QUANTITATIVE STATEMENTS ABOUT THRESHOLDS BEYOND WHICH UNACCEPTABLE ENVIRONMENTAL CHANGE WILL OCCUR?** 565

Summary 566

Study Questions 567

Appendix A-1

Glossary G-1

Notes N-1

Photo Credits P-1

Index I-1

CHAPTER 1

Key Themes in Environmental Sciences

Lions are a tourist attraction at Amboseli National Reserve in southern Kenya, and are a valuable resource. Massi people are beginning to help protect them, rather than hunt or poison them as they have traditionally done.

LEARNING OBJECTIVES

Certain themes are basic to environmental science. After reading this chapter, you should understand . . .

- That people and nature are intimately connected;
- Why rapid human population growth is the fundamental environmental issue;
- What sustainability is, and why we must learn to sustain our environmental resources;
- How human beings affect the environment of the entire planet;
- Why urban environments need attention;
- Why solutions to environmental problems involve making value judgments, based on scientific knowledge;
- What the precautionary principle is and why it is important.

CASE STUDY

Amboseli National Reserve: A Story of Change

Amboseli National Reserve in southern Kenya is home to the Maasai people, who are nomadic some of the time and raise cattle. The reserve is also a major tourist destination, where people from around the world can experience Africa and wild animals, such as lions and elephants. Today, environmental change and the future of tourism are being threatened in the area. We will consider long-term change and the more recent management of lions that may result in their local extinction.

Environmental change is often caused by a complex web of interactions among living things and between living things and their environment. In seeking to determine what caused a particular change, the most obvious answer may not be the right answer. Amboseli National Reserve is a case in point. In the short span of a few decades, this reserve, located at the foot of Mount Kilimanjaro (Figure 1.1), underwent a significant environmental change.

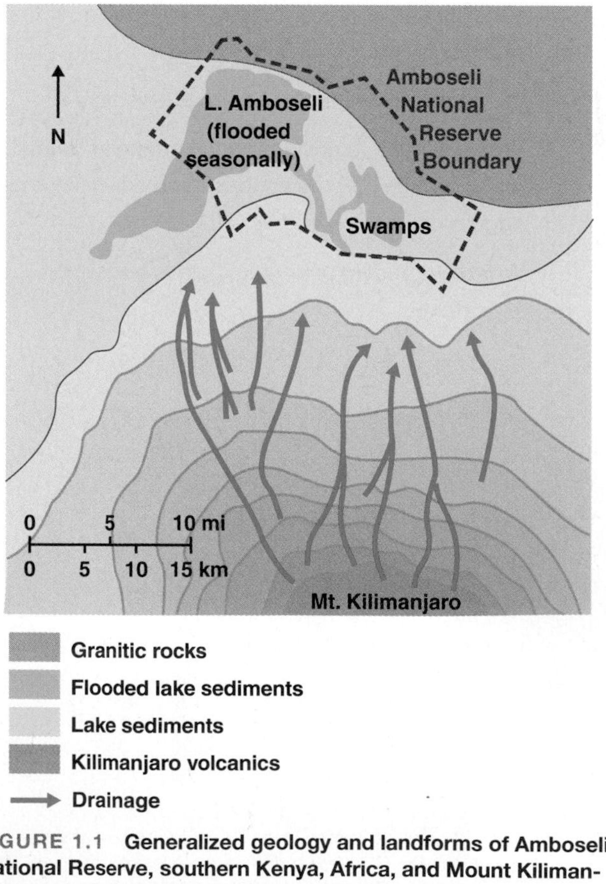

FIGURE 1.1 Generalized geology and landforms of Amboseli National Reserve, southern Kenya, Africa, and Mount Kilimanjaro. (*Source:* T. Dunn and L.B. Leopold, *Water in Environmental Planning* [San Francisco: Freeman, 1978].)

An understanding of physical, biological, and human-use factors—and how these factors are linked—is needed to explain what happened.

Before the mid-1950s, fever-tree woodlands—mostly acacia trees and associated grasses and shrubs—dominated the land and provided habitat for mammals that lived in these open woodlands, such as kudu, baboons, vervet monkeys, leopards, and impalas. Then, beginning in the 1950s and accelerating in the 1960s, these woodlands disappeared and were replaced by short grass and brush, which provided habitat for typical plains animals, such as zebras and wildebeest. Since the mid-1970s, Amboseli has remained a grassland with scattered brush and few trees.

Loss of the woodland habitat was initially blamed on overgrazing of cattle by the Maasai people (Figure 1.2) and damage to the trees from elephants (Figure 1.3). Environmental scientists eventually rejected these hypotheses as the main causes of the environmental change. Their careful work showed that changes in rainfall and soils were the primary culprits, rather than people or elephants.[1, 2] How did they arrive at this explanation?

During recent decades, the mean daily temperature rose dramatically, and annual rainfall increased but continued to vary from year to year by a factor of four, though with no regular pattern.[1, 2] Increased rainfall is generally associated with an increased abundance of trees, unlike what happened at Amboseli.

Why did scientists reject the overgrazing and elephant-damage hypothesis as the sole explanation for changes in Amboseli? Investigators were surprised to note that most dead trees were in an area that had been free of cattle since 1961, which was before the major decline in the woodland environment. Furthermore, some of the woodlands that suffered the least decline had the highest density of people and cattle. These observations suggested that overgrazing by cattle was not responsible for loss of the trees.

Elephant damage was thought to be a major factor because elephants had stripped bark from more than 83% of the trees in some areas and had pushed over some younger, smaller trees. However, researchers concluded that elephants played only a secondary role in changing the habitat. As the density of fever trees and other woodland plants decreased, the incidence of damage caused by elephants increased. In other words, elephant damage interacted with some other, primary factor in changing the habitat.[1]

FIGURE 1.2 **Maasai people grazing cattle in Amboseli National Reserve, Kenya.** Grazing was prematurely blamed for loss of fever-tree woodlands.

Figure 1.1 shows the boundary of the reserve and the major geologic units. The park is centered on an ancient lakebed, remnants of which include the seasonally flooded Lake Amboseli and some swampland. Mount Kilimanjaro is a well-known volcano, composed of alternating layers of volcanic rock and ash deposits. Rainfall that reaches the slopes of Mount Kilimanjaro infiltrates the volcanic material (becomes groundwater) and moves slowly down the slopes to saturate the ancient lakebed, eventually emerging at springs in the swampy, seasonally flooded land. The groundwater becomes saline (salty) as it percolates through the lakebed, since the salt stored in the lakebed sediments dissolves easily when the sediments are wet.

Because a lot of land has been transformed to agricultural uses, the slopes of Mount Kilimanjaro above Amboseli have less forest cover than they did 25 years ago. The loss of trees exposed dark soils that absorb solar energy, and this could cause local warming and drier conditions. In addition, there had been a significant decrease in snow and ice cover on the high slopes and summit of the mountain. Snow and ice reflect sunlight. As snow and ice decrease and dark rock is exposed, more solar energy is absorbed at the surface, warming it. Therefore, decreased snow and ice might cause some local warming.[3]

Research on rainfall, groundwater history, and soils suggested that the area is very sensitive to changing amounts of rainfall. During dry periods, the salty groundwater sinks lower into the earth, and the soil near the surface has a relatively low salt content. The fever trees grow well in the nonsalty soil. During wet periods, the groundwater rises closer to the surface, bringing with it salt, which invades the root zones of trees and kills them. The groundwater level rose as much as 3.5 m (11.4 ft) in response to unusually wet years in the 1960s. Analysis of the soils confirmed that the tree stands that suffered the most damage were those growing in highly saline soils. As the trees died, they were replaced by salt-tolerant grasses and low brush.[1, 2]

Evaluation of the historical record—using information from Maasai herders recorded by early European explorers—and of fluctuating lake levels in other East African lakes suggested that before 1890 there had been another period of above-normal rainfall and loss of woodland environment. Thus, the scientists concluded that cycles of greater and lesser rainfall change hydrology and soil conditions, which in turn change the plant and animal life of the area.[1] Cycles of wet and dry periods can be expected to continue, and associated with these will be changes in the soils, distribution of plants, and abundance and types of animals present.[1]

Management by the Maasai is proving difficult. Tourists want to see wild lions, but the lions sometimes kill and eat Maasai cattle, so the Maasai are killing the

FIGURE 1.3 **Elephant feeding on a yellow-bark acacia tree.** Elephant damage to trees is considered a factor in loss of woodland habitat in Amboseli National Reserve. However, elephants probably play a relatively minor role compared with oscillations in climate and groundwater conditions.

FIGURE 1.4 **Dead lions poisoned by a cheap agriculture pesticide.**

lions. Spearing, a Maasai passage to manhood, remains the dominant way to do it: In recent years, of 20 lions killed, 17 were speared and 3 were poisoned (Figure 1.4).[4] The poison also kills other animals that scavenge cattle, such as hyenas and vultures. Programs to pay the Maasai for cattle lost to lions have problems, so the killing continues. Over 100 lions have been killed in the past ten years, and in spite of declining lion populations, the killing is still increasing.[5] If it doesn't stop, lions may become locally extinct in the reserve, which will damage tourism which brings much needed cash to the reserve. As a result, some Massi are now protecting lions and thus the tourist income (see opening photograph). It may come down to a value judgment: lions on the one hand and cattle and people on the other. The lions may also be threatened by a loss of grasslands if the climate continues to change and becomes drier. Such a change favors woodlands, wherein the lion's natural prey, such as zebras and wildebeest, are replaced by kudu, impalas, monkeys, and baboons.

The Amboseli story illustrates that many environmental factors operate together, and that causes of change can be subtle and complex. The story also illustrates how environmental scientists attempt to work out sequences of events that follow a particular change. At Amboseli, rainfall cycles change hydrology and soil conditions, which in turn change the vegetation and animals of the area, and these in turn impact the people living there. To understand what happens in natural ecosystems, we can't just look for an answer derived from a single factor. We have to look at the entire environment and all of the factors that together influence what happens to life. In this chapter, we discuss some of the fundamental concepts of studying the environment in terms of several key themes that we will revisit at the end of each chapter.

1.1 Major Themes of Environmental Science

The study of environmental problems and their solutions has never been more important. Modern society in 2009 is hooked on oil. Production has declined, while demand has grown, and the population of the world has been increasing by more than 70 million each year. The emerging energy crisis is producing an economic crisis, as the prices of everything produced from oil (fertilizer, food, and fuel) rise beyond what some people can afford to pay. Energy and economic problems come at a time of unprecedented environmental concerns, from the local to global level.

At the beginning of the modern era—in A.D. 1—the number of people in the world was probably about 100 million, one-third of the present population of the United States. In 1960 the world contained 3 billion people. Our population has more than doubled in the last 40 years, to 6.8 billion people today. In the United States, population increase is often apparent when we travel. Urban traffic snarls, long lines to enter national parks, and difficulty getting tickets to popular attractions are all symptoms of a growing population. If recent human population growth rates continue, our numbers could reach 9.4 billion by 2050. The problem is that the Earth has not grown any larger, and the abundance of its resources has not increased—in many cases, quite the opposite. How, then, can Earth sustain all these people? And what is the maximum number of people that could live on Earth, not just for a short time but *sustained* over a long period?

Estimates of how many people the planet can support range from 2.5 billion to 40 billion (a population not possible with today's technology). Why do the estimates vary so widely? Because the answer depends on what quality of life people are willing to accept. Beyond a threshold world population of about 4–6 billion, the quality of life declines. How many people the Earth can sustain depends on *science and values* and is also a question about *people and nature*. The more people we pack onto the Earth, the less room and resources there are for wild animals and plants, wilderness, areas for recreation, and other aspects of nature—and the faster Earth's resources will be used. The answer also depends on how the people are distributed on the Earth—whether they are concentrated mostly in cities or spread evenly across the land.

Although the environment is complex and environmental issues seem sometimes to cover an unmanageable number of topics, the science of the environment comes down to the central topics just mentioned: the human population, urbanization, and sustainability within a global perspective. These issues have to be evaluated in light of the interrelations between people and nature, and the answers ultimately depend on both science and nature.

This book therefore approaches environmental science through six interrelated themes:

- *Human population growth* (the environmental problem).
- *Sustainability* (the environmental goal).
- *A global perspective* (many environmental problems require a global solution).
- *An urbanizing world* (most of us live and work in urban areas).
- *People and nature* (we share a common history with nature).
- *Science and values* (science provides solutions; which ones we choose are in part value judgments).

You may ask, "If this is all there is to it, what is in the rest of this book?" (See A Closer Look 1.1.) The answer

A CLOSER LOOK 1.1

A Little Environmental History

A brief historical explanation will help clarify what we seek to accomplish. Before 1960, few people had ever heard the word *ecology*, and the word *environment* meant little as a political or social issue. Then came the publication of Rachel Carson's landmark book, *Silent Spring* (Boston: Houghton Mifflin, 1960, 1962). At about the same time, several major environmental events occurred, such as oil spills along the coasts of Massachusetts and southern California, and highly publicized threats of extinction of many species, including whales, elephants, and songbirds. The environment became a popular issue.

As with any new social or political issue, at first relatively few people recognized its importance. Those who did found it necessary to stress the problems—to emphasize the negative—in order to bring public attention to environmental concerns. Adding to the limitations of the early approach to environmental issues was a lack of scientific knowledge and practical know-how. Environmental sciences were in their infancy. Some people even saw science as part of the problem.

The early days of modern environmentalism were dominated by confrontations between those labeled "environmentalists" and those labeled "anti-environmentalists." Stated in the simplest terms, environmentalists believed that the world was in peril. To them, economic and social development meant destruction of the environment and ultimately the end of civilization, the extinction of many species, and perhaps the extinction of human beings. Their solution was a new worldview that depended only secondarily on facts, understanding, and science. In contrast, again in simplest terms, the anti-environmentalists believed that whatever the environmental effects, social and economic health and progress were necessary for people and civilization to prosper. From their perspective, environmentalists represented a dangerous and extreme view with a focus on the environment to the detriment of people, a focus they thought would destroy the very basis of civilization and lead to the ruin of our modern way of life.

Today, the situation has changed. Public-opinion polls now show that people around the world rank the environment among the most important social and political issues. There is no longer a need to prove that environmental problems are serious.

We have made significant progress in many areas of environmental science (although our scientific understanding of the environment still lags behind our need to know). We have also begun to create legal frameworks for managing the environment, thus providing a new basis for addressing environmental issues. The time is now ripe to seek truly lasting, more rational solutions to environmental problems.

lies with the old saying "The devil is in the details." The solution to specific environmental problems requires specific knowledge. The six themes listed above help us see the big picture and provide a valuable background. The opening case study illustrates linkages among the themes, as well as the importance of details.

In this chapter we introduce the six themes with brief examples, showing the linkages among them and touching on the importance of specific knowledge that will be the concern of the rest of the book. We start with human population growth.

1.2 Human Population Growth

Our Rapid Population Growth

The most dramatic increase in the history of the human population occurred in the last part of the 20th century and continues today into the early 21st century. As mentioned, in merely the past 40 years the human population of the world more than doubled, from 2.5 billion to about 6.8 billion. Figure 1.5 illustrates this population explosion, sometimes referred to as the "population bomb." The figure shows that the expected decrease in population in the developed regions (for example, the U.S. and Western Europe) is more than offset by rapid population growth in the developing regions (for example, Africa, India, and South America).

Human population growth is, in some important ways, *the* underlying issue of the environment. Much current environmental damage is directly or indirectly the result of the very large number of people on Earth and our rate of increase. As you will see in Chapter 4, where we consider the human population in more detail, for most of human history the total population was small and the average long-term rate of increase was low relative to today's growth rate.[6,7]

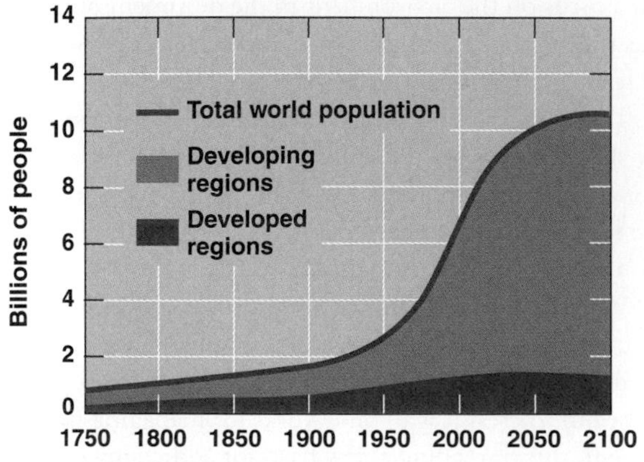

FIGURE 1.5 Population growth in developed and developing nations, 1750 projected to 2100.

Although it is customary to think of the population as increasing continuously without declines or fluctuations, the growth of the human population has not been a steady march. For example, great declines occurred during the time of the Black Death in the 14th century. At that time, entire towns were abandoned, food production declined, and in England one-third of the population died within a single decade.[8]

Famine and Food Crisis

Famine is one of the things that happen when a human population exceeds its environmental resources. Famines have occurred in recent decades in Africa. In the mid-1970s, following a drought in the Sahel region, 500,000 Africans starved to death and several million more were permanently affected by malnutrition.[9] Starvation in African nations gained worldwide attention some ten years later, in the 1980s.[10, 11]

Famine in Africa has had multiple interrelated causes. One, as suggested, is drought. Although drought is not new to Africa, the size of the population affected by drought is new. In addition, deserts in Africa appear to be spreading, in part because of changing climate but also because of human activities. Poor farming practices have increased erosion, and deforestation may be helping to make the environment drier. In addition, the control and destruction of food have sometimes been used as a weapon in political disruptions (Figure 1.6). Today, malnutrition contributes to the death of about 6 million children per year. Low- and middle-income countries suffer the most from malnutrition, as measured by low weight for age (underweight, as shown in Figure 1.7).[12]

Famines in Africa illustrate another key theme: people and nature. People affect the environment, and the environment affects people. The environment affects agriculture, and agriculture affects the environment. Human population growth in Africa has severely stretched the capacity of the land to provide sufficient food and has threatened its future productivity.

The emerging global food crisis in the first decade of the 21st century has not been caused by war or drought but by rising food costs. The cost of basic items, such as rice, corn, and wheat, has risen to the point where low- and moderate-income countries are experiencing a serious crisis. In 2007 and 2008, food riots occurred in many locations, including Mexico, Haiti, Egypt, Yemen, Bangladesh, India, and Sudan (Figure 1.8). The rising cost of oil used to produce food (in fertilizer, transportation, working fields, etc.) and the conversion of some corn production to biofuels have been blamed. This situation involves yet another key theme: science and values. Scien-

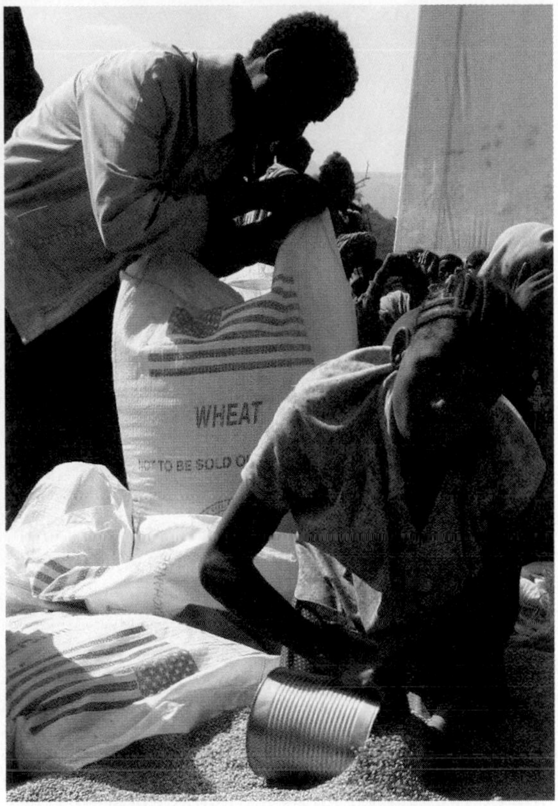

FIGURE 1.6 Science and values. Social conditions affect the environment, and the environment affects social conditions. Political disruption in Somalia (illustrated by a Somalian boy with a gun, left photo) interrupted farming and food distribution, leading to starvation. Overpopulation, climate change, and poor farming methods also lead to starvation, which in turn promotes social disruption. Famine has been common in parts of Africa since the 1980s, as illustrated by gifts of food from aid agencies.

tific knowledge has led to increased agricultural production and to a better understanding of population growth and what is required to conserve natural resources. With this knowledge, we are forced to confront a choice: Which is more important, the survival of people alive today or conservation of the environment on which future food production and human life depend?[13]

Answering this question demands *value judgments* and the information and knowledge with which to make such judgments. For example, we must determine whether we can continue to increase agricultural production without destroying the very environment on which agriculture and, indeed, the persistence of life on Earth depend. Put another way, a technical, scientific investigation provides a basis for a value judgment.

The human population continues to grow, but humans' effects on the environment are growing even faster.[14] People cannot escape the laws of population growth (this is discussed in several chapters). The broad science-and-values question is: What will we do about the increase in our own species and its impact on our planet and on our future?

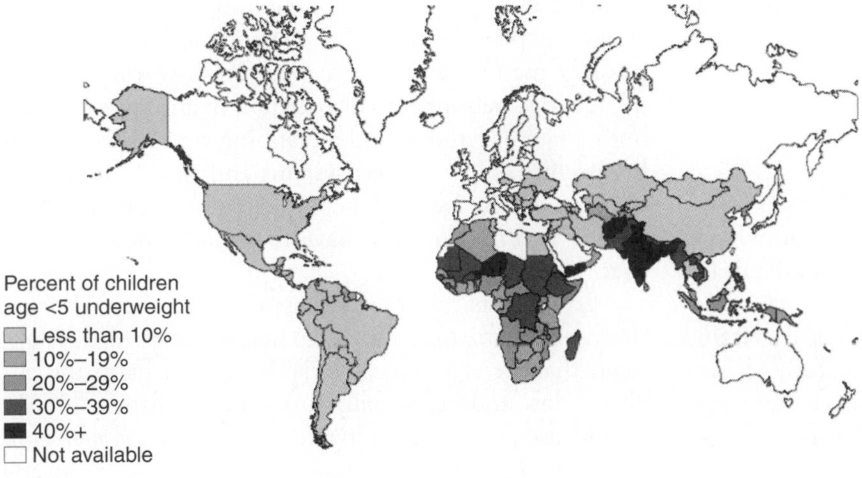

FIGURE 1.7 Underweight children under the age of 5 by region. Most are in low- and middle-income countries. (Source: World Population Data Sheet [Washington, DC: Population Reference Bureau, 2007. Accessed 5/19/08 @www.prb.org].)

(a) (b)

FIGURE 1.8 Food riots over the rising cost of food in 2007. (a) Haiti and **(b)** Bangladesh.

1.3 Sustainability and Carrying Capacity

The story of recent famines and food crises brings up one of the central environmental questions: What is the maximum number of people the Earth can sustain? That is, what is the sustainable human carrying capacity of the Earth? Much of this book will deal with information that helps answer this question. However, there is little doubt that we are using many renewable environmental resources faster than they can be replenished—in other words, we are using them *unsustainably*. In general, we are using forests and fish faster than they can regrow, and we are eliminating habitats of endangered species and other wildlife faster than they can be replenished. We are also extracting minerals, petroleum, and groundwater without sufficient concern for their limits or the need to recycle them. As a result, there is a shortage of some resources and a probability of more shortages in the future. Clearly, we must learn how to sustain our environmental resources so that they continue to provide benefits for people and other living things on our planet.

Sustainability: The Environmental Objective

The environmental catchphrase of the 1990s was "saving our planet." Are all life and the environments on which life depends really in danger? Will we leave behind a dead planet?

In the long view of planetary evolution, it is certain that planet Earth will survive us. Our sun is likely to last another several billion years, and if all humans became extinct in the next few years, life would still flourish here on Earth. The changes we have made—in the landscape, the atmosphere, the waters—would last for a few hundred or thousands of years but in a modest length of time would be erased by natural processes. What we are concerned with, as environmentalists, is the quality of the *human* environment on Earth, for us today and for our children.

Environmentalists agree that sustainability must be achieved, but we are unclear about how to achieve it, in part because the word is used to mean different things, often leading to confusion that causes people to work at cross-purposes. **Sustainability** has two formal scientific meanings with respect to environment: (1) *sustainability of resources*, such as a species of fish from the ocean, a kind of tree from a forest, coal from mines; and (2) *sustainability of an ecosystem*. Strictly speaking, harvesting a resource at a certain rate is sustainable if we can continue to harvest that resource at that same rate for some specified time well into the future. An ecosystem is sustainable if it can continue its primary functions for a specified time in the future. (Economists refer to the specified time in the future as a "planning time horizon.") Commonly, in discussions about environmental problems, the time period is not specified and is assumed to be very long—mathematically an infinite planning time, but in reality as long as it could possibly matter to us. For conservation of the environment and its resources to be based on quantitative science, both a rate of removal and a planning time horizon must be specified. However, ecosystems and species are always undergoing change, and a completely operational definition of *sustainability* will have to include such variation over time.

Economists, political scientists, and others also use the term *sustainability* in reference to types of development that are economically viable, do not harm the environment, and are socially just (fair to all people). We should also point out that the term *sustainable growth* is an oxymoron (i.e., a contradictory term) because any steady growth (fixed-percentage growth per year) produces large numbers in modest periods of time (see Exponential Growth in Chapter 3).

One of the environmental paradigms of the 21st century will be sustainability, but how will it be attained? Economists have begun to consider what is known as the *sustainable global economy:* the careful management and wise use of the planet and its resources, analogous to the management of money and goods. Those focusing on a sustainable global economy generally agree that under present conditions the global economy is *not* sustainable. Increasing numbers of people have resulted in so much pollution of the land, air, and water that the ecosystems that people depend on are in danger of collapse. What, then, are the attributes of a sustainable economy in the information age?[15]

- Populations of humans and other organisms living in harmony with the natural support systems, such as air, water, and land (including ecosystems).
- An energy policy that does not pollute the atmosphere, cause climate change (such as global warming), or pose unacceptable risk (a political or social decision).
- A plan for renewable resources—such as water, forests, grasslands, agricultural lands, and fisheries—that will not deplete the resources or damage ecosystems.
- A plan for nonrenewable resources that does not damage the environment, either locally or globally, and ensures that a share of our nonrenewable resources will be left to future generations.
- A social, legal, and political system that is dedicated to sustainability, with a democratic mandate to produce such an economy.

Recognizing that population is the environmental problem, we should keep in mind that a sustainable global economy will not be constructed around a completely stable global population. Rather, such an economy will take into account that the size of the human population will fluctuate within some stable range necessary to maintain healthy relationships with other components of the environment. To achieve a sustainable global economy, we need to do the following:[15]

- Develop an effective population-control strategy. This will, at least, require more education of people, since literacy and population growth are inversely related.
- Completely restructure our energy programs. A sustainable global economy is probably impossible if it is based on the use of fossil fuels. New energy plans will be based on an integrated energy policy, with more emphasis on renewable energy sources (such as solar and wind) and on energy conservation.
- Institute economic planning, including a tax structure that will encourage population control and wise use of resources. Financial aid for developing countries is absolutely necessary to narrow the gap between rich and poor nations.
- Implement social, legal, political, and educational changes that help to maintain a quality local, regional, and global environment. This must be a serious commitment that all the people of the world will cooperate with.

Moving toward Sustainability: Some Criteria

Stating that we wish to develop a sustainable future acknowledges that our present practices are not sustainable. Indeed, continuing on our present paths of overpopulation, resource consumption, and pollution will not lead to sustainability. We will need to develop new concepts that will mold industrial, social, and environmental interests into an integrated, harmonious system. In other words, we need to develop a new paradigm, an alternative to our present model for running society and creating wealth.[16] The new paradigm might be described as follows.[17]

- *Evolutionary rather than revolutionary.* Developing a sustainable future will require an evolution in our values that involves our lifestyles as well as social, economic, and environmental justice.
- *Inclusive, not exclusive.* All peoples of Earth must be included. This means bringing all people to a higher standard of living in a sustainable way that will not compromise our environment.
- *Proactive, not reactive.* We must plan for change and for events such as human population problems, resource shortages, and natural hazards, rather than waiting for them to surprise us and then reacting. This may sometimes require us to apply the Precautionary Principle, which we discuss with science and values (Section 1.7).
- *Attracting, not attacking.* People must be attracted to the new paradigm because it is right and just. Those who speak for our environment should not take a hostile stand but should attract people to the path of sustainability through sound scientific argument and appropriate values.
- *Assisting the disadvantaged, not taking advantage.* This involves issues of environmental justice. All people have the right to live and work in a safe, clean environment. Working people around the globe need to receive a living wage—wages sufficient to support their families. Exploitation of workers to reduce the costs of manufacturing goods or growing food diminishes us all.

(a) (b)

FIGURE 1.9 How many people do we want on Earth? (a) Streets of Calcutta; **(b)** Davis, California.

The Carrying Capacity of the Earth

Carrying capacity is a concept related to sustainability. It is usually defined as the maximum number of individuals of a species that can be sustained by an environment without decreasing the capacity of the environment to sustain that same number in the future.

There are limits to the Earth's potential to support humans. If we used Earth's total photosynthetic potential with present technology and efficiency to support 6.8 billion people, Earth could support a human population of about 15 billion. However, in doing this, we would share our land with very little else.[18, 19] When we ask "What is the maximum number of people that Earth can sustain?" we are asking not just about Earth's carrying capacity but also about sustainability.

As we pointed out, what we consider a "desirable human carrying capacity" depends in part on our values (Figure 1.9). Do we want those who follow us to live short lives in crowded conditions, without a chance to enjoy Earth's scenery and diversity of life? Or do we hope that our descendants will have a life of high quality and good health? Once we choose a goal regarding the quality of life, we can use scientific information to understand what the sustainable carrying capacity might be and how we might achieve it.

1.4 A Global Perspective

Our actions today are experienced worldwide. Because human actions have begun to change the environment all over the world, the next generation, more than the present generation, will have to take a global perspective on environmental issues (Figure 1.10).

Recognition that civilization can change the environment at a global level is relatively recent. As we discuss in detail in later chapters, scientists now believe that emissions of modern chemicals are changing the ozone layer high in the atmosphere. Scientists also believe that burning fossil fuels increases the concentration of greenhouse gases in the atmosphere, which may change Earth's climate. These atmospheric changes suggest that the actions of many groups of people, at many locations, affect the environment of the entire world.[20] Another new idea explored in later chapters is that not only human life but also nonhuman life affects the environment of our whole planet and has changed it over the course of several billion years. These two new ideas have profoundly affected our approach to environmental issues.

Awareness of the global interactions between life and the environment has led to the development of the **Gaia hypothesis**. Originated by British chemist James Lovelock and American biologist Lynn Margulis, the Gaia hypothesis (discussed in Chapter 3) proposes that over the

FIGURE 1.10 Earth from space. Isolated from other planets, Earth is "home," the only habitat we have.

history of life on Earth, life has profoundly changed the global environment, and that these changes have tended to improve the chances for the continuation of life. Because life affects the environment at a global level, the environment of our planet is different from that of a lifeless one.

1.5 An Urban World

In part because of the rapid growth of the human population and in part because of changes in technology, we are becoming an urban species, and our effects on the environment are more and more the effects of urban life (Figure 1.11a). Economic development leads to urbanization; people move from farms to cities and then perhaps to suburbs. Cities and towns get larger, and because they are commonly located near rivers and along coastlines, urban sprawl often overtakes the agricultural land of river floodplains, as well as the coastal wetlands, which are important habitats for many rare and endangered species. As urban areas expand, wetlands are filled in, forests cut down, and soils covered over with pavement and buildings.

In developed countries, about 75% of the population live in urban areas and 25% in rural areas, but in developing countries only 40% of the people are city dwellers. By 2008, for the first time, more than half of the people on Earth lived in urban areas, and it is estimated that by 2025 almost two-thirds of the population—5 billion people—will live in cities. Only a few urban areas had populations

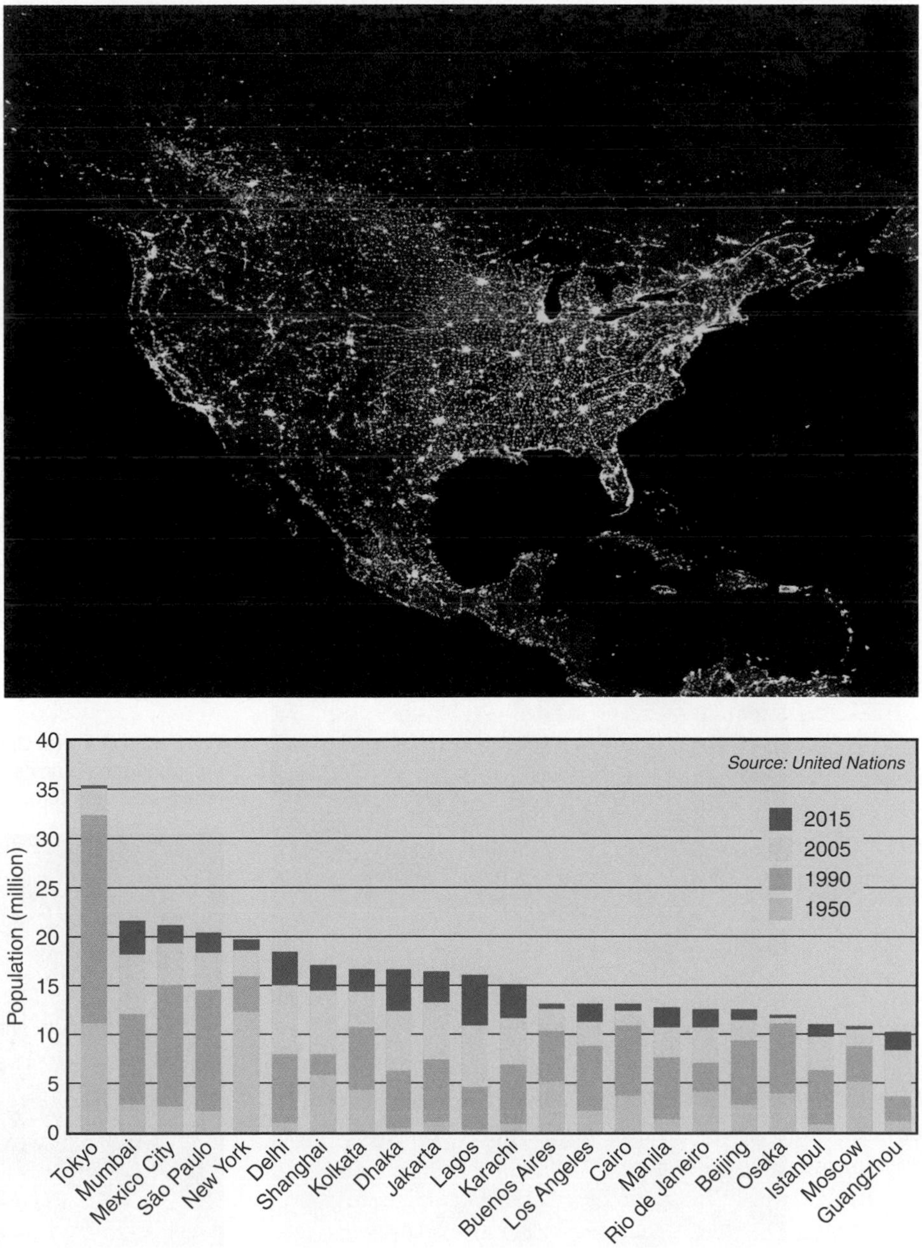

FIGURE 1.11 (a) An urban world and a global perspective. When the United States is viewed at night from space, the urban areas show up as bright lights. The number of urban areas reflects the urbanization of our nation. **(b) Megacities by 2015.** (*Source:* Data from United Nations Population Division, World Urbanization 2005, and *State of the World 2007*. World Watch Institute.)

FIGURE 1.12 An aerial photo of Los Angeles shows the large extent of a megacity.

over 4 million in 1950. In 1999 Tokyo, Japan, was the world's largest city, with a population of about 12 million, and by 2015 Tokyo will likely still be the world's largest city, with a projected population of 28.9 million. The number of **megacities**—urban areas with at least 10 million inhabitants—increased from 2 (New York City and London) in 1950 to 22 (including Los Angeles and New York City) in 2005 (Figures 1.11b and 1.12). Most megacities are in the developing world, and it is estimated that by 2015 most megacities will be in Asia.[21, 22]

In the past, environmental organizations often focused on nonurban issues—wilderness, endangered species, and natural resources, including forests, fisheries, and wildlife. Although these will remain important issues, in the future we must place more emphasis on urban environments and their effects on the rest of the planet.

1.6 People and Nature

Today we stand at the threshold of a major change in our approach to environmental issues. Two paths lie before us. One path is to assume that environmental problems are the result of human actions and that the solution is simply to stop these actions. Based on the notion, popularized some 40 years ago, that people are separate from nature, this path has led to many advances but also many failures. It has emphasized confrontation and emotionalism and has been characterized by a lack of understanding of basic facts about the environment and how natural ecological systems function, often basing solutions instead on political ideologies and ancient myths about nature.

The second path begins with a scientific analysis of an environmental controversy and leads from there to cooperative problem solving. It accepts the connection between people and nature and offers the potential for long-lasting, successful solutions to environmental problems. One purpose of this book is to take the student down the second pathway.

People and nature are intimately integrated. Each affects the other. We depend on nature in countless ways. We depend on nature directly for many material resources, such as wood, water, and oxygen. We depend on nature indirectly through what are called public-service functions. For example, soil is necessary for plants and

(a)

FIGURE 1.13 **(a)** Cross section of a soil; **(b)** earthworms are among the many soil animals important to maintaining soil fertility and structure.

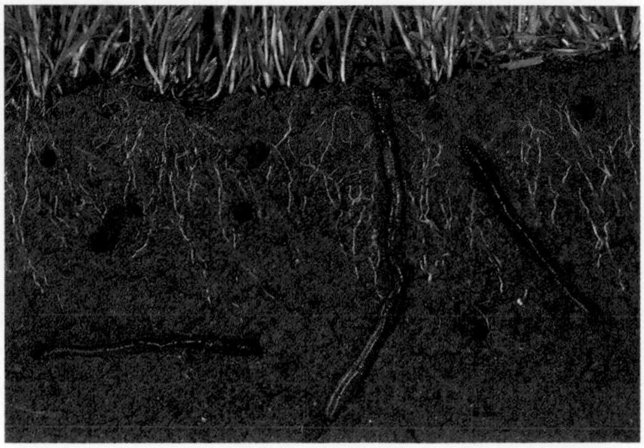

(b)

FIGURE 1.14 Land cleared by African elephants, Tsavo National Park, Kenya.

FIGURE 1.15 People and nature. We feel safe around a campfire—a legacy from our Pleistocene ancestors?

therefore for us (Figure 1.13); the atmosphere provides a climate in which we can live; the ozone layer high in the atmosphere protects us from ultraviolet radiation; trees absorb some air pollutants; wetlands can cleanse water. We also depend on nature for beauty and recreation—the needs of our inner selves—as people always have.

We in turn affect nature. For as long as we have had tools, including fire, we have changed nature, often in ways that we like and have considered "natural." One can argue that it is natural for organisms to change their environment. Elephants topple trees, changing forests to grasslands, and people cut down trees and plant crops (Figure 1.14). Who is to say which is more natural? In fact, few organisms do *not* change their environment.

People have known this for a long time, but the idea that people might change nature to their advantage was unpopular in the last decades of the 20th century. At that time, the word *environment* suggested something separate—"out there"—implying that people were not part of nature. Today, environmental sciences are showing us how people and nature connect, and in what ways this is beneficial to both.

With growing recognition of the environment's importance, we are becoming more Earth-centered. We seek to spend more time in nature for recreation and spiritual activities. We accept that we have evolved on and with the Earth and are not separate from it. Although we are evolving fast, we remain genetically similar to people who lived more than 100,000 years ago. Do you ever wonder why we like to go camping, to sit around a fire at night roasting marshmallows and singing, or exchanging scary stories about bears and mountain lions (Figure 1.15)? More than ever, we understand and celebrate our union with nature as we work toward sustainability.

Most people recognize that we must seek sustainability not only of the environment but also of our economic activities, so that humanity and the environment can persist together. The dichotomy of the 20th century is giving way to a new unity: the idea that a sustainable environment and a sustainable economy may be compatible, that people and nature are intertwined, and that success for one involves success for the other.

1.7 Science and Values

Deciding what to do about an environmental problem involves both values and science, as we have already seen. We must choose what we want the environment to be. But to make this choice, we must first know what is possible. That requires knowing the scientific data and understanding its implications. Scientists rely on critical thinking. Critical scientific thinking is disciplined, using intellectual standards, effective communication, clarity, and commitment to developing scientific knowledge and skills. It leads to conclusions, generalizations, and, sometimes, scientific theories and even scientific laws. Taken together, these comprise a body of beliefs that, at the present time, account for all known observations about a particular phenomenon. Some of the intellectual standards are as follows:

Selected Intellectual Standards

- ***Clarity:*** If a statement is unclear, you can't tell whether it is relevant or accurate.
- ***Accuracy:*** Is a statement true? Can it be checked? To what extent does a measurement agree with the accepted value?
- ***Precision:*** The degree of exactness to which something is measured. Can a statement be more specific, detailed, and exact?

- ***Relevance:*** How well is a statement connected to the problem at hand?
- ***Depth:*** Did you deal with the complexities of a question?
- ***Breadth:*** Did you consider other points of view or look at it from a different perspective?
- ***Logic:*** Does a conclusion make sense and follow from the evidence?
- ***Significance:*** Is the problem an important one? Why?
- ***Fairness:*** Are there any vested interests, and have other points of view received attention?

Modified after R. Paul, and L. Elder, *Critical Thinking* (Dillon Beach, CA: The Foundation for Critical Thinking, 2003).

Once we know our options, we can select from among them. What we choose is determined by our values. An example of a value judgment regarding the world's human environmental problem is the choice between the desire of an individual to have many children and the need to find a way to limit the human population worldwide.

After we have chosen a goal based on knowledge and values, we have to find a way to attain that goal. This step also requires knowledge. And the more technologically advanced and powerful our civilization, the more knowledge is required. For example, current fishing methods enable us to harvest very large numbers of chinook salmon from the Columbia River, and public demand for salmon encourages us to harvest as many as possible. To determine whether chinook salmon are sustainable, we must know how many there are now and how many there have been in the past. We must also understand the processes of birth and growth for this fish, as well as its food requirements, habitat, life cycle, and so forth—all the factors that ultimately determine the abundance of salmon in the Columbia River.

Consider, in contrast, the situation almost two centuries ago. When Lewis and Clark first made an expedition to the Columbia, they found many small villages of Native Americans who depended in large part on the fish in the river for food (Figure 1.16). The human population was small, and the methods of fishing were simple. The maximum number of fish the people could catch probably posed no threat to the salmon, so these people could fish without scientific understanding of numbers and processes. (This example does not suggest that prescientific societies lacked an appreciation for the idea of sustainability. On the contrary, many so-called primitive societies held strong beliefs about the limits of harvests.)

The Precautionary Principle

Science and values come to the forefront when we think about what action to take about a perceived environmental problem for which the science is only partially known. This is often the case because all science is preliminary and subject to analysis of new data, ideas, and tests of hypotheses. Even with careful scientific research, it can be difficult, even impossible, to prove with absolute certainty how relationships between human activities and other physical and biological processes lead to local and global environmental problems, such as global warming, depletion of ozone in the upper atmosphere, loss of biodiversity, and declining resources. For this reason, in 1992 the Rio Earth Summit on Sustainable Development listed as one of its principles what we now call the **Precautionary Principle**. Basically, it says that when there is a threat of serious, perhaps even irreversible, environmental damage, we should not wait for scientific proof before taking precautionary steps to prevent potential harm to the environment.

The Precautionary Principle requires critical thinking about a variety of environmental concerns, such as the manufacture and use of chemicals, including pesticides, herbicides, and drugs; the use of fossil fuels and nuclear energy; the conversion of land from one use to another (for example, from rural to urban); and the management of wildlife, fisheries, and forests.[23]

FIGURE 1.16 Native Americans fishing for salmon on the Columbia River.

FIGURE 1.17 The city of San Francisco, with its scenic bayside environment, has adopted the Precautionary Principle.

One important question in applying the Precautionary Principle is how much scientific evidence we should have before taking action on a particular environmental problem. The principle recognizes the need to evaluate all the scientific evidence we have and to draw provisional conclusions while continuing our scientific investigation, which may provide additional or more reliable data. For example, when considering environmental health issues related to the use of a pesticide, we may have a lot of scientific data, but with gaps, inconsistencies, and other scientific uncertainties. Those in favor of continuing to use that pesticide may argue that there isn't enough proof of its danger to ban it. Others may argue that absolute proof of safety is necessary before a new pesticide is used. Those advocating the Precautionary Principle would argue that we should continue to investigate but, to be on the safe side, should not wait to take cost-effective precautionary measures to prevent environmental damage or health problems. What constitutes a cost-effective measure? Certainly we would need to examine the benefits and costs of taking a particular action versus taking no action. Other economic analyses may also be appropriate.[23, 24]

The Precautionary Principle is emerging as a new tool for environmental management and has been adopted by the city of San Francisco (Figure 1.17) and the European Union. There will always be arguments over what constitutes sufficient scientific knowledge for decision making. Nevertheless, the Precautionary Principle, even though it may be difficult to apply, is becoming a common part of environmental analysis with respect to environmental protection and environmental health issues. It requires us to think ahead and predict potential consequences before they occur. As a result, the Precautionary Principle is a *proactive*, rather than *reactive*, tool—that is, we can use it when we see real trouble coming, rather than reacting after the trouble arises.

Placing a Value on the Environment

How do we place a value on any aspect of our environment? How do we choose between two different concerns? The value of the environment is based on eight justifications: utilitarian (materialistic), ecological, aesthetic, recreational, inspirational, creative, moral, and cultural.

The **utilitarian justification** is that some aspect of the environment is valuable because it benefits individuals economically or is directly necessary to human survival. For example, conserving lions in Africa as part of tourism provides a livelihood for local people.

The **ecological justification** is that an ecosystem is necessary for the survival of some species of interest to us, or that the system itself provides some benefit. For example, a mangrove swamp (a type of coastal wetland) provides habitat for marine fish, and although we do not eat mangrove trees, we may eat the fish that depend on them. Also, the mangroves are habitat for many noncommercial species, some endangered. Therefore, conservation of the mangrove is important ecologically. Another example: Burning coal and oil adds greenhouse gases to the atmosphere, which may lead to a climate change that could affect the entire Earth. Such ecological reasons form a basis for the conservation of nature that is essentially enlightened self-interest.

Aesthetic and **recreational justifications** have to do with our appreciation of the beauty of nature and our desire to get out and enjoy it. For example, many people find wilderness scenery beautiful and would rather live in a world with wilderness than without it. One way we enjoy nature's beauty is to seek recreation in the outdoors.

The aesthetic and recreational justifications are gaining a legal basis. The state of Alaska acknowledges that sea otters have an important recreational role in that people enjoy watching and photographing them in a wilderness setting. And there are many other examples of the aesthetic importance of the environment. When people mourn the death of a loved one, they typically seek out places with grass, trees, and flowers; thus we use these to beautify our graveyards. Conservation of nature can be based on its benefits to the human spirit, our "inner selves" (*inspirational justification*). Nature is also often an aid to human creativity (the *creative justification*). The creativity of artists and poets, among others, is often inspired by their contact with nature. But while nature's aesthetic, recreational, and inspirational value is a widespread reason that people enjoy nature, it is rarely used in formal environmental arguments, perhaps in the belief that they might seem superficial justifications for conserving nature. In fact, however, beauty in their surroundings is of profound importance to people. Frederick Law Olmsted, the great American landscape planner, argued that plantings of vegetation provide medical, psychological, and social benefits and are essential to city life.[18]

Moral justification has to do with the belief that various aspects of the environment have a right to exist and that it is our moral obligation to help them, or at least allow them, to persist. Moral arguments have been extended to many nonhuman organisms, to entire ecosystems, and even to inanimate objects. The historian Roderick Nash, for example, wrote an article entitled "Do Rocks Have Rights?" that discusses such moral justification,[29] and the United Nations General Assembly World Charter for Nature, signed in 1982, states that species have a moral right to exist.

Cultural justification refers to the fact that different cultures have many of the same values but also some different values with respect to the environment. This may also be in terms of specifics of a particular value. All cultures may value nature, but, depending on their religious beliefs, may value it in different degrees of intensity. For example, Buddhist monks when preparing ground for a building may pick up and move disturbed eathhworms, something few others would do. Different cultures integrate nature into their towns, cities, and homes in different ways depending on their view of nature.

Analysis of environmental values is the focus of a new discipline, known as environmental ethics. Another concern of environmental ethics is our obligation to future generations: Do we have a moral obligation to leave the environment in good condition for our descendants, or are we at liberty to use environmental resources to the point of depletion within our own lifetimes?

CRITICAL THINKING ISSUE

Easter Island

The story of Easter Island has been used as an example of how people may degrade the environment as they grow in number, until eventually their overuse of the environment results in the collapse of the society. This story has been challenged by recent work. We will present what is known, and you should examine the case history critically. To help with this issue, look back to the list of intellectual standards useful in critical thinking.

Easter Island's history spans approximately 800 to 1,500 years and illustrates the importance of science and the sometimes irreversible consequences of human population growth and the introduction of a damaging exotic species, accompanied by depletion of resources necessary for survival. Evidence of the island's history is based on detailed studies by earth scientists and social scientists who investigated the anthropological record left in the soil where people lived and the sediment in ponds where pollen from plants that lived at different times was deposited. The goals of the studies were to estimate the number of people, their diet, and their use of resources. This was linked to studies of changes in vegetation, soils, and land productivity.

Easter Island lies about 3,700 km west of South America and 4,000 km from Tahiti (Figure 1.18a), where the people may have come from. The island is small, about 170 km^2, with a rough triangular shape and an inactive volcano at each corner. The elevation is less than about 500 m (1,500 ft) (Figure 1.18b), too low to hold clouds like those in Hawaii that bring rain. As a result, water resources are limited. When Polynesian people first reached it about 800–1,500 years ago, they colonized a green island covered with rich soils and forest. The small group of settlers grew rapidly, to perhaps over 10,000 people, who eventually established a complex society that was spread among a number of small villages. They raised crops and chickens, supplementing their diet with fish from the sea. They used the island's trees to build their homes and to build boats. They also carved massive 8-meter-high statues from volcanic rock and moved them into place at various parts of the island using tree trunks as rollers (Figure 1.18b, c).

When Europeans first reached Easter Island in 1722, the only symbols of the once-robust society were the statues. A study suggested that the island's population had collapsed in just a few decades to about 2,000 people because they had used up (degraded) the isolated island's limited resource base.[25, 26]

At first there were abundant resources, and the human population grew fast. To support their growing population, they cleared more and more land for agriculture and cut more trees for fuel, homes, and boats—and for moving the statues into place. Some of the food plants they brought to the island didn't survive, possibly because the voyage was too long or the climate unsuitable for them. In particular, they did not have the breadfruit tree, a nutritious starchy food source, so they relied more heavily on other crops, which required clearing more land for planting. The island was also relatively dry, so it is likely that fires for clearing land got out of control sometimes and destroyed even more forest than intended.[25, 26]

The cards were stacked against the settlers to some extent—but they didn't know this until too late. Other islands of similar size that the Polynesians had settled did not suffer forest depletion and fall into ruin.[25, 26] This isolated island, however, was more sensitive to change. As the forests were cut down, the soils, no longer protected by forest cover, were lost to erosion. Loss of the soils reduced agricultural productivity, but the biggest loss was the trees. Without wood to build homes and boats, the people were forced to live in caves and could no longer venture out into the ocean for fish.[25]

These changes did not happen overnight—it took more than 1,000 years for the expanding population to deplete its resources. Loss of the forest was irreversible: Because it led to loss of soil, new trees could not grow to replace the forests. As resources grew scarcer, wars between the villages became common, as did slavery, and perhaps even cannibalism.

Easter Island is small, but its story is a dark one that suggests what can happen when people use up the resources of an isolated area. We note, however, that some aspects of the above history of Easter Island have recently been challenged. New data suggest that people first arrived about 800 years ago, not 1,500; thus, much less time was available for people to degrade the land.[27, 28] Deforestation certainly played a role in the loss of trees, and the rats that arrived with the Polynesians were evidently responsible for eating seeds of the palm trees, preventing regeneration. According to the alternative explanation of the island's demise, the Polynesian people on the island at the time of European contact in 1722 numbered about 3,000; this may have been close to the maximum reached around the year 1350. Contact with Europeans introduced new diseases and enslavement, which reduced the population to about 100 by the late 1870s.[27]

Easter Island, also called Rapa Nui, was annexed by Chile in 1888. Today, about 3,000 people live on the island. Tourism is the main source of income; about 90% of the island is grassland, and thin, rocky soil is common. There have been reforestation projects, and about 5% of the island is now forested, mostly by eucalyptus plantations in the central part of the island. There are also fruit trees in some areas.

As more of the story of Easter Island emerges from scientific and social studies, the effects of resource exploitation, invasive rats, and European contact will become clearer, and the environmental lessons of the collapse will lead to a better understanding of how we can sustain our global human culture. However, the primary lesson is that *limited resources can support only a limited human population.*

Like Easter Island, our planet Earth is isolated in our solar system and universe and has limited resources. As a result, the world's growing population is facing the problem of how to conserve those resources. We know it takes a while before environmental damage begins to show, and we know that some environmental damage may be irreversible. We are striving to develop plans to ensure that our natural resources, as well as the other living things we share our planet with, will not be damaged beyond recovery.[29]

Critical Thinking Questions

1. What are the main lessons to take from Easter Island's history?
2. People may have arrived at Easter Island 1,500 years ago or later, perhaps 800 years ago. Does the timing make a significant difference in the story? How?
3. Assuming that an increasing human population, introduction of invasive rats, loss of trees, the resulting soil erosion, and, later, introduced European diseases led to collapse of the society, can Easter Island be used as a model for what could happen to Earth? Why? Why not?

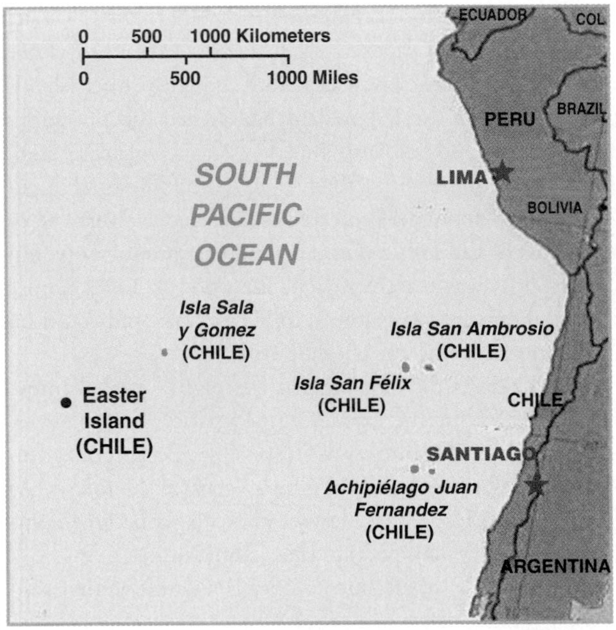

(a)

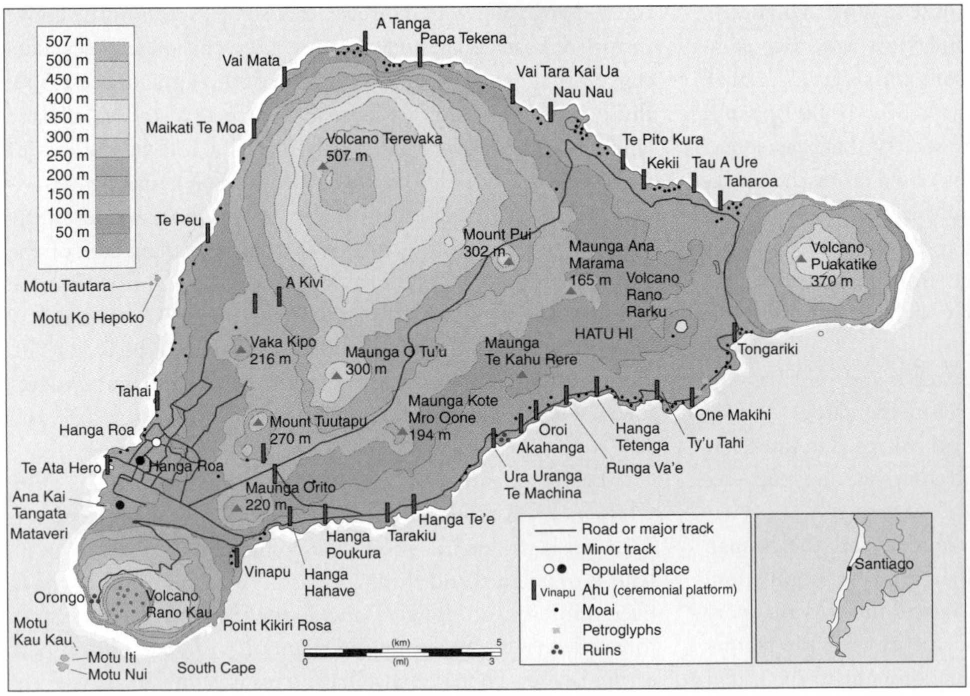

(b)

(c)

FIGURE 1.18 Easter Island, collapse of a society. (a) Location of Easter Island in the Pacific Ocean, several thousand kilometers west of South America; **(b)** map of Easter Island showing the three major volcanoes that anchor the three corners of the small island; and **(c)** large statues carved from volcanic rock before the collapse of a society with several thousand people.

SUMMARY

- Six themes run through this text: the urgency of the population issue; the importance of urban environments; the need for sustainability of resources; the importance of a global perspective; people and nature; and the role of science and values in the decisions we face.
- People and nature are intertwined. Each affects the other.
- The human population grew at a rate unprecedented in history in the 20th century. Population growth is the underlying environmental problem.
- Achieving sustainability, the environmental goal, is a long-term process to maintain a quality environment for future generations. Sustainability is becoming an important environmental paradigm for the 21st century.
- The combined impact of technology and population multiplies the impact on the environment.
- In an increasingly urban world, we must focus much of our attention on the environments of cities and the effects of cities on the rest of the environment.
- Determining Earth's carrying capacity for people and levels of sustainable harvests of resources is difficult but crucial if we are to plan effectively to meet our needs in the future. Estimates of Earth's carrying capacity for people range from 2.5 to 40 billion, but about 15 billion is the upper limit with today's technology. The differences in capacity have to do with the quality of life projected for people—the poorer the quality of life, the more people can be packed onto the Earth.
- Awareness of how people at a local level affect the environment globally gives credence to the Gaia hypothesis. Future generations will need a global perspective on environmental issues.
- Placing a value on various aspects of the environment requires knowledge and understanding of the science, but also depends on our judgments about the uses and aesthetics of the environment and on our moral commitments to other living things and to future generations.
- The Precautionary Principle is emerging as a powerful new tool for environmental management.

REEXAMINING THEMES AND ISSUES

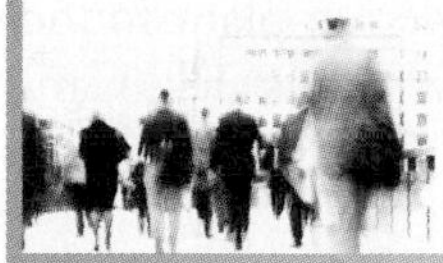

Human Population

What is more important: the quality of life of people alive today or the quality of life of future generations?

Sustainability

What is more important: abundant resources today—as much as we want and can obtain—or the availability of these resources for future generations?

Global Perspective

What is more important: the quality of your local environment or the quality of the global environment—the environment of the entire planet?

Urban World

What is more important: human creativity and innovation, including arts, humanities, and science, or the persistence of certain endangered species? Must this always be a trade-off, or are there ways to have both?

People and Nature

If people have altered the environment for much of the time our species has been on Earth, what then is "natural"?

Science and Values

Does nature know best, so that we never have to ask what environmental goal we should seek, or do we need knowledge about our environment, so that we can make the best judgments given available information?

KEY TERMS

aesthetic justification **15**
carrying capacity **10**
cultural justification **16**
ecological justification **15**
Gaia hypothesis **10**
megacities **12**
moral justification **16**
Precautionary Principle **14**
recreational justification **15**
sustainability **8**
utilitarian justification **15**

STUDY QUESTIONS

1. Why is there a convergence of energy, economics, and environment?

2. In what ways do the effects on the environment of a resident of a large city differ from the effects of someone living on a farm? In what ways are the effects similar?

3. Programs have been established to supply food from Western nations to starving people in Africa. Some people argue that such programs, which may have short-term benefits, actually increase the threat of starvation in the future. What are the pros and cons of international food relief programs?

4. Why is there an emerging food crisis that is different from any in the past?

5. Which of the following are global environmental problems? Why?
(a) Growth of the human population.
(b) Furbish's lousewort, a small flowering plant found in the state of Maine and in New Brunswick, Canada. It is so rare that it has been seen by few people and is considered endangered.
(c) The blue whale, listed as an endangered species under the U.S. Marine Mammal Protection Act.
(d) A car that has air-conditioning.
(e) Seriously polluted harbors and coastlines in major ocean ports.

6. How could you determine the carrying capacity of Earth?

7. Is it possible that sometime in the future all the land on Earth will become one big city? If not, why not? To what extent does the answer depend on the following:
(a) global environmental considerations
(b) scientific information
(c) values

FURTHER READING

Botkin, D.B., ***No Man's Garden: Thoreau and a New Vision for Civilization and Nature*** (Washington, DC: Island Press, 2000). Discusses many of the central themes of this textbook, with special emphasis on values and science and on an urban world. Henry David Thoreau's life and works illustrate approaches that can help us deal with modern environmental issues.

Botkin, D.B., ***Discordant Harmonies: A New Ecology for the 21st Century*** (New York: Oxford University Press, 1990). An analysis of the myths that underlie attempts to solve environmental issues.

Leopold, A., ***A Sand County Almanac*** (New York: Oxford University Press, 1949). Perhaps, along with Rachel Carson's *Silent Spring,* one of the most influential books of the post–World War II and pre–Vietnam War era about the value of the environment. Leopold defines and explains the land ethic and writes poetically about the aesthetics of nature.

Lutz, W., ***The Future of World Population*** (Washington, DC: Population Reference Bureau, 1994). A summary of current information on population trends and future scenarios of fertility, mortality, and migration.

Montgomery, D.K., ***Dirt: The Erosion of Civilizations*** (Berkeley, CA: University of California Press, 2007).

Nash, R.F., ***The Rights of Nature: A History of Environmental Ethics*** (Madison: University of Wisconsin Press, 1988). An introduction to environmental ethics.

CHAPTER 2

Science as a Way of Knowing: Critical Thinking about the Environment

Environmental science poses challenges to traditional science, as these students taking a field course in ecology are finding out. No data were available to tell them about the age of the forests or the grasses growing on the dunes. It's also more difficult to apply the scientific method when you are working in the field rather than in the controlled environment of a laboratory.

LEARNING OBJECTIVES

Science is a process of refining our understanding of nature through continual questioning and active investigation. It is more than a collection of facts to be memorized. After reading this chapter, you should understand that . . .

- Thinking about environmental issues requires thinking scientifically;
- We acquire scientific knowledge of the natural world through observations. The conclusions that we draw from these observations can be stated as hypotheses, theories, and scientific "laws." If they can be disproved, they are scientific. If they can't, they are not;
- Scientific understanding is not fixed; it changes over time as new data, observations, theories, and tests become available;
- Deductive and inductive reasoning are different, and we need to use both in scientific thinking;
- Every measurement involves some degree of approximation—that is, uncertainty—and a measurement without a statement about its degree of uncertainty is meaningless;
- Technology, the application of scientific knowledge, is not science, but science and technology interact, stimulating growth in each other;
- Decision-making about environmental issues involves society, politics, culture, economics, and values, as well as scientific information;
- Environmental scientific findings often get politicized when the use of scientific information is guided by a political goal and only data supporting that goal are selected;
- Forms of life seem so incredible and so well fitted to their environment that we wonder how they have come about. This question leads us to seek to understand different ways of knowing.

CASE STUDY

Birds at Mono Lake: Applying Science to Solve an Environmental Problem

Mono Lake is a large salt lake in California, just east of the Sierra Nevada and across these mountains from Yosemite National Park (Figure 2.1). More than a million birds use the lake; some feed and nest there, some stop on their migrations to feed. Within the lake, brine shrimp and brine fly larvae grow in great abundance, providing food for the birds. The shrimp and fly larvae, in turn, feed on algae and bacteria that grow in the lake (Figure 2.2).

The lake persisted for thousands of years in a desert climate because streams from the Sierra Nevada—fed by mountain snow and rain—flowed into it. But in the 1940s the city of Los Angeles diverted all stream water—beautifully clear water—to provide 17% of the water supply for the city. The lake began to dry out. It covered 60,000 acres in the 1940s, but only 40,000 by the 1980s.

Environmental groups expressed concern that the lake would soon become so salty and alkaline that all the brine shrimp and flies—food for the birds—would die, the birds would no longer be able to nest or feed there, and the beautiful lake would become a hideous eyesore—much like what happened to the Aral Sea in Asia. The Los Angeles Department of Water and Power argued that everything would be all right because rain falling directly on the lake and water flowing underground would provide ample water for the lake. People were unconvinced. "Save Mono Lake" became a popular bumper sticker in California, and the argument about the future of the lake raged for more than a decade.

Scientific information was needed to answer key questions: Without stream input, how small would the lake become? Would it really become too salty and alkaline for the shrimp, fly larvae, algae, and bacteria? If so, when?

The state of California set up a scientific panel to study the future of Mono Lake. The panel discovered that two crucial pieces of knowledge necessary to answer these questions had not been studied: the size and shape of the basin of the lake (so one could determine the lake's volume and, from this, how its salinity and alkalinity would change) and the rate at which water evaporated from the lake (to determine whether and how fast the lake would become too dry to sustain life within it). New research was commissioned that answered these questions. The answers: By about the turn of the 21st century the lake would become so small that it would be too salty for the shrimp, fly larvae, algae, and bacteria.[1]

With this scientific information in hand, the courts decided that Los Angeles would have to stop the removal of water that flowed into Mono Lake. By 2008 the lake still had not recovered to the level required by the courts, indicating that diversion of water had been undesirable for the lake and its ecosystem.

Scientific information had told Californians what would happen, when it would likely happen, and what management approaches were possible. Science was essential to finding a solution that would work. But ultimately

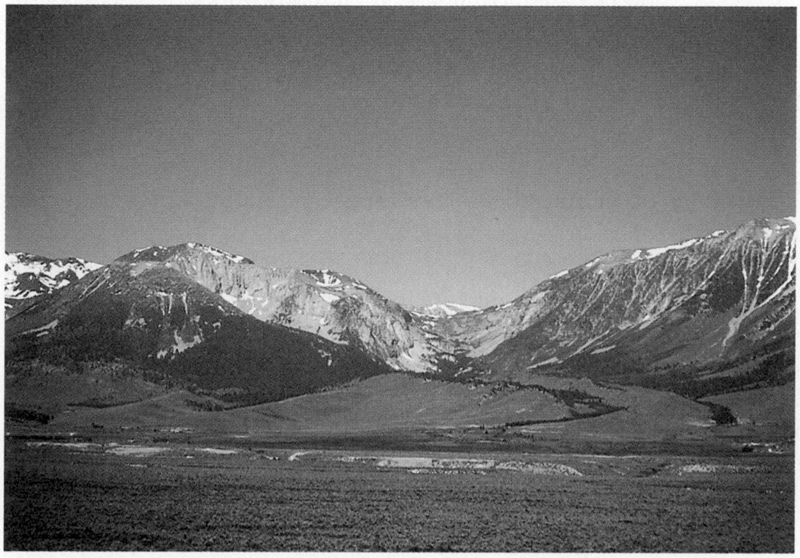

FIGURE 2.1 Mono Lake's watershed below the beautiful east slope of the Sierra Nevada. Streams flowing into the lake are visible as winding blue lines on the lower slopes. The lake and its sandy beaches form the flatlands in the mid-distance.

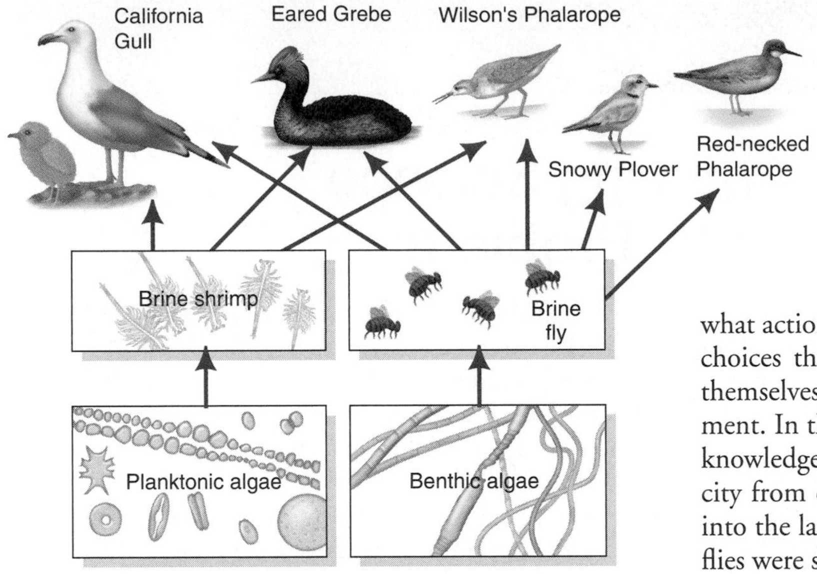

FIGURE 2.2 The Mono Lake food chain. The arrows show who feeds on whom. Just five species of birds are the top predators. This lake is one of the world's simpler ecosystems.

what actions to take, given this scientific knowledge, were choices that depended on values people held regarding themselves, their wants and desires, and the environment. In the end, decisions based on values and scientific knowledge were made by the courts, which stopped the city from diverting any of the stream waters that flowed into the lake. The birds, scenery, brine shrimp, and brine flies were saved.[2]

2.1 Understanding What Science Is—and What It Isn't

As the Mono Lake case study illustrates, modern civilization depends on science. The complexity of environmental sciences raises two fundamental questions: How does science differ from other ways of knowing? And how can we use science to answer practical questions about our effects on nature and what actions we should take to solve environmental problems?

Thinking about the environment is as old as our first human ancestors. Before humans developed the technology to deal with their environment, their very survival depended on knowledge of it. The environment also plays a crucial role in the development of each of us; normal human development does not occur in the absence of environmental stimuli.

However, thinking *scientifically* about the environment is only as old as science itself. Science had its roots in the ancient civilizations of Babylonia and Egypt, where observations of the environment were carried out primarily for practical reasons, such as planting crops, or for religious reasons, such as using the positions of the planets and stars to predict human events. Ancient precursors of science differed from modern science in that they did not distinguish between science and technology, nor between science and religion.

These distinctions first appeared in classical Greek science. Because of their general interest in ideas, the Greeks developed a more theoretical approach to science, in which knowledge for its own sake became the primary goal. At the same time, their philosophical approach began to move science away from religion and toward philosophy.

Modern science is usually considered to have begun toward the end of the 16th and the beginning of the 17th centuries with the development of the **scientific method** by Gilbert (magnets), Galileo (physics of motion), and Harvey (circulation of blood). Earlier classical scientists had asked "Why?" in the sense of "For what purpose?" But these three made important discoveries by asking "How?" in the sense of "How does it work?" Galileo also pioneered in the use of numerical observations and mathematical models. The scientific method, which quickly proved very successful in advancing knowledge, was first described explicitly by Francis Bacon in 1620. Although not a practicing scientist himself, Bacon recognized the importance of the scientific method, and his writings did much to promote scientific research.[3]

Our cultural heritage, therefore, gives us two ways of thinking about the environment: the kind of thinking we do in everyday life and the kind of thinking scientists try to do (Table 2.1). There are crucial differences between these two ways of thinking, and ignoring these differences can lead to invalid conclusions and serious errors in making critical decisions about the environment.

We can look at the world from many points of view, including religious, aesthetic, and moral. They are not science, however, because they are based ultimately on faith, beliefs, and cultural and personal choices, and are not open to disproof in the scientific sense. The distinction between a scientific statement and a nonscientific statement is not a value judgment—there is no implication that science is the only "good" kind of knowledge. The distinction is simply a philosophical one about kinds of knowledge and logic. Each way of viewing the world gives us a different way of perceiving and of making sense of our world, and each is valuable to us.

Table 2.1 KNOWLEDGE IN EVERYDAY LIFE COMPARED WITH KNOWLEDGE IN SCIENCE

FACTOR IN	EVERYDAY LIFE AND IN	SCIENCE
Goal	To lead a satisfying life (implicit)	To know, predict, and explain (explicit)
Requirements	Context-specific knowledge; no complex series of inferences; can tolerate ambiguities and lack of precision	General knowledge; complex, logical sequences of inferences, must be precise and unambiguous
Resolution of questions	Through discussion, compromise, consensus	Through observation, experimentation, logic
Understanding	Acquired spontaneously through interacting with world and people; criteria not well defined	Pursued deliberately; criteria clearly specified
Validity	Assumed, no strong need to check; based on observations, common sense, tradition, authorities, experts, social mores, faith	Must be checked; based on replications, converging evidence, formal proofs, statistics, logic
Organization of knowledge	Network of concepts acquired through experience; local, not integrated	Organized, coherent, hierarchical, logical; global, integrated
Acquisition of knowledge	Perception, patterns, qualitative; subjective	Plus formal rules, procedures, symbols, statistics, mental models; objective
Quality control	Informal correction of errors	Strict requirements for eliminating errors and making sources of error explicit

Source: Based on F. Reif and J.H. Larkin, "Cognition in Scientific and Everyday Domains: Comparison and Learning Implications," *Journal of Research in Science Teaching* 28(9), pp. 733–760. Copyright © 1991 by National Association for Research in Science Teaching. Reprinted by permission of John Wiley & Sons.

Science as a Way of Knowing

Science is a process, a way of knowing. It results in conclusions, generalizations, and sometimes scientific theories and even scientific laws. *Science begins with questions arising from curiosity about the natural world*, such as: How many birds nest at Mono Lake? What species of algae live in the lake? Under what conditions do they live?

Modern science does not deal with things that cannot be tested by observation, such as the ultimate purpose of life or the existence of a supernatural being. Science also does not deal with questions that involve values, such as standards of beauty or issues of good and evil—for example, whether the scenery at Mono Lake is beautiful. On the other hand, the statement that "more than 50% of the people who visit Mono Lake find the scenery beautiful" is a hypothesis (discussed later) that can be tested by public-opinion surveys and can be treated as a scientific statement if the surveys confirm it.

Disprovability

Here's the key to science: It is generally agreed today that the essence of the scientific method is **disprovability** (see Figure 2.3, a diagram that will be helpful throughout this chapter). A statement can be termed "scientific" if someone can state a method of disproving it. If no one can think of such a test, then the statement is said to be nonscientific. Consider, for example, the crop circles discussed in A Closer Look 2.1. One Web site says that some people believe the crop circles are a "spiritual nudge . . . designed to awaken us to our larger context and milieu, which is none other than our collective earth soul." Whether or not this is true, it does not seem open to disproof.

Science is a process of discovery—a continuing process whose essence is change in ideas. The fact that scientific ideas change is frustrating. Why can't scientists agree on what is the best diet for people? Why is a chemical considered dangerous in the environment for a while and then determined not to be? Why do scientists in one decade consider forest fires undesirable disturbances and in a later decade decide forest fires are natural and in fact important? Are we causing global warming or not? And on and on. Can't scientists just find out the truth and give us the final word on all these questions once and for all, and agree on it?

The answer is no—because science is a continuing adventure during which scientists make better and better approximations of how the world works. Sometimes changes in ideas are small, and the major context remains the same. Sometimes a science undergoes a fundamental revolution in ideas.

Science makes certain assumptions about the natural world: that events in the natural world follow patterns that can be understood through careful observation and scientific analysis, which we will describe later; and that these basic patterns and the rules that describe them are the same throughout the universe.

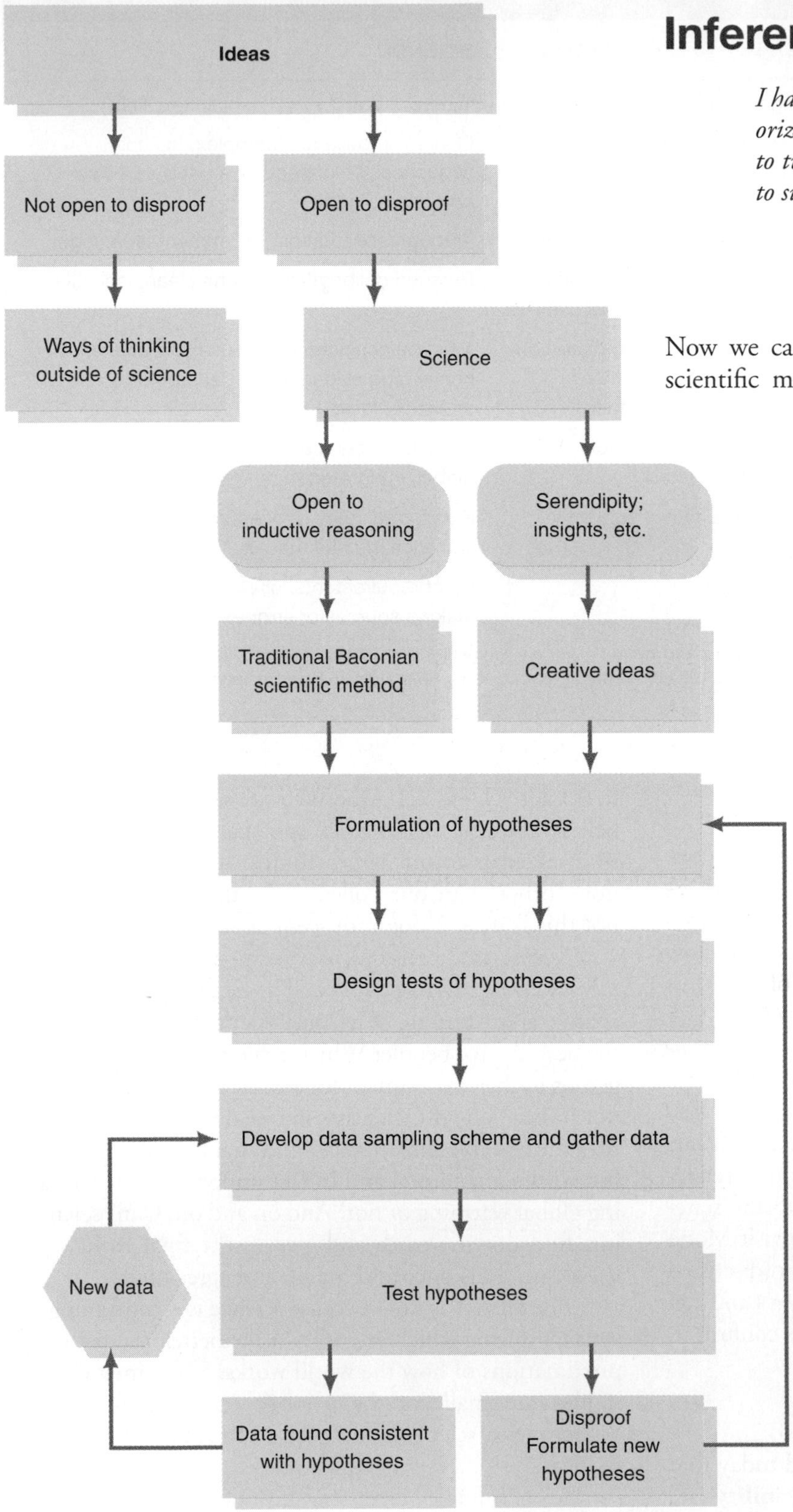

FIGURE 2.3 Schematic diagram of the scientific method. This diagram shows the steps in the scientific method, both traditional and nontraditional, as explained in the text.

2.2 Observations, Facts, Inferences, and Hypotheses

I have no data yet. It is a capital mistake to theorize before one has data. Insensibly one begins to twist facts to suit theories, instead of theories to suit facts.

—Sherlock Holmes,
in Sir Arthur Conan Doyle's
A Scandal in Bohemia

Now we can turn to the specific characteristics of the scientific method. (The steps in the scientific method are shown in Table 2.2.) It is important to distinguish between observations and inferences. **Observations**, the basis of science, may be made through any of the five senses or by instruments that measure beyond what we can sense. **Inferences** are generalizations that arise from a set of observations. When everyone or almost everyone agrees with what is observed about a particular thing, the inference is often called a **fact**.

We might *observe* that a substance is a white, crystalline material with a sweet taste. We might *infer* from these observations alone that the substance is sugar. Before this inference can be accepted as fact, however, it must be subjected to further tests. Confusing observations with inferences and accepting untested inferences as facts are kinds of sloppy thinking described as "Thinking makes it so." When scientists wish to test an inference, they convert it into a **hypothesis**, which is a statement that can be disproved. The hypothesis continues to be accepted until it is disproved.

For example, a scientist is trying to understand how a plant's growth will change with the amount of light it receives. She proposed a hypothesis that a plant can use only so much light and no more—it can be "saturated" by an abundance of light. She measures the rate of photosynthesis at a variety of light intensities. The rate of photosynthesis is called the **dependent variable** because it is affected by, and in this sense depends on, the amount of light, which is called the **independent variable**. The independent variable is also sometimes called a **manipulated variable**

Table 2.2 STEPS IN THE SCIENTIFIC METHOD (TERMS USED HERE ARE DEFINED IN THE TEXT.)

1. Make observations and develop a question about the observations.
2. Develop a tentative answer to the question—a hypothesis.
3. Design a controlled experiment to test the hypothesis (implies identifying and defining independent and dependent variables).
4. Collect data in an organized form, such as a table.
5. Interpret the data visually (through graphs), quantitatively (using statistical analysis) and/or by other means.
6. Draw a conclusion from the data.
7. Compare the conclusion with the hypothesis and determine whether the results support or disprove the hypothesis.
8. If the hypothesis is consistent with observations in some limited experiments, conduct additional experiments to test it further. If the hypothesis is rejected, make additional observations and construct a new hypothesis.

because it is deliberately changed, or manipulated, by the scientist. The dependent variable is then referred to as a **responding variable**—one that responds to changes in the manipulated variable. These values are referred to as *data* (singular: *datum*). They may be numerical, **quantitative data**, or nonnumerical, **qualitative data**. In our example, qualitative data would be the species of a plant; quantitative data would be the tree's mass in grams or the diameter in centimeters. The result of the scientist's observations: The hypothesis is confirmed: The rate of photosynthesis increases to a certain level and does not go higher at higher light intensities (Figure 2.4).

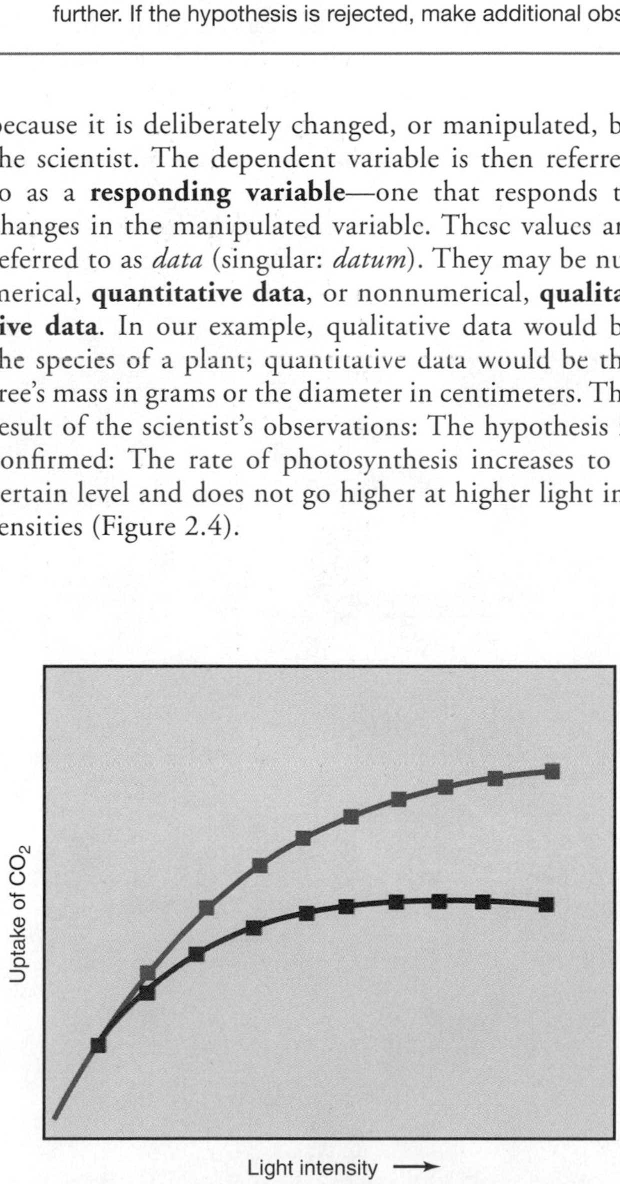

FIGURE 2.4 **Dependent and independent variables: Photosynthesis as affected by light.** In this diagram, photosynthesis is represented by carbon dioxide (CO_2) uptake. Light is the independent variable, uptake is the dependent variable. The blue and red lines represent two plants with different responses to light.

Controlling Variables

In testing a hypothesis, a scientist tries to keep all relevant **variables** constant except for the independent and dependent variables. This practice is known as *controlling variables*. In a **controlled experiment**, the experiment is compared to a standard, or control—an exact duplicate of the experiment except for the one variable being tested (the **independent variable**). Any difference in outcome (dependent variable) between the experiment and the control can be attributed to the effect of the independent variable.

An important aspect of science, but one frequently overlooked in descriptions of the scientific method, is the need to define or describe variables in exact terms that all scientists can understand. The least ambiguous way to define or describe a variable is in terms of what one would have to do to duplicate the measurement of that variable. Such definitions are called **operational definitions**. Before carrying out an experiment, both the independent and dependent variables must be defined operationally. Operational definitions allow other scientists to repeat experiments exactly and to check on the results reported.

Science is based on **inductive reasoning**, also called *induction*: It begins with specific observations and then extends to generalizations, which may be disproved by testing them. If such a test cannot be devised, then we cannot treat the generalization as a scientific statement. Although new evidence can disprove existing scientific theories, science can never provide absolute proof of the truth of its theories.

The Nature of Scientific Proof

One source of serious misunderstanding about science is the use of the word *proof*, which most students encounter in mathematics, particularly in geometry. Proof in mathematics and logic involves reasoning from initial definitions and assumptions. If a conclusion follows logically from these assumptions, or premises, we say it is proven. This process is known as **deductive reasoning**. An example of deductive reasoning is the following syllogism, or series of logically connected statements:

Premise: A straight line is the shortest distance between two points.

Premise: The line from A to B is the shortest distance between points A and B.

Conclusion: Therefore, the line from A to B is a straight line.

Note that the conclusion in this syllogism follows directly from the premises.

Deductive proof does not require that the premises be true, only that the reasoning be foolproof. Statements that are logically valid but untrue can result from false premises, as in the following example (Figure 2.5):

Premise: Humans are the only toolmaking organisms.

Premise: The woodpecker finch uses tools.

Conclusion: Therefore, the woodpecker finch is a human being.

In this case, the concluding statement must be true if both of the preceding statements are true. However, we know that the conclusion is not only false but ridiculous. If the second statement is true (which it is), then the first cannot be true.

The rules of deductive reasoning govern only the process of moving from premises to conclusion. *Science, in contrast, requires not only logical reasoning but also correct premises.* Returning to the example of the woodpecker finch, to be scientific the three statements should be expressed conditionally (that is, with reservation):

If humans are the only toolmaking organisms

and

the woodpecker finch is a toolmaker,

then

the woodpecker finch is a human being.

When we formulate generalizations based on a number of observations, we are engaging in inductive reasoning. To illustrate: One of the birds that feeds at Mono Lake is the eared grebe. The "ears" are a fan of golden feathers that occur behind the eyes of males during the breeding season. Let us define birds with these golden feather fans as eared grebes (Figure 2.6). If we always observe that the breeding male grebes have this feather fan, we may make the inductive statement "All male eared grebes have golden feathers during the breeding season." What we really mean is "All of the male eared grebes *we*

FIGURE 2.5 A woodpecker finch in the Galápagos Islands uses a twig to remove insects from a hole in a tree, demonstrating tool use by nonhuman animals. Because science is based on observations, its conclusions are only as true as the premises from which they are deduced.

FIGURE 2.6 Male eared grebe in breeding season.

have seen in the breeding season have golden feathers." We never know when our very next observation will turn up a bird that is like a male eared grebe in all ways except that it lacks these feathers in the breeding season. This is not impossible; it could occur somewhere due to a mutation.

Proof in inductive reasoning is therefore very different from proof in deductive reasoning. When we say something is proven in induction, what we really mean is that it has a very high degree of probability. Probability is a way of expressing our certainty (or uncertainty)—our estimation of how good our observations are, how confident we are of our predictions.

Theory in Science and Language

A common misunderstanding about science arises from confusion between the use of the word *theory* in science and its use in everyday language. A **scientific theory** is a grand scheme that relates and explains many observations and is supported by a great deal of evidence. In contrast, in everyday usage a theory can be a guess, a hypothesis, a prediction, a notion, a belief. We often hear the phrase "It's just a theory." That may make sense in everyday conversation but not in the language of science. In fact, theories have tremendous prestige and are considered the greatest achievements of science.[3]

Further misunderstanding arises when scientists use the word *theory* in several different senses. For example, we may encounter references to a currently accepted, widely supported theory, such as the theory of evolution by natural selection; a discarded theory, such as the theory of inheritance of acquired characteristics; a new theory, such as the theory of evolution of multicellular organisms by symbiosis; and a model dealing with a specific or narrow area of science, such as the theory of enzyme action.[4]

One of the most important misunderstandings about the scientific method pertains to the relationship between research and theory. Theory is usually presented as growing out of research, but in fact theories also guide research. When a scientist makes observations, he or she does so in the context of existing theories. At times, discrepancies between observations and accepted theories become so great that a scientific revolution occurs: The old theories are discarded and are replaced with new or significantly revised theories.[5]

Knowledge in an area of science grows as more hypotheses are supported. Ideally, scientific hypotheses are continually tested and evaluated by other scientists, and this provides science with a built-in self-correcting feedback system. This is an important, fundamental feature of the scientific method. If you are told that scientists have reached a consensus about something, you want to check carefully to see if this feedback process has been used correctly and is still possible. If not, what began as science can be converted to ideology—a way that certain individuals, groups, or cultures may think despite evidence to the contrary.

Models and Theory

Scientists use accumulated knowledge to develop explanations that are consistent with currently accepted hypotheses. Sometimes an explanation is presented as a model. A **model** is "a deliberately simplified construct of nature."[6] It may be a physical working model, a pictorial model, a set of mathematical equations, or a computer simulation. For example, the U.S. Army Corps of Engineers has a physical model of San Francisco Bay. Open to the public to view, it is a miniature in a large aquarium with the topography of the bay reproduced to scale and with water flowing into it in accordance with tidal patterns. Elsewhere, the Army Corps develops mathematical equations and computer simulations, which are models and attempt to explain some aspects of such water flow.

As new knowledge accumulates, models may no longer be consistent with observations and may have to be revised or replaced, with the goal of finding models more consistent with nature.[5] Computer simulation of the atmosphere has become important in scientific analysis of the possibility of global warming. Computer simulation is becoming important for biological systems as well, such as simulations of forest growth (Figure 2.7).

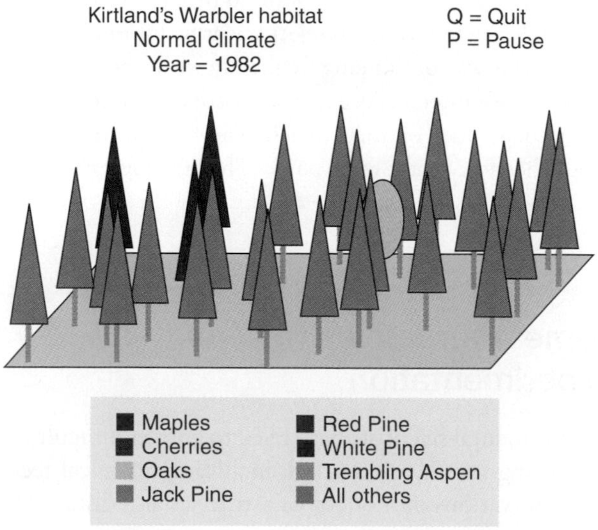

FIGURE 2.7 A computer simulation of forest growth. Shown here is a screen display of individual trees whose growth is forecast year by year, depending on environmental conditions. In this computer run, only three types of trees are present. This kind of model is becoming increasingly important in environmental sciences. (*Source: JABOWA-II* by D.B. Botkin. Copyright © 1993, 2009 by D.B. Botkin.)

A CLOSER LOOK 2.1

The Case of the Mysterious Crop Circles

For 13 years, circular patterns appeared "mysteriously" in grainfields in southern England (Figure 2.8). Proposed explanations included aliens, electromagnetic forces, whirlwinds, and pranksters. The mystery generated a journal and a research organization headed by a scientist, as well as a number of books, magazines, and clubs devoted solely to crop circles. Scientists from Great Britain and Japan brought in scientific equipment to study the strange patterns. Then, in September 1991, two men confessed to having created the circles by entering the fields along paths made by tractors (to disguise their footprints) and dragging planks through the fields. When they made their confession, they demonstrated their technique to reporters and some crop-circle experts.[4, 5]

Despite their confession, some people still believe that the crop circles were caused by something else, and crop-circle organizations not only still exist but also now have Web sites. One report published on the World Wide Web in 2003 stated that "strange orange lightning" was seen one evening and that crop circles appeared the next day.[7, 8]

How is it that so many people, including some scientists, still take those English crop circles seriously? Probably some of these people misunderstand the scientific method and used it incorrectly—and some simply want to believe in a mysterious cause and therefore chose to reject scientific information. We run into this way of thinking frequently with environmental issues. People often believe that some conclusions or some action is good, based on their values. They wish it were so, and decide therefore that it must be so. The false logic here can be phrased: *If it sounds good, it must be good, and if it must be good, we must make it happen.*

(a)

(b)

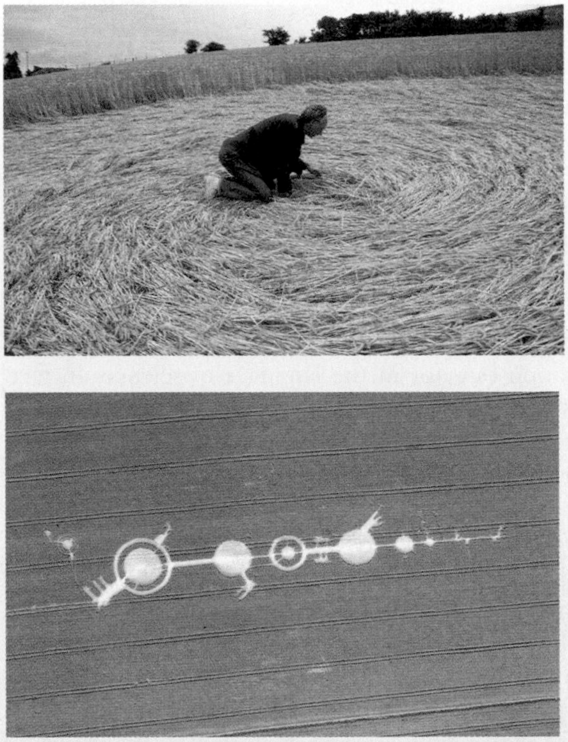

FIGURE 2.8 **(a)** A crop circle close up at the Vale of Pewsey in southern England in July 1990. **(b)** Crop circles seen from the air make distinctive patterns.

Some Alternatives to Direct Experimentation

Environmental scientists have tried to answer difficult questions using several approaches, including historical records and observations of modern catastrophes and disturbances.

Historical Evidence

Ecologists have made use of both human and ecological historical records. A classic example is a study of the history of fire in the Boundary Waters Canoe Area (BWCA) of Minnesota, 1 million acres of boreal forests, streams, and lakes well known for recreational canoeing.

Murray ("Bud") Heinselman had lived near the BWCA for much of his life and was instrumental in having it declared a wilderness area. A forest ecological scientist, Heinselman set out to determine the past patterns of fires in this wilderness. Those patterns are important in maintaining the wilderness. If the wilderness has been characterized by fires of a specific frequency, then one can argue that this frequency is necessary to maintain the area in its most "natural" state.

Heinselman used three kinds of historical data: written records, tree-ring records, and buried records (fossil and prefossil organic deposits). Trees of the boreal forests, like most trees that are conifers or angiosperms (flowering plants),

FIGURE 2.9 Cross section of a tree showing fire scars and tree rings. Together these allow scientists to date fires and to average the time between fires.

produce annual growth rings. If a fire burns through the bark of a tree, it leaves a scar, just as a serious burn leaves a scar on human skin. The tree grows over the scar, depositing a new growth ring for each year. (Figure 2.9 shows fire scars and tree rings on a cross section of a tree.) By examining cross sections of trees, it is possible to determine the date of each fire and the number of years between fires. From written and tree-ring records, Heinselman found that the frequency of fires had varied over time but that since the 17th century the BWCA forests had burned, on average, once per century. Furthermore, buried charcoal dated using carbon-14 revealed that fires could be traced back more than 30,000 years.[9]

The three kinds of historical records provided important evidence about fire in the history of the BWCA. At the time Heinselman did his study, the standard hypothesis was that fires were bad for forests and should be suppressed. The historical evidence provided a disproof of this hypothesis. It showed that fires were a natural and an integral part of the forest and that the forest had persisted with fire for a very long time. Thus, the use of historical information meets the primary requirement of the scientific method—the ability to disprove a statement. Historical evidence is a major source of data that can be used to test scientific hypotheses in ecology.

Modern Catastrophes and Disturbances as Experiments

Sometimes a large-scale catastrophe provides a kind of modern ecological experiment. The volcanic eruption of Mount St. Helens in 1980 supplied such an experiment, destroying vegetation and wildlife over a wide area. The recovery of plants, animals, and ecosystems following this explosion gave scientists insights into the dynamics of ecological systems and provided some surprises. The main surprise was how quickly vegetation recovered and wildlife returned to parts of the mountain. In other ways, the recovery followed expected patterns in ecological succession (see Chapters 5 and 12).

It is important to point out that the greater the quantity and the better the quality of ecological data prior to such a catastrophe, the more we can learn from the response of ecological systems to the event. This calls for careful monitoring of the environment.

Uncertainty in Science

In science, when we have a fairly high degree of confidence in our conclusions, we often forget to state the degree of certainty or uncertainty. Instead of saying, "There is a 99.9% probability that . . . ," we say, "It has been proved that . . ." Unfortunately, many people interpret this as a deductive statement, meaning the conclusion is absolutely true, which has led to much misunderstanding about science. Although science begins with observations and therefore inductive reasoning, deductive reasoning is useful in helping scientists analyze whether conclusions based on inductions are logically valid. *Scientific reasoning combines induction and deduction*—different but complementary ways of thinking.

Leaps of Imagination and Other Nontraditional Aspects of the Scientific Method

What we have described so far is the classic scientific method. Scientific advances, however, often happen somewhat differently. They begin with instances of insight—leaps of imagination that are then subjected to the stepwise inductive process. And some scientists have made major advances by being in the right place at the right time, noticing interesting oddities, and knowing how to put these clues together. For example, penicillin was discovered "by accident" in 1928 when Sir Alexander Fleming was studying the pus-producing bacterium *Staphylococcus aureus*. When a culture of these bacteria was accidentally contaminated by the green fungus *Penicillium notatum*, Fleming noticed that the bacteria did not grow in areas of the culture where the fungus grew. He isolated the mold, grew it in a fluid medium, and found that it produced a substance that killed many of the bacteria that caused diseases. Eventually this discovery led other scientists to develop an injectable agent to treat diseases. *Penicillium notatum* is a common mold found on stale bread. No doubt many others had seen it, perhaps even noticing that other strange growths on bread did not overlap with *Penicillium notatum*. But it took Fleming's knowledge and observational ability for this piece of "luck" to occur.

2.3 Measurements and Uncertainty

A Word about Numbers in Science

We communicate scientific information in several ways. The written word is used for conveying synthesis, analysis, and conclusions. When we add numbers to our analysis, we obtain another dimension of understanding that goes beyond qualitative understanding and synthesis of a problem. Using numbers and statistical analysis allows us to visualize relationships in graphs and make predictions. It also allows us to analyze the strength of a relationship and in some cases discover a new relationship.

People in general put more faith in the accuracy of measurements than do scientists. Scientists realize that all measurements are only approximations, limited by the accuracy of the instruments used and the people who use them. Measurement uncertainties are inevitable; they can be reduced but never completely eliminated. For this reason, *a measurement is meaningless unless it is accompanied by an estimate of its uncertainty.*

Consider the loss of the *Challenger* space shuttle in 1986, the first major space shuttle accident, which appeared to be the result of the failure of rubber O-rings that were supposed to hold sections of rockets together. Imagine a simplified scenario in which an engineer is given a rubber O-ring used to seal fuel gases in a space shuttle. The engineer is asked to determine the flexibility of the O-rings under different temperature conditions to help answer two questions: At what temperature do the O-rings become brittle and subject to failure? And at what temperature(s) is it unsafe to launch the shuttle? After doing some tests, the engineer says that the rubber becomes brittle at –1°C (30°F). So, can you assume it is safe to launch the shuttle at 0°C (32°F)?

At this point, you do not have enough information to answer the question. You assume that the temperature data may have some degree of uncertainty, but you have no idea how great a degree. Is the uncertainty ±5°C, ±2°C, or ±0.5°C? To make a reasonably safe and economically sound decision about whether to launch the shuttle, you must know the amount of uncertainty of the measurement.

Dealing with Uncertainties

There are two sources of uncertainty. One is the real variability of nature. The other is the fact that every measurement has some error. Measurement uncertainties and other errors that occur in experiments are called **experimental errors.** Errors that occur consistently, such as those resulting from incorrectly calibrated instruments, are **systematic errors.**

Scientists traditionally include a discussion of experimental errors when they report results. Error analysis often leads to greater understanding and sometimes even to important discoveries. For example, scientists discovered the eighth planet in our solar system, Neptune, when they investigated apparent inconsistencies—observed "errors"—in the orbit of the seventh planet, Uranus.

We can reduce measurement uncertainties by improving our measurement instruments, standardizing measurement procedures, and using carefully designed experiments and appropriate statistical procedures. Even then, however, uncertainties can never be completely eliminated. Difficult as it is for us to live with uncertainty, that is the nature of nature, as well as the nature of measurement and of science. Our awareness of these uncertainties should lead us to read reports of scientific studies critically, whether they appear in science journals or in popular magazines and newspapers. (See A Closer Look 2.2.)

Accuracy and Precision

A friend inherited some land on an island off the coast of Maine. However, the historical records were unclear about the land's boundaries, and to sell any portion of the land, he first had to determine where his neighbor's land ended and his began. There were differences of opinion about this. In fact, some people said one boundary went right through the house, which would have caused a lot of problems! Clearly what was needed was a good map that everybody could agree on, so our friend hired a surveyor to determine exactly where the boundaries were.

The original surveyor's notes from the early 19th century had vague guidelines, such as "beginning at the mouth of Marsh brook on the Eastern side of the bars at a stake and stones. . . thence running South twenty six rods to a stake & stones. . . ." Over time, of course, the shore, the brook, its mouth, and the stones had moved and the stakes had disappeared. The surveyor was clear about the total distance (a rod, by the way, is an old English measure equal to 16.5 feet or 5.02 meters), but "South" wasn't very specific. So where and in exactly which direction was the true boundary? (This surveyor's method was common in early-19th-century New England. One New Hampshire survey during that time began with "Where you and I were standing yesterday . . ." Another began, "Starting at the hole in the ice [on the pond] . . .").

The 21st-century surveyor who was asked to find the real boundary used the most modern equipment—laser and microwave surveying transits, GPS devices—so he knew where the line he measured went to in millimeters. He could remeasure his line and come within

A CLOSER LOOK 2.2

Measurement of Carbon Stored in Vegetation

A number of people have suggested that a partial solution to global warming might be a massive worldwide program of tree planting. Trees take carbon dioxide (an important greenhouse gas) out of the air in the process of photosynthesis. And because trees live a long time, they can store carbon for decades, even centuries. But how much carbon can be stored in trees and in all perennial vegetation? Many books and reports published during the past 20 years contained numbers representing the total stored carbon in Earth's vegetation, but all were presented without any estimate of error (Table 2.3). Without an estimate of that uncertainty, the figures are meaningless, yet important environmental decisions have been based on them.

Recent studies have reduced error by replacing guesses and extrapolations with scientific sampling techniques similar to those used to predict the outcomes of elections. Even these improved data would be meaningless, however, without an estimate of error. The new figures show that the earlier estimates were three to four times too large, grossly overestimating the storage of carbon in vegetation and therefore the contribution that tree planting could make in offsetting global warming.

Table 2.3 ESTIMATES OF ABOVEGROUND BIOMASS IN NORTH AMERICAN BOREAL FOREST

SOURCE	BIOMASS[a] (kg/m^2)	CARBON[b] (kg/m^2)	TOTAL BIOMASS[c] (10^9 metric tons)	TOTAL CARBON[c] (10^9 metric tons)
This study[d]	4.2 ± 1.0	1.9 ± 0.4	22 ± 5	9.7 ± 2
Previous estimates[e]				
1	17.5	7.9	90	40
2	15.4	6.9	79	35
3	14.8	6.7	76	34
4	12.4	5.6	64	29
5	5.9	2.7	30	13.8

Source: D.B. Botkin and L. Simpson, "The First Statistically Valid Estimate of Biomass for a Large Region," *Biogeochemistry* 9 (1990): 161–274. Reprinted by permission of Klumer Academic, Dordrecht, The Netherlands.

[a]Values in this column are for total aboveground biomass. Data from previous studies giving total biomass have been adjusted using the assumption that 23% of the total biomass is in below-ground roots. Most references use this percentage; Leith and Whittaker use 17%. We have chosen to use the larger value to give a more conservative comparison.

[b]Carbon is assumed to be 45% of total biomass following R.H. Whittaker, *Communities and Ecosystems* (New York: Macmillan, 1974).

[c]Assuming our estimate of the geographic extent of the North American boreal forest: 5,126,427 km^2 (324,166 mi^2).

[d]Based on a statistically valid survey; aboveground woodplants only.

[e]Lacking estimates of error: Sources of previous estimates by number (1) G.J. Ajtay, P. Ketner, and P. Duvigneaud, "Terrestial Primary Production and Phytomass," in B. Bolin, E.T. Degens, S. Kempe, and P. Ketner, eds., *The Global Carbon Cycle* (New York: Wiley, 1979), pp. 129–182. (2) R.H. Whittaker and G.E. Likens, "Carbon in the Biota," in G.M. Woodwell and E.V. Pecam, eds., *Carbon and the Biosphere* (Springfield, VA: National Technical Information Center, 1973), pp. 281–300. (3) J.S. Olson, H.A. Pfuderer, and Y.H. Chan, *Changes in the Global Carbon Cycle and the Biosphere*, ORNL/EIS-109 (Oak Ridge, TN: Oak Ridge National Laboratory, 1978). (4) J.S. Olson, I.A. Watts, and L.I. Allison, *Carbon in Live Vegetation of Major World Ecosystems*, ORNL-5862 (Oak Ridge, TN: Oak Ridge National Laboratory, 1983). (5) G.M. Bonnor, *Inventory of Forest Biomass in Canada* (Petawawa, Ontario: Canadian Forest Service, Petawawa National Forest Institute, 1985).

millimeters of his previous location. But because the original starting point couldn't be determined within many meters, the surveyor didn't know where the true boundary line went; it was just somewhere within 10 meters or so of the line he had surveyed. So the end result was that even after this careful, modern, high-technology survey, nobody really knew where the original boundary lines went. Scientists would say that the modern surveyor's work was precise but not accurate. *Accuracy* refers to what we know; *precision* to how well we measure. With such things as this land survey, this is an important difference.

Accuracy also has another, slightly different scientific meaning. In some cases, certain measurements have been made very carefully by many people over a long period, and accepted values have been determined. In that kind of situation, *accuracy* means the extent to which a measurement agrees with the accepted value. But as before, *precision* retains its original meaning, the degree of exactness with which a quantity is measured. In the case of the land in Maine, we can say that the new measurement had no accuracy in regard to the previous ("accepted") value.

Although a scientist should make measurements as precisely as possible, this friend's experience with surveying his land shows us that it is equally important not to report measurements with more precision than they warrant. Doing so conveys a misleading sense of both precision and accuracy.

2.4 Misunderstandings about Science and Society

Science and Decision Making

Like the scientific method, the process of making decisions is sometimes presented as a series of steps:

1. Formulate a clear statement of the issue to be decided.
2. Gather the scientific information related to the issue.
3. List all alternative courses of action.
4. Predict the positive and negative consequences of each course of action and the probability that each consequence will occur.
5. Weigh the alternatives and choose the best solution.

Such a procedure is a good guide to rational decision making, but it assumes a simplicity not often found in real-world issues. It is difficult to anticipate all the potential consequences of a course of action, and unintended consequences are at the root of many environmental problems. Often the scientific information is incomplete and even controversial. For example, the insecticide DDT causes eggshells of birds that feed on insects to be so thin that unhatched birds die. When DDT first came into use, this consequence was not predicted. Only when populations of species such as the brown pelican became seriously endangered did people become aware of it.

In the face of incomplete information, scientific controversies, conflicting interests, and emotionalism, how can we make sound environmental decisions? We need to begin with the scientific evidence from all relevant sources and with estimates of the uncertainties in each. Avoiding emotionalism and resisting slogans and propaganda are essential to developing sound approaches to environmental issues. Ultimately, however, environmental decisions are policy decisions negotiated through the political process. Policymakers are rarely professional scientists; generally, they are political leaders and ordinary citizens. Therefore, the scientific education of those in government and business, as well as of all citizens, is crucial.

Science and Technology

Science is often confused with technology. As noted earlier, science is a search for understanding of the natural world, whereas technology is the application of scientific knowledge in an attempt to benefit people. Science often leads to technological developments, just as new technologies lead to scientific discoveries. The telescope began as a technological device, such as an aid to sailors, but when Galileo used it to study the heavens, it became a source of new scientific knowledge. That knowledge stimulated the technology of telescope-making, leading to the production of better telescopes, which in turn led to further advances in the science of astronomy.

Science is limited by the technology available. Before the invention of the electron microscope, scientists were limited to magnifications of 1,000 times and to studying objects about the size of one-tenth of a micrometer. (A micrometer is 1/1,000,000 of a meter, or 1/1,000 of a millimeter.) The electron microscope enabled scientists to view objects far smaller by magnifying more than 100,000 times. The electron microscope, a basis for new science, was also the product of science. Without prior scientific knowledge about electron beams and how to focus them, the electron microscope could not have been developed.

Most of us do not come into direct contact with science in our daily lives; instead, we come into contact with the products of science—technological devices such as computers, iPods, and microwave ovens. Thus, people tend to confuse the products of science with science itself. As you study science, it will help if you keep in mind the distinction between science and technology.

Science and Objectivity

One myth about science is the myth of objectivity, or value-free science—the notion that scientists are capable of complete objectivity independent of their personal values and the culture in which they live, and that science deals only with objective facts. Objectivity is certainly a goal of scientists, but it is unrealistic to think they can be totally free of influence by their social environments and personal values. It would be more realistic to admit that scientists do have biases and to try to identify these biases rather than deny or ignore them. In some ways, this situation is similar to that of measurement error: It is inescapable, and we can best deal with it by recognizing it and estimating its effects.

To find examples of how personal and social values affect science, we have only to look at recent controversies about environmental issues, such as whether or not to adopt more stringent automobile emission standards. Genetic engineering, nuclear power, global warming, and the preservation of threatened or endangered species involve conflicts among science, technology, and society. When we function as *scientists* in society, we want to explain the results of science objectively. As citizens who are not scientists, we want scientists to always be objective and tell us the truth about their scientific research.

That science is not entirely value-free should not be taken to mean that fuzzy thinking is acceptable in science. It is still important to think critically and logically about science and related social issues. Without the high standards of evidence held up as the norm for science, we run the risk of accepting unfounded ideas about the world. When we confuse what we would like to believe with what we have the evidence to believe, we have a weak basis for making critical environmental decisions that could have far-reaching and serious consequences.

The great successes of science, especially as the foundation for so many things that benefit us in modern technological societies—from cell phones to CAT scans to space exploration—give science and scientists a societal authority that makes it all the more difficult to know when a scientist might be exceeding the bounds of his or her scientific knowledge. It may be helpful to realize that scientists play three roles in our society: first, as researchers simply explaining the results of their work; second, as almost priestlike authorities who often seem to speak in tongues the rest of us can't understand; and third, as what we could call expert witnesses. In this third role, they will discuss broad areas of research that they are familiar with and that are within their field of study, but about which they may not have done research themselves. Like an expert testifying in court, they are basically saying to us, "Although I haven't done this particular research myself, my experience and knowledge suggest to me that . . ."

The roles of researcher and expert witness are legitimate as long as it is clear to everybody which role a scientist is playing. Whether you want a scientist to be your authority about everything, within science and outside of science, is a personal and value choice.

In the modern world, there is another problem about the role of scientists and science in our society. Science has been so potent that it has become fundamental to political policies. As a result, science can become politicized, which means that rather than beginning with objective inquiry, people begin with a belief about something and pick and choose only the scientific evidence that supports that belief. This can even be carried to the next step, where research is funded only if it fits within a political or an ethical point of view.

Scientists themselves, even acting as best they can *as* scientists, can be caught up in one way of thinking when the evidence points to another. These scientists are said to be working under a certain paradigm, a particular theoretical framework. Sometimes their science undergoes a paradigm shift: New scientific information reveals a great departure from previous ways of thinking and from previous scientific theories, and it is difficult, after working within one way of thinking, to recognize that some or all of their fundamentals must change. Paradigm shifts happen over and over again in science and lead to exciting and often life-changing results for us. The discovery and understanding of electricity are examples, as is the development of quantum mechanics in physics in the early decades of the 20th century.

We can never completely escape biases, intentional and unintentional, in fundamental science, its interpretation, and its application to practical problems, but understanding the nature of the problems that can arise can help us limit this misuse of science. The situation is complicated by legitimate scientific uncertainties and differences in scientific theories. It is hard for us, as citizens, to know when scientists are having a legitimate debate about findings and theories, and when they are disagreeing over personal beliefs and convictions that are outside of science. Because environmental sciences touch our lives in so many ways, because they affect things that are involved with choices and values, and because these sciences deal with phenomena of great complexity, the need to understand where science can go astray is especially important.

Science, Pseudoscience, and Frontier Science

Some ideas presented as scientific are in fact not scientific, because they are inherently untestable, lack empirical support, or are based on faulty reasoning or poor scientific methodology, as illustrated by the case of the mysterious crop circles (A Closer Look 2.1). Such ideas are referred to as pseudoscientific (the prefix *pseudo-* means false).

CRITICAL THINKING ISSUE

How Do We Decide What to Believe about Environmental Issues?

When you read about an environmental issue in a newspaper or magazine, how do you decide whether to accept the claims made in the article? Are they based on scientific evidence, and are they logical? Scientific evidence is based on observations, but media accounts often rely mainly on inferences (interpretations) rather than evidence. Distinguishing inferences from evidence is an important first step in evaluating articles critically. Second, it is important to consider the source of a statement. Is the source a reputable scientific organization or publication? Does the source have a vested interest that might bias the claims? When sources are not named, it is impossible to judge the reliability of claims. If a claim is based on scientific evidence presented logically from a reliable, unbiased source, it is appropriate to accept the claim tentatively, pending further information. Practice your critical evaluation skills by reading the article below and answering the critical thinking questions.

Critical Thinking Questions

1. What is the major claim made in the article?
2. What evidence does the author present to support the claim?
3. Is the evidence based on observations, and is the source of the evidence reputable and unbiased?
4. Is the argument for the claim, whether or not based on evidence, logical?
5. Would you accept or reject the claim?
6. Even if the claim were well supported by evidence based on good authority, why would your acceptance be only tentative?

CLUE FOUND IN DEFORMED FROG MYSTERY

BY MICHAEL CONLON

Reuters News Agency (as printed in the *Toronto Star*)
November 6, 1996

A chemical used for mosquito control could be linked to deformities showing up in frogs across parts of North America, though the source of the phenomenon remains a mystery. "We're still at the point where we've got a lot of leads that we're trying to follow but no smoking gun," says Michael Lannoo of Ball State University in Muncie, Ind. "There are an enormous number of chemicals that are being applied to the environment and we don't understand what the breakdown products of these chemicals are," says Lannoo, who heads the U.S. section of the worldwide Declining Amphibian Population Task Force.

He says one suspect chemical was methoprene, which produces a breakdown product resembling retinoic acid, a substance important in development. "Retinoic acid can produce in the laboratory all or a majority of the limb deformities that we're seeing in nature," he says. "That's not to say that's what's going on. But it is the best guess as to what's happening." Methoprene is used for mosquito control, among other things, Lannoo says.

Both the decline in amphibian populations and the deformities are of concern because frogs and related creatures are considered "sentinel" species that can provide early warnings of human risk. The skin of amphibians is permeable and puts them at particular risk to agents in the water.

Lannoo says limb deformities in frogs had been reported as far back as 1750, but the rate of deformities showing up today was unprecedented in some species. Some were showing abnormalities that affected more than half of the population of a species living in certain areas, he adds. He says he doubted that a parasite believed to have been the cause of some deformities in frogs in California was to blame for similar problems in Minnesota and nearby states. Deformed frogs have been reported in Minnesota, Wisconsin, Iowa, South Dakota, Missouri, California, Texas, Vermont, and Quebec. The deformities reported have included misshapen legs, extra limbs, and missing or misplaced eyes.

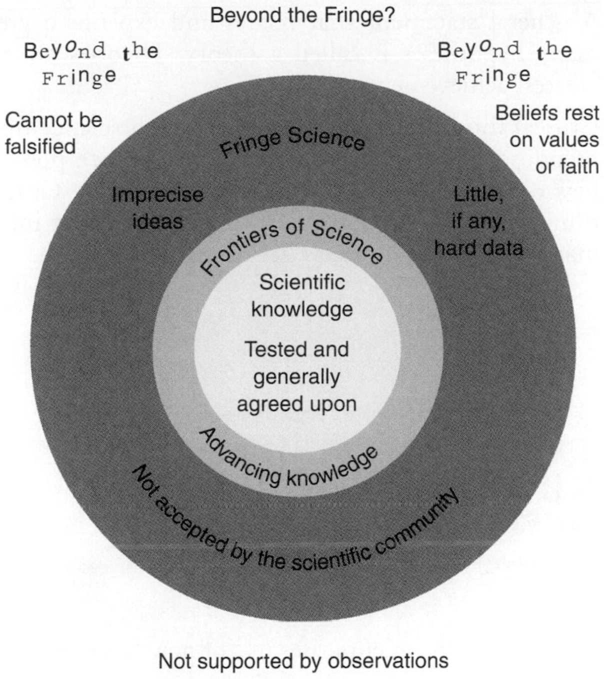

FIGURE 2.10 **Beyond the fringe? A diagrammatic view of different kinds of knowledge and ideas.**

Pseudoscientific ideas arise from various sources. With more research, however, some of the frontier ideas may move into the realm of accepted science, and new ideas will take their place at the advancing frontier (Figure 2.10).[10] Research may not support other hypotheses at the frontier, and these will be discarded. Accepted science may merge into frontier science, which in turn may merge into farther-out ideas, or fringe science. Really wild ideas may be considered beyond the fringe.

2.5 Environmental Questions and the Scientific Method

Environmental sciences deal with especially complex systems and include a relatively new set of sciences. Therefore, the process of scientific study has not always neatly followed the formal scientific method discussed earlier in this chapter. Often, observations are not used to develop formal hypotheses. Controlled laboratory experiments have been the exception rather than the rule. Much environmental research has been limited to field observations of processes and events that have been difficult to subject to controlled experiments.

Environmental research presents several obstacles to following the classic scientific method. The long time frame of many ecological processes relative to human lifetimes, professional lifetimes, and lengths of research grants poses problems for establishing statements that can in practice be subject to disproof. What do we do if a theoretical disproof through direct observation would take a century or more? Other obstacles include difficulties in setting up adequate experimental controls for field studies, in developing laboratory experiments of sufficient complexity, and in developing theory and models for complex systems. Throughout this text, we present differences between the "standard" scientific method and the actual approach that has been used in environmental sciences.

SUMMARY

- Science is one path to critical thinking about the natural world. Its goal is to gain an understanding of how nature works. Decisions on environmental issues must begin with an examination of the relevant scientific evidence. However, environmental decisions also require careful analysis of economic, social, and political consequences. Solutions will reflect religious, aesthetic, and ethical values as well.
- Science is an open-ended process of finding out about the natural world. In contrast, science lectures and texts are usually summaries of the answers arrived at through this process, and science homework and tests are exercises in finding the right answer. Therefore, students often perceive science as a body of facts to be memorized, and they view lectures and texts as authoritative sources of absolute truths about the world.
- Science begins with careful observations of the natural world, from which scientists formulate hypotheses. Whenever possible, scientists test hypotheses with controlled experiments.
- Although the scientific method is often taught as a prescribed series of steps, it is better to think of it as a general guide to scientific thinking, with many variations.
- We acquire scientific knowledge through inductive reasoning, basing general conclusions on specific observations. Conclusions arrived at through induction can never be proved with certainty. Thus, because of the inductive nature of science, it is possible to disprove hypotheses but not possible to prove them with 100% certainty.

- Measurements are approximations that may be more or less exact, depending on the measuring instruments and the people who use them. A measurement is meaningful when accompanied by an estimate of the degree of uncertainty, or error.
- Accuracy in measurement is the extent to which the measurement agrees with an accepted value. Precision is the degree of exactness with which a measurement is made. A precise measurement may not be accurate. The estimate of uncertainty provides information on the precision of a measurement.
- A general statement that relates and explains a great many hypotheses is called a theory. Theories are the greatest achievements of science.
- Critical thinking can help us distinguish science from pseudoscience. It can also help us recognize possible bias on the part of scientists and the media. Critical thinking involves questioning and synthesizing information rather than merely acquiring information.

REEXAMINING THEMES AND ISSUES

Global Perspective

The global perspective on environment arises out of new findings in environmental science.

Urban World

Our increasingly urbanized world is best understood with the assistance of scientific investigation.

People and Nature

Solutions to environmental problems require both values and knowledge. Understanding the scientific method is especially important if we are going to understand the connection between values and knowledge, and the relationship between people and nature. Ultimately, environmental decisions are policy decisions, negotiated through the political process. Policymakers often lack sufficient understanding of the scientific method, leading to false conclusions. Uncertainty is part of the nature of measurement and science. We must learn to accept uncertainty as part of our attempt to conserve and use our natural resources.

Science and Values

This chapter summarizes the scientific method, which is essential to analyzing and solving environmental problems and to developing sound approaches to sustainability.

KEY TERMS

controlled experiment **27**
deductive reasoning **28**
dependent variable **26**
disprovability **25**
experimental errors **32**
fact **26**
hypothesis **26**
independent variable **26**
inductive reasoning **27**
inferences **26**
manipulated variable **26**
model **29**
observations **26**
operational definitions **27**
qualitative data **27**
quantitative data **27**
responding variable **27**
scientific method **24**
scientific theory **29**
systematic errors **32**
variables **27**

STUDY QUESTIONS

1. Which of the following are scientific statements and which are not? What is the basis for your decision in each case?
   (a) The amount of carbon dioxide in the atmosphere is increasing.
   (b) Condors are ugly.
   (c) Condors are endangered.
   (d) Today there are 280 condors.
   (e) Crop circles are a sign from Earth to us that we should act better.
   (f) Crop circles can be made by people.
   (g) The fate of Mono Lake is the same as the fate of the Aral Sea.

2. What is the logical conclusion of each of the following syllogisms? Which conclusions correspond to observed reality?
   (a) All men are mortal. Socrates is a man.
   Therefore____________________________
   (b) All sheep are black. Mary's lamb is white.
   Therefore____________________________
   (c) All elephants are animals. All animals are living beings.
   Therefore____________________________

3. Which of the following statements are supported by deductive reasoning and which by inductive reasoning?
   (a) The sun will rise tomorrow.
   (b) The square of the hypotenuse of a right triangle is equal to the sum of the squares of the other two sides.
   (c) Only male deer have antlers.
   (d) If $A = B$ and $B = C$, then $A = C$.
   (e) The net force acting on a body equals its mass times its acceleration.

4. The accepted value for the number of inches in a centimeter is 0.3937. Two students mark off a centimeter on a piece of paper and then measure the distance using a ruler (in inches). Student A finds the distance equal to 0.3827 in., and student B finds it equal to 0.39 in. Which measurement is more accurate? Which is more precise? If student B measured the distance as 0.3900 in., what would be your answer?

5. (a) A teacher gives five students each a metal bar and asks them to measure the length. The measurements obtained are 5.03, 4.99, 5.02, 4.96, and 5.00 cm. How can you explain the variability in the measurements? Are these systematic or random errors?
   (b) The next day, the teacher gives the students the same bars but tells them that the bars have contracted because they have been in the refrigerator. In fact, the temperature difference would be too small to have any measurable effect on the length of the bars. The students' measurements, in the same order as in part (a), are 5.01, 4.95, 5.00, 4.90, and 4.95 cm. Why are the students' measurements different from those of the day before? What does this illustrate about science?

6. Identify the independent and dependent variables in each of the following:
   (a) Change in the rate of breathing in response to exercise.
   (b) The effect of study time on grades.
   (c) The likelihood that people exposed to smoke from other people's cigarettes will contract lung cancer.

7. (a) Identify a technological advance that resulted from a scientific discovery.
   (b) Identify a scientific discovery that resulted from a technological advance.
   (c) Identify a technological device you used today. What scientific discoveries were necessary before the device could be developed?

8. What is fallacious about each of the following conclusions?
   (a) A fortune cookie contains the statement "A happy event will occur in your life." Four months later, you find a $100 bill. You conclude that the prediction was correct.
   (b) A person claims that aliens visited Earth in prehistoric times and influenced the cultural development of humans. As evidence, the person points to ideas among many groups of people about beings who came from the sky and performed amazing feats.
   (c) A person observes that light-colored animals almost always live on light-colored surfaces, whereas dark forms of the same species live on dark surfaces. The person concludes that the light surface causes the light color of the animals.
   (d) A person knows three people who have had fewer colds since they began taking vitamin C on a regular basis. The person concludes that vitamin C prevents colds.

9. Find a newspaper article on a controversial topic. Identify some loaded words in the article—that is, words that convey an emotional reaction or a value judgment.

10. Identify some social, economic, aesthetic, and ethical issues involved in a current environmental controversy.

FURTHER READING

American Association for the Advancement of Science (AAAS), ***Science for All Americans*** (Washington, DC: AAAS, 1989). This report focuses on the knowledge, skills, and attitudes a student needs in order to be scientifically literate.

Botkin, D.B., ***No Man's Garden: Thoreau and a New Vision for Civilization and Nature*** (Washington, DC: Island Press, 2001). The author discusses how science can be applied to the study of nature and to problems associated with people and nature. He also discusses science and values.

Grinnell, F., ***The Scientific Attitude*** (New York: Guilford, 1992). The author uses examples from biomedical research to illustrate the processes of science (observing, hypothesizing, experimenting) and how scientists interact with each other and with society.

Kuhn, Thomas S., ***The Structure of Scientific Revolutions*** (Chicago: University of Chicago Press, 1996). This is a modern classic in the discussion of the scientific method, especially regarding major transitions in new sciences, such as environmental sciences.

McCain, G., and E.M. Segal, ***The Game of Science*** (Monterey, CA: Brooks/Cole, 1982). The authors present a lively look into the subculture of science.

Sagan, C., ***The Demon-Haunted World*** (New York: Random House, 1995). The author argues that irrational thinking and superstition threaten democratic institutions and discusses the importance of scientific thinking to our global civilization.

CHAPTER 3

The Big Picture: Systems of Change

The Missouri River at Sioux City is a complex system of water, sediment, animals, and fish, all affected by the city and its processes of runoff and river flood control, as well as upstream human intervention in the flow of the river.

LEARNING OBJECTIVES

In this book we discuss a wide range of phenomena. One thing that links them is that they are all part of complex systems. Systems have well-defined properties. Understanding these properties, common to so much of the environment, smooths our way to achieving an understanding of all aspects of environmental science. Changes in systems may occur naturally or may be induced by people, but a key to understanding these systems is that change in them is natural. After reading this chapter you should understand . . .

- Why solutions to many environmental problems involve the study of systems and rates of change;
- What feedback is, the difference between positive and negative feedback, and how these are important to systems;
- The difference between open and closed systems and between static and dynamic systems, and know which kind is characteristic of the environment and of life;
- What residence time is and how it is calculated;
- The principle of uniformitarianism and how it can be used to anticipate future changes;
- The principle of environmental unity and why it is important in studying environmental problems;
- Some helpful ways to think about systems when trying to solve environmental problems that arise from complex natural systems;
- What a stable system is and how this idea relates to the prescientific idea of a balance of nature.

CASE STUDY

Trying to Control Flooding of the Wild Missouri River

The Missouri River drains one-sixth of the United States (excluding Alaska and Hawaii) and flows for more than 3,200 km (2,000 miles). After the land along the Missouri was settled by Europeans, and after large towns and cities were built on the land near the river, flooding of the Missouri became a major problem. The "wild Missouri" became famous in history and folklore for its great fluctuations, its flows and droughts, and as the epitome of unpredictability in nature. One settler said that the Missouri "makes farming as fascinating as gambling. You never know whether you are going to harvest corn or catfish."[1,2]

Two of the river's great floods were in 1927 and 1993 (Figure 3.1). After the 1927 flood, the federal government commissioned the Army Corps of Engineers to build six major dams on the river (Figure 3.2). (The attempt to control the river's flow also included many other alterations of the river, such as straightening the channel and building levees.) Of the six dams, the three largest were built upstream, and each of their reservoirs was supposed to hold the equivalent of an entire year's average flow. The three smaller, downstream dams were meant to serve as safety valves to control the flow more precisely.

The underlying idea was to view the Missouri as a large plumbing system that needed management. When rainfall was sparse in the huge watershed of the river, the three upstream dams were supposed to be able to augment the flow for up to three years, ensuring a constant and adequate supply of water for irrigation and personal use. In flood years, the six dams were supposed to be able to store the dangerous flow, so that the water could be released slowly, the floods controlled, and the flow once again constant. In addition, levees—narrow ridges of higher ground—were built along the river and into it to protect the settled land along the river from floodwaters not otherwise contained. But these idealistic plans did not stop the Missouri from flooding in 1993 (Figures 3.1 and 3.3).

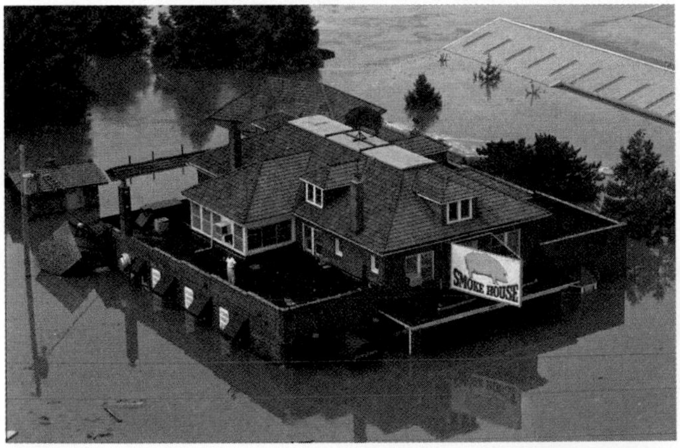

FIGURE 3.1 St. Louis, Missouri, during the 1993 flood of the Missouri River. No matter how hard we try to keep this huge river flowing at a fixed rate, neither flooding nor in drought, we always seem to fail. So it is when we try to tame most natural ecological and environmental systems that are naturally dynamic and always changing.

FIGURE 3.2 The six major dams on the Missouri River. (Based on drawing by Gary Pound from Daniel B. Botkin, *Passage of Discovery: The American Rivers Guide to the Missouri River of Lewis and Clark* [New York: Perigee Books, a division of Penguin-Putnam, 1999].)

Taking the large view, standing way back from the river, this perception of the Missouri River was akin to thinking about it as one huge lake (Figure 3.4) into which water flowed, then drained downstream and out at its mouth at St. Louis, Missouri, into the Mississippi, which carried the waters to New Orleans, Louisiana, and out into the Gulf of Mexico. The hope was that the Missouri River could be managed the way we manage our bathwater—keeping it at a constant level by always matching the outflow down the drain with inflow from the spigot. This is a perception of the river as a system held in *steady state,* a term we will define shortly.

Before there were permanent settlements along the river—both by American Indians and by Europeans—the Missouri's flooding didn't matter. Nomadic peoples could move away from the river when it flooded during the rainy seasons, and wildlife and vegetation generally benefited from the variations in water flow, as will be explained in later chapters. Only with modern civilization

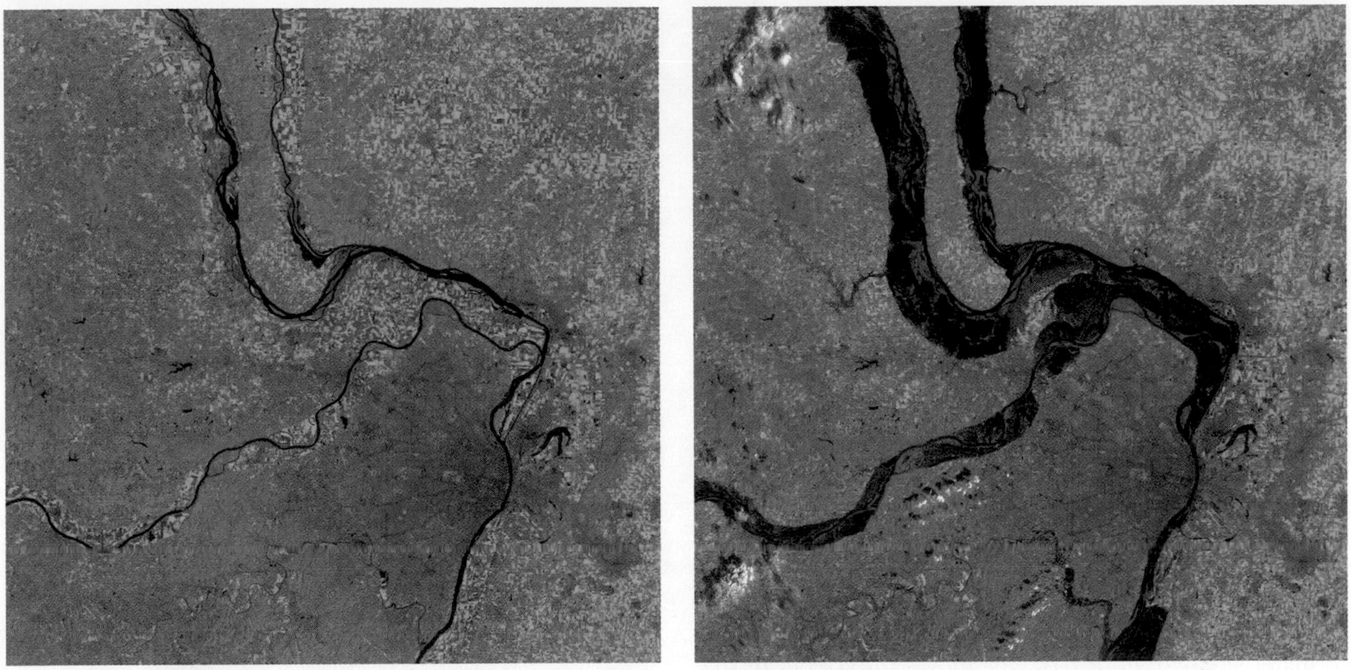

(a) (b)

FIGURE 3.3 Satellite image of the Missouri River at St. Louis before the flood in 1991 (left) and during the 1993 flood. The dark area is water.

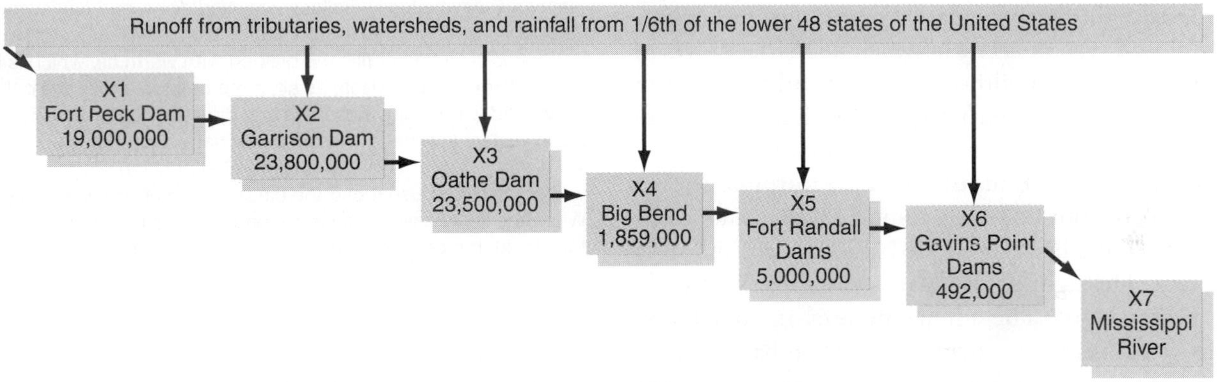

FIGURE 3.4 Imagine the Missouri River as one large lake (composed of the series of dams showed in Figure 3.2) whose water level is controlled. The water level remains constant as water flows into the lake (Fort Peck Dam) at the same rate as water flows out. If more water comes in, more leaves (Gavins Point Dam); if less water comes in, less flows out, and the water level remains at the spillway level. The number inside each box is the dam's maximum storage in acre-feet. The average annual water flow for the Missouri River is 25 million acre-feet (the amount reaching its mouth where it meets the Mississippi at St. Louis, MO)

did it become important to force a huge natural system like the Missouri to flow in steady state.

Unfortunately, people who lived along the Missouri River in 1993 learned that sometimes plans that looked good on paper did not succeed. The Missouri was just too wild and unpredictable—too non-steady-state, to use a systems-analysis term that we will define shortly—for people to control, no matter how great their efforts. The big flood of 1993 breached many levees and affected many lives (Figures 3.1 and 3.3).

The attempt to control the flow of the Missouri is just one of many examples of natural ecological and environmental systems that people thought could be engineered, controlled, tamed, and made to do what they wanted. To understand the environment and people's relation to it, it is necessary to take a systems view, and that is the purpose of this chapter. Once you have read this chapter, you will have one of the foundations for the study of all environmental systems. To understand what happens in natural ecosystems, we can't just look for an answer derived from a single factor. We have to look at the entire system and all of the factors that, together, influence what happens to life.

3.1 Basic Systems Concepts

A **system** is a set of components, or parts, that function together as a whole. A single organism, such as your body, is a system, as are a sewage-treatment plant, a city, and a river. On a much different scale, the entire Earth is a system. In a broader sense, a system is any part of the universe you can isolate in thought (in your brain or on your computer) or, indeed, physically, for the purpose of study. Key systems concepts that we will explain are (1) how a system is connected to the rest of the environment; (2) how matter and energy flow between parts of a system; (3) whether a system is static or dynamic—whether it changes over time; (4) average residence time—how long something stays within a system or part of a system; (5) feedback—how the output from a system can affect its inputs; and (6) linear and nonlinear flows.

In its relation to the rest of the environment, a system can be open or closed. In an **open system**, some energy or material (solid, liquid, or gas) moves into or out of the system. The ocean is an open system with regard to water because water moves into the ocean from the atmosphere and out of the ocean into the atmosphere. In a **closed system**, no such transfers take place. For our purposes, a **materially closed system** is one in which no matter moves in and out of the system, although energy and information can move across the system's boundaries. Earth is a materially closed system (for all practical purposes).

Systems respond to **inputs** and have **outputs**. For example, think of your body as a complex system and imagine you are hiking in Yellowstone National Park and see a grizzly bear. The sight of the bear is an input. Your body reacts to that input: The adrenaline level in your blood goes up, your heart rate increases, and the hair on your head and arms may rise. Your response—perhaps to move slowly away from the bear—is an output.

Static and Dynamic Systems

A **static system** has a fixed condition and tends to remain in that exact condition. A **dynamic system** changes, often continually, over time. A birthday balloon attached to a pole is a static system in terms of space—it stays in one place. A hot-air balloon is a simple dynamic system in terms of space—it moves in response to the winds, air density, and controls exerted by a pilot (Figure 3.5a and b). An important kind of static system is one with **classical stability**. Such a system has a constant condition, and if it is disturbed from that condition, it returns to it once the disturbing factor is removed. The pendulum of an old-fashioned grandfather clock is an example of classical stability. If you push it, the pendulum moves back and forth for a while, but then friction gradually dissipates the energy you just gave it and the pendulum comes to rest

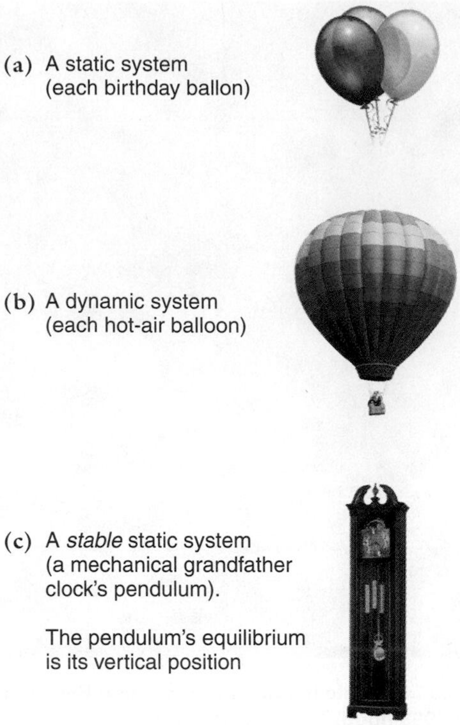

FIGURE 3.5 Static and dynamic systems. (a) *A static system* (each birthday balloon). Balloons are tied down and can't move vertically. **(b)** *A dynamic system* (each hot-air balloon). Hot air generated by a heater fills the balloon with warm air, which is lighter than outside air, so it rises; as air in the balloon cools, the balloon sinks, and winds may move it in any direction. **(c)** *A classical stable static system* (the pendulum on a mechanical grandfather clock). The pendulum's equilibrium is its vertical position. The pendulum will move if you push it or if the clock's mechanism is working. When the source of energy is no longer active (you forgot to wind the clock), the pendulum will come to rest exactly where it started.

exactly where it began. This rest point is known as the **equilibrium** (Figure 3.5c).

We will see that the classic interpretation of populations, species, ecosystems, and Earth's entire biosphere has been to assume that each is a stable, static system. But the more these ecological systems are studied scientifically, the clearer it becomes that these are dynamic systems, always changing *and always requiring change.* An important practical question that keeps arising in many environmental controversies is whether we want to, and should, force ecological systems to be static if and when they are naturally dynamic. You will find this question arising in many of the chapters in this book.

Open Systems

With few exceptions, all real systems that we deal with in the environment are open to the flow of matter, energy, and information. (For all practical purposes, as we noted earlier, Earth as a planet is a materially closed system.) An important distinction for open systems is whether they are steady-state or non-steady-state. In a

steady-state system, the inputs (of anything of interest) are equal to the outputs, so the amount stored within the system is constant. An idealized example of a steady-state system is a dam and lake into which water enters from a river and out of which water flows (Figure 3.4). If the water input equals the water output and evaporation is not considered, the water level in the lake does not change, and so, in regard to water, the lake is in a steady state. (Additional characteristics of systems are discussed in A Closer Look 3.1.)

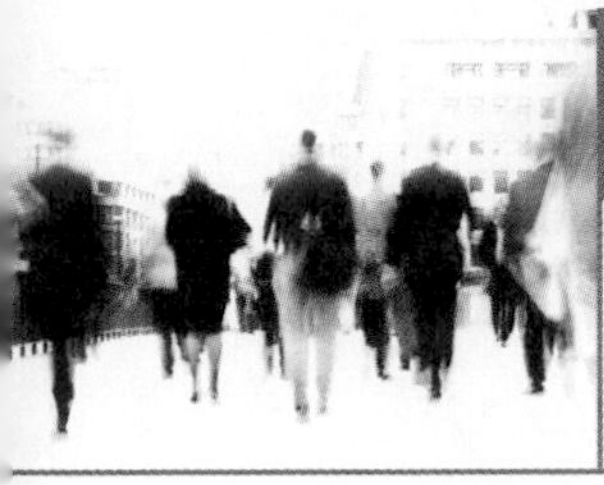

A CLOSER LOOK 3.1

Simple Systems

A simple way to think about a system is to view it as a series of compartments (also called "reservoirs," and we will use these terms interchangeably), each of which can store a certain amount of something you are interested in, and each of which receives input from other compartments and transfers some of its stored material to other compartments (Figure 3.6a).

The general equation is

$$I = O \pm \Delta S$$

where I is input into a compartment; O is output, and ΔS is change in storage. This equation defines a budget for what is being considered. For example, if your checking account has $1,000 in it (no interest rate) and you earn $500 per month at the bookstore, input is $500 per month. If you spend $500 per month, the amount in your account will be $1,000 at the end of the month (no change in storage). If you spend less than $500 per month, your account will grow ($+\Delta S$). If you spend more than $500 per month, the amount of money in your account will decrease ($-\Delta S$).

An environmental water engineer could use this kind of systems diagram (Figure 3.6a) to plan the size of the various dams to be built on the Missouri River, taking into account the desired total storage among the dams (Figure 3.4) and the role of each dam in managing the river's flow (refer back to the opening case study and also see Figure 3.7). In Figure 3.7, the amount stored in a dam's reservoir is listed as Xn, where X is the amount of water stored and n is the number of the compartment. (In this case the dams are numbered in order from upstream to downstream.) Water flows from the environment—tributaries, watersheds, and direct rainfall—into each of the reservoirs, and each is connected to the adjacent reservoirs by the river. Finally, all of the Missouri's water flows into the Mississippi, which carries it to the Gulf of Mexico.

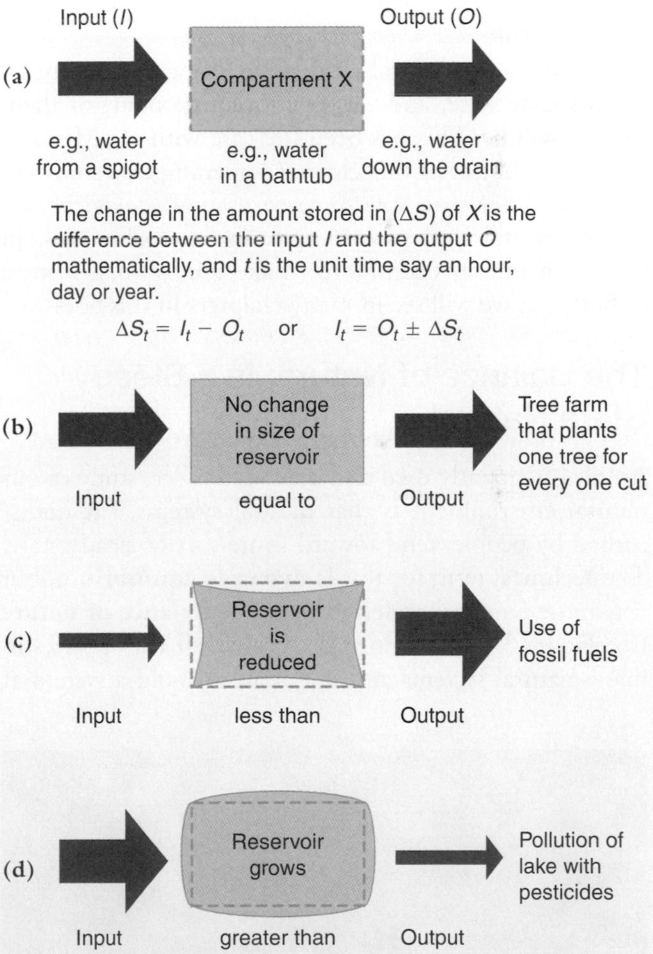

FIGURE 3.6 **(a)** General equation for ways in which a compartment of some material can change. (*Source*: Modified from P.R. Ehrlich, A.H. Ehrlich, and J.P. Holvren, *Ecoscience: Population, Resources, Environment*, 3rd ed. [San Francisco: W.H. Freeman, 1977].) Row **(b)** represents steady-state conditions; rows **(c)** and **(d)** are examples of negative and positive changes in storage.

For this water-flow system, we can make a complete flow diagram. This kind of diagramming helps us to think about and do a scientific analysis of many environmental problems, so you will find such diagrams throughout this book.

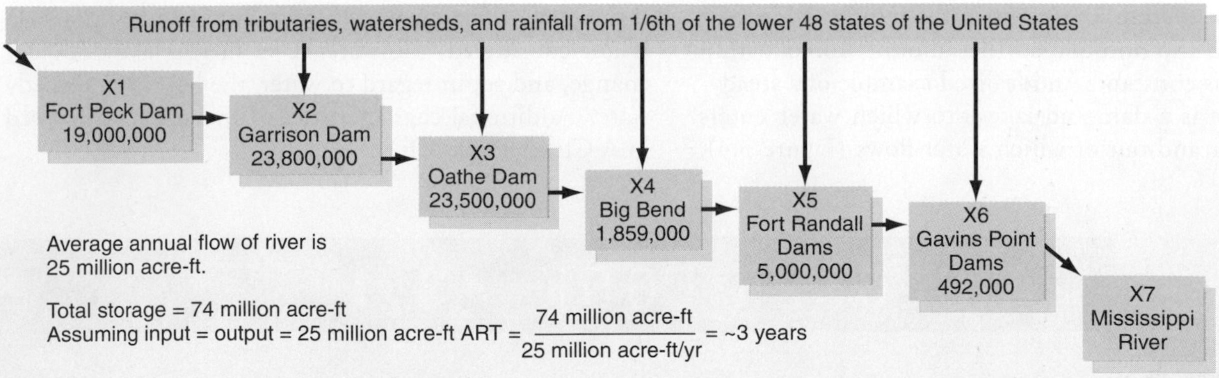

FIGURE 3.7 The Missouri River and its dams viewed as a systems flow chart. The number inside each box is the dam's maximum storage in acre-feet, where one acre-foot is the volume of water that would cover one acre to a depth of 1 foot (1,233 m^3). The average annual water flow for the Missouri River is 25 million acre-feet (the amount reaching its mouth where it meets the Mississippi at St. Louis, Missouri).

Often we want real systems in the environment to be in steady state, and we try to manage many of them so they will be. This has been the case with the Missouri River, the subject of this chapter's opening case study. As with that river, attempts to force natural ecological and environmental systems into a steady state often fail. In fact, such attempts commonly make things worse instead of better, as we will see in many chapters in this book.

The Balance of Nature: Is a Steady State Natural?

An idea frequently used and defended in the study of our natural environment is that natural systems, left undisturbed by people, tend toward some sort of steady state. The technical term for this is **dynamic equilibrium**, but it is more familiarly referred to as the **balance of nature** (see Figure 3.8). Certainly, negative feedback operates in many natural systems and may tend to hold a system at equilibrium. Nevertheless, we need to ask how often the equilibrium model really applies.[3]

FIGURE 3.8 The balance of nature. This painting, *Morning in the Tropics* by Frederic Edwin Church, illustrates the idea of the balance of nature and a dynamic steady state, with everything stationary and still, unchanging.

If we examine natural ecological systems or **ecosystems** (simply defined here as communities of organisms and their nonliving environment in which nutrients and other chemicals cycle and energy flows) in detail and over a variety of time frames, it is evident that a steady state is seldom attained or maintained for very long. Rather, systems are characterized not only by human-induced disturbances but also by natural disturbances (sometimes large-scale ones called natural disasters, such as floods and wildfires). Thus, changes over time can be expected. In fact, studies of such diverse systems as forests, rivers, and coral reefs suggest that disturbances due to natural events, such as storms, floods, and fires, are necessary for the maintenance of those systems, as we will see in later chapters. The environmental lesson is that systems change naturally. If we are going to manage systems for the betterment of the environment, we need to gain a better understanding of how they change.[3,4]

Residence Time

By using rates of change or input-output analysis of systems, we can derive an **average residence time**—how long, on average, a unit of something of interest to us will remain in a reservoir. This is obviously important, as in the case of how much water can be stored for how long in one of the reservoirs on the Missouri River. To compute the average residence time (assuming input is equal to output), we divide the total volume of stored water in the series of dams (Figures 3.4 and 3.7) by the average rate of transfer through the system (Figure 3.7) For example, suppose a university has 10,000 students, and each year 2,500 freshmen start and 2,500 seniors graduate. The average residence time for students is 10,000 divided by 2,500, or four years.

Average residence time has important implications for environmental systems. A system such as a small lake with an inlet and an outlet and a high transfer rate of water has a short residence time for water. On the one hand, from our point of view, that makes the lake especially vulnerable to change because change can happen quickly. On the other hand, any pollutants soon leave the lake.

In large systems with a slow rate of transfer of water, such as oceans, water has a long residence time, and such systems are thus much less vulnerable to quick change. However, once polluted, large systems with slow transfer rates are difficult to clean up. (See Working It Out 3.1.)

WORKING IT OUT 3.1 Average Residence Time (ART)

The average residence time (ART) is the ratio of the size of a reservoir of some material—say, the amount of water in a reservoir—to the rate of its transfer through the reservoir. The equation is

$$ART = S/F$$

where S is the size of the reservoir and F is the rate of transfer.

For example, we can calculate the average residence time for water in the Gavins Point Dam (see Figure 3.7), the farthest downstream of all the dams on the Missouri River, by realizing that the average flow into and out of the dam is about 25 million acre-feet (31 km^3) a year, and that the dam stores about 492,000 acre-feet (0.6 km^3). This suggests that the average residence time in the dam is only about seven days:

$$ART = S/F = 0.6\ km^3\ per\ year\ (31\ km^3\ per\ year)$$

$$S/F = 0.019/year\ (about\ 7\ days)$$

If the total flow were to go through Garrison Dam, the largest of the dams, the residence time would be 347 days, almost a year.

The ART for a chemical element or compound is important in evaluating many environmental problems. For example, knowing the ART of a pollutant in the air, water, or soil gives us a more quantitative understanding of that pollutant, allows us to evaluate the extent to which the pollutant acts in time and space, and helps us to develop strategies to reduce or eliminate the pollutant.

Figure 3.9 shows a map of Big Lake, a hypothetical reservoir impounded by a dam. Three rivers feed a combined 10 m^3/sec (2,640 gal/sec) of water into the lake, and the outlet structure releases an equal 10 m^3/sec. In this simplified example, we will assume that evaporation of water from the lake is negligible. A water pollutant, MTBE (methyl tertiary—butyl ether), is also present in the lake. MTBE is added to gasoline to help reduce emissions of carbon monoxide. MTBE readily dissolves in water and so travels with it. It is toxic; in small concentrations of 20–40 μg/l (thousandths of grams per liter) in water, it smells like turpentine and is nauseating to some people. Concern over MTBE in California led to a decision to stop adding it to gasoline. The sources of MTBE in "Big Lake" are urban runoff from Bear City gasoline stations, gasoline spills on land or in the lake, and gasoline engines used by boats on the lake.

We can ask several questions concerning the water and MTBE in Big Lake.

1. What is the ART of water in the lake?
2. What is the amount of MTBE in the lake, the rate (amount per time) at which MTBE is being put into the lake, and the ART of MTBE in the lake? Because the water and MTBE move together, their

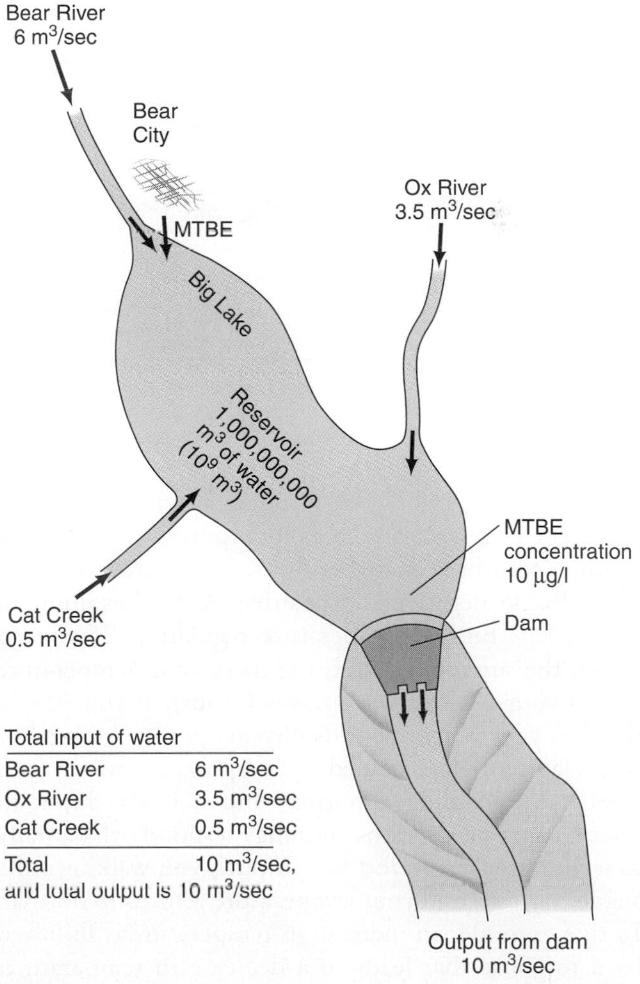

FIGURE 3.9 Idealized diagram of a lake system with MTBE contamination.

ARTs should be the same. We can test this.

ART of Water in Big Lake

For these calculations, use multiplication factors and conversions in Appendixes B and C at the end of this book.

$$ART_{water} = \frac{S}{F} = ART_{water} = \frac{1{,}000{,}000{,}000\, m^3}{10\, m^3/sec}$$

$$or\ \frac{10^9\, m^3}{10\, m^3/sec}$$

The units m^3 cancel out and

$$ART = 100{,}000{,}000\ sec\ or\ 10^8\ sec$$

Convert 10^8 sec to years:

$$\frac{seconds}{year} = \frac{60\ sec}{1\ minute} \times \frac{60\ minutes}{1\ hour} \times \frac{24\ hours}{1\ day} \times \frac{365\ days}{1\ year}$$

Canceling units and multiplying, there are 31,536,000 sec/year, which is

$$3.1536 \times 10^7\ sec/year$$

Then the ART for Big Lake is

$$\frac{100{,}000{,}000\, sec}{31{,}536{,}000\, sec/yr}\ or\ \frac{10^8\ sec}{3.1536 \times 10^7\ sec/yr}$$

Therefore the ART for water in Big Lake is 3.17/years.

ART of MTBE in Big Lake

The concentration of MTBE in water near the dam is measured as 10 μg/l. Then the total amount of MTBE in the lake (size of reservoir or pool of MTBE) is the product of volume of water in the lake and concentration of MTBE:

$$10^9\ m^3 \times \frac{10^3\, 1}{m^3} \times \frac{10\, \mu g}{1} = 10^{13}\, \mu g\ or\ 10^7\ g$$

which is 10^4 kg, or 10 metric tons, of MTBE.

The output of water from Big Lake is 10 m^3/sec, and this contains 10 μg/l of MTBE; the transfer rate of MTBE (g/sec) is

$$MTBE/sec = \frac{10\, m^3}{sec} \times \frac{10^3\, 1}{m^3} \times \frac{10\, \mu g}{1} \times \frac{10^6}{\mu g}$$

$$= 0.1\, g/sec$$

Because we assume that input and output of MTBE are equal, the input is also 0.1 *g/sec*.

$$ART_{MTBE} = \frac{S}{F} = \frac{10^7 g}{0.1 g/sec} = 10^8\ sec,\ \text{or}\ 3.17\ years$$

Thus, as we suspected, the ARTs of the water and MTBE are the same. This is because MTBE is dissolved in the water. If it attached to the sediment in the lake, the ART of the MTBE would be much longer. Chemicals with large reservoirs or small rates of transfer tend to have long ARTs. In this exercise we have calculated the ART of water in Big Lake as well as the input, total amount, and ART of MTBE.

Feedback

Feedback occurs when the output of a system (or a compartment in a system) affects its input. Changes in the output "feed back" on the input. There are two kinds of feedback: negative and positive. A good example of feedback is human temperature regulation. If you go out in the sun and get hot, the increase in temperature affects your sensory perceptions (input). If you stay in the sun, your body responds physiologically: Your pores open, and you are cooled by evaporating water (you sweat). The cooling is output, and it is also input to your sensory perceptions. You may respond behaviorally as well: Because you feel hot (input), you walk into the shade (output) and your temperature returns to normal. In this example, an increase in temperature is followed by a response that leads to a decrease in temperature. This is an example of negative feedback, in which an increase in output now leads to a later *decrease* in output. **Negative feedback** is self-regulating, or stabilizing. It is the way that steady-state systems can remain in a constant condition.

Positive feedback occurs when an increase in output leads to a further *increase* in output. A fire starting in a forest provides an example of positive feedback. The wood may be slightly damp at the beginning and so may not burn readily. Once a fire starts, wood near the flame dries out and begins to burn, which in turn dries out a greater quantity of wood and leads to a larger fire. The larger the fire, the faster more wood becomes dry and the more rapidly the fire grows. Positive feedback, sometimes called a "vicious cycle," is destabilizing.

Environmental damage can be especially serious when people's use of the environment leads to positive feedback. For example, off-road vehicles—including bicycles—may cause positive feedback to soil erosion (Figure 3.10).

(a)

(c)

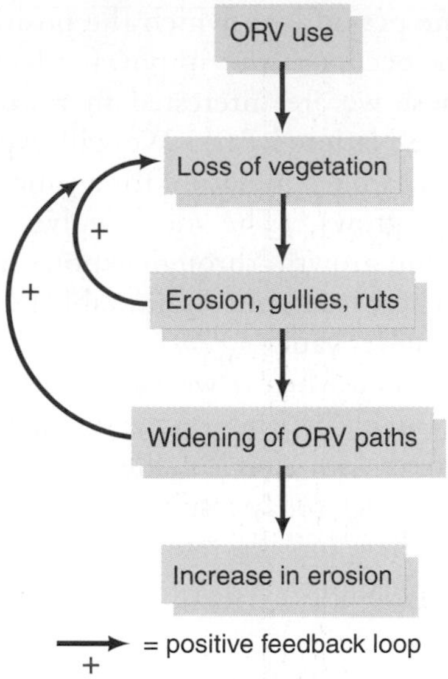

(b)

FIGURE 3.10 How off-road vehicles **(a)** create positive feedback on soil erosion **(b)** and **(c)**.

The vehicles' churning tires are designed to grip the earth, but they also erode the soil and uproot plants. Without vegetation, the soil erodes faster, exposing even more soil (positive feedback). As more soil is exposed, rainwater more easily carves out ruts and gullies (more positive feedback). Drivers of off-road vehicles then avoid the ruts and gullies by driving on adjacent sections that are not as eroded, thus widening paths and further increasing erosion (more positive feedback). The gullies themselves increase erosion because they concentrate runoff and have steep side slopes. Once formed, gullies tend to get longer, wider, and deeper, causing additional erosion (even more positive feedback). Eventually, an area of intensive off-road vehicle use may become a wasteland of eroded paths and gullies. Positive feedback has made the situation increasingly worse.

Some systems have both positive and negative feedbacks, as can occur, for example, for the human population in large cities (Figure 3.11). Positive feedback on the population size may occur when people perceive greater opportunities in cities and move there hoping for a higher standard of living. As more people move to cities, opportunities may increase, leading to even more migration to cities. Negative feedback can then occur when crowding increases air and water pollution, disease, crime, and discomfort. These negatives encourage some people to migrate from the cities to rural areas, reducing the city's population.

Practicing your critical-thinking skills, you may ask, "Is negative feedback generally desirable, and is positive feedback generally undesirable?" Reflecting on this question, we can see that although negative feedback is self-regulating, it may in some instances not be

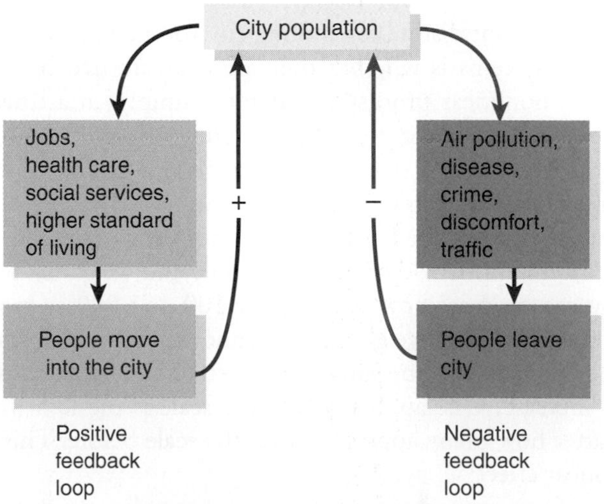

FIGURE 3.11 **Potential positive and negative feedback loops for changes of human population in large cities.** The left side of the figure shows that as jobs increase and health care and the standard of living improve, migration and the city population increase. Conversely, the right side of the figure shows that increased air pollution, disease, crime, discomfort, and traffic tend to reduce the city population. (*Source:* Modified from M. Maruyama, the second cybernetics: Deviation-amplifying mutual causal processes, *American Scientist* 51 [1963]:164–670. Reprinted by permission of *American Scientist* magazine of Sigma Xi, The Scientific Research Society.)

desirable. The period over which the positive or negative feedback occurs is the important factor. For example, suppose we are interested in restoring wolves to Yellowstone National Park. We will expect positive feedback in the wolf population for a time as the number of wolves grows. (The more wolves, the greater their population growth, through exponential growth.) Positive feedback, for a time, is desirable because it produces a change we want.

We can see that whether we view positive or negative feedback as desirable depends on the system and potential changes. Nevertheless, some of the major environmental problems we face today result from positive feedback mechanisms. These include resource use and growth of the human population.

3.2 System Responses: Some Important Kinds of Flows[4]

Within systems, there are certain kinds of flows that we come across over and over in environmental science. (Note that **flow** is an amount transferred; we also refer to the **flux**, which is the rate of transfer per unit time.) Because these are so common, we will explain a few of them here.

Linear and Nonlinear Flows

An important distinction among environmental and ecological systems is whether they are characterized by linear or nonlinear processes. Put most simply, in a **linear process**, if you add the same amount of anything to a compartment in a system, the change will always be the same, no matter how much you have added before and no matter what else has changed about the system and its environment. If you harvest one apple and weigh it, then you can estimate how much 10 or 100 or 1,000 or more of the apples will weigh—adding another apple to a scale does not change the amount by which the scale shows an increase. One apple's effect on a scale is the same, no matter how many apples were on the scale before. This is a linear effect.

Many important processes are **nonlinear**, which means that the effect of adding a specific amount of something changes depending on how much has been added before. If you are very thirsty, one glass of water makes you feel good and is good for your health. Two glasses may also be helpful. But what about 100 glasses? Drinking more and more glasses of water leads quickly to diminishing returns and eventually to water's becoming a poison.

Lag Time

Many responses to environmental inputs (including human population change; pollution of land, water, and air; and use of resources) are nonlinear and may involve delays, which we need to recognize if we are to understand and solve environmental problems. For example, when you add fertilizer to help a tree grow, it takes time for it to enter the soil and be used by the tree.

Lag time is the delay between a cause and the appearance of its effect. (This is also referred to as the time between a stimulus and the appearance of a response.) If the lag time is long, especially compared to human lifetimes (or attention spans or our ability to continue measuring and monitoring), we can fail to recognize the change and know what is the cause and what is the effect. We can also come to believe that a possible cause is not having a detrimental effect, when in reality the effect is only delayed. For example, logging on steep slopes can increase the likelihood and rate of erosion, but in comparatively dry environments this may not become apparent until there is heavy rain, which might not occur until a number of years afterward. If the lag time is short, cause and effect are easier to identify. For example, highly toxic gas released from a chemical plant will likely have rapid effects on the health of people living nearby.

With an understanding of input and output, positive and negative feedback, stable and unstable systems, and systems at steady state, we have a framework for interpreting some of the changes that may affect systems.

Selected Examples of System Responses

Although environmental science deals with very complex phenomena, there are recurring relationships that we can represent with a small number of graphs that show how one part of a system responds to inputs from another part. These graphs include responses of individual organisms, responses of populations and species, responses of entire ecosystems and then large units of the **biosphere**, the planetary system that includes and sustains life, such as how the atmosphere responds to the burning of fossil fuels. Each of these graphs has a mathematical equation that can explain the curve, but it is the shape of the graph and what that shape represents that are key to understanding environmental systems. These curves represent, in one manifestation or another, the fundamental dynamics found in these systems. The graphs show (1) a straight line (linear); (2) the positive exponential; (3) the negative exponential; (4) the logistic curve; and (5) the saturation (Michaelis-Menton) curve. An example of each is shown in Figures 3.12 to 3.15.

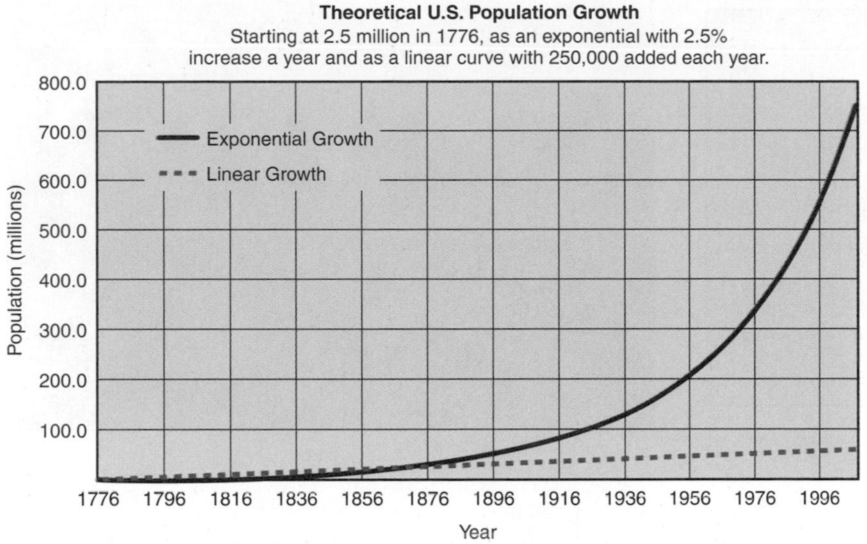

FIGURE 3.12 Curves 1 and 2: linear and positive exponential. This graph shows theoretical growth of the population of the United States, starting with the 2.5 million people estimated to have been here in 1776 and growing as an exponential and a linear curve. Even though the linear curve adds 250,000 people a year—10% of the 1776 population—it greatly lags the exponential by the beginning of the 20th century, reaching fewer than 100 million people today, while the exponential would have exceeded our current population.

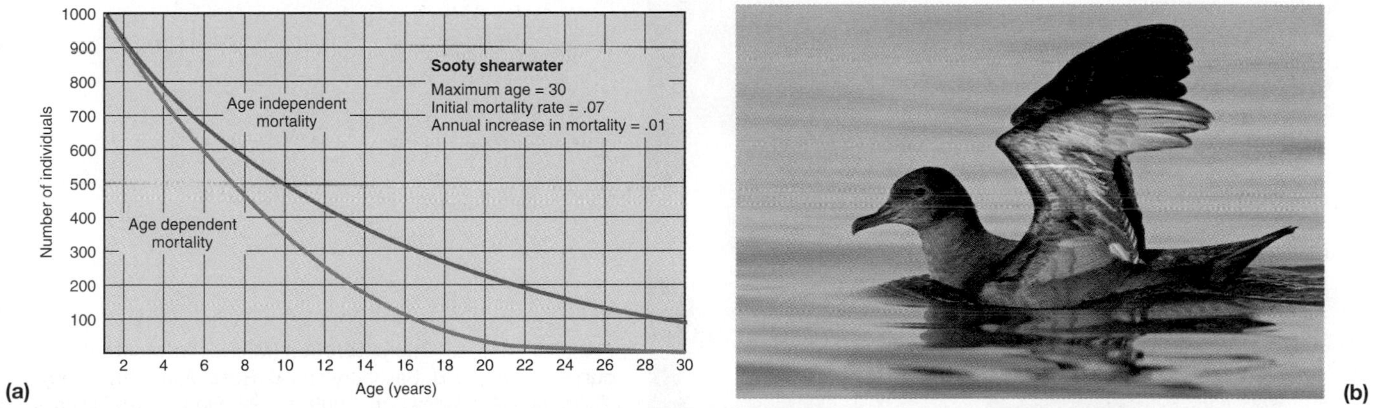

FIGURE 3.13 Negative exponential. Example: the decline in a population of a species of birds when there are no births and the mortality rate is 7% per year. The upper curve is a pure negative exponential. (*Source*: D.B. Botkin and R.S. Miller, 1974, Mortality rates and survival of birds, *American Nat.* 108: 181–192.)

Figure 3.12 shows both a linear relation and a positive exponential relation. A linear relation is of the form $y = a + bx$, where a is the y intercept (in this case, o) and b is the slope of the line (change in y to change in x, where y is the vertical axis and x the horizontal). The form of the positive exponential curve is $y = ax^b$, where a is the y intercept (in this case o) and b is the slope. However, b is a positive exponent (power). (See A Closer Look 3.2, Exponential Growth.)

Figure 3.13 shows two examples of negative exponential relations. Figure 3.14 is the logistic curve, which often has the shape of a lazy *S*; the logistic carrying k is the population eventually reached or approached, based on environmental factors. The saturation (Michaelis-Menton) curve (Figure 3.15) shows initial fast change, followed by a leveling off at saturation. At the point of saturation, the net CO_2 fixed (for soybean)is at a light-intensity value of about 3,000. As light intensity increases above about 3,000, net fixed CO_2 is nearly constant (that is, fixed CO_2 saturates at light intensity of 3,000 and does not change if intensity increases).

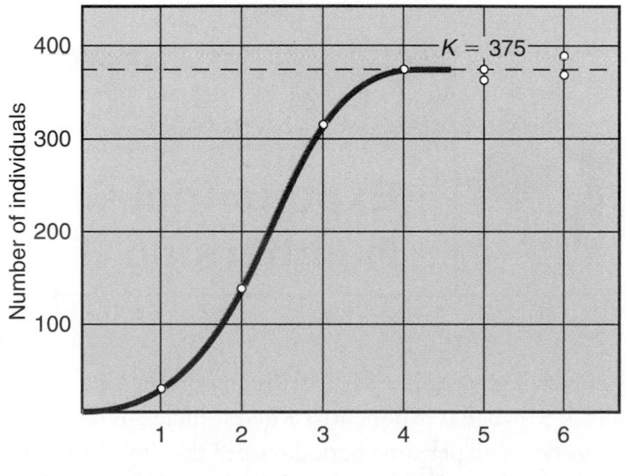

FIGURE 3.14 (a) The logistic curve. Growth of a population of a microorganism in a laboratory test tube under constant conditions with a constant supply of food. (From G.F. Gause, *The Struggle for Existence.*) The logistic carrying capacity is *k*. If you take a population of such bacteria into a laboratory and grow them under constant conditions, you might get the population to change according to the curve above, as Gause did in the 1930s with other microorganisms. (*Source:* D.B. Botkin, *Discordant Harmonies: A New Ecology for the 21st Century* [New York: Oxford University Press, 1990].)

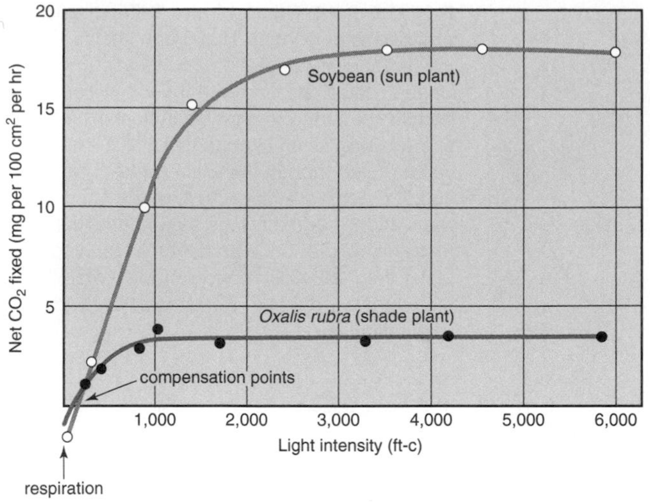

(a)

(b)

(c)

FIGURE 3.15 (a) The saturation (Michaelis-Menton) curve. (*Source:* F.B. Salisbury and C. Ross, *Plant Physiology* [Belmont, CA: Wadsworth, 1969, p. 292, Figure 14-9.] Data from R. Bohning and C. Burnside, 1956, *American Journal of Botany* 43:557].); **(b)** *Glycine max* (soybeans); **(c)** *Oxalis rubra* (shade plant).

A CLOSER LOOK 3.2

Exponential Growth Defined, and Putting Some Numbers on It

Exponential growth is a particularly important kind of feedback. Change is exponential when it increases or decreases at a constant rate per time period, rather than by a constant amount. For instance, suppose you have \$1,000 in the bank and it grows at 10% per year. The first year, \$100 in interest is added to your account. The second year, you earn more, \$110, because you earn 10% on a higher total amount of \$1,100. The greater the amount, the greater the interest earned, so the money increases by larger and larger amounts. When we plot data in which exponential growth is occurring, the curve we obtain is said to be **J**-shaped. It looks like a skateboard ramp, starting out nearly flat and then rising steeply.

Two important qualities of exponential growth are (1) the rate of growth measured as a percentage and (2) the doubling time in years. The **doubling time** is the time necessary for the quantity being measured to double. A useful rule is that the doubling time is approximately equal to 70 divided by the annual percentage growth rate. Working It Out 3.2 describes exponential growth calculations and explains why 70 divided by the annual growth rate is the doubling time.

WORKING IT OUT 3.2 Exponential Growth

If the quantity of something (say, the number of people on Earth) increases or decreases at a fixed fraction per unit of time, whose symbol is k (for example, $k = +0.02$ per year), then the quantity is changing exponentially. With positive k, we have exponential growth. With negative k, we have exponential decay.

The growth rate R is defined as the percent change per unit of time—that is, $k = R/100$. Thus, if $R = 2\%$ per year, then $k = +0.02$ per year. The equation to describe exponential growth is

$$N = N e^{kt}$$

where N is the future value of whatever is being evaluated; N_0 is present value; e, the base of natural logarithms, is a constant 2.71828; k is as defined above; and t is the number of years over which the growth is to be calculated.

This equation can be solved using a simple hand calculator, and a number of interesting environmental questions can then be answered. For example, assume that we want to know what the world population is going to be in the year 2020, given that the population in 2003 is 6.3 billion and the population is growing at a constant rate of 1.36% per year ($k = 0.0136$). We can estimate N, the world population for the year 2020, by applying the preceding equation:

$$\begin{aligned} N &= (6.3 \times 10^9) \times e^{(0.0136 \times 17)} \\ &= 6.3 \times 10^9 \times e^{0.2312} \\ &= 6.3 \times 10^9 \times 2.718^{0.231} \\ &= 7.94 \times 10^9, \text{ or } 7.94 \text{ billion people} \end{aligned}$$

The doubling time for a quantity undergoing exponential growth (i.e., increasing by 100%) can be calculated by the following equation:

$$2N_T = N_0\, e^{kTd}$$

where T_d is the doubling time.

Take the natural logarithm of both sides.

$$\ln 2 = kT_d \quad \text{and} \quad T_d = \ln 2/k$$

Then, remembering that $k = R/100$,

$$\begin{aligned} T_d &= 0.693/(R/100) \\ &= 100(0.693)/R \\ &= 69.3/R, \textit{ or about } 70/R \end{aligned}$$

This result is our general rule—that the doubling time is approximately 70 divided by the growth rate. For example, if $R = 10\%$ per year, then $T = 7$ years.

3.3 Overshoot and Collapse

Figure 3.16 shows the relationship between carrying capacity (maximum population possible without degrading the environment necessary to support the population) and the human population. The carrying capacity starts out being much higher than the human population, but if a population grows exponentially (see Working It Out 3.2), it eventually exceeds—**overshoots**—the carrying capacity. This ultimately results in the **collapse** of a population to some lower level, and the carrying capacity may be reduced as well. In this case, the lag time is the period of exponential growth of a population before it exceeds the carrying capacity. A similar scenario may be posited for harvesting species of fish or trees.

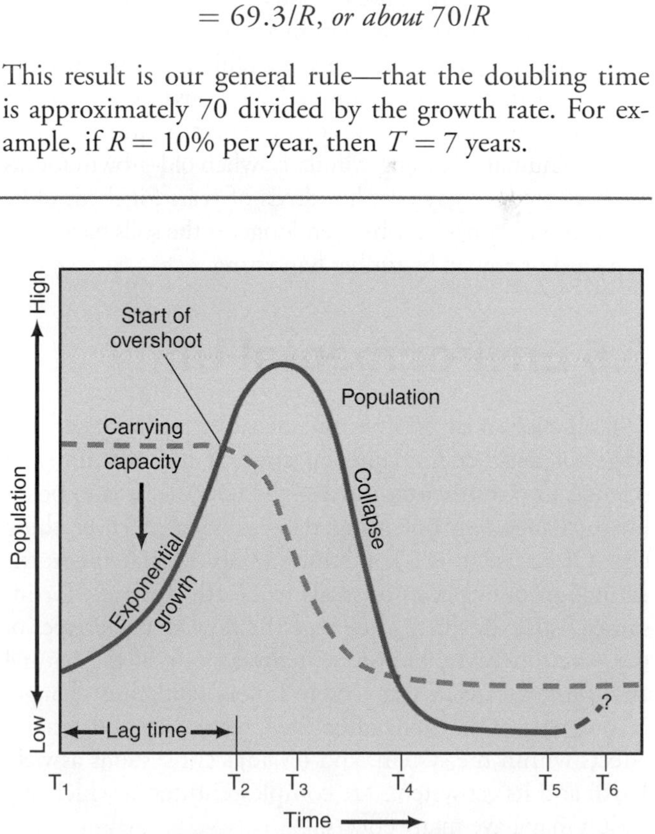

FIGURE 3.16 The concept of overshoot. A population starts out growing exponentially, but as this growth cannot continue indefinitely, it reaches a peak, then declines sharply. Sometimes the population is assumed to have a carrying capacity, which is the maximum number possible, and if the population's habitat is damaged by too great an abundance, the carrying capacity also decreases. (*Source*: Modified after D.H. Meadows and others, 1992.)

3.4 Irreversible Consequences

Adverse consequences of environmental change do not necessarily lead to irreversible consequences. Some do, however, and these lead to particular problems. When we talk about irreversible consequences, we mean

FIGURE 3.17 Timber harvest (clear-cut) can result in soil erosion. Once soil is removed, it can take such a long time for it to rebuild that the damage may be viewed as irreversible on a human time scale.

consequences that may not be easily rectified on a human scale of decades or a few hundred years.

Good examples of this are soil erosion and the harvesting of old-growth forest (Figure 3.17). With soil erosion, there may be a long lag time until the soil erodes to the point where crops no longer have their roots in active soil that has the nutrients necessary to produce a successful crop. But once the soil is eroded, it may take hundreds or thousands of years for new soil to form, and so the consequences are irreversible in terms of human planning. Similarly, when old-growth forests are harvested, it may take hundreds of years for them to be restored. Lag times may be even longer if the soils have been damaged or eroded by timber harvesting.

3.5 Environmental Unity

Our discussion of positive and negative feedback sets the stage for another fundamental concept in environmental science: **environmental unity**—the idea that it is impossible to change only one thing; everything affects everything else. Of course, this is something of an overstatement; the extinction of a species of snails in North America, for instance, is hardly likely to change the flow characteristics of the Amazon River. However, many aspects of the natural environment are in fact closely linked, and thus changes in one part of a system often have secondary and tertiary effects within the system, and on adjacent systems as well. Earth and its ecosystems are complex entities in which any action may have many effects.

We will find many examples of environmental unity throughout this book. Urbanization illustrates it. When cities, such as Chicago and Indianapolis, were developed in the eastern and midwestern United States, the clearing of forests and prairies and the construction of buildings and paved streets increased surface-water runoff and soil erosion, which in turn affected the shape of river channels—some eroded soil was deposited on the bottom of the channel, reducing channel depth and increasing flood hazard. Increased fine sediment made the water muddy, and chemicals from street and yard runoff polluted the stream.[5,6] These changes affected fish and other life in the river, as well as terrestrial wildlife that depended on the river. The point here is that land-use conversion can set off a series of changes in the environment, and each change is likely to trigger additional changes.

3.6 Uniformitarianism

Uniformitarianism is the idea that geological and biological processes that occur today are the same kinds of processes that occurred in the past, and vice versa. Thus, the present is the key to the past, and the past the key to the future. For example, we use measurements of the current rate of erosion of soils and bedrock by rivers and streams to calculate the rate at which this happened in the past and to estimate how long it took for certain kinds of deposits to develop. If a deposit of gravel and sand found at the top of a mountain is similar to stream gravels found today in an adjacent valley, we may infer by uniformitarianism that a stream once flowed in a valley where the mountaintop is now. The concept of uniformitarianism helps explain the geologic and evolutionary history of Earth.

Uniformitarianism was first suggested in 1785 by the Scottish scientist James Hutton, known as the father of geology. Charles Darwin was impressed by the concept, and it pervades his ideas on biological evolution. Today, uniformitarianism is considered one of the fundamental principles of the biological and Earth sciences.

Uniformitarianism does not demand or even suggest that the magnitude and frequency of natural processes remain constant, only that the processes themselves continue. For the past several billion years, the continents, oceans, and atmosphere have been similar to those of today. We assume that the physical and biological processes that form and modify the Earth's surface have not changed significantly over this period. To be useful from an environmental standpoint, the principle of uniformitarianism has to be more than a key to the past; we must turn it around and say that a study of past and present processes is the key to the future. That is, we can assume that in the future the same physical and biological processes will operate, although the rates will vary as the environment is influenced by natural change and human activity. Geologically short-lived landforms, such as beaches (Figure 3.18) and lakes, will continue to appear and disappear in response to storms, fires, volcanic eruptions, and earthquakes. Extinctions of animals and plants will continue, in spite of, as well as because of, human activity.

Obviously, some processes do not extend back through all of geologic time. For example, the early Earth atmosphere did not contain free oxygen. Early photo-

FIGURE 3.18 This beach on the island of Bora Bora, French Polynesia, is an example of a geologically short-lived landform, vulnerable to rapid change from storms and other natural processes.

synthetic bacteria converted carbon dioxide in the atmosphere to hydrocarbons and released free oxygen; before life, this process did not occur. But the process began a long time ago—3.5 billion years ago—and as long as there are photosynthetic organisms, this process of carbon dioxide uptake and oxygen release will continue.

Knowledge of uniformitarianism is one way that we can decide what is "natural" and ascertain the characteristics of nature undisturbed by people. One of the environmental questions we ask repeatedly, in many contexts, is whether human actions are consistent with the processes of the past. If not, we are often concerned that these actions will be harmful. We want to improve our ability to predict what the future may bring, and uniformitarianism can assist in this task.

3.7 Earth as a System

The discussion in this chapter sets the stage for a relatively new way of looking at life and the environment—a global perspective, thinking about our entire planet's life-supporting and life-containing system. This is known as Earth systems science, and it has become especially important in recent years, with concerns about climate change (see Chapter 20).

Our discussion of Earth as a system—life in its environment, the biosphere, and ecosystems—leads us to the question of how much life on Earth has affected our planet. In recent years, the **Gaia hypothesis**—named for Gaia, the Greek goddess Mother Earth—has become a hotly debated subject.[7] The hypothesis states that life manipulates the environment for the maintenance of life. For example, some scientists believe that algae floating near the surface of the ocean influence rainfall at sea and the carbon dioxide content of the atmosphere, thereby significantly affecting the global climate. It follows, then, that the planet Earth is capable of physiological self-regulation.

The idea of a living Earth can be traced back at least to Roman times in the writing of Lucretius.[3] James Hutton, whose theory of uniformitarianism was discussed earlier, stated in 1785 that he believed Earth to be a superorganism, and he compared the cycling of nutrients from soils and rocks in streams and rivers to the circulation of blood in an animal.[7] In this metaphor, the rivers are the arteries and veins, the forests are the lungs, and the oceans are the heart of Earth.

The Gaia hypothesis is really a series of hypotheses. The first is that life, since its inception, has greatly affected the planetary environment. Few scientists would disagree. The second hypothesis asserts that life has altered Earth's environment in ways that have allowed life to persist. Certainly, there is some evidence that life has had such an effect on Earth's climate. A popularized extension of the Gaia hypothesis is that life *deliberately* (consciously) controls the global environment. Few scientists accept this idea.

The extended Gaia hypothesis may have merit in the future, however. We have become conscious of our effects on the planet, some of which influence future changes in the global environment. Thus, the concept that we can consciously make a difference in the future of our planet is not as extreme a view as many once thought. The future status of the human environment may depend in part on actions we take now and in coming years. This aspect of the Gaia hypothesis exemplifies the key theme of thinking globally, which was introduced in Chapter 1.

3.8 Types of Change

Change comes in several forms. Some changes brought on by human activities involve rather slow processes—at least from our point of view—with cumulative effects. For example, in the middle of the 19th century, people began to clear-cut patches of the Michigan forests. It was commonly believed that the forests were so large that it would be impossible to cut them all down before they grew back just as they were. But with many people logging in different, often isolated areas, it took less than 100 years for all but about 100 hectares to be clear-cut.

Another example: With the beginning of the Industrial Revolution, people in many regions began to burn fossil fuels, but only since the second half of the 20th century have the possible global effects become widely evident. Many fisheries appear capable of high harvests for many years. But then suddenly, at least from our perspective—sometimes within a year or a few years—an entire species of fish suffers a drastic decline. In such cases, long-term damage can be done. It has been difficult to recognize when harvesting fisheries is overharvesting and, once it has started, figuring out what can be done to enable a fishery to recover in time for fishermen to continue making a living. A famous example of this was the harvesting of anchovies off the coast of Peru. Once the largest fish catch in the world, within a few years the fish numbers declined so greatly that commercial harvest was threatened. The same thing has happened with the fisheries of Georges Banks and the Grand Banks in the Atlantic Ocean.

You can see from these few examples that environmental problems are often complex, involving a variety of linkages among the major components and within each component, as well as linear and exponential change, lag times, and the possibility of irreversible consequences.

As stated, one of our goals in understanding the role of human processes in environmental change is to help manage our global environment. To accomplish this goal, we need to be able to predict changes, but as the examples above demonstrate, prediction poses great challenges. Although some changes are anticipated, others come as a surprise. As we learn to apply the principles of environmental unity and uniformitarianism more skillfully, we will be better able to anticipate changes that would otherwise have been surprises.

CRITICAL THINKING ISSUE

Is the Gaia Hypothesis Science?

According to the Gaia hypothesis, Earth and all living things form a single system with interdependent parts, communication among these parts, and the ability to self-regulate. Are the Gaia hypothesis and its component hypotheses science, fringe science, or pseudoscience? Is the Gaia hypothesis anything more than an attractive metaphor? Does it have religious overtones? Answering these questions is more difficult than answering similar questions about, say, crop circles, described in Chapter 2. Analyzing the Gaia hypothesis forces us to deal with some of our most fundamental ideas about science and life.

Critical Thinking Questions

1. What are the main hypotheses included in the Gaia hypothesis?
2. What kind of evidence would support each hypothesis?
3. Which of the hypotheses can be tested?
4. Is each hypothesis science, fringe science, or pseudoscience?
5. Some scientists have criticized James E. Lovelock, who formulated the Gaia hypothesis, for using the term *Gaia*. Lovelock responds that it is better than referring to a "biological cybernetic system with homeostatic tendencies." What does this phrase mean?
6. What are the strengths and weaknesses of the Gaia hypothesis?

SUMMARY

- A system is a set of components or parts that function together as a whole. Environmental studies deal with complex systems, and solutions to environmental problems often involve understanding systems and their rates of change.
- Systems respond to inputs and have outputs. Feedback is a special kind of system response, where the output affects the input. Positive feedback, in which increases in output lead to increases in input, is destabilizing, whereas negative feedback, in which increases in output lead to decreases in input, tends to stabilize or encourage more constant conditions in a system.
- Relationships between the input (cause) and output (effect) of systems may be linear, exponential, or represented by a logistic curve or a saturation curve .
- The principle of environmental unity, simply stated, holds that everything affects everything else. It emphasizes linkages among parts of systems.
- The principle of uniformitarianism can help predict future environmental conditions on the basis of the past and the present.
- Although environmental and ecological systems are complex, much of what happens with them can be characterized by just a few response curves or equations: the straight line, the exponential, the logistic, and the saturation curves.
- Exponential growth, long lag times, and the possibility of irreversible change can combine to make solving environmental problems difficult.
- Change may be slow, fast, expected, unexpected, or chaotic. One of our goals is to learn to better recognize change and its consequences in order to better manage the environment.

REEXAMINING THEMES AND ISSUES

Human Population

Due partly to a variety of positive-feedback mechanisms, Earth's human population is increasing. Of particular concern are local or regional increases in population density (the number of people per unit area), which strain resources and lead to human suffering.

Sustainability

Negative feedback is stabilizing. If we are to have a sustainable human population and use our resources sustainably, then we need to put in place a series of negative feedbacks within our agricultural, urban, and industrial systems.

Global Perspective

This chapter introduced Earth as a system. One of the most fruitful areas for environmental research remains the investigation of relationships between physical and biological processes on a global scale. More of these relationships must be discovered if we are to solve environmental problems related to such issues as potential global warming, ozone depletion, and disposal of toxic waste.

Urban World

The concepts of environmental unity and uniformitarianism are particularly applicable to urban environments, where land-use changes result in a variety of changes that affect physical and biochemical processes.

People and Nature

People and nature are linked in complex ways in systems that are constantly changing. Some changes are not related to human activity, but many are—and human-caused changes from local to global in scale are accelerating.

Science and Values

Our discussion of the Gaia hypothesis reminds us that we still know very little about how our planet works and how physical, biological, and chemical systems are linked. What we do know is that we need more scientific understanding. This understanding will be driven, in part, by the value we place on our environment and on the well-being of other living things.

KEY TERMS

average residence time **46**
balance of nature **46**
biosphere **50**
classical stability **44**
closed system **44**
doubling time **52**
dynamic equilibrium **46**
dynamic system **44**
ecosystem **46**
environmental unity **54**
equilibrium **44**
exponential growth **52**
feedback **48**
flow **50**
flux **50**
Gaia hypothesis **55**
input **44**
lag time **50**
linear process **50**
materially closed system **44**
negative feedback **48**
nonlinear process **50**
open system **44**
output **44**
overshoot and collapse **53**
positive feedback **48**
static system **44**
steady-state system **45**
system **44**
uniformitarianism **54**

STUDY QUESTIONS

1. What is the difference between positive and negative feedback in systems? Provide an example of each.
2. What is the main point concerning exponential growth? Is exponential growth good or bad?
3. Why is the idea of equilibrium in systems somewhat misleading in regard to environmental questions? Is it ever possible to establish a balance of nature?
4. Why is the average residence time important in the study of the environment?
5. Is the Gaia hypothesis a true statement of how nature works, or is it simply a metaphor? Explain.
6. How might you use the principle of uniformitarianism to help evaluate environmental problems? Is it possible to use this principle to help evaluate the potential consequences of too many people on Earth?
7. Why does overshoot occur, and what could be done to anticipate and avoid it?

FURTHER READING

Botkin, D.B., M. Caswell, J.E. Estes, and A. Orio, eds., ***Changing the Global Environment: Perspectives on Human Involvement*** (NewYork: Academic Press, 1989). One of the first books to summarize the effects of people on nature; it includes global aspects and uses satellite remote sensing and advanced computer technologies.

Bunyard, P., ed., ***Gaia in Action: Science of the Living Earth*** (Edinburgh: Floris Books, 1996). This book presents investigations into implications of the Gaia hypothesis.

Lovelock, J., ***The Ages of Gaia: A Biography of Our Living Earth*** (New York: Norton, 1995). This small book explains the Gaia hypothesis, presenting the case that life very much affects our planet and in fact may regulate it for the benefit of life.

CHAPTER 4

The Human Population and the Environment

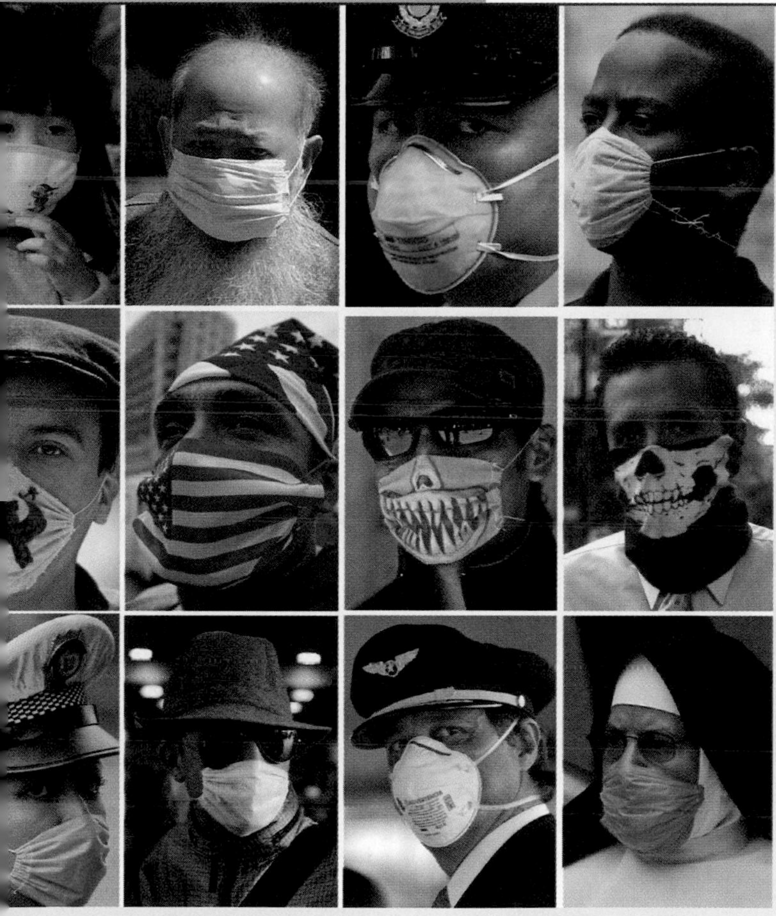

In 2009, people around the world wore masks to protect themselves against swine flu.

LEARNING OBJECTIVES

The human population has been growing rapidly for centuries. What is happening and, most important, what *will* happen to all of us and our planet if this continues? After reading this chapter, you should understand that . . .

- Ultimately, there can be no long-term solutions to environmental problems unless the human population stops increasing;
- Two major questions about the human population are (1) what controls its rate of growth and (2) how many people Earth can sustain;
- Modern medical practices and improvements in sanitation, control of disease-spreading organisms, and supplies of human necessities have lowered death rates and accelerated the net rate of human population growth;
- Although the death rate has declined, so more people live longer, the rapid increase in the human population has occurred with little or no change in the maximum lifetime of an individual, which is still less than 120 years;
- In general, countries with a high standard of living have moved more quickly to a lower birth rate than have countries with a low standard of living;
- Although we cannot predict with absolute certainty what the future human carrying capacity of Earth will be, an understanding of human population dynamics can help us make useful forecasts;
- The principles of population dynamics discussed in this chapter apply to populations of all species and will be useful throughout this book.

CASE STUDY

Pandemics and World Population Growth

On April 14, 2009, the Mexican government reported the first case of a new strain of flu. A genetic combination of flu found in pigs, birds, and people, it was immediately called "swine flu" but formally referred to as flu strain A (H1N1). Because this was a new strain, little natural resistance to it could be expected, and it thus might cause a worldwide disease outbreak—the kind known as a **pandemic**. Indeed, this flu traveled rapidly. By May 1, it had spread to 11 nations.[1]

Nations responded quickly. The government of Hong Kong quarantined a major hotel where one guest from Mexico was diagnosed with the flu. The Mexican government provided open access to information and declared a special "holiday" in Mexico City to prevent the spread of the disease there.

Even so, by mid-May 2009 the disease had spread to 33 nations, causing almost 6,500 cases but few deaths (Figure 4.1). Although it had become widespread rapidly, concerns about swine flu had greatly diminished because it appeared to be a rather mild form of the disease and quick responses seemed to have mostly contained it. Concerns remained, however, that it might spread to the Southern Hemisphere and, during its winter, mutate to a more virulent form, then return to the Northern Hemisphere in the winter of 2009 as a greater threat.

Because this strain of flu did not become a full-blown pandemic and seemed relatively mild, it is easy to believe that nations overreacted. But the failure of this flu to spread more widely appears due in large part to the rapid and widespread response. And the history of recent new diseases—particularly West Nile virus and SARS—supported such a response.

Before 1999, West Nile virus occurred in Africa, West Asia, and the Middle East, but not in the New World. Related to encephalitis, West Nile virus is spread by mosquitoes, which bite infected birds, ingest the virus, and then bite people. It reached the Western Hemisphere through infected birds and has now been found in more than 25 species of birds native to the United States, including crows, the bald eagle, and the black-capped chickadee—a common visitor to bird feeders in the U.S. Northeast. Fortunately, in human beings this disease has lasted only a few days and has rarely caused severe symptoms.[2] By 2007, more than 3,600 people in the United States had contracted this disease, most in California and Colorado, with 124 fatalities.[3] But the speed with which it spread led to concerns about other possible new pandemics.

Four years earlier, in February 2003, the sudden occurrence of a new disease, severe acute respiratory syndrome (SARS), had demonstrated that modern transportation and the world's huge human population could lead to the rapid spread of epidemic diseases. Jet airliners daily carry vast numbers of people and goods around the world. The disease began in China, perhaps spread from some wild animal to human beings. China had become much more open to foreign travelers, with more than 90 million visitors in a recent year.[4] By late spring 2003, SARS had spread to two dozen countries; more than 8,000 people were affected and 774 died. Quick action, led by the World Health Organization (WHO), contained the disease.[5]

And behind all of this is the knowledge of the 1918 world flu virus, which is estimated to have killed as many as 50 million people in one year, probably more than any other single epidemic in human history. It spread around the world in the autumn, striking otherwise healthy young adults in particular. Many died within hours! By the spring of 1919, the virus had virtually disappeared.[6]

Although outbreaks of the well-known traditional epidemic diseases have declined greatly during the past century in industrialized nations, there is now concern that the incidence of pandemics may increase due to several factors. One is that as the human population grows, people live in new habitats, where previously unknown diseases occur. Another is that strains of disease organisms have developed resistance to antibiotics and other modern methods of control.

A broader view of why diseases are likely to increase comes from an ecological and evolutionary perspective (which will be explained in later chapters). Stated simply, the more than 6.6 billion people on Earth constitute a great resource and opportunity for other species; it is naive to think that other species will not take advantage of this huge and easily accessible host. From this perspective, the future promises more diseases rather than fewer. This is a new perspective. In the mid-20th century it was easy to believe that modern medicine would eventually cure all diseases and that most people would live the maximum human life span. It is generally believed, and often forecast, that the human population will simply continue increasing, without any decline. But with increased crowding and its many effects on the environment, there is also concern that the opposite might happen, that our species might suffer a large, if temporary, dieback. This leads us to consider how populations change over time and space, especially our own populations, and this is the subject of the present chapter.

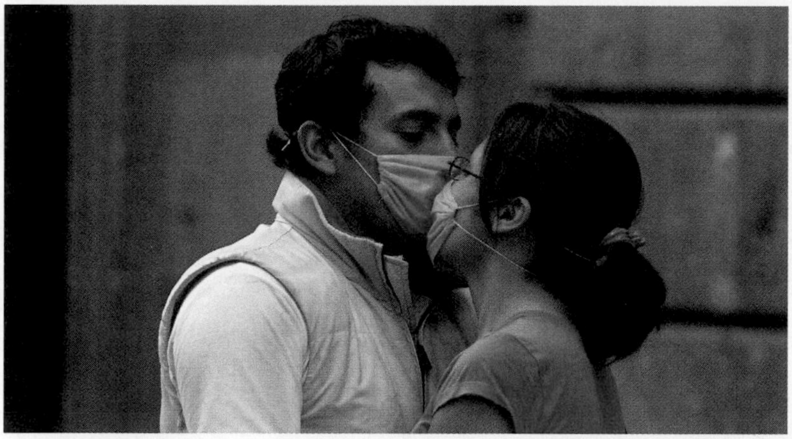

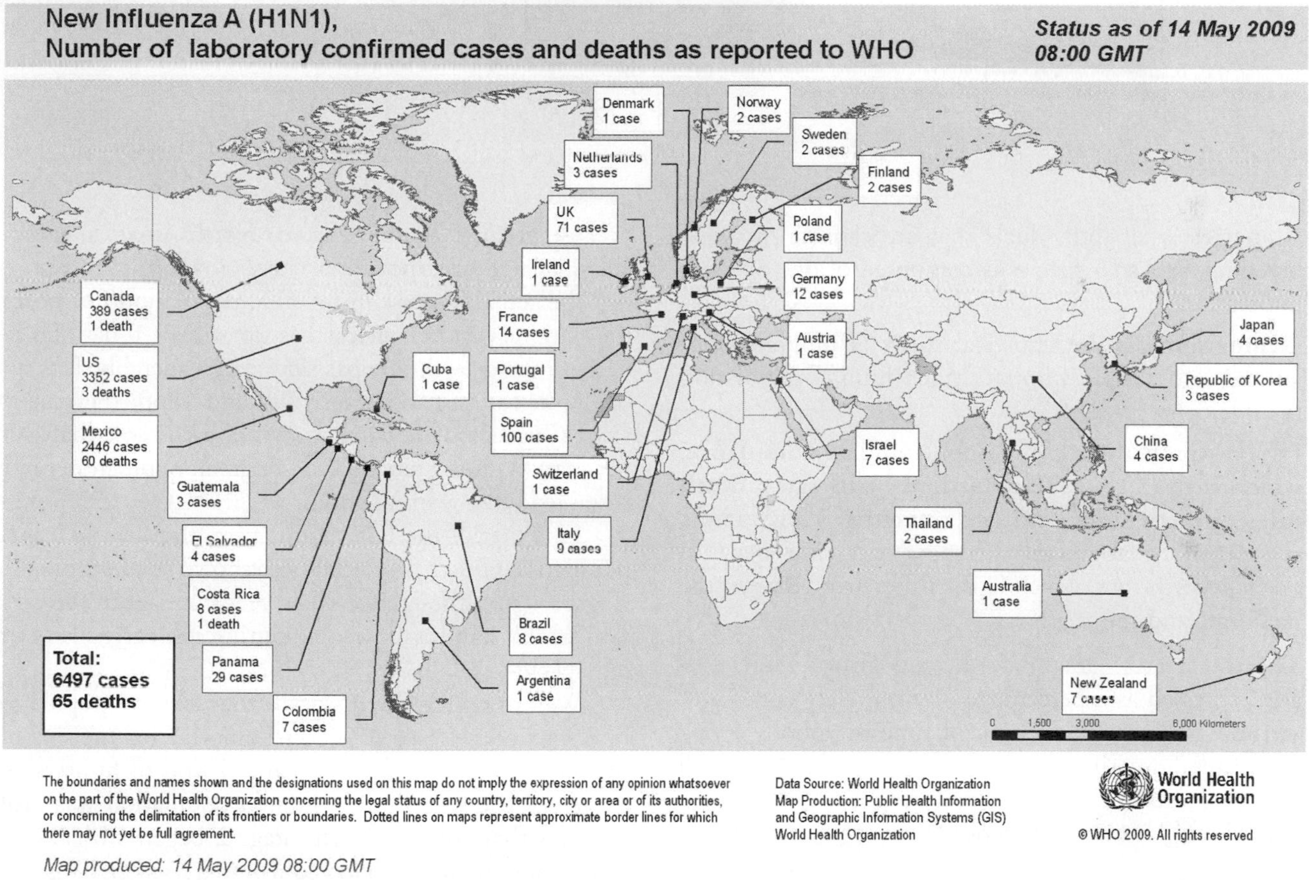

FIGURE 4.1 **(a)** A couple try to take appropriate measures to protect against swine flu. **(b)** Map of the flu's distribution by mid-May 2009. (*Source*: **(b)** World Health Organization).

4.1 Basic Concepts of Population Dynamics

One of the most important properties of living things is that their abundances change over time and space. This is as true for our own species as it is for all others, including those that directly or indirectly affect our lives—for example, by providing our food, or materials for our shelter, or causing diseases and other problems—and those that we just like having around us or knowing that they exist.

In this chapter we focus on the human population because it is so important to all environmental problems, but the concepts we discuss here are useful for the populations of all species, and we will use these concepts throughout this book. You should also familiarize yourself with the following definitions and ideas:

- **Population dynamics** is the general study of population changes.
- A **population** is a group of individuals of the same species living in the same area or interbreeding and sharing genetic information.

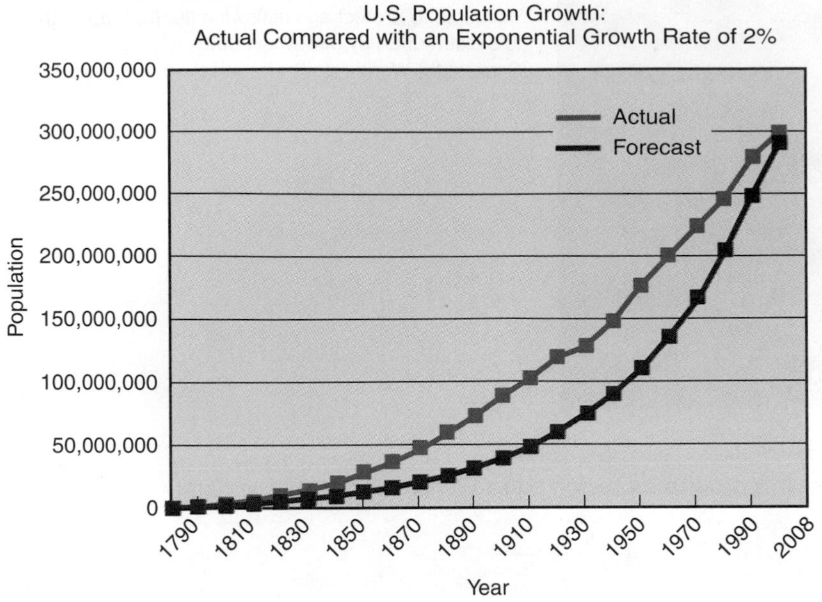

FIGURE 4.2 U.S. population, 1790 to 2008. The actual population growth is shown compared to an exponential curve with an annual growth rate of 2%. (*Source*: Data from U.S. Census Bureau, *US Historical Population Information*.)

- A **species** is all individuals that are capable of interbreeding, and so a species is composed of one or more populations.
- **Demography** is the statistical study of human populations, and people who study the human population include demographers.
- Five key properties of any population are **abundance**, which is the size of a population; **birth rates**; **death rates**; **growth rates**; and **age structure.** How rapidly a population's abundance changes over time depends on its growth rate, which is the difference between the birth rate and the death rate.
- The three rates—birth, death, and growth—are usually expressed as a percentage of a population per unit of time. For people, the unit of time is typically a year or greater. Sometimes these rates are expressed as actual numbers within a population during a specified time. (See Useful Human-Population Terms in Section 4.1.)

Let us begin with the population of the United States, which has grown rapidly since European settlement (Figure 4.2).

The Human Population as an Exponential Growth Curve

It is common to say that human populations, like that of the United States, grow at an **exponential rate**, which means that the annual growth rate is a constant percentage of the population (see Chapter 3). But Figure 4.2 shows that for much of the nation's history the population has grown at a rate that exceeds an exponential. The annual growth rate has changed over time, increasing in the early years, in part because of large immigrations to North America, and decreasing later. An exponential curve growing at 2% per year lags the actual increase in the U.S. population for most of the nation's history but catches up with it today. That is because the growth rate has slowed considerably. It is now 0.6%; in contrast, between 1790 and 1860, the year the Civil War began, the population increased more than 30% per year! (This is a rate that for a human population can be sustained only by immigration.)

Like that of the U.S. population, the world's human population growth is typically also shown as an exponential (Figure 4.3), although we know very little about the variation in the number of people during the early history of our species.

We can divide the history of our species' population into four phases (see A Closer Look 4.1 for more about this history). In Stage 1, the early period of hunters and gatherers, the world's total human population was probably less than a few million. Stage 2 began with the rise of agriculture, which allowed a much greater density of people and the first major increase in the human population. Stage 3, the Industrial Revolution in the late 18th and early 19th centuries, saw improvements in health care and the food supply, which led to a rapid increase in the human population. The growth rate of the world's human population, like that of the early population of the United States, increased but varied during the first part of the 20th century, peaking in 1965–1970 at 2.1% because of improved health care and food production. Stage 4 began around the late 20th century. In this stage, population growth slowed in wealthy, industrialized nations, and although it has continued to increase rapidly in many poorer, less developed nations, globally the growth rate is declining and is now approximately 1.2%.[8]

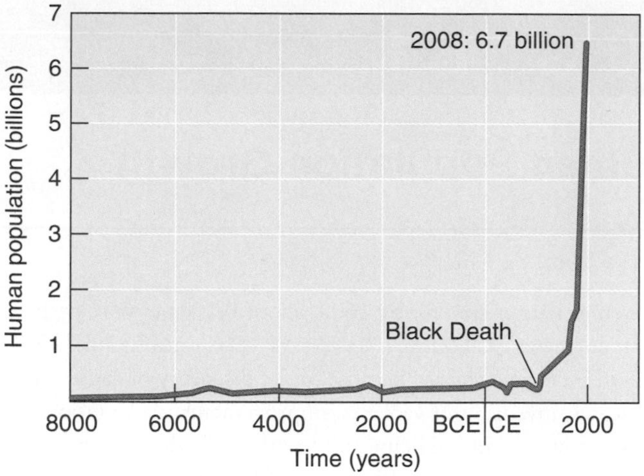

FIGURE 4.3 **Human Population Growth.** It took thousands of years for the human population to reach 1 billion (in 1800) but only 130 years to reach 2 billion (1930). It only took 130 years to reach 3 billion (1960), 15 years to reach 4 billion (1975), 12 years to reach 5 billion (1987), and 12 years to year 6 billion (1999). (*Source*: Reprinted with permission of John Wiley & Sons, Inc.)

Usually in discussions of population dynamics, birth, death, and growth rates are expressed as percentages (the number per 100 individuals). But because the human population is so huge, percentages are too crude a measure, so it is common to state these rates in terms of the number per 1,000, which is referred to as the crude rate. Thus we have the *crude birth rate*, *crude death rate*, and *crude growth rate*. More specifically, here is a list of terms that are used frequently in discussions of human population change and will be useful to us in this book from time to time.

Table 4.1 USEFUL HUMAN-POPULATION TERMS

Crude birth rate: number of births per 1,000 individuals per year; called "crude" because population age structure is not taken into account.

Crude death rate: number of deaths per 1,000 individuals per year.

Crude growth rate: net number added per 1,000 individuals per year; also equal to the crude birth rate minus crude death rate.

Fertility: pregnancy or the capacity to become pregnant or to have children.

General fertility rate: number of live births expected in a year per 1,000 women aged 15–49, considered the childbearing years.

Total fertility rate (TFR): the average number of children expected to be born to a woman throughout her childbearing years.

Age-specific birth rate: number of births expected per year among a fertility-specific age group of women in a population.The fertility-specific age group is, in theory, all ages of women that could have children. In practice, it is typically assumed to be all women between 15 and 49 years old.

Cause-specific death rate: the number of deaths from one cause per 100,000 total deaths.

Morbidity: a general term meaning the occurrence of disease and illness in a population.

Incidence: with respect to disease, the number of people contracting a disease during a specific time period, usually measured per 100 people.

Prevalence: with respect to a disease, the number of people afflicted at a particular time.

Case fatality rate: the percentage of people who die once they contract a disease.

Rate of natural increase (RNI): the birth rate minus the death rate, implying an annual rate of population growth not including migration.

Doubling time: the number of years it takes for a population to double, assuming a constant rate of natural increase.

Infant mortality rate: the annual number of deaths of infants under age 1 per 1,000 live births.

Life expectancy at birth: the average number of years a newborn infant can expect to live given current mortality rates.

GNP per capita: gross national product (GNP), which includes the value of all domestic and foreign output.

(*Source:* C. Haub and D. Cornelius, *World Population Data Sheet* [Washington, DC: Population Reference Bureau, 1998].)

A CLOSER LOOK 4.1

A Brief History of Human Population Growth

STAGE 1. Hunters and Gatherers: From the first evolution of humans to the beginning of agriculture.[7]

Population density: About 1 person per 130–260 km^2 in the most habitable areas.

Total human population: As low as one-quarter million, less than the population of modern small cities like Hartford, Connecticut, and certainly fewer than the number of people—commonly a few million—who now live in many of our largest cities.

Average rate of growth: The average annual rate of increase over the entire history of human population is less than 0.00011% per year.

STAGE 2. Early, Preindustrial Agriculture: Beginning sometime between 9000 B.C. and 6000 B.C. and lasting until approximately the 16th century.

Population density: With the domestication of plants and animals and the rise of settled villages, human population density increased greatly, to about 1 or 2 people/km^2 or more, beginning a second period in human population history.

Total human population: About 100 million by A.D. 1 and 500 million by A.D. 1600.

Average rate of growth: Perhaps about 0.03%, which was high enough to increase the human population from 5 million in 10,000 B.C. to about 100 million in A.D. 1. The Roman Empire accounted for about 54 million. From A.D. 1 to A.D. 1000, the population increased to 200–300 million.

STAGE 3. The Machine Age: Beginning in the 16th century.

Some experts say that this period marked the transition from agricultural to literate societies, when better medical care and sanitation were factors in lowering the death rate.

Total human population: About 900 million in 1800, almost doubling in the next century and doubling again (to 3 billion) by 1960.

Average rate of growth: By 1600, about 0.1% per year, with rate increases of about 0.1% every 50 years until 1950. This rapid increase occurred because of the discovery of causes of diseases, invention of vaccines, improvements in sanitation, other advances in medicine and health, and advances in agriculture that led to a great increase in the production of food, shelter, and clothing.

STAGE 4. The Modern Era: Beginning in the mid-20th century.

Total human population: Reaching and exceeding 6.6 billion.

Average rate of growth: The growth rate of the human population reached 2% in the middle of the 20th century and has declined to 1.2%.[8]

How Many People Have Lived on Earth?

Before written history, there were no censuses. The first estimates of population in Western civilization were attempted in the Roman era. During the Middle Ages and the Renaissance, scholars occasionally estimated the number of people. The first modern census was taken in 1655 in the Canadian colonies by the French and the British.[9] The first series of regular censuses by a country began in Sweden in 1750, and the United States has taken a census every decade since 1790. Most countries began counting their populations much later. The first Russian census, for example, was taken in 1870. Even today, many countries do not take censuses or do not do so regularly. The population of China has only recently begun to be known with any accuracy. However, studying modern primitive peoples and applying principles of ecology can give us a rough idea of the total number of people who may have lived on Earth.

Summing all the values, including those since the beginning of written history, about 50 billion people are estimated to have lived on Earth.[10] If so, then, surprisingly, the more than 6.6 billion people alive today represent more than 10% of all of the people who have ever lived.

4.2 Projecting Future Population Growth

With human population growth a central issue, it is important that we develop ways to forecast what will happen to our population in the future. One of the simplest approaches is to calculate the doubling time.

Exponential Growth and Doubling Time

Recall from Chapter 3 and from the preceding list of Useful Human Population Terms that **doubling time**, a concept used frequently in discussing human population growth, is the time required for a population to double in size (see Working It Out 4.1). The standard way to estimate doubling time is to assume that the population is growing exponentially and then divide 70 by the annual growth rate stated as a percentage. (Dividing into 70 is a consequence of the mathematics of exponential growth.)

The doubling time based on exponential growth is very sensitive to the growth rate—it changes quickly as the growth rate changes (Figure 4.4). A few examples demonstrate this sensitivity. With a current population growth of 1.0%, the United States has a doubling time of 70 years. In contrast, the current growth rate of Nicaragua is 2.0%, giving that nation a doubling time of 35 years. Sweden, with an annual rate of about 0.2%, has a doubling time of 350 years. The world's most populous country, China, has a growth rate of 0.6% and a 117-year doubling time.[11]

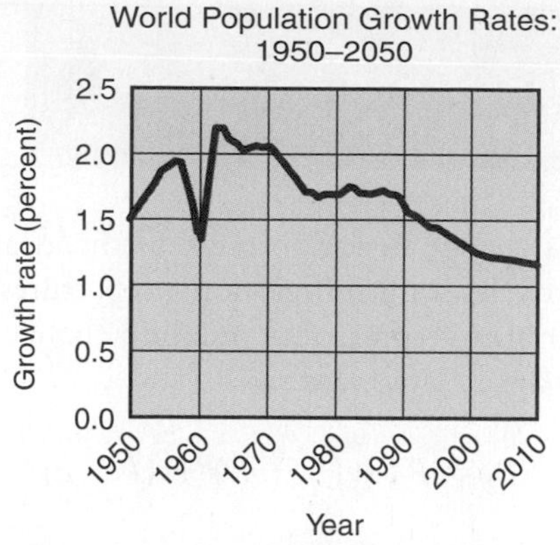

FIGURE 4.5 The annual growth rate of the world's population has been declining since the 1960s.

The world's population growth rate peaked in the 1960s at about 2.2% and is now about 1.1% (Figure 4.5). If the growth rate had continued indefinitely at the 1960s peak, the world population would have doubled in 32 years. At today's rate, it would double in 64 years.

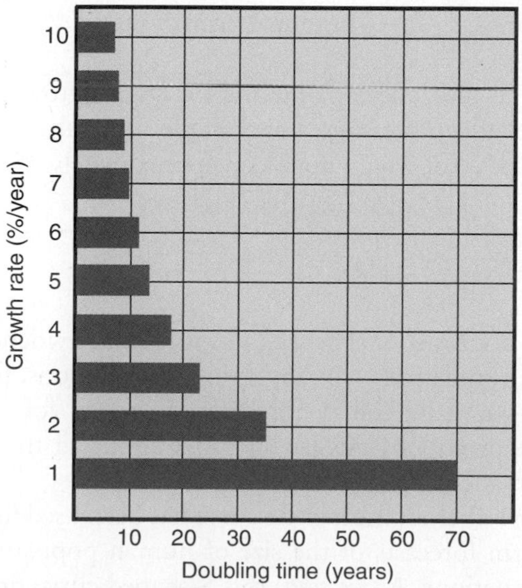

FIGURE 4.4 Doubling time changes rapidly with the population growth rate. Because the world's population is increasing at a rate between 1 and 2%, we expect it to double within the next 35 to 70 years.

Human Population as a Logistic Growth Curve

An exponentially growing population theoretically increases forever. However, on Earth, which is limited in size, this is not possible, as Thomas Henry Malthus pointed out in the 18th century (see A Closer Look 4.2). Eventually the population would run out of food and space and become increasingly vulnerable to catastrophes, as we are already beginning to observe. Consider, a population of 100 increasing at 5% per year would grow to 1 billion in less than 325 years. If the human population had increased at this rate since the beginning of recorded history, it would now exceed all the known matter in the universe.

If a population cannot increase forever, what changes in the population can we expect over time? One of the first suggestions made about population growth is that it would follow a smooth S-shaped curve known as the **logistic growth curve** (see Chapter 3). This was first suggested in 1838 by a European scientist, P. F. Verhulst, as a theory for the growth of animal populations. It has been applied widely to the growth of many animal populations, including those important in wildlife management, endangered species and those in fisheries (see Chapter 13), as well as the human population.

WORKING IT OUT 4.1 Forecasting Population Change

Populations change in size through births, deaths, immigration (arrivals), and emigration (departures). We can write a formula to represent population change in *terms of actual numbers in a population*:

$$P_2 = P_1 + (B - D) + (I - E)$$

where P_1 is the number of individuals in a population at time 1, P_2 is the number of individuals in that population at some later time 2, B is the number of births in the period from time 1 to time 2, D is the number of deaths from time 1 to time 2, I is the number entering as immigrants, and E is the number leaving as emigrants.

So far we have expressed population change in terms of total numbers in the population. We can also express these as rates, including the birth rate (number born divided by the total number in the population), death rate (number dying divided by the total number in the population), and growth rate (change in the population divided by the total number in the population). (In this section, we will use lowercase letters to represent a rate, uppercase letters to represent total amounts.)

Ignoring for the moment immigration and emigration, how rapidly a population changes depends on the growth rate, g, which is the difference between the birth rate and the death rate (see earlier list of useful terms). For example, in 1999 the crude death rate, d, in the United States was 9, meaning that 9 of every 1,000 people died each year. (The same information expressed as a percentage is a rate of 0.9%.) In 1999 the crude birth rate, b, in the United States was 15.[12] The crude growth rate is the net change—the birth rate minus the death rate. Thus the crude growth rate, g, in the United States in 1999 was 6. For every 1,000 people at the beginning of 1999, there were 1,006 at the end of the year.

Continuing for the moment to ignore immigration and emigration, we can state that how rapidly a population grows depends on the difference between the birth rate and the death rate. The growth rate of a population is then

$$g = (B - D)/N \text{ or } g = G/N$$

Note that in all these cases, the units are numbers per unit of time.

It is important to be consistent in using the population at the beginning, middle, or end of the period. Usually, the number at the beginning or the middle is used. Consider an example: There were 19,700,000 people in Australia in mid-2002, and 394,000 births from 2002 to 2003. The birth rate, b, calculated against the mid-2002 population was 394,000/19,700,000, or 2%. During the same period, there were 137,900 deaths; the death rate, d, was 137,900/19,700,000, or 0.7%. The growth rate, g, was (394,000 – 137,900)/19,700,000, or 1.3%.[13]

Recall from Chapter 3 that doubling time—the time it takes a population to reach twice its present size—can be estimated by the formula

$$T = 70/\text{annual growth rate}$$

where T is the doubling time and the annual growth rate is expressed as a percentage. For example, a population growing 2% per year would double in approximately 35 years.

A logistic population would increase exponentially only temporarily. After that, the rate of growth would gradually decline (i.e., the population would increase more slowly) until an upper population limit, called the **logistic carrying capacity**, was reached. Once that had been reached, the population would remain at that number.

Although the logistic growth curve is an improvement over the exponential, it too involves assumptions that are unrealistic for humans and other mammals. Both the exponential and logistic assume a constant environment and a homogeneous population—one in which all individuals are identical in their effects on each other. In addition to these two assumptions, the logistic assumes a constant carrying capacity, which is also unrealistic in most cases, as we will discuss later. There is, in short, little evidence that human populations—or any animal populations, for that matter—actually follow this growth curve, for reasons that are pretty obvious if you think about all the things that can affect a population.[14]

Nevertheless, the logistic curve has been used for most long-term forecasts of the size of human populations in specific nations. As we said, this S-shaped curve first rises steeply upward and then changes slope, curving toward the horizontal carrying capacity. The point at which the curve changes is the **inflection point**, and until a population has reached this point, we cannot project its final logistic size. The human population had not yet made the bend around the inflection point, but forecasters typically dealt with this problem by assuming that the population was just reaching the inflection point at the time the forecast was made. This standard practice inevitably led to a great underestimate of the maximum population. For example, one of the first projections of the upper limit of the U.S. population, made in the 1930s, assumed that the inflection point had been reached then. That assumption resulted in an estimate that the final population of the United States would be approximately 200 million.[15]

Fortunately for us, Figure 4.5 suggests that our species' growth rate has declined consistently since the 1960s, as we noted before, and therefore we can make projections using the logistic, assuming that we have passed the inflection point. The United Nations has made a series of projections based on current birth rates and death rates and assumptions about how these rates will change. These projections form the basis for the curves presented in Figure 4.6. The logistic projections assume that (1) mortality will fall everywhere and level off when female life expectancy reaches 82 years; (2) fertility will reach replacement levels everywhere between 2005 and 2060; and (3) there will be no worldwide catastrophe. This approach projects an equilibrium world population of 10.1–12.5 billion.[16] Developed nations would experience population growth from 1.2 billion today to 1.9 billion, but populations in developing nations would increase from 4.5 billion to 9.6 billion. Bangladesh (an area the size of Wisconsin) would reach 257 million; Nigeria, 453 million; and India, 1.86 billion. In these projections, the developing countries contribute 95% of the increase.[14]

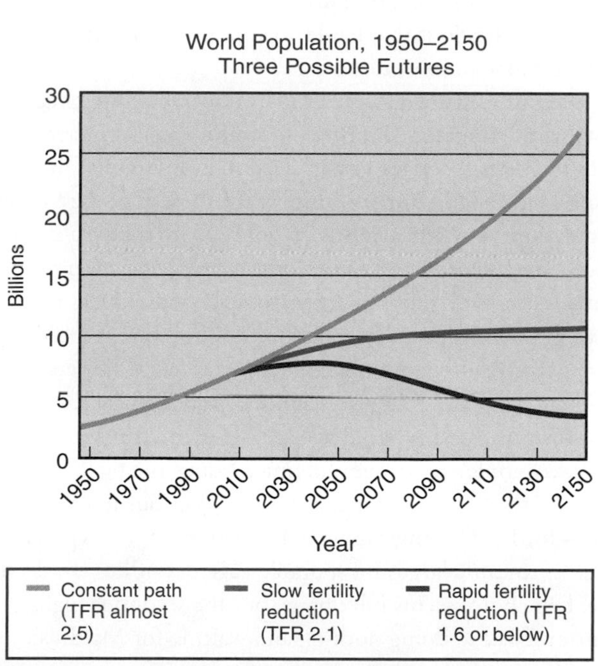

FIGURE 4.6 U.N. projections of world population growth based on the logistic curve and using different total fertility rates (the expected number of children a woman will have during her life—see Chapter 3. The constant path assumes the 1998 growth rate will continue unchanged, resulting in an exponential increase. The slow-fertility-reduction path assumes the world's fertility will decline, reaching replacement level by 2050, so the world's population will stabilize at about 11 billion by the 22nd century. The rapid-fertility reduction path assumes the total fertility rate will go into decline in the 21st century, with the population peaking at 7.7 billion in 2050 and dropping to 3.6 billion by 2150. These are theoretical curves. The total fertility rate has remained high and is now 2.7. (*Source:* U.S. Census Bureau, *Global Population Profile: 2002*, Table A-10,. p.1. http://www.census.gov/ipc/prod/wp02/tabA-10.pdf).

4.3 Age Structure

As we noted earlier, the two standard methods for forecasting human population growth—the exponential and the logistic—ignore all characteristics of the environment and in that way are seriously incomplete. A more comprehensive approach would take into account the effects of the supply of food, water, and shelter; the prevalence of diseases; and other factors that can affect birth and death rates. But with long-lived organisms like ourselves, these environmental factors have different effects on different age groups, and so the next step is to find a way to express how a population is divided among ages. This is known as the population age structure, which is the proportion of the population of each age group. The age structure of a population affects current and future birth rates, death rates, and growth rates; has an impact on the environment; and has implications for current and future social and economic conditions.

We can picture a population's age structure as a pile of blocks, one for each age group, with the size of each block representing the number of people in that group (Figure 4.7). Although age structures can take many shapes, four general types are most important to our discussion: a pyramid, a column, an inverted pyramid (top-heavy), and a column with a bulge. The pyramid age structure occurs in a population that has many young people and a high death rate at each age—and therefore a high birth rate, characteristic of a rapidly growing population and also of a population with a relatively short average lifetime. A column shape occurs where the birth rate and death rate are low and a high percentage of the population is elderly. A bulge occurs if some event in the past caused a high birth rate or death rate for some age group but not others. An inverted pyramid occurs when a population has more older than younger people.

Age structure varies considerably by nation (Figure 4.7) and provides insight into a population's history, its current status, and its likely future. Kenya's pyramid-shaped age structure reveals a rapidly growing population heavily weighted toward youth. In developing

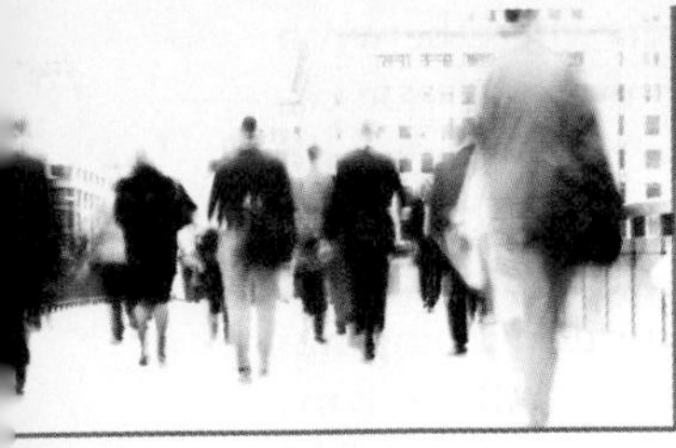

A CLOSER LOOK 4.2

The Prophecy of Malthus

Almost 200 years ago, the English economist Thomas Malthus eloquently stated the human population problem. His writings have gone in and out of fashion, and some people think his views may be out-of-date, but in 2008 Malthus was suddenly back on the front page, the focus of major articles in the *New York Times*[17] and the *Wall Street Journal*, among other places. Perhaps this is because recent events—from natural catastrophes in Asia to rising prices for oil, food, and goods in general—suggest that the human population problem really is a problem.

Malthus based his argument on three simple premises: [18]

- **Food is necessary for people to survive.**
- **"Passion between the sexes is necessary and will remain nearly in its present state"—so children will continue to be born.**
- **The power of population growth is infinitely greater than the power of Earth to produce subsistence.**

Malthus reasoned that it would be impossible to maintain a rapidly multiplying human population on a finite resource base. His projections of the ultimate fate of humankind were dire, as dismal a picture as that painted by today's most extreme pessimists. The power of population growth is so great, he wrote, that "premature death must in some shape or other visit the human race. The vices of mankind are active and able ministers of depopulation, but should they fail, sickly seasons, epidemics, pestilence and plague, advance in terrific array, and sweep off their thousands and ten thousands." Should even these fail, he said, "gigantic famine stalks in the rear, and with one mighty blow, levels the population with the food of the world."

Malthus's statements are quite straightforward. From the perspective of modern science, they simply point out that in a finite world nothing can grow or expand forever, not even the population of the smartest species ever to live on Earth. Critics of Malthus continue to point out that his predictions have yet to come true, that whenever things have looked bleak, technology has provided a way out, allowing us to live at greater densities. Our technologies, they insist, will continue to save us from a Malthusian fate, so we needn't worry about human population growth. Supporters of Malthus respond by reminding them of the limits of a finite world.

Who is correct? Ultimately, in a finite world, Malthus must be correct about the final outcome of unchecked growth. He may have been wrong about the timing; he did not anticipate the capability of technological changes to delay the inevitable. But although some people believe that Earth can support many more people than it does now, in the long run there must be an upper limit. The basic issue that confronts us is this: How can we achieve a constant world population, or at least halt the increase in population, in a way most beneficial to most people? This is undoubtedly one of the most important questions that has ever faced humanity, and it is coming home to roost now.

Recent medical advances in our understanding of aging, along with the potential of new biotechnology to increase both the average longevity and maximum lifetime of human beings, have major implications for the growth of the human population. As medical advances continue to take place, the death rate will drop and the growth rate will rise even more. Thus, a prospect that is positive from the individual's point of view—a longer, healthier, and more active life—could have negative effects on the environment. We will therefore ultimately face the following choices: Stop medical research into chronic diseases of old age and other attempts to increase people's maximum lifetime; reduce the birth rate; or do neither and wait for Malthus's projections to come true—for famine, environmental catastrophes, and epidemic diseases to cause large and sporadic episodes of human death. The first choice seems inhumane, but the second is highly controversial, so doing nothing and waiting for Malthus's projections may be what actually happens, a future that nobody wants. For the people of the world, this is one of the most important issues concerning science and values and people and nature.

countries today, about 34% of the populations are under 15 years of age. Such an age structure indicates that the population will grow very rapidly in the future, when the young reach marriage and reproductive ages, and it suggests that the future for such a nation requires more jobs for the young. This type of age structure has many other social implications that go beyond the scope of this book.

In contrast, the age structure of the United States is more like a column, showing a population with slow growth, while Japan's top-heavy pyramid shows a nation with declining growth.[8] The U.S. age struc-

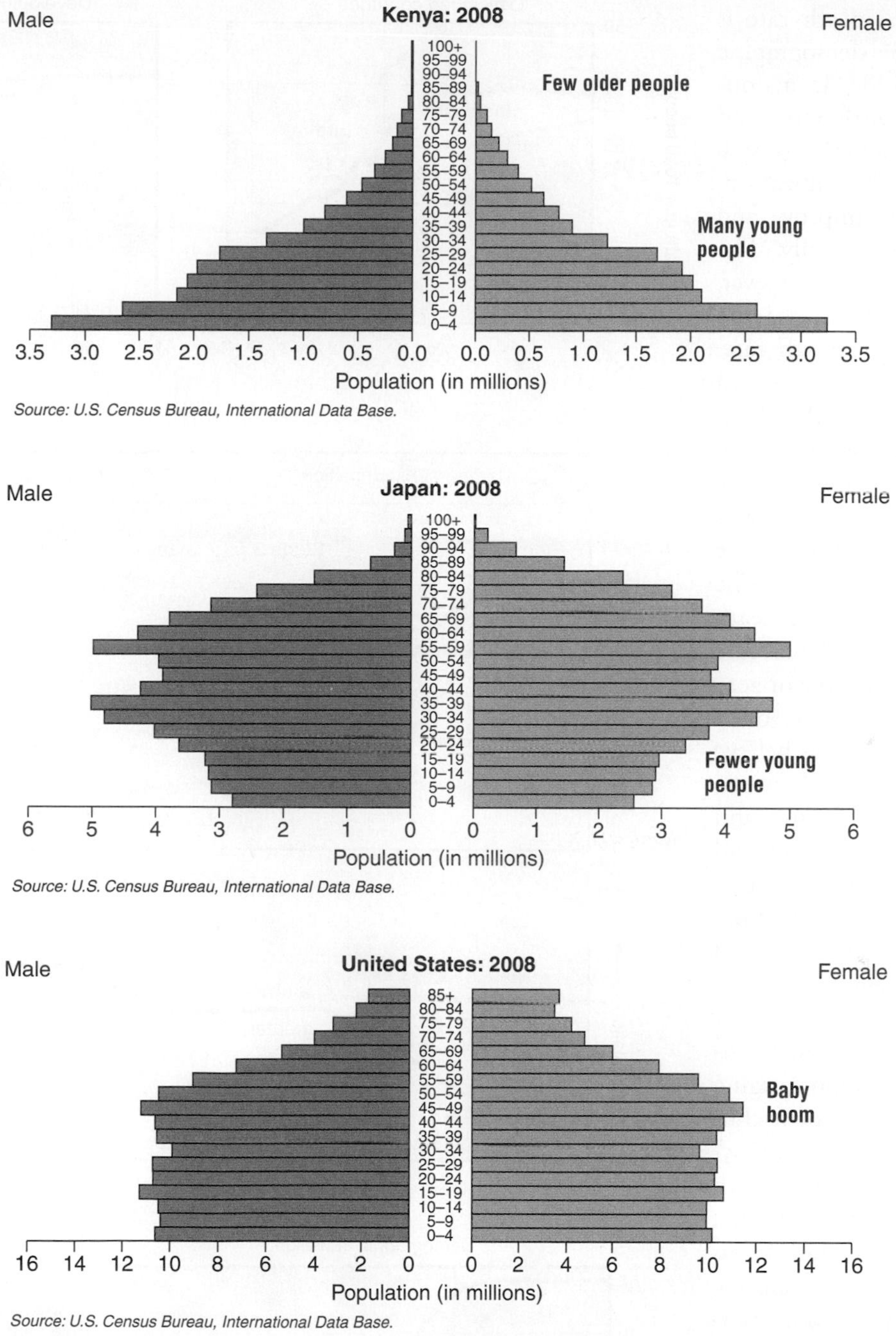

FIGURE 4.7 **Age structure of Kenya, the United States, and Japan, 2008.** The bars to the left are males; those to the right are females. (*Source*: U.S. Bureau of the Census.)

ture also shows the baby boom that occurred in the United States after World War II; a great increase in births from 1946 through 1964 forms a pulse in the population that can be seen as a bulge in the age structure, especially of those aged 45–55 in 2008. At each age, the baby boomers increased demand for social and economic resources; for example, schools were crowded when the baby boomers were of primary- and secondary-school age.

4.4 The Demographic Transition

The **demographic transition** is a three-stage pattern of change in birth rates and death rates that has occurred during the process of industrial and economic development of Western nations. It leads to a decline in population growth.

A decline in the death rate is the first stage of the demographic transition (Figure 4.8).[7] In a nonindustrial country, birth rates and death rates are high, and the growth rate is low. With industrialization, health and sanitation improve and the death rate drops rapidly. The birth rate remains high, however, and the population enters Stage II, a period with a high growth rate. Most European nations passed through this period in the 18th and 19th centuries. As education and the standard of living increase and as family-planning methods become more widely used, the population reaches Stage III. The birth rate drops toward the death rate, and the growth rate therefore declines, eventually to a low or zero growth rate. However, the birth rate declines only if families believe there is a direct connection between future economic well-being and funds spent on the education and care of their young. Such families have few children and put all their resources toward the education and well-being of those few.

Historically, parents have preferred to have large families. Without other means of support, aging parents can depend on grown children for a kind of "social security," and even young children help with many kinds of hunting, gathering, and low-technology farming. Unless there is a change in attitude among parents—unless they see more benefits from a few well-educated children than from many poorer children—nations face a problem in making the transition from Stage II to Stage III (see Figure 4.8c).

Some developed countries are approaching Stage III, but it is an open question whether developing nations will make the transition before a serious population crash occurs. *The key point here is that the demographic transition will take place only if parents come to believe that having a small family is to their*

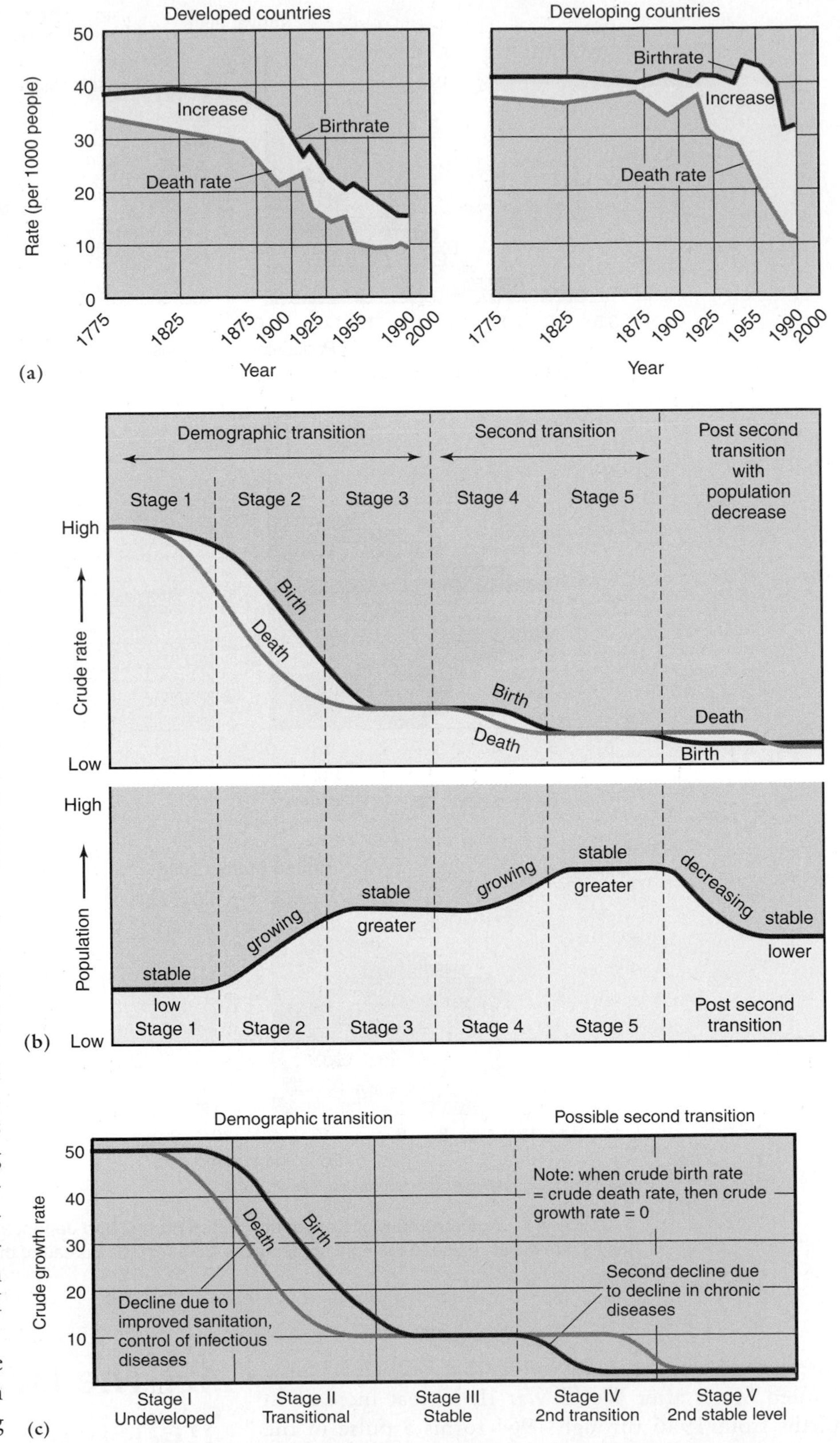

FIGURE 4.8 The demographic transition: (a) Theoretical, including possible fourth and fifth stages that might take place in the future; **(b)** the resulting relative change in population; **(c)** the change in birth rates and death rates from 1775 to 2000 in developed and developing countries. (*Source:* M.M. Kent and K.A. Crews, *World Population: Fundamentals of Growth* [Washington, DC: Population Reference Bureau, 1990]. Copyright 1990 by the Population Reference Bureau, Inc. Reprinted by permission.)

benefit. Here we again see the connection between science and values. Scientific analysis can show the value of small families, but this knowledge must become part of cultural values to have an effect.

Potential Effects of Medical Advances on the Demographic Transition

Although the demographic transition is traditionally defined as consisting of three stages, advances in treating chronic health problems such as heart disease can lead a Stage III country to a second decline in the death rate. This could bring about a second transitional phase of population growth (Stage IV), in which the birth rate would remain the same while the death rate fell. A second stable phase of low or zero growth (Stage V) would be achieved only when the birth rate declined even further to match the decline in the death rate. Thus, there is danger of a new spurt of growth even in industrialized nations that have passed through the standard demographic transition.

4.5 Longevity and Its Effect on Population Growth

The **maximum lifetime** is the genetically determined maximum possible age to which an individual of a species *can* live. **Life expectancy** is the average number of years an individual *can expect* to live given the individual's present age. Technically, life expectancy is an age-specific number: Each age class within a population has its own life expectancy. For general comparison, however, we use the life expectancy at birth.

Life expectancy is much higher in developed, more prosperous nations. Nationally, the highest life expectancy is 84 years, in the tiny nation of Macau. Of the major nations, Japan has the highest life expectancy, 82.1 years. Sixteen other nations have a life expectancy of 80 years or more: Singapore, Hong Kong, Australia, Canada, France, Guernsey, Sweden, Switzerland, Israel, Anguilla, Iceland, Bermuda, Cayman Islands, New Zealand, Gibraltar, and Italy. The United States, one of the richest countries in the world, ranks 50th among nations in life expectancy, at 78 years. China has a life expectancy of just over 73 years; India just over 69 years. Swaziland has the lowest of all nations at 32 years. The ten nations with the shortest life expectancies are all in Africa.[19] Not surprisingly, there is a relationship between per capita income and life expectancy.

A surprising aspect of the second and third periods in the history of human population is that population growth occurred with little or no change in the maximum lifetime. What changed were birth rates, death rates, population growth rates, age structure, and average life expectancy.

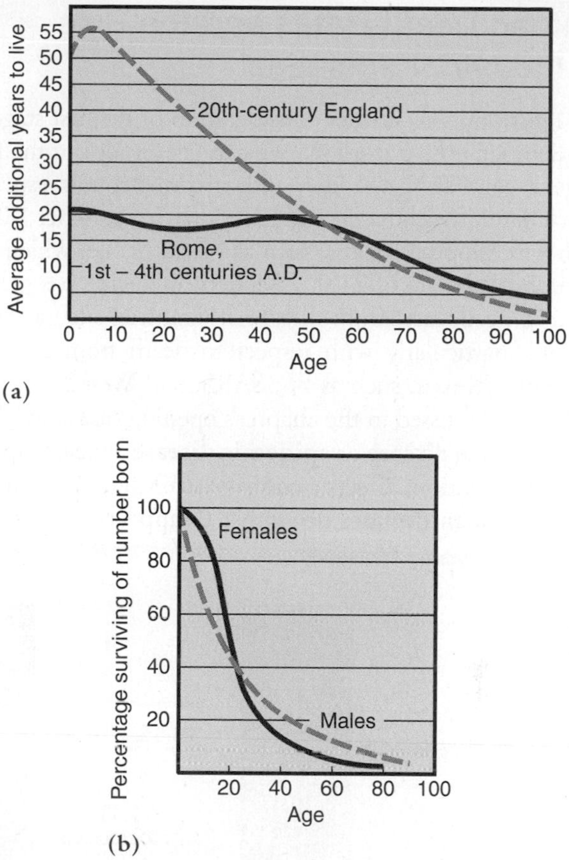

FIGURE 4.9 **(a). Life expectancy in ancient Rome and 20th-century England**. This graph shows the average number of years one could expect to live after reaching a given age: for example, a 10-year-old in England could expect to live about 55 more years; a 10-year-old in Rome, about 20 more years. Life expectancy was greater in 20th-century England than in ancient Rome until about age 55. An 80-year-old Roman could expect to live longer than an 80-year-old Briton. Data for Romans is reconstructed from ages on tombstones. **(b)**. Approximate survivorship curve for Rome for the first four centuries C.E. The percentage surviving drops rapidly in the early years, reflecting high mortality rates for children in ancient Rome. Females had a slightly higher survivorship rate until age 20, after which males had a slightly higher rate. (*Source*: Modified from G.E. Hutchinson, *An Introduction to Population Ecology* [New Haven, CT: Yale University Press, 1978]. Copyright 1978 by Yale University Press. Used by permission.)

Ages at death, from information carved on tombstones, tell us that the chances of a 75-year-old living to age 90 were greater in ancient Rome than they are today in England (Figure 4.9). These also suggest that death rates were much higher in Rome than in 20th-century England. In ancient Rome, the life expectancy of a 1-year-old was about 22 years, while in 20th-century England it was about 50 years. Life expectancy in 20th-century England was greater than in ancient Rome for all ages until age 55, after which it appears to have been higher for ancient Romans than for 20th-century Britons. This suggests that many hazards of modern life may be concentrated more on the aged. Pollution-induced diseases are one factor in this change.

Human Death Rates and the Rise of Industrial Societies

We return now to further consideration of the first stage in the demographic transition. We can get an idea of the first stage by comparing a modern industrialized country, such as Switzerland, which has a crude death rate of 8.59 per 1,000, with a developing nation, such as Sierra Leone, which has a crude death rate of 21.9.[20] Modern medicine has greatly reduced death rates from disease in countries such as Switzerland, particularly with respect to death from acute or epidemic diseases, such as flu, SARS, and West Nile virus, which we discussed in the chapter's opening case study.

An **acute disease** or **epidemic disease** appears rapidly in the population, affects a comparatively large percentage of it, and then declines or almost disappears for a while, only to reappear later. Epidemic diseases typically are rare but have occasional outbreaks during which a large proportion of the population is infected. A **chronic disease**, in contrast, is always present in a population, typically occurring in a relatively small but relatively constant percentage of the population. Heart disease, cancer, and stroke are examples.

The great decrease in the percentage of deaths due to acute or epidemic diseases can be seen in a comparison of causes of deaths in Ecuador in 1987 and in the United States in 1900, 1987, and 1998 (Figure 4.10).[21] In Ecuador, a developing nation, acute diseases and those listed as "all others" accounted for about 60% of mortality in 1987. In the United States in 1987, these accounted for only 20% of mortality. Chronic diseases account for about 70% of mortality in the modern United States. In

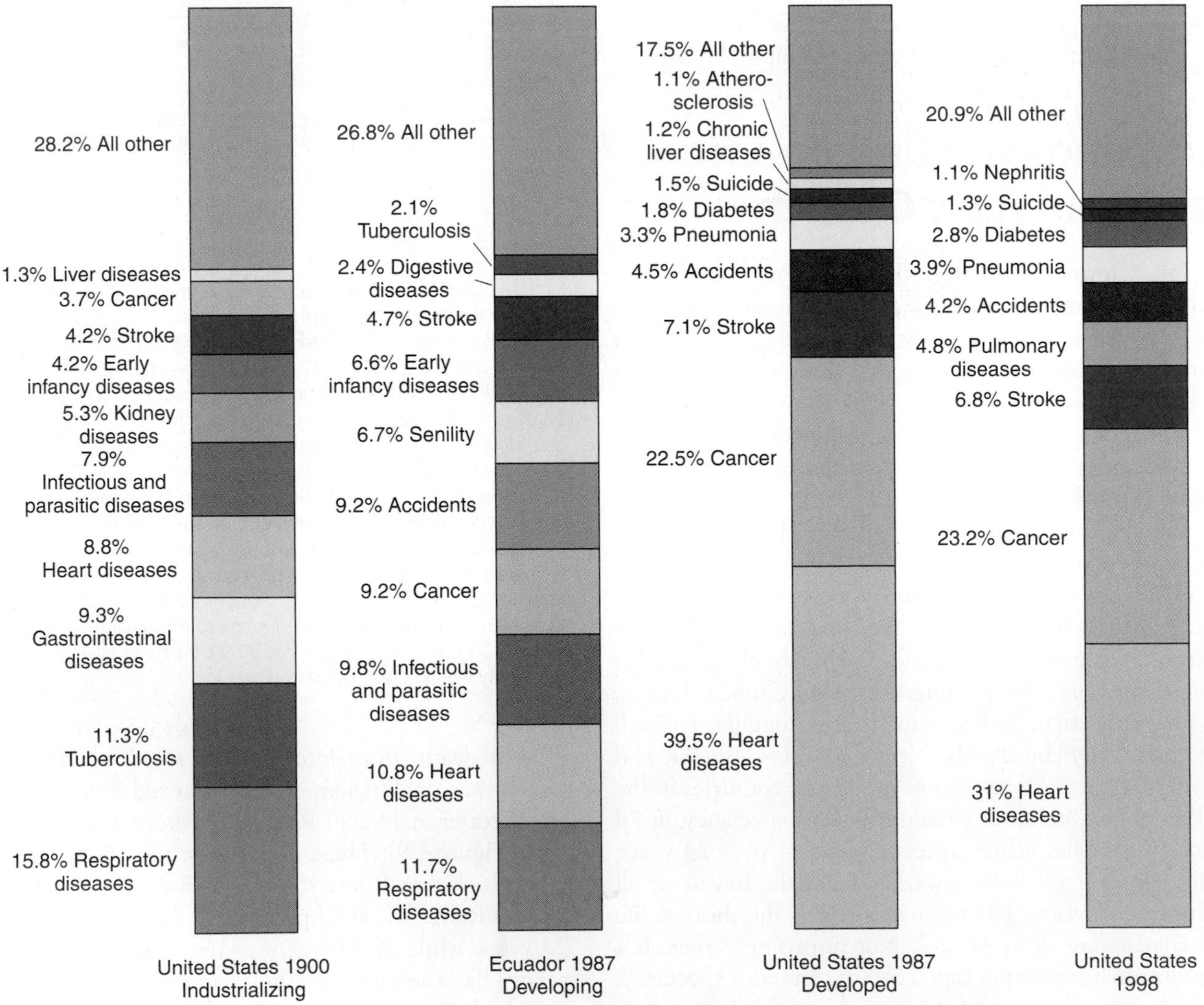

FIGURE 4.10 **Causes of mortality in industrializing, developing, and industrialized nations.** (Sources: U.S. 1900, Ecuador 1987, and U.S. 1987 data from M.M. Kent and K. A. Crews, *World Population: Fundamentals of Growth* [Washington, DC: Population Reference Bureau, 1990]. Copyright 1990 by the Population Reference Bureau, Inc. Reprinted by permission. *National Vital Statistics Report* 48 [11], July 24, 2000.)

contrast, chronic diseases accounted for less than 20% of the deaths in the United States in 1900 and about 33% in Ecuador in 1987. Ecuador in 1987, then, resembled the United States of 1900 more than it resembled the United States of either 1987 or 1998.

4.6 The Human Population's Effects on the Earth

The danger that the human population poses to the environment is the result of two factors: the number of people and the environmental impact of each person. When there were few people on Earth and limited technology, the human impact was primarily local. Even so, people have affected the environment for a surprisingly long time. It started with the use of fire to clear land, and it continued, new research shows, with large effects on the environment by early civilizations. For example, the Mayan temples in South America, standing now in the midst of what were recently believed to be ancient rain forests, actually stood in large areas of farmed land cleared by the Maya. Large areas of North America were modified by American Indians, who used fire for a variety of reasons and modified the forests of the eastern United States.[22] The problem now is that there are so many people and our technologies are so powerful that our effects on the environment are even more global and significant. This could cause a negative feedback—the more people, the worse the environment; the worse the environment, the fewer people.

The simplest way to characterize the total impact of the human population on the environment is to multiply the average impact of an individual by the total number of individuals,[23] or

$$T = P \times I$$

where P is the population size—the number of people—and I is the average environmental impact per person. Of course, the impact per person varies widely, within the same nation and also among nations. The average impact of a person who lives in the United States is much greater than the impact of a person who lives in a low-technology society. But even in a poor, low-technology nation like Bangladesh, the sheer number of people leads to large-scale environmental effects.

Modern technology increases the use of resources and enables us to affect the environment in many new ways, compared with hunters and gatherers or people who farmed with simple wooden and stone tools. For example, before the invention of chlorofluorocarbons (CFCs), which are used as propellants in spray cans and as coolants in refrigerators and air conditioners, we were not causing depletion of the ozone layer in the upper atmosphere. Similarly, before we started driving automobiles, there was much less demand for steel, little demand for oil, and much less air pollution. These linkages between people and the global environment illustrate the global theme and the people-and-nature theme of this book.

The population-times-technology equation reveals a great irony involving two standard goals of international aid: improving the standard of living and slowing overall human population growth. Improving the standard of living increases the total environmental impact, countering the environmental benefits of a decline in population growth.

4.7 The Human Carrying Capacity of Earth

What is the **human carrying capacity** of Earth—that is, how many people can live on Earth at the same time? The answer depends on what quality of life people desire and are willing to accept.

As we have made clear in this chapter, on our finite planet the human population will eventually be limited by some factor or combination of factors. We can group limiting factors into those that affect a population during the year in which they become limiting (short-term factors), those whose effects are apparent after one year but before ten years (intermediate-term factors), and those whose effects are not apparent for ten years (long-term factors). Some factors fit into more than one category, having, say, both short-term and intermediate-term effects.

An important *short-term* factor is the disruption of food distribution in a country, commonly caused by drought or by a shortage of energy for transporting food.

Intermediate-term factors include desertification; dispersal of certain pollutants, such as toxic metals, into waters and fisheries; disruption in the supply of nonrenewable resources, such as rare metals used in making steel alloys for transportation machinery; and a decrease in the supply of firewood or other fuels for heating and cooking.

Long-term factors include soil erosion, a decline in groundwater supplies, and climate change. A decline in resources available per person suggests that we may already have exceeded Earth's long-term human carrying capacity. For example, wood production peaked at 0.67 m^3/person (0.88 yd^3/person) in 1967, fish production at 5.5 kg/person (12.1 lb/person) in 1970, beef at 11.81 kg/person (26.0 lb/person) in 1977, mutton at 1.92 kg/person (4.21 lb/person) in 1972, wool at

0.86 kg/person (1.9 lb/person) in 1960, and cereal crops at 342 kg/person (754.1 lb/person) in 1977.[24] Before these peaks were reached, per capita production of each resource had grown rapidly.

Since the rise of the modern environmental movement in the second half of the 20th century, much attention has focused on estimating the human carrying capacity of Earth—the total number of people that our planet could support indefinitely. This estimation has typically involved three methods. One method, which we have already discussed, is to simply extrapolate from past growth, assuming that the population will follow an S-shaped logistic growth curve and gradually level off (Figure 4.6).

The second method can be referred to as the packing-problem approach. This method simply considers how many people might be packed onto Earth, not taking into sufficient account the need for land and oceans to provide food, water, energy, construction materials, the need to maintain biological diversity, and the human need for scenic beauty. This approach, which could also be called the standing-room-only approach, has led to very high estimates of the total number of people that might occupy Earth—as many as 50 billion.

More recently, a philosophical movement has developed at the other extreme. Known as deep ecology, this third method makes sustaining the biosphere the primary moral imperative. Its proponents argue that the whole Earth is necessary to sustain life, and therefore everything else must be sacrificed to the goal of sustaining the biosphere. People are considered active agents of destruction of the biosphere, and therefore the total number of people should be greatly reduced.[25] Estimates based on this rationale for the desirable number of people vary greatly, from a few million up.

Between the packing-problem approach and the deep-ecology approach are a number of options. It is possible to set goals in between these extremes, but each of these goals is a value judgment, again reminding us of one of this book's themes: *science and values*. What constitutes a desirable quality of life is a value judgment. The perception of what is desirable will depend in part on what we are used to, and this varies greatly. For example, in the United States, New Jersey has only a half acre (0.22 ha) per person, while Wyoming, the most sparsely populated of the lower 48 states, has 116 acres (47.2 ha) per person. For comparison, New York City's Manhattan Island has 71,000 people per square mile, which works out to an area of about 20 × 20 feet per person. Manhattanites manage to live comfortably by using not just the land area but also the airspace to a considerable height. Still, it's clear that people used to living in Wyoming and people living in New Jersey or in Manhattan skyscrapers are likely to have very different views on what is a desirable population density.

Moreover, what quality of life is possible depends not just on the amount of space available but also on technology, which in turn is affected by science. Scientific understanding also tells us what is required to meet each quality-of-life level. The options vary. If all the people of the world were to live at the same level as those of the United States, with our high resource use, then the carrying capacity would be comparatively low. If all the people of the world were to live at the level of those in Bangladesh, with all of its risks as well as its poverty and its heavy drain on biological diversity and scenic beauty, the carrying capacity would be much higher.

In summary, the acceptable carrying capacity is not simply a scientific issue; it is an issue combining science and values, within which science plays two roles. First, by leading to new knowledge, which in turn leads to new technology, it makes possible both a greater impact per individual on Earth's resources and a higher density of human beings. Second, scientific methods can be used to forecast a probable carrying capacity once a goal for the average quality of life, in terms of human values, is chosen. In this second use, science can tell us the implications of our value judgments, but it cannot provide those value judgments.

4.8 Can We Achieve Zero Population Growth?

We have surveyed several aspects of population dynamics. The underlying question is: Can we achieve **zero population growth**—a condition in which the human population, on average, neither increases nor decreases? Much of environmental concern has focused on how to lower the human birth rate and decrease our population growth. As with any long-lived animal population, our species could take several possible approaches to achieving zero population growth. Here are a few.

Age of First Childbearing

The simplest and one of the most effective means of slowing population growth is to delay the age of first childbearing.[26] As more women enter the workforce and as education levels and standards of living rise, this delay occurs naturally. Social pressures that lead to deferred marriage and childbearing can also be effective (Figure 4.11).

FIGURE 4.11 As more and more women enter the workforce and establish professional careers, the average age of first childbearing tends to rise. The combination of an active lifestyle that includes children is illustrated here by the young mother jogging with her child in Perth, Australia.

Typically, countries where early marriage is common have high population growth rates. In South Asia and in Sub-Saharan Africa, about 50% of women marry between the ages of 15 and 19, and in Bangladesh women marry on average at age 16. In Sri Lanka, however, the average age for marriage is 25. The World Bank estimates that if Bangladesh adopted Sri Lanka's marriage pattern, families could average 2.2 fewer children.[26] For many countries, raising the marriage age could account for 40–50% of the drop in fertility required to achieve zero population growth.

Birth Control: Biological and Societal

Another simple way to lower the birth rate is breast feeding, which can delay resumption of ovulation after childbirth.[27] Women in a number of countries use this deliberately as a birth-control method—in fact, according to the World Bank, in the mid-1970s breast feeding provided more protection against conception in developing countries than did family-planning programs.[26]

Family planning is still emphasized, however.[28] Traditional methods range from abstinence to the use of natural agents to induced sterility. Modern methods include the birth-control pill, which prevents ovulation through control of hormone levels; surgical techniques for permanent sterility; and mechanical devices. Contraceptive devices are used widely in many parts of the world, especially in East Asia, where data show that 78% of women use them. In Africa, only 18% of women use them; in Central and South America, the numbers are 53% and 62%, respectively.[26] Abortion is also widespread and is one of the most important birth-control methods in terms of its effects on birth rates—approximately 46 million abortions are performed each year.[29] However, although now medically safe in most cases, abortion is one of the most controversial methods from a moral perspective.

National Programs to Reduce Birth Rates

Reducing birth rates requires a change in attitude, knowledge of the means of birth control, and the ability to afford these means. As we have seen, a change in attitude can occur simply with a rise in the standard of living. In many countries, however, it has been necessary to provide formal family-planning programs to explain the problems arising from rapid population growth and to describe the ways that individuals will benefit from reduced population growth. These programs also provide information about birth-control methods and provide access to these methods.[30] Which methods to promote and use involves social, moral, and religious beliefs, which vary from country to country.

The first country to adopt an official population policy was India in 1952. Few developing countries had official family-planning programs before 1965. Since 1965, many such programs have been introduced, and the World Bank has lent $4.2 billion to more than 80 countries to support "reproductive" health projects.[26,31] Although most countries now have some kind of family-planning program, effectiveness varies greatly.

A wide range of approaches have been used, from simply providing more information to promoting and providing means for birth control, offering rewards, and imposing penalties. Penalties usually take the form of taxes. Ghana, Malaysia, Pakistan, Singapore, and the Philippines have used a combination of methods, including limits on tax allowances for children and on maternity benefits. Tanzania has restricted paid maternity leave for women to a frequency of once in three years. Singapore does not take family size into account in allocating government-built housing, so larger families are more crowded. Singapore also gives higher priority in school admission to children from smaller families. Some countries, including Bangladesh, India, and Sri Lanka, have paid people to be voluntarily sterilized. In Sri Lanka, this practice has applied only to families with two children, and only when a voluntary statement of consent is signed.

CRITICAL THINKING ISSUE

Will the Demographic Transition Hold in the United States?

Earlier in this chapter, we presented the idea of the demographic transition and suggested that it has occurred in developed nations and may continue in the future. But we also noted that improvements in health care can further decrease death rates, which is something everybody wants to see happen but which will increase the human population growth rate, even in technologically developed nations. Recently, Robert Engelman, a vice president of the Worldwatch Institute of Washington, DC, proposed another problem for the demographic transition—an increase in birth rates in nations such as the United States.[32] The accompanying text box has selections from Engelman's article. Using the material in the chapter, the quotes from Engelman here, and any other information you would like to introduce, present an argument either for or against the following: Growth rates will continue to decline in technologically developed nations, leading toward zero population growth.

Robert Engelman, Vice President of the Worldwatch Institute, "World Population Growth: Fertile Ground for Uncertainty," 2008.

Although the average woman worldwide is giving birth to fewer children than ever before, an estimated 136 million babies were born in 2007. Global data do not allow demographers to be certain that any specific year sets a record for births, but this one certainly came close. The year's cohort of babies propelled global population to an estimated 6.7 billion by the end of 2007.

The seeming contradiction between smaller-than-ever families and near-record births is easily explained. The number of women of childbearing age keeps growing and global life expectancy at birth continues to rise. These two trends explain why population continues growing despite declines in family size. There were 1.7 billion women aged 15 to 49 in late 2007, compared with 856 million in 1970. The average human being born today can expect to live 67 years, a full decade longer than the average newborn could expect in 1970.

Only the future growth of the reproductive-age population is readily predictable, however: all but the youngest of the women who will be in this age group in two decades are already alive today. But sustaining further declines in childbearing and increases in life expectancy will require continued efforts by governments to improve access to good health care, and both trends could be threatened by environmental or social deterioration. The uncertain future of these factors makes population growth harder to predict than most people realize.

SUMMARY

- The human population is often referred to as the underlying environmental issue because much current environmental damage results from the very high number of people on Earth and their great power to change the environment.
- Throughout most of our history, the human population and its average growth rate were small. The growth of the human population can be divided into four major phases. Although the population has increased in each phase, the current situation is unprecedented.
- Countries whose birth rates have declined have experienced a demographic transition marked by a decline in death rates followed by a decline in birth rates. In contrast, many developing nations have undergone a great decline in their death rates but still have very high birth rates. It remains an open question whether some of these nations will be able to achieve a lower birth rate before reaching disastrously high population levels.
- The maximum population Earth can sustain and how large a population will ultimately be attained by human beings are controversial questions. Standard estimates

suggest that the human population will reach 10–16 billion before stabilizing.

- How the human population might stabilize, or be stabilized, raises questions concerning science, values, people, and nature.
- One of the most effective ways to lower a population's growth rate is to lower the age of first childbearing. This approach also involves relatively few societal and value issues.

REEXAMINING THEMES AND ISSUES

Human Population

Our discussion in this chapter reemphasizes the point that there can be no long-term solution to our environmental problems unless the human population stops growing at its present rate. This makes the problem of human population a top priority.

Sustainability

As long as the human population continues to grow, it is doubtful that our other environmental resources can be made sustainable.

Global Perspective

Although the growth rate of the human population varies from nation to nation, the overall environmental effects of the rapidly growing human population are global. For example, the increased use of fossil fuels in Western nations since the beginning of the Industrial Revolution has affected the entire world. The growing demand for fossil fuels and their increasing use in developing nations are also having a global effect.

Urban World

One of the major patterns in the growth of the human population is the increasing urbanization of the world. Cities are not self-contained but are linked to the surrounding environment, depending on it for resources and affecting environments elsewhere.

People and Nature

As with any species, the growth rate of the human population is governed by fundamental laws of population dynamics. We cannot escape these basic rules of nature. People greatly affect the environment, and the idea that human population growth is *the* underlying environmental issue illustrates the deep connection between people and nature.

Science and Values

The problem of human population exemplifies the connection between values and knowledge. Scientific and technological knowledge has helped us cure diseases, reduce death rates, and thereby increase growth of the human population. Our ability today to forecast human population growth provides a great deal of useful knowledge, but what we do with this knowledge is hotly debated around the world because values are so important in relation to birth control and family size.

KEY TERMS

abundance **62**
acute disease **72**
age structure **62**
birth rate **62**
chronic disease **72**
death rate **62**
demographic transition **69**
demography **62**
doubling time **65**
epidemic disease **72**
exponential rate **62**
growth rate **62**
human carrying capacity **73**
inflection point **66**
life expectancy **71**
logistic carrying capacity **66**
logistic growth curve **65**
maximum lifetime **71**
pandemic **60**
population **61**
population dynamics **61**
species **62**
zero population growth **74**

STUDY QUESTIONS

1. Refer to three forecasts for the future of the world's human population in Figure 4.6. Each forecast makes a different assumption about the future total fertility rate: that the rate remains constant; that it decreases slowly and smoothly; and that it decreases rapidly and smoothly. Which of these do you think is realistic? Explain why.
2. Why is it important to consider the age structure of a human population?
3. Three characteristics of a population are the birth rate, growth rate, and death rate. How has each been affected by (a) modern medicine, (b) modern agriculture, and (c) modern industry?
4. What is meant by the statement "What is good for an individual is not always good for a population"?
5. Strictly from a biological point of view, why is it difficult for a human population to achieve a constant size?
6. What environmental factors are likely to increase the chances of an outbreak of an epidemic disease?
7. To which of the following can we attribute the great increase in human population since the beginning of the Industrial Revolution: changes in human (a) birth rates, (b) death rates, (c) longevity, or (d) death rates among the very old? Explain.
8. What is the demographic transition? When would one expect replacement-level fertility to be achieved—before, during, or after the demographic transition?
9. Based on the history of human populations in various countries, how would you expect the following to change as per capita income increased: (a) birth rates, (b) death rates, (c) average family size, and (d) age structure of the population? Explain.

FURTHER READING

Barry, J.M., ***The Great Influenza: The Story of the Deadliest Pandemic in History*** (New York: Penguin Books, paperback, 2005). Written for the general reader but praised by such authorities as the *New England Journal of Medicine,* this book discusses the connection between politics, public health, and pandemics.

Cohen, J.E., ***How Many People Can the Earth Support?*** (New York: Norton, 1995). A detailed discussion of world population growth, Earth's human carrying capacity, and factors affecting both.

Ehrlich, P.R., and A.H. Ehrlich, ***One with Nineveh: Politics, Consumption, and the Human Future*** (Washington, DC: Island Press, 2004). An extended discussion of the effects of the human population on the world's resources, and of Earth's carrying capacity for our species. Ehrlich's 1968 book, *The Population Bomb* (New York: Ballantine Books), played an important role in the beginning of the modern environmental movement, and for this reason can be considered a classic.

Livi-Bacci, Massimo, A *Concise History of World Population* (Hoboken, NJ: Wiley-Blackwell, paperback, 2001). A well-written introduction to the field of human demography.

McKee, J.K., *Sparing Nature: The Conflict between Human Population Growth and Earth's Biodiversity* (New Brunswick, NJ: Rutgers University Press, 2003). One of the few recent books about human populations.

CHAPTER 5

Ecosystems: Concepts and Fundamentals

Ecotourists see the first stages in ecological succession at Doñana National Park, Spain—plants that can germinate and grow in sandy soil recently deposited by the winds. Doñana National Park is one of the major stopovers for birds migrating from Europe to Africa, and is one of Europe's most important wildlife parks.

LEARNING OBJECTIVES

Life on Earth is sustained by ecosystems, which vary greatly but have certain attributes in common. After reading this chapter, you should understand . . .

- Why the ecosystem is the basic system that supports life and allows it to persist;
- What food chains, food webs, and trophic levels are;
- What ecosystem chemical cycling is;
- What the ecological community is;
- How to determine the boundaries of an ecosystem;
- How species affect one another indirectly through their ecological community;
- How ecosystems recover from disturbances through ecological succession;
- Whether ecosystems are generally in a steady state.

CASE STUDY

Sea Otters, Sea Urchins, and Kelp: Indirect Effects of Species on One Another

Sea otters, the lovable animals often shown lying faceup among kelp as they eat shellfish, play an important role in their ecosystems. Although they feed on a variety of shellfish, sea otters especially like sea urchins. Sea urchins, in turn, feed on kelp, large brown algae that form undersea "forests" and provide important habitat for many species that require kelp beds for reproduction, places to feed, or havens from predators. Sea urchins graze along the bottoms of the beds, feeding on the base of kelp, called *holdfasts*, which attach the kelp to the bottom. When holdfasts are eaten through, the kelp floats free and dies. Sea urchins thus can clear kelp beds—clear-cutting, so to speak.

While sea otters affect the abundance of kelp, their influence is indirect (Figure 5.1)—they neither feed on kelp nor protect individual kelp plants from attack by sea urchins. But sea otters reduce the number of sea urchins. With fewer sea urchins, less kelp is destroyed. With more kelp, there is more habitat for many other species; so sea otters indirectly increase the diversity of species.[1,2] This is called a **community effect**, and the otters are referred to as **keystone species** in their ecological community and ecosystem.

Sea otters originally occurred throughout a large area of the Pacific coasts, from northern Japan northeastward along the Russian and Alaskan coasts, and southward along the coast of North America to Morro Hermoso in Baja California and to Mexico.[4] But sea otters also like to eat abalone, and this brings them into direct conflict with people, since abalone is a prized seafood for us, too. They also have one of the finest furs in the world and were brought almost to extinction by commercial hunting for their fur during the 18th and 19th centuries. By the end of the 19th century, there were too few otters left to sustain commercial fur hunters.

Several small populations survived and have increased since then, so that today sea otters number in the hundreds of thousands—3,000 in California, 14,000 in southeastern Alaska, and the rest elsewhere in Alaska. According to the Marine Mammal Center, approximately 2,800 sea otters live along the coast of California,[5] a few hundred in Washington State and British Columbia, and about 100,000 worldwide, including Alaska and the coast of Siberia.[6]

Legal protection of the sea otter by the U.S. government began in 1911 and continues under the U.S. Marine Mammal Protection Act of 1972 and the Endangered Species Act of 1973. Today, however, the sea otter continues

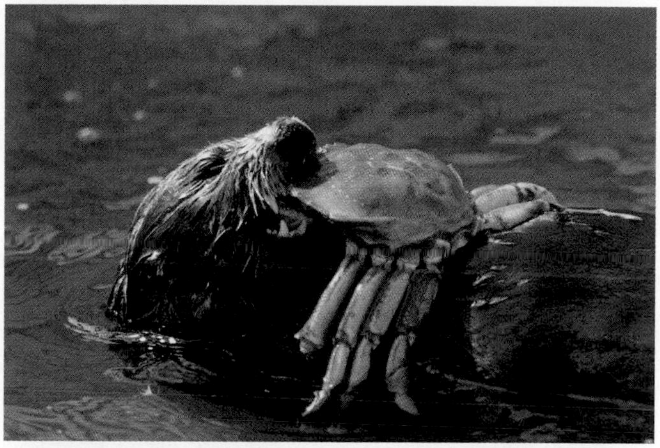

(a)

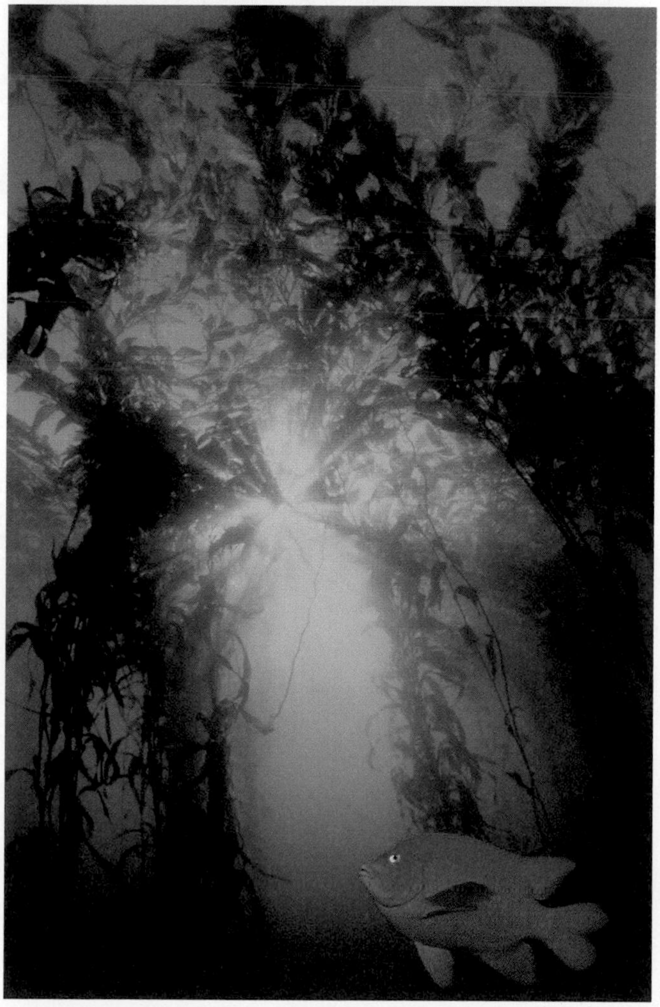

(b)

(c)

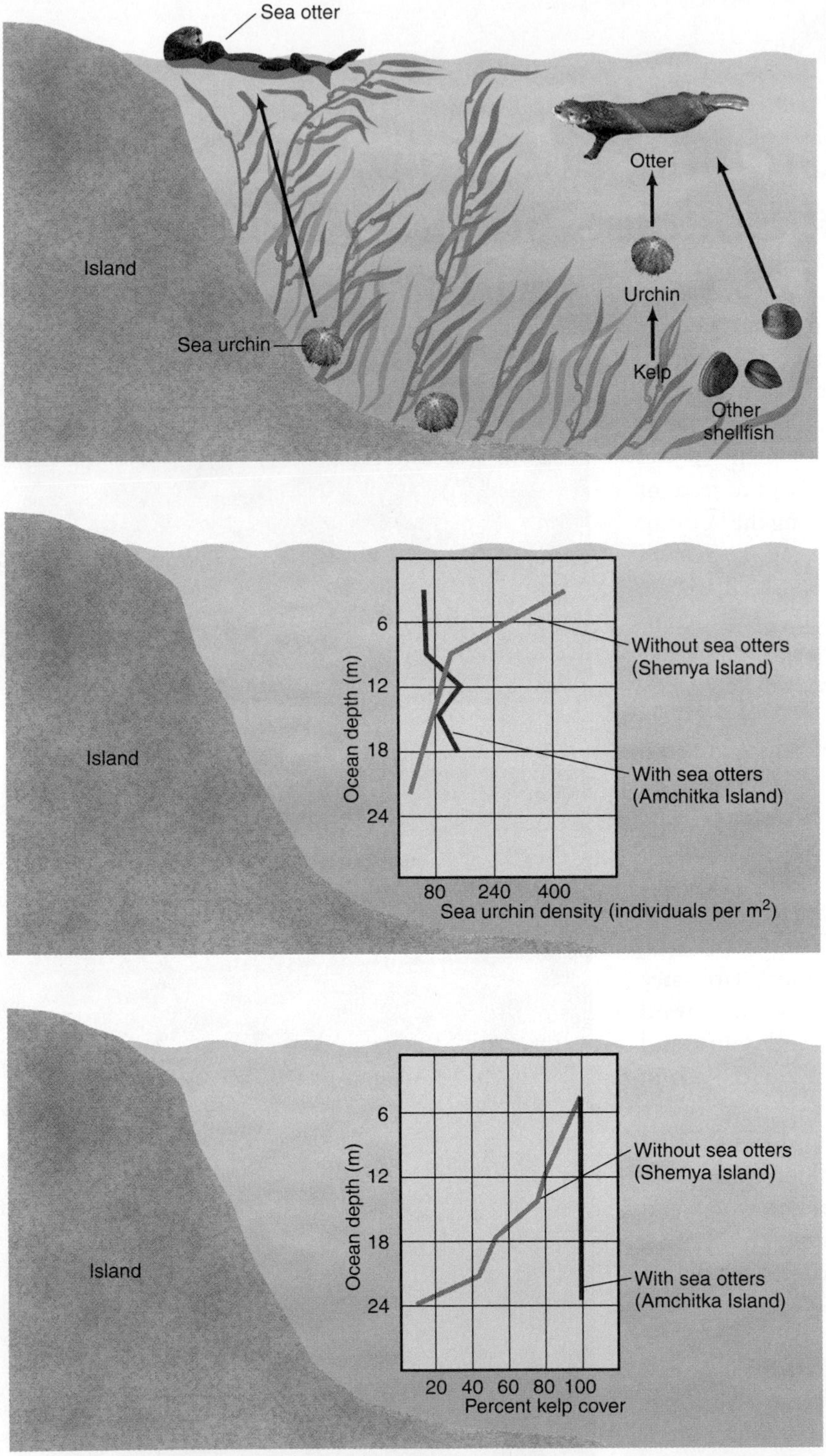

FIGURE 5.1 The effect of sea otters on kelp. (a) Sea otter eating a crab. Sea otters feed on shellfish, including sea urchins. Sea urchins feed on kelp. Where sea otters are abundant, as on Amchitka Island in the Aleutian Islands, there are few sea urchins and kelp beds are abundant (**b** and **c**). At nearby Shemya Island, which lacks sea otters, sea urchins are abundant and there is little kelp.[3] Experimental removal of sea urchins has led to an increase in kelp.[2]

to be a focus of controversy. On the one hand, fishermen argue that the sea otter population has recovered—so much so that they now interfere with commercial fishing because they take large numbers of abalone.[7] On the other hand, conservationists argue that community and ecosystem effects of sea otters make them necessary for the persistence of many oceanic species, and that there are still not enough sea otters to maintain this role at a satisfactory level. Thus, sea otters' indirect effects on many other species have practical consequences. They also demonstrate certain properties of ecosystems and ecological communities that are important to us in understanding how life persists, and how we may be able to help solve certain environmental problems.

5.1 The Ecosystem: Sustaining Life on Earth

We tend to associate life with individual organisms, for the obvious reason that it is individuals that are alive. But sustaining life on Earth requires more than individuals or even single populations or species. Life is sustained by the interactions of many organisms functioning together, interacting through their physical and chemical environments. We call this an **ecosystem**. Sustained life on Earth, then, is a characteristic of ecosystems, not of individual organisms or populations.[8] As the opening case study about sea otters illustrates, to understand important environmental issues—such as conserving endangered species, sustaining renewable resources, and minimizing the effects of toxic substances—we must understand the basic characteristics of ecosystems.

Basic Characteristics of Ecosystems

Ecosystems have several fundamental characteristics, which we can group as *structure* and *processes.*

Ecosystem Structure

An ecosystem has two major parts: nonliving and living. The nonliving part is the physical-chemical environment, including the local atmosphere, water, and mineral soil (on land) or other substrate (in water). The living part, called the **ecological community**, is the set of species interacting within the ecosystem.

Ecosystem Processes

Two basic kinds of processes must occur in an ecosystem: a cycling of chemical elements and a flow of energy. These processes are necessary for all life, but no single species can carry out all necessary chemical cycling and energy flow alone. That is why we said that sustained life on Earth is a characteristic of ecosystems, not of individuals or populations. At its most basic, an ecosystem consists of several species and a fluid medium—air, water, or both (Figure 5.2). Ecosystem energy flow places a fundamental limit on the abundance of life. Energy flow is a difficult subject, which we will discuss in Section 5.4.

Ecosystem chemical cycling is complex as well, and for that reason we have devoted a separate chapter (Chapter 6) to chemical cycling within ecosystems and throughout the entire Earth's biosphere. Briefly, 21 chemical elements are required by at least some form of life, and each chemical element required for growth and reproduction must be available to each organism at the right time, in the right amount, and in the right ratio relative to other elements. These chemical elements must also be recycled—converted to a reusable form: Wastes are converted into food, which is converted into wastes, which must be converted once again into food, with the cycling going on indefinitely if the ecosystem is to remain viable.

For complete recycling of chemical elements to take place, several species must interact. In the presence of light, green plants, algae, and photosynthetic bacteria produce sugar from carbon dioxide and water. From sugar and inorganic compounds, they make many organic compounds, including proteins and woody tissue. But no green plant, algae, or photosynthetic bacteria can decompose woody tissue back to its original inorganic compounds. Other forms of life—primarily bacteria and fungi—can decompose organic matter. But they cannot produce their own food; instead, they obtain energy and chemical nutrition from the dead tissues on which they feed. In an ecosystem, chemical elements recycle, but energy flows one way, into and out of the system, with a small fraction of it stored, as we will discuss later in this chapter.

To repeat, theoretically, at its simplest, an ecosystem consists of at least one species that produces its own food from inorganic compounds in its environment and another species that decomposes the wastes of the first species, plus a fluid medium—air, water, or both (Figure 5.2). But the reality is never as simple as that.

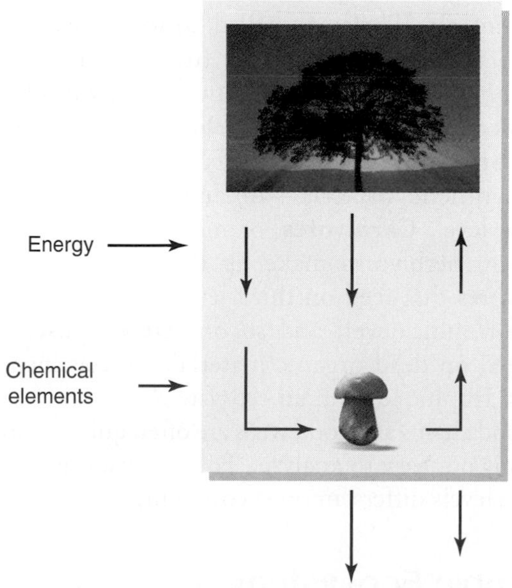

FIGURE 5.2 An idealized minimum ecosystem. Energy flows through an ecosystem one way. A small amount is stored within the system. As chemical elements cycle, there is some small loss, depending on the characteristics of the ecosystem. The size of the arrows in the figure approximates the amount of flow.

5.2 Ecological Communities and Food Chains

In practice, ecologists define the term *ecological community* in two ways. One method defines the community as a set of *interacting* species found in the same place and functioning together, thus enabling life to persist. That is essentially the definition we used earlier. A problem with this definition is that it is often difficult in practice to know the entire set of interacting species. Ecologists therefore may use a practical or an operational definition, in which the community consists of all the species found in an area, whether or not they are known to interact. Animals in different cages in a zoo could be called a community according to this definition.

One way that individuals in a community interact is by feeding on one another. Energy, chemical elements, and some compounds are transferred from creature to creature along **food chains**, the linkage of who feeds on whom. The more complex linkages are called **food webs**. Ecologists group the organisms in a food web into trophic levels. A **trophic level** (from the Greek word *trephein*, meaning to nourish, thus the "nourishing level") consists of all organisms in a food web that are the same number of feeding levels away from the original energy source. The original source of energy in most ecosystems is the sun. In other cases, it is the energy in certain inorganic compounds.

Green plants, algae, and certain bacteria produce sugars through the process of **photosynthesis**, using only energy from the sun and carbon dioxide (CO_2) from the air. They are called **autotrophs**, from the words *auto* (self) and *trephein* (to nourish), thus "self-nourishing," and are grouped into the first trophic level. All other organisms are called **heterotrophs**. Of these, **herbivores**—organisms that feed on plants, algae, or photosynthetic bacteria—are members of the second trophic level. **Carnivores**, or meat-eaters, that feed directly on herbivores make up the third trophic level. Carnivores that feed on third-level carnivores are in the fourth trophic level, and so on. **Decomposers**, those that feed on dead organic material, are classified in the highest trophic level in an ecosystem.

Food chains and food webs are often quite complicated and thus not easy to analyze. For starters, the number of trophic levels differs among ecosystems.

A Simple Ecosystem

One of the simplest natural ecosystems is a hot spring, such as those found in geyser basins in Yellowstone National Park, Wyoming.[9] They are simple because few organisms can live in these severe environments. In and near the center of a spring, water is close to the boiling point, while at the edges, next to soil and winter snow, water is much cooler. In addition, some springs are very acidic and others are very alkaline; either extreme makes a harsh environment.

Photosynthetic bacteria and algae make up the spring's first trophic level. In a typical alkaline hot spring, the hottest waters, between 70° and 80°C (158–176°F), are colored bright yellow-green by photosynthetic blue-green bacteria. One of the few kinds of photosynthetic organisms that can survive at those temperatures, these give the springs the striking appearance for which they are famous (Figure 5.3). In slightly cooler waters, 50° to 60°C (122–140°F), thick mats of other kinds of bacteria and algae accumulate, some becoming 5 cm thick (Figures 5.3 and 5.4).

Ephydrid flies make up the second (herbivore) trophic level. Note that they are the only genus on that entire trophic level, so stressful is the environment, and they live only in the cooler areas of the springs. One of these species, *Ephydra bruesi*, lays bright orange-pink egg masses on stones and twigs that project above the mat. These larvae feed on the bacteria and algae.

The third (carnivore) trophic level is made up of a dolichopodid fly, which feeds on the eggs and larvae of the herbivorous flies, and dragonflies, wasps, spiders,

FIGURE 5.3 **One of the many hot springs in Yellowstone National Park.** The bright yellowish-green color comes from photosynthetic bacteria, one of the few kinds of organisms that can survive in the hot temperatures and chemical conditions of the springs.

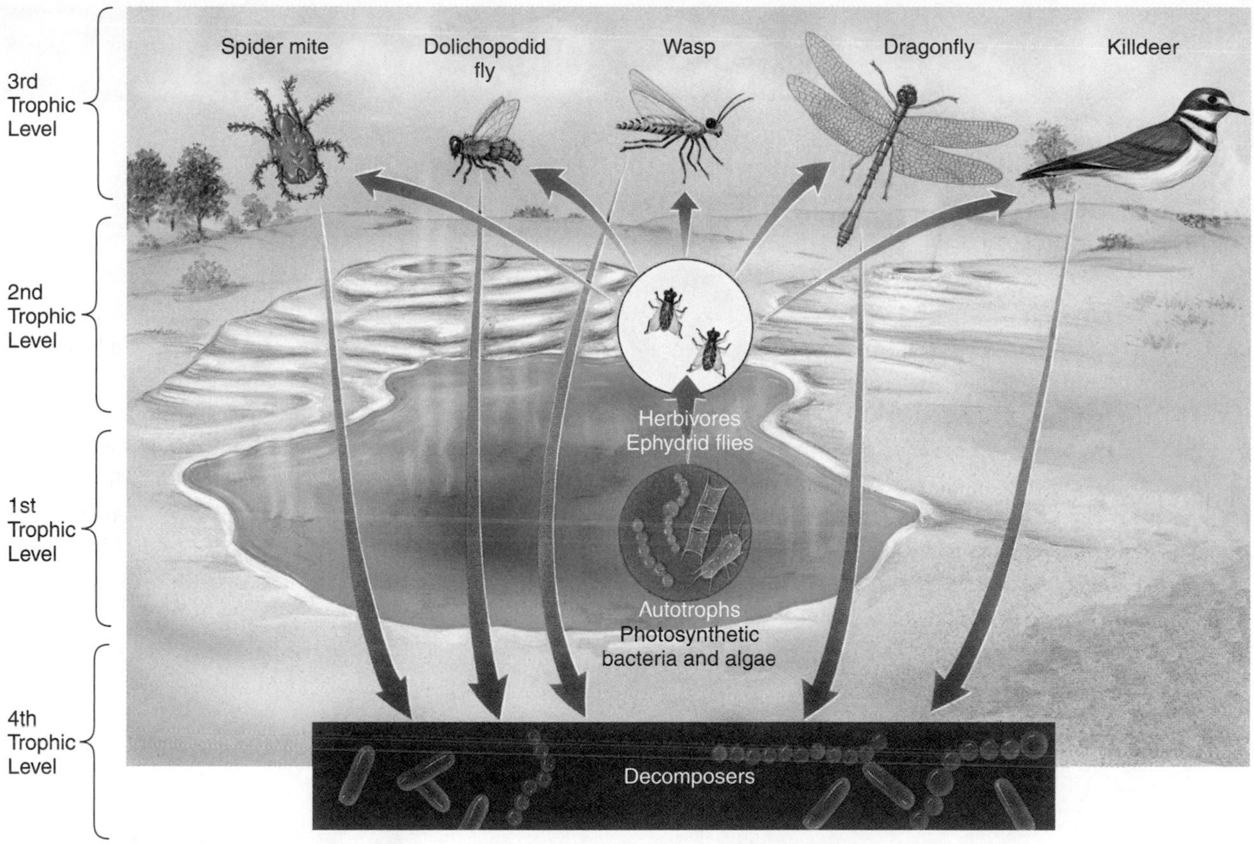

FIGURE 5.4 Food web of a Yellowstone National Park hot spring. Even though this is one of the simplest ecological communities in terms of the numbers of species, a fair number are found. About 20 species in all are important in this ecosystem.

tiger beetles, and one species of bird, the killdeer, that feeds on the ephydrid flies. (Note that the killdeer is a carnivore of the hot springs but also feeds widely in other ecosystems. An interesting question, little addressed in the ecological scientific literature, is how we should list this partial member of the food web: Should there be a separate category of "casual" members? What do you think?)

In addition to their other predators, the herbivorous ephydrid flies have parasites. One is a red mite that feeds on the flies' eggs and travels by attaching itself to the adult flies. Another is a small wasp that lays its eggs within the fly larvae. These are also on the third trophic level.

Wastes and dead organisms of all trophic levels are fed on by decomposers, which in the hot springs are primarily bacteria. These form the fourth trophic level.

The entire hot-springs community of organisms—photosynthetic bacteria and algae, herbivorous flies, carnivores, and decomposers—is maintained by two factors: (1) sunlight, which provides usable energy for the organisms; and (2) a constant flow of hot water, which provides a continual new supply of chemical elements required for life and a habitat in which the bacteria and algae can persist.

An Oceanic Food Chain

In oceans, food webs involve more species and tend to have more trophic levels than they do in a terrestrial ecosystem. In a typical **pelagic** (open-ocean) **ecosystem** (Figure 5.6), microscopic single-cell planktonic algae and planktonic photosynthetic bacteria are in the first trophic level. Small invertebrates called *zooplankton* and some fish feed on the algae and photosynthetic bacteria, forming the second trophic level. Other fish and invertebrates feed on these herbivores and form the third trophic level. The great baleen whales filter seawater for food, feeding primarily on small herbivorous zooplankton (mostly crustaceans), and thus the baleen whales are also in the third level. Some fish and marine mammals, such as killer whales, feed on the predatory fish and form higher trophic levels.

Food Webs Can Be Complex: The Food Web of the Harp Seal

In the abstract or in extreme environments like a hot spring, a diagram of a food web and its trophic levels may seem simple and neat. In reality, however, most food webs are complex. One reason for the complexity

A CLOSER LOOK 5.1

Land and Marine Food Webs

A Terrestrial Food Web

An example of terrestrial food webs and trophic levels is shown in Figure 5.5 for an eastern temperate woodland of North America. The first trophic level, autotrophs, includes grasses, herbs, and trees. The second trophic level, herbivores, includes mice, an insect called the pine borer, and other animals (such as deer) not shown here. The third trophic level, carnivores, includes foxes and wolves, hawks and other predatory birds, spiders, and predatory insects. People, too, are involved as **omnivores** (eaters of both plants and animals), feeding on several trophic levels. In Figure 5.5, people would be included in the fourth trophic level, the highest level in which they would take part. Decomposers, such as bacteria and fungi, feed on wastes and dead organisms of all trophic levels. Decomposers are also shown here on the fourth level. (Here's another interesting question: Should we include people *within* this ecosystem's food web? That would place us within nature. Or should we place people outside of the ecosystem, thus separate from nature?)

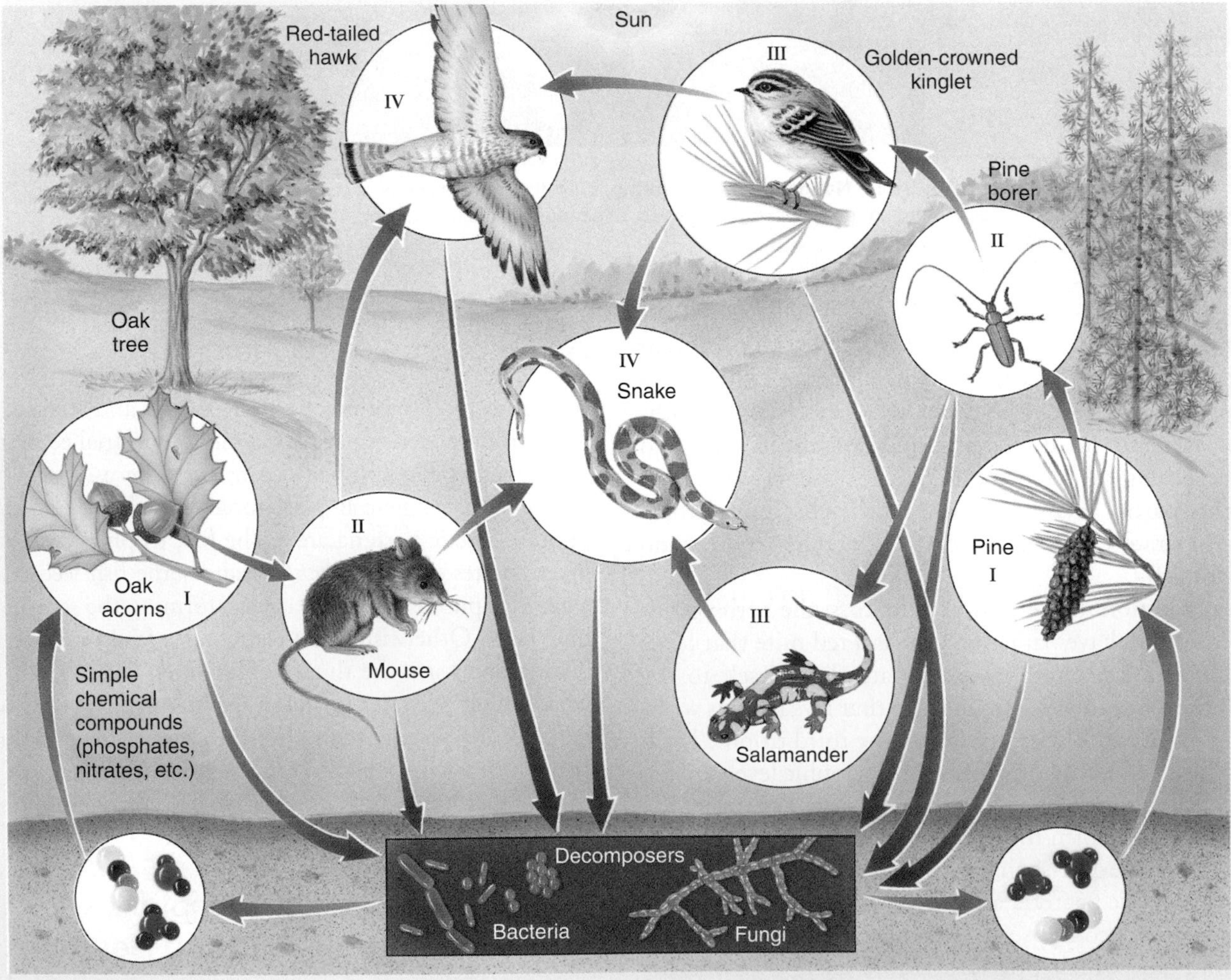

FIGURE 5.5 A typical temperate forest food web.

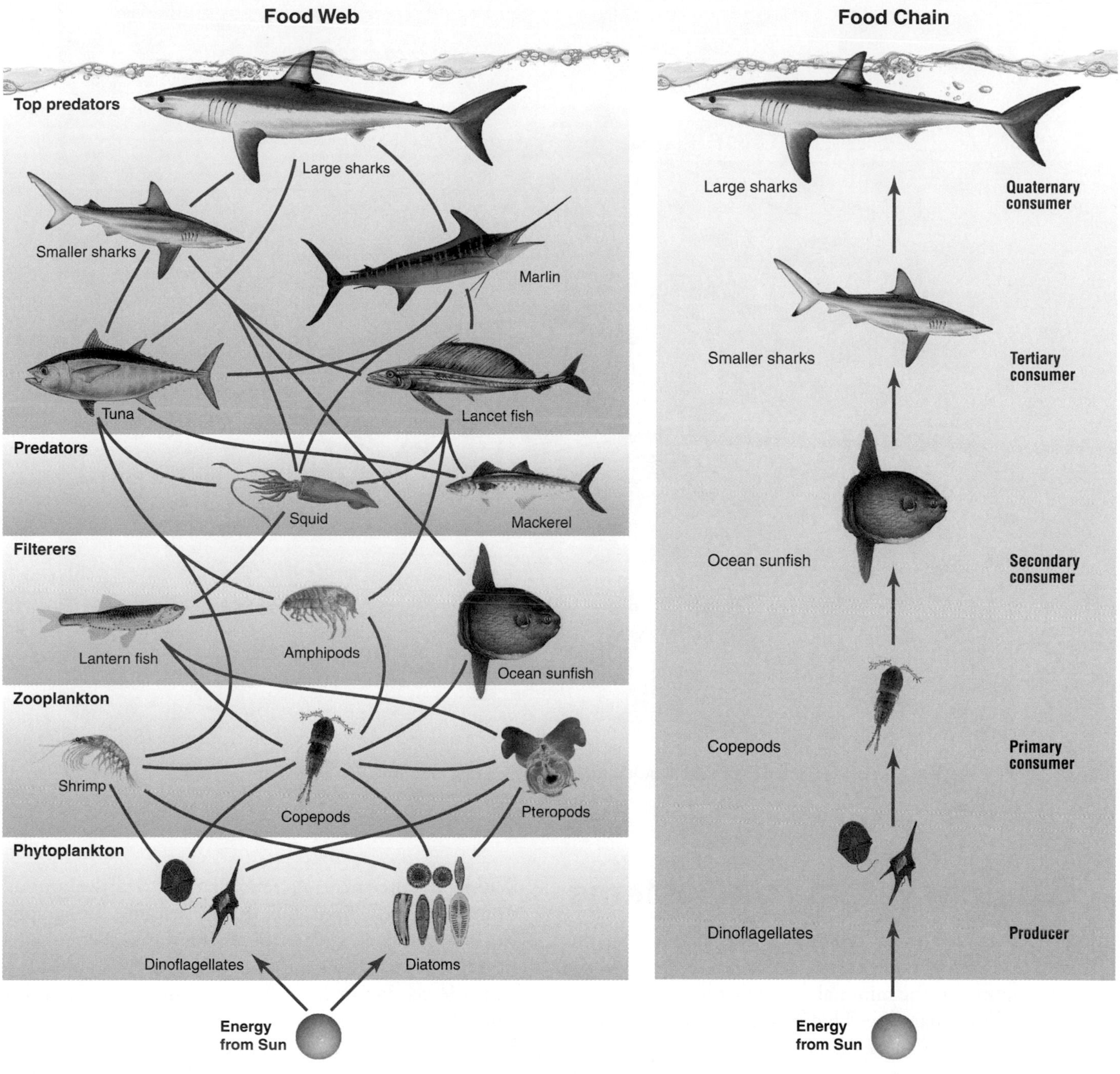

FIGURE 5.6 An oceanic food web. (*Source*: NOAA)

is that many creatures feed on several trophic levels. For example, consider the food web of the harp seal (Figure 5.7). This is a species of special interest because large numbers of the pups are harvested each year in Canada for their fur, giving rise to widespread controversy over the humane treatment of animals even though the species is not endangered (there are more than 5 million harp seals.)[10] This controversy is one reason that the harp seal has been well studied, so we can show its complex food web.

The harp seal is shown at the fifth level.[11] It feeds on flatfish (fourth level), which feed on sand launces (third level), which feed on euphausiids (second level), which feed on phytoplankton (first level). But the harp seal actually feeds at several trophic levels, from the second through the fourth. Thus, it feeds on predators of some of its prey and therefore competes with some of its own prey.[12] A species that feeds on several trophic levels is typically classified as belonging to the trophic level above the highest level from which it feeds. Thus, we place the harp seal on the fifth trophic level.

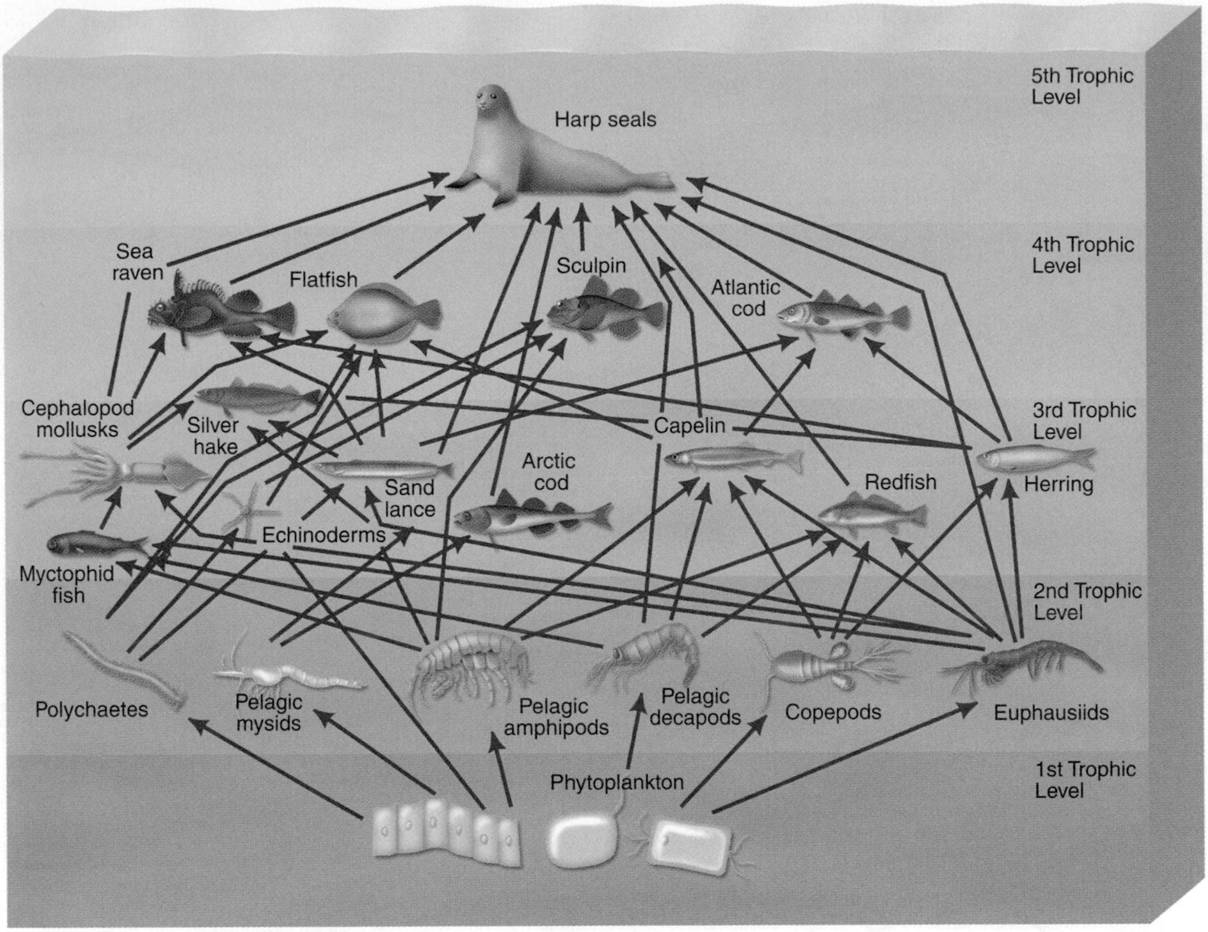

FIGURE 5.7 **Food web of the harp seal showing how complex a real food web can be.**

5.3 Ecosystems as Systems

Ecosystems are open systems: Energy and matter flow into and out of them (see Chapter 3). As we have said, an ecosystem is the minimal entity that has the properties required to sustain life. This implies that an ecosystem is real and important and therefore that we should be able to find one easily. However, ecosystems vary greatly in structural complexity and in the clarity of their boundaries. Sometimes their borders are well defined, such as between a lake and the surrounding countryside (Figure 5.8a). But sometimes the transition from one ecosystem to another is gradual—for example, the transition from deciduous to boreal forest on the slopes of Mt. Washington, N.H. (Figure 5.8b), and in the subtle gradations from grasslands to savannas in East Africa and from boreal forest to tundra in the Far North, where the trees thin out gradually and setting a boundary is difficult, and usually arbitrary.

A commonly used practical delineation of the boundary of an ecosystem on land is the **watershed**. Within a watershed, all rain that reaches the ground from any source flows out in one stream. Topography (the lay of the land) determines the watershed. When a watershed is used to define the boundaries of an ecosystem, the ecosystem is unified in terms of chemical cycling. Some classic experimental studies of ecosystems have been conducted on forested watersheds in U.S. Forest Service experimental areas, including the Hubbard Brook experimental forest in New Hampshire (Figure 5.9) and the Andrews experimental forest in Oregon. In other cases, the choice of an ecosystem's boundary may be arbitrary. For the purposes of scientific analysis, this is okay as long as this boundary is used consistently for any calculation of the exchange of chemicals and energy and the migration of organisms. Let us repeat the primary point: What all ecosystems have in common is not a particular physical size or shape but the processes we have mentioned—the flow of energy and the cycling of chemical elements, which give ecosystems the ability to sustain life.

FIGURE 5.8 **Ecosystem boundaries. (a)** Sometimes the transition from one ecosystem to another is sharp and distinct, as in the transition from lake to forest at Lake Moraine in Banff National Park, Alberta, Canada. **(b)** Sometimes the transition is indistinct and arbitrary, as in the transition from wetlands to uplands in a northern forest as seen from Mount Washington, New Hampshire. Within the distant scene are uplands and wetlands.

FIGURE 5.9 The V-shaped logged area in this picture is the site of the famous Hubbard Brook Ecosystem Study. Here, a watershed defines the ecosystem, and the V shape is an entire watershed cut as part of the experiment.

5.4 Biological Production and Ecosystem Energy Flow

All life requires energy. Energy is the ability to do work, to move matter. As anyone who has dieted knows, our weight is a delicate balance between the energy we take in through our food and the energy we use. What we do not use and do not pass on, we store. Our use of energy, and whether we gain or lose weight, follows the laws of physics. This is true not only for people but also for all populations of living things, for all ecological communities and ecosystems, and for the entire biosphere.

Ecosystem energy flow is the movement of energy through an ecosystem from the external environment through a series of organisms and back to the external environment. It is one of the fundamental processes common to all ecosystems. Energy enters an ecosystem by two pathways: energy fixed by organisms and moving through food webs within an ecosystem; and heat energy that is transferred by air or water currents or by convection through soils and sediments and warms living things. For instance, when a warm air mass passes over a forest, heat energy is transferred from the air to the land and to the organisms.

Energy is a difficult and an abstract concept. When we buy electricity, what are we buying? We cannot see it or feel it, even if we have to pay for it. At first glance, and as we think about it with our own diets, energy flow seems simple enough: We take energy in and use it, just like machines do—our automobiles, cell phones, and so on. But if we dig a little deeper into this subject, we discover a philosophical importance: We learn what distinguishes Earth's life and life-containing systems from the rest of the universe.

Although most of the time energy is invisible to us, with infrared film we can see the differences between warm and cold objects, and we can see some things about energy flow that affect life. With infrared film, warm objects appear red, and cool objects blue. Figure

5.10 shows birch trees in a New Hampshire forest, both as we see them, using standard film, and with infrared film. The infrared film shows tree leaves bright red, indicating that they have been warmed by the sun and are absorbing and reflecting energy, whereas the white birch bark remains cooler. The ability of tree leaves to absorb energy is essential; it is this source of energy that ultimately supports all life in a forest. Energy flows through life, and energy flow is a key concept.

The Laws of Thermodynamics and the Ultimate Limit on the Abundance of Life

When we discuss ecosystems, we are talking about some of the fundamental properties of life and of the ecological systems that keep life going. A question that frequently arises both in basic science and when we want to produce a lot of some kind of life—a crop,

FIGURE 5.10 Making energy visible. Top: A birch forest in New Hampshire as we see it, using normal photographic film **(a)** and the same forest photographed with infrared film **(b)**. Red color means warmer temperatures; the leaves are warmer than the surroundings because they are heated by sunlight. Bottom: A nearby rocky outcrop as we see it, using normal photographic film **(c)** and the same rocky outcrop photographed with infrared film **(d)**. Blue means that a surface is cool. The rocks appear deep blue, indicating that they are much cooler than the surrounding trees.

WORKING IT OUT 5.1 Some Chemistry of Energy Flow

For Those Who Make Their Own Food (autotrophs)

Photosynthesis—the process by which autotrophs make sugar from sunlight, carbon dioxide, and water—is:

$$6CO_2 + 6H_2O + \text{energy} = C_6H_{12}O_6 + 6O_2$$

Chemosynthesis takes place in certain environments. In chemosynthesis, the energy in hydrogen sulfide (H_2S) is used by certain bacteria to make simple organic compounds. The reactions differ among species and depend on characteristics of the environment (Figure 5.11).

Net production for autotrophs is given as

$$NPP = GPP - R_a$$

where *NPP* is net primary production, *GPP* is gross primary production, and R_a is the respiration of autotrophs.

For Those Who Do Not Make Their Own Food (heterotrophs)

Secondary production of a population is given as

$$NSP = B_2 - B_1$$

where *NSP* is net secondary production, B_2 is the biomass (quantity of organic matter) at time 2, and B_1 is the biomass at time 1. (See discussion of biomass in Section 5.5.) The change in biomass is the result of the addition of weight of living individuals, the addition of newborns and immigrants, and loss through death and emigration. The biological use of energy occurs through respiration, most simply expressed as

$$C_6H_{12}O_6 + 6O_2 = 6CO_2 + 6H_2O + \text{Energy}$$

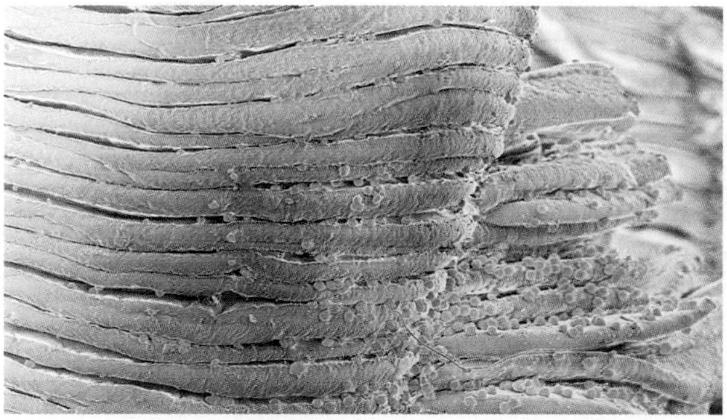

FIGURE 5.11 Deep-sea vent chemosynthetic bacteria.

biofuels, pets—is: *What ultimately limits the amount of organic matter in living things that can be produced anywhere, at any time, forever on the Earth or anywhere in the universe?*

We ask this question when we are trying to improve the production of some form of life. We want to know: How closely do ecosystems, species, populations, and individuals approach this limit? Are any of these near to being as productive as possible?

The answers to these questions, which are at the same time practical, scientifically fundamental, and philosophical, lie in the laws of thermodynamics. The **first law of thermodynamics**, known as the *law of conservation of energy* states that in any physical or chemical change, energy is neither created nor destroyed but merely changed from one form to another. (See A Closer Look 5.1.) This seems to lead us to a confusing, contradictory answer—it seems to say that we don't need to take in any energy at all! If the total amount of energy is always conserved—if it remains constant—then why can't we just recycle energy inside our bodies? The famous 20th-century physicist Erwin Schrödinger asked this question in a wonderful book entitled *What Is Life?* He wrote:

> In some very advanced country (I don't remember whether it was Germany or the U.S.A. or both) you could find menu cards in restaurants indicating, in addition to the price, the energy content of every dish. Needless to say, taken literally, this is . . . absurd. For an adult organism the energy content is as stationary as the material content. Since, surely, any calorie is worth as much as any other calorie, one cannot see how a mere exchange could help.[13]

Schrödinger was saying that, according to the first law of thermodynamics, we should be able to recycle energy in our bodies and never have to eat anything. Similarly, we can ask: Why can't energy be recycled in ecosystems and in the biosphere?

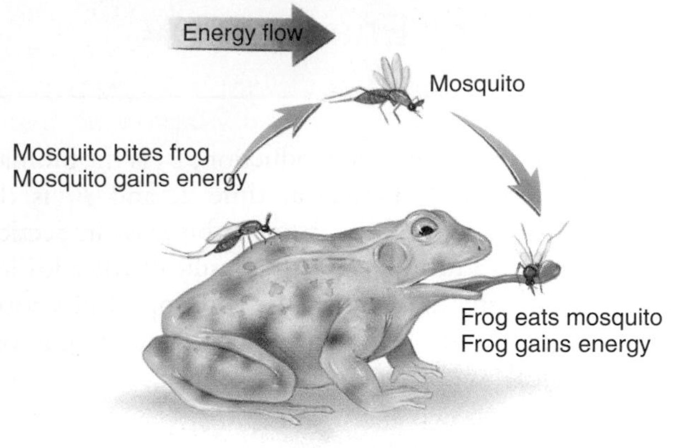

FIGURE 5.12 **An impossible ecosystem.** Energy always changes from a more useful, more highly organized form to a less useful, disorganized form. That is, energy cannot be completely recycled to its original state of organized, high-quality usefulness. For this reason, the mosquito–frog system will eventually stop when not enough useful energy is left. (There is also a more mundane reason: Only female mosquitoes require blood, and then only in order to reproduce. Mosquitoes are otherwise herbivorous.)

Let us imagine how that might work, say, with frogs and mosquitoes. Frogs eat insects, including mosquitoes. Mosquitoes suck blood from vertebrates, including frogs. Consider an imaginary closed ecosystem consisting of water, air, a rock for frogs to sit on, frogs, and mosquitoes. In this system, the frogs get their energy from eating the mosquitoes, and the mosquitoes get their energy from biting the frogs (Figure 5.12). Such a closed system would be a biological perpetual-motion machine. It could continue indefinitely without an input of any new material or energy. This sounds nice, but unfortunately it is impossible. Why? The general answer is found in the *second* law of thermodynamics, which addresses how energy changes in form.

To understand why we cannot recycle energy, imagine a closed system (a system that receives no input after the initial input) containing a pile of coal, a tank of water, air, a steam engine, and an engineer (Figure 5.13). Suppose the engine runs a lathe that makes furniture. The engineer lights a fire to boil the water, creating steam to run the engine. As the engine runs, the heat from the fire gradually warms the entire system.

When all the coal is completely burned, the engineer will not be able to boil any more water, and the engine will stop. The average temperature of the system is now higher than the starting temperature. The energy that was in the coal is dispersed throughout the entire system, much of it as heat in the air. Why can't the engineer recover all that energy, recompact it, put it under the boiler, and run the engine? The answer is in the **second law of thermodynamics**. Physicists have discovered that *no use of energy in the real (not theoretical) world can ever be 100% efficient.* Whenever useful work is done, some energy is inevitably converted to heat. Collecting all the energy dispersed in this closed system would require more energy than could be recovered.

Our imaginary system begins in a highly organized state, with energy compacted in the coal. It ends in a less organized state, with the energy dispersed throughout the system as heat. The energy has been degraded, and the system is said to have undergone a decrease in order. The

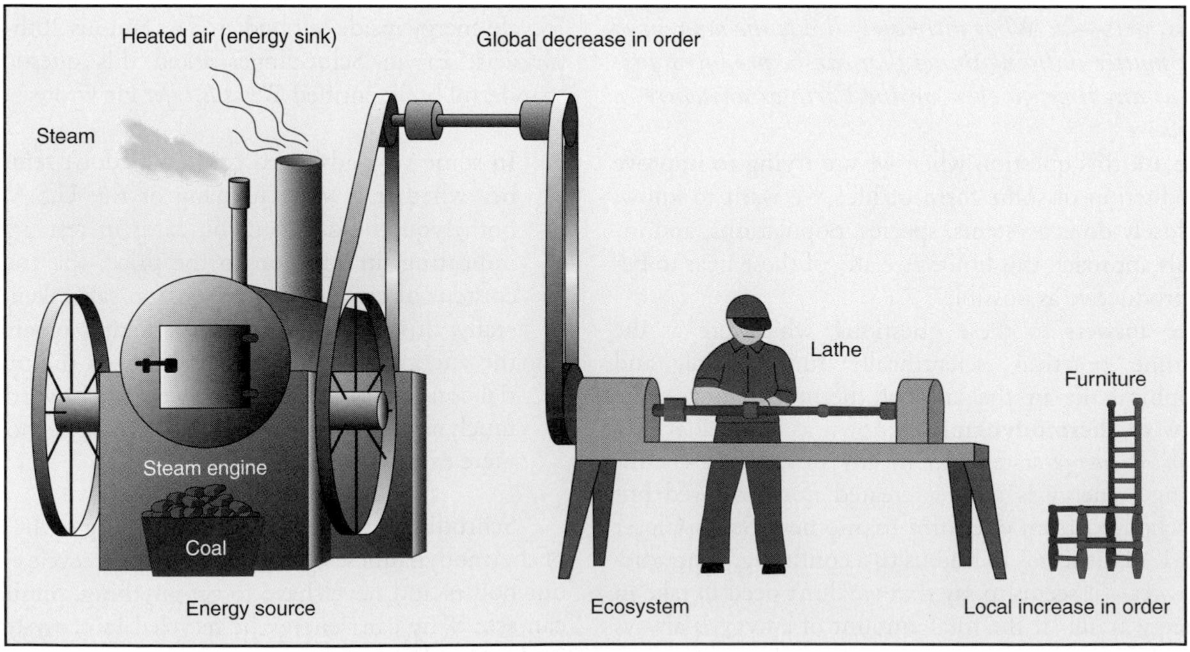

FIGURE 5.13 **A system closed to the flow of energy.**

measure of the decrease in order (the disorganization of energy) is called **entropy**. The engineer did produce some furniture, converting a pile of lumber into nicely ordered tables and chairs. The system had a local increase of order (the furniture) at the cost of a general increase in disorder (the state of the entire system). All energy of all systems tends to flow toward states of increasing entropy.

The second law of thermodynamics gives us a new understanding of a basic quality of life. *It is the ability to create order on a local scale that distinguishes life from its nonliving environment.* This ability requires obtaining energy in a usable form, and that is why we eat. This principle is true for every ecological level: individual, population, community, ecosystem, and biosphere. Energy must continually be added to an ecological system in a usable form. Energy is inevitably degraded into heat, and this heat must be released from the system. If it is not released, the temperature of the system will increase indefinitely. The net flow of energy through an ecosystem, then, is a one-way flow.

Based on what we have said about the energy flow through an ecosystem, we can see that an ecosystem must lie between a source of usable energy and a sink for degraded (heat) energy. The ecosystem is said to be an *intermediate system* between the energy source and the energy sink. The energy source, ecosystem, and energy sink together form a thermodynamic system. The ecosystem can undergo an increase in order, called a *local increase*, as long as the entire system undergoes a decrease in order, called a global decrease. (Note that *order* has a specific meaning in thermodynamics: Randomness is disorder; an ordered system is as far from random as possible.) To put all this simply, creating local order involves the production of organic matter. Producing organic matter requires energy; organic matter stores energy.

With these fundamentals in mind, we can turn to practical and empirical scientific problems, but this requires that we agree how to measure biological production. To complicate matters, there are several measurement units involved, depending on what people are interested in.

5.5 Biological Production and Biomass

The total amount of organic matter in any ecosystem is called its **biomass**. Biomass is increased through biological production (growth). Change in biomass over a given period is called *production*. **Biological production** is the capture of usable energy from the environment to produce organic matter (or organic compounds). This capture is often referred to as energy "fixation," and it is often said that the organism has "fixed" energy. There are two kinds of production, gross and net. **Gross production** is the increase in stored energy before any is used; **net production** is the amount of newly acquired energy stored after some energy has been used. When we use energy, we "burn" a fuel through repiration. The difference between gross and net production is like the difference between a person's gross and net income. Your gross income is the amount you are paid. Your net income is what you have left after taxes and other fixed costs. Respiration is like the expenses that are required in order for you to do your work.

Measuring Biomass and Production

Three measures are used for biomass and biological production: the quantity of organic material (biomass), energy stored, and carbon stored. We can think of these measures as the currencies of production. Biomass is usually measured as the amount per unit surface area—for example, as grams per square meter (g/m^2) or metric tons per hectare (MT/ha). Production, a rate, is the change per unit area in a unit of time—for example, grams per square meter per year. (Common units of measure of production are given in the Appendix.)

The production carried out by autotrophs is called **primary production**; that of heterotrophs is called **secondary production**. As we have said, most autotrophs make sugar from sunlight, carbon dioxide, and water in a process called photosynthesis, which releases free oxygen (see Working It Out 5.1 and 5.2). Some autotrophic bacteria can derive energy from inorganic sulfur compounds; these bacteria are referred to as **chemoautotrophs**. Such bacteria live in deep-ocean vents, where they provide the basis for a strange ecological community. Chemoautotrophs are also found in muds of marshes, where there is no free oxygen.

Once an organism has obtained new organic matter, it can use the energy in that organic matter to do things: to move, to make new compounds, to grow, to reproduce, or to store it for future uses. The use of energy from organic matter by most heterotrophic and autotrophic organisms is accomplished through respiration. In respiration, an organic compound combines with oxygen to release energy and produce carbon dioxide and water (see Working It Out 5.2). The process is similar to the burning of organic compounds but takes place within cells at much lower temperatures through enzyme-mediated reactions. *Respiration is the use of biomass to release energy that can be used to do work.* Respiration returns to the environment the carbon dioxide that had been removed by photosynthesis.

WORKING IT OUT 5.2 Gross and Net Production

The production of biomass and its use as a source of energy by autotrophs include three steps:

1. An organism produces organic matter within its body.
2. It uses some of this new organic matter as a fuel in respiration.
3. It stores some of the newly produced organic matter for future use.

The first step, production of organic matter before use, is called *gross production.* The amount left after utilization is called *net production.*

Net production = Gross production − Respiration

The gross production of a tree—or any other plant—is the total amount of sugar it produces by photosynthesis before any is used. Within living cells in a green plant, some of the sugar is oxidized in respiration. Energy is used to convert sugars to other carbohydrates; those carbohydrates to amino acids; amino acids to proteins and to other tissues, such as cell walls and new leaf tissue. Energy is also used to transport material within the plant to roots, stems, flowers, and fruits. Some energy is lost as heat in the transfer. Some is stored in these other parts of the plant for later use. For woody plants like trees, some of this storage includes new wood laid down in the trunk, new buds that will develop into leaves and flowers the next year, and new roots.

Equations for Production, Biomass, and Energy Flow

We can write a general relation between biomass (B) and net production (NP):

$$B_2 = B_1 + NP$$

where B_2 is the biomass at the end of the time period, B_1 is the amount of biomass at the beginning of the time period, and NP is the change in biomass during the time period. Thus,

$$NP = B_2 - B_1$$

General production equations are given as

$$\mathrm{GP} = \mathrm{NP} + \mathrm{R}$$

$$\mathrm{NP} = \mathrm{GP} - \mathrm{R}$$

where GP is gross production, NP is net production, and R is respiration.

Several units of measures are used in the discussion of biological production and energy flow: calories when people talk about food content; watt-hours when people talk about biofuels; and kilojoules in standard international scientific notation. To make matters even more complicated, there are two kinds of calories: the standard calorie, which is the heat required to heat one gram of water from 15.5°C to 16.5°C, and the kilocalorie, which is 1,000 of the little calories. Even more confusing, when people discuss the energy content of food and diets, they use *calorie* to mean the kilocalorie. Just remember: The calorie you see on the food package is the big calorie, the kilocalorie, equal to 1,000 little calories. Almost nobody uses the "little" calorie, regardless of what they call it. To compare, an average apple contains about 100 Kcal or 116 watt-hours. This means that an apple contains enough energy to run a 100-watt bulb for 1 hour and 9 minutes. For those of you interested in your diet and weight, a Big Mac contains 576 Kcal, which is 669 watt-hours, enough to keep that 100-watt bulb burning for 6 hours and 41 minutes.

The average energy stored in vegetation is approximately 5 Kcal/gram (21 kilojoules per gram [kJ/g]). The energy content of organic matter varies. Ignoring bone and shells, woody tissue contains the least energy per gram, about 4 Kcal/g (17 kJ/g); fat contains the most, about 9 Kcal/g (38 kJ/g); and muscle contains approximately 5–6 Kcal/g (21–25 kJ/g). Leaves and shoots of green plants have about 5 Kcal/g (21–23 kJ/g); roots have about 4.6 Kcal/g (19 kJ/g).[2]

5.6 Energy Efficiency and Transfer Efficiency

How efficiently do living things use energy? This is an important question for the management and conservation of all biological resources. We would like biological resources to use energy efficiently—to produce a lot of biomass from a given amount of energy. This is also important for attempts to sequester carbon by growing trees and other perennial vegetation to remove carbon dioxide from the atmosphere and store it in living and dead organic matter (see Chapter 20).

As you learned from the second law of thermodynamics, no system can be 100% efficient. As energy flows through a food web, it is degraded, and less and less is usable. Generally, the more energy an organism gets, the more it has for its own use. However, organisms differ in how efficiently they use the energy they obtain. A more efficient organism has an advantage over a less efficient one.

Efficiency can be defined for both artificial and natural systems: machines, individual organisms, populations, trophic levels, ecosystems, and the biosphere. **Energy efficiency** is defined as the ratio of output to input, and it is usually further defined as the amount of useful work obtained from some amount of available energy. *Efficiency* has different meanings to different users. From the point of view of a farmer, an efficient corn crop is one that converts a great deal of solar energy to sugar and uses little of that sugar to produce stems, roots, and leaves. In other words, the most efficient crop is the one that has the most harvestable energy left at the end of the season. A truck driver views an efficient truck as just the opposite: For him, an efficient truck uses as much energy as possible from its fuel and stores as little energy as possible (in its exhaust). When we view organisms as food, we define *efficiency* as the farmer does, in terms of energy storage (net production from available energy). When we are energy users, we define *efficiency* as the truck driver does, in terms of how much useful work we accomplish with the available energy.

Consider the use of energy by a wolf and by one of its principal prey, moose. The wolf needs energy to travel long distances and hunt, and therefore it will do best if it uses as much of the energy in its food as it can. For itself, a highly energy-efficient wolf stores almost nothing. But from its point of view, the best moose would be one that used little of the energy it took in, storing most of it as muscle and fat, which the wolf can eat. Thus what is efficient depends on your perspective.

A common ecological measure of energy efficiency is called *food-chain efficiency*, or *trophic-level efficiency*, which is the ratio of production of one trophic level to the production of the next-lower trophic level. This efficiency is never very high. Green plants convert only 1–3% of the energy they receive from the sun during the year to new plant tissue. The efficiency with which herbivores convert the potentially available plant energy into herbivorous energy is usually less than 1%, as is the efficiency with which carnivores convert herbivores into carnivorous energy. In natural ecosystems, the organisms in one trophic level tend to take in much less energy than the potential maximum available to them, and they use more energy than they store for the next trophic level. At Isle Royale National Park, an island in Lake Superior, wolves feed on moose in a natural wilderness. A pack of 18 wolves kills an average of one moose approximately every 2.5 days,[14] which gives wolves a trophic-level efficiency of about 0.01%.

The rule of thumb for ecological trophic energy efficiency is that more than 90% (usually much more) of all energy transferred between trophic levels is lost as heat. Less than 10% (approximately 1% in natural ecosystems) is fixed as new tissue. In highly managed ecosystems, such as ranches, the efficiency may be greater. But even in such systems, it takes an average of 3.2 kg (7 lb) of vegetable matter to produce 0.45 kg (1 lb) of edible meat. Cattle are among the least efficient producers, requiring around 7.2 kg (16 lb) of vegetable matter to produce 0.45 kg (1 lb) of edible meat. Chickens are much more efficient, using approximately 1.4 kg (3 lb) of vegetable matter to produce 0.45 kg (1 lb) of eggs or meat. Much attention has been paid to the idea that humans should eat at a lower trophic level in order to use resources more efficiently. (See Critical Thinking Issue, Should People Eat Lower on the Food Chain?)

5.7 Ecological Stability and Succession

Ecosystems are dynamic: They change over time both from external (environmental) forces and from their internal processes. It is worth repeating the point we made in Chapter 3 about dynamic systems: The classic interpretation of populations, species, ecosystems, and Earth's entire biosphere has been to assume that each is a stable, static system. But the more we study these ecological systems, the clearer it becomes that these are dynamic systems, always changing *and always requiring change.* Curiously, they persist while undergoing change. We say "curiously" because in our modern technological society we are surrounded by mechanical and electronic systems that stay the same in most characteristics and are designed to do so. We don't expect our car or television or cell phone to shrink or get larger and then smaller again; we don't expect that one component will get bigger or smaller over time. If anything like this were to happen, those systems would break.

Ecosystems, however, not only change but also then recover and overcome these changes, and life continues on. It takes some adjustment in our thinking to accept and understand such systems.

When disturbed, ecosystems can recover through **ecological succession** if the damage is not too great. We can classify ecological succession as either primary or secondary. **Primary succession** is the establishment and development of an ecosystem where one did not exist previously. Coral reefs that form on lava emitted from a volcano and cooled in shallow ocean waters are examples of primary succession. So are forests that develop on new lava flows, like those released by the volcano on the big island of Hawaii (Figure 5.14a), and forests that develop at the edges of retreating glaciers (Figure 5.14b).

Secondary succession is reestablishment of an ecosystem after disturbances. In secondary succession, there are remnants of a previous biological community, including such things as organic matter and seeds. A coral reef that has been killed by poor fishing practices, pollution, climate change, or predation, and then

FIGURE 5.14 Primary succession. (a) Forests developing on new lava flows in Hawaii and **(b)** at the edge of a retreating glacier.

recovers, is an example of secondary succession.[15] Forests that develop on abandoned pastures or after hurricanes, floods, or fires are also examples of secondary succession.

Succession is one of the most important ecological processes, and the patterns of succession have many management implications (discussed in detail in Chapter 10). We see examples of succession all around us. When a house lot is abandoned in a city, weeds begin to grow. After a few years, shrubs and trees can be found; secondary succession is taking place. A farmer weeding a crop and a homeowner weeding a lawn are both fighting against the natural processes of secondary succession.

Patterns in Succession

Succession follows certain general patterns. When ecologists first began to study succession, they focused on three cases involving forests: (1) on dry sand dunes along the shores of the Great Lakes in North America; (2) in a northern freshwater bog; and (3) in an abandoned farm field. These were particularly interesting because each demonstrated a repeatable pattern of recovery, and each tended to produce a late stage that was similar to the late stages of the others.

Dune Succession

Sand dunes are continually being formed along sandy shores and then breached and destroyed by storms. In any of the Great Lakes states, soon after a dune is formed on the shores of one of the Great Lakes, dune grass invades. This grass has special adaptations to the unstable dune. Just under the surface, it puts out runners with sharp ends (if you step on one, it will hurt). The dune grass rapidly forms a complex network of underground runners, crisscrossing almost like a coarsely woven mat. Above the ground, the green stems carry out photosynthesis, and the grasses grow. Once the dune grass is established, its runners stabilize the sand, and seeds of other plants have a better chance of germinating. The seeds germinate, the new plants grow, and an ecological community of many species begins to develop. The plants of this early stage tend to be small, grow well in bright light, and withstand the harsh environment—high temperatures in summer, low temperatures in winter, and intense storms.

Slowly, larger plants, such as eastern red cedar and eastern white pine, are able to grow on the dunes. Eventually, a forest develops, which may include species such as beech and maple. A forest of this type can persist for many years, but at some point a severe storm breaches even these heavily vegetated dunes, and the process begins again (Figure 5.15).

FIGURE 5.15 Dune succession on the shores of Lake Michigan. Dune-grass shoots appear scattered on the slope, where they emerge from underground runners.

Bog Succession

A **bog** is an open body of water with surface inlets—usually small streams—but no surface outlet. As a result, the waters of a bog are quiet, flowing slowly if at all. Many bogs that exist today originated as lakes that filled depressions in the land, which in turn were created by glaciers during the Pleistocene ice age. Succession in a northern bog, such as the Livingston Bog in Michigan (Figure 5.16), begins when a sedge (a grasslike herb) puts out floating runners (Figure 5.17a, b). These runners form a complex, matlike network similar to that formed by dune grass. The stems of the sedge grow on the runners and carry out photosynthesis. Wind blows particles onto the mat, and soil, of a kind, develops. Seeds of other plants, instead of falling into the water, land on the mat and can germinate. The floating mat becomes thicker as small shrubs and trees, adapted to wet environments, grow. In the North, these include species of the blueberry family.

FIGURE 5.16 **Livingston Bog,** a famous bog in the northern part of Michigan's lower peninsula.

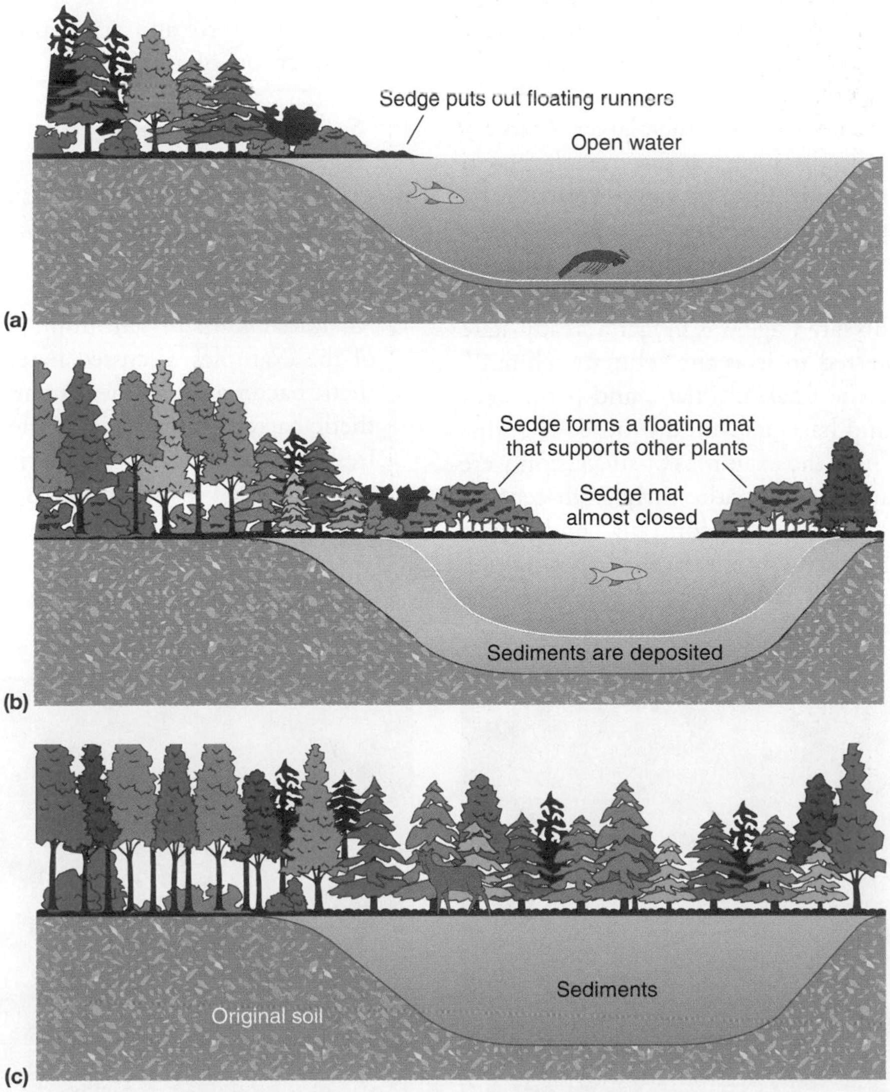

FIGURE 5.17 **Diagram of bog succession.** Open water **(a)** is transformed through formation of a floating mat of sedge and deposition of sediments **(b)** into wetland forest **(c)**.

The bog also fills in from the bottom as streams carry fine particles of clay into it (Figure 5.17b, c). At the shore, the floating mat and the bottom sediments meet, forming a solid surface. But farther out, a "quaking bog" occurs. You can walk on this mat; and if you jump up and down, all the plants around you bounce and shake, because the mat is really floating. Eventually, as the bog fills in from the top and the bottom, trees grow that can withstand wetter conditions—such as northern cedar, black spruce, and balsam fir. The formerly open-water bog becomes a wetland forest. If the bog is farther south, it may eventually be dominated by beech and maple, the same species that dominate the late stages of the dunes.

Old-Field Succession

In the northeastern United States, a great deal of land was cleared and farmed in the 18th and 19th centuries. Today, much of this land has been abandoned for farming and allowed to grow back to forest (Figure 5.18). The first plants to enter the abandoned farmlands are small plants adapted to the harsh and highly variable conditions of a clearing—a wide range of temperatures and precipitation. As these plants become established, other, larger plants enter. Eventually, large trees grow, such as sugar maple, beech, yellow birch, and white pine, forming a dense forest.

Since these three different habitats—one dry (the dunes), one wet (the bog), and one in between (the old field)—tend to develop into similar forests, early ecologists believed that this late stage was in fact a steady-state condition. They referred to it as the "climatic climax," meaning that it was the final, ultimate, and permanent stage to which all land habitats would proceed if undisturbed by people. Thus, the examples of succession were among the major arguments in the early 20th century that nature did in fact achieve a constant condition, a steady state, and there actually was a balance of nature. We know today that this is not true, a point we will return to later.

Coral Reef Succession

Coral reefs (Figure 5.19) are formed in shallow warm waters by corals, small marine animals that live in colonies and are members of the phylum Coelenterata, which also includes sea anemones and jellyfishes. Corals have a whorl of tentacles surrounding the mouth, and feed by catching prey, including planktonic algae, as it passes by. The corals settle on a solid surface and produce a hard polyp of calcium carbonate (in other words, limestone). As old individuals die, this hard material becomes the surface on which new individuals establish themselves. In addition to the coelenterates, other limestone-shell-forming organisms—algae, corals, snails, urchins—live and die on the reef and are glued together primarily by a kind of algae.[16] Eventually a large and complex structure results involving many other species, including autotrophs and heterotrophs, creating one of the most species-diverse of all kinds of ecosystems. Highly valued for this diversity, for production of many edible fish, for the coral itself (used in various handicrafts and arts), and for recreation, coral reefs attract lots of attention.

Succession, in Sum

Even though the environments are very different, these four examples of ecological succession—dune, bog, old field, and coral reef—have common elements found in most ecosystems:

1. An initial kind of autotroph (green plants in three of the examples discussed here; algae and photosynthetic bacteria in marine systems; algae and photosynthetic bacteria, along with some green plants in some freshwater and near-shore marine systems). These are typically small in stature and specially adapted to the unstable conditions of their environment.

FIGURE 5.18 **Old-growth eastern deciduous forest.**

FIGURE 5.19 **Hawaiian coral reef.**

2. A second stage with autotrophs still of small stature, rapidly growing, with seeds or other kinds of reproductive structures that spread rapidly.
3. A third stage in which larger autotrophs—like trees in forest succession—enter and begin to dominate the site.
4. A fourth stage in which a mature ecosystem develops.

Although we list four stages, it is common practice to combine the first two and speak of early-, middle-, and late-successional stages. The stages of succession are described here in terms of autotrophs, but similarly adapted animals and other life-forms are associated with each stage. We discuss other general properties of succession later in this chapter.

Species characteristic of the early stages of succession are called pioneers, or **early-successional species**. They have evolved and are adapted to the environmental conditions in early stages of succession. In terrestrial ecosystems, vegetation that dominates late stages of succession, called **late-successional species**, tends to be slower-growing and longer-lived, and can persist under intense competition with other species. For example, in terrestrial ecosystems, late-successional vegetation tends to grow well in shade and have seeds that, though not as widely dispersing, can persist a rather long time. Typical **middle-successional species** have characteristics in between the other two types.

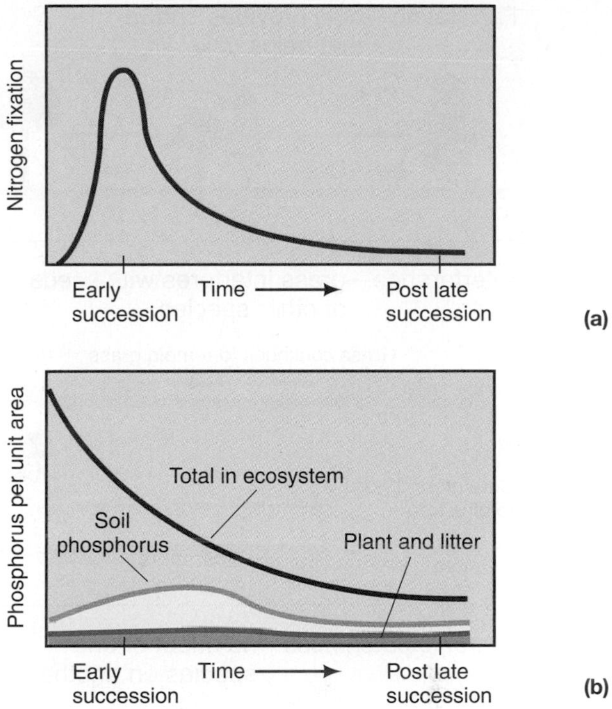

FIGURE 5.20 **(a)** Hypothesized changes in soil nitrogen during the course of soil development. **(b)** Change in total soil phosphorus over time with soil development. (P.M. Vitousek and P.S. White, Process studies in forest succession, in D.C. West, H.H. Shugart, and D.B. Botkin, eds., [New York: Springer-Verlag, 1981], Figure 17.1, p. 269.)

5.8 Chemical Cycling and Succession

One of the important effects of succession is a change in the storage of chemical elements necessary for life. On land, the storage of chemical elements essential for plant growth and function (including nitrogen, phosphorus, potassium, and calcium) generally increases during the progression from the earliest stages of succession to middle succession (Figure 5.20). There are three reasons for this:

Increased storage. Organic matter, living or dead, stores chemical elements. As long as there is an increase in organic matter within the ecosystem, there will be an increase in the storage of chemical elements.

Increased rate of uptake. For example, in terrestrial ecosystems. many plants have root nodules containing bacteria that can assimilate atmospheric nitrogen, which is then used by the plant in a process known as *nitrogen fixation.*

Decreased rate of loss. The presence of live and dead organic matter helps retard erosion. Both organic and inorganic soil can be lost to erosion by wind and water. Vegetation and, in certain marine and freshwater ecosystems, large forms of algae tend to prevent such losses and therefore increase total stored material.

Ideally, chemical elements could be cycled indefinitely in ecosystems, but in the real world there is always some loss as materials are moved out of the system by wind and water. As a result, ecosystems that have persisted continuously for the longest time are less fertile than those in earlier stages. For example, where glaciers melted back thousands of years ago in New Zealand, forests developed, but the oldest areas have lost much of their fertility and have become shrublands with less diversity and biomass. The same thing happened to ancient sand dune vegetation in Australia.[17]

5.9 How Species Change Succession

Early-successional species can affect what happens later in succession in three ways: through (1) facilitation, (2) interference, or (3) life history differences (Figure 5.21).[18,19]

Facilitation

In facilitation, an earlier-successional species changes the local environment in ways that make it suitable for another species that is characteristic of a later successional stage. Dune and bog succession illustrate facilitation. The first plant species—dune grass and floating sedge—prepare the way for other species to grow. Facilitation is common in tropical rain forests,[20] where early-successional species

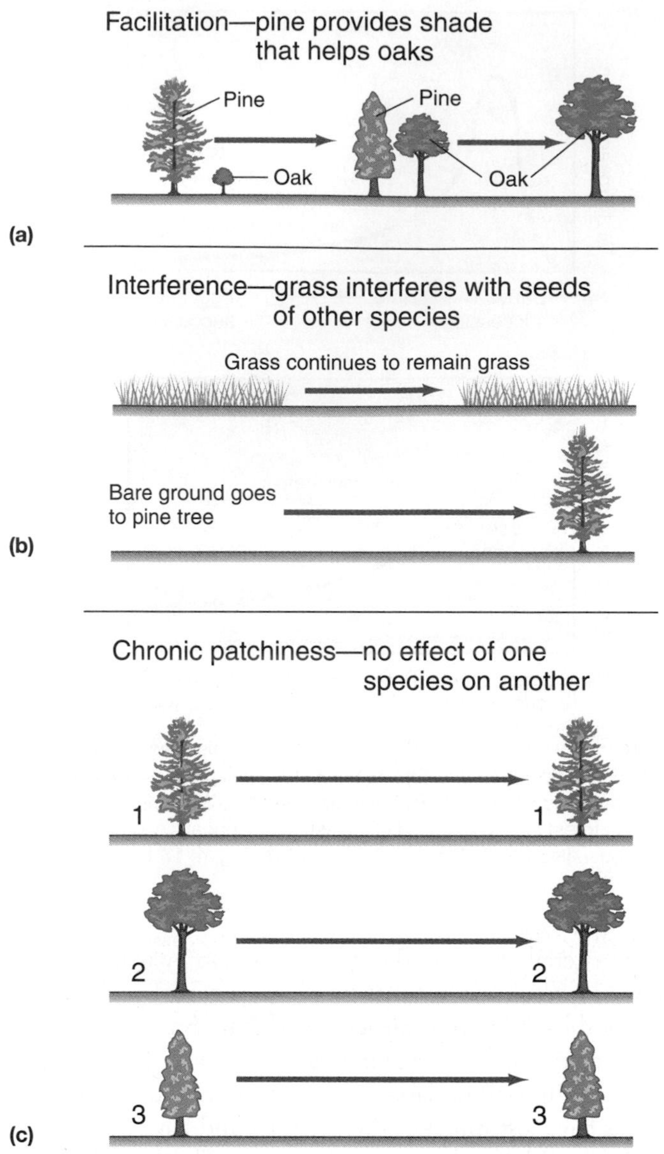

FIGURE 5.21 **Interaction among species during ecological succession. (a) Facilitation.** As Henry David Thoreau observed in Massachusetts more than 100 years ago, pines provide shade and act as "nurse trees" for oaks. Pines do well in openings. If there were no pines, few or no oaks would survive. Thus the pines facilitate the entrance of oaks. **(b) Interference.** Some grasses that grow in open areas form dense mats that prevent seeds of trees from reaching the soil and germinating. **(c) Chronic patchiness.** Earlier-entering species neither help nor interfere with other species; instead, as in a desert, the physical environment dominates.

speed the reappearance of the microclimatic conditions that occur in a mature forest. Because of the rapid growth of early-successional plants, after only 14 years the temperature, relative humidity, and light intensity at the soil surface in tropical forests can approximate those of a mature rain forest.[21] Once these conditions are established, species adapted to deep forest shade can germinate and persist.

Facilitation also occurs in coral reefs, mangrove swamps along ocean shores, kelp beds along cold ocean shores such as the Pacific coast of the United States and Canada's Pacific Northwest, and in shallow marine benthic areas where the water is relatively calm and large algae can become established.

Knowing the role of facilitation can be useful in the restoration of damaged areas. Plants that facilitate the presence of others should be planted first. On sandy areas, for example, dune grasses can help hold the soil before we attempt to plant shrubs or trees.

Interference

In contrast to facilitation, interference refers to situations where an earlier-successional species changes the local environment so it is *unsuitable* to another species characteristic of a later-successional stage. Interference is common, for example, in American tall-grass prairies, where prairie grasses like little bluestem form a mat of living and dead stems so dense that seeds of other plants cannot reach the ground and therefore do not germinate. Interference does not last forever, however. Eventually, some breaks occur in the grass mat—perhaps from surface-water erosion, the death of a patch of grass from disease, or removal by fire. Breaks in the grass mat allow seeds of trees to germinate. For example, in the tall-grass prairie, seeds of red cedar can then reach the ground. Once started, red cedar soon grows taller than the grasses, shading them so much that they cannot grow. More ground is open, and the grasses are eventually replaced.

The same pattern occurs in some Asian tropical rain forests. The grass, *Imperata*, forms stands so dense that seeds of later-successional species cannot reach the ground. *Imperata* either replaces itself or is replaced by bamboo, which then replaces itself. Once established, *Imperata* and bamboo appear able to persist for a long time. Once again, when and if breaks occur in the cover of these grasses, other species can germinate and grow, and a forest eventually develops.

Life History Differences

In this case, changes in the time it takes different species to establish themselves give the appearance of a succession caused by species interactions, but it is not. In cases where no species interact through succession, the result is termed **chronic patchiness**. Chronic patchiness is characteristic of highly disturbed environments and highly stressful ones in terms of temperature, precipitation, or chemical availability, such as deserts. For example, in the warm deserts of California, Arizona, and Mexico, the major shrub species grow in patches, often consisting of mature individuals with few seedlings. These patches tend to persist for long periods until there is a disturbance.[22] Similarly, in highly polluted environments, a sequence of species replacement may not occur. Chronic patchiness also describes planktonic ecological communities and their ecosystems, which occur in the constantly moving waters of the upper ocean and the upper waters of ponds, lakes, rivers, and streams.

CRITICAL THINKING ISSUE

Should People Eat Lower on the Food Chain?

The energy content of a food chain is often represented by an energy pyramid, such as the one shown here in Figure 5.22a for a hypothetical, idealized food chain. In an energy pyramid, each level of the food chain is represented by a rectangle whose area is more or less proportional to the energy content of that level. For the sake of simplicity, the food chain shown here assumes that each link in the chain has only one source of food.

Assume that if a 75 kg (165 lb) person ate frogs (and some people do!), he would need 10 a day, or 3,000 a year (approximately 300 kg, or 660 lb). If each frog ate 10 grasshoppers a day, the 3,000 frogs would require 9 million grasshoppers a year to supply their energy needs, or approximately 9,000 kg (19,800 lb) of grasshoppers. A horde of grasshoppers of that size would require 333,000 kg (732,600 lb) of wheat to sustain them for a year.

As the pyramid illustrates, the energy content decreases at each higher level of the food chain. The result is that the amount of energy at the top of a pyramid is related to the number of layers the pyramid has. For example, if people fed on grasshoppers rather than frogs, each person could probably get by on 100 grasshoppers a day. The 9 million grasshoppers could support 300 people for a year, rather than only one. If, instead of grasshoppers, people ate wheat, then 333,000 kg of wheat could support 666 people for a year.

This argument is often extended to suggest that people should become herbivores (vegetarians) and eat directly from the lowest level of all food chains, the autotrophs. Consider, however, that humans can eat only parts of some plants. Herbivores can eat some parts of plants that humans cannot eat, and some plants that humans cannot eat at all. When people eat these herbivores, more of the energy stored in plants becomes available for human consumption.

The most dramatic example of this is in aquatic food chains. Because people cannot digest most kinds of algae, which are the base of most aquatic food chains, they depend on eating fish that eat algae and fish that eat other fish. So if people were to become entirely herbivorous, they would be excluded from many food chains. In addition, there are major areas of Earth where crop production damages the land but grazing by herbivores does not. In those cases, conservation of soil and biological diversity lead to arguments that support the use of grazing animals for human food. This creates an environmental issue: How low on the food chain should people eat?

Critical Thinking Questions

1. Why does the energy content decrease at each higher level of a food chain? What happens to the energy lost at each level?
2. The pyramid diagram uses mass as an indirect measure of the energy value for each level of the pyramid. Why is it appropriate to use mass to represent energy content?
3. Using the average of 21 kilojoules (kJ) of energy to equal 1 g of completely dried vegetation (see Working It Out 5.2) and assuming that wheat is 80% water, what is the energy content of the 333,000 kg of wheat shown in the pyramid?
4. Make a list of the environmental arguments for and against an entirely vegetarian diet for people. What might be the consequences for U.S. agriculture if everyone in the country began to eat lower on the food chain?
5. How low do you eat on the food chain? Would you be willing to eat lower? Explain.

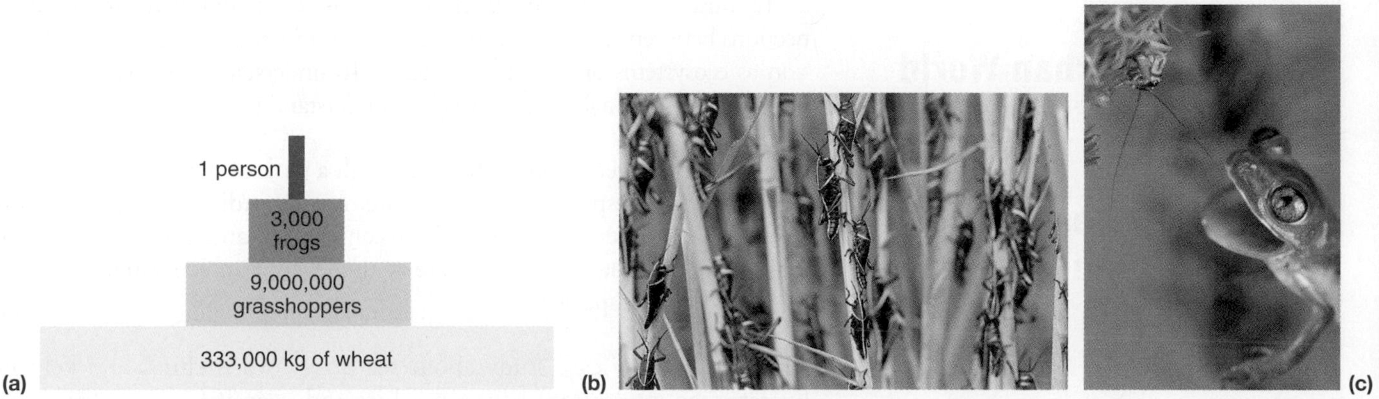

FIGURE 5.22 **(a)** Energy pyramid. **(b)** Grasshoppers. **(c)** Frogs that eat grasshoppers.

SUMMARY

- An ecosystem is the simplest entity that can sustain life. At its most basic, an ecosystem consists of several species and a fluid medium (air, water, or both). The ecosystem must sustain two processes—the cycling of chemical elements and the flow of energy.
- The living part of an ecosystem is the ecological community, a set of species connected by food webs and trophic levels. A food web or food chain describes who feeds on whom. A trophic level consists of all the organisms that are the same number of feeding steps from the initial source of energy.
- Community-level effects result from indirect interactions among species, such as those that occur when sea otters influence the abundance of sea urchins.
- Ecosystems are real and important, but it is often difficult to define the limits of a system or to pinpoint all the interactions that take place. Ecosystem management is considered key to the successful conservation of life on Earth.
- Energy flows one way through an ecosystem; the second law of thermodynamics places a limit on the abundance of productivity of life and requires the one-way flow.
- Chemical elements cycle, and in theory could cycle forever, but in the real world there is always some loss.
- Ecosystems recover from changes through ecological succession, which has repeatable patterns.
- Ecosystems are non-steady-state systems, undergoing changes all the time and requiring change.

REEXAMINING THEMES AND ISSUES

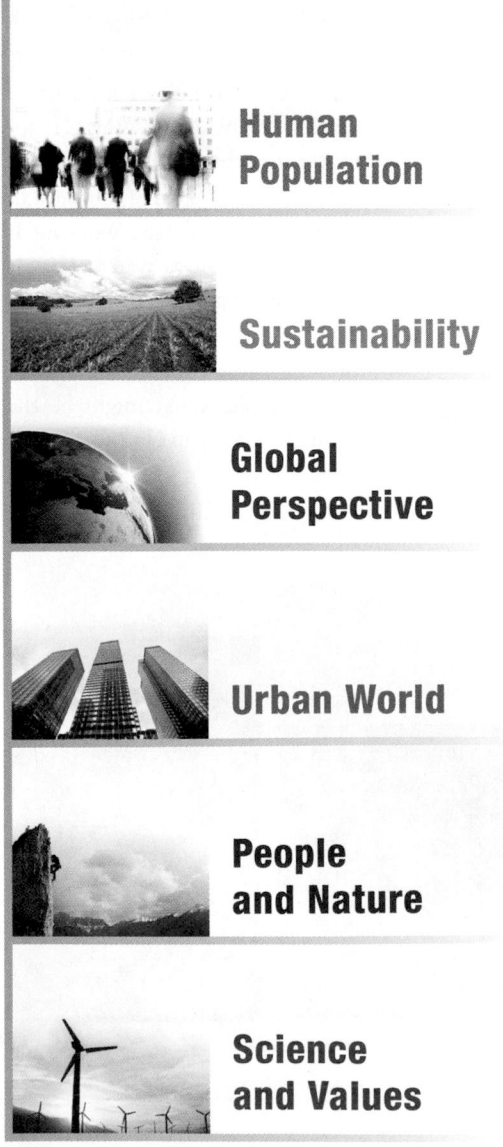

Human Population

The human population depends on many ecosystems that are widely dispersed around the globe. Modern technology may appear to make us independent of these natural systems. In fact, though, the more connections we establish through modern transportation and communication, the more kinds of ecosystems we depend on. Therefore, the ecosystem concept is one of the most important we will learn about in this book.

Sustainability

The ecosystem concept is at the heart of managing for sustainability. When we try to conserve species or manage living resources so that they are sustainable, we must focus on their ecosystem and make sure that it continues to function.

Global Perspective

Our planet has sustained life for approximately 3.5 billion years. To understand how Earth as a whole has sustained life for such a long time, we must understand the ecosystem concept because the environment at a global level must meet the same basic requirements as those of any local ecosystem.

Urban World

Cities are embedded in larger ecosystems. But like any life-supporting system, a city must meet basic ecosystem needs. This is accomplished through connections between cities and surrounding environments. Together, these function as ecosystems or sets of ecosystems. To understand how we can create pleasant and sustainable cities, we must understand the ecosystem concept.

People and Nature

The feelings we get when we hike through a park or near a beautiful lake are as much a response to an ecosystem as to individual species. This illustrates the deep connection between people and ecosystems. Also, many effects we have on nature are at the level of an ecosystem, not just on an individual species.

Science and Values

The introductory case study about sea otters, sea urchins, and kelp illustrates the interactions between values and scientific knowledge about ecosystems. Science can tell us how organisms interact. This knowledge confronts us with choices. Do we want abalone to eat, sea otters to watch, kelp forests for biodiversity? The choices we make depend on our values.

KEY TERMS

autotrophs **84**
biological production **93**
biomass **93**
bog **97**
carnivores **84**
chemoautotrophs **93**
chronic patchiness **100**
community effect **81**
decomposers **84**
early-successional species **99**
ecological community **83**
ecological succession **95**
ecosystem **83**
ecosystem energy flow **89**
energy efficiency **95**
entropy **93**
first law of thermodynamics **91**
food chains **84**
food webs **84**
gross production **93**
herbivores **84**
heterotrophs **84**
keystone species **81**
late-successional species **99**
middle-successional species **99**
net production **93**
omnivores **86**
pelagic ecosystem **85**
photosynthesis **84**
primary production **93**
primary succession **95**
secondary production **93**
secondary succession **95**
second law of thermodynamics **92**
succession **96**
trophic level **84**
watershed **88**

STUDY QUESTIONS

1. Farming has been described as managing land to keep it in an early stage of succession. What does this mean, and how is it achieved?
2. Redwood trees reproduce successfully only after disturbances (including fire and floods), yet individual redwood trees may live more than 1,000 years. Is redwood an early- or late-successional species?
3. What is the difference between an ecosystem and an ecological community?
4. What is the difference between the way energy puts limits on life and the way phosphorus does so?
5. Based on the discussion in this chapter, would you expect a highly polluted ecosystem to have many species or few species? Is our species a keystone species? Explain.
6. Keep track of the food you eat during one day and make a food chain linking yourself with the sources of those foods. Determine the biomass (grams) and energy (kilocalories) you have eaten. Using an average of 5 kcal/g, then using the information on food packaging or assuming that your net production is 10% efficient in terms of the energy intake, how much additional energy might you have stored during the day? What is your weight gain from the food you have eaten?
7. Which of the following are ecosystems? Which are ecological communities? Which are neither?
   (a) Chicago
   (b) A 1,000-ha farm in Illinois
   (c) A sewage-treatment plant
   (d) The Illinois River
   (e) Lake Michigan

FURTHER READING

Modern Studies

Botkin, D.B., *No Man's Garden: Thoreau and a New Vision for Civilization and Nature* (Washington, DC: Island Press, 2001).

Chapin, F. Stuart III, Harold A. Mooney, Melissa C. Chapin, and Pamela Matson, *Principles of Terrestrial Ecosystem Ecology* (New York: Springer, 2004, paperback). Kaiser et al., *Marine Ecology: Processes, Systems, and Impacts* (New York: Oxford University Press, 2005).

Some Classic Studies and Books

Blum, H.F., *Time's Arrow and Evolution* (New York: Harper & Row, 1962). A very readable book discussing how life is connected to the laws of thermodynamics and why this matters.

Bormann, F.H., and G.E. Likens, *Pattern and Process in a Forested Ecosystem,* 2nd ed. (New York: Springer-Verlag, 1994). A synthetic view of the northern hardwood ecosystem, including its structure, function, development, and relationship to disturbance.

Gates, D.M., *Biophysical Ecology* (New York: Springer-Verlag, 1980). A discussion about how energy in the environment affects life.

Morowitz, H.J., *Energy Flow in Biology* (Woodbridge, CT: Oxbow, 1979). The most thorough and complete discussion available about the connection between energy and life, at all levels, from cells to ecosystems to the biosphere.

Odum, Eugene, and G.W. Barrett, *Fundamentals of Ecology* (Duxbury, MA: Brooks/Cole, 2004). Odum's original textbook was a classic, especially in providing one of the first serious introductions to ecosystem ecology. This is the latest update of the late author's work, done with his protégé.

Schrödinger, E. (ed. Roger Penrose), *What Is Life?: With Mind and Matter and Autobiographical Sketches (Canto)* (Cambridge: Cambridge University Press, 1992). The original statement about how the use of energy differentiates life from other phenomena in the universe. Easy to read and a classic.

CHAPTER 6

The Biogeochemical Cycles

As a result of biogeochemical cycles marine life is plentiful in the Santa Barbara Channel of southern California.

LEARNING OBJECTIVES

Life is composed of many chemical elements, which have to be available in the right amounts, the right concentrations, and the right ratios to one another. If these conditions are not met, then life is limited. The study of chemical availability and biogeochemical cycles—the paths chemicals take through Earth's major systems—is important in solving many environmental problems. After reading this chapter, you should understand . . .

- What the major biogeochemical cycles are;
- How life, over the Earth's history, has greatly altered chemical cycles;
- The major factors and processes that control biogeochemical cycles;
- Why some chemical elements cycle quickly and some slowly;
- How each major component of Earth's global system (the atmosphere, waters, solid surfaces, and life) is involved and linked with biogeochemical cycles;
- How the biogeochemical cycles most important to life, especially the carbon cycle, generally operate;
- How humans affect biogeochemical cycles.

CASE STUDY

Methane and Oil Seeps: Santa Barbara Channel

The Santa Barbara Channel off the shore of southern and central California is home to numerous species, including such marine mammals as dolphins, sea otters, elephant seals, sea lions, harbor seals, and blue, humpback, and gray whales; many birds, including brown pelicans; and a wide variety of fish. The channel is also a region with large oil and gas resources that have been exploited by people for thousands of years.[1,2,3] For centuries, Native Americans who lived along the shoreline collected tar from oil seeps to seal baskets and the planks of their seagoing canoes. During the last century, oil wells on land and from platforms anchored on the seabed have been extracting oil and gas. Oil and gas are hydrocarbons, and as such are part of the global carbon cycle that involves physical, geological, biological, and chemical processes.

The story of oil and gas in the Santa Barbara Channel begins 6–18 million years ago with the deposition of a voluminous amount of fine sediment, enriched with planktonic microorganisms whose bodies sank to the ocean floor and were buried. (*Planktonic* refers to small floating algae and animals.) Over geologic time, the sediment was transformed into sedimentary rock, and the organic material was transformed by heat and pressure into oil and gas. About a million or so years ago, tectonic uplift and fracturing forced the oil and gas toward the surface. Oil and gas seepage has reached the surface for at least 120,000 years and perhaps more than half a million years.

Some of the largest seeps of oil and natural gas (primarily methane) are offshore of the University of California, Santa Barbara, at Coal Oil Point, where about 100 barrels of oil and approximately 57,000 m^3 (2 million cubic feet) of gas are released per day (Figures 6.1 and 6.2). To put the amount of oil in perspective, the 1989 *Exxon Valdez* tanker accident in Prince William Sound released about 250,000 barrels of oil. Thus, the oil seeping from the Coal Oil Point area alone equals one *Exxon Valdez* accident every seven years. This is a tremendous amount of oil to be added to the marine environment.

Sudden emissions of gases create small pits on the seafloor. The gas rises as clouds of bubbles clearly visible at the surface (Figure 6.2b and c). Once at the surface, the oil and gas form slicks that are transported by marine currents and wind. On the seafloor, the heaviest materials form mounds of tar several meters or more in diameter.[3] Some of the thicker tar washes up on local beaches, sometimes covering enough of the water and beach to stick to the bare skin of walkers and swimmers. Tar may be found on beaches for several kilometers to the east.

FIGURE 6.1 **Coal Oil Point, Santa Barbara**, the location of large offshore oil and gas seeps on one of America's most beautiful coastlines. Active oil and gas seeps are located from near the shore to just past offshore platform Holly that has many pumping oil wells.

The emitted hydrocarbon gases contribute to air pollution in the Santa Barbara area. Once in the atmosphere, they interact with sunlight to produce smog, much like the smog produced by hydrocarbon emissions from automobiles in Los Angeles. If all the methane ended up in the atmosphere as hydrocarbons, the contribution to air pollution in Santa Barbara County would be about double the emission rate from all on-road vehicles in Santa Barbara County.

Fortunately for us, seawater has a tremendous capacity to take up the methane, and bacteria in the ocean feed on the methane, releasing carbon dioxide (Figure 6.2a). The ocean and its bacteria thus take care of about half the methane moving up from the seeps. Thanks to microbial decomposition of the methane, only about 1% of the methane that is dissolved in the seawater is emitted into the atmosphere.[1, 2]

Even so, in recent years people have taken action to further control the oil and gas seeps at Coal Oil Point. Two steel seep tents (each 30 m by 30 m) have been placed over some of the methane seeps, and the gas is collected and moved to the shore through pipelines, for use as natural gas. Furthermore, the pumping of oil from a single well from a nearby platform with many wells apparently has reduced emissions of methane and oil from the seeps. What drives methane emission is pressure from below, and pumping from the wells evidently reduces that pressure.

The lesson from the methane and oil seeps at Coal Oil Point is twofold: first, that this part of the carbon cycle is a complex linkage of physical, biological, and chemical processes; and second, that human activity may also play a role. These two concepts will be a recurring theme in our discussion of the major biogeochemical cycles that concern us today.

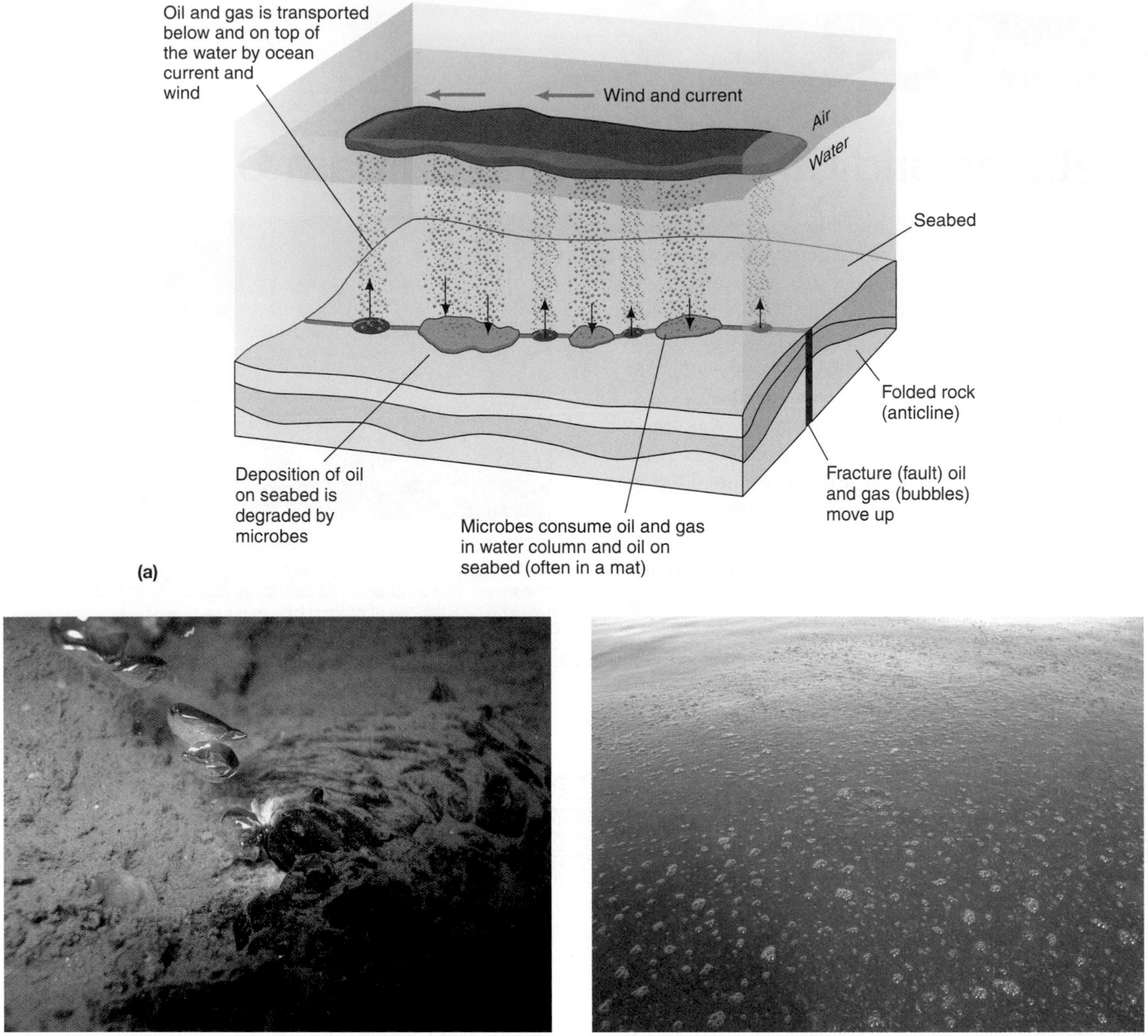

FIGURE 6.2 **(a)** Idealized diagram of physical, chemical, and biological processes with shallow methane and oil seeps; **(b)** small bubbles of methane (~1 cm) from a seep at Coal Oil Point on the seabed; and **(c)** methane bubbles (~1 cm) at the surface. (*Photographs courtesy of David Valentine.*)

6.1 Earth Is a Peculiar Planet

Our planet, Earth, is unique, at least to the extent that we have explored the cosmos. In our solar system, and in the Milky Way galaxy to the extent that we have observed it, Earth is the only body that has the combination of four characteristics: liquid water; water at its triple point (gas, liquid, and solid phases at the same time); plate tectonics; and life (Figure 6.3). (Recent space probes to the moons of Jupiter and Saturn suggest that there may be liquid water on a few of these and perhaps also an equivalent of plate tectonics. And recent studies of Mars suggest that liquid water has broken through to the surface on occasion in the past, causing Earthlike water erosion.)

The above discussion leads to consideration of the history of Earth over billions of years. This has prompted some geologists to propose "big history"—to link contemporary history with geologic history, perhaps even going back all the way to the *Big Bang*

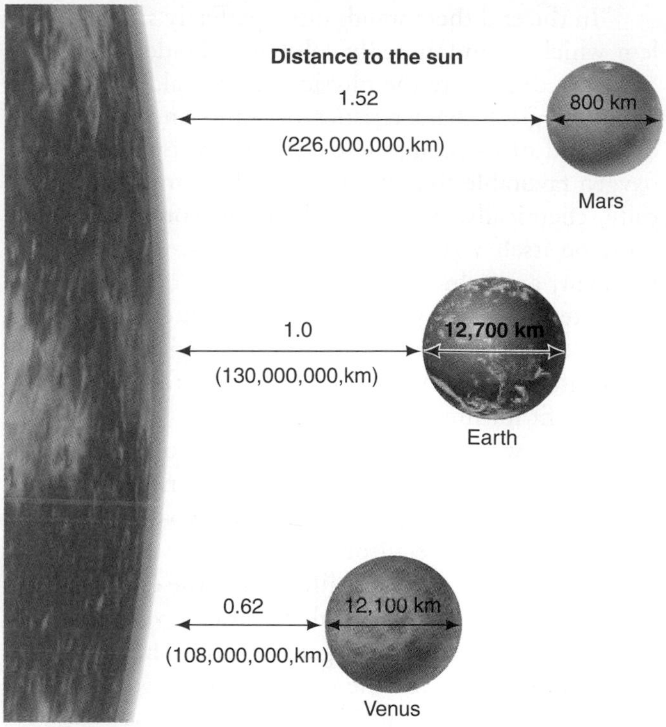

Atmosphere	Venus	Earth	Mars
Carbon dioxide	98%	0.03%	96%
Hydrogen	1.9%	73%	2.7%
Oxygen	Trace	21%	0.13%
Argon	0.1%	1%	2%
Total Pressure (bars)	90	1	0.00
Surface temperature	447°C	13°C	−53°C

FIGURE 6.3 Venus, Earth, and Mars. These three planets had a common origin and should be similar. They are within a factor of 2 in size and distance from the sun, and the atmospheres of Mars and Venus are similar in chemical makeup. Earth's atmosphere, however, is very different.

12 billion years ago, when our universe was born.[4,5] The main regimes of big history include cosmos, Earth, and life. To this, in the context of environmental science, we add human history.[4,5]

Space Travelers and Our Solar System

Life changes the cycling of chemical elements on Earth and has done so for several billion years.[6] To begin to examine this intriguing effect of life at a global level, it is useful to imagine how travelers from another solar system might perceive our planet. Imagine these space travelers approaching our solar system. They find that their fuel is limited and that of the four inner planets, only two, the second (Venus) and the fourth (Mars), are on their approach path. By chance, and because of differences in the orbits of the planets, the first (Mercury) and the third (Earth) are both on the opposite side of the sun, not easily visible for their instruments to observe or possible for their spacecraft to approach closely. However, they can observe Mars and Venus as they fly by them, and from those observations hypothesize about the characteristics of the planet whose orbit is between those two—Earth (Figure 6.4).

The space travelers' instruments tell them that the atmospheres of Venus and Mars are primarily carbon dioxide, with some ammonia (nitrogen combined with hydrogen) and trace amounts of nitrogen, oxygen, argon, and the other "noble gases"—that is, elements like argon that form few compounds. Since the space travelers understand how solar systems originate, they know that the inner planets are formed by the gathering together of particles as a result of gravitational force. Therefore, they believe that the second, third, and fourth planets should have a similar composition, and this leads them to believe that it is reasonable to assume the third planet will have an atmosphere much like that of Venus and Mars.

FIGURE 6.4 Our solar system with the planets (Pluto is not classified as a planet) shown from NASA space probes. Imagine travel to this system from another and wondering what the third planet was like.

Suppose a later space flight from the same solar system visits ours once again, but this time it is able to approach Earth. Knowing the results of the previous voyage, the new travelers are surprised to discover that Earth's atmosphere is entirely different from those of Venus and Mars. It is composed primarily (78%) of free (molecular) nitrogen (N_2), with about 20% oxygen, a trace of carbon dioxide and other gases, and some argon. What has caused this great difference? Because they are trained in science and in the study of life in the universe, and because they come from a planet that has life, these space travelers recognize the cause immediately: *Earth must contain life.* Life changes its planet's atmosphere, oceans, and upper surfaces. Even without seeing life directly, they know from its atmosphere that Earth is a "living" planet.

The great 20th-century ecologist G. Evelyn Hutchinson described this phenomenon succinctly. The strangest characteristic of Earth's surface, he wrote, is that it is not in thermodynamic equilibrium, which is what would happen if you were able to carry out a giant experiment in which you took Earth, with its atmosphere, oceans, and solid surfaces, and put it into a closed container sealed against the flow of energy and matter. Eventually the chemistry of the air, water, and rocks would come into a chemical and physical fixed condition where the energy was dispersed as heat and there would be no new chemical reactions. Physicists will tell you that everything would be at the lowest energy level, that matter and energy would be dispersed randomly, and nothing would be happening. This is called the **thermodynamic equilibrium**. In this giant experiment, this equilibrium would resemble that in the atmospheres of Mars and Venus, and Earth's atmosphere would be very different from the way it is now.

Life on Earth acts as a pump to keep the atmosphere, ocean, and rocks far from a thermodynamic equilibrium. The highly oxygenated atmosphere is so far from a thermodynamic equilibrium that it is close to an explosive combination with the organic matter on the Earth. James Lovelock, the originator of the Gaia hypothesis, has written that if the oxygen concentration in the atmosphere rose a few percentage points, to around 22% or higher, fires would break out spontaneously in dead wood on Earth's surface.[6] It's a controversial idea, but it suggests how close the present atmosphere is to a violent disequilibrium.[7]

The Fitness of the Environment[8]

Early in the 20th century, a scientist named Lawrence Henderson wrote a book with a curious title: *The Fitness of the Environment.*[9] In this book, Henderson observed that the environment on Earth was peculiarly suited to life. The question was, how did this come about? Henderson sought to answer this question in two ways: first, by examining the cosmos and seeking an answer in the history of the universe and in fundamental characteristics of the universe; second, by examining the properties of Earth and trying to understand how these may have come about.

"In the end there stands out a perfectly simple problem which is undoubtedly soluble," Henderson wrote. "In what degree are the physical, chemical, and general meteorological characteristics of water and carbon dioxide and of the compounds of carbon, hydrogen, and oxygen favorable to a mechanism which must be physically, chemically, and physiologically complex, which must be itself well regulated in a well-regulated environment, and which must carry on an active exchange of matter and energy with that environment?" In other words, to what extent are the nonbiological properties of the global environment favorable to life? And why is Earth so fit for life?

Today, we can give partial answers to Henderson's question. The answers involve recognizing that "environmental fitness" is the result of a two-way process. Life evolved in an environment conducive for that to occur, and then, over time, life altered the environment at a global level. These global alterations were originally problems for existing organisms, but they also created opportunities for the evolution of new life-forms adapted to the new conditions.

The Rise of Oxygen

The fossil record provides evidence that before about 2.3 billion years ago Earth's atmosphere was very low in oxygen (anoxic), much closer to the atmospheres of Mars and Venus. The evidence for this exists in water-worn grains of pyrite (iron sulfide, FeS_2), which appear in sedimentary rocks formed before 2.3 billion years ago. Today, when pure iron gets into streams, it is rapidly oxidized because there is so much oxygen in the atmosphere, and the iron forms sediments of iron oxides (what we know familiarly as rusted iron). If there were similar amounts of oxygen in the ancient waters, these ancient deposits would not have been pyrite—iron combined with sulfur—but would have been oxidized, just as they are today. This tells us that Earth's ancient Precambrian atmosphere and oceans were low in oxygen.

The ancient oceans had a vast amount of dissolved iron, which is much more soluble in water in its unoxidized state. Oxygen released into the oceans combined with the dissolved iron, changing it from a more soluble to a less soluble form. No longer dissolved in the water, the iron settled (precipitated) to the bottom of the oceans and became part of deposits that slowly were turned into rock. Over millions of years, these deposits formed the thick bands of iron ore that are mined today all around Earth, with notable deposits found today from Minnesota to Australia. That was the major time when the great iron ore deposits, now mined, were formed. It is intriguing to realize that very ancient Earth history affects our economic and environmental lives today.

FIGURE 6.5 **Banded-iron formations. Photograph of the Grand Canyon showing red layers that contain oxidized iron below layers of other colors that lack the deposited iron.**

The most convincing evidence of an oxygen-deficient atmosphere is found in ancient chemical sediments called *banded-iron formations* (Figure 6.5).

These sediments were laid down in the sea, a sea that must have been able to carry dissolved iron—something it can't do now because oxygen precipitates the iron. If the ancient sea lacked free oxygen, then oxygen must also have been lacking in the atmosphere; otherwise, simple diffusion would have brought free oxygen into the ocean from the atmosphere.

How did the atmosphere become high enough in oxygen to change iron deposits from unoxidized to oxidized? The answer is that life changed the environment at a global level, adding free oxygen and removing carbon dioxide, first within the oceans, then in the atmosphere. This came about as a result of the evolution of photosynthesis. In photosynthesis, you will recall, carbon dioxide and water are combined, in the presence of light, to form sugar and free oxygen. But in early life, oxygen was a toxic waste that was eliminated from the cell and emitted into the surrounding environment. Scientists calculate that before oxygen started to accumulate in the air, 25 times the present-day amount of atmospheric oxygen had been neutralized by reducing agents such as dissolved iron. It took about 2 billion years for the unoxidized iron in Earth's oceans to be used up.

Life Responds to an Oxygen Environment

Early in Earth's history, from 4.6 billion years ago until about 2.3 billion years ago, was the oxygen-deficient phase of life's history. Some of the earliest photosynthetic organisms (3.4 billion years ago) were ancestors of bacteria that formed mats in shallow water (Figure 6.6). Photosynthesis became well established by about 1.9 billion years ago, but until sufficient free oxygen was in the atmosphere, organisms could get their energy only by

(a) (b)

FIGURE 6.6 **Stromatolites.** Among the earliest photosynthetic organisms were bacteria that formed these large mounds called stromatolites **(a)**, fossils of which, shown here, have been dated as old as 3.4 billion years. The bacteria grew in long filaments—cells connected to one another in a line—and these formed layers that were infiltrated by sand and clay that washed down from the land into the shallow waters of the bays where this kind of photosynthetic bacteria lived (they took oxygen from water). Over time, the combination of sediments and living and dead bacterial cells formed large mounds. Similar bacteria, if not exactly the same species, still live, and form the same kind of mounds shown here in Shark Bay, Australia **(b)**. The ancient stromatolites were among the organisms that eventually created an oxygen-rich atmosphere.

fermentation, and the only kinds of organisms were bacteria and their relatives called **prokaryotes**. These have a simpler internal cell structure than that of the cells of our bodies and other familiar forms of life, known as **eukaryotes** (Figure 6.7).

Without oxygen, organisms cannot completely "burn" organic compounds. Instead, they can get some energy from what we call fermentation, whose waste products are carbon dioxide and alcohol. Alcohol is a high-energy compound that, for the cell in an oxygenless atmosphere, is a waste product that has to be gotten rid of. Fermentation's low-energy yield to the organism puts limitations on the anaerobic cell. For example:

- Anaerobic cells must be small because a large surface-to-volume ratio is required to allow rapid diffusion of food in and waste out.
- Anaerobic cells have trouble keeping themselves supplied with energy. They cannot afford to use energy to maintain **organelles**—specialized cell parts that function like the organs of multicelled organisms. This means that all anaerobic bacteria were prokaryotes lacking specialized organelles, including a nucleus.
- Prokaryotes need free space around them; crowding interferes with the movement of nutrients and water into and out of the cell. Therefore, they live singly or strung end-to-end in chains. They cannot form three-dimensional structures. They are life restricted to a plane.

For other forms of life to evolve and persist, bacteria had to convert Earth's atmosphere to one high in oxygen. Once the atmosphere became high in oxygen, complete respiration was possible, and organisms could have much more complex structures, with three-dimensional bodies, and could use energy more efficiently and rapidly. Thus the presence of free oxygen, a biological product, made possible the evolution of eukaryotes, organisms with more structurally complex cells and bodies, including the familiar animals and plants. Eukaryotes can do the following:

- Use oxygen for respiration, and because oxidative respiration is much more efficient (providing more energy) than fermentation, eukaryotes do not require as large a surface-to-volume ratio as anaerobic cells do, so eukaryote cells are larger.
- Maintain a nucleus and other organelles because of their superior metabolic efficiency.
- Form three-dimensional colonies of cells. Unlike prokaryotes, aerobic eukaryotes are not inhibited by crowding, so they can exist close to each other. This made possible the complex, multicellular body structures of animals, plants, and fungi.

Animals, plants, and fungi first evolved about 700 million to 500 million years ago. Inside their cells, DNA, the genetic material, is concentrated in a nucleus rather than distributed throughout the cell, and the cell contains organelles such as mitochondria which process energy. With the appearance of eukaryotes and the growth of an oxygenated atmosphere, the **biosphere**—our planet's system that includes and sustains all life—started to change rapidly and to influence more processes on Earth.

In sum, life affected Earth's surface—not just the atmosphere, but the oceans, rocks, and soils—and it still does so today. Billions of years ago, Earth presented a habitat where life could originate and flourish. Life, in turn, fundamentally changed the characteristics of the planet's surface, providing new opportunities for life and ultimately leading to the evolution of us.

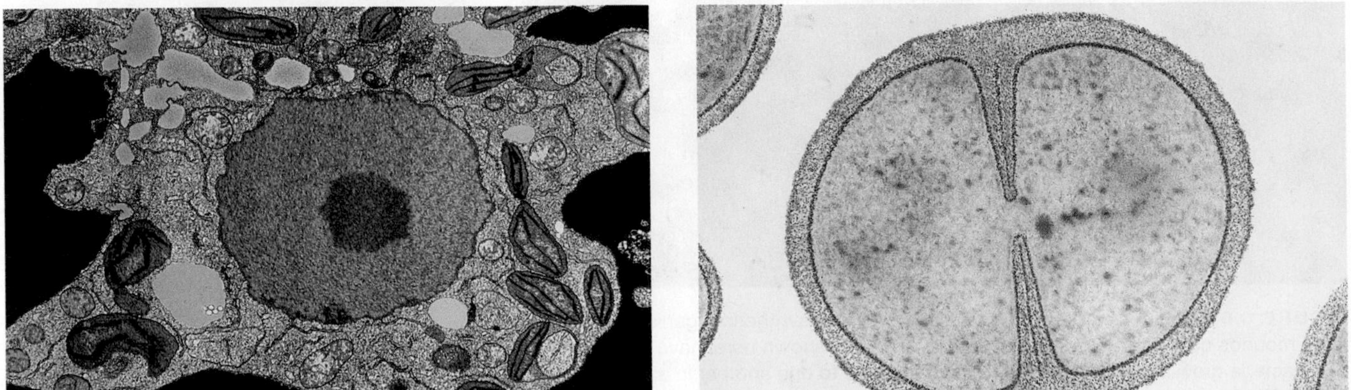

(a) (b)

FIGURE 6.7 **Prokaryotes and eukaryotes.** Photomicrograph of **(a)** bacterial (Prokaryote)cell and **(b)** a bacterial (prokaryote) cell. From these images you can see that the eukaryotic cell has a much more complex structure, including many organelles.

6.2 Life and Global Chemical Cycles

All living things are made up of chemical elements (see Appendix D for a discussion of matter and energy), but of the more than 103 known chemical elements, only 24 are required by organisms (see Figure 6.8). These 24 are divided into the **macronutrients**, elements required in large amounts by all life, and **micronutrients**, elements required either in small amounts by all life or in moderate amounts by some forms of life and not at all by others. (*Note*: For those of you unfamiliar with the basic chemistry of the elements, we have included an introduction in Appendix E, which you might want to read now before proceeding with the rest of this chapter.)

The macronutrients in turn include the "big six" elements that are the fundamental building blocks of life: carbon, hydrogen, nitrogen, oxygen, phosphorus, and sulfur. Each one plays a special role in organisms. Carbon is the basic building block of organic compounds; along with oxygen and hydrogen, carbon forms carbohydrates. Nitrogen, along with these other three, makes proteins. Phosphorus is the "energy element"—it occurs in compounds called ATP and ADP, important in the transfer and use of energy within cells.

Other macronutrients also play specific roles. Calcium, for example, is the structure element, occurring in bones and teeth of vertebrates, shells of shellfish, and wood-forming cell walls of vegetation. Sodium and potassium are important to nerve-signal transmission. Many of the metals required by living things are necessary for specific enzymes. (An enzyme is a complex organic compound that acts as a catalyst—it causes or speeds up chemical reactions, such as digestion.)

For any form of life to persist, chemical elements must be available at the right times, in the right amounts, and in the right concentrations. When this does not happen, a chemical can become a **limiting factor**, preventing the growth of an individual, a population, or a species, or even causing its local extinction.

Chemical elements may also be toxic to some life-forms and ecosystems. Mercury, for example, is toxic even in low concentrations. Copper and some other

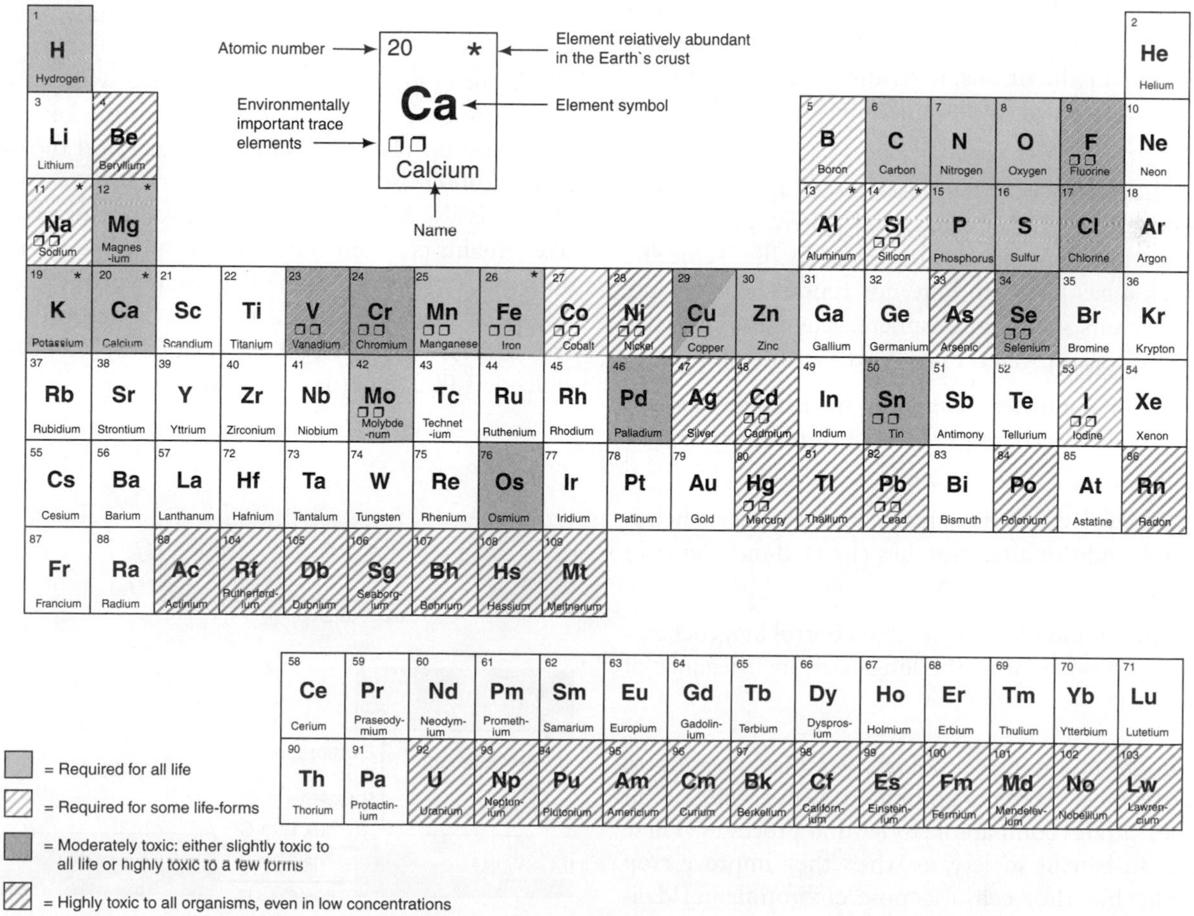

FIGURE 6.8 The Periodic Table of the Elements. The elements in green are required by all life; those in hatched green are micronutrients—required in very small amounts by all life-forms or required by only some forms of life. Those that are moderately toxic are in hatched red, and those that are highly toxic are solid red.

elements are required in low concentrations for life processes but are toxic in high concentrations.

Finally, some elements are neutral for life. Either they are chemically inert, such as the noble gases (for example, argon and neon), which do not react with other elements, or they are present on Earth in very low concentrations.

6.3 General Aspects of Biogeochemical Cycles

A **biogeochemical cycle** is the complete path a chemical takes through the four major components, or reservoirs, of Earth's system: atmosphere, hydrosphere (oceans, rivers, lakes, groundwaters, and glaciers), lithosphere (rocks and soils), and biosphere (plants and animals). A biogeochemical cycle is *chemical* because it is chemicals that are cycled, *bio-* because the cycle involves life, and *geo-* because a cycle may include atmosphere, water, rocks, and soils. Although there are as many biogeochemical cycles as there are chemicals, certain general concepts hold true for these cycles.

- Some chemical elements, such as oxygen and nitrogen, cycle quickly and are readily regenerated for biological activity. Typically, these elements have a gas phase and are present in the atmosphere and/or easily dissolved in water and carried by the hydrologic cycle (discussed later in the chapter).
- Other chemical elements are easily tied up in relatively immobile forms and are returned slowly, by geologic processes, to where they can be reused by life. Typically, they lack a gas phase and are not found in significant concentrations in the atmosphere. They also are relatively insoluble in water. Phosphorus is an example.
- Most required nutrient elements have a light atomic weight. The heaviest required micronutrient is iodine, element 53.
- Since life evolved, it has greatly altered biogeochemical cycles, and this alteration has changed our planet in many ways.
- The continuation of processes that control biogeochemical cycles is essential to the long-term maintenance of life on Earth.

Through modern technology, we have begun to transfer chemical elements among air, water, and soil, in some cases at rates comparable to natural processes. These transfers can benefit society, as when they improve crop production, but they can also pose environmental dangers, as illustrated by the opening case study. To live wisely with our environment, we must recognize the positive and negative consequences of altering biogeochemical cycles.

The simplest way to visualize a biogeochemical cycle is as a box-and-arrow diagram of a system (see the discussion of systems in Chapter 3), with the boxes representing places where a chemical is stored (*storage compartments*) and the arrows representing pathways of transfer (Figure 6.9a). In this kind of diagram, the **flow** is the amount moving from one compartment to another, whereas the **flux** is the rate of transfer—the amount per unit time—of a chemical that enters or leaves a storage compartment. The **residence time** is the average time that an atom is stored in a compartment. The donating compartment is a **source**, and the receiving compartment is a **sink**.

A biogeochemical cycle is generally drawn for a single chemical element, but sometimes it is drawn for a compound—for example, water (H_2O). Figure 6.9b shows the basic elements of a biogeochemical cycle for water, represented about as simply as it can be, as three compartments: water stored temporarily in a lake (compartment B); entering the lake from the atmosphere (compartment A) as precipitation and from the land around the lake as runoff (compartment C). It leaves the lake through evaporation to the atmosphere or as runoff via a surface stream or subsurface flows. We diagrammed the Missouri River in the opening case study of Chapter 3 in this way.

As an example, consider a salt lake with no transfer out except by evaporation. Assume that the lake contains 3,000,000 m^3 (106 million ft^3) of water and the evaporation is 3,000 m^3/day (106,000 ft^3/day). Surface runoff into the lake is also 3,000 m^3/day, so the volume of water in the lake remains constant (input = output). We can calculate the average residence time of the water in the lake as the volume of the lake divided by the evaporation rate (rate of transfer), or 3,000,000 m^3 divided by 3,000 m^3/day, which is 1,000 days (or 2.7 years).

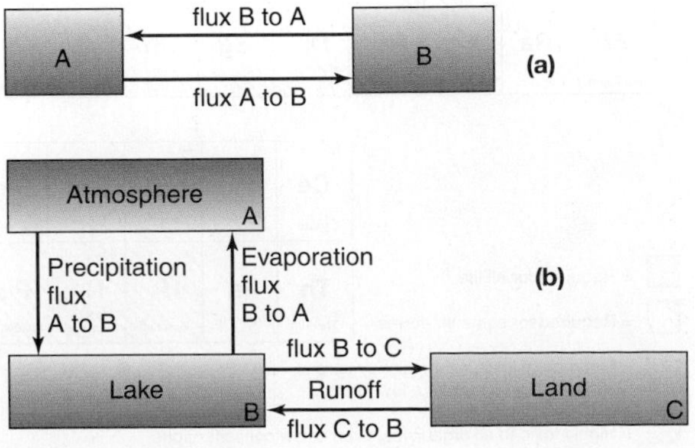

FIGURE 6.9 **(a)** A unit of a biogeochemical cycle viewed as a systems diagram; **(b)** a highly simplified systems diagram of the water cycle.

6.4 The Geologic Cycle

Throughout the 4.6 billion years of Earth's history, rocks and soils have been continuously created, maintained, changed, and destroyed by physical, chemical, and biological processes. This is another illustration that the biosphere is a dynamic system, not in steady state. Collectively, the processes responsible for formation and change of Earth materials are referred to as the **geologic cycle** (Figure 6.10). The geologic cycle is best described as a group of cycles: tectonic, hydrologic, rock, and biogeochemical. (We discuss the last cycle separately because it requires lengthier examination.)

The Tectonic Cycle

The **tectonic cycle** involves the creation and destruction of Earth's solid outer layer, the *lithosphere*. The lithosphere is about 100 km (60 mi) thick on average and is broken into several large segments called *plates*, which are moving relative to one another (Figure 6.11). The slow movement of these large segments of Earth's outermost rock shell is referred to as **plate tectonics**. The plates "float" on denser material and move at rates of 2 to 15 cm/year (0.8 to 6.9 in./year), about as fast as your fingernails grow. The tectonic cycle is driven by forces originating deep within the earth. Closer to the surface, rocks are deformed by spreading plates, which produce ocean basins, and by collisions of plates, which produce mountain ranges and island-arc volcanoes.

Plate tectonics has important environmental effects. Moving plates change the location and size of continents, altering atmospheric and ocean circulation and thereby altering climate. Plate movement has also created ecological islands by breaking up continental areas. When this happens, closely related life-forms are isolated from one another for millions of years, leading to the evolution of new species. Finally, boundaries

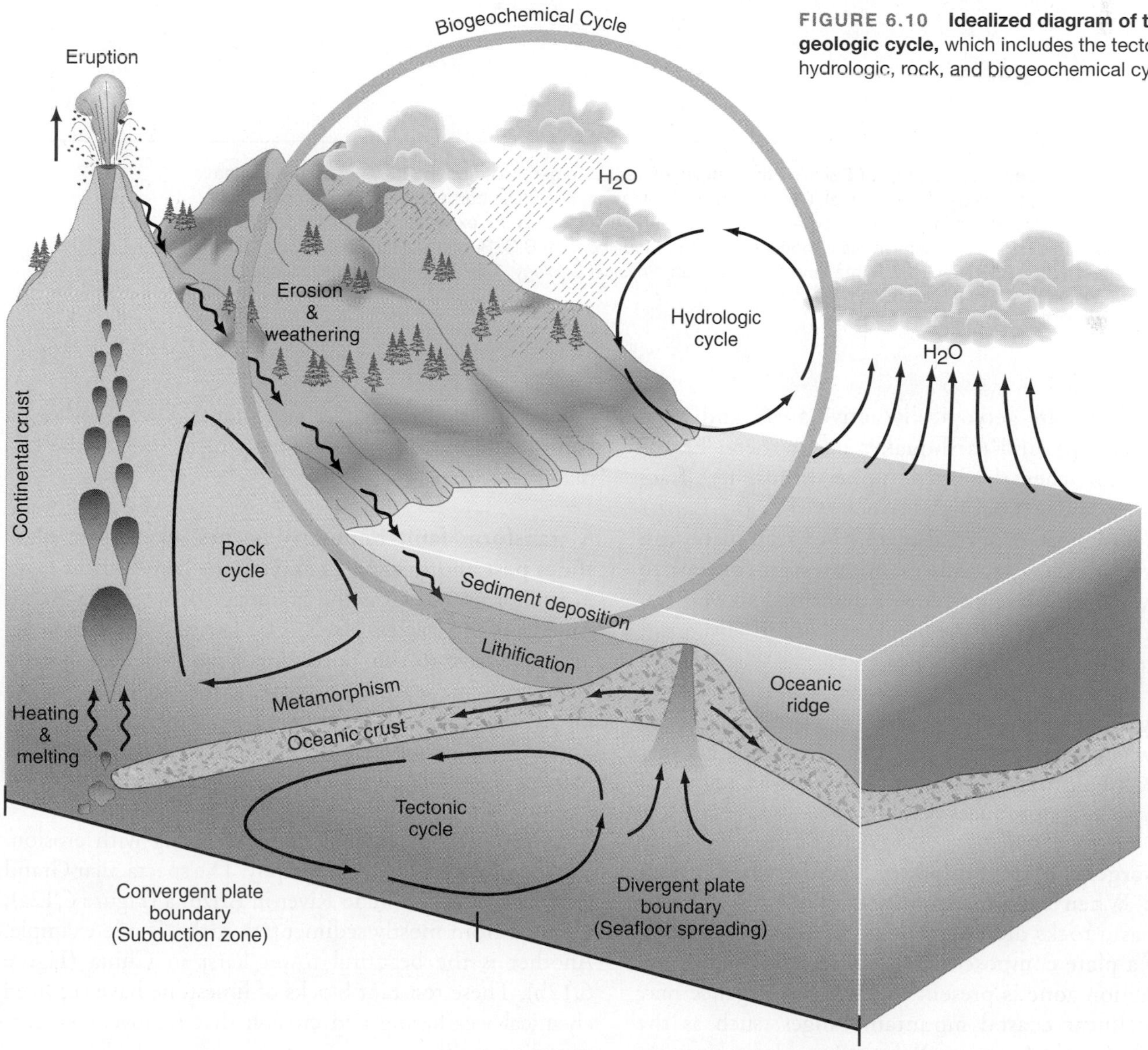

FIGURE 6.10 Idealized diagram of the geologic cycle, which includes the tectonic, hydrologic, rock, and biogeochemical cycles.

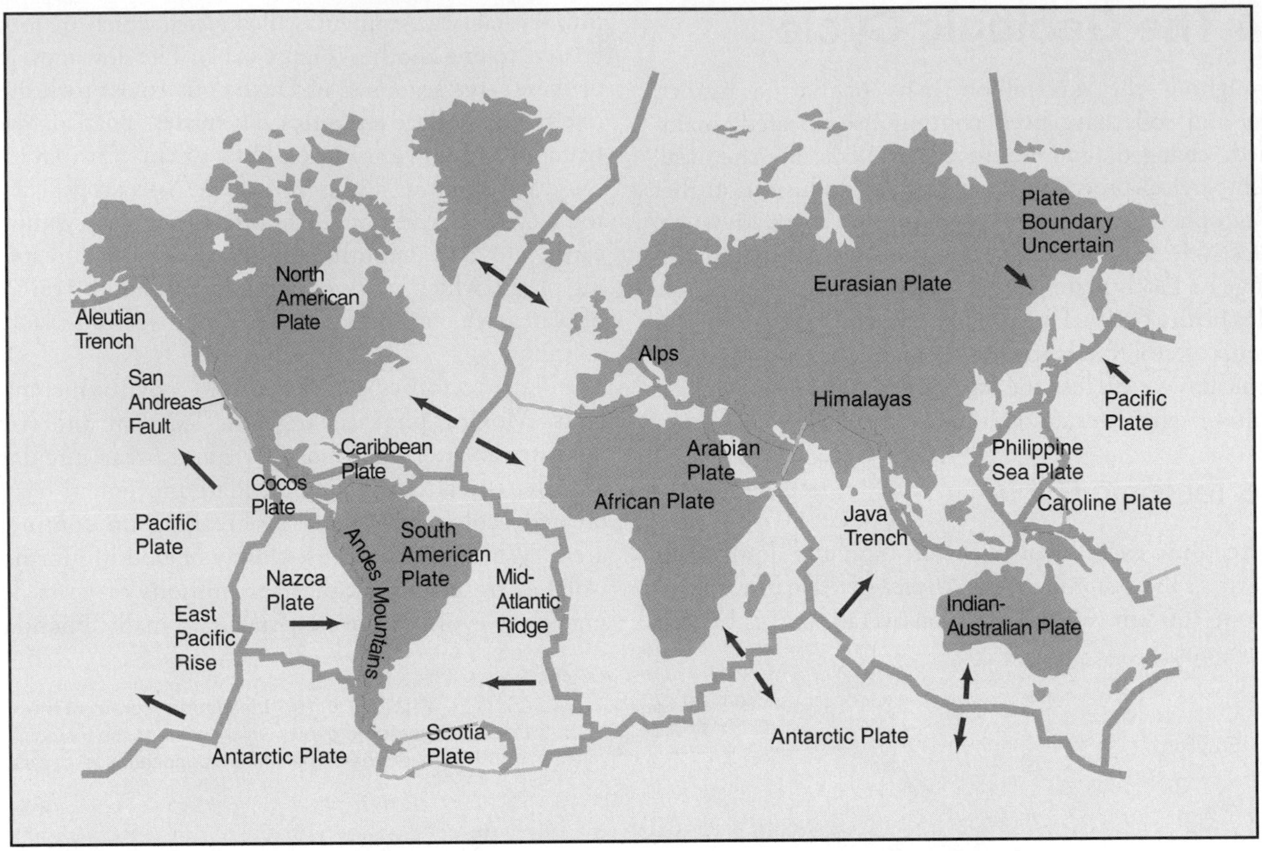

FIGURE 6.11 **Generalized map of Earth's lithospheric plates.** Divergent plate boundaries are shown as heavy lines (for example, the Mid-Atlantic Ridge). Convergent boundaries are shown as barbed lines (for example, the Aleutian trench). Transform fault boundaries are shown as yellow, thinner lines (for example, the San Andreas Fault). Arrows indicate directions of relative plate motions. (*Source*: Modified from B.C. Burchfiel, R.J. Foster, E.A. Keller, W.N. Melhorn, D.G. Brookins, L.W. Mintz, and H.V. Thurman, *Physical Geology: The Structures and Processes of the Earth* [Columbus, Ohio: Merrill, 1982].)

between plates are geologically active areas, and most volcanic activity and earthquakes occur there. Earthquakes occur when the brittle upper lithosphere fractures along faults (fractures in rock within the Earth's crust). Movement of several meters between plates can occur within a few seconds or minutes, in contrast to the slow, deeper plate movement described above.

Three types of plate boundaries occur: divergent, convergent, and transform faults.

A divergent plate boundary occurs at a spreading ocean ridge, where plates are moving away from one another and new lithosphere is produced. This process, known as *seafloor spreading*, produces ocean basins.

A convergent plate boundary occurs when plates collide. When a plate composed of relatively heavy ocean-basin rocks dives (subducts) beneath the leading edge of a plate composed of lighter continental rocks, a subduction zone is present. Such a convergence may produce linear coastal mountain ranges, such as the Andes in South America. When two plates that are both composed of lighter continental rocks collide, a continental mountain range may form, such as the Himalayas in Asia.

A transform fault boundary occurs where one plate slides past another. An example is the San Andreas Fault in California, which is the boundary between the North American and Pacific plates. The Pacific plate is moving north, relative to the North American plate, at about 5 cm/year (2 in./year). As a result, Los Angeles is moving slowly toward San Francisco, about 500 km (300 mi) north. If this continues, in about 10 million years San Francisco will be a suburb of Los Angeles.

Uplift and subsidence of rocks, along with erosion, produce Earth's varied topography. The spectacular Grand Canyon of the Colorado River in Arizona (Figure 6.12a), sculpted from mostly sedimentary rocks, is one example. Another is the beautiful tower karst in China (Figure 6.12b). These resistant blocks of limestone have survived chemical weathering and erosion that removed the surrounding rocks.

(a)

(b)

FIGURE 6.12 Plate tectonics and landscapes. (a) In response to slow tectonic uplift of the region, the Colorado River has eroded through the sedimentary rocks of the Colorado plateau to produce the spectacular Grand Canyon. The river in recent years has been greatly modified by dams and reservoirs above and below the canyon. Sediment once carried to the Gulf of California is now deposited in reservoirs. The dam stores sediments, and some of the water released is from the deeper and thus cooler parts of the reservoir, so water flowing out of the dam and down through the Grand Canyon is clearer and colder than it used to be. Fewer sandbars are created; this and the cooler water change which species of fish are favored. Thus this upstream dam has changed the hydrology and environment of the Colorado River in the Grand Canyon. **(b)** This landscape in the People's Republic of China features tower karst, steep hills or pinnacles composed of limestone. The rock has been slowly dissolving through chemical weathering. The pinnacles and hills are remnants of the weathering and erosion processes.

The Hydrologic Cycle

The **hydrologic cycle** (Figure 6.13) is the transfer of water from the oceans to the atmosphere to the land and back to the oceans. It includes evaporation of water from the oceans; precipitation on land; evaporation from land; transpiration of water by plants; and runoff from streams, rivers, and subsurface groundwater. Solar energy drives the hydrologic cycle by evaporating water from oceans, freshwater bodies, soils, and vegetation. Of the total 1.3 billion km^3 of water on Earth, about 97% is in oceans and about 2% is in glaciers and ice caps; 0.76% is shallow groundwater: 0.013% is in lakes and rivers; and only 0.001% is in the atmosphere. Although water on land and in the atmosphere accounts for only a small fraction of the water on Earth, this water is important in moving chemicals,

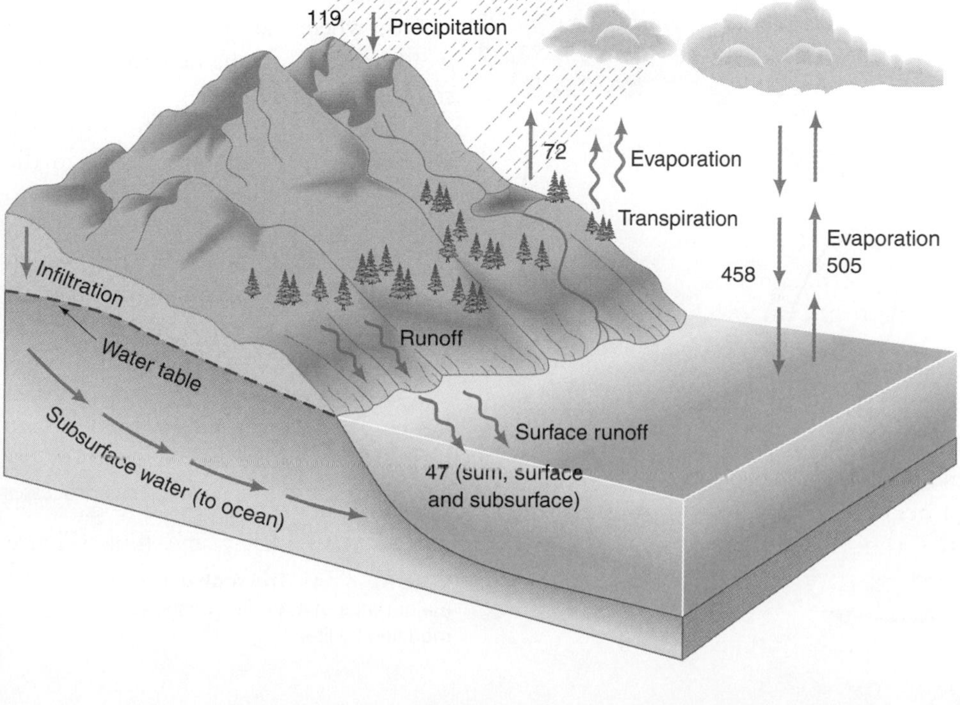

FIGURE 6.13 The hydrologic cycle, showing the transfer of water (thousands of km^3/yr) from the oceans to the atmosphere to the continents and back to the oceans again. (*Source*: From P.H. Gleick, *Water in Crisis* [New York: Oxford University Press, 1993].)

sculpting landscape, weathering rocks, transporting sediments, and providing our water resources.

The rates of transfer of water from land to the ocean are relatively low, and the land and oceans are somewhat independent in the water cycle because most of the water that evaporates from the ocean falls back into the ocean as precipitation, and most of the water that falls as precipitation on land comes from evaporation of water from land, as shown in Figure 6.13. Approximately 60% of precipitation on land evaporates each year back to the atmosphere, while the rest, about 40%, returns to the ocean as surface and subsurface runoff. The distribution of water is far from uniform on the land, and this has many environmental and ecological effects, which we discuss in Chapter 8 (on biological diversity), Chapters 19 and 20 (on water and climate), and Chapter 22 (urban environments).

At the regional and local levels, the fundamental hydrologic unit of the landscape is the *drainage basin* (also called a *watershed* or *catchment*). As explained in Chapter 5, a watershed is the area that contributes surface runoff to a particular stream or river. The term is used in evaluating the hydrology of an area (such as the stream flow or runoff from slopes) and in ecological research and biological conservation. Watersheds are best categorized by drainage basin area, and further by how many streams flow into the final, main channel. A first-order watershed is drained by a single small stream; a second-order watershed includes streams from first-order watersheds, and so on. Drainage basins vary greatly in size, from less than a hectare (2.5 acres) for a first-order watershed to millions of square kilometers for major rivers like the Missouri, Amazon, and Congo. A watershed is usually named for its main stream or river, such as the Mississippi River drainage basin.

The Rock Cycle

The **rock cycle** consists of numerous processes that produce rocks and soils. The rock cycle depends on the tectonic cycle for energy, and on the hydrologic cycle for water. As shown in Figure 6.14, rock is classified as igneous, sedimentary, or metamorphic. These three types of rock are involved in a worldwide recycling process. Internal heat from the tectonic cycle produces *igneous rocks* from molten material (magma) near the surface, such as lava from volcanoes. When magma crystalized deep in the earth the igneous rock granite was formed. These new rocks weather when exposed at the surface. Water in cracks of rocks expands when it freezes, breaking the rocks apart. This physical weathering makes smaller particles of rock from bigger ones, producing sediment, such as gravel, sand, and silt. Chemical weathering occurs, too, when the weak acids in water dissolve chemicals from rocks. The sediments and dissolved chemicals are then transported by water, wind, or ice (glaciers).

Weathered materials that accumulate in *depositional basins*, such as the oceans, are compacted by overlying sediments and converted to *sedimentary rocks*. The process of creating rock by compacting and cementing particles is called *lithification*. Sedimentary rocks buried at sufficient depths (usually tens to hundreds of kilometers) are altered by heat, pressure, or chemically active fluids and transformed into *metamorphic rocks*. Later, plate tectonics uplift may bring these deeply buried rocks to the surface, where they, too, are subjected to weathering, producing new sediment and starting the cycle again.

You can see in Figure 6.14 that life processes play an important role in the rock cycle by adding organic carbon to rocks. The addition of organic carbon produces rocks such as limestone, which is mostly calcium carbonate (the material of seashells and bones), as well as fossil fuels, such as coal.

Our discussion of geologic cycles has emphasized tectonic, hydrologic, and rock-forming processes. We can now begin to integrate biogeochemical processes into the picture.

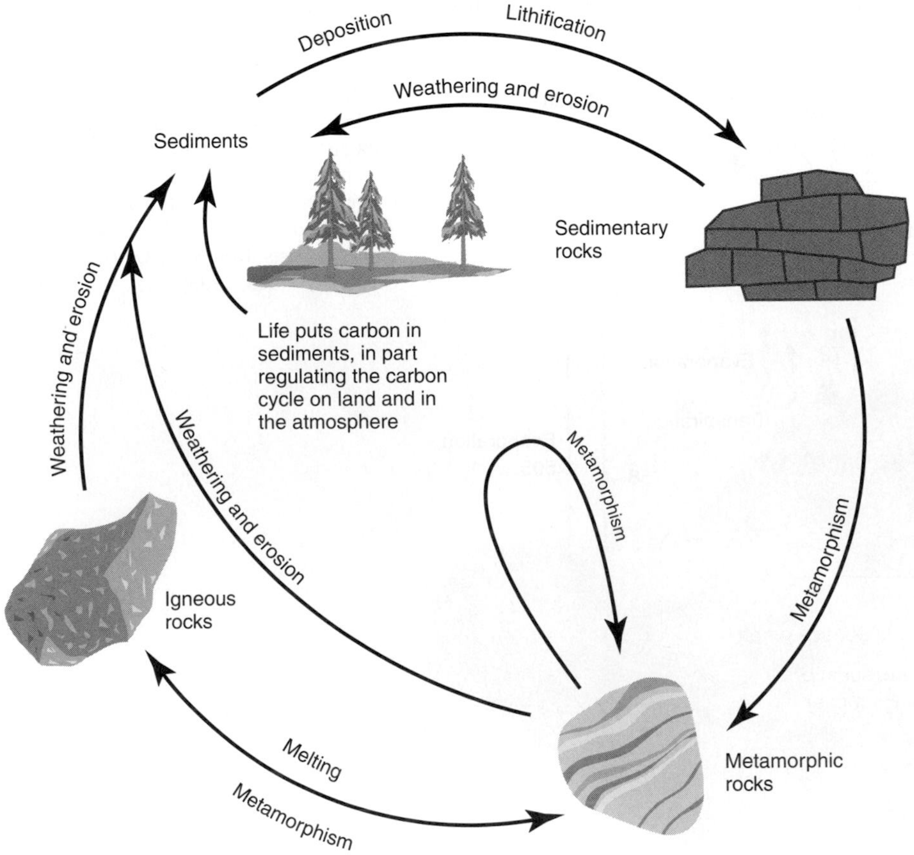

FIGURE 6.14 **The rock cycle** and major paths of material transfer as modified by life.

6.5 Some Major Global Biogeochemical Cycles

With Figure 6.14's basic diagram in mind, we can now consider complete chemical cycles, though still quite simplified. Each chemical element has its own specific cycle, but all the cycles have certain features in common (Figure 6.15).

The Carbon Cycle

Carbon is the basic building block of life and the element that anchors all organic substances, from coal and oil to DNA (deoxyribonucleic acid), the compound that carries genetic information. Although of central importance to life, carbon is not one of the most abundant elements in Earth's crust. It contributes only 0.032% of the weight of the crust, ranking far behind oxygen (45.2%), silicon (29.5%), aluminum (8.0%), iron (5.8%), calcium (5.1%), and magnesium (2.8%).[10, 11]

The major pathways and storage reservoirs of the **carbon cycle** are shown in Figure 6.16. This diagram is simplified to show the big picture of the carbon cycle. Details are much more complex. [12]

- **Oceans and land ecosystems:** In the past half-century, ocean and land ecosystems have removed about 3.1 ± 0.5 GtC/yr, which is approximately 45% of the carbon emitted from burning fossil fuels during that period.
- **Land-use change:** Deforestation and decomposition of what is cut and left, as well as burning of forests to make room for agriculture in the tropics, are the main reasons 2.2 ± 0.8 GtC/yr is added to the atmosphere. A small flux of carbon (0.2 ± 0.5 GtC/yr) from hot tropical areas is pulled from the atmosphere by growing forests. In other words, when considering land-use change, deforestation is by far the dominant process.
- **Residual land sink:** The observed net uptake of CO_2 from the atmosphere (see Figure 6.16) by land ecosystems suggests there must be a sink for carbon in land ecosystems that has not been adequately identified. The sink is large, at 2 to 3 GtC/yr, with large uncertainty (± 1.7GtC/yr). Thus, our understanding of the carbon cycle is not yet complete.

Carbon has a gaseous phase as part of its cycle, occurring in the atmosphere as carbon dioxide (CO_2) and methane (CH_4), both greenhouse gases. Carbon enters

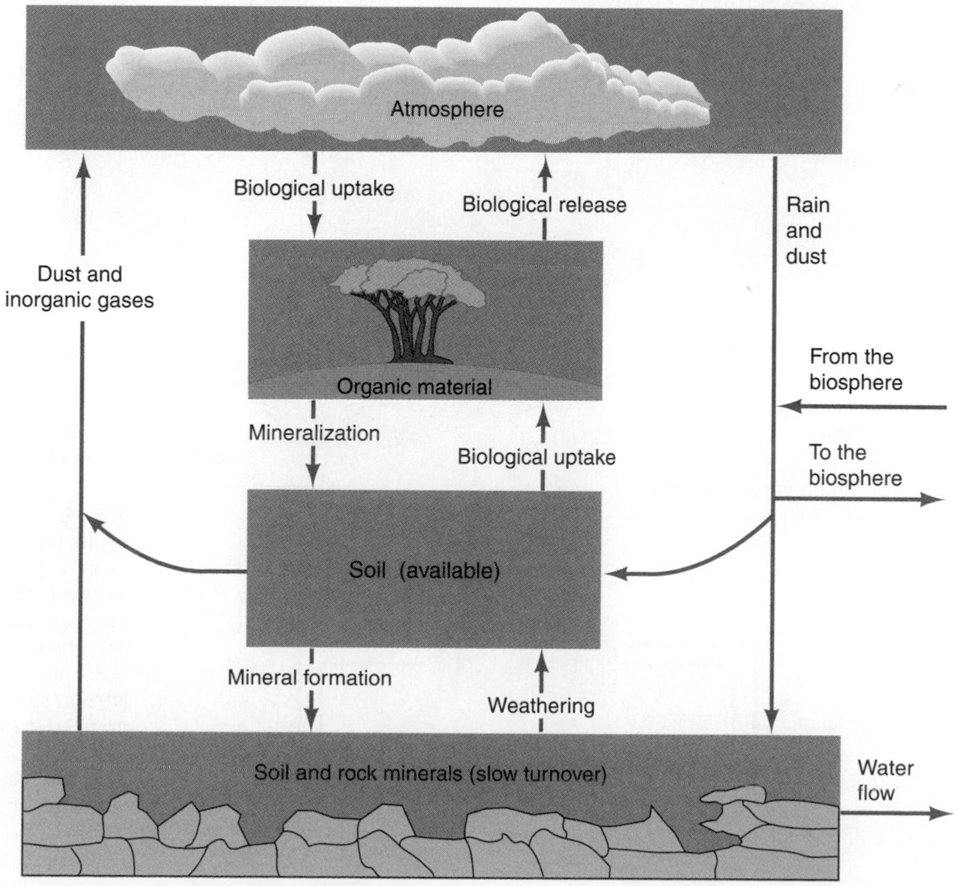

FIGURE 6.15 Basic biogeochemical cycle.

the atmosphere through the respiration of living things, through fires that burn organic compounds, and by diffusion from the ocean. It is removed from the atmosphere by photosynthesis of green plants, algae, and photosynthetic bacteria and enters the ocean from the atmosphere by the simple diffusion of carbon dioxide. The carbon dioxide then dissolves, some of it remaining in that state and the rest converting to carbonate ($CO_3^{=}$) and bicarbonate (HCO_3^-). Marine algae and photosynthetic bacteria obtain the carbon dioxide they use from the water in one of these three forms.

Carbon is transferred from the land to the ocean in rivers and streams as dissolved carbon, including organic compounds, and as organic particulates (fine particles of organic matter) and seashells and other forms of calcium carbonate ($CaCO_3$). Winds, too, transport small organic particulates from the land to the ocean. Rivers and streams transfer a relatively small fraction of the total global carbon flux to the oceans. However, on the local and regional scale, input of carbon from rivers to nearshore areas, such as deltas and salt marshes, which are often highly biologically productive, is important.

Carbon enters the **biota**—the term for all life in a region—through photosynthesis and is returned to the atmosphere or waters by respiration or by wildfire. When an organism dies, most of its organic material decomposes into inorganic compounds, including carbon dioxide. Some carbon may be buried where there is not sufficient

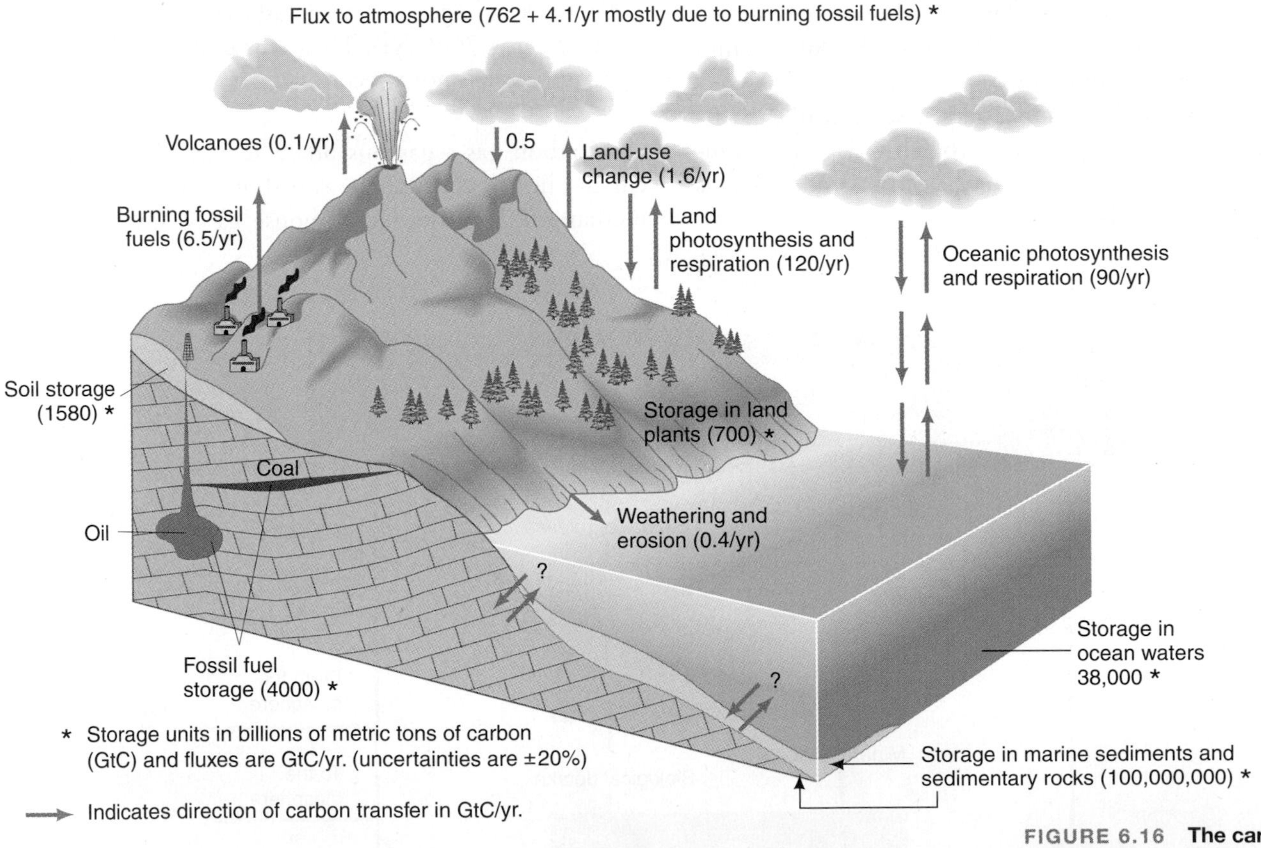

(a)

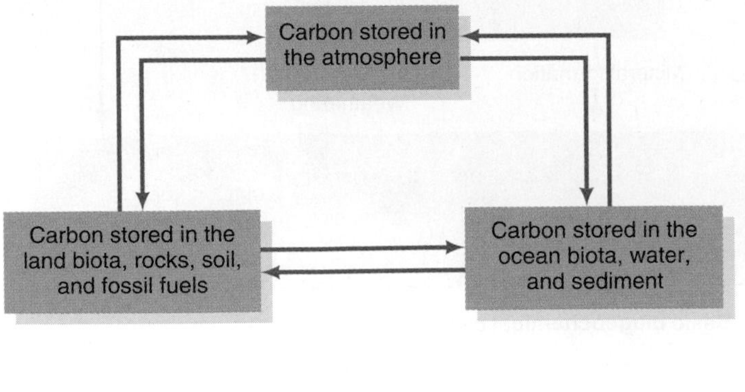

(b)

FIGURE 6.16 The carbon cycle. (a) Generalized global carbon cycle. **(b)** Parts of the carbon cycle simplified to illustrate the cyclic nature of the movement of carbon. (*Source*: Modified from G. Lambert, *La Recherche* 18 [1987]:782–83, with some data from R. Houghton, *Bulletin of the Ecological Society of America* 74, no. 4 [1993]: 355–356, and R. Houghton, *Tellus* 55B, no. 2 [2003]: 378–390), and IPCC, The Physical Science Basis: Working Group I. Contribution to the *Fourth Assessment Report* [New York: Cambridge University Press, 2007].)

oxygen to make this conversion possible or where the temperatures are too cold for decomposition. In these locations, organic matter is stored. Over years, decades, and centuries, storage of carbon occurs in wetlands, including parts of floodplains, lake basins, bogs, swamps, deep-sea sediments, and near-polar regions. Over longer periods (thousands to several million years), some carbon may be buried with sediments that become sedimentary rocks. This carbon is transformed into fossil fuels. Nearly all of the carbon stored in the lithosphere exists as sedimentary rocks, mostly carbonates, such as limestone, much of which has a direct biological origin.

The cycling of carbon dioxide between land organisms and the atmosphere is a large flux. Approximately 15% of the total carbon in the atmosphere is taken up by photosynthesis and released by respiration on land annually. Thus, as noted, life has a large effect on the chemistry of the atmosphere.

Because carbon forms two of the most important greenhouse gases—carbon dioxide and methane—much research has been devoted to understanding the carbon cycle, which will be discussed in Chapter 20 about the atmosphere and climate change.

The Carbon–Silicate Cycle

Carbon cycles rapidly among the atmosphere, oceans, and life. However, over geologically long periods, the cycling of carbon becomes intimately involved with the cycling of silicon. The combined carbon–silicate cycle is therefore of geologic importance to the long-term stability of the biosphere over periods that exceed half a billion years.[12]

The **carbon–silicate cycle** begins when carbon dioxide in the atmosphere dissolves in the water to form weak carbonic acid (H_2CO_3) that falls as rain (Figure 6.17). As the mildly acidic water migrates through the ground, it chemically weathers (dissolves) rocks and facilitates the erosion of Earth's abundant silicate-rich rocks. Among other products, weathering and erosion release calcium ions (Ca^{++}) and bicarbonate ions (HCO_3^-). These ions enter the groundwater and surface waters and eventually are transported to the ocean. Calcium and bicarbonate ions make up a major portion of the chemical load that rivers deliver to the oceans.

Tiny floating marine organisms use the calcium and bicarbonate to construct their shells. When these organisms die, the shells sink to the bottom of the ocean, where they accumulate as carbonate-rich sediments. Eventually, carried by moving tectonic plates, they enter a subduction zone (where the edge of one continental plate slips under the edge of another). There they are subjected to increased heat, pressure, and partial melting. The resulting magma releases carbon dioxide, which rises in volcanoes and is released into the atmosphere. This process provides a lithosphere-to-atmosphere flux of carbon.

The long-term carbon–silicate cycle (Figure 6.17) and the short-term carbon cycle (Figure 6.16) interact to affect levels of CO_2 and O_2 in the atmosphere. For example, the burial of organic material in an oxygen-poor environment amounts to a net increase of photosynthesis (which produces O_2) over respiration (which produces CO_2). Thus, if burial of organic carbon in oxygen-poor environments increases, the concentration of atmospheric

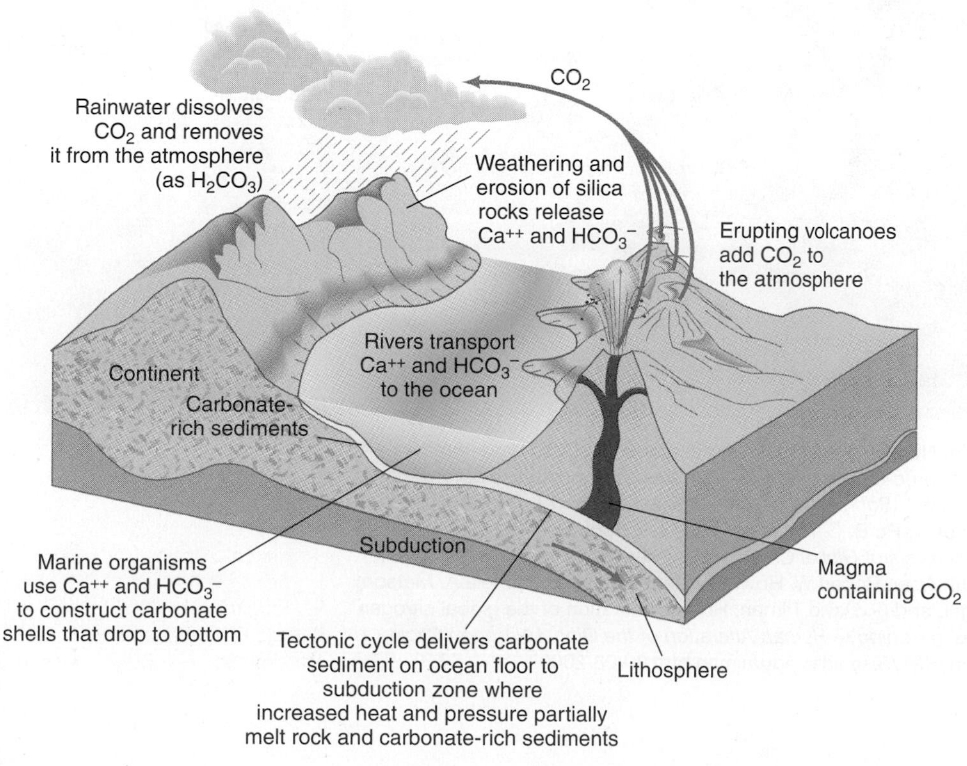

FIGURE 6.17 An idealized diagram showing the carbon–silicate cycle. (*Source*: Modified from J.E. Kasting, O.B. Toon, and J.B. Pollack, How climate evolved on the terrestrial planets, *Scientific American* 258 [1988]:2.)

oxygen will increase. Conversely, if more organic carbon escapes burial and is oxidized to produce CO_2, then the CO_2 concentration in the atmosphere will increase.[13]

The Nitrogen Cycle

Nitrogen is essential to life in proteins and DNA. As we discussed at the beginning of this chapter, free or diatonic nitrogen (N_2 uncombined with any other element) makes up approximately 78% of Earth's atmosphere. However, no organism can use molecular nitrogen directly. Some organisms, such as animals, require nitrogen in an organic compound. Others, including plants, algae, and bacteria, can take up nitrogen either as the nitrate ion (NO_3^-) or the ammonium ion (NH_4^+). Because nitrogen is a relatively unreactive element, few processes convert molecular nitrogen to one of these compounds. Lightning oxidizes nitrogen, producing nitric oxide. In nature, essentially all other conversions of molecular nitrogen to biologically useful forms are conducted by bacteria.

The **nitrogen cycle** is one of the most important and most complex of the global cycles (Figure 6.18). The process of converting inorganic, molecular nitrogen in the atmosphere to ammonia or nitrate is called **nitrogen fixation**. Once in these forms, nitrogen can be used on land by plants and in the oceans by algae. Bacteria, plants, and algae then convert these inorganic nitrogen compounds into organic ones through chemical reactions, and the nitrogen becomes available in ecological food chains. When organisms die, bacteria convert the organic compounds containing nitrogen back to ammonia, nitrate, or molecular nitrogen, which enters the atmosphere. The process of releasing fixed nitrogen back to molecular nitrogen is called **denitrification**.

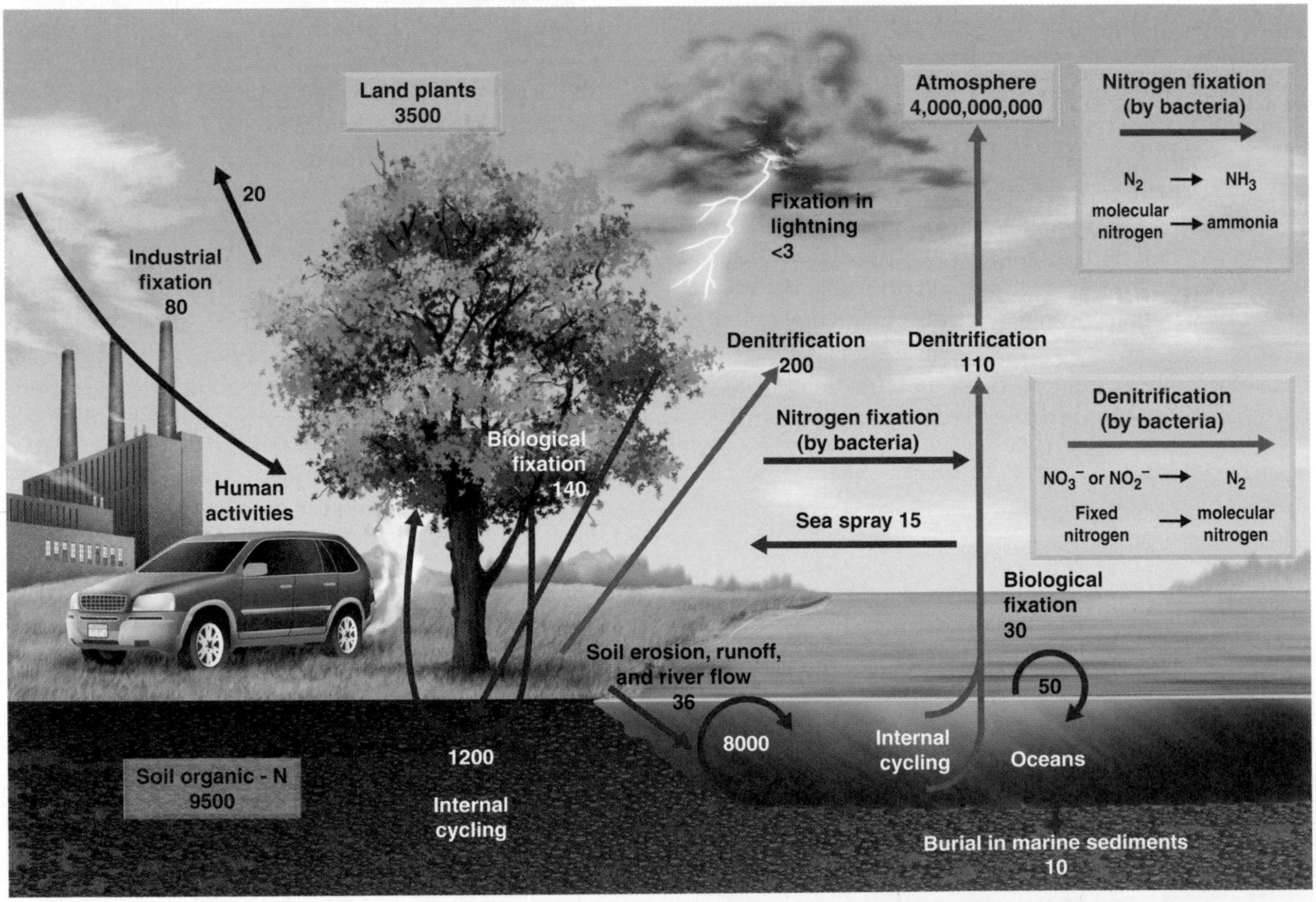

FIGURE 6.18 **The global nitrogen cycle.** Numbers in boxes indicate amounts stored, and numbers with arrows indicate annual flux, in millions of metric tons of nitrogen. Note that the industrial fixation of nitrogen is nearly equal to the global biological fixation. (*Source*: Data from R. Söderlund and T. Rosswall, in *The Handbook of Environmental Chemistry*, Vol. 1, Pt. B, O. Hutzinger, ed. [New York: Springer-Verlag, 1982]; W.H. Schlosinger, *Biogeochemistry: An Analysis of Global Change* [San Diego: Academic Press, 1997], p. 386; and Peter M. Vitousek, Chair, John Aber, Robert W. Howarth, Gene E. Likens, Pamela A. Matson, David W. Schindler, William H. Schlesinger, and G. David Tilman, Human alteration of the global nitrogen cycle: Causes and consequences, *Issues in Ecology—Human Alteration of the Global Nitrogen Cycle*, Ecological Society of America publication http://esa.sdsc.edu/tilman.htm 30/08/2000.)

Thus, all organisms depend on nitrogen-converting bacteria. Some organisms, including termites and ruminant (cud-chewing) mammals, such as cows, goats, deer, and bison, have evolved symbiotic relationships with these bacteria. For example, the roots of the pea family have nodules that provide a habitat for the bacteria. The bacteria obtain organic compounds for food from the plants, and the plants obtain usable nitrogen. Such plants can grow in otherwise nitrogen-poor environments. When these plants die, they contribute nitrogen-rich organic matter to the soil, improving the soil's fertility. Alder trees, too, have nitrogen-fixing bacteria in their roots. These trees grow along streams, and their nitrogen-rich leaves fall into the streams and increase the supply of organic nitrogen to freshwater organisms.

In terms of availability for life, nitrogen lies somewhere between carbon and phosphorus. Like carbon, nitrogen has a gaseous phase and is a major component of Earth's atmosphere. Unlike carbon, however, it is not very reactive, and its conversion depends heavily on biological activity. Thus, the nitrogen cycle is not only essential to life but also primarily driven by life.

In the early part of the 20th century, scientists invented industrial processes that could convert molecular nitrogen into compounds usable by plants. This greatly increased the availability of nitrogen in fertilizers. Today, industrial fixed nitrogen is about 60% of the amount fixed in the biosphere and is a major source of commercial nitrogen fertilizer.[14]

Although nitrogen is required for all life, and its compounds are used in many technological processes and in modern agriculture, nitrogen in agricultural runoff can pollute water, and many industrial combustion processes and automobiles that burn fossil fuels produce nitrogen oxides that pollute the air and play a significant role in urban smog (see Chapter 21).

The Phosphorus Cycle

Phosphrus, one of the "big six" elements required in large quantities by all forms of life, is often a limiting nutrient for plant and algae growth. We call it the "energy element" because it is fundamental to a cell's use of energy, and therefore to the use of energy by all living things. Phosphorus is in DNA, which carries the genetic material of life. It is an important ingredient in cell membranes.

The **phosphorus cycle** is significantly different from the carbon and nitrogen cycles. Unlike carbon and nitrogen, phosphorus does not have a gaseous phase on Earth; it is found in the atmosphere only in small particles of dust (Figure 6.19). In addition, phosphorus tends to form compounds that are relatively insoluble in water, so phosphorus is not readily weathered chemically. It does occur commonly in an oxidized state as phosphate, which combines with calcium, potassium, magnesium, or iron to form minerals. All told, however, the rate of transfer of phosphorus in Earth's system is slow compared with that of carbon or nitrogen.

Phosphorus enters the biota through uptake as phosphate by plants, algae, and photosynthetic bacteria. It is recycled locally in life on land nearly 50 times before being transported by weathering and runoff. Some phosphorus is inevitably lost to ecosystems on the land. It is transported by rivers to the oceans, either in a water-soluble form or as suspended particles. When it finally reaches the ocean, it may be recycled about 800 times before entering marine sediments to become part of the rock cycle. Over tens to hundreds of millions of years, the sediment is transformed into sedimentary rocks, after which it may eventually be returned to the land by uplift, weathering, and erosion.[15]

Ocean-feeding birds, such as the brown pelican, provide an important pathway in returning phosphrus from the ocean to the land. These birds feed on small fish, especially anchovies, which in turn feed on tiny ocean plankton. Plankton thrive where nutrients, such as phosphorus, are present. Areas of rising oceanic currents known as upwellings are such places. Upwellings occur near continents where the prevailing winds blow offshore, pushing surface waters away from the land and allowing deeper waters to rise and replace them. Upwellings carry nutrients, including phosphorus, from the depths of the oceans to the surface.

The fish-eating birds nest on offshore islands, where they are protected from predators. Over time, their nesting sites become covered with their phosphorus-laden excrement, called guano. The birds nest by the thousands, and deposits of guano accumulate over centuries. In relatively dry climates, guano hardens into a rocklike mass that may be up to 40 m (130 ft) thick. The guano results from a combination of biological and nonbiological processes. Without the plankton, fish, and birds, the phosphorus would have remained in the ocean. Without the upwellings, the phosphorus would not have been available.

Guano deposits were once major sources of phosphorus for fertilizers. In the mid-1800s, as much as 9 million metric tons per year of guano deposits were shipped to London from islands near Peru (Figure 6.20). Today, most phosphorus fertilizers come from the mining of phosphate-rich sedimentary rocks containing fossils of marine animals. The richest phosphate mine in the world is Bone Valley, 40 km east of Tampa, Florida. But 10–15 million years ago Bone Valley was the bottom of a shallow sea where marine invertebrates lived and died.[16]

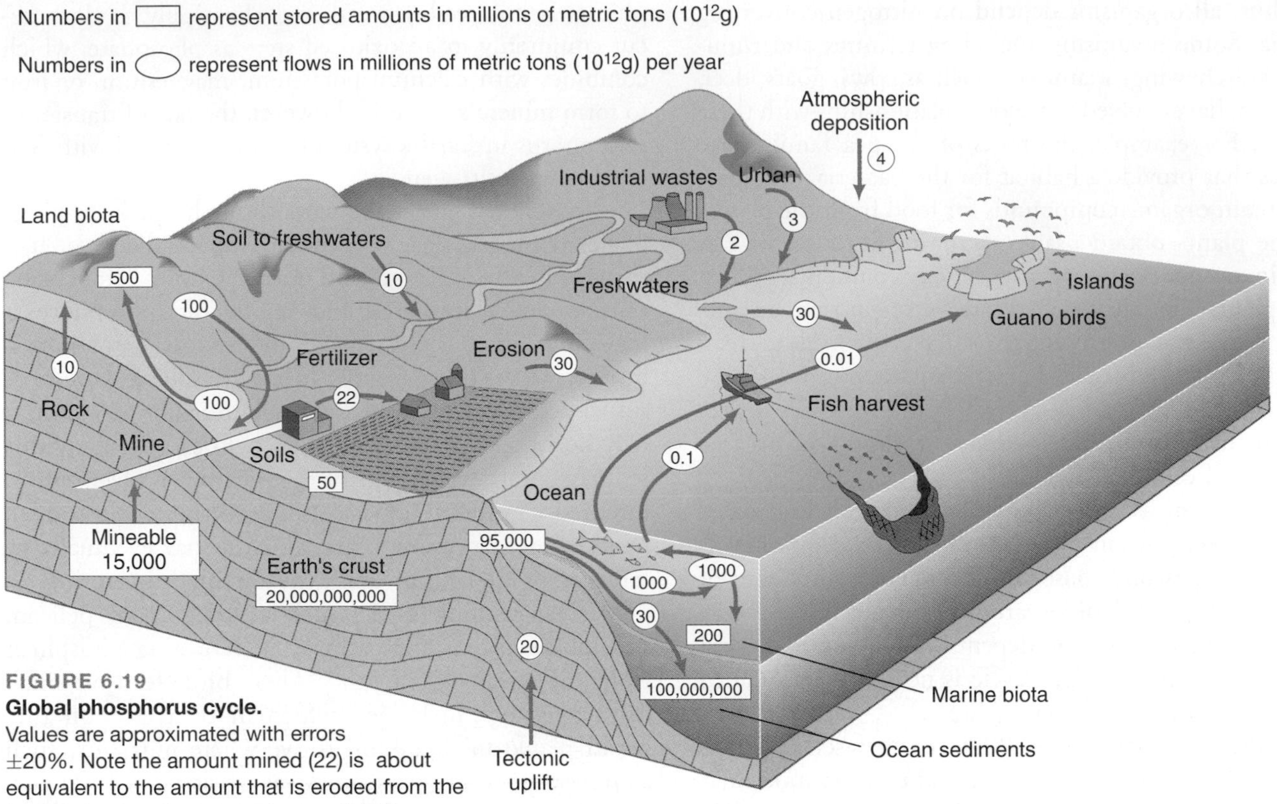

FIGURE 6.19
Global phosphorus cycle.
Values are approximated with errors ±20%. Note the amount mined (22) is about equivalent to the amount that is eroded from the land and enters the oceans by runoff (25).
(*Sources*: Data from 5 mil, 2000 *Phosphorus in the environment: Natural flows and human interference*. Annual Review of Environment and Resources 25:53–88.)

(a) (b)

FIGURE 6.20 **Guano Island**, Peru. For centuries the principal source of phosphorus fertilizer was guano deposits from seabirds. The birds feed on fish and nest on small islands. Their guano accumulates in this dry climate over centuries, forming rocklike deposits that continue to be mined commercially for phosphate fertilizers. On the Peruvian Ballestas Islands **(a)** seabirds (in this case, Incan terns) nest, providing some of the guano, and **(b)** sea lions haul out and rest on the rocklike guano.

Through tectonic processes, the valley was slowly uplifted, and in the 1880s and 1890s phosphate ore was discovered there. Today, Bone Valley provides about 20% of the world's phosphate (Figure 6.21).

About 80% of phosphorus is produced in four countries: the United States, China, South Africa, and Morocco.[17,18] The global supply of phosphorus that can be extracted economically is about 15 billion tons (15,000 million tons). Total U.S. reserves are estimated at 1.2 billion metric tons. In 2009, in the United States, approximately 30.9 million tons of marketable phosphorus rocks valued at $3.5 billion were obtained by removing more than 120 million tons of rocks from mines. Most of the U.S. phosphorus, about 85%, came from Florida and North Carolina, the rest from Utah and Idaho.[19] All of our industrialized agriculture—most of the food produced in the United States—depends on phosphorus for fertilizers that comes from just four states!

Phosphorus may become much more difficult to obtain in the next few decades. According to the U.S. Geological Survey, in 2007 the price of phosphate rock "jumped dramatically worldwide owing to increased agricultural demand and tight supplies," and by 2009 "the average U.S. price was more than double that of 2007," reaching as much as $500 a ton in some parts of the world.[20]

One fact is clear: Without phosphorus, we cannot produce food. Thus, declining phosphorus resources will harm the global food supply and affect all of the world's economies. Extraction continues to increase as the expanding human population demands more food and as we grow more corn for biofuel. However, if the price of phosphorus rises as high-grade deposits dwindle, phosphorus from lower-grade deposits can be mined at a profit. Florida is thought to have as much as 8 billion metric tons of phosphorus that might eventually be recovered if the price is right.

Mining, of course, may have negative effects on the land and ecosystems. For example, in some phosphorus mines, huge pits and waste ponds have scarred the landscape, damaging biologic and hydrologic resources. Balancing the need for phosphorus with the adverse environmental impacts of mining is a major environmental issue. Following phosphate extraction, land disrupted by open-pit phosphate mining, shown in Figure 6.21 is reclaimed to pastureland, as mandated by law.

As with nitrogen, an overabundance of phosphorus causes environmental problems. In bodies of water, from ponds to lakes and the ocean, phosphorus can promote unwanted growth of photosynthetic bacteria. As the algae proliferate, oxygen in the water may be depleted. In oceans, dumping of organic materials high in nitrogen and phosphorus has produced several hundred "dead zones." collectively covering about 250,000 km^2. Although this is an area almost as large as Texas, it represents less than 1% of the area of the Earth's oceans (335,258,000 km^2).

What might we do to maintain our high agriculture production but reduce our need for newly mined phosphate? Among the possibilities:

- Recycle human waste in the urban environment to reclaim phosphorus and nitrogen.
- Use wastewater as a source of fertilizer, rather than letting it end up in waterways.
- Recycle phosphorus-rich animal waste and bones for use in fertilizer.
- Further reduce soil erosion from agricultural lands so that more phosphorus is retained in the fields for crops.
- Apply fertilizer more efficiently so less is immediately lost to wind and water erosion.
- Find new phosphorus sources and more efficient and less expensive ways to mine it.
- Use phosphorus to grow food crops rather than biofuel crops.

We have focused on the biogeochemical cycles of three of the macronutrients, illustrating the major kinds of biogeochemical cycles—those with and those without an atmospheric component—but obviously this is just an introduction about methods that can be applied to all elements required for life and especially in agriculture.

FIGURE 6.21 **A large open-pit phosphate mine in Florida (similar to Bone Valley), with piles of waste material.** The land in the upper part of the photograph has been reclaimed and is being used for pasture.

CRITICAL THINKING ISSUE

How Are Human Activities Linked to the Phosphorus and Nitrogen Cycles?

Scientists estimate that nitrogen deposition to Earth's surface will double in the next 25 years and that the use of phosphorus will also increase greatly as we attempt to feed a few billion more people in coming decades. The natural rate of nitrogen fixation is estimated to be 140 teragrams (Tg) of nitrogen a year (1 teragram = 1 million metric tons). Human activities—such as the use of fertilizers, draining of wetlands, clearing of land for agriculture, and burning of fossil fuels—are causing additional nitrogen to enter the environment. Currently, human activities are responsible for more than half of the fixed nitrogen that is deposited on land. Before the 20th century, fixed nitrogen was recycled by bacteria, with no net accumulation. Since 1900, however, the use of commercial fertilizers has increased exponentially (Figure 6.22). Nitrates and ammonia from burning fossil fuels have increased about 20% in the last decade or so. These inputs have overwhelmed the denitrifying part of the nitrogen cycle and the ability of plants to use fixed nitrogen.

Nitrate ions, in the presence of soil or water, may form nitric acid. With other acids in the soil, nitric acid can leach out chemicals important to plant growth, such as magnesium and potassium. When these chemicals are depleted, more toxic ones, such as aluminum, may be released, damaging tree roots. Acidification of soil by nitrate ions is also harmful to organisms. When toxic chemicals wash into streams, they can kill fish. Excess nitrates in rivers and along coasts can cause algae to overgrow, damaging ecosystems. High levels of nitrates in drinking water from streams or groundwater contaminated by fertilizers are a health hazard.[21, 22, 23, 24]

The nitrogen, phosphorus, and carbon cycles are linked because nitrogen is a component of chlorophyll, the molecule that plants use in photosynthesis. Phosphorus taken up by plants enters the food chain and, thus, the carbon cycle. It is an irreplaceable ingredient in life. Because nitrogen is a limiting factor on land, it has been predicted that rising levels of global nitrogen may increase plant growth. Recent studies have suggested, however, that a beneficial effect from increased nitrogen would be short-lived. As plants use additional nitrogen, some other factor, such as phosphorus, will become limiting. When that occurs, plant growth will slow, and so will the uptake of carbon dioxide. More research is needed to understand the interactions between carbon and the phosphorus and nitrogen cycles and to be able to predict the long-term effects of human activities.

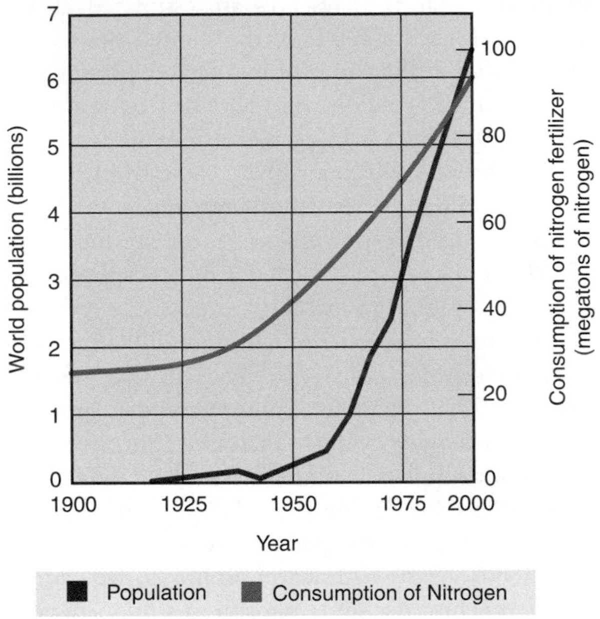

FIGURE 6.22 The use of nitrogen fertilizers has increased greatly. (*Source*: Modified from Rhodes, D. 2009. Purdue University Department of Horticulture & Landscape Architecture.)

Critical Thinking Questions

1. The supply of phosphorus from mining is a limited resource. In the U.S., extraction is decreasing, and the price is rising dramatically. Do you think phosphorus can be used sustainably? How? If not, what are the potential consequences for agriculture?
2. Do you think phosphorus use should be governed by an international body? Why? Why not?
3. Compare the rate of human contributions to nitrogen fixation with the natural rate.
4. How does the change in fertilizer use relate to the change in world population? Why?
5. Develop a diagram to illustrate the links between the phosphorus, nitrogen, and carbon cycles.
6. Make a list of ways in which we could modify our activities to reduce our contributions to the phosphorus and nitrogen cycles.
7. Should phosphorus and nitrogen be used to produce corn as a biofuel (alcohol)? Why? Why not?

SUMMARY

- Biogeochemical cycles are the major way that elements important to Earth processes and life are moved through the atmosphere, hydrosphere, lithosphere, and biosphere.
- Biogeochemical cycles can be described as a series of reservoirs, or storage compartments, and pathways, or fluxes, between reservoirs.
- In general, some chemical elements cycle quickly and are readily regenerated for biological activity. Elements whose biogeochemical cycles include a gaseous phase in the atmosphere tend to cycle more rapidly.
- Life on Earth has greatly altered biogeochemical cycles, creating a planet with an atmosphere unlike those of any others known, and especially suited to sustain life.
- Every living thing, plant or animal, requires a number of chemical elements. These chemicals must be available at the appropriate time and in the appropriate form and amount.
- Chemicals can be reused and recycled, but in any real ecosystem some elements are lost over time and must be replenished if life in the ecosystem is to persist. Change and disturbance of natural ecosystems are the norm. A steady state, in which the net storage of chemicals in an ecosystem does not change with time, cannot be maintained.
- Our modern technology has begun to alter and transfer chemical elements in biogeochemical cycles at rates comparable to those of natural processes. Some of these activities are beneficial to society, but others create problems, such as pollution by nitrogen and phosphorus
- To be better prepared to manage our environment, we must recognize both the positive and the negative consequences of activities that transfer chemical elements, and we must deal with them appropriately.
- Biogeochemical cycles tend to be complex, and Earth's biota has greatly altered the cycling of chemicals through the air, water, and soil. Continuation of these processes is essential to the long-term maintenance of life on Earth.
- There are many uncertainties in measuring either the amount of a chemical in storage or the rate of transfer between reservoirs.

REEXAMINING THEMES AND ISSUES

Human Population

Through modern technology, we are transferring some chemical elements through the air, water, soil, and biosphere at rates comparable to those of natural processes. As our population increases, so does our use of resources and so do these rates of transfer. This is a potential problem because eventually the rate of transfer for a particular chemical may become so large that pollution of the environment results.

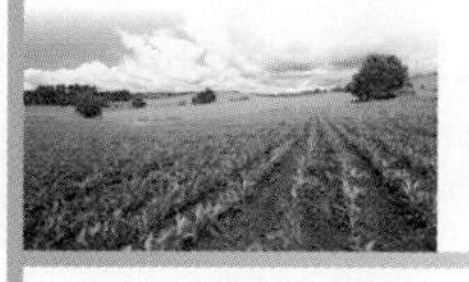

Sustainability

If we are to sustain a high-quality environment, the major biogeochemical cycles must transfer and store the chemicals necessary to maintain healthy ecosystems. That is one reason why understanding biogeochemical cycles is so important. For example, the release of sulfur into the atmosphere is degrading air quality at local to global levels. As a result, the United States is striving to control these emissions.

Global Perspective

The major biogeochemical cycles discussed in this chapter are presented from a global perspective. Through ongoing research, scientists are trying to better understand how major biogeochemical cycles work. For example, the carbon cycle and its relationship to the burning of fossil fuels and the storage of carbon in the biosphere and oceans are being intensely investigated. Results of these studies are helping us to develop strategies for reducing carbon emissions. These strategies are implemented at the local level, at power plants, and in cars and trucks that burn fossil fuels.

Urban World

Our society has concentrated the use of resources in urban regions. As a result, the release of various chemicals into the biosphere, soil, water, and atmosphere is often greater in urban centers, resulting in biogeochemical cycles that cause pollution problems.

People and Nature

Humans, like other animals, are linked to natural processes and nature in complex ways. We change ecosystems through land-use changes and the burning of fossil fuels, both of which change biogeochemical cycles, especially the carbon cycle that anchors life and affects Earth's climate.

Science and Values

Our understanding of biogeochemical cycles is far from complete. There are large uncertainties in the measurement of fluxes of chemical elements—nitrogen, carbon, phosphorus, and others. We are studying biogeochemical cycles because understanding them will help us to solve environmental problems. Which problems we address first will reflect the values of our society.

KEY TERMS

biogeochemical cycle **112**
biosphere **110**
biota **118**
carbon cycle **117**
carbon–silicate cycle **119**
denitrification **120**
eukaryote **110**
flow **112**
flux **112**
geologic cycle **113**
hydrologic cycle **115**
limiting factor **111**
macronutrients **111**
micronutrients **111**
nitrogen cycle **120**
nitrogen fixation **120**
organelle **110**
phosphorus cycle **121**
plate tectonics **113**
prokaryote **110**
residence time **112**
rock cycle **116**
sink **112**
source **112**
tectonic cycle **113**
thermodynamic equilibrium **108**

STUDY QUESTIONS

1. Why is an understanding of biogeochemical cycles important in environmental science? Explain your answer, using two examples.
2. What are some of the general rules that govern biogeochemical cycles, especially the transfer of material?
3. Identify the major aspects of the carbon cycle and the environmental concerns associated with it.
4. What are the differences in the geochemical cycles for phosphorus and nitrogen, and why are the differences important in environmental science?
5. What are the major ways that people have altered biogeochemical cycles?
6. If all life ceased on Earth, how quickly would the atmosphere become like that of Venus and Mars? Explain.

FURTHER READING

Lane, Nick, *Oxygen: The Molecule That Made the World* (Oxford: Oxford University Press, 2009).

Lovelock, J., *The Ages of Gaia: A Biography of the Earth* (Oxford: Oxford University Press, 1995).

Schlesinger, W.H., *Biogeochemistry: An Analysis of Global Change*, 2nd ed. (San Diego: Academic Press, 1997). This book provides a comprehensive and up-to-date overview of the chemical reactions on land, in the oceans, and in the atmosphere of Earth.

CHAPTER 7

Dollars and Environmental Sense: Economics of Environmental Issues

A New England common illustrates the tragedy of the Commons, one of the key ideas of Environmental Economics.

LEARNING OBJECTIVES

Why do people value environmental resources? To what extent are environmental decisions based on economics? Other chapters in this text explain the causes of environmental problems and discuss technical solutions. The scientific solutions, however, are only part of the answer. This chapter introduces some basic concepts of environmental economics and shows how these concepts help us understand environmental issues. After reading this chapter, you should understand . . .

- How the perceived future value of an environmental benefit affects our willingness to pay for it now;
- What "externalities" are and why they matter;
- How much risk we should be willing to accept for the environment and ourselves;
- How we can place a value on environmental intangibles, such as landscape beauty.

CASE STUDY

Cap, Trade, and Carbon Dioxide

We hear a lot in the news these days about a public and congressional debate over "cap-and-trade" and the control of carbon dioxide emissions. The question our society is wrestling with is this: Assuming that carbon dioxide, as a greenhouse gas, can be treated like any other air pollutant legally and economically, how can its emissions best be controlled? "Best" in this context means reducing human-induced carbon dioxide emissions as much as possible, doing so in the least expensive way, and in a way that is fair to all participants. Among the most commonly discussed methods are the following:

Tax on emitters: The government levies a tax based on the quantity of pollution emitted.

Legal emissions limit: The law will limit the amount of emissions allowed from each source—individual, corporation, facility (e.g., a single power plant), or government organization. The control is applied individually, emitter by emitter, with each assigned a maximum.

Cap-and-trade: First, the government decides how much of a particular pollutant will be permitted, either as a *total amount* in the environment or as the *total amount emitted into the environment per year*. (This is one way that cap-and-trade differs from legal emissions limits, which are set per emitter.) Next, the limit is divided up among the sources of the pollutant, but the owners of those sources can trade among themselves. (In contrast, legal emissions limits allow no trading among participants.)

The rationale behind the tax on emitters is twofold. First of all, it raises tax money, which in theory could be used to find ways to reduce and better control the pollution. Second, it is supposed to discourage businesses from emitting pollution. But critics of a direct tax on emitters say it doesn't work and is bad for business. They say businesses simply pass on the cost of the taxes in their prices, so the tax burden is on the consumer but the likely economic result is fewer sales.

Critics of legal emissions limits say that since government sets the limits, this could be arbitrary and place an unfair burden on certain businesses. It also requires extensive, costly monitoring, placing a further burden on society's economics.

The potential problems of direct taxes and emissions limits led to the idea of cap-and-trade. Here's how it works. Suppose you own a coal-fired power plant and the government gives you a certain number of carbon allowances—tons of carbon dioxide you will be allowed to emit into the air each year. These allowances come to you as "ration coupons," something like food stamps, and you can either "spend" them yourself by emitting the amount of pollutant each stamp permits, or sell them to someone else. If you decide to build a solar power plant that replaces your coal plant, you can sell your pollution allowances to a power company that is still using coal and is emitting more than its allowed amount. That company could then increase its emissions. In theory, both you and the other company make money—you by simply selling your credits, the other company by not having to build a completely new power plant.

Does cap-and-trade work? The Environmental Protection Agency (EPA) now has three decades of experience with attempts to control air pollutants and has found that cap-and-trade works very well—for example, in reducing acid rain resulting from sulfur dioxide emissions from power plants (Figure 7.1).[1] Proponents of cap-and-trade argue that it places the benefits and disbenefits of pollution control directly in the hands of the polluters, rather than passing them on to consumers (which a tax usually ends up doing). Also, it keeps the activity in a kind of free market, rather than forcing specific actions on individuals, and in this way minimizes government interference.

Opponents of cap-and-trade argue that it is really just a tax in disguise, that its end result is the same as a direct tax—the corporations that produce electricity from fossil fuel will be burdened with a huge tax and put at a disadvantage versus corporations that turn to alternative energy sources. Advocates of cap-and-trade say that this is just the point—that the whole idea is to encourage our society to move away from fossil fuels, and that this has proved to be an efficient way to do it. Opponents counter that the net result will be a burden on everybody and that the average family's energy bill could go up an estimated $1,500 a year.[2] Proponents of cap-and-trade cite its success with acid rain from sulfur dioxide emissions (Figure 7.1), but critics say that, unlike sulfur dioxide, carbon dioxide isn't really a pollutant in the usual legal sense, and that because carbon dioxide is a global problem, cap-and-trade can't work without unusual treaties among nations.

You can see from the cap-and-trade example that what seems at first glance a simple and straightfor-

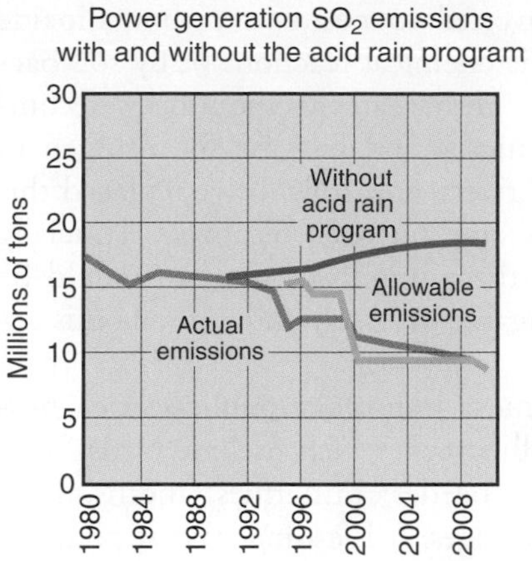

(a)

(b)

FIGURE 7.1 Cap-and-trade helped the EPA reduce acid rain caused by sulfur dioxide emissions (a) from power plants like this coal-fired power plant in Arizona (b).

ward solution can turn out to be much more difficult. Deciding whether cap-and-trade can be a good way to reduce carbon dioxide emissions into the atmosphere depends in part on scientific knowledge, but also on economic analyses. So it is with most environmental issues.

7.1 Overview of Environmental Economics

The history of modern environmental law can be traced back to the 1960s and the beginnings of the modern social and political movement we know as environmentalism. Its foundation is the "three E's": ecology, engineering, and economics.[3] Although in this environmental science textbook we will devote most of our time to the scientific basis—ecology, geology, climatology, and all the other sciences involved in environmental analysis—economics underlies much of the discussion. It is always a factor in finding solutions that work, are efficient, and are fair. This is why we are devoting one of our early overview chapters to environmental economics.

Environmental economics is not simply about money; it is about how to persuade people, organizations, and society at large to act in a way that benefits the environment, keeping it as free as possible of pollution and other damage, keeping our resources sustainable, and accomplishing these goals within a democratic framework. Put most simply, environmental economics focuses on two broad areas: controlling pollution and environmental damage in general, and sustaining renewable resources—forests, fisheries, recreational lands, and so forth. Environmental economists also explore the reasons why people don't act in their own best interests when it comes to the environment. Are there rational explanations for what seem to be irrational choices? If so, and if we can understand them, perhaps we can do something about them. What we do, what we can do, and how we do it are known collectively as **policy instruments**.

Environmental decision-making often, perhaps even usually, involves analysis of tangible and intangible factors. In the language of economics, a **tangible factor** is one you can touch, buy, and sell. A house lost in a mudslide due to altering the slope of the land is an example of a tangible factor. For economists, an **intangible factor** is one you can't touch directly, but you value it, as with the beauty of the slope before the mudslide. Of the two, the intangibles are obviously more difficult to deal with because they are harder to measure and to value economically. Nonetheless, evaluation of intangibles is becoming more important. As you will see in later chapters, huge amounts of money and resources are involved in economic decisions about both tangible and intangible aspects of the environment: There are the costs of pollution and the loss of renewable resources, and there are the costs of doing something about these problems.

In every environmental matter, there is a desire on the one hand to maintain individual freedom of choice, and on the other to achieve a specific social goal. In ocean fishing, for example, we want to allow every individual to choose whether or not to fish, but we want to prevent everyone from fishing at the same time and bringing fish species to extinction. This interplay between private good and public good is at the heart of environmental issues.

In this chapter we will examine some of the basic issues in environmental economics: the environment as a commons; risk-benefit analysis; valuing the future; and why people often do not act in their own best interest.

7.2 Public-Service Functions of Nature

A complicating factor in maintaining clean air, soils, and water, and sustaining our renewable resources is that ecosystems do some of this without our help. Forests absorb particulates, salt marshes convert toxic compounds to nontoxic forms, wetlands and organic soils treat sewage. These are called the **public-service functions** of nature. Economists refer to the ecological systems that provide these benefits as **natural capital**.

The atmosphere performs a public service by acting as a large disposal site for toxic gases. And carbon monoxide is eventually converted to nontoxic carbon dioxide either by inorganic chemical reactions or by soil bacteria. Bacteria also clean water in the soil by decomposing toxic chemicals, and bacteria fix nitrogen in the oceans, lakes, rivers, and soils. If we replaced this function by producing nitrogen fertilizers artificially and transporting them ourselves, the cost would be immense—but, again, we rarely think about this activity of bacteria.

FIGURE 7.2 **Public-service functions of living things.** Wild creatures and natural ecosystems carry out tasks that are important for our survival and would be extremely expensive for us to accomplish by ourselves. For example, bees pollinate millions of flowers important for food production, timber supply, and aesthetics. As a result, beekeeping is a commercial enterprise, with rewards and risks, as shown in this photograph.

Among the most important public-service providers are the pollinators, which include birds, bats, ants, bees, wasps, beetles, butterflies, moths, flies, mosquitoes, and midges. It is estimated that pollinating animals pollinate about $15 billion worth of crops grown on 2 million acres in the United States,[4, 5] that about one bite in three of the food you eat depends on pollinators, and that their total economic impact can reach $40 billion a year (Figure 7.2).[6] The cost of pollinating these crops by hand would be exorbitant, so a pollutant that eliminated bees would have large indirect economic consequences. We rarely think of this benefit of bees, but it has received wide attention in recent years because of a disease called Colony Collapse Disorder (CCD), which affected food costs, agricultural practices, and many companies that provide bees to pollinate crops.[7]

Public-service functions of living things are estimated to provide between $3 trillion and $33 trillion in benefits to human beings and other forms of life per year.[8] However, current estimates are only rough approximations because the value is difficult to measure.

7.3 The Environment as a Commons

Often people use a natural resource without regard for maintaining that resource and its environment in a renewable state—that is, they don't concern themselves with that resource's sustainability. At first glance, this seems puzzling, but economic analysis suggests that the profit motive, by itself, will not always lead a person to act in the best interests of the environment.

One reason has to do with what the ecologist Garrett Hardin called "the tragedy of the commons."[9] When a resource is shared, an individual's personal share of profit from its exploitation is usually greater than his or her share of the resulting loss. A second reason has to do with the low growth rate, and therefore low productivity, of a resource.

A **commons** is land (or another resource) owned publicly, with public access for private uses. The term *commons* originated from land owned publicly in

English and New England towns and set aside so that all the farmers of the town could graze their cattle. Sharing the grazing area worked as long as the number of cattle was low enough to prevent overgrazing. It would seem that people of goodwill would understand the limits of a commons. But take a dispassionate view and think about the benefits and costs to each farmer as if it were a game. Phrased simply, each farmer tries to maximize personal gain and must periodically consider whether to add more cattle to the herd on the commons. The addition of one cow has both a positive and a negative value. The positive value is the benefit when the farmer sells that cow. The negative value is the additional grazing by the cow. The personal profit from selling a cow is greater than the farmer's share of the loss caused by the degradation of the commons. Therefore, the short-term successful game plan is always to add another cow.

Since individuals will act to increase use of the common resource, eventually the common grazing land is so crowded with cattle that none can get adequate food and the pasture is destroyed. In the short run, everyone seems to gain, but in the long run, everyone loses. This applies generally: Complete freedom of action in a commons inevitably brings ruin to all. The implication seems clear: Without some management or control, all natural resources treated like a commons will inevitably be destroyed.

How can we deal with the tragedy of the commons? It is only a partially solved problem. As several scientists wrote recently, "No single broad type of ownership—government, private or community—uniformly succeeds or fails to halt major resource deterioration." Still, in trying to solve this puzzle, economic analysis can be helpful.

There are many examples of commons, both past and present. In the United States, 38% of forests are on publicly owned lands; as such, these forests are commons. Resources in international regions, such as ocean fisheries away from coastlines, and the deep-ocean seabed, where valuable mineral deposits lie, are international commons not controlled by any single nation.

The Arctic sea ice is a commons (Figure 7.3), as is most of the continent of Antarctica, although there are some national territorial claims, and international negotiations have continued for years about conserving Antarctica and about the possible use of its resources.

The atmosphere, too, is a commons, both nationally and internationally. Consider the possibility of global warming. Individuals, corporations, public utilities, motor vehicles, and nations add carbon dioxide to the air by burning fossil fuels. Just as Garrett Hardin suggested, people tend to respond by benefiting themselves (burning more fossil fuel) rather than by benefiting the commons (burning less fossil fuel). The picture here is quite mixed, however, with much ongoing effort to bring cooperation to this common issue.

FIGURE 7.3 Arctic sea ice and polar bears, which live in many areas of the Arctic, are part of a commons.

In the 19th century, burning wood in fireplaces was the major source of heating in the United States (and fuel wood is still the major source of heat in many nations). Until the 1980s, a wood fire in a fireplace or woodstove was considered a simple good, providing warmth and beauty. People enjoyed sitting around a fire and watching the flames—an activity with a long history in human societies. But in the 1980s, with increases in populations and vacation homes in states such as Vermont and Colorado, home burning of wood began to pollute air locally. Especially in valley towns surrounded by mountains, the air became fouled, visibility declined, and there was a potential for ill effects on human health and the environment. Several states, including Vermont, have had programs offering rebates to buyers of newer, lower-polluting woodstoves.[10] The local air is a commons, and its overuse required a societal change.

Recreation is a problem of the commons—overcrowding of national parks, wilderness areas, and other nature–recreation areas. An example is Voyageurs National Park in northern Minnesota. The park, within North America's boreal-forest biome, includes many lakes and islands and is an excellent place for fishing, hiking, canoeing, and viewing wildlife. Before the area became a national park, it was used for motorboating, snowmobiling, and hunting; a number of people in the region made their living from tourism based on these kinds of recreation. Some environmental groups argue that Voyageurs National Park is ecologically fragile and needs to be legally designated a U.S. wilderness area to protect it from overuse and from the adverse effects of motorized vehicles. Others argue that the nearby million-acre Boundary Waters Canoe Area provides ample wilderness, that Voyageurs can withstand a moderate level of hunting and motorized transportation,

FIGURE 7.4 **Voyageurs National Park** in northern Minnesota has many lakes well suited to recreational boating. But what kind of boating—what kinds of motors, what size boats—is a long-running controversy. In a commons such as a national park, these are the kinds of conflicts that arise over intangible value (such as scenic beauty) and tangible value (such as the opportunity for boat owners and guides to make a living). Here we see a guided tour on a motorized boat.

and that these uses should be allowed. At the heart of this conflict is the problem of the commons, which in this case can be summed up as follows:

- What is the appropriate public use of public lands?
- Should all public lands be open to all public uses?
- Should some public lands be protected from people?

At present, the United States has a policy of different uses for different lands. In general, national parks are open to the public for many kinds of recreation, whereas designated wildernesses have restricted visitorship and kinds of uses (Figure 7.4).

7.4 Low Growth Rate and Therefore Low Profit as a Factor in Exploitation

We said earlier that the second reason individuals tend to overexploit natural resources held in common is the low growth rate of many biological resources.[11] For example, one way to view whales economically is to consider them solely in terms of whale oil. Whale oil, a marketable product, and the whales alive in the ocean, can be thought of as the capital investment of the industry.

From an economic point of view, how can whalers get the best return on their capital? Keeping in mind that whale populations, like other populations, increase only if there are more births than deaths, we will examine two approaches: *resource sustainability* and *maximum profit*. If whalers adopt a simple, one-factor resource-sustainability policy, they will harvest only the net biological productivity each year (the number by which the population increased). Barring disease or disaster, this will maintain the total abundance of whales at its current level and keep the whalers in business indefinitely. If, on the other hand, they choose to simply maximize immediate profit, they will harvest all the whales now, sell the oil, get out of the whaling business, and invest their profits.

Suppose they adopt the first policy. What is the maximum gain they can expect? Whales, like other large, long-lived creatures, reproduce slowly, with each female typically giving birth to a calf every three or four years. Thus, the total net growth of a whale population is likely to be no more than 5% per year and probably more like 3%. This means that if all the oil in the whales in the oceans today represented a value of $100 million, then the most the whalers could expect to take in each year would be no more than 5% of this amount, or $5 million. Until the 2008 economic recession, 5% interest was considered a modest, even poor, rate of return on one's money. And meanwhile the whalers would have to pay for the upkeep of ships and other equipment, salaries of employees, and interest on loans—all of which would decrease profit.

However, if whalers opted for the second policy and harvested all the whales, they could invest the money from the oil. Although investment income varies, even a conservative return on their investment of $100 million would likely yield millions of dollars annually, and since they would no longer be hunting whales, this would be clear profit, without the costs of paying a crew, maintaining ships, buying fuel, marketing the oil, and so on.

Clearly, if one considers only direct profit, it makes sense to adopt the second policy: Harvest all the whales, invest the money, and relax. And this seems to have been the case for those who hunted bowhead whales in the 19th and early 20th centuries (Figures 7.5 and 7.6).[12] Whales simply are not a highly profitable long-term investment under the resource-sustainability policy. From a

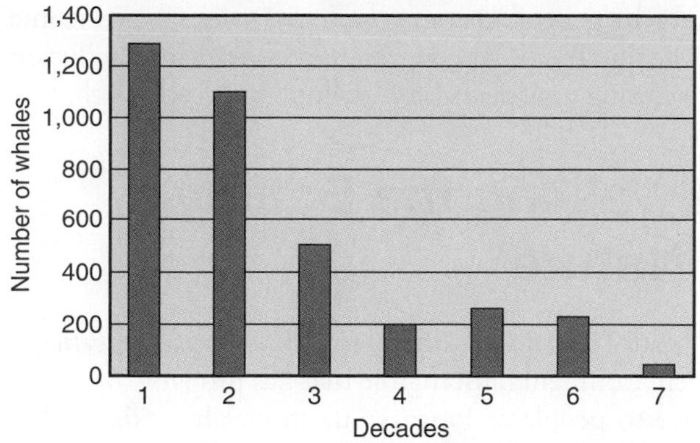

FIGURE 7.5 **Bowhead whales caught and killed by Yankee whalers from 1849 to 1914.** The number killed, shown for each decade, declined rapidly, indicating that the whale population was unable to reproduce at a rate that could replace the initial large catches, yet the whalers kept killing at a nonsustainable rate, as economics would predict. (*Source*: Redrawn from J.R. Bockstoce and D.B. Botkin, *The Historical Status and Reduction of the Western Arctic Bowhead Whale* (Balaena mysticetus) *Population by the Pelagic Whaling Industry, 1849–1914*. Final report to the U.S. National Marine Fisheries Service by the Old Dartmouth Historical Society, 1980; and J.R. Bockstoce, D.B. Botkin, A. Philp, B.W. Collins, and J.C. George, The geographic distribution of bowhead whales in the Bering, Chukchi, and Beaufort seas: Evidence from whaleship records, 1849–1914, *Marine Fisheries Review* 67(3) [2007]:1–43.)

FIGURE 7.6 **Bowhead whale baleen (their modified teeth)** on a dock in San Francisco in the late 19th century, when these flexible plates were needed for women's corsets and other uses where strength and flexibility were important. Baleen and whale oil were the two commercial products obtained from bowheads. When baleen was replaced by new forms of steel, the baleen market disappeared. Commercial bowhead whale hunting ended with the beginning of World War I. However, whale oil remained listed as a strategic material by the U.S. Department of Defense for decades afterward because of its lubricating qualities.

tangible economic perspective, without even getting into the intangible ethical and environmental concerns, it is no wonder that there are fewer and fewer whaling companies and that companies left the whaling business when their ships became old and inefficient. Few nations support whaling; those that do have stayed with whaling for cultural reasons. For example, whaling is important to the Eskimo culture, so some harvest of bowheads takes place in Alaska; and whale meat is a traditional Japanese and Norwegian food, so these countries continue to harvest whales for this reason.

Scarcity Affects Economic Value

The relative scarcity of a necessary resource is another factor to consider in resource use, because this affects its value and therefore its price. For example, if a whaler lived on an isolated island where whales were the only food and he had no communication with other people, then his primary interest in whales would be as a way for him to stay alive. He couldn't choose to sell off all whales to maximize profit, since he would have no one to sell them to. He might harvest at a rate that would maintain the whale population. Or, if he estimated that his own life expectancy was only about ten years, he might decide that he could take a chance on consuming whales beyond their ability to reproduce. Cutting it close to the line, he might try to harvest whales at a rate that would cause them to become extinct at the same time that he would. "You can't take it with you" would be his attitude.

If ships began to land regularly at this island, he could leave, or he could trade and begin to benefit from some of the future value of whales. If ocean property rights existed, so he could "own" the whales that lived within a certain distance of his island, then he might consider the economic value of owning this right to the whales. He could sell rights to future whales, or mortgage against them, and thus reap the benefits during his lifetime from whales that could be caught after his death. Causing the extinction of whales would not be necessary.

From this example, we see that policies that seem ethically good may not be the most profitable for an individual. We must think beyond the immediate, direct economic advantages of harvesting a resource. Economic analysis clarifies how an environmental resource should be used, what is perceived as its intrinsic value and therefore its price. And this brings us to the question of externalities.

7.5 Externalities

One gap in our thinking about whales, an environmental economist would say, is that we must be concerned with externalities in whaling. An **externality**, also called an **indirect cost**, is often not recognized by producers as part of their costs and benefits, and therefore not normally accounted for in their cost-revenue analyses.[11] Put simply, externalities are costs or benefits that don't show up in the price tag.[13] In the case of whaling, externalities include the loss of revenue to whale-watching tourist boats and the loss of the ecological role that whales play in marine ecosystems. Classically, economists agree that the only way for a consumer to make a rational decision is by comparing the true costs—including externalities—against the benefits the consumer seeks.

Air and water pollution provide good examples of externalities. Consider the production of nickel from ore at the Sudbury, Ontario, smelters, which has serious environmental effects. Traditionally, the economic costs associated with producing commercially usable nickel from an ore were only the **direct costs**—that is, those borne by the producer in obtaining, processing, and distributing a product—passed directly on to the user or purchaser. In this case, direct costs include purchasing the ore, buying energy to run the smelter, building the plant, and paying employees. The externalities, however, include costs associated with degradation of the environment from the plant's emissions. For example, prior to implementation of pollution control, the Sudbury smelter destroyed vegetation over a wide area, which led to increased erosion. Although air emissions from smelters have been substantially reduced and restoration efforts have initiated a slow recovery of the area, pollution remains a problem, and total recovery of the local ecosystem may take a century or more.[14] There are costs associated with the value of trees and soil, with restoring vegetation and land to a productive state.

Problem number one: What is the true cost of clean air over Sudbury? Economists say that there is plenty of disagreement about the cost, but that everyone agrees that it is larger than zero. In spite of this, clean air and water are traded and dealt with in today's world as if their value were zero. How do we get the value of clean air and water and other environmental benefits to be recognized socially as greater than zero? In some cases, we can determine the dollar value. We can evaluate water resources for power or other uses based on the amount of flow of the rivers and the quantity of water storage in rivers and lakes. We can evaluate forest resources based on the number, types, and sizes of trees and their subsequent yield of lumber. We can evaluate mineral resources by estimating how many metric tons of economically valuable mineral material exist at particular locations. Quantitative evaluation of the tangible natural resources—such as air, water, forests, and minerals—prior to development or management of a particular area is now standard procedure.

Problem number two: Who should bear the burden of these costs? Some suggest that environmental and ecological costs should be included in costs of production through taxation or fees. The expense would be borne by the corporation that benefits directly from the sale of the resource (nickel in the case of Sudbury) or would be passed on in higher sales prices to users (purchasers) of nickel. Others suggest that these costs be shared by the entire society and paid for by general taxation, such as a sales tax or income tax. The question is whether it is better to finance pollution control using tax dollars or a "polluter pays" approach.

7.6 Valuing the Beauty of Nature

The beauty of nature—often termed *landscape aesthetics*—is an environmental intangible that has probably been important to people as long as our species has existed. We know it has been important since people have written, because the beauty of nature is a continuous theme in literature and art. Once again, as with forests cleaning the air, we face the difficult question: How do we arrive at a price for the beauty of nature? The problem is even more complicated because among the kinds of scenery we enjoy are many modified by people. For example, the open farm fields in Vermont improved the view of the mountains and forests in the distance, so when farming declined in the 1960s, the state began to provide tax incentives for farmers to keep their fields open and thereby help the tourism economy (Figure 7.7).

One of the perplexing problems of aesthetic evaluation is personal preference. One person may appreciate a high mountain meadow far removed from civilization; a second person may prefer visiting with others on a patio at a trailhead lodge; a third may prefer to visit a city park; and a fourth may prefer the austere beauty of a desert. If we are going to consider aesthetic factors in environmental analysis, we must develop a method of aesthetic evaluation that allows for individual differences—another yet unsolved topic.

One way the intangible value of landscape beauty is determined is by how much people are willing to pay for it, and how high a price people will pay for land with a beautiful view, compared with the price of land without a view. As apartment dwellers in any big city will tell you,

FIGURE 7.7 **How much is a beautiful scene worth?** Consider, for example, this view in New Hampshire looking west to the Connecticut River and Vermont. Is landscape beauty an externality?

the view makes a big difference in the price of their unit. For example, in mid-2009, the *New York Times* listed two apartments, both with two bedrooms, for sale in the same section of Manhattan, one without a view for $850,000 and one with a wonderful view of the Hudson River estuary for $1,315,000 (Figure 7.8).

Some philosophers suggest that there are specific characteristics of landscape beauty and that we can use these characteristics to help us set the value of intangibles. Some suggest that the three key elements of landscape beauty are coherence, complexity, and mystery—mystery in the form of something seen in part but not completely, or not completely explained. Other philosophers suggest that the primary aesthetic qualities are unity vividness, and variety.[15] *Unity* refers to the quality or wholeness of the perceived landscape—not as an assemblage but as a single, harmonious unit. *Vividness* refers to that quality of landscape that makes a scene visually striking; it is related to intensity, novelty, and clarity. People differ in what they believe are the key qualities of landscape beauty, but again, almost everyone would agree that the value is greater than zero.

FIGURE 7.8 A view from a New York City apartment greatly increased its price compared with similar apartments without a view.

7.7 How Is the Future Valued?

The discussion about whaling—explaining why whalers may not find it advantageous to conserve whales—reminds us of the old saying "A bird in the hand is worth two in the bush." In economic terms, a profit now is worth much more than a profit in the future. This brings up another economic concept important to environmental issues: the future value of anything compared with its present value.

Suppose you are dying of thirst in a desert and meet two people. One offers to sell you a glass of water now, and the other offers to sell you a glass of water if you can be at the well tomorrow. How much is each glass worth? If you believe you will die today without water, the glass of water today is worth all your money, and the glass tomorrow is worth nothing. If you believe you can live another day without water, but will die in two days, you might place more value on tomorrow's glass than on today's, since it will gain you an extra day—three rather than two.

In practice, things are rarely so simple and distinct. We know we aren't going to live forever, so we tend to value personal wealth and goods more if they are available now than if they are promised in the future. This evaluation is made more complex, however, because we are accustomed to thinking of the future—to planning a nest egg for retirement or for our children. Indeed, many people today argue that we have a debt to future generations and must leave the environment in at least as good a condition as we found it. These people would argue that the future environment is not to be valued less than the present one (Figure 7.9).

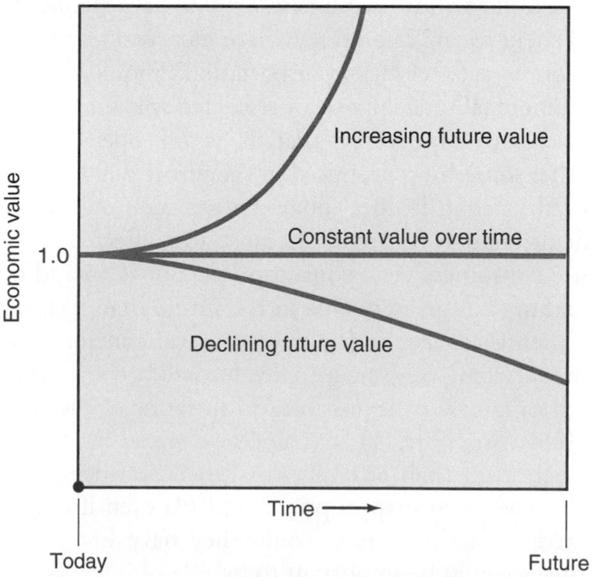

FIGURE 7.9 Economic value as a function of time—a way of comparing the value of having something now with the value of having it in the future. A negative value means that there is more value attached to having something in the present than having it in the future. A positive value means that there is more value attached to having something in the future than having it today.

Since the future existence of whales and other endangered species has value to those interested in biological conservation, the question arises: Can we place a dollar value on the *future* existence of anything? The future value depends on how far into the future you are talking about. The future times associated with some important global environmental topics, such as stratospheric ozone depletion and global warming, extend longer than a century. This is because chlorofluorocarbons (CFCs) have such a long residence time in the atmosphere and because of the time necessary to realize benefits from changing energy policy to offset global climate change.

Another aspect of future versus present value is that spending on the environment can be viewed as diverting resources from alternative forms of productive investment that will be of benefit to future generations. (This assumes that spending on the environment is not itself a productive investment.)

A further issue is that as we get wealthier, the value we place on many environmental assets (such as wilderness areas) increases dramatically. Thus, if society continues to grow in wealth over the next century as it has over the past century, the environment will be worth far more to our great-grandchildren than it was to our great-grandparents, at least in terms of willingness to pay to protect it. The implication—which complicates this topic even more—is that conserving resources and environment for the future is tantamount to taking from the poor today and giving to the possibly rich in the future. To what extent should we ask the average American today to sacrifice now for richer great-great-grandchildren? How can we know the future usefulness of today's sacrifices? Put another way, what would you have liked your ancestors in 1900 to have sacrificed for our benefit today? Should they have increased research and development on electric transportation? Should they have saved more tall-grass prairie or restricted whaling?

Economists observe that it is an open question whether something promised in the future will have more value then than it does today. Future economic value is difficult enough to predict because it is affected by how future consumers view consumption. But if, in addition, something has greater value in the future than it does today, then that leads to the mathematical conclusion that in the very long run, the future value will become infinite, which of course is impossible. So in terms of the future, the basic issues are (1) that since we are so much richer and better off than our ancestors, their sacrificing for us might have been inappropriate; and (2) even if they had wanted to sacrifice, how would they have known what sacrifices would be important to us?

As a general rule, one answer to the thorny questions about future value is: Do not throw away or destroy something that cannot be replaced if you are not sure of its future value. For example, if we do not fully understand the value of the wild relatives of potatoes that grow in Peru but do know that their genetic diversity might be helpful in developing future strains of potatoes, then we ought to preserve those wild strains.

7.8 Risk-Benefit Analysis

Death is the fate of all individuals, and almost every activity in life involves some risk of death or injury. How, then, do we place a value on saving a life by reducing the level of a pollutant? This question raises another important area of environmental economics: **risk-benefit analysis**, in which the riskiness of a present action in terms of its possible outcomes is weighed against the benefit, or value, of the action. Here, too, difficulties arise.

With some activities, the relative risk is clear. It is much more dangerous to stand in the middle of a busy highway than to stand on the sidewalk, and hang gliding has a much higher mortality rate than hiking. The effects of pollutants are often more subtle, so the risks are harder to pinpoint and quantify. Table 7.1 gives the lifetime risk of death associated with a variety of activities and some forms of pollution. In looking at the table, remember that since the ultimate fate of everyone is death, the total lifetime risk of death from all causes must be 100%. So if you are going to die of something and you smoke a pack of cigarettes a day, you have 8 chances in 100 that your death will be a result of smoking. At the same time, your risk of death from driving an automobile is 1 in 100. Risk tells you the chance of an event but not its timing. So you might smoke all you want and die from the automobile risk first.

One of the striking things about Table 7.1 is that death from outdoor environmental pollution is comparatively low—even compared to the risks of drowning or of dying in a fire. This suggests that the primary reason we value lowering air pollution is not to lengthen our lives but to improve the quality of our lives. Considering people's great interest in air pollution today, the quality of life must be much more important than is generally recognized. We are willing to spend money on improving that quality rather than just extending our lives. Another striking observation in this table is that natural *indoor* air pollution is much more deadly than most outdoor air pollution—unless, of course, you live at a toxic-waste facility.

It is commonly believed that future discoveries will help to decrease various risks, perhaps eventually allowing us to approach a zero-risk environment. But complete elimination of risk is generally either technologically impossible or prohibitively expensive. Societies differ in their views of what constitutes socially, psychologically, and ethically acceptable levels of risk for any cause of death or injury, but we can make some generalizations about the acceptability of various risks. One factor is the number of people affected. Risks that affect a small population (such as employees at nuclear power plants) are usually more

Table 7.1 RISK OF DEATH FROM VARIOUS CAUSES

CAUSE	RESULT	RISK OF DEATH (PER LIFETIME)	LIFETIME RISK OF DEATH (%)	COMMENT
Cigarette smoking (pack a day)	Cancer, effect on heart, lungs, etc.	8 in 100	8.0%	
Breathing radon-containing air in the home	Cancer	1 in 100	1.0%	Naturally occurring
Automobile driving		1 in 100	1.0%	
Death from a fall		4 in 1,000	0.4%	
Drowning		3 in 1,000	0.3%	
Fire		3 in 1,000	0.3%	
Artificial chemicals in the home	Cancer	2 in 1,000	0.2%	Paints, cleaning agents, pesticides
Sunlight exposure	Melanoma	2 in 1,000	0.2%	Of those exposed to sunlight
Electrocution		4 in 10,000	0.04%	
Air outdoors in an industrial area		1 in 10,000	0.01%	
Artificial chemicals in water		1 in 100,000	0.001%	
Artificial chemicals in foods		less than 1 in 100,00	0.001%	
Airplane passenger (commercial airline)		less than 1 in 1,000,000	0.00010%	

Source: From *Guide to Environmental Risk* (1991), U.S. EPA Region 5 Publication Number 905/91/017.

acceptable than those that involve all members of a society (such as risk from radioactive fallout).

In addition, novel risks appear to be less acceptable than long-established or natural risks, and society tends to be willing to pay more to reduce such risks. For example, in the late part of the 20th century, France spent about $1 million each year to reduce the likelihood of one air-traffic death but only $30,000 for the same reduction in automobile deaths.[16] Some argue that the greater safety of commercial air travel versus automobile travel is in part due to the relatively novel fear of flying compared with the more ordinary fear of death from a road accident. That is, because the risk is newer to us and thus less acceptable, we are willing to spend more per life to reduce the risk from flying than to reduce the risk from driving.

People's willingness to pay for reducing a risk also varies with how essential and desirable the activity associated with the risk is. For example, many people accept much higher risks for athletic or recreational activities than they would for transportation- or employment-related activities (see Table 7.1). People volunteer to climb Mt. Everest even though many who have attempted it have died, but the same people could be highly averse to risking death in a train wreck or commercial airplane crash. The risks associated with playing a sport or using transportation are assumed to be inherent in the activity. The risks to human health from pollution may be widespread and linked to a large number of deaths. But although risks from pollution are often unavoidable and unseen, people want a lesser risk from pollution than from, say, driving a car or playing a sport.

In an ethical sense, it is impossible to put a value on a human life. However, it is possible to determine how much people are willing to pay for a certain amount of risk reduction or a certain probability of increased longevity. For example, a study by the Rand Corporation considered measures that would save the lives of heart-attack victims, including increasing ambulance services and initiating pretreatment screening programs. According to the study, which identified the likely cost per life saved and people's willingness to pay, people favored government spending of about $32,000 per life saved, or $1,600 per year of longevity. Although information is incomplete, it is possible to estimate the cost of extending lives in terms of dollars per person per year for various actions (Figure 7.10 and Table 7.1). For example, on the basis of direct effects on human health, it costs more to increase longevity by reducing air pollution than to directly reduce deaths by adding a coronary-ambulance system.

Such a comparison is useful as a basis for decision-making. Clearly, though, when a society chooses to reduce air pollution, many factors beyond the direct, measurable health benefits are considered. Pollution not only directly affects our health but also causes ecological and aesthetic damage, which can indirectly affect human health (see Section 7.4). We might want to choose a slightly higher risk of death in a more pleasant environment rather than increase the chances of living longer in a poor environment—spend money to clean up the air rather than increase ambulance services to reduce deaths from heart attacks.

Comparisons like these may make you uncomfortable. But like it or not, we cannot avoid making choices of this kind. The issue boils down to whether we should improve the quality of life for the living or extend life expectancy regardless of the quality of life.[17]

The degree of risk is an important concept in our legal processes. For example, the U.S. Toxic Substances Control Act states that no one may manufacture a new chemical substance or process a chemical substance for a new use without obtaining clearance from the EPA. The Act establishes procedures for estimating the hazard to the environment and to human health of any new chemical before its use becomes widespread. The EPA examines the data provided and judges the degree of risk associated with all aspects of the production of the new chemical or process, including extraction of raw materials, manufacturing, distribution, processing, use, and disposal. The chemical can be banned or restricted in either manufacturing or use if the evidence suggests that it will pose an unreasonable risk to human health or to the environment.

But what is unreasonable?[18] This question brings us back to Table 7.1 and makes us realize that deciding what is "unreasonable" involves judgments about the quality of life as well as the risk of death. The level of acceptable pollution (and thus risk) is a social-economic-environmental trade-off. Moreover, the level of acceptable risk changes over time in society, depending on changes in scientific knowledge, comparison with risks from other causes, the expense of decreasing the risk, and the social and psychological acceptability of the risk.

When adequate data are available, it is possible to take scientific and technological steps to estimate the level of risk and, from this, to estimate the cost of reducing risk and compare the cost with the benefit. However, what constitutes an acceptable risk is more than a scientific or technical issue. The acceptability of a risk involves ethical and psychological attitudes of individuals and society. We must therefore ask several questions: What risk from a particular pollutant is acceptable? How much is a given reduction in risk from that pollutant worth to us? How much will each of us, as individuals or collectively as a society, be willing to pay for a given reduction in that risk?

The answers depend not only on facts but also on societal and personal values. What must also be factored into the equation is that the costs of cleaning up pollutants and polluted areas and the costs of restoration programs can be minimized, or even eliminated, if a recognized pollutant is controlled initially. The total cost of pollution control need not increase indefinitely.

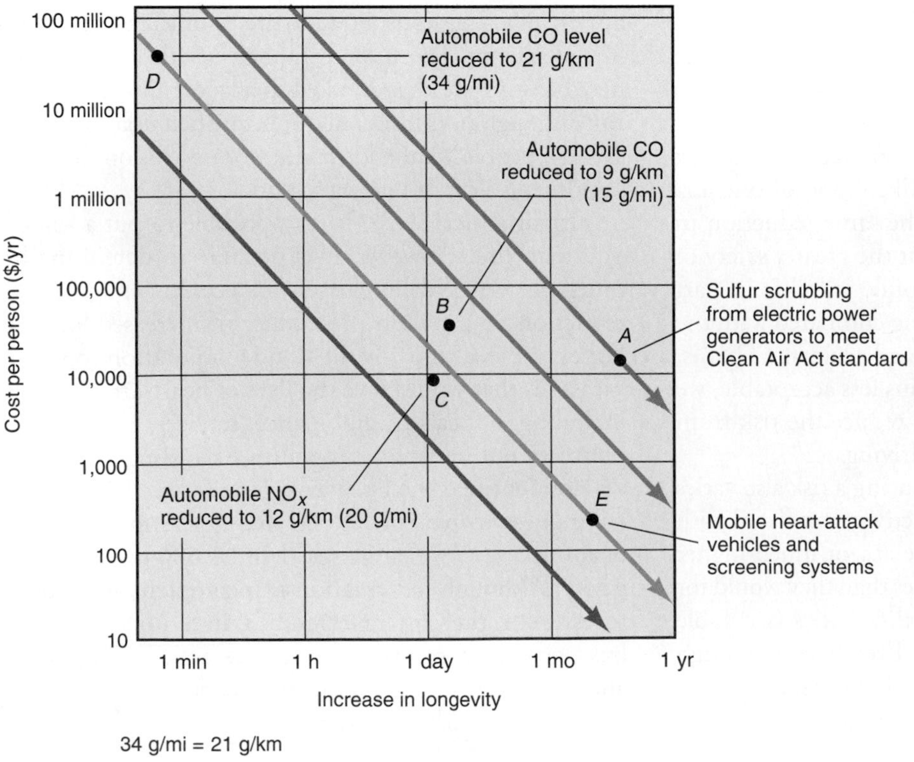

FIGURE 7.10 The cost of extending a life in dollars per year is one way to rank the effectiveness of various efforts to reduce pollutants. This graph shows that reducing sulfur emissions from power plants to the Clean Air Act level **(A)** would extend a human life 1 year at a cost of about $10,000. Similar restrictions applied to automobile emissions **(B, C)** would increase lifetimes by 1 day. More stringent automobile controls would be much more expensive **(D)**; mobile units and screening programs for heart problems would be much cheaper **(E)**. This graph represents only one step in an environmental analysis. (*Source:* Based on R. Wilson, Risk-benefit analysis for toxic chemicals, *Ecotoxicology and Environmental Safety* 4 [1980]: 370–83.)

CRITICAL THINKING ISSUE

Georges Bank: How Can U.S. Fisheries Be Made Sustainable?

The opening case study discussed several ways that economists help make policy. Ocean fishing in Georges Bank—a large, shallow area between Cape Cod, Massachusetts, and Cape Sable Island in Nova Scotia, Canada—illustrates different ways of making a policy work. Both overfishing and pollution have been blamed for the alarming decline in groundfish (cod, haddock, flounder, redfish, pollack, hake) off the northeastern coast of the United States and Canada. Governments' attempts to regulate fishing have generated bitter disputes with fishermen, many of whom contend that restrictions on fishing make them scapegoats for pollution problems. The controversy has become a classic battle between short-term economic interests and long-term environmental concerns.

The oceans outside of national territorial waters are *commons*—open to free use by all—and thus the fish and mammals that live in them are common resources. What is a common resource may change over time, however. The move by many nations to define international waters as beginning 325 kilometers (200 miles) from their coasts has turned some fisheries that used to be completely open common resources into national resources open only to domestic fishermen.

In fisheries, there have been four main management options:

1. Establish total catch quotas for the entire fishery and allow anybody to fish until the total is reached.
2. Issue a restricted number of licenses but allow each licensed fisherman to catch many fish. (This is equivalent to the ***legal emissions limit*** explained in the opening case study.)
3. Tax the catch (the fish brought in) or the effort (the cost of ships, fuel, and other essential items). (This is equivalent to the ***tax on emitters*** in the opening case study.)
4. Allocate fishing rights—that is, assign each fisherman a transferable and salable quota (see the ***cap-and-trade*** option in the opening case study.)

With total-catch quotas, the fishery is closed when the quota is reached. Whales, Pacific halibut, tropical tuna, and anchovies have been regulated in this way. Although regulating the total catch can be done in a way that helps the fish, it tends to increase the number of fishermen and encourage them to buy larger and larger boats. The end result is a hardship for fishermen—huge boats usable for only a brief time each year. When Alaska tried this, all of the halibut were caught in a few days, with the result that restaurants no longer had halibut available for most of the year. This undesirable result led to a change in policy: The total-catch approach was replaced by the sale of licenses.

Issues relating to U.S. fisheries are hardly new. In the early 1970s, fishing was pretty open, but in 1977, in response to concerns about overfishing in U.S. waters by foreign factory ships, the U.S. government extended the nation's coastal waters from 12 to 200 miles (from 19 to 322 km). To encourage domestic fishermen, the National Marine Fisheries Service provided loan guarantees for replacing older vessels and equipment with newer boats carrying high-tech equipment for locating fish. During this same period, demand for fish increased as Americans became more concerned about cholesterol in red meat. Consequently, the number of fishing boats, the number of days at sea, and fishing efficiency increased sharply, and 50–60% of the populations of some species were landed each year.

The international battle over Georges Bank led to a consideration by the International Court of Justice in The Hague. This court's 1984 decision intensified competition. Overfishing continued, and in 1992 Canada was forced to suspend all cod fishing to save the stock from complete annihilation. Later that year, Canada prohibited fishing at certain times and in certain areas on Georges Bank, mandated minimum net sizes, and set quotas on the catch.

These measures were intended to cut the fishing effort in half by 1997. A limited number of fishing permits were issued, limiting the number of days at sea and number of trips for harvesting certain species. High-tech monitoring equipment ensured compliance. Still, things got worse. Recently, portions of Georges Bank were closed indefinitely to fishing for some fish species, including yellowtail, cod, and haddock.

In the spring of 2009, fishermen suggested that the limit on individual fishermen be replaced by a group quota, a variation on Management Option 4 (above). Fishermen would work together in groups called "sectors," and each sector could take a set percentage of the annual catch of one species. This approach is being used elsewhere in U.S. waters. It places fewer restrictions on individual fishermen, such as limiting each one's number of trips or days at sea.

Recent economic analysis suggests that taxes taking into account the cost of externalities (such as water pollution from motorboat oil) can work to the best advantage of fishermen and fish. Allocating a transferable and salable quota to each fisherman produces similar results. However, after decades of trying to find a way to regulate fishing so that Georges Bank becomes a sustainable fishery, nothing has worked well. The fisheries remain in trouble.

Critical Thinking Questions

1. Which of the policy options described above attempt to convert the fishing industry from a commons system to private ownership? How might these measures help prevent overfishing? Is it right to institute private ownership of public resources?

2. Thinking over the choices discussed in this chapter, what policy option do you think has the best chance of sustaining the fisheries on Georges Bank? Explain your answer.
3. What approach to future value (approximately) do each of the following people assume for fish?

   Fisherman: If you don't get it now, someone else will.
   Fisheries manager: By sacrificing now, we can do something to protect fish stocks.
4. Develop a list of the environmental and economic advantages and disadvantages of ITQs. Would you support instituting ITQs in New England? Explain why or why not.
5. Do you think it is possible to reconcile economic and environmental interests in the case of the New England fishing industry? If so, how? If not, why not?

SUMMARY

- Economic analysis can help us understand why environmental resources have been poorly conserved in the past and how we might more effectively achieve conservation in the future.
- Economic analysis is applied to two different kinds of environmental issues: the use of desirable resources (fish in the ocean, oil in the ground, forests on the land) and the minimization of pollution.
- Resources may be common property or privately controlled. The kind of ownership affects the methods available to achieve an environmental goal. There is a tendency to overexploit a common-property resource and to harvest to extinction nonessential resources whose innate growth rate is low, as suggested in Hardin's "tragedy of the commons."
- Future worth compared with present worth can be an important determinant of the level of exploitation.
- The relation between risk and benefit affects our willingness to pay for an environmental good.
- Evaluation of environmental intangibles, such as landscape aesthetics, is becoming more common in environmental analysis. Such evaluation can be used to balance the more traditional economic evaluation and to help separate facts from emotion in complex environmental problems.
- Societal methods to achieve an environmental goal include moral suasion, direct controls, market processes, and government investment. Many kinds of controls have been applied to pollution and the use of desirable resources.

REEXAMINING THEMES AND ISSUES

Human Population

The tragedy of the commons will worsen as human population density increases because more and more individuals will seek personal gain at the expense of community values. For example, more and more individuals will try to make a living from harvesting natural resources. How people can use resources while at the same time conserving them requires an understanding of environmental economics.

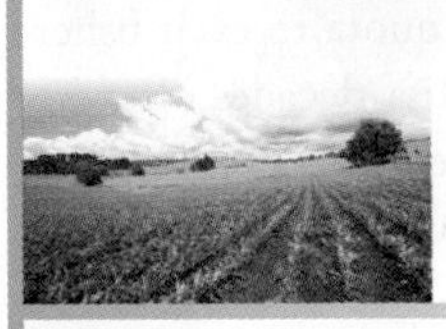

Sustainability

From this chapter, we learn why people sometimes are not interested in sustaining an environmental resource from which they make a living. When the goal is simply to maximize profits, it is sometimes a rational decision to liquidate an environmental resource and put the money gained into a bank or another investment, to avoid such liquidation, we need to understand economic externalities and intangible values.

Global Perspective

Solutions to global environmental issues, such as global warming, require that we understand the different economic interests of developed and developing nations. These can lead to different economic policies and different valuation of global environmental issues.

Urban World

The tragedy of the commons began with grazing rights in small villages. As the world becomes increasingly urbanized, the pressure to use public lands for private economic gain is likely to increase. An understanding of environmental economics can help us find solutions to urban environmental problems.

People and Nature

This chapter brings us to the heart of the matter: How do we value the environment, and when can we attach a monetary value to the benefits and costs of environmental actions? People are intimately involved with nature. While we seek rational methods to put a value on nature, the values we choose often derive from intangible benefits, such as an appreciation of the beauty of nature.

Science and Values

One of the central questions of environmental economics concerns how to develop equivalent economic valuation for tangible and intangible factors. For example, how can we compare the value of timber with the beauty people attach to the scenery, trees intact? How can we compare the value of a dam that provides irrigation water and electrical power on the Columbia River with the scenery without the dam, and the salmon that could inhabit that river?

KEY TERMS

commons **130**
direct costs **134**
environmental economics **129**
externality **133**
indirect cost **133**
intangible factor **129**
natural capital **130**
policy instruments **129**
public-service functions **130**
risk-benefit analysis **136**
tangible factor **129**

STUDY QUESTIONS

1. What is meant by the term *the tragedy of the commons?* Which of the following are the result of this tragedy?
(a) The fate of the California condor
(b) The fate of the gray whale
(c) The high price of walnut wood used in furniture

2. What is meant by risk-benefit analysis?

3. Cherry and walnut are valuable woods used to make fine furniture. Basing your decision on the information in the following table, which would you invest in? (*Hint:* Refer to the discussion of whales in this chapter.)
(a) A cherry plantation
(b) A walnut plantation
(c) A mixed stand of both species
(d) An unmanaged woodland where you see some cherry and walnut growing

Species	*Longevity*	*Maximum Size*	*Maximum Value*
Walnut	400 years	1 m	$15,000/tree
Cherry	100 years	1 m	$10,000/tree

4. Bird flu is spread in part by migrating wild birds. How would you put a value on (a) the continued existence of one species of these wild birds; (b) domestic chickens important for food but also a major source of the disease; (c) control of the disease for human health? What relative value would you place on each (that is, which is most important and which least)? To what extent would an economic analysis enter into your valuation?

5. Which of the following are intangible resources? Which are tangible?
(a) The view of Mount Wilson in California
(b) A road to the top of Mount Wilson
(c) Porpoises in the ocean
(d) Tuna in the ocean
(e) Clean air

6. What kind of future value is implied by the statement "Extinction is forever"? Discuss how we might approach providing an economic analysis for extinction.

7. Which of the following can be thought of as commons in the sense meant by Garrett Hardin? Explain your choice.
   (a) Tuna fisheries in the open ocean
   (b) Catfish in artificial freshwater ponds
   (c) Grizzly bears in Yellowstone National Park
   (d) A view of Central Park in New York City
   (e) Air over Central Park in New York City

FURTHER READING

Daly, H.E., and J. Farley, *Ecological Economics: Principles and Applications* (Washington, DC: Island Press, 2003). Discusses an interdisciplinary approach to the economics of environment.

Goodstein, E.S., *Economics and the Environment*, 3rd ed. (New York: Wiley, 2000).

Hanley, Nick, Jason Shogren, and Ben White, *Environmental Economics in Theory and Practice*, 2nd ed., paperback (New York: Macmillan, 2007).

Hardin, G., Tragedy of the Commons, *Science* 162: 1243–1248, 1968. One of the most cited papers in both science and social science, this classic work outlines the differences between individual interest and the common good.

CHAPTER 8

Biological Diversity and Biological Invasions

Yellow dragon disease—*huanglongbing* in Chinese, "citrus greening" in the United States, represented in China by this iconic symbol for the disease, is a growing threat worldwide to citrus crops. The reason this is a threat has to do with patterns of biodiversity and how people interface with these.

LEARNING OBJECTIVES

Biological diversity has become one of the "hot-button" environmental topics—there is a lot of news about endangered species, loss of biodiversity, and its causes. This chapter provides a basic scientific introduction that will help you understand the background to this news, the causes of and solutions to species loss, and the problems that arise when we move species around the globe. Interest in the variety of life on Earth is not new; people have long wondered how the amazing diversity of living things on Earth came to be. This diversity has developed through biological evolution and is affected by interactions among species and by the environment. After reading this chapter, you should understand . . .

- How biological evolution works—how mutation, natural selection, migration, and genetic drift lead to evolution of new species;
- Why people value biological diversity;
- How people affect biological diversity: by eliminating, reducing, or altering habitats; harvesting; introducing new species where they had not lived before; and polluting the environment;
- When and how biological diversity is important to ecosystems—how it may affect biological production, energy flow, chemical cycling, and other ecosystem processes;
- What major environmental problems are associated with biological diversity;
- Why so many species have been able to evolve and persist;
- The concepts of the ecological niche and habitat;
- The theory of island biogeography;
- How species invade new habitats, and when this can be beneficial and when harmful.

CASE STUDY

Citrus Greening

In 2005 a tiny fruit fly that carries and disperses a bacterial disease of citrus plants arrived in the United States from China (Figure 8.1). This disease, known as "citrus greening" or Chinese *huanglongbing* (yellow dragon disease) had been extending its range and had reached India, many African countries, and Brazil (Figure 8.2). Wherever the fly and the bacteria have gone, citrus crops have failed. The bacteria interfere with the flow of organic compounds in the phloem (the living part of the plant's bark). The larvae of the fruit fly sucks juices from the tree, inadvertently injecting the bacteria. Winds blow the adult flies from one tree to another, making control of the fly difficult. According to the U.S. Department of Agriculture (USDA), citrus greening is the most severe new threat to citrus plants in the United States and might end commercial orange production in Florida. Many Florida counties are under a quarantine that prevents citrus plants from being moved from one area to another.[1]

Introductions of new species into new habitats have occurred as long as life has existed on Earth. And beginning with the earliest human travelers, our ancestors have moved species around the world, sometimes on purpose, sometimes unknowingly. Polynesians brought crops, pigs, and many other animals and plants from one Pacific Island to another as they migrated and settled widely before A.D. 1000. The intentional spread of crop plants around the world has been one of the primary reasons that our species has been able to survive in so many habitats and has grown to such a huge number. But if invasion by species is as old as life, and often beneficial to people, why is it also the source of so many environmental problems? The answers lie in this chapter.

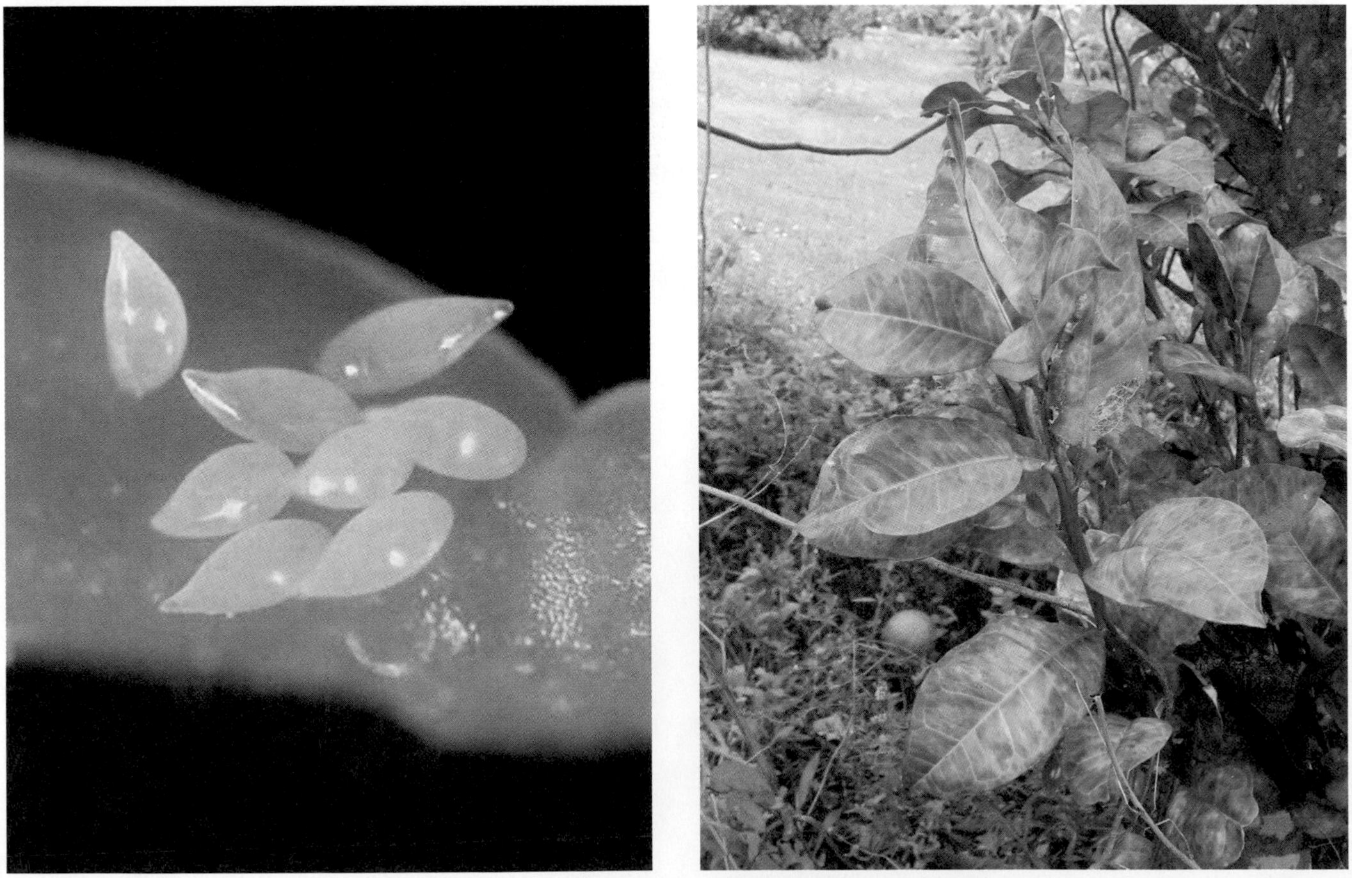
(a) (b)

FIGURE 8.1 **(a) Larvae of *Diaphorina citri*, the tiny fly that spreads citrus greening bacteria (*Candidatus* Liberibacter *asiaticus*); (b) the disease yellows leaves, turns fruit greenish brown, and eventually kills the tree.**

FIGURE 8.2 **Where citrus greening has spread from its origin in China.** The yellow shows the disease locations. (*Source*: USDA.)

8.1 What Is Biological Diversity?

Biological diversity refers to the variety of life-forms, commonly expressed as the number of species or the number of genetic types in an area. (We remind you of the definitions of *population* and *species* in Chapter 4: A **population** is a group of individuals of the same species living in the same area or interbreeding and sharing genetic information. A **species** is all individuals that are capable of interbreeding. A species is made up of populations.)

Conservation of biological diversity gets lots of attention these days. One day we hear about polar bears on the news, the next day something about wolves or salmon or elephants or whales. What should we do to protect these species that mean so much to people? What do we need to do about biological diversity in general—all the life-forms, whether people enjoy them or not? And is this a scientific issue or not? Is it even partially scientific?

That's what this chapter is about. It introduces the scientific concepts concerning biological diversity, explains the aspects of biological diversity that have a scientific base, distinguishes the scientific aspects from the nonscientific ones, and thereby provides a basis for you to evaluate the biodiversity issues you read about.

Why Do People Value Biodiversity?

Before we discuss the scientific basis of biodiversity and the role of science in its conservation, we should consider why people value it. There are nine primary reasons: utilitarian; public-service; ecological; moral; theological; aesthetic; recreational; spiritual; and creative.[2]

Utilitarian means that a species or group of species provides a product that is of direct value to people. *Public-service* means that nature and its diversity provide some service, such as taking up carbon dioxide or pollinating flowers, that is essential or valuable to human life and would be expensive or impossible to do ourselves. *Ecological* refers to the fact that species have roles in their ecosystems, and that some of these are necessary for the persistence of their ecosystems, perhaps even for the persistence of all life. Scientific research tells us which species have such ecosystem roles. The *moral* reason for valuing biodiversity is the belief that species have a right to exist, independent of their value to people. The *theological* reason refers to the fact that some religions value nature and its diversity, and a person who subscribes to that religion supports this belief.

The last four reasons for valuing nature and its diversity—aesthetic, recreational, spiritual, and creative—have to do with the intangible (nonmaterial) ways that nature and its diversity benefit people (see Figure 8.3). These four are often lumped together, but we separate them here. *Aesthetic* refers to the beauty of nature, including the variety of life. *Recreational* is self-explanatory—people enjoy getting out into nature, not just because it is beautiful to look at but because it provides us with healthful activities that we enjoy. *Spiritual* describes the way contact with nature and its diversity often moves people, an uplifting often perceived as a religious experience. *Creative* refers to the fact that artists, writers, and musicians find stimulation for their creativity in nature and its diversity.

Science helps us determine what are utilitarian, public-service, and ecosystem functions of biological diversity, and scientific research can lead to new utilitarian benefits from biological diversity. For example, medical research led to the discovery and development of paclitaxel (trade name Taxol), a chemical found in the Pacific yew and now used widely in chemotherapy treatment of certain cancers. (Ironically, this discovery led at first to the harvest of this endangered tree species, creating an environmental controversy until the compound could be made artificially.)

The rise of the scientific and industrial age brought a great change in the way that people valued nature. Long ago, for example, when travel through mountains was arduous, people struggling to cross them were probably not particularly interested in the scenic vistas. But around the time of the Romantic poets, travel through the Alps became easier, and suddenly poets began to appreciate the "terrible joy" of mountain scenery. Thus scientific knowledge indirectly influences the nonmaterial ways that people value biological diversity.

FIGURE 8.3 People have long loved the diversity of life. Here, a late-15th-century Dutch medieval tapestry, *The Hunting of the Unicorn* (now housed in The Cloisters, part of the New York City's Metropolitan Museum of Art), celebrates the great diversity of life. Except for the mythological unicorn, all the plants and animals shown, including frogs and insects, are familiar to naturalists today and are depicted with great accuracy.

8.2 Biological Diversity Basics

Biological diversity involves the following concepts:

- ***Genetic diversity***: the total number of genetic characteristics of a specific species, subspecies, or group of species. In terms of genetic engineering and our new understanding of DNA, this could mean the total base-pair sequences in DNA; the total number of genes, active or not; or the total number of active genes.
- ***Habitat diversity***: the different kinds of habitats in a given unit area.
- ***Species diversity***, which in turn has three qualities:
  species richness—the total number of species;
  species evenness—the relative abundance of species; and
  species dominance—the most abundant species.

To understand the differences between species richness, species evenness, and species dominance, imagine two ecological communities, each with 10 species and 100 individuals, as illustrated in Figure 8.4. In the first community (Figure 8.4a), 82 individuals belong to a single species, and the remaining nine species are represented by two individuals each. In the second community (Figure 8.4b), all the species are equally abundant; each therefore has 10 individuals. Which community is more diverse?

At first, one might think that the two communities have the same species diversity because they have the same number of species. However, if you walked through both communities, the second would appear more diverse. In the first community, most of the time you would see individuals only of the dominant species (elephants in Figure 8.4a); you probably wouldn't see many of the other species at all.

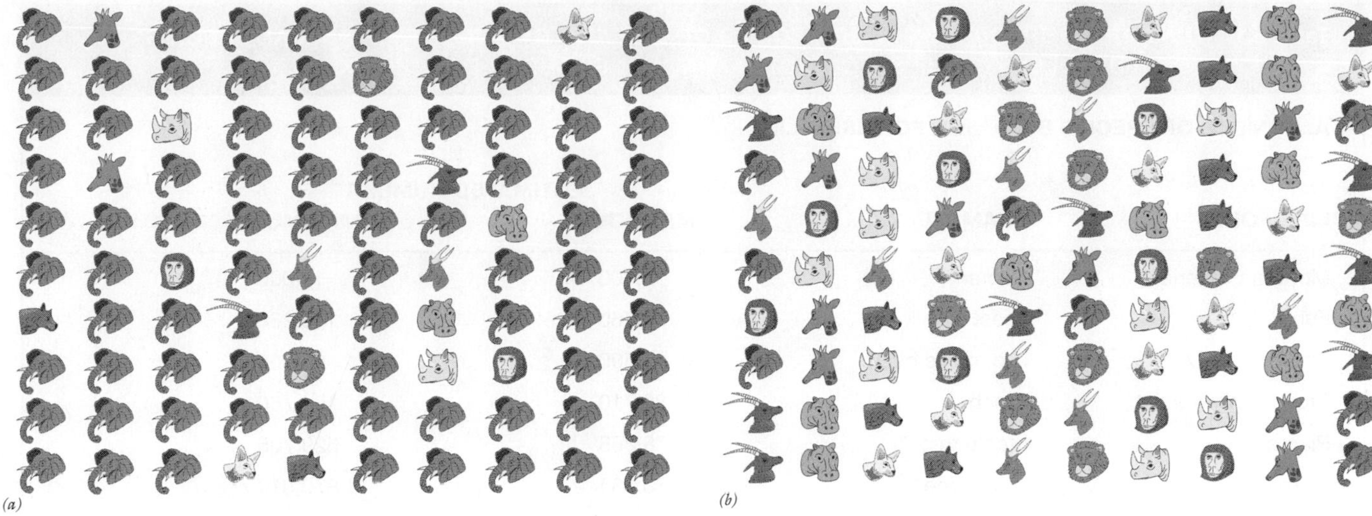

FIGURE 8.4 **Diagram illustrating the difference between species evenness, which is the relative abundance of each species, and species richness, which is the total number of species.** Figures **(a)** and **(b)** have the same number of species but different relative abundances. Lay a ruler across each diagram and count the number of species the edge crosses. Do this several times, and determine how many species are diagram **(a)** and diagram **(b)**. See text for explanation of results.

The first community would appear to have relatively little diversity until it was subjected to careful study, whereas in the second community even a casual visitor would see many of the species in a short time. You can test the probability of encountering a new species in either community by laying a ruler down in any direction on Figures 8.4a and 8.4b and counting the number of species that it touches.

As this example suggests, merely counting the number of species is not enough to describe biological diversity. Species diversity has to do with the relative chance of seeing species as much as it has to do with the actual number present. Ecologists refer to the total number of species in an area as **species richness**, the relative abundance of species as **species evenness**, and the most abundant species as **dominant**.

The Number of Species on Earth

Many species have come and gone on Earth. But how many exist today? Some 1.5 million species have been named, but available estimates suggest there may be almost 3 million (Table 8.1), and some biologists believe the number will turn out to be much, much larger. No one knows the exact number because new species are discovered all the time, especially in little-explored areas such as tropical savannas and rain forests.

For example, in the spring of 2008, an expedition sponsored by Conservation International and led by scientists from Brazilian universities discovered 14 new species in or near Serra Geral do Tocantins Ecological Station, a 716,000-hectare (1.77-million-acre) protected area in the Cerrado, a remote tropical savanna region of Brazil, said to be one of the world's most biodiverse areas. They found eight new fish, three new reptiles, one new amphibian, one new mammal, and one new bird.

In Laos, a new bird, the barefaced bulbul, was discovered in 2009 (Figure 8.5) and five new mammals have been discovered since 1992: (1) the spindle-horned oryx (which is not only a new species but also represents a previously unknown genus); (2) the small black muntjak; (3) the giant muntjak (the muntjak, also known as "barking deer," is a small deer; the giant muntjak is so called because it has large antlers); (4) the striped hare (whose nearest relative lives in Sumatra); and (5) a new species of civet cat. That such a small country with a long history of human occupancy would have so many mammal species previously unknown to science—and some of these were not all that small—suggests how little we still know about the total biological diversity on Earth. But as scientists we must act from what we know, so in this book we will focus on the 1.5 million species identified and named so far (see Table 8.1).

All living organisms are classified into groups called *taxa,* usually on the basis of their evolutionary relationships or similarity of characteristics. (Carl Linnaeus, a Swedish physician and biologist, who lived from 1707 to 1778, was the originator of the classification system and played a crucial role in working all this out. He explained this system in his book *Systema Naturae.*)

The hierarchy of these groups (from largest and most inclusive to smallest and least inclusive) begins with a domain or kingdom. In the recent past, scientists classified life into five kingdoms: animals, plants, fungi, protists, and bacteria. Recent evidence from the

Table 8.1 NUMBER OF SPECIES BY MAJOR FORMS OF LIFE AND BY NUMBER OF ANIMAL SPECIES (FOR A DETAILED LIST OF SPECIES BY TAXONOMIC GROUP, SEE APPENDIX.)

A. NUMBER OF SPECIES BY MAJOR FORMS OF LIFE

		ESTIMATED NUMBER	
LIFE-FORM	EXAMPLE	MINIMUM	MAXIMUM
Monera/Bacteria	Bacteria	4,800	10,000
Fungi	Yeast	71,760	116,260
Lichens	Old man's beard	13,500	13,500
Prostista/Protoctist	Ameba	80,710	194,760
Plantae	Maple tree	478,365	529,705
Animalia	Honeybee	873,084	1,870,019
Total		1,522,219	2,734,244

B. NUMBER OF ANIMAL SPECIES

ANIMALS			
Insecta	Honeybees	668,050	1,060,550
Chondrichthyes	Sharks, rays, etc.	750	850
Osteichthyes	Bony fish	20,000	30,000
Amphibia	Amphibians	200	4,800
Reptilia	Reptiles	5,000	7,000
Aves	Birds	8,600	9,000
Mammalia	Mammals	4,000	5,000
Animal total	Total	873,084	1,870,019

fossil record and studies in molecular biology suggest that it may be more appropriate to describe life as existing in three major domains, one called Eukaryota or Eukarya, which includes animals, plants, fungi, and protists (mostly single-celled organisms); Bacteria; and Archaea.[3] As you learned in Chapter 6, Eukarya cells include a nucleus and other small, organized features called organelles; Bacteria and Archaea do not. (Archaea used to be classified among Bacteria, but they have substantial molecular differences that suggest ancient divergence in heritage—see Chapter 6, Figure 6.7.)

FIGURE 8.5 The barefaced bulbul, discovered in Laos in 2009, shows us once again that there are still species of animals and plants unknown to science.

The plant kingdom is made up of divisions, whereas the animal kingdom is made up of phyla (singular: phylum). A phylum or division is, in turn, made up of classes, which are made up of orders, which are made up of families, which are made up of genera (singular: genus), which are made up of species.

Some argue that the most important thing about biological diversity is the total number of species, and that the primary goal of biological conservation should be to maintain that number at its current known maximum. An interesting and important point to take away from Table 8.1 is that most of the species on Earth are insects (somewhere between 668,000 and more than 1 million) and plants (somewhere between 480,000 and 530,000), and also that there are many species of fungi (about 100,000) and protists (about 80,000 to almost 200,000). In contrast, our own kind, the kind of animals most celebrated on television and in movies, mammals, number a meager 4,000 to 5,000, about the same as reptiles. When it comes to numbers of species on Earth, our kind doesn't seem to matter much—we amount to about half a percent of all animals. If the total number in a species were the only gauge of a species' importance, we wouldn't matter.

8.3 Biological Evolution

The first big question about biological diversity is: How did it all come about? Before modern science, the diversity of life and the adaptations of living things to their environment seemed too amazing to have come about by chance. The great Roman philosopher and writer Cicero put it succinctly: "Who cannot wonder at this harmony of things, at this symphony of nature which seems to will the well-being of the world?" He concluded that "everything in the world is marvelously ordered by divine providence and wisdom for the safety and protection of us all."[4] The only possible explanation seemed to be that this diversity was created by God (or gods).

With the rise of modern science, however, other explanations became possible. In the 19th century, Charles Darwin found an explanation that became known as biological evolution. **Biological evolution** refers to the change in inherited characteristics of a population from generation to generation. It can result in new species—populations that can no longer reproduce with members of the original species but can (and at least occasionally do) reproduce with each other. Along with self-reproduction, biological evolution is one of the features that distinguish life from everything else in the universe. (The others are carbon-based, organic-compound-based, self-replicating systems.)

The word *evolution* in the term *biological evolution* has a special meaning. Outside biology, *evolution* is used broadly to mean the history and development of something. For example, book reviewers talk about the evolution of a novel's plot, meaning how the story unfolds. Geologists talk about the evolution of Earth, which simply means Earth's history and the geologic changes that have occurred over that history. Within biology, however, the term has a more specialized meaning. Biological evolution is a one-way process: Once a species is extinct, it is gone forever. You can run a machine, such as a mechanical grandfather clock, forward and backward, but when a new species evolves, it cannot evolve backward into its parents.

Our understanding of evolution today owes a lot to the modern science of molecular biology and the practice of genetic engineering, which are creating a revolution in how we think about and deal with species. At present, scientists have essentially the complete DNA code for a number of species, including the bacterium *Haemophilus influenzae*; the malaria parasite; its carrier the malaria mosquito;[5] the fruit fly (*Drosophila*); a nematode *C. elegans*, (a very small worm that lives in water); yeast; a small weed plant, thale cress (*Arabidopsis thaliana*); and ourselves—humans. Scientists focused on these species either because they are of great interest to us or because they are relatively easy to study, having either few base pairs (the nematode worm) or having already well-known genetic characteristics (the fruit fly).

According to the theory of biological evolution, new species arise as a result of competition for resources and the differences among individuals in their adaptations to environmental conditions. Since the environment continually changes, which individuals are best adapted changes too. As Darwin wrote, "Can it be doubted, from the struggle each individual has to obtain subsistence, that any minute variation in structure, habits, or instincts, adapting that individual better to the new [environmental] conditions, would tell upon its vigor and health? In the struggle it would have a better chance of surviving; and those of its offspring that inherited the variation, be it ever so slight, would also have a better chance."

Sounds plausible, but how does this evolution occur? Through four processes: mutation, natural selection, migration, and genetic drift.

The Four Key Processes of Biological Evolution

Mutation

Mutations are changes in genes. Contained in the chromosomes within cells, each **gene** carries a single piece of inherited information from one generation to the next, producing a **genotype**, the genetic makeup that is characteristic of an individual or a group.

Genes are made up of a complex chemical compound called deoxyribonucleic acid (DNA). DNA in turn is made up of chemical building blocks that form a code, a kind of alphabet of information. The DNA alphabet consists of four letters that stand for specific nitrogen-containing compounds, called bases, which are combined in pairs: (A) adenine, (C) cytosine, (G) guanine, and (T) thymine. Each gene has a set of the four base pairs, and how these letters are combined in long strands determines the genetic "message" interpreted by a cell to produce specific compounds.

The number of base pairs that make up a strand of DNA varies. To make matters more complex, some base pairs found in DNA are nonfunctional—they are not active and do not determine any chemicals produced by the cell. Furthermore, some genes affect the activity of others, turning those other genes on or off. And creatures such as ourselves have genes that limit the number of times a cell can divide, and thus determine the individual's maximum longevity.

When a cell divides, the DNA is reproduced and each new cell gets a copy. But sometimes an error in reproduction changes the DNA and thereby changes the inherited characteristics. Such errors can arise from various causes. Sometimes an external agent comes in contact with DNA and alters it. Radiation, such as X rays and gamma rays, can break the DNA apart or change its chemical structure. Certain chemicals, also, can change DNA. So can viruses. When DNA changes in any of these ways, it is said to have undergone **mutation**.

In some cases, a cell or an offspring with a mutation cannot survive (Figure 8.6a and b). In other cases, the mutation simply adds variability to the inherited characteristics (Figure 8.6c). But in still other cases, individuals with mutations are so different from their parents that they cannot reproduce with normal offspring of their species, so a new species has been created.

Natural Selection

When there is variation within a species, some individuals may be better suited to the environment than others. (Change is not always for the better. Mutation can result in a new species whether or not that species is better adapted than its parent species to the environment.) Organisms whose biological characteristics make them better able to survive and reproduce in their environment leave more offspring than others. Their descendants form a larger proportion of the next generation and are more "fit" for the environment. This process of increasing the proportion of offspring is called **natural selection**. Which inherited characteristics lead to more offspring depends on the specific characteristics of an environment, and as the environment changes over time, the characteristics' "fit" will also change. In summary, natural selection involves four primary factors:

- Inheritance of traits from one generation to the next and some variation in these traits—that is, genetic variability.
- Environmental variability.
- Differential reproduction (differences in numbers of offspring per individual), which varies with the environment.
- Influence of the environment on survival and reproduction.

Natural selection is illustrated in A Closer Look 8.1, which describes how the mosquitoes that carry malaria develop a resistance to DDT and how the microorganism that causes malaria develops a resistance to quinine, a treatment for the disease.

As explained before, when natural selection takes place over a long time, a number of characteristics can change. The accumulation of these changes may become so great that the present generation can no longer reproduce with individuals that have the original DNA structure, resulting in a new species.

Ironically, the *loss* of geographic isolation can also lead to a new species. This can happen when one population of a species migrates into a habitat already occupied by another population of that species, thereby changing gene frequency in that habitat. Such a change can result, for example, from the migration of seeds of flowering plants blown by wind or carried in the fur of mammals. If the seed lands in a new habitat, the environment may be different enough to favor genotypes not as favored by natural selection in the parents' habitat. Natural selection, in combination with geographic isolation and subsequent migration, can thus lead to new dominant genotypes and eventually to new species.

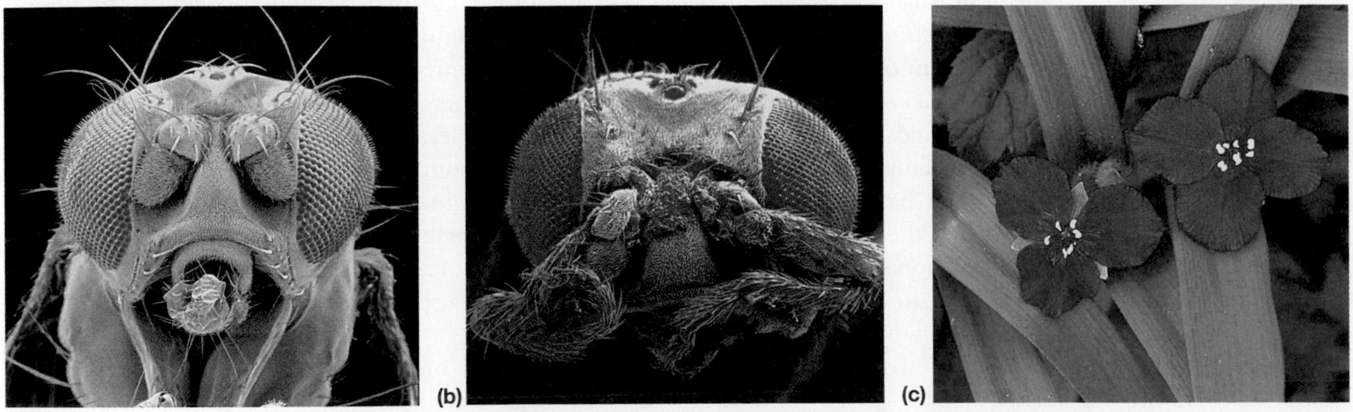

(a) (b) (c)

FIGURE 8.6 (a) A normal fruit fly, (b) a fruit fly with an antennae mutation, and **(c) *Tradescantia*, a small flowering plant used in the study of effects of mutagens.** The color of stamen hairs in the flower (pink versus clear) is the result of a single gene and changes when that gene is mutated by radiation or certain chemicals, such as ethylene chloride.

A CLOSER LOOK 8.1

Natural Selection: Mosquitoes and the Malaria Parasite

Malaria poses a great threat to 2.4 billion people—over one-third of the world's population—living in more than 90 countries, most of them in the tropics. In the United States, in 2003, Palm Beach County, Florida, experienced a small but serious malaria outbreak, and of particular concern is that the malaria was the result of bites from local mosquitoes, not brought in by travelers from nations where malaria is a continual problem. Worldwide, an estimated 300–400 million people are infected each year, and 1.1 million of them die (Figure 8.7).[6] It is the fourth largest cause of death of children in developing nations—in Africa alone, more than 3,000 children die daily from this disease.[7] Once thought to be caused by filth or bad air (hence the name *malaria*, from the Latin for "bad air"), malaria is actually caused by parasitic microbes (four species of the protozoon *Plasmodium*). These microbes affect and are carried by *Anopheles* mosquitoes, which then transfer the protozoa to people. One solution to the malaria problem, then, would be the eradication of *Anopheles* mosquitoes.

By the end of World War II, scientists had discovered that the pesticide DDT was extremely effective against *Anopheles* mosquitoes. They had also found chloroquine highly effective in killing *Plasmodium* parasites. (Chloroquine is an artificial derivative of quinine, a chemical from the bark of the quinine tree that was an early treatment for malaria.) In 1957 the World Health Organization (WHO) began a $6 billion campaign to rid the world of malaria using a combination of DDT and chloroquine.

(a)

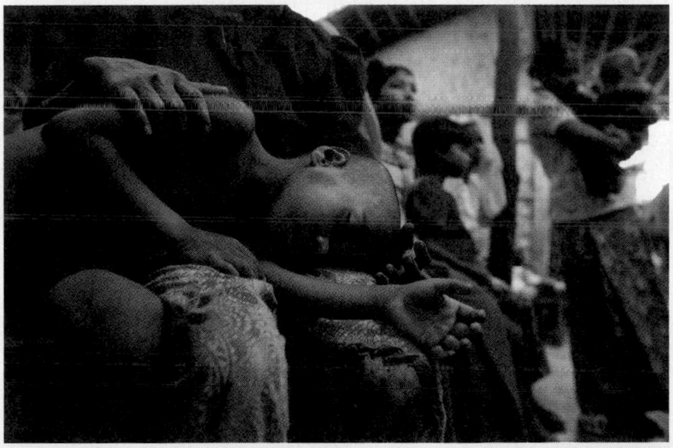

(b)

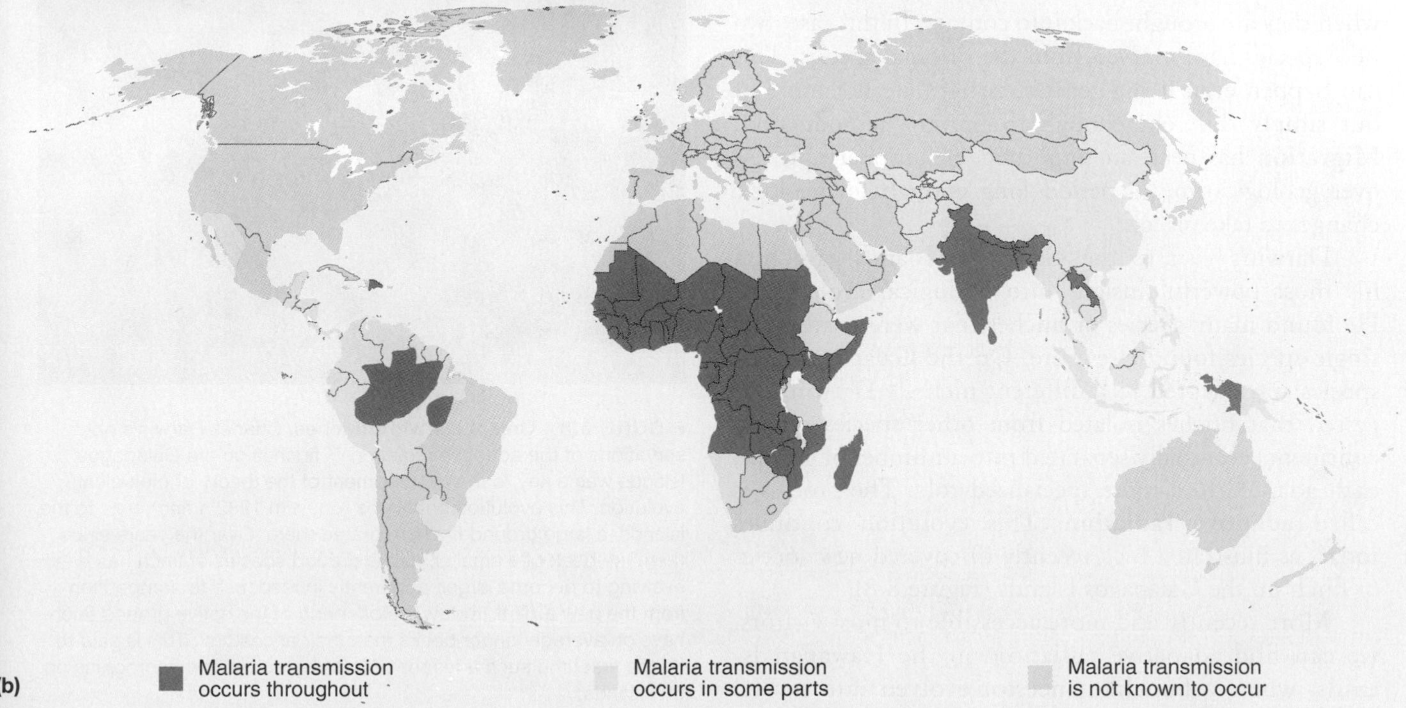

FIGURE 8.7 **(a)** A child with malaria and **(b)** where malaria primarily occurs today. [*Source:* **(b)** U.S. Centers for Disease Control, http://www.cdc.gov/Malaria/distribution_epi/distribution.htm.]

At first, the strategy seemed successful. By the mid-1960s, malaria was nearly gone or had been eliminated from 80% of the target areas. However, success was short-lived. The mosquitoes began to develop a resistance to DDT, and the protozoa became resistant to chloroquine. In many tropical areas, the incidence of malaria worsened. For example, the WHO program had reduced the number of cases in Sri Lanka from 1 million to only 17 by 1963, but by 1975, 600,000 cases had been reported, and the actual number is believed to be four times higher. Worldwide, in 2006 (most recent data available) there were 247 million cases of malaria, resulting in 881,000 deaths. The mosquitoes' resistance to DDT became widespread, and resistance of the protozoa to chloroquine was found in 80% of the 92 countries where malaria was a major killer.[8]

The mosquitoes and the protozoa developed this resistance through natural selection. When they were exposed to DDT and chloroquine, the susceptible individuals died; they left few or no offspring, and any offspring they left were susceptible. The most resistant survived and passed their resistant genes on to their offspring. Thus, a change in the environment—the human introduction of DDT and chloroquine—caused a particular genotype to become dominant in the populations.

A practical lesson from this experience is that if we set out to eliminate a disease-causing species, we must attack it completely at the outset and destroy all the individuals before natural selection leads to resistance. But sometimes this is impossible, in part because of the natural genetic variation in the target species. Since the drug chloroquine is generally ineffective now, new drugs have been developed to treat malaria. However, these second- and third-line drugs will eventually become unsuccessful, too, as a result of the same process of biological evolution by natural selection. This process is speeded up by the ability of the *Plasmodium* to rapidly mutate. In South Africa, for example, the protozoa became resistant to mefloquine immediately after the drug became available as a treatment.

An alternative is to develop a vaccine against the *Plasmodium* protozoa. Biotechnology has made it possible to map the genetic structure of these malaria-causing organisms. Scientists are currently mapping the genetic structure of *P. falciparum*, the most deadly of the malaria protozoa, and expect to finish within several years. With this information, they expect to create a vaccine containing a variety of the species that is benign in human beings but produces an immune reaction.[9] In addition, scientists are mapping the genetic structure of *Anopheles gambiae*, the carrier mosquito. This project could provide insight into genes, which could prevent development of the malaria parasite within the mosquito. In addition, it could identify genes associated with insecticide resistance and provide clues to developing a new pesticide.

Migration and Geographic Isolation

Sometimes two populations of the same species become geographically isolated from each other for a long time. During that time, the two populations may change so much that they can no longer reproduce together even when they are brought back into contact. In this case, two new species have evolved from the original species. This can happen even if the genetic changes are not more fit but simply different enough to prevent reproduction. **Migration** has been an important evolutionary process over geologic time (a period long enough for geologic changes to take place).

Darwin's visit to the Galápagos Islands gave him his most powerful insight into biological evolution.[10] He found many species of finches that were related to a single species found elsewhere. On the Galápagos, each species was adapted to a different niche.[11] Darwin suggested that finches isolated from other species on the continents eventually separated into a number of groups, each adapted to a more specialized role. The process is called **adaptive radiation**. This evolution continues today, as illustrated by a recently discovered new species of finch on the Galápagos Islands (Figure 8.8).

More recently and more accessible to most visitors, we can find adapative radiation on the Hawaiian Islands, where a finchlike ancestor evolved into several

FIGURE 8.8 **One of Darwin's finches.** Charles Darwin's observations of the adaptive radiation of finches on the Galápagos Islands was a key to the development of the theory of biological evolution. This evolution continues today. In 1982 a finch new to the islands, a large ground finch, migrated there. Over the years since then, the beak of a smaller, native ground species of finch has been evolving to become larger, apparently in response to competition from the new arrival. Today, the offspring of the native ground finch have on average longer beaks than their ancestors'. This is said to be the first time such a response has been observed in progress on the islands.

species, including fruit and seed eaters, insect eaters, and nectar eaters, each with a beak adapted for its specific food (Figure 8.9).[12]

Genetic Drift

Genetic drift refers to changes in the frequency of a gene in a population due not to mutation, selection, or migration, but simply to chance. One way this happens is through the **founder effect**. The founder effect occurs when a small number of individuals are isolated from a larger population; they may have much less genetic variation than the original species (and usually do), and the characteristics that the isolated population has will be affected by chance. In the founder effect and genetic drift, individuals may not be better adapted to the environment—in fact, they may be more poorly adapted or neutrally adapted. Genetic drift can occur in any small population and may present conservation problems when it is by chance isolated from the main population.

For example, bighorn sheep live in the mountains of the southwestern deserts of the United States and Mexico. In the summer, these sheep feed high up in the mountains, where it is cooler, wetter, and greener. Before high-density European settlement of the region, the sheep could move freely and sometimes migrated from one mountain to another by descending into the valleys and crossing them in the winter. In this way, large numbers of sheep interbred. With the development of cattle ranches and other human activities, many populations of bighorn sheep could no longer migrate among the mountains by crossing the valleys. These sheep became isolated in very small groups—commonly, a dozen or so—and chance may play a large role in what inherited characteristics remain in the population.

This happened to a population of bighorn sheep on Tiburón Island in Mexico, which was reduced to 20 animals in 1975 but increased greatly to 650 by 1999. Because of the large recovery, this population has been used to repopulate other bighorn sheep habitats in northern Mexico. But a study of the DNA shows that the genetic variability is much less than in other populations in Arizona. Scientists who studied this population suggest that

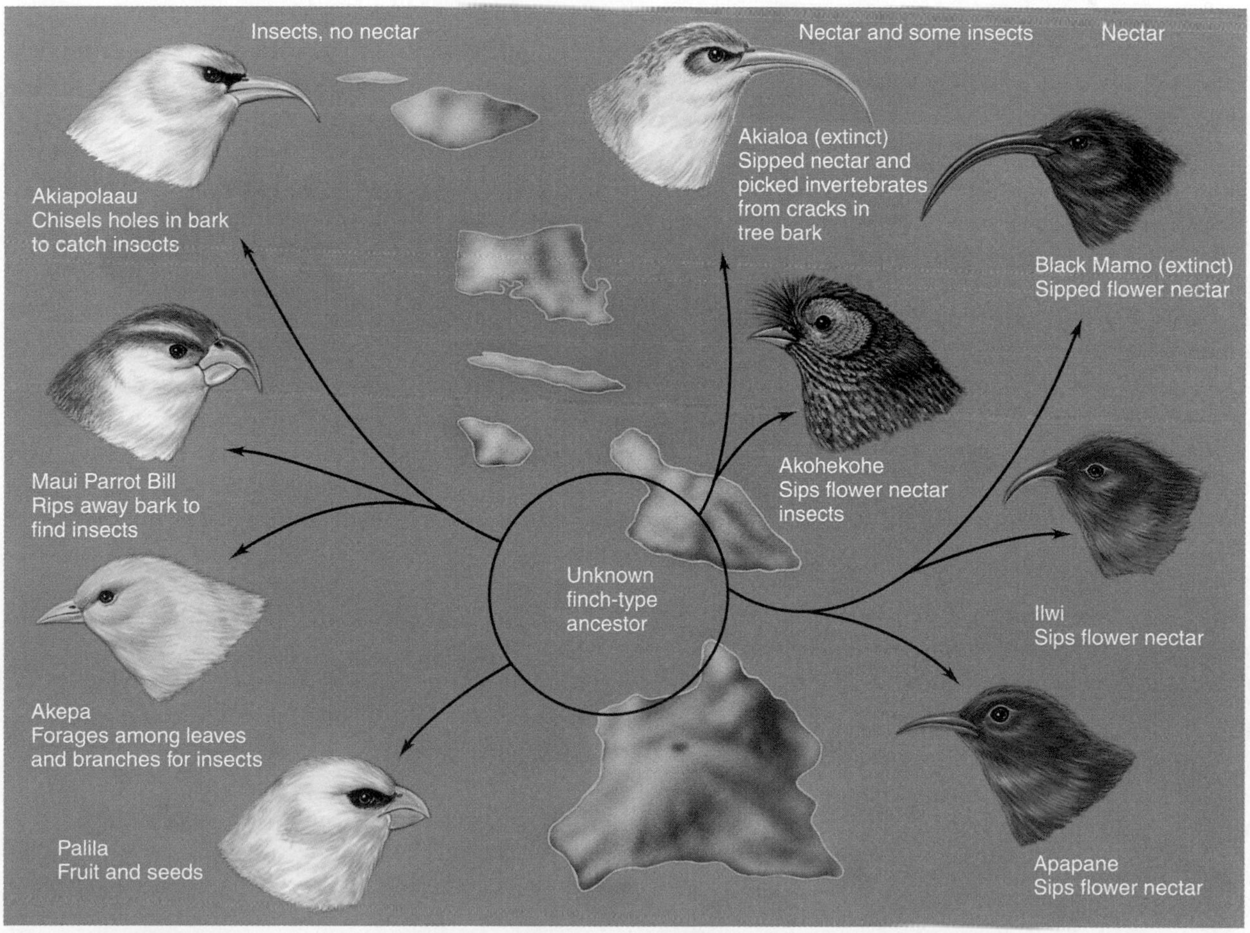

FIGURE 8.9 **Evolutionary divergence among honeycreepers in Hawaii.** Sixteen species of birds, each with a beak specialized for its food, evolved from a single ancestor. Nine of the species are shown here. The species evolved to fit ecological niches that, on the North American continent, had been filled by other species not closely related to the ancestor. (*Source:* From C.B. Cox, I.N. Healey, and P.D. Moore, *Biogeography* [New York: Halsted, 1973].)

individuals from other isolated bighorn sheep populations should be added to any new transplants to help restore some of the greater genetic variation of the past.[13]

Biological Evolution as a Strange Kind of Game

Biological evolution is so different from other processes that it is worthwhile to spend some extra time exploring the topic. There are no simple rules that species must follow to win or even just to stay in the game of life. Sometimes when we try to manage species, we assume that evolution will follow simple rules. But species play tricks on us; they adapt or fail to adapt over time in ways that we did not anticipate. Such unexpected outcomes result from our failure to fully understand how species have evolved in relation to their ecological situations. Nevertheless, we continue to hope and plan as if life and its environment will follow simple rules. This is true even for the most recent work in genetic engineering.

Complexity is a feature of evolution. Species have evolved many intricate and amazing adaptations that have allowed them to persist. It is essential to realize that these adaptations have evolved not in isolation but in the context of relationships to other organisms and to the environment. The environment sets up a situation within which evolution, by natural selection, takes place. The great ecologist G.E. Hutchinson referred to this interaction in the title of one of his books, *The Ecological Theater and the Evolutionary Play*. Here, the ecological situation—the condition of the environment and other species—is the theater and the scenery within which natural selection occurs, and natural selection results in a story of evolution played out in that theater—over the history of life on Earth.[14] These features of evolution are another reason that life and ecosystems are not simple, linear, steady-state systems (see Chapter 3).

In summary, the theory of biological evolution tells us the following about biodiversity:

- Since species have evolved and do evolve, and since some species are also always becoming extinct, biological diversity is always changing, and which species are present in any one location can change over time.
- Adaptation has no rigid rules; species adapt in response to environmental conditions, and complexity is a part of nature. We cannot expect threats to one species to necessarily be threats to another.
- Species and populations do become geographically isolated from time to time, and undergo the founder effect and genetic drift.
- Species are always evolving and adapting to environmental change. One way they get into trouble—become endangered—is when they do not evolve fast enough to keep up with the environment.

8.4 Competition and Ecological Niches

Why there are so many species on Earth has become a key question since the rise of modern ecological and evolutionary sciences. In the next sections we discuss the answers. They partly have to do with how species interact. Speaking most generally, they interact in three ways: competition, in which the outcome is negative for both; symbiosis, in which the interaction benefits both participants; and predation–parasitism, in which the outcome benefits one and is detrimental to the other.

The Competitive Exclusion Principle

The **competitive exclusion principle** supports those who argue that there should be only a few species. It states that *two species with exactly the same requirements cannot coexist in exactly the same habitat.* Garrett Hardin expressed the idea most succinctly: "Complete competitors cannot coexist."[15]

This is illustrated by the introduction of the American gray squirrel into Great Britain. It was introduced intentionally because some people thought it was attractive and would be a pleasant addition to the landscape. About a dozen attempts were made, the first perhaps as early as 1830 (Figure 8.10). By the 1920s, the American gray squirrel was well established in Great Britain, and in the 1940s and 1950s its numbers expanded greatly. It competes with the native red squirrel and is winning—there are now about 2.5 million gray squirrels in Great Britain, and only 140,000 red squirrels, most them in Scotland, where the gray squirrel is less abundant.[16] The two species have almost exactly the same habitat requirements.

One reason for the shift in the balance of these species may be that in the winter the main source of food for red squirrels is hazelnuts, while gray squirrels prefer acorns. Thus, red squirrels have a competitive advantage in areas with hazelnuts, and gray squirrels have the advantage in oak forests. When gray squirrels were introduced, oaks were the dominant mature trees in Great Britain; about 40% of the trees planted were oaks. But that is not the case today. This difference in food preference may allow the coexistence of the two, or perhaps not.

The competitive exclusion principle suggests that there should be very few species. We know from our discussions of ecosystems (Chapter 5) that food webs have at least four levels—producers, herbivores, carnivores, and decomposers. Suppose we allowed for several more levels of carnivores, so that the average food web had six levels. Since there are about 20 major kinds of ecosystems, one would guess that the total number of winners on Earth would be only 6×20, or 120 species.

(a)

(b)

FIGURE 8.10 **(a) British red squirrel,** which is being outcompeted by the **(b) American gray squirrel** introduced into Great Britain.

Being a little more realistic, we could take into account adaptations to major differences in climate and other environmental aspects within kinds of ecosystems. Perhaps we could specify 100 environmental categories: cold and dry; cold and wet; warm and dry; warm and wet; and so forth. Even so, we would expect that within each environmental category, competitive exclusion would result in the survival of only a few species. Allowing six species per major environmental category would result in only 600 species.

That just isn't the case. How did so many different species survive, and how do so many coexist? Part of the answer lies in the different ways in which organisms interact, and part of the answer lies with the idea of the ecological niche.

Niches: How Species Coexist

The **ecological niche** concept explains how so many species can coexist, and this concept is introduced most easily by experiments done with a small, common insect—the flour beetle (*Tribolium*), which, as its name suggests, lives on wheat flour. Flour beetles make good experimental subjects because they require only small containers of wheat flour to live and are easy to grow (in fact, too easy; if you don't store your flour at home properly, you will find these little beetles happily eating in it).

The flour beetle experiments work like this: A specified number of beetles of two species are placed in small containers of flour—each container with the same number of beetles of each species. The containers are then maintained at various temperature and moisture levels—some are cool and wet, others warm and dry. Periodically, the beetles in each container are counted. This is very easy. The experimenter just puts the flour through a sieve that lets the flour through but not the beetles. Then the experimenter counts the number of beetles of each species and puts the beetles back in their container to eat, grow, and reproduce for another interval. Eventually, one species always wins—some of its individuals continue to live in the container while the other species goes extinct. So far, it would seem that there should be only one species of *Tribolium*. But which species survives depends on temperature and moisture. One species does better when it is cold and wet, the other when it is warm and dry (Figure 8.11).

Curiously, when conditions are in between, sometimes one species wins and sometimes the other, seemingly randomly; but invariably one persists while the second becomes extinct. So the competitive exclusion principle holds for these beetles. Both species can survive in a complex environment—one that has cold and wet habitats as well as warm and dry habitats. In no location, however, do the species coexist.

The little beetles provide us with the key to the coexistence of many species. Species that require the same resources can coexist by using those resources under different environmental conditions. So it is habitat *complexity* that allows complete competitors—and not-so-complete competitors—to coexist because they avoid competing with each other.[17]

The flour beetles are said to have the same ecologically functional *niche,* which means they have the same *profession*—eating flour. But they have different *habitats.* Where a species lives is its habitat, but what it does for a living (its profession) is its ecological niche.[18] Suppose you have a neighbor who drives a school bus. Where your neighbor lives and works—your town—is his habitat. What your neighbor does—drive a bus—is his niche. Similarly, if someone says, "Here comes a wolf," you think not only of a creature that inhabits the northern forests (its habitat) but also of a predator that feeds on large mammals (its niche).

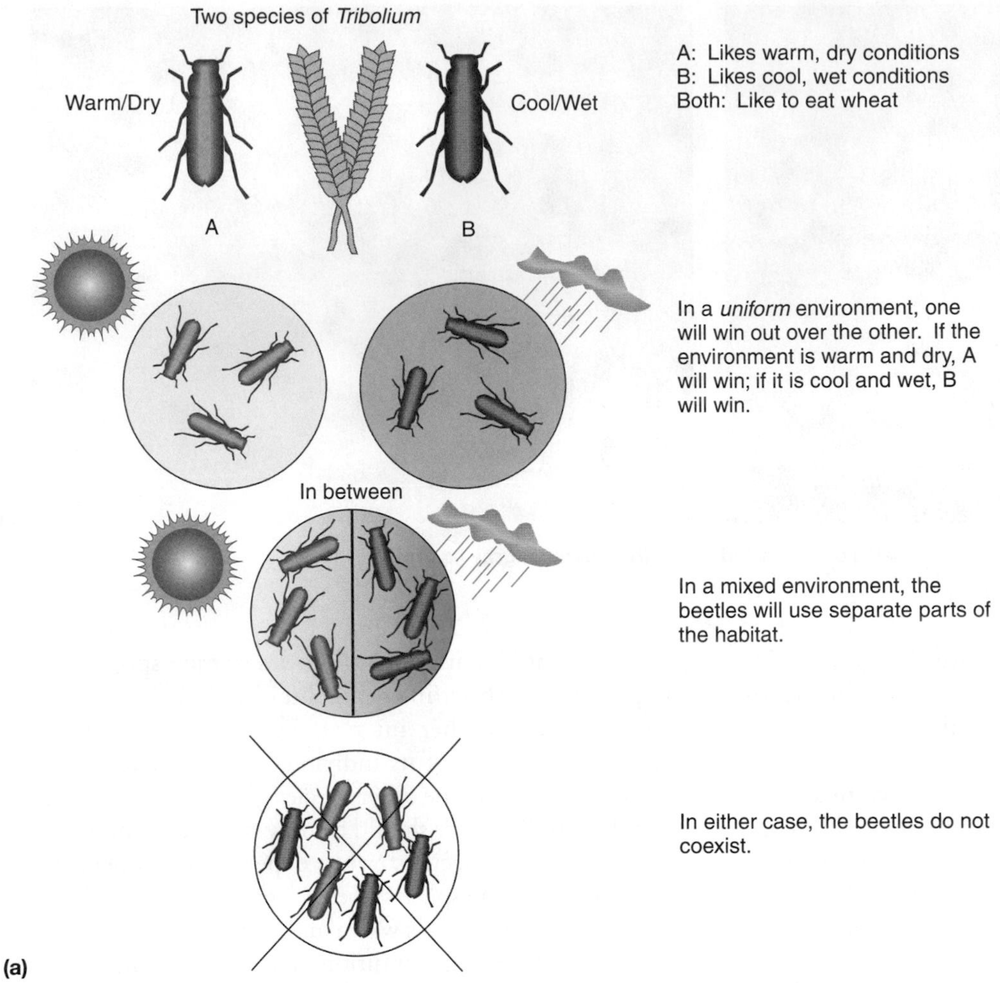
Two species of Tribolium
Warm/Dry
Cool/Wet
A
B
A: Likes warm, dry conditions
B: Likes cool, wet conditions
Both: Like to eat wheat
In a uniform environment, one will win out over the other. If the environment is warm and dry, A will win; if it is cool and wet, B will win.
In between
In a mixed environment, the beetles will use separate parts of the habitat.
In either case, the beetles do not coexist.
(a)

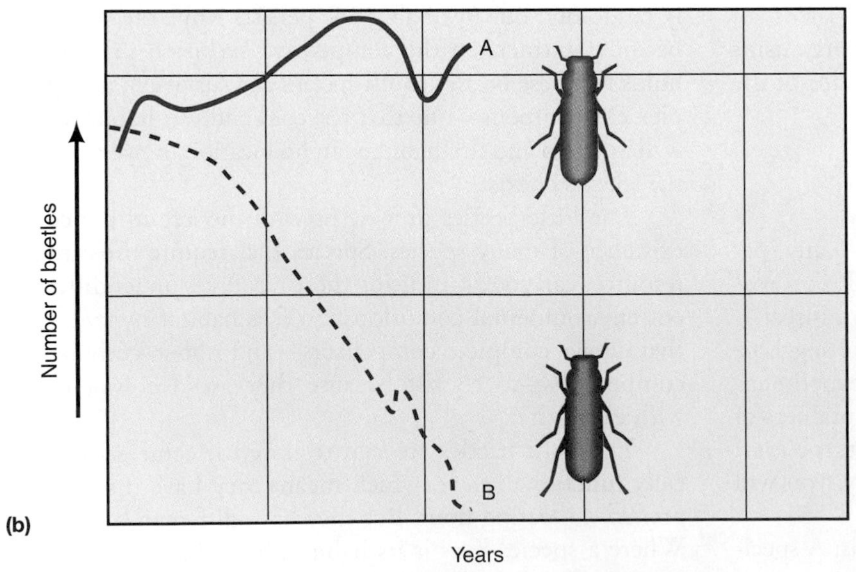
A
B
Number of beetles
Years
(b)

FIGURE 8.11 A classical experiment with flour beetles. Two species of flour beetles are placed in small containers of flour. Each container is kept at a specified temperature and humidity. Periodically, the flour is sifted, and the beetles are counted and then returned to their containers. Which species persists is observed and recorded. **(a)** The general process illustrating competitive exclusion in these species; **(b)** results of a specific, typical experiment under warm, dry conditions.

Understanding the niche of a species is useful in assessing the impact of land development or changes in land use. Will the change remove an essential requirement for some species' niche? A new highway that makes car travel easier might eliminate your neighbor's bus route (an essential part of his habitat) and thereby eliminate his profession (or niche). Other things could also eliminate this niche. Suppose a new school were built and all the children could now walk to school. A school bus driver would not be needed; this niche would no longer exist in your town. In the same way, cutting a forest may drive away prey and eliminate the wolf's niche.

Measuring Niches

An ecological niche is often described and measured as the set of all environmental conditions under which a species can persist and carry out its life functions.[19] It is illustrated by the distribution of two species of flatworm that live on the bottom of freshwater streams. A study of two species of these small worms in Great Britain found that some streams contained one species, some the other, and still others both.[17]

The stream waters are cold at their source in the mountains and become progressively warmer as they flow downstream. Each species of flatworm occurs within a specific range of water temperatures. In streams where species A occurs alone, it is found from 6° to 17°C (42.8°–62.6°F) (Figure 8.12a). Where species B occurs alone, it is found from 6° to 23°C (42.8°–73.4°F) (Figure 8.12b). When they occur in the same stream, their temperature ranges are much narrower. Species A lives in the upstream sections, where the temperature ranges from 6° to 14°C (42.8°–57.2°F), and species B lives in the warmer downstream areas, where temperatures range from 14° to 23°C (57.2°–73.4°F) (Figure 8.12c).

The temperature range in which species A occurs when it has no competition from B is called its *fundamental temperature niche*. The set of conditions under which it persists in the presence of B is called its *realized temperature niche*. The flatworms show that species divide up their habitat so that they use resources from different parts of it. Of course, temperature is only one aspect of the environment. Flatworms also have requirements relating to the acidity of the water and other factors. We could create graphs for each of these factors, showing the range within which A and B occurred. The collection of all those graphs would constitute the complete Hutchinsonian description of the niche of a species.

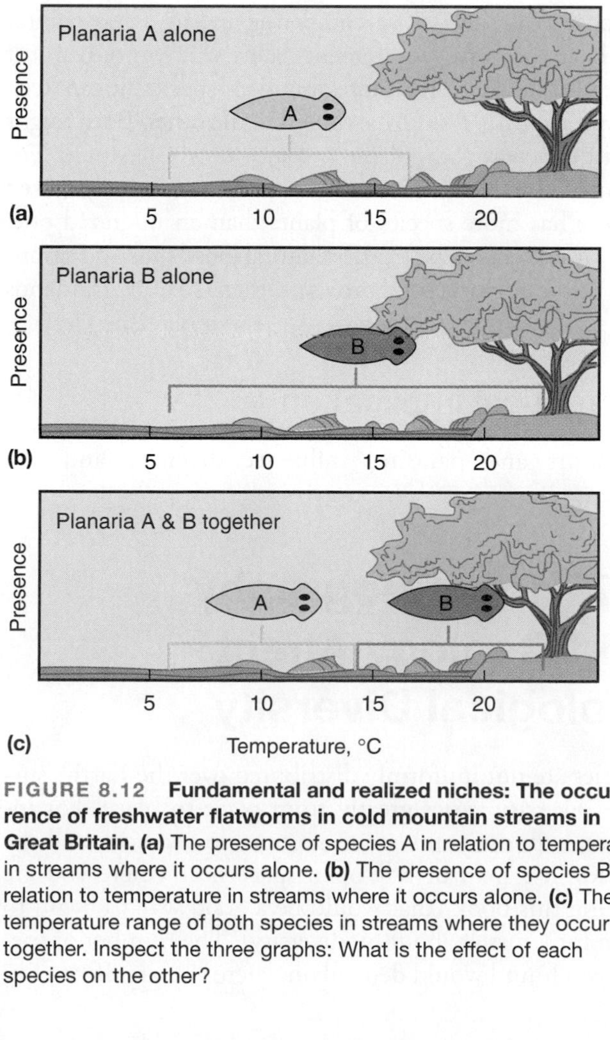

FIGURE 8.12 Fundamental and realized niches: The occurrence of freshwater flatworms in cold mountain streams in Great Britain. (a) The presence of species A in relation to temperature in streams where it occurs alone. **(b)** The presence of species B in relation to temperature in streams where it occurs alone. **(c)** The temperature range of both species in streams where they occur together. Inspect the three graphs: What is the effect of each species on the other?

A Practical Implication

From the discussion of the competitive exclusion principle and the ecological niche, we learn something important about the conservation of species: If we want to conserve a species in its native habitat, we must make sure that all the requirements of its niche are present. Conservation of endangered species is more than a matter of putting many individuals of that species into an area. All the life requirements for that species must also be present—we have to conserve not only a population but also its habitat and its niche.

8.5 Symbiosis

Our discussion up to this point might leave the impression that species interact mainly through competition—by interfering with one another. But symbiosis is also important. This term is derived from a Greek word meaning "living together." In ecology, **symbiosis** describes a relationship between two organisms that is beneficial to both and enhances each organism's chances of persisting. Each partner in symbiosis is called a **symbiont**.

Symbiosis is widespread and common; most animals and plants have symbiotic relationships with other species. We, too, have symbionts—microbiologists tell us that about 10% of our body weight is actually the weight of symbiotic microorganisms that live in our intestines. They help our digestion, and we provide a habitat that supplies all their needs; both we and they benefit. We become aware of this intestinal community when it changes—for example, when we take antibiotics that kill some of these organisms, changing the balance of that community, or when we travel to a foreign country and ingest new strains of bacteria. Then we suffer a well-known traveler's malady, gastrointestinal upset.

Another important kind of symbiotic interaction occurs between certain mammals and bacteria. A reindeer on the northern tundra may appear to be alone but carries with it many companions. Like domestic cattle, the reindeer is a ruminant, with a four-chambered stomach (Figure 8.13) teeming with microbes (a billion per cubic centimeter). In this partially closed environment, the respiration of microorganisms uses up the oxygen ingested by the reindeer while eating. Other microorganisms digest cellulose, take nitrogen from the air in the stomach, and make proteins. The bacterial species that digest the parts of the vegetation that the reindeer cannot digest itself (in particular, the cellulose and lignins of cell walls in woody tissue) require a peculiar environment: They can survive only in an environment without oxygen. One of the few places on Earth's surface where such an environment exists is the inside of a ruminant's stomach.[20] The bacteria and the reindeer are symbionts,

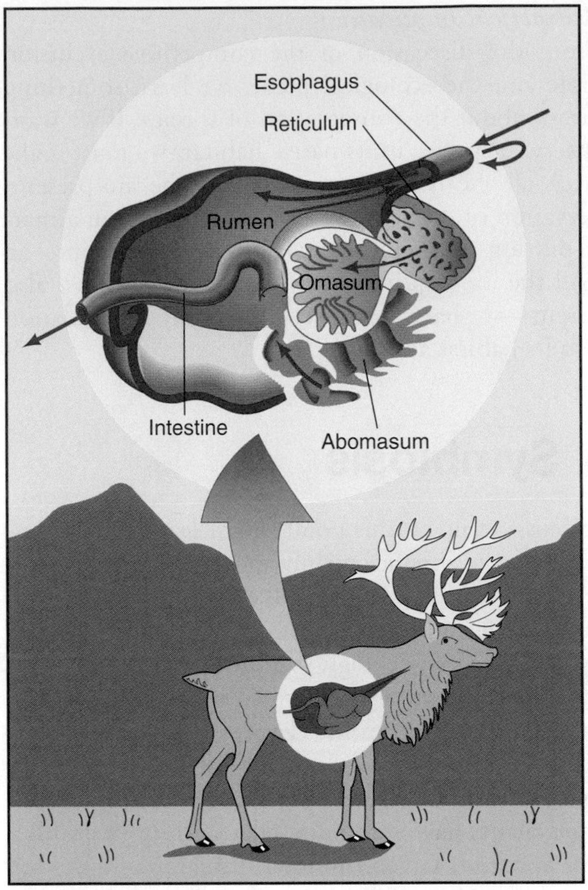

FIGURE 8.13 The stomach of a reindeer illustrates complex symbiotic relationships. For example, in the rumen, bacteria digest woody tissue the reindeer could not otherwise digest. The result is food for the reindeer and food and a home for the bacteria, which could not survive in the local environment outside.

each providing what the other needs, and neither could survive without the other. They are therefore called **obligate symbionts**.

Crop plants illustrate another kind of symbiosis. Plants depend on animals to spread their seeds and have evolved symbiotic relationships with them. That's why fruits are so eatable; it's a way for plants to get their seeds spread, as Henry David Thoreau discussed in his book *Faith in a Seed*.

A Broader View of Symbiosis

So far we have discussed symbiosis in terms of physiological relationships between organisms of different species. But symbiosis is much broader, and includes social and behavioral relationships that benefit both populations. Consider, for example, dogs and wolves. Wolves avoid human beings and have long been feared and disliked by many peoples, but dogs have done very well because of the behavioral connection with people. Being friendly, helpful, and companionable to people has made dogs very abundant. This is another kind of symbiosis.

A Practical Implication

We can see that symbiosis promotes biological diversity, and that if we want to save a species from extinction, we must save not only its habitat and niche but also its symbionts. This suggests another important point that will become more and more evident in later chapters: *The attempt to save a single species almost invariably leads us to conserve a group of species, not just a single species or a particular physical habitat.*

8.6 Predation and Parasitism

Predation–parasitism is the third way in which species interact. *In ecology, a predator–parasite relation is one that benefits one individual (the predator or parasite) and is negative for the other (the prey or host).* **Predation** is when an organism (a predator) feeds on other live organisms (prey), usually of another species. **Parasitism** is when one organism (the parasite) lives on or within another (the host) and depends on it for existence but makes no useful contribution to it and may in fact harm it.

Predation can increase the diversity of prey species. Think again about the competitive exclusion principle. Suppose two species are competing in the same habitat and have the same requirements. One will win out. But if a predator feeds on the more abundant species, it can keep that prey species from overwhelming the other. Both might persist, whereas without the predator only one would. For example, some studies have shown that a moderately grazed pasture has more species of plants than an ungrazed one. The same seems to be true for natural grasslands and savannas. Without grazers and browsers, then, African grasslands and savannas might have fewer species of plants.

A Practical Implication

Predators and parasites influence diversity and can increase it.

8.7 How Geography and Geology Affect Biological Diversity

Species are not uniformly distributed over the Earth's surface; diversity varies greatly from place to place. For instance, suppose you were to go outside and count all the species in a field or any open space near where you are reading this book (that would be a good way to begin to learn for yourself about biodiversity). The number of species you found would depend on where you are. If you live in northern Alaska or Canada, Scandinavia, or Siberia, you would probably find a significantly smaller number of species than if you live in the tropical areas of Brazil, Indonesia, or central Africa. Variation in diversity is partially a question of latitude—in general, greater diversity occurs at lower latitudes. Diversity also varies within local areas. If you count species in the relatively sparse environment of an abandoned city lot, for example, you will find quite a different number than if you count species in an old, long-undisturbed forest.

The species and ecosystems that occur on the land change with soil type and topography: slope, aspect (the direction the slope faces), elevation, and nearness to a drainage basin. These factors influence the number and kinds of plants, and the kinds of plants in turn influence the number and kinds of animals.

Such a change in species can be seen with changes in elevation in mountainous areas like the Grand Canyon and the nearby San Francisco Mountains of Arizona (Figure 8.14). Although such patterns are easiest to see in vegetation, they occur for all organisms.

Some habitats harbor few species because they are stressful to life, as a comparison of vegetation in two areas of Africa illustrates. In eastern and southern Africa, well-drained, sandy soils support diverse vegetation, including many species of *Acacia* and *Combretum* trees, as well as many grasses. In contrast, woodlands on the very heavy clay soils of wet areas near rivers, such as the Sengwa River in Zimbabwe, consist almost exclusively of a single species called *Mopane*. Very heavy clay soils store water and prevent most oxygen from reaching roots. As a result, only tree species with very shallow roots survive.

Moderate environmental disturbance can also increase diversity. For example, fire is a common disturbance in many forests and grasslands. Occasional light fires produce a mosaic of recently burned and unburned areas. These patches favor different kinds of species and increase overall diversity. Table 8.2 shows some of the major influences on biodiversity. Of course, people also affect diversity. In general, urbanization, industrialization, and agriculture decrease diversity, reducing the number of habitats and simplifying habitats. (See, for example, the effects of agriculture on habitats, discussed in Chapter 11.) In addition, we intentionally favor specific species and manipulate populations for our own purposes—for example, when a person plants a lawn or when a farmer plants a single crop over a large area.

Most people don't think of cities as having any beneficial effects on biological diversity. Indeed, the development of cities tends to reduce biological diversity. This is partly because cities have typically been located at good sites for travel, such as along rivers or near oceans, where biological diversity is often high. However, in recent years we have begun to realize that cities can contribute in important ways to the conservation of biological diversity.

Wallace's Realms: Biotic Provinces

As we noted, biological diversity differs among continents, in terms of both total species diversity and the particular species that occur. This large-scale difference has long fascinated naturalists and travelers, many of whom have discovered strange, new (for them) animals and plants as they have traveled between continents. In 1876 the great

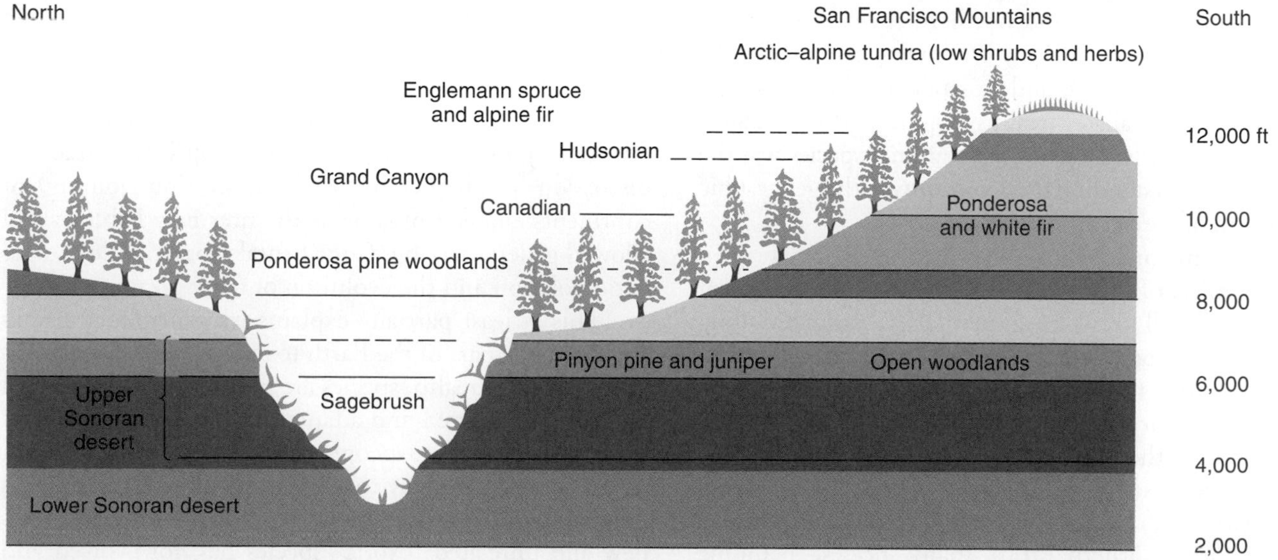

FIGURE 8.14 Change in the relative abundance of a species over an area or a distance is referred to as an *ecological gradient*. Such a change can be seen with changes in elevation in mountainous areas. The altitudinal zones of vegetation in the Grand Canyon of Arizona and the nearby San Francisco Mountains are shown. (*Source:* From C.B. Hunt, *Natural Regions of the United States and Canada* [San Francisco: W.H. Freeman, 1974].)

Table 8.2 SOME MAJOR FACTORS THAT INCREASE AND DECREASE BIOLOGICAL DIVERSITY

A. FACTORS THAT TEND TO INCREASE DIVERSITY
1. A physically diverse habitat
2. Moderate amounts of disturbance (such as fire or storm in a forest or a sudden flow of water from a storm into a pond).
3. A small variation in environmental conditions (temperature, precipitation, nutrient supply, etc.).
4. High diversity at one trophic level increases the diversity at another trophic level. (Many kinds of trees provide habitats for many kinds of birds and insects.)
5. An environment highly modified by life (e.g., a rich organic soil).
6. Middle stages of succession.
7. Evolution.
B. FACTORS THAT TEND TO DECREASE DIVERSITY
1. Environmental stress.
2. Extreme environments (conditions near the limit of what living things can withstand).
3. A severe limitation in the supply of an essential resource.
4. Extreme amounts of disturbance.
5. Recent introduction of exotic species (species from other areas).
6. Geographic isolation (being on a real or ecological island).

British biologist Alfred Russel Wallace (co-discoverer of the theory of biological evolution with Charles Darwin) suggested that the world could be divided into six biogeographic regions on the basis of fundamental features of the animals found in those areas.[21] He referred to these regions as realms and named them Nearctic (North America), Neotropical (Central and South America), Palaearctic (Europe, northern Asia, and northern Africa), Ethiopian (central and southern Africa), Oriental (the Indian subcontinent and Malaysia), and Australian. These have become known as Wallace's realms (Figure 8.15). Recognition of these worldwide patterns in animal species was the first step in understanding **biogeography**—the geographic distribution of species.

In each major biogeographic area (Wallace's realm), certain families of animals are dominant, and animals of these families fill the ecological niches. Animals filling a particular ecological niche in one realm are of different genetic stock from those filling the same niche in the other realms. For example, bison and pronghorn antelope are among the large mammalian herbivores in North America. Rodents such as the capybara fill the same niches in South America, and kangaroos fill them in Australia. In central and southern Africa, many species, including giraffes and antelopes, fill these niches.

This is the basic concept of Wallace's realms, and it is still considered valid and has been extended to all life-forms,[22] including plants (Figure 8.15b)[23] and invertebrates. These realms are now referred to as "biotic provinces."[24] A **biotic province** is a region inhabited by a characteristic set of taxa (species, families, orders), bounded by barriers that prevent the spread of those distinctive kinds of life to other regions and the immigration of foreign species.[10] So in a biotic province, organisms share a common genetic heritage but may live in a variety of environments as long as they are genetically isolated from other regions.

Biotic provinces came about because of continental drift, which is caused by plate tectonics and has periodically joined and separated the continents (see the discussion in Chapter 6).[25] The unification (joining) of continents enabled organisms to enter new habitats and allowed genetic mixing. Continental separation led to genetic isolation and the evolution of new species.

This at least partially explains why introducing species from one part of the Earth to another can cause problems. Within a realm, species are more likely to be related and to have evolved and adapted in the same place for a long time. But when people bring home a species from far away, they are likely to be introducing a species that is unrelated, or only distantly related, to native species. This new and unrelated "exotic" species has not evolved and adapted in the presence of the home species, so ecological and evolutionary adjustments are yet to take place. Sometimes an introduction brings in a superior competitor.

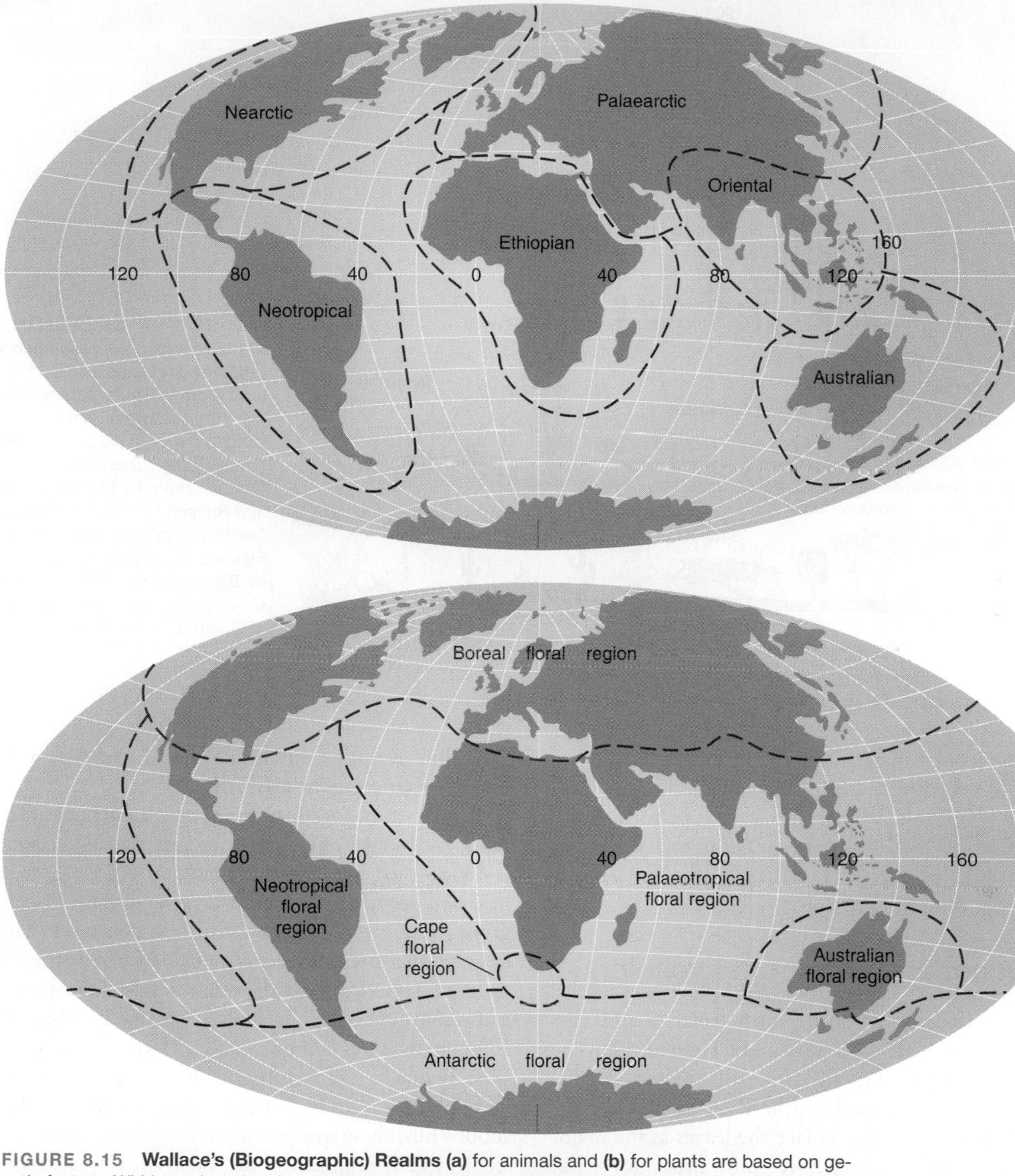

FIGURE 8.15 Wallace's (Biogeographic) Realms (a) for animals and **(b)** for plants are based on genetic factors. Within each realm, the vertebrates are in general more closely related to each other than to vertebrates filling similar niches in other realms; similarly, plants within a realm are more closely related to each other than to plants of other realms.

Biomes

A **biome** is a kind of ecosystem, such as a desert, a tropical rain forest, or a grassland. The same biome can occur on different continents because similar environments provide similar opportunities for life and similar constraints. As a result, similar environments lead to the evolution of organisms similar in form and function (but not necessarily in genetic heritage or internal makeup) and similar ecosystems. This is known as *the rule of climatic similarity*. The close relationship between environment and kinds of life-forms is shown in Figure 8.16.

In sum, the difference between a biome and a biotic province is that a biotic province is based on who is related to whom, while a biome is based on niches and habitats. In general, species within a biotic province are

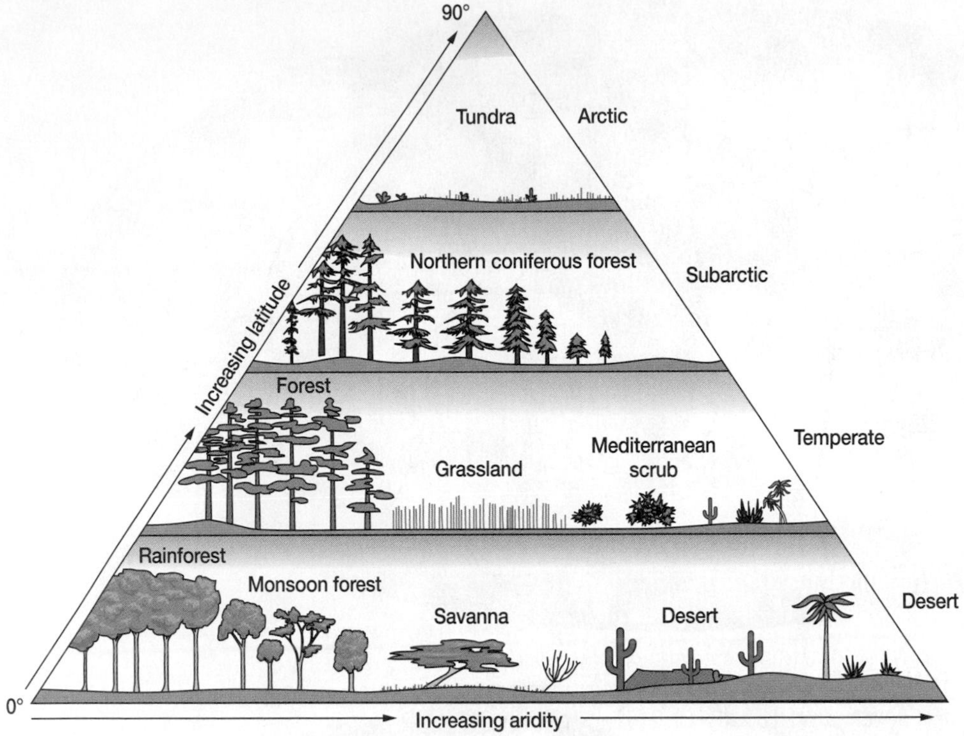

FIGURE 8.16 **Simplified diagram of the relationship between precipitation and latitude and Earth's major land biomes.** Here, latitude serves as an index of average temperature, so latitude can be replaced by average temperature in this diagram. (*Source:* Harm de Blij, Peter O. Muller, and Richard S. Williams, *Physical Geography of the Global Environment*, Figure 27-4, p. 293, edited by Harm de Blij, copyright 2004 by Oxford University Press. Used by permission of Oxford University Press.)

more closely related to each other than to species in other provinces. In two different biotic provinces, the same ecological niche will be filled with species that perform a specific function and may look very similar to each other but have quite different genetic ancestries. In this way, a biotic province is an evolutionary unit.

Convergent and Divergent Evolution

Plants that grow in deserts of North America and East Africa illustrate the idea of a biome (see Figure 8.17). The Joshua tree and saguaro cactus of North America and the giant Euphorbia of East and Southern Africa are tall, have succulent green stems that replace the leaves as the major sites of photosynthesis, and have spiny projections, but these plants are not closely related. The Joshua tree is a member of the agave family, the saguaro is a member of the cactus family, and the Euphorbia is a member of the spurge family. The ancestral differences between these look-alike plants can be found in their flowers, fruits, and seeds, which change the least over time and thus provide the best clues to the genetic history of a species. Geographically isolated for 180 million years, these plants have been subjected to similar climates, which imposed similar stresses and opened up similar ecological opportunities. On both continents, desert plants evolved to adapt to these stresses and potentials, and have come to look alike and prevail in like habitats. Their similar shapes result from evolution in similar desert climates, a process known as **convergent evolution**.

Another important process that influences life's geography is **divergent evolution**. In this process, a population is divided, usually by geographic barriers. Once separated into two populations, each evolves separately, but the two groups retain some characteristics in common. It is now believed that the ostrich (native to Africa), the rhea (native to South America), and the emu (native to Australia) have a common ancestor but evolved separately (Figure 8.18). In open savannas and grasslands, a large bird that can run quickly but feed efficiently on small seeds and insects has certain advantages over other organisms seeking the same food. Thus, these species maintained the same characteristics in widely separated areas. Both convergent and divergent evolution increase biological diversity.

People make use of convergent evolution when they move decorative and useful plants around the world. Cities that lie in similar climates in different parts of the world now share many of the same decorative plants. Bougainvillea, a spectacularly bright flowering shrub originally native to Southeast Asia, decorates cities as distant from each other as Los Angeles and the capital of Zimbabwe. In New York City and its outlying suburbs, Norway maple from Europe and the tree of heaven and gingko tree from China grow with native species such as sweet gum, sugar maple, and pin oak. People intentionally introduced the Asian and European trees.

(a)

(b)

(c)

FIGURE 8.17 Joshua Tree of North America and giant Euphorbia of Africa illustrate convergent evolution. Given sufficient time and similar climates in different areas, species similar in shape will tend to occur. The Joshua tree **(a)** and saguaro cactus **(b)** of North America look similar to the giant Euphorbia **(c)** of East Africa. But these plants are not closely related. Their similar shapes result from evolution under similar climates, a process known as *convergent evolution*.

(a) (b) (c)

FIGURE 8.18 Divergent evolution. Three large, flightless birds evolved from a common ancestor but are now found in widely separated regions: **(a)** the ostrich in Africa, **(b)** the rhea in South America, and **(c)** the emu in Australia.

8.8 Invasions, Invasive Species, and Island Biogeography

Ever since Darwin's voyage on *The Beagle,* which took him to the Galápagos Islands, biologists have been curious about how biological diversity can develop on islands: Do any rules govern this process? How do such invasions happen? And how is biological diversity affected by the size of and distance to a new habitat? E.O. Wilson and R. MacArthur established a theory of island biogeography that sets forth major principles about biological invasion of new habitats,[26] and as it turns out, the many jokes and stories about castaways on isolated islands have a basis in fact.

- Islands have fewer species than continents.
- The two sources of new species on an island are migration from the mainland and evolution of new species in place.
- The smaller the island, the fewer the species, as can be seen in the number of reptiles and amphibians in various West Indian islands (Figure 8.20).
- The farther the island is from a mainland (continent), the fewer the species (Figure 8.19).[27]

Clearly, the farther an island is from the mainland, the harder it will be for an organism to travel the distance, and the smaller the island, the less likely that it will be found by individuals of any species. In addition, the smaller the island, the fewer individuals it can support. Small islands tend to have fewer habitat types, and some habitats on a small island may be too small to support a population large enough to have a good chance of surviving for a long time. Generally, the smaller the population, the greater its risk of extinction. It might be easily extinguished by a storm, flood, or other catastrophe or disturbance, and every species is subject to the risk of extinction by predation, disease (parasitism), competition, climatic change, or habitat alteration.

A final generalization about island biogeography is that over a long time, an island tends to maintain a rather constant number of species, which is the result of the rate at which species are added minus the rate at which they become extinct. These numbers follow the curves shown in Figure 8.20. For any island, the number of species of a particular life-form can be predicted from the island's size and distance from the mainland.

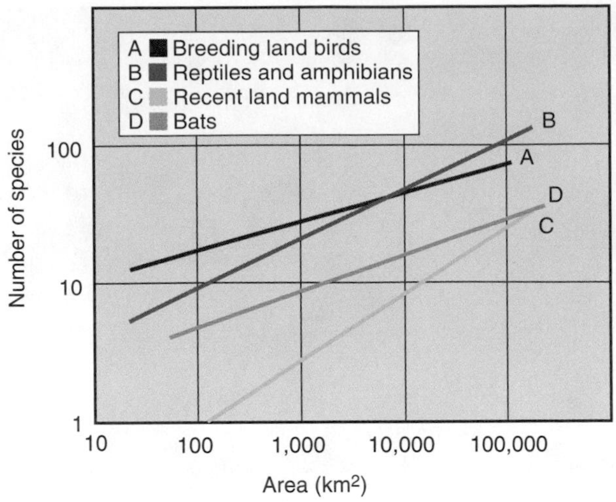

FIGURE 8.20 Islands have fewer species than do mainlands. The larger the island, the greater the number of species. This general rule is shown by a graph of the number of species of birds, reptiles and amphibians, recent land mammals, and bats for islands in the Caribbean. (Modified from B. Wilcox, ed., [Gland, Switzerland: IUCN, 1988].)

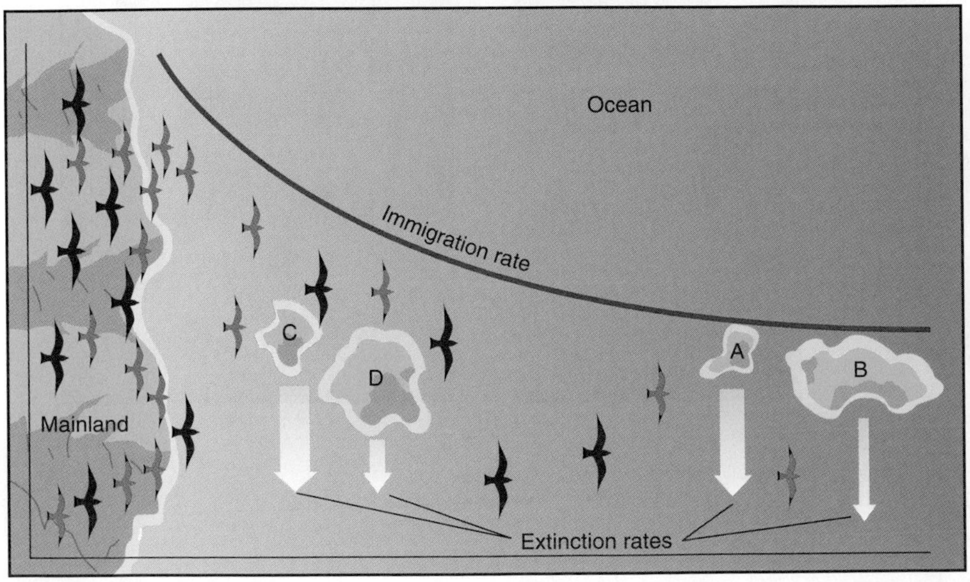

FIGURE 8.19 Idealized relation of an island's size, distance from the mainland, and number of species. The nearer an island is to the mainland, the more likely it is to be found by an individual, and thus the higher the rate of immigration. The larger the island, the larger the population it can support and the greater the chance of persistence of a species—small islands have a higher rate of extinction. The average number of species therefore depends on the rate of immigration and the rate of extinction. Thus, a small island near the mainland may have the same number of species as a large island far from the mainland. The thickness of the arrow represents the magnitude of the rate. (*Source:* Modified from R.H. MacArthur and E.O. Wilson, *The Theory of Island Biogeography* [Princeton, NJ: Princeton University Press, 1967].)

The concepts of island biogeography apply not just to real islands in an ocean but also to ecological islands. An **ecological island** is a comparatively small habitat separated from a major habitat of the same kind. For example, a pond in the Michigan woods is an ecological island relative to the Great Lakes that border Michigan. A small stand of trees within a prairie is a forest island. A city park is also an ecological island. Is a city park large enough to support a population of a particular species? To know whether it is, we can apply the concepts of island biogeography.

Biogeography and People

Benefits of Biological Invasions

We have seen that biogeography affects biological diversity. Changes in biological diversity in turn affect people and the living resources on which we depend. These effects extend from individuals to civilizations. For example, the last ice ages had dramatic effects on plants and animals and thus on human beings. Europe and Great Britain have fewer native species of trees than other temperate regions of the world. Only 30 tree species are native to Great Britain (that is, they were present prior to human settlement), although hundreds of species grow there today.

Why are there so few native species in Europe and Great Britain? Because of the combined effects of climate change and the geography of European mountain ranges. In Europe, major mountain ranges run east–west, whereas in North America and Asia the major ranges run north–south. During the past 2 million years, Earth has experienced several episodes of continental glaciation, when glaciers several kilometers thick expanded from the Arctic over the landscape. At the same time, glaciers formed in the mountains and expanded downward. Trees in Europe, caught between the ice from the north and the ice from the mountains, had few refuges, and many species became extinct. In contrast, in North America and Asia, as the ice advanced, tree seeds could spread southward, where they became established and produced new plants. Thus, the tree species "migrated" southward and survived each episode of glaciation.[16]

Since the rise of modern civilization, these ancient events have had many practical consequences. As we mentioned earlier, soon after Europeans discovered North America, they began to bring exotic North American species of trees and shrubs into Europe and Great Britain. These exotic imports were used to decorate gardens, homes, and parks and formed the basis of much of the commercial forestry in the region. For example, in the famous gardens of the Alhambra in Granada, Spain, Monterey cypress from North America are grown as hedges and cut in elaborate shapes. In Great Britain and Europe, Douglas fir and Monterey pine are important commercial timber trees today. These are only two examples of how knowledge of biogeography—enabling people to predict what will grow where based on climatic similarity—has been used for both aesthetic and economic benefits.

Why Invasive Species Are a Serious Problem Today

The ease and speed of long-distance travel have led to a huge rate of introductions, with invasive pests (including disease-causing microbes) arriving from all around the world both intentionally and unintentionally (Table 8.3 and Figure 8.21). Table 8.3 shows the number of plant pests intercepted by various means in 2007 by the U.S. government. The majority of interceptions—42,003—were at airports, ten times more than maritime interceptions, which before the jet age would have accounted for

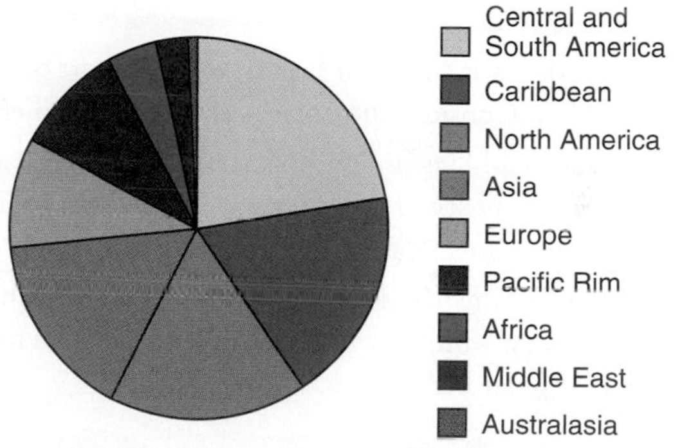

FIGURE 8.21 Where invasive pests are coming from. (*Source:* USDA.)

Table 8.3 REPORTABLE PLANT PEST INTERCEPTIONS, 2007

PLANT PEST INTERCEPTIONS	NUMBER	PERCENT
Airport	42,003	61%
Express carrier	6	0.01%
Inspection station	2,763	4.01%
Land border	14,394	20.91%
Maritime	4,518	6.56%
Other government programs	86	0.12%
Pre-departure	4,869	7.07%
Rail	16	0.02%
USPS Mail	184	0.27%
Total	**68,839**	**100%**

Source: McCullough, D. G., T. T. Works, J. F. Cavey, A. M. Liebold, and D. Marshall. 2006. Interceptions of nonindigenous-plant pests at U.S. ports of entry and border crossings over a 17-year period. *Biological Invasions* 8: 611–630.)

CRITICAL THINKING ISSUE

Polar Bears and the Reasons People Value Biodiversity

In 2008, the U.S. Endangered Species Act listed polar bears as a threatened species. Worldwide, an estimated 20,000 to 25,000 polar bears roam the Arctic, hunting ringed and bearded seals, their primary food. About 5,000 of these polar bears live within the United States. Refer back to the reasons that people value biodiversity. Read up on polar bears (we've listed some sources below) and decide which of these reasons apply to this species. In particular, consider the following questions.

Critical Thinking Questions

1. As a top predator, is the polar bear a necessary part of its ecosystem? (*Hint:* Consider the polar bear's ecological niche.)
2. Do the Inuit who live among polar bears value them as part of the Arctic diversity of life? (This will take some additional study on your part.)
3. Of the nine reasons we discussed earlier for conserving biological diversity, which ones are the primary reasons for conservation of polar bears?

Some Additional Sources of Information about Polar Bears

U.S. Department of Interior Ruling on the Polar Bear: http://alaska.fws.gov/fisheries/mmm/polarbear/pdf/Polar_Bear_Final_Rule.pdf

About Polar Bears as a Species and Their Habitat and Requirements: Polar Bears: Proceedings of the 14th Working Meeting of the IUCN/SSC Polar Bear Specialist Group, June 20–24, 2005, Seattle, Washington. Available at the International Union for the Conservation of Nature (IUCN) website: http://www.iucnredlist.org/search/details.php/22823/summ

Global Warming and Polar Bears: A.E. Derocher, Nicholas J. Lunn, and Ian Stirling, 2004, Polar bears in a warming climate. *Integrative and Comparative Biology*, 44:13–176.

most of them. Passenger ships arrive at fewer locations and far less frequently than do commercial aircraft today. Bear in mind that the 42,003 were just those intercepted—no doubt many passed undetected—and that these are only for pests of plants, not for such things as zebra mussels dumped into American waters from cargo ships. According to the USDA, the present situation is not completely controllable.

Another major avenue of species invasions has been the international trade in exotic pets, like the Burmese python. Many of these are released outdoors when they get to be too big and too much trouble for their owners.

The upshot of this is that we can expect the invasion of species to continue in large numbers, and some will cause problems not yet known in the United States.

SUMMARY

- Biological evolution—the change in inherited characteristics of a population from generation to generation—is responsible for the development of the many species of life on Earth. Four processes that lead to evolution are mutation, natural selection, migration, and genetic drift.
- Biological diversity involves three concepts: genetic diversity (the total number of genetic characteristics), habitat diversity (the diversity of habitats in a given unit area), and species diversity. Species diversity, in turn, involves three ideas: species richness (the total number of species), species evenness (the relative abundance of species), and species dominance (the most abundant species).
- About 1.4 million species have been identified and named. Insects and plants make up most of these species. With further explorations, especially in tropical areas, the number of identified species, especially of invertebrates and plants, will increase.
- Species engage in three basic kinds of interactions: competition, symbiosis, and predation–parasitism. Each type of interaction affects evolution, the persistence of species, and the overall diversity of life. It is important to understand that organisms have evolved together, so predator, parasite, prey, competitor, and symbiont have adjusted to one another. Human interventions frequently upset these adjustments.

- The competitive exclusion principle states that two species that have exactly the same requirements cannot coexist in exactly the same habitat; one must win. The reason more species do not die out from competition is that they have developed a particular niche and thus avoid competition.
- The number of species in a given habitat is determined by many factors, including latitude, elevation, topography, severity of the environment, and diversity of the habitat. Predation and moderate disturbances, such as fire, can actually increase the diversity of species. The number of species also varies over time. Of course, people affect diversity as well.

REEXAMINING THEMES AND ISSUES

Human Population

The growth of human populations has decreased biological diversity. If the human population continues to grow, pressures will continue on endangered species, and maintaining existing biological diversity will be an ever-greater challenge.

Sustainability

Sustainability involves more than just having many individuals of a species. For a species to persist, its habitat must be in good condition and must provide that species' life requirements. A diversity of habitats enables more species to persist.

Global Perspective

For several billion years, life has affected the environment on a global scale. These global effects have in turn affected biological diversity. Life added oxygen to the atmosphere and removed carbon dioxide, thereby making animal life possible.

Urban World

People have rarely thought about cities as having any beneficial effects on biological diversity. However, in recent years there has been a growing realization that cities can contribute in important ways to the conservation of biological diversity. This topic will be discussed in Chapter 22.

People and Nature

People have always treasured the diversity of life, but we have been one of the main causes of the loss in diversity.

Science and Values

Perhaps no environmental issue causes more debate, is more central to arguments over values, or has greater emotional importance to people than biological diversity. Concern about specific endangered species has been at the heart of many political controversies. Resolving these conflicts and debates will require a clear understanding of the values at issue, as well as knowledge about species and their habitat requirements and the role of biological diversity in life's history on Earth.

KEY TERMS

adaptive radiation 152
biogeography 160
biological diversity 145
biological evolution 149
biome 161
biotic province 160
competitive exclusion principle 154
convergent evolution 162
divergent evolution 162
dominant species 147
ecological island 165
ecological niche 155
founder effect 153
gene 149
genetic drift 153
genotype 149
migration 152
mutation 150
natural selection 150
obligate symbionts 158
parasitism 158
population 145
predation 158
species 145
species evenness 147
species richness 147
symbiont 157
symbiosis 157

STUDY QUESTIONS

1. Why do introduced species often become pests?
2. On which of the following planets would you expect a greater diversity of species? (a) a planet with intense tectonic activity; (b) a tectonically dead planet. (Remember that *tectonics* refers to the geologic processes involving the movement of tectonic plates and continents, processes that lead to mountain building and so forth.)
3. You are going to conduct a survey of national parks. What relationship would you expect to find between the number of species of trees and the size of each park?
4. A city park manager has run out of money to buy new plants. How can the park's labor force alone be used to increase the diversity of (a) trees and (b) birds in the park?
5. A plague of locusts visits a farm field. Soon after, many kinds of birds arrive to feed on the locusts. What changes occur in animal dominance and diversity? Begin with the time before the locusts arrive and end after the birds have been present for several days.
6. What will happen to total biodiversity if (a) the emperor penguin becomes extinct? (b) the grizzly bear becomes extinct?
7. What is the difference between a habitat and a niche?
8. More than 600 species of trees grow in Costa Rica, most of them in the tropical rain forests. What might account for the coexistence of so many species with similar resource needs?
9. Which of the following can lead to populations that are less adapted to the environment than were their ancestors?
(a) Natural selection
(b) Migration
(c) Mutation
(d) Genetic drift

FURTHER READING

Botkin, D.B., *No Man's Garden: Thoreau and a New Vision for Civilization and Nature* (Washington, DC: Island Press, 2001). Discusses why people have valued biological diversity from both a scientific and cultural point of view.

Charlesworth, B., and C. Charlesworth, *Evolution: A Very Short Introduction* (Oxford: Oxford University Press, 2003).

Darwin, C.A., *The Origin of Species by Means of Natural Selection, or the Preservation of Proved Races in the Struggle for Life* (London: Murray, 1859., reprinted variously). A book that marked a revolution in the study and understanding of biotic existence.

Dawkins, Richard, *The Selfish Gene* (New York: Oxford University Press; 3rd edition, 2008). Now considered a classic in the discussion of biological evolution for those who are not specialists in the field.

Leveque, C., and J. Mounolou, *Biodiversity* (New York: John Wiley, 2003).

Margulis, L., K.V. Schwartz, M. Dolan, K. Delisle, and C. Lyons, *Diversity of Life: The Illustrated Guide to the Five Kingdoms* (Sudbury, MA: Jones & Bartlett, 1999).

Novacek, M.J. ed., *The Biodiversity Crisis: Losing What Counts* (An American Museum of Natural History Book. New York: New Press, 2001).

Wacey, David, *Early Life on Earth: A Practical Guide* (New York: Springer, 2009). A new and thoughtful book about the beginnings of life on our planet.

CHAPTER 9

Ecological Restoration

The Florida Everglades from the air looks like a field of water islands and grasslike water plants. One of the largest ecological restoration projects is an attempt to help this national park.

LEARNING OBJECTIVES

Ecological restoration is the part of ecosystem management that deals with the recovery of ecosystems that have been damaged by human activities. It is a relatively new field. In this chapter, we explore the concepts of ecological restoration. After reading this chapter, you should understand . . .

- What it means to "restore" an ecosystem, since ecological systems are always changing;
- The main goals of restoration ecology;
- Basic approaches, methods, and limits of restoration;
- The general principles and processes of restoration ecology;
- The role of adaptive management in restoration;
- The criteria used to judge the success of restoration.

CASE STUDY

The Florida Everglades

The Florida Everglades is one of the nation's most valuable ecological treasures, listed by the United Nations as a World Heritage Site. The Everglades is also interconnected with a large area of the rest of Florida, beginning with a series of small lakes near Orlando, Florida (near the center of the state), and extending southward to the Florida Bay. The area south of Lake Okeechobee—about 175 km (110 mi) south of Orlando—is a long, wide system of shallow wetlands with slow-moving water. You can imagine the Everglades as a very wide, grass-filled, slow-moving, shallow river. Everglades National Park is at the very southern end of the system and extends out into Florida Bay to about the Florida Keys. Much of the flow of the Everglades is funneled through a location known as Shark Slough, and the velocity of flow of the lower Everglades, while still slow, is greater than that to the north.

Tourists from all over the world come to the Everglades to see its unusual landscape and wildlife. It is home to more than 11,000 species of plants, several hundred species of birds, and numerous species of fish and mammals. It is the last remaining habitat for approximately 70 threatened or endangered species or subspecies, including the Florida manatee (a subspecies of the West Indian manatee), the Florida panther (a subspecies of the American mountain lion), and the American crocodile. Unfortunately, in the past century, much of the Everglades has been drained for agriculture and urban development; today only 50% of the original wetlands remain (Figure 9.1a and b).

Restoration of the Everglades is complicated by agriculture and urbanization—several million people live in south Florida. All—agriculture, people, and the National Park—compete for the water resources. One of the major issues is to somehow arrive at a plan that will ensure long-term sustainability and quality of the water supply for the Everglades and also supply water for agriculture, towns, and cities.[1] This plan will involve the following:

- Reducing the area in agriculture and/or the water use per hectare in agriculture.
- Reducing the flow of agricultural fertilizers and pesticides from farmland into the Everglades.
- Managing land development that encroaches upon the Everglades.
- Managing access to the Everglades by people.
- Removing introduced exotic species that are dangerous to people, or threaten native species with extinction, or disrupt the presettlement ecosystems.

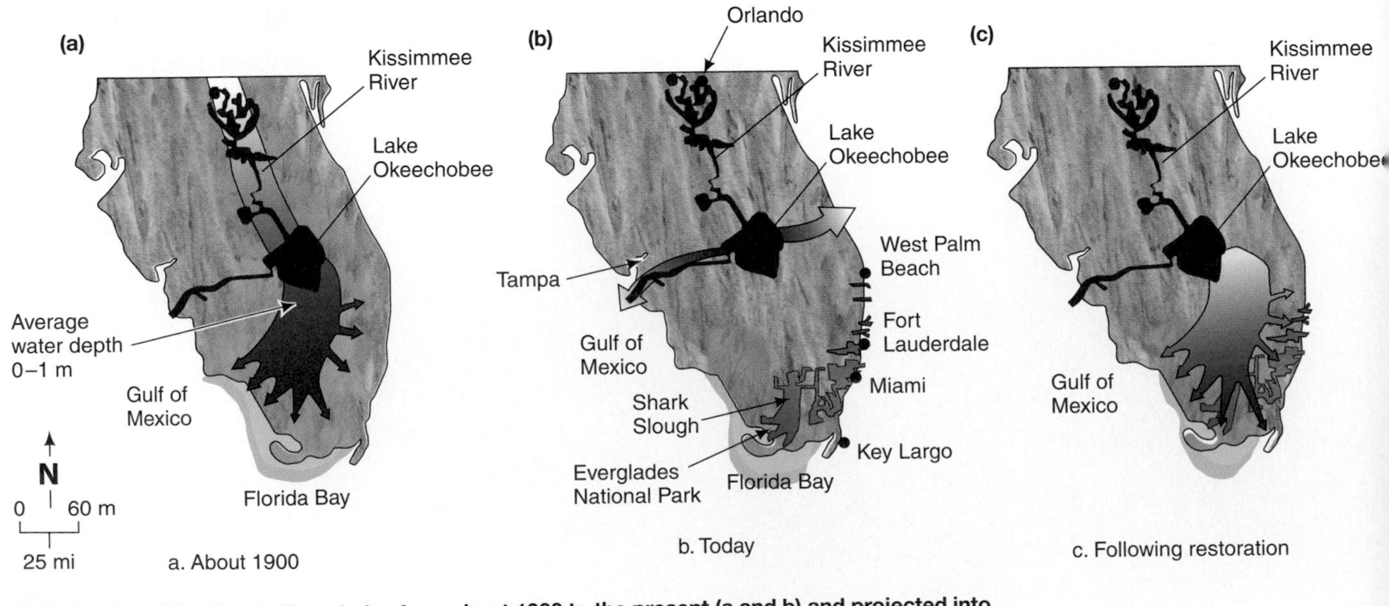

FIGURE 9.1 The Florida Everglades from about 1900 to the present (a and b) and projected into the future following restoration (c). Water flow (both surface and ground water) is shown in blue. Land development and water diversions changed the dominant flow from north-south to east-west. (*Sources*: Modified from the U.S. Geological Survey and Comprehensive Everglades Restoration Plan.)

- Developing scientific methods and theory to better predict the possible consequences of changes to the geologic, hydrologic, and biologic parts of the Everglades as restoration goes forward.
- Restoring the original large-scale water flow in south Florida—that grass-filled, slow-moving, shallow river. (Today the flow of water in the northern part of the Everglades is controlled by a complex system of canals, locks, and levees for a variety of purposes, including agriculture, flood control, and water supply.)

The Everglades restoration is the largest wetland-restoration project in the world. Known as the Comprehensive Everglades Restoration Plan, the project was developed by a number of government agencies, both local and federal, and is slated to continue over a 30-year period at a total cost of about $10 billion. To date, about $2.4 billion has been spent. Land acquired in 2008 that will be removed from agriculture and returned as part of the Everglades will enlarge the park's area by approximately 728 km^2 (281 mi^2). The acquired land will allow restoration of the hydrology of the Everglades in a much more significant way, as it is in the northern part of the system, where much of the water was used for agricultural purposes. Goals for the Florida Everglades include the following:

- Restoration of a more natural water flow (Figure 9.1c).
- Recovery of native and endangered species.
- Removal of invasive exotic species that are causing problems to the ecosystems and to people.
- Restoration of water quality, especially control of nutrients from agricultural and urban areas.
- Habitat restoration for wildlife that use the Everglades.

Progress to date has been significant. The amount of pollutants flowing into the Everglades from a variety of sources has been reduced by about 50%. Thousands of hectares have been treated to remove invasive, exotic species, including the Brazilian pepper tree and tilapia (a fish from Africa). In recent years, the Burmese python, which can grow to over 7 m (21 ft) and weigh over 100 kg (220 lbs), has become established in the Everglades (Figure 9.2). The snakes are an exotic, invasive species introduced by people who purchased small pythons as pets and released them into the wild when they became too large and difficult to keep. They have established a rapidly growing population in the Everglades (and elsewhere in Florida) and are becoming a top predator there, with increasing capacity to deplete populations of native birds and mammals. Recently, a large python killed and attempted to swallow a seven-foot alligator! A program is now under way to attempt to control the growing population of pythons (estimated to exceed 100,000), but it's unlikely that they can be completely eradicated.

FIGURE 9.2 Burmese python, an exotic, invasive species. Released by people who purchased them as pets, it has established a large population in the Everglades.

A purpose of this chapter is to provide an understanding of what is possible and what can be done to restore damaged ecosystems.

9.1 What Is Ecological Restoration?

Ecological restoration is defined as providing assistance to the recovery of an ecosystem that has been degraded, damaged, or destroyed.[2] As such, ecological restoration is applied science and derives from the science of restoration ecology. Some general principles for restoration are listed below.

- Ecosystems are dynamic, not static (change and natural disturbance are expected).
- No simple set of rules will be applicable to a specific restoration project.
- Adaptive management, using the best science, is necessary for restoration to succeed.
- Careful consideration of ecosystems (life), geology (rocks, soils), and hydrology (water) plays an important role in all restoration projects.

Of particular importance is the principle that ecosystems are dynamic, not static—that is, they are always changing and subject to natural disturbance. Any restoration plan must consider disturbance and how resilient the restored system will be. Also important is **adaptive management**, the application of science to the management process. Hypotheses may be tested, and, as restoration goes forward, flexible plans may be developed to accommodate change. Probably the most common projects involve river restoration and the restoration of freshwater and coastal wetlands.[3]

9.2 Goals of Restoration: What Is "Natural"?

If an ecosystem passes naturally through many different states and all of them are "natural," and if the change itself, caused by wildfire, flood, and windstorm, is natural, then what is its natural state? And how can restoration that involves such disturbance occur without damage to human life and property? Can we restore an ecological system to any one of its past states and claim that this is natural and successful restoration?

In Chapters 3 and 5, we discussed the ideas of steady-state and non-steady-state ecological systems. We argued that until the second half of the 20th century the predominant belief in Western civilization was that any natural area—a forest, a prairie, an intertidal zone—left undisturbed by people achieved a single condition that would persist indefinitely. This condition, as mentioned in Chapter 3, has been known as the balance of nature. The major tenets of a belief in the balance of nature are as follows:

- Left undisturbed, nature achieves a permanency of form and structure that persists indefinitely.
- If it is disturbed and the disturbance is removed, nature returns to exactly the same permanent state.
- In this permanent state of nature, there is a "great chain of being," with a place for each creature (a habitat and a niche) and each creature in its appropriate place.

These ideas have their roots in Greek and Roman philosophies about nature, but they have played an important role in modern environmentalism as well. In the early 20th century, ecologists formalized the belief in the balance of nature. At that time, people thought that wildfires were always detrimental to wildlife, vegetation, and natural ecosystems. *Bambi*, a 1942 Walt Disney movie, expressed this belief, depicting a fire that brought death to friendly animals. In the United States, Smokey Bear is a well-known symbol used for many decades by the U.S. Forest Service to warn visitors to be careful with fire and avoid setting wildfires. The message is that wildfires are always harmful to wildlife and ecosystems.

All of this suggests a belief that the balance of nature does in fact exist. But if that were true, the answer to the question "restore to what?" would be simple: restore to the original, natural, permanent condition. The way to do it would be simple, too: Get out of the way and let nature take its course. Since the second half of the 20th century, though, ecologists have learned that nature is not constant, and that forests, prairies—all ecosystems—undergo change. Moreover, since change has been a part of natural ecological systems for millions of years, many species have adapted to change. Indeed, many require specific kinds of change in order to survive. This means that we can restore ecosystem processes (flows of energy, cycling of chemical elements) and help populations of endangered and threatened species increase on average, but the abundances of species and conditions of ecosystems will change over time as they are subjected to internal and external changes, and following the process of succession discussed in Chapter 5.

Dealing with change—natural and human-induced— poses questions of human values as well as science. This is illustrated by wildfires in forests, grasslands, and shrublands, which can be extremely destructive to human life and property. From 1990 to 2009, three wildfires that started in chaparral shrubland in Santa Barbara, California, burned about 1,000 homes. The wildfire hazard can be minimized but not eliminated. Scientific understanding tells us that fires are natural, and that some species require them. But whether we choose to allow fires to burn, or even light fires ourselves, is a matter of values. **Restoration ecology** depends on science to discover what used to be, what is possible, what an ecosystem or species requires to persist, and how different goals can be achieved. But selecting goals for restoration is a matter of human values.

Some possible goals of restoration are listed in Table 9.1. Which state we attempt to restore a landscape to (pre-industrial to modern) depends on more-specific goals and possibilities that, again, are linked to values. For example, restoring the Florida Everglades to a pre-industrial state is not possible or desirable (a value) given the present land and water use that supports the people of Florida. The goal instead is to improve biodiversity, water flow through the Everglades, and water quality (see opening Case Study).

Table 9.1 SOME POSSIBLE RESTORATION GOALS

GOAL	APPROACH
1. Pre-industrial	Maintain ecosystems as they were in A.D. 1500
2. Presettlement (e.g., of North America)	Maintain ecosystems as they were about A.D. 1492
3. Preagriculture	Maintain ecosystems as they were about 5000 B.C.
4. Before any significant impact of human beings	Maintain ecosystems as they were about 10,000 B.C.
5. Maximum production	Independent of a specific time
6. Maximum diversity	Independent of a specific time
7. Maximum biomass	Independent of old growth
8. Preserve a specific endangered species	Whatever stage it is adapted to
9. Historical range of variation	Create the future like the known past

9.3 What Is Usually Restored?

Ecosystems of all types have undergone degradation and need restoration. However, certain kinds of ecosystems have undergone especially widespread loss and degradation and are therefore a focus of attention today. Table 9.2 gives examples of ecosystems that are commonly restored.

Attention has focused on forests, wetlands, and grasslands, especially the North American prairie; streams and rivers and the riparian zones alongside them; lakes; beaches; and habitats of threatened and endangered species. Also included are areas that people desire to restore for aesthetic and moral reasons, showing once again that restoration involves values. In this section, we briefly discuss the restoration of rivers and streams, wetlands, and prairies.

Rivers, Streams, and Wetlands Restoration: Some Examples

Rivers and streams and wetlands probably are restored more frequently than any other systems. Thousands of streams have been degraded by urbanization, agriculture, timber harvesting, and channelization (shortening, widening, and even paving over or confining the channel to culverts). In North America, large areas of both freshwater and coastal wetlands have been greatly altered during the past 200 years. It is estimated that California, for example, has lost more than 90% of its wetlands, both freshwater and coastal, and that the total wetland loss for the United States is about 50%. Not only the United States has suffered; wetlands around the world are affected.

Table 9.2 SELECTED EXAMPLES OF RESTORATION PROJECTS

SYSTEM	OBJECTIVE
Rivers/Streams	Improve biodiversity, water quality, bank stability. Very common practice across the U.S.
Coastal Wetlands	Improve biodiversity and water quality, store water, provide a buffer to erosion from storms to inland areas. Very common practice along all U.S. coastal areas.
Freshwater Wetlands	Improve biodiversity and water quality, store water, and, for river systems, reduce the flood hazard.
Beaches	Sustain beaches and their ecosystems. Most often involve sand supply.
Sand Dunes	Improve biodiversity in both inland and coastal areas.
Landscape	Increase biodiversity and conserve endangered species. Often a very complex process.
Land Disturbed by Mining	Reestablish desired ecosystems, reduce erosion, and improve water quality.

Rivers and Streams

One of the largest and most expensive restoration projects in the United States is the restoration of the Kissimmee River in Florida. This river was channelized, or straightened, by the U.S. Army Corps of Engineers to provide ship passage through Florida. However, although the river and its adjacent ecosystems were greatly altered, shipping never developed, and now several hundred million dollars must be spent to put the river back as it was before. The task includes restoring the meandering flow of the river and replacing the soil layers in the order in which they had lain on the bottom of the river prior to channelization.[4]

The Kissimmee at one time had an unusual hydrology because it inundated its floodplain for prolonged periods (Figure 9.3). The floodplain and river supported a biologically diverse ecosystem, including wetland plants, wading birds, waterfowl, fish, and other wildlife. Few people lived in the Kissimmee basin before about 1940, and the land use was mostly agricultural. Due to rapid development and growth in the past 50 years and a growing flood hazard as a result of inappropriate land use, people asked the federal government to design a flood-control plan for southern Florida. The channelization of the Kissimmee River occurred between 1962 and 1971 as part of the flood-control plan. About two-thirds of the floodplain was drained, and a straight canal was excavated. Turning the meandering river into a straight canal degraded the river ecosystem and greatly reduced the wetlands and populations of birds, mammals, and fish (Figure 9.4).

FIGURE 9.3 The Kissimmee River before channelization.

FIGURE 9.4 The Kissimmee River after channelization that produced a wide, straight ditch.

Criticism of the loss of the river ecosystem went on for years, finally leading to the current restoration efforts. The purpose of the restoration is to return part of the river to its historical meandering riverbed and wide floodplain. Before-and-after photos and specifics of the restoration plan are shown in Figure 9.5 and include restoring as much as possible of the historical biodiversity and ecosystem function; re-creating patterns of wetland plant communities as they existed before channelization; reestablishing prolonged flooding of the floodplain; and re-creating a river floodplain environment and connection to the main river similar to the way it used to be.[4]

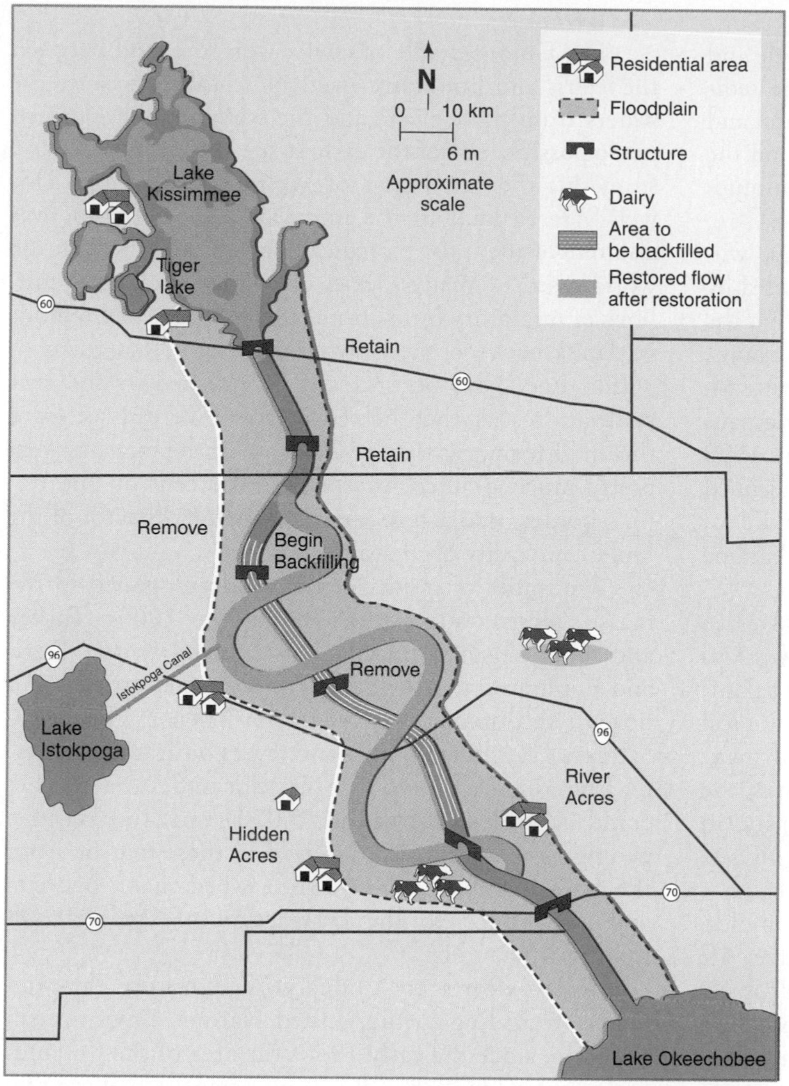

FIGURE 9.5 **Map showing part of the Kissimmee River restoration plan.** A major objective is to re-create the historical river floodplain environment that is wet much of the year, improving biodiversity and ecosystem function. (*Source*: Modified from South Florida Water Management District.)

The cost of the restoration of the Kissimmee River is several times greater than it was to channelize it. Thus, the decision to go forward with the restoration reflects the high value that people in south Florida place on conserving biological diversity and providing for recreational activities in a more natural environment.

Yellowstone National Park is another interesting case. It is the site of an unanticipated stream restoration resulting from the reintroduction of wolves. Wolves were eliminated from Yellowstone by the 1920s and were introduced back into the park in 1995–1996. Initially, 66 wolves were introduced, and by 2007 the wolf population had grown to about 1,500, with about 171 in Yellowstone itself (Figures 9.6 and 9.7), 700–800 in Idaho (outside the park), and the rest in other areas.

Mountain streams in Yellowstone generally consist of stream channels, beds, and banks composed of silt, sand, gravel, and bedrock. Cool, clean water is supplied via the hydrologic and geologic system from snowmelt and rain that infiltrate the rocks and soil to seep into streams. The water supports life in the stream, including fish and other organisms, as well as streamside vegetation adjacent to stream channels. The riparian vegetation is very different from vegetation on adjacent uplands. Stream bank vegetation helps retard erosion of the banks and thus the amount of sediment that enters the stream.

Riparian vegetation, such as cottonwood and willow trees, is also a popular food source for animals, such as elk. Extensive browsing

The restoration project was authorized by the U.S. Congress in 1992, in partnership with the South Florida Water Management District and the U.S. Army Corps of Engineers. It is an ongoing project, and by 2001 approximately 12 km of the nearly straight channel had been restored to a meandering channel with floodplain wetlands about 24 km long. As a result, water again flowed through a meandering channel and onto the floodplain, wetland vegetation was reestablished, and birds and other wildlife returned. The potential flood hazard is being addressed. Some of the structures that control flooding will be removed, and others will be maintained. Flood protection is a main reason the entire river will not be returned to what it was before channelization.

FIGURE 9.6 **Wolves in Yellowstone National Park.**

dramatically reduces the abundance of riparian plants, damaging the stream environment by reducing shade and by increasing bank erosion, which introduces fine sediment into the water. Fine sediment, such as fine sand and silt, not only degrades water quality but also may fill the spaces between gravel particles or seal the bed with mud, damaging fish and aquatic insect habitat.[5]

Before the wolves arrived in the mid-1990s, willows and other streamside plants were nearly denuded by browsing elk. It soon became apparent, however, that the wolves were most successful in hunting elk along streams, where the elk had to negotiate the complex, changing topography. The elk responded by avoiding the dangerous stream environment. Over a four-year period, from 1998 to 2002, the number of willows eaten by elk declined greatly, and the riparian vegetation recovered. As it did so, the stream channel and banks also recovered and became more productive for fish and other animals.

In sum, although the reintroduction of wolves to Yellowstone is controversial, the wolves are a *keystone species*—a species that, even if not overly abundant, plays an important role in maintaining an ecological community. By hunting elk and scaring them away from the streams, wolves improve the stream banks, the water quality, and the broader ecologic community (in this case, the stream ecosystem). The result is a higher-quality stream environment.[5]

Still, the debate about wolf introductions is complex. In Yellowstone, just over 90% of wolf prey is elk; there are far fewer bison, deer, and other animals. Land-use issues associated with grazing for cattle and sheep are more difficult to assess. How we choose to manage wolf populations will reflect both science and values.[5, 6]

FIGURE 9.7 **Wolf hunting elk in Yellowstone National Park.**

Wetlands

The famous cradle of civilization, the land between the Tigris and Euphrates rivers, is so called because the waters from these rivers and the wetlands they formed made possible one of the earliest sites of agriculture, and from this the beginnings of Western civilization. This well-watered land in the midst of a major desert was also one of the most biologically productive areas in the world, used by many species of wildlife, including millions of migratory birds. Ironically, the huge and famous wetlands between these two rivers, land that today is within Iraq, have been greatly diminished by the very civilization that they helped create. "We can see from the satellite images that by 2000, all of the marshes were pretty much drained, except for 7 percent on the Iranian border," said Dr. Curtis Richardson, director of the Duke University Wetland Center.[7]

A number of events of the modern age led to the marsh's destruction. Beginning in the 1960s, Turkey and Syria began to build dams upriver, in the Tigris and Euphrates, to provide irrigation and electricity, and now these number more than 30. Then in the 1980s Saddam Hussein had dikes and levees built to divert water from the marshes so that oil fields under the marshes could be drilled. For at least 5,000 years, the Ma'adan people—the Marsh Arabs—lived in these marshes. But the Iran–Iraq War (1980–1988) killed many of them and also added to the destruction of the wetlands (Figure 9.8a and b).[8]

Today efforts are under way to restore the wetlands. According to the United Nations Environment Program, since the early 1970s the area of the wetlands has increased by 58%.[9] But some scientists believe that there has been little improvement, and the question remains: Can ecosystems be restored once people have seriously changed them?

Prairie Restoration

Tallgrass prairie is also being restored. Prairies once occupied more land in the United States than any other kind of ecosystem. Today, only a few small remnants of prairie exist. Prairie restoration is of two kinds. In a few places, one can still find original prairie that has never been plowed. Here, the soil structure is intact, and restoration is simpler. One of the best known of these areas is the Konza Prairie near Manhattan, Kansas. In other places, where the land has been plowed, restoration is more complicated. Nevertheless, the restoration of prairies has gained considerable attention in recent decades, and restoration of prairie on previously plowed and farmed land is occurring in many midwestern states. The Allwine Prairie, within the city limits of Omaha,

(a) 

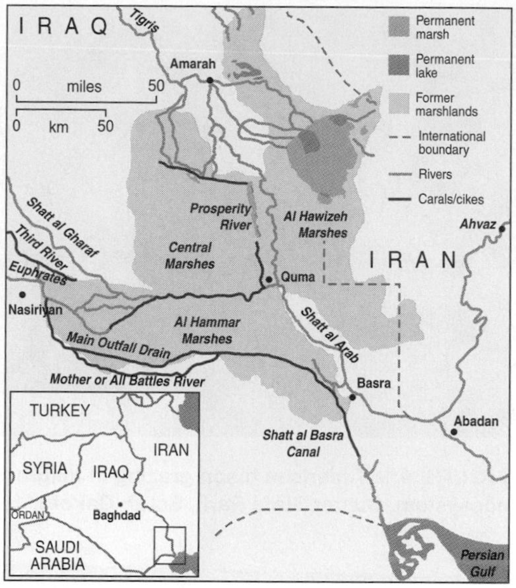

(b)

FIGURE 9.8 **(a)** A Marsh Arab village in the famous wetlands of Iraq, said to be one of the places where Western civilization originated. The people in this picture are among an estimated 100,000 Ma'adan people who now live in their traditional marsh villages, many having returned recently. These marshes are among the most biologically productive areas on Earth. **(b)** Map of the Fertile Crescent, where the Marsh Arabs live, called the cradle of civilization. It is the land between the Tigris (the eastern river) and Euphrates (the western river) in what is now Iraq. Famous cities of history, such as Nineveh, developed here, made possible by the abundant water and good soil. The gray area shows the known original extent of the marshes, the bright green their present area.

Nebraska, has been undergoing restoration from farm to prairie for many years. Prairie restoration has also been taking place near Chicago.

About 10% of North American original tallgrass prairie remains in scattered patches from Nebraska east to Illinois and from southern Canada south to Texas (in the Great Plains physiographic province of the United States). A peculiarity of prairie history is that although most prairie land was converted to agriculture, this was not done along roads and railroads; thus, long, narrow strips of unplowed native prairie remain on these rights-of-way. In Iowa, for example, prairie once covered more than 80% of the state—11 million hectares (28 million acres). More than 99.9% of the prairie land has been converted to other uses, primarily agriculture, but along roadsides there are 242,000 hectares (600,000 acres) of prairie—more than in all of Iowa's county, state, and federal parks. These roadside and railway stretches of prairie provide some of the last habitats for native plants, and restoration of prairies elsewhere in Iowa is making use of these habitats as seed sources.[5]

Studies suggest that the species diversity of tallgrass prairie has declined as a result of land-use changes that have led to the loss or fragmentation of habitat.[9] For example, human-induced changes that nearly eliminated bison (another keystone species) from the prairie greatly changed the community structure and biodiversity of the ecosystem. Scientists evaluating the effects of grazing suggest that, managed properly, grazing by bison can restore or improve biodiversity of tallgrass prairie[9] (Figure 9.9). The effect of grazing by cattle is not as clear. Range managers have for years maintained that cattle grazing is good for ecosystems, but such grazing must be carefully managed. Cattle are not bison. Bison tend to range over a wider area and in a different pattern. Cattle tend to stay longer in a particular area and establish grazing trails with denuded vegetation. However, both cattle and bison, if too many of them are left too long in too small an area, will cause extensive damage to grasses.

Fire is another important factor in tallgrass prairies. Spring fires enhance the growth of the dominant tall grasses that bison prefer. Tallgrass prairie is a mixture of taller grasses, which prefer warm, dry environments, and other grasses, forbs, and woody plants, which prefer cooler, wetter conditions. The tall grasses often dominate and, if not controlled by fire and grazing, form a thick cover (canopy) that makes it more difficult for shorter plants to survive. Grazing opens the canopy and allows more light to reach closer to the ground and support more species. This increases biodiversity. Long ago, fires set by lightning and/or people helped keep the bison's grazing lands from turning into forests. Today, ecological restoration

FIGURE 9.9 **American bison grazing in tallgrass prairie ecosystem, Custer State Park, South Dakota.**

has attempted to use controlled burns to remove exotic species and woody growth (trees). However, fire alone is not sufficient in managing or restoring prairie ecosystem biodiversity. Moderate grazing is the hypothetical solution. Grazing of bison on degraded grassland will have negative impacts, but moderate grazing by bison or cattle on "healthy prairies" may work.

One of the newest threats to tallgrass prairie ecosystems is atmospheric nitrogen from automobile emissions. Nitrogen helps some species, but too much of it causes problems for tallgrass prairie ecosystems, whose diversity and productivity are significantly influenced by the availability of nitrogen. Fire and millions of grazing bison regulated nitrogen availability during prehistoric and pre-automobile times.

A CLOSER LOOK 9.1

Island Fox on Santa Cruz Island

The island fox is found on the Channel Islands, eight islands off the coast of Southern California (Figure 9. 10a). The fox evolved over the past 20,000 years into a separate species of its recent ancestors, the California gray fox. Due to isolation on the islands, as the island fox evolved, it became smaller, and today it is about the size of a house cat (Figure 9.10b).[10]

Island fox most likely reached the islands off the coast of Santa Barbara about 20,000 years ago, when sea levels were more than 120 meters lower than they are today and the distance to the mainland was much shorter, increasing the likelihood of animals' reaching the offshore environment. At that time, this consisted of one large island known as Santa Rosae. By the time Native Americans arrived, about 12,000 years ago, the island fox had become well established. Native Americans evidently kept the foxes as pets, and some burial sites suggest that foxes were, in fact, buried with their owners. The island

(a)

(b)

FIGURE 9.10 **(a) The Santa Barbara Channel and Channel Islands; (b) Island fox.**

(a)

(b)

FIGURE 9.11 **(a) Golden eagle eating an island fox on Santa Cruz Island and (b) a bald eagle hunting fish.**

fox in the Channel Islands lived to ages unheard of for mainland gray foxes. Many of them lived more than 10 years, and a few even about 12 years. A number of them became blind in their old age, from either cataracts or accident, and could be seen feeding on beaches and other areas despite their handicap.

A subspecies of the island fox evolved on the six islands on which they are found today, and until fairly recently they had no natural enemies. But in the 1990s, the populations of Island fox on several islands suddenly plummeted. On San Miguel Island, for example, a population of approximately 400 foxes in 1994 was reduced to about 15 in only five years. Similar declines occurred on Santa Rosa and Santa Cruz islands. At first it seemed that some disease must be spreading rapidly through the fox population. On Catalina Island, in fact, an occurrence of canine distemper did lead to a decline in the number of foxes on that island. On other islands, particularly in the Santa Barbara Channel, the cause was not so easily determined.

Ecologists eventually solved the mystery by discovering that foxes were being killed and eaten by golden eagles (Figure 9.11a), which had only recently arrived on the islands after the apparent demise of the islands' bald eagles (Figure 9.11b). The bald eagles primarily eat fish and hadn't bothered the foxes. Bald eagles are also territorial and kept golden eagles off the islands. It is believed that the bald eagles became endangered because the use of DDT in the 1970s and later led to increasing concentrations of the pesticide in fish that bald eagles ate, causing their eggshells to become too soft to protect the embryos. The golden eagles moved in and colonized the islands in the 1990s, apparently attracted by the amount of food they could easily obtain from their daylight hunting, as well as by the absence of bald eagles. The golden eagles found young feral pigs much to their liking and evidently also found island foxes to be easy targets.

Remains of island foxes have been found in eagle nests, and it is now generally agreed that the golden eagles are responsible for the decline in the fox populations. In fact, of 21 fox carcasses studied on Santa Cruz Island in the 1990s, 19 were apparently victims of golden eagle predation.[11,12]

To conserve the island fox on the three Channel Islands in Santa Barbara Channel, which are part of Channel Islands National Park, a management program was developed. The plan has five steps:[12]

1. Capture remaining island foxes and place them in protected areas.
2. Begin a captive breeding program to rebuild the fox populations.
3. Capture golden eagles and transfer them to the mainland. (The idea is to put them in suitable habitat far from the islands so they will not return.)
4. Reintroduce bald eagles into the island ecosystem. (It is hoped that the birds will establish territories and that this will prevent the return of golden eagles.)
5. Remove populations of feral pigs, which attract golden eagles to the islands.

This five-step program has been put into effect; foxes are now being bred in captivity at several sites, and new kits have been born. By 2009, golden eagles on Santa Cruz Island had been mostly removed, the pigs had also been removed, bald eagles had been reintroduced, and foxes raised in captivity had been released. The historical average population of fox on the islands is 1,400. By 2004 fewer than 100 were present, but by 2009 there were about 700—a remarkable recovery. If all the steps necessary to save the island fox are successful, then the island fox will again take its place as one of the keystone species on the Channel Islands.

9.4 Applying Ecological Knowledge to Restore Heavily Damaged Lands and Ecosystems

An example of how ecological succession can aid in the restoration (termed **reclamation** for land degraded by mining) of heavily damaged lands is the ongoing effort to undo mining damage in Great Britain, where some mines have been used since medieval times and approximately 55,000 hectares (136,000 acres) have been damaged. Recently, programs have been initiated to remove toxic pollutants from the mines and mine tailings, to restore these damaged lands to useful biological production, and to restore the attractiveness of the landscape.[13]

One area damaged by a long history of mining lies within the British Peak District National Park, where lead has been mined since the Middle Ages and waste tailings are as much as 5 m (16.4 ft) deep. The first attempts to restore this area used a modern agricultural approach: heavy application of fertilizers and planting of fast-growing agricultural grasses to revegetate the site rapidly. These grasses quickly green on the good soil of a level farm field, and it was hoped that, with fertilizer, they would do the same in this situation. But after a short period of growth, the grasses died. The soil, leached of its nutrients and lacking organic matter, continued to erode, and the fertilizers that had been added were soon leached away by water runoff. As a result, the areas were shortly barren again.

When the agricultural approach failed, an ecological approach was tried, using knowledge about ecological succession. Instead of planting fast-growing but vulnerable agricultural grasses, ecologists planted slow-growing native grasses, known to be adapted to mineral-deficient soils and the harsh conditions that exist in cleared areas. In choosing these plants, the ecologists relied on their observations of what vegetation first appeared in areas of Great Britain that had undergone succession naturally.[14] The result of the ecological approach has been the successful restoration of damaged lands (Figure 9.12).

FIGURE 9.12 Long-abandoned lead-mining area in Great Britain undergoing restoration (upper half of photograph). Restoration includes planting early-successional native grasses adapted to low-nutrient soil with little physical structure. The bottom hall of the photo shows the unrestored area.

FIGURE 9.13 Restoration of a limestone quarry on Vancouver Island, Canada, in 1921 resulted in a tourist attraction known as Butchart Gardens that draws about 1 million visitors each year.

Heavily damaged landscapes can be found in many places. Restoration similar to that in Great Britain is done in the United States, Canada, and many other places to reclaim lands damaged by strip mining. Butchart Gardens, an open-pit limestone quarry on Vancouver Island, Canada, is an early-20th-century example of mine restoration. The quarry, a large, deep excavation where Portland cement was produced, was transformed into a garden that attracts a large number of visitors each year (Figure 9.13). The project was the vision of one person, the wife of the mine owner. One person can make all the difference!

9.5 Criteria Used to Judge the Success of Restoration

Criteria used to evaluate whether a specific restoration has been successful, and, if so, how successful, will vary, depending on details of the project and the target (reference) ecosystem to which the restoration is compared. Criteria used to judge the success of the Everglades restoration (with issues of endangered species) will be much different from criteria used to evaluate the success of **naturalization** of an urban stream to produce a greenbelt. However, some general criteria apply to both.[2]

- The restored ecosystem has the general structure and process of the target (reference) ecosystem.

- The physical environment (hydrology, soils, rocks) of the restored ecosystem is capable of sustainably supporting the stability of the system.
- The restored ecosystem is linked with and appropriately integrated into the larger landscape community of ecosystems.
- Potential threats to the stability of the restored ecosystem have been minimized to an acceptable level of risk.
- The restored ecosystem is sufficiently adapted to normally withstand expected disturbances that characterize the environment, such as windstorms or fire.
- The restored ecosystem is, as nearly as possible, as self-sustaining as the target (reference) ecosystem. It therefore undergoes the natural range of variation over time and space; otherwise it cannot be self-sustaining.

In the final analysis, restoration, broadly defined to include naturalization, is successful if it improves the environment and the well-being of the people who are linked to the environment. An example is development of city parks that allow people to better communicate with nature.

CRITICAL THINKING ISSUE

How Can We Evaluate Constructed Ecosystems?

What happens when restoring damaged ecosystems is not an option? In such cases, those responsible for the damage may be required to establish alternative ecosystems to replace the damaged ones. An example involved some saltwater wetlands on the coast of San Diego County, California.

In 1984, construction of a flood-control channel and two projects to improve interstate freeways damaged an area of saltwater marsh. The projects were of concern because California had lost 91% of its wetland area since 1943, and the few remaining coastal wetlands were badly fragmented. In addition, the damaged area provided habitat for three endangered species: the California least tern, the light-footed clapper rail, and a plant called the salt-marsh bird's beak. The California Department of Transportation, with funding from the Army Corps of Engineers and the Federal Highway Administration, was required to compensate for the damage by constructing new areas of marsh in the Sweetwater Marsh National Wildlife Refuge. To meet these requirements, eight islands, known as the Connector Marsh, with a total area of 4.9 hectares, were constructed in 1984. An additional 7-hectare area, known as Marisma de Nación, was established in 1990. Goals for the constructed marsh, which were established by the U.S. Fish and Wildlife Service, included the following:[14]

1. Establishment of tide channels with sufficient fish to provide food for the California least tern.
2. Establishment of a stable or increasing population of salt-marsh bird's beak for three years.
3. Selection of the Pacific Estuarine Research Laboratory (PERL) at San Diego State University to monitor progress on the goals and conduct research on the constructed marsh. In 1997, PERL reported that goals for the least tern and bird's beak had been met, but that attempts to establish a habitat suitable for the rail had been only partially successful (see Table 9.3).

During the past decade, PERL scientists have conducted extensive research on the constructed marsh to determine the reasons for its limited success. They found that rails live, forage, and nest in cordgrass more than 60 centimeters tall. Nests are built of dead cordgrass attached to stems of living cordgrass so that the nests can remain above the water as it rises and falls. If the cordgrass is too short, the nests are not high enough to avoid being washed out during high tides.[14]

Researchers suggested that the coarse soil used to construct the marsh did not retain the amount of nitrogen needed for cordgrass to grow tall. Adding nitrogen-rich fertilizer to the soil resulted in taller plants in the constructed marsh, but only if the fertilizer was added on a continuing basis.

Another problem is that the diversity and numbers of large invertebrates, which are the major food source of the rails, are lower in the constructed marsh than in natural marshes. PERL researchers suspect that this, too, is linked to low nitrogen levels. Because nitrogen stimulates the growth of algae and plants, which provide food for small invertebrates, and these in turn provide food for larger invertebrates, low nitrogen can affect the entire food chain.[15, 16, 17]

Table 9.3 GOALS, PROGRESS, AND STATUS AS OF 2006

SPECIES	MITIGATION GOALS	PROGRESS IN MEETING REQUIREMENTS	STATUS AS OF 2006
California	Tidal channels with 75% of the fish species and 75% of the number of fish found in natural channels	Met standards	FWS recommended change from endangered to threatened
Salt-marsh bird's beak	Through reintroduction, at least 5 patches (20 plants each) that remain stable or increase for 3 years	Did not succeed on constructed islands but an introduced population on natural Sweetwater Marsh thrived for 3 years (reached 140,000 plants); continue to monitor because plant is prone to dramatic fluctuations in population	Still listed as endangered
Light-footed clappeer rail	Seven home ranges (82 ha), each having tidal channels with:	Constructed	Still listed as endangered; in 2005, eight captive-raised birds were released
	a. Forage Species equal to 75% of the invertebrate species and 75% of the number of invertebrates in natural areas	Met standards	
	b. High marsh areas for rails to find refuge during high tides	Sufficient in 1996 but two home ranges fell short in 1997	
	c. Low marsh for nesting with 50% coverage by tall cordgrass **d.** Population of tall cordgrass that is self-sustaining for 3 years	All home ranges met low marsh acreage requirement and all but one met cordgrass requirement and six lacked suffcient tall cordgrass Plant height can be increased with continual use of fertilizer but tall cordgrass is not self-sustaining	

Note: FWS stands for U.S. Fish and Wildlife Service

Critical Thinking Questions

1. Make a diagram of the food web in the marsh showing how the clapper rail, cordgrass, invertebrates, and nitrogen are related.
2. The headline of an article about the Sweetwater Marsh project in the April 17, 1998, issue of *Science* declared, "Restored Wetlands Flunk Real-World Test." Based on the information you have about the project, would you agree or disagree with this judgment? Explain your answer.
3. How do you think one can decide whether a constructed ecosystem is an adequate replacement for a natural ecosystem?
4. The term *adaptive management* refers to the use of scientific research in ecosystem management. In what ways has adaptive management been used in the Sweetwater Marsh project? What lessons from the project could be used to improve similar projects in the future?

SUMMARY

- Ecological restoration is the process of helping degraded ecosystems to recover and become more self-sustaining, and therefore able to pass through their natural range of conditions.
- Overarching goals of ecological restoration are to help transform degraded ecosystems into sustainable ecosystems and to develop new relationships between the natural and human-modified environments.
- Adaptive management, which applies science to the restoration process, is necessary if restoration is to be successful.
- Restoration of damaged ecosystems is a relatively new emphasis in environmental sciences that is developing into a new field. Restoration involves a combination of human activities and natural processes. It is also a social activity.
- Disturbance, change, and variation in the environment are natural, and ecological systems and species have evolved in response to these changes. These natural variations must be part of the goals of restoration.

REEXAMINING THEMES AND ISSUES

Human Population

If we degrade ecosystems to the point where their recovery from disturbance is slowed or they cannot recover at all, then we have reduced the local carrying capacity of those areas for human beings. For this reason, an understanding of the factors that determine ecosystem restoration is important to developing a sustainable human population.

Sustainability

Heavily degraded land, such as land damaged by pollution or overgrazing, loses the capacity to recover. By helping degraded ecosystems to recover, we promote sustainability. Ecological principles are useful in restoring ecosystems and thereby achieving sustainability.

Global Perspective

Each degradation of the land takes place locally, but such degradation has been happening around the world since the beginnings of civilization. Ecosystem degradation is therefore a global issue.

Urban World

In cities, we generally eliminate or damage the processes of succession and the ability of ecosystems to recover. As our world becomes more and more urban, we must learn to maintain these processes within cities, as well as in the countryside. Ecological restoration is an important way to improve city life.

People and Nature

Restoration is one of the most important ways that people can compensate for their undesirable effects on nature.

Science and Values

Because ecological systems naturally undergo changes and exist in a variety of conditions, there is no single "natural" state for an ecosystem. Rather, there is the process of succession, with all of its stages. In addition, there are major changes in the species composition of ecosystems over time. While science can tell us what conditions are possible and have existed in the past, which ones we choose to promote in any location is a question of values. Values and science are intimately integrated in ecological restoration.

KEY TERMS

adaptive management **172**
naturalization **180**
reclamation **180**
restoration ecology **172**

STUDY QUESTIONS

1. Develop a plan to restore an abandoned field in your town to natural vegetation for use as a park. The following materials are available: bales of hay; artificial fertilizer; and seeds of annual flowers, grasses, shrubs, and trees.
2. Oil has leaked for many years from the gasoline tanks of a gas station. Some of the oil has oozed to the surface. As a result, the gas station has been abandoned and revegetation has begun to occur. What effects would you expect this oil to have on the process of succession?
3. Refer to the Everglades in the opening case study. Assume there is no hope of changing water diversion from the upstream area that feeds water to the Everglades. Develop a plan to restore the Everglades, assuming the area of wetlands will decrease by another 30% as more water is diverted for people and agriculture in the next 20 years.
4. How can adaptive management best be applied to restoration projects?

FURTHER READING

Botkin, D.B., ***Discordant Harmonies: A New Ecology for the 21st Century*** (New York: Oxford University Press, 1992).

Botkin, D.B., ***No Man's Garden: Thoreau and a New Vision for Civilization and Nature*** (Washington, DC: Island Press, 2001).

Falk, Donald A., and Joy B. Zedler, ***Foundations of Restoration Ecology*** (Washington, DC: Island Press, 2005). A new and important book written by two of the world's experts on ecological restoration.

Higgs, E., ***Nature by Design: People, Natural Process, and Ecological Restoration*** (Cambridge, MA: MIT Press, 2003). A book that discusses the broader perspective on ecological restoration, including philosophical aspects.

Society for Ecological Restoration International Science and Policy Working Group, ***The SER International Primer on Ecological Restoration,*** 2004. www.ser.org. A good handbook on everything you need to know about ecological restoration.

CHAPTER 10

Environmental Health, Pollution, and Toxicology

Houston, Texas, Ship Channel, where many oil refineries are located.

LEARNING OBJECTIVES

Serious health problems may arise from toxic substances in water, air, soil, and even the rocks on which we build our homes. After reading this chapter, you should understand . . .

- How the terms *toxin, pollution, contamination, carcinogen, synergism*, and *biomagnification* are used in environmental health;
- What the classifications and characteristics are of major groups of pollutants in environmental toxicology;
- Why there is controversy and concern about synthetic organic compounds, such as dioxin;
- Whether we should be concerned about exposure to human-produced electromagnetic fields;
- What the dose-response concept is, and how it relates to LD-50, TD-50, ED-50, ecological gradients, and tolerance;
- How the process of biomagnification works, and why it is important in toxicology;
- Why the threshold effects of environmental toxins are important;
- What the process of risk assessment in toxicology is, and why such processes are often difficult and controversial.

CASE STUDY

Toxic Air Pollution and Human Health: Story of a Southeast Houston Neighborhood

Manchester is a neighborhood in southeast Houston, Texas, that is nearly surrounded by oil refineries and petrochemical plants. Residents and others have long noted the peculiar and not so pleasant smells of the area, but only recently have health concerns been raised. The neighborhood is close to downtown Houston, and the houses are relatively inexpensive, the streets safe. It was a generally positive neighborhood except for the occasional complaints about nosebleeds, coughing, and acidic smoke smells. Over a period of years, the number of oil refineries, petrochemical plants, and waste-disposal sites grew along what is known as the Houston Ship Channel (see opening photograph).[1]

Cancer is the second leading cause of death of U.S. children who are not linked to a known health risk before being stricken by the disease. Investigations into childhood cancer from air pollution are few in number but now include exposure to benzene and 1,3-butadiene, commonly referred to simply as butadiene.[1]

Benzene is a colorless toxic liquid that evaporates into the air. Exposure to benzene has a whole spectrum of possible consequences for people, such as drowsiness, dizziness, and headaches; irritation to eyes, skin, and respiratory tract; and loss of consciousness at high levels of exposure. Long-term (chronic) exposure through inhalation can cause blood disorders, including reduced numbers of red blood cells (anemia), in industrial settings. Inhalation has reportedly resulted in reproductive problems for women and, in tests on animals, adverse effects on the developing fetus. In humans, occupational exposure to benzene is linked to increased incidence of leukemia (a cancer of the tissues that form white blood cells). The many potential sources of exposure to benzene include tobacco smoke and evaporating gasoline at service stations. Of particular concern are industrial sources; for example, the chemical is released when gasoline is refined from oil.

The chemical 1,3-butadiene is a colorless gas with a mild gasoline-like odor. One way it is produced is as a by-product of refining oil. Health effects from this toxin are fairly well known and include both acute and chronic problems. Some of the acute problems are irritation of the eyes, throat, nose, and lungs. Possible chronic health effects of exposure to 1,3-butadiene include cancer, disorders of the central nervous system, damage to kidneys and liver, birth defects, fatigue, lowered blood pressure, headache, nausea, and cancer.[1,2] While there is controversy as to whether exposure to 1,3-butadiene causes cancer in people, more definitive studies of animals (rats and mice) exposed to the toxin have prompted the Environmental Protection Agency to classify 1,3-butadiene as a known human carcinogen.[1,2]

Solving problems related to air toxins in the Houston area has not been easy. First of all, the petrochemical facilities along the Houston Ship Channel were first established decades ago, during World War II, when the area was nearly unpopulated; since then, communities such as Manchester have grown up near the facilities. Second, the chemical plants at present are not breaking state or federal pollution laws. Texas is one of the states that have not established air standards for toxins emitted by the petrochemical industry. Advocates of clean air argue that the chemical industry doesn't own the air and doesn't have the right to contaminate it. People in the petrochemical industry say they are voluntarily reducing emissions of some of the chemicals known to cause cancer. Butadiene emissions have in fact decreased significantly in the last several years, but this is not much comfort to parents who believe their child contracted leukemia as a result of exposure to air toxins. Some people examining the air toxins released along Houston's Ship Channel have concluded that although further reducing emissions would be expensive, we have the technology to do it. Petrochemical companies are taking steps to reduce the emissions and the potential health risks associated with them, but more may be necessary.

A recent study set out to study neighborhoods (census tracts near the Ship Channel) with the highest levels of benzene and 1,3-butadiene in the air and evaluate whether these neighborhoods had a higher incidence of childhood lymphohematopoietic cancer. After adjusting for sex, ethnicity, and socioeconomic status, the study found that census tracts with the highest exposure to benzene had higher rates of leukemia.[1] The study concluded that elevated exposure to benzene and 1,3-butadiene may contribute to increased rates of childhood leukemia, but the possible link between the air pollution and disease needs further exploration.

The case history of the Houston Ship Channel, oil refineries, and disease is a complex problem for several reasons:

1. Disease seldom has a one-cause/one-effect relationship.
2. Data on air-pollution exposure are difficult to collect and link to a population of people who are moving around and have different responses to exposure to chemicals.
3. It is difficult to definitively link health problems to toxic air pollutants.
4. There have been few other studies with which the Houston study can be compared.

In this chapter we will explore selected aspects of exposure to toxins in the environment and real and potential health consequences to people and ecosystems.

10.1 Some Basics

As members of Earth's biological community, humans have a place in the biosphere—dependent on complex interrelations among the biosphere, atmosphere, hydrosphere, and lithosphere. We are only beginning to inquire into and gain a basic understanding of the total range of environmental factors that affect our health and well-being. As we continue our exploration of minute quantities of elements in soil, rocks, water, and air in relation to regional and global patterns of climate and earth science, we are making important discoveries about how these factors influence death rates and the incidence of disease. Incidence of a particular disease varies significantly from one area to another,[3,4] and some of the variability is the result of geologic, hydrologic, biologic, and chemical factors linked to Earth's climate system.

Disease—impairment of an individual's well-being and ability to function—is often due to poor adjustment between the individual and the environment. Disease occurs on a continuum—between a state of health and a state of disease is a *gray zone* of suboptimal health, a state of imbalance. In the gray zone, a person may not be diagnosed with a specific disease but may not be healthy.[5] There are many gray zones in environmental health, such as the many possible states of suboptimal health resulting from exposure to man-made chemicals, including pesticides; food additives, such as coloring, preservatives, and artificial saturated fat, some of which alter the chemical structure of food; exposure to tobacco smoke; exposure to air pollutants, such as ozone; exposure to chemicals in gasoline and in many household cleaners; and exposure to heavy metals, such as mercury or lead. As a result of exposure to chemicals in the environment from human activity, we may be in the midst of an epidemic of chronic disease that is unprecedented in human history.[5]

As noted in the opening case study, disease seldom has a one-cause/one-effect relationship with the environment. Rather, the incidence of a disease depends on several factors, including the physical environment, biological environment, and lifestyle. Linkages between these factors are often related to other factors, such as local customs and the level of industrialization. More primitive societies that live directly off the local environment are usually plagued by different environmental health problems than those in an urban society. For example, industrial societies have nearly eliminated such diseases as cholera, dysentery, and typhoid.

People are often surprised to learn that the water we drink, the air we breathe, the soil in which we grow crops, and the rocks on which we build our homes and workplaces may affect our chances of experiencing serious health problems and diseases (although, as suggested, direct relationships between the environment and disease are difficult to establish). At the same time, the environmental factors that contribute to disease—soil, rocks, water, and air—can also influence our chances of living longer, more productive lives.

Many people believe that soil, water, and air in a so-called natural state must be good, and that if human activities have changed or modified them, they have become contaminated, polluted, and therefore bad.[6] This is by no means the entire story; many natural processes—including dust storms, floods, and volcanic processes—can introduce materials harmful to people and other living things into the soil, water, and air.

A tragic example occurred on the night of August 21, 1986, when there was a massive natural release of carbon dioxide (CO_2) gas from Lake Nyos in Cameroon, Africa. The carbon dioxide was probably initially released from volcanic vents at the bottom of the lake and accumulated there over time. Pressure of the overlying lake water normally kept the dissolved gas down at the bottom, but the water was evidently agitated by a slide or small earthquake, and the bottom water moved upward. When the CO_2 gas reached the surface of the lake, it was released quickly into the air. But because CO_2 gas is heavier than air, it flowed downhill from the lake and settled in nearby villages, killing many animals and more than 1,800 people by asphyxiation (Figure 10.1).

It was estimated that a similar event could recur within about 20 years, assuming that carbon dioxide continued to be released at the bottom of the lake.[7] Fortunately, a hazard-reduction project funded by the U.S. Office of Foreign Disaster Assistance (scheduled to be completed early in the 21st century) includes inserting pipes into the

(a)

(b)

FIGURE 10.1 **(a)** In 1986, Lake Nyos in Cameroon, Africa, released carbon dioxide that moved down the slopes of the hills to settle in low places, asphyxiating animals and people. **(b).** Animals asphyxiated by carbon dioxide.

bottom of Lake Nyos, then pumping the gas-rich water to the surface, where the CO_2 gas is safely discharged into the atmosphere. In 2001, a warning system was installed, and one degassing pipe released a little more CO_2 than was seeping naturally into the lake. Recent data suggest that the single pipe now there barely keeps ahead of the CO_2 that continues to enter the bottom, so the lake's 500,000 tons of built-up gas have dropped only 6%. At this rate, it could take 30 to 50 years to make Lake Nyos safe. In the meantime, there could be another eruption.[8]

Terminology

What do we mean when we use the terms *pollution, contamination, toxin,* and *carcinogen?* A polluted environment is one that is impure, dirty, or otherwise unclean. The term **pollution** refers to an unwanted change in the environment caused by the introduction of harmful materials or the production of harmful conditions (heat, cold, sound). **Contamination** has a meaning similar to that of *pollution* and implies making something unfit for a particular use through the introduction of undesirable materials—for example, the contamination of water by hazardous waste. The term **toxin** refers to substances (pollutants) that are poisonous to living things. **Toxicology** is the science that studies toxins or suspected toxins, and toxicologists are scientists in this field. A **carcinogen** is a toxin that increases the risk of cancer. Carcinogens are among the most feared and regulated toxins in our society.

An important concept in considering pollution problems is **synergism**, the interaction of different substances, resulting in a total effect that is greater than the sum of the effects of the separate substances. For example, both sulfur dioxide (SO_2) and coal dust particulates are air pollutants. Either one taken separately may cause adverse health effects, but when they combine, as when SO_2 adheres to the coal dust, the dust with SO_2 is inhaled deeper than SO_2 alone and causes greater damage to lungs. Another aspect of synergistic effects is that the body may be more sensitive to a toxin if it is simultaneously subjected to other toxins.

Pollutants are commonly introduced into the environment by way of **point sources**, such as smokestacks (see A Closer Look 10.1), pipes discharging into waterways, a small stream entering the ocean (Figure 10.2), or accidental spills. **Area sources**, also called *nonpoint sources*, are more diffused over the land and include urban runoff and **mobile sources**, such as automobile exhaust. Area sources are difficult to isolate and correct because the problem is often widely dispersed over a region, as in agricultural runoff that contains pesticides.

FIGURE 10.2 This southern California urban stream flows into the Pacific Ocean at a coastal park. The stream water often carries high counts of fecal coliform bacteria. As a result, the stream is a point source of pollution for the beach, which is sometimes closed to swimming following runoff events.

Measuring the Amount of Pollution

How the amount or concentration of a particular pollutant or toxin present in the environment is reported varies widely. The amount of treated wastewater entering Santa Monica Bay in the Los Angeles area is a big number, reported in millions of gallons per day. Emission of nitrogen and sulfur oxides into the air is also a big number, reported in millions of tons per year. Small amounts of pollutants or toxins in the environment, such as pesticides, are reported in units as parts per million (ppm) or parts per billion (ppb). It is important to keep in mind that the concentration in ppm or ppb may be by volume, mass, or weight. In some toxicology studies, the units used are milligrams of toxin per kilogram of body mass (1 mg/kg is equal to 1 ppm). Concentration may also be recorded as a percentage. For example, 100 ppm (100 mg/kg) is equal to 0.01%. (How many ppm are equal to 1%?)

When dealing with water pollution, units of concentration for a pollutant may be milligrams per liter (mg/L) or micrograms per liter (μg/L). A milligram is one-thousandth of a gram, and a microgram is one-millionth of a gram. For water pollutants that do not cause significant change in the density of water (1 g/cm^3), a pollutant concentration of 1 mg/L is approximately equivalent to 1 ppm. Air pollutants are commonly measured in units such as micrograms of pollutant per cubic meter of air (μg/m^3).

Units such as ppm, ppb, or μg/m^3 reflect very small concentrations. For example, if you were to use 3 g (one-tenth of an ounce) of salt to season popcorn in order to have salt at a concentration of 1 ppm by weight of the popcorn, you would have to pop approximately 3 metric tons of kernels!

10.2 Categories of Pollutants

A partial classification of pollutants by arbitrary categories is presented below. We discuss examples of other pollutants in other parts of the book.

Infectious Agents

Infectious diseases—spread by the interactions between individuals and by the food, water, air, soil, and animals we come in contact with—constitute some of the oldest health problems that people face. Today, infectious diseases have the potential to pose rapid threats, both local and global, by spreading in hours via airplane travelers. Terrorist activity may also spread diseases. Inhalation anthrax caused by a bacterium sent in powdered form in envelopes through the mail killed several people in 2001. New diseases are emerging, and previous ones may emerge again. Although we have cured many diseases, we have no known reliable vaccines for others, such as HIV, hantavirus, and dengue fever.

The H1N1 flu pandemic (widespread outbreak of a disease) that became apparent in 2009 started in Mexico and has spread around the world. The complete origin of H1N1 remains unknown, but it has genetic markers of two swine flues, a human flu, and an avian (bird) flu. As we live closer together, nearer large numbers of animals such as chickens and pigs in large industrial farms and tightly confined animals in smaller farms, the probability of a disease crossing from animals to humans increases. People working closely with pigs have an increased risk of contracting swine flu.

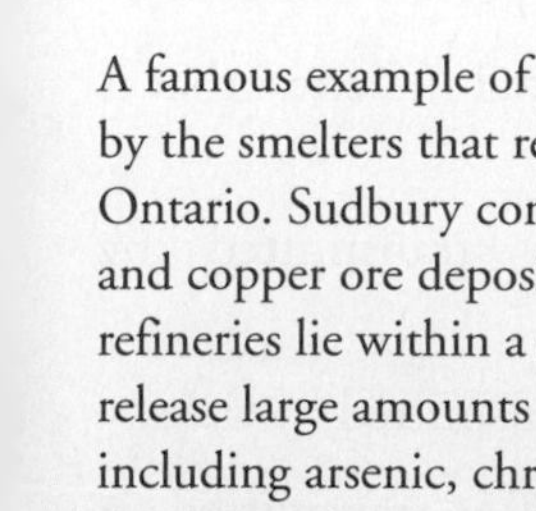

A CLOSER LOOK 10.1

Sudbury Smelters: A Point Source

A famous example of a point source of pollution is provided by the smelters that refine nickel and copper ores at Sudbury, Ontario. Sudbury contains one of the world's major nickel and copper ore deposits. A number of mines, smelters, and refineries lie within a small area. The smelter stacks used to release large amounts of particulates containing toxic metals—including arsenic, chromium, copper, nickel, and lead—into the atmosphere, much of which was then deposited locally in the soil. In addition, because the areas contained a high percentage of sulfur, the emissions included large amounts of sulfur dioxide (SO_2). During its peak output in the 1960s, this complex was the largest single source of SO_2 emissions in North America, emitting 2 million metric tons per year.

As a result of the pollution, nickel contaminated soils up to 50 km (about 31 mi) from the stacks. The forests that once surrounded Sudbury were devastated by decades of acid rain (produced from SO_2 emissions) and the deposition of particulates containing heavy metals. An area of approximately 250 km^2 (96 mi^2) was nearly devoid of vegetation, and damage to forests in the region has been visible over an area

FIGURE 10.3 (a) Lake St. Charles, Sudbury, Ontario, prior to restoration. Note high stacks (smelters) in the background and lack of vegetation in the foreground, resulting from air pollution (acid and heavy-metal deposition). **(b) Recent photo showing regrowth and restoration.**

of approximately 3,500 km^2 (1,350 mi^2); see Figure 10.3a. To control emissions from Sudbury, the Ontario government set standards to reduce emissions to less than 365,000 tons per year by 1994. The goal was achieved by reducing production from the smelters and by treating the emissions to reduce pollution.[9]

Reducing emissions from Sudbury has allowed surrounding areas to recover from the pollution (Figure 10.3b). Species of trees once eradicated from some areas have begun to grow again. Recent restoration efforts have included planting over 7 million trees and 75 species of herbs, moss, and lichens—all of which have contributed to the increase of biodiversity. Lakes damaged by acid precipitation in the area are rebounding and now support populations of plankton and fish.[9]

The case of the Sudbury smelters provides a positive example of emphasizing the key theme of thinking globally but acting locally to reduce air pollution. It also illustrates the theme of science and values: Scientists and engineers can design pollution-abatement equipment, but spending the money to purchase the equipment reflects what value we place on clean air.

Environmentally Transmitted Infectious Disease

Diseases that can be controlled by manipulating the environment, such as by improving sanitation or treating water, are classified as environmental health concerns. Although there is great concern about the toxins and carcinogens produced in industrial society today, the greatest mortality in developing countries is caused by environmentally transmitted infectious disease. In the United States, thousands of cases of waterborne illness and food poisoning occur each year. These diseases can be spread by people; by mosquitoes and fleas; or by contact with contaminated food, water, or soil. They can also be transmitted through ventilation systems in buildings. The following are some examples of environmentally transmitted infectious diseases:

- Legionellosis, or Legionnaires' disease, which often occurs where air-conditioning systems have been contaminated by disease-causing organisms.
- Giardiasis, a protozoan infection of the small intestine, spread via food, water, or person-to-person contact.
- Salmonella, a food-poisoning bacterial infection that is spread via water or food.
- Malaria, a protozoan infection transmitted by mosquitoes.
- Lyme borreliosis (Lyme disease), transmitted by ticks.
- Cryptosporidiosis, a protozoan infection transmitted via water or person-to-person contact (see Chapter 19).[10]
- Anthrax, spread by terrorist activity.

We sometimes hear about epidemics in developing nations. An example is the highly contagious Ebola virus in Africa, which causes external and internal bleeding and kills 80% of those infected. We may tend to think of such epidemics as problems only for developing nations, but this may give us a false sense of security. True, monkeys and bats spread Ebola, but the origin of the virus in the tropical forest remains unknown. Developed countries, where outbreaks may occur in the future, must learn from the developing countries' experiences. To accomplish this and avoid potential global tragedies, more funds must be provided for the study of infectious diseases in developing countries.

Toxic Heavy Metals

The major **heavy metals** (metals with relatively high atomic weight; see Chapter 6) that pose health hazards to people and ecosystems include mercury, lead, cadmium, nickel, gold, platinum, silver, bismuth, arsenic, selenium, vanadium, chromium, and thallium. Each of these elements may be found in soil or water not contaminated by people, each has uses in our modern industrial society, and each is also a by-product of the mining, refining, and use of other elements. Heavy metals often have direct physiological toxic effects. Some are stored or incorporated in living tissue, sometimes permanently. Heavy metals tend to be stored (accumulating with time) in fatty body tissue. A little arsenic each day may eventually result in a fatal dose—the subject of more than one murder mystery.

The quantity of heavy metals in our bodies is referred to as the *body burden*. The body burden of toxic heavy elements for an average human body (70 kg) is about 8 mg of antimony, 13 mg of mercury, 18 mg of arsenic, 30 mg of cadmium, and 150 mg of lead. The average body burden of lead (for which we apparently have no biological need) is about twice that of the others combined, reflecting our heavy use of this potentially toxic metal.

Mercury, thallium, and lead are very toxic to people. They have long been mined and used, and their toxic properties are well known. Mercury, for example, is the "Mad Hatter" element. At one time, mercury was used to stiffen felt hats, and because mercury damages the brain, hatters in Victorian England were known to act peculiarly. Thus, the Mad Hatter in Lewis Carroll's *Alice in Wonderland* had real antecedents in history.

Toxic Pathways

Chemical elements released from rocks or human processes can become concentrated in people (see Chapter 6) through many pathways (Figure 10.4). These pathways may involve what is known as **biomagnification**—the accumulation or increasing concentration of a substance in living tissue as it moves through a food web (also known as *bioaccumulation*). For example, cadmium, which increases the risk of heart disease, may enter the environment via ash from burning coal. The cadmium in coal is in very low concentrations (less than 0.05 ppm). However, after coal is burned in a power plant, the ash is collected in a solid form and disposed of in a landfill. The landfill is covered with soil and revegetated. The low concentration of cadmium in the ash and soil is taken into the plants as they grow, but the concentration of cadmium in the plants is three to five times greater than the concentration in the ash. As the cadmium moves through the food chain, it becomes more and more concentrated. By the

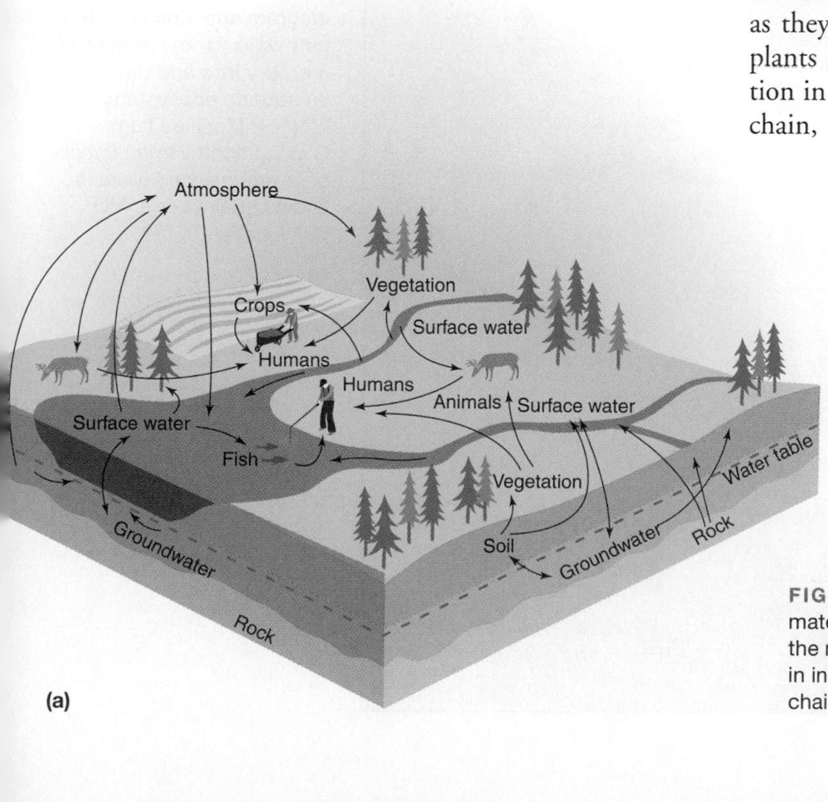

(a)

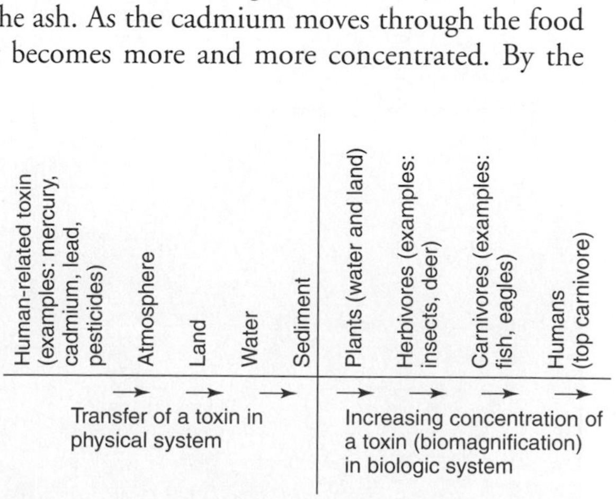

(b)

FIGURE 10.4 **(a)** Potential complex pathways for toxic materials through the living and nonliving environment. Note the many arrows into humans and other animals, sometimes in increasing concentrations as they move through the food chain **(b)**.

time it is incorporated into the tissue of people and other carnivores, the concentration is approximately 50 to 60 times the original concentration in the coal.

Mercury in aquatic ecosystems offers another example of biomagnification. Mercury is a potentially serious pollutant of aquatic ecosystems such as ponds, lakes, rivers, and the ocean. Natural sources of mercury in the environment include volcanic eruptions and erosion of natural mercury deposits, but we are most concerned with human input of mercury into the environment by, for example, burning coal in power plants, incinerating waste, and processing metals such as gold. Rates of input of mercury into the environment through human processes are poorly understood. However, it is believed that human activities have doubled or tripled the amount of mercury in the atmosphere, and it is increasing at about 1.5% per year.[11]

A major source of mercury in many aquatic ecosystems is deposition from the atmosphere through precipitation. Most of the deposition is of inorganic mercury (Hg^{++}, ionic mercury). Once this mercury is in surface water, it enters into complex biogeochemical cycles and a process known as *methylation* may occur. Methylation changes inorganic mercury to methyl mercury $[CH_3Hg]^+$ through bacterial activity. Methyl mercury is much more toxic than inorganic mercury, and it is eliminated more slowly from animals' systems. As the methyl mercury works its way through food chains, biomagnification occurs, resulting in higher concentrations of methyl mercury farther up the food chain. In short, big fish that eat little fish contain higher concentrations of mercury than do smaller fish and the aquatic insects that the fish feed on.

Selected aspects of the mercury cycle in aquatic ecosystems are shown in Figure 10.5. The figure emphasizes the input side of the cycle, from deposition of inorganic mercury through formation of methyl mercury, biomagnification, and sedimentation of mercury at the bottom of a pond. On the output side of the cycle, the mercury that enters fish may be taken up by animals that eat the fish; and sediment may release mercury by a variety of processes, including resuspension in the water, where eventually the mercury enters the food chain or is released into the atmosphere through volatilization (conversion of liquid mercury to a vapor form).

Biomagnification also occurs in the ocean. Because large fish, such as tuna and swordfish, have elevated mercury levels, we are advised to limit our consumption of these fish, and pregnant women are advised not to eat them at all.

The threat of mercury poisoning is widespread. Millions of young children in Europe, the United States, and other industrial countries have mercury levels that exceed health standards.[12] Even children in remote areas of the far north are exposed to mercury through their food chain.

During the 20th century, several significant incidents of methyl mercury poisoning were recorded. One, in Minamata Bay, Japan, involved the industrial release of methyl mercury (see A Closer Look 10.2). Another, in Iran, involved a methyl mercury fungicide used to treat wheat seeds. In each of these cases, hundreds of people were killed and thousands were permanently damaged.[11]

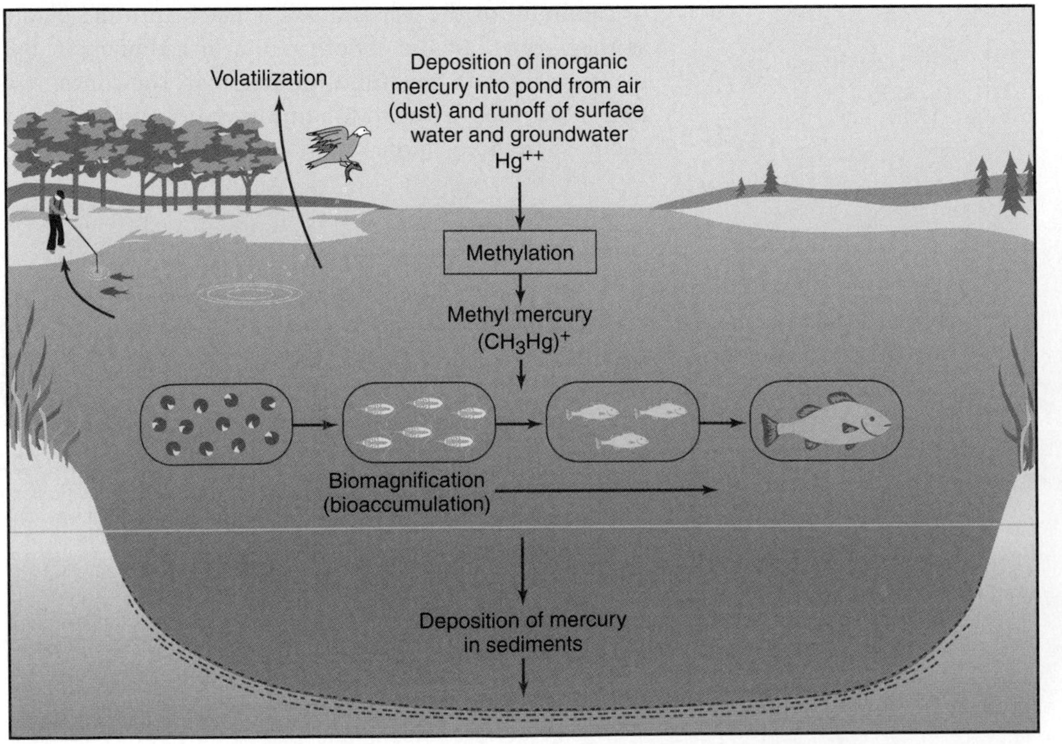

FIGURE 10.5 Idealized diagram showing selected pathways for movement of mercury into and through an aquatic ecosystem. (*Source*: Modified from G.L. Waldbott, *Health Effects of Environmental Pollutants*, 2nd ed. [Saint Louis, MO: C.V. Mosby, 1978].)

The cases in Minamata Bay and Iran involved local-exposure mercury. What is being reported in the Arctic, however, emphasizes mercury at the global level, in a region far from emission sources of the toxic metal. The Inuit people in Quanea, Greenland, live above the Arctic Circle, far from any roads and 45 minutes by helicopter from the nearest outpost of modern society. Nevertheless, they are some of the most chemically contaminated people on Earth, with as much as 12 times more mercury in their blood than is recommended in U.S. guidelines. The mercury gets to the Inuit from the industrialized world by way of what they eat. The whale, seal, and fish they eat contain mercury that is further concentrated in the tissue and blood of the people. The process of increasing concentrations of mercury farther up the food chain is an example of biomagnification.[12]

What needs to be done to stop mercury toxicity at the local to global level is straightforward. The answer is to reduce emissions of mercury by capturing it before emission or by using alternatives to mercury in industry. Success will require international cooperation and technology transfer to countries such as China and India, which, with their tremendous increases in manufacturing, are the world's largest users of mercury today.[12]

Organic Compounds

Organic compounds are carbon compounds produced naturally by living organisms or synthetically by industrial processes. It is difficult to generalize about the environmental and health effects of artificially produced organic compounds because there are so many of them, they have so many uses, and they can produce so many different kinds of effects.

Synthetic organic compounds are used in industrial processes, pest control, pharmaceuticals, and food additives. We have produced over 20 million synthetic chemicals, and new ones are appearing at a rate of about 1 million per year! Most are not produced commercially, but up to 100,000 chemicals are now being used, or have been used in the past. Once used and dispersed in the environment, they may become a hazard for decades or even hundreds of years.

A CLOSER LOOK 10.2

Mercury and Minamata, Japan

In the Japanese coastal town of Minamata, on the island of Kyushu, a strange illness began to occur in the middle of the 20th century. It was first recognized in birds that lost their coordination and fell to the ground or flew into buildings, and in cats that went mad, running in circles and foaming at the mouth.[13] The affliction, known by local fishermen as the "disease of the dancing cats," subsequently affected people, particularly families of fishermen. The first symptoms were subtle: fatigue, irritability, headaches, numbness in arms and legs, and difficulty in swallowing. More severe symptoms involved the sensory organs; vision was blurred and the visual field was restricted. Afflicted people became hard of hearing and lost muscular coordination. Some complained of a metallic taste in their mouths; their gums became inflamed, and they suffered from diarrhea. Lawsuits were brought, and approximately 20,000 people claimed to be affected. In the end, according to the Japanese government, almost 3,000 people were affected and almost 1,800 died. Those affected lived in a small area, and much of the protein in their diet came from fish from Minamata Bay.

A vinyl chloride factory on the bay used mercury in an inorganic form in its production processes. The mercury was released in waste that was discharged into the bay. Mercury forms few organic compounds, and it was believed that the mercury, though poisonous, would not get into food chains. But the inorganic mercury released by the factory was converted by bacterial activity in the bay into methyl mercury, an organic compound that turned out to be much more harmful. Unlike inorganic mercury, methyl mercury readily passes through cell membranes. It is transported by the red blood cells throughout the body, and it enters and damages brain cells.[14] Fish absorb methyl mercury from water 100 times faster than they absorb inorganic mercury. (This was not known before the epidemic in Japan.) And once absorbed, methyl mercury is retained two to five times longer than is inorganic mercury.

In 1982, lawsuits were filed by plaintiffs affected by the mercury. Twenty-two years later, in 2004—almost 50 years after the initial poisonings—the government of Japan agreed to a settlement of $700,000.

Harmful effects of methyl mercury depend on a variety of factors, including the amount and route of intake, the duration of exposure, and the species affected. The effects of the mercury are delayed from three weeks to two months from the time of ingestion. If mercury intake ceases, some symptoms may gradually disappear, but others are difficult to reverse.[14]

The mercury episode at Minamata illustrates four major factors that must be considered in evaluating and treating toxic environmental pollutants.

Individuals vary in their response to exposure to the same dose, or amount, of a pollutant. Not everyone in Minamata responded in the same way; there were variations even among those most heavily exposed. Because we cannot predict exactly how any single individual will respond, we need to find a way to state an expected response of a particular percentage of individuals in a population.

Pollutants may have a threshold—that is, a level below which the effects are not observable and above which the effects become apparent. Symptoms appeared in individuals with concentrations of 500 ppb of mercury in their bodies; no measurable symptoms appeared in individuals with significantly lower concentrations.

Some effects are reversible. Some people recovered when the mercury-filled seafood was eliminated from their diet.

The chemical form of a pollutant, its activity, and its potential to cause health problems may be changed markedly by ecological and biological processes. In the case of mercury, its chemical form and concentration changed as the mercury moved through the food webs.

Sources: Mary Kugler, R.N. Thousands poisoned, disabled, and killed. About.com. Created October 23, 2004. About.com Health's Disease and Condition content is reviewed by our Medical Review Board. Also, BBC News, "Japan remembers mercury victims." http://news.bbc.co.uk/go/pr/fr/-/2/hi/asia-pacific/4959562.stm Published 2006/05/01 15:03:11 GMT ©BBC MM VIII.

Persistent Organic Pollutants

Some synthetic compounds are called **persistent organic pollutants**, or **POPs**. Many were first produced decades ago, when their harm to the environment was not known, and they are now banned or restricted (see Table 10.1 and A Closer Look 10.3). POPs have several properties that define them:[15]

- They have a carbon-based molecular structure, often containing highly reactive chlorine.
- Most are manufactured by people—that is, they are synthetic chemicals.
- They are persistent in the environment—they do not easily break down in the environment.
- They are polluting and toxic.
- They are soluble in fat and likely to accumulate in living tissue.
- They occur in forms that allow them to be transported by wind, water, and sediments for long distances.

For example, consider polychlorinated biphenyls (PCBs), which are heat-stable oils originally used as an insulator in electric transformers.[15] A factory in Alabama manufactured PCBs in the 1940s, shipping them to a General Electric factory in Massachusetts. They were put in insulators and mounted on poles in thousands of locations. The transformers deteriorated over time. Some were damaged by lightning, and others were damaged or destroyed during demolition. The PCBs leaked into the soil or were carried by surface runoff into streams and rivers. Others combined with dust, were transported by wind around the world, and were deposited in ponds, lakes, or rivers, where they entered the food chain. First the PCBs entered algae. Insects ate the algae and were in turn eaten by shrimp and fish. In each stage up the food web, the concentration of PCBs increased. Fish are caught and eaten, passing the PCBs on to people, where they are concentrated in fatty tissue and mother's milk.

Table 10.1 SELECTED COMMON PERSISTENT ORGANIC POLLUTANTS (POPs)

CHEMICAL	EXAMPLE OF USE
Aldrin[a]	Insecticide
Atrazine[b]	Herbicide
DDT[a]	Insecticide
Dieldrin[a]	Insecticide
Endrin[c]	Insecticide
PCBs[a]	Liquid insulators in electric transformers
Dioxins	By-product of herbicide production

[a] Banned in the United States and many other countries.

[b] Degrades in the environment. It is persistent when reapplied often.

[c] Restricted or banned in many countries.

Source: Data in part from Anne Platt McGinn, "Phasing Out Persistent Organic Pollutants," in Lester R. Brown et al., *State of the World 2000* (New York: Norton, 2000).

A CLOSER LOOK 10.3

Dioxin: How Dangerous Is It?

Dioxin, a persistent organic pollutant, or POP, may be one of the most toxic man-made chemicals in the environment. The history of the scientific study of dioxin and its regulation illustrates the interplay of science and values.

Dioxin is a colorless crystal made up of oxygen, hydrogen, carbon, and chlorine. It is classified as an organic compound because it contains carbon. About 75 types of dioxin and dioxinlike compounds are known; they are distinguished from one another by the arrangement and number of chlorine atoms in the molecule.

Dioxin is not normally manufactured intentionally but is a by-product of chemical reactions, including the combustion of compounds containing chlorine in the production of herbicides.[16] In the United States, there are a variety of sources for dioxinlike compounds (specifically, chlorinated dibenzo-*p*-dioxin, or CDD, and chlorinated dibenzofurans, or CDF). These compounds are emitted into the air through such processes as incineration of municipal waste (the major source), incineration of medical waste, burning of gasoline and diesel fuels in vehicles, burning of wood as a fuel, and refining of metals such as copper.

The good news is that releases of CDDs and CDFs decreased about 75% from 1987 to 1995. However, we are only beginning to understand the many sources of dioxin emissions into the air, water, and land and the linkages and rates of transfer from dominant airborne transport to deposition in water, soil, and the biosphere. In too many cases, the amounts of dioxins emitted are based more on expert opinion than on high-quality data, or even on limited data.[17]

Studies of animals exposed to dioxin suggest that some fish, birds, and other animals are sensitive to even small amounts. As a result, it can cause widespread damage to wildlife, including birth defects and death. However, the concentration at which it poses a hazard to human health is still controversial. Studies suggest that workers exposed to high concentrations of dioxin for longer than a year have an increased risk of dying of cancer.[18]

The Environmental Protection Agency (EPA) has classified dioxin as a known human carcinogen, but the decision is controversial. For most of the exposed people, such as those eating a diet high in animal fat, the EPA puts the risk of developing cancer between 1 in 1,000 and 1 in 100. This estimate represents the highest possible risk for individuals who have had the greatest exposure. For most people, the risk will likely be much lower.[19] The EPA has set an acceptable intake of dioxin at 0.006 pg per kilogram of body weight per day (1 pg $= 10^{-12}$ g; see Appendix for prefixes and multiplication factors). This level is deemed too low by some scientists, who argue that the acceptable intake ought to be 100 to 1,000 times higher, or approximately 1 to 10 pg per day.[18] The EPA believes that setting the level this much higher could result in health effects.

The dioxin problem became well known in 1983 when Times Beach, Missouri, a river town just west of Saint Louis with a population of 2,400, was evacuated and purchased for $36 million by the government. The evacuation and purchase occurred after the discovery that oil sprayed on the town's roads to control dust contained dioxin, and that the entire area had been contaminated. Times Beach was labeled a dioxin ghost town (Figure 10.6). The buildings were bulldozed, and all that was left was a grassy and woody area enclosed by a barbed-wire-topped chain-link fence. The evacuation has since been viewed by some scientists (including the person who ordered the evacuation) as a government overreaction to a perceived dioxin hazard. Following clean up, trees were planted and today Times Beach is part of Route 66 State Park and a bird refuge.

The controversy about the toxicity of dioxin is not over.[20-23] Some environmental scientists argue that the regulation of dioxin must be tougher, whereas the industries producing the chemical argue that the dangers of exposure are exaggerated.

FIGURE 10.6 Soil samples from Times Beach, Missouri, thought to be contaminated by dioxin.

Hormonally Active Agents (HAAs)

HAAs are also POPs. An increasing body of scientific evidence indicates that certain chemicals in the environment, known as **hormonally active agents (HAAs)**, may cause developmental and reproductive abnormalities in animals, including humans (see A Closer Look 10.4). HAAs include a wide variety of chemicals, such as some herbicides, pesticides, phthalates (compounds found in many chlorine-based plastics), and PCBs. Evidence in support of the hypothesis that HAAs are interfering with the growth and development of organisms comes from studies of wildlife in the field and laboratory studies of human diseases, such as breast, prostate, and ovarian cancer, as well as abnormal testicular development and thyroid-related abnormalities.[24]

Studies of wildlife include evidence that alligator populations in Florida that were exposed to pesticides, such as DDT, have genital abnormalities and low egg production. Pesticides have also been linked to reproductive problems in several species of birds, including gulls, cormorants, brown pelicans, falcons, and eagles. Studies are ongoing on Florida panthers; they apparently have abnormal ratios of sex hormones, and this may be affecting their reproductive capability. In sum, the studies of major disorders in wildlife have centered on abnormalities, including thinning of birds' eggshells, decline in populations of various animals and birds, reduced viability of offspring, and changes in sexual behavior.[25]

With respect to human diseases, much research has been done on linkages between HAAs and breast cancer by exploring relationships between environmental estrogens and cancer. Other studies are ongoing to understand relationships between PCBs and neurological behavior that results in poor performance on standard intelligence tests. Finally, there is concern that exposure of people to phthalates that are found in plastics containing chlorine is also causing problems. Consumption of phthalates in the United States is considerable, with the highest exposure in women of childbearing age. The products being tested as the source of contamination include perfumes and other cosmetics, such as nail polish and hairspray.[25]

In sum, there is good scientific evidence that some chemical agents, in sufficient concentrations, will affect human reproduction through endocrine and hormonal disruption. The human endocrine system is of primary importance because it is one of the two main systems (the other is the nervous system) that regulate and control growth, development, and reproduction. The human endocrine system consists of a group of hormone-secreting glands, including the thyroid, pancreas, pituitary, ovaries (in women), and testes (in men). The bloodstream transports the hormones to virtually all parts of the body, where they act as chemical messengers to control growth and development of the body.[24]

The National Academy of Sciences completed a review of the available scientific evidence concerning HAAs and recommends continued monitoring of wildlife and human populations for abnormal development and reproduction. Furthermore, where wildlife species are known to be experiencing declines in population associated with abnormalities, experiments should be designed to study the phenomena with respect to chemical contamination. For people, the recommendation is for additional studies to document the presence or absence of associations between HAAs and human cancers. When associations are discovered, the causality is investigated in the relationship between exposure and disease, and indicators of susceptibility to disease of certain groups of people by age and sex.[25]

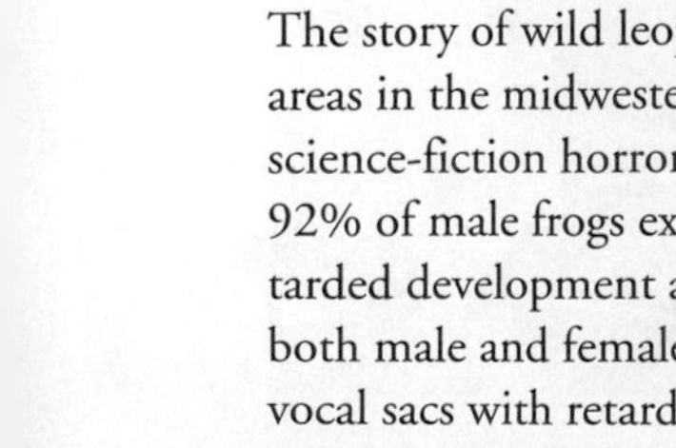

A CLOSER LOOK 10.4

Demasculinization and Feminization of Frogs

The story of wild leopard frogs (Figure 10.7) from a variety of areas in the midwestern United States sounds something like a science-fiction horror story. In affected areas, between 10 and 92% of male frogs exhibit gonadal abnormalities, including retarded development and hermaphroditism, meaning they have both male and female reproductive organs. Other frogs have vocal sacs with retarded growth. Since their vocal sacs are used to attract female frogs, these frogs are less likely to mate.

What is apparently causing some of the changes in male frogs is exposure to atrazine, the most widely used herbicide in the United States today. The chemical is a weed killer, used primarily in agricultural areas. The region of the United States with the highest frequency (92%) of sex reversal of male frogs is in Wyoming, along the North Platte River. Although the region is not near any large agricultural activity, and the use of atrazine there is not particularly significant, hermaphrodite frogs are common there because the North Platte River flows from areas in Colorado where atrazine is commonly used.

The amount of atrazine released into the environment of the United States is estimated at approximately 7.3 million kg (16 million lbs) per year. The chemical degrades in the environment, but the degradation process is longer than the application cycle. Because of its continual application every year, the waters of the Mississippi River basin, which drains about 40% of the lower

FIGURE 10.7 Wild leopard frogs in America have been affected by man-made chemicals (the herbicide atrazine) in the environment.

United States, discharge approximately 0.5 million kg (1.2 million lbs) of atrazine per year to the Gulf of Mexico. Atrazine easily attaches to dust particles and has been found in rain, fog, and snow. As a result, it has contaminated groundwater and surface water in regions where it isn't used. The EPA states that up to 3 parts per billion (ppb) of atrazine in drinking water is acceptable, but at this concentration it definitely affects frogs that swim in the water. Other studies around the world have confirmed this. For example, in Switzerland, where atrazine is banned, it commonly occurs with a concentration of about 1 ppb, and that is sufficient to change some male frogs into females. In fact, atrazine can apparently cause sex change in frogs at concentrations as low as one-thirteenth of the level set by the EPA for drinking water.

Of particular interest and importance is the process that causes the changes in leopard frogs. We begin the discussion with the endocrine system, composed of glands that secrete hormones such as testosterone and estrogen directly into the bloodstream, which carries them to parts of the body where they regulate and control growth and sexual development. Testosterone in male frogs is partly responsible for development of male characteristics. The atrazine is believed to switch on a gene that turns testosterone into estrogen, a female sex hormone. It's the hormones, not the genes, that actually regulate the development and structure of reproductive organs.

Frogs are particularly vulnerable during their early development, before and as they metamorphose from tadpoles into adult frogs. This change occurs in the spring, when atrazine levels are often at a maximum in surface water. Apparently, a single exposure to the chemical may affect the frog's development. Thus, the herbicide is known as a hormone disrupter.

In a more general sense, substances that interact with the hormone systems of an organism, whether or not they are linked to disease or abnormalities, are known as hormonally active agents (HAAs). These HAAs are able to trick the organism's body (in this case, the frog's) into believing that the chemicals have a role to play in its functional development. An analogy you might be more familiar with is a computer virus that fools the computer into accepting it as part of the system by which the computer works. Similar to computer viruses, the HAAs interact with an organism and the mechanisms for regulating growth and development, thus disrupting normal growth functions.

What happens when HAAs—in particular, hormone disrupters (such as pesticides and herbicides)—are introduced into the system is shown in Figure 10.8. Natural hormones produced by the body send chemical messages to cells, where receptors for the hormone molecules are found on the outside and inside of cells. These natural hormones then transmit instructions to the cells' DNA, eventually directing development and growth. We now know that chemicals, such as some pesticides and herbicides, can also bind to the receptor molecules and either mimic or obstruct the role of the natural hormones. Thus, hormonal disrupters may also be known as HAAs.[24–28]

The story of wild leopard frogs in America dramatizes the importance of carefully evaluating the role of man-made chemicals in the environment. Populations of frogs and other amphibians are declining globally, and much research has been directed toward understanding why. Studies to evaluate past or impending extinctions of organisms often center on global processes such as climate change, but the story of leopard frogs leads us down another path, one associated with our use of the natural environment. It also raises a number of more disturbing questions: Are we participating in an unplanned experiment on how man-made chemicals, such as herbicides and pesticides, might transform the bodies of living beings, perhaps even people? Are these changes in organisms limited to only certain plants and animals, or are they a forerunner of what we might expect in the future on a much broader scale? Perhaps we will look back on this moment of understanding as a new beginning in meaningful studies that will answer some of these important questions.

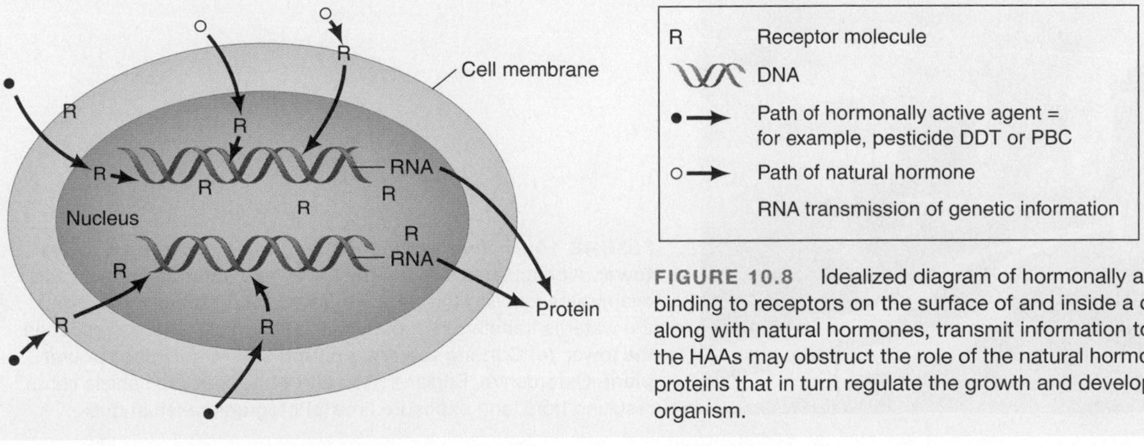

FIGURE 10.8 Idealized diagram of hormonally active agents (HAAs) binding to receptors on the surface of and inside a cell. When HAAs, along with natural hormones, transmit information to the cells' DNA, the HAAs may obstruct the role of the natural hormones that produce proteins that in turn regulate the growth and development of an organism.

Nuclear Radiation

Nuclear radiation is introduced here as a category of pollution. We discuss it in detail in Chapter 17, in conjunction with nuclear energy. We are concerned about nuclear radiation because excessive exposure is linked to serious health problems, including cancer. (See Chapter 21 for a discussion of radon gas as an indoor air pollutant.)

Thermal Pollution

Thermal pollution, also called *heat pollution*, occurs when heat released into water or air produces undesirable effects. Heat pollution can occur as a sudden, acute event or as a long-term, chronic release. Sudden heat releases may result from natural events, such as brush or forest fires and volcanic eruptions, or from human activities, such as agricultural burning.

The major sources of chronic heat pollution are electric power plants that produce electricity in steam generators and release large amounts of heated water into rivers. This changes the average water temperature and the concentration of dissolved oxygen (warm water holds less oxygen than cooler water), thereby changing a river's species composition (see the discussion of eutrophication in Chapter 19). Every species has a temperature range within which it can survive and an optimal temperature for living. For some species of fish, the range is small, and even a small change in water temperature is a problem. Lake fish move away when the water temperature rises more than about 1.5°C above normal; river fish can withstand a rise of about 3°C.

Heating river water can change its natural conditions and disturb the ecosystem in several ways. Fish spawning cycles may be disrupted, and the fish may have a heightened susceptibility to disease. Warmer water also causes physical stress in some fish, making them easier for predators to catch, and warmer water may change the type and abundance of food available for fish at various times of the year.

There are several solutions to chronic thermal discharge into bodies of water. The heat can be released into the air by cooling towers (Figure 10.9), or the heated water can be temporarily stored in artificial lagoons until it

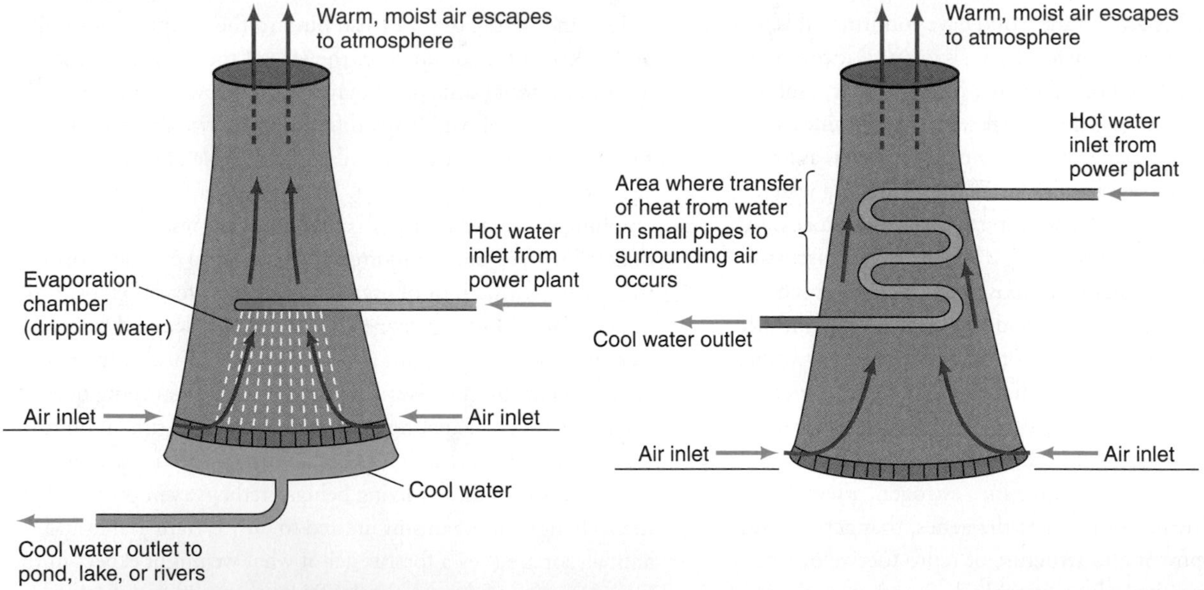

(c)

FIGURE 10.9 **Two types of cooling towers. (a) Wet cooling tower.** Air circulates through the tower; hot water drips down and evaporates, cooling the water. **(b) Dry cooling tower.** Heat from the water is transferred directly to the air, which rises and escapes the tower. **(c) Cooling towers emitting steam at Didcot power plant**, Oxfordshire, England. Red and white lines are vehicle lights resulting from long exposure time (photograph taken at dusk).

cools down to normal temperatures. Some attempts have been made to use the heated water to grow organisms of commercial value that require warmer water. Waste heat from a power plant can also be captured and used for a variety of purposes, such as warming buildings (see Chapter 14 for a discussion of cogeneration).

Particulates

Particulates here refer to small particles of dust (including soot and asbestos fibers) released into the atmosphere by many natural processes and human activities. Modern farming and the burning of oil and coal add considerable amounts of particulates to the atmosphere, as do dust storms, fires (Figure 10.10), and volcanic eruptions. The 1991 eruptions of Mount Pinatubo in the Philippines were the largest volcanic eruptions of the 20th century, explosively hurling huge amounts of volcanic ash, sulfur dioxide, and other volcanic material and gases as high as 30 km (18.6 mi) into the atmosphere. Eruptions can have a significant impact on the global environment and are linked to global climate change and stratospheric ozone depletion (see Chapters 21 and 22). In addition, many chemical toxins, such as heavy metals, enter the biosphere as particulates. Sometimes, nontoxic particulates link with toxic substances, creating a synergetic threat. (See discussion of particulates in Chapter 21.)

Asbestos

Asbestos is a term for several minerals that take the form of small, elongated particles, or fibers. Industrial use of asbestos has contributed to fire prevention and has provided protection from the overheating of materials. Asbestos is also used as insulation for a variety of other purposes. Unfortunately, however, excessive contact with asbestos has led to asbestosis (a lung disease caused by inhaling asbestos) and to cancer in some industrial workers. Experiments with animals have demonstrated that asbestos can cause tumors if the fibers are embedded in lung tissue.[29] The hazard related to certain types of asbestos under certain conditions is considered so serious that extraordinary steps have been taken to reduce the use of asbestos or ban it outright. The expensive process of asbestos removal from old buildings (particularly schools) in the United States is one of those steps.

FIGURE 10.10 Fires in Indonesia in 1997 caused serious air pollution. The person here is wearing a surgical mask in an attempt to breathe cleaner air.

There are several types of asbestos, and they are not equally hazardous. Most commonly used in the United States is white asbestos, which comes from the mineral chrysolite. It has been used to insulate pipes, floor and ceiling tiles, and brake linings of automobiles and other vehicles. Approximately 95% of the asbestos that is now in place in the United States is of the chrysolite type. Most of this asbestos was mined in Canada, and environmental health studies of Canadian miners show that exposure to chrysolite asbestos is not particularly harmful. However, studies involving another type of asbestos, known as crocidolite asbestos (blue asbestos), suggest that exposure to this mineral can be very hazardous and evidently does cause lung disease. Several other types of asbestos have also been shown to be harmful.[29]

A great deal of fear has been associated with nonoccupational exposure to chrysolite asbestos in the United States. Tremendous amounts of money have been spent to remove it from homes, schools, public buildings, and other sites, even though no asbestos-related disease has been recorded among those exposed to chrysolite in nonoccupational circumstances. It is now thought that much of the removal was unnecessary and that chrysolite asbestos doesn't pose a significant health hazard. Additional research into health risks from other varieties of asbestos is necessary to better understand the potential problem and to outline strategies to eliminate potential health problems.

For example, from 1979 to 1998 a strip mine near Libby, Montana, produced vermiculite (a natural mineral) that was contaminated (commingled) with a fibrous form of the mineral tremolite, classified as an asbestos. People in Libby were exposed to asbestos by workers in the mines (occupational exposure) who brought it home on clothes. Libby is in a valley with very poor ventilation, allowing the asbestos particles to settle out over everything. The EPA has documented hundreds of asbestos-related cases of disease, including many deaths. Asbestos mortality in Libby was much higher than expected, compared to the United States as a whole and to other parts of Montana. In 2009 the EPA declared Libby a public-health emergency. Medical care is being provided, and plans for cleanup of the now closed mine and Libby are under way.[30]

Electromagnetic Fields

Electromagnetic fields (EMFs) are part of everyday urban life. Cell phones, electric motors, electric transmission lines for utilities, and our electrical appliances—toasters, electric blankets, computers, and so forth—all produce magnetic fields. There is currently a controversy over whether these fields produce a health risk.

Early on, investigators did not believe that magnetic fields were harmful, because fields drop off quickly with distance from the source, and the strengths of the fields that most people come into contact with are relatively weak. For example, the magnetic fields generated by power transmission lines or by a computer terminal are normally only about 1% of Earth's magnetic field; directly below power lines, the electric field induced in the body is about what the body naturally produces within cells.[31]

Several early studies, however, concluded that children exposed to EMFs from power lines have an increased risk of contracting leukemia, lymphomas, and nervous-system cancers.[32] Investigators concluded that children so exposed are about one and a half to three times more likely to develop cancer than children with very low exposure to EMFs, but the results were questioned because of perceived problems with the research design (problems of sampling, tracking children, and estimating exposure to EMFs).

A later study analyzed more than 1,200 children, approximately half of them suffering from acute leukemia. It was necessary to estimate residential exposure to magnetic fields generated by power lines near the children's present and former homes. That study, the largest such investigation to date, found no association between childhood leukemia and measured exposure to magnetic fields.[31, 32]

In other studies, electric utility workers' exposure to magnetic fields has been compared with the incidence of brain cancer and leukemia. One study concluded that the association between exposure to magnetic fields and both brain cancer and leukemia is not strong and not statistically significant.[33]

Saying that data are not statistically significant is another way of stating that the relationship between exposure and disease cannot be reasonably established given the database that was analyzed. It does not mean that additional data in a future study will not find a statistically significant relationship. Statistics can predict the strength of the relationship between variables, such as exposure to a toxin and the incidence of a disease, but statistics cannot prove a cause-and-effect relationship between them.

In sum, despite the many studies that have evaluated relationships between cancer (brain, leukemia, and breast) and exposure to magnetic fields in our modern urban environment, the jury is still out.[34, 35] There seems to be some indication that magnetic fields cause health problems for children,[36, 37] but the risks to adults (with the exception of utility workers) appear relatively small and difficult to quantify.[38-41]

Noise Pollution

Noise pollution is unwanted sound. Sound is a form of energy that travels as waves. We hear sound because our ears respond to sound waves through vibrations of the eardrum. The sensation of loudness is related to the intensity of the energy carried by the sound waves and is measured in decibels (dB). The threshold for human hearing is 0 dB; the average sound level in the interior of a home is

Table 10.2 EXAMPLES OF SOUND LEVELS

SOUND SOURCE	INTENSITY OF SOUND (dB)	HUMAN PERCEPTION
Threshold of hearing	0	
Rustling of leaf	10	Very quiet
Faint whisper	20	Very quiet
Average home	45	Quiet
Light traffic (30 m away)	55	Quiet
Normal conversation	65	Quiet
Chain saw (15 m away)	80	Moderately loud
Jet aircraft flyover at 300 m	100	Very loud
Rock music concert	110	Very loud
Thunderclap (close)	120	Uncomfortably loud
Jet aircraft takeoff at 100 m	125	Uncomfortably loud
	140	Threshold of pain
Rocket engine (close)	180	Traumatic injury

Source:

about 45 dB; the sound of an automobile, about 70 dB; and the sound of a jet aircraft taking off, about 120 dB (see Table 10.2). A tenfold increase in the strength of a particular sound adds 10 dB units on the scale. An increase of 100 times adds 20 units.[13] The decibel scale is logarithmic—it increases exponentially as a power of 10. For example, 50 dB is 10 times louder than 40 dB and 100 times louder than 30 dB.

Environmental effects of noise depend not only on the total energy but also on the sound's pitch, frequency, and time pattern and length of exposure to the sound. Very loud noises (more than 140 dB) cause pain, and high levels can cause permanent hearing loss. Human ears can take sound up to about 60 dB without damage or hearing loss. Any sound above 80 dB is potentially dangerous. The noise of a lawn mower or motorcycle will begin to damage hearing after about eight hours of exposure. In recent years, there has been concern about teenagers (and older people, for that matter) who have suffered some permanent loss of hearing following extended exposure to amplified rock music (110 dB). At a noise level of 110 dB, damage to hearing can occur after only half an hour. Loud sounds at the workplace are another hazard. Even noise levels below the hearing-loss level may still interfere with human communication and may cause irritability. Noise in the range of 50–60 dB is sufficient to interfere with sleep, producing a feeling of fatigue upon awakening.

Voluntary Exposure

Voluntary exposure to toxins and potentially harmful chemicals is sometimes referred to as exposure to personal pollutants. The most common of these are tobacco, alcohol, and other drugs. Use and abuse of these substances have led to a variety of human ills, including death and chronic disease; criminal activity, such as reckless driving and manslaughter; loss of careers; street crime; and the straining of human relations at all levels.

10.3 General Effects of Pollutants

Almost every part of the human body is affected by one pollutant or another, as shown in Figure 10.11a. For example, lead and mercury (remember the Mad Hatter) affect the brain; arsenic, the skin; carbon monoxide, the heart; and fluoride, the bones. Wildlife is affected as well. Locations in the body where pollutants may affect humans and wildlife are shown in Figure 10.11b; effects of pollutants on wildlife populations are listed in Table 10.3.

The lists of potential toxins and affected body sites for humans and other animals in Figure 10.11 may be somewhat misleading. For example, chlorinated hydrocarbons, such as dioxin, are stored in the fat cells of animals, but they cause damage not only to fat cells but to the entire organism through disease, damaged skin, and birth defects. Similarly, a toxin that affects the brain, such as mercury, causes a wide variety of problems and symptoms, as illustrated in the Minamata, Japan, example (discussed in A Closer Look 10.2). The value of Figure 10.11 is in helping us to understand in general the adverse effects of excess exposure to chemicals.

Concept of Dose and Response

Five centuries ago, the physician and alchemist Paracelsus wrote that "everything is poisonous, yet nothing is poisonous." By this he meant, essentially, that too much of any substance can be dangerous, yet in an extremely

Table 10.3 EFFECTS OF POLLUTANTS ON WILDLIFE

EFFECT ON POPULATION	EXAMPLES OF POLLUTANTS
Changes in abundance	Arsenic, asbestos, cadmium, fluoride, hydrogen sulfide, nitrogen oxides, particulates, sulfur oxides, vanadium, POPs[a]
Changes in distribution	Fluoride, particulates, sulfur oxides, POPs
Changes in birth rates	Arsenic, lead, POPs
Changes in death rates	Arsenic, asbestos, beryllium, boron, cadmium, fluoride, hydrogen sulfide, lead, particulates, selenium, sulfur oxides, POPs
Changes in growth rates	Boron, fluoride, hydrochloric acid, lead, nitrogen oxides, sulfur oxides, POPs

[a] Pesticides, PCBs, hormonally active agents, dioxin, and DDT are examples (see Table 10.1).

Source: J.R. Newman, *Effects of Air Emissions on Wildlife*, U.S. Fish and Wildlife Service, 1980. Biological Services Program, National Power Plant Team, FWS/OBS-80/40, U.S. Fish and Wildlife Service, Washington, DC.

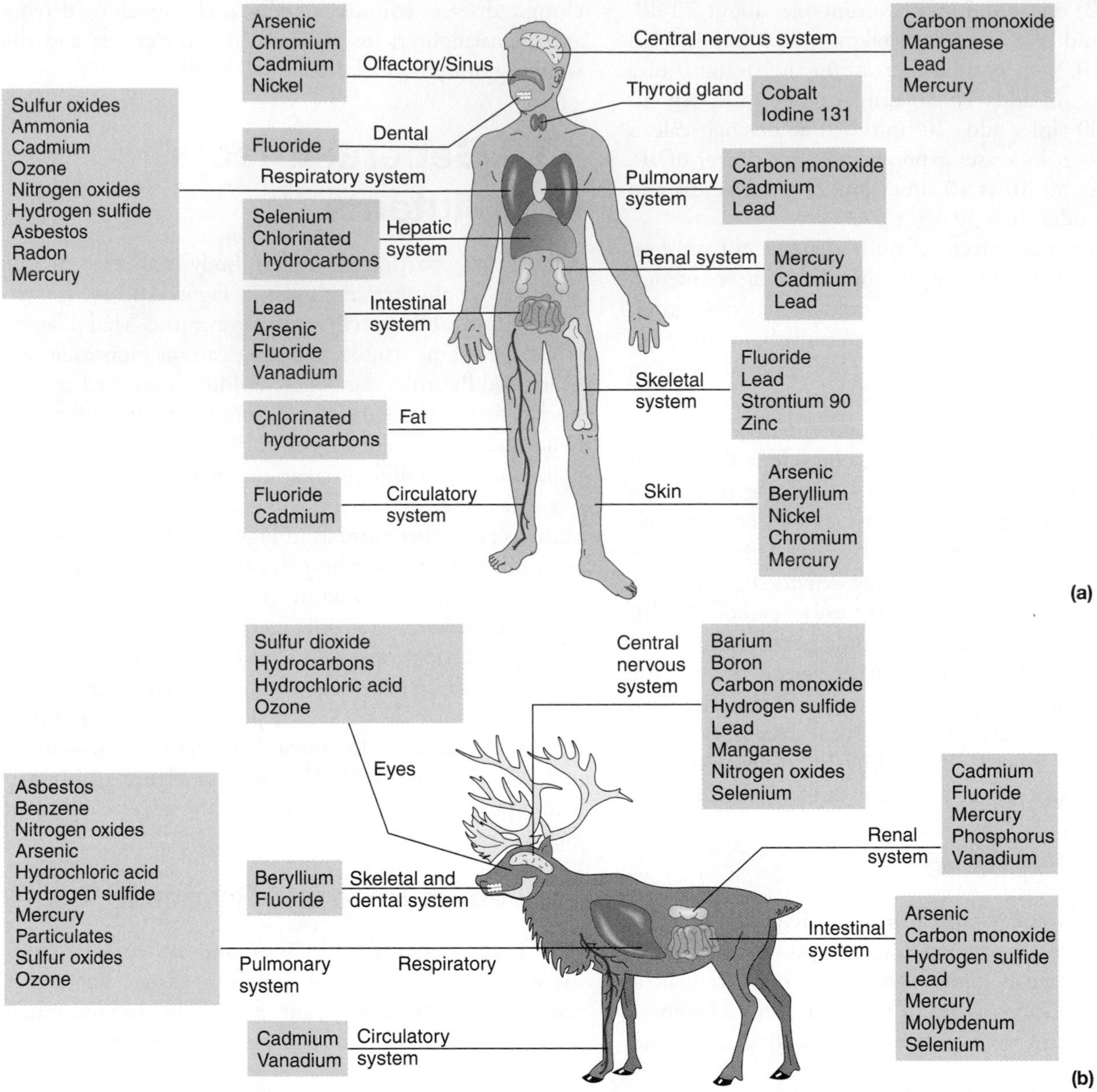

FIGURE 10.11 **(a)** Effects of some major pollutants in human beings. **(b)** Known sites of effects of some major pollutants in wildlife.

small amount can be relatively harmless. Every chemical element has a spectrum of possible effects on a particular organism. For example, selenium is required in small amounts by living things but may be toxic or increase the probability of cancer in cattle and wildlife when it is present in high concentrations in the soil. Copper, chromium, and manganese are other chemical elements required in small amounts by animals but toxic in higher amounts.

It was recognized many years ago that the effect of a certain chemical on an individual depends on the dose. This concept, termed **dose response**, can be represented by a generalized dose-response curve, such as that shown in Figure 10.12. When various concentrations of a chemical present in a biological system are plotted against the effects on the organism, two things are apparent: Relatively large concentrations are toxic and even lethal (points *D*, *E*, and *F* in Figure 10.12), but trace concentrations may actually be beneficial for life (between points *A* and *D*). The dose-response curve forms a plateau of optimal concentration and maximum benefit between two points (*B* and *C*). Points *A*, *B*, *C*, *D*, *E*, and *F* in Figure 10.12 are important thresholds in the dose-response curve. Unfortunately, the amounts at which points *E* and *F* occur are known only for a few substances for a few organisms, including people, and the very important point *D* is all but unknown. Doses that are beneficial, harmful, or lethal may differ widely for different organisms and are difficult to characterize.

Fluorine provides a good example of the general dose-response concept. Fluorine forms fluoride compounds that prevent tooth decay and promote development of a healthy bone structure. Relationships between the concentration of fluoride (in a compound of fluorine, such

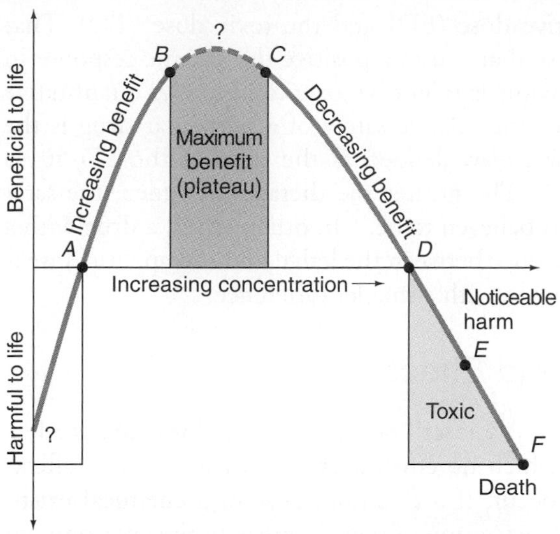

FIGURE 10.12 Generalized dose-response curve. Low concentrations of a chemical may be harmful to life (below point *A*). As the concentration of the chemical increases from *A* to *B*, the benefit to life increases. The maximum concentration that is beneficial to life lies within the benefit plateau (*B–C*). Concentrations greater than this plateau provide less and less benefit (*C–D*) and will harm life (*D–F*) as toxic concentrations are reached. Increased concentrations above the toxic level may result.

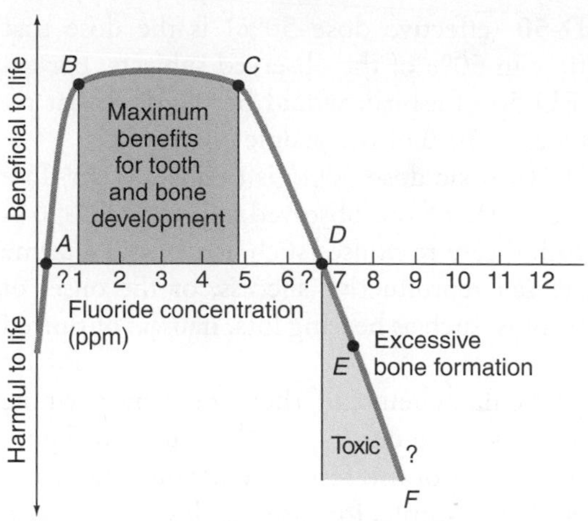

FIGURE 10.13 General dose-response curve for fluoride, showing the relationship between fluoride concentration and physiological benefit.

as sodium fluoride, NaF) and health show a specific dose-response curve (Figure 10.13). The plateau for an optimal concentration of fluoride (point *B* to point *C*) to reduce dental caries (cavities) is from about 1 ppm to just less than 5 ppm. Levels greater than 1.5 ppm do not significantly decrease tooth decay but do increase the occurrence of tooth discoloration. Concentrations of 4–6 ppm reduce the prevalence of osteoporosis, a disease characterized by loss of bone mass; and toxic effects are noticed between 6 and 7 ppm (point *D* in Figure 10.13).

Dose-Response Curve (LD-50, ED-50, and TD-50)

Individuals differ in their response to chemicals, so it is difficult to predict the dose that will cause a response in a particular individual. It is more practical to predict instead what percentage of a population will respond to a specific dose of a chemical.

For example, the dose at which 50% of the population dies is called the lethal dose 50, or LD-50. The **LD-50** is a crude approximation of a chemical's toxicity. It is a gruesome index that does not adequately convey the sophistication of modern toxicology and is of little use in setting a standard for toxicity. However, the LD-50 determination is required for new synthetic chemicals as a way of estimating their toxic potential. Table 10.4 lists, as examples, LD-50 values in rodents for selected chemicals.

Table 10.4 APPROXIMATE LD–50 VALUES (FOR RODENTS) FOR SELECTED AGENTS

AGENT	LD-50(mg/kg)[a]
Sodium chloride (table salt)	4,000
Ferrous sulfate (to treat anemia)	1,520
2,4-D (a weed killer)	368
DDT (an insecticide)	135
Caffeine (in coffee)	127
Nicotine (in tobacco)	24
Strychnine sulfate (used to kill certain pests)	3
Botulinum toxin (in spoiled food)	0.00001

[a] Milligrams per kilogram of body mass (termed mass weight, although it really isn't a weight) administered by mouth to rodents. Rodents are commonly used in such evaluations, in part because they are mammals (as we are), are small, have a short life expectancy, and their biology is well known.

Source: H.B. Schiefer, D.C. Irvine, and S.C. Buzik, *Understanding Toxicology* (New York: CRC Press, 1997).

The **ED-50** (effective dose 50%) is the dose that causes an effect in 50% of the observed subjects. For example, the ED-50 of aspirin would be the dose that relieves headaches in 50% of the people observed.[42]

The **TD-50** (toxic dose 50%) is defined as the dose that is toxic to 50% of the observed subjects. TD-50 is often used to indicate responses such as reduced enzyme activity, decreased reproductive success, or the onset of specific symptoms, such as hearing loss, nausea, or slurred speech.

For a particular chemical, there may be a whole family of dose-response curves, as illustrated in Figure 10.14. Which dose is of interest depends on what is being evaluated. For example, for insecticides we may wish to know the dose that will kill 100% of the insects exposed; therefore, LD-95 (the dose that kills 95% of the insects) may be the minimum acceptable level. However, when considering human health and exposure to a particular toxin, we often want to know the LD-0—the maximum dose that does not cause any deaths.[42] For potentially toxic compounds, such as insecticides that may form a residue on food or food additives, we want to ensure that the expected levels of human exposure will have no known toxic effects. From an environmental perspective, this is important because of concerns about increased risk of cancer associated with exposure to toxic agents.[42]

For drugs used to treat a particular disease, the efficiency of the drug as a treatment is of paramount importance. In addition to knowing what the effective dose (ED-50) is, it is important to know the drug's relative safety. For example, there may be an overlap between the effective dose (ED) and the toxic dose (TD). That is, the dose that causes a positive therapeutic response in some individuals might be toxic to others. A quantitative measure of the relative safety of a particular drug is the *therapeutic index*, defined as the ratio of the LD-50 to the ED-50. The greater the therapeutic index, the safer the drug is believed to be.[43] In other words, a drug with a large difference between the lethal and therapeutic dose is safer than one with a smaller difference.

Threshold Effects

Recall from A Closer Look 10.2 that a **threshold** is a level below which no effect occurs and above which effects begin to occur. If a threshold dose of a chemical exists, then a concentration of that chemical in the environment below the threshold is safe. If there is no threshold dose, then even the smallest amount of the chemical has some negative effect (Figure 10.15).

Whether or not there is a threshold for environmental toxins is an important environmental issue. For example, the U.S. Federal Clean Water Act originally stated a goal to reduce to zero the discharge of pollutants into water. The goal implies there is no such thing as a threshold—that no level of toxin will be legally permitted. However, it is unrealistic to believe that zero discharge of a water pollutant can be achieved or that we can reduce to zero the concentration of chemicals shown to be carcinogenic.

A problem in evaluating thresholds for toxic pollutants is that it is difficult to account for synergistic effects. Little

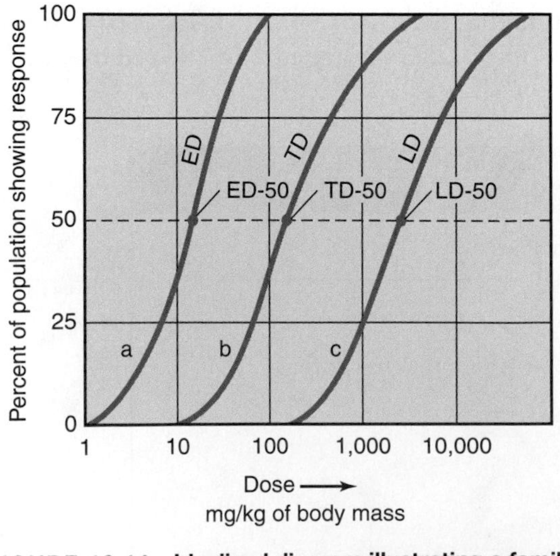

FIGURE 10.14 Idealized diagram illustrating a family of dose-response curves for a specific drug: ED (effective dose), TD (toxic dose), and LD (lethal dose). Notice the overlap for some parts of the curves. For example, at ED-50, a small percentage of the people exposed to that dose will suffer a toxic response, but none will die. At TD-50, about 1% of the people exposed to that dose will die.

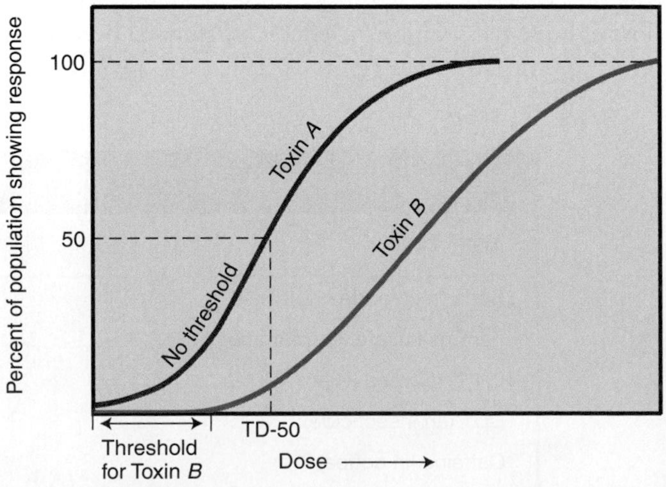

FIGURE 10.15 In this hypothetical toxic dose-response curve, toxin *A* has no threshold; even the smallest amount has some measurable effect on the population. The TD-50 for toxin *A* is the dose required to produce a response in 50% of the population. Toxin *B* has a threshold (flat part of curve) where the response is constant as the dose increases. After the threshold dose is exceeded, the response increases.

is known about whether or how thresholds might change if an organism is exposed to more than one toxin at the same time or to a combination of toxins and other chemicals, some of which are beneficial. Exposures of people to chemicals in the environment are complex, and we are only beginning to understand and conduct research on the possible interactions and consequences of multiple exposures.

Ecological Gradients

Dose response differs among species. For example, the kinds of vegetation that can live nearest to a toxic source are often small plants with relatively short lifetimes (grasses, sedges, and weedy species usually regarded as pests) that are adapted to harsh and highly variable environments. Farther from the toxic source, trees may be able to survive. Changes in vegetation with distance from a toxic source define the **ecological gradient**.

Ecological gradients may be found around smelters and other industrial plants that discharge pollutants into the atmosphere from smokestacks. For example, ecological gradient patterns can be observed in the area around the smelters of Sudbury, Ontario, discussed earlier in this chapter (see A Closer Look 10.1). Near the smelters, an area that was once forest was a patchwork of bare rock and soil occupied by small plants.

Tolerance

The ability to resist or withstand stress from exposure to a pollutant or harmful condition is referred to as **tolerance**. Tolerance can develop for some pollutants in some populations, but not for all pollutants in all populations. Tolerance may result from behavioral, physiological, or genetic adaptation.

Behavioral tolerance results from changes in behavior. For example, mice learn to avoid traps.

Physiological tolerance results when the body of an individual adjusts to tolerate a higher level of pollutant. For example, in studies at the University of California Environmental Stress Laboratory, students were exposed to ozone (O_3), an air pollutant often present in large cities (Chapter 21). The students at first experienced symptoms that included irritation of eyes and throat and shortness of breath. However, after a few days, their bodies adapted to the ozone, and they reported that they believed they were no longer breathing ozone-contaminated air, even though the concentration of O_3 stayed the same. This phenomenon explains why some people who regularly breathe polluted air say they do not notice the pollution. Of course, it does not mean that the ozone is doing no damage; it is, especially to people with existing respiratory problems. There are many mechanisms for physiological tolerance, including *detoxification*, in which the toxic chemical is converted to a nontoxic form, and internal transport of the toxin to a part of the body where it is not harmful, such as fat cells.

Genetic tolerance, or adaptation, results when some individuals in a population are naturally more resistant to a toxin than others. They are less damaged by exposure and more successful in breeding. Resistant individuals pass on the resistance to future generations, who are also more successful at breeding. Adaptation has been observed among some insect pests following exposure to some chemical pesticides. For example, certain strains of malaria-causing mosquitoes are now resistant to DDT, and some organisms that cause deadly infectious diseases have become resistant to common antibiotic drugs, such as penicillin.

Acute and Chronic Effects

Pollutants can have acute and chronic effects. *An acute effect* is one that occurs soon after exposure, usually to large amounts of a pollutant. A *chronic effect* occurs over a long period, often from exposure to low levels of a pollutant. For example, a person exposed all at once to a high dose of radiation may be killed by radiation sickness soon after exposure (an acute effect). However, that same total dose received slowly in small amounts over an entire lifetime may instead cause mutations and lead to disease or affect the person's DNA and offspring (a chronic effect).

10.4 Risk Assessment

Risk assessment in this context can be defined as the process of determining potential adverse health effects of exposure to pollutants and potentially toxic materials (recall the discussion of measurements and methods of science in Chapter 2). Such an assessment generally includes four steps:[44]

1. *Identification of the hazard.* This consists of testing materials to determine whether exposure is likely to cause health problems. One method used is to investigate populations of people who have been previously exposed. For example, to understand the toxicity of radiation produced from radon gas, researchers studied workers in uranium mines. Another method is to perform experiments to test effects on animals, such as mice, rats, or monkeys. This method has drawn increasing criticism from groups who believe such experiments are unethical. Another approach is to try to understand how a particular chemical works at the molecular level on cells. For example, research has been

done to determine how dioxin interacts with living cells to produce an adverse response. After quantifying the response, scientists can develop mathematical models to assess dioxin's risk.[18, 19] This relatively new approach might also be applicable to other potential toxins that work at the cellular level.

2. *Dose-response assessment.* This next step involves identifying relationships between the dose of a chemical (therapeutic drug, pollutant, or toxin) and the health effects on people. Some studies involve administering fairly high doses of a chemical to animals. The effects, such as illness, or symptoms, such as rashes or tumor development, are recorded for varying doses, and the results are used to predict the response in people. This is difficult, and the results are controversial for several reasons:
   - The dose that produces a particular response may be very small and subject to measurement errors.
   - There may be arguments over whether thresholds are present or absent.
   - Experiments on animals such as rats, mice, or monkeys may not be directly applicable to humans.
   - The assessment may rely on probability and statistical analysis. Although statistically significant results from experiments or observations are accepted as evidence to support an argument, statistics cannot establish that the substance tested *caused* the observed response.

3. *Exposure assessment.* This step evaluates the intensity, duration, and frequency of human exposure to a particular chemical pollutant or toxin. The hazard to society is directly proportional to the total population exposed. The hazard to an individual is generally greater closer to the source of exposure. Like dose-response assessment, exposure assessment is difficult, and the results are often controversial, in part because of difficulties in measuring the concentration of a toxin in doses as small as parts per million, billion, or even trillion. Some questions that exposure assessment attempts to answer follow:
   - How many people were exposed to concentrations of a toxin thought to be dangerous?
   - How large an area was contaminated by the toxin?
   - What are the ecological gradients for exposure to the toxin?
   - How long were people exposed to a particular toxin?

4. *Risk characterization:* The goal of this final step is to delineate health risk in terms of the magnitude of the health problem that might result from exposure to a particular pollutant or toxin. To do this, it is necessary to identify the hazard, complete the dose-response assessment, and evaluate the exposure assessment, as has been outlined. This step involves all the uncertainties of the prior steps, and results are again likely to be controversial.

In sum, *risk assessment* is difficult, costly, and controversial. Each chemical is different, and there is no one method of determining responses of humans to specific EDs or TDs. Toxicologists use the scientific method of hypothesis-testing with experiments (see Chapter 2) to predict how specific doses of a chemical may affect humans. Warning labels listing potential side effects of a specific medication are required by law, and these warnings result from toxicology studies to determine a drug's safety. Finally, risk assessment requires making scientific judgments and formulating actions to help minimize health problems related to human exposure to environmental pollutants and toxins.

The process of *risk management* integrates the assessment of risk with technical, legal, political, social, and economic issues.[18, 19] The toxicity of a particular material is often open to debate. For example, there is debate as to whether the risk from dioxin is linear. That is, do effects start at minimum levels of exposure and gradually increase, or is there a threshold exposure beyond which health problems occur? (See A Closer Look 10.3.)[18, 19, 29] It is the task of people in appropriate government agencies assigned to manage risk to make judgments and decisions based on the risk assessment and then to take appropriate actions to minimize the hazard resulting from exposure to toxins. This might involve invoking the precautionary principle discussed in Chapter 1.

CRITICAL THINKING ISSUE

Is Lead in the Urban Environment Contributing to Antisocial Behavior?

Lead is one of the most common toxic metals in our inner-city environments, and it may be linked to delinquent behavior in children. Lead is found in all parts of the urban environment (air, soil, older pipes, and some paint, for example) and in biological systems, including people (Figure 10.16). There is no apparent biological need for lead, but it is sufficiently concentrated in the blood and bones of children living in inner cities to cause health and behavior problems. In some populations, over 20% of the children have blood concentrations of lead that are higher than those believed safe.[45]

Lead affects nearly every system of the body. Thus, acute lead toxicity may cause a variety of symptoms, including anemia, mental retardation, palsy, coma, seizures, apathy, uncoordination, subtle loss of recently acquired skills, and bizarre behavior.[46,47] Lead toxicity is particularly a problem for young children, who are more apt than adults to put things in their mouths and apparently are also more susceptible to lead poisoning. In some children the response to lead poisoning is aggressive, difficult-to-manage behavior.[45-48]

The occurrence of lead toxicity or lead poisoning has cultural, political, and sociological implications. Over 2,000 years ago, the Roman Empire produced and used tremendous amounts of lead for a period of several hundred years. Production rates were as high as 55,000 metric tons per year. Romans had a wide variety of uses for lead. Lead was used in pots in which grapes were crushed and processed into a syrup for making wine, in cups and goblets from which wine was drunk, and as a base for cosmetics and medicines. In the homes of Romans wealthy enough to have running water, lead was used to make the pipes that carried the water. It has been argued that lead poisoning among the upper class in Rome was partly responsible for Rome's decline. Lead poisoning probably resulted in widespread stillbirths, deformities, and brain damage. Studies analyzing the lead content of bones of ancient Romans tend to support this hypothesis.[49]

FIGURE 10.16 The lead in urban soils (a legacy of our past use of lead in gasoline) is still concentrated where children are likely to play. Lead-based paint in older buildings, such as these in New York, also remains a hazard to young children, who sometimes ingest flakes of paint.

The occurrence of lead in glacial ice cores from Greenland has also been studied. Glaciers have an annual growth layer of ice. Older layers are buried by younger layers, allowing us to identify the age of each layer. Researchers drill glaciers, taking continuous samples of the layers. The samples look like long, solid rods of glacial ice and are called *cores*. Measurements of lead in these cores show that lead concentrations during the Roman period, from approximately 500 B.C. to A.D. 300, are about four times higher than before and after this period. This suggests that the mining and smelting of lead in the Roman Empire added small particles of lead to the atmosphere that eventually settled out in the glaciers of Greenland.[49]

Lead toxicity, then, seems to have been a problem for a long time. Now, an emerging, interesting, and potentially significant hypothesis is that, in children, even lead concentrations below the levels known to cause physical damage may be associated with an increased potential for antisocial, delinquent behavior. This is a testable hypothesis. (See Chapter 2 for a discussion of hypotheses.) If the hypothesis is correct, then some of our urban crime may be traced to environmental pollution!

A recent study in children aged 7 to 11 measured the amount of lead in bones and compared it with data concerning behavior over a four-year period. Even taking into account such factors as maternal intelligence, socioeconomic status, and quality of child rearing, the study concluded that an above-average concentration of lead in children's bones was associated with an increased risk of attention-deficit disorder, aggressive behavior, and delinquency.[45]

Critical Thinking Questions

1. What is the main point of the discussion about lead in the bones of children and children's behavior?
2. What are the main assumptions of the argument? Are they reasonable?
3. What other hypotheses might be proposed to explain the behavior?

SUMMARY

- Disease is an imbalance between an organism and the environment. Disease seldom has a one-cause/one-effect relationship, and there is often a gray zone between the state of health and the state of disease.
- Pollution produces an impure, dirty, or otherwise unclean state. Contamination means making something unfit for a particular use through the introduction of undesirable materials.
- Toxic materials are poisonous to people and other living things; toxicology is the study of toxic materials.
- A concept important in studying pollution problems is synergism, whereby actions of different substances produce a combined effect greater than the sum of the effects of the individual substances.
- How we measure the amount of a particular pollutant introduced into the environment or the concentration of that pollutant varies widely, depending on the substance. Common units for expressing the concentration of pollutants are parts per million (ppm) and parts per billion (ppb). Air pollutants are commonly measured in units such as micrograms of pollutant per cubic meter of air ($\mu g/m^3$).
- Categories of environmental pollutants include toxic chemical elements (particularly heavy metals), organic compounds, nuclear radiation, heat, particulates, electromagnetic fields, and noise.
- Organic compounds of carbon are produced by living organisms or synthetically by people. Artificially produced organic compounds may have physiological, genetic, or ecological effects when introduced into the environment. The potential hazards of organic compounds vary: Some are more readily degraded in the environment than others; some are more likely to undergo biomagnification; and some are extremely toxic, even at very low concentrations. Organic compounds of serious concern include persistent organic pollutants, such as pesticides, dioxin, PCBs, and hormonally active agents.
- The effect of a chemical or toxic material on an individual depends on the dose. It is also important to determine tolerances of individuals, as well as acute and chronic effects of pollutants and toxins.
- Risk assessment involves identifying the hazard, assessing the exposure and the dose response, and characterizing the possible results.

REEXAMINING THEMES AND ISSUES

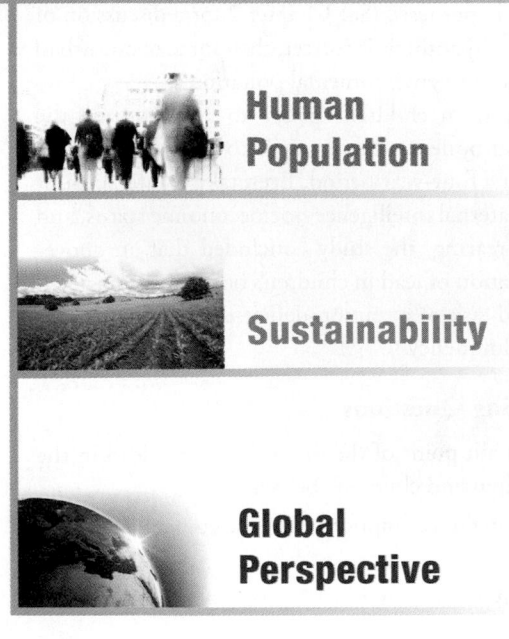

Human Population

As the total population and population density increase, the probability that more people will be exposed to hazardous materials increases as well. Finding acceptable ways to dispose of hazardous substances also becomes more difficult as populations increase and people live closer to industrial areas and waste-disposal sites.

Sustainability

Ensuring that future generations inherit a relatively unpolluted, healthy environment remains a challenging problem. Sustainable development requires that our use of chemicals and other materials not damage the environment.

Global Perspective

Releasing toxins into the environment can cause global patterns of contamination or pollution, particularly when a toxin or contaminant enters the atmosphere, surface water, or oceans and becomes widely dispersed. For example, pesticides, herbicides, and heavy metals emitted into the atmosphere in the midwestern United States may be transported by winds and deposited on glaciers in polar regions.

Urban World

Industrial processes in urban areas concentrate potentially toxic materials that may be inadvertently, accidentally, or deliberately released into the environment. Human exposure to a variety of pollutants—including lead, asbestos, particulates, organic chemicals, radiation, and noise—is often greater in urban areas.

People and Nature

Feminization of frogs and other animals from exposure to human-produced, hormonally active agents (HAAs) is an early warning or red flag that we are disrupting some basic aspects of nature. We are performing unplanned experiments on nature, and the consequences to us and other living organisms with which we share the environment are poorly understood. Control of HAAs seems an obvious candidate for application of the precautionary principle, discussed in Chapter 1.

Science and Values

Because we value both human and nonhuman life, we are interested in learning all we can about the risks of exposing living things to chemicals, pollutants, and toxins. Unfortunately, our knowledge of risk assessment is often incomplete, and the dose response for many chemicals is poorly understood. What we decide to do about exposure to toxic chemicals reflects our values. Increased control of toxic materials in homes and the work environment is expensive. To reduce environmental hazards at worksites in other countries, are we willing to pay more for the goods those workers manufacture?

KEY TERMS

area sources **188**
asbestos **199**
biomagnification **191**
carcinogen **188**
contamination **188**
disease **187**
dose response **202**
ecological gradient **205**
ED-50 **204**
electromagnetic fields (EMFs) **199**
heavy metals **191**
hormonally active agents (HAAs) **196**
LD-50 **203**
mobile sources **188**
noise pollution **200**
organic compounds **193**
particulates **199**
persistent organic pollutants (POPs) **194**
point sources **188**
pollution **188**
risk assessment **205**
synergism **188**
synthetic organic compounds **193**
TD-50 **204**
thermal pollution **198**
threshold **204**
tolerance **205**
toxicology **188**
toxin **188**

STUDY QUESTIONS

1. Do you think the hypothesis that some crime is caused in part by environmental pollution is valid? Why? Why not? How might the hypothesis be further tested? What are the social ramifications of the tests?
2. What kinds of life-forms would most likely survive in a highly polluted world? What would be their general ecological characteristics?
3. Some environmentalists argue that there is no such thing as a threshold for pollution effects. What do they mean? How would you determine whether it was true for a specific chemical and a specific species?
4. What is biomagnification, and why is it important in toxicology?

5. You are lost in Transylvania while trying to locate Dracula's castle. Your only clue is that the soil around the castle is known to have an unusually high concentration of the heavy metal arsenic. You wander in a dense fog, able to see only the ground a few meters in front of you. What changes in vegetation warn you that you are nearing the castle?
6. Distinguish between acute effects and chronic effects of pollutants.
7. Design an experiment to test whether tomatoes or cucumbers are more sensitive to lead pollution.
8. Why is it difficult to establish standards for acceptable levels of pollution? In giving your answer, consider physical, climatological, biological, social, and ethical reasons.
9. A new highway is built through a pine forest. Driving along the highway, you notice that the pines nearest the road have turned brown and are dying. You stop at a rest area and walk into the woods. One hundred meters away from the highway, the trees seem undamaged. How could you make a crude dose–response curve from direct observations of the pine forest? What else would be necessary to devise a dose–response curve from direct observation of the forest? What else would be necessary to devise a dose–response curve that could be used in planning the route of another highway?
10. Do you think your personal behavior is placing you in the gray zone of suboptimal health? If so, what can you do to avoid chronic disease in the future?

FURTHER READING

Amdur, M., J. Doull, and C.D. Klaasen, eds., *Casarett & Doull's Toxicology: The Basic Science of Poisons,* 4th ed. (Tarrytown, NY: Pergamon, 1991). A comprehensive and advanced work on toxicology.

Carson, R., *Silent Spring* (Boston: Houghton Mifflin, 1962). A classic book on problems associated with toxins in the environment.

Schiefer, H.B., D.G. Irvine, and S.C. Buzik, *Understanding Toxicology: Chemicals, Their Benefits and Risks* (Boca Raton, FL: CRC Press, 1997). A concise introduction to toxicology as it pertains to everyday life, including information about pesticides, industrial chemicals, hazardous waste, and air pollution.

Travis, C.C., and H.A. Hattemer-Frey. "Human Exposure to Dioxin," *The Science of the Total Environment* 104: 97–127, 1991. An extensive technical review of dioxin accumulation and exposure.

CHAPTER 11

Agriculture, Aquaculture, and the Environment

Modern agriculture uses modern technology (here a computer controls more than 100 center-pivot sprinklers) but continues to depend on the environment in major ways (here the need for water to grow wheat, alfalfa, potatoes, and melons along the Columbia River near Hermiston, Oregon).

LEARNING OBJECTIVES

The big question about farming and the environment is: Can we produce enough food to feed Earth's growing human population, and do this sustainably? The major agricultural challenges facing us today are to increase the productivity of the land, acre by acre, hectare by hectare; to distribute food adequately around the world; to decrease the negative environmental effects of agriculture; and to avoid creating new kinds of environmental problems as agriculture advances. After reading this chapter, you should understand . . .

- How agroecosystems differ from natural ecosystems;
- What role limiting factors play in determining crop yield;
- How the growing human population, the loss of fertile soils, and the lack of water for irrigation can lead to future food shortages worldwide;
- The relative importance of food *distribution* and food *production*;
- Why some lands are best used for grazing, but how overgrazing can damage land;
- How alternative agricultural methods—including integrated pest management, no-till agriculture, mixed cropping, and other methods of soil conservation—can provide major environmental benefits;
- That genetic modification of crops could improve food production and benefit the environment, but perhaps also could create new environmental problems.

CASE STUDY

Biofuels and Banana Chips: Food Crops vs. Fuel Crops

In 2007 Alfred Smith, a farmer in Garland, North Carolina, was feeding his pigs trail mix, banana chips, yogurt-covered raisins, dried papaya, and cashews, according to an article in the *Wall Street Journal.*[1] The pigs were on this diet, Mr. Smith says, because the demand for the biofuel ethanol, produced from corn and other crops, had driven up prices of feed (the largest cost of raising livestock) to the point where it became cheaper to feed his animals our snack food. In 2007 he bought enough trail mix to feed 5,000 hogs, saving $40,000. Other farmers in the U.S. Midwest were feeding their pigs and cattle cookies, licorice, cheese curls, candy bars, french fries, frosted Mini-Wheats, and Reese's Peanut Butter Cups. Near Hershey, Pennsylvania, farmers were getting waste cocoa and candy trimmings from the Hershey Company and feeding it to their cattle. Their problem has been caused by competition with crops grown directly to be turned into fuels (Figure 11.1).

FIGURE 11.1 Harvest of experimental oilseed crops at Piedmont Biofuels farm in Moncure, North Carolina.[2,3] Interest in and enthusiasm about biofuels are growing among some operating small farms, like the members of the Piedmont Biofuels Cooperative in North Carolina. Will biofuel agriculture be the wave of the future and prove environmentally sound and sustainable?

This raises the fundamental question about agriculture and the environment: For many decades world food production has exceeded demand. But the demand for food crops is growing rapidly due to a rapid rise in the standards of living of many people and continued human population growth, both of which have also increased the demand for fuels and thus competition with crops grown for biofuels. Can we find ways to feed all the people of the world without undue damage to the environment?

World food prices rose almost 40% in 2007. According to the United Nations Food and Agriculture Organization (UNFAO), in February 2008 corn prices had risen 25% and the price of wheat was 80% higher than a year before.[4] Wheat prices have reached record levels, doubling the average cost of just a few years ago, and stocks of wheat (the amount stored for future sale and use) are reaching a 30-year low, in part because of Australian droughts.[5] The FAO predicts a world food crisis, with 36 countries currently facing food crises and

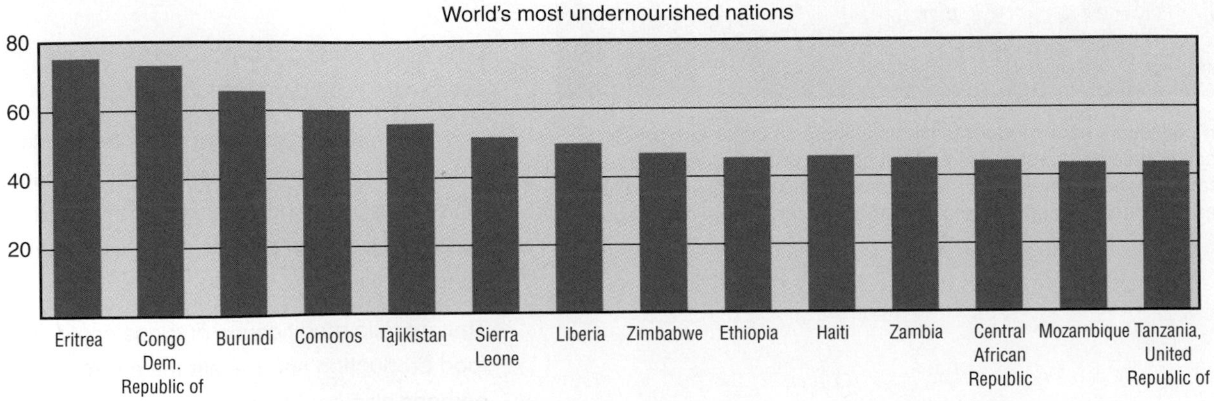

FIGURE 11.2 The world's most undernourished peoples, according to the United Nations Food and Agriculture Organization. This graph shows the 14 nations with the greatest percentage of undernourished people. Of these, 10 are in Africa. In the New World, only Haiti makes this group.[7]

Africa leading the list with 21 nations. In the Republic of the Congo, three-quarters of the population is undernourished, and in 25 nations at least one-third of the population is undernourished, according to the UNFAO (Figure 11.2).[6]

Suddenly, people of the world need a great increase in food production, in part because of the expanding human population, in part because of competition between crops for food and crops for fuel, and in part because of droughts, floods, and other environmental impacts that have decreased agricultural production. What can be done to meet the world's growing need for food and, second to that, its need for fuel? Can agricultural production increase? By how much? And at what costs, environmental and economic?

11.1 An Ecological Perspective on Agriculture

Farming creates novel ecological conditions, referred to as **agroecosystems.** Agroecosystems differ from natural ecosystems in six ways (Figure 11.3).

Ecological succession is halted to keep the agroecosystem in an early-successional state (see Chapter 5). Most crops are early-successional species, which means that they grow fast, spread their seeds widely and rapidly, and do best when sunlight, water, and chemical nutrients in the soil are abundant. Under natural conditions, crop species would eventually be replaced by later-successional

FIGURE 11.3 **How farming changes an ecosystem.** It converts complex ecosystems of high structural and species diversity to a monoculture of uniform structure, and greatly modifies the soil. See text for additional information about the agricultural effects on ecosystems.

plants. Slowing or stopping natural ecological succession requires time and effort on our part.

Biological diversity and food chains are simplified. The focus is on monoculture, one plant species rather than many. Large areas are planted with a single species or even a single strain or subspecies, such as a single hybrid of corn. The downside of monoculture is that it makes the entire crop vulnerable to attack by a single disease or a single change in environmental conditions. Repeated planting of a single species can reduce the soil content of certain essential elements, reducing overall soil fertility.

Crops are planted in neat rows and fields. These simple geometric layouts make life easy for pests because the crop plants have no place to hide. In natural ecosystems, many different species of plants grow mixed together in complex patterns, so it is harder for pests to find their favorite victims.

Agroecosystems require plowing, which is unlike any natural soil disturbance—nothing in nature repeatedly and regularly turns over the soil to a specific depth. Plowing exposes the soil to erosion and damages its physical structure, leading to a decline in organic matter and a loss of chemical elements.

They may include genetically modified crops.

FIGURE 11.4 **Plowing rich prairie soil in South Dakota, around 1916**, a historical photograph from the Library of Congress. The prairie had never been turned over like this.

The Plow Puzzle

There is nothing in nature like a plow, and thus there are big differences between the soils of a forest or grassland and the soils of land that has been plowed and used for crops for several thousand years. These differences were observed and written about by one of the originators of the modern study of the environment, George Perkins Marsh. Born in Vermont in the 19th century, Marsh became the U.S. ambassador to Italy and Egypt. While in Italy, he was so struck by the differences in the soils of the forests of his native Vermont and the soils that had been farmed for thousands of years on the Italian peninsula that he made this a major theme in his landmark book, *Man and Nature*, published in 1864. The farmland he observed in Italy had once been forests. While the soil in Vermont was rich in organic matter and had definite layers, the soil of Italian farmland had little organic matter and lacked definite layers.

Here's the plow puzzle: One would expect that farming that caused such major modification (see Figure 11.4) would eventually make the soils unsustainable, at least in terms of crop production, but much of the farmland in Italy and France, in China, and elsewhere, has been in continuous use since pre-Roman times and is still highly productive. How can this be? And what has been the long-term effect of such agriculture on the environment? The American Dust Bowl seemed to demonstrate how destructive plowing could be (see A Closer Look 11.1).

Deepening the plow puzzle, since the end of World War II, mechanized farming has seriously damaged more than 1 billion hectares (2.47 billion acres) of land. That's about 10.5% of the world's best soil, equal to the combined area of China and India. In addition, overgrazing and deforestation have damaged approximately 9 million hectares (22 million acres) to the point where recovery will be difficult; restoration of the rest will require serious actions.[8]

In the United States, since European settlement, about one-third of the topsoil has been lost, making 80 million hectares (198 million acres) unproductive or only marginally productive.[8] For now, think about this puzzle. We will discuss solutions to it later.

11.2 Can We Feed the World?

Can we produce enough food to feed Earth's growing human population? The answer has a lot to do with the environment and our treatment of it: Can we grow crops sustainably, so that both crop production and agricultural ecosystems remain viable? Can we produce this food without seriously damaging other ecosystems that receive the wastes of agriculture? And to these concerns we must now add: Can we produce all this

Table 11.1 LAND, PEOPLE AND AGRICULTURE, 2006

LOCATION	TOTAL LAND AREA (sq km)	HUMAN POPULATION (MILLIONS)	PEOPLE PER AREA	CROP AREA (sq km)	CROP AREA PER PERSON (sq km)	CROP LAND AS % OF TOTAL LAND
Asia	30,988,970	3,823	123.37	16,813,750	0.044	54%
Africa	29,626,570	850	28.69	11,460,700	0.135	39%
N. and C. America	21,311,580	507	23.79	6,189,030	0.122	29%
S. America	17,532,370	936	53.39	5,842,850	0.062	33%
Europe	22,093,160	362	16.39	4,836,410	0.134	22%
Australia	7,682,300	19	2.47	4,395,000	2.313	57%
World	130,043,970	6,301	48.45	49,734,060	0.079	38%

Source: FAO Statistics 2006 http://faostat.fao.org/faostat/

Note: Data are available for crops until 2003; hence, some population values in this table will differ from those elsewhere in the chapter, which are for 2005.

food and also grow crops used only to produce fuels? To answer these questions, let us begin by considering how crops grow and how productive they can be.

A surprisingly large percentage of the world's land area is in agriculture: approximately 38% (excluding Antarctica), an area about the size of South and North America combined and enough to make agriculture a human-induced biome (Table 11.1 and Figure 11.5).[9]

The history of agriculture is a series of human attempts to overcome environmental limitations and problems. Each new solution has created new environmental problems, which in turn have required their own solutions. Thus, in seeking to improve agricultural systems, we should expect some undesirable side effects and be ready to cope with them.

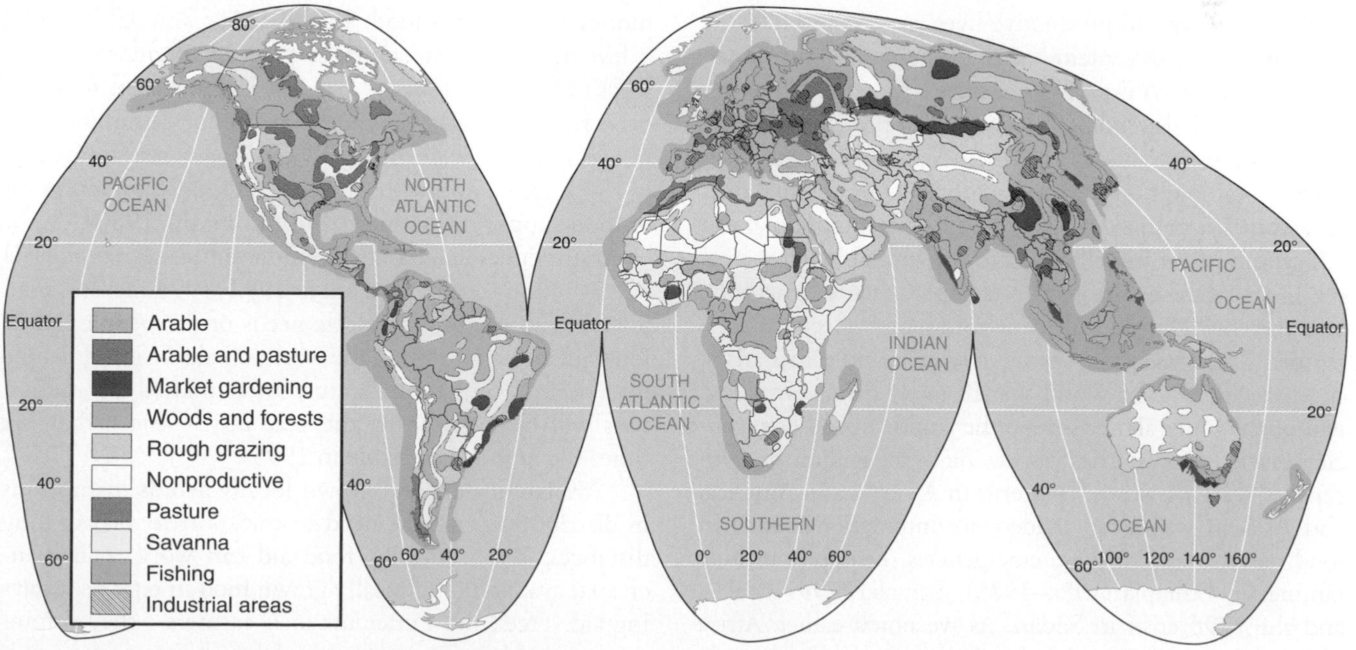

FIGURE 11.5 World land use showing arable (farmable) land. The percentage of land in agriculture varies considerably among continents, from 22% of the land in Europe to 57% in Australia. In the United States, cropland occupies 18% of the land and an additional 26% is used for pasture and rangeland, so agriculture uses 44% of the land. (*Source*: Phillips Atlas.)

The world's food supply is also greatly influenced by social disruptions and social attitudes, which affect the environment and in turn affect agriculture. In Africa, social disruptions since 1960 have included more than 20 major wars and more than 100 coups.[10] Such social instability makes sustained agricultural yields difficult—indeed, it makes any agriculture difficult if not impossible.[11] So does variation in weather, the traditional bane of farmers.[12]

How We Starve

People "starve" in two ways: undernourishment and malnourishment. World food production must provide adequate nutritional quality, not just total quantity. **Undernourishment** results from insufficient calories in available food, so that one has little or no ability to work or even move and eventually dies from the lack of energy. **Malnourishment** results from a lack of specific chemical components of food, such as proteins, vitamins, or other essential chemical elements. Widespread undernourishment manifests itself as famines that are obvious, dramatic, and fast-acting. Malnourishment is long term and insidious. Although people may not die outright, they are less productive than normal and can suffer permanent impairment and even brain damage.

Among the major problems of undernourishment are marasmus, which is progressive emaciation caused by a lack of protein and calories; kwashiorkor, which results from a lack of sufficient protein in the diet and in infants leads to a failure of neural development and thus to learning disabilities (Figure 11.6); and chronic hunger, when people have enough food to stay alive but not enough to lead satisfactory and productive lives (see Figure 11.7).

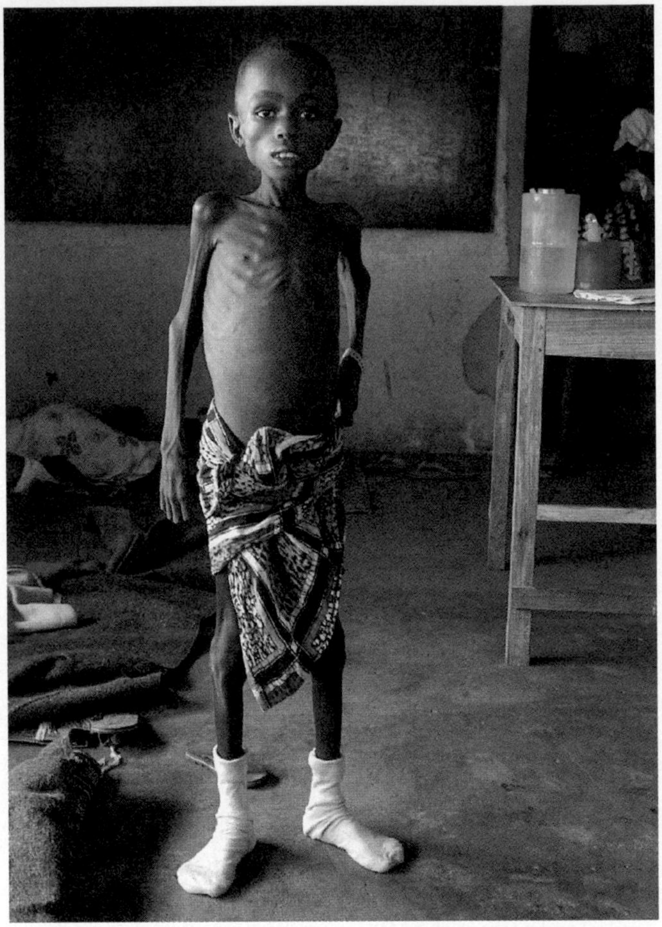

FIGURE 11.6 **Photograph of a child suffering from kwashiorkor.**

The supply of protein has been the major nutritional-quality problem. Animals are the easiest protein food source for people, but depending on animals for protein raises several questions of values. These include ecological ones (Is it better to eat lower on the food chain?), environmental ones (Do domestic animals erode soil faster than crops do?), and ethical ones (Is it morally right to eat animals?). How people answer these questions affects approaches to agriculture and thereby influences the environmental effects of agriculture. Once again, the theme of science and values arises.

Since the end of World War II, rarely has a year passed without a famine somewhere in the world.[11] Food emergencies affected 34 countries worldwide at the end of the 20th century. Varying weather patterns in Africa, Latin America, and Asia, as well as an inadequate international trade in food, contributed to these emergencies. Examples include famines in Ethiopia (1984–1985), Somalia (1991–1993), and the 1998 crisis in Sudan. As we noted earlier, Africa remains the continent with the most acute food shortages, due to adverse weather and civil strife.[11]

A common remedy is food aid among nations, where one nation provides food to another or gives or lends money to purchase food. In the 1950s and 1960s, only a few industrialized countries provided food aid, using stocks of surplus food. A peak in international food aid occurred in the 1960s, when a total of 13.2 million tons per year of food were given. A world food crisis in the early 1970s raised awareness of the need for greater attention to food supply and stability. But during the 1980s, donor commitments totaled only 7.5 million tons. A record level of 15 million tons of food aid in 1992–1993 met less than 50% of the minimum caloric needs of the people fed. If food aid alone is to bring the world's malnourished people to a desired nutritional status, an estimated 55 million tons will be required by the year 2010—more than six times the amount available in 1995.[13]

Availability of food grown locally avoids disruptions in distribution and the need to transport food over long distances. But ironically, food aid can work against increased availability of locally grown food in regions receiving aid. Free food undercuts local farmers—they cannot compete with it. The only complete solution to famine is to develop long-term, sustainable, local agriculture. The old saying "Give a man a fish and feed him for a day; teach him to fish and feed him for life" is true.

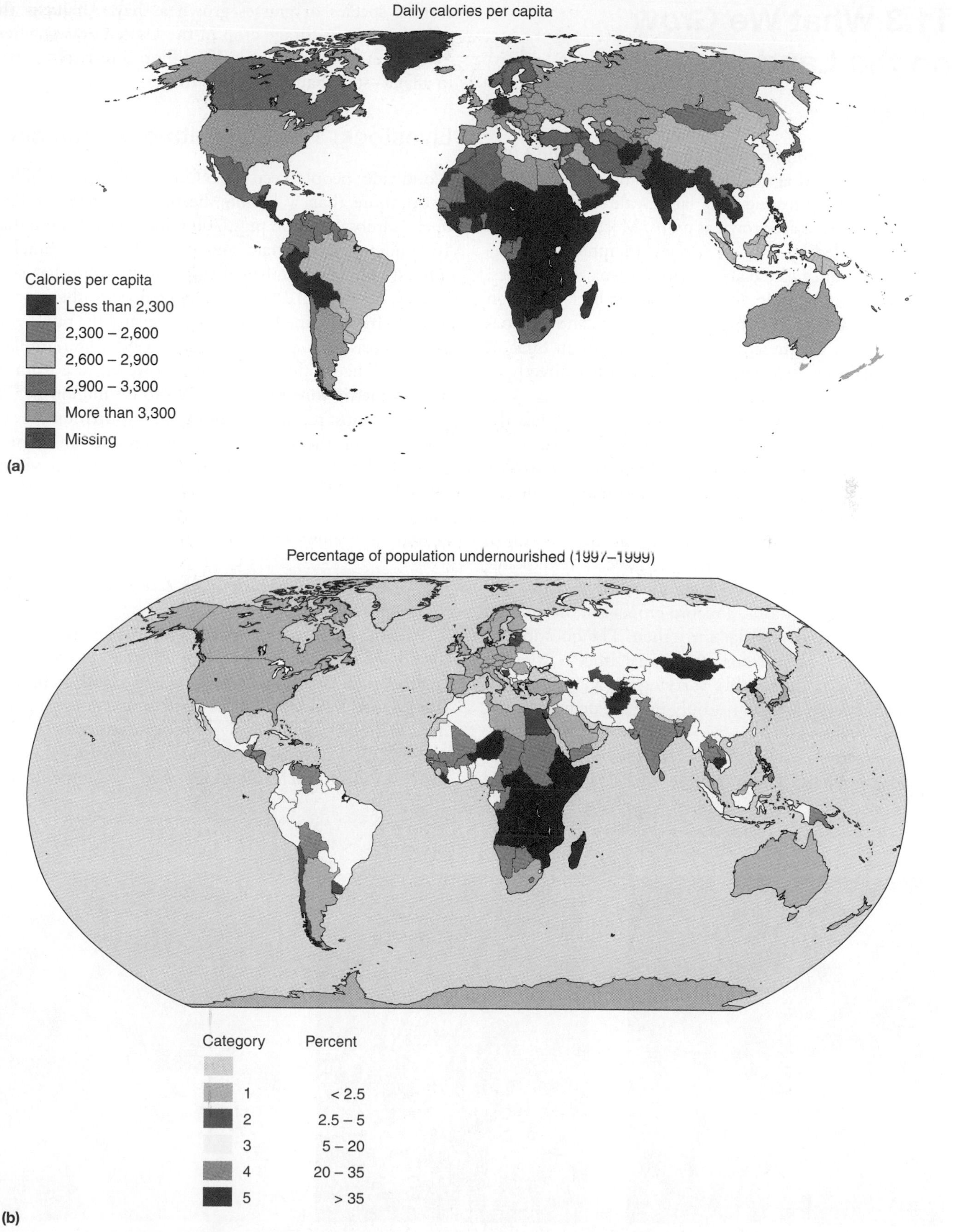

FIGURE 11.7 **(a) Daily intake of calories worldwide. (b) Where people are undernorished.** The percentage is the portion of the country's total population that is undernourished. (*Source*: World Resources Institute Web site: http://www.wri.org/.)

11.3 What We Grow on the Land

Crops

Of Earth's half-million plant species, only about 3,000 have been used as agricultural crops and only 150 species have been cultivated on a large scale. In the United States, 200 species are grown as crops. Most of the world's food is provided by only 14 species. In approximate order of importance, these are wheat, rice, maize, potatoes, sweet potatoes, manioc, sugarcane, sugar beet, common beans, soybeans, barley, sorghum, coconuts, and bananas (Figure 11.8). Of these, six provide more than 80% of the total calories that people consume either directly or indirectly.[12]

There is a large world trade in small grains. Only the United States, Canada, Australia, and New Zealand are major exporters (see Figure 11.9); the rest of the world's nations are net importers. World small-grain production increased greatly in the second half of the 20th century, from 0.8 billion metric tons in 1961 to 1 billion in 1966, and doubled to 2 billion in 1996, a remarkable increase in 30 years. In 2005, world small-grain production was 2.2 billion tons, a record crop.[2] But production has remained relatively flat since then. The question we must ask, and cannot answer at this time, is whether this means that the world's carrying capacity for small grains has been reached or simply that the demand is not growing (Figure 11.10).

Some crops, called *forage*, are grown as food for domestic animals. These include alfalfa, sorghum, and various species of grasses grown as hay. Alfalfa is the most important forage crop in the United States, where 14 million hectares (about 30 million acres) are planted in alfalfa—one-half the world's total.

Livestock: The Agriculture of Animals

Worldwide, people keep 14 billion chickens, 1.3 billion cattle, more than 1 billion sheep, more than a billion ducks, almost a billion pigs, 700 million goats, more than 160 million water buffalo, and about 18 million camels.[18] Interestingly, the number of cattle in the world has risen slightly, by about 0.2% in the past ten years; the number of sheep has remained about the same; and the number of goats increased from 660 million in 1995 to 807 million in 2005. The production of beef, however, rose from 57 million metric tons (MT) in 1995 to 63 million MT in 2005 (the most recent date for which information is available). During the same period, the production of meat from chickens increased greatly, from 46 million MT to 70 million MT, and meat from pigs increased from 80 million MT to more than 100 million MT.[2] These are important food sources and have a major impact on the land.

Grazing on Rangelands: An Environment Benefit or Problem?

Traditional herding practices and industrialized production of domestic animals have different effects on the environment. Most cattle live on rangeland or pasture. **Rangeland** provides food for grazing and browsing animals without plowing and planting. **Pasture** is plowed, planted, and harvested to provide forage for animals. More than 34 million square kilometers (km^2) are in permanent

(a) (b) (c)

FIGURE 11.8 **Among the world's major crops are (a) wheat, (b) rice, and (c) soybeans.** See text for a discussion of the relative importance of these three.

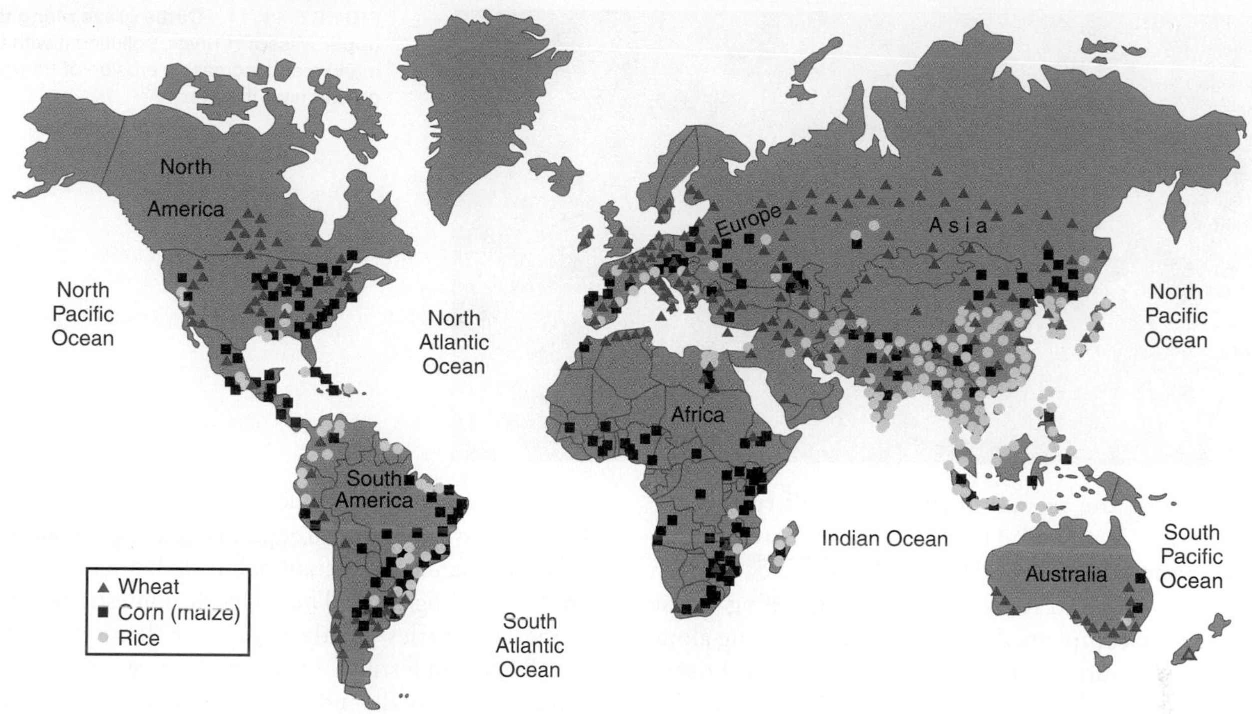

FIGURE 11.9 Geographic distribution of world production of a few major small-grain crops.

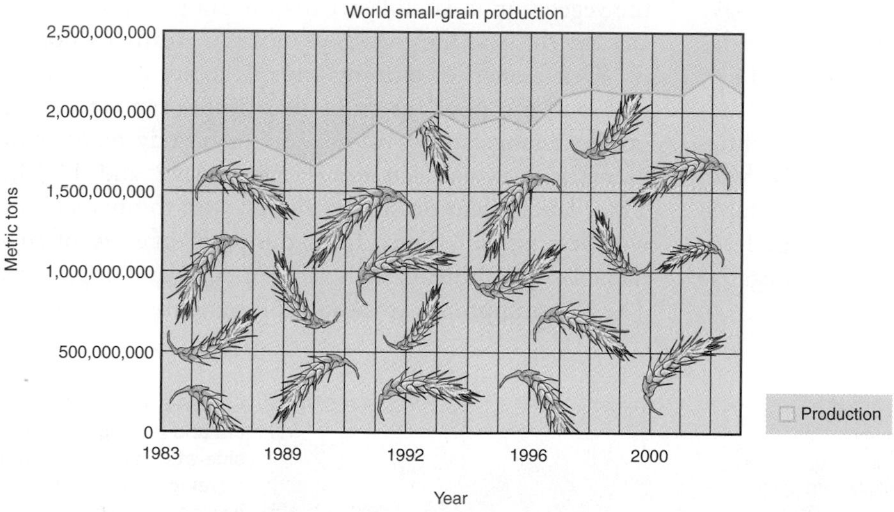

FIGURE 11.10 World small-grain production since 1983. (*Source*: FAO statistics FAOSTATS Web site.)

pasture worldwide—an area larger than the combined sizes of Canada, the United States, Mexico, Brazil, Argentina, and Chile.[2]

Almost half of Earth's land area is used as rangeland, and about 30% of Earth's land is arid rangeland, land easily damaged by grazing, especially during drought. Much of the world's rangeland is in poor condition from overgrazing. In the United States, where more than 99% of rangeland is west of the Mississippi River, rangeland conditions have improved since the 1930s, especially in upland areas. However, land near streams and the streams themselves continue to be heavily affected by grazing.

Grazing cattle trample stream banks and release their waste into stream water. Therefore, maintaining a high-quality stream environment requires that cattle be fenced behind a buffer zone. The upper Missouri River is famous for its beautiful "white cliffs," but private lands along the river that are used to graze cattle take away from the scenic splendor. The large numbers of cattle that come down to the Missouri River to drink damage

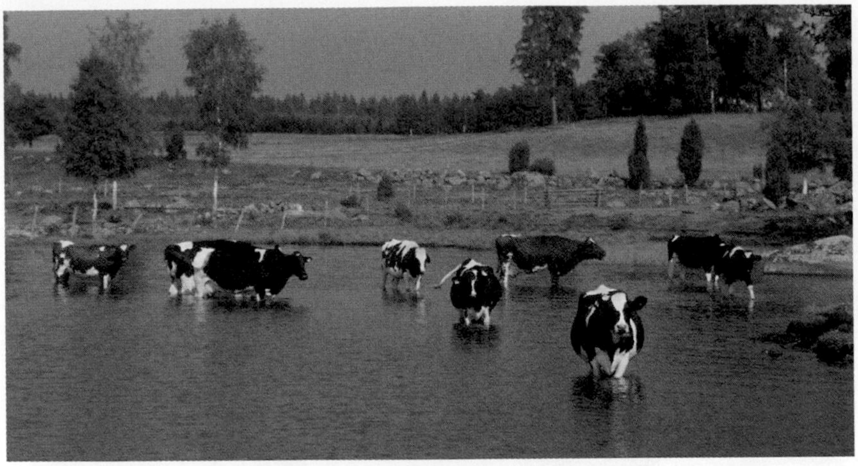

FIGURE 11.11 Cattle graze along the upper Missouri River, polluting it with their manure and increasing erosion of trampled ground near the river.

the land along the river, and the river itself runs heavy with manure (Figure 11.11). These effects extend to an area near a federally designated wild and scenic portion of the upper Missouri River, and tourists traveling on the Missouri have complained. In recent years, fencing along the upper Missouri River has increased, with small openings to allow cattle to drink, but otherwise restricting what they can do to the shoreline.

In modern industrialized agriculture, cattle are initially raised on open range and then transported to feedlots, where they are fattened for market. Feedlots have become widely known in recent years as sources of local pollution. The penned cattle are often crowded and are fed grain or forage that is transported to the feedlot. Manure builds up in large mounds and pollutes local streams when it rains. Feedlots are popular with meat producers because they are economical for rapid production of good-quality meat. However, large feedlots require intense use of resources and have negative environmental effects.

Traditional herding practices, by comparison, chiefly affect the environment through overgrazing. Goats are especially damaging to vegetation, but all domestic herbivores can destroy rangeland. The effect of domestic herbivores on the land varies greatly with their density relative to rainfall and soil fertility. At low to moderate densities, the animals may actually aid growth of aboveground vegetation by fertilizing soil with their manure and stimulating plant growth by clipping off plant ends in grazing, just as pruning stimulates plant growth. But at high densities, the vegetation is eaten faster than it can grow; some species are lost, and the growth of others is greatly reduced.

One benefit of farming animals rather than crops is that land too poor for crops that people can eat can be excellent rangeland, with grasses and woody plants that domestic livestock can eat (Figures 11.12 and 11.13). These lands occur on steeper slopes, with thinner soils or with less rainfall. Thus, from the point of view of sustainable agriculture, there is value in rangeland or pasture. The wisest approach to sustainable agriculture involves a

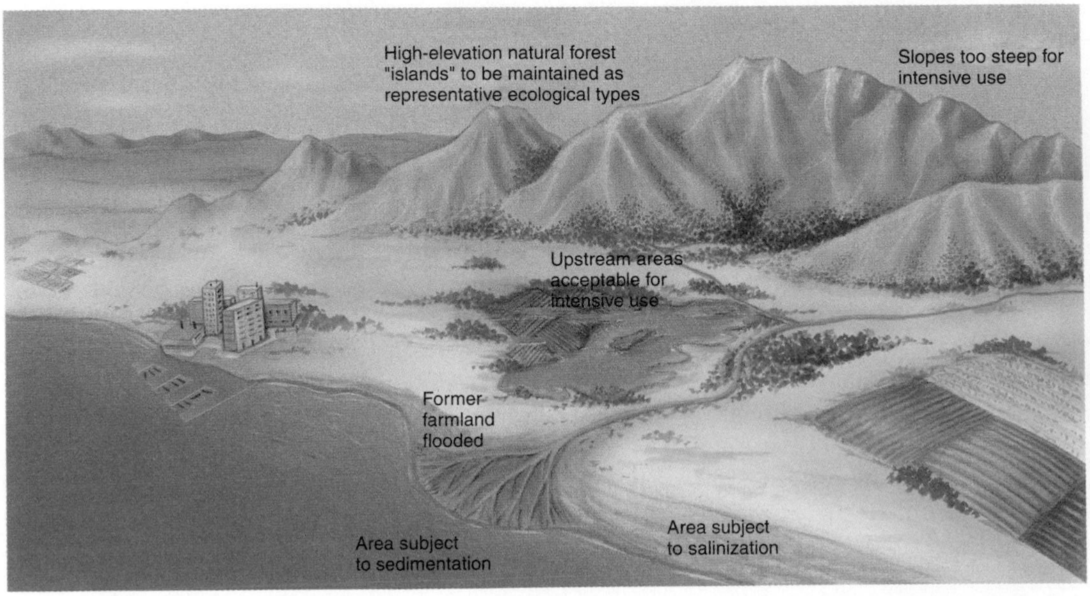

FIGURE 11.12 Physical and ecological considerations in watershed development—such as slope, elevation, floodplain, and river delta location—limit land available for agriculture.

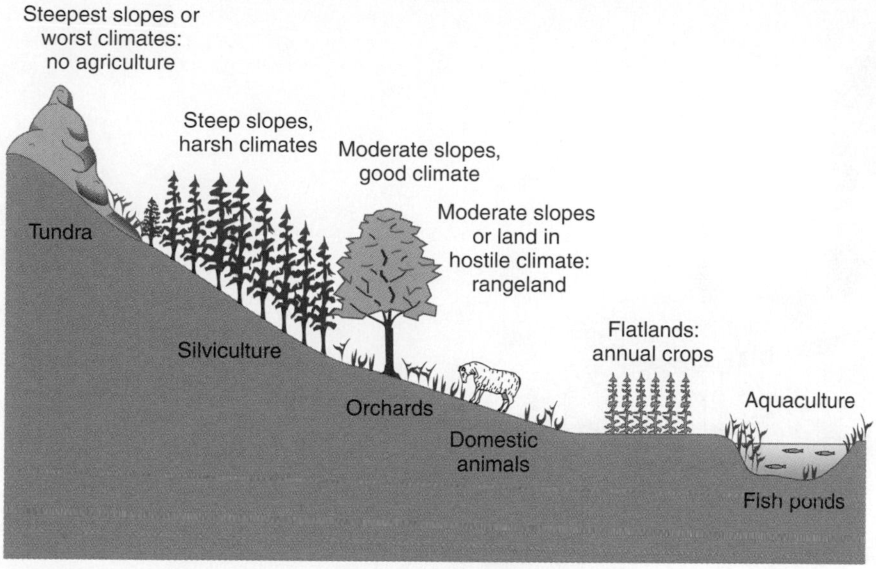

FIGURE 11.13 **Land unsuitable for crop production on a sustainable basis can be used for grazing and for other purposes.**

combination of different kinds of land use: using the best agricultural lands for crops, using poorer lands for pasture and rangeland.

11.4 Soils

To most of us, soils are just "dirt." But in fact soils are a key to life on the land, affecting life and affected by it. Soils develop for a very long time, perhaps thousands of years, and if you look at them closely, they are quite remarkable. You won't find anything like Earth soil on Mars or Venus or the moon. The reason is that water and life have greatly altered the land surface.

Geologically, soils are earth materials modified over time by physical, chemical, and biological processes into a series of layers. Each kind of soil has its own chemical composition. If you dig carefully into a soil so that you leave a nice, clean vertical cut, you will see the soil's layers. In a northern forest, a soil is dark at the top, then has a white powdery layer, pale as ash, then a brightly colored layer, usually much deeper than the white one and typically orangish. Below that is a soil whose color is close to that of the bedrock (which geologists call "the parent material," for obvious reasons). We call the layers *soil horizons.* The soil horizons shown in Figure 11.14 are not necessarily all present in any one soil. Very young soils may have only an upper *A* horizon over a *C* horizon, whereas mature soils may have nearly all the horizons shown.

Rainwater is slightly acid (it has a pH of about 5.5) because it has some carbon dioxide from the air dissolved in it, and this forms carbonic acid, a mild acid. As a result, when rainwater moves down into the soil, iron, calcium, magnesium, and other nutritionally important elements are leached from the upper horizons (*A* and *E*) and may be deposited in a lower horizon (*B*). The upper horizons are usually full of life and are viewed by ecologists as complex ecosystems, or ecosystem units (horizons *O* and *A*).

Within soil, decomposition is the name of the game as fungi, bacteria, and small animals live on what plants and animals on the surface produce and deposit. Bacteria and fungi, the great chemical factories of the biosphere, decompose organic compounds from the surface. Soil animals, such as earthworms, eat leaves, twigs, and other remains, breaking them into smaller pieces that are easier for the fungi and bacteria to process. In this way, the earthworms and other soil animals affect the rate of chemical reactions in the soil. There are also predators on soil animals, so there is a soil ecological food chain.

Soil *fertility* is the capacity of a soil to supply nutrients necessary for plant growth. Soils that have formed on geologically young materials are often nutrient-rich. Soils in humid areas and tropics may be heavily leached and relatively nutrient-poor due to the high rainfall. In such soils, nutrients may be cycled through the organic-rich upper horizons; and if forest cover is removed, reforestation may be very difficult. Soils that accumulate certain clay minerals in semiarid regions may swell when they get wet and shrink as they dry out, cracking roads, walls, buildings, and other structures. Expansion and contraction of soils in the United States cause billions of dollars' worth of property damage each year.

Coarse-grained soils, especially those composed primarily of sand, are particularly susceptible to erosion by water and wind. Sand and gravel have relatively large spaces between grains, so water moves through them quickly. Soils with small clay particles retain water well and retard the movement of water. Soils with a mixture of clay and sand can retain water well enough for plant growth but also drain well. Soils with a high percentage of organic

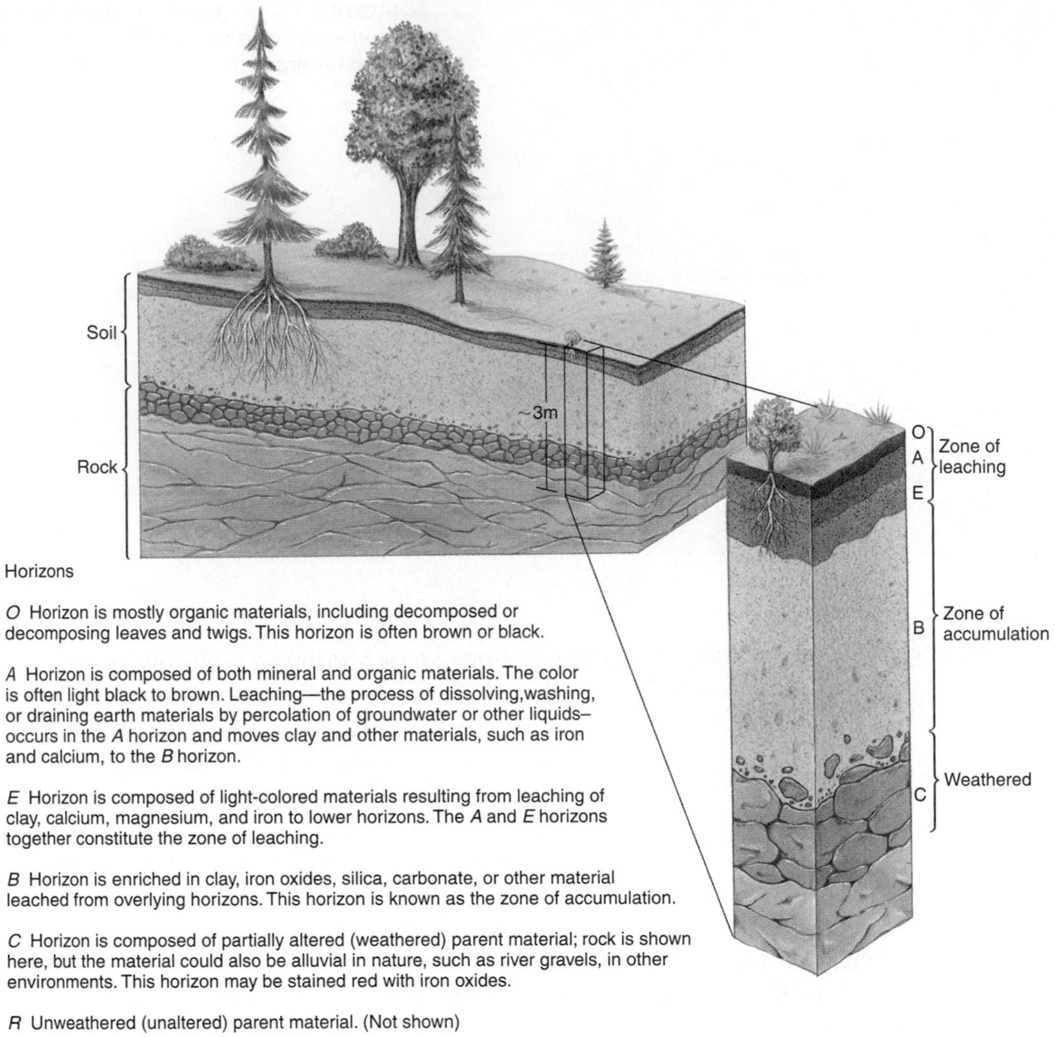
Soil
Rock
~3m
O
A
E
Zone of leaching
B
Zone of accumulation
C
Weathered
Horizons
O Horizon is mostly organic materials, including decomposed or decomposing leaves and twigs. This horizon is often brown or black.
A Horizon is composed of both mineral and organic materials. The color is often light black to brown. Leaching—the process of dissolving,washing, or draining earth materials by percolation of groundwater or other liquids–occurs in the *A* horizon and moves clay and other materials, such as iron and calcium, to the *B* horizon.
E Horizon is composed of light-colored materials resulting from leaching of clay, calcium, magnesium, and iron to lower horizons. The *A* and *E* horizons together constitute the zone of leaching.
B Horizon is enriched in clay, iron oxides, silica, carbonate, or other material leached from overlying horizons. This horizon is known as the zone of accumulation.
C Horizon is composed of partially altered (weathered) parent material; rock is shown here, but the material could also be alluvial in nature, such as river gravels, in other environments. This horizon may be stained red with iron oxides.
R Unweathered (unaltered) parent material. (Not shown)

FIGURE 11.14 **Idealized diagram of a soil, showing soil horizons.**

matter also retain water and chemical nutrients for plant growth. It is an advantage to have good drainage, so a coarse-grained soil is a good place to build your house. If you are going to farm, you'll do best in a loam soil, which has a mixture of particle sizes.

Restoring Our Soils

Part of the answer to the plow puzzle that we discussed earlier in this chapter is the application of organic and inorganic fertilizer. Another part of the answer is general improvements in how plowing is done and where.

Because of improved farming practices, soil erosion has slowed 40% in the United States. According to the U.S. Department of Agriculture, "water (sheet & rill) erosion on cropland dropped from 4.0 tons per acre per year in 1982 to 2.6 tons per acre per year in 2003; wind erosion rates dropped from 3.3 to 2.1 tons per acre per year."[14]

One outstanding example is the drainage area of Coon Creek, Wisconsin, an area of 360 km^2 that has been heavily farmed for more than a century. This stream's watershed was the subject of a detailed study in the 1930s by the U.S. Soil Conservation Service, and was then restudied in the 1970s and 1990s. Measurements at these three times showed that soil erosion was only 6% of what it had been in the 1930s.[15] The bad news is that, even so, the soil is eroding faster than new soil is being generated.[16]

Fertilizers

Traditionally, farmers combated the decline in soil fertility by using organic fertilizers, such as animal manure, which improve both chemical and physical characteristics of soil. But organic fertilizers have drawbacks, especially under intense agriculture on poor soils. In such situations, they do not provide enough of the chemical elements needed to replace what is lost.

A CLOSER LOOK 11.1

The Great American Dust Bowl

Soil erosion became a national issue in the United States in the 1930s, when intense plowing, combined with a major drought, loosened the soil over large areas. The soil blew away, creating dust storms that buried automobiles and houses, destroyed many farms, impoverished many people, and led to a large migration of farmers from Oklahoma and other western and midwestern states to California. The human tragedies of the Dust Bowl were made famous by John Steinbeck's novel *The Grapes of Wrath,* later a popular movie starring Henry Fonda (Figure 11.15).

The land that became the Dust Bowl had been part of America's great prairie, where grasses rooted deep, creating a heavily organic soil a meter or more down. The dense cover provided by grass stems and the anchoring power of roots protected the soil from the erosive forces of water and wind. When the plow turned over those roots, the soil was exposed directly to sun, rain, and wind, which further loosened the soil.

FIGURE 11.15 **The Dust Bowl.** Poor agricultural practices and a major drought created the Dust Bowl, which lasted about ten years during the 1930s. Heavily plowed lands lacking vegetative cover blew away easily in the dry winds, creating dust storms and burying houses.

The development of industrially produced fertilizers, commonly called "chemical" or "artificial" fertilizers, was a major factor in greatly increasing crop production in the 20th century. One of the most important advances was the invention of industrial processes to convert molecular nitrogen gas in the atmosphere to nitrate that can be used directly by plants. Phosphorus, another biologically important element, is mined, usually from a fossil source that was biological in origin, such as deposits of bird guano on islands used for nesting (Figure 11.16). The scientific-industrial age brought with it mechanized mining of phosphates and their long-distance transport, which, at a cost, led to short-term increases in soil fertility. Nitrogen, phosphorus, and other elements are combined in proportions appropriate for specific crops in specific locations.

Limiting Factors

Crops require about 20 chemical elements. These must be available in the right amounts, at the right times, and in the right proportions to each other. It is customary to divide these life-important chemical elements into two groups, macronutrients and micronutrients. A **macronutrient** is a chemical element required by all living things in relatively large amounts. Macronutrients are sulfur, phosphorus, magnesium, calcium, potassium, nitrogen, oxygen, carbon, and hydrogen. A **micronutrient** is a chemical element required in small amounts—either in extremely small amounts by all forms of life or in moderate to small amounts for some forms of life. Micronutrients are often rarer metals, such as molybdenum, copper, zinc, manganese, and iron.

High-quality agricultural soil has all the chemical elements required for plant growth and also has a physical structure that lets both air and water move freely through the soil and yet retain water well. The best agricultural soils have a high organic content and a mixture of sediment particle sizes. Lowland rice grows in flooded ponds and requires a heavy, water-saturated soil, while watermelons grow best in very sandy soil. Soils rarely have everything a crop needs. The question for a farmer is: What needs to be added or done to make a soil more productive for a crop? The traditional answer is that, at any time, just one factor is limiting. If that **limiting factor** can be improved, the soil will be more productive; if that single factor is not improved, nothing else will make a difference.

The idea that some single factor determines the growth and therefore the presence of a species is known as **Liebig's law of the minimum**, after Justus von Liebig, a 19th-century agriculturalist credited with first stating this idea. A general statement of Liebig's law is: The growth of a plant is affected by one limiting factor at a time—the one whose availability is the least in comparison to the needs of a plant.

FIGURE 11.16 **Boobies on a guano island stand on centuries of bird droppings.** The birds feed on ocean fish and nest on islands. In dry climates, their droppings accumulate, becoming a major source of phosphorus for agriculture for centuries.

Striking cases of soil-nutrient limitations have been found in Australia—which has some of the oldest soils in the world—on land that has been above sea level for many millions of years, during which time severe leaching has taken place. Sometimes trace elements are required in extremely small amounts. For example, it is estimated that in certain Australian soils adding an ounce of molybdenum to a field increases the yield of grass by 1 ton/year. The idea of a limiting growth factor, originally used in reference to crop plants, has been extended by ecologists to include all life requirements for all species in all habitats.

If Liebig were always right, then environmental factors would always act one by one to limit the distribution of living things. But there are exceptions. For example, nitrogen is a necessary part of every protein, and proteins are essential building blocks of cells. Enzymes, which make many cell reactions possible, contain nitrogen. A plant given little nitrogen and phosphorus might not make enough of the enzymes involved in taking up and using phosphorus. Increasing nitrogen to the plant might therefore increase the plant's uptake and use of phosphorus. If this were so, the two elements would have a **synergistic effect**, in which a change in the availability of one resource affects the response of an organism to some other resource.

So far we have discussed the effects of chemical elements when they are in short supply. But it is also possible to have too much of a good thing—most chemical elements become toxic in concentrations that are too high. As a simple example, plants die when they have too little water but also when they are flooded, unless they have specific adaptations to living in water. So it is with chemical elements required for life.

11.5 Controlling Pests

From an ecological point of view, pests are undesirable competitors, parasites, or predators. The major agricultural pests are insects that feed mainly on the live parts of plants, especially leaves and stems; nematodes (small worms), which live mainly in the soil and feed on roots and other plant tissues; bacterial and viral diseases; weeds (plants that compete with the crops); and vertebrates (mainly rodents and birds) that feed on grain or fruit. Even today, with modern technology, the total losses from all pests are huge; in the United States, pests account for an estimated loss of one-third of the potential harvest and about one-tenth of the harvested crop. Preharvest losses are due to competition from weeds, diseases, and herbivores; postharvest losses are largely due to herbivores.[17]

Because a farm is maintained in a very early stage of ecological succession and is enriched by fertilizers and water, it is a good place not only for crops but also for other early-successional plants. These noncrop and therefore undesirable plants are what we call weeds. A weed is just a plant in a place we do not want it to be. There are about 30,000 species of weeds, and in any year a typical farm field is infested with between 10 and 50 of them. Some weeds can have a devastating effect on crops. For example, the production of soybeans is reduced by 60% if a weed called cocklebur grows three individuals per meter (one individual per foot).[18]

Pesticides

Before the Industrial Revolution, farmers could do little to prevent pests except remove them or use farming methods that tended to decrease their density. Pre-industrial farmers planted aromatic herbs and other vegetation that repels insects.

The scientific industrial revolution brought major changes in agriculture pest control, which we can divide into four stages:

Stage 1: Broad-Spectrum Inorganic Toxins
With the beginning of modern science-based agriculture, people began to search for chemicals that would reduce the abundance of pests. Their goal was a "magic bullet"—a chemical (referred to as a narrow-spectrum pesticide) that would have a single target, just one pest, and not affect anything else. But this proved elusive. The earliest pesticides were simple inorganic compounds that were widely toxic. One of the earliest was arsenic, a chemical element toxic to all life, including people. It was certainly effective in killing pests, but it killed beneficial organisms as well and was very dangerous to use.

Stage 2: Petroleum-Based Sprays and Natural Plant Chemicals (1930s on)

Many plants produce chemicals as a defense against disease and herbivores, and these chemicals are effective pesticides. Nicotine, from the tobacco plant, is the primary agent in some insecticides still widely used today. However, although natural plant pesticides are comparatively safe, they were not as effective as desired.

Stage 3: Artificial Organic Compounds

Artificial organic compounds have created a revolution in agriculture, but they have some major drawbacks (see Risk–Benefit Analysis in Chapter 7). One problem is secondary pest outbreaks, which occur after extended use (and possibly because of extended use) of a pesticide. Secondary pest outbreaks can come about in two ways: (1) Reducing one target species reduces competition with a second species, which then flourishes and becomes a pest, or (2) the pest develops resistance to the pesticides through evolution and natural selection, which favor those who have a greater immunity to the chemical. Resistance has developed to many pesticides. For example, Dasanit (fensulfothion), an organophosphate first introduced in 1970 to control maggots that attack onions in Michigan, was originally successful but is now so ineffective that it is no longer used for that crop.

Some artificial organic compounds, such as DDT, are broad-spectrum, but more effective than natural plant chemicals. However, they also have had unexpected environmental effects. For example, aldrin and dieldrin have been widely used to control termites as well as pests on corn, potatoes, and fruits. Dieldrin is about 50 times as toxic to people as DDT. These chemicals are designed to remain in the soil and typically do so for years. Therefore, they have spread widely.

World use of pesticides exceeds 2.5 billion kg (5 billion pounds), and in the United States it exceeds 680 million kg (1,200 million pounds) (Figure 11.17). The total amount paid for these pesticides is $32 billion worldwide and $11 billion in the United States.[12] But the magic bullet has remained elusive. Once applied, these chemicals may decompose in place or may be blown by the wind or transported by surface and subsurface waters, meanwhile continuing to decompose. Sometimes the initial breakdown products (the first, still complex chemicals produced from the original pesticides) are toxic, as is the case with DDT. Eventually, the toxic compounds are decomposed to their original inorganic or simple, nontoxic organic compounds, but for some chemicals this can take a very long time.

Public-health standards and environmental-effects standards have been established for some of these compounds. The United States Geological Survey has established a network for monitoring 60 sample watersheds throughout the nation. These are medium-size watersheds, not the entire flow from the nation's major rivers. One such watershed is that of the Platte River, a major tributary of the Missouri River.

The most common herbicides used for growing corn, sorghum, and soybeans along the Platte River were alachlor, atrazine, cyanazine, and metolachlor, all organonitrogen herbicides. Monitoring of the Platte near Lincoln, Nebraska, suggested that during heavy spring runoff, concentrations of some herbicides might be reaching or exceeding established public-health standards. But this research is just beginning, and it is difficult to reach definitive conclusions as to whether present concentrations are causing harm in public water supplies or to wildlife, fish, algae in freshwater, or vegetation. Advances in knowledge give us much more information, on a more regular basis, about how much of many artificial compounds are in our waters, but we are still unclear about their environmental effects. A wider and better program to monitor pesticides in water and soil is important to provide a sound scientific basis for dealing with pesticides.

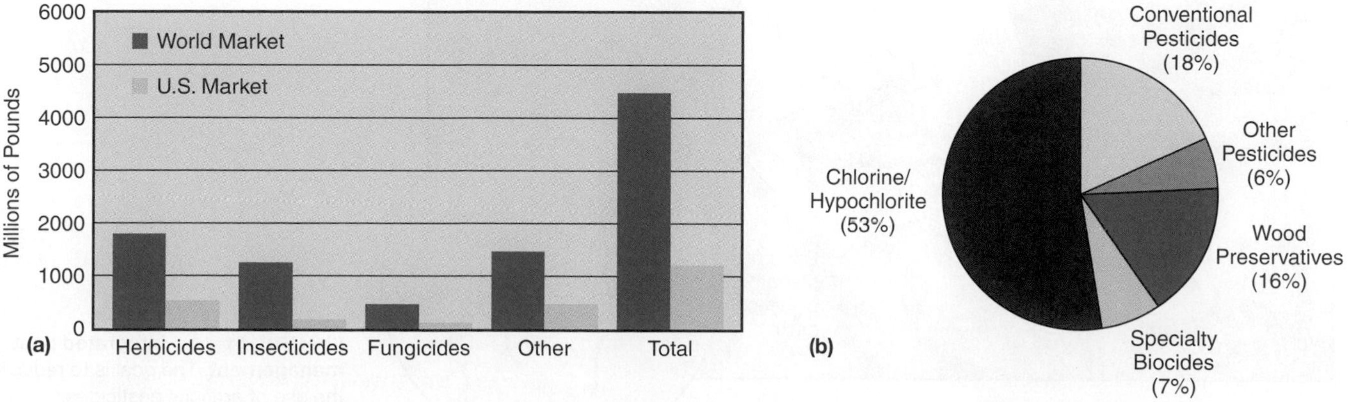

FIGURE 11.17 World use of pesticides. (a) Total amounts, **(b)** percentage by main type. (*Sources*: EPA [2006]."2000–2001Pesticide Market Esimates: Usage.")

Stage 4: Integrated Pest Management and Biological Control

Integrated pest management (IPM) uses a combination of methods, including biological control, certain chemical pesticides, and some methods of planting crops (Figure 11.18). A key idea underlying IPM is that the goal can be control rather than complete elimination of a pest. This is justified for several reasons. Economically, it becomes more and more expensive to eliminate a greater and greater percentage of a pest, while the value of ever-greater elimination, in terms of crops to sell, becomes less and less. This suggests that it makes economic sense to eliminate only enough to provide benefit and leave the rest. In addition, allowing a small but controlled portion of a pest population to remain does less damage to ecosystems, soils, water, and air.

Integrated pest management also moves away from monoculture growing in perfectly regular rows. Studies have shown that just the physical complexity of a habitat can slow the spread of parasites. In effect, a pest, such as a caterpillar or mite, is trying to find its way through a maze. If the maze consists of regular rows of nothing but what the pest likes to eat, the maze problem is easily solved by the dumbest of animals. But if several species, even two or three, are arranged in a more complex pattern, pests have a hard time finding their prey.

No-till or low-till agriculture is another feature of IPM because this helps natural enemies of some pests to build up in the soil, whereas plowing destroys the habitats of these enemies.

Biological control includes using one species that is a natural enemy of another. One of the most effective is the bacterium ***Bacillus thuringiensis***, known as BT, which causes a disease that affects caterpillars and the larvae of other insect pests. Spores of BT are sold commercially—you can buy them at your local garden store and use them in your home garden. BT has been one of the most important ways to control epidemics of gypsy moths, an introduced moth whose larvae periodically strip most of the leaves from large areas of forests in the eastern United States. BT has proved safe and effective—safe because it causes disease only in specific insects and is harmless to people and other mammals, and because, as a natural biological "product," its presence and its decay are nonpolluting.

Another group of effective biological-control agents are small wasps that are parasites of caterpillars. Control of the oriental fruit moth, which attacks a number of fruit crops, is an example of IPM biological control. The moth was found to be a prey of a species of wasp, *Macrocentrus ancylivorus*, and introducing the wasp into fields helped control the moth. Interestingly, in peach fields the wasp was more effective when strawberry fields were nearby. The strawberry fields provided an alternative habitat for the wasp, especially important for overwintering.[9] As this example shows, spatial complexity and biological diversity also become parts of the IPM strategy.

In the list of biological-control species we must not forget ladybugs, which are predators of many pests. You can buy these, too, at many garden stores and release them in your garden.

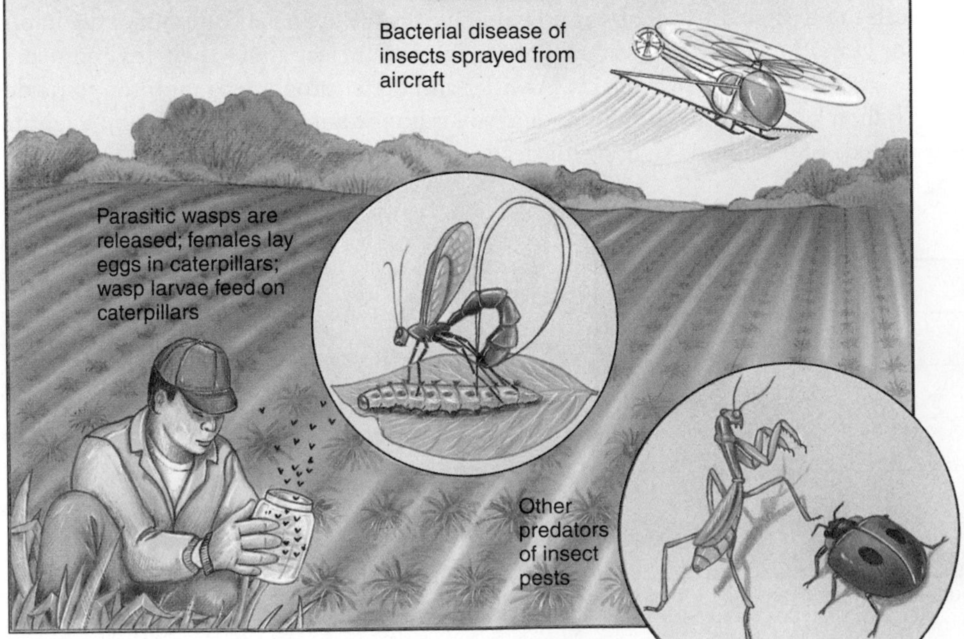
Bacterial disease of insects sprayed from aircraft
Parasitic wasps are released; females lay eggs in caterpillars; wasp larvae feed on caterpillars
Other predators of insect pests

FIGURE 11.18 Integrated pest management. The goal is to reduce the use of artificial pesticides, reduce costs, and efficiently control pests.

Another biological control uses sex pheromones, chemicals released by most species of adult insects (usually the female) to attract members of the opposite sex. In some species, pheromones have been shown to be effective up to 4.3 km (2.7 mi) away. These chemicals have been identified, synthesized, and used as bait in insect traps, in insect surveys, or simply to confuse the mating patterns of the insects involved.

While biological control works well, it has not solved all problems with agricultural pests. As for artificial pesticides, although they are used in integrated pest management, they are used along with the other techniques, so the application of these pesticides can be sparing and specific. This would also greatly reduce the costs to farmers for pest control.

11.6 The Future of Agriculture

Today, there are three major technological approaches to agriculture. One is modern mechanized agriculture, where production is based on highly mechanized technology that has a high demand for resources—including land, water, and fuel—and makes little use of biologically-based technologies. Another approach is resource-based—that is, agriculture based on biological technology and conservation of land, water, and energy. An offshoot of this second approach is organic food production—growing crops without artificial chemicals (including pesticides) or genetic engineering but instead using ecological control methods. The third approach is genetic engineering.

In mechanized agriculture, production is determined by economic demand and limited by that demand, not by resources. In resource-based agriculture, production is limited by environmental sustainability and the availability of resources, and economic demand usually exceeds production.

With these methods in mind, we can consider what can be done to help crop production keep pace with human population growth. Here are some possibilities.

Increased Production per Acre

Some agricultural scientists and agricultural corporations believe that production per unit area will continue to increase, partially through advances in genetically modified crops. This new methodology, however, raises some important potential environmental problems, which we will discuss later. Furthermore, increased production in the past has depended on increased use of water and fertilizers. Water is a limiting factor in many parts of the world and will become a limiting factor in more areas in the future.

Increased Farmland Area

A United Nations Food and Agriculture Organization conference held in early 2008 considered the coming world food crisis. It was reported that 23 million hectares (more than 46 million acres) of farmland had been withdrawn from production in Eastern Europe and the Commonwealth of Independent States (CIS) region, especially in countries such as Kazakhstan, Russia, and Ukraine, and that at least half—13 million hectares (15 million acres)—could be readily put back into production with little environmental effect.[19] This would be like adding all the farmland in Iowa, Illinois, and Indiana.[20]

The need for additional farmland brings up, once again, the problem that agrifuels bring to agriculture: taking land away from food production to produce fuels instead.

New Crops and Hybrids

Since there are so many plant species, perhaps some yet unused ones could provide new sources of food and grow in environments little used for agriculture. Those interested in conserving biological diversity urge a search for such new crops on the grounds that this is one utilitarian justification for the conservation of species. It is also suggested that some of these new crops may be easier on the environment and therefore more likely to allow sustainable agriculture. But it may be that over the long history of human existence those species that are edible have already been found, and the number is small. Research is under way to seek new crops or plants that have been eaten locally but whose potential for widespread, intense cultivation has not been tested.

Among the likely candidates for new crops are amaranth for seeds and leaves; *Leucaena*, a legume useful for animal feed; and triticale, a synthetic hybrid of wheat and rye.[21] A promising source of new crops is the desert; none of the 14 major crops are plants of arid or semiarid regions, yet there are vast areas of desert and semidesert. The United States has 200,000 million hectares (about 500,000 million acres) of arid and semiarid rangeland. In Africa, Australia, and South America, the areas are even greater. Several species of plants can be grown commercially under arid conditions, allowing us to use a biome for agriculture that has been little used in this way in the past. Examples of these species are guayule (a source of rubber), jojoba (for oil), bladderpod (for oil from seeds), and gumweed (for resin). Jojoba, a native shrub of the American Sonoran Desert, produces an extremely fine oil, remarkably resistant to bacterial degradation, which is useful in cosmetics and as a fine lubricant. Jojoba is now grown commercially in Australia, Egypt, Ghana, Iran, Israel, Jordan, Mexico, Saudi Arabia, and the United States.[22] Although these examples are not food crops, they release other lands, now used to produce similar products, to grow food.

FIGURE 11.19 Experimental rice plots at the International Rice Research Institute, Philippines, showing visual crop variation based on the use of fertilizers.

Among the most successful developments of new hybrids have been those developed by the **green revolution**, the name attached to post–World War II programs that have led to the development of new strains of crops with higher yields, better resistance to disease, or better ability to grow under poor conditions. These crops include superstrains of rice (at the International Rice Research Institute in the Philippines) (see Figure 11.19) and strains of maize with improved disease resistance (at the International Maize and Wheat Improvement Center in Mexico).

Better Irrigation

Drip irrigation—from tubes that drip water slowly—greatly reduces the loss of water from evaporation and increases yield. However, it is expensive and thus most likely to be used in developed nations or nations with a large surplus of hard currency—in other words, in few of the countries where hunger is most severe.

Organic Farming

Organic farming is typically considered to have three qualities: It is more like natural ecosystems than monoculture; it minimizes negative environmental impacts; and the food that results from it does not contain artificial compounds. According to the U.S. Department of Agriculture (USDA), organic farming has been one of the fastest-growing sectors in U.S. agriculture, although it still occupies a small fraction of U.S. farmland and contributes only a small amount of agriculture income. By the end of the 20th century it amounted to about $6 billion—much less than the agricultural production of California.

In the 1990s, the number of organic milk cows rose from 2,300 to 12,900, and organic layer hens increased from 44,000 to more than 500,000. In the United States only 0.01% of the land planted in corn and soybeans used certified organic farming systems in the mid-1990s; about 1% of dry peas and tomatoes were grown organically, and about 2% of apples, grapes, lettuce, and carrots. On the high end, nearly one-third of U.S. buckwheat, herb, and mixed vegetable crops were grown under organic farming conditions.[23] USDA certification of organic farming became mandatory in 2002. After that, organic cropland that was listed as certified more than doubled, and the number of farmers certifying their products rose 40%. In the United States today, more than 1.3 million acres are certified as organic. There are about 12,000 organic farmers in the United States, and the number is growing 12% per year.[24]

Eating Lower on the Food Chain

Some people believe it is ecologically unsound to use domestic animals as food, on the grounds that eating each step farther up a food chain leaves much less food to eat per acre. This argument is as follows: No organism is 100% efficient; only a fraction of the energy in food taken in is converted to new organic matter. Crop plants may convert 1–10% of sunlight to edible food, and cows may convert only 1–10% of hay and grain to meat. Thus, the same area could produce 10 to 100 times more vegetation than meat per year. This holds true for the best agricultural lands, which have deep, fertile soils on level ground.

11.7 Genetically Modified Food: Biotechnology, Farming, and Environment

The discovery that DNA is the universal carrier of genetic information has led to development and use of genetically modified crops, which has given rise to new environmental controversies as well as a promise of increased agricultural production.

Genetic engineering in agriculture involves several different practices, which we can group as follows: (1) faster and more efficient ways to develop new hybrids; (2) introduction of the "terminator gene"; and (3) transfer of genetic properties from widely divergent kinds of life. These three practices have quite different potentials and problems. We need to keep in mind a general rule of environmental actions, the rule of natural change: If actions we take are similar in kind and frequency to natural changes, then the effects on the environment are likely to be benign. This is because species have had a long time to evolve and adapt to these changes. In contrast, changes that are novel—that do not occur in nature—are more likely to have negative or undesirable environmental effects, both direct and indirect. We can apply this rule to the three categories of genetically engineered crops.

The jury is out as to whether the benefits of genetically modified crops will outweigh undesirable effects. As with many new technologies of the industrial age, application has preceded environmental investigation and understanding, and the widespread use of **genetically modified crops** (GMCs) is under way before the environmental effects are well understood. The challenge for environmental science is to gain an understanding of environmental effects of GMCs quickly.

New Hybrids

The development of hybrids within a species is a natural phenomenon, and the development of hybrids of major crops, especially of small grains, has been a major factor in the great increase in productivity of 20th-century agriculture. So, strictly from an environmental perspective, genetic engineering to develop hybrids within a species is likely to be as benign as the development of agricultural hybrids has been with conventional methods.

There is an important caveat, however. Some people are concerned that the great efficiency of genetic modification methods may produce "superhybrids" that are so productive they can grow where they are not wanted and become pests. There is also concern that some of the new hybrid characteristics could be transferred by interbreeding with closely related weeds. This could inadvertently create a "superweed" whose growth, persistence, and resistance to pesticides would make it difficult to control. Another environmental concern is that new hybrids might be developed that could grow on more and more marginal lands. Raising crops on such marginal lands might increase erosion and sedimentation and lead to decreased biological diversity in specific biomes. Still another potential problem is that "superhybrids" might require much more fertilizer, pesticide, and water. This could lead to greater pollution and the need for more irrigation.

On the positive side, genetic engineering could lead to hybrids that require less fertilizer, pesticide, and water. For example, right now only legumes (peas and their relatives) have symbiotic relationships with bacteria and fungi that allow them to fix nitrogen. Attempts are under way to transfer this capability to other crops, so that more kinds of crops would enrich the soil with nitrogen and require much less external application of nitrogen fertilizer.

The Terminator Gene

The **terminator gene** makes seeds from a crop sterile. This is done for environmental and economic reasons. In theory, it prevents a genetically modified crop from spreading. It also protects the market for the corporation that developed it: Farmers cannot avoid purchasing seeds by using some of their crops' hybrid seeds the next year. But this poses social and political problems. Farmers in less-developed nations, and governments of nations that lack genetic-engineering capabilities, are concerned that the terminator gene will allow the United States and a few of its major corporations to control the world food supply. Concerned observers believe that farmers in poor nations must be able to grow next year's crops from their own seeds because they cannot afford to buy new seeds every year. This is not directly an environmental problem, but it can become an environmental problem indirectly by affecting total world food production, which then affects the human population and how land is used in areas that have been in agriculture.

Transfer of Genes from One Major Form of Life to Another

Most environmental concerns have to do with the third kind of genetic modification of crops: the transfer of genes from one major kind of life to another. This is a novel effect and, as we have explained, therefore more likely to have undesirable results. In several cases, in fact, this type of genetic modification has affected the environment in unforeseen and undesirable ways. Perhaps the best-known involves potatoes and corn, caterpillars that eat these crops, a disease of caterpillars that controls these pests, and an endangered species, monarch butterflies. Here is what happened.

As discussed earlier, the bacterium *Bacillus thuringiensis* is a successful pesticide that causes a disease in many caterpillars. With the development of biotechnology, agricultural scientists studied the bacteria and discovered the toxic chemical and the gene that caused its production within the bacteria. This gene was then transferred to potatoes and corn so that the biologically engineered plants produced their own pesticide. At first, this was believed to be a constructive step in pest control because it was no longer necessary to spray a pesticide. However, the genetically engineered potatoes and corn produced the toxic BT substance in every cell—not just in the leaves that the caterpillars ate, but also in the potatoes and corn sold as food, in the flowers, and in the pollen. This has a potential, not yet demonstrated, to create problems for species that are not intended targets of the BT (Figure 11.20).

A strain of rice has been developed that produces beta-carotene, important in human nutrition. The rice thus has added nutritional benefits that are particularly valuable for the poor of the world who depend on rice as a primary food. The gene that enables rice to make beta-carotene comes from daffodils, but the modification actually required the introduction of four specific genes and would likely be impossible without genetic-engineering techniques. That is, genes were transferred between plants that would not exchange genes in nature. Once again, the rule of natural change suggests that we should monitor such actions carefully. Indeed, although the genetically modified rice appears to have beneficial effects, the government of India has refused to allow it to be grown in that country.[15]

There is much concern worldwide about the political, social, and environmental effects of genetic modification of crops. This is a story in process, one that will change rapidly in the next few years. You can check on these fast-moving events on the textbook's Web site.

(a)

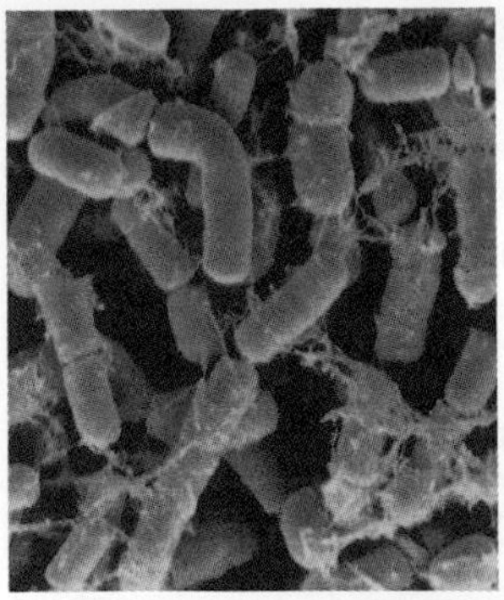

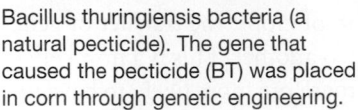

Bacillus thuringiensis bacteria (a natural pecticide). The gene that caused the pecticide (BT) was placed in corn through genetic engineering.

(b)

BT corn contains its own pesticide in every cell of the plant.

(c)

Pollen from the BT corn is also toxic and when it lands on milkweed, monarch butterflies that eat the milkweed may die.

FIGURE 11.20 The flow of the BT toxin from bacteria **(a)** to corn through genetic engineering **(b)** and the possible ecological transfer of toxic substances to monarch butterflies **(c)**.

11.8 Aquaculture

In contrast to food obtained on land, we still get most of our marine and freshwater food by hunting. Hunting wild fish has not been sustainable (see Chapter 13), and thus **aquaculture**, the farming of this important source of protein in both marine and freshwater habitats, is growing rapidly and could become one of the major ways to provide food of high nutritional quality. Popular aquacultural animals include carp, tilapia, oysters, and shrimp, but in many nations other species are farm-raised and culturally important, such as yellowtail (important in Japan and perhaps just one of several species); crayfish (United States); eels and minnows (China); catfish (southern and midwestern United States); salmon (Canada, Chile, Norway, and the United States); trout (United States); plaice, sole, and the Southeast Asian milkfish (Great Britain); mussels (Canada, France, Spain, and Southeast Asian countries); and sturgeon (Ukraine). A few species—trout and carp—have been subject to genetic breeding programs.[25]

Although relatively new in the United States, aquaculture has a long history elsewhere, especially in China, where at least 50 species are grown, including finfish, shrimp, crab, other shellfish, sea turtles, and sea cucumbers (not a vegetable but a marine animal).[14] In the Szechuan area of China, fish are farmed on more than 100,000 hectares (about 250,000 acres) of flooded rice fields. This is an ancient practice that can be traced back to a treatise on fish culture written by Fan Li in 475 B.C.[14] In China and other Asian countries, farmers often grow several species of fish in the same pond, exploiting their different ecological niches. Ponds developed mainly for carp, a bottom-feeding fish, also contain minnows, which feed at the surface on leaves added to the pond.

Aquaculture can be extremely productive on a per-area basis, in part because flowing water brings food from outside into the pond or enclosure. Although the area of Earth that can support freshwater aquaculture is small, we can expect this kind of aquaculture to increase and become a more important source of protein.

Sometimes fishponds use otherwise wasted resources, such as fertilized water from treated sewage. Other fishponds exist in natural hot springs (Idaho) or use water warmed by being used to cool electric power plants (Long Island, New York; Great Britain).[14]

Mariculture, the farming of ocean fish, though producing a small part of the total marine fish catch, has grown rapidly in the last decades and will likely continue to do so. Oysters and mussels are grown on rafts lowered into the ocean, a common practice in the Atlantic Ocean in Portugal and in the Mediterranean in such nations as France. These animals are filter feeders—they obtain food from water that moves past them. Because a small raft is exposed to a large volume of water, and thus a large volume of food, rafts can be extremely productive. Mussels grown on rafts in bays of Galicia, Spain, produce 300 metric tons per hectare, whereas public harvesting grounds of wild shellfish in the United States yield only about 10 kg/ha (that's just a hundredth of a metric ton).[14] Oysters and mussels are grown on artificial pilings in the intertidal zone in the state of Washington (Figure 11.21).

FIGURE 11.21 **An oyster farm in Poulsbo, Washington.** Oysters are grown on artificial pilings in the intertidal zone.

Some Negatives

Although aquaculture has many benefits and holds great promise for our food supply, it also causes environmental problems. Fishponds and marine fish kept in shallow enclosures connected to the ocean release wastes from the fish and chemicals such as pesticides, polluting local environments. In some situations, aquaculture can damage biological diversity. This is a concern with salmon aquaculture in the Pacific Northwest, where genetic strains not native to a stream are grown and some are able to mix with wild populations and breed. Problems with salmon aquaculture demonstrate the need for improved methods and greater care about environmental effects.

CRITICAL THINKING ISSUE

Will There Be Enough Water to Produce Food for a Growing Population?

Between 2000 and 2025, scientists estimate, the world population will increase from 6.6 billion to 7.8 billion, approximately double what it was in 1974. To keep pace with the growing population, the United Nations Food and Agriculture Organization predicts, that food production will have to double by 2025, and so will the amount of water consumed by food crops. Will the supply of freshwater be able to meet this increased demand, or will the water supply limit global food production?

Growing crops consume water through transpiration (loss of water from leaves as part of the photosynthetic process) and evaporation from plant and soil surfaces. The volume of water consumed by crops worldwide—including rainwater and irrigated water—is estimated at 3,200 billion m^3 per year. An almost equal amount of water is used by other plants in and near agricultural fields. Thus, it takes 7,500 billion m^3 per year of water to supply crop ecosystems around the world (see Table 11.2). Grazing and pastureland account for another 5,800 billion m^3, and evaporation from irrigated water another 500 billion m^3, for a total of 13,800 billion m^3 of water per year for food production, or 20% of the water evaporated and transpired worldwide. By 2025, therefore, humans will be appropriating almost half of all the water available to life on land for growing food for their own use. Where will the additional water come from?

Table 11.2 ESTIMATED WATER REQUIREMENTS OF FOOD AND FORAGE CROPS

CROP	LITERS/KG
Potatoes	500
Wheat	900
Alfalfa	900
Sorghum	1,110
Corn	1,400
Rice	1,912
Soybeans	2,000
Broiler chicken	3,500*
Beef	100,000*

* Includes water used to raise feed and forage.

Source: D. Pimentel et al., "Water Resources: Agriculture, the Environment, and Society," *Bioscience* 4, no. 2 [February 1997]: 100.

Although the amount of rainwater cannot be increased, it can be used more efficiently through farming methods such as terracing, mulching, and contouring. Forty percent of the global food harvest now comes from irrigated land, and some scientists estimate that the volume of irrigation water available to crops will have to triple by 2025—to a volume equaling 24 Nile rivers or 110 Colorado rivers.[26] A significant saving of water can therefore come from more efficient irrigation methods, such as improved sprinkler systems, drip irrigation, night irrigation, and surge flow.

Surge flow is the intermittent application of water along furrows—on and off periods of water flow at constant or variable intervals. Often, this can completely irrigate a crop in much less time and therefore wastes much less water than does constant irrigation, which allows much more time for water to evaporate. Surge flow is also useful for young plants, which need only a small amount of water.

Additional water could be diverted from other uses to irrigation, but this might not be as easy as it sounds because of competing needs for water. For example, if water were provided to the 1 billion people in the world who currently lack drinking and household water, less would be available for growing crops. And the new billions of people to be added to the world population in the next decades will also need water. People already use 54% of the world's runoff. Increasing this to more than 70%, as will be required to feed the growing population, may result in a loss of freshwater ecosystems, decline in world fisheries, and extinction of aquatic species.

In many places, groundwater and aquifers are being used faster than they are being replaced—a process that is unsustainable in the long run. Many rivers are already so heavily used that they release little or no water to the ocean. These include the Ganges and most other rivers in India, the Huang He (Yellow River) in China, the Chao Phraya in Thailand, the Amu Darya and Syr Darya in the Aral Sea basin, and the Nile and Colorado rivers.

Two hundred years ago, Thomas Malthus put forth the proposition that population grows more rapidly than the ability of the soil to grow food and that at some time the human population will outstrip the food supply. Malthus might be surprised to know that by applying science and technology to agriculture, food production has so far kept pace with population growth. For example, between 1950 and 1995, the world population increased 122% while grain productivity increased 141%. Since 1995, however, grain production has slowed down (see Figure 11.22), and the question remains whether Malthus will be proved right in the 21st century. Will science and technology be able to solve the problem of water supply for growing food for people, or will water prove a limiting factor in agricultural production?

Critical Thinking Questions

1. How might dietary changes in developed countries affect water availability?
2. How might global warming affect estimates of the amount of water needed to grow crops in the 21st century?
3. Withdrawing water from aquifers faster than the replacement rate is sometimes referred to as "mining water." Why do you think this term is used?
4. Many countries in warm areas of the world are unable to raise enough food, such as wheat, to supply their populations. Consequently, they import wheat and other grains. How is this equivalent to importing water?
5. Malthusians are those who believe that sooner or later, unless population growth is checked, there will not be enough food for the world's people. Anti-Malthusians believe that technology will save the human race from a Malthusian fate. Analyze the issue of water supply for agriculture from both points of view.

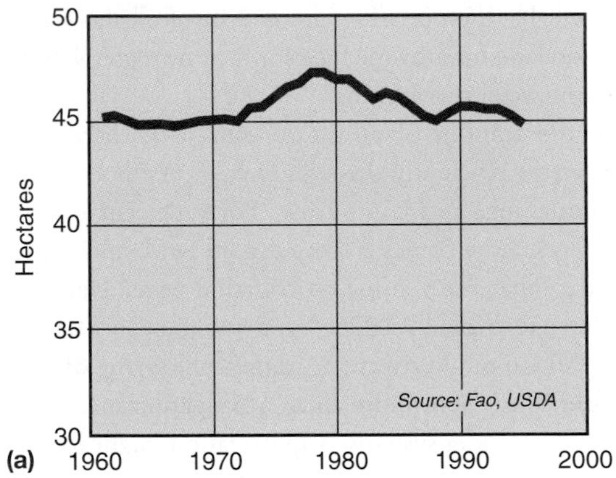

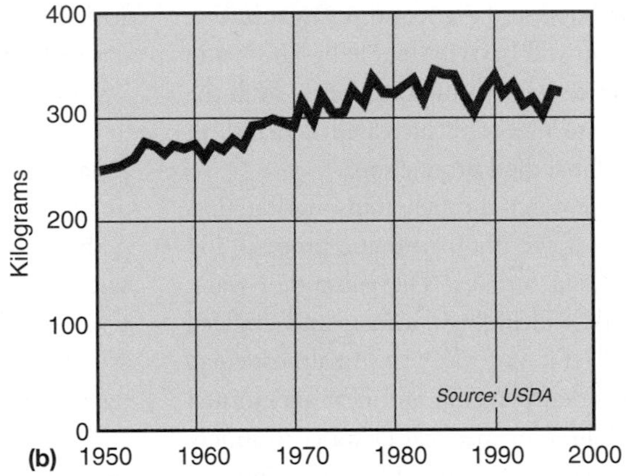

FIGURE 11.22 (a) World irrigated area per thousand people, 1961–95. (*Source*: L.R. Brown, M. Renner, and C. Flavin, *Vital Signs: 1998* [New York: Norton, 1998], p. 47.) **(b)** World grain production per person, 1950–1977. (*Source*: Brown, Renner, and Flavin, *Vital Signs*, p. 29.)

SUMMARY

- Agriculture changes the environment; the more intense the agriculture, the greater the changes.
- From an ecological perspective, agriculture is an attempt to keep an ecosystem in an early-successional stage.
- Farming greatly simplifies ecosystems, creating short and simple food chains, growing a single species or genetic strain in regular rows in large areas, reducing species diversity and reducing the organic content and overall fertility of soils.
- These simplifications open farmed land to predators and parasites, increased soil loss, erosion, and, thus, downstream sedimentation, and increased pollution of soil and water with pesticides, fertilizers, and heavy metals concentrated by irrigation.
- The history of agriculture can be viewed as a series of attempts to overcome environmental limitations and problems. Each new solution has created new environmental problems, which have in turn required their own solutions.
- The Industrial Revolution and the rise of agricultural sciences have revolutionized agriculture in two areas—one ecological and the other genetic—with many benefits and some serious drawbacks.
- Modern fertilizers, irrigation methods, and hybridization have greatly increased the yield per unit area. Modern chemistry has led to the development of a wide variety of pesticides that have reduced, though not eliminated, the loss of crops to weeds, diseases, and herbivores, but these have also had undesirable environmental effects. In the future, pest control will be dominated by integrated pest management.

- Most 20th-century agriculture has relied on machinery and the use of abundant energy, with relatively little attention paid to the loss of soils, the limits of groundwater, and the negative effects of chemical pesticides.
- Overgrazing has severely damaged lands. It is important to properly manage livestock, including using appropriate lands for grazing and keeping livestock at a sustainable density.

REEXAMINING THEMES AND ISSUES

Human Population

Agriculture is the world's oldest and largest industry; more than one-half of all the people in the world still live on farms. Because the production, processing, and distribution of food alter the environment, and because of the size of the industry, large effects on the environment are unavoidable.

Sustainability

Alternative agricultural methods appear to offer the greatest hope of sustaining agricultural ecosystems and habitats over the long term, but more tests and better methods are needed. As the experience with European agriculture shows, crops can be produced on the same lands for thousands of years as long as sufficient fertilizers and water are available; however, the soils and other aspects of the original ecosystem are greatly changed—these are not sustained. In agriculture, production can be sustained, but the ecosystem may not be.

Global Perspective

Agriculture has numerous global effects. It changes land cover, affecting climate at regional and global levels, increasing carbon dioxide in the atmosphere, and adding to the buildup of greenhouse gases, which in turn affects climate. Fires to clear land for agriculture may significantly affect the climate by adding small particulates to the atmosphere. Genetic modification is a new global issue that has not only environmental but also political and social effects.

Urban World

The agricultural revolution makes it possible for fewer and fewer people to produce more and more food and leads to greater productivity per acre. Freed from dependence on farming, people flock to cities, which leads to increased urban effects on the land. Thus, agricultural effects on the environment indirectly extend to the cities.

People and Nature

Farming is one of the most direct and large-scale ways that people affect nature. Our own sustainability, as well as the quality of our lives, depends heavily on how we farm.

Science and Values

Human activities have seriously damaged one-fourth of the world's total land area, impacting one-sixth of the world's population (about 1 billion people). Overgrazing, deforestation, and destructive farming practices have caused so much damage that recovery in some areas will be difficult, and restoration of the rest will require serious actions. A major value judgment we must make in the future is whether our societies will allocate funds to restore these damaged lands. Restoration requires scientific knowledge, both about present conditions and about actions required for restoration. Will we seek this knowledge and pay for it?

KEY TERMS

agroecosystem **213**
aquaculture **230**
biological control **226**
genetically modified crops **229**
green revolution **228**
integrated pest management **226**
Liebig's law of the minimum **223**
limiting factor **223**
macronutrient **223**
malnourishment **216**
mariculture **230**
micronutrient **223**
monoculture **214**
organic farming **228**
pasture **218**
rangeland **218**
synergistic effect **224**
terminator gene **229**
undernourishment **216**

STUDY QUESTIONS

1. Design an integrated pest management scheme for a small vegetable garden in a city lot behind a house. How would this scheme differ from integrated pest management used on a large farm? What aspects of IPM could not be used? How might the artificial structures of a city be put to use to benefit IPM?
2. Under what conditions might grazing cattle be sustainable when growing wheat is not? Under what conditions might a herd of bison provide a sustainable supply of meat when cows might not?
4. Pick one of the nations in Africa that has a major food shortage. Design a program to increase its food production. Discuss how reliable that program might be given the uncertainties that nation faces.
5. Should genetically modified crops be considered acceptable for "organic" farming?
6. You are about to buy your mother a bouquet of 12 roses for Mother's Day, but you discover that the roses were genetically modified to give them a more brilliant color and to produce a natural pesticide through genetic energy. Do you buy the flowers? Explain and justify your answer based on the material presented in this chapter.
7. A city garbage dump is filled, and it is suggested that the area be turned into a farm. What factors in the dump might make it a good area to farm, and what might make it a poor area to farm?
8. You are sent into the Amazon rain forest to look for new crop species. In what kinds of habitats would you look? What kinds of plants would you look for?

FURTHER READING

Borgstrom, G., ***The Hungry Planet: The Modern World at the Edge of Famine*** (New York: Macmillan, 1965). A classic book by one of the leaders of agricultural change.

Cunfer, G., ***On the Great Plains: Agriculture and Environment*** (College Station: Texas A&M University Press, 2005). Uses the history of European agriculture applied to the American Great Plains as a way to discuss the interaction between nature and farming.

Manning, R., ***Against the Grain: How Agriculture Has Hijacked Civilization*** (New York: North Point Press, 2004). An important iconoclastic book in which the author attributes many of civilization's ills—from war to the spread of disease—to the development and use of agriculture. In doing so, he discusses many of the major modern agricultural issues.

Mazoyer, Marcel, and Laurence Roudar, ***A History of World Agriculture: From the Neolithic Age to the Current Crisis*** (New York: Monthly Review Press, 2006). By two French professors of agriculture, this book argues that the world is about to reach a new farming crisis, which can be understood from the history of agriculture.

McNeely, J.A., and S.J. Scherr, ***Ecoagriculture*** (Washington, DC: Island Press, 2003). Smil, V., *Feeding the World* (Cambridge, MA: MIT Press, 2000).

Seymour, John, and Deirdre Headon, eds., ***The Self-sufficient Life and How to Live It*** (Cambridge: DK ADULT, 2003). Ever think about becoming a farmer and leading an independent life? This book tells you how to do it. It is an interesting, alternative way to learn about agriculture. The book is written for a British climate, but the messages can be applied generally.

Terrence, J. Toy, George R. Foster, and Kenneth G. Renard, ***Soil Erosion: Processes, Prediction, Measurement, and Control*** (New York: John Wiley, 2002).

CHAPTER 12

Landscapes: Forests, Parks, and Wilderness

A long line of trucks in Malaysia carrying logs from tropical rain forests. As land ownership changes in the United States, American corporations have purchased more and more forestland in less-developed parts of the world.

LEARNING OBJECTIVES

Forests and parks are among our most valued resources. Their conservation and management require that we understand landscapes—a larger view that includes populations, species, and groups of ecosystems connected together. After reading this chapter, you should understand . . .

- What ecological services are provided by landscapes of various kinds;
- The basic principles of forest management, including its historical context;
- The basic conflicts over forest ownership.
- The basic principles of park management;
- The roles that parks and nature preserves play in the conservation of wilderness.

CASE STUDY

Jamaica Bay National Wildlife Refuge: Nature and the Big City

The largest bird sanctuary in the northeastern United States is—surprise!—in New York City. It is the Jamaica Bay Wildlife Refuge, covering more than 9,000 acres—14 square miles of land, 20,000 acres in total, within view of Manhattan's Empire State Building (see Figure 12.1). Jamaica Bay is run by the National Park Service, and you can get there by city bus or subway.[1] More than 300 bird species have been seen there, including the glossy ibis, common farther south, and the curlew sandpiper, which breeds in northern Siberia. Clearly, this wildlife refuge, like the city itself, is a major transportation crossroads. In fact, it is one of the major stopovers on the Atlantic bird migration flyway.

We are not as likely to think of viewing nature near a big city as we are to think of taking a trip far away to wilderness, but as more and more of us become urban dwellers, parks and preserves within easy reach of cities are going to become more important. Also, cities like New York usually lie at important crossroads, not just for people but for wildlife, as illustrated by Jamaica Bay's many avian visitors.

In the 19th century, this bay was a rich source of shellfish, but these were fished out and their habitats destroyed by urban development of many kinds. And like so many other natural areas, parks, and preserves, Jamaica Bay Wildlife Refuge has troubles. The estuary that it is part of is today only half the size it was in colonial times, and the refuge's salt marshes are disappearing at a rate that alarms conservationists. Some of the wetlands have been filled, some shorelines bulkheaded to protect developments, and channels dredged. A lot of marshland disappeared with the building of Kennedy International Airport, just a few miles away. The salt marshes and brackish waters of the bay are also damaged by a large

(a)

(b)

FIGURE 12.1 Jamaica Bay Wildlife Refuge, New York City. **(a)** The largest wildlife refuge in the northeastern United States is within view of New York City's Empire State Building. It's a surprisingly good place for birdwatching, since it is used by 325 species of birds. **(b) This map of the Jamaica Bay Wildlife Refuge shows how near the refuge is to Manhattan Island.**

flow of freshwater from treated sewage. Contrary to what you may think, the only difficulty with this water is that it is fresh, which is a problem to the bay's ecosystems.

Help may be on the way. A watershed protection plan has been written, and there is growing interest in this amazing refuge. The good news is that plentiful wildlife viewing is within a commuter's trip for more than 10 million people. Still, natural areas like the wetlands and bay near New York City and the forests and prairies throughout North America present a conflict. On the one hand, they have been valued for the profits to be made from developing the land for other uses. On the other hand, people value and want to preserve the wildlife and vegetation, the natural ecosystems, for all the reasons discussed in Chapter 7 on biological diversity.

In the 17th century, when the first Europeans arrived in what is now New York City and Long Island, they found a landscape already occupied by the Lenape Indians, who farmed, hunted, fished, and made trails that ran from Manhattan to Jamaica Bay.[2] Much of the land, especially land extending north along the Hudson River, was forested, and the forests, too, were occupied and used for their resources by the Lenape and other Indians. The dual uses of landscapes were already established: They were both harvested for many resources and appreciated for their beauty and variety.

Although since then the entire landscape has been heavily altered, those dual uses of the land are still with us and give rise to conflicts about which should dominate.

In this chapter we look at various kinds of landscapes: parks, nature preserves, and especially forests, a major kind of landscape that is harvested for commercial products but is also considered important for biological conservation. Which use to emphasize—harvest, or preservation and aesthetic appreciation—underlies all the environmental issues about landscapes. We will talk about these kinds of natural resources and how to conserve and manage them while benefiting from them in many ways.

12.1 Forests and Forestry

How People Have Viewed Forests

Forests have always been important to people; indeed, forests and civilization have always been closely linked. Since the earliest civilizations—in fact, since some of the earliest human cultures—wood has been one of the major building materials and the most readily available and widely used fuel. Forests provided materials for the first boats and the first wagons. Even today, nearly half the people in the world depend on wood for cooking, and in many developing nations wood remains the primary heating fuel.[3]

At the same time, people have appreciated forests for spiritual and aesthetic reasons. There is a long history of sacred forest groves. When Julius Caesar was trying to conquer the Gauls in what is now southern France, he found the enemy difficult to defeat on the battlefield, so he burned the society's sacred groves to demoralize them—an early example of psychological warfare. In the Pacific Northwest, the great forests of Douglas fir provided the Indians with many practical necessities of life, from housing to boats, but they were also important to them spiritually.

Today, forests continue to benefit people and the environment indirectly through what we call *public-service functions*. Forests retard erosion and moderate the availability of water, improving the water supply from major watersheds to cities. Forests are habitats for endangered species and other wildlife. They are important for recreation, including hiking, hunting, and bird and wildlife viewing. At regional and global levels, forests may also be significant factors affecting the climate.

Forestry

Forestry has a long history as a profession. The professional growing of trees is called **silviculture** (from *silvus*, Latin for "forest," and *cultura*, for "cultivate"). People have long practiced silviculture, much as they have grown crops, but forestry developed into a science-based activity and into what we today consider a profession in the late 19th and early 20th centuries. The first modern U.S. professional forestry school was established at Yale University around the turn of the 20th century, spurred by growing concerns about the depletion of America's living resources. In the early days of the 20th century, the goal of silviculture was generally to maximize the yield in the harvest of a single resource. The ecosystem was a minor concern, as were nontarget, noncommercial species and associated wildlife.

In this chapter, we approach forestry as professionals who make careful use of science and whose goals are the conservation and preservation of forests and the sustainability of timber harvest and of forest ecosystems. Unfortunately, these goals sometimes conflict with the goals of others.

Modern Conflicts over Forestland and Forest Resources

What is the primary purpose of national forests? A national source of timber? The conservation of living resources? Recreation?

Who should own and manage our forests and their resources? The people? Corporations? Government agencies?

In the past decade a revolution has taken place as to who owns America's forests, and this has major implications for how, and how well, our forests will be managed, conserved, sustained, and used in the future. The state of Maine illustrates the change. About 80% of forestland owned by industrial forest companies was sold in that state between 1994 and 2000. Most of it (60%) was purchased by timber investment management organizations (TIMOs). The rest was sold to nongovernment entities, primarily conservation and environmental organizations.

Industrial forest companies, such as International Paper and Weyerhaeuser, owned the forestland, harvested the timber and planned how to do it, and made products from it. They employed professional foresters, and the assumption within the forest industry was that the profession of forestry and the science on which it was based played an important role in improving harvests and maintaining the land. Although timber companies' practices were often heavily criticized by environmental groups, both sides shared a belief in sound management of forests, and in the 1980s and 1990s the two sides made many attempts to work together to improve forest ecosystem sustainability.

In contrast, TIMOs are primarily financial investors who view forestland as an opportunity to profit by buying and selling timber. It is unclear how much sound forestry will be practiced on TIMO-owned land, but there is less emphasis on professional forestry and forest science,[4] and far fewer professional foresters have been employed. The danger is that forestland viewed only as a commercial commodity will be harvested and abandoned once the resource is used. If this happens, it will be the exact opposite of what most people involved in forestry, both in the industry and in conservation groups, hoped for and thought was possible throughout the 20th century.

Meanwhile, funding for forest research by the U.S. Forest Service has also been reduced. Our national forests, part of our national heritage, may also be less well managed and therefore less well conserved in the future.

How could this have come about? It is an ironic result of political and ideological activities. Ultimately, the conflict between industrial forestry and environmental conservation seems to have led timberland owners to decide it was less bothersome and less costly to just sell off forestland, buy wood from whomever owned it, and let them deal with the consequences of land use. Consistent with this rationale, much forest ownership by organizations in the United States has moved offshore, to places with fewer environmental constraints and fewer and less powerful environmental groups. This change should be all the more worrisome to those interested in environmental conservation because it has happened without much publicity and

FIGURE 12.2 The dual human uses of forests. This temperate rain forest on Vancouver Island illustrates the beauty of forests. Its tree species are also among those most desired for commercial timber production.

is relatively little known by the general public except where forestry is a major livelihood, as it is in the state of Maine.

In sum, then, modern conflicts about forests center on the following questions:

- Should a forest be used only as a resource to provide materials for people and civilization, or should a forest be used only to conserve natural ecosystems and biological diversity (see Figure 12.2), including specific endangered species?
- Can a forest serve some of both of these functions at the same time and in the same place?
- Can a forest be managed sustainably for either use? If so, how?
- What role do forests play in our global environment, such as climate?
- When are forests habitats for specific endangered species?
- When and where do we need to conserve forests for our water supply?

World Forest Area and Global Production and Consumption of Forest Resources

At the beginning of the 21st century, approximately 26% of Earth's surface was forested—about 3.8 billion hectares (15 million square miles) (Figure 12.3).[5] This works out to about 0.6 hectares (about 1 acre) per person. The forest area is up from 3.45 billion hectares (13.1 million square

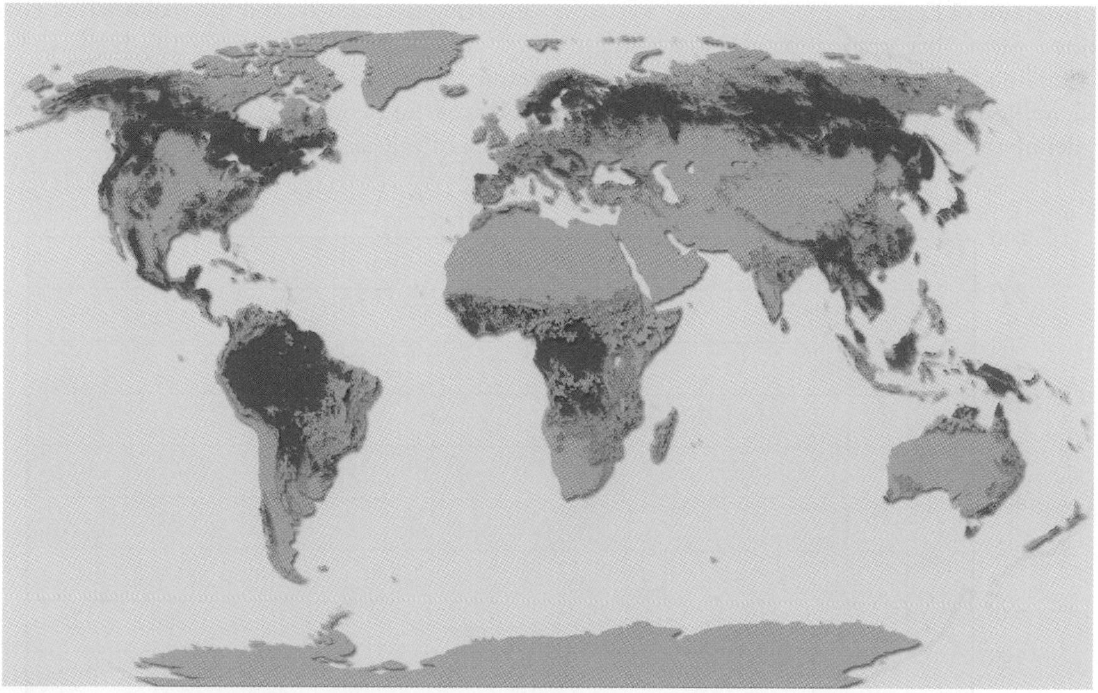

FIGURE 12.3 Forests of the world. (*Source:* Food and Agriculture Organization of the United Nations, Viale delle Terme di Caracalla, 00153 Rome, Italy.)

miles, or 23% of the land area) estimated in 1990, but down from 4 billion hectares (15.2 million square miles, or 27%) in 1980.

Countries differ greatly in their forest resources, depending on the suitability of their land and climate for tree growth and on their history of land use and deforestation. Ten nations have two-thirds of the world's forests. In descending order, these are the Russian Federation, Brazil, Canada, the United States, China, Australia, the Democratic Republic of the Congo, Indonesia, Angola, and Peru (Figure 12.4).

Developed countries account for 70% of the world's total production and consumption of industrial wood products; developing countries produce and consume about 90% of wood used as firewood. Timber for construction, pulp, and paper makes up approximately 90% of the world timber trade; the rest consists of hardwoods used for furniture, such as teak, mahogany, oak, and maple. North America is the world's dominant supplier. Total global production/consumption is about 1.5 billion m^3 annually. To think of this in terms easier to relate to, a cubic meter of timber is a block of wood 1 meter thick on each side. A billion cubic meters would be a block of wood 1 meter (39 inches) thick in a square 1,000 km (621 miles) long on each side. This is a distance greater than that between Washington, DC, and Atlanta, Georgia, and longer than the distance between San Diego and Sacramento, California. The great pyramid of Giza, Egypt, has a volume of more than 2.5 million cubic meters, so the amount of timber consumed in a year would fill 600 great pyramids of Egypt.

The United States has approximately 304 million hectares (751 million acres) of forests, of which 86 million hectares (212 million acres) are considered commercial-grade forest, defined as forest capable of producing at least 1.4 m^3/ha (20 ft^3/acre) of wood per year.[6] Commercial timberland occurs in many parts of the United States. Nearly 75% is in the eastern half of the country (about equally divided between the North and South); the rest is in the West (Oregon, Washington, California, Montana, Idaho, Colorado, and other Rocky Mountain states) and in Alaska.

In the United States, 56% of forestland is privately owned, 33% is federal land, 9% is state land, and 3% is on county and town land.[7] Publicly owned forests are primarily in the Rocky Mountain and Pacific Coast states on sites of poor quality and high elevation (Figure 12.5).[8] In contrast, worldwide most forestland (84%) is said to be publicly owned, although information is spotty.[9]

In the last several decades, world trade in timber does not appear to have grown much, if at all, based on the information reported by nations to the United Nations Food and Agriculture Organization. Thus, the amount traded annually (about 1.5 billion m^3, as mentioned earlier) is a reasonable estimate of the total present world demand for the 6.6 billion people on Earth, at their present standards of living. The fundamental questions are whether and how Earth's forests can continue to produce at least this amount of timber for an indefinite period, and whether and how they can produce even more as the world's human population continues to grow and as standards of living rise worldwide. Keep in mind, all of this has to happen while forests continue to perform their other functions, which include public-service functions, biological conservation functions, and functions involving the aesthetic and spiritual needs of people.

In terms of the themes of this book, the question is: How can forest production be sustainable while meeting the needs of people *and* nature? The answer involves science and values.

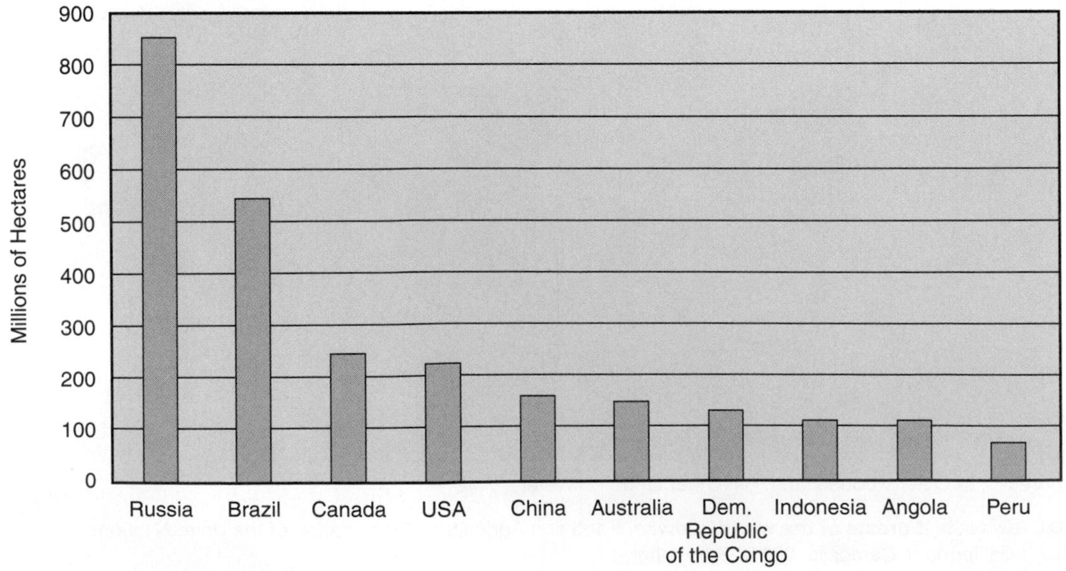

FIGURE 12.4 **Countries with the largest forest areas.** (*Source*: Data from www.mapsofworld.com)

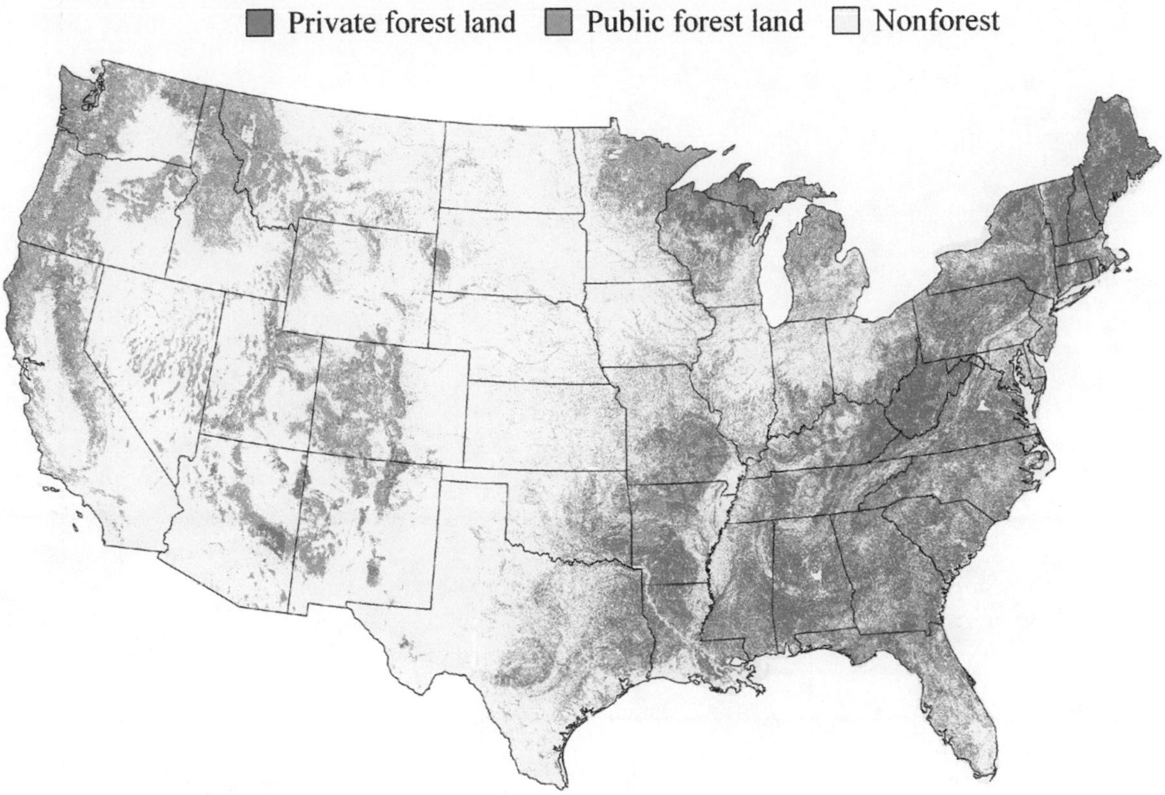

FIGURE 12.5 **Forest ownership in the lower 48 states of the United States in 2008.** (*Source:* U.S. Forest Service, Northern Research Station.)

As we mentioned, wood is a major energy source in many parts of the world. Some 63% of all wood produced in the world, or 2.1 million m^3, is used for firewood. Firewood provides 5% of the world's total energy use,[10] 2% of total commercial energy in developed countries, but 15% of the energy in developing countries, and is the major source of energy for most countries of sub-Saharan Africa, Central America, and continental Southeast Asia.[11]

As the human population grows, the use of firewood increases. In this situation, management is essential, including management of woodland stands (an informal term that foresters use to refer to groups of trees) to improve growth. However, well-planned management of firewood stands has been the exception rather than the rule.

How Forests Affect the Whole Earth

Trees affect the earth by evaporating water, slowing erosion, and providing habitat for wildlife (see Figure 12.6). Trees can also affect climate. Indeed, vegetation of any kind can affect the atmosphere in four ways, and since forests cover so much of the land, they can play an especially important role in the biosphere (Figure 12.7):

1. By changing the color of the surface and thus the amount of sunlight reflected and absorbed.
2. By increasing the amount of water transpired and evaporated from the surface to the atmosphere.
3. By changing the rate at which greenhouse gases are released from Earth's surface into the atmosphere.
4. By changing "surface roughness," which affects wind speed at the surface.

In general, vegetation warms the Earth by making the surface darker, so it absorbs more sunlight and reflects less. The contrast is especially strong between the dark needles of conifers and winter snow in northern forests and between the dark green of shrublands and the yellowish soils of many semiarid climates. Vegetation in general and forests in particular tend to evaporate more water than bare surfaces. This is because the total surface area of the many leaves is many times larger than the area of the soil surface.

FIGURE 12.6 **A forested watershed,** showing the effects of trees in evaporating water, retarding erosion, and providing wildlife habitat.

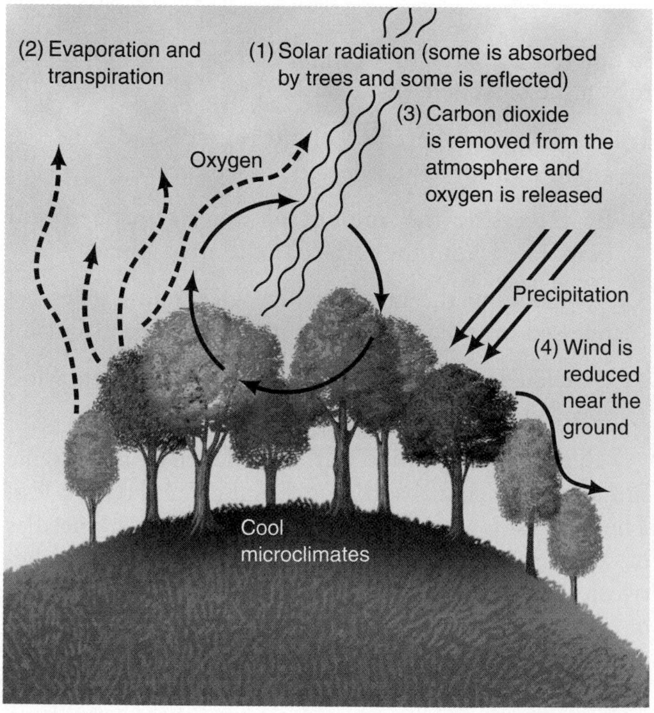

FIGURE 12.7 **Four ways that a forest (or a vegetated area) can affect the atmosphere:** (1) Some solar radiation is absorbed by vegetation and some is reflected, changing the local energy budget, compared to a nonforest environment; (2) evaporation and transpiration from plants, together called *evapotranspiration*, transfers water to the atmosphere; (3) photosynthesis by trees releases oxygen into the atmosphere and removes carbon dioxide, a greenhouse gas, cooling the temperature of the atmosphere; and (4) near-surface wind is reduced because the vegetation—especially trees—produces roughness near the ground that slows the wind.

Is this increased evaporation good or bad? That depends on one's goals. Increasing evaporation means that less water runs off the surface. This reduces erosion. Although increased evaporation also means that less water is available for our own water supply and for streams, in most situations the ecological and environmental benefits of increased evaporation outweigh the disadvantages.

The Ecology of Forests

Each species of tree has its own niche (see Chapter 5) and is thus adapted to specific environmental conditions. For example, in boreal forests, one of the determinants of a tree niche is the water content of the soil. White birch grows well in dry soils; balsam fir in well-watered sites; and northern white cedar in bogs (Figure 12.8).

Another determinant of a tree's niche is its tolerance of shade. Some trees, such as birch and cherry, can grow only in the bright sun of open areas and are therefore found in clearings and called "shade-intolerant." Other species, such as sugar maple and beech, can grow in deep shade and are called "shade-tolerant."

Most of the big trees of the western United States require open, bright conditions and certain kinds of disturbances in order to germinate and survive the early stages of their lives. These trees include coastal redwood, which wins in competition with other species only if both fires and floods occasionally occur; Douglas fir, which begins its growth in openings; and the giant sequoia, whose seeds will germinate only on bare, mineral soil—where there is a thick layer of organic mulch, the sequoia's seeds cannot reach the surface and will die before they can germinate. Some trees are adapted to early stages of succession, where sites are open and there is bright sunlight. Others are adapted to later stages of succession, where there is a high density of trees (see the discussion of ecological succession in Chapter 5).

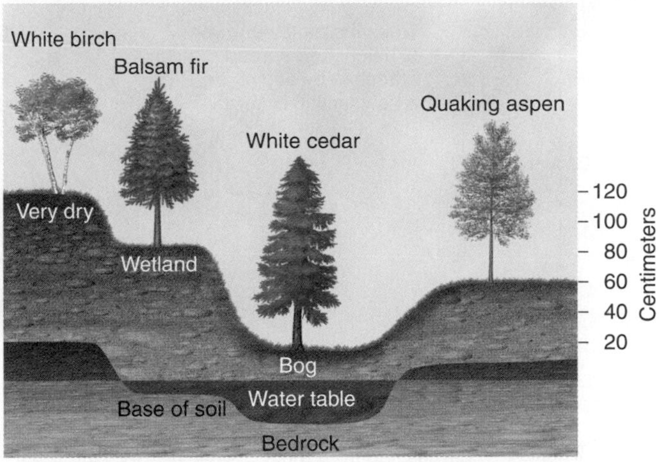

FIGURE 12.8 Some characteristics of tree niches. Tree species have evolved to be adapted to different kinds of environments. In northern boreal forests, white birch grows on dry sites (and early-successional sites); balsam fir grows in wetter soils, up to wetlands; and white cedar grows in even the wetter sites of northern bogs.

Understanding the niches of individual tree species helps us to determine where we might best plant them as a commercial crop, and where they might best contribute to biological conservation or to landscape beauty.

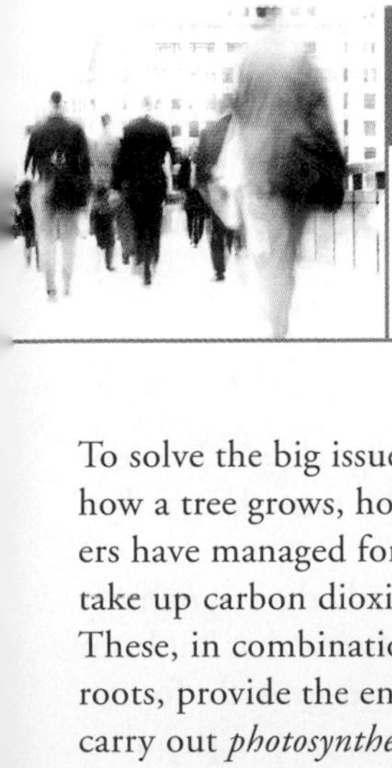

A CLOSER LOOK 12.1

The Life of a Tree

To solve the big issues about forestry, we need to understand how a tree grows, how an ecosystem works, and how foresters have managed forestland (Figure 12.9). Leaves of a tree take up carbon dioxide from the air and absorb sunlight. These, in combination with water transported up from the roots, provide the energy and chemical elements for leaves to carry out *photosynthesis*. Through photosynthesis, the leaves convert carbon dioxide and water into a simple sugar and molecular oxygen. This simple sugar is then combined with other chemical elements to provide all the compounds that the tree uses.

Tree roots take up water, along with chemical elements dissolved in the water and small inorganic compounds, such as the nitrate or ammonia necessary to make proteins. Often the process of extracting minerals and compounds from the soil is aided by symbiotic relationships between the tree

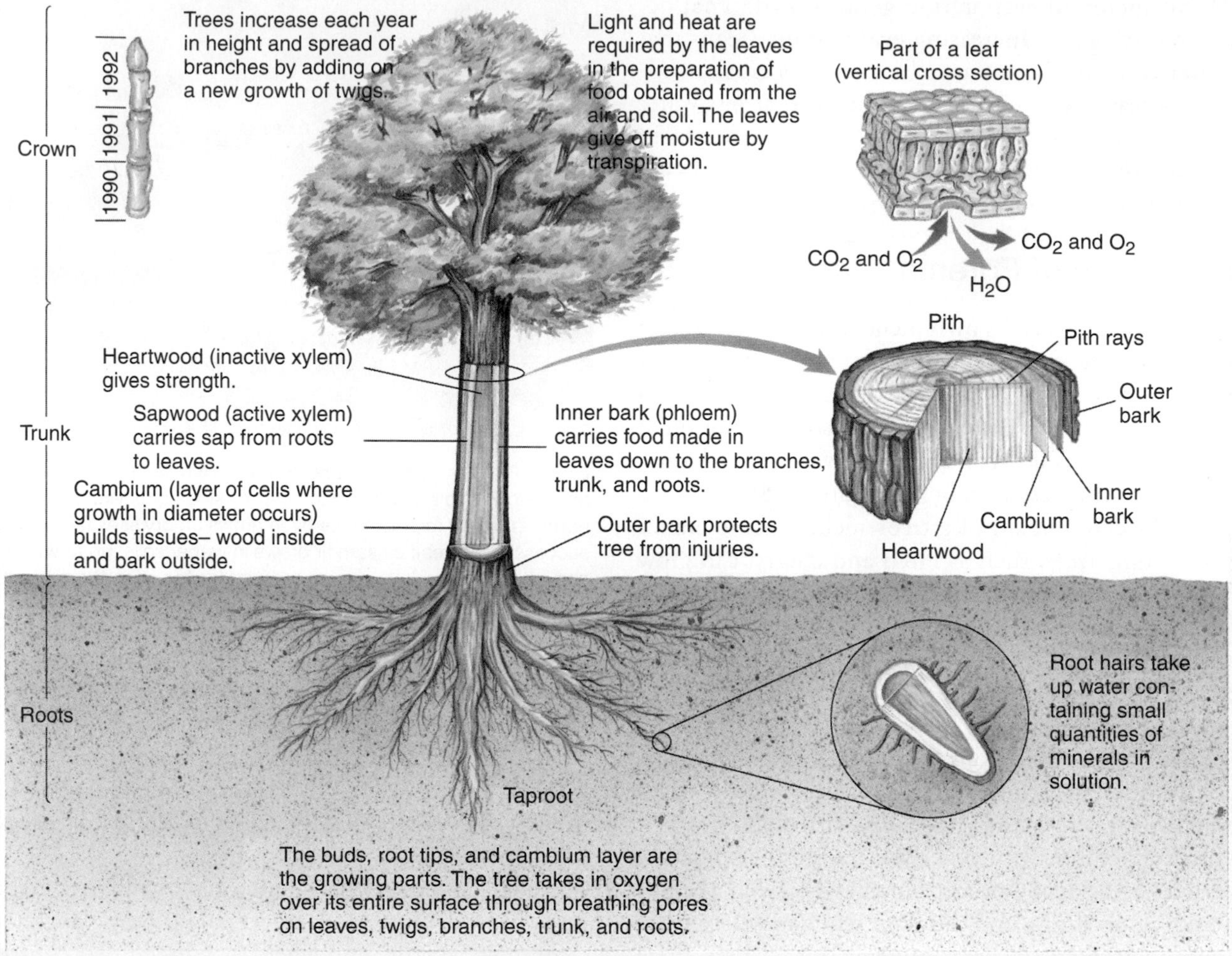
Trees increase each year in height and spread of branches by adding on a new growth of twigs.
1990
1991
1992
Crown
Light and heat are required by the leaves in the preparation of food obtained from the air and soil. The leaves give off moisture by transpiration.
Part of a leaf (vertical cross section)
CO_2 and O_2
CO_2 and O_2
H_2O
Pith
Pith rays
Outer bark
Inner bark
Cambium
Heartwood
Heartwood (inactive xylem) gives strength.
Trunk
Sapwood (active xylem) carries sap from roots to leaves.
Cambium (layer of cells where growth in diameter occurs) builds tissues– wood inside and bark outside.
Inner bark (phloem) carries food made in leaves down to the branches, trunk, and roots.
Outer bark protects tree from injuries.
Roots
Taproot
Root hairs take up water containing small quantities of minerals in solution.
The buds, root tips, and cambium layer are the growing parts. The tree takes in oxygen over its entire surface through breathing pores on leaves, twigs, branches, trunk, and roots.

FIGURE 12.9 How a tree grows. (*Source*: C.H. Stoddard, *Essentials of Forestry Practice*, 3rd ed. [New York: Wiley, 1978].)

roots and fungi. Tree roots release sugars and other compounds that are food for the fungi, and the fungi benefit the tree as well.

Leaves and roots are connected by two transportation systems. Phloem, on the inside of the living part of the bark, transports sugars and other organic compounds down to stems and roots. Xylem, farther inside (Figure 12.9), transports water and inorganic molecules upward to the leaves. Water is transported upward by a sun-powered pump—that is, sunlight provides energy to pump the water up the tree by heating leaves so they evaporate water. Water from below is then pulled upward to replace water that evaporated.

Forest Management

A Forester's View of a Forest

Traditionally, foresters have managed trees locally in stands. Trees in a **stand** are usually of the same species or group of species and often at the same successional stage. Stands can be small (half a hectare) to medium size (several hundred hectares) and are classified by foresters on the basis of tree composition. The two major kinds of commercial stands are *even-aged stands*, where all live trees began growth from seeds and roots germinating the same year, and *uneven-aged stands*, which have at least three distinct age classes. In even-aged stands, trees are approximately the same height but differ in girth and vigor.

A forest that has never been cut is called a *virgin forest* or sometimes an **old-growth forest**. A forest that has been cut and has regrown is called a **second-growth forest**. Although the term old-growth forest has gained popularity in several well-publicized disputes about forests, it is not a scientific term and does not yet have an agreed-on, precise meaning. Another important management term is **rotation time**, the time between cuts of a stand.

Foresters and forest ecologists group the trees in a forest into the **dominants** (the tallest, most common, and most vigorous), **codominants** (fairly common, sharing the canopy or top part of the forest), **intermediate** (forming a layer of growth below dominants), and **suppressed** (growing in the understory). The productivity of a forest varies according to soil fertility, water supply, and local climate. Foresters classify sites by **site quality**, which is the maximum timber crop the site can produce in a given time. Site quality can decline with poor management.

Although forests are complex and difficult to manage, one advantage they have over many other ecosystems is that trees provide easily obtained information that can be a great help to us. For example, the age and growth rate of trees can be measured from tree rings. In temperate and boreal forests, trees produce one growth ring per year.

Harvesting Trees

Managing forests that will be harvested can involve removing poorly formed and unproductive trees (or selected other trees) to permit larger trees to grow faster, planting genetically controlled seedlings, controlling pests and diseases, and fertilizing the soil. Forest geneticists breed new strains of trees just as agricultural geneticists breed new strains of crops. There has been relatively little success in controlling forest diseases, which are primarily fungal.

Harvesting can be done in several ways. **Clear-cutting** (Figure 12.10) is the cutting of all trees in a stand at the same time. Alternatives to clear-cutting are selective cutting, strip-cutting, shelterwood cutting, and seed-tree cutting.

In **selective cutting**, individual trees are marked and cut. Sometimes smaller, poorly formed trees are selectively removed, a practice called **thinning**. At other times, trees of specific species and sizes are removed. For example, some forestry companies in Costa Rica cut only some of the largest mahogany trees, leaving less valuable trees to help maintain the ecosystem and permitting some of the large mahogany trees to continue to provide seeds for future generations.

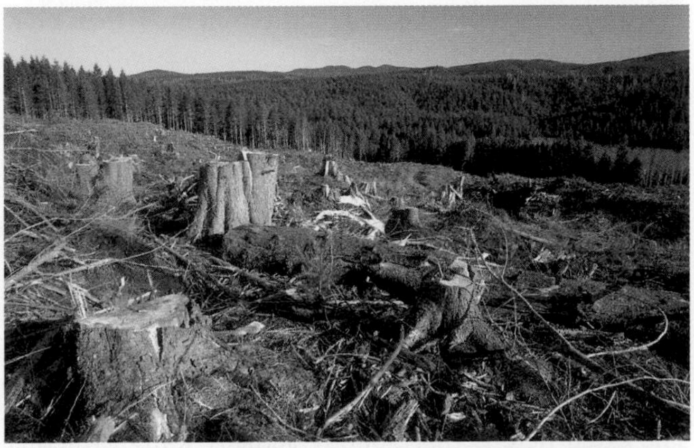

FIGURE 12.10 **A clear-cut forest in western Washington.**

In **strip-cutting**, narrow rows of forest are cut, leaving wooded corridors whose trees provide seeds. Strip-cutting offers several advantages, such as protection against erosion.

Shelterwood cutting is the practice of cutting dead and less desirable trees first, and later cutting mature trees. As a result, there are always young trees left in the forest.

Seed-tree cutting removes all but a few seed trees (mature trees with good genetic characteristics and high seed production) to promote regeneration of the forest.

Scientists have tested the effects of clear-cutting, which is one of the most controversial forest practices.[12, 13, 14] For example, in the U.S. Forest Service Hubbard Brook experimental forest in New Hampshire, an entire watershed was clear-cut, and herbicides were applied to prevent regrowth for two years.[14] The results were dramatic. Erosion increased, and the pattern of water runoff changed substantially. The exposed soil decayed more rapidly, and the concentrations of nitrates in the stream water exceeded public-health standards. In another experiment, at the U.S. Forest Service H.J. Andrews experimental forest in Oregon, a forest where rainfall is high (about 240 cm, or 94 in., annually), clear-cutting greatly increased the frequency of landslides, as did the construction of logging roads.[15]

Clear-cutting also changes chemical cycling in forests and can open the way for the soil to lose chemical elements necessary for life. Exposed to sun and rain, the ground becomes warmer. This accelerates the process of decay, with chemical elements, such as nitrogen, converted more rapidly to forms that are water-soluble and thus readily lost in runoff during rains (Figure 12.11).[16]

The Forest Service experiments show that clear-cutting can be a poor practice on steep slopes in areas of moderate to heavy rainfall. The worst effects of clear-cutting resulted from the logging of vast areas of North America during the 19th and early 20th centuries. Clear-cutting on such a large scale is neither necessary nor desirable for the best timber production. However, where the ground is level or slightly sloped, where rainfall is moderate, and where the desirable species require open areas for growth, clear-cutting on an appropriate spatial scale may be a useful way to regenerate desirable species. The key here is that clear-cutting is neither all good nor all bad for timber production or forest ecosystems. Its use must be evaluated on a case-by-case basis, taking into account the size of cuts, the environment, and the available species of trees.

Plantations

Sometimes foresters grow trees in a **plantation**, which is a stand of a single species, typically planted in straight rows (Figure 12.12). Usually plantations are fertilized, sometimes by helicopter, and modern machines harvest rapidly—some remove the entire tree, root and all.

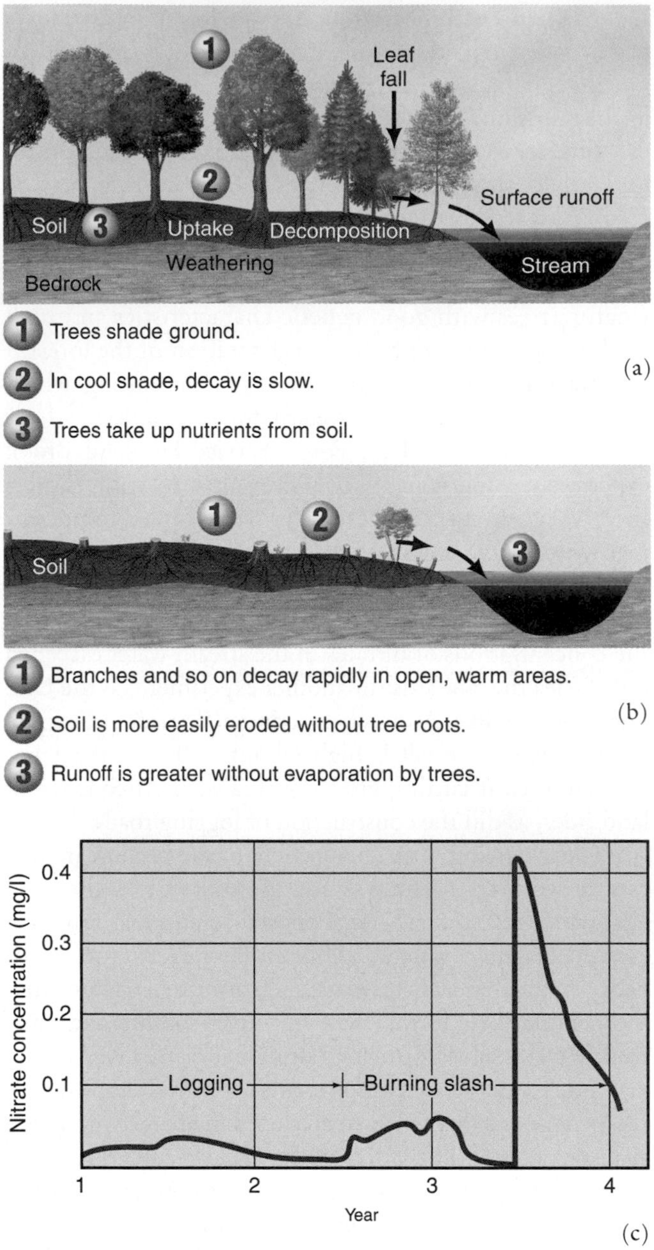

FIGURE 12.11 Effects of clear-cutting on forest chemical cycling. Chemical cycling **(a)** in an old-growth forest and **(b)** after clear-cutting. **(c)** Increased nitrate concentration in streams after logging and the burning of slash (leaves, branches, and other tree debris). (*Source*: adapted from R.L. Fredriksen, "Comparative Chemical Water Quality—Natural and Disturbed Streams Following Logging and Slash Burning," in *Forest Land Use and Stream Environment* [Corvallis: Oregon State University, 1971], pp. 125–137.)

FIGURE 12.12 A modern forest plantation in Queensland, Australia. Note that the trees are evenly spaced and similar, if not identical, in size.

In short, plantation forestry is a lot like modern agriculture. Intensive management like this is common in Europe and parts of the northwestern United States and offers an important alternative solution to the pressure on natural forests. If plantations were used where forest production was high, then a comparatively small percentage of the world's forestland could provide all the world's timber. For example, high-yield forests produce 15–20 m^3/ha/yr. According to one estimate, if plantations were put on timberland that could produce at least 10 m^3/ha/yr, then 10% of the world's forestland could provide enough timber for the world's timber trade.[17] This could reduce pressure on old-growth forests, on forests important for biological conservation, and on forestlands important for recreation.

Can We Achieve Sustainable Forestry?

There are two basic kinds of ecological sustainability: (1) sustainability of the harvest of a specific resource that grows within an ecosystem; and (2) sustainability of the entire ecosystem—and therefore of many species, habitats, and environmental conditions. For forests, this translates into sustainability of the harvest of timber and sustainability of the forest as an ecosystem. Although sustainability has long been discussed in forestry, we don't have enough scientific data to show that sustainability of either kind has been achieved in forests in more than a few unusual cases.

Certification of Forest Practices

If the data do not indicate whether a particular set of practices has led to sustainable forestry, what can be done? The general approach today is to compare the actual practices of specific corporations or government agencies with practices that are believed to be consistent with sustainability. This has become a formal process called **certification of forestry**, and there are organizations whose main function is to certify forest practices. The catch here is that nobody actually knows whether the beliefs are correct and therefore whether the practices will turn out to be sustainable. Since trees take a long time to grow, and a series of harvests is necessary to prove sustainability, the proof lies in the future. Despite this limitation, certification of forestry is becoming common. As practiced today, it is as much an art or a craft as it is a science.

Worldwide concern about the need for forest sustainability has led to international programs for certifying forest practices, as well as to attempts to ban imports of wood produced from purportedly unsustainable forest practices. Some European nations have banned the import of certain tropical woods, and some environmental organizations have led demonstrations in support of such bans. However, there is a gradual movement away from calling certified forest practices "sustainable," instead referring to "well-managed forests" or "improved management."[19, 20] And some scientists have begun to call for a new forestry that includes a variety of practices that they believe increase the likelihood of sustainability.

Most basic is accepting the dynamic characteristics of forests—that to remain sustainable over the long term, a forest may have to change in the short term. Some of the broader, science-based concerns are spoken of as a group—the need for ecosystem management and a landscape context. Scientists point out that any application of a certification program creates an experiment and should be treated accordingly. Therefore, any new programs that claim to provide sustainable practices must include, for comparison, control areas where no cutting is done and must also include adequate scientific monitoring of the status of the forest ecosystem.

Deforestation

Deforestation is believed to have increased erosion and caused the loss of an estimated 562 million hectares (1.4 billion acres) of soil worldwide, with an estimated annual loss of 5–6 million hectares.[21] Cutting forests in one country affects other countries. For example, Nepal, one of the most mountainous countries in the world, lost more than half its forest cover between 1950 and 1980. This destabilized soil, increasing the frequency of landslides, amount of runoff, and sediment load in streams. Many Nepalese streams feed rivers that flow into India (Figure 12.13). Heavy flooding in India's Ganges Valley has caused about a billion dollars' worth of property damage a year and is blamed on the loss of large forested watersheds in Nepal and other countries.[20] Nepal continues to lose forest cover at a rate of about 100,000 hectares (247,000 acres) per year. Reforestation efforts replace less than 15,000 hectares (37,050 acres) per year. If present trends continue, little forestland will remain in Nepal, thus permanently exacerbating India's flood problems.[19, 20]

Because forests cover large, often remote areas that are little visited or studied, information is lacking on which to determine whether the world's forestlands are expanding or shrinking, and precisely how fast and how much. Some experts argue that there is a worldwide net increase in forests because large areas in the temperate zone, such as the eastern and midwestern United States, were cleared in the 19th and early 20th centuries and are now regenerating. Only recently have programs begun to obtain accurate estimates of the distribution and abundance of forests, and these suggest that past assessments overestimated forest biomass by 100 to 400%.[22]

On balance, we believe that the best estimates are those suggesting that the rate of deforestation in the 21st century is 7.3 million hectares a year—an annual loss equal to the size of Panama. The good news is that this is 18% less than the average annual loss of 8.9 million hectares in the 1990s.[23]

(a)

(b)

FIGURE 12.13 **(a) Planting pine trees on the steep slopes in Nepal to replace entire forests that were cut.** The dark green in the background is yet-uncut forest, and the contrast between foreground and background suggests the intensity of clearing that is taking place. **(b) The Indus River in northern India carries a heavy load of sediment**, as shown by the sediments deposited within and along the flowing water and by the color of the water itself. This scene, near the headwaters, shows that erosion takes place at the higher reaches of the river.

(a) (b)

FIGURE 12.14 **(a)** A satellite image showing clearings in the tropical rain forests in the Amazon in Brazil. The image is in false infrared. Rivers appear black, and the bright red is the leaves of the living rain forest. The straight lines of other colors, mostly light blue to gray, are of deforestation by people extending from roads. Much of the clearing is for agriculture. The distance across the image is about 100 km (63 mi). **(b)** An intact South American rain forest with its lush vegetation of many species and a complex vertical structure. This one is in Peru.

History of Deforestation

Forests were cut in the Near East, Greece, and the Roman Empire before the modern era. Removal of forests continued northward in Europe as civilization advanced. Fossil records suggest that prehistoric farmers in Denmark cleared forests so extensively that early-successional weeds occupied large areas. In medieval times, Great Britain's forests were cut, and many forested areas were eliminated. With colonization of the New World, much of North America was cleared.[24]

The greatest losses in the present century have taken place in South America, where 4.3 million acres have been lost on average per year since 2000 (Figure 12.14). Many of these forests are in the tropics, mountain regions, or high latitudes, places difficult to exploit before the advent of modern transportation and machines. The problem is especially severe in the tropics because of rapid human population growth. Satellite images provide a new way to detect deforestation (Figure 12.14a).

Causes of Deforestation

Historically, the two most common reasons people cut forests are to clear land for agriculture and settlement and to use or sell timber for lumber, paper products, or fuel. Logging by large timber companies and local cutting by villagers are both major causes of deforestation. Agriculture is a principal cause of deforestation in Nepal and Brazil and was one of the major reasons for clearing forests in New England during the first settlement by Europeans. A more subtle cause of the loss of forests is indirect deforestation—the death of trees from pollution or disease.

If global warming occurs as projected by global climate models, indirect forest damage might occur over large regions, with major die-offs in many areas and major shifts in the areas of potential growth for each species of tree due to altered combinations of temperature and rainfall.[25] The extent of this effect is controversial. Some suggest that global warming would merely change the location of forests, not their total area or production. However, even if a climate conducive to forest growth were to move to new locations, trees would have to reach these areas. This would take time because changes in the geographic distribution of trees depend primarily on seeds blown by the wind or carried by animals. In addition, for production to remain as high as it is now, climates that meet the needs of forest trees would have to occur where the soils also meet these needs. This combination of climate and soils occurs widely now but might become scarcer with large-scale climate change.

12.2 Parks, Nature Preserves, and Wilderness

As suggested by this chapter's opening case study about Jamaica Bay Wildlife Refuge, governments often protect landscapes from harvest and other potentially destructive uses by establishing parks, nature preserves, and legally designated wilderness areas. So do private organizations, such as the Nature Conservancy, the Southwest Florida Nature Conservancy, and the Land

Trust of California, which purchase lands and maintain them as nature preserves. Whether government or private conservation areas succeed better in reaching the goals listed in Table 12.1 is a matter of considerable controversy.

Parks, natural areas, and wilderness provide benefits within their boundaries and can also serve as migratory corridors between other natural areas. Originally, parks were established for specific purposes related to the land within the park boundaries (discussed later in this chapter). In the future, the design of large landscapes to serve a combination of land uses—including parks, preserves, and wilderness—needs to become more important and a greater focus of discussion.

What's the Difference between a Park and a Nature Preserve?

A park is an area set aside for use by people. A nature preserve, although it may be used by people, has as its primary purpose the conservation of some resource, typically a biological one. Every park or preserve is an ecological island of one kind of landscape surrounded by a different kind of landscape, or several different kinds. Ecological and physical islands have special ecological qualities, and concepts of island biogeography are used in the design and management of parks. Specifically, the size of the park and the diversity of habitats determine the number of species that can be maintained there. Also, the farther the park is from other parks or sources of species, the fewer species are found. Even the shape of a park can determine what species can survive within it.

One of the important differences between a park and a truly natural wilderness area is that a park has definite boundaries. These boundaries are usually arbitrary from an ecological viewpoint and have been established for political, economic, or historical reasons unrelated to the natural ecosystem. In fact, many parks have been developed on areas that would have been considered wastelands, useless for any other purpose. Even where parks or preserves have been set aside for the conservation of some species, the boundaries are usually arbitrary, and this has caused problems.

For example, Lake Manyara National Park in Tanzania, famous for its elephants, was originally established with boundaries that conflicted with elephant habits. Before this park was established, elephants spent part of the year feeding along a steep incline above the lake. At other times of the year, they would migrate down to the valley floor, depending on the availability of food and water. These annual migrations were necessary for the elephants to obtain food of sufficient nutritional quality throughout the year. However, when the park was established, farms that were laid out along its northern border crossed the traditional pathways of the elephants. This had two negative effects. First, elephants came into direct conflict with farmers. Elephants crashed through farm fences, eating corn and other crops and causing general disruption. Second, whenever the farmers succeeded in keeping elephants out, the animals were cut off from reaching their feeding ground near the lake.

When it became clear that the park boundaries were arbitrary and inappropriate, the boundaries were adjusted to include the traditional migratory routes. This eased the conflicts between elephants and farmers.

Table 12.1 GOALS OF PARKS, NATURE PRESERVES, AND WILDERNESS AREAS

Parks are as old as civilization. The goals of park and nature-preserve management can be summarized as follows:

1. Preservation of unique geological and scenic wonders of nature, such as Niagara Falls and the Grand Canyon
2. Preservation of nature without human interference (preserving wilderness for its own sake)
3. Preservation of nature in a condition thought to be representative of some prior time (e.g., the United States prior to European settlement)
4. Wildlife conservation, including conservation of the required habitat and ecosystem of the wildlife
5. Conservation of specific endangered species and habitats
6. Conservation of the total biological diversity of a region
7. Maintenance of wildlife for hunting
8. Maintenance of uniquely or unusually beautiful landscapes for aesthetic reasons
9. Maintenance of representative natural areas for an entire country
10. Maintenance for outdoor recreation, including a range of activities from viewing scenery to wilderness recreation (hiking, cross-country skiing, rock climbing) and tourism (car and bus tours, swimming, downhill skiing, camping)
11. Maintenance of areas set aside for scientific research, both as a basis for park management and for the pursuit of answers to fundamental scientific questions
12. Provision of corridors and connections between separated natural areas

A CLOSER LOOK 12.2

A Brief History of Parks Explains Why Parks Have Been Established

The French word *parc* once referred to an enclosed area for keeping wildlife to be hunted. Such areas were set aside for the nobility and excluded the public. An example is Coto Doñana National Park on the southern coast of Spain. Originally a country home of nobles, today it is one of Europe's most important natural areas, used by 80% of birds migrating between Europe and Africa (Figure 12.16).

The first major *public* park of the modern era was Victoria Park in Great Britain, authorized in 1842. The concept of a *national* park, whose purposes would include protection of nature as well as public access, originated in North America in the 19th century.[26] The world's first national park was Yosemite National Park in California (Figure 12.15), made a park by an act signed by President Lincoln in 1864. The term *national park*, however, was not used until the establishment of Yellowstone in 1872.

The purpose of the earliest national parks in the United States was to preserve the nation's unique, awesome landscapes—a purpose that Alfred Runte, a historian of national parks, refers to as "monumentalism." In the 19th century, Americans considered their national parks a contribution to civilization equivalent to the architectural treasures of the Old World and sought to preserve them as a matter of national pride.[27]

In the second half of the 20th century, the emphasis of park management became more ecological, with parks established both to conduct scientific research and to maintain examples of representative natural areas. For instance, Zimbabwe established Sengwa National Park (now called Matusadona National Park) solely for scientific research. It has no tourist areas, and tourists are not generally allowed; its purpose is the study of natural ecosystems with as little human interference as possible so that the principles of wildlife and wilderness management can be better formulated and understood. Other national parks in the countries of eastern and southern Africa—including those of Kenya, Uganda, Tanzania, Zimbabwe, and South Africa—have been established primarily for viewing wildlife and for biological conservation.

In recent years, the number of national parks throughout the world has increased rapidly. The law establishing national parks in France was first enacted in 1960. Taiwan had no national parks prior to 1980 but now has six. In the United States, the area in national and state parks has expanded from less than 12 million hectares (30 million acres) in 1950 to nearly 83.6 million acres today, with much of the increase due to the establishment of parks in Alaska.[28]

FIGURE 12.15 The famous main valley of Yosemite National Park.

(a)

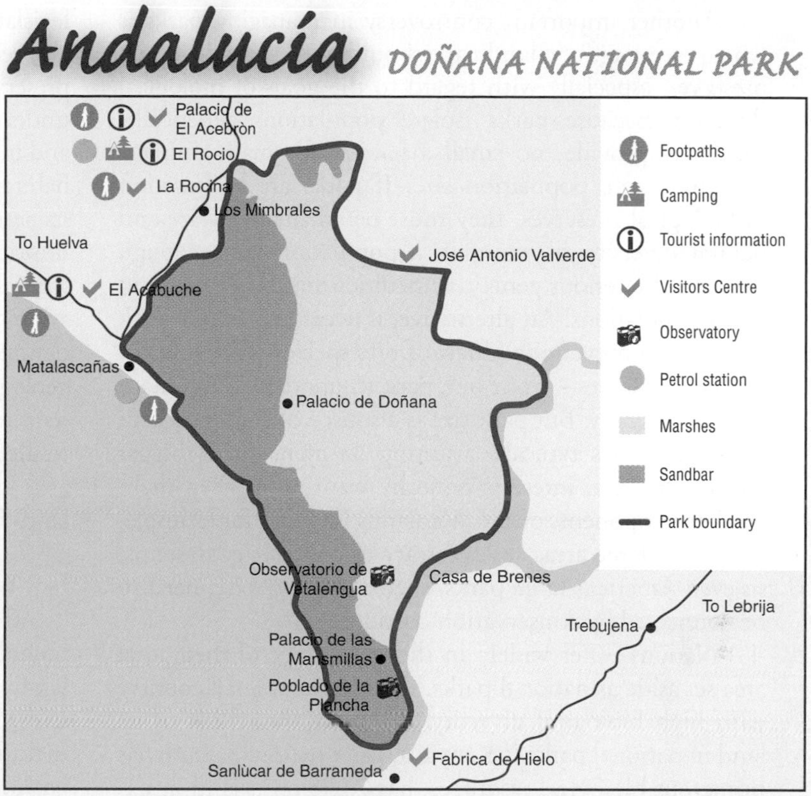

(b)

FIGURE 12.16 **(a) Flamingos are among the many birds that use Coto Doñana National Park,** a major stopover on the Europe-to-Africa flyway. **(b)** Map of Coto Doñana National Park, Spain (*Source*: Colours of Spain. World Heritage Sites http://www.coloursofspain.com/travelguidedetail/17/andalucia_andalusia/world_heritage_sites_donana_national_park/)

Conserving representative natural areas of a country is an increasingly common goal of national parks. For example, the goal of New Zealand's national park planning is to include at least one area representative of each major ecosystem of the nation, from seacoast to mountain peak. In some cases, such as Spain's Coto Doñana National Park, national parks are among the primary resting grounds of major bird flyways (Figure 12.16) or play other crucial roles in conservation of biodiversity.

Conflicts Relating to Parks

Size, Access, and Types of Activities

Major conflicts over parks generally have to do with their size and what kinds and levels of access and activities will be available. The idea of a national, state, county, or city park is well accepted in North America, but conflicts arise over what kinds of activities and what intensity of activities should be allowed in parks. Often, biological conservation and the needs of individual species require limited human access, but, especially in beautiful areas desirable for recreation, people want to go there. As a recent example, travel into Yellowstone National Park by snowmobile in the winter has become popular, but this has led to noise and air pollution and has marred the experience of the park's beauty for many visitors. In 2003 a federal court determined that snowmobile use should be phased out in this park.

Alfred Runte explained the heart of the conflict. "This struggle was not against Americans who like their snowmobiles, but rather against the notion that anything goes in the national parks," he said. "The courts have reminded us that we have a different, higher standard for our national parks. Our history proves that no one loses when beauty wins. We will find room for snowmobiles, but just as important, room without them, which is the enduring greatness of the national parks."[29]

Many of the recent conflicts relating to national parks have concerned the use of motor vehicles. Voyageurs National Park in northern Minnesota, established in 1974—fairly recently compared with many other national parks—occupies land that was once used by a variety of recreational vehicles and provided livelihoods for hunting and fishing guides and other tourism businesses. These people felt that restricting motor-vehicle use would destroy their livelihoods. Voyageurs National Park has 100 miles of snowmobile trails and is open to a greater variety of motor-vehicle recreation than Yellowstone.[30]

Interactions Between People and Wildlife

While many people like to visit parks to see wildlife, some wildlife, such as grizzly bears in Yellowstone National Park, can be dangerous. There has been conflict in the past between conserving the grizzly and making the park as open as possible for recreation.

How Much Land Should Be in Parks?

Another important controversy in managing parks is what percentage of a landscape should be in parks or nature preserves, especially with regard to the goals of biological diversity. Because parks isolate populations genetically, they may provide too small a habitat for maintaining a minimum safe population size. If parks are to function as biological preserves, they must be adequate in size and habitat diversity to maintain a population large enough to avoid the serious genetic difficulties that can develop in small populations. An alternative, if necessary, is for a park manager to move individuals of one species—say, lions in African preserves—from one park to another to maintain genetic diversity. But park size is a source of conflicts, with conservationists typically wanting to make parks bigger and commercial interests typically wanting to keep them smaller. Proponents of the Wildlands Projects, for example, argue that large areas are necessary to conserve ecosystems, so even America's large parks, such as Yellowstone, need to be connected by conservation corridors.

Nations differ widely in the percentage of their total area set aside as national parks. Costa Rica, a small country with high biological diversity, has more than 12% of its land in national parks.[31] Kenya, a larger nation that also has numerous biological resources, has 7.6% of its land in national parks.[32] In France, an industrialized nation in which civilization has altered the landscape for several thousand years, only 0.7% of the land is in the nation's six national parks. However, France has 38 regional parks that encompass 11% (5.9 million hectares) of the nation's area.

The total amount of protected natural area in the United States is more than 104 million hectares (about 240 million acres), approximately 11.2% of the total U.S. land area.[33] However, the states differ greatly in the percentage of land set aside for parks, preserves, and other conservation areas. The western states have vast parks, whereas the six Great Lakes states (Michigan, Minnesota, Illinois, Indiana, Ohio, and Wisconsin), covering an area approaching that of France and Germany combined, allocate less than 0.5% of their land to parks and less than 1% to designated wilderness.[34]

12.3 Conserving Wilderness

What It Is, and Why It Is of Growing Importance

As a modern legal concept, **wilderness** is an area undisturbed by people. The only people in a wilderness are visitors, who do not remain. The conservation of wilderness is a new idea introduced in the second half of the 20th century. It is one that is likely to become more important as the human population increases and the effects of civilization become more pervasive throughout the world.

The U.S. Wilderness Act of 1964 was landmark legislation, marking the first time anywhere that wilderness was recognized by national law as a national treasure to be preserved. Under this law, wilderness includes "an area of undeveloped Federal land retaining its primeval character and influence, without permanent improvements or human habitation, which is protected and managed so as to preserve its natural conditions." Such lands are those in which (1) the imprint of human work is unnoticeable, (2) there are opportunities for solitude and for primitive and unconfined recreation, and (3) there are at least 5,000 acres. The law also recognizes that these areas are valuable for ecological processes, geology, education, scenery, and history. The Wilderness Act required certain maps and descriptions of wilderness areas, resulting in the U.S. Forest Service's Roadless Area Review and Evaluation (RARE I and RARE II), which evaluated lands for inclusion as legally designated wilderness.

Where You'll Find It and Where You Won't

Countries with a significant amount of wilderness include New Zealand, Canada, Sweden, Norway, Finland, Russia, and Australia; some countries of eastern and southern Africa; many countries of South America, including parts of the Brazilian and Peruvian Amazon basin; the mountainous high-altitude areas of Chile and Argentina; some of the remaining interior tropical forests of Southeast Asia; and the Pacific Rim countries (parts of Borneo, the Philippines, Papua New Guinea, and Indonesia). In addition, wilderness can be found in the polar regions, including Antarctica, Greenland, and Iceland.

Many countries have no wilderness left to preserve. In the Danish language, the word for wilderness has even disappeared, although that word was important in the ancestral languages of the Danes.[32] Switzerland is a country in which wilderness is not a part of preservation. For example, a national park in Switzerland lies in view of the Alps—scenery that inspired the English romantic poets of the early 19th century to praise what they saw as wilderness and to attach the adjective *awesome* to what they saw. But the park is in an area that has been heavily exploited for such activities as mining and foundries since the Middle Ages. All the forests are planted.[32]

The Wilderness Experience: Natural vs. Naturalistic

In a perhaps deeper sense, wilderness is an idea and an ideal that can be experienced in many places, such as Japanese gardens, which might occupy no more than a few hundred square meters. Henry David Thoreau distinguished between "wilderness" and "wildness." He thought of wilderness as a physical place and wildness as a state of mind. During his travels through the Maine woods in the 1840s, he concluded that wilderness was an interesting place to visit but not to live in. He preferred long walks through the woods and near swamps around his home in Concord, Massachusetts, where he was able to experience a *feeling* of wildness. Thus,

Thoreau raised a fundamental question: Can one experience true wildness only in a huge area set aside as a wilderness and untouched by human actions, or can wildness be experienced in small, heavily modified and, though not entirely natural, *naturalistic* landscapes, such as those around Concord in the 19th century?[31]

As Thoreau suggests, small, local, naturalistic parks may have more value than some of the more traditional wilderness areas as places of solitude and beauty. In Japan, for instance, there are roadless recreation areas, but they are filled with people. One two-day hiking circuit leads to a high-altitude marsh where people can stay in small cabins. Trash is removed from the area by helicopter. People taking this hike experience a sense of wildness.

In some ways, the answer to the question raised by Thoreau is highly personal. We must discover for ourselves what kind of natural or naturalistic place meets our spiritual, aesthetic, and emotional needs. This is yet another area in which one of our key themes, science and values, is evident.

Conflicts in Managing Wilderness

The legal definition of *wilderness* has given rise to several controversies. The wilderness system in the United States began in 1964 with 3.7 million hectares (9.2 million acres) under U.S. Forest Service control. Today, the United States has 633 legally designated wilderness areas, covering 44 million hectares (106 million acres)—more than 4% of the nation. Another 200 million acres meet the legal requirements and could be protected by the Wilderness Act. Half of this area is in Alaska, including the largest single area, Wrangell–St. Elias (Figure 12.17), covering 3.7 million hectares (9 million acres).[33, 35]

Those interested in developing the natural resources of an area, including mineral ores and timber, have argued that the rules are unnecessarily stringent, protecting too much land from exploitation when there is plenty of wilderness elsewhere. Those who wish to conserve additional wild areas have argued that the interpretation of the U.S. Wilderness Act is too lenient and that mining and logging are inconsistent with the wording of the Act. These disagreements are illustrated by the argument over drilling in the Arctic National Wildlife Refuge, a dispute that reemerged with the rising price of petroleum.

FIGURE 12.17 Wrangell–St. Elias Wilderness Area, Alaska, designated in 1980 and now covering 9,078,675 acres. As the photograph suggests, this vast area gives a visitor a sense of wilderness as a place where a person is only a visitor and human beings seem to have no impact.

The notion of managing wilderness may seem paradoxical—is it still wilderness if we meddle with it? In fact, though, with the great numbers of people in the world today, even wilderness must be defined, legally set aside, and controlled. We can view the goal of managing wilderness in two ways: in terms of the wilderness itself and in terms of people. In the first instance, the goal is to preserve nature undisturbed by people. In the second, the purpose is to provide people with a wilderness experience.

Legally designated wilderness can be seen as one extreme in a spectrum of environments to manage. The spectrum ranges from wilderness rarely disturbed by anyone to preserves in which some human activities are allowed to be visible—parks designed for outdoor recreation, forests for timber production and various kinds of recreation, hunting preserves, and urban parks—and finally, at the other extreme, open-pit mines. You can think of many stages in between on this spectrum.

Wilderness management should involve as little direct action as possible, so as to minimize human influence. This also means, ironically, that one of the necessities is to control human access so that a visitor has little, if any, sense that other people are present.

Consider, for example, the Desolation Wilderness Area in California, consisting of more than 24,200 hectares (60,000 acres), which in one year had more than 250,000 visitors. Could each visitor really have a wilderness experience there, or was the human carrying capacity of the wilderness exceeded? This is a subjective judgment. If, on one hand, all visitors saw only their own companions and believed they were alone, then the actual number of visitors did not matter for each visitor's wilderness experience. On the other hand, if every visitor found the solitude ruined by strangers, then the management failed, no matter how few people visited.

Wilderness designation and management must also take into account adjacent land uses. A wilderness next to a garbage dump or a power plant spewing smoke is a contradiction in terms. Whether a wilderness can be adjacent to a high-intensity campground or near a city is a more subtle question that must be resolved by citizens.

Today, those involved in wilderness management recognize that wild areas change over time and that these changes should be allowed to occur as long as they are natural. This is different from earlier views that nature undisturbed was unchanging and should be managed so that it did not change. In addition, it is generally argued now that in choosing what activities can be allowed in a wilderness, we should emphasize activities that depend on wilderness (the experience of solitude or the observation of shy and elusive wildlife) rather than activities that can be enjoyed elsewhere (such as downhill skiing).

Another source of conflict is that wilderness areas frequently contain economically important resources, including timber, fossil fuels, and mineral ores. There has been heated debate about whether wilderness areas should be open to the extraction of these.

Still another controversy involves the need to study wilderness versus the desire to leave wilderness undisturbed. Those in favor of scientific research in the wilderness argue that it is necessary for the conservation of wilderness. Those opposed argue that scientific research contradicts the purpose of a designated wilderness as an area undisturbed by people. One solution is to establish separate research preserves.

CRITICAL THINKING ISSUE

Can Tropical Forests Survive in Bits and Pieces?

Although tropical rain forests occupy only about 7% of the world's land area, they provide habitat for at least half of the world's species of plants and animals. Approximately 100 million people live in rain forests or depend on them for their livelihood. Tropical plants provide products such as chocolate, nuts, fruits, gums, coffee, wood, rubber, pesticides, fibers, and dyes. Drugs for treating high blood pressure, Hodgkin's disease, leukemia, multiple sclerosis, and Parkinson's disease have been made from tropical plants, and medical scientists believe many more are yet to be discovered.

In the United States, most of the interest in tropical rain forests has focused on Brazil, whose forests are believed to have more species than any other geogaphic area. Estimates of destruction in the Brazilian rain forest range from 6 to 12%, but numerous studies have shown that deforested area alone does not adequately measure habitat destruction because surrounding habitats are also affected (refer back to Figure 12.14a). For example, the more fragmented a forest is, the more edges there are, and the greater the impact on the living organisms. Such edge effects vary depending on the species, the characteristics of the land surrounding the forest fragment, and the distance between fragments. For example, a forest surrounded by farmland is more deeply affected than one surrounded by abandoned land in which secondary growth presents a more gradual transition between forest and deforested areas. Some insects, small mammals, and many birds find only 80 m (262.5 ft) to be a barrier to movement from one fragment to another, whereas one small marsupial has been found to cross distances of 250 m (820.2 ft). Corridors between forested areas also help to offset the negative effects of deforestation on plants and animals of the forest.

Critical Thinking Questions

1. Look again at Figure 12.14a, the satellite image of part of the Brazilian rain forest. You are asked to make a plan that will allow 50% of the area to be cut, and the rest established as a national park. Make a design for how you think this would best be done, taking into account conservation of biological diversity, the difficulty of travel in tropical rain forests, and the needs of local people to make a living. In your plan, the areas to be harvested will not change over time once the design is in place.
2. You are asked to create a park like the one in question 1, taking into account that the forested areas cut for timber will be allowed to regenerate and during that time, until actual harvest, could be used for recreation. Modify your design to take that into account.
3. The forest fragments left uncut in Figure 12.14 are sometimes compared with islands. What are some ways in which this is an appropriate comparison? Some ways in which it is not?

SUMMARY

- In the past, land management for harvesting resources and conserving nature was mostly local, with each parcel of land considered independently.
- Today, a landscape perspective has developed, and lands used for harvesting resources are seen as part of a matrix that includes lands set aside for the conservation of biological diversity and for landscape beauty.
- Forests are among civilization's most important renewable resources. Forest management seeks a sustainable harvest and sustainable ecosystems. Because examples of success-

ful sustainable forestry are rare, "certification of sustainable forestry" has developed to determine which methods appear most consistent with sustainability and then compare the management of a specific forest with those standards.

- Given their rapid population growth, continued use of firewood as an important fuel in developing nations is a major threat to forests. It is doubtful that these nations can implement successful management programs in time to prevent serious damage to their forests and severe effects on their people.
- Clear-cutting is a major source of controversy in forestry. Some tree species require clearing to reproduce and grow, but the scope and method of cutting must be examined carefully in terms of the needs of the species and the type of forest ecosystem.
- Properly managed plantations can relieve pressure on forests.
- Managing parks for biological conservation is a relatively new idea that began in the 19th century. The manager of a park must be concerned with its shape and size. Parks that are too small or the wrong shape may have too small a population of the species for which the park was established and thus may not be able to sustain the species.
- A special extreme in conservation of natural areas is the management of wilderness. In the United States, the 1964 Wilderness Act provided a legal basis for such conservation. Managing wilderness seems a contradiction—trying to make sure it will be undisturbed by people requires interference to limit user access and to maintain the natural state, so an area that is not supposed to be influenced by people actually is.
- Parks, nature preserves, wilderness areas, and actively harvested forests affect one another. The geographic pattern of these areas on a landscape, including corridors and connections among different types, is part of the modern approach to biological conservation and the harvest of forest resources.

REEXAMINING THEMES AND ISSUES

Human Population

Forests provide essential resources for civilization. As the human population grows, there will be greater and greater demand for these resources. Because forest plantations can be highly productive, we are likely to place increasing emphasis on them as a source of timber. This would free more forestland for other uses.

Sustainability

Sustainability is the key to conservation and management of wild living resources. However, sustainable harvests have rarely been achieved for timber production, and sustained ecosystems in harvested forests are even rarer. Sustainability must be the central focus for forest resources in the future.

Global Perspective

Forests are global resources. A decline in the availability of forest products in one region affects the rate of harvest and economic value of these products in other regions. Biological diversity is also a global resource. As the human population grows, the conservation of biological diversity is likely to depend more and more on legally established parks, nature preserves, and wilderness areas.

Urban World

We tend to think of cities as separated from living resources, but urban parks are important in making cities pleasant and livable; if properly designed, they can also help to conserve wild living resources.

People and Nature

Forests have provided essential resources, and often people have viewed them as perhaps sacred but also dark and scary. Today, we value wilderness and forests, but we rarely harvest forests sustainably. Thus, the challenge for the future is to reconcile our dual and somewhat opposing views so that we can enjoy both the deep meaningfulness of forests and their important resources.

Science and Values

Many conflicts over parks, nature preserves, and legally designated wilderness areas also involve science and values. Science tells us what is possible and what is required in order to conserve both a specific species and total biological diversity. But what society desires for such areas is, in the end, a matter of values and experience, influenced by scientific knowledge.

KEY TERMS

certification of forestry **246**
clear-cutting **245**
codominants **245**
dominants **245**
intermediate **245**
old-growth forest **244**
plantation **245**
rotation time **244**
second-growth forest **244**
seed-tree cutting **245**
selective cutting **245**
shelterwood cutting **245**
silviculture **238**
site quality **245**
stand **244**
strip-cutting **245**
suppressed **245**
thinning **245**
wilderness **252**

STUDY QUESTIONS

1. What environmental conflicts might arise when a forest is managed for the multiple uses of (a) commercial timber, (b) wildlife conservation, and (c) a watershed for a reservoir? In what ways could management for one use benefit another?
2. What arguments could you offer for and against the statement "Clear-cutting is natural and necessary for forest management"?
3. Can a wilderness park be managed to supply water to a city? Explain your answer.
4. A park is being planned in rugged mountains with high rainfall. What are the environmental considerations if the purpose of the park is to preserve a rare species of deer? If the purpose is recreation, including hiking and hunting?
5. What are the environmental effects of decreasing the rotation time (accelerating the rate of cutting) in forests from an average of 60 years to 10 years? Compare these effects for (a) a woodland in a dry climate on a sandy soil and (b) a rain forest.
6. In a small but heavily forested nation, two plans are put forward for forest harvests. In Plan A, all the forests to be harvested are in the eastern part of the nation, while all the forests of the West are set aside as wilderness areas, parks, and nature preserves. In Plan B, small areas of forests to be harvested are distributed throughout the country, in many cases adjacent to parks, preserves, and wilderness areas. Which plan would you choose? Note that in Plan B, wilderness areas would be smaller than in Plan A.
7. The smallest legally designated wilderness in the United States is Pelican Island, Florida (Figure 12.18), covering 5 acres. Do you think this can meet the meaning of *wilderness* and the intent of the Wilderness Act?

FURTHER READING

Botkin, D.B., ***No Man's Garden: Thoreau and a New Vision for Civilization and Nature*** (Washington, DC: Island Press, 2001).

Hendee, J.C., ***Wilderness Management: Stewardship and Protection of Resources and Values*** (Golden, CO: Fulcrum Publishing, 2002). Considered the classic work on this subject.

Kimmins, J.P., ***Forest Ecology*****, 3rd ed.** (Upper Saddle River, NJ: Prentice Hall, 2003). A textbook that applies recent developments in ecology to the practical problems of managing forests.

Runte, A., ***National Parks: The American Experience*** (Lincoln: Bison Books of the University of Nebraska, 1997). The classic book about the history of national parks in America and the reasons for their development.

CHAPTER 13

Wildlife, Fisheries, and Endangered Species

Brown pelicans, once endangered because of DDT, have come back in abundance and are common along both the Atlantic and Pacific coasts. This pelican is fishing from a breakwater on the Atlantic coast of Florida.

LEARNING OBJECTIVES

Wildlife, fish, and endangered species are among the most popular environmental issues today. People love to see wildlife; many people enjoy fishing, make a living from it, or rely on fish as an important part of their diet; and since the 19th century the fate of endangered species has drawn public attention. You would think that by now we would be doing a good job of conserving and managing these kinds of life, but often we are not. This chapter tells you how we are doing and how we can improve our conservation and management of wildlife, fisheries, and endangered species. After reading this chapter, you should understand . . .

- Why people want to conserve wildlife and endangered species;
- The importance of habitat, ecosystems, and landscape in the conservation of endangered species;
- Current causes of extinction;
- Steps we can take to achieve sustainability of wildlife, fisheries, and endangered species;
- The concepts of species persistence, maximum sustainable yield, the logistic growth curve, carrying capacity, optimum sustainable yield, and minimum viable populations.

CASE STUDY

Stories Told by the Grizzly Bear and the Bison

The grizzly bear and the American bison illustrate many of the general problems of conserving and managing wildlife and endangered species. In Chapter 2 we pointed out that the standard scientific method sometimes does not seem suited to studies in the environmental sciences. This is also true of some aspects of wildlife management and conservation. Several examples illustrate the needs and problems.

The Grizzly Bear

A classic example of wildlife management is the North American grizzly bear. An endangered species, the grizzly has been the subject of efforts by the U.S. Fish and Wildlife Service to meet the requirements of the U.S. Endangered Species Act, which include restoring the population of these bears.

The grizzly became endangered as a result of hunting and habitat destruction. It is arguably the most dangerous North American mammal, famous for unprovoked attacks on people, and has been eliminated from much of its range for that reason. Males weigh as much as 270 kg (600 pounds), females as much as 160 kg (350 pounds). When they rear up on their hind legs, they are almost 3 m (8 ft) tall. No wonder they are frightening (see Figure 13.1). Despite this, or perhaps because of it, grizzlies intrigue people, and watching grizzlies from a safe distance has become a popular recreation.

At first glance, restoring the grizzly seems simple enough. But then that old question arises: Restore to what? One answer is to restore the species to its abundance at the time of the European discovery and settlement of North America. But it turns out that there is very little historical information about the abundance of the grizzly at that time, so it is not easy to determine how many (or what density per unit area) could be considered a "restored" population. We also lack a good estimate of the grizzly's present abundance, and thus we don't know how far we will have to take the species to "restore" it to some hypothetical past abundance. Moreover, the grizzly is difficult to study—it is large and dangerous and tends to be reclusive. The U.S. Fish and Wildlife Service attempted to count the grizzlies in Yellowstone National Park by installing automatic flash cameras that were set off when the grizzlies took a bait. This seemed a good idea, but the grizzlies didn't like the cameras and destroyed them,[1] so we still don't have a good scientific estimate of their present number. The National Wildlife Federation lists 1,200 in the contiguous states, 32,000 in Alaska, and about 25,000 in Canada, but these are crude estimates.[2]

FIGURE 13.1 **Grizzly bear.** Records of bear sightings by Lewis and Clark have been used to estimate their population at the beginning of the 19th century.

How do we arrive at an estimate of a population that existed at a time when nobody thought of counting its members? Where possible, we use historical records, as discussed in Chapter 2. We can obtain a crude estimate of the grizzly's abundance at the beginning of the 19th century from the journals of the Lewis and Clark expedition. Lewis and Clark did not list the numbers of most wildlife they saw; they simply wrote that they saw "many" bison, elk, and so forth. But the grizzlies were especially dangerous and tended to travel alone, so Lewis and Clark noted each encounter, stating the exact number they met. On that expedition, they saw 37 grizzly bears over a distance of approximately 1,000 miles (their records were in miles).[1]

Lewis and Clark saw grizzlies from near what is now Pierre, South Dakota, to what is today Missoula, Montana. A northern and southern geographic limit to the grizzly's range can be obtained from other explorers. Assuming Lewis and Clark could see a half-mile to each side of their line of travel on average, the density of the bears was approximately 3.7 per 100 square miles. If we estimate that the bears' geographic range was 320,000 square miles in the mountain and western plains states, we arrive at a total population of 320,000 x 0.37, or about 12,000 bears.

Suppose we phrase this as a hypothesis: "The number of grizzly bears in 1805 in what is now the United States was 12,000." Is this open to disproof? Not without time travel. Therefore, it is not a scientific statement; it can only be taken as an educated guess or, more formally, an assumption or premise. Still, it has some basis in historical documents, and it is better than no information, since we have few alternatives to determine what used to be. We can use this assumption to create a plan to restore the grizzly to that abundance. But is this the best approach?

Another approach is to ask what is the minimum viable population of grizzly bears—forget completely about what might have been the situation in the past and make use of modern knowledge of population dynamics and genetics, along with food requirements and potential production of that food. Studies of existing populations of brown and grizzly bears suggest that only populations larger than 450 individuals respond to protection with rapid growth.[2] Using this approach, we could estimate how many bears appear to be a "safe" number—that is, a number that carries small risk of extinction and loss of genetic diversity. More precisely, we could phrase this statement as "How many bears are necessary so that the probability that the grizzly will become extinct in the next ten years [or some other period that we consider reasonable for planning] is less than 1% [or some other percentage that we would like]?"

With appropriate studies, this approach could have a scientific basis. Consider a statement of this kind phrased as a hypothesis: "A population of 450 bears [or some other number] results in a 99% chance that at least one mature male and one mature female will be alive ten years from today." We can disprove this statement, but only by waiting for ten years to go by. Although it is a scientific statement, it is a difficult one to deal with in planning for the present.

The American Bison

Another classic case of wildlife management, or mismanagement, is the demise of the American bison, or buffalo (Figure 13.2a). The bison was brought close to extinction in the 19th century for two reasons: They were hunted because coats made of bison hides had become fashionable in Europe, and they also were killed as part of warfare against the Plains peoples (Figure 13.2b). U.S. Army Colonel R.I. Dodge was quoted in 1867 as saying, "Kill every buffalo you can. Every buffalo dead is an Indian gone."[1]

Unlike the grizzly bear, the bison has recovered, in large part because ranchers have begun to find them profitable to raise and sell for meat and other products. Informal estimates, including herds on private and public ranges, suggest there are 200,000–300,000, and bison are said to occur in every state in the United States, including Hawaii, a habitat quite different from their original Great Plains home range.[3] About 20,000 roam wild on public lands in the United States and Canada.[4]

How many bison were there before European settlement of the American West? And how low did their numbers drop? Historical records provide insight. In 1865 the U.S. Army, in response to Indian attacks in the fall of 1864, set fires to drive away the Indians and the buffalo, killing vast numbers of animals.[5] The speed with which bison were almost eliminated was surprising—even to many of those involved in hunting them.

Many early writers tell of immense herds of bison, but few counted them. One exception was General Isaac I. Stevens, who, on July 10, 1853, was surveying for the transcontinental railway in North Dakota. He and his men climbed a high hill and saw "for a great distance ahead every square mile" having "a herd of buffalo upon it." He wrote that "their number was variously estimated by the members of the party—some as high as half a million. I do not think it any exaggeration to

(a)

(b)

FIGURE 13.2 **(a) A bison ranch in the United States.** In recent years, interest in growing bison ranches has increased greatly. In part, the goal is to restore bison to a reasonable percentage of its numbers before the Civil War. In part, bison are ranched because people like them. In addition, there is a growing market for bison meat and other products, including cloth made from bison hair. **(b) Painting of a buffalo hunt** by George Catlin in 1832–1833 at the mouth of the Yellowstone River.

set it down at 200,000."[1] In short, his estimate of just one herd was about the same number of bison that exist in total today!

One of the better attempts to estimate the number of buffalo in a herd was made by Colonel R. I. Dodge, who took a wagon from Fort Zarah to Fort Larned on the Arkansas River in May 1871, a distance of 34 miles. For at least 25 of those miles, he found himself in a "dark blanket" of buffalo. He estimated that the mass of animals he saw in one day totaled 480,000. At one point, he and his men traveled to the top of a hill from which he estimated that he could see six to ten miles, and from that high point there appeared to be a single solid mass of buffalo extending over 25 miles. At ten animals per acre, not a particularly high density, the herd would have numbered 2.7–8.0 million animals.[1]

In the fall of 1868, "a train traveled 120 miles between Ellsworth and Sheridan, Wyoming, through a continuous, browsing herd, packed so thick that the engineer had to stop several times, mostly because the buffaloes would scarcely get off the tracks for the whistle and the belching smoke."[5] That spring, a train had been delayed for eight hours while a single herd passed "in one steady, unending stream." We can use accounts like this one to set bounds on the possible number of animals seen. At the highest extreme, we can assume that the train bisected a circular herd with a diameter of 120 miles. Such a herd would cover 11,310 square miles, or more than 7 million acres. If we suppose that people exaggerated the density of the buffalo, and there were only ten per acre, this single herd would still have numbered 70 million animals!

Some might say that this estimate is probably too high, because the herd would more likely have formed a broad, meandering, migrating line rather than a circle. The impression remains the same—there were huge numbers of buffalo in the American West even as late as 1868, numbering in the tens of millions and probably 50 million or more. Ominously, that same year, the Kansas Pacific Railroad advertised a "Grand Railway Excursion and Buffalo Hunt."[5] Some say that many hunters believed the buffalo could never be brought to extinction because there were so many. The same was commonly believed about all of America's living resources throughout the 19th century.

We tend to view environmentalism as a social and political movement of the 20th century, but it is said that after the Civil War there were angry protests in every legislature over the slaughter of buffalo. In 1871 the U.S. Biological Survey sent George Grinnell to survey the herds along the Platte River. He estimated that only 500,000 buffalo remained there and that at the then-current rate of killing, the animals would not last long. As late as the spring of 1883, a herd of an estimated 75,000 crossed the Yellowstone River near Miles City, Montana, but fewer than 5,000 reached the Canadian border.[5] By the end of that year—only 15 years after the Kansas Pacific train was delayed for eight hours by a huge herd of buffalo—only a thousand or so buffalo could be found, 256 in captivity and about 835 roaming the plains. A short time later, there were only 50 buffalo wild on the plains.

Today, more and more ranchers are finding ways to maintain bison, and the market for bison meat and other bison products is growing, along with an increasing interest in reestablishing bison herds for aesthetic, spiritual, and moral reasons. The history of the bison once again raises the question of what we mean by "restore" a population. Even with our crude estimates of original abundances, the numbers would have varied from year to year. So we would have to "restore" bison not to a single number independent of the ability of its habitat to support the population, but to some range of abundances. How do we approach that problem and estimate the range?

13.1 Traditional Single-Species Wildlife Management

Wildlife, fisheries, and endangered species are considered together in this chapter because they have a common history of exploitation, management, and conservation, and because modern attempts to manage and conserve them follow the same approaches. Although any form of life, from bacteria and fungi to flowering plants and animals, can become endangered, concern about endangered species has tended to focus on wildlife. We will maintain that focus, but we ask you to remember that the general principles apply to all forms of life.

Attempts to apply science to the conservation and management of wildlife and fisheries, and therefore to endangered species, began around the turn of the 20th century and viewed each species as a single population in isolation.

Carrying Capacity and Sustainable Yields

The classical, early-20th-century idea of wildlife and fisheries was formalized in the S-shaped logistic growth curve (Figure 13.3), which we discussed in Chapter 4.

As explained in that chapter, the logistic growth curve assumes that changes in the size of a population are simply the result of the population's size in relation to a maximum, called the *carrying capacity*. The logistic characterizes the population only by its total size, nothing else—not the ratio of young to old, healthy to sick, males to females, and the environment doesn't appear at all; it is just assumed to be constant. (See the accompanying box Key Characteristics of a Logistic Population.) The carrying capacity is defined simply as the maximum population that can be sustained indefinitely. Implicit in this definition is the idea that if the population exceeds the carrying capacity, it will damage its environment and/or its own ability to grow and reproduce, and therefore the population will decline.

Two management goals resulted from these ideas and the logistic equation: For a species that we intend to harvest, the goal was maximum sustainable yield (MSY); for a species that we wish to conserve, the goal was to have that species reach, and remain at, its carrying capacity. **Maximum sustainable yield** is defined as the maximum growth rate (measured either as a net increase in the number of individuals or in biomass over a specified time period) that the population could sustain indefinitely. The **maximum sustainable-yield population** is defined as the population size at which the maximum growth rate occurs. More simply, the population was viewed as a factory that could keep churning out exactly the same quantity of a product year after year.

Key Characteristics of a Logistic Population

- The population exists in an environment assumed to be constant.
- The population is small in relation to its resources and therefore grows at a nearly exponential rate.
- Competition among individuals in the population slows the growth rate.
- The greater the number of individuals, the greater the competition and the slower the rate of growth.
- Eventually, a point is reached, called the **logistic carrying capacity**, at which the number of individuals is just sufficient for the available resources.
- At this level, the number of births in a unit time equals the number of deaths, and the population is constant.
- A population can be described simply by its total number.
- Therefore, all individuals are equal.

Today a broader view is developing. It acknowledges that a population exists within an ecological community and within an ecosystem, and that the environment is always, or almost always, changing (including human-induced changes). Therefore, the population you are interested in interacts with many others, and the size and condition of those can affect the one on which you are focusing. With this new understanding, the harvesting goal is to harvest sustainably, removing in each time period the maximum number of individuals (or maximum biomass) that can be harvested indefinitely without diminishing either the population that particularly interests you or its ecosystem. The preservation goal for a threatened or an endangered species becomes more open, with more choices. One is to try to keep the population at its carrying capacity. Another is to sustain a **minimum viable population**, which is the estimated smallest population that can maintain itself and its genetic variability indefinitely. A third option, which leads to a population size somewhere between the other two, is the **optimum sustainable population**.

In the logistic curve, the greatest production occurs when the population is exactly one-half of the carrying capacity (see Figure 13.3). This is nifty because it makes everything seem simple—all you have to do is figure out the carrying capacity and keep the population at one-half of it. But what seems simple can easily become troublesome. Even if the basic assumptions of the logistic curve were true, which they are not, the slightest overestimate of carrying capacity, and therefore MSY, would lead to overharvesting, a decline in production, and a decline in the abundance of the species. If a population is harvested as if it were actually at one-half its carrying capacity, then unless a logistic population is actually maintained at exactly that number, its growth will decline. Since it is almost impossible to maintain a wild population at some exact number, the approach is doomed from the start.

One important result of all of this is that a *logistic population is stable in terms of its carrying capacity*—it will return to that number after a disturbance. If the population grows beyond its carrying capacity, deaths exceed births, and the population declines back to the carrying capacity. If the population falls below the carrying capacity, births exceed deaths, and the population increases. Only if the population is exactly at the carrying capacity do births exactly equal deaths, and then the population does not change.

Despite its limitations, the logistic growth curve was used for all wildlife, especially fisheries and including endangered species, throughout much of the 20th century.

The term **carrying capacity** as used today has three definitions. The first is the carrying capacity as defined by the logistic growth curve, the *logistic carrying capacity* already discussed. The second definition contains the same idea but is not dependent on that specific equation. It states that the

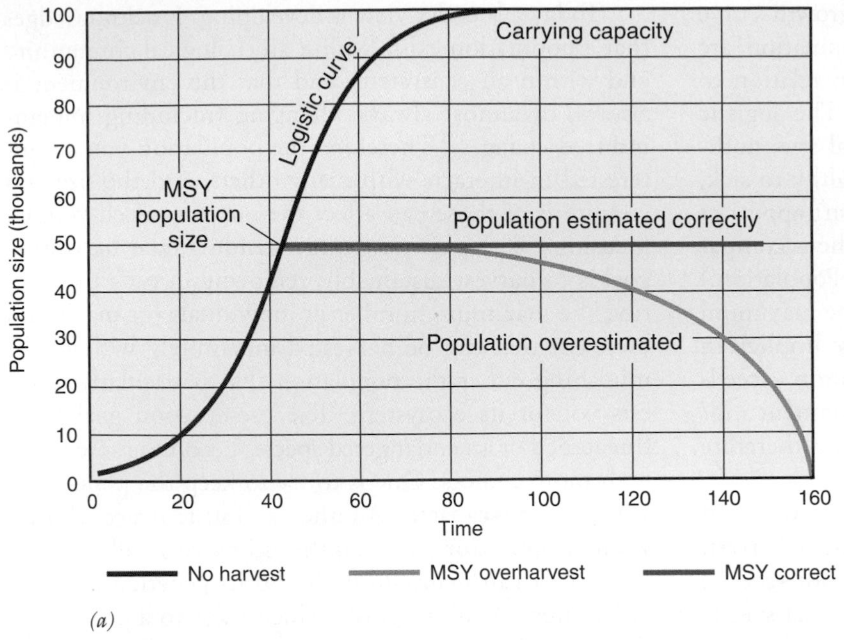

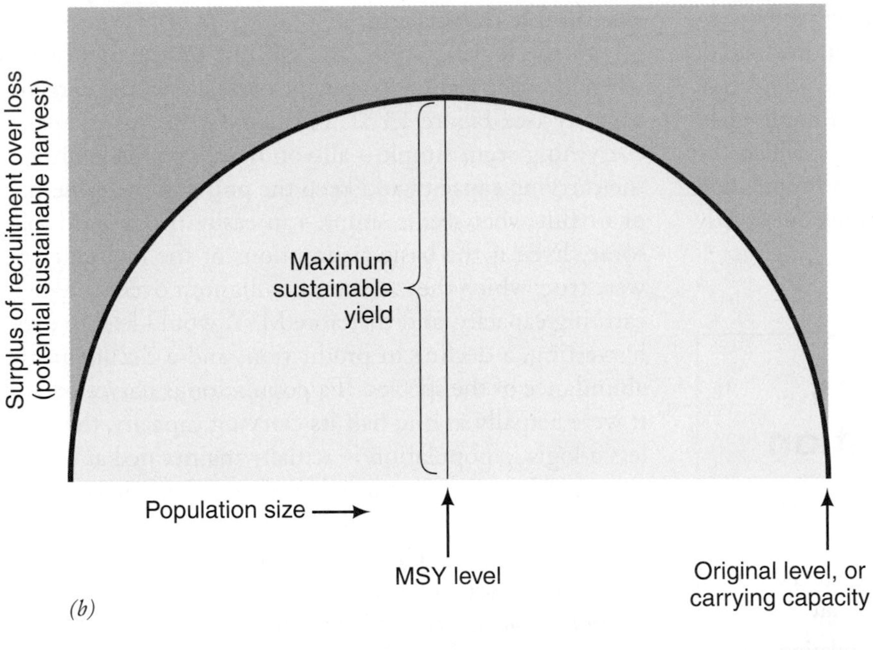

FIGURE 13.3 **(a) The logistic growth curve,** showing the carrying capacity and the maximum sustainable yield (MSY) population (where the population size is one-half the carrying capacity). The figure shows what happens to a population when we assume it is at MSY and it is not. Suppose a population grows according to the logistic curve from a small number to a carrying capacity of 100,000 with an annual growth rate of 5%. The correct maximum sustainable yield would be 50,000. When the population reaches exactly the calculated maximum sustainable yield, the population continues to be constant. But if we make a mistake in estimating the size of the population (for example, if we believe that it is 60,000 when it is only 50,000), then the harvest will always be too large, and we will drive the population to extinction. **(b) Another view of a logistic population.** Growth in the population here is graphed against population size. Growth peaks when the population is exactly at one-half the carrying capacity. This is a mathematical consequence of the equation for the curve. It is rarely, if ever, observed in nature.

carrying capacity is an abundance at which a population can sustain itself without any detrimental effects that would lessen the ability *of that species*—in the abstract, treated as separated from all others—to maintain that abundance. The third, the *optimum sustainable population*, already discussed, leads to a population size between the other two.

An Example of Problems with the Logistic Curve

Suppose you are in charge of managing a deer herd for recreational hunting in one of the 50 U.S. states. Your goal is to maintain the population at its MSY level, which, as you can see from Figure 13.3, occurs at exactly one-half of the carrying capacity. At this abundance, the population increases by the greatest number during any time period.

To accomplish this goal, first you have to determine the logistic carrying capacity. You are immediately in trouble because, first, only in a few cases has the carrying capacity ever been determined by legitimate scientific methods (see Chapter 2) and, second, we now know that the carrying capacity varies with changes in the environment. The procedure in the past was to estimate the carrying capacity by *nonscientific* means and then attempt to maintain the population at one-half that level. This method requires accurate counts each year. It also requires that the environment not vary, or, if it does, that it vary in a way that does not affect the population. Since these conditions cannot be met, the logistic curve has to fail as a basis for managing the deer herd.

An interesting example of the staying power of the logistic growth curve can be found in the United States Marine Mammal Protection Act of 1972. This act states that its primary goal is to conserve "the health and stability of marine ecosystems," which is part of the modern approach, and so the act seems to be off to a good start. But then the act states that the secondary goal is to maintain an "optimum sustainable population" of marine mammals. What is this? The wording of the act allows two interpretations. One is the logistic carrying capacity, and the other is the MSY population level of the logistic growth curve. So the act takes us back to square one, the logistic curve.

13.2 Improved Approaches to Wildlife Management

The U.S. Council on Environmental Quality (an office within the executive branch of the federal government), the World Wildlife Fund of the United States, the Ecological Society of America, the Smithsonian Institution, and the International Union for the Conservation of Nature (IUCN) have proposed four principles of wildlife conservation:

- A safety factor in terms of population size, to allow for limitations of knowledge and imperfections of procedures. An interest in harvesting a population should not allow the population to be depleted to some theoretical minimum size.
- Concern with the entire community of organisms and all the renewable resources, so that policies developed for one species are not wasteful of other resources.
- Maintenance of the ecosystem of which the wildlife are a part, minimizing risk of irreversible change and long-term adverse effects as a result of use.
- Continual monitoring, analysis, and assessment. The application of science and the pursuit of knowledge about the wildlife of interest and its ecosystem should be maintained and the results made available to the public.

These principles broaden the scope of wildlife management from a narrow focus on a single species to inclusion of the ecological community and ecosystem. They call for a safety net in terms of population size, meaning that no population should be held at exactly the MSY level or reduced to some theoretical minimum abundance. These new principles provide a starting point for an improved approach to wildlife management.

Time Series and Historical Range of Variation

As illustrated by the opening case study about the American buffalo and grizzly bears, it is best to have an estimate of population over a number of years. This set of estimates is called a **time series** and could provide us with a measure of the **historical range of variation**—the known range of abundances of a population or species over some past time interval. Such records exist for few species. One is the American whooping crane (Figure 13.4), America's tallest bird, standing about 1.6 m (5 ft) tall. Because this species became so rare and because it migrated as a single flock, people began counting the total population in the late 1930s. At that time, they saw only 14 whooping cranes. They counted not only the total number but also the number born that year. The difference between these two numbers gives the number dying each year as well. And from this time series, we can estimate the probability of extinction.

The first estimate of the probability of extinction based on the historical range of variation, made in the early 1970s, was a surprise. Although the birds were few, the probability of extinction was less than one in a billion.[6] How could this number be so low? Use of the historical range of variation carries with it the assumption that causes of variation in the future will be only those that occurred during the historical period. For the whooping cranes, one catastrophe—such as a long, unprecedented drought on the wintering grounds—could cause a population decline not observed in the past.

Not that the whooping crane is without threats—current changes in its environment may not be simply repeats of what happened in the past. According to Tom Stehn, Whooping Crane Coordinator for the U.S. Fish and Wildlife Service, the whooping crane population that winters in Aransas National Wildlife Refuge, Texas, and summers in Wood Buffalo National Park, Canada,

> *reached a record population of 270 at Aransas in December, 2008. The number would have been substantially higher but for the loss of 34 birds that left Aransas in the spring, 2008 and failed to return in the fall. Faced with food shortages from an "exceptional" drought that hammered Texas, record high mortality during the 2008–09 winter of 23 cranes (8.5% of the flock) left the AWBP at 247 in the spring, 2009. Total flock mortality for the 12 months following April, 2008 equaled 57 birds (21.4% of the flock). The refuge provided supplemental feed during the 2008–09 winter to provide some cranes with additional calories. Two whooping cranes failed to migrate north, but survived the hot and dry 2009 Aransas summer.*

A below-average 2009 production year in Canada with 22 fledged chicks from 62 nests was half the production of the previous summer and is expected to result in a break-even year for the AWBP. Threats to the flock including land and water development in Texas, the spread of black mangrove on the wintering grounds, and wind farm construction in the migration corridor all remained unabated in 2009.[7]

Even with this limitation, this method provides invaluable information. Unfortunately, at present, mathematical estimates of the probability of extinction have been done for just a handful of species. The good news is that the wild whooping cranes on the main flyway have continued to increase and in 2008 numbered 274. The total wild population (all flyways) was 382, and there were 162 in captive flocks, for a total of 534.[8, 5]

Age Structure as Useful Information

An additional key to successful wildlife management is monitoring of the population's age structure (see Chapter 4), which can provide many different kinds of information. For example, the age structures of the catch of salmon from the Columbia River in Washington for two different periods, 1941–1943 and 1961–1963, were quite different. In the first period, most of the catch (60%) consisted of four-year-olds; the three-year-olds and 5-year-olds each made up about 15% of the population. Twenty years later, in 1961 and 1962, half the catch consisted of 3-year-olds, the number of 5-year-olds had declined to about 8%, and the total catch had declined considerably. During the period 1941–1943, 1.9 million fish were caught. During the second period, 1961–1963, the total catch dropped to 849,000, just 49% of the total caught in the earlier period. The shift in catch toward younger ages, along with an overall decline in catch, suggests that the fish were being exploited to a point at which they were not reaching older ages. Such a shift in the age structure of a harvested population is an early sign of overexploitation and of a need to alter allowable catches.

Harvests as an Estimate of Numbers

Another way to estimate animal populations is to use the number harvested. Records of the number of buffalo killed were neither organized nor all that well kept, but they were sufficient to give us some idea of the number taken. In 1870, about 2 million buffalo were killed. In 1872, one company in Dodge City, Kansas, handled 200,000 hides. Estimates based on the sum of reports from such companies, together with guesses at how many

(a)

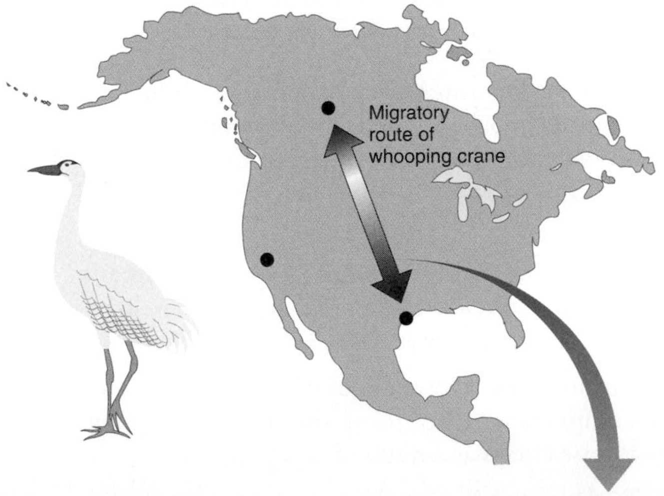

(b

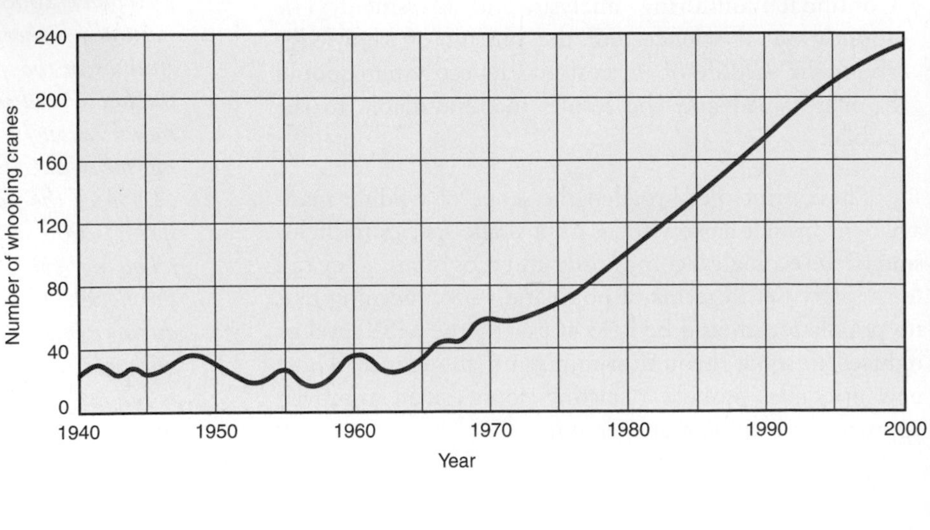

(c

FIGURE 13.4 **The whooping crane** **(a)** is one of many species that appear to have always been rare. Rarity does not necessarily lead to extinction, but a rare species, especially one that has undergone a rapid and large decrease in abundance, needs careful attention and assessment as to threatened or endangered status; **(b) migration route**; and **(c) change in population from 1940 to 2000.**

animals were likely taken by small operators and not reported, suggest that about 1.5 million hides were shipped in 1872 and again in 1873.[5] In those years, buffalo hunting was the main economic activity in Kansas. The Indians were also killing large numbers of buffalo for their own use and for trade. Estimates range to 3.5 million buffalo killed per year, nationwide, during the 1870s.[9] The bison numbered at least in the low millions.

Still another way harvest counts are used to estimate previous animal abundance is the **catch per unit** effort. This method assumes that the same effort is exerted by all hunters/harvesters per unit of time, as long as they have the same technology. So if you know the total time spent in hunting/harvesting and you know the catch per unit of effort, you can estimate the total population. This method leads to rather crude estimates with a large observational error; but where there is no other source of information, it can offer unique insights.

An interesting application of this method is the reconstruction of the harvest of the bowhead whale and, from that, an estimate of the total bowhead population. Taken traditionally by Eskimos, the bowhead was the object of "Yankee," or American, whaling from 1820 until the beginning of World War I. (See A Closer Look 13.3 later in this chapter for a general discussion of marine mammals.) Every ship's voyage was recorded, so we know essentially 100% of all ships that went out to catch bowheads. In addition, on each ship a daily log was kept, with records including sea conditions, ice conditions, visibility, number of whales caught, and their size in terms of barrels of oil. Some 20% of these logbooks still exist, and their entries have been computerized. Using some crude statistical techniques, it was possible to estimate the abundance of the bowhead in 1820 at 20,000, plus or minus 10,000. Indeed, it was possible to estimate the total catch of whales and the catch for each year—and therefore the entire history of the hunting of this species.

13.3 Fisheries

Fish are important to our diets—they provide about 16% of the world's protein and are especially important protein sources in developing countries. Fish provide 6.6% of food in North America (where people are less interested in fish than are people in most other areas), 8% in Latin America, 9.7% in Western Europe, 21% in Africa, 22% in central Asia, and 28% in the Far East.

Fishing is an international trade, but a few countries dominate: Japan, China, Russia, Chile, and the United States are among the major fisheries nations. And commercial fisheries are concentrated in relatively few areas of the world's oceans (Figure 13.5). Continental shelves, which make up only 10% of the oceans, provide more than 90% of the fishery harvest. Fish are abundant where their food is abundant and, ultimately, where there is high production of algae at the base of the food chain. Algae are most abundant in areas with relatively high concentrations of the chemical elements necessary for life, particularly nitrogen and phosphorus. These areas occur most commonly along the continental shelf, particularly in regions of wind-induced upwellings and sometimes quite close to shore.

The world's total fish harvest has increased greatly since the middle of the 20th century. The total harvest was 35 million metric tons (MT) in 1960. It more than doubled in just 20 years (an annual increase of about 3.6%) to 72 million MT in 1980 and has since grown to 132,000 MT, but seems to be leveling off.[10] The total global fish harvest doubled in 20 years because of increases in the number of boats, improvements in technology, and especially increases in aquaculture production, which also more than doubled between 1992 and 2001, from about 15 million MT to more than 37 million MT. Aquaculture presently provides more than 20% of all fish harvested, up from 15% in 1992.

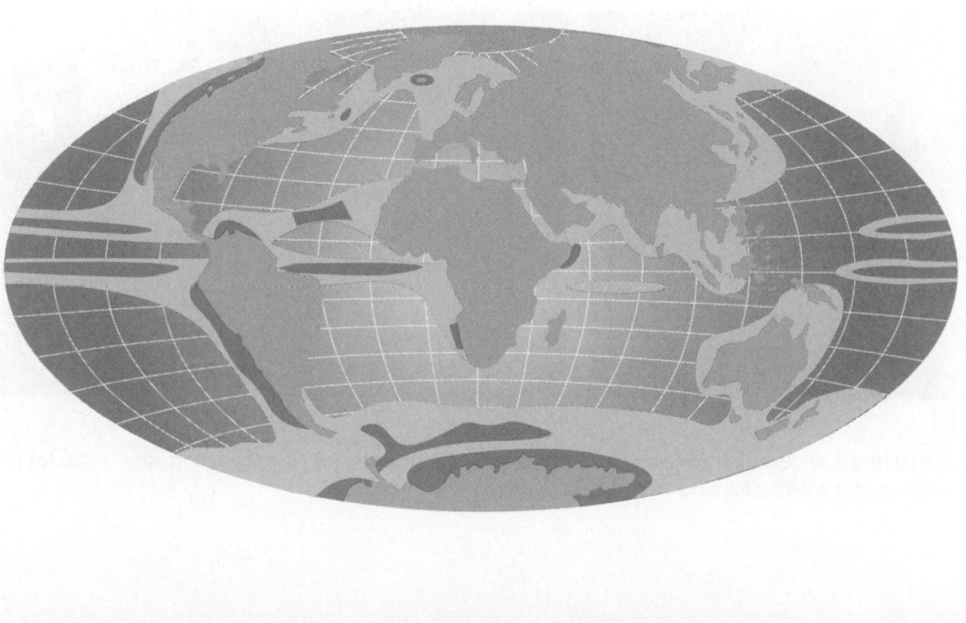

FIGURE 13.5 The world's major fisheries. Red areas are major fisheries; the darker the red, the greater the harvest and the more important the fishery. Most major fisheries are in areas of ocean upwellings, where currents rise, bringing nutrient-rich waters up from the depths of the ocean. Upwellings tend to occur near continents.[11]

Table 13.1 WORLD FISHERIES CATCH

KIND	HARVEST (MILLIONS OF METRIC TONS)	PERCENT	ACCUMULATED PERCENTAGE
Herring, sardines, and anchovies	25	19.23%	19.23%
Carp and relatives	15	11.54%	30.77%
Cod, hake, and haddock	8.6	6.62%	37.38%
Tuna and their relatives	6	4.62%	42.00%
Oysters	4.2	3.23%	45.23%
Shrimp	4	3.08%	48.31%
Squid and octopus	3.7	2.85%	51.15%
Other mollusks	3.7	2.85%	54.00%
Clams and relatives	3	2.31%	56.31%
Tilapia	2.3	1.77%	58.08%
Scallops	1.8	1.38%	59.46%
Mussels and relatives	1.6	1.23%	60.69%
Subtotal	78.9	60.69%	
TOTAL ALL SPECIES	130	100%	

Source: National Oceanic & Atmospheric Administration World

Scientists estimate that there are 27,000 species of fish and shellfish in the oceans. People catch many of these species for food, but only a few kinds provide most of the food—anchovies, herrings, and sardines account for almost 20% (Table 13.1).

In summary, new approaches to wildlife conservation and management include (1) historical range of abundance; (2) estimation of the probability of extinction based on historical range of abundance; (3) use of age-structure information; and (4) better use of harvests as sources of information. These, along with an understanding of the ecosystem and landscape context for populations, are improving our ability to conserve wildlife.

Although the total marine fisheries catch has increased during the past half-century, the effort required to catch a fish has increased as well. More fishing boats with better and better gear sail the oceans (Figure 13.6). That is why the total catch can increase while the total population of a fish species declines.

(a) (b) (c)

FIGURE 13.6 Some modern commercial fishing methods. (a) Trawling using large nets; **(b)** longlines have caught a swordfish; **(c)** workers on a factory ship.

The Decline of Fish Populations

Evidence that the fish populations were declining came from the catch per unit effort. A unit of effort varies with the kind of fish sought. For marine fish caught with lines and hooks, the catch rate generally fell from 6–12 fish caught per 100 hooks—the success typical of a previously unexploited fish population—to 0.5–2.0 fish per 100 hooks just ten years later (Figure 13.7). These observations suggest that fishing depletes fish quickly—about an 80% decline in 15 years. Many of the fish that people eat are predators, and on fishing grounds the biomass of large predatory fish appears to be only about 10% of pre-industrial levels. These changes indicate that the biomass of most major commercial fish has declined to the point where we are mining, not sustaining, these living resources.

Species suffering these declines include codfish, flatfishes, tuna, swordfish, sharks, skates, and rays. The North Atlantic, whose Georges Bank and Grand Banks have for centuries provided some of the world's largest fish harvests, is suffering. The Atlantic codfish catch was 3.7 million MT in 1957, peaked at 7.1 million MT in 1974, declined to 4.3 million MT in 2000, and climbed slightly to 4.7 in 2001.[12] European scientists called for a total ban on cod fishing in the North Atlantic, and the European Union came close to accepting this call, stopping just short of a total ban and instead establishing a 65% cut in the allowed catch for North Sea cod for 2004 and 2005.[12] (See also A Closer Look 13.1.)

Scallops in the western Pacific show a typical harvest pattern, starting with a very low harvest in 1964 at 200 MT, increasing rapidly to 5,887 MT in 1975, declining to 1,489 in 1974, rising to about 7,670 MT in 1993, and then declining to 2,964 in 2002.[13] Catch of tuna and their relatives peaked in the early 1990s at about 730,000 MT and fell to 680,000 MT in 2000, a decline of 14% (Figure 13.7).

Chesapeake Bay, America's largest estuary, was another of the world's great fisheries, famous for oysters and crabs and as the breeding and spawning ground for bluefish, sea bass, and many other commercially valuable species (Figure 13.8). The bay, 200 miles long and 30 miles wide, drains an area of more than 165,000 km^2 from New York State to Maryland and is fed by 48 large rivers and 100 small ones.

Adding to the difficulty of managing the Chesapeake Bay fisheries, food webs are complex. Typical of marine food webs, the food chain of the bluefish that spawns and breeds in Chesapeake Bay shows links to a number of other species, each requiring its own habitat within the space and depending on processes that have a variety of scales of space and time (Figure 13.9).

Furthermore, Chesapeake Bay is influenced by many factors on the surrounding lands in its watershed—runoff from farms, including chicken and turkey farms, that are highly polluted with fertilizers and pesticides; introductions of exotic species; and direct alteration of habitats from fishing and the development of shoreline homes. There is also the varied salinity of the bay's waters—freshwater inlets from rivers and streams, seawater from the Atlantic, and brackish water resulting from the mixture of these.

Just determining which of these factors, if any, are responsible for a major change in the abundance of any fish species is difficult enough, let alone finding a solution that is economically feasible and maintains traditional levels of fisheries employment in the bay. The Chesapeake Bay's fisheries resources are at the limit of what environmental sciences can deal with at this time. Scientific theory remains inadequate, as do observations, especially of fish abundance.

Ironically, this crisis has arisen for one of the living resources most subjected to science-based management. How could this have happened? First, management has been based largely on the logistic growth curve, whose problems we have discussed. Second, fisheries are an open resource, subject to the problems of Garrett Hardin's

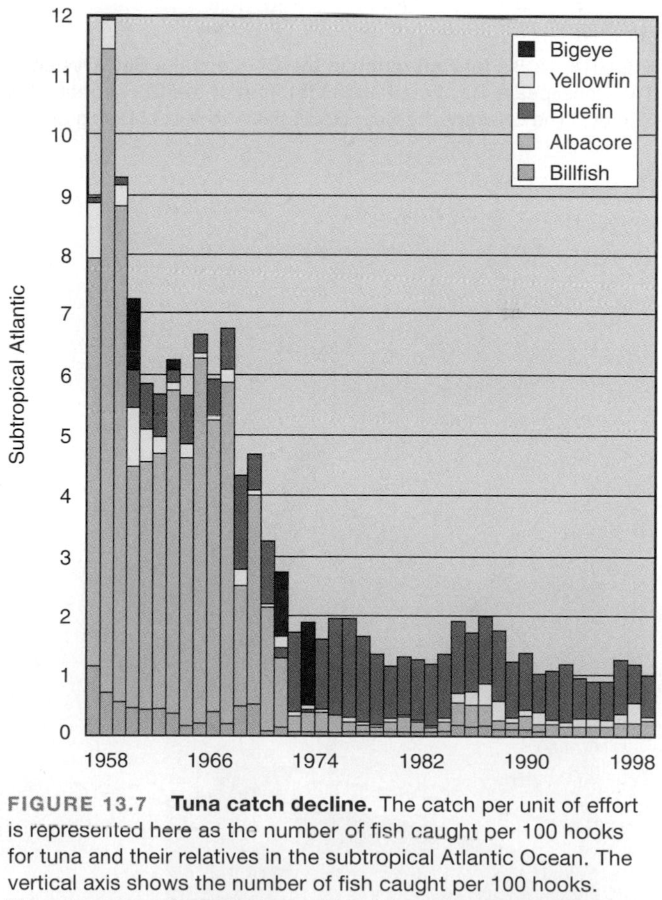

FIGURE 13.7 Tuna catch decline. The catch per unit of effort is represented here as the number of fish caught per 100 hooks for tuna and their relatives in the subtropical Atlantic Ocean. The vertical axis shows the number of fish caught per 100 hooks. The catch per unit of effort was 12 in 1958, when heavy modern industrial fishing for tuna began, and declined rapidly to about 2 by 1974. This pattern occurred worldwide in all the major fishing grounds for these species. (*Source*: Ransom A. Meyers and Boris Worm, "Rapid Worldwide Depletion of Predatory Fish Communities," *Nature* [May 15, 2003].)

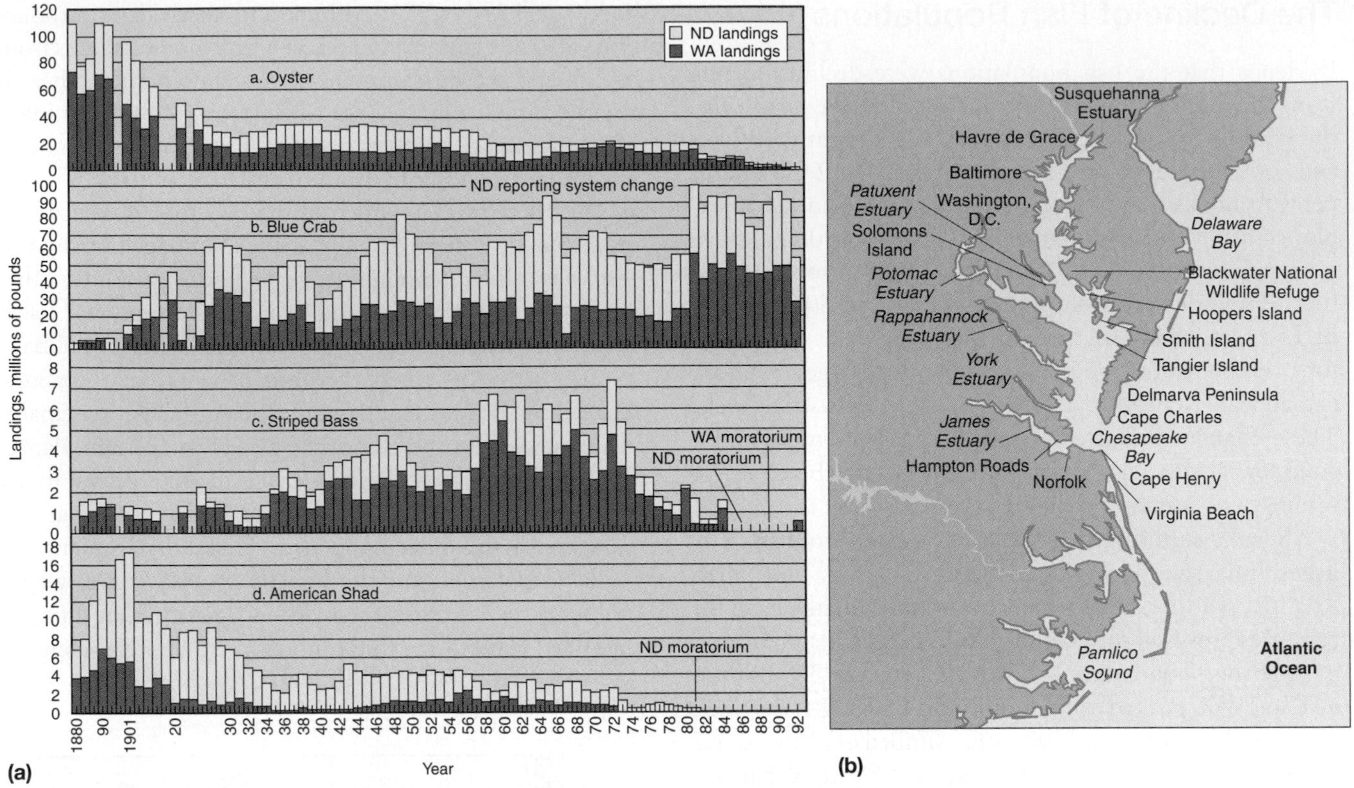

FIGURE 13.8 **(a) Fish catch in the Chesapeake Bay**. Oysters have declined dramatically. (*Source:* The Chesapeake Bay Foundation.) **(b) Map of the Chesapeake Bay Estuary**. (*Source:* U.S. Geological Survey, "The Chesapeake Bay: Geological Product of Rising Sea Level," 1998.)

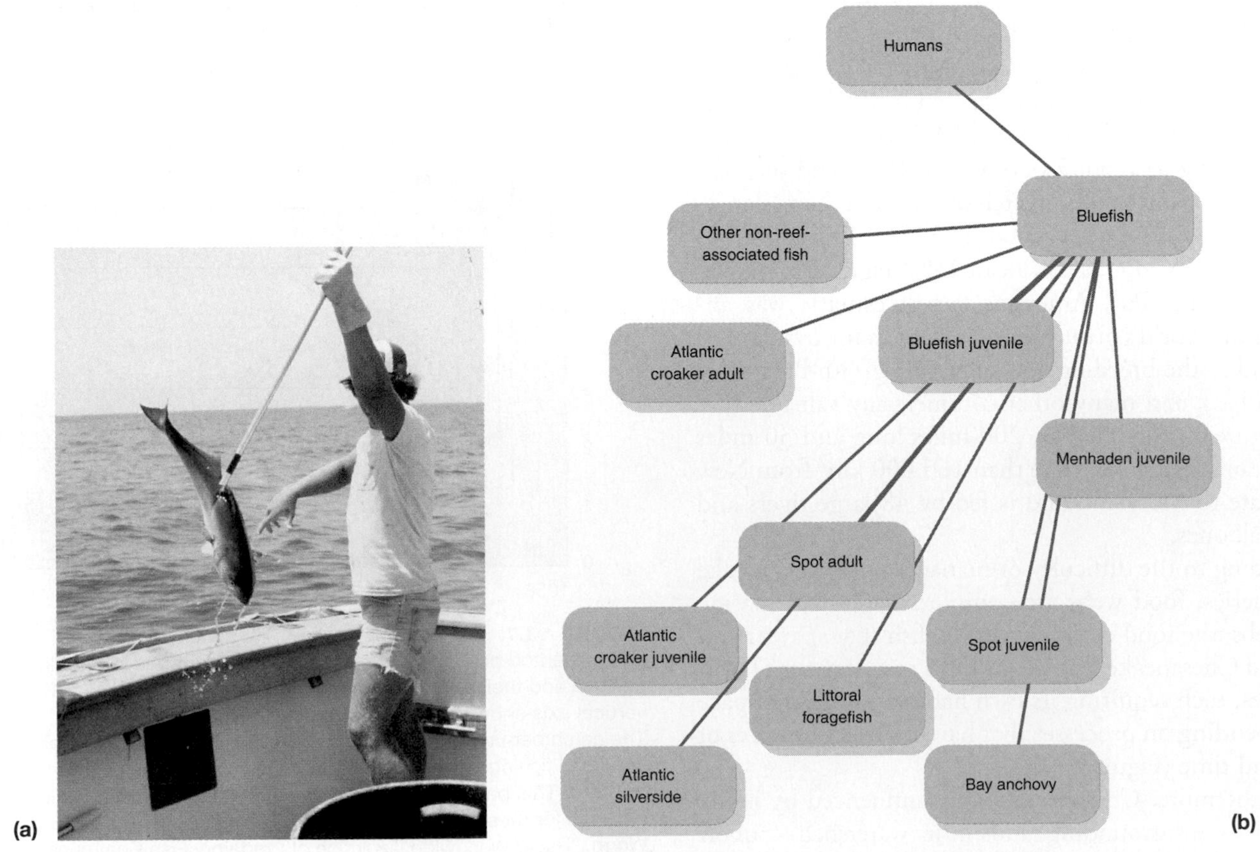

FIGURE 13.9 **(a) Bluefish; (b) food chain of the bluefish in Chesapeake Bay** (*Source*: Chesapeake Bay Foundation.)

"tragedy of the commons," discussed in Chapter 7. In an open resource, often in international waters, the numbers of fish that may be harvested can be limited only by international treaties, which are not tightly binding. Open resources offer ample opportunity for unregulated or illegal harvest, or harvest contrary to agreements.

Exploitation of a new fishery usually occurs before scientific assessment, so the fish are depleted by the time any reliable information about them is available. Furthermore, some fishing gear is destructive to the habitat. Ground-trawling equipment destroys the ocean floor, ruining habitat for both the target fish and its food. Longline fishing kills sea turtles and other nontarget surface animals. Large tuna nets have killed many dolphins that were hunting the tuna.

In addition to highlighting the need for better management methods, the harvest of large predators raises questions about ocean ecological communities, especially whether these large predators play an important role in controlling the abundance of other species.

Human beings began as hunter-gatherers, and some hunter-gatherer cultures still exist. Wildlife on the land used to be a major source of food for hunter-gatherers. It is now a minor source of food for people of developed nations, although still a major food source for some indigenous peoples, such as the Eskimo. In contrast, even developed nations are still primarily hunter-gatherers in the harvesting of fish (see the discussion of aquaculture in Chapter 11).

Can Fishing Ever Be Sustainable?

Suppose you went into fishing as a business and expected reasonable growth in that business in the first 20 years. The world's ocean fish catch rose from 39 million MT in 1964 to 68 million MT in 2003, an average of 3.8% per year for a total increase of 77%.[15] From a business point of view, even assuming all fish caught are sold, that is not considered rapid sales growth. But it is a heavy burden on a living resource.

There is a general lesson to be learned here: Few wild biological resources can sustain a harvest large enough to meet the requirements of a growing business. Although when the overall economy is poor, such as during the economic downturn of 2008–2009, these growth rates begin to look pretty good, most wild biological resources really aren't a good business over the long run. We learned this lesson also from the demise of the bison, discussed earlier, and it is true for whales as well (see A Closer Look 13.3). There have been a few exceptions, such as the several hundred years of fur trading by the Hudson's Bay Company in northern Canada. However, past experience suggests that economically beneficial sustainability is unlikely for most wild populations.

With that in mind, we note that farming fish—aquaculture, discussed in Chapter 11—has been an important source of food in China for centuries and is an increasingly important food source worldwide. But aquaculture can create its own environmental problems (see A Closer Look 13.1).

A CLOSER LOOK 13.1

King Salmon Fishing Season Canceled: Can We Save Them from Extinction?

On May 1, 2008, Secretary of Commerce Carlos M. Gutierrez declared "a commercial fishery failure for the West Coast salmon fishery due to historically low salmon returns," and ordered that salmon fishing be closed. It was an unprecedented decision, the first time since California and Oregon became states, because experts decided that numbers of salmon on the Sacramento River had dropped drastically. The decision was repeated in April 2009, halting all king salmon catch off of California but allowing a small catch off of Oregon, and also allowing a small salmon season on the Sacramento River from mid-November to December 31.[12] Figure 13.10 shows an example of the counts that influenced the decision. These counts were done at the Red Bluff Dam, an irrigation dam for part of the flow of the Sacramento River, completed in 1966. There the fish could be observed as they traversed a fish ladder. Between 1967 and 1969, an average of more than 85,000 adult king salmon crossed the dam. In 2007, there were fewer than 7,000.[13]

While the evidence from Red Bluff Dam seems persuasive, there are several problems. First, the number of salmon varies greatly from year to year, as you can see from the graph. Second, there is no single, consistent way that all the salmon on the Sacramento River system are counted, and in some places the counts are much more ambiguous, suggesting that the numbers may not have dropped severely, as shown in Figure 13.11.

Lacking the best long-term observations, those in charge of managing the salmon decided to err on the side of caution:

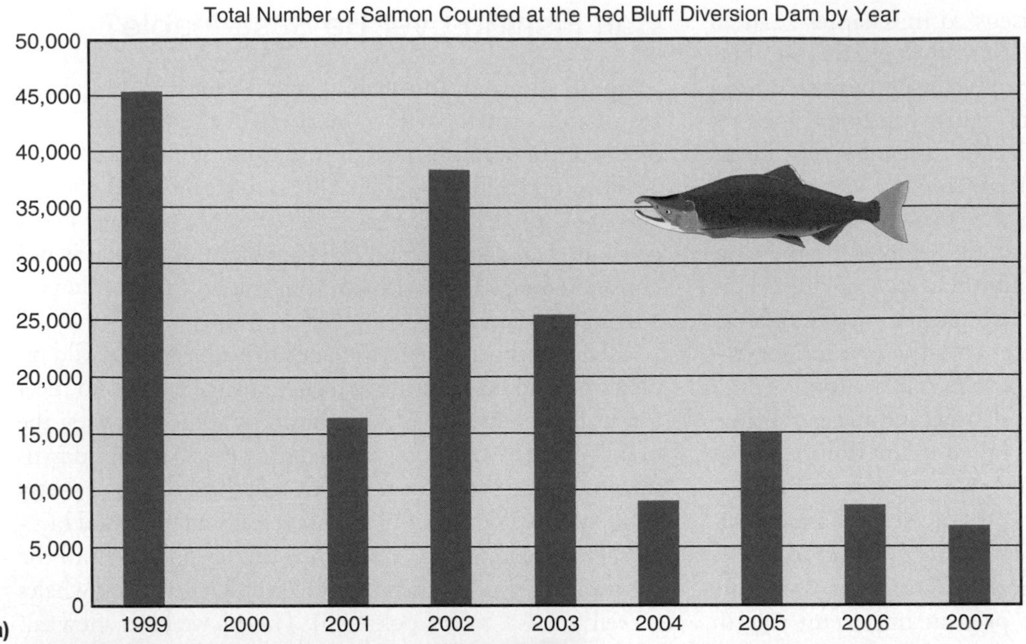

(a)

(b)

FIGURE 13.10 Salmon on the Sacramento River. (a) Counted at the Red Bluff Diversion Dam, Red Bluff, California, between May 14 and September 15 each year as they traverse a fish ladder. (*Source*: http://www.rbuhsd.k12.ca.us/~mpritcha/salmoncount.html); **(b)** Red Bluff Dam photo.

for the fish. As California governor Arnold Schwarzenegger said, "These restrictions will have significant impacts to California's commercial and recreational ocean salmon and Central Valley in-river recreation salmon fisheries and will result in severe economic losses throughout the State, including an estimated $255 million economic impact and the loss of an estimated 2,263 jobs."[14]

It was a difficult choice and typical of the kinds of decisions that arise over the conservation of wildlife, fisheries, and endangered species. Far too often, the data necessary for the wisest planning and decisions are lacking. As we learned in earlier chapters, populations and their environment are always changing, so we can't assume there is one single, simple number that represents the natural state of affairs. Look at the graph of the counts of salmon at Red Bluff Dam. What would you consider an average number of salmon? Is there such a thing? What decision would you have made?

This chapter provides the background necessary to conserve and manage these kinds of life, and raises the important questions that our society faces about wildlife, fisheries, and endangered species in the next decades.

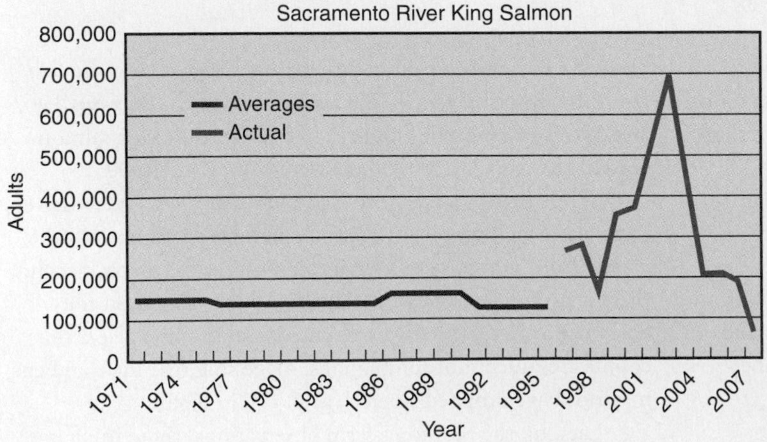

(a)

(b)

FIGURE 13.11 Less precise estimates of king salmon for the entire Sacramento River.[16]

In sum, fish are an important food, and world harvests of fish are large, but the fish populations on which the harvests depend are generally declining, easily exploited, and difficult to restore. We desperately need new approaches to forecasting acceptable harvests and establishing workable international agreements to limit catch. This is a major environmental challenge, needing solutions within the next decade.

13.4 Endangered Species: Current Status

When we say that we want to save a species, what is it that we really want to save? There are four possible answers:

- A wild creature in a wild habitat, as a symbol to us of wilderness.
- A wild creature in a managed habitat, so the species can feed and reproduce with little interference and so we can see it in a naturalistic habitat.
- A population in a zoo, so the genetic characteristics are maintained in live individuals.
- Genetic material only—frozen cells containing DNA from a species for future scientific research.

Which of these goals we choose involves not only science but also values. People have different reasons for wishing to save endangered species—utilitarian, ecological, cultural, recreational, spiritual, inspirational, aesthetic, and moral (see A Closer Look 13.2). Policies and actions differ widely, depending on which goal is chosen.

We have managed to turn some once-large populations of wildlife, including fish, into endangered species. With the expanding public interest in rare and endangered species, especially large mammals and birds, it is time for us to turn our attention to these species.

First some facts. The number of species of animals listed as threatened or endangered rose from about 1,700 in 1988 to 3,800 in 1996 and 5,188 in 2004, the most recent assessment by the International Union for the Conservation of Nature (IUCN).[16] The IUCN's *Red List of Threatened Species* reports that about 20% of all known species of mammals are at risk of extinction, as are 12% of known birds, 4% of known reptiles, 31% of amphibians, and 3% of fish, primarily freshwater fish (see Table 13.2).[18] The *Red List* also estimates that 33,798 species of vascular plants (the familiar kind of plants—trees, grasses, shrubs, flowering herbs), or 12.5% of those known, have recently become extinct or endangered.[17] It lists more than 8,000 plants that are threatened, approximately 3% of all plants.[18]

Table 13.2 NUMBER OF THREATENED SPECIES

LIFE-FORM	NUMBER THREATENED	PERCENT OF SPECIES KNOWN
Vertebrates	5,188	9
Mammals	1,101	20
Birds	1,213	12
Reptiles	304	4
Amphibians	1,770	31
Fish	800	3
Invertebrates	1,992	0.17
Insects	559	0.06
Mollusks	974	1
Crustaceans	429	1
Others	30	0.02
Plants	8,321	2.89
Mosses	80	0.5
Ferns and "Allies"	140	1
Gymnosperms	305	31
Dicots	7,025	4
Monocots	771	1
Total Animals and Plants	**31,002**	**2%**

Source: IUCN Red List www.iuenredlist.org/info/table/table1 (2004)

What does it mean to call a species "threatened" or "endangered"? The terms can have strictly biological meanings, or they can have legal meanings. The U.S. Endangered Species Act of 1973 defines *endangered species* as "any species which is in danger of extinction throughout all or a significant portion of its range other than a species of the Class Insecta determined by the Secretary to constitute a pest whose protection under the provisions of this Act would present an overwhelming and overriding risk to man." In other words, if certain insect species are pests, we want to be rid of them. It is interesting that insect pests can be excluded from protection by this legal definition, but there is no mention of disease-causing bacteria or other microorganisms.

Threatened species, according to the Act, "means any species which is likely to become an endangered species within the foreseeable future throughout all or a significant portion of its range."

A CLOSER LOOK 13.2

Reasons for Conserving Endangered Species—and All Life on Earth

Important reasons for conserving endangered species are of two types: those having to do with tangible qualities and those dealing with intangible ones (see Chapter 7 for an explanation of tangible and intangible qualities). The tangible ones are utilitarian and ecological. The intangible are aesthetic, moral, recreational, spiritual, inspirational, and cultural.[18]

Utilitarian Justification

Many of the arguments for conserving endangered species, and for conserving biological diversity in general, have focused on the utilitarian justification: that many wild species have proved useful to us and many more may yet prove useful now or in the future, and therefore we should protect every species from extinction.

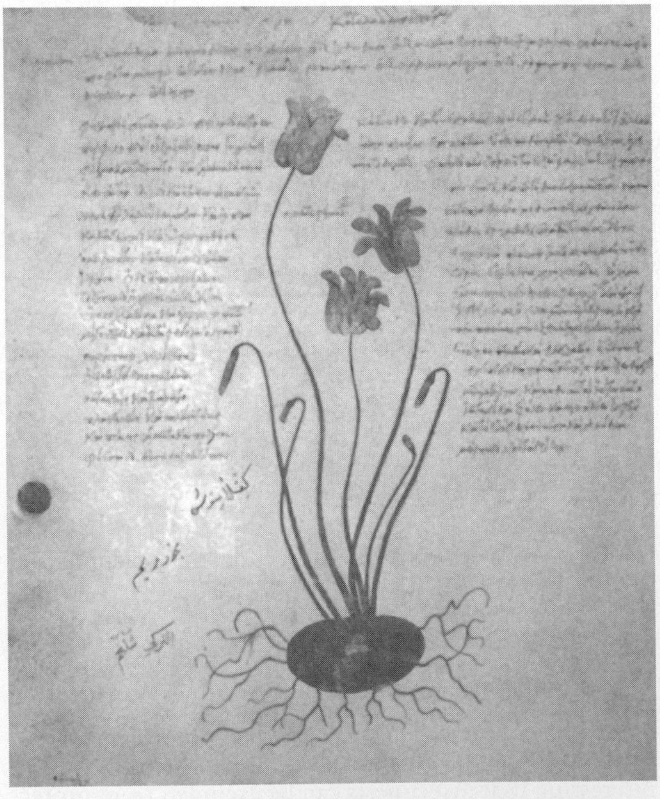

FIGURE 13.12 **Sowbread** (*Sow cyclamen*), a small flowering plant, was believed useful medically at least 1,500 years ago, when this drawing of it appeared in a book published in Constantinople. Whether or not it is medically useful, the plant illustrates the ancient history of interest in medicinal plants. (*Source*: James J. O'Donnell, *The Ruin of the Roman Empire* [New York: ECCO (HarperCollins), 2008], from *Materia Medica* by Dioscorides.)

One example is the need to conserve wild strains of grains and other crops because disease organisms that attack crops evolve continually, and as new disease strains develop, crops become vulnerable. Crops such as wheat and corn depend on the continued introduction of fresh genetic characteristics from wild strains to create new, disease-resistant genetic hybrids. Related to this justification is the possibility of finding new crops among the many species of plants (see Chapter 11).

Another utilitarian justification is that many important chemical compounds come from wild organisms. Medicinal use of plants has an ancient history, going back into human prehistory. For example, a book titled *Materia Medica*, about the medicinal use of plants, was written in the 6th century A.D. in Constantinople by a man named Dioscorides (Figure 13.12).[19] To avoid scurvy, Native Americans advised early European explorers to chew on the bark of eastern hemlock trees (*Tsuga canadensis*); we know today that this was a way to get a little vitamin C.

Digitalis, an important drug for treating certain heart ailments, comes from purple foxglove, and aspirin is a derivative of willow bark. A more recent example was the discovery of a cancer-fighting chemical, paclitaxel, in the Pacific yew tree (genus name *Taxus*; hence the trade name Taxol). Well-known medicines derived from tropical forests include anticancer drugs from rosy periwinkles, steroids from Mexican yams, antihypertensive drugs from serpentwood, and antibiotics from tropical fungi.[20] Some 25% of prescriptions dispensed in the United States today contain ingredients extracted from vascular plants,[21] and these represent only a small fraction of the estimated 500,000 existing plant species. Other plants and organisms may produce useful medical compounds that are as yet unknown.

Scientists are testing marine organisms for use in pharmaceutical drugs. Coral reefs offer a promising area of study for such compounds because many coral-reef species produce toxins to defend themselves. According to the National Oceanic and Atmospheric Administration (NOAA), "Creatures found in coral ecosystems are important sources of new medicines being developed to induce and ease labor; treat cancer, arthritis, asthma, ulcers, human bacterial infections, heart disease, viruses, and other diseases; as well as sources of nutritional supplements, enzymes, and cosmetics."[22, 23]

Some species are also used directly in medical research. For example, the armadillo, one of only two animal species (the other is us) known to contract leprosy, is used to study cures for that disease. Other animals, such as horseshoe crabs and barnacles, are important because of physiologically active compounds they make. Still others may have similar uses as yet unknown to us.

Tourism provides yet another utilitarian justification. Ecotourism is a growing source of income for many countries. Ecotourists value nature, including its endangered species, for aesthetic or spiritual reasons, but the result can be utilitarian.

Ecological Justification

When we reason that organisms are necessary to maintain the functions of ecosystems and the biosphere, we are using an ecological justification for conserving these organisms. Individual species, entire ecosystems, and the biosphere provide public-service functions essential or important to the persistence of life, and as such they are indirectly necessary for our survival. When bees pollinate flowers, for example, they provide a benefit to us that would be costly to replace with human labor. Trees remove certain pollutants from the air; and some soil bacteria fix nitrogen, converting it from molecular nitrogen in the atmosphere to nitrate and ammonia that can be taken up by other living things. That some such functions involve the entire biosphere reminds us of the global perspective on conserving nature and specific species.

Aesthetic Justification

An aesthetic justification asserts that biological diversity enhances the quality of our lives by providing some of the most beautiful and appealing aspects of our existence. Biological diversity is an important quality of landscape beauty. Many organisms—birds, large land mammals, and flowering plants, as well as many insects and ocean animals—are appreciated for their beauty. This appreciation of nature is ancient. Whatever other reasons Pleistocene people had for creating paintings in caves in France and Spain, their paintings of wildlife, done about 14,000 years ago, are beautiful. The paintings include species that have since become extinct, such as mastodons. Poetry, novels, plays, paintings, and sculpture often celebrate the beauty of nature. It is a very human quality to appreciate nature's beauty and is a strong reason for the conservation of endangered species.

Moral Justification

Moral justification is based on the belief that species have a right to exist, independent of our need for them; consequently, in our role as global stewards, we are obligated to promote the continued existence of species and to conserve biological diversity. This right to exist was stated in the U.N. General Assembly World Charter for Nature, 1982. The U.S. Endangered Species Act also includes statements concerning the rights of organisms to exist. Thus, a moral justification for the conservation of endangered species is part of the intent of the law.

Moral justification has deep roots within human culture, religion, and society. Those who focus on cost-benefit analyses tend to downplay moral justification, but although it may not seem to have economic ramifications, in fact it does. As more and more citizens of the world assert the validity of moral justification, more actions that have economic effects are taken to defend a moral position.

The moral justification has grown in popularity in recent decades, as indicated by the increasing interest in the deep-ecology movement. Arne Næss, one of its principal philosophers, explains: "The right of all the forms [of life] to live is a universal right which cannot be quantified. No single species of living being has more of this particular right to live and unfold than any other species."[24]

Cultural Justification

Certain species, some threatened or endangered, are of great importance to many indigenous peoples, who rely on these species of vegetation and wildlife for food, shelter, tools, fuel, materials for clothing, and medicine. Reduced biological diversity can severely increase the poverty of these people. For poor indigenous people who depend on forests, there may be no reasonable replacement except continual outside assistance, which development projects are supposed to eliminate. Urban residents, too, share in the benefits of biological diversity, even if these benefits may not be apparent or may become apparent too late.

Other Intangible Justifications: Recreational, Spiritual, Inspirational

As any mountain biker, scuba diver, or surfer will tell you, the outdoors is great for recreation, and the more natural, the better. Beyond improving muscle tone and cardiovascular strength, many people find a spiritual uplifting and a connectedness to nature from the outdoors, especially where there is a lot of diversity of living things. It has inspired poets, novelists, painters, and even scientists.

13.5 How a Species Becomes Endangered and Extinct

Extinction is the rule of nature (see the discussion of biological evolution in Chapter 7). **Local extinction** means that a species disappears from a part of its range but persists elsewhere. **Global extinction** means a species can no longer be found anywhere. Although extinction is the ultimate fate of all species, the rate of extinctions has varied greatly over geologic time and has accelerated since the Industrial Revolution.

From 580 million years ago until the beginning of the Industrial Revolution, about one species per year, on average, became extinct. Over much of the history of life on Earth, the rate of evolution of new species equaled or slightly exceeded the rate of extinction. The average longevity of a species has been about 10 million years.[25] However, as discussed in Chapter 7, the fossil record suggests that there have been periods of catastrophic losses of species and other periods of rapid evolution of new species (see Figures 13.13 and 13.14), which some refer to as "punctuated extinctions." About 250 million years ago, a mass extinction occurred in which approximately 53% of marine animal species disappeared; and about 65 million years ago, most of the dinosaurs became extinct. Interspersed with the episodes of mass extinctions, there seem to have been periods of hundreds of thousands of years with comparatively low rates of extinction.

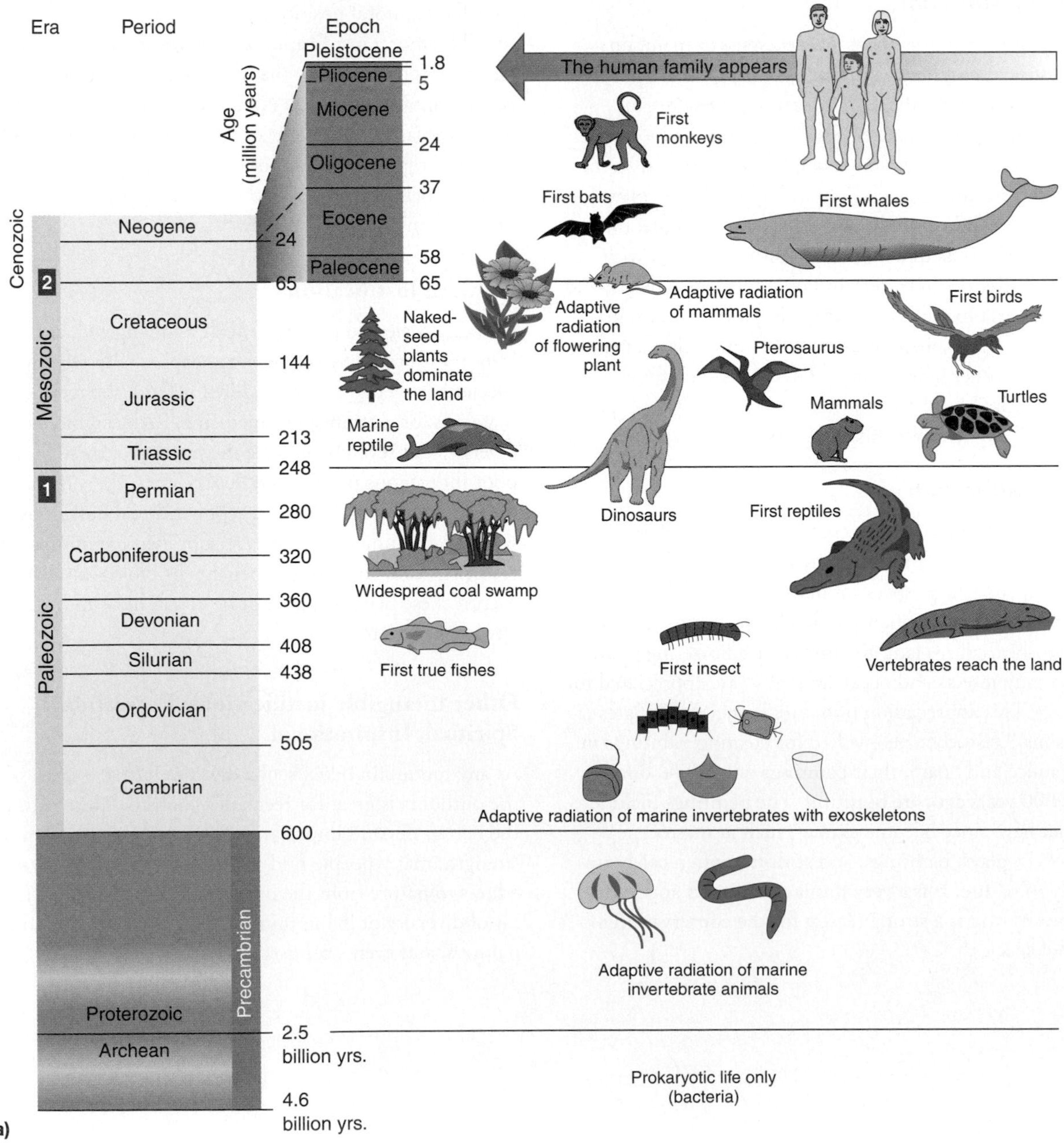

(a)

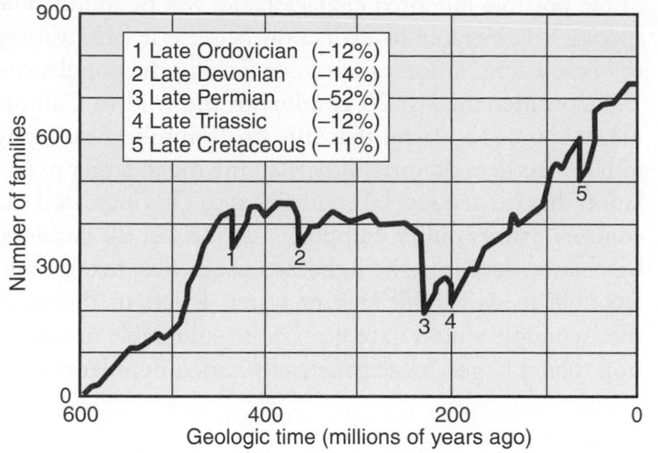

(b)

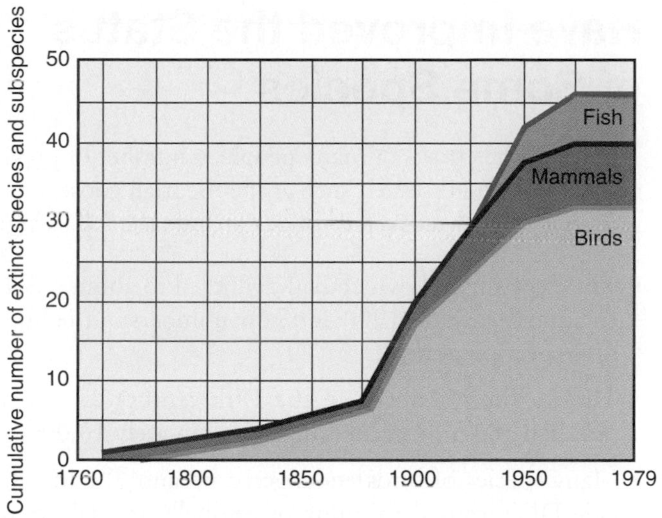

(c)

FIGURE 13.13 **(a) A brief diagrammatic history of evolution and extinction of life on Earth.** There have been periods of rapid evolution of new species and episodes of catastrophic losses of species. Two major catastrophes were the Permian loss, which included 52% of marine animals, as well as land plants and animals, and the Cretaceous loss of dinosaurs. **(b) Graph of the number of families of marine animals in the fossil records,** showing long periods of overall increase in the number of families punctuated by brief periods of major declines. **(c) Extinct vertebrate species and subspecies, 1760–1979.** The number of species becoming extinct increases rapidly after 1860. Note that most of the increase is due to the extinction of birds. (*Sources*: [a] D.M. Raup, "Diversity Crisis in the Geological Past," in E.O. Wilson, ed., *Biodiversity* [Washington, DC: National Academy Press, 1988], p. 53; derived from S.M. Stanley, *Earth and Life through Time* [New York: W.H. Freeman, 1986]. Reprinted with permission. [b] D.M. Raup and J.J. Sepkoski Jr., "Mass Extinctions in the Marine Fossil Record," *Science* 215 [1982]:1501–1502. [c] Council on Environmental Quality; additional data from B. Groombridge England: IUCN, 1993].)

An intriguing example of punctuated extinctions occurred about 10,000 years ago, at the end of the last great continental glaciation. At that time, massive extinctions of large birds and mammals occurred: 33 genera of large mammals—those weighing 50 kg (110 lb) or more—became extinct, whereas only 13 genera had become extinct in the preceding 1 or 2 million years (Figure 13.13a). Smaller mammals were not as affected, nor were marine mammals. As early as 1876, Alfred Wallace, an English biological geographer, noted that "we live in a zoologically impoverished world, from which all of the hugest, and fiercest, and strangest forms have recently disappeared." It has been suggested that these sudden extinctions coincided with the arrival, on different continents at different times, of Stone Age people and therefore may have been caused by hunting.[26]

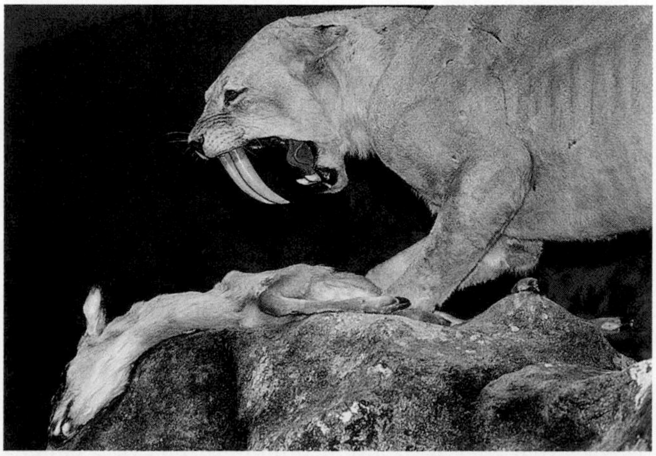

FIGURE 13.14 **Artist's restoration of an extinct saber-toothed cat with prey.** The cat is an example of one of the many large mammals that became extinct about 10,000 years ago.

Causes of Extinction

Causes of extinction are usually grouped into four risk categories: population risk, environmental risk, natural catastrophe, and genetic risk. *Risk* here means the chance that a species or population will become extinct owing to one of these causes.

Population Risk

Random variations in population rates (birth rates and death rates) can cause a species in low abundance to become extinct. This is termed *population risk*. For example, blue whales swim over vast areas of ocean. Because whaling once reduced their total population to only several hundred individuals, the success of individual blue whales in finding mates probably varied from year to year. If in one year most whales were unsuccessful in finding mates, then births could be dangerously low. Such random variation in populations, typical among many species, can occur without any change in the environment. It is a risk especially to species that consist of only a single population in one habitat. Mathematical models of population growth can help calculate the population risk and determine the minimum viable population size.

Environmental Risk

Population size can be affected by changes in the environment that occur from day to day, month to month, and year to year, even though the changes are not severe enough to be considered environmental catastrophes. Environmental risk involves variation in the physical or biological environment, including variations in predator, prey, symbiotic species, or competitor species. Some species are so rare and isolated that such normal variations can lead to their extinction.

For example, Paul and Anne Ehrlich described the local extinction of a population of butterflies in the Colorado mountains.[27] These butterflies lay their eggs in the unopened buds of a single species of lupine (a member of the legume family), and the hatched caterpillars feed on the flowers. One year, however, a very late snow and freeze killed all the lupine buds, leaving the caterpillars without food and causing local extinction of the butterflies. Had this been the only population of that butterfly, the entire species would have become extinct.

Natural Catastrophe

A sudden change in the environment that is not caused by human action is a natural catastrophe. Fires, major storms, earthquakes, and floods are natural catastrophes on land; changes in currents and upwellings are ocean catastrophes. For example, the explosion of a volcano on the island of Krakatoa in Indonesia in 1883 caused one of recent history's worst natural catastrophes. Most of the island was blown to bits, bringing about local extinction of most life-forms there.

Genetic Risk

Detrimental change in genetic characteristics, not caused by external environmental changes, is called *genetic risk*. Genetic changes can occur in small populations from reduced genetic variation and from genetic drift and mutation (see Chapter 8). In a small population, only some of the possible inherited characteristics will be found. The species is vulnerable to extinction because it lacks variety or because a mutation can become fixed in the population.

Consider the last 20 condors in the wild in California. It stands to reason that this small number was likely to have less genetic variability than the much larger population that existed several centuries ago. This increased the condors' vulnerability. Suppose that the last 20 condors, by chance, had inherited characteristics that made them less able to withstand lack of water. If left in the wild, these condors would have been more vulnerable to extinction than a larger, more genetically varied population.

13.6 The Good News: We Have Improved the Status of Some Species

Thanks to the efforts of many people, a number of previously endangered species, such as the Aleutian goose, have recovered. Other recovered species include the following:

- The elephant seal, which had dwindled to about a dozen animals around 1900 and now numbers in the hundreds of thousands.
- The sea otter, reduced in the 19th century to several hundred and now numbering approximately 10,000.
- Many species of birds endangered because the insecticide DDT caused thinning of eggshells and failure of reproduction. With the elimination of DDT in the United States, many bird species recovered, including the bald eagle, brown pelican, white pelican, osprey, and peregrine falcon.
- The blue whale, thought to have been reduced to about 400 when whaling was still actively pursued by a number of nations. Today, 400 blue whales are sighted annually in the Santa Barbara Channel along the California coast, a sizable fraction of the total population.
- The gray whale, which was hunted to near-extinction but is now abundant along the California coast and in its annual migration to Alaska.

Since the U.S. Endangered Species Act became law in 1973, 13 species within the United States have officially recovered (Table 13.3), according to the U.S. Fish and Wildlife Service, which has also "delisted" from protection of the Act 9 species because they have gone extinct, and 17 because they were listed in error or because it was decided that they were not a unique species or a genetically significant unit within a species. (The Act allows listing of subspecies so genetically different from the rest of the species that they deserve protection.)[28]

Table 13.3 RECOVERED SPECIES IN THE UNITED STATES

	DATE SPECIES		
N	FIRST LISTED	DATE DELISTED	SPECIES NAME
1	7/27/1979	6/4//1987	Alligator, American (Alligator mississippiensis)
2	9/17/1980	8/27/2002	Cinquefoil, Robbins' (Potentilla robbinsiana)
3	7/24/2003	7/24/2003	Deer, Columbian white-tailed Douglas County DPS (Odocoileus Virginianus leucurus)
4	3/11/1967	7/9/2007	Eagle, bald lower 48 states (Haliaeetus leucocephalus)
5	6/2/1970	8/25/1999	Falcon, Amercian peregrine (Falco peregrinus anatum)
6	6/2/1970	10/5/1994	Falcon, Arctic peregrine (Falco peregrinus tundrius)
7	3/11/1967	3/20/2001	Goose, Aleutian Canada (Branta Canadensis leucopareia)
8	6/2/1970	2/4/1985	Pelican brown U.S. Atlantic coast, FL, AL (Pelecanus occidentails)
9	7/1/1985	8/26/2008	Squirrel, Virginia northern flying (Glaucomys sabrinus fuscus)
10	5/22/1997	8/18/2005	Sunflower, Eggert's (Helianthus eggertii)
11	6/16/1994	6/16/1994	Whale, gray except where listed (Eschrichtius robustus)
12	3/28/2008	4/2/2009	Wolf, gray Northern Rocky Mountain DPS (Canis lupus)
13	7/19/1990	10/7/2003	Woolly-star, Hoover's (Eriastrum hooveri)

Source: U.S. Fish and Wildlife Service, Delisting Report, http://ecos.fws.gov/tess_public/DelistingReport.do.

A CLOSER LOOK 13.3

Conservation of Whales and Other Marine Mammals

Fossil records show that all marine mammals were originally inhabitants of the land. During the last 80 million years, several separate groups of mammals returned to the oceans and underwent adaptations to marine life. Each group of marine mammals shows a different degree of transition to ocean life. Understandably, the adaptation is greatest for those that began the transition longest ago. Some marine mammals—such as dolphins, porpoises, and great whales—complete their entire life cycle in the oceans and have organs and limbs that are highly adapted to life in the water; they cannot move on the land. Others, such as seals and sea lions, spend part of their time on shore.

Cetaceans

Whales

Whales fit into two major categories: baleen and toothed (Figure 13.15a and b). The sperm whale is the only great whale that is toothed; the rest of the toothed group are smaller whales, dolphins, and porpoises. The other great whales, in the baleen group, have highly modified teeth that look like giant combs and act as water filters. Baleen whales feed by filtering ocean plankton.

Drawings of whales have been dated as early as 2200 B.C.[29] Eskimos used whales for food and clothing as long ago as 1500 B.C. In the 9th century, whaling by Norwegians was reported by travelers whose accounts were written down in the court of the English king Alfred. The earliest whale hunters killed these huge mammals from the shore or from small boats near shore, but gradually whale hunters ventured farther out. In the 11th and 12th centuries, Basques hunted the Atlantic right whale from open boats in the Bay of Biscay, off the western coast of France. Whales were brought ashore for processing, and the boats returned to land once the search for whales was finished.

The Industrial Revolution eventually made whaling pelagic—almost entirely conducted on the open sea. Whalers searched for whales from ships that remained at sea for long periods, and the whales were brought on board and processed

(a) (b)

FIGURE 13.15 **(a) A sperm whale; (b) a blue whale.**

Table 13.4 ESTIMATES OF THE NUMBER OF WHALES

Whales are difficult to count, and the wide range of estimates indicates this difficulty. Of the baleen whales, the most numerous are the smallest—the minke and the pilot. The only toothed great whale, the sperm whale, is thought to be relatively numerous, but actual counts of this species have been made only in comparatively small areas of the ocean.

	RANGE OF ESTIMATES	
SPECIES	MINIMUM	MAXIMUM
Blue	400	1,400
Bowhead	6,900	9,200
Fin	27,700	82,000
Gray	21,900	32,400
Humpback	5,900	16,800
Minke	510,000	1,140,000
Pilot	440,000	1,370,000
Sperm	200,000	1,500,000

Source: International Whaling Commission, August 29, 2006, http://www.iwcoffice.org/conservation/estimate.htm. The estimate for sperm whale is from U.S. NOAA http://www.nmfs.noaa.gov/pr/species/mammals/cetaceans/spermwhale.htm

there by newly invented furnaces and boilers for extracting whale oil at sea. With these inventions, whaling grew as an industry. American fleets developed in the 18th century in New England, and by the 19th century the United States dominated the industry, providing most of the whaling ships and even more of the crews.[30, 31]

Whales provided many 19th-century products. Whale oil was used for cooking, lubrication, and lamps. Whales provided the main ingredients for the base of perfumes. The elongated teeth (whale-bone or baleen) that enable baleen whales to strain the ocean waters for food are flexible and springy and were used for corset stays and other products before the invention of inexpensive steel springs.

Although the 19th-century whaling ships were made famous by novels such as *Moby Dick*, more whales were killed in the 20th century than in the 19th. The resulting worldwide decline in most species of whales made this a global environmental issue.

Conservationists have been concerned about whales for many years. Attempts to control whaling began with the League of Nations in 1924. The first agreement, the Convention for the Regulation of Whaling, was signed by 21 countries in 1931. In 1946 a conference in Washington, DC, initiated the International Whaling Commission (IWC), and in 1982 the IWC established a moratorium on commercial whaling. Currently, 12 of approximately 80 species of whales are protected.[32]

The IWC has played a major role in reducing (almost eliminating) the commercial harvesting of whales. Since its formation, no whale species has become extinct, the total take of whales has decreased, and harvesting of whale species considered endangered has ceased. Endangered species protected from hunting have had a mixed history (see Table 13.4). Blue whales appear to have recovered somewhat but remain rare and endangered. Gray whales are now relatively abundant, numbering about 26,000.[34] However, global climate change, pollution, and ozone depletion now pose greater risks to whale populations than does whaling.

The establishment of the IWC was not only vitally important to whales but also a major landmark in wildlife conservation. It was one of the first major attempts by a group of nations to agree on a reasonable harvest of a biological resource. The annual meeting of the IWC has become a forum for discussing international conservation, working out basic concepts of maximum and optimum sustainable yields, and formulating a scientific basis for commercial harvesting. The

IWC demonstrates that even an informal commission whose decisions are accepted voluntarily by nations can function as a powerful force for conservation.

In the past, each marine mammal population was treated as if it were isolated, had a constant supply of food, and was subject only to the effects of human harvesting. That is, its growth was assumed to follow the logistic curve. We now realize that management policies for marine mammals must be expanded to include ecosystem concepts and the understanding that populations interact in complex ways.

The goal of marine mammal management is to prevent extinction and maintain large population sizes rather than to maximize production. For this reason, the Marine Mammal Protection Act, enacted by the United States in 1972, has as its goal an optimum sustainable population (OSP) rather than a maximum or an optimum sustainable yield. An OSP is the largest population that can be sustained indefinitely without deleterious effects on the ability of the population or its ecosystem to continue to support that same level.

Some of the great whales remain rare.

Dolphins and Other Small Cetaceans

Among the many species of small "whales" are dolphins and porpoises, more than 40 species of which have been hunted commercially or have been killed inadvertently by other fishing efforts.[33] A classic case is the inadvertent catch of the spinner, spotted, and common dolphins of the eastern Pacific. Because these carnivorous, fish-eating mammals often feed with yellowfin tuna, a major commercial fish, more than 7 million dolphins have been netted and killed inadvertently in the past 40 years.[34]

The U.S. Marine Mammal Commission and commercial fishermen have cooperated in seeking ways to reduce dolphin mortality. Research into dolphin behavior helped in the design of new netting procedures that trapped far fewer dolphins. The attempt to reduce dolphin mortality illustrates cooperation among fishermen, conservationists, and government agencies and indicates the role of scientific research in managing renewable resources.

13.7 Can a Species Be Too Abundant? If So, What to Do?

All marine mammals are protected in the United States by the Federal Marine Mammal Protection Act of 1972, which has improved the status of many marine mammals. Sometimes, however, we succeed too well in increasing the populations of a species. Case in point: Sea lions now number more than 190,000 and have become so abundant as to be local problems.[35, 36] In San Francisco Harbor and in Santa Barbara Harbor, for example, sea lions haul out and sun themselves on boats and pollute the water with their excrement near shore. In one case, so many hauled out on a sailboat in Santa Barbara Harbor that they sank the boat, and some of the animals were trapped and drowned.

Mountain lions, too, have become locally overabundant. In the 1990s, California voters passed an initiative that protected the endangered mountain lion but contained no provisions for managing the lion if it became overabundant, unless it threatened human life and property. Few people thought the mountain lion could ever recover enough to become a problem, but in several cases in recent years mountain lions have attacked and even killed people. Current estimates suggest there may be as many as 4,000 to 6,000 in California.[37] These attacks become more frequent as the mountain lion population grows and as the human population grows and people build houses in what was mountain lion habitat.

13.8 How People Cause Extinctions and Affect Biological Diversity

People have become an important factor in causing species to become threatened, endangered, and finally extinct. We do this in several ways:

- By intentional hunting or harvesting (for commercial purposes, for sport, or to control a species that is considered a pest).
- By disrupting or eliminating habitats.
- By introducing exotic species, including new parasites, predators, or competitors of a native species.
- By creating pollution.

People have caused extinctions over a long time, not just in recent years. The earliest people probably caused extinctions through hunting. This practice continues, especially for specific animal products considered valuable, such as elephant ivory and rhinoceros horns. When people learned to use fire, they began to change habitats over large areas. The development of agriculture and the rise of civilization led to rapid deforestation and other habitat changes. Later, as people explored new areas, the introduction of exotic species became a greater cause of extinction (see Chapter 8), especially after Columbus's voyage to the New World, Magellan's circumnavigation of the globe, and the resulting spread of European civilization and technology. The introduction of thousands of novel chemicals into the environment made pollution an increasing cause of extinction in the 20th century, and pollution control has proved a successful way to help species.

The IUCN estimates that 75% of the extinctions of birds and mammals since 1600 have been caused by human beings. Hunting is estimated to have caused 42% of the extinctions of birds and 33% of the extinctions of mammals. The current extinction rate among most groups of mammals is estimated to be 1,000 times greater than the extinction rate at the end of the Pleistocene epoch.[22]

13.9 Ecological Islands and Endangered Species

The history of the Kirtland's warbler illustrates that a species may inhabit "ecological islands," which the isolated jack-pine stands of the right age range are for that bird. Recall from our discussion in Chapter 8 that an ecological island is an area that is biologically isolated, so that a species living there cannot mix (or only rarely mixes) with any other population of the same species (Figure 13.16). Mountaintops and isolated ponds are ecological islands. Real geographic islands may also be ecological islands. Insights gained from studies of the biogeography of islands have important implications for the conservation of endangered species and for the design of parks and preserves for biological conservation.

Almost every park is a biological island for some species. A small city park between buildings may be an island for trees and squirrels. At the other extreme, even a large national park is an ecological island. For example, the Masai Mara Game Reserve in the Serengeti Plain, which stretches from Tanzania to Kenya in East Africa, and other great wildlife parks of eastern and southern Africa are becoming islands of natural landscape surrounded by human settlements. Lions and other great cats exist in these parks as isolated populations, no longer able to roam completely freely and to mix over large areas. Other examples are islands of uncut forests left by logging operations, and oceanic islands, where intense fishing has isolated parts of fish populations.

How large must an ecological island be to ensure the survival of a species? The size varies with the species but can be estimated. Some islands that seem large to us are too small for species we wish to preserve. For example, a preserve was set aside in India in an attempt to reintroduce the Indian lion into an area where it had been eliminated by hunting and by changing patterns of land use. In 1957, a male and two females were introduced into a 95 km^2 (36 mi^2) preserve in the Chakia forest, known as the Chandraprabha Sanctuary. The introduction was carried out carefully and the population was counted annually. There were four lions in 1958, five in 1960, seven in 1962, and eleven in 1965, after which they disappeared and were never seen again.

Why did they go? Although 95 km^2 seems large to us, male Indian lions have territories of 130 km^2

FIGURE 13.16 **Ecological islands: (a)** Central Park in New York City; **(b)** a mountaintop in Arizona where there are bighorn sheep; **(c)** an African wildlife park.

(50 mi^2), within which females and young also live. A population that could persist for a long time would need a number of such territories, so an adequate preserve would require 640–1,300 km^2 (247–500 mi^2). Various other reasons were also suggested for the disappearance of the lions, including poisoning and shooting by villagers. But regardless of the immediate cause, a much larger area was required for long-term persistence of the lions.

13.10 Using Spatial Relationships to Conserve Endangered Species

The red-cockaded woodpecker (Figure 13.17a) is an endangered species of the American Southeast, numbering approximately 15,000.[38] The woodpecker makes its nests in old dead or dying pines, and one of its foods is the pine bark beetle (Figure 13.17b). To conserve this species of woodpecker, these pines must be preserved. But the old pines are home to the beetles, which are pests to the trees and damage them for commercial logging. This presents an intriguing problem: How can we maintain the woodpecker and its food (which includes the pine bark beetle) and also maintain productive forests?

The classic 20th-century way to view the relationship among the pines, the bark beetle, and the woodpecker would be to show a food chain (see Chapter 6). But this alone does not solve the problem for us. A newer approach is to consider the habitat requirements of the pine bark beetle and the woodpecker. Their requirements are somewhat different, but if we overlay a map of one's habitat requirements over a map of the other's, we can compare the co-occurrence of habitats. Beginning with such maps, it becomes possible to design a landscape that would allow the maintenance of all three—pines, beetles, and birds.

(a)

(b)

FIGURE 13.17 (a) Endangered red-cockaded woodpecker, and (b) the pine bark beetle, food for the woodpecker.

CRITICAL THINKING ISSUE

Should Wolves Be Reestablished in the Adirondack Park?

With an area slightly over 24,000 km^2, the Adirondack Park in northern New York is the largest park in the lower 48 states. Unlike most parks, however, it is a mixture of private (60%) and public (40%) land and is home to 130,000 people. When European colonists first came to the area, it was, like much of the rest of North America, inhabited by gray wolves. By 1960, wolves had been exterminated in all of the lower 48 states except for northern Minnesota. The last official sighting of a wolf in the Adirondacks was in the 1890s.

Although the gray wolf was not endangered globally—there were more than 60,000 in Canada and Alaska—it was one of the first animals listed as endangered under the 1973 Endangered Species Act. As required, the U.S. Fish and Wildlife Service developed a plan for recovery that included protection of the existing population and reintroduction of wolves to wilderness areas. The recovery plan would be considered a success if survival of the Minnesota wolf population was assured and at least one other population of more than 200 wolves had been established at least 320 km from the Minnesota population.

Under the plan, Minnesota's wolf population increased, and some wolves from that population, as well as others from southern Canada, dispersed into northern Michigan and Wisconsin, each of which had populations of approximately 100 wolves in 1998. Also, 31 wolves from Canada were introduced into Yellowstone National Park in 1995, and that population grew to over 100. By the end of 1998, it seemed fairly certain that the criteria for removing the wolf from the Endangered Species list would soon be met.

In 1992, when the results of the recovery plan were still uncertain, the Fish and Wildlife Service proposed to investigate the possibility of reintroducing wolves to northern Maine and the Adirondack Park. A survey of New York State residents in 1996 funded by Defenders of Wildlife found that 76% of people living in the park supported reintroduction. However, many residents and organizations within the park vigorously opposed reintroduction and questioned the validity of the survey. Concerns focused primarily on the potential dangers to people, livestock, and pets and the possible impact on the deer population. In response to the public outcry, Defenders of Wildlife established a citizens' advisory committee that initiated two studies by outside experts, one on the social and economic aspects of reintroduction and another on whether there were sufficient prey and suitable habitat for wolves.

Wolves prey primarily on moose, deer, and beaver. Moose have been returning to the Adirondacks in recent years and now number about 40, but this is far less than the moose population in areas where wolves are successfully reestablishing. Beaver are abundant in the Adirondacks, with an estimated population of over 50,000. Because wolves feed on beaver primarily in the spring and the moose population is small, the main food source for Adirondack wolves would be deer.

Deer thrive in areas of early-successional forest and edge habitats, both of which have declined in the Adirondacks as logging has decreased on private forestland and has been eliminated altogether on public lands. Furthermore, the Adirondacks are at the northern limit of the range for white-tailed deer, where harsh winters can result in significant mortality. Deer density in the Adirondacks is estimated at 3.25 per square kilometer, fewer than in the wolf habitat in Minnesota, which also has 8,500 moose. If deer were the only prey available, wolves would kill between 2.5% and 6.5% of the deer population, while hunters take approximately 13% each year. Determining whether there is a sufficient prey base to support a population of wolves is complicated by the fact that coyotes have moved into the Adirondacks and occupy the niche once filled by wolves. Whether wolves would add to the deer kill or replace coyotes, with no net impact on the deer population, is difficult to predict.

An area of 14,000 km^2 in various parts of the Adirondack Park meets criteria established for suitable wolf habitat, but this is about half of the area required to maintain a wolf population for the long term. Based on the average deer density and weight, as well as the food requirements of wolves, biologists estimate that this habitat could support about 155 wolves. However, human communities are scattered throughout much of the park, and many residents are concerned that wolves would not remain on public land and would threaten local residents as well as back-country hikers and hunters. Also, private lands around the edges of the park, with their greater density of deer, dairy cows, and people, could attract wolves.

Critical Thinking Questions

1. Who should make decisions about wildlife management, such as returning wolves to the Adirondacks—scientists, government officials, or the public?
2. Some people advocate leaving the decision to the wolves—that is, waiting for them to disperse from southern Canada and Maine into the Adirondacks. Study a map of the northeastern United States and southeastern Canada. What do you think is the likelihood of natural recolonization of the Adirondacks by wolves?
3. Do you think wolves should be reintroduced to the Adirondack Park? If you lived in the park, would that affect your opinion? How would removal of the wolf from the Endangered Species list affect your opinion?
4. Some biologists recently concluded that wolves in Yellowstone and the Great Lakes region belong to a different subspecies, the Rocky Mountain timber wolf, from those that formerly lived in the northeastern United States, the eastern timber wolf. This means that the eastern timber wolf is still extinct in the lower 48 states. Would this affect your opinion about reintroducing wolves into the Adirondacks?

SUMMARY

- Modern approaches to management and conservation of wildlife use a broad perspective that considers interactions among species as well as the ecosystem and landscape contexts.
- To successfully manage wildlife for harvest, conservation, and protection from extinction, we need certain quantitative information about the wildlife population, including measures of total abundance and of births and deaths, preferably recorded over a long period. The age structure of a population can also help. In addition, we have to characterize the habitat in quantitative terms. However, it is often difficult to obtain these data.

- A common goal of wildlife conservation today is to "restore" the abundance of a species to some previous number, usually a number thought to have existed prior to the influence of modern technological civilization. Information about long-ago abundances is rarely available, but sometimes we can estimate numbers indirectly—for example, by using the Lewis and Clark journals to reconstruct the 1805 population of grizzly bears, or using logbooks from whaling ships. Adding to the complexity, wildlife abundances vary over time even in natural systems uninfluenced by modern civilization. Also, historical information often cannot be subjected to formal tests of disproof and therefore does not by itself qualify as scientific. Adequate information exists for relatively few species.
- Another approach is to seek a minimum viable population, a carrying capacity, or an optimal sustainable population or harvest based on data that can be obtained and tested today. This approach abandons the goal of restoring a species to some hypothetical past abundance.
- The good news is that many formerly endangered species have been restored to an abundance that suggests they are unlikely to become extinct. Success is achieved when the habitat is restored to conditions required by a species. The conservation and management of wildlife present great challenges but also offer great rewards of long-standing and deep meaning to people.

REEXAMINING THEMES AND ISSUES

Human Population

Human beings are a primary cause of species extinctions today and also contributed to extinctions in the past. Nonindustrial societies have caused extinction by such activities as hunting and the introduction of exotic species into new habitats. With the age of exploration in the Renaissance and with the Industrial Revolution, the rate of extinctions accelerated. People altered habitats more rapidly and over greater areas. Hunting efficiency increased, as did the introduction of exotic species into new habitats. As the human population grows, conflicts over habitat between people and wildlife increase. Once again, we find that the human population problem underlies this environmental issue.

Sustainability

At the heart of issues concerning wild living resources is the question of sustainability of species and their ecosystems. One of the key questions is whether we can sustain these resources at a constant abundance. In general, it has been assumed that fish and other wildlife that are hunted for recreation, such as deer, could be maintained at some constant, highly productive level. Constant production is desirable economically because it would provide a reliable, easily forecast income each year. But despite attempts at direct management, few wild living resources have remained at constant levels. New ideas about the intrinsic variability of ecosystems and populations lead us to question the assumption that such resources can or should be maintained at constant levels.

Global Perspective

Although the final extinction of a species takes place in one locale, the problem of biological diversity and the extinction of species is global because of the worldwide increase in the rate of extinction and because of the growth of the human population and its effects on wild living resources.

Urban World

We have tended to think of wild living resources as existing outside of cities, but there is a growing recognition that urban environments will be more and more important in conserving biological diversity. This is partly because cities now occupy many sensitive habitats around the world, such as coastal and inland wetlands. It is also because appropriately designed parks and backyard plantings can provide habitats for some endangered species. As the world becomes increasingly urbanized, this function of cities will become more important.

People and Nature

Wildlife, fish, and endangered species are popular issues. We seem to have a deep feeling of connectedness to many wild animals—we like to watch them, and we like to know that they still exist even when we cannot see them. Wild animals have always been important symbols to people, sometimes sacred ones. The conservation of wildlife of all kinds is therefore valuable to our sense of ourselves, both as individuals and as members of a civilization.

Science and Values

The reasons that people want to save endangered species begin with human values, including values placed on the continuation of life and on the public-service functions of ecosystems. Among the most controversial environmental issues in terms of values are the conservation of biological diversity and the protection of endangered species. Science tells us what is possible with respect to conserving species, which species are likely to persist, and which are not. Ultimately, therefore, our decisions about where to focus our efforts in sustaining wild living resources depend on our values.

KEY TERMS

carrying capacity **261**
catch per unit effort **265**
global extinction **274**
historical range of variation **263**
local extinction **274**
logistic carrying capacity **261**
maximum sustainable yield **261**
maximum-sustainable-yield population **261**
minimum viable population **261**
optimum sustainable population **261**
time series **263**

STUDY QUESTIONS

1. Why are we so unsuccessful in making rats an endangered species?
2. What have been the major causes of extinction (a) in recent times and (b) before people existed on Earth?
3. Refer back to the introductory case study about the American buffalo and grizzly bear. The U.S. Fish and Wildlife Service suggested three key indicators of the status of the grizzly bear: (1) sufficient reproduction to offset the existing levels of human-caused mortality, (2) adequate distribution of breeding animals throughout the area, and (3) a limit on total human-caused mortality. Are these indicators sufficient to assure the recovery of this species? What would you suggest instead?
4. This chapter discussed eight justifications for preserving endangered species. Which of them apply to the following? (You can decide that none apply.)
   (a) the black rhinoceros of Africa
   (b) the Furbish lousewort, a rare small flowering plant of New England, seen by few people
   (c) an unnamed and newly discovered beetle from the Amazon rain forest
   (d) smallpox
   (e) wild strains of potatoes in Peru
   (f) the North American bald eagle
5. Locate an ecological island close to where you live and visit it. Which species are most vulnerable to local extinction?
6. Oysters were once plentiful in the waters around New York City. Create a plan to restore them to numbers that could be the basis for commercial harvest.
7. Using information available in libraries, determine the minimum area required for a minimum viable population of the following:
   (a) domestic cats
   (b) cheetahs
   (c) the American alligator
   (d) swallowtail butterflies
8. Both a ranch and a preserve will be established for the North American bison. The goal of the ranch owner is to show that bison can be a better source of meat than introduced cattle and at the same time have a less detrimental effect on the land. The goal of the preserve is to maximize the abundance of the bison. How will the plans for the ranch and preserve differ, and how will they be similar?

FURTHER READING

Botkin, D.B., *No Man's Garden: Thoreau and a New Vision for Civilization and Nature* (Washington, DC: Island Press, 2001). A work that discusses deep ecology and its implications for biological conservation, as well as reasons for the conservation of nature, both scientific and beyond science.

Caughley, G., and A.R.E. Sinclair, *Wildlife Ecology and Management* (London: Blackwell Scientific, 1994). A valuable textbook based on new ideas of wildlife management.

"Estimating the Abundance of Sacramento River Juvenile Winter Chinook Salmon with Comparisons to Adult Escapement," Final Report Red Bluff Research Pumping Plant Report Series: Volume 5. Prepared by: U.S. Fish and Wildlife Service, Red Bluff Fish and Wildlife Office, 10950 Tyler Road, Red Bluff, CA 96080. Prepared for: U.S. Bureau of Reclamation, Red Bluff Fish Passage Program, P.O. Box 159, Red Bluff, CA 96080, July.

MacKay, R., *The Penguin Atlas of Endangered Species: A Worldwide Guide to Plants and Animals* (New York: Penguin, 2002). A geographic guide to endangered species.

Pauly, D., J. Maclean, and J.L. Maclean, *The State of Fisheries and Ecosystems in the North Atlantic Ocean* (Washington, DC: Island Press, 2002).

Schaller, George B., and Lu Zhi (photographers), ***Pandas in the Wild: Saving an Endangered Species*** (New York: Aperture, 2002). A photographic essay about scientists attempting to save an endangered species.

CHAPTER 14

Energy: Some Basics

Manhattan Island from New Jersey during the blackout of August 14, 2003. During rush hour, millions of people walked dark streets to go home.

LEARNING OBJECTIVES

Understanding the basics of what energy is, as well as the sources and uses of energy, is essential for effective energy planning. After reading this chapter, you should understand . . .

- That energy is neither created nor destroyed but is transformed from one kind to another;
- Why, in all transformations, energy tends to go from a more usable to a less usable form;
- What energy efficiency is and why it is always less than 100%;
- That people in industrialized countries consume a disproportionately large share of the world's total energy, and how efficient use and conservation of energy can help us make better use of global energy resources;
- Why some energy planners propose a business-as-usual approach to energy (based on large power plants using fossil fuels, especially coal), and others a new approach (based on more disseminated and renewable energy sources), and why both of these approaches have positive and negative points;
- Why moving toward global sustainable energy planning with integrated energy planning is an important goal;
- What elements are needed to develop integrated energy planning.

CASE STUDY

National Energy Policy: From Coast-to-Coast Energy Crisis to Promoting Energy Independence

The most serious blackout in U.S. history occurred on August 14, 2003. New York City, along with eight states and parts of Canada, suddenly lost electric power at about 4:00 p.m. More than 50 million people were affected, some trapped in elevators or electric trains underground. People streamed into the streets of New York, unsure whether or not the power failure was due to a terrorist attack. Power was restored within 24 hours to most places, but the event was an energy shock that demonstrated our dependence on aging power distribution systems and centralized electric power generation. Terrorists had nothing to do with the blackout, but the event caused harm, anxiety, and financial loss to millions of people.

Seven presidents of the United States since the mid-1970s have attempted to address energy problems and how to become independent of foreign energy sources. The Energy Policy Act of 2005, passed by Congress and signed into law by President George W. Bush in the summer of 2005, has been followed by heated debate about energy policy in the 21st century. A number of topics related to energy are being discussed, including the American Clean Energy and Security Act of 2009, which took a serious step toward energy self-sufficiency in the United States.

The 2009 Act has four parts: (1) *clean energy,* which involves renewable energy, sequestration of carbon, development of clean fuels and vehicles, and a better electricity transmission grid; (2) *energy efficiency,* for buildings, homes, transportation, and utilities; (3) *reduction of carbon dioxide and other greenhouse gases associated with global warming,* including programs to reduce global warming by reducing emissions of carbon dioxide in coming years; and (4) *making the transition to a clean energy economy,* including economic incentives for development of green-energy jobs, exporting clean technology, increasing domestic competitiveness, and finding ways to adapt to global warming.

Today we are more dependent than ever on imported oil. We import about 65% of our oil, often from countries that do not particularly like us. This presents a security risk. Since the 1970s, U.S. consumption of gasoline (for which most oil is used) has risen 50%, while domestic production of oil has dropped by nearly one-half, due in part to a dramatic 50% reduction in Alaska's oil production since the late 1980s. One result was that gasoline prices rose to a peak of $4 per gallon in 2008.

Natural gas has followed a similar pattern with respect to production and consumption since the late 1980s. New power plants today use natural gas as the desired fuel because it is cleaner-burning, resulting in fewer pollutants, and the United States has abundant potential supplies. The problem with natural gas will be to bring production in line with consumption in the future.

Energy planning at the national level in the first five years of the 21st century was marked by an ongoing debate about future supplies of fossil fuels, including coal, oil, and natural gas. Planning objectives have centered on providing a larger supply of coal, natural gas, and, to a lesser extent, oil. Planners concluded that if the United States is to meet electricity demands by the year 2020, over 1,000 new power plants will have to be constructed. When we work out the numbers, this means building about 60 per year between now and 2020—more than one new facility per week!

The key to energy planning is a diversity of energy sources with a better mix of fossil fuels and alternative sources that must eventually replace them. What is apparent is that in the first decades of the 21st century we are going to be continually plagued by dramatic price changes in energy and accompanying shortages. This pattern will continue until we become much more independent from foreign energy sources. Using our remaining fossil fuels, particularly the cleaner fuels such as natural gas, will represent a transitional phase to more sustainable sources. What is really necessary is a major program to develop sources such as wind and solar much more vigorously than has been done in the past or, apparently, will be done in the next few years. If we are unable to make the transition as world production of petroleum peaks and declines, then we will face an energy crisis unsurpassed in our history.

The United States faces serious energy problems. Energy policy, from local to global, has emerged as a central economic concern, national security issue, and environmental question.[1] How we respond to energy issues will largely define who and what we are and will become in the 21st century. With this in mind, in this chapter we explore some of the basic principles associated with what energy is, how much energy we consume, and how we might manage energy for the future.

14.1 Outlook for Energy

Energy crises are nothing new. People have faced energy problems for thousands of years, as far back as the early Greek and Roman cultures.

Energy Crises in Ancient Greece and Rome

The climate in Greece's coastal areas 2,500 years ago was characterized by warm summers and cool winters, much as it is today. To warm their homes in winter, the Greeks used small, charcoal-burning heaters that were not very efficient. Since charcoal is produced from burning wood, wood was their primary source of energy, as it is today for half the world's people.

By the 5th century B.C., fuel shortages had become common, and much of the forested land in many parts of Greece was depleted of firewood. As local supplies diminished, it became necessary to import wood from farther away. Olive groves became sources of fuel; olive wood was turned into charcoal for burning, reducing a valuable resource. By the 4th century B.C., the city of Athens had banned the use of olive wood for fuel.

At about this time, the Greeks began to build their houses facing south, designing them so that the low winter sun entered the houses, providing heat, and the higher summer sun was partially blocked, cooling the houses. Recent excavations of ancient Greek cities suggest that large areas were planned so that individual homes could make maximum use of solar energy, which was a logical answer to their energy problem.[2]

The use of wood in ancient Rome is somewhat analogous to the use of oil and gas in the United States today. The homes of wealthy Romans about 2,000 years ago had central heating that burned as much as 125 kg (275 lb) of wood every hour. Not surprisingly, local wood supplies were exhausted quickly, and the Romans had to import wood from outlying areas, eventually from as far away as 1,600 km (about 1,000 mi).[2]

The Romans turned to solar energy for the same reasons as the Greeks but with much broader application and success. They used glass windows to increase the effectiveness of solar heat, developed greenhouses to raise food during the winter, and oriented large public bathhouses (some accommodated up to 2,000 people) to use passive solar energy (Figure 14.1). The Romans believed that sunlight in bathhouses was healthy, and it also saved greatly on fuel costs. The use of solar energy in ancient Rome was widespread and resulted in laws to protect a person's right to solar energy. In some areas, it was illegal for one person to construct a building that shaded another's.[2]

The ancient Greeks and Romans experienced an energy crisis in their urban environments. In turning to solar energy, they moved toward what today we call *sustainability*. We are on that same path today as fossil fuels become scarce.

FIGURE 14.1 Roman bathhouse (lower level) in the town of Bath, England. The orientation of the bathhouse and the placement of windows are designed to maximize the benefits of passive solar energy.

Energy Today and Tomorrow

The energy situation facing the United States and the world today is in some ways similar to that faced by the early Greeks and Romans. The use of wood in the United States peaked in the 1880s, when the use of coal became widespread. The use of coal, in turn, began to decline after 1920, when oil and gas started to become available. Today, we are facing the global peak of oil production, which is expected by about 2020. Fossil fuel resources, which took millions of years to form, may be essentially exhausted in just a few hundred years.

The decisions we make today will affect energy use for generations. Should we choose complex, centralized energy-production methods, or simpler and widely dispersed methods, or a combination of the two? Which energy sources should be emphasized? Which uses of energy should be emphasized for increased efficiency? How can we develop a sustainable energy policy? There are no easy answers.

The use of fossil fuels, especially oil, improved sanitation, medicine, and agriculture, helping to make possible the global human population increase that we have discussed in other chapters. Many of us are living longer, with a higher standard of living, than people before us. However, burning fossil fuels imposes growing environmental costs, ranging from urban pollution to a change in the global climate.

One thing certain about the energy picture for tomorrow is that it will involve living with uncertainty when it comes to energy availability and cost. The sources of energy and the patterns of energy use will undoubtedly change. We can expect problems, with growing demand and insufficient supply leading to higher costs. Supplies will continue to be regulated and could be disrupted. Oil embargoes could cause significant economic impact in the United States and other countries, and a war or revolution in a petroleum-producing country would significantly reduce petroleum exports.

It is clear that we need to rethink our entire energy policy in terms of sources, supply, consumption, and environmental concerns. We can begin by understanding basic facts about what energy is.

14.2 Energy Basics

The concept of energy is somewhat abstract: You cannot see it or feel it, even though you have to pay for it.[3] To understand energy, it is easiest to begin with the idea of a force. We all have had the experience of exerting force by pushing or pulling. The strength of a force can be measured by how much it accelerates an object.

What if your car stalls going up a hill and you get out to push it uphill to the side of the road (Figure 14.2)? You apply a force against gravity, which would otherwise cause the car to roll downhill. If the brake is on, the brakes, tires, and bearings may heat up from friction. The longer the distance over which you exert force in pushing the car, the greater the change in the car's position and the greater the amount of heat from friction in the brakes, tires, and bearings. In physicists' terms, exerting the force over the distance moved is work. That is, **work** is the product of a force times a distance. Conversely, energy is the ability to do work. Thus, if you push hard but the car doesn't move, you have exerted a force but have not done any work (according to the definition), even if you feel very tired and sweaty.[3]

In pushing your stalled car, you have moved it against gravity and caused some of its parts (brakes, tires, bearings) to be heated. These effects have something in common: They are forms of energy. You have converted chemical energy in your body to the energy of motion of the car (kinetic energy). When the car is higher on the hill, the potential energy of the car has been increased, and friction produces heat energy.

Energy is often converted or transformed from one kind to another, but the total energy is always conserved. The principle that energy cannot be created or destroyed but is always conserved is known as the **first law of thermodynamics.** Thermodynamics is the science that keeps track of energy as it undergoes various transformations from one type to another. We use the first law to keep track of the quantity of energy.[4]

To illustrate the conservation and conversion of energy, think about a tire swing over a creek (Figure 14.3). When the tire swing is held in its highest position, it is not moving. It does contain stored energy, however, owing to its position. We refer to the stored energy as *potential energy*. Other examples of potential energy are the gravitational energy in water behind a dam; the chemical energy in coal, fuel oil, and gasoline, as well as in the fat in your body; and nuclear energy, which is related to the forces binding the nuclei of atoms.[3]

The tire swing, when released from its highest position, moves downward. At the bottom (straight down), the speed of the tire swing is greatest, and no potential energy remains. At this point, all the swing's energy is the energy of motion, called *kinetic energy*. As the tire swings back and forth, the energy continuously changes between the two forms, potential and kinetic. However, with each swing, the tire slows down a little more and goes a little less high because of friction created by the movement of the tire and rope through air and friction at the pivot where the rope is tied to the tree. The friction slows the swing, generating *heat energy*, which is energy from random motion of atoms and molecules. Eventually, all the energy is converted to heat and emitted to the environment, and the swing stops.[3]

The example of the swing illustrates the tendency of energy to dissipate and end up as heat. Indeed, physicists have found that it is possible to change all the gravitational energy in a tire swing (a type of pendulum) to heat. However, it is impossible to change all the

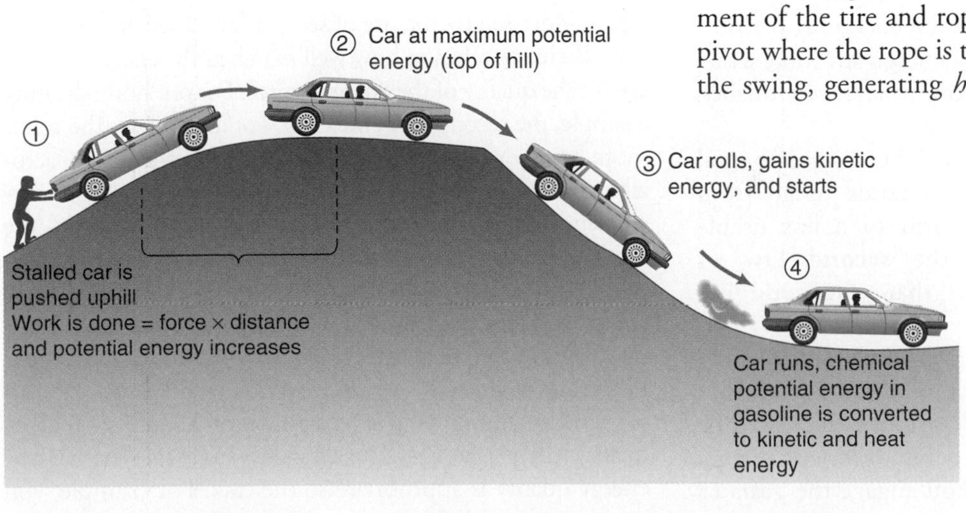
② Car at maximum potential energy (top of hill)
①
③ Car rolls, gains kinetic energy, and starts
④
Stalled car is pushed uphill
Work is done = force × distance and potential energy increases
Car runs, chemical potential energy in gasoline is converted to kinetic and heat energy

FIGURE 14.2 **Some basic energy concepts**, including potential energy, kinetic energy, and heat energy.

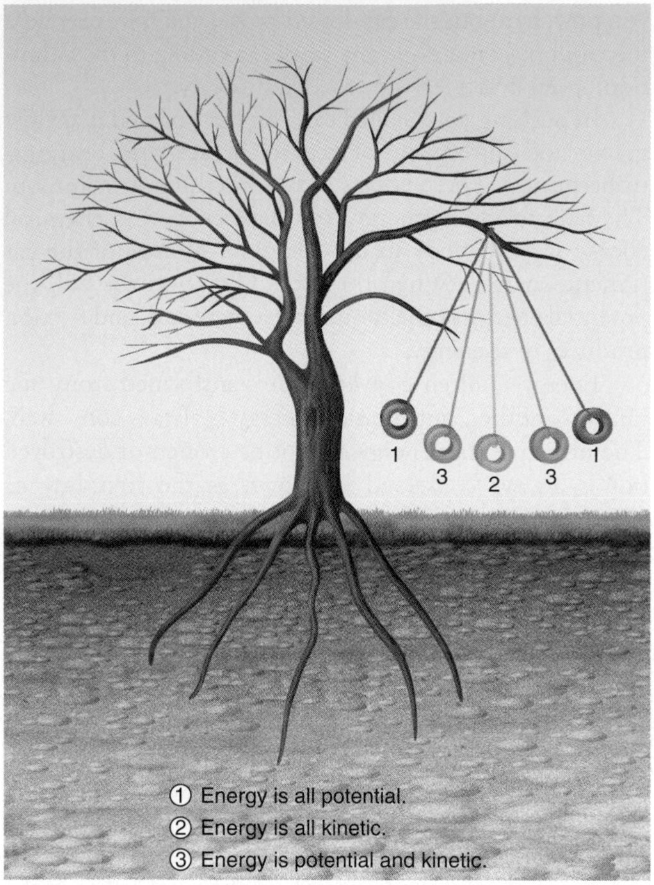
1
3
2
3
1
① Energy is all potential.
② Energy is all kinetic.
③ Energy is potential and kinetic.

FIGURE 14.3 **Diagram of a tire swing**, illustrating the relation between potential and kinetic energy.

heat energy thus generated back into potential energy. Energy is conserved in the tire swing. When the tire swing finally stops, all the initial gravitational potential energy has been transformed by way of friction to heat energy. If the same amount of energy, in the form of heat, were returned to the tire swing, would you expect the swing to start again? The answer is no! What, then, is used up? It is not energy because energy is always conserved. What is used up is the energy *quality*—the availability of the energy to perform work. The higher the quality of the energy, the more easily it can be converted to work; the lower the energy quality, the more difficult to convert it to work.

This example illustrates another fundamental property of energy: Energy always tends to go from a more usable (higher-quality) form to a less usable (lower-quality) form. This is the **second law of thermodynamics**, and it means that when you use energy, you lower its quality.

Let's return to the example of the stalled car, which you have now pushed to the side of the road. Having pushed the car a little way uphill, you have increased its potential energy. You can convert this to kinetic energy by letting it roll back downhill. You engage the gears to restart the car. As the car idles, the potential chemical energy (from the gasoline) is converted to waste heat energy and other energy forms, including electricity to charge the battery and play the radio.

Why can't we collect the wasted heat and use it to run the engine? Again, as the second law of thermodynamics tells us, once energy is degraded to low-quality heat, it can never regain its original availability or energy grade. When we refer to low-grade heat energy, we mean that relatively little of it is available to do useful work. High-grade energy, such as that of gasoline, coal, or natural gas, has high potential to do useful work. The biosphere continuously receives high-grade energy from the sun and radiates low-grade heat to the depths of space.[3, 4]

14.3 Energy Efficiency

Two fundamental types of energy efficiencies are derived from the first and second laws of thermodynamics: first-law efficiency and second-law efficiency. **First-law efficiency** deals with the amount of energy without any consideration of the quality or availability of the energy. It is calculated as the ratio of the actual amount of energy delivered where it is needed to the amount of energy supplied to meet that need. Expressions for efficiencies are given as fractions; multiplying the fraction by 100 converts it to a percentage. As an example, consider a furnace system that keeps a home at a desired temperature of 18°C (65°F) when the outside temperature is 0°C (32°F). The furnace, which burns natural gas, delivers 1 unit of heat energy to the house for every 1.5 units of energy extracted from burning the fuel. That means it has a first-law efficiency of 1 divided by 1.5, or 67% (see Table 14.1 for other examples).[4] The "unit" of energy for our furnace is arbitrary for the purpose of discussion; we also could use the British thermal unit (Btu) or some other units (see A Closer Look 14.1).

First-law efficiencies are misleading because a high value suggests (often incorrectly) that little can be done to save energy through additional improvements in efficiency. This problem is addressed by the use of second-law efficiency. **Second-law efficiency** refers to how well matched the energy end use is with the quality of the energy source. For our home-heating example, the second-law efficiency would compare the minimum energy necessary to heat the home to the energy actually used by the gas furnace. If we calculated the second-law efficiency (which is beyond the scope of this discussion), the result might be 5%—much lower than the first-law efficiency of 67%.[4] (We will see why later.) Table 14.1 also lists some second-law efficiencies for common uses of energy.

Values of second-law efficiency are important because low values indicate where improvements in energy technology and planning may save significant amounts of high-quality energy. Second-law efficiency tells us whether the energy quality is appropriate to the task. For example, you could use a welder's acetylene blowtorch to light a candle, but a match is much more efficient (and safer as well).

Table 14.1 EXAMPLES OF FIRST- AND SECOND-LAW EFFICIENCIES

ENERGY (END USE)	FIRST-LAW EFFICIENCY (%)	WASTE HEAT (%)	SECOND-LAW EFFICIENCY (%)	POTENTIAL FOR SAVINGS
Incandescent lightbulb	5	95		
Fluorescent light	20	80		
Automobile	20-25	75-80	10	Moderate
Power plants (electric); fossil fuel and nuclear	30-40	60-70	30	Low to moderate
Burning fossil fuels (used directly for heat)	65	35		
Water heating			2	Very high
Space heating and cooling			6	Very high
All energy (U.S.)	50	50	10-15	High

A CLOSER LOOK 14.1

Energy Units

When we buy electricity by the kilowatt-hour, what are we buying? We say we are buying energy, but what does that mean? Before we go deeper into the concepts of energy and its uses, we need to define some basic units.

The fundamental energy unit in the metric system is the *joule;* 1 joule is defined as a force of 1 newton* applied over a distance of 1 meter. To work with large quantities, such as the amount of energy used in the United States in a given year, we use the unit *exajoule,* which is equivalent to 10^{18} (a billion billion) joules, roughly equivalent to 1 quadrillion, or 10^{15}, Btu, referred to as a *quad.* To put these big numbers in perspective, the United States today consumes approximately 100 exajoules (or quads) of energy per year, and world consumption is about 425 exajoules (quads) annually.

In many instances, we are particularly interested in the rate of energy use, or *power,* which is energy divided by time. In the metric system, power may be expressed as joules per second, or *watts* (W); 1 joule per second is equal to 1 watt. When larger power units are required, we can use multipliers, such as *kilo*-(thousand), *mega*- (million), and *giga*-(billion). For example, a modern nuclear power plant's electricity production rate is 1,000 megawatts (MW) or 1 gigawatt (GW).

Sometimes it is useful to use a hybrid energy unit, such as the watt-hour (Wh); remember, energy is power multiplied by time. Electrical energy is usually expressed and sold in *kilowatt-hours* (kWh, or 1,000 Wh). This unit of energy is 1,000 W applied for 1 hour (3,600 seconds), the equivalent energy of 3,600,000 J (3.6 MJ).

The average estimated electrical energy in kilowatt-hours used by various household appliances over a period of a year is shown in Table 14.2. The total energy used annually is the power rating of the appliance multiplied by the time the appliance was actually used. The appliances that use most of the electrical energy are water heaters, refrigerators, clothes driers, and washing machines. A list of common household appliances and the amounts of energy they consume is useful in identifying the ones that might help save energy through conservation or improved efficiency.

* A newton (N) is the force necessary to produce an acceleration of 1 m per sec (m/s^2) to a mass of 1 kg.

Table 14.2 POWER USE OF TYPICAL HOUSEHOLD APPLIANCES IN WATTS (W)

APPLIANCE	POWER (W)
Clock	2
Coffee maker	900–1200
Clothes washer	350–500
Clothes dryer	1800–5000
Dishwasher	1200–2400 (using the drying feature greatly increases energy consumption)
Electric blanket - *Single/Double*	60/100
Fans	
Ceiling	65–175
Window	55–250
Furnace	750
Whole house	240–750
Hair dryer	1200–1875
Heater (*portable*)	750–1500
Clothes iron	1000–1800
Microwave oven	750–1100
Personal computer	
CPU - awake/asleep	120/30 or less
Monitor - awake/asleep	150/30 or less
Laptop	50
Radio (*stereo*)	70–400
Refrigerator (*frost-free, 16 cubic feet*)	725
Televisions (color)	
19"	65–110
36" = 133 W	
53"–61" Projection	170
Flat screen	120
Toaster	800–1400
Toaster oven	1225
VCR/DVD	17–21/20–25
Vacuum cleaner	1000–1440
Water heater (*40 gallon*)	4500–5500
Water pump (*deep well*)	250–1100
Water bed (*with heater, no cover*)	120–380

*You can use this formula below to estimate an appliance's energy use:

Wattage × Hours Used Per Day ÷ 1000 = Daily Kilowatt-hour (kWh) consumption: remember 1kW = 1,000 watts, which is why we divide by 1,000

Multiply by the number of days you use a particular appliance during the year for the annual energy consumption.

You can calculate the annual cost to run an appliance by multiplying the kWh per year by your local utility's rate per kWh consumed.

Example: Personal Computer and Monitor: (120 + 150 Watts × 4 hours/day × 365 days/year) ÷ 1000 ≅ 394 kWh × 11 cents/kWh (approx national average) ≅ $43.34/year

*Note: To estimate the number of hours that a refrigerator actually operates at its maximum wattage, divide the total time the refrigerator is plugged in by three. Refrigerators, although turned "on" all the time, cycle on and off as needed to maintain interior temperatures.

Source: Modified from U.S. Department of Energy. Your Home. accessed January 27, 2010 at http://www.energysavers.gov

We are now in a position to understand why the second-law efficiency is so low (5%) for the house-heating example discussed earlier. This low efficiency implies that the furnace is consuming too much high-quality energy in carrying out the task of heating the house. In other words, the task of heating the house requires heat at a relatively low temperature, near 18°C (65°F), not heat with temperatures in excess of 1,000°C (1,832°F), such as is generated inside the gas furnace. Lower-quality energy, such as solar energy, could do the task and yield a higher second-law efficiency because there is a better match between the required energy quality and the house-heating end use. Through better energy planning, such as matching the quality of energy supplies to the end use, higher second-law efficiencies can be achieved, resulting in substantial savings of high-quality energy.

Examination of Table 14.1 indicates that electricity-generating plants have nearly the same first-law and second-law efficiencies. These generating plants are examples of heat engines. A heat engine produces work from heat. Most of the electricity generated in the world today comes from *heat engines* that use nuclear fuel, coal, gas, or other fuels. Our own bodies are examples of heat engines, operating with a capacity (power) of about 100 watts and fueled indirectly by solar energy. (See A Closer Look 14.1 for an explanation of watts and other units of energy.) The internal combustion engine (used in automobiles) and the steam engine are additional examples of heat engines. A great deal of the world's energy is used in heat engines, with profound environmental effects, such as thermal pollution, urban smog, acid rain, and global warming.

The maximum possible efficiency of a heat engine, known as *thermal efficiency*, was discovered by the French engineer Sadi Carnot in 1824, before the first law of thermodynamics was formulated.[5] Modern heat engines have thermal efficiencies that range between 60 and 80% of their ideal Carnot efficiencies. Modern 1,000-megawatt (MW) electrical generating plants have thermal efficiencies ranging between 30 and 40%; that means at least 60–70% of the energy input to the plant is rejected as waste heat. For example, assume that the electric power output from a large generating plant is 1 unit of power (typically 1,000 MW). Producing that 1 unit of power requires 3 units of input (such as burning coal) at the power plant, and the entire process produces 2 units of waste heat, for a thermal efficiency of 33%. The significant number here is the waste heat, 2 units, which amounts to twice the actual electric power produced.

Electricity may be produced by large power plants that burn coal or natural gas, by plants that use nuclear fuel, or by smaller producers, such as geothermal, solar, or wind sources (see Chapters 15, 16, and 17). Once produced, the electricity is fed into the grid, which is the network of power lines, or the distribution system. Eventually it reaches homes, shops, farms, and factories, where it provides light and heat and also drives motors and other machinery used by society. As electricity moves through the grid, losses take place. The wires that transport electricity (power lines) have a natural resistance to electrical flow. Known as *electrical resistivity*, this resistance converts some of the electric energy in the transmission lines to heat energy, which is radiated into the environment surrounding the lines.

14.4 Energy Sources and Consumption

People living in industrialized countries make up a relatively small percentage of the world's population but consume a disproportionate share of the total energy consumed in the world. For example, the United States, with only 5% of the world's population, uses approximately 20% of the total energy consumed in the world. There is a direct relationship between a country's standard of living (as measured by gross national product) and energy consumption per capita.

After the peak in oil production, expected in 2020–2050, oil and gasoline will be in shorter supply and more expensive. Before then, use of these fuels may be curtailed in an effort to lessen global climate change. As a result, within the next 30 years both developed and developing countries will need to find innovative ways to obtain energy. In the future, affluence may be related as closely to more efficient use of a wider variety of energy sources as it is now to total energy consumption.

Fossil Fuels and Alternative Energy Sources

Today, approximately 90% of the energy consumed in the United States is derived from petroleum, natural gas, and coal. Because they originated from plant and animal material that existed millions of years ago, they are called fossil fuels. They are forms of stored solar energy that are part of our geologic resource base, and they are essentially nonrenewable. Other sources of energy—geothermal, nuclear, hydropower, and solar, among others—are referred to as *alternative* energy sources because they may serve as alternatives to fossil fuels in the future. Some of them, such as solar and wind, are not depleted by consumption and are known as *renewable energy* sources.

The shift to alternative energy sources may be gradual as fossil fuels continue to be used, or it could be accelerated by concern about potential environmental effects of burning fossil fuels. Regardless of which path we take, one thing is certain: Fossil fuels are finite. It took millions of years to form them, but they will be depleted in only a few hundred years of human history. Using even the most optimistic predictions, the fossil fuel epoch that started with the Industrial Revolution will represent only about 500 years of human history. Therefore, although fossil fuels have been extremely significant in the development of modern civilization, their use will be a brief event in the span of human history.[6, 7]

Energy Consumption in the United States

Energy consumption in the United States from 1980 and projected to 2030 is shown in Figure 14.4. The figure dramatically illustrates our ongoing dependence on the three major fossil fuels: coal, natural gas, and petroleum. From approximately 1950 through the late 1970s, energy consumption soared, from about 30 exajoules to 75 exajoules. (Energy units are defined in A Closer Look 14.1.) Since about 1980, energy consumption has risen by only about 25 exajoules. This is encouraging because it suggests that policies promoting energy-efficiency improvements (such as requiring new automobiles to be more fuel-efficient and buildings to be better insulated) have been at least partially successful.

What is not shown in the figure, however, is the huge energy loss. For example, energy consumption in the United States in 1965 was approximately 50 exajoules, of which only about half was used effectively. Energy losses were about 50% (the number shown earlier in Table 14.1 for all energy). In 2009, energy consumption in the United States was about 100 exajoules, and again about 50% was lost in conversion processes. Energy losses in 2009 were about equal to total U.S. energy consumption in 1965! The largest energy losses are associated with the production of electricity and with transportation, mostly through the use of heat engines, which produce waste heat that is lost to the environment.

Another way to examine energy use is to look at the generalized energy flow of the United States by end use for a particular year (Figure 14.5). In 2008 we imported considerably more oil than we produced (we import about 65% of the oil we use), and our energy consumption was fairly evenly distributed in three sectors: residential/commercial, industrial, and transportation. It is clear that we remain dangerously vulnerable to changing world conditions affecting the production and delivery of crude oil. We need to evaluate the entire spectrum of potential energy sources to ensure that sufficient energy will be available in the future, while sustaining environmental quality.

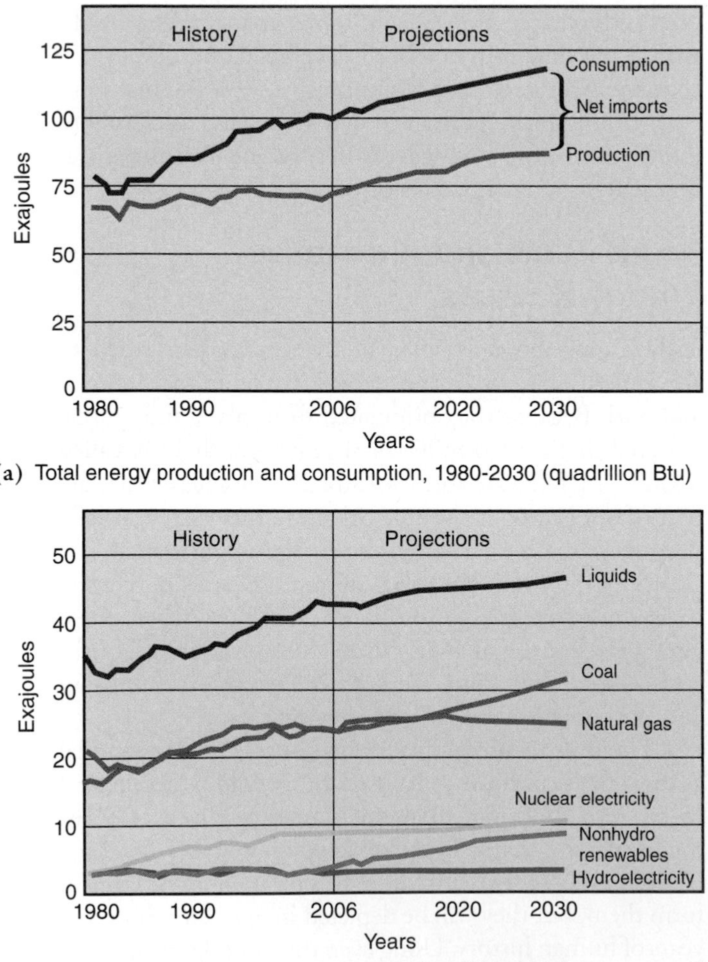

(a) Total energy production and consumption, 1980-2030 (quadrillion Btu)

(b) Energy consumption by fuel, 1980-2030 (quadrillion Btu)

FIGURE 14.4 U.S. energy from 1980 and projected to 2030. **(a)** Total consumption and production; **(b)** consumption by source. (*Source:* Department of Energy, Energy Information Agency, *Annual Report 2008*.) These forecasts are conservative in terms of expected increases in alternative energy.

14.5 Energy Conservation, Increased Efficiency, and Cogeneration

There is a movement to change patterns of energy consumption in the United States through such measures as conservation, improved energy efficiency, and cogeneration. **Conservation** of energy refers simply to using less energy and adjusting our energy needs and uses to minimize the amount of high-quality energy necessary for a given task.[8] Increased **energy efficiency** involves designing equipment to yield more energy output from a given amount of energy input (first-law efficiency) or better matches between energy source and end use (second-law efficiency). **Cogeneration** includes a number of processes designed to capture and use waste heat, rather than simply releasing it into the atmosphere, water, or other parts of the environment as a thermal pollutant. In other words, we design energy systems and power plants to provide energy more than once[9]—that is, to use it a second time, at a lower temperature, but possibly to use it in more than one way as well.

An example of cogeneration is the *natural gas combined cycle power plant* that produces electricity in two ways: gas cycle and steam cycle. In the gas cycle, the natural gas fuel is burned in a gas turbine to produce electricity. In the steam cycle, hot exhaust from the gas turbine is used to create steam that is fed into a steam generator to produce additional electricity. The combined cycles capture waste heat from the gas cycle, nearly doubling the efficiency of the power plant from about 30 to 50–60%. Energy conservation is particularly attractive because it provides more than a one-to-one savings. Remember that it takes 3 units of fuel such as coal to produce 1 unit of power such as electricity (two-thirds is waste heat). Therefore, not using (conserving) 1 unit of power saves 3 units of fuel!

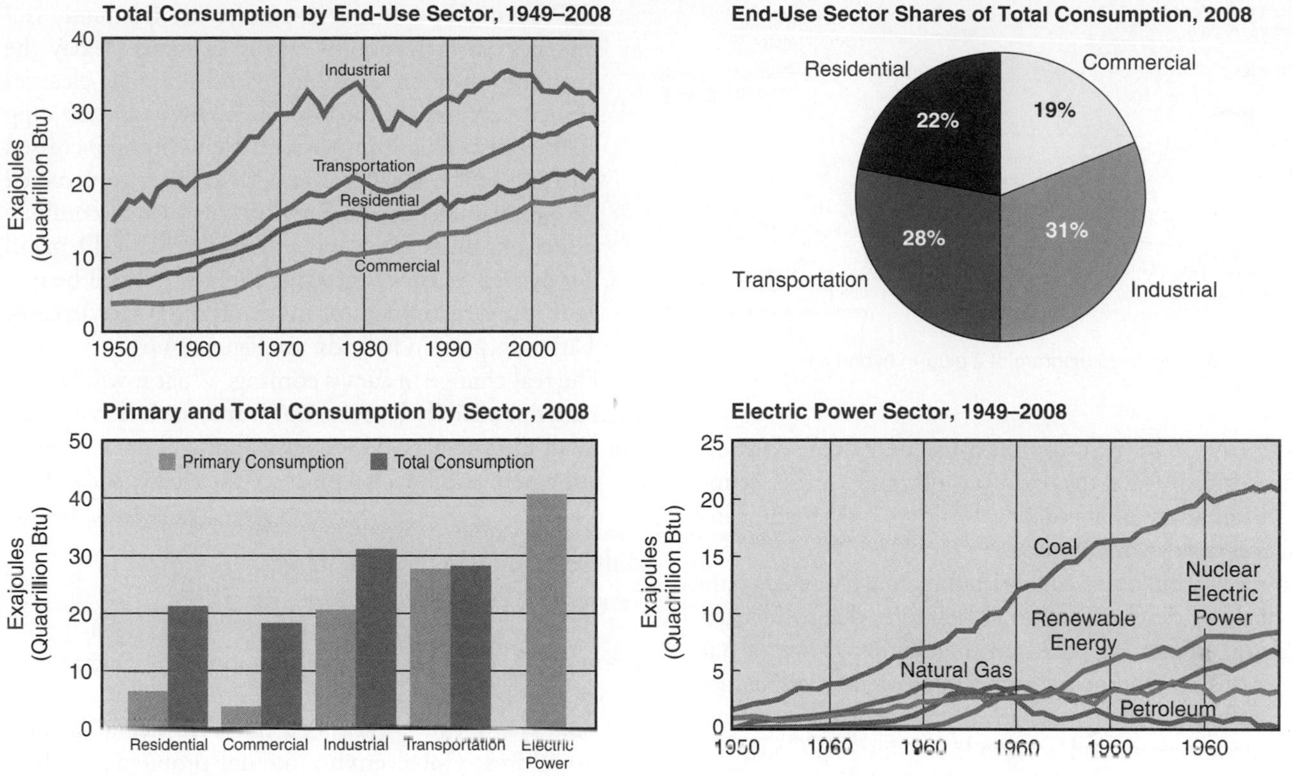

FIGURE 14.5 Energy consumption in the United States by sector (approximate). Total primary consumption is the amount of fossil and renewable fuels consumed. Total consumption refers to fossil and renewable fuels consumed plus electricity used. (*Source:* U.S. Energy Information Administration, *Annual Energy Review, 2008.*)

These three concepts—energy conservation, energy efficiency, and cogeneration—are all interlinked. For example, when big, coal-burning power stations produce electricity, they may release large amounts of heat into the atmosphere. Cogeneration, by using that waste heat, can increase the overall efficiency of a typical power plant from 33% to as much as 75%, effectively reducing losses from 67 to 25%. Cogeneration also involves generating electricity as a by-product of industrial processes that produce steam as part of their regular operations. Optimistic energy forecasters estimate that eventually we may meet approximately one-half the electrical power needs of industry through cogeneration.[8, 9] Another source has estimated that cogeneration could provide more than 10% of the power capacity of the United States.

The average first-law efficiency of only 50% (Table 14.1) illustrates that large amounts of energy are currently lost in producing electricity and in transporting people and goods. Innovations in how we produce energy for a particular use can help prevent this loss, raising second-law efficiencies. Of particular importance will be energy uses with applications below 100°C (212°F), because a large portion of U.S. energy consumption for uses below 300°C, or 572°F, is for space heating and water heating.

In considering where to focus our efforts to improve energy efficiency, we need to look at the total energy-use picture. In the United States, space heating and cooling of homes and offices, water heating, industrial processes (to produce steam), and automobiles account for nearly 60% of the total energy use, whereas transportation by train, bus, and airplane accounts for only about 5%. Therefore, the areas we should target for improvement are building design, industrial energy use, and automobile design. We note, however, that debate continues as to how much efficiency improvements and conservation can reduce future energy demands and the need for increased energy production from traditional sources, such as fossil fuel.

Building Design

A spectrum of possibilities exists for increasing energy efficiency and conservation in residential buildings. For new homes, the answer is to design and construct homes that require less energy for comfortable living.[8] For example, we can design buildings to take advantage of passive solar potential, as did the early Greeks and Romans and the Native American cliff dwellers. (Passive solar energy systems collect solar heat without using moving parts.) Windows and overhanging structures can be positioned so that the overhangs shade the windows from solar energy in summer, thereby keeping the house cool, while allowing winter sun to penetrate the windows and warm the house.

The potential for energy savings through architectural design for older buildings is extremely limited. The position of the building on the site is already established, and reconstruction and modifications are often not

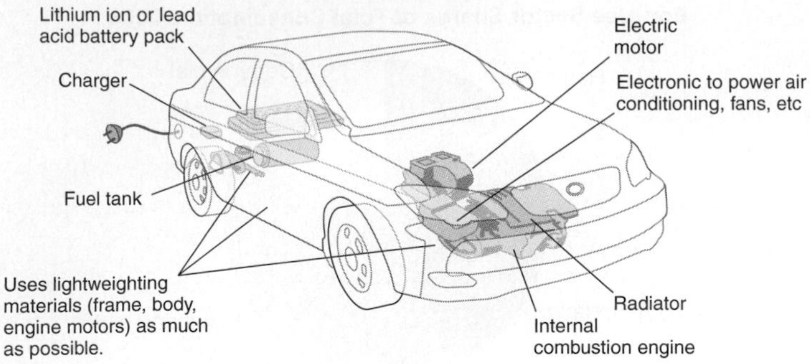

FIGURE 14.6 Idealized diagram of a plug-in hybrid car.

cost-effective. The best approach to energy conservation for these buildings is insulation, caulking, weather stripping, installation of window coverings and storm windows, and regular maintenance.

Ironically, buildings constructed to conserve energy are more likely to develop indoor air pollution due to reduced ventilation. In fact, air pollution is emerging as one of our most serious environmental problems. Potential difficulties can be reduced by better designs for air-circulation systems that purify indoor air and bring in fresh, clean air. Construction that incorporates environmental principles is more expensive owing to higher fees for architects and engineers, as well as higher initial construction costs. Nevertheless, moving toward improved design of homes and residential buildings to conserve energy remains an important endeavor.

Industrial Energy

The rate of increase in energy use (consumption) leveled off in the early 1970s. Nevertheless, industrial production of goods (automobiles, appliances, etc.) continued to grow significantly. Today, U.S. industry consumes about one-third of the energy produced. The reason we have had higher productivity with lower growth of energy use is that more industries are using cogeneration and more energy-efficient machinery, such as motors and pumps designed to use less energy.[8, 10]

Automobile Design

The development of fuel-efficient automobiles has steadily improved during the last 30 years. In the early 1970s, the average U.S. automobile burned approximately 1 gallon of gas for every 14 miles traveled. By 1996, the miles per gallon (mpg) had risen to an average of 28 for highway driving and as high as 49 for some automobiles.[11] Fuel consumption rates did not improve much from 1996 to 1999. In 2004, many vehicles sold were SUVs and light trucks with fuel consumption of 10–20 mpg. A loophole in regulations permits these vehicles to have poorer fuel consumption than conventional automobiles.[11] As a result of higher gasoline prices, sales of larger SUVs declined in 2006, but smaller SUVs remain popular as consumers are apparently sacrificing size for economy (up to a point). Today the fuel consumption of some hybrid (gasoline-electric) vehicles exceeds 90 mpg on the highway and 60 mpg in the city. This improvement stems from increased fuel efficiency; smaller cars with engines constructed of lighter materials; and hybrid cars, which combine a fuel-burning engine and an electric motor. Demand for hybrid vehicles is growing rapidly and will be met with the development of more advanced rechargeable batteries (plug-in hybrids; see Figure 14.6).

The real change in cars is coming. What it will be and when are not entirely known, but it may be a transformation to all-electric cars. Miles per gallon will not be the issue, but where and how we produce the electricity will be.

Values, Choices, and Energy Conservation

A potentially effective method of conserving energy is to change our behavior by using less energy. This involves our values and the choices we make to act at a local level to address global environmental problems, such as human-induced warming caused by burning fossil fuels. For example, we make choices as to how far we commute to school or work and what method of transport we use to get there. Some people commute more than an hour by car to get to work, while others ride a bike, walk, or take a bus or train. Other ways of modifying behavior to conserve energy include the following:

- Using carpools to travel to and from work or school
- Purchasing a hybrid car (gasoline-electric)
- Turning off lights when leaving rooms
- Taking shorter showers (conserves hot water)
- Putting on a sweater and turning down the thermostat in winter
- Using energy-efficient compact fluorescent lightbulbs
- Purchasing energy-efficient appliances
- Sealing drafts in buildings with weather stripping and caulk
- Better insulating your home
- Washing clothes in cold water whenever possible
- Purchasing local foods rather than foods that must be brought to market from afar
- Reducing standby power for electronic devices and appliances by using power strips and turning them off when not in use

What other ways of modifying your behavior would help conserve energy?

14.6 Sustainable-Energy Policy

Energy policy today is at a crossroads. One path leads to the "business-as-usual" approach—find greater amounts of fossil fuels, build larger power plants, and go on using energy as freely as we always have. The business-as-usual path is more comfortable—it requires no new thinking; no realignment of political, economic, or social conditions; and little anticipation of coming reductions in oil production.

People heavily invested in the continued use of fossil fuels and nuclear energy often favor the traditional path. They argue that much environmental degradation around the world has been caused by people who have been forced to use local resources, such as wood, for energy, leading to the loss of plant and animal life and increasing soil erosion. They argue that the way to solve these environmental problems is to provide cheap, high-quality energy, such as fossil fuels or nuclear energy.

In countries like the United States, with sizable resources of coal and natural gas, people supporting the business-as-usual path argue that we should exploit those resources while finding ways to reduce their environmental impact. According to these proponents, we should (1) let the energy industry develop the available energy resources and (2) let industry, free from government regulations, provide a steady supply of energy with less total environmental damage.

The previous U.S. energy plan, suggested by then President George W. Bush, was largely a business-as-usual proposal: Find and use more coal, oil, and natural gas; use more nuclear power; and build more than 1,000 new fossil fuel plants in the next 20 years. Energy conservation and development of alternative energy sources, while encouraged, were not considered of primary importance.

A visionary path for energy policy was suggested more than 30 years ago by Amory Lovins.[12] That path focuses on energy alternatives that emphasize energy quality and are renewable, flexible, and environmentally more benign than those of the business-as-usual path. As defined by Lovins, these alternatives have the following characteristics:

- They rely heavily on renewable energy resources, such as sunlight, wind, and biomass (wood and other plant material).
- They are diverse and are tailored for maximum effectiveness under specific circumstances.
- They are flexible, accessible, and understandable to many people.
- They are matched in energy quality, geographic distribution, and scale to end-use needs, increasing second-law efficiency.

Lovins points out that people are not particularly interested in having a certain amount of oil, gas, or electricity delivered to their homes; they are interested in having comfortable homes, adequate lighting, food on the table, and energy for transportation.[12] According to Lovins, only about 5% of end uses require high-grade energy, such as electricity. Nevertheless, a lot of electricity is used to heat homes and water. Lovins shows that there is an imbalance in using nuclear reactions at extremely high temperatures and in burning fossil fuels at high temperatures simply to meet needs where the necessary temperature increase may be only a few 10s of degrees. He considers such large discrepancies wasteful and a misallocation of high-quality energy.

Energy for Tomorrow

The availability of energy supplies and the future demand for energy are difficult to predict because the technical, economic, political, and social assumptions underlying predictions are constantly changing. In addition, seasonal and regional variations in energy consumption must also be considered. For example, in areas with cold winters and hot, humid summers, energy consumption peaks during the winter months (from heating) and again in the summer (from air-conditioning). Regional variations in energy consumption are significant. For example, in the United States as a whole, the transportation sector uses about one-fourth of the energy consumed. However, in California, where people often commute long distances to work, about one-half of the energy is used for transportation, more than double the national average. Energy sources, too, vary by region. For example, in the eastern and southwestern United States, the fuel of choice for power plants is often coal, but power plants on the West Coast are more likely to burn oil or natural gas or use hydropower from dams to produce electricity.

Future changes in population densities, as well as intensive conservation measures, will probably alter existing patterns of energy use. This might involve a shift to more reliance on alternative (particularly renewable) energy sources.[13, 14] Energy consumption in the United States in the year 2050 may be about 160 exajoules. What will be the energy sources for the anticipated growth in energy consumption? Will we follow our past policy of business as usual (coal, oil, nuclear), or will we turn more to alternative energy sources (wind, solar, geothermal)? What is clear is that the mix of energy sources in 2030 will be different from today's and more diversified.[13-15]

All projections of specific sources and uses of energy in the future must be considered speculative. Perhaps most speculative of all is the idea that we really can meet most of our energy needs with alternative, renewable energy sources in the next several decades. From an energy viewpoint, the next 20 to 30 years, as we move through the maximum production of petroleum, will be crucial to the United States and to the rest of the industrialized world.

The energy decisions we make in the very near future will greatly affect both our standard of living and our quality of life. From an optimistic point of view, we have the necessary information and technology to ensure a bright, warm, lighted, and mobile future. But time may be running out, and we need action now. We can continue to take things as they come and live with the results of our present dependence on fossil fuels, or we can build a sustainable energy future based on careful planning, innovative thinking, and a willingness to move from our dependence on petroleum.

U.S. energy policy for the 21st century is being discussed seriously, and significant change in policy is likely. Some of the recommendations are as follows:

- Promote conventional energy sources: Use more natural gas to reduce our reliance on energy from foreign countries.
- Encourage alternative energy: Support and subsidize wind, solar, geothermal, hydrogen, and biofuels (ethanol and biodiesel).
- Provide for energy infrastructure: Ensure that electricity is transmitted over dependable, modern infrastructure.
- Promote conservation measures: Set higher efficiency standards for buildings and for household products. Require that waste heat from power generation and industrial processes be used to produce electricity or other products. Recommend stronger fuel-efficiency standards for cars, trucks, and SUVs. Provide tax credits for installing energy-efficient windows and appliances in homes and for purchasing fuel-efficient hybrids or clean-diesel vehicles.
- Carefully evaluate the pros and cons of nuclear power, which can generate large amounts of electricity without emitting greenhouse gases, but has serious negatives as well.
- Promote research: Develop new alternative energy sources; find new, innovative ways to improve existing coal plants and to help construct cleaner coal plants; determine whether it is possible to extract vast amounts of oil trapped in oil shale and tar sands without harming the environment; and develop pollution-free, electric automobiles.

Which of the above points will become policy in future years is not known, but parts of the key ideas will move us toward sustainable energy.

Integrated, Sustainable Energy Management The concept of **integrated energy management** recognizes that no single energy source can provide all the energy required by the various countries of the world.[16] A range of options that vary from region to region will have to be employed. Furthermore, the mix of technologies and sources of energy will involve both fossil fuels and alternative, renewable sources.

A basic goal of integrated energy management is to move toward **sustainable energy development** that is implemented at the local level. Sustainable energy development would have the following characteristics:

- It would provide reliable sources of energy.
- It would not destroy or seriously harm our global, regional, or local environments.
- It would help ensure that future generations inherit a quality environment with a fair share of the Earth's resources.

To implement sustainable energy development, leaders in various regions of the world will need energy plans based on local and regional conditions. The plans will integrate the desired end uses for energy with the energy sources that are most appropriate for a particular region and that hold potential for conservation and efficiency. Such plans will recognize that preserving resources can be profitable and that degradation of the environment and poor economic conditions go hand in hand.[16] In other words, degradation of air, water, and land resources depletes assets and ultimately will lower both the standard of living and the quality of life. A good energy plan recognizes that energy demands can be met in environmentally preferred ways and is part of an aggressive environmental policy whose goal is a quality environment for future generations. The plan should do the following:[16]

- Provide for sustainable energy development.
- Provide for aggressive energy efficiency and conservation.
- Provide for diversity and integration of energy sources.
- Develop and use the "smart grid" to optimally manage energy flow on the scale of buildings to regions.
- Provide for a balance between economic health and environmental quality.
- Use second-law efficiencies as an energy policy tool—that is, strive to achieve a good balance between the quality of an energy source and end uses for that energy.

An important element of the plan involves the energy used for automobiles. This builds on policies of the past 30 years to develop hybrid vehicles that use both an electric motor and an internal combustion engine, and to improve fuel technology to reduce both fuel consumption and emission of air pollutants. Finally, the plan should factor in the marketplace through pricing that reflects the economic cost of using the fuel, as well as its cost to the environment. In sum, the plan should be an integrated energy-management statement that moves toward sustainable development. Those who develop such plans recognize that a diversity of energy supplies will be necessary and that the key components are (1) improvements in energy efficiency and conservation and (2) matching energy quality to end uses.[16]

The global pattern of ever-increasing energy consumption led by the United States and other nations cannot be sustained without a new energy paradigm that includes changes in human values, not just a breakthrough in technology. Choosing to own lighter, more fuel-efficient automobiles and living in more energy-efficient homes is consistent with a sustainable energy system that focuses on providing and using energy to improve human welfare. A sustainable energy paradigm establishes and maintains multiple linkages among energy production, energy consumption, human well-being, and environmental quality.[17] It might also involve using smaller generating facilities that are more widely distributed (see A Closer Look 14.2).

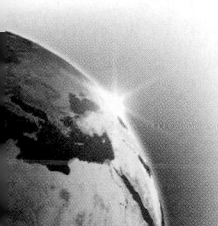

A CLOSER LOOK 14.2

Micropower

It is likely that sustainable energy management will include the emerging concept of **micropower**—smaller, distributed systems for production of electricity. Such systems are not new; the inventor Thomas Edison evidently anticipated that electricity-generating systems would be dispersed. By the late 1890s, many small electrical companies were marketing and building power plants, often located in the basements of businesses and factories. These early plants evidently used cogeneration principles, since waste heat was reused for heating buildings.[18] Imagine if we had followed this early model: Homes would have their own power systems, power lines wouldn't snake through our neighborhoods, and we could replace older, less efficient systems as we do refrigerators.

Instead, in the 20th century U.S. power plants grew larger. By the 1930s, industrializing countries had set up utility systems based on large-scale central power plants, as diagrammed in Figure 14.7a. Today, however, we are again evaluating the merits of distributive power systems, as shown in Figure 14.7b.

Large, centralized power systems are consistent with the hard path, while the distributive power system is more aligned with the soft path. Micropower devices rely heavily on renewable energy sources such as wind and sunlight, which feed into the electric grid system, as shown in Figure 14.7b. Use of micropower systems in the future is being encouraged because they are reliable and are associated with less environmental damage than are large fossil-fuel-burning power plants.[18]

Uses for micropower are emerging in both developed and developing countries. In countries that lack a centralized power-generating capacity, small-scale electrical power generation from solar and wind has become the most economical option. In nations with a high degree of industrialization, micropower may emerge as a potential replacement for aging electric power plants. For micropower to be a significant factor in energy production, a shift in policies and regulations to allow micropower devices to be more competitive with centralized generation of electrical power will be required. Regardless of the obstacles that micropower devices face, distributive power systems will probably play an important role in achieving our goal of integrated, sustainable energy management for the future.

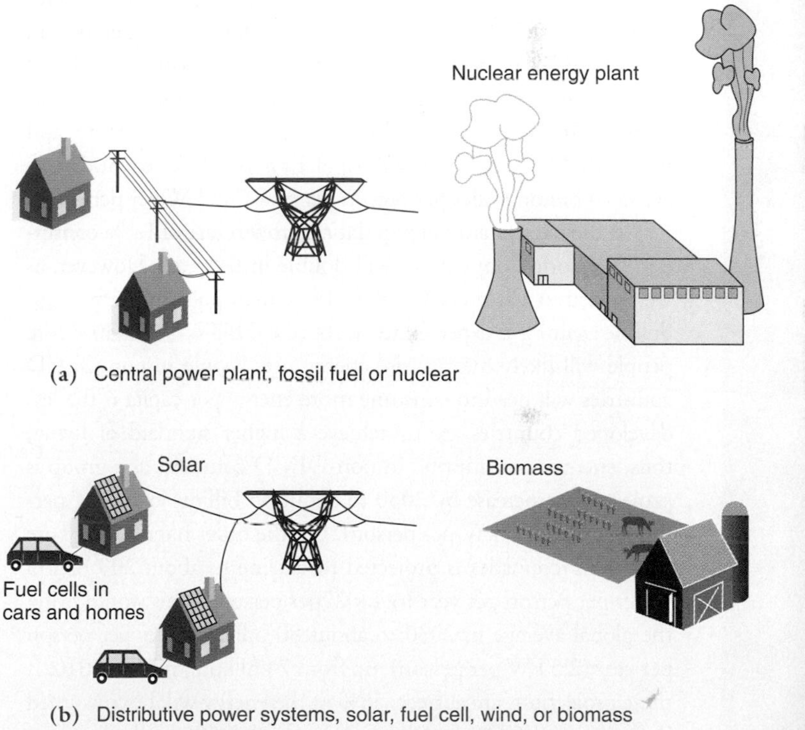

FIGURE 14.7 Idealized diagram comparing **(a)** a centralized power system, such as those used in industrial developed countries today, with **(b)** a distributive power system based on generating electricity from biomass, wind, solar, and other sources, all of which feed into the transmission and distribution system. (*Source*: Modified from S. Dunn, *Micropower, the Next Electrical Era*, Worldwatch Paper 151 [Washington, DC: Worldwatch Institute, 2000].)

CRITICAL THINKING ISSUE

Use of Energy Today and in 2030

Note: Before proceeding with this exercise, refer back to A Closer Look 4.1 to be sure you are comfortable with the units and big numbers.

The Organization for Economic Cooperation and Development (OECD) is a group of 30 countries, 27 of which are classified by the World Bank as having high-income economies. Non-OECD members are not all low-income countries, but many are. The developing countries (all of which are non-OECD) have most of the world's 6.8 billion people and are growing in population faster than the more affluent countries. The average rate of energy use in 2010 for an individual in non-OECD countries is 46 billion joules per person per year (1.5 kW per person), whereas for the OECD countries it is 210 billion joules per person per year (6.7 kW per person). In other words, people in OECD countries use about 4.5 times more energy per person than those in non-OECD countries. In 2010 each group—OECD and non-OECD—used about 250 EJ (1 EJ is 10^{18} J). The world average is 74 billion joules per person per year (2.3 kW per person).[19]

If the current annual population growth rate of 1.1% continues, the world's population will double in 64 years. However, as we learned in Chapters 1 and 4, the human population may not double again. It is expected to be about 8.5 billion by 2030. More people will likely mean more energy use. People in non-OECD countries will need to consume more energy per capita if the less developed countries are to achieve a higher standard of living; thus, energy consumption in non-OECD countries as a group is projected to increase by 2030 to about 55 billion joules per person per year (1.7 kW per person). On the other hand, energy use in OECD countries is projected to decline to about 203 billion joules per person per year (6.4 kW per person). This would bring the global average in 2030 to about 80 billion joules per person per year (2.5 kW per person), up from 74 billion joules in 2010. If these projections are correct, 58% of the energy will be consumed in the non-OECD countries, compared with 50% today.

With worldwide average energy use of 2.3 kW per person in 2010, the 6.8 billion people on Earth use about 16 trillion watts annually. A projected population of 8.5 billion in 2030 with an estimated average per capita energy use rate of 2.5 kW would use about 21 trillion watts annually, an increase of about 33% from today.[19]

A realistic goal is for annual per capita energy use to remain about 2.5 kW, with the world population peaking at 8.5 billion people by the year 2030. If this goal is to be achieved, non-OECD countries will be able to increase their populations by no more than about 50% and their energy use by about 70%; OECD nations can increase their population by only a few percent and will have to reduce their energy use slightly.

Critical Thinking Questions

1. Using only the data presented in this exercise, how much energy, in exajoules, did the world use in 2010 and what would you project global energy use to be in 2030?
2. The average person emits as heat 100 watts of power (the same as a 100 W bulb). If we assume that 25% of it is emitted by the brain, how much energy does your brain emit as heat in a year? Calculate this in joules and kWh. What is the corresponding value for all people today, and how does that value compare with world energy use per year? Can this help explain why a large, crowded lecture hall (independent of the professor pontificating) might get warm over an hour?
3. Can the world supply one-third more energy by 2030 without unacceptable environmental damage? How?
4. What would the rate of energy use be if all people on Earth had a standard of living supported by energy use of 10 kW per person, as in the United States today? How do these totals compare with the present energy-use rate worldwide?
5. In what specific ways could energy be used more efficiently in the United States? Make a list of the ways and compare your list with those of your classmates. Then compile a class list.
6. In addition to increasing efficiency, what other changes in energy consumption might be required to provide an average energy-use rate in 2030 of 6.4 kW per person in OECD countries?
7. Would you view the energy future in 2030 as a continuation of the business-as-usual approach with more large, centralized energy production based on fossil fuels, or a softer path, with more use of alternative, distributed energy sources? Justify your view.

SUMMARY

- The first law of thermodynamics states that energy is neither created nor destroyed but is always conserved and is transformed from one kind to another. We use the first law to keep track of the quantity of energy.
- The second law of thermodynamics tells us that as energy is used, it always goes from a more usable (higher-quality) form to a less usable (lower-quality) form.

- Two fundamental types of energy efficiency are derived from the first and second laws of thermodynamics. In the United States today, first-law efficiencies average about 50%, which means that about 50% of the energy produced is returned to the environment as waste heat. Second-law efficiencies average 10–15%, so there is a high potential for saving energy through better matching of the quality of energy sources with their end uses.
- Energy conservation and improvements in energy efficiency can have significant effects on energy consumption. It takes three units of a fuel such as oil to produce one unit of electricity. As a result, each unit of electricity conserved or saved through improved efficiency saves three units of fuel.
- There are arguments for both the business-as-usual path and changing to a new path. The first path has a long history of success and has produced the highest standard of living ever experienced. However, present sources of energy (based on fossil fuels) are causing serious environmental degradation and are not sustainable (especially with respect to conventional oil). A second path, based on alternative energy sources that are renewable, decentralized, diverse, and flexible, provides a better match between energy quality and end use, and emphasizes second-law efficiencies.
- The transition from fossil fuels to other energy sources requires sustainable, integrated energy management. The goal is to provide reliable sources of energy that do not cause serious harm to the environment and ensure that future generations will inherit a quality environment.

REEXAMINING THEMES AND ISSUES

Human Population

The industrialized and urbanized countries produce and use most of the world's energy. As societies change from rural to urban, energy demands generally increase. Controlling the increase of human population is an important factor in reducing total demand for energy (total demand is the product of average demand per person and number of people).

Sustainability

It will be impossible to achieve sustainability in the United States if we continue with our present energy policies. The present use of fossil fuels is not sustainable. We need to rethink the sources, uses, and management of energy. Sustainability is the central issue in our decision to continue on the hard path or change to the soft path.

Global Perspective

Understanding global trends in energy production and consumption is important if we are to directly address the global impact of burning fossil fuels with respect to air pollution and global warming. Furthermore, the use of energy resources greatly influences global economics, as these resources are transported and utilized around the world.

Urban World

A great deal of the total energy demand is in urban regions, such as Tokyo, Beijing, London, New York, and Los Angeles. How we choose to manage energy in our urban regions greatly affects the quality of urban environments. Burning cleaner fuels results in far less air pollution. This has been observed in several urban regions, such as London. Burning of coal in London once caused deadly air pollution; today, natural gas and electricity heat homes, and the air is cleaner. Burning coal in Beijing continues to cause significant air pollution and health problems for millions of people living there.

People and Nature

Our development and use of energy are changing nature in significant ways. For example, burning fossil fuels is changing the composition of the atmosphere, particularly through the addition of carbon dioxide. The carbon dioxide is contributing to the warming of the atmosphere, water, and land (see Chapter 20 for details). A warmer Earth is, in turn, changing the climates of some regions and affecting weather patterns and the intensity of storms.

Science and Values

Public-opinion polls consistently show that people value a quality environment. In response, energy planners are evaluating how to use our present energy resources more efficiently, practice energy conservation, and reduce adverse environmental effects of energy consumption. Science is providing options in terms of energy sources and uses; our choices will reflect our values.

KEY TERMS

- cogeneration **294**
- conservation **294**
- energy efficiency **294**
- first-law efficiency **290**
- first law of thermodynamics **289**
- integrated energy management **298**
- micropower **299**
- second-law efficiency **290**
- second law of thermodynamics **290**
- sustainable energy development **298**
- work **289**

STUDY QUESTIONS

1. What evidence supports the notion that, although present energy problems are not the first in human history, they are unique in other ways?
2. How do the terms *energy*, *work*, and *power* differ in meaning?
3. Compare and contrast the potential advantages and disadvantages of a major shift from hard-path to soft-path energy development.
4. You have just purchased a 100-hectare wooded island in Puget Sound. Your house is built of raw timber and is uninsulated. Although the island receives some wind, trees over 40 m tall block most of it. You have a diesel generator for electric power, and hot water is produced by an electric heater run by the generator. Oil and gas can be brought in by ship. What steps would you take in the next five years to reduce the cost of the energy you use with the least damage to the island's natural environment?
5. How might better matching of end uses with potential sources yield improvements in energy efficiency?
6. Complete an energy audit of the building you live in, then develop recommendations that might lead to lower utility bills.
7. How might plans using the concept of integrated energy management differ for the Los Angeles area and the New York City area? How might both of these plans differ from an energy plan for Mexico City, which is quickly becoming one of the largest urban areas in the world?
8. A recent energy scenario for the United States suggests that in the coming decades energy sources might be natural gas (10%), solar power (30%), hydropower (20%), wind power (20%), biomass (10%), and geothermal energy (10%). Do you think this is a likely scenario? What would be the major difficulties and points of resistance or controversy?

FURTHER READING

Botkin, D.B. 2010. *Powering the Future.* Pearson Education Inc. Upper Saddle River, N.J. An up-to-date Summary of energy sources and planning for the future.

Lindley, D. The energy should always work twice. *Nature* 458, no. 7235 (2009):138–141. A good paper on cogeneration.

Lovins, A.B., *Soft Energy Paths: Towards a Durable Peace* (New York: Harper & Row. 1979). A classic energy book.

McKibben, B. Energizing America. *Sierra* 92, no. 1(2007):30–38; 112–113. A good recent summary of energy for the future.

Miller, P. Saving energy. *National Geographic* 251, no. 3 (2009): 60–81. Many ways to conserve energy.

Wald, M.L. 2009. The power of renewables. *Scientific American* 300, no. 3 (2009):57–61. A good summary of renewable energy.

CHAPTER 15

Fossil Fuels and the Environment

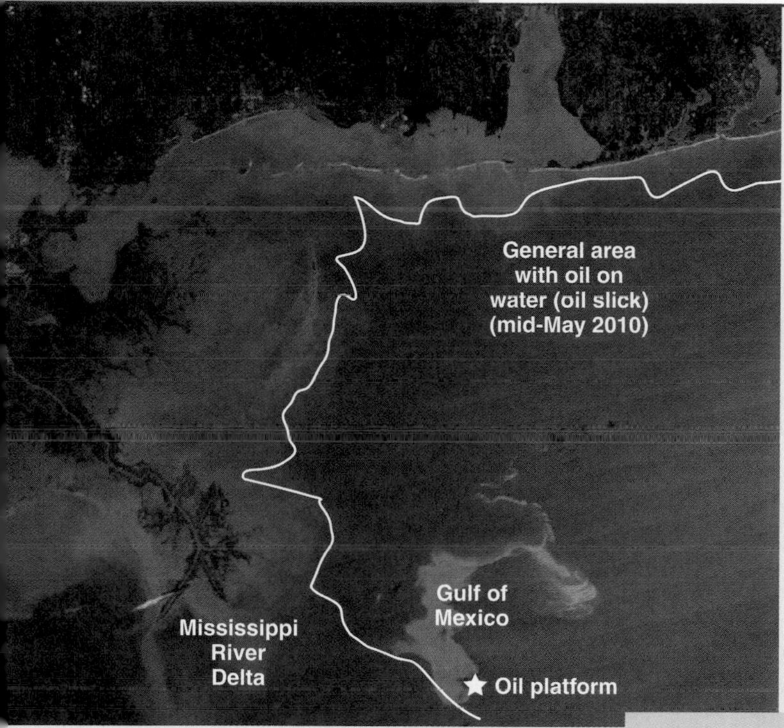

LEARNING OBJECTIVES

We rely almost completely on fossil fuels—oil, natural gas, and coal—for our energy needs. However, these are nonrenewable resources, and their production and use have a variety of serious environmental impacts. After reading this chapter, you should understand . . .

- Why we may have serious, unprecedented supply problems with oil and gasoline within the next 20 to 50 years;
- How oil, natural gas, and coal form;
- What the environmental effects are of producing and using oil, natural gas, and coal.

Oil is the most important fossil fuel today, but working in oil fields has never been easy. Drilling for oil is difficult, dangerous and potentially damaging to the environment. On April 20, 2010 a blowout on the platform Deepwater Horizon in the Gulf of Mexico while drilling in water about 1.6 km (1 mile) deep occurred. Explosive gas rose up the well and exploded. Eleven men were killed, the platform destroyed, and oil leaked for 3 months at a rate of 36,000 to 60,000 barrels of oil per day (1 barrel is 42 gallons).By mid-July about 5 million barrels of oil were in the water column or on the surface and moving with the currents. The high altitude image shows the extent of the spill, off the coast of Louisiana, The smaller photo is of a wetland contaminated with oil. The oil by mid-July had washed up in variable amounts as thick oil or tar balls on beaches and coastal wetlands from Texas to Florida.The spill is the largest in U.S. history, far exceeding the Exxon Valdez in Alaska (1989). A temporary cap on the well first stopped the leak on July 17. Relief wells drilled "killed" the well.

CASE STUDY

Peak Oil: Are We Ready for It?

People in the wealthier countries have grown prosperous and lived longer during the past century as a result of abundant low-cost energy in the form of crude oil. The benefits of oil are undeniable, but so are the potential problems they create, from air and water pollution to climate change. In any case, we are about to learn what life will be like with less, more expensive oil. The question is no longer whether the peak in production will come, but when it will come and what the consequences to a society's economics and politics will be.[1] The peak, or **peak oil**, is the time when one-half of Earth's oil has been exploited.

The global history of oil in terms of rate of discovery and consumption is shown in Figure 15.1. Notice that in 1940 five times as much oil was discovered as was consumed; by 1980 the amount discovered equaled the amount consumed; and in the year 2000 the consumption of oil was three times the amount discovered. Obviously, the trend is not sustainable.

The concept of peak oil production is shown in Figure 15.2. We aren't sure what the peak production will be, but let's assume it will be about 40–50 billion barrels (bbl) per year and that the peak will arrive sometime between 2020 and 2050. In 2004 the growth rate for oil production was 3.4%. Moving from the present production rate of about 31 billion barrels per year (85 million bbl per day) to 50 billion barrels in a few decades is an optimistic estimate that may not be realized. Several oil company executives believe that even 40 billion barrels per year will be difficult. For the past several years, production has been flat, at about 30 billion barrels per year, leading some to believe that the peak is close.[2, 3]

When production peaks, and if demand increases, a gap between production and demand will result. If demand exceeds supply, the cost will rise, as it did in 2008. The price of a barrel of oil doubled from 2007 to mid-2008, and a gallon of gasoline in the United States approached $5 (Figure 15.3), causing a lot of anxiety for consumers. However, the latter part of 2008 saw the cost of oil drop more than 50% from its earlier high, and gasoline prices fell below $2 a gallon. The price was about $40/barrel by April 2009, but rose again, to about $65/barrel, by July. The instability in the cost of oil and gasoline in the first years of the 21st century reflects uncertainty about supplies because of wars and the delivery/refining processes.

We have time now to prepare for the eventual peak and to use the fossil fuels we have more carefully during the time of transition to other energy sources. If we have not prepared for the peak, then disruption to society is likely. In the best scenario, the transition from oil will not occur until we have cost-competitive alternatives in place.[2,4] Alternatives for liquid fuels include conservation (using less); producing massive amounts of biofuel from corn, sugarcane, and other plants; turning our vast coal reserves into liquid fuel; and developing other conventional sources of oil, including tar sands and oil shale. With the exception of conservation, all these have potentially significant environmental consequences. We will return to the concept of peak oil in the Critical Thinking exercise at the end of the chapter.

The approaching peak in oil production is a wake-up call, reminding us that although we will not run out of oil,

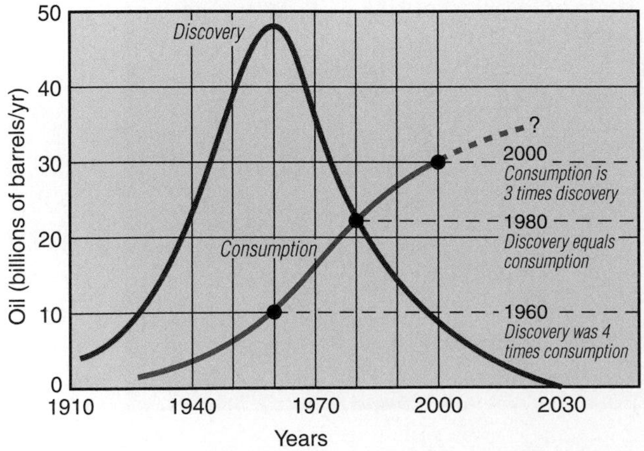

FIGURE 15.1 Discovery of oil peaked around 1960, and consumption exceeded discovery by 1980. *Source:* Modified after K. Alekett, "Oil: A Bumpy Road Ahead," *World Watch 19*, no. 1(2006):10–12.)

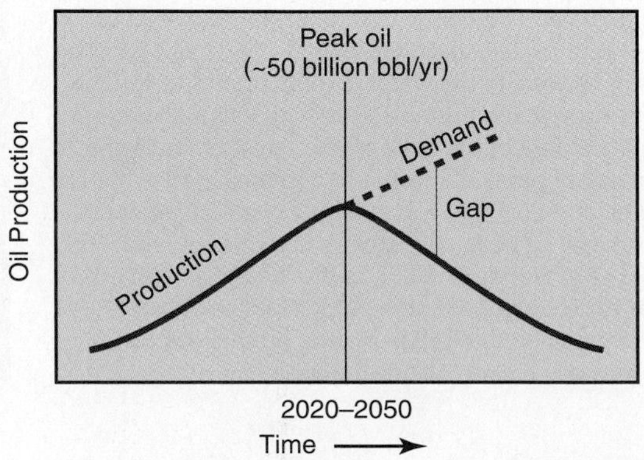

FIGURE 15.2 Idealized diagram of world oil production and peak between 2020 and 2050. When production cannot meet demand, a gap (shortage) develops.

FIGURE 15.3 A California gas station in 2008, at the height of gas pricing.

it will become much more expensive, and that there will be supply problems as demand increases by about 50% in the next 30 years. The peak in world oil production, when it arrives, will be unlike any problem we have faced in the past. The human population will increase by several billion in the coming decades, and countries with growing economies, such as China and India, will consume more oil. China, in fact, expects to double its import of oil in the next five years! Clearly, the social, economic, and political ramifications of peak oil will be enormous. Planning now for ways to conserve oil and transition to alternative energy sources will be critical in the coming decades.[4] We cannot afford to leave the age of oil until alternatives are firmly in place. The remainder of this chapter will discuss the various fossil fuels and their uses.

15.1 Fossil Fuels

Fossil fuels are forms of stored solar energy. Plants are solar energy collectors because they can convert solar energy to chemical energy through photosynthesis (see Chapter 6). The main fossil fuels used today were created from incomplete biological decomposition of dead organic matter (mostly land and marine plants). Buried organic matter that was not completely oxidized was converted by chemical reactions over hundreds of millions of years to oil, natural gas, and coal. Biological and geologic processes in various parts of the geologic cycle produce the sedimentary rocks in which we find these fossil fuels.[5, 6]

The major fossil fuels—crude oil, natural gas, and coal—are our primary energy sources; they provide approximately 90% of the energy consumed worldwide (Figure 15.4). World energy consumption grew about 1.4% in 2008. The largest increase, 7.2%, was in China. In the United States, consumption dropped about 2.8%. Most of the global increase (75%) was due to burning coal in China.[7] Globally, oil and natural gas provide 70 to 80% of the primary energy. Two exceptions are Asia, which uses a lot of coal, and the Middle East, where oil and gas provide nearly all of the energy. In this chapter, we focus primarily on these major fossil fuels. We also briefly discuss two other fossil fuels, oil shale and tar

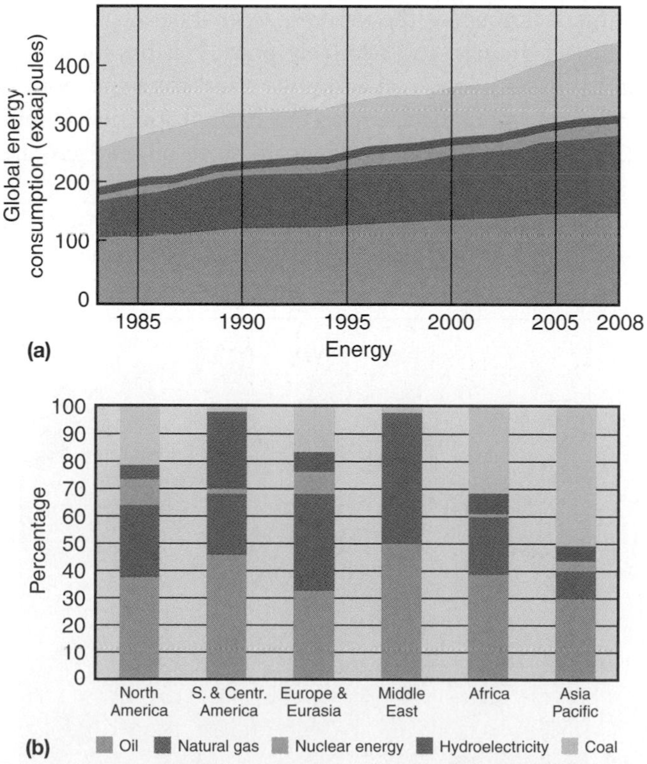

FIGURE 15.4 **(a)** World energy consumption (in exajoules) by primary source from 1983 to 2008; **(b)** world energy consumption by primary sources in 2008. (*Source*: *BP Statistical Review of World Energy 2010.* BP p.l.c.)

sands, that may become increasingly important as oil, gas, and coal reserves are depleted.

15.2 Crude Oil and Natural Gas

Most geologists accept the hypothesis that **crude oil** (petroleum) and **natural gas** are derived from organic materials (mostly plants) that were buried with marine or lake sediments in what are known as *depositional basins.* Oil and gas are found primarily along geologically young tectonic belts at plate boundaries, where large depositional basins are more likely to occur (see Chapter 6). However, there are exceptions, such as in Texas, the Gulf of Mexico, and the North Sea, where oil has been discovered in depositional basins far from active plate boundaries.

The source material, or *source rock*, for oil and gas is fine-grained (less than 1/16 mm, or 0.0025 in., in diameter), organic-rich sediment buried to a depth of at least 500 m (1,640 ft), where it is subjected to increased heat and pressure. The elevated temperature and pressure initiate the chemical transformation of the sediment's organic material into oil and gas. The pressure compresses the sediment; this, along with the elevated temperature in the source rock, initiates the upward migration of the oil and gas, which are relatively light, to a lower-pressure environment (known as the *reservoir rock*). The reservoir rock is coarser-grained and relatively porous (it has more and larger spaces between the grains). Sandstone and porous limestone, which have a relatively high proportion (about 30%) of empty space in which to store oil and gas, are common reservoir rocks.

As mentioned, oil and gas are light; if their upward mobility is not blocked, they will escape to the atmosphere. This explains why oil and gas are not generally found in geologically old rocks. Oil and gas in rocks older than about 0.5 billion years have had ample time to migrate to the surface, where they have either vaporized or eroded away.[6]

The oil and gas fields from which we extract resources are places where the natural upward migration of the oil and gas to the surface is interrupted or blocked by what is known as a *trap* (Figure 15.5). The rock that helps form the trap, known as the *cap rock*, is usually a very fine-grained sedimentary rock, such as shale, composed of silt and clay-sized particles. A favorable rock structure, such as an anticline (arch-shaped fold) or a fault (fracture in the rock along which displacement has occurred), is necessary to form traps, as shown in Figure 15.5. The important concept is that the combination of favorable rock structure and the presence of a cap rock allow deposits of oil and gas to accumulate in the geologic environment, where they are then discovered and extracted.[6]

Petroleum Production

Production wells in an oil field recover oil through both primary and enhanced methods. *Primary production* involves simply pumping the oil from wells, but this method can recover only about 25% of the petroleum in the reservoir. To increase the amount of oil recovered to about 60%, enhanced methods are used. In *enhanced* recovery, steam, water, or chemicals, such as carbon dioxide or nitrogen gas, are injected into the oil reservoir to push the oil toward the wells, where it can be more easily recovered by pumping.

Next to water, oil is the most abundant fluid in the upper part of the Earth's crust. Most of the known, proven oil reserves, however, are in a few fields. *Proven* oil reserves are

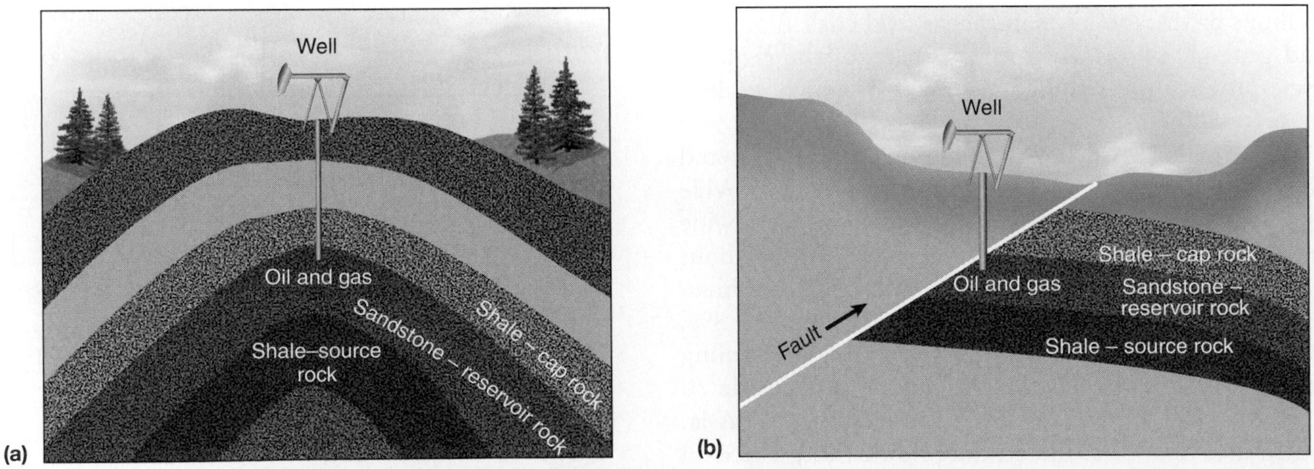

FIGURE 15.5 **Two types of oil and gas traps: (a)** anticline and **(b)** fault.

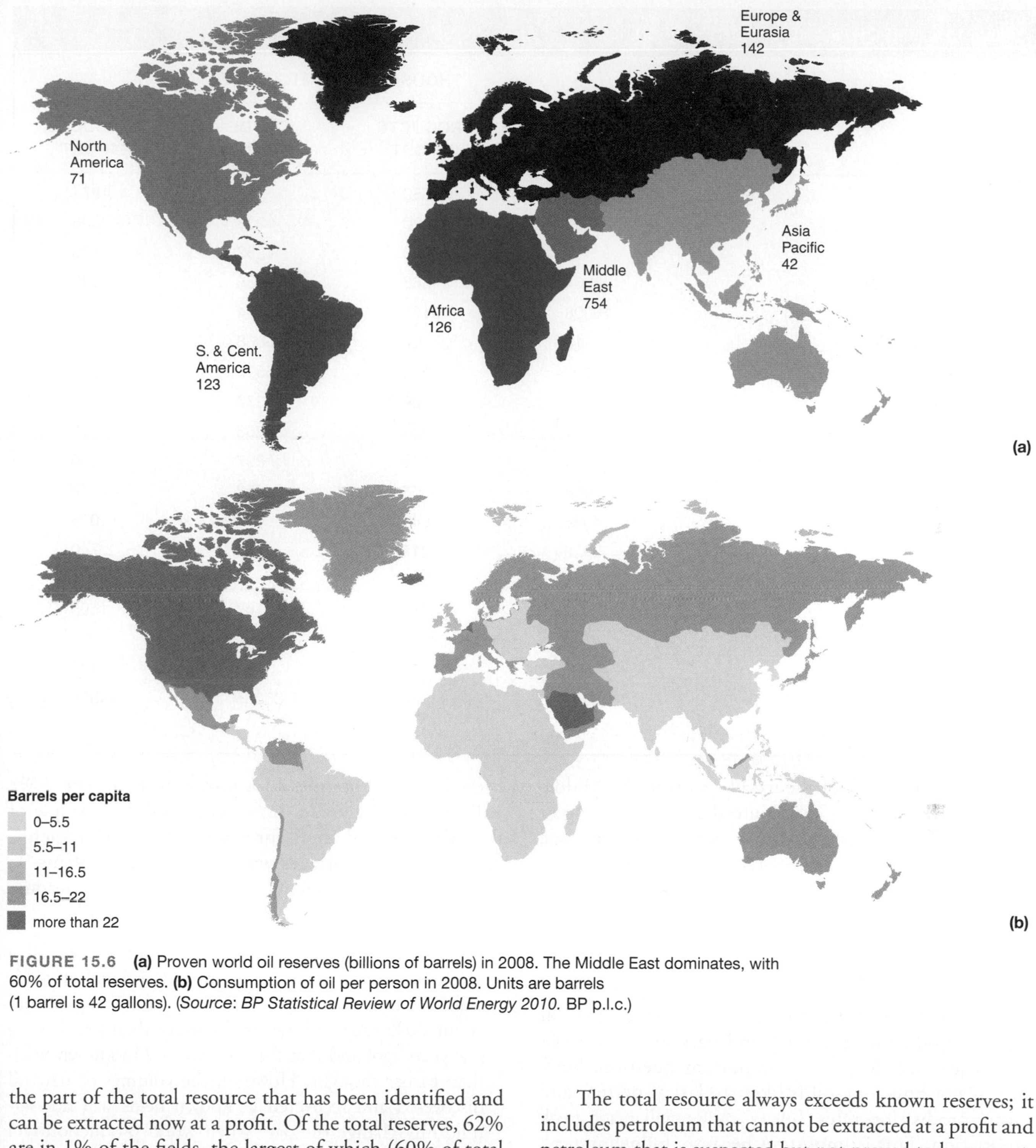

FIGURE 15.6 **(a)** Proven world oil reserves (billions of barrels) in 2008. The Middle East dominates, with 60% of total reserves. **(b)** Consumption of oil per person in 2008. Units are barrels (1 barrel is 42 gallons). (*Source*: *BP Statistical Review of World Energy 2010.* BP p.l.c.)

the part of the total resource that has been identified and can be extracted now at a profit. Of the total reserves, 62% are in 1% of the fields, the largest of which (60% of total known reserves) are in the Middle East (Figure 15.6a). The consumption of oil per person is shown in Figure 15.6b. Notice the domination of energy use in North America. Although new oil and gas fields have recently been and continue to be discovered in Alaska, Mexico, South America, and other areas of the world, the present known world reserves may be depleted in the next few decades.

The total resource always exceeds known reserves; it includes petroleum that cannot be extracted at a profit and petroleum that is suspected but not proved to be present. Several decades ago, the amount of oil that ultimately could be recovered (the total resource) was estimated to be about 1.6 trillion barrels. Today, that estimate is just over 2 trillion barrels.[8] The increases in proven reserves of oil in the last few decades have primarily been due to discoveries in the Middle East, Venezuela, Kazakhstan, and other areas.

Table 15.1 IMPORTS AND EXPORTS OF OIL, 2010

	THOUSAND BARRELS DAILY			
	CRUDE IMPORT	PRODUCTS IMPORT	CRUDE EXPORT	PRODUCTS EXPORT
US	8893	2550	44	1871
Canada	785	320	1938	538
Mexico	9	439	1282	168
S. & Cent. America	504	863	2588	1137
Europe	10308	3177	464	1523
Former Soviet Union	18	67	6868	2197
Middle East	140	219	16510	1916
North Africa	369	209	2232	528
West Africa	1	254	4263	110
East & Southern Africa	439	119	297	6
Australasia	458	358	258	42
China	4086	1041	94	614
India	2928	217	1.9	740
Japan	3545	738	–	345
Singapore	930	1668	47	1505
Other Asia Pacific	4590	2667	807	1252
Unidentified*	–	18	311	430
Total World	38005	14925	38005	14925

Source: *BP Statistical Review of World Energy* (2010.B.P. p.l.c.)

Because so much of the world's oil is in the Middle East, oil revenues have flowed into that area, resulting in huge trade imbalances. Table 15.1 shows the major trade for oil. The United States imports oil from Venezuela, the Middle East, Africa, Mexico, Canada, and Europe. Japan is dependent on oil from the Middle East and Africa.

Oil in the 21st Century

Recent estimates of proven oil reserves suggest that, at present production rates, oil and natural gas will last only a few decades.[7, 9] The important question, however, is not how long oil is likely to last at present and future production rates, but when we will reach peak production. This is important because, following peak production, less oil will be available, leading to shortages and price shocks. World oil production, as mentioned in the opening case study, is likely to peak between the years 2020 and 2050, within the lifetime of many people living today.[10] Even those who think peak oil production in the near future is a myth acknowledge that the peak is coming and that we need to be prepared.[2] Whichever projections are correct, there is a finite amount of time (a few decades or perhaps a bit longer) left in which to adjust to potential changes in lifestyle and economies in a post-petroleum era.[10] We will never entirely run out of crude oil, but people of the world depend on oil for nearly 40% of their energy, and significant shortages will cause major problems.[11]

Consider the following argument that we are heading toward a potential crisis in availability of crude oil:

- We are approaching the time when approximately 50% of the total crude oil available from traditional oil fields will have been consumed.[10] Recent studies suggest that about 20% more oil awaits discovery than predicted a few years ago, and that there is more oil in known fields than earlier thought. However, the volumes of new oil discovered and recovered in known fields will not significantly change the date when world production will peak and a decline in production will begin.[9] This point is controversial. Some experts believe that modern technology for exploration, drilling, and recovery of oil will ensure an adequate supply of oil for the distant future.[8]
- Proven reserves are about 1.3 trillion barrels.[7] It is estimated that approximately 2 trillion barrels of crude oil may ultimately be recovered from remaining oil resources. World production today is about 31 billion barrels per year (85 million barrels per day), and we are using what is left very rapidly.[2]

- Today, for every three barrels of oil we consume, we are finding only one barrel.[11] In other words, output is three times higher than input. However, this could improve in the future.[8]
- Forecasts that predict a decline in oil production are based on the estimated amount of oil that may ultimately be recoverable (2 trillion barrels, nearly two times today's proven reserves), along with projections of new discoveries and rates of future consumption. As already mentioned, it has been estimated that the peak in world crude oil production, about 40 billion bbl/yr, will occur between the years 2020 and 2050.[1,2,9,11] The production of 40 billion bbl/yr is about a 30% increase over 2007. Whether you think this increase is optimistic or pessimistic depends on your view of past oil history, in which oil has survived several predicted shortages, or beliefs that the peak is inevitable sooner than later.[1, 2] Most oil experts believe peak oil is only a few decades away.
- It is expected that U.S. production of oil as we know it now will end by about 2090 and that world production of oil will be nearly exhausted by 2100.[11]

Table 15.1 suggests that world exports of oil are about one-half of world production. We conclude that the other half is often used in the country that produced it. A prospect perhaps as significant as peak oil may be the time when exporting nations no longer have significant oil to export. This will occur in different exporting countries at different times and is sure to cause problems with global supply and demand.

What is an appropriate response to the likelihood that oil production will likely decline in the mid-21st century? First, we need an improved educational program to inform or remind people and governments of the potential depletion of crude oil and the consequences of shortages. Presently, many people seem to be in denial. Planning and appropriate action are necessary to avoid military confrontation (we have already had one oil war), food shortages (oil is used to make the fertilizers modern agriculture depends on), and social disruption. Before significant oil shortages occur, we need to develop alternative energy sources, such as solar energy and wind power, and perhaps rely more on nuclear energy. This is a proactive response to a potentially serious situation.

Natural Gas

We have only begun to seriously search for natural gas and to utilize this resource to its full potential. One reason for the slow start is that natural gas is transported primarily by pipelines, and only in the last few decades have these pipelines been constructed in large numbers. In fact, until recently, natural gas found with petroleum was often simply burned off as waste; in some cases this practice continues.[12]

The worldwide estimate of recoverable natural gas is about 185 trillion cubic meters (m^3), which, at the current rate of world consumption, will last approximately 60 years.[7] Considerable natural gas was recently discovered in the United States, and at present U.S. consumption levels (0.7 trillion m^3 in 2008) this resource is expected to last at least 30 years. Furthermore, new supplies are being found in surprisingly large amounts, particularly at greater depths than those at which oil is found. Optimistic estimates of the total resource suggest that, at current rates of consumption, the supply may last approximately 100 years.[5]

This possibility has important implications. Natural gas is considered a clean fuel; burning it produces fewer pollutants than does burning oil or coal, so it causes fewer environmental problems than do the other fossil fuels. As a result, it is being considered as a possible transition fuel from other fossil fuels (oil and coal) to alternative energy sources, such as solar power, wind power, and hydropower.

Despite the new discoveries and the construction of pipelines, long-term projections for a steady supply of natural gas are uncertain. The supply is finite, and at present rates of consumption it is only a matter of time before the resources are depleted.

Coal-Bed Methane

The processes responsible for the formation of coal include partial decomposition of plants buried by sediments that slowly convert the organic material to coal. This process also releases a lot of methane (natural gas) that is stored within the coal. The methane is actually stored on the surfaces of the organic matter in the coal, and because coal has many large internal surfaces, the amount of methane for a given volume of rock is something like seven times more than could be stored in gas reservoirs associated with petroleum. The estimated amount of coal-bed methane in the United States is more than 20 trillion cubic meters, of which about 3 trillion cubic meters could be recovered economically today with existing technology. At current rates of consumption in the United States, this represents about a five-year supply of methane.[13]

Two areas within the nation's coalfields that are producing methane are the Wasatch Plateau in Utah and the Powder River Basin in Wyoming. The Powder River Basin is one of the world's largest coal basins, and presently an energy boom is occurring in Wyoming, producing an "energy rush." The technology to recover coal-bed methane is a young one, but it is developing quickly. As of early 2003, approximately 10,000 shallow wells were producing methane in the Powder River Basin, and some say there will eventually be about 100,000 wells. The big advantage of the coal-bed methane wells is that they only need to be drilled to shallow depths (about 100 m, or a few hundred feet). Drilling can be done with conventional water-well technology, and the cost is about $100,000 per well, compared to several million dollars for an oil well.[14]

Coal-bed methane is a promising energy source that comes at a time when the United States is importing vast amounts of energy and attempting to evaluate a transition from fossil fuels to alternative fuels. However, coal-bed methane presents several environmental concerns, including (1) disposal of large volumes of water produced when the methane is recovered and (2) migration of methane, which may contaminate groundwater or migrate into residential areas.

A major environmental benefit of burning coal-bed methane, as well as methane from other sources, is that its combustion produces a lot less carbon dioxide than does the burning of coal or petroleum. Furthermore, production of methane gas prior to mining coal reduces the amount of methane that would be released into the atmosphere. Both methane and carbon dioxide are strong greenhouse gases that contribute to global warming. However, because methane produces a lot less carbon dioxide, it is considered one of the main transitional fuels from fossil fuels to alternative energy sources.

Of particular environmental concern in Wyoming is the safe disposal of salty water that is produced with the methane (the wells bring up a mixture of methane and water that contains dissolved salts from contact with subsurface rocks). Often, the water is reinjected into the subsurface, but in some instances the water flows into surface drainages or is placed in evaporation ponds.[13]

Some of the environmental conflicts that have arisen are between those producing methane from wells and ranchers trying to raise cattle on the same land. Frequently, the ranchers do not own the mineral rights; and although energy companies may pay fees for the well, the funds are not sufficient to cover damage resulting from producing the gas. The problem results when the salty water produced is disposed of in nearby streams. When ranchers use the surface water to irrigate crops for cattle, the salt damages the soils, reducing crop productivity. Although it has been argued that ranching is often a precarious economic venture, and that ranchers have in fact been saved by the new money from coal-bed methane, many ranchers oppose coal-bed methane production without an assurance that salty waters will be safely disposed of.

People are also concerned about the sustainability of water resources as vast amounts of water are removed from the groundwater aquifers. In some instances, springs have been reported to have dried up after coal-bed methane extraction in the area.[14] In other words, the "mining" of groundwater for coal-bed methane extraction will remove water that has perhaps taken hundreds of years to accumulate in the subsurface environment.

Another concern is the migration of methane away from the well sites, possibly to nearby urban areas. The problem is that unlike the foul-smelling variety in homes, methane in its natural state is odorless as well as explosive. For example, in the 1970s an urban area near Gallette, Wyoming, had to be evacuated because methane was migrating into homes from nearby coal mines.

Finally, coal-bed methane wells, with their compressors and other equipment, have caused people living a few hundred meters away to report serious and distressing noise pollution.[14]

In sum, coal-bed methane is a tremendous source of energy and relatively clean-burning, but its extraction must be closely evaluated and studied to minimize environmental degradation.

Black Shale (tight) Natural Gas

According to the U.S. Geological Survey, Black Devonian shale over 350 million years old buried a kilometer or so beneath northern Appalachia, contains about 500 trillion cubic feet of natural gas (mostly methane) of which10% or more may be ultimately recovered. The methane is distributed throughout the black shale as an unconventional gas resource compared to gas fields where the methane is in rock pockets often associated with oil. A very large area including parts of Ohio, New York, Pennsylvania, Virginia and Kentucky. Recovery of the methane is costly, because deep wells that turn at depth to a horizontal position are necessary to extract the gas. Water and other chemicals are used to fracture the rocks (hydrofracturing) to recover the gas. An energy rush is now occurring to develop the recovery of tight natural gas. Hundreds of gas wells have already been permitted in Pennsylvania alone. There is concern that drilling and hydrofracturing could result in water pollution, because the fluids used to fracture the rock must be recovered from wells and disposed of before gas production starts. There is also concern that contaminate water could migrate upward and leak from wells to pollute water supplies. The city of New York is very concerned that drilling may contaminated their water supply in upstate New York.[15]

Methane Hydrates

Beneath the seafloor, at depths of about 1,000 meters, there exist deposits of **methane hydrate**, a white, ice-like compound made up of molecules of methane gas (CH_4), molecular "cages" of frozen water. The methane has formed as a result of microbial digestion of organic matter in the sediments of the seafloor and has become trapped in these ice cages. Methane hydrates in the oceans were discovered over 30 years ago and are widespread in both the Pacific and Atlantic oceans. Methane hydrates are also found on land; the first ones discovered were in permafrost areas of Siberia and North America, where they are known as marsh gas.[16]

Methane hydrates in the ocean occur where deep, cold seawater provides high pressure and low temperatures. They are not stable at lower pressure and warmer temperatures. At a water depth of less than about 500 m, methane hydrates decompose rapidly, freeing methane gas from the ice cages to move up as a flow of methane bubbles (like rising helium balloons) to the surface and the atmosphere.

In 1998 researchers from Russia discovered the release of methane hydrates off the coast of Norway. During the release, scientists documented plumes of methane gas as tall as 500 m being emitted from methane hydrate deposits on the seafloor. It appears that there have been large emissions of methane from the sea. The physical evidence includes fields of depressions, looking something like bomb craters, that pockmark the seafloor near methane hydrate deposits. Some of the craters are as large as 30 m deep and 700 m in diameter, suggesting that they were produced by rapid, if not explosive, eruptions of methane.

Methane hydrates in the marine environment are a potential energy resource with approximately twice as much energy as all the known natural gas, oil, and coal deposits on Earth.[16] Methane hydrates are particularly attractive to countries such as Japan that rely exclusively on foreign oil and coal for their fossil fuel needs. Unfortunately, mining methane hydrates will be a difficult task, at least for the near future. The hydrates tend to be found along the lower parts of the continental slopes, where water is often deeper than 1 km. The deposits themselves extend into the seafloor sediments another few hundred meters. Drilling rigs have more problems operating safely at these depths, and developing a way to produce the gas and transport it to land will be challenging.

(a) (b)

FIGURE 15.7 Drilling for oil in **(a)** the Sahara Desert of Algeria; **(b)** the Cook Inlet of southern Alaska.

The Environmental Effects of Oil and Natural Gas

Recovering, refining, and using oil—and to a lesser extent natural gas—cause well-known, documented environmental problems, such as air and water pollution, acid rain, and global warming. People have benefited in many ways from abundant, inexpensive energy, but at a price to the global environment and human health.

Recovery

Development of oil and gas fields involves drilling wells on land or beneath the seafloor (Figure 15.7).

Possible environmental impacts on land include the following:

- Use of land to construct pads for wells, pipelines, and storage tanks and to build a network of roads and other production facilities.
- Pollution of surface waters and groundwater from (1) leaks from broken pipes or tanks containing oil or other oil-field chemicals and (2) salty water (brine) brought to the surface in large volumes with the oil. The brine is toxic and may be disposed of by evaporation in lined pits, which may leak. Alternatively, it may be disposed of by pumping it into the ground, using deep disposal wells outside the oil fields. However, disposal wells may pollute groundwater.
- Accidental release of air pollutants, such as hydrocarbons and hydrogen sulfide (a toxic gas).
- Land subsidence (sinking) as oil and gas are withdrawn.
- Loss or disruption of and damage to fragile ecosystems, such as wetlands or other unique landscapes. This is the center of the controversy over the development of petroleum resources in pristine environments such as the Arctic National Wildlife Refuge in Alaska (see A Closer Look 15.1).

Environmental impacts associated with oil production in the marine environment include the following:

- Oil seepage into the sea from normal operations or large spills from accidents, such as blowouts or pipe ruptures (see photograph opening this chapter). The very serious oil spill in the Gulf of Mexico (April 20, 2010) began from a blowout when equipment designed to prevent a blowout for this well drilled in very deep water (over 1.5 km,1 mile) failed to operate properly. The platform was destroyed by a large explosion and 11 oil workers were killed. By the middle of May oil started to make landfall in Louisiana and other areas, and fishing was shut down over a large area. The oil spread despite the many and continuing efforts (chemical dispersants, skimming and burning among others) to deter the oil from spreading. The spill was the largest and potentially most damaging in U.S. history (see Chapter 19 for a discussion of other oil spills and Chapter 24 for a more detailed case study of the 2010 Gulf spill).
- Release of drilling muds (heavy liquids injected into the borehole during drilling to keep the hole open). These contain heavy metals, such as barium, which may be toxic to marine life.
- Aesthetic degradation from the presence of offshore oil-drilling platforms, which some people consider unsightly.

Refining

Refining crude oil and converting it to products also has environmental impacts. At refineries, crude oil is heated so that its components can be separated and collected (this process is called *fractional distillation*). Other industrial processes then make products such as gasoline and heating oil.

Refineries may have accidental spills and slow leaks of gasoline and other products from storage tanks and pipes. Over years of operation, large amounts of liquid hydrocarbons may be released, polluting soil and groundwater below the site. Massive groundwater-cleaning projects have been required at several West Coast refineries.

Crude oil and its distilled products are used to make fine oil, a wide variety of plastics, and organic chemicals used by society in huge amounts. The industrial processes involved in producing these chemicals have the potential to release a variety of pollutants into the environment.

Delivery and Use

Some of the most extensive and significant environmental problems associated with oil and gas occur when the fuel is delivered and consumed. Crude oil is mostly transported on land in pipelines or across the ocean by tankers; both methods present the danger of oil spills. For example, a bullet from a high-powered rifle punctured the Trans-Alaska Pipeline in 2001, causing a small but damaging oil spill. Strong earthquakes may pose a problem for pipelines in the future, but proper engineering can minimize earthquake hazard. The large 2002 Alaskan earthquake ruptured the ground by several meters where it crossed the Trans-Alaska Pipeline. The pipeline's design prevented damage to the pipeline and to the environment. Although most effects of oil spills are relatively short-lived (days to years), marine spills have killed thousands of seabirds, spoiled beaches for decades (especially beneath the surface of gravel beaches), and caused loss of tourist and fishing revenues (see Chapter 19).

A CLOSER LOOK 15.1

The Arctic National Wildlife Refuge: To Drill or Not to Drill

The Arctic National Wildlife Refuge (ANWR) on the North Slope of Alaska is one of the few pristine wilderness areas remaining in the world (Figure 15.8). The U.S. Geological Survey estimates that the refuge contains about 3 billion barrels

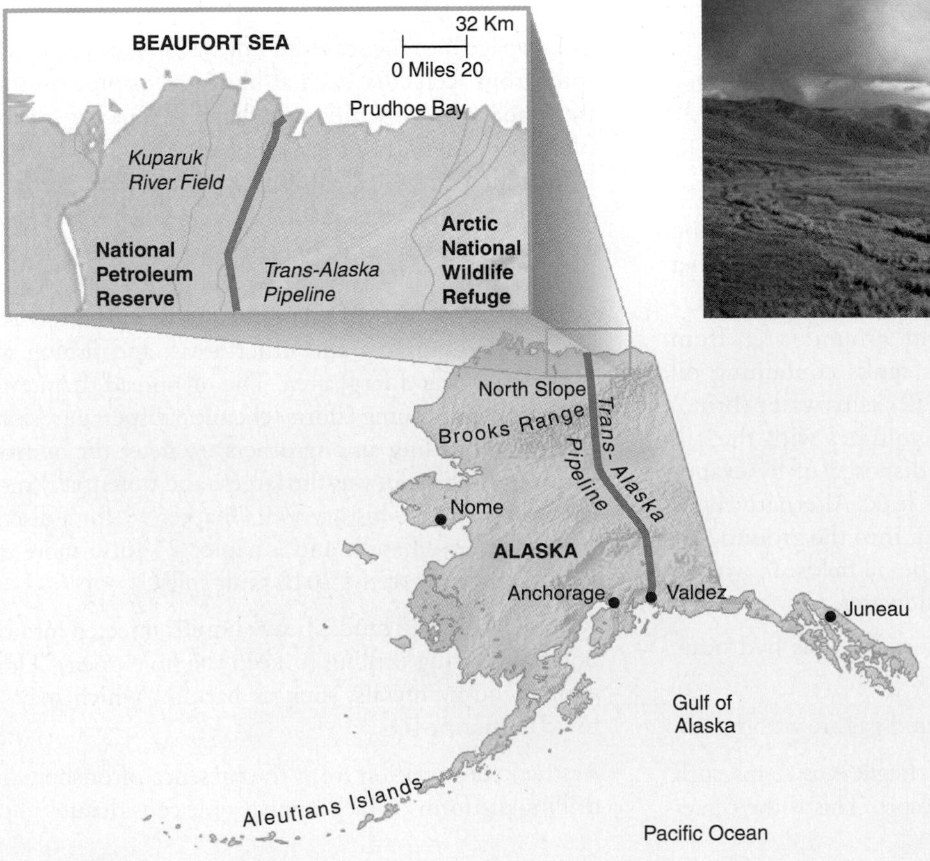

FIGURE 15.8 The Arctic National Wildlife Refuge, on Alaska's North Slope, is valued for its scenery, wildlife, and oil.

of recoverable oil. The United States presently consumes about 20 million barrels of oil per day, so ANWR could provide about a six-month supply if that were the only oil we used. Spread out to supply 1 million barrels per day, the supply at ANWR would last about eight years. According to the oil industry, several times more oil than that can be recovered. The oil industry has long argued in favor of drilling for oil in the ANWR, but the idea has been unpopular for decades among many members of the public and the U.S. government, and no drilling has been permitted. Former president George W. Bush favored drilling in the ANWR, which renewed the controversy over this issue. President Barack Obama is opposed to such drilling.

Arguments in Favor of Drilling in the ANWR

- The United States needs the oil, and it will help us to be more independent of imported oil.
- The unprecedented price increase in oil in 2008 has provided a big economic incentive to develop our domestic oil reserves.
- New oil facilities will bring jobs and dollars to Alaska.
- New exploration tools to evaluate the subsurface for oil pools require far fewer exploratory wells.
- New drilling practices have much less impact on the environment (Figure 15.9a and b). These include (1) constructing roads of ice in the winter that melt in the summer instead of constructing permanent roads; (2) elevating pipelines to allow for animal migration (Figure 15.9c); (3) drilling in various directions from a central location, thus minimizing land needed for wells; and (4) disposing of fluid oil-field wastes by putting them back into the ground to minimize surface pollution.
- The land area affected will be small relative to the total area.

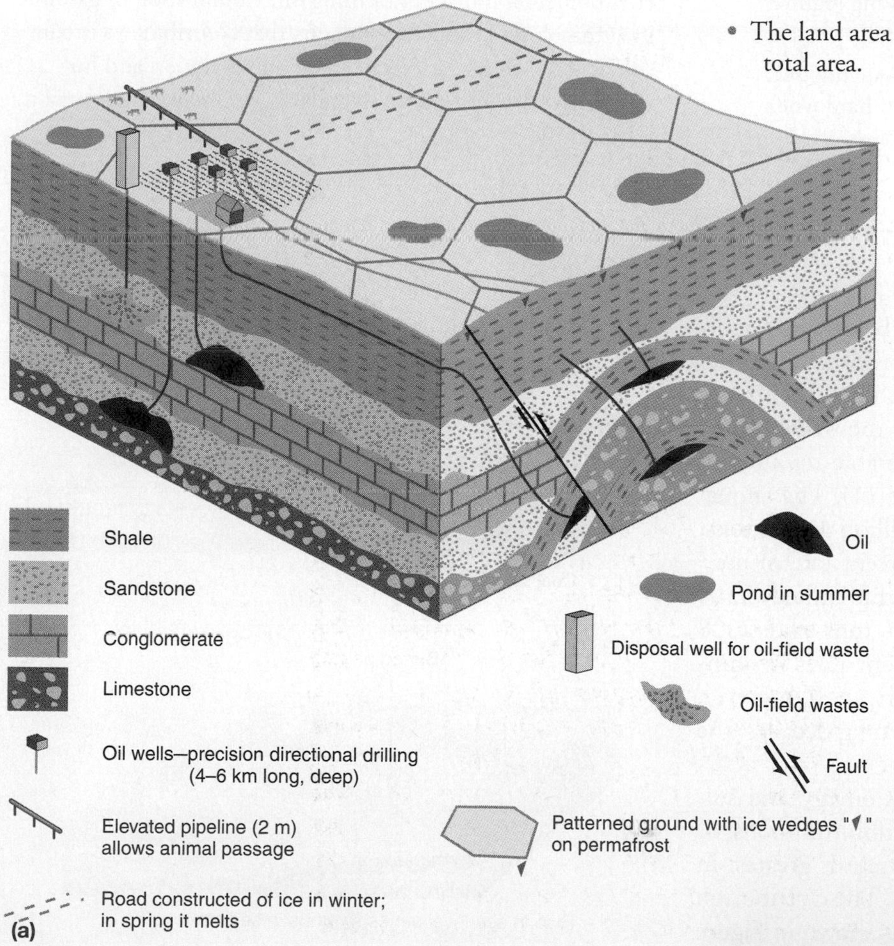

FIGURE 15.9 **(a)** Those in favor of ANWR drilling argue that new technology can reduce the impact of developing oil fields in the Arctic: Wells are located in a central area and use directional drilling; roads are constructed of ice in the winter, melting to become invisible in summer; pipelines are elevated to allow animals, in this case caribou, to pass through the area; and oil-field and drilling wastes are injected deep underground. (See text for arguments against drilling.) **(b)** Oil wells being drilled on frozen ground (tundra), North Slope of Alaska. **(c)** Caribou passing under a pipeline near the Arctic National Wildlife Refuge in Alaska.

- Many Alaskans want the drilling to proceed and point out that oil drilling for 30 years on the North Slope of Alaska (Prudhoe Bay) has not harmed animals or the environment.

Arguments against Drilling in the ANWR

- Advances in technology are irrelevant to the question of whether or not the ANWR should be drilled. Some wilderness should remain wilderness! Drilling will forever change the pristine environment of the North Slope.
- Even with the best technology, oil exploration and development will impact the ANWR. Intensive activity, even in winter on roads constructed of ice, will probably disrupt wildlife.
- Ice roads are constructed from water from the tundra ponds. To build a road 1 km (0.63 mi) long requires about 3,640 m^3 (1 million gallons) of water.
- Heavy vehicles used in exploration permanently scar the ground—even if the ground is frozen hard when the vehicles travel across the open tundra.
- Accidents may occur in even the best facilities.
- Oil development is inherently damaging because it involves a massive industrial complex of people, vehicles, equipment, pipelines, and support facilities.

Our decision about drilling in the Alaska National Wildlife Refuge will reflect both science and values at a basic level. New technology will lessen the environmental impact of oil drilling. How we value the need for energy from oil compared with how we value preservation of a pristine wilderness area will determine our action: to drill or not to drill. The debate is not over yet. The oil crisis of 2008 that drove the price of gasoline to nearly $5 per gallon placed ANWR oil development in the spotlight again.

Air pollution is perhaps the most familiar and serious environmental impact of burning oil. Combustion of gasoline in automobiles produces pollutants that contribute to urban smog. (The adverse effects of smog on vegetation and human health are well documented and are discussed in detail in Chapter 21.)

15.3 Coal

Partially decomposed vegetation, when buried in a sedimentary environment, may be slowly transformed into the solid, brittle, carbonaceous rock we call **coal**. This process is shown in Figure 15.10. Coal is by far the world's most abundant fossil fuel, with a total recoverable resource of about 825 billion metric tons (Figure 15.11). The annual world consumption of coal is about 7 billion metric tons, sufficient for about 120 years at the current rate of use.[7] There are about 18,500 coal mines in the United States with combined reserves of 262 billion tons and 2008 production of 1.2 billion tons. At present rates of mining, U.S. reserves will last nearly 250 years.[7, 17] If, however, consumption of coal increases in the coming decades, the resource will not last nearly as long.[18]

Coal is classified, depending on its energy and sulfur content, as anthracite, bituminous, subbituminous, or lignite (see Table 15.2). The energy content is greatest in anthracite coal and lowest in lignite coal. The distribution of coal in the contiguous United States is shown in Figure 15.12.

The sulfur content of coal is important because low-sulfur coal emits less sulfur dioxide (SO_2) and is therefore more desirable as a fuel for power plants. Most low-sulfur coal in the United States is the relatively low-grade, low-energy lignite and subbituminous coal found west of the Mississippi River. Power plants on the East

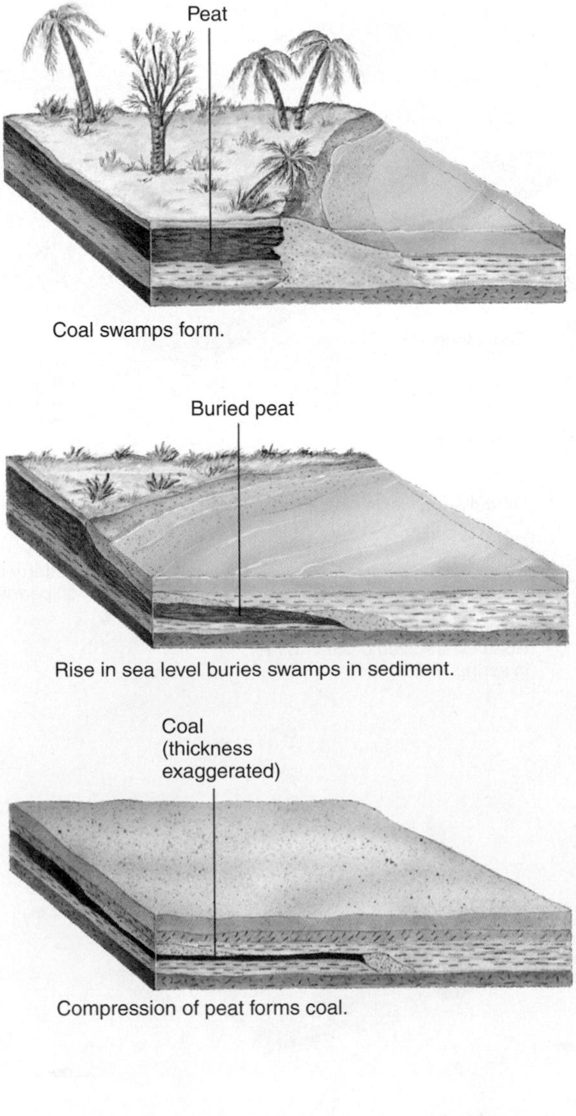
Peat
Coal swamps form.
Buried peat
Rise in sea level buries swamps in sediment.
Coal
(thickness
exaggerated)
Compression of peat forms coal.

FIGURE 15.10 Processes by which buried plant debris (peat) is transformed into coal.

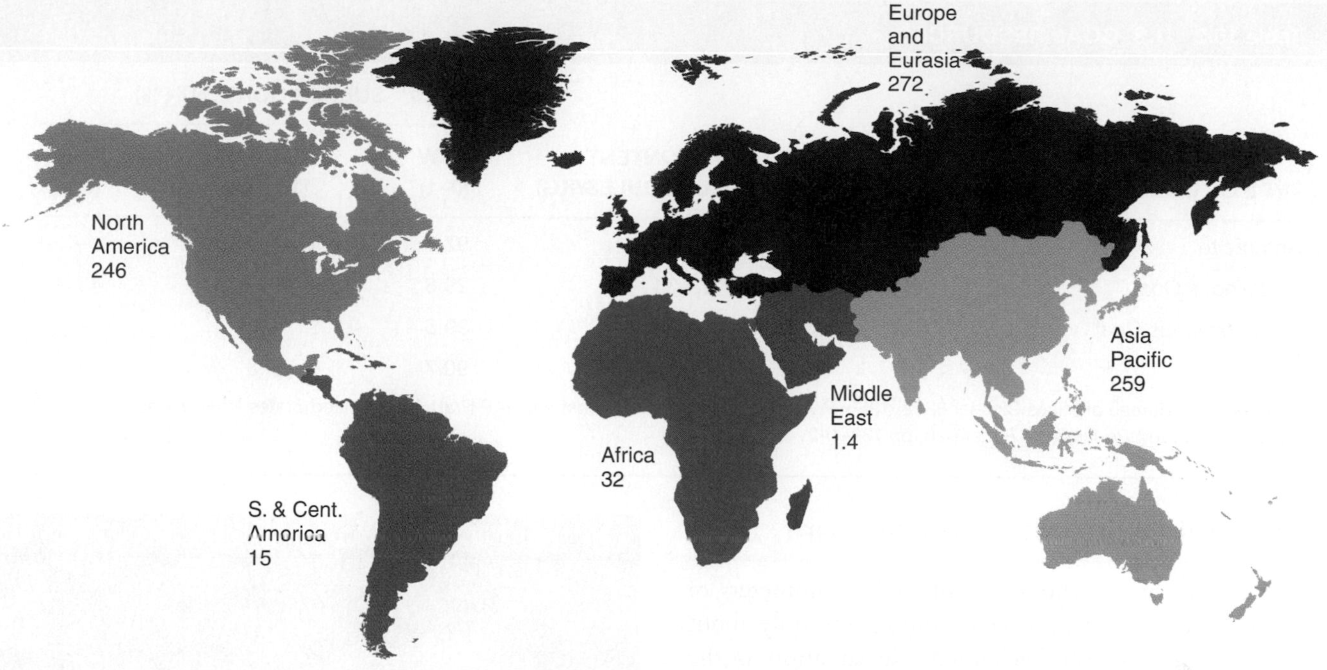

FIGURE 15.11 World coal reserves (billions of tons) in 2008. (*Source*: *BP Statistical Review of World Energy* 2009.B.P. p.l.c.).

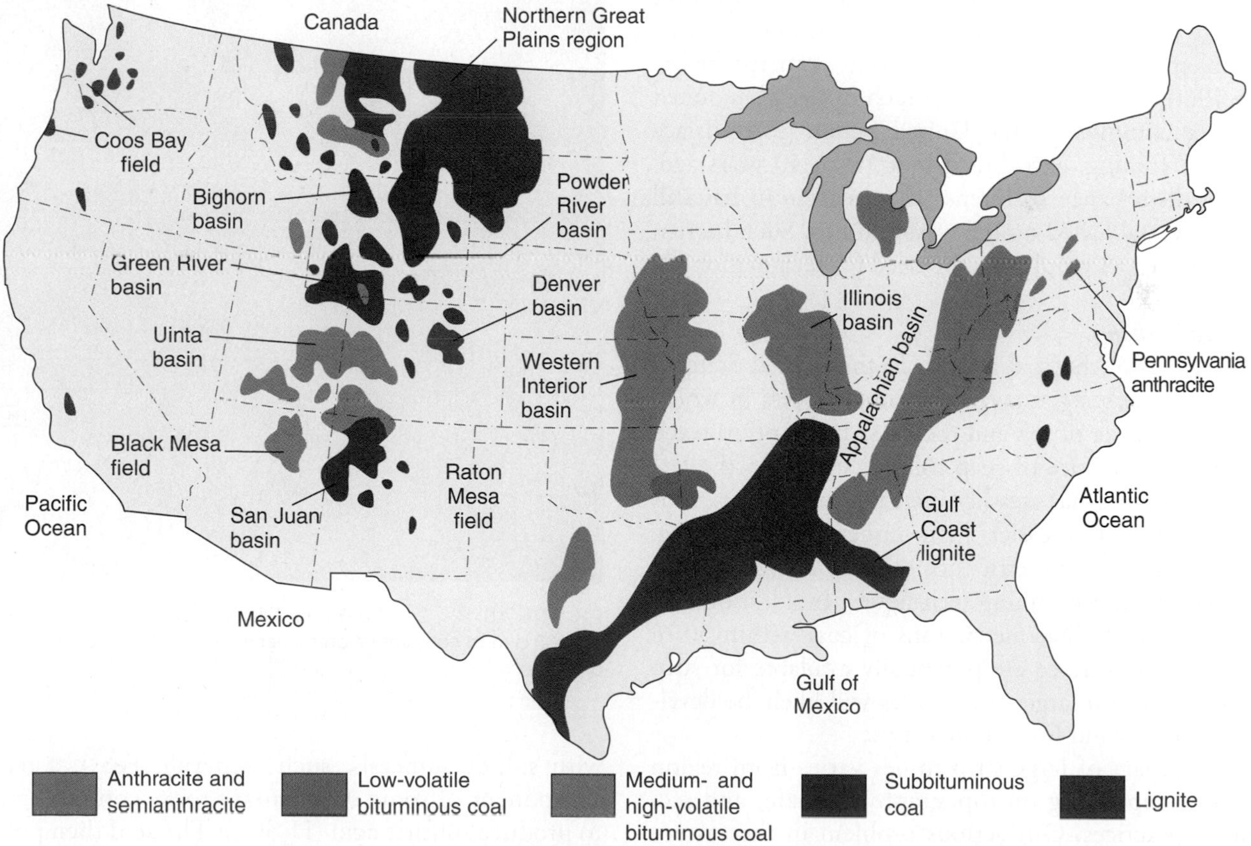

FIGURE 15.12 Coal areas of the contiguous United States. This is a highly generalized map, and numerous relatively small occurrences of coal are not shown. (*Source*: S. Garbini and S.P. Schweinfurth, *U.S. Geological Survey Circular 979, 1986.)*

Coast treat the high-sulfur coal mined in their own region to lower its sulfur content before, during, or after combustion and, thus, avoid excessive air pollution. Although it is expensive, treating coal to reduce pollution may be more economical than transporting low-sulfur coal from the western states.

Table 15.2 U.S. COAL RESOURCES

TYPE OF COAL	RELATIVE RANK	ENERGY OF CONTENT (MILLIONS OF JOULES/KG)	SULFUR CONTENT (%) LOW (0–1)	MEDIUM (1.1–3.0)	HIGH (3 +)
Anthracite	1	30–34	97.1	2.9	-
Bituminous Coal	2	23–34	29.8	26.8	43.4
Subbituminous Coal	3	16–23	99.6	0.4	-
Lignite	4	13–16	90.7	9.3	-

Sources: U.S. Bureau of Mines Circular 8312, 1966: P. Averitt, "Coal" in D.A. Brobst and W.P Pratt, eds., United States Mineral Resources, *U.S. Geological Survey, Professional Paper* 820, pp 133–142.

Coal Mining and the Environment

In the United States, thousands of square kilometers of land have been disturbed by coal mining, and only about half this land has been reclaimed. Reclamation is the process of restoring and improving disturbed land, often by re-forming the surface and replanting vegetation (see Chapter 9). Unreclaimed coal dumps from open-pit mines are numerous and continue to cause environmental problems. Because little reclamation occurred before about 1960, and mining started much earlier, abandoned mines are common in the United States. One surface mine in Wyoming, abandoned more than 40 years ago, caused a disturbance so intense that vegetation has still not been reestablished on the waste dumps. Such barren, ruined landscapes emphasize the need for reclamation.[18]

Strip Mining

Over half of the coal mining in the United States is done by *strip mining*, a surface mining process in which the overlying layer of soil and rock is stripped off to reach the coal. The practice of strip mining started in the late 19th century and has steadily increased because it tends to be cheaper and easier than underground mining. More than 40 billion metric tons of coal reserves are now accessible to surface mining techniques. In addition, approximately 90 billion metric tons of coal within 50 m (165 ft) of the surface are potentially available for strip mining. More and larger strip mines will likely be developed as the demand for coal increases.

The impact of large strip mines varies from region to region, depending on topography, climate, and reclamation practices. One serious problem in the eastern United States that get abundant rainfall is *acid mine drainage*—the drainage of acidic water from mine sites (see Chapter 19). Acid mine drainage occurs when surface water (H_2O) infiltrates the spoil banks (rock debris left after the coal is removed). The water reacts chemically with sulfide minerals, such as pyrite (FeS_2), a natural component of some sedimentary rocks containing coal, to produce sulfuric acid (H_2SO_4). The acid then pollutes streams and groundwater (Figure15.13). Acid water also drains from underground mines and from roads cut in areas where coal and pyrite are abundant, but the problem of acid mine drainage is magnified when large areas of disturbed material remain exposed to surface waters.

FIGURE 15.13 Tar Creek near Miami, Oklahoma, runs orange in 2003 due to contamination by heavy metals from acid mine drainage.

FIGURE 15.14 Strip coal mine in Wyoming. The land in the foreground is being mined, and the green land in the background has been reclaimed after mining.

Acid mine drainage from active mines can be minimized by channeling surface runoff or groundwater before it enters a mined area and diverting it around the potentially polluting materials. However, diversion is not feasible in heavily mined regions where spoil banks from unreclaimed mines may cover hundreds of square kilometers. In these areas, acid mine drainage will remain a long-term problem.

Water problems associated with mining are not as pronounced in arid and semiarid regions as they are in wetter regions, but the land may be more sensitive to activities related to mining, such as exploration and road building. In some arid areas of the western and southwestern United States, the land is so sensitive that tire tracks can remain for years. (Indeed, wagon tracks from the early days of the westward migration reportedly have survived in some locations.) To complicate matters, soils are often thin, water is scarce, and reclamation work is difficult.

Strip mining has the potential to pollute or damage water, land, and biological resources. However, good reclamation practices can minimize the damage (Figure 15.14). Reclamation practices required by law necessarily vary by site. Some of the principles of reclamation are illustrated in the case history of a modern coal mine in Colorado (see A Closer Look 15.2).

Large surface coal mining is almost always controversial. One of the most controversial has been the Black Mesa Mine in Arizona. The mine is in the Black Mesa area of the Hopi Reservation and was the only supplier of coal to the very large 1.5-MW Mohave Generating Station, a power plant at Laughlin, Nevada (144 km, or 90 mi, southeast of Las Vegas). The coal was delivered to the plant by a 440-km (275-mi) pipeline that transported slurry (crushed coal and water). The pipeline used over 1 billion gallons of water pumped from the ground per year—water that nourishes sacred springs and water for irrigation. Both the mine and the power plant suspended operation on December 31, 2005.

Mountaintop Removal

Coal mining in the Appalachian Mountains of West Virginia is a major component of the state's economy. However, there is growing environmental concern about a strip-mining technique known as "mountaintop removal" (Figure 15.16). This technique is very effective in obtaining coal as it levels the tops of mountains. But as mountaintops are destroyed, valleys are filled with waste rock and other mine waste, and the flood hazard increases as toxic wastewater is stored behind coal-waste sludge dams. Several hundred mountains have been destroyed, and by 2103 over 3,840 km (2,400 mi) of stream channels will likely have been damaged or destroyed.[19]

In October 2000, one of the worst environmental disasters in the history of mining in the Appalachian Mountains occurred in southeastern Kentucky. About 1 million cubic meters (250 million gallons) of toxic, thick black coal sludge, produced when coal is processed, was released into the environment. Part of the bottom of the impoundment (reservoir) where the sludge was being stored collapsed, allowing the sludge to enter an abandoned mine beneath the impoundment. The abandoned mine had openings to the surface, and sludge emerging from the mine flowed across people's yards and roads into a stream of the Big Sandy River drainage. About 100 km (65 mi) of stream was severely contaminated, killing several hundred thousand fish and other life in the stream.

Mountaintop removal also produces voluminous amounts of coal dust that settles on towns and fields, polluting the land and causing or exacerbating lung diseases, including asthma. Protests and complaints by communities in the path of mining were formerly ignored but are now getting more attention from state mining boards. As people become better educated about mining laws, they are more effective in confronting mining companies to get them to reduce potential adverse consequences of mining. However, much more needs to be done.

Those in favor of mountaintop mining emphasize its value to the local and regional economy. They further argue that only the mountaintops are removed, leaving most of the mountain, with only the small headwater streams filled with mining debris. They go on to say that

A CLOSER LOOK 15.2

The Trapper Mine

The Trapper Mine on the western slope of the Rocky Mountains in northern Colorado is a good example of a new generation of large coal strip mines. The operation, in compliance with mining laws, is designed to minimize environmental degradation, including damage to land farming and to grazing of livestock and big game. A landslide in 2006 impacted 102 hectares (200 acres) of the mine and caused a rethinking of the landslide hazard at the mine.

Over a 35-year period, the mine will produce 68 million metric tons of coal from the 20–24 km (50–60 mi^2) site, to be delivered to a 1,300-MW power plant adjacent to the mine. Today the mine produces about 2 million tons of coal per year, enough power for about half a million homes. Four coal seams, varying in thickness from about 1 to 4 m (3.3–13.1 ft), will be mined. The seams are separated by layers of rock, called overburden (rocks without coal), and there is additional overburden above the top seam of coal. The depth of the overburden varies from 0 to about 50 m (165 ft).

A number of steps are involved in the actual mining. First, bulldozers and scrapers remove the vegetation and topsoil from an area up to 1.6 km long and 53 m wide (1 mi by 175 ft), and the soil is stockpiled for reuse. Then the overburden is removed with a 23-m^3 (800-ft) dragline bucket. Next, the exposed coal beds are drilled and blasted to fracture the coal, which is removed with a backhoe and loaded onto trucks (Figure 15.15). Finally, the cut is filled, the topsoil replaced, and the land either planted with a crop or returned to rangeland.

At the Trapper Mine, the land is reclaimed without artificially applying water. Precipitation (mostly snow) is about 35 cm/year (about 14 in./year), which is sufficient to reestablish vegetation if there is adequate topsoil. That reclamation is possible at this site emphasizes an important point about reclamation: It is site-specific—what works at one location may not be applicable to other areas. Drainage of reclaimed land has been improved.

Water and air quality are closely monitored at the Trapper Mine. Surface water is diverted around mine pits, and groundwater is intercepted while pits are open. "Settling basins," constructed downslope from the pit, allow suspended solids in the water to settle out before the water is discharged into local streams. Although air quality at the mine can be degraded by dust from the blasting, hauling, and grading of the coal, the dust is minimized by regular sprinkling of water on the dirt roads. Recently, drainage from reclaimed lands has been redesigned to use a sinuous channel rather than a straighter one to slow down sediment deposition, reduce erosion, and minimize maintenance of the channels.

Reclamation at the Trapper Mine has been successful during the first years of operation. In fact, the U.S. Department of the Interior named it one of the best examples of mine reclamation. Although reclamation increases the cost of the coal as much as 50%, it will pay off in the long-range productivity of the land as it is returned to farming and grazing. Wildlife also thrives; the local elk population has significantly increased, and the reclaimed land is home to sharp-tailed grouse, a threatened species.

On the one hand, it might be argued that the Trapper Mine is unique in its combination of geology, hydrology, and topography, which has allowed for successful reclamation. To some extent this is true, and perhaps the Trapper Mine presents an overly optimistic perspective on mine reclamation compared with sites that have less favorable conditions. On the other hand, the success of the mine operation demonstrates that with careful site selection and planning, the development of energy resources can be compatible with other land uses.

(a)

(b)

FIGURE 15.15 **(a)** Mining an exposed coal bed at the Trapper Mine, in Colorado; and **(b)** the land during restoration following mining. Topsoil (lower right) is spread prior to planting of vegetation.

FIGURE 15.16 Mountaintop mining in West Virginia has been criticized as damaging to the environment as vegetation is removed, stream channels are filled with rock and sediment, and the land is changed forever.

the mining, following reclamation, produces flat land for a variety of uses, such as urban development, in a region where flat land is mostly on floodplains with fewer potential uses.

Since the adoption of the Surface Mining Control and Reclamation Act of 1977, the U.S. government has required that mined land be restored to support its premining use. The regulations also prohibit mining on prime agricultural land and give farmers and ranchers the opportunity to restrict or prohibit mining on their land, even if they do not own the mineral rights. Reclamation includes disposing of wastes, contouring the land, and replanting vegetation.

Reclamation is often difficult and unlikely to be completely successful. In fact, some environmentalists argue that reclamation success stories are the exception and that strip mining should not be allowed in the semiarid southwestern states because reclamation is uncertain in that fragile environment.

Underground Mining

Underground mining accounts for approximately 40% of the coal mined in the United States and poses special risks both for miners and for the environment. The dangers to miners have been well documented over the years in news stories, books, and films. Hazards include mine shaft collapses (cave-ins), explosions, fires, and respiratory illnesses, especially the well-known black lung disease, which is related to exposure to coal dust, which has killed or disabled many miners over the years.

Some of the environmental problems associated with underground mining include the following:

- Acid mine drainage and waste piles have polluted thousands of kilometers of streams (see Chapter 19).
- Land subsidence can occur over mines. Vertical subsidence occurs when the ground above coal mine tunnels collapses, often leaving a crater-shaped pit at the surface (Figure 15.17). Coal-mining areas in Pennsylvania and West Virginia, for example, are well known for serious subsidence problems. In recent years, a parking lot and crane collapsed into a hole over a coal mine in Scranton, Pennsylvania; and damage from subsidence caused condemnation of many buildings in Fairmont, West Virginia.
- Coal fires in underground mines, either naturally caused or deliberately set, may belch smoke and hazardous fumes, causing people in the vicinity to suffer from a variety of respiratory diseases. For example, in Centralia, Pennsylvania, a trash fire set in 1961 lit nearby underground coal seams on fire. They are still burning today and have turned Centralia into a ghost town.

Transporting Coal

Transporting coal from mining areas to large population centers where energy is needed is a significant environmental issue. Although coal can be converted at the

FIGURE 15.17 Subsidence below coal mines in the Appalachian coal belt.

production site to electricity, synthetic oil, or synthetic gas, these alternatives have their own problems. Power plants for converting coal to electricity require water for cooling, and in semiarid coal regions of the western United States there may not be sufficient water. Furthermore, transmitting electricity over long distances is inefficient and expensive (see Chapter 14). Converting coal to synthetic oil or gas also requires a huge amount of water, and the process is expensive.[20, 21]

Freight trains and coal-slurry pipelines (designed to transport pulverized coal mixed with water) are options to transport the coal itself over long distances. Trains are typically used, and will continue to be used, because they provide relatively low-cost transportation compared with the cost of constructing pipelines. The economic advantages of slurry pipelines are tenuous, especially in the western United States, where large volumes of water to transport the slurry are not easily available.

The Future of Coal

The burning of coal produces nearly 50% of the electricity used and about 25% of the total energy consumed in the United States today. Coal accounts for nearly 90% of the fossil fuel reserves in the United States, and we have enough coal to last at least several hundred years. However, there is serious concern about burning that coal. Giant power plants that burn coal as a fuel to produce electricity in the United States are responsible for about 70% of the total emissions of sulfur dioxide, 30% of the nitrogen oxides, and 35% of the carbon dioxide. (The effects of these pollutants are discussed in Chapter 21.)

Legislation as part of the Clean Air Amendments of 1990 mandated that sulfur dioxide emissions from coal-burning power plants be eventually cut by 70–90%, depending on the sulfur content of the coal, and that nitrogen oxide emissions be reduced by about 2 million metric tons per year. As a result of this legislation, utility companies are struggling with various new technologies designed to reduce emissions of sulfur dioxide and nitrogen oxides from burning coal. Options being used or developed include the following:[21, 22]

- Chemical and/or physical cleaning of coal prior to combustion.
- Producing new boiler designs that permit a lower temperature of combustion, reducing emissions of nitrogen oxides.
- Injecting material rich in calcium carbonate (such as pulverized limestone or lime) into the gases produced by the burning of coal. This practice, known as **scrubbing**, removes sulfur dioxides. In the scrubber—a large, expensive component of a power plant—the carbonate reacts with sulfur dioxide, producing hydrated calcium sulfite as sludge. The sludge has to be collected and disposed of, which is a major problem.
- Converting coal at power plants into a gas (syngas, a methane-like gas) before burning. This technology is being tested and may become commercial by 2013 at the Polk Power Station in Florida. The syngas, though cleaner-burning than coal, is still more polluting than natural gas.
- Converting coal to oil: We have known how to make oil (gasoline) from coal for decades. Until now, it has been thought too expensive. South Africa is doing this now, producing over 150,000 barrels of oil per day from coal, and China in 2009 finished construction of a plant in Mongolia. In the United States, we could produce 2.5 million barrels per day, which would require about 500 million tons of coal per year by 2020. There are environmental consequences, however, as superheating coal to produce oil generates a lot of carbon dioxide (CO_2), the major greenhouse gas.
- Educating consumers about energy conservation and efficiency to reduce the demand for energy and, thus, the amount of coal burned and emissions released.
- Developing zero-emission coal-burning electric power plants. Emissions of particulates, mercury, sulfur dioxides, and other pollutants would be eliminated by physical and chemical processes. Carbon dioxide would be eliminated by injecting it deep into the earth or using a chemical process to sequester it (tie it up) with calcium or magnesium as a solid. The concept of zero emission is in the experimental stages of development.

The bottom line is that as oil prices rise, coal is getting a lot of attention in the attempt to find ways to lessen the economic shock . The real shortages of oil and gas may still be a few years away, but when they come, they will put pressure on the coal industry to open more and larger mines in both the eastern and western coal beds of the United States. Increased use of coal will have significant environmental impacts for several reasons.

First, more and more land will be strip-mined and will therefore require careful and expensive restoration.

Second, unlike oil and gas, burning coal, as already mentioned, produces large amounts of air pollutants. It also creates ash, which can be as much as 20% of the coal burned; boiler slag, a rocklike cinder produced in the furnace; and calcium sulfite sludge, produced from removing sulfur through scrubbing. Coal-burning power plants in the United States today produce about 90 million tons of these materials per year. Calcium sulfite from scrubbing can be used to make wallboard (by converting calcium sulfite to calcium sulfate, which is gypsum) and other products. Gypsum is being produced for wallboard this way in Japan and Germany, but the United States can make wallboard less expensively from abundant natural gypsum deposits. Another waste product, boiler slag, can be used for fill along railroad tracks and at construction projects. Nevertheless, about 75% of the combustion products of burning coal in the United States today end up in waste piles or landfills.

Third, handling large quantities of coal through all stages (mining, processing, shipping, combustion, and final disposal of ash) could have adverse environmental effects. These include aesthetic degradation, noise, dust, and—most significant from a health standpoint—release of toxic or otherwise harmful trace elements into the water, soil, and air. For example, in late December 2008, the retaining structure of an ash pond at the Kingston Fossil Plant in Tennessee failed, releasing a flood of ash and water that destroyed several homes, ruptured a gas line, and polluted a river.[23]

All of these negative effects notwithstanding, it seems unlikely that the United States will abandon coal in the near future because we have so much of it and have spent so much time and money developing coal resources. Some suggest that we should now promote the use of natural gas in preference to coal because it burns so much cleaner, but that raises the valid concern that we might then become dependent on imports of natural gas. Regardless, it remains a fact that coal is the most polluting of all the fossil fuels.

Allowance Trading

An innovative approach to managing U.S. coal resources and reducing pollution is **allowance trading**, through which the Environmental Protection Agency grants utility companies tradable allowances for polluting: One allowance is good for one ton of sulfur dioxide emissions per year. In theory, some companies wouldn't need all their allowances because they use low-sulfur coal or new equipment and methods that have reduced their emissions. Their extra allowances could then be traded and sold by brokers to utility companies that are unable to stay within their allocated emission levels. The idea is to encourage competition in the utility industry and reduce overall pollution through economic market forces.[22]

Some environmentalists are not comfortable with the concept of allowance trading. They argue that although buying and selling may be profitable to both parties in the transaction, it is less acceptable from an environmental viewpoint. They believe that companies should not be able to buy their way out of taking responsibility for pollution problems.

15.4 Oil Shale and Tar Sands

Oil shale and tar sands play a minor role in today's mix of available fossil fuels, but they may be more significant in the future, when traditional oil from wells becomes scarce.

Oil Shale

Oil shale is a fine-grained sedimentary rock containing organic matter (kerogen). When heated to 500°C (900°F) in a process known as *destructive distillation*, oil shale yields up to nearly 60 liters (14 gallons) of oil per ton of shale. If not for the heating process, the oil would remain in the rock. The oil from shale is one of the so-called **synfuels** (from the words *synthetic* and *fuel*), which are liquid or gaseous fuels derived from solid fossil fuels. The best-known sources of oil shale in the United States are found in the Green River formation, which underlies approximately 44,000 km^2 (17,000 mi^2) of Colorado, Utah, and Wyoming.

Total identified world oil shale resources are estimated to be equivalent to about 3 trillion barrels of oil. However, evaluation of the oil grade and the feasibility of economic recovery with today's technology is not complete. Oil shale resources in the United States amount to about 2 trillion bbl of oil, or two-thirds of the world total. Of this, 90%, or 1.8 trillion bbl, is located in the Green River oil shales. The total oil that could be removed from U.S. oil shale deposits is about 100 billion barrels. This exceeds the oil reserves of the Middle East! But extraction is not easy, and environmental impacts would be serious.[24, 25]

The environmental impact of developing oil shale varies with the recovery technique used. Both surface and subsurface mining techniques have been considered. Surface mining is attractive to developers because nearly 90% of the shale oil can be recovered, compared with less than 60% by underground mining. However, waste disposal is a major problem with either surface or subsurface mining. Both require that oil shale be processed, or *retorted* (crushed and heated), at the surface. The volume of waste will exceed the original volume of shale mined by 20–30% because crushed rock has pore spaces and thus more volume than the solid rock had. (If you doubt this, pour some concrete into a milk carton, remove it when it hardens, and break it into small pieces with a hammer. Then try to put the pieces back into the carton.) Thus, the mines from which the shale is removed will not be able to accommodate all the waste, and its disposal will become a problem.[12]

Although it is much more expensive to extract a barrel of oil from shale than to pump it from a well, interest in oil shale was heightened by an oil embargo in 1973 and by fear of continued shortages of crude oil. In the 1980s through the mid-1990s, however, plenty of cheap oil was available, so oil-shale development was put on the back burner. Today, when it is clear that we will face oil shortages in the future, we are seeing renewed interest in oil shale,[25] and it is clear that any steep increases in oil prices will likely heighten this interest. This would result in significant environmental, social, and economic impacts in the oil shale areas, including rapid urbanization to house a large workforce, construction of industrial facilities, and increased demand on water resources.

Tar Sands

Tar sands are sedimentary rocks or sands impregnated with tar oil, asphalt, or bitumen. Petroleum cannot be recovered from tar sands by pumping wells or other usual commercial methods because the oil is too viscous (thick) to flow easily. Oil in tar sands is recovered by first mining the sands—which are very difficult to remove—and then washing the oil out with hot water. It takes about two tons of tar sand to produce one barrel of oil.

About 19% of U.S. oil imports come from Canada, and about one-half of this is from tar sands.[26] Some 75% of the world's known tar sand deposits are in the Athabasca Tar Sands near Alberta, Canada. The total Canadian resource that lies beneath approximately 78,000 km^2 (30,116 mi^2) of land is about 300 billion barrels that might be recovered. About half of this (173 billion barrels) can be economically recovered today.[26] Production of the Athabasca Tar Sands is currently about 1.2 million barrels of synthetic crude oil per day. Production will likely increase to about 3 million barrels per day in the next decade or so.

In Alberta, tar sand is mined in a large open-pit mine (Figure 15.18). The mining process is complicated by the fragile native vegetation, a water-saturated mat known as a muskeg swamp—a kind of wetland that is difficult to remove except when frozen. The mining of the tar sands does have environmental consequences, ranging from a rapid increase in the human population in mining areas to the need to reclaim the land disturbed by the mining. Restoration of this fragile, naturally frozen (permafrost) environment is difficult. There is also a waste-disposal problem because the mined sand material, like the mined oil shale just discussed, has a greater volume than the unmined material. The land surface can be up to 20 m (66 ft) higher after mining than it was originally. The Canadian approach is to require that the land be returned not to its original use but to some equivalent use, such as turning what was forestland into grazing land.[27]

FIGURE 15.18 Mining tar sands north of Fort McMurray in Alberta, Canada. The large shovel-bucket holds about 100 tons of tar sand. It takes about two tons of tar sand to produce one barrel of oil.

CRITICAL THINKING ISSUE

What Will Be the Consequences of Peak Oil?

The summer of 2008 brought record oil prices. By late May the price had risen to $133 per barrel. Each barrel has 42 gallons. That is $3.17 per gallon of oil, before being shipped, refined, and taxed! Consider the following:[28]

1. Oil doubled in price from 2007 to 2008, then dropped to about $70 per barrel as demand declined. The price remains very uncertain.
2. Grain production increased from about 1.8 billion tons to 2.15 billion tons per year from 2002 to 2008.
3. The global food price index rose about 30% from 2007 to 2008. Sugar prices rose about 40% and grain prices about 90%. Wheat that cost about $375 per ton in 2006 soared to more than $900 in 2008. Large discount stores in the United States in 2008 set limits on the amount of rice that a person could purchase.
4. World biofuel (biodiesel and bioethanol) production increased from about 6.4 billion gallons per year in 2000 to just over 20 billion gallons in 2008, and probably will continue to rise rapidly. In the late 1990s the United States used about 5% of corn production for biofuel (ethanol); in 2009 it used 25% for this purpose.
5. Our worldwide safety net of grain stocks on hand declined from about 525 million tons in 2000 to about 300 million tons in 2008.

Critical Thinking Questions

Assume oil prices will remain unstable:

1. Examine Figure 15.19. What are the main points you can conclude from reading the graph? (*Hint:* Look closely at the shape of the curves and the labeling.) Summarize your thoughts.
2. How is the above information linked? (*Hint:* Make linkages between Figure 15.19 and points 1–5 above.) Summarize your thoughts.
3. What will be the differences in the potential economic and environmental impacts on countries ranging from poor to rich?
4. Should the United States stop producing corn for biofuel?
5. Will famine in the future be due to rising food prices? Why? Why not?
6. Do you think food riots may lead to civil wars in some countries?
7. What solutions can you offer to minimize the impacts of increasing energy costs tied to food production and cost?

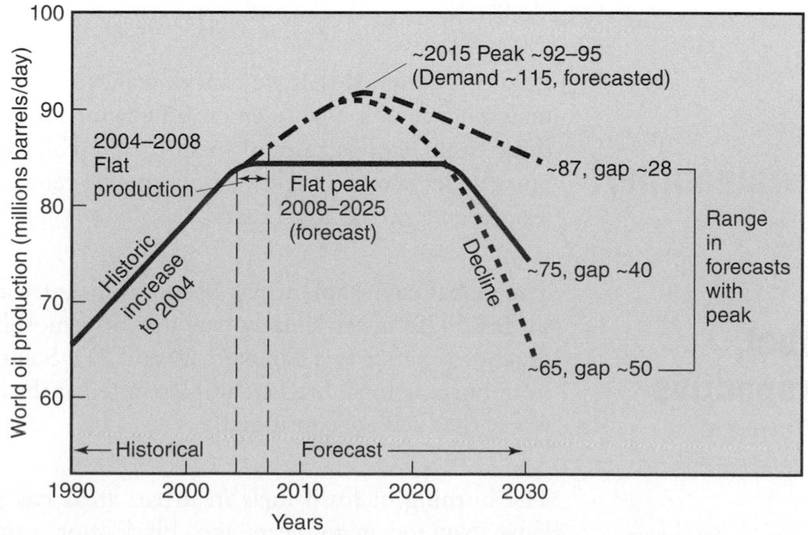

FIGURE 15.19 Peak oil with two scenarios: a long, flat peak with production as it is today; and a sharper peak in about 2015. In either case, there are significant shortages (gap between production and demand). (*Source*: Data from P. Roberts, "Tapped Out," *National Geographic* 213, no. 6 (2008):86–91.)

SUMMARY

- The United States has an energy problem caused by dependence on fossil fuels, especially oil. Maximum global production (peak oil) is expected between 2020 and 2050, followed by a decline in production. The challenge is to plan now for the decline in oil supply and a shift to alternative energy sources.
- Fossil fuels are forms of stored solar energy. Most are created from the incomplete biological decomposition of dead and buried organic material that is converted by complex chemical reactions in the geologic cycle.
- Because fossil fuels are nonrenewable, we will eventually have to develop other sources to meet our energy demands. We must decide when the transition to alternative fuels will occur and what the impacts of the transition will be.
- Environmental impacts related to oil and natural gas include those associated with exploration and development (damage to fragile ecosystems, water pollution, air pollution, and waste disposal); those associated with refining and processing (pollution of soil, water, and air); and those associated with burning oil and gas for energy to power automobiles, produce electricity, run industrial machinery, heat homes, and so on (air pollution).
- Coal is an energy source that is particularly damaging to the environment. The environmental impacts of mining, processing, transporting, and using coal are many. Mining coal can cause fires, subsidence, acid mine drainage, and difficulties related to land reclamation. Burning coal can release air pollutants, including sulfur dioxide and carbon dioxide, and produces a large volume of combustion products and by-products, such as ash, slag, and calcium sulfite (from scrubbing). The environmental objective for coal is to develop a zero-emission power plant.

REEXAMINING THEMES AND ISSUES

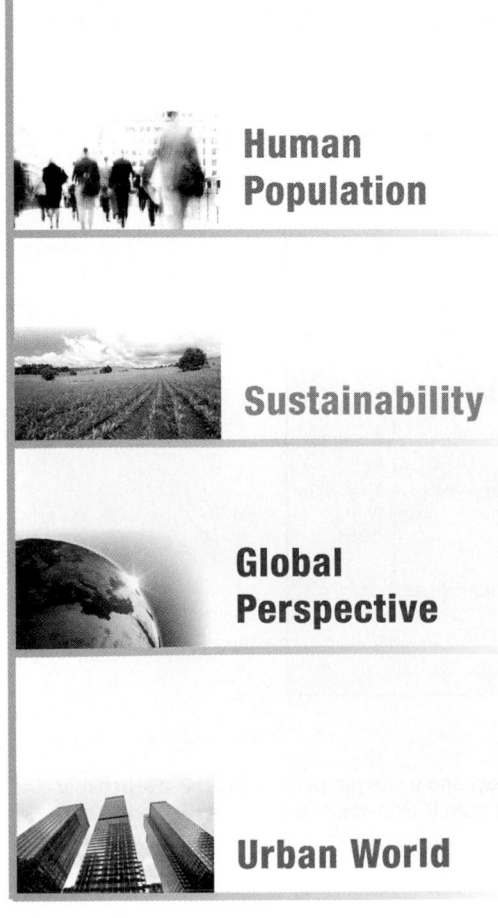

Human Population

As the human population (particularly in developed countries, such as the United States) has increased, so has the total impact from the use of fossil fuels. Total impact is the impact per person times the total number of people. Reducing the impact will require that all countries adopt a new energy paradigm emphasizing use of the minimum energy needed to complete a task (end use) rather than the current prodigious overuse of energy.

Sustainability

It has been argued that we cannot achieve sustainable development and maintenance of a quality environment for future generations if we continue to increase our use of fossil fuels. Achieving sustainability will require wider use of a variety of alternative renewable energy sources and less dependence on fossil fuels.

Global Perspective

The global environment has been significantly affected by burning fossil fuels. This is particularly true for the atmosphere, where fast-moving processes operate (see Chapters 20 and 21). Solutions to global problems from burning fossil fuels are implemented at the local and regional levels, where the fuels are consumed.

Urban World

The burning of fossil fuels in urban areas has a long history of problems. Not too many years ago, black soot from burning coal covered the buildings of most major cities of the world, and historical pollution events killed thousands of people. Today, we are striving to improve our urban environments and reduce urban environmental degradation from burning fossil fuels.

People and Nature

Our exploration, extraction, and use of fossil fuels have changed nature in fundamental ways—from the composition of the atmosphere and the disturbance of coal mines to the pollution of ground and surface waters. Some people still buy giant SUVs supposedly to connect with nature, but using them causes more air pollution than do automobiles and, if used off-road, often degrades nature.

Science and Values

Scientific evidence of the adverse effects of burning fossil fuels is well documented. The controversy over their use is linked to our values. Do we value burning huge amounts of fossil fuels to increase economic growth more highly than we value living in a quality environment? Economic growth is possible without damaging the environment; developing a sustainable energy policy that doesn't harm the environment is possible with present technology. What is required are changes in values and lifestyle that are linked to energy production and use, human well-being, and environmental quality.

KEY TERMS

allowance trading **321**
coal **314**
crude oil **306**
fossil fuels **305**
methane hydrate **310**
natural gas **306**
oil shale **321**
peak oil **304**
scrubbing **320**
synfuels **321**
tar sands **322**

STUDY QUESTIONS

1. Assuming that oil production will peak in about 2020 and then decline about 3% per year, when will production be half of that in 2020? What could be the consequences, both good and bad? Why? How might negative consequences be avoided?
2. Compare the potential environmental consequences of burning oil, burning natural gas, and burning coal.
3. What actions can you personally take to reduce consumption of fossil fuels?
4. What environmental and economic problems could result from a rapid transition from fossil fuels to alternative sources?
5. The transition from wood to fossil fuels took about 100 years. How long do you think the transition from fossil fuels to alternative energy sources will take? What will determine the time of transition?
6. What are some of the technical solutions to reducing air-pollutant emissions from burning coal? Which are best? Why?
7. What do you think of the idea of allowance trading as a potential solution to reducing pollution from burning coal?
8. Do you think we can develop a zero-emission coal-burning power plant? What about for natural gas?
9. Discuss how the rising cost of energy is linked to food supply and environmental problems.
10. What are the ethical issues associated with the energy problems? Is a child born in 2050 more important than a child today? Why or why not?

FURTHER READING

Boyl, G., B. Everett, and J. Ramage, ***Energy Systems and Sustainability*** (Oxford (UK): Oxford University Press, 2003). An excellent discussion of fossil fuel.

British Petroleum Company, ***B.P. Statistical Review of World Energy*** (London: British Petroleum Company, 2010). Good, up-to-date statistics on fossil fuels.

CHAPTER 16

Alternative Energy and the Environment

This wind farm in Tarifa, Spain, has a generating capacity of 5 MW produced by just 12 wind turbines.

LEARNING OBJECTIVES

Alternatives to fossil fuels and nuclear energy include biofuels, solar energy, water power, wind power, and geothermal energy. Some of these are already being used, and efforts are under way to develop others. After reading this chapter, you should understand . . .

- The advantages and disadvantages of each kind of alternative energy;
- What passive, active, and photovoltaic solar energy systems are;
- What may be the important fuels of the future;
- Why water power is unlikely to become more important in the future;
- Why wind power has tremendous potential, and how its development and use could affect the environment;
- Whether biofuels are likely to help us move away from fossil fuels;
- What geothermal energy is, and how developing and using it affect the environment.

CASE STUDY

Using Wind Power in New Ways for an Old Application

On March 14, 2008, a new kind of sailing ship, the MV *Beluga SkySails,* completed its maiden voyage of 11,952 nautical miles, sailing from Bremen, Germany, to Venezuela and back carrying heavy industrial equipment (Figure 16.1). *SkySails* is novel in two ways. First of all, it did not use a set of fixed masts with traditional sails that had to be monitored constantly and their settings changed with each variation in the wind, thereby requiring either a large crew or very sophisticated and expensive equipment. Second, wind provided only part of the power—on the maiden voyage 20%—the rest coming from standard marine diesel engines.

Instead, this ship flew a huge kite that spread out over more than 160 square yards. Flying high above the ship, it caught more reliable winds aloft than occur at the surface, and required only a flexible cable to tether it to the ship. Diesel engines provided the steady energy; the kite-sail helped when wind was available, reducing fuel costs by $1,000 a day. With this saving, and also saving the expense of a large crew, *Beluga SkySails* was an economic success. At the end of the voyage, the ship's captain, Lutz Heldt, said, "We can once again actually 'sail' with cargo ships, thus opening a new chapter in the history of commercial shipping."

This ship takes another important step in the use of alternative energy, integrating it with another source. Also important is that the kite works *with* natural forces, unlike a traditional sailing ship's rigid masts, which had to withstand the force of the winds. Therefore, the *SkySails* is much less likely to be damaged by storms. These design elements can be helpful as we think through a major transition from fossil fuels to alternative energy.

FIGURE 16.1 Wind power is becoming so popular that it is even making a comeback as a way to propel ships. Here a new kind of sail, actually a cable-tethered kite, helps pull a new ship, the *Beluga SkySails*, through the ocean. (*Source: SkySails* website March 31, 2008. First retrofit SkySails-System aboard a cargo vessel—Initial operation successful SkySails pilot project on board the MS "Michael A." of the WESSELS Reederei GmbH & Co.KG based in Haren/Ems http://www.skysails.info/index.php?id=64&L=1&tx_ttnews[tt_news]=98&tx_ttnews[backP id]=6&cHash=c1a209e350 and http://www.skysails.info/index.php?id=64&L=1&tx_ttnews[tt_news]=104&tx_ttnews[back Pid]=6&cHash=db100ad2b6)

16.1 Introduction to Alternative Energy Sources

As we all know, the primary energy sources today are fossil fuels—they supply approximately 90% of the energy consumed by people. All other sources are considered **alternative energy** and are divided into renewable energy and nonrenewable energy. **Nonrenewable alternative energy** sources include nuclear energy (discussed in Chapter 17) and deep-earth geothermal energy (the energy from the Earth's geological processes). This kind of geothermal energy is considered nonrenewable for the most part because heat can be extracted from Earth faster than it is naturally replenished—that is, output exceeds input (see Chapter 3). Nuclear energy is nonrenewable because it requires a mineral fuel mined from Earth.

The **renewable energy** sources are solar; freshwater (hydro); wind; ocean; low-density, near-surface geothermal; and biofuels. Low-density, near-surface geothermal

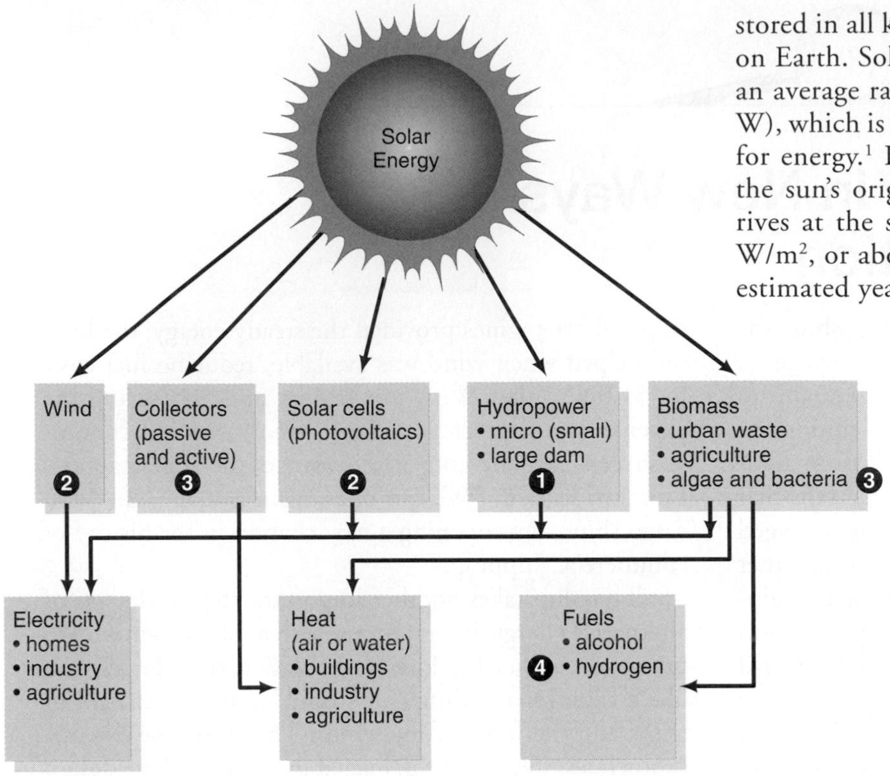

❶ Produces most electricity from renewable solar energy

❷ Rapidly growing, strong potential (Wind and solar are growing at 30% per year.)

❸ Used today; important energy source

❹ Potentially a very important fuel to transition from fossil fuels

FIGURE 16.2 Routes of various types of renewable solar energy.

is simply solar energy stored by soil and rock near the surface. It is widespread and easily obtained and is renewed by the sun. Biofuels are made from biomass (crops, wood, and so forth). Renewable energy sources are often discussed as a group because they all derive from the sun's energy (Figure 16.2). We consider them renewable because they are regenerated by the sun within a time period useful to people.

The total energy we may be able to extract from alternative energy sources is enormous. For example, the estimated recoverable energy from solar energy is about 75 times as much as all the people of the world use each year. The estimated recoverable energy from wind alone is comparable to current global energy consumption.

16.2 Solar Energy

The total amount of solar energy reaching Earth's surface is tremendous. For example, on a global scale, ten weeks of solar energy is roughly equal to the energy stored in all known reserves of coal, oil, and natural gas on Earth. Solar energy is absorbed at Earth's surface at an average rate of 90,000 terawatts (1 TW equals 10^{12} W), which is about 7,000 times the total global demand for energy.[1] In the United States, on average, 13% of the sun's original energy entering the atmosphere arrives at the surface (equivalent to approximately 177 W/m^2, or about 16 W/ft^2, on a continuous basis). The estimated year-round availability of solar energy in the United States is shown in Figure 16.3. However, solar energy is site-specific, and detailed observation of a potential site is necessary to evaluate the daily and seasonal variability of its solar energy potential.[2]

Solar energy may be used by passive or active solar systems. **Passive solar energy systems** do not use mechanical pumps or other active technologies to move air or water. Instead, they typically use architectural designs that enhance the absorption of solar energy (Figure 16.4). Since the rise of civilization, many societies have used passive solar energy (see Chapter 22). Islamic architects, for example, have traditionally used passive solar energy in hot climates to cool buildings.

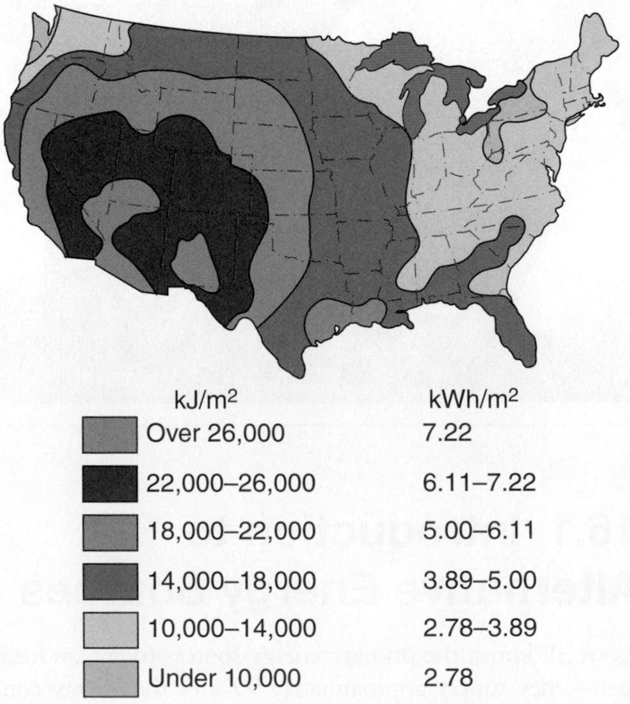

FIGURE 16.3 Estimated solar energy for the contiguous United States. (*Source*: Modified from Solar Energy Research Institute, 1978.)

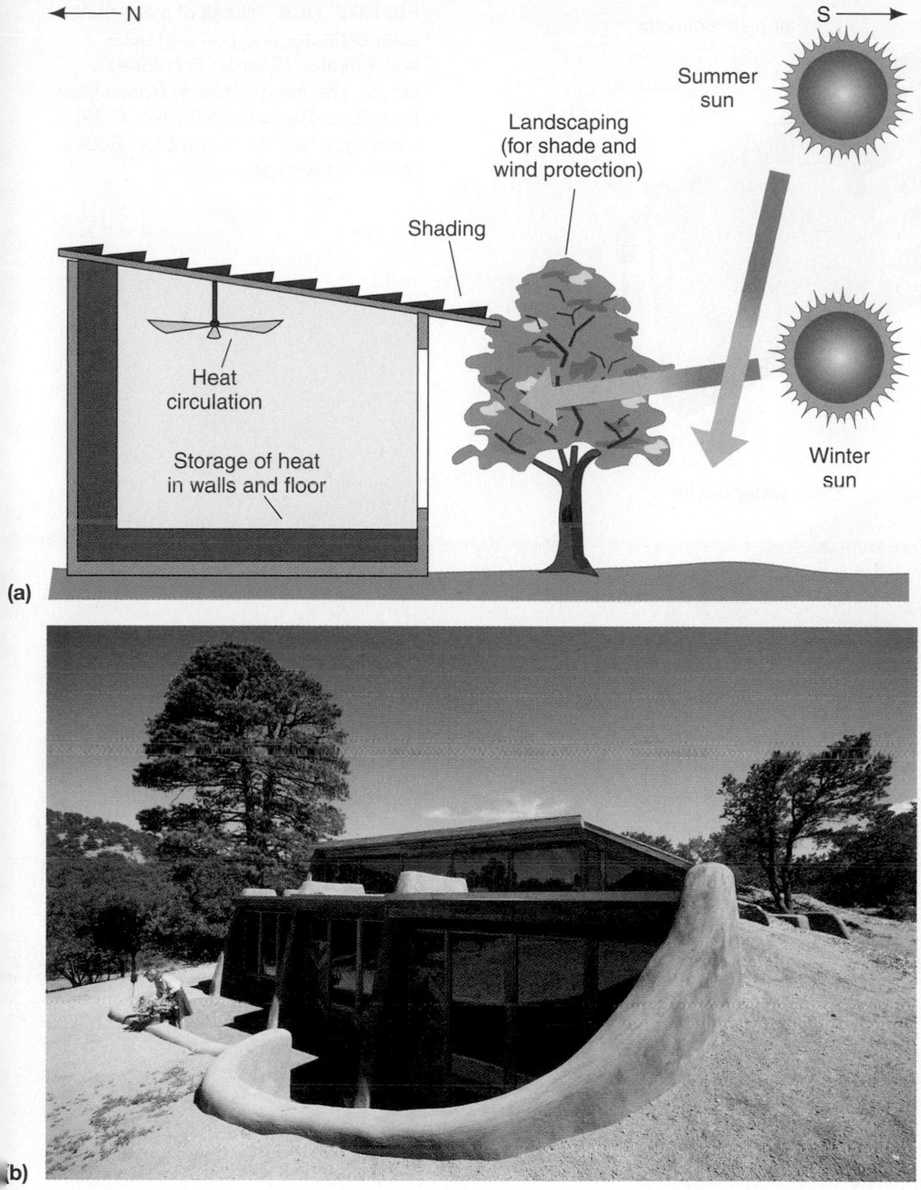

FIGURE 16.4 **(a)** Essential elements of passive solar design. High summer sunlight is blocked by the overhang, but low winter sunlight enters the south-facing window. Deciduous trees shade the building in the summer and allow sunlight to warm it in the winter. These are best located on the southside in the northern hemisphere. Other features are designed to facilitate the storage and circulation of passive solar heat. **(b)** The design of this home uses passive solar energy. Sunlight enters through the windows and strikes a specially designed masonry wall that is painted black. The masonry wall heats up and radiates this heat, warming the house during the day and into the evening. The house is deep set into the ground, which provides winter and summer insolation. It is in a dry climate with few deceduous trres; hence conifers provide some shading. (*Source*: Moran, Morgan, and Wiersma, *Introduction to Environmental Science* [New York: Freeman, 1986]. Copyright by W.H. Freeman & Company. Reprinted with permission.)

Passive Solar Energy

Thousands of buildings in the United States—not just in the sunny Southwest but in other parts of the country, such as New England—now use passive solar systems. Passive solar energy promotes cooling in hot weather and retaining heat in cold weather. Methods include (1) overhangs on buildings to block summer (high-angle) sunlight but allow winter (low-angle) sunlight to penetrate and warm rooms; (2) building a wall that absorbs sunlight during the day and radiates heat that warms the room at night; (3) planting deciduous trees on the sunny side of a building. In summer, these shade and cool the building; in the winter, with the leaves gone, they let the sunlight in.

Passive solar energy also provides natural lighting to buildings through windows and skylights. Modern window glass can have a special glazing that transmits visible light, blocks infrared, and provides insulation.

Active Solar Energy

Active solar energy systems require mechanical power, such as electric pumps, to circulate air, water, or other fluids from solar collectors to a location where the heat is stored and then pumped to where the energy is used.

Solar Collectors

Solar collectors to provide space heating or hot water are usually flat, glass-covered plates over a black background where a heat-absorbing fluid (water or some other liquid) is circulated through tubes (Figure 16.5). Solar radiation enters the glass and is absorbed by the

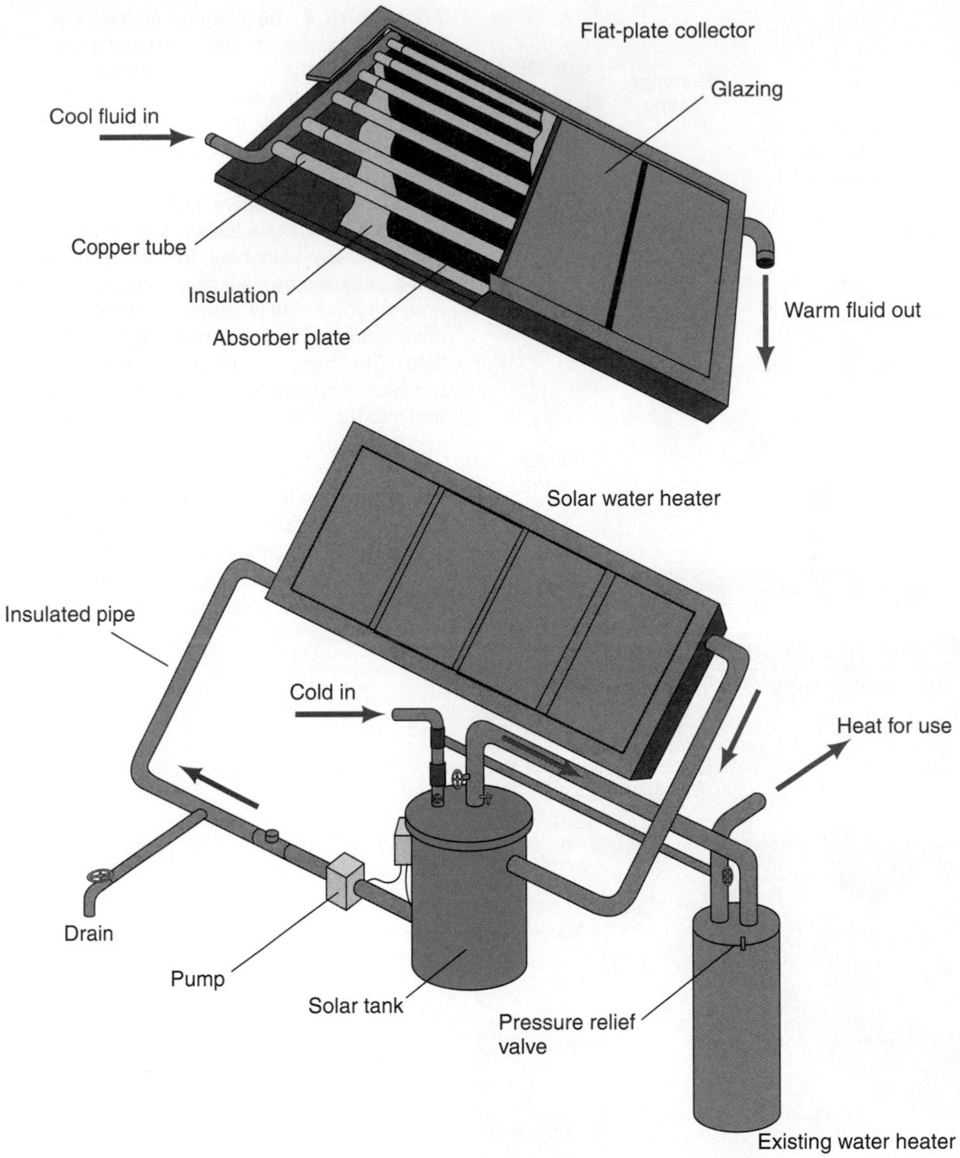

FIGURE 16.5 Detail of a flat-plate solar collector and pumped solar water heater. (*Source*: Farallones Institute, *The Integral Urban House* [San Francisco: Sierra Club Books, 1979]. Copyright 1979 by Sierra Club Books. Reprinted with permission.)

black background. Heat is emitted from the black material, heating the fluid in the tubes.

A second type of solar collector, the evacuated tube collector, is similar to the flat-plate collector, except that each tube, along with its absorbing fluid, passes through a larger tube that helps reduce heat loss. The use of solar collectors is expanding very rapidly; the global market grew about 50% from 2001 to 2004. In the United States, solar water-heating systems generally pay for themselves in only four to eight years.[3]

Photovoltaics

Photovoltaics convert sunlight directly into electricity (Figure 16.6). The systems use solar cells, also called *photovoltaic cells,* made of thin layers of semiconductors (silicon or other materials) and solid-state electronic components with few or no moving parts.

Photovoltaics are the world's fastest growing source of energy, with a growth rate of about 35% per year (doubling every two years). In the United States, the amount of photovoltaics shipped increased 90% between 2007 and 2008, with an average increase of 64% a year since

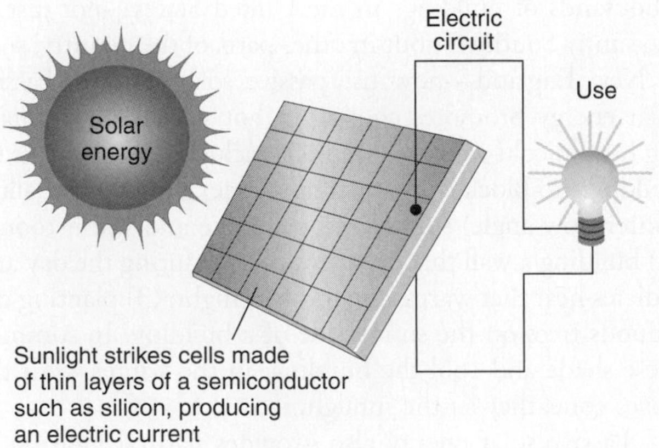

FIGURE 16.6 Idealized diagram illustrating how photovoltaic solar cells work.

(a)

(b)

FIGURE 16.7 **(a)** Panels of photovoltaic cells are used here to power a small refrigerator to keep vaccines cool. The unit is designed to be carried by camels to remote areas in Chad. **(b)** Photovoltaics are used to power emergency telephones along a highway on the island of Tenerife in the Canary Islands. [*Source:* **(a)** H. Gruyaeart/Magnum Photos. Inc. **(b)** Ed Keller]

2005.[4] The photovoltaic industry is expected to grow to $30 billion by 2010.[4]

Solar cell technology is advancing rapidly. While a few decades ago they converted only about 1 or 2% of sunlight into electricity, today they convert as much as 20%. The cells are constructed in standardized modules and encapsulated in plastic or glass, which can be combined to produce systems of various sizes so that power output can be matched to the intended use. Electricity is produced when sunlight strikes the cell. The different electronic properties of the layers cause electrons to flow out of the cell through electrical wires.

Large photovoltaic installations may be connected to an electrical grid. Off-the-grid applications can be large or small and include powering satellites and space vehicles, and powering electric equipment, such as water-level sensors, meteorological stations, and emergency telephones in remote areas (Figure 16.7).

Off-the-grid photovoltaics are emerging as a major contributor to developing countries that can't afford to build electrical grids or large central power plants that burn fossil fuels. One company in the United States is manufacturing photovoltaic systems that power lights and televisions at an installed cost of less than $400 per household.[5] About half a million homes, mostly in villages not linked to a countrywide electrical grid, now receive their electricity from photovoltaic cells.[6]

Solar Thermal Generators

Solar thermal generators focus sunlight onto water-holding containers. The water boils and is used to run such machines as conventional steam-driven electrical generators. These include solar power towers, shown in Figure 16.8. The first large-scale test of using sunlight to boil water and using the steam to run an electric

FIGURE 16.8 Solar power tower at Barstow, California. Sunlight is reflected and concentrated at the central collector, where the heat is used to produce steam to drive turbines and generate electric power. [*Source:* Photograph by Daniel B. Botkin.]

(a)

FIGURE 16.9 **(a)** Acciona solar thermal power plant, south of Las Vegas, which uses more than 180,000 parabolic mirrors to concentrate sunlight onto pipes containing a fluid that is heated above 300°C. The heated fluid, in turn, boils water, and the steam runs a conventional electrical generating turbine. **(b)** Diagram illustrating how such a system works. (*Source*: Courtesy of LUZ International.)

(b)

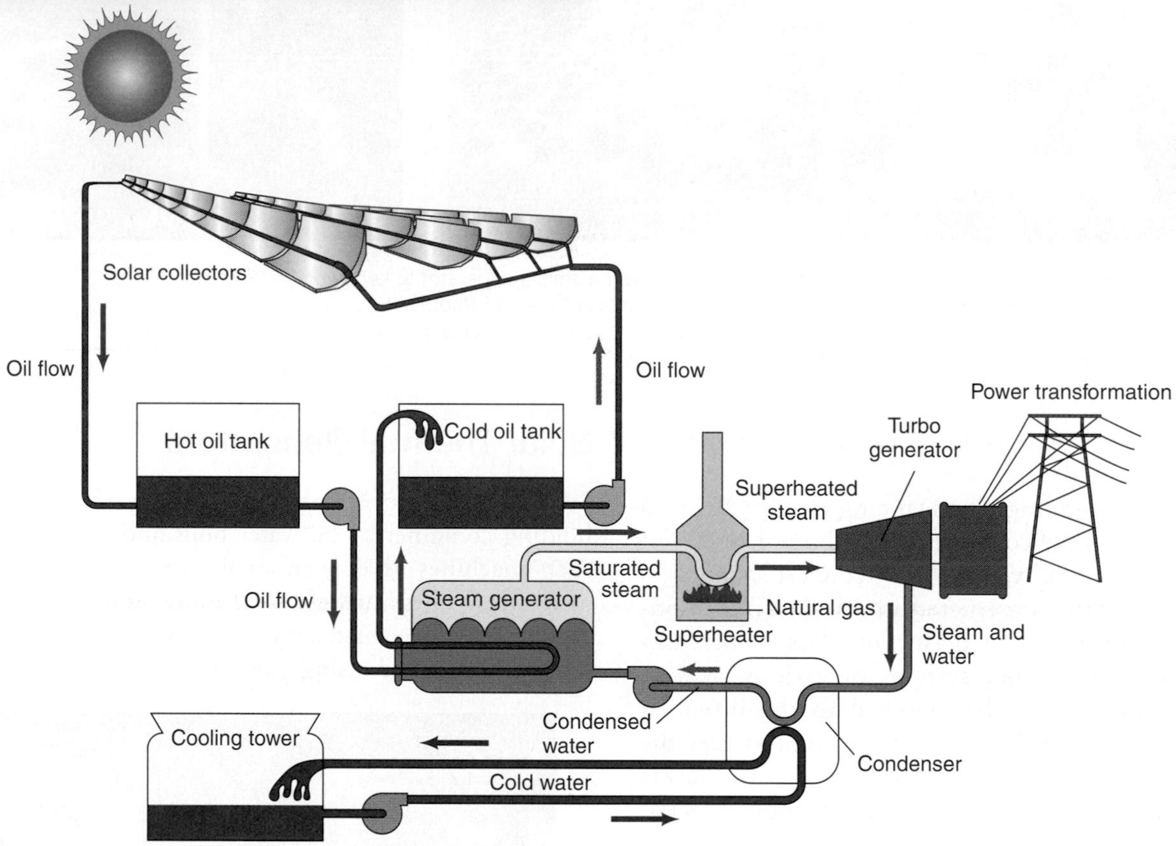

generator was "Solar One," funded by the U.S. Department of Energy. It was built in 1981 by Southern California Edison and operated by that company along with the Los Angeles Department of Water & Power and the California Energy Commission. Sunlight was concentrated onto the top of the tower by 1,818 large mirrors (each about 20 feet in diameter) that were mechanically linked to each other and tracked the sun. At the end of 1999, this power tower was shut down, in part because the plant was not economically competitive with other sources of electricity. New solar thermal generators are being built with very large output (Figure 16.9).

More recently, solar devices that heat a liquid and produce electricity from steam have used many mirrors without a tower, each mirror concentrating sunlight onto a pipe containing the liquid (as shown in Figures 16.8 and 16.9). This is a simpler system and has been considered cheaper and more reliable.[2]

Solar Energy and the Environment

The use of solar energy generally has a relatively low impact on the environment, but there are some environmental concerns nonetheless. One concern is the large variety of metals, glass, plastics, and fluids used in the manufacture and use of solar equipment. Some of these substances may cause environmental problems through production and by accidental release of toxic materials.

16.3 Converting Electricity from Renewable Energy into a Fuel for Vehicles

An obvious question about solar energy, as well as energy from wind, oceans, and freshwater, is how to convert this energy into a form that we can easily transport and can use to power our motor vehicles. Basically, there are two choices: Store the electricity in batteries and use electrical vehicles, or transfer the energy in the electricity to a liquid or gaseous fuel. Of the latter, the simplest is hydrogen.

An electrical current can be used to separate water into hydrogen and oxygen. Hydrogen can power fuel cells (see A Closer Look 16.1). As the hydrogen is combined again with oxygen, electrons flow between negative and positive poles—that is, an electric current is generated. Hydrogen can be produced using solar and other renewable energy sources and, like natural gas, can be transported in pipelines and stored in tanks. Furthermore, it is a clean fuel; the combustion of hydrogen produces water, so it does not contribute to global warming, air pollution, or acid rain. Hydrogen gas may be an important fuel of the future.[7] It is also possible to do additional chemical conversions, combining hydrogen with the carbon in carbon dioxide to produce methane (a primary component of natural gas) and then combining that with oxygen to produce ethanol, which can also power motor vehicles.

A CLOSER LOOK 16.1

Fuel Cells—An Attractive Alternative

Even if we weren't going to run out of fossil fuels, we would still have to deal with the fact that burning fossil fuels, particularly coal and fuels used in internal combustion engines (cars, trucks, ships, and locomotives), causes serious environmental problems. This is why we are searching not only for alternative energy sources but for those that are environmentally benign ways to convert a fuel to useable ways to generate power.[8] One promising technology uses fuel cells, which produce fewer pollutants, are relatively inexpensive, and have the potential to store and produce high-quality energy.

Fuel cells are highly efficient power-generating systems that produce electricity by combining fuel and oxygen in an electrochemical reaction. (They are not sources of energy but a way to convert energy to a useful form.) Hydrogen is the most common fuel type, but fuel cells that run on methanol, ethanol, and natural gas are also available. Traditional generating technologies require combustion of fuel to generate heat, then convert that heat into mechanical energy (to drive pistons or turbines), and convert the mechanical energy into electricity. With fuel cells, however, chemical energy is converted directly into electricity, thus increasing second-law efficiency (see Chapter 17) while reducing harmful emissions.

The basic components of a hydrogen-burning fuel cell are shown in Figure 16.10. Both hydrogen and oxygen are added to the fuel cell in an electrolyte solution. The hydrogen and oxygen remain separated from one another, and a platinum membrane prevents electrons from flowing directly to the positive side of the fuel cell. Instead, they are routed through an external circuit,

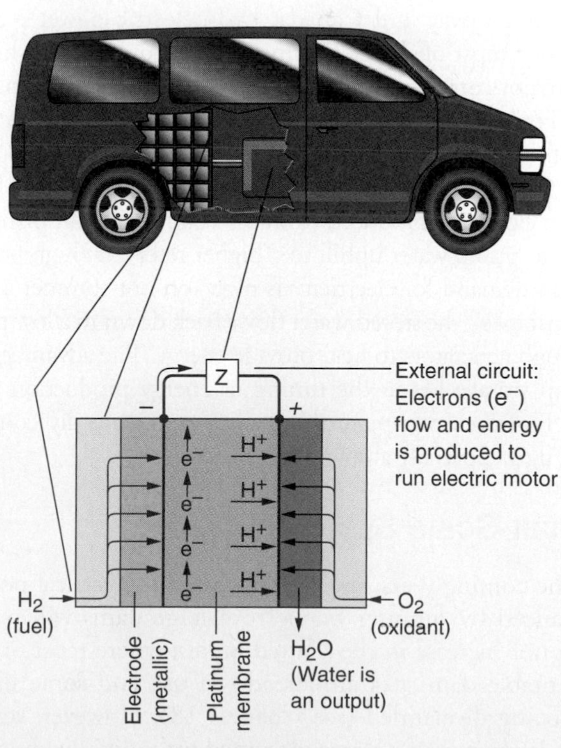

FIGURE 16.10 Idealized diagram showing how a fuel cell works and its application to power a vehicle.

and along the way from the negative to the positive electrode, they are diverted into an electrical motor, supplying current to keep the motor running.[9, 10] To maintain this reaction, hydrogen and oxygen are added as needed. When hydrogen is used in a fuel cell, the only waste product is water.[11]

Fuel cells are efficient and clean, and they can be arranged in a series to produce the appropriate amount of energy for a particular task. In addition, the efficiency of a fuel cell is largely independent of its size and energy output. They can also be used to store energy to be used as needed.

Fuel cells are used in many locations. For example, they power buses at Los Angeles International Airport and in Vancouver. They also power a few buses in east San Francisco Bay[13] and provide heat and power at Vandenberg Air Force Base in California.[12] But this experimental technology is expensive, with buses estimated to cost more than $1 million each.[14]

Technological improvements in producing hydrogen are certain, and the fuel price of hydrogen may be substantially lower in the future.[15]

16.4 Water Power

Water power is a form of stored solar energy that has been successfully harnessed since at least the time of the Roman Empire. Waterwheels that convert water power to mechanical energy were turning in Western Europe in the Middle Ages. During the 18th and 19th centuries, large waterwheels provided energy to power grain mills, sawmills, and other machinery in the United States.

Today, hydroelectric power plants use the water stored behind dams. In the United States, hydroelectric plants generate about 80,000 MW of electricity—about 10% of the total electricity produced in the nation. In some countries, such as Norway and Canada, hydroelectric power plants produce most of the electricity used. Figure 16.11a shows the major components of a hydroelectric power station.

Hydropower can also be used to store energy produced by other means, through the process of pump storage (Figure 16.11b and c). During times when demand for power is low, excess electricity produced from oil, coal, or nuclear plants is used to pump water uphill to a higher reservoir (high pool). When demand for electricity is high (on hot summer days, for instance), the stored water flows back down to a low pool through generators to help provide energy. The advantage of pump storage lies in the timing of energy production and use. However, pump storage facilities are generally considered ugly, especially at low water times.

Small-Scale Systems

In the coming years, the total amount of electrical power produced by running water from large dams will probably not increase in the United States, where most of the acceptable dam sites are already in use and some dams are being dismantled (see Chapter 18). However, small-scale hydropower systems, designed for individual homes, farms, or small industries, may be more common in the future. These small systems, known as microhydropower systems, have power output of less than 100 kW.[16]

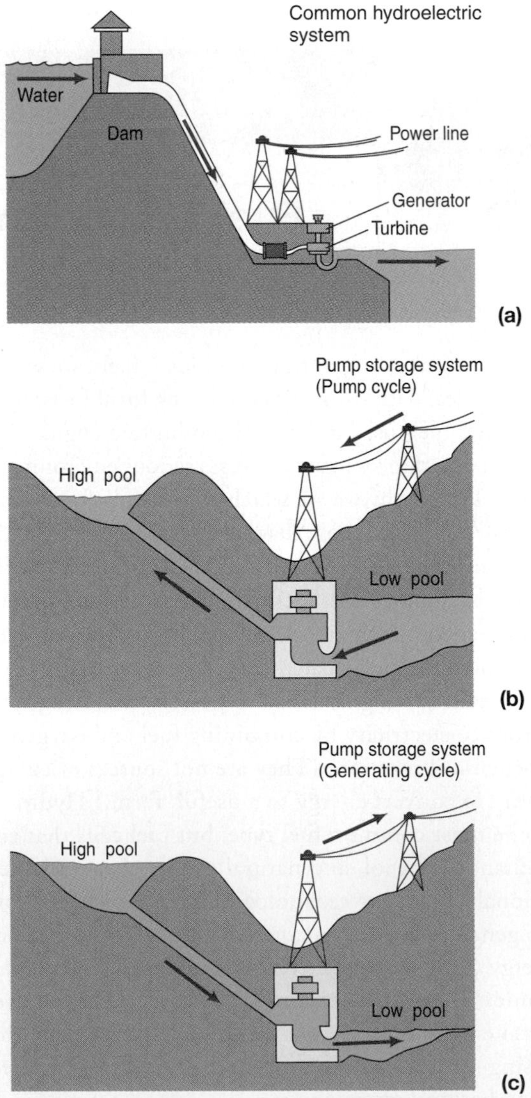

FIGURE 16.11 **(a)** Basic components of a hydroelectric power station. **(b)** A pump storage system. During light power load, water is pumped from low pool to high pool. **(c)** During peak power load, water flows from high pool to low pool through a generator. (*Source*: Modified from the Council on Environmental Quality, *Energy Alternatives: A Comparative Analysis* [Norman: University of Oklahoma Science and Policy Program, 1975].)

Many locations have potential for producing small-scale electrical power, either through small dams or by placing turbines in the free-flowing waters of a river. This is particularly true in mountainous areas, where energy from stream water is often available. Microhydropower development is site-specific, depending on local regulations, economic situations, and hydrologic limitations. Hydropower can be used to generate either electrical power or mechanical power to run machinery. Its use may help reduce the high cost of importing energy and may also help small operations become more independent of local utility providers. However, these systems can interfere with freshwater ecosystems, including fish habitats, and can take away from landscape beauty.

Because small dams can damage stream environments by blocking fish passage and changing downstream flow, careful consideration must be given to their construction. A few small dams cause little environmental degradation beyond the specific sites, but a large number of dams can have an appreciable impact on a region. (This principle applies to many forms of technology and development. The impact of a single development may be nearly negligible over a broad region; but as the number of such developments increases, the total impact may become significant.)

Water Power and the Environment

Water power is clean power in that it requires no burning of fuel, does not pollute the atmosphere, produces no radioactive or other waste, and is efficient. However, there are environmental prices to pay (see Chapter 21):

- Large dams and reservoirs flood large tracts of land that could have had other uses. For example, towns and agricultural lands may be lost.
- Dams block the migration of some fish, such as salmon, and the dams and reservoirs greatly alter habitats for many kinds of fish.
- Dams trap sediment that would otherwise reach the sea and eventually replenish the sand on beaches.
- Reservoirs with large surface areas increase evaporation of water compared to pre-dam conditions. In arid regions, evaporative loss of water from reservoirs is more significant than in more humid regions.
- For a variety of reasons, many people do not want to turn wild rivers into a series of lakes.

For all these reasons, and because many good sites for dams already have one, the growth of large-scale water power in the future (with the exception of a few areas, including Africa, South America, and China) appears limited. Indeed, in the United States there is an emerging social movement to remove dams. Hundreds of dams, especially those with few useful functions, are being considered for removal, and a few have already been removed (see Chapters 18 and 19). The U.S. Department of Energy forecasts that electrical generation from large hydropower dams will decrease significantly.

16.5 Ocean Energy

A lot of energy is involved in the motion of waves, currents, and tides in oceans. Many have dreamed of harnessing this energy, but it's not easy, for the obvious reasons that ocean storms are destructive and ocean waters are corrosive. The most successful development of energy from the ocean has been **tidal power**. Use of the water power of ocean tides can be traced back to the Roman occupation of Great Britain around Julius Caesar's time, when the Romans built a dam that captured tidal water and let it flow out through a waterwheel. By the 10th century, tides were used once again to power coastal mills in Britain.[17] However, only in a few places with favorable topography—such as the north coast of France, the Bay of Fundy in Canada, and the northeastern United States—are the tides strong enough to produce commercial electricity. The tides in the Bay of Fundy have a maximum range of about 15 m (49 ft). A minimum range of about 8 m (26 ft) appears necessary with present technology for development of tidal power.

To harness tidal power, a dam is built across the entrance to a bay or an estuary, creating a reservoir. As the tide rises (flood tide), water is initially prevented from entering the bay landward of the dam. Then, when there is sufficient water (from the oceanside high tide) to run the turbines, the dam is opened, and water flows through it into the reservoir (the bay), turning the blades of the turbines and generating electricity. When the bay is filled, the dam is closed, stopping the flow and holding the water in that reservoir. When the tide falls (ebb tide), the water level in the reservoir is higher than that in the ocean. The dam is then opened to run the turbines (which are reversible), and electric power is produced as the water is let out of the reservoir. Figure 16.12 shows the Rance tidal power plant

FIGURE 16.12 Tidal power station on the river Rance near Saint-Malo, France.

on the north coast of France. Constructed in the 1960s, it is the first and largest modern tidal power plant and has remained in operation since. The plant at capacity produces about 240,000 kW from 24 power units spread out across the dam. At the Rance power plant, most electricity is produced from the ebb tide, which is easier to control.

Tidal power, too, has environmental impacts. The dam changes the hydrology of a bay or an estuary, which can adversely affect the vegetation and wildlife. The dam restricts upstream and downstream passage of fish, and the periodic rapid filling and emptying of the bay as the dam opens and closes with the tides rapidly changes habitats for birds and other organisms.

16.6 Wind Power

Wind power, like solar power, has evolved over a long time. From early Chinese and Persian civilizations to the present, wind has propelled ships and has driven windmills to grind grain and pump water. In the past, thousands of windmills in the western United States were used to pump water for ranches. More recently, wind has been used to generate electricity. The trouble is, wind tends to be highly variable in time, place, and intensity.[18]

Basics of Wind Power

Winds are produced when differential heating of Earth's surface creates air masses with differing heat contents and densities. The potential for energy from the wind is large, and thus wind "prospecting" has become an important endeavor. On a national scale, regions with the greatest potential are the Pacific Northwest coastal area, the coastal region of the northeastern United States, and a belt within the Great Plains extending from northern Texas through the Rocky Mountain states and the Dakotas (Figure 16.13). Other windy sites include mountain areas in North Carolina and the northern Coachella Valley in Southern California. A site with average wind velocity of about 18 kilometers per hour (11 mph) or greater is considered a good prospect for wind energy development, although starting speeds for modern wind turbines can be considerably lower.[19]

In any location, the wind's direction, velocity, and duration may be quite variable, depending on local topography and temperature differences in the atmosphere. For example, wind velocity often increases over hilltops and when wind is funneled through a mountain pass (Figure 16.14). The increase in wind velocity over a mountain is due to a vertical convergence of wind, whereas in a pass the increase is partly due to a horizontal convergence. Because the shape of a mountain or a pass is often related to the local or regional geology, prospecting for wind energy is a geologic as well as a geographic and meteorological task. The wind energy potential of a region or site is determined by instruments that measure and monitor over time the strength, direction, and duration of the wind.

Significant improvements in the size of windmills and the amount of power they produce occurred from the late 1800s to the present, when many European countries and the United States became interested in

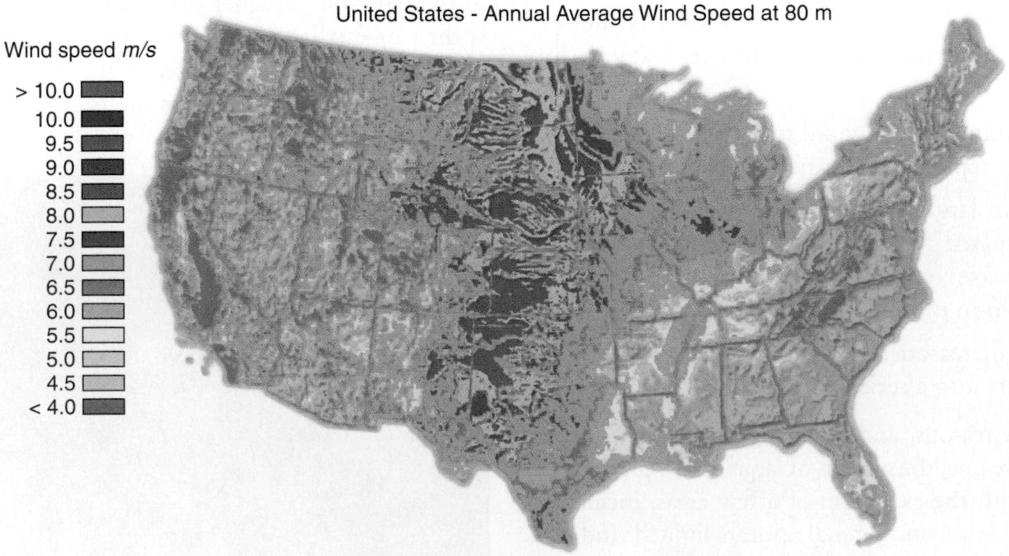

FIGURE 16.13 Wind energy potential in the United States.

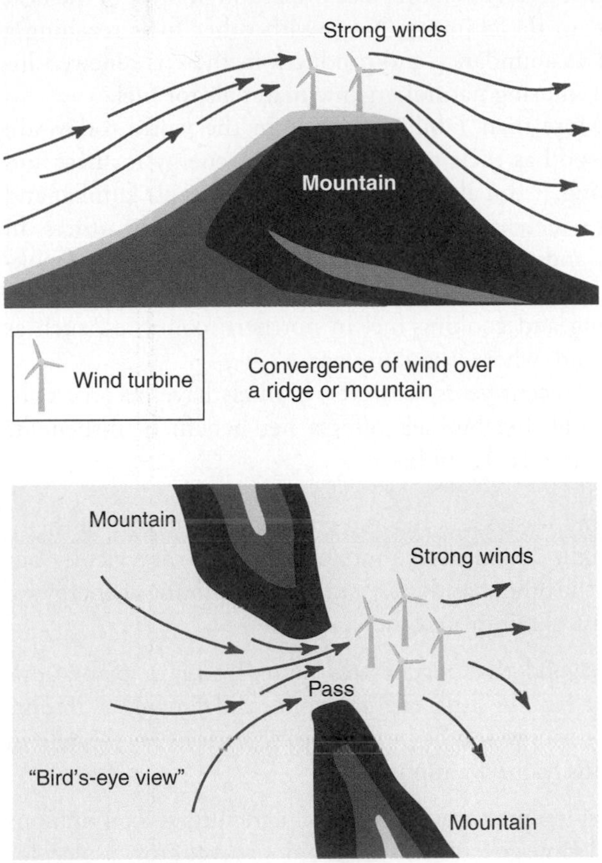

FIGURE 16.14 Idealized diagram showing how wind energy is concentrated by topography.

large-scale generators driven by wind. In the United States in the first half of the 20th century, thousands of small, wind-driven generators were used on farms. Most of these small windmills generated approximately 1 kW of power, which is much too little to be considered for central power-generation needs. Interest in wind power declined for several decades prior to the 1970s because of the abundance of cheap fossil fuels. Since the 1980s, interest in building windmills has revived greatly.

Today, wind energy is the cheapest form of alternative energy. Electricity produced from wind often costs less than that from natural gas and coal. Worldwide, wind power's generating capacity exceeded 120,000 megawatts in 2008. The United States has just taken over first place in having the largest installed wind energy capacity, now ahead of Germany, which had been the leader for a number of years, followed by Spain, China, India, and Italy.[20] U.S. wind power capacity reached 34,000 MW in 2007 (the most recent year for which data are available), increasing 33% a year on average since 2003.[21] Modern wind turbines are big, as much as 70 m (230 feet) high, as tall as a 23-story building, and have a generating capacity of more than 1 million watts—enough electricity for 500 modern U.S. homes (Figure 16.15).[22]

Much of the electricity produced from these large turbines is connected to the electrical grid, and many of them are installed in what are called "wind farms," which are providing large amounts of electricity. One of the biggest in the United States is the Horse Hollow Wind Energy Center near Abilene, Texas, owned and operated by Florida Power & Light. It has 421 wind turbines with a total generating capacity of 735 megawatts, enough to meet the electricity needs of approximately 220,000 homes, and enough for all domestic use for a city the size of Austin, Texas. (To put this in perspective, one large fossil fuel plant or nuclear power plant produces about 1,000 MW.) The Horse Hollow wind turbines are spread widely across approximately 47,000 acres, and the land is used for both ranching and energy production.[23]

Wind Power and the Environment

Wind energy does have a few disadvantages:

- Wind turbines can kill birds. (Birds of prey, such as hawks and falcons, are particularly vulnerable.)
- Wind turbines and wind farms may degrade an area's scenery.

However, although wind farms must often compete with other land uses, in many cases wind turbines can share land used for farms, military bases, and other facilities. Everything considered, wind energy has a relatively low environmental impact.

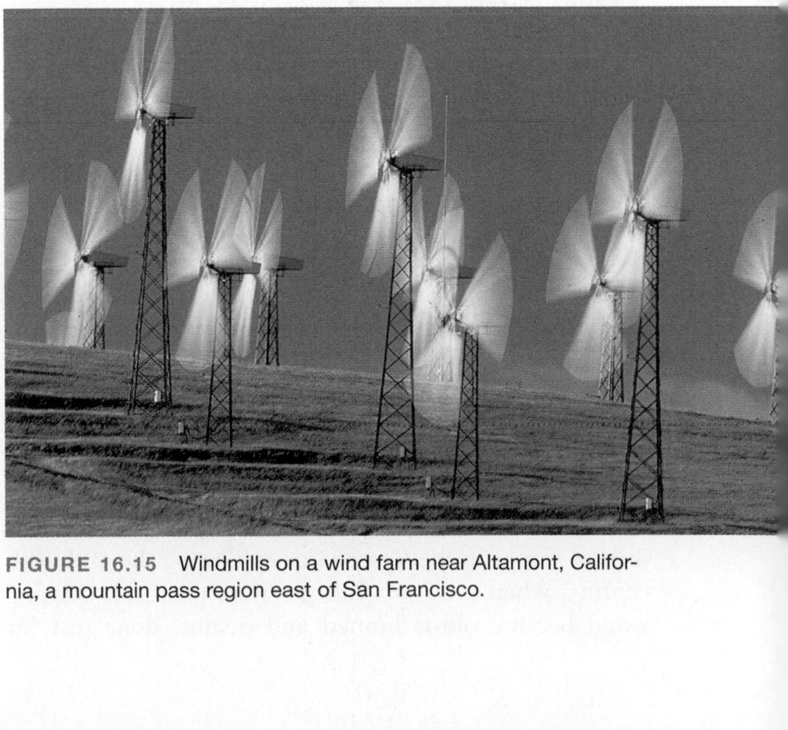

FIGURE 16.15 Windmills on a wind farm near Altamont, California, a mountain pass region east of San Francisco.

The Future of Wind Power

Wind power's continued use is certain. As noted earlier, the use of wind energy has been growing approximately 33% per year, nearly ten times faster than oil use. It is believed that there is sufficient wind energy in Texas, South Dakota, and North Dakota to satisfy the electricity needs of the entire United States. Consider the implications for nations such as China. China burns tremendous amounts of coal at a heavy environmental cost that includes exposing millions of people to hazardous air pollution. In rural China, exposure to the smoke from burning coal in homes has increased the rate of lung cancer by a factor of nine or more. China could probably double its current capacity to generate electricity with wind alone![24]

Wind energy supplies 1.5% of the world's demand for electricity, and its growth rate of more than 30% indicates that it could be a major supplier of power in the relatively near future.[25] One scenario suggests that wind power could supply 10% of the world's electricity in the coming decades and, in the long run, could provide more energy than hydropower, which today supplies approximately 20% of the electricity in the world. The wind energy industry has created thousands of jobs in recent years. Worldwide, more than half a million people are employed in wind energy, and according to the World Wind Energy Association, wind energy "creates many more jobs than centralized, nonrenewable energy sources."[26] Technology, meanwhile, is producing more efficient wind turbines, thereby reducing the price of wind power. All told, wind power is becoming a major investment opportunity.

16.7 Biofuels

Biofuel is energy recovered from biomass (organic matter). We can divide biofuels into three groups: firewood, organic wastes, and crops grown to be converted into liquid fuels.

Biofuels and Human History

Biomass is the oldest fuel used by humans. Our Pleistocene ancestors burned wood in caves to keep warm and cook food. Biofuels remained a major source of energy throughout most of the history of civilization. When North America was first settled, there was more wood fuel than could be used. Forests often were cleared for agriculture by girdling trees (cutting through the bark all the way around the base of a tree) to kill them and then burning the forests.

Until the end of the 19th century, wood was the major fuel source in the United States. During the mid-20th century, when coal, oil, and gas were plentiful, burning wood became old-fashioned and quaint, done just for pleasure in an open fireplace even though most of the heat went up the chimney. Now, with other fuels reaching a limit in abundance and production, there is renewed interest in using natural organic materials for fuel.

More than 1 billion people in the world today still use wood as their primary source of energy for heat and cooking. [13] But although firewood is the most familiar and most widely used biomass fuel, there are many others. In India and other countries, cattle dung is burned for cooking. Peat, a form of compressed dead vegetation, provides heating and cooking fuel in northern countries, such as Scotland, where it is abundant.

In recent years, however, biofuels have become controversial. Do biofuels offer a net benefit or disbenefit? (See Table 16.1.) In brief:

- Using wastes as a fuel is a good way to dispose of them. Making them takes more energy than they yield—but on the other hand, they reduce the amount of energy we must obtain from other sources.
- Firewood that regenerates naturally or in plantations that require little energy input will remain an important energy source in developing nations and locally in industrialized nations.
- Despite pressure from some agricultural corporations and some governments to promote crops grown solely for conversion into liquid fuels (called agrifuels), at present these are poor sources of energy. Most scientific research shows that producing agrifuels takes more energy than they yield. In some cases, there appears to be a net benefit, but the energy produced per unit of land area is low, much lower than can be obtained from solar and wind.

What it boils down to is that photosynthesis, though a remarkable natural process, is less efficient than modern photovoltaic cells in converting sunlight to electricity. Some algae and bacteria appear to provide a net energy benefit and can yield ethanol directly, but production of ethanol from these sources is just beginning and is experimental.[27]

Biofuels and the Environment

The conversion of farmland from food crops to biofuels appears to be one of the main reasons that food prices have risen rapidly worldwide, and that worldwide food production no longer exceeds demand. It also has environmental effects.

Biofuel agriculture competes for water with all other uses, and the main biofuel crops require heavy use of artificial fertilizers and pesticides.

Biofuels are supposed to reduce the production of greenhouse gases, but when natural vegetation is removed to grow biofuel crops, the opposite may be the case. The environmental organization Friends of the Earth says

Table 16.1 SELECTED EXAMPLES OF BIOMASS ENERGY SOURCES, USES, AND PRODUCTS

SOURCES	EXAMPLES	USES/PRODUCTS	COMMENT
Forest Products	Wood, chips	Direct burning,[a] charcoal[b]	Major source today in developing countries
Agriculture residues	Coconut husks, sugarcane waste, corncobs, peanut shells	Direct burning	Minor source
Energy crops	Sugarcane, corn, sorghum	Ethanol (alcohol)[c], gasification[d]	Ethanol is major source of fuel in Brazil for automobiles
Algae and bacteria	Special farms	Experimental	
Trees	Palm oil	Biodiesel	Fuel for vehicles
Animal residues	Manure	Methane[e]	Used to run farm machinery
Urban waste	Waste paper, organic household waste	Direct burning of methane from wastewater treatment or from landfills[f]	Minor source

[a]Principal biomass conversion.

[b]Secondary product from burning wood.

[c]Ethanol is an alcohol produced by fermentation, which uses yeast to convert carbohydrates into alcohol in fermentation chambers (distillery).

[d]Biogas from gasification is a mixture of methane and carbon dioxide produced by pyroclytic technology, Which is a thermochemical process that breaks down solid biomass into an oil-like liquid and almost pure carbon char.

[e]Methane is produced by anaerobic fermentation in a digester.

[f]Naturally produced in landfills by anaerobic fermentation.

that as much as 8% of the world's annual CO_2 emissions can be attributed to draining and deforesting peatlands in Southeast Asia to create palm plantations. The organization estimates that in Indonesia alone 44 million acres have been cleared for these plantations, an area equal to more than 10% of all the cropland in the United States, as large as Oklahoma and larger than Florida.[28]

The use of biofuels can pollute the air and degrade the land. For most of us, the smell of smoke from a single campfire is part of a pleasant outdoor experience. Under certain weather conditions, however, woodsmoke from many campfires or chimneys in narrow valleys can lead to air pollution. The use of biomass as fuel places pressure on an already heavily used resource. A worldwide shortage of firewood is adversely affecting natural areas and endangered species. For example, the need for firewood has threatened the Gir Forest in India, the last remaining habitat of the Indian lion (not to be confused with the Indian tiger). The world's forests will also shrink and in some cases vanish if our need for their products exceeds their productivity.

Biofuels do have some potential benefits. One is that certain kinds of crops, such as nuts produced by trees, may provide a net energy benefit in environments that are otherwise not suited to the growth of food crops. For example, some remote mountainous areas of China may become productive of biofuels. But this is not commonly the case.[29]

Another environmental plus is that combustion of biofuels generally releases fewer pollutants, such as sulfur dioxide and nitrogen oxides, than does combustion of coal and gasoline. This is not always the case for burning urban waste, however. Although plastics and hazardous materials are removed before burning, some inevitably slip through the sorting process and are burned, releasing air pollutants, including heavy metals. There is a conflict in our society as to whether it is better, in terms of the environment, to burn urban waste to recover energy and risk some increase in air pollution, or dump these wastes in landfills, which can then pollute soil and groundwater.

16.8 Geothermal Energy

There are two kinds of **geothermal energy**: deep-earth, high-density; and shallow-earth, low-density. The first makes use of energy within the Earth. The second is a form of solar energy: When the sun warms the surface soils, water, and rocks, some of this heat energy is gradually transmitted down into the ground.

The first kind of geothermal energy—deep-earth, high-density—is natural heat from the interior of the Earth. It is mined and then used to heat buildings and generate electricity. The idea of harnessing Earth's internal heat goes back more than a century. As early as 1904, geothermal power was used in Italy. Today, Earth's natural internal heat is being used to generate electricity in 21 countries, including Russia, Japan, New Zealand, Iceland, Mexico, Ethiopia, Guatemala, El Salvador, the Philippines, and the United States. Total worldwide production is approaching 9,000 MW (equivalent to nine large, modern coal-burning or nuclear power plants)—double the amount in 1980. Some 40 million people today receive their electricity from geothermal energy at a cost competitive with that of other energy sources. In El Salvador, geothermal energy is supplying 25% of the total electric energy used. However, at the global level, geothermal energy accounts for less than 0.15% of the total energy supply.[30]

This kind of geothermal energy may be considered a nonrenewable energy source when rates of extraction are greater than rates of natural replenishment. However, geothermal energy has its origin in the natural heat production within Earth, and only a small fraction of the vast total resource base is being used today. Although most geothermal energy production involves tapping high-heat sources, people are also using the low-temperature geothermal energy of groundwater in some applications.

FIGURE 16.16 Geysers Geothermal Field, north of San Francisco, California, is the largest geothermal power operation in the world and produces energy directly from steam.

Geothermal Systems

The average heat flow from the interior of the Earth is very low, about 0.06 watts per square meter (W/m^2). This amount is trivial compared with the 177 W/m^2 from sunlight at the surface, but in some areas the heat flow is high enough to be useful.[31] For the most part, high heat flow occurs mostly at plate tectonic boundaries (see Chapter 5), including oceanic ridge systems (divergent plate boundaries) and areas where mountains are being uplifted and volcanic islands are forming (convergent plate boundaries). You can see the effects of such natural heat flow at Yellowstone National Park. On the basis of geologic criteria, several types of hot geothermal systems (with temperatures greater than about 80°C, or 176°F) have been defined.

Some communities in the United States, including Boise, Idaho, and Klamath Falls, Oregon, are using deep-earth, high-density geothermal heating systems. A common type of geothermal system uses hydrothermal convection, where the circulation of steam and/or hot water transfers heat from the depths to the surface. An example is the Geysers Geothermal Field, 145 km (90 mi) north of San Francisco. It is the largest geothermal power operation in the world (Figure 16.16), producing about 1,000 MW of electrical energy. At the Geysers, the hot water is maintained in part by injecting treated wastewater from urban areas into hot rocks.

The second kind of geothermal energy—shallow-earth, low-density—is at much lower temperatures than geothermal sources and is used not to produce electricity but for heating buildings and swimming pools, and for heating soil to boost crop production in greenhouses. The potential for this kind of energy is huge, and it is cheap to obtain.[32] Such systems are extensively used in Iceland.

It may come as a surprise to learn that most groundwater can be considered a source of shallow-earth, low-density geothermal energy. It is geothermal because the normal heat flow from the sun to the Earth keeps the temperature of groundwater, at a depth of 100 m (320 ft), at about 13°C (55°F). This is cold for a shower but warmer than winter temperatures in much of the United States, where it can help heat a house. In warmer regions, with summer temperatures of 30–35°C (86–95°F), groundwater at 13°C is cool and thus can be used to provide air conditioning, as it does today in coastal Florida and elsewhere. In summer, heat can be transferred from the warm air in a building to the cool groundwater. In winter, when the outdoor temperature is below about 4°C (40°F), heat can be transferred from the groundwater to the air in the building, reducing the need for heating from other sources. "Heat pumps" for this kind of heat transfer are used in warm locations such as Florida and as far north as Juneau, Alaska, but are limited by extreme temperatures and can't function in the cold winters of northern New Hampshire or interior Alaska, such as in Fairbanks. (Juneau, on the coast, has a much more moderate climate because of the influence of the ocean waters.)

Geothermal Energy and the Environment

Deep-earth, high-density, geothermal energy development produces considerable thermal pollution from its hot wastewaters, which may be saline and highly corrosive. Other environmental problems associated with this kind of geothermal energy use include on-site noise, emissions of gas, and disturbance of the land at drilling sites, disposal sites, roads and pipelines, and power plants. The good news is that the use of deep-earth, high-density geothermal energy releases almost 90% less carbon dioxide and sulfur dioxide than burning coal releases to produce the same amount of electricity.[33] Furthermore, development of geothermal energy does not require large-scale transportation of raw materials or refining of chemicals, as development of fossil fuels does. Nor does geothermal energy produce the atmospheric pollutants associated with burning fossil fuels or the radioactive waste associated with nuclear energy.

Even so, deep-earth, high-density geothermal power is not always popular. For instance, on the island of Hawaii, where active volcanoes provide abundant near-surface heat, some argue that the exploration and development of geothermal energy degrade the tropical forest as developers construct roads, build facilities, and drill wells. In Hawaii, geothermal energy also raises religious and cultural issues. Some people, for instance, are offended by the use of the "breath and water of Pele," the volcano goddess, to make electricity. This points out the importance of being sensitive to the values and cultures of people where development is planned.

The Future of Geothermal Energy

By 2008, the United States produced only 7,500 MW of geothermal energy.[34] Globally, the likelihood is that high-density, deep-earth geothermal can be only a minor contributor to world energy demand, but that low-density, shallow-earth geothermal can be a major source of alternative and renewable energy.[35]

CRITICAL THINKING ISSUE

Should Wind Turbines Be Installed in Nantucket Sound?

In 2001, Cape Wind Associates proposed the Cape Wind Project, which would install 130 wind turbines, each higher than the Statue of Liberty, on floating offshore platforms distributed in Nantucket Sound (Figure 16.17), between Nantucket Island and the coast of Massachusetts. The 130 turbines would cover more than 62 square kilometers (24 square miles) and therefore would not be very close together. The turbines are expected to generate enough electricity for 420,000 homes, but although originally proposed in 2001, work on the $1 billion project has yet to begin. The federal government has received more than 40,000 comments about it, the greatest number ever about an alternative energy project,[31] and in response has held 50 public hearings.

Fishermen have expressed concern about possible effects on fish habitats, including spawning grounds. Shippers have expressed concern about possible dangers to commercial shipping. And some residents of Cape Cod, Martha's Vineyard, and Nantucket have opposed the project on aesthetic grounds—that the wind turbines will spoil the scenery.[36] The late senator Edward M. Kennedy, Democrat of Massachusetts, was one of the leaders in the fight against it, as is his nephew, Robert Kennedy Jr.[37]

Nantucket Sound is about 30 km (19 miles) wide between Nantucket Island and the mainland, while visibility at ground level is only about 19 kilometers (12 miles— limited by the curvature of the Earth). Thus, the turbines would be scattered points in the distance from either shore. They would, however, be visible from ships and boats passing through the channel, such as the yachts of wealthy residents on holiday.

It is worth noting that in the meantime, a Dutch company, Blue H, USA LLC, announced plans in March 2008 to put 120 turbines on floats 23 miles south of Martha's Vineyard and 45 miles west of New Bedford, farther away from the view of residents.[38] Further complicating matters, the National Park Service says that Nantucket Sound is eligible for listing on the National Register of Historic Places. And two tribes, the Mashpee Wampanoag of Cape Cod and the Aquinnah Wampanoag

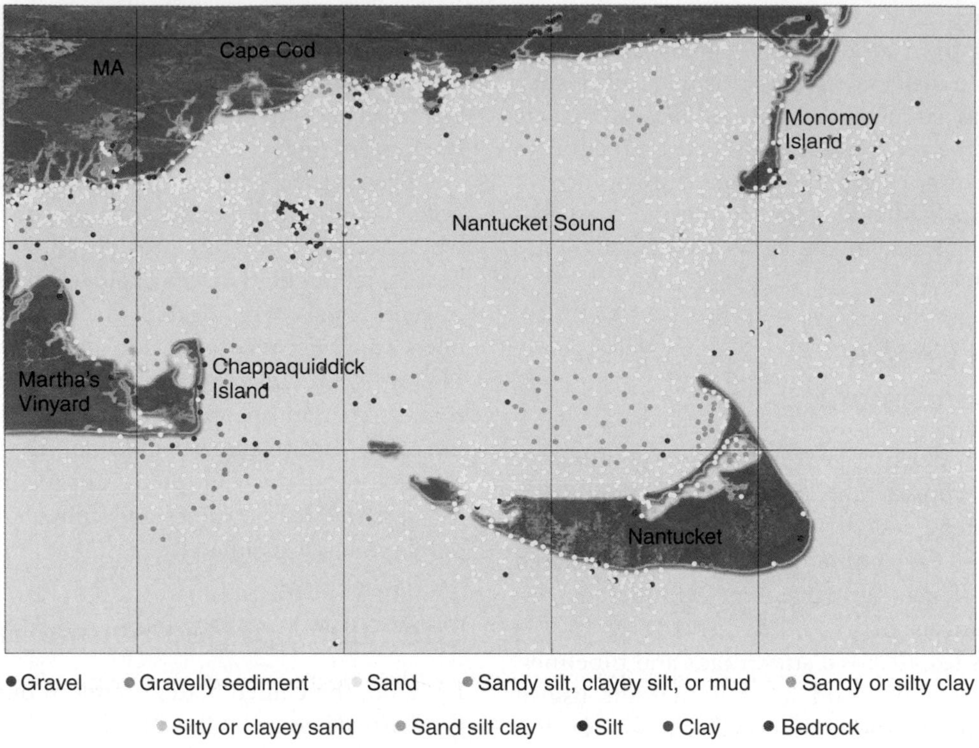

FIGURE 16.17 Map of Nantucket Sound.

of Martha's Vineyard, make two claims against the development: that it might be built over ancestral burial grounds and might interfere with a ritual that requires an unobstructed view of the sunrise.[39]

Critical Thinking Questions

1. Which of the arguments against the Cape Wind Project do you think is most justified? Explain your answer.
   a. The effect on scenic beauty
   b. The effect on fisheries
   c. The traditional cultural practices of Native Americans
2. Since another project would place 120 turbines farther away from the view of residents, what would justify continuation of the Cape Wind Project?
3. Do you think the Cape Wind Project could be important as a precedent for other offshore wind energy projects in the United States? Why or why not?
4. On balance, do you support the Cape Wind Project or oppose it? Explain your answer.

SUMMARY

- The use of renewable alternative energy sources, such as wind and solar energy, is growing rapidly. These energy sources do not cause air pollution, health problems, or climate changes. They offer our best chance to replace fossil fuels and develop a sustainable energy policy.
- Passive solar energy systems often involve architectural designs that enhance absorption of solar energy without requiring mechanical power or moving parts.
- Some active solar energy systems use solar collectors to heat water for homes.
- Systems to produce heat or electricity include power towers and solar farms.
- Photovoltaics convert sunlight directly into electricity.
- Hydrogen gas may be an important fuel of the future, especially when used in fuel cells.
- Water power today provides about 10% of the total electricity produced in the United States. Except in some developing nations, good sites for large dams are already in use. Water power is clean, but there is an environmental price to pay in terms of disturbance of ecosystems, sediment trapped in reservoirs, loss of wild rivers, and loss of productive land.
- Wind power has tremendous potential as a source of electrical energy in many parts of the world. Many

utility companies are using wind power as part of energy production or as part of long-term energy planning. Environmental impacts of wind installations include loss of land, killing of birds, and degradation of scenery.

- Biofuels are of three kinds: firewood, wastes, and crops grown to produce fuels. Firewood has been important historically and will remain so in many developing nations and many rural parts of developed nations. Burning wastes is a good way to dispose of them, and their energy is a useful by-product. Crops grown solely as biofuels appear to be a net energy sink or of only marginal benefit, at considerable environmental costs, including competition for land, water, and fertilizers, and the use of artificial pesticides. At present they are not a good option.
- Geothermal energy, natural heat from Earth's interior, can be used as an energy source. The environmental effects of developing geothermal energy relate to specific site conditions and the type of heat—steam, hot water, or warm water. Environmental impacts may involve on-site noise, industrial scars, emission of gas, and disposal of saline or corrosive waters.

REEXAMINING THEMES AND ISSUES

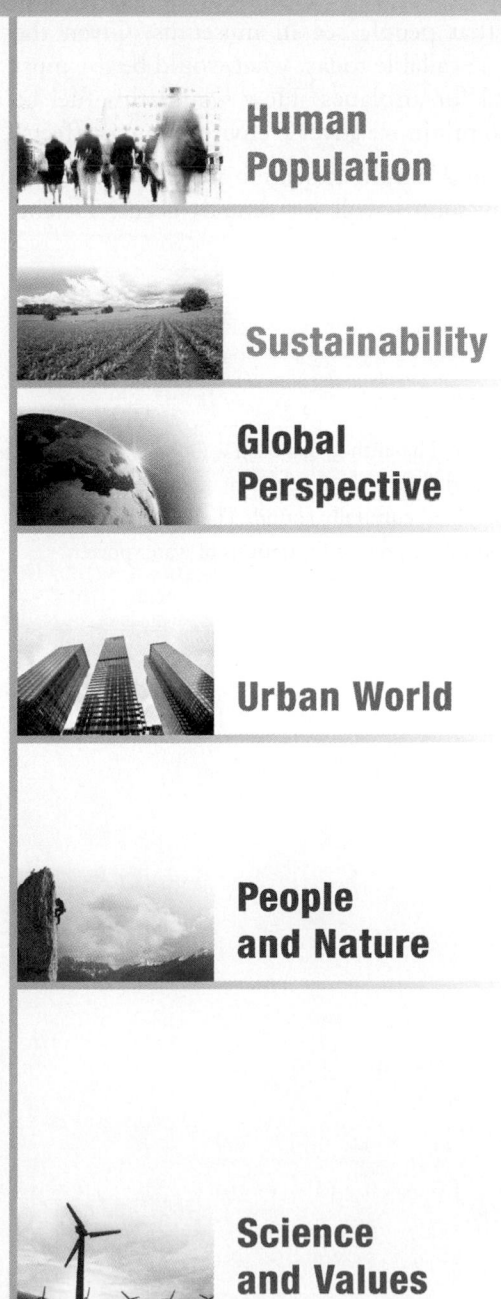

Human Population

As the human population continues to increase, so does global demand for energy. Environmental problems from increased use of fossil fuels could be minimized by controlling population growth, increasing conservation efforts, and using alternative, renewable energy sources that do not harm the environment.

Sustainability

Use of fossil fuels is not sustainable. To plan for energy sustainability, we need to rely more on alternative energy sources that are naturally renewable and do not damage the environment. To do otherwise is antithetical to the concept of sustainability.

Global Perspective

To evaluate the potential of alternative energy sources, we need to understand global Earth systems and identify regions likely to produce high-quality alternative energy that could be used in urban regions of the world.

Urban World

Alternative renewable energy sources have a future in our urban environments. For example, the roofs of buildings can be used for passive solar collectors or photovoltaic systems. Patterns of energy consumption can be regulated through use of innovative systems such as pump storage to augment production of electrical energy when demand is high in urban areas.

People and Nature

Many environmentalists perceive alternative energy sources, such as solar and wind, as being linked more closely with nature than are fossil fuels, nuclear energy, or even water power. This is because solar and wind energy development requires less human modification of the environment. Solar and wind energy allow us to live more in harmony with the environment, and thus we feel more connected to the natural world.

Science and Values

We are seriously considering alternative energy today because we value environmental quality and energy independence and want to plan for a future when we run out of fossil fuels. Recognizing that burning fossil fuels creates many serious environmental problems and that petroleum will soon become less available, we are trying to increase our scientific knowledge and improve our technology to meet our energy needs for the future while minimizing environmental damage. Our present science and technology can lead to a sustainable energy future, but we will need to change our values and our behavior to achieve it.

KEY TERMS

active solar energy systems **329**
alternative energy **327**
biofuel **338**
fuel cells **333**
geothermal energy **339**
nonrenewable alternative energy **327**
passive solar energy systems **328**
photovoltaics **330**
renewable energy **327**
solar collectors **329**
tidal power **335**
water power **334**
wind power **336**

STUDY QUESTIONS

1. What types of government incentives might encourage use of alternative energy sources? Would their widespread use affect our economic and social environment?
2. Your town is near a large river that has a nearly constant water temperature of about 15°C (60°F). Could the water be used to cool buildings in the hot summers? How? What would be the environmental effects?
3. Which has greater future potential for energy production, wind or water power? Which causes more environmental problems? Why?
4. What are some of the problems associated with producing energy from biomass?
5. It is the year 2500, and natural oil and gas are rare curiosities that people see in museums. Given the technologies available today, what would be the most sensible fuel for airplanes? How would this fuel be produced to minimize adverse environmental effects?
6. When do you think the transition from fossil fuels to other energy sources will (or should) occur? Defend your answer.

FURTHER READING

Botkin, D.B., ***Powering the Future: A Scientist's Guide to Energy Independence*** (Indianapolis, IN: Pearson FT Press, 2010).

Boyle, G., ***Renewable Energy*** (New York: Oxford University Press, 2004, paperback).

Scudder, T., ***The Future of Large Dams: Dealing with Social, Environmental, Institutional and Political Costs*** (London: Earthscan, 2006). The author has been a member of a World Bank team that tried to make a new dam and reservoir in Laos environmentally and culturally sound. The book is one of the best summaries of the present limitations of water power.

CHAPTER 17

Nuclear Energy and the Environment

Indian Point Energy Center is a two-unit (originally three-unit) nuclear power plant installation on the eastern shore of the Hudson River within 24 miles of New York City. It must be relicensed, and, as the lower photograph shows, this is creating a major controversy about whether such an installation should be near tens of millions of people.

LEARNING OBJECTIVES

As one of the alternatives to fossil fuels, nuclear energy generates a lot of controversy. After reading this chapter, you should understand . . .

- What nuclear fission is and the basic components of a nuclear power plant;
- Nuclear radiation and its three major types;
- How radioisotopes affect the environment, and the major pathways of radioactive materials in the environment.
- The relationships between radiation doses and health;
- The advantages and disadvantages of nuclear power;
- What the future of nuclear power is likely to be.

CASE STUDY

Indian Point: Should a Nuclear Power Installation Operate Near One of America's Major Cities?

In 1962, after a series of contentious public hearings, Consolidated Edison began operating the first of three nuclear reactors at Indian Point, on the eastern shore of the Hudson River in Buchanan, New York, 38 km (24 mi) north of New York City. Indian Point's second and third reactors began operating in 1974 and 1976, respectively. The first unit had major problems and was finally shut down in 1974. The second and third have been operating since then, but their licenses run out in 2013 and 2015, respectively, and under U.S. law nuclear power plants must be relicensed. All three units are owned by Entergy Nuclear Northeast, a subsidiary of Entergy Corporation.

Twenty million people live within 80 km (50 miles) of this power plant, and this causes considerable concern. Joan Leary Matthews, a lawyer for the New York State Department of Environmental Conservation, said that "whatever the chances of a failure at Indian Point, the consequences could be catastrophic in ways that are almost too horrific to contemplate."[1]

The federal Nuclear Regulatory Commission (NRC) announced the beginning of the relicensing process on May 2, 2007. By 2008 the relicensing of the plant had become a regional controversy, opposed by the New York State government, Westchester County (where the plant is located), and a number of nongovernmental environmental organizations. The plant has operated for almost 50 years, so what's the problem?

There have been some: In 1980, one of the plant's two units filled with water (an operator's mistake). In 1982, the same unit's steam generator piping leaked and released radioactive water. In 1999, it shut down unexpectedly, but operators didn't realize it until the next day, when the batteries that automatically took over ran down. In April 2007, a transformer burned in the second unit, radioactive water leaked into groundwater, and the source of the leak was difficult to find. Most recently, in 2009, a leak in the cooling system allowed 100,000 gallons of water to escape from the main system. Uneasiness about the plant's location increased after the terror attack on September 11, 2001. One of the hijacked jets flew close to the plant, and diagrams of unspecified nuclear plants in the United States have since been found in al Qaeda hideouts in Afghanistan.[2]

Proponents of nuclear power say these are minor problems, and there has been no major one. As far as they can tell, the plant is safe. The Energy Policy Act of 2005 promoted nuclear energy, and the Obama administration is moving ahead with federal funding of nuclear power plants. For 2010, the administration has allocated $18.5 billion for new "next-generation" nuclear power plants. Others, however, such as New York State's attorney general Andrew Cuomo, believe the location is just too dangerous, and he has asked the Nuclear Regulatory Commission to deny Indian Point's relicensing, saying that it has "a long and troubling history of problems."

The conflict at Indian Point illustrates the worldwide debate about nuclear energy. Growing concern about fossil fuels has led to calls for increased use of nuclear power despite unanswered questions and unsolved problems regarding its use. This chapter provides a basis for you to decide whether nuclear power could be, and should be, a bigger supplier of energy in the future. We begin with the basics about the nature of nuclear energy, then go on to explore nuclear reactors, radiation, accidents, waste management, and the future of nuclear power.

17.1 Current Role of Nuclear Power Plants in World Energy Production

Today, nuclear power provides about 17% of the world's electricity and 4.8% of the total energy. In the United States, 104 nuclear power plants produce about 20% of the country's electricity and about 8% of the total energy used (Figure 17.1).[3] Worldwide, there are 436 operating nuclear power plants.[4] Nations differ greatly in the amount of energy they obtain from these plants. France ranks first, with about 80% of its electricity produced by nuclear energy (Table 17.1). The United States ranks tenth in the percentage of electricity it obtains from nuclear power plants.

Most of the world's nuclear power plants are in North America, Western Europe, Russia, China, and In-

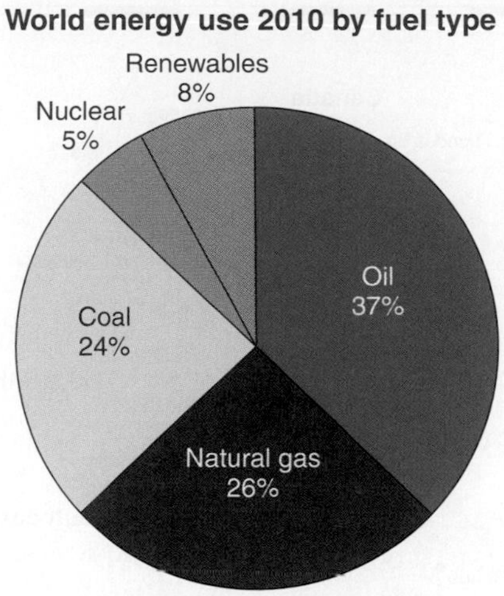

FIGURE 17.1 World energy use. (*Source*: D.B. Botkin, 2010.)

Table 17.1 LEADING NATIONS IN THE USE OF NUCLEAR ENERGY

COUNTRY	% TOTAL ELECTRICITY	GENERATION (MILLION KWH)
France	78%	368,188
Belgium	60%	41,927
Sweden	43%	61,395
Spain	36%	56,060
S. Korea	36%	58,138
Ukraine	33%	75,243
Germany	29%	153,476
Japan	28%	249,256
United Kingdom	28%	89,353
United States	19%	610,365
Canada	18%	94,823
Russia	12%	119,186
World Totals*	18%	2,167,515

(*Source*: D.B. Botkin, 2010.)

dia (Figure 17.2). Most of the U.S. nuclear power plants are in the eastern half of the nation (Figure 17.3). The very few west of the Mississippi River are in Washington, California, Arizona, Nebraska, Kansas, and Texas. The last nuclear plant to be completed in the United States went on line in 1996. However, since the early 1990s, U.S. nuclear plants have added over 23,000 MW, equivalent to the output of 23 large fossil fuel–burning power plants. The electricity produced from nuclear power plants increased 33% between 1980 and 2001, because only two thirds of their capacity was used in 1980, but this increased to more than 90% by 2002. Even if all these power plants operated at only 66% of their capacity, this would be the equivalent of building four new nuclear power plants.

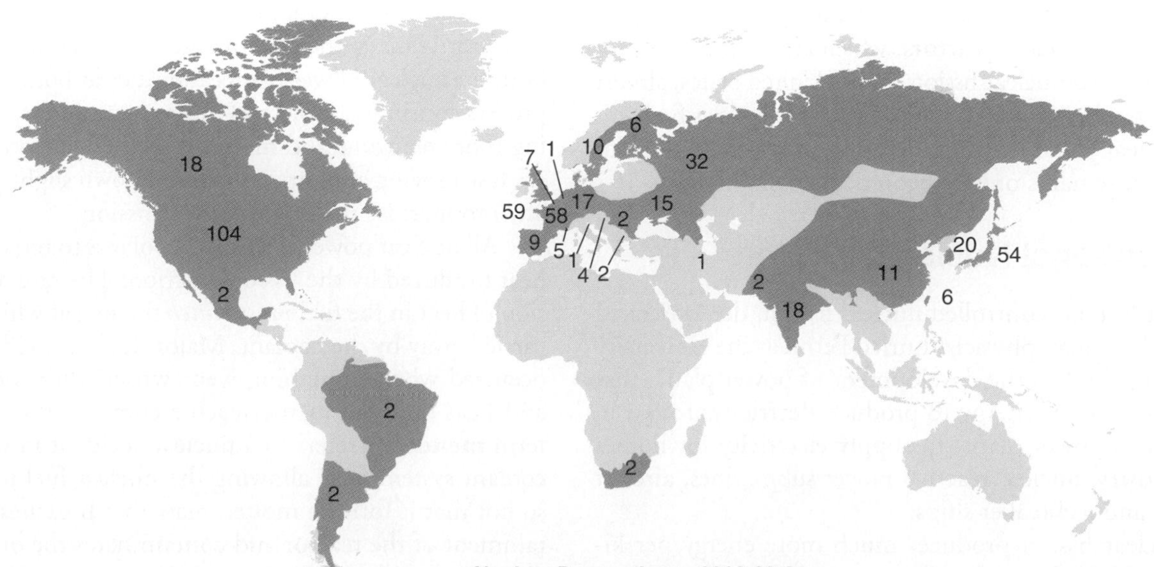

FIGURE 17.2 Where major nuclear power plants are worldwide. (*Source*: Informationskreis KernEnergie, Berlin.)

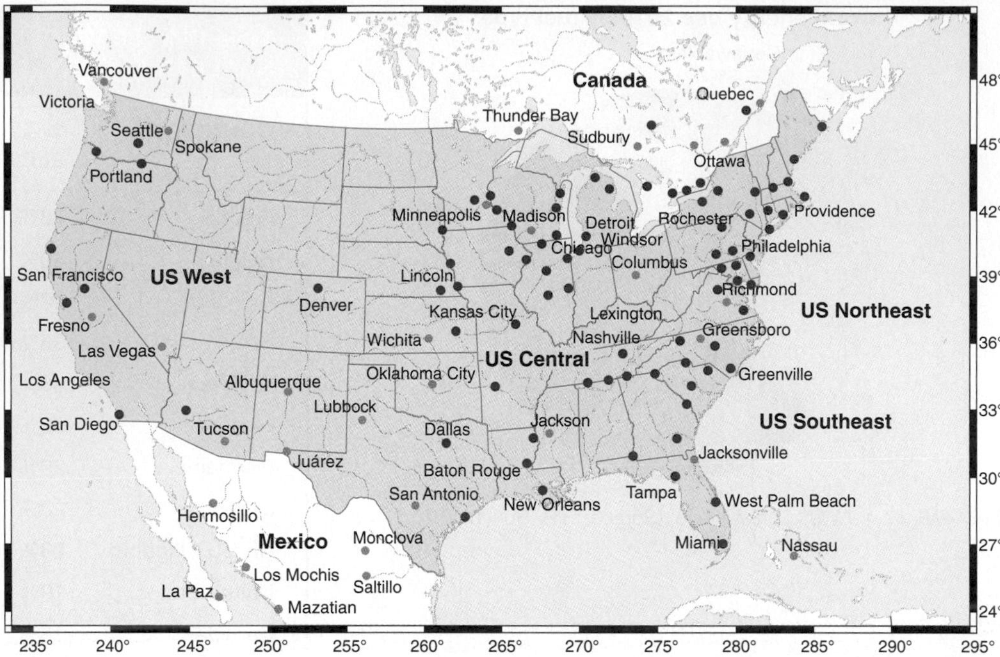

FIGURE 17.3 Where nuclear power plants are in the United States. (*Source*: http://www.insc.anl.gov/pwrmaps/)

17.2 What Is Nuclear Energy?

Hard as it may be to believe, **nuclear energy** is the energy contained in an atom's nucleus. Two nuclear processes can be used to release that energy to do work: fission and fusion. Nuclear **fission** is the splitting of atomic nuclei, and nuclear **fusion** is the fusing, or combining, of atomic nuclei. A by-product of both fission and fusion is the release of enormous amounts of energy. (Radiation and related terms are explained in A Closer Look 17.1. You may also wish to review the discussion of matter and energy in Chapter 14's A Closer Look 14.1.)

Nuclear energy for commercial use is produced by splitting atoms in **nuclear reactors**, which are devices that produce controlled nuclear fission. In the United States, almost all of these reactors use a form of uranium oxide as fuel.

Nuclear *fusion*, despite decades of research to try to develop it, remains only a theoretical possibility.

Conventional Nuclear Reactors

The first human-controlled nuclear fission, demonstrated in 1942 by Italian physicist Enrico Fermi at the University of Chicago, led to the development of power plants that could use nuclear energy to produce electricity. Today, in addition to power plants to supply electricity for homes and industry, nuclear reactors power submarines, aircraft carriers, and icebreaker ships.

Nuclear fission produces much more energy per kilogram of fuel than other fuel-requiring sources, such as biomass and fossil fuels. For example, 1 kilogram (2.2 lb) of uranium oxide produces about the same amount of heat as 16 metric tons of coal.

Three types—isotopes—of uranium occur in nature: uranium-238, which accounts for approximately 99.3% of all natural uranium; uranium-235, which makes up about 0.7%; and uranium-234, about 0.005%. However, uranium-235 is the only naturally occurring fissionable (or *fissile*) material and is therefore essential to the production of nuclear energy. A process called *enrichment* increases the concentration of uranium-235 from 0.7% to about 3%. This enriched uranium is used as fuel.

The spontaneous decay of uranium atoms emits neutrons. Fission reactors split uranium-235 by neutron bombardment. This releases more neutrons than it took to create the first splitting (Figure 17.4). These released neutrons strike other uranium-235 atoms, releasing still more neutrons, other kinds of radiation, fission products, and heat. This is the "chain reaction" that is so famous, both for nuclear power plants and nuclear bombs— as the process continues, more and more uranium is split, releasing more neutrons and more heat. The neutrons released are fast-moving and must be slowed down slightly (*moderated*) to increase the probability of fission.

All nuclear power plants use coolants to remove excess heat produced by the fission reaction. The rate of generation of heat in the fuel *must match* the rate at which heat is carried away by the coolant. Major nuclear accidents have occurred when something went wrong with the balance and heat built up in the reactor core.[5] The well-known term **meltdown** refers to a nuclear accident in which the coolant system fails, allowing the nuclear fuel to become so hot that it forms a molten mass that breaches the containment of the reactor and contaminates the outside environment with radioactivity.

The nuclear steam-supply system includes heat exchangers (which extract heat produced by fission) and primary coolant loops and pumps (which circulate the

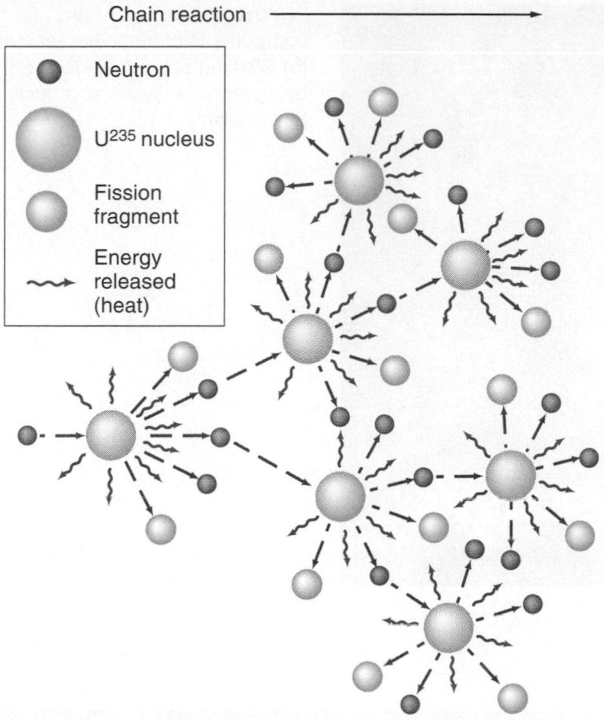

FIGURE 17.4 Fission of uranium-235. A neutron strikes the U-235 nucleus, producing fission fragments and free neutrons and releasing heat. The released neutrons may then strike other U-235 atoms, releasing more neutrons, fission fragments, and energy. As the process continues, a chain reaction develops.

coolant through the reactor). The heat is used to boil water, releasing steam that runs conventional steam-turbine electrical generators (Figure 17.5). In most common reactors, ordinary water is used as the coolant as well as the moderator. Reactors that use ordinary water are called "light water reactors" because there is also "heavy water," which combines deuterium with oxygen.[6]

Most reactors now in use consume more fissionable material than they produce and are known as **burner reactors**. Figure 17.6 shows the main components of a reactor: the core (consisting of fuel and moderator), control rods, coolant, and reactor vessel. The core is enclosed in the heavy, stainless-steel reactor vessel; then, for safety and security, the entire reactor is contained in a reinforced-concrete building.

In the reactor core, fuel pins—enriched uranium pellets in hollow tubes (3–4 m long and less than 1 cm, or 0.4 in., in diameter)—are packed together (40,000 or more in a reactor) in fuel subassemblies. A minimum fuel concentration is necessary to keep the reactor *critical*—that is, to achieve a self-sustaining chain reaction.

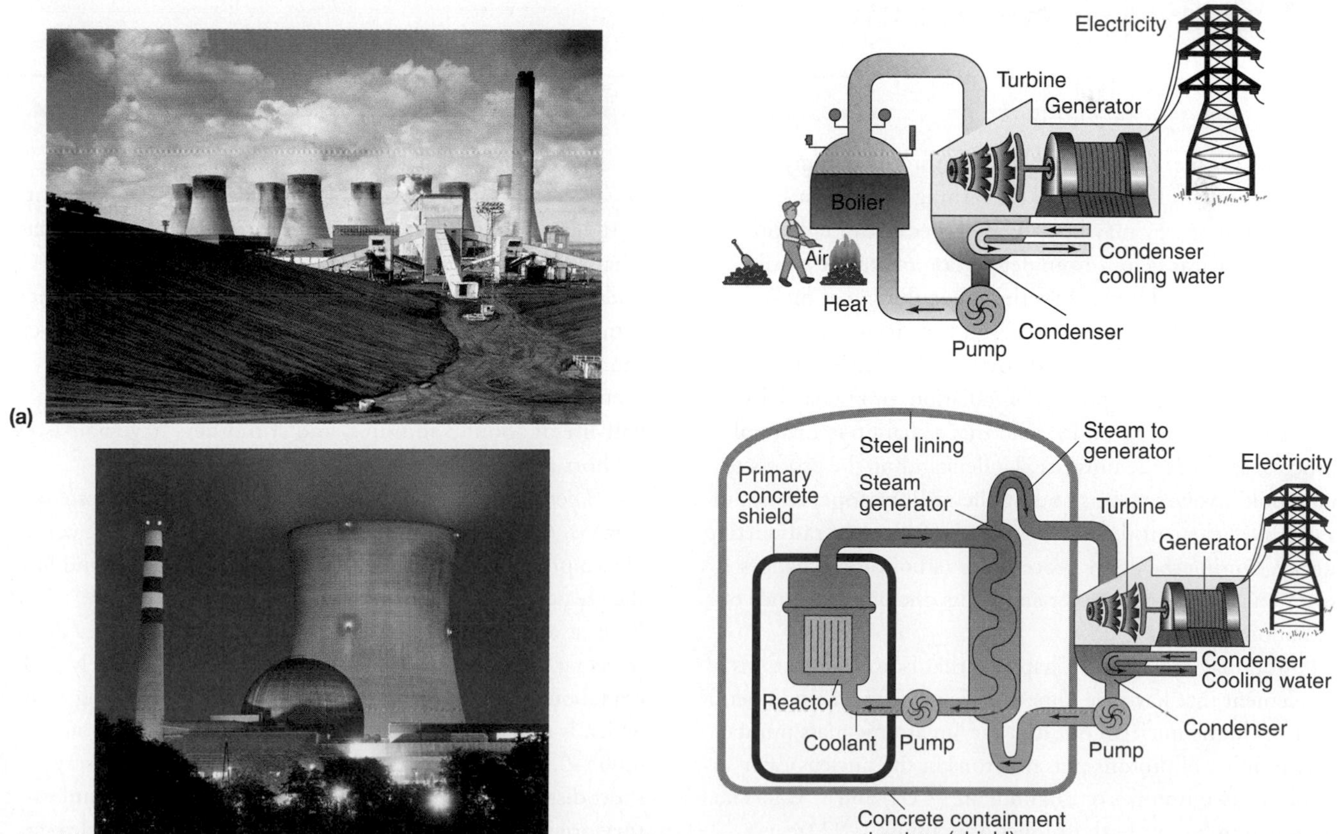

FIGURE 17.5 Comparison of **(a)** a fossil-fuel power plant and **(b)** a nuclear power plant with a boiling-water reactor. Notice that the nuclear reactor has exactly the same function as the boiler in the fossil-fuel power plant. The coal-burning plant **(a)** is Ratcliffe-on-Saw, in Nottinghamshire, England, and the nuclear power station **(b)** is in Leibstadt, Switzerland. (*Source*: American Nuclear Society, *Nuclear Power and the Environment*, 1973.)

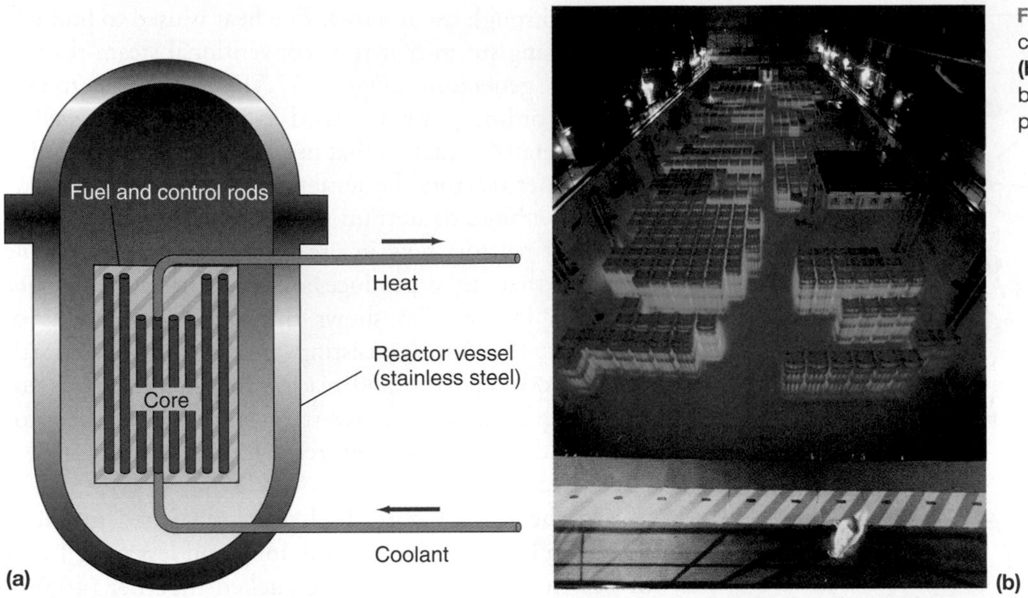

FIGURE 17.6 **(a)** Main components of a nuclear reactor. **(b)** Glowing spent fuel elements being stored in water at a nuclear power plant.

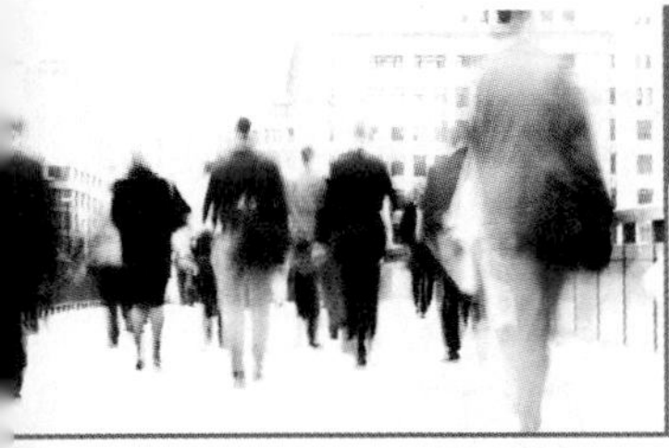

A CLOSER LOOK 17.1

Radioactive Decay

To many people, radiation is a subject shrouded in mystery. They feel uncomfortable with it, learning from an early age that nuclear energy may be dangerous because of radiation and that nuclear fallout from detonation of atomic bombs can cause widespread human suffering. One thing that makes radiation scary is that we cannot see it, taste it, smell it, or feel it. In this closer look, we try to demystify some aspects of radiation by discussing the process of radiation, or radioactivity.

First, we need to understand that radiation is a natural process, as old as the universe. Understanding the process of radiation involves understanding the **radioisotope**, a form of a chemical element that spontaneously undergoes **radioactive decay**. During the decay process, the radioisotope changes from one isotope to another and emits one or more kinds of radiation (Figure 17.7).

You may recall from Chapter 6 that isotopes are atoms of an element that have the same atomic number (the number of protons in the nucleus) but that vary in atomic mass number (the number of protons plus neutrons in the nucleus). For example, two isotopes of uranium are $^{235}U_{92}$ and $^{238}U_{92}$. The atomic number for both isotopes of uranium is 92 (revisit Figure 6.8); however, the atomic mass numbers are 235 and 238. The two different uranium isotopes may be written as uranium-235 and uranium-238 or ^{235}U and ^{238}U.

An important characteristic of a radioisotope is its *half-life*, the time required for one-half of a given amount of the isotope to decay to another form. Uranium-235 has a half-life of 700 million years, a very long time indeed! Radioactive carbon-14 has a half-life of 5,570 years, which is in the intermediate range, and radon-222 has a relatively short half-life of 3.8 days. Other radioactive isotopes have even shorter half-lives; for example, polonium-218 has a half-life of about 3 minutes, and still others have half-lives as short as a fraction of a second.

There are three major kinds of nuclear radiation: *alpha particles, beta particles*, and *gamma rays*. An alpha particle consists of two protons and two neutrons (a helium nucleus) and has the greatest mass of the three types of radiation (Figure 17.7a). Because alpha particles have a relatively high mass, they do not travel far. In air, alpha particles can travel approximately 5–8 cm (about 2–3 in.) before they stop. However, in living tissue, which is much denser than air, they can travel only about 0.005–0.008 cm (0.002–0.003 in.). Because this is a very short distance, they can't cause damage to living cells unless they originate very close to the cells. Also, alpha particles can be stopped by a sheet or so of paper.

Beta particles are electrons and have a mass of 1/1,840 of a proton. Beta decay occurs when one of the protons or

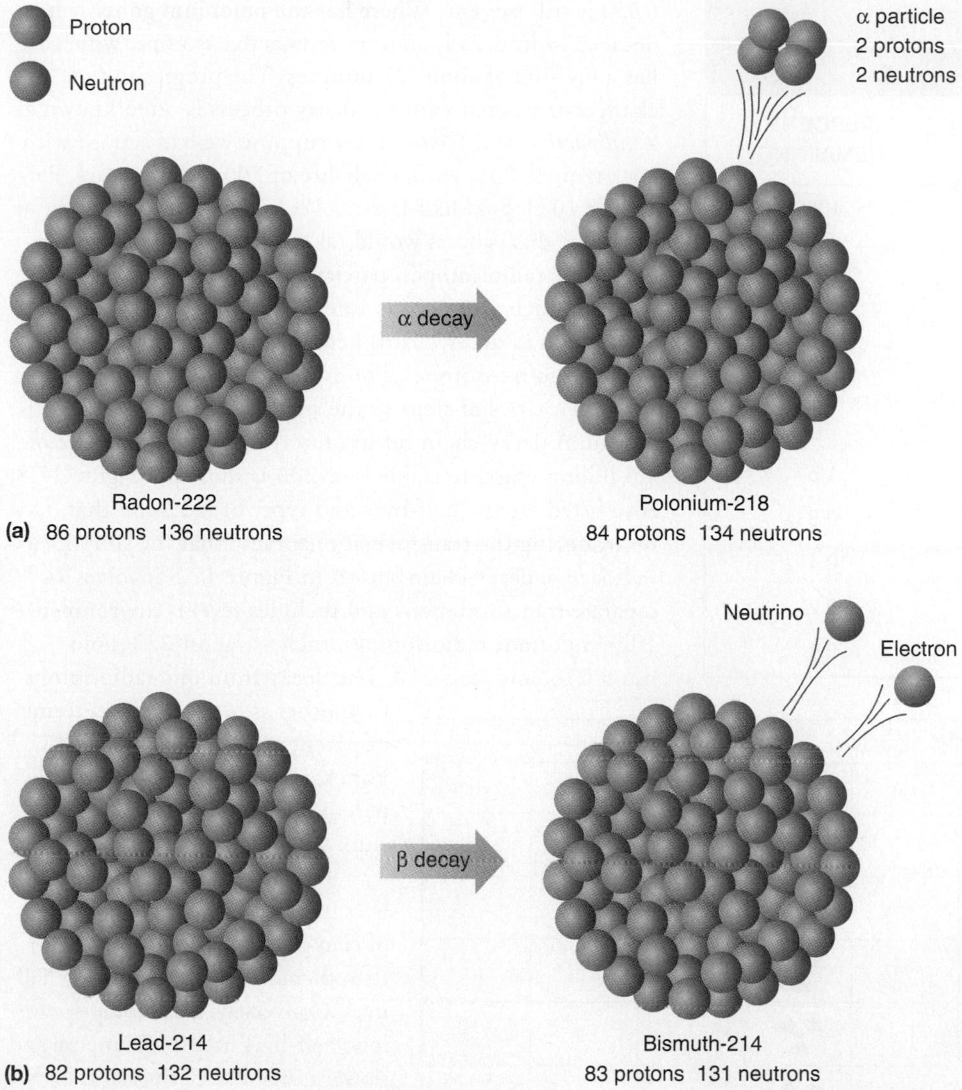

FIGURE 17.7 (Idealized diagrams showing **(a)** alpha and **(b)** beta decay processes. (*Source*: D. J. Brenner, *Radon: Risk and Remedy* [New York: Freeman, 1989]. Copyright 1989 by W.H. Freeman & Company. Reprinted with permission.)

neutrons in the nucleus of an isotope spontaneously changes. What happens is that a proton turns into a neutron, or a neutron is transformed into a proton (Figure 17.7b). As a result of this process, another particle, known as a *neutrino*, is also ejected. A neutrino is a particle with no *rest mass* (the mass when the particle is at rest with respect to an observer).[8] Beta particles travel farther through air than the more massive alpha particles but are blocked by even moderate shielding, such as a thin sheet of metal (aluminum foil) or a block of wood.

The third and most penetrating type of radiation comes from *gamma decay*. When gamma decay occurs, a gamma ray, a type of electromagnetic radiation, is emitted from the isotope. Gamma rays are similar to X-rays but are more energetic and penetrating; and of all types of radiation, they travel the longest average distance. Protection from gamma rays requires thick shielding, such as about a meter of concrete or several centimeters of lead.

Each radioisotope has its own characteristic emissions: Some emit only one type of radiation; others emit a mixture. In addition, the different types of radiation have different toxicities (potential to harm or poison). In terms of human health and the health of other organisms, alpha radiation is most toxic when inhaled or ingested. Because alpha radiation is stopped within a very short distance by living tissue, much of the damaging radiation is absorbed by the tissue. When alpha-emitting isotopes are stored in a container, however, they are relatively harmless. Beta radiation has intermediate toxicity, although most beta radiation is absorbed by the body when a beta emitter is ingested. Gamma emitters are dangerous inside or outside the body; but when they are ingested, some of the radiation passes out of the body.

Each radioactive isotope has its own half-life. Isotopes with very short half-lives are present only briefly, whereas those with long half-lives remain in the environment for long periods. Table 17.2 illustrates the general pattern for decay in terms of the elapsed half-lives and the fraction remaining. For example, suppose we start with 1g polonium-218 with a half-life of approximately 3 minutes. After an elapsed time of 3 minutes, 50% of the polonium-218 remains. After 5 elapsed half-lives, or 15 minutes, only 3% is still present; and after 10 elapsed half-lives (30 minutes),

Table 17.2 GENERALIZED PATTERN OF RADIOACTIVE DECAY

ELAPSED HALF-LIFE	FRACTION REMAINING	PERCENT REMAINING
0	—	100
1	1/2	50
2	1/4	25
3	1/8	13
4	1/16	6
5	1/32	3
6	1/64	1.5
7	1/128	0.8
8	1/256	0.4
9	1/512	0.2
10	1/1024	0.1

0.1% is still present. Where has the polonium gone? It has decayed to lead-214, another radioactive isotope, which has a half-life of about 27 minutes. The progression of changes associated with the decay process is often known as a *radioactive decay chain.* Now suppose we had started with 1 g uranium-235, with a half-life of 700 million years. Following 10 elapsed half-lives, 0.1% of the uranium would be left—but this process would take 7 billion years.

Some radioisotopes, particularly those of very heavy elements such as uranium, undergo a series of radioactive decay steps (a decay chain) before finally becoming stable, nonradioactive isotopes. For example, uranium decays through a series of steps to the stable nonradioactive isotope of lead. A decay chain for uranium-238 (with a half-life of 4.5 billion years) to stable lead-206 is shown in Figure 17.8. Also listed are the half-lives and types of radiation that occur during the transformations. Note that the simplified radioactive decay chain shown in Figure 17.8 involves 14 separate transformations and includes several environmentally important radioisotopes, such as radon-222, polonium-218, and lead-210. The decay from one radioisotope to another is often stated in terms of parent and daughter products. For example, uranium-238 is the parent of daughter product thorium-234.

Radioisotopes with short half-lives initially have a more rapid rate of change (nuclear transformation) than do radioisotopes with long half-lives. Conversely, radioisotopes with long half-lives have a less intense and slower initial rate of nuclear transformation but may be hazardous much longer.[9]

To sum up, when considering radioactive decay, two important facts to remember are (1) the half-life and (2) the type of radiation emitted.

Radioactive Elements	Radiation Emitted			Half-life		
	Alpha	Beta	Gamma	Minutes	Days	Years
Uranium-238	☢		☢			4.5 billion
Thorium-234		☢	☢		24.1	
Protactinium-234		☢	☢	1.2		
Uranium-234	☢		☢			247,000
Thorium-230	☢		☢			80,000
Radium-226	☢		☢			1,622
Radon-222	☢				3.8	
Polonium-218	☢	☢		3.0		
Lead-214		☢	☢	26.8		
Bismuth-214		☢	☢	19.7		
Polonium-214	☢			0.00016 (sec)		
Lead-210		☢	☢			22
Bismuth-210		☢			5.0	
Polonium-210	☢		☢		138.3	
Lead-206	None			Stable		

FIGURE 17.8 Uranium-238 decay chain. (*Source*: F. Schroyer, ed., *Radioactive Waste*, 2nd printing [American Institute of Professional Geologists, 1985].)

A stable fission chain reaction in the core is maintained by controlling the number of neutrons that cause fission. Control rods, which contain materials that capture neutrons, are used to regulate the chain reaction. As the control rods are moved out of the core, the chain reaction increases; as they are moved into the core, the reaction slows. Full insertion of the control rods into the core stops the fission reaction.[7]

17.3 Nuclear Energy and the Environment

The **nuclear fuel cycle** begins with the mining and processing of uranium, its transportation to a power plant, its use in controlled fission, and the disposal of radioactive waste. Ideally, the cycle should also include the reprocessing of spent nuclear fuel, and it must include the decommissioning of power plants. Since much of a nuclear power plant becomes radioactive over time from exposure to radioisotopes, disposal of radioactive wastes eventually involves much more than the original fuel.

Throughout this cycle, radiation can enter and affect the environment (Figure 17.9).

Problems with the Nuclear Fuel Cycle

- Uranium mines and mills produce radioactive waste that can expose mining workers and the local environment to radiation. Radioactive dust produced at mines and mills can be transported considerable distances by wind and water, so pollution can be widespread. Tailings—materials removed by mining but not processed—are generally left at the site, but in some instances radioactive mine tailings were used in foundations and other building materials, contaminating dwellings.
- Uranium-235 enrichment and the fabrication of fuel assemblies also produce radioactive waste that must be carefully handled and disposed of.
- Site selection and construction of nuclear power plants in the United States are highly controversial. The environmental review process is extensive and expensive, often centering on hazards related to such events as earthquakes.
- The power plant or reactor is the site most people are concerned about because it is the most visible part of the cycle. It is also the site of past accidents, including partial meltdowns that have released harmful radiation into the environment.
- The United States does not reprocess spent fuel from reactors to recover uranium and plutonium at this time. However, many problems are associated with the handling and disposal of nuclear waste, as discussed later in this chapter.
- Waste disposal is controversial because no one wants a nuclear waste disposal facility nearby. The problem is that no one has yet figured out how to isolate nuclear waste for the millions of years that it remains hazardous.

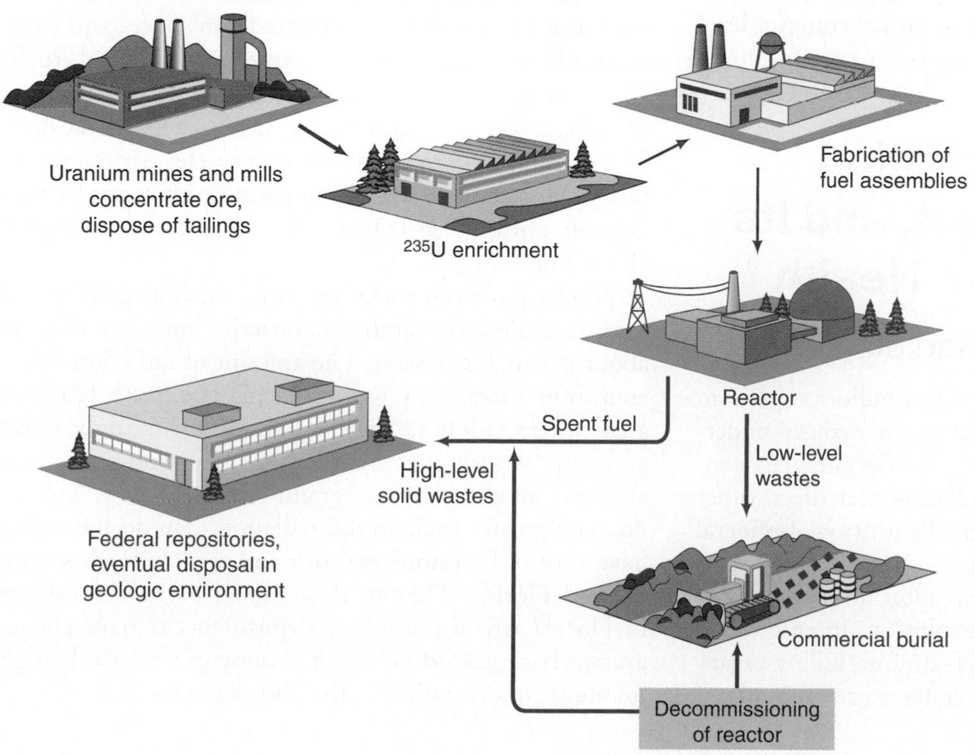

FIGURE 17.9 Idealized diagram showing the nuclear fuel cycle for the U.S. nuclear energy industry. Disposal of tailings, which because of their large volume may be more toxic than high-level waste, was treated casually in the past. (*Source*: Office of Industry Relations, The Nuclear Industry, 1974.)

- Nuclear power plants have a limited lifetime, usually estimated at only several decades, but decommissioning a plant (removing it from service) or modernizing it is a controversial part of the cycle and one with which we have little experience. For one thing, like nuclear waste, contaminated machinery must be safely disposed of or securely stored indefinitely.

Decommissioning or refitting a nuclear plant will be very expensive (perhaps several hundred million dollars) and is an important aspect of planning for the use of nuclear power. It will cost more to dismantle a nuclear reactor than to build it. At present, as we saw in this chapter's opening case study, power companies are filing to extend the licenses of several nuclear power plants that were originally slated to be decommissioned and taken down.

In addition to the above list of hazards in transporting and disposing of radioactive material, there are potential hazards in supplying other nations with reactors. Terrorist activity and the possibility of irresponsible people in governments add risks that are not present in any other form of energy production. For example, Kazakhstan inherited a large nuclear weapons testing facility, covering hundreds of square kilometers, from the former Soviet Union. The soil in several sites contains "hot spots" of plutonium that pose a serious problem of toxic contamination. The facility also poses a security problem. There is international concern that this plutonium could be collected and used by terrorists to produce "dirty" bombs (conventional explosives that disperse radioactive materials). There may even be enough plutonium to produce small nuclear bombs.

Nuclear energy may indeed be one answer to some of our energy needs, but with nuclear power comes a level of responsibility not required by any other energy source.

17.4 Nuclear Radiation in the Environment, and Its Effects on Human Health

Ecosystem Effects of Radioisotopes

As explained in A Closer Look 17.1, a radioisotope is an isotope of a chemical element that spontaneously undergoes radioactive decay. Radioisotopes affect the environment in two ways: by emitting radiation that affects other materials and by entering the normal pathways of mineral cycling and ecological food chains.

The explosion of a nuclear weapon does damage in many ways. At the time of the explosion, intense radiation of many kinds and energies is sent out, killing organisms directly. The explosion generates large amounts of radioactive isotopes, which are dispersed into the environment. Nuclear bombs exploding in the atmosphere produce a huge cloud that sends radioisotopes directly into the stratosphere, where the radioactive particles are widely dispersed by winds. Atomic fallout—the deposit of these radioactive materials around the world—was an environmental problem in the 1950s and 1960s, when the United States, the former Soviet Union, China, France, and Great Britain were testing and exploding nuclear weapons in the atmosphere.

The pathways of some of these isotopes illustrate the second way in which radioactive materials can be dangerous in the environment: They can enter ecological food chains (Figure 17.10). Let's consider an example. One of the radioisotopes emitted and sent into the stratosphere by atomic explosions was cesium-137. This radioisotope was deposited in relatively small concentrations but was widely dispersed in the Arctic region of North America. It fell on reindeer moss, a lichen that is a primary winter food of the caribou. A strong seasonal trend in the levels of cesium-137 in caribou was discovered; the level was highest in winter, when reindeer moss was the principal food, and lowest in summer. Eskimos who obtained a high percentage of their protein from caribou ingested the radioisotope by eating the meat, and their bodies concentrated the cesium. The more that members of a group depended on caribou as their primary source of food, the higher the level of the isotope in their bodies.

People are exposed to a variety of radiation sources from the sky, the air, and the food we eat (Figure 17.11). We receive natural background radiation from cosmic rays entering Earth's atmosphere from space, and from naturally occurring radioisotopes in soil and rock. The average American receives about 2 to 4 mSv/yr. Of this, about 1 to 3 mSv/yr, or 50–75%, is natural. The differences are primarily due to elevation and geology. More cosmic radiation from outer space (which delivers about 0.3–1.3 mSv/yr) is received at higher elevations.

Radiation from rocks and soils (such as granite and organic shales) containing radioactive minerals delivers about 0.3 to 1.2 mSv/yr. The amount of radiation delivered from rocks, soils, and water may be much larger in areas where radon gas (a naturally occurring radioactive gas) seeps into homes. As a result, mountain states that also have an abundance of granitic rocks, such as Colorado, have greater background radiation than do states that have a lot of limestone bedrock and are low in elevation, such as Florida. Despite this general pattern, locations in Florida where phosphate deposits occur have above-average background radiation because of a relatively high uranium concentration in the phosphate rocks.[10]

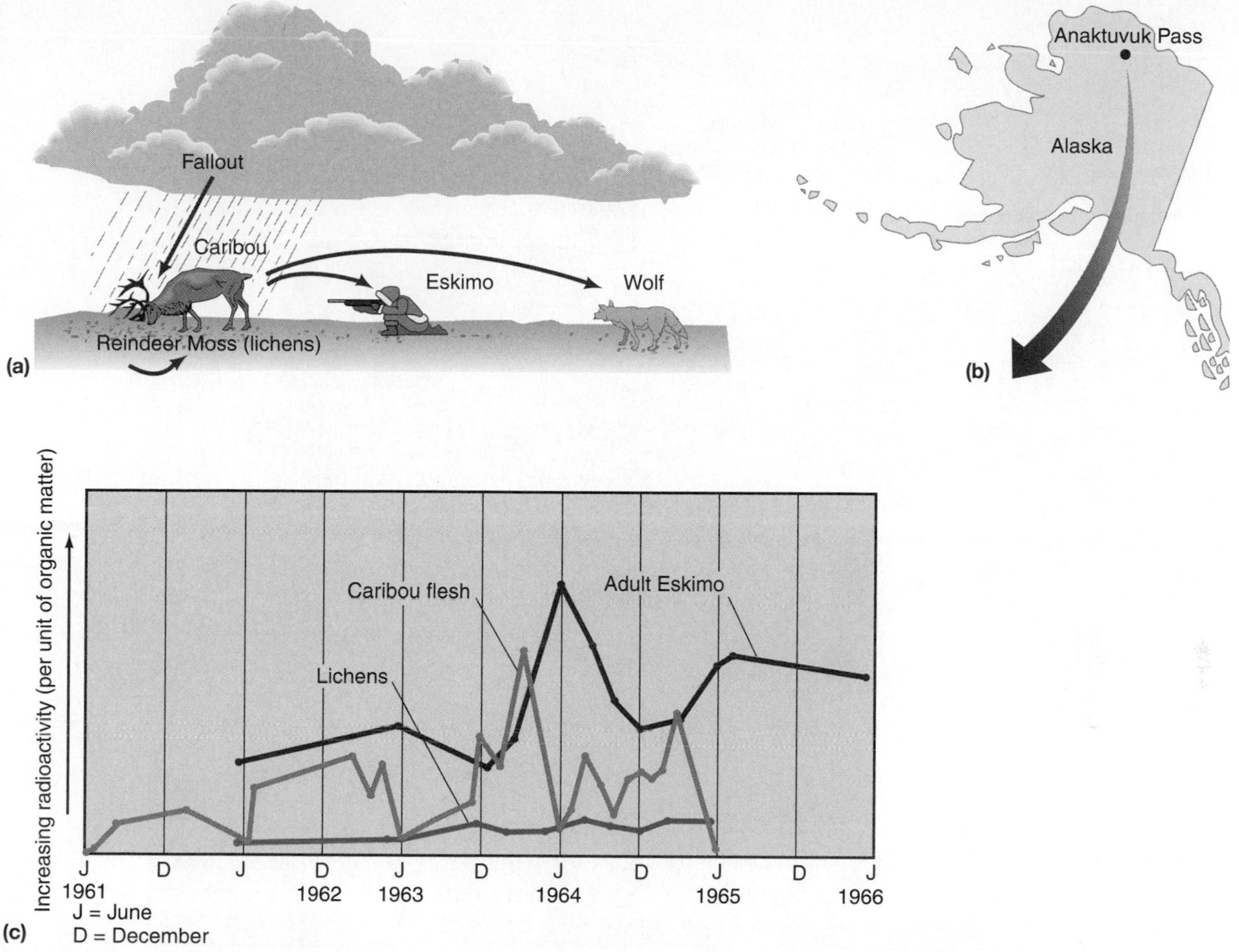

FIGURE 17.10 Cesium-137 released into the atmosphere by atomic bomb tests was part of the fallout deposited on soil and plants. **(a)** The cesium fell on lichens, which were eaten by caribou. The caribou were in turn eaten by Eskimos. **(b)** Measurements of cesium were taken in the lichens, caribou, and Eskimo in the Anaktuvuk Pass of Alaska. **(c)** The cesium was concentrated by the food chain. Peaks in concentrations occurred first in the lichens, then in the caribou, and last in the Eskimos. (*Source*: [c] W.G. Hanson, "Cesium-137 in Alaskan Lichens, Caribou, and Eskimos," *Health Physics* 13 [1967]: 383–389. Copyright 1967 by Pergamon Press. Reprinted with permission.)

The amount of radiation we receive from our own bodies and other people is about 1.35 mSv/yr. Two sources are naturally occurring radioactive potassium-40 and carbon-14, which are present in our bodies and produce about 0.35 mSv/yr. Potassium is an important electrolyte in our blood, and one isotope of potassium (potassium-40) has a very long half-life. Although potassium-40 makes up only a very small percentage of the total potassium in our bodies, it is present in all of us. In short, we are all slightly radioactive, and if you choose to share your life with another person, you are also exposing yourself to a little bit more radiation.

To understand the effects of radiation, you need to be acquainted with the units used to measure radiation and the amount or dose of radiation that may cause a health problem. These are explained in A Closer Look 17.2.

Sources of low-level radiation from our modern technology include X-rays for medical and dental purposes, which may deliver an average of 0.8–0.9 mSv/yr; nuclear weapons testing, approximately 0.04 mSv/yr; the burning of fossil fuels, such as coal, oil, and natural gas, 0.03 mSv/yr; and nuclear power plants (under normal operating conditions), 0.002 mSv/yr.[12]

Your occupation and lifestyle can affect the annual dose of radiation you receive. If you fly at high altitudes in jet aircraft, you receive an additional small dose of radiation—about 0.05 mSv for each flight across the

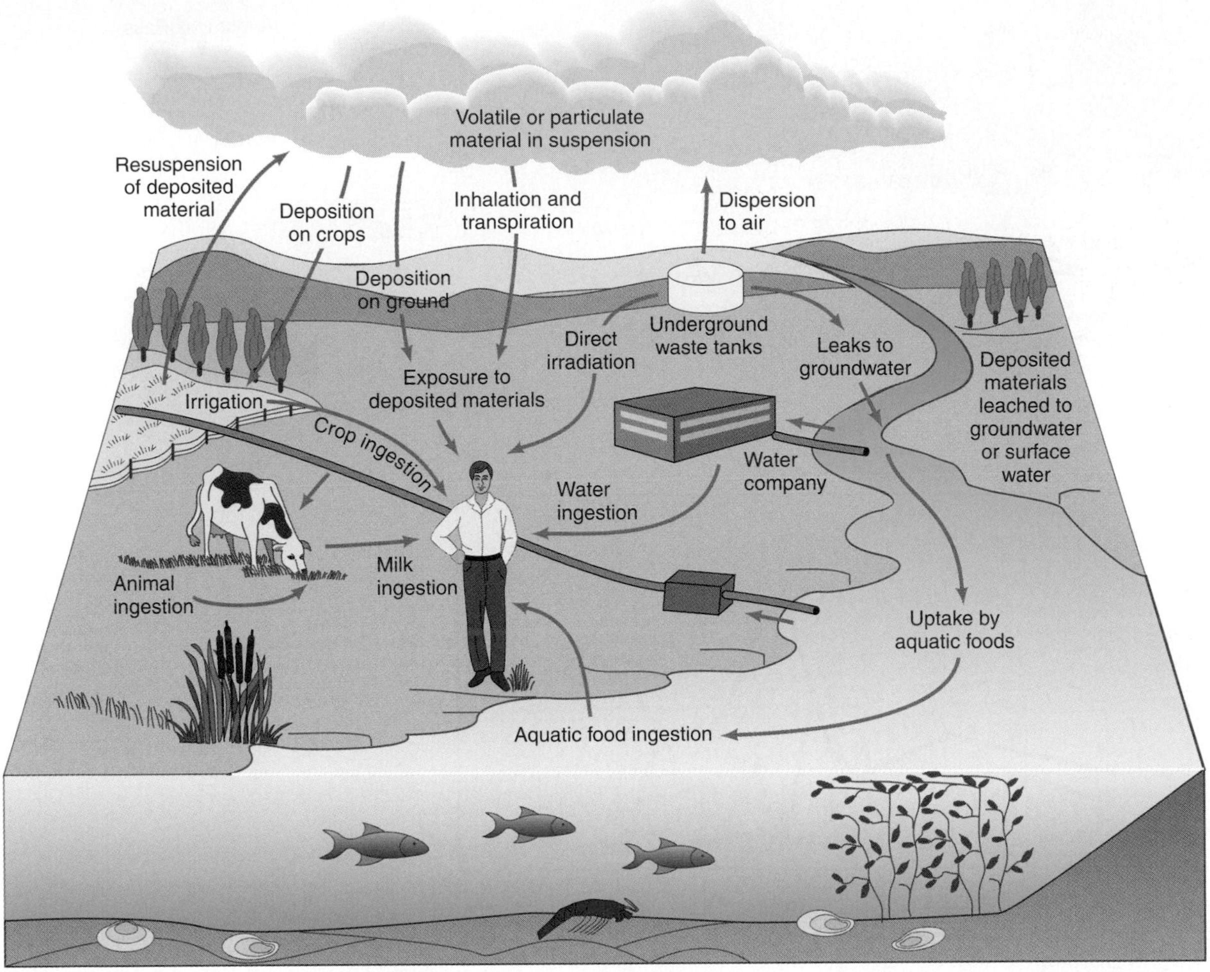

FIGURE 17.11 How radioactive substances reach people. (*Source*: F. Schroyer, ed., *Radioactive Waste*, 2nd printing [American Institute of Professional Geologists, 1985].)

A CLOSER LOOK 17.2

Radiation Units and Doses

The units used to measure radioactivity are complex and somewhat confusing. Nevertheless, a modest acquaintance with them is useful in understanding and talking about radiation's effects on the environment.

A commonly used unit for radioactive decay is the *curie* (Ci), a unit of radioactivity defined as 37 billion nuclear transformations per second. The curie is named for Marie Curie and her husband, Pierre, who discovered radium in the 1890s. They also discovered polonium, which they named after Marie's homeland, Poland. The harmful effects of radiation were not known at that time, and both Marie Curie and her daughter died of radiation-induced cancer.[11] Her laboratory (Figure 17.12) is still contaminated today.

In the International System (SI) of measurement, the unit commonly used for radioactive decay is the *becquerel* (Bq), which is one radioactive decay per second. Units of measurement often used in discussions of radioactive isotopes, such as radon-222, are becquerels per cubic meter and *picocuries* per liter (pC/l). A picocurie is one-trillionth (10^{-12}) of a curie. Becquerels per cubic meter or picocuries

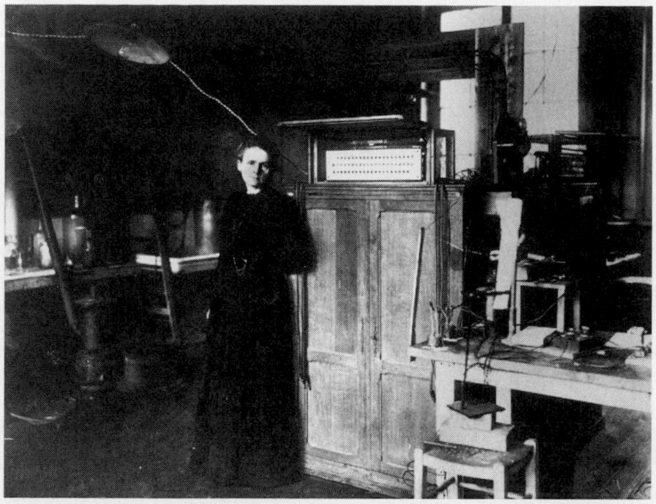

FIGURE 17.12 Marie Curie in her laboratory.

per liter are therefore measures of the number of radioactive decays that occur each second in a cubic meter or liter of air.

When dealing with the environmental effects of radiation, we are most interested in the actual dose of radiation delivered by radioactivity. That dose is commonly measureds in terms of *rads* (rd) and *rems*. In the International System, the corresponding units are *grays* (Gy) and *sieverts* (Sv). Rads and grays are the units of the absorbed dose of radiation; 1 gray is equivalent to 100 rads. Rems and sieverts are units of equivalent dose, or effective equivalent dose, where 1 sievert is 100 rems. The energy retained by living tissue that has been exposed to radiation is called the *radiation absorbed dose*, which is where the term *rad* comes from. Because different types of radiation have different penetrations and thus cause different degrees of damage to living tissue, the rad is multiplied by a factor known as the *relative biological effectiveness* to produce the rem or sievert units. When very small doses of radioactivity are being considered, the millirem (mrem) or millisievert (mSv)—that is, one-thousandth (0.001) of a rem or sievert—is used. For gamma rays, the unit commonly used is the roentgen ®, or, in SI units, coulombs per kilogram (C/kg).

United States. If you work at a nuclear power plant, you can receive up to about 3 mSv/yr. Living next door to a nuclear power plant adds 0.01 mSv/year, and sitting on a bench watching a truck carrying nuclear waste pass by would add 0.001 mSv to your annual exposure. Sources of radiation are summarized in Figure 17.13*a*, assuming an annual total of 3 mSv/yr.[13, 14] The amount of radiation received at certain job sites, such as nuclear power plants and laboratories where X-rays are produced, is closely monitored. At such locations, personnel wear badges that indicate the dose of radiation received.

Figure 17.13 shows some of the common sources of radiation to which we are exposed. Notice that exposure to radon gas can equal what people were exposed to as a result of the Chernobyl nuclear power accident, which occurred in the Soviet Union in 1986. In other words, in some homes, people are exposed to about the same radiation as that experienced by the people evacuated from the Chernobyl area.

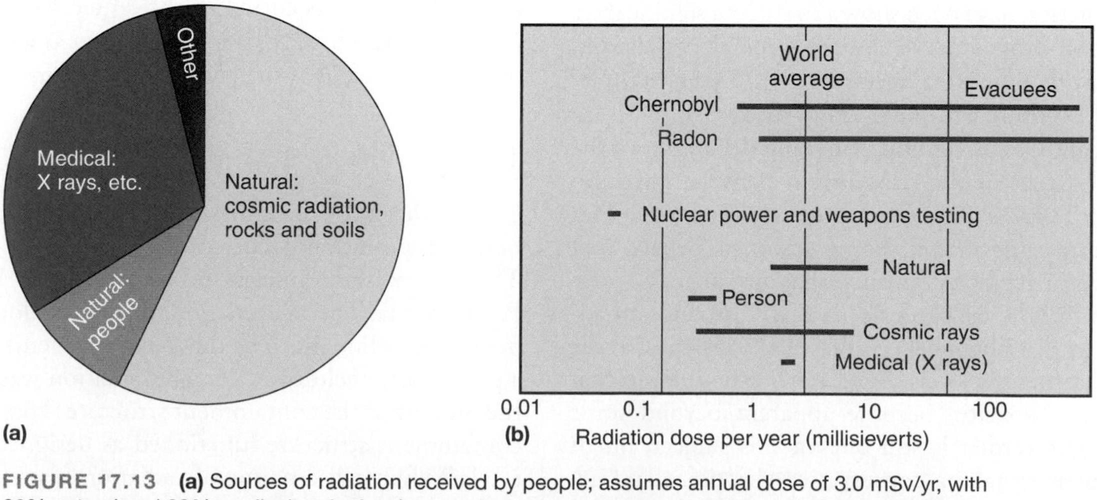

FIGURE 17.13 **(a)** Sources of radiation received by people; assumes annual dose of 3.0 mSv/yr, with 66% natural and 33% medical and other (occupational, nuclear weapons testing, television, air travel, smoke detectors, etc.). (*Sources*: U.S. Department of Energy, 1999; *New Encyclopedia Britannica*, 1997. Radiation V26, p. 487.) **(b)** Range in annual radiation dose to people from major sources. (*Source*: Data in part from A.V. Nero Jr., "Controlling Indoor Air Pollution," *Scientific American* 258[5] [1998]: 42–48.)

Radiation Doses and Health

The most important question in studying radiation exposure in people is: At what point does the exposure or dose becomes a hazard to health? (See again A Closer Look 17.2.) Unfortunately, there are no simple answers to this seemingly simple question. We do know that a dose of about 5,000 mSv (5 sieverts) is considered lethal to 50% of people exposed to it. Exposure to 1,000–2,000 mSv is sufficient to cause health problems, including vomiting, fatigue, potential abortion of pregnancies of less than two months' duration, and temporary sterility in males. At 500 mSv, physiological damage is recorded. The maximum allowed dose of radiation per year for workers in industry is 50 mSv, approximately 30 times the average natural background radiation we all receive.[15] For the general public, the maximum permissible annual dose (for infrequent exposure) is set in the United States at 5 mSv, about three times the annual natural background radiation.[16] For continuous or frequent exposure, the limit for the general public is 1 mSv.

Most information about the effects of high doses of radiation comes from studies of people who survived the atomic bomb detonations in Japan at the end of World War II. We also have information about people exposed to high levels of radiation in uranium mines, workers who painted watch dials with luminous paint containing radium, and people treated with radiation therapy for disease.[17] Starting around 1917 in New Jersey, approximately 2,000 young women were employed painting watch dials with luminous paint. To maintain a sharp point on their brushes, they licked them and as a result were swallowing radium, which was in the paint. By 1924, dentists in New Jersey were reporting cases of jaw rot; and within five years radium was known to be the cause. Many of the women died of anemia or bone cancer.[18]

Workers in uranium mines who were exposed to high levels of radiation have been shown to suffer a significantly higher rate of lung cancer than the general population. Studies show that there is a delay of 10 to 25 years between the time of exposure and the onset of disease.

Although there is vigorous, ongoing debate about the nature and extent of the relationship between radiation exposure and cancer mortality, most scientists agree that radiation can cause cancer. Some scientists believe that there is a linear relationship, such that any increase in radiation beyond the background level will produce an additional hazard. Others believe that the body can handle and recover from low levels of radiation exposure but that health effects (toxicity) become apparent beyond some threshold. The verdict is still out on this subject, but it seems prudent to take a conservative viewpoint and accept that there may be a linear relationship. Unfortunately, chronic health problems related to low-level exposure to radiation are neither well known nor well understood.

Radiation has a long history in the field of medicine. Drinking waters that contain radioactive materials goes back to Roman times. By 1899, the adverse effects of radiation had been studied and were well known; and in that year, the first lawsuit for malpractice in using X-rays was filed. Because science had shown that radiation could destroy human cells, however, it was a logical step to conclude that drinking water containing radioactive material such as radon might help fight diseases such as stomach cancer. In the early 1900s it became popular to drink water containing radon, and the practice was supported by doctors, who stated that there were no known toxic effects. Although we now know that was incorrect, radiotherapy, which uses radiation to kill cancer cells in humans, has been widely and successfully used for a number of years.[19]

17.5 Nuclear Power Plant Accidents

Although the chance of a disastrous nuclear accident is estimated to be very low, the probability that an accident will occur increases with every reactor put into operation. According to the U.S. Nuclear Regulatory Commission's performance goal for a single reactor, the probability of a large-scale core meltdown in any given year should be no greater than 0.01%—one chance in 10,000. However, if there were 1,500 nuclear reactors (about three and a half times the present world total), a meltdown could be expected (at the low annual probability of 0.01%) every seven years. This is clearly an unacceptable risk.[20] Increasing safety by about 10 times would result in lower, more manageable risk, but the risk would still be appreciable because the potential consequences remain large.

Next, we discuss the two most well-known nuclear accidents, which occurred at the Three Mile Island and Chernobyl reactors. It is important to understand that these serious accidents resulted in part from human error.

Three Mile Island

One of the most dramatic events in the history of U.S. radiation pollution occurred on March 28, 1979, at the Three Mile Island nuclear power plant near Harrisburg, Pennsylvania. The malfunction of a valve, along with human errors (thought to be the major problem), resulted in a partial core meltdown. Intense radiation was released to the interior of the containment structure. Fortunately, the containment structure functioned as designed, and only a relatively small amount of radiation was released into the environment. Average exposure from the radiation emitted into the atmosphere has been estimated at 1 mSv, which is low in terms of the amount required to cause acute toxic effects. Average exposure to radiation in the surrounding area is estimated to have been approximately 0.012 mSv, which is only about 1% of the natural background radiation that people receive. However, radiation levels were much higher near the site. On the third day after the accident, 12 mSv/hour were measured at ground level near the site. By comparison, the average American receives about 2 mSv/year from natural radiation.

Because the long-term chronic effects of exposure to low levels of radiation are not well understood, the effects of Three Mile Island exposure, though apparently small, are difficult to estimate. However, the incident revealed many potential problems with the way U.S. society dealt with nuclear power. Historically, nuclear power had been considered relatively safe, so the state of Pennsylvania was unprepared to deal with the accident. For example, there was no state bureau for radiation help, and the state Department of Health did not have a single book on radiation medicine (the medical library had been dismantled two years earlier for budgetary reasons). One of the major impacts of the incident was fear, yet there was no state office of mental health, and no staff member from the Department of Health was allowed to sit in on important discussions following the accident.[21]

Chernobyl

Lack of preparedness to deal with a serious nuclear power plant accident was dramatically illustrated by events that began unfolding on Monday morning, April 28, 1986. Workers at a nuclear power plant in Sweden, frantically searching for the source of elevated levels of radiation near their plant, concluded that it was not their installation that was leaking radiation; rather, the radioactivity was coming from the Soviet Union by way of prevailing winds. When confronted, the Soviets announced that an accident had occurred at a nuclear power plant at Chernobyl two days earlier, on April 26. This was the first notice to the world of the worst accident in the history of nuclear power generation.

It is speculated that the system that supplied cooling waters for the Chernobyl reactor failed as a result of human error, causing the temperature of the reactor core to rise to over 3,000°C (about 5,400°F), melting the uranium fuel, setting fire to the graphite surrounding the fuel rods that were supposed to moderate the nuclear reactions, and causing explosions that blew off the top of the building over the reactor. The fires produced a cloud of radioactive particles that rose high into the atmosphere. There were 237 confirmed cases of acute radiation sickness, and 31 people died of radiation sickness.[22]

In the days following the accident, nearly 3 billion people in the Northern Hemisphere received varying amounts of radiation from Chernobyl. With the exception of the 30-km (19-mi) zone surrounding Chernobyl, the world human exposure was relatively small. Even in Europe, where exposure was highest, it was considerably less than the natural radiation received during one year.[23]

In that 30-km zone, approximately 115,000 people were evacuated, and as many as 24,000 people were estimated to have received an average radiation dose of 0.43 Sv (430 mSv).

It was expected, based on results from Japanese A-bomb survivors, that approximately 122 spontaneous leukemias would occur during the period from 1986 through 1998.[24] Surprisingly, as of late 1998, there was no significant increase in the incidence of leukemia, even among the most highly exposed people. However, an increased incidence of leukemia could still become manifest in the future.[25] Meanwhile, studies have found that since the accident the number of childhood thyroid cancer cases per year has risen steadily in Belarus, Ukraine, and the Russian Federation, the three countries most affected by Chernobyl. A total of 1,036 thyroid cancer cases have been diagnosed in children under 15 in the region. These cancer cases are believed to be linked to the released radiation from the accident, but other factors, such as environmental pollution, may also play a role. It is predicted that a few percent of the roughly 1 million children exposed to the radiation eventually will contract thyroid cancer. Outside the 30 km zone, the increased risk of contracting cancer is very small and not likely to be detected from an ecological evaluation.

To date, 4,000 deaths can be directly attributed to the Chernobyl accident, and according to one estimate, Chernobyl will ultimately be responsible for approximately 16,000 to 39,000 deaths. Proponents of nuclear power point out that this is fewer than the number of deaths caused each year by burning coal.[26, 27]

Vegetation within 7 km of the power plant was either killed or severely damaged by the accident. Pine trees examined in 1990 around Chernobyl showed extensive tissue damage and still contained radioactivity. The distance between annual rings (a measure of tree growth) had decreased since 1986.

Scientists returning to the evacuated zone in the mid-1990s found, to their surprise, thriving and expanding animal populations. Species such as wild boar, moose, otters, waterfowl, and rodents seemed to be enjoying a population boom in the absence of people. The wild boar population had increased tenfold since the evacuation. These animals may be paying a genetic price for living within the contaminated zone, but so far the benefit of excluding humans apparently outweighs the negatives associated with radioactive contamination. The area now resembles a wildlife reserve.

In areas surrounding Chernobyl, radioactive materials continue to contaminate soils, vegetation, surface water, and groundwater, presenting a hazard to plants

FIGURE 17.14 Guard halting entry of people into the forbidden zone evacuated in 1986 as a result of the Chernobyl nuclear accident.

and animals. The evacuation zone may be uninhabitable for a very long time unless some way is found to remove the radioactivity (Figure 17.14). For example, the city of Prypyat, 5 km from Chernobyl, which had a population of 48,000 prior to the accident, is a "ghost city." It is abandoned, with blocks of vacant apartment buildings and rusting vehicles. Roads are cracking and trees are growing as new vegetation transforms the urban land back to green fields.

The final story of the world's most serious nuclear accident is yet to completely unfold.[28] Estimates of the total cost of the Chernobyl accident vary widely, but it will probably exceed $200 billion.

Although the Soviets were accused of not paying attention to reactor safety and of using outdated equipment, people are still wondering if such an accident could happen again elsewhere. Because more than 400 nuclear power plants are producing power in the world today, the answer has to be yes. It is difficult to get an exact account of nuclear power plant accidents that have released radiation into the environment since the first nuclear power plants were built in the 1960s. This is partly because of differences in what is considered a significant radiation emission. As best as can be estimated, there appear to have been 20 to 30 such incidents worldwide—at least that is the range of numbers released to the public. Therefore, although Chernobyl is the most serious nuclear accident to date, it certainly was not the first and is unlikely to be the last. Although the probability of a serious accident is very small at a particular site, the consequences may be great, perhaps posing an unacceptable risk to society. This is really not so much a scientific issue as a political one involving values.

Advocates of nuclear power argue that nuclear power is safer than other energy sources, that many more deaths are caused by air pollution from burning fossil fuels than by nuclear accidents. For example, the 16,000 deaths that might eventually be attributed to Chernobyl are fewer than the number of deaths caused each year by air pollution from burning coal.[29] Those arguing against nuclear power say that as long as people build nuclear power plants and manage them, there will be the possibility of accidents. We can build nuclear reactors that are safer, but people will continue to make mistakes, and accidents will continue to happen.

17.6 Radioactive-Waste Management

Examination of the nuclear fuel cycle (refer back to Figure 17.9) illustrates some of the sources of waste that must be disposed of as a result of using nuclear energy to produce electricity. Radioactive wastes are by-products of using nuclear reactors to generate electricity. The U.S. Federal Energy Regulatory Commission (FERC) defines three categories of radioactive waste: mine tailings, low-level, and high-level. Other groups list a fourth category: transuranic wastes.[30] In the western United States, more than 20 million metric tons of abandoned tailings will continue to produce radiation for at least 100,000 years.

Low-Level Radioactive Waste

Low-level radioactive waste contains radioactivity in such low concentrations or quantities that it does not present a significant environmental hazard if properly handled. Low-level waste includes a wide variety of items, such as residuals or solutions from chemical processing; solid or liquid plant waste, sludges, and acids; and slightly contaminated equipment, tools, plastic, glass, wood, and other materials.[31]

Low-level waste has been buried in near-surface burial areas in which the hydrologic and geologic conditions were thought to severely limit the migration of radioactivity.[32] However, monitoring has shown that several U.S. disposal sites for low-level radioactive waste have not adequately protected the environment, and leaks of liquid waste have polluted groundwater. Of the original six burial sites, three closed prematurely by 1979 due to unexpected leaks, financial problems, or loss of license, and as of 1995 only two remaining government low-level nuclear-waste repositories were still operating in the United States, one in Washington and the other in South Carolina. In addition, a private facility in Utah, run by Envirocare, accepts low-level waste. Construction of new burial sites, such as the Ward Valley site in southeastern California, has been met with strong public opposition, and controversy continues as to whether low-level radioactive waste can be disposed of safely.[33]

Transuranic Waste

As noted earlier, it is useful to also list separately **transuranic waste**, which is waste contaminated by manmade radioactive elements, including plutonium, americum, and einsteineum, that are heavier than uranium and are produced in part by neutron bombardment of uranium in reactors. Most transuranic waste is industrial trash, such as clothing, rags, tools, and equipment, that has been contaminated. The waste is low-level in terms of its intensity of radioactivity, but plutonium has a long half-life and must be isolated from the environment for about 250,000 years. Most transuranic waste is generated from the production of nuclear weapons and, more recently, from cleanup of former nuclear weapons facilities.

Some nuclear weapons transuranic wastes (as of 2000) are being transported to a disposal site near Carlsbad, New Mexico, and to date more than 5,000 shipments have been delivered.[34] The waste is isolated at a depth of 655 m (2,150 ft) in salt beds (rock salt) that are several hundred meters thick (Figure 17.15). Rock salt at the New Mexico site has several advantages:[35, 36, 37]

- The salt is about 225 million years old, and the area is geologically stable, with very little earthquake activity.
- The salt has no flowing groundwater and is easy to mine. Excavated rooms in the salt, about 10 m wide and 4 m high, will be used for disposal.
- Rock salt flows slowly into mined openings. The waste-filled spaces in the storage facility will be naturally closed by the slow-flowing salt in 75 to 200 years, sealing the waste.

The New Mexico disposal site is important because it is the first geologic disposal site for radioactive waste in the United States. As a pilot project, it will be evaluated very carefully. Safety is the primary concern. Procedures have been established to transport the waste to the disposal site as safely as possible and place it underground in the disposal facility. Because the waste will be hazardous for many thousands of years, it was decided that warnings had to be created that would be understandable to future peoples no matter what their cultures and languages. But of course it is unclear today whether any such sign will actually communicate anything to people thousands of years from now.[38]

High-Level Radioactive Waste

High-level radioactive waste consists of commercial and military spent nuclear fuel; uranium and plutonium derived from military reprocessing; and other radioactive nuclear weapons materials. It is extremely toxic, and a sense of urgency surrounds its disposal as the total volume of spent fuel accumulates. At present, in the United States, tens of thousands of metric tons of high-level waste are being stored at more than a hundred sites in 40 states. Seventy-two of the sites are commercial nuclear reactors.[39, 40, 41]

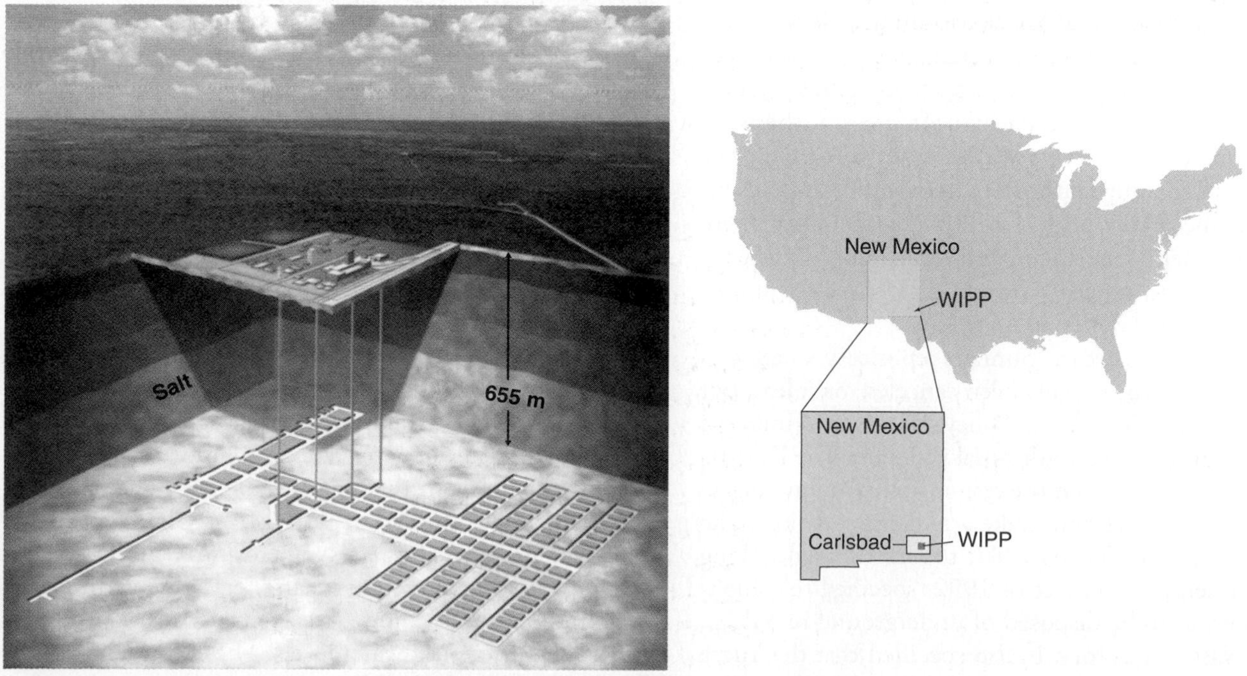

FIGURE 17.15 Waste isolation pilot plant (WIPP) in New Mexico for disposal of transuranic waste. (*Source*: U.S. Department of Energy, 1999.)

These storage arrangements are at best a temporary solution, and serious problems with radioactive waste have occurred where it is being stored. Although improvements in storage tanks and other facilities will help, eventually some sort of disposal program must be initiated. Some scientists believe the geologic environment can best provide safe containment of high-level radioactive waste. Others disagree and have criticized proposals for long-term underground disposal of high-level radioactive waste. A comprehensive geologic disposal development program should have the following objectives:[42]

- Identification of sites that meet broad geologic criteria, including ground stability and slow movement of groundwater with long flow paths to the surface.
- Intensive subsurface exploration of possible sites to positively determine geologic and hydrologic characteristics.
- Predictions of the behavior of potential sites based on present geologic and hydrologic situations and assumptions about future changes in climate, groundwater flow, erosion, ground movements, and other variables.
- Evaluation of risk associated with various predictions.
- Political decision making based on risks acceptable to society.

What Should the United States Do with Its Nuclear Wastes?

For decades in the United States, the focal point for debates over nuclear wastes has been the plan to bury them deep in the earth at Yucca Mountain, Nevada. But the Obama administration rejected that plan, and Secretary of Energy Steven Chu has set up a blue ribbon panel to consider the alternatives. At present, there are 70,000 tons of radioactive wastes from nuclear power plants, and federally authorized temporary storage facilities for these are said to be full. That is to say, there is no government-sanctioned and locally approved place to put any more nuclear wastes. Yet they continue to build up.

Why was the Yucca Mountain repository so controversial, and why has it finally been canceled, or at least put on hold? The Nuclear Waste-Policy Act of 1982 initiated a high-level nuclear-waste-disposal program. The Department of Energy was given the responsibility to investigate several potential sites and make a recommendation. The 1982 act was amended in 1987; the amendment, along with the Energy Power Act of 1992, specified that high-level waste was to be disposed of underground in a deep, geologic waste repository. It also specified that the Yucca Mountain site in Nevada was to be the only site evaluated. Costs to build the facility reached $77 billion, but no nuclear wastes have ever been sent there.[43]

Evaluation of the safety and utility of a new waste repository would have to consider factors such as the following:

- The probability and consequences of volcanic eruptions.
- Earthquake hazard.
- Estimation of changes in the storage environment over long periods.
- Estimation of how long the waste may be contained and the types and rates of radiation that may escape from deteriorated waste containers.
- How heat generated by the waste may affect moisture in and around the repository and the design of the repository.
- Characterization of groundwater flow near the repository.
- Identification and understanding of major geochemical processes that control the transport of radioactive materials.

One of the problems is just transporting the present amount of nuclear waste from power plants to any repository. According to previous U.S. government plans, beginning in 2010 some 70,000 tons of highly radioactive nuclear waste were going to be moved across the country to Yucca Mountain, Nevada, by truck and train, one to six trainloads or truck convoys every day for 24 years. These train and truck convoys would have to be heavily guarded against terrorism and protected as much as possible against accidents.

Extensive scientific evaluations of the Yucca Mountain site have been carried out.[44] Use of this site remains controversial and is generating considerable resistance from the state and people of Nevada as well as from scientists not confident of the plan. Some of the scientific questions at Yucca Mountain have concerned natural processes and hazards that might allow radioactive materials to escape, such as surface erosion, groundwater movement, earthquakes, and volcanic eruptions. In 2002, Congress voted to submit a license of application for Yucca Mountain to the Nuclear Regulatory Commission.

A major question about the disposal of high-level radioactive waste is this: How credible are extremely long-range geologic predictions—those covering several thousand to a few *million* years?[45] Unfortunately, there is no easy answer to this question because geologic processes vary over both time and space. Climates change over long periods, as do areas of erosion, deposition, and groundwater activity. For example, large earthquakes even thousands of kilometers from a site may permanently change groundwater levels. The earthquake record for most of the United States extends back only a few hundred years; therefore, estimates of future earthquake activity are tenuous at best.

The bottom line is that geologists can suggest sites that have been relatively stable in the geologic past, but they cannot absolutely guarantee future stability. This means that policymakers (not geologists) need to evaluate the uncertainty of predictions in light of pressing political, economic, and social concerns.[46] In the end, the geologic environment may be deemed suitable for safe containment of high-level radioactive waste, but care must be taken to ensure that the best possible decisions are made on this important and controversial issue.

17.7 The Future of Nuclear Energy

The United States would need 1,000 new nuclear power plants of the same design and efficiency as existing nuclear plants to completely replace fossil fuels. The International Atomic Energy Agency, which promotes nuclear energy, says a total of just 4.7 million tons of "identified" conventional uranium stock can be mined economically. If we switched from fossil fuels to nuclear today, that uranium would run out in four years. Even the most optimistic estimate of the quantity of uranium ore would last only 29 years.[47]

Nevertheless, nuclear energy as a power source for electricity is now being seriously evaluated. Its advocates argue that nuclear power is good for the environment because (1) it does not contribute to potential global warming through release of carbon dioxide (see Chapter 20) and (2) it does not cause the kinds of air pollution or emit precursors (sulfates and nitrates) that cause acid rain (see Chapter 21). They also argue that developing breeder reactors for commercial use would greatly increase the amount of fuel available for nuclear plants, that nuclear power plants are safer than other means of generating power, and that we should build many more nuclear power plants in the future. Their argument assumes that if we standardize nuclear reactors and make them safer and smaller, nuclear power could provide much of our electricity in the future,[48] although the possibility of accidents and the disposal of spent fuel remain concerns.

The argument against nuclear power is based on political and economic considerations as well as scientific uncertainty about safety issues. Opponents emphasize, as we pointed out earlier, that more than half the U.S. population lives within 75 miles of one of the nation's 104 nuclear power plants. They also argue, correctly, that converting from coal-burning plants to nuclear power plants for the purpose of reducing carbon dioxide emissions would require an enormous investment in nuclear power to make a real impact. Furthermore, they say, given that safer nuclear reactors are only just being developed, there will be a time lag, so nuclear power is unlikely to have a real impact on environmental problems—such as air pollution, acid rain, and potential global warming—before at least the year 2050.

Furthermore, uranium ore to fuel conventional nuclear reactors is limited. The International Nuclear Energy Association estimates that at the 2004 rate of use, there would be 85 years of uranium fuel from known reserves, but if nations attempt to build many new power plants in the next decade, known reserves of uranium ore would be used up much more quickly.[49] Nuclear power can thus be a long-term energy source only through the development of breeder reactors.

Another argument against nuclear power is that some nations may use it as a path to nuclear weapons. Reprocessing used nuclear fuel from a power plant produces plutonium that can be used to make nuclear bombs. There is concern that rogue nations with nuclear power could divert plutonium to make weapons, or may sell plutonium to others, even terrorists, who would make nuclear weapons.[50]

Until 2001, proponents of nuclear energy were losing ground. Nearly all energy scenarios were based on the expectation that nuclear power would grow slowly or perhaps even decline in coming years. Since the Chernobyl accident, many European countries have been reevaluating the use of nuclear power, and in most instances the number of nuclear power plants being built has significantly declined. Germany, which gets about one-third of its electricity from nuclear power, has decided to shut down all nuclear power plants in the next 25 years as they become obsolete.

There is also a problem with present nuclear technology: Today's light-water reactors use uranium very inefficiently; only about 1% of it generates electricity, and the other 99% ends up as waste heat and radiation. Therefore, our present reactors are part of the nuclear-waste problem and not a long-term solution to the energy problem.

One design philosophy that has emerged in recent decades in the nuclear industry is to build less complex, smaller reactors that are safer. Large nuclear power plants, which produce about 1,000 MW of electricity, require an extensive set of pumps and backup equipment to ensure that adequate cooling is available to the reactor. Smaller reactors can be designed with cooling systems that work by gravity and thus are less vulnerable to pump failure caused by power loss. Such cooling systems are said to have *passive stability*, and the reactors are said to be *passively safe*. Another approach is the use of helium gas to cool reactors that have specially designed fuel capsules capable of withstanding temperatures as high as 1,800°C (about 3,300°F). The idea is to design the fuel assembly so that it can't hold enough fuel to reach this temperature and thus can't experience a core meltdown.

One way for nuclear power to be sustainable for at least hundreds of years would be to use a process known as *breeding*. **Breeder reactors** are designed to produce new nuclear fuel by transforming waste or lower-grade uranium into fissionable material. Although proponents of nuclear energy suggest that breeder reactors are the future of nuclear power, only a few are known to be operating anywhere in the world. Bringing breeder reactors online to produce safe nuclear power will take planning, research, and advanced reactor development. Also, fuel for the breeder reactors will have to be recycled because reactor fuel must be replaced every few years. What is needed is a new type of breeder reactor comprising an entire system that includes reactor, fuel cycle (especially fuel recycling and reprocessing), and less production of waste. Such a reactor appears possible but will require redefining our national energy policy and turning energy production in new directions. It remains to be seen whether this will happen.

Possible New Kinds of Nuclear Power Plants

New Kinds of Fission Reactors

Several new designs for conventional nonbreeder fission nuclear power plants are in development and the object of widespread discussion. Among these are the Advanced Boiling Water Reactor, the High Temperature Gas Reactor, and the Pebble Reactor.[51, 52] None are yet installed or operating anywhere in the world. The general goals of these designs are to increase safety, energy efficiency, and ease of operation. Some are designed to shut down automatically if there is any failure in the cooling system, rather than require the action of an operator. Although proponents of nuclear power believe these will offer major advances, it will be years, perhaps decades, until even one of each kind achieves commercial operation, so planning for the future cannot depend on them.

Fusion Reactors

In contrast to fission, which involves splitting heavy nuclei (such as uranium), fusion involves combining the nuclei of light elements (such as hydrogen) to form heavier ones (such as helium). As fusion occurs, heat energy is released. Nuclear fusion is the source of energy in our sun and other stars.

In a hypothetical fusion reactor, two isotopes of hydrogen—deuterium and tritium—are injected into the reactor chamber, where the necessary conditions for fusion are maintained. Products of the deuterium–tritium (DT) fusion include helium, producing 20% of the energy released, and neutrons, producing 80%.

Several conditions are necessary for fusion to take place. First, the temperature must be extremely high (approximately 100 million degrees Celsius for DT fusion). Second, the density of the fuel elements must be sufficiently high. At the temperature necessary for fusion, nearly all atoms are stripped of their electrons, forming a *plasma*—an electrically neutral material consisting of positively charged nuclei, ions, and negatively charged electrons. Third, the plasma must be confined long enough to ensure that the energy released by the fusion reactions exceeds the energy supplied to maintain the plasma.

The potential energy available when and if fusion reactor power plants are developed is nearly inexhaustible. One gram of DT fuel (from a water and lithium fuel supply) has the energy equivalent of 45 barrels of oil. Deuterium can be extracted economically from ocean water, and tritium can be produced in a reaction with lithium in a fusion reactor. Lithium can be extracted economically from abundant mineral supplies.

Many problems remain to be solved before nuclear fusion can be used on a large scale. Research is still in the first stage, which involves basic physics, testing of possible fuels (mostly DT), and magnetic confinement of plasma.

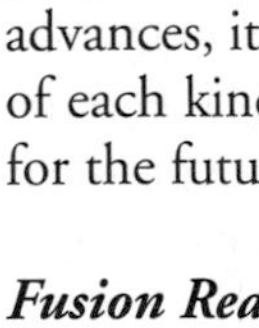

CRITICAL THINKING ISSUE

Should the United States Increase or Decrease the Number of Nuclear Power Plants?

There are two contradictory political movements regarding nuclear power plants in the United States. The federal government has supported an increase in the number of plants. The G. W. Bush administration did, and the Obama administration has allocated $18.5 for new "next-generation" nuclear power plants. But in February 2010, the Vermont Senate voted to prevent relicensing of the Yankee Power Plant, the state's only nuclear plant, after its current license expires in 2012. In 2009 the power plant leaked radioactive tritium into groundwater, and the plant's owners have been accused of misleading state regulators about underground pipes that carry cooling water at the plant.[53] Also, as you saw in this chapter's opening case study, there are major political pressures at the state, county, and local level to prevent the relicensing of Indian Point Power Plant near New York City.

Critical Thinking Questions

1. Refer to the map of nuclear power plants in the United States (Figure 17.4) and to other material in this chapter. Taking into account safety and the problem of transporting large amounts of electricity long distances, choose three locations that you consider appropriate for new nuclear power plants. Or, if you believe there should be none, present your argument for that conclusion. (Be as specific as possible about the locations—include the state and, if possible, name the nearest city.) In answering this question, you can take into account information from other chapters you have read.
2. Should new nuclear power plants be licensed now and built as soon as possible using existing and proven designs? Or would you propose putting off any new nuclear power plants until one of the safer and more efficient designs has been proved—let's say, two decades from now?
3. Which do you believe is the greater environmental problem facing the United States: global warming or the dangers of nuclear power plants? Explain your answer.

SUMMARY

- Nuclear fission is the process of splitting an atomic nucleus into smaller fragments. As fission occurs, energy is released. The major components of a fission reactor are the core, control rods, coolant, and reactor vessel.
- Nuclear radiation occurs when a radioisotope spontaneously undergoes radioactive decay and changes into another isotope.
- The three major types of nuclear radiation are alpha, beta, and gamma.
- Each radioisotope has its own characteristic emissions. Different types of radiation have different toxicities; and in terms of the health of humans and other organisms, it is important to know the type of radiation emitted and the half-life.
- The nuclear fuel cycle consists of mining and processing uranium, generating nuclear power through controlled fission, reprocessing spent fuel, disposing of nuclear waste, and decommissioning power plants. Each part of the cycle is associated with characteristic processes, all with different potential environmental problems.
- The present burner reactors (mostly light-water reactors) use uranium-235 as a fuel. Uranium is a nonrenewable resource mined from the Earth. If many more burner reactors were constructed, we would face fuel shortages. Nuclear energy based on burning uranium-235 in light-water reactors is thus not sustainable. For nuclear energy to be sustainable, safe, and economical, we will need to develop breeder reactors.
- Radioisotopes affect the environment in two major ways: by emitting radiation that affects other materials, and by entering ecological food chains.
- Major environmental pathways by which radiation reaches people include uptake by fish ingested by people, uptake by crops ingested by people, inhalation from air, and exposure to nuclear waste and the natural environment.
- The dose response for radiation is fairly well established. We know the dose–response for higher exposures, when illness or death occurs. However, there are vigorous debates about the health effects of low-level exposure to radiation and what relationships exist between exposure and cancer. Most scientists believe that radiation can cause cancer. But, Ironically, radiation can be used to kill cancer cells, as in radiotherapy treatments.
- We have learned from accidents at nuclear power plants that it is difficult to plan for the human factor. People make mistakes. We have also learned that we are not as prepared for accidents as we would like to think. Some believe that people are not ready for the responsibility of nuclear power. Others believe that we can design much safer power plants where serious accidents are impossible.
- Transuranic nuclear waste is now being disposed of in salt beds—the first disposal of radioactive waste in the geologic environment in the United States.
- There is a consensus that high-level nuclear waste may be safely disposed of in the geologic environment. The problem has been to locate a site that is safe and not objectionable to the people who make the decisions and to those who live in the region.
- Nuclear power is again being seriously evaluated as an alternative to fossil fuels. On the one hand, it has advantages: It emits no carbon dioxide, will not contribute to global warming or cause acid rain, and can be used to produce alternative fuels such as hydrogen. On the other hand, people are uncomfortable with nuclear power because of waste-disposal problems and possible accidents.

REEXAMINING THEMES AND ISSUES

Human Population

As the human population has increased, so has demand for electrical power. In response, a number of countries have turned to nuclear energy. The California energy crisis has caused many people in the United States to rethink the value of nuclear energy. Though relatively rare, accidents at nuclear power plants such as Chernobyl have exposed people to increased radiation, and there is considerable debate over potential adverse effects of that radiation. The fact remains that as the world population increases, and if the number of nuclear power plants increases, the total number of people exposed to a potential release of toxic radiation will increase as well.

Sustainability

Some argue that sustainable energy will require a return to nuclear energy because it doesn't contribute to a variety of environmental problems related to burning fossil fuels. However, for nuclear energy to significantly contribute to sustainable energy development, we cannot depend on burner reactors that will quickly use Earth's uranium resources; rather, development of safer breeder reactors will be necessary.

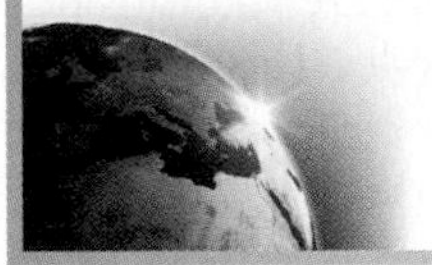

Global Perspective

Use of nuclear energy fits into our global management of the entire spectrum of energy sources. In addition, testing of nuclear weapons has spread radioactive isotopes around the entire planet, as have nuclear accidents. Radioactive isotopes that enter rivers and other waterways may eventually enter the oceans of the world, where oceanic circulation may further disperse and spread them.

Urban World

Development of nuclear energy is a product of our technology and our urban world. In some respects, it is near the pinnacle of our accomplishments in terms of technology.

People and Nature

Nuclear reactions are the source of heat for our sun and are fundamental processes of the universe. Nuclear fusion has produced the heavier elements of the universe. Our use of nuclear reactions in reactors to produce useful energy is a connection to a basic form of energy in nature. However, abuse of nuclear reactions in weapons could damage or even destroy nature on Earth.

Science and Values

We have a good deal of knowledge about nuclear energy and nuclear processes. Still, people remain suspicious and in some cases frightened by nuclear power—in part because of the value we place on a quality environment and our perception that nuclear radiation is toxic to that environment. As a result, the future of nuclear energy will depend in part on how much risk is acceptable to society. It will also depend on research and development to produce much safer nuclear reactors.

KEY TERMS

breeder reactors **363**
burner reactors **349**
fission **348**
fusion **348**
high-level radioactive waste **361**
low-level radioactive waste **360**
meltdown **348**
nuclear energy **348**
nuclear fuel cycle **353**
nuclear reactors **348**
radioactive decay **350**
radioisotope **350**
transuranic waste **361**

STUDY QUESTIONS

1. If exposure to radiation is a natural phenomenon, why are we worried about it?
2. What is a radioisotope, and why is knowing its half-life important?
3. What is the normal background radiation that people receive? Why is it variable?
4. What are the possible relationships between exposure to radiation and adverse health effects?
5. What processes in our environment may result in radioactive substances reaching people?
6. Suppose it is recommended that high-level nuclear waste be disposed of in the geologic environment of the region in which you live. How would you go about evaluating potential sites?
7. Are there good environmental reasons to develop and build new nuclear power plants? Discuss both sides of the issue.

FURTHER READING

Botkin, D.B., ***Powering the Future: A Scientist's Guide to Energy Independence*** (Indianapolis: Pearson FT Press, 2010).

Hore Lacy, Ian, ***Nuclear Energy in the 21st Century*** (New York: Academic Press, 2006). A pro-nuclear power plant book.

World Nuclear Association, "Waste Management in the Nuclear Fuel Cycle." www.world nuclear.org/info/inf04.html. A major international organization's review of this problem.

CHAPTER 18

Water Supply, Use, and Management

Great blue heron and young in Wakodahatchee Wetlands near Palm Beach, Florida.

LEARNING OBJECTIVES

Although water is one of the most abundant resources on Earth, water management involves many important issues and problems. After reading this chapter, you should understand . . .

- Why water is one of the major resource issues of the 21st century;
- What a water budget is, and why it is useful in analyzing water-supply problems and potential solutions;
- What groundwater is, and what environmental problems are associated with its use;
- How water can be conserved at home, in industry, and in agriculture;
- Why sustainable water management will become more difficult as the demand for water increases;
- The concepts of virtual water and a water footprint and their link to water management and conservation;
- The environmental impacts of water projects such as dams;
- What a wetland is, how wetlands function, and why they are important;
- Why we are facing a growing global water shortage linked to our food supply.

CASE STUDY

Palm Beach County, Florida: Water Use, Conservation, and Reuse

The southeastern United States–experienced one of the worst droughts on record from 2006 to 2008. Although March of 2008 brought significant rainfall to south Florida, it was not sufficient to end the shortage that had built up over several years. Hurricane Fran brought another 15–30 cm (6–12 in.) to south Florida in August 2008, relieving drought conditions. Water shortages during the drought in Palm Beach County led to water restrictions and water rules. For example, lawns could be watered and cars washed only once a week on a Saturday or Sunday, depending on whether your home address was an odd or even number.

Even with such rules, there were water-use problems because people use very different amounts of water. Palm Beach County and its famous resort city of Palm Beach have some large estates that use huge quantities of water. During one year of the ongoing drought, one estate of about 14 acres (6 hectares) reportedly used an average of 57,000 gallons per day—about as much as a modest single-family home in Palm Beach County uses in an entire year. Some landowners continue to use very large amounts of water during a drought, while others choose to conserve water and let their lawns go brown.[1]

Although that drought has ended, it highlighted the need to plan for projected greater shortages in the future. To this end, Florida has turned to water-conservation projects, including the use of reclaimed and purified water from wastewater-treatment plants. Florida has several hundred water-recycling projects, making it a national leader in water reuse, and Palm Beach County is a leader in south Florida. Water-conservation measures include installing low-flow showers and toilets in homes, businesses, and public buildings; limiting lawn watering and car washing; and promoting landscaping that uses less water.

The county has reclaimed approximately 9 million gallons of water per day, distributing it to parks, golf courses, and homes by way of separate water pipes painted purple (the color for reclaimed water). In addition, over 1 million gallons a day of highly treated wastewater are sent to Wakodahatchee Wetlands (see opening photograph), constructed (human-made) wetlands of approximately 25 hectares. In the Seminole language, *wakodahatchee* means "created water." The wetlands function as giant filters where wetland plants and soil use and reduce the concentration of nitrogen and phosphorus in the water and thus further treat the water. A second, larger wetland in Palm Beach County, the Green Cay Wetlands, constructed from about 50 hectares of farmland, receives over 1 million gallons of treated wastewater per day. Both are contributing to the fresh water resource base of south Florida.

Using reclaimed water has some significant benefits: (1) people who use it for private lawns or golf courses save money because the reclaimed water is less expensive; (2) reclaimed water used on lawns, golf courses, and parks has traces of nitrogen and phosphorus, which are types of fertilizer, (3) reclaimed water leaves more fresh drinking water available to the rest of the community; and (4) constructed wetlands that accept treated wastewater help the natural environment by creating wildlife habitat as well as green space in which people can walk, bird-watch, and generally enjoy a more natural setting (see Figure 18.1).[2]

Water is a critical, limited, resource in many regions on Earth. As a result, water is one of the major resource issues of the 21st century. This chapter discusses our water resources in terms of supply, use, management, and sustainability. It also addresses important environmental concerns related to water: wetlands, dams and reservoirs, channelization, and flooding.

FIGURE 18.1 Boardwalk for viewing the Wakodahatchee Wetlands near Palm Beach, Florida.

18.1 Water

To understand water as a necessity, as a resource, and as a factor in the pollution problem, we must understand its characteristics, its role in the biosphere, and its role in sustaining life. Water is a unique liquid; without it, life as we know it is impossible. Consider the following:

- Compared with most other common liquids, water has a high capacity to absorb and store heat. Its capacity to hold heat has important climatic significance. Solar energy warms the oceans, storing huge amounts of heat. The heat can be transferred to the atmosphere, developing hurricanes and other storms. The heat in warm oceanic currents, such as the Gulf Stream, warms Great Britain and Western Europe, making these areas much more hospitable for humans than would otherwise be possible at such high latitudes.
- Water is the universal solvent. Because many natural waters are slightly acidic, they can dissolve a great variety of compounds, ranging from simple salts to minerals, including sodium chloride (common table salt) and calcium carbonate (calcite) in limestone rock. Water also reacts with complex organic compounds, including many amino acids found in the human body.
- Compared with other common liquids, water has a high surface tension, a property that is extremely important in many physical and biological processes that involve moving water through, or storing water in, small openings or pore spaces.
- Water is the only common compound whose solid form is lighter than its liquid form. (It expands by about 8% when it freezes, becoming less dense.) That is why ice floats. If ice were heavier than liquid water, it would sink to the bottom of the oceans, lakes, and rivers. If water froze from the bottom up, shallow seas, lakes, and rivers would freeze solid. All life in the water would die because cells of living organisms are mostly water, and as water freezes and expands, cell membranes and walls rupture. If ice were heavier than water, the biosphere would be vastly different from what it is, and life, if it existed at all, would be greatly altered[3].
- Sunlight penetrates water to variable depths, permitting photosynthetic organisms to live below the surface.

A Brief Global Perspective

The water-supply problem, in brief, is that we are facing a growing global water shortage that is linked to our food supply. We will return to this important concept at the end of the chapter, following a discussion of water use, supply, and management.

A review of the global hydrologic cycle, introduced in Chapter 6, is important here. The main process in the cycle is the global transfer of water from the atmosphere to the land and oceans and back to the atmosphere (Figure 18.2). Table 18.1 lists the relative amounts of water in the major storage compartments of the cycle. Notice that more than 97% of Earth's water is in the oceans; the next-largest storage compartment, the ice caps and glaciers, accounts for another 2%. Together, these sources account for more than 99% of the total water, and both are generally unsuitable for human use because of salinity (seawater) and location (ice caps and glaciers). Only about 0.001% of the total water on Earth is in the atmosphere at any one time. However, this relatively small amount of water in the global water cycle, with an average atmosphere residence time of only about nine days, produces all our freshwater resources through the process of precipitation.

Table 18.1 THE WORLD'S WATER SUPPLY (SELECTED EXAMPLES)

LOCATION	SURFACE AREA (KM^2)	WATER VOLUME (KM^3)	PERCENTAGE OF TOTAL WATER	ESTIMATED AVERAGE RESIDENCE TIME OF WATER
Oceans	361,000,000	1,230,000,000	97.2	Thousands of years
Atmosphere	510,000,000	12,700	0.001	9 days
Rivers and streams	–	1,200	0.0001	2 weeks
Groundwater (shallow to depth of 0.8 km)	130,000,000	4,000,000	0.31	Hundreds to many thousands of years
Lakes (freshwater)	855,000	123,000	0.01	Tens of years
Ice caps and glaciers	28,200,000	28,600,000	2.15	Tens of thousands of years and longer

Source: U.S. Geological Survey

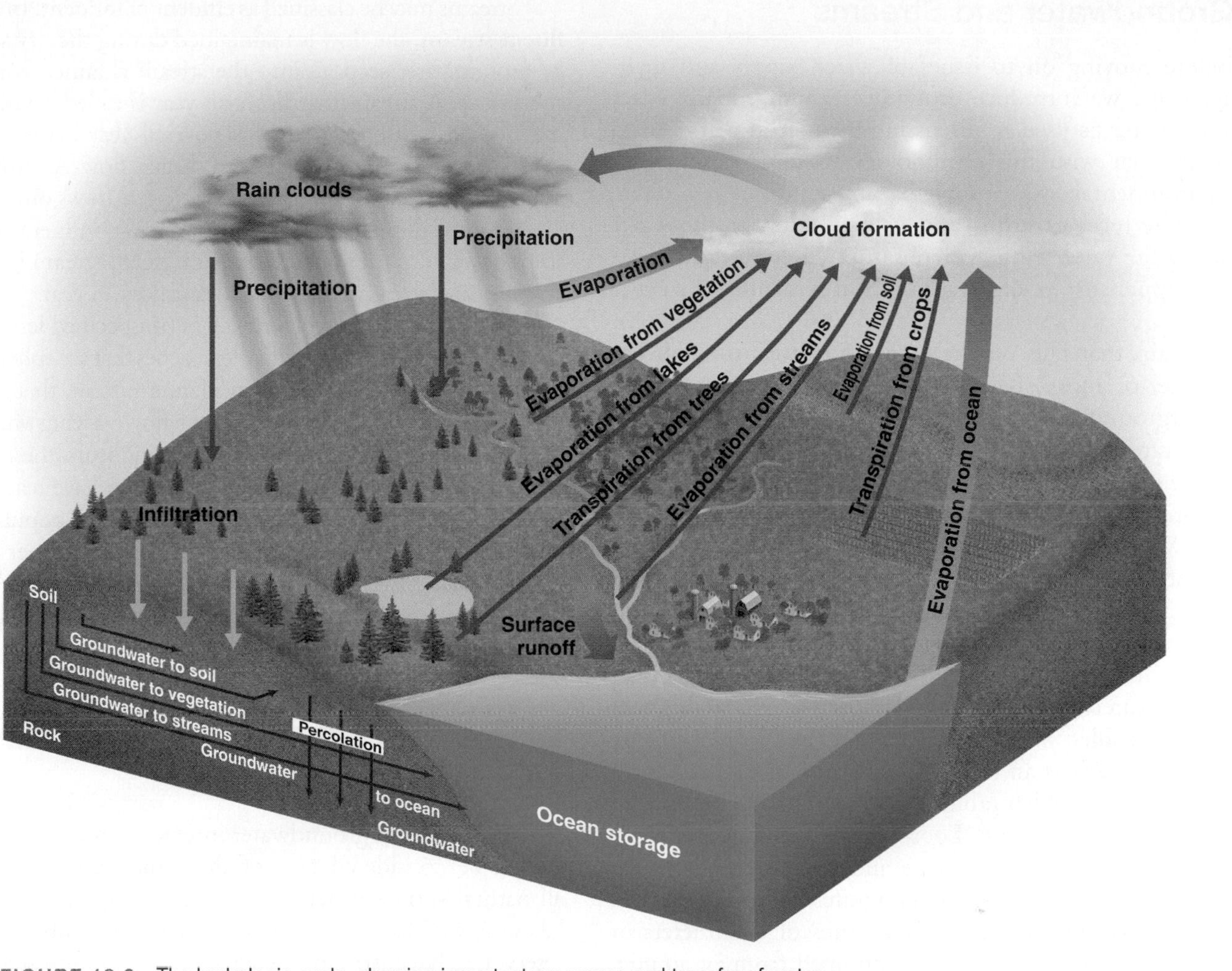

FIGURE 18.2 The hydrologic cycle, showing important processes and transfer of water. (*Source:* Modified from Council on Environment Quality and Department of State, *The Global 2000 Report to the President*, vol. 2 [Washington, DC].)

Water can be found in either liquid, solid, or gaseous form at a number of locations at or near Earth's surface. Depending on the specific location, the water's residence time may vary from a few days to many thousands of years (see Table 18.1). However, as mentioned, more than 99% of Earth's water in its natural state is unavailable or unsuitable for beneficial human use. Thus, the amount of water for which all the people, plants, and animals on Earth compete is much less than 1% of the total.

As the world's population and industrial production of goods increase, the use of water will also accelerate. The global per capita use of water in 1975 was about 700 m^3/year, or 2,000 gallons/day (185,000 gal/yr), and the total human use of water was about 3,850 km^3/year (about 10^{15} gal/yr). Today, world use of water is about 6,000 km^3/yr (about 1.58×10^{15} gal/yr), which is a significant fraction of the naturally available freshwater.

Compared with other resources, water is used in very large quantities. In recent years, the total mass (or weight) of water used on Earth per year has been approximately 1,000 times the world's total production of minerals, including petroleum, coal, metal ores, and nonmetals. Where it is abundant and readily available, water is generally a very inexpensive resource. In places where it is not abundant, such as the southwestern United States, the cost of water has been kept artificially low by government subsidies and programs.

Because the quantity and quality of water available at any particular time are highly variable, water shortages have occurred, and they will probably occur with increasing frequency, sometimes causing serious economic disruption and human suffering.[4] In the Middle East and northern Africa, scarce water has led to harsh exchanges and threats between countries and could even lead to war. The U.S. Water Resources Council estimates that water use in the United States by the year 2020 may exceed surface-water resources by 13%.[4] Therefore, an important question is, How can we best manage our water resources, use, and treatment to maintain adequate supplies?

Groundwater and Streams

Before moving on to issues of water supply and management, we introduce groundwater and surface water and the terms used in discussing them. You will need to be familiar with this terminology to understand many environmental issues, problems, and solutions.

The term **groundwater** usually refers to the water below the water table, where saturated conditions exist. The upper surface of the groundwater is called the *water table*.

Rain that falls on the land evaporates, runs off the surface, or moves below the surface and is transported underground. Locations where surface waters move into (infiltrate) the ground are known as *recharge zones*. Places where groundwater flows or seeps out at the surface, such as springs, are known as *discharge zones* or *discharge points*.

Water that moves into the ground from the surface first seeps through pore spaces (empty spaces between soil particles or rock fractures) in the soil and rock known as the *vadose zone*. This area is seldom saturated (not all pore spaces are filled with water). The water then enters the groundwater system, which is saturated (all of its pore spaces are filled with water).

An *aquifer* is an underground zone or body of earth material from which groundwater can be obtained (from a well) at a useful rate. Loose gravel and sand with lots of pore space between grains and rocks or many open fractures generally make good aquifers. Groundwater in aquifers usually moves slowly at rates of centimeters or meters per day. When water is pumped from an aquifer, the water table is depressed around the well, forming a cone of *depression*. Figure 18.3 shows the major features of a groundwater and surface-water system.

Streams may be classified as effluent or influent. In an **effluent stream**, the flow is maintained during the dry season by groundwater seepage into the stream channel from the subsurface. A stream that flows all year is called a *perennial stream*. Most perennial streams flow all year because they constantly receive groundwater to sustain flow. An **influent stream** is entirely above the water table and flows only in direct response to precipitation. Water from an influent stream seeps down into the subsurface. An influent stream is called an *ephemeral* stream because it doesn't flow all year.

A given stream may have reaches (unspecified lengths of stream) that are perennial and other reaches that are ephemeral. It may also have reaches, known as intermittent, that have a combination of influent and effluent flow varying with the time of year. For example, streams flowing from the mountains to the sea in Southern California often have reaches in the mountains that are perennial, supporting populations of trout or endangered southern steelhead, and lower intermittent reaches that transition to ephemeral reaches. At the coast, these streams may receive fresh or salty groundwater and tidal flow from the ocean to become a perennial lagoon.

Interactions between Surface Water and Groundwater

Surface water and groundwater interact in many ways and should be considered part of the same resource. Nearly all natural surface-water environments, such as rivers and lakes, as well as man-made water environments, such as reservoirs, have strong linkages with groundwater. For example, pumping groundwater from wells may reduce stream flow, lower lake levels, or change the quality of surface water. Reducing effluent stream flow by lowering

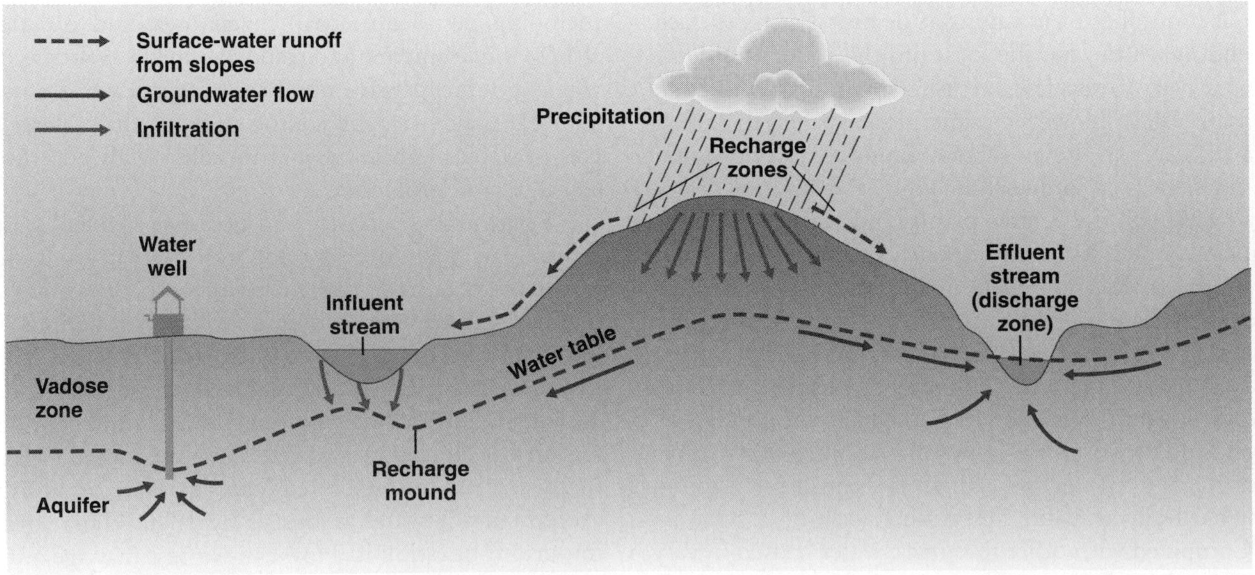

FIGURE 18.3 Groundwater and surface-water flow system.

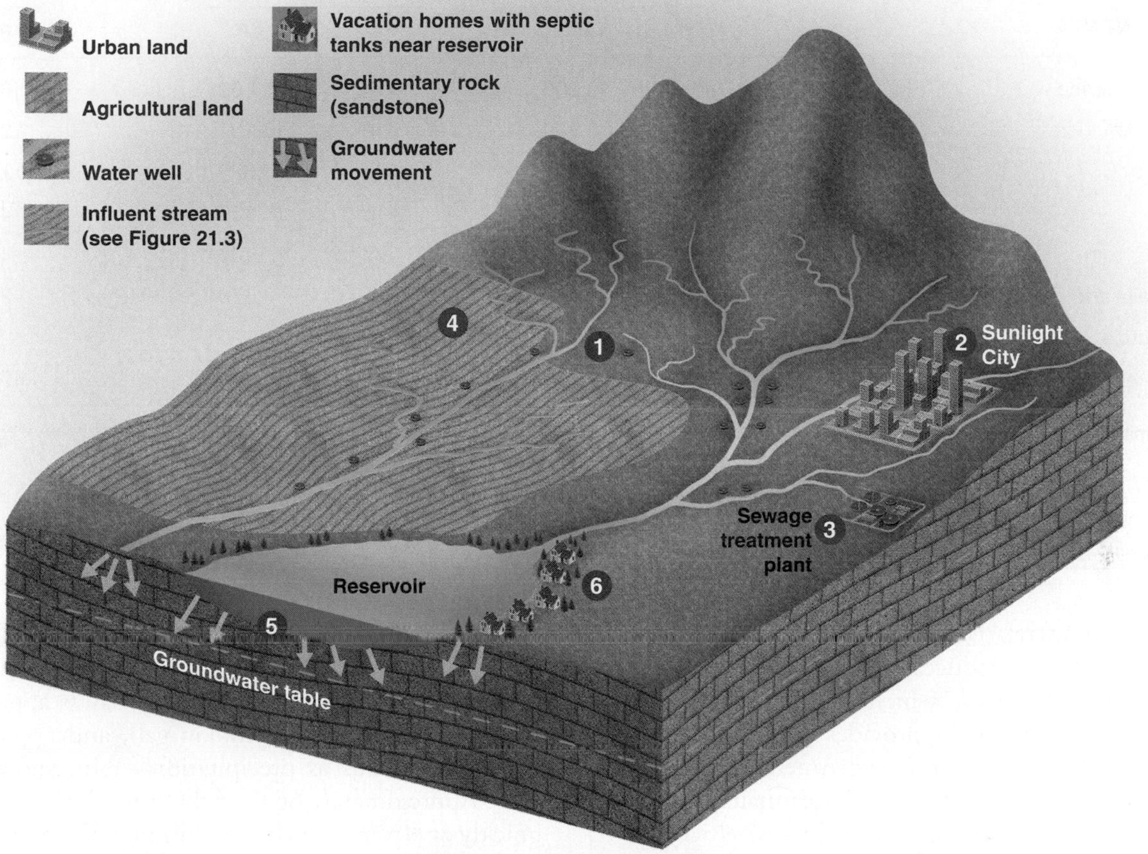

FIGURE 18.4 Idealized diagram illustrating some interactions between surface water and groundwater for a city in a semiarid environment with adjacent agricultural land and reservoir. (1) Water pumped from wells lowers the groundwater level. (2) Urbanization increases runoff to streams. (3) Sewage treatment discharges nutrient-rich waters into stream, groundwater, and reservoir. (4) Agriculture uses irrigation waters from wells, and runoff to stream from fields contains nutrients from fertilizers. (5) Water from the reservoir is seeping down to the groundwater. (6) Water from septic systems for homes is seeping down through the soil to the groundwater.

the groundwater level may change a perennial stream into an intermittent influent stream. Similarly, withdrawing surface water by diverting it from streams and rivers can deplete groundwater or change its quality. Diverting surface waters that recharge groundwaters may increase concentrations of dissolved chemicals in the groundwater because dissolved chemicals in the groundwater will no longer be diluted by infiltrated surface water. Finally, pollution of groundwater may result in polluted surface water, and vice versa.[5]

Selected interactions between surface water and groundwater in a semiarid urban and agricultural environment are shown in Figure 18.4. Urban and agricultural runoff increases the volume of water in the reservoir. Pumping groundwater for agricultural and urban uses lowers the groundwater level. The quality of surface water and groundwater is reduced by urban and agricultural runoff, which adds nutrients from fertilizers, oil from roads, and nutrients from treated wastewaters to streams and groundwater.

18.2 Water Supply: A U.S. Example

The water supply at any particular point on the land surface depends on several factors in the hydrologic cycle, including the rates of precipitation, evaporation, transpiration (water in vapor form that directly enters the atmosphere from plants through pores in leaves and stems), stream flow, and subsurface flow. A concept useful in understanding water supply is the **water budget**, a model that balances the inputs, outputs, and storage of water in a system. Simple annual water budgets (precipitation – evaporation = runoff) for North America and other continents are shown in Table 18.2. The total average annual water yield (runoff) from Earth's rivers is approximately 47,000 km^3 (1.2×10^{16} gal), but its distribution is far from uniform (see Table 18.2). Some

Table 18.2 ANNUAL WATER BUDGETS FOR THE CONTINENTS[a]

	PRECIPITATION		EVAPORATION		RUNOFF
CONTINENTAL	mm/yr	km^3	mm/yr	km^3	km^3/yr
North America	756	18,300	418	10,000	8,180
South America	1,600	28,400	910	16,200	12,200
Europe	790	8,290	507	5,320	2,970
Asia	740	32,200	416	18,100	14,100
Africa	740	22,300	587	17,700	4,600
Australia and Oceania	791	7,080	511	4,570	2,510
Antarctica	165	2,310	0	0	2,310
Earth (entire land area)	800	119,000	485	72,000	47,000[b]

[a] Precipitation – evaporation = runoff.

[b] Surface runoff is 44,800; groundwater runoff is 2,200.

Source: I. A. Shiklomanov, "World Fresh Water Resources," in P. H. Gleick, ed., *Water in Crisis* (New York: Oxford University Press, 1993), pp. 3–12.

runoff occurs in relatively uninhabited regions, such as Antarctica, which produces about 5% of Earth's total runoff. South America, which includes the relatively uninhabited Amazon basin, provides about 25% of Earth's total runoff. Total runoff in North America is about two-thirds that of South America. Unfortunately, much of the North American runoff occurs in sparsely settled or uninhabited regions, particularly in the northern parts of Canada and Alaska.

The daily water budget for the contiguous United States is shown in Figure 18.5. The amount of water vapor passing over the United States every day is approximately 152,000 million m^3 (40 trillion gal), and approximately 10% of this falls as precipitation—rain, snow, hail, or sleet. Approximately 66% of the precipitation evaporates quickly or is transpired by vegetation. The remaining 34% enters the surface water or groundwater storage systems, flows to the oceans or across the nation's boundaries, is

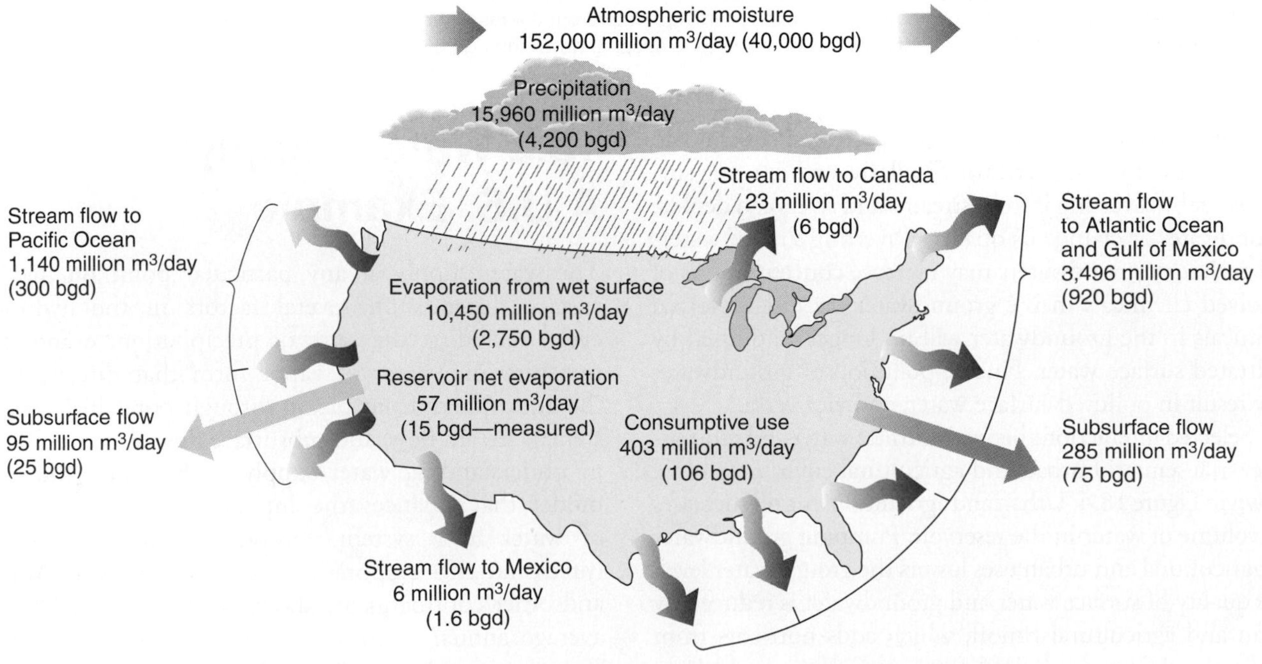

FIGURE 18.5 Water budget for the United States (bgd = billion gallons per day). (*Source:* Water Resources Council, *The Nation's Water Resources 1975–2000* [Washington, DC: Water Resources Council, 1978].)

used by people, or evaporates from reservoirs. Owing to natural variations in precipitation that cause either floods or droughts, only a portion of this water can be developed for intensive uses (only about 50% is considered available 95% of the time).[4]

Precipitation and Runoff Patterns

To put all this information in perspective, consider just the water in the Missouri River. In an average year, enough water flows down the Missouri River to cover 25 million acres a foot deep—8.4 trillion gallons. The average water use in the United States is about 100 gallons a day per person—very high compared to the rest of the world. People in Europe use about half that amount, and in some regions, such as sub-Saharan Africa, people make do with 5 gallons a day. At 100 gallons use a day, the Missouri's flow is enough to provide water for domestic and public use in the United States for about 230 million people. With a little water conservation and reduction in per capita use, the Missouri could provide enough water for all the people in the United States, so great is its flow. Not that people would actually use the Missouri's water that way, but you can stand on the shore of the Missouri, where the river flows under a major highway bridge, and get an idea of just how much water it would take to supply all those people.

In developing water budgets for water resources management, it is useful to consider annual precipitation and runoff patterns. Potential problems with water supply can be predicted in areas where average precipitation and runoff are relatively low, such as the arid and semiarid parts of the southwestern and Great Plains regions of the United States. Surface-water supply can never be as high as the average annual runoff because not all runoff can be successfully stored, due to evaporative losses from river channels, ponds, lakes, and reservoirs. Water shortages are common in areas that have naturally low precipitation and runoff, coupled with strong evaporation. In such areas, rigorous conservation practices are necessary to help ensure an adequate supply of water.[4]

Droughts

Because of large annual and regional variations in stream flow, even areas with high precipitation and runoff may periodically suffer from droughts. For example, recent dry years in the western United States produced serious water shortages. Fortunately for the more humid eastern United States, stream flow there tends to vary less than in other regions, and drought is less likely.[5] Nevertheless, summertime droughts in the southeastern United States in the early 21st century are causing hardships and billions of dollars of damage from Georgia to Florida (see opening case study).

Groundwater Use and Problems

Nearly half the people in the United States use groundwater as a primary source of drinking water. It accounts for approximately 20% of all water used. Fortunately, the total amount of groundwater available in the United States is enormous. In the contiguous United States, the amount of shallow groundwater within 0.8 km (about 0.5 mi) of the surface is estimated at 125,000 to 224,000 km^3 (3.3×10^{16} to 5.9×10^{16} gal). To put this in perspective, the lower estimate of the amount of shallow groundwater is about equal to the total discharge of the Mississippi River during the last 200 years. However, the high cost of pumping limits the total amount of groundwater that can be economically recovered.[4]

In many parts of the country, groundwater withdrawal from wells exceeds natural inflow. In such cases of **overdraft**, we can think of water as a nonrenewable resource that is being *mined*. This can lead to a variety of problems, including damage to river ecosystems and land subsidence. Groundwater overdraft is a serious problem in the Texas–Oklahoma–High Plains area (which includes much of Kansas and Nebraska and parts of other states), as well as in California, Arizona, Nevada, New Mexico, and isolated areas of Louisiana, Mississippi, Arkansas, and the South Atlantic region.

In the Texas–Oklahoma–High Plains area, the overdraft amount per year is approximately equal to the natural flow of the Colorado River for the same period.[4] The Ogallala Aquifer (also called the High Plains Aquifer), which is composed of water-bearing sands and gravels that underlie an area of about 400,000 km^2 from South Dakota into Texas, is the main groundwater resource in this area. Although the aquifer holds a tremendous amount of groundwater, it is being used in some areas at a rate up to 20 times higher than the rate at which it is being naturally replaced. As a result, the water table in many parts of the aquifer has declined in recent years (Figure 18.6), causing yields from wells to decrease and energy costs for pumping the water to rise. The most severe water-depletion problems in the Ogallala Aquifer today are in locations where irrigation was first used in the 1940s. There is concern that eventually a significant portion of land now being irrigated will be returned to dryland farming as the resource is used up.

Some towns and cities in the High Plains are also starting to have water-supply problems. Along the Platte River in northern Kansas there is still plenty of water, and groundwater levels are high (Figure 18.6). Farther south, in southwest Kansas and the panhandle in western Texas, where water levels have declined the most, supplies may last only another decade or so. In Ulysses, Kansas (population 6,000), and Lubbock, Texas (population 200,000), the situation is already getting serious. South of Ulysses, Lower Cimarron Springs, which was a famous water hole along a dry part of

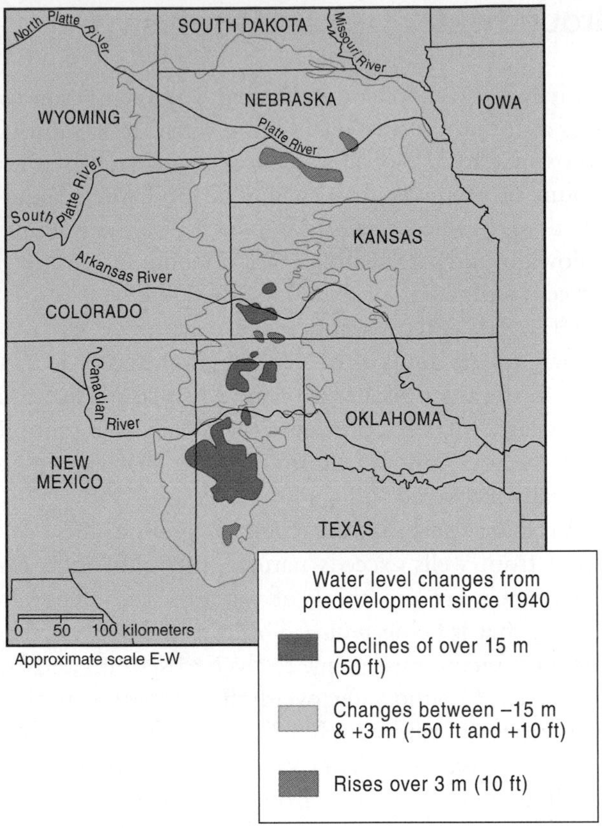

FIGURE 18.6 Groundwater-level changes as a result of pumping in the Texas–Oklahoma–High Plains region. (*Source:* U.S. Geological Survey.)

the Santa Fe Trail, dried up decades ago due to pumping groundwater. It was a symptom of what was coming. Both Ulysses and Lubbock are now facing water shortages and will need to spend millions of dollars to find alternative sources.

Desalination as a Water Source

Seawater is about 3.5% salt; that means each cubic meter of seawater contains about 40 kg (88 lb) of salt. **Desalination**, a technology for removing salt from water, is being used at several hundred plants around the world to produce water with reduced salt. To be used as a freshwater resource, the salt content must be reduced to about 0.05%. Large desalination plants produce 20,000–30,000m^3 (about 5–8 million gal) of water per day. Today, about 15,000 desalination plants in over 100 countries are in operation, and improving technology is significantly lowering the cost of desalination.

Even so, desalinated water costs several times as much as traditional water supplies in the United States. Desalinated water has a *place value*, which means that the price rises quickly with the transport distance and the cost of moving water from the plant. Because the various processes that remove the salt require large amounts of energy, the cost of the water is also tied to ever-increasing energy costs. For these reasons, desalination will remain an expensive process, used only when alternative water sources are not available.

Desalination also has environmental impacts. Discharge of very salty water from a desalination plant into another body of water, such as a bay, may locally increase salinity and kill some plants and animals. The discharge from desalination plants may also cause wide fluctuations in the salt content of local environments, which may damage ecosystems.

18.3 Water Use

In discussing water use, it is important to distinguish between off-stream and in-stream uses. **Off-stream use** refers to water removed from its source (such as a river or reservoir) for use. Much of this water is returned to the source after use; for example, the water used to cool industrial processes may go to cooling ponds and then be discharged to a river, lake, or reservoir. **Consumptive use** is an off-stream use in which water is consumed by plants and animals or used in industrial processes. The water enters human tissue or products or evaporates during use and is not returned to its source.[4]

In-stream use includes the use of rivers for navigation, hydroelectric power generation, fish and wildlife habitats, and recreation. These multiple uses usually create controversy because each requires different conditions. For example, fish and wildlife require certain water levels and flow rates for maximum biological productivity. These levels and rates will differ from those needed for hydroelectric power generation, which requires large fluctuations in discharges to match power needs. Similarly, in-stream uses of water for fish and wildlife will likely conflict with requirements for shipping and boating. Figure 18.7 demonstrates some of these conflicting demands on a graph that shows optimal discharge for various uses

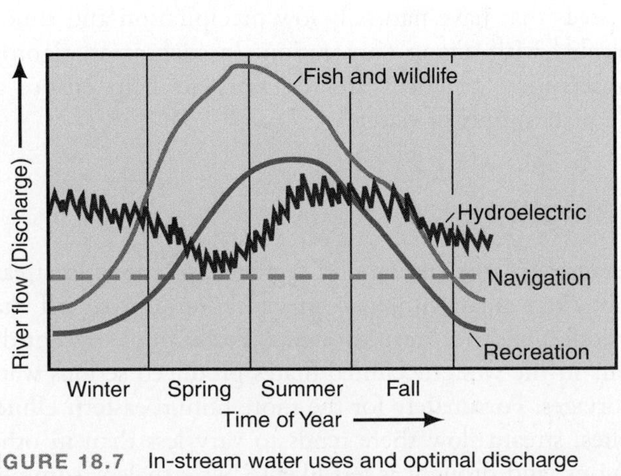

FIGURE 18.7 In-stream water uses and optimal discharge (volume of water flowing per second) for each use. Discharge is the amount of water passing by a particular location and is measured in cubic meters per second. Obviously, all these needs cannot be met simultaneously.

throughout the year. In-stream water use for navigation is optimal at a constant fairly high discharge. Some fish, however, prefer higher flows in the spring for spawning.

One problem for off-stream use is how much water can be removed from a stream or river without damaging the stream's ecosystem. This is an issue in the Pacific Northwest, where fish, such as steelhead trout and salmon, are on the decline partly because diversions for agricultural, urban, and other uses have reduced stream flow to the point where fish habitats are damaged.

The Aral Sea in Kazakhstan and Uzbekistan provides a wake-up call regarding the environmental damage that can be caused by diverting water for agriculture. Diverting water from the two rivers that flow into the Aral Sea has transformed one of the largest bodies of inland water in the world from a vibrant ecosystem into a dying sea. The present shoreline is surrounded by thousands of square kilometers of salt flats that formed as the sea's surface area shrank about 90% in the past 50 years (Figures 18.8 and 18.9). The volume of the sea was reduced by more than 50%, and the salt content increased to more than twice that of seawater, causing fish kills, including sturgeon, an important component of the economy. Dust raised by winds from the dry salt flats is producing a regional air-pollution problem, and the climate in the region has changed as the moderating effect of the sea has been reduced. Winters have grown colder and summers warmer. Fishing centers, such as Muynak in the south and Aralsk to the north that were once on the shore of the sea, are now many kilometers inland (Figure 18.10). Loss of fishing, along with a decline in tourism, has damaged the local economy.

A restoration of the small northern port of the Aral Sea is ongoing. A low, long dam was constructed across the lakebed just south of where the Syr Darya River enters the lake (see Figure 18.8). Conservation of water and the construction of the dam are producing dramatic improvement to the northern port of the lake, and some fishing is returning there. The future of the lake has improved, but great concern remains.[6]

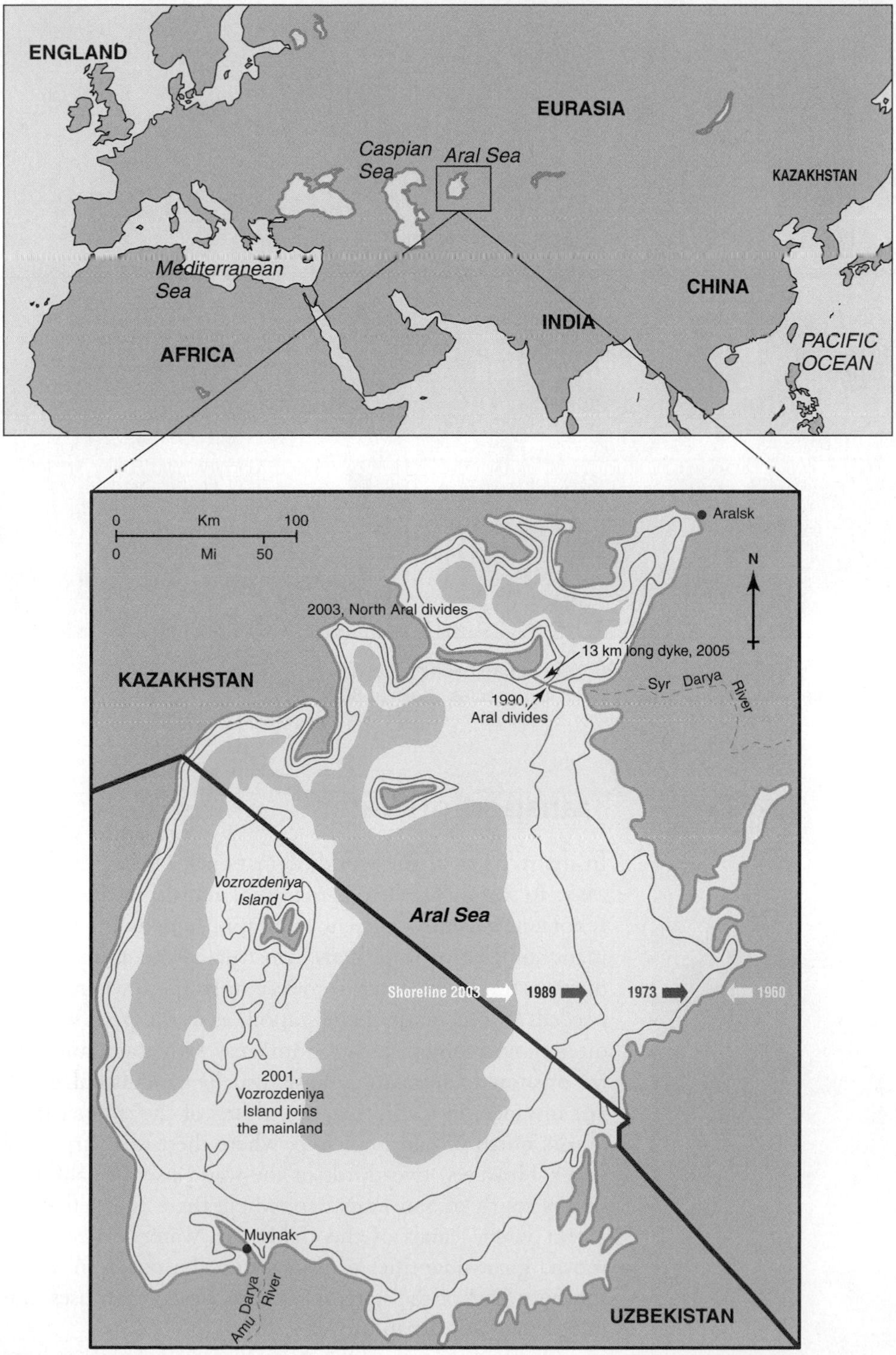

FIGURE 18.8 The Aral Sea from 1960 to 2003. A strong dike (dam), 13 km long, was constructed in 2005, and the northern lake increased in area by 18% and in depth by 2 km by 2007. (Modified after *unimaps.com 2004*.)

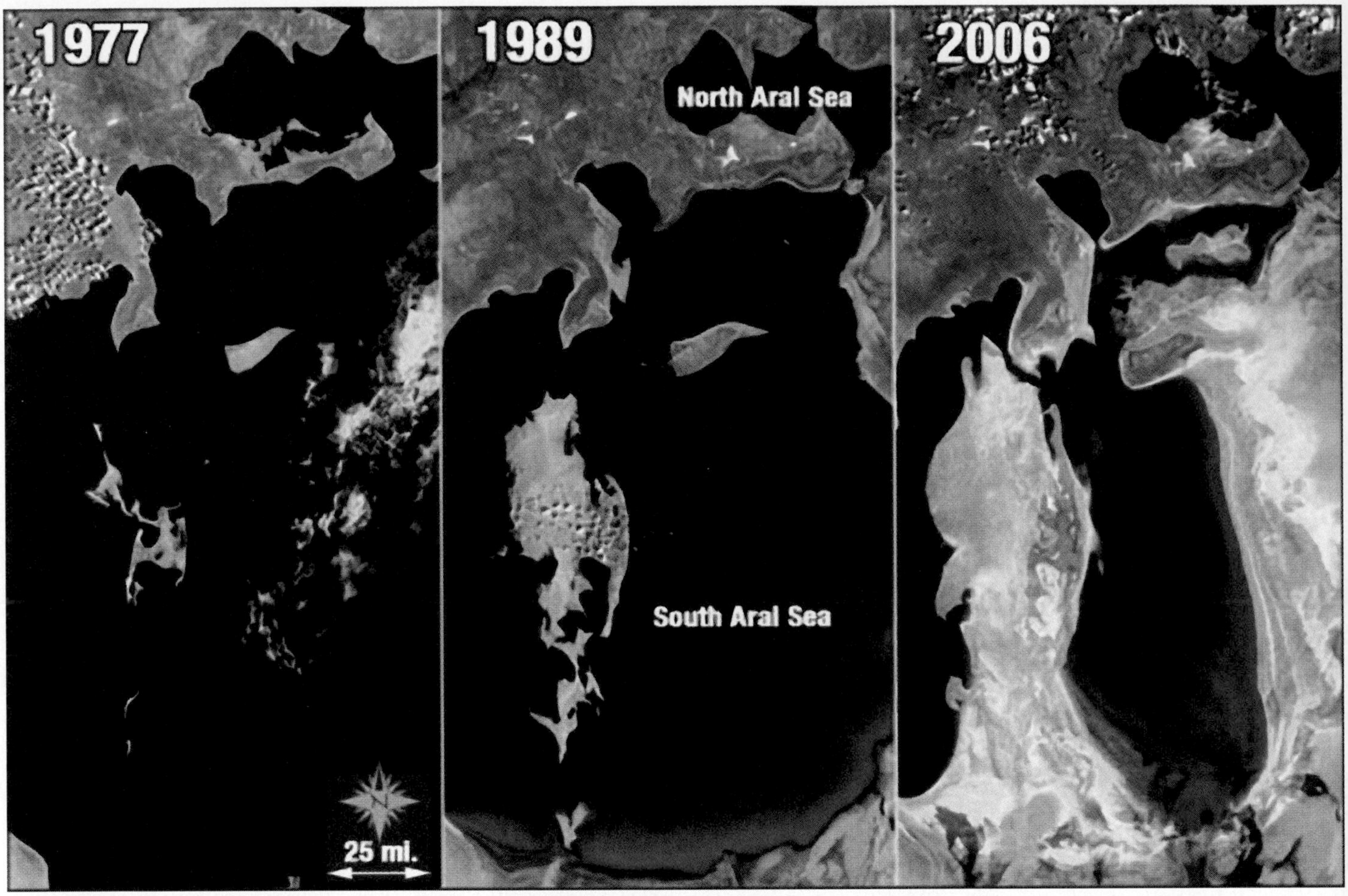

FIGURE 18.9 Three images of the Aral Sea from 1977 to 2006. By 2006, the sea had been reduced to about 10% of its original size. Wetlands around the sea were reduced by 85%; fish species declined 80%, and birds 50%.

FIGURE 18.10 Ships grounded in the dry seabed as the fishing industry collapsed.

Transport of Water

In many parts of the world, demands are being made on rivers to supply water to agricultural and urban areas. This is not a new trend—ancient civilizations, including the Romans and Native Americans, constructed canals and aqueducts to transport water from distant rivers to where it was needed. In our modern civilization, as in the past, water is often moved long distances from areas with abundant rainfall or snow to areas of high use (usually agricultural areas). For instance, in California, two-thirds of the state's runoff occurs north of San Francisco, where there is a surplus of water. However, two-thirds of the water use in California occurs south of San Francisco, where there is a deficit. In recent years, canals of the California Water Project have moved great quantities of water from the northern to the southern part of the state, mostly for agricultural uses, but increasingly for urban uses as well.

On the opposite coast, New York City has imported water from nearby areas for more than 100 years. Water use and supply in New York City show a repeating pattern. Originally, local groundwater, streams, and the

Hudson River itself were used. However, as the population increased and the land was paved over, surface waters were diverted to the sea rather than percolating into the soil to replenish groundwater. Furthermore, what water did infiltrate the soil was polluted by urban runoff. Water needs in New York exceeded local supply, and in 1842 the first large dam was built.

As the city rapidly expanded from Manhattan to Long Island, water needs increased. The shallow aquifers of Long Island were at first a source of drinking water, but this water was used faster than the infiltration of rainfall could replenish it. At the same time, the groundwater became contaminated with urban and agricultural pollutants and from saltwater seeping in underground from the ocean. (The pollution of Long Island groundwater is explored in more depth in the next chapter.) Further expansion of the population created the same pattern: initial use of groundwater; pollution, salinization, and overuse of the resource. A larger dam was built in 1900 about 30 miles north of New York City, at Croton-on-Hudson, and later on new, larger dams farther and farther upstate, in forested areas.

From a broader perspective, the cost of obtaining water for large urban centers from far-off sources, along with competition for available water from other sources and users, will eventually place an upper limit on the water supply of New York City. As shortages develop, stronger conservation measures are implemented, and the cost of water increases. As with other resources, as the water supply shrinks and demand for water rises, so does its price. If the price goes high enough, costlier sources may be developed—for example, pumping from deeper wells or desalinating.

Some Trends in Water Use

Trends in freshwater withdrawals and human population for the United States from 1950 to 2005 (the most recent data available) are shown in Figure 18.11. You can see that during that period, withdrawal of surface water far exceeded withdrawal of groundwater. In addition, withdrawals of both surface water for human uses and groundwater increased between 1950 and 1980, reaching a total maximum of approximately 375,000 million gal/day. However, after 1980, water withdrawals decreased and leveled off. It is encouraging that water withdrawals decreased after 1980 while the U.S. population continued to increase. This suggests that we have improved our water management and water conservation.[7]

Trends in freshwater withdrawals by water-use categories for the United States from 1950 to 2005 (most recent data available) are shown in Figure 18.12. Examination of this graph suggests the following:

1. The major uses of water were for irrigation and the thermoelectric industry. Excluding thermoelectric use, agriculture accounted for 65% of total withdrawals in 2005.
2. The use of water for irrigation by agriculture increased about 68% from 1950 to 1980. It decreased and leveled off from about 1985 to 2005, due in part to better irrigation efficiency, crop type, and higher energy costs.
3. Water use by the thermoelectric industry decreased slightly, beginning in 1980, and has stabilized since 1985 due to recirculating water for cooling in closed-loop systems. During the same period, electrical generation from power plants increased by more than 10 times.

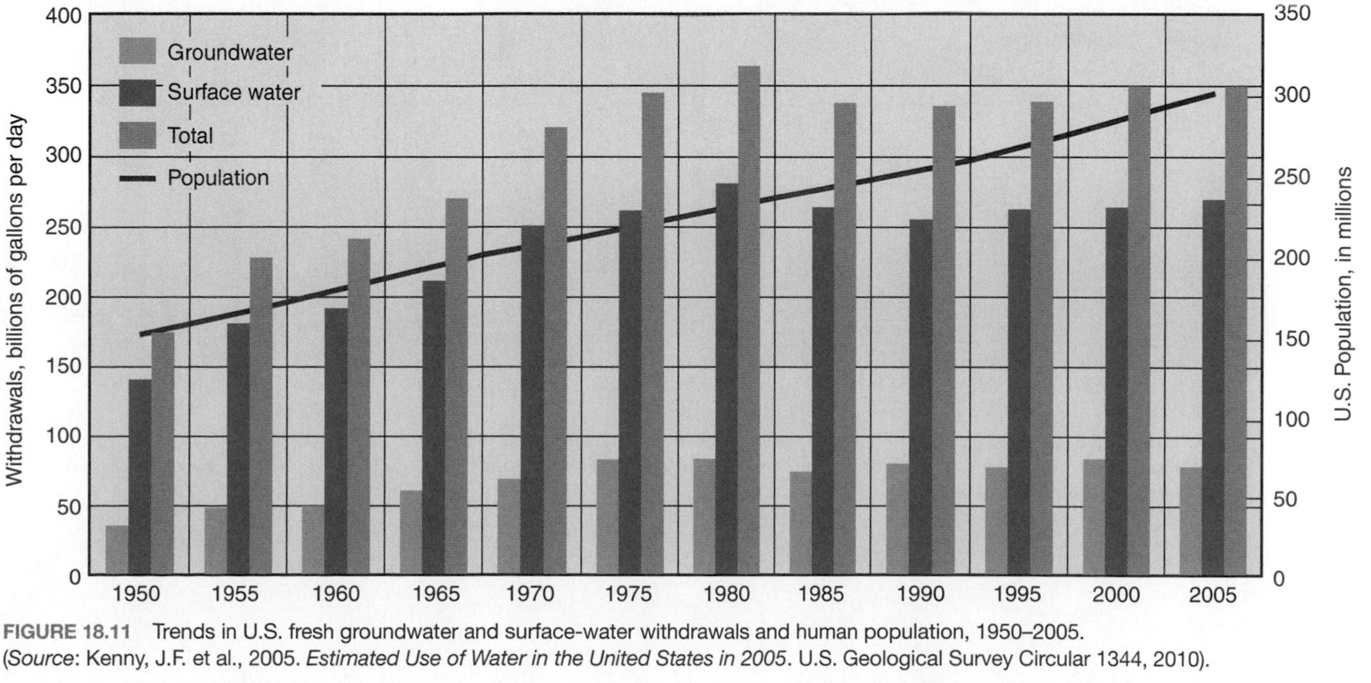

FIGURE 18.11 Trends in U.S. fresh groundwater and surface-water withdrawals and human population, 1950–2005. (*Source*: Kenny, J.F. et al., 2005. *Estimated Use of Water in the United States in 2005*. U.S. Geological Survey Circular 1344, 2010).

4. Use of water for public and rural supplies continued to increase through the period 1950– 2005, presumably due to the increase in human population.[7, 8]

18.4 Water Conservation

Water conservation is the careful use and protection of water resources. It involves both the quantity of water used and its quality. Conservation is an important component of sustainable water use. Because the field of water conservation is changing rapidly, it is expected that a number of innovations will reduce the total withdrawals of water for various purposes, even though consumption will continue to increase.[4]

Agricultural Use

Improved irrigation (Figure 18.13) could reduce agricultural withdrawals by 20 to 30%. Because agriculture is the biggest water user, this would be a huge savings. Suggestions for agricultural conservation include the following:

- Price agricultural water to encourage conservation (subsidizing water encourages overuse).
- Use lined or covered canals that reduce seepage and evaporation.
- Use computer monitoring and schedule release of water for maximum efficiency.
- Integrate the use of surface water and groundwater to more effectively use the total resource. That is, irrigate with surplus surface water when it is abundant, and also use surplus surface water to recharge groundwater aquifers, using specially designed infiltration ponds or injection wells. When surface water is in short supply, use more groundwater.
- Irrigate when evaporation is minimal, such as at night or in the early morning.
- Use improved irrigation systems, such as sprinklers or drip irrigation, that apply water to crops more effectively.
- Improve land preparation for water application—that is, improve the soil so that more water sinks in and less runs off. Where applicable, use mulch to help retain water around plants.
- Encourage the development of crops that require less water or are more salt-tolerant, so that less periodic flooding of irrigated land is necessary to remove accumulated salts in the soil.

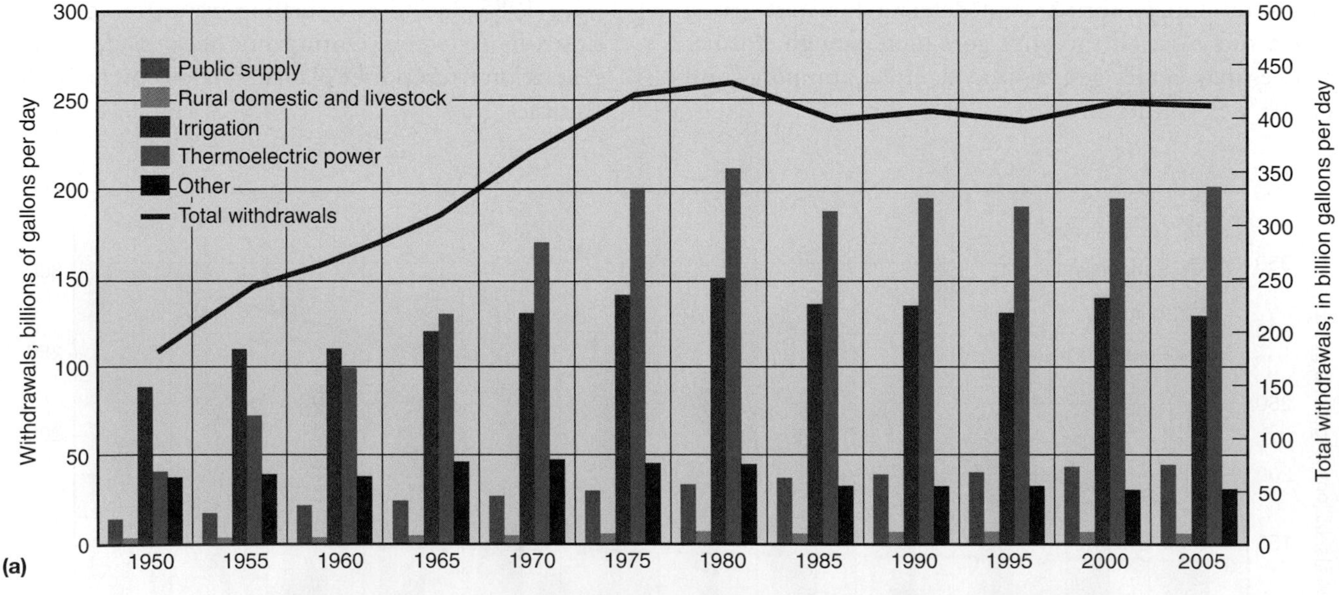

FIGURE 18.12 **(a)** Trends in total U.S. water withdrawals by water-use category (1950–2005). (*Source*: Kenny J. F. et al., 2010, *Estimated Use of Water in the United States in 2005.* U.S. Geological Survey Circular 1344); **(b)** U.S. water use in 2005 (by percent).

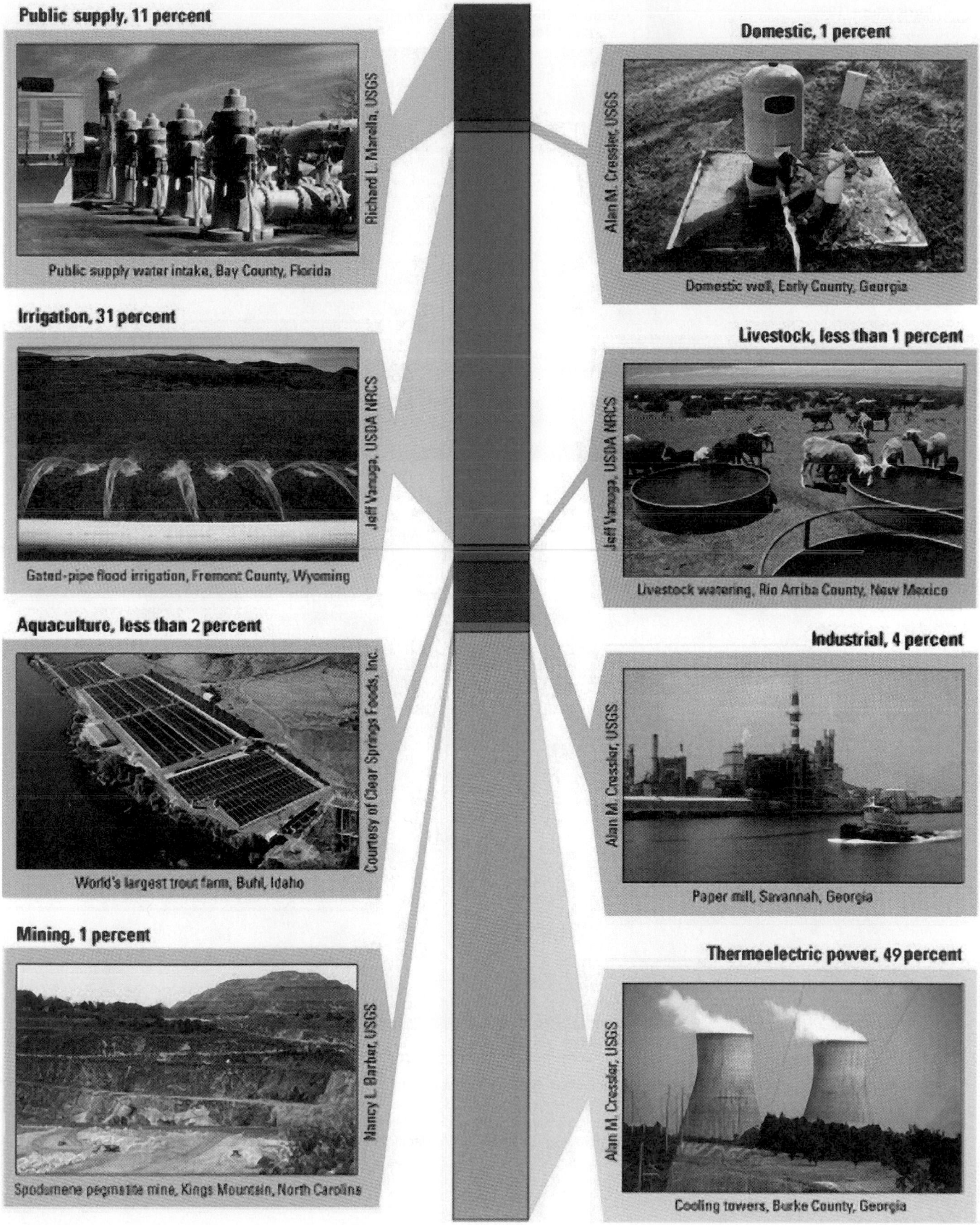
Public supply, 11 percent
Richard L. Marella, USGS
Public supply water intake, Bay County, Florida
Irrigation, 31 percent
Jeff Vanuga, USDA NRCS
Gated-pipe flood irrigation, Fremont County, Wyoming
Aquaculture, less than 2 percent
Courtesy of Clear Springs Foods, Inc.
World's largest trout farm, Buhl, Idaho
Mining, 1 percent
Nancy L. Barber, USGS
Spodumene pegmatite mine, Kings Mountain, North Carolina
Domestic, 1 percent
Alan M. Cressler, USGS
Domestic well, Early County, Georgia
Livestock, less than 1 percent
Jeff Vanuga, USDA NRCS
Livestock watering, Rio Arriba County, New Mexico
Industrial, 4 percent
Alan M. Cressler, USGS
Paper mill, Savannah, Georgia
Thermoelectric power, 49 percent
Alan M. Cressler, USGS
Cooling towers, Burke County, Georgia

(b)

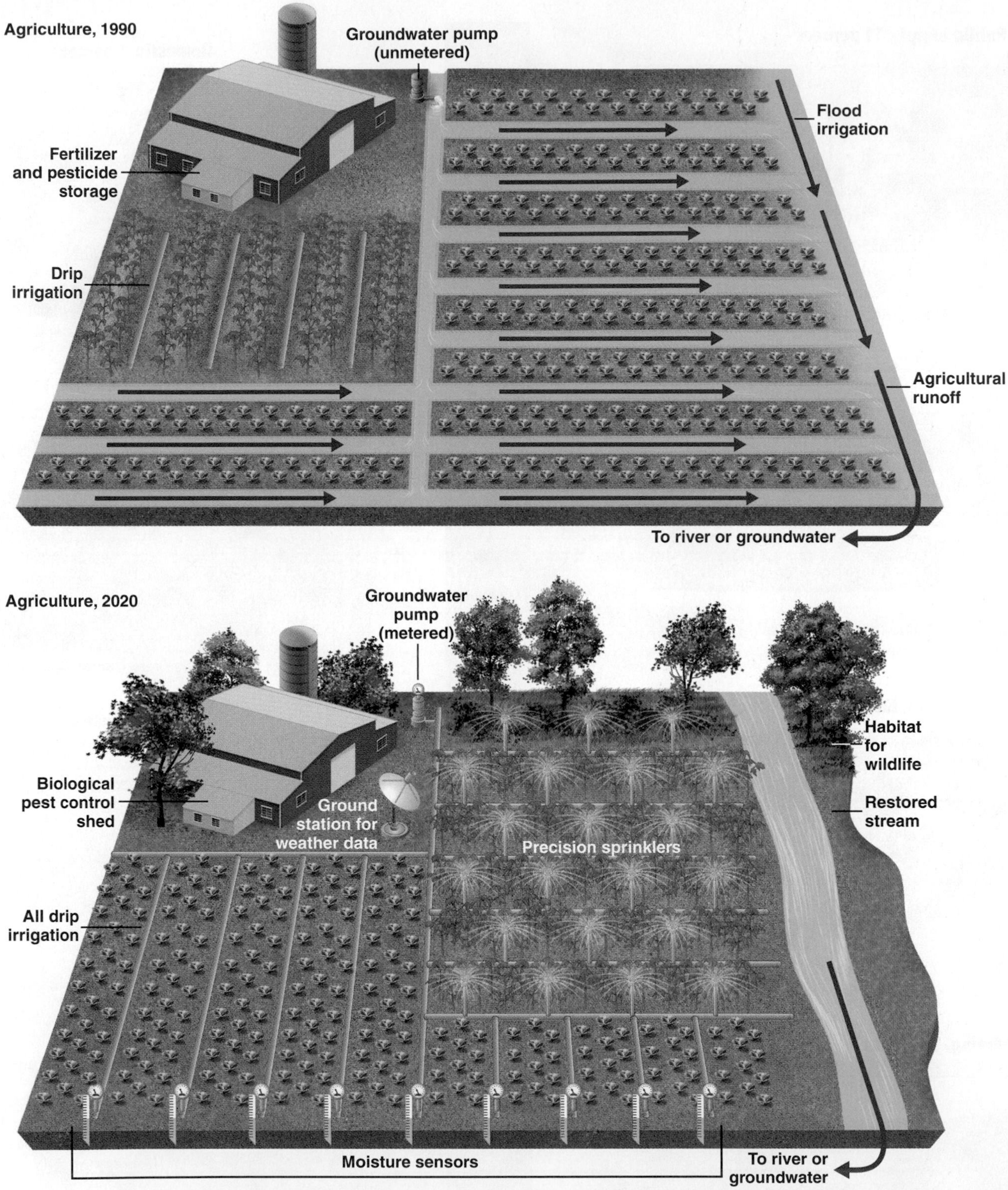

FIGURE 18.13 Comparison of agricultural practices in 1990 with what they might be by 2020. The improvements call for a variety of agricultural procedures, including biological pest control, more efficient irrigation, and restoration of water resources and wildlife habitat. (*Source:* P.H. Gleick, P. Loh, S.V. Gomez, and J. Morrison, *California Water 2020, a Sustainable Vision* [Oakland, CA: Pacific Institute for Studies in Development, Environment, and Security, 1995].)

Public Supply and Domestic Use

Domestic use of water accounts for only about 12% of total national water withdrawals. However, because public supply water use is concentrated in urban areas, it may pose major local problems in areas where water is periodically or often in short supply. The population of the United States continues to grow, and many urban areas in the United States are experiencing or will experience the impact of population growth on water supply. For example:

- Southern California, in particular San Diego and Los Angeles, is growing rapidly, and its water needs are quickly exceeding local supplies. As a result, the city of San Diego has negotiated with farmers to the east, in the Imperial Valley, to purchase water for urban areas. The city is also building desalination plants and considering raising the height of dams so more water can be stored for urban uses. Southern California has long imported water from the Sierra Nevada to the north. If climate change brings less snow and more rain, that supply may become more variable because snow melts slowly and thus serves as a water source for a longer time than rain, which quickly runs off. In the expectation of more rain than snow, plans for what is called the Inland Feeder Project include a series of large-diameter tunnels to quickly deliver large volumes of water from northern California to Southern California during periods of rapid runoff. They will be used to fill local reservoirs and groundwater basins, providing water during dry periods and emergencies.
- In Denver, city officials, fearing future water shortages, are proposing strict conservation measures that include limits on water use for landscaping and the amount of grass that can be planted around new homes.
- Chicago, one of the fastest-growing urban areas in the United States and located on the shore of Lake Michigan, one of the largest sources of freshwater in the world, reports groundwater-depletion problems. Water shortages in outlying urban areas may become apparent by 2020.
- Tampa, Florida, fearing shortages of freshwater because of its continuing growth, began operating a desalination plant in 2003 that produces approximately 25 million gallons of water daily.
- Atlanta, Georgia, another fast-growing urban area in the United States, expects increased demand on its water supplies as a result and is exploring ways to meet that demand.
- New York City, which imports water from the upstate Catskill Mountains, periodically has water shortages during droughts. The city placed water restrictions on its more than 9 million citizens in 2002.

What is clear from these examples is that while there is no shortage of water in the United States or the world, there are local and regional shortages, particularly in large, growing urban areas in the semiarid western and southwestern United States.[9]

Most water in homes is used in the bathroom and for washing laundry and dishes. Domestic water use can be substantially reduced at a relatively small cost by the following measures:

- In semiarid regions, replace lawns with decorative gravel and native plants.
- Use more efficient bathroom fixtures, such as low-flow toilets that use 1.6 gallons or less per flush rather than the standard 5 gallons, and low-flow showerheads that still deliver sufficient water.
- Flush only when really necessary.
- Turn off water when not absolutely needed for washing, brushing teeth, shaving, and so on.
- Fix all leaks quickly. Dripping pipes, faucets, toilets, or garden hoses waste water. A small drip can waste several liters per day; multiply this by millions of homes with a leak, and a large volume of water is lost.
- Purchase dishwashers and laundry machines that minimize water use.
- Take a long bath rather than a long shower.
- Don't hose sidewalks and driveways; sweep them.
- Consider using gray water (from showers, bathtubs, sinks, and washing machines) to water vegetation. The gray water from laundry machines is easiest to use, as it can be easily diverted before entering a drain.
- Water lawns and plants in the early morning, late afternoon, or at night to reduce evaporation.
- Use drip irrigation and place water-holding mulch around garden plants.
- Plant drought-resistant vegetation that requires less water.
- Learn how to read the water meter to monitor for unobserved leaks and record your conservation successes.
- Use reclaimed water (see opening case study).

In addition, local water districts should encourage water pricing policies that make water use more expensive for those who exceed some baseline amount determined by the number of people in a home and the size of the property.

Industry and Manufacturing Use

Water conservation by industry can be improved. For instance, water use for steam generation of electricity could be reduced 25 to 30% by using cooling towers that require less or no water (as has often been done in the United States). Manufacturing and industry could curb water use by increasing in-plant treatment and recycling water and by developing new equipment and processes that require less water.[4]

18.5 Sustainability and Water Management

Because water is essential to sustain life and maintain ecological systems necessary for human survival, it plays important roles in ecosystem support, economic development, cultural values, and community well-being. Managing water use for sustainability is thus important in many ways.

Sustainable Water Use

From a supply and management perspective, **sustainable water use** can be defined as use of water resources in a way that allows society to develop and flourish in an indefinite future without degrading the various components of the hydrologic cycle or the ecological systems that depend on it. Some general criteria for water-use sustainability are as follows.[10]

- Develop enough water resources to maintain human health and well-being.
- Provide sufficient water resources to guarantee the health and maintenance of ecosystems.
- Ensure basic standards of water quality for the various users of water resources.
- Ensure that people do not damage or reduce the long-term renewability of water resources.
- Promote the use of water-efficient technology and practice.
- Gradually eliminate water-pricing policies that subsidize inefficient use of water.

Groundwater Sustainability

The concept of sustainability, by definition, implies a long-term perspective. With groundwater resources, effective management for sustainability requires an even longer time frame than for other renewable resources. Surface waters, for example, may be replaced over a relatively short time, whereas replacement of groundwater may take place slowly over many years. The effects of pumping groundwater faster than it is being replenished—drying up of springs, weaker stream flow—may not be noticed until years after pumping begins. The long-term approach to sustainability with respect to groundwater is basically not to take out more than is going in; to keep monitoring input and adjusting output accordingly.[11]

Water Management

Maintaining a water supply is a complex issue that will become more difficult as demand for water increases in the coming years. The problem will be especially challenging in the southwestern United States and other semiarid and arid parts of the world where water is in short supply or soon will be. Options for minimizing problems include finding alternative water supplies and managing existing supplies better. In some areas, finding new supplies is so unlikely that people are seriously considering some literally far-fetched water sources, such as towing icebergs to coastal regions where freshwater is needed. It seems apparent that water will become much more expensive in the future; and if the price is right, many innovative programs are possible.

A number of municipalities are using the *variable-water-source approach*. The city of Santa Barbara, California, for example, has developed a variable-water-source approach that uses several interrelated measures to meet present and future demand. Details of the plan (shown in Figure 18.14) include importing state water, developing new sources, using reclaimed water, and instituting a permanent conservation program. In addition, there is a desalination plant near the ocean and a wastewater-treatment plant (see Figure 18.14) that is in long-term storage but could be brought online if needed. In essence, this seaside community has developed a master water plan.

A Master Plan for Water Management

Luna Leopold, a famous U.S. hydrologist, suggests that a new philosophy of water management is needed, one based on geologic, geographic, and climatic factors, as well as on the traditional economic, social, and political factors. He argues that the management of water resources cannot be successful as long as it is perceived only from an economic and political standpoint.

The essence of Leopold's water-management philosophy is that surface water and groundwater are both subject to natural flux with time. In wet years, there is plenty of surface water, and the near-surface groundwater is replenished. But we must have in place, and ready to

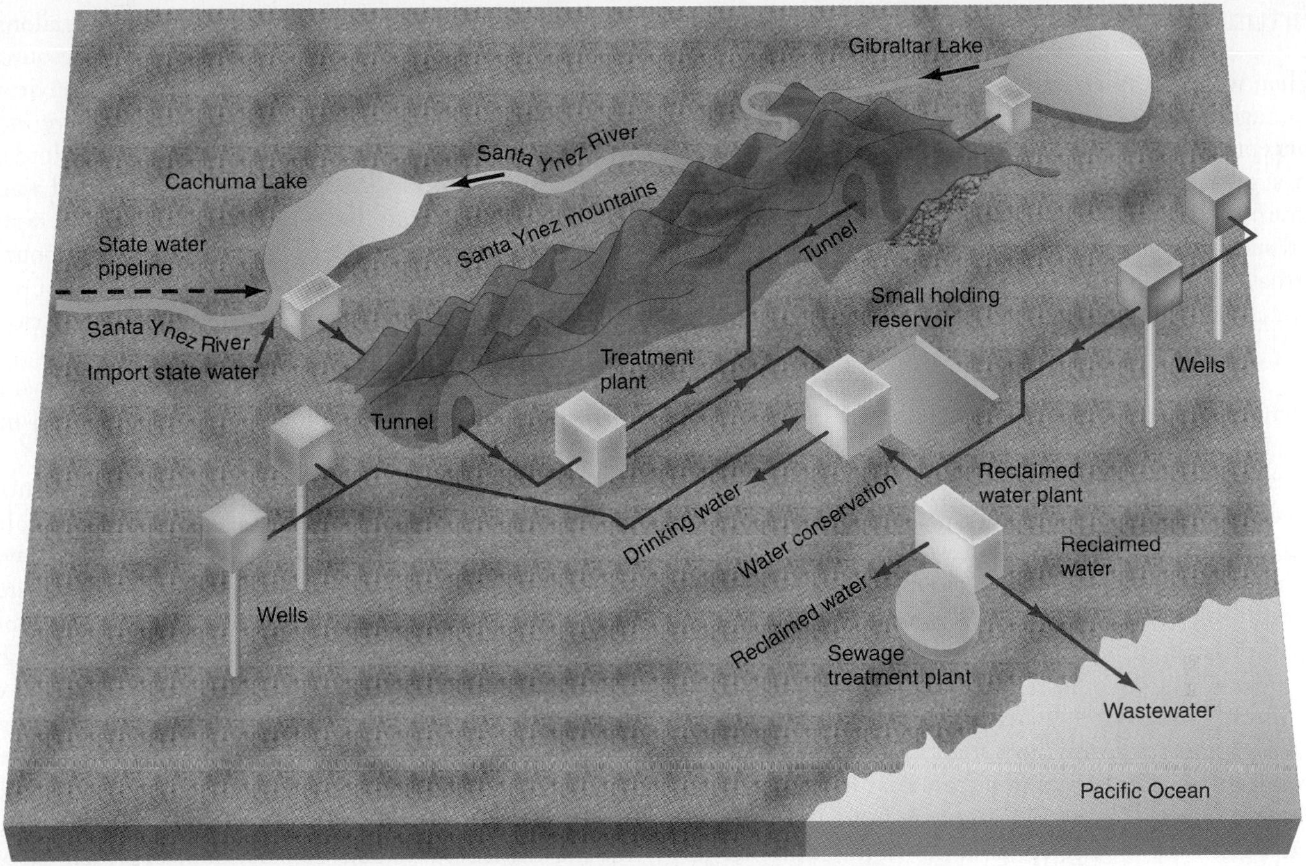

FIGURE 18.14 Schematic drawing of a variable source model (present and future) for water supply for the city of Santa Barbara, California. (*Source:* Santa Barbara City Council, data from 2009.)

use, specific plans to supply water on an emergency basis to minimize hardships in dry years, which we must expect even though we can't accurately predict them.

For example, subsurface waters in various locations in the western United States are too deep to be economically pumped from wells, or else are of marginal quality. These waters may be isolated from the present hydrologic cycle and therefore not subject to natural recharge, but might be used when the need is great if wells have been drilled and connected to existing water lines so as to be ready when the need arises.

Another possible emergency plan might involve the treatment of wastewater. Its reuse on a regular basis is expensive, but advance planning to reuse treated wastewater during emergencies is a wise decision.

Finally, we should develop plans to use surface water when available, and not be afraid to use groundwater as needed in dry years. During wet years, natural recharge as well as artificial recharge (pumping excess surface water into the ground) will replenish the groundwater. This water-management plan recognizes that excesses and deficiencies in water are natural and can be planned for.[12]

Water Management and the Environment

Many agricultural and urban areas depend on water delivered from nearby (and in some cases not-so-nearby) sources. Delivering the water requires a system for water storage and routing by way of canals and aqueducts from reservoirs. As a result, dams are built, wetlands may be modified, and rivers may be channelized to help control flooding—all of which usually generates a good deal of controversy.

The days of developing large projects in the United States without environmental and public review have passed. Resolving water-development issues now involves input from a variety of government and public groups, which may have very different needs and concerns. These range from agricultural groups that see water development as critical for their livelihood to groups primarily concerned with wildlife and wilderness preservation. It is a positive sign that the various parties with interests in water issues are encouraged—and in some cases required—to meet and communicate their desires and concerns. Below we discuss some of these concerns: wetlands, dams, channelization, and flooding.

Virtual Water

When we think of water resources, we generally think of drainage basins or groundwater reservoirs. An emerging concept is that we can also think about water resources on a global scale in terms of what is known as **virtual water**: the amount of water necessary to produce a product, such as an automobile, or a crop such as rice.[13-15] The virtual water content is measured at the place where the product is produced or the crop grown. It is called "virtual" because the water content in the product or crop is very small compared with the amount of water used to produce it.[14]

The amount of virtual water necessary for crops and animals is surprisingly large and variable. A few years ago, the question of how much water is required to produce a cup of coffee was asked. The answer is not trivial. Coffee is an important crop for many countries and the major social drink in much of the world. Many a romance has been initiated with the question "Would you like a cup of coffee?"

How much water is necessary to produce a cup of coffee requires knowing how much water is necessary to produce the coffee berries (that contain the bean) and the roasted coffee. The question is complicated by the fact that water used to raise coffee varies from location to location, as does the yield of berries. Much of the water in coffee-growing areas is free; it comes from rain. However, that doesn't mean the water has no value. People are usually surprised to learn that it takes about 140 liters (40 gallons) of water to produce one cup of coffee. The amount of water that is needed to produce a ton of a crop varies from a low of about 175 m^3 for sugarcane to 1,300 m^3 for wheat, 3,400 m^3 for white rice, and 21,000 m^3 for roasted coffee. For the meat we eat, the amount per ton is 3,900 m^3 for chicken, 4,800 m^3 for pork, and 15,500 m^3 for beef.[14]

The United States produces food that is exported around the world. The concept of virtual water shows that people consuming imported U.S. crops in Western Europe directly affect the regional water resources of the United States. Similarly, our consumption of imported foods—such as cantaloupes grown in Mexico, or blueberries from Chile—affect the regional water supply and groundwater resources of the countries that grew and exported them.

The concept of virtual water is useful in water-resource planning from the local to global scale. A country with an arid climate and restricted water resources can choose between developing those resources for agriculture or for other water uses—for example, to support wetland ecosystems or a growing human population. Since the average global amount of water necessary to produce a ton of white rice is about 3,400 m^3 (nearly 900,000 gallons), growing rice in countries with abundant water resources makes sense. For countries with a more arid environment, it might be prudent to import rice and save local and regional water resources for other purposes. Jordan, for example, imports about 7 billion m^3 of virtual water per year by importing foods that requires a lot of water to produce. As a result, Jordan withdraws only about 1 billion m^3 of water per year from its own water resources. Egypt, on the other hand, has the Nile River and imports only about one-third as much water (virtual water) as it withdraws from its domestic supply. Egypt has a goal of water independence and is much less dependent on imported virtual water than is Jordan.[14]

Examination of global water resources and potential global water conservation is an important part of sustaining our water supply. For example, by trading virtual water, the international trade markets reduce agriculture's global water use by about 5%.[14] Figure 18.15 shows net virtual water budgets (balances) for major trades. The balance is determined by import minus export in km^3 of virtual water, where 1 km^3 is 10^9 (one billion) m^3. For example, when the United States and Canada export wheat and other products to Mexico and Eastern Europe, a lot of virtual water is exported, explaining the negative balance for the United States and Canada, both of which export more virtual water than they import. On the other hand, countries that import a lot of food and other products have a positive balance because their imports of virtual water exceed their exports.

The concept of virtual water has three major uses to society and the world: [16]

- It promotes efficient use of water from a local to global scale. Trading virtual water can conserve global water resources by producing products that require a lot of water in places where water is abundant and can be efficiently used. When those products are exported to places where water is scarce or difficult to use efficiently, water is conserved and real water savings are realized.
- It offers countries and regions an opportunity to enjoy greater water security. Virtual water can be thought of as an alternative, additional water supply that, from a political point of view, can increase security and help solve geopolitical problems between nations.
- It helps us to understand relationships between water-consumption patterns and their environmental, economic, and political impacts. Knowing the virtual water content of the products we produce and where and how they are produced increases our awareness of water demand and ways to realize water savings.

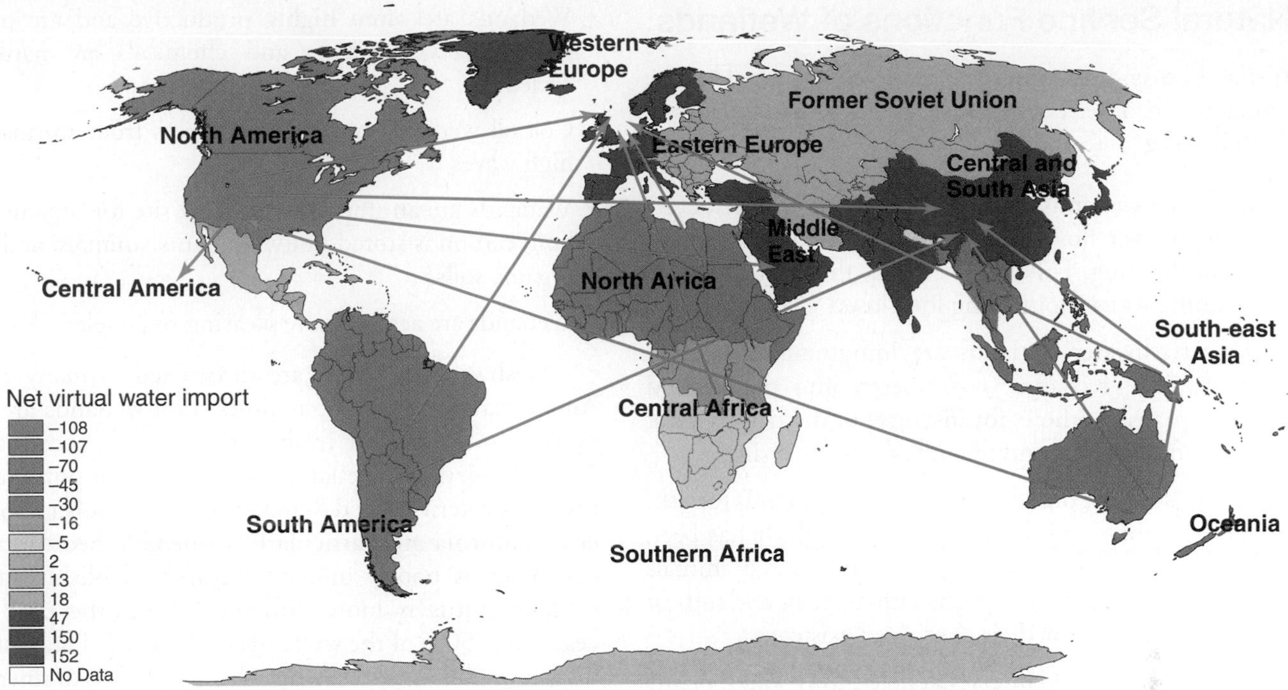

FIGURE 18.15 Virtual-water balances and transfers (6m^3 of water). 1 Gm^3 is 1 billion cubic meters. (*Source:* A.Y. Hoekstra, ed., 2003, Virtual water trade. Proceedings of the International Export Meeting on Virtual Water Trade. Value of Water Research Report Series 12. IHE Delft. The Netherlands.)

Water Footprint

The water footprint is the total volume of freshwater used to produce the products and services used by an individual, community, country, or region. The footprint is generally expressed as the volume of water used per year and is divided into three components:[14]

- *Green water*, defined as precipitation that contributes to water stored in soils. This is the water consumed by crops (consumptive use) that evaporates or transpires from plants we cultivate.
- *Blue water*, defined as surface and groundwater. This is used to produce our goods and services (consumptive use).
- *Gray water,* defined as water polluted by the production of goods and services and rendered not available for other uses. The volume of gray water use has been estimated by calculating the amount of water required to dilute pollutants to the point that the water quality is acceptable and consistent with water quality standards.

The concept of virtual water, when linked to the water footprint, provides new tools to better manage water resources sustainably. An objective is to work toward water conservation and ultimately water self-sufficiency.

18.6 Wetlands

Wetlands is a comprehensive term for landforms such as salt marshes, swamps, bogs, prairie potholes, and vernal pools (shallow depressions that seasonally hold water). Their common feature is that they are wet at least part of the year and, as a result, have a particular type of vegetation and soil. Figure 18.16 shows several types of wetlands.

Wetlands may be defined as areas inundated by water or saturated to a depth of a few centimeters for at least a few days per year. Three major characteristics in identifying wetlands are hydrology, or wetness; type of vegetation; and type of soil. Of these, hydrology is often the most difficult to define because some freshwater wetlands may be wet for only a few days a year. The duration of inundation or saturation must be sufficient for the development of wetland soils, which are characterized by poor drainage and lack of oxygen, and for the growth of specially adapted vegetation.

Natural Service Functions of Wetlands

Wetland ecosystems may serve a variety of natural service functions for other ecosystems and for people, including the following:

- Freshwater wetlands are a natural sponge for water. During high river flow, they store water, reducing downstream flooding. Following a flood, they slowly release the stored water, nourishing low flows.
- Many freshwater wetlands are important as areas of groundwater recharge (water seeps into the ground from a prairie pothole, for instance) or discharge (water seeps out of the ground in a marsh fed by springs).
- Wetlands are one of the primary nursery grounds for fish, shellfish, aquatic birds, and other animals. It has been estimated that as many as 45% of endangered animals and 26% of endangered plants either live in wetlands or depend on them for their continued existence.[17]
- Wetlands are natural filters that help purify water; plants in wetlands trap sediment and toxins.
- Wetlands are often highly productive and are places where many nutrients and chemicals are naturally cycled.
- Coastal wetlands buffer inland areas from storms and high waves.
- Wetlands are an important storage site for organic carbon; carbon is stored in living plants, animals, and rich organic soils.
- Wetlands are aesthetically pleasing to people.

Freshwater wetlands are threatened in many areas. An estimated 1% of the nation's total wetlands are lost every two years, and freshwater wetlands account for 95% of this loss. Wetlands such as prairie potholes in the midwestern United States and vernal pools in Southern California are particularly vulnerable because their hydrology is poorly understood and establishing their wetland status is more difficult.[18] Over the past 200 years, over 50% of the wetlands in the United States have disappeared because they have been diked or drained for agriculture or filled for urban or industrial development. Perhaps as much as 90% of the freshwater wetlands have disappeared.

Although most coastal marshes are now protected in the United States, the extensive salt marshes at many of the nation's major estuaries, where rivers entering the ocean widen and are influenced by tides, have been modified or lost. These include deltas and estuaries of major rivers,

(a)

(b)

(c)

FIGURE 18.16 Several types of wetlands: **(a)** aerial view of part of the Florida Everglades at a coastal site; **(b)** cypress swamp water surface covered with a floating mat of duckweed, northeast Texas; and **(c)** aerial view of farmlands encroaching on prairie potholes, North Dakota.

such as the Mississippi, Potomac, Susquehanna (Chesapeake Bay), Delaware, and Hudson.[19] The San Francisco Bay estuary, considered the estuary most modified by human activity in the United States today, has lost nearly all its marshlands to leveeing and filling (Figure 18.17).[19] Modifications result not only from filling and diking but also from loss of water. The freshwater inflow has been reduced by more than 50%, dramatically changing the hydrology of the bay in terms of flow characteristics and water quality. As a result of the modifications, the plants and animals in the bay have changed as habitats for fish and wildfowl have been eliminated.[19]

The delta of the Mississippi River includes some of the major coastal wetlands of the United States and the world. Historically, coastal wetlands of southern Louisiana were maintained by the flooding of the Mississippi River, which delivered water, mineral sediments, and nutrients to the coastal environment. The mineral sediments contributed to the vertical accretion (building up) of wetlands. The nutrients enhanced growth of wetland plants, whose coarse, organic components (leaves, stems, roots) also accreted. These accretion processes counter processes that naturally submerge the wetlands, including a slow rise in sea level and subsidence (sinking) due to compaction. If the rates of submergence of wetlands exceed the rates of accretion, then the area of open water increases, and the wetlands are reduced.

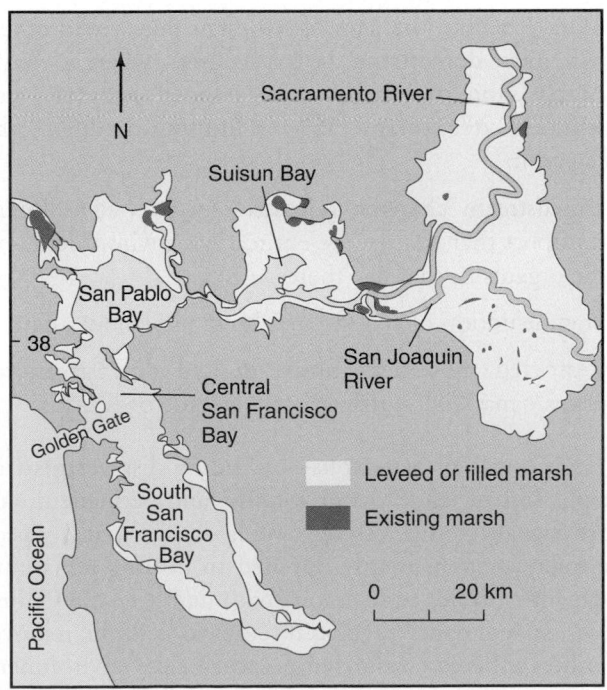

FIGURE 18.17 Loss of marshlands in the San Francisco Bay estuary from about 1850 to the present. (*Sources:* T.J. Conomos, ed., *San Francisco, the Urbanized Estuary* [San Francisco: American Association for the Advancement of Science, 1979]; F.H. Nichols, J.E. Cloern, S.N. Luoma, and D.H. Peterson, "The Modification of an Estuary," *Science* 231 [1986]: 567–573. Copyright 1986 by the American Association for the Advancement of Science.)

Today, levees line the lower Mississippi River, confining the river and directing floodwaters, mineral sediments, and nutrients into the Gulf of Mexico, rather than into the coastal wetlands. Deprived of water, sediments, and nutrients, in a coastal environment where the sea level is rising, the coastal wetlands are being lost. The global sea level is rising 1 to 2 mm/yr as a result of processes that began at the end of the last ice age: the melting of glaciers and expansion of ocean waters as they warm. Regional and local subsidence in the Mississippi delta region combines with the global rise in sea level to produce a relative sea-level rise of about 12 mm/yr. To keep the coastal wetlands from declining, the rate of vertical accretion would thus need to be about 13 mm/yr. Currently, natural vertical accretion is only about 5 to 8 mm/yr.[20]

Most people agree that wetlands are valuable and productive for fish and wildlife. But wetlands are also valued as potential lands for agriculture, mineral exploitation, and building sites. Wetland management is drastically in need of new incentives for private landowners (who own the majority of several types of wetlands in the United States) to preserve wetlands rather than fill them in and develop the land.[18] Management strategies must also include careful planning to maintain the water quantity and quality necessary for wetlands to flourish or at least survive. Unfortunately, although laws govern the filling and draining of wetlands, no national wetland policy for the United States is in place. Debate continues as to what constitutes a wetland and how property owners should be compensated for preserving wetlands.[17, 21]

Restoration of Wetlands

A related management issue is wetlands restoration. A number of projects have attempted to restore wetlands, with varied success. The most important factor to be considered in most freshwater marsh restoration projects is the availability of water. If water is present, wetland soils and vegetation will likely develop. The restoration of salt marshes is more difficult because of the complex interactions among the hydrology, sediment supply, and vegetation that allow salt marshes to develop. Careful studies of relationships between the movement of sediment and the flow of water in salt marshes is providing information crucial to restoration, which makes successful reestablishment of salt marsh vegetation more likely. The restoration of wetlands has become an important topic in the United States because of the mitigation requirement related to environmental impact analysis, as set forth in the National Environmental Policy Act of 1969. According to this requirement, if wetlands are destroyed or damaged by a particular project, the developer must obtain or create additional wetlands at another site to compensate.[17] Unfortunately, the state of the art of restoration is not adequate to ensure that specific restoration projects will be successful.[22]

Constructing wetlands for the purpose of cleaning up agricultural runoff is an idea being implemented in areas with extensive agricultural runoff. Wetlands have a natural ability to remove excess nutrients, break down pollutants, and cleanse water. A series of wetlands are being created in Florida to remove nutrients (especially phosphorus) from agricultural runoff and thus help restore the Everglades to more natural functioning. The Everglades are a huge wetland ecosystem that functions as a wide, shallow river flowing south through southern Florida to the ocean. Fertilizers applied to farm fields north of the Everglades make their way directly into the Everglades by way of agricultural runoff, disrupting the ecosystem. (Phosphorus enrichment causes undesired changes in water quality and aquatic vegetation; see the discussion of eutrophication in the next chapter.) The man-made wetlands are designed to intercept and hold the nutrients so that they do not enter and damage the Everglades.[23]

In southern Louisiana, restoration of coastal wetlands includes the application of treated wastewater, which adds nutrients, nitrogen, and phosphorous to accelerate plant growth. As plants grow, organic debris (stems, leaves, and so forth) builds up on the bottom and causes the wetland to grow vertically. This growth helps offset a relative rise in sea level, maintaining and restoring the wetland.[20]

18.7 Dams and the Environment

Dams and their reservoirs generally are designed to be multifunctional. People who propose the construction of dams and reservoirs point out that they may be used for recreational activities and for generating electricity, as well as providing flood control and ensuring a more stable water supply. However, it is often difficult to reconcile these various uses at a given site. For example, water demands for agriculture might be high during the summer, resulting in a drawdown of the reservoir and leaving extensive mudflats or an exposed bank area subject to erosion (Figure 18.18). Recreational users find the low water level and the mudflats aesthetically displeasing. Also, high demand for water may cause quick changes in lake levels, which could interfere with wildlife (particularly fish) by damaging or limiting spawning opportunities. Another consideration is that dams and reservoirs tend to give a false sense of security to those living below them. Dams may fail. Flooding may originate from tributary rivers that enter the main river below a dam, and dams cannot be guaranteed to protect people against floods larger than those for which they have been designed.

FIGURE 18.18 Erosion along the shoreline of a reservoir in central California following release of water, exposing bare banks.

The environmental effects of dams are considerable and include the following:

- Loss of land, cultural resources, and biological resources in the reservoir area.
- Potential serious flood hazard, should larger dams and reservoirs fail.
- Storage behind the dam of sediment that would otherwise move downstream to coastal areas, where it would supply sand to beaches. The trapped sediment also reduces water storage capacity, limiting the life of the reservoir.
- Downstream changes in hydrology and in sediment transport that change the entire river environment and the organisms that live there.
- Fragmentation of ecosystems above and below a dam.
- Restricted movement upstream and downstream of organic material, nutrients, and aquatic organisms.

For a variety of reasons—including displacement of people, loss of land, loss of wildlife, and permanent, adverse changes to river ecology and hydrology—many people today are vehemently opposed to turning remaining rivers into a series of reservoirs with dams. In the United States, several dams have been removed, and the removal of others is being considered because of the environmental damage they are causing. In contrast, China recently constructed the world's largest dam. Three Gorges Dam, on the Yangtze River (Figure 18.19), has drowned cities, farm fields, important archaeological sites, and highly scenic gorges and has displaced approximately 2 million people from their homes. In the river, rare freshwater dolphins called *Baiji* (by legend the reincarnated 3rd-century

FIGURE 18.19 Three Gorges on the Yangtze River is a landscape of high scenic value. Shown here is the Wu Gorge, near Wushan, one of the gorges flooded by the water in the reservoir.

Chinese princess who symbolizes peace, prosperity, and love) are functionally extinct. A few may still exist, but scientists believe recovery is not possible. On land, habitats were fragmented and isolated as mountaintops became islands in the reservoir.

The dam, which is approximately 185 m high and more than 1.6 km wide, produces a reservoir nearly 600 km long. Raw sewage and industrial pollutants that are discharged into the river are deposited in the reservoir, and there is concern that the reservoir will become seriously polluted. Since the reservoir has been filling for several years, the banks are becoming saturated, increasing the landslide hazard. Large ships make matters worse by generating waves that increase shoreline erosion and cause the rock slopes and shoreline homes to vibrate and shake. Some older homes are thought to be unsafe due to the landslide hazard that has evidently increased since the reservoir began filling. In addition, the Yangtze River has a high sediment load, and it is feared that damage to deepwater shipping harbors will eventually occur at the upstream end of the reservoir, where sediments now being deposited are producing shallower water.

The dam may also give people living downstream a false sense of security. Should the dam fail, downstream cities, such as Wushan, with a population of several million people, might be submerged, with catastrophic loss of life.[24] The dam may also encourage further development in flood-prone areas, which will be damaged or lost if the dam and reservoir are unable to hold back floods in the future. If this happens, loss of property and life from flooding may be greater than if the dam had not been built. Contributing to this problem is the dam's location in a seismically active region where earthquakes and large landslides have been common in the past.

A positive attribute of the giant dam and reservoir is the capacity to produce about 18,000 MW of electricity, the equivalent of about 18 large coal-burning power plants. As pointed out in earlier discussions, pollution from coal burning is a serious problem in China. Some of the dam's opponents have pointed out, however, that a series of dams on tributaries to the Yangtze River could have produced similar electric power without causing environmental damage to the main river.[25]

The Glen Canyon Dam on the Colorado River was completed in 1963. From a hydrologic viewpoint, the Colorado River has been changed by the dam. The river has been tamed. The higher flows have been reduced, the average flow has increased, and the flow changes often because of fluctuating needs to generate electrical power. Changing the hydrology of the river has also changed other aspects, including the rapids, the distribution of sediments that form sandbars (called beaches by rafters; Figure 18.20), and the vegetation near the water's edge.[26]

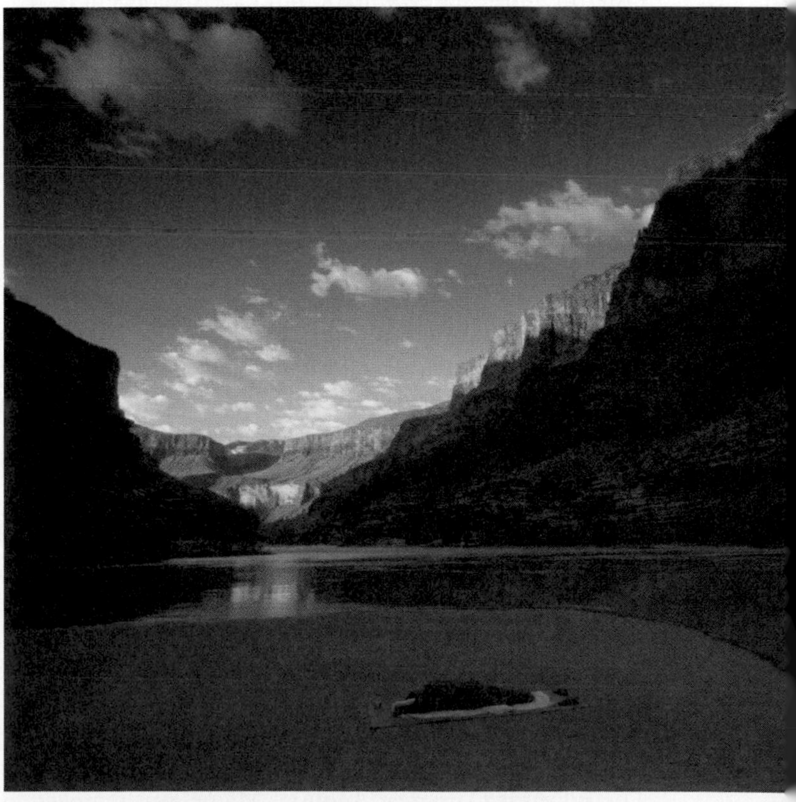

FIGURE 18.20 A sandbar in the Colorado River below Glen Canyon Dam in the Grand Canyon. Releases of relatively large flows in recent years are designed to help maintain the sandbar.

The sandbars, valuable wildlife habitats, shrank in size and number following construction of the dam because sediment that would have moved downstream to nourish them was trapped in the reservoir. All these changes affect the Grand Canyon, which is downstream from the dam. In an effort to restore part of the sand flow and maintain sandbars, releases from the dam are periodically increased to more closely match natural pre-dam flows.[27, 28] The higher flows have helped maintain sandbars by mobilizing sand, but cannot be expected to restore the river to pre-dam conditions.

There is little doubt that if our present water-use practices continue, we will need additional dams and reservoirs, and some existing dams will be heightened to increase water storage. However, there are few acceptable sites for new dams, and conflicts over the construction of additional dams and reservoirs are bound to occur. Water developers may view a canyon dam site as a resource for water storage, whereas others may view that canyon as a wilderness area and recreation site for future generations. The conflict is common because good dam sites are often sites of high-quality scenic landscape.

Dams also have an economic aspect: They are expensive to build and operate, and they are often constructed with federal tax dollars in the western United States, where they provide inexpensive subsidized water for agriculture. This has been a point of concern to some taxpayers in the eastern United States, who do not have the benefit of federally subsidized water. Perhaps a different pricing structure for water would encourage conservation, and fewer new dams and reservoirs would be needed.

Removal of Dams

Many dams in the United States have outlived their original purposes or design lives. These are generally small structures that now are viewed as a detriment to the ecological community of the river. The dams fragment river ecosystems by producing an upstream-of-dam environment and a downstream environment. Often, the structure blocks upstream migration of threatened or endangered fish species (for example, salmon in the Pacific Northwest). The perceived solution is to remove the dam and restore the river's more natural hydrology and ecological functions. However, removal must be carefully planned and executed to ensure that the removal doesn't itself cause ecological problems, such as sudden release of an unacceptable amount of sediment to the river or contaminated sediment.[29]

A large number of U.S. dams (mostly small ones) have been removed or are in the planning stages for removal. The Edwards Dam near Augusta, Maine, was removed in 1999, opening about 29 km (18 mi) of river habitat to migrating fish, including Atlantic salmon, striped bass, shad, alewives, and Atlantic sturgeon. The Kennebec River came back to life as millions of fish migrated upstream for the first time in 160 years.[30]

The Marmot Dam on the Sandy River in northwest Oregon was removed in 2007 (Figure 18.21). The dam was 15 m (45 ft) high, 50 m (150 ft) wide, and was filled with 750,000 cubic meters of sand and gravel. The removal was a scientific experiment and provided useful information for future removal projects. Salmon again swim up the river and spawn. People are kayaking in stretches where the river hadn't been run by a boat in almost 100 years.[31]

The Elwha and Glines Canyon dams, constructed in the early 18th century on the Elwha River in Washington State's Puget Sound, are scheduled for removal beginning in 2012. The largest of the two is the Glines Canyon Dam, which is about 70 m (210 ft) high (the highest dam ever removed). The Elwha headwater is in Olympic National Park, and prior to the dams' construction it supported large salmon and steelhead runs. Denied access to almost their entire spawning habitat, fish populations there have declined greatly. Large runs of fish had also brought nutrients from the ocean to the river and landscape. Bears, birds, and other animals used to eat the salmon and transfer nutrients to the forest ecosystem. Without the salmon, both wildlife and forest suffer. The dams also keep sediment from reaching the sea. Denied sediment, beaches at the river's mouth have eroded, causing the loss of clam beds. The dams will be removed in stages to minimize downstream impacts from the release of sediment. With the dams removed, the river will flow freely for the first time in a century, and it is hoped that the ecosystem will recover and the fish will return in greater numbers.[32]

The Matilija Dam, completed by California's Ventura County in 1948, is about 190 m (620 ft) wide and 60 m (200 ft) high. The structure is in poor condition, with leaking, cracked concrete and a reservoir nearly filled with sediment. The dam serves no useful purpose and blocks endangered southern steelhead trout from their historical spawning grounds. The sediment trapped behind the dam also reduces the natural nourishment of sand on beaches and increases coastal erosion.

The removal process began with much fanfare in October 2000, when a 27 m (90 ft) section was removed from the top of the dam. The entire removal process may take years, after scientists have determined how to safely remove the sediment stored behind the dam. If released quickly, it could damage the downstream river environment, filling pools and killing river organisms such as fish, frogs, and salamanders. If the sediment can be slowly released in a more natural manner, the downstream damage can be minimized.

(a) (b)

FIGURE 18.21 The concrete Marmot Dam before **(a)** and during **(b)** removal in 2007.

The cost of the dam in 1948 was about $300,000. The cost to remove the dam and sediment will be more than ten times that amount.

The perception of dams as permanent edifices, similar to the pyramids of Egypt, has clearly changed. What is learned from studying the removal of the Edwards Dam in Maine, the Marmot Dam in Oregon, and the Matilija Dam in California will be useful in planning other dam-removal projects. The studies will also provide important case histories to evaluate ecological restoration of rivers after removal of dams. In sum, removing dams is simple in concept, but involves complex problems relating to sediment and water. It provides an opportunity to restore ecosystems, but with that opportunity comes responsibility.[31]

18.8 Global Water Shortage Linked to Food Supply

As a capstone to this chapter, we present the hypothesis that we are facing a growing water shortage linked to our food supply. This is potentially a very serious problem. In the past few years, we have begun to realize that isolated water shortages are apparently indicators of a global pattern.[33] At numerous locations on Earth, both surface water and groundwater are being stressed and depleted:

- Groundwater in the United States, China, India, Pakistan, Mexico, and many other countries is being mined (used faster than it is being renewed) and is therefore being depleted.
- Large bodies of water—for example, the Aral Sea—are drying up (see earlier Figures 18.8–10).
- Large rivers, including the Colorado in the United States and the Yellow in China, do not deliver any water to the ocean in some seasons or years. Others, such as the Nile in Africa, have had their flow to the ocean greatly reduced.

Water demand during the past half-century has tripled as the human population more than doubled. In the next half-century, the human population is expected to grow by another 2 to 3 billion. There is growing concern that there won't be enough water to grow the food to feed the 8–9 billion people expected to be inhabiting the planet by the year 2050. Therefore, a food shortage linked to water resources seems a real possibility. The problem is that our increasing use of groundwater and surface water for irrigation has allowed for increased food production—mostly crops such as rice, corn, and soybeans. These same water resources are being depleted, and as water shortages for an agricultural region occur, food shortages may follow. Water is also linked to energy because irrigation water is often pumped from groundwater. As the cost of energy rises, so does the cost of food and the difficulty of purchasing food, especially in poor countries. This scenario in 2007-2008 resulted in a number of food riots in over 30 countries including Haiti, Mexico and west Bengal.

The way to avoid food shortages caused by water-resource depletion is clear: We need to control human population growth and conserve and sustain water resources. In this chapter, we have outlined a number of ways to conserve, manage, and sustain water. The good news is that a solution is possible—but it will take time, and we need to be proactive, taking steps now before significant food shortages develop. For all the reasons discussed, one of the most important and potentially serious resource issues of the 21st century is water supply and management.

CRITICAL THINKING ISSUE

What Is Your Water Footprint?

The concept of an environmental footprint has been developed in recent years to become a part of environmental science. The footprint, as it has evolved, is a quantitative measure of how our use of natural resources is affecting the environment. There are three main environmental footprints:

- The ecological footprint is a measure of the biologically productive space, measured in hectares, that an individual, community, or country uses.
- The carbon footprint is a measure of the amount of greenhouse gases produced and emitted into the atmosphere by the activities of an individual, group of people, or nation.
- The water footprint is a measure of water use by an individual, community, or country, measured in cubic meters of water per year.

Here, we are concerned with the water footprint, which measures direct water use by people and as such involves virtual water. The water footprint is somewhat different from the other two footprints in that water is a variable resource in terms of where it is found on Earth and its availability.

Water is the most abundant fluid in the crust of the Earth, and we use lots of it for drinking and washing our clothes, dishes, and ourselves. However, even more water is used in producing the goods and the services that our society wishes. The water footprint of an individual, group of people, or even a country is defined as the total volume of freshwater, in cubic meters, used per year to produce the goods or services that the individual, group, or country uses.[34]

The water footprint is closely related to our society's consumption of crops, goods, and services. As such, it is also a measure of environmental stress. The basic assumption is that there is a finite amount of water, and the more we use, the less is available for ecosystems and other purposes. Also, the more water we use for industrial processes, the likelier we are to use more of our other resources, such as energy and materials. The general idea with the water footprint is that by examining it we can better understand how we actually use water, which could lead to better water management and conservation. The water footprint includes both direct and indirect uses of water. Of the two, the indirect or virtual water is often much larger than the amount of water directly consumed (taken up in tissue)—for example, for personal consumption or growing crops.

Calculating the water footprint is a quantitative exercise based on gathering extensive data about how much water is used for agriculture, industry, and other activities. Based on water-use data and analysis, average amounts of water use can be estimated, and these data are used to generate an individual's water footprint. To estimate your personal water footprint, you can go to a water-footprint calculator, such as that on the Web site *waterfootprint.org*. In this Critical Thinking exercise, go to that Web site and use the water-footprint calculator for two scenarios: (1) the quick individual water footprint; and (2) the extended individual water footprint. For income, put in the amount of money you spend per year for your college education, plus whatever other money you earn. The quick individual water-footprint calculation involves very few variables, whereas the extended calculator includes a number of variables relating to how you eat and your personal lifestyle.

After you have calculated your personal water footprint, try some experiments by substituting at least three or four other countries for the country in which you live. These countries

Table 18.3 WATER FOOTPRINTS FOR SELECTED COUNTRIES (ARRANGED FROM LOWEST TO HIGHEST WATER FOOTPRINT PER PERSON)

COUNTRY	APPROXIMATE POPULATION (MILLIONS)[a]	TOTAL WATER FOOTPRINT (GM^3/YR)[b]	WATER FOOTPRINT PER PERSON (M^3/YR)	GROSS DOMESTIC PRODUCT PER PERSON (1,000S US$)[c]
Afghanistan	26	17.3	660	0.4
China	1258	883.4	702	3
South Africa	42	39.5	931	6
India	1007	987.4	980	1
Egypt	63	69.5	1097	2
Japan	127	146.1	1153	37
United Kingdom	59	73.1	1245	44
Mexico	97	140.2	1441	10
Switzerland	7	12.1	1682	67
France	59	110.2	1875	46
Spain	40	94.0	2325	35
USA	280	696	2483	47

[a] Population (approx.) in 2000

[b] $1Gm^3$=1 billion cubic meters

[c] Gross domestic product (GDP) per person in 2008. International Monetary Fund 2008. *Source:* National Water Footprints: Statistics. Accessed August 25, 2009 at www.waterfootprint.org.

should range from some that are wealthier than the United States (such as Switzerland) and some that are poorer (such as China). By "rich" and "poor," we are not referring to absolutes of happiness or any other measure, but only the median income for an individual (see Table 18.3).

Critical Thinking Questions

1. How well do you think the variables in the extended individual footprint characterize your water use?
2. Do you think the water footprint you calculated is a useful concept to better understand water resources?
3. In evaluating your individual water footprint living in the United States versus several other countries, what is actually controlling the footprint that you produced? Why is individual income or GDP per person apparently so important?
4. Has calculating your personal water footprint led you to a better understanding of some of the components of water use? What could you do to reduce your water footprint?

SUMMARY

- Water is a liquid with unique characteristics that have made life on Earth possible.
- Although it is one of the most abundant and important renewable resources on Earth, more than 99% of Earth's water is unavailable or unsuitable for beneficial human use because of its salinity or location.
- The pattern of water supply and use at any particular point on the land surface involves interactions and linkages among the biological, hydrological, and rock cycles. To evaluate a region's water resources and use patterns, a water budget is developed to define the natural variability and availability of water.
- It is expected that during the next several decades the total water withdrawn from streams and groundwater in the United States will decrease slightly, but that the consumptive use will increase because of greater demands from our growing population and industry.

- Water withdrawn from streams competes with instream needs, such as maintaining fish and wildlife habitats and navigation, and may therefore cause conflicts.
- Groundwater use has led to a variety of environmental problems, including loss of vegetation along watercourses, and land subsidence.
- Because agriculture is the biggest user of water, conservation of water in agriculture has the most significant effect on sustainable water use. However, it is also important to practice water conservation at the personal level in our homes and to price water in a way that encourages conservation and sustainability.
- There is a need for a new philosophy in water-resource management that considers sustainability and uses creative alternatives and variable sources. A master plan must include normal sources of surface water and groundwater, conservation programs, and use of reclaimed water.
- Development of water supplies and facilities to more efficiently move water may cause considerable environmental degradation; construction of dams and reservoirs should be considered carefully in light of potential environmental impacts.
- Removal of dams as a way to reconnect river ecosystems is becoming more common.
- The concepts of virtual water and the water footprint are becoming important in managing water resources at the regional to global level.
- Wetlands serve a variety of functions at the ecosystem level that benefit other ecosystems and people.
- We are facing a growing global water shortage linked to the food supply.
- Water supply and management is one of the major resource issues of the 21st century.

REEXAMINING THEMES AND ISSUES

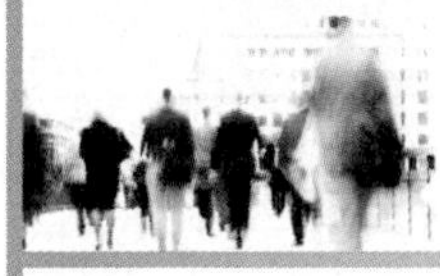

Human Population

As the human population has increased, so has the demand for water resources. As a result, we must be more careful in managing Earth's water resources, particularly near urban centers.

Sustainability

Our planet's water resources are sustainable, provided we manage them properly and they are not overused, polluted, or wasted. This requires good water-management strategies. We believe that the move toward sustainable water use must begin now if we are to avoid conflicts in the future. Principles of water management presented in this chapter help delineate what needs to be done.

Global Perspective

The water cycle is one of the major global geochemical cycles. It is responsible for the transfer and storage of water on a global scale. Fortunately, the total abundance of water on Earth is not a problem. However, ensuring that it is available when and where it is needed in a sustainable way *is* a problem.

Urban World

Although urban areas consume only a small portion of the water resources used by people, it is in urban areas that shortages are often most apparent. Thus, the concepts of water management and water conservation are critical in urban areas.

People and Nature

To many people, water is an icon of nature. Waves crashing on a beach, water flowing in a river or over falls or shimmering in lakes have inspired poets and countless generations of people to connect with nature.

Science and Values

Conflicts result from varying values related to water resources. We value natural areas such as wetlands and free-running rivers, but we also want water resources and protection from hazards such as flooding. As a result, we must learn to align more effectively with nature to minimize natural hazards, maintain a high-quality water resource, and provide the water necessary for the ecosystems of our planet. The relatively high, controlled releases of the Colorado River discussed in this chapter are examples of new river-management practices that embody both scientific knowledge and social values—in this case, scientific understanding of river processes and the wish to sustain the Colorado as a vibrant, living river.

KEY TERMS

consumptive use **376**
desalination **376**
effluent stream **372**
groundwater **372**
influent stream **372**
in-stream use **376**
off-stream use **376**
overdraft **375**
sustainable water use **384**
virtual water **386**
water budget **373**
water conservation **380**
wetlands **387**

STUDY QUESTIONS

1. If water is one of our most abundant resources, why are we concerned about its availability in the future?
2. Which is more important from a national point of view, conservation of water use in agriculture or in urban areas? Why?
3. Distinguish between in-stream and off-stream uses of water. Why is in-stream use controversial?
4. What are some important environmental problems related to groundwater use?
5. How might your community better manage its water resources?
6. Discuss how the concept of virtual water is related to water conservation and management at the global level.
7. What are some of the major environmental impacts associated with the construction of dams? How might these be minimized?
8. What are the most important factors in planning to remove a dam?
9. How can we reduce or eliminate the growing global water shortage? Do you believe the shortage is related to our food supply? Why? Why not?
10. Why is water such an important resource issue?

FURTHER READING

Gleick. P.H, Global freshwater resources: Soft-path solutions for the 21st century, *Science* 302(2003):1524–1528.

Gleick, P.H., *The World's Water 2000–2001* (Washington, DC.: Island Press, 2000).

Hoekstra, A.Y., and A.K. Chapagain, *Globalization of Water* (Malden, MA: Blackwell Publishing, 2008).

James, W., and J. Neimczynowicz, eds., *Water, Development and the Environment* (Boca Raton, FL: CRC Press, 1992). Covers problems with water supplies imposed by a growing population, including urban runoff, pollution and water quality, and management of water resources.

La Riviere, J.W.M. Threats to the world's water, *Scientific American* 261, no. 3 (1989): 80–84. Summary of water supply and demand and threats to continued supply.

Spulber, N., and A. Sabbaghi, *Economics of Water Resources: From Regulation to Privatization* (London: Kluwer Academic, 1994). Discussions of water supply and demand, pollution and its ecological consequences, and water on the open market.

Twort, A.C., F.M. Law, F.W.Crowley, and D.D.Ratnayaka, *Water Supply,* 4th ed. (London: Edward Arnold, 1994). Good coverage of water topics from basic hydrology to water chemistry, and water use, management, and treatment.

Wheeler, B.D., S.C. Shaw, W.J. Fojt, and R.A.Robertson, *Restoration of Temperate Wetlands* (New York: Wiley, 1995). Discussions of wetland restoration around the world.

CHAPTER 19

Water Pollution and Treatment

Dredging began in 2009 to remove contaminant sediment from the Hudson River in New York.

LEARNING OBJECTIVES

Degradation of our surface-water and groundwater resources is a serious problem. Although all of its effects are not yet fully known, we can and should begin taking steps to treat water and to minimize pollution. After reading this chapter, you should understand . . .

- What constitutes water pollution and what the major categories of pollutants are;
- Why the lack of disease-free drinking water is the primary water-pollution problem in many locations around the world;
- How point and nonpoint sources of water pollution differ;
- What biochemical oxygen demand is, and why it is important;
- What eutrophication is, why it is an ecosystem effect, and how human activity can cause it;
- Why sediment pollution is a serious problem;
- What acid mine drainage is, and why it is a problem;
- How urban processes can cause shallow-aquifer pollution;
- What the various methods of wastewater treatment are, and why some are environmentally preferable to others;
- Which environmental laws protect water resources and ecosystems.

CASE STUDY

America's "First River": A Success Story

The Hudson River is sometimes referred to as America's first river. It is named after Henry Hudson, who sailed up the river from the Atlantic in 1609 looking for a route to Asia. Native Americans, however, called the river by a name meaning "flows in two directions." This is because, for much of its lower course, the river ebbs and flows with the ocean tides, flowing upstream, then downstream, and sometimes part of the river is flowing up while part is flowing down. The total length of the river from the Adirondack Mountains to the Atlantic Ocean is just over 480 kilometers, but it's the lower 160 kilometers that has gained the most attention. Before emptying into New York Harbor, the lower Hudson flows past New Jersey communities such as Fort Lee, Union City, and Hoboken on its western shore and Mt Vernon, Yonkers, and Manhattan Island on its eastern shore. Farther upstream, about 50 to 150 km from Manhattan, the river flows through the scenic Hudson River Highlands. The high hills that border the river have a core of ancient hard igneous and metamorphic rock that erosion has carved into beautiful scenery, such as Storm King Mountain.

The story of the Hudson River and the environment go back to the 1800s. The nation's first military academy, West Point, was established along the river in 1802, and during the War of 1812 industrial activity sprang up along the river. A foundry opened near West Point to manufacture such products as cannonballs, pipes, and railroad engines. The foundry closed after about 100 years, but other factories became established along the river, including the Anaconda Wire and Cable Company. That plant closed in 1974, leaving behind a legacy of toxic pollution that helped turn the Hudson River into an environmental battleground in the 20th century between people who revered the river and those deemed responsible for making it unsafe to swim in its waters and making the fish that lived in it unsafe to eat.[1] In one of the nation's earliest battles to eliminate water pollutants, activists and others in the early 1970s sued the company, which was fined about $200,000, a very large fine for pollution violations at that time.

Those who thought that from then on the river would be clean again for future generations were mistaken: From around 1950 to 1977, General Electric discharged (dumped) over a million pounds of polychlorinated biphenyls (PCBs) into the Hudson River from two manufacturing plants. PCBs are highly stable man-made chemical compounds produced by combining chlorine and biphenyl (an organic compound). Because they are good electrical insulators and were considered safe, PCBs were widely used to prevent fires in electric transformers and capacitors. GE operated the plants for decades. Then, in the early 1970s, the story of the catastrophe unfolding in the Hudson River became common knowledge. Commercial fishing for striped bass and other fish was banned in 1976, and fishermen blamed GE for destroying a fishing industry, a way of life, and a culture that had been going on for centuries in the river valley.

PCBs were found to cause liver disease and are a suspected carcinogen in humans and a known carcinogen in other animals. PCBs were found to be persistent in the environment and entered the food chain to damage the river ecosystem, especially fish and invertebrates. As a result, they were banned in the United States in 1977. During this period, environmentalism became important and federal water laws were passed, including the Clean Water Act of 1972. In that year the U.S. government started dealing with hazardous waste and passed the Resource Conservation and Recovery Act, followed a few years later by the Comprehensive Environmental Response Compensation and Liability Act of 1980, which established the so-called Superfund to clean up several hundred of the most hazardous sites in the country.

At the top of the list was the Hudson River, and in 1983 a roughly 300-kilometer reach of the river was classified as the largest and one of the most serious Superfund sites in the country. The new law also changed the way the federal government dealt with industries that polluted the environment. Companies became responsible for their previous pollution of Superfund sites and liable for cleanup.

Today, however, over 100 tons of PCBs are still in the Hudson River sediments, with concentrations thousands of times greater than what is considered safe. PCBs accumulate in food chains in a process known as bioaccumulation. For example, organisms in the sediment contain a concentration of PCBs that increases as these organisms are eaten by fish, and these concentrations may increase further when the fish are eaten by predators such as eagles or people. Health advisories were issued in the mid-1970s and these remain in place today, warning women and children not to eat fish from the Hudson River.

A battle raged on as to what should be done about the PCBs. The two major alternatives were either to dredge areas where PCB concentrations are particularly high or just let natural processes in the river clean up the PCBs. The second option assumed that the sources of PCBs have been nearly eliminated and that the river would naturally cleanse itself of approximately half of the PCBs in three or four years.[2,3] Those arguing for dredging said the pollution was far too great to leave the cleanup to natural processes, and that dredging would greatly shorten the time necessary for the river to clean itself. To dredge or not to dredge was a several-hundred-million-dollar question. General Electric spent millions in an attempt to avoid spending hundreds of millions to clean up the river by dredging. The company said dredging would just stir up the PCBs in the riverbed, moving them up into the water and thus into the food chain.

The issue was settled in 2001: General Electric would have to pay several hundred million dollars to clean up the river by dredging. The work on mapped PCB hotspots began in 2009, using barges to dredge the contaminated sediment from the river bottom and place it in hopper barges. From there it is sent to a processing facility, and then finally transported by train about 3,000 kilometers to a waste-disposal site in west Texas. The cleanup is expected to take until 2015, ending an era of water pollution and toxic legacy in the Hudson.

The lower Hudson River Valley is urbanizing. There are more parking lots and cars than ever before. There is concern that runoff from streets and parking lots and urban houses will lead to a new wave of urban pollution. On the other hand, more and more people are experiencing the Hudson River in very positive ways. Numerous river groups focus their time and effort on cleaning up the river and promoting activities such as boating, hiking, and bird-watching. Parks of all sizes are being established at scenic sites along the river, factories have been removed, making room for some of the new parks, and there is an attempt to join these together in a greenbelt that would stretch many miles up the river from Manhattan. Some of the river culture from times past is reappearing.

In sum, the future of the Hudson River seems secure as progress in its cleanup and preservation continues. Some say that modern environmentalism was born on the Hudson River, one of the few American rivers to be designated an American Heritage River. Many people made important personal sacrifices to their careers, reputations, and livelihoods to protect the river.

These people truly revered the Hudson River and were in the forefront of fighting to protect our natural environment. An organization known as Scenic Hudson led the fight to protect the river. It was joined in 1969 by Clearwater, which included the folksinger, activist, and environmentalist Pete Seeger. Clearwater built a sloop that took people up and down the river, educating them on environmental concerns, fighting to control and eliminate pollution, and encouraging river restoration. Both Scenic Hudson and Clearwater remain active today.

The story of PCB pollution is a powerful reminder that individuals and groups can make a difference in correcting past environmental errors and working toward sustainability.

19.1 Water Pollution

Water pollution refers to degradation of water quality. In defining pollution, we generally look at the intended use of the water, how far the water departs from the norm, its effects on public health, or its ecological impacts. From a public-health or ecological view, a pollutant is any biological, physical, or chemical substance that, in an identifiable excess, is known to be harmful to desirable living organisms. Water pollutants include heavy metals, sediment, certain radioactive isotopes, heat, fecal coliform bacteria, phosphorus, nitrogen, sodium, and other useful (even necessary) elements, as well as certain pathogenic bacteria and viruses. In some instances, a material may be considered a pollutant to a particular segment of the population, although it is not harmful to other segments. For example, excessive sodium as a salt is not generally harmful, but it may be harmful to people who must restrict salt intake for medical reasons.

Today, the world's primary water-pollution problem is a lack of clean, disease-free drinking water. In the past, epidemics (outbreaks) of waterborne diseases such as cholera have killed thousands of people in the United States. Fortunately, we have largely eliminated epidemics of such diseases in the United States by treating drinking water prior to consumption. This certainly is not the case worldwide, however. Every year, several billion people are exposed to waterborne diseases. For example, an epidemic of cholera occurred in South America in the early 1990s, and outbreaks of waterborne diseases continue to be a threat even in developed countries.

Many different processes and materials may pollute surface water or groundwater. Some of these are listed in Table 19.1. All segments of society—urban, rural, industrial, agricultural, and military—may contribute to the problem of water pollution. Most of it results from runoff and leaks or seepage of pollutants into surface water or groundwater. Pollutants are also transported by air and deposited in bodies of water.

Increasing population often results in the introduction of more pollutants into the environment as well as greater demands on finite water resources.[4] As a result, we can expect sources of drinking water in some locations to be degraded in the future.[5,6]

Table 19.1 SOME SOURCES AND PROCESSES OF WATER POLLUTION

SURFACE WATER	GROUNDWATER
Urban runoff (oil, chemicals, organic matter, etc.) (U, I, M) Agricultural runoff (oil, metals, fertilizers, pesticides, etc.) (A)	Leaks from waste-disposal sites (chemicals, radioactive materials, etc.) (I, M)
Accidental spills of chemicals including oil (U, R, I, A, M)	Leaks from buried tanks and pipes (gasoline, oil, etc.) (I, A, M)
Radioactive materials (often involving truck or train accidents) (I, M)	Seepage from agricultural activities (nitrates, heavy metals, pesticides, herbicides, etc.) (A)
Runoff (solvents, chemicals, etc.) from industrial sites (factories, refineries, mines, etc.) (I, M)	Saltwater intrusion into coastal aquifers (U, R, I, M) Seepage from cesspools and septic systems (R)
Leaks from surface storage tanks or pipelines (gasoline, oil, etc.) (I, A, M)	Seepage from acid-rich water from mines (I) Seepage from mine waste piles (I)
Sediment from a variety of sources, including agricultural lands and construction sites (U, R, I, A, M)	Seepage of pesticides, herbicide nutrients, and so on from urban areas (U)
Air fallout (particles, pesticides, metals, etc.) into rivers, lakes, oceans (U, R, I, A, M)	Seepage from accidental spills (e.g., train or truck accidents) (I, M) Inadvertent seepage of solvents and other chemicals including radioactive materials from industrial sites or small businesses (I, M)

Key: U = urban; R = rural; I = industrial; A = agricultural; M = military.

The U.S. Environmental Protection Agency has set thresholds limiting the allowable levels for some (but not all) drinking-water pollutants. Because it is difficult to determine the effects of exposure to low levels of pollutants, thresholds have been set for only a small fraction of the more than 700 identified drinking-water contaminants. If the pollutant exceeds an established threshold, then the water is unsatisfactory for a particular use. Table 19.2 lists selected pollutants included in the national drinking-water standards for the United States.

Water withdrawn from surface or groundwater sources is treated by filtering and chlorinating before distribution to urban users. Sometimes it is possible to use the natural environment to filter the water as a service function, saving treatment cost (see A Closer Look 19.1).

The following sections focus on several water pollutants to emphasize principles that apply to pollutants in general. (See Table 19.3 for categories and examples of water pollutants.) Before proceeding to our discussion of pollutants, however, we first consider biochemical oxygen demand and dissolved oxygen. Dissolved oxygen is not a pollutant but rather is needed for healthy aquatic ecosystems.

Table 19.2 NATIONAL DRINKING-WATER STANDARDS

CONTAMINANT	MAXIMUM CONTAMINANT LEVEL (MG/L)
Inorganics	
Arsenic	0.05
Cadmium	0.01
Lead	0.015 action level[a]
Mercury	0.002
Selenium	0.01
Organic chemicals	
Pesticides	
Endrin	0.0002
Lindane	0.004
Methoxychlor	0.1
Herbicides	
2,4-D	0.1
2,4,S-TP	0.01
Silvex	0.01
Volatile organic chemicals	
Benzene	0.005
Carbon tetrachloride	0.005
Trichloroethylene	0.005
Vinyl chloride	0.002
Microbiological organisms	
Fecal coliform bacteria	1 cell/100 ml

[a] Action level is related to the treatment of water to reduce lead to a safe level. There is no maximum contaminant level for lead.

Source: U.S. Environmental Protection Agency.

A CLOSER LOOK 19.1

What Is the Value of Clean Water to New York City?

The forest of the Catskill Mountains in upstate New York (Figure 19.1) provides water to about 9 million people in New York City. The total contributing area in the forest is about 5,000 km^2 (2,000 square miles), of which the city of New York owns less than 8%. The water from the Catskills has historically been of high quality and in fact was once regarded as one of the largest municipal water supplies in the United States that did not require extensive filtering. Of course, what we are talking about here is industrial filtration plants, where the water enters from reservoirs and groundwater and is then treated before being dispersed to users.

In the past, the water from the Catskills has been filtered very effectively by natural processes. When rain or melting snow drips from trees or melts on slopes in the spring, some of it infiltrates the soil and moves down into the rocks below as groundwater. Some emerges to feed streams that flow into reservoirs. During its journey, the water enters into a number of physical and chemical processes that naturally treat and filter the water. These natural-service functions that the Catskill forest ecosystem provides to the people of New York were taken for granted until about the 1990s, when it became apparent that the water supply was becoming vulnerable to pollution from uncontrolled development in the watershed.

A particular concern was runoff from buildings and streets, as well as seepage from septic systems that treat wastewater from homes and buildings, partly by allowing it to seep through soil.

The city is also concerned that drilling for natural gas, that uses water and contaminates the groundwater, could damage surface water resources (see Chapter 15). Drilling for natural gas in the watershed supplying New York City was virtual banned in 2010.

The Environmental Protection Agency has warned that unless the water quality improved, New York City would have to build a water treatment plant to filter the water. The cost of such a facility was estimated at \$6–8 billion, with an annual operating expense of several hundred million dollars. As an alternative, New York City chose to attempt to improve the water quality at the source. The city built a sewage treatment plant upstate in the Catskill Mountains at a cost of about \$2 billion. This seems very expensive but was about one-third the cost of building the treatment plant to filter water. Thus, the city chose to invest in the "natural capital" of the forest, hoping that it will continue its natural service function of providing clean water. It will probably take several decades to tell whether New York City's gamble will work in the long term.[4]

There have been unanticipated benefits from maintaining the Catskill Mountain forest ecosystem. These benefits come from recreational activities, particularly trout fishing, which is a multibillion-dollar enterprise in upstate New York. In addition to the trout fishermen are people wanting to experience the Catskill Mountains through hiking, winter sports, and wildlife observation, such as bird-watching.

You might wonder why the city has been successful in its initial attempt to maintain high-quality water when it owns only about 8% of the land the water comes from. The reason is that the city has offered farmers, homeowners, and other people living in the forest financial incentives to maintain high-quality water resources. Although the amount of money is not large, it is sufficient to provide a sense of stewardship among the landowners, and they are attempting to abide by guidelines that help protect water quality.

New York isn't the only U.S. city that has chosen to protect watersheds to produce clean, high-quality drinking water rather than constructing and maintaining expensive water treatment plants. Others include Boston, Massachusetts; Seattle, Washington; and Portland, Oregon.

The main point of this story is that we mustn't undervalue the power of natural ecosystems to provide a variety of important services, including improved water and air quality.[4]

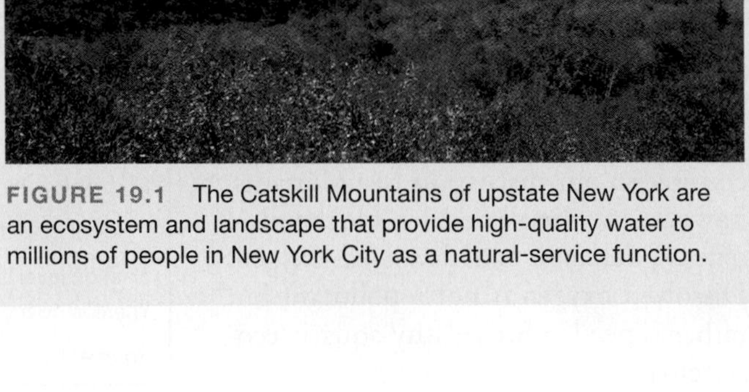

FIGURE 19.1 The Catskill Mountains of upstate New York are an ecosystem and landscape that provide high-quality water to millions of people in New York City as a natural-service function.

Table 19.3 CATEGORIES OF WATER POLLUTANTS

POLLUTANT CATEGORY	EXAMPLES OF SOURCES	COMMENTS
Dead organic matter	Raw sewage, agricultural waste, urban garbage	Produces biochemical oxygen demand and diseases.
Pathogens	Human and animal excrement and urine	Examples: Recent cholera epidemics in South America and Africa; 1993 epidemic of cryptosporidiosis in Milwaukee, Wisconsin. See discussion of fecal coliform bacteria in Section 22.3.
Drugs	Urban wastewater, painkillers, birth control pills, antidepressants, antibiotics	Pharmaceuticals flushed through our sewage treatment plants are contaminating our rivers and groundwater. Hormone residues or hormone mimickers are thought to be causing genetic problems in aquatic animals.
Organic chemicals	Agricultural use of pesticides and herbicides (Chapter 11); industrial processes that produce dioxin (Chapter 10)	Potential to cause significant ecological damage and human health problems. Many of these chemicals pose hazardous-waste problems (Chapter 23).
Nutrients	Phosphorus and nitrogen from agricultural and urban land use (fertilizers) and wastewater from sewage treatment	Major cause of artificial eutrophication. Nitrates in groundwater and surface waters can cause pollution and damage to ecosystems and people.
Heavy metals	Agricultural, urban, and industrial use of mercury, lead, selenium, cadmium, and so on (Chapter 10)	Example: Mercury from industrial processes that is discharged into water (Chapter 10). Heavy metals can cause significant ecosystem damage and human health problems
Acids	Sulfuric acid (H_2SO_4) from coal and some metal mines; industrial processes that dispose of acids improperly	Acid mine drainage is a major water pollution problem in many coal mining areas, damaging ecosystems and spoiling water resources.
Sediment	Runoff from construction sites, agricultural runoff, and natural erosion	Reduces water quality and results in loss of soil resources.
Heat (thermal pollution)	Warm to hot water from power plants and other industrial facilities	Causes ecosystem disruption (Chapter 10).
Radioactivity	Contamination by nuclear power industry, military, and natural sources (Chapter 17)	Often related to storage of radioactive waste. Health effects vigorously debated (Chapters 10 and 17).

19.2 Biochemical Oxygen Demand (BOD)

Dead organic matter in streams decays. Bacteria carrying out this decay use oxygen. If there is enough bacterial activity, the oxygen in the water available to fish and other organisms can be reduced to the point where they may die. A stream with low oxygen content is a poor environment for fish and most other organisms. A stream with an inadequate oxygen level is considered polluted for organisms that require dissolved oxygen above the existing level.

The amount of oxygen required for biochemical decomposition processes is called the **biological** or **biochemical oxygen demand (BOD)**. BOD is commonly used in water-quality management (Figure 19.2a). It measures the amount of oxygen consumed by microorganisms as they break down organic matter within small water samples, which are analyzed in a laboratory. BOD is routinely measured at discharge points into surface water, such as at wastewater treatment plants. At treatment plants, the BOD of the incoming sewage water from sewer lines is measured, as is water from locations both upstream and downstream of the plant. This allows comparison of upstream, or background, BOD, with the BOD of the water being discharged by the plant.

Dead organic matter—which produces BOD—enters streams and rivers from natural sources (such as dead leaves from a forest) as well as from agricultural runoff and urban sewage. Approximately 33% of all BOD in streams results from agricultural activities. However, urban areas, particularly those with older, combined sewer systems (in which stormwater runoff and urban sewage share the same line), also considerably increase BOD in streams. This is because during times of high flow, when sewage treatment plants are unable to handle the total volume of water, raw sewage mixed with storm runoff overflows and is discharged untreated into streams and rivers.

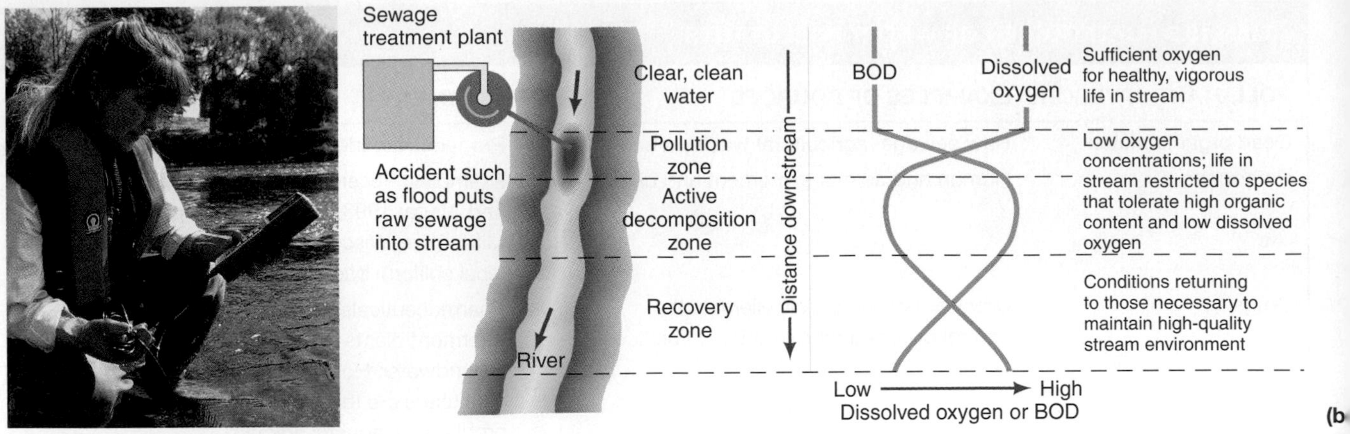

FIGURE 19.2 **(a)** Pollution-control officer measuring oxygen content of the River Severn near Shrewsbury, England. **(b)** The relationship between dissolved oxygen and biochemical oxygen demand (BOD) for a stream after the input of sewage.

When BOD is high, as suggested earlier, the *dissolved oxygen content* of the water may become too low to support life in the water. The U.S. Environmental Protection Agency defines the threshold for a water-pollution alert as a dissolved oxygen content of less than 5 mg/l of water. Figure 19.2b illustrates the effect of high BOD on dissolved oxygen content in a stream when raw sewage is introduced as a result of an accidental spill. Three zones are identified:

1. *A pollution zone,* where a high BOD exists. As waste decomposes, microorganisms use the oxygen, decreasing the dissolved oxygen content of the water.
2. *An active decomposition zone,* where the dissolved oxygen reaches a minimum owing to rapid biochemical decomposition by microorganisms as the organic waste is transported downstream.
3. *A recovery zone,* where dissolved oxygen increases and BOD is reduced because most of the oxygen-demanding organic waste from the input of sewage has decomposed and natural stream processes are replenishing the water's dissolved oxygen. For example, in quickly moving water, the water at the surface mixes with air, and oxygen enters the water.

All streams have some ability to degrade organic waste. Problems result when the stream is overloaded with oxygen-demanding waste, overpowering the stream's natural cleansing function.

19.3 Waterborne Disease

As mentioned earlier, the primary water-pollution problem in the world today is the lack of clean drinking water. Each year, particularly in less-developed countries, several billion people are exposed to waterborne diseases whose effects vary in severity from an upset stomach to death. As recently as the early 1990s, epidemics of cholera, a serious waterborne disease, caused widespread suffering and death in South America.

In the United States, we tend not to think much about waterborne illness. Although historically epidemics of waterborne disease killed thousands of people in U.S. cities, such as Chicago, public-health programs have largely eliminated such epidemics by treating drinking water to remove disease-carrying microorganisms and not allowing sewage to contaminate drinking-water supplies. As we will see, however, North America is not immune to **outbreaks**—or sudden occurrences—of waterborne disease.

Fecal Coliform Bacteria

Because it is difficult to monitor disease-carrying organisms directly, we use the count of **fecal coliform bacteria** as a standard measure and indicator of disease potential. The presence of fecal coliform bacteria in water indicates that fecal material from mammals or birds is present, so organisms that cause waterborne diseases may be present as well. Fecal coliform bacteria are usually (but not always) harmless bacteria that normally inhabit the intestines of all animals, including humans, and are present in all their waste. The EPA's threshold for swimming water is not more than 200 cells of fecal coliform bacteria per 100 ml of water; if fecal coliform is above the threshold level, the water is considered unfit for swimming (Figure 19.3). Water with *any* fecal coliform bacteria is unsuitable for drinking.

One type of fecal coliform bacteria, *Escherichia coli*, or *E. coli 0157*, has caused human illness and death. In the U.S., there are about 73,000 cases and 60 deaths per year from *E. coli 0157*. Outbreaks have resulted from eating contaminated meat (fecal transmission from humans and other animals) and drinking contaminated juices or water.[7-10] Table 19.4 lists some recent outbreaks of disease resulting from *E. coli 0157*.

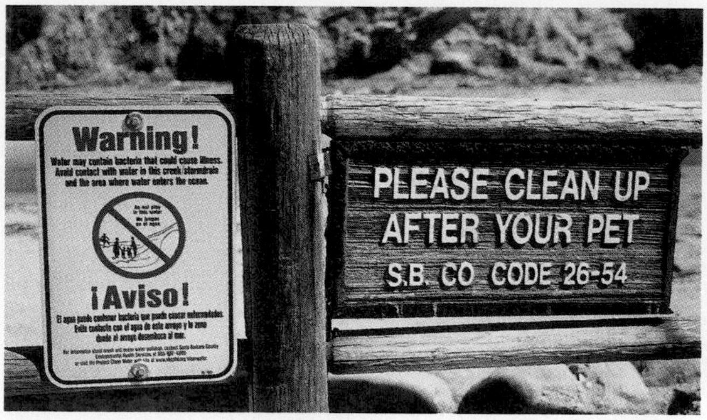

FIGURE 19.3 This beach in Southern California is occasionally closed as a result of contamination by bacteria.

FIGURE 19.4 Cattle feedlot in Colorado. High numbers of cattle in small areas have the potential to pollute both surface water and groundwater because of runoff and infiltration of urine.

19.4 Nutrients

Two important nutrients that cause water-pollution problems are phosphorus and nitrogen, and both are released from sources related to land use. Stream waters on forested land have the lowest concentrations of phosphorus and nitrogen because forest vegetation efficiently removes phosphorus and nitrogen. In urban streams, concentrations of these nutrients are greater because of fertilizers, detergents, and products of sewage treatment plants. Often, however, the highest concentrations of phosphorus and nitrogen are found in agricultural areas, where the sources are fertilized farm fields and feedlots (Figure 19.4). Over 90% of all nitrogen added to the environment by human activity comes from agriculture.

Eutrophication

Eutrophication is the process by which a body of water develops a high concentration of nutrients, such as nitrogen and phosphorus (in the forms of nitrates and phosphates). The nutrients increase the growth of aquatic plants in general, as well as production of photosynthetic blue-green bacteria and algae. Algae may form surface mats that shade the water and block light to algae below the surface, greatly reducing photosynthesis. The bacteria and algae die, and as they decompose, BOD increases, reducing the water's oxygen content, sometimes to the point where other organisms, such as fish, will die.[11, 12] They die not from phosphorus poisoning but from a chain of events that started with

Table 19.4 RECENT OUTBREAKS OF *E. COLI 0157* INFECTIONS[1]

YEAR	WHERE	SOURCE	COMMENT
1993	Washington State (fast-food restaurant)	Meat	5 children died; several hundred illnesses
1998	Georgia (water park)	Water in park pools	26 illnesses in children
1998	Town in Wyoming	Water supply	1 death
2000	Walkerton, Canada	Water supply	5 deaths
2006	23 states	Spinach	5 deaths; over 100 illnesses
2007	Hawaii (restaurant)	Lettuce	several illnesses (mostly tourists)
2009	Across the U.S.	Peanut butter	several deaths; several hundred illnesses
2009	29 states	Raw cookie dough	65 illnesses
2010	Several states, especially California, Colorado and N. Carolina	Raw eggs	at least 1,500 illnesses; 500 million eggs recalled

[a] *E. coli* 0157, a strain of *E. coli* bacteria, has been responsible for many human illnesses and deaths. *E. coli 0157* produces strong toxins in humans that may lead to bloody diarrhea, dehydration, kidney failure, and death.

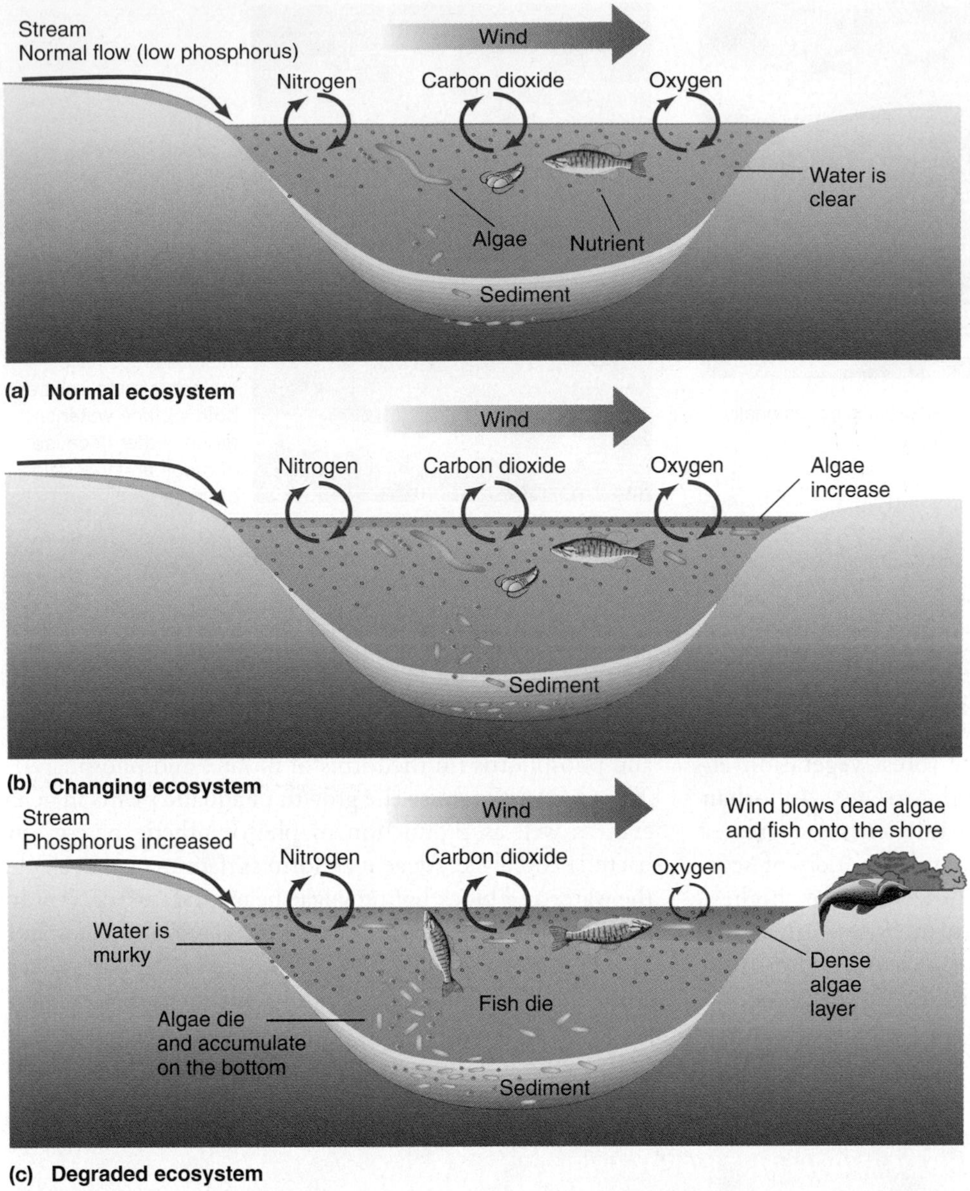

FIGURE 19.5 The eutrophication of a lake. **(a)** In an oligotrophic, or low-nutrient, lake, the abundance of green algae is low, the water clear. **(b)** Phosphorus is added to streams and enters the lake. Algae growth is stimulated, and a dense layer forms. **(c)** The algae layer becomes so dense that the algae at the bottom die. Bacteria feed on the dead algae and use up the oxygen. Finally, fish die from lack of oxygen.

the input of phosphorus and affected the whole ecosystem. The unpleasant effects result from the interactions among different species, the effects of the species on chemical elements in their environment, and the condition of the environment (the body of water and the air above it). This is what we call an **ecosystem effect**.

The process of eutrophication of a lake is shown in Figure 19.5. A lake that has a naturally high concentration of the chemical elements required for life is called a *eutrophic lake.* A lake with a relatively low concentration of chemical elements required by life is called an *oligotrophic lake*. The water in oligotrophic lakes is clear and pleasant for swimmers and boaters and has a relatively low abundance of life. Eutrophic lakes have an abundance of life, often with mats of algae and bacteria and murky, unpleasant water.

When eutrophication is accelerated by human processes that add nutrients to a body of water, we say that **cultural eutrophication** is occurring. Problems associated with the artificial eutrophication of bodies of water are not restricted to lakes (see A Closer Look 19.2). In recent years, concern has grown about the outflow of sewage from urban areas into tropical coastal waters and cultural eutrophication on coral reefs.[13, 14] For example, parts of the famous Great Barrier Reef of Australia, as well as some reefs that fringe the Hawaiian Islands, are being damaged by eutrophication.[15, 16] The damage to corals occurs as nutrient input stimulates algal growth on the reef, which smothers the coral.

The solution to artificial eutrophication is fairly straightforward and involves ensuring that high concentrations of nutrients from human sources do not enter lakes and other bodies of water. This can be accomplished by using phosphate-free detergents, controlling nitrogen-rich runoff from agricultural and urban lands, disposing of or reusing treated wastewater, and using more advanced water treatment methods, such as special filters and chemical treatments that remove more of the nutrients.

A CLOSER LOOK 19.2

Cultural Eutrophication in the Gulf of Mexico

Each summer, a so-called dead zone develops off the nearshore environment of the Gulf of Mexico, south of Louisiana. The zone varies in size from about 13,000 to 18,000 km^2 (5,000 to 7,000 mi^2), an area about the size of the small country of Kuwait or the state of New Jersey. Within the zone, bottom water generally has low concentrations of dissolved oxygen (less than 2 mg/l; a water-pollution alert occurs if the concentration of dissolved oxygen is less than 5 mg/l). Shrimp and fish can swim away from the zone, but bottom dwellers such as shellfish, crabs, and snails are killed. Nitrogen is believed to be the most significant cause of the dead zone (Figure 19.6).

The low concentration of oxygen occurs because the nitrogen causes cultural eutrophication. Algae bloom, and as the algae die and sink, their decomposition depletes the oxygen in the water. The source of nitrogen is believed to be in one of the richest, most productive agricultural regions of the world—the Mississippi River drainage basin.

The Mississippi River drains about 3 million km^2, which is about 40% of the land area of the lower 48 states. The use of nitrogen fertilizers in that area greatly increased beginning in the mid-20th century but leveled off in the 1980s and 1990s. The level of nitrogen in the river has also leveled off, suggesting that the dead zone may have reached its maximum size. This gives us time to study the cultural eutrophication problem carefully and make sound decisions to reduce or eliminate it.

We can partially reduce the amount of nitrogen (nitrates) reaching the Gulf of Mexico via the Mississippi River if we do the following:[12]

- Use fertilizers more effectively and efficiently.
- Restore and create river wetlands between farm fields and streams and rivers, particularly in areas known to contribute high amounts of nitrogen. The wetland plants use nitrogen, lowering the amount that enters the river.
- Implement nitrogen-reduction processes at wastewater treatment plants for towns, cities, and industrial facilities.
- Implement better flood control in the upper Mississippi River to confine floodwaters to floodplains, where nitrogen can be used by riparian vegetation.

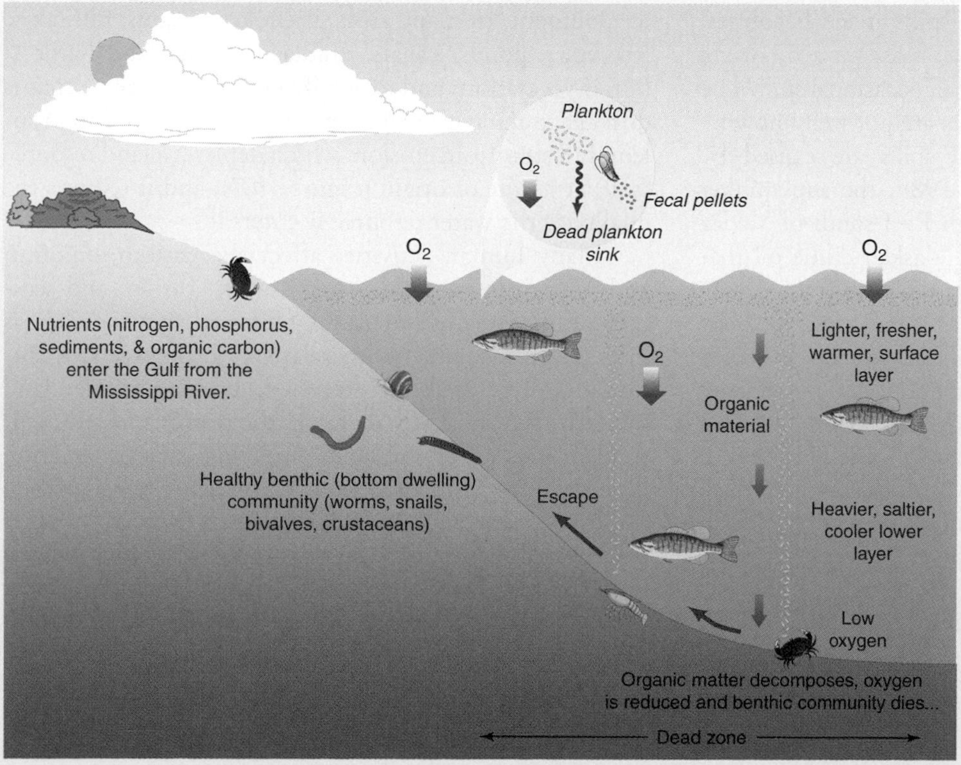

FIGURE 19.6 Idealized drawing showing some of the processes in the dead zone. Low oxygen from cultural eutrophication produces the dead zone. (*Source*: Modified after U.S. Environmental Protection Agency, www.epa.gov, accessed May, 30, 2005).

- Divert floodwater from the Mississippi to backwaters and coastal wetlands of the Mississippi River Delta. At present, levees in the delta push river waters directly into the gulf. Plants in coastal wetlands will use the nitrogen, reducing the concentration that reaches the Gulf of Mexico.

Better agricultural practices could reduce the amount of nitrogen reaching the Mississippi by up to 20%. But this would require reducing fertilizer use by about 20%, which farmers say would harm productivity. Still, restoring and creating river wetlands and riparian forests hold the promise of reducing nitrogen input to the river by up to 40%. This would require some combination of about 10 million hectares (24 million acres) of wetlands and forest, which is about 3.4% of the Mississippi River Basin.[11] That is a lot of land!

There is no easy solution to cultural eutrophication in the Gulf of Mexico. Clearly, however, we need to reduce the amount of nitrogen entering the Gulf. Also needed is a more detailed understanding of the nitrogen cycle within the Mississippi River basin and Delta. Gaining this understanding will require monitoring nitrogen and developing mathematical models of sources, sinks, and rates of nitrogen transfer. With an improved understanding of the nitrogen cycle, a management strategy to reduce or eliminate the dead zone can be put in place.

The dead zone in the Gulf of Mexico is not unique in the world. Other dead zones exist offshore of Europe, China, Australia, South America, and the northeastern United States. In all, about 150 dead zones in the oceans of the world have been observed. Most are much smaller than the zone in the Gulf of Mexico.

As with the Gulf, the other dead zones are due to oxygen depletion resulting from nitrogen, mostly from agricultural runoff. A few result from industrial pollution or runoff from urban areas, especially untreated sewage.

19.5 Oil

Oil discharged into surface water—usually in the ocean but also on land and in rivers—has caused major pollution problems. Several large oil spills from underwater oil drilling have occurred in recent years (for example the 2010 spill in the Gulf of Mexico, discussed in detail in Chapter 24). However, although spills make headlines, normal shipping activities probably release more oil over a period of years than is released by the occasional spill. The cumulative impacts of these releases are not well known.

Some of the best-known oil spills are caused by tanker accidents. On March 24, 1989, the supertanker *Exxon Valdez* ran aground on Bligh Reef south of Valdez in Prince William Sound, Alaska. Alaskan crude oil that had been delivered to the *Valdez* through the Trans-Alaska Pipeline poured out of the vessel's ruptured tanks at about 20,000 barrels per hour. The tanker was loaded with about 1.2 million barrels of oil, and about 250,000 barrels (11 million gal) entered the sound. The spill could have been larger than it was, but fortunately some of the oil in the tanker was offloaded (pumped out) into another vessel. Even so, the *Exxon Valdez* spill produced an environmental shock that resulted in passage of the Oil Pollution Act of 1990 and a renewed evaluation of cleanup technology.

The long-term effects of large oil spills are uncertain. We know that the effects can last several decades; toxic levels of oil have been identified in salt marshes 20 years after a spill.[17, 18]

19.6 Sediment

Sediment consisting of rock and mineral fragments—ranging from gravel particles greater than 2 mm in diameter to finer sand, silt, clay, and even finer colloidal particles—can produce a *sediment pollution* problem. In fact, by volume and mass, sediment is our greatest water pollutant. In many areas, it chokes streams; fills lakes, reservoirs, ponds, canals, drainage ditches, and harbors; buries vegetation; and generally creates a nuisance that is difficult to remove. Sediment pollution is a twofold problem: It results from erosion, which depletes a land resource (soil) at its site of origin (Figure 19.7), and it reduces the quality of the water resource it enters.[19]

Many human activities affect the pattern, amount, and intensity of surface water runoff, erosion, and sedimentation. Streams in naturally forested or wooded areas may be nearly stable, with relatively little excessive erosion or sedimentation. However, converting forested land to agriculture generally increases the runoff and sediment yield or erosion of the land. Applying soil-conservation procedures to farmland can minimize but not eliminate soil loss. The change from agricultural, forested, or rural land to highly urbanized land has even more dramatic effects. But although the construction phase of urbanization can produce large quantities of sediment, sediment production and soil erosion can be minimized by on-site erosion control.[20]

FIGURE 19.7 Massive erosion has produced this large, canyon-size gully, near Rockport Washington (see person in white, looks like a dot, at the top of gully). Note area of clearcutting (timber harvesting on right) and on mountain in background. Clearcutting is a practice that may result in accelerated soil erosion.

19.7 Acid Mine Drainage

Acid mine drainage is water with a high concentration of sulfuric acid (H_2SO_4) that drains from mines—mostly coal mines but also metal mines (copper, lead, and zinc). Coal and the rocks containing coal are often associated with a mineral known as fool's gold or pyrite (FeS_2), which is iron sulfide. When the pyrite, which may be finely disseminated in the rock and coal, comes into contact with oxygen and water, it weathers. A product of the chemical weathering is sulfuric acid. In addition, pyrite is associated with metallic sulfide deposits, which, when weathered, also produce sulfuric acid. The acid is produced when surface water or shallow groundwater runs through or moves into and out of mines or tailings (Figure 19.8). If the acidic water runs off to a natural stream, pond, or lake, significant pollution and ecological damage may result. The acidic water is toxic to the plants and animals of an aquatic ecosystem; it damages biological productivity, and fish and other aquatic life may die. Acidic water can also seep into and pollute groundwater.

Acid mine drainage is produced by complex geochemical and microbial reactions. The general equation is as follows:

$$4\,FeS_2 + 15\,O_2 + 14\,H_2O \rightarrow 4\,Fe\,(OH)_3 + 8\,H_2SO_4$$

Pyrite + Oxygen + Water → Ferric Hydroxide + Sulfuric Acid

Acid mine drainage is a significant water-pollution problem in Wyoming, Indiana, Illinois, Kentucky, Tennessee, Missouri, Kansas, and Oklahoma, and is probably the most significant water-pollution problem in West Virginia, Maryland, Pennsylvania, Ohio, and Colorado. The total impact is significant because thousands of kilometers of streams have been damaged.

Even abandoned mines can cause serious problems. Subsurface mining for sulfide deposits containing lead and zinc began in the tristate area of Kansas, Oklahoma, and Missouri in the late 19th century and ended in some areas in the 1960s. When the mines were operating, they were kept dry by pumping out the groundwater that seeped in. However, since the mining ended, some of them have flooded and overflowed into nearby creeks, polluting the creeks with acidic water. The problem was so severe in the Tar Creek area of Oklahoma that it was at one time designated by the U.S. Environmental Protection Agency as the nation's worst hazardous-waste site.

One solution being used in Tar Creek and other areas is passive treatment methods that use naturally occurring chemical and/or biological reactions in controlled environments to treat acid mine drainage. The simplest and least expensive method is to divert acidic water to an open limestone channel, where it reacts with crushed limestone and the acid is neutralized. A general reaction that neutralizes the acid is

$$H_2SO_4 + CaCO_3 \rightarrow CaSO_4 + H_2O + CO_2$$

Sulfuric Acid + Calcium Carbonate (crushed limestone) → Calcium Sulfate + Water + Carbon Dioxide

Another solution is to divert the acidic water to a bioreactor (an elongated trough) containing sulfate-reducing bacteria and a bacteria nutrient to encourage bacterial growth. The sulfate-reducing bacteria are held in cells that have a honeycomb structure, forcing the acidic water to follow a tortuous path through the bacteria-laden cells of the reactor. Complex biochemical reactions between the acidic

FIGURE 19.8 Aerial view of an acid mine drainage holding pond adjacent to an iron mine (located in the mountains of southwestern Colorado).

water and bacteria in the reactor produce metal sulfides and in the process reduce the sulfuric acid content of the water. Both methods result in cleaner water with a lower concentration of acid being released into the environment.

19.8 Surface-Water Pollution

Pollution of surface water occurs when too much of an undesirable or harmful substance flows into a body of water, exceeding that body of water's natural ability to remove it, dilute it to a harmless concentration, or convert it to a harmless form.

Water pollutants, like other pollutants, are categorized as being emitted from point or nonpoint sources (see Chapter 10). **Point sources** are distinct and confined, such as pipes from industrial and municipal sites that empty into streams or rivers (Figure 19.9). In general, point source pollutants from industries are controlled through on-site treatment or disposal and are regulated by permit. Municipal point sources are also regulated by permit. In older cities in the northeastern and Great Lakes areas of the United States, most point sources are outflows from combined sewer systems. As mentioned earlier, such systems combine stormwater flow with municipal wastewater. During heavy rains, urban storm runoff may exceed the capacity of the sewer system, causing it to overflow and deliver pollutants to nearby surface waters.

FIGURE 19.9 This pipe is a point source of chemical pollution from an industrial site entering a river in England.

Nonpoint sources, such as runoff, are diffused and intermittent and are influenced by factors such as land use, climate, hydrology, topography, native vegetation, and geology. Common urban nonpoint sources include runoff from streets or fields; such runoff contains all sorts of pollutants, from heavy metals to chemicals and sediment. Rural sources of nonpoint pollution are generally associated with agriculture, mining, or forestry. Nonpoint sources are difficult to monitor and control.

Reducing Surface-Water Pollution

From an environmental view, two approaches to dealing with surface-water pollution are (1) to reduce the sources and (2) to treat the water to remove pollutants or convert them to forms that can be disposed of safely. Which option is used depends on the specific circumstances of the pollution problem. Reduction at the source is the environmentally preferable way of dealing with pollutants. For example, air-cooling towers, rather than water-cooling towers, may be used to dispose of waste heat from power plants, thereby avoiding thermal pollution of water. The second method—water treatment—is used for a variety of pollution problems. Water treatments include chlorination to kill microorganisms such as harmful bacteria, and filtering to remove heavy metals.

There is a growing list of success stories in the treatment of water pollution. One of the most notable is the cleanup of the Thames River in Great Britain. For centuries, London's sewage had been dumped into that river, and there were few fish to be found downstream in the estuary. In recent decades, however, improved water treatment has led to the return of a number of species of fish, some not seen in the river for centuries.

Many large cities in the United States—such as Boston, Miami, Cleveland, Detroit, Chicago, Portland, and Los Angeles—grew on the banks of rivers, but the rivers were often nearly destroyed by pollution and concrete. Today, there are grassroots movements all around the country dedicated to restoring urban rivers and adjacent lands as greenbelts, parks, and other environmentally sensitive developments. For example, the Cuyahoga River in Cleveland, Ohio, was so polluted by 1969 that sparks from a train ignited oil-soaked wood in the river, setting the surface of the river on fire! The burning of an American river became a symbol for a growing environmental consciousness. The Cuyahoga River today is cleaner and no longer flammable—from Cleveland to Akron, it is a beautiful greenbelt (Figure 19.10). The greenbelt changed part of the river from a sewer into a valuable public resource and focal point for economic and environmental renewal.[21] However, in downtown Cleveland and Akron, the river remains an industrial stream, and parts remain polluted.

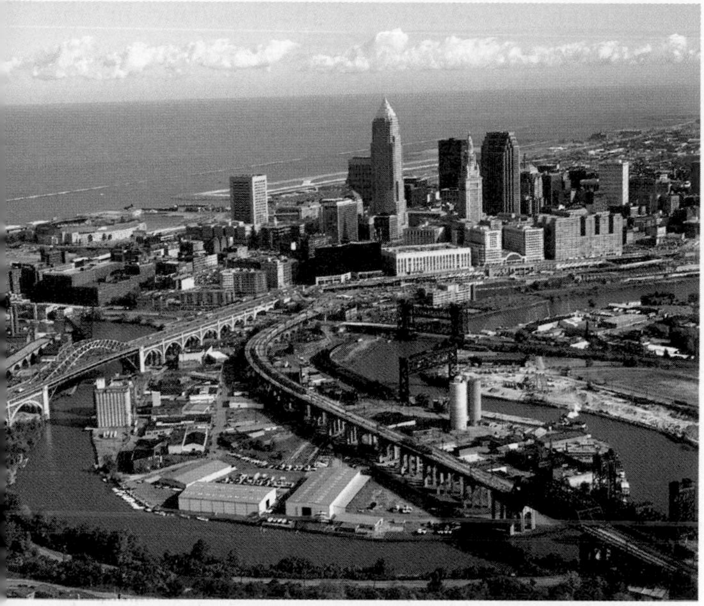

FIGURE 19.10 The Cuyahoga River (lower left) flows toward Cleveland, Ohio, and the Erie Canal (lower right) is in the Cuyahoga National Park. The skyline is that of industrial Cleveland.

Two of the newer techniques are nanotechnology and urban-runoff naturalization. **Nanotechnology** uses extremely small material particles (10^{-9}m size, about 100,000 times thinner than human hair) designed for a number of purposes. Some nano particles can capture heavy metals such as lead, mercury, and arsenic from water. The nano particles have a tremendous surface area to volume. One cubic centimeter of particles has a surface area exceeding a football field and can take up over 50% of its weight in heavy metals.[22]

FIGURE 19.12 Bioswales collect runoff from Manzaneta Village Dormitory Complex at the University of California, Santa Barbara. **(a)** Plants in bioswales **(b)** help filter water and remove nutrients, reducing cultural eutrophication.

Urban-runoff naturalization is an emerging bioengineering technology to treat urban runoff before it reaches streams, lakes, or the ocean. One method is to create a "closed-loop" local landscape that does not allow runoff to leave a property. Plants may be located as "rain gardens" below downspouts, and parking-lot drainage is directed to plants instead of the street (Figure 19.11).[23] Runoff from five large building complexes such as Manzaneta Village at the University of California, Santa Barbara, can be directed to engineered wetlands (bioswales) where wetland plants remove contaminants before water is discharged into the campus lagoon and then the ocean. Removing nutrients has helped reduce cultural eutrophication of the lagoon (Figure 19.12).

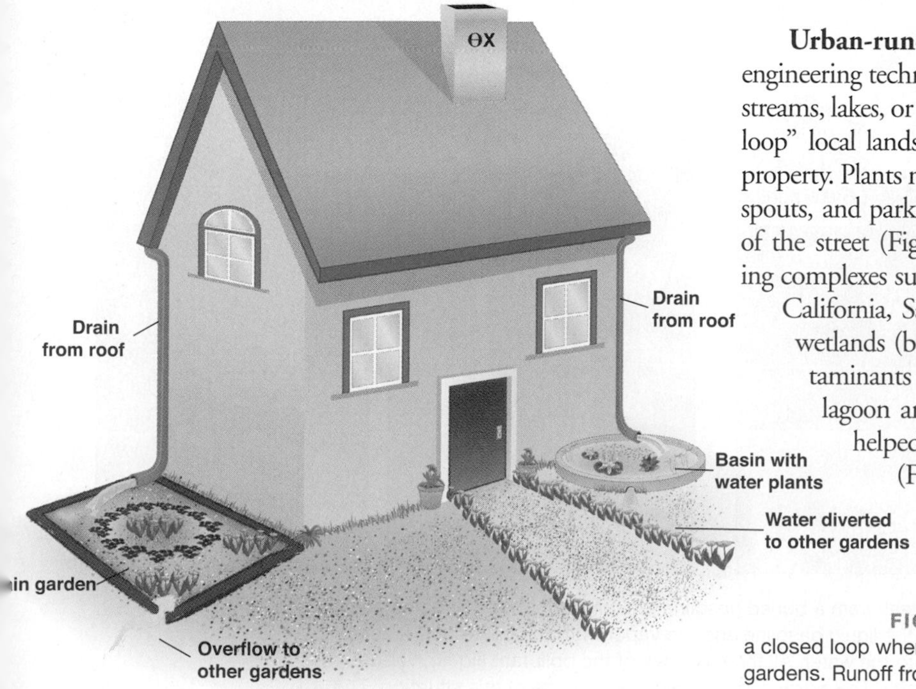

FIGURE 19.11 Water from roof runoff is part of a closed loop where water remains on the site and is used in rain gardens. Runoff from parking areas is diverted to other gardens.

19.9 Groundwater Pollution

Approximately half of all people in the United States today depend on groundwater as their source of drinking water. (Water for domestic use in the United States is discussed in A Closer Look 19.3.) People have long believed that groundwater is, in general, pure and safe to drink. In fact, however, groundwater can be easily polluted by any one of several sources (see Table 19.1), and the pollutants, though very toxic, may be difficult to recognize. (Groundwater processes were discussed in Section 19.1, and you may wish to review them.)

In the United States today, only a small portion of the groundwater is known to be seriously contaminated. However, as mentioned earlier, the problem may become worse as human population pressure on water resources increases. Our realization of the extent of the problem is growing as the testing of groundwater becomes more common. For example, Atlantic City and Miami are two eastern cities threatened by polluted groundwater that is slowly migrating toward their wells.

It is estimated that 75% of the 175,000 known waste-disposal sites in the United States may be spewing plumes of hazardous chemicals that are migrating into groundwater resources. Because many of the chemicals are toxic or are suspected carcinogens, it appears that we have inadvertently been conducting a large-scale experiment on how people are affected by chronic low-level exposure to potentially harmful chemicals. The final results of the experiment will not be known for many years.[24]

The hazard presented by a particular groundwater pollutant depends on several factors, including the concentration or toxicity of the pollutant in the environment and the degree of exposure of people or other organisms to the pollutants.[25] (See the section on risk assessment in Chapter 10.)

Principles of Groundwater Pollution: An Example

Some general principles of groundwater pollution are illustrated by an example. Pollution from leaking underground gasoline tanks belonging to automobile service stations is a widespread environmental problem that no one thought very much about until only a few years ago. Underground tanks are now strictly regulated. Many thousands of old, leaking tanks have been removed, and the surrounding soil and groundwater have been treated to remove the gasoline. Cleanup can be a very expensive process, involving removal and disposal of soil (as a hazardous waste) and treatment of the water using a process known as vapor extraction (Figure 19.13). Treatment may also be accomplished underground by microorganisms that consume the gasoline. This is known as **bioremediation** and is much less expensive than removal, disposal, and vapor extraction.

Pollution from leaking buried gasoline tanks emphasizes some important points about groundwater pollutants:

- Some pollutants, such as gasoline, are lighter than water and thus float on the groundwater.
- Some pollutants have multiple phases: liquid, vapor, and dissolved. Dissolved phases chemically combine with the groundwater (e.g., salt dissolves into water).

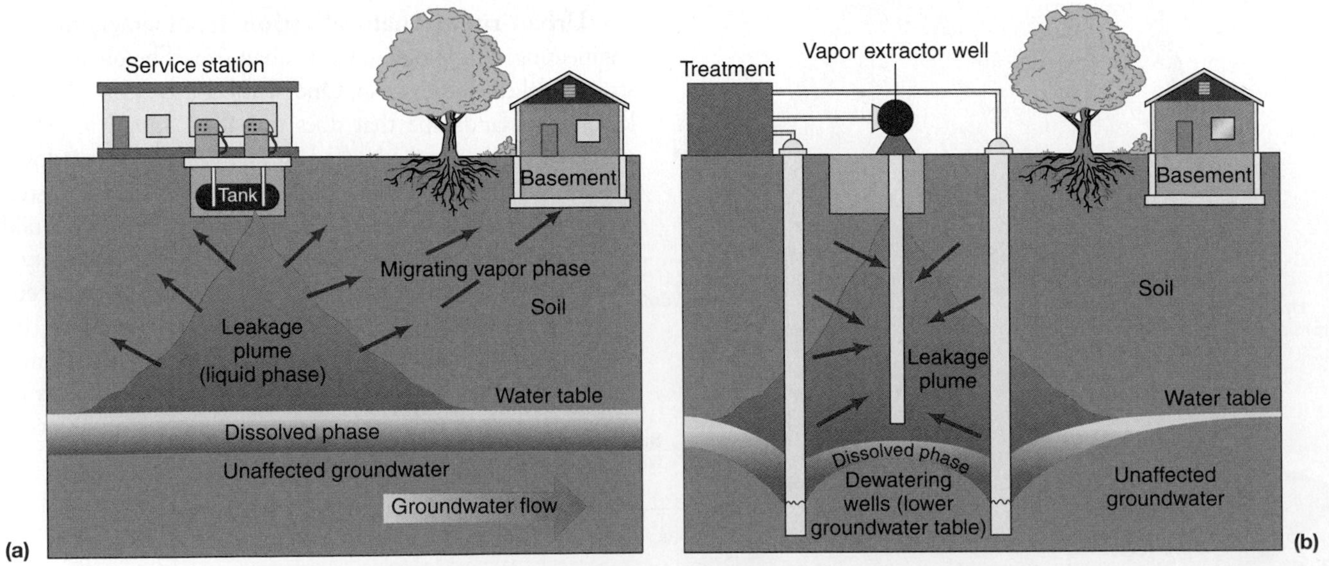

FIGURE 19.13 Diagram illustrating **(a)** a leak from a buried gasoline tank and **(b)** possible remediation using a vapor extractor system. Notice that the liquid gasoline and the vapor from the gasoline are above the water table; a small amount dissolves into the water. All three phases of the pollutant (liquid, vapor, and dissolved) float on the denser groundwater. The extraction well takes advantage of this situation. The function of the dewatering wells is to pull the pollutants in where the extraction is most effective. (*Source*: Courtesy of the University of California Santa Barbara Vadose Zone Laboratory and David Springer.)

- Some pollutants are heavier than water and sink or move downward through groundwater. Examples of sinkers include some particulates and cleaning solvents. Pollutants that sink may become concentrated deep in groundwater aquifers.
- The method used to treat or eliminate a water pollutant must take into account the physical and chemical properties of the pollutant and how these interact with surface water or groundwater. For example, the extraction well for removing gasoline from a groundwater resource (Figure 19.13) takes advantage of the fact that gasoline floats on water.
- Because cleanup or treatment of water pollutants in groundwater is very expensive, and because undetected or untreated pollutants may cause environmental damage, the emphasis should be on preventing pollutants from entering groundwater in the first place.

Groundwater pollution differs in several ways from surface-water pollution. Groundwater often lacks oxygen, a situation that kills aerobic types of microorganisms (which require oxygen-rich environments) but may provide a happy home for anaerobic varieties (which live in oxygen-deficient environments). The breakdown of pollutants that occurs in the soil and in material a meter or so below the surface does not occur readily in groundwater. Furthermore, the channels through which groundwater moves are often very small and variable. Thus, the rate of movement is low in most cases, and the opportunity for dispersion and dilution of pollutants is limited.

Long Island, New York

Another example—that of Long Island, New York—illustrates several groundwater pollution problems and how they affect people's water supply. Two counties on Long Island, New York (Nassau and Suffolk), with a population of several million people, depend entirely on groundwater. Two major problems with the groundwater in Nassau County are intrusion of saltwater and shallow-aquifer contamination.[27] Saltwater intrusion, where subsurface salty water migrates to wells being pumped, is a problem in many coastal areas of the world. The general movement of groundwater under natural conditions for Nassau County is illustrated in Figure 19.14. Salty groundwater is restricted from migrating inland by the large wedge of freshwater moving beneath the island.

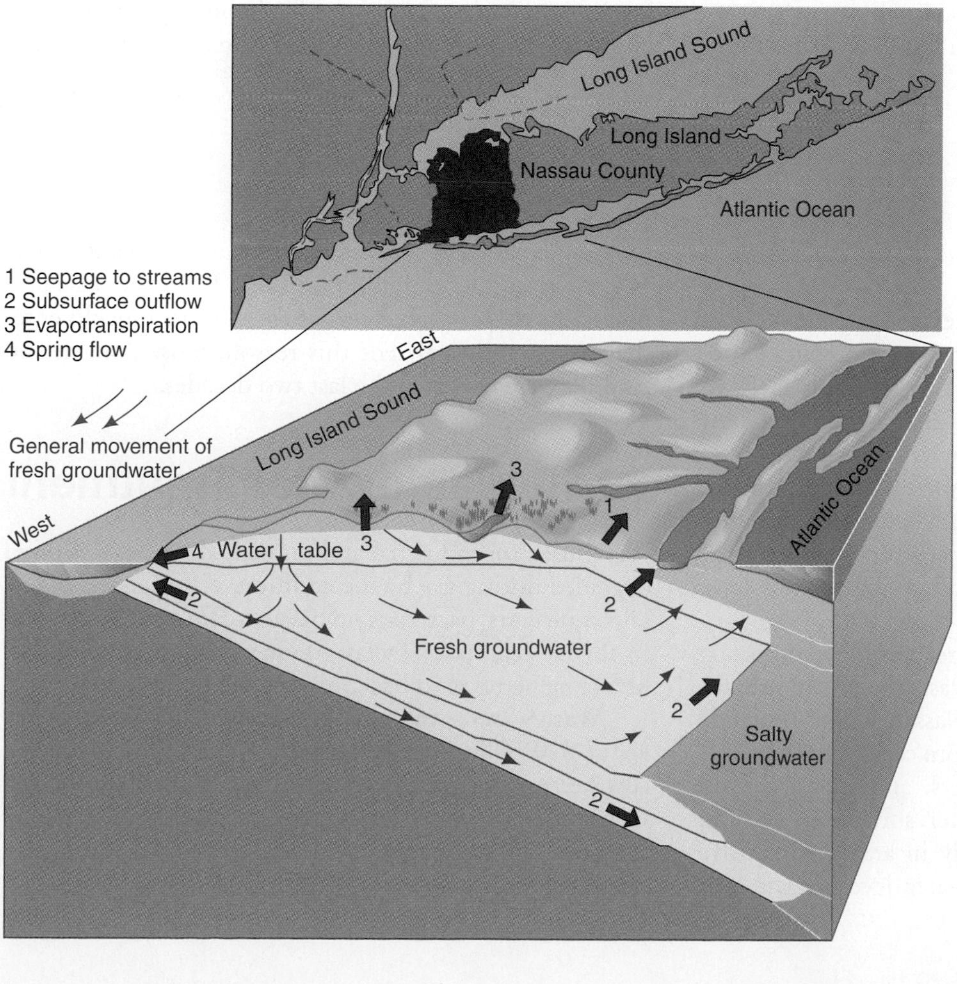

FIGURE 19.14 The general movement of fresh groundwater for Nassau County, Long Island. (*Source*: G.L. Foxworth, Nassau County, Long Island, New York, "Water Problems in Humid County," in G.D. Robinson and A.M. Spieke, eds., *Nature to Be Commanded*, U.S. Geological Survey Professional Paper 950, 1978, pp. 55–68.)

A CLOSER LOOK 19.3

Water for Domestic Use: How Safe Is It?

Water for domestic use in the United States is drawn from surface waters and groundwater. Although some groundwater sources are high quality and need little or no treatment, most are treated to conform to national drinking water standards (revisit Table 19.2).

Before treatment, water is usually stored in reservoirs or special ponds. Storage allows for solids, such as fine sediment and organic matter, to settle out, improving the clarity of water. The water is then run through a water plant, where it is filtered and chlorinated before it is distributed to individual homes. Once in people's homes, it may be further treated. For example, many people run their tap water through readily available charcoal filters before using it for drinking and cooking.

A growing number of people prefer not to drink tap water and instead drink bottled water. As a result, bottled water has become a multibillion-dollar industry. A lot of bottled water is filtered tap water delivered in plastic containers, and health questions have arisen regarding toxins leaching from the plastic, especially if bottles are left in the sun. Hot plastics can leach many more chemicals into the water than cool plastic. In any case, the plastic bottles should be used only once, then recycled.[26]

Some people prefer not to drink water that contains chlorine or that runs through metal pipes. Furthermore, water supplies vary in clarity, hardness (concentration of calcium and magnesium), and taste; and the water available locally may not be to some people's liking. A common complaint about tap water is a chlorine taste, which may be detectable at chlorine concentrations as low as 0.2–0.4 mg/l. People may also fear contamination by minute concentrations of pollutants.

The drinking water in the United States is among the safest in the world. There is no doubt that treating water with chlorine has nearly eliminated waterborne diseases, such as typhoid and cholera, which previously caused widespread suffering and death in the developed world and still do in many parts of the world. However, we need to know much more about the long-term effects of exposure to low concentrations of toxins in our drinking water. How safe is the water in the United States? It's much safer than it was 100 years ago, but low-level contamination (below what is thought dangerous) of organic chemicals and heavy metals is a concern that requires continued research and evaluation.

Notice also that the aquifers are layered, with those closest to the surface being the most salty.

In spite of the huge quantities of water in Nassau County's groundwater system, intensive pumping in recent years has lowered water levels by as much as 15 m (50 ft) in some areas. As groundwater is removed near coastal areas, the subsurface outflow to the ocean decreases, allowing saltwater to migrate inland. Saltwater intrusion has become a problem for south shore communities, which now must pump groundwater from a deeper aquifer, below and isolated from the shallow aquifers that have saltwater-intrusion problems.

The most serious groundwater problem on Long Island is shallow-aquifer pollution associated with urbanization. Sources of pollution in Nassau County include urban runoff, household sewage from cesspools and septic tanks, salt used to de-ice highways, and industrial and solid waste. These pollutants enter surface waters and then migrate downward, especially in areas of intensive pumping and declining groundwater levels.[27] Landfills for municipal solid waste have been a significant source of shallow-aquifer pollution on Long Island because pollutants (garbage) placed on sandy soil can quickly enter shallow groundwater. For this reason, most Long Island landfills were closed in the last two decades.

19.10 Wastewater Treatment

Water used for industrial and municipal purposes is often degraded during use by the addition of suspended solids, salts, nutrients, bacteria, and oxygen-demanding material. In the United States, by law, these waters must be treated before being released back into the environment.

Wastewater treatment—sewage treatment—costs about $20 billion per year in the United States, and the cost keeps rising, but it will continue to be big business. Conventional wastewater treatment includes septic-tank disposal systems in rural areas and centralized wastewater treatment plants in cities. Recent, innovative approaches include applying wastewater to the land and renovating and reusing wastewater. We discuss the conventional methods in this section and some newer methods in later sections.

Septic-Tank Disposal Systems

In many rural areas, no central sewage systems or wastewater treatment facilities are available. As a result, individual septic-tank disposal systems, not connected to sewer systems, continue to be an important method of sewage disposal in rural areas as well as outlying areas of cities. Because not all land is suitable for a septic-tank disposal system, an evaluation of each site is required by law before a permit can be issued. An alert buyer should make sure that the site is satisfactory for septic-tank disposal before purchasing property in a rural setting or on the fringe of an urban area where such a system is necessary.

The basic parts of a septic-tank disposal system are shown in Figure 19.15. The sewer line from the house leads to an underground septic tank in the yard. The tank is designed to separate solids from liquid, digest (biochemically change) and store organic matter through a period of detention, and allow the clarified liquid to discharge into the drain field (absorption field) from a piping system through which the treated sewage seeps into the surrounding soil. As the wastewater moves through the soil, it is further treated by the natural processes of oxidation and filtering. By the time the water reaches any freshwater supply, it should be safe for other uses.

Sewage drain fields may fail for several reasons. The most common causes are failure to pump out the septic tank when it is full of solids, and poor soil drainage, which allows the effluent to rise to the surface in wet weather. When a septic-tank drain field does fail, pollution of groundwater and surface water may result. Solutions to septic-system problems include siting septic tanks on well-drained soils, making sure systems are large enough, and practicing proper maintenance.

Wastewater Treatment Plants

In urban areas, wastewater is treated at specially designed plants that accept municipal sewage from homes, businesses, and industrial sites. The raw sewage is delivered to the plant through a network of sewer pipes. Following treatment, the wastewater is discharged into the surface-water environment (river, lake, or ocean) or, in some limited cases, used for another purpose, such as crop irrigation. The main purpose of standard treatment plants is to break down and reduce the BOD and kill bacteria with chlorine. A simplified diagram of the wastewater treatment process is shown in Figure 19.16.

Wastewater treatment methods are usually divided into three categories: **primary treatment**, **secondary treatment**, and **advanced wastewater treatment**. Primary and secondary treatments are required by federal law for all municipal plants in the United States. However, treatment plants may qualify for a waiver exempting them from secondary treatment if installing secondary treatment facilities poses an excessive financial burden. Where secondary treatment is not sufficient to protect the quality of the surface water into which the treated water is discharged—for example, a river with endangered fish species that must be protected—advanced treatment may be required.[28]

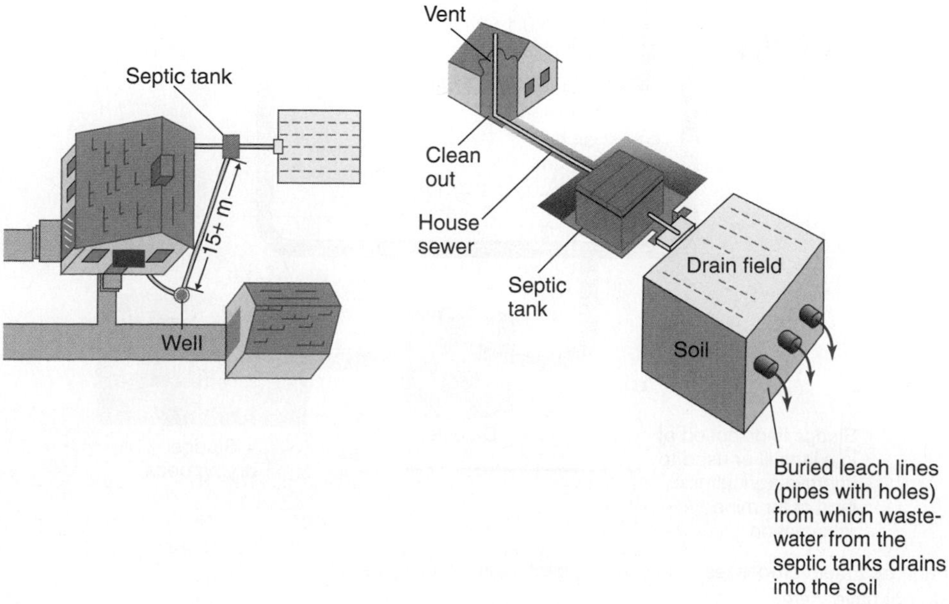

FIGURE 19.15 Septic-tank sewage system and location of the drain field with respect to the house and well. (*Source*: Based on Indiana State Board of Health.)

Primary Treatment

Incoming raw sewage enters the plant from the municipal sewer line and first passes through a series of screens to remove large floating organic material. The sewage next enters the "grit chamber," where sand, small stones, and grit are removed and disposed of. From there, it goes to the primary sedimentation tank, where particulate matter settles out to form sludge. Sometimes, chemicals are used to help the settling process. The sludge is removed and transported to the "digester" for further processing. Primary treatment removes approximately 30 to 40% of BOD by volume from the wastewater, mainly in the form of suspended solids and organic matter.[28]

Secondary Treatment

There are several methods of secondary treatment. The most common treatment is known as *activated sludge*, because it uses living organisms—mostly bacteria. In this procedure, the wastewater from the primary sedimentation tank enters the aeration tank (Figure 19.16), where it is mixed with air (pumped in) and with some of the sludge from the final sedimentation tank. The sludge contains aerobic bacteria that consume organic material (BOD) in the waste. The wastewater then enters the final sedimentation tank, where sludge settles out. Some of this "activated sludge," rich in bacteria, is recycled and mixed again in the aeration tank with air and new, incoming wastewater acting as a starter. The bacteria are used again and again. Most of the sludge from the final sedimentation tank, however, is transported to the sludge digester. There, along with sludge from the primary sedimentation tank, it is treated by anaerobic bacteria (bacteria that can live and grow without oxygen), which further degrade the sludge by microbial digestion.

Methane gas (CH_4) is a product of the anaerobic digestion and may be used at the plant as a fuel to run equipment or to heat and cool buildings. In some cases, it is burned off. Wastewater from the final sedimentation tank is next disinfected, usually by chlorination, to eliminate disease-causing organisms. The treated wastewater is then discharged into a river, lake, or ocean (see A Closer Look 19.4), or in some limited cases used to irrigate farmland. Secondary treatment removes about 90% of BOD that enters the treatment plant in the sewage.[28]

The sludge from the digester is dried and disposed of in a landfill or applied to improve soil. In some instances, treatment plants in urban and industrial areas contain many pollutants, such as heavy metals, that are not removed in the treatment process. Sludge from these plants is too polluted to use in the soil, and sludge must

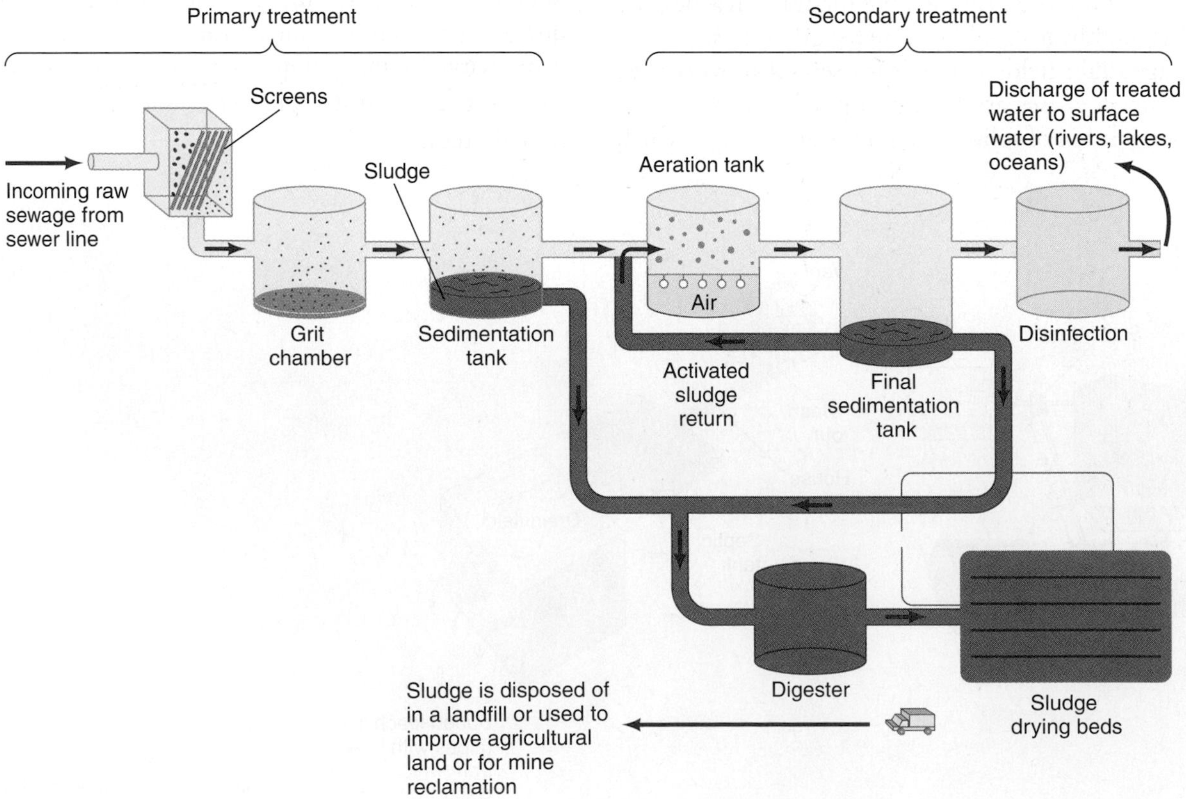

FIGURE 19.16 Diagram of sewage treatment processes. The use of digesters is relatively new, and many older treatment plants do not have them.

be disposed of. Some communities, however, require industries to pretreat sewage to remove heavy metals before the sewage is sent to the treatment plant; in these instances, the sludge can be more safely used for soil improvement.

Advanced Wastewater Treatment

As noted above, primary and secondary treatments do not remove all pollutants from incoming sewage. Some additional pollutants, however, can be removed by adding more treatment steps. For example, phosphates and nitrates, organic chemicals, and heavy metals can be removed by specifically designed treatments, such as sand filters, carbon filters, and chemicals applied to assist in the removal process.[28] Treated water is then discharged into surface water or may be used for irrigating agricultural lands or municipal properties, such as golf courses, city parks, and grounds surrounding wastewater treatment plants.

Advanced wastewater treatment is used when it is particularly important to maintain good water quality. For example, if a treatment plant discharges treated wastewater into a river and there is concern that nutrients remaining after secondary treatment may cause damage to the river ecosystem (eutrophication), advanced treatment may be used to reduce the nutrients.

Chlorine Treatment

As mentioned, chlorine is frequently used to disinfect water as part of wastewater treatment. Chlorine is very effective in killing the pathogens responsible for outbreaks of serious waterborne diseases that have killed many thousands of people. However, a recently discovered potential is that chlorine treatment also produces minute quantities of chemical by-products, some of which are potentially hazardous to people and other animals. For example, a recent study in Britain revealed that in some rivers, male fish sampled downstream from wastewater treatment plants had testes containing both eggs and sperm. This is likely related to the concentration of sewage effluent and the treatment method used.[30] Evidence also suggests that these by-products in the water may pose a risk of cancer and other human health effects. The degree of risk is controversial and currently being debated.[31]

A CLOSER LOOK 19.4

Boston Harbor: Cleaning Up a National Treasure

The city of Boston is steeped in early American history. Samuel Adams and Paul Revere immediately come to mind when considering the late 1700s, when the colonies were struggling for freedom from Britain. In 1773, Samuel Adams led a group of patriots who boarded three British ships and dumped their cargo of tea into Boston Harbor. The patriots were protesting what they considered an unfair tax on tea, and the event came to be known as the Boston Tea Party. The tea they dumped overboard did not pollute the harbor, but the growing city and the dumping of all sorts of waste eventually did.

Late in the 20th century, after more than 200 years of using Boston Harbor as a disposal site for dumping sewage, sewer overflows during storms, and treated wastewater into Massachusetts Bay, the courts demanded that measures be taken to clean up the bay. Studies concluded that the harbor had become polluted because waste being placed there moved into a small, shallow part of Massachusetts Bay, and despite vigorous tidal action between the harbor and the bay, the flushing time is about one week. It was decided that relocating the areas of waste discharge (called "outfalls") farther offshore, where the water is deeper and currents are stronger, would lower the pollution levels in Boston Harbor.

Moving the wastewater outfalls offshore was definitely a step in the right direction, but the long-term solution to protecting the marine ecosystem from pollutants will require additional measures. Even when placed farther offshore, in deeper water with greater circulation, pollutants will eventually accumulate and cause environmental damage. As a result, any long-term solution must include source reduction of pollutants. To this end, the Boston Regional Sewage Treatment Plan included a new treatment plant designed to significantly reduce the levels of pollutants discharged into the bay. This acknowledges that dilution by itself cannot solve the urban waste-management problem. Moving the sewage outfall offshore, when combined with source reduction of pollutants, is a positive example of what can be done to better manage our waste and reduce environmental problems.[29]

19.11 Land Application of Wastewater

The practice of applying wastewater to the land arose from the fundamental belief that waste is simply a resource out of place. Land application of untreated human waste was practiced for hundreds if not thousands of years before the development of wastewater treatment plants, which have sanitized the process by reducing BOD and using chlorination.

Many sites around the United States are now recycling wastewater, and the technology for wastewater treatment is rapidly evolving. An important question is: Can we develop environmentally preferred, economically viable wastewater treatment plants that are fundamentally different from those in use today? An idea for such a plant, called a resource-recovery wastewater treatment plant, is shown in Figure 19.17. The term *resource recovery* here refers to the production of resources, including methane gas (which can be burned as a fuel), as well as ornamental plants and flowers that have commercial value.[32]

Wastewater and Wetlands

Wastewater is being applied successfully to natural and constructed wetlands at a variety of locations.[33–35] Natural or man-made wetlands can be effective in treating the following water-quality problems:

- municipal wastewater from primary or secondary treatment plants (BOD, pathogens, phosphorus, nitrate, suspended solids, metals)
- stormwater runoff (metals, nitrate, BOD, pesticides, oils)
- industrial wastewater (metals, acids, oils, solvents)
- agricultural wastewater and runoff (BOD, nitrate, pesticides, suspended solids)
- mining waters (metals, acidic water, sulfates)
- groundwater seeping from landfills (BOD, metals, oils, pesticides)

Using wetlands to treat wastewater is particularly attractive to communities that find it difficult to purchase traditional wastewater treatment plants. For example,

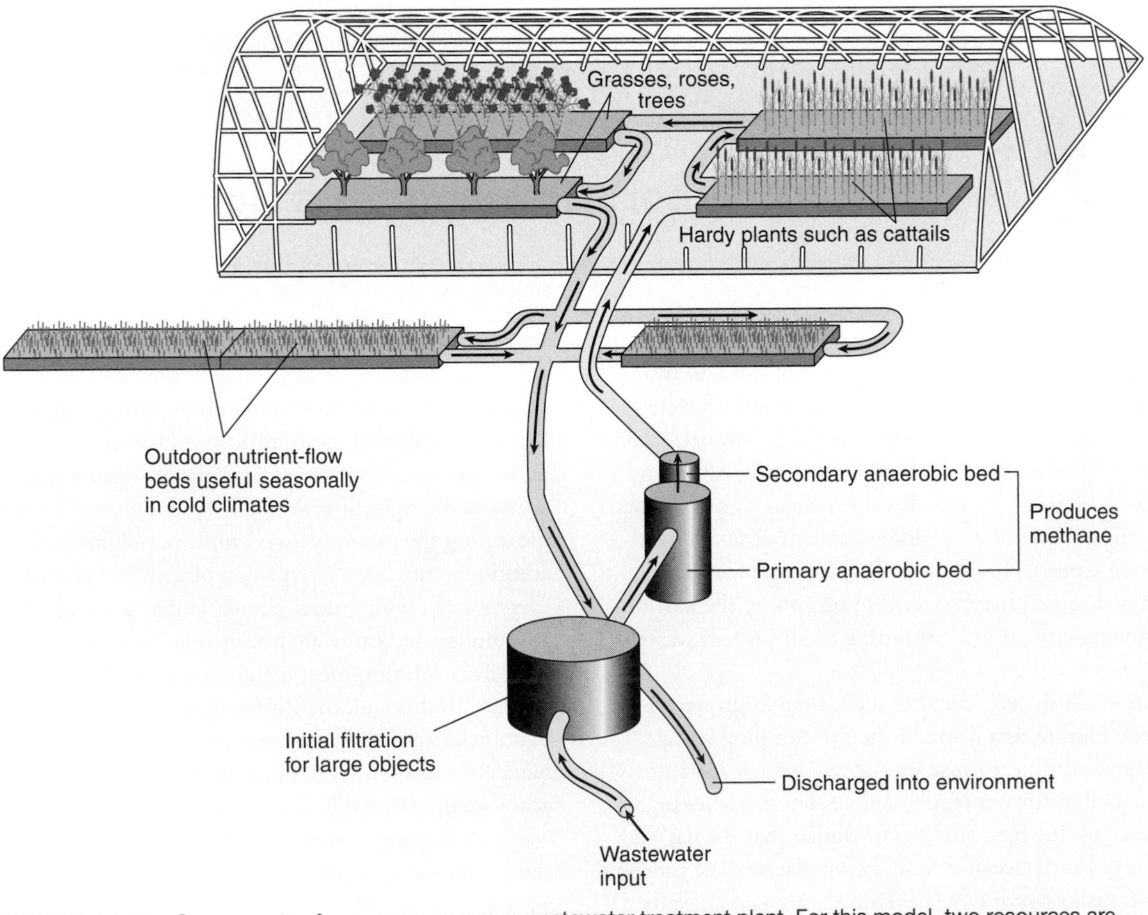

FIGURE 19.17 Components of a resource-recovery wastewater treatment plant. For this model, two resources are recovered: methane, which can be burned to produce energy from the anaerobic beds; and ornamental plants, which can be sold. (*Source*: Based on W.J. Jewell, "Resource-Recovery Wastewater Treatment," *American Scientist* [1994] 82:366–375.)

the city of Arcata, in northern California, makes use of a wetland as part of its wastewater treatment system. The wastewater comes mostly from homes, with minor inputs from the numerous lumber and plywood plants in Arcata. It is treated by standard primary and secondary methods, then chlorinated and dechlorinated before being discharged into Humboldt Bay.[33]

Louisiana Coastal Wetlands

The state of Louisiana, with its abundant coastal wetlands, is a leader in the development of advanced treatment using wetlands after secondary treatment (Figure 19.18). Wastewater rich in nitrogen and phosphorus, applied to coastal wetlands, increases the production of wetland plants, thereby improving water quality as these nutrients are used by the plants. When the plants die, their organic material (stems, leaves, roots) causes the wetland to grow vertically (or accrete), partially offsetting wetland loss due to sea-level rise.[36] There are also significant economic savings in applying treated wastewater to wetlands, because the financial investment is small compared with the cost of advanced treatment at conventional treatment plants. Over a 25-year period, a savings of about \$40,000 per year is likely.[35]

In sum, the use of isolated wetlands, such as those in coastal Louisiana, is a practical way to improve water quality in small, widely dispersed communities in the coastal zone. As water-quality standards are tightened, wetland wastewater treatment will become a viable, effective alternative that is less costly than traditional treatment.[36, 37]

Phoenix, Arizona: Constructed Wetlands

Wetlands can be constructed in arid regions to treat poor-quality water. For example, at Avondale, Arizona, near Phoenix, a wetland treatment facility for agricultural wastewater is sited in a residential community (Figure 19.19). The facility is designed to eventually treat about 17,000 m^3/day (4.5 million gal/day) of water. Water entering the facility has nitrate (NO_3) concentrations as high as 20 mg/l. The artificial wetlands contain naturally occurring bacteria that reduce the nitrate to below the maximum contaminant level of 10 mg/l. Following treatment, the water flows by pipe to a recharge basin on the nearby Agua Fria River, where it seeps into the ground to become a groundwater resource. The cost of the wetland treatment facility was about \$11 million, about half the cost of a more traditional treatment facility.

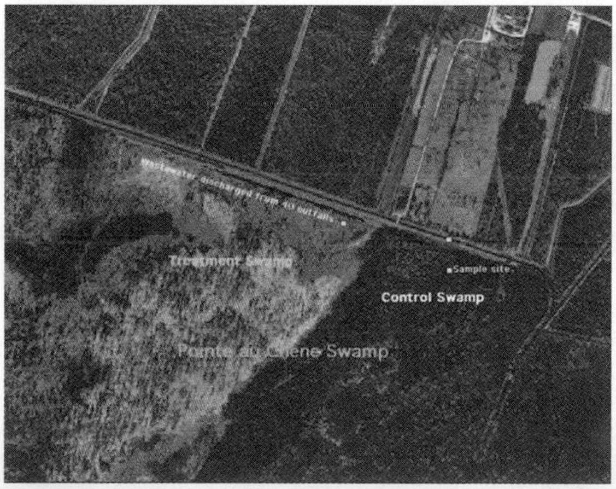

(a)

(c)

(b)

FIGURE 19.18 **(a)** Wetland Pointe au Chene Swamp, three miles south of Thibodaux, Louisiana, receives wastewater; **(b)** one of the outfall pipes delivering wastewater; and **(c)** ecologists doing field work at the Pointe au Chene Swamp to evaluate the wetland.

(a)

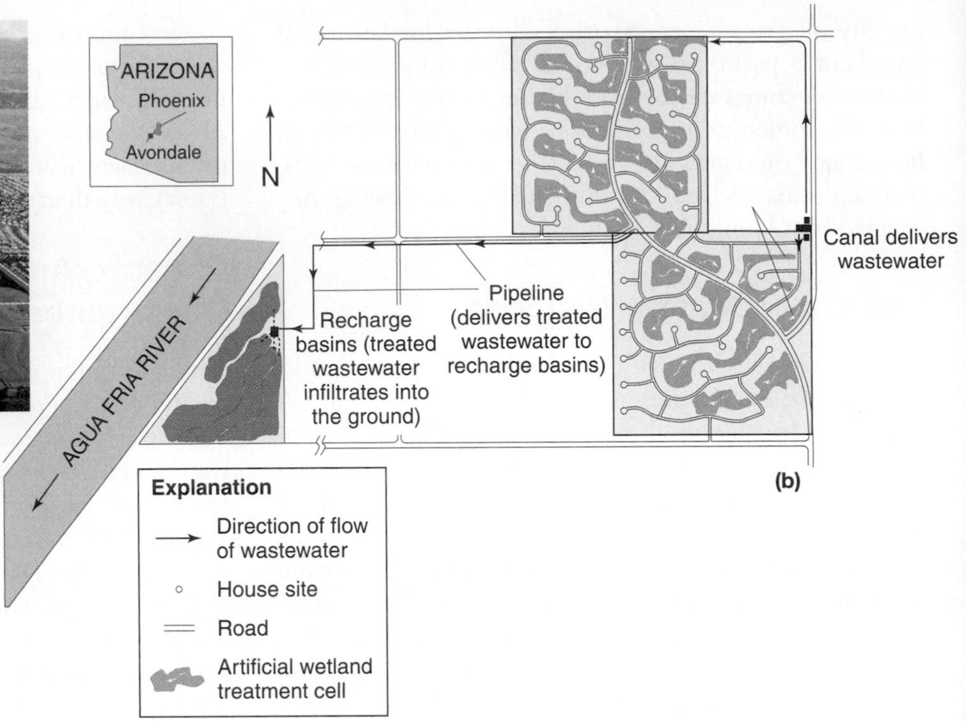

(b)

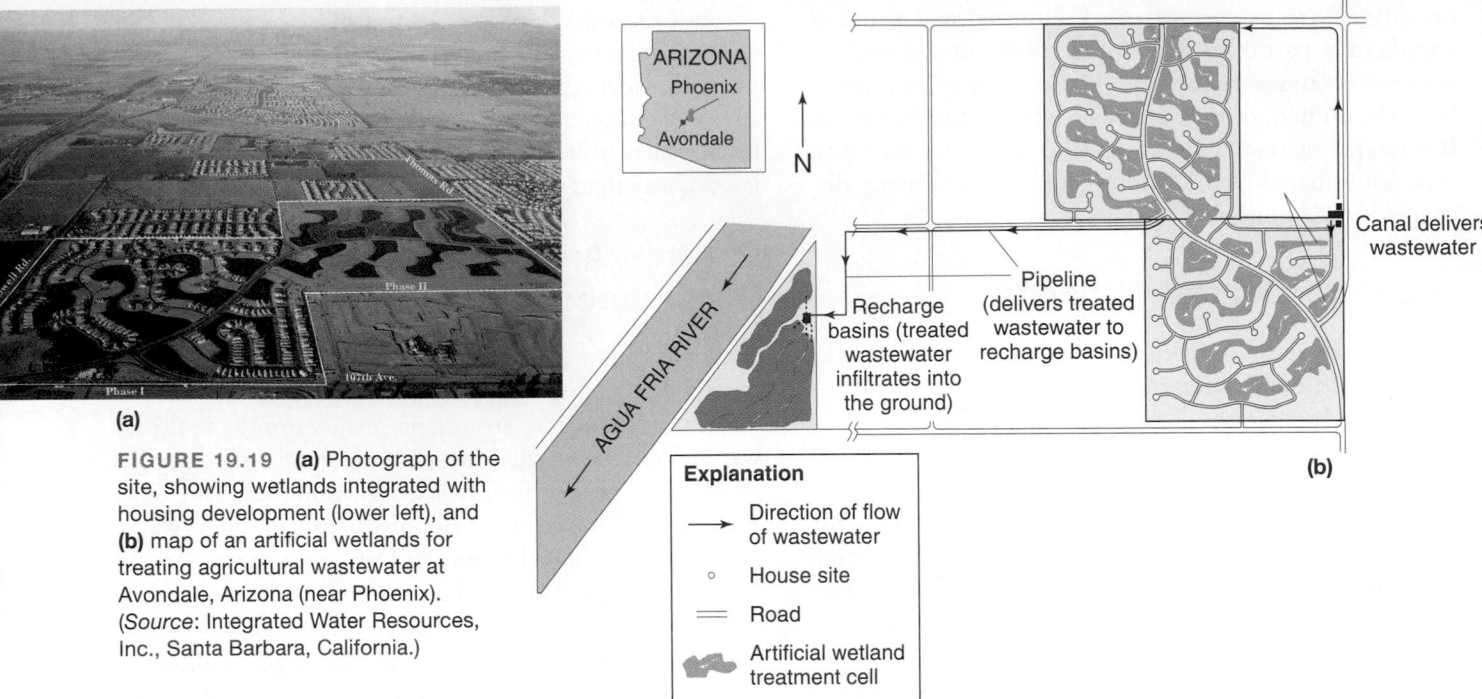

FIGURE 19.19 **(a)** Photograph of the site, showing wetlands integrated with housing development (lower left), and **(b)** map of an artificial wetlands for treating agricultural wastewater at Avondale, Arizona (near Phoenix). (*Source*: Integrated Water Resources, Inc., Santa Barbara, California.)

19.12 Water Reuse

Water reuse can be inadvertent, indirect, or direct. *Inadvertent water reuse* results when water is withdrawn, treated, used, treated, and returned to the environment, followed by further withdrawals and use. Inadvertent water reuse is very common and a fact of life for millions of people who live along large rivers. Many sewage treatment plants are located along rivers and discharge treated water into the rivers. Downstream, other communities withdraw, treat, and consume the water.

Several risks are associated with inadvertent reuse:

1. Inadequate treatment facilities may deliver contaminated or poor-quality water to downstream users.
2. Because the fate of all disease-causing viruses during and after treatment is not completely known, the health hazards of treated water remain uncertain.
3. Every year, new and potentially hazardous chemicals are introduced into the environment. Harmful chemicals are often difficult to detect in the water; and if they are ingested in low concentrations over many years, their effects on people may be difficult to evaluate.[33]

Indirect water reuse is a planned endeavor. For example, in the United States, several thousand cubic meters of treated wastewater per day have been applied to numerous sites to recharge groundwater and then reused for agricultural and municipal purposes.

Direct water reuse refers to use of treated wastewater that is piped directly from a treatment plant to the next user. In most cases, the water is used in industry, in agricultural activity, or for irrigating golf courses, institutional grounds (such as university campuses), and parks. Direct water reuse is growing rapidly and is the norm for industrial processes in factories. In Las Vegas, Nevada, new resort hotels that use a great deal of water for fountains, rivers, canals, and lakes are required to treat wastewater and reuse it (Figure 19.20). Because of perceived risks and negative cultural attitudes toward using treated wastewater, there has been little direct reuse of water for human consumption, except in emergencies. However, that is changing in Orange County, California, where an ambitious program to reuse treated wastewater is under way. The program processes 70 million gallons a day by injecting treated wastewater into the groundwater system to be further filtered underground. The water is then pumped out, further treated, and used in homes and businesses.[38]

FIGURE 19.20 Water reuse at a Las Vegas, Nevada, resort hotel.

19.13 Conditions of Stream Ecosystems in the United States

Assessment of stream ecosystem conditions in the United States has been an important research goal since passage of the Clean Water Act of 1977. Until recently, no straightforward way to do this had been seriously attempted, so the condition of small streams that can be waded was all but unknown. That void is now partly filled by recent studies aimed at providing a credible, broad-scale assessment of small streams in the United States. A standardized field collection of data was an important step, and the data at each site include the following:

- measurement of stream channel morphology and habitat characteristic
- measurement of the streamside and near-stream vegetation, known as the *riparian vegetation*
- measurement of water chemistry
- measurement of the assemblage and composition of the stream environment (biotic environment)

The evaluation includes two key biological indicators: (1) an index of how pristine a stream ecosystem is and (2) an index that represents a loss of biodiversity. Results of the study are shown in Figure 19.21. The top graph is for the entire United States, while the three lower graphs are done on a regional basis. The ratings range from poor conditions—that is, those most disturbed by environmental stress—to good conditions that mostly correspond with undisturbed stream systems. Streams with poor quality are most numerous in the northeastern part of the United States, as well as in the midsection of the country. The percentage of stream miles in good condition is considerably higher in the West.

This is not surprising, given the extent of stream-channel modifications and changes in land use in the eastern half of the country compared to the western half. Western states tend to have more mountains and more areas of natural landscape that have not been modified by agriculture and other human activities. However, streams in the West were deemed to have a higher risk of future degradation than those in other areas because there are more pristine streams to measure changes against and because there are more high-quality stream ecosystem conditions to potentially be degraded in the West than there are in other parts of the country.[39]

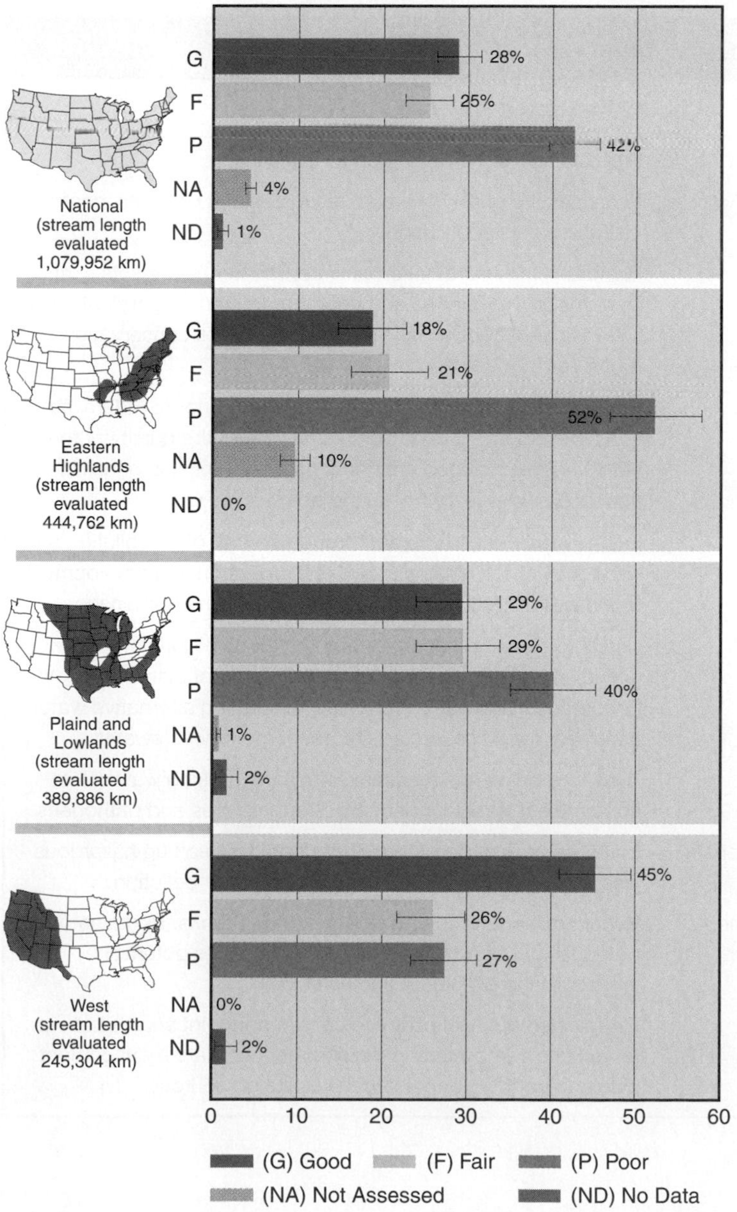

19.14 Water Pollution and Environmental Law

Environmental law, the branch of law dealing with conservation and use of natural resources and control of pollution, is very important as we debate environmental issues and make decisions about how best to protect our environment. In the United States, laws at the federal, state, and local levels address these issues.

FIGURE 19.21 Condition of ecosystems of small streams that can be waded in the U.S. (*Source*: Modified after J.M. Faustini et al. 2009, "Assessing Stream Ecosystem Condition in the United States." *EOS, Transactions, American Geophysical Union* 36: 309–310).

Federal laws to protect water resources go back to the Refuse Act of 1899, which was enacted to protect navigable streams, rivers, and lakes from pollution. Table 19.5 lists major federal laws that have a strong water-resource/pollution component. Each of these major pieces of legislation has had a significant impact on water-quality issues. Many federal laws have been passed with the purpose of cleaning up or treating pollution problems or treating wastewater. However, there has also been a focus on preventing pollutants from entering water. Prevention has the advantage of avoiding environmental damage and costly cleanup and treatment.

From the standpoint of water pollution, the mid-1990s in the United States was a time of debate and controversy. In 1994, Congress attempted to rewrite major environmental laws, including the Clean Water Act (1972, amended in 1977). The purpose was to give industry greater flexibility in choosing how to comply with environmental regulations concerning water pollution. Industry interests favored proposed new regulations that, in their estimation, would be more cost-effective without causing increased environmental degradation. Environmentalists, on the other hand, viewed attempts to rewrite the Clean Water Act as a giant step backward in the nation's fight to clean up our water resources. Apparently, Congress had incorrectly assumed it knew the public's values on this issue. Survey after survey has established that there is strong support for a clean environment in the United States and that people are willing to pay to have clean air and clean water. Congress has continued to debate changes in environmental laws, but little has been resolved.[40]

Table 19.5 FEDERAL WATER LEGISLATION

DATE	LAW	OVERVIEW
1899	Refuse Act	Protects navigable water from pollution
1956	Federal Water and Pollution Control Act	Enhances the quality of water resources and prevents, controls, and abates water pollution.
1958	Fish and Wildlife Coordination Act	Mandates the coordination of water resources projects such as dams, power plants, and flood control must coordinate with U.S Fish and Wildlife Service to enact wildlife conservation measures
1969	National Environmental Policy Act	Requires environmental impact statement prior to federal actions (development) that significantly affect the quality of the environment. Included are dams and reservoirs, channelization, power plants, bridges, and so on
1970	Water Quality Improvement Act	Expands power of 1956 act through control of oil pollution and hazardous pollutants and provides for research and development to eliminate pollution in Great Lakes and acid mine drainage.
1972 (amended in 1977)	Federal Water Pollution Control Act (Clean Water Act)	Seeks to clean up nation's water. Provides billions of dollars in federal grants for sewage treatment plants. Encourages innovative technology, including alternative water treatment methods and aquifer recharge of wastewater.
1974	Federal Safe Drinking Water Act	Aims to provide all Americans with safe drinking water. Sets contaminant levels for dangerous substances and pathogens
1980	Comprehensive Environmental Response, Compensation, and Liability Act	Established revolving fund (Superfund) to clean up hazardous waste disposal sites, reducing ground water pollution.
1984	Hazardous and Solid Waste Amendments to the Resource Conservation and Recovery Act	Regulates underground gasoline storage tanks. Reduces potential for gasoline storage tanks. Reduces potential for gasoline to pollute groundwater
1987	Water Quality Act	Established national policy to control nonpoint sources of water pollution. Important in development of state management plants to control nonpoint water pollution sources.

CRITICAL THINKING ISSUE

Is Water Pollution from Pig Farms Unavoidable?

Hurricane Floyd struck the Piedmont area of North Carolina in September 1999. The killer storm took a number of lives while flooding many homes and forcing some 48,000 people into emergency shelters. The storm had another, more unusual effect as well. Floodwaters containing thousands of dead pigs, along with their feces and urine, flowed through schools, churches, homes, and businesses. The stench was reportedly overwhelming, and the count of pig carcasses may have been as high as 30,000. The storm waters had overlapped and washed out over 38 pig lagoons with as much as 950 million liters (250 million gal) of liquid pig waste, which ended up in flooded creeks, rivers, and wetlands. In all, something like 250 large commercial pig farms flooded out, drowning hogs whose floating carcasses had to be collected and disposed of (Figure 19.22).

Prior to Hurricane Floyd, the pig farm industry in North Carolina had been involved in a scandal reported by newspapers and television—and even by *60 Minutes*. North Carolina has a long history of hog production, and the population of pigs swelled from about 2 million in 1990 to nearly 10 million in 1997. At that time, North Carolina became the second-largest pig-farming state in the nation.[41] As the number of large commercial pig farms grew, the state allowed the hog farmers to build automated and very confining farms housing hundreds or thousands of pigs. There were no restrictions on farm location, and many farms were constructed on floodplains.

Each pig produces approximately 2 tons of waste per year. The North Carolina herd was producing approximately 20 million tons of waste a year, mostly manure and urine, which was flushed out of the pig barns and into open, unlined lagoons about the size of football fields. Favorable regulations, along with the availability of inexpensive waste disposal systems (the lagoons), were responsible for the tremendous growth of the pig population in North Carolina in the 1990s.

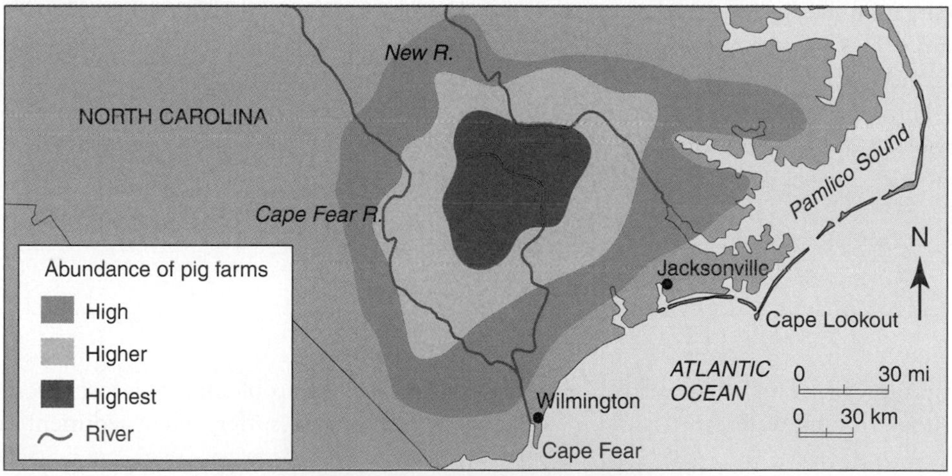

(a)

(b)

FIGURE 19.22 North Carolina's "Bay of Pigs." **(a)** Map of areas flooded by Hurricane Floyd in 1999 with relative abundance of pig farms. **(b)** Collecting dead pigs near Boulaville, North Carolina. The animals were drowned when floodwaters from the Cape Fear River inundated commercial pig farms

After the hurricane, mobile incinerators were moved into the hog region to burn the carcasses, but there were so many that hog farmers had to bury some animals in shallow pits. The pits were supposed to be at least 1 meter deep, and dry, but there wasn't always time to find dry ground, and for the most part the pits were dug and filled on floodplains. As these pig carcasses rot, bacteria will leak into the groundwater and surface water for some appreciable time.

An early warning occurred in 1995, when a pig-waste lagoon failed and sent approximately 950 million liters (250 million gal) of concentrated pig feces down the New River past the city of Jacksonville, South Carolina, and into the New River estuary. The spill's adverse effects on marine life lasted for approximately three months.

The lesson to be learned from North Carolina's so-called Bay of Pigs is that we are vulnerable to environmental catastrophes caused by large-scale industrial agriculture. Economic growth and production of livestock must be carefully planned to anticipate problems, and waste-management facilities must be designed so as not to pollute local streams, rivers, and estuaries.

Was the lesson learned in North Carolina? The pig farmers had powerful friends in government and big money. Incredible as it may seem, following the hurricane, the farmers asked for $1 billion in grants to help repair and replace the pig facilities, including waste lagoons, destroyed by the hurricane. Furthermore, they asked for exemptions from the Clean Water Act for a period of six months so that waste from the pig lagoons could be discharged directly into streams.[42] This was not allowed.

With regard to future management, considering that North Carolina is frequently struck by hurricanes, barring pig operations from floodplains seems obvious. However, this is only the initial step. The whole concept of waste lagoons needs to be rethought and alternative waste-management practices put into effect if pollution of surface waters and groundwaters is to be avoided. To this end, North Carolina in 2007 passed legislation to ban construction or expansion of new waste lagoons and encouraged pig farms to treat pig waste to extract methane (gas) as an energy source. Other methods of on-site treatment to reduce organic matter and nutrients is ongoing

North Carolina's pig problem led to the formation of what is called the "Hog Roundtable," a coalition of civic, health, and environmental groups with the objective of controlling industrial-scale pig farming. Its efforts, with others, resulted in a mandate to phase out pig-waste lagoons and expand regulations to require buffers between pig farms and surface waters and water wells. The coalition also halted construction of a proposed slaughterhouse that would have allowed more pig farms to be established.

Critical Thinking Questions

1. Can future pollution from large pig farms in areas with recurring hurricane hazards be eliminated or minimized? If so, how?
2. Do you think the pollution caused by pig farm flooding as a result of hurricanes is a natural event, a so-called act of God? Pig farmers blamed the hurricane for the water pollution. Are they right, or are people responsible?
3. Do you think the actions of the Hog Roundtable can succeed over the long term in minimizing environmental problems caused by large pig farms?
4. Discuss the moral and ethical issues of industrial-scale agriculture that confines large numbers of animals, often in small spaces. Is there a better way to produce our food? What are alternatives?

SUMMARY

- The primary water-pollution problem in the world today is the lack of disease-free drinking water.
- Water pollution is degradation of quality that renders water unusable for its intended purpose.
- Major categories of water pollutants include disease-causing organisms, dead organic material, heavy metals, organic chemicals, acids, sediment, heat, and radioactivity.
- Sources of pollutants may be point sources, such as pipes that discharge into a body of water, or nonpoint sources, such as runoff, which are diffused and intermittent.
- Eutrophication of water is a natural or human-induced increase in the concentration of nutrients, such as phosphorus and nitrogen, required for living things. A high concentration of such nutrients may cause a population explosion of photosynthetic bacteria. As the bacteria die and decay, the concentration of dissolved oxygen in the water is lowered, leading to the death of fish.
- Sediment is a twofold problem: Soil is lost through erosion, and water quality suffers when sediment enters a body of water.
- Acid mine drainage is a serious water-pollution problem because when water and oxygen react with sulfide minerals that are often associated with coal or metal sulfide deposits, they form sulfuric acid. Acidic water draining from mines or tailings pollutes streams and other bodies of water, damaging aquatic ecosystems and degrading water quality.
- Urban processes—for example, waste disposal in landfills, application of fertilizers, and dumping of chemicals such as motor oil and paint—can contribute to shallow-aquifer contamination. Overpumping of aquifers near the ocean may cause saltwater, found below the freshwater, to rise closer to the surface, contaminating the water resource by a process called saltwater intrusion.

- Wastewater treatment at conventional treatment plants includes primary, secondary, and, occasionally, advanced treatment. In some locations, natural ecosystems, such as wetlands and soils, are being used as part of the treatment process.
- Water reuse is the norm for millions of people living along rivers where sewage treatment plants discharge treated wastewater back into the river. People who withdraw river water downstream are reusing some of the treated wastewater.
- Industrial reuse of water is the norm for many factories.
- Deliberate use of treated wastewater for irrigating agricultural lands, parks, golf courses, and the like is growing rapidly as demand for water increases.
- Cleanup and treatment of both surface water and groundwater pollution are expensive and may not be completely successful. Furthermore, environmental damage may result before a pollution problem i s identified and treated. Therefore, we should continue to focus on preventing pollutants from entering water, which is a goal of much water-quality legislation.

REEXAMINING THEMES AND ISSUES

Human Population

We state in this chapter that the number one water-pollution problem in the world today is the lack of disease-free drinking water. This problem is likely to get worse in the future as the number of people, particularly in developing countries, continues to increase. As population increases, so does the possibility of continued water pollution from a variety of sources relating to agricultural, industrial, and urban activities.

Sustainability

Any human activity that leads to water pollution—such as the building of pig farms and their waste facilities on floodplains—is antithetical to sustainability. Groundwater is fairly easy to pollute and, once degraded, may remain polluted for a long time. Therefore, if we wish to leave a fair share of groundwater resources to future generations, we must ensure that these resources are not polluted, degraded, or made unacceptable for use by people and other living organisms.

Global Perspective

Several aspects of water pollution have global implications. For example, some pollutants may enter the atmosphere and be transported long distances around the globe, where they may be deposited and degrade water quality. Examples include radioactive fallout from nuclear reactor accidents or experimental detonation of nuclear devices. Waterborne pollutants from rivers and streams may enter the ocean and circulate with marine waters around the ocean basins of the world.

Urban World

Urban areas are centers of activities that may result in serious water pollution. A broad range of chemicals and disease-causing organisms are present in large urban areas and may enter surface waters and groundwaters. An example is bacterial contamination of coastal waters, resulting in beach closures. Many large cities have grown along the banks of streams and rivers, and the water quality of those streams and rivers is often degraded as a result. There are positive signs that some U.S. cities are viewing their rivers as valuable resources, with a focus on environmental and economic renewal. Thus, rivers flowing through some cities are designated as greenbelts, with parks and trail systems along river corridors. Examples include New York City; Cleveland, Ohio; San Antonio, Texas; Corvallis, Oregon; and Sacramento and Los Angeles, California.

People and Nature

Polluting our water resources endangers people and ecosystems. When we dump our waste in rivers, lakes, and oceans, we are doing what other animals have done for millions of years—it is natural. For example, a herd of hippopotamuses in a small pool may pollute the water with their waste, causing problems for other living things in the pond. The difference is that we understand that dumping our waste damages the environment, and we know how to reduce our impact.

Science and Values

It is clear that the people of the United States place a high value on the environment and, in particular, on critical resources such as water. Attempts to weaken water-quality standards are viewed negatively by the public. There is also a desire to protect water resources necessary for the variety of ecosystems found on Earth. This has led to research and development aimed at finding new technologies to reduce, control, and treat water pollution. Examples include development of new wastewater treatments and support of laws and regulations that protect water resources.

KEY TERMS

acid mine drainage **409**
advanced wastewater treatment **415**
biological or biochemical oxygen demand (BOD) **403**
bioremediation **412**
cultural eutrophication **406**
ecosystem effect **406**
environmental law **421**
eutrophication **405**
fecal coliform bacteria **404**
nanotechnology **411**
nonpoint sources **410**
outbreaks **404**
point sources **410**
primary treatment **415**
urban-runoff naturalization **411**
secondary treatment **415**
wastewater treatment **414**
water reuse **420**

STUDY QUESTIONS

1. Do you think outbreaks of waterborne diseases will be more common or less common in the future? Why? Where are outbreaks most likely to occur?
2. What was learned from the *Exxon Valdez* oil spill that might help reduce the number of future spills and their environmental impact?
3. What is meant by the term *water pollution*, and what are several major processes that contribute to water pollution?
4. Compare and contrast point and nonpoint sources of water pollution. Which is easier to treat, and why?
5. What is the twofold effect of sediment pollution?
6. In the summer, you buy a house with a septic system that appears to function properly. In the winter, effluent discharges at the surface. What could be the environmental cause of the problem? How could the problem be alleviated?
7. Describe the major steps in wastewater treatment (primary, secondary, advanced). Can natural ecosystems perform any of these functions? Which ones?
8. In a city along an ocean coast, rare waterbirds inhabit a pond that is part of a sewage treatment plant. How could this have happened? Is the water in the sewage pond polluted? Consider this question from the birds' point of view and from your own.
9. How does water that drains from coal mines become contaminated with sulfuric acid? Why is this an important environmental problem?
10. What is eutrophication, and why is it an ecosystem effect?
11. How safe do you believe the drinking water is in your home? How did you reach your conclusion? Are you worried about low-level contamination by toxins in your water? What could be the sources of contamination?

12. Do you think our water supply is vulnerable to terrorist attacks? Why? Why not? How could potential threats be minimized?

13. Would you be willing to use treated wastewater in your home for personal consumption, as they are doing in Orange County, California? Why? Why not?

14. How would you design a system to capture runoff where you live before it enters a storm drain?

FURTHER READING

Borner, H., ed., *Pesticides in Ground and Surface Water.* Vol. 9 of *Chemistry of Plant Protection* (New York: Springer-Verlag, 1994). Essays on the fate and effects of pesticides in surface water and groundwater, including methods to minimize water pollution from pesticides.

Dunne, T., and L.B. Leopold, *Water and Environmental Planning* (San Francisco: W.H. Freeman, 1978). A great summary and detailed examination of water resources and problems.

Hester, R.E., and R.M. Harrison, eds., *Agricultural Chemicals and the Environment* (Cambridge: Royal Society of Chemistry, Information Services, 1996). A good source of information about the impact of agriculture on the environment, including eutrophication and the impact of chemicals on water quality.

Manahan, S.E., *Environmental Chemistry* (Chelsea, MI: Lewis, 1991). A detailed primer on the chemical processes pertinent to a broad array of environmental problems, including water pollution and treatment.

Newman, M.C., *Quantitative Methods in Aquatic Ecotoxicology* (Chelsea, MI: Lewis, 1995). Up-to-date text on the fate, effects, and measurement of pollutants in aquatic ecosystems.

Nichols, C. "Trouble at the Waterworks," *The Progressive* 53 (1989): 33–35. A concise report on the problem of tainted water supplies in the United States.

Rao, S.S., ed., *Particulate Matter and Aquatic Contaminants* (Chelsea, MI.: Lewis, 1993). Coverage of the biological, microbiological, and ecotoxicological principles associated with interaction between suspended particulate matter and contaminants in aquatic environments.

CHAPTER 20

The Atmosphere, Climate, and Global Warming

The years of the Medieval Warm Period, from about A.D. 950 to 1250, were good times for the people in Western Europe. Harvests were good, cultures flourished, the population expanded, and great cathedrals were built. In the southwestern United States, Mexico, and Central America, the same period brought persistent droughts that contributed to the collapse of some civilizations, including the Maya.

LEARNING OBJECTIVES

Earth's atmosphere—the layer of gases surrounding the Earth—is a complex, dynamic system that is changing continuously. After reading this chapter, you should understand . . .

- The basic composition and structure of the atmosphere;
- How the processes of atmospheric circulation and climate work;
- How the climate has changed over the Earth's history;
- What the term *greenhouse gas* means and what the major greenhouse gases are;
- What global warming is and what major kinds of evidence point to it;
- What effects global warming might have and how we can adjust to them.

CASE STUDY

What Does History Tell Us about Global Warming's Potential Consequences for People?

During an approximate 300-year period from A.D. 950 to 1250, Earth's surface was considerably warmer than what climatologists today call normal (meaning the average surface temperature during the past century or some shorter interval, such as 1960–1990). This warm time is known as the Medieval Warm Period (MWP). With all the concerns today about climate change, perhaps we can learn some lessons from that time. Since weather records were not kept then, we do not have a global picture of what it was like. What we do know is that parts of the world, in particular Western Europe and the Atlantic, may have been warmer some of the time than they were in the last decade of the 20th century. However on a global basis the MWP was not as warm as it is today.

In Western Europe, it was a time of flourishing culture and activity, as well as expansion of the population; a time when harvests were plentiful, people generally prospered, and many of Europe's grand cathedrals were constructed.[1,2] Sea temperatures evidently were warmer, and there was less sea ice. Viking explorers from Scandinavia traveled widely in the Far North and established settlements in Iceland, Greenland, and even briefly in North America. Near the end of the 10th century, Erik the Red, the famous Viking explorer, arrived at Greenland with his ships and set up settlements that flourished for several hundred years. The settlers were able to raise domestic animals and grow a variety of crops that had never before been cultivated in Greenland (Figure 20.1). During the same warm period, Polynesian people in the Pacific, taking advantage of winds flowing throughout the Pacific, were able to sail to and colonize islands over vast areas of the Pacific, including Hawaii.[2]

FIGURE 20.1 Remains of a Viking settlement in Greenland from the Medieval Warm Period.

While some prospered in Western Europe and the Pacific during the Medieval Warm Period, other cultures appear to have been less fortunate. Associated with the warming period were long, persistent droughts (think human-generational length) that appear to have been partially responsible for the collapse of sophisticated cultures in North and Central America. The collapses were not sudden but occurred over a period of many decades, and in some cases the people just moved away. These included the people living near Mono Lake on the eastern side of the Sierra Nevada in California, the Chacoan people in what is today Chaco Canyon in New Mexico, and the Mayan civilization in the Yucatán of southern Mexico and Central America.

The Medieval Warm Period was followed by the Little Ice Age (LIA), which lasted from approximately mid-1400 to 1700. The cooling made life more difficult for people in Western Europe and North America. Crop failures occurred in Western Europe, and some mountain glaciers in the Swiss Alps advanced to the extent that they filled valleys and destroyed villages. Areas to the north that had enjoyed abundant crop production were under ice.[3] The population was devastated by the Black Plague, whose effects may have been exacerbated by poor nutrition as a result of crop failures and by the damp and cold that reached out across Europe and even to Iceland by about 1400.

Travel and trade became difficult in the Far North. Eventually, the Viking colonies in North America were abandoned and those in Greenland declined greatly. Part of the reason for the abandonment in North America, and particularly in Newfoundland, was that the Vikings may not have been able to adapt to the changing conditions, as did the Inuit peoples living there. As times became tough, the two cultures collided, and the Vikings, despite their fierce reputation, were less able than the Inuit to adapt to the cooling climate.

We do not know what caused the Medieval Warm Period, and the details about it are obscured by insufficient climate data to help us estimate temperatures during that period. We do know that it was relatively warm (in Western Europe). We can't associate the warming 1000 years ago with burning of fossil fuels. This suggests that more than one factor can cause warming. In this chapter we will explore climate dynamics so you can better understand what may be the causes of climate change and what might be the best estimates of how it could affect life on Earth and civilizations.[1–5]

20.1 Fundamental Global Warming Questions

The modern concern about global warming arose from two kinds of observations. The first, shown in Figure 20.2, is of the average surface temperature of the Earth from 1850 to the present. This graph shows an increase beginning in the 1930s and accelerating, especially after 1960, when the increase was about 0.2°C per decade.

The second kind of key observation is the measurement of carbon dioxide concentrations in the atmosphere. Of these, the best-known were made on Mauna Loa Mountain, Hawaii, by Charles Keeling and are now known as the Keeling Curve (Figure 20.3). Taken at 3,500 m (11,500 ft) on an island far from most human activities, these measurements provide an excellent estimate of the background condition of the atmosphere.

Here are the fundamental questions about global warming:

- What is the origin of known periods of rapid warming in the geologic record? This fundamental question is the subject of intense ongoing research and is not yet solved.
- Is the present rapid warming unprecedented or at least so rare that many living things will not be able to respond successfully to it?
- To what extent, have people caused it?
- What are likely to be the effects on people?
- What are likely to be the effects on all life on Earth?
- How can we make forecasts about it and other kinds of climate change?
- What can we do to minimize potential negative effects?

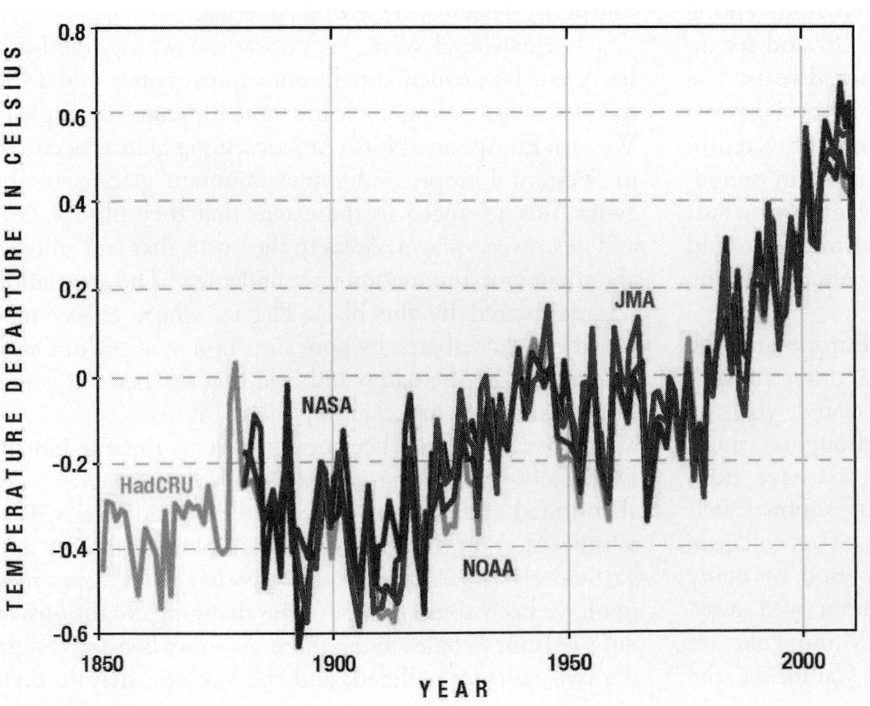

FIGURE 20.2 The temperature difference between the average at the end of the 19th century and the years between 1860 and today. This graph shows the difference between calculated world surface temperatures for each year and the average at the end of the 19th century. Temperature departure refers to changes in mean global temperature from some standard such as 1951–1980. Climatologists studying climate change prefer, in general, to look at the difference between temperatures at one time compared to another, rather than the actual temperature, for a variety of technical reasons. (*Source:* Hadley Meteorological Center, Great Britain.) http://www.metoffice.gov.uk/corporate/pressoffice/myths/2.html)

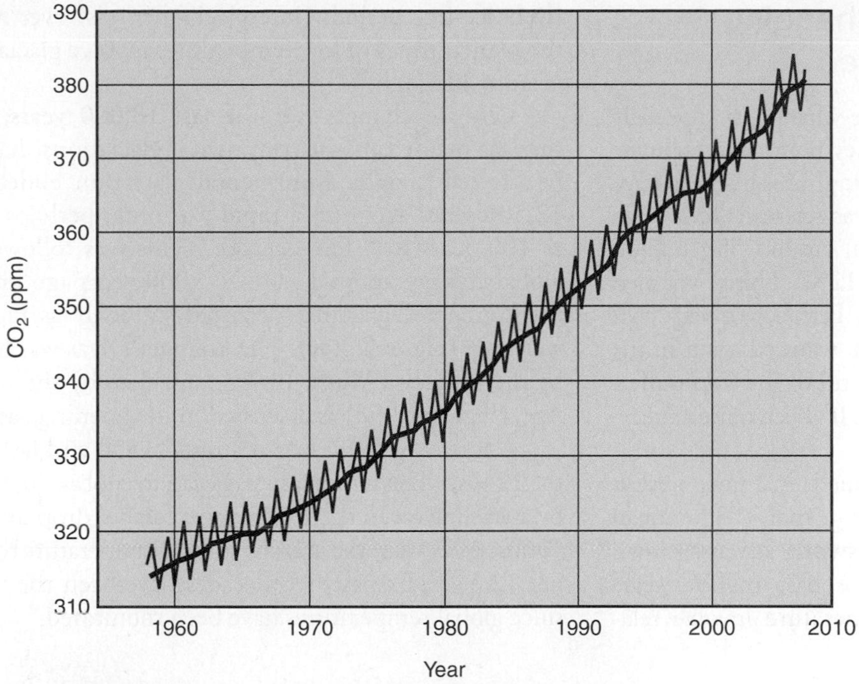

(a)

(b)

FIGURE 20.3 **(a)** Carbon dioxide concentrations in the air above Mauna Loa, Hawaii; **(b)** the NOAA Observatory on Mauna Loa, where these measurements were made. (*Sources:* **(a)** *Encyclopedia of Earth,* http://www.eoearth.org/article/Climate_change; taken from C.D. Keeling and T.P. Whorf, 2005. "Atmospheric CO_2 records from sites in the SIO air sampling network," in *Trends: A Compendium of Data on Global Change.* Carbon Dioxide Information Analysis Center, Oak Ridge National Laboratory, U.S. Department of Energy, Oak Ridge, TN).

20.2 Weather and Climate

Weather is what's happening now or over some short time period—this hour, today, this week—in the atmosphere near the ground: its temperature, pressure, cloudiness, precipitation, winds. **Climate** is the average weather and usually refers to average weather conditions over long periods, at least seasons, but more often years or decades. When we say it's hot and humid in New York today or raining in Seattle, we are speaking of weather. When we say Los Angeles has cool, wet winters and warm, dry summers, we are referring to the Los Angeles climate.

Since climates are characteristic of certain latitudes (and other factors that we will discuss later), they are classified mainly by latitude—tropical, subtropical, midlatitudinal (continental), sub-Arctic (continental), and Arctic—but also by wetness/dryness, such as humid continental, Mediterranean, monsoon, desert, and tropical wet–dry (Figure 20.4). Recall from the discussion of biogeography in Chapter 7 that similar climates produce similar kinds of ecosystems. Therefore, knowing the climate, we can make pretty good predictions about what kinds of life we will find there and what kinds could survive there if introduced.

The Climate Is Always Changing at a Variety of Time Scales

Answering questions about climate change is especially complicated because—and this is a key point about climate and life—the climate is always changing. This has been happening as far back in Earth's history as scientists have been able to study. The Precambrian Era, around 550 million years ago, averaged a relatively cool 12°C. Things warmed up to about 22°C in the Cambrian Period, got very cool in the Ordovician/Silurian transition, warmed again in the Devonian, cooled a lot again at the end of the Carboniferous, and warmed again in the Triassic. It's been quite a rollercoaster ride.

Climate changes have continued in more recent times—"recent" geologically speaking, that is. The mean annual temperature of Earth has swung up and down by several degrees Celsius over the past million years (Figure 20.5). Times of high temperature involve relatively ice-free periods (interglacial periods) over much of the planet; times of low temperature involve glacial events (Figure 20.5 a, b).[6-8]

Climate change over the last 18,000 years, during the last major time of continental glaciations, has greatly affected people. Continental glaciation ended about 12,500 years ago with a rapid warming, perhaps as brief as 100 years to a few decades.[7] This was followed by a global cooling about 11,000–13,000 years ago known as the Younger Diyas that occurred suddenly as Earth was warming (Figure 20.5c). The Younger Diyas was followed by the Medieval Warm Period, and then by the Little Ice Age, (Figure 12.5d) as discussed in the opening case study.

A warming trend began around 1850 and lasted until the 1940s, when temperatures began to cool again, followed by a leveling off in the 1950s and a further drop during the 1960s. After that, the average surface temperature rose (Figure 12.5e). The past two decades have been the warmest since global temperatures have been monitored.[6, 8]

FIGURE 20.4 The climates of the world and some of the major climate types in terms of characteristic precipitation and temperature conditions. (*Source:* Modified from W.M. Marsh and J. Dozier, *Landscape* [New York: John Wiley & Sons, 1981. Reprinted with permission of John Wiley & Sons, Inc.]).

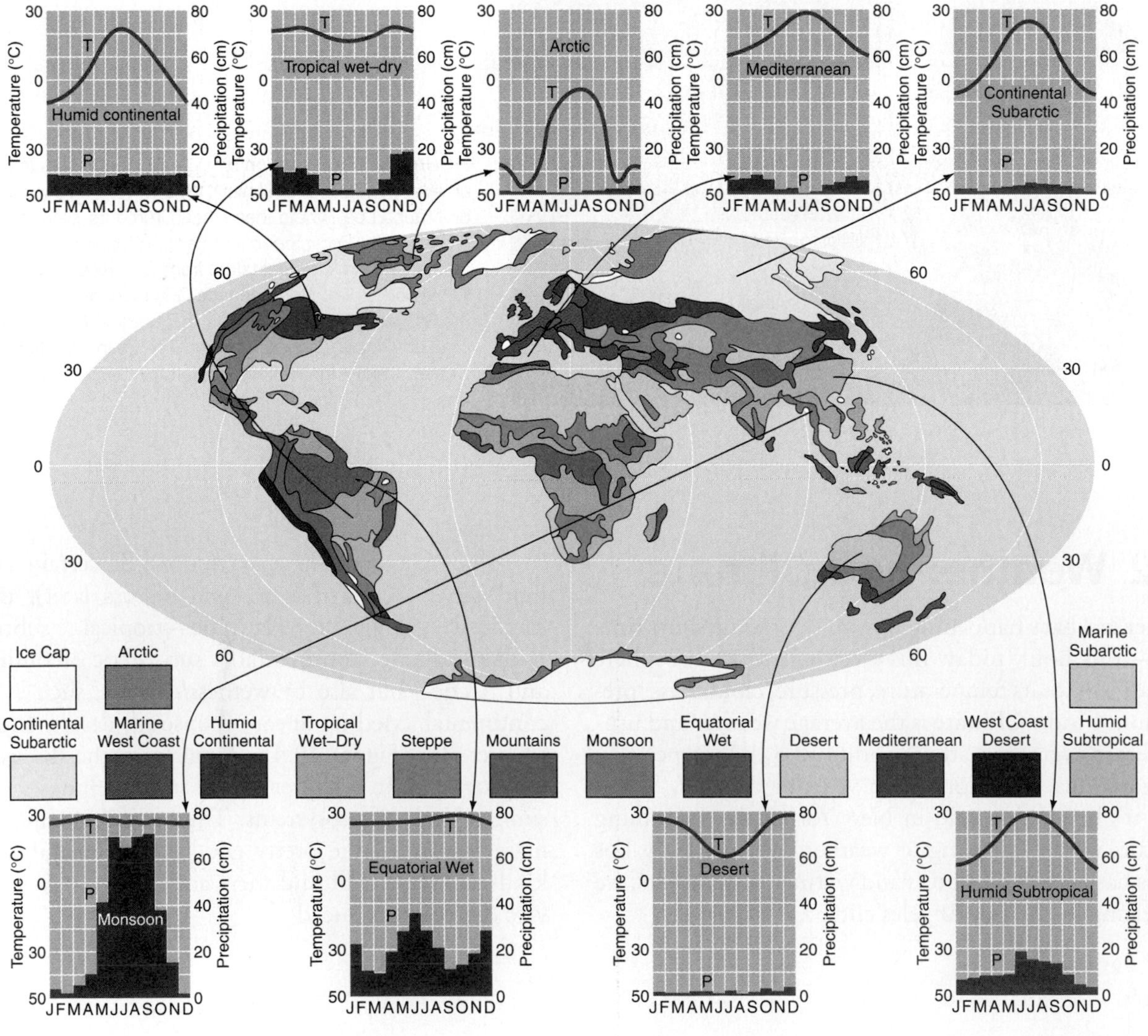

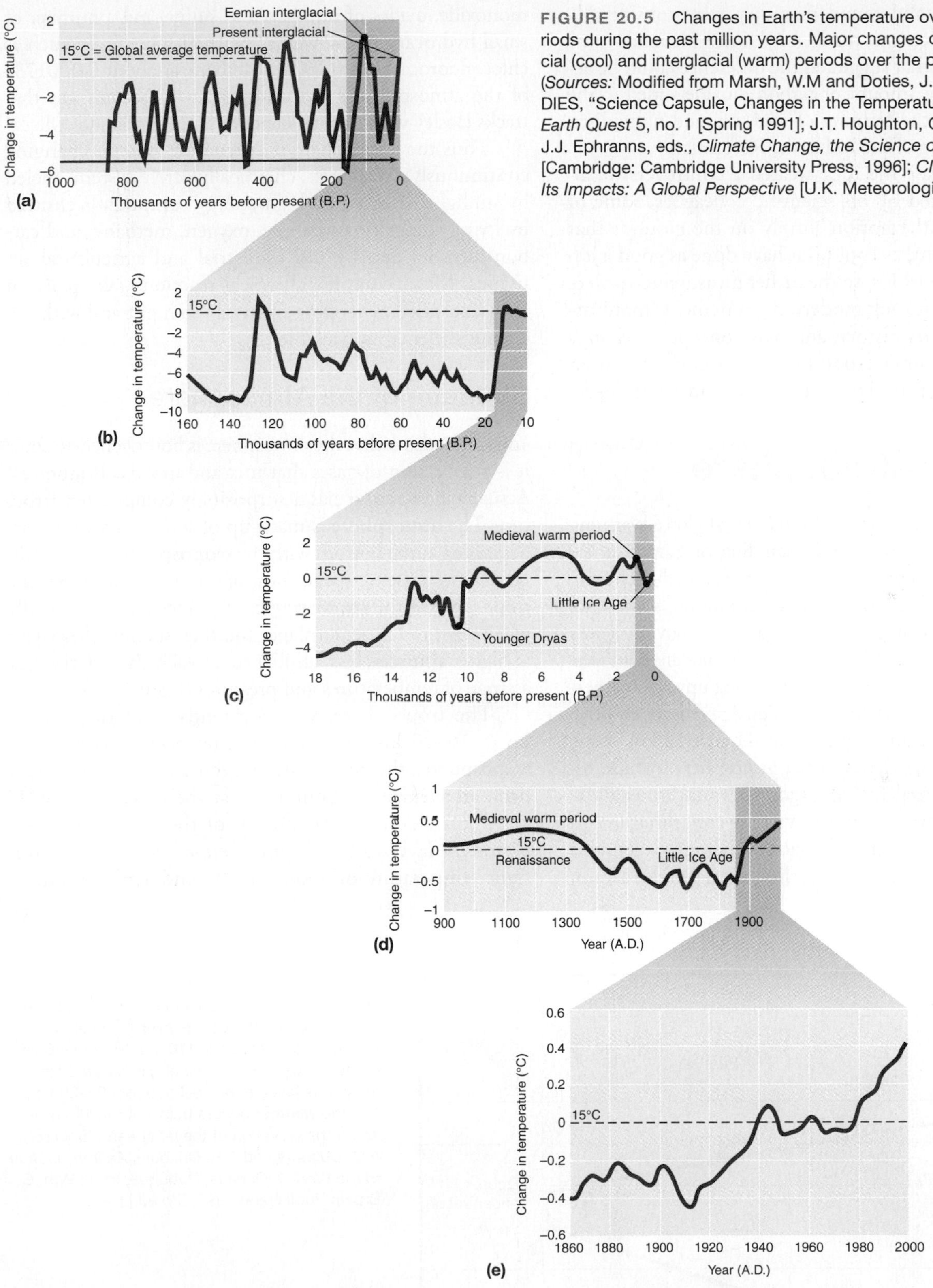

FIGURE 20.5 Changes in Earth's temperature over varying time periods during the past million years. Major changes correspond to glacial (cool) and interglacial (warm) periods over the past 800,000 years. (*Sources:* Modified from Marsh, W.M and Doties, J., 1981; and UCAR/DIES, "Science Capsule, Changes in the Temperature of the Earth," *Earth Quest* 5, no. 1 [Spring 1991]; J.T. Houghton, G.L. Jenkins, and J.J. Ephranns, eds., *Climate Change, the Science of Climate Change* [Cambridge: Cambridge University Press, 1996]; *Climate Change and Its Impacts: A Global Perspective* [U.K. Meteorological Office, 1997].

20.3 The Origin of the Global Warming Issue

That burning fossil fuels might enhance the levels of **greenhouse gases**—gases that warm the Earth's surface—was first proposed in the early 19th century, about half a century after the discovery of carbon dioxide, oxygen, and the other gases that make up the atmosphere. But well into the 20th century most scientists did not take the idea of global warming seriously. It just seemed impossible that people could be affecting the entire planet. For example, in 1938 the scientist Gary Stewart Callendar studied measurements of carbon dioxide concentration in the atmosphere taken in the 19th century and found that they were considerably lower than measurements made at his time.[9] He made some other calculations that suggested the difference could be accounted for by the amount of carbon dioxide added to the atmosphere from the burning of coal, oil, and natural gas since the beginning of the Industrial Revolution.

As it has today, the idea created a controversy. Callendar was attacked by his scientific colleagues, some of whom dismissed the notion simply on the grounds that 19th-century scientists could not have done as good a job as scientists in the 1930s, so the earlier measurements were likely inaccurate. It took modern measurement, monitoring, study of Earth's history, and new concepts to change the way scientists understood life and its environment at a global level. We turn now to that new understanding.

20.4 The Atmosphere

To understand and answer the fundamental global warming questions, we need a basic understanding of Earth's atmosphere. This atmosphere is the thin layer of gases that envelop Earth. These gases are almost always in motion, sometimes rising, sometimes falling, most of the time moving across Earth's surface. The atmosphere's gas molecules are held near to the Earth's surface by gravity and pushed upward by thermal energy—heating—of the molecules. Approximately 90% of the weight of the atmosphere is in the first 12 km above Earth's surface. Major gases in the atmosphere include nitrogen (78%), oxygen (21%), argon (0.9%), carbon dioxide (0.03%), and water vapor in varying concentrations in the lower few kilometers. The atmosphere also contains trace amounts of methane ozone, hydrogen sulfide, carbon monoxide, oxides of nitrogen and sulfur, and a number of small hydrocarbons, as well as synthetic chemicals, such as chlorofluorocarbons (CFCs). Methane at about 0.00017% of the atmosphere is emerging as an important gas that tracks closley with climate change (more so than CO_2).

Thus the atmosphere is a dynamic system, changing continuously. It is a vast, chemically active system, fueled by sunlight, affected by high-energy compounds emitted by living things (for example, oxygen, methane, and carbon dioxide) and by our industrial and agricultural activities. Many complex chemical reactions take place in the atmosphere, changing from day to night and with the chemical elements available.

Structure of the Atmosphere

You might think that the atmosphere is homogeneous, since it is a collection of gases that mix and move continuously. Actually, however, it has a surprisingly complicated structure. The **atmosphere** is made up of several vertical layers, beginning at the bottom with the **troposphere,** most familiar to us because we spend most of our lives in it. Above the troposphere is the **stratosphere,** which we visit occasionally when we travel by jet airplane, and then several other layers at higher altitudes, less familiar to us, each characterized by a range of temperatures and pressures (Figure 20.6).

The troposphere, which extends from the ground up to 10–20 km, is where weather occurs. Within the troposphere, the temperature decreases with elevation, from an average of about 17°C at the surface to –60°C at 12 km elevation. At the top of the troposphere is a boundary layer called the *tropopause,* which has a constant temperature of about –60°C and acts as a lid, or

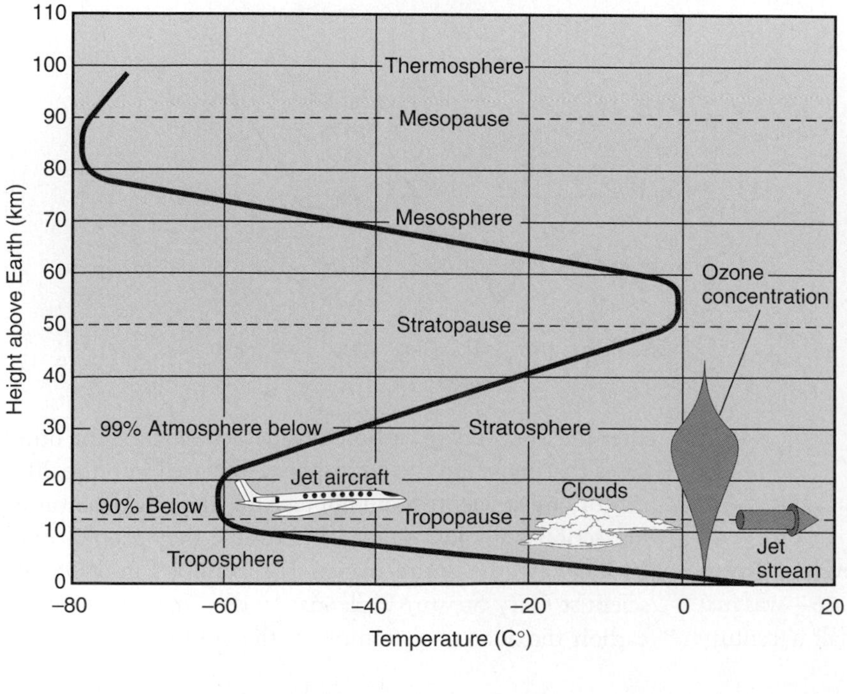

FIGURE 20.6 An idealized diagram of the structure of the atmosphere showing temperature profile and the ozone layer of the atmosphere to an altitude of 110 km. Note that 99% of the atmosphere (by weight) is below 30 km, the ozone layer is thickest at about 25–30 km, and the weather occurs below about 11 km—about the elevation of the jet stream. (*Source:* A.C. Duxbury and A.B. Duxbury, *An Introduction to the World's Oceans* [Dubuque, Iowa: Wm. C. Brown Publishers, 1997, 5th ed.].)

cold trap, on the troposphere because it is where almost all remaining water vapor condenses.

Another important layer for life is the *stratospheric ozone layer,* which extends from the tropopause to an elevation of approximately 40 km (25 mi), with a maximum concentration of ozone above the equator at about 25–30 km (16–19 mi) (Figure 20.6). Stratospheric ozone (O_3) protects life in the lower atmosphere from receiving harmful doses of ultraviolet radiation (see Chapter 21).

Atmospheric Processes: Temperature, Pressure, and Global Zones of High and Low Pressure

Two important qualities of the atmosphere are *pressure* and *temperature.* Pressure is force per unit area. Atmospheric pressure is caused by the weight of overlying atmospheric gases on those below and therefore decreases with altitude. At sea level, atmospheric pressure is 10^5 N/m^2 (newtons per square meter) (14.7 lb/in). We are familiar with this as **barometric pressure**, which the weatherman gives to us in units that are the height to which a column of mercury is raised by that pressure. We are also familiar with low- and high-pressure systems in the atmosphere. When the air pressure is low, air tends to rise, cooling as it rises and condensing its water vapor; it is therefore characterized by clouds and precipitation. When air pressure is high, it is moving downward, which warms the air, changing the condensed water drops in clouds to vapor; therefore high-pressure systems are clear and sunny.

Temperature, familiar to us as the relative warmth or coldness of materials, is a measure of thermal energy, which is the *kinetic* energy—the motion of atoms and molecules in a substance.

Water vapor content is another important characteristic of the lower atmosphere. It varies from less than 1% to about 4% by volume, depending on air temperature, air pressure, and availability of water vapor from the surface.

The atmosphere moves because of the Earth's rotation and differential heating of Earth's surface and atmosphere. These produce global patterns that include prevailing winds and latitudinal belts of low and high air pressure from the equator to the poles. Three cells of atmospheric circulation (Hadley cells) are present in

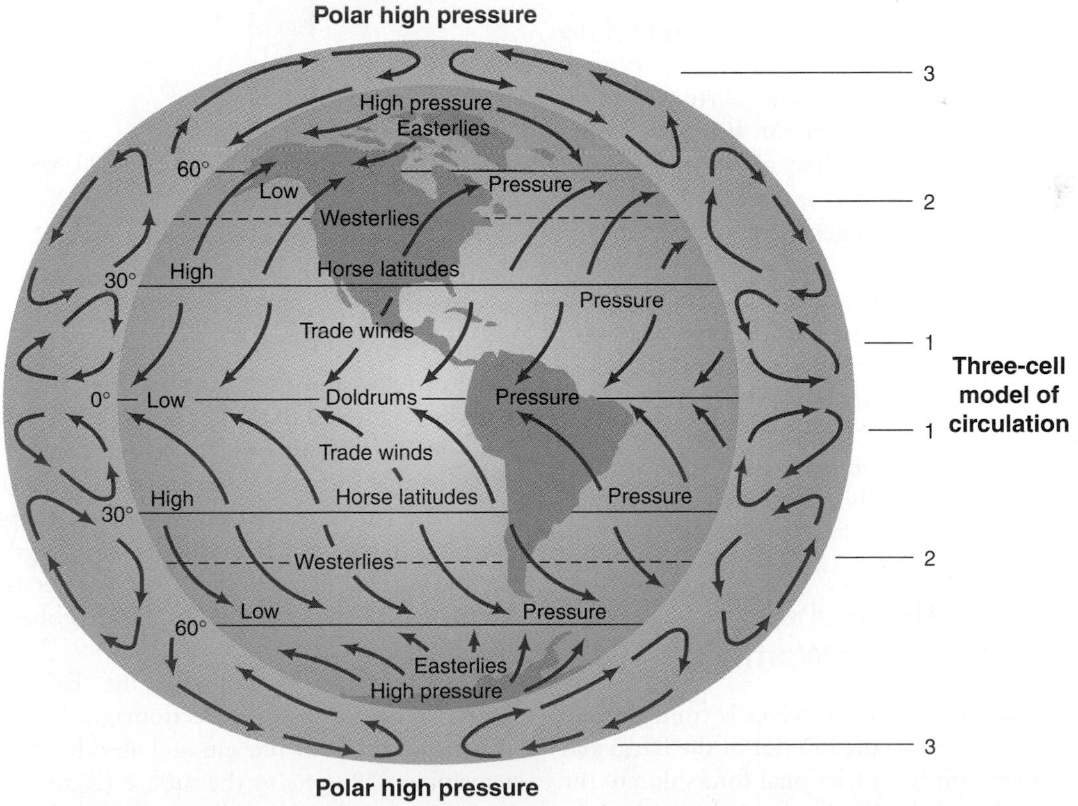

FIGURE 20.7 Generalized circulation of the atmosphere. The heating of the surface of the Earth is uneven, producing pressure differences (warm air is less dense than cooler air). There is rising warm air at the equator and sinking cool air at the poles. With rotation of Earth three cells of circulating air are formed in each hemisphere (called Hadley Cells after George Hadley who first proposed a model of atmospheric circulation in 1735). (*Source:* Samuel J. Williamson, *Fundamentals of Air Pollution*, Figure 5.5 [Reading, MS: Addison-Wesley, 1973]. Reprinted with permission of Addison-Wesley.)

each hemisphere (see Figure 20.7 for more details). In general belts of low air pressure develop at the equator where the air is warmed most of the time during the day by the sun. The heated air rises, creating an area of low pressure and a cloudy and rainy climate (cell 1 Figure 20.7). This air then moves to higher latitudes (toward the poles), and because it is cooler at higher elevations and because sunlight is less intense at the higher latitudes. By the time the air that was heated at the equator reaches about 30° latitude, it has cooled enough to become heavier, and it descends, creating a region of high pressure, with its characteristic sunny skies and low rainfall, forming a latitude belt where many of the world's deserts are found. Then the air that descended at 30° latitude moves poleward along the surface warms and rises again, creating another region of generally low pressure around 50° to 60° (cell 2, Figure 20.7) latitude and once again becoming a region of clouds and precipitation. Atmospheric orientation in cell 3 (Figure 20.7) moves air toward the poles at higher elevation and toward the equator along the surface. Sinking cool air at the poles produces the polar high-pressure zones at both poles. At the most basis level warm air rises at the equator moves toward the poles where it sinks after going through (cell 2) and return flow is along the surface of Earth toward the equator.

Of course, the exact locations of the areas of rising (low-pressure) and falling (high-pressure) air vary with the season, as the sun's position moves north and south relative to the Earth's surface. You can begin to understand that what would seem at first glance to be just a simple container of gases has complicated patterns of movement and that these change all the time for a variety of reasons.

The latitudinal belts (cells) just described have names, most of which came about during the days of sailing ships. Such names include the "doldrums," regions at the equator with little air movement; "trade winds," northeast and southeast winds important when clipper ships moved the world's goods; and "horse latitudes," two belts centered about 30° north and south of the equator with descending air and high pressure.

Energy and the Atmosphere: What Makes the Earth Warm

Almost all the energy the Earth receives is from the sun (a small amount comes from the interior of the Earth and an even smaller amount from frictional forces due to the moon revolving around the Earth). Sunlight comes in a wide range of electromagnetic radiation, from very long radio waves to much shorter infrared waves, then shorter wavelengths of visible light, even shorter wavelengths

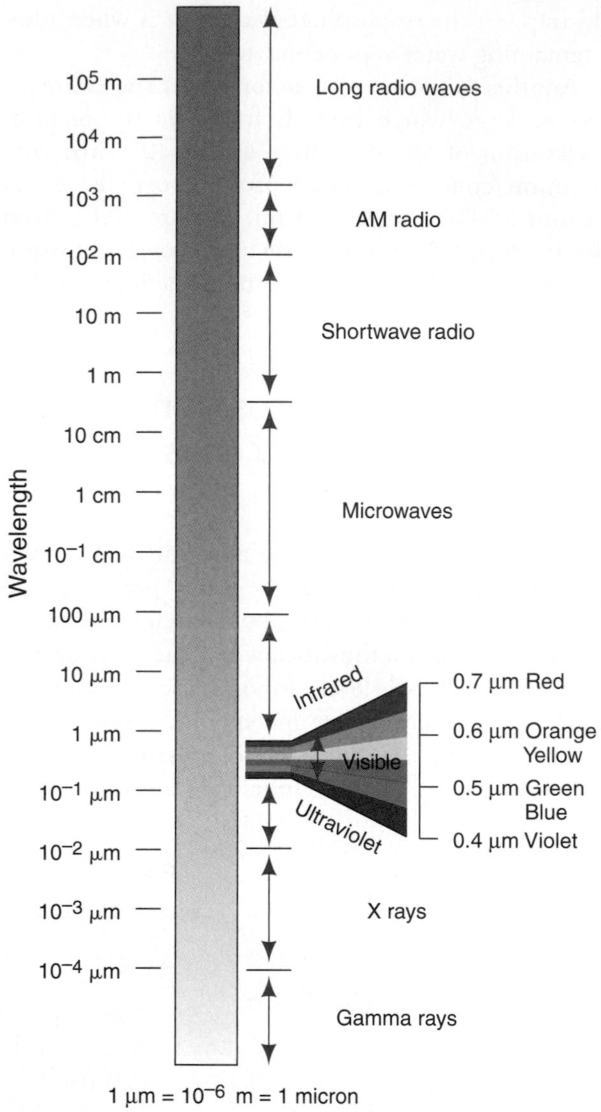

FIGURE 20.8 Kinds of electromagnetic radiation the Earth receives.

of ultraviolet, and then on to shorter and shorter wavelengths (Figure 20.8).

Most of the sun's radiation that reaches the Earth is in the visible and near infrared wavelengths (Figure 20.9), while the Earth, much cooler, radiates energy mostly in the far infrared, which has longer wavelengths. (The hotter the surface of any object, the shorter the dominant wavelengths. That's why a hot flame is blue and a cooler flame red.)

Under typical conditions, the Earth's atmosphere reflects about 30% of the electromagnetic (radiant) energy that comes in from the sun and absorbs about 25%. The remaining 45% gets to the surface (Figure 20.10). As the surface warms up, it radiates more energy back to the atmosphere, which absorbs some of it. The warmed atmosphere radiates some of its energy upward into outer space and some downward to the Earth's surface.

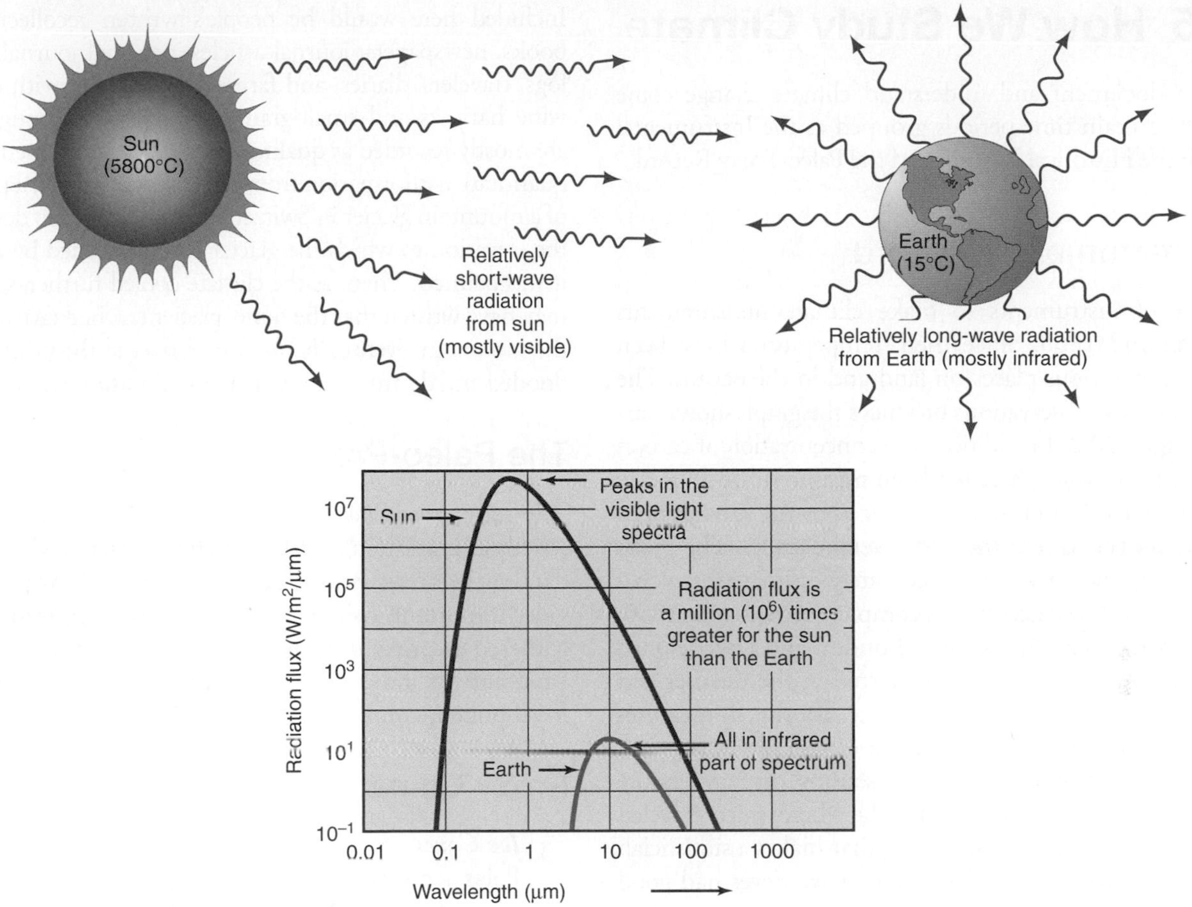

FIGURE 20.9 The sun, much hotter than the Earth, mostly emits energy in the visible and near infrared. The cooler Earth emits energy mostly in the far (longer-wavelength) infrared.

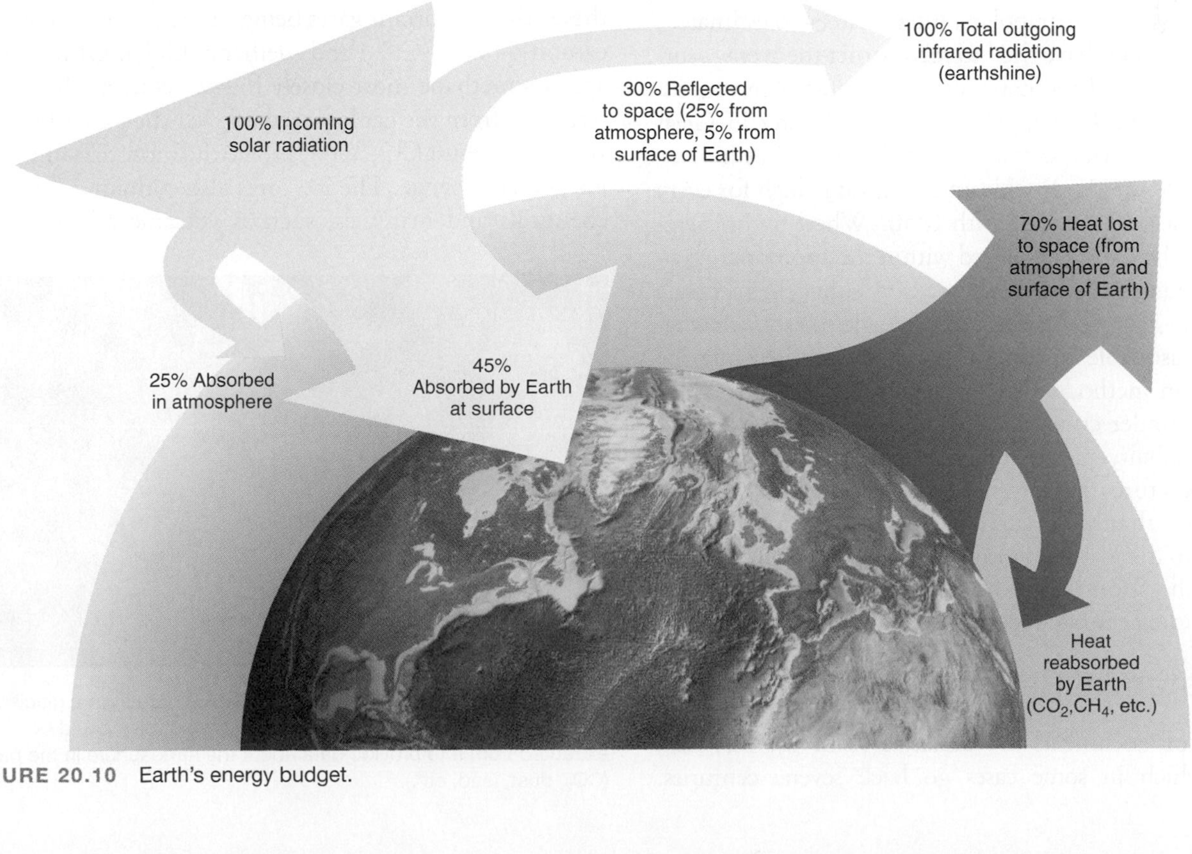

FIGURE 20.10 Earth's energy budget.

20.5 How We Study Climate

Data to document and understand climate change come from three main time periods grouped as the Instrumental Record; the Historical Record; and the Paleo-Proxy Record.[3, 10]

The Instrumental Record

The use of instruments to make climate measurements began around 1860. Since then, temperatures have been measured at various places on land and in the oceans. The average of these observations produces the graph shown earlier in Figure 20.2. In addition, the concentration of carbon dioxide in the atmosphere has been measured continuously since about 1957, and energy produced by the sun has been carefully measured over the past several decades. The problem in using these records to accurately estimate the global average is that few places have a complete record since 1850, and the places that do—such as London and Philadelphia, where people were especially interested in the weather and where scientists could readily make continual measurements—are not very representative of the global average. Until the advent of satellite remote sensing, air temperature over the oceans was measured only where ships traveled (which is not the kind of sampling that makes a statistician happy), and many of Earth's regions have never had good long-term, ground-based temperature measurements. As a result, when we want or need to know what the temperatures were like in the 19th century, before carbon dioxide concentrations began to rise from the burning of fossil fuels, experts seek ways to extrapolate, interpolate, and estimate.

Several groups have tried to reconstruct the average surface temperature of the Earth using available observations. For example, the Hadley Meteorological Center in Great Britain created a data set that divides Earth's surface into areas that are 5° longitude wide and 5° latitude high for every month of each year starting with 1850. Where historical records exist, these data are placed within the appropriate geographic rectangle. Where they are not, either the rectangle is left empty or various attempts are made to estimate what might be reasonable for that location and time. Recently, the extrapolation methods used to make these reconstructions have come under criticism, and today there is controversy over the reliability and usefulness of such attempts.

Temperature measurement has improved greatly in recent years thanks to such devices as ocean platforms with automatic weather-monitoring equipment, coordinated by the World Meteorological Organization. Thus, we have more accurate records since about 1960.

The Historical Record

A variety of documents are available from the historical records, which in some cases go back several centuries. Included here would be people's written recollections in books, newspapers, journal articles, personal journals, ships' logs, travelers' diaries, and farmers' logs, along with dates of wine harvests and small-grain harvests.[3, 10] Although these are mostly recorded as qualitative data, we can sometimes get quantitative information from them. For example, a painting of a mountain glacier in Switzerland can be used to determine the elevation to which the glacier had descended by the year it was painted. Then, as the climate cooled further, someone may have written that the same glacier reached farther down the mountain, eventually blocking a river in the valley, which flooded and destroyed a town, whose elevation is also known.

The Paleo-Proxy Record

The term *proxy data* refers to scientific data that are not strictly climatic in nature but can be correlated with climate data, such as temperature of the land or sea. Proxy data provides important insights into climate change. Information gathered as proxy data includes natural records of climate variability as indicated by tree rings, sediments, ice cores, fossil pollen, corals, and carbon-14 (^{14}C).[10]

Proxy Climate Records

Ice Cores

Polar ice caps and mountain glaciers have an accumulation record of snow that has been transformed into glacial ice over hundreds to thousands of years. Ice cores often contain small bubbles of air deposited at the time of the snow, and we can measure the atmospheric gases in these. Two important gases being measured in ice cores are carbon-dioxide (CO_2) and methane (CH_4). Of the two it appears methane most closely follows climate change determined from the geologic record over the past 1,000,000 years. As a result CO_2 and CH_4 are the most relevant proxy for climate change. The ice cores also contain a variety of chemicals and materials, such as volcanic ash and dust,

FIGURE 20.11 Scientist examining an ice core from a glacier. The core was stored in a freezer so that ice bubbles could be extracted from it to provide data about the atmosphere in the past (CO_2, dust, lead, etc.).

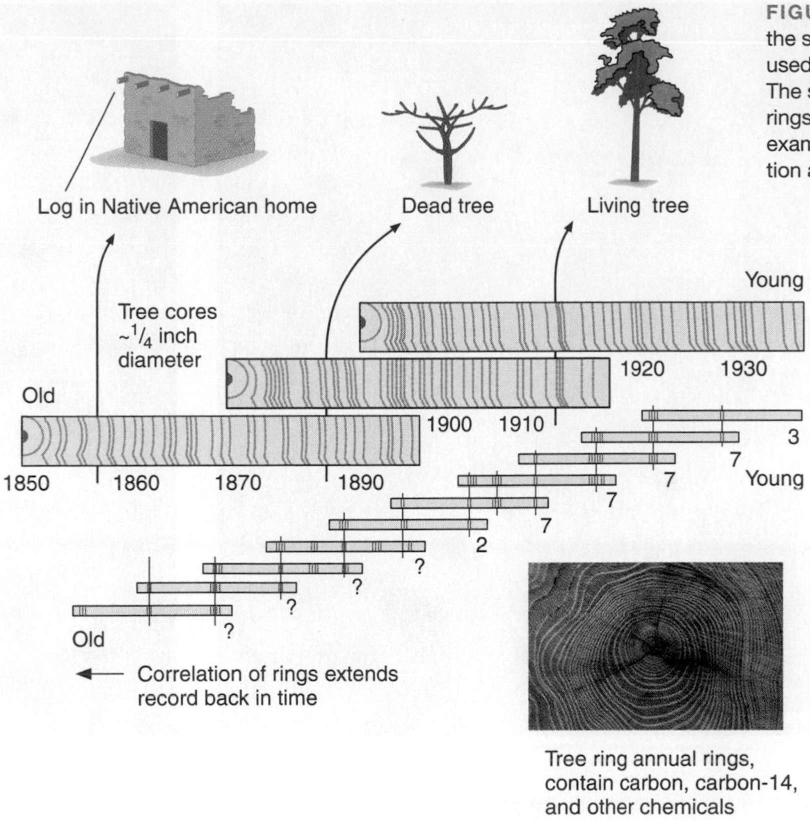

FIGURE 20.12 Dendrochronology, the study of tree growth rings, can be used as an indicator of past climate. The spacing (relative volume of wood) of rings and the isotopic content (^{14}C, for example) of wood can provide information about past rainfall and solar activity.

which may provide additional insights into possible causes of climate change. Ice cores are obtained by drilling into the ice (Figure 20.11). The age of glacial ice back to about 800,000 years is estimated by correlating ice accumulation rates linked to the geologic record of climate change from other proxy sources

Tree Rings

The growth of trees is influenced by climate, both temperature and precipitation. Many trees put on one growth ring per year, and patterns in the tree rings—their width, density, and isotopic composition—tell us something about the variability of the climate. When conditions are good for growth, a ring is wide; when conditions are poor, the ring is narrow. Tree-ring chronology, known as dendrochronology, has produced a proxy record of climate that extends back over 10,000 years (Figure 20.12).

Sediments

Biological material, including pollen from plants, is deposited on the land and stored for very long periods in lake, bog, and pond sediments and, once transported downstream to the coast, in the oceans. Samples may be taken of very small fossils and of chemicals in the sediments, and these may be interpreted to study past climates and extend our knowledge back hundreds of thousand years (Figure 20.13). Pollen is useful because (1) the quantity of pollen is an indicator of the relative abundance of each plant species; (2) the pollen can be dated, and since the grains are preserved in sedimentary layers that also might be dated, we can develop a chronology; and (3) based on the types of plants found at different times, we can construct a climatic history.

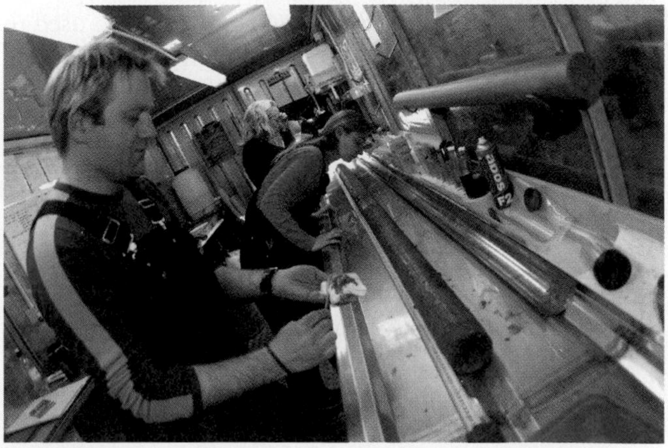

FIGURE 20.13 Scientist examining a sediment core taken by drilling into the seafloor.

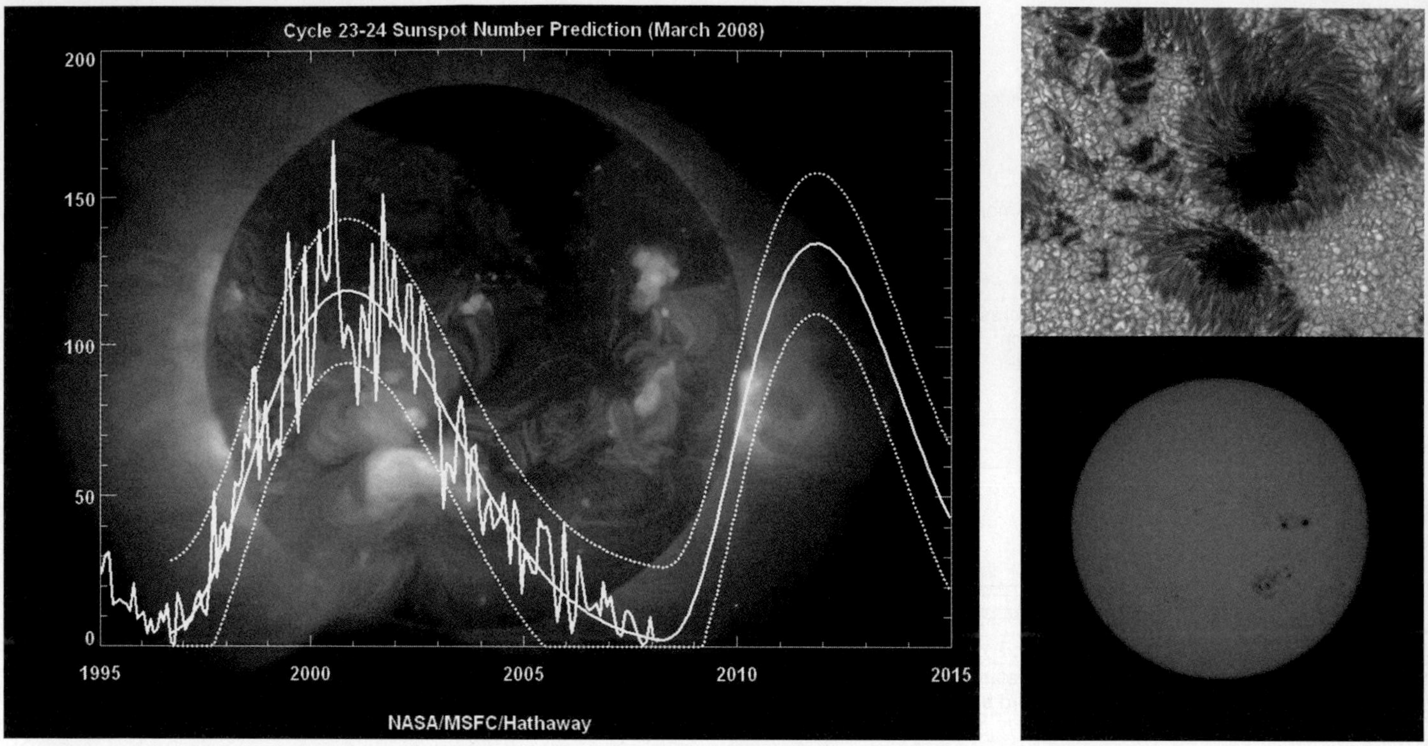

(a) (b)

FIGURE 20.14 **(a)** Sunspot cycle with number of sunspots per year, from 1995 to 2009 and **(b)** photos of sunspots in 2003. The large sunspots are about 150,000 km across (for comparison, Earth's diameter is about 12,750 km). The sunspots are the dark areas, and each is surrounded by an orange ring and then a hot gold color. Notice the great variety in the number of sunspots from year to year in the 11-year cycle (over 150 in 2001 and very few in 2008 and 2009). (*Source:* NOAA Images.)

Sediments recovered by drilling in the bottom of the ocean basin provide some of the very strongest evidence of past climate change.

Corals

Corals have hard skeletons composed of calcium carbonate ($CaCO_3$), a mineral extracted by the corals from seawater. The carbonate contains isotopes of oxygen, as well as a variety of trace metals, which have been used to determine the temperature of the water in which the coral grew. The growth of corals has been dated directly with a variety of dating techniques over short time periods of coral growth thereby revealing the chronology of climate change over variable time periods.

Carbon-14

Radioactive carbon-14 (^{14}C) is produced in the upper atmosphere by the collision of cosmic rays and nitrogen-14 (^{14}N). Cosmic rays come from outer space; those the Earth receives are predominantly from the sun. The abundance of cosmic rays varies with the number of sunspots, so called because they appear as dark areas on the sun (Figure 20.14). The frequency of sunspots has been accurately measured for decades and observed by people for nearly 1,000 years. As sunspot activity increases, more energy from the sun reaches Earth. There is an associated solar wind, which produces ionized particles consisting mostly of protons and electrons, emanating from the sun. The radioactive ^{14}C is taken up by photosynthetic organisms—green plants, algae, and some bacteria—and stored in them. If these materials become part of sediments (see above), the year at which they were deposited can be estimated from the decay rate of the ^{14}C.

The record of ^{14}C in the atmosphere has been correlated with tree-ring chronology. Each ring of wood of known age contains carbon, and the amount of ^{14}C can be measured. Then, given the climatic record, it may be correlated with ^{14}C, and that correlation has been shown to be very strong.[10]

Thus, we can examine the output of the sun, going back thousands of years, by studying tree rings and the carbon-14 they contain. This connects to our opening case study about the Medieval Warm Period. Based on these records, it appears that the production of solar energy was slightly higher around A.D. 1000, during the Medieval Warm Period, and slightly lower during the Little Ice Age that followed several hundred years later and lasted from A.D. 1300 to 1850.

20.6 The Greenhouse Effect

Each gas in the atmosphere has its own absorption spectrum—which wavelengths it absorbs and which it transmits. Certain gases in Earth's atmosphere are especially strong absorbers in the infrared and therefore absorb radiation emitted by the warmed surfaces of the Earth. Warmed by this, the gases re-emit this radiation. Some of it reaches back to the surface, making Earth warmer than it otherwise would be. The process by which the heat is trapped is not the same as in a greenhouse (air in a closed greenhouse has restricted circulation and will heat up). Still, in trapping heat this way, the gases act a little like the glass panes, which is why it is called the **greenhouse effect.** The major greenhouse gases are water vapor, carbon dioxide, methane, some oxides of nitrogen, and chlorofluorocarbons (CFCs). The greenhouse effect is a natural phenomenon that occurs on Earth and on other planets in our solar system. Most natural greenhouse warming is due to water in the atmosphere—water vapor and small particles of water in the atmosphere produce about 85% and 12%, respectively, of the total greenhouse warming.

How the Greenhouse Effect Works

Figure 20.15 is a highly idealized diagram of some important aspects of the greenhouse effect. The arrows labeled "energy input" represent the energy from the sun absorbed at or near Earth's surface. The arrows labeled "energy output" represent energy emitted from the upper atmosphere and Earth's surface, which balances input, consistent with Earth's energy balance. The highly contorted lines near the surface of the Earth represent the absorption of infrared radiation (IR) occurring there and producing the 15°C (59°F) near-surface temperature. Following many scatterings and absorptions and reemissions, the infrared radiation emitted from levels near the top of the atmosphere (troposphere) corresponds to a temperature of approximately 18°C (0°F). The one output arrow that goes directly through Earth's atmosphere represents radiation emitted through what is called the atmospheric window (Figure 20.16). The atmospheric window, centered on a wavelength of 10 m, is a region of wavelengths (8–12 m) where outgoing radiation from Earth is not absorbed well by natural greenhouse gases (water vapor and carbon dioxide). Anthropogenic CFCs do absorb in this region, however, and CFCs significantly contribute to the greenhouse effect in this way.

Let us look more closely at the relation of the greenhouse effect to Earth's energy balance, which was introduced in a simple way. The figure showed that, of the simple solar radiation, approximately 30% is reflected back to space from the atmosphere as shortwave solar radiation, while 70% is absorbed by Earth's surface and atmosphere. The 70% that is absorbed is eventually reemitted as infrared radiation into space. The sum of the reflected solar radiation and the outgoing infrared radiation balances with the energy arriving from the sun.

This simple balance becomes much more complicated when we consider exchanges of infrared radiation within the atmosphere and Earth surface. In some instances, these internal radiation fluxes (rates of transfer) may have magnitudes greater than the amount of energy entering Earth's atmospheric system from the sun, as shown in Figure 20.17. A major contributor to the fluxes is the greenhouse effect. At first glance, you might think it would be impossible to have internal radiation fluxes greater than the total amount of incoming solar radiation (shown as 100 units in Figure 20.17). It is possible because the infrared radiation is reabsorbed and readmitted many times in the

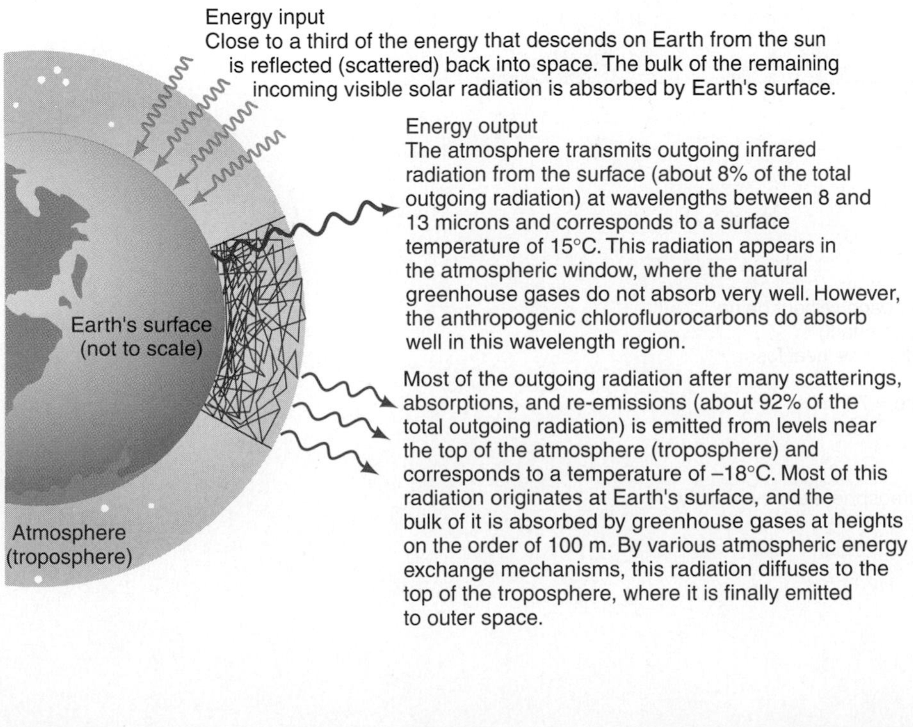

Energy input
Close to a third of the energy that descends on Earth from the sun is reflected (scattered) back into space. The bulk of the remaining incoming visible solar radiation is absorbed by Earth's surface.
Energy output
The atmosphere transmits outgoing infrared radiation from the surface (about 8% of the total outgoing radiation) at wavelengths between 8 and 13 microns and corresponds to a surface temperature of 15°C. This radiation appears in the atmospheric window, where the natural greenhouse gases do not absorb very well. However, the anthropogenic chlorofluorocarbons do absorb well in this wavelength region.
Earth's surface (not to scale)
Most of the outgoing radiation after many scatterings, absorptions, and re-emissions (about 92% of the total outgoing radiation) is emitted from levels near the top of the atmosphere (troposphere) and corresponds to a temperature of –18°C. Most of this radiation originates at Earth's surface, and the bulk of it is absorbed by greenhouse gases at heights on the order of 100 m. By various atmospheric energy exchange mechanisms, this radiation diffuses to the top of the troposphere, where it is finally emitted to outer space.
Atmosphere (troposphere)

FIGURE 20.15 Idealized diagram showing the greenhouse effect. Incoming visible solar radiation is absorbed by Earth's surface, to be reemitted in the infrared region of the electromagnetic spectrum. Most of this reemitted infrared radiation is absorbed by the atmosphere, maintaining the greenhouse effect. (*Source*: Developed by M.S. Manalis and E.A. Keller, 1990.)

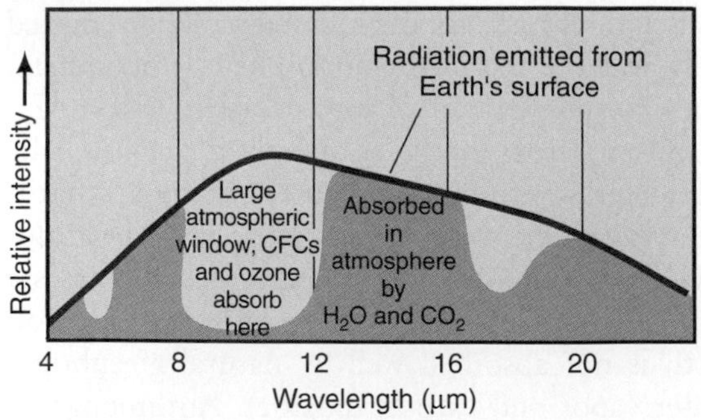

FIGURE 20.16 What the major greenhouse gases absorb in the Earth's atmosphere. Earth's surface radiates mostly in the infrared, which is the range of electromagnetic energy shown here. Water and carbon dioxide absorb heavily in some wavelengths within this range, making them major greenhouse gases. The other greenhouse gases, including methane, some oxides of nitrogen, CFCs, and ozone, absorb smaller amounts but in wavelengths not absorbed by water and carbon dioxide. (*Source*: Modified from T.G. Spiro and W.M. Stigliani, *Environmental Science in Perspective* [Albany: State University of New York Press, 1980].)

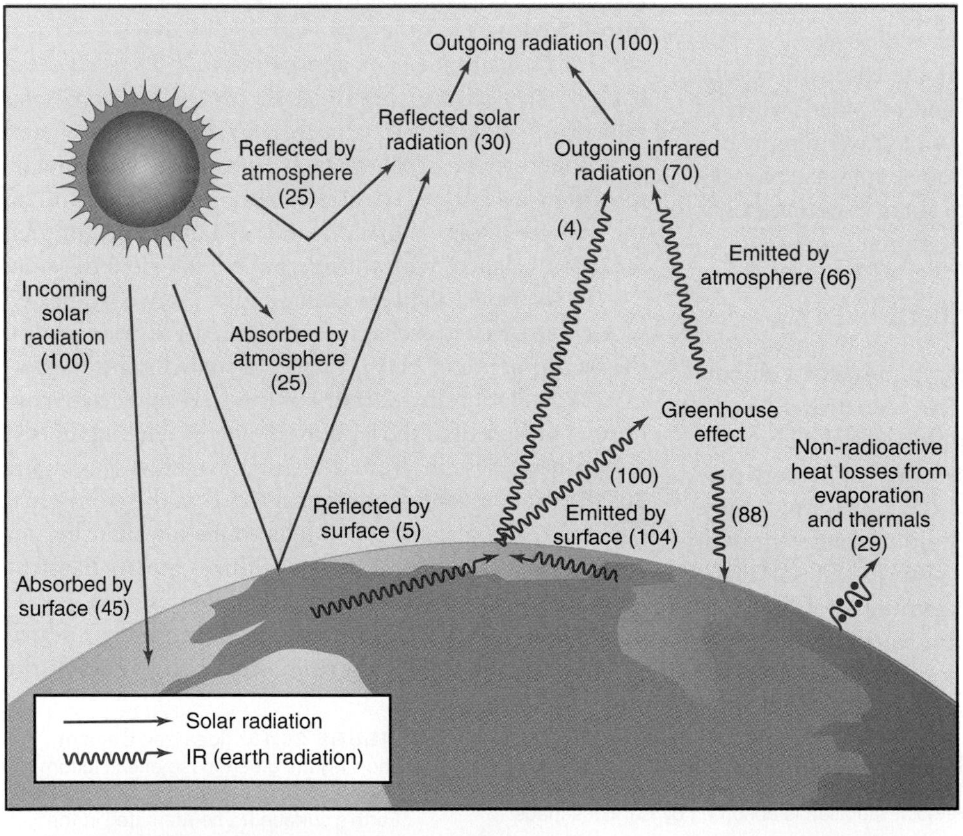

FIGURE 20.17 Idealized diagram showing Earth's energy balance and the greenhouse effect. Incoming solar radiation is arbitrarily set at 100 units, and this is balanced by outgoing radiation of 100 units. Notice that some of the fluxes (rates of transfer) of infrared radiation (IR) are greater than 100, reflecting the role of the greenhouse effect. Some of these fluxes are explained in the diagram. (*Source*: Modified from D.I. Hartmann, *Global Physical Climatology, International Geophysics Series*, vol. 56 [New York: Academic Press. 1994] and S. Schneider, "Climate Modeling," *Scientific American* 256 No. 5.)

•Total incoming solar radiation = 100 units

•Total absorbed by surface = 133 units
  45 from solar radiation (shortwave)
  88 from greenhouse effect IR (infrared)

•Total emitted by surface = 133 units
  104 IR (of this, only 4 units pass directly to space without being absorbed or re-emitted in greenhouse effect)
  29 from evaporation and thermals (non-radioactive heat loss)

•Total IR emitted by upper atmosphere to space = 70 units
  66 units emitted by atmosphere
  4 units emitted by surface

•The 25 units of solar radiation absorbed by atmosphere are eventually emitted as IR (part of the 66 units)

•Total outgoing radiation = 100 units
  70 IR
  30 reflected solar radiation

atmosphere, resulting in high internal fluxes. For example, in terms of the figure, the amount of IR absorbed at Earth's surface from the greenhouse effect is approximately 88 units, which is about twice the amount of shortwave solar radiation (45 units) absorbed by Earth's surface. Despite the large internal fluxes, the overall energy balance remains the same. At the top of the atmosphere, the net downward solar radiation (70 units, 45 units + 25 units) balances the outgoing IR from the top of the atmosphere (70 units).

The important point here is recognizing the strength of the greenhouse effect. For example, notice in the figure that of the 104 units of IR emitted by the surface of Earth, only 4 go directly to the upper atmosphere and are emitted. The rest is reabsorbed and reemitted by greenhouse gases. Of these, 88 units are directed downward to Earth and 66 units upward to the upper atmosphere.

All this may sound somewhat complicated, but if you read and study the points mentioned in Figures 20.15 to 20.17 carefully and work through the balances of the various parts of the energy fluxes, you will gain a deeper understanding of why the greenhouse effect is so important. The greenhouse effect keeps Earth's lower atmosphere approximately 33°C warmer than it would otherwise be and performs other important service functions as well. For example, without the strong downward emission of IR from the greenhouse effect, the land surface would cool much faster at night and warm much more quickly during the day. In sum, the greenhouse effect helps to limit temperature swings from day to night and maintain relatively comfortable surface temperatures. It is, then, not the greenhouse effect itself but the changes in greenhouse gases that are a concern.

20.7 The Major Greenhouse Gases

The major anthropogenic greenhouse gases are listed in Table 20.1. The table also lists the recent rate of increase for each gas and its relative contribution to the anthropogenic greenhouse effect.

Carbon Dioxide

Current estimates suggest that approximately 200 billion metric tons of carbon in the form of carbon dioxide (CO_2) enter and leave Earth's atmosphere each year as a result of a number of biological and physical processes: 50 to 60% of the anthropogenic greenhouse effect is attributed to this gas. Measurements of carbon dioxide trapped in air bubbles in the Antarctic ice sheet suggest that 160,000 years before the Industrial Revolution the atmospheric concentration of carbon dioxide varied from approximately 200 to 300 ppm.[12] The highest level or concentration of carbon dioxide in the atmosphere, other than today's, occurred during the major interglacial period about 125,000 years ago.

About 140 years ago, just before the major use of fossil fuels began as part of the Industrial Revolution, the atmospheric concentration of carbon dioxide was approximately 280 ppm.[13] Since then, and especially in the past few decades, the concentration of CO_2 in the atmosphere has grown rapidly. Today, the CO_2 concentration is about 392 ppm, and at its current rate of increase of about 0.5% per year, the level may rise to approximately 450 ppm by the year 2050—more than 1.5 times the preindustrial level.[13]

Table 20.1 MAJOR GREENHOUSE GASES

TRACE GASES	RELATIVE CONTRIBUTION (%)	GROWTH RATE (%/YR)
CFC	15[a]-25[b]	5
CH_4	12[a]-20[b]	0.4[c]
O_3(troposphere)	8[d]	0.5
N_2O	5[d]	0.2
Total	40–50	
Contribution of CO_2	50–60	0.3[e]–0.5[d,f]

[a]W. A. Nierenberg, "Atmospheric CO_2: Causes, Effects, and Options," *Chemical Engineering Progress* 85, no.8 (August 1989): 27

[b]J. Hansen, A. Lacis, and M. Prather, "Greenhouse Effect of Chlorofluorocarbons and Other Trace Gases," *Journal of Geophysical Research* 94 (November 20, 1989): 16, 417.

[c]Over the past 200 yrs.

[d]H. Rodha, "A Comparison of the Contribution of Various Gases to the Greenhouse Effect," *Science* 248 (1990): 1218, Table 2.

[e]W. W. Kellogg, "Economic and Political Implications of Climate Change," paper presented at Conference on Technology-based Confidence Building: Energy and Environment, University of California, Los Alamos National Laboratory, July 9–14, 1989.

[f]H. Abelson, "Uncertainties about Global Warming," *Science* 247 (March 30,1990): 1529.

Methane

The concentration of methane (CH_4) in the atmosphere more than doubled in the past 200 years and is thought to contribute approximately 12 to 20% of the anthropogenic greenhouse effect.[14, 15] Certain bacteria that can live only in oxygenless atmospheres produce methane and release it. These bacteria live in the guts of termites and the intestines of ruminant mammals, such as cows, which produce methane as they digest woody plants. These bacteria also live in oxygenless parts of freshwater wetlands, where they decompose vegetation, releasing methane as a decay product. Methane is also released with seepage from oil fields and seepage from methane hydrates (see Chapter 15).

Our activities also release methane. These activities include landfills (the major methane source in the United States), the burning of biofuels, production of coal and natural gas, and agriculture, such as raising cattle and cultivating rice. (Methane is also released by anaerobic activity in flooded lands where rice is grown.) As with carbon dioxide, there are important uncertainties in our understanding of the sources and sinks of methane in the atmosphere.

Chlorofluorocarbons

Chlorofluorocarbons (CFCs) are inert, stable compounds that have been used in spray cans as aerosol propellants and in refrigerators. The rate of increase of CFCs in the atmosphere in the recent past was about 5% per year, and it has been estimated that approximately 15 to 25% of the anthropogenic greenhouse effect may be related to CFCs. Because they affect the stratospheric ozone layer and also play a role in the greenhouse effect, the United States banned their use as propellants in 1978. In 1987, 24 countries signed the Montreal Protocol to reduce and eventually eliminate production of CFCs and accelerate the development of alternative chemicals. As a result of the treaty, production of CFCs was nearly phased out by 2000.

Potential global warming from CFCs is considerable because they absorb in the atmospheric window, as explained earlier, and each CFC molecule may absorb hundreds or even thousands of times more infrared radiation emitted from Earth than is absorbed by a molecule of carbon dioxide. Furthermore, because CFCs are highly stable, their residence time in the atmosphere is long. Even though their production was drastically reduced, their concentrations in the atmosphere will remain significant (although lower than today's) for many years, perhaps for as long as a century.[16, 17] (CFCs are discussed in Chapter 21, which examines stratospheric ozone depletion.)

Nitrous Oxide

Nitrous oxide (N_2O) is increasing in the atmosphere and probably contributes as much as 5% of the anthropogenic greenhouse effect.[18] Anthropogenic sources of nitrous oxide include agricultural application of fertilizers and the burning of fossil fuels. This gas, too, has a long residence time; even if emissions were stabilized or reduced, elevated concentrations of nitrous oxide would persist for at least several decades.

20.8 Climate Change and Feedback Loops

Part of the reason climate change is so complex is that there can be many positive and negative feedback loops. Only a few possible feedback loops of many are discussed here. Negative feedbacks are self-inhacing and help stabilize a system. Positive feedbacks are self-regulating, so a greater change now will result in an even greater change in the future. This is a simplistic statement about feedback and climate because some changes are associated with both positive and negative feedback. With this caveat stated, we will discuss positive and negative feedbacks with respect to climate change. We discussed feedback in Chapter 3; you may wish to review those concepts.

Here are some feedback loops that have been suggested for climate change.[19]

Possible Negative Feedback Loops for Climate Change

- As global warming occurs, the warmth and additional carbon dioxide could stimulate algae growth. This, in turn, could absorb carbon dioxide, reducing the concentration of CO_2 in the atmosphere and cooling Earth's climate.
- Increased CO_2 concentration with warming might similarly stimulate growth of land plants, leading to increased CO_2 absorption and reducing the greenhouse effect.
- If polar regions receive more precipitation from warmer air carrying more moisture, the increasing snowpack and ice buildup could reflect solar energy away from Earth's surface, causing cooling.
- Increases in water evaporation with warming from the ocean and the land could lead to cloudier conditions (the water vapor condenses), and the clouds would reflect more sunlight and cool the surface.

Possible Positive Feedback Loops for Climate Change

- The warming Earth increases water evaporation from the oceans, adding water vapor to the atmosphere. Water vapor is a major greenhouse gas that, as it increases, causes additional warming. If more clouds form from the increaced water vapor, and more solar radiation is reflected this would cause cooling as discussed with negative feedback above. Thus water vapor is associated with both positive and negative feedback. This makes study of clouds and global climate change complex.
- The warming Earth could melt a large amount of permafrost at high latitudes, which would in turn release the greenhouse gas methane, a by-product of decomposition of organic material in the melted permafrost layer. This would cause additional warming.
- Replacing some of the summer snowpack or glacial ice with darker vegetation and soil surfaces decreases the albedo (reflectivity) increasing the absorption of solar energy, further warming surface. This is a powerful positive feedback explaining, in part, why the Arctic is warming faster than at lower latitudes.
- In warming climates, people use more air-conditioning and thus more fossil fuels. The resulting increase in carbon dioxide could lead to additional global warming.

Since negative and positive feedback can occur simultaneously in the atmosphere, the dynamics of climate change are all the more complex. Research is ongoing to better understand negative feedback processes associated with clouds and their water vapor.

20.9 Causes of Climate Change

Not until the 19th century did scientists begin to understand that climate changed greatly over long periods and included times of continental glaciations. The realization that there had been glacial and interglacial episodes began in 1815 when a Swiss peasant, Jean-Paul Perraudin, suggested to a Swiss civil engineer, Ignaz Venetz-Sitten, that some features of mountain valleys, including the boulders and soil debris, were due to glaciers that in a previous time had extended down the slopes beyond their present limits. Impressed with these observations, Venetz-Sitten spoke before a natural history society at Lucerne in 1821 and suggested that the glaciers had at some previous time extended considerably beyond their present range.

At first, he wasn't taken seriously—in fact, the famous 19th-century geologist Louis Agassiz traveled to the Alps to refute these ideas. But once Agassiz saw the evidence, he changed his mind and formulated a theory of continental glaciation. The evidence was debris—rocks and soils—at the edges of existing mountain glaciers and the same kinds of deposits at lower elevations. Agassiz realized that only glaciers could have produced the kinds of debris now far below the ice. It was soon recognized that glaciers had covered vast areas in Great Britain, mainland Europe, and North America.[20]

This began the search for an answer to a puzzling question: Why does the climate change, and change so drastically? One of the most important insights was achieved in the 1920s by Milutin Milankovitch, a Serbian astronomer who looked at long-term climate records and began to think about what might correlate with these records. Look at Figure 20.18 and you will see that cycles of about 100,000 years are apparent; these seem to be divided as well into shorter cycles of about 20,000 to 40,000 years.

Milankovitch Cycles

Milankovitch realized that the explanation might have to do with the way the Earth revolved on its axis and rotated around the sun. Our spinning Earth is like a wobbling top following an elliptical orbit around the sun. Three kinds of changes occur.

First, the wobble means that the Earth is unable to keep its poles at a constant angle in relation to the sun (Figure 20.18a). Right now, the North Pole points to Polaris, the North Star, but this changes as the planet wobbles. The wobble makes a complete cycle in 26,000 years.

Second, the tilt of Earth's axis varies over a period of 41,000 years (Figure 20.18b).

Third, the elliptical orbit around the sun also changes. Sometimes it is a more extreme ellipse; at other times it is closer to a circle (Figure 20.18c), and this occurs over 100,000 years.

The combination of these changes leads to periodic changes in the amount and distribution of sunlight reaching the Earth. Sometimes the wobble causes the Northern Hemisphere to be tilted toward the sun (Northern Hemisphere summertime) when the Earth is closest to the sun. At other times, the opposite occurs—the Northern Hemisphere is tipped away from the sun (northern wintertime) when the Earth is closest to the sun. Milankovitch showed that these variations correlated with the major glacial and interglacial periods (Figure 20.18). They are now called Milankovitch cycles.[21]

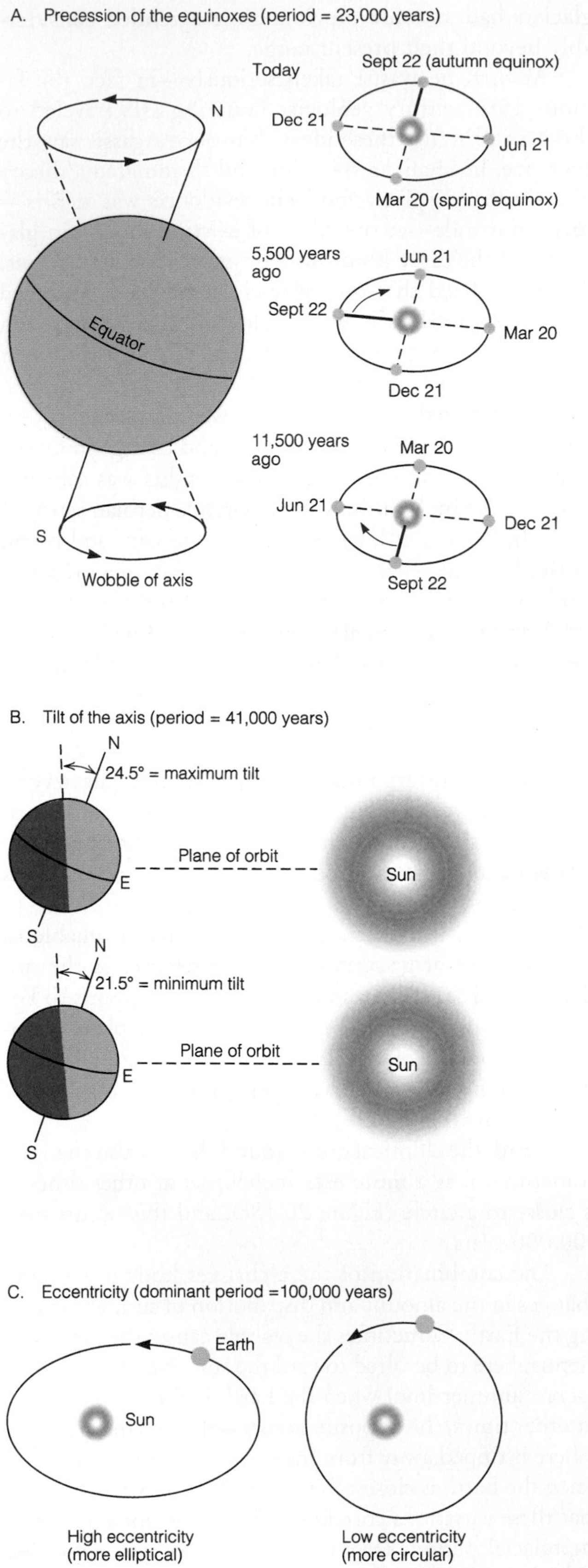

FIGURE 20.18 The Earth wobbles, changes its tilt, and has an elliptical orbit that changes as well. (*Source:* Skinner, Porter, and Botkin, *The Blue Planet* [John Wiley], 2nd edition, p. 335.)

Milankovich attempted explain ice ages through changes in solar radiation reaching Earth. His contribution is very significant. While Milankovitch cycles are consistent with the timing of variations in glacial and inter glacial change, they were not intented to not account for all the large-scale climatic variations in the geologic record. It is perhaps best to think of these cycles as response of climate to orbital variations.

Once Earth receives energy from the sun, Earth's surface features affect the climate. These earthly factors that affect, and are in turn affected by, regional and global temperature changes include warmer ice-sheet temperatures; changes in vegetation; changes in atmospheric gases, such as carbon dioxide, methane, and nitrous oxide; and particulates and aerosols. Volcanoes inject aerosols into the upper atmosphere, where they reflect sunlight and cool the Earth's surface.

Solar Cycles

As we discussed earlier, the sun goes through cycles too, sometimes growing hotter, sometimes colder. Today, solar intensity is observed directly with telescopes and other instruments. Variations in the sun's intensity in the past can be determined because hotter and cooler sun periods emit different amounts of **radionuclides**—atoms with unstable nuclei that undergo radioactive decay (such as beryllium-10 and carbon-14), which are trapped in glacial ice and can then be measured. As we mentioned earlier, evaluation of these radionuclides in ice cores from glaciers reveals that during the Medieval Warm Period, from approximately A.D. 950 to 1250, the amount of solar energy reaching Earth was relatively high, and that minimum solar activity occurred during the 14th century, coincident with the beginning of the Little Ice Age. Thus, it appears that the variability of solar energy input explains a small part of the Earth's climatic variability.[22, 23] Since about 1880 solar input has increased about 0.5% while CO_2 has increased about 33%. Solar input in the Arctic has closely followed annual surface temperature. Since 1960, CO_2 increase in the atmosphere has been about 25% in close agreement with Arctic surface temperature increase. Thus in the past 50 years CO_2 appears to be a dominant factor in increasing surface temperature in the Arctic as well as the entire Earth.[24] That is recent warming cannot be explained by solar activity.

Atmospheric Transparency Affects Climate and Weather

How transparent the atmosphere is to the radiation coming to it, from both the sun and Earth's surface, affects the temperature of the Earth. Dust and aerosols absorb light, cooling the Earth's surfaces. Volcanoes and large forest fires put dust into the atmosphere, as do various human activities, such as plowing large areas. Each gas compound has its own absorption spectrum (the electromagnetic radiation that is absorbed by a gas as it passes through the atmosphere). Thus the chemical and physical composition of the atmosphere can make things warmer or cooler.

The Surface of Earth and Albedo (reflectivity) Affects

Albedo is the reflectivity of an object that is measured as the percentage of incoming radiation that is reflected. For examples the approximate albedos are: Earth (as a whole) is 30%, clouds depending on type and thickness are 40–90%, fresh snow is 85%, glacial ice depending on soil rock cover is 20–40%, a pine forest is 10% dark rock is 5–15%, dry sand is 40%, and a grass-covered meadow is 15%.

A dark rock surface exposed near the North Pole absorbs more of the sunlight it receives than it reflects in the summer, warming the surface and the air passing over it. When a glacier spreads out and covers that rock, it reflects more of the incoming sunlight than the darker rock cooling both the surface and the air that comes in contact with it.

Vegetation also affects the climate and weather in the same way. If vegetation is a darker color than the soil, it warms the surface. If it is a lighter color than the soil, it cools the surface. Now you know why if you walk barefoot on dark asphalt on a hot day you feel the heat radiating from the surface (you may burn the bottom of your feet).

Roughness of the Earth's Surface Affects the Atmosphere

Above a completely smooth surface, air flows smoothly—a flow called "laminar." A rough surface causes air to become turbulent—to spin, rotate, reverse, and so forth. Turbulent air gives up some of the energy in its motion (its kinetic energy), and that energy is turned into heat. This affects the weather above. Forests are a much rougher surface than smooth rock or glaciers, so in this way, too, vegetation affects weather and climate.

The Chemistry of Life Affects the Atmosphere

The emission and uptake of chemicals by living things affect the weather and climate, as we will discuss in detail in the next section. Thus, a planet with water vapor, liquid water, frozen water, and living things has a much more complex energy-exchange system than a lifeless, waterless planet. This is one reason (of many) why it is difficult to forecast climate change.

Climate Forcing

It can be helpful to view climate change in terms of ***climate forcing***—defined as an imposed perturbation of Earth's energy balance.[8] The major forcings associated with the glaciations are shown in Figure 20.19a. Factors that affect and are in turn affected by regional global temperature changes include higher ice-sheet temperatures; changes in vegetation; changes in atmospheric gases, such as carbon dioxide, methane, and nitrous oxide; and changes in sunlight intensity. Aerosols, such as those ejected by a volcano into the upper atmosphere, reflect sunlight and cool the Earth's surface. Changes in sunlight intensity are caused both by variations in sunlight brightness and by changes in Earth's orbit. Total energy forcing of the Last Glacial Maximum (LGM), around 22,000 years ago, when glaciers were at their thickest and sea levels their lowest, is calculated to have been about 6.6 ± 1.5 W/m^2. Thus, the average 5°C lowering equates to 3/4°C per W/m^2. The units are power per unit area.

Climate forcing during the industrial age is shown on Figure 20-19b. Positive forcings cause warm and negative forcings cause cooling. In recent decades human caused forcings have dominated over natural forcings. Total forcing is 1.6 ± 0.1 W/m^2, consistent with the observed rise in global surface air temperature over the past few decades.

Forcings operate by changing the properties of both the atmosphere and the surface that feed back into climate. As ice sheets grow, they reflect more incoming solar radiation, which enhances cooling. Changes in the amount of area covered by vegetation and the kinds of vegetation change reflectivity and absorption of solar energy and the uptake and release of atmospheric gases. Atmospheric gases, such as carbon dioxide, methane, and nitrous oxide, also play important roles. For now, it is more important to recognize that climate forcing is related to Earth's energy balance and as such is an important quantitative tool with which to evaluate global change in the geologic past, as well as global warming, which refers to the more recent rise in global surface air temperatures over the past few decades.[25]

As a way to visualize forcing, consider a checking account that you are free to use but earns no interest use. Assume you initially deposit $1,000, and each year you deposit $500 and write checks for $500. You do this for many years, and at the end of that time you still have $1,000 in your account. The point is that for any system, when input equals output for some material (in this case, dollars), the amount in the system remains constant. In our bank account example, if we increased the total amount in the account by only $3 per year (a 0.3% increase per year), it would double to $2,000 over a period of about 233 years. In short, a small imbalance over many years can cause significant change.

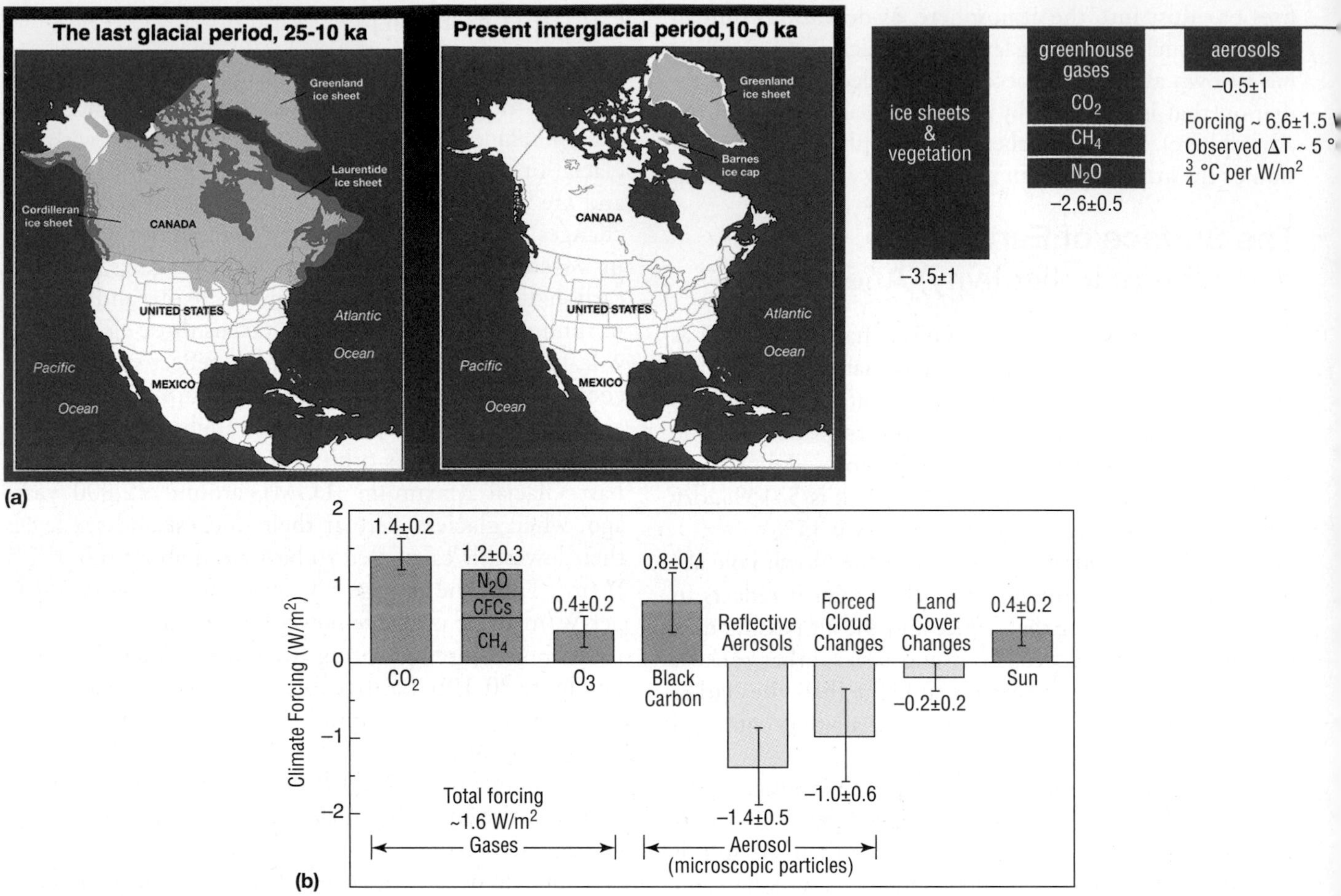

- Increases of greenhouse gases (except O_3) are known from observations and bubbles of air trapped in ice sheels. The increase of CO_2 from 285 parts per million (ppm) in 1850 to 392 in 2010 is accurate to about 5 ppm. The conversion of this gas change to a climate forcing (1.4 W/m^2), from calculation of the infrared opacity, adds about 10% to the uncertainty.
- Increase of CH_4 since 1850, including its effect on stratospheric H_2O and tropospheric O_3, causes a climate forcing about half as large as that by CO_2. Main sources of CH_4 include landfills, coal mining, leaky natural gas lines, increasing ruminant (cow) population, rice cultivation, and waste management. Growth rate of CH_4 has slowed in recent years.
- Tropospheric O_3 is increasing. The U.S. and Europe have reduced O_3 precursor emissions (hydrocarbons) in recent years, but increased emissions are occurring in the developing world.
- Block carbon ("soot"), a product of incomplete combustion, is visible in the exhaust of diesel trucks. It is also produced by biofuels and outdoor biomass burning. Black carbon aerosols are not well measured, and their climate forcing is estimated from measurements of total aerosol absorption. The forcing includes the effect of soot in reducing the reflectance of snow and ice.
- Human-made reflective aerosols include sulfates, nitrates, organic carbon, and soil dust. Sources include burning fossil fuel and agricultural activities. Uncertainty in the forcing by reflective aerosols is at least 35%.
- Indirect effects of aerosols on cloud properties are difficult to compute, but satellite measurements of the correlation of aerosol and cloud properties are consistent with the estimated net forcing of −1 W/m^2, with uncertainty of at least 50%.

FIGURE 20.19 **(a)** Climate forcing during the last major glaciations about 22,000 years ago was 6.6 ± 1.5 W/m^2, which produced a drop in global lower atmospheric temperature **(b)** climate forcing during industrial age. (*Source*: **(a)** USGS and NASA **(b)** modified from Hansen, J. 2003. *Can we defuse the global warming time bomb*? Edited version of the presentation to the Council on Environmental Quality. June 12. Washington DC, also *Natural Science* http://www.naturalscience.com)

20.10 The Oceans and Climate Change

The oceans play an important role in climate because two-thirds of the Earth is covered by water. Moreover, water has the highest heat-storage capacity of any compound, so a very large amount of heat energy can be stored in the world's oceans. There is a complex, dynamic, and ongoing relationship between the oceans and the atmosphere. If carbon dioxide increases in the atmosphere, it will also increase in the oceans, and, over time the oceans can absorb a very large quantity of CO_2. This can cause seawater to become more acidic ($H_2O + CO_2 \rightarrow H_2CO_3$) as carbonic acid increases.

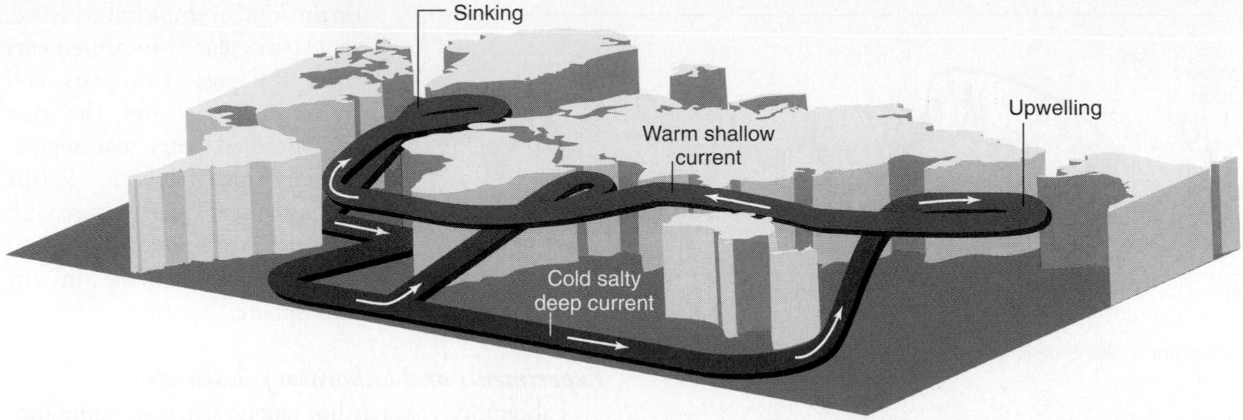

FIGURE 20.20 Idealized diagram of the oceanic conveyor belt. The actual system is more complex, but in general the warm surface water (red) is transported westward and northward (increasing in salinity because of evaporation) to near Greenland, where it cools from contact with cold Canadian air. As the surface water becomes denser, it sinks to the bottom and flows south, then east to the Pacific, then north, where upwelling occurs in the North Pacific. The masses of sinking and upwelling waters balance, and the total flow rate is about 20 million m^3/sec. The heat released to the atmosphere from the warm water keeps Northern Europe 5°C—10°C warmer than if the oceanic conveyor belt were not present. (*Source*: Modified from W. Broker, "Will Our Ride into the Greenhouse Future Be a Smooth One?" *Geology Today* 7, no. 5 [1997]:2–6.)

Part of what may drive the climate system and its changes is the "ocean conveyor belt"—a global circulation of ocean waters characterized by strong northward movement of upper warm waters of the Gulf Stream in the Atlantic Ocean. The temperature of these waters is approximately 12°–13°C when they arrive near Greenland, and they are cooled in the North Atlantic to a temperature of 2°–4°C (Figure 20.20).[25] As the water cools, it becomes saltier and denser, causing it to sink to the bottom. The cold, deep current flows southward, then eastward, and finally northward in the Pacific Ocean. Upwelling in the North Pacific starts the warm, shallow current again. The flow in this conveyor-belt current is huge—20 million m^3/sec, about equal to 100 Amazon rivers.

If the ocean conveyor belt were to shut down, some major changes might occur in the climates of some regions. Western Europe would cool but probably not experience extreme cold or icebound conditions.[26]

The ocean currents of the world have oscillations related to changes in water temperature, air pressure, storms, and weather over periods of a year or so to decades. They occur in the North Pacific, South Pacific, Indian, and North Atlantic oceans and can influence the climate. The Pacific Decadal Oscillation (PDO) for the North Pacific from 1900 to about 2010 is shown in Figure 20.21a. Natural oscillations of the ocean linked to the atmosphere can produce warmer or cooler periods of a few years to a decade or so. The effect of the oscillations can be ten times as strong (in a given year) as long-term warming that we have observed over the past century—larger, over a period of a few decades, than human-induced climate change. By comparison, the annual increase in warming due mostly to human activity is about two-hundredths of a degree Celsius per year.[27] Some scientists attribute the cool winter of 2009–2010 to natural ocean–atmosphere oscillations, and also suggest that these caused a cool year in 1911 that froze Niagara Falls. The more famous El Niño oscillations that occur in the Pacific Ocean are connected to large-scale but short-term changes in weather.

El Niño and Climate

A curious and historically important climate change linked to variations in ocean currents is the Southern Oscillation, known informally as El Niño. From the time of early Spanish settlement of the west coast of South America, people observed a strange event that occurred about every seven years. Usually starting around Christmas (hence the Spanish name El Niño, referring to the little Christ Child), the ocean waters would warm up, fishing would become poor, and seabirds would disappear.

Under normal conditions, there are strong vertical, rising currents, called upwelling, off the shore of Peru. These are caused by prevailing winds coming westward off the South American Continent, which move the surface water away from the shore and allow cold water to rise from the depths, along with important nutrients that promote the growth of algae (the base of the food chain) and thus produce lots of fish. Seabirds feed on those fish and live in great numbers, nesting on small islands just offshore.

El Niño occurs when those cold upwellings weaken or stop rising altogether. As a result, nutrients decline, algae grow poorly, and so do the fish, which either die, fail to reproduce, or move away. The seabirds, too, either leave or die. Because rainfall follows warm water eastward during El Niño years, there are high rates of precipitation and flooding in Peru, while droughts and fires are common in Australia and Indonesia. Because warm ocean water provides an atmospheric heat source, El Niño changes global atmospheric circulation, which causes changes in weather in regions that are far removed from the tropical Pacific.[28]

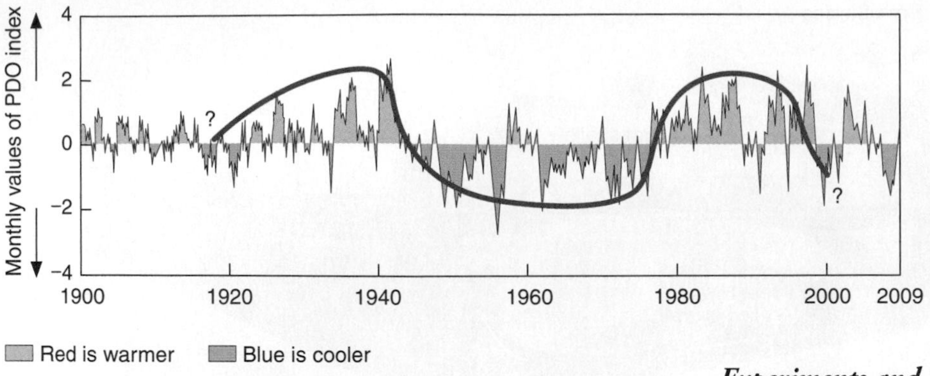

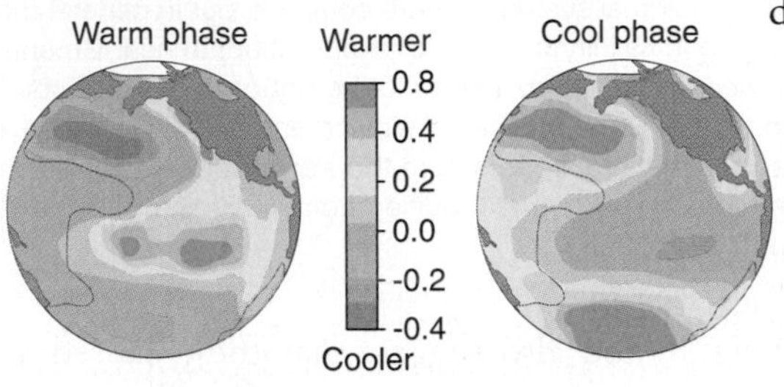

FIGURE 20.21 **(a)** The Pacific Decadal Oscillation (PDO) from 1900 to 2009. Three oscillations are clear (heavy line); **(b)** PDO warm and cool phases with characteristic changes in water temperature (°C) in the North Pacific and along the coast of the Pacific Northwest. (*Source*: Modified after http://jisao.washington.edu. Accessed March 26, 2010.)

20.11 Forecasting Climate Change

Concerns about global warming have to do with the future of the climate. This presents a problem because predicting the future has always been difficult and because people who make predictions have often been wrong. For climate and its effects on living things, there are two approaches to forecasting the future: empirical and theoretical.

Past Observations and Laboratory Research

Past Observations

We discussed this first kind of empirical approach in Section 20.5 in terms of how past climates are reconstructed. The use of empirical records is based on the idea of **uniformitarianism**—the idea that processes occurring in the past occur today and that processes occurring today occurred in the past. Therefore, the past is the key to the present and future. This leads to an "if–then" way of thinking about the future: *If* climate change in the past correlated with change of a certain factor, *then* perhaps a change in that factor will lead to a similar climate change in the future. The argument that human actions are leading to global warming is heavily based on this kind of empirical evidence, in particular measurements from the past 150 years and proxy evidence over the past few hundred years that suggest relationships between Earth's average surface temperature with both the concentations of carbon dioxide and methane in the atmosphere.[15]

Experiments and Laboratory Research

Laboratory research has taught scientists some fundamental things about the cause and effect of climate change. For example, the understanding that carbon dioxide absorbs in specific infrared wavelengths that are different from those of the other gases in the atmosphere comes from a long history of laboratory studies of the air around us, beginning with the work of one of the first modern chemical scientists, the Englishman Joseph Priestley (1733–1804). In the 1770s he did experiments with plants, mice, and candles that were in closed glass containers. He found that if a mouse was kept in a jar by itself, it soon died; but if there was a plant in the jar, the mouse lived. He also found that a plant grew better in a jar in which a mouse had died than it did when by itself. He put a mint plant and a lighted candle in a glass jar, closed the jar, and the candle soon went out. He left the closed jar for a month, and when he came back he lit the candle without opening the jar (focusing sunlight on the candle's wick) and it burned again. Obviously, the mint plant had somehow changed the air in the jar, as had the mouse.

In this way, Priestley discovered that animals and plants change the atmosphere. It wasn't long afterward that oxygen (given off, of course by the green plant) and carbon dioxide (given off by the mouse and the candle) were identified and their light-absorption spectra were determined. Without this kind of study, we wouldn't know that there were such things as greenhouse gases.[29]

Computer Simulations

Scientists have been trying to use mathematics to predict the weather since the beginning of the 20th century. They began by trying to forecast the weather a day in advance, using the formal theory of how the atmosphere functioned. The first person to try this eventually went off to fight in World War I and never completed the forecast. By the early 1970s, computers had gotten fast enough and models sophisticated enough to forecast the next day's weather in two days—not much help in practice, but a start. At least they knew whether their forecast of yesterday's weather had been right!

Computers are much faster today, and the major theoretical method used today to forecast climate change is a group of computer models called **general circulation**

models (GCMs). Mathematically, these are deterministic differential equation models. The dominant computer models of Earth's climate are all based on the general idea shown in Figure 20.22. The atmosphere is divided into rectangular solids, each a few kilometers high and several kilometers north and south. For each of these, the flux of energy and matter is calculated for each of the adjacent cells. Since there are many cells, and each cell has six sides, a huge number of calculations have to be made. Determining how well these GCMs work is a major challenge because the real test is what the future brings.

Many such computer simulations are in use around the world, but all are very similar mathematically. They all use deterministic differential equations to calculate the rate of exchange of energy and matter among the atmospheric cells. They are all steady-state models, meaning that for any given set of input information about the climate at the beginning, the result will always be the same—there is no chance or randomness involved. These models assume that the climate is in a steady state except for specific perturbations, especially those belived to be caused by human activities. Thus an assumption of these models, and a necessary outcome, is that the climate, if left to itself, will be in balance, in a steady state. This is unlike the real world's global environmental systems, which are inherently non-steady-state—always changing, as we saw at the beginning of this chapter.

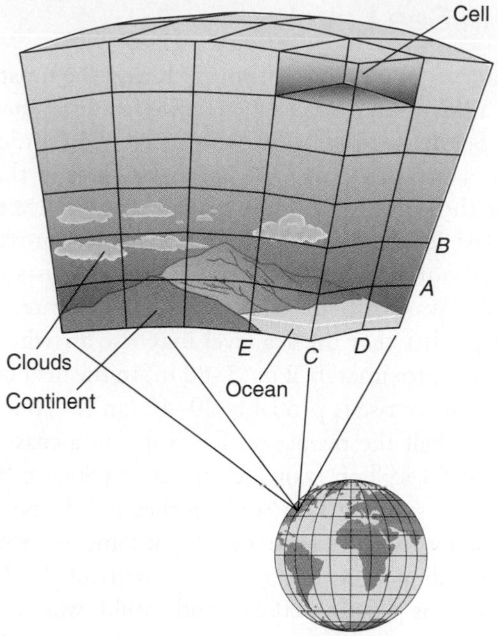

FIGURE 20.22 Idealized diagram of how the huge, general circulation models (models of the entire Earth's climate) are viewed by the computers that run the programs. The atmosphere is divided into hundreds of rectangular cells, and the flow of energy and material is calculated for the transfer between each boundary of every cell to its adjacent cells.

20.12 Potential Rates of Global Climate Change

The global average temperature since 1900 has risen by about 0.8° C (1.5°F).[31] Global surface temperature has risen about 0.2 °C (0.36° F) per decade in the past 30 years.[8] The warmest year since direct surface air temperature has been measured was 2005 (but 2005 will likely be surpassed by 2010 when all data is in). Virtually tied for second were 2002, 2003, 2006, 2007 and 2009. The decade 2001–2010 will be the warmest on record.

According to current GCMs, if the concentrations of all greenhouse gases and aerosols had been kept constant at year 2000 levels, warming of about 0.1° C per decade would be expected.[30, 31] Based on current and expected rates of CO_2 release by human activities, it is estimated that by 2030 the concentration of carbon dioxide in the atmosphere will be double the pre–Industrial Revolution level. If so, the GCMs forecast that the average global temperature will rise approximately 1°–2°C (2°–4°F), with greater increases toward the poles.[31]

In recent decades, the surface air temperature has risen more in some polar regions, in part because of positive feedback. As snow and ice melt, solar energy that used to be reflected outward by ice and snow is now absorbed by vegetation and water, resulting in enhanced warming. This is termed **polar amplification**.

20.13 Potential Environmental, Ecological, and Human Effects of Global Warming

Changes in River Flow

With a continuation of global warming, melting of glacial ice and reductions in snow cover are anticipated to accelerate throughout the twenty-first century. This is also projected to reduce water availability and hydropower potential, and change the seasonality of flows in regions supplied by meltwater from major mountain ranges (e.g., Hindu-Kush, Himalaya, Andes), where more than one-sixth of the world population currently lives.[31, 32]

California, which depends on snowmelt from the Sierra Nevada for water to irrigate one of the richest agriculture regions in the world, will have problems storing water in reservoirs if these forecasts became true. Rainfall will likely increase, but there will be less snowpack with warming. Runoff, will be more rapid than if snow slowly melts. As a result, reservoirs will fill sooner and more water will escape to the Pacific Ocean. Lower runoff is projected for much of Mexico, South America, southern Europe, India, southern Africa, and Australia.

Rise in Sea Level

The sea level reached a minimum during the most recent glacial maximum. Since then, the sea level has risen slowly. Sea level rises from two causes: (1) Liquid water expands as it warms; and (2) ice sheets on land that melt increase the amount of water in the oceans. Since the end of the last ice age, the sea level has risen approximately 23 cm (about 1 foot) per century. Climatologists forecast that global warming could about double that rate. Various models predict that the sea level may rise anywhere from 20 cm to approximately 2 m (8–80 in) in the next century; the most likely rise is probably 20–40 cm (8–16 in).[31]

About half the people on Earth live in a coastal zone, and about 50 million people each year experience flooding due to storm surges. As the sea level rises and the population increases, more and more people become vulnerable to coastal flooding. The rising sea level particularly threatens island nations (Figure 20.23) and could worsen coastal erosion on open beaches, making structures more vulnerable to damage from waves. This could lead to further investments to protect cities in the coastal zone by constructing seawalls, dikes, and other structures to control erosion. Groundwater supplies for coastal communities could also be threatened by saltwater intrusion (see Chapter 19). In short, coastal erosion is a difficult problem that is very expensive to deal with. In many cases, it is best to allow erosion to take place naturally where feasible and defend against coastal erosion only where absolutely necessary.

Glaciers and Sea Ice

The amount of ice on the Earth's surface changes in complicated ways. A major concern is whether global warming will lead to a great decline in the volume of water stored as ice, especially because melting of glacial ice raises the mean sea level and because mountain glaciers are often significant sources of water for lower-elevation ecosystems. At present, many more glaciers in North America, Europe, and other areas are retreating, than are advancing (Figure 20.24). In the Cascades of the Pacific Northwest and the Alps in Switzerland and Italy, retreats are accelerating. For example, on Mt. Baker in the Northern Cascades of Washington, all eight glaciers on the mountain were advancing in 1976. Today all eight are retreating.[32] If present trends continue, all glaciers in Glacier National Park in Montana could be gone by 2030 and most glaciers in the European Alps could be gone by the end of the century.[33]

Not all melting of glacial ice is due to global warming. For example, the study of decrease in the glacier ice on Mt. Kilimanjaro in Africa shows that the primary cause of the ice loss is not melting. The glaciers of Kilimanjaro formed during African Humid Period about 4,000 to 11,000 years ago. Although there have been wet periods since then—notably in the nineteenth century, which appears to have led to a secondary increase in ice—condition have generally been drier.[34]

Since they were first observed in 1912, the glaciers of Kilimanjaro have decreased in area by about 80%. The ice is disappearing not from warmer temperatures at the top of the mountain, which are almost always below freezing, but because less snowfall is occurring and ice is being depleted by solar radiation and sublimation (ice is transformed from solid state to water vapor without melting). More arid conditions in the past century led to air that contained less moisture and thus favored sublimation. This may be due to land use changes from native vegetation to agriculture. Much of the ice depletion had occurred by the mid-1950s.[34]

In addition to many glaciers melting back, the Northern Hemisphere sea ice coverage in September, the time of the ice minimum, has declined an average of 10.7% per decade since satellite remote sensing became possible in the 1970s (Figure 20.25). If present trends were to continue, the Arctic Ocean might be seasonally

(a)

(b)

FIGURE 20.23 The world's smallest nation, Tuvalu, may succumb to sea-level rise. Tuvalu consists of nine coral islands in the South Pacific, with a total area smaller than Manhattan, and its highest elevation above sea level is 4.5 meters. Sea levels have been rising since the end of the last ice age, a natural response. But global warming could accelerate this rise, making the 12,000 citizens of Tuvalu the world's first sea-level-rise refugees.

FIGURE 20.24 **(a)** The thinning of selected glaciers (m^2) since 1977 (National Geographic Maps); **(b)** Muir Glacier in 1941 and 2004. The glacier retreated over 12 km (7 mi) and has thinned by over 800 m (2625 ft). *Source:* National Snow and Ice Data Center W.O. Field (1941) and Molina (2004).

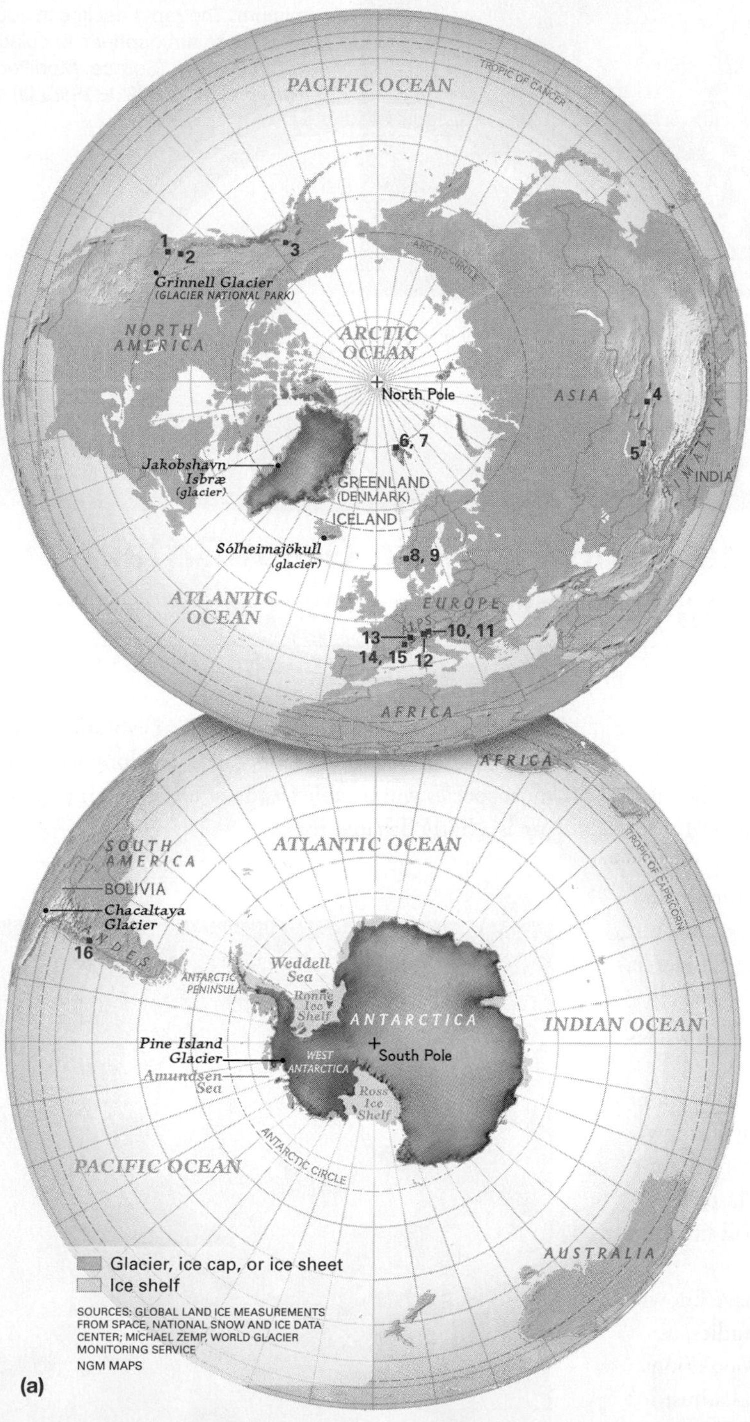

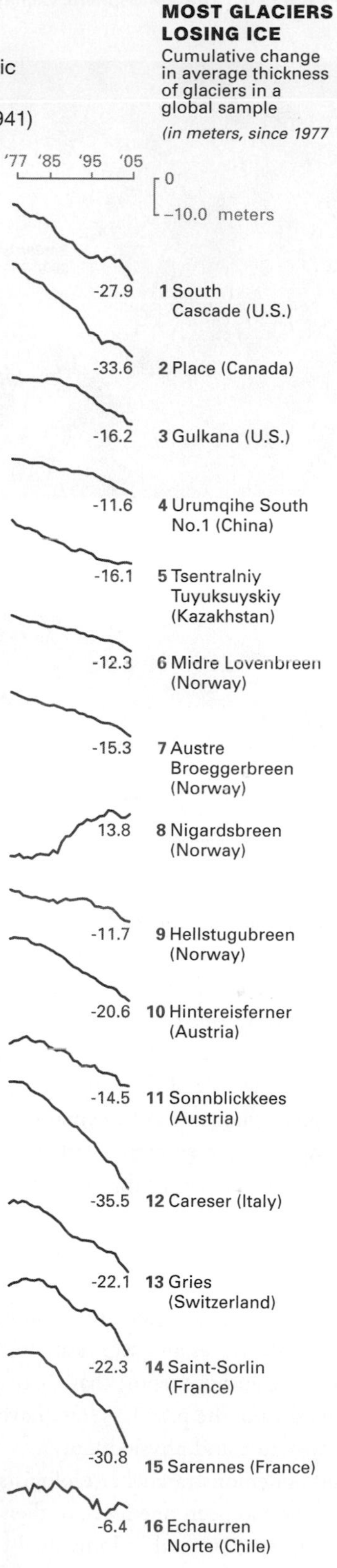

(b)

1941 2004

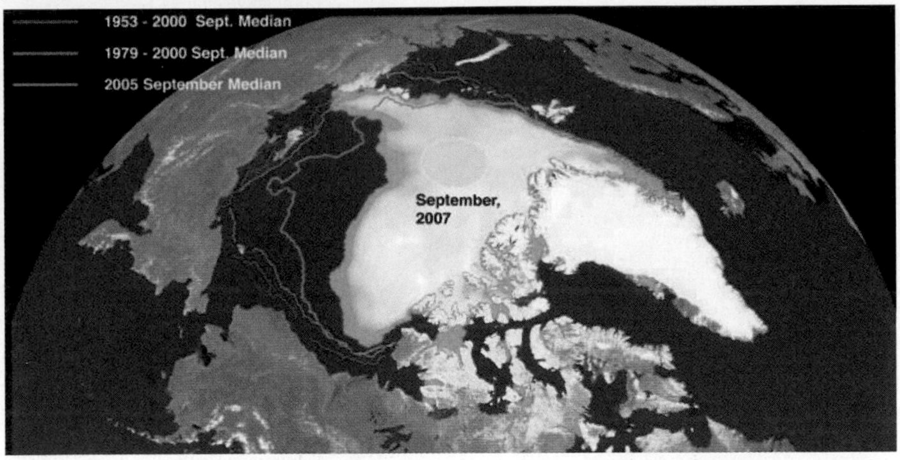

FIGURE 20.25 Satellite observations, which began in 1977, show that Arctic sea ice reached a minimum in September 2007 and has increased since then. The sea ice coverage varies greatly between summer and winter, with July marking the summer minimum. The rapid decline in 2007 was partly due to atmospheric circulation that favored melting. (*Source*: Modified after Stroever et al., 2008. EOS 89 [2] 13–14.)

A CLOSER LOOK 20.1

Some Animals and Plants in Great Britain Are Adjusting to Global Warming

Two of the longest studies of animals and plants in Great Britain show that at least some are adjusting to recent and rapid climate change. The first is a 47-year study of the bird *Parus major*. This study, one of the longest for any bird species, shows that these birds are responding behaviorally to rapid climate change. It's the case of the early bird gets the worm. A species of caterpillar that is one of the main foods of this bird during egg laying has been emerging earlier as the climate has warmed. In response, females of this bird species are laying their eggs an average of two weeks earlier (Figure 20.26). Both birds and caterpillars are doing okay so far.[39]

The second study, one of the longest experiments about how vegetation responds to temperature and rainfall, shows that long-lived small grasses and sedges are highly resistant to climate change. The authors report that changes in temperature and rainfall during the past 13 years "have had little effect on vegetation structure and physiognomy".[40]

These studies demonstrate what ecologists have known for a long time and has been one focus of their studies, as described in Chapters 5, 6, and 7: Individuals, populations, and species have evolved with, are adapted to, and adjust to environmental change, including climate change. However, as we learned from the niche concept, each species persists within a certain range of each environmental condition, so there are limits to the adjustment any one species can make over a short time. Larger changes require biological evolution, which for long-lived animals and plants can take a long time. Whether most species will be able to adjust fast enough to global warming is a hotly debated topic.

FIGURE 20.26 This pretty bird, *Parus major*, a native of Great Britain, is adjusting to rapid climate change.

ice-free by 2030.[35] On the other hand, the central ice cap on Antarctica has grown during the same time. Satellite measurement from 1992 to 2003 suggests the East Antarctica ice sheet increased in mass by about 50 billion tons per year during the period of measurement.[36] As Earth warms, more snow falls on Antarctica.

Changes in sea ice involve more than total area; also involved is the depth of the ice and the age of the ice. The newer the ice, the thinner it is, and therefore the smaller amount of water that is frozen.

The rate of melting of the Greenland ice sheet has doubled since about 1998. As melting produces surface water, it flows into the interior to the base of the ice sheet, causing the ice to flow faster, further destabilizing the ice sheet.[33] It is clear that the polar regions are complex regions on Earth. Changing patterns of ocean and atmosphere circulation in the Arctic and Antarctic regions influence everything from snowfall to melting of glacial and sea ice and movement of glacial ice.[36, 37]

Satellite observations of sea ice became possible in the 1970s . Since then, Northern Hemisphere sea ice reached an observed minimum area covered in September 2007 (Figure 20.25). However, Arctic sea ice has increased since 2007,[19] and so has the central ice cap on Antarctica. Satellite measurement from 1992 to 2003 suggests the East Antarctica ice sheet increased in mass by about 50 billion tons per year during the period of measurement.[36] Changes in sea ice involve more than total area; also involved are the depth of the ice and the age of the ice. The newer the ice, the thinner it is, and therefore the smaller amount of water that is frozen.

FIGURE 20.27 Black guillemots, medium-size birds also called sea pigeons, nest in the Far North (blue in the map above), including Cooper Island, Alaska.

Changes in Biological Diversity

Some of the greatest uncertainties about the consequence of global warming have to do with changes in biodiversity. This is because organisms are complex and so their responses to change can be complex. Warming is one change, but others—such as availability of nutrients, relations with other organisms (predator and prey), and competition for habitat and niches in ecosystems—also affect biodiversity. Because we lack adequate theoretical models to link specific climate changes to specific changes in overall biodiversity, our best insights come from empirical evidence. Surprisingly few species went extinct as a result of climate change during the past 2.5 million years, even though the amount of changes was about the same as that forecast for today and the next few decades.[38] Warming will certainly change some areas, and plants and animals will experience stress. Many will adapt, as apparently occurred during the Medieval Warm Period (see A Closer Look 20.1). For example, polar bears were undoubtedly stressed during this period but did not become extinct.

On the other hand, black guillemots, birds that nest on Cooper Island, Alaska, illustrate the concerns some scientists have about global warming and certain species (Figure 20.27). The abundance of this species has declined since ternperature increases in the 1990s caused the sea ice to recede farther from Cooper Island each spring. The parent birds feed on Arctic cod found under the sea ice and must then return to the nest to feed their chicks, who are not yet mature enough to survive on their own. For the parents to do this, the distance from feeding grounds to nest must be less than about 30 km, but in recent years the ice in the spring has been receding as much as 250 km from the island. As a result, the black guillemots on the island have lost an important source of food. The future of black guillemots on Cooper Island depends on future springtime weather. Too warm and the birds may disappear; Too cold and there may be too few snow-free days for breeding, in which case they also will disappear.

Agricultural Productivity

Globally, agricultural production will likely increase in some regions and decline in others.[19] In the Northern Hemisphere, some of the more northern areas, such as Canada and Russia, may become more productive. Although global warming might move North America's prime farming climate north from the Midwestern United States to the region of Saskatchewan, Canada, the U.S. loss would not simply be translated into a gain for Canada. Saskatchewan would have the optimum climate for growing, but Canadian soils are thinner and less fertile than the prairie-formed soils of the U.S. Midwest. Therefore, a climate shift could have serious negative effects on midlatitude food production. Meanwhile, lands in the southern part of the Northern Hemisphere may become more arid. Prolonged drought as a result of future warming as evidently occurred during the Medieval warming period (see opening case study) with loss of agricultural productivity could be one of the serious impacts of global warming.

Human Health Effects

Like other biological and ecological responses, the effects of global warrning on human health are difficult to forecast. The IPCC *Climate Change 2007: Synthesis Report* is cautious about these possible effects, stating only that one needs to be thinking about "some aspects of human health, such as excess heat-related mortality in Europe, changes in infectious disease vectors in parts of Europe, and earlier onset of and increases in seasonal production of allergenic pollen in Northern Hemisphere high and mid-latitudes."[31] Some have suggested that global warming might increase the incidence of malaria. However, this has been shown not to be the case in past and present circumstances because temperature alone is not a good correlate for malaria.[41] The same has been found for tick-borne encephalitis, another disease that some thought might increase from global warming.[42]

20.14 Adjusting to Potential Global Warming

People can adjust to the threat of global warming in two ways:

- ***Adapt***: Learn to live with future global climate change over the next 20 years because there is warming in the pipeline from greenhouse gases already emitted.
- ***Mitigate***: Work to reduce the emissions of greenhouse gases and take actions to reduce the undesirable effects of a global warming.

How can carbon dioxide emissions be reduced? Increasing energy conservation and efficiency, along with the use of alternative energy sources, can reduce emissions of carbon dioxide. Rebalancing our use of fossil fuels so that we burn more natural gas would also be helpful because natural gas releases 28% less carbon per unit of energy than does oil and 50% less than coal.[43, 44] Conservation strategies to reduce CO_2 emissions include greater use of mass transit and less use of automobiles; providing larger economic incentives to energy-efficient technology; setting higher fuel-economy standards for cars, trucks, and buses; and establishing higher standards of energy efficiency.

Because clearing forests for agriculture reduces storage of carbon dioxide, protecting the world's forests would help reduce the threat of global warming, as would reforestation.[45]

Geologic (rock) sequestration is another way to reduce the amount of carbon dioxide that would otherwise enter the atmosphere. The idea is to capture carbon dioxide from power plants and industrial smokestacks and inject it into deep subsurface geologic reservoirs. Geologic environments suitable for carbon sequestration are sedimentary rocks that contain saltwater, and sedimentary rocks at the sites of depleted oil and gas fields. To significantly mitigate the adverse effects of CO_2 emissions that result in global warming, we need to sequester approximately 2 gigatons of CO_2 per year.[46]

The process of carbon sequestration involves compressing carbon dioxide and changing it to a mixture of both liquid and gas, then injecting it deep underground. Individual injection projects can sequester approximately 1 million tons of CO_2 per year. A carbon-sequestration project is under way in Norway beneath the North Sea. The carbon dioxide from a large natural-gas production facility is injected approximately 1 km into sedimentary rocks below a natural-gas field. The project, begun in 1996, injects about 1 million tons of CO_2 every year, and it is estimated that the entire reservoir can hold up to about 600 billion tons of CO_2—about as much as is likely to be produced from all of Europe's fossil-fuel plants in the next several hundred years. Sequestering carbon dioxide beneath the North Sea is expensive, but it saves the company from paying carbon dioxide taxes for emissions into the atmosphere.

Pilot projects to demonstrate the potential of sequestering CO_2 in sedimentary rocks have been initiated in Texas beneath depleted oil fields. The storage potential at sites in Texas and Louisiana is immense: One estimate is that 200–250 billion tons of CO_2 could be sequestered in this region.[47]

International Agreements to Mitigate Global Warming

There are several approaches to seeking international agreements to limit greenhouse-gas emissions. One major approach is the international agreement in which each nation agrees to some specific limit on emissions. Another major approach is carbon trading in which a nation agrees to cap its carbon emissions at a certain total amount and then issues emission permits to its corporations and other entities, allowing each to emit a certain quantity. These permits can be traded: For example, a power company that wants to build a new fossil-fuel power plant might trade permits with a company that does reforestation, based on estimates of the amount of CO_2 the power plant would release and of an area of forest that could take up that amount. One of the most important programs of this kind is the European Climate Exchange. Carbon trading in the U.S. has come under criticism from both sides of the debate of what to do about potential global warming. That is should we use "cap and trade or not"? Those in favor argue we need to control CO_2 emissions to be proactive and reduce potential adverse inputs at global warming. Those opposed to cap and trade say the economic impact of reducing emissions and changing energy policy is too expensive and will result in economic disaster.

Attempts to establish international treaties limiting greenhouse-gas emissions began in 1988 at a major scientific conference on global warming held in Toronto, Canada. Scientists recommended a 20% reduction in carbon dioxide emissions by 2005. The meeting was a catalyst for scientists to work with politicians to initiate international agreements for reducing emissions of greenhouse gases.

In 1992, at the Earth Summit in Rio de Janeiro, Brazil, a general blueprint for reducing global emissions was suggested. Some in the United States, however, objected that the reductions in CO_2 emissions would be too costly. Agreements from the Earth Summit did not include legally binding limits. After the meetings in Rio de Janeiro, governments worked to strengthen a climate-control treaty that included specific limits on the amounts of greenhouse gases that each industrialized country could emit into the atmosphere.

Legally binding emission limits were discussed in Kyoto, Japan, in December 1997, but specific aspects of the agreement divided the delegates. The United States eventually agreed to cut emissions to about 7% below 1990 levels, but that was far short of the reductions suggested by leading global warmng scientists, who recommended reductions of 60–80% below 1990 levels. A "Kyoto Protocol" resulted from this meeting, was signed by 166 nations, and became a formal international treaty in February 2006.

In July 2008, the leaders of the G-8 nations, meeting in Japan, agreed to "consider and adopt" reductions of at least 50% in greenhouse gas emissions as part of a new U.N. treaty to be discussed in Copenhagen in 2009. This was the first time the United States agreed in principle to such a reduction (in practice, the United States has not gone along with it).

The United States, with 5% of the world's population, emits about 20% of the world's atmospheric carbon dioxide. The fast-growing economies of China and India are rapidly increasing their CO_2 emissions and are not bound by the Kyoto Protocol. California, which by itself is twelfth in the world in CO_2 emissions, passed legislation in 2006 to reduce emissions by 25% by 2020. Some have labeled the action a "job killer" but environmentalists point out that the legislation will bring opportunity and new jobs to the state. California is often a leader, and other states are considering how to control greenhouse gases. The U.S. (as a mid 2010) had not agreed to any international agreements to address climate change. New energy bills to reduce greenhouse gas emissions and turn to alternative energy to reduce our dependency on fossil fuels have been stopped in Congress. Failure to address global change will compromise our ability to be proactive and require our response to be reactive as change occurs. This is not effective environmental planning.

CRITICAL THINKING ISSUE

What Is Valid Science in the Global Warming Debate?

Modern concerns about global warming began in the 1960s with a scientific inquiry into the possibility that human activities might be able to affect the Earth's climate. But the issue touched on so many aspects of civilization, as well as the activities of individuals, that by 1990 global warming had become an intensely debated political and ideological subject

In November 2009, 1,000 e-mails and 3,000 other documents from computers at the University of East Anglia's Climatic Research Unit were hacked and released to the public. Some people claimed that the e-mails revealed efforts by IPCC scientists to exclude views of others, to withhold scientific data, and to tamper with data (a claim now refuted). In response, some defenders of the IPCC made a request to the U.S. government, under the Freedom of Information Act, to gain access to the e-mails of scientists who were said to oppose global warming. The overall result was to further transform what should have been a scientific inquiry into an even more highly politicized and ideological debate.

In Chapter 2 we discussed the scientific method. It may have seemed somewhat academic at the beginning of this book, but with the current chapter's topic you can see that understanding what is and what is not legitimate science—and what are and are not legitimate scientific findings—can affect many aspects of our lives, from international economics and trade to individual choices about how to travel. Therefore, it is important to think about the scientific method within the context of global warming. If you have the interest, you can investigate this topic broadly, but for our purposes consider the following:

A 2007 IPCC report stated that Himalayan glaciers were in danger of melting as soon as 2035. This report quickly gained wide acceptance. Some glaciers in the central Himalayas are in fact melting fast. The western larger glaciers, however, are growing, as warming has increased snowfall. In 2009, the IPCC withdrew the assertion after it was revealed that the source turned out not to be a peer-reviewed scientific article but a report of the World Wildlife Fund, which in turn was based on an article in a popular magazine in India, whose author later disavowed his offhand speculation.[48]

Scientific study suggest that the huge Himalayan ice fields would take more than 300 years to melt if global warming occurred over the next century as currently forecast by general circulation models. Others that feed water to places like India are melting faster and could impact water resources in less than 100 years.

Critical Thinking Questions

1. If you were in charge of IPCC, overseeing the report writing of many scientists, what precautions would you take to ensure that only the best scientific information got into the publications?
2. Give a general description of the kinds of data and other scientific analysis that could be used to determine the rate at which huge areas of mountain glaciers could melt. (You don't have to learn all about glaciers to answer this question, just consider the information in this chapter about what affects the Earth's climate.)
3. In what ways could laboratory research help you study how fast or how slowly glaciers might melt?
4. Why do you think so much of the climate change debate has moved from the scientific arena to the political arena? What are implications of this shift?

SUMMARY

- The atmosphere, a layer of gases that envelops Earth, is a dynamic system that is constantly changing. A great number of complex chemical reactions take place in the atmosphere, and atmospheric circulation takes place on a variety of scales, producing the world's weather and climates.
- Nearly all the compounds found in the atmosphere either are produced primarily by biological activity or are greatly affected by life.
- Major climate changes have occurred throughout the Earth's history. Of special interest to us is that periodic glacial and interglacial episodes have characterized the Earth since the evolution of our species.
- During the past 1,000 years, several warming and cooling trends have affected civilizations.
- During the past 100 years, the mean global surface air temperature has risen by about 0.8°C. About 0.5°C of this increase has occurred since about 1960.
- Water vapor, carbon dioxide, methane, some oxides of nitrogen, and CFCs are the major greenhouse gases. The vast majority of the greenhouse effect is produced by water vapor, a natural constituent of the atmosphere. Carbon dioxide and other greenhouse gases also occur naturally in the atmosphere. However, especially since the Industrial Revolution, human activity has added substantial amounts of carbon dioxide to the atmosphere, along with such greenhouse gases as methane and CFCs.
- Climate models suggest that a doubling of carbon dioxide concentration in the atmosphere could raise the mean global temperature 1°–2°C in the next few decades and 1.5°C–4.5°C by the end of this century.
- Many complex positive feedback and negative feedback cycles affect the atmosphere. Natural cycles, solar forcing, aerosol forcing, particulate forcing from volcanic eruptions, and El Niño events also affect the temperature of Earth.
- There are concerns based on scientific evidence that global warming is leading to changes in climate patterns, rise in sea level, melting of glaciers, and changes in the biosphere. A potential threat from future warming, as in the Medieval Warm Period, is the occurrence of prolonged drought that would compromise our food supply.
- Adjusting to global warming includes learning to live with the changes and attempting to mitigate warming by reducing emissions of greenhouse gases.

REEXAMINING THEMES AND ISSUES

Human Population

Burning of fossil fuels and trees has increased emissions of carbon dioxide into the atmosphere. As the human population increases and standards of living rise, the demand for energy increases; and, as long as fossil fuels are used, greenhouse gases will also increase.

Sustainability

Through our emissions of greenhouse gases, we are conducting global experiments, the final results of which are difficult to predict. As a result, achieving sustainability in the future will be more difficult. If we do not know in detail what the consequences or magnitude of human-induced climate change will be, then it is difficult to predict how we might achieve sustainable development for future generations.

Global Perspective

Global warming is a global problem.

Urban World

If sea levels rise as climate models forecast coastal cities will be affected by higher storm surges. Rising air temperatures can accentuate urban heat-island effects, making life in cities more unpleasant in the summer. If global warming reduces the availability of freshwater, cities will feel the impact.

People and Nature

Our ancestors adapted to natural climate change over the past million years. During that period, Earth experienced glacial and interglacial periods that were colder and warmer than today. Burning fossil fuels has led to human-induced climate changes different in scope from past human effects.

Science and Values

Responding to global warming requires choices based on value judgments. Scientific information, especially geologic data and the instrumental (historic) record along with modern computer simulations, is providing a solid foundation for the belief that global warming is happening. The extent to which scientific information of this kind is accepted involves value decisions.

KEY TERMS

atmosphere **434**
barometric pressure **435**
climate **431**
climate forcing **447**
general circulation models (GCMs) **451**
greenhouse effect **441**
greenhouse gases **433**
polar amplification **451**
radionuclides **446**
stratosphere **434**
troposphere **434**
uniformitarianism **450**
weather **431**

STUDY QUESTIONS

1. Summarize the scientific data that global warming as a result of human activity is occuring.
2. What is the composition of Earth's atmosphere, and how has life affected the atmosphere during the past several billion years?
3. What is the greenhouse effect? What is its importance to global climate?
4. What is an anthropogenic greenhouse gas? Discuss the various anthropogenic greenhouse gases in terms of their potential to cause global warming.
5. What are some of the major negative feedback cycles and positive feedback cycles that might increase or decrease global warming?
6. In terms of the effects of global warming, do you think that a change in climate patterns and storm frequency and intensity is likely to be more serious than a global rise in sea level? Illustrate your answer with specific problems and areas where the problems are likely to occur.
7. How would you refute or defend the statement that the best adjustment to global warming is to do little or nothing and learn to live with change?

FURTHER READING

Fay, J. A., and D. Golumb, *Energy and the Environment* (New York: Oxford University Press, 2002). See Chapter 10 on global warming.

IPCC, *Climate Change 2007. The Physical Science Basis* (New York: Cambridge University Press, 2007). A report by the international panel that was awarded the Nobel Prize for its work on global warming

Lovejoy, T.E., and Lee Hannah, *Climate Change and Biodiversity* (New Haven, CT: Yale University Press, 2005). Discusses, continent by continent, what has happened to biodiversity in the past when climate has changed.

Rohli, R.V., and A. J. Vega, *Climatology* (Sudbury, MA: Jones & Bartlett, 2008). An introduction to the basic science of how the atmosphere works.

Weart, S.R., *The Discovery of Global Warming* (Cambridge, MA: Harvard University Press, 2008). Tells how the possibility of global warming was discovered, and the history of the controversies about it.

CHAPTER 21

Air Pollution

Shown here is an aerial view of Dallas, Texas and the greater Kansas City metropolitan area are both participating in the new initiative called Sustainable Skylines, through the U.S. Environmental Protection Agency.

LEARNING OBJECTIVES

The atmosphere has always been a sink—a place of deposition and storage—for gaseous and particulate wastes. When the amount of waste entering an area of the atmosphere exceeds the atmosphere's ability to disperse or break down the pollutants, problems result. After reading this chapter, you should understand . . .

- The two major ways that pollution affects living things: by direct contact down here and by altering the atmosphere above us;
- Why air pollution from human activities, combined with meteorological conditions, may exceed the atmosphere's natural ability to remove wastes;
- What the major categories and sources of air pollutants are;
- How "acid rain" is produced and how its environmental impacts might be minimized;
- Why air quality standards are important;
- Why the economics of air pollution is controversial and difficult;
- What the major indoor air pollutants are, where they come from, and why they cause some of our most serious environmental health problems;
- "Green buildings" and other major strategies for controlling and minimizing indoor air pollution;
- The "ozone hole" and the science of ozone depletion.

CASE STUDY

Sustainable Skylines: Dallas and Kansas City

Sustainable Skylines is an initiative that has been launched by the Environmental Protection Agency (EPA). Its objective is to achieve sustainable air quality by reducing the six major air pollutants, as well as other toxic air pollutants and greenhouse gases. Cities that participate in the program are encouraged to integrate energy, land use, transportation, and air quality planning in order to achieve measurable improvements within a three-year period. As of 2009, two cities participated—Dallas, Texas, and Kansas City, Kansas and Missouri (the greater Kansas City metropolitan area). The EPA hopes to have ten cities invested in the program by 2010. Among the projects included in a particular sustainable Skyline venture are the following:

- Reducing emissions from landscape equipment by improved irrigation of lawns and turf management, as well as retrofitting small off-road equipment to achieve reduced emissions of air pollutants.
- Reducing vehicle emissions by increasing public transportation and reducing the distances traveled in vehicles.
- Replacing existing taxis with "green taxis" that emit far less pollution.
- Encouraging "**green buildings**" with healthier interior environments and landscaping that benefit the local external environment.
- Reducing emissions from idling vehicles and retrofitting diesel engines to reduce emissions.
- Programs to encourage planting trees in the city to develop a tree canopy in as many areas as possible.

Each city that participates in the Sustainable Skylines Program will have its own local programs and policies, developed in collaboration with the city's inhabitants and city leaders, along with public and private partners. For example, in Dallas, Texas, the description of activities has the goal of helping to reduce the urban "heat island" effect. Urban areas are often warmer than surrounding areas due to the abundance of equipment and lights, as well as surfaces that absorb heat. Cities with little vegetation also have less evaporative cooling. This is a particular problem in Dallas, which has a naturally warm climate much of the year. As a result, the goal of the Sustainable Skylines Program for Dallas is to increase the number of shaded surfaces and green vegetated surfaces of roofs and surrounding buildings in order to reduce the heat island effect and cool the city.

In the greater Kansas City area, the objectives are to encourage a variety of sustainable environmental projects with social benefits. They plan to address such issues as transportation, energy, land use, resource efficiency, green buildings, and air quality, with a focus on projects that will result in cleaner, healthier air for this large urban area.

21.1 Air Pollution in the Lower Atmosphere

A Brief Overview

As the fastest-moving fluid medium in the environment, the atmosphere has always been one of the most convenient places to dispose of unwanted materials. The atmosphere has been a sink for waste disposal ever since we first used fire, and people have long recognized the existence of atmospheric pollutants, both natural and human-induced. Leonardo da Vinci wrote in 1550 that a blue haze formed from materials emitted into the atmosphere by trees. What he had observed is a natural photochemical smog from hydrocarbons given off by living trees. This haze, whose cause is still not fully understood, gave rise to the name Smoky Mountains for the range in the southeastern United States.

The phenomenon of acid rain was first described in the 17th century, and by the 18th century it was known that plants in London were damaged by air pollution. Beginning with the Industrial Revolution in the 18th century, air pollution became more noticeable. The word *smog* was introduced by a physician at a public-health conference in 1905 to denote poor air quality resulting from a mixture of smoke and fog.

Stationary and Mobile Sources of Air Pollution

The two major categories of air pollution sources are stationary sources and mobile sources. **Stationary sources** have a relatively fixed location and include point sources, fugitive sources, and area sources.

- *Point sources*, discussed in Chapter 10, emit pollutants from one or more controllable sites, such as power-plant smokestacks (Figure 21.1).
- *Fugitive sources* generate air pollutants from open areas exposed to wind. Examples include burning for agricultural purposes (Figure 21.2), as well as dirt roads, construction sites, farmlands, storage piles, surface mines, and other exposed areas.
- *Area sources*, also discussed in Chapter 10, are well-defined areas within which are several sources of air pollutants—for example, small urban communities, areas of intense industrialization within urban complexes, and agricultural areas sprayed with herbicides and pesticides.

Mobile sources of air pollutants include automobiles, trucks, buses, aircraft, ships, trains, and anything else that pollutes as it moves from place to place.[2]

FIGURE 21.2 Burning sugarcane fields, Maui, Hawaii—an example of a fugitive source of air pollution.

General Effects of Air Pollution

Air pollution affects many aspects of our environment, including its visual qualities, vegetation, animals, soils, water quality, natural and artificial structures, and human health. Air pollutants affect visual resources by discoloring the atmosphere and by reducing visual range and atmospheric clarity. We cannot see as far in polluted air, and what we do see has less color contrast. These effects were once limited to cities but now extend to some wide-open spaces of the United States. For example, near the Four Corners, where New Mexico, Arizona, Colorado, and Utah meet, emissions from two large fossil-fuel-burning power plants have altered visibility in a region where visibility used to be 80 km (50 mi) from a mountaintop on a clear day.[1] The power plants are two of the largest pollution sources in the U.S.

FIGURE 21.1 This steel mill in Beijing, China, is a major source of air pollution.

Air pollution's numerous effects on vegetation include damage to leaves, needles, and fruit; reduced or suppressed growth; increased susceptibility to diseases, pests, and adverse weather; and disruption of reproductive processes.[1, 2]

Air pollution is a significant factor in the human death rate in many large cities. For example, it has been estimated that in Athens, Greece, the number of deaths is several times higher on days when the air is heavily polluted; and in Hungary, where air pollution has been a serious problem in recent years, it may contribute to as many as 1 in 17 deaths. The United States is certainly not immune to health problems related to air pollution. The most polluted air in the nation is in the Los Angeles urban area, where millions of people are exposed to it. An estimated 175 million people live in areas of the United States where exposure to air pollution contributes to lung disease, which causes more than 300,000 deaths per year. Air pollution in the United States is directly responsible for annual health costs of over \$50 billion. In China, whose large cities have serious air pollution problems, mostly from burning coal, the health cost is now about \$50 billion per year and may rise to about \$100 billion per year by 2020.

Air pollutants can affect our health in several ways, depending on the dose or concentration and other factors, including individual susceptibility (see the discussion of dose–response in Chapter 10). Some of the primary effects are cancer, birth defects, eye and respiratory system irritation, greater susceptibility to heart disease, and aggravation of chronic diseases, such as asthma and emphysema. People suffering from respiratory diseases are the most likely to be affected. Healthy people tend to acclimate to pollutants, but this is a physiological tolerance; as explained in Chapter 10, it doesn't mean that the pollutants are doing no harm (Figure 21.3).

It is worth noting here that many air pollutants have *synergistic effects*—that is, the combined effects are greater than the sum of the separate effects. For example, sulfate and nitrate may attach to small particles in the air, facilitating their inhalation deep into lung tissue. There, they may do greater damage than a combination of the two pollutants would be expected to, based on their separate effects. This phenomenon has obvious health consequences; consider joggers breathing deeply and inhaling particulates as they run along the streets of a city. The effects of air pollutants on vertebrate animals in general include impairment of the respiratory system; damage to eyes, teeth, and bones; increased susceptibility to disease, parasites, and other stress-related environmental hazards; decreased availability of food sources (such as vegetation affected by air pollutants); and reduced ability for successful reproduction.[2]

Air-pollution deposits can also make soil and water toxic. In addition, soils may be leached of nutrients by pollutants that form acids. Air pollution's effects on man-made structures include discoloration, erosion, and decomposition of building materials (see the discussion of acid rain later in this chapter).

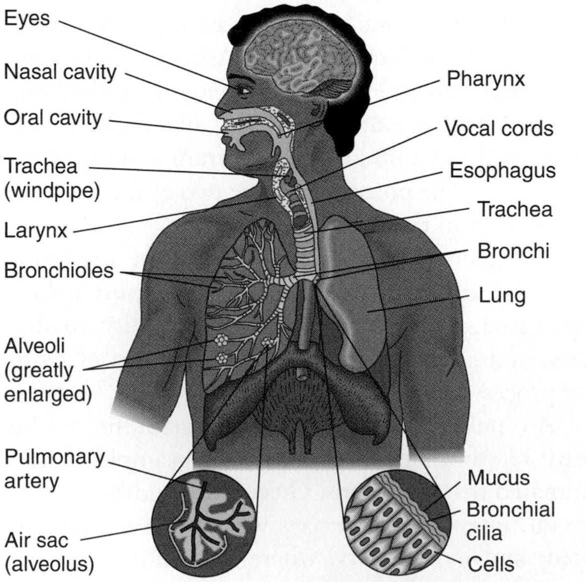

FIGURE 21.3 Idealized diagram showing some of the parts of the human body (brain, cardiovascular system, and pulmonary system) that can be damaged by common air pollutants. The most severe health risks from normal exposures are related to particulates. Other substances of concern include carbon monoxide, photochemical oxidants, sulfur dioxide, and nitrogen oxides. Toxic chemicals and tobacco smoke also can cause chronic or acute health problems.

The Major Air Pollutants

Nearly 200 air pollutants are recognized and assessed by the EPA and listed in the Clean Air Act. They can be classified as primary or secondary. **Primary pollutants** are emitted directly into the air. They include particulates, sulfur dioxide, carbon monoxide, nitrogen oxides, and hydrocarbons. **Secondary pollutants** are produced by reactions between primary pollutants and normal atmospheric compounds. For example, ozone forms over urban areas through reactions of primary pollutants, sunlight, and natural atmospheric gases. Thus, ozone is a secondary pollutant.

The major air pollutants occur either as particulate matter (PM) or in gaseous forms. Particulates are very small particles of solid or liquid substances and may be organic or inorganic. Gaseous pollutants include sulfur dioxide (SO_2), nitrogen oxides (NO_x), carbon monoxide (CO), ozone (O_3) and volatile organic compounds (VOCs), such as hydrocarbons (compounds containing only carbon and hydrogen that include petroleum products), hydrogen sulfide (H_2S), and hydrogen fluoride (HF).

The primary pollutants that account for nearly all air-pollution problems are carbon monoxide (58%), volatile organic compounds (11%), nitrogen oxides (15%), sulfur oxides (13%), and particulates (3%). In the United States today, about 140 million metric tons of these substances enter the atmosphere from human-related processes. If these pollutants were uniformly distributed in the atmosphere, the concentration would be only a few parts per million by weight. Unfortunately, pollutants are not uniformly distributed but tend to be produced, released, and concentrated locally or regionally—for example, in large cities.

In addition to pollutants from human sources, our atmosphere contains many pollutants of natural origin, such as sulfur dioxide from volcanic eruptions; hydrogen sulfide from geysers and hot springs, as well as from biological decay in bogs and marshes; ozone in the lower atmosphere as a result of unstable meteorological conditions, such as violent thunderstorms; a variety of particles from wildfires and windstorms;[1] and natural hydrocarbon seeps, such as La Brea Tar Pits in Los Angeles.

The data in Table 21.1 suggest that, except for sulfur and nitrogen oxides, natural emissions of air pollutants exceed human-produced emissions. Nevertheless, it is the human component that is most abundant in urban areas and leads to the most severe problems for human health.

Criteria Pollutants

The six most common pollutants are called **criteria pollutants** because the EPA has set specific limits on the levels of these six and they are responsible for most of our air-pollution problems. The six are sulfur dioxide, nitrogen oxides, carbon monoxide, ozone, particulates, and lead.

Sulfur Dioxide

Sulfur dioxide (SO_2) is a colorless and odorless gas normally present at Earth's surface in low concentrations. A significant feature of SO_2 is that once emitted into the atmosphere, it can be converted into fine particulate sulfate (SO_4) and removed from the atmosphere by wet or dry deposition. The major anthropogenic (human) source of sulfur dioxide is the burning of fossil fuels, mostly coal in power plants (see Table 21.1). Another major source comprises a variety of industrial processes, ranging from petroleum refining to the production of paper, cement, and aluminum.[1-4]

Adverse effects of sulfur dioxide depend on the dose or the concentrations (see Chapter 10) and include injury or death to animals and plants, as well as corrosion of paint and metals. Crops, such as alfalfa, cotton, and barley, are especially susceptible. Sulfur dioxide can severely damage the lungs of people and other animals, especially in the sulfate form. It is also an important precursor to acid rain (see A Closer Look 21.1).[1-4]

U.S. emission rates of SO_2 from 1970 to 2007 are shown in Table 21.2. Emissions peaked at about 32 million tons in the early 1970s and since then have fallen 60%, to about 13 million tons, as a result of effective emission controls.

Nitrogen Oxides

Although nitrogen oxides (NO_x) occur in many forms in the atmosphere, they are emitted largely as nitric oxide (NO) and nitrogen dioxide (NO_2), and only these two forms are subject to emission regulations. The more important of the two is NO_2, a yellow-brown to reddish-brown gas. A major concern with NO_2 is that it may be converted by complex reactions in the atmosphere to an ion, NO_3^{2-}, within small water particulates, impairing visibility. As mentioned earlier, both NO and NO_2 are major contributors to smog, and NO_2 is also a major contributor to acid rain (see A Closer Look 21.1). Nitrogen oxides contribute to nutrient enrichment and eutrophication of water in ponds, lakes, rivers, and the ocean (see Chapter 19). Nearly all NO_2 is emitted

Table 21.1 MAJOR NATURAL AND HUMAN-PRODUCED COMPONENTS OF SELECTED AIR POLLUTANTS

AIR POLLUTANTS	EMISSIONS (% OF TOTAL) NATURAL	HUMAN-PRODUCED	MAJOR SOURCES OF HUMAN-PRODUCED COMPONENTS	PERCENT
Particulates	85	15	Fugitive (mostly dust)	85
			Industrial processes	7
			Combustion of fuels (stationary sources)	8
Sulfur oxides (SO_x)	50	50	Combustion of fuels (stationary sources, mostly coal)	84
			Industrial processes	9
Carbon monoxide (CO)	91	9	Transportation (automobiles)	54
Nitrogen dioxide (NO_2)		Nearly all	Transportation (mostly automobiles)	37
			Combustion of fuels (stationary sources, mostly natural gas and coal)	38
Ozone (O_3)	A secondary pollutant derived from reaction with sunlight NO_2, and oxygen (O_2)		Concentration present depends on reaction in lower atmosphere involving hydrocarbons and thus automobile exhaust	
Hydrocarbons (HC)	84	16	Transportation (automobiles)	27
			Industrial processes	7

Table 21.2 U.S. EMISSIONS OF CRITERIA POLLUTANTS FROM 1970–2007

	MILLIONS OF TONS PER YEAR							
	1970	1980	1985	1990	1995	2000	2005	2007
Carbon Monoxide (CO)	200	178	170	144	120	102	89	81
Lead	ND	0.074	0.023	0.005	0.004	0.002	0.003	0.002
Nitrogen Oxides (NO_x)	~27	27	26	25	25	22	19	17
Volatile Organic Compounds (VOC)	~30	30	27	23	22	17	15	15
Particulate Matter (PM)								
PM_{10}	ND	6	4	3	3	2	2	2
$PM_{2.5}$		ND	ND	2	2	2	1	1
Sulfur Dioxide (SO_2)	32	26	23	23	19	16	15	13
Totals	ND	267	250	220	191	161	141	129

Notes:

1. In 1985 and 1996 EPA refined its methods for estimating emissions. Between 1970 and 1975, EPA revised its methods for estimating PM emissions.
2. The estimates for 2002 are from 2002 NEI v2; the estimates for 2003 and beyond are preliminary and based on 2002 NEI v2.
3. No data (ND)

Source: Environmental Protection Agency, 2008. Air Trends accessed June 10, 2008 @ www.epa.gov.

from anthropogenic sources. The two main sources are automobiles and power plants that burn fossil fuels.[1, 2]

Nitrogen oxides have various effects on people, including irritation of eyes, nose, throat, and lungs and increased susceptibility to viral infections, including influenza (which can cause bronchitis and pneumonia).[1, 2] Dissolved in water, nitrogen oxides form acids that can harm vegetation. But when the oxides are converted to nitrates, they can promote plant growth.

U.S. emission rates of NO_x from 1970 to 2007 are shown in Table 21.2. Emissions are primarily from combustion of fuels in power plants and vehicles. They have been reduced by about 30% since 1980.

Carbon Monoxide

Carbon monoxide (CO) is a colorless, odorless gas that, even at very low concentrations, is extremely toxic to humans and other animals. The high toxicity results from a physiological effect: Carbon monoxide and hemoglobin have a strong natural attraction for one another; if there is any carbon monoxide in the vicinity, the hemoglobin in our blood will take it up nearly 250 times faster than it will oxygen and carry mostly carbon monoxide, rather than oxygen, from the atmosphere to the internal organs. Effects range from dizziness and headaches to death. Many people have been accidentally asphyxiated by carbon monoxide from incomplete combustion of fuels in campers, tents, and houses. Carbon monoxide is particularly hazardous to people with heart disease, anemia, or respiratory disease. It may also cause birth defects, including mental retardation and impaired fetal growth. Its effects tend to be worse at higher altitudes, where oxygen levels are lower. Detectors (similar to smoke detectors) are now commonly used to warn people if CO in a building reaches a dangerous level.

Approximately 90% of the carbon monoxide in the atmosphere comes from natural sources. The other 10% comes mainly from fires, automobiles, and other sources of incomplete burning of organic compounds, but these are easily concentrated locally, especially by enclosures, so this 10% causes most of the health problems. Emissions of CO peaked in the early 1970s at about 200 million metric tons and declined 60% to about 81 million metric tons by 2007 (Table 21.2). This significant reduction stemmed largely from cleaner-burning engines despite an increased number of vehicles.

Ozone and Other Photochemical Oxidants

Photochemical oxidants are secondary pollutants arising from atmospheric interactions of nitrogen dioxide and sunlight. Ozone, of primary concern here, is a form of oxygen in which three atoms of oxygen occur together rather than the normal two. A number of other photochemical oxidants, known as PANs (peroxyacyl nitrates), occur with photochemical smog.

Ozone is relatively unstable and releases its third oxygen atom readily, so it oxidizes or burns things more readily and at lower concentrations than does normal oxygen. Released into the air or produced in the air, ozone may injure living things. However, since these include bacteria and other organisms, it is sometimes used for sterilizing purposes—for example, bubbling ozone gas through water is one way to purify water.

Ozone in the lower atmosphere is a secondary pollutant produced on bright, sunny days in areas where there is significant primary pollution. The major sources of ozone, as well as other oxidants, are automobiles and industrial processes that release nitrogen dioxide by burning fossil fuels. Because of the nature of its formation, ozone is difficult to regulate and thus is the pollutant whose health standard is most frequently exceeded in U.S. urban areas.[5, 6]

The adverse environmental effects of ozone and other oxidants, like those of other pollutants, depend in part on the dose or concentration of exposure and include damage to plants and animals, as well as to materials, such as rubber, paint, and textiles. Ozone's effects on plants can be subtle. At very low concentrations, it can slow growth without visible injury. At higher concentrations, it kills leaf tissue and, if pollutant levels remain high, whole plants. The death of white pine trees along highways in New England is believed to be due in part to ozone pollution. In animals, including people, ozone causes various kinds of damage, especially to the eyes and respiratory system. Many millions of Americans are often exposed to ozone levels that damage cell walls in lungs and airways. Tissue reddens and swells, and cellular fluids seep into the lungs. Eventually, the lungs lose elasticity and are more susceptible to bacterial infection, and scars and lesions may form in the airways. Even young, healthy people may be unable to breathe normally, and on especially polluted days breathing may be shallow and painful. Ground-level ozone decreased by 9% from 1990 to 2007.[1, 2, 6]

While too much ozone causes problems down here, too little of it has become a problem in the stratosphere. Because of the effect of sunlight on normal oxygen, ozone forms a natural layer high in the stratosphere that protects us from harmful ultraviolet radiation from the sun. However, the emission of certain chemicals in the lower atmosphere has led to serious ozone depletion in the stratosphere. We will discuss the ozone-depletion story later in the chapter. Suffice it to say here that it is becoming an environmental success story at the global level.

Particulate Matter: PM_{10}, $PM_{2.5}$, and Ultrafine Particles

Particulate matter (PM) is made up of tiny particles. The term *particulate matter* is used for varying mixtures of particles suspended in the air we breathe, but in regulations these are divided into three categories: PM_{10}, particles up to 10 micrometers (μm) in diameter; $PM_{2.5}$, particles between 2.5 and 0.18 microns; and UP, **ultrafine particles** smaller than 0.18 micrometers in diameter, released into the air by vehicles on streets and freeways. For comparison, the diameter of a human hair is about 60 to 150 μm (Figure 21.4).

Nearly all industrial processes, as well as the burning of fossil fuels, release particulates into the atmosphere. Farming, too, adds considerable particulate matter to the atmosphere, as do windstorms in areas with little vegetation and volcanic eruptions. Particles are everywhere, and high concentrations and/or specific types of particles pose a serious danger to human health, including aggravation of cardiovascular and respiratory diseases. Major particulates include asbestos (especially dangerous, discussed in detail in Chapter 10)[2] and small particles of heavy metals, such as arsenic, copper, lead, and zinc, which are usually emitted from smelters and other industrial facilities. Particulates can reduce visibility and affect climate (see Chapter 20).[2] Much particulate matter is easily visible as smoke, soot, or dust; other particulate matter is not easily visible.

Fine particles—$PM_{2.5}$ and smaller—are easily inhaled into the lungs, where they can be absorbed into the bloodstream or remain embedded for a long time. Among the most significant of these particles are sulfates and nitrates. As already explained, these are mostly secondary pollutants produced in the atmosphere by chemical reactions between normal atmospheric constituents and sulfur dioxide and nitrogen oxides. These reactions are important in the formation of sulfuric and nitric acids in the atmosphere and are further discussed when we consider acid rain.[1, 2]

Ultrafine particles (UP), released into the air by motor vehicles, are so small that they cannot be easily filtered and can enter the bloodstream. Rich in organic compounds and other reactive chemicals, they may be the most hazardous components of air pollution, especially with respect to heart disease. They evidently can contribute to inflammation (cell and tissue damage by oxidation), reducing the protective quality of "good" cholesterol and leading to plaque buildup in the arteries that can result in heart attack and stroke. Those most at risk are the young, the elderly, and individuals living near a freeway, or exercising near heavy traffic, or spending a lot of time in traffic (sitting in slow-moving traffic can roughly triple your short-term risk of a heart attack). The risk to an individual is very small, but when millions of people are exposed to a small risk, large numbers are affected. The prudent approach is to limit your exposure. For example, avoid jogging or bike riding near heavy traffic for extended periods.[3]

Particulate matter is measured as *total suspended particulates* (TSPs). Values for TSPs tend to be much higher in large cities in developing countries, such as Mexico,

China, and India, than in developed countries, such as Japan and the United States.

Particulates affect human health, ecosystems, and the biosphere. In the United States, particulate air pollution is estimated to contribute to the death of 60,000 people annually.[7] Studies estimate that 2 to 9% of human mortality in cities is associated with particulate pollution, and that the mortality risk is about 15 to 25% higher in cities with the highest levels of fine particulate pollution.[8] Particulates are linked to both lung cancer and bronchitis (see Figure 21.4) and are especially hazardous to the elderly and to people who have respiratory problems, such as asthma. There is a direct relationship between particulate pollution and increased hospital admissions for respiratory distress.

Dust raised by road building and plowing not only makes breathing more difficult for animals (including humans) but also can be deposited on green plants, interfering with absorption of carbon dioxide and oxygen and release of water (transpiration). On a larger scale, particulates associated with large construction projects—such as housing developments, shopping centers, and industrial parks—may injure or kill plants and animals and damage surrounding areas, changing species composition, altering food chains, and thereby affecting ecosystems. The terrorist attacks that destroyed the Twin Towers in New York City on September 11, 2001, sent huge amounts of particles of all sizes into the air, causing serious health problems that continue even today in people who were exposed to it.

Modern industrial processes have greatly increased the total amount of suspended particulates in Earth's atmosphere. These particulates block sunlight and can cause **global dimming**, a gradual reduction in the solar energy that reaches Earth's surface. Global dimming cools the atmosphere and has lessened the global warming that has been predicted. Its effects are most apparent in the midlatitudes of the Northern Hemisphere, especially over urban regions or where jet air traffic is more common. Jet plane exhaust emits particulates high in the atmosphere. That this could affect the climate was suggested in 2001, when civil air traffic was shut down for two days after the September 11 attacks in New York. During those two days, the daily temperature range over the United States was about 1°C higher than usual.[9] Of course, this may have been just a coincidence.

Table 21.2 shows that anthropogenic emissions of PM_{10} in the United States from 1970 to 2007 declined by about two-thirds (66%).

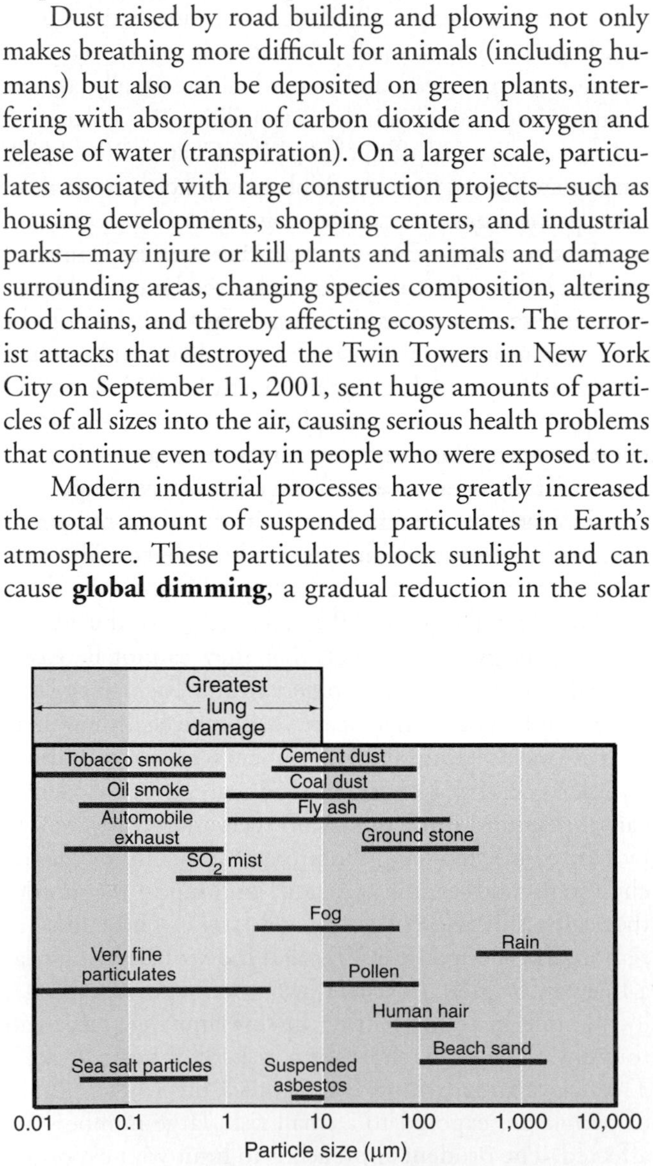

FIGURE 21.4 Sizes of selected particulates. The shaded area shows the size range that produces the greatest lung damage. (*Source:* Modified from Fig. 7–8, p. 244 in *Chemistry, Man and Environmental Change: An Integrated Approach*, by J. Calvin Giddings. Copyright © 1973 by J. Calvin Giddings. Reprinted by permission of Harper Collins Publishers, Inc.).

Lead

Lead (a heavy metal) is an important constituent of automobile batteries and many other industrial products. Leaded gasoline (still used in some countries) helps protect engines and promotes more effective fuel consumption. However, the lead is emitted into the air with the exhaust and has thereby been spread widely around the world, reaching high levels in soils and waters along roadways. Once released, lead can be transported through the air as particulates to be taken up by plants through the soil or deposited directly on their leaves. Thus, it enters terrestrial food chains. When lead is carried by streams and rivers, deposited in quiet waters, or transported to oceans or lakes, it is taken up by aquatic organisms and enters aquatic food chains. Lead is toxic to wildlife and people. It can damage the nervous system, impair learning, and reduce IQ and memory. In children it can also contribute to behavioral problems. (Recall that this is the subject of the Critical Thinking section in Chapter 10.) In adults it can contribute to cardiovascular and kidney disease, as well as anemia.[1, 2]

Lead reaches Greenland as airborne particulates and in seawater and is stored in glacial ice. The concentration of lead in Greenland glaciers was essentially zero in A.D. 800 and reached measurable levels with the beginning of the Industrial Revolution in the mid-18th century. The lead content of the glacial ice increased steadily from 1750 until about 1950, when there was a sudden upsurge in the rate of lead accumulation, reflecting rapid growth in the use of leaded gasoline. The accumulation of lead in Greenland's ice illustrates that our use of heavy metals in the 20th century reached the point of affecting the entire biosphere.

Lead has now been removed from nearly all gasoline in the United States, Canada, and much of Europe. In the United States, lead emissions have declined about 98% since the early 1980s (Table 21.2). The reduction and eventual elimination of lead in gasoline are a good start in reducing levels of anthropogenic lead in the biosphere.

A CLOSER LOOK 21.1

Acid Rain

Acid rain is precipitation in which the pH is below 5.6. The pH, a measure of acidity and alkalinity, is the negative logarithm of the concentration of the hydrogen ion (H^+). Because the pH scale is logarithmic, a pH value of 3 is 10 times more acidic than a pH value of 4 and 100 times more acidic than a pH value of 5. Automobile battery acid has a pH value of 1. Many people are surprised to learn that all rainfall is slightly acidic; water reacts with atmospheric carbon dioxide to produce weak carbonic acid. Thus, pure rainfall has a pH of about 5.6, where 1 is highly acidic and 7 is neutral (see Figure 21.5). (Natural rainfall in tropical rain forests has been observed in some instances to have a pH of less than 5.6; this is probably related to acid precursors emitted by the trees.)

Acid rain includes both wet (rain, snow, fog) and dry (particulate) acidic depositions. The depositions occur near and downwind of areas where the burning of fossil fuels produces major emissions of sulfur dioxide (SO_2) and nitrogen oxides (NO_x). Although these oxides are the primary contributors to acid rain, other acids are also involved. An example is hydrochloric acid emitted from coal-fired power plants.

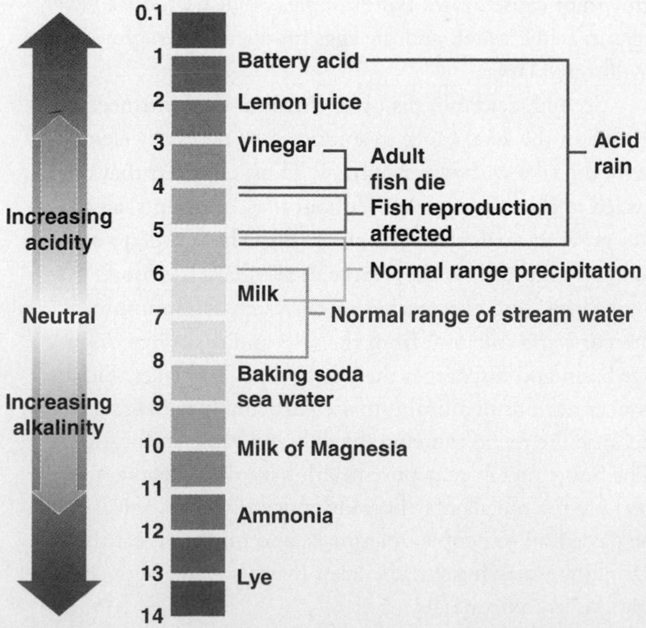

FIGURE 21.5 The pH scale shows the levels of acidity in various fluids. The scale ranges from less than 1 to 14, with 7 being neutral: pHs lower than 7 are acidic; pHs greater than 7 are alkaline (basic). Acid rain can be very acidic and harmful to the environment. (*Source:* http:/ga.water.usgs.gov/edu/phdiagram.html. Accessed August 12, 2005.)

Acid rain has likely been a problem at least since the beginning of the Industrial Revolution. In recent decades, however, it has gained more and more attention, and today it is a major, global environmental problem affecting all industrial countries. In the United States, nearly all of the eastern states are affected, as well as West Coast urban centers, such as Seattle, San Francisco, and Los Angeles. The problem is also of great concern in Canada, Germany, Scandinavia, and Great Britain. Developing countries that rely heavily on coal, such as China, are facing serious acid rain problems as well.

Causes of Acid Rain

As we have said, sulfur dioxide and nitrogen oxides are the major contributors to acid rain. Amounts of these substances emitted into the environment in the United States are shown in Table 21.1. As shown earlier in Table 21.2, emissions of SO_2 peaked in the 1970s and declined to about 13 million metric tons per year by 2007; and nitrogen oxides leveled off at about 25 million metric tons per year in the mid-1980s and had dropped to 17 million metric tons by 2007.

In the atmosphere, reactions with oxygen and water vapor transform SO_2 NO_x into sulfuric and nitric acids, which may travel long distances with prevailing winds and be deposited as acid precipitation—rainfall, snow, or fog (Figure 21.6). Sulfate and nitrate particles may also be deposited directly on the surface of the land as dry deposition and later be activated by moisture to become sulfuric and nitric acids.

Again, sulfur dioxide is emitted primarily by stationary sources, such as power plants that burn fossil fuels, whereas nitrogen oxides are emitted by both stationary and mobile sources, such as automobiles. Approximately 80% of sulfur dioxide and 65% of nitrogen oxides in the United States come from states east of the Mississippi River.

Sensitivity to Acid Rain

Geology, climate, vegetation, and soil help determine the effects of acid rain, because these differ widely in their "buffers"—chemicals that can neutralize acids. Sensitive areas are those in which the type of bedrock (such as granite) or soils (such as those consisting largely of sand) cannot buffer acid input. Limestone bedrock provides the best buffering because it is made up mainly of calcium carbonate ($CaCO_3$), the mineral known as calcite. Calcium carbonate reacts with the hydrogen in the water and neutralizes the acid.

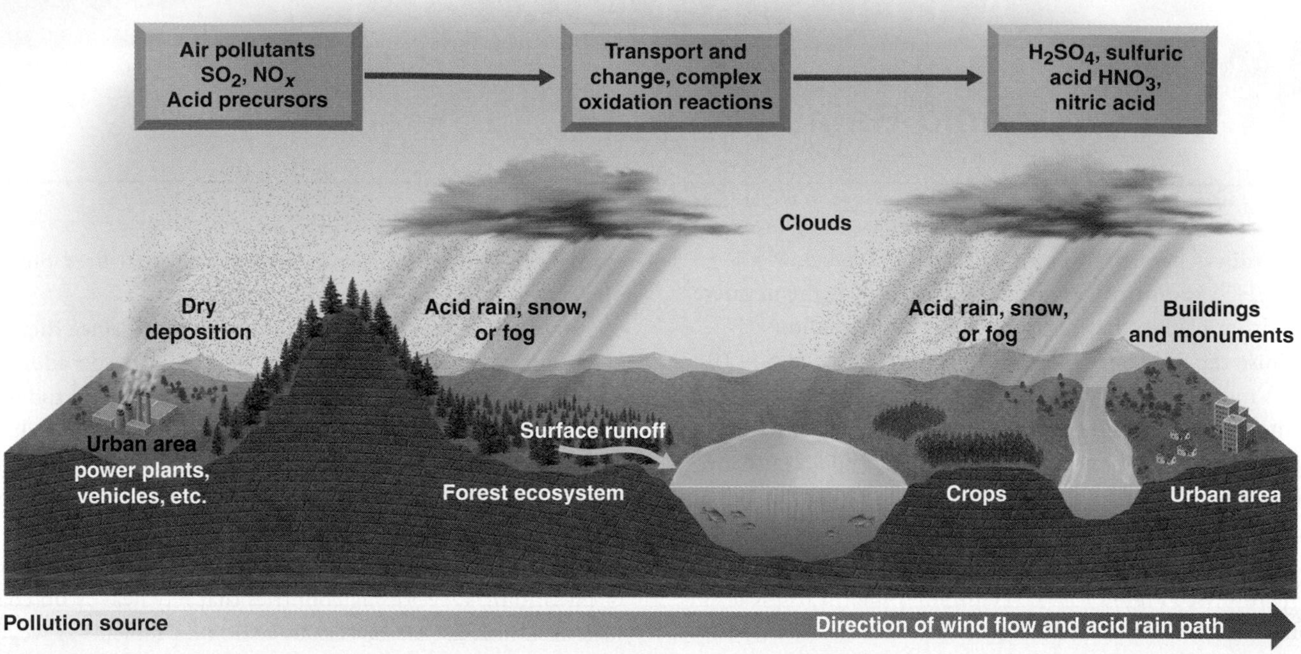

FIGURE 21.6 Idealized diagram showing selected aspects of acid rain formation and paths.

Soils may lose their fertility when exposed to acid rain, either because nutrients are leached out by acid water or because the acid in the soil releases elements that are toxic to plants.

Acid Rain's Effects on Forest Ecosystems

It has long been suspected that acid precipitation damages trees. Studies in Germany led scientists to cite acid rain and other air pollution as the cause of death for thousands of acres of evergreen trees in Bavaria. Similar studies in the Appalachian Mountains of Vermont (where many soils are naturally acidic) suggest that in some locations half the red spruce trees have died in recent years. Some high-elevation forests of the Appalachian Mountains, including the Great Smoky Mountains and Shenandoah National Park, have been impacted by acid rain, acid fog, and dry deposition of acid. Symptoms include slowed tree growth, leaves and needles that turn brown and fall off, and in extreme cases the death of trees. The acid rain does not directly kill trees; rather, it weakens them as essential nutrients are leached from soils or stripped from leaves by acid fog. Acidic rainfall also may release toxic chemicals, such as aluminum, that damage trees.[10]

Acid Rain's Effects on Lake Ecosystems

Records from Scandinavian lakes show an increase in acidity accompanied by a decrease in fish. The increased acidity has been traced to acid rain caused by industrial processes in other countries, particularly Germany and Great Britain. Thousands of lakes, ponds, and streams in the eastern United States are sensitive to acidification, including the Adirondacks and Catskill Mountains of New York State and others in the Midwest and in the mountains of the Western U.S. Little Echo Pond in Franklin, New York, is one of the most acidic lakes, with a measured pH of 4.2.[10]

Acid rain affects lake ecosystems in three ways. First, it damages aquatic species (fish, amphibians, and crayfish) directly by disrupting their life processes in ways that limit growth or cause death. For example, crayfish produce fewer eggs in acidic water, and the eggs produced often grow into malformed larvae.

Second, acid rain dissolves chemical elements necessary for life in the lake. Once in solution, the necessary elements leave the lake with water outflow. Thus, elements that once cycled in the lake are lost. Without these nutrients, algae do not grow, animals that feed on the algae have little to eat, and animals that feed on these animals also have less food.[10, 11]

Third, acid rain leaches metals, such as aluminum, lead, mercury, and calcium, from the soils and rocks in a drainage basin and discharges them into rivers and lakes. Elevated concentrations of aluminum are particularly damaging to fish because the metal can clog the gills and cause suffocation. The heavy metals may pose health hazards to people, too, because the metals may become concentrated in fish and then be passed on to people, mammals, and birds that eat the fish. Drinking water from acidic lakes may also have high concentrations of toxic metals.

Not all lakes are vulnerable to acidification. Acid is neutralized in waters with a high carbonate content (in the form of the ion HCO_3). Therefore, lakes on limestone or other rocks rich in calcium or magnesium carbonates can readily buffer river and lake water against acids. Lakes with high concentrations of such elements are called hard water lakes. Lakes on sand or igneous rocks, such as granite, tend to lack sufficient buffering to neutralize acids and are more susceptible to acidification.[12]

Acid Rain's Effects on Human Society

Acid rain damages not only our forests and lakes but also many building materials, including steel, galvanized steel, paint, plastics, cement, masonry, and several types of rock, especially limestone, sandstone, and marble. Classical buildings on the Acropolis in Athens and in other cities show considerable decay (chemical weathering) that accelerated in the 20th century as a result of air pollution. The problem has grown to such an extent that buildings require restoration, and the protective coatings on statues and other monuments must be replaced quite frequently, at a cost of billions of dollars a year. Particularly important statues in Greece and other areas have been removed and placed in protective glass containers, leaving replicas standing in their former outdoor locations for tourists to view.[11]

Stone decays about twice as rapidly in cities as it does in less urban areas. The damage comes mainly from acid rain and humidity in the atmosphere, as well as from corrosive groundwater.[15] This implies that measuring rates of stone decay will tell us something about changes in the acidity of rain and groundwater in different regions and ages. It is now possible, where the ages of stone buildings and other structures are known, to determine whether the acid rain problem has changed over time.

Control of Acid Rain

We know what causes acid precipitation—the solution is what we are struggling with. One solution to lake acidification is rehabilitation by the periodic addition of lime, as has been done in New York State, Sweden, and Ontario. This solution is not satisfactory over a long period, however, because the continuing effort is expensive. A better approach is to target the components of acid rain, the emissions of sulfur dioxide and nitrogen oxides. As noted, sulfur dioxide emissions in the United States are down about 60% since 1970—a big improvement that is significantly reducing acid rain. Emissions were lowered by a market-based SO_2 cap-and-trade program of the U.S. Environmental Protection Agency's Acid Rain Program, by which utilities receive pollution allowances that they can trade or sell if they lower emissions from their power plants (see Chapter 7 for a discussion of cap and trade).[14]

Air Toxics

Toxic air pollutants, or **air toxics**, are among those pollutants known or suspected to cause cancer and other serious health problems after either long-term or short-term exposure. The most serious exposure to air toxics occurs in California and New York, with Oregon, Washington, DC, and New Jersey making up the rest of the top five. States with the cleanest air include Montana, Wyoming, and South Dakota.

Air toxics include gases, metals, and organic chemicals that are emitted in relatively small volumes. They cause respiratory, neurological, reproductive, or immune diseases, and some may be carcinogenic. A 2006 EPA report estimated that the average risk of cancer from exposure to air toxics is about 1 in 21,000. The assessment concluded that benzene poses the most significant risk for cancer, accounting for 25% of the average individual cancer risk from all air toxics. Again, the effect on an individual's health depends on a number of factors, including duration and frequency of exposure, toxicity of the chemical, concentration of the pollutant the individual is exposed to, and method of exposure, as well as an individual's general health.[15]

Among the more than 150 known toxic air pollutants are hydrogen sulfide, hydrogen fluoride, various chlorine gases, benzene, methanol, and ammonia. In 2006 the EPA released an assessment of the national health risk from air toxics. It focused on exposure from breathing the pollutants; it did not address other ways people are exposed to them.

Standards and regulations established for more than 150 air toxics are expected to reduce annual emissions from 1990 levels. Even though vehicle miles will likely increase significantly by 2020, emissions of gaseous air toxics (such as benzene) from vehicles on highways are projected to decline about 80% from 1990 levels. Following are several examples of air toxics.

Hydrogen Sulfide

Hydrogen sulfide (H_2S) is a highly toxic corrosive gas, easily identified by its rotten-egg odor. Hydrogen sulfide is produced from natural sources, such as geysers, swamps, and bogs, and from human sources, such as petroleum refineries and metal smelters. The potential effects of hydrogen sulfide include functional damage to plants and health problems ranging from toxicity to death for humans and other animals.[4]

Hydrogen Fluoride

Hydrogen fluoride (HF) is a gas released by some industrial activities, such as aluminum production, coal gasification, and burning of coal in power plants. Hydrogen fluoride is extremely toxic; even a small concentration (as low as 1 ppb) may cause problems for plants and animals. It is potentially dangerous to grazing animals because some forage plants can become toxic when exposed to this gas.[1]

Mercury

Mercury is a heavy metal released into the atmosphere by coal-burning power plants, other industrial processes, and mining. Natural processes—such as volcanic eruptions and evaporation from soil, wetlands, and oceans—also release mercury into the air. Its toxicity to people is well documented and includes neurological and development damage, as well as damage to the brain, liver, and kidneys. Mercury from the atmosphere may be deposited in rivers, ponds, lakes, and the ocean, where it accumulates through biomagnification and both wildlife and people are exposed to it.[2]

Volatile Organic Compounds

Volatile organic compounds (VOCs) include a variety of organic compounds. Some of these compounds are used as solvents in industrial processes, such as dry cleaning, degreasing, and graphic arts. Hydrocarbons (compounds of hydrogen and carbon) comprise one group of VOCs. Thousands of hydrocarbons exist, including natural gas, or methane (CH_4); butane (C_4H_{10}); and propane (C_3H_8). Analysis of urban air has identified many hydrocarbons, and their potential adverse effects are numerous. Some are toxic to plants and animals, and others may be converted to harmful compounds through complex chemical changes that occur in the atmosphere. Some react with sunlight to produce photochemical smog.

Globally, our activities produce only about 15% of hydrocarbon emissions. In the United States, however, nearly half the hydrocarbons entering the atmosphere are emitted from anthropogenic sources. The largest of these sources in the United States is automobiles. Anthropogenic sources are particularly abundant in urban regions. However, in some southeastern U.S. cities, such as Atlanta, Georgia, natural emissions (in Atlanta's case, apparently from trees) probably exceed those from automobiles and other human sources.[3]

Like emissions of sulfur dioxide and nitrogen oxide, VOCs peaked in the 1970s and have been reduced by 50% (Table 21.2) thanks to effective government-mandated emission controls for automobiles.

Methyl Isocyanate

Some chemicals are so toxic that extreme care must be taken to ensure they do not enter the environment. This was demonstrated on December 3, 1984, when a toxic liquid from a pesticide plant leaked, vaporized, and formed a deadly cloud of gas that settled over a 64 km^2 area of Bhopal, India. The gas leak lasted less than an hour; yet more than 2,000 people were killed and more than 15,000 injured. The colorless gas that resulted from the leak was methyl isocyanate (C_2H_3NO), which causes severe irritation (burns on contact) to eyes, nose, throat, and lungs. Breathing the gas in concentrations of only a few parts per million (ppm) causes violent coughing, swelling of the lungs, bleeding, and death. Less exposure can cause a variety of problems, including blindness.

Methyl isocyanate is an ingredient of a common pesticide known in the United States as Sevin, as well as two other insecticides used in India. An industrial plant in West Virginia also makes the chemical. Small leaks not leading to major accidents occurred there both before and after the catastrophe in Bhopal.

Clearly, chemicals that can cause widespread injury and death should not be stored near large population centers. In addition, chemical plants should have reliable accident-prevention equipment, as well as personnel trained to control and prevent problems.

Benzene

Benzene (C_6H_6) is a gasoline additive and an important industrial solvent. Generally, it is produced when carbon-rich materials, such as oil and gasoline, undergo incomplete combustion. It is also a component of cigarette smoke. Automobiles, trucks, airplanes, trains, and farm machinery are major sources of environmental benzene.[15]

Acrolein

Acrolein (CH_2CHCHO) is a volatile hydrocarbon that is extremely irritating to the eyes, nose, and respiratory system in general. It is produced by manufacturing processes that involve combustion of petroleum fuels and is a component of cigarette smoke.[15]

Variability of Air Pollution

Pollution problems vary greatly among the different regions of the world and even within just the United States. For example, as noted earlier, in the Los Angeles basin and many U.S. cities, nitrogen oxides and hydrocarbons are particularly troublesome because they combine in the presence of sunlight to form photochemical smog. Most of the nitrogen oxides and hydrocarbons are emitted from automobiles and other mobile sources. In other U.S. regions, such as Ohio and the Great Lakes region, air quality also suffers from emissions of sulfur dioxide and particulates from industry and from coal-burning power plants, which are point sources.

Air pollution also varies with the time of year. For example, smog is usually a problem in the summer, when there is a lot of sunlight. Particulates are a problem in dry months, when wildfires are likely, and during months when the wind blows across the desert. For example, drought and heat in August of 2010 resulted in wildfires in Russia that produced a thick hazardous smoke and resulting very poor air quality in Moscow. The combination of heat and air pollution at the height of the pollution event killed about 700 people per day in Moscow.

Pollution from particulates is a problem in arid regions, where there is little vegetation and the wind easily picks up and transports fine dust. Las Vegas, Nevada, the fastest-growing urban area in the United States in the 1990s, now has some of the most polluted air in the southwestern United States. The brown haze over Las Vegas is due mostly to the nearly 80,000 metric tons of PM_{10} that enter the air in that region from the desert environment. About 60% of the dust comes from new construction sites, dirt roads, and vacant land. The rest is natural windblown dust. Las Vegas also has a carbon monoxide problem from vehicles, but it is the particulates that are causing concern, possibly leading to future EPA sanctions and growth restrictions.

Haze from Afar

Air pollution has become global and is not limited to urban areas. One example of this is Alaska's North Slope, a vast strip of land approximately 200 km (125 mi) wide that many consider to be one of the few unspoiled wilderness areas left on Earth. It seems logical to assume that air quality in the Arctic environments of Alaska would be pristine, except perhaps near areas where petroleum is being vigorously developed. However, ongoing studies suggest that the North Slope has an air-pollution problem that originates in Eastern Europe and Eurasia.

It is suspected that pollutants from burning fossil fuels in Eurasia are transported via the jet stream, at speeds that may exceed 400 km/hr (250 mi/hr), northeast over the North Pole to the North Slope of Alaska. There, they slow, stagnate, and produce a reddish-brown air mass known as "Arctic haze." The concentrations of air pollutants, including oxides of sulfur and nitrogen, are comparable to those of some eastern U.S. cities, such as Boston. Air quality problems in remote areas, such as Alaska, have significance as we try to understand air pollution at the global level.[16]

A curious global event occurred in the spring of 2001 when a white haze consisting of dust from Mongolia and industrial particulate pollutants arrived in North America. The haze affected one-fourth of the United States and could be seen from Canada to Mexico. In the United States, pollution levels from the haze alone were as high as two-thirds of federal health limits and caused respiratory problems. The haze demonstrates that pollution from Asia is carried by winds across the Pacific Ocean. Today we know from satellite observation that air pollutants transported by winds from East Asia to North America account for about 15% of the total pollutants originating from the United States and Canada.[17]

Urban Air Pollution: Chemical and Atmospheric Processes

Now that we have introduced and discussed the various types of air pollutants. This preparation allows for a more detailed discussion of the processes and chemistry of urban **smog**.

There are two major types of urban smog: photochemical smog, sometimes called L.A.-type smog or brown air; and sulfurous smog, sometimes referred to as London-type smog, gray air, or industrial smog. **Sulfurous smog** is produced primarily by the burning of coal or oil at large power plants. Sulfur oxides and particulates combine under certain conditions to produce a concentrated sulfurous smog. **Photochemical smog** is directly related to automobile use.

Figure 21.7 shows a characteristic pattern in the way nitrogen oxides, hydrocarbons, and oxidants (mostly ozone) vary during a typically smoggy day in Southern California. Early in the morning, when commuter traffic begins to build up, concentrations of nitrogen oxide (NO) and hydrocarbons begin to increase. At the same time, nitrogen dioxide (NO_2) may decrease because sunlight breaks it down to produce NO plus atomic oxygen (NO + O). The atomic oxygen (O) is then free to combine with molecular oxygen (O_2) to form ozone (O_3). As a result, the concentration of ozone also increases after sunrise. Shortly thereafter, oxidized hydrocarbons react with NO to increase the concentration of NO_2 by midmorning. This causes the NO concentration to decrease and allows ozone to build up, producing a midday peak in ozone and a minimum in NO. As the smog develops, visibility may be greatly reduced as light is scattered by the pollutants. Figure 21.8 shows Los Angeles on a clear day, in sharp contrast to the way the city looks on a smoggy day.

What are the chances that a deadly smog will occur somewhere in the world? Unfortunately, the answer is all too good, given the amount of air pollution in some large cities. Beijing, for example, might be a candidate; the city uses an immense amount of coal, and coughing is so pervasive that residents often refer to it as the "Beijing cough." Another likely candidate is Mexico City, which has one of the worst air-pollution problems anywhere in the world today.

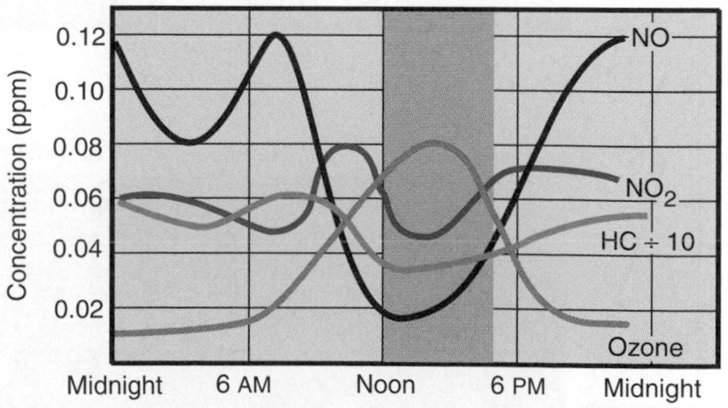

FIGURE 21.7 Development of photochemical smog over the Los Angeles area on a typical warm day.

(a) (b)

FIGURE 21.8 The city of Los Angeles on **(a)** a clear day and **(b)** a smoggy day.

Wherever multiple sources emit air pollutants over a wide area, air pollution can develop. Whether it does or not depends on topography and meteorological conditions, which can determine whether air pollution is a nuisance or a major health problem. The primary adverse effects are damage to green plants and aggravation of chronic illnesses. Most of these effects are due to relatively low concentrations of pollutants over a long period. Periods of pollution generally do not directly cause numerous deaths. Serious pollution events (disasters) can develop over a period of days and lead to increases in illnesses and deaths.

In the lower atmosphere, restricted circulation associated with an atmospheric inversion may lead to pollution events. An **atmospheric inversion** occurs when warmer air lies above cooler air and there is little wind. The air stays still both vertically and horizontally, so any pollutant emissions stay there and build up. Figure 21.9 shows two types of atmospheric inversion that may contribute to air-pollution problems. In the upper diagram, which is somewhat analogous to the situation in the Los Angeles area, descending warm air forms a semipermanent inversion layer. Because the mountains act as a barrier to the pollution, polluted air moving in response to the sea breeze

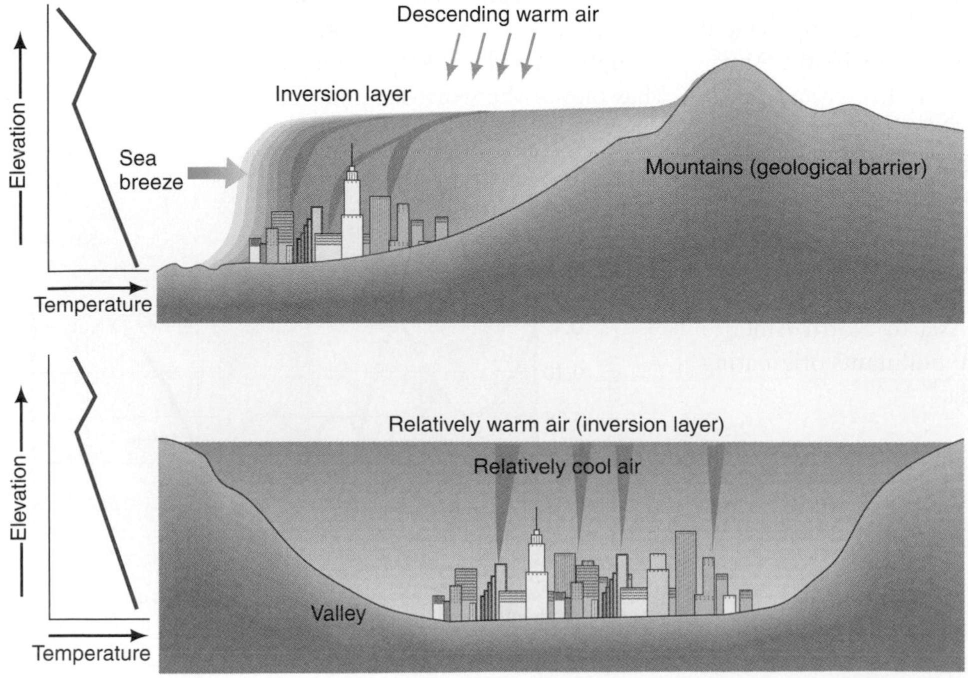

FIGURE 21.9 Two causes of the development of atmospheric inversion, which may aggravate air-pollution problems.

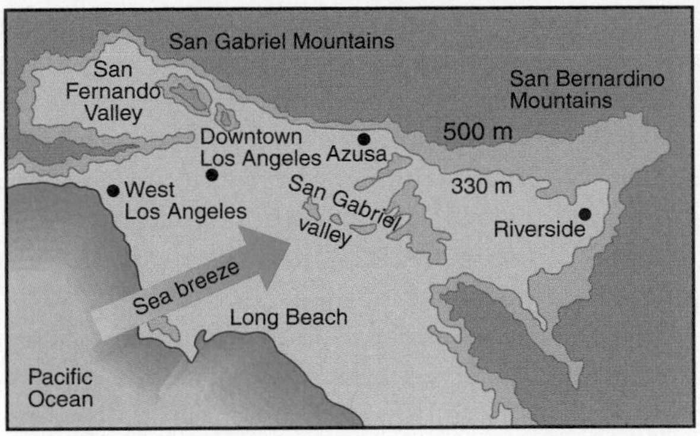

FIGURE 21.10 Part of Southern California showing the Los Angeles basin (south coast air basin). (*Source*: Modified from S.J. Williamson, *Fundamentals of Air Pollution* [Reading, MA: Addison-Wesley, 1973].)

and other processes tends to move up canyons, where it is trapped. The air pollution that develops occurs primarily in summer and fall.

The lower part of Figure 21.9 shows a valley with relatively cool air overlain by warm air. This type of inversion can occur when cloud cover associated with a stagnant air mass develops over an urban area. Incoming solar radiation is blocked by the clouds, which reflect and absorb some of the solar energy and are warmed. On the ground or near Earth's surface, the air cools. If there is moisture in the air (humidity), then, as the air cools, the dew point (the temperature at which water vapor condenses) is reached, and fog may form. Because the air is cold, people burn more fuel to heat their homes and factories, so more pollutants are delivered into the atmosphere. As long as the stagnant conditions exist, the pollutants will build up. It was this mechanism that caused the deadly 1952 London smog that killed about 4,000 people over a one week period December 4 to 10.

Cities in a valley or topographic bowl surrounded by mountains are more susceptible to smog problems than are cities in open plains. Surrounding mountains and the occurrence of temperature inversions prevent pollutants from being dispersed by winds and weather systems. The production of air pollution is particularly well documented for Los Angeles, which has mountains surrounding part of the urban area and lies within a region where the air lingers, allowing pollutants to build up (Figure 21.10).

In sum, the potential for air pollution in urban areas is determined by the following:

- The rate of emission of pollutants per unit area.
- The distance that an air mass moves downwind through a city.
- The average speed of the wind.
- The elevation to which potential pollutants can be thoroughly mixed by naturally moving air in the lower atmosphere (Figure 21.11).[18]

The concentration of air pollutants is directly proportional to the first two factors: As either the emission rate or downwind travel distance through an urban area increases, so will the concentration of pollutants in the air. Again, the Los Angeles basin is a good example (see Figure 21.10). If there is a wind from the ocean, as is generally the case, coastal areas will experience much less air pollution than inland areas. Assuming a constant rate of emission of air pollutants, the air mass will collect more and more pollutants as it moves through the urban area; the inversion layer acts as a lid for the pollutants. However, near a geological barrier, such as a mountain, there may be a chimney effect in which the pollutants spill over the top of the mountain (see Figures 21.10 and 21.11). This has been noticed in the Los Angeles basin, where pollutants may climb several thousand meters, damaging mountain pines and other vegetation and spoiling the air of mountain valleys.

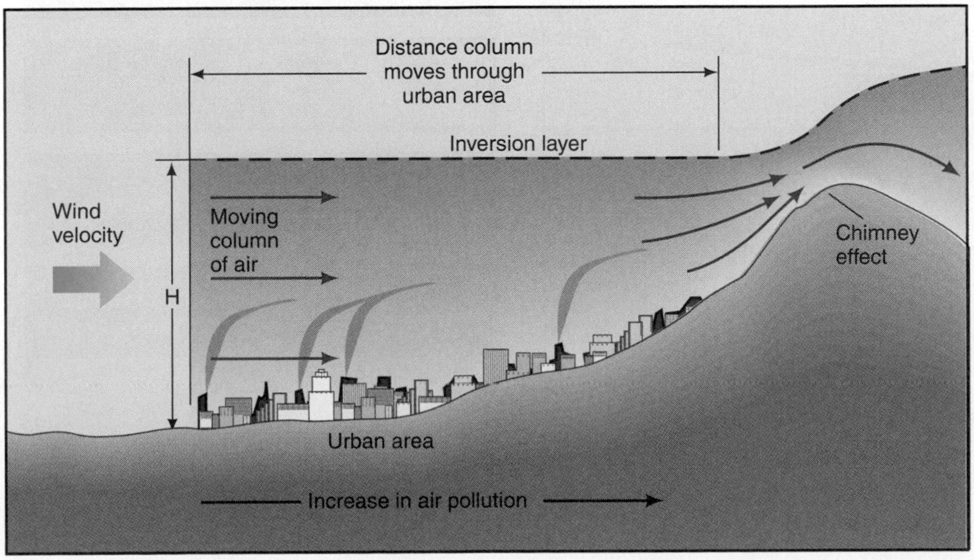

FIGURE 21.11 The higher the wind velocity and the thicker the mixing layer (shown here as H), the less air pollution. The greater the emission rate and the longer the downwind length of the city, the more air pollution. The chimney effect allows polluted air to move over a mountain and down into an adjacent valley.

City air pollution diminishes with increases in the third and fourth factors, which are meteorological: the wind velocity and the height of mixing. The stronger the wind and the higher the mixing layer, the lower the pollution.

Future Trends for Urban Air Pollution

The United States

What does the future hold for U.S. urban areas with respect to air pollution? The optimistic view is that urban air quality will continue to improve as it has in the past 40 years because we know so much about the sources of air pollution and have developed effective ways to reduce it (see Table 21.2). In recent years, as the U.S. population, gross domestic product, and energy consumption have increased, emissions of the major pollutants have decreased (Figure 21.12).[2]

Despite improvements, air pollution in the United States remains a serious problem in many parts of the country. The Los Angeles urban area, for example, still has the worst air quality in the United States. Southern California is coming to grips with the problem, and the people studying air pollution there understand that further pollution abatement will require massive efforts. There are encouraging signs of improvement. For example, from the 1950s to the present, the peak level of ozone (considered one of the best indicators of air pollution) has declined, even though the population nearly tripled and the number of motor vehicles quadrupled during this period. Nevertheless, exposure to ozone in Southern California remains the nation's worst. Even if all the aforementioned controls in urban areas are implemented, air quality will continue to be a significant problem in coming decades, particularly if the urban population continues to increase.

We have focused on air pollution in Southern California because its air quality is especially poor. However, most large and not-so-large U.S. cities have poor air quality for a significant part of the year. With the exception of the Pacific Northwest, no U.S. region is free from air pollution and its health effects.[6]

Developing Countries

The pessimistic view is that population pressures and environmentally unsound policies and practices will dictate what happens in many developing parts of the world, and the result will be poorer air quality. They often don't have the financial base necessary to fight air pollution and are more concerned with finding ways to house and feed their growing populations.

Consider Mexico City. With a population of about 25 million, Mexico City is one of the four largest urban areas in the world. Cars, buses, industry, and power plants in the city emit hundreds of thousands of metric

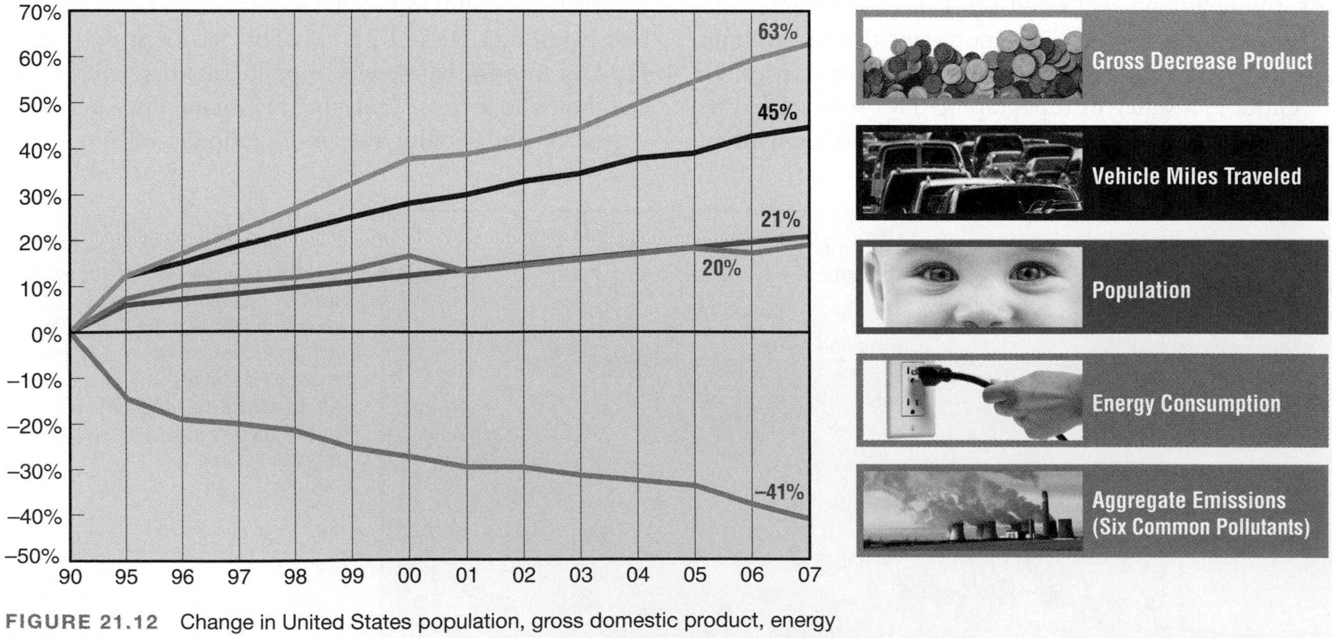

FIGURE 21.12 Change in United States population, gross domestic product, energy consumption, and aggregate emission of the six common air pollutants (ground-level ozone, particulates, lead, nitrogen dioxide, sulfur dioxide, and carbon monoxide) from 1990 to 2007. (*Source*: U.S. Environmental Protection Agency, 2008, air quality trends through 2007. www.epa.gov.

tons of pollutants into the atmosphere each year. The city is at an elevation of about 2,255 m (7,400 ft) in a natural basin surrounded by mountains, a perfect situation for a severe air pollution problem. It is becoming a rare day in Mexico City when the mountains can be seen. Headaches, irritated eyes, and sore throats are common when the pollution settles in, and physicians report a steady increase in respiratory diseases. They advise parents to take their children out of the city permanently. The people in Mexico City do not need to be told they have an air-pollution problem; it is all too apparent. However, developing a successful strategy to improve the quality of the air is difficult.[19]

21.2 Controlling Common Pollutants of the Lower Atmosphere

The most reasonable ways to control the most common air pollutants in our cities include reducing emissions, capturing them before they reach the atmosphere, and removing them from the atmosphere. From an environmental viewpoint, reducing emissions through energy efficiency and conservation (such as burning less fuel) is the preferred strategy, with clear advantages over all other approaches (see Chapters 14 to 16). Here, we discuss control of selected air pollutants.

Particulates

Particulates emitted from fugitive, point, or area stationary sources are much easier to control than the very small particulates of primary or secondary origin released from mobile sources, such as automobiles. As we learn more about these very small particles, we will have to devise new methods to control them.

A variety of "settling chambers" or collectors are used to control emissions of coarse particulates from power plants and industrial sites (point or area sources) by providing a mechanism that causes particles in gases to settle out in a location where they can be collected for disposal in landfills. In recent decades, we have made great strides in controlling particulates, such as ash, from power plants and industry.

Automobiles

Controlling such pollutants as carbon monoxide, nitrogen oxides, and hydrocarbons in urban areas is best achieved by pollution-control measures for automobiles. Control of these materials will also limit ozone formation in the lower atmosphere, since, as you have learned, ozone forms through reactions with nitrogen oxides and hydrocarbons in the presence of sunlight.

Nitrogen oxides from automobile exhausts are controlled by recirculating exhaust gas and diluting the air-to-fuel mixture burned in the engine. Dilution lowers the temperature of combustion and decreases the oxygen concentration in the burning mixture so that it produces fewer nitrogen oxides. Unfortunately, the same process increases hydrocarbon emissions. Nevertheless, exhaust recirculation to reduce nitrogen oxide emissions has been common practice in the United States for more than 20 years.[20]

The exhaust system's catalytic converter is the device most commonly used to reduce carbon monoxide and hydrocarbon emissions from automobiles. In the converter, oxygen from outside air is introduced, and exhaust gases from the engine are passed over a catalyst, typically platinum or palladium. Two important chemical reactions occur: Carbon monoxide is converted to carbon dioxide; and hydrocarbons are converted to carbon dioxide and water.

Other approaches to reducing air pollution from vehicles include reducing the number and types of cars on roads; developing cleaner fuels through use of fuel additives and reformulation; and requiring more fuel-efficient motor vehicles, such as those with electric engines and hybrid cars that have both an electric engine and an internal combustion engine.

Sulfur Dioxide

Sulfur dioxide emissions have been reduced by using abatement measures before, during, or after combustion. Technology to clean up coal so that it will burn more cleanly is already available. Although removing the sulfur makes fuel more expensive, the expense must be balanced against the long-term consequences of burning high-sulfur coal. Switching from high-sulfur coal to low-sulfur coal seems an obvious way to reduce emissions of sulfur dioxide, and in some regions this will work. Unfortunately, however, most of the naturally low-sulfur coal in the United States is in the western part of the country, whereas most coal is burned in the East, so transportation is an issue and using low-sulfur coal is a solution only in cases where it is economically feasible.

Sulfur emissions can also be reduced by washing coal. When finely ground coal is washed with water, iron sulfide (mineral pyrite) settles out because of its relatively high density. But this is ineffective for removing organic sulfur bound up with carbonaceous material, and it is expensive.

Another option is *coal gasification*, which converts relatively high-sulfur coal to a gas in order to remove the sulfur.

The gas is quite clean and can be transported relatively easily, augmenting supplies of natural gas. True, it is still fairly expensive compared with gas from other sources, but its price may become more competitive in the future.

Desulfurization, or **scrubbing** (Figure 21.13), removes sulfur from stationary sources such as power plants. This technology was developed in the 1970s in the United States in response to passage of the Clean Air Act. However, the technology was not initially implemented in the United States; instead, regulators chose to allow plants to disperse pollutants through very tall smokestacks. This worsened the regional acid-rain problem.

Nearly all scrubbers (90%) used at coal-burning power plants in the United States are wet scrubbers that use a lot of water and produce a wet end product. Wet scrubbing is done after coal is burned. The SO_2 rich gases are treated with a slurry (a watery mixture) of lime (calcium oxide, CaO) or limestone (calcium carbonate, $CaCO_3$). The sulfur oxides react with the calcium to form calcium sulfite, which is collected and then usually disposed of in a landfill. [21]

In West Virginia, a coal mine, power plant, and synthetic gypsum plant located close to each other joined forces in 2008 to produce electric energy, recover sulfur dioxide, and produce high-quality wallboard (sheetrock) for the construction industry. The power plant benefits by selling the raw gypsum (from scrubbers) rather than paying to dispose of it in a landfill. The wallboard plant is right next to the power plant and uses gypsum that does not have to be mined from earth.[22]

Air Pollution Legislation and Standards

Clean Air Act Amendments of 1990

The Clean Air Act Amendments of 1990 are comprehensive regulations enacted by the U.S. Congress that address the problems of acid rain, toxic emissions, ozone depletion, and automobile exhaust. In dealing with acid rain, the amendments establish limits on the maximum permissible emissions of sulfur dioxide from utility companies burning coal. The goal of the legislation—to reduce such emissions by about 50%, to 10 million tons a year, by 2000—was more than achieved (refer back to Table 21.2).

An innovative aspect of the legislation is the incentives it offers to utility companies to reduce emissions of sulfur dioxide. As explained earlier here and discussed in detail in Chapter 7, the incentives are marketable permits (allowances) that allow companies to buy and sell the right to pollute[23] (see A Closer Look 21.1). The 1990 amendments also call for reducing emissions of nitrogen dioxides by approximately 2 million tons from the 1980 level. The actual reduction has been 10 million tons—an air- pollution success story!

Ambient Air Quality Standards

Air quality standards are important because they are tied to emission standards that attempt to control the concentrations of various pollutants in the atmosphere. The many countries that have developed air quality standards include France, Japan, Israel, Italy, Canada, Germany, Norway, and the United States. National Ambient Air Quality Standards (NAAQS) for the United States, defined to comply with the Clean Air Act, are shown in Table 21.3. Tougher standards were set for ozone and $PM_{2.5}$ in recent years to reduce adverse health effects on children and elderly people, who are most susceptible to air pollution. The new standards are saving the lives of thousands and improving the health of hundreds of thousands of children. The ozone standard was significantly strengthened in

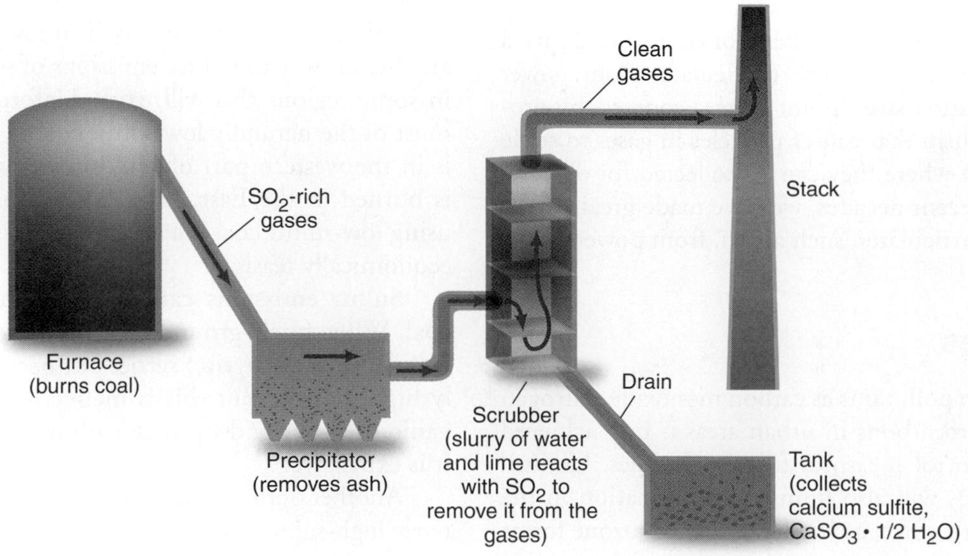

FIGURE 21.13 Scrubber used to remove sulfur oxides from the gases emitted by tall stacks.

2008. The change is expected to result in health benefits of more than $15 billion per year.

Air Quality Index

In the United States, the Air Quality Index (AQI) (Table 21.4) is used to describe air pollution on a given day. For example, air quality in urban areas is often reported as good, moderate, unhealthy for sensitive groups, unhealthy, very unhealthy, or hazardous, corresponding to a color code of the Air Quality Index. The AQI is determined by measuring the concentration of five major pollutants: particulate matter, sulfur dioxide, carbon monoxide, ozone, and nitrogen dioxide. An AQI value greater than 100 is unhealthy. In most U.S. cities, AQI values range between 0 and 100. Values above 100 are generally recorded for a particular city only a few times a year, but some cities with serious air-pollution problems may exceed an AQI of 100 many times a year. In a typical year, AQI values above 200 (for all U.S. sites) are rare, and those above 300 are very rare. In large cities outside the United States with dense human populations and numerous uncontrolled sources of pollution, AQIs greater than 200 are frequent.

The Cost of Controlling Outdoor Air Pollution

The cost of outdoor air pollution control varies widely from one industry to another. For example, the cost for incremental control in a fossil-fuel-burning utility is a few hundred dollars per additional ton of particulates removed. For an aluminum refinery, the cost to remove an additional ton of particulates may be as much as several thousand dollars. Some economists would argue that it is wise to raise the standards for utilities and relax them, or at least not raise them, for aluminum plants. This would lead to more cost-efficient pollution control while maintaining good air quality. However, the geographic distribution of various facilities will determine the trade-offs possible.[23, 24]

Economic analysis of air pollution is not simple. There are many variables, some of which are hard to quantify. We do know the following:

- With increasing air pollution controls, the capital cost for technology to control air pollution increases.
- As the controls for air pollution increase, the loss from pollution damages decreases.

Table 21.3 U.S. NATIONAL AMBIENT AIR QUALITY STANDARDS (NAAQS)

POLLUTANT	STANDARD VALUE[a]		STANDARD TYPE
Carbon monoxide (CO)			
8-hour average	9 ppm	(10 mg/m³)	Primary[c]
1-hour average	35 ppm	(40 mg/m³)	Primary
Nitrogen dioxide (NO_2)			
Annual arithmetic mean	0.053 ppm	(100 μg/m³)	Primary and secondary[d]
Ozone (O_3)			
8-hour average	0.075 ppm	(147 μg/m³)	Primary and secondary
Lead (Pb)			
Quarterly average	1.5 μg/m³		Primary and secondary
Particulate (PM 10) *Particles with diameters of 10 micrometers or less*			
Annual arithmetic mean	50 μg/m³		Primary and secondary
24-hour average	150 μg/m³		Primary and secondary
Particulate (PM 2.5)[b] *Particles with diameters of 2.5 micrometers or less*			
Annual arithmetic mean	15 μg/m³		Primary and secondary
24-hour average	65 μg/m³		Primary and secondary
Sulfur dioxide (SO_2)			
Annual arithmetic mean	0.03 ppm	(80 μg/m³)	Primary
24-hour average	0.14 ppm	(365 μg/m³)	Primary
3-hour average	0.50 ppm	(1300 μg/m³)	Secondary

[a] Parenthetical value is an approximately equivalent concentration.

[b] The ozone 8-hour standard and the PM 2.5 standards are included for information only. A 1999 federal court ruling blocked implementation of these standards, which the EPA proposed in 1997. EPA has asked the U.S. Supreme Court to reconsider that decision. (Note: In March 2001, the Court ruled in favor of the EPA, and the new standards are expected to take effect within a few years.)

[c] Primary standards set limits to protect public health, including the health of sensitive populations such as asthmatics, children, and the elderly.

[d] Secondary standards set limits to protect public welfare, including protection against decreased visibility and damage to animals, crops, vegetation, and buildings.

Source: U.S. Environmental Protection Agency.

Table 21.4 AIR QUALITY INDEX (AQI) AND HEALTH CONDITIONS

INDEX VALUES	DESCRIPTOR	CAUTIONARY STATEMENT	GENERAL ADVERSE HEALTH EFFECTS	ACTION LEVEL (AQI)[a]
0–50	Good	None	None	None
51–100	Moderate	Unusually sensitive people should consider limiting prolonged outdoor exertion.	Very few symptoms[b] for the most susceptible people[c]	None
101–150	Unhealthy for sensitive groups	Active children and adults, and people with respiratory disease, such as asthma, should limit prolonged outdoor exertion.	Mild aggravation of symptoms in susceptible people, few symptoms for healthy people	None
151–199	Unhealthy	Active children and adults, and people with respiratory disease, such as asthma, should avoid prolonged outdoor exertion; everyone else, especially children, should limit prolonged outdoor exertion.	Mild aggravation of symptoms in susceptible people, irritation symptoms for healthy people	None
200–300	Very unhealthy	Active children and adults, and people with respiratory disease, such as asthma, should avoid outdoor exertion; everyone else, especially children, should limit outdoor exertion.	Significant aggravation of symptoms in susceptible people, widespread symptoms in healthy people	Alert (200+)
Over 300	Hazardous	*Everyone* should avoid outdoor exertion.	300–400: Widespread symptoms in healthy people 400–500: Premature onset of some diseases Over 500: Premature death of ill and elderly people; healthy people experience symptoms that affect normal activity	Warning (300+) Emergency (400+)

[a] Triggers preventative action by state or local officials.

[b] Symptoms include eye, nose, and throat irritation; chest pain; breathing difficulty.

[c] Susceptible people are young, old, and ill people and people with lung or heart disease.

AQI 51–100	Health advisories for susceptible individuals.
AQI 101–150	Health advisories for all.
AQI 151–200	Health advisories for all.
AQI 200+	Health advisories for all; triggers an alert; activities that cause pollution might be restricted.
AQI 300	Health advisories to all; triggers a warning; probably would require power plant operations to be reduced and carpooling to be used.
AQI 400+	Health advisories for all; triggers an emergency; cessation of most industrial and commercial activities, including power plants; nearly all private use of vehicles prohibited.

Source: U.S. Environmental Protection Agency.

- The total cost of air pollution is the cost of pollution control plus the environmental damages of the pollution.

Although the cost of pollution-abatement technology is fairly well known, it is difficult to accurately determine the loss from pollution damages, particularly when considering health problems and damage to vegetation, including food crops. For example, exposure to air pollution may cause or aggravate chronic respiratory diseases in people, at a very high cost. A recent study of the health benefits of cleaning up the air quality in the Los Angeles basin estimated that the annual cost of air pollution in the basin is 1,600 lives and about $10 billion.[25] Air pollution also leads to loss of revenue from people who choose not to visit some areas, such as Los Angeles and Mexico City, because of known air-pollution problems.[26, 27]

With our discussion of the more traditional outdoor air pollutants behind us, we turn now to ozone depletion in the stratosphere. This will be followed by a discussion of indoor air pollution, which has emerged as a serious environmental problem.

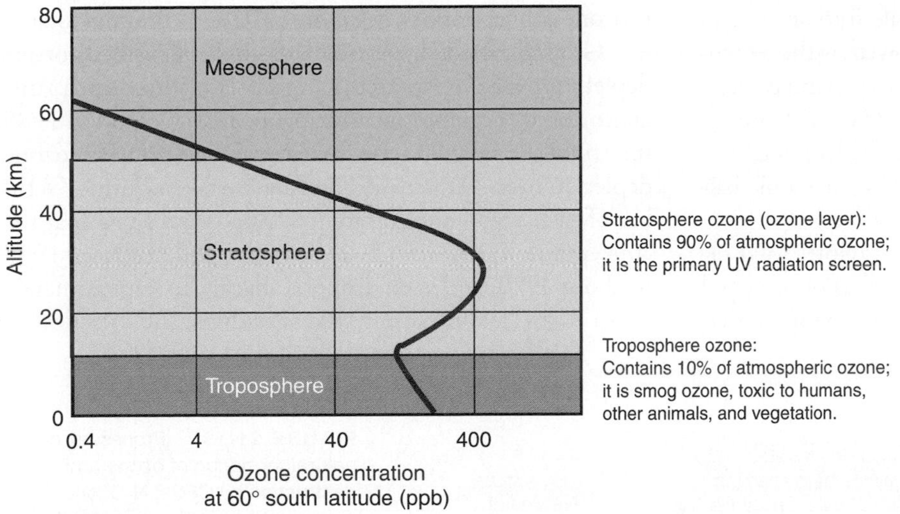

(a)

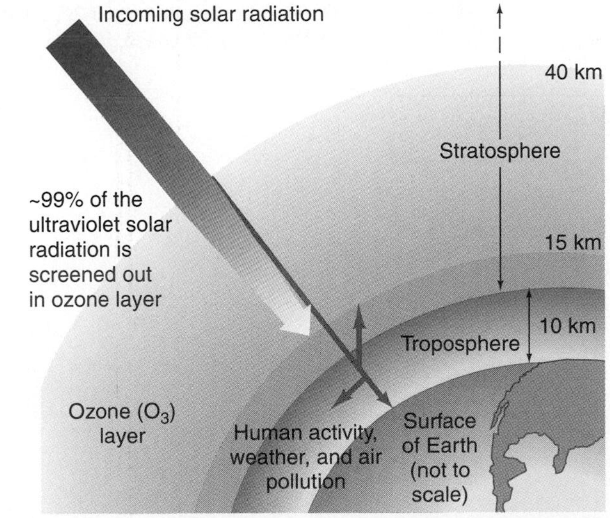

(b)

FIGURE 21.14 **(a)** Structure of the atmosphere and ozone concentration. **(b)** Reduction of the potentially most biologically damaging ultraviolet radiation by ozone in the stratosphere. (*Source:* Ozone concentrations modified from R.T. Watson, "Atmospheric Ozone," in J.G. Titus, ed., *Effects of Change in Stratospheric Ozone and Global Climate*, vol. 1, *Overview*, p. 70 (U.S. Environmental Protection Agency).

21.3 High-Altitude (Stratospheric) Ozone Depletion

The serious problem of ozone depletion in the stratosphere (about 9 to 25 km above Earth's surface) starts down here in the lower atmosphere.

About 21% of the air we breathe at sea level is *diatomic* oxygen (O_2), which is two oxygen atoms bonded together. **Ozone (O_3)** is a *triatomic* form of oxygen in which three atoms of oxygen are bonded. Ozone is a strong oxidant and reacts chemically with many materials in the atmosphere. In the lower atmosphere, as we have discussed, ozone is a pollutant produced by photochemical reactions involving sunlight, nitrogen oxides, hydrocarbons, and diatomic oxygen. In the stratosphere, however, ozone plays an entirely different role, protecting us from ultraviolet radiation.

Ultraviolet Radiation and Ozone

The ozone layer in the stratosphere is often called the **ozone shield** because it absorbs most of the potentially hazardous ultraviolet radiation that enters Earth's atmosphere from the sun. Ultraviolet radiation has wavelengths between 0.1 and 0.4 μm and is subdivided into ultraviolet A (UVA), ultraviolet B (UVB), and ultraviolet C (UVC). Ultraviolet radiation with a wavelength of less than about 0.3 μm can be very hazardous to life. If much of this radiation reached Earth's surface, it would injure or kill most living things.[28, 29]

Ultraviolet C (UVC) has the shortest wavelength and is the most energetic of the three types. It has enough energy to break down diatomic oxygen (O_2) in the stratosphere into two oxygen atoms, each of which may combine with an O_2 molecule to create ozone. Ultraviolet C is strongly absorbed in the stratosphere, and negligible amounts reach Earth's surface.[28,29]

Ultraviolet A (UVA) radiation has the longest wavelength and the least energy of the three types. UVA can cause some damage to living cells, is not affected by stratospheric ozone, and is transmitted to the surface of Earth.[28]

Ultraviolet B (UVB) radiation is energetic and strongly absorbed by stratospheric ozone. In fact, ozone is the only known gas that absorbs UVB. Thus, depletion of ozone in the stratosphere allows more UVB to reach the Earth. Because UVB radiation is known to be hazardous to living things,[28-30] this increase in UVB is the hazard we are talking about when we discuss the problem of ozone depletion in the stratosphere.

The structure of the atmosphere and concentrations of ozone are shown in Figure 21.14. Approximately 90% of the ozone in the atmosphere is in the stratosphere, ranging from about 15 km to 40 km (9 to 25 mi) in altitude, with peak concentrations of about 400 ppb. The altitude of peak concentration varies from about 30 km (19 mi) near the equator to about 15 km (9 mi) in polar regions.[28]

Processes that produce ozone in the stratosphere are illustrated in Figure 21.15. The first step in ozone production is *photodissociation*—intense ultraviolet radiation (UVC) breaks an oxygen molecule (O_2) into two oxygen atoms. These atoms then react with another oxygen molecule to form two ozone molecules. Ozone, once produced, may absorb UVC radiation, which breaks the ozone molecule into an oxygen molecule and an oxygen atom. This is followed by the recombination of the oxygen atom with another oxygen molecule to re-form into ozone. As part of this process, UVC radiation is converted to heat energy in the stratosphere. Natural conditions that prevail in the stratosphere result in a dynamic balance between the creation and destruction of ozone.

In sum, approximately 99% of all ultraviolet solar radiation (all UVC and most UVB) is absorbed or screened out in the ozone layer. The absorption of ultraviolet radiation by ozone is a natural service function of the ozone shield and protects us from the potentially harmful effects of ultraviolet radiation.

Measuring Stratospheric Ozone

Scientists first measured the concentration of atmospheric ozone in the 1920s from the ground, using an instrument known as a Dobson ultraviolet spectrometer. The Dobson unit (DU) is still commonly used to measure the ozone concentrations; 1 DU equals a concentration of 1 ppb O_3. Today, we have a record of ozone concentrations spanning about 50 years. Most of the measurement stations are in the midlatitudes, and the accuracy of the data varies with the levels of quality control.[28] Satellite measurements of atmospheric ozone concentrations began in 1970 and continue today.

Ground-based measurements first identified ozone depletion over the Antarctic. Members of the British Antarctic Survey began to measure ozone in 1957, and in 1985 they published the first data that suggested significant ozone depletion over Antarctica. The data are taken during October of each year—the Antarctic spring—and show that the concentration of ozone hovered around 300 DU from 1957 to about 1970, and then dropped sharply, to approximately 140 DU by 1986. Despite the variations, the direction of

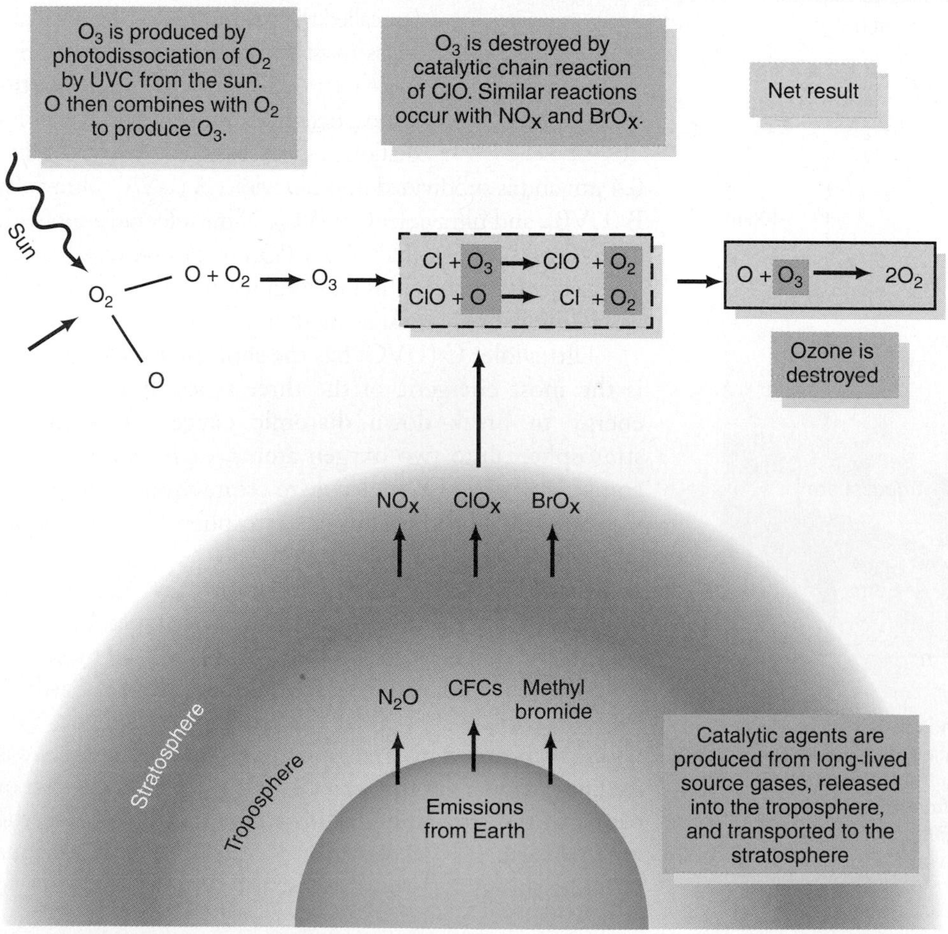

FIGURE 21.15 Processes of natural formation of ozone and destruction by CFCs, N_2O, and methyl bromide. (*Source:* Modified from NASA-GSFC, "Stratospheric Ozone," accessed August 22, 2000, at http://see.gsfc.nasa.gov.)

change, with minor exceptions, is clear: Ozone concentrations in the stratosphere during the Antarctic spring have been decreasing since the mid-1970s.[31-34] The depletion in ozone was dubbed the *ozone hole*. There is no actual hole in the ozone shield where all the ozone is depleted; rather, the term describes a relative depletion in the concentration of ozone that occurs during the Antarctic spring.

Ozone Depletion and CFCs

The hypothesis that ozone in the stratosphere is being depleted by **chlorofluorocarbons (CFCs)** was first suggested in 1974 by Mario Molina and F. Sherwood Rowland.[34] This hypothesis, based mostly on physical and chemical properties of CFCs and knowledge about atmospheric conditions, was immediately controversial and vigorously debated by scientists, companies producing CFCs, and other interested parties.[35-36] The major features of the Molina and Rowland hypothesis are as follows:[28, 29]

- CFCs emitted in the lower atmosphere by human activity are very stable and nonreactive in the lower atmosphere and therefore have a very long residence time (about 100 years). No significant sinks for CFCs are known, with the possible exception of soils, which evidently do remove an unknown amount of CFCs from the atmosphere at Earth's surface.[36]
- Because of their long residence time in the lower atmosphere, and because the lower atmosphere is very fluid, the CFCs eventually disperse, wander upward, and enter the stratosphere. Once they reach altitudes above most of the stratospheric ozone, they may be destroyed by the highly energetic solar ultraviolet radiation. This releases chlorine, a highly reactive atom.
- The reactive chlorine may then enter into reactions that deplete ozone in the stratosphere.
- Ozone depletion allows an increased amount of UVB radiation to reach Earth. Ultraviolet B is a cause of human skin cancers and is also believed to be harmful to the human immune system.

Simplified Stratospheric Chlorine Chemistry

CFCs are considered responsible for most of the ozone depletion. Let us look more closely at how this occurs.

Earlier, we noted that there are no tropospheric sinks for CFCs. That is, the processes that remove most chemicals in the lower atmosphere—destruction by sunlight, rain-out, and oxidation—do not break down CFCs because CFCs are transparent to sunlight, are essentially insoluble, and are nonreactive in the oxygen-rich lower atmosphere.[37] Indeed, the fact that CFCs are nonreactive in the lower atmosphere was one reason they were attractive for use as propellants.

When CFCs wander to the upper part of the stratosphere, however, reactions do occur. Highly energetic ultraviolet radiation (UVC) splits up the CFC, releasing chlorine. When this happens, the following two reactions can take place:[37]

$$(1)\ Cl + O_3 \rightarrow ClO + O_2$$
$$(2)\ ClO + O \rightarrow Cl + O_2$$

These two equations define a chemical cycle that can deplete ozone (Figure 21.15). In the first reaction, chlorine combines with ozone to produce chlorine monoxide. which, in the second reaction, combines with monatomic oxygen to produce chlorine again. The chlorine can then enter another reaction with ozone and cause additional ozone depletion. This series of reactions is what is known as a *catalytic chain reaction*. Because the chlorine is not removed but reappears as a product of the second reaction, the process may be repeated over and over again. It has been estimated that each chlorine atom may destroy approximately 100,000 molecules of ozone in one or two years before the chlorine is finally removed from the stratosphere through other chemical reactions and rain-out.[37] The significance of these reactions is apparent when we realize how many metric tons of CFCs have been emitted into the atmosphere.

It should be noted that what actually happens chemically in the stratosphere is considerably more complex than the two equations shown here. The atmosphere is essentially a chemical soup in which a variety of processes related to aerosols and clouds take place (some of these are addressed in the discussion of the ozone hole). Nevertheless, these equations show us the basic chemical chain reaction that occurs in the stratosphere to deplete ozone.

The catalytic chain reaction just described can be interrupted through storage of chlorine in other compounds in the stratosphere. Two possibilities are as follows:

1. Ultraviolet light breaks down CFCs to release chlorine, which combines with ozone to form chlorine monoxide (ClO), as already described. This is the first reaction discussed. The chlorine monoxide may then react with nitrogen dioxide (NO_2) to form a chlorine nitrate ($ClONO_2$). If this reaction occurs, ozone depletion is minimal. The chlorine nitrate, however, is only a temporary reservoir for chlorine. The compound may be destroyed, and the chlorine released again.
2. Chlorine released from CFCs combine with methane (CH_4) to form hydrochloric acid (HCl). The hydrochloric acid may then diffuse downward. If it enters the troposphere, rain may remove it, thus removing the chlorine from the ozone-destroying chain reaction. This is the ultimate end for most chlorine atoms in the stratosphere. However, while the hydrochloric acid molecule is in the stratosphere, it may be destroyed by incoming solar radiation, releasing the chlorine for additional ozone depletion.

It has been estimated that the chlorine chain reaction that destroys ozone may be interrupted by the processes just described as many as 200 times while a chlorine atom is in the stratosphere.[28, 38]

It is important to remember that for the Southern Hemisphere and under natural conditions, the highest concentration of ozone is in the polar regions (about 60° south latitude) and the lowest near the equator. At first this may seem strange because ozone is produced in the stratosphere by solar energy, and there is more solar energy near the equator. But although much of the world's ozone is produced near the equator, the ozone in the stratosphere moves from the equator toward the poles with global air-circulation patterns.[32]

In part as a result of ozone depletion, concentrations of ozone have declined in both northern and southern temperate latitudes. While remaining relatively constant at the equator, ozone has been significantly reduced in the Antarctic since the 1970s. Massive destruction of ozone in the Antarctic constitutes the "ozone hole." [28]

The Antarctic Ozone Hole

Since the Antarctic ozone hole was first reported in 1985, it has captured the interest of many people around the world. Every year since then, ozone depletion has been observed in the Antarctic in October, the spring season there. Because the thickness of the ozone layer above the Antarctic in springtime has been declining since the mid-1970s, the geographic area covered by the ozone hole has grown from a million or so square kilometers in the late 1970s and early 1980s to about 29 million square kilometers by 1995—about the size of North America in 2000. It has since stabilized as the ozone concentration has ceased its steep decline.[31, 39]

Polar Stratospheric Clouds

The minimum concentration of ozone in the Antarctic since 1980 has varied from about 50% to 70% of that in the 1970s. Polar stratospheric clouds over the Antarctic appear to be one of the causes of this variation. Observed for at least the past hundred years about 20 km (12 mi) above the polar regions, the clouds have an eerie beauty and an iridescent glow, reminiscent of mother-of-pearl.[38] They form during the polar winter (called the polar night because the tilt of Earth's axis limits sunlight). During the polar winter, the Antarctic air mass is isolated from the rest of the atmosphere and circulates about the pole in what is known as the Antarctic *polar vortex*. The vortex forms as the isolated air mass cools, condenses, and descends.[31,32]

Clouds form in the vortex when the air mass reaches a temperature between 195 K and 190 K (–78° to –83°C; –108° to –117°F). At these very low temperatures, small sulfuric acid particles (approximately 01. μm) freeze and serve as seed particles for nitric acid (HNO_3). These clouds are called Type I polar stratospheric clouds. If temperatures drop below 190 K (–83°C; –117°F), water vapor condenses around some of the earlier-formed Type I cloud particles, forming Type II polar stratospheric clouds, which contain larger particles. Type II polar stratospheric clouds are the ones with the mother-of-pearl color.

During the formation of polar stratospheric clouds, nearly all the nitrogen oxides in the air mass are converted to the clouds as nitric acid particles, which grow heavy and descend below the stratosphere, leaving very little nitrogen oxide in the vicinity of the clouds.[28, 38, 39] This facilitates ozone-depleting reactions that may ultimately reduce stratospheric ozone in the polar vortex by as much as 1% to 2% per day in the early spring, when sunlight returns to the polar region (Figure 21.16).

An idealized diagram of the polar vortex that forms over Antarctica is shown in Figure 21.16*a*. The Ozone-depleting reactions within the vortex are illustrated in Figure 21.16*b*. As shown, in the dark Antarctic winter almost all available nitrogen oxides are tied up on the edges of particles in the polar stratospheric clouds or have settled out. Hydrochloric acid and chlorine nitrate (the two important sinks of chlorine) act on particles of polar stratospheric clouds to form dimolecular chlorine (Cl_2) and nitric acid through the following reaction:[40]

$$HCl + ClONO_2 \rightarrow Cl_2 + HNO_3$$

In the spring, when sunlight returns and breaks apart chlorine (Cl_2), the ozone-depleting reactions discussed earlier occur. Nitrogen oxides are absent from the Antarctic stratosphere in the spring, so the chlorine cannot be sequestered to form chlorine nitrate, one of its major sinks, and remains free to destroy ozone. In the early Antarctic spring, these ozone-depleting reactions can be rapid, producing the 50% reduction in ozone observed in recent years. Ozone depletion in the Antarctic vortex ceases later in spring as the environment warms and the polar stratospheric clouds disappear, releasing nitrogen back into the atmosphere, where it can combine with chlorine and thus be removed from ozone-depleting reactions. Stratospheric ozone concentrations then increase as ozone-rich air masses again migrate to the polar region.

A weaker, shorter polar vortex forms over the North Pole area and can lead to ozone depletion of as much as 30–40%. When the vortex breaks up, it can send ozone-deficient air masses southward to drift over areas of Europe and North America.[41]

Environmental Effects of Ozone Depletion

Ozone depletion damages some food chains on land and in the oceans and is dangerous to people, increasing the incidence of skin cancers and cataracts and suppressing immune systems.[41, 42] A 1% decrease

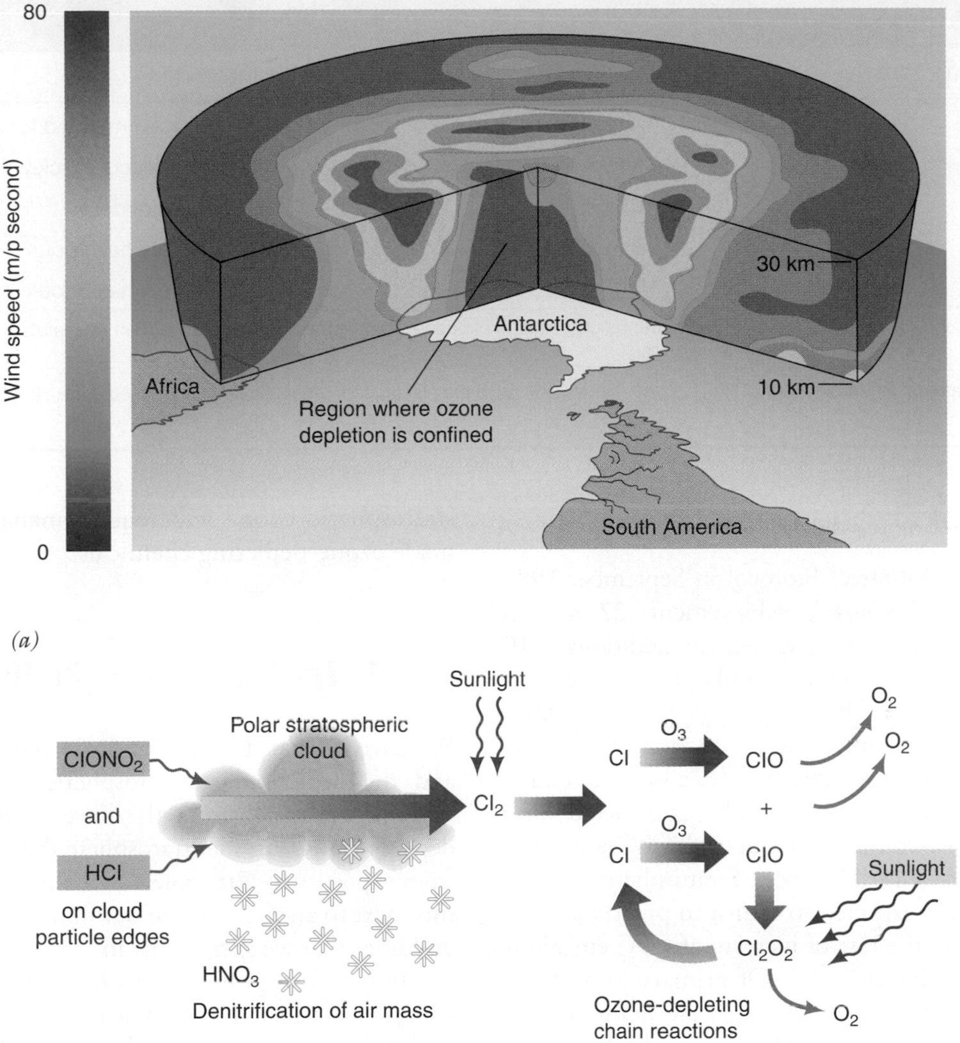

FIGURE 21.16 **(a)** Idealized diagram of the Antarctic polar vortex and **(b)** the role of polar stratospheric clouds in the ozone-depletion chain reaction. (*Source*: Based on O.B. Toon and R.P. Turco, "Polar Stratospheric Clouds and Ozone Depletion," *Scientific American*, 264, no. 6 [1991]: 68–74.)

in ozone can cause a 1–2% increase in UVB radiation and a 2% increase in skin cancer.[43] Because skin cancers have increased globally, health-conscious people today are replacing tanning oils with sunblocks and hats, and newspapers in the United States now provide the **Ultraviolet (UV) Index** (Table 21.5). Developed by the National Weather Service and EPA, the index predicts UV intensity on a scale from 1 to 11+. Some news agencies also use the index to recommend the level of sunblock. It is speculated that the incidence of skin cancer due to ozone depletion will rise until about 2060 and then decline as the ozone shield recovers as a result of controls on CFC emissions.[44, 45]

You can lower your risk of skin cancer and other skin damage from UV exposure by taking a few simple precautions:

- Limit exposure to the sun between 10 A.M. and 4 P.M., the hours of intense solar radiation, and stay in the shade when possible.
- Use a sunscreen with an SPF of at least 30 (but remember that protection diminishes with increased exposure), or use clothing to cover up.
- Wear UV-protective sunglasses.
- Avoid tanning salons and sun lamps.
- Consult the UV Index before going out.

A simple guideline: If your shadow is longer than you are, such as in the evening or early morning, UV exposure is relatively low. If your shadow is shorter than you are, you are in the part of the day with highest UV exposure.

Table 21.5 ULTRAVIOLET (UV) INDEX FOR HUMAN EXPOSURE

EXPOSURE CATEGORY	UV INDEX	COMMENT
Low	< 2	Sunblock recommended for all exposure
Moderate	3 to 5	Sunburn can occur quickly
High	6 to 7	Potentially hazardous
Very high	8 to 10	Potentially very hazardous
Extreme	11	Potentially very hazardous

Note: At moderate exposure to UV, sunburn can occur quickly, at high exposure, fair-skinned people may burn in 10 minutes or less of exposure.

Source: Modified after U.S. Environmental Protection Agency 2004 (with the National Weather Service). Accessed June 16, 2004 at www.epa.gov.

The Future of Ozone Depletion

The signing of the Montreal Protocol in September 1987 was an important diplomatic achievement: 27 nations signed the agreement originally, and an additional 119 signed later. The protocol outlined a plan to eventually reduce global emissions of CFCs to 50% of 1986 emissions. It originally called for eliminating production of CFCs by 1999, but the period was shortened because of scientific evidence that stratospheric ozone was being depleted faster than predicted. An eventual phase-out of all CFC consumption is part of the Montreal Protocol. Stratospheric concentrations of CFCs are expected to return to pre-1980 levels by about 2050, and the rate of increase of CFC emissions has already been reduced.[31, 46, 47] Of primary importance is developing substitutes for CFCs that are both safe and effective. Hydrofluorocarbons (HFCs) are the long-term substitute for CFCs because they do not contain chlorine.

However, a troubling aspect of ozone depletion is that if the manufacture, use, and emission of all ozone-depleting chemicals were to stop today, the problem would not go away—because millions of metric tons of those chemicals are now in the lower atmosphere, working their way up to the stratosphere. Several CFCs have atmospheric lifetimes of 75–140 years. Thus, an estimated 35% of the CFC-12 molecules in the atmosphere will likely still be there in 2100, and approximately 15% in 2200.[28] In addition, some 10–15% of the CFC molecules manufactured in recent years have not yet been admitted to the atmosphere because they remain in foam insulation, air-conditioning units, and refrigerators.[28] Nevertheless, indicators suggest that growth in the concentrations of CFCs has been slowed and in some cases reversed, and recovery of ozone should be noticeable by 2020 or later.[31]

Today by necessity, we are adapting to ozone depletion by learning to live with higher levels of exposure to ultraviolet radiation. (For example using sunblock, wearing hats and avoiding direct mid day solar radiation.) In the long term, achieving a sustainable level of stratospheric ozone will require management of man-made ozone-depleting chemicals.

21.4 Indoor Air Pollution

We have discussed air pollution in the lower atmosphere and the depletion of stratospheric ozone by chemical emissions that rise from the lower atmosphere to cause depletion of O_3 in the stratosphere that produces a hazard from exposure to ultravioleted radiation from the sun. We turn next to air pollution in our homes, schools, and other buildings that we spend time in.

Indoor air pollution from fires for cooking and heating has affected human health for thousands of years. A detailed autopsy of a 4th-century Native American woman, frozen shortly after death, revealed that she suffered from **black lung disease** from breathing very polluted air over many years. The pollutants included hazardous particles from lamps that burned seal and whale blubber.[48] This same disease has long been recognized as a major health hazard for underground coal miners and has been called "coal miners' disease." As recently as the mid-1970s, black lung disease was estimated to be responsible for about 4,000 deaths each year in the United States.[49]

People today spend between 70% and 90% of their time in enclosed places—homes, workplaces, automobiles, restaurants, and so forth—but only recently have we begun to fully study the indoor environment and how pollution of that environment affects our health. The World Health Organization has estimated that as many as one in three people may be working in a building that causes them to become sick, and as many as 20% of public schools in the United States have problems related to indoor air quality. The EPA considers indoor air pollution one of the most significant environmental health hazards people face in the modern workplace.[50]

Hurricane Katrina in 2005 (see Chapter 22) left a great number of people homeless. In response, the Federal Emergency Management Agency (FEMA) provided thousands of trailers for people to live in. That sounded like a great idea until complaints started to come in about health problems of people living in the trailers. A study by the Centers for Disease Control and Prevention (CDC) confirmed that the mobile homes suffered from indoor air pollution by formaldehyde in their construction materials. Formaldehyde is a chemical widely used in the manufacture of building materials, as well as a number of other products. It is considered a probable human carcinogen (a substance that causes or promotes cancer). Common symptoms of exposure to formaldehyde include irritation of the skin, nose, throat, and eyes. People with asthma may be more sensitive to the chemical, and their symptoms may be worse. Since discovery of the high levels of formaldehyde in mobile homes in late 2007, plans have gone forward to remove the remaining people, particularly those experiencing symptoms of formaldehyde toxicity.[51, 52]

The history of formaldehyde in the mobile homes provided to Katrina victims is a sad legacy of the entire way our federal government responded to Hurricane Katrina and its aftermath. It is also important because it brings to the public consciousness the potential problems of indoor air pollution, which is often more significant than outdoor air pollution.

Sources of Indoor Air Pollution

The sources of indoor air pollution are incredibly varied (Figure 21.17) and can arise from both human activities and natural processes. Two common pollutants are shown in Figure 21.18. Other common indoor air pollutants, together with guidelines for allowable exposure, are listed in Table 21.6.

1. Heating, ventilation, and air-conditioning systems may be sources of indoor air pollutants, including molds and bacteria, if filters and equipment are not maintained properly. Gas and oil furnaces release carbon monoxide, nitrogen dioxide, and particles.
2. Restrooms may have a variety of indoor air pollutants, including secondhand smoke, and also molds and fungi due to humid conditions.
3. Furniture and carpets often contain toxic chemicals (formaldehyde, organic solvents, asbestos) that may be released over time in buildings.
4. Coffee machines, fax machines, computers, and printers can release particles and chemicals, including ozone (O_3), which is highly oxidizing.
5. Pesticides can contaminate buildings with cancer-causing chemicals.
6. Fresh-air intake that is poorly located—for example, above a loading dock or first-floor restaurant exhaust fan—can bring in air pollutants.
7. People who smoke indoors, perhaps in restaurants or offices, pollute the indoor environment, and even people who smoke outside buildings, particularly near open or revolving doors, may cause pollution as the smoke (secondhand smoke) is drawn into and up through the building by the chimney effect.
8. Remodeling, painting, and other such activities often bring a variety of chemicals and materials into a building. Fumes from such activities may enter the building's heating, ventilation, and air-conditioning system, causing widespread pollution.
9. A variety of cleaning products and solvents used in offices and other parts of buildings contain harmful chemicals whose fumes may circulate throughout a building.
10. People can increase carbon dioxide levels; they can emit bioeffluents and spread bacterial and viral contaminants.
11. Loading docks can be sources of organics from garbage containers, of particulates, and of carbon monoxide from vehicles.
12. Radon gas can seep into a building from soil; rising damp (water), which facilitates the growth of molds, can enter foundations and rise up walls.
13. Dust mites and molds can live in carpets and other indoor places.
14. Pollen can come from inside and outside sources.

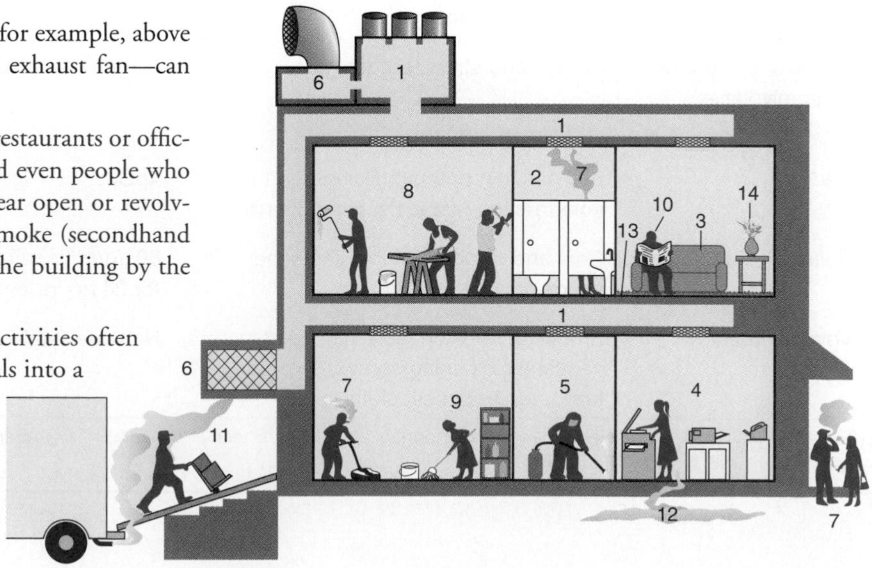

FIGURE 21.17 Some potential sources of indoor air pollution.

Table 21.6 SOURCES, CONCENTRATIONS, OCCURRENCES, AND POSSIBLE HEALTH EFFECTS OF INDOOR AIR POLLUTANTS

POLLUTANT	SOURCE	GUIDELINES (DOSE OR CONCENTRATIONS)	POSSIBLE HEALTH EFFECTS
Asbestos	Fireproofing; insulation, vinyl floor, and cement products; vehicle brake linings	0.2 fibers/mL for fibers larger than 5 μm	Skin irritation, lung cancer
Biological aerosols/ microorganisms	Infectious agents, bacteria in heating, ventilation, and air-conditioning systems: allergens	None available	Diseases, weakened immunity
Carbon dioxide	Motor vehicles, gas appliances, smoking	1,0000 ppm	Dizziness, headaches, nausea
Carbon monoxide	Motor vehicles, kerosene and gas space heaters, gas and wood stoves, fireplaces; smoking	10,000 $\mu g/m^2$ for 8 hours; 40,000 $\mu g/m^3$ for 1 hour	Dizziness, headaches, nausea, death
Formaldehyde	Foam insulation; plywood, particleboard, ceiling tile, paneling, and other construction materials	120 $\mu g/m^3$	Skin irritant, carcinogen
Inhalable particulates	Smoking, fireplaces, dust, combustion sources (wildfires, burning trash, etc.)	55-110 $\mu g/m^3$ annual; 350 $\mu g/m^2$ for 1 hour	Respiratory and mucous irritant, carcinogen
Inorganic particulates			
Nitrates	Outdoor air	None available	
Sulfates	Outdoor air	4 $\mu g/m^3$ annual; 12 $\mu g/m^3$ for 24 hours	
Metal particulates			
Arsenic	Smoking, pesticides, rodent poisons	None available	Toxic, carcinogen
Cadmium	Smoking, fungicides	2 $\mu g/m^3$ for 24 hours	
Lead	Automobile exhaust	1.5 $\mu g/m^3$ for 3 months	
Mercury	Old fungicides; fossil fuel combustion	2 $\mu g/m^3$ for 24 hours	
Nitrogen dioxide	Gas and kerosene space heaters, gas stoves, vehicular exhaust	100 $\mu g/m^3$ annual	Respiratory and mucous irritant
Ozone	Photocopying machines, electrostatic air cleaners, outdoor air	235 $\mu g/m^3$ for 1 hour	Respiratory irritant causes fatigue
Pesticides and other semivolatile organics	Sprays and strips, outdoor air	5 $\mu g/m^3$ for chlordane	Possible carcinogens
Radon	Soil gas that enters buildings, construction materials, groundwater	4pCi/L	Lung cancer
Sulfur dioxide	Coal and oil combustion, kerosene space heaters, outside air	80 $\mu g/m^3$ annual; 365 $\mu g/m^3$ for 24 hours	Respiratory and mucous irritant
Volatile organics	Smoking, cooking, solvents, paints, varnishes, cleaning sprays, carpets, furniture, draperies, clothing	None available	Possibe carcinogens

Source: N. L. Nagda, H. E. Rector, and M. D. Koontz, 1987; M. C. Baechler et al., 1991; E. J. Bardana Jr. and A. Montaro (eds.), 1997; M. Meeker,1996; D. W. Moffatt, 1997.

(a) 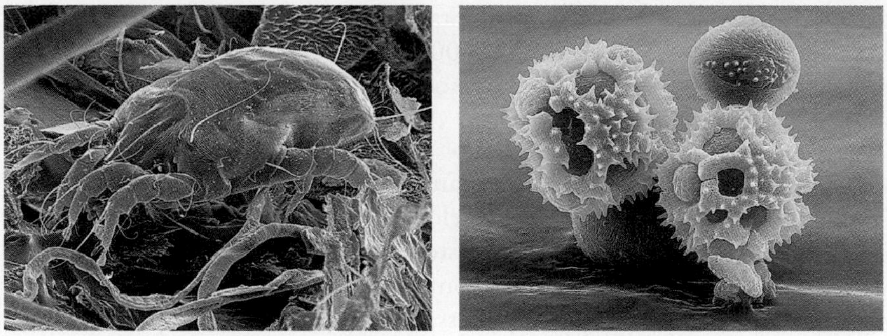 (b)

FIGURE 21.18 **(a)** This dust mite (magnified about 140 times) is an eight-legged relative of spiders. It feeds on human skin in household dust and lives in materials such as fabrics on furniture. Dead dust mites and their excrement can cause allergic reactions and asthma attacks in some people. **(b)** Microscopic pollen grains that in large amounts may be visible as a brown or yellow powder. The pollen shown here is from dandelions and horse chestnuts.

Many products and processes used in our homes and workplaces are sources of pollution. Other air pollutants—such as carbon monoxide, particulates, nitrogen dioxide, radon, and carbon dioxide—may enter a building by infiltration, either through cracks and other openings in the foundations and walls or by way of ventilation systems, and are generally found in much higher concentrations indoors than outdoors (see Figure 21.19). The reason is somewhat ironic: The steps we have taken to make our homes and offices energy-efficient often trap pollutants inside. Two of the best ways to conserve energy in homes and other buildings are to increase insulation and decrease infiltration of outside air. But windows that don't open and extensive caulking and weather stripping, while reducing energy consumption, also reduce natural ventilation. With less natural ventilation, we must depend more on the ventilation systems that are part of heating and air-conditioning systems.

Pathways, Processes, and Driving Forces

Both natural and human processes create differential pressures that move air and contaminants from one area of a building to another. Areas of high pressure may develop on the windward side of a building, whereas pressure is lower on the leeward, or protected, side. As a result, air is drawn into a building from the windward side. Opening and closing doors produces pressure differentials that cause air to move within buildings. Wind, too, can affect the movement of air in a building, particularly if the structure is leaky.[53]

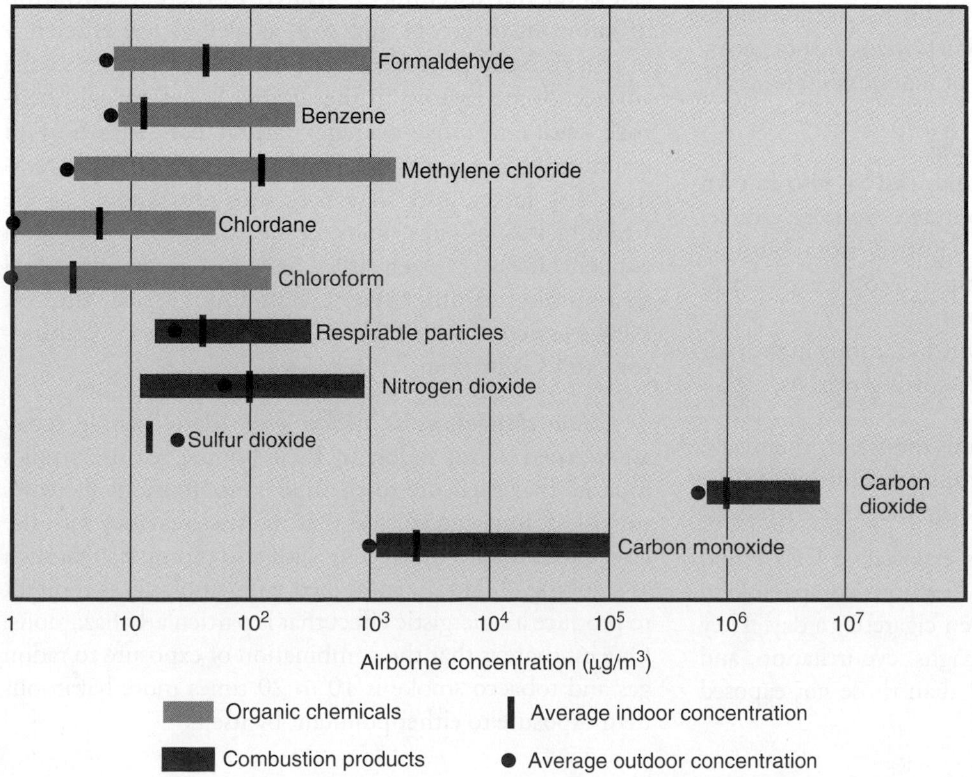

FIGURE 21.19 Concentrations of common indoor air pollutants compared with outdoor concentrations plotted on a log scale, that is, $10^2 = 100$; $10^3 = 1{,}000$, $10^4 = 10{,}000$, etc. (*Source:* A.V. Nero Jr., "Controlling Indoor Air Pollution," *Scientific American*, 258, no. 5 [1998]: 42–48.)

A **chimney effect (or stack effect)** occurs when the indoor and outdoor temperatures differ. Warm air rises within a building. If the indoor air is warmer than the outdoor air, then as the warmer air rises to the building's upper levels, it is replaced in the lower levels by outdoor air drawn in through various openings—windows, doors, cracks in the foundations and walls. Because air is so fluid, the possible interactions between the driving forces and the building are complex, and the distribution of potential air contaminants and pollutants can be extensive. One outcome is that people in various parts of a building may complain about the air quality even if they are widely separated from each other and from potential sources of pollution.[53]

Heating, Ventilation, and Air-Conditioning Systems

Heating, ventilation, and air-conditioning systems are designed to provide a comfortable indoor environment. Their design depends on a number of variables, including the activity of people in the building, air temperature and humidity, and air quality. If the heating, ventilation, and air-conditioning system is designed correctly and functions properly, it will maintain a comfortable temperature and adequate ventilation (using outdoor air), and also remove common air pollutants via exhaust fans and filters.[53]

Regardless of the type of system used in a home or other building, its effectiveness depends on the proper design of the equipment for that building, proper installation, and correct maintenance and operating procedures. Indoor air pollution may result if any one of these factors concentrates pollutants from the many possible sources. Filters plugged or contaminated with fungi, bacteria, or other potentially infectious agents can cause serious problems. In addition, as we see later in this chapter, ventilation systems are not generally designed to reduce some types of indoor pollution.[53, 54]

Environmental Tobacco Smoke

Environmental tobacco smoke (ETS), also known as *secondhand smoke*, comes from two sources: smoke exhaled by smokers and smoke emitted from burning tobacco in cigarettes, cigars, or pipes. People exposed to ETS are referred to as *passive smokers*.[55]

ETS is the most widely known hazardous indoor air pollutant. It is hazardous for the following reasons:[55, 56]

- Tobacco smoke contains several thousand chemicals, many of which are irritants. Examples include NOx, CO, hydrogen cyanide, and about 40 carcinogenic chemicals.
- Studies of nonsmoking workers exposed to ETS found that they have impaired airway functions comparable to that caused by smoking up to ten cigarettes a day. They suffer more illnesses, such as coughs, eye irritation, and colds, and lose more work time than those not exposed to ETS.
- In the United States, about 3,000 deaths from lung cancer and 40,000 deaths from heart disease a year are thought to be associated with ETS.

The number of smokers in the United States has declined, but there are still about 40 million. The rate is higher in the developing world, where health warnings are few or nonexistent. Smoking is extremely addictive because tobacco contains nicotine, a highly addictive substance. Nevertheless, education and social pressure have persuaded some thoughtful people to quit smoking and to encourage others to keep trying.

Radon Gas

It has become apparent over the past few decades that radon gas—colorless, odorless, and tasteless—may be a significant environmental health problem in the United States.[57, 58] **Radon** comes from natural processes, not from human activities. It is a naturally occurring radioactive gas that is a product of the radioactive-decay chain from uranium to stable lead. Radon-222, which has a half-life of 3.8 days, is emitted during the radioactive decay of radium-226. Radon decays with emission of an alpha particle to polonium-218, which has a half-life of approximately 3 minutes. (The discussion of radiation, radiation units, radiation doses, and health problems related to radiation in Chapter 17 will help you understand the following discussion.)

Geology and radon gas. The concentration of radon gas that reaches the surface of the Earth and thus can enter our dwellings is related to the concentration of radon in the rocks and soil, as well as the efficiency of the transfer processes from the rocks or soil to the surface. Some regions in the United States contain bedrock with an above-average natural concentration of uranium. A large area that includes parts of Pennsylvania, New Jersey, and New York—an area known as the Reading Prong—has many homes with elevated radon concentrations.[57] Such areas have also been identified in a number of other states, including Florida, Illinois, New Mexico, South Dakota, North Dakota, Washington, and California.

How dangerous is radon gas? Many people today are worried about radon in their homes because studies indicate that exposure to elevated concentrations increases the risk of lung cancer, and that the risk increases with the level and duration of exposure and also certain habits, such as smoking.[58] Radon, combined with smoking, is thought to produce a synergistic effect that is particularly hazardous. One estimate is that the combination of exposure to radon gas and tobacco smoke is 10 to 20 times more hazardous than exposure to either pollutant by itself.[58]

The Environmental Protection Agency (EPA) estimates that 14,000 lung cancer deaths per year in the United States are related to exposure to radon and its daughter products (products that result from its radioactive decay), primarily polonium-218. (The estimate actually ranges from 7,000 to 30,000.) By comparison, approximately 140,000 people die of lung cancer in the United States each year. If these estimates are correct—and they are controversial—approximately 10% of the lung cancer deaths in the United States can be attributed to radon. Exposure to radon has also been linked to other forms of cancer, such as melanoma (a deadly skin cancer) and leukemia, but, again, such linkages are highly controversial.[59, 60] The link between radon and cancer is mostly based on studies of uranium miners, a group of people exposed to high concentrations of radon in mines.

If the estimated risks from radon are anywhere close to the actual risk, then the hazard is a large one. The U.S. Surgeon General has stated that "indoor radon gas is a national health problem." The risks posed by radon are thought to be hundreds of times higher than risks from outdoor pollutants in air and water. Such pollutants are generally regulated to reduce the risk of premature death and disease to less than 0.001%. Risks from some indoor pollutants, such as organic chemicals, may be as high as 0.1%.[61] These risks still are very small compared with the risk for radon. For example, people who live in homes for about 20 years with an average concentration of radon of about 25 pCi/L are estimated to have a 1 to 2% chance of contracting lung cancer.[58, 61]

How does radon enter homes and other buildings? Radon enters homes and other buildings in three main ways: (1) It migrates up from soil and rock into basements and lower floors; (2) dissolved in groundwater, it is pumped into wells and then into homes; and (3) radon-contaminated materials, such as building blocks, are used in construction. It is difficult to estimate how many homes in the United States may have elevated concentrations of radon. The EPA estimates that about 7% have elevated radon levels and recommends that all homes and schools be tested. The test is simple and inexpensive.

Symptoms of Indoor Air Pollution

People living or working in particular indoor environments may react to pollutants in different ways: Some are particularly susceptible to indoor air pollution; some report different symptoms from the same pollutant; and some report symptoms that turn out not to stem from air pollution.

A wide variety of symptoms can result from exposure to indoor air pollutants (see Table 21.7). Some chemical pollutants can cause nosebleeds, chronic sinus infections, headaches, and irritation of the skin or eyes, nose, and throat. More-serious problems include loss of balance and memory, chronic fatigue, difficulty in speaking, and allergic reactions, including asthma.

Sick Buildings

An entire building can be considered "sick" because of environmental problems. There are two types of "sick" buildings:

- Buildings with identifiable problems, such as toxic molds or bacteria known to cause disease. The diseases are known as *building-related illnesses* (BRI).
- Buildings with **sick building syndrome (SBS)**, where the symptoms people report cannot be traced to any known cause.

A sick building's indoor environment appears to be unhealthy in that a number of people in the building report adverse health effects that they believe are related to the amount of time they spend in the building. Their complaints may range from funny odors to more-serious symptoms, such as headaches, dizziness, nausea, and so forth. In addition, an unusual number of people in the building may feel sick, or may have contracted a serious disease, such as cancer.[62-64]

In many cases, it is difficult to establish what may be causing the sick building syndrome. It has sometimes been found to be related to poor management and low worker morale, rather than to toxins in the building. When the occupants of a building report adverse health effects and a study does not detect the cause, a number of other things may be happening:[53]

21.5 Controlling Indoor Air Pollution

As much as $250 billion per year might be saved by decreasing illnesses and increasing productivity through improving the work environment.[50] A good starting point would be environmental legislation requiring certain indoor air quality standards. At a minimum, these should include increasing the inflow of fresh air through ventilation. In Europe, systems of filters and pumps in many office buildings circulate air three times as frequently as is typical in the United States. Many building codes in Europe require that workers have access to fresh air (windows) and natural light. Unfortunately, no similar codes

Table 21.7 SOME SYMPTOMS OF INDOOR AIR POLLUTION

SYMPTOMS	ETS[a]	COMBUSTION PRODUCTS[b]	BIOLOGIC POLLUTANTS[c]	VOCS[d]	HEAVY METALS[e]	SBS[f]
Respiratory						
Inflammation of mucous membranes of the nose, nasal congestion	Yes	Yes	Yes	Yes	No	Yes
Nosebleed	No	No	No	Yes	No	Yes
Cough	Yes	Yes	Yes	Yes	No	Yes
Wheezing, worsening asthma	Yes	Yes	No	Yes	No	Yes
Labored breathing	Yes	No	Yes	No	No	Yes
Severe lung disease	Yes	Yes	Yes	No	No	Yes
Other						
Irritation of mucous membranes of eyes	Yes	Yes	Yes	Yes	No	Yes
Headache or dizziness	Yes	Yes	Yes	Yes	Yes	Yes
Lethargy, fatigue, malaise	No	Yes	Yes	Yes	Yes	Yes
Nausea, vomiting, anorexia	No	Yes	Yes	Yes	Yes	No
Cognitive impairment, personality change	No	Yes	No	Yes	Yes	Yes
Rashes	No	No	Yes	Yes	Yes	No
Fever, chills	No	No	Yes	No	Yes	No
Abnormal heartbeat	Yes	Yes	No	No	Yes	No
Retinal hemorrhage	No	Yes	No	No	No	No
Muscle pain, cramps	No	No	No	Yes	No	Yes
Hearing loss	No	No	No	Yes	No	No

[a] Environmental tobacco smoke.

[b] Combustion products Include particles, NO_x, CO, and CO_2 .

[c] Biologic pollutants include molds, dust mites, pollen, bacteria, and viruses.

[d] Volatile organic compounds, including formaldehyde and solvents.

[e] Heavy metals include lead and mercury.

[f] Sick building syndrome.

Source: Modified from American Lung Association, Environmental Protection Agency, and American Medical Association, "Indoor Air Pollution—An introduction for Health Professionals," 523-217/81322 (Washington, D.C.: GPO, 1994).

exist for U.S. workers, and many buildings use central air-conditioning with windows permanently sealed.[50]

You might think that heating, ventilating, and air-conditioning systems, operating properly and well maintained, will ensure good indoor air quality, but in fact these systems are not designed to maintain all aspects of air quality. For example, commonly used ventilation systems do not generally reduce radon gas. Other strategies include source removal, source modification, and air cleaning.[62]

Education also plays an important role in developing strategies to reduce indoor air pollution; it enables people to make informed decisions about exposure to chemicals, such as paints and solvents, and about strategies to avoid potentially hazardous conditions in the home and workplace.[62] At one level, this may involve deciding not to install unvented or poorly vented appliances. A surprising (and tragic) number of people are killed each year by carbon monoxide poisoning due to poor ventilation in homes, campers, and tents. Educated people are also more aware of their legal rights with respect to product liability and safety.

Making Homes and Other Buildings Radon Resistant

Protecting new homes from potential radon problems is straightforward and relatively inexpensive. It is also easy to upgrade an older home to reduce radon. The techniques vary according to the type of foundation the structure has. The basic strategy is to prevent radon from entering a home (usually sealing entry points) and ensure that radon is removed from the site (this generally involves designing a ventilation system).[65, 66]

Designing Buildings to Minimize Indoor Air Pollution

There is a movement under way in the United States and the world to create buildings specifically designed to provide a healthful indoor environment for their occupants. The basic objectives of the design are to minimize indoor air pollutants; ensure that fresh air is supplied and circulated; manage moisture to avoid problems such as mold; reduce energy use; use materials whose origin is environmentally benign and can be recycled, as much as possible; create as pleasing a working environment as possible; use vegetation planted on roofs and wherever else possible to take up carbon dioxide, release oxygen, and add to the general pleasantness of the working environment.

CRITICAL THINKING ISSUE

Should Carbon Dioxide Be Regulated along with Other Major Air Pollutants?

The six common pollutants, sometimes called the *criteria pollutants*, are ozone, particulate matter, lead, nitrogen dioxide, carbon monoxide, and sulfur dioxide. These pollutants have a long history with the EPA, and major efforts have been made to reduce them in the lower atmosphere over the United States. This effort has been largely successful—all of them have been significantly reduced since 1990.

In 2009, the EPA suggested that we add carbon dioxide to this list. Two years earlier, the U.S. Supreme Court had ordered the EPA to make a scientific review of carbon dioxide as an air pollutant that could possibly endanger public health and welfare. Following that review, the EPA announced that greenhouse gases pose a threat to public health and welfare. This proclamation makes it possible that greenhouse gases, especially carbon dioxide, will be regulated by the Clean Air Act, which regulates most other serious air pollutants. The EPA's conclusion that greenhouse gases harm or endanger public health and welfare is based primarily on the role these gases play in climate change. The analysis states that the impacts include, but are not limited to, increased drought that will impact agricultural productivity; more intense rainfall, leading to a greater flood hazard; and increased frequency of heat waves that affect human health.

The next step in adding carbon dioxide and other greenhouse gasses, such as methane, to the list of pollutants regulated by the EPA will be a series of public hearings and feedback from a variety of people and agencies. Some people oppose listing carbon dioxide as an air pollutant because, first of all, it is a nutrient and stimulates plant growth; and, second, it does not directly affect human health in most cases (the exception being carbon dioxide emitted by volcanic eruption and other volcanic activity, which can be extremely toxic).

Critical Thinking Questions

After going over the information concerning global climate change and the role of carbon dioxide in causing change, consider the following questions:

1. Do you think carbon dioxide, along with other greenhouse gases, should be controlled under the Clean Air Act? Why? Why not?
2. Assuming carbon dioxide and other greenhouse gases are to be controlled under the Clean Air Act, what sorts of programs might be used for such control? For example, the control of sulfur dioxide was primarily through a cap-and-trade program where the total amount of emissions were set, and companies bought and sold shares of allowed pollution up to the cap.
3. If the United States can curtail emissions of carbon dioxide under the Clean Air Act, how effective will this be in, say, reducing the global concentration of carbon dioxide to about 350 parts per million given what other countries are likely to do in the future with respect to emissions and given that the concentration today is about 390 parts per million?

SUMMARY

- There are two main kinds of air pollutants: primary and secondary. Primary pollutants are emitted directly into the air: particulates, sulfur dioxide, carbon monoxide, nitrogen oxides, and hydrocarbons. Secondary pollutants are produced through reactions between primary pollutants and other atmospheric compounds. Ozone is a secondary pollutant that forms over urban areas through photochemical reactions between primary pollutants and natural atmospheric gases.
- There are also two kinds of sources: stationary and mobile. Stationary sources have a relatively fixed position and include point sources, area sources, and fugitive sources.
- Meteorological conditions—in particular, restricted circulation in the lower atmosphere due to temperature inversion—greatly determine whether or not polluted air is a problem in an urban area.
- Pollution-control methods are tailored to specific pollution sources and types and vary from settling chambers for particulates to scrubbers that remove sulfur before it enters the atmosphere.
- Emissions of air pollutants in the United States are decreasing, but in large urban areas of developing countries it remains a serious problem.
- The concentration of atmospheric ozone has been measured for more than 70 years. Concentrations in the stratosphere have declined since the mid-1970s, allowing more ultraviolet radiation to reach the lower atmosphere, where it can damage living things.
- In 1974, Mario Molina and F. Sherwood Rowland hypothesized that stratospheric ozone might be depleted by emissions of chlorofluorocarbons (CFCs) into the lower atmosphere. Major features of the hypothesis are that CFCs are very stable and have a long residence time in the atmosphere. Eventually they reach the stratosphere, where they may be destroyed by solar ultraviolet radiation, releasing chlorine. The chlorine may then enter into a catalytic chain reaction that depletes ozone in the stratosphere.
- Banning chemicals that deplete stratospheric ozone is a step in the right direction. However, millions of tons of these are now in the lower atmosphere and working their way up, so even if all production, use, and emission of these chemicals stopped today, the problem would continue for a long time. The good news is that concentrations of CFCs in the atmosphere have apparently peaked and are now static or in slow decline.
- Possible sources of indoor air pollution are construction materials, furnishings, types of equipment used for heating and cooling, as well as natural processes that allow gases to seep into buildings.
- Indoor concentrations of air pollutants are generally greater than outdoor concentrations of the same pollutants.
- Ventilation is commonly used to control indoor air pollution, but tighter construction impedes ventilation, and many popular ventilation systems do not reduce certain types of indoor air pollutants.
- The most common natural process that affects interior air quality is the "chimney" or "stack effect" that occurs when the indoor and outdoor environments differ in temperature.
- People react to indoor air pollution in different ways, and so reported symptoms may vary.
- In some cases, reported symptoms have nothing to do with air pollution.
- Controlling indoor air pollution involves several strategies, including ventilation, source removal, source modification, and air-cleaning equipment, as well as education.

REEXAMINING THEMES AND ISSUES

Human Population

Population growth will exacerbate air pollution problems. As the number of people increases, so does the use of resources, many of which are related to emissions of air pollutants. This may be partially offset in developed countries, where the per capita emissions of air pollutants have been reduced in recent years.

Sustainability

Ensuring that future generations inherit a quality environment is an important objective of sustainability. Thus, it is vital that we develop technology that minimizes air pollution.

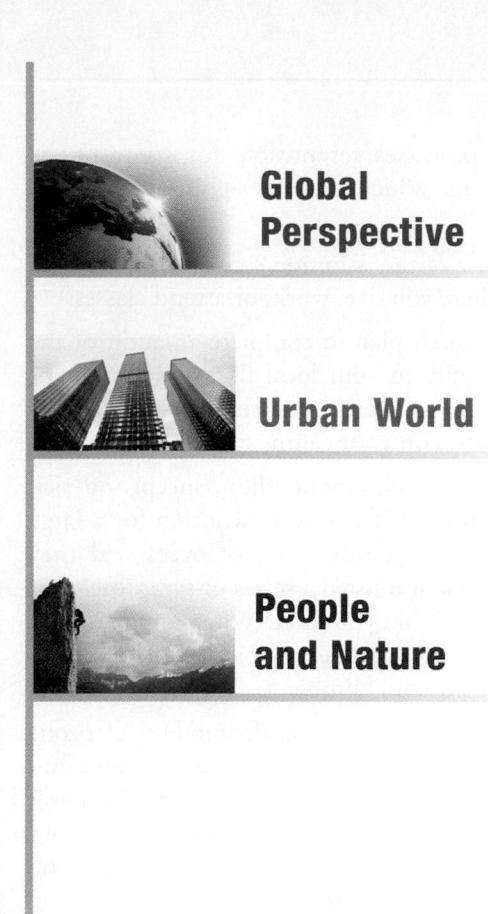

Global Perspective

Atmospheric processes and atmospheric pollution occur on regional and global scales. Pollutants emitted into the atmosphere at a particular site may join the global circulation pattern and spread throughout the world, and pollutants emitted from urban or agricultural areas may be dispersed to pristine areas far removed from human activities. Therefore, an understanding of global atmospheric processes is critical to finding solutions to many air pollution problems, including acid deposition.

Urban World

Cities and urban corridors are sites of intense human activity, and many of these activities contribute to air pollution. Some large cities have such severe air pollution problems that the health and lives of people are being affected.

People and Nature

Although we think of nature as unspoiled, in reality nature can be toxic. This is especially true with regard to air pollution—for example, the vast majority of particulates and carbon monoxide are generated by volcanic eruption and wildfire. Even hydrocarbons have local sources, such as seeps, which, in areas such as offshore Goleta, California, emit a significant amount of hydrocarbons that contribute to smog.

Science and Values

The science and technology necessary to reduce air pollution are well known; what we do with these tools involves a value judgment. It is clear that people value a high-quality environment, and clean air is at the top of the list. The developed countries have an obligation to take a leadership role in finding ways to use resources while minimizing air pollution. Of particular importance is finding methods and technologies that will allow for reducing air pollution while stimulating economies. What is considered waste in one part of the urban-industrial complex may be used as resources in another part. This idea is at the heart of what is sometimes called industrial ecology. The discovery, understanding, and management of ozone-depleting chemicals is an environmental success, reflecting the value of the environment.

KEY TERMS

acid rain **469**
air toxics **471**
atmospheric inversion **474**
black lung disease **486**
chimney effect (or stack effect) **490**
chlorofluorocarbons (CFCs) **483**
criteria pollutants **465**
environmental tobacco smoke (ETS) **490**
global dimming **468**
green building **462**
mobile sources **463**
ozone (O_3) **481**
ozone shield **481**
photochemical smog **473**
primary pollutants **464**
radon **490**
scrubbing **478**
secondary pollutants **464**
sick building syndrome (SBS) **491**
smog **473**
stationary sources **463**
sulfurous smog **473**
Ultraviolet (UV) Index **485**
ultrafine particles **467**

STUDY QUESTIONS

1. Since the amount of pollution emitted into the air is a very small fraction of the total material in the atmosphere, why do we have air-pollution problems?
2. What are the differences between primary and secondary pollutants?
3. Carefully examine Figure 21.11, which shows a column of air moving through an urban area, and Figure 21.7, which shows relative concentrations of pollutants that develop on a typical warm day in Los Angeles. What linkages between the information in these two figures might be important in trying to identify and learn more about potential air pollution in an area?
4. Why is acid deposition a major environmental problem, and how can it be minimized?
5. Why will air-pollution abatement strategies in developed countries probably be much different in terms of methods, process, and results than air-pollution abatement strategies in developing countries?
6. In a highly technological society, is it possible to have 100% clean air? Is it likely?
7. Study Figure 21.14 carefully and discuss how the information in parts (a) and (b) are linked and related to the chapter in general. Apply Critical Thinking skills.
8. Discuss the processes responsible for stratospheric ozone depletion. Which are most significant? Where? Why?
9. What are some of the common sources of indoor air pollutants where you live, work, or attend classes?
10. Develop a research plan to complete an audit of the indoor air quality in your local library. How might that research plan differ from a similar audit for the science buildings on your campus?
11. What do you think about the concept of sick building syndrome? If you were working for a large corporation and a number of employees said they were getting sick and listed a series of symptoms and problems, how would you react? What could you do? Play the role of the administrator and develop a plan to look at the potential problem.
12. Suppose that next year our understanding of ozone depletion is changed by the discovery that concentrations of stratospheric ozone have natural cycles and that lower concentrations in recent years have resulted not from our activities but from natural processes. How would you put all the information in this chapter into perspective? Would you think science had let you down?

FURTHER READING

Boubel, R.W., D.L. Fox, D.B. Turner, and A.C. Stern, *Fundamentals of Air Pollution,* 4th ed. (New York: Academic, 2008). A thorough book covering the sources, mechanisms, effects, and control of air pollution.

Brenner, D.J., *Radon: Risk and Remedy* (New York: Freeman, 1989). A wonderful book about the hazard of radon gas. It covers everything from the history of the problem to what was happening in 1989, as well as solutions.

Christie, M., *The Ozone Layer: A Philosophy of Science Perspective* (Cambridge: Cambridge University Press, 2000). A complete look at the history of the ozone hole, from the first discovery of its existence to more recent studies of the hole over Antarctica.

Hamill, P., and O.B. Toon. "Polar Stratospheric Clouds and the Ozone Hole," *Physics Today* 44, no. 12 (1991): 34–42. A good review of the ozone problem and important chemical and physical processes related to ozone depletion.

Reid, S., *Ozone and Climate Change: A Beginner's Guide* (Amsterdam: Gordon & Breach Science Publishers, 2000). A look at the science behind the ozone hole and future predictions, written for a general audience to make the science understandable.

Rowland, F.S. "Stratospheric Ozone Depletion by Chlorofluorocarbons." *AMBIO* 19, no. 6–7 (1990):281–292. An excellent summary of stratospheric ozone depletion, it discusses some of the major issues.

Wang, L.,"Paving out Pollution," *Scientific American,* February 2002, p. 20. Discussion of an innovative approach to reducing air pollution.

CHAPTER 22

Urban Environments

New Orleans suffering right after Hurricane Katrina.

LEARNING OBJECTIVES

Because the world is becoming increasingly urbanized, it is important to learn how to improve urban environments—to make cities more pleasant and healthier places to live, and reduce undesirable effects on the environment. After reading this chapter, you should understand . . .

- How to view a city from an ecosystem perspective;
- How location and site conditions determine the success, importance, and longevity of a city;
- How cities have changed with changes in technology and in ideas about city planning;
- How a city changes its own environment and affects the environment of surrounding areas, and how we can plan cities to minimize some of these effects;
- How trees and other vegetation not only beautify cities but also provide habitats for animals, and how we can alter the urban environment to encourage wildlife and discourage pests;
- How cities can be designed to promote biological conservation and become pleasant environments for people;
- The fundamental choices we face in deciding what kind of future we want and what the role of cities will be in that future.

CASE STUDY

New York's High Line Park in the Sky

From the late 19th century to the 1930s, trains ran on street level in New York City, carrying goods such as meat to the meatpacking district near the Hudson River. This was dangerous and caused accidents. In the 1930s, a public–private project built an elevated railroad 30 feet above the streets, making the streets safer, if darker and less attractive.

By the 1980s the economy had changed, and the trains were no longer needed and so stopped running. The elevated rail line that had served Manhattan's meatpacking district near the Hudson River was just another urban eyesore. Although the narrow, rusty steel stairs that led up to them were closed off and hard to find, a few New Yorkers did manage to climb up and discovered that nature was taking over—the open ground between and along the tracks was undergoing secondary succession, with trees, shrubs, tall grasses, and flowering plants growing wild. Following the rails above the city streets, you could get a secret taste of what Henry David Thoreau would have called "wildness"—the feeling of nature—even within one of the world's largest cities.

Some people proposed tearing down the useless rail line to let more light reach the streets beneath, but another group of citizens had a different idea: to turn it into a park. They formed "Friends of the High Line" in 1999 and got city approval. In 2009 the eagerly anticipated first section of the High Line Park opened, carefully planned by professional landscape designers who strove to keep the natural, uncultivated look by using many of the same plant species that had been growing wild on their own.

Although New York City has many parks, including the very large Central Park, its millions of residents need more places to relax, stroll, sit, and enjoy nature. The park will ultimately be a mile and a half long, and as soon as the first section opened, it was filled with people (Figure 22.1), strolling, enjoying the views of the city on one side and the river on the other, relaxing on wheeled wooden lounges mounted on the tracks. A lot of imagination had helped bring nature to the city, and city people to nature.[1]

(a)

(c)

(b)

FIGURE 22.1 A portion of the newly opened first section of New York City's High Line **(a)** as originally designed for use as an elevated freight railway within Manhattan and **(b)** as it is today, an urban park, ultimately a mile and a half long when completed, planted with many of the species that grew wild after the railway was abandoned **(c)** the HighLine as an elevated with Field.

22.1 City Life

In the past, the emphasis of environmental action has most often been on wilderness, wildlife, endangered species, and the impact of pollution on natural landscapes outside cities. Now it is time to turn more of our attention to city environments. In the development of the modern environmental movement in the 1960s and 1970s, it was fashionable to consider everything about cities bad and everything about wilderness good. Cities were viewed as polluted, dirty, lacking in wildlife and native plants, and artificial—therefore bad. Wilderness was viewed as unpolluted, clean, teeming with wildlife and native plants, and natural—therefore good.

Although it was fashionable to disdain cities, many people live in urban environments and have suffered directly from their decline. According to the United Nations Environment Program, in 1950 fewer than a third of the people of the world lived in a town or city, while today almost half of the world's population is urban, and the forecasts are that in just 20 years—by 2030—almost two-thirds of the people will live in cities and towns.[2] (See Chapter 4.)

In the United States, about 75% of the population live in urban areas and about 25% in rural areas. Perhaps even more striking, half of all Americans live in one of the 39 cities with populations over 1 million.[3] However, in the past decade more people have moved out of the largest cities in the United States than have moved into them. The New York, Los Angeles, Chicago, and San Francisco/Oakland metropolitan areas each averaged a net loss of more than 60,000 people a year. Chicago's Cook County lost a half million people between 2000 and 2004.[4, 5, 6, 7] Today approximately 45% of the world's population live in cities, and it is projected that 62% of the population will live in cities by the year 2025.[6] Economic development leads to urbanization; 75% of people in developed countries live in cities, but only 38% of people in the poorest of the developing countries are city dwellers.[7]

Megacities—huge metropolitan areas with more than 8 million residents—are cropping up more and more. In 1950 the world had only two: the New York City and nearby urban New Jersey metropolitan area (12.2 million residents altogether) and greater London (12.4 million). By 1975, Mexico City, Los Angeles, Tokyo, Shanghai, and São Paulo, Brazil, had joined this list. By 2002, the most recent date for which data are available, 30 urban areas had more than 8 million people.[5]

Yet comparatively little public concern has focused on urban ecology. Many urban people see environmental issues as outside their realm, but the reality is just the opposite: City dwellers are at the center of some of the most important environmental issues. People are realizing that city and wilderness are inextricably connected. We cannot fiddle in the wilderness while our Romes burn from sulfur dioxide and nitrogen oxide pollution. Fortunately, we are experiencing a rebirth of interest in urban environments and urban ecology. The National Science Foundation has added two urban areas, Baltimore and Phoenix, to its Long-Term Ecological Research Program, a program that supports research on, and long-term monitoring of, specific ecosystems and regions.

In the future, most people will live in cities. In most nations, most urban residents will live in the country's single largest city. For most people, living in an environment of good quality will mean living in a city that is managed carefully to maintain that environmental quality.

22.2 The City as a System

We need to analyze a city as the ecological system that it is—but of a special kind. Like any other life-supporting system, a city must maintain a flow of energy, provide necessary material resources, and have ways of removing wastes. These ecosystem functions are maintained in a city by transportation and communication with outlying areas. A city is not a self-contained ecosystem; it depends on other cities and rural areas. A city takes in raw materials from the surrounding countryside: food, water, wood, energy, mineral ores—everything that a human society uses. In turn, the city produces and exports material goods and, if it is a truly great city, also exports ideas, innovations, inventions, arts, and the spirit of civilization. A city cannot exist without a countryside to support it. As was said half a century ago, city and country, urban and rural, are one thing—one connected system of energy and material flows—not two things (see Figure 22.2).

As a consequence, if the environment of a city declines, almost certainly the environment of its surroundings will also decline. The reverse is also true: If the environment around a city declines, the city itself will be threatened. Some people suggest, for example, that the ancient Native American settlement in Chaco Canyon, Arizona, declined after the environment surrounding it either lost soil fertility from poor farming practices or suffered a decline in rainfall.

Cities also export waste products to the countryside, including polluted water, air, and solids. The average city resident in an industrial nation annually uses (directly or indirectly) about 208,000 kg (229 tons) of water, 660 kg (0.8 ton) of food, and 3,146 kg (3.5 tons) of fossil fuels and produces 1,660,000 kg (1,826 tons) of sewage, 660 kg (0.8 ton) of solid wastes, and 200 kg (440 lb) of air pollutants. If these are exported without care, they pollute the countryside, reducing its ability to provide necessary resources for the city and making life in the surroundings less healthy and less pleasant.

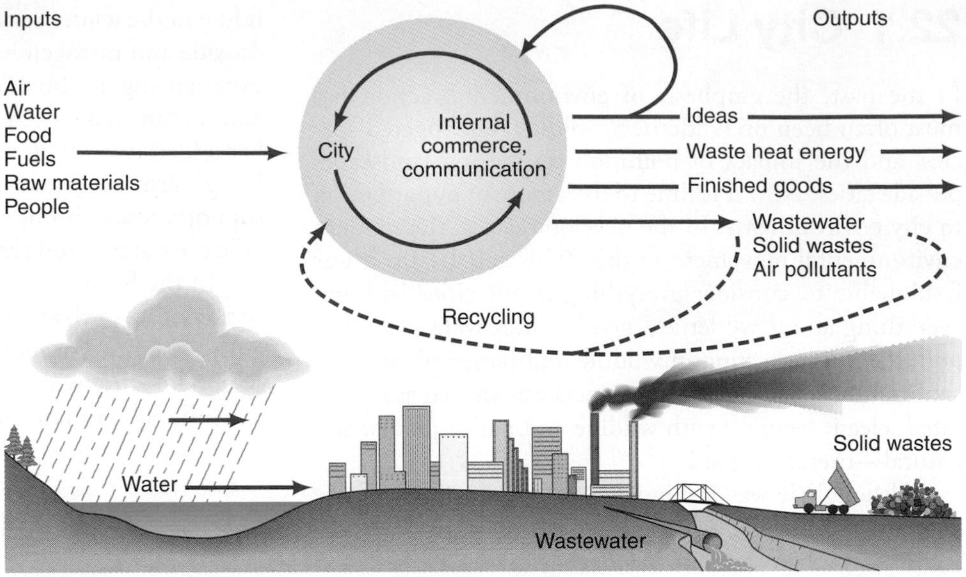

FIGURE 22.2 The city as a system with flows of energy and materials. A city must function as part of a city–countryside ecosystem, with an input of energy and materials, internal cycling, and an output of waste heat energy and material wastes. As in any natural ecosystem, recycling of materials can reduce the need for input and the net output of wastes.

Given such dependencies and interactions between city and surroundings, it's no wonder that relationships between people in cities and in the countryside have often been strained. Why, country dwellers want to know, should they have to deal with the wastes of those in the city? The answer is that many of our serious environmental problems occur at the interface between urban and rural areas. People who live outside but near a city have a vested interest in maintaining both a good environment for that city and maintaining a good system for managing the city's resources. The more concentrated the human population, the more land is available for other uses, including wilderness, recreation, conservation of biological diversity, and production of renewable resources. So cities benefit wilderness, rural areas, and so forth.

With the growing human population, we can imagine two futures. In one, cities are pleasing and livable, use resources from outside the city in a sustainable way, minimize pollution of the surrounding countryside, and allow room for wilderness, agriculture, and forestry. In the other future, cities continue to be seen as environmental negatives and are allowed to decay from the inside. People flee to grander and more expansive suburbs that occupy much land, and the poor who remain in the city live in an unhealthy and unpleasant environment. Without care for the city, its technological structure declines and it pollutes even more than in the past. Trends in both directions appear to be occurring.

In light of all these concerns, this chapter describes how a city can fit within, use, and avoid destroying the ecological systems on which it depends, and how the city itself can serve human needs and desires as well as environmental functions. With this information, you will have the foundation for making decisions, based on science and on what you value, about what kind of urban-rural landscape you believe will provide the most benefits for people and nature.

22.3 The Location of Cities: Site and Situation

Here is an idea that our modern life, with its rapid transportation and its many electronic tools, obscures: Cities are not located at random but develop mainly because of local conditions and regional benefits. In most cases they grow up at crucial transportation locations—an aspect of what is called the city's situation—and at a good site, one that can be readily defended, with good building locations, water supplies, and access to resources. The primary exceptions are cities that have been located primarily for political reasons. Washington, DC, for example, was located to be near the geographic center of the area of the original 13 states; but the site was primarily swampland, and nearby Baltimore provided the major harbor of the region.

The Importance of Site and Situation

The location of a city is influenced primarily by the **site**, which is the summation of all the environmental features of that location; and the **situation**, which is the placement of the city with respect to other areas. A good site includes a geologic substrate suitable for buildings, such as a firm rock base and well-drained soils that are above the water table; nearby supplies of drinkable water; nearby lands suitable for agriculture; and forests. Sometimes, however, other factors—such as the importance of creating a port city—can compensate for a poor geological site, as long as people are able to build an artificial foundation for the city and maintain that foundation despite nature's attempts to overwhelm it.

Cities influence and are influenced by their environment. The environment of a city affects its growth, success, and importance—and can also provide the seeds of its destruction. All cities are so influenced, and those who plan, manage, and live in cities must be aware of all aspects of the urban environment.

The environmental situation is especially important with respect to transportation and defense. Waterways, for example, are important for transportation. Before railroads, automobiles, and airplanes, cities depended on water for transportation, so most early cities—including all the important cities of the Roman Empire—were on or near waterways. Waterways continue to influence the locations of cities; most major cities of the eastern United States are situated either at major ocean harbors, like New Orleans (see A Closer Look 22.1), or at the fall line on major rivers.

A **fall line** on a river occurs where there is an abrupt drop in elevation of the land, creating waterfalls (Figure 22.3), typically where streams pass from harder, more erosion-resistant rocks to softer rocks. In eastern North America the major fall line occurs at the transition from the granitic and metamorphic bedrock that forms the Appalachian Mountains to the softer, more easily eroded and more recent sedimentary rocks. In general, the transition from major mountain range bedrock to another bedrock forms the primary fall line on continents.

Cities have frequently been established at fall lines, especially the major continental fall lines, for a number

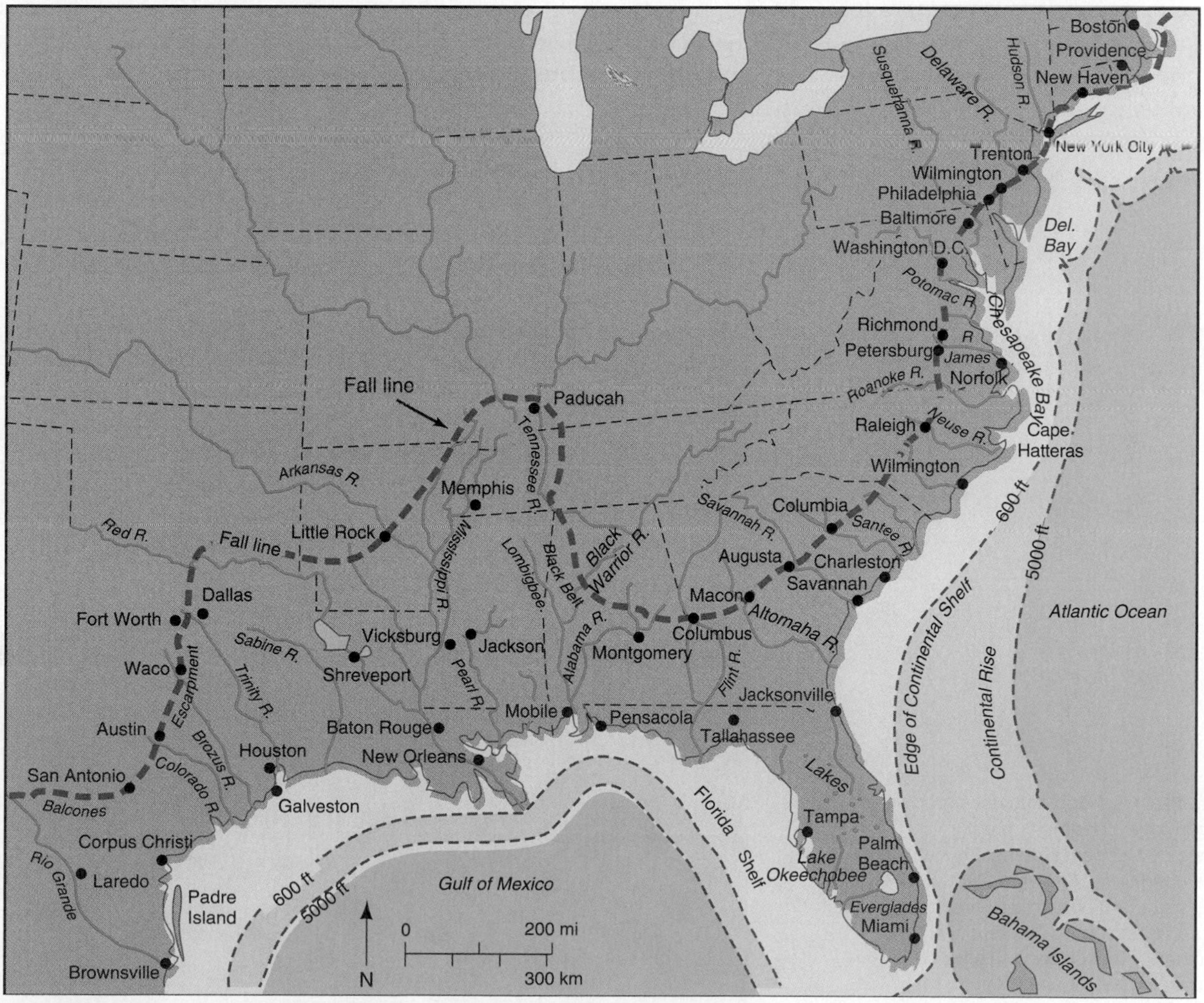

FIGURE 22.3 The fall line. Most major cities of the eastern and southern United States lie either at the sites of harbors or along a fall line (shown by the dashed line in the figure), which marks locations of waterfalls and rapids on major rivers. This is one way the location of cities is influenced by the characteristics of the environment. (*Source*: C.B. Hunt, *Natural Regions of the United States and Canada* [San Francisco: Freeman, 1974]. Copyright 1974 by W.H. Freeman & Co.)

of reasons. Fall lines provide waterpower, an important source of energy in the 18th and 19th centuries, when the major eastern cities of the United States were established or rose to importance. At that time, the fall line was the farthest inland that larger ships could navigate; and just above the fall line was the farthest downstream that the river could be easily bridged. Not until the development of steel bridges in the late 19th century did it become practical to span the wider regions of a river below the fall line. The proximity of a city to a river has another advantage: River valleys have rich, water-deposited soils that are good for agriculture. In early times, rivers also provided an important means of waste disposal, which today has become a serious problem.

Cities also are often founded at other kinds of crucial locations, growing up around a market, a river crossing, or a fort. Newcastle, England, and Budapest, Hungary, are located at the lowest bridging points on their rivers. Other cities, such as Geneva, are located where a river enters or leaves a major lake. Some well-known cities are at the confluence of major rivers: Saint Louis lies at the confluence of the Missouri and Mississippi rivers; Manaus (Brazil), Pittsburgh (Pennsylvania), Koblenz (Germany), and Khartoum (Sudan) are at the confluence of several rivers. Many famous cities are at crucial defensive locations, such as on or adjacent to easily defended rock outcrops. Examples include Edinburgh, Athens, and Salzburg. Other cities and municipalities are situated on peninsulas—for example, Istanbul and Monaco. Cities also frequently arise close to a mineral resource, such as salt (Salzburg, Austria), metals (Kalgoorlie, Australia), or medicated waters and thermal springs (Spa, Belgium; Bath, Great Britain; Vichy, France; and Saratoga Springs, New York).

When a successful city grows and spreads over surrounding terrain, its original purpose may be obscured. Its original market or fort may have evolved into a square or a historical curiosity. In most cases, though, cities originated where the situation provided a natural meeting point for people.

An ideal location for a city has both a good site and a good situation, but such a place is difficult to find. Paris is perhaps one of the best examples of a perfect location for a city—one with both a good site and a good situation. Paris began on an island more than 2,000 years ago, the situation providing a natural moat for defense and waterways for transportation. Surrounding countryside, a fertile lowland called the Paris basin, affords good

(a)

(b)

FIGURE 22.4 **(a)** Geologic, topographic, and hydrologic conditions greatly influence how successful the city can be. If these conditions, known collectively as the city's site, are poor, much time and effort are necessary to create a livable environment. New Orleans has a poor site but an important situation. **(b)** In contrast, New York City's Manhattan is a bedrock island rising above the surrounding waters, providing a strong base for buildings and a soil that is sufficiently above the water table so that flooding and mosquitoes are much less of a problem.

A CLOSER LOOK 22.1

Should We Try to Restore New Orleans?

On August 29, 2005, Hurricane Katrina roared, slammed, and battered its way into New Orleans with 192 km/hr (120 mph) winds. Its massive storm surges breached the levees that had protected many of the city's residents from the Gulf Coast's waters, flooding 80% of the city and an estimated 40% of the houses (opening photo). With so many people suddenly homeless and such major damage (Figure 22.5), the New Orleans mayor, Ray Nagin, ordered a first-time-ever complete evacuation of the city, an evacuation that became its own disaster. Some estimates claimed that 80% of the 1.3 million residents of the greater New Orleans metropolitan area evacuated.

By the time it was over, Katrina was the most costly hurricane in the history of the United States—between $75 billion and $100 billion, in addition to an estimated $200 billion in lost business revenue. A year after the hurricane, much of the damage remained, and even today much of New Orleans is not yet restored. Citizens remain frustrated by the lack of progress on many fronts.[8] An estimated 50,000 homes will have to be demolished. Many former residents are still living elsewhere, scattered across the nation. The storm affected the casino and entertainment industry, as many of the Gulf Coast's casinos were destroyed or sustained considerable damage. New Orleans also was home to roughly 115,000 small businesses, many of which will likely never reopen.

The problem with New Orleans is that it is built in the wetlands at the mouth of the Mississippi River, and much of it is below sea level (Figure 22.6). Although a port at the mouth of the Mississippi River has always been an important location for a city, there just wasn't a great place to build that city. The original development, the French Quarter, was just barely above sea level, about the best that could be found.

Hurricane Katrina was rated a Category 3 hurricane, but discussions about protecting the city from future storms focus on an even worse scenario, a Category 5 hurricane with winds up to 249 km/hr (155 mph). As of 2009, 68,000 homes remained abandoned.[9] Clearly, New Orleans requires an expensive improvement in its site if it is to survive at all. And if it does, will it be restored to its former glory and importance? Will it continue in any fashion, even as a mere shadow of its former self? Fortunately, the city has many residents who love it and are working hard to restore it.

To know how to rebuild the city, to decide whether this is worth doing, and to forecast whether such a restoration is likely, we have to understand the ecology of cities, how cities fit into the environment, the complex interplay between a city and its surroundings, and how a city acts as an environment for its residents.

a) (b)

FIGURE 22.5 **(a)** Aerial photograph of New Orleans skyline before Hurricane Katrina struck. The Super Dome is near the center left. **(b)** A similar view of New Orleans, but after Hurricane Katrina struck on August 31, 2005. The widespread flooding of the city from the hurricane is visible.

FIGURE 22.6 Map of New Orleans showing how much of the city is below sea level. The city began with the French Quarter, which is above sea level. As the population grew and expanded, levees were built to keep the water out, and the city became an accident waiting to happen. (*Source*: Tim Vasquez/Weather Graphics)

local agricultural land and other natural resources. New Orleans on the other hand, is an example of a city with an important situation but, as Hurricane Katrina made abundantly clear, a poor site (Figure 22.4).

Site Modification

Site is provided by the environment, but technology and environmental change can alter a site for better or worse. People can improve the site of a city and have done so when the situation of the city made it important and when its citizens could afford large projects. An excellent situation can sometimes compensate for a poor site. However, improvements are almost always required to the site so the city can persist.

Changes in a site over time can have adverse effects on a city. For example, Bruges, Belgium, developed as an important center for commerce in the 13th century because its harbor on the English Channel permitted trade with England and other European nations. By the 15th century, however, the harbor had seriously silted in, and the limited technology of the time did not make dredging possible (Figure 22.7). This problem, combined with political events, led to a decline in the importance of Bruges—a decline from which it never recovered. Nevertheless, today, Bruges still lives, a beautiful city with many fine examples of medieval architecture. Ironically, that these buildings were never replaced with modern ones makes Bruges a modern tourist destination.

Ghent, Belgium, and Ravenna, Italy, are other examples of cities whose harbors silted in. As human effects on the environment bring about global change, there may be rapid, serious changes in the sites of many cities.

FIGURE 22.7 Bruges, Belgium, was once an important seaport, but over the years sand that the ocean deposited in the harbor left the city far inland. Today, Bruges is a beautiful historic city, though no longer important for commerce.

22.4 An Environmental History of Cities

The Rise of Towns

The first cities emerged on the landscape thousands of years ago, during the New Stone Age, with the development of agriculture, which provided enough food to sustain a city.[8] In this first stage, the number of city dwellers per square kilometer was much higher than the number of people in the surrounding countryside, but the density was still too low to cause rapid, serious disturbance to the land. In fact, the waste from city dwellers and their animals was an important fertilizer for the surrounding farmlands. In this stage, the city's size was restricted by the primitive means of transporting food and necessary resources into the city and removing waste.[10]

The Urban Center

In the second stage, more efficient transportation made possible the development of much larger urban centers. Boats, barges, canals, and wharves, as well as roads, horses, carriages, and carts, enabled cities to rise up and thrive farther from agricultural areas. Ancient Rome, originally dependent on local produce, became a city fed by granaries in Africa and the Near East.

The population of a city is limited by how far a person can travel in one day to and from work and by how many people can be packed into an area (density). In the second stage, the internal size of a city was limited by pedestrian travel. A worker had to be able to walk to work, do a day's work, and walk home the same day. The density of people per square kilometer was limited by architectural techniques and primitive waste disposal. These cities never exceeded a population of 1 million, and only a few approached this size, most notably Rome and some cities in China.

The Industrial Metropolis

The Industrial Revolution allowed greater modification of the environment than had been possible before. Three technological advances that had significant effects on the city environment were improved medicine and sanitation, which led to the control of many diseases, and improved transportation.

Modern transportation makes a larger city possible. Workers can live farther from their place of work and commerce, and communication can extend over larger areas. Air travel has freed cities even more from the traditional limitation of situation. We now have thriving urban areas where previously transportation was poor: in the Far North (Fairbanks, Alaska) and on islands (Honolulu). These changes increase city dwellers' sense of separateness from their natural environment.

Subways and commuter trains have also led to the development of suburbs. In some cities, however, the negative effects of urban sprawl have prompted many people to return to the urban centers or to smaller, satellite cities surrounding the central city. The drawbacks of suburban commuting and the destruction of the landscape in suburbs have brought new appeal to the city center.

The Center of Civilization

We are at the beginning of a new stage in the development of cities. With modern telecommunications, people can work at home or at distant locations. Perhaps, as telecommunication frees us from the necessity for certain kinds of commercial travel and related activities, the city can become a cleaner, more pleasing center of civilization.

An optimistic future for cities requires a continued abundance of energy and material resources, which are certainly not guaranteed, and wise use of these resources. If energy resources are rapidly depleted, modern mass transit may fail, fewer people will be able to live in suburbs, and the cities will become more crowded. Reliance on coal and wood will increase air pollution. Continued destruction of the land within and near cities could compound transportation problems, making local production of food impossible. The future of our cities depends on our ability to plan and to use our resources wisely.

22.5 City Planning and the Environment

If people live in densely populated cities, ways must be found to make urban life healthy and pleasant and to keep the cities from polluting the very environment that their population depends on. City planners have found many ways to make cities pleasing environments: developing parks and connecting cities to rivers and nearby mountains in environmentally and aesthetically pleasing ways. City planning has a long and surprising history, with the paired goals of defense and beauty. Long experience in city planning, combined with modern knowledge from environmental sciences, can make cities of the future healthier and more satisfying to people and better integrated within the environment. Beautiful cities are not only healthy but also attract more people, relieving pressure on the countryside.

A city can never be free of environmental constraints, even though its human constructions give us a false sense of security. Lewis Mumford, a historian of cities, wrote, "Cities give us the illusion of self-sufficiency and independence and of the possibility of physical continuity without conscious renewal."[8,11] But this security is only an illusion.

A danger in city planning is the tendency to totally transform the features of a city center from natural to artificial—to completely replace grass and soil with pavement, gravel, houses, and commercial buildings, creating an impression that civilization has dominated the environment. Ironically, the artificial aspects of the city that make it seem so independent of the rest of the world actually make it more dependent on its rural surroundings for all resources. Although such a city appears to its inhabitants to grow stronger and more independent, it actually becomes more fragile.[8]

City Planning for Defense and Beauty

Many cities in history grew without any conscious plan. However, **city planning**—formal, conscious planning for new cities in modern Western civilization—can be traced back as far as the 15th century. Sometimes cities have been designed for specific social purposes, with little consideration of the environment. In other cases the environment and its effect on city residents have been major planning considerations.

Defense and beauty have been two dominant themes in formal city planning (see A Closer Look 22.2). We can think of these two types of cities as fortress cities and park cities. The ideas of the fortress city and the park city influenced the planning of cities in North America. The importance of aesthetic considerations is illustrated in the plan of Washington, DC, designed by Pierre Charles L'Enfant. L'Enfant mixed a traditional rectangular grid pattern of streets (which can be traced back to the Romans) with broad avenues set at angles. The goal was to create a beautiful city with many parks, including small ones at the intersections of avenues and streets. This design has made Washington, DC, one of the most pleasant cities in the United States.

The City Park

Parks have become more and more important in cities. A significant advance for U.S. cities was the 19th-century planning and construction of Central Park in New York City, the first large public park in the United States. The park's designer, Frederick Law Olmsted, was one of the most important modern experts on city planning. He took site and situation into account and attempted to blend improvements to a site with the aesthetic qualities of the city.[12]

Central Park is an example of "design with nature," a term coined much later, and its design influenced other U.S. city parks. For Olmsted, the goal of a city park was to provide psychological and physiological relief from city life through access to nature and beauty. Vegetation was one of the keys to creating beauty in the park, and Olmsted carefully considered the opportunities and limitations of topography, geology, hydrology, and vegetation.

In contrast to the approach of a preservationist, who might simply have strived to return the area to its natural, wild state, Olmsted created a naturalistic environment, keeping the rugged, rocky terrain but putting ponds where he thought they were desirable. To add variety, he constructed "rambles, " walkways that were densely planted and followed circuitous patterns. He created a "sheep meadow" by using explosives to flatten the terrain. In the southern part of the park, where there were flat meadows, he created recreational areas. To meet the needs of the city, he built transverse roads through the park and also created depressed roadways that allowed traffic to cross the park without detracting from the vistas seen by park visitors.

Olmsted has remained a major figure in American city planning, and the firm he founded continued to be important in city planning into the 20th century. His skill in creating designs that addressed both the physical and aesthetic needs of a city is further illustrated by his work in Boston. Boston's original site had certain advantages: a narrow peninsula with several hills that could be easily defended, a good harbor, and a good water supply. But as Boston grew, demand increased for more land for buildings, a larger area for docking ships, and a better water supply. The need to control ocean floods and to dispose of solid

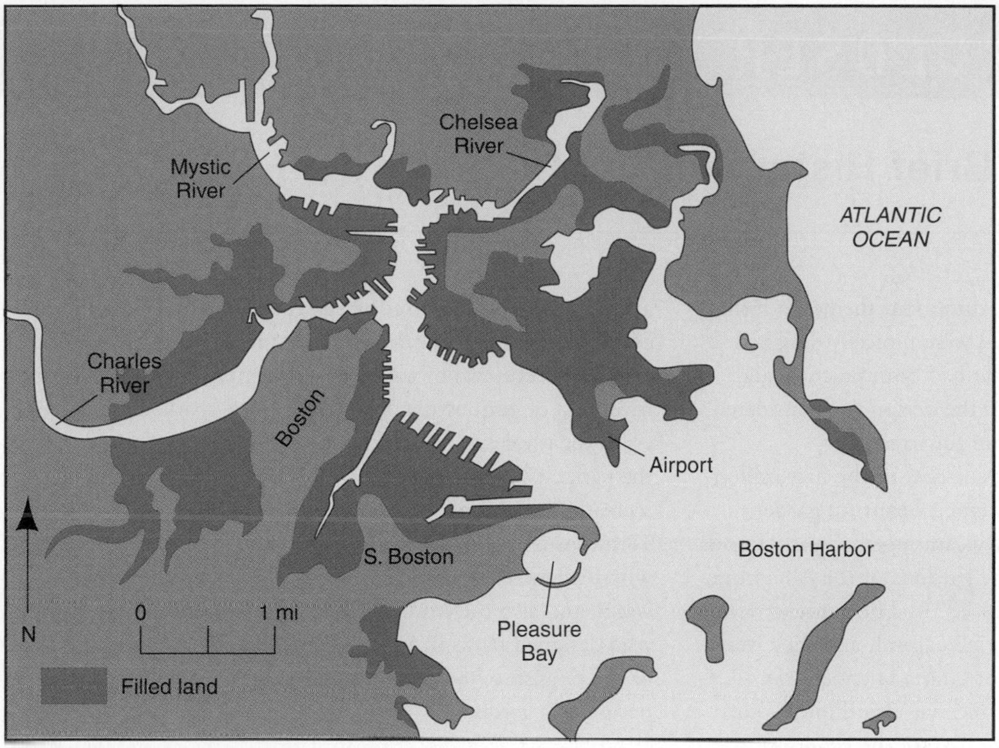

FIGURE 22.8 Nature integrated into a city plan. Boston has been modified over time to improve the environment and provide more building locations. This map of Boston shows land filled in to provide new building sites as of 1982. Although such landfill allows for expansion of the city, it can also create environmental problems, which then must be solved. (*Source*: A.W. Spirn, *The Granite Garden: Urban Nature and Human Design* [New York: Basic Books, 1984].)

and liquid wastes grew as well. Much of the original tidal flats area, which had been too wet to build on and too shallow to navigate, had been filled in (Figure 22.8). Hills had been leveled and the marshes filled with soil. The largest project had been the filling of Back Bay, which began in 1858 and continued for decades. Once filled, however, the area had suffered from flooding and water pollution.

Olmsted's solution to these problems was a water-control project called the "fens." His goal was to "abate existing nuisances" by keeping sewage out of the streams and ponds and building artificial banks for the streams to prevent flooding—and to do this in a natural-looking way. His solution included creating artificial watercourses by digging shallow depressions in the tidal flats, following meandering patterns like natural streams; setting aside other artificial depressions as holding ponds for tidal flooding; restoring a natural salt marsh planted with vegetation tolerant of brackish water; and planting the entire area to serve as a recreational park when not in flood. He put a tidal gate on the Charles River—Boston's major river—and had two major streams diverted directly through culverts into the Charles so that they flooded the fens only during flood periods. He reconstructed the Muddy River primarily to create new, accessible landscape. The result of Olmsted's vision was that control of water became an aesthetic addition to the city. The blending of several goals made the development of the fens a landmark in city planning. Although to the casual stroller it appears to be simply a park for recreation, the area serves an important environmental function in flood control.

Parks near rivers and the ocean are receiving more and more attention. For example, New York City is spending several hundred million dollars to build the Hudson River Park along the Hudson River, where previously abandoned docks and warehouses littered the shoreline and barred public access to the river.

An extension of the park idea was the "garden city," a term coined in 1902 by Ebenezer Howard. Howard believed that city and countryside should be planned together. A **garden city** was one that was surrounded by a **greenbelt**, a belt of parkways, parks, or farmland. The idea was to locate garden cities in a set connected by greenbelts, forming a system of countryside and urban landscapes. The idea caught on, and garden cities were planned and developed in Great Britain and the United States. Greenbelt, Maryland, just outside Washington, DC, is one of these cities, as is Lecheworth, England. Howard's garden city concept, like Olmsted's use of the natural landscape in designing city parks, continues to be a part of city planning today.

A CLOSER LOOK 22.2

A Brief History of City Planning

Defense and beauty have been two dominant themes in formal city planning. Ancient Roman cities were typically designed along simple geometric patterns that had both practical and aesthetic benefits. The symmetry of the design was considered beautiful but was also a useful layout for streets.

During the height of Islamic culture, in the first millennium, Islamic cities typically contained beautiful gardens, often within the grounds of royalty. Among the most famous urban gardens in the world are the gardens of the Alhambra, a palace in Granada, Spain (Figure 22.9). The gardens were created when this city was a Moorish capital, and they were maintained after Islamic control of Granada ended in 1492. Today, as a tourist attraction that receives 2 million visitors a year, the Alhambra gardens demonstrate the economic benefits of aesthetic considerations in city planning. They also illustrate that making a beautiful park a specific focus in a city benefits the city environment by providing relief from the city itself.

After the fall of the Roman Empire, the earliest planned towns and cities in Europe were walled fortress cities designed for defense. But even in these instances, city planners considered the aesthetics of the town. In the 15th century, one such planner, Leon Battista Alberti, argued that large and important towns should have broad and straight streets; smaller, less fortified towns should have winding streets to increase their beauty. He also advocated the inclusion of town squares and recreational areas, which continue to be important considerations in city planning.[13] One of the most successful of these walled cities is Carcassonne, in southern France, now the third most visited tourist site in that country. Today, walled cities have become major tourist attractions, again illustrating the economic benefits of good aesthetic planning in urban development.

The usefulness of walled cities essentially ended with the invention of gunpowder. The Renaissance sparked an interest in the ideal city, which in turn led to the development of the park city. A preference for gardens and parks, emphasizing recreation, developed in Western civilization in the 17th and 18th centuries. It characterized the plan of Versailles, France, with its famous formal parks of many sizes and tree-lined walks, and also the work of the Englishman Capability Brown, who designed parks in England and was one of the founders of the English school of landscape design, which emphasized naturalistic gardens.

FIGURE 22.9 Planned beauty. The Alhambra gardens of Granada, Spain, illustrate how vegetation can be used to create beauty within a city.

22.6 The City as an Environment

A city changes the landscape, and because it does, it also changes the relationship between biological and physical aspects of the environment. Many of these changes were discussed in earlier chapters as aspects of pollution, water management, or climate. You may find some mentioned again in the following sections, generally with a focus on how effective city planning can reduce the problems.

The Energy Budget of a City

Like any ecological and environmental system, a city has an "energy budget." The city exchanges energy with its environment in the following ways: (1) absorption and reflection of solar energy, (2) evaporation of water, (3) conduction of air, (4) winds (air convection), (5) transport of fuels into the city and burning of fuels by people in the city, and (6) convection of water (subsurface and surface stream flow). These in turn affect the climate in the city, and the city may affect the climate in the nearby surroundings, a possible landscape effect.

The Urban Atmosphere and Climate

Cities affect the local climate; as the city changes, so does its climate (see Chapter 20). Cities are generally less windy than nonurban areas because buildings and other structures obstruct the flow of air. But city buildings also channel the wind, sometimes creating local wind tunnels with high wind speeds. The flow of wind around one building is influenced by nearby buildings, and the total wind flow through a city is the result of the relationships among all the buildings. Thus, plans for a new building must take into account its location among other buildings as well as its shape. In some cases, when this has not been done, dangerous winds around tall buildings have blown out windows, as happened to the John Hancock Building in Boston on January 20, 1973, a famous example of the problem.

A city also typically receives less sunlight than the countryside because of the particulates in the atmosphere over cities—often over ten times more particulates than in surrounding areas.[15] Despite reduced sunlight, a city is a heat island, warmer than surrounding areas, for two reasons: (1) the burning of fossil fuels and other industrial and residential activities and (2) a lower rate of heat loss, partly because buildings and paving materials act as solar collectors (Figure 22.10).[14]

Solar Energy in Cities

Until modern times, it was common to use solar energy, through what is called today *passive solar energy*, to help heat city houses. Cities in ancient Greece, Rome, and China were designed so that houses and patios faced south and passive solar energy applications were accessible to each household.[19] The 20th century in America and Europe was a major exception to this approach because cheap and easily accessible fossil fuels led people to forget certain fundamental lessons. Today, the industrialized nations are beginning to appreciate the importance of solar energy once again. Solar photovoltaic devices that convert sunlight to electricity are becoming a common sight in many cities, and some cities have enacted solar energy ordinances that make it illegal to shade another property owner's building in such a way that it loses solar heating capability. (See Chapter 16 for a discussion of solar energy.)

Water in the Urban Environment

Modern cities affect the water cycle, in turn affecting soils and consequently plants and animals in the city. Because city streets and buildings prevent water infiltration, most rain runs off into storm sewers. The streets and sidewalks also add to the heat island effect by preventing water in the soil from evaporating to the atmosphere, a process that cools natural ecosystems. Chances of flooding increase both within the city and downstream outside the city. New, ecological methods of managing stormwater can alleviate these problems by controlling the speed and quality of water running off pavements and into streams. For example, a plan for the central library's parking lot in Alexandria, Virginia, includes wetland vegetation and soils that temporarily absorb runoff from the parking lot, remove some of the pollutants, and slow down the water flow (Figure 22.11).

Most cities have a single underground sewage system. During times of no rain or light rain, this system handles only sewage. But during periods of heavy rain, the runoff is mixed with the sewage and can exceed the capacity

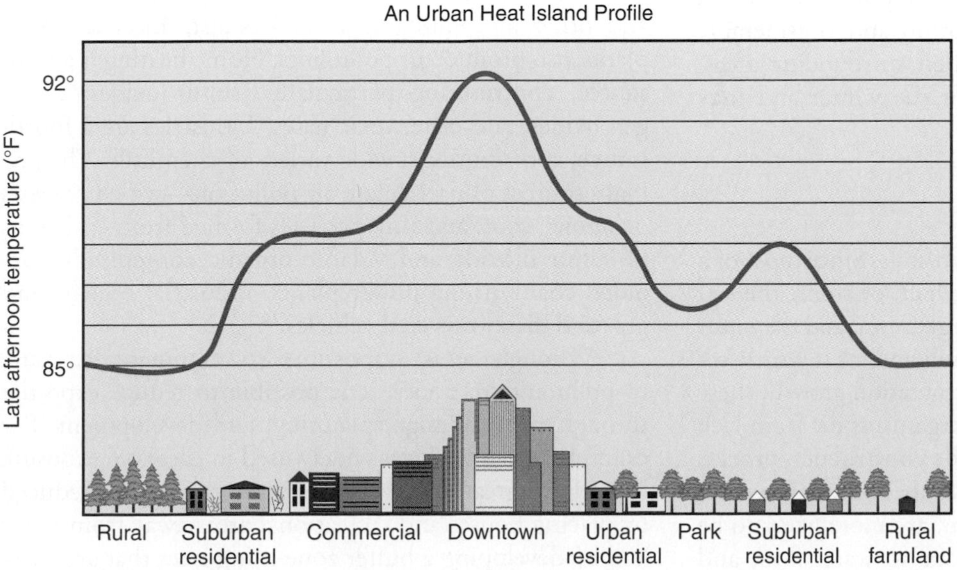

FIGURE 22.10 A typical urban heat island profile. The graph shows temperature changes correlated with the density of development and trees. (*Source*: Andrasko and Huang, in H. Akbari et al., *Cooling Our Communities: A Guidebook on Tree Planting and Light-Colored Surfacing* [Washington, DC: U.S. EPA Office of Policy Analysis, 1992].)

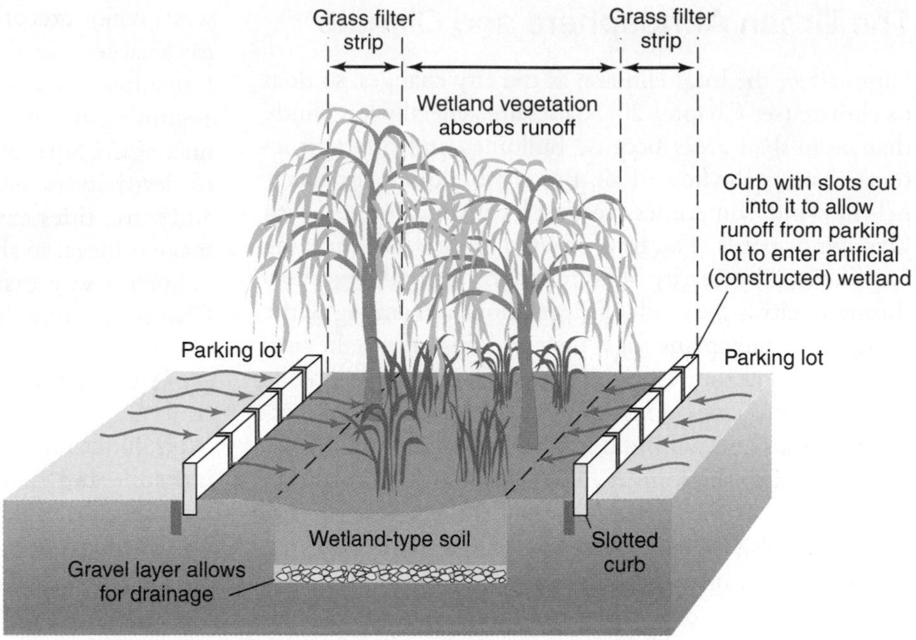

FIGURE 22.11 Planned for better drainage. A plan for the Alexandria, Virginia, central library parking lot includes wetland vegetation and soils that temporarily absorb runoff from the parking lot (see arrows). The landscape architecture firm of Rhodeside & Harwell planned the project. (*Source*: Modified after Rhodeside & Harwell Landscape Architects.)

of sewage-treatment plants, causing sewage to be released downstream without sufficient treatment. In most cities that already have such systems, the expense of building a completely new and separate runoff system is prohibitive, so other solutions must be found. One city that avoids this problem is Woodlands, Texas. It was designed by the famous landscape architect Ian McHarg, who originated the phrase "design with nature," the subject of A Closer Look 22.3.[15]

Because of reduced evaporation, midlatitude cities generally have lower relative humidity (2% lower in winter to 8% lower in summer) than the surrounding countryside. At the same time, cities can have higher rainfall than their surroundings because dust above a city provides particles for condensation of water vapor. Some urban areas have 5–10% more precipitation and considerably more cloud cover and fog than their surrounding areas. Fog is particularly troublesome in the winter and may impede ground and air traffic.

Soils in the City

A modern city has a great impact on soils. Since most of a city's soil is covered by cement, asphalt, or stone, the soil no longer has its natural cover of vegetation, and the natural exchange of gases between the soil and air is greatly reduced. No longer replenished by vegetation growth, these soils lose organic matter, and soil organisms die from lack of food and oxygen. In addition, the construction process and the weight of the buildings compact the soil, which restricts water flow. City soils, then, are more likely to be compacted, waterlogged, impervious to water flow, and lacking in organic matter.

Pollution in the City

In a city, everything is concentrated, including pollutants. City dwellers are exposed to more kinds of toxic chemicals in higher concentrations and to more human-produced noise, heat, and particulates than are their rural neighbors (see Chapter 15). This environment makes life riskier—in fact, lives are shortened by an average of one to two years in the most polluted cities in the United States. The city with the greatest number of early deaths is Los Angeles, with an estimated 5,973 early deaths per year, followed by New York with 4,024, Chicago with 3,479, Philadelphia with 2,590, and Detroit with 2,123.

Some urban pollution comes from motor vehicles, which emit nitrogen oxides, ozone, carbon monoxide, and other air pollutants from exhaust. Electric power plants also produce air pollutants. Home heating is a third source, contributing particulates, sulfur oxides, nitrogen oxides, and other toxic gases. Industries are a fourth source, contributing a wide variety of chemicals. The primary sources of particulate air pollution—which consists of smoke, soot, and tiny particles formed from emissions of sulfur dioxide and volatile organic compounds—are older, coal-burning power plants, industrial boilers, and gas- and diesel-powered vehicles.[16]

Although it is impossible to eliminate exposure to pollutants in a city, it is possible to reduce exposure through careful design, planning, and development. For example, when lead was widely used in gasoline, exposure to lead was greater near roads. Exposure could be reduced by placing houses and recreational areas away from roads and by developing a buffer zone using trees that are resistant to the pollutant and that absorb pollutants.

22.7 Bringing Nature to the City

As we saw in this chapter's opening case study about New York City's High Line, a practical problem is how to bring nature to the city—how to make plants and animals part of a city landscape (see A Closer Look 22.3). This has evolved into several specialized professions, including urban forestry (whose professionals are often called tree wardens), landscape architecture, city planning and management, and civil engineering specializing in urban development. Most cities have an urban forester on the payroll who determines the best sites for planting trees and the tree species best suited to those environments. These professionals take into account climate, soils, and the general influences of the urban setting, such as the shade imposed by tall buildings and the pollution from motor vehicles.

Cities and Their Rivers

Traditionally, rivers have been valued for their usefulness in transportation and as places to dump wastes and therefore not places of beauty or recreation. The old story was that a river renewed and cleaned itself every mile or every 3 miles (depending on who said it). That may have been relatively correct when there was one person or one family per linear river mile, but it is not for today's cities, with their high population densities and widespread use and dumping of modern chemicals.

Kansas City, Missouri, at the confluence of the Kansas and Missouri rivers, illustrates the traditional disconnect between a city and its river. The Missouri River's floodplain provides a convenient transportation corridor, so the south shore is dominated by railroads, while downtown the north shore forms the southern boundary of the city's airport. Except for a small riverfront park, the river has little place in this city as a source of recreation and relief for its citizens or in the conservation of nature.

The same used to be true of the Hudson River in New York, but that river has undergone a major cleanup since the beginning of the project *Clearwater,* led in part by folksinger Pete Seeger and also by activities of the city's Hudson River Foundation and Metropolitan Waterfront Alliance. Not only is the river cleaner, but an extensive Hudson River Park is being completed, transforming Manhattan's previously industrial and uninviting riverside into a beautifully landscaped and inviting park (Figure 22.12) extending for miles from the southern end of Manhattan to near the George Washington Bridge.

The throngs of sunbathers, picnickers, older people, young couples, and parents with children relaxing on the grass and enjoying the river views are proof of city dwellers' need for contact with nature. And a lesson we are learning is that for cities on rivers, one way to bring nature to the city is to connect the city to its river.

A CLOSER LOOK 22.3

Design with Nature

The new town of Woodlands, a suburb of Houston, Texas, is an example of professional planning. Woodlands was designed so that most houses and roads were on ridges; the lowlands were left as natural open space. The lowlands provide areas for temporary storage of floodwater and, because the land is unpaved, allow rain to penetrate the soil and recharge the aquifer for Houston. Preserving the natural lowlands has other environmental benefits as well. In this region of Texas, low-lying wetlands are habitats for native wildlife, such as deer. Large, attractive trees, such as magnolias, grow here, providing food and habitat for birds. The innovative city plan has economic as well as aesthetic and conservational benefits. It is estimated that a conventional drainage system would have cost $14 million more than the amount spent to develop and maintain the wetlands.[21]

A kind of soil important in modern cities is the soil that occurs on **made lands**—lands created from fill, sometimes as waste dumps of all kinds, sometimes to create more land for construction. The soils of made lands are different from those of the original landscape. They may be made of all kinds of trash, from newspapers to bathtubs, and may contain some toxic materials. The fill material is unconsolidated, meaning that it is loose material without rock structure. Thus, it is not well suited to be a foundation for buildings. Fill material is particularly vulnerable to earthquake tremors and can act somewhat like a liquid and amplify the effects of the earthquake on buildings. However, some made lands have been turned into well-used parks. For example, a marina park in Berkeley, California, is built on a solid-waste landfill. It extends into San Francisco Bay, providing public access to beautiful scenery, and is a windy location, popular for kite flying and family strolls. (See Chapter 23 for more information about solid-waste disposal.)

FIGURE 22.12 The newly built Hudson River Park on Manhattan's West Side illustrates the changing view of rivers and the improved use of riverfronts for recreation and urban landscape beauty.

FIGURE 22.13 Paris was one of the first modern cities to use trees along streets to provide beauty and shade, as shown in this picture of the famous Champs-Elysées.

Vegetation in Cities

Trees, shrubs, and flowers add to the beauty of a city. Plants fill different needs in different locations. Trees provide shade, which reduces the need for air-conditioning and makes travel much more pleasant in hot weather. In parks, vegetation provides places for quiet contemplation; trees and shrubs can block some of the city sounds, and their complex shapes and structures create a sense of solitude. Plants also provide habitats for wildlife, such as birds and squirrels, which many urban residents consider pleasant additions to a city.

The use of trees in cities has expanded since the Renaissance. In earlier times, trees and shrubs were set apart in gardens, where they were viewed as scenery but not experienced as part of ordinary activities. Street trees were first used in Europe in the 18th century; among the first cities to line streets with trees were London and Paris (Figure 22.13). In many cities, trees are now considered an essential element of the urban visual scene, and major cities have large tree-planting programs. In New York City, for example, 11,000 trees are planted each year, and in Vancouver, Canada, 4,000 are planted each year.[17] Trees are also increasingly used to soften the effects of climate near houses. In colder climates, rows of conifers planted to the north of a house can protect it from winter winds. Deciduous trees to the south can provide shade in the summer, reducing requirements for air-conditioning, yet allowing sunlight to warm the house in the winter (Figure 22.14).

Cities can even provide habitat for endangered plants. For example, Lakeland, Florida, uses endangered plants in local landscaping with considerable success. However, it is necessary to select species carefully because vegetation in cities must be able to withstand special kinds of stress, such as compacted soils, poor drainage, and air pollution. Because trees along city streets are often surrounded by cement, and because the soils tend to be compacted and drain poorly, the root systems are likely to suffer from extremes of drought on the one hand and soil saturation (immediately following or during a rainstorm) on the other. The solution to this particular problem is to specially prepare streets and sidewalks for tree growth. A tree-planting project was completed for the World Bank Building in Washington, DC, in 1996. Special care was taken to provide good growing conditions for trees, including aeration, irrigation, and adequate drainage so that the soils did not become waterlogged. The trees continue to grow and remain healthy.[23]

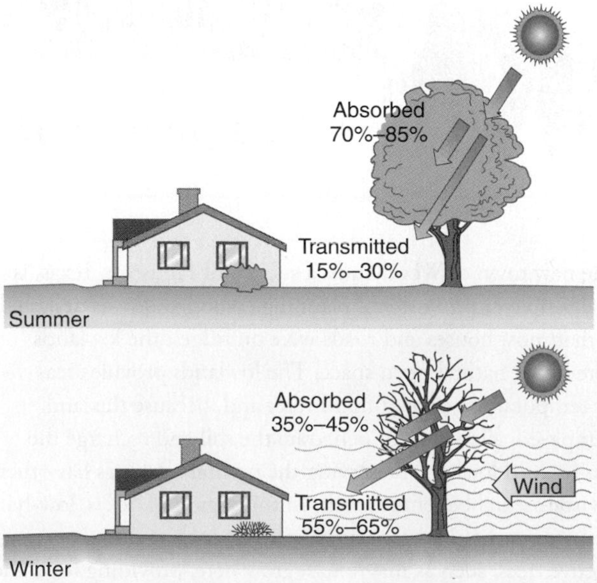

FIGURE 22.14 Trees cool homes. Trees can improve the microclimate near a house, protecting the house from winter winds and providing shade in the summer while allowing sunlight through in the winter. (*Source*: J. Huang and S. Winnett, in H. Akbari et al., *Cooling Our Communities: A Guidebook on Tree Planting and Light-Colored Surfacing* [Washington, DC: U.S. EPA Office of Policy Analysis, 1992].)

Many species of trees and plants are very sensitive to air pollution and will not thrive in cities. The eastern white pine of North America, for example, is extremely sensitive to ozone pollution and does not do well in cities with heavy motor-vehicle traffic or along highways. Dust, too, can interfere with the exchange of oxygen and carbon dioxide necessary for photosynthesis and respiration of the trees. City trees also suffer direct damage from pets, from the physical impact of bicycles, cars, and trucks, and from vandalism. Trees subject to such stresses are more susceptible to attacks by fungus diseases and insects. The lifetime of trees in a city is generally shorter than in their natural woodland habitats unless they are given considerable care.

Some species of trees are more useful and successful in cities than are others. An ideal urban tree would be resistant to all forms of urban stress, have a beautiful form and foliage, and produce no messy fruit, flowers, or leaf litter that required removal. In most cities, in part because of these requirements, only a few tree species are used for street planting. However, reliance on one or a few species results in ecologically fragile urban planting, as we learned when Dutch elm disease spread throughout the eastern United States, destroying urban elms and leaving large stretches of a city treeless. It is prudent to use a greater diversity of trees to avoid the effects of insect infestations and tree diseases.[18]

Cities, of course, have many recently disturbed areas, including abandoned lots and the medians in boulevards and highways. Disturbed areas provide habitat for early-successional plants, including many that we call "weeds," which are often introduced (exotic) plants, such as European mustard. Therefore, wild plants that do particularly well in cities are those characteristic of disturbed areas and of early stages in ecological succession (see Chapter 5). City roadsides in Europe and North America have wild mustards, asters, and other early-successional plants.

Urban "Wilds": The City as Habitat for Wildlife and Endangered Species

We don't associate wildlife with cities—indeed, with the exception of some birds and small, docile mammals such as squirrels, most wildlife in cities are considered pests. But there is much more wildlife in cities, a great deal of it unnoticed. In addition, there is growing recognition that urban areas can be modified to provide habitats for wildlife that people can enjoy. This can be an important method of biological conservation.[19, 20]

We can divide city wildlife into the following categories: (1) species that cannot persist in an urban environment and disappear; (2) those that tolerate an urban environment but do better elsewhere; (3) those that have adapted to urban environments, are abundant there, and are either neutral or beneficial to human beings; and (4) those that are so successful they become pests.

Cooper's hawks probably belong in the third category. They are doing pretty well in Tucson, Arizona, a city of 900,000 people. Although this hawk is a native of the surrounding Sonoran Desert, some of them are nesting in groves of trees within the city. Nest success in 2005 was 84%, between two-thirds and three-quarters of the juvenile hawks that left the nest were still alive six months later, and the population is increasing (Figure 23.15). Scientists studying the hawk in Tucson concluded that "urbanized landscape can provide high-quality habitat."[21]

Cities can even be home to rare or endangered species. Peregrine falcons once hunted pigeons above the streets of Manhattan. Unknown to most New Yorkers, the falcons nested on the ledges of skyscrapers and dived on their prey in an impressive display of predation. The falcons disappeared when DDT and other organic pollutants caused a thinning of their eggshells and a failure in reproduction, but they have been reintroduced into the city. The first reintroduction into New York City took place in 1982, and today 32 falcons are living there.[22] The reintroduction of peregrine falcons illustrates an important recent trend: the growing understanding that city environments can assist in the conservation of nature, including the conservation of endangered species.

In sum, cities are a habitat, albeit artificial. They can provide all the needs—physical structures and necessary resources such as food, minerals, and water—for many plants and animals. We can identify ecological food chains in cities, as shown in Figure 22.16 for insect-eating birds and for a fox. These can occur when areas cleared of buildings and abandoned begin to recover and are in an early stage of ecological succession. For some species, cities' artificial structures are sufficiently like their original habitat to be home.[23] Chimney swifts, for example, which once lived in hollow trees, are now common in chimneys and other vertical shafts, where they glue their nests to the walls with saliva. A city can easily have more chimneys per square kilometer than a forest has hollow trees.

FIGURE 22.15 Cooper's hawks, like this one, live, nest, and breed in the city of Tucson, Arizona.

Cities also have natural habitats in parks and preserves. In fact, modern parks provide some of the world's best wildlife habitats. In New York City's Central Park, approximately 260 species of birds have been observed—100 in a single day. Urban zoos, too, play an important role in conserving endangered species, and the importance of parks and zoos will increase as truly wild areas shrink.

Finally, cities that are seaports often have many species of marine wildlife at their doorsteps. New York City's waters include sharks, bluefish, mackerel, tuna, striped bass, and nearly 250 other species of fish.[24]

City environments can contribute to wildlife conservation in a number of ways. Urban kitchen gardens—backyard gardens that provide table vegetables and decorative plants—can be designed to provide habitats. For instance, these gardens can include flowers that provide nectar for threatened or endangered hummingbirds. Rivers and their riparian zones, ocean shorelines, and wooded parks can provide habitats for endangered species and ecosystems. For example, prairie vegetation, which once occupied more land area than any other vegetation type in the United States, is rare today, but one restored prairie exists within the city limits of Omaha, Nebraska. (Some urban nature preserves are not accessible to the public or offer only limited access, as is the case with the prairie preserve in Omaha.)

Urban drainage structures can also be designed as wildlife habitats. A typical urban runoff design depends on concrete-lined ditches that speed the flow of water from city streets to lakes, rivers, or the ocean. However, as with Boston's Back Bay design, discussed earlier, these features can be planned to maintain or create stream and marsh habitats, with meandering waterways and storage areas that do not interfere with city processes. Such areas can become habitats for fish and mammals (Figure 22.17). Modified to promote wildlife, cities can provide urban corridors that allow wildlife to migrate along their natural routes.[25] Urban corridors also help to prevent some of the effects of ecological islands (see Chapter 8) and are increasingly important to biological conservation.

Animal Pests

Pests are familiar to urban dwellers. The most common city pests are cockroaches, fleas, termites, rats, pigeons, and (since banning DDT) bedbugs, but there are many more, especially species of insects. In gardens and parks, pests include insects, birds, and mammals that feed on fruit and vegetables and destroy foliage of shade trees and plants. Pests compete with people for food and spread diseases. Indeed, before modern sanitation and medicine, such diseases played a major role in limiting human population density in cities. Bubonic plague is spread by fleas found on rodents; mice and rats in cities

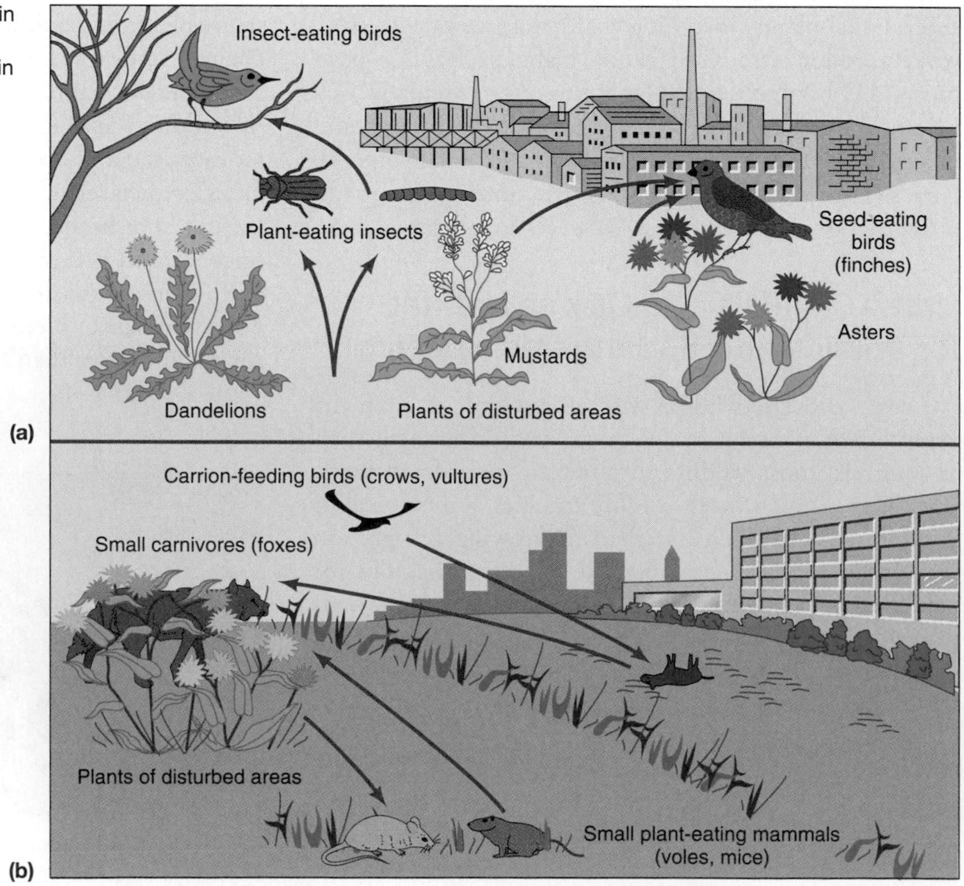

FIGURE 22.16 **(a)** An urban food chain based on plants of disturbed places and insect herbivores. **(b)** An urban food chain based on roadkill.

promoted the spread of the Black Death. Bubonic plague continues to be a health threat in cities—the World Health Organization reports several thousand cases a year.[26] Poor sanitation and high population densities of people and rodents set up a situation where the disease can strike.

An animal is a pest to people when it is in an undesired place at an undesirable time doing an unwanted thing. A termite in a woodland helps the natural regeneration of wood by hastening decay and speeding the return of chemical elements to the soil, where they are available to living plants. But termites in a house are pests because they threaten the house's physical structure.

Animals that do well enough in cities to become pests have certain characteristics in common. They are generalists in their food choice, so they can eat what we eat (including the leftovers we throw in the trash), and they have a high reproductive rate and a short average lifetime.

Controlling Pests

We can best control pests by recognizing how they fit their natural ecosystem and identifying the things that control them in nature. People often assume that the only way to control animal pests is with poisons, but there are limitations to this approach. Early poisons used in pest control were generally also toxic to people and pets (see Chapter 11). Another problem is that reliance on one toxic compound can cause a species to develop a resistance to it, which can lead to rebound—a renewed increase in that pest's population. A pesticide used once and spread widely will greatly reduce the population of the pest. However, when the pesticide loses its effectiveness, the pest population can increase rapidly as long as habitat is suitable and food plentiful. This is what happened when an attempt was made to control Norway rats in Baltimore.

One of the keys to controlling pests is to eliminate their habitats. For example, the best way to control rats is to reduce the amount of open garbage and eliminate areas to hide and nest. Common access areas used by rats are the spaces within and between walls and the openings between buildings where pipes and cables enter. Houses can be constructed to restrict access by rats. In older buildings, we can seal areas of access.

FIGURE 22.17 How water drainage systems in a city can be modified to provide wildlife habitat. In the community on the right, concrete-lined ditches speed runoff and have little value to fish and wildlife. In the community on the left, the natural stream and marsh were preserved; water is retained between rains, and an excellent habitat is provided. (*Source*: D.L. Leedly and L.W. Adams, A *Guide to Urban Wildlife Management* [Columbia, MD: National Institute for Urban Wildlife, 1984], pp. 20–21.)

CRITICAL THINKING ISSUE

How Can Urban Sprawl Be Controlled?

As the world becomes increasingly urbanized, individual cities are growing in area as well as population. Residential areas and shopping centers move into undeveloped land near cities, impinging on natural areas and creating a chaotic, unplanned human environment. "Urban sprawl" has become a serious concern in communities all across the United States. According to the U.S. EPA, in a recent six-month period approximately 5,000 people left Baltimore City to live in suburbs, with the result that nearly 10,000 acres of forests and farmlands were converted to housing. At this rate, the state of Maryland could use as much land for development in the next 25 years as it has used in the entire history of the state.[27] In the past ten years, 22 states have enacted new laws to try to control urban sprawl.

The city of Boulder, Colorado, has been in the forefront of this effort since 1959, when it created the "blue line"—a line at an elevation of 1,761 m (the city itself is at 1,606 m) above which it would not extend city water or sewer services. Boulder's citizens felt, however, that the blue line was insufficient to control development and maintain the city's scenic beauty in the face of rapid population growth. (Boulder's population had grown in the decade before 1959 from 29,000 to 66,000 and reached 96,000 by 1998.) To prevent uncontrolled development in the area between the city and the blue line, in 1967 Boulder began to use a portion of the city sales tax to purchase land, creating a 10,800-hectare greenbelt around the city proper.

In 1976 Boulder went one step further and limited increases in new residences to 2% a year. Two years later, recognizing that planned development requires a regional approach, the city and surrounding Boulder County adopted a coordinated development plan. By the early 1990s, it had become apparent that further growth control was needed for nonresidential building. The plan that the city finally adopted reduced the allowable density of many commercial and industrial properties, in effect limiting jobs rather than limiting building space.

Boulder's methods to limit the size of its population have worked. The most recent census (2002) showed that the population had increased by a mere 2,000 people and totaled just a little more than 94,000.

The benefits of Boulder's controlled-growth initiatives have been a defined urban–rural boundary; rational, planned development; protection of sensitive environmental areas and scenic vistas; and large areas of open space within and around the city for recreation. And in spite of its growth-control measures, Boulder's economy has remained strong. However, restraints on residential growth forced many people who found jobs in Boulder to seek affordable housing in adjoining communities, where populations ballooned. The population of Superior, Colorado, for example, grew from 225 in 1990 to 9,000 in 2000. Further, as commuting workers—40,000 a day—tried to get to and from their jobs in Boulder, traffic congestion and air pollution increased. In addition, because developers had built housing but not stores in the outlying areas, shoppers flocked into Boulder's downtown mall. When plans for a competing mall in the suburbs were finally announced, however, Boulder officials worried about the loss of revenue if the new mall drew shoppers away from the city. At the same time, sprawl from Denver (only 48 km from Boulder), as well as its infamous "brown cloud" of polluted air, began to spill out along the highway connecting the two communities.

Critical Thinking Questions

1. Is a city an open or a closed system (see Chapter 3)? Use examples from the case of Boulder to support your answer.
2. As Boulder takes steps to limit growth, it becomes an even more desirable place to live, which subjects it to even greater growth pressures. What ways can you suggest to avoid such a positive-feedback loop?
3. Some people in Boulder think the next step is to increase residential density within the city. Do you think people living there will accept this plan? What are the advantages and disadvantages of increasing density?
4. To some, the story of Boulder is the saga of a heroic battle against commercial interests that would destroy environmental resources and a unique quality of life. To others, it is the story of an elite group building an island of prosperity and the good life for themselves. Which do you think it is?

SUMMARY

- As an urban society, we must recognize the city's relation to the environment. A city influences and is influenced by its environment and is an environment itself.
- Like any other life-supporting system, a city must maintain a flow of energy, provide necessary material resources, and have ways of removing wastes. These functions are accomplished through transportation and communication with outlying areas.
- Because cities depend on outside resources, they developed only when human ingenuity resulted in modern agriculture and thus excess food production. The history of cities divides into four stages: (1) the rise of

towns; (2) the era of classic urban centers; (3) the period of industrial metropolises; and (4) the age of mass telecommunication, computers, and new forms of travel.

- Locations of cities are strongly influenced by environment. It is clear that cities are not located at random but in places of particular importance and environmental advantage. A city's site and situation are both important.
- A city creates an environment that is different from surrounding areas. Cities change local climate; they are commonly cloudier, warmer, and rainier than surrounding areas.
- In general, life in a city is riskier because of higher concentrations of pollutants and pollutant-related diseases.
- Cities favor certain animals and plants. Natural habitats in city parks and preserves will become more important as wilderness shrinks.
- Trees are an important part of urban environments, but cities place stresses on trees. Especially important are the condition of urban soils and the supply of water for trees.
- Cities can help to conserve biological diversity, providing habitat for some rare and endangered species.
- As the human population continues to increase, we can envision two futures: one in which people are dispersed widely throughout the countryside and cities are abandoned except by the poor; and another in which cities attract most of the human population, freeing much landscape for conservation of nature, production of natural resources, and public-service functions of ecosystems.

REEXAMINING THEMES AND ISSUES

Human Population

As the world's human population increases, we are becoming an increasingly urbanized species. Present trends indicate that in the future, most citizens of most nations will live in their country's single largest city. Thus, concern about urban environments will be increasingly important.

Cities contain the seeds of their own destruction: The very artificiality of a city gives its inhabitants the sense that they are independent of their surrounding environment. But the opposite is the case: The more artificial a city, the more it depends on its surrounding environment for resources and the more susceptible it becomes to major disasters unless this susceptibility is recognized and planned for. The keys to sustainable cities are an ecosystem approach to urban planning and a concern with the aesthetics of urban environments.

Sustainability

Cities depend on the sustainability of all renewable resources and must therefore recognize that they greatly affect their surrounding environments. Urban pollution of rivers that flow into an ocean can affect the sustainability of fish and fisheries. Urban sprawl can have destructive effects on endangered habitats and ecosystems, including wetlands. At the same time, cities designed to support vegetation and some wildlife can contribute to the sustainability of nature.

Global Perspective

The great urban centers of the world produce global effects. As an example, because people are concentrated in cities and because many cities are located at the mouths of rivers, most major river estuaries of the world are severely polluted.

Urban World

The primary message of this chapter is that Earth is becoming urbanized and that environmental science must deal more and more with urban issues.

People and Nature

It has been a modern tendency to focus environmental conservation efforts on wilderness, large parks, and preserves outside of cities. Meanwhile, city environments have been allowed to decay. As the world becomes increasingly urbanized, however, a change in values is necessary. If we are serious about conserving biological diversity, we must assign greater value to urban environments. The more pleasant city environments are, and the more recreation people can find in them, the less pressure there will be on the countryside.

Science and Values

Modern environmental sciences tell us much that we can do to improve the environments of cities and the effects of cities on their environments. What we choose to do with this knowledge depends on our values. Scientific information can suggest new options, and we can select among these for the future of our cities, depending on our values.

KEY TERMS

city planning **506**
fall line **501**
garden city **507**
greenbelt **507**
made lands **511**
site **500**
situation **500**

STUDY QUESTIONS

1. Should we try to save New Orleans or just give up and move the port at the mouth of the Mississippi River elsewhere? Explain your answer in terms of environment and economics.

2. Which of the following cities are most likely to become ghost towns in the next 100 years? In answering this question, use your knowledge of changes in resources, transportation, and communications.
(a) Honolulu, Hawaii
(b) Fairbanks, Alaska
(c) Juneau, Alaska
(d) Savannah, Georgia
(e) Phoenix, Arizona

3. Some futurists picture a world that is one giant biospheric city. Is this possible? If so, under what conditions?

4. The ancient Greeks said that a city should have only as many people as can hear the sound of a single voice. Would you apply this rule today? If not, how would you plan the size of a city?

5. You are the manager of Central Park in New York City and receive the following two offers. Which would you approve? Explain your reasons.
(a) A gift of $1 billion to plant trees from all the eastern states.
(b) A gift of $1 billion to set aside half the park to be forever untouched, thus producing an urban wilderness.

6. Your state asks you to locate and plan a new town. The purpose of the town is to house people who will work at a wind farm—a large area of many windmills, all linked to produce electricity. You must first locate the site for the wind farm and then plan the town. How would you proceed? What factors would you take into account?

7. Visit your town center. What changes, if any, would make better use of the environmental location? How could the area be made more livable?

8. In what ways does air travel alter the location of cities? The value of land within a city?

9. You are put in charge of ridding your city's parks of slugs, which eat up the vegetable gardens rented to residents. How would you approach controlling this pest?

10. It is popular to suggest that in the Information Age people can work at home and live in the suburbs and the countryside, so cities are no longer necessary. List five arguments for and five arguments against this point of view.

FURTHER READING

Beveridge, C.E., and P. Rocheleau, ***Frederick Law Olmsted: Designing the American Landscape*** (New York: Rizzoli International, 1995). The most important analysis of the work of the father of landscape architecture.

Howard, E., ***Garden Cities of Tomorrow*** (Cambridge, MA: MIT Press, 1965, reprint). A classic work of the 19th century that has influenced modern city design, as in Garden City, New Jersey, and Greenbelt, Maryland. It presents a methodology for designing cities with the inclusion of parks, parkways, and private gardens.

McHarg, I.L., ***Design with Nature*** (New York: Wiley, 1995). A classic book about cities and environment.

Ndubisi, F., ***Ecological Planning: A Historical and Comparative Synthesis*** (Baltimore: Johns Hopkins University Press, 2002). An important discussion of an ecological approach to cities.

CHAPTER 23

Materials Management

College student on campus texting on a smartphone. These phones are e-waste when disposed of, but the plastic and metals in them can be recycled for a profit.

LEARNING OBJECTIVES

The waste-management concept of "dilute and disperse" (for example, dumping waste into a river) is a holdover from our frontier days, when we mistakenly believed that land and water were limitless resources. We next attempted to "concentrate and contain" waste in disposal sites—which also proved to pollute land, air, and water. We are now focusing on *managing materials* to reduce environmental degradation associated with resource use and eventually eliminate waste entirely. Finally, we are getting it right! After reading this chapter, you should understand . . .

- That the standard of living in modern society is related in part to the availability of natural resources;
- The importance of resources to society;
- The differences between mineral resources and reserves;
- The factors that control the environmental impact of mineral exploitation;
- How wastes generated from the use of mineral resources affect the environment;
- The social impacts of mineral exploitation;
- How sustainability may be linked to the way we use nonrenewable minerals.
- The emerging concept of *materials management* and how to achieve it;
- The advantages and disadvantages of each of the major methods that constitute integrated waste management;
- The various methods of managing hazardous chemical waste;
- The problems related to ocean dumping and why they will likely persist for some time.

CASE STUDY

Treasures of the Cell Phone

The number of people who use cell phones in the United States has risen from about 5 million in 1990 to nearly 200 million today. In 2009 more than 1 billion cell phones were sold worldwide, about half of them in Asia and Japan. Along with calls, text messaging, and video, cell phones have connected us as never before (see opening photograph). Cell phones are commonly replaced every two to three years as new features and services become available—witness the iPhone's popularity in 2008 when the new phones came out. Each cell phone is small, but the millions of phones retired each year in the United States collectively contain a treasure chest of valuable metals worth over $300 million, not counting the cost of recycling (Table 23.1). Worldwide, their value probably exceeds a billion dollars, but although the money potentially available is attractive, a very small percentage of discarded cell phones are recycled. Most end up stored in our closets or disposed of at municipal solid-waste facilities.

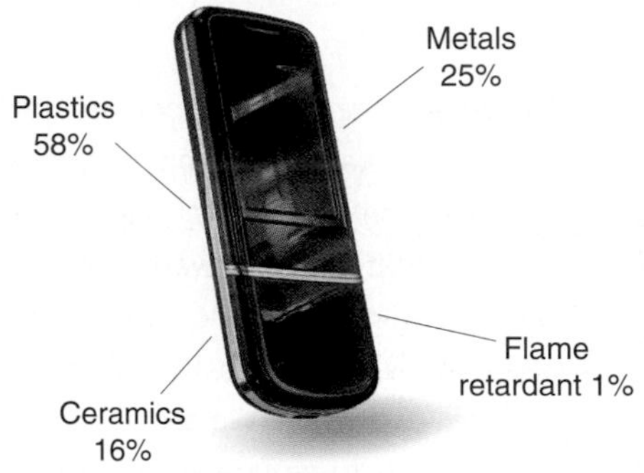

The life cycle of a cell phone is shown in Figure 23.1 and is typical of most electronic waste (e-waste).[1] The primary reason more e-waste is not recycled is that we lack a simple, effective, small-scale, inexpensive way to do it. We also need to better educate people about the environmental value of recycling and to offer more attractive financial incentives to do it. Some states (California,

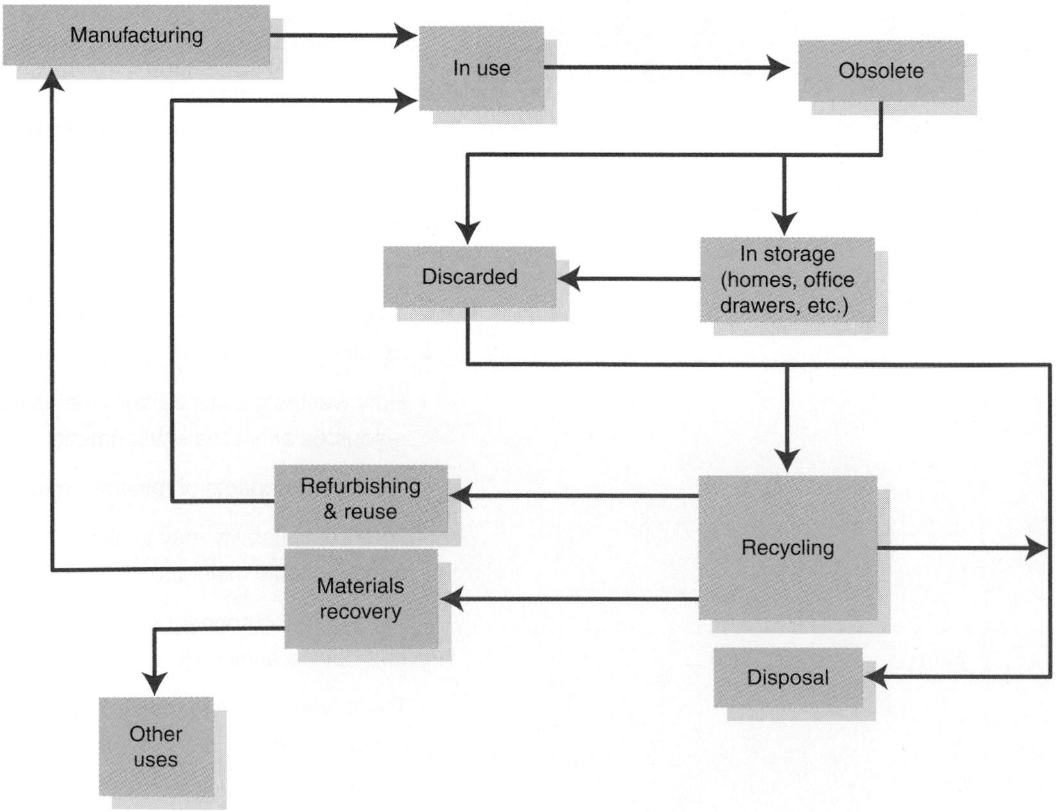

FIGURE 23.1 Composition and life cycle of a cell phone. (*Source*: Modified from D.E. Sullivan, 2006, "Recycled Cell Phones—A Treasure Trove of Valuable Metals," U.S. Geological Survey Fact Sheet 2006–3097.)

Table 23.1 METAL CONTENT AND VALUE OF U.S. CELL PHONES, NOT COUNTING COST TO RECYCLE

	METAL CONTENT AND VALUE ESTIMATED FOR A TYPICAL CELL PHONE		METAL CONTENT AND VALUE FOR 500 MILLION OBSOLETE CELL PHONES IN STORAGE IN 2005	
METAL	WT (g)	VALUE	WT (t)	VALUE
Copper	16	$0.03	7,900	$17 million
Silver	0.35	$0.06	178	$31 million
Gold	0.034	$0.40	17	$199 million
Palladium	0.015	$0.13	7.4	$63 million
Platinum	0.00034	$0.01	0.18	$3.9 million
Total		$0.63	8,102	$314 million

Source: Modified from Sullivan, D.E., 2006. Recycled cell phones—A treasure trove of valuable metals. U.S. Geological Survey Fast Sheet 2006–3097.

for example) have laws that require building the recycling costs into the prices of products.

Our failure to manage cell phones and other e-waste reminds us that we have failed in the past 50 years to move from a throwaway, waste-oriented society to a society that sustains natural resources through improved materials management. In some cases we are moving in that direction by producing less waste and recycling more discarded products. With this in mind, in this chapter we introduce concepts of waste management applied to urban waste, hazardous chemical waste, and waste in the marine environment.

23.1 The Importance of Resources to Society

Modern society depends on the availability of both **renewable resources** (air, surface water, some groundwater, plants, animals, and some energy sources) and **nonrenewable** resources (soil, some groundwater, oil, coal, and most minerals).[2-4] What partially differentiates renewable from nonrenewable resources is their availability in a human time framework. Air and water, along with biological resources such as fish and crops, are regularly replenished as long as the processes that renew them continue to operate at an adequate rate. Nonrenewable resources, such as oil and minerals, even those that are being replenished by Earth processes today, are not being replenished in a time frame useful to people. Thus, strategies to use resources sustainably are linked to specific resources. We can sustain water resources by careful water management (see Chapter 18), but sustaining minerals or oil requires strategies linked more to conservation, recycling, reuse, and substitution than to management of next year's supply delivered by Earth processes.

Many products made from both renewable and nonrenewable resources are found and consumed in a typical American home (see Figure 23.2 for nonrenewable minerals used in a home office). Consider this morning's breakfast (food is a renewable resource). You probably drank from a glass made primarily of sand; ate from dishes made of clay; flavored your food with salt mined from Earth; ate fruit grown with the aid of fertilizers, such as potassium carbonate (potash) and phosphorus; and used utensils made of stainless steel, which comes from processing iron ore and other minerals. While eating your tasty renewable resources, you may have viewed the news on a television or computer screen, listened to music on your iPod, or made appointments using your cell phone. All these electronic items are made from metals and petroleum.

Resources are vital to people, and the standard of living increases with their availability in useful forms. Indeed, the availability of resources is one measure of a society's wealth. Those who have been most successful in locating and extracting or importing and using resources have grown and prospered. Without resources to grow food, construct buildings and roads, and manufacture everything from computers to televisions to automobiles, modern technological civilization as we know it would not be possible. For example, to maintain our standard of living in the United States, each person requires about 10 tons of nonfuel minerals per year.[5] We use other resources, such as food and water, in much greater amounts.

1. Computer—Includes gold, silica, nickel, aluminum, zinc, iron, petroleum products and about thirty other minerals.
2. Pencil—Includes graphite and clays.
3. Telephone—Includes copper, gold and petroleum products.
4. Books—Includes limestone and clays.
5. Pens—Includes limestone, mica, petroleum products, clays, silica and talc.
6. Film—Includes petroleum products and silver.
7. Camera—Includes silica, zinc, copper, aluminum and petroleum products
8. Chair—Includes aluminum and petroleum products.
9. Television—Includes aluminum, copper, iron, nickel, silica, rare earth, and strontium.
10. Stereo—Includes gold, iron, nickel, beryllium and petroleum products.
11. Compact Disc—Includes aluminum and petroleum products.
12. Metal Chest—Includes iron and nickel. The brass trim is made of copper and zinc.
13. Carpet—Includes limestone, petroleum products and selenium.
14. Drywall—Includes gypsum clay, vermiculite, calcium carbonate and micas.
15. Geologic Map—Includes clays, petroleum products, mineral pigments.
16. Concrete Foundation—Includes limestone, clays, sand and gravel
17. Paint-mineral Pigments—Includes pigments (such as iron, zinc and titanium).
18. Cosmetics—Includes mineral chemicals.

FIGURE 23.2 Mineral products used in a home office. (*Source*: Modified from S.J. Kropschot and K.M. Johnson, 2006. U.S. 65. Mineral Resources Program. USGS Circular 1289. Menlo Park, CA)

23.2 Materials Management: What It Is

Materials management has the visionary environmental goal of sustainably obtaining and using renewable and nonrenewable resources. This goal can be pursued in the following ways: [6]

- Eliminate subsidies for extracting virgin materials such as minerals, oil, and timber.
- Establish "green building" incentives that encourage the use of recycled-content materials and products in new construction.
- Assess financial penalties for production that uses poor materials-management practices.
- Provide financial incentives for industrial practices and products that benefit the environment by enhancing sustainability (for example, by reducing waste production and using recycled materials).
- Provide more incentives for people, industry, and agriculture to develop materials-management programs that eliminate or reduce waste by using it as raw material for other products.

Materials management in the United States today is beginning to influence where industries are located. For example, because approximately 50% of the steel produced in the nation now comes from scrap, new steel mills are no longer located near resources such as coal and iron ore. New steel mills are now found in a variety of places, from California to North Carolina and Nebraska; their resource is the local supply of scrap steel. Because they are starting with scrap metal, the new industrial facilities use far less energy and cause much less pollution than older steel mills that must start with virgin iron ore.[7]

Similarly, the recycling of paper is changing where new paper mills are constructed. In the past, mills were built near forested areas where the timber for paper production was being logged. Today, they are being built near cities that have large supplies of recycled paper. New Jersey, for example, has 13 paper mills using recycled paper and 8 steel "mini-mills" producing steel from scrap metal. What is remarkable is that New Jersey has little forested land and no iron mines. Resources for the paper and steel mills come from materials already in use, exemplifying the power of materials management.[7]

We have focused on renewable resources in previous parts of this book (Chapter 11, agriculture; Chapter 12, forests; Chapter 13, wildlife; Chapter 18, water; and Chapter 21, air). We discussed nonrenewable resources with respect to fossil fuels in Chapter 15. The remainder of this chapter will discuss other nonrenewable mineral resources and how to sustain them as long as possible by intelligent waste management.

23.3 Mineral Resources

Minerals can be considered a very valuable, nonrenewable heritage from the geologic past. Although new deposits are still forming from Earth processes, these processes are producing new deposits too slowly to be of use to us today or anytime soon. Also, because mineral deposits are generally in small, hidden areas, they must be discovered, and unfortunately most of the easy-to-find deposits have already been discovered and exploited. Thus, if modern civilization were to vanish, our descendants would have a harder time finding rich mineral deposits than we did. It is interesting to speculate that they might mine landfills for metals thrown away by our civilization. Unlike biological resources, minerals cannot be easily managed to produce a sustained yield; the supply is finite. Recycling and conservation will help, but, eventually, the supply will be exhausted.

How Mineral Deposits Are Formed

Metals in mineral form are generally extracted from naturally occurring, unusually high concentrations of Earth materials. When metals are concentrated in such high amounts by geologic processes, **ore deposits** are formed. The discovery of natural ore deposits allowed early peoples to exploit copper, tin, gold, silver, and other metals while slowly developing skills in working with metals.

The origin and distribution of mineral resources is intimately related to the history of the biosphere and to the entire geologic cycle (see Chapter 6). Nearly all aspects and processes of the geologic cycle are involved to some extent in producing local concentrations of useful materials. Earth's outer layer, or crust, is silica-rich, made up mostly of rock-forming minerals containing silica, oxygen, and a few other elements. The elements are not evenly distributed in the crust: Nine elements account for about 99% of the crust by weight (oxygen, 45.2%; silicon, 27.2%; aluminum, 8.0%; iron, 5.8%; calcium, 5.1%; magnesium, 2.8%; sodium, 2.3%; potassium, 1.7%; and titanium, 0.9%). In general, the remaining elements are found in trace concentrations.

The ocean, covering nearly 71% of Earth, is another reservoir for many chemicals other than water. Most elements in the ocean have been weathered from crustal rocks on the land and transported to the oceans by rivers. Others are transported to the ocean by wind or glaciers. Ocean water contains about 3.5% dissolved solids, mostly chlorine (55.1% of the dissolved solids by weight). Each cubic kilometer of ocean water contains about 2.0 metric tons of zinc, 2.0 metric tons of copper, 0.8 metric ton of tin, 0.3 metric ton of silver, and 0.01 metric ton of gold. These concentrations are low compared with those in the crust, where corresponding values (in metric tons/km^3) are zinc, 170,000; copper, 86,000; tin, 5,700; silver, 160; and gold, 5. After rich crustal ore deposits are depleted, we will be more likely to extract metals from lower-grade deposits or even from common rock than from ocean water, unless mineral-extraction technology becomes more efficient.

Why do the minerals we mine occur in deposits—with anomalously high local concentrations? Planetary scientists now believe that all the planets in our solar system were formed by the gravitational attraction of the forming sun, which brought together the matter dispersed around it. As the mass of the proto-Earth increased, the material condensed and was heated by the process. The heat was sufficient to produce a molten liquid core, consisting primarily of iron and other heavy metals, which sank toward the center of the planet. When molten rock material known as *magma* cools, heavier minerals that crystallize (solidify) early may slowly sink toward the bottom of the magma, whereas lighter minerals that crystallize later are left at the top. Deposits of an ore of chromium, called chromite, are thought to be formed in this way. When magma containing small amounts of carbon is deeply buried and subjected to very high pressure during slow cooling (crystallization), diamonds (which are pure carbon) may be produced (Figure 23.3).[8, 9]

Earth's crust formed from generally lighter elements and is a mixture of many different kinds. The elements in the crust are not uniformly distributed because geologic processes (such as volcanic activity, plate tectonics, and sedimentary processes), as well as some biological processes, selectively dissolve, transport, and deposit elements and minerals.

FIGURE 23.3 Diamond mine near Kimberley, South Africa. This is the largest hand-dug excavation in the world.

Sedimentary processes related to the transport of sediments by wind, water, and glaciers often concentrate materials in amounts sufficient for extraction. As sediments are transported, running water and wind help segregate them by size, shape, and density. This sorting is useful to people. The best sand or sand and gravel deposits for construction, for example, are those in which the finer materials have been removed by water or wind. Sand dunes, beach deposits, and deposits in stream channels are good examples. The sand and gravel industry amounts to several billion dollars annually and, in terms of the total volume of materials mined, is one of the largest nonfuel mineral industries in the United States.[5]

Rivers and streams that empty into oceans and lakes carry tremendous quantities of dissolved material from the weathering of rocks. Over geologic time, a shallow marine basin may be isolated by tectonic activity that uplifts its boundaries, or climate variations, such as the ice ages, may produce large inland lakes with no outlets. As these basins and lakes eventually dry up, the dissolved materials drop out of solution and form a wide variety of compounds, minerals, and rocks that have important commercial value.[10]

Biological processes form some mineral deposits, such as phosphates and iron ore deposits. The major iron ore deposits exist in sedimentary rocks that were formed more than 2 billion years ago.[10] Although the processes are not fully understood, it appears that major deposits of iron stopped forming when the atmospheric concentration of oxygen reached its present level.[11]

Organisms, too, form many kinds of minerals, such as the calcium minerals in shells and bones. Some of these minerals cannot be formed inorganically in the biosphere. Thirty-one biologically produced minerals have been identified.[12]

Weathering, the chemical and mechanical decomposition of rock, concentrates some minerals in the soil, such as native gold and oxides of aluminum and iron. (The more soluble elements, such as silica, calcium, and sodium, are selectively removed by soil and biological processes.) If sufficiently concentrated, residual aluminum oxide forms an ore of aluminum known as bauxite. Important nickel and cobalt deposits are also found in soils developed from iron- and magnesium-rich igneous rocks.

23.4 Figuring Out How Much Is Left

Estimating how much is left of our valuable and nonrenewable mineral resources will help us estimate how long they are likely to last at our present rate of use and motivate us to do everything we can to sustain them as long as possible for future generations. We can begin by looking at the classification of minerals as *resources* and *reserves*.

Mineral Resources and Reserves

Mineral **resources** are broadly defined as known concentrations of elements, chemical compounds, minerals, or rocks. Mineral **reserves** are concentrations that at the time of evaluation can be legally and economically extracted as a commodity that can be sold at a profit (Figure 23.4).

The main point here is that *resources are not reserves.* An analogy from a student's personal finances may help clarify this point. A student's reserves are liquid assets, such as money in the bank, whereas the student's resources include the total income the student can expect to earn during his or her lifetime. This distinction is often critical to the student in school because resources that may become available in the future cannot be used to pay this month's bills.[6] For planning purposes, it is important to continually reassess all components of a total resource, considering new technology, the probability of geologic discovery, and shifts in economic and political conditions.[13]

Availability and Use of Our Mineral Resources

Earth's mineral resources can be divided into broad categories according to their use: elements for metal production and technology, building materials, minerals for the chemical industry, and minerals for agriculture. Metallic minerals can be further classified by their abundance. Abundant metals include iron, aluminum, chromium, manganese, titanium, and magnesium. Scarce metals include copper, lead, zinc, tin, gold, silver, platinum, uranium, mercury, and molybdenum.

Some minerals, such as salt (sodium chloride), are necessary for life. Primitive peoples traveled long distances

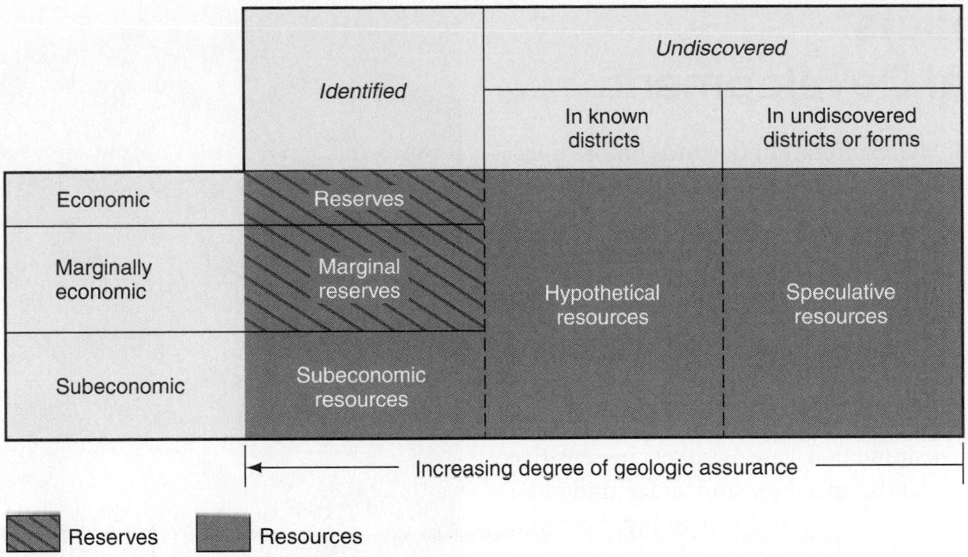

FIGURE 23.4 Classification of mineral resources used by the U.S. Geological Survey and the U.S. Bureau of Mines. (*Source: Principles of a Resource Preserve Classification for Minerals*, U.S. Geological Survey Circular 831, 1980.)

to obtain salt when it was not locally available. Other minerals are desired or considered necessary to maintain a particular level of technology.

When we think about minerals, we usually think of metals; but with the exception of iron, the predominant minerals are not metallic. Consider the annual world consumption of a few selected elements. Sodium and iron are used at a rate of approximately 100–1,000 million metric tons per year; and nitrogen, sulfur, potassium, and calcium at a rate of approximately 10–100 million metric tons per year, primarily as soil conditioners or fertilizers. Elements such as zinc, copper, aluminum, and lead have annual world consumption rates of about 3–10 million metric tons, and gold and silver are consumed at annual rates of 10,000 metric tons or less. Of the metallic minerals, iron makes up 95% of all the metals consumed; and nickel, chromium, cobalt, and manganese are used mainly in alloys of iron (as in stainless steel).

The basic issue associated with mineral resources is not actual exhaustion or extinction but the cost of maintaining an adequate stock by mining and recycling. At some point, the costs of mining exceed the worth of material. When the availability of a particular mineral becomes limited, there are four possible solutions:

1. Find more sources.

2. Recycle and reuse what has already been obtained.

3. Reduce consumption.

4. Find a substitute.

Which choice or combination of choices is made depends on social, economic, and environmental factors.

U.S. Supply of Mineral Resources

Domestic supplies of many mineral resources in the United States are insufficient for current use and must be supplemented by imports from other nations. For example, the United States imports many of the minerals needed for its complex military and industrial system, called strategic minerals (such as bauxite, manganese, graphite, cobalt, strontium, and asbestos). Of particular concern is the possibility that the supply of a much-desired or much-needed mineral will be interrupted by political, economic, or military instability in the supplying nation.

That the United States—along with many other countries—depends on a steady supply of imports to meet its domestic demand for them does not necessarily mean that sufficient kinds and amounts can't be mined domestically. Rather, it suggests economic, political, or environmental reasons that make it easier, more practical, or more desirable to import the material. This has resulted in political alliances that otherwise would be unlikely. Industrial countries often need minerals from countries whose policies they don't necessarily agree with; as a result they make political concessions, on human rights and other issues, that they would not otherwise make.[3]

Moreover, the fact remains that mineral resources are limited, and this raises important questions. How long will a particular resource last? How much short-term or long-term environmental deterioration are we willing to accept to ensure that resources are developed in a particular area? How can we make the best use of available resources?

23.5 Impacts of Mineral Development

The impact of mineral exploitation depends on ore quality, mining procedures, local hydrologic conditions, climate, rock types, size of operation, topography, and many more interrelated factors. In addition, our use of mineral resources has a significant social impact.

Environmental Impacts

Exploration for mineral deposits generally has a minimal impact on the environment if care is taken in sensitive areas, such as arid lands, marshes, and areas underlain by permafrost. Mineral mining and processing, however, generally have a considerable impact on land, water, air, and living things. Furthermore, as it becomes necessary to use ores of lower and lower grades, the environmental effects tend to worsen. One example is the asbestos fibers in the drinking water of Duluth, Minnesota, from the disposal of waste from mining low-grade iron ore.

A major practical issue is whether open-pit or underground mines should be developed in an area. As you saw in our earlier discussion of coal mining in Chapter 15, there are important differences between the two kinds of mining.[2] The trend in recent years has been away from subsurface mining and toward large, open-pit mines, such as the Bingham Canyon copper mine in Utah (Figure 23.5). The Bingham Canyon mine is one of the world's largest man-made excavations, covering nearly 8 km^2 (3 mi^2) to a maximum depth of nearly 800 m (2,600 ft).

Surface mines and quarries today cover less than 0.5% of the total area of the United States, but even though their impacts are local, numerous local occurrences will eventually constitute a larger problem. Environmental degradation tends to extend beyond the immediate vicinity of a mine. Large mining operations remove material in some areas and dump waste in others, changing topography. At the very least, severe aesthetic degradation is the result. In addition, dust may affect the air quality, even though care is taken to reduce it by sprinkling water on roads and on other sites that generate dust.

A potential problem with mineral resource development is the possible release of harmful trace elements into the environment. Water resources are particularly vulnerable even if drainage is controlled and sediment pollution is reduced (see Chapter 15 for more about this, including a discussion of acid mine drainage). The white streaks in Figure 23.6 are mineral deposits apparently leached from tailings from a zinc mine in Colorado. Similar-looking deposits may cover rocks in rivers for many kilometers downstream from some mining areas.

FIGURE 23.5 Aerial photograph of Bingham Canyon Copper Pit, Utah. It is one of the largest artificial excavations in the world.

FIGURE 23.6 Tailings from a lead, zinc, and silver mine in Colorado. White streaks on the slope are mineral deposits apparently leached from the tailings.

Mining-related physical changes in the land, soil, water, and air indirectly affect the biological environment. Plants and animals killed by mining activity or by contact with toxic soil or water are some of the direct impacts. Indirect impacts include changes in nutrient cycling, total biomass, species diversity, and ecosystem stability. Periodic or accidental discharge of low-grade pollutants through failure of barriers, ponds, or water diversions, or through the breaching of barriers during floods, earthquakes, or volcanic eruptions, also may damage local ecological systems to some extent.

Social Impacts

The social impacts of large-scale mining result from the rapid influx of workers into areas unprepared for growth. This places stress on local services, such as water supplies, sewage and solid-waste disposal systems, and also on schools, housing, and nearby recreation and wilderness areas. Land use shifts from open range, forest, and agriculture to urban patterns. Construction and urbanization affect local streams through sediment pollution, reduced water quality, and increased runoff. Air quality suffers as a result of more vehicles, construction dust, and power generation.

Perversely, closing down mines also has adverse social impacts. Nearby towns that have come to depend on the income of employed miners can come to resemble the well-known "ghost towns" of the old American West. The price of coal and other minerals also directly affects the livelihood of many small towns. This is especially evident in the Appalachian Mountain region of the United States, where coal mines have closed partly because of lower prices for coal and partly because of rising mining costs. One of the reasons mining costs are rising is the increased level of environmental regulation of the mining industry. Of course, regulations have also helped make mining safer and have facilitated land reclamation. Some miners, however, believe the regulations are not flexible enough, and there is some truth to their arguments. For example, some mined areas might be reclaimed for use as farmland now that the original hills have been leveled. Regulations, however, may require the restoration of the land to its original hilly state, even though hills make inferior farmland.

Minimizing the Environmental Impact of Mineral Development

Minimizing the environmental impacts of mineral development requires consideration of the entire cycle of mineral resources shown in Figure 23.7. This diagram reveals that waste is produced by many components of the cycle. In fact, the major environmental impacts of mineral use are related to waste products. Waste produces pollution that may be toxic to people, may harm natural ecosystems and the biosphere, and may be aesthetically displeasing. Waste may attack and degrade air, water, soil, and living things. Waste also depletes nonrenewable mineral resources and, when simply disposed of, provides no offsetting benefits for human society.

Environmental regulations at the federal, state, and local levels address pollution of air and water by all aspects of the mineral cycle, and may also address reclamation of land used for mining minerals. Today, in the United States, approximately 50% of the land used by the mining industry has been reclaimed.

Minimizing the environmental effects of mining takes several interrelated paths:[3]

- *Reclaiming* areas disturbed by mining (see A Closer Look 23.1).
- *Stabilizing soils* that contain metals to minimize their release into the environment. Often this requires placing contaminated soils in a waste facility.
- *Controlling air emissions* of metals and other materials from mining areas.
- *Treating contaminated water before it can leave a mining site or treating contaminated water that has left a mining site.*
- *Treating waste onsite and offsite.* Minimizing onsite and offsite problems by controlling sediment, water, and air pollution through good engineering and conservation practices is an important goal. Of particular interest is the development of biotechnological processes such as biooxidation, bioleaching, and biosorption, the bonding of waste to microbes, as well as genetic engineering of microbes. These practices have enormous potential for both extracting metals and minimizing environmental degradation. At several sites, for example, constructed wetlands use acid-tolerant plants to remove metals from mine wastewaters and neutralize acids by biological activity.[14] The Homestake Gold Mine in South Dakota uses biooxidation to convert contaminated water from the mining operation into substances that are environmentally safe; the process uses bacteria that have a natural ability to oxidize cyanide to harmless nitrates.[15]
- *Practicing the three R's of waste management.* That is, **R**educe the amount of waste produced, **R**euse waste as much as possible, and maximize **R**ecycling opportunities. Wastes from some parts of the mineral cycle, for example, may themselves be considered ores because they contain materials that might be recycled to provide energy or other products.[16–18]

We will look at the three R's in greater detail in Section 23.7, Integrated Waste Management.

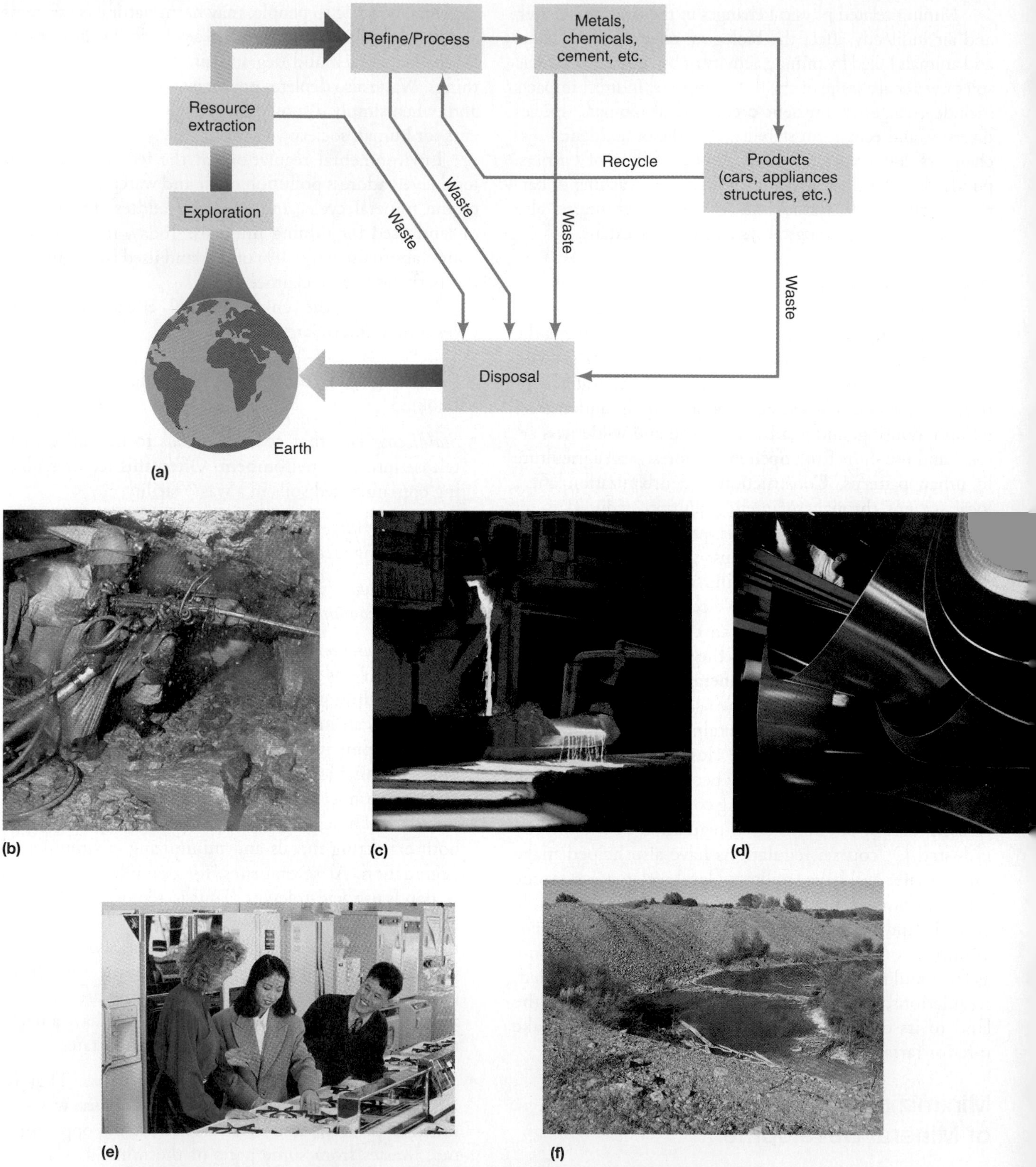

FIGURE 23.7 **(a)** Simplified flowchart of the resource cycle: **(b)** mining gold in South Africa; **(c)** copper smelter, Montana; **(d)** sheets of copper for industrial use; **(e)** appliances made in part from metals; and **(f)** disposal of mining waste from a Montana gold mine into a tailings pond.

A CLOSER LOOK 23.1

Golden, Colorado: Open-Pit Mine Becomes a Golf Course

The city of Golden, Colorado, has an award-winning golf course on land that, for about 100 years, was an open-pit mine (quarry) excavated in limestone rock (Figure 23.8). The mine produced clay for making bricks from clay layers between limestone beds. Over the life of the mine, the clay was used as a building material at many sites, including prominent buildings in the Denver area, such as the Colorado Governor's Mansion. The mine site included unsightly pits with vertical limestone walls as well as a landfill for waste disposal. However, it had spectacular views of the Rocky Mountain foothills. Today the limestone cliffs with their exposed plant and dinosaur fossils have been transformed into golf greens, fairways, and a driving range. The name Fossil Trace Golf Club reflects its geologic heritage. The course includes trails to fossil locations and also has channels, constructed wetlands, and three lakes that store floodwater runoff, helping to protect Golden from flash floods. The reclamation project started with a grassroots movement by the people of Golden to have a public golf course. The reclamation is now a moneymaker for the city and demonstrates that mining sites can not only be reclaimed, but also be transformed into valuable property.

FIGURE 23.8 This award-winning golf course in Golden, Colorado, was for a century an open-pit mine (quarry) for clay to produce bricks.

23.6 Materials Management and Our Waste

History of Waste Disposal

During the first century of the Industrial Revolution, the volume of waste produced in the United States was relatively small and could be managed using the concept of "dilute and disperse." Factories were located near rivers because the water provided a number of benefits, including easy transport of materials by boat, enough water for processing and cooling, and easy disposal of waste into the river. With few factories and a sparse population, dilute and disperse was sufficient to remove the waste from the immediate environment.[19]

As industrial and urban areas expanded, the concept of dilute and disperse became inadequate, and a new concept, "concentrate and contain," came into use. It has become apparent, however, that containment was, and is, not always achieved. Containers, whether simple trenches excavated in the ground or metal drums and tanks, may leak or break and allow waste to escape. Health hazards resulting from past waste-disposal practices have led to the present situation, in which many people have little confidence in government or industry to preserve and protect public health.[20]

In the United States and many other parts of the world, people are facing a serious solid-waste disposal problem. Basically, we are producing a great deal of waste and don't have enough acceptable space for disposing of it. It has been estimated that within the next few years approximately half the cities in the United States may run out of landfill space. Philadelphia, for example, is essentially out of landfill space now and is bargaining with other states on a monthly or yearly basis to dispose of its trash. The Los Angeles area has landfill space for only about ten more years.

To say we are actually running out of space for landfills isn't altogether accurate—land used for landfills is minute compared to the land area of the United States. Rather, existing sites are being filled, and it is difficult to site new landfills. After all, no one wants to live near a waste-disposal site, be it a sanitary landfill for municipal waste, an incinerator that burns urban waste, or a hazardous-waste disposal operation for chemical materials.

This attitude is widely known as NIMBY ("not in my backyard").

The environmentally correct concept with respect to waste management is to consider wastes as resources out of place. Although we may not soon be able to reuse and recycle all waste, it seems apparent that the increasing cost of raw materials, energy, transportation, and land will make it financially feasible to reuse and recycle more resources and products. Moving toward this objective is moving toward an environmental view that there is no such thing as waste. Under this concept, waste would not exist because it would not be produced—or, if produced, would be a resource to be used again. This is referred to as the "zero waste" movement.

Zero waste is the essence of what is known as **industrial ecology**, the study of relationships among industrial systems and their links to natural systems. Under the principles of industrial ecology, our industrial society would function much as a natural ecosystem functions. Waste from one part of the system would be a resource for another part.[21]

Until recently, zero waste production was considered unreasonable in the waste-management arena. However, it is catching on. The city of Canberra, Australia, may be the first community to propose a zero waste plan. Thousands of kilometers away, in the Netherlands, a national waste-reduction goal of 70 to 90% has been set. How this goal is to be met is not entirely clear, but a large part of the planning involves taxing waste in all its various forms, from smokestack emissions to solids delivered to landfills. Already, in the Netherlands, pollution taxes have nearly eliminated discharges of heavy metals into waterways. At the household level, the government is considering programs—known as "pay as you throw"—that would charge people by the volume of waste they produce. Taxing waste, including household waste, motivates people to produce less of it.[22]

Of particular importance to waste management is the growing awareness that many of our waste-management programs involve moving waste from one site to another, not really managing it. For example, waste from urban areas may be placed in landfills; but eventually these landfills may cause new problems by producing methane gas or noxious liquids that leak from the site and contaminate the surrounding areas. Managed properly, however, methane produced from landfills is a resource that can be burned as a fuel (an example of industrial ecology).

In sum, previous notions of waste disposal are no longer acceptable, and we are rethinking how we deal with materials, with the objective of eliminating the concept of waste entirely. In this way, we can reduce the consumption of minerals and other virgin materials, which depletes our environment, and live within our environment more sustainably.[21]

23.7 Integrated Waste Management

The dominant concept today in managing waste is known as **integrated waste management (IWM)**, which is best defined as a set of management alternatives that includes *reuse, source reduction, recycling, composting, landfill,* and *incineration.*[20]

Reduce, Reuse, Recycle

The ultimate objective of the three R's of IWM is to reduce the amount of urban and other waste that must be disposed of in landfills, incinerators, and other waste-management facilities. Study of the *waste stream* (the waste produced) in areas that use IWM technology suggests that the amount (by weight) of urban refuse disposed of in landfills or incinerated can be reduced by at least 50% and perhaps as much as 70%. A 50% reduction by weight could be achieved by (1) source reduction, such as packaging better designed to reduce waste (10% reduction); (2) large-scale composting programs (10% reduction); and (3) recycling programs (30% reduction).[20]

As this list indicates, recycling is a major player in reducing the urban waste stream. Metals such as iron, aluminum, copper, and lead have been recycled for many years and are still being recycled today. The metal from almost all of the millions of automobiles discarded annually in the United States is recycled.[16, 17] The total value of recycled metals is about $50 billion. Iron and steel account for approximately 90% by weight and 40% by total value of recycled metals. Iron and steel are recycled in such large volumes for two reasons. First, the market for iron and steel is huge, and as a result there is a large scrap-collection and scrap-processing industry. Second, an enormous economic and environmental burden would result from failure to recycle because over 50 million tons of scrap iron and steel would have to be disposed of annually.[17, 18]

Today in the United States we recycle over 30% of our total municipal solid waste, up 10% from 25 years ago. This amounts to 99% of automobile batteries, 63% of steel cans, 71% office type papers, 48% of aluminum cans, 35% of tires, 28% of glass containers, and about 30% of various plastic containers (Figure 23.9).[23] This is encouraging news. Can recycling actually reduce the waste stream by 50%? Recent work suggests that the 50% goal is reasonable. In fact, it has been reached in some parts of the United States, and the potential upper limit for recycling is considerably higher. It is estimated that as much as 80 to 90% of the U.S. waste stream might be recovered through what is known as "intensive recycling."[24] A pilot study involving 100 families in East Hampton, New York,

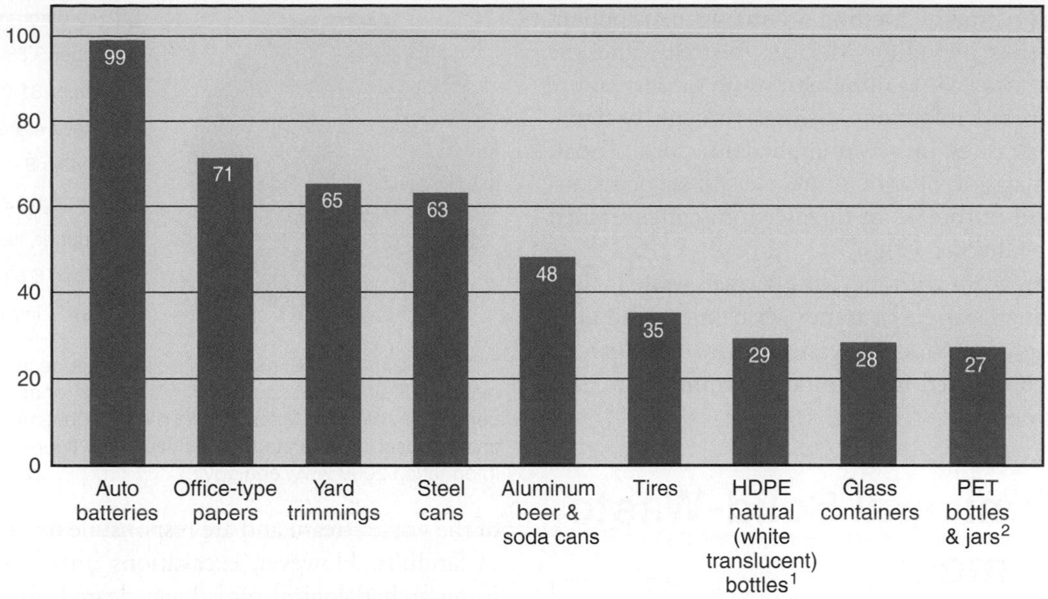

[1] HDPE is high-density polyethylene produced from ethylene to make blow-molded bottles.

[2] PET is a type of plastic labled with a recycling number code (in a triangle) on the bottom of the bottle.

FIGURE 23.9 Recycling rates of selected materials from municipal solid waste in 2008 for the United States. (*Source:* Municipal solid waste generation, recycling, and disposal in the United States: facts and figures 2008. Basic Information 2009. www.epa.gov.)

achieved a level of 84%. More realistic for many communities is partial recycling, which targets specific materials, such as glass, aluminum cans, plastic, organic material, and newsprint. Partial recycling can provide a significant reduction, and in many places it is approaching or even exceeding 50%.[25, 26]

Recycling is simplified with **single-stream recycling,** in which paper, plastic, glass, and metals are not separated before collection; the waste is commingled in one container and separated later at recycling centers. This is more convenient for homeowners, reduces the cost of collection, and increases the rate of recycling. Thus, single-stream recycling is growing rapidly.

Public Support for Recycling

An encouraging sign of public support for the environment is the increased willingness of industry and business to support recycling on a variety of scales. For example, fast-food restaurants are using less packaging and providing onsite bins for recycling paper and plastic. Groceries and supermarkets are encouraging the recycling of plastic and paper bags by providing bins for their collection, and some offer inexpensive reusable canvas shopping bags instead of disposables. Companies are redesigning products so that they can be more easily disassembled after use and the various parts recycled. As this idea catches on, small appliances, such as electric frying pans and toasters, may be recycled rather than ending up in landfills. The automobile industry is also responding by designing automobiles with coded parts so that they can be more easily disassembled (by professional recyclers) and recycled, rather than left to become rusting eyesores in junkyards.

On the consumer front, people are now more likely to purchase products that can be recycled or that come in containers that are more easily recycled or composted. Many consumers have purchased small home appliances that crush bottles and aluminum cans, reducing their volume and facilitating recycling. The entire arena is rapidly changing, and innovations and opportunities will undoubtedly continue.

As with many other environmental solutions, implementing the IWM concept successfully can be a complex undertaking. In some communities where recycling has been successful, it has resulted in glutted markets for recycled products, which has sometimes required temporarily stockpiling or suspending the recycling of some items. It is apparent that if recycling is to be successful, markets and processing facilities will also have to be developed to ensure that recycling is a sound financial venture as well as an important part of IWM.

Recycling of Human Waste

The use of human waste, or "night soil," on croplands is an ancient practice. In Asia, recycling of human waste has a long history. Chinese agriculture was sustained for thousands of years through collection of human waste, which was spread over agricultural fields. The practice grew, and by the early 20th century the land application of sewage was a primary disposal method in many metropolitan areas in countries including Mexico, Australia, and the United States.[27] Early uses of human waste for agriculture occasionally spread infectious diseases through bacteria, viruses, and parasites in waste applied to crops. Today, with the globalization of agriculture, we still see occasional warnings and outbreaks of disease from contaminated vegetables (see Chapter 19).

A major problem with recycling human waste is that, along with human waste, thousands of chemicals and metals flow through our modern waste stream. Even garden waste that is composted may contain harmful chemicals, such as pesticides.[27]

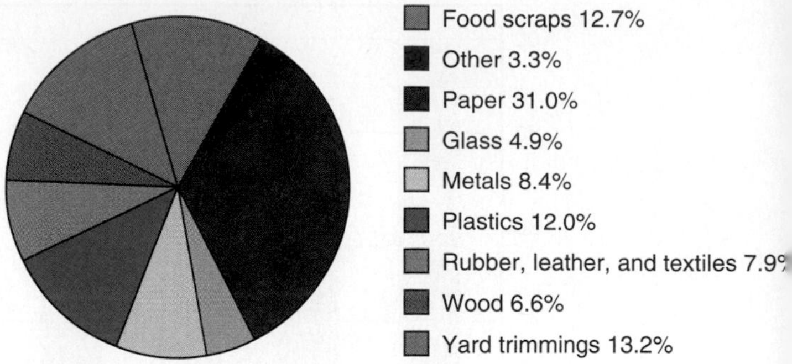

FIGURE 23.10 U.S. municipal solid-waste generation before recycling in 2008 was about 250 million tons, or about 4.6 lbs (2 kg) per person. (*Source*: Municipal solid waste generation, recycling, and disposal in the United States: facts and figures 2008_Basic Information 2009.www.epa.gov.)

23.8 Municipal Solid-Waste Management

Municipal solid-waste management continues to be a problem in the United States and other parts of the world. In many areas, particularly in developing countries, waste-management practices are inadequate. These practices, which include poorly controlled open dumps and illegal roadside dumping, can spoil scenic resources, pollute soil and water, and pose health hazards.

Illegal dumping is a social problem as much as a physical one because many people are simply disposing of waste as inexpensively and as quickly as possible, perhaps not seeing their garbage as an environmental problem. If nothing else, this is a tremendous waste of resources, since much of what is dumped could be recycled or reused. In areas where illegal dumping has been reduced, the keys have been awareness, education, and alternatives. Education programs teach people about the environmental problems of unsafe, unsanitary dumping of waste, and funds are provided for cleanup and for inexpensive collection and recycling of trash at sites of origin.

We look next at the composition of solid waste in the United States and then go on to describe specific disposal methods: onsite disposal, composting, incineration, open dumps, and sanitary landfills.

Composition of Solid Waste

The average content of unrecycled solid waste likely to end up at a disposal site in the United States is shown in Figure 23.10. It is no surprise that paper is by far the most abundant component. However, considerable variation can be expected, based on factors such as land use, economic base, industrial activity, climate, and time of year.

People have many misconceptions about our waste stream.[28] With all the negative publicity about fast-food packaging, polystyrene foam, and disposable diapers, many people assume that these make up a large percentage of the waste stream and are responsible for the rapid filling of landfills. However, excavations into modern landfills using archaeological tools have cleared up some misconceptions. We now know that fast-food packaging accounts for only about 0.25% of the average landfill; disposable diapers, approximately 0.8%; and polystyrene products about 0.9%.[29] Paper is a major constituent in landfills, perhaps as much as 50% by volume and 40% by weight. The largest single item is newsprint, which accounts for as much as 18% by volume.[29] Newsprint is one of the major items targeted for recycling because big environmental dividends can be expected. However (and this is a value judgment), the need to deal with the major waste products doesn't mean that we need not cut down on our use of disposable diapers, polystyrene, and other paper products. In addition to creating a need for disposal, these products are made from resources that might be better managed.

Onsite Disposal

A common onsite disposal method in urban areas is the garbage-disposal device installed in the wastewater pipe under the kitchen sink to grind garbage and flush it into the sewer system. This effectively reduces the amount of handling and quickly removes food waste. What's left of it is transferred to sewage-treatment plants, where solids remaining as sewage sludge still must be disposed of.[30, 31]

Composting

Composting is a biochemical process in which organic materials, such as lawn clippings and kitchen scraps, decompose to a rich, soil-like material. The process involves rapid partial decomposition of moist solid organic waste by aerobic organisms. Although simple backyard compost piles may come to mind, large-scale composting as a waste-management option is generally carried out in the controlled environment of mechanical digesters. This technique is popular in Europe and Asia, where intense farming creates a demand for compost. However, a major drawback of composting is the necessity of separating organic material from other waste. Therefore, it is probably economically advantageous only where organic material is collected separately from other waste. Another negative is that composting plant debris previously treated with herbicides may produce a compost toxic to some plants. Nevertheless, composting is an important component of IWM, and its contribution continues to grow.[30, 31]

Incineration

Incineration burns combustible waste at temperatures high enough (900°–1,000°C, or 1,650°–1,830°F) to consume all combustible material, leaving only ash and noncombustibles to dispose of in a landfill. Under ideal conditions, incineration may reduce the volume of waste by 75–95%.[31] In practice, however, the actual decrease in volume is closer to 50% because of maintenance problems as well as waste-supply problems. Besides reducing a large volume of combustible waste to a much smaller volume of ash, incineration has another advantage: It can be used to supplement other fuels and generate electrical power.

Incineration of urban waste is not necessarily a clean process; it may produce air pollution and toxic ash. In the United States, for example, incineration is apparently a significant source of environmental dioxin, a carcinogenic toxin (see Chapter 10).[32] Smokestacks from incinerators also may emit oxides of nitrogen and sulfur, which lead to acid rain; heavy metals, such as lead, cadmium, and mercury; and carbon dioxide, which is related to global warming.

In modern incineration facilities, smokestacks fitted with special devices trap pollutants, but the process of pollutant abatement is expensive. The plants themselves are expensive, and government subsidization may be needed to aid in their establishment. Evaluation of the urban waste stream suggests that an investment of $8 billion could build enough incinerators in the United States to burn approximately 25% of the solid waste that is generated. However, a similar investment in source reduction, recycling, and composting could divert as much as 75% of the nation's urban waste stream away from landfills.[24]

The economic viability of incinerators depends on revenue from the sale of the energy produced by burning the waste. As recycling and composting increase, they will compete with incineration for their portion of the waste stream, and sufficient waste (fuel) to generate a profit from incineration may not be available. The main conclusion that can be drawn based on IWM principles is that a combination of reusing, recycling, and composting could reduce the volume of waste requiring disposal at a landfill by at least as much as incineration.[24]

Open Dumps (Poorly Controlled Landfills)

In the past, solid waste was often disposed of in open dumps (now called landfills), where refuse was piled up and left uncovered. Thousands of open dumps have been closed in recent years, and new open dumps are banned in the United States and many other countries. Nevertheless, many are still being used worldwide (Figure 23.11).[31]

Sanitary Landfills

A **sanitary landfill** (also called a municipal solid-waste landfill) is designed to concentrate and contain refuse without creating a nuisance or hazard to public health or safety. The idea is to confine the waste to the smallest practical area, reduce it to the smallest practical volume, and cover it with a layer of compacted soil at the end of each day of operation, or more frequently if necessary. Covering the waste is what makes the landfill sanitary. The compacted layer restricts (but does not eliminate) continued access to the waste by insects, rodents, and other animals, such as seagulls. It also isolates the refuse, minimizing the amount of surface water seeping into it and the amount of gas escaping from it.[33]

Leachate

The most significant hazard from a sanitary landfill is pollution of groundwater or surface water. If waste buried in a landfill comes into contact with water percolating down from the surface or with groundwater moving laterally through the refuse, **leachate**—noxious, mineralized liquid capable of transporting bacterial pollutants—is

FIGURE 23.11 Urban garbage dump in Rio de Janeiro, Brazil. At this site, people are going through the waste and recycling materials that can be reused or resold. This activity is all too common in dumps for large cities in the developing world. In some cases several thousand scavengers, including children, sift through tons of burning garbage to collect cans and bottles.

produced.[34] For example, two landfills dating from the 1930s and 1940s on Long Island, New York, have produced subsurface leachate trails (plumes) several hundred meters wide that have migrated kilometers from the disposal site. The nature and strength of the leachate produced at a disposal site depend on the composition of the waste, the amount of water that infiltrates or moves through the waste, and the length of time that infiltrated water is in contact with the refuse.[31]

Site Selection

The siting of a sanitary landfill is very important and must take into consideration a number of factors, including topography, location of the groundwater table, amount of precipitation, type of soil and rock, and location of the disposal zone in the surface water and groundwater flow system. A favorable combination of climatic, hydrologic, and geologic conditions helps to ensure reasonable safety in containing the waste and its leachate.[35] The best sites are in arid regions, where disposal conditions are relatively safe because little leachate is produced. In a humid environment, some leachate is always produced; therefore, an acceptable level of leachate production must be established to determine the most favorable sites in such environments. What is acceptable varies with local water use, regulations, and the ability of the natural hydrologic system to disperse, dilute, and otherwise degrade the leachate to harmless levels.

Elements of the most desirable site in a humid climate with moderate to abundant precipitation are shown in Figure 23.12. The waste is buried above the water table in relatively impermeable clay and silt that water cannot easily move through. Any leachate therefore remains in the vicinity of the site and degrades by natural filtering action and chemical reactions between clay and leachate.[36, 37]

Siting waste-disposal facilities also involves important social considerations. Often, planners choose sites where they expect minimal local resistance or where they perceive land to have little value. Waste-disposal facilities are frequently located in areas where residents tend to have low socioeconomic status or belong to a particular racial or ethnic group. The study of social issues in siting waste facilities, chemical plants, and other such facilities is an emerging field known as **environmental justice**.[38, 39]

Monitoring Pollution in Sanitary Landfills

Once a site is chosen for a sanitary landfill and before filling starts, monitoring the movement of groundwater should begin. Monitoring involves periodically taking samples of water and gas from specially designed monitoring wells. Monitoring the movement of leachate and gases should continue as long as there is any possibility of pollution and is particularly important after the site is completely filled and permanently covered. Continued monitoring is necessary because a certain amount of settling always occurs after a landfill is completed; and if small depressions form, surface water may collect, infiltrate, and produce leachate. Monitoring and proper maintenance of an abandoned landfill reduce its pollution potential.[33]

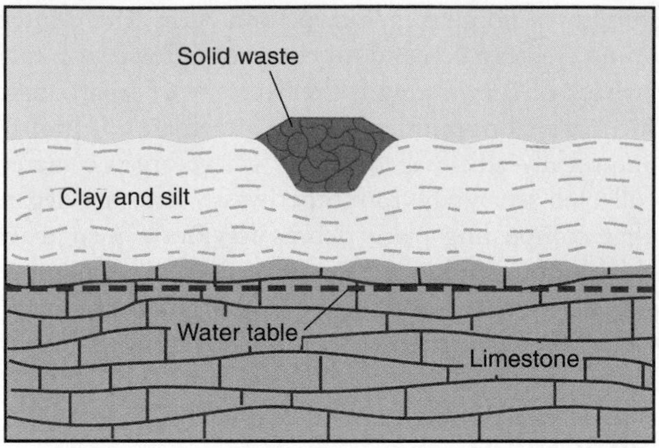

FIGURE 23.12 The most desirable landfill site in a humid environment. Waste is buried above the water table in a relatively impermeable environment. (*Source*: W.J. Schneider, *Hydraulic Implications of Solid-Waste Disposal*, U.S. Geological Survey Circular 601F, 1970.)

How Pollutants Can Enter the Environment from Sanitary Landfills

Pollutants from a solid-waste disposal site can enter the environment through as many as eight paths (Figure 23.13):[40]

1. Methane, ammonia, hydrogen sulfide, and nitrogen gases can be produced from compounds in the waste and the soil and can enter the atmosphere.
2. Heavy metals, such as lead, chromium, and iron, can be retained in the soil.
3. Soluble materials, such as chloride, nitrate, and sulfate, can readily pass through the waste and soil to the groundwater system.
4. Overland runoff can pick up leachate and transport it into streams and rivers.
5. Some plants (including crops) growing in the disposal area can selectively take up heavy metals and other toxic materials. These materials are then passed up the food chain as people and animals eat the plants.
6. If plant residue from crops left in fields contains toxic substances, these substances return to the soil.
7. Streams and rivers may become contaminated by waste from groundwater seeping into the channel (3) or by surface runoff (4).
8. Wind can transport toxic materials to other areas.

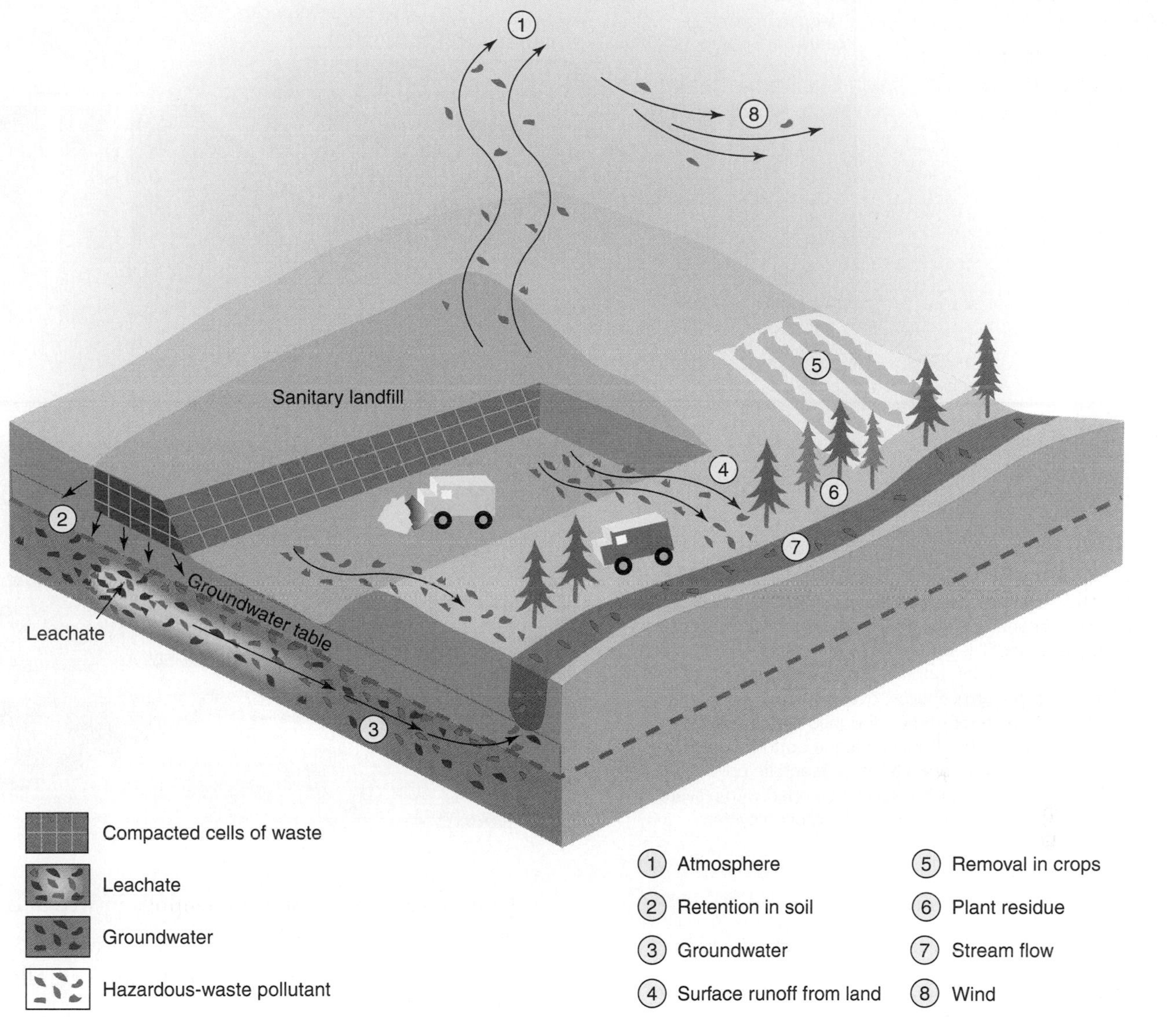

FIGURE 23.13 Idealized diagram showing eight paths that pollutants from a sanitary landfill site may follow to enter the environment.

Modern sanitary landfills are engineered to include multiple barriers: clay and plastic liners to limit the movement of leachate; surface and subsurface drainage to collect leachate; systems to collect methane gas from decomposing waste; and groundwater monitoring to detect leaks of leachate below and adjacent to the landfill. A thorough monitoring program considers all eight possible paths by which pollutants enter the environment. In practice, however, monitoring seldom includes all pathways. It is particularly important to monitor the zone above the water table to identify potential pollution before it reaches and contaminates groundwater, where correction would be very expensive. Figure 23.14 shows (a) an idealized diagram of a landfill that uses the multiple-barrier approach and (b) a photograph of a landfill site under construction.

Federal Legislation for Sanitary Landfills

New landfills that opened in the United States after 1993 must comply with stricter requirements under the Resource Conservation and Recovery Act of 1980. The legislation, as its title states, is intended to strengthen and standardize the design, operation, and monitoring of sanitary landfills. Landfills that cannot comply with regulations face closure. However, states may choose between two options: (1) comply with federal standards or (2) seek EPA approval of solid-waste management plans.

The federal standards include the following:

- Landfills may not be sited on floodplains, wetlands, earthquake zones, unstable land, or near airports (birds drawn to landfill sites are a hazard to aircraft).

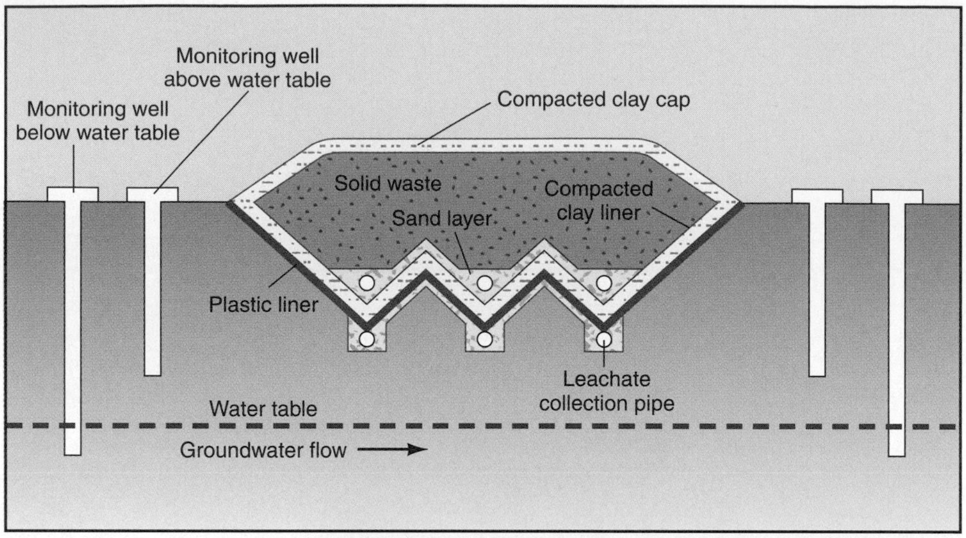

FIGURE 23.14 **(a)** Idealized diagram of a solid-waste facility (sanitary landfill) illustrating multiple-barrier design, monitoring system, and leachate collection system. **(b)** Rock Creek landfill under construction in Calaveras County, California. This municipal solid-waste landfill is underlain by a compacted clay liner, exposed in the center left portion of the photograph. The darker slopes, covered with gravel piles, overlie the compacted clay layer. These form a vapor barrier designed to keep moisture in the clay so it won't crack. Trenches at the bottom of the landfill are lined with plastic and are part of the leachate collection system for the landfill. The landfill is also equipped with a system to monitor the water below the leachate collection system.

- Landfills must have liners.
- Landfills must have a leachate collection system.
- Landfill operators must monitor groundwater for many specified toxic chemicals.
- Landfill operators must meet financial assurance criteria to ensure that monitoring continues for 30 years after the landfill is closed.

EPA approval of a state's landfill program allows greater flexibility:

- Groundwater monitoring may be suspended if the landfill operator can demonstrate that hazardous constituents are not migrating from the landfill.
- Alternative types of daily cover over the waste may be used.
- Alternative groundwater-protection standards are allowed.
- Alternative schedules for documentation of groundwater monitoring are allowed.
- Under certain circumstances, landfills in wetlands and fault zones are allowed.
- Alternative financial assurance mechanisms are allowed.

Given the added flexibility, it appears advantageous for states to develop EPA-approved waste-management plans.

Reducing the Waste that Ends Up in a Landfill

Most of the municipal solid waste we generate is from our homes, and over 50% of it could be diverted from the landfill by the 3 R's of waste management: reduce, reuse, and recycle. Diversion may eventually be increased to as much as 85% through improved waste management. In other words, the life of the landfill can be extended by keeping more waste out of the landfill through conservation and recycling, or by turning waste, even waste that is presently buried, into a source of clean energy. The latter involves first removing materials that can be recycled, then linking noncombustion thermal or biochemical processes with the

Table 23.2 ACTIONS YOU CAN TAKE TO REDUCE THE WASTE YOU GENERATE

Keep track of the waste you personally generate: Know how much waste you produce. This will make you conscious of how to reduce it.

Recycle as much as is possible and practical: Take your cans, glass, and paper to a recycling center or use curbside pickup. Take your hazardous materials such as batteries, cell phones, computers, paint, used oil, and solvents to a hazardous waste collection site.

Reduce packaging: Whenever possible buy your food items in bulk or concentrated form.

Use durable products: Choose automobiles, light bulbs, furniture, sports equipment, and tools that will last a longer time.

Reuse products: Some things may be used several times. For example, you can reuse boxes and shipping "bubble wrap" to ship packages.

Purchase products made from recycled material: Many bottles, cans, boxes, containers, cartons, carpets, clothing, floor tiles, and other products are made from recycled material. Select these whenever you can.

Purchase products designed for ease in recycling: Products as large as automobiles along with many other items are being designed with recycling in mind. Apply pressure to manufacturers to produce items that can be easily recycled.

Source: Modified from U.S. Environmental Protection Agency. Accessed April 21, 2006 at www.epa.gov.

remaining solid waste to produce electricity and alternative fuels (for example, biodiesel). The less waste in the landfill, the less potential for pollution of ground and surface water, along with the important fringe benefit of green energy.

The average waste per person in the United States increased from about 1 kg (2.2 lb) per day in 1960 to 2 kg (4.5 lb) per day in 2008. This is an annual growth rate of about 1.5% per year and is not sustainable because the doubling time for waste production is only a few decades. The 236 million tons we produced in 2003 would be close to 500 million tons by 2050, and we are already having big waste-management problems today. Table 23.2 lists some of the many ways you could reduce the waste you generate. What other ways can you think of?

23.9 Hazardous Waste

So far in this chapter we have discussed integrated waste management and materials management for the everyday waste stream from homes and businesses. We now consider the important topic of hazardous waste.

Creation of new chemical compounds has proliferated in recent years. In the United States, approximately 1,000 new chemicals are marketed each year, and about 70,000 chemicals are currently on the market. Although many have been beneficial to people, approximately 35,000 chemicals used in the United States are classified as definitely or potentially hazardous to people or ecosystems if they are released into the environment as waste—and unfortunately, a lot of it is.

The United States currently produces about 700 million metric tons of hazardous chemical waste per year, referred to more commonly as **hazardous waste**. About 70% of it is generated east of the Mississippi River, and about half of the total by weight is generated by chemical-products industries. The electronics industry (see A Closer Look 23.2 for discussion of e-waste) and petroleum and coal products industries each contribute about 10%.[41-43] Hazardous waste may also enter the environment when buildings are destroyed by events such as fires and hurricanes, releasing paints, solvents, pesticides, and other chemicals that were stored in them, or when debris from damaged buildings is later burned or buried. As a result, collection of such chemicals after natural disasters is an important goal in managing hazardous materials.

In the mid-20th century, as much as half the total volume of hazardous waste produced in the United States was indiscriminately dumped.[42] Some was illegally dumped on public or private lands, a practice called midnight dumping. Buried drums of illegally dumped hazardous waste have been discovered at hundreds of sites by contractors constructing buildings and roads. Cleanup has been costly and has delayed projects.[41]

The case of Love Canal is a well-known hazardous-waste horror story. In 1976, in a residential area near Niagara Falls, New York, trees and gardens began to die. Rubber on tennis shoes and bicycle tires disintegrated. Puddles of toxic substances began to ooze through the soil. A swimming pool popped from its foundation and floated in a bath of chemicals.

The story of Love Canal started in 1892 when William Love excavated a canal 8 km (5 mi) long as part of the development of an industrial park. The development didn't need the canal when inexpensive electricity arrived, so the uncompleted canal remained unused for decades

A CLOSER LOOK 23.2

"e-waste": A Growing Environmental Problem

Hundreds of millions of computers and other electronic devices—such as cell phones, iPods, televisions, and computers games—are discarded every year. The average life of a computer is about three years, and it is not manufactured with recycling in mind. That is changing in the United States. as the cost to recycle TV and computer screens is charged to their manufacturers.

When we take our electronic waste, called **e-waste**, to a location where computers are turned in, we assume that it will be handled properly, but this is too often not what happens. In the United States, which helped start the technology revolution and produces most of the e-waste, its eventual disposal may cause serious environmental problems. The plastic housing for computers, for example, may produce toxins when burned. Computer parts also have small amounts of heavy metals—including gold, tin, copper, cadmium, and mercury—that are harmful and may cause cancer if inhaled, ingested, or absorbed through the skin. At present, many millions of computers are disposed of by what is billed as recycling, but the EPA has no official process to ensure that this e-waste won't cause future problems. In fact, most of these computers are being exported under the label of "recycling" to countries such as Nigeria and China.

China's largest e-waste facility is in Guiyu, near Hong Kong. People in the Guiyu area process more than 1 million tons of e-waste each year with little thought to the potential toxicity of the material the workers are handling (Figure 23.15). In the United States, computers cannot be recycled profitably without charging the people who dump them a fee. Even with that, many U.S. firms ship their e-waste out of the country, where greater profits are possible. The revenue to the Guiyu area is about $1 million per year, so the central government is reluctant to regulate the activity. Workers at locations where computers are disassembled may be unaware that some of the materials they are handling are toxic and that they thus have a hazardous occupation. Altogether, in the Guiyu area, more than 5,000 family-run facilities specialize in scavenging e-waste for raw materials. While doing this, they are exposing themselves to a variety of toxins and potential health problems.

FIGURE 23.15 e-waste being processed in China—a hazardous occupation.

To date, the United States has not made a proactive attempt to regulate the computer industry so that less waste is produced. In fact, the United States is the only major nation that did not ratify an international agreement that restricts and bans exports of hazardous e-waste.[43]

Our current ways of handling e-waste are not sustainable, and the value we place on a quality environment should include the safe handling and recycling of such waste. Hopefully, that is the path we will take in the future. There are positive signs. Some companies are now processing e-waste to reclaim metals such as gold and silver. Others are designing computers that use less toxic materials and are easier to recycle. The European Union is taking a leadership role in requiring more responsible management of e-waste.

and became a dump for wastes. From 1920 to 1952, some 20,000 tons of more than 80 chemicals were dumped into the canal. In 1953 the Hooker Chemical Company—which produced the insecticide DDT as well as an herbicide and chlorinated solvents, and had dumped chemicals into the canal—was pressured to donate the land to the city of Niagara Falls for $1.00. The city knew that chemical wastes were buried there, but no one expected any problems. Eventually, several hundred homes and an elementary school were built on and near the site, and for years everything seemed fine. Then, in 1976–1977, heavy rains and snows triggered a number of events, making Love Canal a household word.[41]

A study of the site identified many substances suspected of being carcinogens, including benzene, dioxin, dichlorethylene, and chloroform. Although officials admitted that little was known about the impact of these chemicals, there was grave concern for people living in the area. Eventually, concern centered on alleged high rates of miscarriages, blood and liver abnormalities, birth defects, and chromosome damage. The government had to destroy about 200 homes and a school, and about 800 families were relocated and reimbursed. After about $400 million was spent on cleaning up the site, the EPA eventually declared the area clean, and about 280 remaining homes were sold.[44] Today, the community around the canal is known as Black Creek Village, and many people live there.

Uncontrolled or poorly controlled dumping of chemical waste has polluted soil and groundwater in several ways:

- In some places, chemical waste is still stored in barrels, either stacked on the ground or buried. The barrels may eventually corrode and leak, polluting surface water, soil, and groundwater.
- When liquid chemical waste is dumped into an unlined lagoon, contaminated water may percolate through soil and rock to the groundwater table.
- Liquid chemical waste may be illegally dumped in deserted fields or even along roads.

Some sites pose particular dangers. The floodplain of a river, for example, is not an acceptable site for storing hazardous waste. Yet, that is exactly what occurred at a site on the floodplain of the River Severn near a village in one of the most scenic areas of England. Several fires at the site in 1999 were followed by a large fire of unknown origin on October 30, 2000. Approximately 200 tons of chemicals, including industrial solvents (xylene and toluene), cleaning solvents (methylene chloride), and various insecticides and pesticides, produced a fireball that rose into the night sky (Figure 23.16). Wind gusts of hurricane strength spread toxic smoke and ash to nearby farmlands and villages, which had to be evacuated. People exposed to the smoke complained of a variety of symptoms, including headaches, stomachaches and vomiting, sore throats, coughs, and difficulty breathing.

A few days later, on November 3, the site flooded (Figure 23.17). The floodwaters interfered with cleanup after the fire and increased the risk of downstream contamination by waterborne hazardous wastes. In one small village, contaminated floodwaters apparently inundated farm fields, gardens, and even homes.[45] Of course, the solution to this problem is to clean up the site and move waste storage to a safer location.

23.10 Hazardous-Waste Legislation

Recognition in the 1970s that hazardous waste was a danger to people and the environment and that the waste was not being properly managed led to important federal legislation in the United States.

Resource Conservation and Recovery Act

Management of hazardous waste in the United States began in 1976 with passage of the Resource Conservation and Recovery Act (RCRA). At the heart of the act is identification of hazardous wastes and their life cycles. The idea was to issue guidelines and assign responsibilities to those who manufacture, transport, and dispose of hazardous waste. This

FIGURE 23.16 On October 30, 2000, fire ravaged a site on the floodplain of the River Severn in England where hazardous waste was being stored. Approximately 200 tons of chemicals burned.

FIGURE 23.17 Flooding on November 3, 2000, followed the large fire at a hazardous-waste storage site on the floodplain of the River Severn in England (see Figure 23.16 at left).

is known as "cradle-to-grave" management. Regulations require stringent record keeping and reporting to verify that the wastes are not a public nuisance or a health problem.

RCRA applies to solid, semisolid, liquid, and gaseous hazardous wastes. It considers a waste hazardous if its concentration, volume, or infectious nature may contribute to serious disease or death or if it poses a significant hazard to people and the environment as a result of improper management (storage, transport, or disposal).[41] The act classifies hazardous wastes in several categories: materials highly toxic to people and other living things; wastes that may ignite when exposed to air; extremely corrosive wastes; and reactive unstable wastes that are explosive or generate toxic gases or fumes when mixed with water.

Comprehensive Environmental Response, Compensation, and Liability Act

In 1980, Congress passed the Comprehensive Environmental Response, Compensation, and Liability Act (CERCLA). It defined policies and procedures for release of hazardous substances into the environment (for example, landfill regulations). It also mandated development of a list of sites where hazardous substances were likely to produce or already had produced the most serious environmental problems and established a revolving fund (*Superfund*) to clean up the worst abandoned hazardous-waste sites. In 1984 and 1986, CERCLA was strengthened by amendments that made the following changes:

- Improved and tightened standards for disposal and cleanup of hazardous waste (for example, requiring double liners, monitoring landfills).
- Banned land disposal of certain hazardous chemicals, including dioxins, polychlorinated biphenyls (PCBs), and most solvents.
- Initiated a timetable for phasing out disposal of all untreated liquid hazardous waste in landfills or surface impoundments.
- Increased the Superfund. The fund was allocated about \$8.5 billion in 1986; Congress approved another \$5.1 billion for fiscal year 1998, which almost doubled the Superfund budget.[46]

The Superfund has had management problems, and cleanup efforts are far behind schedule. Unfortunately, the funds available are not sufficient to pay for decontaminating all targeted sites. Furthermore, present technology may not be sufficient to treat all abandoned waste-disposal sites; it may be necessary to simply try to confine waste at those sites until better disposal methods are developed. It seems apparent that abandoned disposal sites are likely to remain problems for some time to come.

Federal legislation has also changed the ways in which real estate business is conducted. For example, there are provisions by which property owners may be held liable for costly cleanup of hazardous waste on their property, even if they did not directly cause the problem. As a result, banks and other lending institutions might be held liable for release of hazardous materials by their tenants.

The Superfund Amendment and Reauthorization Act (SARA) of 1986 permits a possible defense against such liability if the property owner completed an **environmental audit** before purchasing the property. Such an audit involves studying past land use at the site, usually determined by analyzing old maps, aerial photographs, and reports. It may also involve drilling and sampling groundwater and soil to determine whether hazardous materials are present. Environmental audits are now completed routinely before purchasing property for development.[46]

In 1990 the U.S. Congress reauthorized hazardous-waste-control legislation. Priorities include:

- Establishing who is responsible (liable) for existing hazardous waste problems.
- When necessary, assisting in or providing funding for cleanup at sites identified as having a hazardous-waste problem.
- Providing measures whereby people who suffer damages from the release of hazardous materials are compensated.
- Improving the required standards for disposal and cleanup of hazardous waste.

23.11 Hazardous-Waste Management: Land Disposal

Management of hazardous chemical waste involves several options, including recycling; onsite processing to recover by-products that have commercial value; microbial breakdown; chemical stabilization; high-temperature decomposition; incineration; and disposal by **secure landfill** (Figure 23.18) or deep-well injection. A number of technological advances have been made in toxic-waste management; as land disposal becomes more expensive, the recent trend toward onsite treatment is likely to continue. However, onsite treatment will not eliminate all hazardous chemical waste; disposal of some waste will remain necessary.

Table 23.3 compares hazardous "waste" reduction technologies for treatment and disposal. Notice that all available technologies cause some environmental disruption. There is no simple solution for all waste-management issues.

Table 23.3 COMPARISON OF HAZARD REDUCTION TECHNOLOGIES

	DISPOSAL		TREATMENT			
PARAMETER COMPARED	LANDFILLS AND IMPOUNDMENTS	INJECTION WELLS	INCINERATION AND OTHER THERMAL DESTRUCTION	HIGH-TEMPERATURE DECOMPOSITION[A]	CHEMICAL STABILIZATION	MICROBIAL BREAKDOWN
Effectiveness: how well it contains or destroys hazardous characteristics	Low for volatiles, high for unsoluble solids	High, for waste compatible with the disposal environment	High	High for many chemicals	High for many metals	High for many metals and some organic waste such as oil
Reliability issues	Siting, construction, and operation Uncertainties: long-term integrity and cover	Site history and geology, well depth, construction, and operation	Monitoring uncertainties with respect to high degree of DRE: surrogate measures, PICs, incinerability[b]	Mobile units; on-site treament avoids hauling risks Operational simplicity	Some inorganics still soluble Uncertain leachate production	Monitoring uncertainties during construction and operation
Environment media most affected	Surface water and groundwater	Surface water and groundwater	Air	Air	Groundwater	Soil, groundwater
Least compatible wastes[c]	Highly toxic, persistent chemicals	Reactive; corrosive; highly toxic, mobile, and persistent	Highly toxic organics, high heavy-metal concentration	Some inorganics	Organics	Highly toxic persistent chemicals
Relative costs	Low to moderate	Low	Moderate to high	Moderate to high	Moderate	Moderate
Resource recovery potential	None	None	Energy and some acids	Energy and some metals	Possible building materials	Some metals

[a]Molten salt, high-temperature fluid well, and plasma arc treatments.

[b]DRE=destruction and removal efficiency; PIC = product of incomplete combustion.

[c]Wastes for which this method may be less effective for reducing exposure, relative to other technologies. Wastes listed do not necessarily denote common usage.

Source: Modified after Council on Environmental Quality, 1983.

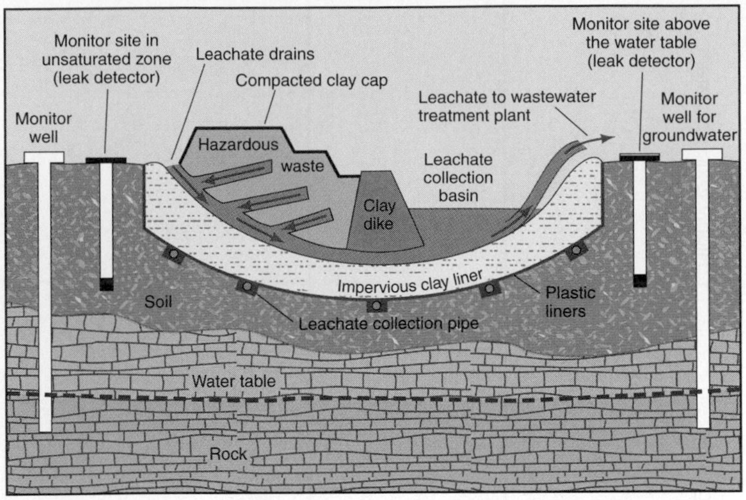

FIGURE 23.18 A secure landfill for hazardous chemical waste. The impervious liners, systems of drains, and leak detectors are integral parts of the system to ensure that leachate does not escape from the disposal site. Monitoring in the unsaturated zone is important and involves periodic collection of soil water.

Direct land disposal of hazardous waste is often not the best initial alternative. The consensus is that even with extensive safeguards and state-of-the-art designs, land disposal alternatives cannot guarantee that the waste will be contained and will not cause environmental disruption in the future. This concern holds true for all land disposal facilities, including landfills, surface impoundments, land application, and injection wells. Pollution of air, land, surface water, and groundwater may result if a land disposal site fails to contain hazardous waste. Pollution of groundwater is perhaps the most significant risk because groundwater provides a convenient route for pollutants to reach people and other living things.

Some of the paths that pollutants may take from land disposal sites to contaminate the environment include leakage and runoff to surface water or groundwater from improperly designed or maintained landfills; seepage, runoff, or air emissions from unlined lagoons; percolation and seepage from failure of surface land application of waste to soils; leaks in pipes or other equipment associated with deep-well injection; and leaks from buried drums, tanks, or other containers.[47-50]

23.12 Alternatives to Land Disposal of Hazardous Waste

Our handling of hazardous chemical waste should be multifaceted. In addition to the disposal methods just discussed, chemical-waste management should include such processes as source reduction, recycling and resource recovery, treatment, and incineration. Recently, it has been argued that these alternatives to land disposal are not being used to their full potential—that is, the volume of waste could be reduced, and the remaining waste could be recycled or treated in some form prior to land disposal of the treatment residues.[51] The advantages of source reduction, recycling, treatment, and incineration include the following:

- Useful chemicals can be reclaimed and reused.
- Treatment may make wastes less toxic and therefore less likely to cause problems in landfills.
- The volume of waste that must eventually be disposed of is reduced.
- Because a reduced volume of waste is finally disposed of, there is less stress on the dwindling capacity of waste-disposal sites.

Although some of the following techniques have been discussed as part of integrated waste management, they have special implications and complications in regard to hazardous wastes.

Source Reduction

The object of source reduction in hazardous-waste management is to reduce the amount of hazardous waste generated by manufacturing or other processes. For example, changes in the chemical processes involved, equipment and raw materials used, or maintenance measures may successfully reduce the amount or toxicity of hazardous waste produced.[51]

Recycling and Resource Recovery

Hazardous chemical waste may contain materials that can be recovered for future use. For example, acids and solvents collect contaminants when they are used in manufacturing processes. These acids and solvents can be processed to remove the contaminants and then be reused in the same or different manufacturing processes.[51]

Treatment

Hazardous chemical waste can be treated by a variety of processes to change its physical or chemical composition and reduce its toxicity or other hazardous characteristics. For example, acids can be neutralized, heavy metals can be separated from liquid waste, and hazardous chemical compounds can be broken up through oxidation.[51]

Incineration

High-temperature incineration can destroy hazardous chemical waste. However, incineration is considered a waste treatment, not a disposal method, because the process produces an ash residue that must itself be disposed of in a landfill. Hazardous waste has also been incinerated offshore on ships, creating potential air pollution and ash-disposal problems in the marine environment—an environment we consider next.

23.13 Ocean Dumping

Oceans cover more than 70% of Earth. They play a part in maintaining our global environment and are of major importance in the cycling of carbon dioxide, which helps regulate the global climate. Oceans are also important in cycling many chemical elements important to life, such as nitrogen and phosphorus, and are a valuable resource because they provide us with such necessities as food and minerals.

It seems reasonable that such an important resource would receive preferential treatment, and yet oceans have long been dumping grounds for many types of waste, including industrial waste, construction debris, urban sewage, and plastics (see A Closer Look 23.3). Ocean dumping contributes to the larger problem of ocean pollution, which has seriously damaged the marine environment and caused a health hazard. Figure 23.19 shows locations in the oceans of the world that are accumulating pollution continuously, or have intermittent pollution problems, or have potential for pollution from ships in the major shipping lanes. Notice that the areas with continual or intermittent pollution are near the shore.

Unfortunately, these are also areas of high productivity and valuable fisheries. Shellfish today often contain organisms that cause diseases such as polio and hepatitis. In the United States, at least 20% of the nation's commercial shellfish beds have been closed (mostly temporarily) because of pollution. Beaches and bays have been closed (again, mostly temporarily) to recreational uses. Lifeless zones in the marine environment have been created. Heavy kills of fish and other organisms have occurred, and profound changes in marine ecosystems have taken place (see Chapter 22).[52, 53]

Marine pollution has a variety of specific effects on oceanic life, including the following:

- Death or retarded growth, vitality, and reproductivity of marine organisms.
- Reduction of dissolved oxygen necessary for marine life, due to increased biochemical oxygen demand.
- Eutrophication caused by nutrient-rich waste in shallow estuaries, bays, and parts of the continental shelf, resulting in oxygen depletion and subsequent killing of algae, which may wash up and pollute coastal areas. (See Chapter 19 for a discussion of eutrophication in the Gulf of Mexico.)
- Habitat change caused by waste-disposal practices that subtly or drastically change entire marine ecosystems.[52]

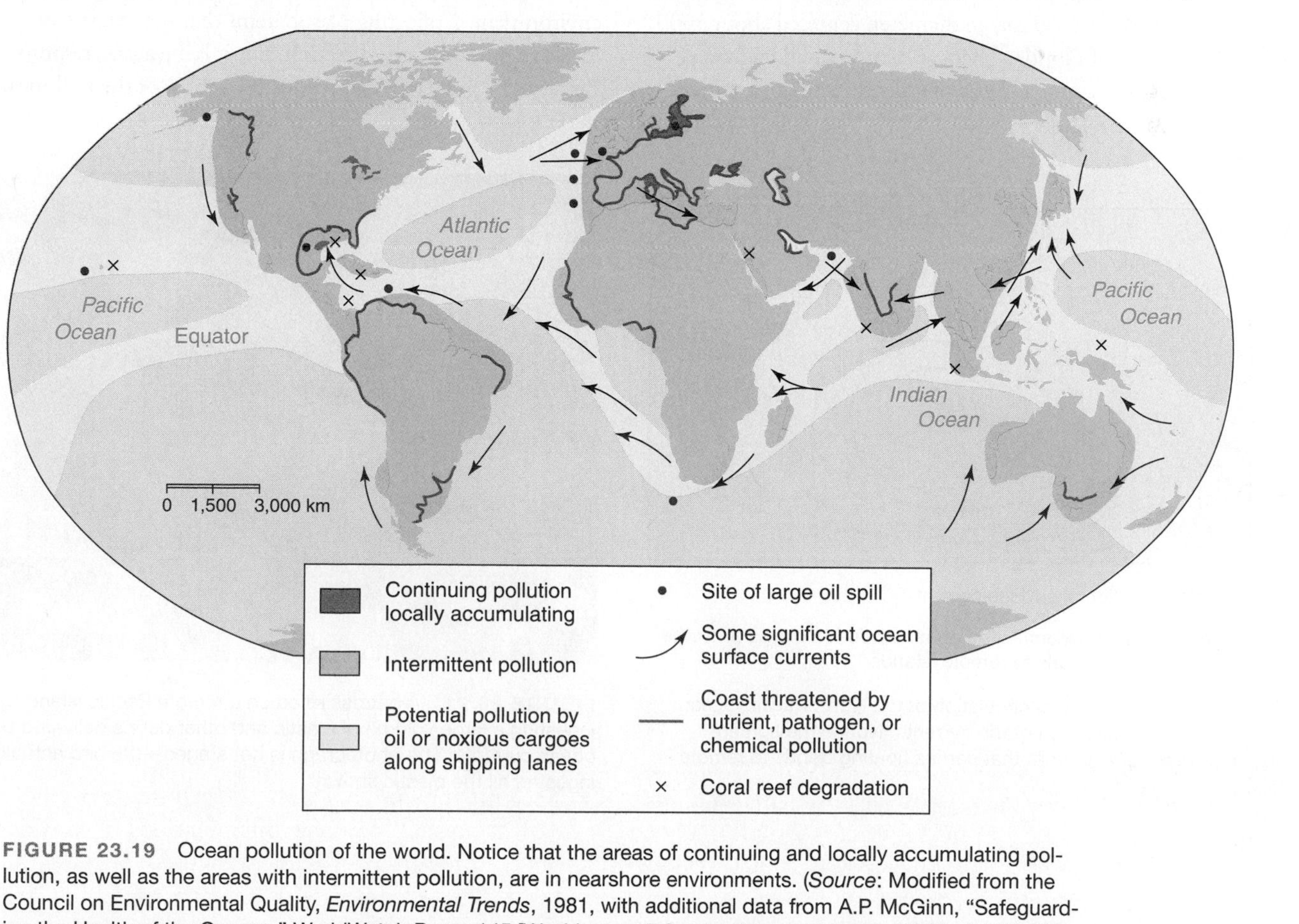

FIGURE 23.19 Ocean pollution of the world. Notice that the areas of continuing and locally accumulating pollution, as well as the areas with intermittent pollution, are in nearshore environments. (*Source*: Modified from the Council on Environmental Quality, *Environmental Trends*, 1981, with additional data from A.P. McGinn, "Safeguarding the Health of the Oceans," WorldWatch Paper 145 [Washington, DC: WorldWatch Institute, 1999], pp. 22–23.)

A CLOSER LOOK 23.3

Plastics in the Ocean

Vast quantities of plastic are used for a variety of products, ranging from beverage containers to cigarette lighters. For decades, people have been dumping plastics into the oceans. Some are dumped by passengers from passing ships; others are dropped as litter along beaches and swept into the water by the tides. Once in the ocean, plastics that float move with the currents and tend to accumulate where currents converge, concentrating the debris. Convergent currents of the Pacific (Figure 23.20) have a whirlpool-like action that concentrates debris near the center of these zones. One such zone is north of the equator, near the northwestern Hawaiian Islands. These islands are so remote that most people would expect them to be unspoiled, even pristine. In fact, however, there are literally hundreds of tons of plastics and other types of human debris on these islands. Recently, the National Oceanographic and Atmospheric Administration collected more than 80 tons of marine debris on Pearl and Hermes Atolls. Plastic debris is also widespread throughout the western North Atlantic Ocean. In the large North Atlantic subtropical gyre about 1200 km (750 mi) in diameter, centered about 1000 km (625 mi) east of Florida.[54] Most plastic is small fragments of a few mm up to about 1/2 size of a penny, and apparently is being digested by microbes.

The island ecosystems include sea turtles, monk seals, and a variety of birds, including albatross. Marine scientist Jean-Michel Cousteau and his colleagues have been studying the problem of plastics on the northwestern Hawaiian Islands, including Midway Island and Kure Atoll. They reported that the beaches of some of the islands and atolls look like a "recycling bin" of plastics. They found numerous cigarette lighters, some with fuel still in them, as well as caps from plastic bottles and all kinds of plastic toys and other debris. Birds on the islands pick up the plastic, attracted to it but not knowing what it is, and eat it. Figure 23.21 shows a dead albatross with debris in its stomach that caused its death. Plastic rings from a variety of products are also ingested by sea turtles and have been found around the snouts of seals, causing them to starve to death. In some areas, the carcasses of albatrosses litter the shorelines.

The solution to the problem is to be more careful about recycling plastic products to ensure they do not enter the marine environment. Collecting plastic items that accumulate on beaches is a step in the right direction, but it is a reactive response. Better to be proactive and reduce the source of the pollution.

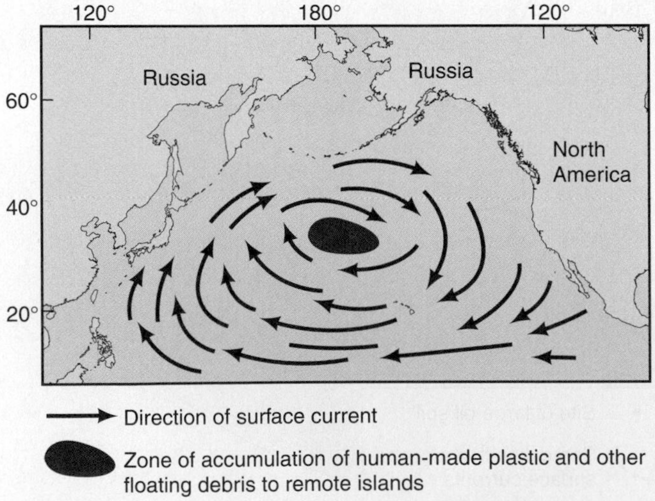

FIGURE 23.20 General circulation of the North Pacific Ocean. Arrows show the direction of the currents. Notice the tightening clockwise spiral pattern that carries floating debris to remote islands.

FIGURE 23.21 Albatross killed on a remote Pacific island by ingesting a large volume of plastic and other debris delivered by ocean currents. The photograph is not staged—the bird actually ingested all the plastic shown!

Marine waters of Europe are in particular trouble, in part because urban and agricultural pollutants have raised concentrations of nutrients in seawater. Blooms (heavy, sudden growth) of toxic algae are becoming more common. For example, in 1988 a bloom was responsible for killing nearly all marine life to a depth of about 15 m (50 ft), in the waterway connecting the North Sea to the Baltic Sea. It is believed that urban waste and agricultural runoff contributed to the toxic bloom.

Although oceans are vast, they are basically giant sinks for materials from continents, and parts of the marine environment are extremely fragile.[53] One area of concern is the *microlayer*, the upper 3 mm of ocean water. The base of the marine food chain consists of planktonic life abundant in the microlayer, and the young of certain fish and shellfish also reside there in the early stages of their life. Unfortunately, these upper few millimeters of the ocean also tend to concentrate pollutants, such as toxic chemicals and heavy metals. One study reported that concentrations of heavy metals—including zinc, lead, and copper—in the microlayer are from 10 to 1,000 times higher than in the deeper waters. It is feared that disproportionate pollution of the microlayer will have especially serious effects on marine organisms.[53] There is also concern that ocean pollution is a threat to some marine ecosystems, such as coral reefs, estuaries. salt marshes, and mangrove swamps.

Marine pollution can also have major impacts on people and society. Contaminated marine organisms, as we mentioned, may transmit toxic elements or diseases to people who eat them. In addition, beaches and harbors polluted by solid waste, oil, and other materials may not only damage marine life but also lose their visual appeal and other amenities. Economic loss is considerable as well. Loss of shellfish from pollution in the United States, for example, amounts to many millions of dollars per year. A great deal of money also is spent cleaning up solid waste, liquid waste, and other pollutants in coastal areas.[51]

23.14 Pollution Prevention

Approaches to waste management are changing. During the first several decades of environmental concern and management (the 1970s and 1980s), the United States approached the problem through government regulations and waste-control measures: chemical, physical, or biological treatment and collection (for eventual disposal), or transformation or destruction of pollutants after they had been generated. This was considered the most cost-effective approach to waste management.

With the 1990s came a growing emphasis on **pollution prevention**—ways to stop generating so much waste, rather than ways to dispose of it or manage it. This approach, which is part of materials management, includes the following:[55]

- Purchasing the proper amount of raw materials so that no excess remains to be disposed of.
- Exercising better control of materials used in manufacturing processes so that less waste is produced.
- Substituting nontoxic chemicals for hazardous or toxic materials currently used.
- Improving engineering and design of manufacturing processes so less waste is produced.

These approaches are often called P-2 approaches, for "pollution prevention." Probably the best way to illustrate the P-2 process is through a case history.[55]

A Wisconsin firm that produced cheese was faced with the disposal of about 2,000 gallons a day of a salty solution generated during the cheese-making process. Initially, the firm spread the salty solution on nearby agricultural lands—common practice for firms that could not discharge wastewater into publicly owned treatment plants. This method of waste disposal, when done incorrectly, caused the level of salts in the soil to rise so much that it damaged crops. As a result, the Department of Natural Resources in Wisconsin placed limitations on this practice.

The cheese firm decided to modify its cheese-making processes to recover salt from the solution and reuse it in production. This involved developing a recovery process that used an evaporator. The recovery process reduced the salty waste by about 75% and at the same time reduced the amount of the salt the company had to purchase by 50%. The operating and maintenance costs for recovery were approximately 3 cents per pound of salt recovered, and the extra cost of the new equipment was recovered in only two months. The firm saved thousands of dollars a year by recycling its salt.

The case history of the cheese firm suggests that rather minor changes can often result in large reductions of waste produced. And this case history is not an isolated example. Thousands of similar cases exist today as we move from the era of recognizing environmental problems, and regulating them at a national level, to providing economic incentives and new technology to better manage materials.[55]

23.15 Sustainable Resource Management

Sustaining renewable resources, such as water, wildlife, crops, and forests, though complex and sometimes difficult to achieve, is fairly easy to understand. Management of the environment must include development of goals and procedures to ensure that what makes a particular resource renewable persists over the long term (numerous generations). We have devoted several chapters in this book to sustainability with respect to renewable resources (water, air, energy, crops, forests, fish, and wildlife). However, simultaneously considering sustainable development and mineral exploitation and use is problematic. This is because, even with the most careful use, nonrenewable mineral resources will eventually be used up, and sustainability is a long-term concept that requires finding ways to assure future generations a fair share of Earth's resources. Recently, it has been argued that, given human ingenuity and sufficient lead time, we can find solutions for sustainable development that incorporate nonrenewable mineral resources.

Human ingenuity is important because often it is not the mineral we need so much as what we use the mineral for. For example, we mine copper and use it to transmit electricity in wires or electronic pulses in telephone wires. It is not the copper itself we desire but the properties of copper that allow these transmissions. We can use fiberglass cables in telephone wires, eliminating the need for copper. Digital cameras have eliminated the need for film development that uses silver. The message is that it is possible to compensate for a nonrenewable mineral by finding new ways to do things. We are also learning that we can use raw mineral materials more efficiently. For example, in the late 1800s when the Eiffel Tower was constructed, 8,000 metric tons of steel were used. Today the tower could be built with only a quarter of that amount.[56]

Finding substitutes or ways to more efficiently use nonrenewable resources generally requires several decades of research and development. A measure of how much time we have for finding solutions to the depletion of nonrenewable reserves is the ***R*-to-*C* ratio**, where R is the known reserves (for example, hundreds of thousands of tons of a metal) and C is the rate of consumption (for example, thousands of tons per year used by people). The R-to-C ratio is often misinterpreted as the time a reserve will last at the present rate of consumption. During the past 50 years, the R-to-C ratios for metals, such as zinc and copper, have fluctuated around 30 years. During that time, consumption of the metals roughly tripled, but we discovered new deposits. Although the R-to-C ratio is a *present* analysis of a dynamic system in which both the amount of reserves and consumption may change over time, it does provide a view of how scarce a particular mineral resource may be. Metals with relatively small ratios can be viewed as being in short supply, and it is those resources for which we should find substitutes through technological innovation.[56]

In sum, we may approach sustainable development and use of nonrenewable mineral resources by developing more efficient ways of mining resources and finding ways to more efficiently use available resources, recycling more and applying human ingenuity to find substitutes for a nonrenewable mineral.

CRITICAL THINKING ISSUE

Can We Make Recycling a More Financially Viable Industry?

There is enthusiastic public support for recycling in the United States today. Many people understand that managing our waste has many advantages to society as a whole and the environment in particular. People like the notion of recycling because they correctly assume they are helping to conserve resources, such as forests, that make up much of the nonurban environment of the planet. Large cities from New York to Los Angeles have initiated recycling programs, but many people are concerned that recycling is not yet "cost-effective."

To be sure, there are success stories, such as a large urban paper mill on New York's Staten Island that recycles more than 1,000 tons of paper per day. It is claimed that this paper mill saves more than 10,000 trees a day and uses only about 10% of the electricity required to make paper from virgin wood processing. On the West Coast, San Francisco has an innovative and ambitious recycling program that diverts nearly 50% of the urban waste from landfills to recycling programs. The city is even talking about the concept of zero waste, hoping to achieve total recycling of waste by 2020. In part, this is achieved by instigating a "pay-as-you-throw-away" approach; businesses and individuals are charged for disposal of garbage but not for materials that are recycled. Materials from the waste of the San Francisco urban area are shipped as far away as China and the Philippines to be recycled into usable products; organic waste is sent to agricultural areas; and metals, such as aluminum, are sent around California and to other states where they are recycled.

To understand some of the issues concerning recycling and its cost, consider the following points:

- The average cost of disposal at a landfill is about $40/ton in the United States, and even at a higher price of about $80/ton it may be cheaper than recycling.
- Landfill fees in Europe range from $200 to $300/ton.
- Europe has been more successful in recycling, in part because countries such as Germany hold manufacturers responsible for disposing of the industrial goods they produce, as well as the packaging.
- In the United States, packaging accounts for approximately one-third of all waste generated by manufacturing.
- The cost to cities such as New York, which must export their waste out of state, is steadily rising and is expected to exceed the cost of recycling within about ten years.
- Placing a 10-cent refundable deposit on all beverage containers except milk would greatly increase the number recycled. For example, states with a deposit system have an average recycling rate of about 70–95% of bottles and cans, whereas states that do not have a refundable-deposit system average less than 30%.
- When people have to pay for trash disposal at a landfill, but are not charged for materials that are recycled—such as paper, plastic, glass, and metals—the success of recycling is greatly enhanced.
- Beverage companies do not particularly favor requiring a refundable deposit for containers. They claim that the additional costs would be several billion dollars, but do agree that recovery rates would be higher, providing a steadier supply of recycled metal, such as aluminum, as well as plastic.
- Education is a big issue with recycling. Many people still don't know which items are recyclable and which are not.
- Global markets for recyclable materials, such as paper and metals, have potential for expansion, particularly for large urban areas on the seacoast, where shipping materials is economically viable. Recycling in the United States today is a $14 billion industry; and if it is done right, it generates new jobs and revenue for participating communities.
- The economic downturn since 2004 has resulted in much lower prices for recycled materials, such as paper, and even aluminum. The drop in demand for recycled materials in 2009 was global.

Critical Thinking Questions

1. What can be done about the global problem of e-waste? Could more be recycled safely?
2. What can be done to help recycling industries become more cost-effective?
3. What are some of the indirect benefits to society and the environment from recycling?
4. Defend or criticize the contention that if we really want to improve the environment by reducing our waste, we have to focus on more than the fact that recycling waste may cost more than dumping it at a landfill.
5. What are the recycling efforts in your community and university, and how could they be improved?
6. Do you think the global economic downturn since 2004 will cause a permanent problem for the recycling industry? Why? Why not?

SUMMARY

- Mineral resources are usually extracted from naturally occurring, anomalously high concentrations of Earth materials. Such natural deposits allowed early peoples to exploit minerals while slowly developing technological skills.
- Mineral resources are not mineral reserves. Unless discovered and developed, resources cannot be used to ease present shortages.
- The availability of mineral resources is one measure of the wealth of a society. Modern technological civilization would not be possible without the exploitation of mineral resources. However, it is important to recognize that mineral deposits are not infinite and that we cannot maintain exponential population growth on a finite resource base.
- The United States and many other affluent nations rely on imports for their supplies of many minerals. As other nations industrialize and develop, such imports may be more difficult to obtain, and affluent countries may have to find substitutes for some minerals or use a smaller portion of the world's annual production.

- The mining and processing of minerals greatly affect the land, water, air, and biological resources and have social impacts as well, including increased demand for housing and services in mining areas.
- Sustainable development and use of nonrenewable resources are not necessarily incompatible. Reducing consumption, reusing, recycling, and finding substitutes are environmentally preferable ways to delay or alleviate possible crises caused by the convergence of a rapidly rising population and a limited resource base.
- The history of waste-disposal practices since the Industrial Revolution has progressed from dilution and dispersion to the concept of integrated waste management (IWM), which emphasizes the three R's: reducing waste, reusing materials, and recycling.
- One goal of the emerging concept of industrial ecology is a system in which the concept of waste doesn't exist because waste from one part of the system would be a resource for another part.
- The most common way to dispose of solid waste is the sanitary landfill. However, around many large cities, space for landfills is hard to find, partly because few people wish to live near a waste-disposal site.
- Hazardous chemical waste is one of the most serious environmental problems in the United States. Hundreds or even thousands of abandoned, uncontrolled disposal sites could be time bombs that will eventually cause serious public health problems. We know that we will continue to produce some hazardous chemical waste. Therefore, it is imperative that we develop and use safe ways to dispose of it.
- Ocean dumping is a significant source of marine pollution. The most seriously affected areas are near shore, where valuable fisheries often exist.
- Pollution prevention (P-2)—identifying and using ways to prevent the generation of waste—is an important emerging area of materials management.

REEXAMINING THEMES AND ISSUES

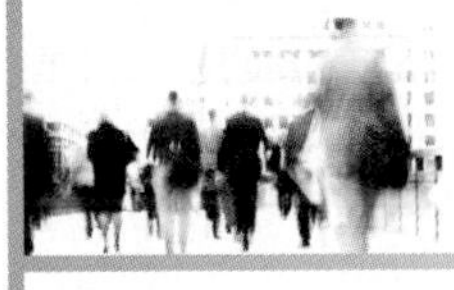

Human Population

Materials-management strategies are inextricably linked to the human population. As the population increases, so does the waste generated. In developing countries where population increase is the most dramatic, increases in industrial output, when linked to poor environmental control, produce or aggravate waste-management problems.

Sustainability

Assuring a quality environment for future generations is closely linked to materials management. Of particular importance here are the concepts of integrated waste management, materials management, and industrial ecology. Carried to their natural conclusion, the ideas behind these concepts would lead to a system in which the issue would no longer be waste management but instead resource management. Pollution prevention (P-2) is a step in this direction.

Global Perspective

Materials management is becoming a global problem. Improper management of materials contributes to air and water pollution and can cause environmental disruption on a regional or global scale. For example, waste generated by large inland cities and disposed of in river systems may eventually enter the oceans and be dispersed by the global circulation patterns of ocean currents. Similarly, soils polluted by hazardous materials may erode, and the particles may enter the atmosphere or water system, to be dispersed widely.

Urban World

Because so much of our waste is generated in the urban environment, cities are a focus of special attention for materials management. Where population densities are high, it is easier to implement the principles behind "reduce, reuse, and recycle." There are greater financial incentives for materials management where waste is more concentrated.

People and Nature

Production of waste is a basic process of life. In nature, waste from one organism is a resource for another. Waste is recycled in ecosystems as energy flows and chemicals cycle. As a result, the concept of waste in nature is much different than that in the human waste stream. In the human system, waste may be stored in facilities such as landfills, where it may remain for long periods, far from natural cycling. Our activities to recycle waste or burn it for energy move us closer to transforming waste into resources. Converting waste into resources brings us closer to nature by causing urban systems to operate in parallel with natural ecosystems.

Science and Values

People today value a quality, pollution-free environment. The way materials have been managed continues to affect health and other environmental problems. An understanding of these problems has resulted in a considerable amount of work and research aimed at reducing or eliminating the impact of resource use. How a society manages its waste is a sign of the maturity of the society and its ethical framework. Accordingly, we have become more conscious of environmental justice issues related to materials management.

KEY TERMS

composting **532**
environmental audit **540**
environmental justice **534**
e-waste **538**
hazardous waste **537**
incineration **533**
industrial ecology **530**
integrated waste management (IWM) **530**
leachate **533**
materials management **522**
nonrenewable resources **521**
ore deposits **523**
pollution prevention **545**
R-to-*C* ratio **546**
renewable resources **521**
reserves **524**
resources **524**
sanitary landfill **533**
secure landfill **540**
single-stream recycling **531**

STUDY QUESTIONS

1. What is the difference between a resource and a reserve?
2. Under what circumstances might sewage sludge be considered a mineral resource?
3. If surface mines and quarries cover less than 0.5% of the land surface of the United States, why is there so much environmental concern about them?
4. A deep-sea diver claims that the oceans can provide all our mineral resources with no negative environmental effects. Do you agree or disagree?
5. What factors determine the availability of a mineral resource?
6. Using a mineral resource involves four general phases: (a) exploration, (b) recovery, (c) consumption, and (d) disposal of waste. Which phase do you think has the greatest environmental effect?
7. Have you ever contributed to the hazardous-waste problem through disposal methods used in your home, school laboratory, or other location? How big a problem do you think such events are? For example, how bad is it to dump paint thinner down a drain?
8. Why is it so difficult to ensure safe land disposal of hazardous waste?
9. Would you approve the siting of a waste-disposal facility in your part of town? If not, why, and where do you think such facilities should be?
10. Why might there be a trend toward onsite disposal rather than land disposal of hazardous waste? Consider the physical, biological, social, legal, and economic aspects of the question.
11. Considering how much waste has been dumped in the nearshore marine environment, how safe is it to swim in bays and estuaries near large cities?

12. Do you think we should collect household waste and burn it in special incinerators to make electrical energy? What problems and what advantages do you see for this method, compared with other waste-management options?

13. Should companies that dumped hazardous waste years ago, when the problem was not understood or recognized, be held liable today for health problems to which their dumping may have contributed?

14. Suppose you found that the home you had been living in for 15 years was atop a buried waste-disposal site. What would you do? What kinds of studies should be done to evaluate potential problems?

FURTHER READING

Allenby, B.R., ***Industrial Ecology: Policy Framework and Implementation*** (Upper Saddle River, NJ: Prentice Hall, 1999). A primer on industrial ecology.

Ashley, S., It's not easy being green. *Scientific American*, April 2002, pp. 32–34. A look at the economics of developing biodegradable products and a little of the chemistry involved.

Brookins, D.G., ***Mineral and Energy Resources*** (Columbus, Ohio: Charles E. Merrill, 1990). A good summary of mineral resources.

Kesler, S.F., ***Mineral Resources, Economics and the Environment*** (Upper Saddle River, NJ: Prentice Hall, 1994). A good book about mineral resources.

Kreith, F. ed., ***Handbook of Solid Waste Management*** (New York: McGraw-Hill, 1994). Thorough coverage of municipal waste management, including waste characteristics, federal and state legislation, source reduction, recycling, and landfilling.

Watts, R.J., ***Hazardous Wastes*** (New York: John Wiley, 1998). A to Z of hazardous wastes.

CHAPTER 24

Our Environmental Future

The *Deepwater Horizon* oil drilling platform on fire April 22, 2010. Eleven workers were killed.

LEARNING OBJECTIVES

The learning objectives of this final chapter are to put it all together and then ask yourself the following questions:

- How can we take what we have learned and apply it to improving the environment?
- How can we think about a future in which we use our environment wisely—an "ecotopia"?
- Can we do this in a way that is good for people, human societies, and for nature?
- Will our environmental future "self-organize," or will it require more laws and formal planning?
- How have environmental laws affected people and environment, and what guidance does that experience provide for us in planning the future?

CASE STUDY

The Oil Spill in the Gulf of Mexico in 2010

America's biggest oil spill began on April 20, 2010, about 66 km (41 miles) south of the Louisiana Coast in the Gulf of Mexico. Everything about the spill was big, very big (Figure 24.1a). It happened on the *Deepwater Horizon,* a floating, semisubmerged drilling platform whose surface was larger than a football field—121m (396 ft) long and 78 m (256 feet) wide. Built in 2001 at a cost of $600 million and owned by Transocean, the platform had previously dug the deepest offshore gas and oil well ever, down 10,685 m (35,055 feet). In February 2010, Transocean began a new job under lease by British Petroleum (BP), drilling in waters 1,500 m (5,000 ft) deep. BP's plan was to use this platform to drill an exploratory well into the bedrock below to a depth of 5,600 m (18,360 feet)—almost 3½ miles into the rock! Pipes descended from the platform through the seawater to the bedrock below, and then drilling began. The wellhead, which sits atop the seafloor, contains devices to control the drilling, to insert drilling fluids (called "muds") into the hole drilled below, and to control the upward flow of oil and gas once those deposits are reached.

On April 20, things went wrong. Methane (natural gas) from the oil and gas deposits that were being drilled into from the platform broke through the wellhead at the surface far below. It rose rapidly, reaching the platform in a short time, starting a fire there at 9:56 P.M. local time, and then causing a major explosion. Eleven of the 126 crew members were killed, and many others were injured; some saved themselves by diving off the collapsing rig into the ocean. The fire was big—so big and bright that people in boats that came to help said it was hard to look at and melted the paint off the boats.

The *Deepwater Horizon* burned for 36 hours and then, on April 22, it sank, and the oil spill began in earnest. At first, the U.S. Coast Guard reported that 8,000 barrels a day were leaking, but it was difficult to determine just how much oil was pouring out thousands of feet below. By July the best estimate was about 60,000 barrels per day. The oil spread widely; by mid-June 2010, medium to heavy amounts of oil had reached more than 160 km (100 miles) east of the platform's position (Figure 24.1*b*).[1]

The total amount of oil spilled by mid-July when the leak was stopped was about 5 million barrels (210 million gallons). At this rate of release, the BP spill equaled the *Exxon Valdez* oil spill (until then the largest spill in U.S. history) every 4½ days.

To put these large numbers in perspective, the average school gymnasium would hold about 1.3 million gallons of oil. Thus, the oil spilled in just the first two months of the BP spill would fill over 100 school gymnasiums.

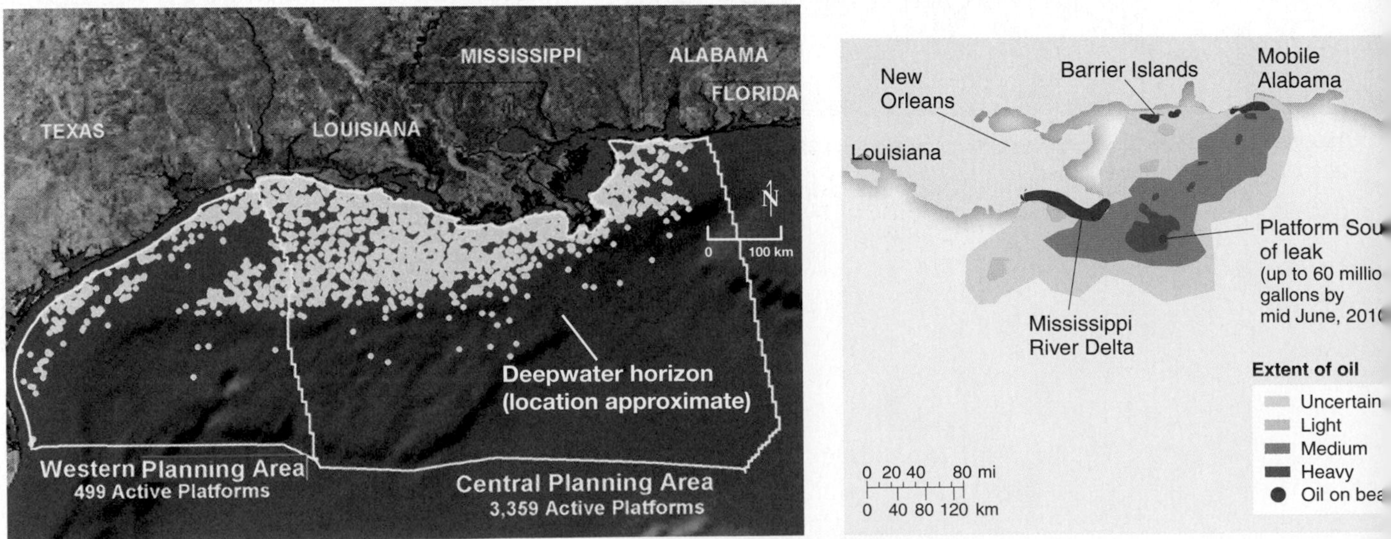

(a) (b)

FIGURE 24.1 **(a)** Active oil platforms (about 4,000) on the northern slope of the Gulf of Mexico. The location of the *Deepwater Horizon* is shown; **(b)** the extent of the Gulf spill as of mid-June 2010. (*Source*: Modified after NOAA.)

How does this compare to other blowouts and oil spills? The largest known blowout on land, which happened in Iran in 1956, involved about 120,000 barrels per day and lasted 3 months before being capped, releasing a total of almost 11 million barrels. A number of other "gushers" in the history of oil drilling released about 100,000 barrels per day. (We discuss other spills on land and offshore in Chapter 15.)

Any way you look at it, the BP spill is a lot of oil released into the Gulf's fragile marine and coastal environments. Spilled oil that remains near the water surface moves with currents and winds, some of it ending up on shorelines, partly covering plants and animals, infiltrating the sediment, and doing other kinds of ecological damage (Figure 24.2). Although most effects of oil spills are relatively short-lived (days to a few years), previous marine spills have killed thousands of seabirds, temporarily spoiled beaches, and caused loss of tourist and fishing revenues. Complicating matters, four species of sea turtles (loggerheads, Kemp's ridley, leatherback, and green) lay their eggs along Gulf State coasts that either have already been reached by the oil or are likely to in the near future. By June 25, 2010, 555 turtles had been found within the spill, 417 of them were dead.[2]

FIGURE 24.2 Oil on land: **(a)** Chandeleur Beach, Louisiana, 2010; **(b)** oil invades a Louisiana coastal wetland marsh in 2010; **(c)** dolphins swimming through some of the BP oil spill; **(d)** Kemp's ridley sea turtle at a rehabilitation center; **(e)** oil-covered seabird in Louisiana.

A large volume of natural gas (methane) has been released with the oil, some dissolved in the oil and some from gas pockets. Eventually, much of the methane in the water is degraded by bacteria, whose increased respiration decreases oxygen levels in the water. Scientists studying the oxygen content of the deep water near the oil rig have found oxygen depletion of 2% to 30% at depths of about 1,000 feet.

The spill is an economic disaster for BP and for the U.S. states along the Gulf coast. If the oil that was spilled in the first 60 days had been obtained and sold, it would have provided BP with about $288 million. But by that time BP had already admitted that the spill had cost the company $1 billion, and BP had agreed to provide the U.S. government with $20 billion to repay those who suffered damage from the spill. Fishermen, for example, have been put out of work because a large area of the Gulf's waters—from Morgan City west of New Orleans, Louisiana, to well east of Panama City, Florida, and south parallel to the Florida Keys, approximately 300 miles east–west and 300 miles north–south—were closed to fishing (Figure 24.3). Some closed areas were opened by August 2010. People in tourist areas lost money, too, because vacationers are choosing other locations where they won't have to contend with oil on beaches. All the supporting businesses for fishing and tourism also suffered.

Will the Gulf recover from the 2010 oil spill? Certainly it will. Scientific studies of previous oil spills show that there is always an immediate (scientists call it an "acute") effect, killing fish, birds, and marine mammals and damaging vegetation and nearshore algae. Over the long run, the oil decomposes and much of it becomes food for bacteria, or nutrients for algae and plants. However, studies of previous oil spills also show that some oil remains even decades after spills, and therefore

FIGURE 24.3 A large part of the commercial fisheries in the Gulf of Mexico were closed to fishing by late June 2010 because of the BP oil spill.

the effects on people, economics, and the environment that people value and enjoy are, from a human perspective, damaged for a long time.

By late August 2010 it was hard to find much floating oil in the Gulf and some saltmarsh plants were showing signs of recovery. Recovery may be quicker in the Gulf than in Alaska because the warm water favors biologic decomposition of the oil, the oil is light and the Gulf is a very large, deep, body of water subject to active surface processes from storm generated wind and waves.

How did the BP spill happen? As with many major environmental disasters, a series of poor decisions were involved, including a failure to take advantage of the safest and best technology. Before the blowout, problems with the well caused workers and others to express concern about being able to prevent an incident in which oil or natural gas would escape. The *Deepwater Horizon* had problems prior to the blowout and received 18 government citations for pollution. The BP wellhead had been fitted with a blowout preventer, but not with remote control or acoustically activated triggers for use in an emergency. The blowout preventer malfunctioned shortly after the heavy drilling mud had been withdrawn from the wellhead (the function of the drilling mud is to help keep oil from moving up the well to the surface). This is considered one of the major mistakes because without the mud and with the blowout-preventer malfunction, there was only water pressure to keep the oil and gas from escaping, and that was a recipe for disaster.

In addition, the response to the oil spill was inadequate. Rather than proactive, it was reactive: Each time something went wrong, there was spur-of-the- moment action. Also lacking was a clear line of authority and responsibility. The drilling was being done offshore by a private corporation, but it had large-scale effects, many of which were on government lands and waters and thus came under government control. Some available technologies that could have been applied were not; others were applied in too limited and tentative a way (Figure 24.4). News reports were rife with speculation by poorly informed people about all sorts of things that might be done, from gathering the oil with hay to blowing up the well with an atomic bomb.

About 6,000 boats were deployed with about 25,000 workers to try to minimize the spread of the spill by collecting it in the sea and on land. Some oil was burned, and chemical dispersants were applied from aircraft as well as at the bottom of the sea where the leak was occurring. These dispersants are chemicals and have environmental impacts themselves. It is known that these chemicals can damage marine ecosystems, but in this situation some scientists considered dispersants the lesser of two evils. Dispersants are being used, but their long-term impact, particularly on the deep-sea bed and in the seawater, is largely un-

(a)

(b)

FIGURE 24.4 Cleaning up an oil spill. **(a)** Boats use booms and skimmers to collect oil during 2010 spill; **(b)** cleaning a Louisiana beach, 2010.

known. This brings up an important point: The science of the deep-ocean basin has not progressed enough to be able to adequately predict the processes there and how they will interact with the oil and dispersants. [3]

The *Deepwater Horizon* was just one of nearly 4,000 other platforms in the Gulf off the coast of the United States. From the perspective of environmental science, what lessons can we take home from the BP oil spill?

First of all, it did not have to happen. Best practices—those that take advantage of the best and safest modern technology, developed from modern science—were not followed.

Second, modern industrialized nations use huge amounts of petroleum, and even with widespread movements away from petroleum, the need will not cease quickly. Therefore, it is essential that oil exploration and development make use of the best available technology and science, including the sciences that inform us about the environmental and ecological effects of an oil spill.

And third, after decades of concern about offshore oil spills, the technologies to deal with their cleanup remain insufficient. What is needed is an oversight program that includes advance planning, early warning, and rapid and sufficient response. Given the huge amount of money spent on energy within the United States and the importance of energy to our nation's standard of living, creativity, and productivity, we can no longer deal with such things as oil spills in a haphazard way.

24.1 Imagine an Ecotopia

Imagine a future in which we use our environment wisely—an "ecotopia." A learning objective of this chapter is to work out, to the best that present information allows, what you think is possible and desirable for this future world, focusing primarily on the United States, and also describe how this might be accomplished. Having read this book, you may well imagine a future in which, for example, we move away from fossil fuels and shift to renewable energy, no longer needing to damage the environment by mining and burning fossil fuels, nor forced to import them from uncertain and unfriendly sources. But which alternative energy sources would you favor?

This may seem an empty academic exercise, but unless we have an idea of what we want, we won't know in which direction to seek our future. Ideas are powerful, as history has proved. Wars have been fought over ideas. Ideas led Europeans to the New World and forged the American democracy. So what seems simply an academic exercise could be a powerful force for the future. It is not difficult today to imagine an "ecotopia"—a world in which the environment, human societies, and individuals are treated well in the present and helped to persist long into the future. But it would be extremely difficult to help it come about. What would that ecotopia be like? Here are a dozen qualities you would probably want to include:

- Since human population growth is the underlying environmental problem, an ecotopia would have to include a human population that had stabilized or even perhaps declined.
- All living resources would be sustainable, as would harvests of those resources.
- There would be enough wilderness and other kinds of natural or naturalistic areas for everyone to have opportunities for recreation and the enjoyment of nature.
- Pollution would be minimized.
- The risk of extinction of many species would be minimized.
- There would be enough functioning ecosystems to handle the public-service functions of ecosystems.

- Representatives of all natural ecosystems would be sustained in their dynamic ecological states.
- Poverty would be alleviated, benefiting both people and environment, because when you are poor it is hard to devote your resources to anything beyond immediate necessities.
- Energy would be abundant but, as much as possible, not cause pollution or otherwise damage land, water, and ecosystems.
- Water would be available to meet the needs of people and natural ecosystems,
- Natural resources, both finite and renewable, would also be available, and recycled where possible.
- Societies would have ample resources to be creative and innovative.

Admittedly, achieving all of this—and/or whatever else you've thought of—will be far from easy.

24.2 The Process of Planning a Future

Both human societies and natural ecosystems are complex systems. One of the questions asked by modern science is the degree to which such systems are *self-organizing*. A seed of a plant, for example, is a self-organizing system: It can develop into a mature plant without any outside planning or rational effort. But plants grown in agriculture are not simply left to their self-organizing abilities. Farmers plan for them and carry out those plans, and in these ways a plant is no longer completely self-organizing.

In our discussion of ecosystems in Chapter 5, we said that an ecosystem is the basic unit that can sustain life, and, in that sense, is necessary for life to persist. To some extent ecosystems show self-organizing characteristics, as in ecological succession, but that process of succession isn't as fixed, neat, and perfect a pattern as the growth of a seed into a mature plant.

One of the major themes of this book is the connection between people and nature. We understand today that human societies are linked to natural ecosystems. To what degree can these linked, complex systems self-organize? In various chapters, we have reviewed some examples that appear as self-organization. For example, as we saw in Chapter 22 (urban environments), cities developed at important transportation centers and where local resources could support a high density of people. In medieval Europe, bridges and other transportation aids developed in response to local needs. People arriving at a river would pay the farmer whose land lay along the river to row them across. Sometimes this would become more profitable than farming, or at least an important addition to the farmer's income. Eventually, he might build a toll bridge. People would congregate naturally at such a crossing and begin to trade. A town would develop.[4] The combination of environment and society led in a self-organizing way to cities.[4]

In contrast, the oil-drilling platform *Deepwater Horizon* was not self-organizing at all. It was imagined, designed, and built by a large manufacturing corporation with a planned purpose: to serve as a floating platform for drilling into difficult oil and gas reserves. It functioned within the laws of the United States and international treaties that affected activities in the Gulf of Mexico. These are external plans and agreements to regulate and control how the complex structure of the *Deepwater Horizon* could be and would be used. The failure of this platform was also not the result of self-organization, but of external (human) decisions.

In a democracy, planning with the environment in mind leads to a tug-of-war between individual freedom and the welfare of society as a whole. On one hand, citizens of a democracy want freedom to do what they want, wherever they want, especially on land that, in Western civilizations, is "owned" by the citizens or where citizens have legal rights to water or other resources. On the other hand, land and resource development and use affect society at large, and in either direct or indirect ways everyone benefits or suffers from a specific development. Society's concerns lead to laws, regulations, bureaucracies, forms to fill out, and limitations on land use.

Our society has formal planning processes for land use. These processes have two qualities: a set of rules (laws, regulations, etc.) requiring forms to be filled out and certain procedures to be followed; and an imaginative attempt to use land and resources in ways that are beautiful, economically beneficial, and sustainable. All human civilizations plan the development and use of land and resources in one way or another—through custom or by fiat of a king or emperor, if not by democratic processes. For thousands of years, experts have created formal plans for cities (see Chapter 22) and for important buildings and other architectural structures, such as bridges.

How can we balance freedom of individual action with effects on society? How can we achieve a sustainable use of Earth's natural resources, making sure that they will still be available for future generations to use and enjoy? In short, the questions are: Who speaks for nature? Who legally represents the environment? The landowner? Society at large? At this time, we have no definitive answers. Planning is a social experiment in which we all participate. Planning occurs at every level of activity, from a garden to a house, a neighborhood, a city park and its surroundings, a village, town, or city, a county, state, or nation. However, the history of our laws provides insight into our modern dilemma.

Issues of environmental planning and review are closely related to how land is used. Land use in the United States is dominated by agriculture and forestry; only a small portion of land (about 3%) is urban. However, rural lands are being converted to nonagricultural uses at about 9,000 km^2 (about 3,500 mi^2) per year. About half the conversion is for wilderness areas, parks, recreational areas, and wildlife refuges; the other half is for urban development, transportation networks, and other facilities. On a national scale, there is relatively little conversion of rural lands to urban uses. But in rapidly growing urban areas, increasing urbanization may be viewed as destroying agricultural land and exacerbating urban environmental problems, and urbanization in remote areas with high scenic and recreational value may be viewed as potentially damaging to important ecosystems.

24.3 Environment and Law: A Horse, a Gun, and a Plan

The legal system of the United States has historical origins in the British common law system—that is, laws derived from custom, judgment, and decrees of the courts rather than from legislation. The U.S. legal system preserved and strengthened British law to protect the individual from society—expressed best perhaps in the frontier spirit of "Just give me a little land, a horse, and a gun and leave me alone." Individual freedom—nearly unlimited discretion to use one's own property as one pleases—was given high priority, and the powers of the federal government were strictly limited.

But there is a caveat: When individual behavior infringed on the property or well-being of others, the common law provided protection through doctrines prohibiting trespass and nuisance. For example, if your land is damaged by erosion or flooding caused by your neighbor's improper management of his land, then you have recourse under common law. If the harm is more widespread through the community, creating a public nuisance, then only the government has the authority to take action—for instance, to limit certain air and water pollution.

The common law provides another doctrine, that of public trust, which both grants and limits the authority of government over certain natural areas of special character. Beginning with Roman law, navigable and tidal waters were entrusted to the government to hold for public use. More generally, "The public trust doctrine makes the government the public guardian of those valuable natural resources which are not capable of self-regeneration and for which substitutes cannot be made by man."[5] For such resources, the government has the strict responsibility of a trustee to provide protection and is not permitted to transfer such properties into private ownership. This doctrine was considerably weakened by the exaltation of private-property rights and by strong development pressures in the United States, but in more recent times it has shown increased vitality, especially concerning the preservation of coastal areas. Here is the basis for much modern environmental law, policy, regulation, and planning: common law with respect to you and your neighbors and the public trust doctrine.

The Three Stages in the History of U.S. Environmental Law

The history of federal legislation affecting land and natural resources occurred in three stages. In the first stage, the goal for public lands was to convert them to private uses. During this phase, Congress passed laws that were not intended to address environmental issues but did affect land, water, minerals, and living resources—and thereby had large effects on the environment. In 1812, Congress established the General Land Office, whose original purpose was to dispose of federal lands. The government disposed of federal lands through the Homestead Act of 1862 and other laws. As an example of Stage 1, in the 19th century the U.S. government granted rights-of-way to railroad companies to promote the development of rapid transportation. In addition to rights-of-way, the federal government granted the railroads every other square mile along each side of the railway line, creating a checkerboard pattern. The square miles in between were kept as federal land and are administered today by the Bureau of Land Management. These lands are difficult to manage for wildlife or vegetation because their artificial boundaries rarely fit the habitat needs of species, especially those of large mammals.

The second stage began in the second half of the 19th century, when Congress began to pass laws that conserved public lands for recreation, scenic beauty, and historic preservation. Late in the 19th century, Americans came to believe that the nation's grand scenery should be protected and that public lands provided benefits, some directly economic, such as rangelands for private ranching.

Federal laws created the National Park Service in the second half of the 19th century in response to Americans' growing interest in their scenic resources. Congress made Yosemite Valley a California state park in 1864 and created Yellowstone National Park in 1872 "as a public park or pleasuring-ground for the benefit and enjoyment of the people."[6] Interest in Indian ruins led soon after to the establishment in 1906 of Mesa Verde National Park, putting into public lands the prehistoric

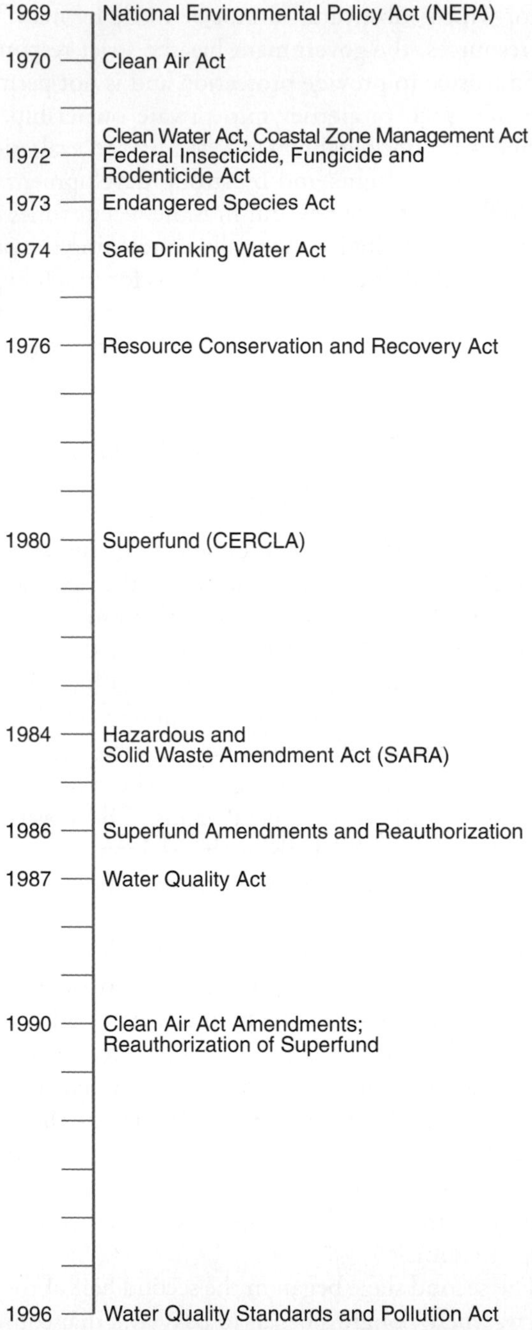

FIGURE 24.5 Major federal environmental legislation and the year enacted. Most of the important environmental legislation was adopted from 1969 to 1996. Some laws were enacted earlier in a much less comprehensive form (e.g., the Clean Air Act in 1963), and most were amended subsequently.

cliff dwellings of early North Americans and at the same time creating national monuments. The National Park System was created by Congress in 1916. Today it consists of 379 areas.

Also in the second stage, the United States Forest Service began in 1898, and President Grover Cleveland appointed Gifford Pinchot to be head of the Division of Forestry, soon renamed the U.S. Forest Service. Pinchot believed that the purpose of national forests was "the art of producing from the forest whatever it can yield for the service of man." The focus was on production of useful products.

Although the term *sustainability* had not yet become popular, in 1937 the federal government passed the Oregon and California Act, which required that timberland in western Oregon be managed to give sustained yields.[7]

In the third stage, Congress enacted laws whose primary purpose was environmental. This stage has antecedents in the 1930s but didn't get going in force until the 1960s and it continues today. The acknowledged need to regulate the use of land and resources has been filled by legislation enacted at all levels of government. In the late 1960s, public awareness and concern in the United States that our environment was deteriorating reached a high level. Congress responded by passing the National Environmental Protection Act (NEPA) in 1969 and a series of other laws in the 1970s (Figure 24.5). Federal laws relating to land management proliferated to the point where they became confusing. By the end of World War II, there were 2,000 laws about managing public lands, often contradicting one another. In 1946 Congress set up the Bureau of Land Management (BLM) to help correct this confusion.

Government regulation of land and resources has also given rise to controversy: How far should the government be allowed to go to protect what appears to be the public good against what have traditionally been private rights and interests? Today, the BLM attempts to balance the traditional uses of public lands—grazing and mining—with the environmental era's interest in outdoor recreation, scenic beauty, and biological conservation. Part of achieving a sustainable future in the United States will be finding a balance among these uses, as well as a balance between the amount of land that should be public and the amount of land that need not be.

24.4 Planning to Provide Environmental Goods and Services

One important experiment of the 20th century was regional planning. In the United States, this means planning across state boundaries. One of the best-known regional plans in the United States began in 1933, when President Franklin D. Roosevelt proposed the establishment of the Tennessee Valley Authority (TVA), a semi-independent agency responsible for promoting economic growth and social well-being for the people throughout parts of seven states, which were economically depressed at the time the authority was established. There had been rampant exploitation of timber and fossil-fuel resources in the region, and the people living there were among the poorest in the country.[8]

Today, the TVA is considered one of the world's best examples of regional planning (Figure 24.6). It is characterized by multidimensional and multilevel planning to manage land and water resources and is involved in the production and regulation of electrical power, as well as flood control, navigation, and outdoor recreation. In the midst of the Great Depression, Roosevelt sought new ways to invigorate the economy, especially in depressed rural areas. He envisioned the TVA as a corporation clothed with the power of government but with the flexibility and initiative of a private enterprise. The TVA granted legal control over land use to a multistate authority of a new kind and posed novel issues of governmental authority. The act creating the TVA contained the following stipulations:

> The unified development and regulation of the Tennessee River system require that no dam, appurtenant works, or other obstruction, affecting navigation, flood control, or public lands or reservations shall be constructed, and thereafter operated or maintained across, along, or in the said river or any of its tributaries until plans for such construction, operation, and maintenance shall have been submitted to and approved by the Board; and the construction, commencement of construction, operation, or maintenance of such structures without such approval is hereby prohibited.[9]

24.5 Planning for Recreation on Public Lands

Today, management of public lands for recreational activities requires planning at a variety of levels, with considerable public input. For example, when a national forest is developing management plans, public meetings are often held to inform people about the planning process and to ask for ideas and suggestions. Maximizing public input promotes better communication between those responsible for managing resources and those using them for recreational purposes.

Government officials and scientists involved in developing plans for public lands are often faced with land-use problems so complex that no easy answers can be found. Nonetheless, because action or inaction today can have serious consequences tomorrow, it is best to have at least some plans to protect and preserve a quality environment for future generations. Plans for many of the national forests and national parks in the United States have been or are being developed, generally taking into account a spectrum of recreational activities and attempting to balance the desires of several user groups.

Severe 1996–1997 winter floods in Yosemite National Park damaged roads, campgrounds, bridges, and other structures. The flood led to a rethinking of the goals and objectives of park management, and one result was that some land claimed by the floods was returned to natural

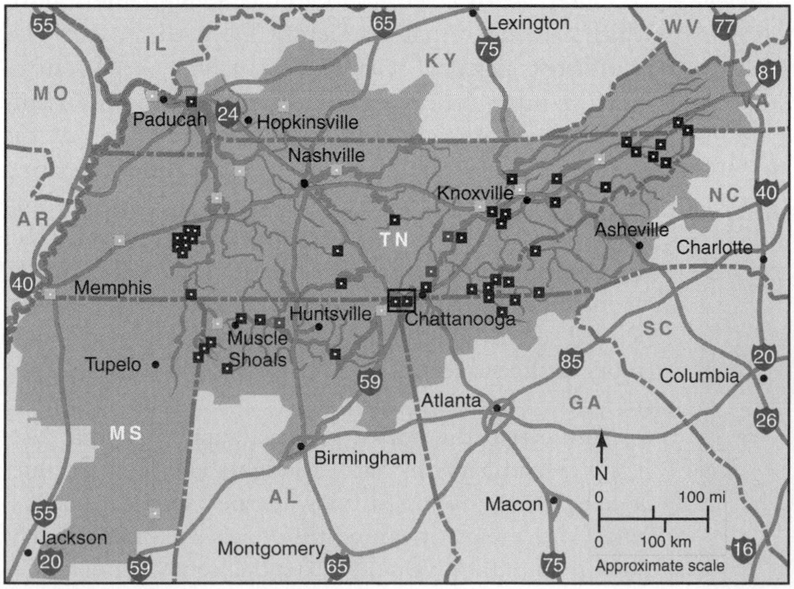

(a)

(b)

FIGURE 24.6 **(a)** A map showing the region encompassed by the TVA (darker area) and one of the major impoundments, the Raccoon Mountain Dam; **(b)** a large reservoir created by one of the TVA dams.

ecosystems. Another result was the elimination of private vehicles in parts of the park. Many other important policies have also been implemented in U.S. forests and parks. For example:

- In wilderness areas, only a limited number of people are admitted.
- In coastal areas, regulations may limit such activities as jet skiing and surfing in swimming areas.
- Regions that are home to endangered species, or to species that may pose a danger to people, may have more stringent regulations governing the activities of visitors. In Yellowstone National Park in Wyoming and Montana, for example, special consideration is given to grizzly bear habitats through controls on where people may venture.

Other recreational activities that are, or may become, subject to increased regulation include hiking, camping, fishing, boating, skiing, snowmobiling, and such recently popularized activities as treasure hunting, which includes panning for gold. At the extremes, certain areas have been set aside for intensive off-road-vehicle use, while other areas have been closed entirely. Activities on government lands can be more easily regulated than those occurring elsewhere. However, park management may be difficult if goals are not clear and natural processes not understood.

Who Stands for Nature? Skiing at Mineral King

Planning for recreational activities on U.S. government lands (including national forests and national parks) is controversial. At the heart of the controversy are two different moral positions, both with wide support in the United States. On one side, some argue that public land must be open to public use, and therefore the resources within those

FIGURE 24.7 Mineral King Valley, now part of Sequoia National Park after nearly 20 years of controversy about the development of a ski resort in the valley.

lands should be available to citizens and corporations for economic benefit. On the other side are those who argue that public lands should serve the needs of society first and individuals second, and that public lands can and must provide for land uses not possible on private lands.

A classic example of this controversy concerned a plan by the Disney Corporation in the 1960s and 1970s to develop a ski resort with a multimillion-dollar complex of recreational facilities on federal land in a part of California's Sierra Nevada called Mineral King Valley (Figure 24.7), which had been considered a wilderness area. The Sierra Club, arguing that such a development would adversely affect the aesthetics of this wilderness, as well as its ecological balance, brought a suit against the government.

The case raised a curious question: If a wrong was being done, who was wronged? Christopher D. Stone, a lawyer, discussed this idea in an article entitled "Should Trees Have Standing? Toward Legal Rights for Natural Objects." The California courts decided that the Sierra Club itself could not claim direct harm from the development, and because the government owned the land but also represented the people, it was difficult to argue that the people in general were wronged. Stone said that the Sierra Club's case might be based, by common-law analogy, on the idea that in some cases inanimate objects have been treated as having legal standing—as, for example, in lawsuits involving ships, where ships have legal standing. Stone suggested that trees should have that legal standing, that although the Sierra Club was not able to claim direct damage to itself, it could argue on behalf of the nonhuman wilderness.

The case was taken to the U.S. Supreme Court, which concluded that the Sierra Club itself did not have a sufficient "personal stake in the outcome of the controversy" to bring the case to court. But in a famous dissenting statement, Justice William O. Douglas addressed the question of legal standing (*standing* is a legal term relating here to the right to bring suit). He proposed establishing a new federal rule that would allow "environmental issues to be litigated before federal agencies or federal courts in the name of the inanimate object about to be despoiled, defaced, or invaded by roads and bulldozers and where injury is the subject of public outrage." In other words, trees would have legal standing.

While trees did not achieve legal standing in that case, it was a landmark in that legal rights and ethical values were explicitly discussed for wilderness and natural systems. This subject in ethics still evokes lively controversy. Should our ethical values be extended to nonhuman, biological communities and even to Earth's life-support system? What position you take will depend in part on your understanding of the characteristics of wilderness, natural systems, and other environmental factors and features, and in part on your values.

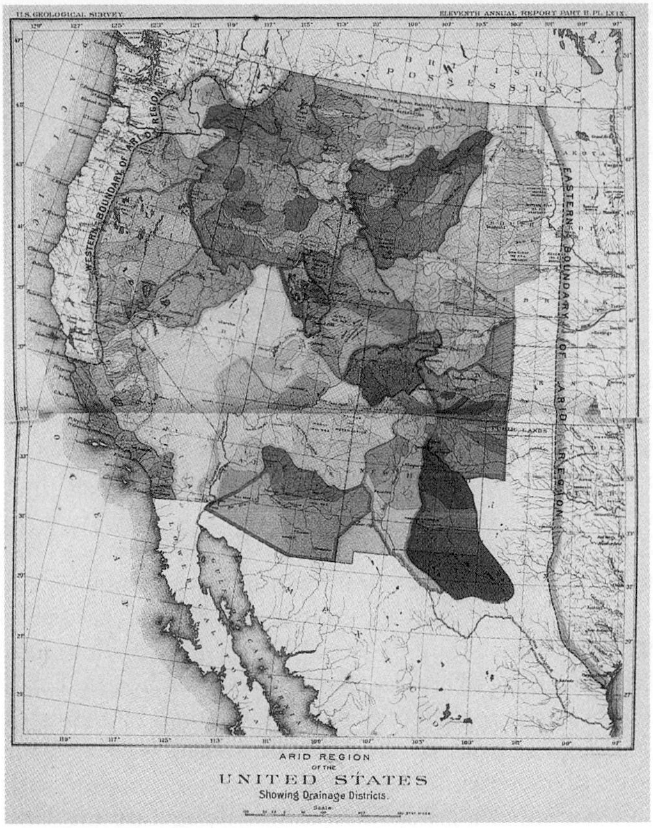

FIGURE 24.8 Powell's map of water in the West.

Mineral King Valley and surrounding peaks of Mineral King, about 6,000 ha (12,600 acres), were transferred from the national forest to Sequoia National Park in September 1978. The transfer ended nearly 20 years of controversy over proposed development of a ski resort.

How Big Should Wildlands Be? Planning a Nation's Landscapes

Recent thinking about the environment has focused on the big picture: What is necessary at a national scale, or at some landscape scale, to achieve our goals? We are not the first to ask this question. John Wesley Powell, the famous one-armed American explorer who was the first to lead men down the Colorado River through the Grand Canyon, observed the dry American West and suggested that the land should be organized around major watersheds rather than laid out for political and social reasons, as the states ultimately were (Figure 24.8). His utopian vision was of a landscape where farmers spent their own money on dams and canals, doing so because the land was organized politically around watersheds. They could use, but not sell, their water. This plan seemed to impose too much control from the top and never happened.[10] Instead, in 1902 Congress passed an act that began the 20th-century construction of large dams and canals funded with federal dollars. Water rights could be sold, and cities like Los Angeles could assert the right to water hundreds of miles away.

While we cannot go back to Powell's vision completely, our society is gradually thinking more and more in terms of planning around large watersheds. This regional approach may help us move closer to the dream of our ecotopia. Modern scientific studies of ecosystems and landscapes also lead to speculation about the best way to conserve biological resources. Some argue that nature can be saved only in the large. A group called the Wildlands Project argues that big predators, referred to as "umbrella species," are keys to ecosystems, and that these predators require large home ranges. The assumption is that big, wide-ranging carnivores offer a wide umbrella of land protection under which many species that are more abundant but smaller and less charismatic find safety and resources.[11] Leaders of the Wildlands Project feel that even the biggest national parks, such as Yellowstone, are not big enough, and that America needs "rewilding." They propose that large areas of the United States be managed around the needs of big predators and that we replan our landscapes to provide a combination of core areas, corridors, and inner and outer buffers (Figure 24.9). No human activities would take place in the core areas, and even in the corridors and buffers human activity would be restricted.

The Wildlands Project has created a major controversy, with some groups seeing the project as a fundamental threat to American democracy. Another criticism of the Wildlands Project is directed at its scientific foundation. These critics say that although some ecological research suggests that large predators may be important, what controls populations in all ecosystems is far from understood. Similarly, the idea of keystone species, central to the rationale of the Wildlands Project, lacks an adequate scientific base.

A related idea that developed in the last two decades was *rewilding*—that is, returning the land that was once American prairie to land without towns and cities, where bison are once again allowed to roam free. As Reed Noss, one of the founders of the Wildlands Project, has written:

> A cynic might describe rewilding as an atavistic obsession with the resurrection of Eden. A more sympathetic critic might label it romantic. We contend, however, that rewilding is simply scientific realism, assuming that our goal is to insure the long-term integrity of the land community. Rewilding with extirpated carnivores and other keystone species is a means as well as an end. The "end" is the moral obligation to protect wilderness and to sustain the remnants of the Pleistocene—animals and plants—not only for our human enjoyment, but because of their intrinsic value.[11]

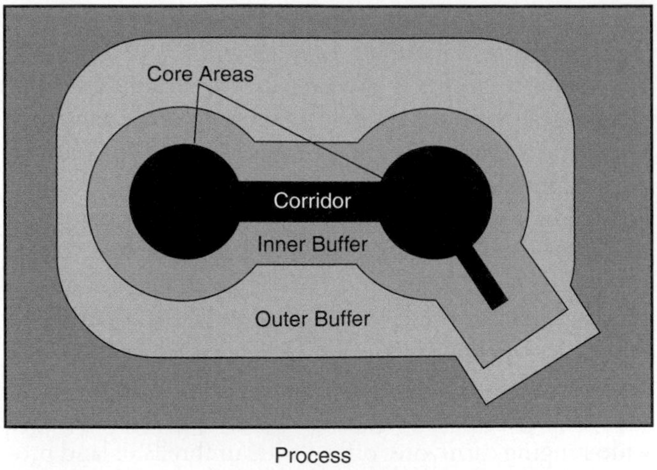

FIGURE 24.9 Wildlands Project diagram of land divisions.

Proposals for the environment of the future thus involve science and values, and people and nature. So what do you want? A vast area of the United States returned to what might be self-functioning ecosystems? Or some open system of conservation that integrates people and allows for more freedom of action? The choices lie with your generation and the next, and tests of those choices' validity are also yours. The implications for the environment and for people are huge.[11, 12]

24.6 How You Can Be an Actor in the Environmental Law Processes

The case of Mineral King raises the question: What is the role of our legal system—laws, courts, judges, lawyers—in achieving environmental goals? The current answer is that environmental groups working through the courts have been a powerful force in shaping the direction of environmental quality control since the early 1970s. Their influence arose in part because the courts, appearing to respond to the national sense of environmental crisis of that time, took a more activist stance and were less willing to defer to the judgment of government agencies. At the same time, citizens were granted unprecedented access to the courts and, through them, to environmental policy.

Citizen Actions

Even without specific legislative authorization for citizens' suits, courts have allowed citizen actions in environmental cases as part of a trend to liberalize standing requirements.[13]

In the 1980s, a new type of environmentalism (which some people would label radical) arose, based in part on the premise that when it comes to the defense of wilderness, there can be no compromise. Methods used by these new environmentalists have included sit-ins to block roads into forest areas where mining or timber harvesting is scheduled; sitting in trees to block timber harvesting; implanting large steel spikes in trees to discourage timber harvesting; and sabotaging equipment, such as bulldozers (a practice known as "ecotage").

Ecotage and other forms of civil disobedience have undoubtedly been responsible for millions of dollars' worth of damage to a variety of industrial activities related to the use of natural resources in wilderness areas. One result of civil disobedience by some environmental groups is that other environmental groups, such as the Sierra Club, are now considered moderate in their approach to protecting the environment. There is no doubt, however, that civil disobedience has been successful in defending the environment in some instances. For example, members of the group Earth First succeeded in halting construction of a road being built to allow timber harvesting in an area of southwestern Oregon. Earth First's tactics included blockading the road by sitting or standing in front of the bulldozers, which slowed the pace of road work considerably. In conjunction with this action, the group filed a lawsuit against the U.S. Forest Service.

Environmentalists are now relying more on the law when arguing for ecosystem protection. The Endangered Species Act has been used as a tool in attempts to halt activities such as timber harvesting and development. Although the presence of an endangered species is rarely responsible for stopping a proposed development, those species are increasingly being used as weapons in attempts to save remaining portions of relatively undisturbed ecosystems.

Mediation

The expense and delay of litigation have led people to seek other ways to resolve disputes. In environmental conflicts, an alternative that has recently received considerable attention is mediation, a negotiation process between the adversaries guided by a neutral facilitator. The task of the mediator is to clarify the issues, help each party understand the position and the needs of the other parties, and attempt to arrive at a compromise whereby each party gains enough to prefer a settlement to the risks and costs of litigation. Often, a citizens' suit, or the possibility that a suit might be filed, gives an environmental group a place at the table in mediation. Litigation, which may delay a project for years, becomes something that can be bargained away in return for concessions from a developer. Some states require mediation as an alternative or prior to litigation in the highly contentious siting of waste-treatment facilities. In Rhode Island, for example, a developer who wishes to construct a hazardous-waste treatment facility must negotiate with representatives of the host community and submit to arbitration of any issues not resolved by negotiation. The costs of the negotiation process are borne by the developer.

A classic example of a situation in which mediation could have saved millions of dollars in legal costs and years of litigation is the Storm King Mountain case, a conflict between a utility company and conservationists. In 1962, the Consolidated Edison Company of New York announced plans for a new hydroelectric project in the Hudson River Highlands, an area with thriving fisheries and also considered to have unique aesthetic value (Figure 24.10). The utility company argued that it needed the new facility, and the environmentalists fought to preserve the landscape and the fisheries. Litigation began with a suit filed in 1965 and ended in 1981 after 16 years of intense courtroom battles that left a paper trail exceeding 20,000 pages. After spending millions of dollars and untold hours, the various parties finally managed to forge an agreement with the assistance of an outside mediator. If they had been able to sit down and talk at an early stage, mediation might have settled the issue much sooner and at much less cost to the parties and to society.[14] The Storm King Mountain case is often cited as a major victory for environmentalists, but the cost was great to both sides.

FIGURE 24.10 Storm King Mountain and the Hudson River Highlands in New York State were the focus of environmental conflict between a utility company and conservationists for nearly 20 years before a dispute about building a power plant was finally resolved by mediation.

24.7 International Environmental Law and Diplomacy

Legal issues involving the environment are difficult enough within a nation; they become extremely complex in international situations. International law is different from domestic law in basic concept because there is no world government with enforcement authority over nations. As a result, international law must depend on the agreement of the parties to bind themselves to behavior that many residents of a particular nation may oppose. Certain issues of multinational concern are addressed by a collection of policies, agreements, and treaties that are loosely called international environmental law. There have been encouraging developments in this area, such as agreements to reduce air pollutants that destroy stratospheric ozone (the Montreal Protocol of 1987 and subsequent discussion and agreements; see Chapter 21).

Antarctica provides a positive example of using international law to protect the environment. Antarctica, a continent of 14 million km^2, was first visited by a Russian ship in 1820, and people soon recognized that the continent contained unique landscapes and life-forms (Figure 24.11). By 1960, a number of countries had claimed parts of Antarctica to exploit mineral and fossil-fuel resources. Then, in 1961, an international treaty was established designating Antarctica a "scientific sanctuary." Thirty years later, in 1991, a major environmental agreement, the Protocol of Madrid, was reached, protecting Antarctica, including islands and seas south of 60° latitude. The continent was designated "nuclear-free," and access to its resources was restricted. This was the first step in conserving Antarctica from territorial claims and establishing the "White Continent" as a heritage for all people on Earth.

Other environmental problems addressed at the international level include persistent organic pollutants (POPs), such as dioxins, DDT, and other pesticides. After several years of negotiations in South Africa and Sweden, 127 nations adopted a treaty in May 2001 to greatly reduce or eliminate the use of toxic chemicals known to contribute to cancer and harm the environment.

24.8 Global Security and Environment

The terrorist attacks on New York City and Washington, DC, on September 11, 2001, brought the realization that the United States—in fact the world—is not as safe as we had assumed. The attacks led to a war on terrorists and their financial and political networks around the world. However, for every terrorist removed, another will fill the void unless the root causes are recognized and eliminated.

Achieving sustainability in the world today has strong political and economic components, but it also has an environmental component. Terrorism comes in part from poverty, overcrowding, disease, and conflicts

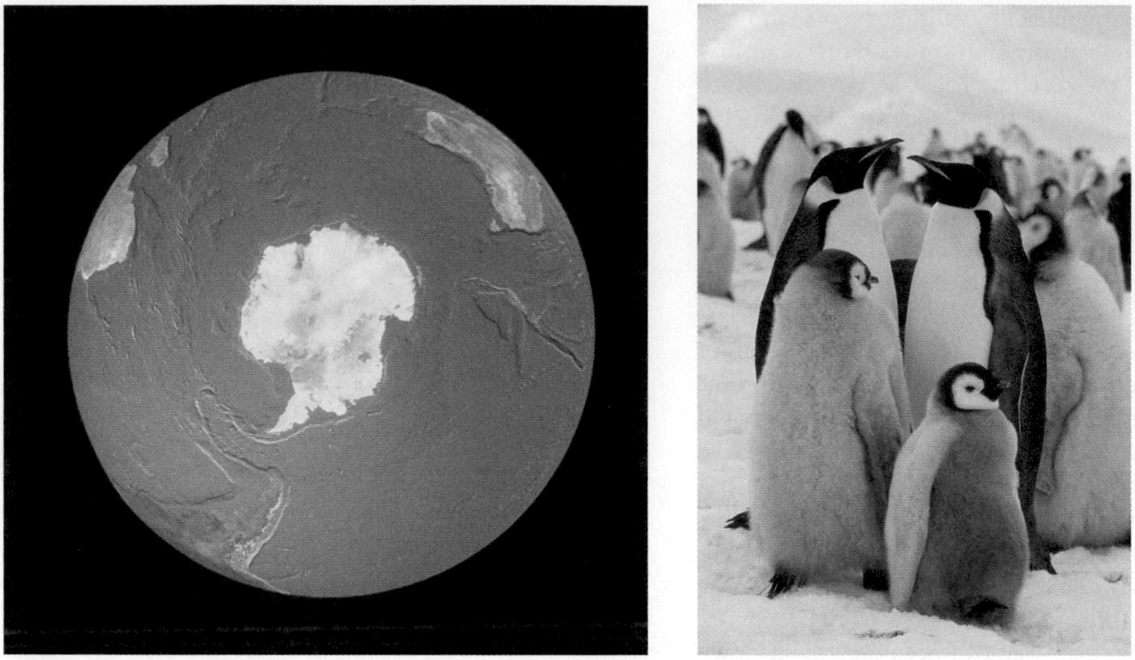

(a) (b)

FIGURE 24.11 International agreements determine environmental practices in Antarctica. (a) Satellite image of Antarctica and surrounding southern oceans; (b) emperor penguins and chicks in Antarctica.

that have environmental significance. Over 1 billion people on Earth today live in poverty with little hope for the future. In some large urban regions, tens of millions of people exist in crowded, unsanitary conditions, with unsafe drinking water and inadequate sewage disposal. In the countryside, rural people in many developing countries are being terrorized and displaced by armed conflicts over the control of valuable resources, such as oil, diamonds, and timber. Examples include oil in Nigeria, Sudan, and Colombia; diamonds in Sierra Leone, Angola, and the Democratic Republic of the Congo; and timber in Cambodia, Indonesia, and Borneo.[15]

The goal of the 1992 Rio Earth Summit on Sustainable Development was to address global environmental problems of both developed and developing countries, with an emphasis on solving conflicts between economic interests and environmental concerns. In many countries today, the gap between the rich and the poor is even wider than it was in the early 1990s. As a result, political, social, and economic security remains threatened, and serious environmental damage from overpopulation and resource exploitation continues. Environmental protection continues to be inadequately funded. Worldwatch Institute reported in 2002 that the United Nations' annual budget for the environment is about $100 million, while the governments of the world are spending $2 billion per day for military purposes.[15]

24.9 Challenges to Students of the Environment

To end this book on an optimistic note—and there *are* reasons to be optimistic—we note that the Earth Summit on Sustainable Development, held in the summer of 2002 in Johannesburg, South Africa, had the following objectives:

- To continue to work toward environmental and social justice for all the people in the world.
- To enhance the development of sustainability.
- To minimize local, regional, and global environmental degradation resulting from overpopulation, deforestation, mining, agriculture, and pollution of the land, water, and air.
- To develop and support international agreements to control global warming and pollutants, and to foster environmental and social justice.

Solving our environmental problems will help build a more secure and sustainable future. This is becoming your charge and responsibility, as you, students of the environment and our future leaders, graduate from colleges and universities. This transfer of knowledge and leadership is a major reason why we wrote this book.

CRITICAL THINKING ISSUE

Is It Possible to Derive Some Quantitative Statements about Thresholds beyond Which Unacceptable Environmental Change Will Occur?

A *threshold* is a condition or level that, if exceeded, will cause a system to change, often from one mode of operation to another, in terms of actual processes or rates of processes. In the environmental literature, thresholds are sometimes spoken of as tipping points, beyond which adverse consequences are likely to occur. Other definitions of a tipping point are: a point when a system (say the global climate) changes from one stable state to another stable state (this is a threshold); and a point where slow small changes over time results in a sudden large change (also a threshold). However, thresholds are not tipping points where change becomes catastrophic and may be irreversible. For example, some believe that if global warming continues past a particular point, say a two degree Celsius rise of temperature, then changes will become more rapid and the consequences of those changes more severe. The purpose of this critical thinking issue is to examine some of these hypotheses in more detail.[16]

In previous chapters, we discussed the major environmental problems related to human population, water, energy, and climate. In discussing human population, we introduced the concept of what Earth's carrying capacity might be. In answering that question, we posed another: "What would we *like* it to be?" It is acknowledged that human population growth is the environmental problem, but at what population level would the degree of environmental degradation become unacceptable to us? Similar limits or thresholds might be introduced for biological productivity; loss of biological diversity; use of nutrients, such as nitrogen and phosphorus; transformation of the land; and our use of freshwater resources. For this list, some scientists have tried to pinpoint thresholds beyond which environmental degradation is unacceptable (a value judgment).

Table 24.1 is based on a paper published in 2009 in the major scientific journal *Nature* and entitled "A Safe Operating Space for Humanity." You should treat these ideas as proposals for discussion, not as truths or facts. The table lists these systems in terms of parameters that may be measured, along with suggested thresholds, which are compared to the present status and also to pre-industrial levels. For example, for human population, a suggested threshold might be 5 billion people—fewer than are on Earth today and 4 billion more than the pre-industrial level of about 1 billion. This 5 billion threshold might be based on the fact that biological productivity, when it was more in balance with human needs, peaked around 1985, when the population was 5 billion people. The arbitrary choice of 5 billion is obviously linked to other factors shown in the table, as they are interrelated. Any specific number for the optimum carrying capacity of the planet will be controversial, but your evaluation will depend on the knowledge you bring to bear and your values.

With respect to climate change, Table 24.1 lists a hypothetical threshold of 350 parts per million for carbon dioxide concentration in the atmosphere, versus the present level of 390 parts per million and the pre-industrial level 280 parts per million. Setting the threshold at 350 parts per million was based on examination of the geologic record, the possible effects of previous climate change, and the likely levels of carbon dioxide in the atmosphere. This table is intended just for the sake of our discussion here. Similarly, the amount of land transformation or water use is also related to our present scientific knowledge.

Looking at Table 24.1 in more detail, we can see that some of the suggested thresholds have already been exceeded, and others have not. However, whether they actually have been exceeded will depend on how much we know about the particular system, whether the consequences are unacceptable, and whether this can be shown with some degree of certainty.

Critical Thinking Questions

1. Do you think it is a valid argument that some sorts of thresholds, or tipping points, exist beyond which unacceptable environmental degradation will occur?
2. Has science satisfactorily answered whether or not these thresholds, or tipping points, can in fact be established?
3. From your reading of *Environmental Science*, can you make other suggestions as to where thresholds or tipping points might be placed?
4. If you are not able to set thresholds, what sorts of studies might be necessary to establish them in the future? Of course, this assumes that the whole concept of thresholds, or tipping points, is a valid approach in environmental science.

Table 24.1 GLOBAL THRESHOLDS THAT, TRANSGRESSED, COULD CAUSE UNACCEPTABLE ENVIRONMENTAL CHANGE [FOR DISCUSSION PURPOSES ONLY; NOT TO BE TAKEN AS FACTS]

SYSTEM	PARAMETER	SUGGESTED THRESHOLD	PRESENT STATUS	PRE-INDUSTRIAL LEVEL
Human population	Billions of people	5.0	6.8	1.0
Climate	Carbon dioxide concentration (parts per million)	350	390	280
Biological productivity	Portion used by humans	0.6	1.2	<0.2
Biodiversity loss (extinction)	Extinction rate (number of species per million species per year)	10	>100	0.1–1.0
Nitrogen use	Amount removed from the air for human use (millions of tons per year)	35	120	0
Phosphorus use	Quantity flowing into the ocean (millions of tons per year)	11	9	-1.0
Land transformation	% of land converted to agriculture	15	12	Low
Global freshwater use	km^3/yr	4,000	2,600	415
Air pollution	Metric tons per year	To be determined	To be determined	To be determined
Water pollution	Metric tons per year	To be determined	To be determined	To be determined

Source: Modified from J. Rockström et al., 2009. "A Safe Operating Space for Humanity," *Nature* 461: 472–475. doi:10.1038/461472a.

SUMMARY

- A fundamental question, continuously debated in a democracy, is the extent to which human societies and their environment can function as self-organizing systems, and how much formal planning—laws and so on—is necessary.
- Both natural ecosystems and human societies are complex systems. The big question is how the interaction among these can lead to the long-term persistence of both, and perhaps even improvements.
- Mistakes are always likely; advance planning, including rapid response, is essential to maintaining the best environment.
- Our environmental laws have grown out of a combination of the English common law—derived from custom and judgment, rather than legislation—and American perspectives on freedom and planning.
- In the 19th and 20th centuries, America experimented with a variety of approaches to conserving nature, some involving laws, some new kinds of plans and organizations. The best combination is yet to be determined.
- International environmental law is proving useful in addressing several important environmental problems, including preservation of resources and pollution abatement.
- Global security, sustainability, and environment are linked in complex ways. Solving environmental problems will improve both sustainability and security.

STUDY QUESTIONS

1. Based on what you have learned in this book and in your studies about environment, what would an "ecotopia" include, in addition to what is mentioned in this chapter? Which of these items, if any, do you think could be achieved during your lifetime?
2. Just how big should a wilderness be?
3. The famous ecologist Garrett Hardin argued that designated wilderness areas should not have provisions for people with handicaps, even though he himself was confined to a wheelchair. He believed that wilderness should be truly natural in the ultimate sense—that is, without any trace of civilization. Argue for or against Garrett Hardin's position. In your argument, consider the "people and nature" theme of this book.
4. How can we balance freedom of individual action with the need to sustain our environment?
5. Visit a local natural or naturalistic place, even a city park, and write down what is necessary for that area to be sustainable in its present uses.
6. Should trees—and other nonhuman organisms—have legal standing? Explain your position on this topic.
7. Since there are no international laws that are binding in the same way that laws govern people within a nation, what can be done to achieve a sustainable environment for world fisheries or other international resources?
8. Do you think the Gulf oil spill could have been prevented? If so, how?
9. Do you think Garrett Hardin is right—that there are some technologies (such as drilling in deep water) that humans are not prepared to adequately address and that there will thus be continued accidents due to human error?

Appendix

A Special Feature: Electromagnetic Radiation (EMR) Laws

Properties of Waves

- Direction of wave propagation

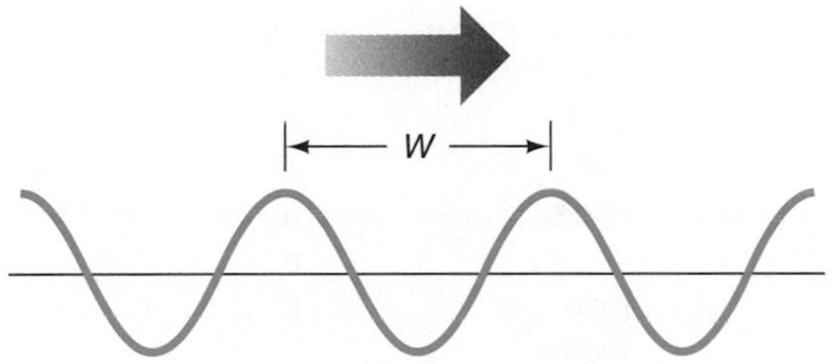

- W = wavelength (distance from one wave crest to the next)
- An EMR wave travels at the speed of light (C) in a vacuum, or about 300,000 km/s (3×10^8 m/s).
- The period T of a wave is the time it takes for a wave to travel a distance of one wavelength W. Then since distance is the product of speed (velocity) and time, W = CT.
- The frequency f of a wave is the number of cycles (each wavelength that passes a point is a cycle) of a wave that pass a particular point per unit time. Frequency f is measured in cycles per second (hertz). The frequency f is the inverse of T: f = 1/T and W = C/f. For example, the period T of a 6000 hertz EMF wave is: 6000 hertz = 1/T and T = 1/6000 S, or 1.7×10^{-4} S. The wavelength W = CT is 3×10^8m/s times 1.7×10^{-4} S, or 5.1×10^4 m, which according to Figure 20.18 is a long radio wave.

Absolute Temperature Scale (kelvin, K)

- Zero is really zero; there are no negative values of K
- Temperature in K = temperature in °C + 273

$$K = °C + 273$$

- Example: water freezes at °C = 0 = 273 K
  water boils at °C = 100 = 373 K

Stefan–Boltzmann Law

- All bodies with a temperature greater than absolute zero radiate EMR. These bodies are called thermal radiators. The amount of energy per second radiated from thermal radiators is called *intensity* and is given by the Stefan–Boltzmann law

$$E = aT^4$$

where E is the energy per second (intensity); T is the absolute temperature; and a is a constant (the nature of this constant involves physical ideas beyond the scope of this text).

- The Stefan–Boltzmann law states that the intensity of EMR coming from a thermal radiator is directly proportional to the fourth power of its absolute temperature.

Wien's Law

$$W_P = a/T$$

where WP is the wavelength of the peak intensity of a thermal radiator; T is temperature in K; and a is a constant. For example, Figure 20.19 shows that WP for the earth is about 10 μm. Wien's law states in a general way that the hotter a substance is, the shorter the wavelength of the emitted predominant electromagnetic radiation. That is, wavelength is inversely proportional to temperature.

B Prefix and Multiplication Factors

Number	10×, Power of 10	Prefix	Symbol
1,000,000,000,000,000,000	10^{18}	exa	E
1,000,000,000,000,000	10^{15}	peta	P
1,000,000,000,000	10^{12}	tera	T
1,000,000,000	10^{9}	giga	G
1,000,000	10^{6}	mega	M
10,000	10^{4}	myria	
1,000	10^{3}	kilo	k
100	10^{2}	hecto	h
10	10^{1}	deca	da
0.1	10^{-1}	deci	d
0.01	10^{-2}	centi	c
0.001	10^{-3}	milli	m
0.000 001	10^{-6}	micro	μ
0.000 000 001	10^{-9}	nano	n
0.000 000 000 001	10^{-12}	pico	p
0.000 000 000 000 001	10^{-15}	femto	f
0.000 000 000 000 000 001	10^{-18}	atto	a

C Common Conversion Gactors

LENGTH

1 yard = 3 ft, 1 fathom = 6 ft

	in	ft	mi	cm	m	km
1 inch (in) =	1	0.083	1.58×10^{-5}	2.54	0.0254	2.54×10^{-5}
1 foot (ft) =	12	1	1.89×10^{-4}	30.48	0.3048	—
1 mile (mi) =	63,360	5,280	1	160,934	1,609	1.609
1 centimeter (cm) =	0.394	0.0328	6.2×10^{-6}	1	0.01	1.0×10^{-5}
1 meter (m) =	39.37	3.281	6.2×10^{-4}	100	1	0.001
1 kilometer (km) =	39,370	3,281	0.6214	100,000	1,000	1

AREA

1 square mi = 640 acres, 1 acre = 43,560 ft^2 = 4046.86 m^2 = 0.4047 ha
1 ha = 10,000 m^2 = 2.471 acres

	in^2	ft^2	mi^2	cm^2	m^2	km^2
1 in^2 =	1	—	—	6.4516	—	—
1 ft^2 =	144	1	—	929	0.0929	—
1 mi^2 =	—	27,878,400	1	—	—	2.590
1 cm^2 =	0.155	—	—	1	—	—
1 m^2 =	1,550	10.764	—	10,000	1	—
1 km^2 =	—	—	0.3861	—	1,000,000	1

VOLUME	in^3	ft^3	yd^3	m^3	qt	liter	barrel	gal (U.S.)
1 in^3 =	1	—	—	—	—	0.02	—	—
1 ft^3 =	1,728	1	—	0.0283	—	28.3	—	7.480
1 yd^3 =	—	27	1	0.76	—	—	—	—
1 m^3 =	61,020	35.315	1.307	1	—	1,000	—	—
1 quart (qt) =	—	—	—	—	1	0.95	—	0.25
1 liter (l) =	61.02	—	—	—	1.06	1	—	0.2642
1 barrel (oil) =	—	—	—	—	168	159.6	1	42
1 gallon (U.S.) =	231	0.13	—	—	4	3.785	0.02	1

Mass and Weight

1 pound = 453.6 grams = 0.4536 kilogram = 16 ounces

1 gram = 0.0353 ounce = 0.0022 pound

1 short ton = 2000 pounds = 907.2 kilograms

1 long ton = 2240 pounds = 1008 kilograms

1 metric ton = 2205 pounds = 1000 kilograms

1 kilogram = 2.205 pounds

Energy and Power[a]

1 kilowatt-hour = 3413 Btus = 860,421 calories

2 Btu = 0.000293 kilowatt-hour = 252 calories = 1055 joules

1 watt = 3.413 Btu/hr = 14.34 calorie/min

1 calorie = the amount of heat necessary to raise the temperature of 1 gram (1 cm^3) of water 1 degree Celsius

1 quadrillion Btu = (approximately) 1 exajoule

1 horsepower = 7.457×10^2 watts

1 joule = 9.481×10^{-4} Btu = 0.239 cal = 2.778×10^{-7} kilowatt-hour

[a]Values from Lange, N. A., 1967, *Handbook of Chemistry*, New York: McGraw-Hill.

Temperature

$F = \frac{9}{5}C + 32$ — F is degrees Fahrenheit. C is degrees Celsius (centigrade).

Fahrenheit		Celsius
32	Freezing of H_20 (Atmospheric Pressure)	0
50		10
68		20
86		30
104		40
122		50
140		60
158		70
176		80
194		90
212	Boiling of H_20 (Atmospheric Pressure)	100

Other Conversion Factors

1 ft^3/sec = 0.0283 m^3/sec = 7.48 gal/sec = 28.32 liter/sec

1 acre-foot = 43,560 ft^3 = 1233 m^3 = 325,829 gal

1 m^3/sec = 35.32 ft^3/sec

1 ft^3/sec for one day = 1.98 acre-feet

1 m/sec = 3.6 km/hr = 2.24 mi/hr

1 ft/sec = 0.682 mi/hr = 1.097 km/hr

1 atmosphere = 14.7 lb(in.$^{-2}$) = 2116 lb(ft^{-2}) = 1.013×10^5 N(m^{-2})

D Geologic Time Scale and Biologic Evolution

Era	Approximate Age in Millions of Years Before Present	Period	Epoch	Life Form
	Less than 0.01		Recent (Holocene)	
	0.01–2	Quaternary	Pleistocene	Humans
	2			
	2–5		Pliocene	
	5–23		Miocene	
Cenozoic	23–35	Tertiary	Oligocene	
	35–56		Eocene	Mammals
	56–65		Paleocene	
	65			
	65–146	Cretaceous		
Mesozoic	146–208	Jurassic		Flying reptiles, birds
	208–245	Triassic		Dinosaurs
	245			
	245–290	Permian		Reptiles
	290–363	Carboniferous		Insects
Paleozoic	363–417	Devonian		Amphibians
	417–443	Silurian		Land plants
	443–495	Ordovician		Fish
	495–545	Cambrian		
	545			
	700			Multicelled organisms
	3,400			One-celled organisms
	4,000	Approximate age of oldest rocks discovered on Earth		
Precambrian				
	4,600	Approximate age of Earth and meteorites		

E Matter, Energy and Chemistry

Matter and Energy

The universe, as we know it, consists of two entities: matter and energy. Matter is the material that makes up our physical and biological environments (you are composed of matter). Energy is the ability to do work. The first law of thermodynamics—also known as the law of conservation of energy or the first energy law—states that energy cannot be created or destroyed but can change from one form to another. This law stipulates that the total amount of energy in the universe does not change.

Our sun produces energy through nuclear reactions at high temperatures and pressures that change mass (a measure of the amount of matter) into energy. At first glance, this may seem to violate the law of conservation of energy. However, this is not the case, because energy and matter are interchangeable. Albert Einstein first described the equivalence of energy and mass in his famous equation $E = mc^2$, where E is energy, m is mass, and c is the velocity of light in a vacuum, such as outer space (approximately 300,000 km/s or 186,000 mi/s). Because the velocity of light squared is a very large number, even a small amount of conversion of mass to energy produces very large amounts of energy. Energy, then, may be thought of as an abstract, mathematical quantity that is always conserved. This means that it is impossible to get something for nothing when dealing with energy; it is impossible to extract more energy from any system than the amount of energy that originally entered the system. In fact, the second law of thermodynamics states that you cannot break even. When energy is changed from one form to another, it always moves from a more useful form to a less useful one. Thus, as energy moves through a real system and is changed from one form to another, energy is conserved, but it becomes less useful.

Basic Chemistry

We turn next to a brief introduction to the basic chemistry of matter, which will help you in understanding biogeochemical cycles. An atom is the smallest part of a chemical element that can take part in a chemical reaction with another atom. An element is a chemical substance composed of identical atoms that cannot be separated by ordinary chemical processes into different substances. Each element is given a symbol. For example, the symbol for the element carbon is C, and that for phosphorus is P.

A model of an atom (Figure E.1) shows three subatomic particles: neutrons, protons, and electrons. The atom is visualized as a central nucleus composed of neutrons with no electrical charge and protons with a positive charge. A cloud of electrons, each with a negative charge, revolves about the nucleus. The number of protons in the nucleus is unique for each element and is the atomic number for that element. For example, hydrogen, H, has one proton in its nucleus, and its atomic number is 1. Uranium has 92 protons in its nucleus, and its atomic number is 92. A list of known elements with their atomic numbers, called the Periodic Table, is shown in Chapter 6 (Figure 6.8)

Electrons, in our model of the atom, are arranged in shells (representing energy levels), and the electrons closest to the nucleus are bound tighter to the atom than those in the outer shells. Electrons have negligible mass compared with neutrons or protons; therefore, nearly the entire mass of an atom is in the nucleus.

The sum of the number of neutrons and protons in the nucleus of an atom is known as the atomic weight. Atoms of the same element always have the same atomic number (the same number of protons in the nucleus), but they can have different numbers of neutrons and, therefore, different atomic weights. Two atoms of the same element with different numbers of neutrons in their nuclei and different atomic weights are known as isotopes of that element. For example, two isotopes of oxygen are ^{16}O and ^{18}O, where 16 and 18 are the atomic weights. Both isotopes have an atomic number of 8, but ^{18}O has two more neutrons than ^{16}O. Such study is proving very useful in learning how Earth works. For example, the study of oxygen isotopes has resulted in a better understanding of how the global climate has changed. This topic is beyond the scope of our present discussion but can be found in many basic textbooks on oceanography.

An atom is chemically balanced in terms of electric charge when the number of protons in the nucleus is equal to the number of electrons. However, an atom may lose or gain electrons, changing the balance in the electrical charge. An atom that has lost or gained electrons is called an ion. An atom that has lost one or more electrons has a net positive charge and is called a cation. For example, the potassium ion K^+ has lost one electron, and the calcium ion Ca^{2+} has lost two electrons. An atom that has gained electrons has a net negative charge and is called an anion. For example, O^{2-} is an anion of oxygen that has gained two electrons.

A compound is a chemical substance composed of two or more atoms of the same or different elements. The smallest unit of a compound is a molecule. For example, each molecule of water, H_2O, contains two atoms of hydrogen and one atom of oxygen, held together by chemical bonds. Minerals that form rocks are compounds, as are most chemical substances found in a solid, liquid, or gaseous state in the environment.

The atoms that constitute a compound are held together by *chemical bonding*. The four main types of chemical bonds are covalent, ionic, Vander Waals, and metallic. It is important to recognize that, when talking about chemical bonding of compounds, we are dealing with a complex subject. Although some compounds have a particular type of bond, many other compounds have more than one type of bond. With this caveat in mind, let's define each type of bond.

Covalent bonds result when atoms share electrons. This sharing takes place in the region between the atoms, and the strength of the bond is related to the number of pairs of electrons that are shared. Some important environmental compounds are held together solely by covalent bonds. These include carbon dioxide (CO_2) and water (H_2O). Covalent bonds are stronger than *ionic bonds*, which form as a result of attraction between cations and anions. An example of an environmentally important compound with ionic bonds is table salt (mineral halite), or sodium chloride (NaCL). Compounds with ionic bonds such as sodium chloride tend to be soluble in water and, thus, dissolve easily, as salt does. Van der Waals bonds are weak forces of attraction between molecules that are not bound to each other. Such bonding is much weaker than either covalent or ionic bonding. For example, the mineral graphite (which is the "lead" in pencils) is black and consists of sheets of carbon atoms that easily part or break from one another because the bonds are of the weak Van der Waals type. Finally, metallic bonds are those in which electrons are *c* shared, as with covalent bonds. However, they differ because, in metallic bonding, the electrons are shared by all atoms of the solid, rather than by specific atoms. As a result, the electrons can flow. For example, the mineral and element gold is an excellent conductor of electricity and can be pounded easily into thin sheets because the electrons have the freedom of movement that is characteristic of metallic bonding.

In summary, in the study of the environment, we are concerned with matter (chemicals) and energy that moves in and between the major components of the Earth system.

An example is the element carbon, which moves through the atmosphere, hydrosphere, lithosphere, and biosphere in a large variety of compounds. These include carbon dioxide (CO_2) and methane (CH_4), which are gases in the atmosphere; sugar ($C_6H_{12}O_6$) in plants and animals; and complex hydrocarbons (compounds of hydrogen and carbon) in coal and oil deposits.

Chemical Reactions

It is important to acknowledge in our discussion of how chemical cycles work that the emphasis is on chemistry. Many chemical reactions occur within and between the living and nonliving portions of ecosystems. A **chemical reaction** is a process in which new chemicals are formed from elements and compounds that undergo a chemical change. For example, a simple reaction between rainwater (H_2O) and carbon dioxide (CO_2) in the atmosphere produces weak carbonic acid (H_2CO_3):

$$H_2O + CO_2 \rightarrow H_2CO_3$$

This weak acid reacts with earth materials, such as rock and soil, to release chemicals into the environment. The released chemicals include calcium, sodium, magnesium, and sulfur, with smaller amounts of heavy metals such as lead, mercury, and arsenic. The chemicals appear in various forms, such as compounds and ions in solution.

Many other chemical reactions determine whether chemicals are available to life. For example, photosynthesis is a series of chemical reactions by which living green plants, with sunlight as an energy source, convert carbon dioxide (CO_2) and water (H_2O) to sugar ($C_6H_{12}O_6$) and oxygen (O_2). The general chemical reaction for photosynthesis is:

$$6CO_2 + 6H_2O \xrightarrow{\textit{sunlight}} C_6H_{12}O_6 + 6O_2$$

Photosynthesis produces oxygen as a by-product, and that is why we have free oxygen in our atmosphere.

After considering the two chemical reactions and applying critical thinking, you may recognize that both reactions combine water and carbon dioxide, but the products are very different: carbonic acid in one combination and a sugar in the other. How can this be so? The answer lies in an important difference between the simple reaction in the atmosphere that takes place to produce carbonic acid and the production of sugar and oxygen in the series of reactions of photosynthesis. Green plants use the energy from the sun, which they absorb through the chemical chlorophyll. Thus, active solar energy is converted to a stored chemical energy in sugar.

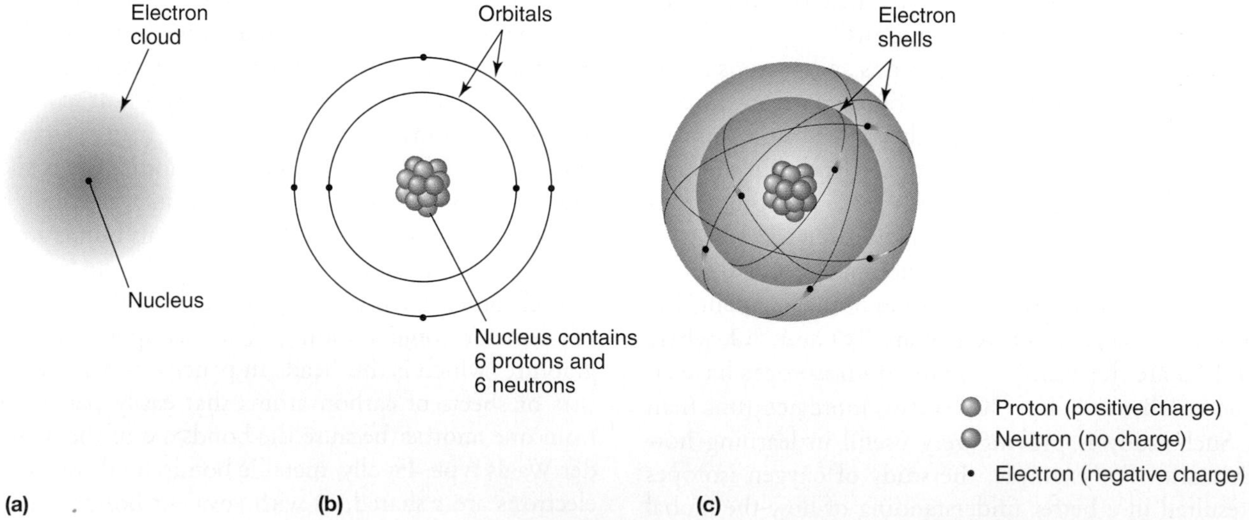

E.1 **(a)** Idealized diagram showing the basic structure of an atom as a nucleus surrounded by a cloud of electrons. **(b)** Conceptualized model of an atom of carbon with six protons and six neutrons in the nucleus and six orbiting electrons in two energy shells. **(c)** Three-dimensional view of (b). Size of nucleus relative to size of electron shells is greatly exaggerated.
[*Source*: After F. Press, and R. Siever *Understanding Earth* (New York: Freeman, 1994).]

Glossary

Abortion rate The estimated number of abortions per 1,000 women aged 15 to 44 in a given year. (Ages 15 to 44 are taken to be the limits of ages during which women can have babies. This, of course, is an approximation, made for convenience.)

Abortion ratio The estimated number of abortions per 1,000 live births in a given year.

Acceptable Risk The risk that individuals, society, or institutions are willing to take.

Acid mine drainage Acidic water that drains from mining areas (mostly coal but also metal mines). The acidic water may enter surface water resources, causing environmental damage.

Acid rain Rain made acid by pollutants, particularly oxides of sulfur and nitrogen. (Natural rainwater is slightly acid owing to the effect of carbon dioxide dissolved in the water.)

Active solar energy systems Direct use of solar energy that requires mechanical power; usually consists of pumps and other machinery to circulate air, water, or other fluids from solar collectors to a heat sink where the heat may be stored.

Acute disease a disease that appears rapidly in the population, affects a comparatively large percentage of it, and then declines or almost disappears for a while, only to reappear later.

Adaptive radiation The process that occurs when a species enters a new habitat that has unoccupied niches and evolves into a group of new species, each adapted to one of these niches.

Advanced wastewater treatment Treatment of wastewater beyond primary and secondary procedures. May include sand filters, carbon filters, or application of chemicals to assist in removing potential pollutants such as nutrients from the wastewater stream.

Aerobic Characterized by the presence of free oxygen.

Aesthetic justification for the conservation of nature An argument for the conservation of nature on the grounds that nature is beautiful and that beauty is important and valuable to people.

Age dependency ratio The ratio of dependent-age people (those unable to work) to working-age people. It is customary to define working-age people as those aged 15 to 65.

Age structure (of a population) A population divided into groups by age. Sometimes the groups represent the actual number of each age in the population; sometimes the groups represent the percentage or proportion of the population of each age.

Agroecosystem An ecosystem created by agriculture. Typically it has low genetic, species, and habitat diversity.

Air quality standards Levels of air pollutants that delineate acceptable levels of pollution over a particular time period. Valuable because they are often tied to emission standards that attempt to control air pollution.

Air toxics Those air pollutants known or suspected to cause cancer and other serious health problems from either long- or short-term exposure.

Allowance trading An approach to managing coal resources and reducing pollution through buying, selling, and trading of allowances to emit pollutants from burning coal. The idea is to control pollution by controlling the number of allowances issued.

Alpha particles One of the major types of nuclear radiation, consisting of two protons and two neutrons (a helium nucleus).

Alternative energy Renewable and nonrenewable energy resources that are alternatives to the fossil fuels.

Anaerobic Characterized by the absence of free oxygen.

Aquaculture Production of food from aquatic habitats.

Aquifer An underground zone or body of earth material from which groundwater can be obtained from a well at a useful rate.

Area sources Sometimes also called nonpoint sources. These are diffuse sources of pollution such as urban runoff or automobile exhaust. These sources include emissions that may be over a broad area or even over an entire region. They are often difficult to isolate and correct because of the widely dispersed nature of the emissions.

Asbestos A term for several minerals that have the form of small elongated particles. Some types of particles are believed to be carcinogenic or to carry with them carcinogenic materials.

Atmosphere Layer of gases surrounding Earth.

Atmospheric inversion A condition in which warmer air is found above cooler air, restricting air circulation; often associated with a pollution event in urban areas.

Autotroph An organism that produces its own food from inorganic compounds and a source of energy. There are photoautotrophs and chemical autotrophs.

Average residence time A measure of the time it takes for a given part of the total pool or reservoir of a particular material in a system to be cycled through the system. When the size of the pool and rate of throughput are constant, average residence time is the ratio of the total size of the pool or reservoir to the average rate of transfer through the pool.

Balance of nature An environmental myth that the natural environment, when not influenced by human activity, will reach a constant status, unchanging over time, referred to as an equilibrium state.

Barrier island An island separated from the mainland by a salt marsh. It generally consists of a multiple system of beach ridges and is separated from other barrier islands by inlets that allow the exchange of seawater with lagoon water.

Becquerel A unit commonly used for radioactive decay in the International System (IS) of measurement.

Beta particles One of the three major kinds of nuclear radiation; electrons that are emitted when one of the protons or neutrons in the nucleus of an isotope spontaneously changes.

Biochemical oxygen demand (BOD) A measure of the amount of oxygen necessary to decompose organic material in a unit volume of water. As the amount of organic waste in water increases, more oxygen is used, resulting in a higher BOD.

Biogeochemical cycle The cycling of a chemical element through the biosphere; its pathways, storage locations, and chemical forms in living things, the atmosphere, oceans, sediments, and lithosphere.

Biogeography The large-scale geographic pattern in the distribution of species, and the causes and history of this distribution.

Biohydromettalurgy Combining biological and mining processes, usually involving microbes to help extract valuable metals such as gold from the ground. May also be used to remove pollutants from mining waste.

Biological control A set of methods to control pest organisms by using natural ecological interactions, including predation, parasitism, and competition. Part of integrated pest management.

Biological diversity Used loosely to mean the variety of life on Earth, but scientifically typically used as to consisting of three components: (1) genetic diversity—the total number of genetic characteristics; (2) species diversity; and (3) habitat or ecosystem diversity—the number of kinds of habitats or ecosystems in a given unit area. Species diversity in turn includes three concepts: *species richness*, *evenness*, and *dominance*.

Biological evolution The change in inherited characteristics of a population from generation to generation, which can result in new species.

Biological production The capture of usable energy from the environment to produce organic compounds in which that energy is stored.

Biomagnification Also called *biological concentration*. The tendency for some substances to concentrate with each trophic level. Organisms preferentially store certain chemicals and excrete others. When this occurs consistently among organisms, the stored chemicals increase as a percentage of the body weight as the material is transferred along a food chain or trophic level. For example, the concentration of DDT is greater in herbivores than in plants and greater in plants than in the nonliving environment.

Biomass The amount of living material, or the amount of organic material contained in living organisms, both as live and dead material, as in the leaves (live) and stem wood (dead) of trees.

Biomass energy The energy that may be recovered from biomass, which is organic material such as plants and animal waste.

Biomass fuel A new name for the oldest fuel used by humans. Organic matter, such as plant material and animal waste, that can be used as a fuel.

Biome A kind of ecosystem. The rain forest is an example of a biome; rain forests occur in many parts of the world but are not all connected to each other.

Bioremediation A method of treating groundwater pollution problems that utilizes microorganisms in the ground to consume or break down pollutants.

Biosphere Has several meanings. One is that part of a planet where life exists. On Earth it extends from the depths of the oceans to the summits of mountains, but most life exists within a few meters of the surface. A second meaning is the planetary system that includes and sustains life, and therefore is made up of the atmosphere, oceans, soils, upper bedrock, and all life.

Biota All the organisms of all species living in an area or region up to and including the biosphere, as in "the biota of the Mojave Desert" or "the biota in that aquarium."

Biotic province A geographic region inhabited by life-forms (species, families, orders) of common ancestry, bounded by barriers that prevent the spread of the distinctive kinds of life to other regions and the immigration of foreign species into that region.

Birth control The number born divided by the total number in the population.

Birth rate The rate at which births occur in a population, measured either as the number of individuals born per unit of time or as the percentage of births per unit of time compared with the total population.

Black lung disease Often called coal miner disease because it is caused by years of inhaling coal dust, resulting in damage to the lungs.

Body burden The amount of concentration of a toxic chemical, especially radionuclides, in an individual.

Breeder reactor A type of nuclear reactor that utilizes between 40% and 70% of its nuclear fuel and converts fertile nuclei to fissile nuclei faster than the rate of fission. Thus breeder reactors actually produce nuclear fuels.

Brines With respect to mineral resources, refers to waters with a high salinity that contain useful materials such as bromine, iodine, calcium chloride, and magnesium.

Buffers Materials (chemicals) that have the ability to neutralize acids. Examples include the calcium carbonate that is present in many soils and rocks. These materials may lessen potential adverse effects of acid rain.

Burner reactors A type of nuclear reactor that consumes more fissionable material than it produces.

Capillary action The rise of water along narrow passages, facilitated and caused by surface tension.

Carbon cycle Biogeochemical cycle of carbon. Carbon combines with and is chemically and biologically linked with the cycles of oxygen and hydrogen that form the major compounds of life.

Carbon monoxide (CO) A colorless, odorless gas that at very low concentrations is extremely toxic to humans and animals.

Carbon-silicate cycle A complex biogeochemical cycle over time scales as long as one-half billion years. Included in this cycle are major geologic processes, such as weathering, transport by ground and surface waters, erosion, and deposition of crustal rocks. The carbonate-silicate cycle is believed to provide important negative feedback mechanisms that control the temperature of the atmosphere.

Carcinogen Any material that is known to produce cancer in humans or other animals.

Carnivores Organisms that feed on other live organisms; usually applied to animals that eat other animals.

Carrying capacity The maximum abundance of a population or species that can be maintained by a habitat or ecosystem without degrading the ability of that habitat or ecosystem to maintain that abundance in the future.

Cash crops Crops grown to be traded in a market.

Catastrophe A situation or event that causes great damage to people, property or society and from which recovery is a long and involved process. Also defined as a very serious disaster.

Catch per unit effort The number of animals caught per unit of effort, such as the number of fish caught by a fishing ship per day. It is used to estimate the population abundance of a species.

Channelization An engineering technique that consists of straightening, deepening, widening, clearing, or lining existing stream channels. The purpose is to control floods, improve drainage, control erosion, or improve navigation. It is a very controversial practice that may have significant environmental impacts.

Chaparral A dense scrubland found in areas with Mediterranean climate (a long warm, dry season and a cooler rainy season).

Chemical reaction The process in which compounds and elements undergo a chemical change to become a new substance or substances.

Chemoautotrophs Autotrophic bacteria that can derive energy from chemical reactions of simple inorganic compounds.

Chemosynthesis Synthesis of organic compounds by energy derived from chemical reactions.

Chimney (or stack) effect A process whereby warmer air rises in buildings to upper levels and is replaced in the lower portion of the building by outdoor air drawn through a variety of openings, such as windows, doors, or cracks in the foundations and walls.

Chlorofluorocarbons (CFCs) Highly stable compounds that have been or are being used in spray cans as aerosol propellants and in refrigeration units (the gas that is compressed and expanded in a cooling unit). Emissions of chlorofluorocarbons have been associated with potential global warming and stratospheric ozone depletion.

Chronic disease A disease that is persistent in a population, typically occurring in a relatively small but constant percentage of the population.

Chronic hunger A condition in which there is enough food available per person to stay alive, but not enough to lead a satisfactory and productive life.

Chronic patchiness A situation where ecological succession does not occur. One species may replace another, or an individual of the first species may replace it, but no overall general temporal pattern is established. Characteristic of harsh environments such as deserts.

Classical stability A system characterized by constant conditions that, if disturbed from those conditions, will return to it once the factor that disturbed the system has been removed.

Clay May refer to a mineral family or to a very fine-grained sediment. It is associated with many environmental problems, such as shrinking and swelling of soils and sediment pollution.

Clean Air Act Amendments of 1990 Comprehensive regulations (federal statute) that address acid rain, toxic emissions, ozone depletion, and automobile exhaust.

Clear-cutting In timber harvesting, the practice of cutting all trees in a stand at the same time.

Climate The representative or characteristic conditions of the atmosphere at particular places on Earth. Climate refers to the average or expected conditions over long periods; weather refers to the particular conditions at one time in one place.

Climatic change Change in mean annual temperature and other aspects of climate over periods of time ranging from decades to hundreds of years to several million years.

Climate forcing An imposed perturbation of Earth energy that balance major climatic forcings associated with global warming. Includes: greenhouse gases, such as carbon dioxide and methane; reflective aerosols in the atmosphere, black carbon. Forcing of the climate system also includes solar activity and Milankovitch Cycles.

Climax stage (or ecological succession) The final stage of ecological succession and therefore an ecological community that continues to reproduce itself over time.

Closed system A system in which there are definite boundaries to mass and energy and thus exchange of these factors with other systems does not occur.

Closed-canopy forest Forests in which the leaves of adjacent trees overlap or touch, so that the trees form essentially continuous cover.

Coal Solid, brittle carbonaceous rock that is one of the world's most abundant fossil fuels. It is classified according to energy content as well as carbon and sulfur content.

Coal gasification Process that converts coal that is relatively high in sulfur to a gas in order to remove the sulfur.

Cogeneration The capture and use of waste heat; for example, using waste heat from a power plant to heat adjacent factories and other buildings.

Cohort All the individuals in a population born during the same time period. Thus all the people born during the year 2005 represent the world human cohort for that year.

Common law Law derived from custom, judgment, or decrees of courts rather than from legislation.

Commons Land that belongs to the public, not to individuals. Historically a part of old English and New England towns where all the farmers could graze their cattle.

Community, ecological A group of populations of different species living in the same local area and interacting with one another. A community is the living portion of an ecosystem.

Community effect (community-level effect) When the interaction between two species leads to changes in the presence or absence of other species or to a large change in abundance of other species, then a community effect is said to have occurred.

Competition The situation that exists when different individuals, populations, or species compete for the same resource(s) and the presence of one has a detrimental effect on the other. Sheep and cows eating grass in the same field are competitors.

Competitive exclusion principle The idea that two populations of different species with exactly the same requirements cannot persist indefinitely in the same habitat—one will always win out and the other will become extinct.

Composting Biochemical process in which organic materials, such as lawn clippings and kitchen scraps, are decomposed to a rich, soil-like material.

Comprehensive plan An official plan adopted by local government formally stating general and long-range policies concerning future development.

Cone of depression A cone-shaped depression in the water table around a well caused by withdrawal by pumping water faster than the water can be replenished by natural groundwater flow.

Conservation With respect to resources such as energy, refers to changing our patterns of use or simply getting by with less. In a pragmatic sense the term means adjusting our needs to minimize the use of a particular resource, such as energy.

Consumptive use A type of off-stream water use. This water is consumed by plants and animals or in industrial processes or evaporates during use. It is not returned to its source.

Contamination The presence of undesirable material that makes something unfit for a particular use.

Continental drift The movement of continents in response to seafloor spreading. The most recent episode of continental drift started about 200 million years ago with the breakup of the supercontinent Pangaea.

Continental shelf The relatively shallow ocean area between the shoreline and the continental slope that extends to approximately a 600-foot (~200 m) water depth surrounding a continent.

Contour plowing Plowing land along topographic contours, as much in a horizontal plane as possible, thereby decreasing the erosion rate.

Controlled experiment A controlled experiment is designed to test the effects of independent variables on a dependent variable by changing only one independent variable at a time. For each variable tested, there are two setups (an experiment and a control) that are identical except for the independent variable being tested. Any difference in the outcome (dependent variable) between the experiment and the control can then be attributed to the effects of the independent variable tested.

Convection The transfer of heat involving the movement of particles; for example, the boiling water in which hot water rises to the surface and displaces cooler water, which moves toward the bottom.

Convergent evolution The process by which species evolve in different places or different times and, although they have different genetic heritages, develop similar external forms and structures as a result of adaptation to similar environments. The similarity in the shapes of sharks and porpoises is an example of convergent evolution.

Convergent plate boundary A boundary between two lithosphere plates in which one plate descends below the other (subduction).

Cosmopolitan species A species with a broad distribution, occurring wherever in the world the environment is appropriate.

Creative justification for the conservation of nature An argument for the conservation of nature on the grounds that people often find sources of artistic and scientific creativity in their contacts with the unspoiled natural world.

Criteria pollutants Are the sixth most common air pollutants: sulfur dioxide, nitrogen oxides, carbon monoxide, ozone and other photochemical oxidants, particulate matter, and lead.

Crop rotation A series of different crops planted successively in the same field, with the field occasionally left fallow, or grown with a cover crop.

Crude oil Naturally occurring petroleum, normally pumped from wells in oil fields. Refinement of crude oil produces most of the petroleum products we use today.

Cultural eutrophication Human-induced eutrophication that involves nutrients such as nitrates or phosphates that cause a rapid increase in the rate of plant growth in ponds, lakes, rivers, or the ocean.

Cultural justification With respect to environmental values refers to the fact that different cultures have many of the same values but differ in others.

Curie Commonly used unit to measure radioactive decay; the amount of radioactivity from 1 gram of radium 226 that undergoes about 37 billion nuclear transformations per second.

Death rate The rate at which deaths occur in a population, measured either as the number of individuals dying per unit time or as the percentage of a population dying per unit time.

Decomposers Organisms that feed on dead organic matter.

Deductive reasoning Drawing a conclusion from initial definitions and assumptions by means of logical reasoning.

Deep-well disposal Method of disposal of hazardous liquid waste that involves pumping the waste deep into the ground below and completely isolated from all freshwater aquifers. A controversial method of waste disposal that is being carefully evaluated.

Demand for food The amount of food that would be bought at a given price if it were available.

Demand-based agriculture Agriculture with production determined by economic demand and limited by that demand rather than by resources.

Demographic transition The pattern of change in birth and death rates as a country is transformed from undeveloped to developed. There are three stages: (1) in an undeveloped country, birth and death rates are high and the growth rate low; (2) the death rate decreases, but the birth rate remains high and the growth rate is high; (3) the birth rate drops toward the death rate and the growth rate therefore also decreases.

Demography The study of populations, especially their patterns in space and time.

Dependent variable See **Variable, dependent.**

Denitrification The conversion of nitrate to molecular nitrogen by the action of bacteria—an important step in the nitrogen cycle.

Density-dependent population effects Factors whose effects on a population change with population density.

Density-independent population effects Changes in the size of a population due to factors that are independent of the population size. For example, a storm that knocks down all trees in a forest, no matter how many there are, is a density-independent population effect.

Desalination The removal of salts from seawater or brackish water so that the water can be used for purposes such as agriculture, industrial processes, or human consumption.

Desertification The process of creating a desert where there was not one before.

Dioxin An organic compound composed of oxygen, hydrogen, carbon, and chlorine. About 75 types are known. Dioxin is not normally manufactured intentionally but is a by-product resulting from chemical reactions in the production of other materials, such as herbicides. Known to be extremely toxic to mammals, its effects on the human body are being intensively studied and evaluated.

Direct costs Costs borne by the producer in obtaining, processing, and distributing a product.

Direct effects With respect to natural hazards, refers to the number of people killed, injured, dislocated, made homeless or otherwise damaged by a hazardous event.

Disaster A hazardous event that occurs over a limited span of time in a defined geographic area. Loss of human life and property damage are significant.

Disprovability The idea that a statement can be said to be scientific if someone can clearly state a method or test by which it might be disproved.

Divergent evolution Organisms with the same ancestral genetic heritage migrate to different habitats and evolve into species with different external forms and structures, but typically continue to use the same kind of habitats. The ostrich and the emu are believed to be examples of divergent evolution.

Divergent plate boundary A boundary between lithospheric plates characterized by the production of new lithosphere; found along oceanic ridges.

Diversity, genetic The total number of genetic characteristics, sometimes of a specific species, subspecies, or group of species.

Diversity, habitat The number of kinds of habitats in a given unit area.

Diversity, species Used loosely to mean the variety of species in an area or on Earth. Technically, it is composed of three components: species richness—the total number of species; species evenness—the relative abundance of species; and species dominance—the most abundant species.

Dobson unit Commonly used to measure the concentration of ozone. One Dobson unit is equivalent to a concentration of 1 ppb ozone.

Dominant species Generally, the species that are most abundant in an area, ecological community, or ecosystem.

Dominants In forestry, the tallest, most numerous, and most vigorous trees in a forest community.

Dose dependency Dependence on the dose or concentration of a substance for its effects on a particular organism.

Dose–response The principle that the effect of a certain chemical on an individual depends on the dose or concentration of that chemical.

Doubling time The time necessary for a quantity of whatever is being measured to double.

Drainage basin The area that contributes surface water to a particular stream network.

Drip irrigation Irrigation of soil through tubes that drip water slowly, greatly reducing the loss of water from direct evaporation and increasing yield.

Drought A period of months or more commonly years of unusually dry weather.

Dynamic equilibrium A steady state of a system that with negative feedback will return to a quasi-equilibrium state following disturbance.

Dynamic system Characterized by a system that changes often and continually over time.

Early-successional species Species that occur only or primarily during early stages of succession.

Earth system science The science of Earth as a system. It includes understanding processes and linkages between the lithosphere, hydrosphere, biosphere, and atmosphere.

Earthquake Generation of earthquake or seismic waves when rocks under stress fracture and break, resulting in displacement along a fault.

Ecological community This term has two meanings. (1) A conceptual or functional meaning: a set of interacting species that occur in the same place (sometimes extended to mean a set that interacts in a way to sustain life). (2) An operational meaning: a set of species found in an area, whether or not they are interacting.

Ecological economics Study and evaluation of relations between humans and the economy with emphasis on long-term health of ecosystems and sustainability.

Ecological gradient A change in the relative abundance of a species or group of species along a line or over an area.

Ecological island An area that is biologically isolated so that a species occurring within the area cannot mix (or only rarely mixes) with any other population of the same species.

Ecological justification for the conservation of nature An argument for the conservation of nature on the grounds that a species, an ecological community, an ecosystem, or Earth's biosphere provides specific functions necessary to the persistence of our life or of benefit to life. The ability of trees in forests to remove carbon dioxide produced in burning fossil fuels is such a public benefit and an argument for maintaining large areas of forests.

Ecological niche The general concept is that the niche is a species' "profession"—what it does to make a living. The term is also used to refer to a set of environmental conditions within which a species is able to persist.

Ecological succession The process of the development of an ecological community or ecosystem, usually viewed as a series of stages—early, middle, late, mature (or climax), and sometimes postclimax. Primary succession is an original establishment; secondary succession is a reestablishment.

Ecology The science of the study of the relationships between living things and their environment.

Ecosystem An ecological community and its local, nonbiological community. An ecosystem is the minimum system that includes and sustains life. It must include at least an autotroph, a decomposer, a liquid medium, a source and sink of energy, and all the chemical elements required by the autotroph and the decomposer.

Ecosystem effect Effects that result from interactions among different species, effects of species on chemical elements in their environment, and conditions of the environment.

Ecosystem energy flow The flow of energy through an ecosystem—from the external environment through a series of organisms and back to the external environment.

Ecotopia A society based on sustainable development and sound environmental planning characterized by a stable human population within the carrying capacity of earth. It is thought of as an ideal state.

Ecotourism Tourism based on an interest in observing of nature.

ED-50 The effective dose, or dose that causes an effect in 50% of the population on exposure to a particular toxicant. It is related to the onset of specific symptoms, such as loss of hearing, nausea, or slurred speech.

Edge effect An effect that occurs following the forming of an ecological island; in the early phases the species diversity along the edge is greater than in the interior. Species escape from the cut area and seek refuge in the border of the forest, where some may last only a short time.

Efficiency The primary definition used in the text is the ratio of output to input. With machines, usually the ratio of work or power produced to the energy or power used to operate or fuel them. With living things, efficiency may be defined as either the useful work done or the energy stored in a useful form compared with the energy taken in.

Efficiency improvements With respect to energy, refers to designing equipment that will yield more energy output from a given amount of energy input.

Effluent Any material that flows outward from something. Examples include wastewater from hydroelectric plants and water discharged into streams from waste-disposal sites.

Effluent stream Type of stream where flow is maintained during the dry season by groundwater seepage into the channel.

El Niño Natural perturbation of the physical earth system that affects global climate. Characterized by development of warm oceanic waters in the eastern part of the tropical Pacific Ocean, a weakening or reversal of the trade winds, and a weakening or even reversal of the equatorial ocean currents. Reoccurs periodically and affects the atmosphere and global temperature by pumping heat into the atmosphere.

Electromagnetic fields (EMFs) Magnetic and electrical fields produced naturally by our planet and also by appliances such as toasters, electric blankets, and computers. There currently is controversy concerning potential adverse health effects related to exposure to EMFs in the workplace and home from such artificial sources as power lines and appliances.

Electromagnetic spectrum All the possible wavelengths of electromagnetic energy, considered as a continuous range. The spectrum includes long wavelength (used in radio transmission), infrared, visible, ultraviolet, X rays, and gamma rays.

Endangered species A species that faces threats that might lead to its extinction in a short time.

Endemic species A species that has evolved in, and lives only within, a specific location. e.g. the California condor is endemic to the Pacific coast of North America.

Energy An abstract concept referring to the ability or capacity to do work.

Energy efficiency Refers to both first-law efficiency and second-law efficiency, where first-law efficiency is the ratio of the actual amount of energy delivered to the amount of energy supplied to meet a particular need, and second-law efficiency is the ratio of the maximum available work needed to perform a particular task to the actual work used to perform that task.

Energy flow The movement of energy through an ecosystem from the external environment through a series of organisms and back to the external environment. It is one of the fundamental processes common to all ecosystems.

Entropy A measure in a system of the amount of energy that is unavailable for useful work. As the disorder of a system increases, the entropy in a system also increases.

Environment All factors (living and nonliving) that actually affect an individual organism or population at any point in the life cycle. *Environment* is also sometimes used to denote a certain set of circumstances surrounding a particular occurrence (environments of deposition, for example).

Environmental audit A process of determining the past history of a particular site, with special reference to the existence of toxic materials or waste.

Environmental economics Economic effects of the environment and how economic processes affect that environment, including its living resources.

Environmental ethics A school, or theory, in philosophy that deals with the ethical value of the environment.

Environmental geology The application of geologic information to environmental problems.

Environmental impact The effects of some action on the environment, particularly action by human beings.

Environmental impact report (EIR) Similar to the environmental impact statement, a report describing potential environmental impacts resulting from a particular project, often at the state level.

Environmental impact statement (EIS) A written statement that assesses and explores possible impacts associated with a particular project that may affect the human environment. The statement is required in the United States by the National Environmental Policy Act of 1969.

Environmental justice The principle of dealing with environmental problems in such a way as to not discriminate against people based upon socioeconomic status, race, or ethnic group.

Environmental law A field of law concerning the conservation and use of natural resources and the control of pollution.

Environmental tobacco smoke Commonly called secondhand smoke from people smoking tobacco.

Environmental unity A principle of environmental sciences that states that everything affects everything else, meaning that a particular course of action could lead to a string of events. Another way of stating this idea is that you can't do only one thing.

Environmentalism A social, political, and ethical movement concerned with protecting the environment and using its resources wisely.

Epidemic disease A disease that appears occasionally in the population, affects a large percentage of it, and declines or almost disappears for a while only to reappear later.

Equilibrium A point of rest. At equilibrium, a system remains in a single, fixed condition and is said to be in equilibrium. Compare with **Steady state**.

Eukaryote An organism whose cells have nuclei and organelles. The eukaryotes include animals, fungi, vegetation, and many single-cell organisms.

Eutrophic Referring to bodies of water having an abundance of the chemical elements required for life.

Eutrophication Increase in the concentration of chemical elements required for living things (for example, phosphorus). Increased nutrient loading may lead to a population explosion of photosynthetic algae and blue-green bacteria that become so thick that light cannot penetrate the water. Bacteria deprived of light beneath the surface die; as they decompose, dissolved oxygen in the lake is lowered and eventually a fish kill may result. Eutrophication of lakes caused by human-induced processes, such as nutrient-rich sewage water entering a body of water, is called cultural eutrophication.

Even-aged stands A forest of trees that began growth in or about the same year.

Evolution, biological The change in inherited characteristics of a population from generation to generation, sometimes resulting in a new species.

Evolution, nonbiological Outside the realm of biology, the term *evolution* is used broadly to mean the history and development of something.

Exotic species Species introduced into a new area, one in which it had not evolved.

Experimental errors There are two kinds of experimental errors, random and systematic. Random errors are those due to chance events, such as air currents pushing on a scale and altering a measurement of weight. In contrast, a miscalibration of an instrument would lead to a systematic error. Human errors can be either random or systematic.

Exponential growth Growth in which the rate of increase is a constant percentage of the current size; that is, the growth occurs at a constant rate per time period.

Exponential growth rate The annual growth rate is a constant percentage of the population.

Externality In economics, an effect not normally accounted for in the cost–revenue analysis.

Extinction Disappearance of a life-form from existence; usually applied to a species.

Facilitation During succession, one species prepares the way for the next (and may even be necessary for the occurrence of the next).

Fact Something that is known based on actual experience and observation.

Fall line The point on a river where there is an abrupt drop in elevation of the land and where numerous waterfalls occur. The line in the eastern United States is located where streams pass from harder to softer rocks.

Fallow A farm field unplanted or allowed to grow with a cover crop without harvesting for at least one season.

Fecal coliform bacteria Bacteria that occur naturally in human intestines and are used as a standard measure of microbial pollution and an indicator of disease potential for a water source.

Feedback A kind of system response that occurs when output of the system also serves as input leading to changes in the system.

First law of thermodynamics The principle that energy may not be created or destroyed but is always conserved.

First-law efficiency The ratio of the actual amount of energy delivered where it is needed to the amount of energy supplied in order to meet that need; expressed as a percentage.

Fission The splitting of an atom into smaller fragments with the release of energy.

Flood Inundation of an area by water, often produced by intense rain storms, melting of snow, storm surges from a hurricane, or tsunami, or failure of a flood-protection structure such as a dam.

Flooding, natural The process whereby waters emerge from their stream channel to cover part of the floodplain. Natural flooding is not a problem until people choose to build homes and other structures on floodplains.

Floodplain Flat topography adjacent to a stream in a river valley that has been produced by the combination of overbank flow and lateral migration of meander bends.

Fluidized-bed combustion A process used during the combustion of coal to eliminate sulfur oxides. Involves mixing finely ground limestone with coal and burning it in suspension.

Flux The rate of transfer of material within a system per unit time.

Food chain The linkage of who feeds on whom.

Food-chain concentration See **Biomagnification**.

Food web A network of who feeds on whom or a diagram showing who feeds on whom. It is synonymous with **food chain**.

Force A push or pull that affects motion. The product of mass and acceleration of a material.

Forcing With respect to global change, processes capable of changing global temperature, such as changes in solar energy emitted from the sun, or volcanic activity.

Fossil fuels Forms of stored solar energy created from incomplete biological decomposition of dead organic matter. Include coal, crude oil, and natural gas.

Fuel cell A device that produces electricity directly from a chemical reaction in a specially designed cell. In the simplest case the cell uses hydrogen as a fuel, to which an oxidant is supplied. The hydrogen is combined with oxygen as if the hydrogen were burned, but the reactants are separated by an electrolyte solution that facilitates the migration of ions and the release of electrons (which may be tapped as electricity).

Fugitive sources Type of stationary air pollution sources that generate pollutants from open areas exposed to wind processes.

Fusion, nuclear Combining of light elements to form heavier elements with the release of energy.

Gaia hypothesis The Gaia hypothesis states (1) that life has greatly altered the Earth's environment globally for more than 3 billion years and continues to do so; and (2) that these changes benefit life (increase its persistence). Some extend this, nonscientifically, to assert that life did it on purpose.

Gamma rays One of the three major kinds of nuclear radiation. A type of electromagnetic radiation emitted from the isotope similar to X-rays but more energetic and penetrating.

Garden city Land planning that considers a city and countryside together.

Gene A single unit of genetic information comprising of a complex segment of the four DNA base-paircompounds.

General circulation model (GCM) Consists of a group of computer models that focus on climate change using a series of equations, often based on conservation of mass and energy.

Genetic drift Changes in the frequency of a gene in a population as a result of chance rather than of mutation, selection, or migration.

Genetic risk Used in discussions of endangered species to mean detrimental change in genetic characteristics not caused by external environmental changes. Genetic changes can occur in small populations from such causes as reduced genetic variation, genetic drift, and mutation.

Genetically modified crops Crop species modified by genetic engineering to produce higher crop yields and increase resistance to drought, cold, heat, toxins, plant pests, and disease.

Genetically modified organisms Organisms created by genetic engineering, the altering of genes or genetic material to produce new organisms or organisms with desired characteristics, or to eliminate undesirable characteristics in organisms.

Geochemical cycles The pathways of chemical elements in geologic processes, including the chemistry of the lithosphere, atmosphere, and hydrosphere.

Geographic Information System (GIS) Technology capable of storing, retrieving, transferring, and displaying environmental data.

Geologic cycle The formation and destruction of earth materials and the processes responsible for these events. The geologic cycle includes the following subcycles: hydrologic, tectonic, rock, and geochemical.

Geometric growth See **Exponential growth**.

Geopressurized systems Geothermal systems that exist when the normal heat flow from the Earth is trapped by impermeable clay layers that act as an effective insulator.

Geothermal energy The useful conversion of natural heat from the interior of Earth.

Global circulation model (GCM) A type of mathematical model used to evaluate global change, particularly related to climatic change. GCMs are very complex and require supercomputers for their operation.

Global dimming The reduction of incoming solar radiation by reflection from suspended particles in the atmosphere and their interaction with water vapor (especially clouds).

Global forecasting Predicting or forecasting future change in environmental areas such as world population, natural resource utilization, and environmental degradation.

Global warming Natural or human-induced increase in the average global temperature of the atmosphere near Earth's surface.

Gravel Unconsolidated, generally rounded fragments of rocks and minerals greater than 2 mm in diameter.

Green building Designing buildings that have a healthy interior environment and landscape that benefits the local external environment as well.

Green plans Long-term strategies for identifying and solving global and regional environmental problems. The philosophical heart of green plans is sustainability.

Green revolution Name attached to post-World War II agricultural programs that have led to the development of new strains of crops with higher yield, better resistance to disease, or better ability to grow under poor conditions.

Greenbelt A belt of recreational parks, farmland, or uncultivated land surrounding or connecting urban communities, forming a system of countryside and urban landscapes.

Greenhouse effect Occurs when water vapor and several other gases warm the Earth's atmosphere by trapping some of the heat radiating from the Earth's atmospheric system.

Greenhouse gases The suite of gases that produce a greenhouse effect, such as carbon dioxide, methane, and water vapor.

Gross production (biology) Production before respiration losses are subtracted.

Groundwater Water found beneath the Earth's surface within the zone of saturation, below the water table.

Growth efficiency Gross production efficiency (P/C), or ratio of the material produced (P = net production) by an organism or population to the material ingested or consumed (C).

Growth rate The net increase in some factor per unit time. In ecology, the growth rate of a population, sometimes measured as the increase in numbers of individuals or biomass per unit time and sometimes as a percentage increase in numbers or biomass per unit time.

Habitat Where an individual, population, or species exists or can exist. For example, the habitat of the Joshua tree is the Mojave Desert of North America.

Half-life The time required for half the amount of a substance to disappear; the average time required for one-half of a radioisotope to be transformed to some other isotope; the time required for one-half of a toxic chemical to be converted to some other form.

Hard path Energy policy based on the emphasis of energy quantity generally produced from large, centralized power plants.

Hazardous waste Waste that is classified as definitely or potentially hazardous to the health of people. Examples include toxic or flammable liquids and a variety of heavy metals, pesticides, and solvents.

Heat energy Energy of the random motion of atoms and molecules.

Heat island Usually, a large city that is warmer air of a city than surrounding areas as a result of increased heat production and decreased heat loss because building and paving materials act as solar collectors.

Heat island effect Urban areas are several degrees warmer than their surrounding areas. During relatively calm periods there is an upward flow of air over heavily developed areas accompanied by a downward flow over nearby greenbelts. This produces an air-temperature profile that delineates the heat island.

Heat pumps Devices that transfer heat from one material to another, such as from groundwater to the air in a building.

Heat wave A period of days of weeks or unusually hot weather. A natural recurring weather phenomenon related to heating of the atmosphere and moving of air masses.

Heavy metals Refers to a number of metals, including lead, mercury, arsenic, and silver (among others); that have a relatively high atomic number (the number of protons in the nucleus of an atom). They are often toxic even at relatively low concentrations, causing a variety of environmental problems.

Herbivore An organism that feeds on an autotroph.

Heterotrophs Organisms that cannot make their own food from inorganic chemicals and a source of energy and therefore live by feeding on other organisms.

High-level radioactive waste Extremely toxic nuclear waste, such as spent fuel elements from commercial reactors.

Historical range of variation The known range of an environmental variable, such as the abundance of a species or the depth of a lake, over some past time interval.

Homeostasis The ability of a cell or organism to maintain a constant environment. Results from negative feedback, resulting in a state of dynamic equilibrium.

Hormonally active agents Chemicals in the environment able to cause reproductive and developmental abnormalities in animals, including humans.

Hot igneous systems Geothermal systems that involve hot, dry rocks with or without the presence of near-surface molten rock.

Human carrying capacity Theoretical estimates of the number of humans who could inhabit Earth at the same time.

Human demography The study of human population characteristics, such as age structure, demographic transition, total fertility, human population and environment relationships, death-rate factors, and standard of living.

Hurricane A tropical storm with circulating winds in excess of 120 km (74 mi) per hour that moves across warm ocean waters of the tropics.

Hutchinsonian niche The idea of a measured niche, the set of environmental conditions within which a species is able to persist.

Hydrocarbons Compounds containing only carbon and hydrogen. These organic compounds include petroleum products, such as crude oil and natural gas.

Hydrochlorofluorocarbons Also known as HCFCs, these are a group of chemicals containing hydrogen, chlorine, fluorine, and carbon, produced as a potential substitute for chlorofluorocarbons (CFCs).

Hydrofluorocarbons Chemicals containing hydrogen, fluorine, and carbons, produced as potential substitutes for chlorofluorocarbons (CFCs).

Hydrologic cycle Circulation of water from the oceans to the atmosphere and back to the oceans by way of evaporation, runoff from streams and rivers, and groundwater flow.

Hydrology The study of surface and subsurface water.

Hydroponics The practice of growing plants in a fertilized water solution on a completely artificial substrate in an artificial environment such as a greenhouse.

Hydrothermal convection systems A type of geothermal energy characterized by circulation of steam and/or hot water that transfers to the surface.

Hypothesis In science, an explanation set forth in a manner that can be tested and disproved. A tested hypothesis is accepted until and unless it has been disproved.

Igneous rocks Rocks made of solidified magma. They are extrusive if they crystallize on the surface of Earth and intrusive if they crystallize beneath the surface.

Incineration Combustion of waste at high temperature, consuming materials and leaving only ash and noncombustibles to dispose of in a landfill.

Independent variable See **Variable, independent.**

Indirect effects In regards to natural hazards, effects disaster. Include donations of money and goods as well as providing shelter for people and paying taxes that will help finance recovery and relief of emotional distress caused by natural hazardous events.

Inductive reasoning Drawing a general conclusion from a limited set of specific observations.

Industrial ecology The process of designing industrial systems to behave more like ecosystems where waste from one part of the system is a resource for another part.

Inference (1) A conclusion derived by logical reasoning from premises and/or evidence (observations or facts), or (2) a conclusion, based on evidence, arrived at by insight or analogy, rather than derived solely by logical processes.

Inflection point The point where a graphed curve or an equation representing that curve changes from convex to concave, or from concave to convex.

Influent stream Type of stream that is everywhere above the groundwater table and flows in direct response to precipitation. Water from the channel moves down to the water table, forming a recharge mound.

Input With respect to basic concepts of systems, refers to material or energy that enters a system.

Inspirational justification for the conservation of nature An argument for the conservation of nature on the grounds that direct experience of nature is an aid to spiritual or mental well-being.

In-stream use A type of water use that includes navigation, generation of hydroelectric power, fish and wildlife habitat, and recreation.

Intangible factor In economics, an intangible factor is one you can't touch directly, but you value it.

Integrated energy management Use of a range of energy options that vary from region to region, including a mix of technology and sources of energy.

Integrated pest management Control of agricultural pests using several methods together, including biological and chemical agents. A goal is to minimize the use of artificial chemicals; another goal is to prevent or slow the buildup of resistance by pests to chemical pesticides.

Integrated waste management (IWM) Set of management alternatives including reuse, source reduction, recycling, composting, landfill, and incineration.

Interference When, during succession, one species prevents the entrance of later-successional species into an ecosystem. For example, some grasses produce such dense mats that seeds of trees cannot reach the soil to germinate. As long as these grasses persist, the trees that characterize later stages of succession cannot enter the ecosystem.

Island arc A curved group of volcanic islands associated with a deep oceanic trench and subduction zone (convergent plate boundary).

Isotope Atoms of an element that have the same atomic number (the number of protons in the nucleus of the atom) but vary in atomic mass number (the number of protons plus neutrons in the nucleus of an atom).

Keystone species A species, such as the sea otter, that has a large effect on its community or ecosystem so that its removal or addition to the community leads to major changes in the abundances of many or all other species.

Kinetic energy The energy of motion. For example, the energy in a moving car that results from the mass of the car traveling at a particular velocity.

Kwashiorkor Lack of sufficient protein in the diet, which leads to a failure of neural development in infants and therefore to learning disabilities.

Lag time The delay in time between the cause and appearance of an effect in a system.

Land application Method of disposal of hazardous waste that involves intentional application of waste material to surface soil. Useful for certain biodegradable industrial waste, such as oil and petroleum waste, and some organic chemical waste.

Land ethic A set of ethical principles that affirm the right of all resources, including plants, animals, and earth materials, to continued existence and, at least in some locations, to continued existence in a natural state.

Landscape perspective The concept that effective management and conservation recognizes that ecosystems, populations, and species are interconnected across large geographic areas.

Landslide Comprehensive term for earth materials moving downslope.

Land-use planning A complex process involving development of a land-use plan to include a statement of land-use issues, goals, and objectives; a summary of data collection and analysis; a land-classification map; and a report that describes and indicates appropriate development in areas of special environmental concern.

Late-successional species Species that occur only or primarily in, or are dominant in, late stages in succession.

Law of the minimum (Liebig's law of the minimum) The concept that the growth or survival of a population is directly related to the single life requirement that is in least supply (rather than due to a combination of factors).

LD-50 A crude approximation of a chemical toxicity defined as the dose at which 50% of the population dies on exposure.

Leachate Noxious, mineralized liquid capable of transporting bacterial pollutants. Produced when water infiltrates through waste material and becomes contaminated and polluted.

Leaching Water infiltration from the surface, dissolving soil materials as part of chemical weathering processes and transporting the dissolved materials laterally or downward.

Lead A heavy metal that is an important constituent of automobile batteries and other industrial products. Lead is a toxic metal capable of causing environmental disruption and health problems to people and other living organisms.

Liebig's law of the minimum See **Law of the minimum**.

Life expectancy The estimated average number of years (or other time period used as a measure) that an individual of a specific age can expect to live.

Limiting factor The single requirement for growth available in the least supply in comparison to the need of an organism. Originally applied to crops but now often applied to any species.

Linear process With respect to systems, refers to the addition or subtraction of anything to a compartment in a system where the amount will always be the same, no matter how much you have added before and what else has changed about the system and the environment. For example, if you collect stones from a particular site and place them in a basket and place one stone per hour, you will have placed 6 stones in 6 hours and 24 in 24 hours, and the change is linear with time.

Lithosphere The outer layer of Earth, approximately 100 km thick, of which the plates that contain the ocean basins and the continents are composed.

Little Ice Age (LIA) A period of approximately 300 years from the mid 1400 to 1700 where the Earth was a bit cooler than it is today. During the Little Ice Age, glaciers expanded in mountainous regions. Cold, wet years during the Little Ice Age, may have contributed to the devastation caused by the Black Plague.

Littoral drift Movement caused by wave motion in nearshore and beach environment.

Local extinction The disappearance of a species from part of its range but continued persistence elsewhere.

Logistic carrying capacity In terms of the logistic curve, the population size at which births equal deaths and there is no net change in the population.

Logistic equation The equation that results in a logistic growth curve; that is, the growth rate $dN/dt = rN[(K - N)/N]$, where r is the intrinsic rate of increase, K is the carrying capacity, and N is the population size.

Logistic growth curve The S-shaped growth curve that is generated by the logistic growth equation. In the logistic, a small population grows rapidly, but the growth rate slows down, and the population eventually reaches a constant size.

Low-level radioactive waste Waste materials that contain sufficiently low concentrations or quantities of radioactivity so as not to present a significant environmental hazard if properly handled.

Luz solar electric generating system Solar energy farms comprising a power plant surrounded by hundreds of solar collectors (curved mirrors) that heat a synthetic oil, which flows through heat exchangers to drive steam turbine generators.

Macronutrients Elements required in large amounts by living things. These include the big six—carbon, hydrogen, oxygen, nitrogen, phosphorus, and sulfur.

Made lands Man-made areas created artificially with fill, sometimes as waste dumps of all kinds and sometimes to make more land available for construction.

Magma A naturally occurring silica melt, a good deal of which is in a liquid state.

Malnourishment The lack of specific components of food, such as proteins, vitamins, or essential chemical elements.

Manipulated variable See **Variable, independent**.

Marasmus Progressive emaciation caused by a lack of protein and calories.

Marginal cost In environmental economics, the cost to reduce one additional unit of a type of degradation; for example, pollution.

Marginal land An area of Earth with minimal rainfall or otherwise limited severely by some necessary factor, so that it is a poor place for agriculture and easily degraded by agriculture. Typically, these lands are easily converted to deserts even when used for light grazing and crop production.

Mariculture Production of food from marine habitats.

Marine evaporites With respect to mineral resources, refers to materials such as potassium and sodium salts resulting from the evaporation of marine waters.

Materially closed system Characterized by a system in which no matter moves in and out of the system, although energy and information may move across the system's boundaries. For example, Earth is a materially closed system for all practical purposes.

Materials management In waste management, methods consistent with the ideal of industrial ecology, making better use of materials and leading to more sustainable use of resources.

Matter Anything that occupies space and has mass. It is the substance of which physical objects are composed.

Maximum lifetime Genetically determined maximum possible age to which an individual of a species can live.

Maximum sustainable population The largest population size that can be sustained indefinitely.

Maximum sustainable yield (MSY) The maximum usable production of a biological resource that can be obtained in a specified time period without decreasing the ability of the resource to sustain that level of production.

Mediation A negotiation process between adversaries, guided by a neutral facilitator.

Medieval Warming Period (MWP) A period of approximately 300 years from A.D. 950 to 1250 when Earth's surface was considerably warmer than the normal that we experience today. The warming was particularly relevant and important in Western Europe and the Atlantic Ocean where the MWP was a time of flourishing culture and activity, as well as expansion of population.

Megacities Urban areas with at least 8 million inhabitants.

Meltdown A nuclear accident in which the nuclear fuel forms a molten mass that breaches the containment of the reactor, contaminating the outside environment with radioactivity.

Methane (CH_4) A molecule of carbon and four hydrogen atoms. It is a naturally occurring gas in the atmosphere, one of the so-called greenhouse gases.

Methane hydrate A white icelike compound made up of molecules of methane gas trapped in "cages" of frozen water in the sediments of the deep seafloor.

Microclimate The climate of a very small local area. For example, the climate under a tree, near the ground within a forest, or near the surface of streets in a city.

Micronutrients Chemical elements required in very small amounts by at least some forms of life. Boron, copper, and molybdenum are examples of micronutrients.

Micropower The production of electricity using smaller distributed systems rather than relying on large central power plants.

Migration The movement of an individual, population, or species from one habitat to another or more simply from one geographic area to another.

Migration corridor Designated passageways among parks or preserves allowing migration of many life-forms among several of these areas.

Mineral A naturally occurring inorganic material with a definite internal structure and physical and chemical properties that vary within prescribed limits.

Mineral resources Elements, chemical compounds, minerals, or rocks concentrated in a form that can be extracted to obtain a usable commodity.

Minimum viable population The minimum number of individuals that have a reasonable chance of persisting for a specified time period.

Missing carbon sink The unknown location of substantial amounts of carbon dioxide released into the atmosphere but apparently not reabsorbed and thus remaining unaccounted for.

Mitigated negative declaration A special type of negative declaration that suggests that the adverse environmental aspects of a particular action may be mitigated through modification of the project in such a way as to reduce the impacts to near insignificance.

Mitigation A process that identifies actions to avoid, lessen, or compensate for anticipated adverse environmental impacts.

Mobile sources Sources of air pollutants that move from place to place; for example, automobiles, trucks, buses, and trains.

Model A deliberately simplified explanation, often physical, mathematical, pictorial, or computer-simulated, of complex phenomena or processes.

Monitoring Process of collecting data on a regular basis at specific sites to provide a database from which to evaluate change. For example, collection of water samples from beneath a landfill to provide early warning should a pollution problem arise.

Monoculture (Agriculture) The planting of large areas with a single species or even a single strain or subspecies in farming.

Moral justification for the conservation of nature An argument for the conservation of nature on the grounds that aspects of the environment have a right to exist, independent of human desires, and that it is our moral obligation to allow them to continue or to help them persist.

Multiple use Literally, using the land for more than one purpose at the same time. For example, forestland can be used to produce commercial timber but at the same time serve as wildlife habitat and land for recreation. Usually multiple use requires compromises and trade-offs, such as striking a balance between cutting timber for the most efficient production of trees at a level that facilitates other uses.

Mutation Stated most simply, a chemical change in a DNA molecule. It means that the DNA carries a different message than it did before, and this change can affect the expressed characteristics when cells or individual organisms reproduce.

Mutualism See **Symbiosis**.

Natural capital Ecological systems that provide public service benefits.

Natural catastrophe Sudden catastrophic change in the environment, not the result of human actions.

Natural gas Naturally occurring gaseous hydrocarbon (predominantly methane) generally produced in association with crude oil or from gas wells; an important efficient and clean-burning fuel commonly used in homes and industry.

Natural hazard Any natural process that is a potential threat to human life and property.

Natural selection A process by which organisms whose biological characteristics better fit them to the environment are represented by more descendants in future generations than those whose characteristics are less fit for the environment.

Nature preserve An area set aside with the primary purpose of conserving some biological resource.

Negative declaration A document that may be filed if an agency has determined that a particular project will not have a significant adverse effect on the environment.

Negative feedback A type of feedback that occurs when the system's response is in the opposite direction of the output. Thus negative feedback is self-regulating.

Net growth efficiency Net production efficiency (P/A), or the ratio of the material produced (P) to the material assimilated (A) by an organism. The material assimilated is less than the material consumed because some food taken in is egested as waste (discharged) and never used by an organism.

Net production (biology) The production that remains after utilization. In a population, net production is sometimes measured as the net change in the numbers of individuals. It is also measured as the net change in biomass or in stored energy. In terms of energy, it is equal to the gross production minus the energy used in respiration.

New forestry The name for a new variety of timber harvesting practices to increase the likelihood of sustainability, including recognition of the dynamic characteristics of forests and of the need for management within an ecosystem context.

Niche (1) The "profession," or role, of an organism or species; or (2) all the environmental conditions under which the individual or species can persist. The fundamental niche is all the conditions under which a species can persist in the absence of competition; the realized niche is the set of conditions as they occur in the real world with competitors.

Nitrogen cycle A complex biogeochemical cycle responsible for moving important nitrogen components through the biosphere and other Earth systems. This is an extremely important cycle because nitrogen is required by all living things.

Nitrogen fixation The process of converting inorganic, molecular nitrogen in the atmosphere to ammonia. In nature it is carried out only by a few species of bacteria, on which all life depends.

Nitrogen oxides Occur in several forms: NO, NO_2, and NO_3. Most important as an air pollutant is nitrogen dioxide, which is a visible yellow brown to reddish brown gas. It is a precursor of acid rain and produced through the burning of fossil fuels.

Noise pollution A type of pollution characterized by unwanted or potentially damaging sound.

Non-linear process Characterized by system operation in which the effect of adding a specific amount of something changes, depending upon how much has been added before.

Nonmarine evaporites With respect to mineral resources, refers to useful deposits of materials such as sodium and calcium bicarbonate, sulfate, borate, or nitrate produced by evaporation of surficial waters on the land, as differentiated from marine waters in the oceans.

Nonpoint sources Pollution sources that are diffused and intermittent and are influenced by factors such as land use, climate, hydrology, topography, native vegetation, and geology.

Nonrenewable energy Energy sources, including nuclear and geothermal, that are dependent on fuels, or a resource that may be used up much faster than it is replenished by natural processes.

Nonrenewable resource A resource that is cycled so slowly by natural Earth processes that once used, it is essentially not going to be made available within any useful time framework.

No-till agriculture A combination of farming practices that includes not plowing the land and using herbicides to keep down weeds.

Nuclear cycle The series of processes that begins with the mining of uranium to be processed and used in nuclear reactors and ends with the disposal of radioactive waste.

Nuclear energy The energy of the atomic nucleus that, when released, may be used to do work. Controlled nuclear fission reactions take place within commercial nuclear reactors to produce energy.

Nuclear fuel cycle Processes involved with producing nuclear power from the mining and processing of uranium to control fission, reprocessing of spent nuclear fuel, decommissioning of power plants, and disposal of radioactive waste.

Nuclear reactors Devices that produce controlled nuclear fission, generally for the production of electric energy.

Obligate symbionts A symbiotic relationship between two organisms in which neither by themselves can exist without the other.

Observations Information obtained through one or more of the five senses or through instruments that extend the senses. For example, some remote sensing instruments measure infrared intensity, which we do not see, and convert the measurement into colors, which we do see.

Ocean thermal conversion Direct utilization of solar energy using part of a natural oceanic environment as a gigantic solar collector.

Off-site effect An environmental effect occurring away from the location of the causal factors.

Off-stream use Type of water use where water is removed from its source for a particular use.

Oil shale A fine-grained sedimentary rock containing organic material known as kerogen. On distillation, it yields significant amounts of hydrocarbons, including oil.

Oil spill The accidental release of oil from a ship transporting oil, an oil pipeline leak, or release of oil from a well during or after drilling.

Old-growth forest A nontechnical term often used to mean a virgin forest (one never cut), but also used to mean a forest that has been undisturbed for a long, but usually unspecified, time.

Oligotrophic Referring to bodies of water having a low concentration of the chemical elements required for life.

Omnivores Organisms that eat both plants and animals.

On-site effect An environmental effect occurring at the location of the causal factors.

Open dump An area where solid waste is disposed of by simply dumping it. It often causes severe environmental problems, such as water pollution, and creates a health hazard. Illegal in the United States and in many other countries around the world.

Open system A type of system in which exchanges of mass or energy occur with other systems.

Open woodlands Areas in which trees are a dominant vegetation form but the leaves of adjacent trees generally do not touch or overlap, so that there are gaps in the canopy. Typically, grasses or shrubs grow in the gaps among the trees.

Operational definitions Definitions that tell you what you need to look for or do in order to carry out an operation, such as measuring, constructing, or manipulating.

Optimal carrying capacity A term that has several meanings, but the major idea is the maximum abundance of a population or species that can persist in an ecosystem without degrading the ability of the ecosystem to maintain (1) that population or species; (2) all necessary ecosystem processes; and (3) the other species found in that ecosystem.

Optimum sustainable population (OSP) The population size that is in some way best for the population, its ecological community, its ecosystem, or the biosphere.

Optimum sustainable yield (OSY) The largest yield of a renewable resource achievable over a long time period without decreasing the ability of the resource, its ecosystem or its environment to maintain this level of yield. OSY differs from maximum sustainable yield (MSY) by taking the ecosystem of a resource into account.

Ore deposits Earth materials in which metals exist in high concentrations, sufficient to be mined.

Organelle Specialized parts of cells that function like organs in multi-celled organisms.

Organic compound A compound of carbon; originally used to refer to the compounds found in and formed by living things.

Organic farming Farming that is more "natural" in the sense that it does not involve the use of artificial pesticides and, more recently, genetically modified crops. In recent years governments have begun to set up legal criteria for what constitutes organic farming.

Output With respect to basic operation of system, refers to material or energy that leaves a particular storage compartment.

Overdraft Groundwater withdrawal when the amount pumped from wells exceeds the natural rate of replenishment.

Overgrazing Exceeding the carrying capacity of land for an herbivore, such as cattle or deer.

Overshoot and collapse Occurs when growth in one part of a system over time exceeds carrying capacity, resulting in a sudden decline in one or both parts of the system.

Ozone (O_3) A form of oxygen in which three atoms of oxygen occur together. It is chemically active and has a short average lifetime in the atmosphere. Forms a natural layer high in the atmosphere (stratosphere) that protects us from harmful ultraviolet radiation from the sun, is an air pollutant when present in the lower atmosphere above the National Air Quality Standards.

Ozone shield Stratospheric ozone layer that absorbs ultraviolet radiation.

Pandemic A worldwide disease outbreak.

Particulate matter Small particles of solid or liquid substances that are released into the atmosphere by many activities, including farming, volcanic eruption, and burning fossil fuels. Particulates affect human health, ecosystems, and the biosphere.

Passive solar energy system Direct use of solar energy through architectural design to enhance or take advantage of natural changes in solar energy that occur throughout the year without requiring mechanical power.

Pasture Land plowed and planted to provide forage for domestic herbivorous animals.

Peak oil Refers to the time in the future when one-half of Earth's oil has been exploited. Peak oil is expected to occur sometime between 2020 and 2050.

Pebble A rock fragment between 4 and 64 mm in diameter.

Pedology The study of soils.

Pelagic ecosystem An ecosystem that occurs in the floating part of an ocean or sea, without any physical connections to the bottom of the ocean or sea.

Pelagic whaling Practice of whalers taking to the open seas and searching for whales from ships that remained at sea for long periods.

Per-capita availability The amount of a resource available per person.

Per-capita demand The economic demand per person.

Per-capita food production The amount of food produced per person.

Permafrost Permanently frozen ground.

Persistent organic pollutants Synthetic carbon-based compounds, often containing chlorine, that do not easily break down in the environment. Many were introduced decades before their harmful effects were fully understood and are now banned or restricted.

Pesticides, broad-spectrum Pesticides that kill a wide variety of organisms. E.g. arsenic, one of the first elements used as a pesticide, is toxic to many life-forms, including people.

Phosphorus cycle A major biogeochemical cycle involving the movement of phosphorus throughout the biosphere and lithosphere. This cycle is important because phosphorus is an essential element for life and often is a limiting nutrient for plant growth.

Photochemical oxidants Result from atmospheric interactions of nitrogen dioxide and sunlight. Most common is ozone (O_3).

Photochemical smog Sometimes called L.A.-type smog or brown air. Directly related to automobile use and solar radiation. Reactions that occur in the development of the smog are complex and involve both nitrogen oxides and hydrocarbons in the presence of sunlight.

Photosynthesis Synthesis of sugars from carbon dioxide and water by living organisms using light as energy. Oxygen is given off as a by-product.

Photovoltaics Technology that converts sunlight directly into electricity using a solid semiconductor material.

Physiographic province A region characterized by a particular assemblage of landforms, climate, and geomorphic history.

Pioneer species Species found in early stages of succession.

Placer deposit A type of ore deposit found in material transported and deposited by agents such as running water, ice, or wind. Examples include gold and diamonds found in stream deposits.

Plantations In forestry, managed forests, in which a single species is planted in straight rows and harvested at regular intervals.

Plate tectonics A model of global tectonics that suggests that the outer layer of Earth, known as the lithosphere, is composed of several large plates that move relative to one another. Continents and ocean basins are passive riders on these plates.

Point sources Sources of pollution such as smokestacks, pipes, or accidental spills that are readily identified and stationary. They are often thought to be easier to recognize and control than are area sources. This is true only in a general sense, as some very large point sources emit tremendous amounts of pollutants into the environment.

Polar amplification Processes in which global warming causes greater temperature increases at polar regions.

Polar stratospheric clouds Clouds that form in the stratosphere during the polar winter.

Polar vortex Arctic air masses that in the winter become isolated from the rest of the atmosphere and circulate about the pole. The vortex rotates counterclockwise because of the rotation of Earth in the Southern Hemisphere.

Policy instruments The means to implement a society's policies. Such instruments include moral suasion (jawboning—persuading people by talk, publicity, and social pressure); direct controls, including regulations; and market processes affecting the price of goods, subsidies, licenses, and deposits.

Pollutant In general terms, any factor that has a harmful effect on living things or their environment.

Pollution The process by which something becomes impure, defiled, dirty, or otherwise unclean.

Pollution prevention Identifying ways to avoid the generation of waste rather than finding ways to dispose of it.

Pool (in a stream) A stream bed produced by scour.

Population A group of individuals of the same species living in the same area or interbreeding and sharing genetic information.

Population age structure The number of individuals or the proportion of the population in each age class.

Population dynamics The causes of changes in population size.

Population momentum or lag effect The continued growth of a population after replacement-level fertility is reached.

Population regulation See **Density-dependent population effects** and **Density-independent population effects.**

Population risk A term used in discussions of endangered species to mean random variation in population—birth rates and death rates—possibly causing species in low abundance to become extinct.

Positive feedback A type of feedback that occurs when an increase in output leads to a further increase in output. This is sometimes known as a vicious cycle, since the more you have, the more you get.

Potential energy Energy that is stored. Examples include the gravitational energy of water behind a dam; chemical energy in coal, fuel oil, and gasoline; and nuclear energy (in the forces that hold atoms together).

Power The amount of energy used per unit of time.

Precautionary principle The idea that even full scientific certainty is not available to prove cause and effect, we should still take cost-effective precautions to solve environmental problems when appears to be a threat of potential serious and irreversible environmental damage.

Predation-parasitism Interaction between individuals of two species in which the outcome benefits one and is detrimental to the other.

Predator An organism that feeds on other live organisms, usually of other species. The term is usually applied to animals that feed on other animals.

Premises In science, initial definitions and assumptions.

Primary pollutants Air pollutants emitted directly into the atmosphere. Included are particulates, sulfur oxides, carbon monoxide, nitrogen oxides, and hydrocarbons.

Primary production See **Production, primary.**

Primary succession The initial establishment and development of an ecosystem.

Primary treatment (of wastewater) Removal of large particles and organic materials from wastewater through screening.

Probability The likelinood that an event will occur.

Production, ecological The amount of increase in organic matter, usually measured per unit area of land surface or unit volume of water, as in grams per square meter (g/m^2). Production is divided into *primary* (that of autotrophs) and *secondary* (that of heterotrophs). It is also divided into *net* (that which remains stored after use) and *gross* (that added before any use).

Production, primary The production by autotrophs.

Production, secondary The production by heterotrophs.

Productivity, ecological The *rate* of production; that is, the amount of increase in organic matter per unit time (for example, grams per meter squared per year).

Prokaryote A kind of organism that lacks a true cell nucleus and has other cellular characteristics that distinguish it from the *eukaryotes*. Bacteria are prokaryotes.

Pseudoscientific Describes ideas that are claimed to have scientific validity but are inherently untestable and/or lack empirical support and/or were arrived at through faulty reasoning or poor scientific methodology.

Public service functions Functions performed by ecosystems that benefit other forms of life in other ecosystems. Examples include the cleansing of the air by trees and removal of pollutants from water by infiltration through the soil.

Public trust Grants and limits the authority of government over certain natural areas of special character.

Qualitative data Data distinguished by qualities or attributes that cannot be or are not expressed as quantities. For example, blue and red are qualitative data about the electromagnetic spectrum.

Quantitative data Data expressed as numbers or numerical measurements. For example, the wavelengths of specific colors of blue and red light (460 and 650 nanometers, re-spectively) are quantitative data about the electromagnetic spectrum.

R-to-C ratio A measure of the time available for finding the solutions to depletion of nonrenewable reserves, where R is the known reserves (for example, hundreds of thousands of tons of a metal) and C is the rate of consumption (for example, thousands of tons per year used by people).

Radiation absorbed dose (RAD) Energy retained by living tissue that has been exposed to radiation.

Radioactive decay A process of decay of radioisotopes that change from one isotope to another and emit one or more forms of radiation.

Radioactive waste Type of waste produced in the nuclear fuel cycle; generally classified as high-level or low-level.

Radioisotope A form of a chemical element that spontaneously undergoes radioactive decay.

Radionuclides Atoms with unstable nuclei that undergo radioactive decay.

Radon A naturally occurring radioactive gas. Radon is colorless, odorless, and tasteless and must be identified through proper testing.

Rangeland Land used for grazing.

Rare species Species with a small total population, or restricted to a small area, but not necessarily declining or in danger of extinction.

Realms (ecological) Major biogeographic regions of Earth in which most animals have some common genetic heritage.

Record of decision Concise statement prepared by the agency planning a proposed project; it outlines the alternatives considered and discusses which alternatives are environmentally preferable.

Recreational justification for the conservation of nature An argument for the conservation of nature on the grounds that direct experience of nature is inherently enjoyable and that the benefits derived from it are important and valuable to people.

Recycle To collect and reuse resources in the waste stream.

Reduce With respect to waste management, refers to practices that will reduce the amount of waste we produce.

Reduce, reuse, and recycle The three Rs of integrated waste management.

Renewable energy Alternative energy sources, such as solar, water, wind, and biomass, that are more or less continuously available in a time framework useful to people.

Renewable resource A resource, such as timber, water, or air, that is naturally recycled or recycled by artificial processes within a time frame useful for people.

Replacement-level fertility The fertility rate required for the population to remain a constant size.

Representative natural areas Parks or preserves set aside to represent presettlement conditions of a specific ecosystem type.

Reserves Known and identified deposits of earth materials from which useful materials can be extracted profitably with existing technology and under present economic and legal conditions.

Resource-based agriculture Agricultural practices that rely on extensive use of resources, so that production is limited by the availability of resources.

Resources Reserves plus other deposits of useful earth materials that may eventually become available.

Respiration The complex series of chemical reactions in organisms that make energy available for use. Water, carbon dioxide, and energy are the products of respiration.

Responding variable See **Variable, dependent.**

Restoration ecology The field within the science of ecology whose goal is to return damaged ecosystems to ones that are functional, sustainable, and more natural in some meaning of this word.

Reuse With respect to waste management, refers to finding ways to reuse products and materials so they need not be disposed of.

Riffle A section of stream channel characterized at low flow by fast, shallow flow. Generally contains relatively coarse bed-load particles.

Risk The product of the probability of an event occurring and the consequences should that event occur.

Risk assessment The process of determining potential adverse environmental health effects to people following exposure to pollutants and other toxic materials. It generally includes four steps: identification of the hazard, dose-response assessment, exposure assessment, and risk characterization.

Risk-benefit analysis In environmental economics, weighing the riskiness of the future against the value we place on things in the present.

Rock (engineering) Any earth material that has to be blasted in order to be removed.

Rock (geologic) An aggregate of a mineral or minerals.

Rock cycle A group of processes that produce igneous, metamorphic, and sedimentary rocks.

Rotation time Time between cuts of a stand or area of forest.

Rule of climatic similarity Similar environments lead to the evolution of organisms similar in form and function (but not necessarily in genetic heritage or internal makeup) and to similar ecosystems.

Ruminants Animals having a four-chambered stomach within which bacteria convert the woody tissue of plants to proteins and fats that, in turn, are digested by the animal. Cows, camels, and giraffes are ruminants; horses, pigs, and elephants are not.

Sand Grains of sediment between 1/16 and 2 mm in diameter; often sediment composed of quartz particles of this size.

Sand dune A ridge or hill of sand formed by wind action.

Sanitary landfill A method of disposal of solid waste without creating a nuisance or hazard to public health or safety. Sanitary landfills are highly engineered structures with multiple barriers and collection systems to minimize environmental problems.

Savanna An area with trees scattered widely among dense grasses.

Scientific method A set of systematic methods by which scientists investigate natural phenomena, including gathering data, formulating and testing hypotheses, and developing scientific theories and laws.

Scientific theory A grand scheme that relates and explains many observations and is supported by a great deal of evidence, in contrast to a guess, a hypothesis, a prediction, a notion, or a belief.

Scoping The process of early identification of important environmental issues that require detailed evaluation.

Scrubbing A process of removing sulfur from gases emitted from power plants burning coal. The gases are treated with a slurry of lime and limestone, and the sulfur oxides react with the calcium to form insoluble calcium sulfides and sulfates that are collected and disposed of.

Second growth A forest that has been logged and regrown.

Secondary enrichment A weathering process of sulfide ore deposits that may concentrate the desired minerals.

Secondary pollutants Air pollutants produced through reactions between primary pollutants and normal atmospheric compounds. An example is ozone that forms over urban areas through reactions of primary pollutants, sunlight, and natural atmospheric gases.

Secondary production See **Production, secondary.**

Secondary succession The reestablishment of an ecosystem where there are remnants of a previous biological community.

Secondary treatment (of wastewater) Use of biological processes to degrade wastewater in a treatment facility.

Second-law efficiency The ratio of the minimum available work needed to perform a particular task to the actual work used to perform that task. Reported as a percentage.

Second law of thermodynamics The law of thermodynamics which states that *no use of energy in the real (not theoretical) world can ever be 100% efficient.*

Secure landfill A type of landfill designed specifically for hazardous waste. Similar to a modern sanitary landfill in that it includes multiple barriers and collection systems to ensure that leachate does not contaminate soil and other resources.

Sediment pollution By volume and mass, sediment is our greatest water pollutant. It may choke streams, fill reservoirs, bury vegetation, and generally create a nuisance that is difficult to remove.

Seed-tree cutting A logging method in which mature trees with good genetic characteristics and high seed production are preserved to promote regeneration of the forest. It is an alternative to clear-cutting.

Seismic Referring to vibrations in Earth produced by earth-quakes.

Selective cutting In timber harvesting, the practice of cutting some, but not all, trees, leaving some on the site. There are many kinds of selective cutting. Sometimes the biggest trees with the largest market value are cut, and smaller trees are left to be cut later. Sometimes the best trees are left to provide seed for future generations. Sometimes trees are left for wildlife habitat and recreation.

Shelterwood cutting A logging method in which dead and less desirable trees are cut first; mature trees are cut later. This ensures that young, vigorous trees will always be left in the forest. It is an alternative to clear-cutting.

Sick building syndrome A condition associated with a particular indoor environment that appears to be unhealthy for the human occupants.

Silicate minerals The most important group of rock-forming minerals.

Silt Sediment between 1/16 and 1/256 mm in diameter.

Silviculture The practice of growing trees and managing forests, traditionally with an emphasis on the production of timber for commercial sale.

Single stream recycling Recycling process in which paper, plastic, glass, and metals are not separated prior to collection.

Sink With respect to systems operation, refers to a component or storage cell within a system that is receiving a material, such as a chemical. The donating compartment is called a source, and there generally is a flux or rate of transfer between the source and sink.

Sinkhole A surface depression formed by the solution of limestone or the collapse over a subterranean void such as a cave.

Site (in relation to cities) Environmental features of a location that influence the placement of a city. For example, New Orleans is built on low-lying muds, which form a poor site, while New York City's Manhattan is built on an island of strong bedrock, an excellent site.

Site quality Used by foresters to mean an estimator of the maximum timber crop the land can produce in a given time.

Situation (in relation to cities) The relative geographic location of a site that makes it a good location for a city. For example, New Orleans has a good situation because it is located at the mouth of the Mississippi River and is therefore a natural transportation junction.

Smog A term first used in 1905 for a mixture of smoke and fog that produced unhealthy urban air. There are several types of fog, including photochemical smog and sulfurous smog.

Soft path Energy policy that relies heavily on renewable energy resources as well as other sources that are diverse, flexible, and matched to the end-use needs.

Soil The top layer of a land surface where the rocks have been weathered to small particles. Soils are made up of inorganic particles of many sizes, from small clay particles to large sand grains. Many soils also include dead organic material.

Soil (in engineering) Earth material that can be removed without blasting.

Soil (in soil science) Earth material modified by biological, chemical, and physical processes such that the material will support rooted plants.

Soil fertility The capacity of a soil to supply the nutrients and physical properties necessary for plant growth.

Soil horizon A layer in soil (A, B, C) that differs from another layer in chemical, physical, and biological properties.

Solar cell (photovoltaic) A device that directly converts light into electricity.

Solar collector A device for collecting and storing solar energy. For example, home water heating is done by flat panels consisting of a glass cover plate over a black background on which water is circulated through tubes. Short-wave solar radiation enters the glass and is absorbed by the black background. As long-wave radiation is emitted from the black material, it cannot escape through the glass, so the water in the circulating tubes is heated, typically to temperatures of 38° to 93°C.

Solar energy Energy from the sun.

Solar pond A shallow pond filled with water and used to generate relatively low-temperature water.

Solar power tower A system of collecting solar energy that delivers the energy to a central location where the energy is used to produce electric power.

Source With respect to storage compartments within a system, such as the atmosphere or land, refers to a compartment that donates to another compartment. The donating compartment is the source; the receiving compartment the sink.

Source reduction A waste-management process to reduce the amounts of materials that must be handled in the waste stream.

Species A group of individuals capable of interbreeding.

Stable equilibrium A condition in which a system will remain if undisturbed and to which it will return when displaced.

Stand An informal term used by foresters to refer to a group of trees.

Stationary sources Air pollution sources that have a relatively fixed location, including point sources, fugitive sources, and area sources.

Steady state When input equals output in a system, there is no net change and the system is said to be in a steady state. A bathtub with water flowing in and out at the same rate maintains the same water level and is in a steady state. Compare with **equilibrium**.

Stratosphere Overlies the troposphere and the atmosphere from approximately 20-70 kilometers above the Earth. The stratosphere contains the higher concentrations of ozone at about 25 kilometers above the Earth known as the ozone layer.

Stress Force per unit area. May be compression, tension, or shear.

Strip cutting In timber harvesting, the practice of cutting narrow rows of forest, leaving wooded corridors.

Strip mining Surface mining in which the overlying layer of rock and soil is stripped off to reach the resource. Large strip mines are some of the world's largest excavations by people.

Subduction A process in which one lithospheric plate descends beneath another.

Subsidence A sinking, settling, or otherwise lowering of parts of crust.

Subsistence crops Crops used directly for food by a farmer or sold locally where the food is used directly.

Succession The process of establishment and development of an ecosystem.

Sulfur dioxide (SO_2) A colorless and odorless gas normally present at Earth's surface in low concentrations. An important precursor to acid rain, its major anthropogenic source is burning fossil fuels.

Sulfurous smog Produced primarily by burning coal or oil at large power plants. Sulfur oxides and particulates combine under certain meteorological conditions to produce a concentrated form of this smog.

Suppressed In forestry, describes tree species growing in the understory, beneath the dominant and intermediate species.

Surface impoundment Method of disposal of some liquid hazardous waste. This method is controversial, and many sites have been closed.

Sustainability Management of natural resources and the environment with the goals of allowing the harvest of resources to remain at or above some specified level, and the ecosystem to retain its functions and structure.

Sustainable development The ability of a society to continue to develop its economy and social institutions and also maintain its environment for an indefinite time.

Sustainable ecosystem An ecosystem that is subject to some human use but at a level that leads to no loss of species or of necessary ecosystem functions.

Sustainable energy development A type of energy management that provides for reliable sources of energy while not causing environmental degradation and while ensuring that future generations will have a fair share of the Earth's resources.

Sustainable forestry Managing a forest so that a resource in it can be harvested at a rate that does not decrease the ability of the forest ecosystem to continue to provide that same rate of harvest indefinitely.

Sustainable harvest An amount of a resource that can be harvested at regular intervals indefinitely.

Sustainable water use Use of water resources that does not harm the environment and provides for the existence of high-quality water for future generations.

Symbiont Each partner in symbiosis.

Symbiosis An interaction between individuals of two different species that benefits both. For example, lichens contain an alga and a fungus that require each other to persist. Sometimes this term is used broadly, so that domestic corn and people could be said to have a symbiotic relationship—domestic corn cannot reproduce without the aid of people, and some people survive because they have corn to eat.

Symbiotic Relationships that exist between different organisms. Mutually beneficial.

Synergism Cooperative action of different substances such that the combined effect is greater than the sum of the effects taken separately.

Synergistic effect When the change in availability of one resource affects the response of an organism to some other resource.

Synfuels Synthetic fuels, which may be liquid or gaseous, derived from solid fuels, such as oil from kerogen in oil shale, or oil and gas from coal.

Synthetic organic compounds Compounds of carbon produced synthetically by human industrial processes, as for example pesticides and herbicides.

System A set of components that are linked and interact to produce a whole. For example, the river as a system is composed of sediment, water, bank, vegetation, fish, and other living things that all together produce the river.

Systematic errors Errors that occur consistently in scientific experiments, such as those resulting from incorrectly calibrated instruments.

Taiga Forest of cold climates of high latitudes and high altitudes, also known as boreal forest.

Tangible factor In economics, something one you can touch, buy and sell.

Tar sands Sedimentary rocks or sands impregnated with tar oil, asphalt, or bitumen.

Taxa Categories that identify groups of living organisms based on evolutionary relationships or similarity of characters.

Taxon A grouping of organisms according to evolutionary relationships.

TD-50 The toxic dose defined as the dose that is toxic to 50% of a population exposed to the toxin.

Tectonic cycle The processes that change Earth's crust, producing external forms such as ocean basins, continents, and mountains.

Terminator gene A genetically modified crop that has a gene to cause the plant to become sterile after the first year.

Tertiary treatment (of wastewater) An advanced form of wastewater treatment involving chemical treatment or advanced filtration. An example is chlorination of water.

Theories Scientific models that offer broad, fundamental explanations of related phenomena and are supported by consistent and extensive evidence.

Thermal (heat) energy The energy of the random motion of atoms and molecules.

Thermal pollution A type of pollution that occurs when heat is released into water or air and produces undesirable effects on the environment.

Thermodynamic equilibrium With respect to systems, is a physical concept of equilibrium where everything is at the lowest energy level in the system, and matter and energy are dispersed randomly.

Thermodynamic system Formed by an energy source, ecosystem, and energy sink, where the ecosystem is said to be an intermediate system between the energy source and the energy sink.

Thermodynamics, first law of See **First law of thermodynamics.**

Thermodynamics, second law of See **Second law of thermodynamics.**

Thinning The timber-harvesting practice of selectively removing only smaller or poorly formed trees.

Threatened species Species experiencing a decline in the number of individuals to a degree that raises concern about the possibility of extinction of that species.

Threshold A point in the operation of a system at which a change occurs. With respect to toxicology, it is a level below which effects are not observable and above which effects become apparent.

Tidal power Energy generated by ocean tides in places where favorable topography allows for construction of a power plant.

Time series The set of estimates of some variable over a number of years.

Tolerance The ability to withstand stress resulting from exposure to a pollutant or other harmful condition.

Tornado A funnel-shaped cloud of violently rotating air that extends downward from large thunder storms to contact the surface of earth.

Total fertility rate (TFR) The average number of children expected to be born to a woman during her lifetime. (Usually defined as the number born to a woman between the ages of 15 and 44, taken conventionally as the lower and upper limit of reproductive ages for women.)

Toxic Harmful, deadly, or poisonous.

Toxicology The science concerned with the study of poisons (or toxins) and their effects on living organisms. The subject also includes the clinical, industrial, economic, and legal problems associated with toxic materials.

Transuranic waste Radioactive waste consisting of human-made radioactive elements heavier than uranium. Includes clothing, rags, tools, and equipment that has been contaminated.

Trophic level In an ecological community, all the organisms that are the same number of food-chain steps from the primary source of energy. For example, in a grassland the green grasses are on the first trophic level, grasshoppers are on the second, birds that feed on grasshoppers are on the third, and so forth.

Trophic level efficiency The ratio of the biological production of one trophic level to the biological production of the next lower trophic level.

Troposphere The atmospheric zone from the surface of the Earth to an altitude of approximately 20 kilometers above the Earth. The troposphere is the zone of the atmosphere we are most familiar with because we spend most of our lives in it.

Tundra The treeless land area in alpine and arctic areas characterized by plants of low stature and including bare areas without any plants and areas covered with lichens, mosses, grasses, sedges, and small flowering plants, including low shrubs.

Ubiquitous species Species that are found almost anywhere on Earth.

Ultraviolet A (UVA) The longest wavelength of ultraviolet radiation (0.32–0.4 micrometers), not affected by stratospheric ozone, and transmitted to the surface of Earth.

Ultraviolet B (UVB) Intermediate-wavelength radiation which is the ozone problem. Wavelengths are approximately 0.28–0.32 micrometers and are the most harmful of the ultraviolet radiation types. Most of this radiation is absorbed by stratospheric ozone, and depletion of ozone has led to increased ultraviolet B radiation reaching Earth.

Ultraviolet C (UVC) The shortest wavelength of the ultraviolet radiation with wavelengths of approximately 0.2–0.28 micrometers. It is the most energetic of the ultraviolet radiation and is absorbed strongly in the atmosphere. Only a negligible amount of Ultraviolet C reaches the surface of Earth.

Ultraviolet (UV) index An index based on the exposure to ultraviolet radiation to humans. Varies from low to extreme and is useful for people wishing recommendation of how much exposure to the sun they should incur and how much sun block to use.

Undernourishment The lack of sufficient calories in available food, so that one has little or no ability to move or work.

Uneven-aged stands A forest stand with at least three distinct age classes.

Unified soil classification system A classification of soils, widely used in engineering practice, based on the amount of coarse particles, fine particles, or organic material.

Uniformitarianism The principle stating that processes that operate today operated in the past. Therefore, observations of processes today can explain events that occurred in the past and leave evidence, for example, in the fossil record or in geologic formations.

Urban dust dome Polluted urban air produced by the combination of lingering air and abundance of particulates and other pollutants in the urban air mass.

Urban forestry The practice and profession of planting and maintaining trees in cities, including trees in parks and other public areas.

Urban-runoff naturalization An emerging bioengineering technology with the objective to treat urban runoff before it reaches streams, lakes, or the ocean.

Utilitarian justification for the conservation of nature An argument for the conservation of nature on the grounds that the environment, an ecosystem, habitat, or species provides individuals with direct economic benefit or is directly necessary to their survival.

Utility Value or worth in economic terms.

UVA **See Ultraviolet A.**

UVB **See Ultraviolet B.**

UVC **See Ultraviolet C.**

Vadose zone Zone or layer above the water table where water may be stored as it moves laterally or down to the zone of saturation. Part of the vadose zone may be saturated part of the time.

Variable, dependent A variable that changes in response to changes in an independent variable; a variable taken as the outcome of one or more other variables.

Variable, independent In an experiment, the variable that is manipulated by the investigator. In an observational study, the variable that is believed by the investigator to affect an outcome, or dependent, variable.

Variable, manipulated See **Variable, independent**.

Variable, responding See **Variable, dependent**.

Virgin forest A forest that has never been cut.

Virtual water The amount of water necessary to produce a product, such as rice or, in industry, an automobile.

Volcanic eruption Extrusion at the surface of Earth of molten rock (magma). May be explosive and violent or less energetic lava flows.

Vulnerable species Another term for *threatened species*—species experiencing a decline in the number of individuals.

Waldsterben German phenomenon of forest death as the result of acid rain, ozone, and other air pollutants.

Wallace's realms Six biotic provinces, or biogeographic regions, divided on the basis of fundamental inherited features of the animals found in those areas, suggested by A. R. Wallace (1876). His realms are Nearctic (North America), Neotropical (Central and South America), Palearctic (Europe, northern Asia, and northern Africa), Ethiopian (central and southern Africa), Oriental (the Indian subcontinent and Malaysia), and Australian.

Wastewater renovation and conservation cycle The practice of applying wastewater to the land. In some systems, treated wastewater is applied to agricultural crops, and as the water infiltrates through the soil layer, it is naturally purified. Reuse of the water is by pumping it out of the ground for municipal or agricultural uses.

Wastewater treatment The process of treating wastewater (primarily sewage) in specially designed plants that accept municipal wastewater. Generally divided into three categories: primary treatment, secondary treatment, and advanced wastewater treatment.

Water budget Inputs and outputs of water for a particular system (a drainage basin, region, continent, or the entire Earth).

Water conservation Practices designed to reduce the amount of water we use.

Water power An alternative energy source derived from flowing water. One of the world's oldest and most common energy sources. Sources vary in size from microhydropower systems to large reservoirs and dams.

Water reuse The use of wastewater following some sort of treatment. Water reuse may be inadvertent, indirect, or direct.

Water table The surface that divides the zone of aeration from the zone of saturation, the surface below which all the pore space in rocks is saturated with water.

Watershed An area of land that forms the drainage of a stream or river. If a drop of rain falls anywhere within a watershed, it can flow out only through that same stream or river.

Weather What is happening in the atmosphere over a short time period or what may be happening now in terms of temperature, pressure, cloudiness, precipitation, and winds. The average of weather over longer periods and regions refers to the climate.

Weathering Changes that take place in rocks and minerals at or near the surface of Earth in response to physical, chemical, and biological changes; the physical, chemical, and biological breakdown of rocks and minerals.

Wetlands A comprehensive term for landforms such as salt marshes, swamps, bogs, prairie potholes, and vernal pools. Their common feature is that they are wet at least part of the year and as a result have a particular type of vegetation and soil. Wetlands form important habitats for many species of plants and animals, while serving a variety of natural service functions for other ecosystems and people.

Wilderness An area unaffected now or in the past by human activities and without a noticeable presence of human beings.

Wildfire Self-sustaining rapid oxidation that releases light, heat, carbon dioxide, and other gases as it moves across the landscape. Also known as a fire in the natural environment that may be initiated by natural processes such as lighting strike or deliberately set by humans.

Wind power Alternative energy source that has been used by people for centuries. More recently, thousands of windmills have been installed to produce electric energy.

Work (physics) Force times the distance through which it acts. When work is done we say energy is expended.

Zero population growth Results when the number of births equals the number of deaths so that there is no net change in the size of the population.

Zone of aeration The zone or layer above the water table in which some water may be suspended or moving in a downward migration toward the water table or laterally toward a discharge point.

Zone of saturation Zone or layer below the water table in which all the pore space of rock or soil is saturated.

Zooplankton Small aquatic invertebrates that live in the sunlit waters of streams, lakes, and oceans and feed on algae and other invertebrate animals.

Photo Credits

CHAPTER 1 Ch Op 1: Image Bank/Getty Images, Inc. Fig. 1.2: Corbis/SuperStock (top). Fig. 1.3: AfriPics.com/Alamy (bottom). Fig. 1.4: Images of Africa Photobank/Alamy. Fig. 1.6a: Peter Turnley/©Corbis (left). Fig. 1.6b: dbimages/Alamy (right). Fig. 1.8a: Thony Belizaire/AFP/Getty Images, Inc. (left). Fig. 1.8b: STRDEL/AFP/Getty Images, Inc. (right). Fig. 1.9a: Mira/Alamy (left). Fig. 1.9b: Comstock/Jupiter Images (right). Fig. 1.10: NASA Earth Observatory (bottom). Fig. 1.11: NASA/©Corbis. Fig. 1.12: Bill Ross/Corbis Images (top left). Fig. 1.13a: Kenneth W. Fink/Photo Researchers, Inc. (bottom left). Fig. 1.13b: © J.P. Ferrero/Jacana/Photo Researchers, Inc. (bottom right). Fig. 1.14: Courtesy Dan Botkin (top left). Fig. 1.15: Exactostock/SuperStock (top right). Fig. 1.16: Natalie Fobes/©Corbis. Fig. 1.17: Phil Dyer/iStockphoto. Fig. 1.18c: Michael Wozniak/iStockphoto.

CHAPTER 2 Ch Op 2: Dan Botkin. Fig. 2.1: Phil Schermeister/©Corbis. Fig. 2.5: Alan Root/Photo Researchers, Inc (left). Fig. 2.6: All Canada Photos/SuperStock (right). Fig. 2.8a: Gideon Mendel//Magnum Photos, Inc. (top). Fig. 2.8b: D. Hudson/©Corbis (bottom). Fig. 2.9: Visuals Unlimited.

CHAPTER 3 Ch Op 3: Lyroky/Alamy. Fig. 3.1: ©AP/Wide World Photos. Fig. 3.2: Goddard Space Flight Center Scientific Visualization Studio./Courtesy NASA (left). Fig. 3.3: Goddard Space Flight Center Scientific Visualization Studio./Courtesy NASA (right). Fig. 3.8: Morning in the Tropics, c.1858 (oil on canvas), Church, Frederic Edwin (1826-1900) / © Walters Art Museum, Baltimore, USA / The Bridgeman Art Library. Fig. 3.10a: Tim Tadder/©AP/Wide World Photos (top). Fig. 3.10c: Ed Keller (bottom). Fig. 3.13: All Canada Photos/Alamy. Fig. 3.15b: AGStockUSA/Alamy (right). Fig. 3.15c: Colin Woodbridge/Alamy (left). Fig. 3.17: Bohemian Nomad Picturemakers/©Corbis Images. Fig. 3.18: Wolfgang Kaehler/Alamy.

CHAPTER 4 Ch Op 4: AFP/Getty Images, Inc. Fig. 4.1a: ALFREDO ESTRELLA/AFP/Getty Images, Inc. (top). Fig. 4.1b: ©World Health Organization (bottom). Fig. 4.11: Jacobs Stock Photography/Photodisc/Getty Images, Inc.

CHAPTER 5 Ch Op 5: Dan Botkin. Fig. 5.1a: Alaska Stock/Alamy (top). Fig. 5.1b: David Nardini/Taxi/Getty Images, Inc. (bottom). Fig. 5.3: Farrell Grehan/Photo Researchers, Inc. Fig. 5.8a: Photographers Choice/Getty Images, Inc. (top left). Fig. 5.8b: Dan Botkin (top right). Fig. 5.9: Hubbard Brook Archives (bottom). Fig. 5.10a: Daniel Botkin (top left). Fig. 5.10b: Daniel Botkin (top right). Fig. 5.10c: Daniel Botkin (bottom left). Fig. 5.10d: Daniel Botkin (bottom right). Fig. 5.11: Mona Lisa Production/Science Photo Library. Fig. 5.14a: Masha Nordbye/Bruce Coleman, Inc. (left). Fig. 5.14b: Robert Harding Picture Library Ltd/Alamy (right). Fig. 5.15: Curlyson Photography/Flickr/Getty Images, Inc. (bottom). Fig. 5.16: Courtesy Daniel Botkin. Fig. 5.18: Michael P. Gadomski/ (left). Fig. 5.19: David B Fleetham/Photolibrary/Getty Images, Inc. (right). Fig. 5.22b: Theo Allofs/Getty Images, Inc. (left). Fig. 5.22c: David Aubrey/©Corbis Images (right).

CHAPTER 6 Ch Op 6: Mr. Eric Zimmerman. Fig. 6.1: Prof. Ed Keller. Fig. 6.2b: Prof. David Valentine (left). Fig. 6.2c: Prof. David Valentine (right). Fig. 6.4: NASA Images. Fig. 6.5: Flirt/SuperStock. Fig. 6.6a: MICHAEL AND PATRICIA FOGDEN/MINDEN PICTURES/NG Image Collection (bottom left). Fig. 6.6b: Georgette Douwma/Photo Researchers (bottom right). Fig. 6.7a: Dr. Patricia Schultz/Peter Arnold, Inc. (left). Fig. 6.7b: Kari Lounatmaa/Photo Researchers (right). Fig. 6.12a: Kim Heacox/Peter Arnold, Inc. (left). Fig. 6.12b: LUIS MARDEN/NG Image Collection (right). Fig. 6.20a: Peter Crighton/Alamy (left). Fig. 6.20b: Markus Renner/Getty Images, Inc. (right). Fig. 6.21: Melissa Farlow/NG Image Collection.

CHAPTER 7 Ch Op 7: Tom Brakefield /SuperStock. Fig. 7.1: Eric Draper/Aurora Photos/. Fig. 7.2: SuperStock. Fig. 7.3: Steve Kazlowski/Danita Delimont. Fig. 7.4: DENNIS ANDERSON/MCT/Landov LLC. Fig. 7.6: Jaime Abecasis / TopFoto/The Image Works. Fig. 7.7: Dan Botkin. Fig. 7.8: Photonica/Getty Images, Inc.

CHAPTER 8 Fig. 8.1a: Courtesy USDA (left). Fig. 8.1b: ©AP/Wide World Photos (right). Fig. 8.3: Rob Crandall/The Image Works. Fig. 8.5: Dr. Ian Woxvold of the University of Melbourne Zoology Dept. Fig. 8.6a: Oliver Meckes/Photo Researchers, Inc (left). Fig. 8.6b: Oliver Meckas/Photo Researchers, Inc (center). Fig. 8.6c: Adam Jones/Photo Researchers, Inc (right). Fig. 8.7a: Paula Bronstein / Getty Images, Inc. Fig. 8.8: All Canada Photos/Alamy. Fig. 8.10a: David Tipling/Getty Images, Inc. (left). Fig. 8.10b: Big Stock Photo (right). Fig. 8.17a: S. Greg Panosian/iStockphoto (top left). Fig. 8.17b: Jim Parkin/iStockphoto (top right). Fig. 8.17c: imagebroker/Alamy (center). Fig. 8.18a: John Carnemolla/iStockphoto (bottom left). Fig. 8.18b: Toni Angermayer/Photo Researchers (bottom center). Fig. 8.18c: Gary Unwin/iStockphoto (bottom right).

CHAPTER 9 Ch Op 9: Franz Marc Frei/Corbis Images. Fig. 9.2: Joe Raedle/Getty Images, Inc. Fig. 9.3: Cameron Davidson /Alamy (left). Fig. 9.4: Cameron Davidson/Alamy (right). Fig. 9.6: age fotostock/SuperStock. Fig. 9.7: Brand X Pictures/Getty Images, Inc. Fig. 9.8a: Nik Wheeler/©Corbis (left). Fig. 9.9: Corbis/SuperStock (top). Fig. 9.10a: Gary Crabbe/Alamy (bottom left). Fig. 9.10b: Ian Shive/Aurora Photos/Corbis Images (bottom right). Fig. 9.11a: Kaido KÑrner/Alamy (left). Fig. 9.11b: Tom Brakefield/©Photolibrary (right). Fig. 9.12: Robert Morris//Alamy (left). Fig. 9.13: Courtesy Ed Keller (right).

CHAPTER 10 Ch Op 10: ©AP/Wide World Photos. Fig. 10.1a: T. Orban/©Corbis (top). Fig. 10.1b: David & Peter Turnley/©Corbis (bottom). Fig. 10.2: Prof. Ed Keller. Fig. 10.3a: Bill Brooks/Masterfile (left). Fig. 10.3b: Greg Taylor/Alamy (right). Fig. 10.6: O. Franken/©Corbis. Fig. 10.7: Stephen Dalton/Photo Researchers. Fig. 10.9c: Martin Bond/Photo Researchers. Fig. 10.10: NANJLA TANJUNG/Reuters/Landov LLC. Fig. 10.16: (c)AP/Wide World Photos.

CHAPTER 11 Ch Op 11: Courtesy USDA. Fig. 11.1: Debbie Roos, North Carolina Cooperative Extension. Fig. 11.4: Bettmann Archive/©Corbis. Fig. 11.6: ©AP/Wide World Photos. Fig. 11.8a: Kevin Morris/Getty Images, Inc. (left). Fig. 11.8b: JC Carton/Bruce Coleman, Inc. (center). Fig. 11.8c: Du an Kosti /iStockphoto (right). Fig. 11.15: ©Corbis (top left). Fig. 11.11: SuperStock. Fig. 11.16: Daniel Valla/Alamy. Fig. 11.19: Bullit Marquez/©AP/Wide World Photos. Fig. 11.20a: Simko/Visuals Unlimited (top left). Fig. 11.20b: JAY DIRECTO/AFP/Getty Images/NewsCom (top right). Fig. 11.20c: Gaertner/Alamy (bottom). Fig. 11.21: Doug Plummer/Photo Researchers (bottom).

CHAPTER 12 Ch Op 12: Courtesy Dan Botkin. Fig. 12.1a: Courtesy Dan Botkin. Fig. 12.2: Radius/SuperStock (top). Fig. 12.1: IndexStock/SuperStock. Fig. 12.12: Global Warming Images/Alamy. Fig. 12.13a: Steve McCurry/Magnum Photos, Inc. (top). Fig. 12.13b: John Warburton-Lee Photography/Alamy (bottom). Fig. 12.14a: Photo Researchers (left). Fig. 12.14b: Gregory G. Dimijian, M.D./Photo Researchers (right). Fig. 12.15: Chee-Onn Leong/iStockphoto. Fig. 12.16a: Courtesy Dan Botkin. Fig. 12.17: Rich Reid/Getty Images, Inc.

CHAPTER 13 Ch Op 13: Courtesy Dan Botkin. Fig. 13.1: PhotoDisc, Inc. Fig. 13.2a: WILLIAM ALBERT ALLARD/NG Image Collection (top). Fig. 13.2b: Buffalo Hunt Chase (colour litho); Catlin, George (1794-1872); colour lithograph; 600 X 414; © Butler Institute of American Art, Youngstown, OH, USA; PERMISSION REQUIRED FOR NON EDITORIAL USAGE; Out of copyright (bottom). Fig. 13.4a: Ken Lucas/Visuals Unlimited. Fig. 13.6a: David South /Alamy (left). Fig. 13.6b: Jeffrey L. Rotman/©Corbis (center). Fig. 13.6c: Jeffrey Rotman/©Corbis (right). Fig. 13.9a: Courtesy NOAA. Fig. 13.10a: US Bureau of Reclamation (top). Fig. 13.11: Corbis/SuperStock, Inc. (bottom). Fig. 13.12: ©NB, Picture Archive, Cod. med. graec, fol. 164v. Fig. 13.14: John Cunningham/Visuals Unlimited. Fig. 13.15a: Francois Gohier/Photo Researchers (left). Fig. 13.15b: Denis Scott/©Corbis (right). Fig. 13.16a: Richard Elliott/Getty Images, Inc. (top). Fig. 13.16b: ©T. ULRICH / ClassicStock/The Image Works (center). Fig. 13.16c: Dianne Blell/Peter Arnold, Inc. (bottom). Fig. 13.17a: Gilbert Grant/Photo Researchers (top). Fig. 13.17b: Photo Researchers (bottom).

CHAPTER 14 Ch Op 14: ©AP/Wide World Photos. Fig. 14.1: Taxi/Getty Images, Inc.

CHAPTER 15 Ch Op 15: NASA Images (top). Ch Op 15-inset: Gerald Herbert/©AP/Wide World Photos (bottom). Fig. 15.3: Courtesy Ed Keller. Fig. 15.7a: George Hunter/Getty Images, Inc. (left). Fig. 15.7b: Ken Graham/Bruce Coleman, Inc. (right). Fig. 15.8: U.S. Fish and Wildlife Service/Liaison Agency, Inc./Getty Images. Fig. 15.9b: ©AP/Wide World Photos (left). Fig. 15.9c: U.S. Fish and Wildlife Service/Getty Images, Inc. (right). Fig. 15.13: ©AP/Wide World Photos. Fig. 15.14: IndexStock/SuperStock. Fig. 15.15a: Prof. Ed Keller (top). Fig. 15.15b: Prof. Ed Keller (bottom). Fig. 15.16: Mandel Ngan/Getty Images (top). Fig. 15.17: William P. Hines/The Scranton Times Tribune Library (bottom). Fig. 15.18: Wally Bauman/Alamy.

CHAPTER 16 Ch Op 16: age fotostock/SuperStock. Fig. 16.1: JORGE SILVA/Reuters/Landov LLC. Fig. 16.4b: JAMES P. BLAIR/NG Image Collection. Fig. 16.7a: H. Gruyaeart//Magnum Photos, Inc. (left). Fig. 16.7b: Prof. Ed Keller (right). Fig. 16.8: Courtesy Dan Botkin (bottom). Fig. 16.9a: Paul Harris/NewsCom. Fig. 16.12: Yann Arthus-Bertrand/©Corbis. Fig. 16.15: Glen Allison/Getty Images, Inc. Fig. 16.16: Kim Steele/Getty Images, Inc.

CHAPTER 17 Ch Op 17a: Courtesy Dan Botkin (top). Ch Op 17b: Courtesy Dan Botkin (bottom). Fig. 17.5a: Graham Finlayson/Getty Images, Inc. (top). Fig. 17.5: Martin Bond/Science Photo Library (bottom). Fig. 17.6: Roger Ressmeyer/Starlight/©Corbis. Fig. 17.12: ©Corbis. Fig. 17.14: Igor Kostin/©Corbis. Fig. 17.15: U.S. Department of Energy.

CHAPTER 18 Ch Op 18: Courtesy Dan Botkin. Fig. 18.1: Courtesy Ed Keller. Fig. 18.9: NASA Images (top). Fig. 18.10: AFP/EPA/NewsCom (bottom). Fig. 18.12b: USGS. Fig. 18.16a: Stephen Krasemann/Stone/Getty Images (top). Fig. 18.16b: Gregory G. Dimijian, M.D./Photo Researchers (bottom left). Fig. 18.16c: Jim Brandenburg/Minden Pictures, Inc. (bottom). Fig. 18.18: Courtesy Ed Keller. Fig. 18.19: Liu Liqun/©Corbis (top). Fig. 18.20: SuperStock (bottom). Fig. 18.21a: ©AP/Wide World Photos (left). Fig. 18.21b: STEVEN NEHL/The Oregonian/Landov LLC (right).

CHAPTER 19 Ch Op 19: Hans Pennink/©AP/Wide World Photos. Fig. 19.1: oote boe/Alamy Images. Fig. 19.2: Ben Osborne/Stone/Getty Images. Fig. 19.3: Courtesy Ed Keller (left). Fig. 19.4: McAllister/Liaison Agency, Inc./Getty Images (right). Fig. 19.7: Bohemian Nomad Picturemakers/©Corbis (top). Fig. 19.8: Thomas Del Brase/Getty Images, Inc. (bottom). Fig. 19.9: Photoshot Holdings Ltd/Alamy. Fig. 19.10: J. BLANK / ClassicStock/The Image Works (left). Fig. 19.12a: Courtesy Ed Keller (top). Fig. 19.12b: Courtesy Ed Keller (bottom). Fig. 19.18a: Courtesy John Day, Louisiana State University (top left). Fig. 19.18b: Courtesy John Day, Louisiana State University (bottom left). Fig. 19.18c: Courtesy John Day, Louisiana Stae University (right). Fig. 19.19: (c)Integrated Water Systmen, Inc (top). Fig. 19.20: Prof. Ed Keller (bottom). Fig. 19.22: John Althouse/Liaison Agency, Inc./Getty Images.

CHAPTER 20 Ch Op 20: Art Archive, The/SuperStock. Fig. 20.1: SuperStock. Fig. 20.3: NOAA. Fig. 20.12: Glasshouse Images/SuperStock. Fig. 20.12: Science Source/Photo Researchers. Fig. 20.13: Roger Ressmeyer/©Corbis. Fig. 20.14a: NASA Images (left). Fig. 20.14b: NASA Images (top). Fig. 20.14c: NASA Images (bottom). Fig. 20.24a: TORSTEN BLACKWOOD/AFP/Getty Images (left). Fig. 20.24b: Ashley Cooper/Alamy (right). Fig. 20.27: age fotostock/SUPERSTOCK. Fig. 20.28: David Tipling/Getty Images.

CHAPTER 21 Ch Op 21a: Larry Dunmire/SuperStock. Ch Op 21b: Comstock/Photolibrary. Fig. 21.1: F. Hoffman/The Image Works (left). Fig. 21.2: Courtesy Ed Keller (right). Fig. 21.8a: iStockphoto (left). Fig. 21.8b: David McNew/Getty Images, Inc. (right). Fig. 21.12a: YinYang/iStockphoto (top left). Fig. 21.12b: Natalia Bratslavsky/iStockphoto (top center). Fig. 21.12c: Lev Dolgatshjov/iStockphoto (top right). Fig. 21.12d: Eric Hood/iStockphoto (bottom left). Fig. 21.12e: oversnap/iStockphoto (bottom right). Fig. 21.18a: Andrew Syred/Photo Researchers (left). Fig. 21.18b: Oliver Meckes/Photo Researchers (right).

CHAPTER 22 Ch Op 22: Irwin Thompson/Melanie Burford/Dallas Morning News/©Corbis. Fig. 22.1a: Bettmann/©Corbis (top left). Fig. 22.1b: Courtesy Dan Botkin (bottom left). Fig. 22.1c: BEBETO MATTHEWS/AP/Wide World Photos (top right). Fig. 22.4a: U. S. Army Corps of Engineers (bottom). Fig. 22.4b: Thinkstock/Getty Images (top). Fig. 22.5a: Alamy Images (left). Fig. 22.5b: Marc Serota/Reuters/Landov LLC (right). Fig. 22.6: Tim Vasquez/Weather Graphics. Fig. 22.7: Photononstop/SUPERSTOCK. Fig. 22.9: Patrick Ward/©Corbis. Fig. 22.12: Courtesy Dan Botkin (left). Fig. 22.13: Robert Holmes/©Corbis (right). Fig. 22.15: Jim Zipp/Photo Researchers, Inc.

CHAPTER 23 Ch Op 23: Joanne Carole/©AP/Wide World Photos. Fig. 23.3: Helen Thompson/Animals Animals Earth Scenes. Fig. 23.5: Bettmann/©Corbis (top). Fig. 23.6: Ed Keller (bottom). Fig. 23.7a: Stone/Getty Images (top left). Fig. 23.7b: Craig Aurness/©Corbis (top center). Fig. 23.7c: © R. Maissonneuve/Publiphoto/Photo Researchers (top right). Fig. 23.7d: Stone/Getty Images (bottom left). Fig. 23.7e: Montana Stock Photography/Alamy (bottom right). Fig. 23.8: Courtesy of Fossil Trace Golf Course. Fig. 23.11: Ricardo Moraes/©AP/Wide World Photos. Fig. 23.14b: Courtesy EPA. Fig. 23.15: Bob Sacha/©Corbis. Fig. 23.16: Courtesy of Gloucester Fire Service & Sandhurst Area Action Group (left). Fig. 23.17: Courtesy of Gloucester Fire Service & Sandhurst Area Action Group (right). Fig. 23.21: Courtesy of Cynthia Vanderlip/Algalita Marine Research Foundation.

CHAPTER 24 Ch Op 24: United States Coast Guard. Fig. 24.1a: NOAA. Fig. 24.2a: NOAA (top left). Fig. 24.2b: Patrick Semansky/AP/Wide World Photos (top right). Fig. 24.2c: NOAA (center left). Fig. 24.2d: NOAA (center right). Fig. 24.2e: Win McNamee /Getty Images, Inc. (bottom). Fig. 24.3: John Dooley/Sipa Press/NewsCom. Fig. 24.4a: Win McNamee/Getty Images, Inc. (top). Fig. 24.4b: Spencer Platt /Getty Images, Inc. (bottom). Fig. 24.6: Tennessee Valley Authority. Fig. 24.7: B. Anthony Stewart/National Geographic/Getty Images, Inc. Fig. 24.8: Library of Congress Prints and Photographs Division. Fig. 24.10: Stone/Getty Images. Fig. 24.11a: Tom Van Sant/Photo Researchers (left). Fig. 24.11b: David W. Hamilton/The Image Bank/Getty Images (right).

Notes

Chapter 1 Notes

1. Western, D., and C. Van Prat. 1973. Cyclical changes in habitat and climate of an East African ecosystem. *Nature* 241(549):104–106.
2. Dunne, T., and L.B. Leopold. 1978. *Water in Environmental Planning.* San Francisco: Freeman.
3. Altmann, J., S.C. Alberts, S.A. Altman, and S.B. Roy. 2002. Dramatic change in local climate patterns in the Amboseli basin, Kenya. *African Journal of Ecology* 40:248–251.
4. Maclennan, S.D., R.J. Groom, D.W. Macdonald, and L.G. Frank. 2009. Evaluation of a compensation scheme to bring about a pastoralist tolerance of lions. *Bio Conserve.* (in Press).
5. Frank, L., S. Maclennan, L. Mazzah, R. Bonham, and T. Hill. 2006. Lion killing in the Amboseli-Tsavo ecosystem, 2001–2006, and its implications for Kenya's lion population. *Report by the Kilimanjaro Lion Conservation Project and Museum of Vertebrate Zoology*, University of California, Berkeley.
6. Deevey, E.S. 1960. The human population. *Scientific American* 203:194–204.
7. Keyfitz, N. 1989. The growing human population. *Scientific American* 261: 118–126.
8. Gottfield, R.S. 1983. *The Black Death: Natural and Human Disaster in Medieval Europe.* New York: Free Press.
9. Field, J.O., ed. 1983. *The Challenge of Famine: Recent Experience, Lessons Learned.* Hartford, CT: Kumarian Press.
10. Glantz, M.H., ed. 1987. *Drought and Hunger in Africa: Denying Famine a Future.* Cambridge: Cambridge University Press.
11. Seavoy, R.E. 1989. *Famine in East Africa: Food Production and Food Politics.* Westport, CT: Greenwood.
12. Levinson, F.J., and Bassett. 2007. *Malnutrition is Still a Major Contributor to Child Deaths.* Washington, DC: Population Reference Bureau.
13. Gower, B.S. 1992. What do we owe future generations? In D.E. Cooper and J.A. Palmers, eds., *The Environment in Question: Ethics and Global Issues*, pp. 1–12. New York: Routledge.
14. Haub, C. 2007. *World Population Data Sheet.* Washington, DC: Population Reference Bureau.
15. Bartlett, A.A. 1997–1998. Reflections on sustainability, population growth, and the environment, revisited. *Renewable Resources Journal* 15(4):6–23.
16. Hawken, P., A. Lovins, and L.H. Lovins. 1999. *Natural Capitalism.* Boston: Little, Brown.
17. Hubbard, B.M. 1998. *Conscious Evolution.* Novato, CA: New World Library.
18. Botkin, D.B., M. Caswell, J.E. Estes, and A. Orio, eds. 1989. *Changing the Global Environment: Perspectives on Human Involvement.* New York: Academic Press.
19. Montgmery, D.R. 2007. *Dirt: The Erosion of Civilizations.* Berkeley: University of California Press.
20. World Resources Institute. 1998. *Teacher's Guide to World Resources: Exploring Sustainable Communities.* Washington, DC: World Resources Institute. http://www.igc.org/wri/wr-98–99/citygrow.htm.
21. World Resources Institute. 1999. *Urban Growth.* Washington, DC: World Resources Institute.
22. Starke, L., ed. 2007. *State of the World: Our Urban Future.* Worldwatch Institute. New York: W. W. Norton.
23. Foster, K.R., P. Vecchia, and M.H. Repacholi. 2000. Science and the Precautionary Principle. *Science* 288:979–981.
24. Easton, T.A., and T.D. Goldfarb, eds. 2003. *Taking Sides, Environmental Issues*, 10th ed. Issue 5: Is the Precautionary Principle a sound basis for international policy? pp. 76–101. Guilford, CT: McGraw-Hill/Dushkin.
25. Rolett, B., and J. Diamond. 2004. Environmental predictors of pre-European deforestation on Pacific Islands. *Nature* 431:443–446.
26. Stokstad, E. 2004. Heaven or hellhole? Islands' destinies were shaped by geography. *Science* 305:1889.
27. Hunt, T.L. 2006. Rethinking the fall of Easter Island. *American Scientist* 94(5): 412–419.
28. Hunt,T.L, and C.P. Lipo. 2008. Evidence for a shorter chronology on Rapa Nui (Easter Island). *Journal of Island and Coastal Archaeology* 3:140–148.
29. Brown, L.R., and C. Flavin. 1999. A new economy for a new century. In L. Star, ed., *State of the World*, pp. 3–21. New York: W.W. Norton.

Chapter 2 Notes

1. Wiens, J.A., D.T. Pattern, and D.B. Botkin. 1993. Assessing ecological impact assessment: Lessons from Mono Lake, California. *Ecological Applications* 3(4): 595—609; and Botkin, D.B., W.S. Broecker, L.G. Everett, J. Shapiro, and J.A. Wiens. 1988. *The Future of Mono Lake* (Report No. 68). Riverside: California Water Resources Center, University of California and http://www.monolake.org; accessed June 17, 2008.
2. Botkin, D.B. 2001. *No Man's Garden: Thoreau and a New Vision for Civilization and Nature.* Washington, DC: Island Press.
3. Lerner, L.S., and W.J. Bennetta. 1988 (April). The treatment of theory in textbooks. *The Science Teacher*, pp. 37–41.
4. Vickers, B., ed. *English Science: Bacon to Newton* (Cambridge English Prose Texts). 1987 (Paperback)
5. Kuhn, T.S. 1970. *The Structure of Scientific Revolutions.* Chicago: University of Chicago Press.
6. Pease, C.M., and J.J. Bull. 1992. Is science logical? *Bioscience* 42:293–298.

7. Schmidt, W.E. 1991 (September 10). "Jovial con men" take credit(?) for crop circles. *New York Times*, p. 81.
8. This information about crop circles is from *Crop Circle News*, http://cropcirclenews.com.
9. Heinselman, H.M. 1973. Fire in the virgin forests of the Boundary Waters Canoe Area, Minnesota. *Journal of Quaternary Research* 3:329–382.
10. Trefil, J.S. 1978. A consumer's guide to pseudoscience. *Saturday Review* 4:16–21.

Chapter 3 Notes

1. Botkin, D.B. 1999. *Passage of Discovery: The American Rivers Guide to the Missouri River of Lewis and Clark*. New York: Perigee Books (a Division of Penguin-Putnam).
2. Botkin, D.B. 2004. *Beyond the Stony Mountains: Nature in the American West from Lewis and Clark to Today*. New York: Oxford University Press.
3. Botkin, D.B. 1990. *Discordant Harmonies: A New Ecology for the 21st Century*. New York: Oxford University Press.
4. The discussion of basic systems responses in from Botkin, D.B. and K. Woods, *Fundamentals of Ecology* (in press).
5. Dunne, T., and L.B. Leopold. 1978. *Water in Environmental Planning*. San Francisco: Freeman.
6. Leach, M.K., and T.J. Givnich. 1996. Ecological determinants of species loss in remnant prairies. *Science* 273:1555–1558.
7. Lovelock, J. 1995. *The Ages of Gaia: A Biography of Our Living Earth*, rev. ed. New York: Norton.

Chapter 3 Critical Thinking Issue References

Barlow, C. 1993. *From Gaia to Selfish Genes.* Cambridge, MA: MIT Press.

Kirchner, J.W. 1989. The Gaia hypothesis: Can it be tested? *Reviews of Geophysics* 27:223–235.

Lovelock, J.E. 1995. *Gaia: A New Look at Life on Earth.* New York: Oxford University Press.

Lyman, F. 1989. What hath Gaia wrought? *Technology Review* 92(5): 55–61.

Resnik, D.B. 1992. Gaia: From fanciful notion to research program. *Perspectives in Biology and Medicine* 35(4):572–582.

Schneider, S.H. 1990. Debating Gaia. *Environment* 32(4):4–9, 29–32.

Chapter 3 Working It Out 3.1 Reference

Bartlett, A. A. 1993. The arithmetic of growth: Methods of calculation. *Population and Environment* 14(4):359–387.

Chapter 4 Notes

1. World Health Organization: http://www.who.int/csr/don/2009_05_14/en/index.html. Accessed May 14, 2009.
2. U.S. Centers for Disease Control. West Nile virus statistics, surveilance, and control. http://www.cdc.gov/ncidod/dvbid/westnile/surv&controlCaseCount07_detailed.htm.
3. U.S. Centers for Disease Control Web site, available at http:// www.cdc.gov/ncidod/sars/factsheet.htm and http://www.cdc.gov/ncidod/dvbid/westnile/qa/overview.htm.
4. Xinhua News Agency, 2002 (December 31). China's cross-border tourism prospers in 2002. From the Population Reference Bureau Web site, available at http://www.prb.org/Template.cfm? Section=PRB&template=/ContentManagement/Content_Display. cfm&ContentID=8661.
5. Graunt, J. (1662). 1973. *Natural and Political Observations Made upon the Bill of Mortality*. London, 662.
6. Barry, J.M. 2005. *The Great Influenza: The Story of the Deadliest Pandemic in History* (New York: Penguin Books, paperback).
7. Keyfitz, N. 1992. Completing the worldwide demographic transition: The relevance of past experience. *Ambio* 21:26–30.
8. U.S. Census Bureau. International Data Base, http://www.census.gov/ipc/www/idb/idbsprd.html. Accessed May 14, 2009.
9. Graunt, J. *Natural and Political Observations Made upon the Bill of Mortality.*
10. Dumond, D.E. 1975. The limitation of human population: A natural history. *Science* 187:713–721.
11. All population growth rate data are from U.S. Census Bureau, http://www.census.gov/ipc/www/idb/country/usportal.html. Accessed May 12, 2009.
12. Population Reference Bureau. 2005. World Population Data Sheet.
13. Ibid.
14. Botkin, D.B. 1990. *Discordant Harmonies: A New Ecology for the 21st Century.* New York: Oxford University Press.
15. Zero Population Growth. 2000. U.S. population. Washington, DC.: Zero Population Growth.
16. World Bank. 1984. World Development Report 1984. New York: Oxford University Press.
17. McNeil, Donald G. Jr. 2008 (June 15). Malthus redux: Is doomsday upon us, again? *New York Times.*
18. Malthus, T.R. (1803). 1992. An essay on the principle of population. Selected and introduced by Donald Winch. Cambridge, England: Cambridge University Press.
19. *Source:* U.S. Census Bureau, International Data Base. http://www.census.gov/cgi-bin/ipc/idbagg.
20. U.S. Census Bureau. International Data Base, http://www.census.gov/ipc/www/idb/idbsprd.html. Accessed May 14, 2009.
21. Bureau of the Census, 1990. *Statistical Abstract of the United States* 1990. Washington, DC: U.S. Department of Commerce.
22. Botkin, D.B. 1990. *Discordant Harmonies: A New Ecology for the 21st Century.* New York: Oxford University Press.
23. Erhlich, P.R. 1971. *The Population Bomb, rev. ed.* New York: Ballantine.
24. Central Intelligence Agency. 1990. *The World Factbook.* Washington, DC: CIA.
25. Naess, Arne. 1989. *Ecology, Community and Lifestyle: Outline of an Ecosophy.* New York: Cambridge University Press.

26. World Bank. 1992. World Development Report. The Relevance of Past Experience. Washington, DC: World Bank.
27. Guz, D., and J. Hobcraft. 1991. Breastfeeding and fertility: A comparative analysis. *Population Studies* 45:91–108.
28. Fathalla, M.F. 1992. Family planning: Future needs, *AMBIO* 21: 84–87.
29. Alan Guttmacher Institute. 1999. Sharing Responsibility: Women, *Society and Abortion Worldwide.* New York: AGI.
30. Xinhua News Agency, March 13, 2002, untitled, available at http://www.16da.org.cn/english/archiveen/28691.htm.
31. Xinhua News Agency, March 13, 2002, untitled, available at http://www.16da.org.cn/english/archiveen/28691.htm.
32. Revkin, A.C. 2008 (March 14). *Earth 2050: Population Unknowable?* **Dot Earth** (a Web page of the *New York Times*). http://dotearth.blogs.nytimes.com/2008/03/14/earth-2050-population-unknowable/?scp=2-b&sq =human+population+growth&st=nyt.

Chapter 5 Notes

1. Kvitek, R.G., J.S. Oliver, A.R. DeGange, and B.S. Anderson. 1992. Changes in Alaskan soft-bottom prey communities along a gradient in sea otter predation. *Ecology* 73:413–428.
2. Paine, R.T. 1969. A note on trophic complexity and community stability. *American Naturalist* 100:65–75.
3. Duggins, D.O. 1980. Kelp beds and sea otters: An experimental approach. *Ecology* 61:447–453.
4. Kenyon, K.W. 1969. The sea otter in the eastern Pacific Ocean. North American Fauna, no. 68. Washington, DC: Bureau of Sports Fisheries and Wildlife, U.S. Department of the Interior.
5. Sea otter 2008 population size from the following: U.S. Geological Survey, Western Ecological Research Center, Santa Cruz Field Station, Spring 2008 Mainland California Sea Otter Survey Results. Brian Hatfield and Tim Tinker, brian_hatfield@usgs.gov, ttinker@usgs.gov June, 24 2008, U.S. Geological Survey, http://www.werc.usgs.gov/otters/ca-surveydata.html.
6. IUCN otter specialist group. 2006. http://www.otterspecialistgroup.org/Species/Enhydra_lutris.html#Distribution. The current worldwide population estimate for *E. lutris* is approximately 108,000.
7. Kvitek et al. Changes in Alaskan soft-bottom prey communities.
8. Morowitz, H.J. 1979. *Energy Flow in Biology.* Woodbridge, CT: Oxbow Press.
9. Brock, T.D. 1967. Life at high temperatures. *Science* 158:1012–1019.
10. NOAA Office of Protected Resources, http://www.nmfs.noaa.gov/pr/species/mammals/pinnipeds/harpseal.htm, accessed June 26, 2009.
11. Lavigne, D.M., W. Barchard, S. Innes, and N.A. Oritsland. 1976. *Pinniped bioenergetics.* ACMRR/MM/SC/12. Rome: United Nations Food and Agriculture Organization.
12. Estes, J.A., and J.F. Palmisano. 1974. Sea otters: Their role in structuring nearshore communities. *Science* 185:1058–1060.
13. Schrödinger, Erwin, 1944. *What Is Life? The Physical Aspect of the Living Cell.* Cambridge: The University Press. Based on lectures delivered under the auspices of the Institute of Trinity College, Dublin, in February 1943. Widely reprinted since then and available as a PDF file on the Internet.
14. Peterson, R.O. 1995. *The Wolves of Isle Royale: A Broken Balance.* Minocqua, WI: Willow Creek Press.
15. Mitchell, W. Colgan. 1987 (December). Coral Reef Recovery on Guam (Micronesia) after Catastrophic Predation by *Acanthaster Planci. Ecology*, 68, No. 68 (6): 1592–1605.
16. Tissot, Brian N. 2005. The Hawaiian reef ecosystem. Hawai'i Coral Reef Network, http://www.coralreefnetwork.com/reefs/ecology/ecology.htm.
17. Botkin, D.B. 1990. *Discordant Harmonies: A New Ecology for the 21st Century.* New York: Oxford University Press.
18. Connell, J.H., and R.O. Slatyer. 1977. Mechanism of succession in natural communities and their role in community stability and organization. *American Naturalist* 111:1119–1144.
19. Pickett, S.T.A., S.L. Collins, and J.J. Armesto. 1987. Models, mechanisms and pathways of succession. *Botanical Review* 53:335–371.
20. Gomez-Pompa, A., and C. Vazquez-Yanes. 1981. Successional studies of a rain forest in Mexico. In D.C. West, H.H. Shugart, and D.B. Botkin eds., *Forest Succession: Concepts and Application,* pp. 246–266. New York: Springer-Verlag.
21. Gorham, E., P.M. Vitousek, and W.A. Reiners. 1979. The regulation of chemical budgets over the course of terrestrial ecosystem succession. *Annual Review Ecology and Systematics* 10:53–84.
22. MacMahon, J.A. 1981. Successional processes: Comparison among biomes with special reference to probable roles of and influences on animals. In D.C. West, H.H. Shugart, and D.B. Botkin, eds., *Forest Succession: Concepts and Application,* pp. 277–304. New York: Springer-Verlag.

Chapter 6 Notes

1. Mau, S., D.L. Valentine, J.F. Clark, J. Reed, R. Camilli, and L. Washburn, 2007. Dissolved methane distributions and air-sea flux in the plume of a massive seep field, Coal Oil Point, California. *Geophysical Research Letters*, 34, L22603, doi:10.1029/2007GL031344, 20.
2. Ding, H., and D.L. Valentine. 2008. Methanotrophic bacteria occupy benthic microbial mats in shallow marine hydrocarbon seeps, Coal Oil Point, California. *Journal of Geophysical Research Biogeosciences* 113 (GI) G01015.
3. Leifer, I., B.P. Luyendyk , J. Boles, and J.F. Clark, 2006. Natural marine seepage blowout: Contribution to atmospheric methane. *Global Biogeochemical Cycles*, 20, GB3008, doi:10.1029/2005GB002668.

4. Christian, D. 2004. *Maps of Time.* Berkeley: University of California Press.
5. Gaddis, J.L. 2002. *The Landscape of History.* New York, Oxford University Press.
6. Lovelock, J. 1995. *The Ages of Gaia: A Biography of the Earth.* Oxford: Oxford University Press.
7. Lane, Nick. 2009. *Oxygen: The Molecule That Made the World.* Oxford: Oxford University Press.
8. This section is based on, and used with permission, Botkin, D.B. 1990. *Discordant Harmonies: A New Ecology for the 21st Century.* New York, Oxford University Press.
9. Henderson, Lawrence J. 1913 reprinted 1958 by the same publisher. *The Fitness of the Environment.* Boston: Beacon Press.
10. Ehrlich, P.R., A.H. Ehrlich, and J.P. Holdren. 1970. *Ecoscience: Population, Resources, Environment.* San Francisco: W.H. Freeman, p. 1051.
11. Post, W.M., T. Peng, W.R. Emanuel, A.W. King, V.H. Dale, and D.L. De Angelis. 1990. The global carbon cycle. *American Scientist* 78:310–326.
12. Kasting, J.F., O.B. Toon, and J.B. Pollack. 1988. How climate evolved on the terrestrial planets. *Scientific American* 258:90–97.
13. Berner, R.A. 1999. A new look at the long-term carbon cycle. *GSA Today* 9(11):2–6.
14. Peter M. Vitousek, Chair, John Aber, Robert W. Howarth, Gene E. Likens, Pamela A. Matson, David W. Schindler, William H. Schlesinger, and G. David Tilman. 2000. Human Alteration of the Global Nitrogen Cycle: Causes and Consequences. *Issues in Ecology—Human Alteration of the Global Nitrogen Cycle.* ESA publication http://esa.sdsc.edu/tilman.htm 30/08/2000.
15. Chameides, W.L., and E.M. Perdue. 1997. *Biogeochemical Cycles.* New York: Oxford University Press.
16. Carter, L.J. 1980. Phosphate: Debate over an essential resource. *Science* 209:4454.
17. Vaccari, D.A. 2009. Phosphorus: A looming crisis. *Scientific American* 300(6):54–59.
18. Smil, V. 2000. Phosphorus in the environment: Natural flows and human interference. *Annual Review of Environment and Resources,* 25:53–88.
19. U.S. Geological Survey. 2009. http://minerals.usgs.gov/minerals/pubs/commodity/phosphate_rock/mcs-2009-phosp.pdf.
20. Ibid.
21. Asner, G.P., T.R. Seastedt, and A.R. Townsend. 1997 (April). The decoupling of terrestrial carbon and nitrogen cycles. *Bioscience* 47 (4):226–234.
22. Hellemans, A. 1998 (February 13). Global nitrogen overload problem grows critical. *Science,* 279:988–989.
23. Smil, V.1997 (July). Global populations and the nitrogen cycle. *Scientific American,* 76–81.
24. Vitousek, P.M., J. Aber, R.W. Howarth, G.E. Likens, P.A. Matson, D.W. Schindler, W.H. Schlesinger, and G.D. Tilman. 1997. Human alteration of the global nitrogen cycle: Causes and consequences. *Issues in Ecology.* http://esa.sdsc.edu/.

Chapter 7 Notes

1. EPA. 2009. *Cap and Trade Essentials.* Adobe pdf file available at http://www.epa.gov/captrade/documents/ctessentials.pdf accessed June 16, 2009.
2. Loris, Nicolas and Ben Lieberman. 2009. *Cap and trade: A handout for corporations and a huge tax on consumers.*" Thomas A. Roe Institute for Economic Policy Studies. Washington, DC: Heritage Foundation.
3. Tarlock, Dan. 1994. The nonequilibrium paradigm in ecology and the partial unraveling of environmental law." Loyola of Los Angeles Law Review **2**(1121): 1–25.
4. Senate Resolution 580 2007, Recognizing the importance of pollinators to ecosystem health and agriculture in the United States and the value of partnership efforts to increase awareness about pollinators and support ... (Agreed to by Senate). http://www.ucs.iastate.edu/mnet/_repository/2005/plantbee/pdf/Senate_Resolution_580.pdf.
5. The North American Pollinator Protection Campaign and the Pollinator Partnership. http://www.pollinator.org. Accessed July 18, 2009.
6. National Biological Information Infrastructure. 2009. http://www.nbii.gov/portal/community/Communities/Ecological_Topics/Pollinators/Pollinator_Species/Invertebrates/. Accessed July 18, 2009.
7. Hills, S. 2008. Study sheds light on bee decline threatening crops. Food Navigator USA.com 20-Aug-2008. http://www.foodnavigator-usa.com/Product-Categories/Fruit-vegetable-nut-ingredients. Accessed July 18, 2009.
8. Costanza, R., et al. 1997. The value of the world's ecosystem services and natural capital. *Nature* 387:253–260.
9. Hardin, G. 1968. The tragedy of the commons. *Science* 162:1243–1248.
10. State of Vermont. http://www.anr.state.vt.us/air/htm/woodstoverebate.htm Accessed July 18, 2009.
11. Clark, C.W. 1973. The economics of overexploitation. *Science* 181:630–634.
12. Bockstoce, J.R., D.B. Botkin, A. Philp, B.W. Collins, and J.C. George, 2007. The geographic distribution of Bowhead Whales in the Bering, Chukchi, and Beaufort seas: Evidence from whaleship records, 1849–1914. 2007 *Marine Fisheries Review* 67 (3): 1–43.
13. Freudenburg, W.R. 2004. Personal communication.
14. Gunn, J.M., ed. 1995. *Restoration and Recovery of an Industrial Region: Progress in Restoring the Smelter-damaged Landscape near Sudbury, Canada.* New York: Springer-Verlag.
15. Litton, R.B. 1972. Aesthetic dimensions of the landscape. In J.V. Krutilla, ed. *Natural Environments.* Baltimore: Johns Hopkins University Press.
16. Schwing, R.C. 1979. Longevity and benefits and costs of reducing various risks *Technological Forecasting and Social Change* 13:333–345.
17. Gori, G.B. 1980. The regulation of carcinogenic hazards. *Science* 208:256–261.
18. Cairns, J., Jr., 1980. Estimating hazard. *BioScience* 20:101–107.

Chapter 8 Notes

1. Dowdy, Alan K. 2008. Current issues in invasive species management. *PowerPoint Presentation* USDA, APHIS Plant Protection and Quarantine, Emergency and Domestic Programs. Provided courtesy of Dowdy.
2. Botkin, D.B. 2001. *No Man's Garden: Thoreau and a New Vision for Civilization and Nature*. Washington, DC: Island Press.
3. Woese, C.R., O. Kandler, and M.L. Wheelis. 1990. Towards a natural system of organisms: Proposals for the domains Archaea, Bacteria, and Eucharya. *Proceedings of the National Academy of Sciences* (USA) 87:4576–4579.
4. Cicero. *The Nature of the Gods* (44 B.C.). Reprinted by Penguin Classics. New York, 1972.
5. *Christian Science Monitor*. 2008. *Complete DNA Coding Opens New Ways to Beat Malaria:* Published: 10/2/2002. © Guardian News & Media 2008 Boston.
6. United Nations World Health Organization. 2003. http:// www. who.int/mediacentre/releases/2003/pr33/en.
7. World Health Organization. 2008. World malaria 2008.
8. World Health Organization. 2008. World malaria 2008; and U.S. Centers for Disease Control. Malaria, April 11, 2007. http://www.cdc.gov/malaria/facts.htm; United Nations World Health Organization. 2003. http://www. who.int/ mediacentre/releases/2003/pr33/en/; World Health Organization. 2000. Overcoming antimicrobial resistance: World Health Report on Infectious Diseases 2000.
9. James, A.A. 1992. Mosquito molecular genetics: The hands that feed bite back. *Science* 257:37–38; Kolata, G., 1984. The search for a malaria vaccine. *Science* 226:679–682; Miller, L.H. 1992. The challenge of malaria. *Science* 257:36–37; World Health Organization. 1999 (June). Using malaria information. News Release No. 59. WHO; World Health Organization. 1999 (July). Sequencing the *Anopheles gambiae genome*. News Release No. 60. WHO.
10. Darwin, C.R.1859. *The Origin of Species by Means of Natural Selection or the Preservation of Favored Races in the Struggle for Life*. London: Murray. (Originally published as: Darwin, C. 1859. *On the Origin of Species by Means of Natural Selection*. London: John Murray.)
11. Grant, P.R. 1986. *Ecology and Evolution of Darwin's Finches*. Princeton, NJ: Princeton University Press.
12. Cox, Call., I.N. Healey, and P.D. Moore. 1973. *Biogeography*. New York: Halsted.
13. Hedrick, P.W., G.A. Gutierrez-Espeleta, and R.N. Lee. 2001. Founder effect in an island population of bighorn sheep. *Molecular Ecology* 10: 851–857.
14. Hutchinson, G.E. 1965. *The Ecological Theater and the Evolutionary Play*. New Haven, CT: Yale University Press.
15. Hardin, G. 1960. The competitive exclusion principle. *Science* 131:1292–1297.
16. British Forestry Commission. http://www.forestry.gov.uk/ forestry/Redsquirrel. May 20, 2008.
17. Miller, R.S. 1967. Pattern and process in competition. *Advances in Ecological Research* 4:1–74.
18. Elton, C.S. 1927. *Animal Ecology*. New York: Macmillan.
19. Hutchinson, G.E. 1958. Concluding remarks. *Cold Spring Harbor Symposium in Quantitative Biology* 22:415–427.
20. Botkin, D.B. 1985. The need for a science of the biosphere. *Interdisciplinary Science Reviews* 10:267–278.
21. Wallace, A.R. 1896. *The Geographical Distribution of Animals*. Vol. 1. New York: Hafner. Reprinted 1962.
22. Pielou, E.C. 1979. *Biogeography*. New York: Wiley.
23. Takhtadzhian, A.L. 1986. *Floristic Regions of the World*. Berkeley: University of California Press; and Udvardy, M. 1975. *A Classification of the Biogeographical Provinces of the World*. IUCN Occasional Paper 18. Morges, Switzerland: IUCN.
24. Lentine, J.W. 1973. Plates and provinces, a theoretical history of environmental discontinuity. In N.F. Hughes, ed., *Organisms and Continents through Time*, pp. 79–92. Special Papers in Paleontology 12.
25. Mather, J.R., and G.A. Yoshioka. 1968. The role of climate in the distribution of vegetation. *Annals of the Association of American Geography*, 58:29–41; and Prentice, I.C., W. Cramer, S.P. Harrison, R. Leemans, R.A. Monserud, and A.M. Solomon. 1992. A global biome model based on plant physiology and dominance, soil properties and climate. *Journal of Biogeography* 19:117–134.
26. Wilson, E.O. and R.H. MacArthur 2001. *The Theory of Island Biogeography*. Princeton, NJ: Princeton University Press.
27. Tallis, J.H. 1991. *Plant Community History*. London: Chapman & Hall.

Chapter 9 Notes

1. Comprehensive Everglades Restoration Plan. www. evergladesplan.org. Accessed March 11, 2006.
2. Society for Ecological Restoration. 2004. The SER international primer on ecological restoration. www.ser.org. Accessed March 11, 2006.
3. Riley, A.L. 1998. *Restoring Streams in Cities*. Washington, DC: Island Press.
4. South Florida Water Management District. Kissimmee River restoration. www.sfwmd.gov. Accessed March 11, 2006.
5. Ripple, J.W., and Robert L. Beschta, 2004. Wolves and the ecology of fear: Can predation risk structure ecosystems? *BioScience* 54(8):755–66.
6. Smith, D.W., R.O. Smith, and D.B. Houston, 2003. Yellowstone after wolves. *BioScience* 53(4):330–340.
7. CNN News, Tuesday, February 22, 2005. Posted: 2:28 PM EST (1928) GMT, http://www.cnn.com/2005/TECH/ science/02/22/iraq.marshes/index.html. Accessed April 10, 2006.
8. UNEP project to help manage and restore the Iraq Marshlands, Iraqi Marshlands Observation System (IMOS). 2006. http://imos.grid.unep.ch.

9. Collins, S.L., A.K. Knapp, J.M. Briggs, J.M. Blair, and E.M. Steinauer, 1998. Modulation of diversity by grazing and mowing in native tallgrass prairie. *Science* 280:745–747.
10. Roemer, G.W., A.D. Smith, D.K. Garcelon, and R.K. Wayne 2001. The behavioral ecology of the Island Fox (*Urocyon littoralis*). *Journal of Zoology* 255:1–14.
11. Roemer, G.W., T.J. Coonan, D.K. Garcelon, J. Dascompte, and L. Laugrin 2001. Feral pigs facilitate hyperpredation by golden eagles and indirectly cause the decline of island fox. *Animal Conservation* 4:307–318.
12. Taylor, P. 2000. Nowhere to run, nowhere to hide. *National Wildlife* 38(5).
13. Wathern, Peter. 1986. Restoring derelict lands in Great Britain. In G. Orians, ed., *Ecological Knowledge and Environmental Problem-solving: Concepts and Case Studies*, pp. 248–274. Washington, DC: National Academy Press.
14. Pacific Estuarine Research Laboratory (PERL). 1997 (November). The status of constructed wetlands at Sweetwater Marsh National Wildlife Refuge. Annual Report to the California Department of Transportation.
15. Malakoff, D. 1998 (April 17). Restored wetlands flunk real-world test. *Science* 280:371–372.
16. Zedler, J.B. 1997. Adaptive management of coastal ecosystems designed to support endangered species. *Ecology Law Quarterly* 24:735–743.
17. Zedler, J.B., and A. Powell. 1993. Problems in managing coastal wetlands: Complexities, compromises, and concerns. *Oceanus* 36(2):19–28.

Chapter 10 Notes

1. Whitworth, K.W., E. Symanski, and A.L. Coker. 2008. Childhood lymphohematopoietic cancer incidence and hazardous air pollutants in southeast Texas, 1995–2004. *Environmental Health Perspectives* 116(11):1576–1580.
2. US Department of Labor. 1,3-Butadiene. Health Effects. Accessed April 28, 2008, www.osha.gov.
3. Selinus, O., ed. 2005. *Essentials of Medical Geology*. Burlington, MA: Elsevier Academic Press.
4. Rubonowitz-Lundun, E., and K.M. Hiscock. 2005. Water hardness and health effects. In O. Selinus, ed., *Essentials of Medical Geology*, Ch. 13. Burlington, MA: Elsevier Academic Press.
5. Han, H. 2005. The grey zone. In C. Dawson and G. Gendreau. *Healing Our Planet, Healing Ourselves*. Santa Rosa, CA: Elite Books.
6. Warren, H.V., and R.E. DeLavault. 1967. A geologist looks at pollution: Mineral variety. *Western Mines* 40:23–32.
7. Evans, W. 1996. Lake Nyos. Knowledge of the fount and the cause of disaster. *Science* 379:21.
8. Krajick, K. 2003. Efforts to tame second African killer Lake Begin. *Science* 379:21.
9. Gunn, J., ed. 1995. *Restoration and Recovery of an Industrial Region: Progress in Restoring the Smelter-damaged Landscape near Sudbury, Canada*. New York: Springer-Verlag.
10. Blumenthal, D.S., and J. Ruttenber. 1995. *Introduction to Environmental Health*, 2nd ed. New York: Springer.
11. U.S. Geological Survey. 1995. *Mercury Contamination of Aquatic Ecosystems*. USGS FS 216–95.
12. Greer, L., M. Bender, P. Maxson, and D. Lennett. 2006. Curtailing mercury's global reach. In L. Starke, ed., *State of the World 2006*, pp. 96–114. New York: Norton.
13. Ehrlich, P.R., A.H. Ehrlich, and J.P. Holdren. 1970. *Ecoscience: Population, Resources, Environment*. San Francisco: Freeman.
14. Waldbott, G.L. 1978. *Health Effects of Environmental Pollutants*, 2nd ed. Saint Louis: Moseby.
15. McGinn, A.P. 2000 (April 1). POPs culture. *World Watch*, pp. 26–36.
16. Carlson, E.A. 1983. International symposium on herbicides in the Vietnam War: An appraisal. *BioScience* 33:507–512.
17. Cleverly, D., J. Schaum, D. Winters, and G. Schweer. 1999. Inventory of sources and releases of dioxin-like compounds in the United States. Paper presented at the 19th International Symposium on Halogenated Environmental Organic Pollutants and POPs, September 12–17, Venice, Italy. Short paper in *Organohalogen Compounds* 41:467–472.
18. Roberts, L. 1991. Dioxin risks revisited. *Science* 251: 624–626.
19. Kaiser, J. 2000. Just how bad is dioxin? *Science* 5473: 1941–1944.
20. Johnson, J. 1995. SAB Advisory Panel rejects dioxin risk characterization. *Environmental Science & Technology* 29:302A.
21. National Research Council Committee on EPA's Exposure and Human Health Reassessment of TCDD and Related Compounds.2006. *Health Risks from Dioxin and Related Compounds*. Washington, DC: National Academy Press.
22. Thomas, V.M., and T.G. Spiro.1996. The U.S. dioxin inventory: Are there missing sources? *Environmental Science &Technology* 30:82A–85A.
23. U.S. Environmental Protection Agency. 1994 (June). Estimating exposure to dioxinlike compounds. Review draft. Office of Research and Development, EPA/600/6–88/005 Ca-c.
24. Krimsky, S. 2001. Hormone disrupters: A clue to understanding the environmental cause of disease. *Environment* 43(5):22–31.
25. Committee on Hormonally Active Agents in the Environment, National Research Council, National Academy of Sciences. 1999. *Hormonally Active Agents in the Environment.* Washington, DC: National Academy Press.
26. Royte, E. 2003. Transsexual frogs. *Discover* 24(2):26–53.
27. Hayes, T.B., K. Haston, M. Tsui, A. Hong, C. Haeffele, and A. Vock. 2002. Hermaphrodites beyond the cornfield. Atrozine-induced testicular oogenesis in leopard frogs (Rana pipiens). *Nature* 419:895–896.
28. Hanes, T.B., et al. 2002. Hermaphroditic demasculinized frogs after exposure to the herbicide atrozine at low ecologi-

cally relevant doses. *Proceedings of the National Academy (PNAS)* 99(8):5476–5480.

29. Ross, M. 1990. Hazards associated with asbestos minerals. In B.R. Doe, ed., *Proceedings of a U.S. Geological Survey Workshop on Environmental Geochemistry*, pp. 175–176. U.S. Geological Survey Circular 1033.
30. Agency for Toxic Substances & Disease Registry. 2002. Mortality from Asbestosis in Libby, Montana, 1979–1998. www.atsdr.cdc.gov.
31. Pool, R. 1990. Is there an EMF-cancer connection? *Science* 249:1096–1098.
32. Linet, M.S., E.E. Hatch, R.A. Kleinerman, L.L. Robison, W.T. Kaune, D.R. Friedman, R.K. Severson, C.M. Haines, C.T. Hartsock, S. Niwa, S. Wacholder, and R.E. Tarone. 1997. Residential exposure to magnetic fields and acute lymphoblastic leukemia in children. *New England Journal of Medicine* 337(1):1–7.
33. Kheifets, L.I., E.S. Gilbert, S.S. Sussman, P. Guaenel, S.D. Sahl, D.A. Savitz, and G. Thaeriault. 1999. Comparative analyses of the studies of magnetic fields and cancer in electric utility workers: Studies from France, Canada, and the United States. *Occupational and Environmental Medicine* 56(8):567–574.
34. Ahlbom, A., E. Cardis, A. Green, M. Linet, D. Savitz, and A. Swerdlow. 2001. Review of the epidemiologic literature on EMF and health. *Environmental Perspectives* 109(6):911–933.
35. World Health Organization, International Agency for Research on Cancer. Volume 80: Non-ionizing radiation, Part 1, Static and extremely low-frequency (ELF) electric and magnetic fields. IARC Working Group on the Evaluation of Carcinogenic Risks to Humans. 2002: Lyon, France.
36. Kleinerman, R.A., W.T. Kaune, E.E. Hatch, et al. 2000. Are children living near high voltage power lines at increased risk of acute lymphocytic leukemia? *American Journal of Epidemiology* 15:512–515.
37. Greenland, S., A.R. Sheppard, W.T. Kaune, C. Poole, and M.A. Kelsh. 2000. A pooled analysis of magnetic fields, wire codes, and childhood leukemia. Childhood Leukemia-EMF Study Group. *Epidemiology* 11(6):624–634.
38. Schoenfeld, E.R., E.S. O'Leary, K. Henderson, et al. 2003. Electromagnetic fields and breast cancer on Long Island: A case-control study. *American Journal of Epidemiology* 158:47–58.
39. Kabat, G.C., E.S. O'Leary, E.R. Schoenfeld, et al. 2003. Electric blanket use and breast cancer on Long Island. *Epidemiology* 14(5):514–520.
40. Kliukiene, J., T. Tynes, and A. Andersen. 2004. Residential and occupational exposures to 50-Hz magnetic fields and breast cancer in women: A population-based study. *American Journal of Epidemiology* 159(9):852–861.
41. Zhu, K., S. Hunter, K. Payne-Wilks, et al. 2003. Use of electric bedding devices and risk of breast cancer in African-American women. *American Journal of Epidemiology* 158: 798–806.
42. Francis, B.M. 1994. *Toxic Substances in the Environment*. New York: John Wiley & Sons.
43. Poisons and poisoning. 1997. *Encyclopedia Britannica*. Vol. 25, p. 913. Chicago: Encyclopedia Britannica.
44. Air Risk Information Support Center (Air RISC), U.S. Environmental Protection Agency. 1989. *Glossary of Terms Related to Health Exposure and Risk Assessment*. EPA/450/_3–88/016. Research Triangle Park, NC.
45. Needleman, H.L., J.A. Riess, M.J. Tobin, G.E. Biesecker, and J.B. Greenhouse. 1996. Bone lead levels and delinquent behavior. *Journal of the American Medical Association* 275:363–369.
46. Centers for Disease Control. 1991. *Preventing Lead Poisoning in Young Children*. Atlanta: Public Health Service, Centers for Disease Control.
47. Goyer, R.A. 1991. Toxic effects of metals. In M.O. Amdur, J. Doull, and C.D. Klaassen, eds., *Toxicology*, pp. 623–680. New York: Pergamon.
48. Bylinsky, G. 1972. Metallic nemesis. In B. Hafen, ed., *Man, Health and Environment*, pp. 174–185. Minneapolis: Burgess.
49. Hong, S., J. Candelone, C.C. Patterson, and C.F. Boutron. 1994. Greenland ice evidence of hemispheric lead pollution two millennia ago by Greek and Roman civilizations. *Science* 265:1841–1843.

Chapter 11 Notes

1. Etter, L. 2007. With corn prices rising, pigs switch to fatty snacks on the menus: Trail mix, cheese curls, tater tots; Farmer Jones's ethanol fix. _*Wall Street Journal*.
2. Farm Photos of the Week, August 1, 2005. Photos by Debbie Roos, Agricultural Extension Agent. Co-op member John Bonitz demonstrates how to catch the seeds that shatter during harvest. He is harvesting mustard, one of the many oilseed crops that can be used to create biodiesel fuel. http://www.ces.ncsu.edu/chatham/ag/SustAg/farmphotoaugust0105.html.
3. The North Carolina Biodiesel Trade Group was started in 2007 and has its own Web site. http://news.biofuels.coop/2008/01/15/north-carolina-biodiesel-trade-group-launched.
4. UN News Central UN predicts rise in global cereal production but warns prices will remain high 14 February 2008, The United Nations Food and *Agriculturel Organization* (FAO) http://www.un.org/apps/news/story.asp?NewsID=25621&Cr=cereal&Cr1=.
5. BBC news about a new UN FAO report February 18th, 2008. http://news.bbc.co.uk/2/hi/business/7148374.stm.
6. UN News Central UN predicts rise in global cereal production but warns prices will remain high 14 February 2008, The United Nations Food and *Agricultural Organization* (FAO) http://www.un.org/apps/news/story.asp?NewsID=25621&Cr=cereal&Cr1=.
7. Vocke, Gary. 2007, Global production shortfalls bring record wheat prices. AmberWaves November 2007.

8. U.S. Department of Agriculture (USDA). 2007. National Resources Inventory 2003 Annual NRI Soil erosion report February 2007. Accessed December 23, 2008 from http://www.nrcs.usda.gov/technical/NRI/2003/SoilErosion mrb.pdf.
9. Land area in agriculture is from United Nations Food and *Agricultural Organization*, Statistics 2006, http://faostat.fao.org/faostat.
10. Information on African wars is from Globalsecurity.org http://www.globalsecurity.org/military/world/war/index.html, accessed June 16, 2008. Information on African coups is from http://www.crisisgroup.org/home/index.cfm?id=4040.
11. Field, J.O., ed. 1993. *The Challenge of Famine: Recent Experience, Lessons Learned.* Hartford, CT.: Kumarian Press.
12. World Food Programme. 1998. *Tackling Hunger in a World Full of Food: Tasks Ahead for Food Aid*; and U.N. FAO. "FAOSTAT" 2003 Web site.
13. UNFAO. 1998 (September). Global information.
14. USDA, National Resources Inventory. 2003. Annual NRI. http://www.nrcs.usda.gov/technical/NRI/2003/nri03eros-mrb.html
15. Trimble, S.W., and P. Crosson. July 2000. U.S. soil erosion rates myth and reality. *Science* 289:248–250; and Trimble, S.W. 2000. Soil conservation and soil erosion in the upper Midwest. *Environmental Review* 7(1):3–9.
16. USDA, National Resources Inventory. 2003. Annual NRI. http://www.nrcs.usda.gov/technical/NRI/2003/ images/eros chart.
17. Lashof, J. C., ed. 1979. *Pest Management Strategies in Crop Protection.* Vol. 1. Washington, D.C.: Office of Technology Assessment, U.S. Congress; and Baldwin, F.L., and P.W. Santelmann. 1980. Weed science in integrated pest management. *BioScience* 30:675–678.
18. Barfield, C.S., and J.L. Stimac. 1980. Pest management: An entomological perspective. *BioScience* 30:683–688.
19. EBRD and FAO call for bold steps to contain soaring food prices, UN FAO http://www.fao.org/newsroom/en/news/2008/1000808/index.html, FAO news room, March 10, 2008.
20. USDA, National Agricultural Statistics Service Research and Development Division. 1997 Census of Agriculture acreage by state for harvested cropland, corn, soybeans, wheat, hay, and cotton" http://www.nass.usda.gov/research/sumpant.htm. Accessed March 10, 2008.
21. USDA. U.S. organic farming emerges in the 1990s. Available at http://www.ers.usda.gov/publications/aib770/aib770.pdf.
22. PEW Biotechnology Initiative, Pew Charitable Trusts. Web site. http://pewagbiotech.org/resources/factsheets/display.php3? FactsheetID.
23. California Agricultural Statistics Service. 1999. California agricultural statistics. Sacramento, Calif.
24. USDA. July 2000. Magriet Caswell interview.
25. USDA. 2003. Alternative Farming Systems Information Center. Available at http://www.nal.usda.gov/afsic/ofp/.
26. IPCC, 2007: Climate Change 2007: Impacts, Adaptation, and Vulnerability. Contribution of Working Group II to the Third Assessment Report of the Intergovernmental Panel on Climate Change. Martin L. Parry, Osvaldo F. Canziani, Jean P. Palutikof, Paul J. van der Linden, and Clair E. Hanson (eds.). Cambridge, UK: Cambridge University Press, 1,000 pp.

Chapter 12 Notes

1. Lloyd, E.C. 2007. Jamaica Bay Watershed Protection Plan Volume I—Regional Profile, New York City Department of Environmental Protection, Emily Lloyd, Commissioner, October 1, 2007.
2. Lloyd, E.C. 2007. Jamaica Bay Watershed Protection Plan Volume I—Regional Profile, New York City Department of Environmental Protection, Emily Lloyd, Commissioner, October 1, 2007.
3. United Nations Food and Agriculture Organization. 2001. Rome: UN FAO. Available at ftp://ftp.fao.org/docrep/fao/003/y0900e/ y0900e02.pdf.
4. Suming Jin. 2006. Effects of forest ownership and change on forest harvest rates, types and trends in northern Maine. *Science Direct* Available online May 2, 2006. http://www.sciencedirect.com/science?_ob=ArticleURL&_udi=B6T6X-4JVTCCJ-4&_user=10&_rdoc=1&_fmt=&_orig=search&_sort=d&_docanchor=&view=c&_searchStrld=1050312920&_rerunOrigin=google&_acct=C000050221&_version=1&_urlVersion=0&_userid=10&md5=5c36bcd396911d59aff4583aec8d9095.
5. United Nations Food and Agriculture Organization. 2001. Rome: UN FAO. ftp://ftp.fao.org/docrep/fao/003/y0900e/y0900e02.pdf.
6. USFS. 2008. Forest Ownership Patterns and Family Forest Highlights from the National Woodland Owner Survey USDA, Northern Research Station.
7. USFS. 2008. Forest Ownership Patterns and Family Forest Highlights from the National Woodland Owner Survey USDA, Northern Research Station.
8. Busby, F.E., et al. 1994. *Rangeland Health: New Methods to Classify Inventory and Monitor Rangelands.* Washington, D.C.: National Academy Press.
9. UNFAO. 2006. Global Forest Resources Assessment 2005: Progress toward Sustainable Forest Management. Rome, United Nations Food and Agriculture Organization. p.122, Table 7.8.
10. World Firewood Supply, World Energy Council website. www. worldenergy.org. April 24, 2006. Accessed 6/01/08.
11. World Resources Institute. 1993. *World Resources 1992–93.* New York: Oxford University Press.
12. Botkin, D.B. 1990. *Discordant Harmonies: A New Ecology for the 21st Century.* NewYork: Oxford University Press.
13. Likens. G.E., F.H. Borman, R.S. Pierce, J.S. Eaton, and N.M. Johnson. 1977. *The Biogeochemistry of a Forested Ecosystem.* New York: Springer-Verlag.

14. The Hubbard Brook ecosystem continues to be one of the most active and long-term ecosystem studies in North America. An example of a recent publication is: Bailey, S. W., D.C. Buso, and G.E. Likens. 2003. Implications of sodium mass balance for interpreting the calcium cycle of a forested ecosystem. *Ecology* 84(2):471–484.
15. Swanson, F.J., and C.T. Dyrness. 1975. Impact of clearcutting and road construction on soil erosion by landslides in the western Cascade Range, Oregon. *Geology* 3:393–396.
16. Fredriksen, R.L. 1971. Comparative chemical water quality—natural and disturbed streams following logging and slash burning. In *Forest Land Use and Stream Environments*, pp. 125–137. Corvallis: Oregon State University.
17. Sedjo, R.A., and D.B. Botkin. 1997. Using forest plantations to spare the natural forest environment 39(10):14–20.
18. http://www.google.com/imgres?imgurl=http://www.scienceimage.csiro.au/index.cfm%3Fevent%3Dsite.image.thumbnail%26id%3D4808%26viewfile%3Df%26divid%3DBU&imgrefurl=http://www.scienceimage.csiro.au/index.cfm%3Fevent%3Dsite.image.detail%26id%3D4808&h=270&w=270&sz=51&tbnid=pRK6RyGx2HvnnM:&tbnh=113&tbnw=113&prev=/images%3Fq%3Dforest%2Bplantation%2Bphoto&usg=__i1a9J6H8DL3gHqq5Nb8pp-VjO_E=&ei=RSjeSufqMZXVlAevoYSoAw&sa=X&oi=image_result&resnum=1&ct=image&ved=0CAsQ9QEwAA
19. Kimmins, H. 1995. Proceedings of the conference on certification of sustainable forestry practices. Malaysia.
20. Jenkins, Michael B. 1999. *The Business of Sustainable Forestry.* Washington, D.C.: Island Press.
21. World Resources Institute. *Disappearing Land: Soil Degradation.* Washington, D.C.: WRI.
22. Manandhar, A. 1997. *Solar Cookers as a Means for Reducing Deforestation in Nepal.* Nepal: Center for Rural Technology.
23. Council on Environmental Quality and U.S. Department of State. 1981. The Global 2000 Report to the President: Entering the Twenty-first Century. Washington, D.C.: Council on Environmental Quality.
24. Perlin, J. 1989. *A Forest Journey: The Role of Wood in the Development of Civilization.* New York: Norton.
25. Botkin, D.B. 1992. Global warming and forests of the Great Lakes states. In J. Schrnandt, ed., *The Regions and Global Warming: Impacts and Response Strategies.* New York: Oxford University Press.
26. Runte, A. 1997. *National Parks: The American Experience.* Lincoln: Bison Books of the University of Nebraska.
27. Runte, A.1997. *National Parks: The American Experience.* Lincoln: Bison Books of the University of Nebraska.
28. National Park Service Web site, http://www/wilderness.net/index.cfm?fuse=NWPS&sec=fastfacts.
29. Quotations from Alfred Runte cited by the Wilderness Society on its Web site, available at http://www.wilderness.org/NewsRoom/ Statement/20031216.cm.
30. Voyageurs National Park Web site, http://www.npsgov/voya/home.htm.
31. Costa Rica's TravelNet.National parks of Costa Rica. 1999. Costa Rica: Costa Rica's TravelNet.
32. Kenyaweb. 1998. National parks and reserves. Kenya.
33. National Park Service Web site, http://www/wilderness.net/index.cfm?fuse=NWPS&sec=fastfacts.
34. The National Wilderness Preservation System. 2006. http://nationalatlas. gov/articles/boundaries/a_nwps.html.
35. Hendee, J.C., G.H. Stankey, and R.C. Lucas. 1978. Wilderness management. *United States Forest Service Misc. Pub.* No. 1365.

Chapter 13 Notes

1. Botkin, D.B. 2004. *Beyond the Stony Mountains: Nature in the American West from Lewis and Clark to Today.* New York: Oxford University Press.
2. Mattson, D.J., and M.W. Reid. 1991. Conservation of the Yellowstone grizzly bear. *Conservation Biology* 5:364–372.
3. The National Zoo. April 10, 2006. http://nationalzoo.si.edu/support/adoptspecies/Animalinfo/biosn/default.cfm.
4. www.bisoncentral.com. April 10, 2006.
5. Haines, F. 1970. *The Buffalo.* New York: Thomas Y. Crowell.
6. Miller, R.S., D.B. Botkin, and R. Mendelssohn. 1974. The whooping crane (*Grus americana*) population of North America, *Biol. Conservation* 6: 106–111.
7. Stehn, Tom. 2009. Whooping crane recovery activities. U.S. Fish and Wildlife Service. Aransas Pass, TX, U.S. Fish and Wildlife Service: 27 pp.
8. Stehn, Tom. 2009. Whooping crane recovery activities. U.S. Fish and Wildlife Service. Aransas Pass, TX, U.S. Fish and Wildlife Service: 27pp.
9. The National Zoo. April 10, 2006. http://nationalzoo.si.edu/support/adoptspecies/Animalinfo/biosn/default.cfm.
10. Martin, P.S. 1963. *The Last 10,000 Years.* Tucson: University of Arizona Press.
11. http://www.defenders.org/wildlife/dolphin/tundolph.html.
12. April 9, 2009, New York Times, National Briefing | West, California: Agency Calls Off Chinook Salmon Fishing Season, by the Associated Press.
13. Specific data obtained from National Marine Fisheries Service. http://swr.nmfs.noaa.gov/biologic.htm.
14. Pacific Fishery Management Council. 2008, February. *Preseason Report I, Stock Abundance Analysis for 2008 Ocean Salmon Fisheries,* Chapter 2.
15. FAO Statistics. 2006. http://www.fao.org/waicent/portal/statistics_en.asp.
16. Species Survival Commission. 2006. *Red List of Threatened Species.* Geneva: IUCN.
17. Botkin, D.B., and L.M. Talbot. 1992. Biological diversity and forests. In N. Sharma, ed., *Contemporary Issues in Forest Management: Policy Implications.* Washington, D.C: World Bank.
18. Botkin, D. B. 2001. *No Man's Garden: Thoreau and a New Vision for Civilization and Nature*, Island Press.

19. O'Donnell, James J. 2008. *The Ruin of the Roman Empire.* New York: ECCO Press (HarperCollins), p. 190.
20. Botkin, D.B. 1990. *Discordant Harmonies: A New Ecology for the 21st Century.* New York: Oxford University Press.
21. Michigan Department of Natural Resources. 2008. Michigan's 2008 Kirtland's Warbler Population Reaches Another Record High.
22. NOAA's Coral Reef Conservation Program: Medicine. http://coralreef.noaa.gov/aboutcorals/values/medicine.
23. Bruckner, A. 2002. Life-Saving Products from Coral Reefs: *Issues in Science and Technology.* Spring Ed.
24. Naess, A. 1989. *Ecology, Community, and Lifestyle.* Cambridge, England: Cambridge University Press.
25. Regan, H.M., R. Lupia, A.N. Drinnan, and M.A. Burgman. 2001. The currency and tempo of extinction. *American Naturalist* 157(1):1–10.
26. Martin, P.S. 1963. *The Last 10,000 Years.* Tucson: University of Arizona Press.
27. http://www.defenders.org/wildlife/dolphin/tundolph.html.
28. U.S. Fish and Wildlife Service Delisting Report http://ecos.fws.gov/tess_public/DelistingReport.do)
29. Friends of the Earth. 1979. *The Whaling Question: The Inquiry by Sir Sidney Frost of Australia.* San Francisco: Friends of the Earth.
30. United Nations Food and Agriculture Organization. 1978. *Mammals in the Seas.* Report of the FAO Advisory Committee on Marine Resources Research, Working Party on Marine Mammals. FAO Fisheries Series 5, vol. 1. Rome: UNFAO.
31. NOAA. 2003. World fisheries. Available at http://www.st.nmfs. gov/st1/fus/current/04_world2002.pdf.
32. United Nations Food and Agriculture Organization. 1978. *Mammals in the Seas.* Report of the FAO Advisory Committee on Marine Resources Research, Working Party on Marine Mammals. FAO Fisheries Series 5, vol. 1. Rome: UNFAO.
33. Stenn, T. 2003. Whooping Crane Report, Arkansas National Wildlife Refuge, December 10, 2003. Available at http://www.birdrockport. com/tom_stehn_whooping_crane__report.htm
34. Regan, H.M., R. Lupia, A.N. Drinnan, and M.A. Burgman. 2001. The currency and tempo of extinction. *American Naturalist* 157(1):1–10.
35. Impacts of California sea lion and Pacific Harbor Seal on Salmonids and west coast ecosystems, U.S Department of Commerce, National Oceanic and Atmospheric Administration, National Marine Fisheries Service, February 10, 1999; p. Appendix-7.
36. NOAA delisted species, http://www.nmfs.noaa.gov/pr/species/ esa.htm#delisted and U.S. Fish and Wildlife Service Threatened and Endangered Species System (TESS).
37. California Department of Fish and Game. http://www.dfg.ca.gov/news/issues/lion/lion_faq.html, mountain lion abundances.
38. U.S. Department of Defense and U.S. Fish and Wildlife Service. 2006. Red-cockaded Woodpecker *(Picoides borealis).* http://www.fws. gov/endangered/pdfs/DoD/RCW_fact_sheet-Aug06.pdf, accessed 12128108.

Chapter 14 Notes

1. McKibben, B. 2007. Energizing America. *Sierra* 92(1): 30–38; 112–113.
2. Butti, K., and J. Perlin. 1980. *A Golden Thread: 2500 Years of Solar Architecture and Technology.* Palo Alto, CA: Cheshire Books.
3. Morowitz, H.J. 1979. *Energy Flow in Biology.* New Haven, CT: Oxbow Press.
4. Ehrlich, P.R., A.H. Ehrlich, and J.P. Holdren. 1970. *Ecoscience: Population, Resources, Environment.* San Francisco: W.H. Freeman.
5. Feynman, R.P., R.B. Leighton, and M. Sands. 1964. *The Feynman Lectures on Physics.* Reading, MA: Addison-Wesley.
6. Alekett, K. 2006. Oil: A bumpy road ahead. *World Watch* 19(1):10–12.
7. Cavanay, R. 2006. Global oil about to peak? A recurring myth. *World Watch* 19(1):13–15.
8. Miller, P. 2009. Saving energy. *National Geographic* 251(3): 60–81.
9. Lindley, D. 2009. The energy should always work twice. *Nature* 458(7235): 138–141.
10. Flavin, C. 1984. *Electricity's Future: The Shift to Efficiency and Small-scale Power.* Worldwatch Paper 61. Washington, DC: Worldwatch Institute.
11. Consumers' Research. 1995. Fuel economy rating: 1996 mileage estimates. *Consumers' Research* 78:22–26.
12. Lovins, A.B. 1979. *Soft Energy Paths: Towards a Durable Peace.* New York: Harper & Row.
13. Duval, J. 2007. The fix. *Sierra* 92(1):40–41.
14. Wald, M.L. 2009. The power of renewables. *Scientific American* 300(3):57–61.
15. Flavin, C. 2008. *Low-Carbon Energy: A Roadmap.* Worldwatch report 178. Worldwatch Institute. Washington, DC.
16. California Energy Commission. 1991.*California's Energy Plan: Biennial Report.* Sacramento, CA.
17. Flavin, C., and S. Dunn. 1999. Reinventing the energy system. In L.R. Brown et al., eds., *State of the World 1999: A Worldwatch Institute Report on Progress toward a Sustainable Society.* New York: W.W. Norton.
18. Dunn, S. 2000. *Micropower, the Next Electrical Era.* Worldwatch Paper 151. Washington, DC: Worldwatch Institute.
19. U.S. Energy Information Administration. International Energy Outlook 2009. Accessed January 18, 2010. www.eia.doe.gov

Chapter 15 Notes

1. Alekett, K. 2006. Oil: A bumpy road ahead. *World Watch* 19(1):10–12.
2. Roberts, P. 2008. Tapped out. *National Geographic* 213(6):86–91

3. Cavanay, R. 2006. Global oil about to peak? A recurring myth. *World Watch* 19(1):13–15.
4. McKibben, B. 2007. Energizing America. *Sierra* 92(1):30–38; 112–113.
5. Van Koevering, T.E., and N.J. Sell. 1986. *Energy: A Conceptual Approach*. Englewood Cliffs, NJ: Prentice-Hall.
6. McCulloh, T.H. 1973. In D.A. Brobst and W.P. Pratt, eds., *Oil and Gas in United States Mineral Resources*, pp. 477–496. U.S. Geological Survey Professional Paper 820.
7. British Petroleum Company. 2009. *B.P. Statistical Review of World Energy 2008*. London: British Petroleum Company.
8. Maugeri, L. 2004. Oil: Never cry wolf—when the petroleum age is far from over. *Science* 304:1114–1115.
9. Kerr, R.A. 2000. USGS optimistic on world oil prospects. *Science* 289:237.
10. Youngquist, W. 1998. Spending our great inheritance. Then what? *Geotimes* 43(7):24–27.
11. Edwards, J.D. 1997. Crude oil and alternative energy production forecast for the twenty-first century: The end of the hydrocarbon era. *American Association of Petroleum Geologists Bulletin* 81(8):1292–1305.
12. Darmstadter, J., H.H. Landsberg, H.C. Morton, and M.J. Coda. 1983. *Energy Today and Tomorrow: Living with Uncertainty*. Englewood Cliffs, NJ: Prentice-Hall.
13. Nuccio, V. 2000. *Coal-bed Methane: Potential Environmental Concerns*. U.S. Geological Survey. USGS Fact Sheet. FS-123-00.
14. Wood, T. 2003 (February 2). Prosperity's brutal price. *Los Angeles Time Magazine*.
15. Milici, R. C. and Swezey, C. S. 2006. Assessment of Appalachian Basin Oil and Gas Resources: Devonian Shale-Middle and Upper Paleozoic Total Petroleum System. *U.S. Geological Survey Open File Report* 2006-1237. Reston, VA.
16. Suess, E., G. Bohrmann, J. Greinert, and E. Lauch. 1999. Flammable ice. *Scientific American* 28(5):76–83.
17. U.S. Energy Administration 2008. Annual Coal Report 2007. Accessed July 20, 2009 at http://www.eia.doe.gov.
18. Rahn, P.H. 1996. *Engineering Geology: An Environmental Approach*, 2d ed. New York: Elsevier.
19. Stewart Burns, S. 2009. Mountain top removal in Central Appalachia. *Southern Spaces*, September 30.
20. Webber, M.E. 2009. Coal-to-liquids: The good, the bad and the ugly. *Earth* 54(4):44–47.
21. Berlin Snell, M. 2007. Can coal be clean? *Sierra* 92(1): 32–33.
22. Corcoran, E. 1991. Cleaning up coal. *Scientific American* 264:106–116.
23. Environmental Protection Agency. 2009. EPAs response to the TVA Kingston Fossil Plant Fly ash release. Accessed March 15, 2009 at http://www.epa.gov/region4/kingston.
24. Knapp, D.H. 1995. Non-OPEC oil supply continues to grow. *Oil & Gas Journal* 93:35–45.
25. Dyni, J.R. 2006. Geology and resources of some world oil-shale deposits. *U.S. Geological Survey Scientific Investigations Report 2005-5294*. Reston, VA.
26. Kunzig, R. 2009. The Canadian oil boom. *National Geographic* 215(3):34–59.
27. EIS. 2008. About tar sands. Accessed May 27, 2008 at http://ostseis.anl.gov.
28. Sachs, J.D. 2008. Surging food prices and global stability. *Scientific American* 298(6):40.

Chapter 16 Notes

1. Eaton, W.W. 1978. Solar energy. In L.C. Ruedisili and M.W. Firebaugh, eds., *Perspectives on Energy*, 2nd ed., pp. 418–436. New York: Oxford University Press.
2. Brown, L.R. 1999 (March–April). Crossing the threshold. *Worldwatch*, pp. 12–22.
3. U.S. Energy Information "Administration, Form EIA-63B 2008 (December). Annual Photovoltaic Module/Cell Manufacturers Survey."
4. Starke, L., ed. 2005. *Vital Signs 2005*. New York: Norton.
5. Berger, J.J. 2000. *Beating the Heat*. Berkeley, CA: Berkeley Hills Books.
6. Demeo, E.M., and P. Steitz. 1990. The U.S. electric utility industry's activities in solar and wind energy. In K.W. Böer, ed., *Advances in Solar Energy*, Vol. 6, pp. 1–218. New York: American Solar Energy Society.
7. Johnson, J.T. 1990 (May). The hot path to solar electricity. *Popular Science*, pp. 82–85.
8. Kartha, S., and P. Grimes. 1994. Fuel cells: Energy conversion for the next century. *Physics Today* 47:54–61.
9. Kartha, S., and P. Grimes. 1994. Fuel cells: Energy conversion for the next century. *Physics Today* 47:54–61.
10. Piore, A. 2002 (April 15). Hot springs eternal: Hydrogen power. *Newsweek*, pp.32H.
11. Demeo, E.M., and P. Steitz. 1990. The U.S. electric utility industry's activities in solar and wind energy. In K.W. Böer, ed., *Advances in Solar Energy*, Vol. 6, pp. 1–218. New York: American Solar Energy Society.
12. Schatz Solar Hydrogen Project. N. D. Pamphlet. Arcata, CA: Humboldt State University.
13. Fuel cell buses. http://www.fuelcells.org/info/charts/buses.pdf, created by Fuel Cells 2000.
14. Fuel Cell Bus Programs Worldwide. http://www.cleanairnet.org/infopool/1411/propertyvalue19516.html.
15. Schatz Solar Hydrogen Project. N. D. Pamphlet. Arcata, CA: Humboldt State University.
16. World Firewood Supply. World Energy Council. 2006 (April 24). www.worldenergy.org.
17. Botkin, D.B., 2010 *Powering the Future: A Scientist's Guide to Energy Independence* (Indianapolis In: Pearson FT Press Science).
18. Showstack, R. 2003. Re-examining potential for geothermal energy in United States. *EOS* 84(23):214.

19. Piore, 2002 (April 15). Hot springs Eternal: Hydrogen power. *newsweek*, P.32H.
20. Anonymous. 2009. *World Wind Energy Report 2008*. Bonn, Germany, World Wind Energy Association. http://www.wwindea.org/home/index.php.
21. Electric Power: Energy Information Administration, Form EIA-923, Power Plant Operations Report, and predecessor forms: Form EIA-906, Power Plant Report, and Form EIA-920, Combined Heat and Power Plant Report.
22. Botkin, D.B. 2010. *Powering the Future: A Scientist's Guide to Energy Independence* (Indianapolis, IN: Pearson FT Press).
23. Top 50 Cities in the U.S. by Population and Rank. Infoplease. © 2000–2007 Pearson Education, publishing as Infoplease. June 17, 2008. http://www.infoplease.com/ipa/A0763098.html.
24. Quinn, R. 1997 (March). Sunlight brightens our energy future. *The World and I,* pp. 156–163.
25. Anonymous. 2009. *World Wind Energy Report 2008*. Bonn, Germany, World Wind Energy Association. http://www.wwindea.org/home/index.php.
26. Anonymous. 2009. *World Wind Energy Report 2008*. Bonn, Germany, World Wind Energy Association.
27. Botkin, 2010.
28. Botkin, 2010.
29. Botkin, 2010.
30. Wright, P. 2000. Geothermal energy. *Geotimes* 45(7):16–18.
31. Wright, P. 2000. Geothermal energy. *Geotimes* 45(7):16–18.
32. Botkin, 2010.
33. Duffield, W.A., J.H. Sass, and M.L. Sorey. 1994. Tapping the Earth's natural heat. U.S. Geological Survey Circular 1125.
34. Botkin, 2010.
35. Johnson, J.T. 1990 (May). The hot path to solar electricity. *Popular Science,* pp. 82–85.
36. Reuters, 2008, (January 15). Wind Farm Clears Hurdle.
37. Galbraith, Kate, 2006 (September 25). Texas is more hospitable than Mass. to wind farms economy, culture fueling a boom, *The Boston Globe.*
38. Cassidy, P. 2008, April 19. Floating wind farm plan dealt blow, *Cape Cod Times.*
39. Zezima, K. 2010. Interior secretary sees little hope for consensus on wind farm. New York Times.

Chapter 17 Notes

1. Rosa, E.A., and R.E. Dunalp. 1994. Nuclear power: Three decades of public opinion. *Public Opinion Quarterly* 58(2):295–324.
2. New York Times Topics > Subjects > I > Indian Point Nuclear Power Plant (NY). Updated: May 6, 2009. http://topics.nytimes.com/top/reference/timestopics/subjects/i/indian_point_nuclear_power_plant_ny/index.html?=8qa&scp=1-spot&sq=indian+point&st=nyt
3. Botkin, D.B., 2010, *Powering the Future: A Scientist's Guide to Energy Independence* (Indianapolis IN: Pearson FT Press).
4. European Nuclear Society. Accessed February 26, 2010 at http://www.euronuclear.org/info/encyclopedia/n/nuclear-power-plant-world-wide.htm.
5. Till, O.E. 1989. Advanced reactor development. *Annals of Nuclear Energy* 16(6):301–305.
6. Churchill, A.A. 1993 (July). Review of WEC Commission: Energy for tomorrow's world. *World Energy Council Journal*, pp. 19–22.
7. Duderstadt, J.J. 1978. Nuclear power generation. In L. C. Ruedisili and M. W. Firebaugh, eds., *Perspectives on Energy*, 2d ed., pp. 249–273. New York: Oxford University Press.
8. Brenner, D.J. 1989. *Radon: Risk and Remedy*. New York: Freeman.
9. Ehrlich, P.R., A.H. Ehrlich, and J.P. Holdren. 1970. *Ecoscience: Population, Resources, Environment*. San Francisco: Freeman.
10. *New Encyclopedia Britannica*. 1997. *Radiation*, 26, p. 487.
11. Brenner,1989.
12. Waldbott, G.L. 1978. *Health Effects of Environmental Pollutants*, 2d ed. St. Louis: C.V. Moseby.
13. U.S. Department of Energy. 1999. Radiation (in) waste isolation pilot plant. Carlsbad, New Mexico. Accessed at www.wipp. carlsbad.nm.us
14. New Encyclopedia Britannica. 1997. *Radiation*, 26, p. 487.
15. Waldbott, G.L. 1978. *Health Effects of Environmental Pollutants*, 2d ed. St. Louis: C.V. Moseby.
16. Ehrlich et al. 1970.
17. University of Maine and Maine Department of Human Services. 1983 (February). Radon in water and air. *Resource Highlights*.
18. Brenner, 1989.
19. Ibid., 1989.
20. Till, O.E. 1989. Advanced reactor development. *Annals of Nuclear Energy* 16(6):301–305.
21. MacLeod, G.K. 1981. Some public health lessons from Three Mile Island: A case study in chaos. *Ambio* 10:18–23.
22. Anspaugh, L.R. , R.J. Catlin, and M. Goldman. 1988. The global impact of the Chernobyl reactor accident. *Science* 242: 1513–1518.
23. Nuclear Energy Agency. 2002. Chernobyl Assessment of Radiological and Health Impacts: 2002 Update of Chernobyl: Ten Years On.
24. Anspaugh, L.R. , R.J. Catlin, and M. Goldman. 1988. The global impact of the Chernobyl reactor accident. *Science* 242: 1513–1518.
25. Nuclear Energy Agency. 2002.
26. U.S. Department of Energy. 1979. *Environmental Development Plan, Magnetic Fusion*. DOE/ER-0052. Washington, DC: U.S. Department of Energy.
27. Cohen, B.L. 1990. *The Nuclear Energy Option: An Alternative for the 90s*. New York: Plenum.
28. Fletcher, M. 2000 (November 14). The last days of Chernobyl. *Times 2* London, pp. 3–5.

29. Cohen, B.L. 1990. *The Nuclear Energy Option: An Alternative for the 90s*. New York: Plenum.
30. U.S. Nuclear Regulatory Commissions Accessed March 17, 2010 at http://www.nrc.gov/reading rm/doc collections/fact sheets/radwaste.html.
31. Office of Industry Relations. 1974. *Development, Growth and State of the Nuclear Industry*. Washington, DC: U.S. Congress, Joint Committee on Atomic Energy.
32. Ibid.
33. Weisman, J. 1996. Study inflames Ward Valley controversy. *Science* 271:1488.
34. DOE Waste Isolation Pilot Plant "WIPP Chronology." Accessed February 26, 2010 http://www.wipp.energy.gov/fctshts/factsheet.htm.
35. Waste Isolation_Pilot_Plant (2047). The Remote Handled Transuranic Waste Program.
36. Weart, W.D., M.T. Rempe, and D.W. Powers. 1998 (October). The waste isolation plant. *Geotimes*.
37. U.S. Department of Energy. 1999. Waste isolation pilot plant, Carlsbad, NM. Accessed at www.wipp.carlsbad.nm.us.
38. Ibid.
39. Roush, W. 1995. Can nuclear waste keep Yucca Mountain dry— and safe? *Science* 270:1761.
40. Hanks, T.C., I.J. Winograd, R.E. Anderson, T.E. Reilly, and E.P. Weeks. 1999. *Yucca Mountain as a Radioactive Waste Repository*. U.S. Geological Survey Circular 1184.
41. *New York Times*. Nevada: A Shift at Yucca Mountain. New York, Associated Press.
42. Ibid.
43. Hanks et al., 1999. *New York Times*. Nevada: A Shift at Yucca Mountain. Nuclear Regulatory Commission. 2000 NRC's high level waste program. Accessed July 18, 2000 at http://www.nrc.gov/NMSS/ DWM/hlw.htm.
44. *New York Times*, Nevada: A shift at Yucca Mountain.
45. Nuclear Regulatory Commission, 2000 NRC's high level waste program.
46. Ibid.
47. Botkin, 2010.
48. Cohen, 1990.
49. Botkin, D.B. The limits of nuclear power 2008 (October 20). *International Herald Tribune*.
50. Starke, L., ed. 2005. *Vital Signs* 2005. New York: W.W. Norton p. 139.
51. U.S. Nuclear Regulatory Commission, 2008. Backgrounder on New Nuclear Plant Designs, Accessed February 24, 2010 at http://www.nrc.gov/reading-rm/doc-collections/fact-sheets/new-nuc-plant-des-bg.html.
52. A.C. Kadak, R.G. Ballinger, T. Alvey, C.W. Kang, P. Owen, A. Smith, M. Wright and X. Yao. 1998. Nuclear Power Plant Design Project: A Response to the Environmental and Economic Challenge Of Global Warming Phase 1 Review of Options & Selection of Technology of Choice. http://web.mit.edu/pebble-bed/background.html
53. Anonymous, 2010. Vermont Senate Votes to Close Nuke Plant in 2012. *New York Times*.

Chapter 18 Notes

1. Frank, R. What drought? Palm Beach. Wall Street Journal online November 16, 2007, accessed April 28, 2008 @ wsj.com
2. Palm Beach County. Reclaimed water. Accessed April 28, 2008 @ www. pbcgov.com.
3. Henderson, L.J. 1913. *The Fitness of the Environment: An Inquiry into the Biological Significance of the Properties of Matter.* New York: Macmillan.
4. Water Resources Council. 1978. The Nation's Water Resources, 1975–2000, Vol. 1. Washington, DC.
5. Winter, T.C.. J.W. Harvey, O.L. Franke, and W.M. Alley. 1998. *Groundwater and Surface Water: A Single Resource.* U.S. Geological Survey Circular 1139.
6. Micklin, P., and Aladin, N.V. 2008. Reclaiming the Aral Sea. *Scientific American*, 298(4):64–71.
7. Kenny, J.F., et al. 2010. *Estimated Use of Water in the United States in 2005.* U.S. Geological Survey Circular 1344.
8. Alexander, G. 1984 (February/March). Making do with less. *National Wildlife*, special report, pp. 11–13.
9. U.S. General Accounting Office. 2003. *Freshwater Supply: States' View of How Federal Agencies Could Help Them Meet the Challenges of Expected Shortages.* Report GAO-03-514.
10. Gfeick, P.H., P. Loh, S.V. Gomez, and J. Morrison. 1995. *California Water 2020, A Sustainable Vision.* Oakland, CA: Pacific Institute for Studies in Development, Environment and Security.
11. Alley, W.M., T.E. Reilly, and O.L. Franke. 1999. *Sustainability of Ground-Water Resources.* U.S. Geological Survey Circular.
12. Leopold, L.B. 1977. A reverence for rivers. *Geology* 5: 429–430.
13. Allen, 1998. Virtual water: A strategic resource, global solutions to regional deficits. *Groundwater* 36(4): 545–546.
14. Hoekstra, A.Y. and A.K. Chapagain. 2008. *Globalization of Water*. Malden MA: Blackwell Publishing.
15. Smil, V. 2008.Water news: Bad, good and virtual. *American Scientist* 96:399–407
16. Hoekstra, A.Y ed. 2003. Virtual water trade: Proceedings of the International Expert Meeting on Virtual Water Trade. Value of Water Research Report Series 12. IHE Delft. The Netherlands.
17. Holloway, M. 1991. High and dry. *Scientific American* 265:16–20.
18. Levinson, M. 1984 (February/March). Nurseries of life. *National Wildlife*, special report, pp. 18–21.
19. Nichols, F.H., J.E. Cloern, S.N. Luoma, and D.H. Peterson. 1986. The modification of an estuary. *Science* 231:567–573.
20. Day, J.W., Jr., J.M. Rybczyk, L. Carboch, W.H. Conner, P. Delgado-Sanchez, R.I. Partt, and A. Westphal. 1998. A review of recent studies of the ecology and economic aspects of the

application of secondary treated municipal effluent to wetlands in southern Louisiana. From L.P. Rozas et al., eds., *Symposium on Recent Research in Coastal Louisiana*, February 3–5, 1998, Louisiana Sea Grant College Program, pp. 1–12.

21. Hileman, B. 1995. Rewrite of Clean Water Act draws praise, fire. *Chemical & Engineering News* 73:8.
22. Kaiser, J. 2001. Wetlands restoration: Recreated wetlands no match for original. *Science* 293:25a.
23. Gurardo, D., M.L. Fink, T.D. Fontaine, S. Newman, M. Chimney, R. Bearxotti, and G. Goforth. 1995. Large-scale constructed wetlands for nutrient removal from stormwater runoff: An Everglades restoration project. *Environmental Management* 19:879–889.
24. Pearce, M. 1995 (January). The biggest dam in the world. *New Scientist*, pp. 25–29.
25. Zich, R. 1997. China's Three Gorges: Before the flood. *National Geographic* 192(3):2–33.
26. Dolan, R., A. Howard, and A. Gallenson. 1974. Man's impact on the Colorado River and the Grand Canyon. *American Scientist* 62:392–401.
27. Hecht, J. 1996. Grand Canyon flood a roaring success. *New Scientist* 151 (2045):8.
28. Wright, S.A., Schmidt, J.C., Melis, T.S., Topping, D.J., and D.M. Rubin. 2008. Is there enough sand? Evaluating the fate of Grand Canyon sandbars. *GSA Today*. 18 (8):4–10.
29. Grant, G. 2001. Dam removal: Panacea or Pandora for rivers. *Hydrological Processes* 15:1531–1532.
30. State of Maine. 2001. A brief history of the Edwards Dam. Accessed January 15, 2000 at http://janus.state.me.us/spo/edwares/_timeline.htm.
31. O'Connor, J., Major, J., and G. Grant. 2008. The dams come down. *Geotimes* 53(3):22–28.
32. American Rivers. Elwha River Restoration accessed March 1, 2006 at www.americanrivers.org.
33. Brown, L.R. 2003. *Plan B: Rescuing a Planet under Stress and a Civilization in Trouble*. New York: Norton.
34. Hoekstra, A.Y. 2009. A comprehensive introduction to water footprints. Accessed August 25, 2009 at http: //www.waterfootprint.org.

Chapter 19 Notes

1. Environmental Protection Agency, American Heritage Rivers, Hudson River. Accessed February 19, 2005 www.epa.gov/rivers/98rivers/hudson.html .
2. Bopp, R.F., and H.J. Simpson, 1989. *Contamination of the Hudson River: The Sediment Record in Contaminated Marine Sediments, an Assessment and Remediation*. Washington, DC: National Academy Press.
3. Wall, G.R., Riva-Murray, K., Phillips, P.J. 1998. and , A.C. 2009. Water quality in the Hudson River Basin. New York and adjacent states, 1992–95.
4. Morrison, J. 2005. How much is clean water worth? *National Wildlife* 43(2):22–28.
5. Gleick, P.H. 1993. An introduction to global fresh water issues. In P. H. Gleick, ed., *Water in Crisis*, pp. 3–120 New York: Oxford University Press.
6. Hileman, B. 1995. Pollution tracked in surface and groundwater. *Chemical & Engineering News* 73:5.
7. Lewis, S.A. 1995. Trouble on tap. *Sierra* 80:54–58.
8. Smith, R.A. 1994. Water quality and health. *Geotimes* 39:19–21.
9. MacKenzie, W.R., et al. 1994. A massive outbreak in Milwaukee of Cryptosporidium infection transmitted through the public water supply. *New England Journal of Medicine* 331:161–167.
10. Kluger, J. 1998. Anatomy of an outbreak. *Time* 152 (5): 56–62.
11. Maugh, T.H. 1979. Restoring damaged lakes. *Science* 203:425–127.
12. Mitch, W.J., J.W. Day, Jr., J.W. Gilliam, P.M. Groffman, D.L. Hey, G.W. Randall, and N. Wang. 2001. The Gulf of Mexico hypoxia— approaches to reducing nitrate in the Mississippi River or reducing a persistent large-scale ecological problem. *Bioscience* (in Press).
13. Hinga, K.R. 1989. Alteration of phosphorus dynamics during experimental eutrophication of enclosed-marine ecosystems. *Marine Pollution Bulletin* 20:624–628.
14. Richmond, R.H. 1993. Coral reefs: Present problems and future concerns resulting from anthopogenic disturbance. *American Zoologist* 33:524–536.
15. Bell, P.R. 1991. Status of eutrophication in the Great Barrier Reef Lagoon. *Marine Pollution Bulletin* 23:89–93.
16. Hunter, C.L., and C.W. Evans. 1995. Coral reefs in Kaneohe Bay, Hawaii: Two centuries of Western influence and two decades of data. *Bulletin of Marine Science* 57:499.
17. Department of Alaska Fish and game 1918. *Alaska Fish and Game* 21(4), Special Issue.
18. Holway, M. 1991. Soiled shores. *Scientific American* 265: 102–106.
19. Robinson, A.R. 1973. Sediment, our greatest pollutant? In RW. Tank, ed., *Focus on Environmental Geology*, pp. 186–192. New York: Oxford University Press.
20. Yorke, T.H. 1975. Effects of sediment control on sediment transport in the northwest branch, Anacostia River basin, Montgomery Country, Maryland. *Journal of Research* 3:481–494.
21. Poole, W. 1996. Rivers run through them. *Land and People* 8:16–21.
22. Haikin, M. 2007. Nanotechnology takes on water pollution. *Business 2.0 Magazine*. Accessed May 30, 2008 @ money.cnn.com.
23. Natural Resources Defense Council. 2008. Mimicking nature to solve a water pollution problem. Accessed May 30, 2008 @ www.nrdc.org.
24. Carey, J. 1984 (February/'March). Is it safe to drink? *National Wildlife*, Special Report pp. 19–2l.
25. Pye, U.I., and R. Patrick. 1983. Groundwater contamination in the United States. *Science* 221:713–718.
26. Newmark, J. 2008, May. Plastic people of the universe. *Discover.*

Better Planet Special Issue, pp. 46–51.

27. Foxworthy, G. L. 1978. Nassau Country, Long Island, New York—Water problems in humid country. In G. D. Robinson and A. M. Spieker, eds., *Nature to Be Commanded,* pp. 55–68. U.S. Geological Survey Professional Paper 950. Washington, DC: U.S. Government Printing Office.
28. Van der Leeden, F., F.L. Troise, and D.K. Tood. 1990. *The Water Encyclopedia,* 2d ed. Chelsea, MI: Lewis Publishers.
29. U.S. Geologic Survey. 1997. Predicting the impact of relocating Boston's sewage outfall. *UCGC Fact Sheet* 185–97.
30. Jobling, S., M. Nolan, C.R. Tyler, G. Brighty, and J.P. Sumpter. 1998. Widespread sexual disturbance in wild fish. *Environmental Science and Technology* 32(17):2498–2506.
31. Environmental Protection Agency, Drinking Water Committee of the Science Advisory Board. 1995. *An SAB Report: Safe Drinking Water. Future Trends and Challenges.* Washington, DC: Environmental Protection Agency.
32. Jewell, W.J. 1994. Resource—recovery wastewater treatment. *American Scientist* 82:366–375.
33. Task Force on Water Reuse. 1989. *Water Reuse: Manual of Practice SM-3.* Alexandria. VA: Water Pollution Control Federation.
34. Kadlec, R.H., and R.L. Knight. 1996. *Treatment Wetlands.* New York: Lewis Publishers.
35. Breaux, A.M., and J.W. Day Jr. 1994. Policy considerations for wetland wastewater treatment in the coastal zone: A case study for Lousiana. *Coastal Management* 22:285–307.
36. Day, J.W., Jr., J.M. Rybczyk, L. Carboch, W.H. Conner, P. Delgado Sanchez, R.T. Pratt, and W. Westphal. 1998. A review of recent studies of the ecology and economic aspects of the application of secondary treated municipal effluent wetlands in southern Louisiana. In L.P. Rozas et al., eds., Symposium on Recent Research in Coastal Louisiana, February 1998, Louisiana Sea Great College Program, pp. 1–12.
37. Breaux, A., S. Fuber, and J. Day 1995. Using natural coastal wetland systems: An economic benefit analyis. *Journal of Environmental Management* 44:285–291.
38. Barone, J. 2008, May. Better Water. *Discover.* Better Planet Special Issue, pp. 31--32.
39. Faustini, J.M., et al.2009. Assessing stream ecosystem condition in the United States. *EOS, Transactions, American Geophysical Union,* 36:309–310.
40. Hileman, B. 1995. Rewrite of Clean Water Act draws praise, fire. *Chemical & Engineering News* 73:8.
41. Mallin, M.A. 2000. Impacts of industrial animal production on rivers and estuaries. *American Scientist* 88(l):26–37.
42. Bowie, P. 2000. No act of God. *The Amichs Journal* 21(4): 16–21.

Chapter 20 Notes

1. Mann, M.E. and 8 others. 2009. Global signatures and dynamical origins of the Little Ice Age and Medieval Climate Anomaly. *Science* 326:1256–1260.
2. Fagan, B.M. 2008.*The Great Warming: Climate Change and the Rise and Fall of Civilizations.* New York: Bloomsbury Press.
3. Le Roy Ladurie, E. (1971)
4. National Research Council. 2006. Surface temperature reconstructions for the last 2,000 years. Washington DC: National Academy Press.
5. Fagan, B 2004. *The Long Summer: How Climate Changed Civilization* NY: Basic Books.
6. Marsh, W.M., and Dozier, J. 1981. *Landscape.* New York: John Wiley & Sons.
7. Crowley, T.J. 2000. Causes of climate change over the past 1000 years. *Science* 289:270–277.
8. Hansen, J., et al. 2005. Efficiency of climate forcrngs. *Joumal of Geophysical Research* 110 (D18104):45P.
9. Callendar, C.S. The artifcial production of carbon dioxiod and its influence on tempeature. *Quarterly Journal of the Royal Meteorological Society* 84(1938):223–237; Can carbon dioxide influence climate? *Weather* 4(1949):310–14; On the Amount of Carbon Dioxide in the Atmosphere Talus 10(1958):243
10. NOAA. 2009. Paleo proxy data. In Introduction to Paleoclimatology. Accessed March 24, 2010. www.ncdc.noaa.gov.
11. Caillon, N., J.P. Severinghouse, et al, 2003. Timing of atmospheric CO_2 and Antarctic temperature changes across termination III. *Science* 299: 1728–1731.
12. Encyclopedia Britannica online. Accessed January2, 2009. http://www.britannica.com/EBchecked/topic-art/174962/69345/Carbon-dioxide-concentrations-in-Earths-atmosphere-plotted-over-the-past.
13. Hansen, J., et al. 2005. Efficiency of climate forcing. *Journal of Geophysical Research* 110 (D18104):45P.
14. Brook, E. 2008, Palaeoclimate – Windows on the greenhouse, *Nature,* 453(7193), 291-292, Doi 10.1038/453291a.
15. Dlugokencky, E.J., L.P. Steele, P.M. Lang, and K.A. Masarie. 1994. The growth rate and distribution of atmospheric methane. *Journal of Geophysical Research* 99(D8): 17021-17043.
16. Hansen, J., A. Lacis, and M. Prather 1989. Greenhouse effect of chlorofluorocarbons and other trace gases. *Journal of Geophysical Research* 94(D13): 16417-16421.
17. Rodhe, H. 1990. A comparison of the contribution of various gases to the greenhouse effect. *Science* 248:1217-1219.
18. Hansen, J., A. Lacis and M. Prather 1989. Greenhouse effect of chlorofluorocarbons and other trace gases. *Journal of Geophysical Research* 94(D13): 16417-16421.
19. IPCC. 2007. Climate Change 2007: The Physical Science Basis: Working Group I Contribution to the Fourth Assessment Report, IPCC. New York: Cambridge University Press, 989P.
20. Botkin, D.B. 1990. *Discordant Harmonies: A New Ecology for the 21st Century.* New York: Oxford University Press.'
21. Bennett, K. 1990. Milankovitch Cycles and their effects on Species in ecological and evolutionary time. *Paleobiology* 16(1):11–12.

22. Foukal, P., C. Frohlich, H. Sprint, and T. M. L. Wigley. 2006. Variations in solar luminosity and their effect on the Earth's climate. *Nature* 443:151-166.
23. Soon, W. 2007. Implications of the secondary role of carbon dioxide and methane forcing in climate change: Past, present and future. *Physical Geography* 28(2): 97-125.
24. IPCC. 2007. Climate change 2007: The Physical science basis: Working Group I Contribution to the Fourth Assessment Report, IPCC. New York: Cambridge University Press, 989P.
25. Broker, W. 1997. Will our ride into the greenhouse future be a smouth one? *Geology Today* 7 no.5: 2-6.
26. Steager, R. 2006. The source of Europe's mild climate. *American Scientist* 94:334-341.
27. Hidore, J. J., Oliver, J.E., Snow M., and Snow, R. 2010. *Climatology*. Upper Saddle River, NJ: Prentice Hall.
28. Jet Propulsion Laboratory. El Niño–When the Pacific Ocean speaks, Earth listens. Accessed September 25, 2008 at wmv.jpl.nasa.gov.
29. Botkin, D.B. 1990. *Discordant Harmonies: A New Ecology for the 21st Century*. New York: Oxford University Press.
30. NOAA. 2009. Global climate change impacts in the United States. Washington, DC: Cambridge University Press. Editors: Thomas R. Karl, NOAA National Climatic Data Conter; Jerry M. Melillo, Marine Biological Laboratory, Thomas C. Peterson, NOAA National Climatic Data Center.
31. IPCC. 2007. Climate change 2007: Synthesis report. Valencia, Spain: IPCC.
32. Pelto, M.S. 1996. Recent Changes in Glacier and Alpine Runoff in the North Cascades, Washington. *Hydrological Processes* 10:1173-80.
33. Appenzeller, T. 2007. The Big Thaw. *National Geographic* 211(6):56-71.
34. Mote. P.W., and G. Kasen, 2007. The shrinking glaciers of Kilimanjaro: can global warming be blamed? *American Scientist* 95(4): 218-25.
35. Stroeve, J., M. Serreze, and S. Drobot. 2008. "Arctic Sea Ice Extent Plummets in 2007. EOS, Transactions." *American Geophysical Union* 89(2): 13–14.
36. Parkinson, C. L. 2008. "Recent Trend Reversals in Arctic Sea Ice Extent—Possible Connections to the North Atlantic Oscillation." *Polar Geography* 31(1):3-14.
37. Morrison, J., J. Wahr, R. Kwok, and Perelta Ferriz. 2006. "Recent trends in Arctic Ocean Mass Distribution Revealed by GRACE." *Geophysical Reserve Letters.* 34:L07602.
38. Botkin, D.B. 2007. Forcasting effects of global warming on biodiversity. *Bio Science* 57(3):227–236.
39. Charmantier, A., Robin H. McCleery, Lionel R. Cole, Chris Perrins, Loeske E. B. Kruuk, Ben C. Sheldon. 2008, "Adaptive Phenotypic Plasticity in Response to Climate Change in a Wild Bird Population." *Science* 320(5877): 800–803.
40. Grime, J. P., Jason D. Fridley, Andrew P. Askew, Ken Thompson, John G. Hodgson, and Chris R. Bennett. 2008. "Long-term Resistance to Simulated Climate Change in an Infertile Grassland." *PNAS* 105(29):10028 10032. Earth System Science Committee (1988). *Earth System Science: A Preview.* Boulder, Colorado: University Corporation for Atmospheric Research.
41. Rogers, D. J., and S. E. Randolph. 2000. "The Global Spread of Malaria in a Future, Warmer World." *Science* 289:1763–1766.
42. Sumilo, D., Asokliene, L., Bormane, A. Vasilenko, V., and Golovijova,I. 2007. "Climate Change Cannot Explain the Upsurge of Tick-Borne Encephalitis in the Baltics." *PLoS ONE* 2(6): e500 doi:10.1371/journa.pone.0000500.
43. Botkin, D.B., 2010. *Powering the Future: A Scientist's Guide to Energy independence.* (Pearson FT Press Science. May 2010.
44. Dunn, S. 2001. "Decarbonizing the Energy Economy." In *World Watch Institute State of the World 2001.* New York: Norton.
45. Rice, C. W. 2002. "Storing Carbon in Soil: Why and How." *Geotimes* 47(1):14–17.
46. Friedman, S. J. 2003. "Storing Carbon in Earth." *Geotimes* 48(3):16–20.
47. Bartlett, K. 2003. "Demonstrating Carbon Sequestration." *Geotimes* 48 (3):22–23. 48 on P66 http://www. juliantrubin.com/bigten/photosynthesisexperiments.html; and a variety of other sources.
48. http://www.IPCC.ch/pdf/presentations/himalanga-statement-20january2010.pdf

Chapter 21 Notes

1. National Park Service. 1984. *Air Resources Management Manual.*
2. U.S. Environmental Protection Agency. 2008. National air quality trends through 2007. www.epa.gov.
3. Araujo, J.A. and 10 others. 2008. Ambient particulate pollutants in the ultrafine range promote early atherosclerosis and systematic oxidative stress. *Circulation Research* 102: 589–596.
4. Godish, T. 1991. *Air Quality*, 2d ed. Chelsea, MI.: Lewis Publishers.
5. Seitz, F., and C. Plepys. 1995. Monitoring air quality in healthy people 2000. *Healthy People 2000: Statistical Notes no. 9.* Atlanta: Centers for Disease Control and Prevention, National Center for Health Statistics.
6. American Lung Association. 2009. *State of the Air 2009.* www. stateofftheair.org
7. Moore, C. 1995. Poisons in the air. *International Wildlife* 25:38–45.
8. Pope, C. A., III, D.V. Bates, and M.E. Raizenne. 1995. Health effects of particulate air pollution: Time for reassessment? *Environmental Health Perspectives* 103:472–480.
9. Travis, D.J., A.M. Carleton, and R.G. Lauritsen 2002. Contrails reduce daily temperature range. *Nature* 419:601.
10. U.S. Environmental Protection Agency. 2008. Acid Rain. www.epa.gov.
11. Canadian Department of the Environment. 1984. *The acid rain story*. Ottawa: Minister of Supply and Services.

12. Lippmann, M., and R.B. Schlesinger. 1979. *Chemical Contamination in the Human Environment.* New York: Oxford University Press.
13. Winkler, E.M. 1998 (September). The complexity of urban stone decay. *Geotimes*, pp. 25–29.
14. U.S. Environmental Protection Agency. 2009. Acid Rain Program SO_2 Allowances Fact Sheet. www.epa.gov.
15. U.S. Environmental Protection Agency 2006. National scale air toxics assessment for 1999. Estimated emissions. Concentrations and risks. Technical Fact Sheet. Accessed April 10, 2006 at www.epa.gov.
16. Tyson, P. 1990. Hazing the Arctic. *Earthwatch* 10:23–29.
17. NASA. 2008. NASA satellite measures pollution from East Asia to North America. www.nasa.gov.
18. Pittock, A.B., L.A. Frakes, D. Jenssen, J.A. Peterson, and J.W. Zillman, eds. 1978. *Climatic Change and Variability: A Southern Perspective.* New York: Cambridge University Press.
19. Blake, D.R., and F.S. Rowland. 1995. Urban leakage of liquefied petroleum gas and its impact on Mexico City air quality. *Science* 269:953.
20. Pountain, D. 1993 (May). Complexity on wheels. *Byte*, pp. 213–220.
21. Moore, C. 1995. Green revolution in the making. *Sierra* 80:50.
22. U.S. Environmental Protection Agency. 2008. Case Study 22. AEP and Centain Teed put environmental process by-products to beneficial uase in wallboard. www.epa.gov.
23. Kolstad, C. D. 2000. *Environmental Economics.* New York: Oxford University Press.
24. Crandall, R. W. 1983. *Controlling Industrial Pollution: The Economics and Politics of Clean Air.* Washington, DC: Brookings Institution.
25. Hall, J.V., A.M. Winer, M.T. Kleinman, F.W. Lurmann, V. Brajer, and S.D. Colome. 1992. Valuing the health benefits of clean air. *Science* 255:812–816.
26. Krupnick, A.J., and P.R. Portney. 1991. Controlling urban air pollution: A benefits-cost assessment. *Science* 252:522–528.
27. Lipfert, F.W., S.C. Morris, R.M. Friedman, and J.M. Lents. 1991. Air pollution benefit-cost assessment. *Science* 253:606.
28. Rowland, F.S. 1990. Stratospheric ozone depletion of chlorofluorocarbons. *AMBIO* 19:281–292.
29. Rowland F.S. 2007. Stratospheric ozone depletion by chlorofluorocarbons. Nobel Lecture for the encyclopedia of Earth. www.eoearth.org.
30. Smith, R.C., B.B. Prezelin, K.S. Baker, R.R. Bidigare, N.P. Boucher, T. Coley, D. Karentz, S. Macintyre, H.A. Matlick, D. Menzies, M. Ondrusek, Z. Wan, and K.J. Waters. 1992. Ozone depletion: Ultraviolet saturation and phytoplankton biology in Antarctic waters. *Science* 255:952–959.
31. NASA. 2009 Ozone Facts: History of the ozone hole. http:// earth observatory.nasa.gov.
32. Hamill, P., and O.B. Toon. 1991. Polar stratospheric clouds and the ozone hole. *Physics Today* 44:34–42.
33. Stolarski, R.S. 1988. The Antarctic ozone hole. *Scientific American* 258:30–36.
34. Molina, M.J., and F.S. Rowland. 1974. Stratospheric sink for chlorofluoromethanes: Chlorine-atom catalyzed distribution of ozone. *Nature* 249:810–812.
35. Brouder, P. 1986 (June). Annals of chemistry in the face of doubt. *New Yorker*, pp. 20–87.
36. Khalil, M.A.K., and R.A. Rasmussen. 1989. The potential of soils as a sink of chlorofluorocarbons and other manmade chlorocarbons. *Geophysical Research Letters* 16:679–682.
37. Rowland, F.S. 1989. Chlorofluorocarbons and the depletion of stratospheric ozone. *American Scientist* 77:36–45.
38. Toon, O.B., and R.P. Turco. 1991. Polar stratospheric clouds and ozone depletion. *Scientific American* 264:68–74.
39. Worldwatch Institute 2008. Vital Signs 2007–2008. New York: W.W. Norton.
40. Webster, C.R., R.D. May, D.W. Toohey, L.M. Avallone, J.G. Anderson, P. Newman, L. Lait, M. Schoeberl, J.W. Elkins, and K.R. Chay. 1983. Chlorine chemistry on polar stratospheric cloud particles in the Arctic winter. *Science* 261:1130–1134.
41. Kerr, R.A. 1992. New assaults seen on Earth's ozone shield. *Science* 255:797–798.
42. Shea, C.P. 1989. Mending the Earth's shield. *World Watch* 2:28–34.
43. Kerr, J.B., and C.T. McElroy. 1993. Evidence for large upward trends of ultraviolet-B radiation linked to ozone depletion. *Science* 262:1032–1034.
44. U.S. Environmental Protection Agency. 2006. Ozone depletion. Accessed April 16, 2006. www.epa.gov.
45. Cutter Information Corp. 1996. Reports discuss present and future state of ozone layer. *Global Environmental Change Report* V, VIII, 21, no. 22, pp. 1–3.
46. Showstack, R. 1998. Ozone layer is on slow road to recovery, new science assessment indicates. *Eos* 79(27):317–318.
47. Spurgeon, D. 1998. Surprising success of the Montreal Protocol. *Nature* 389(6648):219.
48. Zimmerman, M.R. 1985. Pathology in Alaskan mummies. *American Scientist* 73:20–25.
49. Ehrlich, P.R., A.H. Ehrlich, and J.P. Holdren. 1970. *Ecoscience: Population, Resources, Environment.* San Francisco: Freeman.
50. Conlin, M. 2000 (June 5). Is your office killing you? *Business Week*, pp. 114–124.
51. U.S. Environmental Protection Agency. Basic information: Formaldehyde. Accessed April 28, 2008 at www.epa.gov
52. Anonymous. FEMA Hurries Hurricane Survivors Out of Toxic Trailers. Environment News Service (ENS). February 15, 2008.
53. U.S. Environmental Protection Agency. 1991. *Building Air Quality: A Guide for Building Owners and Facility Managers.* EPA/400/ 1–91/033, DHHS (NIOSH) Pub. No. 91–114. Washington, DC: Environmental Protection Agency.

54. Zummo, S.M., and M.H. Karol. 1996. Indoor air pollution: Acute adverse health effects and host susceptibility. *Environmental Health* 58:25–29.
55. Godish, T. 1997. *Air Quality,* 3d ed. Boca Raton, FL: Lewis Publishers.
56. O'Reilly, J.T., P. Hagan, R. Gots, and A. Hedge. 1998. *Keeping Buildings Healthy.* New York: Wiley.
57. Brenner, D.J. 1989. *Radon: Risk and Remedy.* New York: W. H. Freeman.
58. U.S. Environmental Protection Agency. 1992. *A Citizen's Guide to Radon: The Guide to Protecting Yourself and Your Family from Radon,* 2d ed. ANR-464. Washington, DC: Environmental Protection Agency.
59. Henshaw, D.L., J.P. Eatough, and R.B. Richardson. 1990. Radon as a causative factor in induction of myeloid leukaemia and other cancers. *The Lancet* 335:1008–1012.
60. Pershagen, G., G. Akerblom, O. Axelson, B. Clavensjo, L. Damber, G. Desai, A. Enflo, F. Lagarde, H. Mellander, M. Svartengren, and G.A. Swedjemark. 1994. Residential radon exposure and lung cancer in Sweden. *New England Journal of Medicine* 330:159–164.
61. Nero, A.V., Jr. 1988. Controlling indoor air pollution. *Scientific American* 258:42–48.
62. Committee on Indoor Air Pollution. 1981. *Indoor Pollutants.* Washington, DC: National Academy Press.
63. Massachusetts Department of Public Health, Bureau of Environmental Health Assessments. 1995. *Symptom Prevalence Survey Related to Indoor Air Concerns at the Registry of Motor Vehicles Building, Ruggles Station.*
64. Horton, W.G.B. 1995. *NOVA: Can Buildings Make You Sick?* Video production. Boston: WGBH.
65. U.S. Environmental Protection Agency. 1986. *Radon Reduction Techniques for Detached Houses: Technical Guidance.* EPA 625/5–86–019. Research Triangle Park, NC: Air and Energy Engineering Research Laboratory, Office of Research and Development, U.S. Environmental Protection Agency.
66. U.S. Environmental Protection Agency. Radon-resistance new construction (RRNC). Accessed April 22, 2006. www.epa.gov.

Chapter 22 Notes

1. The official Web site of the High Line and Friends of the High Line, http://www.thehighline.org/about/high-line-history.
2. UNEP, http://www.unep.org/web/2005/english/information_Material/facts.asp.
3. World Resources Institute, http://archive.wri.org/item_detail.cfm?id=2795&section=climate&pages=pubs_content_text&z=?
4. EPA. May 11, 2006. What is urban sprawl? http://www.epa.gov/maia/html/sprawl.html.
5. Butler, Rhett. 2003. *World's Largest Urban Areas [Ranked by Urban Area Population, The World Gazetteer,* http://www.mongabay.com citiesurban01.htm.
6. Neubauer, D. 2004 (September 24). Mixed blessings of the megacities. *Yale Global Online Magazine,* http://yalelobal.–yale.edu.
7. Haub, C., and D. Cornelius. 2000. *World Population Data Sheet.* Washington, DC: Population Reference Bureau.
8. Nossiter, Adam. 2008. Big plans are slow to bear fruit in New Orleans. *New York Times,* April 1, 2008.
9. Richard Fausset. 2009. Defying economy, New Orleans keeps rebuilding. *Los Angeles Times,* April 03, 2009. http://artices. latimes.com/2009/apr/03/nation/na–rebuilding–new–orleans3.
10. Leibbrand, K. 1970. *Transportation and Town Planning.* Translated by N. Seymer. Cambridge, MA.: MIT Press.
11. Mumford, L. 1972. The natural history of urbanization. In R.L. Smith, ed., *The Ecology of Man: An Ecosystem Approach,* pp. 140–152. New York: Haper & Row.
12. The material about Olmsted is primarily from Beveridge, C.E. 2001. *The Papers of Frederick Law Olmsted,* Vols. 1–6 and supplementary series vol. 1; 1997–2001. Baltimore: Johns Hopkins University Press; and Beveridge, C. 1998. *Frederick Law Olmsted: Designing the American Landscape.* Universe.
13. Detwyler, T.R., and M.G. Marcus, eds.1972. *Urbanization and the Environment: The Physical Geography of the City.* North Scituate, MA: Duxbury Press.
14. Butti, K., and J. Perlin. 1980. *A Golden Thread: 2500 Years of Solar Architecture and Technology.* New York: Cheshire.
15. McHarg, I.L. 1971. *Design with Nature.* Garden City, NY: Doubleday.
16. Ford, A.B., and O. Bialik. 1980. Air pollution and urban factors in relation to cancer mortality. *Archives of Environmental Health* 35:350–359.
17. Nadel, I.B., C.H. Oberlander, and L.R. Bohm. 1977. *Trees in the City.* New York: Pergamon.
18. Dreistadt, S.H., D.L. Dahlsten, and G.W. Frankie. 1990. Urban forests and insect ecology. *Bioscience* 40:192–198.
19. Leedly, D.L., and L.W. Adams. 1984. *A Guide to Urban Wildlife Management.* Columbia, MD: National Institute for Urban Wildlife.
20. Tylka, D. 1987. Critters in the city. *American Forests* 93:61–64.
21. Mannan, R.W., Robert J. Steidl, and Clint W. Boal. 2008. Identifying habitat sinks: A case study of Cooper's hawk in an urban environment. *Urban Ecosystems* II:141–148.
22. Department of Environmental Protection, New York. 2006. Peregrine falcons in New York City. http://www.nyc.gov/html/dep/–html/news/falcon.html.
23. Mannan, Steidl, and Boal, Identifying habitat sinks.
24. Tylka, Critters in the city.
25. Adams, L.W., and L.E. Dove. 1989. Wildlife reserves and corridors in the urban environment. Columbia, MD: National Institute for Urban Wildlife.
26. http://www.responsiblewildifemanagement.org/bubo–nic–plague.htm.

27. U.S. Environmental Protection Agency. 2006 (May 11). http://www.epa.gov/maia/html/sprawl.html.

Chapter 23 Notes

1. Sullivan, D.E. 2006. *Recycled cell phones—A treasure trove of valuable metals.* U.S. Geological Survey Fact Sheet 2006–3097.
2. Kropschot, S.J., and K.M. Johnson. 2006. U.S. Geological Survey Circular 1289. Menlo Park, CA.
3. Hudson, T.L., F.D. Fox, and G.S. Plumlee. 1999. *Metal Mining and the Environment.* Alexandria, VA: American Geological Institute.
4. McKelvey, V.E. 1973. Mineral resource estimates and public policy. In D.A. Brobst and W.P. Pratt, eds., *United States Mineral Resources*, pp. 9–19. U.S. Geological Survey Professional Paper 820.
5. U.S. Department of the Interior, Bureau of Mines. 1993. *Mineral Commodity Summaries*, 1993. I 28.149:993. Washington, DC: U.S. Department of the Interior.
6. McGreery, P. 1995. Going for the goals: Will states hit the wall? *Waste Age* 26:68–76.
7. Brown, L.R. 1999 (March/April). Crossing the threshold. *World Watch*, pp. 12–22.
8. Meyer, H.O.A. 1985. Genesis of diamond: A mantle saga. *American Mineralogist* 70:344–355.
9. Kesler, S.F. 1994. *Mineral Resources, Economics, and the Environment.* New York: Macmillan.
10. Awramik, S.A. 1981. The pre-Phanerozoic biosphere—three billion years of crises and opportunities. In M.H. Nitecki, ed., *Biotic Crises in Ecological and Evolutionary Time*, pp. 83–102. Spring Systematics Symposium. New York: Academic Press.
11. Margulis, L., and J.E. Lovelock. 1974. Biological modulation of the Earth's atmosphere. *Icarus* 21:471–489.
12. Lowenstam, H.A. 1981. Minerals formed by organisms. *Science* 211:1126–1130.
13. Brobst, D.A., W.P. Pratt, and V.E. McKelvey, 1973. *Summary of United States mineral resources.* U.S. Geological Survey Circular 682.
14. Jeffers, T.H. 1991 (June). Using microorganisms to recover metals. *Minerals Today*. Washington, DC: U.S. Department of Interior, Bureau of Mines, pp. 14–18.
15. Haynes, B.W. 1990 (May). Environmental technology research. *Minerals Today*. Washington, DC: U.S. Bureau of Mines, pp. 13–17.
16. Sullivan, P.M., M.H. Stanczyk, and M.J. Spendbue. 1973. *Resource Recovery from Raw Urban Refuse.* U.S. Bureau of Mines Report of Investigations 7760.
17. Davis, F.F. 1972 (May). Urban ore. *California Geology*, pp. 99–112.
18. U.S. Geological Survey. 2010. *Minerals Yearbook* 2007—*Recycling Metals*. Accessed April 3, 2010 at http://minerals.usgs.gov.
19. Galley, J.E. 1968. Economic and industrial potential of geologic basins and reservoir strata. In J.E. Galley, ed., *Subsurface Disposal in Geologic Basins: A Study of Reservoir Strata*, pp. 1–19. American Association of Petroleum Geologists Memoir 10. Tulsa, OK.: American Association of Petroleum Geologists.
20. Relis, P., and A. Dominski. 1987. *Beyond the Crisis: Integrated Waste Management.* Santa Barbara, CA: Community Environmental Council.
21. Allenby, B.R. 1999. *Industrial Ecology: Policy Framework and Implementation.* Upper Saddle River, NJ: Prentice-Hall.
22. Garner, G., and P. Sampat. 1999 (May). Making things last: Reinventing of material culture. *The Futurist*, pp. 24–28.
23. U.S. Environmental Protection Agency. *Municipal solid waste.* Accessed April 21, 2006 at www.epa.gov.
24. Relis, P., and H. Levenson. 1998. *Discarding Solid Waste as We Know It: Managing Materials in the 21st Century*. Santa Barbara, CA: Community Environmental Council.
25. Young, J.E. 1991. Reducing waste-saving materials. In L.R. Brown, ed., *State of the World*, 1991, pp. 39–55. New York: W.W. Norton.
26. Steuteville, R. 1995. The state of garbage in America: Part I. *BioCycle* 36:54.
27. Gardner, G. 1998 (January/February). Fertile ground or toxic legacy? *World Watch*, pp. 28–34.
28. Rathje, W.L., and C. Murphy. 1992. Five major myths about garbage, and why they're wrong. *Smithsonian* 23:113–122.
29. Rathje, W.L. 1991. Once and future landfills. *National Geographic* 179(5):116–134.
30. U.S. Environmental Protection Agency. 2002. *Solid waste management: A local challenge with global impacts.* Accessed May 26, 2010 at www.epa.gov.
31. Schneider, W.J. 1970. *Hydrologic implications of solid-waste disposal.* 135(22). U.S. Geological Survey Circular 601F. Washington, DC: U.S. Geological Survey.
32. Thomas, V.M., and T.G. Spiro. 1996. The U.S. dioxin inventory: Are there missing sources? *Environmental Science & Technology* 30:82A-85A.
33. Turk, L.J. 1970. Disposal of solid wastes—acceptable practice or geological nightmare? In *Environmental Geology,* pp. 1–42. Washington, DC: American Geological Institute Short Course, American Geological Institute.
34. Hughes, G.M. 1972. Hydrologic considerations in the siting and design of landfills. *Environmental Geology Notes*, no. 51. Urbana: Illinois State Geological Survey.
35. Bergstrom, R.E. 1968. Disposal of wastes: Scientific and administrative considerations. *Environmental Geology Notes*, no 20. Urbana: Illinois State Geological Survey.
36. Cartwright, K., and Sherman, F.B. 1969. Evaluating sanitary landfill sites in Illinois. *Environmental Geology Notes*, no. 27. Urbana: Illinois State Geological Survey.
37. Rahn, P.H. 1996. *Engineering Geology*, 2nd ed. Upper Saddle River, NJ: Prentice-Hall.

38. Bullard, R.D. 1990. *Dumping in Dixie: Race, Class and Environmental Quality*. Boulder, CO: Westview Press.
39. Sadd, J.L., J.T. Boer, M Foster, Jr., and L.D. Snyder 1997. Addressing environmental justice: Demographics of hazardous waste in Los Angeles County. *Geology Today* 7(8):18–19.
40. Walker, W.H. 1974 Monitoring toxic chemical pollution from land disposal sites in humid regions. *Ground Water* 12:213–218.
41. Watts, R.J. 1998. *Hazardous Wastes*. New York: John Wiley & Sons.
42. Wilkes, A.S. 1980. *Everybody's Problem: Hazardous Waste*. SW-826. Washington, DC: U.S. Environmental Protection Agency, Office of Water and Waste Management.
43. Harder, B. 2005 (November 8). Toxic e-waste is cashed in poor nations. *National Geographic News*.
44. Elliot, J. 1980. Lessons from Love Canal. *Journal of the American Medical Association* 240:2033–2034, 2040.
45. Whittell, G. 2000 (November 29). Poison in paradise. *(London) Times* 2, p. 4.
46. U.S. Environmental Protection Agency. 2010. Summary of the Comprehensive Environmental Response, Compensation, and Liability (Superfund). Accessed March 19, 2010 at www.epa.gov.
47. Bedient, P.B., H.S. Rifai, and C J. Newell. 1994. *Ground Water Contamination*. Englewood Cliffs, NJ: Prentice-Hall.
48. Huddleston, R.L. 1979. Solid-waste disposal: Land farming. *Chemical Engineering* 86:119–124.
49. McKenzie, G.D., and W.A. Pettyjohn. 1975. Subsurface waste management. In G.D. McKenzie and R.O. Utgard, eds., *Man and His Physical Environment: Readings in Environmental Geology*, 2nd ed., pp. 150–156. Minneapolis, MN: Burgess Publishing.
50. National Research Council, Committee on Geological Sciences 1972. *The Earth and Human Affairs*. San Francisco: Canfield Press.
51. Cox, C. 1985. The Buried Threat: Getting Away from Land Disposal of Hazardous Waste. No. 115–5. California Senate Office of Research.
52. Council on Environmental Quality. 1970. Ocean Dumping: A National Policy: A Report to the President. Washington, DC: U.S. Government Printing Office.
53. Lenssen, N. 1989 (July–August). The ocean blues. *World Watch*, pp. 26–35.
54. Woods Hole Oceanographic Institute, 2010. Plastic particles permeate the Atlantic. *Oceanus* Sept. 2, 2010 at www.whoi.edu
55. U.S. Environmental Protection Agency. 2000. Forward pollution protection: The future look of environmental protection. Accessed August 12, 2000, at http://www.epa.gov/p2/p2case.htm#num4.
56. Wellmar, F.W., and M. Kosinowoski. 2003. Sustainable development and the use of non renewable sources. *Geotimes* 48(12):14–17.

Chapter 24 Notes

1. NOAA Website http://response.restoration.noaa.gov/dwh.php?entry_id=809, accessed June 27, 2010.
2. NOAA Website http://response.restoration.noaa.gov/dwh.php?entry_id=809, accessed June 27, 2010.
3. NOAA.2010. Deepwater Horizon incident, Gulf of Mexico. at http://response.restoration.noaa.gov.
4. Jusserand, J. 1897. *English Wayfaring Life in the Middle Ages* (XIVth Century). London: T. Fisher Unwin.
5. Cohen, B.S. 1970. The Constitution, the public trust doctrine and the environment. *Utah Law Review* 388.
6. From http://www.cr.nps.gov/history/hisnps/NPSHistory/briefhistory.htm.
7. Bureau of Land Management facts, on the BLM Website.
8. Steiner, F. 1983. Regional planning: Historic and contemporary examples. *Landscape Planning* 10:297–315.
9. Section 26 of the TVA Act, http://www.tva.gov/
10. From http://www.npr.org/programs/atc/features/2003/aug/water/part1.html. See also http://www.library.unt.edu/gpo/powell/publications.htm.
11. Forman, D. 2004. *Rewilding North America: A vision of conservation for the 21st century*. Washington, D.C.: Island Press.
12. From http://www.eco.freedom.org/el/20031001/joyce.shtml.
13. Yannacone, V.J., Jr., B.S. Cohen, and S.G. Davison. 1972. Environmental rights and remedies. *Lawyers Co-operative Pub*., pp. 39–46.
14. Bacow, L.S., and M. Wheeler. 1984. *Environmental Dispute Resolution*. New York: Plenum Press.
15. Renner, M. 2002. Breaking the link between resources and repression. In *Worldwatch Institute State of the World 2002*. New York: Norton.
16. Rockström, J., et al., 2009. A safe operating space for humanity. *Nature* 461: 472–475.

Index

Entries followed by an f may be found in figures, those followed by a t may be found in tables.

A

Abandoned mines, 409–410
Abundance, population, 62
Abundance of life, 90–93
Acacia trees, 159
Acciona solar thermal power plant, Las Vegas, Nevada, 332f
Accuracy, measurement, 32–34
Acid mine drainage, 316–317, 316f, 409–410, 409f
Acid rain, 469–471, 469f
 causes of, 469
 control of, 471
 forest ecosystems and, 470
 formation and paths of, 470f
 human society and, 471
 lake ecosystems and, 470–471
 sensitivity to, 469–470
 stone decay and, 471
Acrolein, 472
Activated sludge, 416
Active decompensation zone, 404
Active solar energy, 329–331. *See also* Solar energy
Acute disease, 72
Acute effects, pollutants, 205
Adaptive management, 172
Adaptive radiation, 152
Adirondack Park, New York, 281–282
Advanced Boiling Water Reactor, 364
Advanced wastewater treatment, 415, 417
Aesthetic degradation, offshore oil drilling platforms and, 311
Aesthetic justifications, environment and, 15
Africa
 contraceptive devices and, 75
 famine in, 6
 Masai Mara Game Reserve, 280
 social disruptions and, 216
 vegetation comparison and, 159
Agassiz, Louis, 445
Age-specific birth rate, 63t
Age structure, population, 62, 67–69, 69f
Agriculture
 agricultural runoff, 390
 biofuels and, 338–339
 controlling pests and, 224–227
 crops, 218, 218f
 ecological perspective on, 213–214
 farmland area and, 227
 feeding world and, 214–217
 food chain and, 228
 future of, 227–228
 genetically modified food, 228–230
 global warming and, 455–456
 increased production per acre, 227
 irrigation and, 228
 livestock and, 218–221
 new crops and hybrids and, 227–228
 organic farming and, 228
 plowing and, 214, 214f
 small-grain production and, 218, 219f
 soils and, 221–224
 water requirements and, 231–232, 231t, 380, 382f
 world land use, 215, 215f, 215t
Agroecosystems, 213–214, 213f
Air-conditioning systems, indoor air pollution and, 490
Air pollution. *See also* Indoor air pollution
 acid rain and, 469–471, 469f
 air toxics and, 471–472
 atmospheric inversion and, 474–475, 474f
 automobiles and, 477
 carbon monoxide and, 466
 chemical and atmospheric processes and, 473–476
 controlling lower atmosphere pollutants, 477–481
 cost of controlling, 479–481
 criteria pollutants and, 465–468, 466t
 developing countries and, 476–477
 externalities and, 133–134
 future trends and, 476m 476f
 general effects of, 463–464
 haze from afar, 473
 high-altitude ozone depletion and, 481–486
 human death rates and, 463–464
 human health and, 464, 464f
 lead and, 468
 legislation/standards and, 478–479
 major pollutants, 465t
 nitrogen oxides and, 465–466
 oil and natural gas recovery and, 311
 overview of, 462
 ozone and, 466–467
 particulate matter and, 467–468, 468f, 477
 photochemical oxidants and, 466–467
 primary pollutants and, 464–465
 secondary pollutants and, 464–465
 soil and water toxicity and, 464
 stationary and mobile sources of, 463
 sulfur dioxide and, 465, 477–478
 synergistic effects and, 464
 variability of, 472–473
 vegetation and, 463
 wind velocity and, 475m 475f
Air Quality Index (AQI), 479, 480f
Air toxics
 acrolein, 472
 benzene, 472
 hydrogen fluoride, 471
 hydrogen sulfide, 471
 mercury, 472
 methyl isocyanate, 472
 volatile organic compounds, 472
Alaska
 Cook Inlet, 311f
 Cooper Island, 455f
 Wrangell-St. Elias Wilderness Area, 253, 253f
Albatross, 544f
Albedo effects, 447
Aldrin, 194t, 225
Alexandria, Virginia, 509, 510f
Alfalfa, 218
Algae, 339t
Algeria, Sahara Desert and, 311f
Alligators, pesticides and, 196
Allowance trading, 321
Allwine Prairie, Nebraska, 176–177
Alpha particles, 350–351, 351f
Alternative energy sources, 293–294
 biofuels, 338–339
 fuel cells, 333–334, 333f
 geothermal energy, 339–342
 nonrenewable alternative energy, 327
 ocean energy and, 335–336, 335f
 renewable energy, 327–328
 routes of, 328f
 solar energy, 328–332
 water power, 334–335
 wind power, 336–338
Amaranth, 227
Ambient air quality standards, 478–479, 479t
Ambroseli National Reserve, Kenya, 2–4, 2f–4f
America Clean Energy and Security Act of 2009, 287
American Bison, 17f, 177–178, 259–260, 259f
American gray squirrel, 154–155, 155f
Amu Darya River, 232
Anaerobic cells, 110
Animal pests, urban environments and, 514–515
Animal residues, energy uses of, 339t
Animal species number, 148t
Anopheles mosquitos, 151–152, 151f
Antarctic, ozone depletion over, 482–484
Antarctica, 131, 131f, 563, 564f
Anthrax, 190
Antisocial behavior, lead toxicity and, 207, 207f
Appalachian Mountains, 317, 319f, 470
Aquaculture, 230–231, 230f
Aquifer, 372
Aral Sea, 232, 377, 377f, 378f
Aransas National Wildlife Refuge, Texas, 263
Arctic National Wildlife Refuge, 312–314, 312f, 313f
Arctic sea ice, 131, 131f, 454–455
Area sources, air pollution and, 463, 463f
Arizona
 Black Mesa Mine, 317
 Grand Canyon, 159, 159f
 Phoenix, 419, 419f
 Tucson, 513, 513f

Armadillo, 273
Artificial fertilizers, 223
Artificial organic compounds, 225
Asbestos, 199, 488t
Atlanta, Georgia, 383
Atmosphere, 434–437
  albedo effects and, 447
  atmospheric inversion, 474–475, 474f
  atmospheric transparency, 446–447
  barometric pressure and, 435–436
  chemistry of life and, 447
  circulation of, 435–436, 435f
  earth's surface roughness and, 447
  energy and, 436, 436f, 437f
  ozone layer and, 434, 434f
  processes of, 435–436
  structure of, 434–435
  temperatures and, 434f, 435–436
  water vapor content and, 435
Atrazine, 194t, 197
Australia
  Canberra, 530
  Great Barrier Reef, 406
  modern forest plantation in, 246f
  Shark Bay, 109f
Automobile design, 296, 296f
Automobiles, air pollution and, 477
Autotrophs, 84
Average residence time, 46–48
Average wind speed, wind power and, 336, 336f

B

Bacillus thuringiensis, 226, 229, 230f
Background radiation, 354–357, 356f, 357f
Bacon, Francis, 24
Bacteria, energy uses of, 339t
Bacterial contamination, water, 404, 405f
Balance of nature, 46, 46f
Banded-iron formations, 109, 109f
Bangladesh
  food riots in, 8f
  population growth rates and, 75
  population policy and, 75
Barefaced bulbul, 148, 148f
Barking deer, 147
Barnacles, 273
Barstow, California, 331f
Basins, depositional, 306
Beauty of nature, valuing, 134–135, 134f
Becquerel (Bq), 356
Beef production, 218
Behavioral tolerance, 205
Beijing, China, 463f
Belgium, Bruges, 505f
Beluga SkySails, 327, 327f
Benzene, 186, 472
Beta carotene, 229
Beta particles, 350–351, 351f
Bighorn sheep, 153–154
Bingham Canyon Copper Pit, Utah, 526f
Bioaccumulation, 191
Biochemical oxygen demand, 403–404, 404f
Biofuels
  environment and, 338–339
  human history and, 338
  sources, uses, and products and, 339t
Biogeochemical cycles
  basic cycle, 117, 117f
  carbon cycle, 117–119, 118f
  carbon-silicate cycle, 119–120, 119f
  Earth's history and, 106–110
  general aspects of, 112
  geologic cycle, 113–116, 113f
  life and global chemical cycles, 111–112
  nitrogen cycle, 120–121, 120f, 124
  phosphorous cycle, 121–124, 122f
  as systems diagram, 112f
Biogeography, 159–160, 161f
Biological aerosols, 488t
Biological control, 226–227
Biological diversity
  aesthetic reasons for valuing, 146
  basics of, 146–149
  biomes and, 161–162
  biotic provinces and, 159–160, 161f
  competitive exclusion principle and, 154–155
  convergent and divergent evolution and, 162–163, 163f
  creative reasons for valuing, 146
  ecological niche and, 155–157
  geography and geology affects on, 158–163
  global warming and, 455
  major influences on, 159, 160f
  moral reasons for valuing, 145
  population and species and, 145
  predation-parasitism and, 158
  recreational reasons for valuing, 146
  species numbers on earth, 147–149, 148f
  spiritual reasons for valuing, 146
  symbiosis and, 157–158
  theological reasons for valuing, 145
  value of, 145–146
Biological evolution, 149–154
  complexity and, 154
  convergent and divergent, 162–163, 163f
  defined, 149
  genetic drift and, 153–154
  key processes of, 149–154
  migration and geographic isolation, 152–153
  mutation and, 149–150
  natural selection and, 150–152
Biological invasions, 164–166, 165f
Biological production, ecosystem energy flow and, 89–93
Biomagnification, 191–192
Biomass, 93–94
Biomass energy sources, 339t
Biomes, 161–162
Bioremediation, 412
Biosphere, 50, 110
Bioswales, 411f
Biota, 118
Biotic provinces, 159–160
Birds, pesticides and, 196
Birth control, zero population growth and, 74–75
Birth rates, 62
Bison, 177–178, 178f
Black guillemots, 455. 455f
Black lung disease, 486
Black Mesa Mine, Arizona, 317
Black plague, 429
Black shale natural gas, 310
Bladderpod, 227
Blue water, 387
Blue whales, 276, 278, 278f
Bluefish, 267, 268f
Body burden, 191
Bog succession, 97–98, 97f
Boiler slag, 321
Boise, Idaho, 346
Bone Valley, Florida, 121–123, 123f
Bora Bora, French Polynesia, 55f
Boreal forests, 33t
Boston, Massachusetts, 417, 506–507, 507f
Bottled water, 414
Boulder, Colorado, 516
Boundaries, ecosystem, 88, 89f
Boundary Waters Canoe Area, Minnesota, 30–31
Bowhead whales, 132–133, 133f, 265
Brazil
  Rio de Janeiro, 533f
  tropical rain forests, 248, 248f, 254
Breeder reactors, 363–364
British Peak District National Park, 180, 180f
British Petroleum (BP), 552–555
British red squirrel, 154–155, 155f
Broad-spectrum inorganic toxins, 224
Brown pelican, 121
Bruges, Belgium, 505f
Building design, indoor air pollution and, 493
Building-related illnesses, 491
Bureau of Land Management (BLM), 558
Burmese python, 171, 171f
Burner reactors, 349
Bush, George W., 287, 297
Butadiene, 186
Butchart Gardens, Vancouver Island, Canada, 180, 180f

C

Cadmium toxicity, 191–192
Calcium sulfite, 321
California
  air pollution and, 473–476
  Barstow, 331f
  beach bacterial contamination, 405f
  California Water Project, 378
  Channel Islands, 178–179, 178f
  constructed ecosystems and, 181–182
  Desolation Wilderness Area, 253
  Geysers Geothermal Field, 340, 340f
  Los Angeles, 12f, 473–475, 473f, 474f
  Matilija Dam, 392
  Mineral King Valley, 560–561, 560f
  Mono Lake, 23–24, 23f–24f
  Red Bluff Dam, 269, 270f
  river flow changes and, 451
  San Andreas Fault, 114
  San Francisco, 15, 15f
  San Francisco Bay estuary, 389, 389f
  Santa Barbara, 172, 384, 385f, 411f
  Santa Barbara Channel, 105–106, 105f–106f, 178–179, 178f
  Santa Cruz Island, 179, 179f
  shoreline erosion and, 390f

Sweetwater Marsh National Wildlife Refuge, 181
Ward Valley, 360
water use and, 383
Yosemite National Park, 250, 250f
Callendar, Gary Stewart, 433–434
Calorie intake, worldwide, 216, 217f
Cameroon, Lake Nyos, 187–188, 188f
Canada
Alberta, 322, 322f
Butchart Gardens, Vancouver Island, 180, 180f
Lake Moraine, 89f
overfishing and, 139
tar sands and, 322, 322f
Wood Buffalo National Park, 263
Canary Islands, Tenerife, 331f
Canberra, Australia, 530
Cancer, 186, 200
Cap-and-trade, carbon dioxide emissions and, 128–129, 129f
Cap rocks, 306
Cape Wind Project, 341–342, 342f
Carbon, stored in vegetation, 33, 33t
Carbon-14 records, 440, 440f
Carbon cycle, 117–119, 118f
Carbon dioxide
air pollution and, 466
emissions, 128–129, 129f
ice core measurements and, 438
as indoor air pollutant, 488t
as major greenhouse gas, 443
massive natural release of, 187–188, 188f
Carbon monoxide, 477, 488t
Carbon sequestration, 456
Carbon-silicate cycle, 119–120, 119f
Carcinogens, 188
Carlsbad, New Mexico, 361
Carnivores, 84
Carnot, Sadi, 293
Carrying capacity, 10, 10f, 66, 260–262, 261
Carson, Rachel, 5
Case fatality rate, 63t
Catalytic chain reactions, 483
Catalytic converters, 477
Catastrophes as experiments, 31
Catch per unit, 265
Catchment, 116
Catskill Mountains, New York, 402, 402f
Cattle, grazing of, 177
Cause-specific death rate, 63t
Cell phones, 520–521, 520f, 521t
Central Park, New York City, 506
Centralia, Pennsylvania, 319
Certification of forest practices, 246–247
Cesium-137 release, 355f
Chain reaction, 348
Challenger space shuttle, 32
Chandeleur Beach, Louisiana, 553f
Chandraprabha Sanctuary, India, 280
Channel Islands, Southern California, 178–179, 178f
Channelization, 174–175, 174f
Chao Phraya River, Thailand, 232
Chemical cycling, 99, 99f
Chemical elements, 111–112, 111f
Chemical fertilizers, 223
Chemical weathering, 116
Chemoautotrophs, 93
Chemosynthesis, 91
Chernobyl nuclear power plant, Soviet Union, 357, 359–360, 360f
Chesapeake Bay fisheries, 267, 268f
Chicago, Illinois, 383
Childbearing age, zero population growth and, 74–75
Children
childhood cancer, 186
death from malnutrition and, 6
electromagnetic fields and, 200
lead toxicity and, 207, 207f
Chimney effect, 490
China
aquaculture and, 230
Beijing, 463f
electronic waste and, 535, 535f
Huang He River, 232
phosphorous reserves and, 123
Three Gorges Dam, 390–391, 391f
tower karst, 114, 115f
wind power potential and, 338
Chlorine, drinking water and, 414
Chlorine chain reactions, 483
Chlorine chemistry, 483–484
Chlorine treatment, wastewater, 417
Chlorofluorocarbons (CFCs), 73, 444, 483, 486
Chloroquine, 151
Chronic disease, 72
Chronic effects, pollutants and, 205
Chronic patchiness, ecological succession and, 99–100, 100f
Circulation, atmospheric, 435–436, 435f
Cities
as an environment, 508–510
animal pests and, 514–515
bringing nature to, 511–515
center of civilization and, 505
city as a system, 499–500, 500f
city planning, 506–508
energy budgets and, 508
environmental history of, 505
fall lines and, 501–502, 501f
ideal locations for, 502–504
industrial metropolis and, 505
life in, 499
location of, 500–504, 502f
parks and, 506–507
pollution and, 510
rivers and, 511
site and situation importance, 500–502
site modification and, 504
soils and, 510
solar energy in, 509
urban atmosphere and climate, 509
urban centers and, 505
vegetation in, 512–513
water and, 509–510
wildlife and endangered species and, 513–514, 515f
Citizens actions, 562
Citrus greening, 144–145, 144f, 145f
Classical stability, 44
Clean Air Act, 478
Clean Water Act, 399, 421–422
Clearcutting, 245, 246f, 409f
Clearwater, 400, 511
Cleveland, Grover, 558
Climate. *See also* Greenhouse effect
albedo effects and, 447
atmospheric transparency and, 446–447
carbon-14 records and, 440, 440f
causes of change in, 445–447
characteristics of, 431
chemistry of life ad, 447
climate forcing, 447, 448f
climatic similarity, 161–162, 162f
computer simulations and, 450–451
coral growth and, 440
Earth's surface and, 447
El Niño and, 449
experiments and laboratory research and, 450
feedback loops and, 444–445
forecasting change and, 450–451
general circulation models and, 451, 451f
historical records and, 438
ice cores and, 438–439, 438f
instrumental records and, 438
laboratory research and, 450
Milankovitch cycles and, 445–446, 446f
negative feedback loops and, 444
oceans and, 448–449, 449f
past observations and, 450
polar amplification and, 451
positive feedback loops and, 445
potential rates of change and, 451
precipitation and temperature conditions and, 431, 432f, 433f
proxy data and, 438–440
sediment samples and, 439–440, 439f
solar cycles and, 446
study of, 438–440
time scales and, 432, 433f
tree ring chronology and, 439, 439f
urban atmosphere and, 509
Closed systems, 44
Coal, 314–321
allowance trading and, 321
converting to oil, 320
future of, 320–321
mountaintop removal and, 317–319, 319f
peat transformation into, 314, 314f
scrubbing, 320
strip mining and, 316–317, 317f
sulfur content of, 314–315
syngas and, 320
transportation of, 319–320
underground mining and, 319
U.S. reserves, 314–315, 315f, 316t
world coal reserves, 315f
Coal-bed methane, 309–310
Coal dust, 317
Coal fires in underground mines, 319
Coal gasification, 477
Coal Oil Point, Santa Barbara, 105–106, 105f–106f
Coastal erosion, 452
Coastal flooding, 452
Coastal wetlands, Louisiana, 419, 419f
Cod fishing, 267

Codominants, tree, 245
Coelenterata, 98
Cogeneration, energy, 294–296
Coliform bacteria, 188, 188f
Colony collapse disorder (CCD), 130
Colorado
  background radiation and, 354
  Boulder, 516
  Denver, 383
  Golden, 529, 529f
  Trapper Mine, 318, 318f
Columbia River, 14, 14f, 264
Combretum trees, 159
Commons, environment as, 130–132
Community effect, 81
Competitive exclusion principle, 154–155
Composting, solid-waste and, 532–533
Comprehensive Environmental Response, Compensation, and Liability Act (CERCLA), 399, 540
Comprehensive Everglades Restoration Plan, 171
Computer simulations, 29, 29f, 450–451
Computer waste, 535
Consolidated Edison Company, 346, 563
Constructed ecosystems, 181–182
Consumptive use, water, 376
Contamination, 188
Controlled experiments, 27
Controlling variables, 27
Conventional nuclear reactors, 348–349
Convergent evolution, 162–163, 163f
Convergent plate boundary, 114
Cook Inlet, Alaska, 311f
Cooling towers, 198–199, 198f
Coon Creek, Wisconsin, 222
Cooper Island, Alaska, 455, 455f
Cooper's hawks, 513, 513f
Copper ore deposits, 189–190
Coral growth, climate study and, 440
Coral reef succession, 98, 98f
Costa Rica, national parks in, 252
Coto Doñana National Park, Spain, 251, 251f
Creative justification, 15
Crop circles, 30, 30f
Crop plants, symbiotic relationships and, 158
Crops, 218, 218f
Crude birth rate, 63, 63t
Crude death rate, 63, 63t
Crude growth rate, 63, 63t
Crude oil transportation, 312
Cryptosporidiosis, 190
Cultural eutrophication, 406–408, 407f
Cultural justification, 16
Cuomo, Andrew, 346
Curie, Marie, 356, 357f
Curie, Pierre, 356
Custer State Park, South Dakota, 178f
Cuyahoga River, Cleveland, Ohio, 410, 411f
Cypress swamp water, 388f

D

Dallas, Texas, 462
Dams
  economic aspect of, 392
  Edwards Dam, Maine, 392, 393f
  Elwha River Dam, Washington, 392
  environmental effects of, 390
  Glen Canyon Dam, Colorado River, 391–392, 391f
  Matilija Dam, California, 392
  Missouri River, 42, 42f
  opposition to, 390
  removal of, 392–393, 393f
  role of, 335
  Sandy River Dam, Washington, 392
  Three Gorges Dam, China, 390–391, 391f
Darwin, Charles, 54, 149, 152
Dasanit, 225
Data types, 27
DDT
  alligator populations and, 196
  endangered birds and, 276
  malaria and, 151–152
  toxicity of, 225
  uses of, 194t
Dead organic matter in streams and rivers, 403–404, 404f
Dead zone, Gulf of Mexico, 407–408, 407f
Death rates, 62, 72–73, 72f
Death risks, 137t
Decision making, science and, 34
Decomposers, 84
Deductive reasoning, 28
Deep-earth, high-density geothermal energy, 340
Deep ecology, 74
Deep-sea vent chemosynthetic bacteria, 91f
Deepwater Horizon oil spill, 202, 311, 552–555, 552f–555f, 556
Deforestation, 247–248, 247f, 248f
Demasculinization, frogs, 196–197, 197f
Demographic transition, 69–71, 70f
Demography, 62
Denitrification, 120
Denver, Colorado, 383
Deoxyribonucleic acid (DNA), 149–150
Dependent variables, 26, 27
Depositional basins, 116, 306
Depression cone, 372
Desalination, 376
Desertification, 73
Desolation Wilderness Area, California, 253
Destructive distillation, 321
Desulfurization, 478
Detoxification, 205
Deuterium-tritium (DT) fusion, 364
Developing countries, air pollution and, 476–477
Diamond mines, 523, 524f
Diaphorina citri, 144–145, 144f, 145f
Didcot power plant, 198f
Dieldrin, 194t, 225
Digitalis, 272
Dioxins, 194t, 195
Direct costs, 134
Direct water reuse, 420
Discharge points, 372
Discharge zones, 372
Disney Corporation, 560
Disprovability, 25
Dissolved oxygen content, water, 404
Distributive power systems, 299, 299f
Disturbances as experiments, 31
Divergent evolution, 162–163, 163f
Divergent plate boundary, 114
Dobson ultraviolet spectrometer, 482
Dodge, Colonel R. I., 259–260
Dodge City, Kansas, 264
Dolphins, 279
Domestic water use, 383
Dominant species, 147
Dominants, tree, 245
Dose response, 202, 203f
Dose-response curve, 202–204, 203f
Doubling time, 52–53, 63t, 65, 65f
Douglas, William O., 560
Drainage basin, 116
Drainage systems, wildlife habitats and, 514, 515f
Drilling muds, 311
Drinking water
  radioactive materials in, 358
  safety of, 414
  standards for, 401, 401t
Droughts, 369, 375
Dry cooling towers, 198–199, 198f
Dune succession, 96, 96f
Dust, 468
Dust Bowl, 223, 223f
Dust mites, 489f
Dynamic equilibrium, 46
Dynamic systems, 44, 44f

E

Early-successional species, 99, 213
Earth
  atmosphere of, 106–110, 107f
  lithospheric plates, 113–114, 114f
  from space, 10f
  as a system, 55
Earth First, 562
Earth Summit in Rio de Janeiro, Brazil, 457
Earthworms, 12f
East Asia, contraceptive devices and, 75
Easter Island, 17–18, 18f
Ebola virus, 191
Ecological communities and food chains, 84–88
Ecological gradients, 159, 159f, 205
Ecological islands, 165, 280–281, 280f
Ecological justification, 15
Ecological niches, 155–157
Ecological restoration
  applying ecological knowledge to, 180
  characteristics of, 171–172
  examples of, 173–179, 173t
  goals of, 172, 173t
  judging success of, 180–181
  prairie restoration, 176–178
  rivers and streams and, 174–176
  wetlands and, 176
Ecological Society of America, 263
Ecological species, 145
Ecological stability, 95–99
Ecological succession, 95–99
  bog succession, 97–98, 97f
  chemical cycling and, 99, 99f
  chronic patchiness and, 100, 100f
  coral reef succession, 98, 98f

dune succession, 96, 96f
early-successional species, 99
facilitation and, 99–100
how species change, 99–100
interference and, 100
late-successional species, 99
life history differences and, 100
middle-successional species, 99
old-field succession, 98, 98f
patterns of, 96–99
Ecological trophic energy efficiency, 95
Ecosystems
basic characteristics of, 83
biological production and biomass, 93
boundaries, 88, 89f
damage, oil and natural gas recovery and, 311
ecological communities and food chains, 84–88
ecological stability and succession, 95–99
ecosystem effect, 406
energy efficiency and transfer efficiency, 94–95
energy flow and, 89–93
gross and net production, 94
idealized minimum ecosystem, 83f
impossible ecosystem, 92, 92f
laws of thermodynamics and, 90–93
making energy visible and, 89–90, 90f
processes, 83
simple ecosystem, 84–85
steady state and, 46
structures, 83
sustainability of, 8
as systems, 88–89
Ecotopia, 555–556
Ecotourism, 273
ED-50 values, 204, 204f
Edison, Thomas, 299
Edwards Dam, Maine, 392, 393f
Effluent streams, 372
Ehrlich, Anne, 275
Ehrlich, Paul, 275
El Niño, 449
El Salvador, geothermal energy and, 340
Electrical resistivity, 293
Electromagnetic fields, 199–200
Electromagnetic radiation, 436, 436f
Electronic waste (e-waste), 520–521, 521t, 535
Elephant seal, 276
Elephants, 2, 13, 13f, 249
Elwha River Dam, Washington, 392
Emu, 163f
Endangered species, 271–282
aesthetic justifications and, 273
becoming, 274–276
cultural justifications and, 273
current status of, 271, 271t
diagrammatic history and, 275f
ecological islands and, 280–281, 280f
ecological justifications and, 273
global extinction, 274
improvements and, 276, 277f
local extinction, 274
moral justifications and, 273
reasons for conserving, 272–273
spatial relationships and, 281
urban environments and, 513–514
utilitarian justifications and, 272–273
Endangered Species Act of 1973, 81, 271, 276
Endrin, 194t
Energy
atmospheric, 436, 436f, 437f
basics of, 289–290, 289f, 290f
city budgets and, 508
conservation of, 294–296
in early Greek and Roman cultures, 288, 288f
efficiency and, 94–95, 290–296
energy flow chemistry, 91
energy pyramid, 101, 101f
energy units, 291
flow equations, 93–94
making visible, 89–90, 90f
quality of, 290
sources and consumption of, 293–294
system closed to, 92, 92f
in world today, 288–289
Energy crops, 339t
Energy Policy Act, 287, 346
Energy Power Act, 362
Engelman, Robert, 76
England
life expectancy and, 71, 71f
Ratcliffe-on-Saw, 349f
River Severn, 404f, 539, 539f
Vale of Pewsey, 30, 30f
Enhanced recovery, petroleum production and, 306
Enrichment, nuclear, 348
Entropy, 93
Environmental economics
environment as a commons, 130–132
externalities and, 133–134
low growth rate and, 132–133
overview of, 129
public-service functions of nature, 130, 130f
risk-benefit analysis and, 136–138, 137t
scarcity and economic value, 133
valuing beauty of nature and, 134–135, 134f
valuing the future and, 135–136, 135f
Environmental fitness, 108
Environmental health. *See also* Pollutants
asbestos and, 199
basics and, 187–188
disease incidence and, 187
electromagnetic fields and, 199–200
environmentally transmitted infectious disease, 190–191
hormonally active agents, 196–197, 197f
infectious agents, 189–190
noise pollution and, 200–201, 200t
nuclear radiation and, 198
organic compounds and, 193
particulates and, 199, 199f
persistent organic pollutants and, 194, 194t
pollutant categories, 189–201
pollution measurement and, 189
terminology and, 188
thermal pollution, 198–199, 198f
toxic heavy metals, 191
toxic pathways and, 191–193, 191f, 192f
voluntary exposure and, 201
Environmental justice, 534
Environmental law, 421–422, 422t, 557–558, 558f
Environmental risk, extinction and, 276
Environmental tobacco smoke, 490
Environmental unity, 54
Environmentally transmitted infectious disease, 190–191
Ephemeral streams, 372
Epidemic disease, 72
Equator, death rates and, 72, 72f
Equilibrium, 44, 44f
Erosion, soil, 222–224
Escherichia coli infection outbreaks, 404, 405t
Eukarya, 148
Eukaryota, 148
Eukaryotes, 110, 110f
Euphorbia, 163f
European Climate Exchange, 457
Eutrophic lakes, 406, 406f
Eutrophication, 405–408, 406f
Evapotranspiration, 241–243, 242f
Everglades, Florida, 170–171, 170f, 388f, 390
Evolution. *See* Biological evolution
Exajoules, 291
Experimental errors, 32
Exponential growth, 52–53, 65
Exponential growth curve, 62–63
Exponential rate, 62
Externalities, 133–134
Extinction, causes of, 275–276, 279–280
Exxon Valdez oil spill, 105, 408, 552

F

Facilitation, ecological succession and, 99–100, 100f
Fall lines, 501–502, 501f
Family planning, zero population growth and, 75
Famine, 6–7, 7f, 216
Farmland area, 227. *See also* Agriculture
Fecal coliform bacteria, 404
Feedback, 48–50
Feedback loops, climate change and, 444–445
Feedlots, 220
Feminization, frogs, 196–197, 197f
Fermi, Enrico, 348
Fertile Crescent, 177f
Fertility, 63t
Fertilizers, 222–223
Fever trees, 3
Finches, 152, 152f
Fine particles, air pollution and, 467
Fire, tallgrass prairies and, 177–178
Fire scars, 31, 31f
First-law efficiency, energy, 290, 291t
First law of thermodynamics, 91, 289
Fish spawning cycles, 198
Fisheries, 265–271
decline in populations, 267–269, 267f
international trade and, 265
management options and, 139
modern commercial methods, 266, 266f

Fisheries *(cont.)*
sustainability of fishing, 269
total global fish harvest, 265, 266t
total species, 266
world's major fisheries, 265f
Fishing rights, allocating, 139
Fission, 348
Fission reactors, 364
Flamingos, 251f
Flatworms, freshwater, 157, 157f
Flemining, Sir Alexander, 31
Florida
background radiation and, 354
Bone Valley, 121–123, 123f
Everglades, 170–171, 170f, 387, 388f, 390
Kissimmee River, 174–175, 174f
Lakeland, 512
Palm Beach County, 369, 369f
Tampa, 383
Flour beetles, 155–156, 156f
Flow, biogeochemical cycle and, 112
Fluorine, 202–203, 203f
Flux, biogeochemical cycle and, 112
Food
distribution disruptions and, 73
emergencies and, 216
global water shortage and, 393–394
international aid and, 216
population growth and, 6–7
prices and, 212
riots and, 6, 8f
Food chain, 84–88
eating lower on, 228
efficiency and, 95
energy content of, 101, 101f
Mono Lake, 23–24, 24f
oceanic, 85
Food webs
defined, 84
of harp seal, 85–88, 88f
oceanic food web, 87f
temperate, 86f
terrestrial, 86
Footprints, water, 387, 394–395
Forage, 218
Forecasting population change, 66
Forestry, 238
Forests. *See also* Trees
acid rain and, 470
atmosphere and, 242f
certification of forest practices, 246–247
countries with largest areas of, 240, 240f
deforestation, 247–248, 247f
developing on lava flows, 96f
dual human uses of, 239, 239f
ecology of, 243
effects of, 241–243
forested watershed, 242f
global warming and, 248
growth, computer simulation of, 29, 29f
harvesting trees and, 245, 245f
management of, 244–246
modern conflicts over, 238–239
old-growth forest, 244
ownership in U.S., 241f
plantations, 245–246, 246f
products, energy uses of, 339t
public-service functions and, 238
rotation time, 244
second-growth forest, 244
site quality and, 245
sustainable forestry, 246–247
tree growth, 243–244, 244f
tree niches, 243, 243f
viewpoint of, 237–238
wood as energy source, 241
world forest area, 239–240, 239f
Formaldehyde, 488t
Fort McMurray, Alberta, Canada, 322, 322f
Fossil fuels
alternatives to, 293–294
black shale natural gas, 310
cap rocks, 306
characteristics of, 305–306
coal, 314–321
coal-bed methane, 309–310
delivery and use, 312
depositional basins and, 306
methane hydrates, 310–311
natural gas, 309
oil and gas traps, 306, 306f
oil in twenty-first century, 208t, 308–309
oil shale, 321–322
petroleum production, 306–307
power plants and, 349f
recovery and, 311, 311f
refinement and, 312
reservoir rock and, 306
source rock and, 306
tar sands, 322, 322f
Founder effect, 153
Fox, 178–179, 178f
Fractional distillation, 312
France
national parks in, 252
Paris, 512f
Rance tidal power station, 335–336, 335f
French Polynesia, Bora Bora, 55f
Freshwater flatworms, 157, 157f
Freshwater wetlands, 388–389
Frogs
demasculinization and feminization of, 196–197, 197f
food chain and, 101, 191f
Frontier science, 35–37
Fruit fly mutation, 150, 150f
Fuel cells, 333–334, 333f
Fugitive sources, air pollution and, 463, 463f
Fundamental niches, 157, 157f
Fusion, 348
Fusion reactors, 364
Future *vs.* present value, 135–136, 135f

G

Gaia hypothesis, 10–11, 55, 108
Galápagos Islands, 28f, 152, 152f
Galileo, 24
Gamma decay, 351
Gamma rays, 350–351, 351f
Ganges River, India, 232
Garden city, 507
Gasoline tank leakage, 412, 412f
General circulation models, 451, 451f
General Electric, 399–400
General fertility rate, 63t
Generators, solar thermal, 331–332, 331f, 332f
Genes, mutations and, 149–150
Genetic diversity, 146
Genetic drift, biological evolution and, 153–154
Genetic risk, extinction and, 276
Genetic tolerance, 205
Genetically modified crops, 214
Genetically modified food
gene transfer and, 229–230, 230f
new hybrids, 229
terminator gene and, 229
Genotypes, 149–150
Geographic isolation, biological evolution and, 152–153
Geologic cycle, 113–116, 113f
hydrologic cycle, 115–116, 115f
rock cycle, 116, 116f
tectonic cycle, 113–114, 114f
Geologic sequestration, 456
Georges Bank, 139
Georgia, Atlanta, 383
Geothermal energy
environment and, 341
future of, 341
geothermal systems, 340–341
types of, 339–340
Geysers Geothermal Field, San Francisco, California, 340, 340f
Ghana, population policy and, 75
Giardiasis, 190
Gir Forest, India, 339
Glacier National Park, Montana, 452
Glaciers, global warming and, 452–455
Glen Canyon Dam, Colorado River, 391–392, 391f
Global chemical cycles, 111–112, 111f
Global dimming, 468
Global extinction, 274
Global perspective, 10–11
Global security and environment, 563–564
Global thresholds, 565–566, 566t
Global warming. *See also* Climate
adjusting to, 456–457
agricultural productivity and, 455–456
animals and plants and, 454
average surface temperature and, 430, 430f
biological diversity and, 455
fundamental questions regarding, 430–431
glaciers and sea ice and, 452–455
human health effects and, 456
indirect forest damage and, 248
international agreements and, 456–457
issue origins and, 433–434
river flow changes and, 451
seal level rises and, 452, 452f
Goats, 220
Golden, Colorado, 529, 529f
Golden eagle, 179, 179f
Granada, Spain, 508, 508f
Grand Canyon, Arizona
banded-iron formations and, 109, 109f

Colorado River, 114, 115f
species abundance and, 159, 159f
Grasshoppers, 101, 191f
Gray water, 387
Gray whales, 276, 278
Gray zone of suboptimal health, 187
Grazing
rangeland, 218–221, 220f
tallgrass prairies and, 177–178
Great Barrier Reef, Australia, 406
Great Britain
American gray squirrel and, 154–155, 155f
global warming and, 454
Hadley Meteorological Center, 438
mining area restoration in, 180, 180f
native species and, 165
Thames River, 410
Victoria Park, 250
Grebe, 28–29, 28f
Green buildings, 462
Green revolution, 228
Green water, 387
Greenbelts, 507
Greenhouse effect
absorption and, 442f
characteristics of, 441–443
Earth's energy balance and, 442f
idealized diagram of, 441f
major greenhouse gases, 443–444, 443t
Greenhouse gases, 433, 443–444, 443t
carbon dioxide, 443
chlorofluorocarbons, 444
methane, 444
nitrous oxide, 444
Greenland
ice sheet, 455
lead accumulation and, 468
Quanea, 193
Grinnell, George, 260
Grit chamber, 416
Grizzly bear, 258–259, 258f
Gross national product (GNP), 63t
Gross production, 93–94
Ground-trawling equipment, ocean floor destruction and, 269
Groundwater
geothermal energy and, 340–341
interactions with surface water, 372–373, 372f, 373f
overdraft, 375
supplies, 73
sustainability, 384
use and problems, 375–376
water pollution sources and processes, 401t
Groundwater pollution
bioremediation and, 412
Long Island, New York and, 413–414, 413f
principles of, 412–413, 413f
waste-disposal sites and, 412
Growth rates, population, 62
Guano Islands, 121, 122f, 223, 224f
Guayule, 227
Gulf of Mexico
active oil platforms in, 552f
cultural eutrophication of, 407–408, 407f
Deepwater Horizon oil spill and, 202, 311, 552–556
marine and coastal environment and, 553–554
Gumweed, 227
Gutierrez, Carlos M., 269
Gypsum, 321

H

Habitat diversity, 146
Hadley cells, 435
Hadley Meteorological Center, Great Britain, 438
Haiti, food riots in, 8f
Half-life, radioisotope, 350–352
Hardin, Garrett, 130
Harp seal, 85–88, 88f
Harvest counts, 264–265
Hawaii
adaptive radiation and, 152–153, 153f
coral reef, 98, 98f
forests developing on lava flows, 96f
Maui, 463f
Mauna Loa Mountain, 430, 431f
Hazardous waste
incineration and, 542–543
land disposal and, 540–542
legislation and, 539–540
Love Canal case and, 537–539
nuclear power reactors and, 353
production of, 537
recycling and resource recovery and, 542
reduction technologies, 541t
River Severn, England, 539, 539f
secure landfill for, 542f
source reduction and, 542
treatment and, 542
Heat energy, 289
Heat engines, 293
Heat pollution, 198–199, 198f
Heating, indoor air pollution and, 490
Heavy metals, 191
Heavy water, 349
Heinselman, Murray, 30–31
Henderson, Lawrence, 108
Herbicides, 225
Herbivores, 101
Herding practices, 220
Heterotrophs, 84, 91
High-altitude ozone depletion, 481–486
Antarctic ozone hole, 484
chlorofluorocarbons and, 483
environmental effects and, 484485
future of, 486
measuring stratospheric ozone, 482–483
polar stratospheric clouds, 484
simplified stratospheric chlorine chemistry, 483–484
ultraviolet radiation and, 481–482, 481f, 482f
High-level radioactive waste, 361–362
High Line Park, New York City, 498, 498f
High Plains Aquifer, 375
High Temperature Gas Reactor, 364
Historical evidence, 30–31
Historical range of variation, 263–264
Historical record, climate and, 438
H.J. Andrews experimental forest, Oregon, 245
H1N1 flu pandemic, 189
Holdfasts, 81
Honeycreepers, 153, 153f
Hormonally active agents, 196–197, 197f
Horse Hollow Wind Energy Center, Texas, 337
Horseshoe crabs, 273
Hot springs, 84–85, 84f, 85f
Household appliances, power use of, 292t
Houston, Texas, 186–187
Houston Ship Channel, 186–187
Howard, Ebenezer, 507
Huang He River, China, 232
Hubbard Brook Ecosystem Study, 89
Hubbard Brook experimental forest, New Hampshire, 245
Hudson River, New York, 399–400, 511, 512f
Human population
age structure and, 67–69
annual growth rate of, 65f
brief history of, 64
carrying capacity and, 73–74
death rates and industrial rise, 72–73, 72f
definitions and terminology and, 61–62, 63t
demographic transition and, 69–71m 79f
doubling time and, 65, 65f
effects on earth, 73
exponential growth and, 65
exponential growth curve and, 62–63
famine and food crisis and, 6–7, 7f
forecasting change in, 66
future growth projections and, 65–67
growth of, 6–8
logistic growth curve and, 65–67
longevity and, 71–73, 71f
Malthus prophecy and, 68
medical advances and, 71
pandemics and, 60–61, 61f
population dynamics and, 61–63
population history of our species, 63f
rapid growth of, 6, 6f
U.N. projections and, 67, 67f
U.S. (1790-2008) population statistics, 62f
zero population growth, 74–75
Human waste recycling, 531–532
Hurricane Katrina, 487, 503–504, 504f
Hussein, Saddam, 176
Hutchinson, G. Evelyn, 108, 154
Hutton, James, 54, 55
Hybrid cars, 296, 296f
Hybrid crops, 227–228, 229
Hydrocarbons, 472
Hydroelectric power. *See* Water power
Hydrogen fluoride, 471
Hydrogen sulfide, 471
Hydrologic cycle, 115–116, 115f, 370, 371f
Hydrology, rainfall and, 2–4
Hypothesis, 26

I

Ice cores, 438–439, 438f
Idaho, Boise, 346

Igneous rocks, 116
Illinois, Chicago, 383
Imagination, leaps of, 31
In-stream use, water, 376–377, 376f
Inadvertent water reuse, 420
Incidence, 63t
Incineration
  hazardous waste and, 542–543
  solid-waste and, 533
Independent variables, 26, 27
India
  Ganges River, 232
  Gir Forest, 339
  Indus River, 247, 247f
  population policy and, 75
Indian Point Energy Center, New York, 345f, 346
Indirect water reuse, 420
Indonesia
  fires in, 199f
  palm plantations and, 339
Indoor air pollution, 486–493
  building design and, 493
  chimney effect and, 490
  control of, 491–493
  environmental tobacco smoke and, 490
  heating, ventilation, and air-conditioning systems and, 490
  pathways, processes, and driving forces, 489–491
  radon gas and, 490–491
  radon-resistant buildings and, 493
  sick buildings and, 491
  sources of, 487–489, 487f, 488t
  symptoms of, 491, 492t
Induction, 27
Inductive reasoning, 27, 28–29
Indus River, India, 247, 247f
Industrial ecology, 530
Industrial energy, 296
Industrial forest companies, 238
Industrial metropolis, 505
Industrial societies, death rates and, 72–73, 72f
Industrial water use, 384
Infant mortality rate, 63t
Infectious agents, 189–190
Inferences, 26
Inflection point, 66
Influent streams, 372
Infrared wavelengths, solar energy and, 436, 437f
Inhalable particulates, 488t
Inputs, systems and, 44
Inspirational justification, 15
Instrumental records, climate study and, 438
Intangible factors, 129
Integrated energy management, 295
Integrated pest management, 226, 226f
Integrated waste management, 530–532
Intellectual standards, 13–14
Interference, ecological succession and, 99–100, 100f
Intermediates, tree, 245
International agreements on global warming, 456–457
International environmental law and diplomacy, 563
International food aid, 216
International Union for the Conservation of Nature (IUCN), 263, 271
Intuit People, Greenland, 193
Invasions, biological, 164–166, 165f
Invasive species, 164–166, 165f
Iowa, prairie in, 177
Iran, methyl mercury poisoning in, 192–193
Iraq-Iran War, 176
Irreversible consequences, 53–54
Irrigated area, world, 232t
Irrigation, 228
Island biogeography, 164–165, 164f
Island fox, 178–179, 178f
Isle Royale National Park, 95
Isotopes, 348

J

Jamaica Bay National Wildlife Refuge, New York City, 236–237, 236f, 237f
Japan, Minamata Bay, 192–194
Jojoba, 227
Jordan, virtual water use and, 386
Joshua trees, 162–163, 163f
Joules, 291

K

Kansas
  Dodge City, 264
  Kansas City, 462
  Konza Prairie, 176
  Platte River, 375
  Ulysses, 375
Kansas City, Kansas and Missouri, 462
Kazakhstan, nuclear weapons testing facility and, 354
Keeling, Charles, 430
Kelp, 81–82, 81f–82f
Kenya
  age structure in, 69f
  Ambroseli National Reserve, 2–4, 2f–4f
  national parks in, 252
Keystone species, 81, 176
Kilowatt-hours, 291
Kimberley, South Africa, 524f
Kinetic energy, 289
King Salmon fishing, 269–270, 270f
Kingston Fossil Plant, Tennessee, 321
Kissimmee River, Florida, 174–175, 174f
Klamath Falls, Oregon, 346
Knowledge, science and, 24–25, 25t
Konza Prairie, Kansas, 176
Kwashiorkor, 216, 216f
Kyoto Protocol, 457

L

Lag time, 50
Lake ecosystems, acid rain and, 470–471
Lake Manyara National Park, Tanzania, 249
Lake Michigan, 96, 96f
Lake Moraine, Banff National Park, Alberta, Canada, 89f
Lake Nyos, Cameroon, 187–188, 188f
Lake St. Charles, Sudbury, Ontario, 189–190, 190f
Lakeland, Florida, 512
Lakes, eutrophication of, 405–408, 406f
Land ecosystems, carbon cycle and, 117
Land subsidence, oil and natural gas recovery and, 311
Land use, world, 215, 215f, 215t
Land-use changes, carbon cycle and, 117
Landscape aesthetics, 134–135, 134f
Laos, new species found in, 147
Las Vegas, Nevada, 420, 420f, 472–473
Last glacial maximum, 447
Late-successional species, 99
Latitude, precipitation and, 161–162, 162f
Law of conservation of energy, 91
Laws of thermodynamics, 90–93
LD-50 values, 203, 203t
Leachate, 533–534
Lead, air pollution and, 468
Lead toxicity, 191, 207, 207f
Leaps of imagination, 31
Legal emissions limits, carbon dioxide and, 128
Legionnaires' disease, 190
Leibstadt, Switzerland, 349f
Leopard frogs, 197
Leopold, Luna, 384–385
Leucaena, 227
Leukemia, 200
Lewis and Clark expedition, 14, 258
Libby, Montana, 199
License restrictions, overfishing and, 139
Liebig, Justus von, 223
Liebig's law of the minimum, 223
Life, cost of extending, 137–138, 138f
Life expectancy, 63t, 71–73, 71f
Life form numbers, 148t
Light water reactors, 349
Limestone quarry restoration, 180, 180f
Limiting factors, 111, 223–224
Linear exponential, 51, 51f
Linear flows, 50
Linear process, 50
Linnaeus, Carl, 147
Lions, 3–4
Lithification, 116
Lithosphere, 113
Lithospheric plates, 113–114, 114f
Little ice age, 429, 432, 440, 446
Livestock, 218–221
Livingston Bog, Michigan, 97f
Local extinction, 274
Local increase, ecosystems and, 93
Logistic carrying capacity, 66, 261
Logistic control curve, 262f
Logistic curve, 51, 51f
Logistic growth curve, human population as, 65–67
Logistic population, 261–262, 262f
Long Island, New York, 413–414, 413f
Longevity, population growth and, 71–73, 71f
Los Angeles, California, 12f, 473–475, 473f, 474f
Louisiana
  Chandeleur Beach, 553f
  coastal wetlands, 390, 419, 419f
  New Orleans, 503–504, 504f
Love Canal, New York, 537–539

Lovelock, James, 10, 108
Lovins, Amory, 297
Low-level radioactive waste, 360
Lubbock, Texas, 375
Lucretius, 55
Lyme disease, 190

M

Ma'adan people, 176, 177f
Maasai people, 2–4
MacArthur, R., 164
Macrocentrus ancylivorus, 226
Macronutrients, 111, 223
Made lands, 511
Maize, 228
Malaria, 151–152, 151f, 190
Malaysia, population policy and, 75
Male eared grebe, 28–29, 28f
Malnourishment, 216, 217f
Malnutrition, 6
Malthus, Thomas, 65, 68, 232
Manipulated variables, 26–27
Manufacturing water use, 384
Margulis, Lynn, 10
Mariculture, 230
Marine mammal conservation, 277–279, 278f
Marine Mammal Protection Act, 263, 270, 279
Marine pollution, 543–545
Marmot Dam, Oregon, 383f, 392
Mars, atmosphere of, 106–107, 107f
Marsh, George Perkins, 214
Marsh Arabs, 176, 177f
Marshes, 388–389
Masai Mara Game Reserve, East Africa, 280
Massachusetts, Boston, 417, 506–507, 507f
Materially closed system, 44
Materials management
    goals of, 522
    hazardous waste and, 537–543
    industry location and, 522–523
    integrated waste management, 530–532
    mineral resources, 523–528
    municipal solid-waste management, 532–537
    ocean dumping and, 543–545
    pollution prevention and, 545
    sustainable resource management and, 546
    waste disposal history and, 529–530
Matilija Dam, California, 392
Matusadona National Park, Zimbabwe, 250
Maui, Hawaii, 463f
Mauna Loa Mountain, Hawaii, 430, 431f
Maximum lifetime, 51
Maximum sustainable yields, 261
McHarg, Ian, 510
Measurements and uncertainty, 32–34
Mediation, 562–563
Medical advances, 71
Medical research, endangered species and, 272–273
Medieval warm period, 429–430, 429f, 432, 440, 446
Mefloquine, 152
Megacities, 11f, 12, 12f, 499
Meltdown, nuclear, 348
Mercury, 472
Mercury toxicity, 191, 192–194, 192f
Metal particulates, 488t
Metamorphic rocks, 116
Methane, 309–310, 416, 438, 444
Methane hydrates, 310–311
Methane seeps, 105–106, 105f–106f
Methyl isocyanate, 472
Methylation, 192
Mexico
    air pollution and, 476–477
    Mexico City, 476–477
    Tiburón Island, 153–154
Mexico City, Mexico, 476–477
Michaelis-Menton curve, 51, 52f
Michigan, Livingston Bog, 97f
Microhydropower systems, 334–335
Micronutrients, 111, 223
Micropower, 299, 299f
Middle-successional species, 99
Migration, biological evolution and, 152–153
Milankovitch, Milutin, 445
Milankovitch cycles, 445–446, 446f
Minamata Bay, Japan, 192–194
Mineral King Valley, California, 560–561, 560f
Mineral products, 521, 522f
Mineral resources, 523–528
    availability and use of, 524–525
    classification of, 525f
    environmental impacts and, 526–527
    mineral deposit formation, 523–524
    minimizing environmental impact and, 527–538
    reserves and, 524
    social impacts and, 427
    U.S. supply of, 525
Minimum viable population, 261
Mining
    area restoration and, 180, 180f
    contaminated water and, 527
    controlling air emissions and, 527
    minimizing environmental effects of, 527–528, 528f
    reclaiming areas and, 527
    stabilizing soils and, 527
    treating wastes and, 527
Minnesota
    Boundary Waters Canoe Area, 30–31
    Voyageurs National Park, 131–132, 132f, 251
Mississippi River delta, 389
Missouri
    Kansas City, 462
    St. Louis, 42–43, 42f
    Times Beach, 195, 195f
Missouri River, 42–43, 42f–43f, 46f, 220, 220f, 375
Mobile sources, pollution, 188, 463
Models, 29
Modern environmentalism, 5
Molina, Mario, 483
Moncure, North Carolina, 212f
Mono Lake, California, 23–24, 23f–24f
Monoculture, 214
Montana
    Glacier National Park, 452
    Libby, 199
Montreal Protocol, 486, 563
Monumentalism, 250
Mopane, 159
Moral justification, 16
Morbidity, 63t
Morocco, phosphorous reserves and, 123
Moscow, Russia, 472
Mosquito-frog system, 92, 92f
Mosquitos, 151–152, 151f
Mount Baker, Washington, 452
Mount Kilimanjaro, Tanzania, 2–3, 452
Mount Pinatubo, Philippines, 199
Mount St. Helens, Washington, 31
Mount Washington, New Hampshire, 89f
Mountain lions, 270
Mountaintop removal, coal, 317–319, 319f
Muds, drilling, 311
Muir Glacier, 453f
Mumford, Lewis, 506
Municipal solid-waste management, 532–537
    composting and, 532–533
    incineration and, 533
    onsite disposal and, 532
    open dumps and, 533, 533f
    sanitary landfills and, 533–536
    solid waste composition, 532, 532f
Muntjak, 147
Mussel farms, 230
Mutation, biological evolution and, 149–150

N

Nanotechnology, 411
Nantucket Sound, 341–342, 342f
Nash, Roderick, 16
National Ambient Air Quality Standards (NAAQS), 478–479, 479t
National drinking-water standards, 401, 401t
National energy policy, 287
National Environmental Policy Act, 389
National Environmental Protection Act (NEPA), 558
National landscape planning, 561–562
National Marine Fisheries Service, 139
National Park Service, 557–558
National parks, 250. *See also* Parks
Native Americans, 14, 14f
Natural capital, 130
Natural catastrophe, extinction and, 276
Natural gas, 309
Natural gas combined cycle power plant, 294
Natural gas recovery, 311, 311f
Natural plant chemicals, 225
Natural selection, 150–152
Naturalistic landscapes, 253
Naturalization, 180–181
Nature preserves, 248–252, 249t
Nebraska
    Allwine Prairie, 176–177
    Omaha, 514
Negative exponential, 51, 51f
Negative feedback, 48–50, 49f
Negative feedback loops, climate change and, 444
Nepal, deforestation and, 247, 247f
Net production, 93–94
Netherlands, waste reduction in, 530

Neutrino, 351
Nevada
    Acciona solar thermal power plant, 332f
    Las Vegas, 420, 420f, 472–473
    Yucca Mountain, 362–363
New crops, 227–228
New Hampshire
    birch forests in, 89–90, 90f
    Hubbard Brook experimental forest, 245
    Mount Washington, 89f
New Mexico, Carlsbad, 361
New Orleans, Louisiana, 503–504, 504f
New York
    Adirondack Park, 281–282
    Catskill Mountains, 402, 402f
    Hudson River, 399–400, 511, 512f
    Indian Point Energy Center, 345f, 346
    Long Island, 413–414, 413f
    Love Canal, 537–539
New York City
    apartment views and, 135, 135f
    Central Park, 506
    clean water and, 402
    High Line Park and, 498, 498f
    Jamaica Bay National Wildlife Refuge, 236–237, 236f, 237f
    water importation and, 378–379
    water use and, 383
Niches, ecological, 155–157, 157f
Nickel ore deposits, 189–190
Nitrogen
    fertilization and, 124, 124f, 223
    fixation, 99, 120
    tallgrass prairie ecosystem and, 178
    water-pollution problems and, 405–408
Nitrogen-converting bacteria, 121
Nitrogen cycle, 120–121, 120f, 124
Nitrogen dioxide, 473, 488t
Nitrogen oxides, 465–466, 477
Nitrous oxides, 444
Noise pollution, 200–201, 200t
Nonlinear flows, 50
Nonpoint sources, pollution, 188, 410
Nonrenewable alternative energy, 327
Nonrenewable resources, 521
North Carolina
    Moncure, 212f
    pig farms and, 421–422, 422t
North Dakota, prairie potholes and, 388f
North Pacific Ocean circulation, 544f
North Platte River, Wyoming, 196–197, 197f
Noss, Reed, 561
Nuclear energy
    arguments against, 363
    breeder reactors and, 363–364
    conventional nuclear reactors, 348–349
    current role of, 346–347
    fuel cycle and, 353–354, 353f
    fusion reactors and, 364
    future of, 363–364
    leading nations using, 347t
    new kinds of fission reactors, 364
    proponents of, 363
    radiation doses and health, 358
    radiation units and doses, 356–357
    radioactive decay and, 350–352
    radioactive-waste management, 360–363
    radioisotopes and, 354–357
    world use of, 347f
    worldwide location of, 347f
Nuclear fission, 348
Nuclear fusion, 348
Nuclear meltdown, 348
Nuclear power plants
    accidents and, 358–360
    construction of, 353
    decommissioning of, 354
    types of, 349f
Nuclear power reactors
    components of, 350f
    lifetimes of, 354
    spent fuel and, 353
    waste disposal and, 353
Nuclear radiation, 198
Nuclear Regulatory Commission (NRC), 346
Nuclear steam-supply system, 348–349
Nuclear Waste Policy Act, 362
Nutrients, water pollution problems and, 405–408

O

Objectivity, 35
Obligate symbionts, 158
Observations, 26
Oceans
    carbon cycle and, 117
    climate change and, 448–449, 449f
    conveyor belt, 449, 449f
    dumping in, 543–545, 543f
    fishing and, 139
    mineral resources and, 523
    ocean energy, 335–336, 335f
    oceanic food chain, 85, 87f
    plastics discarded in, 544, 544f
Off-stream use, water, 376–377
Off-the-grid photovoltaics, 331
Ogallala Aquifer, 375
Ohio, Cuyahoga River, 410, 411f
Oil
    converting coal to, 320
    embargoes and, 289
    imports and exports of, 308t
    proven reserves, 306–307, 307f
    recovery of, 311, 311f
    refineries and, 312
    seepage of, 105–106, 105f–106f, 311
    twenty-first century and, 308–309
    water pollution and, 408
Oil and gas traps, 306, 306f
Oil Pollution Act, 408
Oil shale, 321–322
Oilseed crops, 212, 212f
Oklahoma, Tar Creek, 316f, 409
Old-field succession, 98, 98f
Old-growth forest, 244
Oligotrophic lakes, 406
Olmsted, Frederick Law, 15, 506–507
Omaha, Nebraska, 514
Omnivores, 86
Onsite disposal, solid-waste and, 532
Ontario, Sudbury, 189–190, 190f
Open dumps, solid-waste and, 533, 533f
Open-pit mines, 123f, 526
Open systems, 44–45
Operational definitions, 27
Optimum sustainable population, 261, 262, 279
Ore deposits, 523
Oregon
    H.J. Andrews experimental forest, 245
    Klamath Falls, 346
    Marmot Dam, 392, 393f
Organelles, 110
Organic compounds, 193
Organic farming, 228
Organization for Economic Cooperation and Development (OECD), 300
Oriental fruit moth, 226
Ostrich, 163f
Outbreaks, waterborne disease, 404
Overabundance, species, 270
Overdraft, 375
Overfishing, 139
Overshoot, carrying capacity and, 53, 53f
Oxygen, 108–110
Oyster farms, 230, 230f
Ozone
    air pollution and, 466–467
    chlorofluorocarbons and, 483
    depletion of, 484–485
    future and, 486
    indoor air pollution and, 488t
    measuring stratospheric ozone, 482–483
    ozone hole, 483
    ozone shield, 481
    ultraviolet radiation and, 481–482, 481f, 482f

P

Pacific decadal oscillation, 449
Pacific Estuarine Research Laboratory (PERL), 181
Packing-problem approach, human population, 74
Paclitaxel, 272
Pakistan, population policy and, 75
Palm Beach County, Florida, 369, 369f
Palm plantations, 339
Pandemics, 60–61, 61f
Parasitism, 158
Paris, France, 512f
Parks, 248–252, 249t
    city, 506–507
    conflicts relating to, 251–252
    history of, 250
    landscape percentage and, 252
    nature preserves compared, 249
    people and wildlife interactions and, 251
    size, access, and activity types, 251
Particulate matter, air pollution and, 199, 199f, 467–468, 468f, 477
Parus major, 454, 454f
Passive smokers, 490
Passive solar energy, 328–329, 329f, 509
Passive stability, nuclear cooling systems and, 363
Past observations, forecasting climate change and, 450
Pastures, 218
Peak oil, 304–305, 304f
Pebble Reactor, 364

Pelagic ecosystem, 85, 87f
Penicillin discovery, 31
Pennsylvania
- Centralia, 319
- Three Mile Island nuclear power plant, 358–359

Peregrine falcons, 513
Perennial streams, 372
Periodic table of elements, 111–112, 111f
Permafrost, 445
Perraudin, Jean-Paul, 445
Persistent organic pollutants, 194, 194t
Peru
- Guano Island, 121, 122f
- tropical rain forests, 248, 248f

Pest control, 224–227
Pesticides, 224–227, 488t
- artificial organic compounds, 225
- broad-spectrum inorganic toxins, 224
- integrated pest management, 226, 226f
- natural plant chemicals, 225
- petroleum-based sprays, 225
- world use of, 225, 225f

Petroleum-based sprays, 225
Petroleum production, 306–307
Pheromones, 227
Philippines
- experimental rice plots and, 228
- Mount Pinatubo, 199
- population policy and, 75

Phoenix, Arizona, 419, 419f
Phosphorous, 223, 224f, 405–408
Phosphorous cycle, 121–124, 122f
Photochemical oxidants, air pollution and, 466–467
Photochemical smog, 473, 473f
Photodissociation, 482
Photosynthesis
- chemistry of, 91
- dependent and independent variables and, 27, 27f
- ecological communities and, 84
- trees and, 242f

Photovoltaics, 330–331, 330f, 331f
Physiological tolerance, 205
Picocuries per liter, 356
Pig farms, 421–422, 422t
Pinchot, Gifford, 558
Pine bark beetle, 281, 281f
Place value, desalinated water and, 375–376
Plankton, 121
Plant pest interceptions, 165–166, 165t
Plantations, forest, 245–246, 246f
Plasma, 364
Plasmodium parasites, 151–152, 151f
Plastics, in ocean, 544, 544f
Plate tectonics, 113–114, 114f, 115f
Platte River, 225, 375
Plowing, 214, 214f
Point sources, pollution, 188, 188f, 410, 410f, 463, 463f
Polar amplification, 451
Polar Bears, 166
Polar night, 484
Polar stratospheric clouds, 484
Polar vortex, 484, 485f
Policy instruments, 129
Pollen grains, 489f
Pollinators, 130, 130f
Pollution
- acute and chronic effects of, 205
- asbestos and, 199
- defined, 188
- dose-response and, 201–204, 203f, 206
- ecological gradients and, 205
- ED-50 values and, 204, 204f
- electromagnetic fields and, 199–200
- exposure assessment and, 206
- general effects of, 201–205
- hazard identification and, 205–206
- hormonally active agents and, 196–197, 197f
- infectious agents and, 189–190
- LD-50 values and, 203, 203t
- measuring amount of, 189
- noise pollution, 200–201, 200t
- nuclear radiation, 198
- organic compounds and, 193
- particulates and, 199, 199f
- persistent organic pollutants, 194, 194t
- risk assessment and, 205–206
- risk characterization and, 206
- sanitary landfills and, 534–535, 536f
- TD-50 values and, 204, 204f
- thermal pollution, 198–199, 198f
- threshold effects and, 204–205, 204f
- tolerance and, 205
- toxic heavy metals and, 191
- toxic pathways and, 191–193, 191f, 192f
- urban, 510
- wildlife effects and, 201, 201t
- zone classification and, 404

Polychlorinated biphenyls (PCBs), 194, 194t, 399–400
Population. *See* Human population
Positive exponential, 51, 51f
Positive feedback, 48–50, 49f, 445
Potassium-40, 355
Potential energy, 289
Powder River Basin, Wyoming, 309
Powell, John Wesley, 561, 561f
Power, energy units and, 291
Prairie potholes, 388f
Prairie restoration, 176–178
Precautionary Principle, 14–15, 16f
Precipitation, 161–162, 162f, 375
Precision, measurement, 32–34
Predation, 158
Prevalence, disease, 63t
Priestley, Joseph, 450
Primary production, autotrophs, 93
Primary production, petroleum, 306
Primary succession, ecosystem, 95, 96f
Primary treatment, wastewater, 415–416
Production, equations for, 94
Production per acre, 227
Prokaryotes, 110, 110f
Protein supply, malnourishment and, 216
Protocol of Madrid, 563
Proven oil reserves, 306–307, 307f
Proxy data, climate, 438–440
Pseudoscience, 35–37
Public land recreation, 559–600
Public-service functions, 130, 130f, 238
Public-service species, 145
Pump storage systems, 334f
Python, Burmese, 171, 171f

Q

Quads, 291
Qualitative data, 27
Quanea, Greenland, 193
Quantitative data, 27
Quarries, 526
Queensland, Australia, 246f

R

R-to-*C* ratio, 546
Radiation
- dosage and health, 358
- electromagnetic, 436, 436f
- exposure to, 354–357, 356f, 357f
- units and doses of, 356–357

Radioactive decay, 350–352, 352f, 352t
Radioactive-waste management
- high-level radioactive waste, 361–362
- low-level radioactive waste, 360
- transuranic waste, 361, 361f
- United States and, 362–363

Radioisotopes, 350, 354–357
Radionuclides, 446
Radon gas, 488t
- danger and, 490–491
- entering homes and buildings, 491
- geology and, 490
- making homes and buildings resistant to, 493

Rainwater, acidity of, 221
Rance Tidal Power Station, France, 335–336, 335f
Rangeland grazing, 218–221, 220f
Rapa Nui. *See* Easter Island
Ratcliffe-on-Saw, England, 349f
Rate of natural increase, 63t
Ration coupons, 128
Realized niches, 157, 157f
Recharge zones, 372
Reclaimed water, 369
Reclamation, 180, 319
Recovered species, 276, 277f
Recovery, oil and natural gas, 311, 311f
Recovery zone, 404
Recycling, 530–531, 531f
- financial viability and, 546–547
- hazardous waste and, 542
- human waste and, 531–532
- public support for, 531
- single-stream recycling, 531

Red Bluff Dam, California, 269, 270f
Red-cockaded woodpecker, 281, 281f
Red List, 271
Refineries, crude oil, 312
Refuse Act, 422
Reindeer, 157, 158f
Removal, dams, 392–393, 393f
Renewable energy, 293–294, 327–328
Renewable resources importance, 521
Reservoir rock, 306
Reservoirs, 335
Residence time, 46–48, 112
Residual land sink, 117

Resource Conservation and Recovery Act, 399, 535, 539–540
Resource recovery, hazardous waste and, 542
Resource sustainability, 8
Respiration, 93
Responding variables, 27
Rest mass, 351
Restoration, ecological. *See* Ecological restoration
Restoration, soils, 222–224
Reuse, waste management and, 530–531
Rewilding, 561
Rice, 218f, 229
Rice superstrains, 228, 228f
Richardson, Curtis, 176
Rio de Janeiro, Brazil, 533f
Rio Earth Summit on Sustainable Development, 14, 564
Riparian vegetation, 421
Risk assessment, 205–206
Risk-benefit analysis, 136–138, 137t
River floodplain environment, 174–175, 174f
River flow changes, global warming and, 451
River restorations, 174–176
River Severn, England, 404f, 539, 539f
River water, heating, 198
Rivers, cities and, 511
Rock cycle, 116, 116f
Rockport, Washington, 409f
Rome, Italy, 71, 71f
Roof runoff, 411, 411f
Roosevelt, Franklin D., 558
Rotation time, forests and, 244
Rowland, F. Sherwood, 483
Rule of climatic similarity, 161–162, 162f
Runoff patterns, 375
Russia, Moscow, 472

S

Sacramento River, 269, 270f
Saguaro cactus, 162–163, 163f
Sahara Desert, Algeria, 311f
St. Louis, Missouri, 42–43, 42f
Salmonella, 190
San Andreas Fault, California, 114
San Francisco, California, 15, 15f, 389, 389f
Sandy River Dam, Washington, 392
Sanitary landfills
 federal legislation for, 535–536
 leachate and, 533–534
 monitoring pollution and, 534
 multiple-barrier approach and, 535, 536f
 pollutants entering environment from, 534–535, 536f
 reducing waste and, 536–537, 537t
 site selection and, 534
Santa Barbara, California, 105–106, 105f–106f, 172, 178–179, 178f, 384, 385f, 411f
Santa Cruz Island, California, 179, 179f
Saturation (Michaelis-Menton) curve, 51, 52f
Scallops harvest, 267
Scarcity, economic value and, 133
Scenic Hudson, 400
Schrödinger, Erwin, 91
Schwarzenegger, Arnold, 270
Science
 decision making and, 34
 objectivity and, 35
 scientific proof, 27–29
 scientific theory, 29
 technology and, 34
 values and, 13–16
Scientific method, 24, 26f, 27t, 37
Scrubbing, 320, 478, 478f
Sea ice, global warming and, 452–455
Sea level, global warming and, 452, 452f
Sea lions, 270
Sea otters, 81–82, 81f–82f, 276
Sea urchins, 81–82, 81f–82f
Seafloor spreading, 114
Second-growth forest, 244
Second-law efficiency, energy, 290–293, 291t
Second law of thermodynamics, 92–93, 290
Secondary production, 93
Secondary succession, 95–96
Secondary treatment, wastewater, 415–417
Secondhand smoke, 490
Secure landfill, hazardous waste and, 540–542
Sediment pollution, 408
Sediment samples, climate study and, 439–440, 439f
Sedimentary processes, 524
Sedimentary rocks, 116
Seed-tree cutting, 245
Seeger, Pete, 400
Selective cutting, trees, 245
Sengwa National Park, Zimbawe, 250
September 11 attacks, 468, 563
Septic-tank disposal systems, 415, 415f
Serra Geral do Tocantins Ecologocial Station, 147
Settling chambers, 477
Severe acute respiratory syndrome (SARS), 60
Sewage drainage fields, 415, 415f
Sewage treatment process diagram, 416f
Sex pheromones, 227
Sex reversal, frogs and, 196–197, 197f
Shallow earth, low-density geothermal energy, 340
Shark Bay, Australia, 109f
Shelterwood cutting, 245
Sick building syndrome, 491
Sierra Club, 560, 562
Silent Spring (Carson), 5
Silviculture, 238
Simple ecosystem, 84–85
Simple systems, 45, 45f
Simplified stratospheric chlorine chemistry, 483–484
Singapore, population policy and, 75
Single-species wildlife management, 260–262
Single-stream recycling, 531
Sink, biogeochemical cycle and, 112
Site quality, forests, 245
Small-grain production, 218, 219f
Smelting, 134, 189–190, 190f
Smithsonian Institution, 263
Smog, 473–476
Soil
 in cities, 510
 coarse-grained soils, 221–222
 cross section of, 12f
 decomposition and, 221
 erosion and, 73
 fertility and, 221
 fertilizers and, 222–223
 horizons and, 221, 222f
 limiting factors and, 223–224
 nitrogen and soil development, 99, 99f
 phosphorous and soil development, 99, 99f
 rainfall and, 2–4
 restoration of, 222–224
 timber harvest and, 54, 54f
 toxicity and air pollution, 464
Solar collectors, 329–330, 330f
Solar cycles, climate change and, 446
Solar energy
 active solar energy, 329–331
 cities and, 509
 environment and, 332
 passive solar energy systems, 328–329, 329f
 photovoltaics, 330–331, 330f, 331f
 solar collectors, 329–330, 330f
 solar thermal generators, 331–332, 331f, 332f
Solar One, 332
Solar radiation, vegetation and, 242f
Solar system, atmospheres and, 106–110, 107f
Solid-waste management. *See* Municipal solid-waste management
Somalia, political disruption and starvation, 7f
Sounds levels, 200t
Source reduction, hazardous waste and, 542
Source rock, 306
South Africa
 Kimberley, 524f
 phosphorous reserves and, 123
South America, contraceptive devices and, 75
South Asia, population growth rates and, 75
South Dakota, Custer State Park, 178f
Soviet Union, Chernobyl nuclear power plant, 357, 359–360, 360f
Sowbread, 272f
Soybeans, 218f
Spain
 Coto Doñana National Park, 251, 251f
 Granada, 508, 508f
 Tarifa, 326f
Spatial relationships, endangered species and, 281
Species
 biological diversity and, 145–146
 evenness and, 146–147, 147f
 numbers of on earth, 147–149, 148f
 overabundance and, 270
 population dynamics and, 62
Species richness, 146–147, 147f
Spent fuel, 353
Sperm whales, 278f
Spindle-horned oryx, 147
Spring fires, tallgrass prairies and, 177–178
Squirrel, American gray, 154–155, 155f
Sri Lanka
 malaria and, 152
 population growth rates and, 75
 population policy and, 75
Stack effect, 490
Starvation, 216, 217f

Static systems, 44, 44f
Stationary sources, air pollution and, 463
Steady state systems, 45–46
Stevens, General Isaac I., 259
Stone, Christopher D, 560
Stone decay, acid rain and, 471
Storm King Mountain case, 563, 563f
Stratosphere, 434–435
Stratospheric ozone layer, 435
Streams
    classification of, 372, 372f
    ecosystem conditions and, 421
    restorations of, 174–176
Strip-cutting, trees, 245
Strip mining, 316–317, 316f, 317f
Striped hare, 147
Stromatolites, 109, 109f
Sub-Saharan Africa, population growth rates and, 75
Subsurface mining, 526
Succession. *See* Ecological succession
Sudbury, Ontario, 189–190, 190f
Sugarcane fields, burning, 463f
Sulfur dioxide, 465, 477–478, 488t
Sulfurous smog, 473
Superfund Amendment and Reauthorization Act (SARA), 540
Superhybrids, 229
Superweeds, 229
Surface -water pollution
    sources and processes and, 401t
Surface mines, 526
Surface Mining Control and Reclamation Act, 319
Surface water
    flow system and, 372, 372f
    groundwater interactions and, 372–373, 373f
Surface-water pollution
    nanotechnology and, 411
    nonpoint sources and, 410
    point sources and, 410, 410f
    reduction of, 410–411
    urban-water naturalization and, 411, 411f
Surge flow, 231
Sustainable energy development, 298
Sustainable-energy policy, 297–299
Sustainable forestry, 246–247
Sustainable global economy, 9
Sustainable growth, 8–9
Sustainable resource management, 546
Sustainable water use, 384
Sustainable yields, 260–262
Sweetwater Marsh National Wildlife Refuge, California, 181
Swine flu, 60–61, 61f
Switzerland, Leibstadt, 349f
Symbionts, 157–158
Symbiosis, 157–158
Synergism, 188
Synergistic effect, 224, 464
Synfuels, 321
Syngas, 320
Synthetic organic compounds, 193
Syr Darya River, 232
System responses
    lag time, 50
    linear and nonlinear flows, 50
    selected examples of, 50–52
Systematic errors, 32
Systems concepts
    average residence time, 46–48
    balance of nature, 46, 46f
    feedback, 48–50
    open systems, 44–45
    residence time, 46–48
    simple systems, 45, 45f
    static and dynamic systems, 44, 44f
    systems defined, 44

T

Tallgrass prairie, 176–178
Tampa, Florida, 383
Tangible factors, 129
Tanzania
    Lake Manyara National Park, 249
    population policy and, 75
Tar Creek, Oklahoma, 316f, 409
Tar sands, 322, 322f
Tarifa, Spain, 326f
Tax on emitters, carbon dioxide emissions and, 128
Taxa, 147
TD-50 values, 204, 204f
Tectonic cycle, 113–114, 114f
Temperate forest food web, 86f
Temperature
    atmospheric, 434f, 435–436
    during past million years, 432, 433f
Tenerife, Canary Islands, 331f
Tennessee, Kingston Fossil Plant, 321
Tennessee Valley Authority (TVA), 558–559, 559f
Terminator gene, 229
Termolite, 199
Terrestrial food web, 86
Terrorism, 563–564
Texas
    Arkansas National Wildlife Refuge, 263
    cypress swamp water in, 388f
    Dallas, 462
    Horse Hollow Wind Energy Center, 337
    Houston, 186–187
    Lubbock, 375
    Woodlands, 510, 511
Thailand, Chao Phraya River, 232
Thallium toxicity, 191
Thames River, Great Britain, 410
Therapeutic index, 204
Thermal efficiency, 293
Thermal pollution, 198–199, 198f
Thermodynamic equilibrium, 108
Thinning, trees, 245
Thoreau, Henry David, 158, 252–253
Threatened species, 271
Three Gorges Dam, China, 390–391, 391f
Three Mile Island nuclear power plant, Pennsylvania, 358–359
Threshold effects, 204–205, 204f
Threshold world population, 5
Thresholds, global, 565–566, 566t
Tiburón Island, Mexico, 153–154
Tidal power, 335–336, 335f
Tigris and Euphrates river system, 176, 177f
Timber harvest, soil erosion and, 54, 54f
Timber investment management organizations, 238
Time, economic value and, 135–136, 135f
Time series, 263–264
Times Beach, Missouri, 195, 195f
Tobacco smoke, 490
Tolerance, 205
Total catch quotas, 139
Total fertility rate, 63t
Total suspended particulates, 467
Toxic heavy metals, 191
Toxic pathways, 191–193, 191f, 192f
Toxic Substances Control Act, 138
Toxicology, 188
Toxins, 188
Trade winds, 436
Tradescantia, 150, 150f
Tragedy of the commons, 130
Trans-Alaska Pipeline, 312, 408
Transfer efficiency, 94–95
Transform fault boundary, 114
Transport, water, 378–379
Transportation, coal, 319–320
Transuranic waste, 361, 361f
Trapper Mine, Colorado, 318, 318f
Traps, oil and gas, 306, 306f
Treatment, hazardous waste and, 542
Trees. *See also* Forests
    clear cutting and, 245, 246f
    codominants, 245
    dominants, 245
    growth of, 243–244, 244f
    harvesting of, 245, 245t
    home cooling and, 512, 512f
    intermediates, 245
    niches, 243, 243f
    seed-tree cutting, 245
    selective cutting of, 245
    shelterwood cutting, 245
    strip-cutting, 245
    suppressed, 245
    thinning of, 245
    tree rings, 31, 31f, 439, 439f
    in urban environments, 512–513
Trophic level, food web, 84
Trophic-level efficiency, 95
Tropical rain forests, 248, 248f, 254
Tropopause, 434
Troposphere, 434–435
Tucson, Arizona, 513, 513f
Tuna catch, 267, 267f
Tuna nets, 269
Tuvalu, 452f

U

Ultrafine particles, air pollution and, 467
Ultraviolet (UV) Index, 485, 486t
Ultraviolet radiation, ozone and, 481–482, 481f, 482f
Ulysses, Kansas, 375
Uncertainty in science, 31–32
Underground mines, 319, 526
Undernourishment, 212–213, 212f, 216, 217f
Underweight children, 7f
Uniformitarianism, 54–55, 450

United States
  coal reserves and, 314–315, 315f, 316t
  death rates and, 72, 72f
  demographic transition and, 76
  desalination and, 376
  droughts and, 375
  energy consumption and, 294, 294f, 295f
  estimated solar energy and, 328, 328f
  future energy consumption needs and, 297
  groundwater use and problems, 375–376
  hydroelectric power and, 334
  mineral resources and, 525
  national parks in, 252
  nuclear energy location and, 347, 348f
  nuclear fuel cycle and, 353–354, 353f
  phosphorous reserves and, 123
  population (1790-2008), 62f
  precipitation and runoff patterns, 375
  radioactive-waste management and, 362–363
  stream ecosystems in, 421, 421f
  twenty-first century energy policy and, 298
  water budget for, 374–375, 374f
  water supply and, 373–376
  water usage trends and, 379–380, 379f
  wind energy and, 336, 336f
Unity, landscape aesthetics and, 135
Uranium-235, 348, 349f, 353
Uranium-238 decay chain, 352f
Uranium mines and mills, 353
Urban environments. *See also* Cities
  bringing nature to, 511–515
  city as a system, 499–500, 500f
  city as an environment, 508–510
  city life, 499
  city planning and the environment, 506–507
  environmental history of, 505
  location of, 500–504, 502f
  made lands, 511
  megacities, 499
  pollution and, 510
  site modification and, 504
  soils, lead toxicity and, 207, 207f
  urban heat island profile, 509, 509f
  urban sprawl, 516
  urban-water naturalization, 411, 411f
  urban world, 11–12, 11f
  waste, energy uses of, 339t
U.S. Council on Environmental Quality, 263
U.S. Forest Service's Roadless Area Review and Evaluation, 252
U.S. Marine Mammal Protection Act, 81
U.S. Office of Foreign Disaster Assistance, 187
U.S. Water Resources Council, 371
U.S. Wilderness Act, 252
Utah
  Bingham Canyon Copper Pit, 526f
  Wasatch Plateau, 309
Utilitarian justification, 15
Utilitarian species, 145

V

Vadose zones, 372
Vale of Pewsey, England, 30, 30f
Variable-water-source approach, 384, 385f
Variables
  controlling, 27
  dependent, 26–27
  independent, 26–27
  manipulated, 26–27
  responding, 27
Vegetation
  affects of, 241–243, 242f
  air pollution and, 463
  carbon stored in, 33, 33t
  cities and, 512–513
Venetz-Sitten, Ignaz, 445
Ventilation, indoor air pollution and, 490
Venus, 106–107, 107f
Verhulst, P.F., 65
Vermiculite, 199
Victoria Park, Great Britain, 250
Vikings, 429–430
Virginia, Alexandria, 509, 510f
Virtual water balances and transfers, 386, 387f
Vividness, landscape aesthetics and, 135
Volatile organics, 472, 488t
Volcanic eruptions, 199
Voyageurs National Park, Minnesota, 131–132, 132f, 251

W

Wallace, Alfred, 160, 275
Wallace's realm, 159–160, 161f
Ward Valley, California, 360
Wasatch Plateau, Utah, 309
Washington
  Columbia River, 264
  Elwha River Dam, 392
  Mt. Baker, 452
  Rockport, 409f
  Sandy River Dam, 392
Wasps, 226
Waste-disposal sites, groundwater pollution and, 412
Waste isolation pilot plant, 361, 361f
Waste stream, 530
Wastewater
  chlorine treatment and, 417
  land application of, 418–419
  primary treatment and, 416
  resource-recovery treatment plants and, 418f
  secondary treatment and, 416–417
  septic-tank disposal systems and, 415, 415f
  treatment of, 414–417
  treatment plants and, 415
  wetlands and, 418–419
Water
  agricultural use and, 380, 382f
  characteristics of, 370
  conservation of, 380–384
  consumptive use of, 376
  dams and, 390–393
  desalination and, 376
  dissolved oxygen content and, 404
  domestic use and, 383
  droughts and, 375
  environment and, 385
  food production requirements and, 231–232, 231t
  food supply and, 393–394
  global perspective on, 370–371, 370t
  groundwater and streams, 372–373, 372f, 373f
  groundwater sustainability, 384
  groundwater use and problems, 375–376
  hydrologic cycle and, 370, 371f
  in-stream use, 376–377, 376f
  industry and manufacturing use and, 384
  management of, 384
  master plan for, 384–385
  off-stream use, 376–377
  precipitation and runoff patterns, 375
  recycling of, 369
  reuse of, 420, 420f
  sustainable water use, 384
  toxicity, air pollution and, 464
  transport of, 378–379
  urban environment and, 509–510
  U.S. water supply, 373–376
  usage trends and, 379–380, 379f
  use of, 376–380
  virtual water balances and transfers, 386, 387f
  water budget, 373–375, 374f, 374t
  water footprints, 387, 394–395
  water table, 372
  wetlands and, 387–390
Water pollution and treatment
  acid mine drainage and, 409–410, 409f
  biochemical oxygen demand and, 403–404, 404f
  domestic use water safety, 414
  environmental law and, 421–422, 422t
  eutrophication and, 405–408, 406f
  externalities and, 133–134
  groundwater pollution, 412–414, 412f
  land application of wastewater, 418–419
  nutrients and, 405–408
  oil and natural gas recovery and, 311
  pig farms and, 423–424, 423f
  primary problems and, 400
  sediment and, 408
  sources and processes and, 400, 401t
  standards and, 401, 401t
  stream ecosystems and, 421, 421f
  surface-water pollution, 410–411, 410f, 411f
  wastewater treatment, 414–417
  water pollutant categories, 403t
  water reuse, 420, 420f
  waterborne disease and, 404
Water power
  environment and, 335
  hydroelectric power station components, 334f
  pump storage systems, 334f
  small-scale systems, 334–335
Water vapor, 435, 445
Waterborne disease, 404
Watersheds
  development and, 220–221, 220f, 221f
  ecosystems and, 88
  forested, 242f
  hydrologic cycle and, 116

Watts, 291, 292t
Weather, 431–433
Weathering, 524
West Nile virus, 60
West Virginia
  Appalachian Mountains, 317, 319f
  sulfur dioxide and, 478
Wet cooling towers, 198–199, 198f
Wetlands, 387–390
  constructed, 419, 419f
  natural service functions of, 388–389
  restoration of, 170–171, 170f, 176, 389–390
  types of, 387, 388f
  wastewater and, 418–419
Whales
  conservation of, 277–279, 278f
  number estimates of, 278t
  oil from, 132–133
Whaling, 132–133, 133f
Wheat, 218f
Whooping crane, 263, 264f
Wild leopard frogs, 196–197, 197f
Wilderness areas
  conflicts in managing, 253–254
  conservation of, 252–253
  goals of, 249t
  growing importance of, 252
  locations of, 252
  natural *vs.* naturalistic, 252–253
Wildfires, 172
Wildlands Projects, 252, 561–562, 562f
Wildlife
  age structure and, 264
  carrying capacity and, 260–262
  harvests and, 264–265
  historical ranges and, 263–264
  improved approaches to, 263–265
  logistic curves and, 262–263, 262f
  pollutants effects on, 201, 201t
  sustainable yields and, 260–262
  time series and, 263–264
  traditional single-species, 260–262
  urban environments and, 513–514
Wilson, E.O., 164
Wind power
  average wind speed, 336, 336f
  basics of, 336–337
  concentration of, 336, 337f
  environment and, 337
  future of, 338
  new uses of, 327, 327f
  wind farms, 337
  wind turbines, 341–342, 342f
Wind velocity, air pollution and, 475, 475f
Windmills, 336–337, 337f
Wisconsin, Coon Creek, 222
Wolves, 95, 175–176, 175f, 176f, 281–282
Wood as energy source, 241
Wood Buffalo National Park, Canada, 263
Wood burning for heating, 131
Wood stoves, 131
Woodlands, Texas, 510, 511
Woodpecker finch, 28f
Work, energy basics and, 289
World coal reserves, 315f
World energy consumption, 305f
World flu virus, 60
World food crisis, 212–213
World Meteorological Organization, 438
World population growth, pandemics and, 60–61, 61f
World Wildlife Fund, 263
World Wind Energy Association, 338
Worldwatch Institute, 76, 564
Wrangell-St. Elias Wilderness Area, Alaska, 253, 253f
Wyoming
  North Platte River, 196–197, 197f
  Powder River Basin, 309
  strip mining in, 317f

Y

Yellow dragon disease, 144–145, 144f, 145f
Yellowstone National Park, 50, 84–85, 84f, 85f, 175–176, 175f, 176f, 340
Yosemite National Park, California, 250, 250f
Younger Diyas, 432
Yucca Mountain, Nevada, 362–363

Z

Zero-emission coal-burning electric power plants, 320
Zero population growth
  age of first childbearing and, 74–75
  birth control and, 74–75
  national programs and, 75
Zero waste production, 530
Zimbawe, Matusadona National Park, 250
Zooplankton, 85

ALGEBRA

ARITHMETIC OPERATIONS

$$a(b + c) = ab + ac \qquad \frac{a}{b} + \frac{c}{d} = \frac{ad + bc}{bd}$$

$$\frac{a + c}{b} = \frac{a}{b} + \frac{c}{b} \qquad \frac{\frac{a}{b}}{\frac{c}{d}} = \frac{a}{b} \times \frac{d}{c} = \frac{ad}{bc}$$

EXPONENTS AND RADICALS

$$x^m x^n = x^{m+n} \qquad \frac{x^m}{x^n} = x^{m-n}$$

$$(x^m)^n = x^{mn} \qquad x^{-n} = \frac{1}{x^n}$$

$$(xy)^n = x^n y^n \qquad \left(\frac{x}{y}\right)^n = \frac{x^n}{y^n}$$

$$x^{1/n} = \sqrt[n]{x} \qquad x^{m/n} = \sqrt[n]{x^m} = (\sqrt[n]{x})^m$$

$$\sqrt[n]{xy} = \sqrt[n]{x}\sqrt[n]{y} \qquad \sqrt[n]{\frac{x}{y}} = \frac{\sqrt[n]{x}}{\sqrt[n]{y}}$$

FACTORING SPECIAL POLYNOMIALS

$$x^2 - y^2 = (x + y)(x - y)$$

$$x^3 + y^3 = (x + y)(x^2 - xy + y^2)$$

$$x^3 - y^3 = (x - y)(x^2 + xy + y^2)$$

BINOMIAL THEOREM

$$(x + y)^2 = x^2 + 2xy + y^2 \qquad (x - y)^2 = x^2 - 2xy + y^2$$

$$(x + y)^3 = x^3 + 3x^2y + 3xy^2 + y^3$$

$$(x - y)^3 = x^3 - 3x^2y + 3xy^2 - y^3$$

$$(x + y)^n = x^n + nx^{n-1}y + \frac{n(n-1)}{2}x^{n-2}y^2 + \cdots + \binom{n}{k}x^{n-k}y^k + \cdots + nxy^{n-1} + y^n$$

where $\binom{n}{k} = \dfrac{n(n-1)\cdots(n-k+1)}{1 \cdot 2 \cdot 3 \cdot \cdots \cdot k}$

QUADRATIC FORMULA

If $ax^2 + bx + c = 0$, then $x = \dfrac{-b \pm \sqrt{b^2 - 4ac}}{2a}$.

INEQUALITIES AND ABSOLUTE VALUE

If $a < b$ and $b < c$, then $a < c$.

If $a < b$, then $a + c < b + c$.

If $a < b$ and $c > 0$, then $ca < cb$.

If $a < b$ and $c < 0$, then $ca > cb$.

If $a > 0$, then

$|x| = a$ means $x = a$ or $x = -a$

$|x| < a$ means $-a < x < a$

$|x| > a$ means $x > a$ or $x < -a$

GEOMETRY

GEOMETRIC FORMULAS

Formulas for area A, circumference C, and volume V:

Triangle

$A = \frac{1}{2}bh$

$= \frac{1}{2}ab\sin\theta$

Circle

$A = \pi r^2$

$C = 2\pi r$

Sector of Circle

$A = \frac{1}{2}r^2\theta$

$s = r\theta$ (θ in radians)

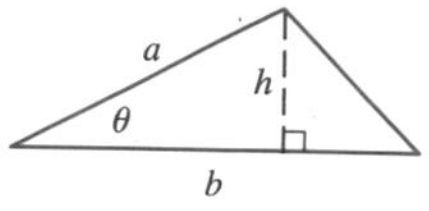

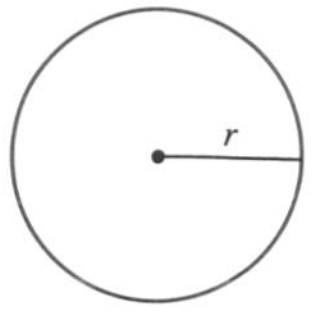

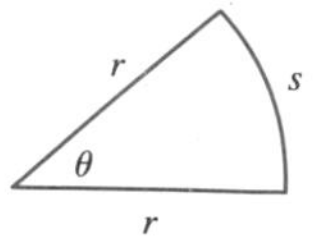

Sphere

$V = \frac{4}{3}\pi r^3$

$A = 4\pi r^2$

Cylinder

$V = \pi r^2 h$

Cone

$V = \frac{1}{3}\pi r^2 h$

$A = \pi r\sqrt{r^2 + h^2}$

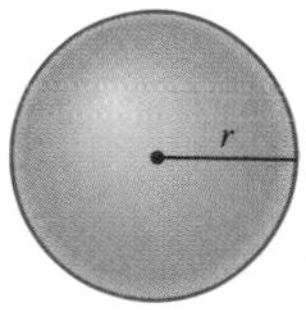

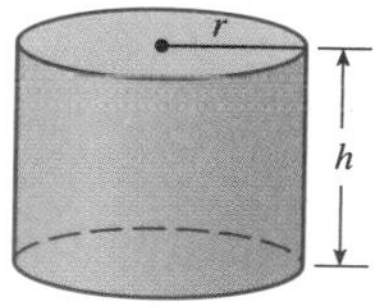

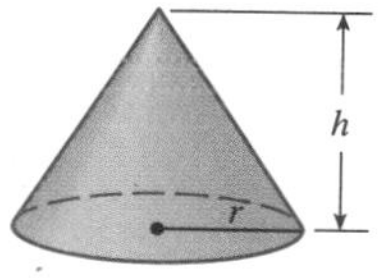

DISTANCE AND MIDPOINT FORMULAS

Distance between $P_1(x_1, y_1)$ and $P_2(x_2, y_2)$:

$$d = \sqrt{(x_2 - x_1)^2 + (y_2 - y_1)^2}$$

Midpoint of $\overline{P_1P_2}$: $\left(\dfrac{x_1 + x_2}{2}, \dfrac{y_1 + y_2}{2}\right)$

LINES

Slope of line through $P_1(x_1, y_1)$ and $P_2(x_2, y_2)$:

$$m = \frac{y_2 - y_1}{x_2 - x_1}$$

Point-slope equation of line through $P_1(x_1, y_1)$ with slope m:

$$y - y_1 = m(x - x_1)$$

Slope-intercept equation of line with slope m and y-intercept b:

$$y = mx + b$$

CIRCLES

Equation of the circle with center (h, k) and radius r:

$$(x - h)^2 + (y - k)^2 = r^2$$

TRIGONOMETRY

ANGLE MEASUREMENT

π radians $= 180°$

$1° = \frac{\pi}{180}$ rad $\qquad$ 1 rad $= \frac{180°}{\pi}$

$s = r\theta$

(θ in radians)

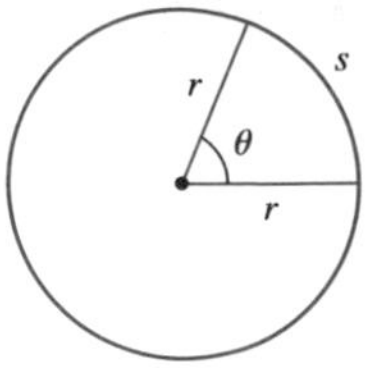

RIGHT ANGLE TRIGONOMETRY

$\sin\theta = \frac{\text{opp}}{\text{hyp}}$ $\qquad$ $\csc\theta = \frac{\text{hyp}}{\text{opp}}$

$\cos\theta = \frac{\text{adj}}{\text{hyp}}$ $\qquad$ $\sec\theta = \frac{\text{hyp}}{\text{adj}}$

$\tan\theta = \frac{\text{opp}}{\text{adj}}$ $\qquad$ $\cot\theta = \frac{\text{adj}}{\text{opp}}$

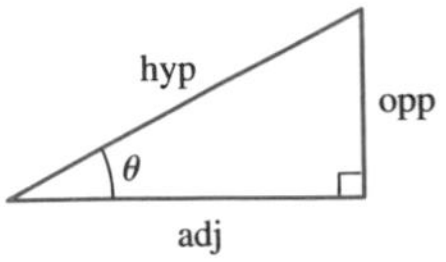

TRIGONOMETRIC FUNCTIONS

$\sin\theta = \frac{y}{r}$ $\qquad$ $\csc\theta = \frac{r}{y}$

$\cos\theta = \frac{x}{r}$ $\qquad$ $\sec\theta = \frac{r}{x}$

$\tan\theta = \frac{y}{x}$ $\qquad$ $\cot\theta = \frac{x}{y}$

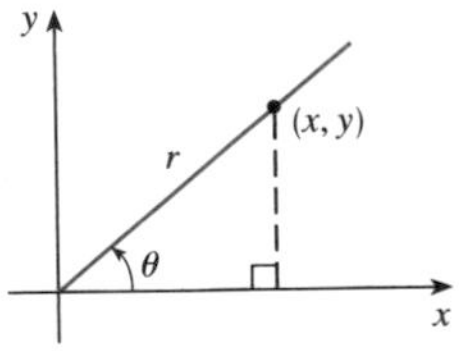

GRAPHS OF TRIGONOMETRIC FUNCTIONS

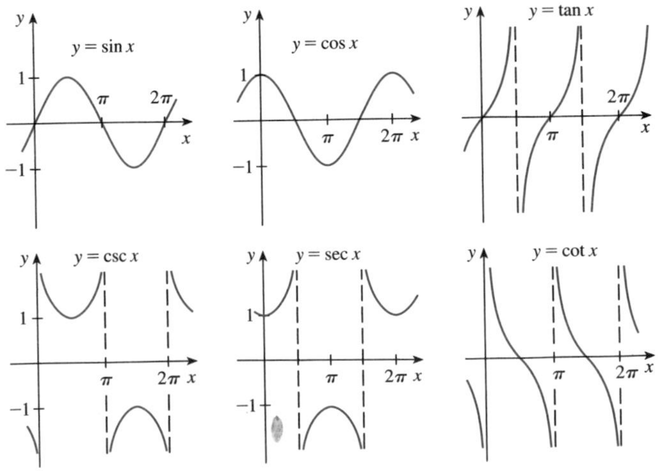

TRIGONOMETRIC FUNCTIONS OF IMPORTANT ANGLES

θ	radians	$\sin\theta$	$\cos\theta$	$\tan\theta$
0°	0	0	1	0
30°	$\pi/6$	$1/2$	$\sqrt{3}/2$	$\sqrt{3}/3$
45°	$\pi/4$	$\sqrt{2}/2$	$\sqrt{2}/2$	1
60°	$\pi/3$	$\sqrt{3}/2$	$1/2$	$\sqrt{3}$
90°	$\pi/2$	1	0	—

FUNDAMENTAL IDENTITIES

$\csc\theta = \frac{1}{\sin\theta}$ $\qquad$ $\sec\theta = \frac{1}{\cos\theta}$

$\tan\theta = \frac{\sin\theta}{\cos\theta}$ $\qquad$ $\cot\theta = \frac{\cos\theta}{\sin\theta}$

$\cot\theta = \frac{1}{\tan\theta}$ $\qquad$ $\sin^2\theta + \cos^2\theta = 1$

$1 + \tan^2\theta = \sec^2\theta$ $\qquad$ $1 + \cot^2\theta = \csc^2\theta$

$\sin(-\theta) = -\sin\theta$ $\qquad$ $\cos(-\theta) = \cos\theta$

$\tan(-\theta) = -\tan\theta$ $\qquad$ $\sin\left(\frac{\pi}{2} - \theta\right) = \cos\theta$

$\cos\left(\frac{\pi}{2} - \theta\right) = \sin\theta$ $\qquad$ $\tan\left(\frac{\pi}{2} - \theta\right) = \cot\theta$

THE LAW OF SINES

$$\frac{\sin A}{a} = \frac{\sin B}{b} = \frac{\sin C}{c}$$

THE LAW OF COSINES

$a^2 = b^2 + c^2 - 2bc\cos A$

$b^2 = a^2 + c^2 - 2ac\cos B$

$c^2 = a^2 + b^2 - 2ab\cos C$

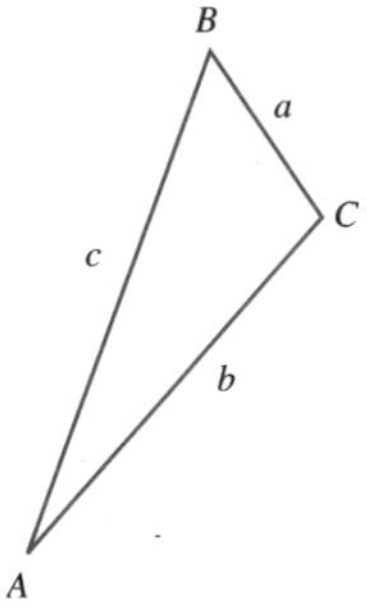

ADDITION AND SUBTRACTION FORMULAS

$\sin(x + y) = \sin x\cos y + \cos x\sin y$

$\sin(x - y) = \sin x\cos y - \cos x\sin y$

$\cos(x + y) = \cos x\cos y - \sin x\sin y$

$\cos(x - y) = \cos x\cos y + \sin x\sin y$

$\tan(x + y) = \frac{\tan x + \tan y}{1 - \tan x\tan y}$

$\tan(x - y) = \frac{\tan x - \tan y}{1 + \tan x\tan y}$

DOUBLE-ANGLE FORMULAS

$\sin 2x = 2\sin x\cos x$

$\cos 2x = \cos^2 x - \sin^2 x = 2\cos^2 x - 1 = 1 - 2\sin^2 x$

$\tan 2x = \frac{2\tan x}{1 - \tan^2 x}$

HALF-ANGLE FORMULAS

$\sin^2 x = \frac{1 - \cos 2x}{2}$ $\qquad$ $\cos^2 x = \frac{1 + \cos 2x}{2}$

Single Variable Calculus

Early Transcendentals for UC Berkeley

Stewart

CENGAGE Learning™

Australia • Brazil • Japan • Korea • Mexico • Singapore • Spain • United Kingdom • United States

CENGAGE Learning™

Single Variable Calculus
Early Transcendentals for UC Berkeley

Stewart

Executive Editor:
Maureen Staudt
Michael Stranz

Senior Project Development Manager:
Linda de Stefano

Marketing Specialist:
Sara Mercurio
Lindsay Shapiro

Production/Manufacturing Manager:
Donna M. Brown

PreMedia Supervisor:
Joel Brennecke

Rights & Permissions Specialist:
Kalina Hintz
Todd Osborne

Cover Image:
Getty Images*

* Unless otherwise noted, all cover images used by Custom Solutions, a part of Cengage Learning, have been supplied courtesy of Getty Images with the exception of the Earthview cover image, which has been supplied by the National Aeronautics and Space Administration (NASA).

ISBN-13: 978-1-4240-5500-5

ISBN-10: 1-4240-5500-8

Cengage Learning
5191 Natorp Boulevard
Mason, Ohio 45040
USA

Printed in the United States of America

Custom Table of Contents

	Preview of Calculus	2
1	Functions and Models	10
2	Limits and Derivatives	82
3	Differentiation Rules	172
4	Applications of Differentiation	270
5	Integrals	354
6	Applications of Integration	414
7	Techniques of Integration	452
8	Further Applications of Integration	524
9	Differential Equations	566
11	Infinite Sequences and Series	674
17	Second-Order Differential Equations	1110
	Appendixes	A1
	Index	A131

CALCULUS

EARLY TRANSCENDENTALS

A PREVIEW OF CALCULUS

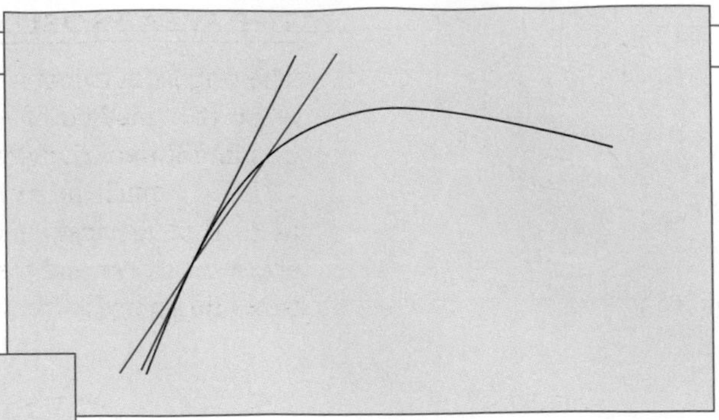

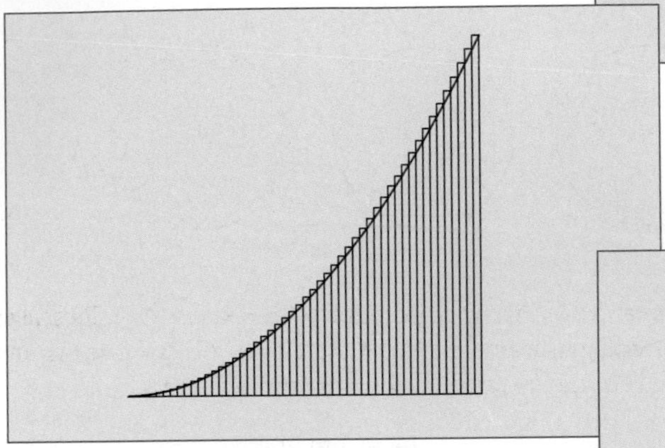

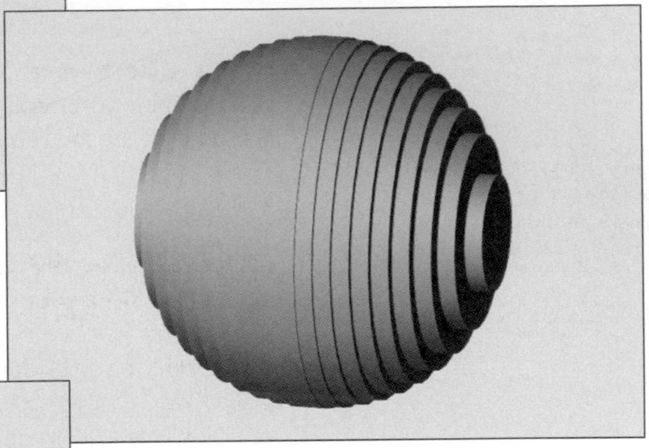

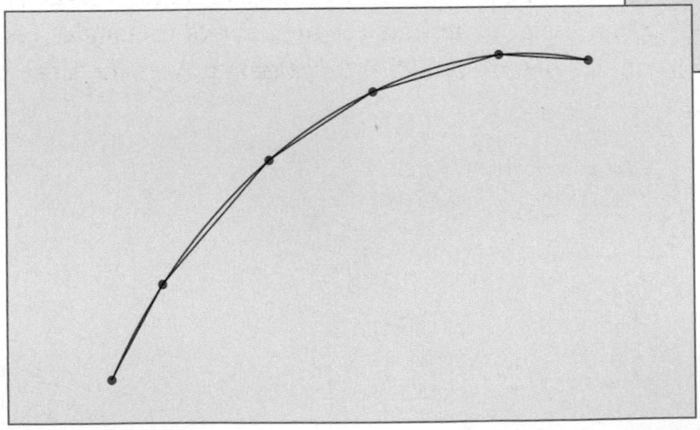

Calculus is fundamentally different from the mathematics that you have studied previously: calculus is less static and more dynamic. It is concerned with change and motion; it deals with quantities that approach other quantities. For that reason it may be useful to have an overview of the subject before beginning its intensive study. Here we give a glimpse of some of the main ideas of calculus by showing how the concept of a limit arises when we attempt to solve a variety of problems.

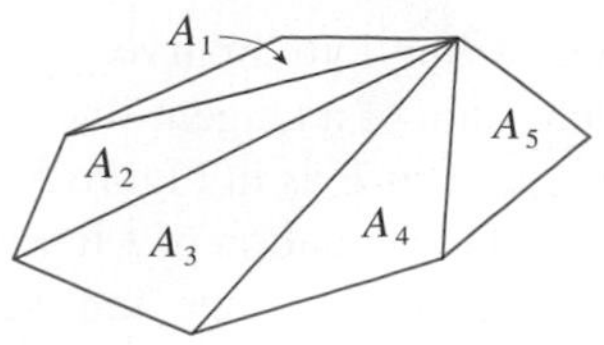

$A = A_1 + A_2 + A_3 + A_4 + A_5$

FIGURE 1

THE AREA PROBLEM

The origins of calculus go back at least 2500 years to the ancient Greeks, who found areas using the "method of exhaustion." They knew how to find the area A of any polygon by dividing it into triangles as in Figure 1 and adding the areas of these triangles.

It is a much more difficult problem to find the area of a curved figure. The Greek method of exhaustion was to inscribe polygons in the figure and circumscribe polygons about the figure and then let the number of sides of the polygons increase. Figure 2 illustrates this process for the special case of a circle with inscribed regular polygons.

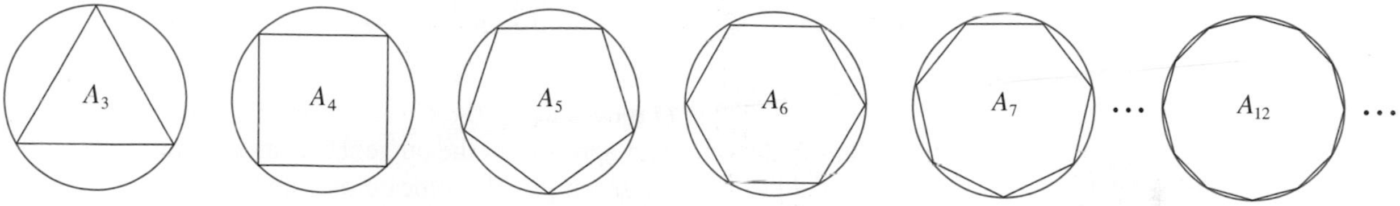

FIGURE 2

Let A_n be the area of the inscribed polygon with n sides. As n increases, it appears that A_n becomes closer and closer to the area of the circle. We say that the area of the circle is the *limit* of the areas of the inscribed polygons, and we write

TEC In the Preview Visual, you can see how inscribed and circumscribed polygons approximate the area of a circle.

$$A = \lim_{n \to \infty} A_n$$

The Greeks themselves did not use limits explicitly. However, by indirect reasoning, Eudoxus (fifth century BC) used exhaustion to prove the familiar formula for the area of a circle: $A = \pi r^2$.

We will use a similar idea in Chapter 5 to find areas of regions of the type shown in Figure 3. We will approximate the desired area A by areas of rectangles (as in Figure 4), let the width of the rectangles decrease, and then calculate A as the limit of these sums of areas of rectangles.

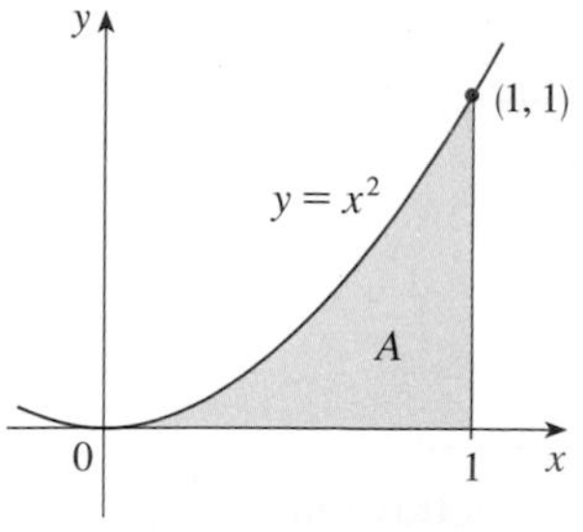

FIGURE 3

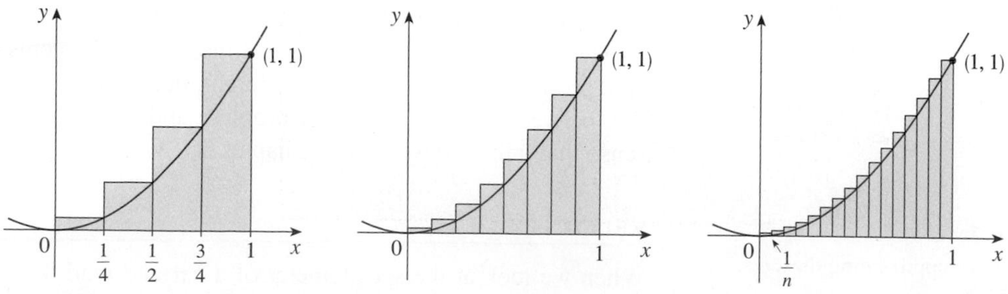

FIGURE 4

The area problem is the central problem in the branch of calculus called *integral calculus*. The techniques that we will develop in Chapter 5 for finding areas will also enable us to compute the volume of a solid, the length of a curve, the force of water against a dam, the mass and center of gravity of a rod, and the work done in pumping water out of a tank.

THE TANGENT PROBLEM

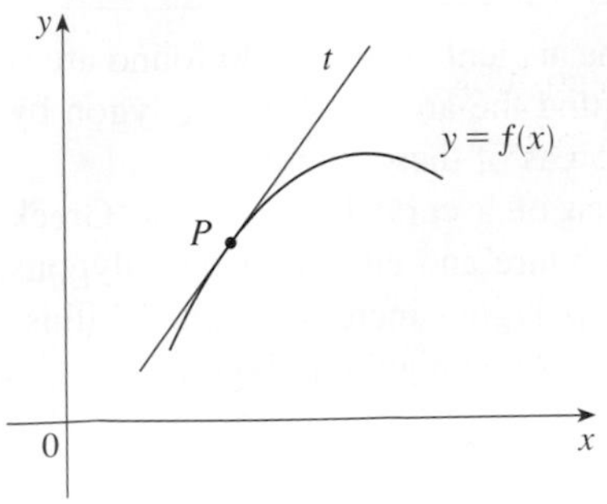

FIGURE 5
The tangent line at P

Consider the problem of trying to find an equation of the tangent line t to a curve with equation $y = f(x)$ at a given point P. (We will give a precise definition of a tangent line in Chapter 2. For now you can think of it as a line that touches the curve at P as in Figure 5.) Since we know that the point P lies on the tangent line, we can find the equation of t if we know its slope m. The problem is that we need two points to compute the slope and we know only one point, P, on t. To get around the problem we first find an approximation to m by taking a nearby point Q on the curve and computing the slope m_{PQ} of the secant line PQ. From Figure 6 we see that

1 $$m_{PQ} = \frac{f(x) - f(a)}{x - a}$$

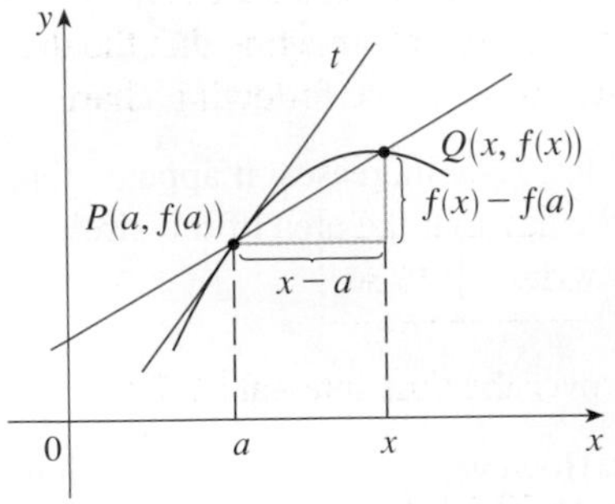

FIGURE 6
The secant line PQ

Now imagine that Q moves along the curve toward P as in Figure 7. You can see that the secant line rotates and approaches the tangent line as its limiting position. This means that the slope m_{PQ} of the secant line becomes closer and closer to the slope m of the tangent line. We write

$$m = \lim_{Q \to P} m_{PQ}$$

and we say that m is the limit of m_{PQ} as Q approaches P along the curve. Since x approaches a as Q approaches P, we could also use Equation 1 to write

2 $$m = \lim_{x \to a} \frac{f(x) - f(a)}{x - a}$$

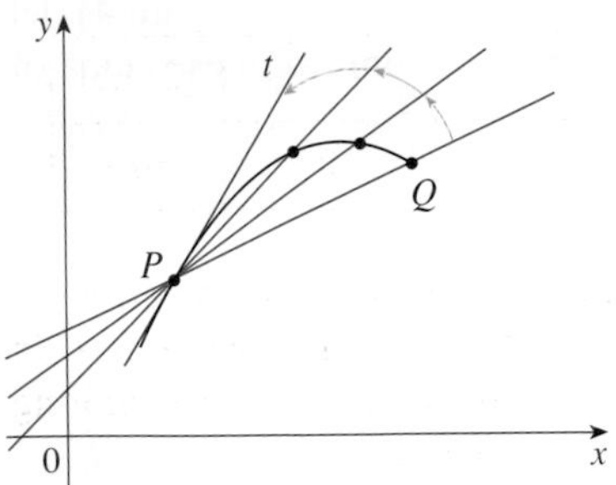

FIGURE 7
Secant lines approaching the tangent line

Specific examples of this procedure will be given in Chapter 2.

The tangent problem has given rise to the branch of calculus called *differential calculus,* which was not invented until more than 2000 years after integral calculus. The main ideas behind differential calculus are due to the French mathematician Pierre Fermat (1601–1665) and were developed by the English mathematicians John Wallis (1616–1703), Isaac Barrow (1630–1677), and Isaac Newton (1642–1727) and the German mathematician Gottfried Leibniz (1646–1716).

The two branches of calculus and their chief problems, the area problem and the tangent problem, appear to be very different, but it turns out that there is a very close connection between them. The tangent problem and the area problem are inverse problems in a sense that will be described in Chapter 5.

VELOCITY

When we look at the speedometer of a car and read that the car is traveling at 48 mi/h, what does that information indicate to us? We know that if the velocity remains constant, then after an hour we will have traveled 48 mi. But if the velocity of the car varies, what does it mean to say that the velocity at a given instant is 48 mi/h?

In order to analyze this question, let's examine the motion of a car that travels along a straight road and assume that we can measure the distance traveled by the car (in feet) at 1-second intervals as in the following chart:

t = Time elapsed (s)	0	1	2	3	4	5
d = Distance (ft)	0	2	9	24	42	71

As a first step toward finding the velocity after 2 seconds have elapsed, we find the average velocity during the time interval $2 \leqslant t \leqslant 4$:

$$\begin{aligned}\text{average velocity} &= \frac{\text{change in position}}{\text{time elapsed}} \\ &= \frac{42 - 9}{4 - 2} \\ &= 16.5 \text{ ft/s}\end{aligned}$$

Similarly, the average velocity in the time interval $2 \leqslant t \leqslant 3$ is

$$\text{average velocity} = \frac{24 - 9}{3 - 2} = 15 \text{ ft/s}$$

We have the feeling that the velocity at the instant $t = 2$ can't be much different from the average velocity during a short time interval starting at $t = 2$. So let's imagine that the distance traveled has been measured at 0.1-second time intervals as in the following chart:

t	2.0	2.1	2.2	2.3	2.4	2.5
d	9.00	10.02	11.16	12.45	13.96	15.80

Then we can compute, for instance, the average velocity over the time interval [2, 2.5]:

$$\text{average velocity} = \frac{15.80 - 9.00}{2.5 - 2} = 13.6 \text{ ft/s}$$

The results of such calculations are shown in the following chart:

Time interval	[2, 3]	[2, 2.5]	[2, 2.4]	[2, 2.3]	[2, 2.2]	[2, 2.1]
Average velocity (ft/s)	15.0	13.6	12.4	11.5	10.8	10.2

The average velocities over successively smaller intervals appear to be getting closer to a number near 10, and so we expect that the velocity at exactly $t = 2$ is about 10 ft/s. In Chapter 2 we will define the instantaneous velocity of a moving object as the limiting value of the average velocities over smaller and smaller time intervals.

In Figure 8 we show a graphical representation of the motion of the car by plotting the distance traveled as a function of time. If we write $d = f(t)$, then $f(t)$ is the number of feet traveled after t seconds. The average velocity in the time interval $[2, t]$ is

$$\text{average velocity} = \frac{\text{change in position}}{\text{time elapsed}} = \frac{f(t) - f(2)}{t - 2}$$

which is the same as the slope of the secant line PQ in Figure 8. The velocity v when $t = 2$ is the limiting value of this average velocity as t approaches 2; that is,

$$v = \lim_{t \to 2} \frac{f(t) - f(2)}{t - 2}$$

and we recognize from Equation 2 that this is the same as the slope of the tangent line to the curve at P.

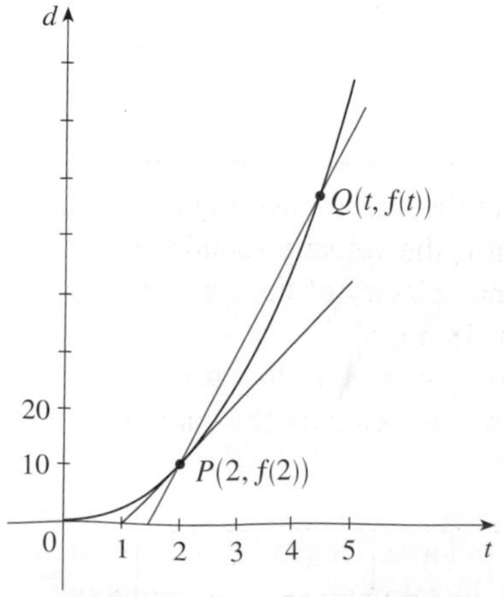

FIGURE 8

Thus, when we solve the tangent problem in differential calculus, we are also solving problems concerning velocities. The same techniques also enable us to solve problems involving rates of change in all of the natural and social sciences.

THE LIMIT OF A SEQUENCE

In the fifth century BC the Greek philosopher Zeno of Elea posed four problems, now known as *Zeno's paradoxes,* that were intended to challenge some of the ideas concerning space and time that were held in his day. Zeno's second paradox concerns a race between the Greek hero Achilles and a tortoise that has been given a head start. Zeno argued, as follows, that Achilles could never pass the tortoise: Suppose that Achilles starts at position a_1 and the tortoise starts at position t_1. (See Figure 9.) When Achilles reaches the point $a_2 = t_1$, the tortoise is farther ahead at position t_2. When Achilles reaches $a_3 = t_2$, the tortoise is at t_3. This process continues indefinitely and so it appears that the tortoise will always be ahead! But this defies common sense.

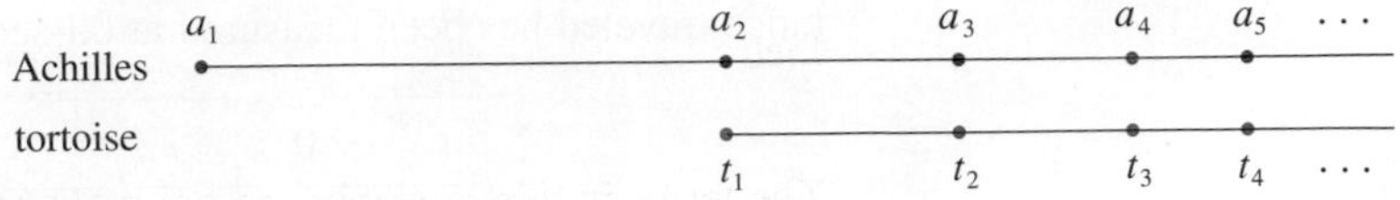

FIGURE 9

One way of explaining this paradox is with the idea of a *sequence.* The successive positions of Achilles $(a_1, a_2, a_3, \ldots)$ or the successive positions of the tortoise $(t_1, t_2, t_3, \ldots)$ form what is known as a sequence.

In general, a sequence $\{a_n\}$ is a set of numbers written in a definite order. For instance, the sequence

$$\left\{1, \tfrac{1}{2}, \tfrac{1}{3}, \tfrac{1}{4}, \tfrac{1}{5}, \ldots\right\}$$

can be described by giving the following formula for the nth term:

$$a_n = \frac{1}{n}$$

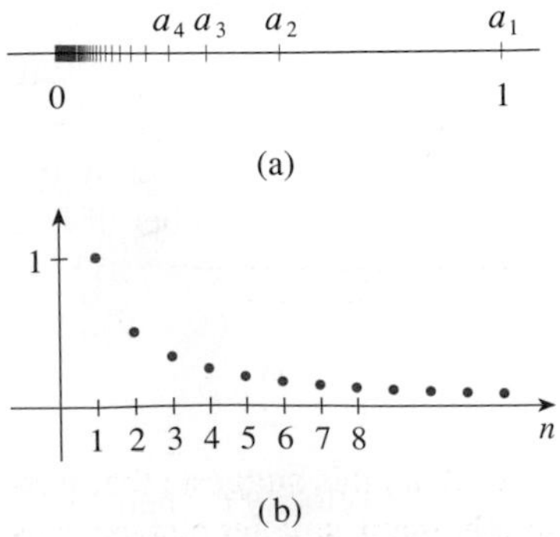

FIGURE 10

We can visualize this sequence by plotting its terms on a number line as in Figure 10(a) or by drawing its graph as in Figure 10(b). Observe from either picture that the terms of the sequence $a_n = 1/n$ are becoming closer and closer to 0 as n increases. In fact, we can find terms as small as we please by making n large enough. We say that the limit of the sequence is 0, and we indicate this by writing

$$\lim_{n \to \infty} \frac{1}{n} = 0$$

In general, the notation

$$\lim_{n \to \infty} a_n = L$$

is used if the terms a_n approach the number L as n becomes large. This means that the numbers a_n can be made as close as we like to the number L by taking n sufficiently large.

The concept of the limit of a sequence occurs whenever we use the decimal representation of a real number. For instance, if

$$a_1 = 3.1$$
$$a_2 = 3.14$$
$$a_3 = 3.141$$
$$a_4 = 3.1415$$
$$a_5 = 3.14159$$
$$a_6 = 3.141592$$
$$a_7 = 3.1415926$$
$$\vdots$$

then $$\lim_{n \to \infty} a_n = \pi$$

The terms in this sequence are rational approximations to π.

Let's return to Zeno's paradox. The successive positions of Achilles and the tortoise form sequences $\{a_n\}$ and $\{t_n\}$, where $a_n < t_n$ for all n. It can be shown that both sequences have the same limit:

$$\lim_{n \to \infty} a_n = p = \lim_{n \to \infty} t_n$$

It is precisely at this point p that Achilles overtakes the tortoise.

THE SUM OF A SERIES

Another of Zeno's paradoxes, as passed on to us by Aristotle, is the following: "A man standing in a room cannot walk to the wall. In order to do so, he would first have to go half the distance, then half the remaining distance, and then again half of what still remains. This process can always be continued and can never be ended." (See Figure 11.)

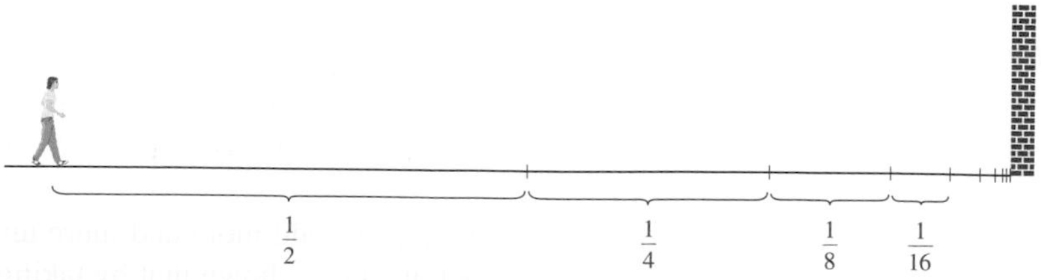

FIGURE 11

Of course, we know that the man can actually reach the wall, so this suggests that perhaps the total distance can be expressed as the sum of infinitely many smaller distances as follows:

3 $$1 = \frac{1}{2} + \frac{1}{4} + \frac{1}{8} + \frac{1}{16} + \cdots + \frac{1}{2^n} + \cdots$$

1

FUNCTIONS AND MODELS

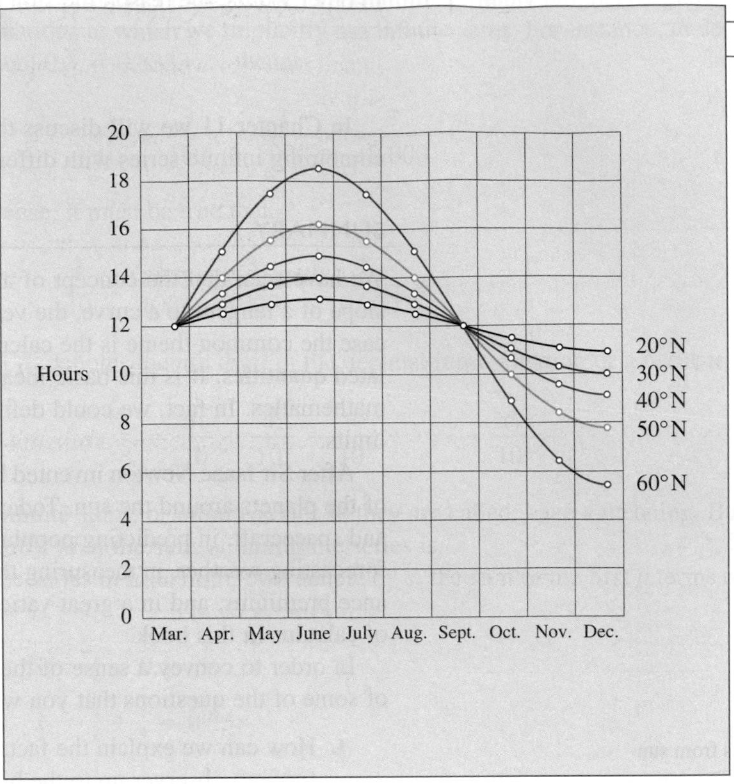

A graphical representation of a function—here the number of hours of daylight as a function of the time of year at various latitudes—is often the most natural and convenient way to represent the function.

The fundamental objects that we deal with in calculus are functions. This chapter prepares the way for calculus by discussing the basic ideas concerning functions, their graphs, and ways of transforming and combining them. We stress that a function can be represented in different ways: by an equation, in a table, by a graph, or in words. We look at the main types of functions that occur in calculus and describe the process of using these functions as mathematical models of real-world phenomena. We also discuss the use of graphing calculators and graphing software for computers.

1.1 FOUR WAYS TO REPRESENT A FUNCTION

Functions arise whenever one quantity depends on another. Consider the following four situations.

A. The area A of a circle depends on the radius r of the circle. The rule that connects r and A is given by the equation $A = \pi r^2$. With each positive number r there is associated one value of A, and we say that A is a *function* of r.

B. The human population of the world P depends on the time t. The table gives estimates of the world population $P(t)$ at time t, for certain years. For instance,

$$P(1950) \approx 2{,}560{,}000{,}000$$

But for each value of the time t there is a corresponding value of P, and we say that P is a function of t.

Year	Population (millions)
1900	1650
1910	1750
1920	1860
1930	2070
1940	2300
1950	2560
1960	3040
1970	3710
1980	4450
1990	5280
2000	6080

C. The cost C of mailing a first-class letter depends on the weight w of the letter. Although there is no simple formula that connects w and C, the post office has a rule for determining C when w is known.

D. The vertical acceleration a of the ground as measured by a seismograph during an earthquake is a function of the elapsed time t. Figure 1 shows a graph generated by seismic activity during the Northridge earthquake that shook Los Angeles in 1994. For a given value of t, the graph provides a corresponding value of a.

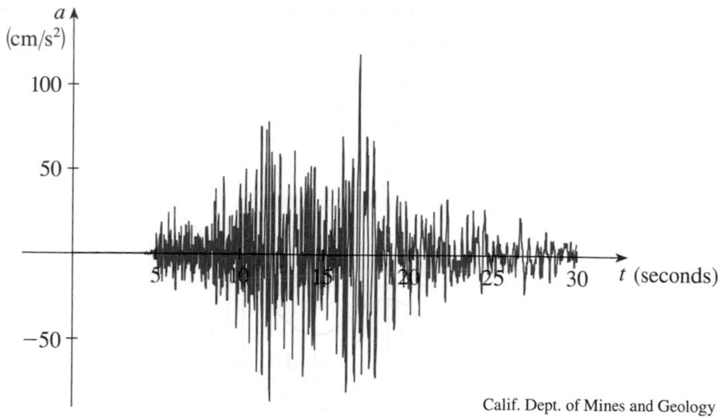

FIGURE 1
Vertical ground acceleration during the Northridge earthquake

Each of these examples describes a rule whereby, given a number (r, t, w, or t), another number (A, P, C, or a) is assigned. In each case we say that the second number is a function of the first number.

> A **function** f is a rule that assigns to each element x in a set D exactly one element, called $f(x)$, in a set E.

We usually consider functions for which the sets D and E are sets of real numbers. The set D is called the **domain** of the function. The number $f(x)$ is the **value of f at x** and is read "f of x." The **range** of f is the set of all possible values of $f(x)$ as x varies throughout the domain. A symbol that represents an arbitrary number in the *domain* of a function f is called an **independent variable**. A symbol that represents a number in the *range* of f is called a **dependent variable**. In Example A, for instance, r is the independent variable and A is the dependent variable.

functions are described more naturally by one method than by another. With this in mind, let's reexamine the four situations that we considered at the beginning of this section.

A. The most useful representation of the area of a circle as a function of its radius is probably the algebraic formula $A(r) = \pi r^2$, though it is possible to compile a table of values or to sketch a graph (half a parabola). Because a circle has to have a positive radius, the domain is $\{r \mid r > 0\} = (0, \infty)$, and the range is also $(0, \infty)$.

Year	Population (millions)
1900	1650
1910	1750
1920	1860
1930	2070
1940	2300
1950	2560
1960	3040
1970	3710
1980	4450
1990	5280
2000	6080

B. We are given a description of the function in words: $P(t)$ is the human population of the world at time t. The table of values of world population provides a convenient representation of this function. If we plot these values, we get the graph (called a *scatter plot*) in Figure 9. It too is a useful representation; the graph allows us to absorb all the data at once. What about a formula? Of course, it's impossible to devise an explicit formula that gives the exact human population $P(t)$ at any time t. But it is possible to find an expression for a function that *approximates* $P(t)$. In fact, using methods explained in Section 1.2, we obtain the approximation

$$P(t) \approx f(t) = (0.008079266) \cdot (1.013731)^t$$

and Figure 10 shows that it is a reasonably good "fit." The function f is called a *mathematical model* for population growth. In other words, it is a function with an explicit formula that approximates the behavior of our given function. We will see, however, that the ideas of calculus can be applied to a table of values; an explicit formula is not necessary.

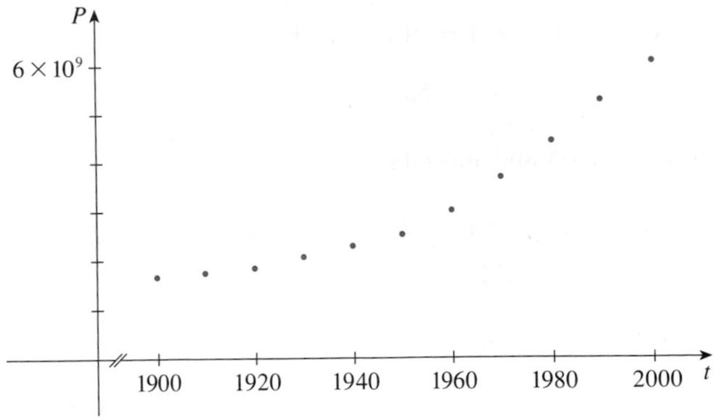

FIGURE 9

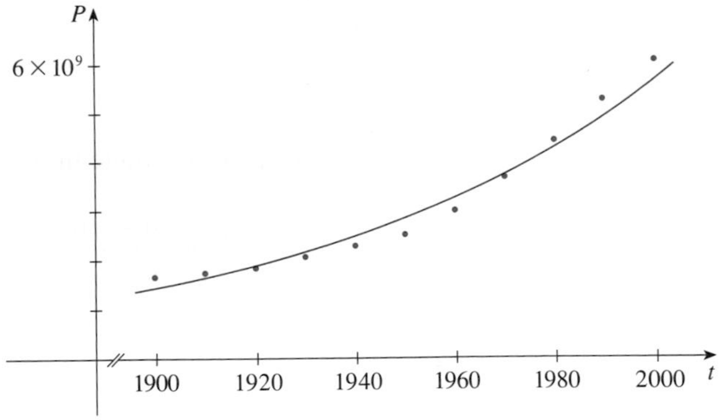

FIGURE 10

■ A function defined by a table of values is called a *tabular* function.

w (ounces)	$C(w)$ (dollars)
$0 < w \leqslant 1$	0.39
$1 < w \leqslant 2$	0.63
$2 < w \leqslant 3$	0.87
$3 < w \leqslant 4$	1.11
$4 < w \leqslant 5$	1.35
⋮	⋮
$12 < w \leqslant 13$	3.27

The function P is typical of the functions that arise whenever we attempt to apply calculus to the real world. We start with a verbal description of a function. Then we may be able to construct a table of values of the function, perhaps from instrument readings in a scientific experiment. Even though we don't have complete knowledge of the values of the function, we will see throughout the book that it is still possible to perform the operations of calculus on such a function.

C. Again the function is described in words: $C(w)$ is the cost of mailing a first-class letter with weight w. The rule that the US Postal Service used as of 2007 is as follows: The cost is 39 cents for up to one ounce, plus 24 cents for each successive ounce up to 13 ounces. The table of values shown in the margin is the most convenient representation for this function, though it is possible to sketch a graph (see Example 10).

D. The graph shown in Figure 1 is the most natural representation of the vertical acceleration function $a(t)$. It's true that a table of values could be compiled, and it is even possible to devise an approximate formula. But everything a geologist needs to know—amplitudes and patterns—can be seen easily from the graph. (The same is true for the patterns seen in electrocardiograms of heart patients and polygraphs for lie-detection.)

In the next example we sketch the graph of a function that is defined verbally.

EXAMPLE 4 When you turn on a hot-water faucet, the temperature T of the water depends on how long the water has been running. Draw a rough graph of T as a function of the time t that has elapsed since the faucet was turned on.

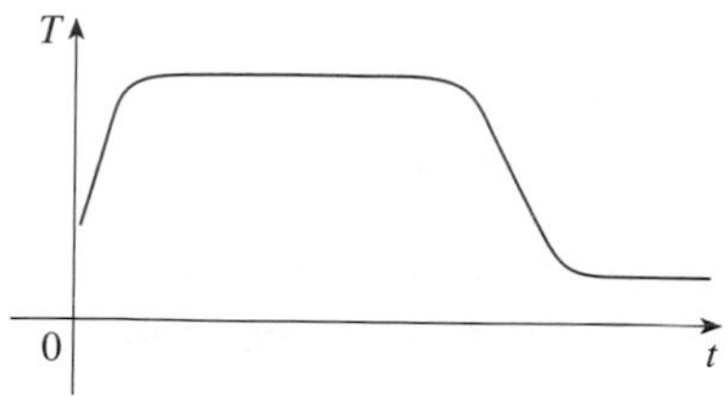

FIGURE 11

SOLUTION The initial temperature of the running water is close to room temperature because the water has been sitting in the pipes. When the water from the hot-water tank starts flowing from the faucet, T increases quickly. In the next phase, T is constant at the temperature of the heated water in the tank. When the tank is drained, T decreases to the temperature of the water supply. This enables us to make the rough sketch of T as a function of t in Figure 11. □

In the following example we start with a verbal description of a function in a physical situation and obtain an explicit algebraic formula. The ability to do this is a useful skill in solving calculus problems that ask for the maximum or minimum values of quantities.

V EXAMPLE 5 A rectangular storage container with an open top has a volume of 10 m^3. The length of its base is twice its width. Material for the base costs \$10 per square meter; material for the sides costs \$6 per square meter. Express the cost of materials as a function of the width of the base.

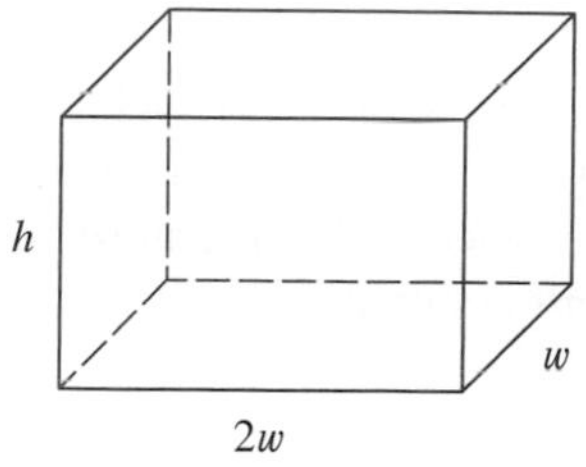

FIGURE 12

SOLUTION We draw a diagram as in Figure 12 and introduce notation by letting w and $2w$ be the width and length of the base, respectively, and h be the height.

The area of the base is $(2w)w = 2w^2$, so the cost, in dollars, of the material for the base is $10(2w^2)$. Two of the sides have area wh and the other two have area $2wh$, so the cost of the material for the sides is $6[2(wh) + 2(2wh)]$. The total cost is therefore

$$C = 10(2w^2) + 6[2(wh) + 2(2wh)] = 20w^2 + 36wh$$

To express C as a function of w alone, we need to eliminate h and we do so by using the fact that the volume is 10 m^3. Thus

$$w(2w)h = 10$$

which gives

$$h = \frac{10}{2w^2} = \frac{5}{w^2}$$

Substituting this into the expression for C, we have

$$C = 20w^2 + 36w\left(\frac{5}{w^2}\right) = 20w^2 + \frac{180}{w}$$

■ In setting up applied functions as in Example 5, it may be useful to review the principles of problem solving as discussed on page 76, particularly *Step 1: Understand the Problem.*

Therefore, the equation

$$C(w) = 20w^2 + \frac{180}{w} \qquad w > 0$$

expresses C as a function of w. □

EXAMPLE 6 Find the domain of each function.

(a) $f(x) = \sqrt{x + 2}$ (b) $g(x) = \dfrac{1}{x^2 - x}$

SOLUTION

■ If a function is given by a formula and the domain is not stated explicitly, the convention is that the domain is the set of all numbers for which the formula makes sense and defines a real number.

(a) Because the square root of a negative number is not defined (as a real number), the domain of f consists of all values of x such that $x + 2 \geqslant 0$. This is equivalent to $x \geqslant -2$, so the domain is the interval $[-2, \infty)$.

(b) Since

$$g(x) = \frac{1}{x^2 - x} = \frac{1}{x(x - 1)}$$

and division by 0 is not allowed, we see that $g(x)$ is not defined when $x = 0$ or $x = 1$. Thus the domain of g is

$$\{x \mid x \neq 0, x \neq 1\}$$

which could also be written in interval notation as

$$(-\infty, 0) \cup (0, 1) \cup (1, \infty)$$ □

The graph of a function is a curve in the xy-plane. But the question arises: Which curves in the xy-plane are graphs of functions? This is answered by the following test.

THE VERTICAL LINE TEST A curve in the xy-plane is the graph of a function of x if and only if no vertical line intersects the curve more than once.

The reason for the truth of the Vertical Line Test can be seen in Figure 13. If each vertical line $x = a$ intersects a curve only once, at (a, b), then exactly one functional value is defined by $f(a) = b$. But if a line $x = a$ intersects the curve twice, at (a, b) and (a, c), then the curve can't represent a function because a function can't assign two different values to a.

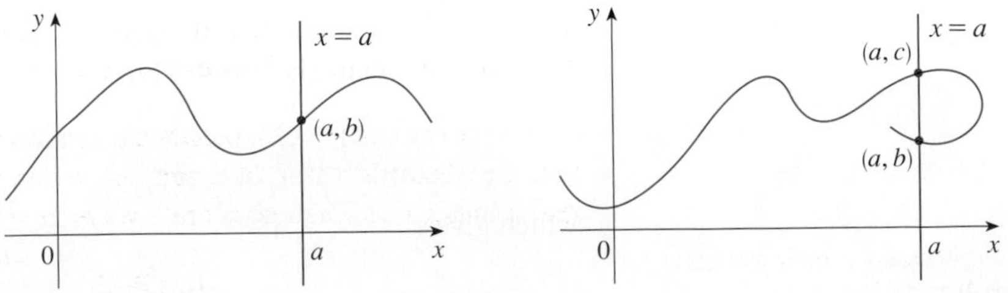

FIGURE 13

For example, the parabola $x = y^2 - 2$ shown in Figure 14(a) on the next page is not the graph of a function of x because, as you can see, there are vertical lines that intersect the parabola twice. The parabola, however, does contain the graphs of *two* functions of x. Notice that the equation $x = y^2 - 2$ implies $y^2 = x + 2$, so $y = \pm\sqrt{x + 2}$. Thus the upper and lower halves of the parabola are the graphs of the functions $f(x) = \sqrt{x + 2}$ [from Example 6(a)] and $g(x) = -\sqrt{x + 2}$. [See Figures 14(b) and (c).] We observe that if we reverse the roles of x and y, then the equation $x = h(y) = y^2 - 2$ does define x as a function of y (with y as the independent variable and x as the dependent variable) and the parabola now appears as the graph of the function h.

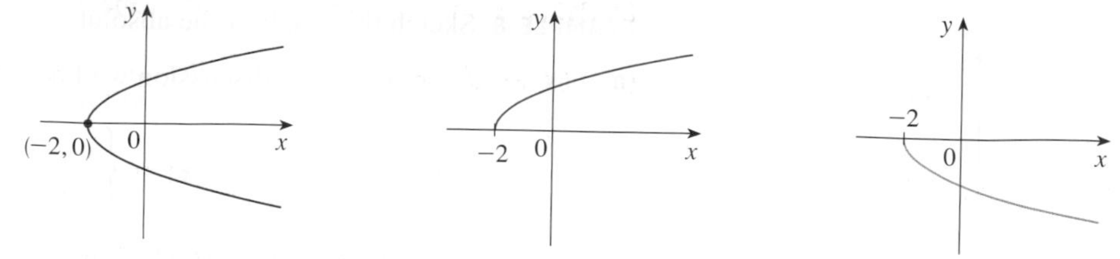

FIGURE 14 (a) $x = y^2 - 2$ (b) $y = \sqrt{x+2}$ (c) $y = -\sqrt{x+2}$

PIECEWISE DEFINED FUNCTIONS

The functions in the following four examples are defined by different formulas in different parts of their domains.

V EXAMPLE 7 A function f is defined by

$$f(x) = \begin{cases} 1 - x & \text{if } x \leqslant 1 \\ x^2 & \text{if } x > 1 \end{cases}$$

Evaluate $f(0)$, $f(1)$, and $f(2)$ and sketch the graph.

SOLUTION Remember that a function is a rule. For this particular function the rule is the following: First look at the value of the input x. If it happens that $x \leqslant 1$, then the value of $f(x)$ is $1 - x$. On the other hand, if $x > 1$, then the value of $f(x)$ is x^2.

Since $0 \leqslant 1$, we have $f(0) = 1 - 0 = 1$.

Since $1 \leqslant 1$, we have $f(1) = 1 - 1 = 0$.

Since $2 > 1$, we have $f(2) = 2^2 = 4$.

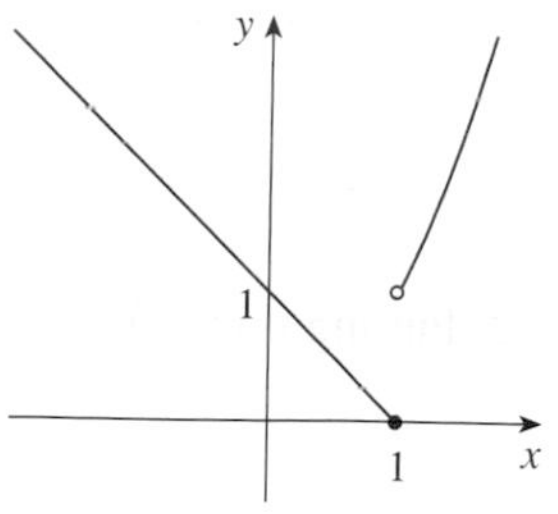

FIGURE 15

How do we draw the graph of f? We observe that if $x \leqslant 1$, then $f(x) = 1 - x$, so the part of the graph of f that lies to the left of the vertical line $x = 1$ must coincide with the line $y = 1 - x$, which has slope -1 and y-intercept 1. If $x > 1$, then $f(x) = x^2$, so the part of the graph of f that lies to the right of the line $x = 1$ must coincide with the graph of $y = x^2$, which is a parabola. This enables us to sketch the graph in Figure 15. The solid dot indicates that the point $(1, 0)$ is included on the graph; the open dot indicates that the point $(1, 1)$ is excluded from the graph. □

The next example of a piecewise defined function is the absolute value function. Recall that the **absolute value** of a number a, denoted by $|a|$, is the distance from a to 0 on the real number line. Distances are always positive or 0, so we have

■ For a more extensive review of absolute values, see Appendix A.

$$|a| \geqslant 0 \qquad \text{for every number } a$$

For example,

$$|3| = 3 \qquad |-3| = 3 \qquad |0| = 0 \qquad |\sqrt{2} - 1| = \sqrt{2} - 1 \qquad |3 - \pi| = \pi - 3$$

In general, we have

$$\boxed{\begin{aligned} |a| &= a && \text{if } a \geqslant 0 \\ |a| &= -a && \text{if } a < 0 \end{aligned}}$$

(Remember that if a is negative, then $-a$ is positive.)

INCREASING AND DECREASING FUNCTIONS

The graph shown in Figure 22 rises from A to B, falls from B to C, and rises again from C to D. The function f is said to be increasing on the interval $[a, b]$, decreasing on $[b, c]$, and increasing again on $[c, d]$. Notice that if x_1 and x_2 are any two numbers between a and b with $x_1 < x_2$, then $f(x_1) < f(x_2)$. We use this as the defining property of an increasing function.

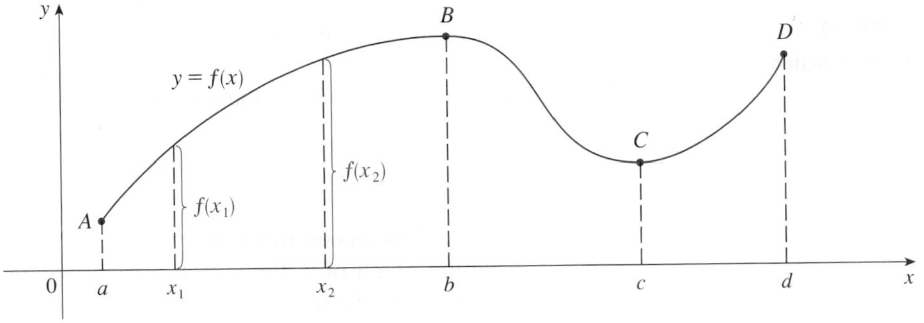

FIGURE 22

> A function f is called **increasing** on an interval I if
>
> $$f(x_1) < f(x_2) \qquad \text{whenever } x_1 < x_2 \text{ in } I$$
>
> It is called **decreasing** on I if
>
> $$f(x_1) > f(x_2) \qquad \text{whenever } x_1 < x_2 \text{ in } I$$

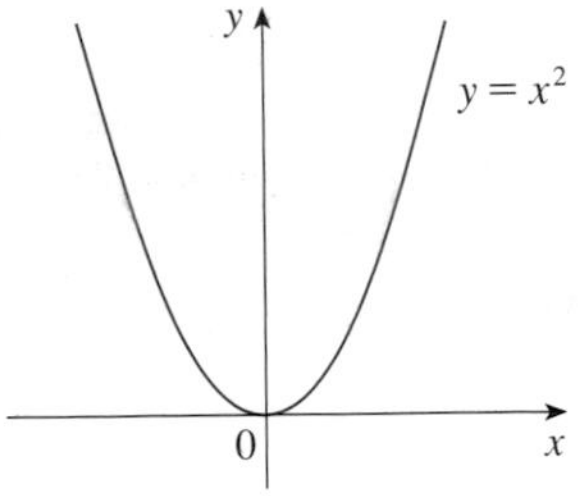

FIGURE 23

In the definition of an increasing function it is important to realize that the inequality $f(x_1) < f(x_2)$ must be satisfied for *every* pair of numbers x_1 and x_2 in I with $x_1 < x_2$.

You can see from Figure 23 that the function $f(x) = x^2$ is decreasing on the interval $(-\infty, 0]$ and increasing on the interval $[0, \infty)$.

1.1 EXERCISES

1. The graph of a function f is given.

(a) State the value of $f(-1)$.

(b) Estimate the value of $f(2)$.

(c) For what values of x is $f(x) = 2$?

(d) Estimate the values of x such that $f(x) = 0$.

(e) State the domain and range of f.

(f) On what interval is f increasing?

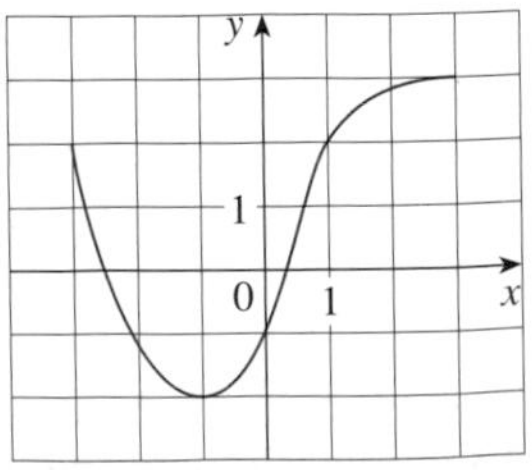

2. The graphs of f and g are given.
   (a) State the values of $f(-4)$ and $g(3)$.
   (b) For what values of x is $f(x) = g(x)$?
   (c) Estimate the solution of the equation $f(x) = -1$.
   (d) On what interval is f decreasing?
   (e) State the domain and range of f.
   (f) State the domain and range of g.

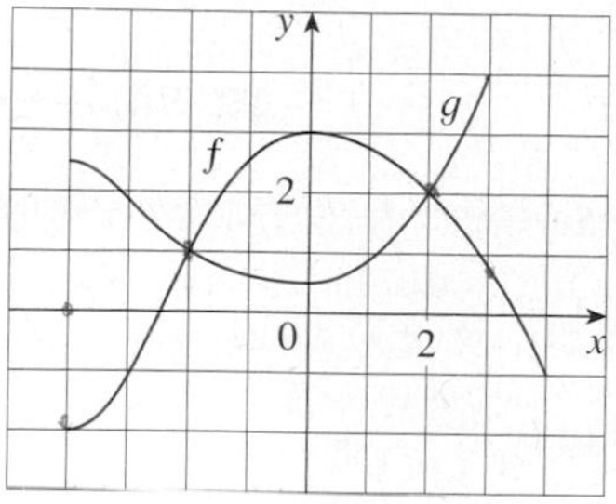

3. Figure 1 was recorded by an instrument operated by the California Department of Mines and Geology at the University Hospital of the University of Southern California in Los Angeles. Use it to estimate the range of the vertical ground acceleration function at USC during the Northridge earthquake.

4. In this section we discussed examples of ordinary, everyday functions: Population is a function of time, postage cost is a function of weight, water temperature is a function of time. Give three other examples of functions from everyday life that are described verbally. What can you say about the domain and range of each of your functions? If possible, sketch a rough graph of each function.

5–8 Determine whether the curve is the graph of a function of x. If it is, state the domain and range of the function.

5.
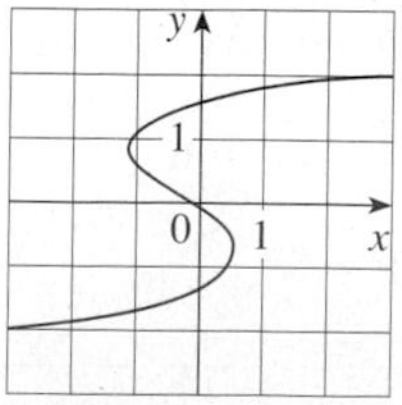

6.
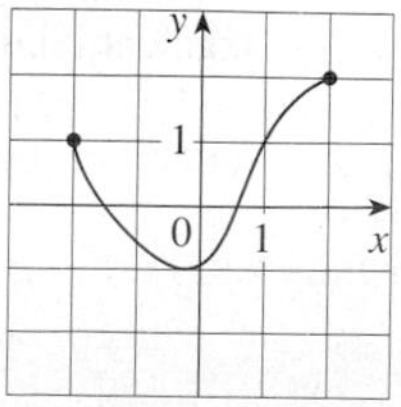

7.
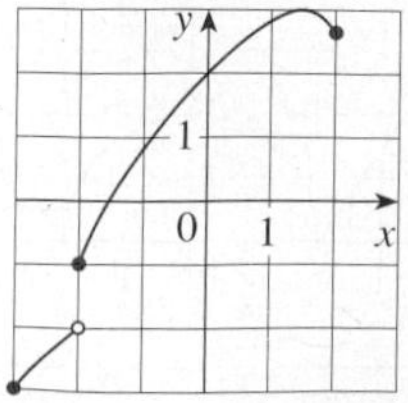

8.
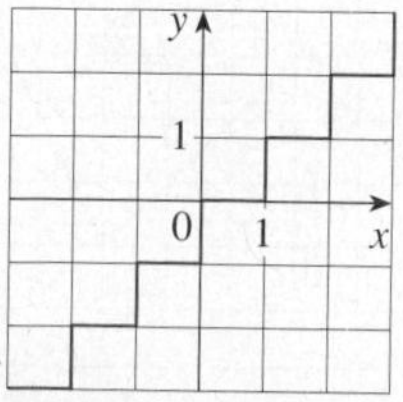

9. The graph shown gives the weight of a certain person as a function of age. Describe in words how this person's weight varies over time. What do you think happened when this person was 30 years old?

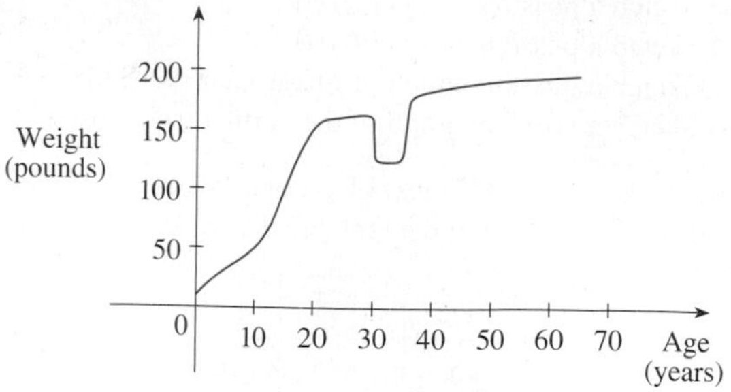

10. The graph shown gives a salesman's distance from his home as a function of time on a certain day. Describe in words what the graph indicates about his travels on this day.

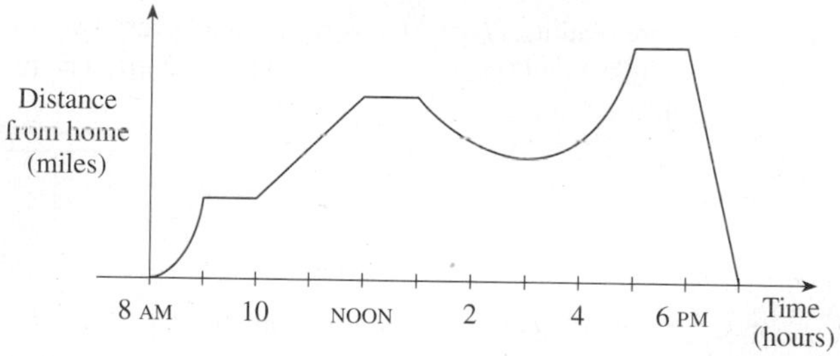

11. You put some ice cubes in a glass, fill the glass with cold water, and then let the glass sit on a table. Describe how the temperature of the water changes as time passes. Then sketch a rough graph of the temperature of the water as a function of the elapsed time.

12. Sketch a rough graph of the number of hours of daylight as a function of the time of year.

13. Sketch a rough graph of the outdoor temperature as a function of time during a typical spring day.

14. Sketch a rough graph of the market value of a new car as a function of time for a period of 20 years. Assume the car is well maintained.

15. Sketch the graph of the amount of a particular brand of coffee sold by a store as a function of the price of the coffee.

16. You place a frozen pie in an oven and bake it for an hour. Then you take it out and let it cool before eating it. Describe how the temperature of the pie changes as time passes. Then sketch a rough graph of the temperature of the pie as a function of time.

17. A homeowner mows the lawn every Wednesday afternoon. Sketch a rough graph of the height of the grass as a function of time over the course of a four-week period.

18. An airplane takes off from an airport and lands an hour later at another airport, 400 miles away. If t represents the time in minutes since the plane has left the terminal building, let $x(t)$ be

the horizontal distance traveled and $y(t)$ be the altitude of the plane.
(a) Sketch a possible graph of $x(t)$.
(b) Sketch a possible graph of $y(t)$.
(c) Sketch a possible graph of the ground speed.
(d) Sketch a possible graph of the vertical velocity.

19. The number N (in millions) of cellular phone subscribers worldwide is shown in the table. (Midyear estimates are given.)

t	1990	1992	1994	1996	1998	2000
N	11	26	60	160	340	650

(a) Use the data to sketch a rough graph of N as a function of t.
(b) Use your graph to estimate the number of cell-phone subscribers at midyear in 1995 and 1999.

20. Temperature readings T (in °F) were recorded every two hours from midnight to 2:00 PM in Dallas on June 2, 2001. The time t was measured in hours from midnight.

t	0	2	4	6	8	10	12	14
T	73	73	70	69	72	81	88	91

(a) Use the readings to sketch a rough graph of T as a function of t.
(b) Use your graph to estimate the temperature at 11:00 AM.

21. If $f(x) = 3x^2 - x + 2$, find $f(2)$, $f(-2)$, $f(a)$, $f(-a)$, $f(a+1)$, $2f(a)$, $f(2a)$, $f(a^2)$, $[f(a)]^2$, and $f(a+h)$.

22. A spherical balloon with radius r inches has volume $V(r) = \frac{4}{3}\pi r^3$. Find a function that represents the amount of air required to inflate the balloon from a radius of r inches to a radius of $r + 1$ inches.

23–26 Evaluate the difference quotient for the given function. Simplify your answer.

23. $f(x) = 4 + 3x - x^2$, $\quad \dfrac{f(3+h) - f(3)}{h}$

24. $f(x) = x^3$, $\quad \dfrac{f(a+h) - f(a)}{h}$

25. $f(x) = \dfrac{1}{x}$, $\quad \dfrac{f(x) - f(a)}{x - a}$

26. $f(x) = \dfrac{x+3}{x+1}$, $\quad \dfrac{f(x) - f(1)}{x - 1}$

27–31 Find the domain of the function.

27. $f(x) = \dfrac{x}{3x - 1}$

28. $f(x) = \dfrac{5x + 4}{x^2 + 3x + 2}$

29. $f(t) = \sqrt{t} + \sqrt[3]{t}$

30. $g(u) = \sqrt{u} + \sqrt{4 - u}$

31. $h(x) = \dfrac{1}{\sqrt[4]{x^2 - 5x}}$

32. Find the domain and range and sketch the graph of the function $h(x) = \sqrt{4 - x^2}$.

33–44 Find the domain and sketch the graph of the function.

33. $f(x) = 5$

34. $F(x) = \frac{1}{2}(x + 3)$

35. $f(t) = t^2 - 6t$

36. $H(t) = \dfrac{4 - t^2}{2 - t}$

37. $g(x) = \sqrt{x - 5}$

38. $F(x) = |2x + 1|$

39. $G(x) = \dfrac{3x + |x|}{x}$

40. $g(x) = \dfrac{|x|}{x^2}$

41. $f(x) = \begin{cases} x + 2 & \text{if } x < 0 \\ 1 - x & \text{if } x \geqslant 0 \end{cases}$

42. $f(x) = \begin{cases} 3 - \frac{1}{2}x & \text{if } x \leqslant 2 \\ 2x - 5 & \text{if } x > 2 \end{cases}$

43. $f(x) = \begin{cases} x + 2 & \text{if } x \leqslant -1 \\ x^2 & \text{if } x > -1 \end{cases}$

44. $f(x) = \begin{cases} x + 9 & \text{if } x < -3 \\ -2x & \text{if } |x| \leqslant 3 \\ -6 & \text{if } x > 3 \end{cases}$

45–50 Find an expression for the function whose graph is the given curve.

45. The line segment joining the points $(1, -3)$ and $(5, 7)$

46. The line segment joining the points $(-5, 10)$ and $(7, -10)$

47. The bottom half of the parabola $x + (y - 1)^2 = 0$

48. The top half of the circle $x^2 + (y - 2)^2 = 4$

49.

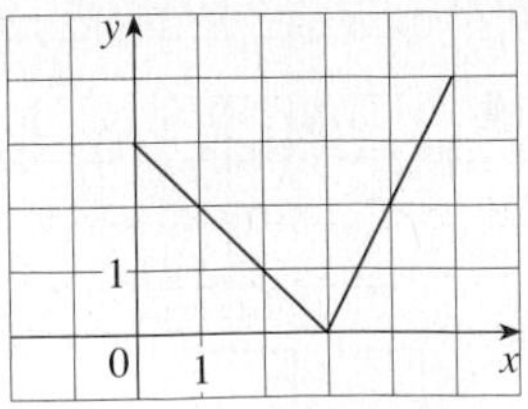

50.

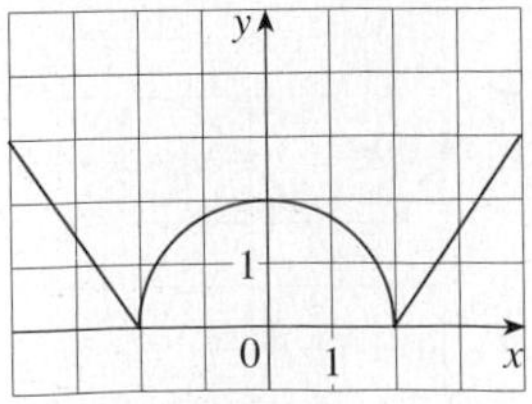

51–55 Find a formula for the described function and state its domain.

51. A rectangle has perimeter 20 m. Express the area of the rectangle as a function of the length of one of its sides.

52. A rectangle has area 16 m^2. Express the perimeter of the rectangle as a function of the length of one of its sides.

53. Express the area of an equilateral triangle as a function of the length of a side.

54. Express the surface area of a cube as a function of its volume.

55. An open rectangular box with volume 2 m^3 has a square base. Express the surface area of the box as a function of the length of a side of the base.

56. A Norman window has the shape of a rectangle surmounted by a semicircle. If the perimeter of the window is 30 ft, express the area A of the window as a function of the width x of the window.

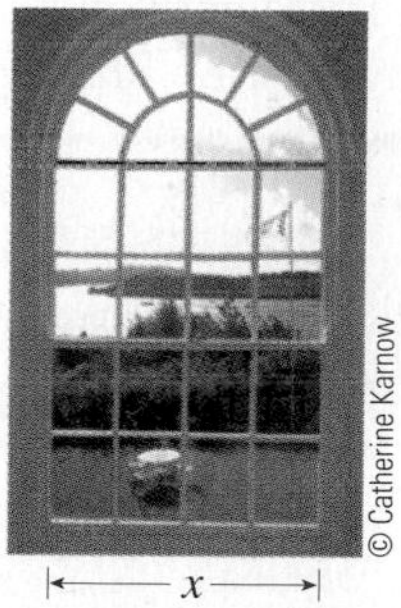

© Catherine Karnow

57. A box with an open top is to be constructed from a rectangular piece of cardboard with dimensions 12 in. by 20 in. by cutting out equal squares of side x at each corner and then folding up the sides as in the figure. Express the volume V of the box as a function of x.

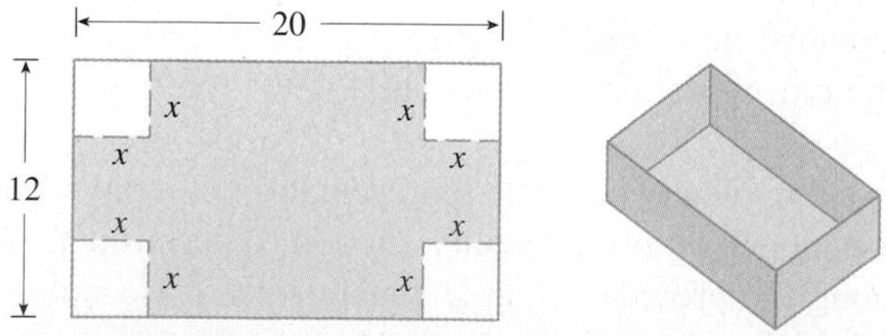

58. A taxi company charges two dollars for the first mile (or part of a mile) and 20 cents for each succeeding tenth of a mile (or part). Express the cost C (in dollars) of a ride as a function of the distance x traveled (in miles) for $0 < x < 2$, and sketch the graph of this function.

59. In a certain country, income tax is assessed as follows. There is no tax on income up to \$10,000. Any income over \$10,000 is taxed at a rate of 10%, up to an income of \$20,000. Any income over \$20,000 is taxed at 15%.

(a) Sketch the graph of the tax rate R as a function of the income I.

(b) How much tax is assessed on an income of \$14,000? On \$26,000?

(c) Sketch the graph of the total assessed tax T as a function of the income I.

60. The functions in Example 10 and Exercises 58 and 59(a) are called *step functions* because their graphs look like stairs. Give two other examples of step functions that arise in everyday life.

61–62 Graphs of f and g are shown. Decide whether each function is even, odd, or neither. Explain your reasoning.

61.

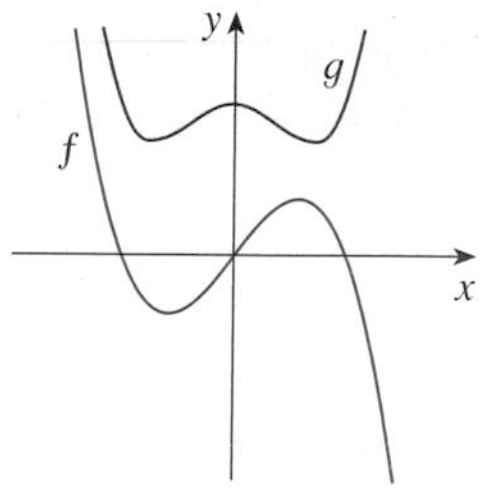

62.

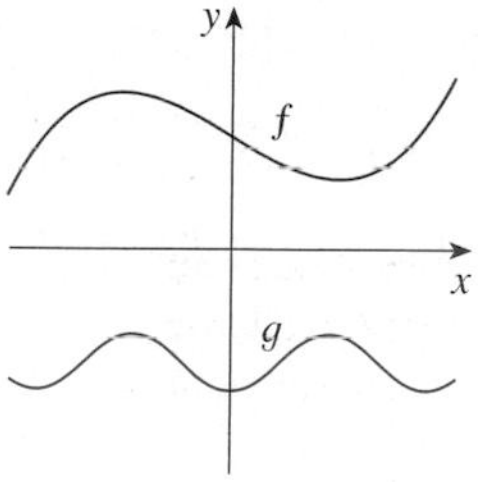

63. (a) If the point (5, 3) is on the graph of an even function, what other point must also be on the graph?

(b) If the point (5, 3) is on the graph of an odd function, what other point must also be on the graph?

64. A function f has domain $[-5, 5]$ and a portion of its graph is shown.

(a) Complete the graph of f if it is known that f is even.

(b) Complete the graph of f if it is known that f is odd.

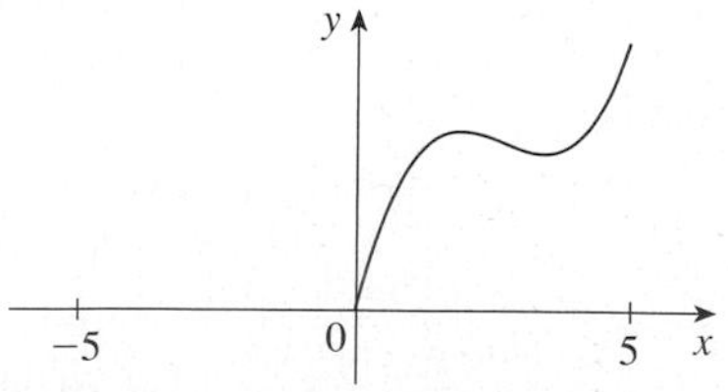

65–70 Determine whether f is even, odd, or neither. If you have a graphing calculator, use it to check your answer visually.

65. $f(x) = \dfrac{x}{x^2 + 1}$

66. $f(x) = \dfrac{x^2}{x^4 + 1}$

67. $f(x) = \dfrac{x}{x + 1}$

68. $f(x) = x|x|$

69. $f(x) = 1 + 3x^2 - x^4$

70. $f(x) = 1 + 3x^3 - x^5$

1.2 MATHEMATICAL MODELS: A CATALOG OF ESSENTIAL FUNCTIONS

A **mathematical model** is a mathematical description (often by means of a function or an equation) of a real-world phenomenon such as the size of a population, the demand for a product, the speed of a falling object, the concentration of a product in a chemical reaction, the life expectancy of a person at birth, or the cost of emission reductions. The purpose of the model is to understand the phenomenon and perhaps to make predictions about future behavior.

Figure 1 illustrates the process of mathematical modeling. Given a real-world problem, our first task is to formulate a mathematical model by identifying and naming the independent and dependent variables and making assumptions that simplify the phenomenon enough to make it mathematically tractable. We use our knowledge of the physical situation and our mathematical skills to obtain equations that relate the variables. In situations where there is no physical law to guide us, we may need to collect data (either from a library or the Internet or by conducting our own experiments) and examine the data in the form of a table in order to discern patterns. From this numerical representation of a function we may wish to obtain a graphical representation by plotting the data. The graph might even suggest a suitable algebraic formula in some cases.

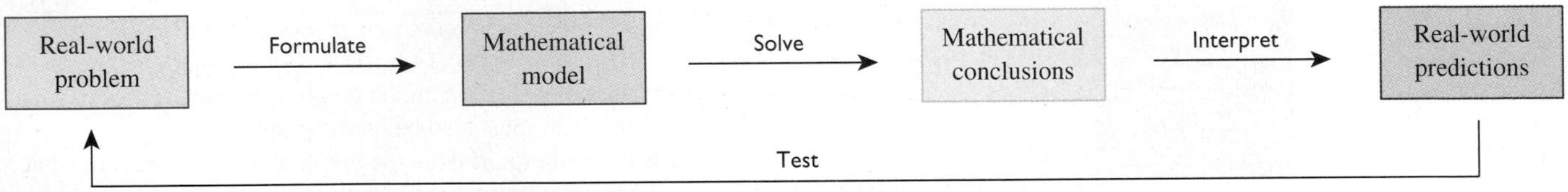

FIGURE 1 The modeling process

The second stage is to apply the mathematics that we know (such as the calculus that will be developed throughout this book) to the mathematical model that we have formulated in order to derive mathematical conclusions. Then, in the third stage, we take those mathematical conclusions and interpret them as information about the original real-world phenomenon by way of offering explanations or making predictions. The final step is to test our predictions by checking against new real data. If the predictions don't compare well with reality, we need to refine our model or to formulate a new model and start the cycle again.

A mathematical model is never a completely accurate representation of a physical situation—it is an *idealization.* A good model simplifies reality enough to permit mathematical calculations but is accurate enough to provide valuable conclusions. It is important to realize the limitations of the model. In the end, Mother Nature has the final say.

There are many different types of functions that can be used to model relationships observed in the real world. In what follows, we discuss the behavior and graphs of these functions and give examples of situations appropriately modeled by such functions.

LINEAR MODELS

■ The coordinate geometry of lines is reviewed in Appendix B.

When we say that y is a **linear function** of x, we mean that the graph of the function is a line, so we can use the slope-intercept form of the equation of a line to write a formula for the function as

$$y = f(x) = mx + b$$

where m is the slope of the line and b is the y-intercept.

A characteristic feature of linear functions is that they grow at a constant rate. For instance, Figure 2 shows a graph of the linear function $f(x) = 3x - 2$ and a table of sample values. Notice that whenever x increases by 0.1, the value of $f(x)$ increases by 0.3. So $f(x)$ increases three times as fast as x. Thus the slope of the graph $y = 3x - 2$, namely 3, can be interpreted as the rate of change of y with respect to x.

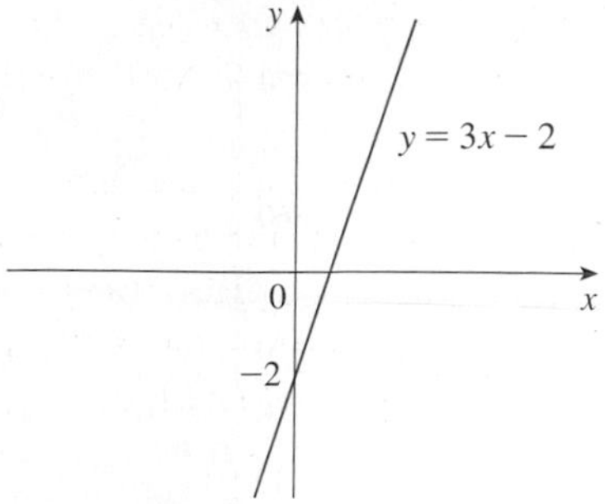

x	$f(x) = 3x - 2$
1.0	1.0
1.1	1.3
1.2	1.6
1.3	1.9
1.4	2.2
1.5	2.5

FIGURE 2

V EXAMPLE 1

(a) As dry air moves upward, it expands and cools. If the ground temperature is 20°C and the temperature at a height of 1 km is 10°C, express the temperature T (in °C) as a function of the height h (in kilometers), assuming that a linear model is appropriate.
(b) Draw the graph of the function in part (a). What does the slope represent?
(c) What is the temperature at a height of 2.5 km?

SOLUTION

(a) Because we are assuming that T is a linear function of h, we can write

$$T = mh + b$$

We are given that $T = 20$ when $h = 0$, so

$$20 = m \cdot 0 + b = b$$

In other words, the y-intercept is $b = 20$.

We are also given that $T = 10$ when $h = 1$, so

$$10 = m \cdot 1 + 20$$

The slope of the line is therefore $m = 10 - 20 = -10$ and the required linear function is

$$T = -10h + 20$$

(b) The graph is sketched in Figure 3. The slope is $m = -10$°C/km, and this represents the rate of change of temperature with respect to height.

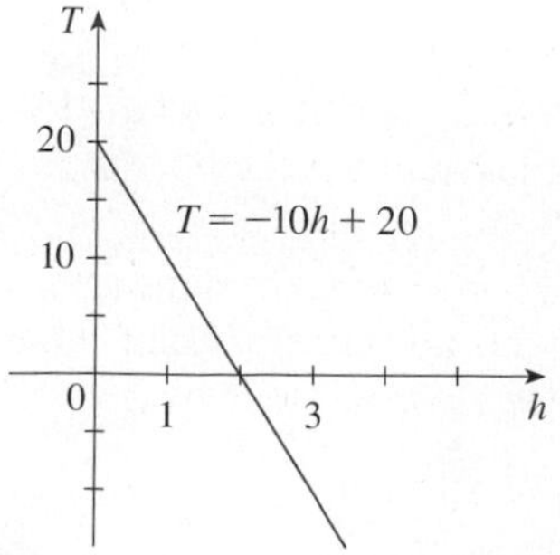

FIGURE 3

(c) At a height of $h = 2.5$ km, the temperature is

$$T = -10(2.5) + 20 = -5°\text{C}$$ □

If there is no physical law or principle to help us formulate a model, we construct an **empirical model**, which is based entirely on collected data. We seek a curve that "fits" the data in the sense that it captures the basic trend of the data points.

EXAMPLE 2 Table 1 lists the average carbon dioxide level in the atmosphere, measured in parts per million at Mauna Loa Observatory from 1980 to 2002. Use the data in Table 1 to find a model for the carbon dioxide level.

SOLUTION We use the data in Table 1 to make the scatter plot in Figure 4, where t represents time (in years) and C represents the CO_2 level (in parts per million, ppm).

TABLE 1

Year	CO_2 level (in ppm)	Year	CO_2 level (in ppm)
1980	338.7	1992	356.4
1982	341.1	1994	358.9
1984	344.4	1996	362.6
1986	347.2	1998	366.6
1988	351.5	2000	369.4
1990	354.2	2002	372.9

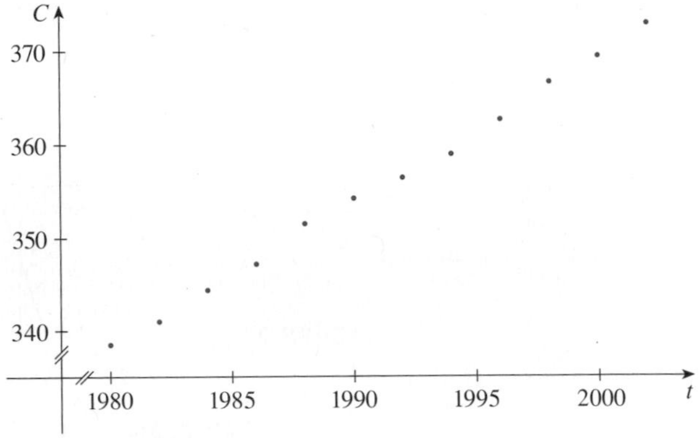

FIGURE 4 Scatter plot for the average CO_2 level

Notice that the data points appear to lie close to a straight line, so it's natural to choose a linear model in this case. But there are many possible lines that approximate these data points, so which one should we use? From the graph, it appears that one possibility is the line that passes through the first and last data points. The slope of this line is

$$\frac{372.9 - 338.7}{2002 - 1980} = \frac{34.2}{22} \approx 1.5545$$

and its equation is

$$C - 338.7 = 1.5545(t - 1980)$$

or

$$C = 1.5545t - 2739.21 \qquad \boxed{1}$$

Equation 1 gives one possible linear model for the carbon dioxide level; it is graphed in Figure 5.

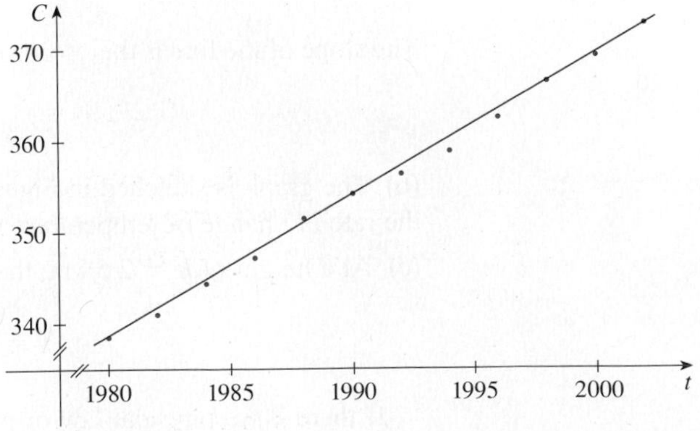

FIGURE 5
Linear model through first and last data points

Although our model fits the data reasonably well, it gives values higher than most of the actual CO_2 levels. A better linear model is obtained by a procedure from statistics

■ A computer or graphing calculator finds the regression line by the method of **least squares**, which is to minimize the sum of the squares of the vertical distances between the data points and the line. The details are explained in Section 14.7.

called *linear regression.* If we use a graphing calculator, we enter the data from Table 1 into the data editor and choose the linear regression command. (With Maple we use the fit[leastsquare] command in the stats package; with Mathematica we use the Fit command.) The machine gives the slope and y-intercept of the regression line as

$$m = 1.55192 \qquad b = -2734.55$$

So our least squares model for the CO_2 level is

$$\boxed{2} \qquad C = 1.55192t - 2734.55$$

In Figure 6 we graph the regression line as well as the data points. Comparing with Figure 5, we see that it gives a better fit than our previous linear model.

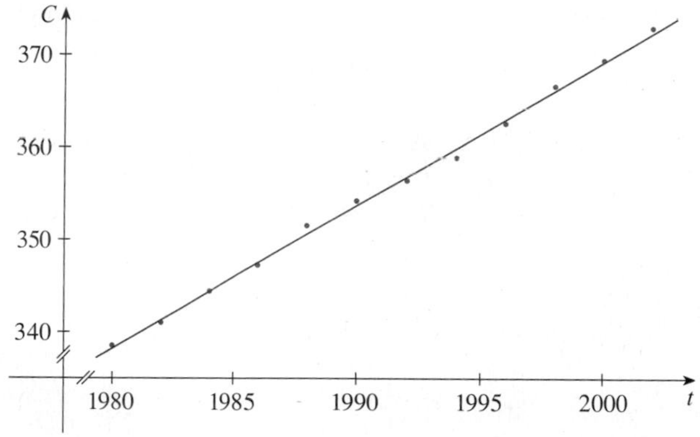

FIGURE 6
The regression line

□

V EXAMPLE 3 Use the linear model given by Equation 2 to estimate the average CO_2 level for 1987 and to predict the level for the year 2010. According to this model, when will the CO_2 level exceed 400 parts per million?

SOLUTION Using Equation 2 with $t = 1987$, we estimate that the average CO_2 level in 1987 was

$$C(1987) = (1.55192)(1987) - 2734.55 \approx 349.12$$

This is an example of *interpolation* because we have estimated a value *between* observed values. (In fact, the Mauna Loa Observatory reported that the average CO_2 level in 1987 was 348.93 ppm, so our estimate is quite accurate.)

With $t = 2010$, we get

$$C(2010) = (1.55192)(2010) - 2734.55 \approx 384.81$$

So we predict that the average CO_2 level in the year 2010 will be 384.8 ppm. This is an example of *extrapolation* because we have predicted a value *outside* the region of observations. Consequently, we are far less certain about the accuracy of our prediction.

Using Equation 2, we see that the CO_2 level exceeds 400 ppm when

$$1.55192t - 2734.55 > 400$$

Solving this inequality, we get

$$t > \frac{3134.55}{1.55192} \approx 2019.79$$

We therefore predict that the CO_2 level will exceed 400 ppm by the year 2019. This prediction is somewhat risky because it involves a time quite remote from our observations. □

POLYNOMIALS

A function P is called a **polynomial** if

$$P(x) = a_n x^n + a_{n-1} x^{n-1} + \cdots + a_2 x^2 + a_1 x + a_0$$

where n is a nonnegative integer and the numbers $a_0, a_1, a_2, \ldots, a_n$ are constants called the **coefficients** of the polynomial. The domain of any polynomial is $\mathbb{R} = (-\infty, \infty)$. If the leading coefficient $a_n \neq 0$, then the **degree** of the polynomial is n. For example, the function

$$P(x) = 2x^6 - x^4 + \tfrac{2}{5}x^3 + \sqrt{2}$$

is a polynomial of degree 6.

A polynomial of degree 1 is of the form $P(x) = mx + b$ and so it is a linear function. A polynomial of degree 2 is of the form $P(x) = ax^2 + bx + c$ and is called a **quadratic function**. Its graph is always a parabola obtained by shifting the parabola $y = ax^2$, as we will see in the next section. The parabola opens upward if $a > 0$ and downward if $a < 0$. (See Figure 7.)

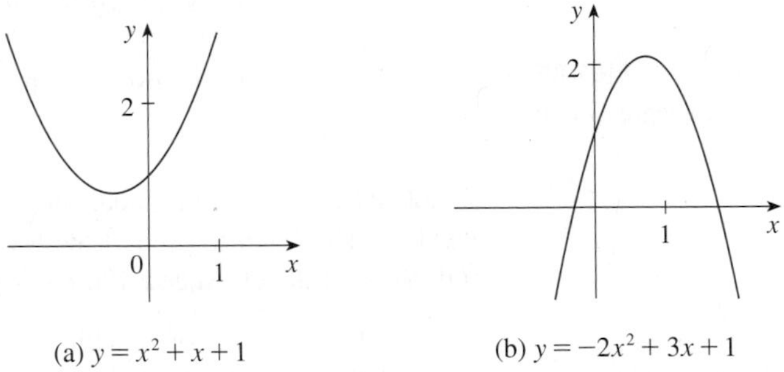

FIGURE 7
The graphs of quadratic functions are parabolas.

A polynomial of degree 3 is of the form

$$P(x) = ax^3 + bx^2 + cx + d \qquad (a \neq 0)$$

and is called a **cubic function**. Figure 8 shows the graph of a cubic function in part (a) and graphs of polynomials of degrees 4 and 5 in parts (b) and (c). We will see later why the graphs have these shapes.

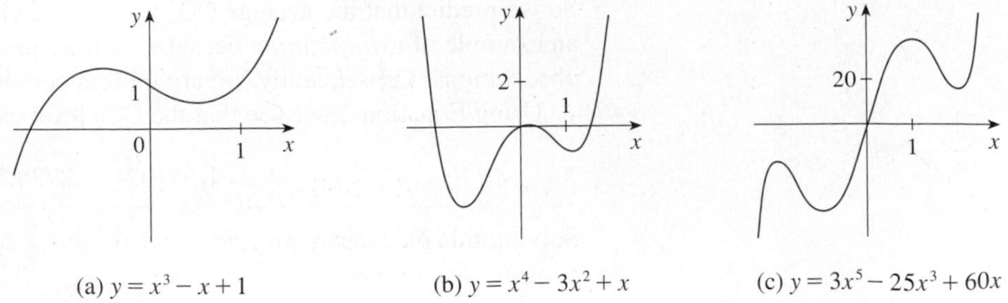

FIGURE 8

Polynomials are commonly used to model various quantities that occur in the natural and social sciences. For instance, in Section 3.7 we will explain why economists often use a polynomial $P(x)$ to represent the cost of producing x units of a commodity. In the following example we use a quadratic function to model the fall of a ball.

TABLE 2

Time (seconds)	Height (meters)
0	450
1	445
2	431
3	408
4	375
5	332
6	279
7	216
8	143
9	61

EXAMPLE 4 A ball is dropped from the upper observation deck of the CN Tower, 450 m above the ground, and its height h above the ground is recorded at 1-second intervals in Table 2. Find a model to fit the data and use the model to predict the time at which the ball hits the ground.

SOLUTION We draw a scatter plot of the data in Figure 9 and observe that a linear model is inappropriate. But it looks as if the data points might lie on a parabola, so we try a quadratic model instead. Using a graphing calculator or computer algebra system (which uses the least squares method), we obtain the following quadratic model:

$$h = 449.36 + 0.96t - 4.90t^2 \quad (3)$$

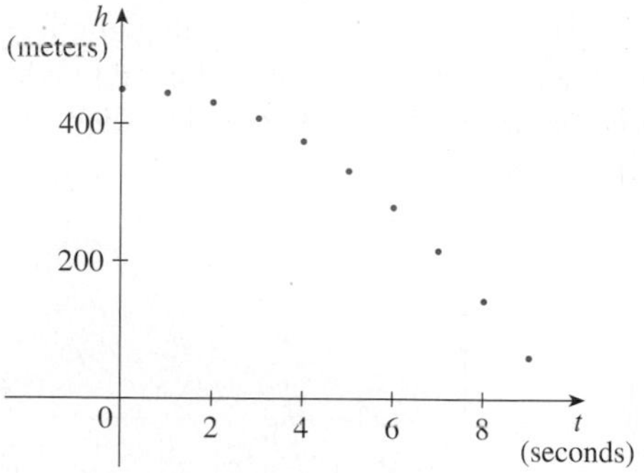

FIGURE 9
Scatter plot for a falling ball

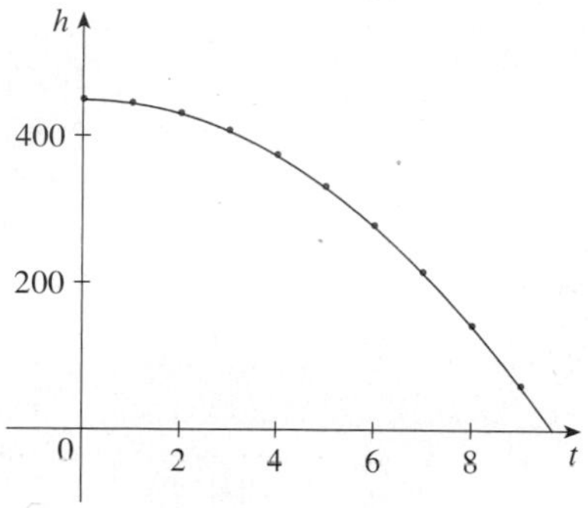

FIGURE 10
Quadratic model for a falling ball

In Figure 10 we plot the graph of Equation 3 together with the data points and see that the quadratic model gives a very good fit.

The ball hits the ground when $h = 0$, so we solve the quadratic equation

$$-4.90t^2 + 0.96t + 449.36 = 0$$

The quadratic formula gives

$$t = \frac{-0.96 \pm \sqrt{(0.96)^2 - 4(-4.90)(449.36)}}{2(-4.90)}$$

The positive root is $t \approx 9.67$, so we predict that the ball will hit the ground after about 9.7 seconds. □

POWER FUNCTIONS

A function of the form $f(x) = x^a$, where a is a constant, is called a **power function**. We consider several cases.

(i) $a = n$, where n is a positive integer

The graphs of $f(x) = x^n$ for $n = 1, 2, 3, 4$, and 5 are shown in Figure 11. (These are polynomials with only one term.) We already know the shape of the graphs of $y = x$ (a line through the origin with slope 1) and $y = x^2$ [a parabola, see Example 2(b) in Section 1.1].

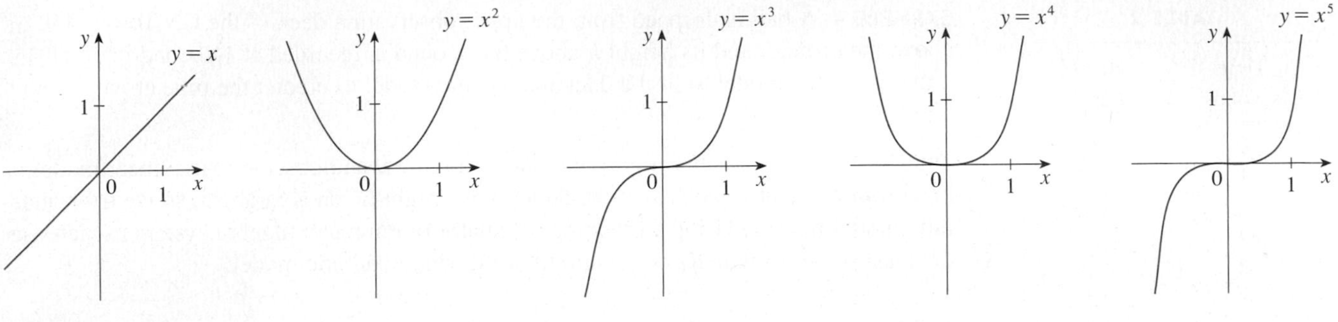

FIGURE 11 Graphs of $f(x) = x^n$ for $n = 1, 2, 3, 4, 5$

The general shape of the graph of $f(x) = x^n$ depends on whether n is even or odd. If n is even, then $f(x) = x^n$ is an even function and its graph is similar to the parabola $y = x^2$. If n is odd, then $f(x) = x^n$ is an odd function and its graph is similar to that of $y = x^3$. Notice from Figure 12, however, that as n increases, the graph of $y = x^n$ becomes flatter near 0 and steeper when $|x| \geqslant 1$. (If x is small, then x^2 is smaller, x^3 is even smaller, x^4 is smaller still, and so on.)

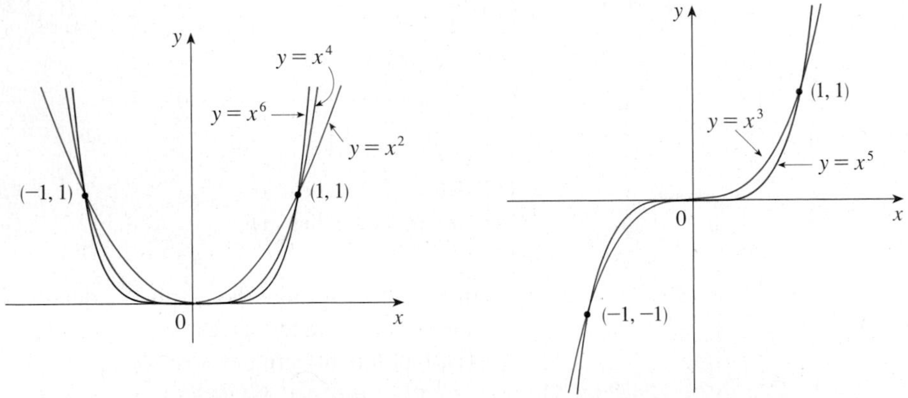

FIGURE 12
Families of power functions

(ii) $a = 1/n$, where n is a positive integer

The function $f(x) = x^{1/n} = \sqrt[n]{x}$ is a **root function**. For $n = 2$ it is the square root function $f(x) = \sqrt{x}$, whose domain is $[0, \infty)$ and whose graph is the upper half of the parabola $x = y^2$. [See Figure 13(a).] For other even values of n, the graph of $y = \sqrt[n]{x}$ is similar to that of $y = \sqrt{x}$. For $n = 3$ we have the cube root function $f(x) = \sqrt[3]{x}$ whose domain is $\mathbb{R}$ (recall that every real number has a cube root) and whose graph is shown in Figure 13(b). The graph of $y = \sqrt[n]{x}$ for n odd ($n > 3$) is similar to that of $y = \sqrt[3]{x}$.

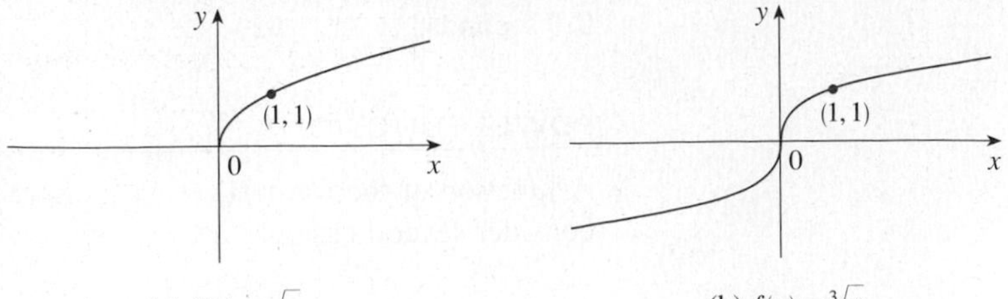

FIGURE 13
Graphs of root functions

(a) $f(x) = \sqrt{x}$ (b) $f(x) = \sqrt[3]{x}$

(iii) $a = -1$

The graph of the **reciprocal function** $f(x) = x^{-1} = 1/x$ is shown in Figure 14. Its graph has the equation $y = 1/x$, or $xy = 1$, and is a hyperbola with the coordinate axes as its asymptotes. This function arises in physics and chemistry in connection with Boyle's Law, which says that, when the temperature is constant, the volume V of a gas is inversely proportional to the pressure P:

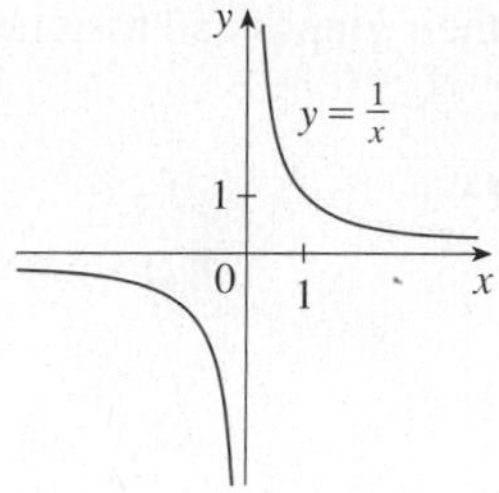

FIGURE 14
The reciprocal function

$$V = \frac{C}{P}$$

where C is a constant. Thus the graph of V as a function of P (see Figure 15) has the same general shape as the right half of Figure 14.

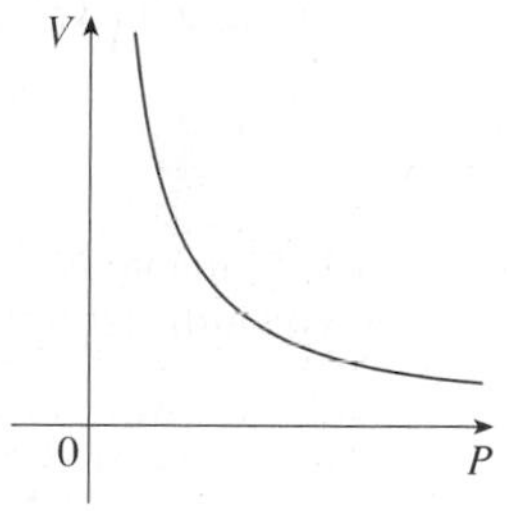

FIGURE 15
Volume as a function of pressure at constant temperature

Another instance in which a power function is used to model a physical phenomenon is discussed in Exercise 26.

RATIONAL FUNCTIONS

A **rational function** f is a ratio of two polynomials:

$$f(x) = \frac{P(x)}{Q(x)}$$

where P and Q are polynomials. The domain consists of all values of x such that $Q(x) \neq 0$. A simple example of a rational function is the function $f(x) = 1/x$, whose domain is $\{x \mid x \neq 0\}$; this is the reciprocal function graphed in Figure 14. The function

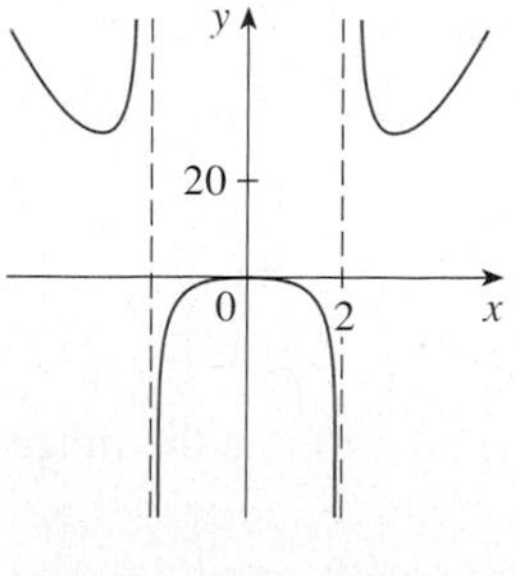

FIGURE 16
$f(x) = \dfrac{2x^4 - x^2 + 1}{x^2 - 4}$

$$f(x) = \frac{2x^4 - x^2 + 1}{x^2 - 4}$$

is a rational function with domain $\{x \mid x \neq \pm 2\}$. Its graph is shown in Figure 16.

ALGEBRAIC FUNCTIONS

A function f is called an **algebraic function** if it can be constructed using algebraic operations (such as addition, subtraction, multiplication, division, and taking roots) starting with polynomials. Any rational function is automatically an algebraic function. Here are two more examples:

$$f(x) = \sqrt{x^2 + 1} \qquad g(x) = \frac{x^4 - 16x^2}{x + \sqrt{x}} + (x - 2)\sqrt[3]{x + 1}$$

When we sketch algebraic functions in Chapter 4, we will see that their graphs can assume a variety of shapes. Figure 17 illustrates some of the possibilities.

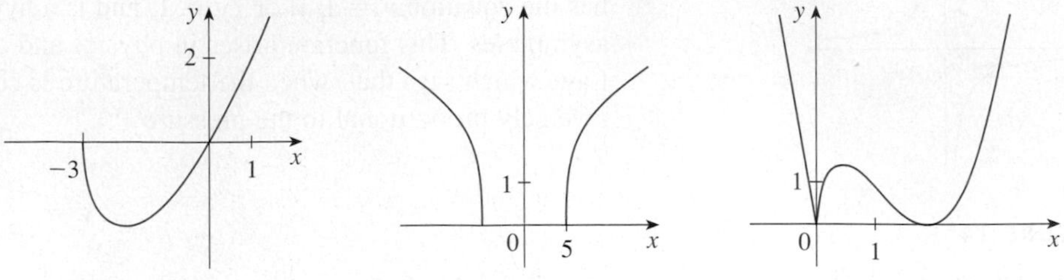

FIGURE 17 (a) $f(x) = x\sqrt{x+3}$ (b) $g(x) = \sqrt[4]{x^2 - 25}$ (c) $h(x) = x^{2/3}(x-2)^2$

An example of an algebraic function occurs in the theory of relativity. The mass of a particle with velocity v is

$$m = f(v) = \frac{m_0}{\sqrt{1 - v^2/c^2}}$$

where m_0 is the rest mass of the particle and $c = 3.0 \times 10^5$ km/s is the speed of light in a vacuum.

TRIGONOMETRIC FUNCTIONS

■ The Reference Pages are located at the front and back of the book.

Trigonometry and the trigonometric functions are reviewed on Reference Page 2 and also in Appendix D. In calculus the convention is that radian measure is always used (except when otherwise indicated). For example, when we use the function $f(x) = \sin x$, it is understood that $\sin x$ means the sine of the angle whose radian measure is x. Thus the graphs of the sine and cosine functions are as shown in Figure 18.

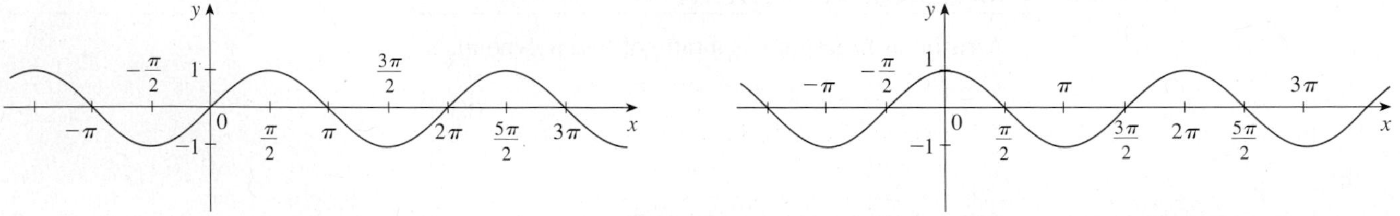

(a) $f(x) = \sin x$ (b) $g(x) = \cos x$

FIGURE 18

Notice that for both the sine and cosine functions the domain is $(-\infty, \infty)$ and the range is the closed interval $[-1, 1]$. Thus, for all values of x, we have

$$-1 \leqslant \sin x \leqslant 1 \qquad -1 \leqslant \cos x \leqslant 1$$

or, in terms of absolute values,

$$|\sin x| \leqslant 1 \qquad |\cos x| \leqslant 1$$

Also, the zeros of the sine function occur at the integer multiples of π; that is,

$$\sin x = 0 \quad \text{when} \quad x = n\pi \quad n \text{ an integer}$$

An important property of the sine and cosine functions is that they are periodic functions and have period 2π. This means that, for all values of x,

$$\sin(x + 2\pi) = \sin x \qquad \cos(x + 2\pi) = \cos x$$

The periodic nature of these functions makes them suitable for modeling repetitive phenomena such as tides, vibrating springs, and sound waves. For instance, in Example 4 in Section 1.3 we will see that a reasonable model for the number of hours of daylight in Philadelphia t days after January 1 is given by the function

$$L(t) = 12 + 2.8 \sin\left[\frac{2\pi}{365}(t - 80)\right]$$

The tangent function is related to the sine and cosine functions by the equation

$$\tan x = \frac{\sin x}{\cos x}$$

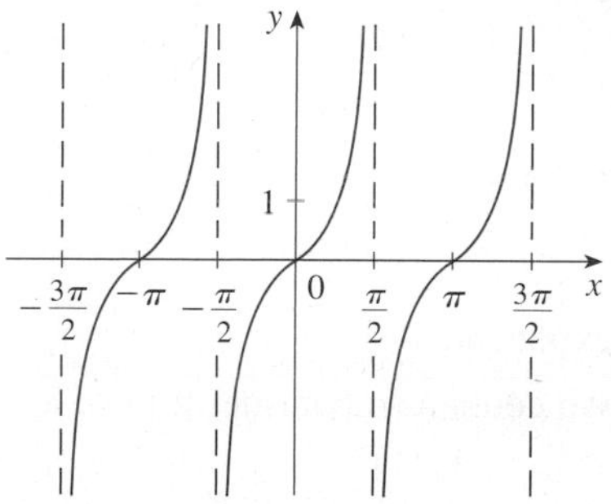

FIGURE 19
$y = \tan x$

and its graph is shown in Figure 19. It is undefined whenever $\cos x = 0$, that is, when $x = \pm\pi/2, \pm 3\pi/2, \ldots$. Its range is $(-\infty, \infty)$. Notice that the tangent function has period π:

$$\tan(x + \pi) = \tan x \qquad \text{for all } x$$

The remaining three trigonometric functions (cosecant, secant, and cotangent) are the reciprocals of the sine, cosine, and tangent functions. Their graphs are shown in Appendix D.

EXPONENTIAL FUNCTIONS

The **exponential functions** are the functions of the form $f(x) = a^x$, where the base a is a positive constant. The graphs of $y = 2^x$ and $y = (0.5)^x$ are shown in Figure 20. In both cases the domain is $(-\infty, \infty)$ and the range is $(0, \infty)$.

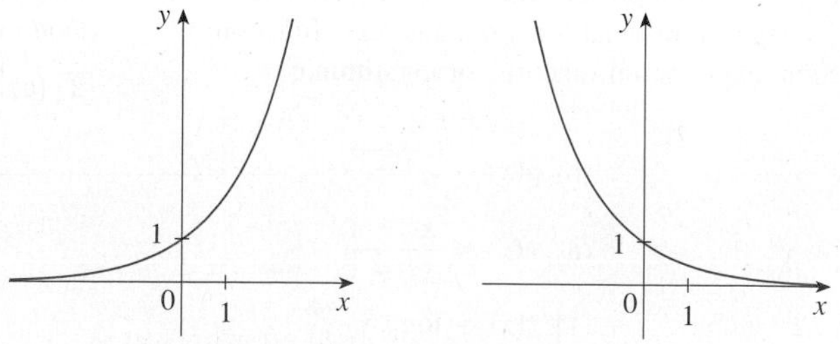

FIGURE 20 (a) $y = 2^x$ (b) $y = (0.5)^x$

Exponential functions will be studied in detail in Section 1.5, and we will see that they are useful for modeling many natural phenomena, such as population growth (if $a > 1$) and radioactive decay (if $a < 1$).

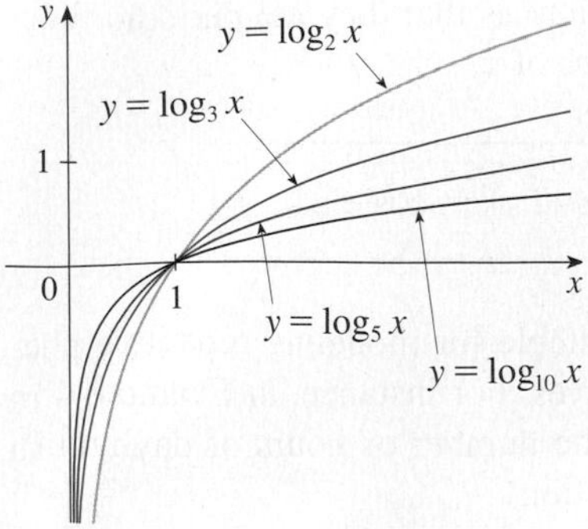

FIGURE 21

LOGARITHMIC FUNCTIONS

The **logarithmic functions** $f(x) = \log_a x$, where the base a is a positive constant, are the inverse functions of the exponential functions. They will be studied in Section 1.6. Figure 21 shows the graphs of four logarithmic functions with various bases. In each case the domain is $(0, \infty)$, the range is $(-\infty, \infty)$, and the function increases slowly when $x > 1$.

TRANSCENDENTAL FUNCTIONS

These are functions that are not algebraic. The set of transcendental functions includes the trigonometric, inverse trigonometric, exponential, and logarithmic functions, but it also includes a vast number of other functions that have never been named. In Chapter 11 we will study transcendental functions that are defined as sums of infinite series.

EXAMPLE 5 Classify the following functions as one of the types of functions that we have discussed.

(a) $f(x) = 5^x$ (b) $g(x) = x^5$

(c) $h(x) = \dfrac{1 + x}{1 - \sqrt{x}}$ (d) $u(t) = 1 - t + 5t^4$

SOLUTION

(a) $f(x) = 5^x$ is an exponential function. (The x is the exponent.)

(b) $g(x) = x^5$ is a power function. (The x is the base.) We could also consider it to be a polynomial of degree 5.

(c) $h(x) = \dfrac{1 + x}{1 - \sqrt{x}}$ is an algebraic function.

(d) $u(t) = 1 - t + 5t^4$ is a polynomial of degree 4. □

1.2 EXERCISES

1–2 Classify each function as a power function, root function, polynomial (state its degree), rational function, algebraic function, trigonometric function, exponential function, or logarithmic function.

1. (a) $f(x) = \sqrt[5]{x}$ (b) $g(x) = \sqrt{1 - x^2}$

(c) $h(x) = x^9 + x^4$ (d) $r(x) = \dfrac{x^2 + 1}{x^3 + x}$

(e) $s(x) = \tan 2x$ (f) $t(x) = \log_{10} x$

2. (a) $y = \dfrac{x - 6}{x + 6}$ (b) $y = x + \dfrac{x^2}{\sqrt{x - 1}}$

(c) $y = 10^x$ (d) $y = x^{10}$

(e) $y = 2t^6 + t^4 - \pi$ (f) $y = \cos\theta + \sin\theta$

3–4 Match each equation with its graph. Explain your choices. (Don't use a computer or graphing calculator.)

3. (a) $y = x^2$ (b) $y = x^5$ (c) $y = x^8$

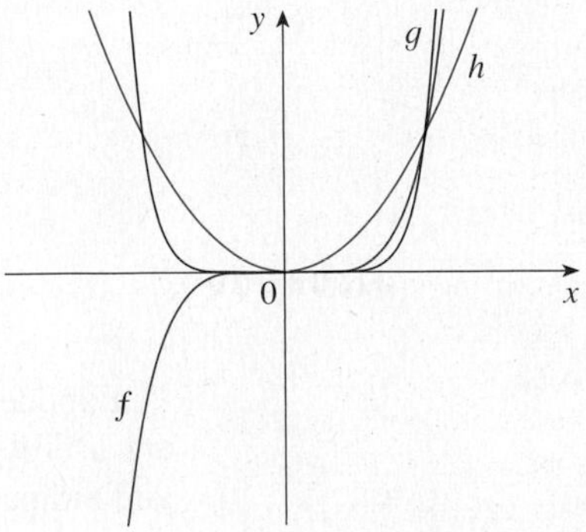

4. (a) $y = 3x$ (b) $y = 3^x$
(c) $y = x^3$ (d) $y = \sqrt[3]{x}$

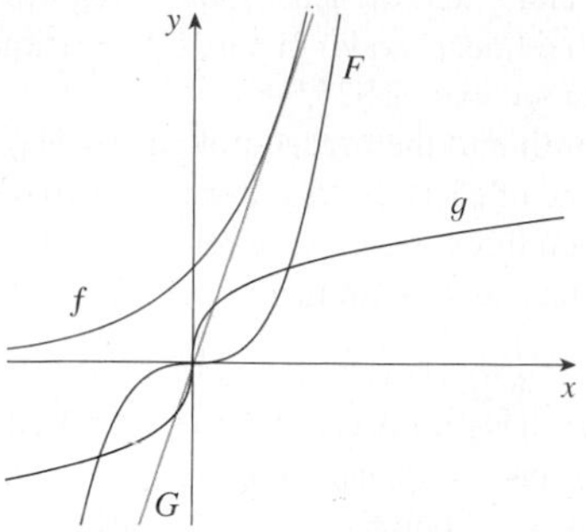

5. (a) Find an equation for the family of linear functions with slope 2 and sketch several members of the family.
(b) Find an equation for the family of linear functions such that $f(2) = 1$ and sketch several members of the family.
(c) Which function belongs to both families?

6. What do all members of the family of linear functions $f(x) = 1 + m(x + 3)$ have in common? Sketch several members of the family.

7. What do all members of the family of linear functions $f(x) = c - x$ have in common? Sketch several members of the family.

8. Find expressions for the quadratic functions whose graphs are shown.

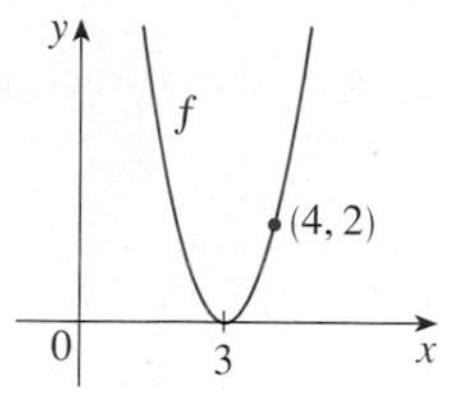

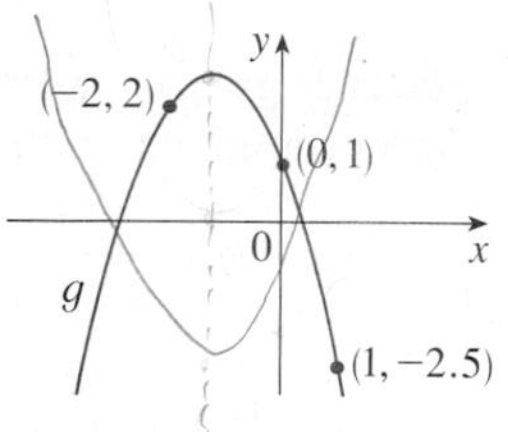

9. Find an expression for a cubic function f if $f(1) = 6$ and $f(-1) = f(0) = f(2) = 0$.

10. Recent studies indicate that the average surface temperature of the earth has been rising steadily. Some scientists have modeled the temperature by the linear function $T = 0.02t + 8.50$, where T is temperature in °C and t represents years since 1900.
(a) What do the slope and T-intercept represent?
(b) Use the equation to predict the average global surface temperature in 2100.

11. If the recommended adult dosage for a drug is D (in mg), then to determine the appropriate dosage c for a child of age a, pharmacists use the equation $c = 0.0417D(a + 1)$. Suppose the dosage for an adult is 200 mg.
(a) Find the slope of the graph of c. What does it represent?
(b) What is the dosage for a newborn?

12. The manager of a weekend flea market knows from past experience that if he charges x dollars for a rental space at the market, then the number y of spaces he can rent is given by the equation $y = 200 - 4x$.
(a) Sketch a graph of this linear function. (Remember that the rental charge per space and the number of spaces rented can't be negative quantities.)
(b) What do the slope, the y-intercept, and the x-intercept of the graph represent?

13. The relationship between the Fahrenheit (F) and Celsius (C) temperature scales is given by the linear function $F = \frac{9}{5}C + 32$.
(a) Sketch a graph of this function.
(b) What is the slope of the graph and what does it represent? What is the F-intercept and what does it represent?

14. Jason leaves Detroit at 2:00 PM and drives at a constant speed west along I-96. He passes Ann Arbor, 40 mi from Detroit, at 2:50 PM.
(a) Express the distance traveled in terms of the time elapsed.
(b) Draw the graph of the equation in part (a).
(c) What is the slope of this line? What does it represent?

15. Biologists have noticed that the chirping rate of crickets of a certain species is related to temperature, and the relationship appears to be very nearly linear. A cricket produces 113 chirps per minute at 70°F and 173 chirps per minute at 80°F.
(a) Find a linear equation that models the temperature T as a function of the number of chirps per minute N.
(b) What is the slope of the graph? What does it represent?
(c) If the crickets are chirping at 150 chirps per minute, estimate the temperature.

16. The manager of a furniture factory finds that it costs \$2200 to manufacture 100 chairs in one day and \$4800 to produce 300 chairs in one day.
(a) Express the cost as a function of the number of chairs produced, assuming that it is linear. Then sketch the graph.
(b) What is the slope of the graph and what does it represent?
(c) What is the y-intercept of the graph and what does it represent?

17. At the surface of the ocean, the water pressure is the same as the air pressure above the water, 15 lb/in^2. Below the surface, the water pressure increases by 4.34 lb/in^2 for every 10 ft of descent.
(a) Express the water pressure as a function of the depth below the ocean surface.
(b) At what depth is the pressure 100 lb/in^2?

18. The monthly cost of driving a car depends on the number of miles driven. Lynn found that in May it cost her \$380 to drive 480 mi and in June it cost her \$460 to drive 800 mi.

(a) Express the monthly cost C as a function of the distance driven d, assuming that a linear relationship gives a suitable model.

(b) Use part (a) to predict the cost of driving 1500 miles per month.

(c) Draw the graph of the linear function. What does the slope represent?

(d) What does the y-intercept represent?

(e) Why does a linear function give a suitable model in this situation?

19–20 For each scatter plot, decide what type of function you might choose as a model for the data. Explain your choices.

19. (a)

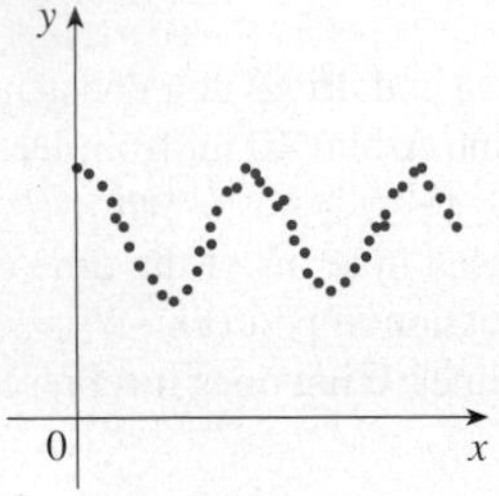

(b)

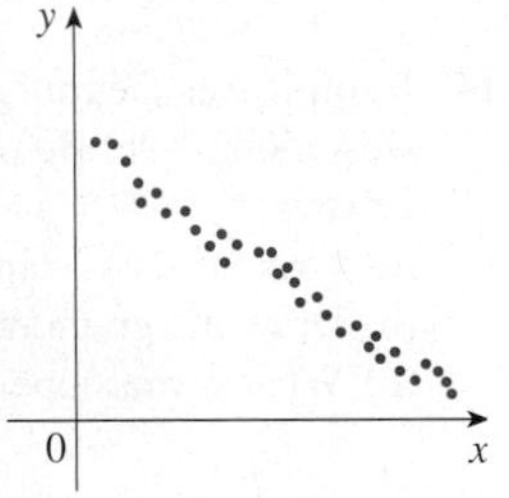

20. (a)

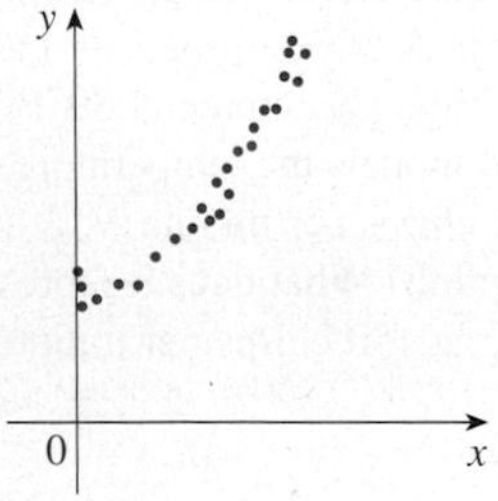

(b)

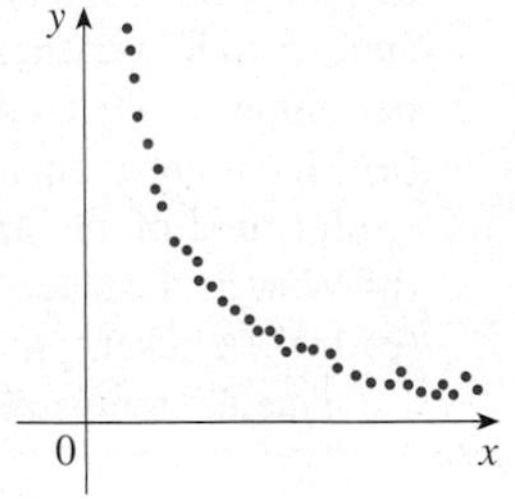

21. The table shows (lifetime) peptic ulcer rates (per 100 population) for various family incomes as reported by the National Health Interview Survey.

Income	Ulcer rate (per 100 population)
\$4,000	14.1
\$6,000	13.0
\$8,000	13.4
\$12,000	12.5
\$16,000	12.0
\$20,000	12.4
\$30,000	10.5
\$45,000	9.4
\$60,000	8.2

(a) Make a scatter plot of these data and decide whether a linear model is appropriate.

(b) Find and graph a linear model using the first and last data points.

(c) Find and graph the least squares regression line.

(d) Use the linear model in part (c) to estimate the ulcer rate for an income of \$25,000.

(e) According to the model, how likely is someone with an income of \$80,000 to suffer from peptic ulcers?

(f) Do you think it would be reasonable to apply the model to someone with an income of \$200,000?

22. Biologists have observed that the chirping rate of crickets of a certain species appears to be related to temperature. The table shows the chirping rates for various temperatures.

Temperature (°F)	Chirping rate (chirps/min)	Temperature (°F)	Chirping rate (chirps/min)
50	20	75	140
55	46	80	173
60	79	85	198
65	91	90	211
70	113		

(a) Make a scatter plot of the data.

(b) Find and graph the regression line.

(c) Use the linear model in part (b) to estimate the chirping rate at 100°F.

23. The table gives the winning heights for the Olympic pole vault competitions in the 20th century.

Year	Height (ft)	Year	Height (ft)
1900	10.83	1956	14.96
1904	11.48	1960	15.42
1908	12.17	1964	16.73
1912	12.96	1968	17.71
1920	13.42	1972	18.04
1924	12.96	1976	18.04
1928	13.77	1980	18.96
1932	14.15	1984	18.85
1936	14.27	1988	19.77
1948	14.10	1992	19.02
1952	14.92	1996	19.42

(a) Make a scatter plot and decide whether a linear model is appropriate.

(b) Find and graph the regression line.

(c) Use the linear model to predict the height of the winning pole vault at the 2000 Olympics and compare with the actual winning height of 19.36 feet.

(d) Is it reasonable to use the model to predict the winning height at the 2100 Olympics?

24. A study by the US Office of Science and Technology in 1972 estimated the cost (in 1972 dollars) to reduce automobile emissions by certain percentages:

Reduction in emissions (%)	Cost per car (in \$)	Reduction in emissions (%)	Cost per car (in \$)
50	45	75	90
55	55	80	100
60	62	85	200
65	70	90	375
70	80	95	600

Find a model that captures the "diminishing returns" trend of these data.

25. Use the data in the table to model the population of the world in the 20th century by a cubic function. Then use your model to estimate the population in the year 1925.

Year	Population (millions)	Year	Population (millions)
1900	1650	1960	3040
1910	1750	1970	3710
1920	1860	1980	4450
1930	2070	1990	5280
1940	2300	2000	6080
1950	2560		

26. The table shows the mean (average) distances d of the planets from the sun (taking the unit of measurement to be the distance from the earth to the sun) and their periods T (time of revolution in years).

Planet	d	T
Mercury	0.387	0.241
Venus	0.723	0.615
Earth	1.000	1.000
Mars	1.523	1.881
Jupiter	5.203	11.861
Saturn	9.541	29.457
Uranus	19.190	84.008
Neptune	30.086	164.784

(a) Fit a power model to the data.

(b) Kepler's Third Law of Planetary Motion states that

"The square of the period of revolution of a planet is proportional to the cube of its mean distance from the sun."

Does your model corroborate Kepler's Third Law?

1.3 NEW FUNCTIONS FROM OLD FUNCTIONS

In this section we start with the basic functions we discussed in Section 1.2 and obtain new functions by shifting, stretching, and reflecting their graphs. We also show how to combine pairs of functions by the standard arithmetic operations and by composition.

TRANSFORMATIONS OF FUNCTIONS

By applying certain transformations to the graph of a given function we can obtain the graphs of certain related functions. This will give us the ability to sketch the graphs of many functions quickly by hand. It will also enable us to write equations for given graphs. Let's first consider **translations**. If c is a positive number, then the graph of $y = f(x) + c$ is just the graph of $y = f(x)$ shifted upward a distance of c units (because each y-coordinate is increased by the same number c). Likewise, if $g(x) = f(x - c)$, where $c > 0$, then the value of g at x is the same as the value of f at $x - c$ (c units to the left of x). Therefore, the graph of $y = f(x - c)$ is just the graph of $y = f(x)$ shifted c units to the right (see Figure 1).

VERTICAL AND HORIZONTAL SHIFTS Suppose $c > 0$. To obtain the graph of

$y = f(x) + c$, shift the graph of $y = f(x)$ a distance c units upward

$y = f(x) - c$, shift the graph of $y = f(x)$ a distance c units downward

$y = f(x - c)$, shift the graph of $y = f(x)$ a distance c units to the right

$y = f(x + c)$, shift the graph of $y = f(x)$ a distance c units to the left

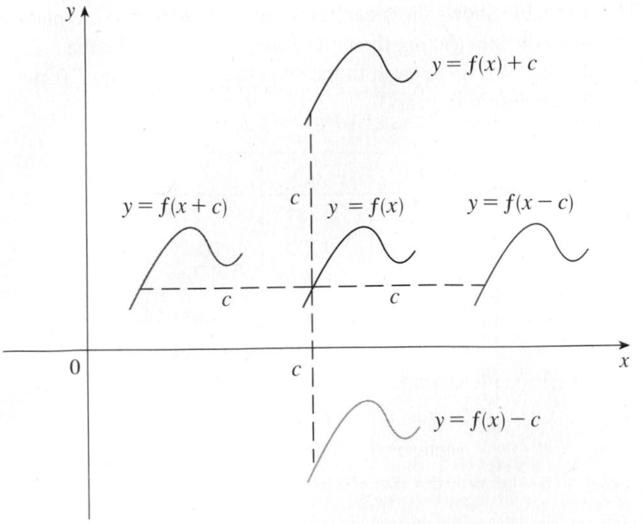

FIGURE 1
Translating the graph of f

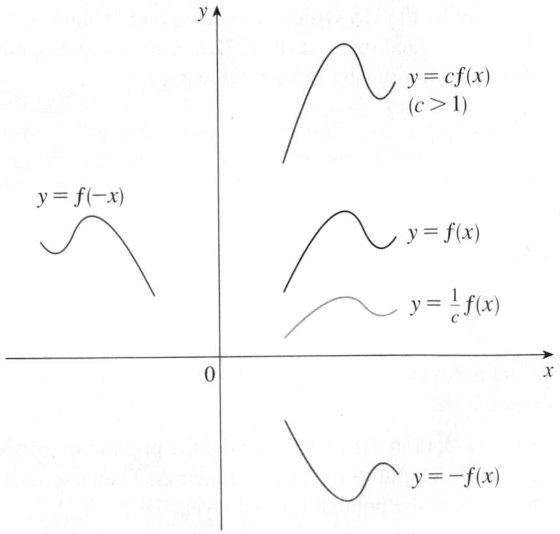

FIGURE 2
Stretching and reflecting the graph of f

Now let's consider the **stretching** and **reflecting** transformations. If $c > 1$, then the graph of $y = cf(x)$ is the graph of $y = f(x)$ stretched by a factor of c in the vertical direction (because each y-coordinate is multiplied by the same number c). The graph of $y = -f(x)$ is the graph of $y = f(x)$ reflected about the x-axis because the point (x, y) is replaced by the point $(x, -y)$. (See Figure 2 and the following chart, where the results of other stretching, compressing, and reflecting transformations are also given.)

VERTICAL AND HORIZONTAL STRETCHING AND REFLECTING Suppose $c > 1$. To obtain the graph of

$y = cf(x)$, stretch the graph of $y = f(x)$ vertically by a factor of c

$y = (1/c)f(x)$, compress the graph of $y = f(x)$ vertically by a factor of c

$y = f(cx)$, compress the graph of $y = f(x)$ horizontally by a factor of c

$y = f(x/c)$, stretch the graph of $y = f(x)$ horizontally by a factor of c

$y = -f(x)$, reflect the graph of $y = f(x)$ about the x-axis

$y = f(-x)$, reflect the graph of $y = f(x)$ about the y-axis

Figure 3 illustrates these stretching transformations when applied to the cosine function with $c = 2$. For instance, in order to get the graph of $y = 2\cos x$ we multiply the y-coordinate of each point on the graph of $y = \cos x$ by 2. This means that the graph of $y = \cos x$ gets stretched vertically by a factor of 2.

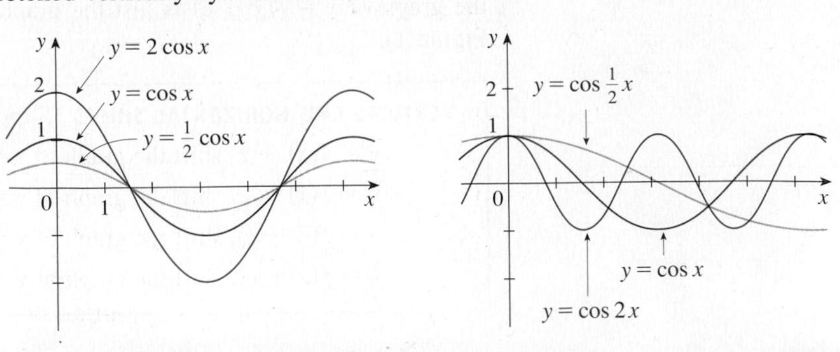

FIGURE 3

V EXAMPLE 1 Given the graph of $y = \sqrt{x}$, use transformations to graph $y = \sqrt{x} - 2$, $y = \sqrt{x - 2}$, $y = -\sqrt{x}$, $y = 2\sqrt{x}$, and $y = \sqrt{-x}$.

SOLUTION The graph of the square root function $y = \sqrt{x}$, obtained from Figure 13(a) in Section 1.2, is shown in Figure 4(a). In the other parts of the figure we sketch $y = \sqrt{x} - 2$ by shifting 2 units downward, $y = \sqrt{x - 2}$ by shifting 2 units to the right, $y = -\sqrt{x}$ by reflecting about the x-axis, $y = 2\sqrt{x}$ by stretching vertically by a factor of 2, and $y = \sqrt{-x}$ by reflecting about the y-axis.

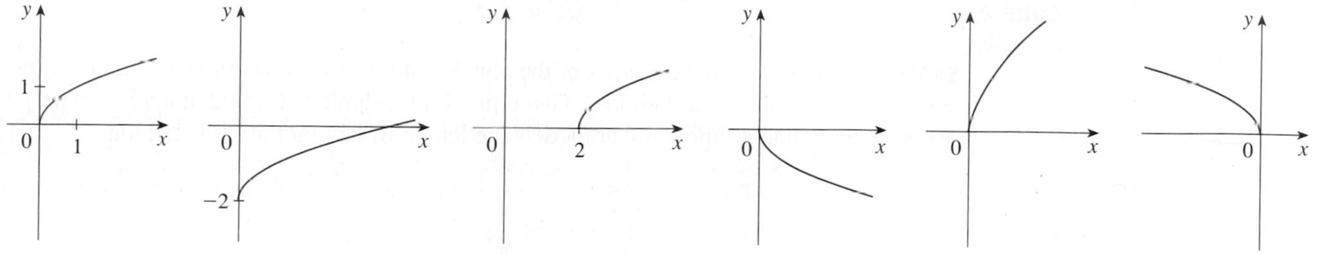

(a) $y = \sqrt{x}$ (b) $y = \sqrt{x} - 2$ (c) $y = \sqrt{x - 2}$ (d) $y = -\sqrt{x}$ (e) $y = 2\sqrt{x}$ (f) $y = \sqrt{-x}$ □

FIGURE 4

EXAMPLE 2 Sketch the graph of the function $f(x) = x^2 + 6x + 10$.

SOLUTION Completing the square, we write the equation of the graph as

$$y = x^2 + 6x + 10 = (x + 3)^2 + 1$$

This means we obtain the desired graph by starting with the parabola $y = x^2$ and shifting 3 units to the left and then 1 unit upward (see Figure 5).

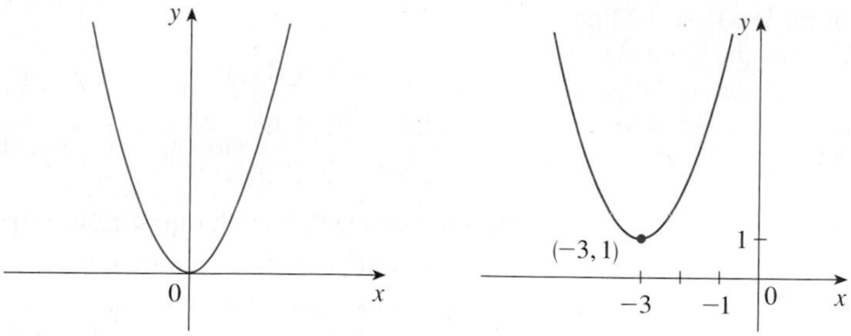

FIGURE 5 (a) $y = x^2$ (b) $y = (x + 3)^2 + 1$ □

EXAMPLE 3 Sketch the graphs of the following functions.
(a) $y = \sin 2x$ (b) $y = 1 - \sin x$

SOLUTION
(a) We obtain the graph of $y = \sin 2x$ from that of $y = \sin x$ by compressing horizontally by a factor of 2 (see Figures 6 and 7). Thus, whereas the period of $y = \sin x$ is 2π, the period of $y = \sin 2x$ is $2\pi/2 = \pi$.

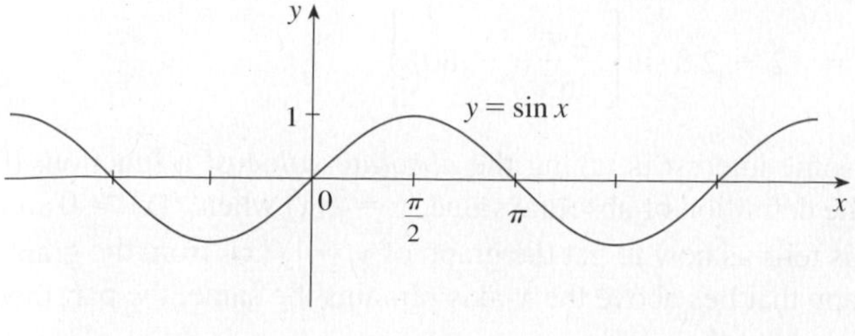

FIGURE 6

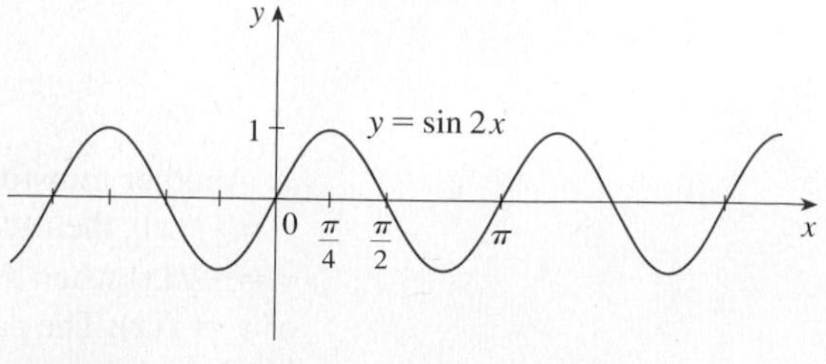

FIGURE 7

(b) To obtain the graph of $y = 1 - \sin x$, we again start with $y = \sin x$. We reflect about the x-axis to get the graph of $y = -\sin x$ and then we shift 1 unit upward to get $y = 1 - \sin x$. (See Figure 8.)

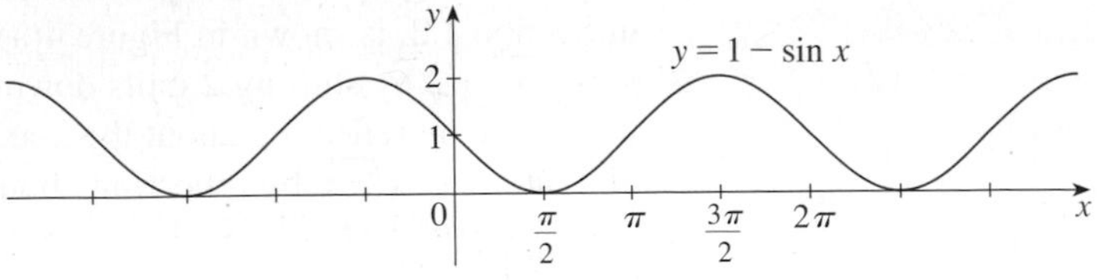

FIGURE 8

□

EXAMPLE 4 Figure 9 shows graphs of the number of hours of daylight as functions of the time of the year at several latitudes. Given that Philadelphia is located at approximately 40°N latitude, find a function that models the length of daylight at Philadelphia.

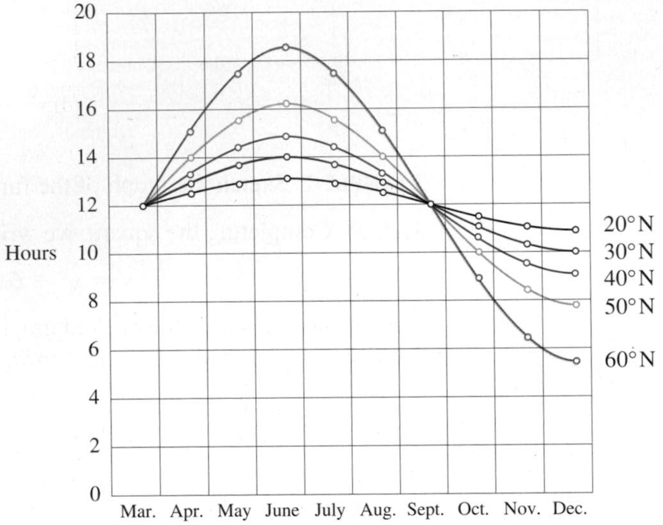

FIGURE 9

Graph of the length of daylight from March 21 through December 21 at various latitudes

Lucia C. Harrison, *Daylight, Twilight, Darkness and Time* (New York: Silver, Burdett, 1935) page 40.

SOLUTION Notice that each curve resembles a shifted and stretched sine function. By looking at the blue curve we see that, at the latitude of Philadelphia, daylight lasts about 14.8 hours on June 21 and 9.2 hours on December 21, so the amplitude of the curve (the factor by which we have to stretch the sine curve vertically) is $\frac{1}{2}(14.8 - 9.2) = 2.8$.

By what factor do we need to stretch the sine curve horizontally if we measure the time t in days? Because there are about 365 days in a year, the period of our model should be 365. But the period of $y = \sin t$ is 2π, so the horizontal stretching factor is $c = 2\pi/365$.

We also notice that the curve begins its cycle on March 21, the 80th day of the year, so we have to shift the curve 80 units to the right. In addition, we shift it 12 units upward. Therefore we model the length of daylight in Philadelphia on the tth day of the year by the function

$$L(t) = 12 + 2.8 \sin\left[\frac{2\pi}{365}(t - 80)\right]$$

□

Another transformation of some interest is taking the *absolute value* of a function. If $y = |f(x)|$, then according to the definition of absolute value, $y = f(x)$ when $f(x) \geq 0$ and $y = -f(x)$ when $f(x) < 0$. This tells us how to get the graph of $y = |f(x)|$ from the graph of $y = f(x)$: The part of the graph that lies above the x-axis remains the same; the part that lies below the x-axis is reflected about the x-axis.

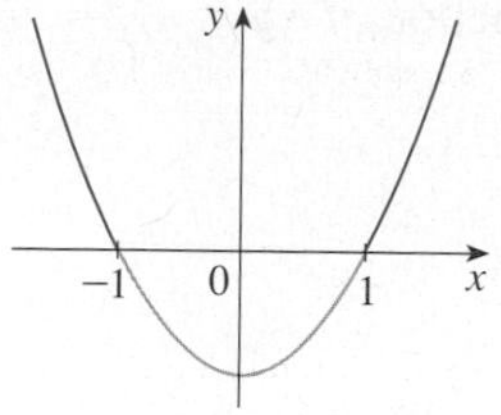

(a) $y = x^2 - 1$

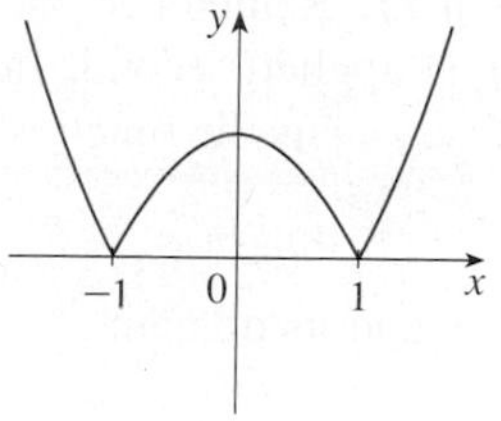

(b) $y = |x^2 - 1|$

FIGURE 10

EXAMPLE 5 Sketch the graph of the function $y = |x^2 - 1|$.

SOLUTION We first graph the parabola $y = x^2 - 1$ in Figure 10(a) by shifting the parabola $y = x^2$ downward 1 unit. We see that the graph lies below the x-axis when $-1 < x < 1$, so we reflect that part of the graph about the x-axis to obtain the graph of $y = |x^2 - 1|$ in Figure 10(b). □

COMBINATIONS OF FUNCTIONS

Two functions f and g can be combined to form new functions $f + g$, $f - g$, fg, and f/g in a manner similar to the way we add, subtract, multiply, and divide real numbers. The sum and difference functions are defined by

$$(f + g)(x) = f(x) + g(x) \qquad (f - g)(x) = f(x) - g(x)$$

If the domain of f is A and the domain of g is B, then the domain of $f + g$ is the intersection $A \cap B$ because both $f(x)$ and $g(x)$ have to be defined. For example, the domain of $f(x) = \sqrt{x}$ is $A = [0, \infty)$ and the domain of $g(x) = \sqrt{2 - x}$ is $B = (-\infty, 2]$, so the domain of $(f + g)(x) = \sqrt{x} + \sqrt{2 - x}$ is $A \cap B = [0, 2]$.

Similarly, the product and quotient functions are defined by

$$(fg)(x) = f(x)g(x) \qquad \left(\frac{f}{g}\right)(x) = \frac{f(x)}{g(x)}$$

The domain of fg is $A \cap B$, but we can't divide by 0 and so the domain of f/g is $\{x \in A \cap B \mid g(x) \neq 0\}$. For instance, if $f(x) = x^2$ and $g(x) = x - 1$, then the domain of the rational function $(f/g)(x) = x^2/(x - 1)$ is $\{x \mid x \neq 1\}$, or $(-\infty, 1) \cup (1, \infty)$.

There is another way of combining two functions to obtain a new function. For example, suppose that $y = f(u) = \sqrt{u}$ and $u = g(x) = x^2 + 1$. Since y is a function of u and u is, in turn, a function of x, it follows that y is ultimately a function of x. We compute this by substitution:

$$y = f(u) = f(g(x)) = f(x^2 + 1) = \sqrt{x^2 + 1}$$

The procedure is called *composition* because the new function is *composed* of the two given functions f and g.

In general, given any two functions f and g, we start with a number x in the domain of g and find its image $g(x)$. If this number $g(x)$ is in the domain of f, then we can calculate the value of $f(g(x))$. The result is a new function $h(x) = f(g(x))$ obtained by substituting g into f. It is called the *composition* (or *composite*) of f and g and is denoted by $f \circ g$ ("f circle g").

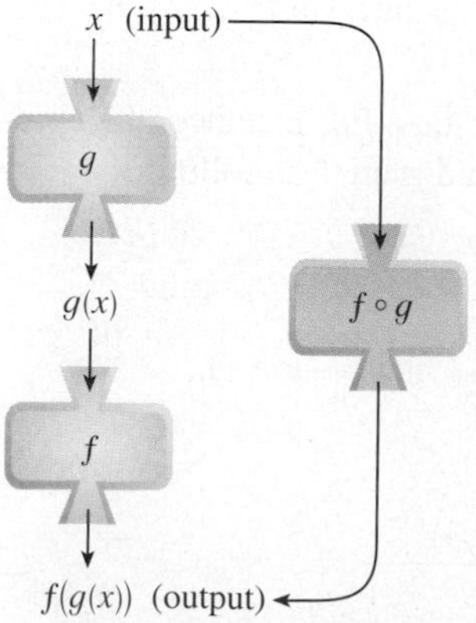

FIGURE 11
The $f \circ g$ machine is composed of the g machine (first) and then the f machine.

> **DEFINITION** Given two functions f and g, the **composite function** $f \circ g$ (also called the **composition** of f and g) is defined by
>
> $$(f \circ g)(x) = f(g(x))$$

The domain of $f \circ g$ is the set of all x in the domain of g such that $g(x)$ is in the domain of f. In other words, $(f \circ g)(x)$ is defined whenever both $g(x)$ and $f(g(x))$ are defined. Figure 11 shows how to picture $f \circ g$ in terms of machines.

EXAMPLE 6 If $f(x) = x^2$ and $g(x) = x - 3$, find the composite functions $f \circ g$ and $g \circ f$.

SOLUTION We have

$$(f \circ g)(x) = f(g(x)) = f(x - 3) = (x - 3)^2$$

$$(g \circ f)(x) = g(f(x)) = g(x^2) = x^2 - 3$$ □

NOTE You can see from Example 6 that, in general, $f \circ g \neq g \circ f$. Remember, the notation $f \circ g$ means that the function g is applied first and then f is applied second. In Example 6, $f \circ g$ is the function that *first* subtracts 3 and *then* squares; $g \circ f$ is the function that *first* squares and *then* subtracts 3.

EXAMPLE 7 If $f(x) = \sqrt{x}$ and $g(x) = \sqrt{2 - x}$, find each function and its domain.
(a) $f \circ g$ (b) $g \circ f$ (c) $f \circ f$ (d) $g \circ g$

SOLUTION

(a) $$(f \circ g)(x) = f(g(x)) = f(\sqrt{2 - x}) = \sqrt{\sqrt{2 - x}} = \sqrt[4]{2 - x}$$

The domain of $f \circ g$ is $\{x \mid 2 - x \geqslant 0\} = \{x \mid x \leqslant 2\} = (-\infty, 2]$.

(b) $$(g \circ f)(x) = g(f(x)) = g(\sqrt{x}) = \sqrt{2 - \sqrt{x}}$$

For $\sqrt{x}$ to be defined we must have $x \geqslant 0$. For $\sqrt{2 - \sqrt{x}}$ to be defined we must have $2 - \sqrt{x} \geqslant 0$, that is, $\sqrt{x} \leqslant 2$, or $x \leqslant 4$. Thus we have $0 \leqslant x \leqslant 4$, so the domain of $g \circ f$ is the closed interval $[0, 4]$.

If $0 \leqslant a \leqslant b$, then $a^2 \leqslant b^2$.

(c) $$(f \circ f)(x) = f(f(x)) = f(\sqrt{x}) = \sqrt{\sqrt{x}} = \sqrt[4]{x}$$

The domain of $f \circ f$ is $[0, \infty)$.

(d) $$(g \circ g)(x) = g(g(x)) = g(\sqrt{2 - x}) = \sqrt{2 - \sqrt{2 - x}}$$

This expression is defined when both $2 - x \geqslant 0$ and $2 - \sqrt{2 - x} \geqslant 0$. The first inequality means $x \leqslant 2$, and the second is equivalent to $\sqrt{2 - x} \leqslant 2$, or $2 - x \leqslant 4$, or $x \geqslant -2$. Thus $-2 \leqslant x \leqslant 2$, so the domain of $g \circ g$ is the closed interval $[-2, 2]$. □

It is possible to take the composition of three or more functions. For instance, the composite function $f \circ g \circ h$ is found by first applying h, then g, and then f as follows:

$$(f \circ g \circ h)(x) = f(g(h(x)))$$

EXAMPLE 8 Find $f \circ g \circ h$ if $f(x) = x/(x + 1)$, $g(x) = x^{10}$, and $h(x) = x + 3$.

SOLUTION

$$(f \circ g \circ h)(x) = f(g(h(x))) = f(g(x + 3))$$

$$= f((x + 3)^{10}) = \frac{(x + 3)^{10}}{(x + 3)^{10} + 1}$$ □

So far we have used composition to build complicated functions from simpler ones. But in calculus it is often useful to be able to *decompose* a complicated function into simpler ones, as in the following example.

EXAMPLE 9 Given $F(x) = \cos^2(x + 9)$, find functions f, g, and h such that $F = f \circ g \circ h$.

SOLUTION Since $F(x) = [\cos(x + 9)]^2$, the formula for F says: First add 9, then take the cosine of the result, and finally square. So we let

$$h(x) = x + 9 \qquad g(x) = \cos x \qquad f(x) = x^2$$

Then

$$(f \circ g \circ h)(x) = f(g(h(x))) = f(g(x + 9)) = f(\cos(x + 9))$$
$$= [\cos(x + 9)]^2 = F(x)$$

□

1.3 EXERCISES

1. Suppose the graph of f is given. Write equations for the graphs that are obtained from the graph of f as follows.
(a) Shift 3 units upward.
(b) Shift 3 units downward.
(c) Shift 3 units to the right.
(d) Shift 3 units to the left.
(e) Reflect about the x-axis.
(f) Reflect about the y-axis.
(g) Stretch vertically by a factor of 3.
(h) Shrink vertically by a factor of 3.

2. Explain how each graph is obtained from the graph of $y = f(x)$.
(a) $y = 5f(x)$ (b) $y = f(x - 5)$
(c) $y = -f(x)$ (d) $y = -5f(x)$
(e) $y = f(5x)$ (f) $y = 5f(x) - 3$

3. The graph of $y = f(x)$ is given. Match each equation with its graph and give reasons for your choices.
(a) $y = f(x - 4)$ (b) $y = f(x) + 3$
(c) $y = \frac{1}{3}f(x)$ (d) $y = -f(x + 4)$
(e) $y = 2f(x + 6)$

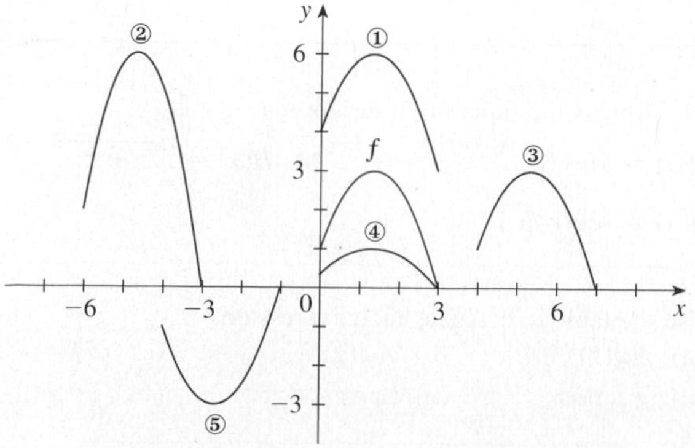

4. The graph of f is given. Draw the graphs of the following functions.
(a) $y = f(x + 4)$ (b) $y = f(x) + 4$
(c) $y = 2f(x)$ (d) $y = -\frac{1}{2}f(x) + 3$

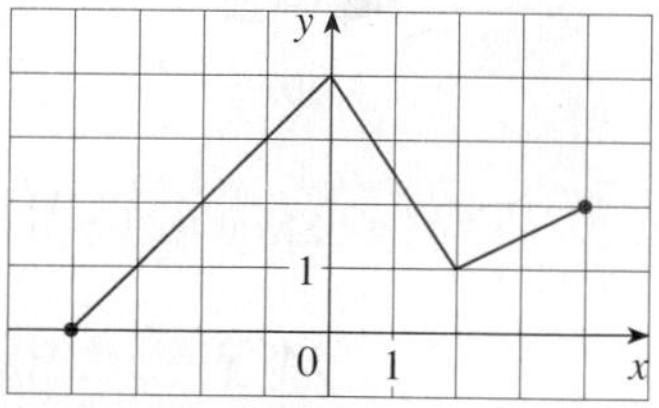

5. The graph of f is given. Use it to graph the following functions.
(a) $y = f(2x)$ (b) $y = f(\frac{1}{2}x)$
(c) $y = f(-x)$ (d) $y = -f(-x)$

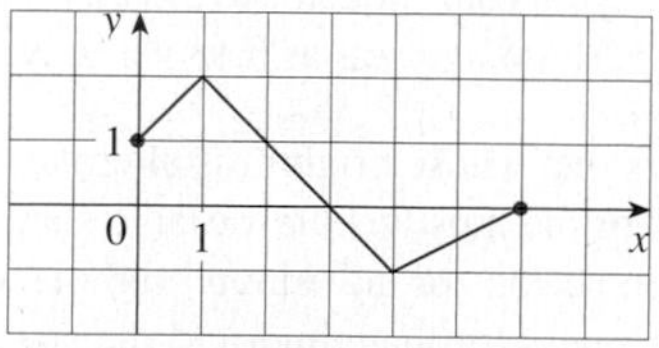

6–7 The graph of $y = \sqrt{3x - x^2}$ is given. Use transformations to create a function whose graph is as shown.

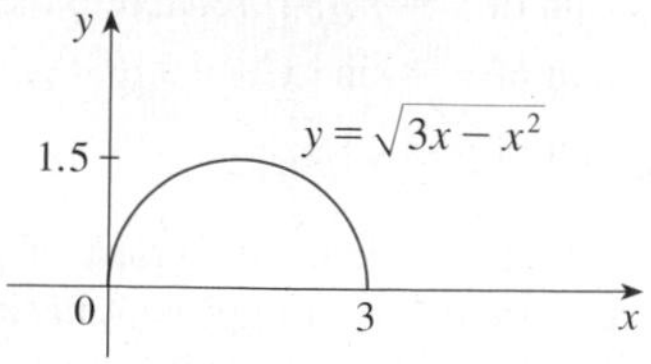

6.

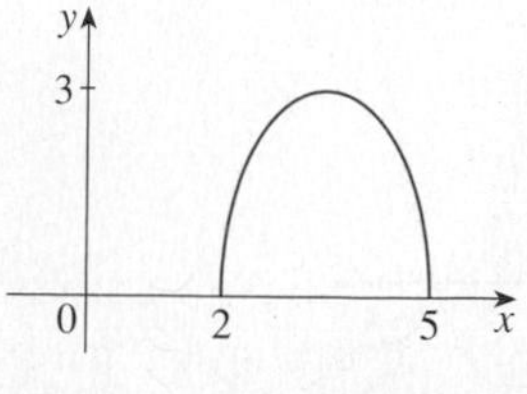

7.

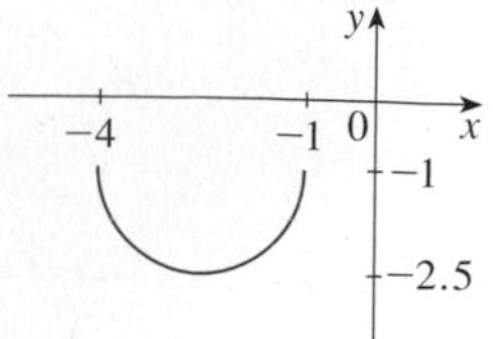

8. (a) How is the graph of $y = 2 \sin x$ related to the graph of $y = \sin x$? Use your answer and Figure 6 to sketch the graph of $y = 2 \sin x$.
(b) How is the graph of $y = 1 + \sqrt{x}$ related to the graph of $y = \sqrt{x}$? Use your answer and Figure 4(a) to sketch the graph of $y = 1 + \sqrt{x}$.

9–24 Graph the function by hand, not by plotting points, but by starting with the graph of one of the standard functions given in Section 1.2, and then applying the appropriate transformations.

9. $y = -x^3$

10. $y = 1 - x^2$

11. $y = (x + 1)^2$

12. $y = x^2 - 4x + 3$

13. $y = 1 + 2 \cos x$

14. $y = 4 \sin 3x$

15. $y = \sin(x/2)$

16. $y = \dfrac{1}{x - 4}$

17. $y = \sqrt{x + 3}$

18. $y = (x + 2)^4 + 3$

19. $y = \frac{1}{2}(x^2 + 8x)$

20. $y = 1 + \sqrt[3]{x - 1}$

21. $y = \dfrac{2}{x + 1}$

22. $y = \dfrac{1}{4} \tan\left(x - \dfrac{\pi}{4}\right)$

23. $y = |\sin x|$

24. $y = |x^2 - 2x|$

25. The city of New Orleans is located at latitude 30°N. Use Figure 9 to find a function that models the number of hours of daylight at New Orleans as a function of the time of year. To check the accuracy of your model, use the fact that on March 31 the sun rises at 5:51 AM and sets at 6:18 PM in New Orleans.

26. A variable star is one whose brightness alternately increases and decreases. For the most visible variable star, Delta Cephei, the time between periods of maximum brightness is 5.4 days, the average brightness (or magnitude) of the star is 4.0, and its brightness varies by ± 0.35 magnitude. Find a function that models the brightness of Delta Cephei as a function of time.

27. (a) How is the graph of $y = f(|x|)$ related to the graph of f?
(b) Sketch the graph of $y = \sin |x|$.
(c) Sketch the graph of $y = \sqrt{|x|}$.

28. Use the given graph of f to sketch the graph of $y = 1/f(x)$. Which features of f are the most important in sketching $y = 1/f(x)$? Explain how they are used.

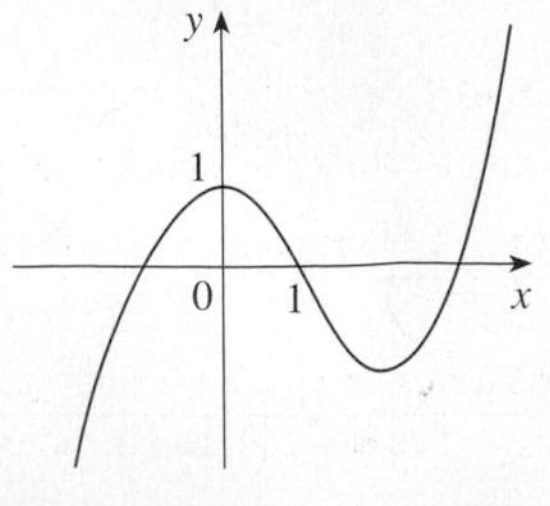

29–30 Find $f + g$, $f - g$, fg, and f/g and state their domains.

29. $f(x) = x^3 + 2x^2$, $g(x) = 3x^2 - 1$

30. $f(x) = \sqrt{3 - x}$, $g(x) = \sqrt{x^2 - 1}$

31–36 Find the functions (a) $f \circ g$, (b) $g \circ f$, (c) $f \circ f$, and (d) $g \circ g$ and their domains.

31. $f(x) = x^2 - 1$, $g(x) = 2x + 1$

32. $f(x) = x - 2$, $g(x) = x^2 + 3x + 4$

33. $f(x) = 1 - 3x$, $g(x) = \cos x$

34. $f(x) = \sqrt{x}$, $g(x) = \sqrt[3]{1 - x}$

35. $f(x) = x + \dfrac{1}{x}$, $g(x) = \dfrac{x + 1}{x + 2}$

36. $f(x) = \dfrac{x}{1 + x}$, $g(x) = \sin 2x$

37–40 Find $f \circ g \circ h$.

37. $f(x) = x + 1$, $g(x) = 2x$, $h(x) = x - 1$

38. $f(x) = 2x - 1$, $g(x) = x^2$, $h(x) = 1 - x$

39. $f(x) = \sqrt{x - 3}$, $g(x) = x^2$, $h(x) = x^3 + 2$

40. $f(x) = \tan x$, $g(x) = \dfrac{x}{x - 1}$, $h(x) = \sqrt[3]{x}$

41–46 Express the function in the form $f \circ g$.

41. $F(x) = (x^2 + 1)^{10}$

42. $F(x) = \sin(\sqrt{x})$

43. $F(x) = \dfrac{\sqrt[3]{x}}{1 + \sqrt[3]{x}}$

44. $G(x) = \sqrt[3]{\dfrac{x}{1 + x}}$

45. $u(t) = \sqrt{\cos t}$

46. $u(t) = \dfrac{\tan t}{1 + \tan t}$

47–49 Express the function in the form $f \circ g \circ h$.

47. $H(x) = 1 - 3^{x^2}$

48. $H(x) = \sqrt[8]{2 + |x|}$

49. $H(x) = \sec^4(\sqrt{x})$

50. Use the table to evaluate each expression.
(a) $f(g(1))$
(b) $g(f(1))$
(c) $f(f(1))$
(d) $g(g(1))$
(e) $(g \circ f)(3)$
(f) $(f \circ g)(6)$

x	1	2	3	4	5	6
$f(x)$	3	1	4	2	2	5
$g(x)$	6	3	2	1	2	3

51. Use the given graphs of f and g to evaluate each expression, or explain why it is undefined.
(a) $f(g(2))$ (b) $g(f(0))$ (c) $(f \circ g)(0)$
(d) $(g \circ f)(6)$ (e) $(g \circ g)(-2)$ (f) $(f \circ f)(4)$

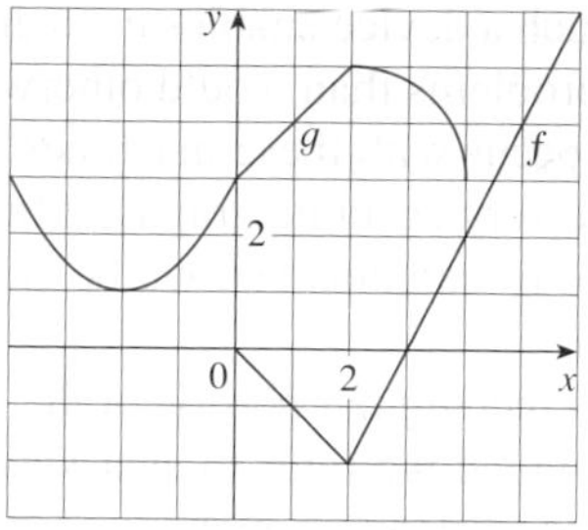

52. Use the given graphs of f and g to estimate the value of $f(g(x))$ for $x = -5, -4, -3, \ldots, 5$. Use these estimates to sketch a rough graph of $f \circ g$.

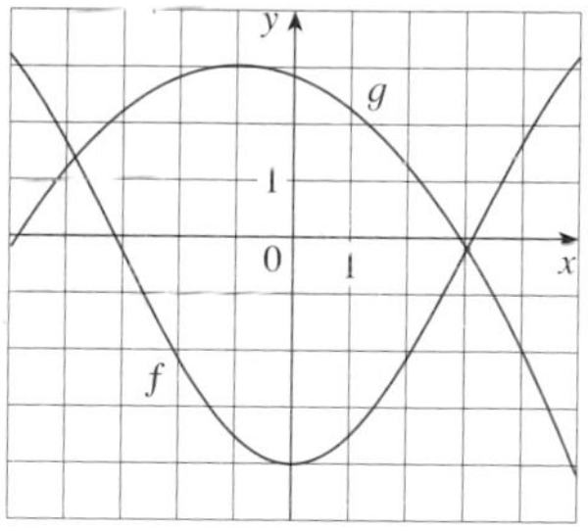

53. A stone is dropped into a lake, creating a circular ripple that travels outward at a speed of 60 cm/s.
(a) Express the radius r of this circle as a function of the time t (in seconds).
(b) If A is the area of this circle as a function of the radius, find $A \circ r$ and interpret it.

54. A spherical balloon is being inflated and the radius of the balloon is increasing at a rate of 2 cm/s.
(a) Express the radius r of the balloon as a function of the time t (in seconds).
(b) If V is the volume of the balloon as a function of the radius, find $V \circ r$ and interpret it.

55. A ship is moving at a speed of 30 km/h parallel to a straight shoreline. The ship is 6 km from shore and it passes a lighthouse at noon.
(a) Express the distance s between the lighthouse and the ship as a function of d, the distance the ship has traveled since noon; that is, find f so that $s = f(d)$.
(b) Express d as a function of t, the time elapsed since noon; that is, find g so that $d = g(t)$.
(c) Find $f \circ g$. What does this function represent?

56. An airplane is flying at a speed of 350 mi/h at an altitude of one mile and passes directly over a radar station at time $t = 0$.
(a) Express the horizontal distance d (in miles) that the plane has flown as a function of t.
(b) Express the distance s between the plane and the radar station as a function of d.
(c) Use composition to express s as a function of t.

57. The **Heaviside function** H is defined by

$$H(t) = \begin{cases} 0 & \text{if } t < 0 \\ 1 & \text{if } t \geqslant 0 \end{cases}$$

It is used in the study of electric circuits to represent the sudden surge of electric current, or voltage, when a switch is instantaneously turned on.
(a) Sketch the graph of the Heaviside function.
(b) Sketch the graph of the voltage $V(t)$ in a circuit if the switch is turned on at time $t = 0$ and 120 volts are applied instantaneously to the circuit. Write a formula for $V(t)$ in terms of $H(t)$.
(c) Sketch the graph of the voltage $V(t)$ in a circuit if the switch is turned on at time $t = 5$ seconds and 240 volts are applied instantaneously to the circuit. Write a formula for $V(t)$ in terms of $H(t)$. (Note that starting at $t = 5$ corresponds to a translation.)

58. The Heaviside function defined in Exercise 57 can also be used to define the **ramp function** $y = ctH(t)$, which represents a gradual increase in voltage or current in a circuit.
(a) Sketch the graph of the ramp function $y = tH(t)$.
(b) Sketch the graph of the voltage $V(t)$ in a circuit if the switch is turned on at time $t = 0$ and the voltage is gradually increased to 120 volts over a 60-second time interval. Write a formula for $V(t)$ in terms of $H(t)$ for $t \leqslant 60$.
(c) Sketch the graph of the voltage $V(t)$ in a circuit if the switch is turned on at time $t = 7$ seconds and the voltage is gradually increased to 100 volts over a period of 25 seconds. Write a formula for $V(t)$ in terms of $H(t)$ for $t \leqslant 32$.

59. Let f and g be linear functions with equations $f(x) = m_1x + b_1$ and $g(x) = m_2x + b_2$. Is $f \circ g$ also a linear function? If so, what is the slope of its graph?

60. If you invest x dollars at 4% interest compounded annually, then the amount $A(x)$ of the investment after one year is $A(x) = 1.04x$. Find $A \circ A$, $A \circ A \circ A$, and $A \circ A \circ A \circ A$. What do these compositions represent? Find a formula for the composition of n copies of A.

61. (a) If $g(x) = 2x + 1$ and $h(x) = 4x^2 + 4x + 7$, find a function f such that $f \circ g = h$. (Think about what operations you would have to perform on the formula for g to end up with the formula for h.)
(b) If $f(x) = 3x + 5$ and $h(x) = 3x^2 + 3x + 2$, find a function g such that $f \circ g = h$.

62. If $f(x) = x + 4$ and $h(x) = 4x - 1$, find a function g such that $g \circ f = h$.

63. (a) Suppose f and g are even functions. What can you say about $f + g$ and fg?
(b) What if f and g are both odd?

64. Suppose f is even and g is odd. What can you say about fg?

65. Suppose g is an even function and let $h = f \circ g$. Is h always an even function?

66. Suppose g is an odd function and let $h = f \circ g$. Is h always an odd function? What if f is odd? What if f is even?

EXAMPLE 4 Graph the function $f(x) = \sin 50x$ in an appropriate viewing rectangle.

SOLUTION Figure 6(a) shows the graph of f produced by a graphing calculator using the viewing rectangle $[-12, 12]$ by $[-1.5, 1.5]$. At first glance the graph appears to be reasonable. But if we change the viewing rectangle to the ones shown in the following parts of Figure 6, the graphs look very different. Something strange is happening.

■ The appearance of the graphs in Figure 6 depends on the machine used. The graphs you get with your own graphing device might not look like these figures, but they will also be quite inaccurate.

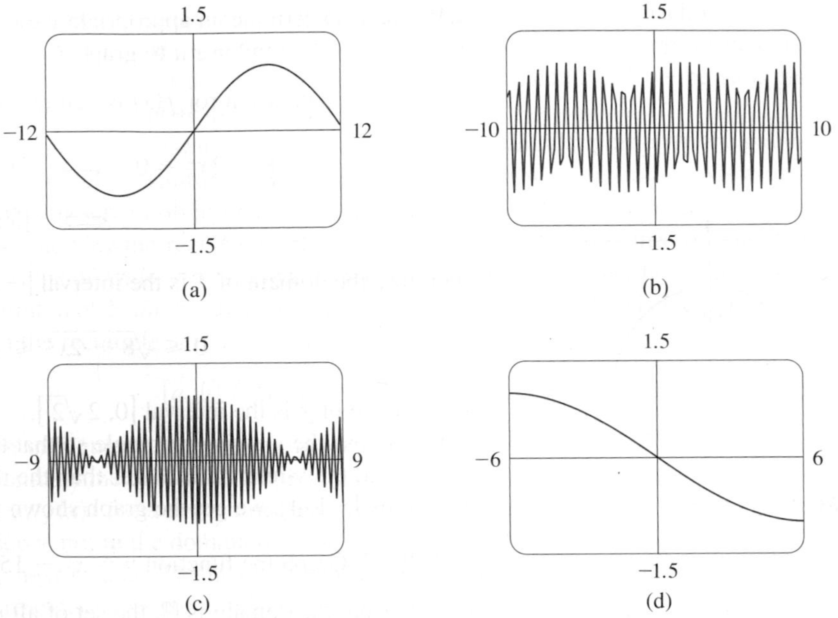

FIGURE 6
Graphs of $f(x) = \sin 50x$ in four viewing rectangles

In order to explain the big differences in appearance of these graphs and to find an appropriate viewing rectangle, we need to find the period of the function $y = \sin 50x$. We know that the function $y = \sin x$ has period 2π and the graph of $y = \sin 50x$ is compressed horizontally by a factor of 50, so the period of $y = \sin 50x$ is

$$\frac{2\pi}{50} = \frac{\pi}{25} \approx 0.126$$

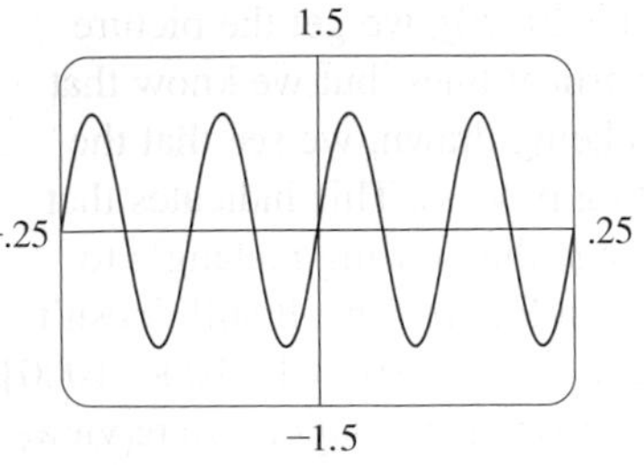

FIGURE 7
$f(x) = \sin 50x$

This suggests that we should deal only with small values of x in order to show just a few oscillations of the graph. If we choose the viewing rectangle $[-0.25, 0.25]$ by $[-1.5, 1.5]$, we get the graph shown in Figure 7.

Now we see what went wrong in Figure 6. The oscillations of $y = \sin 50x$ are so rapid that when the calculator plots points and joins them, it misses most of the maximum and minimum points and therefore gives a very misleading impression of the graph. □

We have seen that the use of an inappropriate viewing rectangle can give a misleading impression of the graph of a function. In Examples 1 and 3 we solved the problem by changing to a larger viewing rectangle. In Example 4 we had to make the viewing rectangle smaller. In the next example we look at a function for which there is no single viewing rectangle that reveals the true shape of the graph.

EXAMPLE 5 Graph the function $f(x) = \sin x + \frac{1}{100} \cos 100x$.

SOLUTION Figure 8 shows the graph of f produced by a graphing calculator with viewing rectangle $[-6.5, 6.5]$ by $[-1.5, 1.5]$. It looks much like the graph of $y = \sin x$, but perhaps with some bumps attached. If we zoom in to the viewing rectangle $[-0.1, 0.1]$ by $[-0.1, 0.1]$, we can see much more clearly the shape of these bumps in Figure 9. The reason for this behavior is that the second term, $\frac{1}{100}\cos 100x$, is very small in comparison with the first term, $\sin x$. Thus we really need two graphs to see the true nature of this function.

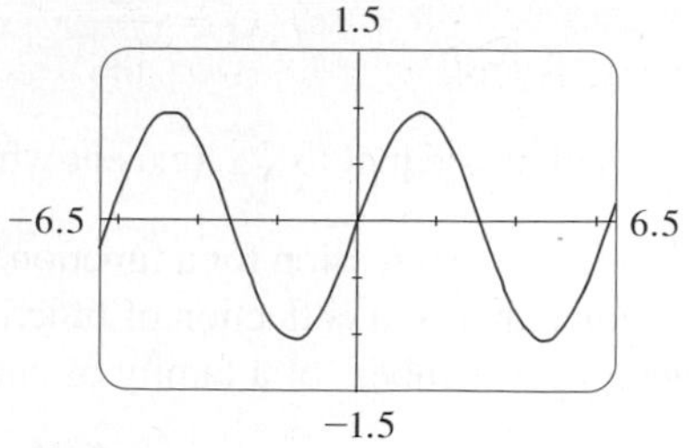

FIGURE 8

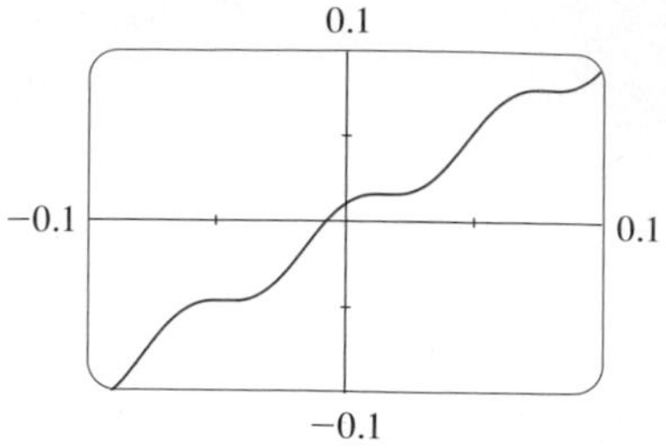

FIGURE 9

□

EXAMPLE 6 Draw the graph of the function $y = \dfrac{1}{1 - x}$.

SOLUTION Figure 10(a) shows the graph produced by a graphing calculator with viewing rectangle $[-9, 9]$ by $[-9, 9]$. In connecting successive points on the graph, the calculator produced a steep line segment from the top to the bottom of the screen. That line segment is not truly part of the graph. Notice that the domain of the function $y = 1/(1 - x)$ is $\{x \mid x \neq 1\}$. We can eliminate the extraneous near-vertical line by experimenting with a change of scale. When we change to the smaller viewing rectangle $[-4.7, 4.7]$ by $[-4.7, 4.7]$ on this particular calculator, we obtain the much better graph in Figure 10(b).

■ Another way to avoid the extraneous line is to change the graphing mode on the calculator so that the dots are not connected.

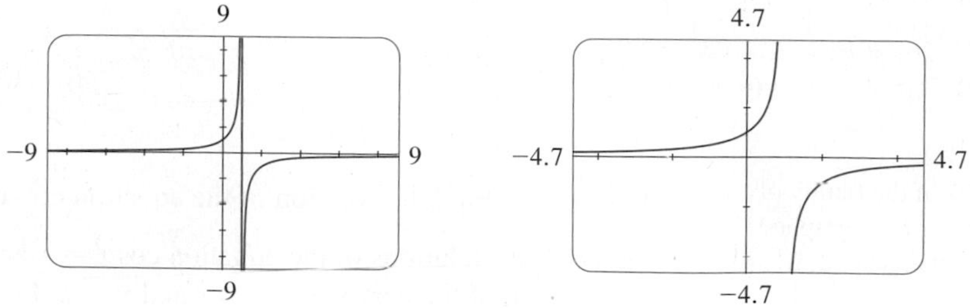

FIGURE 10 (a) (b)

□

EXAMPLE 7 Graph the function $y = \sqrt[3]{x}$.

SOLUTION Some graphing devices display the graph shown in Figure 11, whereas others produce a graph like that in Figure 12. We know from Section 1.2 (Figure 13) that the graph in Figure 12 is correct, so what happened in Figure 11? The explanation is that some machines compute the cube root of x using a logarithm, which is not defined if x is negative, so only the right half of the graph is produced.

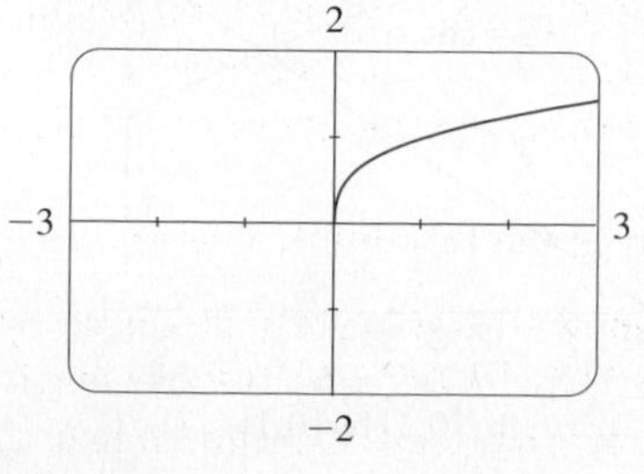

FIGURE 11

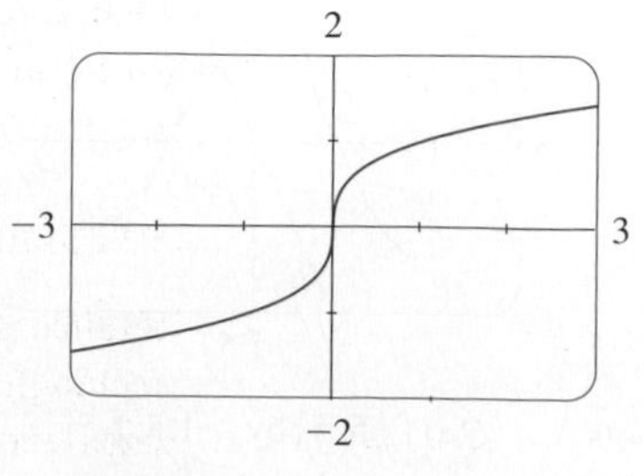

FIGURE 12

You should experiment with your own machine to see which of these two graphs is produced. If you get the graph in Figure 11, you can obtain the correct picture by graphing the function

$$f(x) = \frac{x}{|x|} \cdot |x|^{1/3}$$

Notice that this function is equal to $\sqrt[3]{x}$ (except when $x = 0$). □

To understand how the expression for a function relates to its graph, it's helpful to graph a **family of functions**, that is, a collection of functions whose equations are related. In the next example we graph members of a family of cubic polynomials.

V EXAMPLE 8 Graph the function $y = x^3 + cx$ for various values of the number c. How does the graph change when c is changed?

SOLUTION Figure 13 shows the graphs of $y = x^3 + cx$ for $c = 2, 1, 0, -1$, and -2. We see that, for positive values of c, the graph increases from left to right with no maximum or minimum points (peaks or valleys). When $c = 0$, the curve is flat at the origin. When c is negative, the curve has a maximum point and a minimum point. As c decreases, the maximum point becomes higher and the minimum point lower.

TEC In Visual 1.4 you can see an animation of Figure 13.

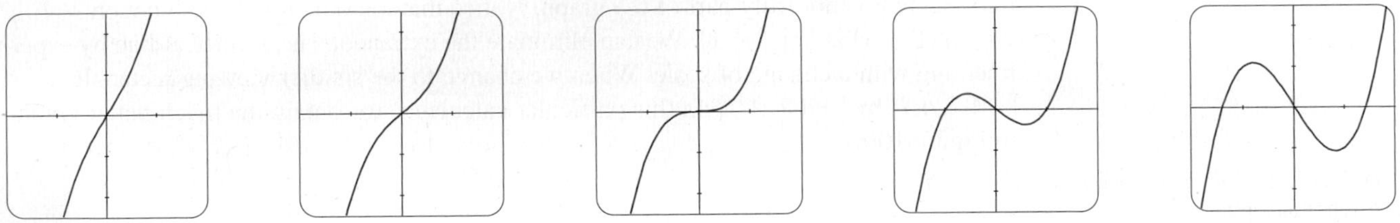

(a) $y = x^3 + 2x$ (b) $y = x^3 + x$ (c) $y = x^3$ (d) $y = x^3 - x$ (e) $y = x^3 - 2x$

□

FIGURE 13
Several members of the family of functions $y = x^3 + cx$, all graphed in the viewing rectangle $[-2, 2]$ by $[-2.5, 2.5]$

EXAMPLE 9 Find the solution of the equation $\cos x = x$ correct to two decimal places.

SOLUTION The solutions of the equation $\cos x = x$ are the x-coordinates of the points of intersection of the curves $y = \cos x$ and $y = x$. From Figure 14(a) we see that there is only one solution and it lies between 0 and 1. Zooming in to the viewing rectangle $[0, 1]$ by $[0, 1]$, we see from Figure 14(b) that the root lies between 0.7 and 0.8. So we zoom in further to the viewing rectangle $[0.7, 0.8]$ by $[0.7, 0.8]$ in Figure 14(c). By moving the cursor to the intersection point of the two curves, or by inspection and the fact that the x-scale is 0.01, we see that the solution of the equation is about 0.74. (Many calculators have a built-in intersection feature.)

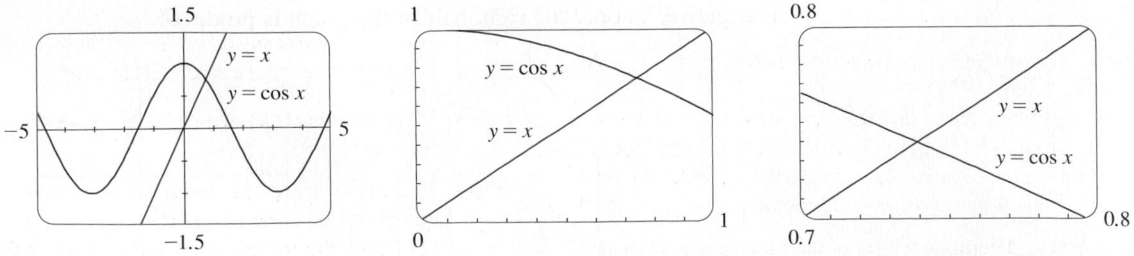

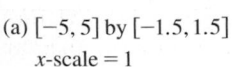

FIGURE 14
Locating the roots of $\cos x = x$

(a) $[-5, 5]$ by $[-1.5, 1.5]$
x-scale = 1

(b) $[0, 1]$ by $[0, 1]$
x-scale = 0.1

(c) $[0.7, 0.8]$ by $[0.7, 0.8]$
x-scale = 0.01

□

1.4 EXERCISES

1. Use a graphing calculator or computer to determine which of the given viewing rectangles produces the most appropriate graph of the function $f(x) = \sqrt{x^3 - 5x^2}$.
(a) $[-5, 5]$ by $[-5, 5]$ (b) $[0, 10]$ by $[0, 2]$
(c) $[0, 10]$ by $[0, 10]$

2. Use a graphing calculator or computer to determine which of the given viewing rectangles produces the most appropriate graph of the function $f(x) = x^4 - 16x^2 + 20$.
(a) $[-3, 3]$ by $[-3, 3]$ (b) $[-10, 10]$ by $[-10, 10]$
(c) $[-50, 50]$ by $[-50, 50]$ (d) $[-5, 5]$ by $[-50, 50]$

3–14 Determine an appropriate viewing rectangle for the given function and use it to draw the graph.

3. $f(x) = 5 + 20x - x^2$
4. $f(x) = x^3 + 30x^2 + 200x$
5. $f(x) = \sqrt[4]{81 - x^4}$
6. $f(x) = \sqrt{0.1x + 20}$
7. $f(x) = x^3 - 225x$
8. $f(x) = \dfrac{x}{x^2 + 100}$
9. $f(x) = \sin^2(1000x)$
10. $f(x) = \cos(0.001x)$
11. $f(x) = \sin\sqrt{x}$
12. $f(x) = \sec(20\pi x)$
13. $y = 10\sin x + \sin 100x$
14. $y = x^2 + 0.02\sin 50x$

---

15. Graph the ellipse $4x^2 + 2y^2 = 1$ by graphing the functions whose graphs are the upper and lower halves of the ellipse.

16. Graph the hyperbola $y^2 - 9x^2 = 1$ by graphing the functions whose graphs are the upper and lower branches of the hyperbola.

17–18 Do the graphs intersect in the given viewing rectangle? If they do, how many points of intersection are there?

17. $y = 3x^2 - 6x + 1$, $y = 0.23x - 2.25$; $[-1, 3]$ by $[-2.5, 1.5]$

18. $y = 6 - 4x - x^2$, $y = 3x + 18$; $[-6, 2]$ by $[-5, 20]$

19–21 Find all solutions of the equation correct to two decimal places.

19. $x^3 - 9x^2 - 4 = 0$
20. $x^3 = 4x - 1$
21. $x^2 = \sin x$

---

22. We saw in Example 9 that the equation $\cos x = x$ has exactly one solution.
(a) Use a graph to show that the equation $\cos x = 0.3x$ has three solutions and find their values correct to two decimal places.
(b) Find an approximate value of m such that the equation $\cos x = mx$ has exactly two solutions.

23. Use graphs to determine which of the functions $f(x) = 10x^2$ and $g(x) = x^3/10$ is eventually larger (that is, larger when x is very large).

24. Use graphs to determine which of the functions $f(x) = x^4 - 100x^3$ and $g(x) = x^3$ is eventually larger.

25. For what values of x is it true that $|\sin x - x| < 0.1$?

26. Graph the polynomials $P(x) = 3x^5 - 5x^3 + 2x$ and $Q(x) = 3x^5$ on the same screen, first using the viewing rectangle $[-2, 2]$ by $[-2, 2]$ and then changing to $[-10, 10]$ by $[-10{,}000, 10{,}000]$. What do you observe from these graphs?

27. In this exercise we consider the family of root functions $f(x) = \sqrt[n]{x}$, where n is a positive integer.
(a) Graph the functions $y = \sqrt{x}$, $y = \sqrt[4]{x}$, and $y = \sqrt[6]{x}$ on the same screen using the viewing rectangle $[-1, 4]$ by $[-1, 3]$.
(b) Graph the functions $y = x$, $y = \sqrt[3]{x}$, and $y = \sqrt[5]{x}$ on the same screen using the viewing rectangle $[-3, 3]$ by $[-2, 2]$. (See Example 7.)
(c) Graph the functions $y = \sqrt{x}$, $y = \sqrt[3]{x}$, $y = \sqrt[4]{x}$, and $y = \sqrt[5]{x}$ on the same screen using the viewing rectangle $[-1, 3]$ by $[-1, 2]$.
(d) What conclusions can you make from these graphs?

28. In this exercise we consider the family of functions $f(x) = 1/x^n$, where n is a positive integer.
(a) Graph the functions $y = 1/x$ and $y = 1/x^3$ on the same screen using the viewing rectangle $[-3, 3]$ by $[-3, 3]$.
(b) Graph the functions $y = 1/x^2$ and $y = 1/x^4$ on the same screen using the same viewing rectangle as in part (a).
(c) Graph all of the functions in parts (a) and (b) on the same screen using the viewing rectangle $[-1, 3]$ by $[-1, 3]$.
(d) What conclusions can you make from these graphs?

29. Graph the function $f(x) = x^4 + cx^2 + x$ for several values of c. How does the graph change when c changes?

30. Graph the function $f(x) = \sqrt{1 + cx^2}$ for various values of c. Describe how changing the value of c affects the graph.

31. Graph the function $y = x^n 2^{-x}$, $x \geqslant 0$, for $n = 1, 2, 3, 4, 5$, and 6. How does the graph change as n increases?

32. The curves with equations

$$y = \frac{|x|}{\sqrt{c - x^2}}$$

are called **bullet-nose curves**. Graph some of these curves to see why. What happens as c increases?

33. What happens to the graph of the equation $y^2 = cx^3 + x^2$ as c varies?

34. This exercise explores the effect of the inner function g on a composite function $y = f(g(x))$.
(a) Graph the function $y = \sin(\sqrt{x})$ using the viewing rectangle $[0, 400]$ by $[-1.5, 1.5]$. How does this graph differ from the graph of the sine function?

(b) Graph the function $y = \sin(x^2)$ using the viewing rectangle $[-5, 5]$ by $[-1.5, 1.5]$. How does this graph differ from the graph of the sine function?

35. The figure shows the graphs of $y = \sin 96x$ and $y = \sin 2x$ as displayed by a TI-83 graphing calculator.

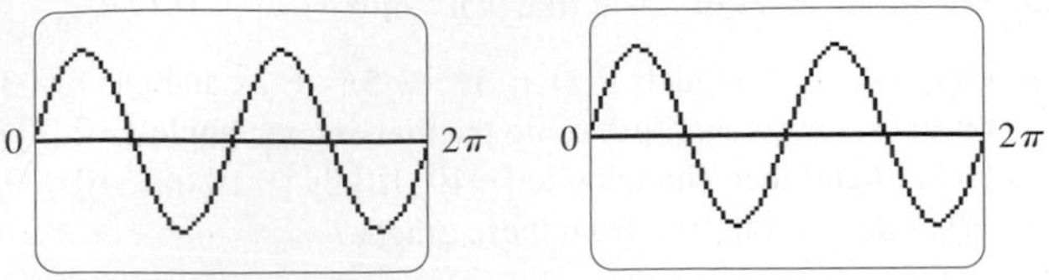

$y = \sin 96x$ $\quad$ $y = \sin 2x$

The first graph is inaccurate. Explain why the two graphs appear identical. [*Hint:* The TI-83's graphing window is 95 pixels wide. What specific points does the calculator plot?]

36. The first graph in the figure is that of $y = \sin 45x$ as displayed by a TI-83 graphing calculator. It is inaccurate and so, to help explain its appearance, we replot the curve in dot mode in the second graph.

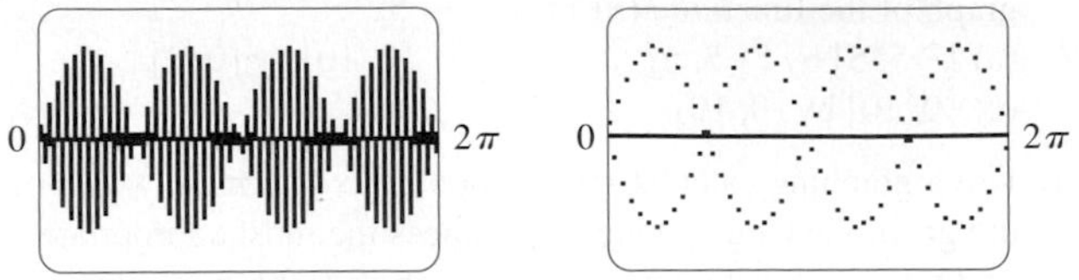

What two sine curves does the calculator appear to be plotting? Show that each point on the graph of $y = \sin 45x$ that the TI-83 chooses to plot is in fact on one of these two curves. (The TI-83's graphing window is 95 pixels wide.)

1.5 EXPONENTIAL FUNCTIONS

The function $f(x) = 2^x$ is called an *exponential function* because the variable, x, is the exponent. It should not be confused with the power function $g(x) = x^2$, in which the variable is the base.

■ In Appendix G we present an alternative approach to the exponential and logarithmic functions using integral calculus.

In general, an **exponential function** is a function of the form

$$f(x) = a^x$$

where a is a positive constant. Let's recall what this means.

If $x = n$, a positive integer, then

$$a^n = \underbrace{a \cdot a \cdot \cdots \cdot a}_{n \text{ factors}}$$

If $x = 0$, then $a^0 = 1$, and if $x = -n$, where n is a positive integer, then

$$a^{-n} = \frac{1}{a^n}$$

If x is a rational number, $x = p/q$, where p and q are integers and $q > 0$, then

$$a^x = a^{p/q} = \sqrt[q]{a^p} = \left(\sqrt[q]{a}\right)^p$$

But what is the meaning of a^x if x is an irrational number? For instance, what is meant by $2^{\sqrt{3}}$ or 5^{π}?

To help us answer this question we first look at the graph of the function $y = 2^x$, where x is rational. A representation of this graph is shown in Figure 1. We want to enlarge the domain of $y = 2^x$ to include both rational and irrational numbers.

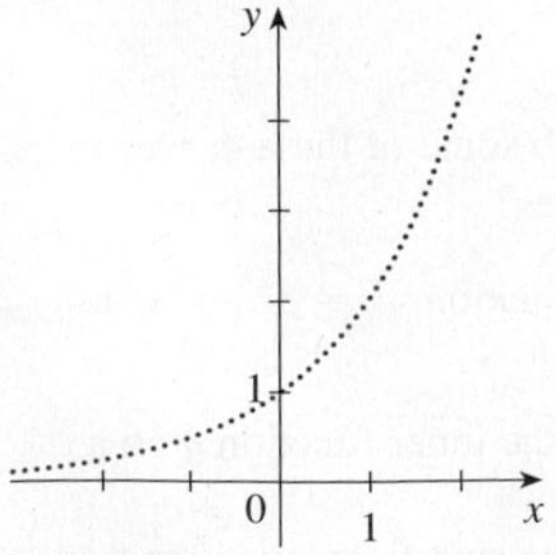

FIGURE 1
Representation of $y = 2^x$, x rational

There are holes in the graph in Figure 1 corresponding to irrational values of x. We want to fill in the holes by defining $f(x) = 2^x$, where $x \in \mathbb{R}$, so that f is an increasing function. In particular, since the irrational number $\sqrt{3}$ satisfies

$$1.7 < \sqrt{3} < 1.8$$

we must have

$$2^{1.7} < 2^{\sqrt{3}} < 2^{1.8}$$

and we know what $2^{1.7}$ and $2^{1.8}$ mean because 1.7 and 1.8 are rational numbers. Similarly, if we use better approximations for $\sqrt{3}$, we obtain better approximations for $2^{\sqrt{3}}$:

$$1.73 < \sqrt{3} < 1.74 \quad \Rightarrow \quad 2^{1.73} < 2^{\sqrt{3}} < 2^{1.74}$$
$$1.732 < \sqrt{3} < 1.733 \quad \Rightarrow \quad 2^{1.732} < 2^{\sqrt{3}} < 2^{1.733}$$
$$1.7320 < \sqrt{3} < 1.7321 \quad \Rightarrow \quad 2^{1.7320} < 2^{\sqrt{3}} < 2^{1.7321}$$
$$1.73205 < \sqrt{3} < 1.73206 \quad \Rightarrow \quad 2^{1.73205} < 2^{\sqrt{3}} < 2^{1.73206}$$
$$\vdots \qquad \vdots \qquad\qquad \vdots \qquad \vdots$$

■ A proof of this fact is given in J. Marsden and A. Weinstein, *Calculus Unlimited* (Menlo Park, CA: Benjamin/Cummings, 1981). For an online version, see

www.cds.caltech.edu/~marsden/volume/cu/CU.pdf

It can be shown that there is exactly one number that is greater than all of the numbers

$$2^{1.7}, \quad 2^{1.73}, \quad 2^{1.732}, \quad 2^{1.7320}, \quad 2^{1.73205}, \quad \ldots$$

and less than all of the numbers

$$2^{1.8}, \quad 2^{1.74}, \quad 2^{1.733}, \quad 2^{1.7321}, \quad 2^{1.73206}, \quad \ldots$$

We define $2^{\sqrt{3}}$ to be this number. Using the preceding approximation process we can compute it correct to six decimal places:

$$2^{\sqrt{3}} \approx 3.321997$$

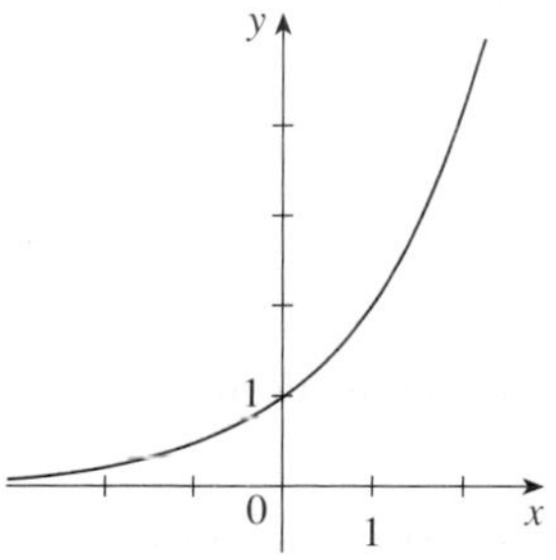

FIGURE 2

$y = 2^x$, x real

Similarly, we can define 2^x (or a^x, if $a > 0$) where x is any irrational number. Figure 2 shows how all the holes in Figure 1 have been filled to complete the graph of the function $f(x) = 2^x$, $x \in \mathbb{R}$.

The graphs of members of the family of functions $y = a^x$ are shown in Figure 3 for various values of the base a. Notice that all of these graphs pass through the same point (0, 1) because $a^0 = 1$ for $a \neq 0$. Notice also that as the base a gets larger, the exponential function grows more rapidly (for $x > 0$).

■ If $0 < a < 1$, then a^x approaches 0 as x becomes large. If $a > 1$, then a^x approaches 0 as x decreases through negative values. In both cases the x-axis is a horizontal asymptote. These matters are discussed in Section 2.6.

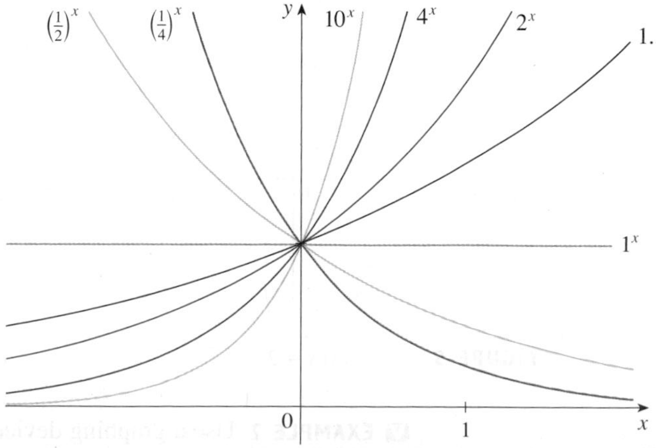

FIGURE 3

You can see from Figure 3 that there are basically three kinds of exponential functions $y = a^x$. If $0 < a < 1$, the exponential function decreases; if $a = 1$, it is a constant; and if $a > 1$, it increases. These three cases are illustrated in Figure 4. Observe that if $a \neq 1$,

The pattern of the data points in Figure 8 suggests exponential growth, so we use a graphing calculator with exponential regression capability to apply the method of least squares and obtain the exponential model

$$P = (0.008079266) \cdot (1.013731)^t$$

Figure 9 shows the graph of this exponential function together with the original data points. We see that the exponential curve fits the data reasonably well. The period of relatively slow population growth is explained by the two world wars and the Great Depression of the 1930s.

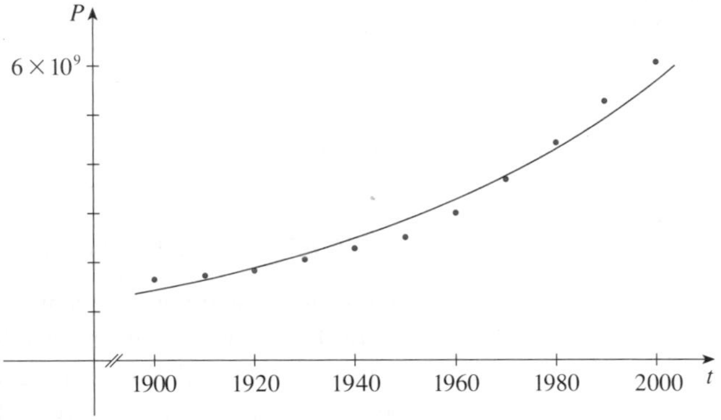

FIGURE 9
Exponential model for population growth

THE NUMBER e

Of all possible bases for an exponential function, there is one that is most convenient for the purposes of calculus. The choice of a base a is influenced by the way the graph of $y = a^x$ crosses the y-axis. Figures 10 and 11 show the tangent lines to the graphs of $y = 2^x$ and $y = 3^x$ at the point (0, 1). (Tangent lines will be defined precisely in Section 2.7. For present purposes, you can think of the tangent line to an exponential graph at a point as the line that touches the graph only at that point.) If we measure the slopes of these tangent lines at (0, 1), we find that $m \approx 0.7$ for $y = 2^x$ and $m \approx 1.1$ for $y = 3^x$.

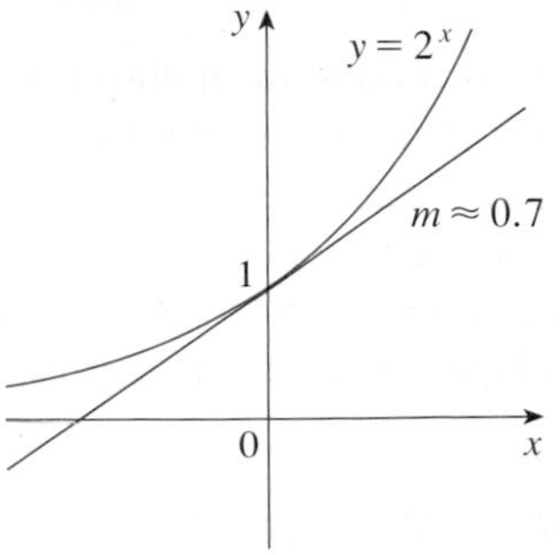

FIGURE 10

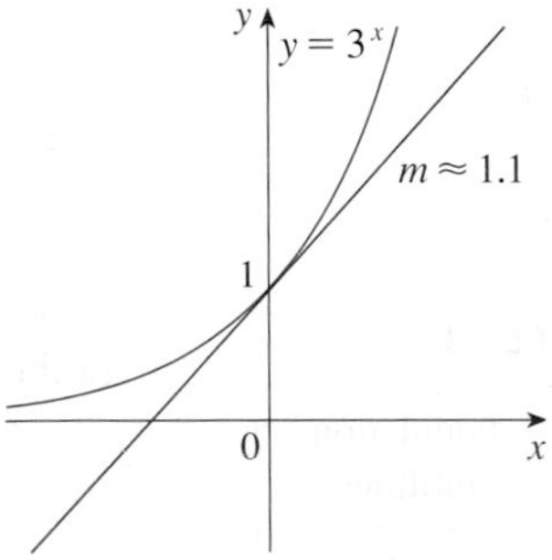

FIGURE 11

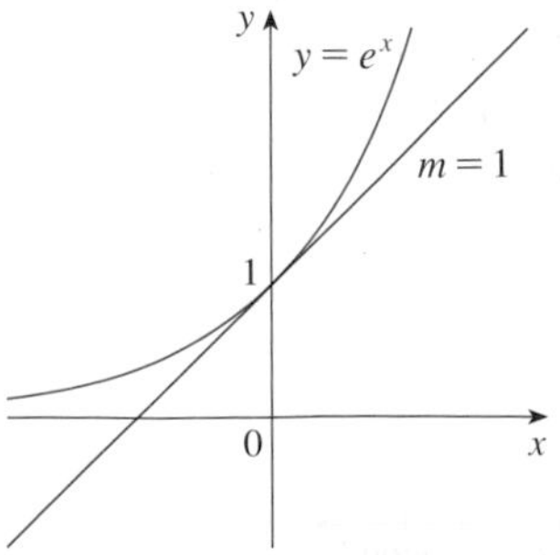

FIGURE 12
The natural exponential function crosses the y-axis with a slope of 1.

It turns out, as we will see in Chapter 3, that some of the formulas of calculus will be greatly simplified if we choose the base a so that the slope of the tangent line to $y = a^x$ at (0, 1) is *exactly* 1. (See Figure 12.) In fact, there *is* such a number and it is denoted by the letter e. (This notation was chosen by the Swiss mathematician Leonhard Euler in 1727, probably because it is the first letter of the word *exponential*.) In view of Figures 10 and 11, it comes as no surprise that the number e lies between 2 and 3 and the graph of $y = e^x$ lies between the graphs of $y = 2^x$ and $y = 3^x$. (See Figure 13.) In Chapter 3 we will see that the value of e, correct to five decimal places, is

$$e \approx 2.71828$$

TEC Module 1.5 enables you to graph exponential functions with various bases and their tangent lines in order to estimate more closely the value of a for which the tangent has slope 1.

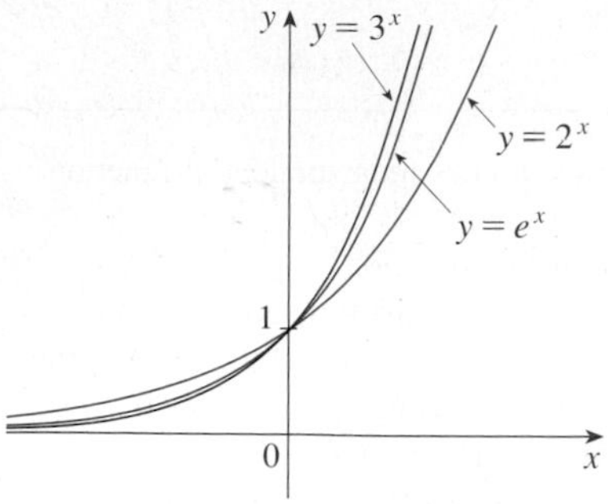

FIGURE 13

V EXAMPLE 3 Graph the function $y = \frac{1}{2}e^{-x} - 1$ and state the domain and range.

SOLUTION We start with the graph of $y = e^x$ from Figures 12 and 14(a) and reflect about the y-axis to get the graph of $y = e^{-x}$ in Figure 14(b). (Notice that the graph crosses the y-axis with a slope of -1). Then we compress the graph vertically by a factor of 2 to obtain the graph of $y = \frac{1}{2}e^{-x}$ in Figure 14(c). Finally, we shift the graph downward one unit to get the desired graph in Figure 14(d). The domain is $\mathbb{R}$ and the range is $(-1, \infty)$.

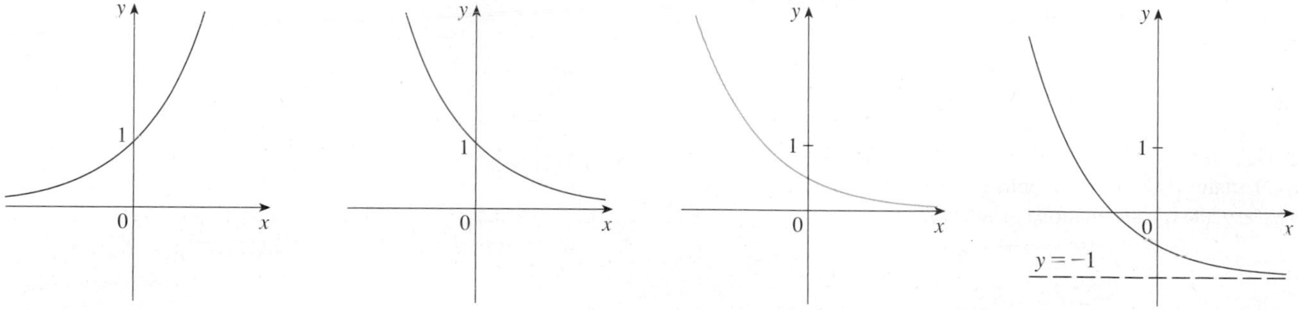

(a) $y = e^x$ (b) $y = e^{-x}$ (c) $y = \frac{1}{2}e^{-x}$ (d) $y = \frac{1}{2}e^{-x} - 1$

FIGURE 14

□

How far to the right do you think we would have to go for the height of the graph of $y = e^x$ to exceed a million? The next example demonstrates the rapid growth of this function by providing an answer that might surprise you.

EXAMPLE 4 Use a graphing device to find the values of x for which $e^x > 1{,}000{,}000$.

SOLUTION In Figure 15 we graph both the function $y = e^x$ and the horizontal line $y = 1{,}000{,}000$. We see that these curves intersect when $x \approx 13.8$. Thus $e^x > 10^6$ when $x > 13.8$. It is perhaps surprising that the values of the exponential function have already surpassed a million when x is only 14.

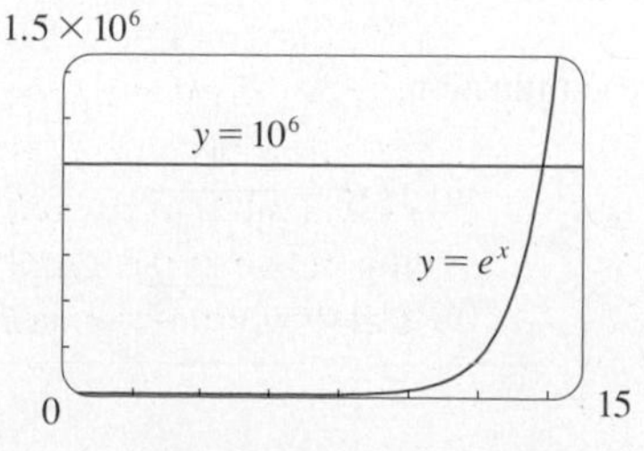

FIGURE 15

□

1.5 EXERCISES

1. (a) Write an equation that defines the exponential function with base $a > 0$.
(b) What is the domain of this function?
(c) If $a \neq 1$, what is the range of this function?
(d) Sketch the general shape of the graph of the exponential function for each of the following cases.
(i) $a > 1$ (ii) $a = 1$ (iii) $0 < a < 1$

2. (a) How is the number e defined?
(b) What is an approximate value for e?
(c) What is the natural exponential function?

3–6 Graph the given functions on a common screen. How are these graphs related?

3. $y = 2^x$, $y = e^x$, $y = 5^x$, $y = 20^x$

4. $y = e^x$, $y = e^{-x}$, $y = 8^x$, $y = 8^{-x}$

5. $y = 3^x$, $y = 10^x$, $y = \left(\frac{1}{3}\right)^x$, $y = \left(\frac{1}{10}\right)^x$

6. $y = 0.9^x$, $y = 0.6^x$, $y = 0.3^x$, $y = 0.1^x$

7–12 Make a rough sketch of the graph of the function. Do not use a calculator. Just use the graphs given in Figures 3 and 12 and, if necessary, the transformations of Section 1.3.

7. $y = 4^x - 3$

8. $y = 4^{x-3}$

9. $y = -2^{-x}$

10. $y = 1 + 2e^x$

11. $y = 1 - \frac{1}{2}e^{-x}$

12. $y = 2(1 - e^x)$

13. Starting with the graph of $y = e^x$, write the equation of the graph that results from
(a) shifting 2 units downward
(b) shifting 2 units to the right
(c) reflecting about the x-axis
(d) reflecting about the y-axis
(e) reflecting about the x-axis and then about the y-axis

14. Starting with the graph of $y = e^x$, find the equation of the graph that results from
(a) reflecting about the line $y = 4$
(b) reflecting about the line $x = 2$

15–16 Find the domain of each function.

15. (a) $f(x) = \dfrac{1}{1 + e^x}$ (b) $f(x) = \dfrac{1}{1 - e^x}$

16. (a) $g(t) = \sin(e^{-t})$ (b) $g(t) = \sqrt{1 - 2^t}$

17–18 Find the exponential function $f(x) = Ca^x$ whose graph is given.

17.

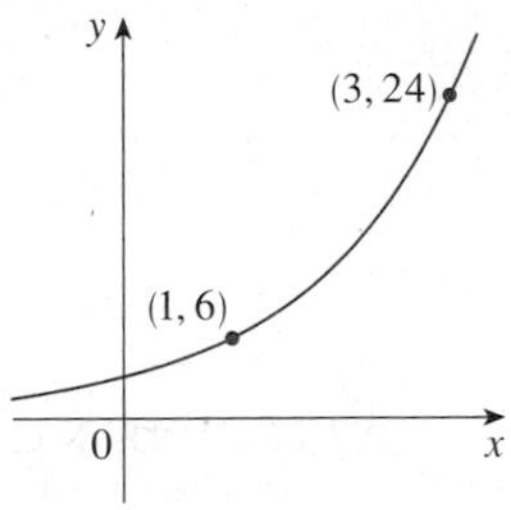

18.

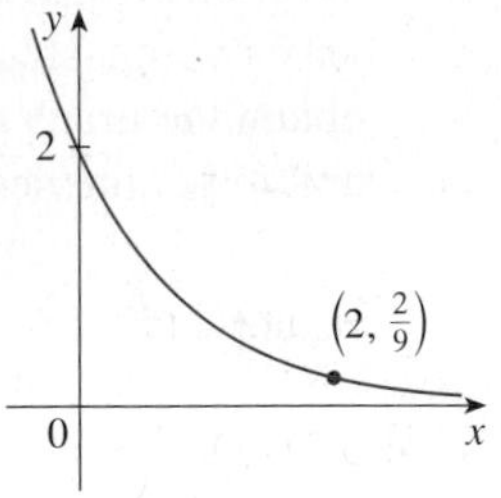

19. If $f(x) = 5^x$, show that

$$\frac{f(x + h) - f(x)}{h} = 5^x\left(\frac{5^h - 1}{h}\right)$$

20. Suppose you are offered a job that lasts one month. Which of the following methods of payment do you prefer?
I. One million dollars at the end of the month.
II. One cent on the first day of the month, two cents on the second day, four cents on the third day, and, in general, 2^{n-1} cents on the nth day.

21. Suppose the graphs of $f(x) = x^2$ and $g(x) = 2^x$ are drawn on a coordinate grid where the unit of measurement is 1 inch. Show that, at a distance 2 ft to the right of the origin, the height of the graph of f is 48 ft but the height of the graph of g is about 265 mi.

22. Compare the functions $f(x) = x^5$ and $g(x) = 5^x$ by graphing both functions in several viewing rectangles. Find all points of intersection of the graphs correct to one decimal place. Which function grows more rapidly when x is large?

23. Compare the functions $f(x) = x^{10}$ and $g(x) = e^x$ by graphing both f and g in several viewing rectangles. When does the graph of g finally surpass the graph of f?

24. Use a graph to estimate the values of x such that $e^x > 1{,}000{,}000{,}000$.

25. Under ideal conditions a certain bacteria population is known to double every three hours. Suppose that there are initially 100 bacteria.
(a) What is the size of the population after 15 hours?
(b) What is the size of the population after t hours?
(c) Estimate the size of the population after 20 hours.
(d) Graph the population function and estimate the time for the population to reach 50,000.

26. A bacterial culture starts with 500 bacteria and doubles in size every half hour.
(a) How many bacteria are there after 3 hours?
(b) How many bacteria are there after t hours?
(c) How many bacteria are there after 40 minutes?
(d) Graph the population function and estimate the time for the population to reach 100,000.

27. Use a graphing calculator with exponential regression capability to model the population of the world with the data from 1950 to 2000 in Table 1 on page 55. Use the model to estimate the population in 1993 and to predict the population in the year 2010.

28. The table gives the population of the United States, in millions, for the years 1900–2000. Use a graphing calculator with exponential regression capability to model the US population since 1900. Use the model to estimate the population in 1925 and to predict the population in the years 2010 and 2020.

Year	Population	Year	Population
1900	76	1960	179
1910	92	1970	203
1920	106	1980	227
1930	123	1990	250
1940	131	2000	281
1950	150		

29. If you graph the function

$$f(x) = \frac{1 - e^{1/x}}{1 + e^{1/x}}$$

you'll see that f appears to be an odd function. Prove it.

30. Graph several members of the family of functions

$$f(x) = \frac{1}{1 + ae^{bx}}$$

where $a > 0$. How does the graph change when b changes? How does it change when a changes?

1.6 INVERSE FUNCTIONS AND LOGARITHMS

Table 1 gives data from an experiment in which a bacteria culture started with 100 bacteria in a limited nutrient medium; the size of the bacteria population was recorded at hourly intervals. The number of bacteria N is a function of the time t: $N = f(t)$.

Suppose, however, that the biologist changes her point of view and becomes interested in the time required for the population to reach various levels. In other words, she is thinking of t as a function of N. This function is called the *inverse function* of f, denoted by f^{-1}, and read "f inverse." Thus $t = f^{-1}(N)$ is the time required for the population level to reach N. The values of f^{-1} can be found by reading Table 1 from right to left or by consulting Table 2. For instance, $f^{-1}(550) = 6$ because $f(6) = 550$.

TABLE 1 N as a function of t

t (hours)	$N = f(t)$ = population at time t
0	100
1	168
2	259
3	358
4	445
5	509
6	550
7	573
8	586

TABLE 2 t as a function of N

N	$t = f^{-1}(N)$ = time to reach N bacteria
100	0
168	1
259	2
358	3
445	4
509	5
550	6
573	7
586	8

Not all functions possess inverses. Let's compare the functions f and g whose arrow diagrams are shown in Figure 1. Note that f never takes on the same value twice (any two inputs in A have different outputs), whereas g does take on the same value twice (both 2 and 3 have the same output, 4). In symbols,

$$g(2) = g(3)$$

but $$f(x_1) \neq f(x_2) \qquad \text{whenever } x_1 \neq x_2$$

Functions that share this property with f are called *one-to-one functions.*

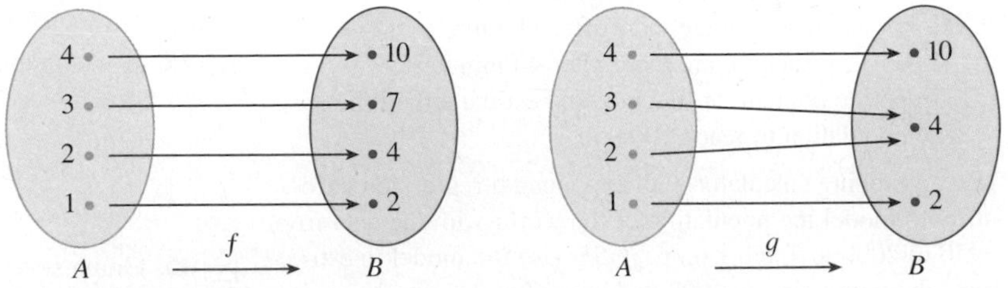

FIGURE 1

f is one-to-one; g is not

■ In the language of inputs and outputs, this definition says that f is one-to-one if each output corresponds to only one input.

> **1 DEFINITION** A function f is called a **one-to-one function** if it never takes on the same value twice; that is,
>
> $$f(x_1) \neq f(x_2) \qquad \text{whenever } x_1 \neq x_2$$

If a horizontal line intersects the graph of f in more than one point, then we see from Figure 2 that there are numbers x_1 and x_2 such that $f(x_1) = f(x_2)$. This means that f is not one-to-one. Therefore we have the following geometric method for determining whether a function is one-to-one.

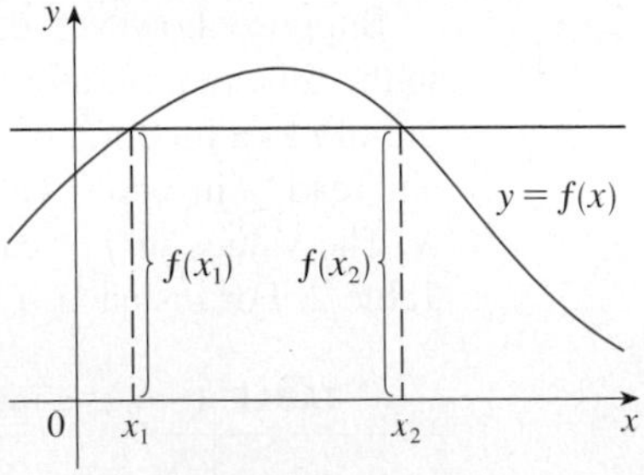

FIGURE 2

This function is not one-to-one because $f(x_1) = f(x_2)$.

> **HORIZONTAL LINE TEST** A function is one-to-one if and only if no horizontal line intersects its graph more than once.

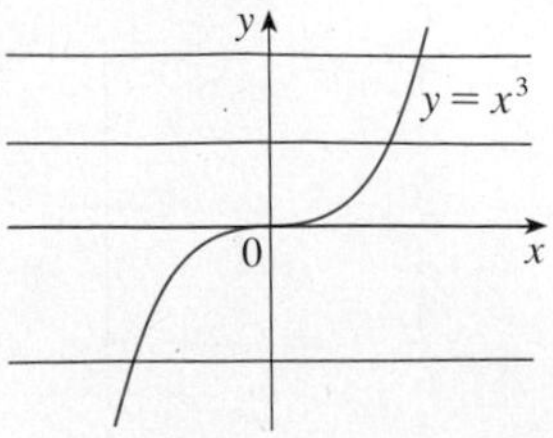

FIGURE 3

$f(x) = x^3$ is one-to-one.

EXAMPLE 1 Is the function $f(x) = x^3$ one-to-one?

SOLUTION 1 If $x_1 \neq x_2$, then $x_1^3 \neq x_2^3$ (two different numbers can't have the same cube). Therefore, by Definition 1, $f(x) = x^3$ is one-to-one.

SOLUTION 2 From Figure 3 we see that no horizontal line intersects the graph of $f(x) = x^3$ more than once. Therefore, by the Horizontal Line Test, f is one-to-one. □

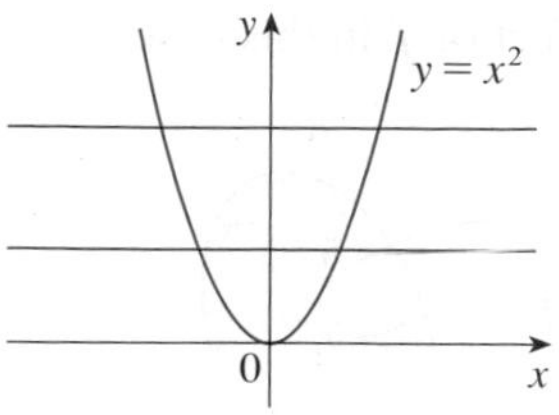

FIGURE 4
$g(x) = x^2$ is not one-to-one.

V EXAMPLE 2 Is the function $g(x) = x^2$ one-to-one?

SOLUTION 1 This function is not one-to-one because, for instance,

$$g(1) = 1 = g(-1)$$

and so 1 and -1 have the same output.

SOLUTION 2 From Figure 4 we see that there are horizontal lines that intersect the graph of g more than once. Therefore, by the Horizontal Line Test, g is not one-to-one. □

One-to-one functions are important because they are precisely the functions that possess inverse functions according to the following definition.

> **2 DEFINITION** Let f be a one-to-one function with domain A and range B. Then its **inverse function** f^{-1} has domain B and range A and is defined by
>
> $$f^{-1}(y) = x \iff f(x) = y$$
>
> for any y in B.

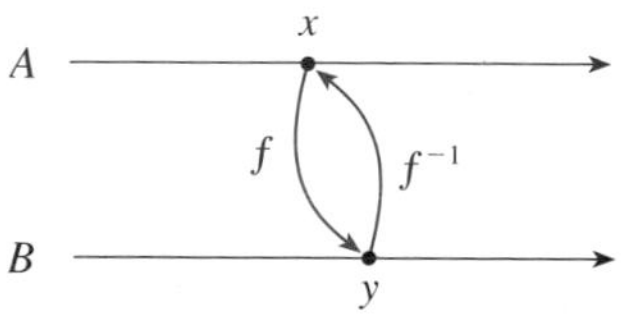

FIGURE 5

This definition says that if f maps x into y, then f^{-1} maps y back into x. (If f were not one-to-one, then f^{-1} would not be uniquely defined.) The arrow diagram in Figure 5 indicates that f^{-1} reverses the effect of f. Note that

$$\text{domain of } f^{-1} = \text{range of } f$$

$$\text{range of } f^{-1} = \text{domain of } f$$

For example, the inverse function of $f(x) = x^3$ is $f^{-1}(x) = x^{1/3}$ because if $y = x^3$, then

$$f^{-1}(y) = f^{-1}(x^3) = (x^3)^{1/3} = x$$

CAUTION Do not mistake the -1 in f^{-1} for an exponent. Thus

$$f^{-1}(x) \quad \text{does } \textit{not} \text{ mean} \quad \frac{1}{f(x)}$$

The reciprocal $1/f(x)$ could, however, be written as $[f(x)]^{-1}$.

V EXAMPLE 3 If $f(1) = 5$, $f(3) = 7$, and $f(8) = -10$, find $f^{-1}(7)$, $f^{-1}(5)$, and $f^{-1}(-10)$.

SOLUTION From the definition of f^{-1} we have

$$f^{-1}(7) = 3 \quad \text{because} \quad f(3) = 7$$

$$f^{-1}(5) = 1 \quad \text{because} \quad f(1) = 5$$

$$f^{-1}(-10) = 8 \quad \text{because} \quad f(8) = -10$$

then we have

$$\boxed{6} \qquad \log_a x = y \iff a^y = x$$

Thus, if $x > 0$, then $\log_a x$ is the exponent to which the base a must be raised to give x. For example, $\log_{10} 0.001 = -3$ because $10^{-3} = 0.001$.

The cancellation equations (4), when applied to the functions $f(x) = a^x$ and $f^{-1}(x) = \log_a x$, become

$$\boxed{7} \qquad \begin{aligned} \log_a(a^x) &= x \quad \text{for every } x \in \mathbb{R} \\ a^{\log_a x} &= x \quad \text{for every } x > 0 \end{aligned}$$

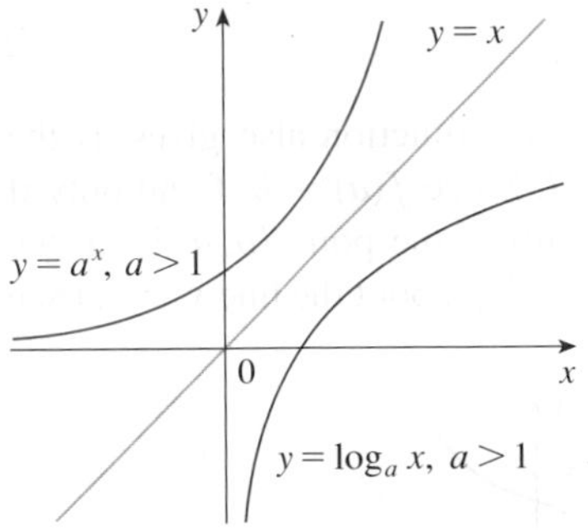

FIGURE 11

The logarithmic function $\log_a$ has domain $(0, \infty)$ and range $\mathbb{R}$. Its graph is the reflection of the graph of $y = a^x$ about the line $y = x$.

Figure 11 shows the case where $a > 1$. (The most important logarithmic functions have base $a > 1$.) The fact that $y = a^x$ is a very rapidly increasing function for $x > 0$ is reflected in the fact that $y = \log_a x$ is a very slowly increasing function for $x > 1$.

Figure 12 shows the graphs of $y = \log_a x$ with various values of the base $a > 1$. Since $\log_a 1 = 0$, the graphs of all logarithmic functions pass through the point $(1, 0)$.

The following properties of logarithmic functions follow from the corresponding properties of exponential functions given in Section 1.5.

LAWS OF LOGARITHMS If x and y are positive numbers, then

1. $\log_a(xy) = \log_a x + \log_a y$

2. $\log_a\left(\dfrac{x}{y}\right) = \log_a x - \log_a y$

3. $\log_a(x^r) = r \log_a x$ (where r is any real number)

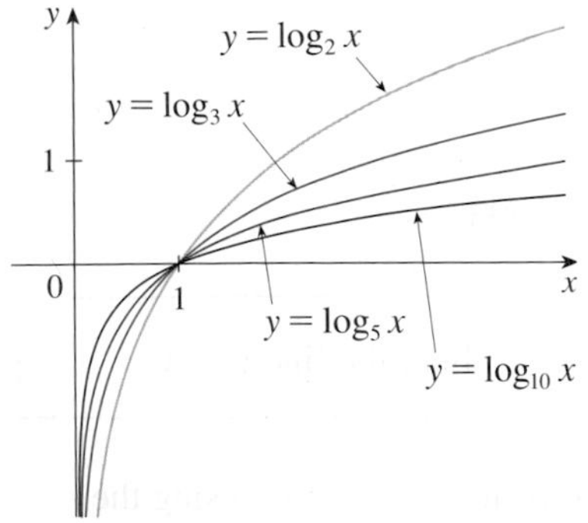

FIGURE 12

EXAMPLE 6 Use the laws of logarithms to evaluate $\log_2 80 - \log_2 5$.

SOLUTION Using Law 2, we have

$$\log_2 80 - \log_2 5 = \log_2\left(\frac{80}{5}\right) = \log_2 16 = 4$$

because $2^4 = 16$. □

NATURAL LOGARITHMS

Of all possible bases a for logarithms, we will see in Chapter 3 that the most convenient choice of a base is the number e, which was defined in Section 1.5. The logarithm with base e is called the **natural logarithm** and has a special notation:

$$\boxed{\log_e x = \ln x}$$

■ **NOTATION FOR LOGARITHMS**
Most textbooks in calculus and the sciences, as well as calculators, use the notation $\ln x$ for the natural logarithm and $\log x$ for the "common logarithm," $\log_{10} x$. In the more advanced mathematical and scientific literature and in computer languages, however, the notation $\log x$ usually denotes the natural logarithm.

If we put $a = e$ and replace $\log_e$ with "ln" in (6) and (7), then the defining properties of the natural logarithm function become

8
$$\ln x = y \iff e^y = x$$

9
$$\ln(e^x) = x \qquad x \in \mathbb{R}$$
$$e^{\ln x} = x \qquad x > 0$$

In particular, if we set $x = 1$, we get

$$\ln e = 1$$

EXAMPLE 7 Find x if $\ln x = 5$.

SOLUTION 1 From (8) we see that

$$\ln x = 5 \qquad \text{means} \qquad e^5 = x$$

Therefore $x = e^5$.

(If you have trouble working with the "ln" notation, just replace it by $\log_e$. Then the equation becomes $\log_e x = 5$; so, by the definition of logarithm, $e^5 = x$.)

SOLUTION 2 Start with the equation

$$\ln x = 5$$

and apply the exponential function to both sides of the equation:

$$e^{\ln x} = e^5$$

But the second cancellation equation in (9) says that $e^{\ln x} = x$. Therefore, $x = e^5$. □

EXAMPLE 8 Solve the equation $e^{5-3x} = 10$.

SOLUTION We take natural logarithms of both sides of the equation and use (9):

$$\ln(e^{5-3x}) = \ln 10$$
$$5 - 3x = \ln 10$$
$$3x = 5 - \ln 10$$
$$x = \tfrac{1}{3}(5 - \ln 10)$$

Since the natural logarithm is found on scientific calculators, we can approximate the solution: to four decimal places, $x \approx 0.8991$. □

we have

$$\sin^{-1}x = y \iff \sin y = x \quad \text{and} \quad -\frac{\pi}{2} \leq y \leq \frac{\pi}{2}$$

⊘ $\sin^{-1}x \neq \dfrac{1}{\sin x}$

Thus, if $-1 \leq x \leq 1$, $\sin^{-1}x$ is the number between $-\pi/2$ and $\pi/2$ whose sine is x.

EXAMPLE 12 Evaluate (a) $\sin^{-1}(\frac{1}{2})$ and (b) $\tan(\arcsin \frac{1}{3})$.

SOLUTION

(a) We have

$$\sin^{-1}(\tfrac{1}{2}) = \frac{\pi}{6}$$

because $\sin(\pi/6) = \frac{1}{2}$ and $\pi/6$ lies between $-\pi/2$ and $\pi/2$.

(b) Let $\theta = \arcsin \frac{1}{3}$, so $\sin \theta = \frac{1}{3}$. Then we can draw a right triangle with angle θ as in Figure 19 and deduce from the Pythagorean Theorem that the third side has length $\sqrt{9 - 1} = 2\sqrt{2}$. This enables us to read from the triangle that

$$\tan(\arcsin \tfrac{1}{3}) = \tan \theta = \frac{1}{2\sqrt{2}}$$ □

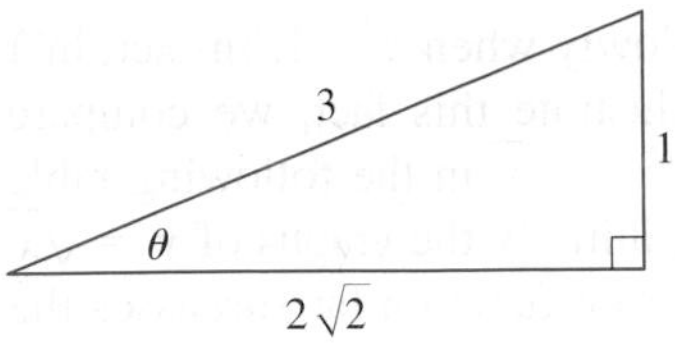

FIGURE 19

The cancellation equations for inverse functions become, in this case,

$$\sin^{-1}(\sin x) = x \quad \text{for } -\frac{\pi}{2} \leq x \leq \frac{\pi}{2}$$

$$\sin(\sin^{-1}x) = x \quad \text{for } -1 \leq x \leq 1$$

The inverse sine function, $\sin^{-1}$, has domain $[-1, 1]$ and range $[-\pi/2, \pi/2]$, and its graph, shown in Figure 20, is obtained from that of the restricted sine function (Figure 18) by reflection about the line $y = x$.

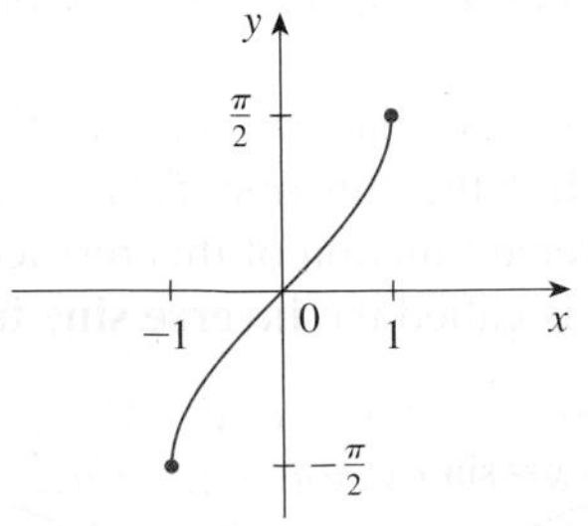

FIGURE 20
$y = \sin^{-1}x = \arcsin x$

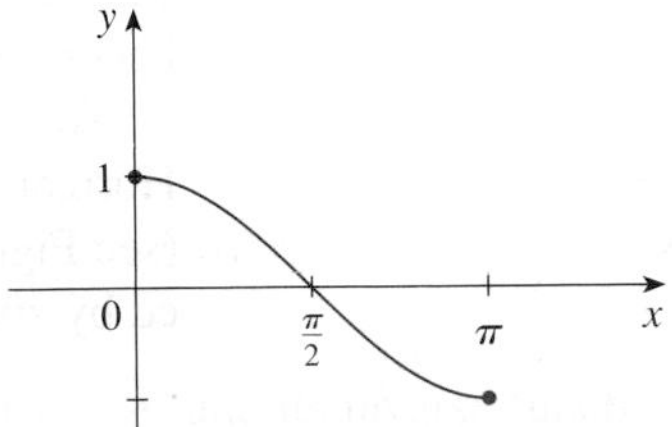

FIGURE 21
$y = \cos x,\ 0 \leq x \leq \pi$

The **inverse cosine function** is handled similarly. The restricted cosine function $f(x) = \cos x$, $0 \leq x \leq \pi$, is one-to-one (see Figure 21) and so it has an inverse function denoted by $\cos^{-1}$ or arccos.

$$\cos^{-1}x = y \iff \cos y = x \quad \text{and} \quad 0 \leq y \leq \pi$$

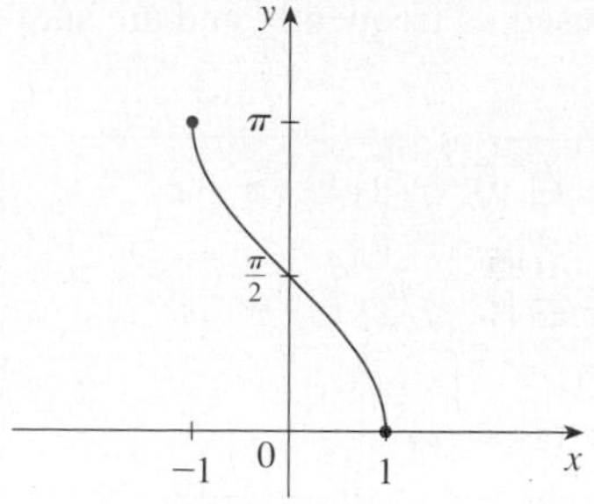

FIGURE 22
$y = \cos^{-1} x = \arccos x$

The cancellation equations are

$$\cos^{-1}(\cos x) = x \quad \text{for } 0 \leqslant x \leqslant \pi$$

$$\cos(\cos^{-1} x) = x \quad \text{for } -1 \leqslant x \leqslant 1$$

The inverse cosine function, $\cos^{-1}$, has domain $[-1, 1]$ and range $[0, \pi]$. Its graph is shown in Figure 22.

The tangent function can be made one-to-one by restricting it to the interval $(-\pi/2, \pi/2)$. Thus the **inverse tangent function** is defined as the inverse of the function $f(x) = \tan x$, $-\pi/2 < x < \pi/2$. (See Figure 23.) It is denoted by $\tan^{-1}$ or arctan.

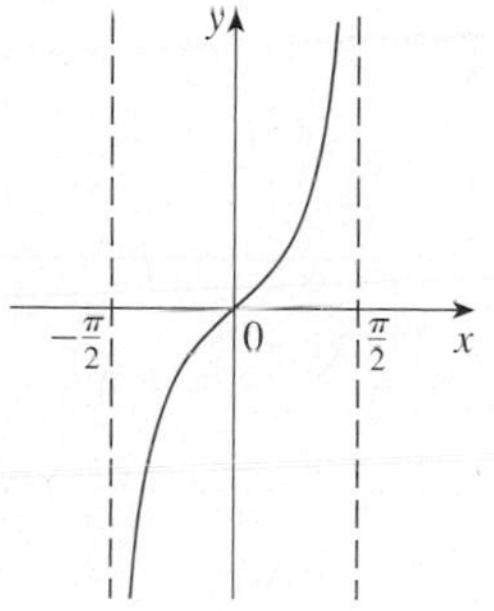

FIGURE 23
$y = \tan x, -\frac{\pi}{2} < x < \frac{\pi}{2}$

$$\tan^{-1} x = y \iff \tan y = x \quad \text{and} \quad -\frac{\pi}{2} < y < \frac{\pi}{2}$$

EXAMPLE 13 Simplify the expression $\cos(\tan^{-1} x)$.

SOLUTION 1 Let $y = \tan^{-1} x$. Then $\tan y = x$ and $-\pi/2 < y < \pi/2$. We want to find $\cos y$ but, since $\tan y$ is known, it is easier to find $\sec y$ first:

$$\sec^2 y = 1 + \tan^2 y = 1 + x^2$$

$$\sec y = \sqrt{1 + x^2} \qquad (\text{since } \sec y > 0 \text{ for } -\pi/2 < y < \pi/2)$$

Thus $$\cos(\tan^{-1} x) = \cos y = \frac{1}{\sec y} = \frac{1}{\sqrt{1 + x^2}}$$

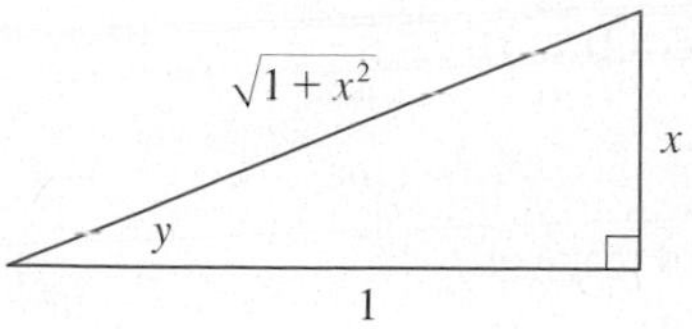

FIGURE 24

SOLUTION 2 Instead of using trigonometric identities as in Solution 1, it is perhaps easier to use a diagram. If $y = \tan^{-1} x$, then $\tan y = x$, and we can read from Figure 24 (which illustrates the case $y > 0$) that

$$\cos(\tan^{-1} x) = \cos y = \frac{1}{\sqrt{1 + x^2}}$$ □

The inverse tangent function, $\tan^{-1} = \arctan$, has domain $\mathbb{R}$ and range $(-\pi/2, \pi/2)$. Its graph is shown in Figure 25.

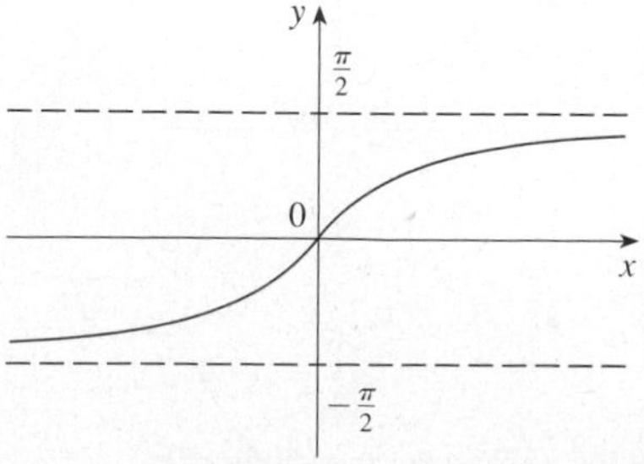

FIGURE 25
$y = \tan^{-1} x = \arctan x$

We know that the lines $x = \pm\pi/2$ are vertical asymptotes of the graph of tan. Since the graph of $\tan^{-1}$ is obtained by reflecting the graph of the restricted tangent function about the line $y = x$, it follows that the lines $y = \pi/2$ and $y = -\pi/2$ are horizontal asymptotes of the graph of $\tan^{-1}$.

The remaining inverse trigonometric functions are not used as frequently and are summarized here.

11 $y = \csc^{-1}x \ (|x| \geqslant 1) \iff \csc y = x \text{ and } y \in (0, \pi/2] \cup (\pi, 3\pi/2]$

$y = \sec^{-1}x \ (|x| \geqslant 1) \iff \sec y = x \text{ and } y \in [0, \pi/2) \cup [\pi, 3\pi/2)$

$y = \cot^{-1}x \ (x \in \mathbb{R}) \iff \cot y = x \text{ and } y \in (0, \pi)$

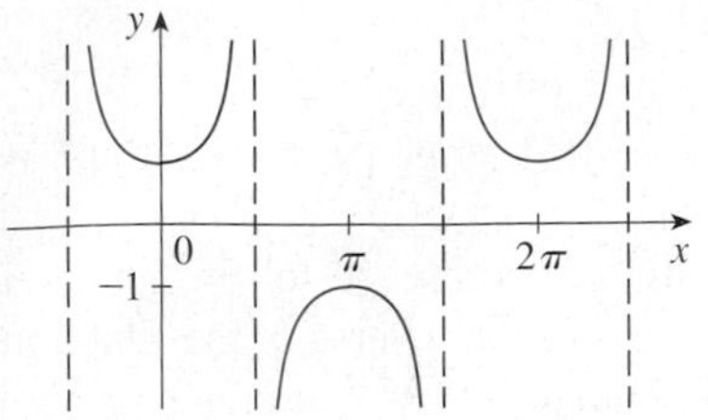

FIGURE 26
$y = \sec x$

The choice of intervals for y in the definitions of $\csc^{-1}$ and $\sec^{-1}$ is not universally agreed upon. For instance, some authors use $y \in [0, \pi/2) \cup (\pi/2, \pi]$ in the definition of $\sec^{-1}$. [You can see from the graph of the secant function in Figure 26 that both this choice and the one in (11) will work.]

1.6 EXERCISES

1. (a) What is a one-to-one function?
(b) How can you tell from the graph of a function whether it is one-to-one?

2. (a) Suppose f is a one-to-one function with domain A and range B. How is the inverse function f^{-1} defined? What is the domain of f^{-1}? What is the range of f^{-1}?
(b) If you are given a formula for f, how do you find a formula for f^{-1}?
(c) If you are given the graph of f, how do you find the graph of f^{-1}?

3–14 A function is given by a table of values, a graph, a formula, or a verbal description. Determine whether it is one-to-one.

3.

x	1	2	3	4	5	6
$f(x)$	1.5	2.0	3.6	5.3	2.8	2.0

4.

x	1	2	3	4	5	6
$f(x)$	1	2	4	8	16	32

5.

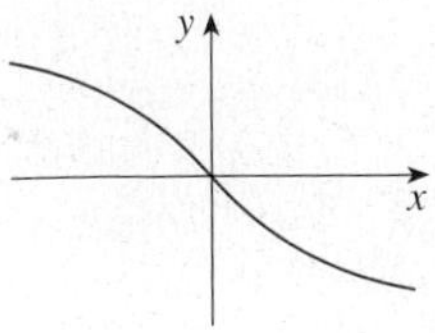

6.

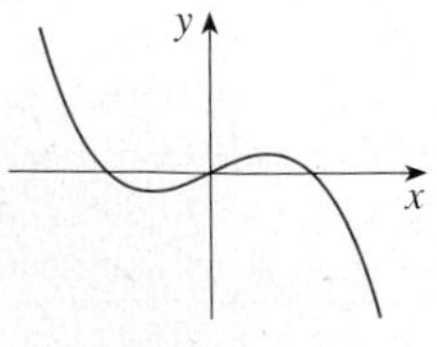

7.

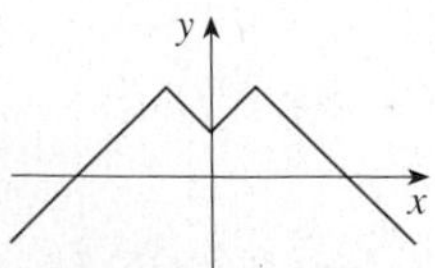

8.

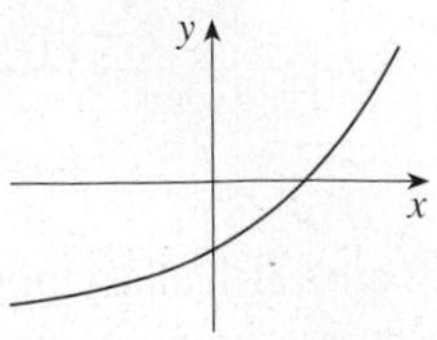

9. $f(x) = x^2 - 2x$

10. $f(x) = 10 - 3x$

11. $g(x) = 1/x$

12. $g(x) = \cos x$

13. $f(t)$ is the height of a football t seconds after kickoff.

14. $f(t)$ is your height at age t.

15. If f is a one-to-one function such that $f(2) = 9$, what is $f^{-1}(9)$?

16. Let $f(x) = 3 + x^2 + \tan(\pi x/2)$, where $-1 < x < 1$.
(a) Find $f^{-1}(3)$.
(b) Find $f(f^{-1}(5))$.

17. If $g(x) = 3 + x + e^x$, find $g^{-1}(4)$.

18. The graph of f is given.
(a) Why is f one-to-one?
(b) What are the domain and range of f^{-1}?
(c) What is the value of $f^{-1}(2)$?
(d) Estimate the value of $f^{-1}(0)$.

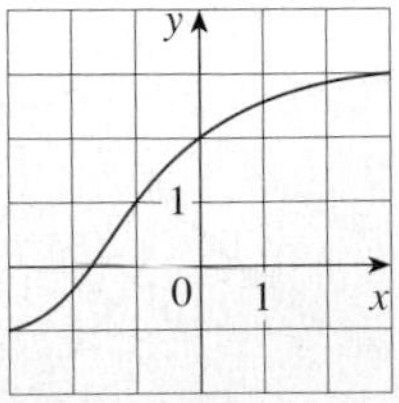

19. The formula $C = \frac{5}{9}(F - 32)$, where $F \geqslant -459.67$, expresses the Celsius temperature C as a function of the Fahrenheit temperature F. Find a formula for the inverse function and interpret it. What is the domain of the inverse function?

20. In the theory of relativity, the mass of a particle with speed v is

$$m = f(v) = \frac{m_0}{\sqrt{1 - v^2/c^2}}$$

where m_0 is the rest mass of the particle and c is the speed of light in a vacuum. Find the inverse function of f and explain its meaning.

21–26 Find a formula for the inverse of the function.

21. $f(x) = \sqrt{10 - 3x}$

22. $f(x) = \dfrac{4x - 1}{2x + 3}$

23. $f(x) = e^{x^3}$

24. $y = 2x^3 + 3$

25. $y = \ln(x + 3)$

26. $y = \dfrac{e^x}{1 + 2e^x}$

27–28 Find an explicit formula for f^{-1} and use it to graph f^{-1}, f, and the line $y = x$ on the same screen. To check your work, see whether the graphs of f and f^{-1} are reflections about the line.

27. $f(x) = x^4 + 1, \quad x \geqslant 0$

28. $f(x) = 2 - e^x$

29–30 Use the given graph of f to sketch the graph of f^{-1}.

29.

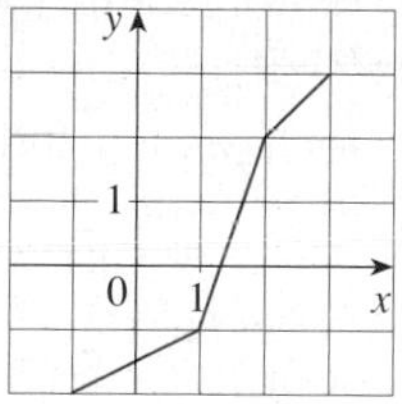

30.

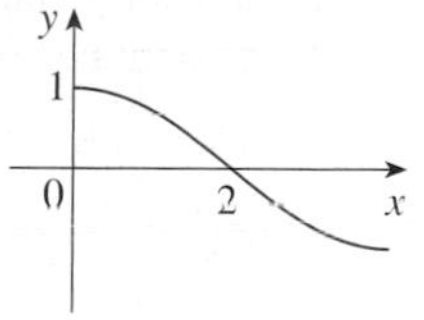

31. (a) How is the logarithmic function $y = \log_a x$ defined?
(b) What is the domain of this function?
(c) What is the range of this function?
(d) Sketch the general shape of the graph of the function $y = \log_a x$ if $a > 1$.

32. (a) What is the natural logarithm?
(b) What is the common logarithm?
(c) Sketch the graphs of the natural logarithm function and the natural exponential function with a common set of axes.

33–36 Find the exact value of each expression.

33. (a) $\log_5 125$ (b) $\log_3 \frac{1}{27}$

34. (a) $\ln(1/e)$ (b) $\log_{10} \sqrt{10}$

35. (a) $\log_2 6 - \log_2 15 + \log_2 20$
(b) $\log_3 100 - \log_3 18 - \log_3 50$

36. (a) $e^{-2\ln 5}$ (b) $\ln\left(\ln e^{e^{10}}\right)$

37–39 Express the given quantity as a single logarithm.

37. $\ln 5 + 5 \ln 3$

38. $\ln(a + b) + \ln(a - b) - 2 \ln c$

39. $\ln(1 + x^2) + \frac{1}{2} \ln x - \ln \sin x$

40. Use Formula 10 to evaluate each logarithm correct to six decimal places.
(a) $\log_{12} 10$ (b) $\log_2 8.4$

41–42 Use Formula 10 to graph the given functions on a common screen. How are these graphs related?

41. $y = \log_{1.5} x, \quad y = \ln x, \quad y = \log_{10} x, \quad y = \log_{50} x$

42. $y = \ln x, \quad y = \log_{10} x, \quad y = e^x, \quad y = 10^x$

43. Suppose that the graph of $y = \log_2 x$ is drawn on a coordinate grid where the unit of measurement is an inch. How many miles to the right of the origin do we have to move before the height of the curve reaches 3 ft?

44. Compare the functions $f(x) = x^{0.1}$ and $g(x) = \ln x$ by graphing both f and g in several viewing rectangles. When does the graph of f finally surpass the graph of g?

45–46 Make a rough sketch of the graph of each function. Do not use a calculator. Just use the graphs given in Figures 12 and 13 and, if necessary, the transformations of Section 1.3.

45. (a) $y = \log_{10}(x + 5)$ (b) $y = -\ln x$

46. (a) $y = \ln(-x)$ (b) $y = \ln |x|$

47–50 Solve each equation for x.

47. (a) $2 \ln x = 1$ (b) $e^{-x} = 5$

48. (a) $e^{2x+3} - 7 = 0$ (b) $\ln(5 - 2x) = -3$

49. (a) $2^{x-5} = 3$ (b) $\ln x + \ln(x - 1) = 1$

50. (a) $\ln(\ln x) = 1$ (b) $e^{ax} = Ce^{bx}$, where $a \neq b$

51–52 Solve each inequality for x.

51. (a) $e^x < 10$ (b) $\ln x > -1$

52. (a) $2 < \ln x < 9$ (b) $e^{2-3x} > 4$

53–54 Find (a) the domain of f and (b) f^{-1} and its domain.

53. $f(x) = \sqrt{3 - e^{2x}}$

54. $f(x) = \ln(2 + \ln x)$

CAS **55.** Graph the function $f(x) = \sqrt{x^3 + x^2 + x + 1}$ and explain why it is one-to-one. Then use a computer algebra system to find an explicit expression for $f^{-1}(x)$. (Your CAS will produce three possible expressions. Explain why two of them are irrelevant in this context.)

CAS **56.** (a) If $g(x) = x^6 + x^4, x \geqslant 0$, use a computer algebra system to find an expression for $g^{-1}(x)$.
(b) Use the expression in part (a) to graph $y = g(x)$, $y = x$, and $y = g^{-1}(x)$ on the same screen.

57. If a bacteria population starts with 100 bacteria and doubles every three hours, then the number of bacteria after t hours is $n = f(t) = 100 \cdot 2^{t/3}$. (See Exercise 25 in Section 1.5.)
(a) Find the inverse of this function and explain its meaning.
(b) When will the population reach 50,000?

58. When a camera flash goes off, the batteries immediately begin to recharge the flash's capacitor, which stores electric charge given by

$$Q(t) = Q_0(1 - e^{-t/a})$$

(The maximum charge capacity is Q_0 and t is measured in seconds.)
(a) Find the inverse of this function and explain its meaning.
(b) How long does it take to recharge the capacitor to 90% of capacity if $a = 2$?

59–64 Find the exact value of each expression.

59. (a) $\sin^{-1}(\sqrt{3}/2)$ (b) $\cos^{-1}(-1)$

60. (a) $\tan^{-1}(1/\sqrt{3})$ (b) $\sec^{-1} 2$

61. (a) $\arctan 1$ (b) $\sin^{-1}(1/\sqrt{2})$

62. (a) $\cot^{-1}(-\sqrt{3})$ (b) $\arccos(-\frac{1}{2})$

63. (a) $\tan(\arctan 10)$ (b) $\sin^{-1}(\sin(7\pi/3))$

64. (a) $\tan(\sec^{-1} 4)$ (b) $\sin(2 \sin^{-1}(\frac{3}{5}))$

65. Prove that $\cos(\sin^{-1} x) = \sqrt{1 - x^2}$.

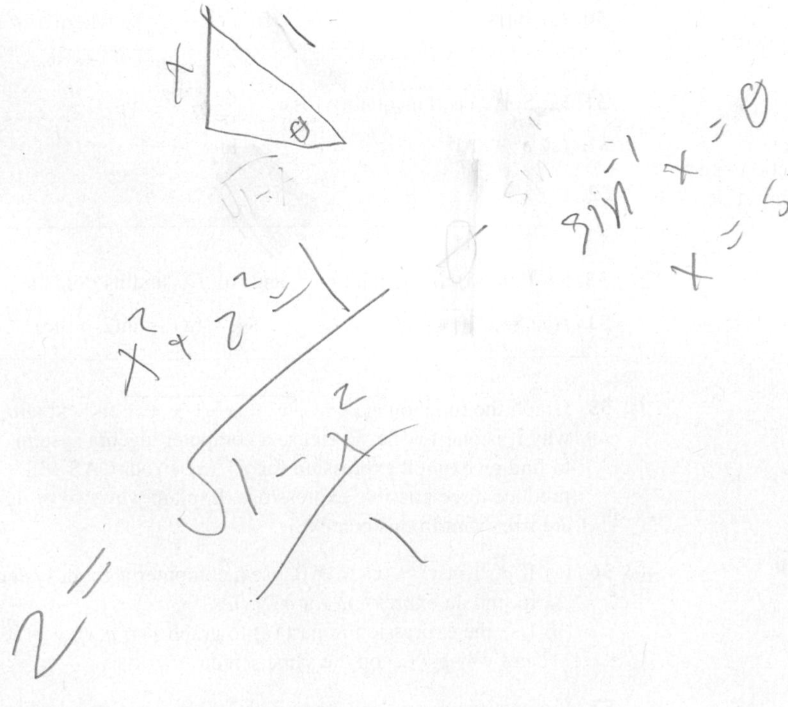

66–68 Simplify the expression.

66. $\tan(\sin^{-1} x)$

67. $\sin(\tan^{-1} x)$

68. $\cos(2 \tan^{-1} x)$

69–70 Graph the given functions on the same screen. How are these graphs related?

69. $y = \sin x,\ -\pi/2 \leqslant x \leqslant \pi/2;\quad y = \sin^{-1} x;\quad y = x$

70. $y = \tan x,\ -\pi/2 < x < \pi/2;\quad y = \tan^{-1} x;\quad y = x$

71. Find the domain and range of the function

$$g(x) = \sin^{-1}(3x + 1)$$

72. (a) Graph the function $f(x) = \sin(\sin^{-1} x)$ and explain the appearance of the graph.
(b) Graph the function $g(x) = \sin^{-1}(\sin x)$. How do you explain the appearance of this graph?

73. (a) If we shift a curve to the left, what happens to its reflection about the line $y = x$? In view of this geometric principle, find an expression for the inverse of $g(x) = f(x + c)$, where f is a one-to-one function.
(b) Find an expression for the inverse of $h(x) = f(cx)$, where $c \neq 0$.

1 REVIEW

CONCEPT CHECK

1. (a) What is a function? What are its domain and range?
   (b) What is the graph of a function?
   (c) How can you tell whether a given curve is the graph of a function?
2. Discuss four ways of representing a function. Illustrate your discussion with examples.
3. (a) What is an even function? How can you tell if a function is even by looking at its graph?
   (b) What is an odd function? How can you tell if a function is odd by looking at its graph?
4. What is an increasing function?
5. What is a mathematical model?
6. Give an example of each type of function.
   (a) Linear function (b) Power function
   (c) Exponential function (d) Quadratic function
   (e) Polynomial of degree 5 (f) Rational function
7. Sketch by hand, on the same axes, the graphs of the following functions.
   (a) $f(x) = x$ (b) $g(x) = x^2$
   (c) $h(x) = x^3$ (d) $j(x) = x^4$
8. Draw, by hand, a rough sketch of the graph of each function.
   (a) $y = \sin x$ (b) $y = \tan x$
   (c) $y = e^x$ (d) $y = \ln x$
   (e) $y = 1/x$ (f) $y = |x|$
   (g) $y = \sqrt{x}$ (h) $y = \tan^{-1}x$
9. Suppose that f has domain A and g has domain B.
   (a) What is the domain of $f + g$?
   (b) What is the domain of fg?
   (c) What is the domain of f/g?
10. How is the composite function $f \circ g$ defined? What is its domain?
11. Suppose the graph of f is given. Write an equation for each of the graphs that are obtained from the graph of f as follows.
   (a) Shift 2 units upward.
   (b) Shift 2 units downward.
   (c) Shift 2 units to the right.
   (d) Shift 2 units to the left.
   (e) Reflect about the x-axis.
   (f) Reflect about the y-axis.
   (g) Stretch vertically by a factor of 2.
   (h) Shrink vertically by a factor of 2.
   (i) Stretch horizontally by a factor of 2.
   (j) Shrink horizontally by a factor of 2.
12. (a) What is a one-to-one function? How can you tell if a function is one-to-one by looking at its graph?
   (b) If f is a one-to-one function, how is its inverse function f^{-1} defined? How do you obtain the graph of f^{-1} from the graph of f?
13. (a) How is the inverse sine function $f(x) = \sin^{-1}x$ defined? What are its domain and range?
   (b) How is the inverse cosine function $f(x) = \cos^{-1}x$ defined? What are its domain and range?
   (c) How is the inverse tangent function $f(x) = \tan^{-1}x$ defined? What are its domain and range?

TRUE-FALSE QUIZ

Determine whether the statement is true or false. If it is true, explain why. If it is false, explain why or give an example that disproves the statement.

1. If f is a function, then $f(s + t) = f(s) + f(t)$.
2. If $f(s) = f(t)$, then $s = t$.
3. If f is a function, then $f(3x) = 3f(x)$.
4. If $x_1 < x_2$ and f is a decreasing function, then $f(x_1) > f(x_2)$.
5. A vertical line intersects the graph of a function at most once.
6. If f and g are functions, then $f \circ g = g \circ f$.
7. If f is one-to-one, then $f^{-1}(x) = \dfrac{1}{f(x)}$.
8. You can always divide by e^x.
9. If $0 < a < b$, then $\ln a < \ln b$.
10. If $x > 0$, then $(\ln x)^6 = 6 \ln x$.
11. If $x > 0$ and $a > 1$, then $\dfrac{\ln x}{\ln a} = \ln \dfrac{x}{a}$.
12. $\tan^{-1}(-1) = 3\pi/4$
13. $\tan^{-1}x = \dfrac{\sin^{-1}x}{\cos^{-1}x}$

PRINCIPLES OF PROBLEM SOLVING

There are no hard and fast rules that will ensure success in solving problems. However, it is possible to outline some general steps in the problem-solving process and to give some principles that may be useful in the solution of certain problems. These steps and principles are just common sense made explicit. They have been adapted from George Polya's book *How To Solve It*.

1 Understand the Problem

The first step is to read the problem and make sure that you understand it clearly. Ask yourself the following questions:

What is the unknown?

What are the given quantities?

What are the given conditions?

For many problems it is useful to

draw a diagram

and identify the given and required quantities on the diagram.

Usually it is necessary to

introduce suitable notation

In choosing symbols for the unknown quantities we often use letters such as a, b, c, m, n, x, and y, but in some cases it helps to use initials as suggestive symbols; for instance, V for volume or t for time.

2 Think of a Plan

Find a connection between the given information and the unknown that will enable you to calculate the unknown. It often helps to ask yourself explicitly: "How can I relate the given to the unknown?" If you don't see a connection immediately, the following ideas may be helpful in devising a plan.

Try to Recognize Something Familiar Relate the given situation to previous knowledge. Look at the unknown and try to recall a more familiar problem that has a similar unknown.

Try to Recognize Patterns Some problems are solved by recognizing that some kind of pattern is occurring. The pattern could be geometric, or numerical, or algebraic. If you can see regularity or repetition in a problem, you might be able to guess what the continuing pattern is and then prove it.

Use Analogy Try to think of an analogous problem, that is, a similar problem, a related problem, but one that is easier than the original problem. If you can solve the similar, simpler problem, then it might give you the clues you need to solve the original, more difficult problem. For instance, if a problem involves very large numbers, you could first try a similar problem with smaller numbers. Or if the problem involves three-dimensional geometry, you could look for a similar problem in two-dimensional geometry. Or if the problem you start with is a general one, you could first try a special case.

Introduce Something Extra It may sometimes be necessary to introduce something new, an auxiliary aid, to help make the connection between the given and the unknown. For instance, in a problem where a diagram is useful the auxiliary aid could be a new line drawn in a diagram. In a more algebraic problem it could be a new unknown that is related to the original unknown.

Take Cases We may sometimes have to split a problem into several cases and give a different argument for each of the cases. For instance, we often have to use this strategy in dealing with absolute value.

Work Backward Sometimes it is useful to imagine that your problem is solved and work backward, step by step, until you arrive at the given data. Then you may be able to reverse your steps and thereby construct a solution to the original problem. This procedure is commonly used in solving equations. For instance, in solving the equation $3x - 5 = 7$, we suppose that x is a number that satisfies $3x - 5 = 7$ and work backward. We add 5 to each side of the equation and then divide each side by 3 to get $x = 4$. Since each of these steps can be reversed, we have solved the problem.

Establish Subgoals In a complex problem it is often useful to set subgoals (in which the desired situation is only partially fulfilled). If we can first reach these subgoals, then we may be able to build on them to reach our final goal.

Indirect Reasoning Sometimes it is appropriate to attack a problem indirectly. In using proof by contradiction to prove that P implies Q, we assume that P is true and Q is false and try to see why this can't happen. Somehow we have to use this information and arrive at a contradiction to what we absolutely know is true.

Mathematical Induction In proving statements that involve a positive integer n, it is frequently helpful to use the following principle.

PRINCIPLE OF MATHEMATICAL INDUCTION Let S_n be a statement about the positive integer n. Suppose that

1. S_1 is true.

2. S_{k+1} is true whenever S_k is true.

Then S_n is true for all positive integers n.

This is reasonable because, since S_1 is true, it follows from condition 2 (with $k = 1$) that S_2 is true. Then, using condition 2 with $k = 2$, we see that S_3 is true. Again using condition 2, this time with $k = 3$, we have that S_4 is true. This procedure can be followed indefinitely.

3 Carry Out the Plan

In Step 2 a plan was devised. In carrying out that plan we have to check each stage of the plan and write the details that prove that each stage is correct.

4 Look Back

Having completed our solution, it is wise to look back over it, partly to see if we have made errors in the solution and partly to see if we can think of an easier way to solve the problem. Another reason for looking back is that it will familiarize us with the method of solution and this may be useful for solving a future problem. Descartes said, "Every problem that I solved became a rule which served afterwards to solve other problems."

These principles of problem solving are illustrated in the following examples. Before you look at the solutions, try to solve these problems yourself, referring to these Principles of Problem Solving if you get stuck. You may find it useful to refer to this section from time to time as you solve the exercises in the remaining chapters of this book.

PRINCIPLES OF PROBLEM SOLVING

EXAMPLE 1 Express the hypotenuse h of a right triangle with area 25 m^2 as a function of its perimeter P.

■ Understand the problem

SOLUTION Let's first sort out the information by identifying the unknown quantity and the data:

Unknown: hypotenuse h

Given quantities: perimeter P, area 25 m^2

■ Draw a diagram

It helps to draw a diagram and we do so in Figure 1.

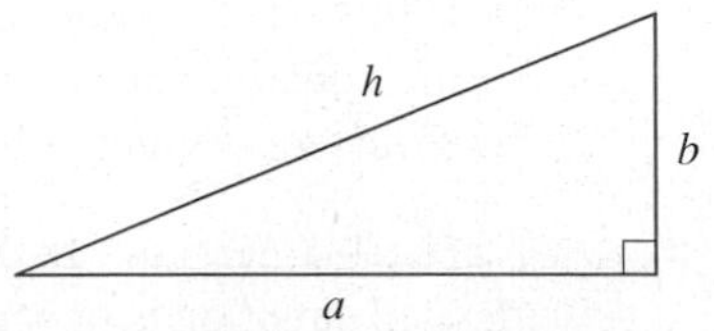

FIGURE 1

■ Connect the given with the unknown

■ Introduce something extra

In order to connect the given quantities to the unknown, we introduce two extra variables a and b, which are the lengths of the other two sides of the triangle. This enables us to express the given condition, which is that the triangle is right-angled, by the Pythagorean Theorem:

$$h^2 = a^2 + b^2$$

The other connections among the variables come by writing expressions for the area and perimeter:

$$25 = \tfrac{1}{2}ab \qquad P = a + b + h$$

Since P is given, notice that we now have three equations in the three unknowns a, b, and h:

$$\boxed{1} \qquad h^2 = a^2 + b^2$$

$$\boxed{2} \qquad 25 = \tfrac{1}{2}ab$$

$$\boxed{3} \qquad P = a + b + h$$

■ Relate to the familiar

Although we have the correct number of equations, they are not easy to solve in a straightforward fashion. But if we use the problem-solving strategy of trying to recognize something familiar, then we can solve these equations by an easier method. Look at the right sides of Equations 1, 2, and 3. Do these expressions remind you of anything familiar? Notice that they contain the ingredients of a familiar formula:

$$(a + b)^2 = a^2 + 2ab + b^2$$

Using this idea, we express $(a + b)^2$ in two ways. From Equations 1 and 2 we have

$$(a + b)^2 = (a^2 + b^2) + 2ab = h^2 + 4(25)$$

From Equation 3 we have

$$(a + b)^2 = (P - h)^2 = P^2 - 2Ph + h^2$$

Thus

$$h^2 + 100 = P^2 - 2Ph + h^2$$

$$2Ph = P^2 - 100$$

$$h = \frac{P^2 - 100}{2P}$$

This is the required expression for h as a function of P. □

As the next example illustrates, it is often necessary to use the problem-solving principle of *taking cases* when dealing with absolute values.

EXAMPLE 2 Solve the inequality $|x - 3| + |x + 2| < 11$.

SOLUTION Recall the definition of absolute value:

$$|x| = \begin{cases} x & \text{if } x \geqslant 0 \\ -x & \text{if } x < 0 \end{cases}$$

It follows that

$$|x - 3| = \begin{cases} x - 3 & \text{if } x - 3 \geqslant 0 \\ -(x - 3) & \text{if } x - 3 < 0 \end{cases}$$

$$= \begin{cases} x - 3 & \text{if } x \geqslant 3 \\ -x + 3 & \text{if } x < 3 \end{cases}$$

Similarly

$$|x + 2| = \begin{cases} x + 2 & \text{if } x + 2 \geqslant 0 \\ -(x + 2) & \text{if } x + 2 < 0 \end{cases}$$

$$= \begin{cases} x + 2 & \text{if } x \geqslant -2 \\ -x - 2 & \text{if } x < -2 \end{cases}$$

■ Take cases

These expressions show that we must consider three cases:

$$x < -2 \qquad -2 \leqslant x < 3 \qquad x \geqslant 3$$

CASE I If $x < -2$, we have

$$|x - 3| + |x + 2| < 11$$
$$-x + 3 - x - 2 < 11$$
$$-2x < 10$$
$$x > -5$$

CASE II If $-2 \leqslant x < 3$, the given inequality becomes

$$-x + 3 + x + 2 < 11$$
$$5 < 11 \qquad \text{(always true)}$$

CASE III If $x \geqslant 3$, the inequality becomes

$$x - 3 + x + 2 < 11$$
$$2x < 12$$
$$x < 6$$

Combining cases I, II, and III, we see that the inequality is satisfied when $-5 < x < 6$. So the solution is the interval $(-5, 6)$. □

In the following example we first guess the answer by looking at special cases and recognizing a pattern. Then we prove it by mathematical induction.

In using the Principle of Mathematical Induction, we follow three steps:

STEP 1 Prove that S_n is true when $n = 1$.

STEP 2 Assume that S_n is true when $n = k$ and deduce that S_n is true when $n = k + 1$.

STEP 3 Conclude that S_n is true for all n by the Principle of Mathematical Induction.

EXAMPLE 3 If $f_0(x) = x/(x + 1)$ and $f_{n+1} = f_0 \circ f_n$ for $n = 0, 1, 2, \ldots$, find a formula for $f_n(x)$.

■ Analogy: Try a similar, simpler problem

SOLUTION We start by finding formulas for $f_n(x)$ for the special cases $n = 1$, 2, and 3.

$$f_1(x) = (f_0 \circ f_0)(x) = f_0(f_0(x)) = f_0\left(\frac{x}{x+1}\right)$$

$$= \frac{\dfrac{x}{x+1}}{\dfrac{x}{x+1} + 1} = \frac{\dfrac{x}{x+1}}{\dfrac{2x+1}{x+1}} = \frac{x}{2x+1}$$

$$f_2(x) = (f_0 \circ f_1)(x) = f_0(f_1(x)) = f_0\left(\frac{x}{2x+1}\right)$$

$$= \frac{\dfrac{x}{2x+1}}{\dfrac{x}{2x+1} + 1} = \frac{\dfrac{x}{2x+1}}{\dfrac{3x+1}{2x+1}} = \frac{x}{3x+1}$$

$$f_3(x) = (f_0 \circ f_2)(x) = f_0(f_2(x)) = f_0\left(\frac{x}{3x+1}\right)$$

■ Look for a pattern

$$= \frac{\dfrac{x}{3x+1}}{\dfrac{x}{3x+1} + 1} = \frac{\dfrac{x}{3x+1}}{\dfrac{4x+1}{3x+1}} = \frac{x}{4x+1}$$

We notice a pattern: The coefficient of x in the denominator of $f_n(x)$ is $n + 1$ in the three cases we have computed. So we make the guess that, in general,

$$f_n(x) = \frac{x}{(n+1)x + 1} \tag{4}$$

To prove this, we use the Principle of Mathematical Induction. We have already verified that (4) is true for $n = 1$. Assume that it is true for $n = k$, that is,

$$f_k(x) = \frac{x}{(k+1)x + 1}$$

Then $\quad f_{k+1}(x) = (f_0 \circ f_k)(x) = f_0(f_k(x)) = f_0\left(\dfrac{x}{(k+1)x+1}\right)$

$$= \frac{\dfrac{x}{(k+1)x+1}}{\dfrac{x}{(k+1)x+1}+1} = \frac{\dfrac{x}{(k+1)x+1}}{\dfrac{(k+2)x+1}{(k+1)x+1}} = \frac{x}{(k+2)x+1}$$

This expression shows that (4) is true for $n = k + 1$. Therefore, by mathematical induction, it is true for all positive integers n. □

PROBLEMS

1. One of the legs of a right triangle has length 4 cm. Express the length of the altitude perpendicular to the hypotenuse as a function of the length of the hypotenuse.

2. The altitude perpendicular to the hypotenuse of a right triangle is 12 cm. Express the length of the hypotenuse as a function of the perimeter.

3. Solve the equation $|2x - 1| - |x + 5| = 3$.

4. Solve the inequality $|x - 1| - |x - 3| \geqslant 5$.

5. Sketch the graph of the function $f(x) = |x^2 - 4|x| + 3|$.

6. Sketch the graph of the function $g(x) = |x^2 - 1| - |x^2 - 4|$.

7. Draw the graph of the equation $x + |x| = y + |y|$.

8. Draw the graph of the equation $x^4 - 4x^2 - x^2y^2 + 4y^2 = 0$.

9. Sketch the region in the plane consisting of all points (x, y) such that $|x| + |y| \leqslant 1$.

10. Sketch the region in the plane consisting of all points (x, y) such that

$$|x - y| + |x| - |y| \leqslant 2$$

11. Evaluate $(\log_2 3)(\log_3 4)(\log_4 5) \cdots (\log_{31} 32)$.

12. (a) Show that the function $f(x) = \ln\left(x + \sqrt{x^2 + 1}\right)$ is an odd function.
(b) Find the inverse function of f.

13. Solve the inequality $\ln(x^2 - 2x - 2) \leqslant 0$.

14. Use indirect reasoning to prove that $\log_2 5$ is an irrational number.

15. A driver sets out on a journey. For the first half of the distance she drives at the leisurely pace of 30 mi/h; she drives the second half at 60 mi/h. What is her average speed on this trip?

16. Is it true that $f \circ (g + h) = f \circ g + f \circ h$?

17. Prove that if n is a positive integer, then $7^n - 1$ is divisible by 6.

18. Prove that $1 + 3 + 5 + \cdots + (2n - 1) = n^2$.

19. If $f_0(x) = x^2$ and $f_{n+1}(x) = f_0(f_n(x))$ for $n = 0, 1, 2, \ldots$, find a formula for $f_n(x)$.

20. (a) If $f_0(x) = \dfrac{1}{2 - x}$ and $f_{n+1} = f_0 \circ f_n$ for $n = 0, 1, 2, \ldots$, find an expression for $f_n(x)$ and use mathematical induction to prove it.
(b) Graph f_0, f_1, f_2, f_3 on the same screen and describe the effects of repeated composition.

Figure 3 illustrates the limiting process that occurs in this example. As Q approaches P along the parabola, the corresponding secant lines rotate about P and approach the tangent line t.

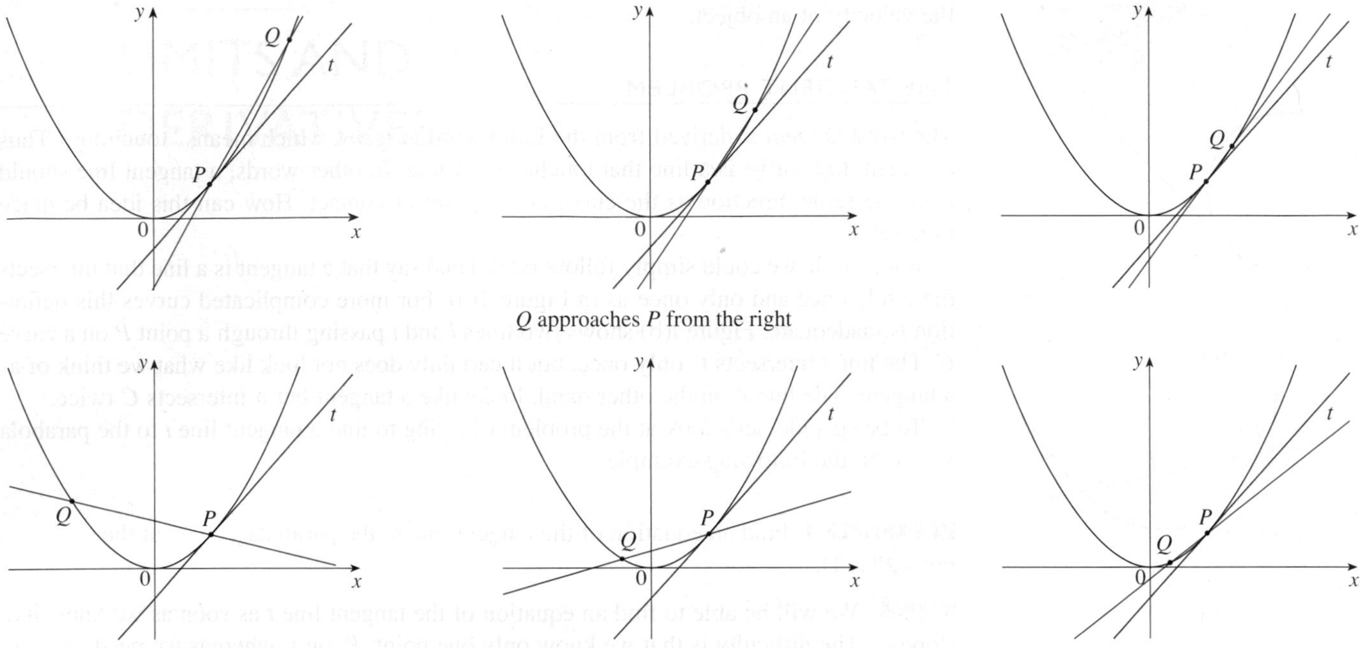

Q approaches P from the right

Q approaches P from the left

FIGURE 3

□

TEC In Visual 2.1 you can see how the process in Figure 3 works for additional functions.

Many functions that occur in science are not described by explicit equations; they are defined by experimental data. The next example shows how to estimate the slope of the tangent line to the graph of such a function.

V EXAMPLE 2 The flash unit on a camera operates by storing charge on a capacitor and releasing it suddenly when the flash is set off. The data in the table describe the charge Q remaining on the capacitor (measured in microcoulombs) at time t (measured in seconds after the flash goes off). Use the data to draw the graph of this function and estimate the slope of the tangent line at the point where $t = 0.04$. [*Note:* The slope of the tangent line represents the electric current flowing from the capacitor to the flash bulb (measured in microamperes).]

t	Q
0.00	100.00
0.02	81.87
0.04	67.03
0.06	54.88
0.08	44.93
0.10	36.76

SOLUTION In Figure 4 we plot the given data and use them to sketch a curve that approximates the graph of the function.

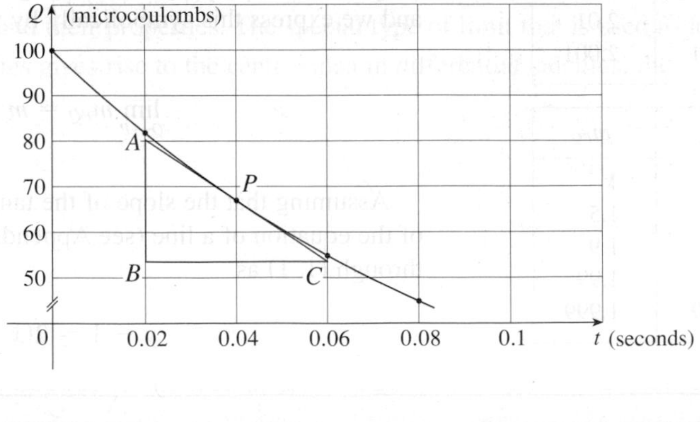

FIGURE 4

Given the points $P(0.04, 67.03)$ and $R(0.00, 100.00)$ on the graph, we find that the slope of the secant line PR is

$$m_{PR} = \frac{100.00 - 67.03}{0.00 - 0.04} = -824.25$$

R	m_{PR}
(0.00, 100.00)	−824.25
(0.02, 81.87)	−742.00
(0.06, 54.88)	−607.50
(0.08, 44.93)	−552.50
(0.10, 36.76)	−504.50

The table at the left shows the results of similar calculations for the slopes of other secant lines. From this table we would expect the slope of the tangent line at $t = 0.04$ to lie somewhere between -742 and -607.5. In fact, the average of the slopes of the two closest secant lines is

$$\tfrac{1}{2}(-742 - 607.5) = -674.75$$

So, by this method, we estimate the slope of the tangent line to be -675.

Another method is to draw an approximation to the tangent line at P and measure the sides of the triangle ABC, as in Figure 4. This gives an estimate of the slope of the tangent line as

$$-\frac{|AB|}{|BC|} \approx -\frac{80.4 - 53.6}{0.06 - 0.02} = -670$$ □

■ The physical meaning of the answer in Example 2 is that the electric current flowing from the capacitor to the flash bulb after 0.04 second is about –670 microamperes.

THE VELOCITY PROBLEM

If you watch the speedometer of a car as you travel in city traffic, you see that the needle doesn't stay still for very long; that is, the velocity of the car is not constant. We assume from watching the speedometer that the car has a definite velocity at each moment, but how is the "instantaneous" velocity defined? Let's investigate the example of a falling ball.

The CN Tower in Toronto is currently the tallest freestanding building in the world.

EXAMPLE 3 Suppose that a ball is dropped from the upper observation deck of the CN Tower in Toronto, 450 m above the ground. Find the velocity of the ball after 5 seconds.

SOLUTION Through experiments carried out four centuries ago, Galileo discovered that the distance fallen by any freely falling body is proportional to the square of the time it has been falling. (This model for free fall neglects air resistance.) If the distance fallen after t seconds is denoted by $s(t)$ and measured in meters, then Galileo's law is expressed by the equation

$$s(t) = 4.9t^2$$

The difficulty in finding the velocity after 5 s is that we are dealing with a single instant of time ($t = 5$), so no time interval is involved. However, we can approximate the desired quantity by computing the average velocity over the brief time interval of a tenth of a second from $t = 5$ to $t = 5.1$:

$$\text{average velocity} = \frac{\text{change in position}}{\text{time elapsed}}$$

$$= \frac{s(5.1) - s(5)}{0.1}$$

$$= \frac{4.9(5.1)^2 - 4.9(5)^2}{0.1} = 49.49 \text{ m/s}$$

The following table shows the results of similar calculations of the average velocity over successively smaller time periods.

Time interval	Average velocity (m/s)
$5 \leqslant t \leqslant 6$	53.9
$5 \leqslant t \leqslant 5.1$	49.49
$5 \leqslant t \leqslant 5.05$	49.245
$5 \leqslant t \leqslant 5.01$	49.049
$5 \leqslant t \leqslant 5.001$	49.0049

It appears that as we shorten the time period, the average velocity is becoming closer to 49 m/s. The **instantaneous velocity** when $t = 5$ is defined to be the limiting value of these average velocities over shorter and shorter time periods that start at $t = 5$. Thus the (instantaneous) velocity after 5 s is

$$v = 49 \text{ m/s}$$ □

You may have the feeling that the calculations used in solving this problem are very similar to those used earlier in this section to find tangents. In fact, there is a close connection between the tangent problem and the problem of finding velocities. If we draw the graph of the distance function of the ball (as in Figure 5) and we consider the points $P(a, 4.9a^2)$ and $Q(a + h, 4.9(a + h)^2)$ on the graph, then the slope of the secant line PQ is

$$m_{PQ} = \frac{4.9(a + h)^2 - 4.9a^2}{(a + h) - a}$$

which is the same as the average velocity over the time interval $[a, a + h]$. Therefore, the velocity at time $t = a$ (the limit of these average velocities as h approaches 0) must be equal to the slope of the tangent line at P (the limit of the slopes of the secant lines).

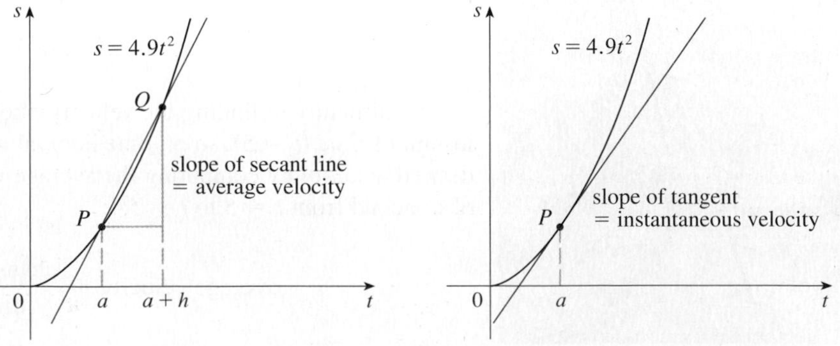

FIGURE 5

Examples 1 and 3 show that in order to solve tangent and velocity problems we must be able to find limits. After studying methods for computing limits in the next five sections, we will return to the problems of finding tangents and velocities in Section 2.7.

2.1 EXERCISES

1. A tank holds 1000 gallons of water, which drains from the bottom of the tank in half an hour. The values in the table show the volume V of water remaining in the tank (in gallons) after t minutes.

t (min)	5	10	15	20	25	30
V (gal)	694	444	250	111	28	0

(a) If P is the point (15, 250) on the graph of V, find the slopes of the secant lines PQ when Q is the point on the graph with $t = 5$, 10, 20, 25, and 30.
(b) Estimate the slope of the tangent line at P by averaging the slopes of two secant lines.
(c) Use a graph of the function to estimate the slope of the tangent line at P. (This slope represents the rate at which the water is flowing from the tank after 15 minutes.)

2. A cardiac monitor is used to measure the heart rate of a patient after surgery. It compiles the number of heartbeats after t minutes. When the data in the table are graphed, the slope of the tangent line represents the heart rate in beats per minute.

t (min)	36	38	40	42	44
Heartbeats	2530	2661	2806	2948	3080

The monitor estimates this value by calculating the slope of a secant line. Use the data to estimate the patient's heart rate after 42 minutes using the secant line between the points with the given values of t.
(a) $t = 36$ and $t = 42$ (b) $t = 38$ and $t = 42$
(c) $t = 40$ and $t = 42$ (d) $t = 42$ and $t = 44$
What are your conclusions?

3. The point $P(1, \frac{1}{2})$ lies on the curve $y = x/(1 + x)$.
(a) If Q is the point $(x, x/(1 + x))$, use your calculator to find the slope of the secant line PQ (correct to six decimal places) for the following values of x:
(i) 0.5 (ii) 0.9 (iii) 0.99 (iv) 0.999
(v) 1.5 (vi) 1.1 (vii) 1.01 (viii) 1.001
(b) Using the results of part (a), guess the value of the slope of the tangent line to the curve at $P(1, \frac{1}{2})$.
(c) Using the slope from part (b), find an equation of the tangent line to the curve at $P(1, \frac{1}{2})$.

4. The point $P(3, 1)$ lies on the curve $y = \sqrt{x - 2}$.
(a) If Q is the point $(x, \sqrt{x - 2})$, use your calculator to find the slope of the secant line PQ (correct to six decimal places) for the following values of x:
(i) 2.5 (ii) 2.9 (iii) 2.99 (iv) 2.999
(v) 3.5 (vi) 3.1 (vii) 3.01 (viii) 3.001
(b) Using the results of part (a), guess the value of the slope of the tangent line to the curve at $P(3, 1)$.
(c) Using the slope from part (b), find an equation of the tangent line to the curve at $P(3, 1)$.
(d) Sketch the curve, two of the secant lines, and the tangent line.

5. If a ball is thrown into the air with a velocity of 40 ft/s, its height in feet t seconds later is given by $y = 40t - 16t^2$.
(a) Find the average velocity for the time period beginning when $t = 2$ and lasting
(i) 0.5 second (ii) 0.1 second
(iii) 0.05 second (iv) 0.01 second
(b) Estimate the instantaneous velocity when $t = 2$.

6. If a rock is thrown upward on the planet Mars with a velocity of 10 m/s, its height in meters t seconds later is given by $y = 10t - 1.86t^2$.
(a) Find the average velocity over the given time intervals:
(i) [1, 2] (ii) [1, 1.5] (iii) [1, 1.1]
(iv) [1, 1.01] (v) [1, 1.001]
(b) Estimate the instantaneous velocity when $t = 1$.

7. The table shows the position of a cyclist.

t (seconds)	0	1	2	3	4	5
s (meters)	0	1.4	5.1	10.7	17.7	25.8

(a) Find the average velocity for each time period:
(i) [1, 3] (ii) [2, 3] (iii) [3, 5] (iv) [3, 4]
(b) Use the graph of s as a function of t to estimate the instantaneous velocity when $t = 3$.

8. The displacement (in centimeters) of a particle moving back and forth along a straight line is given by the equation of motion $s = 2 \sin \pi t + 3 \cos \pi t$, where t is measured in seconds.
(a) Find the average velocity during each time period:
(i) [1, 2] (ii) [1, 1.1]
(iii) [1, 1.01] (iv) [1, 1.001]
(b) Estimate the instantaneous velocity of the particle when $t = 1$.

9. The point $P(1, 0)$ lies on the curve $y = \sin(10\pi/x)$.
(a) If Q is the point $(x, \sin(10\pi/x))$, find the slope of the secant line PQ (correct to four decimal places) for $x = 2$, 1.5, 1.4, 1.3, 1.2, 1.1, 0.5, 0.6, 0.7, 0.8, and 0.9. Do the slopes appear to be approaching a limit?
(b) Use a graph of the curve to explain why the slopes of the secant lines in part (a) are not close to the slope of the tangent line at P.
(c) By choosing appropriate secant lines, estimate the slope of the tangent line at P.

2.2 THE LIMIT OF A FUNCTION

Having seen in the preceding section how limits arise when we want to find the tangent to a curve or the velocity of an object, we now turn our attention to limits in general and numerical and graphical methods for computing them.

Let's investigate the behavior of the function f defined by $f(x) = x^2 - x + 2$ for values of x near 2. The following table gives values of $f(x)$ for values of x close to 2, but not equal to 2.

x	$f(x)$	x	$f(x)$
1.0	2.000000	3.0	8.000000
1.5	2.750000	2.5	5.750000
1.8	3.440000	2.2	4.640000
1.9	3.710000	2.1	4.310000
1.95	3.852500	2.05	4.152500
1.99	3.970100	2.01	4.030100
1.995	3.985025	2.005	4.015025
1.999	3.997001	2.001	4.003001

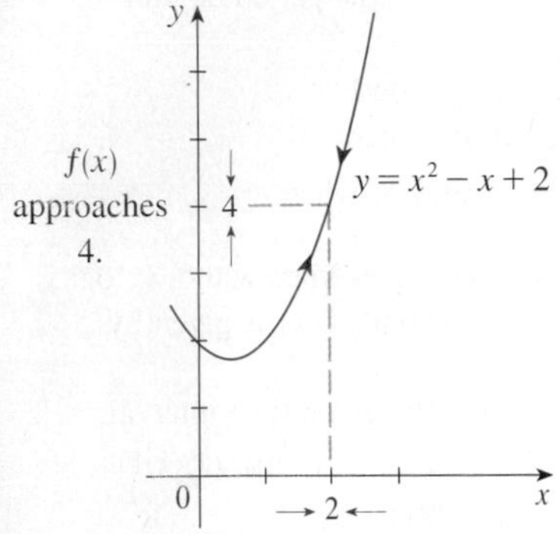

As x approaches 2,

FIGURE 1

From the table and the graph of f (a parabola) shown in Figure 1 we see that when x is close to 2 (on either side of 2), $f(x)$ is close to 4. In fact, it appears that we can make the values of $f(x)$ as close as we like to 4 by taking x sufficiently close to 2. We express this by saying "the limit of the function $f(x) = x^2 - x + 2$ as x approaches 2 is equal to 4." The notation for this is

$$\lim_{x \to 2} (x^2 - x + 2) = 4$$

In general, we use the following notation.

1 DEFINITION We write

$$\lim_{x \to a} f(x) = L$$

and say "the limit of $f(x)$, as x approaches a, equals L"

if we can make the values of $f(x)$ arbitrarily close to L (as close to L as we like) by taking x to be sufficiently close to a (on either side of a) but not equal to a.

Roughly speaking, this says that the values of $f(x)$ tend to get closer and closer to the number L as x gets closer and closer to the number a (from either side of a) but $x \neq a$. (A more precise definition will be given in Section 2.4.)

An alternative notation for

$$\lim_{x \to a} f(x) = L$$

is $$f(x) \to L \quad \text{as} \quad x \to a$$

which is usually read "$f(x)$ approaches L as x approaches a."

Notice the phrase "but $x \neq a$" in the definition of limit. This means that in finding the limit of $f(x)$ as x approaches a, we never consider $x = a$. In fact, $f(x)$ need not even be defined when $x = a$. The only thing that matters is how f is defined *near a*.

Figure 2 shows the graphs of three functions. Note that in part (c), $f(a)$ is not defined and in part (b), $f(a) \neq L$. But in each case, regardless of what happens at a, it is true that $\lim_{x \to a} f(x) = L$.

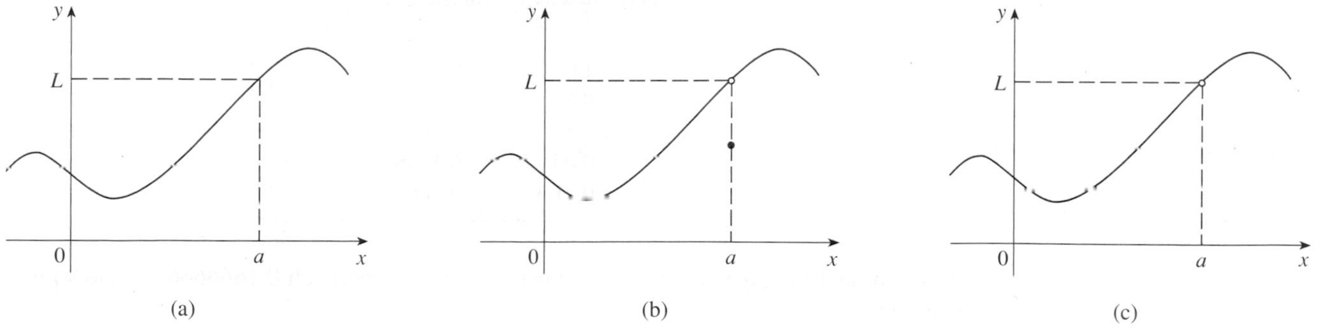

FIGURE 2 $\lim_{x \to a} f(x) = L$ in all three cases

EXAMPLE 1 Guess the value of $\lim_{x \to 1} \dfrac{x - 1}{x^2 - 1}$.

SOLUTION Notice that the function $f(x) = (x - 1)/(x^2 - 1)$ is not defined when $x = 1$, but that doesn't matter because the definition of $\lim_{x \to a} f(x)$ says that we consider values of x that are close to a but not equal to a.

$x < 1$	$f(x)$
0.5	0.666667
0.9	0.526316
0.99	0.502513
0.999	0.500250
0.9999	0.500025

$x > 1$	$f(x)$
1.5	0.400000
1.1	0.476190
1.01	0.497512
1.001	0.499750
1.0001	0.499975

The tables at the left give values of $f(x)$ (correct to six decimal places) for values of x that approach 1 (but are not equal to 1). On the basis of the values in the tables, we make the guess that

$$\lim_{x \to 1} \frac{x - 1}{x^2 - 1} = 0.5$$ □

Example 1 is illustrated by the graph of f in Figure 3. Now let's change f slightly by giving it the value 2 when $x = 1$ and calling the resulting function g:

$$g(x) = \begin{cases} \dfrac{x - 1}{x^2 - 1} & \text{if } x \neq 1 \\ 2 & \text{if } x = 1 \end{cases}$$

This new function g still has the same limit as x approaches 1. (See Figure 4.)

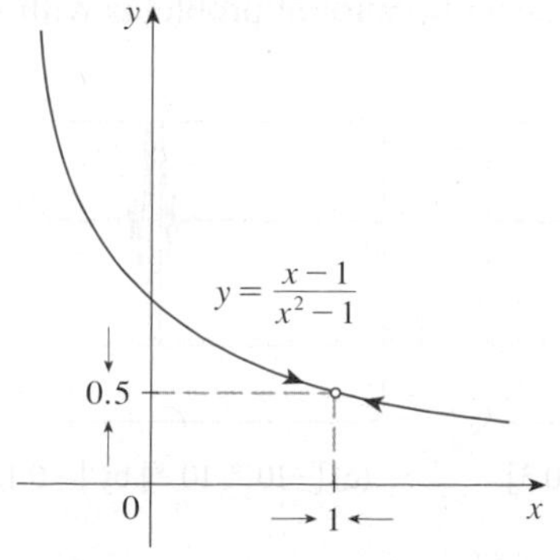

FIGURE 3

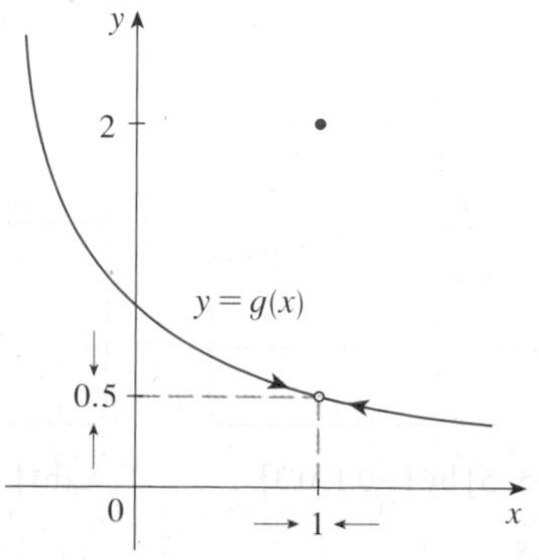

FIGURE 4

The dashed lines near the y-axis indicate that the values of $\sin(\pi/x)$ oscillate between 1 and -1 infinitely often as x approaches 0. (See Exercise 39.)

Since the values of $f(x)$ do not approach a fixed number as x approaches 0,

$$\lim_{x\to 0} \sin\frac{\pi}{x} \quad \text{does not exist}$$ □

x	$x^3 + \dfrac{\cos 5x}{10,000}$
1	1.000028
0.5	0.124920
0.1	0.001088
0.05	0.000222
0.01	0.000101

x	$x^3 + \dfrac{\cos 5x}{10,000}$
0.005	0.00010009
0.001	0.00010000

EXAMPLE 5 Find $\lim_{x\to 0}\left(x^3 + \dfrac{\cos 5x}{10,000}\right)$.

SOLUTION As before, we construct a table of values. From the first table in the margin it appears that

$$\lim_{x\to 0}\left(x^3 + \frac{\cos 5x}{10,000}\right) = 0$$

But if we persevere with smaller values of x, the second table suggests that

$$\lim_{x\to 0}\left(x^3 + \frac{\cos 5x}{10,000}\right) = 0.000100 = \frac{1}{10,000}$$

Later we will see that $\lim_{x\to 0} \cos 5x = 1$; then it follows that the limit is 0.0001. □

⊘ Examples 4 and 5 illustrate some of the pitfalls in guessing the value of a limit. It is easy to guess the wrong value if we use inappropriate values of x, but it is difficult to know when to stop calculating values. And, as the discussion after Example 2 shows, sometimes calculators and computers give the wrong values. In the next section, however, we will develop foolproof methods for calculating limits.

V EXAMPLE 6 The Heaviside function H is defined by

$$H(t) = \begin{cases} 0 & \text{if } t < 0 \\ 1 & \text{if } t \geqslant 0 \end{cases}$$

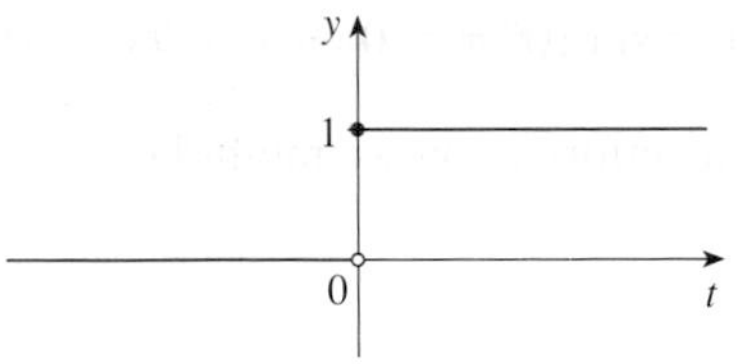

FIGURE 8

[This function is named after the electrical engineer Oliver Heaviside (1850–1925) and can be used to describe an electric current that is switched on at time $t = 0$.] Its graph is shown in Figure 8.

As t approaches 0 from the left, $H(t)$ approaches 0. As t approaches 0 from the right, $H(t)$ approaches 1. There is no single number that $H(t)$ approaches as t approaches 0. Therefore, $\lim_{t\to 0} H(t)$ does not exist. □

ONE-SIDED LIMITS

We noticed in Example 6 that $H(t)$ approaches 0 as t approaches 0 from the left and $H(t)$ approaches 1 as t approaches 0 from the right. We indicate this situation symbolically by writing

$$\lim_{t\to 0^-} H(t) = 0 \qquad \text{and} \qquad \lim_{t\to 0^+} H(t) = 1$$

The symbol "$t \to 0^-$" indicates that we consider only values of t that are less than 0. Likewise, "$t \to 0^+$" indicates that we consider only values of t that are greater than 0.

2 DEFINITION We write

$$\lim_{x \to a^-} f(x) = L$$

and say the **left-hand limit of $f(x)$ as x approaches a** [or the **limit of $f(x)$ as x approaches a from the left**] is equal to L if we can make the values of $f(x)$ arbitrarily close to L by taking x to be sufficiently close to a and x less than a.

Notice that Definition 2 differs from Definition 1 only in that we require x to be less than a. Similarly, if we require that x be greater than a, we get "the **right-hand limit of $f(x)$ as x approaches a** is equal to L" and we write

$$\lim_{x \to a^+} f(x) = L$$

Thus the symbol "$x \to a^+$" means that we consider only $x > a$. These definitions are illustrated in Figure 9.

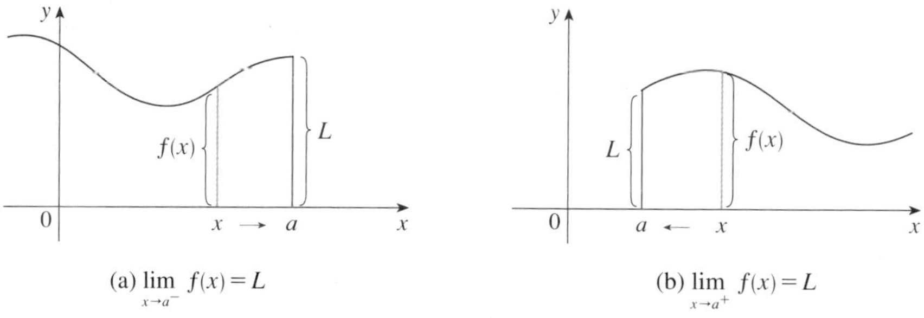

FIGURE 9

By comparing Definition 1 with the definitions of one-sided limits, we see that the following is true.

3 $\lim_{x \to a} f(x) = L$ if and only if $\lim_{x \to a^-} f(x) = L$ and $\lim_{x \to a^+} f(x) = L$

EXAMPLE 7 The graph of a function g is shown in Figure 10. Use it to state the values (if they exist) of the following:

(a) $\lim_{x \to 2^-} g(x)$ (b) $\lim_{x \to 2^+} g(x)$ (c) $\lim_{x \to 2} g(x)$

(d) $\lim_{x \to 5^-} g(x)$ (e) $\lim_{x \to 5^+} g(x)$ (f) $\lim_{x \to 5} g(x)$

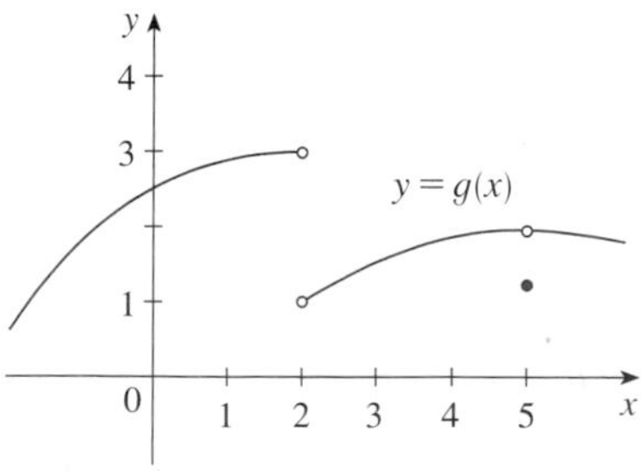

FIGURE 10

SOLUTION From the graph we see that the values of $g(x)$ approach 3 as x approaches 2 from the left, but they approach 1 as x approaches 2 from the right. Therefore

$$\text{(a)} \lim_{x \to 2^-} g(x) = 3 \quad \text{and} \quad \text{(b)} \lim_{x \to 2^+} g(x) = 1$$

(c) Since the left and right limits are different, we conclude from (3) that $\lim_{x \to 2} g(x)$ does not exist.

The graph also shows that

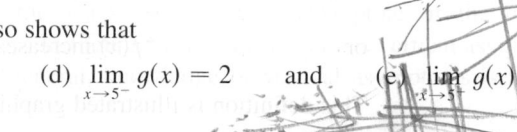

$$\text{(d)} \lim_{x \to 5^-} g(x) = 2 \quad \text{and} \quad \text{(e)} \lim_{x \to 5^+} g(x) = 2$$

EXAMPLE 9 Find $\lim_{x \to 3^+} \frac{2x}{x - 3}$ and $\lim_{x \to 3^-} \frac{2x}{x - 3}$.

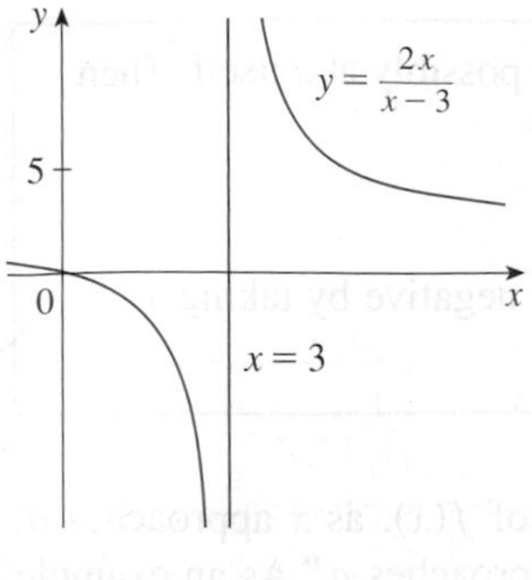

FIGURE 15

SOLUTION If x is close to 3 but larger than 3, then the denominator $x - 3$ is a small positive number and $2x$ is close to 6. So the quotient $2x/(x - 3)$ is a large *positive* number. Thus, intuitively, we see that

$$\lim_{x \to 3^+} \frac{2x}{x - 3} = \infty$$

Likewise, if x is close to 3 but smaller than 3, then $x - 3$ is a small negative number but $2x$ is still a positive number (close to 6). So $2x/(x - 3)$ is a numerically large *negative* number. Thus

$$\lim_{x \to 3^-} \frac{2x}{x - 3} = -\infty$$

The graph of the curve $y = 2x/(x - 3)$ is given in Figure 15. The line $x = 3$ is a vertical asymptote. □

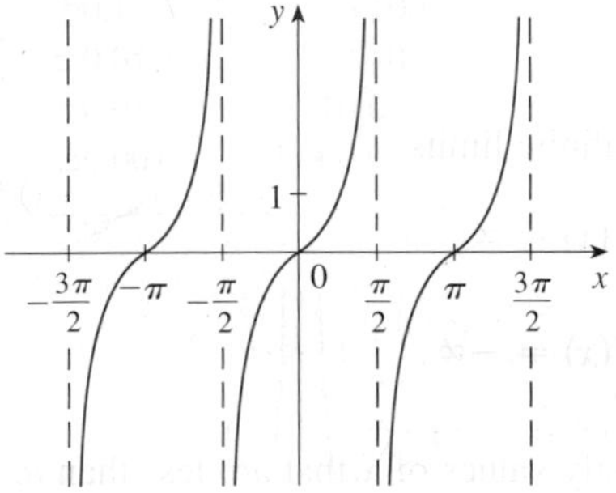

FIGURE 16
$y = \tan x$

EXAMPLE 10 Find the vertical asymptotes of $f(x) = \tan x$.

SOLUTION Because

$$\tan x = \frac{\sin x}{\cos x}$$

there are potential vertical asymptotes where $\cos x = 0$. In fact, since $\cos x \to 0^+$ as $x \to (\pi/2)^-$ and $\cos x \to 0^-$ as $x \to (\pi/2)^+$, whereas $\sin x$ is positive when x is near $\pi/2$, we have

$$\lim_{x \to (\pi/2)^-} \tan x = \infty \qquad \text{and} \qquad \lim_{x \to (\pi/2)^+} \tan x = -\infty$$

This shows that the line $x = \pi/2$ is a vertical asymptote. Similar reasoning shows that the lines $x = (2n + 1)\pi/2$, where n is an integer, are all vertical asymptotes of $f(x) = \tan x$. The graph in Figure 16 confirms this. □

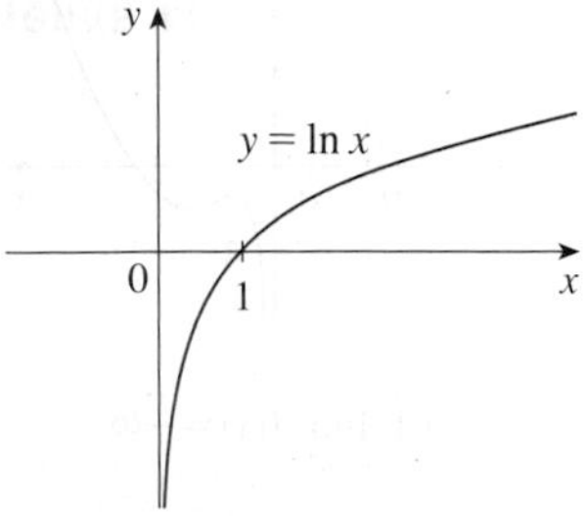

FIGURE 17
The y-axis is a vertical asymptote of the natural logarithmic function.

Another example of a function whose graph has a vertical asymptote is the natural logarithmic function $y = \ln x$. From Figure 17 we see that

$$\boxed{\lim_{x \to 0^+} \ln x = -\infty}$$

and so the line $x = 0$ (the y-axis) is a vertical asymptote. In fact, the same is true for $y = \log_a x$ provided that $a > 1$. (See Figures 11 and 12 in Section 1.6.)

2.2 EXERCISES

1. Explain in your own words what is meant by the equation

$$\lim_{x \to 2} f(x) = 5$$

Is it possible for this statement to be true and yet $f(2) = 3$? Explain.

2. Explain what it means to say that

$$\lim_{x \to 1^-} f(x) = 3 \qquad \text{and} \qquad \lim_{x \to 1^+} f(x) = 7$$

In this situation is it possible that $\lim_{x \to 1} f(x)$ exists? Explain.

3. Explain the meaning of each of the following.

(a) $\lim_{x\to -3} f(x) = \infty$ (b) $\lim_{x\to 4^+} f(x) = -\infty$

4. For the function f whose graph is given, state the value of each quantity, if it exists. If it does not exist, explain why.

(a) $\lim_{x\to 0} f(x)$ (b) $\lim_{x\to 3^-} f(x)$ (c) $\lim_{x\to 3^+} f(x)$

(d) $\lim_{x\to 3} f(x)$ (e) $f(3)$

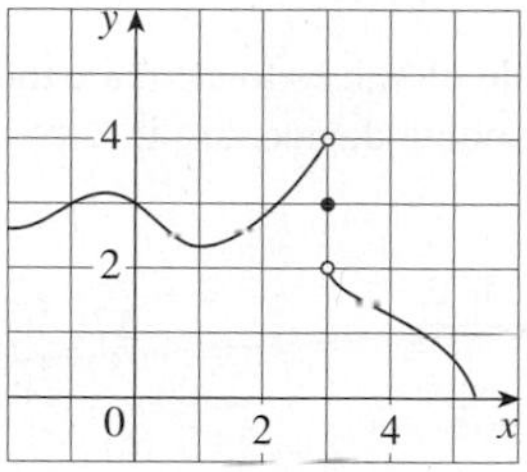

5. Use the given graph of f to state the value of each quantity, if it exists. If it does not exist, explain why.

(a) $\lim_{x\to 1^-} f(x)$ (b) $\lim_{x\to 1^+} f(x)$ (c) $\lim_{x\to 1} f(x)$

(d) $\lim_{x\to 5} f(x)$ (e) $f(5)$

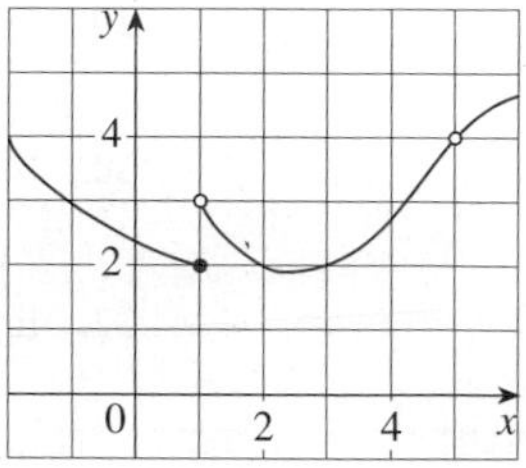

6. For the function h whose graph is given, state the value of each quantity, if it exists. If it does not exist, explain why.

(a) $\lim_{x\to -3^-} h(x)$ (b) $\lim_{x\to -3^+} h(x)$ (c) $\lim_{x\to -3} h(x)$

(d) $h(-3)$ (e) $\lim_{x\to 0^-} h(x)$ (f) $\lim_{x\to 0^+} h(x)$

(g) $\lim_{x\to 0} h(x)$ (h) $h(0)$ (i) $\lim_{x\to 2} h(x)$

(j) $h(2)$ (k) $\lim_{x\to 5^+} h(x)$ (l) $\lim_{x\to 5^-} h(x)$

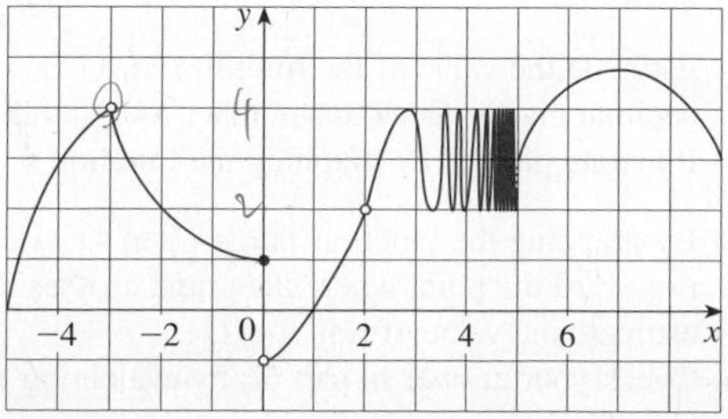

7. For the function g whose graph is given, state the value of each quantity, if it exists. If it does not exist, explain why.

(a) $\lim_{t\to 0^-} g(t)$ (b) $\lim_{t\to 0^+} g(t)$ (c) $\lim_{t\to 0} g(t)$

(d) $\lim_{t\to 2^-} g(t)$ (e) $\lim_{t\to 2^+} g(t)$ (f) $\lim_{t\to 2} g(t)$

(g) $g(2)$ (h) $\lim_{t\to 4} g(t)$

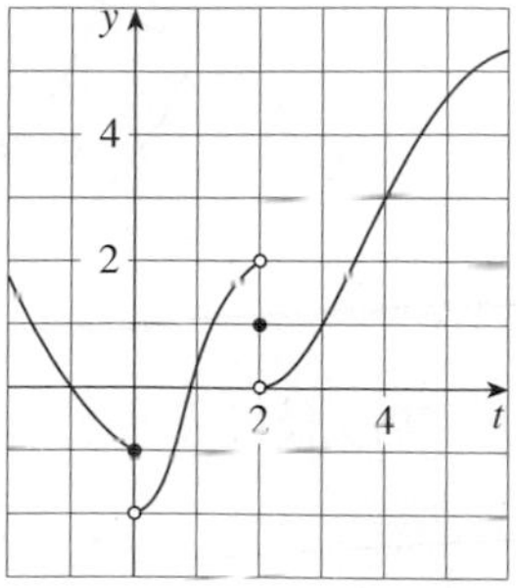

8. For the function R whose graph is shown, state the following.

(a) $\lim_{x\to 2} R(x)$ (b) $\lim_{x\to 5} R(x)$

(c) $\lim_{x\to -3^-} R(x)$ (d) $\lim_{x\to -3^+} R(x)$

(e) The equations of the vertical asymptotes.

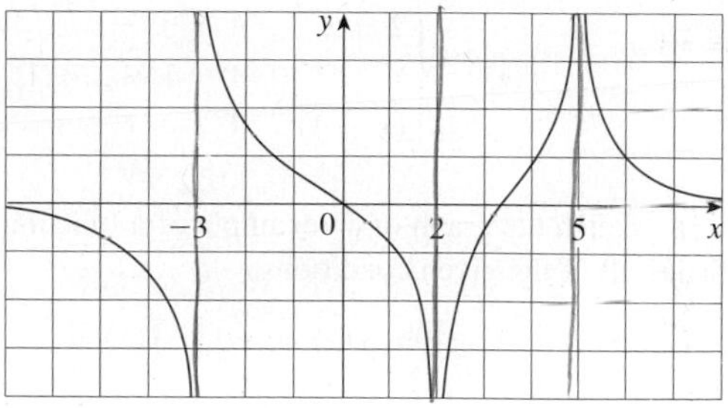

9. For the function f whose graph is shown, state the following.

(a) $\lim_{x\to -7} f(x)$ (b) $\lim_{x\to -3} f(x)$ (c) $\lim_{x\to 0} f(x)$

(d) $\lim_{x\to 6^-} f(x)$ (e) $\lim_{x\to 6^+} f(x)$

(f) The equations of the vertical asymptotes.

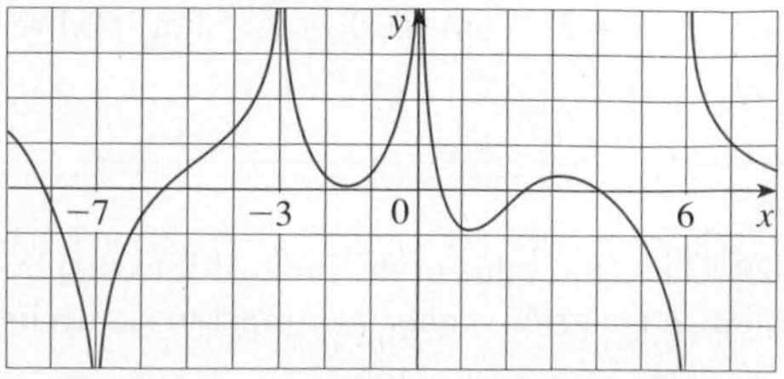

10. A patient receives a 150-mg injection of a drug every 4 hours. The graph shows the amount $f(t)$ of the drug in the bloodstream after t hours. Find

$$\lim_{t \to 12^-} f(t) \quad \text{and} \quad \lim_{t \to 12^+} f(t)$$

and explain the significance of these one-sided limits.

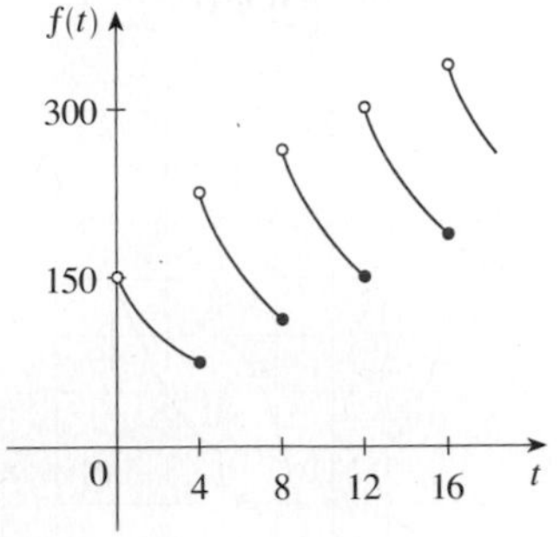

11. Use the graph of the function $f(x) = 1/(1 + e^{1/x})$ to state the value of each limit, if it exists. If it does not exist, explain why.

(a) $\lim_{x \to 0^-} f(x)$ (b) $\lim_{x \to 0^+} f(x)$ (c) $\lim_{x \to 0} f(x)$

12. Sketch the graph of the following function and use it to determine the values of a for which $\lim_{x \to a} f(x)$ exists:

$$f(x) = \begin{cases} 2 - x & \text{if } x < -1 \\ x & \text{if } -1 \leqslant x < 1 \\ (x - 1)^2 & \text{if } x \geqslant 1 \end{cases}$$

13–16 Sketch the graph of an example of a function f that satisfies all of the given conditions.

13. $\lim_{x \to 1^-} f(x) = 2, \quad \lim_{x \to 1^+} f(x) = -2, \quad f(1) = 2$

14. $\lim_{x \to 0^-} f(x) = 1, \quad \lim_{x \to 0^+} f(x) = -1, \quad \lim_{x \to 2^-} f(x) = 0,$
$\lim_{x \to 2^+} f(x) = 1, \quad f(2) = 1, \quad f(0)$ is undefined

15. $\lim_{x \to 3^+} f(x) = 4, \quad \lim_{x \to 3^-} f(x) = 2, \quad \lim_{x \to -2} f(x) = 2,$
$f(3) = 3, \quad f(-2) = 1$

16. $\lim_{x \to 1} f(x) = 3, \quad \lim_{x \to 4^-} f(x) = 3, \quad \lim_{x \to 4^+} f(x) = -3,$
$f(1) = 1, \quad f(4) = -1$

17–20 Guess the value of the limit (if it exists) by evaluating the function at the given numbers (correct to six decimal places).

17. $\lim_{x \to 2} \dfrac{x^2 - 2x}{x^2 - x - 2}$, $x = 2.5, 2.1, 2.05, 2.01, 2.005, 2.001,$
$1.9, 1.95, 1.99, 1.995, 1.999$

18. $\lim_{x \to -1} \dfrac{x^2 - 2x}{x^2 - x - 2}$,
$x = 0, -0.5, -0.9, -0.95, -0.99, -0.999,$
$-2, -1.5, -1.1, -1.01, -1.001$

19. $\lim_{x \to 0} \dfrac{e^x - 1 - x}{x^2}$, $x = \pm 1, \pm 0.5, \pm 0.1, \pm 0.05, \pm 0.01$

20. $\lim_{x \to 0^+} x \ln(x + x^2)$, $x = 1, 0.5, 0.1, 0.05, 0.01, 0.005, 0.001$

21–24 Use a table of values to estimate the value of the limit. If you have a graphing device, use it to confirm your result graphically.

21. $\lim_{x \to 0} \dfrac{\sqrt{x + 4} - 2}{x}$

22. $\lim_{x \to 0} \dfrac{\tan 3x}{\tan 5x}$

23. $\lim_{x \to 1} \dfrac{x^6 - 1}{x^{10} - 1}$

24. $\lim_{x \to 0} \dfrac{9^x - 5^x}{x}$

25–32 Determine the infinite limit.

25. $\lim_{x \to -3^+} \dfrac{x + 2}{x + 3}$

26. $\lim_{x \to -3^-} \dfrac{x + 2}{x + 3}$

27. $\lim_{x \to 1} \dfrac{2 - x}{(x - 1)^2}$

28. $\lim_{x \to 5^-} \dfrac{e^x}{(x - 5)^3}$

29. $\lim_{x \to 3^+} \ln(x^2 - 9)$

30. $\lim_{x \to \pi^-} \cot x$

31. $\lim_{x \to 2\pi^-} x \csc x$

32. $\lim_{x \to 2^-} \dfrac{x^2 - 2x}{x^2 - 4x + 4}$

33. Determine $\lim_{x \to 1^-} \dfrac{1}{x^3 - 1}$ and $\lim_{x \to 1^+} \dfrac{1}{x^3 - 1}$

(a) by evaluating $f(x) = 1/(x^3 - 1)$ for values of x that approach 1 from the left and from the right,

(b) by reasoning as in Example 9, and

(c) from a graph of f.

34. (a) Find the vertical asymptotes of the function

$$y = \frac{x^2 + 1}{3x - 2x^2}$$

(b) Confirm your answer to part (a) by graphing the function.

35. (a) Estimate the value of the limit $\lim_{x \to 0} (1 + x)^{1/x}$ to five decimal places. Does this number look familiar?

(b) Illustrate part (a) by graphing the function $y = (1 + x)^{1/x}$.

36. (a) By graphing the function $f(x) = (\tan 4x)/x$ and zooming in toward the point where the graph crosses the y-axis, estimate the value of $\lim_{x \to 0} f(x)$.

(b) Check your answer in part (a) by evaluating $f(x)$ for values of x that approach 0.

37. (a) Evaluate the function $f(x) = x^2 - (2^x/1000)$ for $x = 1$, 0.8, 0.6, 0.4, 0.2, 0.1, and 0.05, and guess the value of

$$\lim_{x \to 0} \left(x^2 - \frac{2^x}{1000} \right)$$

(b) Evaluate $f(x)$ for $x = 0.04$, 0.02, 0.01, 0.005, 0.003, and 0.001. Guess again.

38. (a) Evaluate $h(x) = (\tan x - x)/x^3$ for $x = 1$, 0.5, 0.1, 0.05, 0.01, and 0.005.

(b) Guess the value of $\displaystyle\lim_{x \to 0} \frac{\tan x - x}{x^3}$.

(c) Evaluate $h(x)$ for successively smaller values of x until you finally reach a value of 0 for $h(x)$. Are you still confident that your guess in part (b) is correct? Explain why you eventually obtained a value of 0. (In Section 4.4 a method for evaluating the limit will be explained.)

(d) Graph the function h in the viewing rectangle $[-1, 1]$ by $[0, 1]$. Then zoom in toward the point where the graph crosses the y-axis to estimate the limit of $h(x)$ as x approaches 0. Continue to zoom in until you observe distortions in the graph of h. Compare with the results of part (c).

39. Graph the function $f(x) = \sin(\pi/x)$ of Example 4 in the viewing rectangle $[-1, 1]$ by $[-1, 1]$. Then zoom in toward the origin several times. Comment on the behavior of this function.

40. In the theory of relativity, the mass of a particle with velocity v is

$$m = \frac{m_0}{\sqrt{1 - v^2/c^2}}$$

where m_0 is the mass of the particle at rest and c is the speed of light. What happens as $v \to c^-$?

41. Use a graph to estimate the equations of all the vertical asymptotes of the curve

$$y = \tan(2 \sin x) \qquad -\pi \leqslant x \leqslant \pi$$

Then find the exact equations of these asymptotes.

42. (a) Use numerical and graphical evidence to guess the value of the limit

$$\lim_{x \to 1} \frac{x^3 - 1}{\sqrt{x} - 1}$$

(b) How close to 1 does x have to be to ensure that the function in part (a) is within a distance 0.5 of its limit?

2.3 CALCULATING LIMITS USING THE LIMIT LAWS

In Section 2.2 we used calculators and graphs to guess the values of limits, but we saw that such methods don't always lead to the correct answer. In this section we use the following properties of limits, called the *Limit Laws,* to calculate limits.

LIMIT LAWS Suppose that c is a constant and the limits

$$\lim_{x \to a} f(x) \qquad \text{and} \qquad \lim_{x \to a} g(x)$$

exist. Then

1. $\displaystyle\lim_{x \to a} [f(x) + g(x)] = \lim_{x \to a} f(x) + \lim_{x \to a} g(x)$

2. $\displaystyle\lim_{x \to a} [f(x) - g(x)] = \lim_{x \to a} f(x) - \lim_{x \to a} g(x)$

3. $\displaystyle\lim_{x \to a} [cf(x)] = c \lim_{x \to a} f(x)$

4. $\displaystyle\lim_{x \to a} [f(x)g(x)] = \lim_{x \to a} f(x) \cdot \lim_{x \to a} g(x)$

5. $\displaystyle\lim_{x \to a} \frac{f(x)}{g(x)} = \frac{\lim_{x \to a} f(x)}{\lim_{x \to a} g(x)} \quad \text{if } \lim_{x \to a} g(x) \neq 0$

These five laws can be stated verbally as follows:

SUM LAW
1. The limit of a sum is the sum of the limits.

DIFFERENCE LAW
2. The limit of a difference is the difference of the limits.

CONSTANT MULTIPLE LAW
3. The limit of a constant times a function is the constant times the limit of the function.

PRODUCT LAW
4. The limit of a product is the product of the limits.

QUOTIENT LAW
5. The limit of a quotient is the quotient of the limits (provided that the limit of the denominator is not 0).

It is easy to believe that these properties are true. For instance, if $f(x)$ is close to L and $g(x)$ is close to M, it is reasonable to conclude that $f(x) + g(x)$ is close to $L + M$. This gives us an intuitive basis for believing that Law 1 is true. In Section 2.4 we give a precise definition of a limit and use it to prove this law. The proofs of the remaining laws are given in Appendix F.

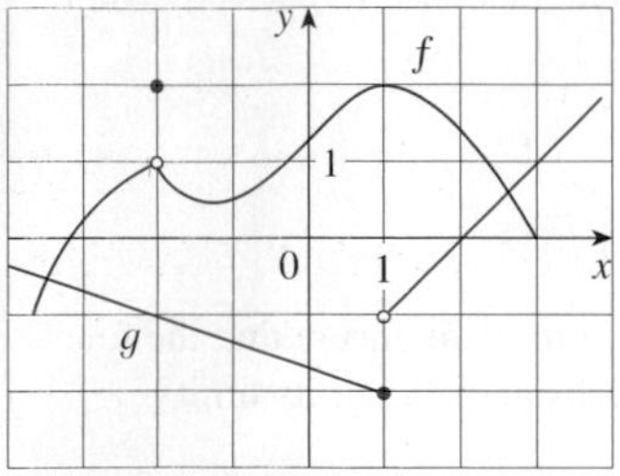

FIGURE 1

EXAMPLE 1 Use the Limit Laws and the graphs of f and g in Figure 1 to evaluate the following limits, if they exist.

(a) $\lim_{x \to -2} [f(x) + 5g(x)]$ (b) $\lim_{x \to 1} [f(x)g(x)]$ (c) $\lim_{x \to 2} \frac{f(x)}{g(x)}$

SOLUTION

(a) From the graphs of f and g we see that

$$\lim_{x \to -2} f(x) = 1 \quad \text{and} \quad \lim_{x \to -2} g(x) = -1$$

Therefore, we have

$$\lim_{x \to -2} [f(x) + 5g(x)] = \lim_{x \to -2} f(x) + \lim_{x \to -2} [5g(x)] \quad \text{(by Law 1)}$$

$$= \lim_{x \to -2} f(x) + 5 \lim_{x \to -2} g(x) \quad \text{(by Law 3)}$$

$$= 1 + 5(-1) = -4$$

(b) We see that $\lim_{x \to 1} f(x) = 2$. But $\lim_{x \to 1} g(x)$ does not exist because the left and right limits are different:

$$\lim_{x \to 1^-} g(x) = -2 \qquad \lim_{x \to 1^+} g(x) = -1$$

So we can't use Law 4 for the desired limit. But we *can* use Law 4 for the one-sided limits:

$$\lim_{x \to 1^-} [f(x)g(x)] = 2 \cdot (-2) = -4 \qquad \lim_{x \to 1^+} [f(x)g(x)] = 2 \cdot (-1) = -2$$

The left and right limits aren't equal, so $\lim_{x \to 1} [f(x)g(x)]$ does not exist.

(c) The graphs show that

$$\lim_{x \to 2} f(x) \approx 1.4 \quad \text{and} \quad \lim_{x \to 2} g(x) = 0$$

Because the limit of the denominator is 0, we can't use Law 5. The given limit does not exist because the denominator approaches 0 while the numerator approaches a nonzero number. □

If we use the Product Law repeatedly with $g(x) = f(x)$, we obtain the following law.

POWER LAW

6. $\lim_{x \to a} [f(x)]^n = \left[\lim_{x \to a} f(x)\right]^n$ where n is a positive integer

In applying these six limit laws, we need to use two special limits:

7. $\lim_{x \to a} c = c$ **8.** $\lim_{x \to a} x = a$

These limits are obvious from an intuitive point of view (state them in words or draw graphs of $y = c$ and $y = x$), but proofs based on the precise definition are requested in the exercises for Section 2.4.

If we now put $f(x) = x$ in Law 6 and use Law 8, we get another useful special limit.

9. $\lim_{x \to a} x^n = a^n$ where n is a positive integer

A similar limit holds for roots as follows. (For square roots the proof is outlined in Exercise 37 in Section 2.4.)

10. $\lim_{x \to a} \sqrt[n]{x} = \sqrt[n]{a}$ where n is a positive integer

(If n is even, we assume that $a > 0$.)

More generally, we have the following law, which is proved in Section 2.5 as a consequence of Law 10.

ROOT LAW

11. $\lim_{x \to a} \sqrt[n]{f(x)} = \sqrt[n]{\lim_{x \to a} f(x)}$ where n is a positive integer

$\left[\text{If } n \text{ is even, we assume that } \lim_{x \to a} f(x) > 0.\right]$

EXAMPLE 2 Evaluate the following limits and justify each step.

(a) $\lim_{x \to 5} (2x^2 - 3x + 4)$ (b) $\lim_{x \to -2} \dfrac{x^3 + 2x^2 - 1}{5 - 3x}$

SOLUTION

(a)
$$\begin{aligned} \lim_{x \to 5} (2x^2 - 3x + 4) &= \lim_{x \to 5} (2x^2) - \lim_{x \to 5} (3x) + \lim_{x \to 5} 4 && \text{(by Laws 2 and 1)} \\ &= 2 \lim_{x \to 5} x^2 - 3 \lim_{x \to 5} x + \lim_{x \to 5} 4 && \text{(by 3)} \\ &= 2(5^2) - 3(5) + 4 && \text{(by 9, 8, and 7)} \\ &= 39 \end{aligned}$$

Some limits are best calculated by first finding the left- and right-hand limits. The following theorem is a reminder of what we discovered in Section 2.2. It says that a two-sided limit exists if and only if both of the one-sided limits exist and are equal.

1 THEOREM $\lim_{x \to a} f(x) = L$ if and only if $\lim_{x \to a^-} f(x) = L = \lim_{x \to a^+} f(x)$

When computing one-sided limits, we use the fact that the Limit Laws also hold for one-sided limits.

■ The result of Example 7 looks plausible from Figure 3.

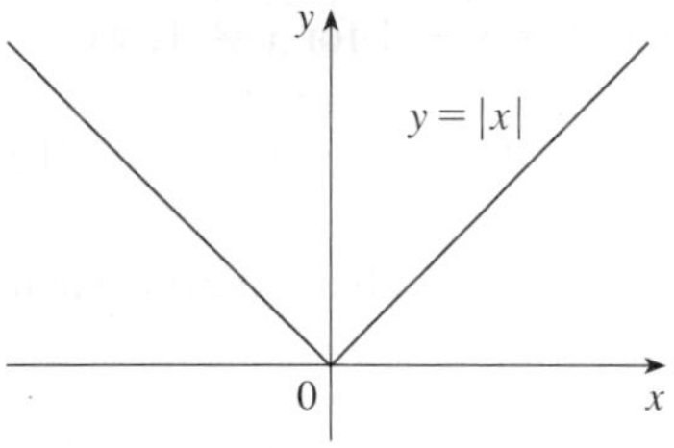

FIGURE 3

EXAMPLE 7 Show that $\lim_{x \to 0} |x| = 0$.

SOLUTION Recall that

$$|x| = \begin{cases} x & \text{if } x \geq 0 \\ -x & \text{if } x < 0 \end{cases}$$

Since $|x| = x$ for $x > 0$, we have

$$\lim_{x \to 0^+} |x| = \lim_{x \to 0^+} x = 0$$

For $x < 0$ we have $|x| = -x$ and so

$$\lim_{x \to 0^-} |x| = \lim_{x \to 0^-} (-x) = 0$$

Therefore, by Theorem 1,

$$\lim_{x \to 0} |x| = 0$$ □

V EXAMPLE 8 Prove that $\lim_{x \to 0} \frac{|x|}{x}$ does not exist.

SOLUTION

$$\lim_{x \to 0^+} \frac{|x|}{x} = \lim_{x \to 0^+} \frac{x}{x} = \lim_{x \to 0^+} 1 = 1$$

$$\lim_{x \to 0^-} \frac{|x|}{x} = \lim_{x \to 0^-} \frac{-x}{x} = \lim_{x \to 0^-} (-1) = -1$$

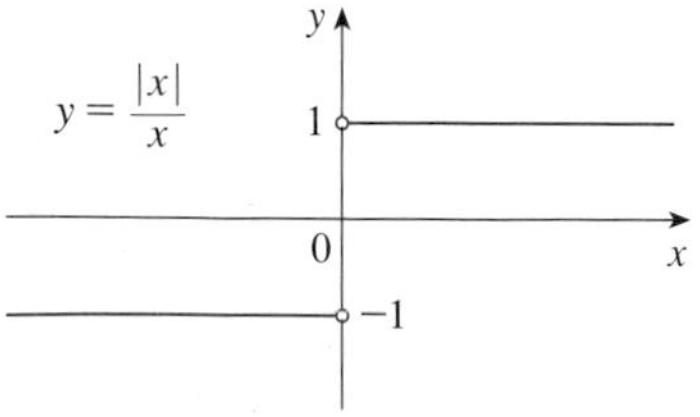

FIGURE 4

Since the right- and left-hand limits are different, it follows from Theorem 1 that $\lim_{x \to 0} |x|/x$ does not exist. The graph of the function $f(x) = |x|/x$ is shown in Figure 4 and supports the one-sided limits that we found. □

EXAMPLE 9 If

$$f(x) = \begin{cases} \sqrt{x - 4} & \text{if } x > 4 \\ 8 - 2x & \text{if } x < 4 \end{cases}$$

determine whether $\lim_{x \to 4} f(x)$ exists.

■ It is shown in Example 3 in Section 2.4 that $\lim_{x \to 0^+} \sqrt{x} = 0$.

SOLUTION Since $f(x) = \sqrt{x - 4}$ for $x > 4$, we have

$$\lim_{x \to 4^+} f(x) = \lim_{x \to 4^+} \sqrt{x - 4} = \sqrt{4 - 4} = 0$$

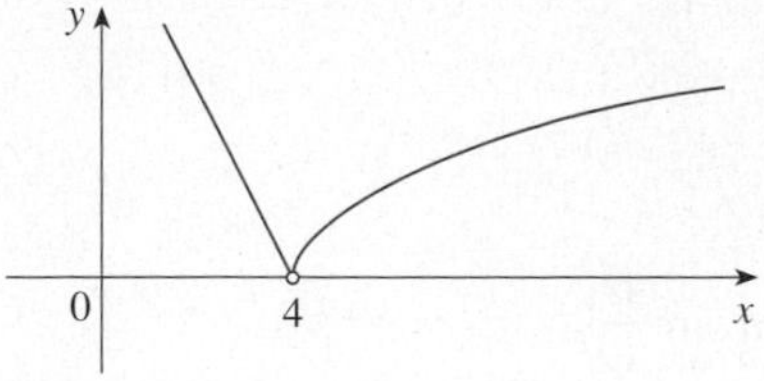

FIGURE 5

Since $f(x) = 8 - 2x$ for $x < 4$, we have

$$\lim_{x \to 4^-} f(x) = \lim_{x \to 4^-} (8 - 2x) = 8 - 2 \cdot 4 = 0$$

The right- and left-hand limits are equal. Thus the limit exists and

$$\lim_{x \to 4} f(x) = 0$$

The graph of f is shown in Figure 5. □

■ Other notations for $[\![x]\!]$ are $[x]$ and $\lfloor x \rfloor$. The greatest integer function is sometimes called the *floor function*.

EXAMPLE 10 The **greatest integer function** is defined by $[\![x]\!]$ = the largest integer that is less than or equal to x. (For instance, $[\![4]\!] = 4$, $[\![4.8]\!] = 4$, $[\![\pi]\!] = 3$, $[\![\sqrt{2}\,]\!] = 1$, $[\![-\frac{1}{2}]\!] = -1$.) Show that $\lim_{x \to 3} [\![x]\!]$ does not exist.

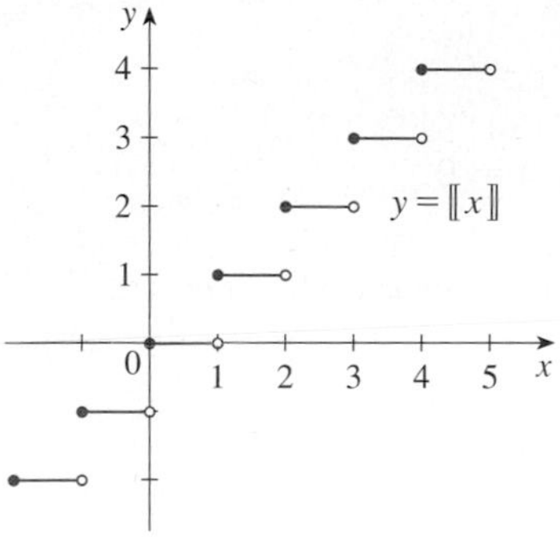

FIGURE 6
Greatest integer function

SOLUTION The graph of the greatest integer function is shown in Figure 6. Since $[\![x]\!] = 3$ for $3 \leqslant x < 4$, we have

$$\lim_{x \to 3^+} [\![x]\!] = \lim_{x \to 3^+} 3 = 3$$

Since $[\![x]\!] = 2$ for $2 \leqslant x < 3$, we have

$$\lim_{x \to 3^-} [\![x]\!] = \lim_{x \to 3^-} 2 = 2$$

Because these one-sided limits are not equal, $\lim_{x \to 3} [\![x]\!]$ does not exist by Theorem 1. □

The next two theorems give two additional properties of limits. Their proofs can be found in Appendix F.

> **2 THEOREM** If $f(x) \leqslant g(x)$ when x is near a (except possibly at a) and the limits of f and g both exist as x approaches a, then
>
> $$\lim_{x \to a} f(x) \leqslant \lim_{x \to a} g(x)$$

> **3 THE SQUEEZE THEOREM** If $f(x) \leqslant g(x) \leqslant h(x)$ when x is near a (except possibly at a) and
>
> $$\lim_{x \to a} f(x) = \lim_{x \to a} h(x) = L$$
>
> then $$\lim_{x \to a} g(x) = L$$

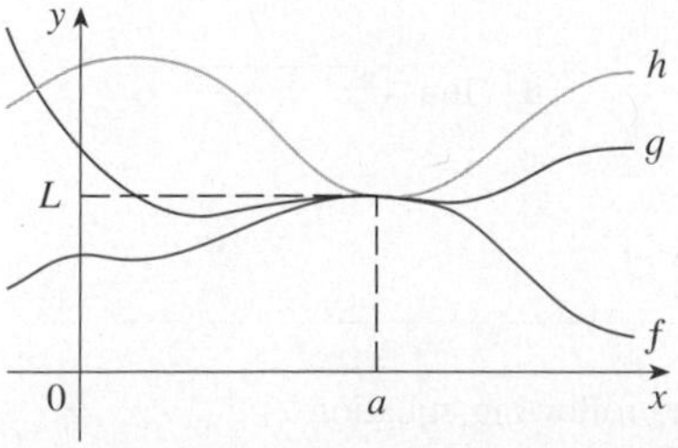

FIGURE 7

The Squeeze Theorem, which is sometimes called the Sandwich Theorem or the Pinching Theorem, is illustrated by Figure 7. It says that if $g(x)$ is squeezed between $f(x)$ and $h(x)$ near a, and if f and h have the same limit L at a, then g is forced to have the same limit L at a.

EXAMPLE 11 Show that $\lim_{x\to 0} x^2 \sin \frac{1}{x} = 0$.

SOLUTION First note that we **cannot** use

$$\lim_{x\to 0} x^2 \sin \frac{1}{x} = \lim_{x\to 0} x^2 \cdot \lim_{x\to 0} \sin \frac{1}{x}$$

because $\lim_{x\to 0} \sin(1/x)$ does not exist (see Example 4 in Section 2.2). However, since

$$-1 \leqslant \sin \frac{1}{x} \leqslant 1$$

we have, as illustrated by Figure 8,

$$-x^2 \leqslant x^2 \sin \frac{1}{x} \leqslant x^2$$

We know that

$$\lim_{x\to 0} x^2 = 0 \quad \text{and} \quad \lim_{x\to 0} (-x^2) = 0$$

Taking $f(x) = -x^2$, $g(x) = x^2 \sin(1/x)$, and $h(x) = x^2$ in the Squeeze Theorem, we obtain

$$\lim_{x\to 0} x^2 \sin \frac{1}{x} = 0$$

□

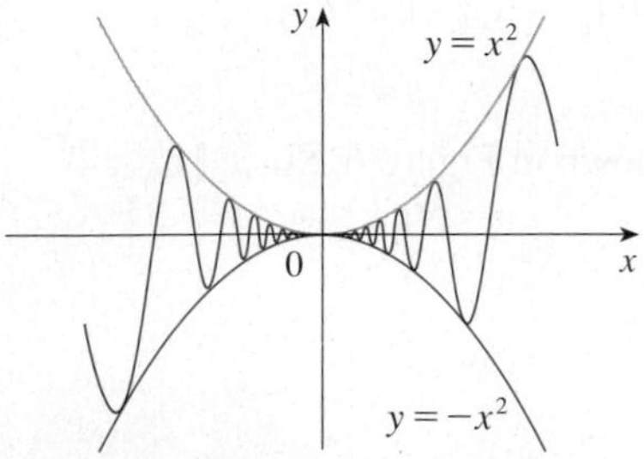

FIGURE 8
$y = x^2 \sin(1/x)$

2.3 EXERCISES

1. Given that

$$\lim_{x\to 2} f(x) = 4 \qquad \lim_{x\to 2} g(x) = -2 \qquad \lim_{x\to 2} h(x) = 0$$

find the limits that exist. If the limit does not exist, explain why.

(a) $\lim_{x\to 2} [f(x) + 5g(x)]$ (b) $\lim_{x\to 2} [g(x)]^3$

(c) $\lim_{x\to 2} \sqrt{f(x)}$ (d) $\lim_{x\to 2} \dfrac{3f(x)}{g(x)}$

(e) $\lim_{x\to 2} \dfrac{g(x)}{h(x)}$ (f) $\lim_{x\to 2} \dfrac{g(x)h(x)}{f(x)}$

2. The graphs of f and g are given. Use them to evaluate each limit, if it exists. If the limit does not exist, explain why.

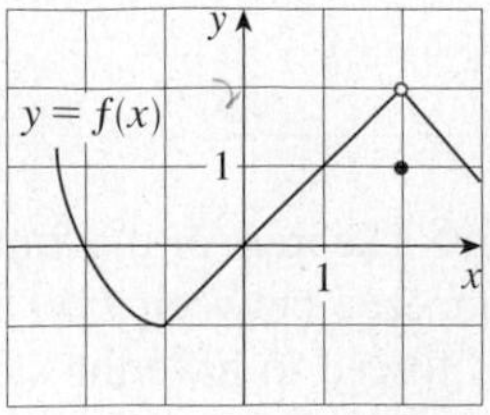

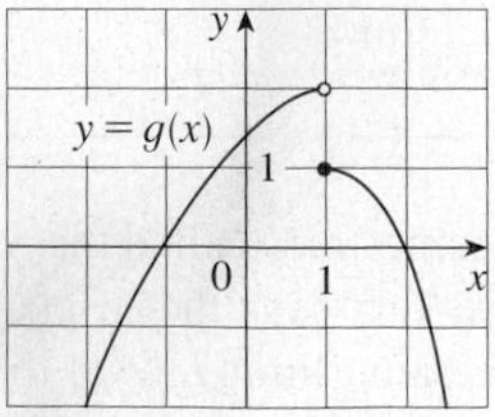

(a) $\lim_{x\to 2} [f(x) + g(x)]$ (b) $\lim_{x\to 1} [f(x) + g(x)]$

(c) $\lim_{x\to 0} [f(x)g(x)]$ (d) $\lim_{x\to -1} \dfrac{f(x)}{g(x)}$

(e) $\lim_{x\to 2} [x^3 f(x)]$ (f) $\lim_{x\to 1} \sqrt{3 + f(x)}$

3–9 Evaluate the limit and justify each step by indicating the appropriate Limit Law(s).

3. $\lim_{x\to -2} (3x^4 + 2x^2 - x + 1)$

4. $\lim_{x\to 2} \dfrac{2x^2 + 1}{x^2 + 6x - 4}$

5. $\lim_{x\to 8} (1 + \sqrt[3]{x})(2 - 6x^2 + x^3)$

6. $\lim_{t\to -1} (t^2 + 1)^3 (t + 3)^5$

7. $\lim_{x\to 1} \left(\dfrac{1 + 3x}{1 + 4x^2 + 3x^4} \right)^3$

8. $\lim_{u\to -2} \sqrt{u^4 + 3u + 6}$

9. $\lim_{x\to 4^-} \sqrt{16 - x^2}$

10. (a) What is wrong with the following equation?

$$\frac{x^2 + x - 6}{x - 2} = x + 3$$

(b) In view of part (a), explain why the equation

$$\lim_{x \to 2} \frac{x^2 + x - 6}{x - 2} = \lim_{x \to 2} (x + 3)$$

is correct.

11–30 Evaluate the limit, if it exists.

11. $\displaystyle \lim_{x \to 2} \frac{x^2 + x - 6}{x - 2}$

12. $\displaystyle \lim_{x \to -4} \frac{x^2 + 5x + 4}{x^2 + 3x - 4}$

13. $\displaystyle \lim_{x \to 2} \frac{x^2 - x + 6}{x - 2}$

14. $\displaystyle \lim_{x \to 4} \frac{x^2 - 4x}{x^2 - 3x - 4}$

15. $\displaystyle \lim_{t \to -3} \frac{t^2 - 9}{2t^2 + 7t + 3}$

16. $\displaystyle \lim_{x \to -1} \frac{x^2 - 4x}{x^2 - 3x - 4}$

17. $\displaystyle \lim_{h \to 0} \frac{(4 + h)^2 - 16}{h}$

18. $\displaystyle \lim_{x \to 1} \frac{x^3 - 1}{x^2 - 1}$

19. $\displaystyle \lim_{x \to -2} \frac{x + 2}{x^3 + 8}$

20. $\displaystyle \lim_{h \to 0} \frac{(2 + h)^3 - 8}{h}$

21. $\displaystyle \lim_{t \to 9} \frac{9 - t}{3 - \sqrt{t}}$

22. $\displaystyle \lim_{h \to 0} \frac{\sqrt{1 + h} - 1}{h}$

23. $\displaystyle \lim_{x \to 7} \frac{\sqrt{x + 2} - 3}{x - 7}$

24. $\displaystyle \lim_{x \to -1} \frac{x^2 + 2x + 1}{x^4 - 1}$

25. $\displaystyle \lim_{x \to -4} \frac{\frac{1}{4} + \frac{1}{x}}{4 + x}$

26. $\displaystyle \lim_{t \to 0} \left(\frac{1}{t} - \frac{1}{t^2 + t} \right)$

27. $\displaystyle \lim_{x \to 16} \frac{4 - \sqrt{x}}{16x - x^2}$

28. $\displaystyle \lim_{h \to 0} \frac{(3 + h)^{-1} - 3^{-1}}{h}$

29. $\displaystyle \lim_{t \to 0} \left(\frac{1}{t\sqrt{1 + t}} - \frac{1}{t} \right)$

30. $\displaystyle \lim_{x \to -4} \frac{\sqrt{x^2 + 9} - 5}{x + 4}$

31. (a) Estimate the value of

$$\lim_{x \to 0} \frac{x}{\sqrt{1 + 3x} - 1}$$

by graphing the function $f(x) = x/(\sqrt{1 + 3x} - 1)$.

(b) Make a table of values of $f(x)$ for x close to 0 and guess the value of the limit.

(c) Use the Limit Laws to prove that your guess is correct.

32. (a) Use a graph of

$$f(x) = \frac{\sqrt{3 + x} - \sqrt{3}}{x}$$

to estimate the value of $\lim_{x \to 0} f(x)$ to two decimal places.

(b) Use a table of values of $f(x)$ to estimate the limit to four decimal places.

(c) Use the Limit Laws to find the exact value of the limit.

33. Use the Squeeze Theorem to show that $\lim_{x \to 0} (x^2 \cos 20\pi x) = 0$. Illustrate by graphing the functions $f(x) = -x^2$, $g(x) = x^2 \cos 20\pi x$, and $h(x) = x^2$ on the same screen.

34. Use the Squeeze Theorem to show that

$$\lim_{x \to 0} \sqrt{x^3 + x^2} \sin \frac{\pi}{x} = 0$$

Illustrate by graphing the functions f, g, and h (in the notation of the Squeeze Theorem) on the same screen.

35. If $4x - 9 \leqslant f(x) \leqslant x^2 - 4x + 7$ for $x \geqslant 0$, find $\lim_{x \to 4} f(x)$.

36. If $2x \leqslant g(x) \leqslant x^4 - x^2 + 2$ for all x, evaluate $\lim_{x \to 1} g(x)$.

37. Prove that $\displaystyle \lim_{x \to 0} x^4 \cos \frac{2}{x} = 0$.

38. Prove that $\displaystyle \lim_{x \to 0^+} \sqrt{x}\, e^{\sin(\pi/x)} = 0$.

39–44 Find the limit, if it exists. If the limit does not exist, explain why.

39. $\displaystyle \lim_{x \to 3} (2x + |x - 3|)$

40. $\displaystyle \lim_{x \to -6} \frac{2x + 12}{|x + 6|}$

41. $\displaystyle \lim_{x \to 0.5^-} \frac{2x - 1}{|2x^3 - x^2|}$

42. $\displaystyle \lim_{x \to -2} \frac{2 - |x|}{2 + x}$

43. $\displaystyle \lim_{x \to 0^-} \left(\frac{1}{x} - \frac{1}{|x|} \right)$

44. $\displaystyle \lim_{x \to 0^+} \left(\frac{1}{x} - \frac{1}{|x|} \right)$

45. The *signum* (or sign) *function*, denoted by sgn, is defined by

$$\operatorname{sgn} x = \begin{cases} -1 & \text{if } x < 0 \\ 0 & \text{if } x = 0 \\ 1 & \text{if } x > 0 \end{cases}$$

(a) Sketch the graph of this function.

(b) Find each of the following limits or explain why it does not exist.

(i) $\lim_{x \to 0^+} \operatorname{sgn} x$ (ii) $\lim_{x \to 0^-} \operatorname{sgn} x$

(iii) $\lim_{x \to 0} \operatorname{sgn} x$ (iv) $\lim_{x \to 0} |\operatorname{sgn} x|$

46. Let

$$f(x) = \begin{cases} 4 - x^2 & \text{if } x \leqslant 2 \\ x - 1 & \text{if } x > 2 \end{cases}$$

(a) Find $\lim_{x \to 2^-} f(x)$ and $\lim_{x \to 2^+} f(x)$.

(b) Does $\lim_{x \to 2} f(x)$ exist?

(c) Sketch the graph of f.

47. Let $\displaystyle F(x) = \frac{x^2 - 1}{|x - 1|}$.

(a) Find

(i) $\lim_{x \to 1^+} F(x)$ (ii) $\lim_{x \to 1^-} F(x)$

(b) Does $\lim_{x \to 1} F(x)$ exist?
(c) Sketch the graph of F.

48. Let

$$g(x) = \begin{cases} x & \text{if } x < 1 \\ 3 & \text{if } x = 1 \\ 2 - x^2 & \text{if } 1 < x \leqslant 2 \\ x - 3 & \text{if } x > 2 \end{cases}$$

(a) Evaluate each of the following limits, if it exists.
(i) $\lim_{x \to 1^-} g(x)$ (ii) $\lim_{x \to 1} g(x)$ (iii) $g(1)$
(iv) $\lim_{x \to 2^-} g(x)$ (v) $\lim_{x \to 2^+} g(x)$ (vi) $\lim_{x \to 2} g(x)$
(b) Sketch the graph of g.

49. (a) If the symbol $[\![\ ]\!]$ denotes the greatest integer function defined in Example 10, evaluate
(i) $\lim_{x \to -2^+} [\![x]\!]$ (ii) $\lim_{x \to -2} [\![x]\!]$ (iii) $\lim_{x \to -2.4} [\![x]\!]$
(b) If n is an integer, evaluate
(i) $\lim_{x \to n^-} [\![x]\!]$ (ii) $\lim_{x \to n^+} [\![x]\!]$
(c) For what values of a does $\lim_{x \to a} [\![x]\!]$ exist?

50. Let $f(x) = [\![\cos x]\!]$, $-\pi \leqslant x \leqslant \pi$.
(a) Sketch the graph of f.
(b) Evaluate each limit, if it exists.
(i) $\lim_{x \to 0} f(x)$ (ii) $\lim_{x \to (\pi/2)^-} f(x)$
(iii) $\lim_{x \to (\pi/2)^+} f(x)$ (iv) $\lim_{x \to \pi/2} f(x)$
(c) For what values of a does $\lim_{x \to a} f(x)$ exist?

51. If $f(x) = [\![x]\!] + [\![-x]\!]$, show that $\lim_{x \to 2} f(x)$ exists but is not equal to $f(2)$.

52. In the theory of relativity, the Lorentz contraction formula

$$L = L_0\sqrt{1 - v^2/c^2}$$

expresses the length L of an object as a function of its velocity v with respect to an observer, where L_0 is the length of the object at rest and c is the speed of light. Find $\lim_{v \to c^-} L$ and interpret the result. Why is a left-hand limit necessary?

53. If p is a polynomial, show that $\lim_{x \to a} p(x) = p(a)$.

54. If r is a rational function, use Exercise 53 to show that $\lim_{x \to a} r(x) = r(a)$ for every number a in the domain of r.

55. If $\lim_{x \to 1} \dfrac{f(x) - 8}{x - 1} = 10$, find $\lim_{x \to 1} f(x)$.

56. If $\lim_{x \to 0} \dfrac{f(x)}{x^2} = 5$, find the following limits.
(a) $\lim_{x \to 0} f(x)$ (b) $\lim_{x \to 0} \dfrac{f(x)}{x}$

57. If

$$f(x) = \begin{cases} x^2 & \text{if } x \text{ is rational} \\ 0 & \text{if } x \text{ is irrational} \end{cases}$$

prove that $\lim_{x \to 0} f(x) = 0$.

58. Show by means of an example that $\lim_{x \to a} [f(x) + g(x)]$ may exist even though neither $\lim_{x \to a} f(x)$ nor $\lim_{x \to a} g(x)$ exists.

59. Show by means of an example that $\lim_{x \to a} [f(x)g(x)]$ may exist even though neither $\lim_{x \to a} f(x)$ nor $\lim_{x \to a} g(x)$ exists.

60. Evaluate $\lim_{x \to 2} \dfrac{\sqrt{6 - x} - 2}{\sqrt{3 - x} - 1}$.

61. Is there a number a such that

$$\lim_{x \to -2} \frac{3x^2 + ax + a + 3}{x^2 + x - 2}$$

exists? If so, find the value of a and the value of the limit.

62. The figure shows a fixed circle C_1 with equation $(x - 1)^2 + y^2 = 1$ and a shrinking circle C_2 with radius r and center the origin. P is the point $(0, r)$, Q is the upper point of intersection of the two circles, and R is the point of intersection of the line PQ and the x-axis. What happens to R as C_2 shrinks, that is, as $r \to 0^+$?

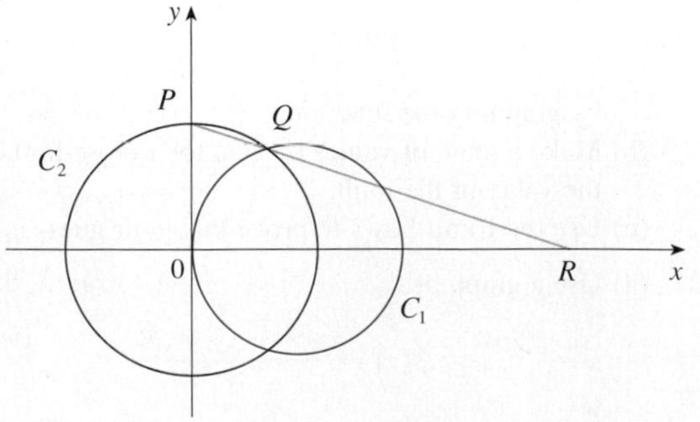

2.4 THE PRECISE DEFINITION OF A LIMIT

The intuitive definition of a limit given in Section 2.2 is inadequate for some purposes because such phrases as "x is close to 2" and "$f(x)$ gets closer and closer to L" are vague. In order to be able to prove conclusively that

$$\lim_{x\to 0}\left(x^3 + \frac{\cos 5x}{10{,}000}\right) = 0.0001 \qquad \text{or} \qquad \lim_{x\to 0}\frac{\sin x}{x} = 1$$

we must make the definition of a limit precise.

To motivate the precise definition of a limit, let's consider the function

$$f(x) = \begin{cases} 2x - 1 & \text{if } x \neq 3 \\ 6 & \text{if } x = 3 \end{cases}$$

Intuitively, it is clear that when x is close to 3 but $x \neq 3$, then $f(x)$ is close to 5, and so $\lim_{x\to 3} f(x) = 5$.

To obtain more detailed information about how $f(x)$ varies when x is close to 3, we ask the following question:

How close to 3 does x have to be so that $f(x)$ differs from 5 by less than 0.1?

The distance from x to 3 is $|x - 3|$ and the distance from $f(x)$ to 5 is $|f(x) - 5|$, so our problem is to find a number δ such that

■ It is traditional to use the Greek letter δ (delta) in this situation.

$$|f(x) - 5| < 0.1 \quad \text{if} \quad |x - 3| < \delta \quad \text{but } x \neq 3$$

If $|x - 3| > 0$, then $x \neq 3$, so an equivalent formulation of our problem is to find a number δ such that

$$|f(x) - 5| < 0.1 \quad \text{if} \quad 0 < |x - 3| < \delta$$

Notice that if $0 < |x - 3| < (0.1)/2 = 0.05$, then

$$|f(x) - 5| = |(2x - 1) - 5| = |2x - 6| = 2|x - 3| < 0.1$$

that is, $$|f(x) - 5| < 0.1 \quad \text{if} \quad 0 < |x - 3| < 0.05$$

Thus an answer to the problem is given by $\delta = 0.05$; that is, if x is within a distance of 0.05 from 3, then $f(x)$ will be within a distance of 0.1 from 5.

If we change the number 0.1 in our problem to the smaller number 0.01, then by using the same method we find that $f(x)$ will differ from 5 by less than 0.01 provided that x differs from 3 by less than $(0.01)/2 = 0.005$:

$$|f(x) - 5| < 0.01 \quad \text{if} \quad 0 < |x - 3| < 0.005$$

Similarly,

$$|f(x) - 5| < 0.001 \quad \text{if} \quad 0 < |x - 3| < 0.0005$$

The numbers 0.1, 0.01, and 0.001 that we have considered are *error tolerances* that we might allow. For 5 to be the precise limit of $f(x)$ as x approaches 3, we must not only be able to bring the difference between $f(x)$ and 5 below each of these three numbers; we must be able to bring it below *any* positive number. And, by the same reasoning, we can! If we write ε (the Greek letter epsilon) for an arbitrary positive number, then we find as before that

$$\boxed{1} \qquad |f(x) - 5| < \varepsilon \quad \text{if} \quad 0 < |x - 3| < \delta = \frac{\varepsilon}{2}$$

This is a precise way of saying that $f(x)$ is close to 5 when x is close to 3 because (1) says that we can make the values of $f(x)$ within an arbitrary distance ε from 5 by taking the values of x within a distance $\varepsilon/2$ from 3 (but $x \neq 3$).

Note that (1) can be rewritten as follows:

$$\text{if} \quad 3 - \delta < x < 3 + \delta \quad (x \neq 3) \quad \text{then} \quad 5 - \varepsilon < f(x) < 5 + \varepsilon$$

and this is illustrated in Figure 1. By taking the values of x ($\neq 3$) to lie in the interval $(3 - \delta, 3 + \delta)$ we can make the values of $f(x)$ lie in the interval $(5 - \varepsilon, 5 + \varepsilon)$.

Using (1) as a model, we give a precise definition of a limit.

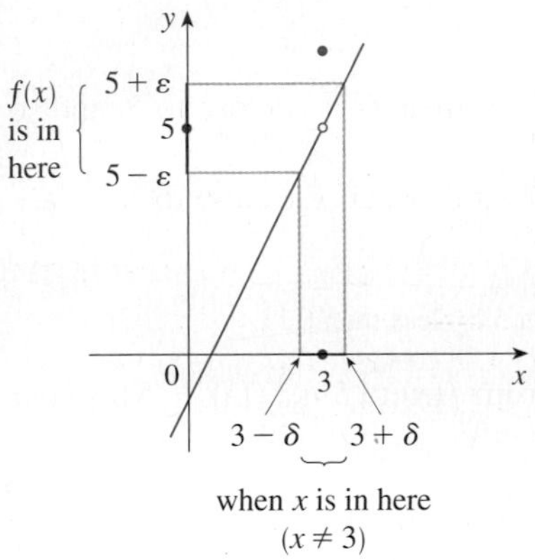

FIGURE 1

2 DEFINITION Let f be a function defined on some open interval that contains the number a, except possibly at a itself. Then we say that the **limit of $f(x)$ as x approaches a is L**, and we write

$$\lim_{x \to a} f(x) = L$$

if for every number $\varepsilon > 0$ there is a number $\delta > 0$ such that

$$\text{if} \quad 0 < |x - a| < \delta \quad \text{then} \quad |f(x) - L| < \varepsilon$$

Since $|x - a|$ is the distance from x to a and $|f(x) - L|$ is the distance from $f(x)$ to L, and since ε can be arbitrarily small, the definition of a limit can be expressed in words as follows:

> $\lim_{x \to a} f(x) = L$ means that the distance between $f(x)$ and L can be made arbitrarily small by taking the distance from x to a sufficiently small (but not 0).

Alternatively,

> $\lim_{x \to a} f(x) = L$ means that the values of $f(x)$ can be made as close as we please to L by taking x close enough to a (but not equal to a).

We can also reformulate Definition 2 in terms of intervals by observing that the inequality $|x - a| < \delta$ is equivalent to $-\delta < x - a < \delta$, which in turn can be written as $a - \delta < x < a + \delta$. Also $0 < |x - a|$ is true if and only if $x - a \neq 0$, that is, $x \neq a$. Similarly, the inequality $|f(x) - L| < \varepsilon$ is equivalent to the pair of inequalities $L - \varepsilon < f(x) < L + \varepsilon$. Therefore, in terms of intervals, Definition 2 can be stated as follows:

> $\lim_{x \to a} f(x) = L$ means that for every $\varepsilon > 0$ (no matter how small ε is) we can find $\delta > 0$ such that if x lies in the open interval $(a - \delta, a + \delta)$ and $x \neq a$, then $f(x)$ lies in the open interval $(L - \varepsilon, L + \varepsilon)$.

We interpret this statement geometrically by representing a function by an arrow diagram as in Figure 2, where f maps a subset of $\mathbb{R}$ onto another subset of $\mathbb{R}$.

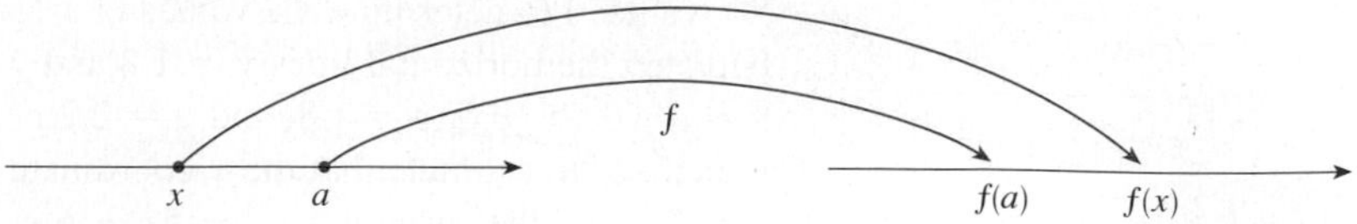

FIGURE 2

The definition of limit says that if any small interval $(L - \varepsilon, L + \varepsilon)$ is given around L, then we can find an interval $(a - \delta, a + \delta)$ around a such that f maps all the points in $(a - \delta, a + \delta)$ (except possibly a) into the interval $(L - \varepsilon, L + \varepsilon)$. (See Figure 3.)

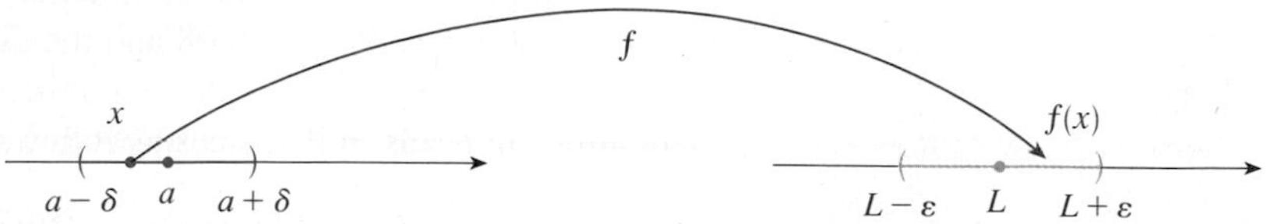

FIGURE 3

Another geometric interpretation of limits can be given in terms of the graph of a function. If $\varepsilon > 0$ is given, then we draw the horizontal lines $y = L + \varepsilon$ and $y = L - \varepsilon$ and the graph of f. (See Figure 4.) If $\lim_{x \to a} f(x) = L$, then we can find a number $\delta > 0$ such that if we restrict x to lie in the interval $(a - \delta, a + \delta)$ and take $x \neq a$, then the curve $y = f(x)$ lies between the lines $y = L - \varepsilon$ and $y = L + \varepsilon$. (See Figure 5.) You can see that if such a δ has been found, then any smaller δ will also work.

It is important to realize that the process illustrated in Figures 4 and 5 must work for *every* positive number ε, no matter how small it is chosen. Figure 6 shows that if a smaller ε is chosen, then a smaller δ may be required.

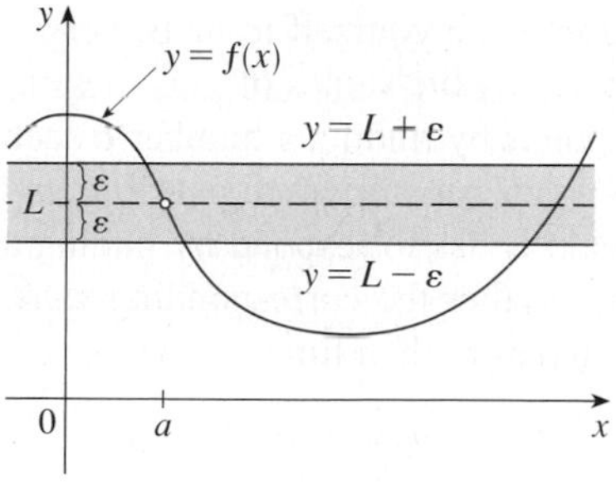

FIGURE 4

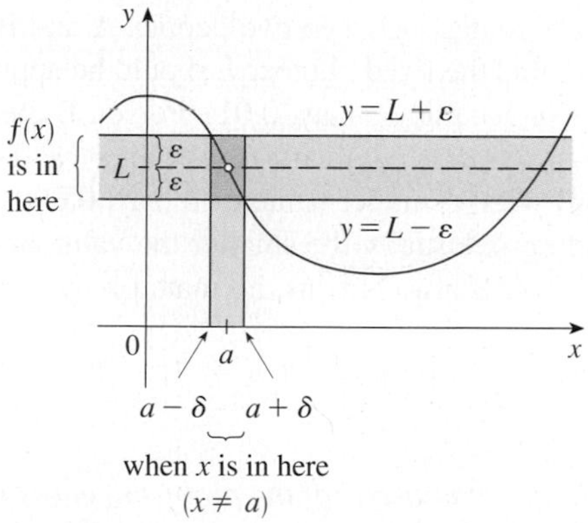

FIGURE 5

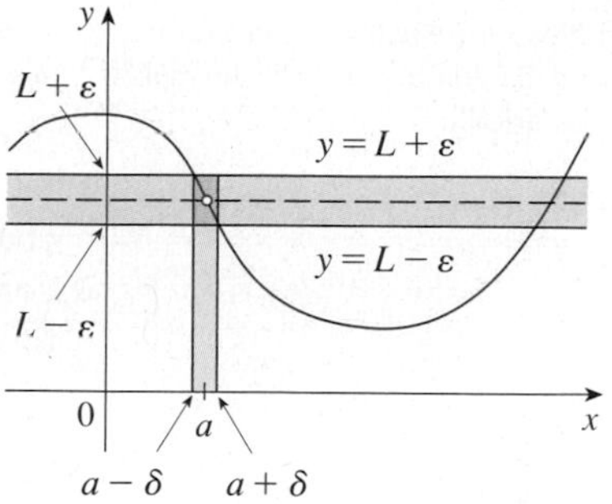

FIGURE 6

EXAMPLE 1 Use a graph to find a number δ such that

$$\text{if} \quad |x - 1| < \delta \quad \text{then} \quad |(x^3 - 5x + 6) - 2| < 0.2$$

In other words, find a number δ that corresponds to $\varepsilon = 0.2$ in the definition of a limit for the function $f(x) = x^3 - 5x + 6$ with $a = 1$ and $L = 2$.

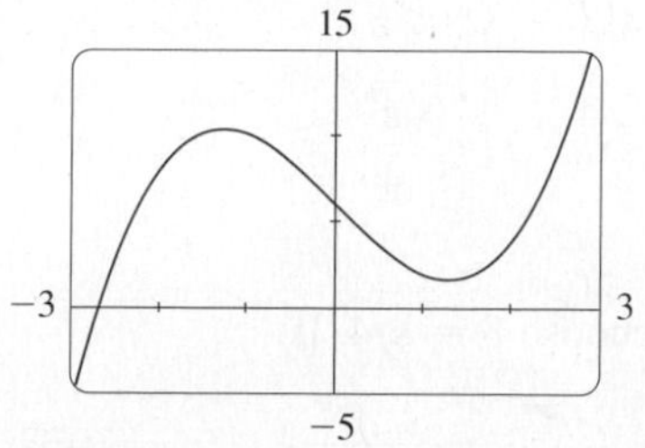

FIGURE 7

SOLUTION A graph of f is shown in Figure 7; we are interested in the region near the point $(1, 2)$. Notice that we can rewrite the inequality

$$|(x^3 - 5x + 6) - 2| < 0.2$$

as

$$1.8 < x^3 - 5x + 6 < 2.2$$

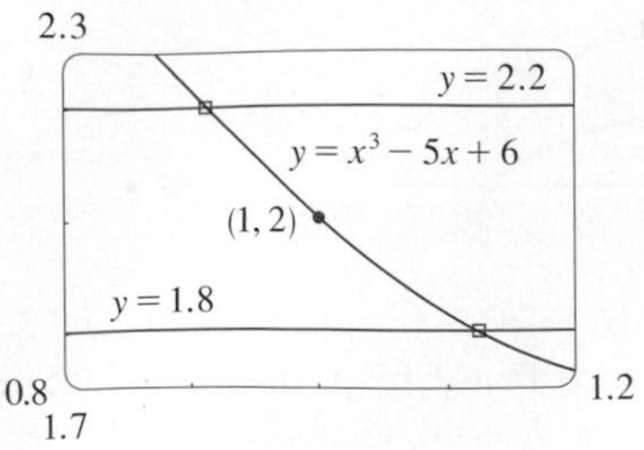

FIGURE 8

So we need to determine the values of x for which the curve $y = x^3 - 5x + 6$ lies between the horizontal lines $y = 1.8$ and $y = 2.2$. Therefore we graph the curves $y = x^3 - 5x + 6$, $y = 1.8$, and $y = 2.2$ near the point $(1, 2)$ in Figure 8. Then we use the cursor to estimate that the x-coordinate of the point of intersection of the line $y = 2.2$ and the curve $y = x^3 - 5x + 6$ is about 0.911. Similarly, $y = x^3 - 5x + 6$ intersects the line $y = 1.8$ when $x \approx 1.124$. So, rounding to be safe, we can say that

$$\text{if} \quad 0.92 < x < 1.12 \quad \text{then} \quad 1.8 < x^3 - 5x + 6 < 2.2$$

This interval $(0.92, 1.12)$ is not symmetric about $x = 1$. The distance from $x = 1$ to the left endpoint is $1 - 0.92 = 0.08$ and the distance to the right endpoint is 0.12. We can choose δ to be the smaller of these numbers, that is, $\delta = 0.08$. Then we can rewrite our inequalities in terms of distances as follows:

$$\text{if} \quad |x - 1| < 0.08 \quad \text{then} \quad |(x^3 - 5x + 6) - 2| < 0.2$$

This just says that by keeping x within 0.08 of 1, we are able to keep $f(x)$ within 0.2 of 2.

Although we chose $\delta = 0.08$, any smaller positive value of δ would also have worked. □

The graphical procedure in Example 1 gives an illustration of the definition for $\varepsilon = 0.2$, but it does not *prove* that the limit is equal to 2. A proof has to provide a δ for *every* ε.

In proving limit statements it may be helpful to think of the definition of limit as a challenge. First it challenges you with a number ε. Then you must be able to produce a suitable δ. You have to be able to do this for *every* $\varepsilon > 0$, not just a particular ε.

TEC In Module 2.4/2.6 you can explore the precise definition of a limit both graphically and numerically.

Imagine a contest between two people, A and B, and imagine yourself to be B. Person A stipulates that the fixed number L should be approximated by the values of $f(x)$ to within a degree of accuracy ε (say, 0.01). Person B then responds by finding a number δ such that if $0 < |x - a| < \delta$, then $|f(x) - L| < \varepsilon$. Then A may become more exacting and challenge B with a smaller value of ε (say, 0.0001). Again B has to respond by finding a corresponding δ. Usually the smaller the value of ε, the smaller the corresponding value of δ must be. If B always wins, no matter how small A makes ε, then $\lim_{x \to a} f(x) = L$.

V EXAMPLE 2 Prove that $\lim_{x \to 3} (4x - 5) = 7$.

SOLUTION

1. *Preliminary analysis of the problem (guessing a value for δ).* Let ε be a given positive number. We want to find a number δ such that

$$\text{if} \quad 0 < |x - 3| < \delta \quad \text{then} \quad |(4x - 5) - 7| < \varepsilon$$

But $|(4x - 5) - 7| = |4x - 12| = |4(x - 3)| = 4|x - 3|$. Therefore, we want

$$\text{if} \quad 0 < |x - 3| < \delta \quad \text{then} \quad 4|x - 3| < \varepsilon$$

that is,
$$\text{if} \quad 0 < |x - 3| < \delta \quad \text{then} \quad |x - 3| < \frac{\varepsilon}{4}$$

This suggests that we should choose $\delta = \varepsilon/4$.

2. *Proof (showing that this δ works).* Given $\varepsilon > 0$, choose $\delta = \varepsilon/4$. If $0 < |x - 3| < \delta$, then

$$|(4x - 5) - 7| = |4x - 12| = 4|x - 3| < 4\delta = 4\left(\frac{\varepsilon}{4}\right) = \varepsilon$$

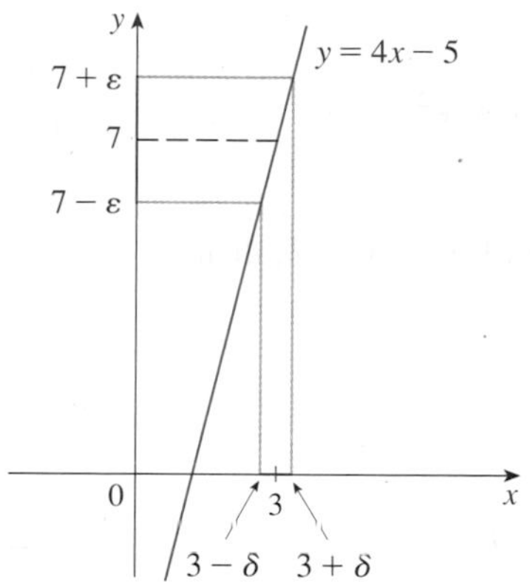

FIGURE 9

Thus

$$\text{if} \quad 0 < |x - 3| < \delta \quad \text{then} \quad |(4x - 5) - 7| < \varepsilon$$

Therefore, by the definition of a limit,

$$\lim_{x \to 3} (4x - 5) = 7$$

This example is illustrated by Figure 9. □

Note that in the solution of Example 2 there were two stages—guessing and proving. We made a preliminary analysis that enabled us to guess a value for δ. But then in the second stage we had to go back and prove in a careful, logical fashion that we had made a correct guess. This procedure is typical of much of mathematics. Sometimes it is necessary to first make an intelligent guess about the answer to a problem and then prove that the guess is correct.

The intuitive definitions of one-sided limits that were given in Section 2.2 can be precisely reformulated as follows.

3 DEFINITION OF LEFT-HAND LIMIT

$$\lim_{x \to a^-} f(x) = L$$

if for every number $\varepsilon > 0$ there is a number $\delta > 0$ such that

$$\text{if} \quad a - \delta < x < a \quad \text{then} \quad |f(x) - L| < \varepsilon$$

4 DEFINITION OF RIGHT-HAND LIMIT

$$\lim_{x \to a^+} f(x) = L$$

if for every number $\varepsilon > 0$ there is a number $\delta > 0$ such that

$$\text{if} \quad a < x < a + \delta \quad \text{then} \quad |f(x) - L| < \varepsilon$$

Notice that Definition 3 is the same as Definition 2 except that x is restricted to lie in the *left* half $(a - \delta, a)$ of the interval $(a - \delta, a + \delta)$. In Definition 4, x is restricted to lie in the *right* half $(a, a + \delta)$ of the interval $(a - \delta, a + \delta)$.

V EXAMPLE 3 Use Definition 4 to prove that $\lim_{x \to 0^+} \sqrt{x} = 0$.

SOLUTION

1. *Guessing a value for* δ. Let ε be a given positive number. Here $a = 0$ and $L = 0$, so we want to find a number δ such that

$$\text{if} \quad 0 < x < \delta \quad \text{then} \quad |\sqrt{x} - 0| < \varepsilon$$

that is,

$$\text{if} \quad 0 < x < \delta \quad \text{then} \quad \sqrt{x} < \varepsilon$$

CAUCHY AND LIMITS

After the invention of calculus in the 17th century, there followed a period of free development of the subject in the 18th century. Mathematicians like the Bernoulli brothers and Euler were eager to exploit the power of calculus and boldly explored the consequences of this new and wonderful mathematical theory without worrying too much about whether their proofs were completely correct.

The 19th century, by contrast, was the Age of Rigor in mathematics. There was a movement to go back to the foundations of the subject—to provide careful definitions and rigorous proofs. At the forefront of this movement was the French mathematician Augustin-Louis Cauchy (1789–1857), who started out as a military engineer before becoming a mathematics professor in Paris. Cauchy took Newton's idea of a limit, which was kept alive in the 18th century by the French mathematician Jean d'Alembert, and made it more precise. His definition of a limit reads as follows: "When the successive values attributed to a variable approach indefinitely a fixed value so as to end by differing from it by as little as one wishes, this last is called the *limit* of all the others." But when Cauchy used this definition in examples and proofs, he often employed delta-epsilon inequalities similar to the ones in this section. A typical Cauchy proof starts with: "Designate by δ and ε two very small numbers; . . ." He used ε because of the correspondence between epsilon and the French word *erreur* and δ because delta corresponds to *différence*. Later, the German mathematician Karl Weierstrass (1815–1897) stated the definition of a limit exactly as in our Definition 2.

INFINITE LIMITS

Infinite limits can also be defined in a precise way. The following is a precise version of Definition 4 in Section 2.2.

> **6 DEFINITION** Let f be a function defined on some open interval that contains the number a, except possibly at a itself. Then
>
> $$\lim_{x \to a} f(x) = \infty$$
>
> means that for every positive number M there is a positive number δ such that
>
> $$\text{if} \quad 0 < |x - a| < \delta \quad \text{then} \quad f(x) > M$$

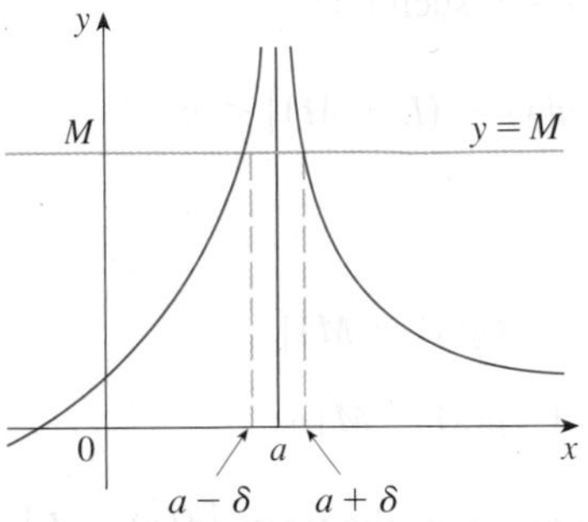

FIGURE 10

This says that the values of $f(x)$ can be made arbitrarily large (larger than any given number M) by taking x close enough to a (within a distance δ, where δ depends on M, but with $x \neq a$). A geometric illustration is shown in Figure 10.

Given any horizontal line $y = M$, we can find a number $\delta > 0$ such that if we restrict x to lie in the interval $(a - \delta, a + \delta)$ but $x \neq a$, then the curve $y = f(x)$ lies above the line $y = M$. You can see that if a larger M is chosen, then a smaller δ may be required.

V EXAMPLE 5 Use Definition 6 to prove that $\lim_{x \to 0} \frac{1}{x^2} = \infty$.

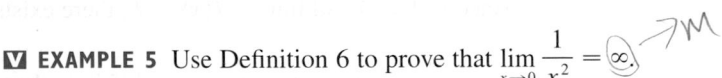

SOLUTION Let M be a given positive number. We want to find a number δ such that

$$\text{if} \quad 0 < |x| < \delta \quad \text{then} \quad 1/x^2 > M$$

But

$$\frac{1}{x^2} > M \iff x^2 < \frac{1}{M} \iff |x| < \frac{1}{\sqrt{M}}$$

So if we choose $\delta = 1/\sqrt{M}$ and $0 < |x| < \delta = 1/\sqrt{M}$, then $1/x^2 > M$. This shows that $1/x^2 \to \infty$ as $x \to 0$. □

Similarly, the following is a precise version of Definition 5 in Section 2.2. It is illustrated by Figure 11.

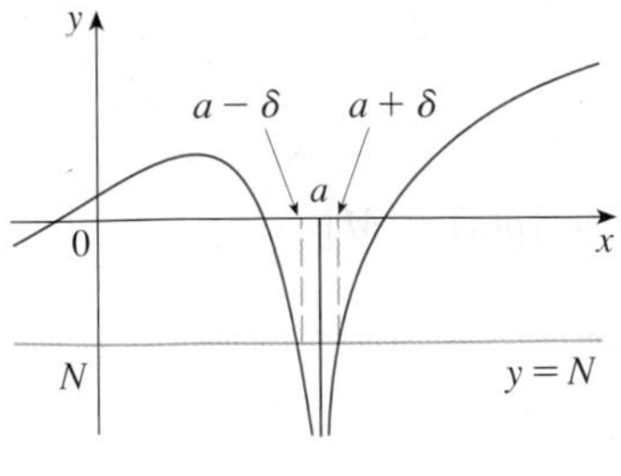

FIGURE 11

> **7 DEFINITION** Let f be a function defined on some open interval that contains the number a, except possibly at a itself. Then
>
> $$\lim_{x \to a} f(x) = -\infty$$
>
> means that for every negative number N there is a positive number δ such that
>
> $$\text{if} \quad 0 < |x - a| < \delta \quad \text{then} \quad f(x) < N$$

2.4 EXERCISES

1. Use the given graph of $f(x) = 1/x$ to find a number δ such that

$$\text{if} \quad |x - 2| < \delta \quad \text{then} \quad \left| \frac{1}{x} - 0.5 \right| < 0.2$$

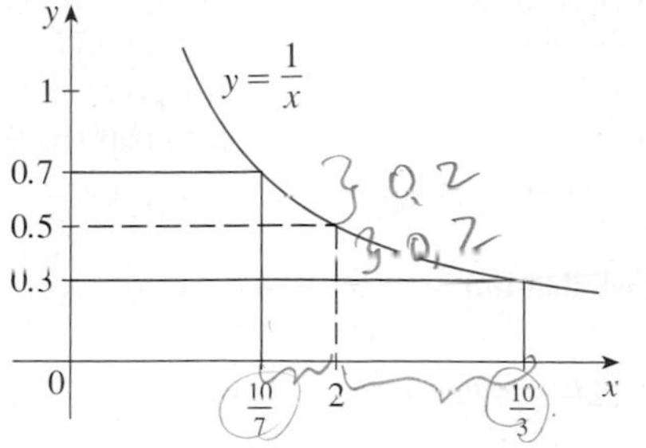

2. Use the given graph of f to find a number δ such that

$$\text{if} \quad 0 < |x - 5| < \delta \quad \text{then} \quad |f(x) - 3| < 0.6$$

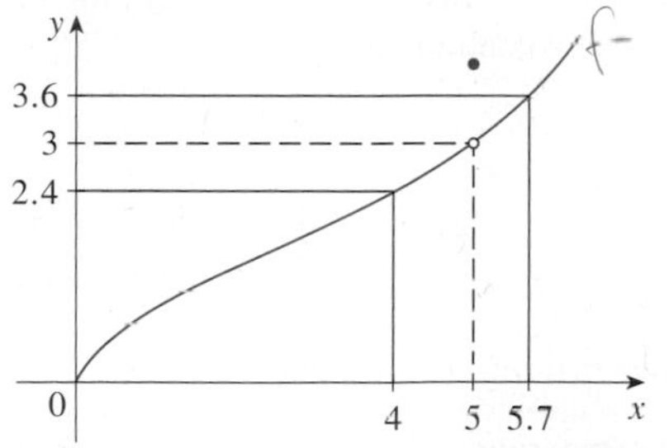

3. Use the given graph of $f(x) = \sqrt{x}$ to find a number δ such that

$$\text{if} \quad |x - 4| < \delta \quad \text{then} \quad |\sqrt{x} - 2| < 0.4$$

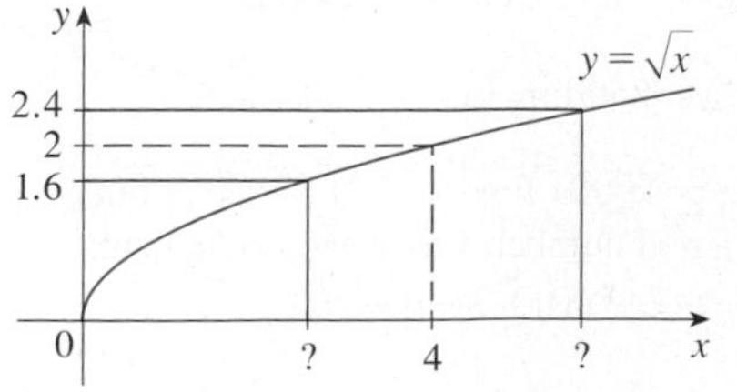

4. Use the given graph of $f(x) = x^2$ to find a number δ such that

$$\text{if} \quad |x - 1| < \delta \quad \text{then} \quad |x^2 - 1| < \tfrac{1}{2}$$

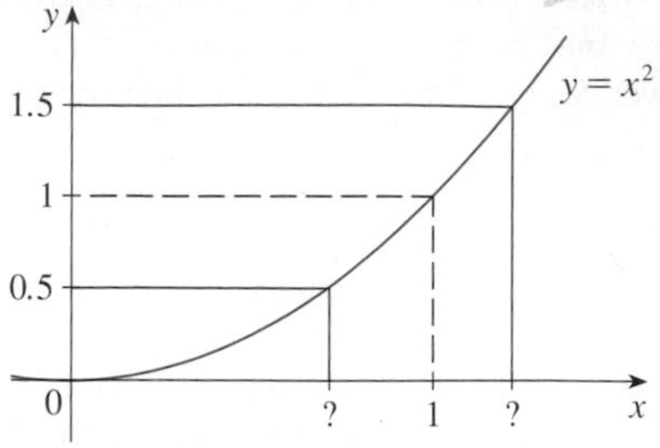

5. Use a graph to find a number δ such that

$$\text{if} \quad \left| x - \frac{\pi}{4} \right| < \delta \quad \text{then} \quad |\tan x - 1| < 0.2$$

6. Use a graph to find a number δ such that

$$\text{if} \quad |x - 1| < \delta \quad \text{then} \quad \left| \frac{2x}{x^2 + 4} - 0.4 \right| < 0.1$$

7. For the limit

$$\lim_{x \to 1} (4 + x - 3x^3) = 2$$

illustrate Definition 2 by finding values of δ that correspond to $\varepsilon = 1$ and $\varepsilon = 0.1$.

8. For the limit

$$\lim_{x \to 0} \frac{e^x - 1}{x} = 1$$

illustrate Definition 2 by finding values of δ that correspond to $\varepsilon = 0.5$ and $\varepsilon = 0.1$.

9. Given that $\lim_{x \to \pi/2} \tan^2 x = \infty$, illustrate Definition 6 by finding values of δ that correspond to (a) $M = 1000$ and (b) $M = 10{,}000$.

10. Use a graph to find a number δ such that

$$\text{if} \quad 5 < x < 5 + \delta \quad \text{then} \quad \frac{x^2}{\sqrt{x - 5}} > 100$$

11. A machinist is required to manufacture a circular metal disk with area 1000 cm^2.

(a) What radius produces such a disk?

(b) If the machinist is allowed an error tolerance of $\pm 5 \text{ cm}^2$ in the area of the disk, how close to the ideal radius in part (a) must the machinist control the radius?

(c) In terms of the ε, δ definition of $\lim_{x \to a} f(x) = L$, what is x? What is $f(x)$? What is a? What is L? What value of ε is given? What is the corresponding value of δ?

12. A crystal growth furnace is used in research to determine how best to manufacture crystals used in electronic components for the space shuttle. For proper growth of the crystal, the temperature must be controlled accurately by adjusting the input power. Suppose the relationship is given by

$$T(w) = 0.1w^2 + 2.155w + 20$$

where T is the temperature in degrees Celsius and w is the power input in watts.

(a) How much power is needed to maintain the temperature at 200°C?

(b) If the temperature is allowed to vary from 200°C by up to ± 1°C, what range of wattage is allowed for the input power?

(c) In terms of the ε, δ definition of $\lim_{x \to a} f(x) = L$, what is x? What is $f(x)$? What is a? What is L? What value of ε is given? What is the corresponding value of δ?

13. (a) Find a number δ such that if $|x - 2| < \delta$, then $|4x - 8| < \varepsilon$, where $\varepsilon = 0.1$.

(b) Repeat part (a) with $\varepsilon = 0.01$.

14. Given that $\lim_{x \to 2}(5x - 7) = 3$, illustrate Definition 2 by finding values of δ that correspond to $\varepsilon = 0.1$, $\varepsilon = 0.05$, and $\varepsilon = 0.01$.

15–18 Prove the statement using the ε, δ definition of limit and illustrate with a diagram like Figure 9.

15. $\lim_{x \to 1} (2x + 3) = 5$

16. $\lim_{x \to -2} \left(\frac{1}{2}x + 3\right) = 2$

17. $\lim_{x \to -3} (1 - 4x) = 13$

18. $\lim_{x \to 4} (7 - 3x) = -5$

19–32 Prove the statement using the ε, δ definition of limit.

19. $\lim_{x \to 3} \frac{x}{5} = \frac{3}{5}$

20. $\lim_{x \to 6} \left(\frac{x}{4} + 3\right) = \frac{9}{2}$

21. $\lim_{x \to 2} \frac{x^2 + x - 6}{x - 2} = 5$

22. $\lim_{x \to -1.5} \frac{9 - 4x^2}{3 + 2x} = 6$

23. $\lim_{x \to a} x = a$

24. $\lim_{x \to a} c = c$

25. $\lim_{x \to 0} x^2 = 0$

26. $\lim_{x \to 0} x^3 = 0$

27. $\lim_{x \to 0} |x| = 0$

28. $\lim_{x \to 9^-} \sqrt[4]{9 - x} = 0$

29. $\lim_{x \to 2} (x^2 - 4x + 5) = 1$

30. $\lim_{x \to 3} (x^2 + x - 4) = 8$

31. $\lim_{x \to -2} (x^2 - 1) = 3$

32. $\lim_{x \to 2} x^3 = 8$

33. Verify that another possible choice of δ for showing that $\lim_{x \to 3} x^2 = 9$ in Example 4 is $\delta = \min\{2, \varepsilon/8\}$.

34. Verify, by a geometric argument, that the largest possible choice of δ for showing that $\lim_{x \to 3} x^2 = 9$ is $\delta = \sqrt{9 + \varepsilon} - 3$.

CAS **35.** (a) For the limit $\lim_{x \to 1} (x^3 + x + 1) = 3$, use a graph to find a value of δ that corresponds to $\varepsilon = 0.4$.

(b) By using a computer algebra system to solve the cubic equation $x^3 + x + 1 = 3 + \varepsilon$, find the largest possible value of δ that works for any given $\varepsilon > 0$.

(c) Put $\varepsilon = 0.4$ in your answer to part (b) and compare with your answer to part (a).

36. Prove that $\lim_{x \to 2} \frac{1}{x} = \frac{1}{2}$.

37. Prove that $\lim_{x \to a} \sqrt{x} = \sqrt{a}$ if $a > 0$.

$$\left[\textit{Hint: } \text{Use } \left|\sqrt{x} - \sqrt{a}\right| = \frac{|x - a|}{\sqrt{x} + \sqrt{a}}.\right]$$

38. If H is the Heaviside function defined in Example 6 in Section 2.2, prove, using Definition 2, that $\lim_{t \to 0} H(t)$ does not exist. [*Hint:* Use an indirect proof as follows. Suppose that the limit is L. Take $\varepsilon = \frac{1}{2}$ in the definition of a limit and try to arrive at a contradiction.]

39. If the function f is defined by

$$f(x) = \begin{cases} 0 & \text{if } x \text{ is rational} \\ 1 & \text{if } x \text{ is irrational} \end{cases}$$

prove that $\lim_{x \to 0} f(x)$ does not exist.

40. By comparing Definitions 2, 3, and 4, prove Theorem 1 in Section 2.3.

41. How close to -3 do we have to take x so that

$$\frac{1}{(x + 3)^4} > 10{,}000$$

42. Prove, using Definition 6, that $\lim_{x \to -3} \frac{1}{(x + 3)^4} = \infty$.

43. Prove that $\lim_{x \to 0^+} \ln x = -\infty$.

44. Suppose that $\lim_{x \to a} f(x) = \infty$ and $\lim_{x \to a} g(x) = c$, where c is a real number. Prove each statement.

(a) $\lim_{x \to a} [f(x) + g(x)] = \infty$

(b) $\lim_{x \to a} [f(x)g(x)] = \infty$ if $c > 0$

(c) $\lim_{x \to a} [f(x)g(x)] = -\infty$ if $c < 0$

2.5 CONTINUITY

We noticed in Section 2.3 that the limit of a function as x approaches a can often be found simply by calculating the value of the function at a. Functions with this property are called *continuous at a*. We will see that the mathematical definition of continuity corresponds closely with the meaning of the word *continuity* in everyday language. (A continuous process is one that takes place gradually, without interruption or abrupt change.)

> **1 DEFINITION** A function f is **continuous at a number *a*** if
>
> $$\lim_{x \to a} f(x) = f(a)$$

■ As illustrated in Figure 1, if f is continuous, then the points $(x, f(x))$ on the graph of f approach the point $(a, f(a))$ on the graph. So there is no gap in the curve.

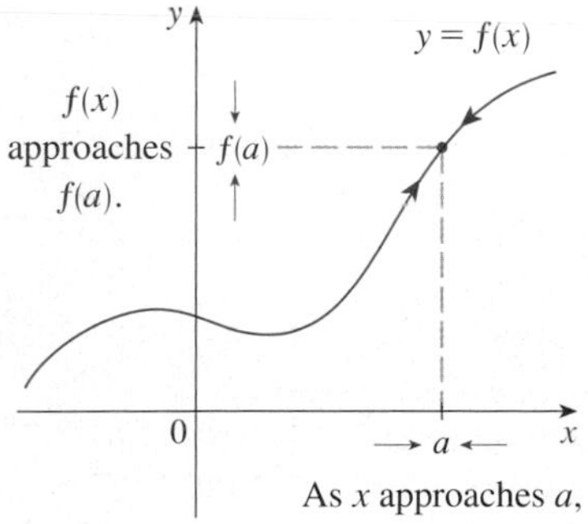

FIGURE 1

Notice that Definition 1 implicitly requires three things if f is continuous at a:

1. $f(a)$ is defined (that is, a is in the domain of f)

2. $\lim_{x \to a} f(x)$ exists

3. $\lim_{x \to a} f(x) = f(a)$

The definition says that f is continuous at a if $f(x)$ approaches $f(a)$ as x approaches a. Thus a continuous function f has the property that a small change in x produces only a small change in $f(x)$. In fact, the change in $f(x)$ can be kept as small as we please by keeping the change in x sufficiently small.

If f is defined near a (in other words, f is defined on an open interval containing a, except perhaps at a), we say that f is **discontinuous at *a*** (or f has a **discontinuity** at a) if f is not continuous at a.

Physical phenomena are usually continuous. For instance, the displacement or velocity of a vehicle varies continuously with time, as does a person's height. But discontinuities do occur in such situations as electric currents. [See Example 6 in Section 2.2, where the Heaviside function is discontinuous at 0 because $\lim_{t \to 0} H(t)$ does not exist.]

Geometrically, you can think of a function that is continuous at every number in an interval as a function whose graph has no break in it. The graph can be drawn without removing your pen from the paper.

EXAMPLE 1 Figure 2 shows the graph of a function f. At which numbers is f discontinuous? Why?

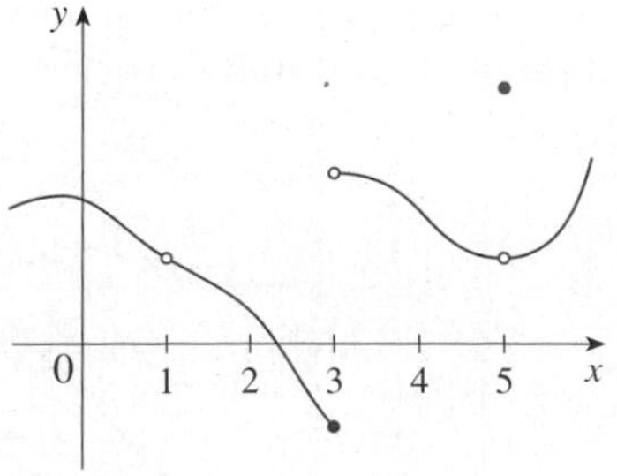

FIGURE 2

SOLUTION It looks as if there is a discontinuity when $a = 1$ because the graph has a break there. The official reason that f is discontinuous at 1 is that $f(1)$ is not defined.

The graph also has a break when $a = 3$, but the reason for the discontinuity is different. Here, $f(3)$ is defined, but $\lim_{x \to 3} f(x)$ does not exist (because the left and right limits are different). So f is discontinuous at 3.

What about $a = 5$? Here, $f(5)$ is defined and $\lim_{x \to 5} f(x)$ exists (because the left and right limits are the same). But

$$\lim_{x \to 5} f(x) \neq f(5)$$

So f is discontinuous at 5. □

Now let's see how to detect discontinuities when a function is defined by a formula.

V EXAMPLE 2 Where are each of the following functions discontinuous?

(a) $f(x) = \dfrac{x^2 - x - 2}{x - 2}$ (b) $f(x) = \begin{cases} \dfrac{1}{x^2} & \text{if } x \neq 0 \\ 1 & \text{if } x = 0 \end{cases}$

(c) $f(x) = \begin{cases} \dfrac{x^2 - x - 2}{x - 2} & \text{if } x \neq 2 \\ 1 & \text{if } x = 2 \end{cases}$ (d) $f(x) = [\![x]\!]$

SOLUTION

(a) Notice that $f(2)$ is not defined, so f is discontinuous at 2. Later we'll see why f is continuous at all other numbers.

(b) Here $f(0) = 1$ is defined but

$$\lim_{x \to 0} f(x) = \lim_{x \to 0} \frac{1}{x^2}$$

does not exist. (See Example 8 in Section 2.2.) So f is discontinuous at 0.

(c) Here $f(2) = 1$ is defined and

$$\lim_{x \to 2} f(x) = \lim_{x \to 2} \frac{x^2 - x - 2}{x - 2} = \lim_{x \to 2} \frac{(x - 2)(x + 1)}{x - 2} = \lim_{x \to 2} (x + 1) = 3$$

exists. But

$$\lim_{x \to 2} f(x) \neq f(2)$$

so f is not continuous at 2.

(d) The greatest integer function $f(x) = [\![x]\!]$ has discontinuities at all of the integers because $\lim_{x \to n} [\![x]\!]$ does not exist if n is an integer. (See Example 10 and Exercise 49 in Section 2.3.) □

Figure 3 shows the graphs of the functions in Example 2. In each case the graph can't be drawn without lifting the pen from the paper because a hole or break or jump occurs in the graph. The kind of discontinuity illustrated in parts (a) and (c) is called **removable** because we could remove the discontinuity by redefining f at just the single number 2. [The function $g(x) = x + 1$ is continuous.] The discontinuity in part (b) is called an **infinite discontinuity**. The discontinuities in part (d) are called **jump discontinuities** because the function "jumps" from one value to another.

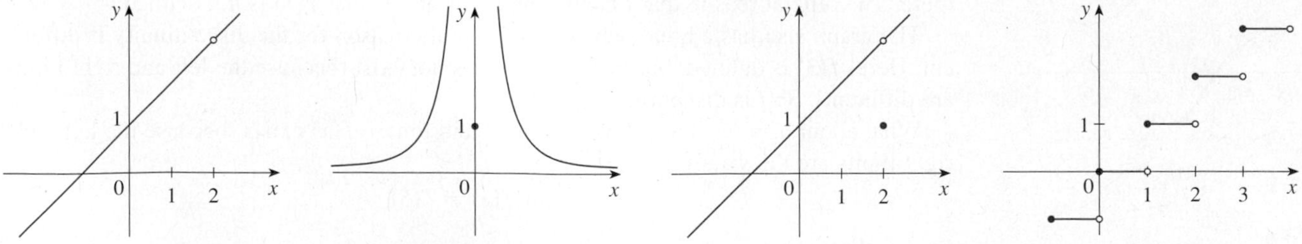

(a) $f(x) = \dfrac{x^2 - x - 2}{x - 2}$ (b) $f(x) = \begin{cases} \dfrac{1}{x^2} & \text{if } x \neq 0 \\ 1 & \text{if } x = 0 \end{cases}$ (c) $f(x) = \begin{cases} \dfrac{x^2 - x - 2}{x - 2} & \text{if } x \neq 2 \\ 1 & \text{if } x = 2 \end{cases}$ (d) $f(x) = [\![x]\!]$

FIGURE 3 Graphs of the functions in Example 2

2 DEFINITION A function f is **continuous from the right at a number a** if

$$\lim_{x \to a^+} f(x) = f(a)$$

and f is **continuous from the left at a** if

$$\lim_{x \to a^-} f(x) = f(a)$$

EXAMPLE 3 At each integer n, the function $f(x) = [\![x]\!]$ [see Figure 3(d)] is continuous from the right but discontinuous from the left because

$$\lim_{x \to n^+} f(x) = \lim_{x \to n^+} [\![x]\!] = n = f(n)$$

but

$$\lim_{x \to n^-} f(x) = \lim_{x \to n^-} [\![x]\!] = n - 1 \neq f(n)$$ □

3 DEFINITION A function f is **continuous on an interval** if it is continuous at every number in the interval. (If f is defined only on one side of an endpoint of the interval, we understand *continuous* at the endpoint to mean *continuous from the right* or *continuous from the left.*)

EXAMPLE 4 Show that the function $f(x) = 1 - \sqrt{1 - x^2}$ is continuous on the interval $[-1, 1]$.

SOLUTION If $-1 < a < 1$, then using the Limit Laws, we have

$$\begin{aligned}
\lim_{x \to a} f(x) &= \lim_{x \to a} \left(1 - \sqrt{1 - x^2}\right) \\
&= 1 - \lim_{x \to a} \sqrt{1 - x^2} && \text{(by Laws 2 and 7)} \\
&= 1 - \sqrt{\lim_{x \to a} (1 - x^2)} && \text{(by 11)} \\
&= 1 - \sqrt{1 - a^2} && \text{(by 2, 7, and 9)} \\
&= f(a)
\end{aligned}$$

Thus, by Definition 1, f is continuous at a if $-1 < a < 1$. Similar calculations show that

$$\lim_{x \to -1^+} f(x) = 1 = f(-1) \qquad \text{and} \qquad \lim_{x \to 1^-} f(x) = 1 = f(1)$$

so f is continuous from the right at -1 and continuous from the left at 1. Therefore, according to Definition 3, f is continuous on $[-1, 1]$.

The graph of f is sketched in Figure 4. It is the lower half of the circle

$$x^2 + (y - 1)^2 = 1$$ □

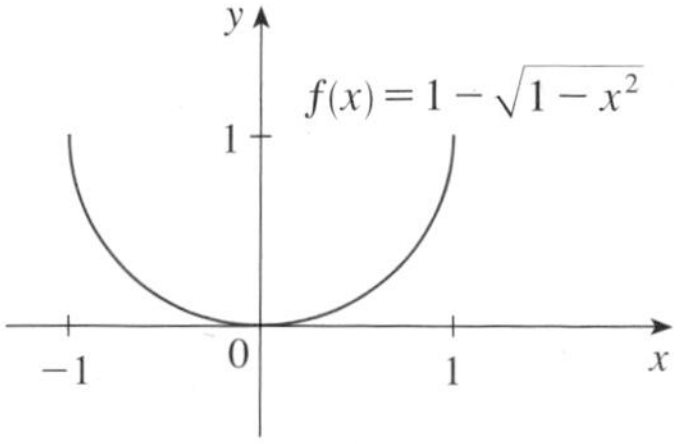

FIGURE 4

Instead of always using Definitions 1, 2, and 3 to verify the continuity of a function as we did in Example 4, it is often convenient to use the next theorem, which shows how to build up complicated continuous functions from simple ones.

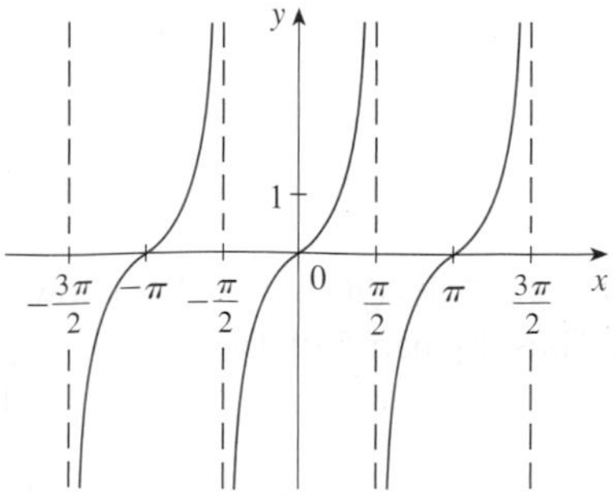

FIGURE 6 $y = \tan x$

is continuous except where $\cos x = 0$. This happens when x is an odd integer multiple of $\pi/2$, so $y = \tan x$ has infinite discontinuities when $x = \pm\pi/2, \pm3\pi/2, \pm5\pi/2$, and so on (see Figure 6).

The inverse function of any continuous one-to-one function is also continuous. (This fact is proved in Appendix F, but our geometric intuition makes it seem plausible: The graph of f^{-1} is obtained by reflecting the graph of f about the line $y = x$. So if the graph of f has no break in it, neither does the graph of f^{-1}.) Thus the inverse trigonometric functions are continuous.

■ The inverse trigonometric functions are reviewed in Section 1.6.

In Section 1.5 we defined the exponential function $y = a^x$ so as to fill in the holes in the graph of $y = a^x$ where x is rational. In other words, the very definition of $y = a^x$ makes it a continuous function on $\mathbb{R}$. Therefore its inverse function $y = \log_a x$ is continuous on $(0, \infty)$.

7 THEOREM The following types of functions are continuous at every number in their domains:

polynomials	rational functions	root functions
trigonometric functions	inverse trigonometric functions	
exponential functions	logarithmic functions	

EXAMPLE 6 Where is the function $f(x) = \dfrac{\ln x + \tan^{-1}x}{x^2 - 1}$ continuous?

SOLUTION We know from Theorem 7 that the function $y = \ln x$ is continuous for $x > 0$ and $y = \tan^{-1}x$ is continuous on $\mathbb{R}$. Thus, by part 1 of Theorem 4, $y = \ln x + \tan^{-1}x$ is continuous on $(0, \infty)$. The denominator, $y = x^2 - 1$, is a polynomial, so it is continuous everywhere. Therefore, by part 5 of Theorem 4, f is continuous at all positive numbers x except where $x^2 - 1 = 0$. So f is continuous on the intervals $(0, 1)$ and $(1, \infty)$. □

EXAMPLE 7 Evaluate $\displaystyle\lim_{x \to \pi} \frac{\sin x}{2 + \cos x}$.

SOLUTION Theorem 7 tells us that $y = \sin x$ is continuous. The function in the denominator, $y = 2 + \cos x$, is the sum of two continuous functions and is therefore continuous. Notice that this function is never 0 because $\cos x \geqslant -1$ for all x and so $2 + \cos x > 0$ everywhere. Thus the ratio

$$f(x) = \frac{\sin x}{2 + \cos x}$$

is continuous everywhere. Hence, by the definition of a continuous function,

$$\lim_{x \to \pi} \frac{\sin x}{2 + \cos x} = \lim_{x \to \pi} f(x) = f(\pi) = \frac{\sin \pi}{2 + \cos \pi} = \frac{0}{2 - 1} = 0$$ □

Another way of combining continuous functions f and g to get a new continuous function is to form the composite function $f \circ g$. This fact is a consequence of the following theorem.

■ This theorem says that a limit symbol can be moved through a function symbol if the function is continuous and the limit exists. In other words, the order of these two symbols can be reversed.

8 THEOREM If f is continuous at b and $\lim_{x \to a} g(x) = b$, then $\lim_{x \to a} f(g(x)) = f(b)$. In other words,

$$\lim_{x \to a} f(g(x)) = f\left(\lim_{x \to a} g(x)\right)$$

Intuitively, Theorem 8 is reasonable because if x is close to a, then $g(x)$ is close to b, and since f is continuous at b, if $g(x)$ is close to b, then $f(g(x))$ is close to $f(b)$. A proof of Theorem 8 is given in Appendix F.

EXAMPLE 8 Evaluate $\lim_{x \to 1} \arcsin\left(\dfrac{1 - \sqrt{x}}{1 - x}\right)$.

SOLUTION Because arcsin is a continuous function, we can apply Theorem 8:

$$\begin{aligned} \lim_{x \to 1} \arcsin\left(\frac{1 - \sqrt{x}}{1 - x}\right) &= \arcsin\left(\lim_{x \to 1} \frac{1 - \sqrt{x}}{1 - x}\right) \\ &= \arcsin\left(\lim_{x \to 1} \frac{1 - \sqrt{x}}{(1 - \sqrt{x})(1 + \sqrt{x})}\right) \\ &= \arcsin\left(\lim_{x \to 1} \frac{1}{1 + \sqrt{x}}\right) \\ &= \arcsin \frac{1}{2} = \frac{\pi}{6} \end{aligned}$$

□

Let's now apply Theorem 8 in the special case where $f(x) = \sqrt[n]{x}$, with n being a positive integer. Then

$$f(g(x)) = \sqrt[n]{g(x)}$$

and

$$f\left(\lim_{x \to a} g(x)\right) = \sqrt[n]{\lim_{x \to a} g(x)}$$

If we put these expressions into Theorem 8, we get

$$\lim_{x \to a} \sqrt[n]{g(x)} = \sqrt[n]{\lim_{x \to a} g(x)}$$

and so Limit Law 11 has now been proved. (We assume that the roots exist.)

9 THEOREM If g is continuous at a and f is continuous at $g(a)$, then the composite function $f \circ g$ given by $(f \circ g)(x) = f(g(x))$ is continuous at a.

This theorem is often expressed informally by saying "a continuous function of a continuous function is a continuous function."

PROOF Since g is continuous at a, we have

$$\lim_{x \to a} g(x) = g(a)$$

Since f is continuous at $b = g(a)$, we can apply Theorem 8 to obtain

$$\lim_{x \to a} f(g(x)) = f(g(a))$$

which is precisely the statement that the function $h(x) = f(g(x))$ is continuous at a; that is, $f \circ g$ is continuous at a. □

V EXAMPLE 9 Where are the following functions continuous?
(a) $h(x) = \sin(x^2)$ (b) $F(x) = \ln(1 + \cos x)$

SOLUTION
(a) We have $h(x) = f(g(x))$, where

$$g(x) = x^2 \qquad \text{and} \qquad f(x) = \sin x$$

Now g is continuous on $\mathbb{R}$ since it is a polynomial, and f is also continuous everywhere. Thus $h = f \circ g$ is continuous on $\mathbb{R}$ by Theorem 9.

(b) We know from Theorem 7 that $f(x) = \ln x$ is continuous and $g(x) = 1 + \cos x$ is continuous (because both $y = 1$ and $y = \cos x$ are continuous). Therefore, by Theorem 9, $F(x) = f(g(x))$ is continuous wherever it is defined. Now $\ln(1 + \cos x)$ is defined when $1 + \cos x > 0$. So it is undefined when $\cos x = -1$, and this happens when $x = \pm\pi, \pm 3\pi, \ldots$. Thus F has discontinuities when x is an odd multiple of π and is continuous on the intervals between these values (see Figure 7). □

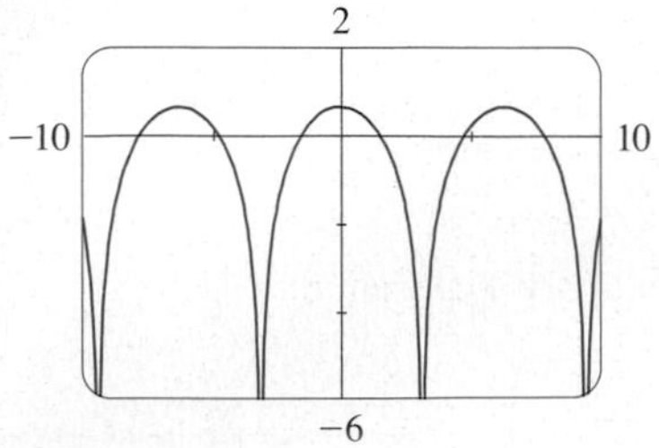

FIGURE 7
$y = \ln(1 + \cos x)$

An important property of continuous functions is expressed by the following theorem, whose proof is found in more advanced books on calculus.

> **10 THE INTERMEDIATE VALUE THEOREM** Suppose that f is continuous on the closed interval $[a, b]$ and let N be any number between $f(a)$ and $f(b)$, where $f(a) \neq f(b)$. Then there exists a number c in (a, b) such that $f(c) = N$.

The Intermediate Value Theorem states that a continuous function takes on every intermediate value between the function values $f(a)$ and $f(b)$. It is illustrated by Figure 8. Note that the value N can be taken on once [as in part (a)] or more than once [as in part (b)].

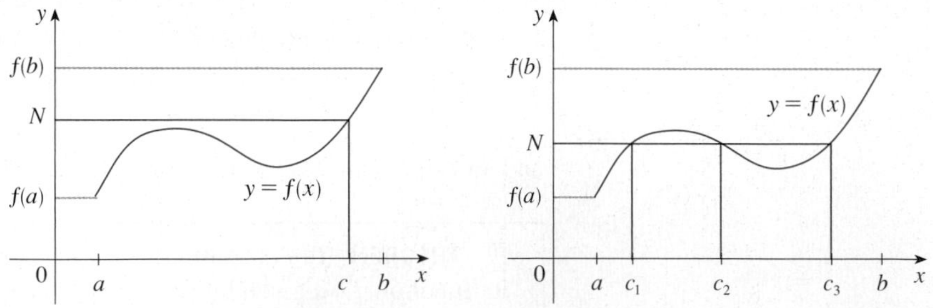

FIGURE 8 (a) (b)

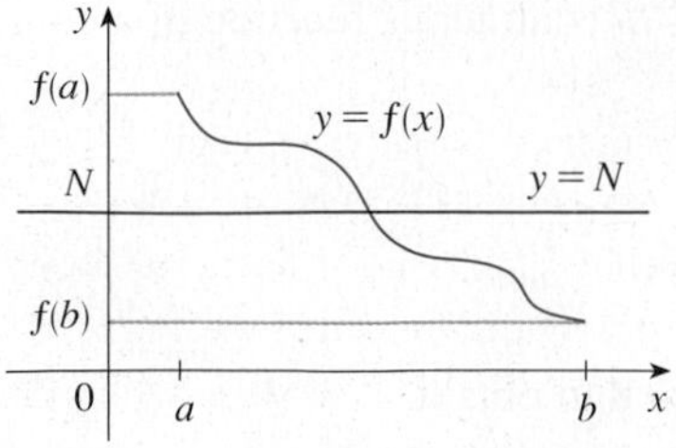

FIGURE 9

If we think of a continuous function as a function whose graph has no hole or break, then it is easy to believe that the Intermediate Value Theorem is true. In geometric terms it says that if any horizontal line $y = N$ is given between $y = f(a)$ and $y = f(b)$ as in Figure 9, then the graph of f can't jump over the line. It must intersect $y = N$ somewhere.

It is important that the function f in Theorem 10 be continuous. The Intermediate Value Theorem is not true in general for discontinuous functions (see Exercise 44).

One use of the Intermediate Value Theorem is in locating roots of equations as in the following example.

EXAMPLE 10 Show that there is a root of the equation

$$4x^3 - 6x^2 + 3x - 2 = 0$$

between 1 and 2.

SOLUTION Let $f(x) = 4x^3 - 6x^2 + 3x - 2$. We are looking for a solution of the given equation, that is, a number c between 1 and 2 such that $f(c) = 0$. Therefore, we take $a = 1$, $b = 2$, and $N = 0$ in Theorem 10. We have

$$f(1) = 4 - 6 + 3 - 2 = -1 < 0$$

and

$$f(2) = 32 - 24 + 6 - 2 = 12 > 0$$

Thus $f(1) < 0 < f(2)$; that is, $N = 0$ is a number between $f(1)$ and $f(2)$. Now f is continuous since it is a polynomial, so the Intermediate Value Theorem says there is a number c between 1 and 2 such that $f(c) = 0$. In other words, the equation $4x^3 - 6x^2 + 3x - 2 = 0$ has at least one root c in the interval $(1, 2)$.

In fact, we can locate a root more precisely by using the Intermediate Value Theorem again. Since

$$f(1.2) = -0.128 < 0 \quad \text{and} \quad f(1.3) = 0.548 > 0$$

a root must lie between 1.2 and 1.3. A calculator gives, by trial and error,

$$f(1.22) = -0.007008 < 0 \quad \text{and} \quad f(1.23) = 0.056068 > 0$$

so a root lies in the interval $(1.22, 1.23)$. □

We can use a graphing calculator or computer to illustrate the use of the Intermediate Value Theorem in Example 10. Figure 10 shows the graph of f in the viewing rectangle $[-1, 3]$ by $[-3, 3]$ and you can see that the graph crosses the x-axis between 1 and 2. Figure 11 shows the result of zooming in to the viewing rectangle $[1.2, 1.3]$ by $[-0.2, 0.2]$.

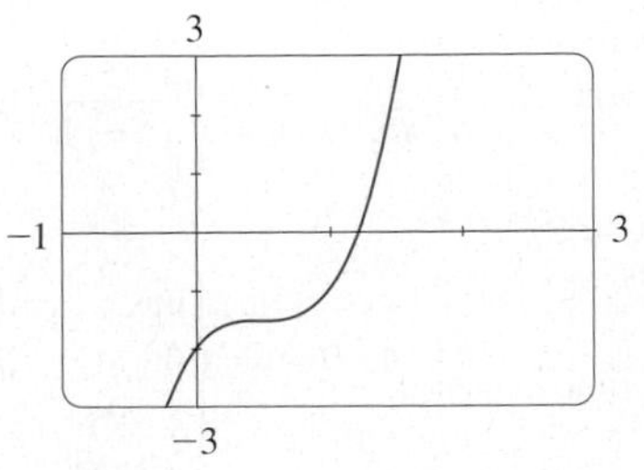

FIGURE 10

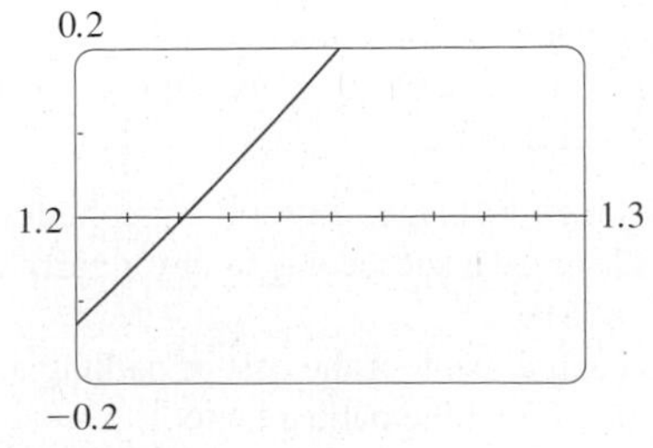

FIGURE 11

In fact, the Intermediate Value Theorem plays a role in the very way these graphing devices work. A computer calculates a finite number of points on the graph and turns on the pixels that contain these calculated points. It assumes that the function is continuous and takes on all the intermediate values between two consecutive points. The computer therefore connects the pixels by turning on the intermediate pixels.

2.5 EXERCISES

1. Write an equation that expresses the fact that a function f is continuous at the number 4.

2. If f is continuous on $(-\infty, \infty)$, what can you say about its graph?

3. (a) From the graph of f, state the numbers at which f is discontinuous and explain why.
(b) For each of the numbers stated in part (a), determine whether f is continuous from the right, or from the left, or neither.

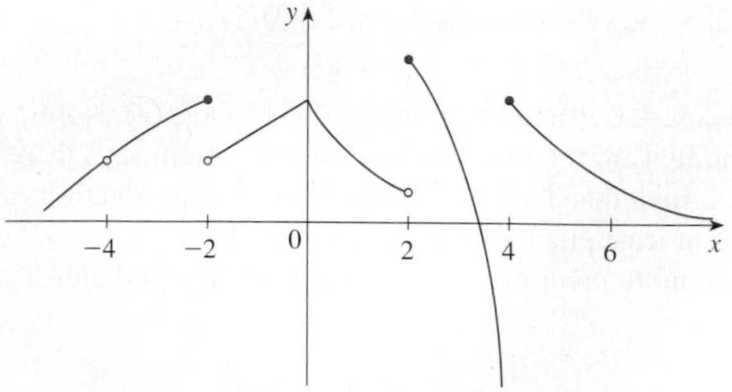

4. From the graph of g, state the intervals on which g is continuous.

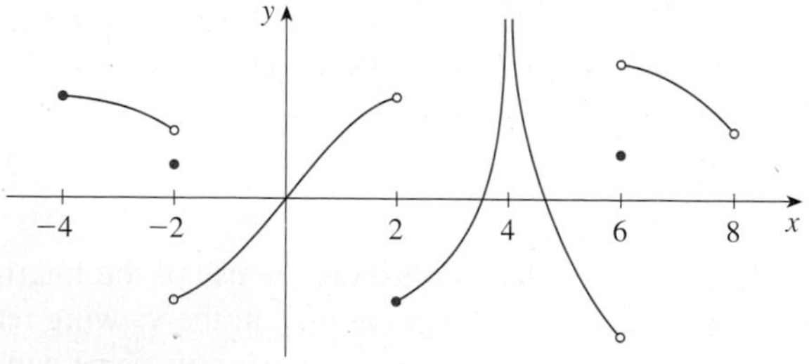

5. Sketch the graph of a function that is continuous everywhere except at $x = 3$ and is continuous from the left at 3.

6. Sketch the graph of a function that has a jump discontinuity at $x = 2$ and a removable discontinuity at $x = 4$, but is continuous elsewhere.

7. A parking lot charges \$3 for the first hour (or part of an hour) and \$2 for each succeeding hour (or part), up to a daily maximum of \$10.
(a) Sketch a graph of the cost of parking at this lot as a function of the time parked there.
(b) Discuss the discontinuities of this function and their significance to someone who parks in the lot.

8. Explain why each function is continuous or discontinuous.
(a) The temperature at a specific location as a function of time
(b) The temperature at a specific time as a function of the distance due west from New York City
(c) The altitude above sea level as a function of the distance due west from New York City
(d) The cost of a taxi ride as a function of the distance traveled
(e) The current in the circuit for the lights in a room as a function of time

9. If f and g are continuous functions with $f(3) = 5$ and $\lim_{x \to 3} [2f(x) - g(x)] = 4$, find $g(3)$.

10–12 Use the definition of continuity and the properties of limits to show that the function is continuous at the given number a.

10. $f(x) = x^2 + \sqrt{7 - x}, \quad a = 4$

11. $f(x) = (x + 2x^3)^4, \quad a = -1$

12. $h(t) = \dfrac{2t - 3t^2}{1 + t^3}, \quad a = 1$

13–14 Use the definition of continuity and the properties of limits to show that the function is continuous on the given interval.

13. $f(x) = \dfrac{2x + 3}{x - 2}, \quad (2, \infty)$

14. $g(x) = 2\sqrt{3 - x}, \quad (-\infty, 3]$

15–20 Explain why the function is discontinuous at the given number a. Sketch the graph of the function.

15. $f(x) = \ln |x - 2| \qquad a = 2$

16. $f(x) = \begin{cases} \dfrac{1}{x - 1} & \text{if } x \neq 1 \\ 2 & \text{if } x = 1 \end{cases} \qquad a = 1$

17. $f(x) = \begin{cases} e^x & \text{if } x < 0 \\ x^2 & \text{if } x \geqslant 0 \end{cases} \qquad a = 0$

18. $f(x) = \begin{cases} \dfrac{x^2 - x}{x^2 - 1} & \text{if } x \neq 1 \\ 1 & \text{if } x = 1 \end{cases} \qquad a = 1$

19. $f(x) = \begin{cases} \cos x & \text{if } x < 0 \\ 0 & \text{if } x = 0 \\ 1 - x^2 & \text{if } x > 0 \end{cases} \qquad a = 0$

20. $f(x) = \begin{cases} \dfrac{2x^2 - 5x - 3}{x - 3} & \text{if } x \neq 3 \\ 6 & \text{if } x = 3 \end{cases} \qquad a = 3$

21–28 Explain, using Theorems 4, 5, 7, and 9, why the function is continuous at every number in its domain. State the domain.

21. $F(x) = \dfrac{x}{x^2 + 5x + 6}$

22. $G(x) = \sqrt[3]{x}\,(1 + x^3)$

23. $R(x) = x^2 + \sqrt{2x - 1}$

24. $h(x) = \dfrac{\sin x}{x + 1}$

25. $L(t) = e^{-5t} \cos 2\pi t$

26. $F(x) = \sin^{-1}(x^2 - 1)$

27. $G(t) = \ln(t^4 - 1)$

28. $H(x) = \cos(e^{\sqrt{x}})$

29–30 Locate the discontinuities of the function and illustrate by graphing.

29. $y = \dfrac{1}{1 + e^{1/x}}$

30. $y = \ln(\tan^2 x)$

31–34 Use continuity to evaluate the limit.

31. $\displaystyle\lim_{x \to 4} \frac{5 + \sqrt{x}}{\sqrt{5 + x}}$

32. $\displaystyle\lim_{x \to \pi} \sin(x + \sin x)$

33. $\displaystyle\lim_{x \to 1} e^{x^2 - x}$

34. $\displaystyle\lim_{x \to 2} \arctan\left(\frac{x^2 - 4}{3x^2 - 6x}\right)$

35–36 Show that f is continuous on $(-\infty, \infty)$.

35. $f(x) = \begin{cases} x^2 & \text{if } x < 1 \\ \sqrt{x} & \text{if } x \geqslant 1 \end{cases}$

36. $f(x) = \begin{cases} \sin x & \text{if } x < \pi/4 \\ \cos x & \text{if } x \geqslant \pi/4 \end{cases}$

37–39 Find the numbers at which f is discontinuous. At which of these numbers is f continuous from the right, from the left, or neither? Sketch the graph of f.

37. $f(x) = \begin{cases} 1 + x^2 & \text{if } x \leqslant 0 \\ 2 - x & \text{if } 0 < x \leqslant 2 \\ (x - 2)^2 & \text{if } x > 2 \end{cases}$

38. $f(x) = \begin{cases} x + 1 & \text{if } x \leqslant 1 \\ 1/x & \text{if } 1 < x < 3 \\ \sqrt{x - 3} & \text{if } x \geqslant 3 \end{cases}$

39. $f(x) = \begin{cases} x + 2 & \text{if } x < 0 \\ e^x & \text{if } 0 \leqslant x \leqslant 1 \\ 2 - x & \text{if } x > 1 \end{cases}$

---

40. The gravitational force exerted by the earth on a unit mass at a distance r from the center of the planet is

$$F(r) = \begin{cases} \dfrac{GMr}{R^3} & \text{if } r < R \\[2ex] \dfrac{GM}{r^2} & \text{if } r \geqslant R \end{cases}$$

where M is the mass of the earth, R is its radius, and G is the gravitational constant. Is F a continuous function of r?

41. For what value of the constant c is the function f continuous on $(-\infty, \infty)$?

$$f(x) = \begin{cases} cx^2 + 2x & \text{if } x < 2 \\ x^3 - cx & \text{if } x \geqslant 2 \end{cases}$$

42. Find the values of a and b that make f continuous everywhere.

$$f(x) = \begin{cases} \dfrac{x^2 - 4}{x - 2} & \text{if } x < 2 \\ ax^2 - bx + 3 & \text{if } 2 < x < 3 \\ 2x - a + b & \text{if } x \geqslant 3 \end{cases}$$

43. Which of the following functions f has a removable discontinuity at a? If the discontinuity is removable, find a function g that agrees with f for $x \neq a$ and is continuous at a.

(a) $f(x) = \dfrac{x^4 - 1}{x - 1}, \quad a = 1$

(b) $f(x) = \dfrac{x^3 - x^2 - 2x}{x - 2}, \quad a = 2$

(c) $f(x) = [\![\sin x]\!], \quad a = \pi$

44. Suppose that a function f is continuous on $[0, 1]$ except at 0.25 and that $f(0) = 1$ and $f(1) = 3$. Let $N = 2$. Sketch two possible graphs of f, one showing that f might not satisfy the conclusion of the Intermediate Value Theorem and one showing that f might still satisfy the conclusion of the Intermediate Value Theorem (even though it doesn't satisfy the hypothesis).

45. If $f(x) = x^2 + 10 \sin x$, show that there is a number c such that $f(c) = 1000$.

46. Suppose f is continuous on $[1, 5]$ and the only solutions of the equation $f(x) = 6$ are $x = 1$ and $x = 4$. If $f(2) = 8$, explain why $f(3) > 6$.

47–50 Use the Intermediate Value Theorem to show that there is a root of the given equation in the specified interval.

47. $x^4 + x - 3 = 0, \quad (1, 2)$

48. $\sqrt[3]{x} = 1 - x, \quad (0, 1)$

49. $\cos x = x, \quad (0, 1)$

50. $\ln x = e^{-x}, \quad (1, 2)$

51–52 (a) Prove that the equation has at least one real root.
(b) Use your calculator to find an interval of length 0.01 that contains a root.

51. $\cos x = x^3$

52. $\ln x = 3 - 2x$

53–54 (a) Prove that the equation has at least one real root.
(b) Use your graphing device to find the root correct to three decimal places.

53. $100e^{-x/100} = 0.01x^2$

54. $\arctan x = 1 - x$

55. Prove that f is continuous at a if and only if

$$\lim_{h \to 0} f(a + h) = f(a)$$

56. To prove that sine is continuous, we need to show that $\lim_{x \to a} \sin x = \sin a$ for every real number a. By Exercise 55 an equivalent statement is that

$$\lim_{h \to 0} \sin(a + h) = \sin a$$

Use (6) to show that this is true.

57. Prove that cosine is a continuous function.

58. (a) Prove Theorem 4, part 3.
(b) Prove Theorem 4, part 5.

59. For what values of x is f continuous?

$$f(x) = \begin{cases} 0 & \text{if } x \text{ is rational} \\ 1 & \text{if } x \text{ is irrational} \end{cases}$$

60. For what values of x is g continuous?

$$g(x) = \begin{cases} 0 & \text{if } x \text{ is rational} \\ x & \text{if } x \text{ is irrational} \end{cases}$$

61. Is there a number that is exactly 1 more than its cube?

62. If a and b are positive numbers, prove that the equation

$$\frac{a}{x^3 + 2x^2 - 1} + \frac{b}{x^3 + x - 2} = 0$$

has at least one solution in the interval $(-1, 1)$.

63. Show that the function

$$f(x) = \begin{cases} x^4 \sin(1/x) & \text{if } x \neq 0 \\ 0 & \text{if } x = 0 \end{cases}$$

is continuous on $(-\infty, \infty)$.

64. (a) Show that the absolute value function $F(x) = |x|$ is continuous everywhere.
(b) Prove that if f is a continuous function on an interval, then so is $|f|$.
(c) Is the converse of the statement in part (b) also true? In other words, if $|f|$ is continuous, does it follow that f is continuous? If so, prove it. If not, find a counterexample.

65. A Tibetan monk leaves the monastery at 7:00 AM and takes his usual path to the top of the mountain, arriving at 7:00 PM. The following morning, he starts at 7:00 AM at the top and takes the same path back, arriving at the monastery at 7:00 PM. Use the Intermediate Value Theorem to show that there is a point on the path that the monk will cross at exactly the same time of day on both days.

2.6 LIMITS AT INFINITY; HORIZONTAL ASYMPTOTES

In Sections 2.2 and 2.4 we investigated infinite limits and vertical asymptotes. There we let x approach a number and the result was that the values of y became arbitrarily large (positive or negative). In this section we let x become arbitrarily large (positive or negative) and see what happens to y.

Let's begin by investigating the behavior of the function f defined by

$$f(x) = \frac{x^2 - 1}{x^2 + 1}$$

as x becomes large. The table at the left gives values of this function correct to six decimal places, and the graph of f has been drawn by a computer in Figure 1.

x	$f(x)$
0	−1
±1	0
±2	0.600000
±3	0.800000
±4	0.882353
±5	0.923077
±10	0.980198
±50	0.999200
±100	0.999800
±1000	0.999998

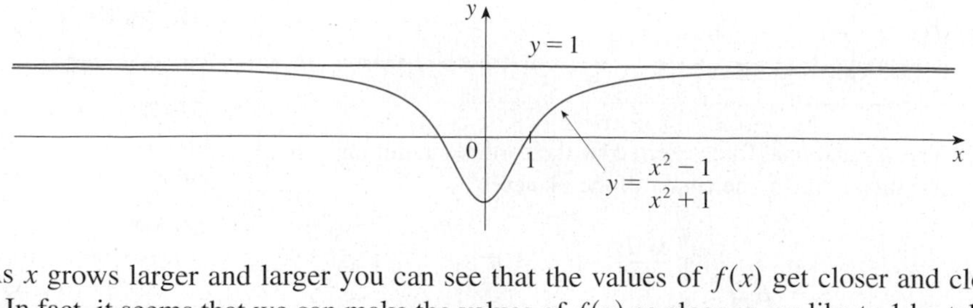

FIGURE 1

As x grows larger and larger you can see that the values of $f(x)$ get closer and closer to 1. In fact, it seems that we can make the values of $f(x)$ as close as we like to 1 by taking x sufficiently large. This situation is expressed symbolically by writing

$$\lim_{x \to \infty} \frac{x^2 - 1}{x^2 + 1} = 1$$

In general, we use the notation

$$\lim_{x \to \infty} f(x) = L$$

to indicate that the values of $f(x)$ become closer and closer to L as x becomes larger and larger.

1 DEFINITION Let f be a function defined on some interval (a, ∞). Then

$$\lim_{x \to \infty} f(x) = L$$

means that the values of $f(x)$ can be made arbitrarily close to L by taking x sufficiently large.

Another notation for $\lim_{x \to \infty} f(x) = L$ is

$$f(x) \to L \quad \text{as} \quad x \to \infty$$

The symbol ∞ does not represent a number. Nonetheless, the expression $\lim\limits_{x \to \infty} f(x) = L$ is often read as

"the limit of $f(x)$, as x approaches infinity, is L"

or "the limit of $f(x)$, as x becomes infinite, is L"

or "the limit of $f(x)$, as x increases without bound, is L"

The meaning of such phrases is given by Definition 1. A more precise definition, similar to the ε, δ definition of Section 2.4, is given at the end of this section.

Geometric illustrations of Definition 1 are shown in Figure 2. Notice that there are many ways for the graph of f to approach the line $y = L$ (which is called a *horizontal asymptote*) as we look to the far right of each graph.

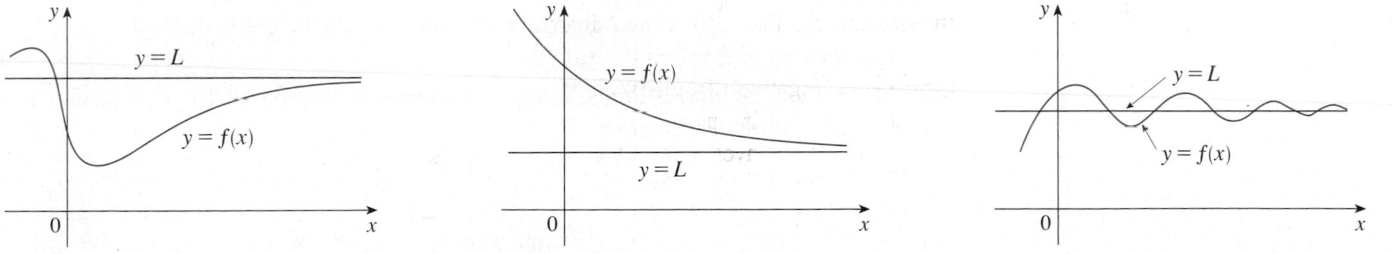

FIGURE 2
Examples illustrating $\lim\limits_{x \to \infty} f(x) = L$

Referring back to Figure 1, we see that for numerically large negative values of x, the values of $f(x)$ are close to 1. By letting x decrease through negative values without bound, we can make $f(x)$ as close as we like to 1. This is expressed by writing

$$\lim_{x \to -\infty} \frac{x^2 - 1}{x^2 + 1} = 1$$

The general definition is as follows.

2 DEFINITION Let f be a function defined on some interval $(-\infty, a)$. Then

$$\lim_{x \to -\infty} f(x) = L$$

means that the values of $f(x)$ can be made arbitrarily close to L by taking x sufficiently large negative.

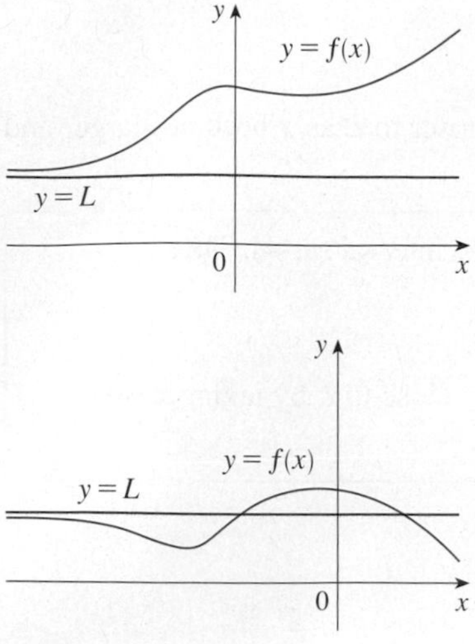

FIGURE 3
Examples illustrating $\lim_{x \to -\infty} f(x) = L$

Again, the symbol $-\infty$ does not represent a number, but the expression $\lim_{x \to -\infty} f(x) = L$ is often read as

"the limit of $f(x)$, as x approaches negative infinity, is L"

Definition 2 is illustrated in Figure 3. Notice that the graph approaches the line $y = L$ as we look to the far left of each graph.

3 DEFINITION The line $y = L$ is called a **horizontal asymptote** of the curve $y = f(x)$ if either

$$\lim_{x \to \infty} f(x) = L \qquad \text{or} \qquad \lim_{x \to -\infty} f(x) = L$$

For instance, the curve illustrated in Figure 1 has the line $y = 1$ as a horizontal asymptote because

$$\lim_{x \to \infty} \frac{x^2 - 1}{x^2 + 1} = 1$$

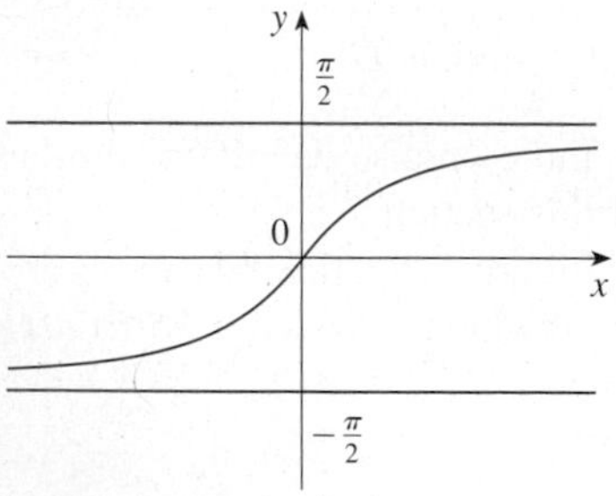

FIGURE 4
$y = \tan^{-1}x$

An example of a curve with two horizontal asymptotes is $y = \tan^{-1}x$. (See Figure 4.) In fact,

$$\boxed{4} \qquad \lim_{x \to -\infty} \tan^{-1}x = -\frac{\pi}{2} \qquad \lim_{x \to \infty} \tan^{-1}x = \frac{\pi}{2}$$

so both of the lines $y = -\pi/2$ and $y = \pi/2$ are horizontal asymptotes. (This follows from the fact that the lines $x = \pm\pi/2$ are vertical asymptotes of the graph of tan.)

EXAMPLE 1 Find the infinite limits, limits at infinity, and asymptotes for the function f whose graph is shown in Figure 5.

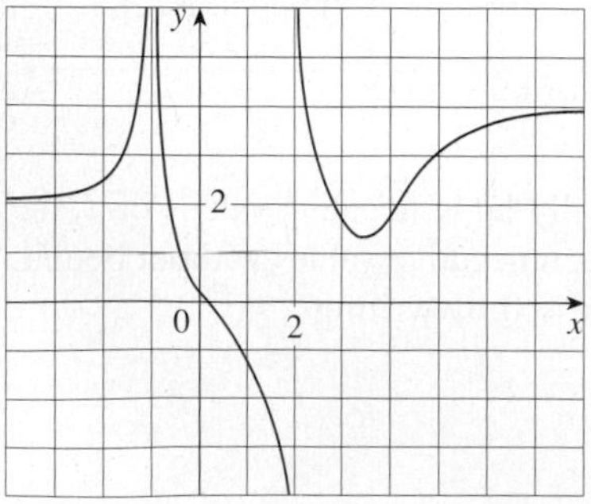

FIGURE 5

SOLUTION We see that the values of $f(x)$ become large as $x \to -1$ from both sides, so

$$\lim_{x \to -1} f(x) = \infty$$

Notice that $f(x)$ becomes large negative as x approaches 2 from the left, but large positive as x approaches 2 from the right. So

$$\lim_{x \to 2^-} f(x) = -\infty \qquad \text{and} \qquad \lim_{x \to 2^+} f(x) = \infty$$

Thus both of the lines $x = -1$ and $x = 2$ are vertical asymptotes.

As x becomes large, it appears that $f(x)$ approaches 4. But as x decreases through negative values, $f(x)$ approaches 2. So

$$\lim_{x \to \infty} f(x) = 4 \qquad \text{and} \qquad \lim_{x \to -\infty} f(x) = 2$$

This means that both $y = 4$ and $y = 2$ are horizontal asymptotes. □

EXAMPLE 2 Find $\lim_{x \to \infty} \frac{1}{x}$ and $\lim_{x \to -\infty} \frac{1}{x}$.

SOLUTION Observe that when x is large, $1/x$ is small. For instance,

$$\frac{1}{100} = 0.01 \qquad \frac{1}{10{,}000} = 0.0001 \qquad \frac{1}{1{,}000{,}000} = 0.000001$$

In fact, by taking x large enough, we can make $1/x$ as close to 0 as we please. Therefore, according to Definition 1, we have

$$\lim_{x \to \infty} \frac{1}{x} = 0$$

Similar reasoning shows that when x is large negative, $1/x$ is small negative, so we also have

$$\lim_{x \to -\infty} \frac{1}{x} = 0$$

It follows that the line $y = 0$ (the x-axis) is a horizontal asymptote of the curve $y = 1/x$. (This is an equilateral hyperbola; see Figure 6.) □

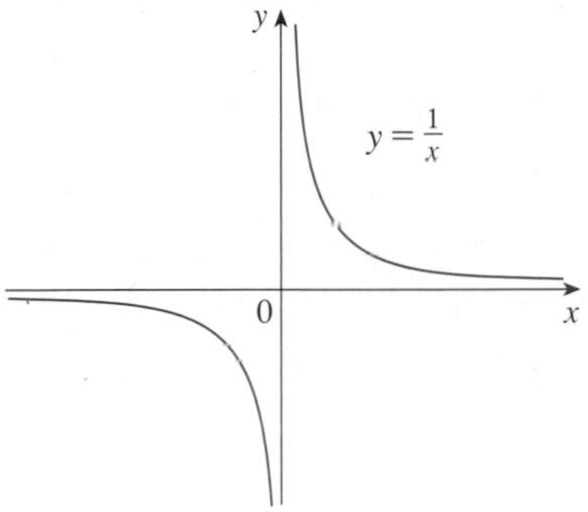

FIGURE 6
$\lim_{x \to \infty} \frac{1}{x} = 0, \ \lim_{x \to -\infty} \frac{1}{x} = 0$

Most of the Limit Laws that were given in Section 2.3 also hold for limits at infinity. It can be proved that the *Limit Laws listed in Section 2.3 (with the exception of Laws 9 and 10) are also valid if "$x \to a$" is replaced by "$x \to \infty$" or "$x \to -\infty$."* In particular, if we combine Laws 6 and 11 with the results of Example 2, we obtain the following important rule for calculating limits.

> **5 THEOREM** If $r > 0$ is a rational number, then
>
> $$\lim_{x \to \infty} \frac{1}{x^r} = 0$$
>
> If $r > 0$ is a rational number such that x^r is defined for all x, then
>
> $$\lim_{x \to -\infty} \frac{1}{x^r} = 0$$

V EXAMPLE 3 Evaluate

$$\lim_{x \to \infty} \frac{3x^2 - x - 2}{5x^2 + 4x + 1}$$

and indicate which properties of limits are used at each stage.

SOLUTION As x becomes large, both numerator and denominator become large, so it isn't obvious what happens to their ratio. We need to do some preliminary algebra.

To evaluate the limit at infinity of any rational function, we first divide both the numerator and denominator by the highest power of x that occurs in the denominator.

(We may assume that $x \neq 0$, since we are interested only in large values of x.) In this case the highest power of x in the denominator is x^2, so we have

$$\lim_{x\to\infty} \frac{3x^2 - x - 2}{5x^2 + 4x + 1} = \lim_{x\to\infty} \frac{\dfrac{3x^2 - x - 2}{x^2}}{\dfrac{5x^2 + 4x + 1}{x^2}} = \lim_{x\to\infty} \frac{3 - \dfrac{1}{x} - \dfrac{2}{x^2}}{5 + \dfrac{4}{x} + \dfrac{1}{x^2}}$$

$$= \frac{\lim_{x\to\infty}\left(3 - \dfrac{1}{x} - \dfrac{2}{x^2}\right)}{\lim_{x\to\infty}\left(5 + \dfrac{4}{x} + \dfrac{1}{x^2}\right)} \qquad \text{(by Limit Law 5)}$$

$$= \frac{\lim_{x\to\infty} 3 - \lim_{x\to\infty} \dfrac{1}{x} - 2 \lim_{x\to\infty} \dfrac{1}{x^2}}{\lim_{x\to\infty} 5 + 4 \lim_{x\to\infty} \dfrac{1}{x} + \lim_{x\to\infty} \dfrac{1}{x^2}} \qquad \text{(by 1, 2, and 3)}$$

$$= \frac{3 - 0 - 0}{5 + 0 + 0} \qquad \text{(by 7 and Theorem 5)}$$

$$= \frac{3}{5}$$

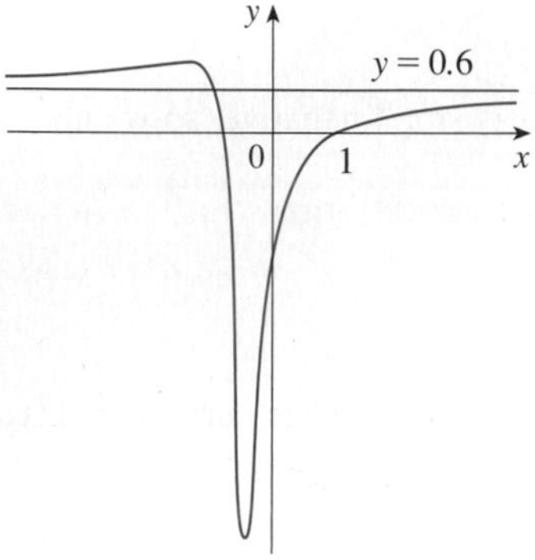

FIGURE 7

$y = \dfrac{3x^2 - x - 2}{5x^2 + 4x + 1}$

A similar calculation shows that the limit as $x \to -\infty$ is also $\frac{3}{5}$. Figure 7 illustrates the results of these calculations by showing how the graph of the given rational function approaches the horizontal asymptote $y = \frac{3}{5}$. □

EXAMPLE 4 Find the horizontal and vertical asymptotes of the graph of the function

$$f(x) = \frac{\sqrt{2x^2 + 1}}{3x - 5}$$

SOLUTION Dividing both numerator and denominator by x and using the properties of limits, we have

$$\lim_{x\to\infty} \frac{\sqrt{2x^2 + 1}}{3x - 5} = \lim_{x\to\infty} \frac{\sqrt{2 + \dfrac{1}{x^2}}}{3 - \dfrac{5}{x}} \qquad \text{(since } \sqrt{x^2} = x \text{ for } x > 0\text{)}$$

$$= \frac{\lim_{x\to\infty} \sqrt{2 + \dfrac{1}{x^2}}}{\lim_{x\to\infty}\left(3 - \dfrac{5}{x}\right)} = \frac{\sqrt{\lim_{x\to\infty} 2 + \lim_{x\to\infty} \dfrac{1}{x^2}}}{\lim_{x\to\infty} 3 - 5 \lim_{x\to\infty} \dfrac{1}{x}} = \frac{\sqrt{2 + 0}}{3 - 5 \cdot 0} = \frac{\sqrt{2}}{3}$$

Therefore the line $y = \sqrt{2}/3$ is a horizontal asymptote of the graph of f.

In computing the limit as $x \to -\infty$, we must remember that for $x < 0$, we have $\sqrt{x^2} = |x| = -x$. So when we divide the numerator by x, for $x < 0$ we get

$$\frac{1}{x}\sqrt{2x^2 + 1} = -\frac{1}{\sqrt{x^2}}\sqrt{2x^2 + 1} = -\sqrt{2 + \frac{1}{x^2}}$$

Therefore

$$\lim_{x\to-\infty}\frac{\sqrt{2x^2+1}}{3x-5}=\lim_{x\to-\infty}\frac{-\sqrt{2+\dfrac{1}{x^2}}}{3-\dfrac{5}{x}}=\frac{-\sqrt{2+\displaystyle\lim_{x\to-\infty}\frac{1}{x^2}}}{3-5\displaystyle\lim_{x\to-\infty}\frac{1}{x}}=-\frac{\sqrt{2}}{3}$$

Thus the line $y=-\sqrt{2}/3$ is also a horizontal asymptote.

A vertical asymptote is likely to occur when the denominator, $3x-5$, is 0, that is, when $x=\frac{5}{3}$. If x is close to $\frac{5}{3}$ and $x>\frac{5}{3}$, then the denominator is close to 0 and $3x-5$ is positive. The numerator $\sqrt{2x^2+1}$ is always positive, so $f(x)$ is positive. Therefore

$$\lim_{x\to(5/3)^+}\frac{\sqrt{2x^2+1}}{3x-5}=\infty$$

If x is close to $\frac{5}{3}$ but $x<\frac{5}{3}$, then $3x-5<0$ and so $f(x)$ is large negative. Thus

$$\lim_{x\to(5/3)^-}\frac{\sqrt{2x^2+1}}{3x-5}=-\infty$$

The vertical asymptote is $x=\frac{5}{3}$. All three asymptotes are shown in Figure 8. □

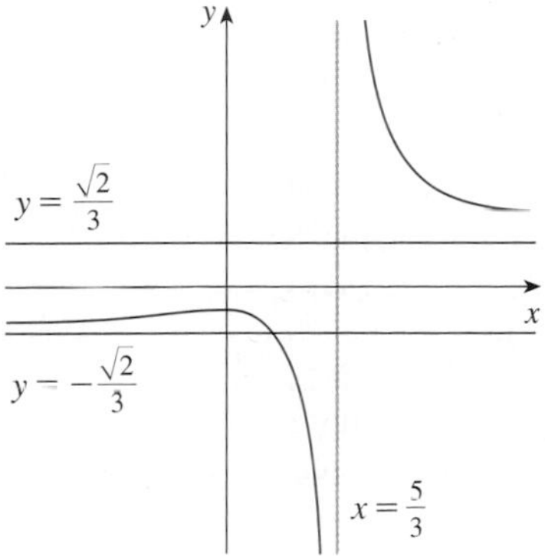

FIGURE 8

$y=\dfrac{\sqrt{2x^2+1}}{3x-5}$

EXAMPLE 5 Compute $\lim_{x\to\infty}(\sqrt{x^2+1}-x)$.

SOLUTION Because both $\sqrt{x^2+1}$ and x are large when x is large, it's difficult to see what happens to their difference, so we use algebra to rewrite the function. We first multiply numerator and denominator by the conjugate radical:

■ We can think of the given function as having a denominator of 1.

$$\lim_{x\to\infty}(\sqrt{x^2+1}-x)=\lim_{x\to\infty}(\sqrt{x^2+1}-x)\frac{\sqrt{x^2+1}+x}{\sqrt{x^2+1}+x}$$
$$=\lim_{x\to\infty}\frac{(x^2+1)-x^2}{\sqrt{x^2+1}+x}=\lim_{x\to\infty}\frac{1}{\sqrt{x^2+1}+x}$$

The Squeeze Theorem could be used to show that this limit is 0. But an easier method is to divide numerator and denominator by x. Doing this and using the Limit Laws, we obtain

$$\lim_{x\to\infty}(\sqrt{x^2+1}-x)=\lim_{x\to\infty}\frac{1}{\sqrt{x^2+1}+x}=\lim_{x\to\infty}\frac{\dfrac{1}{x}}{\dfrac{\sqrt{x^2+1}+x}{x}}$$
$$=\lim_{x\to\infty}\frac{\dfrac{1}{x}}{\sqrt{1+\dfrac{1}{x^2}}+1}=\frac{0}{\sqrt{1+0}+1}=0$$

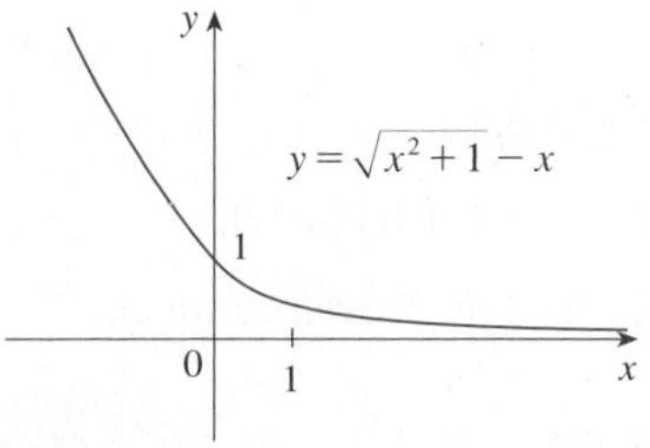

FIGURE 9

Figure 9 illustrates this result. □

The graph of the natural exponential function $y=e^x$ has the line $y=0$ (the x-axis) as a horizontal asymptote. (The same is true of any exponential function with base $a>1$.) In fact, from the graph in Figure 10 and the corresponding table of values, we see that

6 $$\lim_{x \to -\infty} e^x = 0$$

Notice that the values of e^x approach 0 very rapidly.

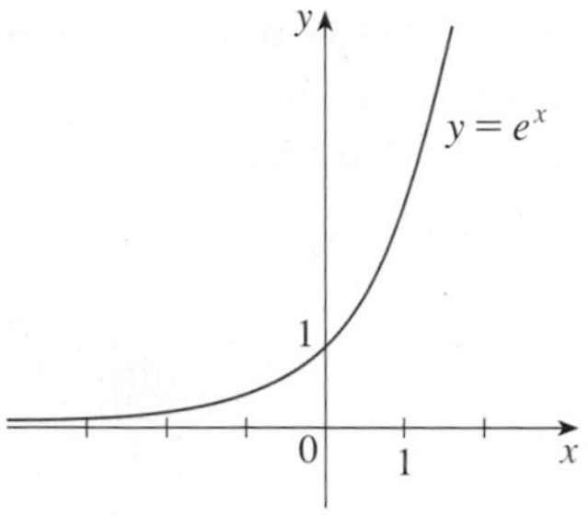

FIGURE 10

x	e^x
0	1.00000
-1	0.36788
-2	0.13534
-3	0.04979
-5	0.00674
-8	0.00034
-10	0.00005

EXAMPLE 6 Evaluate $\lim_{x \to 0^-} e^{1/x}$.

■ The problem-solving strategy for Example 6 is *introducing something extra* (see page 76). Here, the something extra, the auxiliary aid, is the new variable t.

SOLUTION If we let $t = 1/x$, we know that $t \to -\infty$ as $x \to 0^-$. Therefore, by (6),

$$\lim_{x \to 0^-} e^{1/x} = \lim_{t \to -\infty} e^t = 0$$

(See Exercise 71.) □

EXAMPLE 7 Evaluate $\lim_{x \to \infty} \sin x$.

SOLUTION As x increases, the values of $\sin x$ oscillate between 1 and -1 infinitely often and so they don't approach any definite number. Thus $\lim_{x \to \infty} \sin x$ does not exist. □

INFINITE LIMITS AT INFINITY

The notation

$$\lim_{x \to \infty} f(x) = \infty$$

is used to indicate that the values of $f(x)$ become large as x becomes large. Similar meanings are attached to the following symbols:

$$\lim_{x \to -\infty} f(x) = \infty \qquad \lim_{x \to \infty} f(x) = -\infty \qquad \lim_{x \to -\infty} f(x) = -\infty$$

EXAMPLE 8 Find $\lim_{x \to \infty} x^3$ and $\lim_{x \to -\infty} x^3$.

SOLUTION When x becomes large, x^3 also becomes large. For instance,

$$10^3 = 1000 \qquad 100^3 = 1{,}000{,}000 \qquad 1000^3 = 1{,}000{,}000{,}000$$

In fact, we can make x^3 as big as we like by taking x large enough. Therefore we can write

$$\lim_{x \to \infty} x^3 = \infty$$

Similarly, when x is large negative, so is x^3. Thus

$$\lim_{x \to -\infty} x^3 = -\infty$$

These limit statements can also be seen from the graph of $y = x^3$ in Figure 11. □

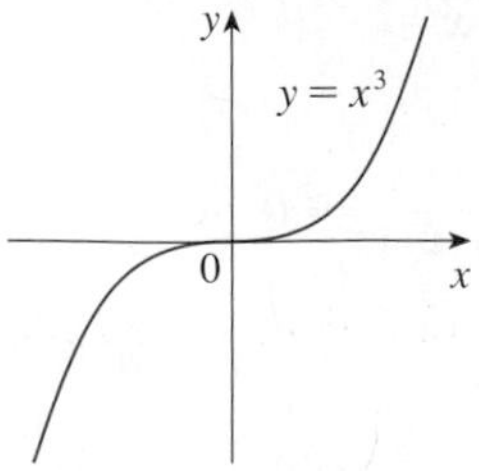

FIGURE 11
$\lim_{x \to \infty} x^3 = \infty, \quad \lim_{x \to -\infty} x^3 = -\infty$

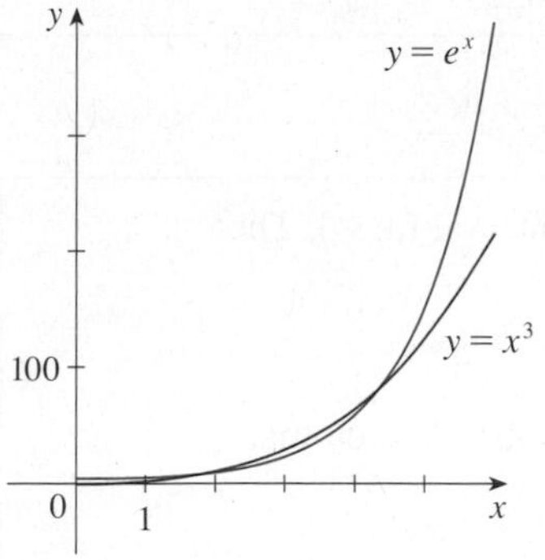

FIGURE 12
e^x is much larger than x^3 when x is large.

Looking at Figure 10 we see that

$$\lim_{x \to \infty} e^x = \infty$$

but, as Figure 12 demonstrates, $y = e^x$ becomes large as $x \to \infty$ at a much faster rate than $y = x^3$.

EXAMPLE 9 Find $\lim_{x \to \infty} (x^2 - x)$.

SOLUTION It would be **wrong** to write

$$\lim_{x \to \infty} (x^2 - x) = \lim_{x \to \infty} x^2 - \lim_{x \to \infty} x = \infty - \infty$$

The Limit Laws can't be applied to infinite limits because ∞ is not a number ($\infty - \infty$ can't be defined). However, we *can* write

$$\lim_{x \to \infty} (x^2 - x) = \lim_{x \to \infty} x(x - 1) = \infty$$

because both x and $x - 1$ become arbitrarily large and so their product does too. □

EXAMPLE 10 Find $\lim_{x \to \infty} \dfrac{x^2 + x}{3 - x}$.

SOLUTION As in Example 3, we divide the numerator and denominator by the highest power of x in the denominator, which is just x:

$$\lim_{x \to \infty} \frac{x^2 + x}{3 - x} = \lim_{x \to \infty} \frac{x + 1}{\dfrac{3}{x} - 1} = -\infty$$

because $x + 1 \to \infty$ and $3/x - 1 \to -1$ as $x \to \infty$. □

The next example shows that by using infinite limits at infinity, together with intercepts, we can get a rough idea of the graph of a polynomial without having to plot a large number of points.

V EXAMPLE 11 Sketch the graph of $y = (x - 2)^4(x + 1)^3(x - 1)$ by finding its intercepts and its limits as $x \to \infty$ and as $x \to -\infty$.

SOLUTION The y-intercept is $f(0) = (-2)^4(1)^3(-1) = -16$ and the x-intercepts are found by setting $y = 0$: $x = 2, -1, 1$. Notice that since $(x - 2)^4$ is positive, the function doesn't change sign at 2; thus the graph doesn't cross the x-axis at 2. The graph crosses the axis at -1 and 1.

When x is large positive, all three factors are large, so

$$\lim_{x \to \infty} (x - 2)^4(x + 1)^3(x - 1) = \infty$$

When x is large negative, the first factor is large positive and the second and third factors are both large negative, so

$$\lim_{x \to -\infty} (x - 2)^4(x + 1)^3(x - 1) = \infty$$

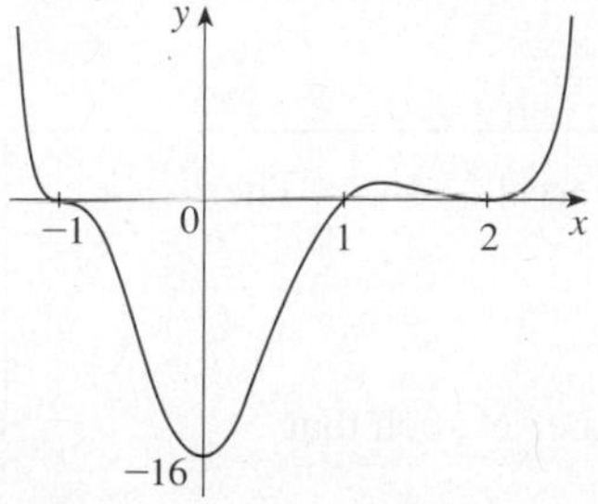

FIGURE 13
$y = (x - 2)^4(x + 1)^3(x - 1)$

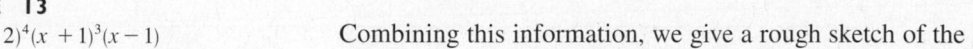

Combining this information, we give a rough sketch of the graph in Figure 13. □

PRECISE DEFINITIONS

Definition 1 can be stated precisely as follows.

7 DEFINITION Let f be a function defined on some interval (a, ∞). Then

$$\lim_{x \to \infty} f(x) = L$$

means that for every $\varepsilon > 0$ there is a corresponding number N such that

$$\text{if} \quad x > N \qquad \text{then} \qquad |f(x) - L| < \varepsilon$$

In words, this says that the values of $f(x)$ can be made arbitrarily close to L (within a distance ε, where ε is any positive number) by taking x sufficiently large (larger than N, where N depends on ε). Graphically it says that by choosing x large enough (larger than some number N) we can make the graph of f lie between the given horizontal lines $y = L - \varepsilon$ and $y = L + \varepsilon$ as in Figure 14. This must be true no matter how small we choose ε. Figure 15 shows that if a smaller value of ε is chosen, then a larger value of N may be required.

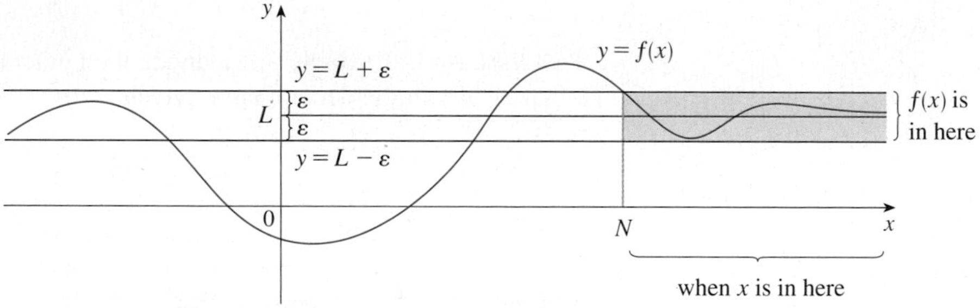

FIGURE 14
$\lim_{x \to \infty} f(x) = L$

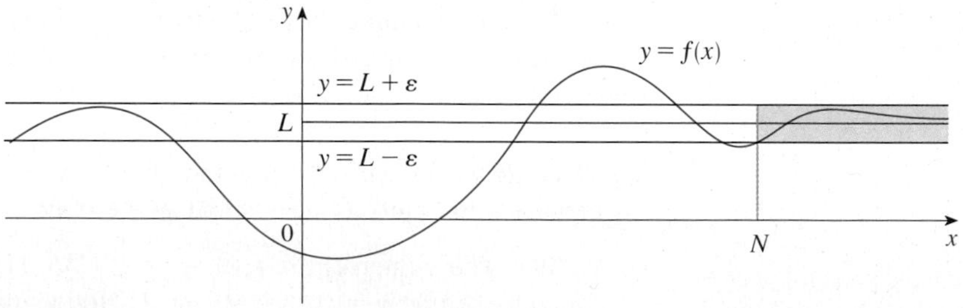

FIGURE 15
$\lim_{x \to \infty} f(x) = L$

Similarly, a precise version of Definition 2 is given by Definition 8, which is illustrated in Figure 16.

8 DEFINITION Let f be a function defined on some interval $(-\infty, a)$. Then

$$\lim_{x \to -\infty} f(x) = L$$

means that for every $\varepsilon > 0$ there is a corresponding number N such that

$$\text{if} \quad x < N \qquad \text{then} \qquad |f(x) - L| < \varepsilon$$

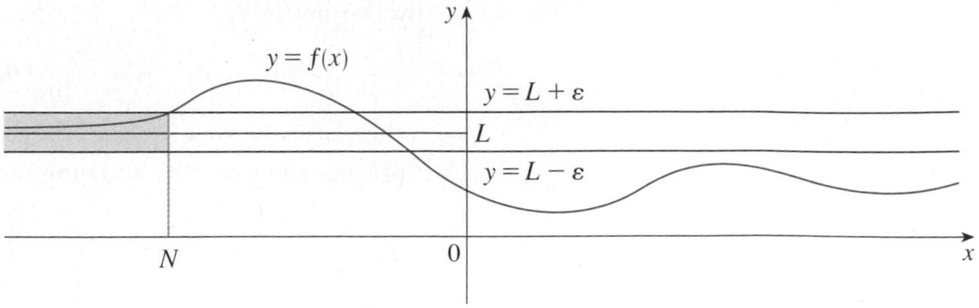

FIGURE 16
$\lim_{x \to -\infty} f(x) = L$

In Example 3 we calculated that

$$\lim_{x \to \infty} \frac{3x^2 - x - 2}{5x^2 + 4x + 1} = \frac{3}{5}$$

In the next example we use a graphing device to relate this statement to Definition 7 with $L = \frac{3}{5}$ and $\varepsilon = 0.1$.

TEC In Module 2.4/2.6 you can explore the precise definition of a limit both graphically and numerically.

EXAMPLE 12 Use a graph to find a number N such that

$$\text{if} \quad x > N \quad \text{then} \quad \left| \frac{3x^2 - x - 2}{5x^2 + 4x + 1} - 0.6 \right| < 0.1$$

SOLUTION We rewrite the given inequality as

$$0.5 < \frac{3x^2 - x - 2}{5x^2 + 4x + 1} < 0.7$$

We need to determine the values of x for which the given curve lies between the horizontal lines $y = 0.5$ and $y = 0.7$. So we graph the curve and these lines in Figure 17. Then we use the cursor to estimate that the curve crosses the line $y = 0.5$ when $x \approx 6.7$. To the right of this number the curve stays between the lines $y = 0.5$ and $y = 0.7$. Rounding to be safe, we can say that

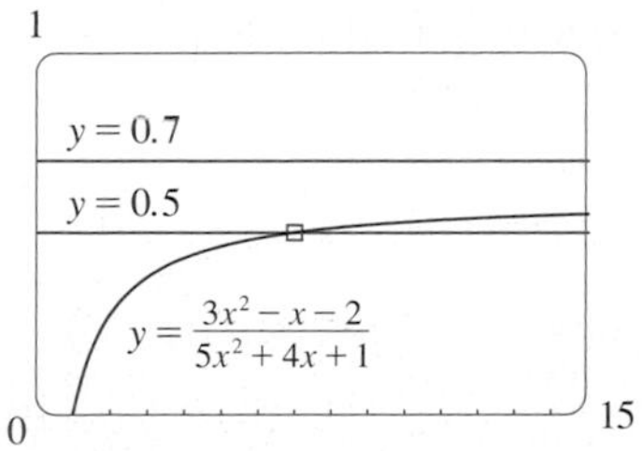

FIGURE 17

$$\text{if} \quad x > 7 \quad \text{then} \quad \left| \frac{3x^2 - x - 2}{5x^2 + 4x + 1} - 0.6 \right| < 0.1$$

In other words, for $\varepsilon = 0.1$ we can choose $N = 7$ (or any larger number) in Definition 7. □

EXAMPLE 13 Use Definition 7 to prove that $\lim_{x \to \infty} \frac{1}{x} = 0$.

SOLUTION Given $\varepsilon > 0$, we want to find N such that

$$\text{if} \quad x > N \quad \text{then} \quad \left| \frac{1}{x} - 0 \right| < \varepsilon$$

In computing the limit we may assume that $x > 0$. Then $1/x < \varepsilon \iff x > 1/\varepsilon$. Let's choose $N = 1/\varepsilon$. So

$$\text{if} \quad x > N = \frac{1}{\varepsilon} \quad \text{then} \quad \left| \frac{1}{x} - 0 \right| = \frac{1}{x} < \varepsilon$$

Therefore, by Definition 7,

$$\lim_{x \to \infty} \frac{1}{x} = 0$$

Figure 18 illustrates the proof by showing some values of ε and the corresponding values of N.

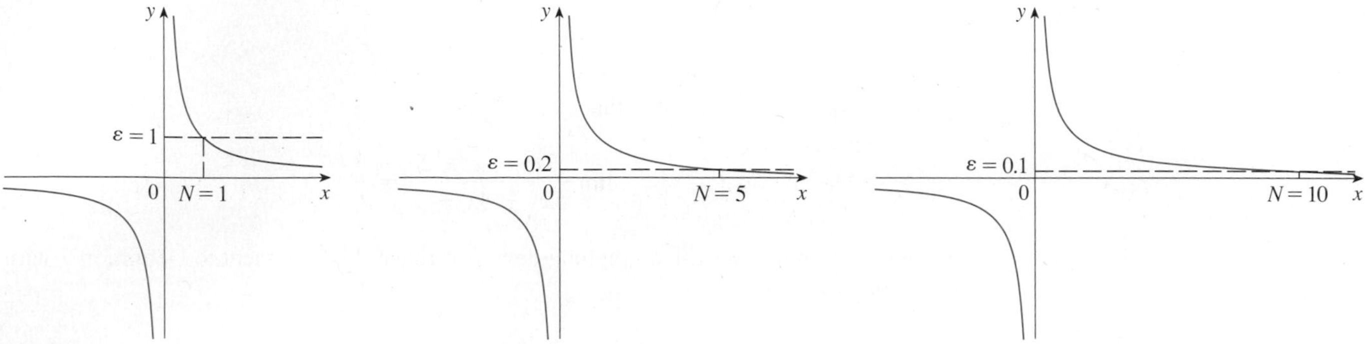

FIGURE 18

□

Finally we note that an infinite limit at infinity can be defined as follows. The geometric illustration is given in Figure 19.

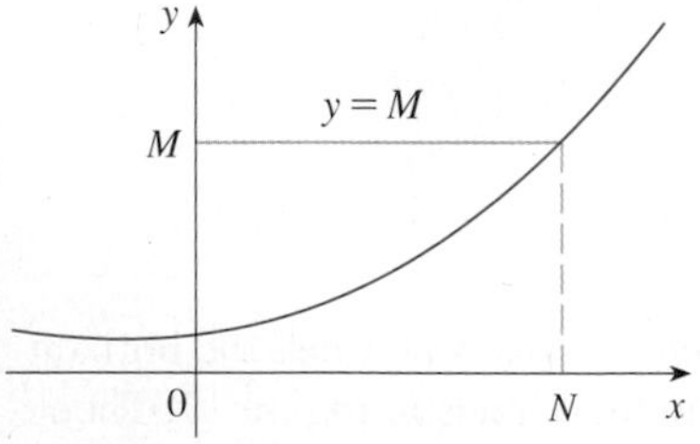

FIGURE 19
$\lim_{x \to \infty} f(x) = \infty$

> **9 DEFINITION** Let f be a function defined on some interval (a, ∞). Then
>
> $$\lim_{x \to \infty} f(x) = \infty$$
>
> means that for every positive number M there is a corresponding positive number N such that
>
> $$\text{if} \quad x > N \qquad \text{then} \qquad f(x) > M$$

Similar definitions apply when the symbol ∞ is replaced by $-\infty$. (See Exercise 70.)

2.6 EXERCISES

1. Explain in your own words the meaning of each of the following.
(a) $\lim_{x \to \infty} f(x) = 5$ (b) $\lim_{x \to -\infty} f(x) = 3$

2. (a) Can the graph of $y = f(x)$ intersect a vertical asymptote? Can it intersect a horizontal asymptote? Illustrate by sketching graphs.
(b) How many horizontal asymptotes can the graph of $y = f(x)$ have? Sketch graphs to illustrate the possibilities.

3. For the function f whose graph is given, state the following.
(a) $\lim_{x \to 2} f(x)$ (b) $\lim_{x \to -1^-} f(x)$
(c) $\lim_{x \to -1^+} f(x)$ (d) $\lim_{x \to \infty} f(x)$
(e) $\lim_{x \to -\infty} f(x)$ (f) The equations of the asymptotes

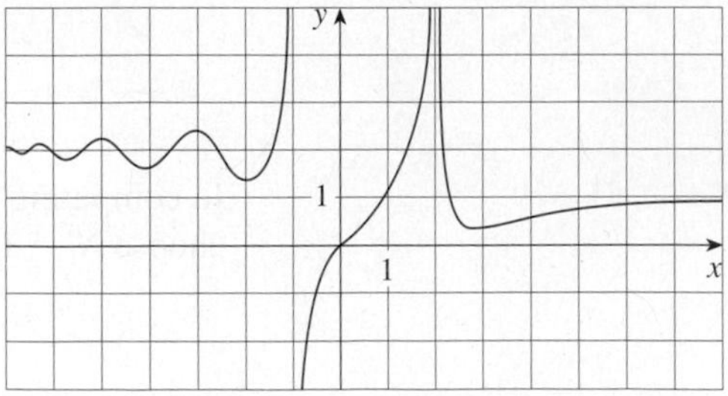

4. For the function g whose graph is given, state the following.

(a) $\lim_{x\to\infty} g(x)$ (b) $\lim_{x\to-\infty} g(x)$

(c) $\lim_{x\to 3} g(x)$ (d) $\lim_{x\to 0} g(x)$

(e) $\lim_{x\to -2^+} g(x)$ (f) The equations of the asymptotes

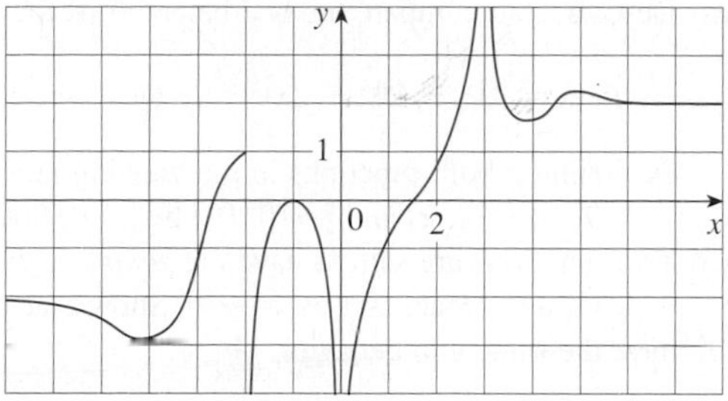

5–10 Sketch the graph of an example of a function f that satisfies all of the given conditions.

5. $f(0) = 0$, $f(1) = 1$, $\lim_{x\to\infty} f(x) = 0$, f is odd

6. $\lim_{x\to 0^+} f(x) = \infty$, $\lim_{x\to 0^-} f(x) = -\infty$, $\lim_{x\to\infty} f(x) = 1$, $\lim_{x\to-\infty} f(x) = 1$

7. $\lim_{x\to 2} f(x) = -\infty$, $\lim_{x\to\infty} f(x) = \infty$, $\lim_{x\to-\infty} f(x) = 0$, $\lim_{x\to 0^+} f(x) = \infty$, $\lim_{x\to 0^-} f(x) = -\infty$

8. $\lim_{x\to -2} f(x) = \infty$, $\lim_{x\to-\infty} f(x) = 3$, $\lim_{x\to\infty} f(x) = -3$

9. $f(0) = 3$, $\lim_{x\to 0^-} f(x) = 4$, $\lim_{x\to 0^+} f(x) = 2$, $\lim_{x\to-\infty} f(x) = -\infty$, $\lim_{x\to 4^-} f(x) = -\infty$, $\lim_{x\to 4^+} f(x) = \infty$, $\lim_{x\to\infty} f(x) = 3$

10. $\lim_{x\to 3} f(x) = -\infty$, $\lim_{x\to\infty} f(x) = 2$, $f(0) = 0$, f is even

11. Guess the value of the limit

$$\lim_{x\to\infty} \frac{x^2}{2^x}$$

by evaluating the function $f(x) = x^2/2^x$ for $x = 0, 1, 2, 3, 4, 5, 6, 7, 8, 9, 10, 20, 50$, and 100. Then use a graph of f to support your guess.

12. (a) Use a graph of

$$f(x) = \left(1 - \frac{2}{x}\right)^x$$

to estimate the value of $\lim_{x\to\infty} f(x)$ correct to two decimal places.

(b) Use a table of values of $f(x)$ to estimate the limit to four decimal places.

13–14 Evaluate the limit and justify each step by indicating the appropriate properties of limits.

13. $\lim_{x\to\infty} \dfrac{3x^2 - x + 4}{2x^2 + 5x - 8}$

14. $\lim_{x\to\infty} \sqrt{\dfrac{12x^3 - 5x + 2}{1 + 4x^2 + 3x^3}}$

15–36 Find the limit.

15. $\lim_{x\to\infty} \dfrac{1}{2x + 3}$

16. $\lim_{x\to\infty} \dfrac{3x + 5}{x - 4}$

17. $\lim_{x\to-\infty} \dfrac{1 - x - x^2}{2x^2 - 7}$

18. $\lim_{y\to\infty} \dfrac{2 - 3y^2}{5y^2 + 4y}$

19. $\lim_{x\to\infty} \dfrac{x^3 + 5x}{2x^3 - x^2 + 4}$

20. $\lim_{t\to-\infty} \dfrac{t^2 + 2}{t^3 + t^2 - 1}$

21. $\lim_{u\to\infty} \dfrac{4u^4 + 5}{(u^2 - 2)(2u^2 - 1)}$

22. $\lim_{x\to\infty} \dfrac{x + 2}{\sqrt{9x^2 + 1}}$

23. $\lim_{x\to\infty} \dfrac{\sqrt{9x^6 - x}}{x^3 + 1}$

24. $\lim_{x\to-\infty} \dfrac{\sqrt{9x^6 - x}}{x^3 + 1}$

25. $\lim_{x\to\infty} \left(\sqrt{9x^2 + x} - 3x\right)$

26. $\lim_{x\to-\infty} \left(x + \sqrt{x^2 + 2x}\right)$

27. $\lim_{x\to\infty} \left(\sqrt{x^2 + ax} - \sqrt{x^2 + bx}\right)$

28. $\lim_{x\to\infty} \cos x$

29. $\lim_{x\to\infty} \dfrac{x + x^3 + x^5}{1 - x^2 + x^4}$

30. $\lim_{x\to\infty} \sqrt{x^2 + 1}$

31. $\lim_{x\to-\infty} (x^4 + x^5)$

32. $\lim_{x\to\infty} \dfrac{x^3 - 2x + 3}{5 - 2x^2}$

33. $\lim_{x\to\infty} \dfrac{1 - e^x}{1 + 2e^x}$

34. $\lim_{x\to\infty} \tan^{-1}(x^2 - x^4)$

35. $\lim_{x\to\infty} (e^{-2x}\cos x)$

36. $\lim_{x\to(\pi/2)^+} e^{\tan x}$

37. (a) Estimate the value of

$$\lim_{x\to-\infty} \left(\sqrt{x^2 + x + 1} + x\right)$$

by graphing the function $f(x) = \sqrt{x^2 + x + 1} + x$.

(b) Use a table of values of $f(x)$ to guess the value of the limit.

(c) Prove that your guess is correct.

38. (a) Use a graph of

$$f(x) = \sqrt{3x^2 + 8x + 6} - \sqrt{3x^2 + 3x + 1}$$

to estimate the value of $\lim_{x\to\infty} f(x)$ to one decimal place.

(b) Use a table of values of $f(x)$ to estimate the limit to four decimal places.

(c) Find the exact value of the limit.

39–44 Find the horizontal and vertical asymptotes of each curve. If you have a graphing device, check your work by graphing the curve and estimating the asymptotes.

39. $y = \dfrac{2x + 1}{x - 2}$

40. $y = \dfrac{x^2 + 1}{2x^2 - 3x - 2}$

41. $y = \dfrac{2x^2 + x - 1}{x^2 + x - 2}$

42. $y = \dfrac{1 + x^4}{x^2 - x^4}$

43. $y = \dfrac{x^3 - x}{x^2 - 6x + 5}$

44. $y = \dfrac{2e^x}{e^x - 5}$

---

45. Estimate the horizontal asymptote of the function

$$f(x) = \frac{3x^3 + 500x^2}{x^3 + 500x^2 + 100x + 2000}$$

by graphing f for $-10 \leqslant x \leqslant 10$. Then calculate the equation of the asymptote by evaluating the limit. How do you explain the discrepancy?

46. (a) Graph the function

$$f(x) = \frac{\sqrt{2x^2 + 1}}{3x - 5}$$

How many horizontal and vertical asymptotes do you observe? Use the graph to estimate the values of the limits

$$\lim_{x \to \infty} \frac{\sqrt{2x^2 + 1}}{3x - 5} \quad \text{and} \quad \lim_{x \to -\infty} \frac{\sqrt{2x^2 + 1}}{3x - 5}$$

(b) By calculating values of $f(x)$, give numerical estimates of the limits in part (a).

(c) Calculate the exact values of the limits in part (a). Did you get the same value or different values for these two limits? [In view of your answer to part (a), you might have to check your calculation for the second limit.]

47. Find a formula for a function f that satisfies the following conditions:

$$\lim_{x \to \pm\infty} f(x) = 0, \quad \lim_{x \to 0} f(x) = -\infty, \quad f(2) = 0,$$

$$\lim_{x \to 3^-} f(x) = \infty, \quad \lim_{x \to 3^+} f(x) = -\infty$$

48. Find a formula for a function that has vertical asymptotes $x = 1$ and $x = 3$ and horizontal asymptote $y = 1$.

49–52 Find the limits as $x \to \infty$ and as $x \to -\infty$. Use this information, together with intercepts, to give a rough sketch of the graph as in Example 11.

49. $y = x^4 - x^6$

50. $y = x^3(x + 2)^2(x - 1)$

51. $y = (3 - x)(1 + x)^2(1 - x)^4$

52. $y = x^2(x^2 - 1)^2(x + 2)$

---

53. (a) Use the Squeeze Theorem to evaluate $\displaystyle\lim_{x \to \infty} \frac{\sin x}{x}$.

(b) Graph $f(x) = (\sin x)/x$. How many times does the graph cross the asymptote?

54. By the *end behavior* of a function we mean the behavior of its values as $x \to \infty$ and as $x \to -\infty$.

(a) Describe and compare the end behavior of the functions

$$P(x) = 3x^5 - 5x^3 + 2x \qquad Q(x) = 3x^5$$

by graphing both functions in the viewing rectangles $[-2, 2]$ by $[-2, 2]$ and $[-10, 10]$ by $[-10{,}000, 10{,}000]$.

(b) Two functions are said to have the *same end behavior* if their ratio approaches 1 as $x \to \infty$. Show that P and Q have the same end behavior.

55. Let P and Q be polynomials. Find

$$\lim_{x \to \infty} \frac{P(x)}{Q(x)}$$

if the degree of P is (a) less than the degree of Q and (b) greater than the degree of Q.

56. Make a rough sketch of the curve $y = x^n$ (n an integer) for the following five cases:

(i) $n = 0$ (ii) $n > 0$, n odd
(iii) $n > 0$, n even (iv) $n < 0$, n odd
(v) $n < 0$, n even

Then use these sketches to find the following limits.

(a) $\displaystyle\lim_{x \to 0^+} x^n$ (b) $\displaystyle\lim_{x \to 0^-} x^n$

(c) $\displaystyle\lim_{x \to \infty} x^n$ (d) $\displaystyle\lim_{x \to -\infty} x^n$

57. Find $\lim_{x \to \infty} f(x)$ if, for all $x > 1$,

$$\frac{10e^x - 21}{2e^x} < f(x) < \frac{5\sqrt{x}}{\sqrt{x - 1}}$$

58. (a) A tank contains 5000 L of pure water. Brine that contains 30 g of salt per liter of water is pumped into the tank at a rate of 25 L/min. Show that the concentration of salt after t minutes (in grams per liter) is

$$C(t) = \frac{30t}{200 + t}$$

(b) What happens to the concentration as $t \to \infty$?

59. In Chapter 9 we will be able to show, under certain assumptions, that the velocity $v(t)$ of a falling raindrop at time t is

$$v(t) = v^*(1 - e^{-gt/v^*})$$

where g is the acceleration due to gravity and v^* is the *terminal velocity* of the raindrop.

(a) Find $\lim_{t \to \infty} v(t)$.

(b) Graph $v(t)$ if $v^* = 1$ m/s and $g = 9.8$ m/s^2. How long does it take for the velocity of the raindrop to reach 99% of its terminal velocity?

60. (a) By graphing $y = e^{-x/10}$ and $y = 0.1$ on a common screen, discover how large you need to make x so that $e^{-x/10} < 0.1$.
(b) Can you solve part (a) without using a graphing device?

61. Use a graph to find a number N such that

$$\text{if} \quad x > N \quad \text{then} \quad \left| \frac{3x^2 + 1}{2x^2 + x + 1} - 1.5 \right| < 0.05$$

62. For the limit

$$\lim_{x \to \infty} \frac{\sqrt{4x^2 + 1}}{x + 1} = 2$$

illustrate Definition 7 by finding values of N that correspond to $\varepsilon = 0.5$ and $\varepsilon = 0.1$.

63. For the limit

$$\lim_{x \to -\infty} \frac{\sqrt{4x^2 + 1}}{x + 1} = -2$$

illustrate Definition 8 by finding values of N that correspond to $\varepsilon = 0.5$ and $\varepsilon = 0.1$.

64. For the limit

$$\lim_{x \to \infty} \frac{2x + 1}{\sqrt{x + 1}} = \infty$$

illustrate Definition 9 by finding a value of N that corresponds to $M = 100$.

65. (a) How large do we have to take x so that $1/x^2 < 0.0001$?
(b) Taking $r = 2$ in Theorem 5, we have the statement

$$\lim_{x \to \infty} \frac{1}{x^2} = 0$$

Prove this directly using Definition 7.

66. (a) How large do we have to take x so that $1/\sqrt{x} < 0.0001$?
(b) Taking $r = \frac{1}{2}$ in Theorem 5, we have the statement

$$\lim_{x \to \infty} \frac{1}{\sqrt{x}} = 0$$

Prove this directly using Definition 7.

67. Use Definition 8 to prove that $\lim_{x \to -\infty} \frac{1}{x} = 0$.

68. Prove, using Definition 9, that $\lim_{x \to \infty} x^3 = \infty$.

69. Use Definition 9 to prove that $\lim_{x \to \infty} e^x = \infty$.

70. Formulate a precise definition of

$$\lim_{x \to -\infty} f(x) = -\infty$$

Then use your definition to prove that

$$\lim_{x \to -\infty} (1 + x^3) = -\infty$$

71. Prove that

$$\lim_{x \to \infty} f(x) = \lim_{t \to 0^+} f(1/t)$$

and

$$\lim_{x \to -\infty} f(x) = \lim_{t \to 0^-} f(1/t)$$

if these limits exist.

2.7 DERIVATIVES AND RATES OF CHANGE

The problem of finding the tangent line to a curve and the problem of finding the velocity of an object both involve finding the same type of limit, as we saw in Section 2.1. This special type of limit is called a *derivative* and we will see that it can be interpreted as a rate of change in any of the sciences or engineering.

TANGENTS

If a curve C has equation $y = f(x)$ and we want to find the tangent line to C at the point $P(a, f(a))$, then we consider a nearby point $Q(x, f(x))$, where $x \neq a$, and compute the slope of the secant line PQ:

$$m_{PQ} = \frac{f(x) - f(a)}{x - a}$$

Then we let Q approach P along the curve C by letting x approach a. If m_{PQ} approaches a number m, then we define the *tangent* t to be the line through P with slope m. (This

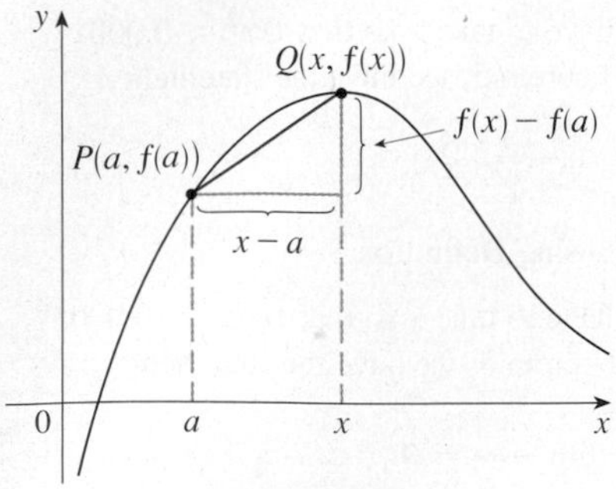

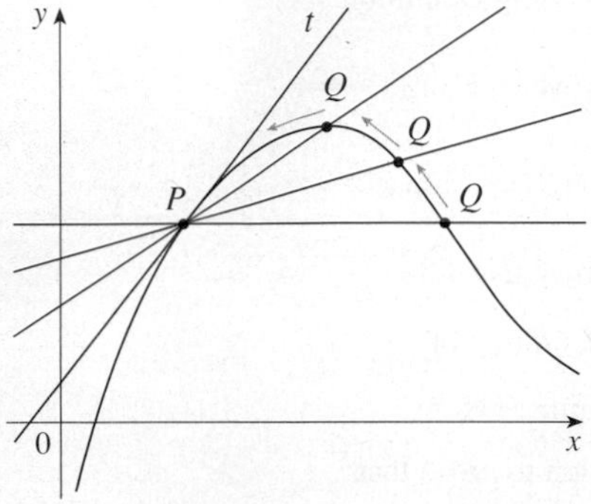

FIGURE 1

amounts to saying that the tangent line is the limiting position of the secant line PQ as Q approaches P. See Figure 1.)

1 DEFINITION The **tangent line** to the curve $y = f(x)$ at the point $P(a, f(a))$ is the line through P with slope

$$m = \lim_{x \to a} \frac{f(x) - f(a)}{x - a}$$

provided that this limit exists.

In our first example we confirm the guess we made in Example 1 in Section 2.1.

EXAMPLE 1 Find an equation of the tangent line to the parabola $y = x^2$ at the point $P(1, 1)$.

SOLUTION Here we have $a = 1$ and $f(x) = x^2$, so the slope is

$$m = \lim_{x \to 1} \frac{f(x) - f(1)}{x - 1} = \lim_{x \to 1} \frac{x^2 - 1}{x - 1}$$

$$= \lim_{x \to 1} \frac{(x - 1)(x + 1)}{x - 1}$$

$$= \lim_{x \to 1} (x + 1) = 1 + 1 = 2$$

■ Point-slope form for a line through the point (x_1, y_1) with slope m:

$$y - y_1 = m(x - x_1)$$

Using the point-slope form of the equation of a line, we find that an equation of the tangent line at $(1, 1)$ is

$$y - 1 = 2(x - 1) \qquad \text{or} \qquad y = 2x - 1$$

□

TEC Visual 2.7 shows an animation of Figure 2.

We sometimes refer to the slope of the tangent line to a curve at a point as the **slope of the curve** at the point. The idea is that if we zoom in far enough toward the point, the curve looks almost like a straight line. Figure 2 illustrates this procedure for the curve $y = x^2$ in Example 1. The more we zoom in, the more the parabola looks like a line. In other words, the curve becomes almost indistinguishable from its tangent line.

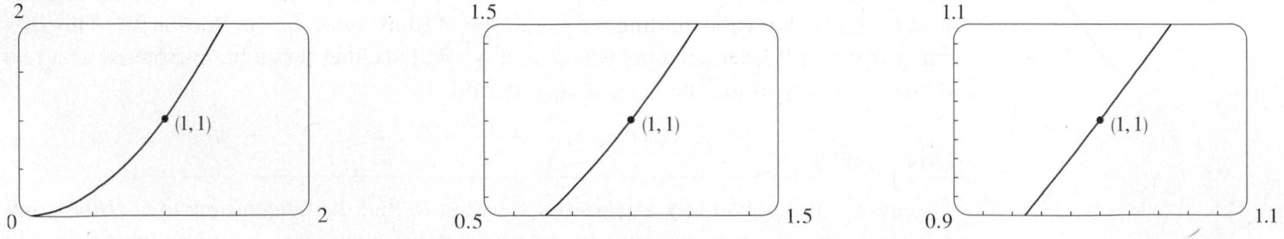

FIGURE 2 Zooming in toward the point (1, 1) on the parabola $y = x^2$

There is another expression for the slope of a tangent line that is sometimes easier to use. If $h = x - a$, then $x = a + h$ and so the slope of the secant line PQ is

$$m_{PQ} = \frac{f(a + h) - f(a)}{h}$$

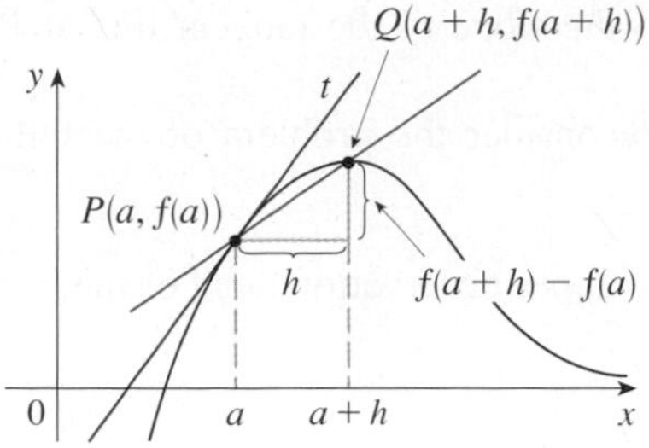

FIGURE 3

(See Figure 3 where the case $h > 0$ is illustrated and Q is to the right of P. If it happened that $h < 0$, however, Q would be to the left of P.)

Notice that as x approaches a, h approaches 0 (because $h = x - a$) and so the expression for the slope of the tangent line in Definition 1 becomes

2

$$m = \lim_{h \to 0} \frac{f(a + h) - f(a)}{h}$$

EXAMPLE 2 Find an equation of the tangent line to the hyperbola $y = 3/x$ at the point (3, 1).

SOLUTION Let $f(x) = 3/x$. Then the slope of the tangent at (3, 1) is

$$m = \lim_{h \to 0} \frac{f(3 + h) - f(3)}{h} = \lim_{h \to 0} \frac{\dfrac{3}{3 + h} - 1}{h} = \lim_{h \to 0} \frac{\dfrac{3 - (3 + h)}{3 + h}}{h}$$

$$= \lim_{h \to 0} \frac{-h}{h(3 + h)} = \lim_{h \to 0} -\frac{1}{3 + h} = -\frac{1}{3}$$

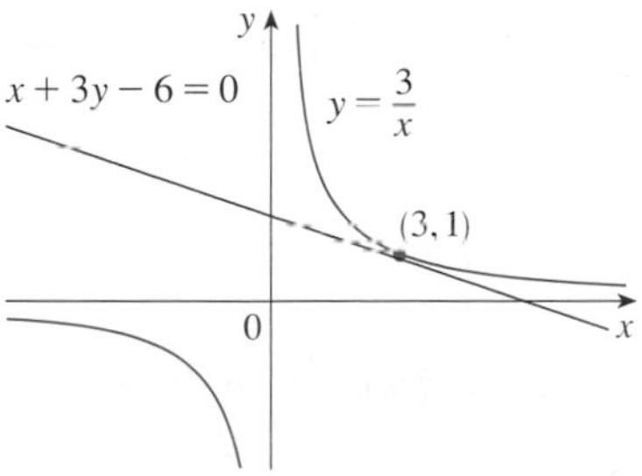

FIGURE 4

Therefore an equation of the tangent at the point (3, 1) is

$$y - 1 = -\tfrac{1}{3}(x - 3)$$

which simplifies to

$$x + 3y - 6 = 0$$

The hyperbola and its tangent are shown in Figure 4. □

VELOCITIES

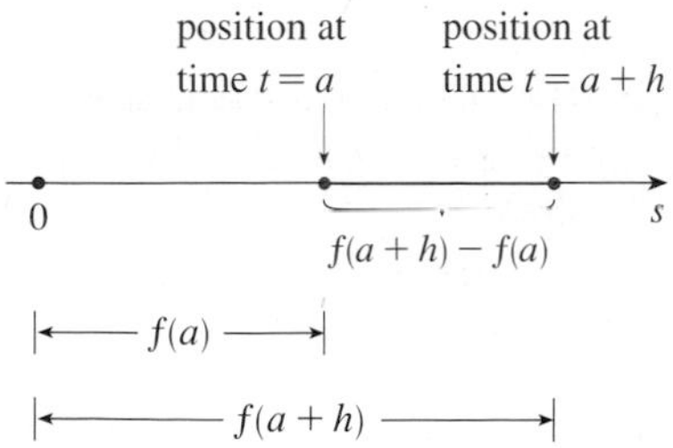

FIGURE 5

In Section 2.1 we investigated the motion of a ball dropped from the CN Tower and defined its velocity to be the limiting value of average velocities over shorter and shorter time periods.

In general, suppose an object moves along a straight line according to an equation of motion $s = f(t)$, where s is the displacement (directed distance) of the object from the origin at time t. The function f that describes the motion is called the **position function** of the object. In the time interval from $t = a$ to $t = a + h$ the change in position is $f(a + h) - f(a)$. (See Figure 5.) The average velocity over this time interval is

$$\text{average velocity} = \frac{\text{displacement}}{\text{time}} = \frac{f(a + h) - f(a)}{h}$$

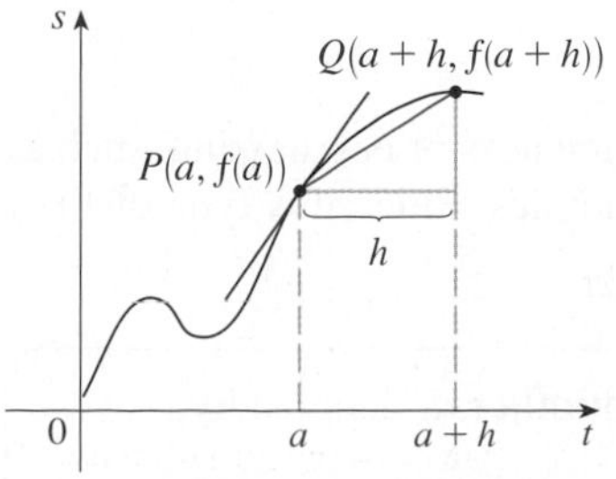

$m_{PQ} = \dfrac{f(a + h) - f(a)}{h}$
= average velocity

FIGURE 6

which is the same as the slope of the secant line PQ in Figure 6.

Now suppose we compute the average velocities over shorter and shorter time intervals $[a, a + h]$. In other words, we let h approach 0. As in the example of the falling ball, we define the **velocity** (or **instantaneous velocity**) $v(a)$ at time $t = a$ to be the limit of these average velocities:

3

$$v(a) = \lim_{h \to 0} \frac{f(a + h) - f(a)}{h}$$

This means that the velocity at time $t = a$ is equal to the slope of the tangent line at P (compare Equations 2 and 3).

Now that we know how to compute limits, let's reconsider the problem of the falling ball.

EXAMPLE 3 Suppose that a ball is dropped from the upper observation deck of the CN Tower, 450 m above the ground.
(a) What is the velocity of the ball after 5 seconds?
(b) How fast is the ball traveling when it hits the ground?

■ Recall from Section 2.1: The distance (in meters) fallen after t seconds is $4.9t^2$.

SOLUTION We will need to find the velocity both when $t = 5$ and when the ball hits the ground, so it's efficient to start by finding the velocity at a general time $t = a$. Using the equation of motion $s = f(t) = 4.9t^2$, we have

$$v(a) = \lim_{h \to 0} \frac{f(a + h) - f(a)}{h} = \lim_{h \to 0} \frac{4.9(a + h)^2 - 4.9a^2}{h}$$

$$= \lim_{h \to 0} \frac{4.9(a^2 + 2ah + h^2 - a^2)}{h} = \lim_{h \to 0} \frac{4.9(2ah + h^2)}{h}$$

$$= \lim_{h \to 0} 4.9(2a + h) = 9.8a$$

(a) The velocity after 5 s is $v(5) = (9.8)(5) = 49$ m/s.

(b) Since the observation deck is 450 m above the ground, the ball will hit the ground at the time t_1 when $s(t_1) = 450$, that is,

$$4.9t_1^2 = 450$$

This gives

$$t_1^2 = \frac{450}{4.9} \qquad \text{and} \qquad t_1 = \sqrt{\frac{450}{4.9}} \approx 9.6 \text{ s}$$

The velocity of the ball as it hits the ground is therefore

$$v(t_1) = 9.8t_1 = 9.8\sqrt{\frac{450}{4.9}} \approx 94 \text{ m/s}$$ □

DERIVATIVES

We have seen that the same type of limit arises in finding the slope of a tangent line (Equation 2) or the velocity of an object (Equation 3). In fact, limits of the form

$$\lim_{h \to 0} \frac{f(a + h) - f(a)}{h}$$

arise whenever we calculate a rate of change in any of the sciences or engineering, such as a rate of reaction in chemistry or a marginal cost in economics. Since this type of limit occurs so widely, it is given a special name and notation.

4 DEFINITION The **derivative of a function f at a number a**, denoted by $f'(a)$, is

$$f'(a) = \lim_{h \to 0} \frac{f(a + h) - f(a)}{h}$$

if this limit exists.

■ $f'(a)$ is read "f prime of a."

If we write $x = a + h$, then we have $h = x - a$ and h approaches 0 if and only if x approaches a. Therefore an equivalent way of stating the definition of the derivative, as we saw in finding tangent lines, is

5
$$f'(a) = \lim_{x \to a} \frac{f(x) - f(a)}{x - a}$$

V EXAMPLE 4 Find the derivative of the function $f(x) = x^2 - 8x + 9$ at the number a.

SOLUTION From Definition 4 we have

$$\begin{aligned} f'(a) &= \lim_{h \to 0} \frac{f(a + h) - f(a)}{h} \\ &= \lim_{h \to 0} \frac{[(a + h)^2 - 8(a + h) + 9] - [a^2 - 8a + 9]}{h} \\ &= \lim_{h \to 0} \frac{a^2 + 2ah + h^2 - 8a - 8h + 9 - a^2 + 8a - 9}{h} \\ &= \lim_{h \to 0} \frac{2ah + h^2 - 8h}{h} = \lim_{h \to 0} (2a + h - 8) \\ &= 2a - 8 \end{aligned}$$

□

We defined the tangent line to the curve $y = f(x)$ at the point $P(a, f(a))$ to be the line that passes through P and has slope m given by Equation 1 or 2. Since, by Definition 4, this is the same as the derivative $f'(a)$, we can now say the following.

The tangent line to $y = f(x)$ at $(a, f(a))$ is the line through $(a, f(a))$ whose slope is equal to $f'(a)$, the derivative of f at a.

If we use the point-slope form of the equation of a line, we can write an equation of the tangent line to the curve $y = f(x)$ at the point $(a, f(a))$:

$$y - f(a) = f'(a)(x - a)$$

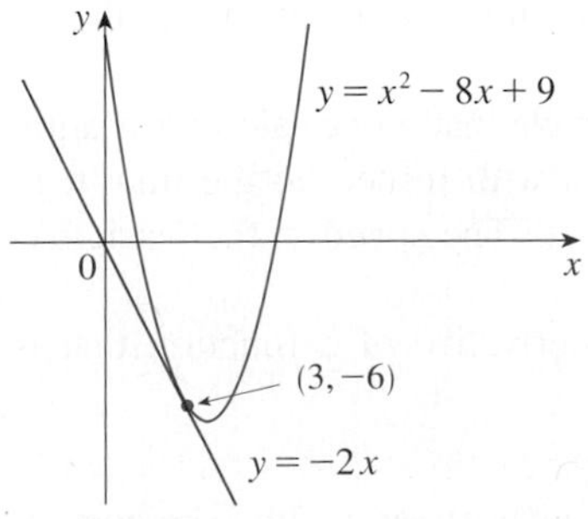

FIGURE 7

V EXAMPLE 5 Find an equation of the tangent line to the parabola $y = x^2 - 8x + 9$ at the point $(3, -6)$.

SOLUTION From Example 4 we know that the derivative of $f(x) = x^2 - 8x + 9$ at the number a is $f'(a) = 2a - 8$. Therefore the slope of the tangent line at $(3, -6)$ is $f'(3) = 2(3) - 8 = -2$. Thus an equation of the tangent line, shown in Figure 7, is

$$y - (-6) = (-2)(x - 3) \qquad \text{or} \qquad y = -2x$$

□

RATES OF CHANGE

Suppose y is a quantity that depends on another quantity x. Thus y is a function of x and we write $y = f(x)$. If x changes from x_1 to x_2, then the change in x (also called the **increment** of x) is

$$\Delta x = x_2 - x_1$$

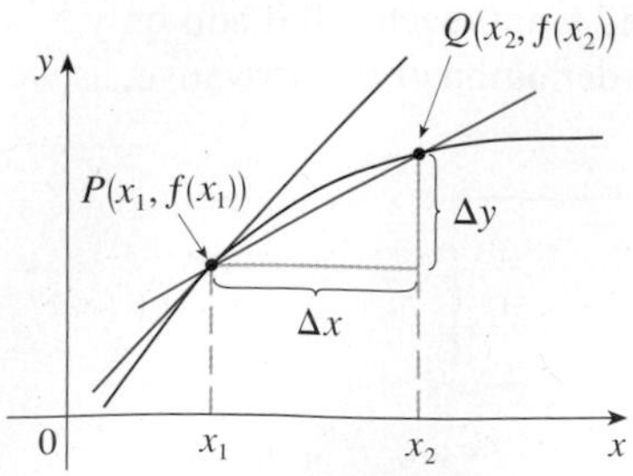

average rate of change = m_{PQ}

instantaneous rate of change = slope of tangent at P

FIGURE 8

and the corresponding change in y is

$$\Delta y = f(x_2) - f(x_1)$$

The difference quotient

$$\frac{\Delta y}{\Delta x} = \frac{f(x_2) - f(x_1)}{x_2 - x_1}$$

is called the **average rate of change of y with respect to x** over the interval $[x_1, x_2]$ and can be interpreted as the slope of the secant line PQ in Figure 8.

By analogy with velocity, we consider the average rate of change over smaller and smaller intervals by letting x_2 approach x_1 and therefore letting Δx approach 0. The limit of these average rates of change is called the (**instantaneous**) **rate of change of y with respect to x** at $x = x_1$, which is interpreted as the slope of the tangent to the curve $y = f(x)$ at $P(x_1, f(x_1))$:

$$\boxed{6} \qquad \text{instantaneous rate of change} = \lim_{\Delta x \to 0} \frac{\Delta y}{\Delta x} = \lim_{x_2 \to x_1} \frac{f(x_2) - f(x_1)}{x_2 - x_1}$$

We recognize this limit as being the derivative $f'(x_1)$.

We know that one interpretation of the derivative $f'(a)$ is as the slope of the tangent line to the curve $y = f(x)$ when $x = a$. We now have a second interpretation:

> The derivative $f'(a)$ is the instantaneous rate of change of $y = f(x)$ with respect to x when $x = a$.

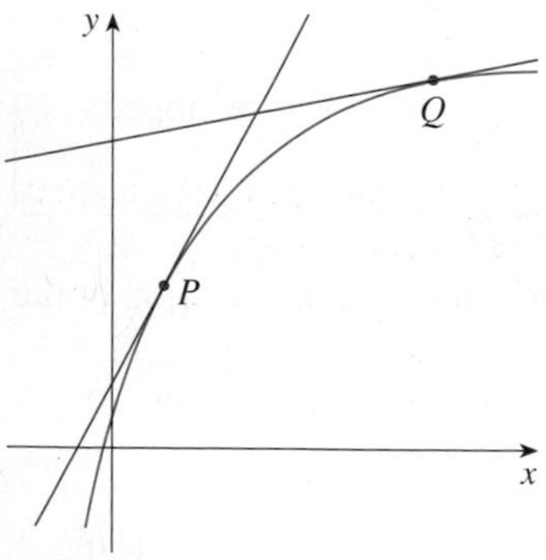

FIGURE 9
The y-values are changing rapidly at P and slowly at Q.

The connection with the first interpretation is that if we sketch the curve $y = f(x)$, then the instantaneous rate of change is the slope of the tangent to this curve at the point where $x = a$. This means that when the derivative is large (and therefore the curve is steep, as at the point P in Figure 9), the y-values change rapidly. When the derivative is small, the curve is relatively flat and the y-values change slowly.

In particular, if $s = f(t)$ is the position function of a particle that moves along a straight line, then $f'(a)$ is the rate of change of the displacement s with respect to the time t. In other words, *$f'(a)$ is the velocity of the particle at time $t = a$*. The **speed** of the particle is the absolute value of the velocity, that is, $|f'(a)|$.

In the next example we discuss the meaning of the derivative of a function that is defined verbally.

V EXAMPLE 6 A manufacturer produces bolts of a fabric with a fixed width. The cost of producing x yards of this fabric is $C = f(x)$ dollars.
(a) What is the meaning of the derivative $f'(x)$? What are its units?
(b) In practical terms, what does it mean to say that $f'(1000) = 9$?
(c) Which do you think is greater, $f'(50)$ or $f'(500)$? What about $f'(5000)$?

SOLUTION
(a) The derivative $f'(x)$ is the instantaneous rate of change of C with respect to x; that is, $f'(x)$ means the rate of change of the production cost with respect to the number of yards produced. (Economists call this rate of change the *marginal cost*. This idea is discussed in more detail in Sections 3.7 and 4.7.)

Because

$$f'(x) = \lim_{\Delta x \to 0} \frac{\Delta C}{\Delta x}$$

the units for $f'(x)$ are the same as the units for the difference quotient $\Delta C/\Delta x$. Since ΔC is measured in dollars and Δx in yards, it follows that the units for $f'(x)$ are dollars per yard.

(b) The statement that $f'(1000) = 9$ means that, after 1000 yards of fabric have been manufactured, the rate at which the production cost is increasing is \$9/yard. (When $x = 1000$, C is increasing 9 times as fast as x.)

Since $\Delta x = 1$ is small compared with $x = 1000$, we could use the approximation

■ Here we are assuming that the cost function is well behaved; in other words, $C(x)$ doesn't oscillate rapidly near $x = 1000$.

$$f'(1000) \approx \frac{\Delta C}{\Delta x} = \frac{\Delta C}{1} = \Delta C$$

and say that the cost of manufacturing the 1000th yard (or the 1001st) is about \$9.

(c) The rate at which the production cost is increasing (per yard) is probably lower when $x = 500$ than when $x = 50$ (the cost of making the 500th yard is less than the cost of the 50th yard) because of economies of scale. (The manufacturer makes more efficient use of the fixed costs of production.) So

$$f'(50) > f'(500)$$

But, as production expands, the resulting large-scale operation might become inefficient and there might be overtime costs. Thus it is possible that the rate of increase of costs will eventually start to rise. So it may happen that

$$f'(5000) > f'(500)$$ □

In the following example we estimate the rate of change of the national debt with respect to time. Here the function is defined not by a formula but by a table of values.

t	$D(t)$
1980	930.2
1985	1945.9
1990	3233.3
1995	4974.0
2000	5674.2

V EXAMPLE 7 Let $D(t)$ be the US national debt at time t. The table in the margin gives approximate values of this function by providing end of year estimates, in billions of dollars, from 1980 to 2000. Interpret and estimate the value of $D'(1990)$.

SOLUTION The derivative $D'(1990)$ means the rate of change of D with respect to t when $t = 1990$, that is, the rate of increase of the national debt in 1990.

According to Equation 5,

$$D'(1990) = \lim_{t \to 1990} \frac{D(t) - D(1990)}{t - 1990}$$

So we compute and tabulate values of the difference quotient (the average rates of change) as follows.

t	$\dfrac{D(t) - D(1990)}{t - 1990}$
1980	230.31
1985	257.48
1995	348.14
2000	244.09

A NOTE ON UNITS

The units for the average rate of change $\Delta D/\Delta t$ are the units for ΔD divided by the units for Δt, namely, billions of dollars per year. The instantaneous rate of change is the limit of the average rates of change, so it is measured in the same units: billions of dollars per year.

From this table we see that $D'(1990)$ lies somewhere between 257.48 and 348.14 billion dollars per year. [Here we are making the reasonable assumption that the debt didn't fluctuate wildly between 1980 and 2000.] We estimate that the rate of increase of the national debt of the United States in 1990 was the average of these two numbers, namely

$$D'(1990) \approx 303 \text{ billion dollars per year}$$

Another method would be to plot the debt function and estimate the slope of the tangent line when $t = 1990$. □

In Examples 3, 6, and 7 we saw three specific examples of rates of change: the velocity of an object is the rate of change of displacement with respect to time; marginal cost is the rate of change of production cost with respect to the number of items produced; the rate of change of the debt with respect to time is of interest in economics. Here is a small sample of other rates of change: In physics, the rate of change of work with respect to time is called *power*. Chemists who study a chemical reaction are interested in the rate of change in the concentration of a reactant with respect to time (called the *rate of reaction*). A biologist is interested in the rate of change of the population of a colony of bacteria with respect to time. In fact, the computation of rates of change is important in all of the natural sciences, in engineering, and even in the social sciences. Further examples will be given in Section 3.7.

All these rates of change are derivatives and can therefore be interpreted as slopes of tangents. This gives added significance to the solution of the tangent problem. Whenever we solve a problem involving tangent lines, we are not just solving a problem in geometry. We are also implicitly solving a great variety of problems involving rates of change in science and engineering.

2.7 EXERCISES

1. A curve has equation $y = f(x)$.
(a) Write an expression for the slope of the secant line through the points $P(3, f(3))$ and $Q(x, f(x))$.
(b) Write an expression for the slope of the tangent line at P.

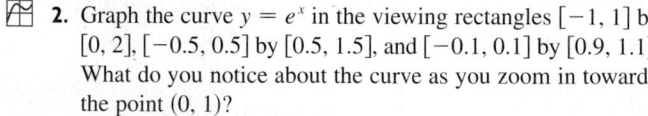

2. Graph the curve $y = e^x$ in the viewing rectangles $[-1, 1]$ by $[0, 2]$, $[-0.5, 0.5]$ by $[0.5, 1.5]$, and $[-0.1, 0.1]$ by $[0.9, 1.1]$. What do you notice about the curve as you zoom in toward the point $(0, 1)$?

3. (a) Find the slope of the tangent line to the parabola $y = 4x - x^2$ at the point $(1, 3)$
(i) using Definition 1 (ii) using Equation 2
(b) Find an equation of the tangent line in part (a).
(c) Graph the parabola and the tangent line. As a check on your work, zoom in toward the point $(1, 3)$ until the parabola and the tangent line are indistinguishable.

4. (a) Find the slope of the tangent line to the curve $y = x - x^3$ at the point $(1, 0)$
(i) using Definition 1 (ii) using Equation 2
(b) Find an equation of the tangent line in part (a).
(c) Graph the curve and the tangent line in successively smaller viewing rectangles centered at $(1, 0)$ until the curve and the line appear to coincide.

5–8 Find an equation of the tangent line to the curve at the given point.

5. $y = \dfrac{x-1}{x-2}$, $(3, 2)$

6. $y = 2x^3 - 5x$, $(-1, 3)$

7. $y = \sqrt{x}$, $(1, 1)$

8. $y = \dfrac{2x}{(x+1)^2}$, $(0, 0)$

9. (a) Find the slope of the tangent to the curve $y = 3 + 4x^2 - 2x^3$ at the point where $x = a$.
(b) Find equations of the tangent lines at the points $(1, 5)$ and $(2, 3)$.
(c) Graph the curve and both tangents on a common screen.

10. (a) Find the slope of the tangent to the curve $y = 1/\sqrt{x}$ at the point where $x = a$.
(b) Find equations of the tangent lines at the points $(1, 1)$ and $(4, \frac{1}{2})$.
(c) Graph the curve and both tangents on a common screen.

11. (a) A particle starts by moving to the right along a horizontal line; the graph of its position function is shown. When is the particle moving to the right? Moving to the left? Standing still?

(b) Draw a graph of the velocity function.

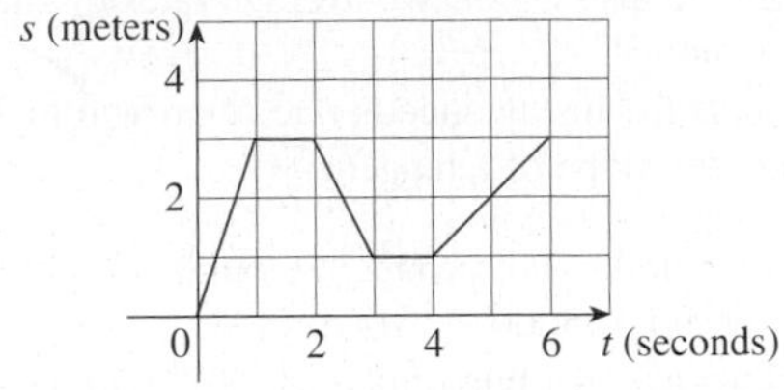

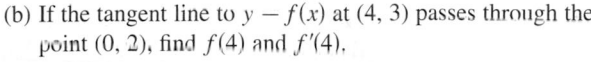

12. Shown are graphs of the position functions of two runners, A and B, who run a 100-m race and finish in a tie.

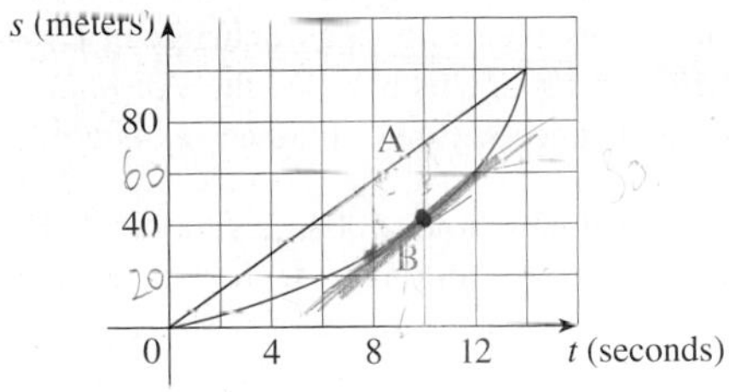

(a) Describe and compare how the runners run the race.
(b) At what time is the distance between the runners the greatest?
(c) At what time do they have the same velocity?

13. If a ball is thrown into the air with a velocity of 40 ft/s, its height (in feet) after t seconds is given by $y = 40t - 16t^2$. Find the velocity when $t = 2$.

14. If a rock is thrown upward on the planet Mars with a velocity of 10 m/s, its height (in meters) after t seconds is given by $H = 10t - 1.86t^2$.
(a) Find the velocity of the rock after one second.
(b) Find the velocity of the rock when $t = a$.
(c) When will the rock hit the surface?
(d) With what velocity will the rock hit the surface?

15. The displacement (in meters) of a particle moving in a straight line is given by the equation of motion $s = 1/t^2$, where t is measured in seconds. Find the velocity of the particle at times $t = a$, $t = 1$, $t = 2$, and $t = 3$.

16. The displacement (in meters) of a particle moving in a straight line is given by $s = t^2 - 8t + 18$, where t is measured in seconds.
(a) Find the average velocity over each time interval:
(i) $[3, 4]$ (ii) $[3.5, 4]$
(iii) $[4, 5]$ (iv) $[4, 4.5]$
(b) Find the instantaneous velocity when $t = 4$.
(c) Draw the graph of s as a function of t and draw the secant lines whose slopes are the average velocities in part (a) and the tangent line whose slope is the instantaneous velocity in part (b).

17. For the function g whose graph is given, arrange the following numbers in increasing order and explain your reasoning:

$$0 \qquad g'(-2) \qquad g'(0) \qquad g'(2) \qquad g'(4)$$

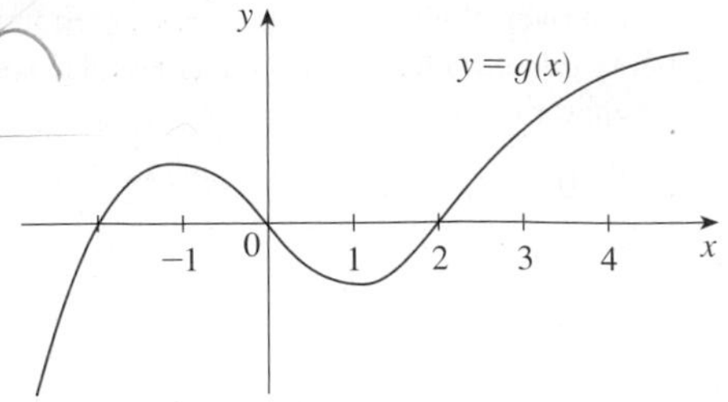

18. (a) Find an equation of the tangent line to the graph of $y = g(x)$ at $x = 5$ if $g(5) = -3$ and $g'(5) = 4$.
(b) If the tangent line to $y = f(x)$ at (4, 3) passes through the point (0, 2), find $f(4)$ and $f'(4)$.

19. Sketch the graph of a function f for which $f(0) = 0$, $f'(0) = 3$, $f'(1) = 0$, and $f'(2) = -1$.

20. Sketch the graph of a function g for which $g(0) = g'(0) = 0$, $g'(-1) = -1$, $g'(1) = 3$, and $g'(2) = 1$.

21. If $f(x) = 3x^2 - 5x$, find $f'(2)$ and use it to find an equation of the tangent line to the parabola $y = 3x^2 - 5x$ at the point (2, 2).

22. If $g(x) = 1 - x^3$, find $g'(0)$ and use it to find an equation of the tangent line to the curve $y = 1 - x^3$ at the point (0, 1).

23. (a) If $F(x) = 5x/(1 + x^2)$, find $F'(2)$ and use it to find an equation of the tangent line to the curve $y = 5x/(1 + x^2)$ at the point (2, 2).
(b) Illustrate part (a) by graphing the curve and the tangent line on the same screen.

24. (a) If $G(x) = 4x^2 - x^3$, find $G'(a)$ and use it to find equations of the tangent lines to the curve $y = 4x^2 - x^3$ at the points (2, 8) and (3, 9).
(b) Illustrate part (a) by graphing the curve and the tangent lines on the same screen.

25–30 Find $f'(a)$.

25. $f(x) = 3 - 2x + 4x^2$

26. $f(t) = t^4 - 5t$

27. $f(t) = \dfrac{2t + 1}{t + 3}$

28. $f(x) = \dfrac{x^2 + 1}{x - 2}$

29. $f(x) = \dfrac{1}{\sqrt{x + 2}}$

30. $f(x) = \sqrt{3x + 1}$

31–36 Each limit represents the derivative of some function f at some number a. State such an f and a in each case.

31. $\displaystyle\lim_{h \to 0} \frac{(1 + h)^{10} - 1}{h}$

32. $\displaystyle\lim_{h \to 0} \frac{\sqrt[4]{16 + h} - 2}{h}$

33. $\displaystyle\lim_{x \to 5} \frac{2^x - 32}{x - 5}$

34. $\displaystyle\lim_{x \to \pi/4} \frac{\tan x - 1}{x - \pi/4}$

35. $\displaystyle\lim_{h \to 0} \frac{\cos(\pi + h) + 1}{h}$

36. $\displaystyle\lim_{t \to 1} \frac{t^4 + t - 2}{t - 1}$

37–38 A particle moves along a straight line with equation of motion $s = f(t)$, where s is measured in meters and t in seconds. Find the velocity and the speed when $t = 5$.

37. $f(t) = 100 + 50t - 4.9t^2$ **38.** $f(t) = t^{-1} - t$

39. A warm can of soda is placed in a cold refrigerator. Sketch the graph of the temperature of the soda as a function of time. Is the initial rate of change of temperature greater or less than the rate of change after an hour?

40. A roast turkey is taken from an oven when its temperature has reached 185°F and is placed on a table in a room where the temperature is 75°F. The graph shows how the temperature of the turkey decreases and eventually approaches room temperature. By measuring the slope of the tangent, estimate the rate of change of the temperature after an hour.

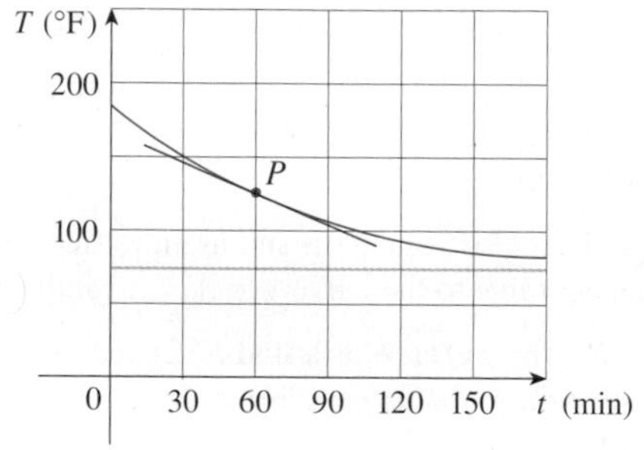

41. The table shows the estimated percentage P of the population of Europe that use cell phones. (Midyear estimates are given.)

Year	1998	1999	2000	2001	2002	2003
P	28	39	55	68	77	83

(a) Find the average rate of cell phone growth
 (i) from 2000 to 2002 (ii) from 2000 to 2001
 (iii) from 1999 to 2000
 In each case, include the units.
(b) Estimate the instantaneous rate of growth in 2000 by taking the average of two average rates of change. What are its units?
(c) Estimate the instantaneous rate of growth in 2000 by measuring the slope of a tangent.

42. The number N of locations of a popular coffeehouse chain is given in the table. (The numbers of locations as of June 30 are given.)

Year	1998	1999	2000	2001	2002
N	1886	2135	3501	4709	5886

(a) Find the average rate of growth
 (i) from 2000 to 2002 (ii) from 2000 to 2001
 (iii) from 1999 to 2000
 In each case, include the units.
(b) Estimate the instantaneous rate of growth in 2000 by taking the average of two average rates of change. What are its units?
(c) Estimate the instantaneous rate of growth in 2000 by measuring the slope of a tangent.

43. The cost (in dollars) of producing x units of a certain commodity is $C(x) = 5000 + 10x + 0.05x^2$.
(a) Find the average rate of change of C with respect to x when the production level is changed
 (i) from $x = 100$ to $x = 105$
 (ii) from $x = 100$ to $x = 101$
(b) Find the instantaneous rate of change of C with respect to x when $x = 100$. (This is called the *marginal cost.* Its significance will be explained in Section 3.7.)

44. If a cylindrical tank holds 100,000 gallons of water, which can be drained from the bottom of the tank in an hour, then Torricelli's Law gives the volume V of water remaining in the tank after t minutes as

$$V(t) = 100{,}000\left(1 - \frac{t}{60}\right)^2 \qquad 0 \leqslant t \leqslant 60$$

Find the rate at which the water is flowing out of the tank (the instantaneous rate of change of V with respect to t) as a function of t. What are its units? For times $t = 0, 10, 20, 30, 40, 50$, and 60 min, find the flow rate and the amount of water remaining in the tank. Summarize your findings in a sentence or two. At what time is the flow rate the greatest? The least?

45. The cost of producing x ounces of gold from a new gold mine is $C = f(x)$ dollars.
(a) What is the meaning of the derivative $f'(x)$? What are its units?
(b) What does the statement $f'(800) = 17$ mean?
(c) Do you think the values of $f'(x)$ will increase or decrease in the short term? What about the long term? Explain.

46. The number of bacteria after t hours in a controlled laboratory experiment is $n = f(t)$.
(a) What is the meaning of the derivative $f'(5)$? What are its units?
(b) Suppose there is an unlimited amount of space and nutrients for the bacteria. Which do you think is larger, $f'(5)$ or $f'(10)$? If the supply of nutrients is limited, would that affect your conclusion? Explain.

47. Let $T(t)$ be the temperature (in °F) in Dallas t hours after midnight on June 2, 2001. The table shows values of this function recorded every two hours. What is the meaning of $T'(10)$? Estimate its value.

t	0	2	4	6	8	10	12	14
T	73	73	70	69	72	81	88	91

48. The quantity (in pounds) of a gourmet ground coffee that is sold by a coffee company at a price of p dollars per pound is $Q = f(p)$.
(a) What is the meaning of the derivative $f'(8)$? What are its units?
(b) Is $f'(8)$ positive or negative? Explain.

49. The quantity of oxygen that can dissolve in water depends on the temperature of the water. (So thermal pollution influences the oxygen content of water.) The graph shows how oxygen solubility S varies as a function of the water temperature T.
(a) What is the meaning of the derivative $S'(T)$? What are its units?
(b) Estimate the value of $S'(16)$ and interpret it.

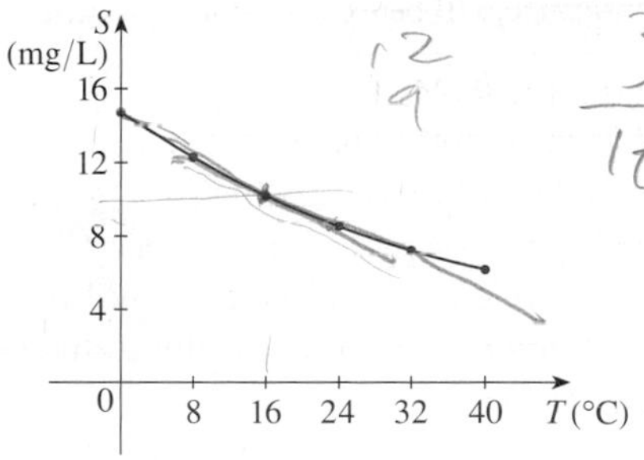

Adapted from *Environmental Science: Living Within the System of Nature*, 2d ed.; by Charles E. Kupchella, © 1989. Reprinted by permission of Prentice-Hall, Inc., Upper Saddle River, NJ.

50. The graph shows the influence of the temperature T on the maximum sustainable swimming speed S of Coho salmon.
(a) What is the meaning of the derivative $S'(T)$? What are its units?
(b) Estimate the values of $S'(15)$ and $S'(25)$ and interpret them.

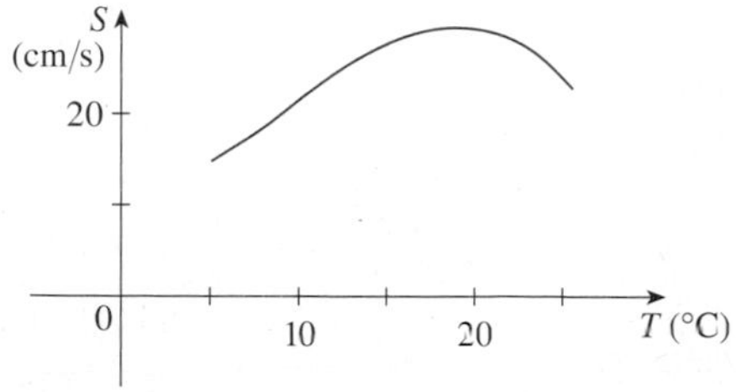

51–52 Determine whether $f'(0)$ exists.

51. $f(x) = \begin{cases} x \sin \dfrac{1}{x} & \text{if } x \neq 0 \\ 0 & \text{if } x = 0 \end{cases}$

52. $f(x) = \begin{cases} x^2 \sin \dfrac{1}{x} & \text{if } x \neq 0 \\ 0 & \text{if } x = 0 \end{cases}$

WRITING PROJECT

EARLY METHODS FOR FINDING TANGENTS

The first person to formulate explicitly the ideas of limits and derivatives was Sir Isaac Newton in the 1660s. But Newton acknowledged that "If I have seen further than other men, it is because I have stood on the shoulders of giants." Two of those giants were Pierre Fermat (1601–1665) and Newton's teacher at Cambridge, Isaac Barrow (1630–1677). Newton was familiar with the methods that these men used to find tangent lines, and their methods played a role in Newton's eventual formulation of calculus.

The following references contain explanations of these methods. Read one or more of the references and write a report comparing the methods of either Fermat or Barrow to modern methods. In particular, use the method of Section 2.7 to find an equation of the tangent line to the curve $y = x^3 + 2x$ at the point $(1, 3)$ and show how either Fermat or Barrow would have solved the same problem. Although you used derivatives and they did not, point out similarities between the methods.

1. Carl Boyer and Uta Merzbach, *A History of Mathematics* (New York: Wiley, 1989), pp. 389, 432.
2. C. H. Edwards, *The Historical Development of the Calculus* (New York: Springer-Verlag, 1979), pp. 124, 132.
3. Howard Eves, *An Introduction to the History of Mathematics*, 6th ed. (New York: Saunders, 1990), pp. 391, 395.
4. Morris Kline, *Mathematical Thought from Ancient to Modern Times* (New York: Oxford University Press, 1972), pp. 344, 346.

2.8 THE DERIVATIVE AS A FUNCTION

In the preceding section we considered the derivative of a function f at a fixed number a:

$$f'(a) = \lim_{h \to 0} \frac{f(a + h) - f(a)}{h} \tag{1}$$

Here we change our point of view and let the number a vary. If we replace a in Equation 1 by a variable x, we obtain

$$f'(x) = \lim_{h \to 0} \frac{f(x + h) - f(x)}{h} \tag{2}$$

Given any number x for which this limit exists, we assign to x the number $f'(x)$. So we can regard f' as a new function, called the **derivative of f** and defined by Equation 2. We know that the value of f' at x, $f'(x)$, can be interpreted geometrically as the slope of the tangent line to the graph of f at the point $(x, f(x))$.

The function f' is called the derivative of f because it has been "derived" from f by the limiting operation in Equation 2. The domain of f' is the set $\{x \mid f'(x) \text{ exists}\}$ and may be smaller than the domain of f.

V EXAMPLE 1 The graph of a function f is given in Figure 1. Use it to sketch the graph of the derivative f'.

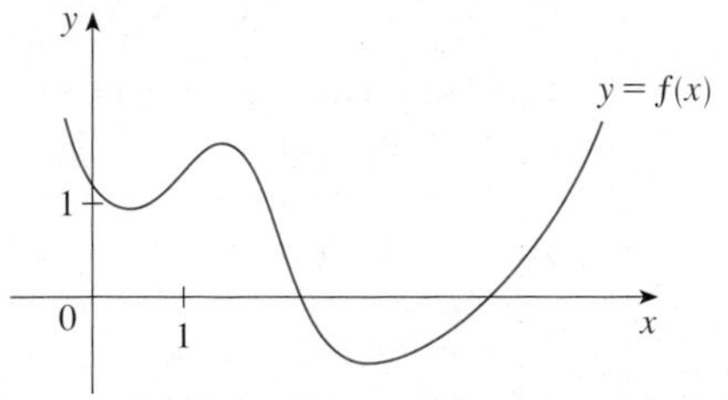

FIGURE 1

SOLUTION We can estimate the value of the derivative at any value of x by drawing the tangent at the point $(x, f(x))$ and estimating its slope. For instance, for $x = 5$ we draw the tangent at P in Figure 2(a) and estimate its slope to be about $\frac{3}{2}$, so $f'(5) \approx 1.5$. This allows us to plot the point $P'(5, 1.5)$ on the graph of f' directly beneath P. Repeating this procedure at several points, we get the graph shown in Figure 2(b). Notice that the tangents at A, B, and C are horizontal, so the derivative is 0 there and the graph of f' crosses the x-axis at the points A', B', and C', directly beneath A, B, and C. Between A and B the tangents have positive slope, so $f'(x)$ is positive there. But between B and C the tangents have negative slope, so $f'(x)$ is negative there.

TEC Visual 2.8 shows an animation of Figure 2 for several functions.

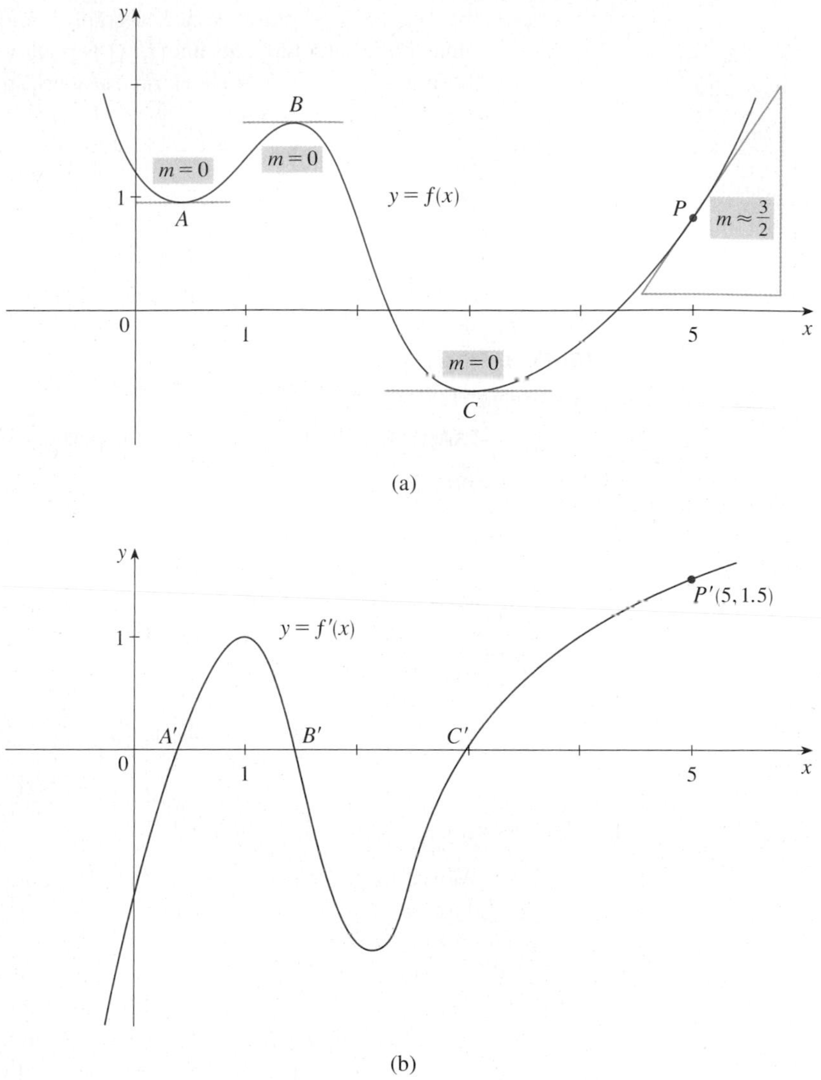

FIGURE 2

□

V EXAMPLE 2

(a) If $f(x) = x^3 - x$, find a formula for $f'(x)$.

(b) Illustrate by comparing the graphs of f and f'.

SOLUTION

(a) When using Equation 2 to compute a derivative, we must remember that the variable is h and that x is temporarily regarded as a constant during the calculation of the limit.

$$f'(x) = \lim_{h \to 0} \frac{f(x+h) - f(x)}{h} = \lim_{h \to 0} \frac{[(x+h)^3 - (x+h)] - [x^3 - x]}{h}$$

$$= \lim_{h \to 0} \frac{x^3 + 3x^2h + 3xh^2 + h^3 - x - h - x^3 + x}{h}$$

$$= \lim_{h \to 0} \frac{3x^2h + 3xh^2 + h^3 - h}{h} = \lim_{h \to 0} (3x^2 + 3xh + h^2 - 1) = 3x^2 - 1$$

(b) We use a graphing device to graph f and f' in Figure 3. Notice that $f'(x) = 0$ when f has horizontal tangents and $f'(x)$ is positive when the tangents have positive slope. So these graphs serve as a check on our work in part (a).

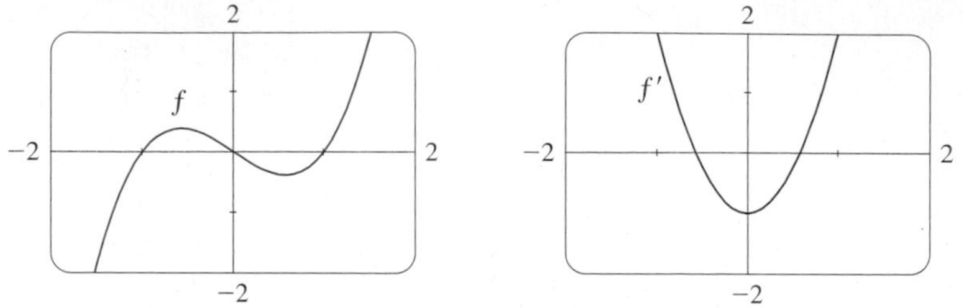

FIGURE 3

EXAMPLE 3 If $f(x) = \sqrt{x}$, find the derivative of f. State the domain of f'.

SOLUTION

$$f'(x) = \lim_{h \to 0} \frac{f(x+h) - f(x)}{h} = \lim_{h \to 0} \frac{\sqrt{x+h} - \sqrt{x}}{h}$$

$$= \lim_{h \to 0} \left(\frac{\sqrt{x+h} - \sqrt{x}}{h} \cdot \frac{\sqrt{x+h} + \sqrt{x}}{\sqrt{x+h} + \sqrt{x}} \right)$$

Here we rationalize the numerator.

$$= \lim_{h \to 0} \frac{(x+h) - x}{h\left(\sqrt{x+h} + \sqrt{x}\right)} = \lim_{h \to 0} \frac{1}{\sqrt{x+h} + \sqrt{x}}$$

$$= \frac{1}{\sqrt{x} + \sqrt{x}} = \frac{1}{2\sqrt{x}}$$

We see that $f'(x)$ exists if $x > 0$, so the domain of f' is $(0, \infty)$. This is smaller than the domain of f, which is $[0, \infty)$.

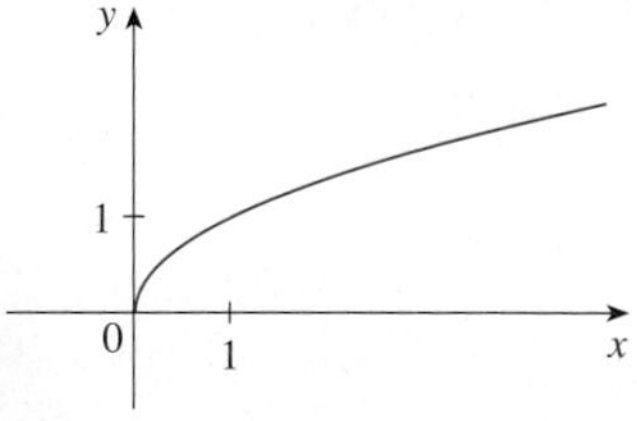

(a) $f(x) = \sqrt{x}$

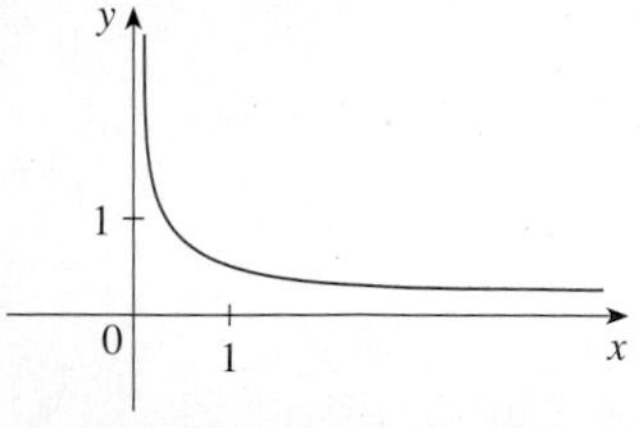

(b) $f'(x) = \dfrac{1}{2\sqrt{x}}$

FIGURE 4

Let's check to see that the result of Example 3 is reasonable by looking at the graphs of f and f' in Figure 4. When x is close to 0, $\sqrt{x}$ is also close to 0, so $f'(x) = 1/\left(2\sqrt{x}\right)$ is very large and this corresponds to the steep tangent lines near $(0, 0)$ in Figure 4(a) and the large values of $f'(x)$ just to the right of 0 in Figure 4(b). When x is large, $f'(x)$ is very small and this corresponds to the flatter tangent lines at the far right of the graph of f and the horizontal asymptote of the graph of f'.

EXAMPLE 4 Find f' if $f(x) = \dfrac{1 - x}{2 + x}$.

SOLUTION

$$f'(x) = \lim_{h \to 0} \frac{f(x+h) - f(x)}{h} = \lim_{h \to 0} \frac{\dfrac{1 - (x+h)}{2 + (x+h)} - \dfrac{1 - x}{2 + x}}{h}$$

$$\frac{\dfrac{a}{b} - \dfrac{c}{d}}{e} = \frac{ad - bc}{bd} \cdot \frac{1}{e}$$

$$= \lim_{h \to 0} \frac{(1 - x - h)(2 + x) - (1 - x)(2 + x + h)}{h(2 + x + h)(2 + x)}$$

$$= \lim_{h \to 0} \frac{(2 - x - 2h - x^2 - xh) - (2 - x + h - x^2 - xh)}{h(2 + x + h)(2 + x)}$$

$$= \lim_{h \to 0} \frac{-3h}{h(2 + x + h)(2 + x)} = \lim_{h \to 0} \frac{-3}{(2 + x + h)(2 + x)} = -\frac{3}{(2 + x)^2}$$

OTHER NOTATIONS

If we use the traditional notation $y = f(x)$ to indicate that the independent variable is x and the dependent variable is y, then some common alternative notations for the derivative are as follows:

$$f'(x) = y' = \frac{dy}{dx} = \frac{df}{dx} = \frac{d}{dx}f(x) = Df(x) = D_x f(x)$$

The symbols D and d/dx are called **differentiation operators** because they indicate the operation of **differentiation**, which is the process of calculating a derivative.

The symbol dy/dx, which was introduced by Leibniz, should not be regarded as a ratio (for the time being); it is simply a synonym for $f'(x)$. Nonetheless, it is a very useful and suggestive notation, especially when used in conjunction with increment notation. Referring to Equation 2.7.6, we can rewrite the definition of derivative in Leibniz notation in the form

$$\frac{dy}{dx} = \lim_{\Delta x \to 0} \frac{\Delta y}{\Delta x}$$

If we want to indicate the value of a derivative dy/dx in Leibniz notation at a specific number a, we use the notation

$$\left.\frac{dy}{dx}\right|_{x=a} \qquad \text{or} \qquad \left.\frac{dy}{dx}\right]_{x=a}$$

which is a synonym for $f'(a)$.

LEIBNIZ

Gottfried Wilhelm Leibniz was born in Leipzig in 1646 and studied law, theology, philosophy and mathematics at the university there, graduating with a bachelor's degree at age 17. After earning his doctorate in law at age 20, Leibniz entered the diplomatic service and spent most of his life traveling to the capitals of Europe on political missions. In particular, he worked to avert a French military threat against Germany and attempted to reconcile the Catholic and Protestant churches.

His serious study of mathematics did not begin until 1672 while he was on a diplomatic mission in Paris. There he built a calculating machine and met scientists, like Huygens, who directed his attention to the latest developments in mathematics and science. Leibniz sought to develop a symbolic logic and system of notation that would simplify logical reasoning. In particular, the version of calculus that he published in 1684 established the notation and the rules for finding derivatives that we use today.

Unfortunately, a dreadful priority dispute arose in the 1690s between the followers of Newton and those of Leibniz as to who had invented calculus first. Leibniz was even accused of plagiarism by members of the Royal Society in England. The truth is that each man invented calculus independently. Newton arrived at his version of calculus first but, because of his fear of controversy, did not publish it immediately. So Leibniz's 1684 account of calculus was the first to be published.

3 DEFINITION A function f is **differentiable at *a*** if $f'(a)$ exists. It is **differentiable on an open interval** (a, b) [or (a, ∞) or $(-\infty, a)$ or $(-\infty, \infty)$] if it is differentiable at every number in the interval.

EXAMPLE 5 Where is the function $f(x) = |x|$ differentiable?

SOLUTION If $x > 0$, then $|x| = x$ and we can choose h small enough that $x + h > 0$ and hence $|x + h| = x + h$. Therefore, for $x > 0$, we have

$$f'(x) = \lim_{h \to 0} \frac{|x + h| - |x|}{h}$$

$$= \lim_{h \to 0} \frac{(x + h) - x}{h} = \lim_{h \to 0} \frac{h}{h} = \lim_{h \to 0} 1 = 1$$

and so f is differentiable for any $x > 0$.

Similarly, for $x < 0$ we have $|x| = -x$ and h can be chosen small enough that $x + h < 0$ and so $|x + h| = -(x + h)$. Therefore, for $x < 0$,

$$f'(x) = \lim_{h \to 0} \frac{|x + h| - |x|}{h}$$

$$= \lim_{h \to 0} \frac{-(x + h) - (-x)}{h} = \lim_{h \to 0} \frac{-h}{h} = \lim_{h \to 0} (-1) = -1$$

and so f is differentiable for any $x < 0$.

For $x = 0$ we have to investigate

$$f'(0) = \lim_{h \to 0} \frac{f(0 + h) - f(0)}{h}$$

$$= \lim_{h \to 0} \frac{|0 + h| - |0|}{h} \qquad \text{(if it exists)}$$

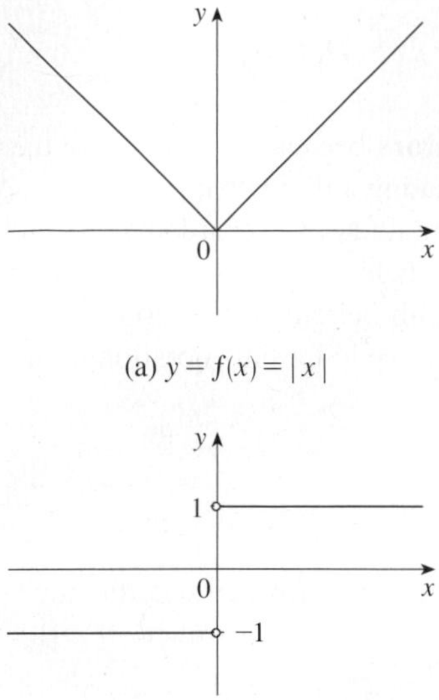

(a) $y = f(x) = |x|$

(b) $y = f'(x)$

FIGURE 5

Let's compute the left and right limits separately:

$$\lim_{h \to 0^+} \frac{|0 + h| - |0|}{h} = \lim_{h \to 0^+} \frac{|h|}{h} = \lim_{h \to 0^+} \frac{h}{h} = \lim_{h \to 0^+} 1 = 1$$

and $$\lim_{h \to 0^-} \frac{|0 + h| - |0|}{h} = \lim_{h \to 0^-} \frac{|h|}{h} = \lim_{h \to 0^-} \frac{-h}{h} = \lim_{h \to 0^-} (-1) = -1$$

Since these limits are different, $f'(0)$ does not exist. Thus f is differentiable at all x except 0.

A formula for f' is given by

$$f'(x) = \begin{cases} 1 & \text{if } x > 0 \\ -1 & \text{if } x < 0 \end{cases}$$

and its graph is shown in Figure 5(b). The fact that $f'(0)$ does not exist is reflected geometrically in the fact that the curve $y = |x|$ does not have a tangent line at (0, 0). [See Figure 5(a).] □

Both continuity and differentiability are desirable properties for a function to have. The following theorem shows how these properties are related.

> **4 THEOREM** If f is differentiable at a, then f is continuous at a.

PROOF To prove that f is continuous at a, we have to show that $\lim_{x \to a} f(x) = f(a)$. We do this by showing that the difference $f(x) - f(a)$ approaches 0 as x approaches a.

The given information is that f is differentiable at a, that is,

$$f'(a) = \lim_{x \to a} \frac{f(x) - f(a)}{x - a}$$

exists (see Equation 2.7.5). To connect the given and the unknown, we divide and multiply $f(x) - f(a)$ by $x - a$ (which we can do when $x \neq a$):

$$f(x) - f(a) = \frac{f(x) - f(a)}{x - a}(x - a)$$

Thus, using the Product Law and (2.7.5), we can write

$$\lim_{x \to a} [f(x) - f(a)] = \lim_{x \to a} \frac{f(x) - f(a)}{x - a}(x - a)$$

$$= \lim_{x \to a} \frac{f(x) - f(a)}{x - a} \cdot \lim_{x \to a} (x - a)$$

$$= f'(a) \cdot 0 = 0$$

To use what we have just proved, we start with $f(x)$ and add and subtract $f(a)$:

$$\begin{aligned}\lim_{x \to a} f(x) &= \lim_{x \to a} [f(a) + (f(x) - f(a))] \\ &= \lim_{x \to a} f(a) + \lim_{x \to a} [f(x) - f(a)] \\ &= f(a) + 0 = f(a)\end{aligned}$$

Therefore f is continuous at a. □

NOTE The converse of Theorem 4 is false; that is, there are functions that are continuous but not differentiable. For instance, the function $f(x) = |x|$ is continuous at 0 because

$$\lim_{x \to 0} f(x) = \lim_{x \to 0} |x| = 0 = f(0)$$

(See Example 7 in Section 2.3.) But in Example 5 we showed that f is not differentiable at 0.

HOW CAN A FUNCTION FAIL TO BE DIFFERENTIABLE?

We saw that the function $y = |x|$ in Example 5 is not differentiable at 0 and Figure 5(a) shows that its graph changes direction abruptly when $x = 0$. In general, if the graph of a function f has a "corner" or "kink" in it, then the graph of f has no tangent at this point and f is not differentiable there. [In trying to compute $f'(a)$, we find that the left and right limits are different.]

Theorem 4 gives another way for a function not to have a derivative. It says that if f is not continuous at a, then f is not differentiable at a. So at any discontinuity (for instance, a jump discontinuity) f fails to be differentiable.

A third possibility is that the curve has a **vertical tangent line** when $x = a$; that is, f is continuous at a and

$$\lim_{x \to a} |f'(x)| = \infty$$

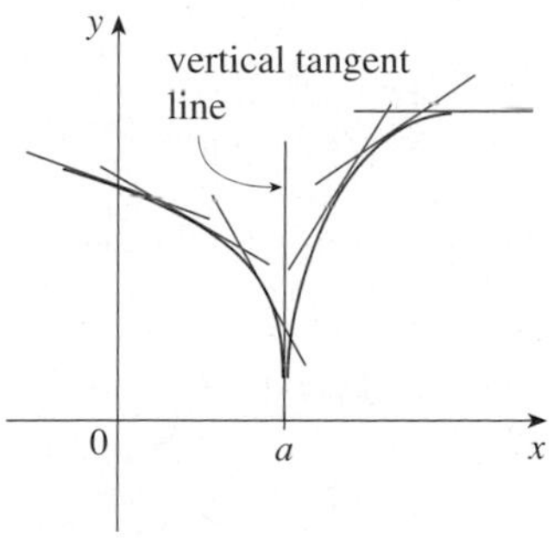

FIGURE 6

This means that the tangent lines become steeper and steeper as $x \to a$. Figure 6 shows one way that this can happen; Figure 7(c) shows another. Figure 7 illustrates the three possibilities that we have discussed.

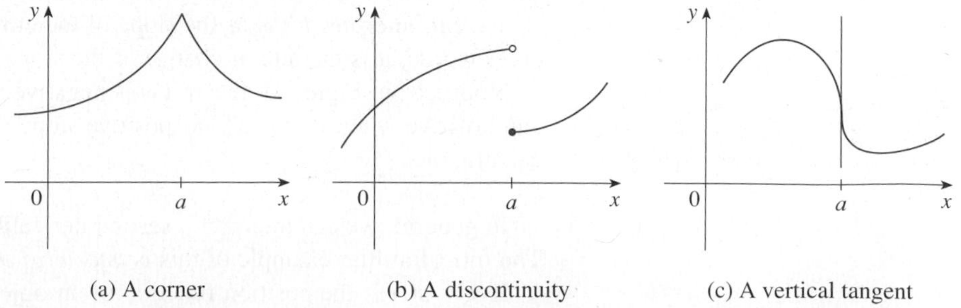

FIGURE 7
Three ways for f not to be differentiable at a

A graphing calculator or computer provides another way of looking at differentiability. If f is differentiable at a, then when we zoom in toward the point $(a, f(a))$ the graph straightens out and appears more and more like a line. (See Figure 8. We saw a specific example of this in Figure 2 in Section 2.7.) But no matter how much we zoom in toward a point like the ones in Figures 6 and 7(a), we can't eliminate the sharp point or corner (see Figure 9).

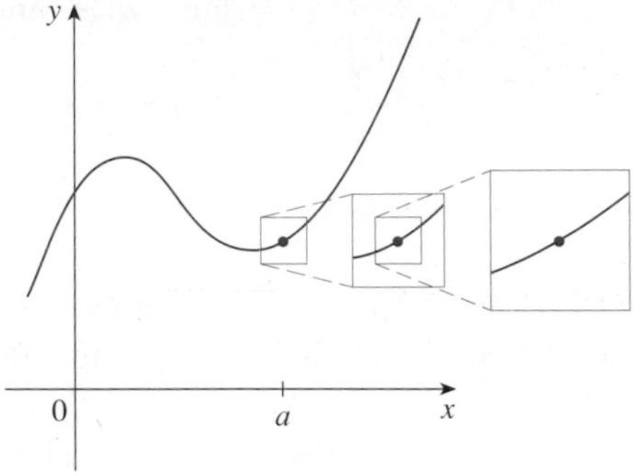

FIGURE 8
f is differentiable at a.

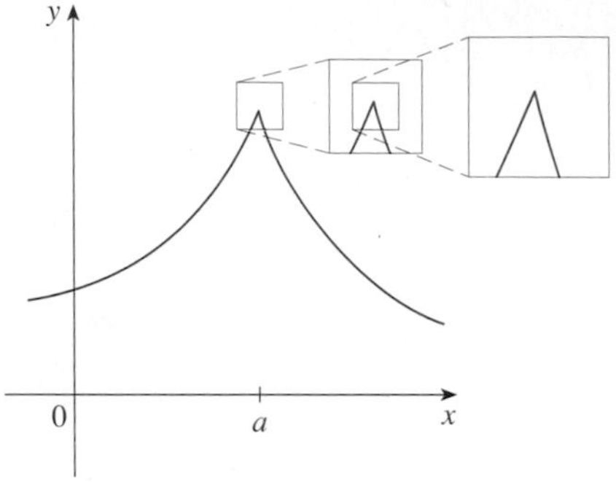

FIGURE 9
f is not differentiable at a.

HIGHER DERIVATIVES

If f is a differentiable function, then its derivative f' is also a function, so f' may have a derivative of its own, denoted by $(f')' = f''$. This new function f'' is called the **second derivative** of f because it is the derivative of the derivative of f. Using Leibniz notation, we write the second derivative of $y = f(x)$ as

$$\frac{d}{dx}\left(\frac{dy}{dx}\right) = \frac{d^2y}{dx^2}$$

EXAMPLE 6 If $f(x) = x^3 - x$, find and interpret $f''(x)$.

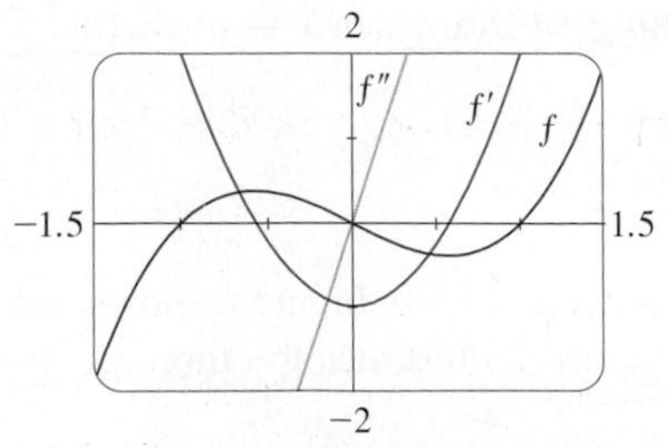

FIGURE 10

SOLUTION In Example 2 we found that the first derivative is $f'(x) = 3x^2 - 1$. So the second derivative is

$$f''(x) = (f')'(x) = \lim_{h\to 0}\frac{f'(x+h) - f'(x)}{h} = \lim_{h\to 0}\frac{[3(x+h)^2 - 1] - [3x^2 - 1]}{h}$$

$$= \lim_{h\to 0}\frac{3x^2 + 6xh + 3h^2 - 1 - 3x^2 + 1}{h} = \lim_{h\to 0}(6x + 3h) = 6x$$

The graphs of f, f', and f'' are shown in Figure 10.

We can interpret $f''(x)$ as the slope of the curve $y = f'(x)$ at the point $(x, f'(x))$. In other words, it is the rate of change of the slope of the original curve $y = f(x)$.

Notice from Figure 10 that $f''(x)$ is negative when $y = f'(x)$ has negative slope and positive when $y = f'(x)$ has positive slope. So the graphs serve as a check on our calculations. □

TEC In Module 2.8 you can see how changing the coefficients of a polynomial f affects the appearance of the graphs of f, f', and f''.

In general, we can interpret a second derivative as a rate of change of a rate of change. The most familiar example of this is *acceleration,* which we define as follows.

If $s = s(t)$ is the position function of an object that moves in a straight line, we know that its first derivative represents the velocity $v(t)$ of the object as a function of time:

$$v(t) = s'(t) = \frac{ds}{dt}$$

The instantaneous rate of change of velocity with respect to time is called the **acceleration** $a(t)$ of the object. Thus the acceleration function is the derivative of the velocity function and is therefore the second derivative of the position function:

$$a(t) = v'(t) = s''(t)$$

or, in Leibniz notation,

$$a = \frac{dv}{dt} = \frac{d^2s}{dt^2}$$

The **third derivative** f''' is the derivative of the second derivative: $f''' = (f'')'$. So $f'''(x)$ can be interpreted as the slope of the curve $y = f''(x)$ or as the rate of change of $f''(x)$. If $y = f(x)$, then alternative notations for the third derivative are

$$y''' = f'''(x) = \frac{d}{dx}\left(\frac{d^2y}{dx^2}\right) = \frac{d^3y}{dx^3}$$

The process can be continued. The fourth derivative f'''' is usually denoted by $f^{(4)}$. In general, the nth derivative of f is denoted by $f^{(n)}$ and is obtained from f by differentiating n times. If $y = f(x)$, we write

$$y^{(n)} = f^{(n)}(x) = \frac{d^ny}{dx^n}$$

EXAMPLE 7 If $f(x) = x^3 - x$, find $f'''(x)$ and $f^{(4)}(x)$.

SOLUTION In Example 6 we found that $f''(x) = 6x$. The graph of the second derivative has equation $y = 6x$ and so it is a straight line with slope 6. Since the derivative $f'''(x)$ is the slope of $f''(x)$, we have

$$f'''(x) = 6$$

for all values of x. So f''' is a constant function and its graph is a horizontal line. Therefore, for all values of x,

$$f^{(4)}(x) = 0$$ □

We can interpret the third derivative physically in the case where the function is the position function $s = s(t)$ of an object that moves along a straight line. Because $s''' = (s'')' = a'$, the third derivative of the position function is the derivative of the acceleration function and is called the **jerk**:

$$j = \frac{da}{dt} = \frac{d^3s}{dt^3}$$

Thus the jerk j is the rate of change of acceleration. It is aptly named because a large jerk means a sudden change in acceleration, which causes an abrupt movement in a vehicle.

We have seen that one application of second and third derivatives occurs in analyzing the motion of objects using acceleration and jerk. We will investigate another application of second derivatives in Section 4.3, where we show how knowledge of f'' gives us information about the shape of the graph of f. In Chapter 11 we will see how second and higher derivatives enable us to represent functions as sums of infinite series.

2.8 EXERCISES

1–2 Use the given graph to estimate the value of each derivative. Then sketch the graph of f'.

1. (a) $f'(-3)$
(b) $f'(-2)$
(c) $f'(-1)$
(d) $f'(0)$
(e) $f'(1)$
(f) $f'(2)$
(g) $f'(3)$

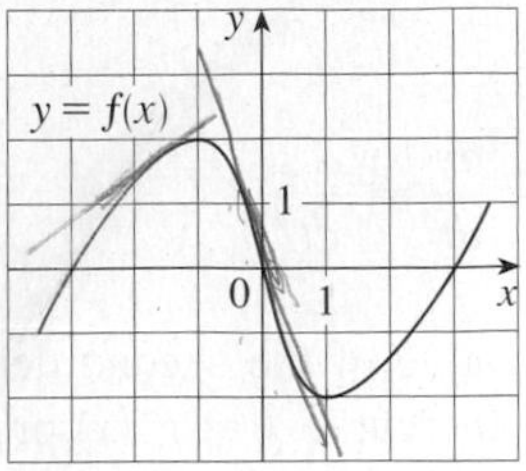

2. (a) $f'(0)$
(b) $f'(1)$
(c) $f'(2)$
(d) $f'(3)$
(e) $f'(4)$
(f) $f'(5)$

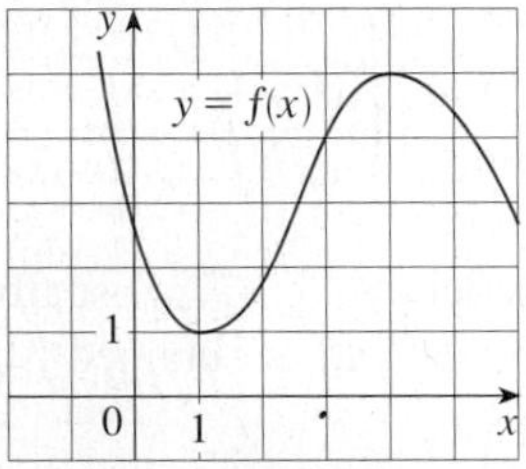

3. Match the graph of each function in (a)–(d) with the graph of its derivative in I–IV. Give reasons for your choices.

(a)
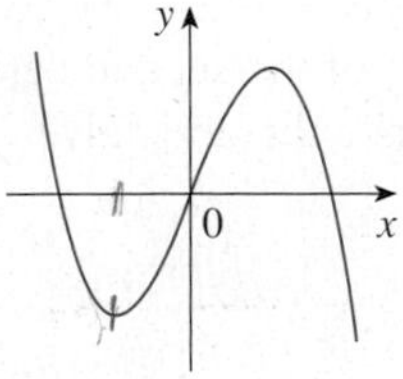

(b)
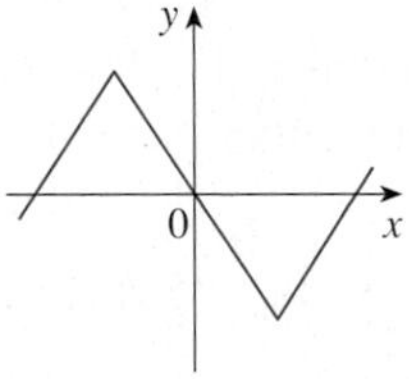

(c)
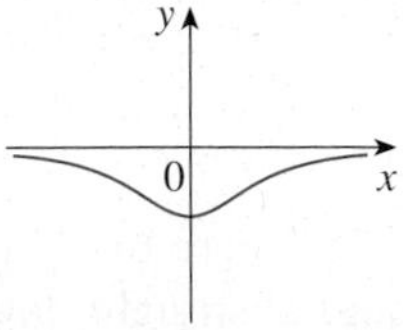

(d)
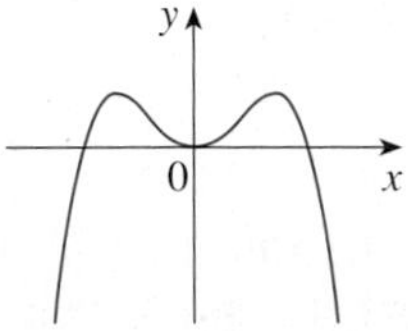

I
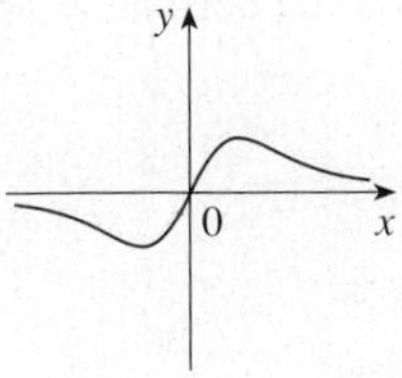

II
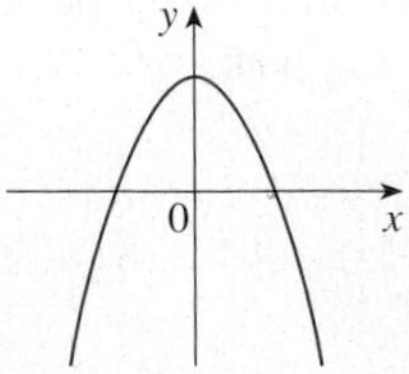

III
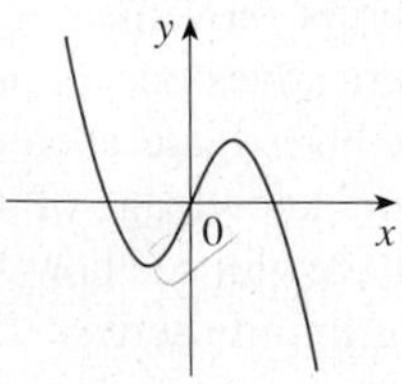

IV
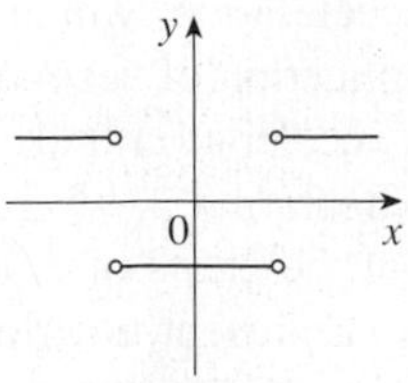

4–11 Trace or copy the graph of the given function f. (Assume that the axes have equal scales.) Then use the method of Example 1 to sketch the graph of f' below it.

4.
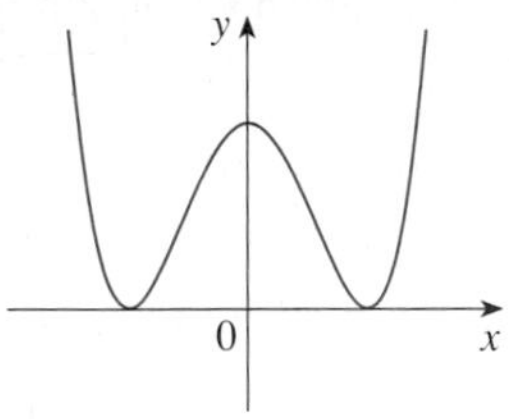

5.
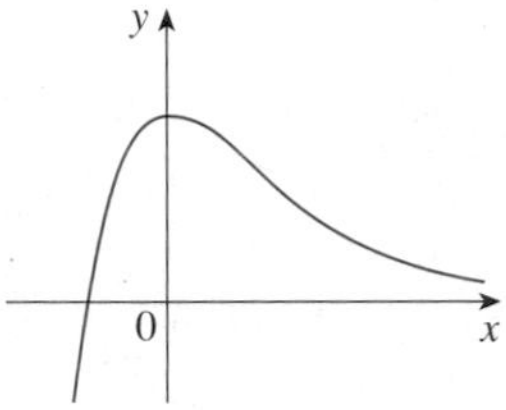

6.
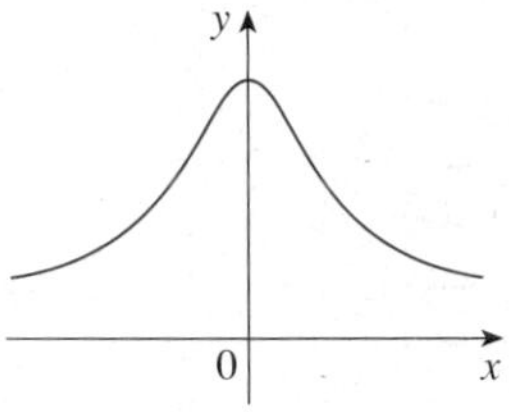

7.
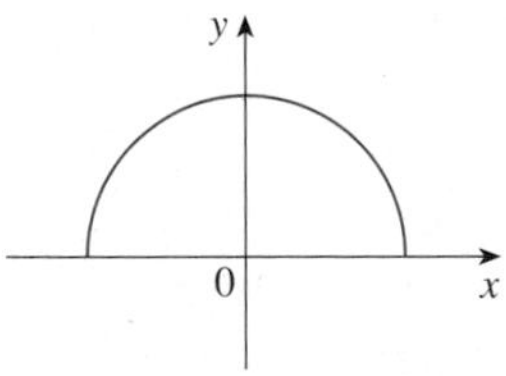

8.
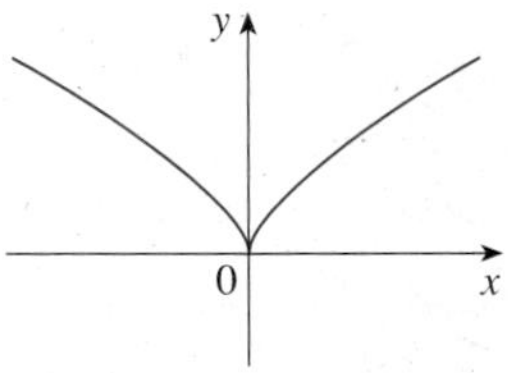

9.
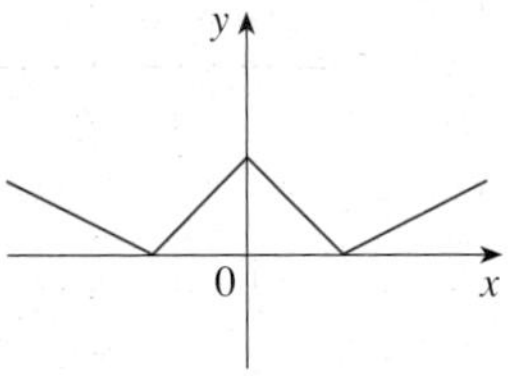

10.
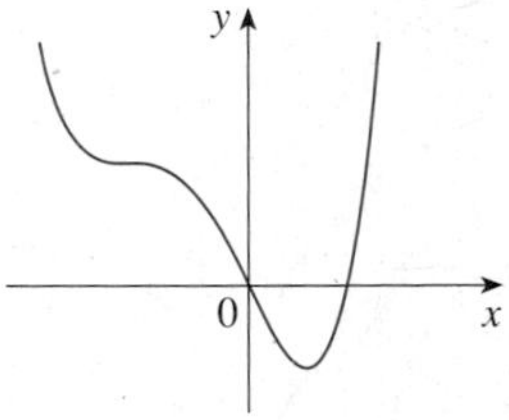

11.
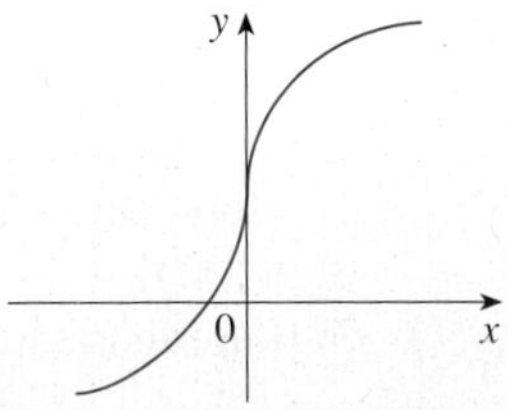

12. Shown is the graph of the population function $P(t)$ for yeast cells in a laboratory culture. Use the method of Example 1 to

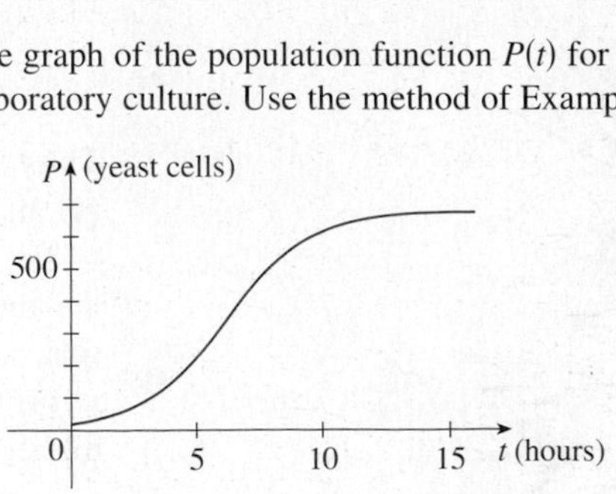

graph the derivative $P'(t)$. What does the graph of P' tell us about the yeast population?

13. The graph shows how the average age of first marriage of Japanese men has varied in the last half of the 20th century. Sketch the graph of the derivative function $M'(t)$. During which years was the derivative negative?

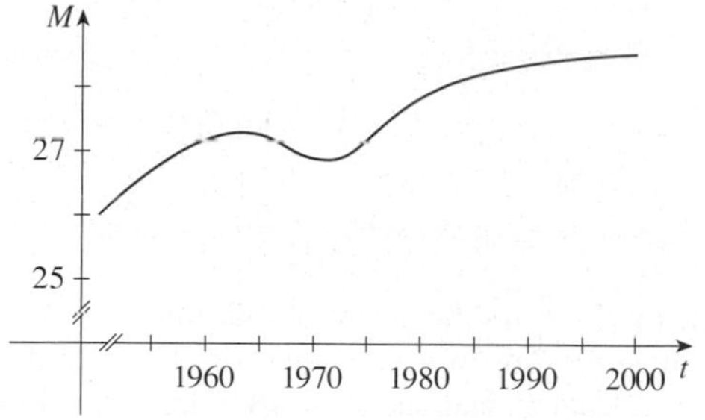

14–16 Make a careful sketch of the graph of f and below it sketch the graph of f' in the same manner as in Exercises 4–11. Can you guess a formula for $f'(x)$ from its graph?

14. $f(x) = \sin x$

15. $f(x) = e^x$

16. $f(x) = \ln x$

17. Let $f(x) = x^2$.
(a) Estimate the values of $f'(0)$, $f'(\frac{1}{2})$, $f'(1)$, and $f'(2)$ by using a graphing device to zoom in on the graph of f.
(b) Use symmetry to deduce the values of $f'(-\frac{1}{2})$, $f'(-1)$, and $f'(-2)$.
(c) Use the results from parts (a) and (b) to guess a formula for $f'(x)$.
(d) Use the definition of a derivative to prove that your guess in part (c) is correct.

18. Let $f(x) = x^3$.
(a) Estimate the values of $f'(0)$, $f'(\frac{1}{2})$, $f'(1)$, $f'(2)$, and $f'(3)$ by using a graphing device to zoom in on the graph of f.
(b) Use symmetry to deduce the values of $f'(-\frac{1}{2})$, $f'(-1)$, $f'(-2)$, and $f'(-3)$.
(c) Use the values from parts (a) and (b) to graph f'.
(d) Guess a formula for $f'(x)$.
(e) Use the definition of a derivative to prove that your guess in part (d) is correct.

19–29 Find the derivative of the function using the definition of derivative. State the domain of the function and the domain of its derivative.

19. $f(x) = \frac{1}{2}x - \frac{1}{3}$

20. $f(x) = mx + b$

21. $f(t) = 5t - 9t^2$

22. $f(x) = 1.5x^2 - x + 3.7$

23. $f(x) = x^3 - 3x + 5$

24. $f(x) = x + \sqrt{x}$

25. $g(x) = \sqrt{1 + 2x}$

26. $f(x) = \dfrac{3 + x}{1 - 3x}$

27. $G(t) = \dfrac{4t}{t + 1}$

28. $g(t) = \dfrac{1}{\sqrt{t}}$

29. $f(x) = x^4$

30. (a) Sketch the graph of $f(x) = \sqrt{6 - x}$ by starting with the graph of $y = \sqrt{x}$ and using the transformations of Section 1.3.
(b) Use the graph from part (a) to sketch the graph of f'.
(c) Use the definition of a derivative to find $f'(x)$. What are the domains of f and f'?
(d) Use a graphing device to graph f' and compare with your sketch in part (b).

31. (a) If $f(x) = x^4 + 2x$, find $f'(x)$.
(b) Check to see that your answer to part (a) is reasonable by comparing the graphs of f and f'.

32. (a) If $f(t) = t^2 - \sqrt{t}$, find $f'(t)$.
(b) Check to see that your answer to part (a) is reasonable by comparing the graphs of f and f'.

33. The unemployment rate $U(t)$ varies with time. The table (from the Bureau of Labor Statistics) gives the percentage of unemployed in the US labor force from 1993 to 2002.

t	$U(t)$	t	$U(t)$
1993	6.9	1998	4.5
1994	6.1	1999	4.2
1995	5.6	2000	4.0
1996	5.4	2001	4.7
1997	4.9	2002	5.8

(a) What is the meaning of $U'(t)$? What are its units?
(b) Construct a table of values for $U'(t)$.

34. Let $P(t)$ be the percentage of Americans under the age of 18 at time t. The table gives values of this function in census years from 1950 to 2000.

t	$P(t)$	t	$P(t)$
1950	31.1	1980	28.0
1960	35.7	1990	25.7
1970	34.0	2000	25.7

(a) What is the meaning of $P'(t)$? What are its units?
(b) Construct a table of estimated values for $P'(t)$.
(c) Graph P and P'.
(d) How would it be possible to get more accurate values for $P'(t)$?

35–38 The graph of f is given. State, with reasons, the numbers at which f is not differentiable.

35.

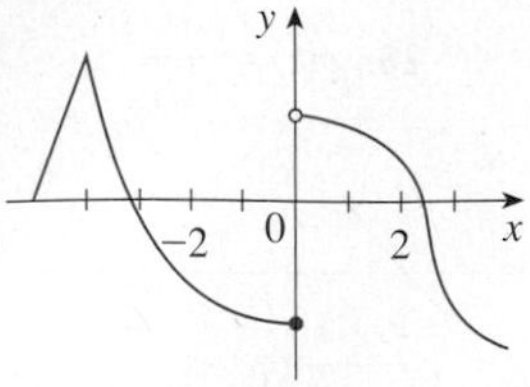

36.

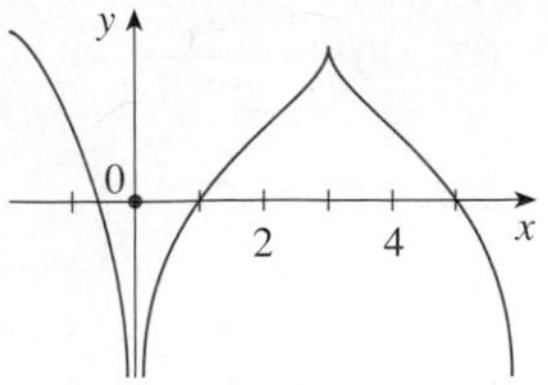

37.

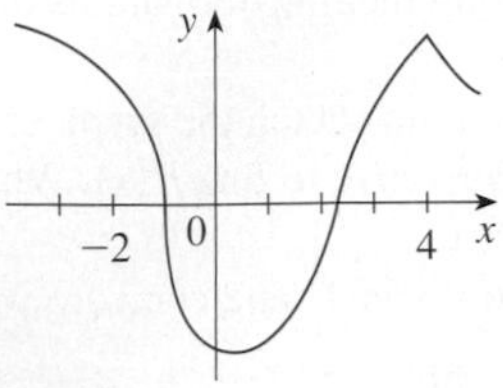

38.

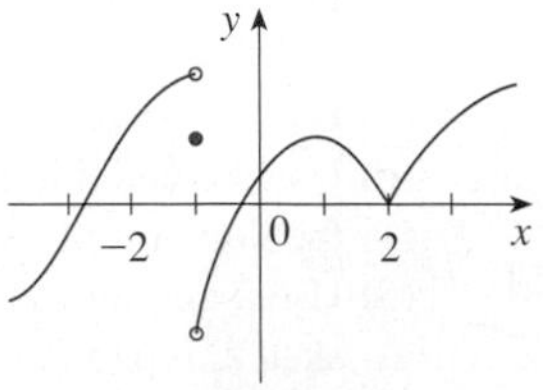

39. Graph the function $f(x) = x + \sqrt{|x|}$. Zoom in repeatedly, first toward the point $(-1, 0)$ and then toward the origin. What is different about the behavior of f in the vicinity of these two points? What do you conclude about the differentiability of f?

40. Zoom in toward the points $(1, 0)$, $(0, 1)$, and $(-1, 0)$ on the graph of the function $g(x) = (x^2 - 1)^{2/3}$. What do you notice? Account for what you see in terms of the differentiability of g.

41. The figure shows the graphs of f, f', and f''. Identify each curve, and explain your choices.

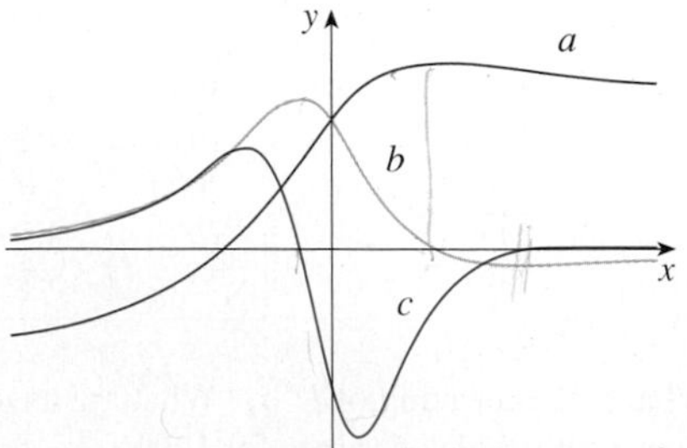

42. The figure shows graphs of f, f', f'', and f'''. Identify each curve, and explain your choices.

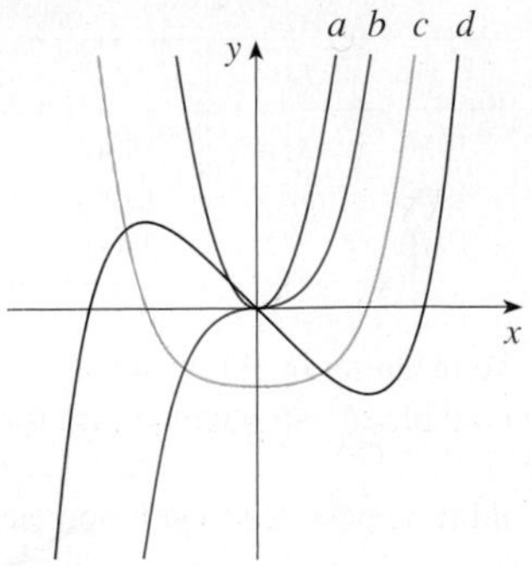

43. The figure shows the graphs of three functions. One is the position function of a car, one is the velocity of the car, and one is its acceleration. Identify each curve, and explain your choices.

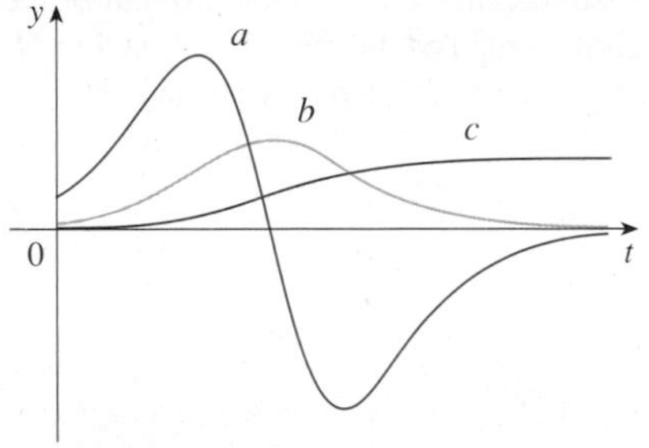

44. The figure shows the graphs of four functions. One is the position function of a car, one is the velocity of the car, one is its acceleration, and one is its jerk. Identify each curve, and explain your choices.

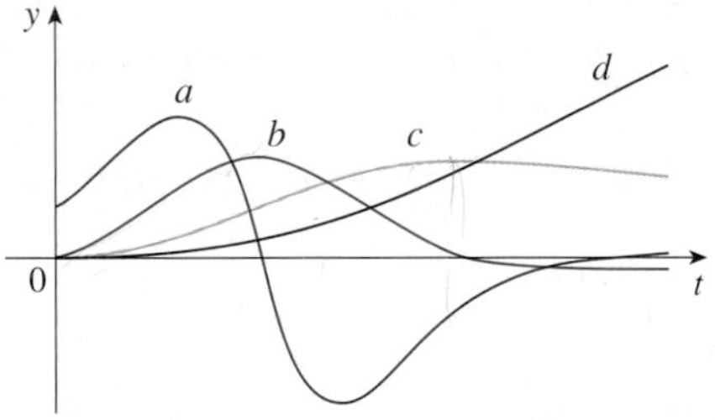

45–46 Use the definition of a derivative to find $f'(x)$ and $f''(x)$. Then graph f, f', and f'' on a common screen and check to see if your answers are reasonable.

45. $f(x) = 1 + 4x - x^2$

46. $f(x) = 1/x$

47. If $f(x) = 2x^2 - x^3$, find $f'(x)$, $f''(x)$, $f'''(x)$, and $f^{(4)}(x)$. Graph f, f', f'', and f''' on a common screen. Are the graphs consistent with the geometric interpretations of these derivatives?

48. (a) The graph of a position function of a car is shown, where s is measured in feet and t in seconds. Use it to graph the velocity and acceleration of the car. What is the acceleration at $t = 10$ seconds?

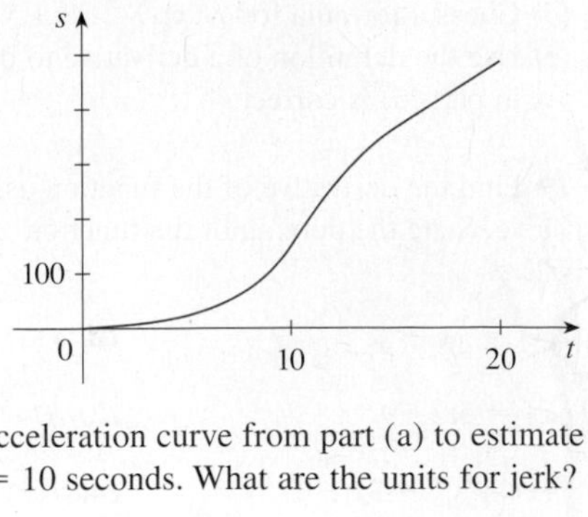

(b) Use the acceleration curve from part (a) to estimate the jerk at $t = 10$ seconds. What are the units for jerk?

49. Let $f(x) = \sqrt[3]{x}$.
(a) If $a \neq 0$, use Equation 2.7.5 to find $f'(a)$.
(b) Show that $f'(0)$ does not exist.
(c) Show that $y = \sqrt[3]{x}$ has a vertical tangent line at (0, 0). (Recall the shape of the graph of f. See Figure 13 in Section 1.2.)

50. (a) If $g(x) = x^{2/3}$, show that $g'(0)$ does not exist.
(b) If $a \neq 0$, find $g'(a)$.
(c) Show that $y = x^{2/3}$ has a vertical tangent line at (0, 0).
(d) Illustrate part (c) by graphing $y = x^{2/3}$.

51. Show that the function $f(x) = |x - 6|$ is not differentiable at 6. Find a formula for f' and sketch its graph.

52. Where is the greatest integer function $f(x) = [\![x]\!]$ not differentiable? Find a formula for f' and sketch its graph.

53 (a) Sketch the graph of the function $f(x) = x|x|$.
(b) For what values of x is f differentiable?
(c) Find a formula for f'.

54. The **left-hand** and **right-hand derivatives** of f at a are defined by

$$f'_-(a) = \lim_{h \to 0^-} \frac{f(a+h) - f(a)}{h}$$

and $$f'_+(a) = \lim_{h \to 0^+} \frac{f(a+h) - f(a)}{h}$$

if these limits exist. Then $f'(a)$ exists if and only if these one-sided derivatives exist and are equal.

(a) Find $f'_-(4)$ and $f'_+(4)$ for the function

$$f(x) = \begin{cases} 0 & \text{if } x \leq 0 \\ 5 - x & \text{if } 0 < x < 4 \\ \dfrac{1}{5 - x} & \text{if } x \geq 4 \end{cases}$$

(b) Sketch the graph of f.
(c) Where is f discontinuous?
(d) Where is f not differentiable?

55. Recall that a function f is called *even* if $f(-x) = f(x)$ for all x in its domain and *odd* if $f(-x) = -f(x)$ for all such x. Prove each of the following.
(a) The derivative of an even function is an odd function.
(b) The derivative of an odd function is an even function.

56. When you turn on a hot-water faucet, the temperature T of the water depends on how long the water has been running.
(a) Sketch a possible graph of T as a function of the time t that has elapsed since the faucet was turned on.
(b) Describe how the rate of change of T with respect to t varies as t increases.
(c) Sketch a graph of the derivative of T.

57. Let ℓ be the tangent line to the parabola $y = x^2$ at the point (1, 1). The *angle of inclination* of ℓ is the angle ϕ that ℓ makes with the positive direction of the x-axis. Calculate ϕ correct to the nearest degree.

2 REVIEW

CONCEPT CHECK

1. Explain what each of the following means and illustrate with a sketch.
(a) $\lim_{x \to a} f(x) = L$
(b) $\lim_{x \to a^+} f(x) = L$
(c) $\lim_{x \to a^-} f(x) = L$
(d) $\lim_{x \to a} f(x) = \infty$
(e) $\lim_{x \to \infty} f(x) = L$

2. Describe several ways in which a limit can fail to exist. Illustrate with sketches.

3. State the following Limit Laws.
(a) Sum Law
(b) Difference Law
(c) Constant Multiple Law
(d) Product Law
(e) Quotient Law
(f) Power Law
(g) Root Law

4. What does the Squeeze Theorem say?

5. (a) What does it mean to say that the line $x = a$ is a vertical asymptote of the curve $y = f(x)$? Draw curves to illustrate the various possibilities.
(b) What does it mean to say that the line $y = L$ is a horizontal asymptote of the curve $y = f(x)$? Draw curves to illustrate the various possibilities.

6. Which of the following curves have vertical asymptotes? Which have horizontal asymptotes?
(a) $y = x^4$
(b) $y = \sin x$
(c) $y = \tan x$
(d) $y = \tan^{-1}x$
(e) $y = e^x$
(f) $y = \ln x$
(g) $y = 1/x$
(h) $y = \sqrt{x}$

7. (a) What does it mean for f to be continuous at a?
(b) What does it mean for f to be continuous on the interval $(-\infty, \infty)$? What can you say about the graph of such a function?

8. What does the Intermediate Value Theorem say?

9. Write an expression for the slope of the tangent line to the curve $y = f(x)$ at the point $(a, f(a))$.

10. Suppose an object moves along a straight line with position $f(t)$ at time t. Write an expression for the instantaneous velocity of the object at time $t = a$. How can you interpret this velocity in terms of the graph of f?

11. If $y = f(x)$ and x changes from x_1 to x_2, write expressions for the following.
(a) The average rate of change of y with respect to x over the interval $[x_1, x_2]$.
(b) The instantaneous rate of change of y with respect to x at $x = x_1$.

12. Define the derivative $f'(a)$. Discuss two ways of interpreting this number.

13. Define the second derivative of f. If $f(t)$ is the position function of a particle, how can you interpret the second derivative?

14. (a) What does it mean for f to be differentiable at a?
(b) What is the relation between the differentiability and continuity of a function?
(c) Sketch the graph of a function that is continuous but not differentiable at $a = 2$.

15. Describe several ways in which a function can fail to be differentiable. Illustrate with sketches.

TRUE-FALSE QUIZ

Determine whether the statement is true or false. If it is true, explain why. If it is false, explain why or give an example that disproves the statement.

1. $\displaystyle\lim_{x\to 4}\left(\frac{2x}{x-4} - \frac{8}{x-4}\right) = \lim_{x\to 4}\frac{2x}{x-4} - \lim_{x\to 4}\frac{8}{x-4}$

2. $\displaystyle\lim_{x\to 1}\frac{x^2+6x-7}{x^2+5x-6} = \frac{\lim_{x\to 1}(x^2+6x-7)}{\lim_{x\to 1}(x^2+5x-6)}$

3. $\displaystyle\lim_{x\to 1}\frac{x-3}{x^2+2x-4} = \frac{\lim_{x\to 1}(x-3)}{\lim_{x\to 1}(x^2+2x-4)}$

4. If $\lim_{x\to 5} f(x) = 2$ and $\lim_{x\to 5} g(x) = 0$, then $\lim_{x\to 5}[f(x)/g(x)]$ does not exist.

5. If $\lim_{x\to 5} f(x) = 0$ and $\lim_{x\to 5} g(x) = 0$, then $\lim_{x\to 5}[f(x)/g(x)]$ does not exist.

6. If $\lim_{x\to 6}[f(x)g(x)]$ exists, then the limit must be $f(6)g(6)$.

7. If p is a polynomial, then $\lim_{x\to b} p(x) = p(b)$.

8. If $\lim_{x\to 0} f(x) = \infty$ and $\lim_{x\to 0} g(x) = \infty$, then $\lim_{x\to 0}[f(x) - g(x)] = 0$.

9. A function can have two different horizontal asymptotes.

10. If f has domain $[0, \infty)$ and has no horizontal asymptote, then $\lim_{x\to\infty} f(x) = \infty$ or $\lim_{x\to\infty} f(x) = -\infty$.

11. If the line $x = 1$ is a vertical asymptote of $y = f(x)$, then f is not defined at 1.

12. If $f(1) > 0$ and $f(3) < 0$, then there exists a number c between 1 and 3 such that $f(c) = 0$.

13. If f is continuous at 5 and $f(5) = 2$ and $f(4) = 3$, then $\lim_{x\to 2} f(4x^2 - 11) = 2$.

14. If f is continuous on $[-1, 1]$ and $f(-1) = 4$ and $f(1) = 3$, then there exists a number r such that $|r| < 1$ and $f(r) = \pi$.

15. Let f be a function such that $\lim_{x\to 0} f(x) = 6$. Then there exists a number δ such that if $0 < |x| < \delta$, then $|f(x) - 6| < 1$.

16. If $f(x) > 1$ for all x and $\lim_{x\to 0} f(x)$ exists, then $\lim_{x\to 0} f(x) > 1$.

17. If f is continuous at a, then f is differentiable at a.

18. If $f'(r)$ exists, then $\lim_{x\to r} f(x) = f(r)$.

19. $\displaystyle\frac{d^2y}{dx^2} = \left(\frac{dy}{dx}\right)^2$

20. The equation $x^{10} - 10x^2 + 5 = 0$ has a root in the interval $(0, 2)$.

EXERCISES

1. The graph of f is given.

(a) Find each limit, or explain why it does not exist.

(i) $\lim_{x\to 2^+} f(x)$ (ii) $\lim_{x\to -3^+} f(x)$

(iii) $\lim_{x\to -3} f(x)$ (iv) $\lim_{x\to 4} f(x)$

(v) $\lim_{x\to 0} f(x)$ (vi) $\lim_{x\to 2^-} f(x)$

(vii) $\lim_{x\to\infty} f(x)$ (viii) $\lim_{x\to -\infty} f(x)$

(b) State the equations of the horizontal asymptotes.

(c) State the equations of the vertical asymptotes.

(d) At what numbers is f discontinuous? Explain.

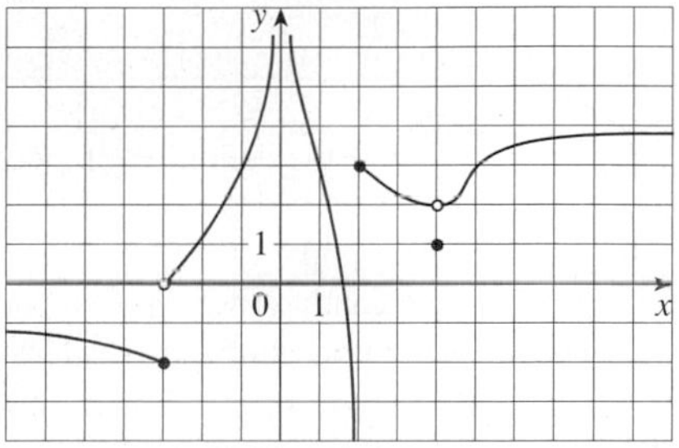

2. Sketch the graph of an example of a function f that satisfies all of the following conditions:

$\lim_{x\to -\infty} f(x) = -2$, $\lim_{x\to\infty} f(x) = 0$, $\lim_{x\to -3} f(x) = \infty$,

$\lim_{x\to 3^-} f(x) = -\infty$, $\lim_{x\to 3^+} f(x) = 2$,

f is continuous from the right at 3

3–20 Find the limit.

3. $\lim_{x\to 1} e^{x^3-x}$

4. $\lim_{x\to 3} \dfrac{x^2-9}{x^2+2x-3}$

5. $\lim_{x\to -3} \dfrac{x^2-9}{x^2+2x-3}$

6. $\lim_{x\to 1^+} \dfrac{x^2-9}{x^2+2x-3}$

7. $\lim_{h\to 0} \dfrac{(h-1)^3+1}{h}$

8. $\lim_{t\to 2} \dfrac{t^2-4}{t^3-8}$

9. $\lim_{r\to 9} \dfrac{\sqrt{r}}{(r-9)^4}$

10. $\lim_{v\to 4^+} \dfrac{4-v}{|4-v|}$

11. $\lim_{u\to 1} \dfrac{u^4-1}{u^3+5u^2-6u}$

12. $\lim_{x\to 3} \dfrac{\sqrt{x+6}-x}{x^3-3x^2}$

13. $\lim_{x\to\infty} \dfrac{\sqrt{x^2-9}}{2x-6}$

14. $\lim_{x\to -\infty} \dfrac{\sqrt{x^2-9}}{2x-6}$

15. $\lim_{x\to \pi^-} \ln(\sin x)$

16. $\lim_{x\to -\infty} \dfrac{1-2x^2-x^4}{5+x-3x^4}$

17. $\lim_{x\to\infty} \left(\sqrt{x^2+4x+1}-x\right)$

18. $\lim_{x\to\infty} e^{x-x^2}$

19. $\lim_{x\to 0^+} \tan^{-1}(1/x)$

20. $\lim_{x\to 1} \left(\dfrac{1}{x-1}+\dfrac{1}{x^2-3x+2}\right)$

21–22 Use graphs to discover the asymptotes of the curve. Then prove what you have discovered.

21. $y = \dfrac{\cos^2 x}{x^2}$

22. $y = \sqrt{x^2+x+1}-\sqrt{x^2-x}$

23. If $2x-1 \leqslant f(x) \leqslant x^2$ for $0 < x < 3$, find $\lim_{x\to 1} f(x)$.

24. Prove that $\lim_{x\to 0} x^2\cos(1/x^2) = 0$.

25–28 Prove the statement using the precise definition of a limit.

25. $\lim_{x\to 2} (14-5x) = 4$

26. $\lim_{x\to 0} \sqrt[3]{x} = 0$

27. $\lim_{x\to 2} (x^2-3x) = -2$

28. $\lim_{x\to 4^+} \dfrac{2}{\sqrt{x-4}} = \infty$

29. Let

$$f(x) = \begin{cases} \sqrt{-x} & \text{if } x < 0 \\ 3-x & \text{if } 0 \leqslant x < 3 \\ (x-3)^2 & \text{if } x > 3 \end{cases}$$

(a) Evaluate each limit, if it exists.

(i) $\lim_{x\to 0^+} f(x)$ (ii) $\lim_{x\to 0^-} f(x)$ (iii) $\lim_{x\to 0} f(x)$

(iv) $\lim_{x\to 3^-} f(x)$ (v) $\lim_{x\to 3^+} f(x)$ (vi) $\lim_{x\to 3} f(x)$

(b) Where is f discontinuous?

(c) Sketch the graph of f.

30. Let

$$g(x) = \begin{cases} 2x-x^2 & \text{if } 0 \leqslant x \leqslant 2 \\ 2-x & \text{if } 2 < x \leqslant 3 \\ x-4 & \text{if } 3 < x < 4 \\ \pi & \text{if } x \geqslant 4 \end{cases}$$

(a) For each of the numbers 2, 3, and 4, discover whether g is continuous from the left, continuous from the right, or continuous at the number.

(b) Sketch the graph of g.

31–32 Show that each function is continuous on its domain. State the domain.

31. $h(x) = xe^{\sin x}$

32. $g(x) = \dfrac{\sqrt{x^2-9}}{x^2-2}$

33–34 Use the Intermediate Value Theorem to show that there is a root of the equation in the given interval.

33. $2x^3 + x^2 + 2 = 0, \quad (-2, -1)$

34. $e^{-x^2} = x, \quad (0, 1)$

---

35. (a) Find the slope of the tangent line to the curve $y = 9 - 2x^2$ at the point $(2, 1)$.
(b) Find an equation of this tangent line.

36. Find equations of the tangent lines to the curve

$$y = \frac{2}{1 - 3x}$$

at the points with x-coordinates 0 and -1.

37. The displacement (in meters) of an object moving in a straight line is given by $s = 1 + 2t + \frac{1}{4}t^2$, where t is measured in seconds.
(a) Find the average velocity over each time period.
(i) $[1, 3]$ (ii) $[1, 2]$
(iii) $[1, 1.5]$ (iv) $[1, 1.1]$
(b) Find the instantaneous velocity when $t = 1$.

38. According to Boyle's Law, if the temperature of a confined gas is held fixed, then the product of the pressure P and the volume V is a constant. Suppose that, for a certain gas, $PV = 800$, where P is measured in pounds per square inch and V is measured in cubic inches.
(a) Find the average rate of change of P as V increases from 200 in^3 to 250 in^3.
(b) Express V as a function of P and show that the instantaneous rate of change of V with respect to P is inversely proportional to the square of P.

39. (a) Use the definition of a derivative to find $f'(2)$, where $f(x) = x^3 - 2x$.
(b) Find an equation of the tangent line to the curve $y = x^3 - 2x$ at the point $(2, 4)$.
(c) Illustrate part (b) by graphing the curve and the tangent line on the same screen.

40. Find a function f and a number a such that

$$\lim_{h \to 0} \frac{(2 + h)^6 - 64}{h} = f'(a)$$

41. The total cost of repaying a student loan at an interest rate of $r\%$ per year is $C = f(r)$.
(a) What is the meaning of the derivative $f'(r)$? What are its units?
(b) What does the statement $f'(10) = 1200$ mean?
(c) Is $f'(r)$ always positive or does it change sign?

42–44 Trace or copy the graph of the function. Then sketch a graph of its derivative directly beneath.

42.

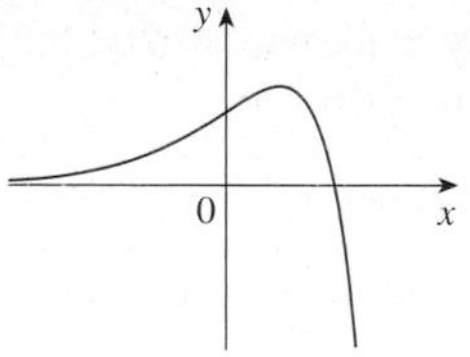

43.

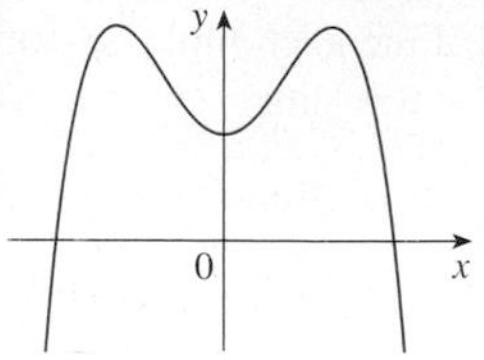

44.

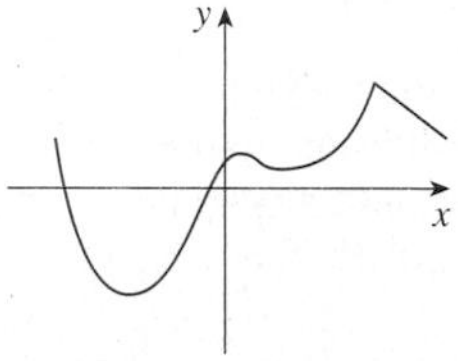

---

45. (a) If $f(x) = \sqrt{3 - 5x}$, use the definition of a derivative to find $f'(x)$.
(b) Find the domains of f and f'.
(c) Graph f and f' on a common screen. Compare the graphs to see whether your answer to part (a) is reasonable.

46. (a) Find the asymptotes of the graph of $f(x) = \dfrac{4 - x}{3 + x}$ and use them to sketch the graph.
(b) Use your graph from part (a) to sketch the graph of f'.
(c) Use the definition of a derivative to find $f'(x)$.
(d) Use a graphing device to graph f' and compare with your sketch in part (b).

47. The graph of f is shown. State, with reasons, the numbers at which f is not differentiable.

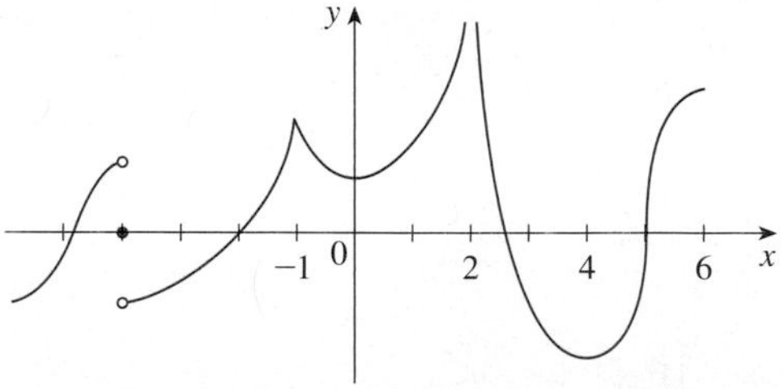

48. The figure shows the graphs of f, f', and f''. Identify each curve, and explain your choices.

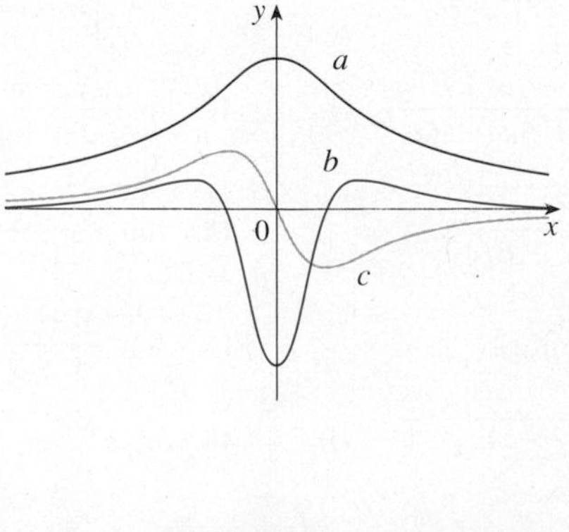

49. Let $C(t)$ be the total value of US currency (coins and banknotes) in circulation at time t. The table gives values of this function from 1980 to 2000, as of September 30, in billions of dollars. Interpret and estimate the value of $C'(1990)$.

t	1980	1985	1990	1995	2000
$C(t)$	129.9	187.3	271.9	409.3	568.6

50. The *total fertility rate* at time t, denoted by $F(t)$, is an estimate of the average number of children born to each woman (assuming that current birth rates remain constant). The graph of the total fertility rate in the United States shows the fluctuations from 1940 to 1990.

(a) Estimate the values of $F'(1950)$, $F'(1965)$, and $F'(1987)$.
(b) What are the meanings of these derivatives?
(c) Can you suggest reasons for the values of these derivatives?

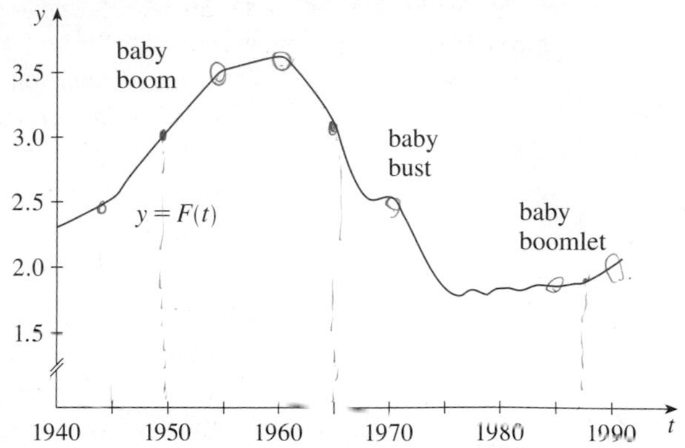

51. Suppose that $|f(x)| \leq g(x)$ for all x, where $\lim_{x \to a} g(x) = 0$. Find $\lim_{x \to a} f(x)$.

52. Let $f(x) = [\![x]\!] + [\![-x]\!]$.

(a) For what values of a does $\lim_{x \to a} f(x)$ exist?
(b) At what numbers is f discontinuous?

PROBLEMS PLUS

In our discussion of the principles of problem solving we considered the problem-solving strategy of *introducing something extra* (see page 76). In the following example we show how this principle is sometimes useful when we evaluate limits. The idea is to change the variable—to introduce a new variable that is related to the original variable—in such a way as to make the problem simpler. Later, in Section 5.5, we will make more extensive use of this general idea.

EXAMPLE 1 Evaluate $\lim_{x \to 0} \frac{\sqrt[3]{1 + cx} - 1}{x}$, where c is a nonzero constant.

SOLUTION As it stands, this limit looks challenging. In Section 2.3 we evaluated several limits in which both numerator and denominator approached 0. There our strategy was to perform some sort of algebraic manipulation that led to a simplifying cancellation, but here it's not clear what kind of algebra is necessary.

So we introduce a new variable t by the equation

$$t = \sqrt[3]{1 + cx}$$

We also need to express x in terms of t, so we solve this equation:

$$t^3 = 1 + cx \qquad x = \frac{t^3 - 1}{c}$$

Notice that $x \to 0$ is equivalent to $t \to 1$. This allows us to convert the given limit into one involving the variable t:

$$\begin{aligned} \lim_{x \to 0} \frac{\sqrt[3]{1 + cx} - 1}{x} &= \lim_{t \to 1} \frac{t - 1}{(t^3 - 1)/c} \\ &= \lim_{t \to 1} \frac{c(t - 1)}{t^3 - 1} \end{aligned}$$

The change of variable allowed us to replace a relatively complicated limit by a simpler one of a type that we have seen before. Factoring the denominator as a difference of cubes, we get

$$\begin{aligned} \lim_{t \to 1} \frac{c(t - 1)}{t^3 - 1} &= \lim_{t \to 1} \frac{c(t - 1)}{(t - 1)(t^2 + t + 1)} \\ &= \lim_{t \to 1} \frac{c}{t^2 + t + 1} = \frac{c}{3} \end{aligned}$$

□

The following problems are meant to test and challenge your problem-solving skills. Some of them require a considerable amount of time to think through, so don't be discouraged if you can't solve them right away. If you get stuck, you might find it helpful to refer to the discussion of the principles of problem solving on page 76.

PROBLEMS

1. Evaluate $\lim_{x \to 1} \frac{\sqrt[3]{x} - 1}{\sqrt{x} - 1}$.

2. Find numbers a and b such that $\lim_{x \to 0} \frac{\sqrt{ax + b} - 2}{x} = 1$.

3. Evaluate $\displaystyle\lim_{x\to 0}\frac{|2x-1|-|2x+1|}{x}$.

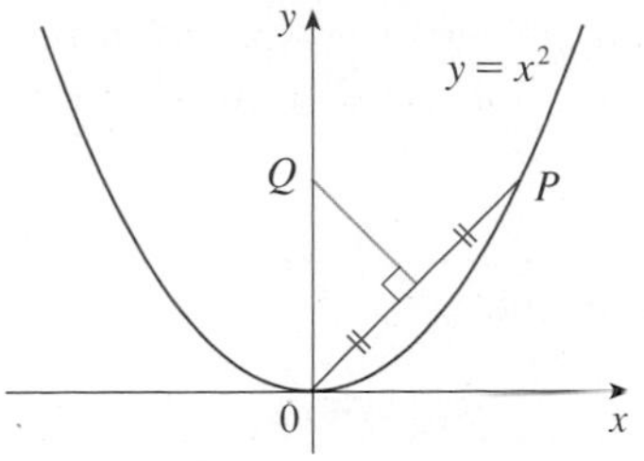

FIGURE FOR PROBLEM 4

4. The figure shows a point P on the parabola $y = x^2$ and the point Q where the perpendicular bisector of OP intersects the y-axis. As P approaches the origin along the parabola, what happens to Q? Does it have a limiting position? If so, find it.

5. If $[\![x]\!]$ denotes the greatest integer function, find $\displaystyle\lim_{x\to\infty}\frac{x}{[\![x]\!]}$.

6. Sketch the region in the plane defined by each of the following equations.
(a) $[\![x]\!]^2 + [\![y]\!]^2 = 1$ (b) $[\![x]\!]^2 - [\![y]\!]^2 = 3$ (c) $[\![x+y]\!]^2 = 1$ (d) $[\![x]\!] + [\![y]\!] = 1$

7. Find all values of a such that f is continuous on $\mathbb{R}$:

$$f(x) = \begin{cases} x+1 & \text{if } x \leqslant a \\ x^2 & \text{if } x > a \end{cases}$$

8. A **fixed point** of a function f is a number c in its domain such that $f(c) = c$. (The function doesn't move c; it stays fixed.)
(a) Sketch the graph of a continuous function with domain $[0, 1]$ whose range also lies in $[0, 1]$. Locate a fixed point of f.
(b) Try to draw the graph of a continuous function with domain $[0, 1]$ and range in $[0, 1]$ that does *not* have a fixed point. What is the obstacle?
(c) Use the Intermediate Value Theorem to prove that any continuous function with domain $[0, 1]$ and range a subset of $[0, 1]$ must have a fixed point.

9. If $\lim_{x\to a}[f(x) + g(x)] = 2$ and $\lim_{x\to a}[f(x) - g(x)] = 1$, find $\lim_{x\to a}[f(x)g(x)]$.

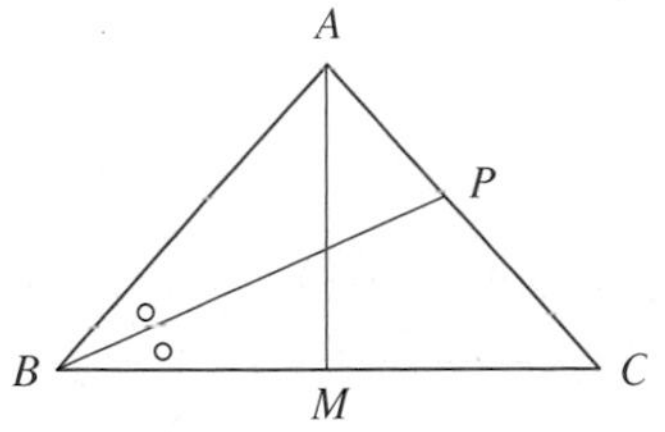

FIGURE FOR PROBLEM 10

10. (a) The figure shows an isosceles triangle ABC with $\angle B = \angle C$. The bisector of angle B intersects the side AC at the point P. Suppose that the base BC remains fixed but the altitude $|AM|$ of the triangle approaches 0, so A approaches the midpoint M of BC. What happens to P during this process? Does it have a limiting position? If so, find it.
(b) Try to sketch the path traced out by P during this process. Then find an equation of this curve and use this equation to sketch the curve.

11. (a) If we start from 0° latitude and proceed in a westerly direction, we can let $T(x)$ denote the temperature at the point x at any given time. Assuming that T is a continuous function of x, show that at any fixed time there are at least two diametrically opposite points on the equator that have exactly the same temperature.
(b) Does the result in part (a) hold for points lying on any circle on the earth's surface?
(c) Does the result in part (a) hold for barometric pressure and for altitude above sea level?

12. If f is a differentiable function and $g(x) = xf(x)$, use the definition of a derivative to show that $g'(x) = xf'(x) + f(x)$.

13. Suppose f is a function that satisfies the equation

$$f(x + y) = f(x) + f(y) + x^2y + xy^2$$

for all real numbers x and y. Suppose also that

$$\lim_{x\to 0}\frac{f(x)}{x} = 1$$

(a) Find $f(0)$. (b) Find $f'(0)$. (c) Find $f'(x)$.

14. Suppose f is a function with the property that $|f(x)| \leqslant x^2$ for all x. Show that $f(0) = 0$. Then show that $f'(0) = 0$.

3

DIFFERENTIATION RULES

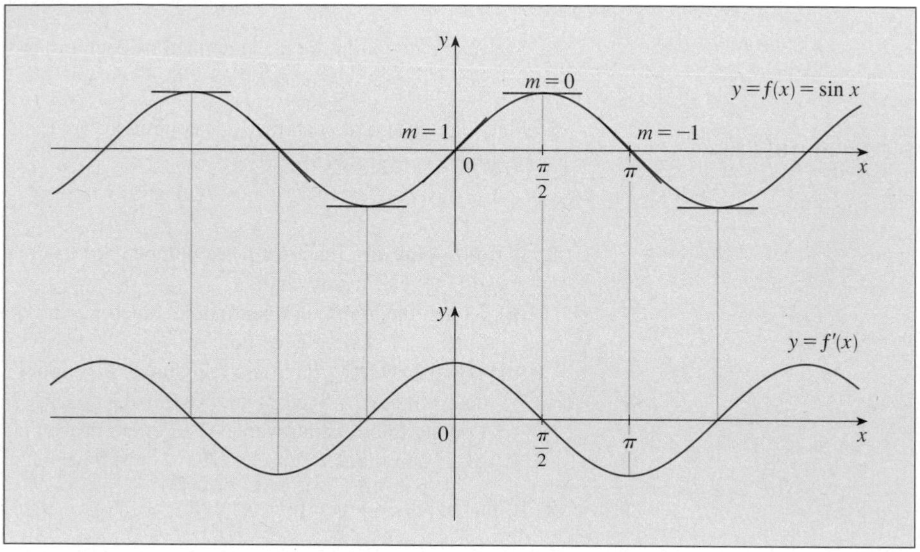

By measuring slopes at points on the sine curve, we get strong visual evidence that the derivative of the sine function is the cosine function.

We have seen how to interpret derivatives as slopes and rates of change. We have seen how to estimate derivatives of functions given by tables of values. We have learned how to graph derivatives of functions that are defined graphically. We have used the definition of a derivative to calculate the derivatives of functions defined by formulas. But it would be tedious if we always had to use the definition, so in this chapter we develop rules for finding derivatives without having to use the definition directly. These differentiation rules enable us to calculate with relative ease the derivatives of polynomials, rational functions, algebraic functions, exponential and logarithmic functions, and trigonometric and inverse trigonometric functions. We then use these rules to solve problems involving rates of change and the approximation of functions.

3.1 DERIVATIVES OF POLYNOMIALS AND EXPONENTIAL FUNCTIONS

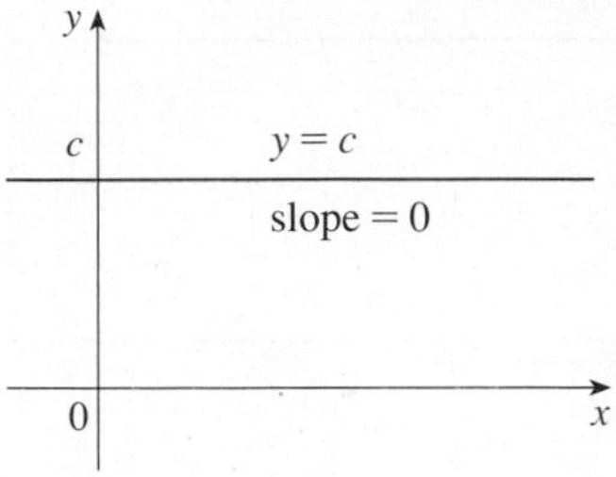

FIGURE 1
The graph of $f(x) = c$ is the line $y = c$, so $f'(x) = 0$.

In this section we learn how to differentiate constant functions, power functions, polynomials, and exponential functions.

Let's start with the simplest of all functions, the constant function $f(x) = c$. The graph of this function is the horizontal line $y = c$, which has slope 0, so we must have $f'(x) = 0$. (See Figure 1.) A formal proof, from the definition of a derivative, is also easy:

$$f'(x) = \lim_{h \to 0} \frac{f(x+h) - f(x)}{h} = \lim_{h \to 0} \frac{c - c}{h} = \lim_{h \to 0} 0 = 0$$

In Leibniz notation, we write this rule as follows.

DERIVATIVE OF A CONSTANT FUNCTION

$$\frac{d}{dx}(c) = 0$$

POWER FUNCTIONS

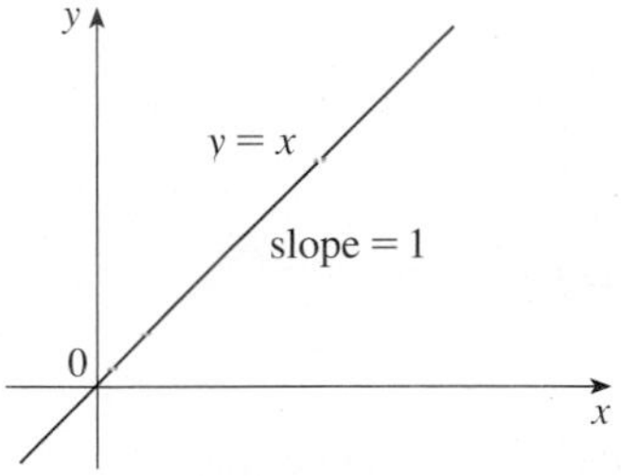

FIGURE 2
The graph of $f(x) = x$ is the line $y = x$, so $f'(x) = 1$.

We next look at the functions $f(x) = x^n$, where n is a positive integer. If $n = 1$, the graph of $f(x) = x$ is the line $y = x$, which has slope 1. (See Figure 2.) So

1 $$\frac{d}{dx}(x) = 1$$

(You can also verify Equation 1 from the definition of a derivative.) We have already investigated the cases $n = 2$ and $n = 3$. In fact, in Section 2.8 (Exercises 17 and 18) we found that

2 $$\frac{d}{dx}(x^2) = 2x \qquad \frac{d}{dx}(x^3) = 3x^2$$

For $n = 4$ we find the derivative of $f(x) = x^4$ as follows:

$$\begin{aligned} f'(x) &= \lim_{h \to 0} \frac{f(x+h) - f(x)}{h} = \lim_{h \to 0} \frac{(x+h)^4 - x^4}{h} \\ &= \lim_{h \to 0} \frac{x^4 + 4x^3h + 6x^2h^2 + 4xh^3 + h^4 - x^4}{h} \\ &= \lim_{h \to 0} \frac{4x^3h + 6x^2h^2 + 4xh^3 + h^4}{h} \\ &= \lim_{h \to 0} (4x^3 + 6x^2h + 4xh^2 + h^3) = 4x^3 \end{aligned}$$

Thus

3 $$\frac{d}{dx}(x^4) = 4x^3$$

Comparing the equations in (1), (2), and (3), we see a pattern emerging. It seems to be a reasonable guess that, when n is a positive integer, $(d/dx)(x^n) = nx^{n-1}$. This turns out to be true.

THE POWER RULE If n is a positive integer, then

$$\frac{d}{dx}(x^n) = nx^{n-1}$$

FIRST PROOF The formula

$$x^n - a^n = (x - a)(x^{n-1} + x^{n-2}a + \cdots + xa^{n-2} + a^{n-1})$$

can be verified simply by multiplying out the right-hand side (or by summing the second factor as a geometric series). If $f(x) = x^n$, we can use Equation 2.7.5 for $f'(a)$ and the equation above to write

$$\begin{aligned} f'(a) &= \lim_{x \to a} \frac{f(x) - f(a)}{x - a} = \lim_{x \to a} \frac{x^n - a^n}{x - a} \\ &= \lim_{x \to a} (x^{n-1} + x^{n-2}a + \cdots + xa^{n-2} + a^{n-1}) \\ &= a^{n-1} + a^{n-2}a + \cdots + aa^{n-2} + a^{n-1} \\ &= na^{n-1} \end{aligned}$$

SECOND PROOF

$$f'(x) = \lim_{h \to 0} \frac{f(x + h) - f(x)}{h} = \lim_{h \to 0} \frac{(x + h)^n - x^n}{h}$$

■ The Binomial Theorem is given on Reference Page 1.

In finding the derivative of x^4 we had to expand $(x + h)^4$. Here we need to expand $(x + h)^n$ and we use the Binomial Theorem to do so:

$$\begin{aligned} f'(x) &= \lim_{h \to 0} \frac{\left[x^n + nx^{n-1}h + \dfrac{n(n-1)}{2}x^{n-2}h^2 + \cdots + nxh^{n-1} + h^n \right] - x^n}{h} \\ &= \lim_{h \to 0} \frac{nx^{n-1}h + \dfrac{n(n-1)}{2}x^{n-2}h^2 + \cdots + nxh^{n-1} + h^n}{h} \\ &= \lim_{h \to 0} \left[nx^{n-1} + \frac{n(n-1)}{2}x^{n-2}h + \cdots + nxh^{n-2} + h^{n-1} \right] \\ &= nx^{n-1} \end{aligned}$$

because every term except the first has h as a factor and therefore approaches 0. □

We illustrate the Power Rule using various notations in Example 1.

EXAMPLE 1

(a) If $f(x) = x^6$, then $f'(x) = 6x^5$.

(b) If $y = x^{1000}$, then $y' = 1000x^{999}$.

(c) If $y = t^4$, then $\dfrac{dy}{dt} = 4t^3$.

(d) $\dfrac{d}{dr}(r^3) = 3r^2$ □

What about power functions with negative integer exponents? In Exercise 61 we ask you to verify from the definition of a derivative that

$$\frac{d}{dx}\left(\frac{1}{x}\right) = -\frac{1}{x^2}$$

We can rewrite this equation as

$$\frac{d}{dx}(x^{-1}) = (-1)x^{-2}$$

and so the Power Rule is true when $n = -1$. In fact, we will show in the next section [Exercise 58(c)] that it holds for all negative integers.

What if the exponent is a fraction? In Example 3 in Section 2.8 we found that

$$\frac{d}{dx}\sqrt{x} = \frac{1}{2\sqrt{x}}$$

which can be written as

$$\frac{d}{dx}(x^{1/2}) = \tfrac{1}{2}x^{-1/2}$$

This shows that the Power Rule is true even when $n = \frac{1}{2}$. In fact, we will show in Section 3.6 that it is true for all real numbers n.

THE POWER RULE (GENERAL VERSION) If n is any real number, then

$$\frac{d}{dx}(x^n) = nx^{n-1}$$

■ Figure 3 shows the function y in Example 2(b) and its derivative y'. Notice that y is not differentiable at 0 (y' is not defined there). Observe that y' is positive when y increases and is negative when y decreases.

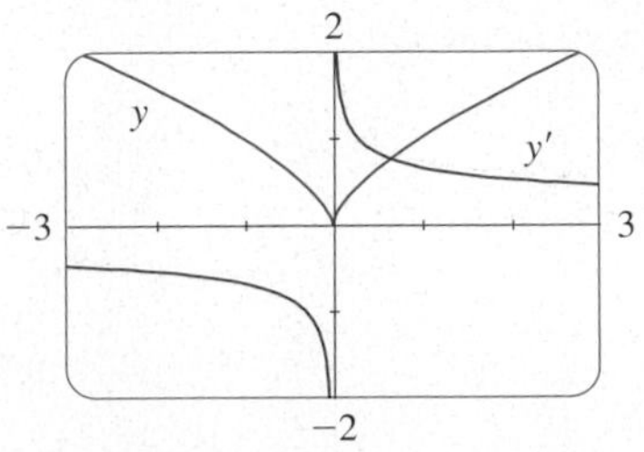

FIGURE 3
$y = \sqrt[3]{x^2}$

EXAMPLE 2 Differentiate:

(a) $f(x) = \dfrac{1}{x^2}$ (b) $y = \sqrt[3]{x^2}$

SOLUTION In each case we rewrite the function as a power of x.

(a) Since $f(x) = x^{-2}$, we use the Power Rule with $n = -2$:

$$f'(x) = \frac{d}{dx}(x^{-2}) = -2x^{-2-1} = -2x^{-3} = -\frac{2}{x^3}$$

(b) $$\frac{dy}{dx} = \frac{d}{dx}\left(\sqrt[3]{x^2}\right) = \frac{d}{dx}(x^{2/3}) = \tfrac{2}{3}x^{(2/3)-1} = \tfrac{2}{3}x^{-1/3}$$ □

The Power Rule enables us to find tangent lines without having to resort to the definition of a derivative. It also enables us to find *normal lines*. The **normal line** to a curve C at a point P is the line through P that is perpendicular to the tangent line at P. (In the study of optics, one needs to consider the angle between a light ray and the normal line to a lens.)

EXAMPLE 3 Find equations of the tangent line and normal line to the curve $y = x\sqrt{x}$ at the point $(1, 1)$. Illustrate by graphing the curve and these lines.

SOLUTION The derivative of $f(x) = x\sqrt{x} = xx^{1/2} = x^{3/2}$ is

$$f'(x) = \tfrac{3}{2}x^{(3/2)-1} = \tfrac{3}{2}x^{1/2} = \tfrac{3}{2}\sqrt{x}$$

So the slope of the tangent line at $(1, 1)$ is $f'(1) = \frac{3}{2}$. Therefore an equation of the tangent line is

$$y - 1 = \tfrac{3}{2}(x - 1) \qquad \text{or} \qquad y = \tfrac{3}{2}x - \tfrac{1}{2}$$

The normal line is perpendicular to the tangent line, so its slope is the negative reciprocal of $\frac{3}{2}$, that is, $-\frac{2}{3}$. Thus an equation of the normal line is

$$y - 1 = -\tfrac{2}{3}(x - 1) \qquad \text{or} \qquad y = -\tfrac{2}{3}x + \tfrac{5}{3}$$

We graph the curve and its tangent line and normal line in Figure 4. □

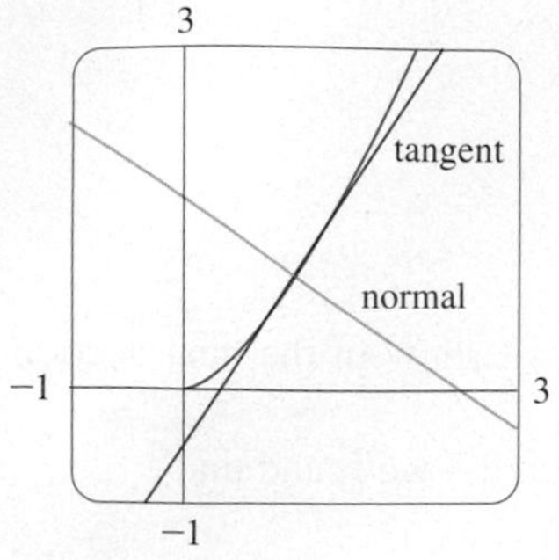

FIGURE 4

NEW DERIVATIVES FROM OLD

When new functions are formed from old functions by addition, subtraction, or multiplication by a constant, their derivatives can be calculated in terms of derivatives of the old functions. In particular, the following formula says that *the derivative of a constant times a function is the constant times the derivative of the function.*

■ GEOMETRIC INTERPRETATION OF THE CONSTANT MULTIPLE RULE

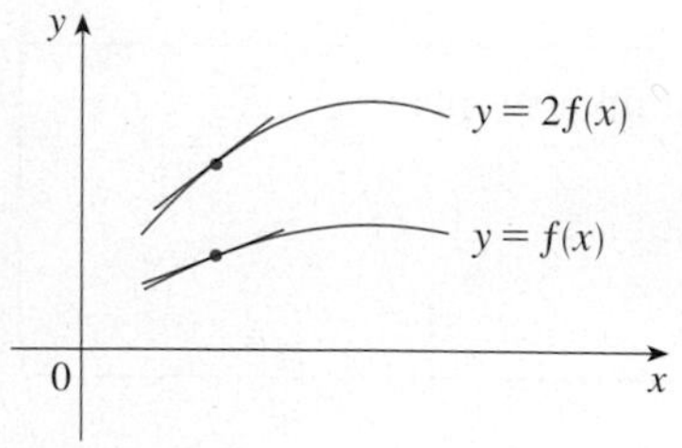

Multiplying by $c = 2$ stretches the graph vertically by a factor of 2. All the rises have been doubled but the runs stay the same. So the slopes are doubled, too.

THE CONSTANT MULTIPLE RULE If c is a constant and f is a differentiable function, then

$$\frac{d}{dx}[cf(x)] = c\frac{d}{dx}f(x)$$

PROOF Let $g(x) = cf(x)$. Then

$$g'(x) = \lim_{h\to 0}\frac{g(x + h) - g(x)}{h} = \lim_{h\to 0}\frac{cf(x + h) - cf(x)}{h}$$

$$= \lim_{h\to 0} c\left[\frac{f(x + h) - f(x)}{h}\right]$$

$$= c\lim_{h\to 0}\frac{f(x + h) - f(x)}{h} \qquad \text{(by Law 3 of limits)}$$

$$= cf'(x)$$ □

EXAMPLE 4

(a) $\dfrac{d}{dx}(3x^4) = 3\dfrac{d}{dx}(x^4) = 3(4x^3) = 12x^3$

(b) $\dfrac{d}{dx}(-x) = \dfrac{d}{dx}[(-1)x] = (-1)\dfrac{d}{dx}(x) = -1(1) = -1$ □

The next rule tells us that *the derivative of a sum of functions is the sum of the derivatives.*

■ Using prime notation, we can write the Sum Rule as

$$(f + g)' = f' + g'$$

THE SUM RULE If f and g are both differentiable, then

$$\frac{d}{dx}[f(x) + g(x)] = \frac{d}{dx}f(x) + \frac{d}{dx}g(x)$$

PROOF Let $F(x) = f(x) + g(x)$. Then

$$\begin{aligned} F'(x) &= \lim_{h \to 0} \frac{F(x+h) - F(x)}{h} \\ &= \lim_{h \to 0} \frac{[f(x+h) + g(x+h)] - [f(x) + g(x)]}{h} \\ &= \lim_{h \to 0} \left[\frac{f(x+h) - f(x)}{h} + \frac{g(x+h) - g(x)}{h} \right] \\ &= \lim_{h \to 0} \frac{f(x+h) - f(x)}{h} + \lim_{h \to 0} \frac{g(x+h) - g(x)}{h} \qquad \text{(by Law 1)} \\ &= f'(x) + g'(x) \end{aligned}$$

□

The Sum Rule can be extended to the sum of any number of functions. For instance, using this theorem twice, we get

$$(f + g + h)' = [(f + g) + h]' = (f + g)' + h' = f' + g' + h'$$

By writing $f - g$ as $f + (-1)g$ and applying the Sum Rule and the Constant Multiple Rule, we get the following formula.

THE DIFFERENCE RULE If f and g are both differentiable, then

$$\frac{d}{dx}[f(x) - g(x)] = \frac{d}{dx}f(x) - \frac{d}{dx}g(x)$$

The Constant Multiple Rule, the Sum Rule, and the Difference Rule can be combined with the Power Rule to differentiate any polynomial, as the following examples demonstrate.

EXAMPLE 5

$$\begin{aligned} &\frac{d}{dx}(x^8 + 12x^5 - 4x^4 + 10x^3 - 6x + 5) \\ &\qquad = \frac{d}{dx}(x^8) + 12\frac{d}{dx}(x^5) - 4\frac{d}{dx}(x^4) + 10\frac{d}{dx}(x^3) - 6\frac{d}{dx}(x) + \frac{d}{dx}(5) \\ &\qquad = 8x^7 + 12(5x^4) - 4(4x^3) + 10(3x^2) - 6(1) + 0 \\ &\qquad = 8x^7 + 60x^4 - 16x^3 + 30x^2 - 6 \end{aligned}$$

□

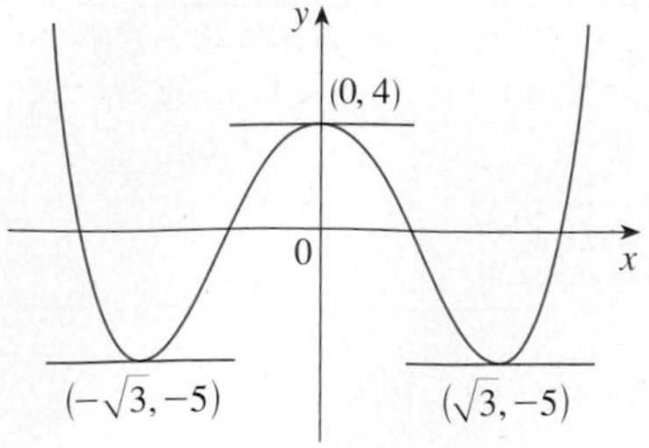

FIGURE 5
The curve $y = x^4 - 6x^2 + 4$ and its horizontal tangents

V EXAMPLE 6 Find the points on the curve $y = x^4 - 6x^2 + 4$ where the tangent line is horizontal.

SOLUTION Horizontal tangents occur where the derivative is zero. We have

$$\frac{dy}{dx} = \frac{d}{dx}(x^4) - 6\frac{d}{dx}(x^2) + \frac{d}{dx}(4)$$

$$= 4x^3 - 12x + 0 = 4x(x^2 - 3)$$

Thus $dy/dx = 0$ if $x = 0$ or $x^2 - 3 = 0$, that is, $x = \pm\sqrt{3}$. So the given curve has horizontal tangents when $x = 0$, $\sqrt{3}$, and $-\sqrt{3}$. The corresponding points are $(0, 4)$, $(\sqrt{3}, -5)$, and $(-\sqrt{3}, -5)$. (See Figure 5.) □

EXAMPLE 7 The equation of motion of a particle is $s = 2t^3 - 5t^2 + 3t + 4$, where s is measured in centimeters and t in seconds. Find the acceleration as a function of time. What is the acceleration after 2 seconds?

SOLUTION The velocity and acceleration are

$$v(t) = \frac{ds}{dt} = 6t^2 - 10t + 3$$

$$a(t) = \frac{dv}{dt} = 12t - 10$$

The acceleration after 2 s is $a(2) = 14 \text{ cm/s}^2$. □

EXPONENTIAL FUNCTIONS

Let's try to compute the derivative of the exponential function $f(x) = a^x$ using the definition of a derivative:

$$f'(x) = \lim_{h \to 0} \frac{f(x + h) - f(x)}{h} = \lim_{h \to 0} \frac{a^{x+h} - a^x}{h}$$

$$= \lim_{h \to 0} \frac{a^x a^h - a^x}{h} = \lim_{h \to 0} \frac{a^x(a^h - 1)}{h}$$

The factor a^x doesn't depend on h, so we can take it in front of the limit:

$$f'(x) = a^x \lim_{h \to 0} \frac{a^h - 1}{h}$$

Notice that the limit is the value of the derivative of f at 0, that is,

$$\lim_{h \to 0} \frac{a^h - 1}{h} = f'(0)$$

Therefore we have shown that if the exponential function $f(x) = a^x$ is differentiable at 0, then it is differentiable everywhere and

4 $$f'(x) = f'(0)a^x$$

This equation says that *the rate of change of any exponential function is proportional to the function itself.* (The slope is proportional to the height.)

h	$\dfrac{2^h - 1}{h}$	$\dfrac{3^h - 1}{h}$
0.1	0.7177	1.1612
0.01	0.6956	1.1047
0.001	0.6934	1.0992
0.0001	0.6932	1.0987

Numerical evidence for the existence of $f'(0)$ is given in the table at the left for the cases $a = 2$ and $a = 3$. (Values are stated correct to four decimal places.) It appears that the limits exist and

$$\text{for } a = 2, \quad f'(0) = \lim_{h \to 0} \frac{2^h - 1}{h} \approx 0.69$$

$$\text{for } a = 3, \quad f'(0) = \lim_{h \to 0} \frac{3^h - 1}{h} \approx 1.10$$

In fact, it can be proved that these limits exist and, correct to six decimal places, the values are

$$\left. \frac{d}{dx}(2^x) \right|_{x=0} \approx 0.693147 \qquad \left. \frac{d}{dx}(3^x) \right|_{x=0} \approx 1.098612$$

Thus, from Equation 4, we have

5 $$\frac{d}{dx}(2^x) \approx (0.69)2^x \qquad \frac{d}{dx}(3^x) \approx (1.10)3^x$$

Of all possible choices for the base a in Equation 4, the simplest differentiation formula occurs when $f'(0) = 1$. In view of the estimates of $f'(0)$ for $a = 2$ and $a = 3$, it seems reasonable that there is a number a between 2 and 3 for which $f'(0) = 1$. It is traditional to denote this value by the letter e. (In fact, that is how we introduced e in Section 1.5.) Thus we have the following definition.

In Exercise 1 we will see that e lies between 2.7 and 2.8. Later we will be able to show that, correct to five decimal places,

$$e \approx 2.71828$$

DEFINITION OF THE NUMBER e

$$e \text{ is the number such that } \lim_{h \to 0} \frac{e^h - 1}{h} = 1$$

Geometrically, this means that of all the possible exponential functions $y = a^x$, the function $f(x) = e^x$ is the one whose tangent line at (0, 1) has a slope $f'(0)$ that is exactly 1. (See Figures 6 and 7.)

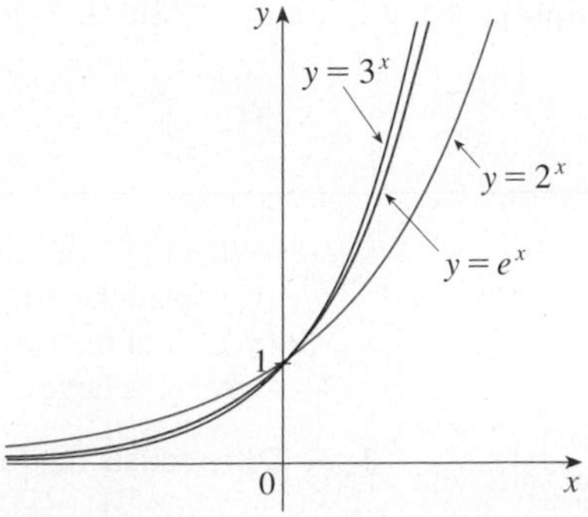

FIGURE 6

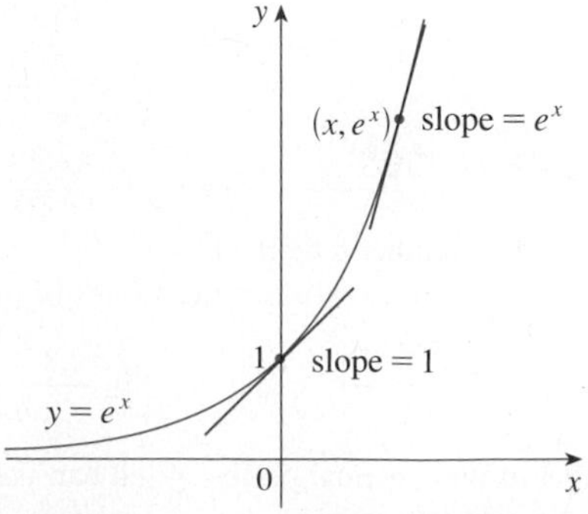

FIGURE 7

If we put $a = e$ and, therefore, $f'(0) = 1$ in Equation 4, it becomes the following important differentiation formula.

TEC Visual 3.1 uses the slope-a-scope to illustrate this formula.

DERIVATIVE OF THE NATURAL EXPONENTIAL FUNCTION

$$\frac{d}{dx}(e^x) = e^x$$

Thus the exponential function $f(x) = e^x$ has the property that it is its own derivative. The geometrical significance of this fact is that the slope of a tangent line to the curve $y = e^x$ is equal to the y-coordinate of the point (see Figure 7).

V EXAMPLE 8 If $f(x) = e^x - x$, find f' and f''. Compare the graphs of f and f'.

SOLUTION Using the Difference Rule, we have

$$f'(x) = \frac{d}{dx}(e^x - x) = \frac{d}{dx}(e^x) - \frac{d}{dx}(x) = e^x - 1$$

In Section 2.8 we defined the second derivative as the derivative of f', so

$$f''(x) = \frac{d}{dx}(e^x - 1) = \frac{d}{dx}(e^x) - \frac{d}{dx}(1) = e^x$$

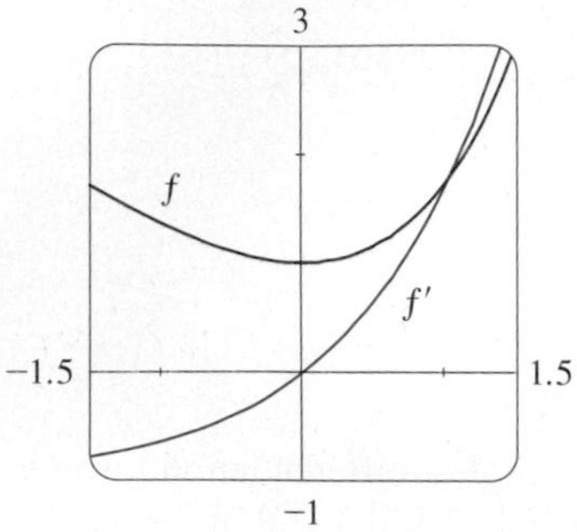

FIGURE 8

The function f and its derivative f' are graphed in Figure 8. Notice that f has a horizontal tangent when $x = 0$; this corresponds to the fact that $f'(0) = 0$. Notice also that, for $x > 0$, $f'(x)$ is positive and f is increasing. When $x < 0$, $f'(x)$ is negative and f is decreasing. □

EXAMPLE 9 At what point on the curve $y = e^x$ is the tangent line parallel to the line $y = 2x$?

SOLUTION Since $y = e^x$, we have $y' = e^x$. Let the x-coordinate of the point in question be a. Then the slope of the tangent line at that point is e^a. This tangent line will be parallel to the line $y = 2x$ if it has the same slope, that is, 2. Equating slopes, we get

$$e^a = 2 \qquad a = \ln 2$$

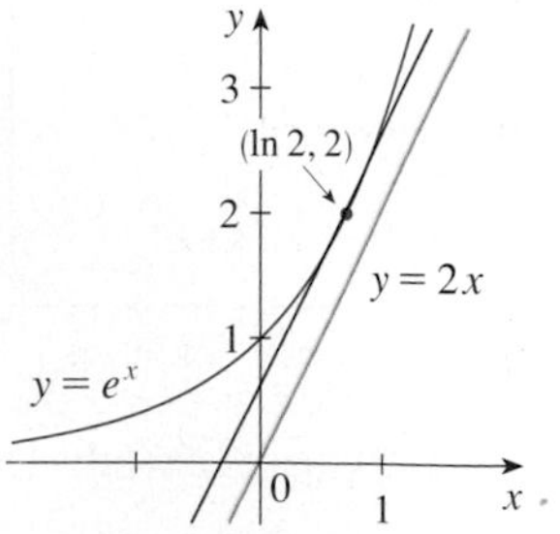

FIGURE 9

Therefore the required point is $(a, e^a) = (\ln 2, 2)$. (See Figure 9.) □

3.1 EXERCISES

1. (a) How is the number e defined?
(b) Use a calculator to estimate the values of the limits

$$\lim_{h \to 0} \frac{2.7^h - 1}{h} \quad \text{and} \quad \lim_{h \to 0} \frac{2.8^h - 1}{h}$$

correct to two decimal places. What can you conclude about the value of e?

2. (a) Sketch, by hand, the graph of the function $f(x) = e^x$, paying particular attention to how the graph crosses the y-axis. What fact allows you to do this?
(b) What types of functions are $f(x) = e^x$ and $g(x) = x^e$? Compare the differentiation formulas for f and g.
(c) Which of the two functions in part (b) grows more rapidly when x is large?

3–32 Differentiate the function.

3. $f(x) = 186.5$

4. $f(x) = \sqrt{30}$

5. $f(t) = 2 - \frac{2}{3}t$

6. $F(x) = \frac{3}{4}x^8$

7. $f(x) = x^3 - 4x + 6$

8. $f(t) = \frac{1}{2}t^6 - 3t^4 + t$

9. $f(t) = \frac{1}{4}(t^4 + 8)$

10. $h(x) = (x - 2)(2x + 3)$

11. $y = x^{-2/5}$

12. $y = 5e^x + 3$

13. $V(r) = \frac{4}{3}\pi r^3$

14. $R(t) = 5t^{-3/5}$

15. $A(s) = -\dfrac{12}{s^5}$

16. $B(y) = cy^{-6}$

17. $G(x) = \sqrt{x} - 2e^x$

18. $y = \sqrt[3]{x}$

19. $F(x) = (\frac{1}{2}x)^5$

20. $f(t) = \sqrt{t} - \dfrac{1}{\sqrt{t}}$

21. $y = ax^2 + bx + c$

22. $y = \sqrt{x}\,(x - 1)$

23. $y = \dfrac{x^2 + 4x + 3}{\sqrt{x}}$

24. $y = \dfrac{x^2 - 2\sqrt{x}}{x}$

25. $y = 4\pi^2$

26. $g(u) = \sqrt{2}\,u + \sqrt{3u}$

27. $H(x) = (x + x^{-1})^3$

28. $y = ae^v + \dfrac{b}{v} + \dfrac{c}{v^2}$

29. $u = \sqrt[5]{t} + 4\sqrt{t^5}$

30. $v = \left(\sqrt{x} + \dfrac{1}{\sqrt[3]{x}}\right)^2$

31. $z = \dfrac{A}{y^{10}} + Be^y$

32. $y = e^{x+1} + 1$

33–34 Find an equation of the tangent line to the curve at the given point.

33. $y = \sqrt[4]{x}$, $(1, 1)$

34. $y = x^4 + 2x^2 - x$, $(1, 2)$

35–36 Find equations of the tangent line and normal line to the curve at the given point.

35. $y = x^4 + 2e^x$, $(0, 2)$

36. $y = (1 + 2x)^2$, $(1, 9)$

37–38 Find an equation of the tangent line to the curve at the given point. Illustrate by graphing the curve and the tangent line on the same screen.

37. $y = 3x^2 - x^3$, $(1, 2)$

38. $y = x - \sqrt{x}$, $(1, 0)$

39–42 Find $f'(x)$. Compare the graphs of f and f' and use them to explain why your answer is reasonable.

39. $f(x) = e^x - 5x$

40. $f(x) = 3x^5 - 20x^3 + 50x$

41. $f(x) = 3x^{15} - 5x^3 + 3$

42. $f(x) = x + \dfrac{1}{x}$

43. (a) Use a graphing calculator or computer to graph the function $f(x) = x^4 - 3x^3 - 6x^2 + 7x + 30$ in the viewing rectangle $[-3, 5]$ by $[-10, 50]$.

(b) Using the graph in part (a) to estimate slopes, make a rough sketch, by hand, of the graph of f'. (See Example 1 in Section 2.8.)

(c) Calculate $f'(x)$ and use this expression, with a graphing device, to graph f'. Compare with your sketch in part (b).

44. (a) Use a graphing calculator or computer to graph the function $g(x) = e^x - 3x^2$ in the viewing rectangle $[-1, 4]$ by $[-8, 8]$.

(b) Using the graph in part (a) to estimate slopes, make a rough sketch, by hand, of the graph of g'. (See Example 1 in Section 2.8.)

(c) Calculate $g'(x)$ and use this expression, with a graphing device, to graph g'. Compare with your sketch in part (b).

45–46 Find the first and second derivatives of the function.

45. $f(x) = x^4 - 3x^3 + 16x$

46. $G(r) = \sqrt{r} + \sqrt[3]{r}$

47–48 Find the first and second derivatives of the function. Check to see that your answers are reasonable by comparing the graphs of f, f', and f''.

47. $f(x) = 2x - 5x^{3/4}$

48. $f(x) = e^x - x^3$

49. The equation of motion of a particle is $s = t^3 - 3t$, where s is in meters and t is in seconds. Find

(a) the velocity and acceleration as functions of t,

(b) the acceleration after 2 s, and

(c) the acceleration when the velocity is 0.

50. The equation of motion of a particle is $s = 2t^3 - 7t^2 + 4t + 1$, where s is in meters and t is in seconds.

(a) Find the velocity and acceleration as functions of t.

(b) Find the acceleration after 1 s.

(c) Graph the position, velocity, and acceleration functions on the same screen.

51. Find the points on the curve $y = 2x^3 + 3x^2 - 12x + 1$ where the tangent is horizontal.

52. For what values of x does the graph of $f(x) = x^3 + 3x^2 + x + 3$ have a horizontal tangent?

53. Show that the curve $y = 6x^3 + 5x - 3$ has no tangent line with slope 4.

54. Find an equation of the tangent line to the curve $y = x\sqrt{x}$ that is parallel to the line $y = 1 + 3x$.

55. Find equations of both lines that are tangent to the curve $y = 1 + x^3$ and are parallel to the line $12x - y = 1$.

56. At what point on the curve $y = 1 + 2e^x - 3x$ is the tangent line parallel to the line $3x - y = 5$? Illustrate by graphing the curve and both lines.

57. Find an equation of the normal line to the parabola $y = x^2 - 5x + 4$ that is parallel to the line $x - 3y = 5$.

58. Where does the normal line to the parabola $y = x - x^2$ at the point $(1, 0)$ intersect the parabola a second time? Illustrate with a sketch.

59. Draw a diagram to show that there are two tangent lines to the parabola $y = x^2$ that pass through the point $(0, -4)$. Find the coordinates of the points where these tangent lines intersect the parabola.

60. (a) Find equations of both lines through the point $(2, -3)$ that are tangent to the parabola $y = x^2 + x$.
(b) Show that there is no line through the point $(2, 7)$ that is tangent to the parabola. Then draw a diagram to see why.

61. Use the definition of a derivative to show that if $f(x) = 1/x$, then $f'(x) = -1/x^2$. (This proves the Power Rule for the case $n = -1$.)

62. Find the nth derivative of each function by calculating the first few derivatives and observing the pattern that occurs.
(a) $f(x) = x^n$ (b) $f(x) = 1/x$

63. Find a second-degree polynomial P such that $P(2) = 5$, $P'(2) = 3$, and $P''(2) = 2$.

64. The equation $y'' + y' - 2y = x^2$ is called a **differential equation** because it involves an unknown function y and its derivatives y' and y''. Find constants A, B, and C such that the function $y = Ax^2 + Bx + C$ satisfies this equation. (Differential equations will be studied in detail in Chapter 9.)

65. Find a cubic function $y = ax^3 + bx^2 + cx + d$ whose graph has horizontal tangents at the points $(-2, 6)$ and $(2, 0)$.

66. Find a parabola with equation $y = ax^2 + bx + c$ that has slope 4 at $x = 1$, slope -8 at $x = -1$, and passes through the point $(2, 15)$.

67. Let

$$f(x) = \begin{cases} 2 - x & \text{if } x \leqslant 1 \\ x^2 - 2x + 2 & \text{if } x > 1 \end{cases}$$

Is f differentiable at 1? Sketch the graphs of f and f'.

68. At what numbers is the following function g differentiable?

$$g(x) = \begin{cases} -1 - 2x & \text{if } x < -1 \\ x^2 & \text{if } -1 \leqslant x \leqslant 1 \\ x & \text{if } x > 1 \end{cases}$$

Give a formula for g' and sketch the graphs of g and g'.

69. (a) For what values of x is the function $f(x) = |x^2 - 9|$ differentiable? Find a formula for f'.
(b) Sketch the graphs of f and f'.

70. Where is the function $h(x) = |x - 1| + |x + 2|$ differentiable? Give a formula for h' and sketch the graphs of h and h'.

71. Find the parabola with equation $y = ax^2 + bx$ whose tangent line at $(1, 1)$ has equation $y = 3x - 2$.

72. Suppose the curve $y = x^4 + ax^3 + bx^2 + cx + d$ has a tangent line when $x = 0$ with equation $y = 2x + 1$ and a tangent line when $x = 1$ with equation $y = 2 - 3x$. Find the values of a, b, c, and d.

73. For what values of a and b is the line $2x + y = b$ tangent to the parabola $y = ax^2$ when $x = 2$?

74. Find the value of c such that the line $y = \frac{3}{2}x + 6$ is tangent to the curve $y = c\sqrt{x}$.

75. Let

$$f(x) = \begin{cases} x^2 & \text{if } x \leqslant 2 \\ mx + b & \text{if } x > 2 \end{cases}$$

Find the values of m and b that make f differentiable everywhere.

76. A tangent line is drawn to the hyperbola $xy = c$ at a point P.
(a) Show that the midpoint of the line segment cut from this tangent line by the coordinate axes is P.
(b) Show that the triangle formed by the tangent line and the coordinate axes always has the same area, no matter where P is located on the hyperbola.

77. Evaluate $\displaystyle\lim_{x \to 1} \frac{x^{1000} - 1}{x - 1}$.

78. Draw a diagram showing two perpendicular lines that intersect on the y-axis and are both tangent to the parabola $y = x^2$. Where do these lines intersect?

79. If $c > \frac{1}{2}$, how many lines through the point $(0, c)$ are normal lines to the parabola $y = x^2$? What if $c \leqslant \frac{1}{2}$?

80. Sketch the parabolas $y = x^2$ and $y = x^2 - 2x + 2$. Do you think there is a line that is tangent to both curves? If so, find its equation. If not, why not?

APPLIED PROJECT — BUILDING A BETTER ROLLER COASTER

Suppose you are asked to design the first ascent and drop for a new roller coaster. By studying photographs of your favorite coasters, you decide to make the slope of the ascent 0.8 and the slope of the drop -1.6. You decide to connect these two straight stretches $y = L_1(x)$ and $y = L_2(x)$ with part of a parabola $y = f(x) = ax^2 + bx + c$, where x and $f(x)$ are measured in feet. For the track to be smooth there can't be abrupt changes in direction, so you want the linear

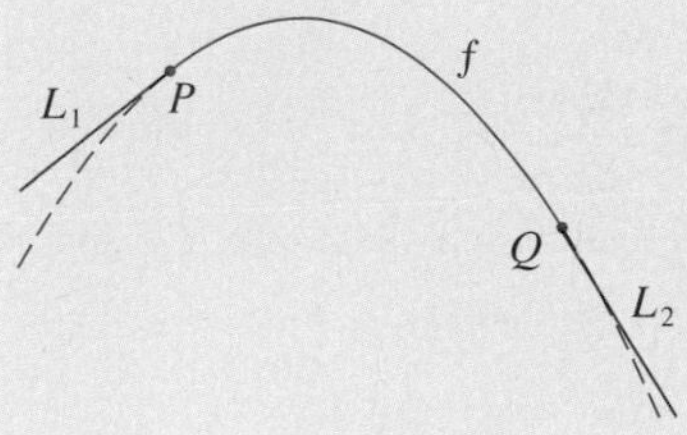

segments L_1 and L_2 to be tangent to the parabola at the transition points P and Q. (See the figure.) To simplify the equations, you decide to place the origin at P.

1. (a) Suppose the horizontal distance between P and Q is 100 ft. Write equations in a, b, and c that will ensure that the track is smooth at the transition points.
(b) Solve the equations in part (a) for a, b, and c to find a formula for $f(x)$.
(c) Plot L_1, f, and L_2 to verify graphically that the transitions are smooth.
(d) Find the difference in elevation between P and Q.

2. The solution in Problem 1 might *look* smooth, but it might not *feel* smooth because the piecewise defined function [consisting of $L_1(x)$ for $x < 0$, $f(x)$ for $0 \leqslant x \leqslant 100$, and $L_2(x)$ for $x > 100$] doesn't have a continuous second derivative. So you decide to improve the design by using a quadratic function $q(x) = ax^2 + bx + c$ only on the interval $10 \leqslant x \leqslant 90$ and connecting it to the linear functions by means of two cubic functions:

$$g(x) = kx^3 + lx^2 + mx + n \qquad 0 \leqslant x < 10$$

$$h(x) = px^3 + qx^2 + rx + s \qquad 90 < x \leqslant 100$$

(a) Write a system of equations in 11 unknowns that ensure that the functions and their first two derivatives agree at the transition points.
CAS (b) Solve the equations in part (a) with a computer algebra system to find formulas for $q(x)$, $g(x)$, and $h(x)$.
(c) Plot L_1, g, q, h, and L_2, and compare with the plot in Problem 1(c).

3.2 THE PRODUCT AND QUOTIENT RULES

The formulas of this section enable us to differentiate new functions formed from old functions by multiplication or division.

THE PRODUCT RULE

By analogy with the Sum and Difference Rules, one might be tempted to guess, as Leibniz did three centuries ago, that the derivative of a product is the product of the derivatives. We can see, however, that this guess is wrong by looking at a particular example. Let $f(x) = x$ and $g(x) = x^2$. Then the Power Rule gives $f'(x) = 1$ and $g'(x) = 2x$. But $(fg)(x) = x^3$, so $(fg)'(x) = 3x^2$. Thus $(fg)' \neq f'g'$. The correct formula was discovered by Leibniz (soon after his false start) and is called the Product Rule.

Before stating the Product Rule, let's see how we might discover it. We start by assuming that $u = f(x)$ and $v = g(x)$ are both positive differentiable functions. Then we can interpret the product uv as an area of a rectangle (see Figure 1). If x changes by an amount Δx, then the corresponding changes in u and v are

$$\Delta u = f(x + \Delta x) - f(x) \qquad \Delta v = g(x + \Delta x) - g(x)$$

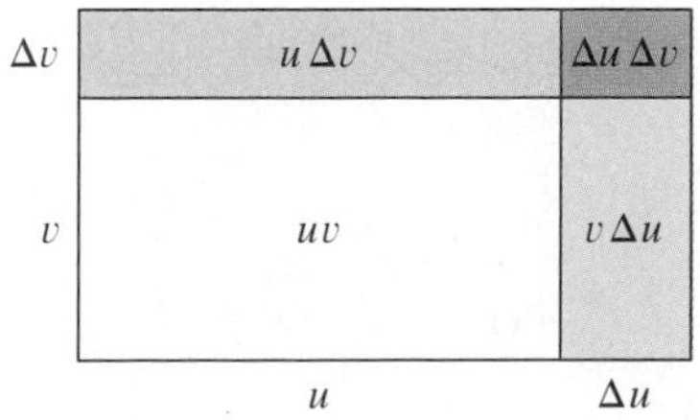

FIGURE 1
The geometry of the Product Rule

and the new value of the product, $(u + \Delta u)(v + \Delta v)$, can be interpreted as the area of the largest rectangle in Figure 1 (provided that Δu and Δv happen to be positive).

The change in the area of the rectangle is

$$\boxed{1} \qquad \Delta(uv) = (u + \Delta u)(v + \Delta v) - uv = u\,\Delta v + v\,\Delta u + \Delta u\,\Delta v$$

$$= \text{the sum of the three shaded areas}$$

If we divide by Δx, we get

$$\frac{\Delta(uv)}{\Delta x} = u\frac{\Delta v}{\Delta x} + v\frac{\Delta u}{\Delta x} + \Delta u\frac{\Delta v}{\Delta x}$$

■ Recall that in Leibniz notation the definition of a derivative can be written as

$$\frac{dy}{dx} = \lim_{\Delta x \to 0} \frac{\Delta y}{\Delta x}$$

If we now let $\Delta x \to 0$, we get the derivative of uv:

$$\begin{aligned}\frac{d}{dx}(uv) &= \lim_{\Delta x \to 0} \frac{\Delta(uv)}{\Delta x} = \lim_{\Delta x \to 0}\left(u\frac{\Delta v}{\Delta x} + v\frac{\Delta u}{\Delta x} + \Delta u\frac{\Delta v}{\Delta x}\right)\\ &= u\lim_{\Delta x \to 0}\frac{\Delta v}{\Delta x} + v\lim_{\Delta x \to 0}\frac{\Delta u}{\Delta x} + \left(\lim_{\Delta x \to 0}\Delta u\right)\left(\lim_{\Delta x \to 0}\frac{\Delta v}{\Delta x}\right)\\ &= u\frac{dv}{dx} + v\frac{du}{dx} + 0 \cdot \frac{dv}{dx}\end{aligned}$$

2 $$\frac{d}{dx}(uv) = u\frac{dv}{dx} + v\frac{du}{dx}$$

(Notice that $\Delta u \to 0$ as $\Delta x \to 0$ since f is differentiable and therefore continuous.)

Although we started by assuming (for the geometric interpretation) that all the quantities are positive, we notice that Equation 1 is always true. (The algebra is valid whether u, v, Δu, and Δv are positive or negative.) So we have proved Equation 2, known as the Product Rule, for all differentiable functions u and v.

■ In prime notation:

$$(fg)' = fg' + gf'$$

THE PRODUCT RULE If f and g are both differentiable, then

$$\frac{d}{dx}[f(x)g(x)] = f(x)\frac{d}{dx}[g(x)] + g(x)\frac{d}{dx}[f(x)]$$

In words, the Product Rule says that *the derivative of a product of two functions is the first function times the derivative of the second function plus the second function times the derivative of the first function.*

EXAMPLE 1
(a) If $f(x) = xe^x$, find $f'(x)$.
(b) Find the nth derivative, $f^{(n)}(x)$.

■ Figure 2 shows the graphs of the function f of Example 1 and its derivative f'. Notice that $f'(x)$ is positive when f is increasing and negative when f is decreasing.

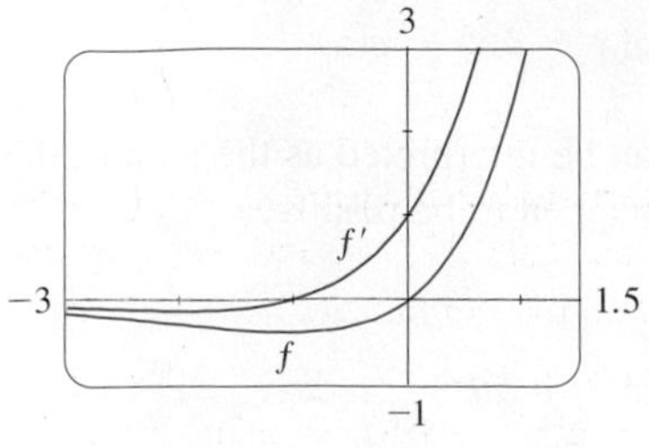

FIGURE 2

SOLUTION
(a) By the Product Rule, we have

$$\begin{aligned}f'(x) &= \frac{d}{dx}(xe^x) = x\frac{d}{dx}(e^x) + e^x\frac{d}{dx}(x)\\ &= xe^x + e^x \cdot 1 = (x+1)e^x\end{aligned}$$

(b) Using the Product Rule a second time, we get

$$\begin{aligned}f''(x) &= \frac{d}{dx}[(x+1)e^x] = (x+1)\frac{d}{dx}(e^x) + e^x\frac{d}{dx}(x+1)\\ &= (x+1)e^x + e^x \cdot 1 = (x+2)e^x\end{aligned}$$

Further applications of the Product Rule give

$$f'''(x) = (x + 3)e^x \qquad f^{(4)}(x) = (x + 4)e^x$$

In fact, each successive differentiation adds another term e^x, so

$$f^{(n)}(x) = (x + n)e^x$$

□

■ In Example 2, a and b are constants. It is customary in mathematics to use letters near the beginning of the alphabet to represent constants and letters near the end of the alphabet to represent variables.

EXAMPLE 2 Differentiate the function $f(t) = \sqrt{t}\,(a + bt)$.

SOLUTION 1 Using the Product Rule, we have

$$\begin{aligned} f'(t) &= \sqrt{t}\,\frac{d}{dt}(a + bt) + (a + bt)\frac{d}{dt}(\sqrt{t}) \\ &= \sqrt{t}\cdot b + (a + bt)\cdot \tfrac{1}{2}t^{-1/2} \\ &= b\sqrt{t} + \frac{a + bt}{2\sqrt{t}} = \frac{a + 3bt}{2\sqrt{t}} \end{aligned}$$

SOLUTION 2 If we first use the laws of exponents to rewrite $f(t)$, then we can proceed directly without using the Product Rule.

$$f(t) = a\sqrt{t} + bt\sqrt{t} = at^{1/2} + bt^{3/2}$$

$$f'(t) = \tfrac{1}{2}at^{-1/2} + \tfrac{3}{2}bt^{1/2}$$

which is equivalent to the answer given in Solution 1. □

Example 2 shows that it is sometimes easier to simplify a product of functions than to use the Product Rule. In Example 1, however, the Product Rule is the only possible method.

EXAMPLE 3 If $f(x) = \sqrt{x}\,g(x)$, where $g(4) = 2$ and $g'(4) = 3$, find $f'(4)$.

SOLUTION Applying the Product Rule, we get

$$\begin{aligned} f'(x) &= \frac{d}{dx}\left[\sqrt{x}\,g(x)\right] = \sqrt{x}\,\frac{d}{dx}[g(x)] + g(x)\frac{d}{dx}\left[\sqrt{x}\right] \\ &= \sqrt{x}\,g'(x) + g(x)\cdot\tfrac{1}{2}x^{-1/2} = \sqrt{x}\,g'(x) + \frac{g(x)}{2\sqrt{x}} \end{aligned}$$

So $$f'(4) = \sqrt{4}\,g'(4) + \frac{g(4)}{2\sqrt{4}} = 2\cdot 3 + \frac{2}{2\cdot 2} = 6.5$$ □

THE QUOTIENT RULE

We find a rule for differentiating the quotient of two differentiable functions $u = f(x)$ and $v = g(x)$ in much the same way that we found the Product Rule. If x, u, and v change by amounts Δx, Δu, and Δv, then the corresponding change in the quotient u/v is

$$\Delta\left(\frac{u}{v}\right) = \frac{u + \Delta u}{v + \Delta v} - \frac{u}{v} = \frac{(u + \Delta u)v - u(v + \Delta v)}{v(v + \Delta v)} = \frac{v\,\Delta u - u\,\Delta v}{v(v + \Delta v)}$$

so

$$\frac{d}{dx}\left(\frac{u}{v}\right) = \lim_{\Delta x \to 0} \frac{\Delta(u/v)}{\Delta x} = \lim_{\Delta x \to 0} \frac{v\dfrac{\Delta u}{\Delta x} - u\dfrac{\Delta v}{\Delta x}}{v(v + \Delta v)}$$

As $\Delta x \to 0$, $\Delta v \to 0$ also, because $v = g(x)$ is differentiable and therefore continuous. Thus, using the Limit Laws, we get

$$\frac{d}{dx}\left(\frac{u}{v}\right) = \frac{v \displaystyle\lim_{\Delta x \to 0} \frac{\Delta u}{\Delta x} - u \lim_{\Delta x \to 0} \frac{\Delta v}{\Delta x}}{v \displaystyle\lim_{\Delta x \to 0} (v + \Delta v)} = \frac{v\dfrac{du}{dx} - u\dfrac{dv}{dx}}{v^2}$$

■ In prime notation:

$$\left(\frac{f}{g}\right)' = \frac{gf' - fg'}{g^2}$$

THE QUOTIENT RULE If f and g are differentiable, then

$$\frac{d}{dx}\left[\frac{f(x)}{g(x)}\right] = \frac{g(x)\dfrac{d}{dx}[f(x)] - f(x)\dfrac{d}{dx}[g(x)]}{[g(x)]^2}$$

In words, the Quotient Rule says that the *derivative of a quotient is the denominator times the derivative of the numerator minus the numerator times the derivative of the denominator, all divided by the square of the denominator.*

The Quotient Rule and the other differentiation formulas enable us to compute the derivative of any rational function, as the next example illustrates.

■ We can use a graphing device to check that the answer to Example 4 is plausible. Figure 3 shows the graphs of the function of Example 4 and its derivative. Notice that when y grows rapidly (near -2), y' is large. And when y grows slowly, y' is near 0.

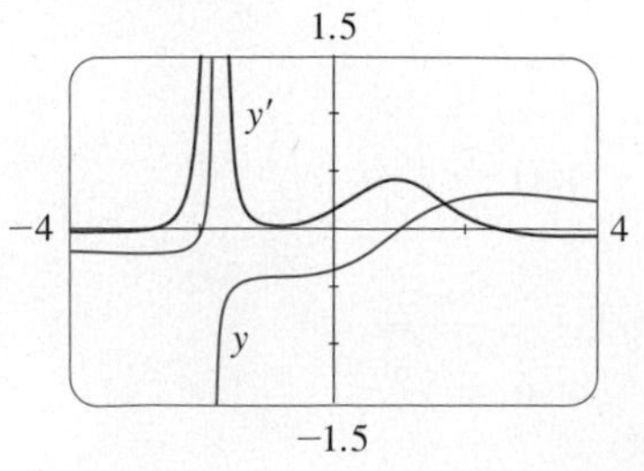

FIGURE 3

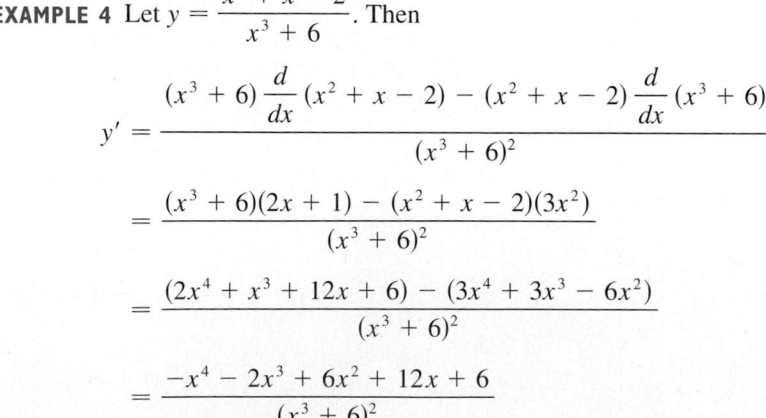

EXAMPLE 4 Let $y = \dfrac{x^2 + x - 2}{x^3 + 6}$. Then

$$y' = \frac{(x^3 + 6)\dfrac{d}{dx}(x^2 + x - 2) - (x^2 + x - 2)\dfrac{d}{dx}(x^3 + 6)}{(x^3 + 6)^2}$$

$$= \frac{(x^3 + 6)(2x + 1) - (x^2 + x - 2)(3x^2)}{(x^3 + 6)^2}$$

$$= \frac{(2x^4 + x^3 + 12x + 6) - (3x^4 + 3x^3 - 6x^2)}{(x^3 + 6)^2}$$

$$= \frac{-x^4 - 2x^3 + 6x^2 + 12x + 6}{(x^3 + 6)^2}$$ □

EXAMPLE 5 Find an equation of the tangent line to the curve $y = e^x/(1 + x^2)$ at the point $\left(1, \tfrac{1}{2}e\right)$.

SOLUTION According to the Quotient Rule, we have

$$\frac{dy}{dx} = \frac{(1 + x^2)\dfrac{d}{dx}(e^x) - e^x\dfrac{d}{dx}(1 + x^2)}{(1 + x^2)^2}$$

$$= \frac{(1 + x^2)e^x - e^x(2x)}{(1 + x^2)^2} = \frac{e^x(1 - x)^2}{(1 + x^2)^2}$$

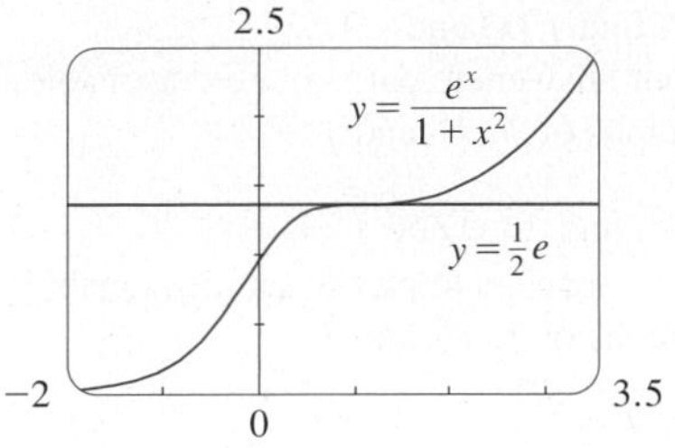

FIGURE 4

So the slope of the tangent line at $(1, \frac{1}{2}e)$ is

$$\left.\frac{dy}{dx}\right|_{x=1} = 0$$

This means that the tangent line at $(1, \frac{1}{2}e)$ is horizontal and its equation is $y = \frac{1}{2}e$. [See Figure 4. Notice that the function is increasing and crosses its tangent line at $(1, \frac{1}{2}e)$.] □

NOTE Don't use the Quotient Rule *every* time you see a quotient. Sometimes it's easier to rewrite a quotient first to put it in a form that is simpler for the purpose of differentiation. For instance, although it is possible to differentiate the function

$$F(x) = \frac{3x^2 + 2\sqrt{x}}{x}$$

using the Quotient Rule, it is much easier to perform the division first and write the function as

$$F(x) = 3x + 2x^{-1/2}$$

before differentiating.

We summarize the differentiation formulas we have learned so far as follows.

TABLE OF DIFFERENTIATION FORMULAS

$\frac{d}{dx}(c) = 0$	$\frac{d}{dx}(x^n) = nx^{n-1}$	$\frac{d}{dx}(e^x) = e^x$
$(cf)' = cf'$	$(f + g)' = f' + g'$	$(f - g)' = f' - g'$
$(fg)' = fg' + gf'$	$\left(\frac{f}{g}\right)' = \frac{gf' - fg'}{g^2}$	

3.2 EXERCISES

1. Find the derivative of $y = (x^2 + 1)(x^3 + 1)$ in two ways: by using the Product Rule and by performing the multiplication first. Do your answers agree?

2. Find the derivative of the function

$$F(x) = \frac{x - 3x\sqrt{x}}{\sqrt{x}}$$

in two ways: by using the Quotient Rule and by simplifying first. Show that your answers are equivalent. Which method do you prefer?

3–26 Differentiate.

3. $f(x) = (x^3 + 2x)e^x$

4. $g(x) = \sqrt{x}\, e^x$

5. $y = \frac{e^x}{x^2}$

6. $y = \frac{e^x}{1 + x}$

7. $g(x) = \frac{3x - 1}{2x + 1}$

8. $f(t) = \frac{2t}{4 + t^2}$

9. $V(x) = (2x^3 + 3)(x^4 - 2x)$

10. $Y(u) = (u^{-2} + u^{-3})(u^5 - 2u^2)$

11. $F(y) = \left(\frac{1}{y^2} - \frac{3}{y^4}\right)(y + 5y^3)$

12. $R(t) = (t + e^t)(3 - \sqrt{t})$

13. $y = \frac{x^3}{1 - x^2}$

14. $y = \frac{x + 1}{x^3 + x - 2}$

15. $y = \frac{t^2 + 2}{t^4 - 3t^2 + 1}$

16. $y = \frac{t}{(t - 1)^2}$

17. $y = (r^2 - 2r)e^r$

18. $y = \frac{1}{s + ke^s}$

19. $y = \dfrac{v^3 - 2v\sqrt{v}}{v}$

20. $z = w^{3/2}(w + ce^w)$

21. $f(t) = \dfrac{2t}{2 + \sqrt{t}}$

22. $g(t) = \dfrac{t - \sqrt{t}}{t^{1/3}}$

23. $f(x) = \dfrac{A}{B + Ce^x}$

24. $f(x) = \dfrac{1 - xe^x}{x + e^x}$

25. $f(x) = \dfrac{x}{x + \dfrac{c}{x}}$

26. $f(x) = \dfrac{ax + b}{cx + d}$

27–30 Find $f'(x)$ and $f''(x)$.

27. $f(x) = x^4e^x$

28. $f(x) = x^{5/2}e^x$

29. $f(x) = \dfrac{x^2}{1 + 2x}$

30. $f(x) = \dfrac{x}{3 + e^x}$

31–32 Find an equation of the tangent line to the given curve at the specified point.

31. $y = \dfrac{2x}{x + 1}$, $(1, 1)$

32. $y = \dfrac{e^x}{x}$, $(1, e)$

33–34 Find equations of the tangent line and normal line to the given curve at the specified point.

33. $y = 2xe^x$, $(0, 0)$

34. $y = \dfrac{\sqrt{x}}{x + 1}$, $(4, 0.4)$

35. (a) The curve $y = 1/(1 + x^2)$ is called a **witch of Maria Agnesi**. Find an equation of the tangent line to this curve at the point $\left(-1, \frac{1}{2}\right)$.

(b) Illustrate part (a) by graphing the curve and the tangent line on the same screen.

36. (a) The curve $y = x/(1 + x^2)$ is called a **serpentine**. Find an equation of the tangent line to this curve at the point $(3, 0.3)$.

(b) Illustrate part (a) by graphing the curve and the tangent line on the same screen.

37. (a) If $f(x) = e^x/x^3$, find $f'(x)$.

(b) Check to see that your answer to part (a) is reasonable by comparing the graphs of f and f'.

38. (a) If $f(x) = x/(x^2 - 1)$, find $f'(x)$.

(b) Check to see that your answer to part (a) is reasonable by comparing the graphs of f and f'.

39. (a) If $f(x) = (x - 1)e^x$, find $f'(x)$ and $f''(x)$.

(b) Check to see that your answers to part (a) are reasonable by comparing the graphs of f, f', and f''.

40. (a) If $f(x) = x/(x^2 + 1)$, find $f'(x)$ and $f''(x)$.

(b) Check to see that your answers to part (a) are reasonable by comparing the graphs of f, f', and f''.

41. If $f(x) = x^2/(1 + x)$, find $f''(1)$.

42. If $g(x) = x/e^x$, find $g^{(n)}(x)$.

43. Suppose that $f(5) = 1$, $f'(5) = 6$, $g(5) = -3$, and $g'(5) = 2$. Find the following values.

(a) $(fg)'(5)$ (b) $(f/g)'(5)$

(c) $(g/f)'(5)$

44. Suppose that $f(2) = -3$, $g(2) = 4$, $f'(2) = -2$, and $g'(2) = 7$. Find $h'(2)$.

(a) $h(x) = 5f(x) - 4g(x)$ (b) $h(x) = f(x)g(x)$

(c) $h(x) = \dfrac{f(x)}{g(x)}$ (d) $h(x) = \dfrac{g(x)}{1 + f(x)}$

45. If $f(x) = e^xg(x)$, where $g(0) = 2$ and $g'(0) = 5$, find $f'(0)$.

46. If $h(2) = 4$ and $h'(2) = -3$, find

$$\frac{d}{dx}\left(\frac{h(x)}{x}\right)\bigg|_{x=2}$$

47. If f and g are the functions whose graphs are shown, let $u(x) = f(x)g(x)$ and $v(x) = f(x)/g(x)$.

(a) Find $u'(1)$. (b) Find $v'(5)$.

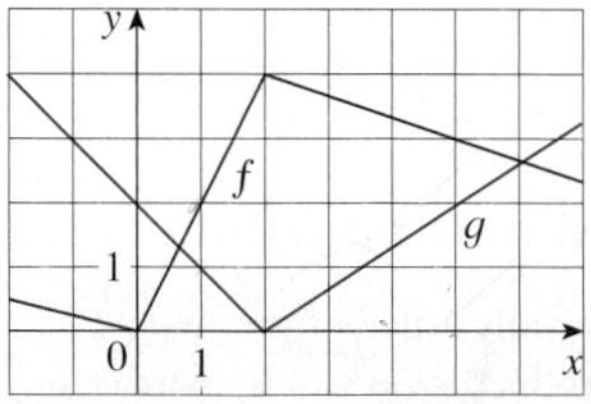

48. Let $P(x) = F(x)G(x)$ and $Q(x) = F(x)/G(x)$, where F and G are the functions whose graphs are shown.

(a) Find $P'(2)$. (b) Find $Q'(7)$.

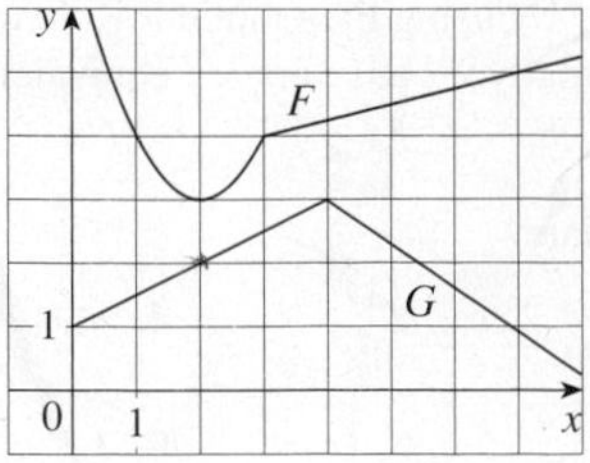

49. If g is a differentiable function, find an expression for the derivative of each of the following functions.

(a) $y = xg(x)$ (b) $y = \dfrac{x}{g(x)}$ (c) $y = \dfrac{g(x)}{x}$

50. If f is a differentiable function, find an expression for the derivative of each of the following functions.

(a) $y = x^2 f(x)$ (b) $y = \dfrac{f(x)}{x^2}$

(c) $y = \dfrac{x^2}{f(x)}$ (d) $y = \dfrac{1 + xf(x)}{\sqrt{x}}$

51. How many tangent lines to the curve $y = x/(x + 1)$ pass through the point $(1, 2)$? At which points do these tangent lines touch the curve?

52. Find equations of the tangent lines to the curve

$$y = \frac{x - 1}{x + 1}$$

that are parallel to the line $x - 2y = 2$.

53. In this exercise we estimate the rate at which the total personal income is rising in the Richmond-Petersburg, Virginia, metropolitan area. In 1999, the population of this area was 961,400, and the population was increasing at roughly 9200 people per year. The average annual income was \$30,593 per capita, and this average was increasing at about \$1400 per year (a little above the national average of about \$1225 yearly). Use the Product Rule and these figures to estimate the rate at which total personal income was rising in the Richmond-Petersburg area in 1999. Explain the meaning of each term in the Product Rule.

54. A manufacturer produces bolts of a fabric with a fixed width. The quantity q of this fabric (measured in yards) that is sold is a function of the selling price p (in dollars per yard), so we can write $q = f(p)$. Then the total revenue earned with selling price p is $R(p) = pf(p)$.

(a) What does it mean to say that $f(20) = 10{,}000$ and $f'(20) = -350$?

(b) Assuming the values in part (a), find $R'(20)$ and interpret your answer.

55. (a) Use the Product Rule twice to prove that if f, g, and h are differentiable, then $(fgh)' = f'gh + fg'h + fgh'$.

(b) Taking $f = g = h$ in part (a), show that

$$\frac{d}{dx}[f(x)]^3 = 3[f(x)]^2 f'(x)$$

(c) Use part (b) to differentiate $y = e^{3x}$.

56. (a) If $F(x) = f(x)g(x)$, where f and g have derivatives of all orders, show that $F'' = f''g + 2f'g' + fg''$.

(b) Find similar formulas for F''' and $F^{(4)}$.

(c) Guess a formula for $F^{(n)}$.

57. Find expressions for the first five derivatives of $f(x) = x^2e^x$. Do you see a pattern in these expressions? Guess a formula for $f^{(n)}(x)$ and prove it using mathematical induction.

58. (a) If g is differentiable, the **Reciprocal Rule** says that

$$\frac{d}{dx}\left[\frac{1}{g(x)}\right] = -\frac{g'(x)}{[g(x)]^2}$$

Use the Quotient Rule to prove the Reciprocal Rule.

(b) Use the Reciprocal Rule to differentiate the function in Exercise 18.

(c) Use the Reciprocal Rule to verify that the Power Rule is valid for negative integers, that is,

$$\frac{d}{dx}(x^{-n}) = -nx^{-n-1}$$

for all positive integers n.

3.3 DERIVATIVES OF TRIGONOMETRIC FUNCTIONS

■ A review of the trigonometric functions is given in Appendix D.

Before starting this section, you might need to review the trigonometric functions. In particular, it is important to remember that when we talk about the function f defined for all real numbers x by

$$f(x) = \sin x$$

it is understood that $\sin x$ means the sine of the angle whose *radian* measure is x. A similar convention holds for the other trigonometric functions cos, tan, csc, sec, and cot. Recall from Section 2.5 that all of the trigonometric functions are continuous at every number in their domains.

If we sketch the graph of the function $f(x) = \sin x$ and use the interpretation of $f'(x)$ as the slope of the tangent to the sine curve in order to sketch the graph of f' (see Exercise 14 in Section 2.8), then it looks as if the graph of f' may be the same as the cosine curve (see Figure 1).

TEC Visual 3.3 shows an animation of Figure 1.

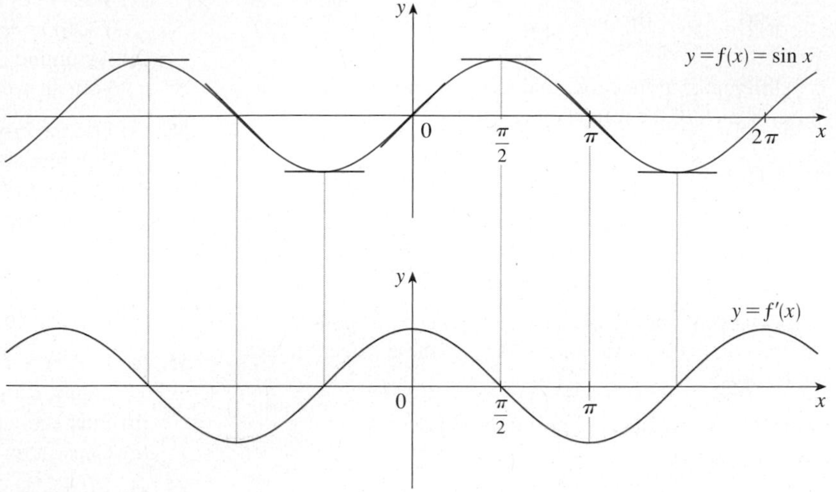

FIGURE 1

Let's try to confirm our guess that if $f(x) = \sin x$, then $f'(x) = \cos x$. From the definition of a derivative, we have

■ We have used the addition formula for sine. See Appendix D.

$$\begin{aligned} f'(x) &= \lim_{h\to 0} \frac{f(x+h) - f(x)}{h} = \lim_{h\to 0} \frac{\sin(x+h) - \sin x}{h} \\ &= \lim_{h\to 0} \frac{\sin x \cos h + \cos x \sin h - \sin x}{h} \\ &= \lim_{h\to 0} \left[\frac{\sin x \cos h - \sin x}{h} + \frac{\cos x \sin h}{h} \right] \\ &= \lim_{h\to 0} \left[\sin x \left(\frac{\cos h - 1}{h} \right) + \cos x \left(\frac{\sin h}{h} \right) \right] \end{aligned}$$

$$\boxed{1} \qquad = \lim_{h\to 0} \sin x \cdot \lim_{h\to 0} \frac{\cos h - 1}{h} + \lim_{h\to 0} \cos x \cdot \lim_{h\to 0} \frac{\sin h}{h}$$

Two of these four limits are easy to evaluate. Since we regard x as a constant when computing a limit as $h \to 0$, we have

$$\lim_{h\to 0} \sin x = \sin x \qquad \text{and} \qquad \lim_{h\to 0} \cos x = \cos x$$

The limit of $(\sin h)/h$ is not so obvious. In Example 3 in Section 2.2 we made the guess, on the basis of numerical and graphical evidence, that

$$\boxed{2} \qquad \boxed{\lim_{\theta\to 0} \frac{\sin \theta}{\theta} = 1}$$

We now use a geometric argument to prove Equation 2. Assume first that θ lies between 0 and $\pi/2$. Figure 2(a) shows a sector of a circle with center O, central angle θ, and

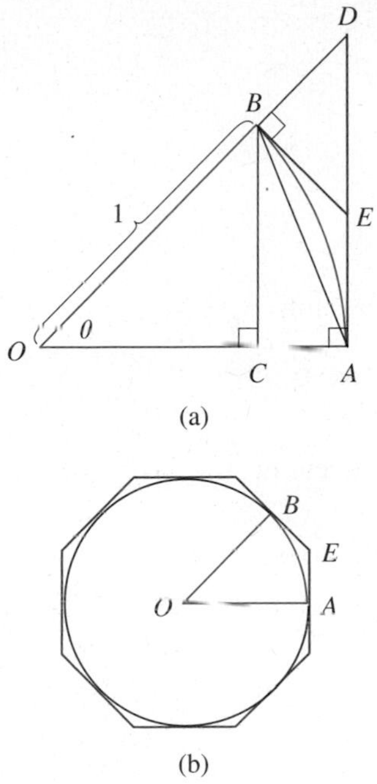

FIGURE 2

radius 1. BC is drawn perpendicular to OA. By the definition of radian measure, we have arc $AB = \theta$. Also $|BC| = |OB| \sin \theta = \sin \theta$. From the diagram we see that

$$|BC| < |AB| < \text{arc } AB$$

Therefore $$\sin \theta < \theta \qquad \text{so} \qquad \frac{\sin \theta}{\theta} < 1$$

Let the tangent lines at A and B intersect at E. You can see from Figure 2(b) that the circumference of a circle is smaller than the length of a circumscribed polygon, and so arc $AB < |AE| + |EB|$. Thus

$$\begin{aligned}\theta = \text{arc } AB &< |AE| + |EB| \\ &< |AE| + |ED| \\ &= |AD| = |OA| \tan \theta \\ &= \tan \theta\end{aligned}$$

(In Appendix F the inequality $\theta \leqslant \tan \theta$ is proved directly from the definition of the length of an arc without resorting to geometric intuition as we did here.) Therefore, we have

$$\theta < \frac{\sin \theta}{\cos \theta}$$

so $$\cos \theta < \frac{\sin \theta}{\theta} < 1$$

We know that $\lim_{\theta \to 0} 1 = 1$ and $\lim_{\theta \to 0} \cos \theta = 1$, so by the Squeeze Theorem, we have

$$\lim_{\theta \to 0^+} \frac{\sin \theta}{\theta} = 1$$

But the function $(\sin \theta)/\theta$ is an even function, so its right and left limits must be equal. Hence, we have

$$\lim_{\theta \to 0} \frac{\sin \theta}{\theta} = 1$$

so we have proved Equation 2.

We can deduce the value of the remaining limit in (1) as follows:

■ We multiply numerator and denominator by $\cos \theta + 1$ in order to put the function in a form in which we can use the limits we know.

$$\begin{aligned}\lim_{\theta \to 0} \frac{\cos \theta - 1}{\theta} &= \lim_{\theta \to 0} \left(\frac{\cos \theta - 1}{\theta} \cdot \frac{\cos \theta + 1}{\cos \theta + 1} \right) = \lim_{\theta \to 0} \frac{\cos^2 \theta - 1}{\theta (\cos \theta + 1)} \\ &= \lim_{\theta \to 0} \frac{-\sin^2 \theta}{\theta (\cos \theta + 1)} = -\lim_{\theta \to 0} \left(\frac{\sin \theta}{\theta} \cdot \frac{\sin \theta}{\cos \theta + 1} \right) \\ &= -\lim_{\theta \to 0} \frac{\sin \theta}{\theta} \cdot \lim_{\theta \to 0} \frac{\sin \theta}{\cos \theta + 1} \\ &= -1 \cdot \left(\frac{0}{1 + 1} \right) = 0 \qquad \text{(by Equation 2)}\end{aligned}$$

3
$$\lim_{\theta \to 0} \frac{\cos \theta - 1}{\theta} = 0$$

If we now put the limits (2) and (3) in (1), we get

$$f'(x) = \lim_{h \to 0} \sin x \cdot \lim_{h \to 0} \frac{\cos h - 1}{h} + \lim_{h \to 0} \cos x \cdot \lim_{h \to 0} \frac{\sin h}{h}$$
$$= (\sin x) \cdot 0 + (\cos x) \cdot 1 = \cos x$$

So we have proved the formula for the derivative of the sine function:

4
$$\frac{d}{dx}(\sin x) = \cos x$$

■ Figure 3 shows the graphs of the function of Example 1 and its derivative. Notice that $y' = 0$ whenever y has a horizontal tangent.

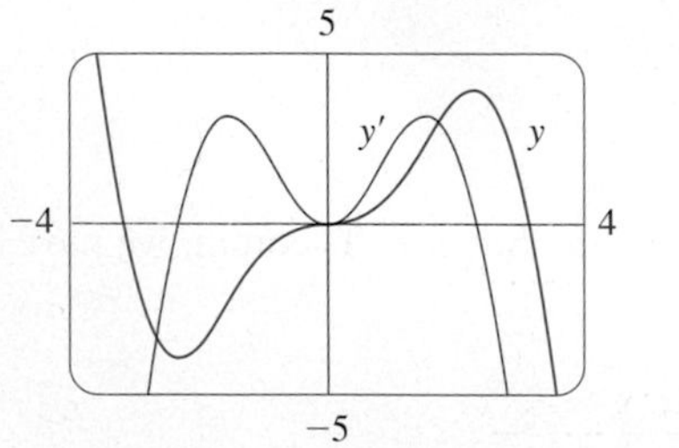

FIGURE 3

V EXAMPLE 1 Differentiate $y = x^2 \sin x$.

SOLUTION Using the Product Rule and Formula 4, we have

$$\frac{dy}{dx} = x^2 \frac{d}{dx}(\sin x) + \sin x \frac{d}{dx}(x^2)$$
$$= x^2 \cos x + 2x \sin x$$ □

Using the same methods as in the proof of Formula 4, one can prove (see Exercise 20) that

5
$$\frac{d}{dx}(\cos x) = -\sin x$$

The tangent function can also be differentiated by using the definition of a derivative, but it is easier to use the Quotient Rule together with Formulas 4 and 5:

$$\frac{d}{dx}(\tan x) = \frac{d}{dx}\left(\frac{\sin x}{\cos x}\right)$$
$$= \frac{\cos x \dfrac{d}{dx}(\sin x) - \sin x \dfrac{d}{dx}(\cos x)}{\cos^2 x}$$
$$= \frac{\cos x \cdot \cos x - \sin x(-\sin x)}{\cos^2 x}$$
$$= \frac{\cos^2 x + \sin^2 x}{\cos^2 x}$$
$$= \frac{1}{\cos^2 x} = \sec^2 x$$

6 $$\frac{d}{dx}(\tan x) = \sec^2 x$$

The derivatives of the remaining trigonometric functions, csc, sec, and cot, can also be found easily using the Quotient Rule (see Exercises 17–19). We collect all the differentiation formulas for trigonometric functions in the following table. Remember that they are valid only when x is measured in radians.

■ When you memorize this table, it is helpful to notice that the minus signs go with the derivatives of the "cofunctions," that is, cosine, cosecant, and cotangent.

DERIVATIVES OF TRIGONOMETRIC FUNCTIONS

$$\frac{d}{dx}(\sin x) = \cos x \qquad \frac{d}{dx}(\csc x) = -\csc x \cot x$$

$$\frac{d}{dx}(\cos x) = -\sin x \qquad \frac{d}{dx}(\sec x) = \sec x \tan x$$

$$\frac{d}{dx}(\tan x) = \sec^2 x \qquad \frac{d}{dx}(\cot x) = -\csc^2 x$$

EXAMPLE 2 Differentiate $f(x) = \dfrac{\sec x}{1 + \tan x}$. For what values of x does the graph of f have a horizontal tangent?

SOLUTION The Quotient Rule gives

$$f'(x) = \frac{(1 + \tan x)\dfrac{d}{dx}(\sec x) - \sec x \dfrac{d}{dx}(1 + \tan x)}{(1 + \tan x)^2}$$

$$= \frac{(1 + \tan x)\sec x \tan x - \sec x \cdot \sec^2 x}{(1 + \tan x)^2}$$

$$= \frac{\sec x\,(\tan x + \tan^2 x - \sec^2 x)}{(1 + \tan x)^2}$$

$$= \frac{\sec x\,(\tan x - 1)}{(1 + \tan x)^2}$$

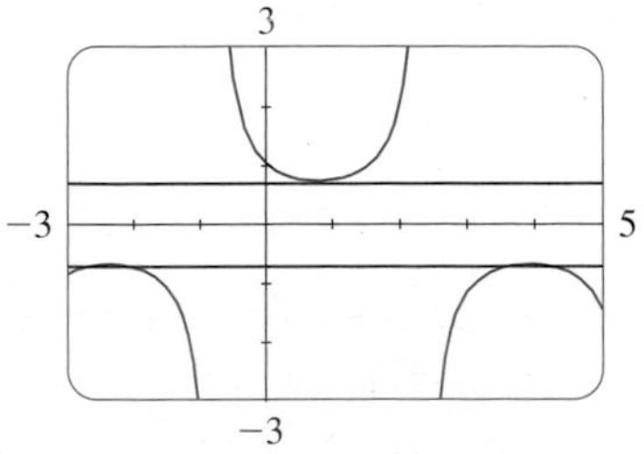

FIGURE 4
The horizontal tangents in Example 2

In simplifying the answer we have used the identity $\tan^2 x + 1 = \sec^2 x$.

Since $\sec x$ is never 0, we see that $f'(x) = 0$ when $\tan x = 1$, and this occurs when $x = n\pi + \pi/4$, where n is an integer (see Figure 4). □

Trigonometric functions are often used in modeling real-world phenomena. In particular, vibrations, waves, elastic motions, and other quantities that vary in a periodic manner can be described using trigonometric functions. In the following example we discuss an instance of simple harmonic motion.

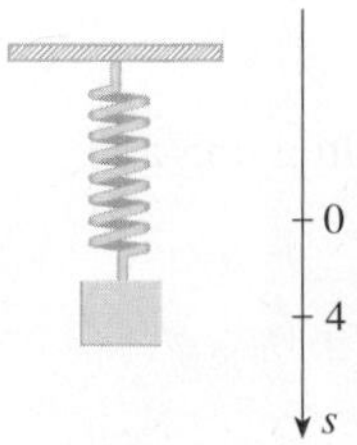

FIGURE 5

EXAMPLE 3 An object at the end of a vertical spring is stretched 4 cm beyond its rest position and released at time $t = 0$. (See Figure 5 and note that the downward direction is positive.) Its position at time t is

$$s = f(t) = 4 \cos t$$

Find the velocity and acceleration at time t and use them to analyze the motion of the object.

SOLUTION The velocity and acceleration are

$$v = \frac{ds}{dt} = \frac{d}{dt}(4\cos t) = 4\frac{d}{dt}(\cos t) = -4\sin t$$

$$a = \frac{dv}{dt} = \frac{d}{dt}(-4\sin t) = -4\frac{d}{dt}(\sin t) = -4\cos t$$

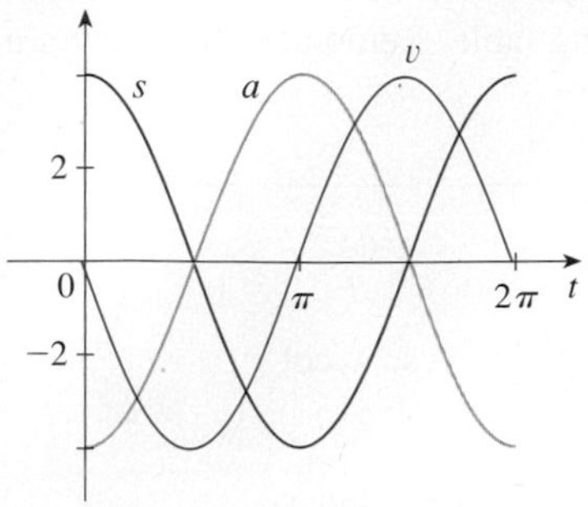

FIGURE 6

The object oscillates from the lowest point ($s = 4$ cm) to the highest point ($s = -4$ cm). The period of the oscillation is 2π, the period of $\cos t$.

The speed is $|v| = 4|\sin t|$, which is greatest when $|\sin t| = 1$, that is, when $\cos t = 0$. So the object moves fastest as it passes through its equilibrium position ($s = 0$). Its speed is 0 when $\sin t = 0$, that is, at the high and low points.

The acceleration $a = -4\cos t = 0$ when $s = 0$. It has greatest magnitude at the high and low points. See the graphs in Figure 6. □

EXAMPLE 4 Find the 27th derivative of $\cos x$.

SOLUTION The first few derivatives of $f(x) = \cos x$ are as follows:

■ Look for a pattern.

$$f'(x) = -\sin x$$
$$f''(x) = -\cos x$$
$$f'''(x) = \sin x$$
$$f^{(4)}(x) = \cos x$$
$$f^{(5)}(x) = -\sin x$$

We see that the successive derivatives occur in a cycle of length 4 and, in particular, $f^{(n)}(x) = \cos x$ whenever n is a multiple of 4. Therefore

$$f^{(24)}(x) = \cos x$$

and, differentiating three more times, we have

$$f^{(27)}(x) = \sin x$$ □

Our main use for the limit in Equation 2 has been to prove the differentiation formula for the sine function. But this limit is also useful in finding certain other trigonometric limits, as the following two examples show.

EXAMPLE 5 Find $\displaystyle\lim_{x\to 0}\frac{\sin 7x}{4x}$.

SOLUTION In order to apply Equation 2, we first rewrite the function by multiplying and dividing by 7:

Note that $\sin 7x \neq 7\sin x$.

$$\frac{\sin 7x}{4x} = \frac{7}{4}\left(\frac{\sin 7x}{7x}\right)$$

If we let $\theta = 7x$, then $\theta \to 0$ as $x \to 0$, so by Equation 2 we have

$$\lim_{x\to 0} \frac{\sin 7x}{4x} = \frac{7}{4} \lim_{x\to 0} \left(\frac{\sin 7x}{7x} \right) = \frac{7}{4} \lim_{\theta\to 0} \frac{\sin\theta}{\theta} = \frac{7}{4} \cdot 1 = \frac{7}{4}$$ □

V EXAMPLE 6 Calculate $\lim_{x\to 0} x \cot x$.

SOLUTION Here we divide numerator and denominator by x:

$$\lim_{x\to 0} x \cot x = \lim_{x\to 0} \frac{x \cos x}{\sin x}$$

$$= \lim_{x\to 0} \frac{\cos x}{\dfrac{\sin x}{x}} = \frac{\lim_{x\to 0} \cos x}{\lim_{x\to 0} \dfrac{\sin x}{x}}$$

$$= \frac{\cos 0}{1} \qquad \text{(by the continuity of cosine and Equation 2)}$$

$$= 1$$ □

3.3 EXERCISES

1–16 Differentiate.

1. $f(x) = 3x^2 - 2\cos x$
2. $f(x) = \sqrt{x}\sin x$
3. $f(x) = \sin x + \frac{1}{2}\cot x$
4. $y = 2\csc x + 5\cos x$
5. $g(t) = t^3 \cos t$
6. $g(t) = 4\sec t + \tan t$
7. $h(\theta) = \csc\theta + e^\theta \cot\theta$
8. $y = e^u(\cos u + cu)$
9. $y = \dfrac{x}{2 - \tan x}$
10. $y = \dfrac{1 + \sin x}{x + \cos x}$
11. $f(\theta) = \dfrac{\sec\theta}{1 + \sec\theta}$
12. $y = \dfrac{1 - \sec x}{\tan x}$
13. $y = \dfrac{\sin x}{x^2}$
14. $y = \csc\theta\,(\theta + \cot\theta)$
15. $f(x) = xe^x \csc x$
16. $y = x^2 \sin x \tan x$

17. Prove that $\dfrac{d}{dx}(\csc x) = -\csc x \cot x$.

18. Prove that $\dfrac{d}{dx}(\sec x) = \sec x \tan x$.

19. Prove that $\dfrac{d}{dx}(\cot x) = -\csc^2 x$.

20. Prove, using the definition of derivative, that if $f(x) = \cos x$, then $f'(x) = -\sin x$.

21–24 Find an equation of the tangent line to the curve at the given point.

21. $y = \sec x, \quad (\pi/3, 2)$
22. $y = e^x \cos x, \quad (0, 1)$
23. $y = x + \cos x, \quad (0, 1)$
24. $y = \dfrac{1}{\sin x + \cos x}, \quad (0, 1)$

25. (a) Find an equation of the tangent line to the curve $y = 2x\sin x$ at the point $(\pi/2, \pi)$.
(b) Illustrate part (a) by graphing the curve and the tangent line on the same screen.

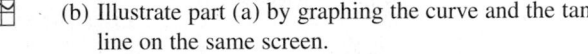

26. (a) Find an equation of the tangent line to the curve $y = \sec x - 2\cos x$ at the point $(\pi/3, 1)$.
(b) Illustrate part (a) by graphing the curve and the tangent line on the same screen.

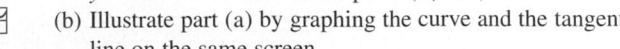

27. (a) If $f(x) = \sec x - x$, find $f'(x)$.
(b) Check to see that your answer to part (a) is reasonable by graphing both f and f' for $|x| < \pi/2$.

28. (a) If $f(x) = e^x \cos x$, find $f'(x)$ and $f''(x)$.
(b) Check to see that your answers to part (a) are reasonable by graphing f, f', and f''.

29. If $H(\theta) = \theta\sin\theta$, find $H'(\theta)$ and $H''(\theta)$.

30. If $f(x) = \sec x$, find $f''(\pi/4)$.

31. (a) Use the Quotient Rule to differentiate the function

$$f(x) = \frac{\tan x - 1}{\sec x}$$

(b) Simplify the expression for $f(x)$ by writing it in terms of $\sin x$ and $\cos x$, and then find $f'(x)$.
(c) Show that your answers to parts (a) and (b) are equivalent.

32. Suppose $f(\pi/3) = 4$ and $f'(\pi/3) = -2$, and let

$$g(x) = f(x) \sin x$$

and $$h(x) = \frac{\cos x}{f(x)}$$

Find (a) $g'(\pi/3)$ and (b) $h'(\pi/3)$.

33. For what values of x does the graph of $f(x) = x + 2 \sin x$ have a horizontal tangent?

34. Find the points on the curve $y = (\cos x)/(2 + \sin x)$ at which the tangent is horizontal.

35. A mass on a spring vibrates horizontally on a smooth level surface (see the figure). Its equation of motion is $x(t) = 8 \sin t$, where t is in seconds and x in centimeters.
(a) Find the velocity and acceleration at time t.
(b) Find the position, velocity, and acceleration of the mass at time $t = 2\pi/3$. In what direction is it moving at that time?

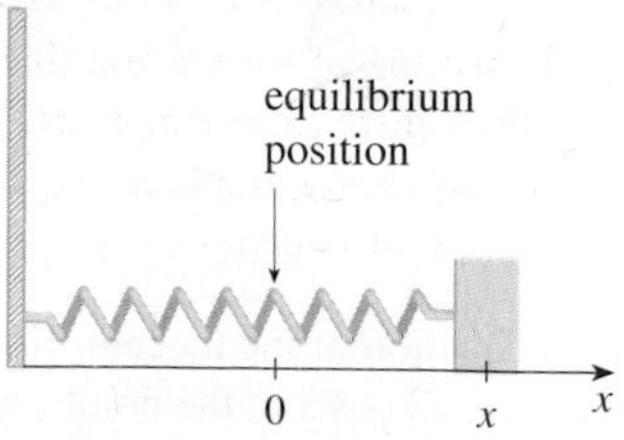

36. An elastic band is hung on a hook and a mass is hung on the lower end of the band. When the mass is pulled downward and then released, it vibrates vertically. The equation of motion is $s = 2 \cos t + 3 \sin t$, $t \geq 0$, where s is measured in centimeters and t in seconds. (Take the positive direction to be downward.)
(a) Find the velocity and acceleration at time t.
(b) Graph the velocity and acceleration functions.
(c) When does the mass pass through the equilibrium position for the first time?
(d) How far from its equilibrium position does the mass travel?
(e) When is the speed the greatest?

37. A ladder 10 ft long rests against a vertical wall. Let θ be the angle between the top of the ladder and the wall and let x be the distance from the bottom of the ladder to the wall. If the bottom of the ladder slides away from the wall, how fast does x change with respect to θ when $\theta = \pi/3$?

38. An object with weight W is dragged along a horizontal plane by a force acting along a rope attached to the object. If the rope makes an angle θ with the plane, then the magnitude of the force is

$$F = \frac{\mu W}{\mu \sin \theta + \cos \theta}$$

where μ is a constant called the *coefficient of friction*.
(a) Find the rate of change of F with respect to θ.
(b) When is this rate of change equal to 0?
(c) If $W = 50$ lb and $\mu = 0.6$, draw the graph of F as a function of θ and use it to locate the value of θ for which $dF/d\theta = 0$. Is the value consistent with your answer to part (b)?

39–48 Find the limit.

39. $\lim_{x \to 0} \frac{\sin 3x}{x}$

40. $\lim_{x \to 0} \frac{\sin 4x}{\sin 6x}$

41. $\lim_{t \to 0} \frac{\tan 6t}{\sin 2t}$

42. $\lim_{\theta \to 0} \frac{\cos \theta - 1}{\sin \theta}$

43. $\lim_{\theta \to 0} \frac{\sin(\cos \theta)}{\sec \theta}$

44. $\lim_{t \to 0} \frac{\sin^2 3t}{t^2}$

45. $\lim_{\theta \to 0} \frac{\sin \theta}{\theta + \tan \theta}$

46. $\lim_{x \to 0} \frac{\sin(x^2)}{x}$

47. $\lim_{x \to \pi/4} \frac{1 - \tan x}{\sin x - \cos x}$

48. $\lim_{x \to 1} \frac{\sin(x - 1)}{x^2 + x - 2}$

49. Differentiate each trigonometric identity to obtain a new (or familiar) identity.

(a) $\tan x = \dfrac{\sin x}{\cos x}$

(b) $\sec x = \dfrac{1}{\cos x}$

(c) $\sin x + \cos x = \dfrac{1 + \cot x}{\csc x}$

50. A semicircle with diameter PQ sits on an isosceles triangle PQR to form a region shaped like a two-dimensional ice-cream cone, as shown in the figure. If $A(\theta)$ is the area of the semicircle and $B(\theta)$ is the area of the triangle, find

$$\lim_{\theta \to 0^+} \frac{A(\theta)}{B(\theta)}$$

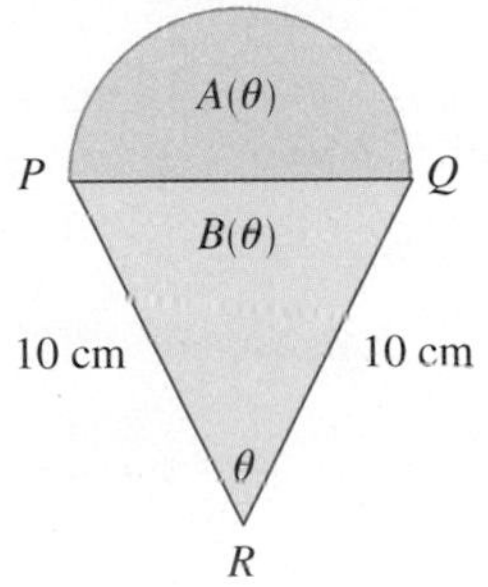

51. The figure shows a circular arc of length s and a chord of length d, both subtended by a central angle θ. Find

$$\lim_{\theta \to 0^+} \frac{s}{d}$$

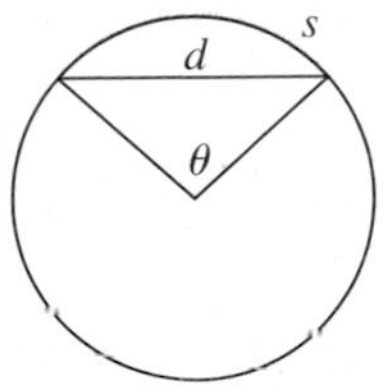

3.4 THE CHAIN RULE

Suppose you are asked to differentiate the function

$$F(x) = \sqrt{x^2 + 1}$$

The differentiation formulas you learned in the previous sections of this chapter do not enable you to calculate $F'(x)$.

■ See Section 1.3 for a review of composite functions.

Observe that F is a composite function. In fact, if we let $y = f(u) = \sqrt{u}$ and let $u = g(x) = x^2 + 1$, then we can write $y = F(x) = f(g(x))$, that is, $F = f \circ g$. We know how to differentiate both f and g, so it would be useful to have a rule that tells us how to find the derivative of $F = f \circ g$ in terms of the derivatives of f and g.

It turns out that the derivative of the composite function $f \circ g$ is the product of the derivatives of f and g. This fact is one of the most important of the differentiation rules and is called the *Chain Rule*. It seems plausible if we interpret derivatives as rates of change. Regard du/dx as the rate of change of u with respect to x, dy/du as the rate of change of y with respect to u, and dy/dx as the rate of change of y with respect to x. If u changes twice as fast as x and y changes three times as fast as u, then it seems reasonable that y changes six times as fast as x, and so we expect that

$$\frac{dy}{dx} = \frac{dy}{du}\frac{du}{dx}$$

THE CHAIN RULE If g is differentiable at x and f is differentiable at $g(x)$, then the composite function $F = f \circ g$ defined by $F(x) = f(g(x))$ is differentiable at x and F' is given by the product

$$F'(x) = f'(g(x)) \cdot g'(x)$$

In Leibniz notation, if $y = f(u)$ and $u = g(x)$ are both differentiable functions, then

$$\frac{dy}{dx} = \frac{dy}{du}\frac{du}{dx}$$

COMMENTS ON THE PROOF OF THE CHAIN RULE Let Δu be the change in u corresponding to a change of Δx in x, that is,

$$\Delta u = g(x + \Delta x) - g(x)$$

Then the corresponding change in y is

$$\Delta y = f(u + \Delta u) - f(u)$$

It is tempting to write

$$\frac{dy}{dx} = \lim_{\Delta x \to 0} \frac{\Delta y}{\Delta x}$$

1 $$= \lim_{\Delta x \to 0} \frac{\Delta y}{\Delta u} \cdot \frac{\Delta u}{\Delta x}$$

$$= \lim_{\Delta x \to 0} \frac{\Delta y}{\Delta u} \cdot \lim_{\Delta x \to 0} \frac{\Delta u}{\Delta x}$$

$$= \lim_{\Delta u \to 0} \frac{\Delta y}{\Delta u} \cdot \lim_{\Delta x \to 0} \frac{\Delta u}{\Delta x}$$

(Note that $\Delta u \to 0$ as $\Delta x \to 0$ since g is continuous.)

$$= \frac{dy}{du}\frac{du}{dx}$$

The only flaw in this reasoning is that in (1) it might happen that $\Delta u = 0$ (even when $\Delta x \neq 0$) and, of course, we can't divide by 0. Nonetheless, this reasoning does at least *suggest* that the Chain Rule is true. A full proof of the Chain Rule is given at the end of this section. □

The Chain Rule can be written either in the prime notation

2 $$(f \circ g)'(x) = f'(g(x)) \cdot g'(x)$$

or, if $y = f(u)$ and $u = g(x)$, in Leibniz notation:

3 $$\frac{dy}{dx} = \frac{dy}{du}\frac{du}{dx}$$

Equation 3 is easy to remember because if dy/du and du/dx were quotients, then we could cancel du. Remember, however, that du has not been defined and du/dx should not be thought of as an actual quotient.

EXAMPLE 1 Find $F'(x)$ if $F(x) = \sqrt{x^2 + 1}$.

SOLUTION 1 (using Equation 2): At the beginning of this section we expressed F as $F(x) = (f \circ g)(x) = f(g(x))$ where $f(u) = \sqrt{u}$ and $g(x) = x^2 + 1$. Since

$$f'(u) = \tfrac{1}{2}u^{-1/2} = \frac{1}{2\sqrt{u}} \qquad \text{and} \qquad g'(x) = 2x$$

we have

$$F'(x) = f'(g(x)) \cdot g'(x)$$

$$= \frac{1}{2\sqrt{x^2 + 1}} \cdot 2x = \frac{x}{\sqrt{x^2 + 1}}$$

SOLUTION 2 (using Equation 3): If we let $u = x^2 + 1$ and $y = \sqrt{u}$, then

$$F'(x) = \frac{dy}{du}\frac{du}{dx} = \frac{1}{2\sqrt{u}}(2x)$$

$$= \frac{1}{2\sqrt{x^2+1}}(2x) = \frac{x}{\sqrt{x^2+1}}$$ □

When using Formula 3 we should bear in mind that dy/dx refers to the derivative of y when y is considered as a function of x (called the *derivative of y with respect to x*), whereas dy/du refers to the derivative of y when considered as a function of u (the derivative of y with respect to u). For instance, in Example 1, y can be considered as a function of x $\left(y = \sqrt{x^2+1}\right)$ and also as a function of u $\left(y = \sqrt{u}\right)$. Note that

$$\frac{dy}{dx} = F'(x) = \frac{x}{\sqrt{x^2+1}} \qquad \text{whereas} \qquad \frac{dy}{du} = f'(u) = \frac{1}{2\sqrt{u}}$$

NOTE In using the Chain Rule we work from the outside to the inside. Formula 2 says that *we differentiate the outer function f [at the inner function g(x)] and then we multiply by the derivative of the inner function.*

$$\frac{d}{dx}\underbrace{f}_{\substack{\text{outer}\\\text{function}}}\underbrace{(g(x))}_{\substack{\text{evaluated}\\\text{at inner}\\\text{function}}} = \underbrace{f'}_{\substack{\text{derivative}\\\text{of outer}\\\text{function}}}\underbrace{(g(x))}_{\substack{\text{evaluated}\\\text{at inner}\\\text{function}}} \cdot \underbrace{g'(x)}_{\substack{\text{derivative}\\\text{of inner}\\\text{function}}}$$

V EXAMPLE 2 Differentiate (a) $y = \sin(x^2)$ and (b) $y = \sin^2 x$.

SOLUTION

(a) If $y = \sin(x^2)$, then the outer function is the sine function and the inner function is the squaring function, so the Chain Rule gives

$$\frac{dy}{dx} = \frac{d}{dx}\underbrace{\sin}_{\substack{\text{outer}\\\text{function}}}\underbrace{(x^2)}_{\substack{\text{evaluated}\\\text{at inner}\\\text{function}}} = \underbrace{\cos}_{\substack{\text{derivative}\\\text{of outer}\\\text{function}}}\underbrace{(x^2)}_{\substack{\text{evaluated}\\\text{at inner}\\\text{function}}} \cdot \underbrace{2x}_{\substack{\text{derivative}\\\text{of inner}\\\text{function}}}$$

$$= 2x\cos(x^2)$$

(b) Note that $\sin^2 x = (\sin x)^2$. Here the outer function is the squaring function and the inner function is the sine function. So

$$\frac{dy}{dx} = \frac{d}{dx}\underbrace{(\sin x)^2}_{\substack{\text{inner}\\\text{function}}} = \underbrace{2}_{\substack{\text{derivative}\\\text{of outer}\\\text{function}}} \cdot \underbrace{(\sin x)}_{\substack{\text{evaluated}\\\text{at inner}\\\text{function}}} \cdot \underbrace{\cos x}_{\substack{\text{derivative}\\\text{of inner}\\\text{function}}}$$

The answer can be left as $2\sin x\cos x$ or written as $\sin 2x$ (by a trigonometric identity known as the double-angle formula). □

■ See Reference Page 2 or Appendix D.

In Example 2(a) we combined the Chain Rule with the rule for differentiating the sine function. In general, if $y = \sin u$, where u is a differentiable function of x, then, by the Chain Rule,

$$\frac{dy}{dx} = \frac{dy}{du}\frac{du}{dx} = \cos u\,\frac{du}{dx}$$

Thus $$\frac{d}{dx}(\sin u) = \cos u \frac{du}{dx}$$

In a similar fashion, all of the formulas for differentiating trigonometric functions can be combined with the Chain Rule.

Let's make explicit the special case of the Chain Rule where the outer function f is a power function. If $y = [g(x)]^n$, then we can write $y = f(u) = u^n$ where $u = g(x)$. By using the Chain Rule and then the Power Rule, we get

$$\frac{dy}{dx} = \frac{dy}{du}\frac{du}{dx} = nu^{n-1}\frac{du}{dx} = n[g(x)]^{n-1}g'(x)$$

4 THE POWER RULE COMBINED WITH THE CHAIN RULE If n is any real number and $u = g(x)$ is differentiable, then

$$\frac{d}{dx}(u^n) = nu^{n-1}\frac{du}{dx}$$

Alternatively, $$\frac{d}{dx}[g(x)]^n = n[g(x)]^{n-1} \cdot g'(x)$$

Notice that the derivative in Example 1 could be calculated by taking $n = \frac{1}{2}$ in Rule 4.

EXAMPLE 3 Differentiate $y = (x^3 - 1)^{100}$.

SOLUTION Taking $u = g(x) = x^3 - 1$ and $n = 100$ in (4), we have

$$\frac{dy}{dx} = \frac{d}{dx}(x^3 - 1)^{100} = 100(x^3 - 1)^{99}\frac{d}{dx}(x^3 - 1)$$
$$= 100(x^3 - 1)^{99} \cdot 3x^2 = 300x^2(x^3 - 1)^{99}$$ □

EXAMPLE 4 Find $f'(x)$ if $f(x) = \dfrac{1}{\sqrt[3]{x^2 + x + 1}}$.

SOLUTION First rewrite f: $$f(x) = (x^2 + x + 1)^{-1/3}$$

Thus $$f'(x) = -\tfrac{1}{3}(x^2 + x + 1)^{-4/3}\frac{d}{dx}(x^2 + x + 1)$$
$$= -\tfrac{1}{3}(x^2 + x + 1)^{-4/3}(2x + 1)$$ □

EXAMPLE 5 Find the derivative of the function

$$g(t) = \left(\frac{t - 2}{2t + 1}\right)^9$$

SOLUTION Combining the Power Rule, Chain Rule, and Quotient Rule, we get

$$g'(t) = 9\left(\frac{t - 2}{2t + 1}\right)^8 \frac{d}{dt}\left(\frac{t - 2}{2t + 1}\right)$$
$$= 9\left(\frac{t - 2}{2t + 1}\right)^8 \frac{(2t + 1) \cdot 1 - 2(t - 2)}{(2t + 1)^2} = \frac{45(t - 2)^8}{(2t + 1)^{10}}$$ □

■ The graphs of the functions y and y' in Example 6 are shown in Figure 1. Notice that y' is large when y increases rapidly and $y' = 0$ when y has a horizontal tangent. So our answer appears to be reasonable.

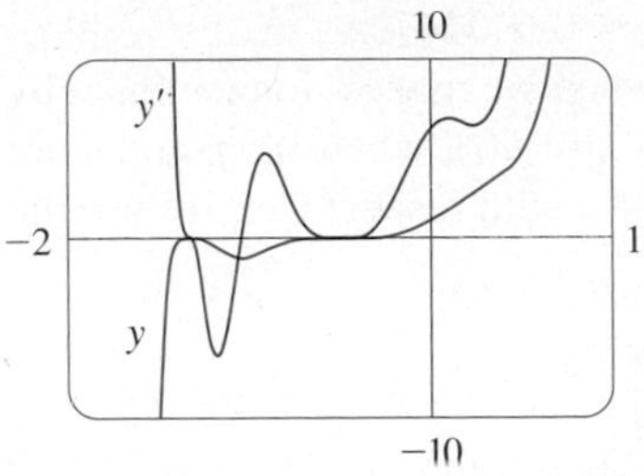

FIGURE 1

EXAMPLE 6 Differentiate $y = (2x + 1)^5(x^3 - x + 1)^4$.

SOLUTION In this example we must use the Product Rule before using the Chain Rule:

$$\begin{aligned}\frac{dy}{dx} &= (2x+1)^5 \frac{d}{dx}(x^3 - x + 1)^4 + (x^3 - x + 1)^4 \frac{d}{dx}(2x+1)^5 \\ &= (2x+1)^5 \cdot 4(x^3 - x + 1)^3 \frac{d}{dx}(x^3 - x + 1) \\ &\quad + (x^3 - x + 1)^4 \cdot 5(2x+1)^4 \frac{d}{dx}(2x+1) \\ &= 4(2x+1)^5(x^3 - x + 1)^3(3x^2 - 1) + 5(x^3 - x + 1)^4(2x+1)^4 \cdot 2\end{aligned}$$

Noticing that each term has the common factor $2(2x + 1)^4(x^3 - x + 1)^3$, we could factor it out and write the answer as

$$\frac{dy}{dx} = 2(2x+1)^4(x^3 - x + 1)^3(17x^3 + 6x^2 - 9x + 3)$$ □

EXAMPLE 7 Differentiate $y = e^{\sin x}$.

SOLUTION Here the inner function is $g(x) = \sin x$ and the outer function is the exponential function $f(x) = e^x$. So, by the Chain Rule,

■ More generally, the Chain Rule gives

$$\frac{d}{dx}(e^u) = e^u \frac{du}{dx}$$

$$\frac{dy}{dx} = \frac{d}{dx}(e^{\sin x}) = e^{\sin x} \frac{d}{dx}(\sin x) = e^{\sin x} \cos x$$ □

We can use the Chain Rule to differentiate an exponential function with any base $a > 0$. Recall from Section 1.6 that $a = e^{\ln a}$. So

$$a^x = (e^{\ln a})^x = e^{(\ln a)x}$$

and the Chain Rule gives

$$\begin{aligned}\frac{d}{dx}(a^x) &= \frac{d}{dx}(e^{(\ln a)x}) = e^{(\ln a)x} \frac{d}{dx}(\ln a)x \\ &= e^{(\ln a)x} \cdot \ln a = a^x \ln a\end{aligned}$$

because $\ln a$ is a constant. So we have the formula

■ Don't confuse Formula 5 (where x is the *exponent*) with the Power Rule (where x is the *base*):

$$\frac{d}{dx}(x^n) = nx^{n-1}$$

5 $$\frac{d}{dx}(a^x) = a^x \ln a$$

In particular, if $a = 2$, we get

6 $$\frac{d}{dx}(2^x) = 2^x \ln 2$$

In Section 3.1 we gave the estimate

$$\frac{d}{dx}(2^x) \approx (0.69)2^x$$

This is consistent with the exact formula (6) because $\ln 2 \approx 0.693147$.

The reason for the name "Chain Rule" becomes clear when we make a longer chain by adding another link. Suppose that $y = f(u)$, $u = g(x)$, and $x = h(t)$, where f, g, and h are differentiable functions. Then, to compute the derivative of y with respect to t, we use the Chain Rule twice:

$$\frac{dy}{dt} = \frac{dy}{dx}\frac{dx}{dt} = \frac{dy}{du}\frac{du}{dx}\frac{dx}{dt}$$

EXAMPLE 8 If $f(x) = \sin(\cos(\tan x))$, then

$$\begin{aligned} f'(x) &= \cos(\cos(\tan x))\frac{d}{dx}\cos(\tan x) \\ &= \cos(\cos(\tan x))[-\sin(\tan x)]\frac{d}{dx}(\tan x) \\ &= -\cos(\cos(\tan x))\sin(\tan x)\sec^2 x \end{aligned}$$

Notice that we used the Chain Rule twice. □

EXAMPLE 9 Differentiate $y = e^{\sec 3\theta}$.

SOLUTION The outer function is the exponential function, the middle function is the secant function and the inner function is the tripling function. So we have

$$\begin{aligned} \frac{dy}{d\theta} &= e^{\sec 3\theta}\frac{d}{d\theta}(\sec 3\theta) \\ &= e^{\sec 3\theta}\sec 3\theta \tan 3\theta \frac{d}{d\theta}(3\theta) \\ &= 3e^{\sec 3\theta}\sec 3\theta \tan 3\theta \end{aligned}$$

□

HOW TO PROVE THE CHAIN RULE

Recall that if $y = f(x)$ and x changes from a to $a + \Delta x$, we defined the increment of y as

$$\Delta y = f(a + \Delta x) - f(a)$$

According to the definition of a derivative, we have

$$\lim_{\Delta x \to 0}\frac{\Delta y}{\Delta x} = f'(a)$$

So if we denote by ε the difference between the difference quotient and the derivative, we obtain

$$\lim_{\Delta x \to 0}\varepsilon = \lim_{\Delta x \to 0}\left(\frac{\Delta y}{\Delta x} - f'(a)\right) = f'(a) - f'(a) = 0$$

But $$\varepsilon = \frac{\Delta y}{\Delta x} - f'(a) \quad \Rightarrow \quad \Delta y = f'(a)\,\Delta x + \varepsilon\,\Delta x$$

If we define ε to be 0 when $\Delta x = 0$, then ε becomes a continuous function of Δx. Thus, for a differentiable function f, we can write

$$\boxed{7} \qquad \Delta y = f'(a)\,\Delta x + \varepsilon\,\Delta x \qquad \text{where} \quad \varepsilon \to 0 \text{ as } \Delta x \to 0$$

and ε is a continuous function of Δx. This property of differentiable functions is what enables us to prove the Chain Rule.

PROOF OF THE CHAIN RULE Suppose $u = g(x)$ is differentiable at a and $y = f(u)$ is differentiable at $b = g(a)$. If Δx is an increment in x and Δu and Δy are the corresponding increments in u and y, then we can use Equation 7 to write

$$\boxed{8} \qquad \Delta u = g'(a)\,\Delta x + \varepsilon_1\,\Delta x = [g'(a) + \varepsilon_1]\,\Delta x$$

where $\varepsilon_1 \to 0$ as $\Delta x \to 0$. Similarly

$$\boxed{9} \qquad \Delta y = f'(b)\,\Delta u + \varepsilon_2\,\Delta u = [f'(b) + \varepsilon_2]\,\Delta u$$

where $\varepsilon_2 \to 0$ as $\Delta u \to 0$. If we now substitute the expression for Δu from Equation 8 into Equation 9, we get

$$\Delta y = [f'(b) + \varepsilon_2][g'(a) + \varepsilon_1]\,\Delta x$$

so $$\frac{\Delta y}{\Delta x} = [f'(b) + \varepsilon_2][g'(a) + \varepsilon_1]$$

As $\Delta x \to 0$, Equation 8 shows that $\Delta u \to 0$. So both $\varepsilon_1 \to 0$ and $\varepsilon_2 \to 0$ as $\Delta x \to 0$. Therefore

$$\frac{dy}{dx} = \lim_{\Delta x \to 0} \frac{\Delta y}{\Delta x} = \lim_{\Delta x \to 0} [f'(b) + \varepsilon_2][g'(a) + \varepsilon_1]$$
$$= f'(b)g'(a) = f'(g(a))g'(a)$$

This proves the Chain Rule. □

3.4 EXERCISES

1–6 Write the composite function in the form $f(g(x))$. [Identify the inner function $u = g(x)$ and the outer function $y = f(u)$.] Then find the derivative dy/dx.

1. $y = \sin 4x$

2. $y = \sqrt{4 + 3x}$

3. $y = (1 - x^2)^{10}$

4. $y = \tan(\sin x)$

5. $y = e^{\sqrt{x}}$

6. $y = \sin(e^x)$

7–46 Find the derivative of the function.

7. $F(x) = (x^4 + 3x^2 - 2)^5$

8. $F(x) = (4x - x^2)^{100}$

9. $F(x) = \sqrt[4]{1 + 2x + x^3}$

10. $f(x) = (1 + x^4)^{2/3}$

11. $g(t) = \dfrac{1}{(t^4 + 1)^3}$

12. $f(t) = \sqrt[3]{1 + \tan t}$

13. $y = \cos(a^3 + x^3)$

14. $y = a^3 + \cos^3 x$

15. $y = xe^{-kx}$

16. $y = 3\cot(n\theta)$

17. $g(x) = (1 + 4x)^5(3 + x - x^2)^8$

18. $h(t) = (t^4 - 1)^3(t^3 + 1)^4$

19. $y = (2x - 5)^4(8x^2 - 5)^{-3}$

20. $y = (x^2 + 1)\sqrt[3]{x^2 + 2}$

21. $y = \left(\dfrac{x^2+1}{x^2-1}\right)^3$

22. $y = e^{-5x}\cos 3x$

23. $y = e^{x\cos x}$

24. $y = 10^{1-x^2}$

25. $F(z) = \sqrt{\dfrac{z-1}{z+1}}$

26. $G(y) = \dfrac{(y-1)^4}{(y^2+2y)^5}$

27. $y = \dfrac{r}{\sqrt{r^2+1}}$

28. $y = \dfrac{e^u - e^{-u}}{e^u + e^{-u}}$

29. $y = \sin(\tan 2x)$

30. $G(y) = \left(\dfrac{y^2}{y+1}\right)^5$

31. $y = 2^{\sin \pi x}$

32. $y = \tan^2(3\theta)$

33. $y = \sec^2 x + \tan^2 x$

34. $y = x \sin \dfrac{1}{x}$

35. $y = \cos\left(\dfrac{1-e^{2x}}{1+e^{2x}}\right)$

36. $f(t) = \sqrt{\dfrac{t}{t^2+4}}$

37. $y = \cot^2(\sin\theta)$

38. $y = e^{k\tan\sqrt{x}}$

39. $f(t) = \tan(e^t) + e^{\tan t}$

40. $y = \sin(\sin(\sin x))$

41. $f(t) = \sin^2(e^{\sin^2 t})$

42. $y = \sqrt{x + \sqrt{x + \sqrt{x}}}$

43. $g(x) = (2ra^{rx} + n)^p$

44. $y = 2^{3^{x^2}}$

45. $y = \cos\sqrt{\sin(\tan \pi x)}$

46. $y = [x + (x + \sin^2 x)^3]^4$

47–50 Find the first and second derivatives of the function.

47. $h(x) = \sqrt{x^2+1}$

48. $y = xe^{cx}$

49. $y = e^{\alpha x}\sin\beta x$

50. $y = e^{e^x}$

51–54 Find an equation of the tangent line to the curve at the given point.

51. $y = (1+2x)^{10}$, $(0, 1)$

52. $y = \sin x + \sin^2 x$, $(0, 0)$

53. $y = \sin(\sin x)$, $(\pi, 0)$

54. $y = x^2 e^{-x}$, $(1, 1/e)$

55. (a) Find an equation of the tangent line to the curve $y = 2/(1+e^{-x})$ at the point $(0, 1)$.
(b) Illustrate part (a) by graphing the curve and the tangent line on the same screen.

56. (a) The curve $y = |x|/\sqrt{2-x^2}$ is called a *bullet-nose curve*. Find an equation of the tangent line to this curve at the point $(1, 1)$.
(b) Illustrate part (a) by graphing the curve and the tangent line on the same screen.

57. (a) If $f(x) = x\sqrt{2-x^2}$, find $f'(x)$.
(b) Check to see that your answer to part (a) is reasonable by comparing the graphs of f and f'.

58. The function $f(x) = \sin(x + \sin 2x)$, $0 \leq x \leq \pi$, arises in applications to frequency modulation (FM) synthesis.
(a) Use a graph of f produced by a graphing device to make a rough sketch of the graph of f'.
(b) Calculate $f'(x)$ and use this expression, with a graphing device, to graph f'. Compare with your sketch in part (a).

59. Find all points on the graph of the function $f(x) = 2\sin x + \sin^2 x$ at which the tangent line is horizontal.

60. Find the x-coordinates of all points on the curve $y = \sin 2x - 2\sin x$ at which the tangent line is horizontal.

61. If $F(x) = f(g(x))$, where $f(-2) = 8$, $f'(-2) = 4$, $f'(5) = 3$, $g(5) = -2$, and $g'(5) = 6$, find $F'(5)$.

62. If $h(x) = \sqrt{4 + 3f(x)}$, where $f(1) = 7$ and $f'(1) = 4$, find $h'(1)$.

63. A table of values for f, g, f', and g' is given.

x	$f(x)$	$g(x)$	$f'(x)$	$g'(x)$
1	3	2	4	6
2	1	8	5	7
3	7	2	7	9

(a) If $h(x) = f(g(x))$, find $h'(1)$.
(b) If $H(x) = g(f(x))$, find $H'(1)$.

64. Let f and g be the functions in Exercise 63.
(a) If $F(x) = f(f(x))$, find $F'(2)$.
(b) If $G(x) = g(g(x))$, find $G'(3)$.

65. If f and g are the functions whose graphs are shown, let $u(x) = f(g(x))$, $v(x) = g(f(x))$, and $w(x) = g(g(x))$. Find each derivative, if it exists. If it does not exist, explain why.
(a) $u'(1)$ (b) $v'(1)$ (c) $w'(1)$

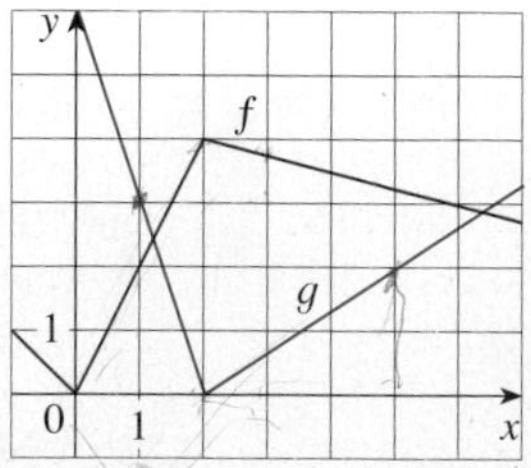

66. If f is the function whose graph is shown, let $h(x) = f(f(x))$ and $g(x) = f(x^2)$. Use the graph of f to estimate the value of each derivative.
(a) $h'(2)$ (b) $g'(2)$

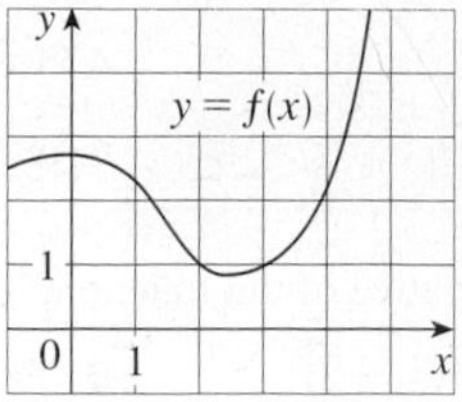

67. Suppose f is differentiable on $\mathbb{R}$. Let $F(x) = f(e^x)$ and $G(x) = e^{f(x)}$. Find expressions for (a) $F'(x)$ and (b) $G'(x)$.

68. Suppose f is differentiable on $\mathbb{R}$ and α is a real number. Let $F(x) = f(x^\alpha)$ and $G(x) = [f(x)]^\alpha$. Find expressions for (a) $F'(x)$ and (b) $G'(x)$.

69. Let $r(x) = f(g(h(x)))$, where $h(1) = 2$, $g(2) = 3$, $h'(1) = 4$, $g'(2) = 5$, and $f'(3) = 6$. Find $r'(1)$.

70. If g is a twice differentiable function and $f(x) = xg(x^2)$, find f'' in terms of g, g', and g''.

71. If $F(x) = f(3f(4f(x)))$, where $f(0) = 0$ and $f'(0) = 2$, find $F'(0)$.

72. If $F(x) = f(xf(xf(x)))$, where $f(1) = 2$, $f(2) = 3$, $f'(1) = 4$, $f'(2) = 5$, and $f'(3) = 6$, find $F'(1)$.

73. Show that the function $y = Ae^{-x} + Bxe^{-x}$ satisfies the differential equation $y'' + 2y' + y = 0$.

74. For what values of r does the function $y = e^{rx}$ satisfy the equation $y'' + 5y' - 6y = 0$?

75. Find the 50th derivative of $y = \cos 2x$.

76. Find the 1000th derivative of $f(x) = xe^{-x}$.

77. The displacement of a particle on a vibrating string is given by the equation

$$s(t) = 10 + \tfrac{1}{4}\sin(10\pi t)$$

where s is measured in centimeters and t in seconds. Find the velocity of the particle after t seconds.

78. If the equation of motion of a particle is given by $s = A\cos(\omega t + \delta)$, the particle is said to undergo *simple harmonic motion.*

(a) Find the velocity of the particle at time t.
(b) When is the velocity 0?

79. A Cepheid variable star is a star whose brightness alternately increases and decreases. The most easily visible such star is Delta Cephei, for which the interval between times of maximum brightness is 5.4 days. The average brightness of this star is 4.0 and its brightness changes by ± 0.35. In view of these data, the brightness of Delta Cephei at time t, where t is measured in days, has been modeled by the function

$$B(t) = 4.0 + 0.35 \sin\left(\frac{2\pi t}{5.4}\right)$$

(a) Find the rate of change of the brightness after t days.
(b) Find, correct to two decimal places, the rate of increase after one day.

80. In Example 4 in Section 1.3 we arrived at a model for the length of daylight (in hours) in Philadelphia on the tth day of the year:

$$L(t) = 12 + 2.8 \sin\left[\frac{2\pi}{365}(t - 80)\right]$$

Use this model to compare how the number of hours of daylight is increasing in Philadelphia on March 21 and May 21.

81. The motion of a spring that is subject to a frictional force or a damping force (such as a shock absorber in a car) is often modeled by the product of an exponential function and a sine or cosine function. Suppose the equation of motion of a point on such a spring is

$$s(t) = 2e^{-1.5t}\sin 2\pi t$$

where s is measured in centimeters and t in seconds. Find the velocity after t seconds and graph both the position and velocity functions for $0 \le t \le 2$.

82. Under certain circumstances a rumor spreads according to the equation

$$p(t) = \frac{1}{1 + ae^{-kt}}$$

where $p(t)$ is the proportion of the population that knows the rumor at time t and a and k are positive constants. [In Section 9.4 we will see that this is a reasonable equation for $p(t)$.]

(a) Find $\lim_{t\to\infty} p(t)$.
(b) Find the rate of spread of the rumor.
(c) Graph p for the case $a = 10$, $k = 0.5$ with t measured in hours. Use the graph to estimate how long it will take for 80% of the population to hear the rumor.

83. A particle moves along a straight line with displacement $s(t)$, velocity $v(t)$, and acceleration $a(t)$. Show that

$$a(t) = v(t)\frac{dv}{ds}$$

Explain the difference between the meanings of the derivatives dv/dt and dv/ds.

84. Air is being pumped into a spherical weather balloon. At any time t, the volume of the balloon is $V(t)$ and its radius is $r(t)$.

(a) What do the derivatives dV/dr and dV/dt represent?
(b) Express dV/dt in terms of dr/dt.

85. The flash unit on a camera operates by storing charge on a capacitor and releasing it suddenly when the flash is set off. The following data describe the charge Q remaining on the capacitor (measured in microcoulombs, μC) at time t (measured in seconds).

t	0.00	0.02	0.04	0.06	0.08	0.10
Q	100.00	81.87	67.03	54.88	44.93	36.76

(a) Use a graphing calculator or computer to find an exponential model for the charge.
(b) The derivative $Q'(t)$ represents the electric current (measured in microamperes, μA) flowing from the capacitor to the flash bulb. Use part (a) to estimate the current when $t = 0.04$ s. Compare with the result of Example 2 in Section 2.1.

86. The table gives the US population from 1790 to 1860.

Year	Population	Year	Population
1790	3,929,000	1830	12,861,000
1800	5,308,000	1840	17,063,000
1810	7,240,000	1850	23,192,000
1820	9,639,000	1860	31,443,000

(a) Use a graphing calculator or computer to fit an exponential function to the data. Graph the data points and the exponential model. How good is the fit?
(b) Estimate the rates of population growth in 1800 and 1850 by averaging slopes of secant lines.
(c) Use the exponential model in part (a) to estimate the rates of growth in 1800 and 1850. Compare these estimates with the ones in part (b).
(d) Use the exponential model to predict the population in 1870. Compare with the actual population of 38,558,000. Can you explain the discrepancy?

CAS **87.** Computer algebra systems have commands that differentiate functions, but the form of the answer may not be convenient and so further commands may be necessary to simplify the answer.
(a) Use a CAS to find the derivative in Example 5 and compare with the answer in that example. Then use the simplify command and compare again.
(b) Use a CAS to find the derivative in Example 6. What happens if you use the simplify command? What happens if you use the factor command? Which form of the answer would be best for locating horizontal tangents?

CAS **88.** (a) Use a CAS to differentiate the function

$$f(x) = \sqrt{\frac{x^4 - x + 1}{x^4 + x + 1}}$$

and to simplify the result.
(b) Where does the graph of f have horizontal tangents?
(c) Graph f and f' on the same screen. Are the graphs consistent with your answer to part (b)?

89. Use the Chain Rule to prove the following.
(a) The derivative of an even function is an odd function.
(b) The derivative of an odd function is an even function.

90. Use the Chain Rule and the Product Rule to give an alternative proof of the Quotient Rule.
[*Hint:* Write $f(x)/g(x) = f(x)[g(x)]^{-1}$.]

91. (a) If n is a positive integer, prove that

$$\frac{d}{dx}(\sin^n x \cos nx) = n \sin^{n-1} x \cos(n+1)x$$

(b) Find a formula for the derivative of $y = \cos^n x \cos nx$ that is similar to the one in part (a).

92. Suppose $y = f(x)$ is a curve that always lies above the x-axis and never has a horizontal tangent, where f is differentiable everywhere. For what value of y is the rate of change of y^5 with respect to x eighty times the rate of change of y with respect to x?

93. Use the Chain Rule to show that if θ is measured in degrees, then

$$\frac{d}{d\theta}(\sin\theta) = \frac{\pi}{180}\cos\theta$$

(This gives one reason for the convention that radian measure is always used when dealing with trigonometric functions in calculus: The differentiation formulas would not be as simple if we used degree measure.)

94. (a) Write $|x| = \sqrt{x^2}$ and use the Chain Rule to show that

$$\frac{d}{dx}|x| = \frac{x}{|x|}$$

(b) If $f(x) = |\sin x|$, find $f'(x)$ and sketch the graphs of f and f'. Where is f not differentiable?
(c) If $g(x) = \sin|x|$, find $g'(x)$ and sketch the graphs of g and g'. Where is g not differentiable?

95. If $y = f(u)$ and $u = g(x)$, where f and g are twice differentiable functions, show that

$$\frac{d^2y}{dx^2} = \frac{d^2y}{du^2}\left(\frac{du}{dx}\right)^2 + \frac{dy}{du}\frac{d^2u}{dx^2}$$

96. If $y = f(u)$ and $u = g(x)$, where f and g possess third derivatives, find a formula for d^3y/dx^3 similar to the one given in Exercise 95.

APPLIED PROJECT

WHERE SHOULD A PILOT START DESCENT?

An approach path for an aircraft landing is shown in the figure on the next page and satisfies the following conditions:

(i) The cruising altitude is h when descent starts at a horizontal distance ℓ from touchdown at the origin.

(ii) The pilot must maintain a constant horizontal speed v throughout descent.

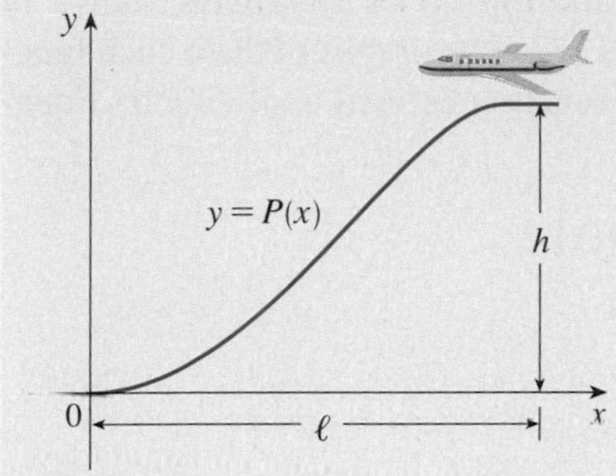

(iii) The absolute value of the vertical acceleration should not exceed a constant k (which is much less than the acceleration due to gravity).

1. Find a cubic polynomial $P(x) = ax^3 + bx^2 + cx + d$ that satisfies condition (i) by imposing suitable conditions on $P(x)$ and $P'(x)$ at the start of descent and at touchdown.

2. Use conditions (ii) and (iii) to show that

$$\frac{6hv^2}{\ell^2} \leqslant k$$

3. Suppose that an airline decides not to allow vertical acceleration of a plane to exceed $k = 860 \text{ mi/h}^2$. If the cruising altitude of a plane is 35,000 ft and the speed is 300 mi/h, how far away from the airport should the pilot start descent?

4. Graph the approach path if the conditions stated in Problem 3 are satisfied.

3.5 IMPLICIT DIFFERENTIATION

The functions that we have met so far can be described by expressing one variable explicitly in terms of another variable—for example,

$$y = \sqrt{x^3 + 1} \qquad \text{or} \qquad y = x \sin x$$

or, in general, $y = f(x)$. Some functions, however, are defined implicitly by a relation between x and y such as

1 $$x^2 + y^2 = 25$$

or

2 $$x^3 + y^3 = 6xy$$

In some cases it is possible to solve such an equation for y as an explicit function (or several functions) of x. For instance, if we solve Equation 1 for y, we get $y = \pm\sqrt{25 - x^2}$, so two of the functions determined by the implicit Equation 1 are $f(x) = \sqrt{25 - x^2}$ and $g(x) = -\sqrt{25 - x^2}$. The graphs of f and g are the upper and lower semicircles of the circle $x^2 + y^2 = 25$. (See Figure 1.)

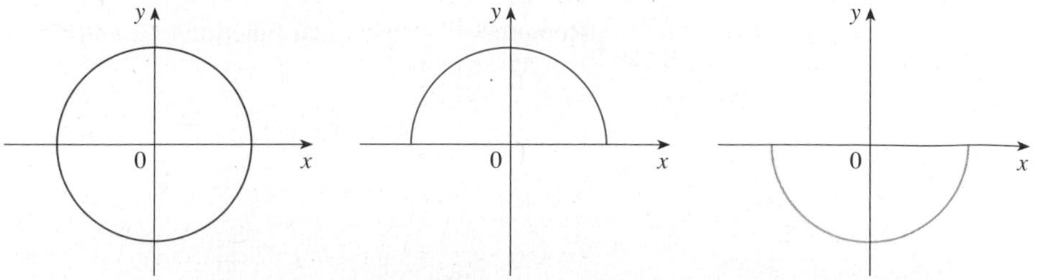

FIGURE 1 (a) $x^2 + y^2 = 25$ (b) $f(x) = \sqrt{25 - x^2}$ (c) $g(x) = -\sqrt{25 - x^2}$

It's not easy to solve Equation 2 for y explicitly as a function of x by hand. (A computer algebra system has no trouble, but the expressions it obtains are very complicated.)

Nonetheless, (2) is the equation of a curve called the **folium of Descartes** shown in Figure 2 and it implicitly defines y as several functions of x. The graphs of three such functions are shown in Figure 3. When we say that f is a function defined implicitly by Equation 2, we mean that the equation

$$x^3 + [f(x)]^3 = 6xf(x)$$

is true for all values of x in the domain of f.

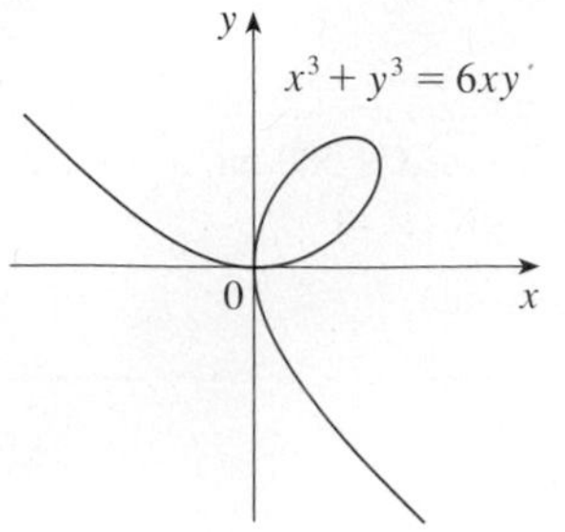

FIGURE 2 The folium of Descartes

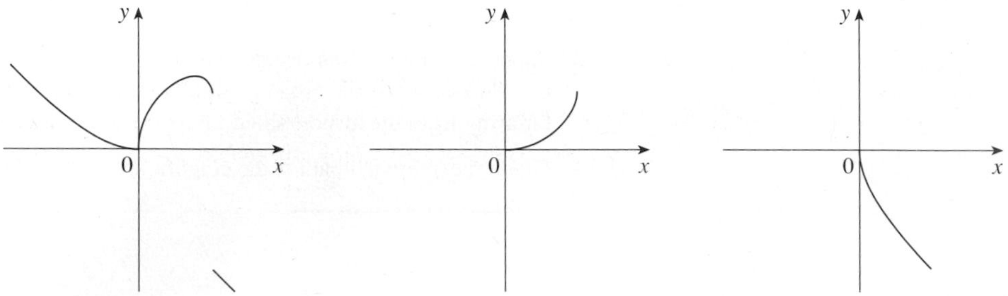

FIGURE 3 Graphs of three functions defined by the folium of Descartes

Fortunately, we don't need to solve an equation for y in terms of x in order to find the derivative of y. Instead we can use the method of **implicit differentiation**. This consists of differentiating both sides of the equation with respect to x and then solving the resulting equation for y'. In the examples and exercises of this section it is always assumed that the given equation determines y implicitly as a differentiable function of x so that the method of implicit differentiation can be applied.

V EXAMPLE 1

(a) If $x^2 + y^2 = 25$, find $\dfrac{dy}{dx}$.

(b) Find an equation of the tangent to the circle $x^2 + y^2 = 25$ at the point $(3, 4)$.

SOLUTION 1

(a) Differentiate both sides of the equation $x^2 + y^2 = 25$:

$$\frac{d}{dx}(x^2 + y^2) = \frac{d}{dx}(25)$$

$$\frac{d}{dx}(x^2) + \frac{d}{dx}(y^2) = 0$$

Remembering that y is a function of x and using the Chain Rule, we have

$$\frac{d}{dx}(y^2) = \frac{d}{dy}(y^2)\frac{dy}{dx} = 2y\frac{dy}{dx}$$

Thus

$$2x + 2y\frac{dy}{dx} = 0$$

Now we solve this equation for dy/dx:

$$\frac{dy}{dx} = -\frac{x}{y}$$

(b) At the point (3, 4) we have $x = 3$ and $y = 4$, so

$$\frac{dy}{dx} = -\frac{3}{4}$$

An equation of the tangent to the circle at (3, 4) is therefore

$$y - 4 = -\tfrac{3}{4}(x - 3) \qquad \text{or} \qquad 3x + 4y = 25$$

SOLUTION 2

(b) Solving the equation $x^2 + y^2 = 25$, we get $y = \pm\sqrt{25 - x^2}$. The point (3, 4) lies on the upper semicircle $y = \sqrt{25 - x^2}$ and so we consider the function $f(x) = \sqrt{25 - x^2}$. Differentiating f using the Chain Rule, we have

$$f'(x) = \tfrac{1}{2}(25 - x^2)^{-1/2} \frac{d}{dx}(25 - x^2)$$

$$= \tfrac{1}{2}(25 - x^2)^{-1/2}(-2x) = -\frac{x}{\sqrt{25 - x^2}}$$

■ Example 1 illustrates that even when it is possible to solve an equation explicitly for y in terms of x, it may be easier to use implicit differentiation.

So $$f'(3) = -\frac{3}{\sqrt{25 - 3^2}} = -\frac{3}{4}$$

and, as in Solution 1, an equation of the tangent is $3x + 4y = 25$. □

NOTE 1 The expression $dy/dx = -x/y$ in Solution 1 gives the derivative in terms of both x and y. It is correct no matter which function y is determined by the given equation. For instance, for $y = f(x) = \sqrt{25 - x^2}$ we have

$$\frac{dy}{dx} = -\frac{x}{y} = -\frac{x}{\sqrt{25 - x^2}}$$

whereas for $y = g(x) = -\sqrt{25 - x^2}$ we have

$$\frac{dy}{dx} = -\frac{x}{y} = -\frac{x}{-\sqrt{25 - x^2}} = \frac{x}{\sqrt{25 - x^2}}$$

V EXAMPLE 2

(a) Find y' if $x^3 + y^3 = 6xy$.

(b) Find the tangent to the folium of Descartes $x^3 + y^3 = 6xy$ at the point (3, 3).

(c) At what point in the first quadrant is the tangent line horizontal?

SOLUTION

(a) Differentiating both sides of $x^3 + y^3 = 6xy$ with respect to x, regarding y as a function of x, and using the Chain Rule on the term y^3 and the Product Rule on the term $6xy$, we get

$$3x^2 + 3y^2y' = 6xy' + 6y$$

or $$x^2 + y^2y' = 2xy' + 2y$$

We now solve for y':

$$y^2y' - 2xy' = 2y - x^2$$
$$(y^2 - 2x)y' = 2y - x^2$$
$$y' = \frac{2y - x^2}{y^2 - 2x}$$

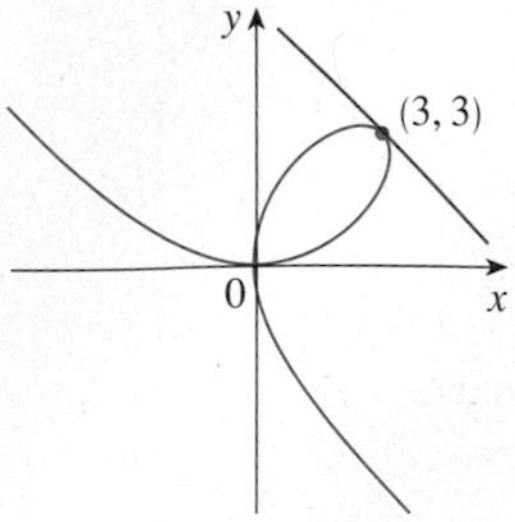

FIGURE 4

(b) When $x = y = 3$,

$$y' = \frac{2 \cdot 3 - 3^2}{3^2 - 2 \cdot 3} = -1$$

and a glance at Figure 4 confirms that this is a reasonable value for the slope at (3, 3). So an equation of the tangent to the folium at (3, 3) is

$$y - 3 = -1(x - 3) \qquad \text{or} \qquad x + y = 6$$

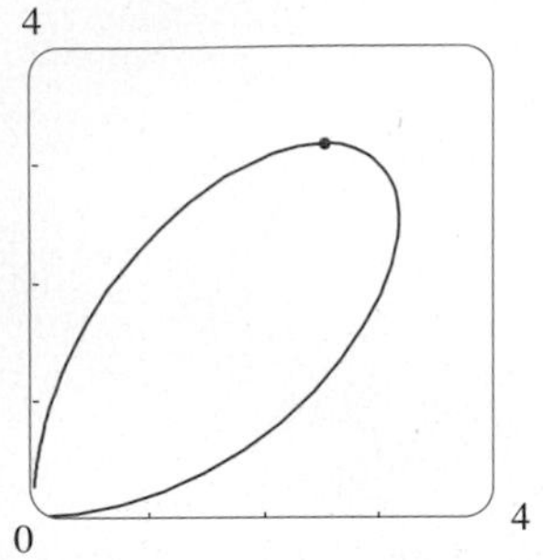

FIGURE 5

(c) The tangent line is horizontal if $y' = 0$. Using the expression for y' from part (a), we see that $y' = 0$ when $2y - x^2 = 0$ (provided that $y^2 - 2x \neq 0$). Substituting $y = \frac{1}{2}x^2$ in the equation of the curve, we get

$$x^3 + \left(\tfrac{1}{2}x^2\right)^3 = 6x\left(\tfrac{1}{2}x^2\right)$$

which simplifies to $x^6 = 16x^3$. Since $x \neq 0$ in the first quadrant, we have $x^3 = 16$. If $x = 16^{1/3} = 2^{4/3}$, then $y = \frac{1}{2}(2^{8/3}) = 2^{5/3}$. Thus the tangent is horizontal at $(2^{4/3}, 2^{5/3})$, which is approximately (2.5198, 3.1748). Looking at Figure 5, we see that our answer is reasonable. □

NOTE 2 There is a formula for the three roots of a cubic equation that is like the quadratic formula but much more complicated. If we use this formula (or a computer algebra system) to solve the equation $x^3 + y^3 = 6xy$ for y in terms of x, we get three functions determined by the equation:

$$y = f(x) = \sqrt[3]{-\tfrac{1}{2}x^3 + \sqrt{\tfrac{1}{4}x^6 - 8x^3}} + \sqrt[3]{-\tfrac{1}{2}x^3 - \sqrt{\tfrac{1}{4}x^6 - 8x^3}}$$

and

$$y = \tfrac{1}{2}\left[-f(x) \pm \sqrt{-3}\left(\sqrt[3]{-\tfrac{1}{2}x^3 + \sqrt{\tfrac{1}{4}x^6 - 8x^3}} - \sqrt[3]{-\tfrac{1}{2}x^3 - \sqrt{\tfrac{1}{4}x^6 - 8x^3}}\right)\right]$$

■ The Norwegian mathematician Niels Abel proved in 1824 that no general formula can be given for the roots of a fifth-degree equation in terms of radicals. Later the French mathematician Evariste Galois proved that it is impossible to find a general formula for the roots of an nth-degree equation (in terms of algebraic operations on the coefficients) if n is any integer larger than 4.

(These are the three functions whose graphs are shown in Figure 3.) You can see that the method of implicit differentiation saves an enormous amount of work in cases such as this. Moreover, implicit differentiation works just as easily for equations such as

$$y^5 + 3x^2y^2 + 5x^4 = 12$$

for which it is *impossible* to find a similar expression for y in terms of x.

EXAMPLE 3 Find y' if $\sin(x + y) = y^2 \cos x$.

SOLUTION Differentiating implicitly with respect to x and remembering that y is a function of x, we get

$$\cos(x + y) \cdot (1 + y') = y^2(-\sin x) + (\cos x)(2yy')$$

(Note that we have used the Chain Rule on the left side and the Product Rule and Chain

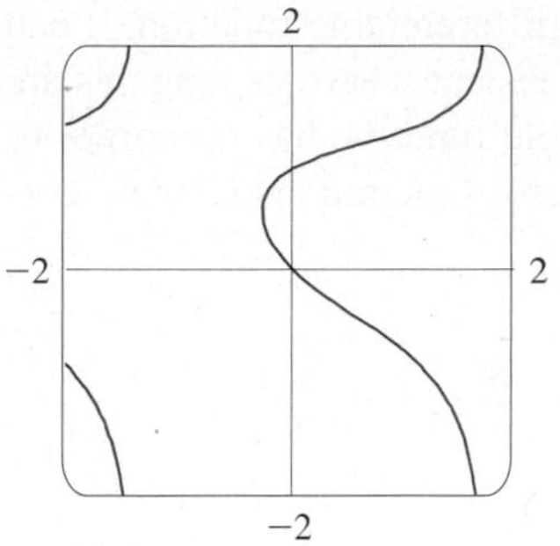

FIGURE 6

Rule on the right side.) If we collect the terms that involve y', we get

$$\cos(x + y) + y^2 \sin x = (2y \cos x)y' - \cos(x + y) \cdot y'$$

So
$$y' = \frac{y^2 \sin x + \cos(x + y)}{2y \cos x - \cos(x + y)}$$

Figure 6, drawn with the implicit-plotting command of a computer algebra system, shows part of the curve $\sin(x + y) = y^2 \cos x$. As a check on our calculation, notice that $y' = -1$ when $x = y = 0$ and it appears from the graph that the slope is approximately -1 at the origin. □

The following example shows how to find the second derivative of a function that is defined implicitly.

EXAMPLE 4 Find y'' if $x^4 + y^4 = 16$.

SOLUTION Differentiating the equation implicitly with respect to x, we get

$$4x^3 + 4y^3y' = 0$$

Solving for y' gives

3
$$y' = -\frac{x^3}{y^3}$$

■ Figure 7 shows the graph of the curve $x^4 + y^4 = 16$ of Example 4. Notice that it's a stretched and flattened version of the circle $x^2 + y^2 = 4$. For this reason it's sometimes called a *fat circle*. It starts out very steep on the left but quickly becomes very flat. This can be seen from the expression

$$y' = -\frac{x^3}{y^3} = -\left(\frac{x}{y}\right)^3$$

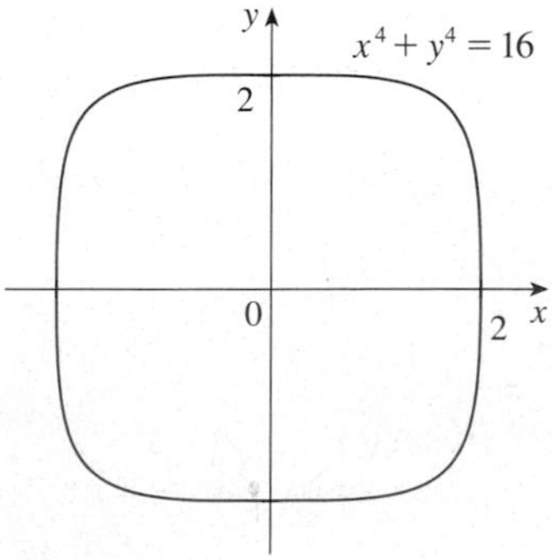

FIGURE 7

To find y'' we differentiate this expression for y' using the Quotient Rule and remembering that y is a function of x:

$$y'' = \frac{d}{dx}\left(-\frac{x^3}{y^3}\right) = -\frac{y^3\,(d/dx)(x^3) - x^3\,(d/dx)(y^3)}{(y^3)^2}$$

$$= -\frac{y^3 \cdot 3x^2 - x^3(3y^2y')}{y^6}$$

If we now substitute Equation 3 into this expression, we get

$$y'' = -\frac{3x^2y^3 - 3x^3y^2\left(-\dfrac{x^3}{y^3}\right)}{y^6}$$

$$= -\frac{3(x^2y^4 + x^6)}{y^7} = -\frac{3x^2(y^4 + x^4)}{y^7}$$

But the values of x and y must satisfy the original equation $x^4 + y^4 = 16$. So the answer simplifies to

$$y'' = -\frac{3x^2(16)}{y^7} = -48\,\frac{x^2}{y^7}$$ □

DERIVATIVES OF INVERSE TRIGONOMETRIC FUNCTIONS

The inverse trigonometric functions were reviewed in Section 1.6. We discussed their continuity in Section 2.5 and their asymptotes in Section 2.6. Here we use implicit differentiation to find the derivatives of the inverse trigonometric functions, assuming that these functions are differentiable. [In fact, if f is any one-to-one differentiable function, it can be proved that its inverse function f^{-1} is also differentiable, except where its tangents are vertical. This is plausible because the graph of a differentiable function has no corner or kink and so if we reflect it about $y = x$, the graph of its inverse function also has no corner or kink.]

Recall the definition of the arcsine function:

$$y = \sin^{-1}x \quad \text{means} \quad \sin y = x \quad \text{and} \quad -\frac{\pi}{2} \leqslant y \leqslant \frac{\pi}{2}$$

Differentiating $\sin y = x$ implicitly with respect to x, we obtain

$$\cos y \frac{dy}{dx} = 1 \qquad \text{or} \qquad \frac{dy}{dx} = \frac{1}{\cos y}$$

Now $\cos y \geqslant 0$, since $-\pi/2 \leqslant y \leqslant \pi/2$, so

$$\cos y = \sqrt{1 - \sin^2 y} = \sqrt{1 - x^2}$$

■ The same method can be used to find a formula for the derivative of *any* inverse function. See Exercise 67.

Therefore $$\frac{dy}{dx} = \frac{1}{\cos y} = \frac{1}{\sqrt{1 - x^2}}$$

$$\boxed{\frac{d}{dx}(\sin^{-1}x) = \frac{1}{\sqrt{1 - x^2}}}$$

■ Figure 8 shows the graph of $f(x) = \tan^{-1}x$ and its derivative $f'(x) = 1/(1 + x^2)$. Notice that f is increasing and $f'(x)$ is always positive. The fact that $\tan^{-1}x \to \pm\pi/2$ as $x \to \pm\infty$ is reflected in the fact that $f'(x) \to 0$ as $x \to \pm\infty$.

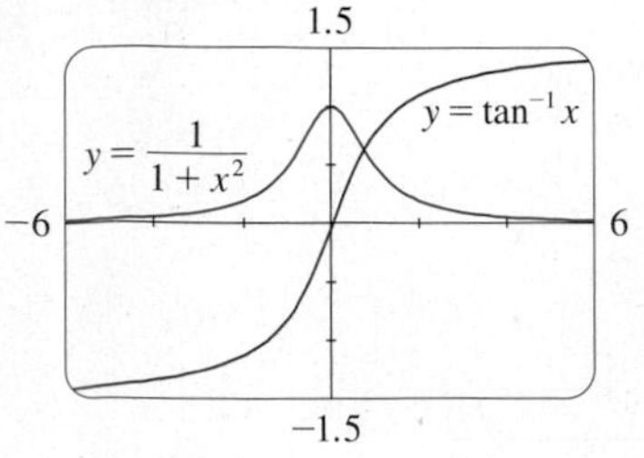

FIGURE 8

The formula for the derivative of the arctangent function is derived in a similar way. If $y = \tan^{-1}x$, then $\tan y = x$. Differentiating this latter equation implicitly with respect to x, we have

$$\sec^2 y \frac{dy}{dx} = 1$$

$$\frac{dy}{dx} = \frac{1}{\sec^2 y} = \frac{1}{1 + \tan^2 y} = \frac{1}{1 + x^2}$$

$$\boxed{\frac{d}{dx}(\tan^{-1}x) = \frac{1}{1 + x^2}}$$

V EXAMPLE 5 Differentiate (a) $y = \dfrac{1}{\sin^{-1}x}$ and (b) $f(x) = x \arctan\sqrt{x}$.

SOLUTION

(a) $$\frac{dy}{dx} = \frac{d}{dx}(\sin^{-1}x)^{-1} = -(\sin^{-1}x)^{-2}\frac{d}{dx}(\sin^{-1}x)$$

$$= -\frac{1}{(\sin^{-1}x)^2\sqrt{1 - x^2}}$$

■ Recall that $\arctan x$ is an alternative notation for $\tan^{-1}x$.

(b) $$f'(x) = x\frac{1}{1 + (\sqrt{x})^2}\left(\tfrac{1}{2}x^{-1/2}\right) + \arctan\sqrt{x}$$

$$= \frac{\sqrt{x}}{2(1 + x)} + \arctan\sqrt{x}$$ □

The inverse trigonometric functions that occur most frequently are the ones that we have just discussed. The derivatives of the remaining four are given in the following table. The proofs of the formulas are left as exercises.

■ The formulas for the derivatives of $\csc^{-1}x$ and $\sec^{-1}x$ depend on the definitions that are used for these functions. See Exercise 58.

DERIVATIVES OF INVERSE TRIGONOMETRIC FUNCTIONS

$$\frac{d}{dx}(\sin^{-1}x) = \frac{1}{\sqrt{1 - x^2}} \qquad \frac{d}{dx}(\csc^{-1}x) = -\frac{1}{x\sqrt{x^2 - 1}}$$

$$\frac{d}{dx}(\cos^{-1}x) = -\frac{1}{\sqrt{1 - x^2}} \qquad \frac{d}{dx}(\sec^{-1}x) = \frac{1}{x\sqrt{x^2 - 1}}$$

$$\frac{d}{dx}(\tan^{-1}x) = \frac{1}{1 + x^2} \qquad \frac{d}{dx}(\cot^{-1}x) = -\frac{1}{1 + x^2}$$

3.5 EXERCISES

1–4
(a) Find y' by implicit differentiation.
(b) Solve the equation explicitly for y and differentiate to get y' in terms of x.
(c) Check that your solutions to parts (a) and (b) are consistent by substituting the expression for y into your solution for part (a).

1. $xy + 2x + 3x^2 = 4$

2. $4x^2 + 9y^2 = 36$

3. $\dfrac{1}{x} + \dfrac{1}{y} = 1$

4. $\cos x + \sqrt{y} = 5$

5–20 Find dy/dx by implicit differentiation.

5. $x^3 + y^3 = 1$

6. $2\sqrt{x} + \sqrt{y} = 3$

7. $x^2 + xy - y^2 = 4$

8. $2x^3 + x^2y - xy^3 = 2$

9. $x^4(x + y) = y^2(3x - y)$

10. $y^5 + x^2y^3 = 1 + ye^{x^2}$

11. $x^2y^2 + x \sin y = 4$

12. $1 + x = \sin(xy^2)$

13. $4 \cos x \sin y = 1$

14. $y \sin(x^2) = x \sin(y^2)$

15. $e^{x/y} = x - y$

16. $\sqrt{x + y} = 1 + x^2y^2$

17. $\sqrt{xy} = 1 + x^2y$

18. $\tan(x - y) = \dfrac{y}{1 + x^2}$

19. $e^y \cos x = 1 + \sin(xy)$

20. $\sin x + \cos y = \sin x \cos y$

21. If $f(x) + x^2[f(x)]^3 = 10$ and $f(1) = 2$, find $f'(1)$.

22. If $g(x) + x \sin g(x) = x^2$, find $g'(0)$.

23–24 Regard y as the independent variable and x as the dependent variable and use implicit differentiation to find dx/dy.

23. $x^4y^2 - x^3y + 2xy^3 = 0$

24. $y \sec x = x \tan y$

25–30 Use implicit differentiation to find an equation of the tangent line to the curve at the given point.

25. $x^2 + xy + y^2 = 3$, $(1, 1)$ (ellipse)

26. $x^2 + 2xy - y^2 + x = 2$, $(1, 2)$ (hyperbola)

27. $x^2 + y^2 = (2x^2 + 2y^2 - x)^2$
$(0, \frac{1}{2})$
(cardioid)

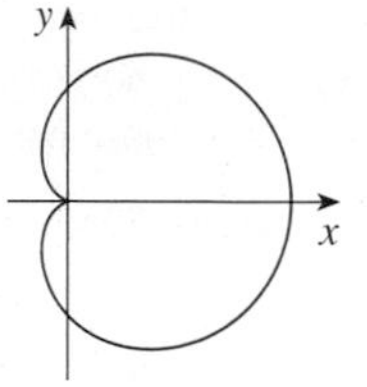

28. $x^{2/3} + y^{2/3} = 4$
$(-3\sqrt{3}, 1)$
(astroid)

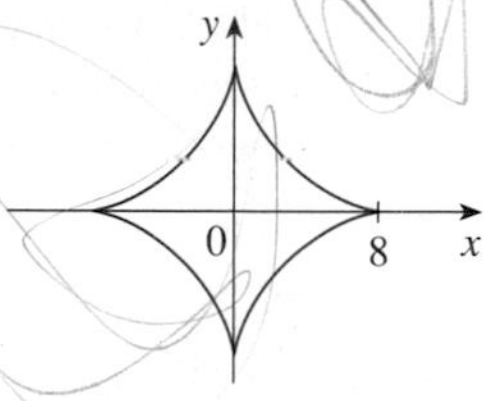

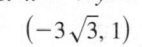

29. $2(x^2 + y^2)^2 = 25(x^2 - y^2)$
$(3, 1)$
(lemniscate)

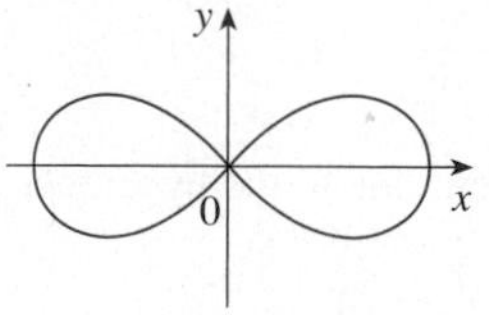

30. $y^2(y^2 - 4) = x^2(x^2 - 5)$
$(0, -2)$
(devil's curve)

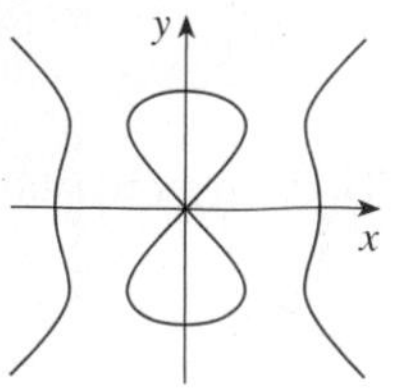

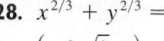

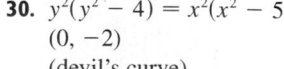

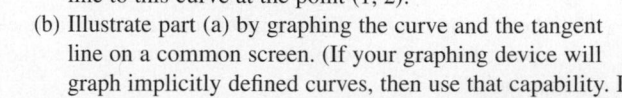

31. (a) The curve with equation $y^2 = 5x^4 - x^2$ is called a **kampyle of Eudoxus**. Find an equation of the tangent line to this curve at the point $(1, 2)$.
(b) Illustrate part (a) by graphing the curve and the tangent line on a common screen. (If your graphing device will graph implicitly defined curves, then use that capability. If

not, you can still graph this curve by graphing its upper and lower halves separately.)

32. (a) The curve with equation $y^2 = x^3 + 3x^2$ is called the **Tschirnhausen cubic**. Find an equation of the tangent line to this curve at the point $(1, -2)$.
(b) At what points does this curve have horizontal tangents?
(c) Illustrate parts (a) and (b) by graphing the curve and the tangent lines on a common screen.

33–36 Find y'' by implicit differentiation.

33. $9x^2 + y^2 = 9$ **34.** $\sqrt{x} + \sqrt{y} = 1$

35. $x^3 + y^3 = 1$ **36.** $x^4 + y^4 = a^4$

CAS **37.** Fanciful shapes can be created by using the implicit plotting capabilities of computer algebra systems.
(a) Graph the curve with equation

$$y(y^2 - 1)(y - 2) = x(x - 1)(x - 2)$$

At how many points does this curve have horizontal tangents? Estimate the x-coordinates of these points.
(b) Find equations of the tangent lines at the points $(0, 1)$ and $(0, 2)$.
(c) Find the exact x-coordinates of the points in part (a).
(d) Create even more fanciful curves by modifying the equation in part (a).

CAS **38.** (a) The curve with equation

$$2y^3 + y^2 - y^5 = x^4 - 2x^3 + x^2$$

has been likened to a bouncing wagon. Use a computer algebra system to graph this curve and discover why.
(b) At how many points does this curve have horizontal tangent lines? Find the x-coordinates of these points.

39. Find the points on the lemniscate in Exercise 29 where the tangent is horizontal.

40. Show by implicit differentiation that the tangent to the ellipse

$$\frac{x^2}{a^2} + \frac{y^2}{b^2} = 1$$

at the point (x_0, y_0) is

$$\frac{x_0 x}{a^2} + \frac{y_0 y}{b^2} = 1$$

41. Find an equation of the tangent line to the hyperbola

$$\frac{x^2}{a^2} - \frac{y^2}{b^2} = 1$$

at the point (x_0, y_0).

42. Show that the sum of the x- and y-intercepts of any tangent line to the curve $\sqrt{x} + \sqrt{y} = \sqrt{c}$ is equal to c.

43. Show, using implicit differentiation, that any tangent line at a point P to a circle with center O is perpendicular to the radius OP.

44. The Power Rule can be proved using implicit differentiation for the case where n is a rational number, $n = p/q$, and $y = f(x) = x^n$ is assumed beforehand to be a differentiable function. If $y = x^{p/q}$, then $y^q = x^p$. Use implicit differentiation to show that

$$y' = \frac{p}{q} x^{(p/q)-1}$$

45–54 Find the derivative of the function. Simplify where possible.

45. $y = \tan^{-1}\sqrt{x}$ **46.** $y = \sqrt{\tan^{-1}x}$

47. $y = \sin^{-1}(2x + 1)$ **48.** $g(x) = \sqrt{x^2 - 1}\,\sec^{-1}x$

49. $G(x) = \sqrt{1 - x^2}\,\arccos x$ **50.** $y = \tan^{-1}\left(x - \sqrt{1 + x^2}\right)$

51. $h(t) = \cot^{-1}(t) + \cot^{-1}(1/t)$ **52.** $F(\theta) = \arcsin\sqrt{\sin\theta}$

53. $y = \cos^{-1}(e^{2x})$ **54.** $y = \arctan\sqrt{\dfrac{1 - x}{1 + x}}$

55–56 Find $f'(x)$. Check that your answer is reasonable by comparing the graphs of f and f'.

55. $f(x) = \sqrt{1 - x^2}\,\arcsin x$ **56.** $f(x) = \arctan(x^2 - x)$

57. Prove the formula for $(d/dx)(\cos^{-1}x)$ by the same method as for $(d/dx)(\sin^{-1}x)$.

58. (a) One way of defining $\sec^{-1}x$ is to say that $y = \sec^{-1}x \iff \sec y = x$ and $0 \leqslant y < \pi/2$ or $\pi \leqslant y < 3\pi/2$. Show that, with this definition,

$$\frac{d}{dx}(\sec^{-1}x) = \frac{1}{x\sqrt{x^2 - 1}}$$

(b) Another way of defining $\sec^{-1}x$ that is sometimes used is to say that $y = \sec^{-1}x \iff \sec y = x$ and $0 \leqslant y \leqslant \pi$, $y \neq 0$. Show that, with this definition,

$$\frac{d}{dx}(\sec^{-1}x) = \frac{1}{|x|\sqrt{x^2 - 1}}$$

59–62 Two curves are **orthogonal** if their tangent lines are perpendicular at each point of intersection. Show that the given families of curves are **orthogonal trajectories** of each other, that is, every curve in one family is orthogonal to every curve in the other family. Sketch both families of curves on the same axes.

59. $x^2 + y^2 = r^2, \quad ax + by = 0$

60. $x^2 + y^2 = ax, \quad x^2 + y^2 = by$

61. $y = cx^2, \quad x^2 + 2y^2 = k$

62. $y = ax^3, \quad x^2 + 3y^2 = b$

63. The equation $x^2 - xy + y^2 = 3$ represents a "rotated ellipse," that is, an ellipse whose axes are not parallel to the coordinate axes. Find the points at which this ellipse crosses the x-axis and show that the tangent lines at these points are parallel.

64. (a) Where does the normal line to the ellipse $x^2 - xy + y^2 = 3$ at the point $(-1, 1)$ intersect the ellipse a second time?

(b) Illustrate part (a) by graphing the ellipse and the normal line.

65. Find all points on the curve $x^2y^2 + xy = 2$ where the slope of the tangent line is -1.

66. Find equations of both the tangent lines to the ellipse $x^2 + 4y^2 = 36$ that pass through the point $(12, 3)$.

67. (a) Suppose f is a one-to-one differentiable function and its inverse function f^{-1} is also differentiable. Use implicit differentiation to show that

$$(f^{-1})'(x) = \frac{1}{f'(f^{-1}(x))}$$

provided that the denominator is not 0.

(b) If $f(4) = 5$ and $f'(4) = \frac{2}{3}$, find $(f^{-1})'(5)$.

68. (a) Show that $f(x) = 2x + \cos x$ is one-to-one.

(b) What is the value of $f^{-1}(1)$?

(c) Use the formula from Exercise 67(a) to find $(f^{-1})'(1)$.

69. The figure shows a lamp located three units to the right of the y-axis and a shadow created by the elliptical region $x^2 + 4y^2 \leqslant 5$. If the point $(-5, 0)$ is on the edge of the shadow, how far above the x-axis is the lamp located?

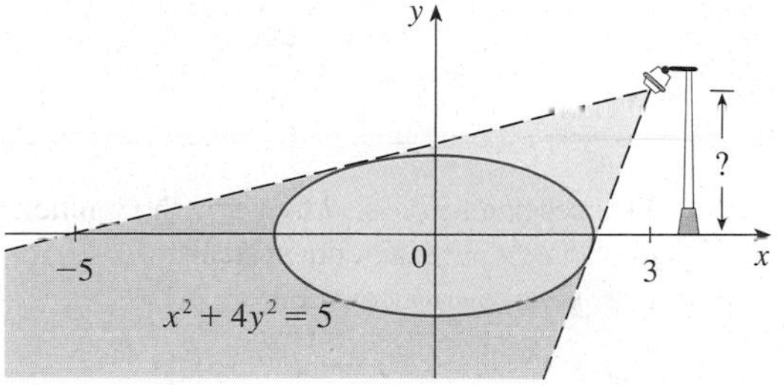

3.6 DERIVATIVES OF LOGARITHMIC FUNCTIONS

In this section we use implicit differentiation to find the derivatives of the logarithmic functions $y = \log_a x$ and, in particular, the natural logarithmic function $y = \ln x$. [It can be proved that logarithmic functions are differentiable; this is certainly plausible from their graphs (see Figure 12 in Section 1.6).]

1
$$\frac{d}{dx}(\log_a x) = \frac{1}{x \ln a}$$

PROOF Let $y = \log_a x$. Then

$$a^y = x$$

■ Formula 3.4.5 says that

$$\frac{d}{dx}(a^x) = a^x \ln a$$

Differentiating this equation implicitly with respect to x, using Formula 3.4.5, we get

$$a^y(\ln a)\frac{dy}{dx} = 1$$

and so

$$\frac{dy}{dx} = \frac{1}{a^y \ln a} = \frac{1}{x \ln a}$$ □

If we put $a = e$ in Formula 1, then the factor $\ln a$ on the right side becomes $\ln e = 1$ and we get the formula for the derivative of the natural logarithmic function $\log_e x = \ln x$:

2
$$\frac{d}{dx}(\ln x) = \frac{1}{x}$$

By comparing Formulas 1 and 2, we see one of the main reasons that natural logarithms (logarithms with base e) are used in calculus: The differentiation formula is simplest when $a = e$ because $\ln e = 1$.

V EXAMPLE 1 Differentiate $y = \ln(x^3 + 1)$.

SOLUTION To use the Chain Rule, we let $u = x^3 + 1$. Then $y = \ln u$, so

$$\frac{dy}{dx} = \frac{dy}{du}\frac{du}{dx} = \frac{1}{u}\frac{du}{dx} = \frac{1}{x^3+1}(3x^2) = \frac{3x^2}{x^3+1}$$ □

In general, if we combine Formula 2 with the Chain Rule as in Example 1, we get

3 $$\frac{d}{dx}(\ln u) = \frac{1}{u}\frac{du}{dx} \qquad \text{or} \qquad \frac{d}{dx}[\ln g(x)] = \frac{g'(x)}{g(x)}$$

EXAMPLE 2 Find $\dfrac{d}{dx}\ln(\sin x)$.

SOLUTION Using (3), we have

$$\frac{d}{dx}\ln(\sin x) = \frac{1}{\sin x}\frac{d}{dx}(\sin x) = \frac{1}{\sin x}\cos x = \cot x$$ □

EXAMPLE 3 Differentiate $f(x) = \sqrt{\ln x}$.

SOLUTION This time the logarithm is the inner function, so the Chain Rule gives

$$f'(x) = \tfrac{1}{2}(\ln x)^{-1/2}\frac{d}{dx}(\ln x) = \frac{1}{2\sqrt{\ln x}}\cdot\frac{1}{x} = \frac{1}{2x\sqrt{\ln x}}$$ □

EXAMPLE 4 Differentiate $f(x) = \log_{10}(2 + \sin x)$.

SOLUTION Using Formula 1 with $a = 10$, we have

$$f'(x) = \frac{d}{dx}\log_{10}(2 + \sin x) = \frac{1}{(2 + \sin x)\ln 10}\frac{d}{dx}(2 + \sin x)$$

$$= \frac{\cos x}{(2 + \sin x)\ln 10}$$ □

■ Figure 1 shows the graph of the function f of Example 5 together with the graph of its derivative. It gives a visual check on our calculation. Notice that $f'(x)$ is large negative when f is rapidly decreasing.

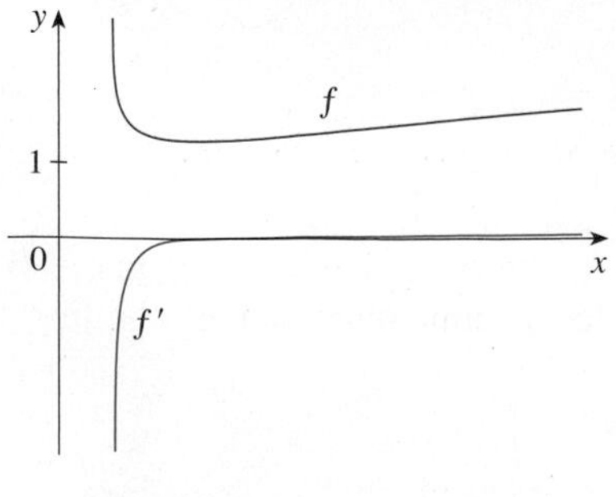

FIGURE 1

EXAMPLE 5 Find $\dfrac{d}{dx}\ln\dfrac{x+1}{\sqrt{x-2}}$.

SOLUTION 1

$$\frac{d}{dx}\ln\frac{x+1}{\sqrt{x-2}} = \frac{1}{\dfrac{x+1}{\sqrt{x-2}}}\frac{d}{dx}\frac{x+1}{\sqrt{x-2}}$$

$$= \frac{\sqrt{x-2}}{x+1}\,\frac{\sqrt{x-2}\cdot 1 - (x+1)\left(\frac{1}{2}\right)(x-2)^{-1/2}}{x-2}$$

$$= \frac{x - 2 - \frac{1}{2}(x+1)}{(x+1)(x-2)} = \frac{x-5}{2(x+1)(x-2)}$$

SOLUTION 2 If we first simplify the given function using the laws of logarithms, then the differentiation becomes easier:

$$\frac{d}{dx} \ln \frac{x+1}{\sqrt{x-2}} = \frac{d}{dx}\left[\ln(x+1) - \tfrac{1}{2}\ln(x-2)\right]$$

$$= \frac{1}{x+1} - \frac{1}{2}\left(\frac{1}{x-2}\right)$$

(This answer can be left as written, but if we used a common denominator we would see that it gives the same answer as in Solution 1.) □

■ Figure 2 shows the graph of the function $f(x) = \ln|x|$ in Example 6 and its derivative $f'(x) = 1/x$. Notice that when x is small, the graph of $y = \ln|x|$ is steep and so $f'(x)$ is large (positive or negative).

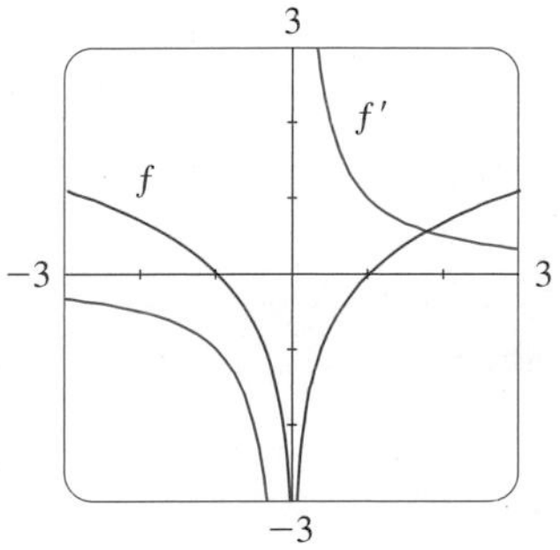

FIGURE 2

V EXAMPLE 6 Find $f'(x)$ if $f(x) = \ln|x|$.

SOLUTION Since

$$f(x) = \begin{cases} \ln x & \text{if } x > 0 \\ \ln(-x) & \text{if } x < 0 \end{cases}$$

it follows that

$$f'(x) = \begin{cases} \dfrac{1}{x} & \text{if } x > 0 \\[2ex] \dfrac{1}{-x}(-1) = \dfrac{1}{x} & \text{if } x < 0 \end{cases}$$

Thus $f'(x) = 1/x$ for all $x \neq 0$. □

The result of Example 6 is worth remembering:

4

$$\frac{d}{dx} \ln|x| = \frac{1}{x}$$

LOGARITHMIC DIFFERENTIATION

The calculation of derivatives of complicated functions involving products, quotients, or powers can often be simplified by taking logarithms. The method used in the following example is called **logarithmic differentiation**.

EXAMPLE 7 Differentiate $y = \dfrac{x^{3/4}\sqrt{x^2+1}}{(3x+2)^5}$.

SOLUTION We take logarithms of both sides of the equation and use the Laws of Logarithms to simplify:

$$\ln y = \tfrac{3}{4}\ln x + \tfrac{1}{2}\ln(x^2+1) - 5\ln(3x+2)$$

Differentiating implicitly with respect to x gives

$$\frac{1}{y}\frac{dy}{dx} = \frac{3}{4}\cdot\frac{1}{x} + \frac{1}{2}\cdot\frac{2x}{x^2+1} - 5\cdot\frac{3}{3x+2}$$

Solving for dy/dx, we get

$$\frac{dy}{dx} = y\left(\frac{3}{4x} + \frac{x}{x^2 + 1} - \frac{15}{3x + 2}\right)$$

■ If we hadn't used logarithmic differentiation in Example 7, we would have had to use both the Quotient Rule and the Product Rule. The resulting calculation would have been horrendous.

Because we have an explicit expression for y, we can substitute and write

$$\frac{dy}{dx} = \frac{x^{3/4}\sqrt{x^2 + 1}}{(3x + 2)^5}\left(\frac{3}{4x} + \frac{x}{x^2 + 1} - \frac{15}{3x + 2}\right)$$ □

STEPS IN LOGARITHMIC DIFFERENTIATION

1. Take natural logarithms of both sides of an equation $y = f(x)$ and use the Laws of Logarithms to simplify.
2. Differentiate implicitly with respect to x.
3. Solve the resulting equation for y'.

If $f(x) < 0$ for some values of x, then $\ln f(x)$ is not defined, but we can write $|y| = |f(x)|$ and use Equation 4. We illustrate this procedure by proving the general version of the Power Rule, as promised in Section 3.1.

THE POWER RULE If n is any real number and $f(x) = x^n$, then

$$f'(x) = nx^{n-1}$$

■ If $x = 0$, we can show that $f'(0) = 0$ for $n > 1$ directly from the definition of a derivative.

PROOF Let $y = x^n$ and use logarithmic differentiation:

$$\ln|y| = \ln|x|^n = n\ln|x| \qquad x \neq 0$$

Therefore

$$\frac{y'}{y} = \frac{n}{x}$$

Hence

$$y' = n\frac{y}{x} = n\frac{x^n}{x} = nx^{n-1}$$ □

⊘ You should distinguish carefully between the Power Rule $[(x^n)' = nx^{n-1}]$, where the base is variable and the exponent is constant, and the rule for differentiating exponential functions $[(a^x)' = a^x \ln a]$, where the base is constant and the exponent is variable.

In general there are four cases for exponents and bases:

1. $\dfrac{d}{dx}(a^b) = 0 \qquad$ (a and b are constants)
2. $\dfrac{d}{dx}[f(x)]^b = b[f(x)]^{b-1}f'(x)$
3. $\dfrac{d}{dx}[a^{g(x)}] = a^{g(x)}(\ln a)g'(x)$
4. To find $(d/dx)[f(x)]^{g(x)}$, logarithmic differentiation can be used, as in the next example.

V EXAMPLE 8 Differentiate $y = x^{\sqrt{x}}$.

SOLUTION 1 Using logarithmic differentiation, we have

$$\ln y = \ln x^{\sqrt{x}} = \sqrt{x}\ln x$$

$$\frac{y'}{y} = \sqrt{x}\cdot\frac{1}{x} + (\ln x)\frac{1}{2\sqrt{x}}$$

$$y' = y\left(\frac{1}{\sqrt{x}} + \frac{\ln x}{2\sqrt{x}}\right) = x^{\sqrt{x}}\left(\frac{2+\ln x}{2\sqrt{x}}\right)$$

■ Figure 3 illustrates Example 8 by showing the graphs of $f(x) = x^{\sqrt{x}}$ and its derivative.

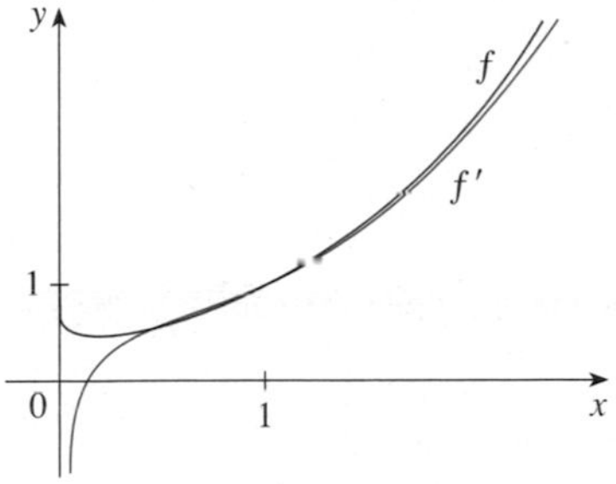

FIGURE 3

SOLUTION 2 Another method is to write $x^{\sqrt{x}} = (e^{\ln x})^{\sqrt{x}}$:

$$\frac{d}{dx}\left(x^{\sqrt{x}}\right) = \frac{d}{dx}\left(e^{\sqrt{x}\ln x}\right) = e^{\sqrt{x}\ln x}\frac{d}{dx}\left(\sqrt{x}\ln x\right)$$

$$= x^{\sqrt{x}}\left(\frac{2+\ln x}{2\sqrt{x}}\right) \qquad \text{(as in Solution 1)}$$ □

THE NUMBER *e* AS A LIMIT

We have shown that if $f(x) = \ln x$, then $f'(x) = 1/x$. Thus $f'(1) = 1$. We now use this fact to express the number e as a limit.

From the definition of a derivative as a limit, we have

$$f'(1) = \lim_{h\to 0}\frac{f(1+h)-f(1)}{h} = \lim_{x\to 0}\frac{f(1+x)-f(1)}{x}$$

$$= \lim_{x\to 0}\frac{\ln(1+x)-\ln 1}{x} = \lim_{x\to 0}\frac{1}{x}\ln(1+x)$$

$$= \lim_{x\to 0}\ln(1+x)^{1/x}$$

Because $f'(1) = 1$, we have

$$\lim_{x\to 0}\ln(1+x)^{1/x} = 1$$

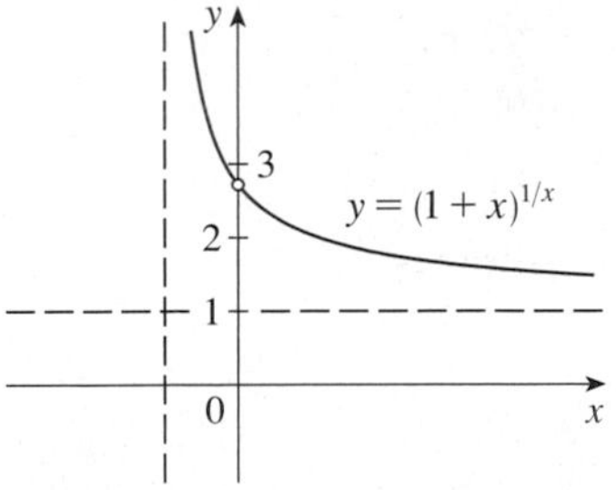

FIGURE 4

Then, by Theorem 2.5.8 and the continuity of the exponential function, we have

$$e = e^1 = e^{\lim_{x\to 0}\ln(1+x)^{1/x}} = \lim_{x\to 0} e^{\ln(1+x)^{1/x}} = \lim_{x\to 0}(1+x)^{1/x}$$

5 $$e = \lim_{x\to 0}(1+x)^{1/x}$$

x	$(1+x)^{1/x}$
0.1	2.59374246
0.01	2.70481383
0.001	2.71692393
0.0001	2.71814593
0.00001	2.71826824
0.000001	2.71828047
0.0000001	2.71828169
0.00000001	2.71828181

Formula 5 is illustrated by the graph of the function $y = (1+x)^{1/x}$ in Figure 4 and a table of values for small values of x. This illustrates the fact that, correct to seven decimal places,

$$e \approx 2.7182818$$

If we put $n = 1/x$ in Formula 5, then $n \to \infty$ as $x \to 0^+$ and so an alternative expression for e is

$$\boxed{6} \qquad e = \lim_{n\to\infty}\left(1 + \frac{1}{n}\right)^n$$

3.6 EXERCISES

1. Explain why the natural logarithmic function $y = \ln x$ is used much more frequently in calculus than the other logarithmic functions $y = \log_a x$.

2–22 Differentiate the function.

2. $f(x) = \ln(x^2 + 10)$

3. $f(x) = \sin(\ln x)$

4. $f(x) = \ln(\sin^2 x)$

5. $f(x) = \log_2(1 - 3x)$

6. $f(x) = \log_5(xe^x)$

7. $f(x) = \sqrt[5]{\ln x}$

8. $f(x) = \ln \sqrt[5]{x}$

9. $f(x) = \sin x \ln(5x)$

10. $f(t) = \dfrac{1 + \ln t}{1 - \ln t}$

11. $F(t) = \ln \dfrac{(2t+1)^3}{(3t-1)^4}$

12. $h(x) = \ln\left(x + \sqrt{x^2 - 1}\right)$

13. $g(x) = \ln\left(x\sqrt{x^2 - 1}\right)$

14. $F(y) = y \ln(1 + e^y)$

15. $f(u) = \dfrac{\ln u}{1 + \ln(2u)}$

16. $y = \dfrac{1}{\ln x}$

17. $y = \ln|2 - x - 5x^2|$

18. $H(z) = \ln\sqrt{\dfrac{a^2 - z^2}{a^2 + z^2}}$

19. $y = \ln(e^{-x} + xe^{-x})$

20. $y = [\ln(1 + e^x)]^2$

21. $y = 2x \log_{10}\sqrt{x}$

22. $y = \log_2(e^{-x}\cos \pi x)$

23–26 Find y' and y''.

23. $y = x^2 \ln(2x)$

24. $y = \dfrac{\ln x}{x^2}$

25. $y = \ln\left(x + \sqrt{1 + x^2}\right)$

26. $y = \ln(\sec x + \tan x)$

27–30 Differentiate f and find the domain of f.

27. $f(x) = \dfrac{x}{1 - \ln(x - 1)}$

28. $f(x) = \dfrac{1}{1 + \ln x}$

29. $f(x) = \ln(x^2 - 2x)$

30. $f(x) = \ln \ln \ln x$

31. If $f(x) = \dfrac{\ln x}{x^2}$, find $f'(1)$.

32. If $f(x) = \ln(1 + e^{2x})$, find $f'(0)$.

33–34 Find an equation of the tangent line to the curve at the given point.

33. $y = \ln\left(xe^{x^2}\right)$, $(1, 1)$

34. $y = \ln(x^3 - 7)$, $(2, 0)$

35. If $f(x) = \sin x + \ln x$, find $f'(x)$. Check that your answer is reasonable by comparing the graphs of f and f'.

36. Find equations of the tangent lines to the curve $y = (\ln x)/x$ at the points $(1, 0)$ and $(e, 1/e)$. Illustrate by graphing the curve and its tangent lines.

37–48 Use logarithmic differentiation to find the derivative of the function.

37. $y = (2x + 1)^5(x^4 - 3)^6$

38. $y = \sqrt{x}\, e^{x^2}(x^2 + 1)^{10}$

39. $y = \dfrac{\sin^2 x \tan^4 x}{(x^2 + 1)^2}$

40. $y = \sqrt[4]{\dfrac{x^2 + 1}{x^2 - 1}}$

41. $y = x^x$

42. $y = x^{\cos x}$

43. $y = x^{\sin x}$

44. $y = \sqrt{x}^{\,x}$

45. $y = (\cos x)^x$

46. $y = (\sin x)^{\ln x}$

47. $y = (\tan x)^{1/x}$

48. $y = (\ln x)^{\cos x}$

49. Find y' if $y = \ln(x^2 + y^2)$.

50. Find y' if $x^y = y^x$.

51. Find a formula for $f^{(n)}(x)$ if $f(x) = \ln(x - 1)$.

52. Find $\dfrac{d^9}{dx^9}(x^8 \ln x)$.

53. Use the definition of derivative to prove that

$$\lim_{x\to 0}\frac{\ln(1 + x)}{x} = 1$$

54. Show that $\lim_{n\to\infty}\left(1 + \dfrac{x}{n}\right)^n = e^x$ for any $x > 0$.

3.7 RATES OF CHANGE IN THE NATURAL AND SOCIAL SCIENCES

We know that if $y = f(x)$, then the derivative dy/dx can be interpreted as the rate of change of y with respect to x. In this section we examine some of the applications of this idea to physics, chemistry, biology, economics, and other sciences.

Let's recall from Section 2.7 the basic idea behind rates of change. If x changes from x_1 to x_2, then the change in x is

$$\Delta x = x_2 - x_1$$

and the corresponding change in y is

$$\Delta y = f(x_2) - f(x_1)$$

The difference quotient

$$\frac{\Delta y}{\Delta x} = \frac{f(x_2) - f(x_1)}{x_2 - x_1}$$

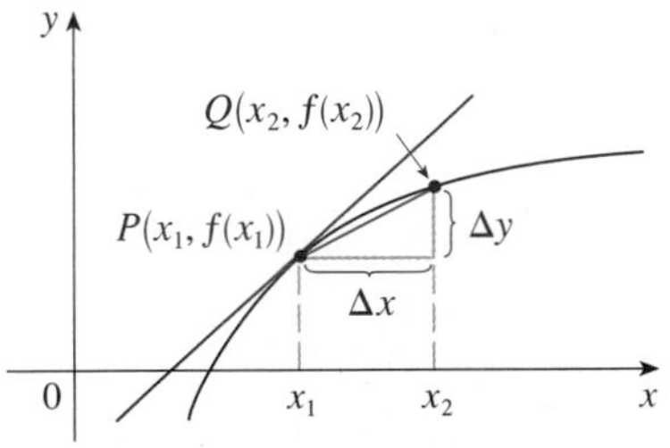

m_{PQ} = average rate of change
$m = f'(x_1)$ = instantaneous rate of change

FIGURE 1

is the **average rate of change of y with respect to x** over the interval $[x_1, x_2]$ and can be interpreted as the slope of the secant line PQ in Figure 1. Its limit as $\Delta x \to 0$ is the derivative $f'(x_1)$, which can therefore be interpreted as the **instantaneous rate of change of y with respect to x** or the slope of the tangent line at $P(x_1, f(x_1))$. Using Leibniz notation, we write the process in the form

$$\frac{dy}{dx} = \lim_{\Delta x \to 0} \frac{\Delta y}{\Delta x}$$

Whenever the function $y = f(x)$ has a specific interpretation in one of the sciences, its derivative will have a specific interpretation as a rate of change. (As we discussed in Section 2.7, the units for dy/dx are the units for y divided by the units for x.) We now look at some of these interpretations in the natural and social sciences.

PHYSICS

If $s = f(t)$ is the position function of a particle that is moving in a straight line, then $\Delta s/\Delta t$ represents the average velocity over a time period Δt, and $v = ds/dt$ represents the instantaneous **velocity** (the rate of change of displacement with respect to time). The instantaneous rate of change of velocity with respect to time is **acceleration**: $a(t) = v'(t) = s''(t)$. This was discussed in Sections 2.7 and 2.8, but now that we know the differentiation formulas, we are able to solve problems involving the motion of objects more easily.

V EXAMPLE 1 The position of a particle is given by the equation

$$s = f(t) = t^3 - 6t^2 + 9t$$

where t is measured in seconds and s in meters.

(a) Find the velocity at time t.
(b) What is the velocity after 2 s? After 4 s?
(c) When is the particle at rest?
(d) When is the particle moving forward (that is, in the positive direction)?
(e) Draw a diagram to represent the motion of the particle.
(f) Find the total distance traveled by the particle during the first five seconds.

(g) Find the acceleration at time t and after 4 s.
(h) Graph the position, velocity, and acceleration functions for $0 \leqslant t \leqslant 5$.
(i) When is the particle speeding up? When is it slowing down?

SOLUTION

(a) The velocity function is the derivative of the position function.

$$s = f(t) = t^3 - 6t^2 + 9t$$

$$v(t) = \frac{ds}{dt} = 3t^2 - 12t + 9$$

(b) The velocity after 2 s means the instantaneous velocity when $t = 2$, that is,

$$v(2) = \left.\frac{ds}{dt}\right|_{t=2} = 3(2)^2 - 12(2) + 9 = -3 \text{ m/s}$$

The velocity after 4 s is

$$v(4) = 3(4)^2 - 12(4) + 9 = 9 \text{ m/s}$$

(c) The particle is at rest when $v(t) = 0$, that is,

$$3t^2 - 12t + 9 = 3(t^2 - 4t + 3) = 3(t - 1)(t - 3) = 0$$

and this is true when $t = 1$ or $t = 3$. Thus the particle is at rest after 1 s and after 3 s.

(d) The particle moves in the positive direction when $v(t) > 0$, that is,

$$3t^2 - 12t + 9 = 3(t - 1)(t - 3) > 0$$

This inequality is true when both factors are positive ($t > 3$) or when both factors are negative ($t < 1$). Thus the particle moves in the positive direction in the time intervals $t < 1$ and $t > 3$. It moves backward (in the negative direction) when $1 < t < 3$.

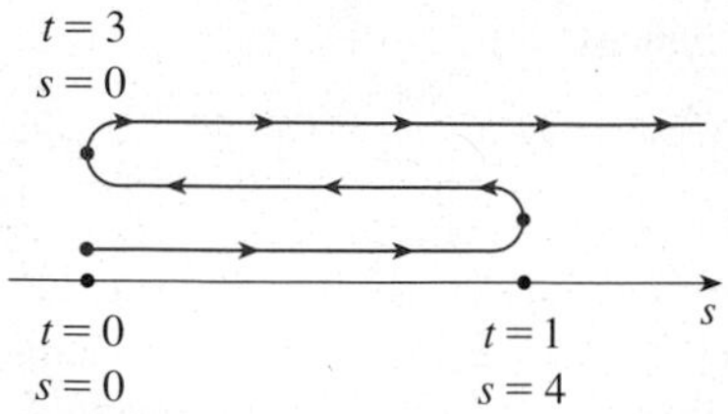

FIGURE 2

(e) Using the information from part (d) we make a schematic sketch in Figure 2 of the motion of the particle back and forth along a line (the s-axis).

(f) Because of what we learned in parts (d) and (e), we need to calculate the distances traveled during the time intervals [0, 1], [1, 3], and [3, 5] separately.

The distance traveled in the first second is

$$|f(1) - f(0)| = |4 - 0| = 4 \text{ m}$$

From $t = 1$ to $t = 3$ the distance traveled is

$$|f(3) - f(1)| = |0 - 4| = 4 \text{ m}$$

From $t = 3$ to $t = 5$ the distance traveled is

$$|f(5) - f(3)| = |20 - 0| = 20 \text{ m}$$

The total distance is $4 + 4 + 20 = 28$ m.

(g) The acceleration is the derivative of the velocity function:

$$a(t) = \frac{d^2s}{dt^2} = \frac{dv}{dt} = 6t - 12$$

$$a(4) = 6(4) - 12 = 12 \text{ m/s}^2$$

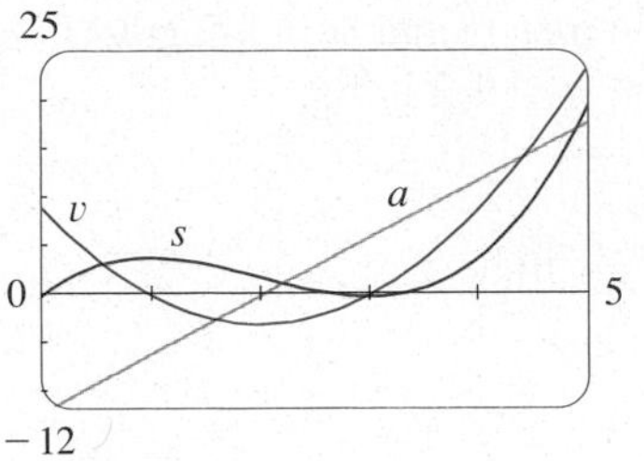

FIGURE 3

(h) Figure 3 shows the graphs of s, v, and a.

(i) The particle speeds up when the velocity is positive and increasing (v and a are both positive) and also when the velocity is negative and decreasing (v and a are both negative). In other words, the particle speeds up when the velocity and acceleration have the same sign. (The particle is pushed in the same direction it is moving.) From Figure 3 we see that this happens when $1 < t < 2$ and when $t > 3$. The particle slows down when v and a have opposite signs, that is, when $0 \leqslant t < 1$ and when $2 < t < 3$. Figure 4 summarizes the motion of the particle.

TEC In Module 3.7 you can see an animation of Figure 4 with an expression for s that you can choose yourself.

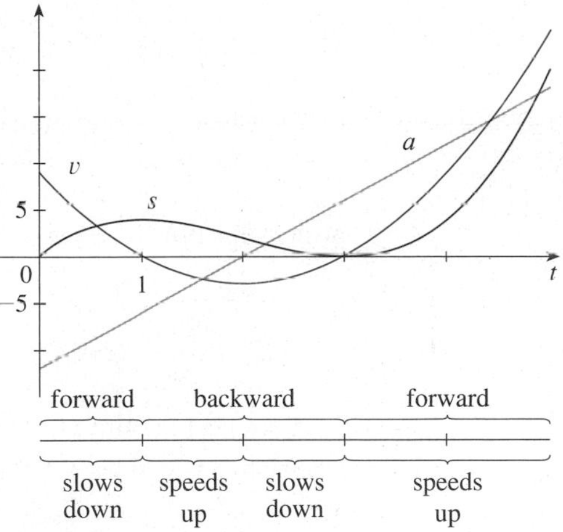

FIGURE 4

□

EXAMPLE 2 If a rod or piece of wire is homogeneous, then its linear density is uniform and is defined as the mass per unit length ($\rho = m/l$) and measured in kilograms per meter. Suppose, however, that the rod is not homogeneous but that its mass measured from its left end to a point x is $m = f(x)$, as shown in Figure 5.

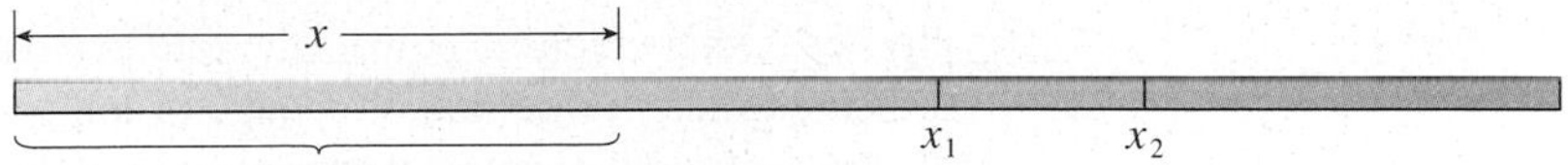

FIGURE 5

The mass of the part of the rod that lies between $x = x_1$ and $x = x_2$ is given by $\Delta m = f(x_2) - f(x_1)$, so the average density of that part of the rod is

$$\text{average density} = \frac{\Delta m}{\Delta x} = \frac{f(x_2) - f(x_1)}{x_2 - x_1}$$

If we now let $\Delta x \to 0$ (that is, $x_2 \to x_1$), we are computing the average density over smaller and smaller intervals. The **linear density** ρ at x_1 is the limit of these average densities as $\Delta x \to 0$; that is, the linear density is the rate of change of mass with respect to length. Symbolically,

$$\rho = \lim_{\Delta x \to 0} \frac{\Delta m}{\Delta x} = \frac{dm}{dx}$$

Thus the linear density of the rod is the derivative of mass with respect to length.

For instance, if $m = f(x) = \sqrt{x}$, where x is measured in meters and m in kilograms, then the average density of the part of the rod given by $1 \leqslant x \leqslant 1.2$ is

$$\frac{\Delta m}{\Delta x} = \frac{f(1.2) - f(1)}{1.2 - 1} = \frac{\sqrt{1.2} - 1}{0.2} \approx 0.48 \text{ kg/m}$$

while the density right at $x = 1$ is

$$\rho = \left.\frac{dm}{dx}\right|_{x=1} = \left.\frac{1}{2\sqrt{x}}\right|_{x=1} = 0.50 \text{ kg/m}$$ □

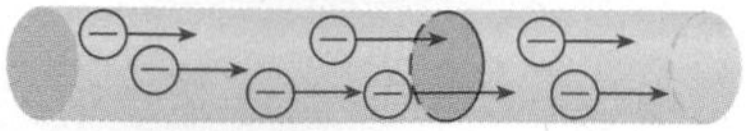

FIGURE 6

V EXAMPLE 3 A current exists whenever electric charges move. Figure 6 shows part of a wire and electrons moving through a shaded plane surface, shaded red. If ΔQ is the net charge that passes through this surface during a time period Δt, then the average current during this time interval is defined as

$$\text{average current} = \frac{\Delta Q}{\Delta t} = \frac{Q_2 - Q_1}{t_2 - t_1}$$

If we take the limit of this average current over smaller and smaller time intervals, we get what is called the **current** I at a given time t_1:

$$I = \lim_{\Delta t \to 0} \frac{\Delta Q}{\Delta t} = \frac{dQ}{dt}$$

Thus the current is the rate at which charge flows through a surface. It is measured in units of charge per unit time (often coulombs per second, called amperes). □

Velocity, density, and current are not the only rates of change that are important in physics. Others include power (the rate at which work is done), the rate of heat flow, temperature gradient (the rate of change of temperature with respect to position), and the rate of decay of a radioactive substance in nuclear physics.

CHEMISTRY

EXAMPLE 4 A chemical reaction results in the formation of one or more substances (called *products*) from one or more starting materials (called *reactants*). For instance, the "equation"

$$2H_2 + O_2 \rightarrow 2H_2O$$

indicates that two molecules of hydrogen and one molecule of oxygen form two molecules of water. Let's consider the reaction

$$A + B \rightarrow C$$

where A and B are the reactants and C is the product. The **concentration** of a reactant A is the number of moles (1 mole $= 6.022 \times 10^{23}$ molecules) per liter and is denoted by [A]. The concentration varies during a reaction, so [A], [B], and [C] are all functions of time (t). The average rate of reaction of the product C over a time interval $t_1 \leqslant t \leqslant t_2$ is

$$\frac{\Delta[\mathrm{C}]}{\Delta t} = \frac{[\mathrm{C}](t_2) - [\mathrm{C}](t_1)}{t_2 - t_1}$$

But chemists are more interested in the **instantaneous rate of reaction**, which is obtained by taking the limit of the average rate of reaction as the time interval Δt approaches 0:

$$\text{rate of reaction} = \lim_{\Delta t \to 0} \frac{\Delta[\mathrm{C}]}{\Delta t} = \frac{d[\mathrm{C}]}{dt}$$

Since the concentration of the product increases as the reaction proceeds, the derivative $d[\mathrm{C}]/dt$ will be positive, and so the rate of reaction of C is positive. The concentrations of the reactants, however, decrease during the reaction, so, to make the rates of reaction of A and B positive numbers, we put minus signs in front of the derivatives $d[\mathrm{A}]/dt$ and $d[\mathrm{B}]/dt$. Since $[\mathrm{A}]$ and $[\mathrm{B}]$ each decrease at the same rate that $[\mathrm{C}]$ increases, we have

$$\text{rate of reaction} = \frac{d[\mathrm{C}]}{dt} = -\frac{d[\mathrm{A}]}{dt} = -\frac{d[\mathrm{B}]}{dt}$$

More generally, it turns out that for a reaction of the form

$$a\mathrm{A} + b\mathrm{B} \to c\mathrm{C} + d\mathrm{D}$$

we have

$$-\frac{1}{a}\frac{d[\mathrm{A}]}{dt} = -\frac{1}{b}\frac{d[\mathrm{B}]}{dt} = \frac{1}{c}\frac{d[\mathrm{C}]}{dt} = \frac{1}{d}\frac{d[\mathrm{D}]}{dt}$$

The rate of reaction can be determined from data and graphical methods. In some cases there are explicit formulas for the concentrations as functions of time, which enable us to compute the rate of reaction (see Exercise 22). □

EXAMPLE 5 One of the quantities of interest in thermodynamics is compressibility. If a given substance is kept at a constant temperature, then its volume V depends on its pressure P. We can consider the rate of change of volume with respect to pressure—namely, the derivative dV/dP. As P increases, V decreases, so $dV/dP < 0$. The **compressibility** is defined by introducing a minus sign and dividing this derivative by the volume V:

$$\text{isothermal compressibility} = \beta = -\frac{1}{V}\frac{dV}{dP}$$

Thus β measures how fast, per unit volume, the volume of a substance decreases as the pressure on it increases at constant temperature.

For instance, the volume V (in cubic meters) of a sample of air at 25°C was found to be related to the pressure P (in kilopascals) by the equation

$$V = \frac{5.3}{P}$$

The rate of change of V with respect to P when $P = 50$ kPa is

$$\left.\frac{dV}{dP}\right|_{P=50} = -\left.\frac{5.3}{P^2}\right|_{P=50}$$

$$= -\frac{5.3}{2500} = -0.00212 \text{ m}^3/\text{kPa}$$

The compressibility at that pressure is

$$\beta = -\left.\frac{1}{V}\frac{dV}{dP}\right|_{P=50} = \frac{0.00212}{\frac{5.3}{50}} = 0.02 \text{ (m}^3/\text{kPa)/m}^3$$ □

BIOLOGY

EXAMPLE 6 Let $n = f(t)$ be the number of individuals in an animal or plant population at time t. The change in the population size between the times $t = t_1$ and $t = t_2$ is $\Delta n = f(t_2) - f(t_1)$, and so the average rate of growth during the time period $t_1 \leqslant t \leqslant t_2$ is

$$\text{average rate of growth} = \frac{\Delta n}{\Delta t} = \frac{f(t_2) - f(t_1)}{t_2 - t_1}$$

The **instantaneous rate of growth** is obtained from this average rate of growth by letting the time period Δt approach 0:

$$\text{growth rate} = \lim_{\Delta t \to 0} \frac{\Delta n}{\Delta t} = \frac{dn}{dt}$$

Strictly speaking, this is not quite accurate because the actual graph of a population function $n = f(t)$ would be a step function that is discontinuous whenever a birth or death occurs and therefore not differentiable. However, for a large animal or plant population, we can replace the graph by a smooth approximating curve as in Figure 7.

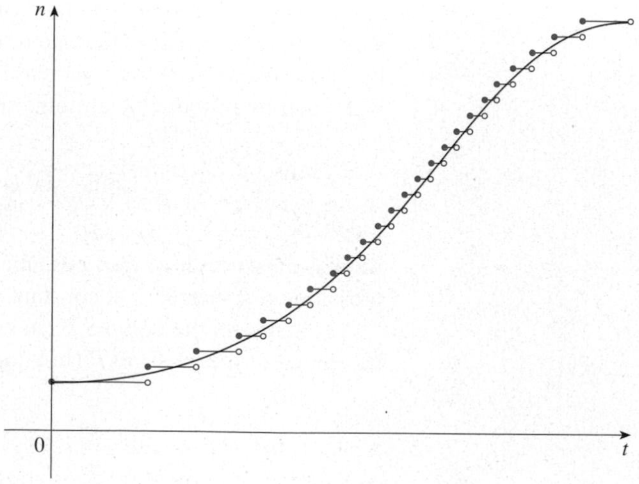

FIGURE 7
A smooth curve approximating a growth function

To be more specific, consider a population of bacteria in a homogeneous nutrient medium. Suppose that by sampling the population at certain intervals it is determined that the population doubles every hour. If the initial population is n_0 and the time t is measured in hours, then

$$f(1) = 2f(0) = 2n_0$$

$$f(2) = 2f(1) = 2^2 n_0$$

$$f(3) = 2f(2) = 2^3 n_0$$

and, in general,

$$f(t) = 2^t n_0$$

The population function is $n = n_0 2^t$.

In Section 3.4 we showed that

$$\frac{d}{dx}(a^x) = a^x \ln a$$

So the rate of growth of the bacteria population at time t is

$$\frac{dn}{dt} = \frac{d}{dt}(n_0 2^t) = n_0 2^t \ln 2$$

For example, suppose that we start with an initial population of $n_0 = 100$ bacteria. Then the rate of growth after 4 hours is

$$\left.\frac{dn}{dt}\right|_{t=4} = 100 \cdot 2^4 \ln 2 = 1600 \ln 2 \approx 1109$$

This means that, after 4 hours, the bacteria population is growing at a rate of about 1109 bacteria per hour. □

EXAMPLE 7 When we consider the flow of blood through a blood vessel, such as a vein or artery, we can model the shape of the blood vessel by a cylindrical tube with radius R and length l as illustrated in Figure 8.

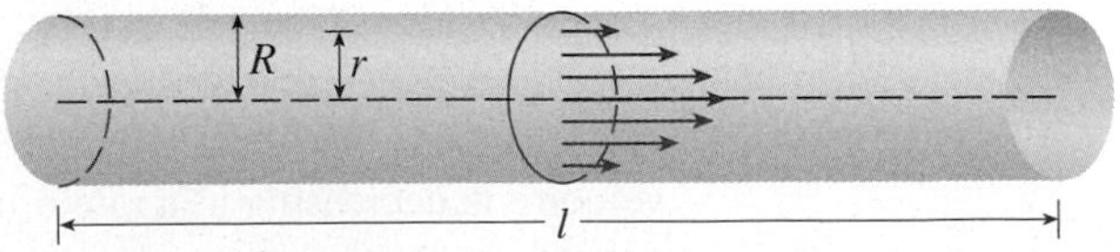

FIGURE 8
Blood flow in an artery

Because of friction at the walls of the tube, the velocity v of the blood is greatest along the central axis of the tube and decreases as the distance r from the axis increases until v becomes 0 at the wall. The relationship between v and r is given by the **law of laminar flow** discovered by the French physician Jean-Louis-Marie Poiseuille in 1840. This law states that

1 $$v = \frac{P}{4\eta l}(R^2 - r^2)$$

where η is the viscosity of the blood and P is the pressure difference between the ends of the tube. If P and l are constant, then v is a function of r with domain $[0, R]$.

■ For more detailed information, see W. Nichols and M. O'Rourke (eds.), *McDonald's Blood Flow in Arteries: Theoretic, Experimental, and Clinical Principles*, 4th ed. (New York: Oxford University Press, 1998).

The average rate of change of the velocity as we move from $r = r_1$ outward to $r = r_2$ is given by

$$\frac{\Delta v}{\Delta r} = \frac{v(r_2) - v(r_1)}{r_2 - r_1}$$

and if we let $\Delta r \to 0$, we obtain the **velocity gradient**, that is, the instantaneous rate of change of velocity with respect to r:

$$\text{velocity gradient} = \lim_{\Delta r \to 0} \frac{\Delta v}{\Delta r} = \frac{dv}{dr}$$

Using Equation 1, we obtain

$$\frac{dv}{dr} = \frac{P}{4\eta l}(0 - 2r) = -\frac{Pr}{2\eta l}$$

For one of the smaller human arteries we can take $\eta = 0.027$, $R = 0.008$ cm, $l = 2$ cm, and $P = 4000$ dynes/cm^2, which gives

$$\begin{aligned} v &= \frac{4000}{4(0.027)2}(0.000064 - r^2) \\ &\approx 1.85 \times 10^4(6.4 \times 10^{-5} - r^2) \end{aligned}$$

At $r = 0.002$ cm the blood is flowing at a speed of

$$\begin{aligned} v(0.002) &\approx 1.85 \times 10^4(64 \times 10^{-6} - 4 \times 10^{-6}) \\ &= 1.11 \text{ cm/s} \end{aligned}$$

and the velocity gradient at that point is

$$\left.\frac{dv}{dr}\right|_{r=0.002} = -\frac{4000(0.002)}{2(0.027)2} \approx -74 \text{ (cm/s)/cm}$$

To get a feeling for what this statement means, let's change our units from centimeters to micrometers (1 cm = 10,000 μm). Then the radius of the artery is 80 μm. The velocity at the central axis is 11,850 μm/s, which decreases to 11,110 μm/s at a distance of $r = 20$ μm. The fact that $dv/dr = -74$ (μm/s)/μm means that, when $r = 20$ μm, the velocity is decreasing at a rate of about 74 μm/s for each micrometer that we proceed away from the center. □

ECONOMICS

V EXAMPLE 8 Suppose $C(x)$ is the total cost that a company incurs in producing x units of a certain commodity. The function C is called a **cost function**. If the number of items produced is increased from x_1 to x_2, then the additional cost is $\Delta C = C(x_2) - C(x_1)$, and the average rate of change of the cost is

$$\frac{\Delta C}{\Delta x} = \frac{C(x_2) - C(x_1)}{x_2 - x_1} = \frac{C(x_1 + \Delta x) - C(x_1)}{\Delta x}$$

The limit of this quantity as $\Delta x \to 0$, that is, the instantaneous rate of change of cost with respect to the number of items produced, is called the **marginal cost** by economists:

$$\text{marginal cost} = \lim_{\Delta x \to 0} \frac{\Delta C}{\Delta x} = \frac{dC}{dx}$$

[Since x often takes on only integer values, it may not make literal sense to let Δx approach 0, but we can always replace $C(x)$ by a smooth approximating function as in Example 6.]

Taking $\Delta x = 1$ and n large (so that Δx is small compared to n), we have

$$C'(n) \approx C(n + 1) - C(n)$$

Thus the marginal cost of producing n units is approximately equal to the cost of producing one more unit [the $(n + 1)$st unit].

It is often appropriate to represent a total cost function by a polynomial

$$C(x) = a + bx + cx^2 + dx^3$$

where a represents the overhead cost (rent, heat, maintenance) and the other terms represent the cost of raw materials, labor, and so on. (The cost of raw materials may be proportional to x, but labor costs might depend partly on higher powers of x because of overtime costs and inefficiencies involved in large-scale operations.)

For instance, suppose a company has estimated that the cost (in dollars) of producing x items is

$$C(x) = 10{,}000 + 5x + 0.01x^2$$

Then the marginal cost function is

$$C'(x) = 5 + 0.02x$$

The marginal cost at the production level of 500 items is

$$C'(500) = 5 + 0.02(500) = \$15/\text{item}$$

This gives the rate at which costs are increasing with respect to the production level when $x = 500$ and predicts the cost of the 501st item.

The actual cost of producing the 501st item is

$$\begin{aligned} C(501) - C(500) &= [10{,}000 + 5(501) + 0.01(501)^2] \\ &\qquad - [10{,}000 + 5(500) + 0.01(500)^2] \\ &= \$15.01 \end{aligned}$$

Notice that $C'(500) \approx C(501) - C(500)$. □

Economists also study marginal demand, marginal revenue, and marginal profit, which are the derivatives of the demand, revenue, and profit functions. These will be considered in Chapter 4 after we have developed techniques for finding the maximum and minimum values of functions.

OTHER SCIENCES

Rates of change occur in all the sciences. A geologist is interested in knowing the rate at which an intruded body of molten rock cools by conduction of heat into surrounding rocks. An engineer wants to know the rate at which water flows into or out of a reservoir. An urban geographer is interested in the rate of change of the population density in a city as the distance from the city center increases. A meteorologist is concerned with the rate of change of atmospheric pressure with respect to height (see Exercise 17 in Section 3.8).

In psychology, those interested in learning theory study the so-called learning curve, which graphs the performance $P(t)$ of someone learning a skill as a function of the training time t. Of particular interest is the rate at which performance improves as time passes, that is, dP/dt.

In sociology, differential calculus is used in analyzing the spread of rumors (or innovations or fads or fashions). If $p(t)$ denotes the proportion of a population that knows a rumor by time t, then the derivative dp/dt represents the rate of spread of the rumor (see Exercise 82 in Section 3.4).

A SINGLE IDEA, MANY INTERPRETATIONS

Velocity, density, current, power, and temperature gradient in physics; rate of reaction and compressibility in chemistry; rate of growth and blood velocity gradient in biology; marginal cost and marginal profit in economics; rate of heat flow in geology; rate of improvement of performance in psychology; rate of spread of a rumor in sociology—these are all special cases of a single mathematical concept, the derivative.

This is an illustration of the fact that part of the power of mathematics lies in its abstractness. A single abstract mathematical concept (such as the derivative) can have different interpretations in each of the sciences. When we develop the properties of the mathematical concept once and for all, we can then turn around and apply these results to all of the sciences. This is much more efficient than developing properties of special concepts in each separate science. The French mathematician Joseph Fourier (1768–1830) put it succinctly: "Mathematics compares the most diverse phenomena and discovers the secret analogies that unite them."

3.7 EXERCISES

1–4 A particle moves according to a law of motion $s = f(t)$, $t \geq 0$, where t is measured in seconds and s in feet.

(a) Find the velocity at time t.
(b) What is the velocity after 3 s?
(c) When is the particle at rest?
(d) When is the particle moving in the positive direction?
(e) Find the total distance traveled during the first 8 s.
(f) Draw a diagram like Figure 2 to illustrate the motion of the particle.
(g) Find the acceleration at time t and after 3 s.
(h) Graph the position, velocity, and acceleration functions for $0 \leq t \leq 8$.
(i) When is the particle speeding up? When is it slowing down?

1. $f(t) = t^3 - 12t^2 + 36t$ **2.** $f(t) = 0.01t^4 - 0.04t^3$

3. $f(t) = \cos(\pi t/4)$, $t \leq 10$ **4.** $f(t) = te^{-t/2}$

5. Graphs of the *velocity* functions of two particles are shown, where t is measured in seconds. When is each particle speeding up? When is it slowing down? Explain.

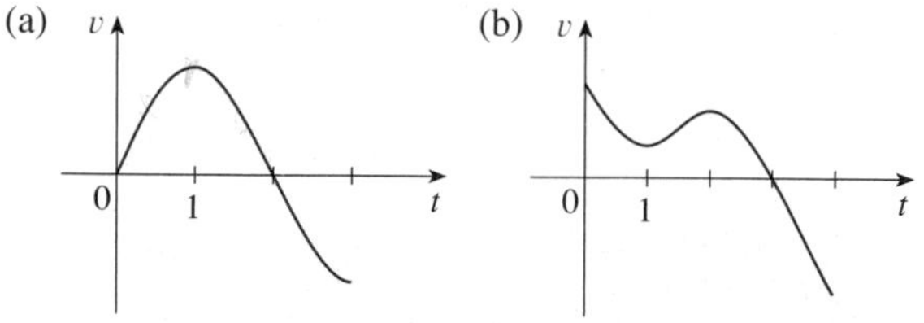

6. Graphs of the *position* functions of two particles are shown, where t is measured in seconds. When is each particle speeding up? When is it slowing down? Explain.

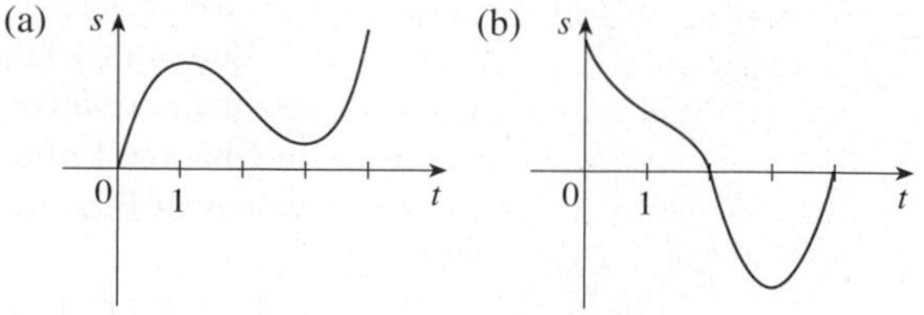

7. The position function of a particle is given by $s = t^3 - 4.5t^2 - 7t$, $t \geq 0$.
(a) When does the particle reach a velocity of 5 m/s?

(b) When is the acceleration 0? What is the significance of this value of t?

8. If a ball is given a push so that it has an initial velocity of 5 m/s down a certain inclined plane, then the distance it has rolled after t seconds is $s = 5t + 3t^2$.
(a) Find the velocity after 2 s.
(b) How long does it take for the velocity to reach 35 m/s?

9. If a stone is thrown vertically upward from the surface of the moon with a velocity of 10 m/s, its height (in meters) after t seconds is $h = 10t - 0.83t^2$.
(a) What is the velocity of the stone after 3 s?
(b) What is the velocity of the stone after it has risen 25 m?

10. If a ball is thrown vertically upward with a velocity of 80 ft/s, then its height after t seconds is $s = 80t - 16t^2$.
(a) What is the maximum height reached by the ball?
(b) What is the velocity of the ball when it is 96 ft above the ground on its way up? On its way down?

11. (a) A company makes computer chips from square wafers of silicon. It wants to keep the side length of a wafer very close to 15 mm and it wants to know how the area $A(x)$ of a wafer changes when the side length x changes. Find $A'(15)$ and explain its meaning in this situation.
(b) Show that the rate of change of the area of a square with respect to its side length is half its perimeter. Try to explain geometrically why this is true by drawing a square whose side length x is increased by an amount Δx. How can you approximate the resulting change in area ΔA if Δx is small?

12. (a) Sodium chlorate crystals are easy to grow in the shape of cubes by allowing a solution of water and sodium chlorate to evaporate slowly. If V is the volume of such a cube with side length x, calculate dV/dx when $x = 3$ mm and explain its meaning.
(b) Show that the rate of change of the volume of a cube with respect to its edge length is equal to half the surface area of the cube. Explain geometrically why this result is true by arguing by analogy with Exercise 11(b).

13. (a) Find the average rate of change of the area of a circle with respect to its radius r as r changes from
(i) 2 to 3 (ii) 2 to 2.5 (iii) 2 to 2.1
(b) Find the instantaneous rate of change when $r = 2$.
(c) Show that the rate of change of the area of a circle with respect to its radius (at any r) is equal to the circumference of the circle. Try to explain geometrically why this is true by drawing a circle whose radius is increased by an amount Δr. How can you approximate the resulting change in area ΔA if Δr is small?

14. A stone is dropped into a lake, creating a circular ripple that travels outward at a speed of 60 cm/s. Find the rate at which the area within the circle is increasing after (a) 1 s, (b) 3 s, and (c) 5 s. What can you conclude?

15. A spherical balloon is being inflated. Find the rate of increase of the surface area ($S = 4\pi r^2$) with respect to the radius r when r is (a) 1 ft, (b) 2 ft, and (c) 3 ft. What conclusion can you make?

16. (a) The volume of a growing spherical cell is $V = \frac{4}{3}\pi r^3$, where the radius r is measured in micrometers ($1\ \mu m = 10^{-6}$ m). Find the average rate of change of V with respect to r when r changes from
(i) 5 to 8 μm (ii) 5 to 6 μm (iii) 5 to 5.1 μm
(b) Find the instantaneous rate of change of V with respect to r when $r = 5$ μm.
(c) Show that the rate of change of the volume of a sphere with respect to its radius is equal to its surface area. Explain geometrically why this result is true. Argue by analogy with Exercise 13(c).

17. The mass of the part of a metal rod that lies between its left end and a point x meters to the right is $3x^2$ kg. Find the linear density (see Example 2) when x is (a) 1 m, (b) 2 m, and (c) 3 m. Where is the density the highest? The lowest?

18. If a tank holds 5000 gallons of water, which drains from the bottom of the tank in 40 minutes, then Torricelli's Law gives the volume V of water remaining in the tank after t minutes as

$$V = 5000\left(1 - \frac{t}{40}\right)^2 \qquad 0 \leqslant t \leqslant 40$$

Find the rate at which water is draining from the tank after (a) 5 min, (b) 10 min, (c) 20 min, and (d) 40 min. At what time is the water flowing out the fastest? The slowest? Summarize your findings.

19. The quantity of charge Q in coulombs (C) that has passed through a point in a wire up to time t (measured in seconds) is given by $Q(t) = t^3 - 2t^2 + 6t + 2$. Find the current when (a) $t = 0.5$ s and (b) $t = 1$ s. [See Example 3. The unit of current is an ampere (1 A = 1 C/s).] At what time is the current lowest?

20. Newton's Law of Gravitation says that the magnitude F of the force exerted by a body of mass m on a body of mass M is

$$F = \frac{GmM}{r^2}$$

where G is the gravitational constant and r is the distance between the bodies.
(a) Find dF/dr and explain its meaning. What does the minus sign indicate?
(b) Suppose it is known that the earth attracts an object with a force that decreases at the rate of 2 N/km when $r = 20{,}000$ km. How fast does this force change when $r = 10{,}000$ km?

21. Boyle's Law states that when a sample of gas is compressed at a constant temperature, the product of the pressure and the volume remains constant: $PV = C$.
(a) Find the rate of change of volume with respect to pressure.

(b) A sample of gas is in a container at low pressure and is steadily compressed at constant temperature for 10 minutes. Is the volume decreasing more rapidly at the beginning or the end of the 10 minutes? Explain.
(c) Prove that the isothermal compressibility (see Example 5) is given by $\beta = 1/P$.

22. If, in Example 4, one molecule of the product C is formed from one molecule of the reactant A and one molecule of the reactant B, and the initial concentrations of A and B have a common value $[A] = [B] = a$ moles/L, then

$$[C] = a^2kt/(akt + 1)$$

where k is a constant.
(a) Find the rate of reaction at time t.
(b) Show that if $x = [C]$, then

$$\frac{dx}{dt} = k(a - x)^2$$

(c) What happens to the concentration as $t \to \infty$?
(d) What happens to the rate of reaction as $t \to \infty$?
(e) What do the results of parts (c) and (d) mean in practical terms?

23. In Example 6 we considered a bacteria population that doubles every hour. Suppose that another population of bacteria triples every hour and starts with 400 bacteria. Find an expression for the number n of bacteria after t hours and use it to estimate the rate of growth of the bacteria population after 2.5 hours.

24. The number of yeast cells in a laboratory culture increases rapidly initially but levels off eventually. The population is modeled by the function

$$n = f(t) = \frac{a}{1 + be^{-0.7t}}$$

where t is measured in hours. At time $t = 0$ the population is 20 cells and is increasing at a rate of 12 cells/hour. Find the values of a and b. According to this model, what happens to the yeast population in the long run?

25. The table gives the population of the world in the 20th century.

Year	Population (in millions)	Year	Population (in millions)
1900	1650	1960	3040
1910	1750	1970	3710
1920	1860	1980	4450
1930	2070	1990	5280
1940	2300	2000	6080
1950	2560		

(a) Estimate the rate of population growth in 1920 and in 1980 by averaging the slopes of two secant lines.
(b) Use a graphing calculator or computer to find a cubic function (a third-degree polynomial) that models the data.
(c) Use your model in part (b) to find a model for the rate of population growth in the 20th century.
(d) Use part (c) to estimate the rates of growth in 1920 and 1980. Compare with your estimates in part (a).
(e) Estimate the rate of growth in 1985.

26. The table shows how the average age of first marriage of Japanese women varied in the last half of the 20th century.

t	$A(t)$	t	$A(t)$
1950	23.0	1980	25.2
1955	23.8	1985	25.5
1960	24.4	1990	25.9
1965	24.5	1995	26.3
1970	24.2	2000	27.0
1975	24.7		

(a) Use a graphing calculator or computer to model these data with a fourth-degree polynomial.
(b) Use part (a) to find a model for $A'(t)$.
(c) Estimate the rate of change of marriage age for women in 1990.
(d) Graph the data points and the models for A and A'.

27. Refer to the law of laminar flow given in Example 7. Consider a blood vessel with radius 0.01 cm, length 3 cm, pressure difference 3000 dynes/cm^2, and viscosity $\eta = 0.027$.
(a) Find the velocity of the blood along the centerline $r = 0$, at radius $r = 0.005$ cm, and at the wall $r = R = 0.01$ cm.
(b) Find the velocity gradient at $r = 0$, $r = 0.005$, and $r = 0.01$.
(c) Where is the velocity the greatest? Where is the velocity changing most?

28. The frequency of vibrations of a vibrating violin string is given by

$$f = \frac{1}{2L}\sqrt{\frac{T}{\rho}}$$

where L is the length of the string, T is its tension, and ρ is its linear density. [See Chapter 11 in D. E. Hall, *Musical Acoustics*, 3d ed. (Pacific Grove, CA: Brooks/Cole, 2002).]
(a) Find the rate of change of the frequency with respect to
(i) the length (when T and ρ are constant),
(ii) the tension (when L and ρ are constant), and
(iii) the linear density (when L and T are constant).
(b) The pitch of a note (how high or low the note sounds) is determined by the frequency f. (The higher the frequency, the higher the pitch.) Use the signs of the derivatives in part (a) to determine what happens to the pitch of a note
(i) when the effective length of a string is decreased by placing a finger on the string so a shorter portion of the string vibrates,
(ii) when the tension is increased by turning a tuning peg,
(iii) when the linear density is increased by switching to another string.

29. The cost, in dollars, of producing x yards of a certain fabric is

$$C(x) = 1200 + 12x - 0.1x^2 + 0.0005x^3$$

(a) Find the marginal cost function.
(b) Find $C'(200)$ and explain its meaning. What does it predict?
(c) Compare $C'(200)$ with the cost of manufacturing the 201st yard of fabric.

30. The cost function for production of a commodity is

$$C(x) = 339 + 25x - 0.09x^2 + 0.0004x^3$$

(a) Find and interpret $C'(100)$.
(b) Compare $C'(100)$ with the cost of producing the 101st item.

31. If $p(x)$ is the total value of the production when there are x workers in a plant, then the *average productivity* of the workforce at the plant is

$$A(x) = \frac{p(x)}{x}$$

(a) Find $A'(x)$. Why does the company want to hire more workers if $A'(x) > 0$?
(b) Show that $A'(x) > 0$ if $p'(x)$ is greater than the average productivity.

32. If R denotes the reaction of the body to some stimulus of strength x, the *sensitivity* S is defined to be the rate of change of the reaction with respect to x. A particular example is that when the brightness x of a light source is increased, the eye reacts by decreasing the area R of the pupil. The experimental formula

$$R = \frac{40 + 24x^{0.4}}{1 + 4x^{0.4}}$$

has been used to model the dependence of R on x when R is measured in square millimeters and x is measured in appropriate units of brightness.
(a) Find the sensitivity.
(b) Illustrate part (a) by graphing both R and S as functions of x. Comment on the values of R and S at low levels of brightness. Is this what you would expect?

33. The gas law for an ideal gas at absolute temperature T (in kelvins), pressure P (in atmospheres), and volume V (in liters) is $PV = nRT$, where n is the number of moles of the gas and $R = 0.0821$ is the gas constant. Suppose that, at a certain instant, $P = 8.0$ atm and is increasing at a rate of 0.10 atm/min and $V = 10$ L and is decreasing at a rate of 0.15 L/min. Find the rate of change of T with respect to time at that instant if $n = 10$ mol.

34. In a fish farm, a population of fish is introduced into a pond and harvested regularly. A model for the rate of change of the fish population is given by the equation

$$\frac{dP}{dt} = r_0\left(1 - \frac{P(t)}{P_c}\right)P(t) - \beta P(t)$$

where r_0 is the birth rate of the fish, P_c is the maximum population that the pond can sustain (called the *carrying capacity*), and β is the percentage of the population that is harvested.
(a) What value of dP/dt corresponds to a stable population?
(b) If the pond can sustain 10,000 fish, the birth rate is 5%, and the harvesting rate is 4%, find the stable population level.
(c) What happens if β is raised to 5%?

35. In the study of ecosystems, *predator-prey models* are often used to study the interaction between species. Consider populations of tundra wolves, given by $W(t)$, and caribou, given by $C(t)$, in northern Canada. The interaction has been modeled by the equations

$$\frac{dC}{dt} = aC - bCW \qquad \frac{dW}{dt} = -cW + dCW$$

(a) What values of dC/dt and dW/dt correspond to stable populations?
(b) How would the statement "The caribou go extinct" be represented mathematically?
(c) Suppose that $a = 0.05$, $b = 0.001$, $c = 0.05$, and $d = 0.0001$. Find all population pairs (C, W) that lead to stable populations. According to this model, is it possible for the two species to live in balance or will one or both species become extinct?

3.8 EXPONENTIAL GROWTH AND DECAY

In many natural phenomena, quantities grow or decay at a rate proportional to their size. For instance, if $y = f(t)$ is the number of individuals in a population of animals or bacteria at time t, then it seems reasonable to expect that the rate of growth $f'(t)$ is proportional to the population $f(t)$; that is, $f'(t) = kf(t)$ for some constant k. Indeed, under ideal conditions (unlimited environment, adequate nutrition, immunity to disease) the mathematical model given by the equation $f'(t) = kf(t)$ predicts what actually happens fairly accurately. Another example occurs in nuclear physics where the mass of a radioactive substance decays at a rate proportional to the mass. In chemistry, the rate of a unimolecular first-order reaction is proportional to the concentration of the substance. In finance, the value of a savings account with continuously compounded interest increases at a rate proportional to that value.

In general, if $y(t)$ is the value of a quantity y at time t and if the rate of change of y with respect to t is proportional to its size $y(t)$ at any time, then

1 $$\frac{dy}{dt} = ky$$

where k is a constant. Equation 1 is sometimes called the **law of natural growth** (if $k > 0$) or the **law of natural decay** (if $k < 0$). It is called a **differential equation** because it involves an unknown function y and its derivative dy/dt.

It's not hard to think of a solution of Equation 1. This equation asks us to find a function whose derivative is a constant multiple of itself. We have met such functions in this chapter. Any exponential function of the form $y(t) = Ce^{kt}$, where C is a constant, satisfies

$$y'(t) = C(ke^{kt}) = k(Ce^{kt}) = ky(t)$$

We will see in Section 9.4 that *any* function that satisfies $dy/dt = ky$ must be of the form $y = Ce^{kt}$. To see the significance of the constant C, we observe that

$$y(0) = Ce^{k \cdot 0} = C$$

Therefore C is the initial value of the function.

2 THEOREM The only solutions of the differential equation $dy/dt = ky$ are the exponential functions

$$y(t) = y(0)e^{kt}$$

POPULATION GROWTH

What is the significance of the proportionality constant k? In the context of population growth, where $P(t)$ is the size of a population at time t, we can write

3 $$\frac{dP}{dt} = kP \qquad \text{or} \qquad \frac{1}{P}\frac{dP}{dt} = k$$

The quantity

$$\frac{1}{P}\frac{dP}{dt}$$

is the growth rate divided by the population size; it is called the **relative growth rate**. According to (3), instead of saying "the growth rate is proportional to population size" we could say "the relative growth rate is constant." Then (2) says that a population with constant relative growth rate must grow exponentially. Notice that the relative growth rate k appears as the coefficient of t in the exponential function Ce^{kt}. For instance, if

$$\frac{dP}{dt} = 0.02P$$

and t is measured in years, then the relative growth rate is $k = 0.02$ and the population grows at a relative rate of 2% per year. If the population at time 0 is P_0, then the expression for the population is

$$P(t) = P_0e^{0.02t}$$

EXAMPLE 1 Use the fact that the world population was 2560 million in 1950 and 3040 million in 1960 to model the population of the world in the second half of the 20th century. (Assume that the growth rate is proportional to the population size.) What is the relative growth rate? Use the model to estimate the world population in 1993 and to predict the population in the year 2020.

SOLUTION We measure the time t in years and let $t = 0$ in the year 1950. We measure the population $P(t)$ in millions of people. Then $P(0) = 2560$ and $P(10) = 3040$. Since we are assuming that $dP/dt = kP$, Theorem 2 gives

$$P(t) = P(0)e^{kt} = 2560e^{kt}$$

$$P(10) = 2560e^{10k} = 3040$$

$$k = \frac{1}{10}\ln\frac{3040}{2560} \approx 0.017185$$

The relative growth rate is about 1.7% per year and the model is

$$P(t) = 2560e^{0.017185t}$$

We estimate that the world population in 1993 was

$$P(43) = 2560e^{0.017185(43)} \approx 5360 \text{ million}$$

The model predicts that the population in 2020 will be

$$P(70) = 2560e^{0.017185(70)} \approx 8524 \text{ million}$$

The graph in Figure 1 shows that the model is fairly accurate to the end of the 20th century (the dots represent the actual population), so the estimate for 1993 is quite reliable. But the prediction for 2020 is riskier.

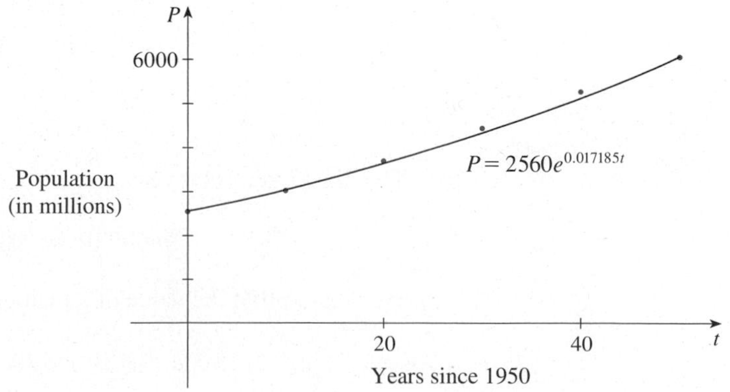

FIGURE 1
A model for world population growth in the second half of the 20th century

□

RADIOACTIVE DECAY

Radioactive substances decay by spontaneously emitting radiation. If $m(t)$ is the mass remaining from an initial mass m_0 of the substance after time t, then the relative decay rate

$$-\frac{1}{m}\frac{dm}{dt}$$

has been found experimentally to be constant. (Since dm/dt is negative, the relative decay rate is positive.) It follows that

$$\frac{dm}{dt} = km$$

where k is a negative constant. In other words, radioactive substances decay at a rate proportional to the remaining mass. This means that we can use (2) to show that the mass decays exponentially:

$$m(t) = m_0 e^{kt}$$

Physicists express the rate of decay in terms of **half-life**, the time required for half of any given quantity to decay.

EXAMPLE 2 The half-life of radium-226 is 1590 years.
(a) A sample of radium-226 has a mass of 100 mg. Find a formula for the mass of the sample that remains after t years.
(b) Find the mass after 1000 years correct to the nearest milligram.
(c) When will the mass be reduced to 30 mg?

SOLUTION
(a) Let $m(t)$ be the mass of radium-226 (in milligrams) that remains after t years. Then $dm/dt = km$ and $y(0) = 100$, so (2) gives

$$m(t) = m(0)e^{kt} = 100e^{kt}$$

In order to determine the value of k, we use the fact that $y(1590) = \frac{1}{2}(100)$. Thus

$$100e^{1590k} = 50 \qquad \text{so} \qquad e^{1590k} = \tfrac{1}{2}$$

and

$$1590k = \ln \tfrac{1}{2} = -\ln 2$$

$$k = -\frac{\ln 2}{1590}$$

Therefore

$$m(t) = 100e^{-(\ln 2)t/1590}$$

We could use the fact that $e^{\ln 2} = 2$ to write the expression for $m(t)$ in the alternative form

$$m(t) = 100 \times 2^{-t/1590}$$

(b) The mass after 1000 years is

$$m(1000) = 100e^{-(\ln 2)1000/1590} \approx 65 \text{ mg}$$

(c) We want to find the value of t such that $m(t) = 30$, that is,

$$100e^{-(\ln 2)t/1590} = 30 \qquad \text{or} \qquad e^{-(\ln 2)t/1590} = 0.3$$

We solve this equation for t by taking the natural logarithm of both sides:

$$-\frac{\ln 2}{1590}t = \ln 0.3$$

Thus

$$t = -1590\frac{\ln 0.3}{\ln 2} \approx 2762 \text{ years}$$

□

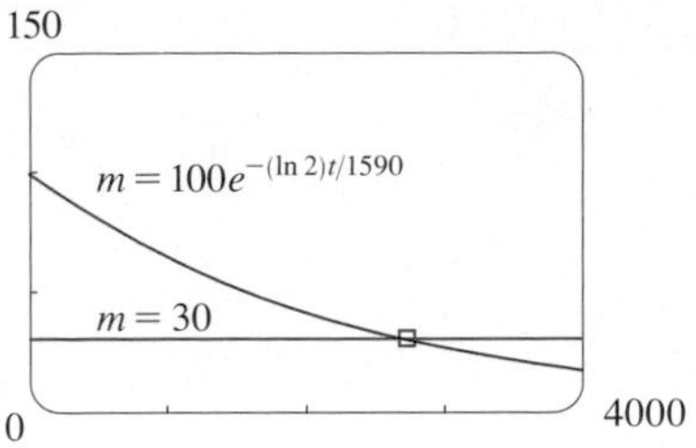

FIGURE 2

As a check on our work in Example 2, we use a graphing device to draw the graph of $m(t)$ in Figure 2 together with the horizontal line $m = 30$. These curves intersect when $t \approx 2800$, and this agrees with the answer to part (c).

NEWTON'S LAW OF COOLING

Newton's Law of Cooling states that the rate of cooling of an object is proportional to the temperature difference between the object and its surroundings, provided that this difference is not too large. (This law also applies to warming.) If we let $T(t)$ be the temperature of the object at time t and T_s be the temperature of the surroundings, then we can formulate Newton's Law of Cooling as a differential equation:

$$\frac{dT}{dt} = k(T - T_s)$$

where k is a constant. This equation is not quite the same as Equation 1, so we make the change of variable $y(t) = T(t) - T_s$. Because T_s is constant, we have $y'(t) = T'(t)$ and so the equation becomes

$$\frac{dy}{dt} = ky$$

We can then use (2) to find an expression for y, from which we can find T.

EXAMPLE 3 A bottle of soda pop at room temperature (72°F) is placed in a refrigerator where the temperature is 44°F. After half an hour the soda pop has cooled to 61°F.
(a) What is the temperature of the soda pop after another half hour?
(b) How long does it take for the soda pop to cool to 50°F?

SOLUTION
(a) Let $T(t)$ be the temperature of the soda after t minutes. The surrounding temperature is $T_s = 44°\text{F}$, so Newton's Law of Cooling states that

$$\frac{dT}{dt} = k(T - 44)$$

If we let $y = T - 44$, then $y(0) = T(0) - 44 = 72 - 44 = 28$, so y satisfies

$$\frac{dy}{dt} = ky \qquad y(0) = 28$$

and by (2) we have

$$y(t) = y(0)e^{kt} = 28e^{kt}$$

We are given that $T(30) = 61$, so $y(30) = 61 - 44 = 17$ and

$$28e^{30k} = 17 \qquad e^{30k} = \tfrac{17}{28}$$

Taking logarithms, we have

$$k = \frac{\ln\left(\frac{17}{28}\right)}{30} \approx -0.01663$$

Thus

$$y(t) = 28e^{-0.01663t}$$

$$T(t) = 44 + 28e^{-0.01663t}$$

$$T(60) = 44 + 28e^{-0.01663(60)} \approx 54.3$$

So after another half hour the pop has cooled to about 54°F.

(b) We have $T(t) = 50$ when

$$44 + 28e^{-0.01663t} = 50$$

$$e^{-0.01663t} = \tfrac{6}{28}$$

$$t = \frac{\ln\left(\frac{6}{28}\right)}{-0.01663} \approx 92.6$$

The pop cools to 50°F after about 1 hour 33 minutes. □

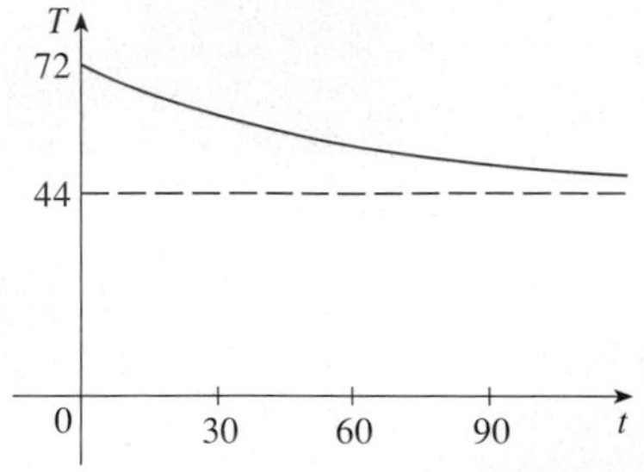

FIGURE 3

Notice that in Example 3, we have

$$\lim_{t \to \infty} T(t) = \lim_{t \to \infty} (44 + 28e^{-0.01663t}) = 44 + 28 \cdot 0 = 44$$

which is to be expected. The graph of the temperature function is shown in Figure 3.

CONTINUOUSLY COMPOUNDED INTEREST

EXAMPLE 4 If \$1000 is invested at 6% interest, compounded annually, then after 1 year the investment is worth \$1000(1.06) = \$1060, after 2 years it's worth \$[1000(1.06)]1.06 = \$1123.60, and after t years it's worth $\$1000(1.06)^t$. In general, if an amount A_0 is invested at an interest rate r ($r = 0.06$ in this example), then after t years it's worth $A_0(1 + r)^t$. Usually, however, interest is compounded more frequently, say, n times a year. Then in each compounding period the interest rate is r/n and there are nt compounding periods in t years, so the value of the investment is

$$A_0\left(1 + \frac{r}{n}\right)^{nt}$$

For instance, after 3 years at 6% interest a \$1000 investment will be worth

$$\$1000(1.06)^3 = \$1191.02 \quad \text{with annual compounding}$$

$$\$1000(1.03)^6 = \$1194.05 \quad \text{with semiannual compounding}$$

$$\$1000(1.015)^{12} = \$1195.62 \quad \text{with quarterly compounding}$$

$$\$1000(1.005)^{36} = \$1196.68 \quad \text{with monthly compounding}$$

$$\$1000\left(1 + \frac{0.06}{365}\right)^{365 \cdot 3} = \$1197.20 \quad \text{with daily compounding}$$

You can see that the interest paid increases as the number of compounding periods (n) increases. If we let $n \to \infty$, then we will be compounding the interest **continuously** and the value of the investment will be

$$A(t) = \lim_{n\to\infty} A_0\left(1 + \frac{r}{n}\right)^{nt} = \lim_{n\to\infty} A_0\left[\left(1 + \frac{r}{n}\right)^{n/r}\right]^{rt}$$

$$= A_0\left[\lim_{n\to\infty}\left(1 + \frac{r}{n}\right)^{n/r}\right]^{rt}$$

$$= A_0\left[\lim_{m\to\infty}\left(1 + \frac{1}{m}\right)^{m}\right]^{rt} \qquad \text{(where } m = n/r\text{)}$$

But the limit in this expression is equal to the number e. (See Equation 3.6.6). So with continuous compounding of interest at interest rate r, the amount after t years is

$$A(t) = A_0e^{rt}$$

If we differentiate this function, we get

$$\frac{dA}{dt} = rA_0e^{rt} = rA(t)$$

which says that, with continuous compounding of interest, the rate of increase of an investment is proportional to its size.

Returning to the example of \$1000 invested for 3 years at 6% interest, we see that with continuous compounding of interest the value of the investment will be

$$A(3) = \$1000e^{(0.06)3} = \$1197.22$$

Notice how close this is to the amount we calculated for daily compounding, \$1197.20. But the amount is easier to compute if we use continuous compounding. □

3.8 EXERCISES

1. A population of protozoa develops with a constant relative growth rate of 0.7944 per member per day. On day zero the population consists of two members. Find the population size after six days.

2. A common inhabitant of human intestines is the bacterium *Escherichia coli*. A cell of this bacterium in a nutrient-broth medium divides into two cells every 20 minutes. The initial population of a culture is 60 cells.
(a) Find the relative growth rate.
(b) Find an expression for the number of cells after t hours.
(c) Find the number of cells after 8 hours.
(d) Find the rate of growth after 8 hours.
(e) When will the population reach 20,000 cells?

3. A bacteria culture initially contains 100 cells and grows at a rate proportional to its size. After an hour the population has increased to 420.
(a) Find an expression for the number of bacteria after t hours.
(b) Find the number of bacteria after 3 hours.
(c) Find the rate of growth after 3 hours.
(d) When will the population reach 10,000?

4. A bacteria culture grows with constant relative growth rate. After 2 hours there are 600 bacteria and after 8 hours the count is 75,000.
(a) Find the initial population.
(b) Find an expression for the population after t hours.

(c) Find the number of cells after 5 hours.
(d) Find the rate of growth after 5 hours.
(e) When will the population reach 200,000?

5. The table gives estimates of the world population, in millions, from 1750 to 2000:

Year	Population	Year	Population
1750	790	1900	1650
1800	980	1950	2560
1850	1260	2000	6080

(a) Use the exponential model and the population figures for 1750 and 1800 to predict the world population in 1900 and 1950. Compare with the actual figures.
(b) Use the exponential model and the population figures for 1850 and 1900 to predict the world population in 1950. Compare with the actual population.
(c) Use the exponential model and the population figures for 1900 and 1950 to predict the world population in 2000. Compare with the actual population and try to explain the discrepancy.

6. The table gives the population of the United States, in millions, for the years 1900–2000.

Year	Population	Year	Population
1900	76	1960	179
1910	92	1970	203
1920	106	1980	227
1930	123	1990	250
1940	131	2000	275
1950	150		

(a) Use the exponential model and the census figures for 1900 and 1910 to predict the population in 2000. Compare with the actual figure and try to explain the discrepancy.
(b) Use the exponential model and the census figures for 1980 and 1990 to predict the population in 2000. Compare with the actual population. Then use this model to predict the population in the years 2010 and 2020.
(c) Graph both of the exponential functions in parts (a) and (b) together with a plot of the actual population. Are these models reasonable ones?

7. Experiments show that if the chemical reaction

$$N_2O_5 \rightarrow 2NO_2 + \tfrac{1}{2}O_2$$

takes place at 45°C, the rate of reaction of dinitrogen pentoxide is proportional to its concentration as follows:

$$-\frac{d[N_2O_5]}{dt} = 0.0005[N_2O_5]$$

(a) Find an expression for the concentration $[N_2O_5]$ after t seconds if the initial concentration is C.
(b) How long will the reaction take to reduce the concentration of N_2O_5 to 90% of its original value?

8. Bismuth-210 has a half-life of 5.0 days.
(a) A sample originally has a mass of 800 mg. Find a formula for the mass remaining after t days.
(b) Find the mass remaining after 30 days.
(c) When is the mass reduced to 1 mg?
(d) Sketch the graph of the mass function.

9. The half-life of cesium-137 is 30 years. Suppose we have a 100-mg sample.
(a) Find the mass that remains after t years.
(b) How much of the sample remains after 100 years?
(c) After how long will only 1 mg remain?

10. A sample of tritium-3 decayed to 94.5% of its original amount after a year.
(a) What is the half-life of tritium-3?
(b) How long would it take the sample to decay to 20% of its original amount?

11. Scientists can determine the age of ancient objects by the method of *radiocarbon dating*. The bombardment of the upper atmosphere by cosmic rays converts nitrogen to a radioactive isotope of carbon, ^{14}C, with a half-life of about 5730 years. Vegetation absorbs carbon dioxide through the atmosphere and animal life assimilates ^{14}C through food chains. When a plant or animal dies, it stops replacing its carbon and the amount of ^{14}C begins to decrease through radioactive decay. Therefore the level of radioactivity must also decay exponentially.

A parchment fragment was discovered that had about 74% as much ^{14}C radioactivity as does plant material on the earth today. Estimate the age of the parchment.

12. A curve passes through the point (0, 5) and has the property that the slope of the curve at every point P is twice the y-coordinate of P. What is the equation of the curve?

13. A roast turkey is taken from an oven when its temperature has reached 185°F and is placed on a table in a room where the temperature is 75°F.
(a) If the temperature of the turkey is 150°F after half an hour, what is the temperature after 45 minutes?
(b) When will the turkey have cooled to 100°F?

14. A thermometer is taken from a room where the temperature is 20°C to the outdoors, where the temperature is 5°C. After one minute the thermometer reads 12°C.
(a) What will the reading on the thermometer be after one more minute?
(b) When will the thermometer read 6°C?

15. When a cold drink is taken from a refrigerator, its temperature is 5°C. After 25 minutes in a 20°C room its temperature has increased to 10°C.
(a) What is the temperature of the drink after 50 minutes?
(b) When will its temperature be 15°C?

16. A freshly brewed cup of coffee has temperature 95°C in a 20°C room. When its temperature is 70°C, it is cooling at a rate of 1°C per minute. When does this occur?

17. The rate of change of atmospheric pressure P with respect to altitude h is proportional to P, provided that the temperature is constant. At 15°C the pressure is 101.3 kPa at sea level and 87.14 kPa at $h = 1000$ m.
(a) What is the pressure at an altitude of 3000 m?
(b) What is the pressure at the top of Mount McKinley, at an altitude of 6187 m?

18. (a) If \$1000 is borrowed at 8% interest, find the amounts due at the end of 3 years if the interest is compounded (i) annually, (ii) quarterly, (iii) monthly, (iv) weekly, (v) daily, (vi) hourly, and (viii) continuously.
(b) Suppose \$1000 is borrowed and the interest is compounded continuously. If $A(t)$ is the amount due after t years, where $0 \leqslant t \leqslant 3$, graph $A(t)$ for each of the interest rates 6%, 8%, and 10% on a common screen.

19. (a) If \$3000 is invested at 5% interest, find the value of the investment at the end of 5 years if the interest is compounded (i) annually, (ii) semiannually, (iii) monthly, (iv) weekly, (v) daily, and (vi) continuously.
(b) If $A(t)$ is the amount of the investment at time t for the case of continuous compounding, write a differential equation and an initial condition satisfied by $A(t)$.

20. (a) How long will it take an investment to double in value if the interest rate is 6% compounded continuously?
(b) What is the equivalent annual interest rate?

3.9 RELATED RATES

If we are pumping air into a balloon, both the volume and the radius of the balloon are increasing and their rates of increase are related to each other. But it is much easier to measure directly the rate of increase of the volume than the rate of increase of the radius.

In a related rates problem the idea is to compute the rate of change of one quantity in terms of the rate of change of another quantity (which may be more easily measured). The procedure is to find an equation that relates the two quantities and then use the Chain Rule to differentiate both sides with respect to time.

V EXAMPLE 1 Air is being pumped into a spherical balloon so that its volume increases at a rate of 100 cm³/s. How fast is the radius of the balloon increasing when the diameter is 50 cm?

■ According to the Principles of Problem Solving discussed on page 76, the first step is to understand the problem. This includes reading the problem carefully, identifying the given and the unknown, and introducing suitable notation.

SOLUTION We start by identifying two things:

the *given information:*

the rate of increase of the volume of air is 100 cm^3/s

and the *unknown*:

the rate of increase of the radius when the diameter is 50 cm

In order to express these quantities mathematically, we introduce some suggestive *notation*:

Let V be the volume of the balloon and let r be its radius.

The key thing to remember is that rates of change are derivatives. In this problem, the volume and the radius are both functions of the time t. The rate of increase of the volume with respect to time is the derivative dV/dt, and the rate of increase of the radius is dr/dt. We can therefore restate the given and the unknown as follows:

$$\textit{Given:} \qquad \frac{dV}{dt} = 100 \text{ cm}^3/\text{s}$$

$$\textit{Unknown:} \qquad \frac{dr}{dt} \quad \text{when } r = 25 \text{ cm}$$

■ The second stage of problem solving is to think of a plan for connecting the given and the unknown.

In order to connect dV/dt and dr/dt, we first relate V and r by the formula for the volume of a sphere:

$$V = \frac{4}{3}\pi r^3$$

In order to use the given information, we differentiate each side of this equation with respect to t. To differentiate the right side, we need to use the Chain Rule:

$$\frac{dV}{dt} = \frac{dV}{dr}\frac{dr}{dt} = 4\pi r^2 \frac{dr}{dt}$$

Now we solve for the unknown quantity:

■ Notice that, although dV/dt is constant, dr/dt is *not* constant.

$$\frac{dr}{dt} = \frac{1}{4\pi r^2}\frac{dV}{dt}$$

If we put $r = 25$ and $dV/dt = 100$ in this equation, we obtain

$$\frac{dr}{dt} = \frac{1}{4\pi(25)^2}100 = \frac{1}{25\pi}$$

The radius of the balloon is increasing at the rate of $1/(25\pi) \approx 0.0127$ cm/s. □

EXAMPLE 2 A ladder 10 ft long rests against a vertical wall. If the bottom of the ladder slides away from the wall at a rate of 1 ft/s, how fast is the top of the ladder sliding down the wall when the bottom of the ladder is 6 ft from the wall?

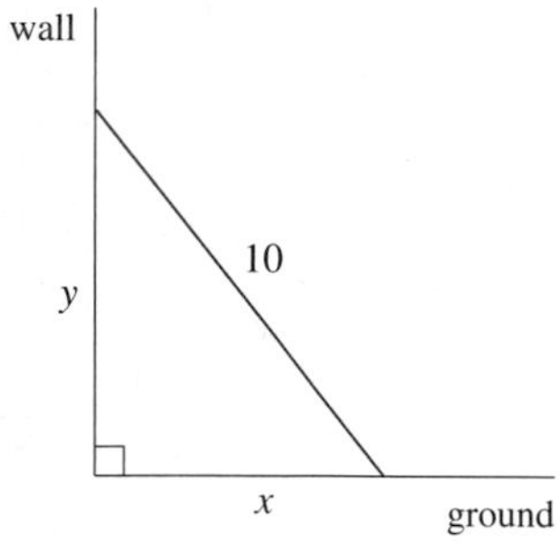

FIGURE 1

SOLUTION We first draw a diagram and label it as in Figure 1. Let x feet be the distance from the bottom of the ladder to the wall and y feet the distance from the top of the ladder to the ground. Note that x and y are both functions of t (time, measured in seconds).

We are given that $dx/dt = 1$ ft/s and we are asked to find dy/dt when $x = 6$ ft (see Figure 2). In this problem, the relationship between x and y is given by the Pythagorean Theorem:

$$x^2 + y^2 = 100$$

Differentiating each side with respect to t using the Chain Rule, we have

$$2x\frac{dx}{dt} + 2y\frac{dy}{dt} = 0$$

and solving this equation for the desired rate, we obtain

$$\frac{dy}{dt} = -\frac{x}{y}\frac{dx}{dt}$$

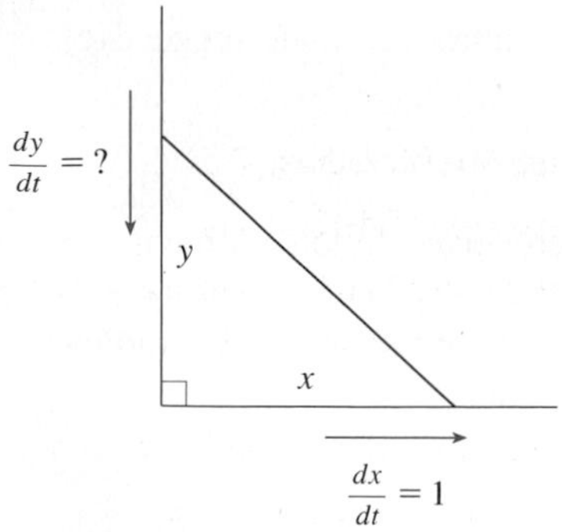

FIGURE 2

When $x = 6$, the Pythagorean Theorem gives $y = 8$ and so, substituting these values and $dx/dt = 1$, we have

$$\frac{dy}{dt} = -\frac{6}{8}(1) = -\frac{3}{4} \text{ ft/s}$$

The fact that dy/dt is negative means that the distance from the top of the ladder to the ground is *decreasing* at a rate of $\frac{3}{4}$ ft/s. In other words, the top of the ladder is sliding down the wall at a rate of $\frac{3}{4}$ ft/s. □

EXAMPLE 3 A water tank has the shape of an inverted circular cone with base radius 2 m and height 4 m. If water is being pumped into the tank at a rate of 2 m^3/min, find the rate at which the water level is rising when the water is 3 m deep.

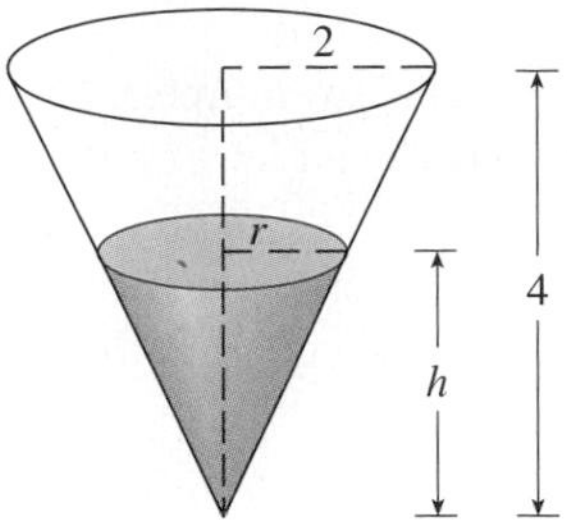

FIGURE 3

SOLUTION We first sketch the cone and label it as in Figure 3. Let V, r, and h be the volume of the water, the radius of the surface, and the height of the water at time t, where t is measured in minutes.

We are given that $dV/dt = 2$ m^3/min and we are asked to find dh/dt when h is 3 m. The quantities V and h are related by the equation

$$V = \tfrac{1}{3}\pi r^2 h$$

but it is very useful to express V as a function of h alone. In order to eliminate r, we use the similar triangles in Figure 3 to write

$$\frac{r}{h} = \frac{2}{4} \qquad r = \frac{h}{2}$$

and the expression for V becomes

$$V = \frac{1}{3}\pi\left(\frac{h}{2}\right)^2 h = \frac{\pi}{12}h^3$$

Now we can differentiate each side with respect to t:

$$\frac{dV}{dt} = \frac{\pi}{4}h^2\frac{dh}{dt}$$

so

$$\frac{dh}{dt} = \frac{4}{\pi h^2}\frac{dV}{dt}$$

Substituting $h = 3$ m and $dV/dt = 2$ m^3/min, we have

$$\frac{dh}{dt} = \frac{4}{\pi(3)^2}\cdot 2 = \frac{8}{9\pi}$$

The water level is rising at a rate of $8/(9\pi) \approx 0.28$ m/min. □

■ Look back: What have we learned from Examples 1–3 that will help us solve future problems?

STRATEGY It is useful to recall some of the problem-solving principles from page 76 and adapt them to related rates in light of our experience in Examples 1–3:

1. Read the problem carefully.
2. Draw a diagram if possible.
3. Introduce notation. Assign symbols to all quantities that are functions of time.
4. Express the given information and the required rate in terms of derivatives.
5. Write an equation that relates the various quantities of the problem. If necessary, use the geometry of the situation to eliminate one of the variables by substitution (as in Example 3).
6. Use the Chain Rule to differentiate both sides of the equation with respect to t.
7. Substitute the given information into the resulting equation and solve for the unknown rate.

The following examples are further illustrations of the strategy.

⊘ **WARNING:** A common error is to substitute the given numerical information (for quantities that vary with time) too early. This should be done only *after* the differentiation. (Step 7 follows Step 6.) For instance, in Example 3 we dealt with general values of h until we finally substituted $h = 3$ at the last stage. (If we had put $h = 3$ earlier, we would have gotten $dV/dt = 0$, which is clearly wrong.)

EXAMPLE 4 Car A is traveling west at 50 mi/h and car B is traveling north at 60 mi/h. Both are headed for the intersection of the two roads. At what rate are the cars approaching each other when car A is 0.3 mi and car B is 0.4 mi from the intersection?

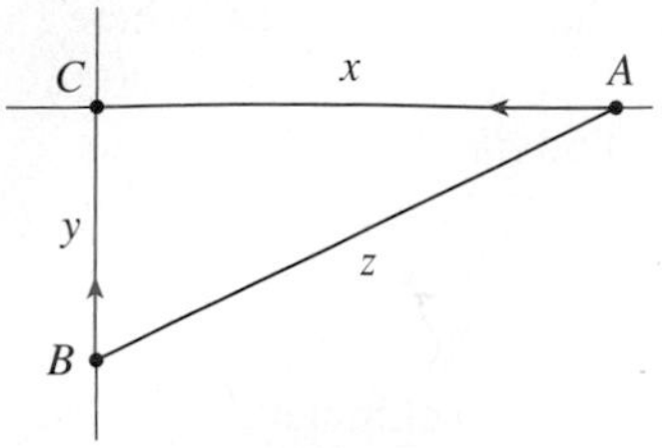

FIGURE 4

SOLUTION We draw Figure 4, where C is the intersection of the roads. At a given time t, let x be the distance from car A to C, let y be the distance from car B to C, and let z be the distance between the cars, where x, y, and z are measured in miles.

We are given that $dx/dt = -50$ mi/h and $dy/dt = -60$ mi/h. (The derivatives are negative because x and y are decreasing.) We are asked to find dz/dt. The equation that relates x, y, and z is given by the Pythagorean Theorem:

$$z^2 = x^2 + y^2$$

Differentiating each side with respect to t, we have

$$2z\frac{dz}{dt} = 2x\frac{dx}{dt} + 2y\frac{dy}{dt}$$

$$\frac{dz}{dt} = \frac{1}{z}\left(x\frac{dx}{dt} + y\frac{dy}{dt}\right)$$

When $x = 0.3$ mi and $y = 0.4$ mi, the Pythagorean Theorem gives $z = 0.5$ mi, so

$$\frac{dz}{dt} = \frac{1}{0.5}[0.3(-50) + 0.4(-60)]$$

$$= -78 \text{ mi/h}$$

The cars are approaching each other at a rate of 78 mi/h. □

EXAMPLE 5 A man walks along a straight path at a speed of 4 ft/s. A searchlight is located on the ground 20 ft from the path and is kept focused on the man. At what rate is the searchlight rotating when the man is 15 ft from the point on the path closest to the searchlight?

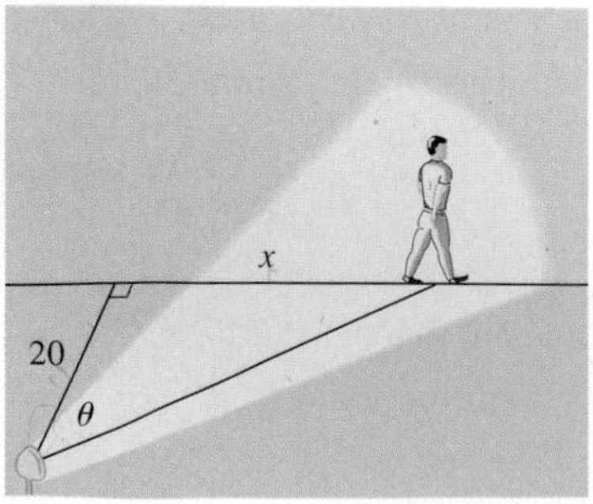

FIGURE 5

SOLUTION We draw Figure 5 and let x be the distance from the man to the point on the path closest to the searchlight. We let θ be the angle between the beam of the searchlight and the perpendicular to the path.

We are given that $dx/dt = 4$ ft/s and are asked to find $d\theta/dt$ when $x = 15$. The equation that relates x and θ can be written from Figure 5:

$$\frac{x}{20} = \tan\theta \qquad x = 20\tan\theta$$

Differentiating each side with respect to t, we get

$$\frac{dx}{dt} = 20\sec^2\theta\,\frac{d\theta}{dt}$$

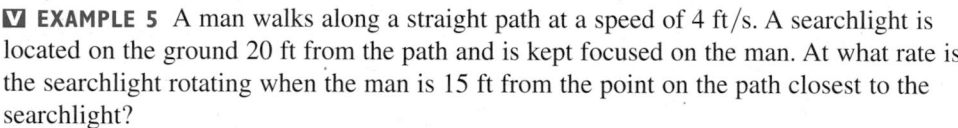

so $$\frac{d\theta}{dt} = \frac{1}{20}\cos^2\theta\,\frac{dx}{dt} = \frac{1}{20}\cos^2\theta\,(4) = \frac{1}{5}\cos^2\theta$$

When $x = 15$, the length of the beam is 25, so $\cos\theta = \frac{4}{5}$ and

$$\frac{d\theta}{dt} = \frac{1}{5}\left(\frac{4}{5}\right)^2 = \frac{16}{125} = 0.128$$

The searchlight is rotating at a rate of 0.128 rad/s. □

3.9 EXERCISES

1. If V is the volume of a cube with edge length x and the cube expands as time passes, find dV/dt in terms of dx/dt.

2. (a) If A is the area of a circle with radius r and the circle expands as time passes, find dA/dt in terms of dr/dt.
(b) Suppose oil spills from a ruptured tanker and spreads in a circular pattern. If the radius of the oil spill increases at a constant rate of 1 m/s, how fast is the area of the spill increasing when the radius is 30 m?

3. Each side of a square is increasing at a rate of 6 cm/s. At what rate is the area of the square increasing when the area of the square is 16 cm^2?

4. The length of a rectangle is increasing at a rate of 8 cm/s and its width is increasing at a rate of 3 cm/s. When the length is 20 cm and the width is 10 cm, how fast is the area of the rectangle increasing?

5. A cylindrical tank with radius 5 m is being filled with water at a rate of 3 m^3/min. How fast is the height of the water increasing?

6. The radius of a sphere is increasing at a rate of 4 mm/s. How fast is the volume increasing when the diameter is 80 mm?

7. If $y = x^3 + 2x$ and $dx/dt = 5$, find dy/dt when $x = 2$.

8. If $x^2 + y^2 = 25$ and $dy/dt = 6$, find dx/dt when $y = 4$.

9. If $z^2 = x^2 + y^2$, $dx/dt = 2$, and $dy/dt = 3$, find dz/dt when $x = 5$ and $y = 12$.

10. A particle moves along the curve $y = \sqrt{1 + x^3}$. As it reaches the point (2, 3), the y-coordinate is increasing at a rate of 4 cm/s. How fast is the x-coordinate of the point changing at that instant?

11–14
(a) What quantities are given in the problem?
(b) What is the unknown?
(c) Draw a picture of the situation for any time t.
(d) Write an equation that relates the quantities.
(e) Finish solving the problem.

11. A plane flying horizontally at an altitude of 1 mi and a speed of 500 mi/h passes directly over a radar station. Find the rate at which the distance from the plane to the station is increasing when it is 2 mi away from the station.

12. If a snowball melts so that its surface area decreases at a rate of 1 cm^2/min, find the rate at which the diameter decreases when the diameter is 10 cm.

13. A street light is mounted at the top of a 15-ft-tall pole. A man 6 ft tall walks away from the pole with a speed of 5 ft/s along a straight path. How fast is the tip of his shadow moving when he is 40 ft from the pole?

14. At noon, ship A is 150 km west of ship B. Ship A is sailing east at 35 km/h and ship B is sailing north at 25 km/h. How fast is the distance between the ships changing at 4:00 PM?

15. Two cars start moving from the same point. One travels south at 60 mi/h and the other travels west at 25 mi/h. At what rate is the distance between the cars increasing two hours later?

16. A spotlight on the ground shines on a wall 12 m away. If a man 2 m tall walks from the spotlight toward the building at a speed of 1.6 m/s, how fast is the length of his shadow on the building decreasing when he is 4 m from the building?

17. A man starts walking north at 4 ft/s from a point P. Five minutes later a woman starts walking south at 5 ft/s from a point 500 ft due east of P. At what rate are the people moving apart 15 min after the woman starts walking?

18. A baseball diamond is a square with side 90 ft. A batter hits the ball and runs toward first base with a speed of 24 ft/s.
(a) At what rate is his distance from second base decreasing when he is halfway to first base?
(b) At what rate is his distance from third base increasing at the same moment?

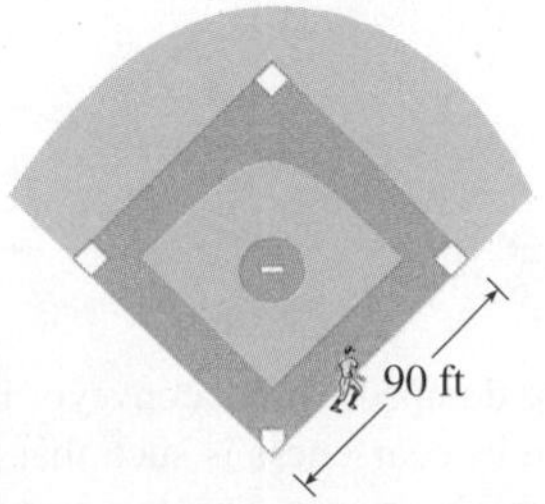

19. The altitude of a triangle is increasing at a rate of 1 cm/min while the area of the triangle is increasing at a rate of 2 cm^2/min. At what rate is the base of the triangle changing when the altitude is 10 cm and the area is 100 cm^2?

20. A boat is pulled into a dock by a rope attached to the bow of the boat and passing through a pulley on the dock that is 1 m higher than the bow of the boat. If the rope is pulled in at a rate of 1 m/s, how fast is the boat approaching the dock when it is 8 m from the dock?

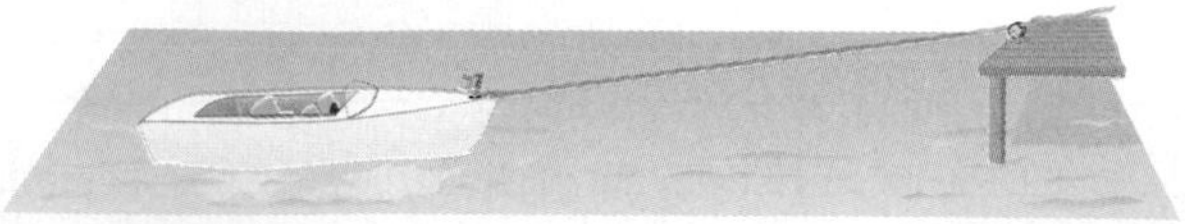

21. At noon, ship A is 100 km west of ship B. Ship A is sailing south at 35 km/h and ship B is sailing north at 25 km/h. How fast is the distance between the ships changing at 4:00 PM?

22. A particle is moving along the curve $y = \sqrt{x}$. As the particle passes through the point (4, 2), its x-coordinate increases at a rate of 3 cm/s. How fast is the distance from the particle to the origin changing at this instant?

23. Water is leaking out of an inverted conical tank at a rate of 10,000 cm^3/min at the same time that water is being pumped into the tank at a constant rate. The tank has height 6 m and the diameter at the top is 4 m. If the water level is rising at a rate of 20 cm/min when the height of the water is 2 m, find the rate at which water is being pumped into the tank.

24. A trough is 10 ft long and its ends have the shape of isosceles triangles that are 3 ft across at the top and have a height of 1 ft. If the trough is being filled with water at a rate of 12 ft^3/min, how fast is the water level rising when the water is 6 inches deep?

25. A water trough is 10 m long and a cross-section has the shape of an isosceles trapezoid that is 30 cm wide at the bottom, 80 cm wide at the top, and has height 50 cm. If the trough is being filled with water at the rate of 0.2 m^3/min, how fast is the water level rising when the water is 30 cm deep?

26. A swimming pool is 20 ft wide, 40 ft long, 3 ft deep at the shallow end, and 9 ft deep at its deepest point. A cross-section is shown in the figure. If the pool is being filled at a rate of 0.8 ft^3/min, how fast is the water level rising when the depth at the deepest point is 5 ft?

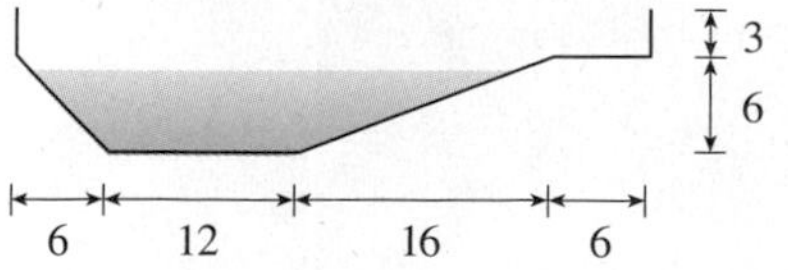

27. Gravel is being dumped from a conveyor belt at a rate of 30 ft^3/min, and its coarseness is such that it forms a pile in the shape of a cone whose base diameter and height are always equal. How fast is the height of the pile increasing when the pile is 10 ft high?

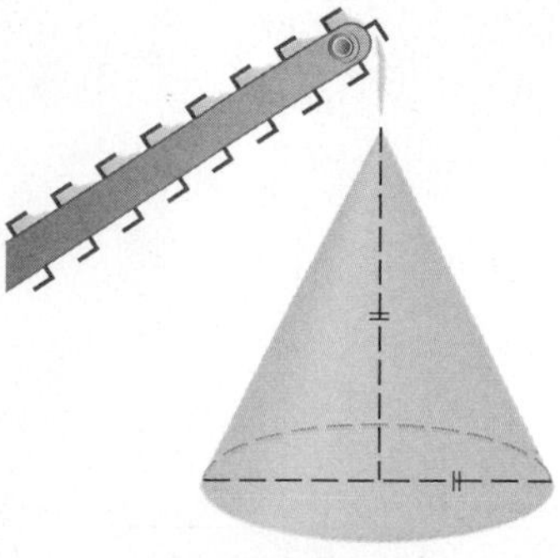

28. A kite 100 ft above the ground moves horizontally at a speed of 8 ft/s. At what rate is the angle between the string and the horizontal decreasing when 200 ft of string has been let out?

29. Two sides of a triangle are 4 m and 5 m in length and the angle between them is increasing at a rate of 0.06 rad/s. Find the rate at which the area of the triangle is increasing when the angle between the sides of fixed length is $\pi/3$.

30. How fast is the angle between the ladder and the ground changing in Example 2 when the bottom of the ladder is 6 ft from the wall?

31. Boyle's Law states that when a sample of gas is compressed at a constant temperature, the pressure P and volume V satisfy the equation $PV = C$, where C is a constant. Suppose that at a certain instant the volume is 600 cm^3, the pressure is 150 kPa, and the pressure is increasing at a rate of 20 kPa/min. At what rate is the volume decreasing at this instant?

32. When air expands adiabatically (without gaining or losing heat), its pressure P and volume V are related by the equation $PV^{1.4} = C$, where C is a constant. Suppose that at a certain instant the volume is 400 cm^3 and the pressure is 80 kPa and is decreasing at a rate of 10 kPa/min. At what rate is the volume increasing at this instant?

33. If two resistors with resistances R_1 and R_2 are connected in parallel, as in the figure, then the total resistance R, measured in ohms (Ω), is given by

$$\frac{1}{R} = \frac{1}{R_1} + \frac{1}{R_2}$$

If R_1 and R_2 are increasing at rates of 0.3 Ω/s and 0.2 Ω/s, respectively, how fast is R changing when $R_1 = 80\ \Omega$ and $R_2 = 100\ \Omega$?

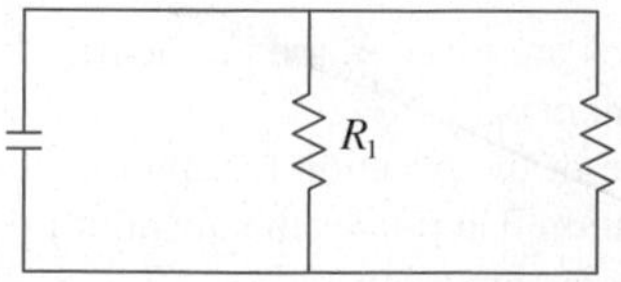

34. Brain weight B as a function of body weight W in fish has been modeled by the power function $B = 0.007W^{2/3}$, where B and W are measured in grams. A model for body weight as a function of body length L (measured in centimeters) is $W = 0.12L^{2.53}$. If, over 10 million years, the average length of a certain species of fish evolved from 15 cm to 20 cm at a constant rate, how fast was this species' brain growing when the average length was 18 cm?

35. Two sides of a triangle have lengths 12 m and 15 m. The angle between them is increasing at a rate of $2°/\text{min}$. How fast is the length of the third side increasing when the angle between the sides of fixed length is $60°$?

36. Two carts, A and B, are connected by a rope 39 ft long that passes over a pulley P (see the figure). The point Q is on the floor 12 ft directly beneath P and between the carts. Cart A is being pulled away from Q at a speed of 2 ft/s. How fast is cart B moving toward Q at the instant when cart A is 5 ft from Q?

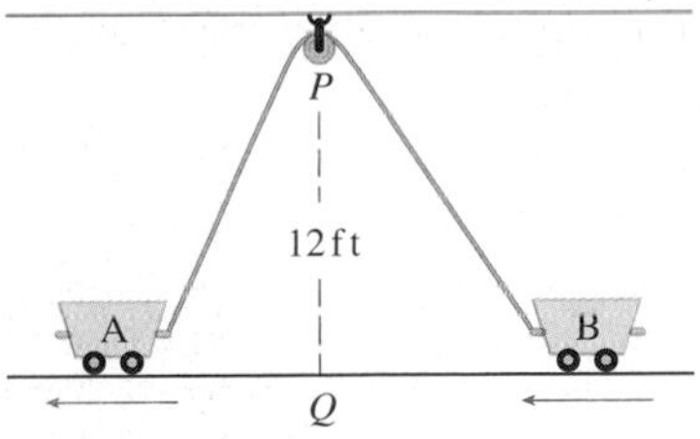

37. A television camera is positioned 4000 ft from the base of a rocket launching pad. The angle of elevation of the camera has to change at the correct rate in order to keep the rocket in sight. Also, the mechanism for focusing the camera has to take into account the increasing distance from the camera to the rising rocket. Let's assume the rocket rises vertically and its speed is 600 ft/s when it has risen 3000 ft.

(a) How fast is the distance from the television camera to the rocket changing at that moment?

(b) If the television camera is always kept aimed at the rocket, how fast is the camera's angle of elevation changing at that same moment?

38. A lighthouse is located on a small island 3 km away from the nearest point P on a straight shoreline and its light makes four revolutions per minute. How fast is the beam of light moving along the shoreline when it is 1 km from P?

39. A plane flies horizontally at an altitude of 5 km and passes directly over a tracking telescope on the ground. When the angle of elevation is $\pi/3$, this angle is decreasing at a rate of $\pi/6$ rad/min. How fast is the plane traveling at that time?

40. A Ferris wheel with a radius of 10 m is rotating at a rate of one revolution every 2 minutes. How fast is a rider rising when his seat is 16 m above ground level?

41. A plane flying with a constant speed of 300 km/h passes over a ground radar station at an altitude of 1 km and climbs at an angle of $30°$. At what rate is the distance from the plane to the radar station increasing a minute later?

42. Two people start from the same point. One walks east at 3 mi/h and the other walks northeast at 2 mi/h. How fast is the distance between the people changing after 15 minutes?

43. A runner sprints around a circular track of radius 100 m at a constant speed of 7 m/s. The runner's friend is standing at a distance 200 m from the center of the track. How fast is the distance between the friends changing when the distance between them is 200 m?

44. The minute hand on a watch is 8 mm long and the hour hand is 4 mm long. How fast is the distance between the tips of the hands changing at one o'clock?

3.10 LINEAR APPROXIMATIONS AND DIFFERENTIALS

We have seen that a curve lies very close to its tangent line near the point of tangency. In fact, by zooming in toward a point on the graph of a differentiable function, we noticed that the graph looks more and more like its tangent line. (See Figure 2 in Section 2.7.) This observation is the basis for a method of finding approximate values of functions.

The idea is that it might be easy to calculate a value $f(a)$ of a function, but difficult (or even impossible) to compute nearby values of f. So we settle for the easily computed values of the linear function L whose graph is the tangent line of f at $(a, f(a))$. (See Figure 1.)

In other words, we use the tangent line at $(a, f(a))$ as an approximation to the curve $y = f(x)$ when x is near a. An equation of this tangent line is

$$y = f(a) + f'(a)(x - a)$$

and the approximation

$$f(x) \approx f(a) + f'(a)(x - a) \tag{1}$$

is called the **linear approximation** or **tangent line approximation** of f at a. The linear

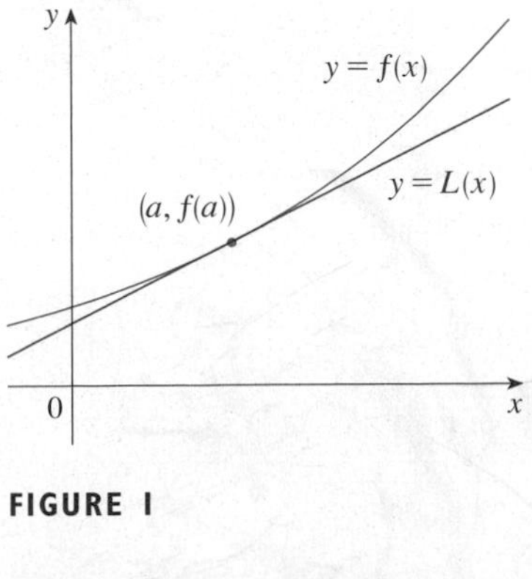

FIGURE 1

function whose graph is this tangent line, that is,

2 $$L(x) = f(a) + f'(a)(x - a)$$

is called the **linearization** of f at a.

V EXAMPLE 1 Find the linearization of the function $f(x) = \sqrt{x + 3}$ at $a = 1$ and use it to approximate the numbers $\sqrt{3.98}$ and $\sqrt{4.05}$. Are these approximations overestimates or underestimates?

SOLUTION The derivative of $f(x) = (x + 3)^{1/2}$ is

$$f'(x) = \tfrac{1}{2}(x + 3)^{-1/2} = \frac{1}{2\sqrt{x + 3}}$$

and so we have $f(1) = 2$ and $f'(1) = \frac{1}{4}$. Putting these values into Equation 2, we see that the linearization is

$$L(x) = f(1) + f'(1)(x - 1) = 2 + \tfrac{1}{4}(x - 1) = \frac{7}{4} + \frac{x}{4}$$

The corresponding linear approximation (1) is

$$\sqrt{x + 3} \approx \frac{7}{4} + \frac{x}{4} \qquad \text{(when } x \text{ is near 1)}$$

In particular, we have

$$\sqrt{3.98} \approx \tfrac{7}{4} + \tfrac{0.98}{4} = 1.995 \qquad \text{and} \qquad \sqrt{4.05} \approx \tfrac{7}{4} + \tfrac{1.05}{4} = 2.0125$$

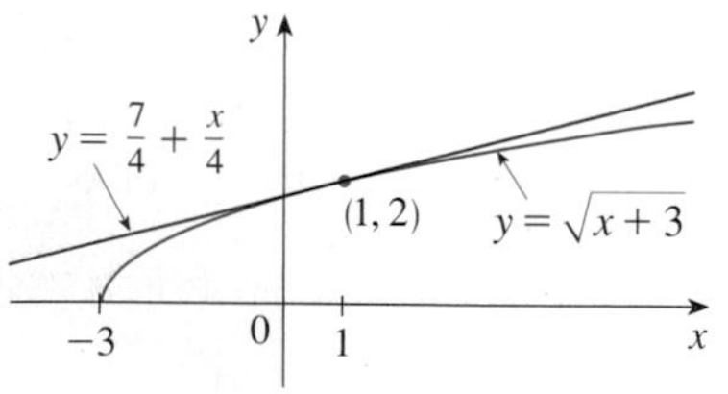

FIGURE 2

The linear approximation is illustrated in Figure 2. We see that, indeed, the tangent line approximation is a good approximation to the given function when x is near 1. We also see that our approximations are overestimates because the tangent line lies above the curve.

Of course, a calculator could give us approximations for $\sqrt{3.98}$ and $\sqrt{4.05}$, but the linear approximation gives an approximation *over an entire interval.* □

In the following table we compare the estimates from the linear approximation in Example 1 with the true values. Notice from this table, and also from Figure 2, that the tangent line approximation gives good estimates when x is close to 1 but the accuracy of the approximation deteriorates when x is farther away from 1.

	x	From $L(x)$	Actual value
$\sqrt{3.9}$	0.9	1.975	1.97484176 . . .
$\sqrt{3.98}$	0.98	1.995	1.99499373 . . .
$\sqrt{4}$	1	2	2.00000000 . . .
$\sqrt{4.05}$	1.05	2.0125	2.01246117 . . .
$\sqrt{4.1}$	1.1	2.025	2.02484567 . . .
$\sqrt{5}$	2	2.25	2.23606797 . . .
$\sqrt{6}$	3	2.5	2.44948974 . . .

How good is the approximation that we obtained in Example 1? The next example shows that by using a graphing calculator or computer we can determine an interval throughout which a linear approximation provides a specified accuracy.

EXAMPLE 2 For what values of x is the linear approximation

$$\sqrt{x+3} \approx \frac{7}{4} + \frac{x}{4}$$

accurate to within 0.5? What about accuracy to within 0.1?

SOLUTION Accuracy to within 0.5 means that the functions should differ by less than 0.5:

$$\left| \sqrt{x+3} - \left(\frac{7}{4} + \frac{x}{4} \right) \right| < 0.5$$

Equivalently, we could write

$$\sqrt{x+3} - 0.5 < \frac{7}{4} + \frac{x}{4} < \sqrt{x+3} + 0.5$$

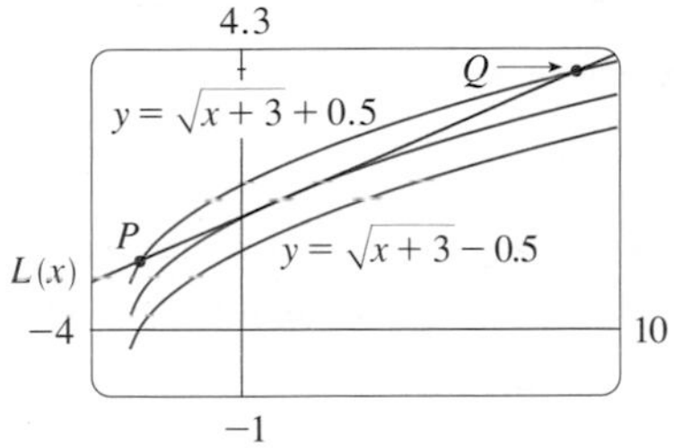

FIGURE 3

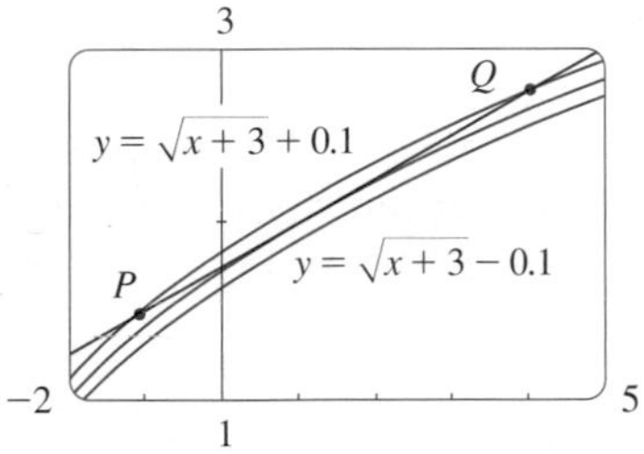

FIGURE 4

This says that the linear approximation should lie between the curves obtained by shifting the curve $y = \sqrt{x+3}$ upward and downward by an amount 0.5. Figure 3 shows the tangent line $y = (7 + x)/4$ intersecting the upper curve $y = \sqrt{x+3} + 0.5$ at P and Q. Zooming in and using the cursor, we estimate that the x-coordinate of P is about -2.66 and the x-coordinate of Q is about 8.66. Thus we see from the graph that the approximation

$$\sqrt{x+3} \approx \frac{7}{4} + \frac{x}{4}$$

is accurate to within 0.5 when $-2.6 < x < 8.6$. (We have rounded to be safe.)

Similarly, from Figure 4 we see that the approximation is accurate to within 0.1 when $-1.1 < x < 3.9$. □

APPLICATIONS TO PHYSICS

Linear approximations are often used in physics. In analyzing the consequences of an equation, a physicist sometimes needs to simplify a function by replacing it with its linear approximation. For instance, in deriving a formula for the period of a pendulum, physics textbooks obtain the expression $a_T = -g \sin \theta$ for tangential acceleration and then replace $\sin \theta$ by θ with the remark that $\sin \theta$ is very close to θ if θ is not too large. [See, for example, *Physics: Calculus,* 2d ed., by Eugene Hecht (Pacific Grove, CA: Brooks/Cole, 2000), p. 431.] You can verify that the linearization of the function $f(x) = \sin x$ at $a = 0$ is $L(x) = x$ and so the linear approximation at 0 is

$$\sin x \approx x$$

(see Exercise 42). So, in effect, the derivation of the formula for the period of a pendulum uses the tangent line approximation for the sine function.

Another example occurs in the theory of optics, where light rays that arrive at shallow angles relative to the optical axis are called *paraxial rays.* In paraxial (or Gaussian) optics, both $\sin \theta$ and $\cos \theta$ are replaced by their linearizations. In other words, the linear approximations

$$\sin \theta \approx \theta \qquad \text{and} \qquad \cos \theta \approx 1$$

are used because θ is close to 0. The results of calculations made with these approximations became the basic theoretical tool used to design lenses. [See *Optics,* 4th ed., by Eugene Hecht (San Francisco: Addison-Wesley, 2002), p. 154.]

In Section 11.11 we will present several other applications of the idea of linear approximations to physics.

DIFFERENTIALS

The ideas behind linear approximations are sometimes formulated in the terminology and notation of *differentials.* If $y = f(x)$, where f is a differentiable function, then the **differential** dx is an independent variable; that is, dx can be given the value of any real number. The **differential** dy is then defined in terms of dx by the equation

■ If $dx \neq 0$, we can divide both sides of Equation 3 by dx to obtain

$$\frac{dy}{dx} = f'(x)$$

We have seen similar equations before, but now the left side can genuinely be interpreted as a ratio of differentials.

$$\boxed{3} \qquad dy = f'(x)\, dx$$

So dy is a dependent variable; it depends on the values of x and dx. If dx is given a specific value and x is taken to be some specific number in the domain of f, then the numerical value of dy is determined.

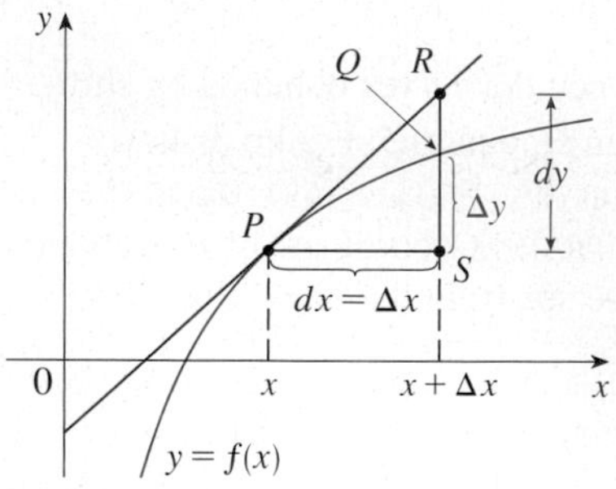

FIGURE 5

The geometric meaning of differentials is shown in Figure 5. Let $P(x, f(x))$ and $Q(x + \Delta x, f(x + \Delta x))$ be points on the graph of f and let $dx = \Delta x$. The corresponding change in y is

$$\Delta y = f(x + \Delta x) - f(x)$$

The slope of the tangent line PR is the derivative $f'(x)$. Thus the directed distance from S to R is $f'(x)\, dx = dy$. Therefore dy represents the amount that the tangent line rises or falls (the change in the linearization), whereas Δy represents the amount that the curve $y = f(x)$ rises or falls when x changes by an amount dx.

EXAMPLE 3 Compare the values of Δy and dy if $y = f(x) = x^3 + x^2 - 2x + 1$ and x changes (a) from 2 to 2.05 and (b) from 2 to 2.01.

SOLUTION

(a) We have

$$f(2) = 2^3 + 2^2 - 2(2) + 1 = 9$$

$$f(2.05) = (2.05)^3 + (2.05)^2 - 2(2.05) + 1 = 9.717625$$

$$\Delta y = f(2.05) - f(2) = 0.717625$$

■ Figure 6 shows the function in Example 3 and a comparison of dy and Δy when $a = 2$. The viewing rectangle is $[1.8, 2.5]$ by $[6, 18]$.

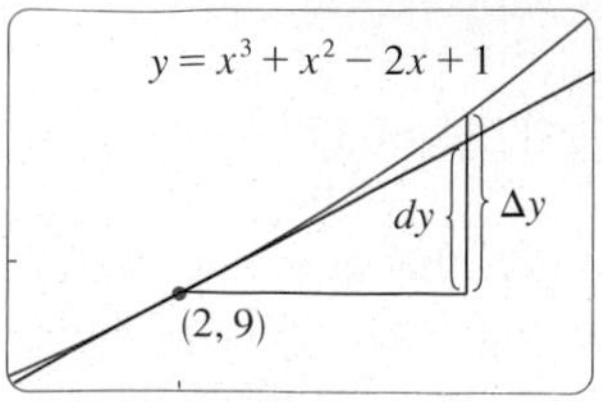

FIGURE 6

In general, $\qquad dy = f'(x)\, dx = (3x^2 + 2x - 2)\, dx$

When $x = 2$ and $dx = \Delta x = 0.05$, this becomes

$$dy = [3(2)^2 + 2(2) - 2]0.05 = 0.7$$

(b) $$f(2.01) = (2.01)^3 + (2.01)^2 - 2(2.01) + 1 = 9.140701$$

$$\Delta y = f(2.01) - f(2) = 0.140701$$

When $dx = \Delta x = 0.01$,

$$dy = [3(2)^2 + 2(2) - 2]0.01 = 0.14$$

□

Notice that the approximation $\Delta y \approx dy$ becomes better as Δx becomes smaller in Example 3. Notice also that dy was easier to compute than Δy. For more complicated functions it may be impossible to compute Δy exactly. In such cases the approximation by differentials is especially useful.

In the notation of differentials, the linear approximation (1) can be written as

$$f(a + dx) \approx f(a) + dy$$

For instance, for the function $f(x) = \sqrt{x + 3}$ in Example 1, we have

$$dy = f'(x)\,dx = \frac{dx}{2\sqrt{x + 3}}$$

If $a = 1$ and $dx = \Delta x = 0.05$, then

$$dy = \frac{0.05}{2\sqrt{1 + 3}} = 0.0125$$

and $$\sqrt{4.05} = f(1.05) \approx f(1) + dy = 2.0125$$

just as we found in Example 1.

Our final example illustrates the use of differentials in estimating the errors that occur because of approximate measurements.

V EXAMPLE 4 The radius of a sphere was measured and found to be 21 cm with a possible error in measurement of at most 0.05 cm. What is the maximum error in using this value of the radius to compute the volume of the sphere?

SOLUTION If the radius of the sphere is r, then its volume is $V = \frac{4}{3}\pi r^3$. If the error in the measured value of r is denoted by $dr = \Delta r$, then the corresponding error in the calculated value of V is ΔV, which can be approximated by the differential

$$dV = 4\pi r^2\,dr$$

When $r = 21$ and $dr = 0.05$, this becomes

$$dV = 4\pi(21)^2 0.05 \approx 277$$

The maximum error in the calculated volume is about 277 cm^3. □

NOTE Although the possible error in Example 4 may appear to be rather large, a better picture of the error is given by the **relative error**, which is computed by dividing the error by the total volume:

$$\frac{\Delta V}{V} \approx \frac{dV}{V} = \frac{4\pi r^2\,dr}{\frac{4}{3}\pi r^3} = 3\,\frac{dr}{r}$$

Thus the relative error in the volume is about three times the relative error in the radius. In Example 4 the relative error in the radius is approximately $dr/r = 0.05/21 \approx 0.0024$ and it produces a relative error of about 0.007 in the volume. The errors could also be expressed as **percentage errors** of 0.24% in the radius and 0.7% in the volume.

3.10 EXERCISES

1–4 Find the linearization $L(x)$ of the function at a.

1. $f(x) = x^4 + 3x^2$, $a = -1$

2. $f(x) = \ln x$, $a = 1$

3. $f(x) = \cos x$, $a = \pi/2$

4. $f(x) = x^{3/4}$, $a = 16$

5. Find the linear approximation of the function $f(x) = \sqrt{1 - x}$ at $a = 0$ and use it to approximate the numbers $\sqrt{0.9}$ and $\sqrt{0.99}$. Illustrate by graphing f and the tangent line.

6. Find the linear approximation of the function $g(x) = \sqrt[3]{1 + x}$ at $a = 0$ and use it to approximate the numbers $\sqrt[3]{0.95}$ and $\sqrt[3]{1.1}$. Illustrate by graphing g and the tangent line.

7–10 Verify the given linear approximation at $a = 0$. Then determine the values of x for which the linear approximation is accurate to within 0.1.

7. $\sqrt[3]{1 - x} \approx 1 - \frac{1}{3}x$

8. $\tan x \approx x$

9. $1/(1 + 2x)^4 \approx 1 - 8x$

10. $e^x \approx 1 + x$

11–14 Find the differential of each function.

11. (a) $y = x^2 \sin 2x$ (b) $y = \ln\sqrt{1 + t^2}$

12. (a) $y = s/(1 + 2s)$ (b) $y = e^{-u} \cos u$

13. (a) $y = \dfrac{u + 1}{u - 1}$ (b) $y = (1 + r^3)^{-2}$

14. (a) $y = e^{\tan \pi t}$ (b) $y = \sqrt{1 + \ln z}$

15–18 (a) Find the differential dy and (b) evaluate dy for the given values of x and dx.

15. $y = e^{x/10}$, $x = 0$, $dx = 0.1$

16. $y = 1/(x + 1)$, $x = 1$, $dx = -0.01$

17. $y = \tan x$, $x = \pi/4$, $dx = -0.1$

18. $y = \cos x$, $x = \pi/3$, $dx = 0.05$

19–22 Compute Δy and dy for the given values of x and $dx = \Delta x$. Then sketch a diagram like Figure 5 showing the line segments with lengths dx, dy, and Δy.

19. $y = 2x - x^2$, $x = 2$, $\Delta x = -0.4$

20. $y = \sqrt{x}$, $x = 1$, $\Delta x = 1$

21. $y = 2/x$, $x = 4$, $\Delta x = 1$

22. $y = e^x$, $x = 0$, $\Delta x = 0.5$

23–28 Use a linear approximation (or differentials) to estimate the given number.

23. $(2.001)^5$

24. $e^{-0.015}$

25. $(8.06)^{2/3}$

26. $1/1002$

27. $\tan 44°$

28. $\sqrt{99.8}$

29–31 Explain, in terms of linear approximations or differentials, why the approximation is reasonable.

29. $\sec 0.08 \approx 1$

30. $(1.01)^6 \approx 1.06$

31. $\ln 1.05 \approx 0.05$

32. Let $f(x) = (x - 1)^2$ $g(x) = e^{-2x}$

and $h(x) = 1 + \ln(1 - 2x)$

(a) Find the linearizations of f, g, and h at $a = 0$. What do you notice? How do you explain what happened?

(b) Graph f, g, and h and their linear approximations. For which function is the linear approximation best? For which is it worst? Explain.

33. The edge of a cube was found to be 30 cm with a possible error in measurement of 0.1 cm. Use differentials to estimate the maximum possible error, relative error, and percentage error in computing (a) the volume of the cube and (b) the surface area of the cube.

34. The radius of a circular disk is given as 24 cm with a maximum error in measurement of 0.2 cm.

(a) Use differentials to estimate the maximum error in the calculated area of the disk.

(b) What is the relative error? What is the percentage error?

35. The circumference of a sphere was measured to be 84 cm with a possible error of 0.5 cm.

(a) Use differentials to estimate the maximum error in the calculated surface area. What is the relative error?

(b) Use differentials to estimate the maximum error in the calculated volume. What is the relative error?

36. Use differentials to estimate the amount of paint needed to apply a coat of paint 0.05 cm thick to a hemispherical dome with diameter 50 m.

37. (a) Use differentials to find a formula for the approximate volume of a thin cylindrical shell with height h, inner radius r, and thickness Δr.

(b) What is the error involved in using the formula from part (a)?

38. One side of a right triangle is known to be 20 cm long and the opposite angle is measured as 30°, with a possible error of $\pm 1°$.

(a) Use differentials to estimate the error in computing the length of the hypotenuse.

(b) What is the percentage error?

39. If a current I passes through a resistor with resistance R, Ohm's Law states that the voltage drop is $V = RI$. If V is constant and R is measured with a certain error, use differentials to show that the relative error in calculating I is approximately the same (in magnitude) as the relative error in R.

40. When blood flows along a blood vessel, the flux F (the volume of blood per unit time that flows past a given point) is proportional to the fourth power of the radius R of the blood vessel:

$$F = kR^4$$

(This is known as Poiseuille's Law; we will show why it is true in Section 8.4.) A partially clogged artery can be expanded by an operation called angioplasty, in which a balloon-tipped catheter is inflated inside the artery in order to widen it and restore the normal blood flow.

Show that the relative change in F is about four times the relative change in R. How will a 5% increase in the radius affect the flow of blood?

41. Establish the following rules for working with differentials (where c denotes a constant and u and v are functions of x).

(a) $dc = 0$ (b) $d(cu) = c\,du$

(c) $d(u + v) = du + dv$ (d) $d(uv) = u\,dv + v\,du$

(e) $d\left(\dfrac{u}{v}\right) = \dfrac{v\,du - u\,dv}{v^2}$ (f) $d(x^n) = nx^{n-1}\,dx$

42. On page 431 of *Physics: Calculus,* 2d ed., by Eugene Hecht (Pacific Grove, CA: Brooks/Cole, 2000), in the course of deriving the formula $T = 2\pi\sqrt{L/g}$ for the period of a pendulum of length L, the author obtains the equation $a_T = -g\sin\theta$ for the tangential acceleration of the bob of the pendulum. He then says, "for small angles, the value of θ in radians is very nearly the value of $\sin\theta$; they differ by less than 2% out to about 20°."

(a) Verify the linear approximation at 0 for the sine function:

$$\sin x \approx x$$

(b) Use a graphing device to determine the values of x for which $\sin x$ and x differ by less than 2%. Then verify Hecht's statement by converting from radians to degrees.

43. Suppose that the only information we have about a function f is that $f(1) = 5$ and the graph of its *derivative* is as shown.

(a) Use a linear approximation to estimate $f(0.9)$ and $f(1.1)$.

(b) Are your estimates in part (a) too large or too small? Explain.

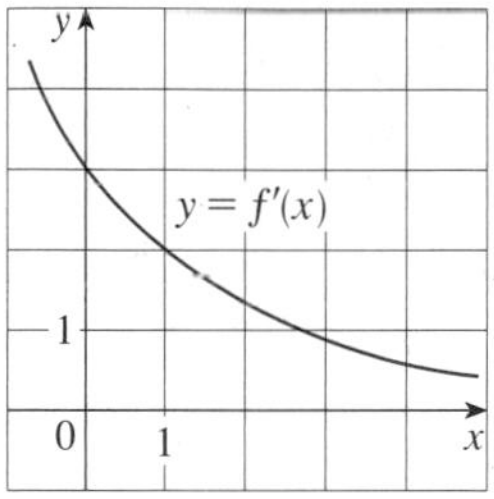

44. Suppose that we don't have a formula for $g(x)$ but we know that $g(2) = -4$ and $g'(x) = \sqrt{x^2 + 5}$ for all x.

(a) Use a linear approximation to estimate $g(1.95)$ and $g(2.05)$.

(b) Are your estimates in part (a) too large or too small? Explain.

LABORATORY PROJECT — TAYLOR POLYNOMIALS

The tangent line approximation $L(x)$ is the best first-degree (linear) approximation to $f(x)$ near $x = a$ because $f(x)$ and $L(x)$ have the same rate of change (derivative) at a. For a better approximation than a linear one, let's try a second-degree (quadratic) approximation $P(x)$. In other words, we approximate a curve by a parabola instead of by a straight line. To make sure that the approximation is a good one, we stipulate the following:

(i) $P(a) = f(a)$ (P and f should have the same value at a.)

(ii) $P'(a) = f'(a)$ (P and f should have the same rate of change at a.)

(iii) $P''(a) = f''(a)$ (The slopes of P and f should change at the same rate at a.)

1. Find the quadratic approximation $P(x) = A + Bx + Cx^2$ to the function $f(x) = \cos x$ that satisfies conditions (i), (ii), and (iii) with $a = 0$. Graph P, f, and the linear approximation $L(x) = 1$ on a common screen. Comment on how well the functions P and L approximate f.

2. Determine the values of x for which the quadratic approximation $f(x) = P(x)$ in Problem 1 is accurate to within 0.1. [*Hint:* Graph $y = P(x)$, $y = \cos x - 0.1$, and $y = \cos x + 0.1$ on a common screen.]

3. To approximate a function f by a quadratic function P near a number a, it is best to write P in the form

$$P(x) = A + B(x - a) + C(x - a)^2$$

Show that the quadratic function that satisfies conditions (i), (ii), and (iii) is

$$P(x) = f(a) + f'(a)(x - a) + \tfrac{1}{2}f''(a)(x - a)^2$$

4. Find the quadratic approximation to $f(x) = \sqrt{x + 3}$ near $a = 1$. Graph f, the quadratic approximation, and the linear approximation from Example 2 in Section 3.10 on a common screen. What do you conclude?

5. Instead of being satisfied with a linear or quadratic approximation to $f(x)$ near $x = a$, let's try to find better approximations with higher-degree polynomials. We look for an nth-degree polynomial

$$T_n(x) = c_0 + c_1(x - a) + c_2(x - a)^2 + c_3(x - a)^3 + \cdots + c_n(x - a)^n$$

such that T_n and its first n derivatives have the same values at $x = a$ as f and its first n derivatives. By differentiating repeatedly and setting $x = a$, show that these conditions are satisfied if $c_0 = f(a)$, $c_1 = f'(a)$, $c_2 = \tfrac{1}{2}f''(a)$, and in general

$$c_k = \frac{f^{(k)}(a)}{k!}$$

where $k! = 1 \cdot 2 \cdot 3 \cdot 4 \cdot \cdots \cdot k$. The resulting polynomial

$$T_n(x) = f(a) + f'(a)(x - a) + \frac{f''(a)}{2!}(x - a)^2 + \cdots + \frac{f^{(n)}(a)}{n!}(x - a)^n$$

is called the ***n*th-degree Taylor polynomial of *f* centered at *a***.

6. Find the 8th-degree Taylor polynomial centered at $a = 0$ for the function $f(x) = \cos x$. Graph f together with the Taylor polynomials T_2, T_4, T_6, T_8 in the viewing rectangle $[-5, 5]$ by $[-1.4, 1.4]$ and comment on how well they approximate f.

3.11 HYPERBOLIC FUNCTIONS

Certain even and odd combinations of the exponential functions e^x and e^{-x} arise so frequently in mathematics and its applications that they deserve to be given special names. In many ways they are analogous to the trigonometric functions, and they have the same relationship to the hyperbola that the trigonometric functions have to the circle. For this reason they are collectively called **hyperbolic functions** and individually called **hyperbolic sine**, **hyperbolic cosine**, and so on.

DEFINITION OF THE HYPERBOLIC FUNCTIONS

$$\sinh x = \frac{e^x - e^{-x}}{2} \qquad \operatorname{csch} x = \frac{1}{\sinh x}$$

$$\cosh x = \frac{e^x + e^{-x}}{2} \qquad \operatorname{sech} x = \frac{1}{\cosh x}$$

$$\tanh x = \frac{\sinh x}{\cosh x} \qquad \coth x = \frac{\cosh x}{\sinh x}$$

The graphs of hyperbolic sine and cosine can be sketched using graphical addition as in Figures 1 and 2.

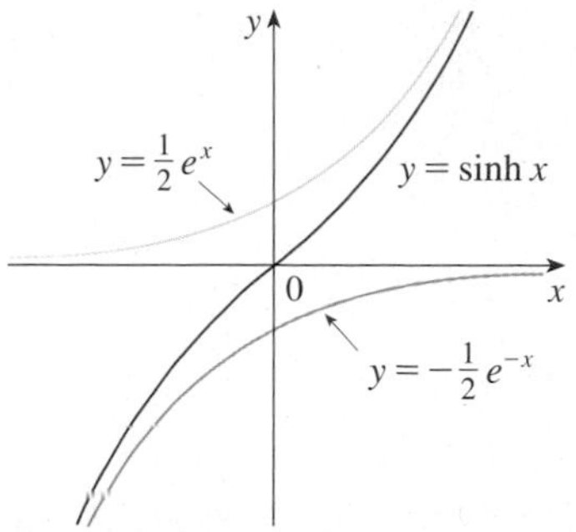

FIGURE 1
$y = \sinh x = \frac{1}{2}e^x - \frac{1}{2}e^{-x}$

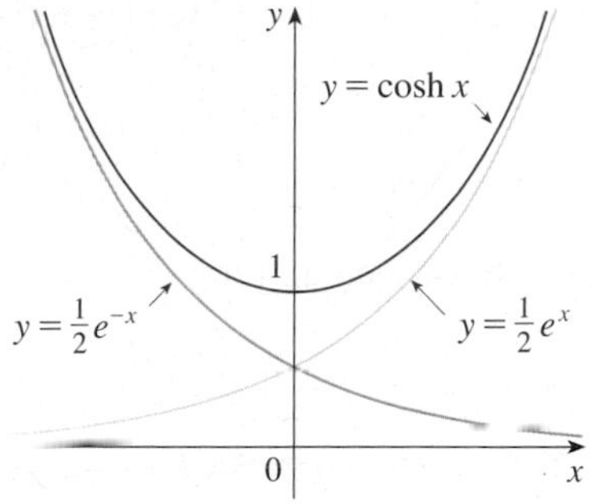

FIGURE 2
$y = \cosh x = \frac{1}{2}e^x + \frac{1}{2}e^{-x}$

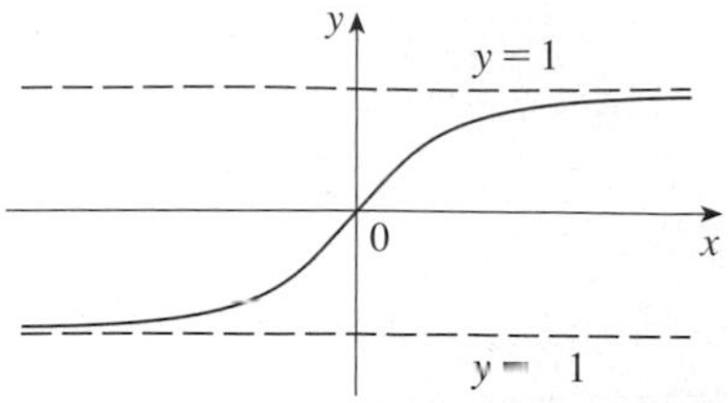

FIGURE 3
$y = \tanh x$

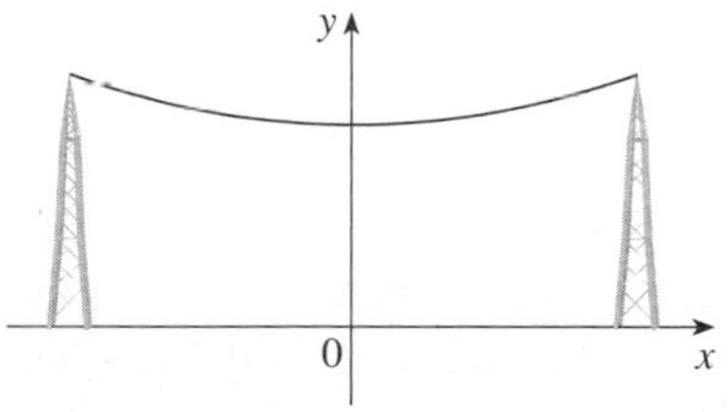

FIGURE 4
A catenary $y = c + a\cosh(x/a)$

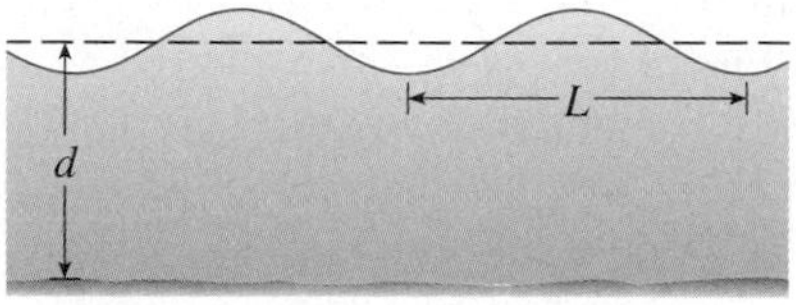

FIGURE 5
Idealized ocean wave

Note that sinh has domain $\mathbb{R}$ and range $\mathbb{R}$, while cosh has domain $\mathbb{R}$ and range $[1, \infty)$. The graph of tanh is shown in Figure 3. It has the horizontal asymptotes $y = \pm 1$. (See Exercise 23.)

Some of the mathematical uses of hyperbolic functions will be seen in Chapter 7. Applications to science and engineering occur whenever an entity such as light, velocity, electricity, or radioactivity is gradually absorbed or extinguished, for the decay can be represented by hyperbolic functions. The most famous application is the use of hyperbolic cosine to describe the shape of a hanging wire. It can be proved that if a heavy flexible cable (such as a telephone or power line) is suspended between two points at the same height, then it takes the shape of a curve with equation $y = c + a\cosh(x/a)$ called a *catenary* (see Figure 4). (The Latin word *catena* means "chain.")

Another application of hyperbolic functions occurs in the description of ocean waves: The velocity of a water wave with length L moving across a body of water with depth d is modeled by the function

$$v = \sqrt{\frac{gL}{2\pi}\tanh\left(\frac{2\pi d}{L}\right)}$$

where g is the acceleration due to gravity. (See Figure 5 and Exercise 49.)

The hyperbolic functions satisfy a number of identities that are similar to well-known trigonometric identities. We list some of them here and leave most of the proofs to the exercises.

HYPERBOLIC IDENTITIES

$$\sinh(-x) = -\sinh x \qquad \cosh(-x) = \cosh x$$

$$\cosh^2 x - \sinh^2 x = 1 \qquad 1 - \tanh^2 x = \text{sech}^2 x$$

$$\sinh(x + y) = \sinh x \cosh y + \cosh x \sinh y$$

$$\cosh(x + y) = \cosh x \cosh y + \sinh x \sinh y$$

The Gateway Arch in St. Louis was designed using a hyperbolic cosine function (Exercise 48).

V EXAMPLE 1 Prove (a) $\cosh^2 x - \sinh^2 x = 1$ and (b) $1 - \tanh^2 x = \operatorname{sech}^2 x$.

SOLUTION

(a)

$$\cosh^2 x - \sinh^2 x = \left(\frac{e^x + e^{-x}}{2}\right)^2 - \left(\frac{e^x - e^{-x}}{2}\right)^2$$

$$= \frac{e^{2x} + 2 + e^{-2x}}{4} - \frac{e^{2x} - 2 + e^{-2x}}{4} = \frac{4}{4} = 1$$

(b) We start with the identity proved in part (a):

$$\cosh^2 x - \sinh^2 x = 1$$

If we divide both sides by $\cosh^2 x$, we get

$$1 - \frac{\sinh^2 x}{\cosh^2 x} = \frac{1}{\cosh^2 x}$$

or

$$1 - \tanh^2 x = \operatorname{sech}^2 x$$

□

The identity proved in Example 1(a) gives a clue to the reason for the name "hyperbolic" functions:

If t is any real number, then the point $P(\cos t, \sin t)$ lies on the unit circle $x^2 + y^2 = 1$ because $\cos^2 t + \sin^2 t = 1$. In fact, t can be interpreted as the radian measure of $\angle POQ$ in Figure 6. For this reason the trigonometric functions are sometimes called *circular* functions.

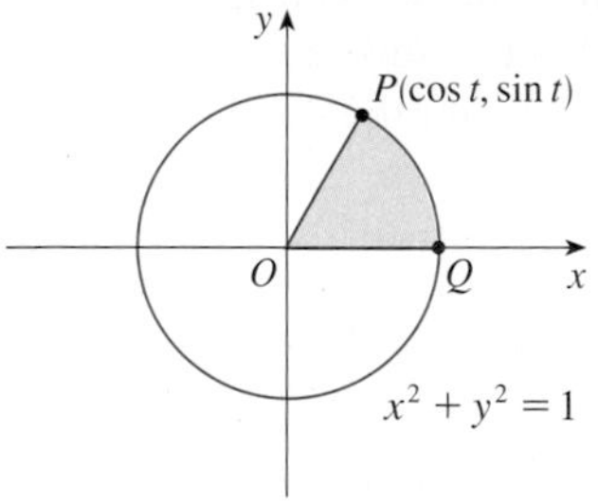

FIGURE 6

Likewise, if t is any real number, then the point $P(\cosh t, \sinh t)$ lies on the right branch of the hyperbola $x^2 - y^2 = 1$ because $\cosh^2 t - \sinh^2 t = 1$ and $\cosh t \geqslant 1$. This time, t does not represent the measure of an angle. However, it turns out that t represents twice the area of the shaded hyperbolic sector in Figure 7, just as in the trigonometric case t represents twice the area of the shaded circular sector in Figure 6.

The derivatives of the hyperbolic functions are easily computed. For example,

$$\frac{d}{dx}(\sinh x) = \frac{d}{dx}\left(\frac{e^x - e^{-x}}{2}\right) = \frac{e^x + e^{-x}}{2} = \cosh x$$

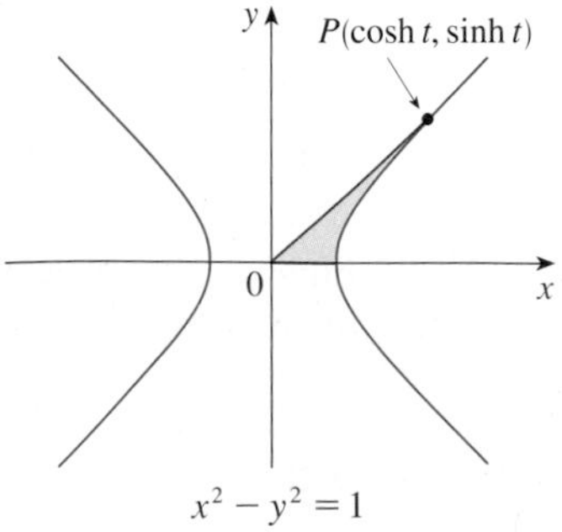

FIGURE 7

We list the differentiation formulas for the hyperbolic functions as Table 1. The remaining proofs are left as exercises. Note the analogy with the differentiation formulas for trigonometric functions, but beware that the signs are different in some cases.

1 DERIVATIVES OF HYPERBOLIC FUNCTIONS

$$\frac{d}{dx}(\sinh x) = \cosh x \qquad \frac{d}{dx}(\operatorname{csch} x) = -\operatorname{csch} x \coth x$$

$$\frac{d}{dx}(\cosh x) = \sinh x \qquad \frac{d}{dx}(\operatorname{sech} x) = -\operatorname{sech} x \tanh x$$

$$\frac{d}{dx}(\tanh x) = \operatorname{sech}^2 x \qquad \frac{d}{dx}(\coth x) = -\operatorname{csch}^2 x$$

EXAMPLE 2 Any of these differentiation rules can be combined with the Chain Rule. For instance,

$$\frac{d}{dx}(\cosh\sqrt{x}) = \sinh\sqrt{x}\cdot\frac{d}{dx}\sqrt{x} = \frac{\sinh\sqrt{x}}{2\sqrt{x}} \qquad \square$$

INVERSE HYPERBOLIC FUNCTIONS

You can see from Figures 1 and 3 that sinh and tanh are one-to-one functions and so they have inverse functions denoted by $\sinh^{-1}$ and $\tanh^{-1}$. Figure 2 shows that cosh is not one-to-one, but when restricted to the domain $[0, \infty)$ it becomes one-to-one. The inverse hyperbolic cosine function is defined as the inverse of this restricted function.

2

$$y = \sinh^{-1}x \iff \sinh y = x$$
$$y = \cosh^{-1}x \iff \cosh y = x \quad \text{and} \quad y \geqslant 0$$
$$y = \tanh^{-1}x \iff \tanh y = x$$

The remaining inverse hyperbolic functions are defined similarly (see Exercise 28).

We can sketch the graphs of $\sinh^{-1}$, $\cosh^{-1}$, and $\tanh^{-1}$ in Figures 8, 9, and 10 by using Figures 1, 2, and 3.

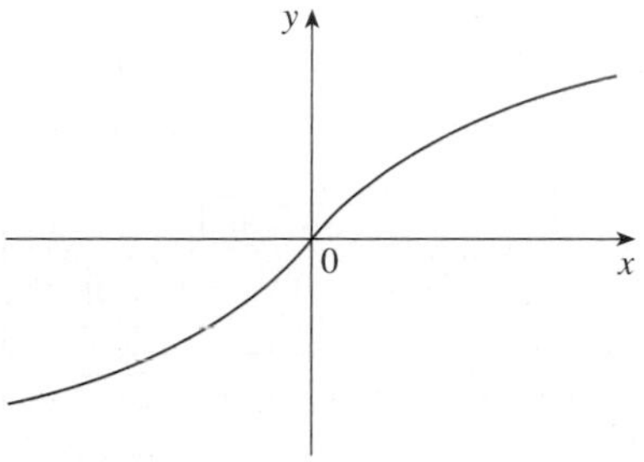

FIGURE 8 $y = \sinh^{-1}x$
domain $= \mathbb{R}$ range $= \mathbb{R}$

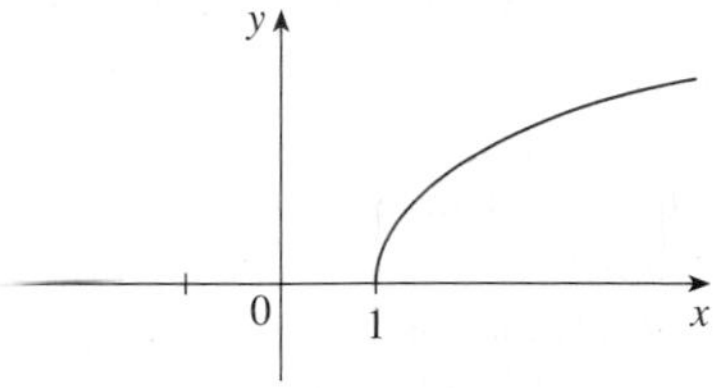

FIGURE 9 $y = \cosh^{-1}x$
domain $= [1, \infty)$ range $= [0, \infty)$

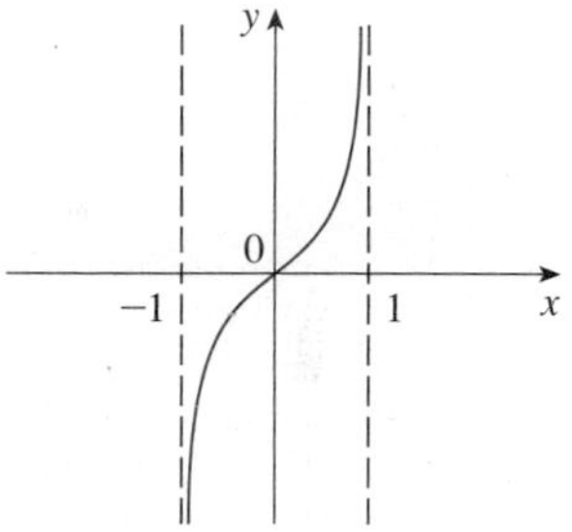

FIGURE 10 $y = \tanh^{-1}x$
domain $= (-1, 1)$ range $= \mathbb{R}$

Since the hyperbolic functions are defined in terms of exponential functions, it's not surprising to learn that the inverse hyperbolic functions can be expressed in terms of logarithms. In particular, we have:

■ Formula 3 is proved in Example 3. The proofs of Formulas 4 and 5 are requested in Exercises 26 and 27.

3 $$\sinh^{-1}x = \ln\left(x + \sqrt{x^2 + 1}\right) \qquad x \in \mathbb{R}$$

4 $$\cosh^{-1}x = \ln\left(x + \sqrt{x^2 - 1}\right) \qquad x \geqslant 1$$

5 $$\tanh^{-1}x = \tfrac{1}{2}\ln\left(\frac{1 + x}{1 - x}\right) \qquad -1 < x < 1$$

EXAMPLE 3 Show that $\sinh^{-1}x = \ln\left(x + \sqrt{x^2 + 1}\right)$.

SOLUTION Let $y = \sinh^{-1}x$. Then

$$x = \sinh y = \frac{e^y - e^{-y}}{2}$$

so $$e^y - 2x - e^{-y} = 0$$

or, multiplying by e^y,

$$e^{2y} - 2xe^y - 1 = 0$$

This is really a quadratic equation in e^y:

$$(e^y)^2 - 2x(e^y) - 1 = 0$$

Solving by the quadratic formula, we get

$$e^y = \frac{2x \pm \sqrt{4x^2 + 4}}{2} = x \pm \sqrt{x^2 + 1}$$

Note that $e^y > 0$, but $x - \sqrt{x^2 + 1} < 0$ $\left(\text{because } x < \sqrt{x^2 + 1}\right)$. Thus the minus sign is inadmissible and we have

$$e^y = x + \sqrt{x^2 + 1}$$

Therefore $$y = \ln(e^y) = \ln\left(x + \sqrt{x^2 + 1}\right)$$

(See Exercise 25 for another method.) □

6 DERIVATIVES OF INVERSE HYPERBOLIC FUNCTIONS

$$\frac{d}{dx}(\sinh^{-1}x) = \frac{1}{\sqrt{1 + x^2}} \qquad \frac{d}{dx}(\text{csch}^{-1}x) = -\frac{1}{|x|\sqrt{x^2 + 1}}$$

$$\frac{d}{dx}(\cosh^{-1}x) = \frac{1}{\sqrt{x^2 - 1}} \qquad \frac{d}{dx}(\text{sech}^{-1}x) = -\frac{1}{x\sqrt{1 - x^2}}$$

$$\frac{d}{dx}(\tanh^{-1}x) = \frac{1}{1 - x^2} \qquad \frac{d}{dx}(\coth^{-1}x) = \frac{1}{1 - x^2}$$

■ Notice that the formulas for the derivatives of $\tanh^{-1}x$ and $\coth^{-1}x$ appear to be identical. But the domains of these functions have no numbers in common: $\tanh^{-1}x$ is defined for $|x| < 1$, whereas $\coth^{-1}x$ is defined for $|x| > 1$.

The inverse hyperbolic functions are all differentiable because the hyperbolic functions are differentiable. The formulas in Table 6 can be proved either by the method for inverse functions or by differentiating Formulas 3, 4, and 5.

V EXAMPLE 4 Prove that $\dfrac{d}{dx}(\sinh^{-1}x) = \dfrac{1}{\sqrt{1 + x^2}}$.

SOLUTION Let $y = \sinh^{-1}x$. Then $\sinh y = x$. If we differentiate this equation implicitly with respect to x, we get

$$\cosh y \frac{dy}{dx} = 1$$

Since $\cosh^2 y - \sinh^2 y = 1$ and $\cosh y \geqslant 0$, we have $\cosh y = \sqrt{1 + \sinh^2 y}$, so

$$\frac{dy}{dx} = \frac{1}{\cosh y} = \frac{1}{\sqrt{1 + \sinh^2 y}} = \frac{1}{\sqrt{1 + x^2}}$$

SOLUTION 2 From Equation 3 (proved in Example 3), we have

$$\begin{aligned}
\frac{d}{dx}(\sinh^{-1}x) &= \frac{d}{dx}\ln\left(x + \sqrt{x^2+1}\right) \\
&= \frac{1}{x + \sqrt{x^2+1}}\frac{d}{dx}\left(x + \sqrt{x^2+1}\right) \\
&= \frac{1}{x + \sqrt{x^2+1}}\left(1 + \frac{x}{\sqrt{x^2+1}}\right) \\
&= \frac{\sqrt{x^2+1} + x}{\left(x + \sqrt{x^2+1}\right)\sqrt{x^2+1}} \\
&= \frac{1}{\sqrt{x^2+1}}
\end{aligned}$$ □

V EXAMPLE 5 Find $\frac{d}{dx}[\tanh^{-1}(\sin x)]$.

SOLUTION Using Table 6 and the Chain Rule, we have

$$\begin{aligned}
\frac{d}{dx}[\tanh^{-1}(\sin x)] &= \frac{1}{1 - (\sin x)^2}\frac{d}{dx}(\sin x) \\
&= \frac{1}{1 - \sin^2 x}\cos x = \frac{\cos x}{\cos^2 x} = \sec x
\end{aligned}$$ □

3.11 EXERCISES

1–6 Find the numerical value of each expression.

1. (a) $\sinh 0$ (b) $\cosh 0$

2. (a) $\tanh 0$ (b) $\tanh 1$

3. (a) $\sinh(\ln 2)$ (b) $\sinh 2$

4. (a) $\cosh 3$ (b) $\cosh(\ln 3)$

5. (a) $\operatorname{sech} 0$ (b) $\cosh^{-1} 1$

6. (a) $\sinh 1$ (b) $\sinh^{-1} 1$

7–19 Prove the identity.

7. $\sinh(-x) = -\sinh x$
(This shows that sinh is an odd function.)

8. $\cosh(-x) = \cosh x$
(This shows that cosh is an even function.)

9. $\cosh x + \sinh x = e^x$

10. $\cosh x - \sinh x = e^{-x}$

11. $\sinh(x + y) = \sinh x \cosh y + \cosh x \sinh y$

12. $\cosh(x + y) = \cosh x \cosh y + \sinh x \sinh y$

13. $\coth^2 x - 1 = \operatorname{csch}^2 x$

14. $\tanh(x + y) = \dfrac{\tanh x + \tanh y}{1 + \tanh x \tanh y}$

15. $\sinh 2x = 2 \sinh x \cosh x$

16. $\cosh 2x = \cosh^2 x + \sinh^2 x$

17. $\tanh(\ln x) = \dfrac{x^2 - 1}{x^2 + 1}$

18. $\dfrac{1 + \tanh x}{1 - \tanh x} = e^{2x}$

19. $(\cosh x + \sinh x)^n = \cosh nx + \sinh nx$
(n any real number)

20. If $\tanh x = \frac{12}{13}$, find the values of the other hyperbolic functions at x.

21. If $\cosh x = \frac{5}{3}$ and $x > 0$, find the values of the other hyperbolic functions at x.

22. (a) Use the graphs of sinh, cosh, and tanh in Figures 1–3 to draw the graphs of csch, sech, and coth.

(b) Check the graphs that you sketched in part (a) by using a graphing device to produce them.

23. Use the definitions of the hyperbolic functions to find each of the following limits.

(a) $\lim_{x\to\infty} \tanh x$ (b) $\lim_{x\to-\infty} \tanh x$

(c) $\lim_{x\to\infty} \sinh x$ (d) $\lim_{x\to-\infty} \sinh x$

(e) $\lim_{x\to\infty} \operatorname{sech} x$ (f) $\lim_{x\to\infty} \coth x$

(g) $\lim_{x\to0^+} \coth x$ (h) $\lim_{x\to0^-} \coth x$

(i) $\lim_{x\to-\infty} \operatorname{csch} x$

24. Prove the formulas given in Table 1 for the derivatives of the functions (a) cosh, (b) tanh, (c) csch, (d) sech, and (e) coth.

25. Give an alternative solution to Example 3 by letting $y = \sinh^{-1}x$ and then using Exercise 9 and Example 1(a) with x replaced by y.

26. Prove Equation 4.

27. Prove Equation 5 using (a) the method of Example 3 and (b) Exercise 18 with x replaced by y.

28. For each of the following functions (i) give a definition like those in (2), (ii) sketch the graph, and (iii) find a formula similar to Equation 3.

(a) csch^{-1} (b) sech^{-1} (c) $\coth^{-1}$

29. Prove the formulas given in Table 6 for the derivatives of the following functions.

(a) $\cosh^{-1}$ (b) $\tanh^{-1}$ (c) csch^{-1}

(d) sech^{-1} (e) $\coth^{-1}$

30–47 Find the derivative. Simplify where possible.

30. $f(x) = \tanh(1 + e^{2x})$

31. $f(x) = x \sinh x - \cosh x$

32. $g(x) = \cosh(\ln x)$

33. $h(x) = \ln(\cosh x)$

34. $y = x \coth(1 + x^2)$

35. $y = e^{\cosh 3x}$

36. $f(t) = \operatorname{csch} t(1 - \ln \operatorname{csch} t)$

37. $f(t) = \operatorname{sech}^2(e^t)$

38. $y = \sinh(\cosh x)$

39. $y = \arctan(\tanh x)$

40. $y = \sqrt[4]{\dfrac{1 + \tanh x}{1 - \tanh x}}$

41. $G(x) = \dfrac{1 - \cosh x}{1 + \cosh x}$

42. $y = x^2 \sinh^{-1}(2x)$

43. $y = \tanh^{-1}\sqrt{x}$

44. $y = x \tanh^{-1}x + \ln\sqrt{1 - x^2}$

45. $y = x \sinh^{-1}(x/3) - \sqrt{9 + x^2}$

46. $y = \operatorname{sech}^{-1}\sqrt{1 - x^2}, \quad x > 0$

47. $y = \coth^{-1}\sqrt{x^2 + 1}$

---

48. The Gateway Arch in St. Louis was designed by Eero Saarinen and was constructed using the equation

$$y = 211.49 - 20.96 \cosh 0.03291765x$$

for the central curve of the arch, where x and y are measured in meters and $|x| \leqslant 91.20$.

(a) Graph the central curve.

(b) What is the height of the arch at its center?

(c) At what points is the height 100 m?

(d) What is the slope of the arch at the points in part (c)?

49. If a water wave with length L moves with velocity v in a body of water with depth d, then

$$v = \sqrt{\frac{gL}{2\pi} \tanh\left(\frac{2\pi d}{L}\right)}$$

where g is the acceleration due to gravity. (See Figure 5.) Explain why the approximation

$$v \approx \sqrt{\frac{gL}{2\pi}}$$

is appropriate in deep water.

50. A flexible cable always hangs in the shape of a catenary $y = c + a\cosh(x/a)$, where c and a are constants and $a > 0$ (see Figure 4 and Exercise 52). Graph several members of the family of functions $y = a\cosh(x/a)$. How does the graph change as a varies?

51. A telephone line hangs between two poles 14 m apart in the shape of the catenary $y = 20\cosh(x/20) - 15$, where x and y are measured in meters.

(a) Find the slope of this curve where it meets the right pole.

(b) Find the angle θ between the line and the pole.

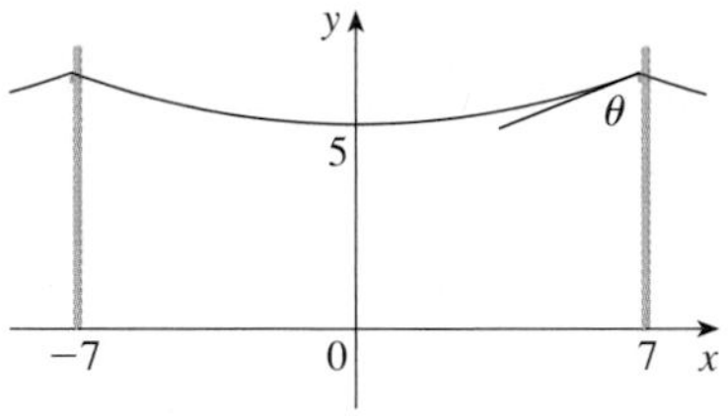

52. Using principles from physics it can be shown that when a cable is hung between two poles, it takes the shape of a curve $y = f(x)$ that satisfies the differential equation

$$\frac{d^2y}{dx^2} = \frac{\rho g}{T}\sqrt{1 + \left(\frac{dy}{dx}\right)^2}$$

where ρ is the linear density of the cable, g is the acceleration due to gravity, and T is the tension in the cable at its lowest point, and the coordinate system is chosen appropriately. Verify that the function

$$y = f(x) = \frac{T}{\rho g}\cosh\left(\frac{\rho g x}{T}\right)$$

is a solution of this differential equation.

53. (a) Show that any function of the form

$$y = A \sinh mx + B \cosh mx$$

satisfies the differential equation $y'' = m^2 y$.

(b) Find $y = y(x)$ such that $y'' = 9y$, $y(0) = -4$, and $y'(0) = 6$.

54. Evaluate $\lim_{x \to \infty} \frac{\sinh x}{e^x}$.

55. At what point of the curve $y = \cosh x$ does the tangent have slope 1?

56. If $x = \ln(\sec \theta + \tan \theta)$, show that $\sec \theta = \cosh x$.

57. Show that if $a \neq 0$ and $b \neq 0$, then there exist numbers α and β such that $ae^x + be^{-x}$ equals either $\alpha \sinh(x + \beta)$ or $\alpha \cosh(x + \beta)$. In other words, almost every function of the form $f(x) = ae^x + be^{-x}$ is a shifted and stretched hyperbolic sine or cosine function.

3 REVIEW

CONCEPT CHECK

1. State each differentiation rule both in symbols and in words.

(a) The Power Rule
(b) The Constant Multiple Rule
(c) The Sum Rule
(d) The Difference Rule
(e) The Product Rule
(f) The Quotient Rule
(g) The Chain Rule

2. State the derivative of each function.

(a) $y = x^n$
(b) $y = e^x$
(c) $y = a^x$
(d) $y = \ln x$
(e) $y = \log_a x$
(f) $y = \sin x$
(g) $y = \cos x$
(h) $y = \tan x$
(i) $y = \csc x$
(j) $y = \sec x$
(k) $y = \cot x$
(l) $y = \sin^{-1} x$
(m) $y = \cos^{-1} x$
(n) $y = \tan^{-1} x$
(o) $y = \sinh x$
(p) $y = \cosh x$
(q) $y = \tanh x$
(r) $y = \sinh^{-1} x$
(s) $y = \cosh^{-1} x$
(t) $y = \tanh^{-1} x$

3. (a) How is the number e defined?

(b) Express e as a limit.

(c) Why is the natural exponential function $y = e^x$ used more often in calculus than the other exponential functions $y = a^x$?

(d) Why is the natural logarithmic function $y = \ln x$ used more often in calculus than the other logarithmic functions $y = \log_a x$?

4. (a) Explain how implicit differentiation works.

(b) Explain how logarithmic differentiation works.

5. (a) Write an expression for the linearization of f at a.

(b) If $y = f(x)$, write an expression for the differential dy.

(c) If $dx = \Delta x$, draw a picture showing the geometric meanings of Δy and dy.

TRUE-FALSE QUIZ

Determine whether the statement is true or false. If it is true, explain why. If it is false, explain why or give an example that disproves the statement.

1. If f and g are differentiable, then

$$\frac{d}{dx}[f(x) + g(x)] = f'(x) + g'(x)$$

2. If f and g are differentiable, then

$$\frac{d}{dx}[f(x)g(x)] = f'(x)g'(x)$$

3. If f and g are differentiable, then

$$\frac{d}{dx}[f(g(x))] = f'(g(x))g'(x)$$

4. If f is differentiable, then $\frac{d}{dx}\sqrt{f(x)} = \frac{f'(x)}{2\sqrt{f(x)}}$.

5. If f is differentiable, then $\frac{d}{dx} f(\sqrt{x}) = \frac{f'(x)}{2\sqrt{x}}$.

6. If $y = e^2$, then $y' = 2e$.

7. $\frac{d}{dx}(10^x) = x10^{x-1}$

8. $\frac{d}{dx}(\ln 10) = \frac{1}{10}$

9. $\frac{d}{dx}(\tan^2 x) = \frac{d}{dx}(\sec^2 x)$

10. $\frac{d}{dx}|x^2 + x| = |2x + 1|$

11. If $g(x) = x^5$, then $\lim_{x \to 2} \frac{g(x) - g(2)}{x - 2} = 80$.

12. An equation of the tangent line to the parabola $y = x^2$ at $(-2, 4)$ is $y - 4 = 2x(x + 2)$.

EXERCISES

1–50 Calculate y'.

1. $y = (x^4 - 3x^2 + 5)^3$

2. $y = \cos(\tan x)$

3. $y = \sqrt{x} + \dfrac{1}{\sqrt[3]{x^4}}$

4. $y = \dfrac{3x - 2}{\sqrt{2x + 1}}$

5. $y = 2x\sqrt{x^2 + 1}$

6. $y = \dfrac{e^x}{1 + x^2}$

7. $y = e^{\sin 2\theta}$

8. $y = e^{-t}(t^2 - 2t + 2)$

9. $y = \dfrac{t}{1 - t^2}$

10. $y = e^{mx} \cos nx$

11. $y = \sqrt{x} \cos \sqrt{x}$

12. $y = (\arcsin 2x)^2$

13. $y = \dfrac{e^{1/x}}{x^2}$

14. $y = \dfrac{1}{\sin(x - \sin x)}$

15. $xy^4 + x^2y = x + 3y$

16. $y = \ln(\csc 5x)$

17. $y = \dfrac{\sec 2\theta}{1 + \tan 2\theta}$

18. $x^2 \cos y + \sin 2y = xy$

19. $y = e^{cx}(c \sin x - \cos x)$

20. $y = \ln(x^2 e^x)$

21. $y = 3^{x \ln x}$

22. $y = \sec(1 + x^2)$

23. $y = (1 - x^{-1})^{-1}$

24. $y = 1/\sqrt[3]{x + \sqrt{x}}$

25. $\sin(xy) = x^2 - y$

26. $y = \sqrt{\sin \sqrt{x}}$

27. $y = \log_5(1 + 2x)$

28. $y = (\cos x)^x$

29. $y = \ln \sin x - \frac{1}{2} \sin^2 x$

30. $y = \dfrac{(x^2 + 1)^4}{(2x + 1)^3(3x - 1)^5}$

31. $y = x \tan^{-1}(4x)$

32. $y = e^{\cos x} + \cos(e^x)$

33. $y = \ln|\sec 5x + \tan 5x|$

34. $y = 10^{\tan \pi\theta}$

35. $y = \cot(3x^2 + 5)$

36. $y = \sqrt{t \ln(t^4)}$

37. $y = \sin\left(\tan \sqrt{1 + x^3}\right)$

38. $y = \arctan\left(\arcsin \sqrt{x}\right)$

39. $y = \tan^2(\sin \theta)$

40. $xe^y = y - 1$

41. $y = \dfrac{\sqrt{x + 1}\,(2 - x)^5}{(x + 3)^7}$

42. $y = \dfrac{(x + \lambda)^4}{x^4 + \lambda^4}$

43. $y = x \sinh(x^2)$

44. $y = \dfrac{\sin mx}{x}$

45. $y = \ln(\cosh 3x)$

46. $y = \ln\left|\dfrac{x^2 - 4}{2x + 5}\right|$

47. $y = \cosh^{-1}(\sinh x)$

48. $y = x \tanh^{-1}\sqrt{x}$

49. $y = \cos\left(e^{\sqrt{\tan 3x}}\right)$

50. $y = \sin^2\left(\cos \sqrt{\sin \pi x}\right)$

51. If $f(t) = \sqrt{4t + 1}$, find $f''(2)$.

52. If $g(\theta) = \theta \sin \theta$, find $g''(\pi/6)$.

53. Find y'' if $x^6 + y^6 = 1$.

54. Find $f^{(n)}(x)$ if $f(x) = 1/(2 - x)$.

55. Use mathematical induction (page 77) to show that if $f(x) = xe^x$, then $f^{(n)}(x) = (x + n)e^x$.

56. Evaluate $\displaystyle\lim_{t \to 0} \frac{t^3}{\tan^3(2t)}$.

57–59 Find an equation of the tangent to the curve at the given point.

57. $y = 4 \sin^2 x$, $(\pi/6, 1)$

58. $y = \dfrac{x^2 - 1}{x^2 + 1}$, $(0, -1)$

59. $y = \sqrt{1 + 4 \sin x}$, $(0, 1)$

60–61 Find equations of the tangent line and normal line to the curve at the given point.

60. $x^2 + 4xy + y^2 = 13$, $(2, 1)$

61. $y = (2 + x)e^{-x}$, $(0, 2)$

62. If $f(x) = xe^{\sin x}$, find $f'(x)$. Graph f and f' on the same screen and comment.

63. (a) If $f(x) = x\sqrt{5 - x}$, find $f'(x)$.
(b) Find equations of the tangent lines to the curve $y = x\sqrt{5 - x}$ at the points $(1, 2)$ and $(4, 4)$.
(c) Illustrate part (b) by graphing the curve and tangent lines on the same screen.
(d) Check to see that your answer to part (a) is reasonable by comparing the graphs of f and f'.

64. (a) If $f(x) = 4x - \tan x$, $-\pi/2 < x < \pi/2$, find f' and f''.
(b) Check to see that your answers to part (a) are reasonable by comparing the graphs of f, f', and f''.

65. At what points on the curve $y = \sin x + \cos x$, $0 \leqslant x \leqslant 2\pi$, is the tangent line horizontal?

66. Find the points on the ellipse $x^2 + 2y^2 = 1$ where the tangent line has slope 1.

67. If $f(x) = (x - a)(x - b)(x - c)$, show that

$$\frac{f'(x)}{f(x)} = \frac{1}{x - a} + \frac{1}{x - b} + \frac{1}{x - c}$$

68. (a) By differentiating the double-angle formula

$$\cos 2x = \cos^2 x - \sin^2 x$$

obtain the double-angle formula for the sine function.
(b) By differentiating the addition formula

$$\sin(x + a) = \sin x \cos a + \cos x \sin a$$

obtain the addition formula for the cosine function.

69. Suppose that $h(x) = f(x)g(x)$ and $F(x) = f(g(x))$, where $f(2) = 3$, $g(2) = 5$, $g'(2) = 4$, $f'(2) = -2$, and $f'(5) = 11$. Find (a) $h'(2)$ and (b) $F'(2)$.

70. If f and g are the functions whose graphs are shown, let $P(x) = f(x)g(x)$, $Q(x) = f(x)/g(x)$, and $C(x) = f(g(x))$. Find (a) $P'(2)$, (b) $Q'(2)$, and (c) $C'(2)$.

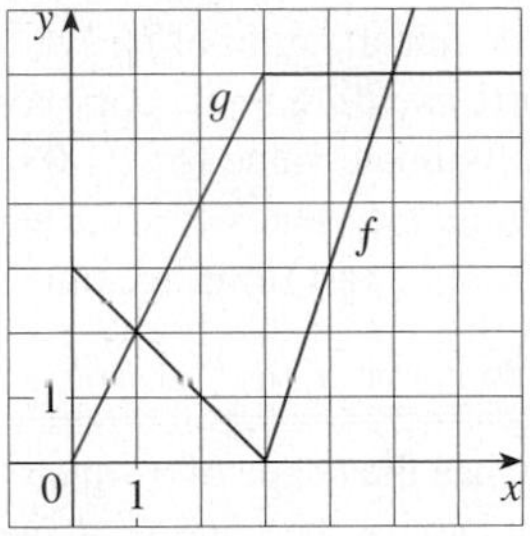

71–78 Find f' in terms of g'.

71. $f(x) = x^2g(x)$

72. $f(x) = g(x^2)$

73. $f(x) = [g(x)]^2$

74. $f(x) = g(g(x))$

75. $f(x) = g(e^x)$

76. $f(x) = e^{g(x)}$

77. $f(x) = \ln|g(x)|$

78. $f(x) = g(\ln x)$

79–81 Find h' in terms of f' and g'.

79. $h(x) = \dfrac{f(x)g(x)}{f(x) + g(x)}$

80. $h(x) = \sqrt{\dfrac{f(x)}{g(x)}}$

81. $h(x) = f(g(\sin 4x))$

82. (a) Graph the function $f(x) = x - 2\sin x$ in the viewing rectangle $[0, 8]$ by $[-2, 8]$.
(b) On which interval is the average rate of change larger: $[1, 2]$ or $[2, 3]$?
(c) At which value of x is the instantaneous rate of change larger: $x = 2$ or $x = 5$?
(d) Check your visual estimates in part (c) by computing $f'(x)$ and comparing the numerical values of $f'(2)$ and $f'(5)$.

83. At what point on the curve $y = [\ln(x + 4)]^2$ is the tangent horizontal?

84. (a) Find an equation of the tangent to the curve $y = e^x$ that is parallel to the line $x - 4y = 1$.
(b) Find an equation of the tangent to the curve $y = e^x$ that passes through the origin.

85. Find a parabola $y = ax^2 + bx + c$ that passes through the point $(1, 4)$ and whose tangent lines at $x = -1$ and $x = 5$ have slopes 6 and -2, respectively.

86. The function $C(t) = K(e^{-at} - e^{-bt})$, where a, b, and K are positive constants and $b > a$, is used to model the concentration at time t of a drug injected into the bloodstream.
(a) Show that $\lim_{t\to\infty} C(t) = 0$.
(b) Find $C'(t)$, the rate at which the drug is cleared from circulation.
(c) When is this rate equal to 0?

87. An equation of motion of the form $s = Ae^{-ct}\cos(\omega t + \delta)$ represents damped oscillation of an object. Find the velocity and acceleration of the object.

88. A particle moves along a horizontal line so that its coordinate at time t is $x = \sqrt{b^2 + c^2t^2}$, $t \geq 0$, where b and c are positive constants.
(a) Find the velocity and acceleration functions.
(b) Show that the particle always moves in the positive direction.

89. A particle moves on a vertical line so that its coordinate at time t is $y = t^3 - 12t + 3$, $t \geq 0$.
(a) Find the velocity and acceleration functions.
(b) When is the particle moving upward and when is it moving downward?
(c) Find the distance that the particle travels in the time interval $0 \leq t \leq 3$.
(d) Graph the position, velocity, and acceleration functions for $0 \leq t \leq 3$.
(e) When is the particle speeding up? When is it slowing down?

90. The volume of a right circular cone is $V = \pi r^2h/3$, where r is the radius of the base and h is the height.
(a) Find the rate of change of the volume with respect to the height if the radius is constant.
(b) Find the rate of change of the volume with respect to the radius if the height is constant.

91. The mass of part of a wire is $x(1 + \sqrt{x})$ kilograms, where x is measured in meters from one end of the wire. Find the linear density of the wire when $x = 4$ m.

92. The cost, in dollars, of producing x units of a certain commodity is

$$C(x) = 920 + 2x - 0.02x^2 + 0.00007x^3$$

(a) Find the marginal cost function.
(b) Find $C'(100)$ and explain its meaning.
(c) Compare $C'(100)$ with the cost of producing the 101st item.

93. A bacteria culture contains 200 cells initially and grows at a rate proportional to its size. After half an hour the population has increased to 360 cells.
(a) Find the number of bacteria after t hours.
(b) Find the number of bacteria after 4 hours.
(c) Find the rate of growth after 4 hours.
(d) When will the population reach 10,000?

94. Cobalt-60 has a half-life of 5.24 years.
(a) Find the mass that remains from a 100-mg sample after 20 years.
(b) How long would it take for the mass to decay to 1 mg?

95. Let $C(t)$ be the concentration of a drug in the bloodstream. As the body eliminates the drug, $C(t)$ decreases at a rate that is proportional to the amount of the drug that is present at the time. Thus $C'(t) = -kC(t)$, where k is a positive number called the *elimination constant* of the drug.
(a) If C_0 is the concentration at time $t = 0$, find the concentration at time t.
(b) If the body eliminates half the drug in 30 hours, how long does it take to eliminate 90% of the drug?

96. A cup of hot chocolate has temperature 80°C in a room kept at 20°C. After half an hour the hot chocolate cools to 60°C.
(a) What is the temperature of the chocolate after another half hour?
(b) When will the chocolate have cooled to 40°C?

97. The volume of a cube is increasing at a rate of $10\ \text{cm}^3/\text{min}$. How fast is the surface area increasing when the length of an edge is 30 cm?

98. A paper cup has the shape of a cone with height 10 cm and radius 3 cm (at the top). If water is poured into the cup at a rate of $2\ \text{cm}^3/\text{s}$, how fast is the water level rising when the water is 5 cm deep?

99. A balloon is rising at a constant speed of 5 ft/s. A boy is cycling along a straight road at a speed of 15 ft/s. When he passes under the balloon, it is 45 ft above him. How fast is the distance between the boy and the balloon increasing 3 s later?

100. A waterskier skis over the ramp shown in the figure at a speed of 30 ft/s. How fast is she rising as she leaves the ramp?

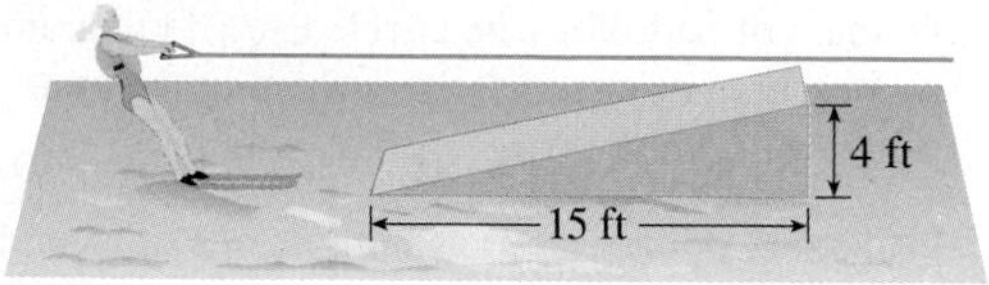

101. The angle of elevation of the sun is decreasing at a rate of 0.25 rad/h. How fast is the shadow cast by a 400-ft-tall building increasing when the angle of elevation of the sun is $\pi/6$?

102. (a) Find the linear approximation to $f(x) = \sqrt{25 - x^2}$ near 3.
(b) Illustrate part (a) by graphing f and the linear approximation.
(c) For what values of x is the linear approximation accurate to within 0.1?

103. (a) Find the linearization of $f(x) = \sqrt[3]{1 + 3x}$ at $a = 0$. State the corresponding linear approximation and use it to give an approximate value for $\sqrt[3]{1.03}$.
(b) Determine the values of x for which the linear approximation given in part (a) is accurate to within 0.1.

104. Evaluate dy if $y = x^3 - 2x^2 + 1$, $x = 2$, and $dx = 0.2$.

105. A window has the shape of a square surmounted by a semicircle. The base of the window is measured as having width 60 cm with a possible error in measurement of 0.1 cm. Use differentials to estimate the maximum error possible in computing the area of the window.

106–108 Express the limit as a derivative and evaluate.

106. $\displaystyle\lim_{x \to 1} \frac{x^{17} - 1}{x - 1}$

107. $\displaystyle\lim_{h \to 0} \frac{\sqrt[4]{16 + h} - 2}{h}$

108. $\displaystyle\lim_{\theta \to \pi/3} \frac{\cos\theta - 0.5}{\theta - \pi/3}$

---

109. Evaluate $\displaystyle\lim_{x \to 0} \frac{\sqrt{1 + \tan x} - \sqrt{1 + \sin x}}{x^3}$.

110. Suppose f is a differentiable function such that $f(g(x)) = x$ and $f'(x) = 1 + [f(x)]^2$. Show that $g'(x) = 1/(1 + x^2)$.

111. Find $f'(x)$ if it is known that

$$\frac{d}{dx}[f(2x)] = x^2$$

112. Show that the length of the portion of any tangent line to the astroid $x^{2/3} + y^{2/3} = a^{2/3}$ cut off by the coordinate axes is constant.

Before you look at the example, cover up the solution and try it yourself first.

EXAMPLE 1 How many lines are tangent to both of the parabolas $y = -1 - x^2$ and $y = 1 + x^2$? Find the coordinates of the points at which these tangents touch the parabolas.

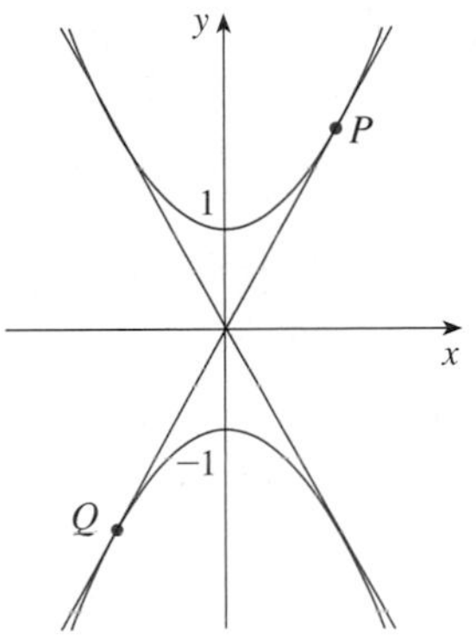

FIGURE 1

SOLUTION To gain insight into this problem, it is essential to draw a diagram. So we sketch the parabolas $y = 1 + x^2$ (which is the standard parabola $y = x^2$ shifted 1 unit upward) and $y = -1 - x^2$ (which is obtained by reflecting the first parabola about the x-axis). If we try to draw a line tangent to both parabolas, we soon discover that there are only two possibilities, as illustrated in Figure 1.

Let P be a point at which one of these tangents touches the upper parabola and let a be its x-coordinate. (The choice of notation for the unknown is important. Of course we could have used b or c or x_0 or x_1 instead of a. However, it's not advisable to use x in place of a because that x could be confused with the variable x in the equation of the parabola.) Then, since P lies on the parabola $y = 1 + x^2$, its y-coordinate must be $1 + a^2$. Because of the symmetry shown in Figure 1, the coordinates of the point Q where the tangent touches the lower parabola must be $(-a, -(1 + a^2))$.

To use the given information that the line is a tangent, we equate the slope of the line PQ to the slope of the tangent line at P. We have

$$m_{PQ} = \frac{1 + a^2 - (-1 - a^2)}{a - (-a)} = \frac{1 + a^2}{a}$$

If $f(x) = 1 + x^2$, then the slope of the tangent line at P is $f'(a) = 2a$. Thus the condition that we need to use is that

$$\frac{1 + a^2}{a} = 2a$$

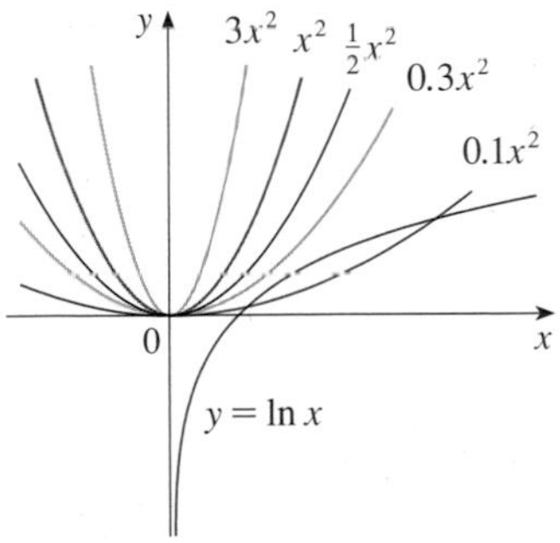

FIGURE 2

Solving this equation, we get $1 + a^2 = 2a^2$, so $a^2 = 1$ and $a = \pm 1$. Therefore the points are $(1, 2)$ and $(-1, -2)$. By symmetry, the two remaining points are $(-1, 2)$ and $(1, -2)$. □

EXAMPLE 2 For what values of c does the equation $\ln x = cx^2$ have exactly one solution?

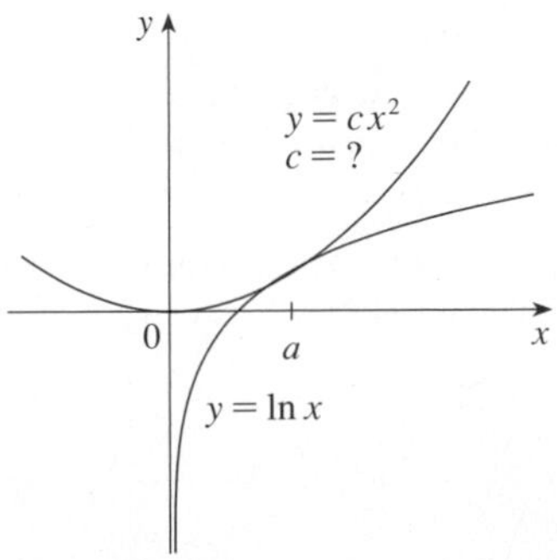

FIGURE 3

SOLUTION One of the most important principles of problem solving is to draw a diagram, even if the problem as stated doesn't explicitly mention a geometric situation. Our present problem can be reformulated geometrically as follows: For what values of c does the curve $y = \ln x$ intersect the curve $y = cx^2$ in exactly one point?

Let's start by graphing $y = \ln x$ and $y = cx^2$ for various values of c. We know that, for $c \neq 0$, $y = cx^2$ is a parabola that opens upward if $c > 0$ and downward if $c < 0$. Figure 2 shows the parabolas $y = cx^2$ for several positive values of c. Most of them don't intersect $y = \ln x$ at all and one intersects twice. We have the feeling that there must be a value of c (somewhere between 0.1 and 0.3) for which the curves intersect exactly once, as in Figure 3.

To find that particular value of c, we let a be the x-coordinate of the single point of intersection. In other words, $\ln a = ca^2$, so a is the unique solution of the given equation. We see from Figure 3 that the curves just touch, so they have a common tangent line when $x = a$. That means the curves $y = \ln x$ and $y = cx^2$ have the same slope when $x = a$. Therefore

$$\frac{1}{a} = 2ca$$

Solving the equations $\ln a = ca^2$ and $1/a = 2ca$, we get

$$\ln a = ca^2 = c \cdot \frac{1}{2c} = \frac{1}{2}$$

Thus $a = e^{1/2}$ and

$$c = \frac{\ln a}{a^2} = \frac{\ln e^{1/2}}{e} = \frac{1}{2e}$$

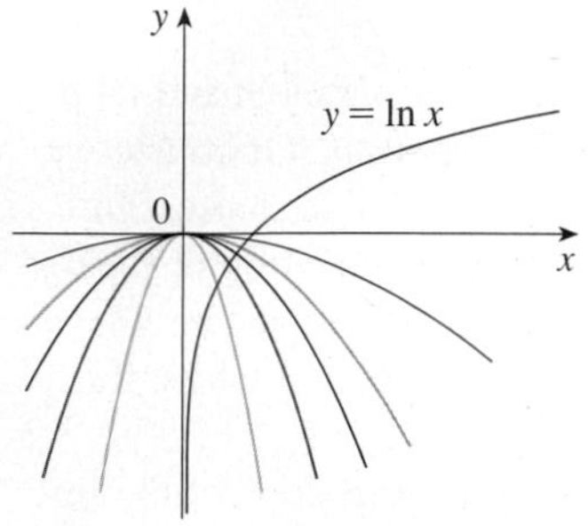

FIGURE 4

For negative values of c we have the situation illustrated in Figure 4: All parabolas $y = cx^2$ with negative values of c intersect $y = \ln x$ exactly once. And let's not forget about $c = 0$: The curve $y = 0x^2 = 0$ is just the x-axis, which intersects $y = \ln x$ exactly once.

To summarize, the required values of c are $c = 1/(2e)$ and $c \leq 0$. □

PROBLEMS

1. Find points P and Q on the parabola $y = 1 - x^2$ so that the triangle ABC formed by the x-axis and the tangent lines at P and Q is an equilateral triangle.

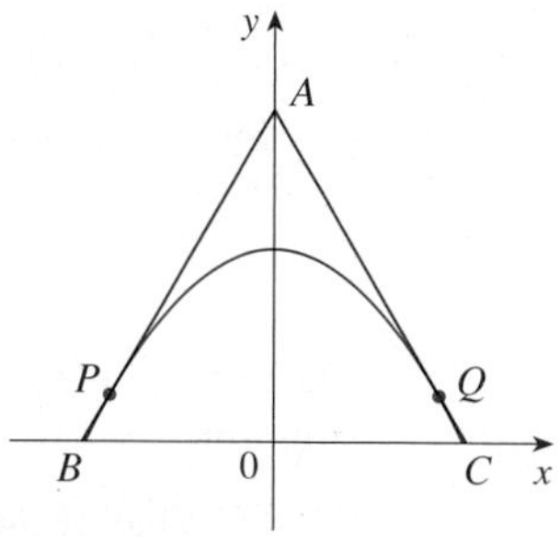

2. Find the point where the curves $y = x^3 - 3x + 4$ and $y = 3(x^2 - x)$ are tangent to each other, that is, have a common tangent line. Illustrate by sketching both curves and the common tangent.

3. Show that the tangent lines to the parabola $y = ax^2 + bx + c$ at any two points with x-coordinates p and q must intersect at a point whose x-coordinate is halfway between p and q.

4. Show that

$$\frac{d}{dx}\left(\frac{\sin^2 x}{1 + \cot x} + \frac{\cos^2 x}{1 + \tan x}\right) = -\cos 2x$$

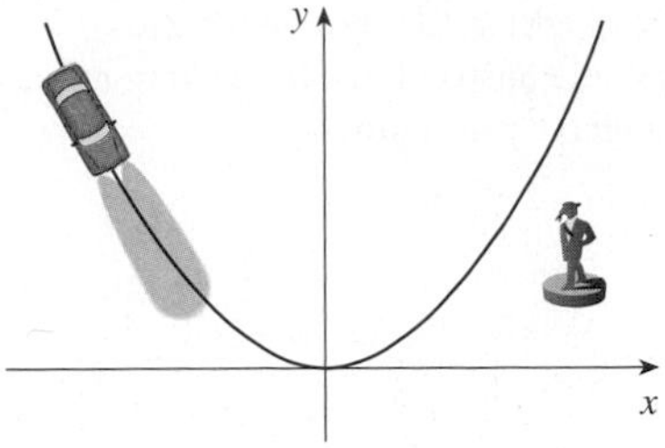

FIGURE FOR PROBLEM 6

5. Show that $\sin^{-1}(\tanh x) = \tan^{-1}(\sinh x)$.

6. A car is traveling at night along a highway shaped like a parabola with its vertex at the origin (see the figure). The car starts at a point 100 m west and 100 m north of the origin and travels in an easterly direction. There is a statue located 100 m east and 50 m north of the origin. At what point on the highway will the car's headlights illuminate the statue?

7. Prove that $\dfrac{d^n}{dx^n}(\sin^4 x + \cos^4 x) = 4^{n-1}\cos(4x + n\pi/2)$.

8. Find the nth derivative of the function $f(x) = x^n/(1 - x)$.

9. The figure shows a circle with radius 1 inscribed in the parabola $y = x^2$. Find the center of the circle.

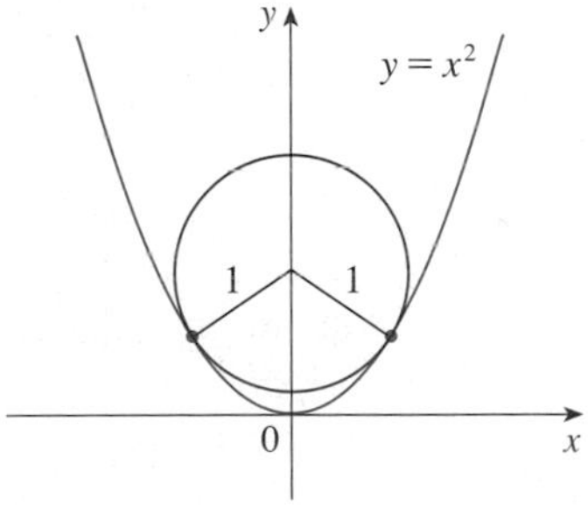

10. If f is differentiable at a, where $a > 0$, evaluate the following limit in terms of $f'(a)$:

$$\lim_{x \to a} \frac{f(x) - f(a)}{\sqrt{x} - \sqrt{a}}$$

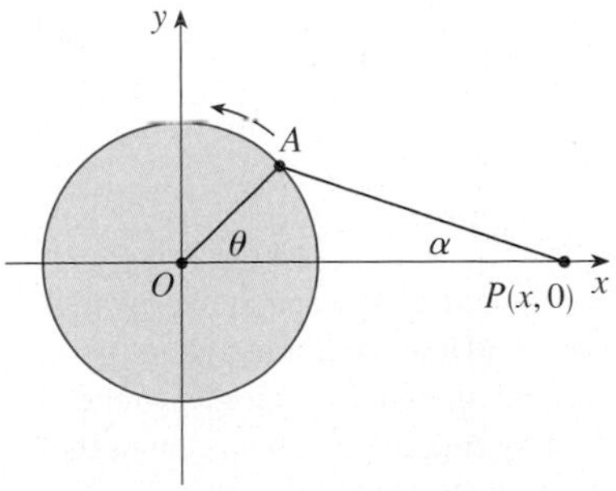

FIGURE FOR PROBLEM 11

11. The figure shows a rotating wheel with radius 40 cm and a connecting rod AP with length 1.2 m. The pin P slides back and forth along the x-axis as the wheel rotates counterclockwise at a rate of 360 revolutions per minute.

(a) Find the angular velocity of the connecting rod, $d\alpha/dt$, in radians per second, when $\theta = \pi/3$.

(b) Express the distance $x = |OP|$ in terms of θ.

(c) Find an expression for the velocity of the pin P in terms of θ.

12. Tangent lines T_1 and T_2 are drawn at two points P_1 and P_2 on the parabola $y = x^2$ and they intersect at a point P. Another tangent line T is drawn at a point between P_1 and P_2; it intersects T_1 at Q_1 and T_2 at Q_2. Show that

$$\frac{|PQ_1|}{|PP_1|} + \frac{|PQ_2|}{|PP_2|} = 1$$

13. Show that

$$\frac{d^n}{dx^n}(e^{ax}\sin bx) = r^n e^{ax}\sin(bx + n\theta)$$

where a and b are positive numbers, $r^2 = a^2 + b^2$, and $\theta = \tan^{-1}(b/a)$.

14. Evaluate $\displaystyle\lim_{x \to \pi} \frac{e^{\sin x} - 1}{x - \pi}$.

15. Let T and N be the tangent and normal lines to the ellipse $x^2/9 + y^2/4 = 1$ at any point P on the ellipse in the first quadrant. Let x_T and y_T be the x- and y-intercepts of T and x_N and y_N be the intercepts of N. As P moves along the ellipse in the first quadrant (but not on the axes), what values can x_T, y_T, x_N, and y_N take on? First try to guess the answers just by looking at the figure. Then use calculus to solve the problem and see how good your intuition is.

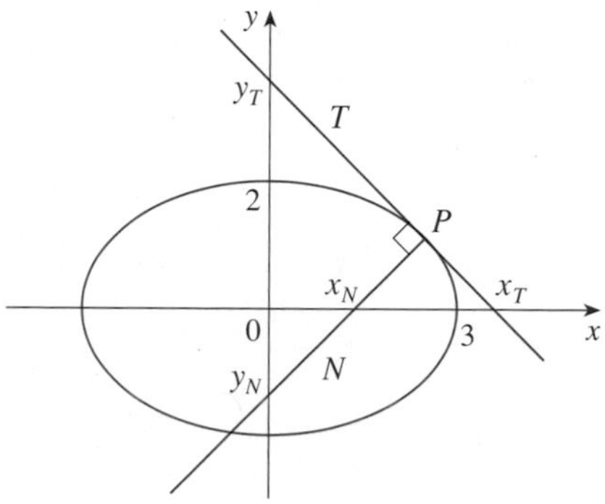

16. Evaluate $\lim_{x \to 0} \dfrac{\sin(3 + x)^2 - \sin 9}{x}$.

17. (a) Use the identity for $\tan(x - y)$ (see Equation 14b in Appendix D) to show that if two lines L_1 and L_2 intersect at an angle α, then

$$\tan \alpha = \frac{m_2 - m_1}{1 + m_1 m_2}$$

where m_1 and m_2 are the slopes of L_1 and L_2, respectively.

(b) The **angle between the curves** C_1 and C_2 at a point of intersection P is defined to be the angle between the tangent lines to C_1 and C_2 at P (if these tangent lines exist). Use part (a) to find, correct to the nearest degree, the angle between each pair of curves at each point of intersection.

(i) $y = x^2$ and $y = (x - 2)^2$

(ii) $x^2 - y^2 = 3$ and $x^2 - 4x + y^2 + 3 = 0$

18. Let $P(x_1, y_1)$ be a point on the parabola $y^2 = 4px$ with focus $F(p, 0)$. Let α be the angle between the parabola and the line segment FP, and let β be the angle between the horizontal line $y = y_1$ and the parabola as in the figure. Prove that $\alpha = \beta$. (Thus, by a principle of geometrical optics, light from a source placed at F will be reflected along a line parallel to the x-axis. This explains why *paraboloids*, the surfaces obtained by rotating parabolas about their axes, are used as the shape of some automobile headlights and mirrors for telescopes.)

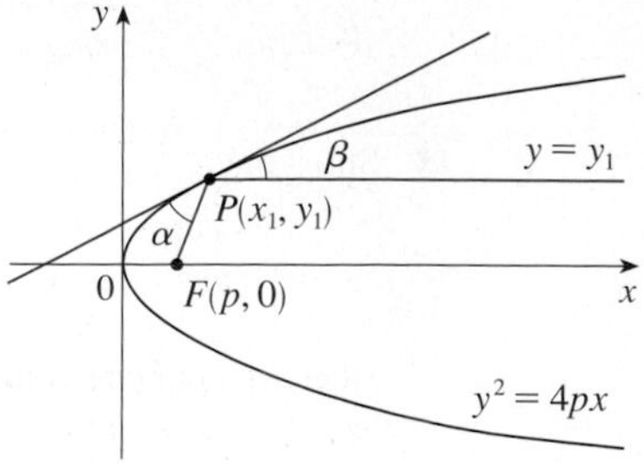

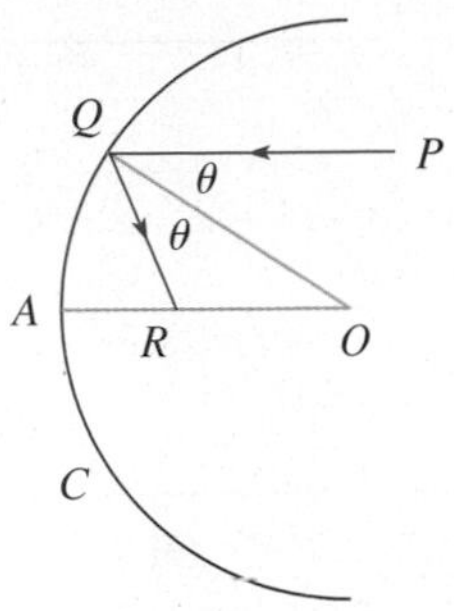

FIGURE FOR PROBLEM 19

19. Suppose that we replace the parabolic mirror of Problem 18 by a spherical mirror. Although the mirror has no focus, we can show the existence of an *approximate* focus. In the figure, C is a semicircle with center O. A ray of light coming in toward the mirror parallel to the axis along the line PQ will be reflected to the point R on the axis so that $\angle PQO = \angle OQR$ (the angle of incidence is equal to the angle of reflection). What happens to the point R as P is taken closer and closer to the axis?

20. If f and g are differentiable functions with $f(0) = g(0) = 0$ and $g'(0) \neq 0$, show that

$$\lim_{x \to 0} \frac{f(x)}{g(x)} = \frac{f'(0)}{g'(0)}$$

21. Evaluate $\lim_{x \to 0} \dfrac{\sin(a + 2x) - 2\sin(a + x) + \sin a}{x^2}$.

CAS **22.** (a) The cubic function $f(x) = x(x - 2)(x - 6)$ has three distinct zeros: 0, 2, and 6. Graph f and its tangent lines at the *average* of each pair of zeros. What do you notice?
(b) Suppose the cubic function $f(x) = (x - a)(x - b)(x - c)$ has three distinct zeros: a, b, and c. Prove, with the help of a computer algebra system, that a tangent line drawn at the average of the zeros a and b intersects the graph of f at the third zero.

23. For what value of k does the equation $e^{2x} = k\sqrt{x}$ have exactly one solution?

24. For which positive numbers a is it true that $a^x \geqslant 1 + x$ for all x?

25. If

$$y = \frac{x}{\sqrt{a^2 - 1}} - \frac{2}{\sqrt{a^2 - 1}} \arctan \frac{\sin x}{a + \sqrt{a^2 - 1} + \cos x}$$

show that $y' = \dfrac{1}{a + \cos x}$.

26. Given an ellipse $x^2/a^2 + y^2/b^2 = 1$, where $a \neq b$, find the equation of the set of all points from which there are two tangents to the curve whose slopes are (a) reciprocals and (b) negative reciprocals.

27. Find the two points on the curve $y = x^4 - 2x^2 - x$ that have a common tangent line.

28. Suppose that three points on the parabola $y = x^2$ have the property that their normal lines intersect at a common point. Show that the sum of their x-coordinates is 0.

29. A *lattice point* in the plane is a point with integer coordinates. Suppose that circles with radius r are drawn using all lattice points as centers. Find the smallest value of r such that any line with slope $\frac{2}{5}$ intersects some of these circles.

30. A cone of radius r centimeters and height h centimeters is lowered point first at a rate of 1 cm/s into a tall cylinder of radius R centimeters that is partially filled with water. How fast is the water level rising at the instant the cone is completely submerged?

31. A container in the shape of an inverted cone has height 16 cm and radius 5 cm at the top. It is partially filled with a liquid that oozes through the sides at a rate proportional to the area of the container that is in contact with the liquid. (The surface area of a cone is πrl, where r is the radius and l is the slant height.) If we pour the liquid into the container at a rate of 2 cm^3/min, then the height of the liquid decreases at a rate of 0.3 cm/min when the height is 10 cm. If our goal is to keep the liquid at a constant height of 10 cm, at what rate should we pour the liquid into the container?

4

APPLICATIONS OF DIFFERENTIATION

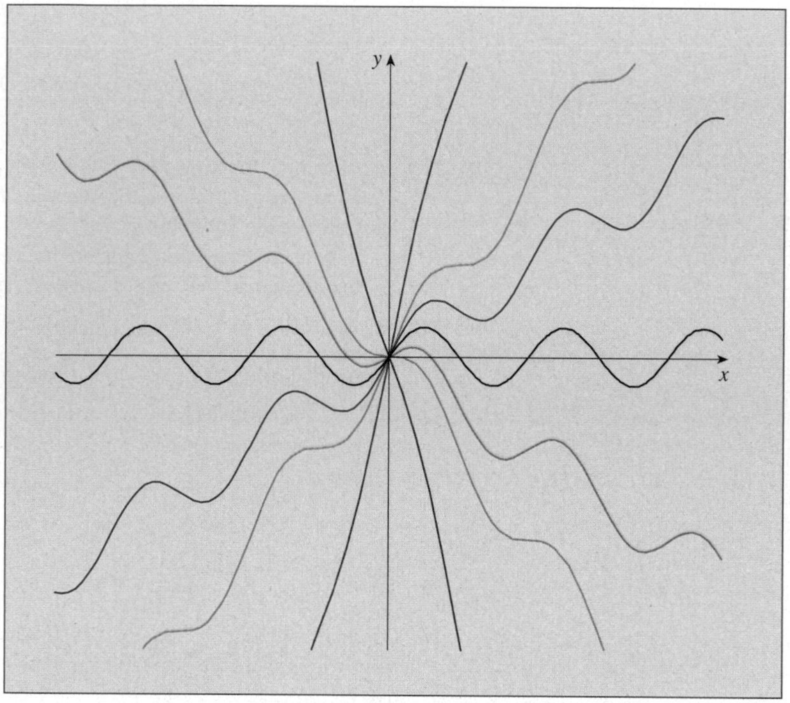

Calculus reveals all the important aspects of graphs of functions. Members of the family of functions $f(x) = cx + \sin x$ are illustrated.

We have already investigated some of the applications of derivatives, but now that we know the differentiation rules we are in a better position to pursue the applications of differentiation in greater depth. Here we learn how derivatives affect the shape of a graph of a function and, in particular, how they help us locate maximum and minimum values of functions. Many practical problems require us to minimize a cost or maximize an area or somehow find the best possible outcome of a situation. In particular, we will be able to investigate the optimal shape of a can and to explain the location of rainbows in the sky.

4.1 MAXIMUM AND MINIMUM VALUES

Some of the most important applications of differential calculus are *optimization problems*, in which we are required to find the optimal (best) way of doing something. Here are examples of such problems that we will solve in this chapter:

- What is the shape of a can that minimizes manufacturing costs?
- What is the maximum acceleration of a space shuttle? (This is an important question to the astronauts who have to withstand the effects of acceleration.)
- What is the radius of a contracted windpipe that expels air most rapidly during a cough?
- At what angle should blood vessels branch so as to minimize the energy expended by the heart in pumping blood?

These problems can be reduced to finding the maximum or minimum values of a function. Let's first explain exactly what we mean by maximum and minimum values.

1 DEFINITION A function f has an **absolute maximum** (or **global maximum**) at c if $f(c) \geqslant f(x)$ for all x in D, where D is the domain of f. The number $f(c)$ is called the **maximum value** of f on D. Similarly, f has an **absolute minimum** at c if $f(c) \leqslant f(x)$ for all x in D and the number $f(c)$ is called the **minimum value** of f on D. The maximum and minimum values of f are called the **extreme values** of f.

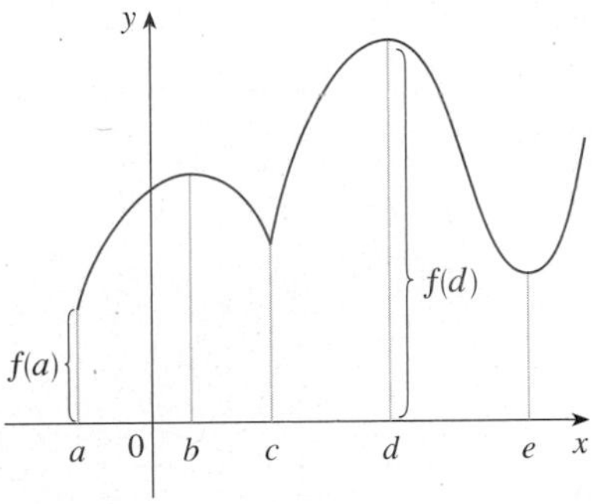

FIGURE 1
Minimum value $f(a)$, maximum value $f(d)$

Figure 1 shows the graph of a function f with absolute maximum at d and absolute minimum at a. Note that $(d, f(d))$ is the highest point on the graph and $(a, f(a))$ is the lowest point. If we consider only values of x near b [for instance, if we restrict our attention to the interval (a, c)], then $f(b)$ is the largest of those values of $f(x)$ and is called a *local maximum value* of f. Likewise, $f(c)$ is called a *local minimum value* of f because $f(c) \leqslant f(x)$ for x near c [in the interval (b, d), for instance]. The function f also has a local minimum at e. In general, we have the following definition.

2 DEFINITION A function f has a **local maximum** (or **relative maximum**) at c if $f(c) \geqslant f(x)$ when x is near c. [This means that $f(c) \geqslant f(x)$ for all x in some open interval containing c.] Similarly, f has a **local minimum** at c if $f(c) \leqslant f(x)$ when x is near c.

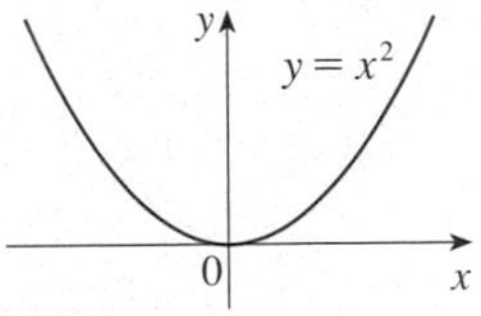

FIGURE 2
Minimum value 0, no maximum

EXAMPLE 1 The function $f(x) = \cos x$ takes on its (local and absolute) maximum value of 1 infinitely many times, since $\cos 2n\pi = 1$ for any integer n and $-1 \leqslant \cos x \leqslant 1$ for all x. Likewise, $\cos(2n + 1)\pi = -1$ is its minimum value, where n is any integer. □

EXAMPLE 2 If $f(x) = x^2$, then $f(x) \geqslant f(0)$ because $x^2 \geqslant 0$ for all x. Therefore $f(0) = 0$ is the absolute (and local) minimum value of f. This corresponds to the fact that the origin is the lowest point on the parabola $y = x^2$. (See Figure 2.) However, there is no highest point on the parabola and so this function has no maximum value. □

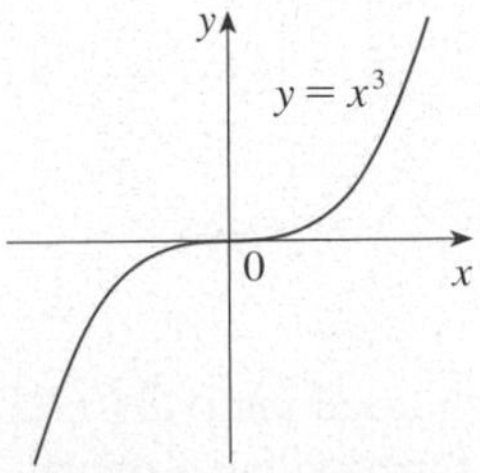

FIGURE 3
No minimum, no maximum

EXAMPLE 3 From the graph of the function $f(x) = x^3$, shown in Figure 3, we see that this function has neither an absolute maximum value nor an absolute minimum value. In fact, it has no local extreme values either. □

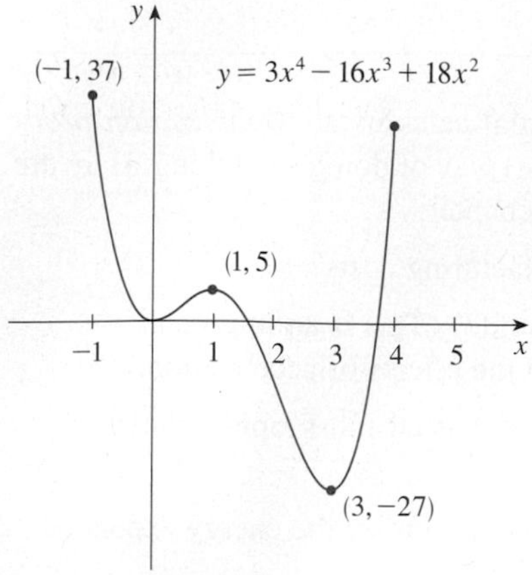

FIGURE 4

V EXAMPLE 4 The graph of the function

$$f(x) = 3x^4 - 16x^3 + 18x^2 \qquad -1 \leqslant x \leqslant 4$$

is shown in Figure 4. You can see that $f(1) = 5$ is a local maximum, whereas the absolute maximum is $f(-1) = 37$. (This absolute maximum is not a local maximum because it occurs at an endpoint.) Also, $f(0) = 0$ is a local minimum and $f(3) = -27$ is both a local and an absolute minimum. Note that f has neither a local nor an absolute maximum at $x = 4$. □

We have seen that some functions have extreme values, whereas others do not. The following theorem gives conditions under which a function is guaranteed to possess extreme values.

> **3 THE EXTREME VALUE THEOREM** If f is continuous on a closed interval $[a, b]$, then f attains an absolute maximum value $f(c)$ and an absolute minimum value $f(d)$ at some numbers c and d in $[a, b]$.

The Extreme Value Theorem is illustrated in Figure 5. Note that an extreme value can be taken on more than once. Although the Extreme Value Theorem is intuitively very plausible, it is difficult to prove and so we omit the proof.

FIGURE 5

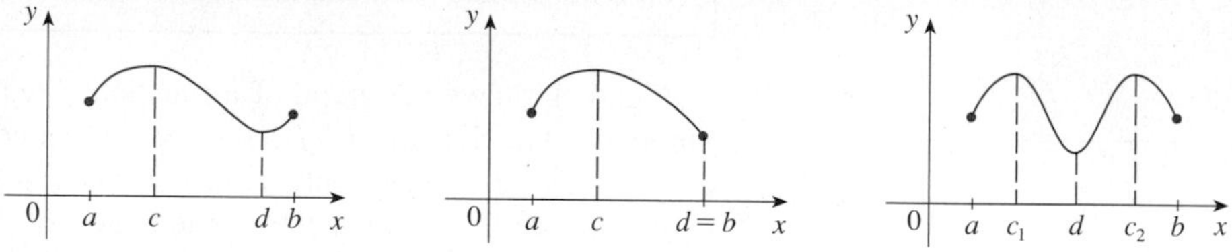

Figures 6 and 7 show that a function need not possess extreme values if either hypothesis (continuity or closed interval) is omitted from the Extreme Value Theorem.

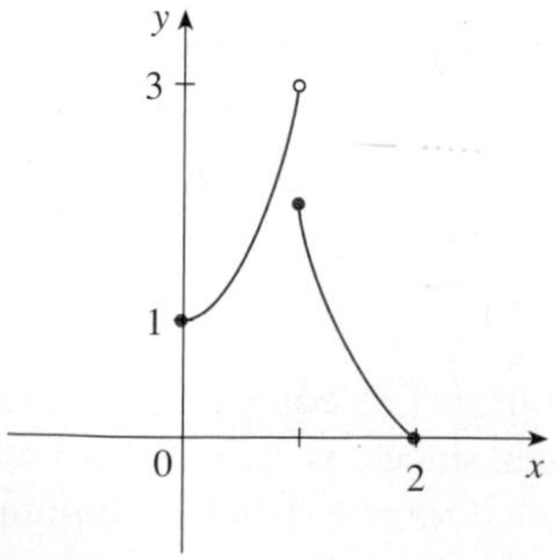

FIGURE 6
This function has minimum value $f(2) = 0$, but no maximum value.

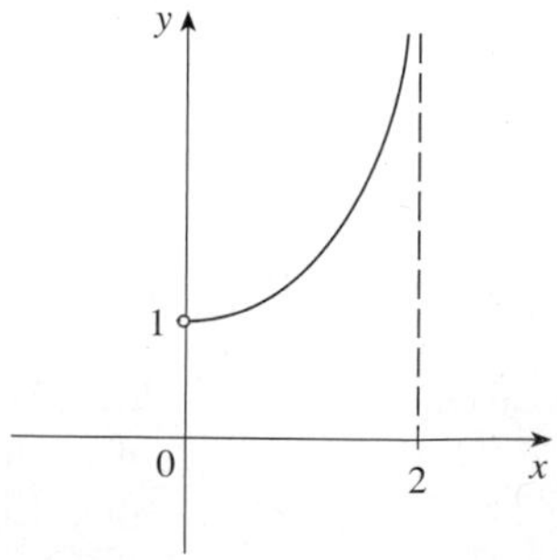

FIGURE 7
This continuous function g has no maximum or minimum.

The function f whose graph is shown in Figure 6 is defined on the closed interval $[0, 2]$ but has no maximum value. (Notice that the range of f is $[0, 3)$. The function takes on values arbitrarily close to 3, but never actually attains the value 3.) This does not contradict the Extreme Value Theorem because f is not continuous. [Nonetheless, a discontinuous function *could* have maximum and minimum values. See Exercise 13(b).]

The function g shown in Figure 7 is continuous on the open interval $(0, 2)$ but has neither a maximum nor a minimum value. [The range of g is $(1, \infty)$. The function takes on arbitrarily large values.] This does not contradict the Extreme Value Theorem because the interval $(0, 2)$ is not closed.

The Extreme Value Theorem says that a continuous function on a closed interval has a maximum value and a minimum value, but it does not tell us how to find these extreme values. We start by looking for local extreme values.

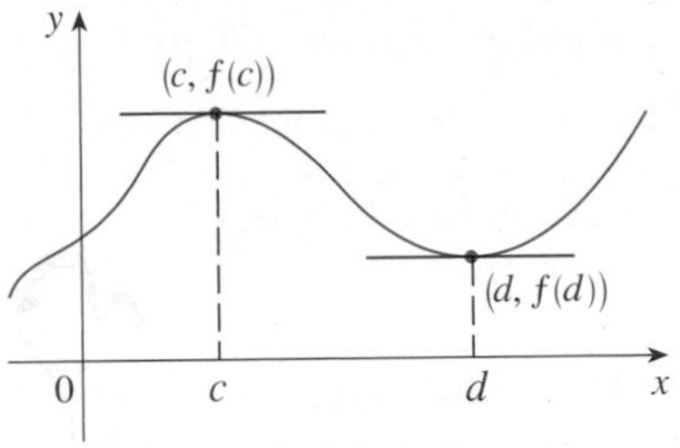

FIGURE 8

Figure 8 shows the graph of a function f with a local maximum at c and a local minimum at d. It appears that at the maximum and minimum points the tangent lines are horizontal and therefore each has slope 0. We know that the derivative is the slope of the tangent line, so it appears that $f'(c) = 0$ and $f'(d) = 0$. The following theorem says that this is always true for differentiable functions.

■ Fermat's Theorem is named after Pierre Fermat (1601–1665), a French lawyer who took up mathematics as a hobby. Despite his amateur status, Fermat was one of the two inventors of analytic geometry (Descartes was the other). His methods for finding tangents to curves and maximum and minimum values (before the invention of limits and derivatives) made him a forerunner of Newton in the creation of differential calculus.

> **4 FERMAT'S THEOREM** If f has a local maximum or minimum at c, and if $f'(c)$ exists, then $f'(c) = 0$.

PROOF Suppose, for the sake of definiteness, that f has a local maximum at c. Then, according to Definition 2, $f(c) \geqslant f(x)$ if x is sufficiently close to c. This implies that if h is sufficiently close to 0, with h being positive or negative, then

$$f(c) \geqslant f(c + h)$$

and therefore

$$f(c + h) - f(c) \leqslant 0 \tag{5}$$

We can divide both sides of an inequality by a positive number. Thus, if $h > 0$ and h is sufficiently small, we have

$$\frac{f(c + h) - f(c)}{h} \leqslant 0$$

Taking the right-hand limit of both sides of this inequality (using Theorem 2.3.2), we get

$$\lim_{h \to 0^+} \frac{f(c + h) - f(c)}{h} \leqslant \lim_{h \to 0^+} 0 = 0$$

But since $f'(c)$ exists, we have

$$f'(c) = \lim_{h \to 0} \frac{f(c + h) - f(c)}{h} = \lim_{h \to 0^+} \frac{f(c + h) - f(c)}{h}$$

and so we have shown that $f'(c) \leqslant 0$.

If $h < 0$, then the direction of the inequality (5) is reversed when we divide by h:

$$\frac{f(c + h) - f(c)}{h} \geqslant 0 \qquad h < 0$$

So, taking the left-hand limit, we have

$$f'(c) = \lim_{h \to 0} \frac{f(c + h) - f(c)}{h} = \lim_{h \to 0^-} \frac{f(c + h) - f(c)}{h} \geqslant 0$$

We have shown that $f'(c) \geqslant 0$ and also that $f'(c) \leqslant 0$. Since both of these inequalities must be true, the only possibility is that $f'(c) = 0$.

We have proved Fermat's Theorem for the case of a local maximum. The case of a local minimum can be proved in a similar manner, or we could use Exercise 76 to deduce it from the case we have just proved (see Exercise 77). □

The following examples caution us against reading too much into Fermat's Theorem. We can't expect to locate extreme values simply by setting $f'(x) = 0$ and solving for x.

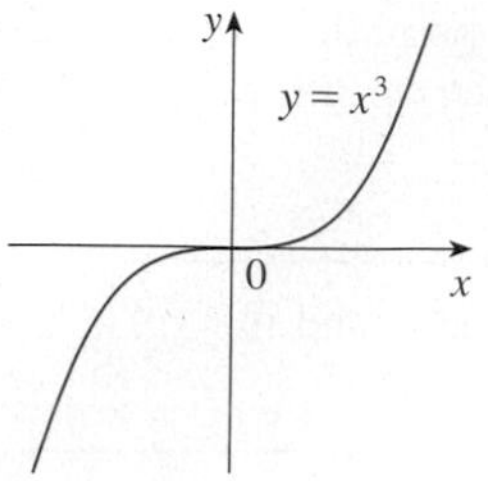

FIGURE 9
If $f(x) = x^3$, then $f'(0) = 0$ but f has no maximum or minimum.

EXAMPLE 5 If $f(x) = x^3$, then $f'(x) = 3x^2$, so $f'(0) = 0$. But f has no maximum or minimum at 0, as you can see from its graph in Figure 9. (Or observe that $x^3 > 0$ for $x > 0$ but $x^3 < 0$ for $x < 0$.) The fact that $f'(0) = 0$ simply means that the curve $y = x^3$ has a horizontal tangent at (0, 0). Instead of having a maximum or minimum at (0, 0), the curve crosses its horizontal tangent there. □

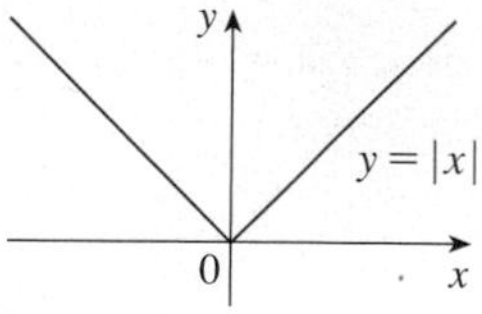

FIGURE 10
If $f(x) = |x|$, then $f(0) = 0$ is a minimum value, but $f'(0)$ does not exist.

EXAMPLE 6 The function $f(x) = |x|$ has its (local and absolute) minimum value at 0, but that value can't be found by setting $f'(x) = 0$ because, as was shown in Example 5 in Section 2.8, $f'(0)$ does not exist. (See Figure 10.) □

WARNING Examples 5 and 6 show that we must be careful when using Fermat's Theorem. Example 5 demonstrates that even when $f'(c) = 0$ there need not be a maximum or minimum at c. (In other words, the converse of Fermat's Theorem is false in general.) Furthermore, there may be an extreme value even when $f'(c)$ does not exist (as in Example 6).

Fermat's Theorem does suggest that we should at least *start* looking for extreme values of f at the numbers c where $f'(c) = 0$ or where $f'(c)$ does not exist. Such numbers are given a special name.

> **6 DEFINITION** A **critical number** of a function f is a number c in the domain of f such that either $f'(c) = 0$ or $f'(c)$ does not exist.

■ Figure 11 shows a graph of the function f in Example 7. It supports our answer because there is a horizontal tangent when $x = 1.5$ and a vertical tangent when $x = 0$.

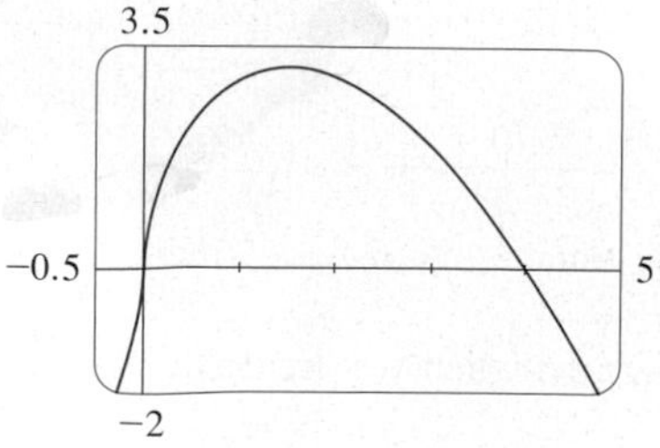

FIGURE 11

EXAMPLE 7 Find the critical numbers of $f(x) = x^{3/5}(4 - x)$.

SOLUTION The Product Rule gives

$$f'(x) = x^{3/5}(-1) + (4 - x)\left(\tfrac{3}{5}x^{-2/5}\right) = -x^{3/5} + \frac{3(4 - x)}{5x^{2/5}}$$

$$= \frac{-5x + 3(4 - x)}{5x^{2/5}} = \frac{12 - 8x}{5x^{2/5}}$$

[The same result could be obtained by first writing $f(x) = 4x^{3/5} - x^{8/5}$.] Therefore $f'(x) = 0$ if $12 - 8x = 0$, that is, $x = \frac{3}{2}$, and $f'(x)$ does not exist when $x = 0$. Thus the critical numbers are $\frac{3}{2}$ and 0. □

In terms of critical numbers, Fermat's Theorem can be rephrased as follows (compare Definition 6 with Theorem 4):

> **7** If f has a local maximum or minimum at c, then c is a critical number of f.

To find an absolute maximum or minimum of a continuous function on a closed interval, we note that either it is local [in which case it occurs at a critical number by (7)] or it occurs at an endpoint of the interval. Thus the following three-step procedure always works.

> **THE CLOSED INTERVAL METHOD** To find the *absolute* maximum and minimum values of a continuous function f on a closed interval $[a, b]$:
>
> **1.** Find the values of f at the critical numbers of f in (a, b).
>
> **2.** Find the values of f at the endpoints of the interval.
>
> **3.** The largest of the values from Steps 1 and 2 is the absolute maximum value; the smallest of these values is the absolute minimum value.

V EXAMPLE 8 Find the absolute maximum and minimum values of the function

$$f(x) = x^3 - 3x^2 + 1 \qquad -\tfrac{1}{2} \leqslant x \leqslant 4$$

SOLUTION Since f is continuous on $\left[-\frac{1}{2}, 4\right]$, we can use the Closed Interval Method:

$$f(x) = x^3 - 3x^2 + 1$$

$$f'(x) = 3x^2 - 6x = 3x(x - 2)$$

Since $f'(x)$ exists for all x, the only critical numbers of f occur when $f'(x) = 0$, that is, $x = 0$ or $x = 2$. Notice that each of these critical numbers lies in the interval $\left(-\frac{1}{2}, 4\right)$. The values of f at these critical numbers are

$$f(0) = 1 \qquad f(2) = -3$$

The values of f at the endpoints of the interval are

$$f\left(-\tfrac{1}{2}\right) = \tfrac{1}{8} \qquad f(4) = 17$$

Comparing these four numbers, we see that the absolute maximum value is $f(4) = 17$ and the absolute minimum value is $f(2) = -3$.

Note that in this example the absolute maximum occurs at an endpoint, whereas the absolute minimum occurs at a critical number. The graph of f is sketched in Figure 12. □

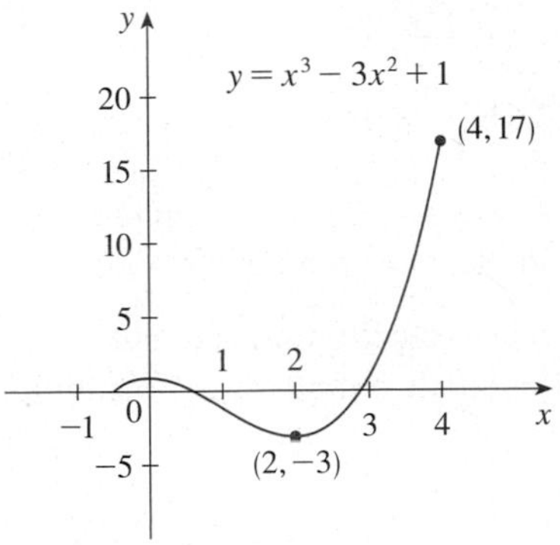

FIGURE 12

If you have a graphing calculator or a computer with graphing software, it is possible to estimate maximum and minimum values very easily. But, as the next example shows, calculus is needed to find the *exact* values.

EXAMPLE 9

(a) Use a graphing device to estimate the absolute minimum and maximum values of the function $f(x) = x - 2\sin x$, $0 \leqslant x \leqslant 2\pi$.

(b) Use calculus to find the exact minimum and maximum values.

SOLUTION

(a) Figure 13 shows a graph of f in the viewing rectangle $[0, 2\pi]$ by $[-1, 8]$. By moving the cursor close to the maximum point, we see that the y-coordinates don't change very much in the vicinity of the maximum. The absolute maximum value is about 6.97 and it occurs when $x \approx 5.2$. Similarly, by moving the cursor close to the minimum point, we see that the absolute minimum value is about -0.68 and it occurs when $x \approx 1.0$. It is

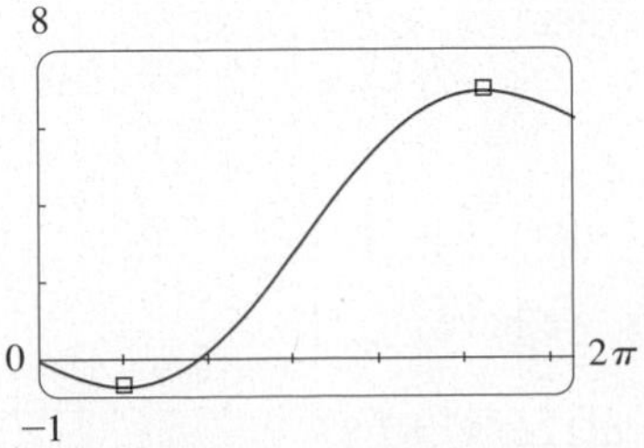

FIGURE 13

possible to get more accurate estimates by zooming in toward the maximum and minimum points, but instead let's use calculus.

(b) The function $f(x) = x - 2\sin x$ is continuous on $[0, 2\pi]$. Since $f'(x) = 1 - 2\cos x$, we have $f'(x) = 0$ when $\cos x = \frac{1}{2}$ and this occurs when $x = \pi/3$ or $5\pi/3$. The values of f at these critical points are

$$f(\pi/3) = \frac{\pi}{3} - 2\sin\frac{\pi}{3} = \frac{\pi}{3} - \sqrt{3} \approx -0.684853$$

and $$f(5\pi/3) = \frac{5\pi}{3} - 2\sin\frac{5\pi}{3} = \frac{5\pi}{3} + \sqrt{3} \approx 6.968039$$

The values of f at the endpoints are

$$f(0) = 0 \qquad \text{and} \qquad f(2\pi) = 2\pi \approx 6.28$$

Comparing these four numbers and using the Closed Interval Method, we see that the absolute minimum value is $f(\pi/3) = \pi/3 - \sqrt{3}$ and the absolute maximum value is $f(5\pi/3) = 5\pi/3 + \sqrt{3}$. The values from part (a) serve as a check on our work. □

NASA

EXAMPLE 10 The Hubble Space Telescope was deployed on April 24, 1990, by the space shuttle *Discovery*. A model for the velocity of the shuttle during this mission, from liftoff at $t = 0$ until the solid rocket boosters were jettisoned at $t = 126$ s, is given by

$$v(t) = 0.001302t^3 - 0.09029t^2 + 23.61t - 3.083$$

(in feet per second). Using this model, estimate the absolute maximum and minimum values of the *acceleration* of the shuttle between liftoff and the jettisoning of the boosters.

SOLUTION We are asked for the extreme values not of the given velocity function, but rather of the acceleration function. So we first need to differentiate to find the acceleration:

$$a(t) = v'(t) = \frac{d}{dt}(0.001302t^3 - 0.09029t^2 + 23.61t - 3.083)$$
$$= 0.003906t^2 - 0.18058t + 23.61$$

We now apply the Closed Interval Method to the continuous function a on the interval $0 \leqslant t \leqslant 126$. Its derivative is

$$a'(t) = 0.007812t - 0.18058$$

The only critical number occurs when $a'(t) = 0$:

$$t_1 = \frac{0.18058}{0.007812} \approx 23.12$$

Evaluating $a(t)$ at the critical number and at the endpoints, we have

$$a(0) = 23.61 \qquad a(t_1) \approx 21.52 \qquad a(126) \approx 62.87$$

So the maximum acceleration is about 62.87 ft/s^2 and the minimum acceleration is about 21.52 ft/s^2. □

4.1 EXERCISES

1. Explain the difference between an absolute minimum and a local minimum.

2. Suppose f is a continuous function defined on a closed interval $[a, b]$.
(a) What theorem guarantees the existence of an absolute maximum value and an absolute minimum value for f?
(b) What steps would you take to find those maximum and minimum values?

3–4 For each of the numbers a, b, c, d, r, and s, state whether the function whose graph is shown has an absolute maximum or minimum, a local maximum or minimum, or neither a maximum nor a minimum.

3.

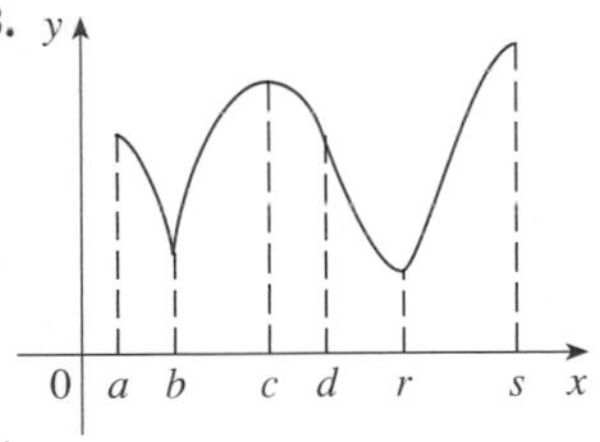

4.

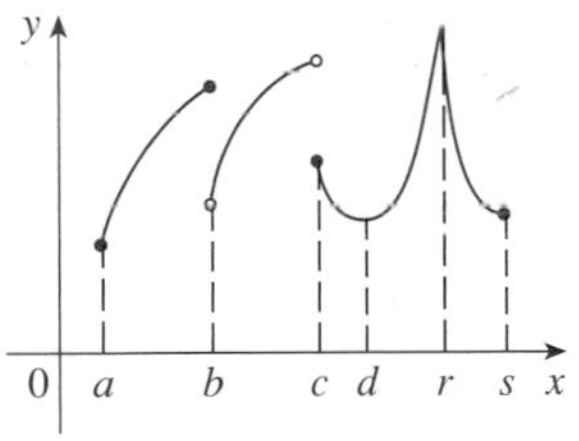

5–6 Use the graph to state the absolute and local maximum and minimum values of the function.

5.

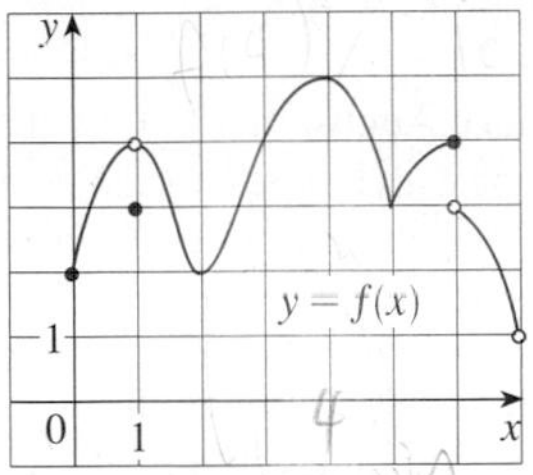

6.

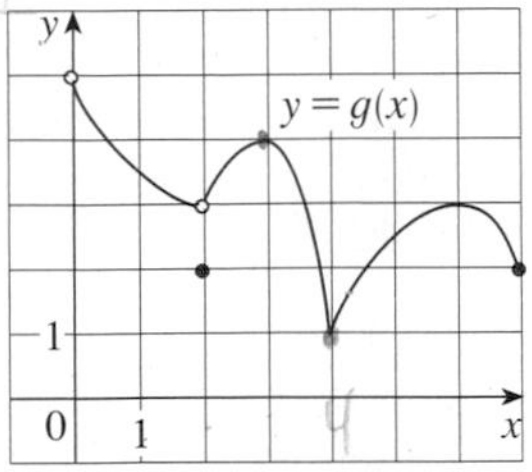

7–10 Sketch the graph of a function f that is continuous on $[1, 5]$ and has the given properties.

7. Absolute minimum at 2, absolute maximum at 3, local minimum at 4

8. Absolute minimum at 1, absolute maximum at 5, local maximum at 2, local minimum at 4

9. Absolute maximum at 5, absolute minimum at 2, local maximum at 3, local minima at 2 and 4

10. f has no local maximum or minimum, but 2 and 4 are critical numbers

11. (a) Sketch the graph of a function that has a local maximum at 2 and is differentiable at 2.
(b) Sketch the graph of a function that has a local maximum at 2 and is continuous but not differentiable at 2.
(c) Sketch the graph of a function that has a local maximum at 2 and is not continuous at 2.

12. (a) Sketch the graph of a function on $[-1, 2]$ that has an absolute maximum but no local maximum.
(b) Sketch the graph of a function on $[-1, 2]$ that has a local maximum but no absolute maximum.

13. (a) Sketch the graph of a function on $[-1, 2]$ that has an absolute maximum but no absolute minimum.
(b) Sketch the graph of a function on $[-1, 2]$ that is discontinuous but has both an absolute maximum and an absolute minimum.

14. (a) Sketch the graph of a function that has two local maxima, one local minimum, and no absolute minimum.
(b) Sketch the graph of a function that has three local minima, two local maxima, and seven critical numbers.

15–28 Sketch the graph of f by hand and use your sketch to find the absolute and local maximum and minimum values of f. (Use the graphs and transformations of Sections 1.2 and 1.3.)

15. $f(x) = 8 - 3x, \quad x \geqslant 1$

16. $f(x) = 3 - 2x, \quad x \leqslant 5$

17. $f(x) = x^2, \quad 0 < x < 2$

18. $f(x) = x^2, \quad 0 < x \leqslant 2$

19. $f(x) = x^2, \quad 0 \leqslant x < 2$

20. $f(x) = x^2, \quad 0 \leqslant x \leqslant 2$

21. $f(x) = x^2, \quad -3 \leqslant x \leqslant 2$

22. $f(x) = 1 + (x + 1)^2, \quad -2 \leqslant x < 5$

23. $f(x) = \ln x, \quad 0 < x \leqslant 2$

24. $f(t) = \cos t, \quad -3\pi/2 \leqslant t \leqslant 3\pi/2$

25. $f(x) = 1 - \sqrt{x}$

26. $f(x) = e^x$

27. $f(x) = \begin{cases} 1 - x & \text{if } 0 \leqslant x < 2 \\ 2x - 4 & \text{if } 2 \leqslant x \leqslant 3 \end{cases}$

28. $f(x) = \begin{cases} 4 - x^2 & \text{if } -2 \leqslant x < 0 \\ 2x - 1 & \text{if } 0 \leqslant x \leqslant 2 \end{cases}$

29–44 Find the critical numbers of the function.

29. $f(x) = 5x^2 + 4x$

30. $f(x) = x^3 + x^2 - x$

31. $f(x) = x^3 + 3x^2 - 24x$

32. $f(x) = x^3 + x^2 + x$

33. $s(t) = 3t^4 + 4t^3 - 6t^2$

34. $g(t) = |3t - 4|$

35. $g(y) = \dfrac{y - 1}{y^2 - y + 1}$

36. $h(p) = \dfrac{p - 1}{p^2 + 4}$

37. $h(t) = t^{3/4} - 2t^{1/4}$

38. $g(x) = \sqrt{1 - x^2}$

39. $F(x) = x^{4/5}(x - 4)^2$

40. $g(x) = x^{1/3} - x^{-2/3}$

41. $f(\theta) = 2\cos\theta + \sin^2\theta$

42. $g(\theta) = 4\theta - \tan\theta$

43. $f(x) = x^2e^{-3x}$

44. $f(x) = x^{-2}\ln x$

45–46 A formula for the *derivative* of a function f is given. How many critical numbers does f have?

45. $f'(x) = 5e^{-0.1|x|}\sin x - 1$

46. $f'(x) = \dfrac{100\cos^2 x}{10 + x^2} - 1$

47–62 Find the absolute maximum and absolute minimum values of f on the given interval.

47. $f(x) = 3x^2 - 12x + 5,\quad [0, 3]$

48. $f(x) = x^3 - 3x + 1,\quad [0, 3]$

49. $f(x) = 2x^3 - 3x^2 - 12x + 1,\quad [-2, 3]$

50. $f(x) = x^3 - 6x^2 + 9x + 2,\quad [-1, 4]$

51. $f(x) = x^4 - 2x^2 + 3,\quad [-2, 3]$

52. $f(x) = (x^2 - 1)^3,\quad [-1, 2]$

53. $f(x) = \dfrac{x}{x^2 + 1},\quad [0, 2]$

54. $f(x) = \dfrac{x^2 - 4}{x^2 + 4},\quad [-4, 4]$

55. $f(t) = t\sqrt{4 - t^2},\quad [-1, 2]$

56. $f(t) = \sqrt[3]{t}\,(8 - t),\quad [0, 8]$

57. $f(t) = 2\cos t + \sin 2t,\quad [0, \pi/2]$

58. $f(t) = t + \cot(t/2),\quad [\pi/4, 7\pi/4]$

59. $f(x) = xe^{-x^2/8},\quad [-1, 4]$

60. $f(x) = x - \ln x,\quad [\tfrac{1}{2}, 2]$

61. $f(x) = \ln(x^2 + x + 1),\quad [-1, 1]$

62. $f(x) = e^{-x} - e^{-2x},\quad [0, 1]$

63. If a and b are positive numbers, find the maximum value of $f(x) = x^a(1 - x)^b$, $0 \leqslant x \leqslant 1$.

64. Use a graph to estimate the critical numbers of $f(x) = |x^3 - 3x^2 + 2|$ correct to one decimal place.

65–68

(a) Use a graph to estimate the absolute maximum and minimum values of the function to two decimal places.

(b) Use calculus to find the exact maximum and minimum values.

65. $f(x) = x^5 - x^3 + 2,\quad -1 \leqslant x \leqslant 1$

66. $f(x) = e^{x^3 - x},\quad -1 \leqslant x \leqslant 0$

67. $f(x) = x\sqrt{x - x^2}$

68. $f(x) = x - 2\cos x,\quad -2 \leqslant x \leqslant 0$

69. Between 0°C and 30°C, the volume V (in cubic centimeters) of 1 kg of water at a temperature T is given approximately by the formula

$$V = 999.87 - 0.06426T + 0.0085043T^2 - 0.0000679T^3$$

Find the temperature at which water has its maximum density.

70. An object with weight W is dragged along a horizontal plane by a force acting along a rope attached to the object. If the rope makes an angle θ with the plane, then the magnitude of the force is

$$F = \frac{\mu W}{\mu\sin\theta + \cos\theta}$$

where μ is a positive constant called the *coefficient of friction* and where $0 \leqslant \theta \leqslant \pi/2$. Show that F is minimized when $\tan\theta = \mu$.

71. A model for the US average price of a pound of white sugar from 1993 to 2003 is given by the function

$$S(t) = -0.00003237t^5 + 0.0009037t^4 - 0.008956t^3 + 0.03629t^2 - 0.04458t + 0.4074$$

where t is measured in years since August of 1993. Estimate the times when sugar was cheapest and most expensive during the period 1993–2003.

72. On May 7, 1992, the space shuttle *Endeavour* was launched on mission STS-49, the purpose of which was to install a new perigee kick motor in an Intelsat communications satellite. The table gives the velocity data for the shuttle between liftoff and the jettisoning of the solid rocket boosters.

Event	Time (s)	Velocity (ft/s)
Launch	0	0
Begin roll maneuver	10	185
End roll maneuver	15	319
Throttle to 89%	20	447
Throttle to 67%	32	742
Throttle to 104%	59	1325
Maximum dynamic pressure	62	1445
Solid rocket booster separation	125	4151

(a) Use a graphing calculator or computer to find the cubic polynomial that best models the velocity of the shuttle for the time interval $t \in [0, 125]$. Then graph this polynomial.

(b) Find a model for the acceleration of the shuttle and use it to estimate the maximum and minimum values of the acceleration during the first 125 seconds.

73. When a foreign object lodged in the trachea (windpipe) forces a person to cough, the diaphragm thrusts upward causing an increase in pressure in the lungs. This is accompanied by a contraction of the trachea, making a narrower channel for the expelled air to flow through. For a given amount of air to escape in a fixed time, it must move faster through the narrower channel than the wider one. The greater the velocity of the airstream, the greater the force on the foreign object. X rays show that the radius of the circular tracheal tube contracts to about two-thirds of its normal radius during a cough. According to a mathematical model of coughing, the velocity v of the airstream is related to the radius r of the trachea by the equation

$$v(r) = k(r_0 - r)r^2 \qquad \tfrac{1}{2}r_0 \leqslant r \leqslant r_0$$

where k is a constant and r_0 is the normal radius of the trachea. The restriction on r is due to the fact that the tracheal wall stiffens under pressure and a contraction greater than $\frac{1}{2}r_0$ is prevented (otherwise the person would suffocate).

(a) Determine the value of r in the interval $\left[\frac{1}{2}r_0, r_0\right]$ at which v has an absolute maximum. How does this compare with experimental evidence?

(b) What is the absolute maximum value of v on the interval?

(c) Sketch the graph of v on the interval $[0, r_0]$.

74. Show that 5 is a critical number of the function

$$g(x) = 2 + (x - 5)^3$$

but g does not have a local extreme value at 5.

75. Prove that the function

$$f(x) = x^{101} + x^{51} + x + 1$$

has neither a local maximum nor a local minimum.

76. If f has a minimum value at c, show that the function $g(x) = -f(x)$ has a maximum value at c.

77. Prove Fermat's Theorem for the case in which f has a local minimum at c.

78. A cubic function is a polynomial of degree 3; that is, it has the form $f(x) = ax^3 + bx^2 + cx + d$, where $a \neq 0$.

(a) Show that a cubic function can have two, one, or no critical number(s). Give examples and sketches to illustrate the three possibilities.

(b) How many local extreme values can a cubic function have?

APPLIED PROJECT

THE CALCULUS OF RAINBOWS

Rainbows are created when raindrops scatter sunlight. They have fascinated mankind since ancient times and have inspired attempts at scientific explanation since the time of Aristotle. In this project we use the ideas of Descartes and Newton to explain the shape, location, and colors of rainbows.

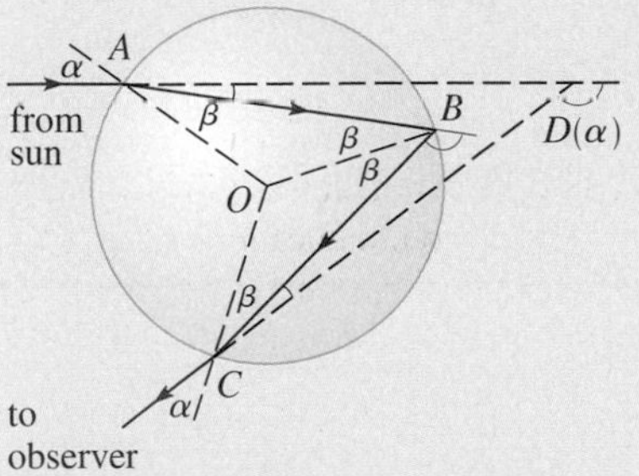

Formation of the primary rainbow

1. The figure shows a ray of sunlight entering a spherical raindrop at A. Some of the light is reflected, but the line AB shows the path of the part that enters the drop. Notice that the light is refracted toward the normal line AO and in fact Snell's Law says that $\sin\alpha = k\sin\beta$, where α is the angle of incidence, β is the angle of refraction, and $k \approx \frac{4}{3}$ is the index of refraction for water. At B some of the light passes through the drop and is refracted into the air, but the line BC shows the part that is reflected. (The angle of incidence equals the angle of reflection.) When the ray reaches C, part of it is reflected, but for the time being we are more interested in the part that leaves the raindrop at C. (Notice that it is refracted away from the normal line.) The *angle of deviation* $D(\alpha)$ is the amount of clockwise rotation that the ray has undergone during this three-stage process. Thus

$$D(\alpha) = (\alpha - \beta) + (\pi - 2\beta) + (\alpha - \beta) = \pi + 2\alpha - 4\beta$$

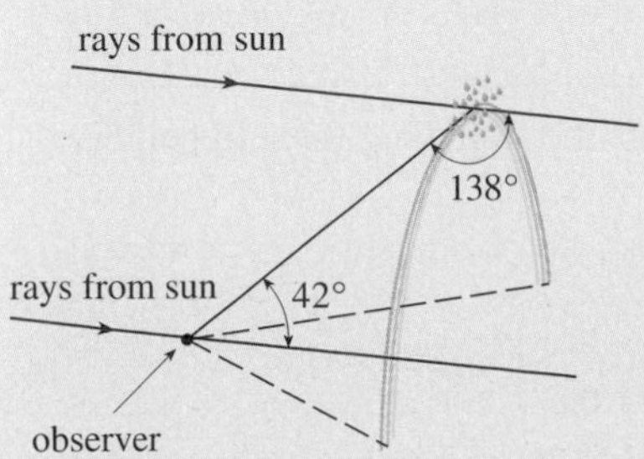

Show that the minimum value of the deviation is $D(\alpha) \approx 138°$ and occurs when $\alpha \approx 59.4°$.

The significance of the minimum deviation is that when $\alpha \approx 59.4°$ we have $D'(\alpha) \approx 0$, so $\Delta D/\Delta\alpha \approx 0$. This means that many rays with $\alpha \approx 59.4°$ become deviated by approximately the same amount. It is the *concentration* of rays coming from near the direction of minimum deviation that creates the brightness of the primary rainbow. The figure at the left shows that the angle of elevation from the observer up to the highest point on the rainbow is $180° - 138° = 42°$. (This angle is called the *rainbow angle*.)

2. Problem 1 explains the location of the primary rainbow, but how do we explain the colors? Sunlight comprises a range of wavelengths, from the red range through orange, yellow, green, blue, indigo, and violet. As Newton discovered in his prism experiments of 1666, the index of refraction is different for each color. (The effect is called *dispersion.*) For red light the refractive index is $k \approx 1.3318$ whereas for violet light it is $k \approx 1.3435$. By repeating the calculation of Problem 1 for these values of k, show that the rainbow angle is about $42.3°$ for the red bow and $40.6°$ for the violet bow. So the rainbow really consists of seven individual bows corresponding to the seven colors.

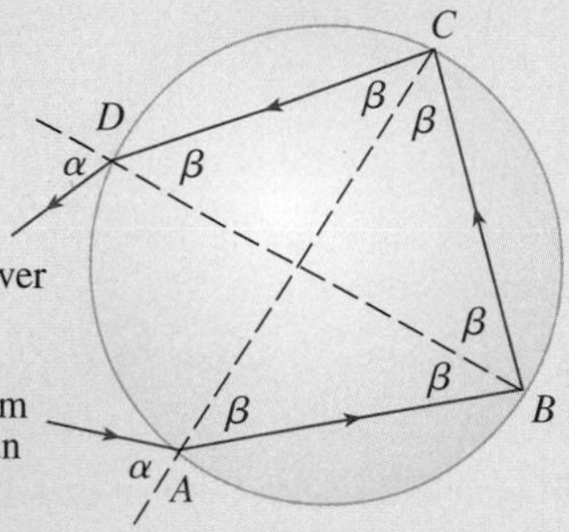

Formation of the secondary rainbow

3. Perhaps you have seen a fainter secondary rainbow above the primary bow. That results from the part of a ray that enters a raindrop and is refracted at A, reflected twice (at B and C), and refracted as it leaves the drop at D (see the figure). This time the deviation angle $D(\alpha)$ is the total amount of counterclockwise rotation that the ray undergoes in this four-stage process. Show that

$$D(\alpha) = 2\alpha - 6\beta + 2\pi$$

and $D(\alpha)$ has a minimum value when

$$\cos\alpha = \sqrt{\frac{k^2 - 1}{8}}$$

Taking $k = \frac{4}{3}$, show that the minimum deviation is about $129°$ and so the rainbow angle for the secondary rainbow is about $51°$, as shown in the figure.

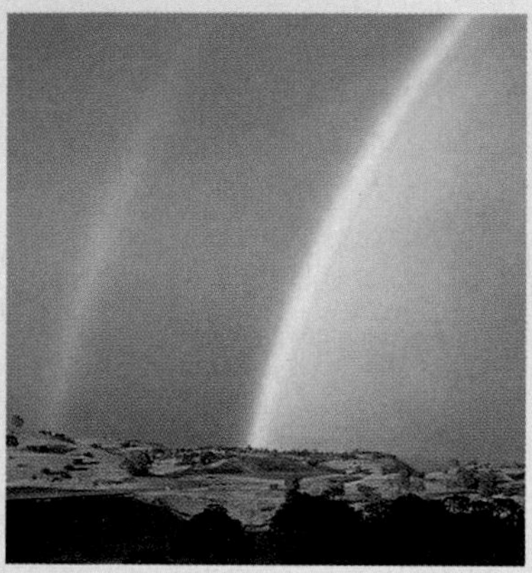

© C. Donald Ahrens

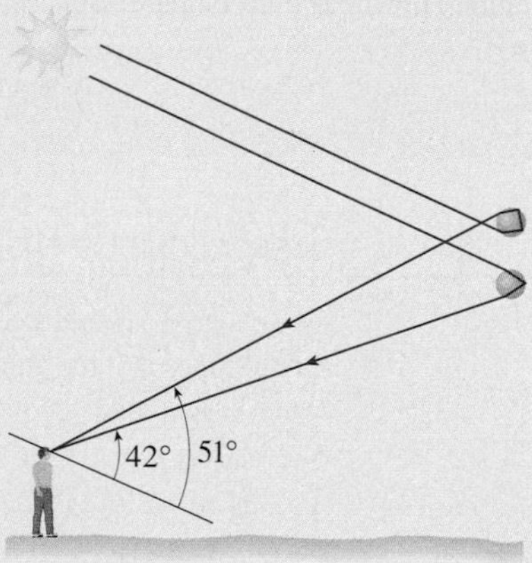

4. Show that the colors in the secondary rainbow appear in the opposite order from those in the primary rainbow.

4.2 THE MEAN VALUE THEOREM

We will see that many of the results of this chapter depend on one central fact, which is called the Mean Value Theorem. But to arrive at the Mean Value Theorem we first need the following result.

■ Rolle's Theorem was first published in 1691 by the French mathematician Michel Rolle (1652–1719) in a book entitled *Méthode pour résoudre les égalitéz*. He was a vocal critic of the methods of his day and attacked calculus as being a "collection of ingenious fallacies." Later, however, he became convinced of the essential correctness of the methods of calculus.

ROLLE'S THEOREM Let f be a function that satisfies the following three hypotheses:

1. f is continuous on the closed interval $[a, b]$.

2. f is differentiable on the open interval (a, b).

3. $f(a) = f(b)$

Then there is a number c in (a, b) such that $f'(c) = 0$.

Before giving the proof let's take a look at the graphs of some typical functions that satisfy the three hypotheses. Figure 1 shows the graphs of four such functions. In each case it appears that there is at least one point $(c, f(c))$ on the graph where the tangent is horizontal and therefore $f'(c) = 0$. Thus Rolle's Theorem is plausible.

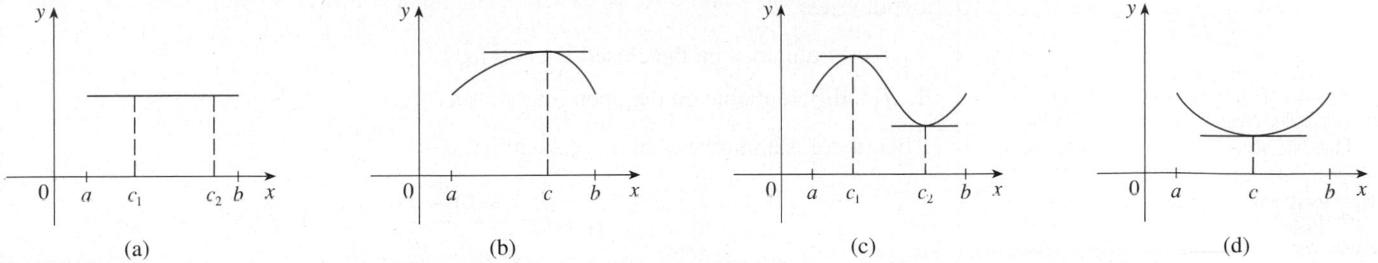

FIGURE 1

■ Take cases

PROOF There are three cases:

CASE I ■ $f(x) = k$, a constant
Then $f'(x) = 0$, so the number c can be taken to be *any* number in (a, b).

CASE II ■ $f(x) > f(a)$ for some x in (a, b) [as in Figure 1(b) or (c)]
By the Extreme Value Theorem (which we can apply by hypothesis 1), f has a maximum value somewhere in $[a, b]$. Since $f(a) = f(b)$, it must attain this maximum value at a number c in the open interval (a, b). Then f has a *local* maximum at c and, by hypothesis 2, f is differentiable at c. Therefore $f'(c) = 0$ by Fermat's Theorem.

CASE III ■ $f(x) < f(a)$ for some x in (a, b) [as in Figure 1(c) or (d)]
By the Extreme Value Theorem, f has a minimum value in $[a, b]$ and, since $f(a) = f(b)$, it attains this minimum value at a number c in (a, b). Again $f'(c) = 0$ by Fermat's Theorem. □

EXAMPLE 1 Let's apply Rolle's Theorem to the position function $s = f(t)$ of a moving object. If the object is in the same place at two different instants $t = a$ and $t = b$, then $f(a) = f(b)$. Rolle's Theorem says that there is some instant of time $t = c$ between a and b when $f'(c) = 0$; that is, the velocity is 0. (In particular, you can see that this is true when a ball is thrown directly upward.) □

EXAMPLE 2 Prove that the equation $x^3 + x - 1 = 0$ has exactly one real root.

■ Figure 2 shows a graph of the function $f(x) = x^3 + x - 1$ discussed in Example 2. Rolle's Theorem shows that, no matter how much we enlarge the viewing rectangle, we can never find a second x-intercept.

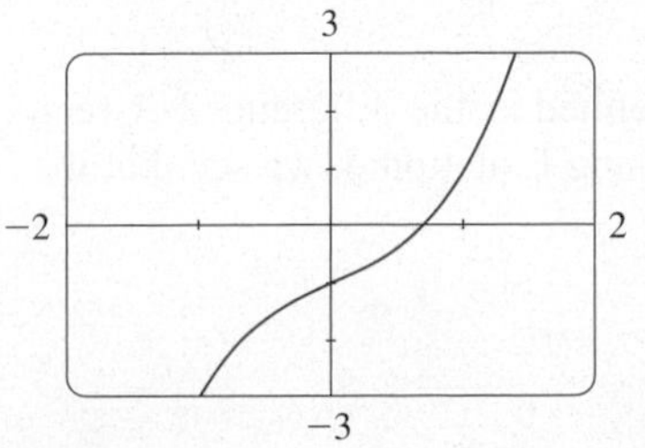

FIGURE 2

SOLUTION First we use the Intermediate Value Theorem (2.5.10) to show that a root exists. Let $f(x) = x^3 + x - 1$. Then $f(0) = -1 < 0$ and $f(1) = 1 > 0$. Since f is a polynomial, it is continuous, so the Intermediate Value Theorem states that there is a number c between 0 and 1 such that $f(c) = 0$. Thus the given equation has a root.

To show that the equation has no other real root, we use Rolle's Theorem and argue by contradiction. Suppose that it had two roots a and b. Then $f(a) = 0 = f(b)$ and, since f is a polynomial, it is differentiable on (a, b) and continuous on $[a, b]$. Thus, by Rolle's Theorem, there is a number c between a and b such that $f'(c) = 0$. But

$$f'(x) = 3x^2 + 1 \geqslant 1 \qquad \text{for all } x$$

(since $x^2 \geqslant 0$) so $f'(x)$ can never be 0. This gives a contradiction. Therefore the equation can't have two real roots. □

Our main use of Rolle's Theorem is in proving the following important theorem, which was first stated by another French mathematician, Joseph-Louis Lagrange.

THE MEAN VALUE THEOREM Let f be a function that satisfies the following hypotheses:

1. f is continuous on the closed interval $[a, b]$.

2. f is differentiable on the open interval (a, b).

Then there is a number c in (a, b) such that

1
$$f'(c) = \frac{f(b) - f(a)}{b - a}$$

or, equivalently,

2
$$f(b) - f(a) = f'(c)(b - a)$$

■ The Mean Value Theorem is an example of what is called an existence theorem. Like the Intermediate Value Theorem, the Extreme Value Theorem, and Rolle's Theorem, it guarantees that there *exists* a number with a certain property, but it doesn't tell us how to find the number.

Before proving this theorem, we can see that it is reasonable by interpreting it geometrically. Figures 3 and 4 show the points $A(a, f(a))$ and $B(b, f(b))$ on the graphs of two differentiable functions. The slope of the secant line AB is

3
$$m_{AB} = \frac{f(b) - f(a)}{b - a}$$

which is the same expression as on the right side of Equation 1. Since $f'(c)$ is the slope of the tangent line at the point $(c, f(c))$, the Mean Value Theorem, in the form given by Equation 1, says that there is at least one point $P(c, f(c))$ on the graph where the slope of the tangent line is the same as the slope of the secant line AB. In other words, there is a point P where the tangent line is parallel to the secant line AB.

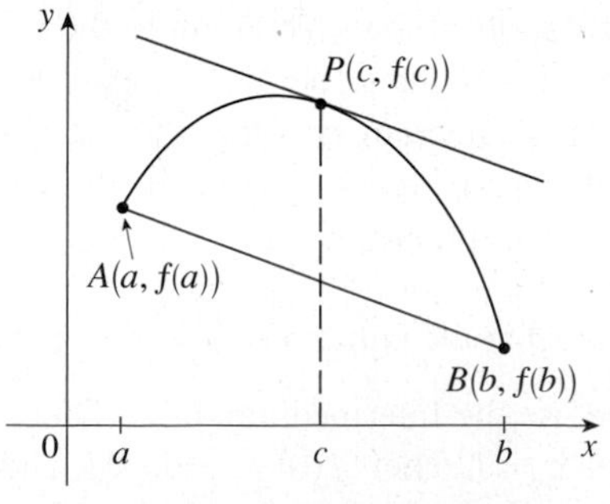

FIGURE 3

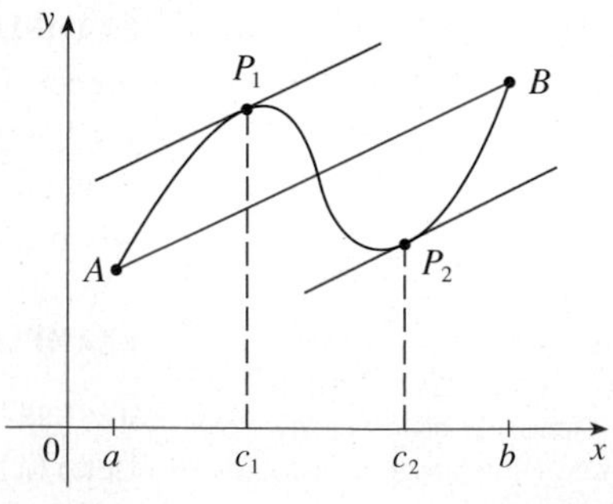

FIGURE 4

PROOF We apply Rolle's Theorem to a new function h defined as the difference between f and the function whose graph is the secant line AB. Using Equation 3, we see that the equation of the line AB can be written as

$$y - f(a) = \frac{f(b) - f(a)}{b - a}(x - a)$$

or as

$$y = f(a) + \frac{f(b) - f(a)}{b - a}(x - a)$$

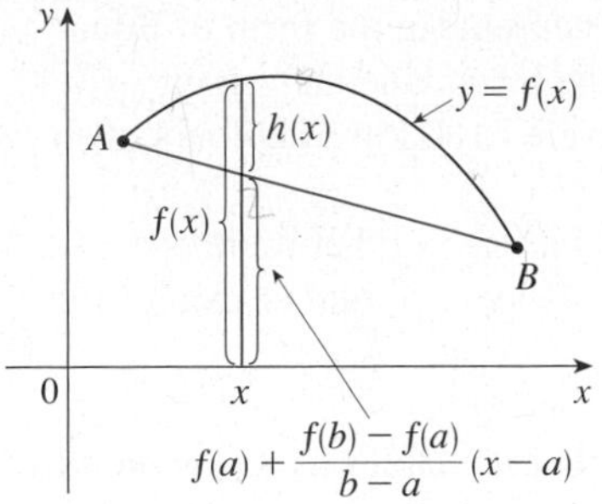

FIGURE 5

So, as shown in Figure 5,

$$\boxed{4} \qquad h(x) = f(x) - f(a) - \frac{f(b) - f(a)}{b - a}(x - a)$$

First we must verify that h satisfies the three hypotheses of Rolle's Theorem.

1. The function h is continuous on $[a, b]$ because it is the sum of f and a first-degree polynomial, both of which are continuous.
2. The function h is differentiable on (a, b) because both f and the first-degree polynomial are differentiable. In fact, we can compute h' directly from Equation 4:

$$h'(x) = f'(x) - \frac{f(b) - f(a)}{b - a}$$

(Note that $f(a)$ and $[f(b) - f(a)]/(b - a)$ are constants.)

3.
$$h(a) = f(a) - f(a) - \frac{f(b) - f(a)}{b - a}(a - a) = 0$$

$$h(b) = f(b) - f(a) - \frac{f(b) - f(a)}{b - a}(b - a)$$

$$= f(b) - f(a) - [f(b) - f(a)] = 0$$

Therefore $h(a) = h(b)$.

LAGRANGE AND THE MEAN VALUE THEOREM

The Mean Value Theorem was first formulated by Joseph-Louis Lagrange (1736–1813), born in Italy of a French father and an Italian mother. He was a child prodigy and became a professor in Turin at the tender age of 19. Lagrange made great contributions to number theory, theory of functions, theory of equations, and analytical and celestial mechanics. In particular, he applied calculus to the analysis of the stability of the solar system. At the invitation of Frederick the Great, he succeeded Euler at the Berlin Academy and, when Frederick died, Lagrange accepted King Louis XVI's invitation to Paris, where he was given apartments in the Louvre and became a professor at the Ecole Polytechnique. Despite all the trappings of luxury and fame, he was a kind and quiet man, living only for science.

Since h satisfies the hypotheses of Rolle's Theorem, that theorem says there is a number c in (a, b) such that $h'(c) = 0$. Therefore

$$0 = h'(c) = f'(c) - \frac{f(b) - f(a)}{b - a}$$

and so

$$f'(c) = \frac{f(b) - f(a)}{b - a}$$

□

V EXAMPLE 3 To illustrate the Mean Value Theorem with a specific function, let's consider $f(x) = x^3 - x$, $a = 0$, $b = 2$. Since f is a polynomial, it is continuous and differentiable for all x, so it is certainly continuous on $[0, 2]$ and differentiable on $(0, 2)$. Therefore, by the Mean Value Theorem, there is a number c in $(0, 2)$ such that

$$f(2) - f(0) = f'(c)(2 - 0)$$

Now $f(2) = 6$, $f(0) = 0$, and $f'(x) = 3x^2 - 1$, so this equation becomes

$$6 = (3c^2 - 1)2 = 6c^2 - 2$$

which gives $c^2 = \frac{4}{3}$, that is, $c = \pm 2/\sqrt{3}$. But c must lie in $(0, 2)$, so $c = 2/\sqrt{3}$. Figure 6 illustrates this calculation: The tangent line at this value of c is parallel to the secant line OB. □

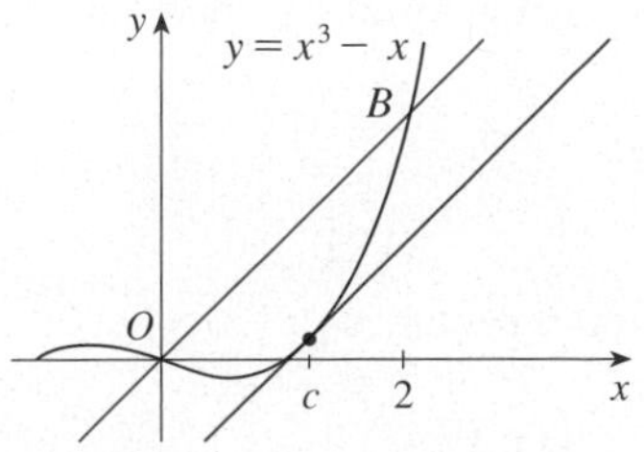

FIGURE 6

V EXAMPLE 4 If an object moves in a straight line with position function $s = f(t)$, then the average velocity between $t = a$ and $t = b$ is

$$\frac{f(b) - f(a)}{b - a}$$

and the velocity at $t = c$ is $f'(c)$. Thus the Mean Value Theorem (in the form of Equation 1) tells us that at some time $t = c$ between a and b the instantaneous velocity $f'(c)$ is equal to that average velocity. For instance, if a car traveled 180 km in 2 hours, then the speedometer must have read 90 km/h at least once.

In general, the Mean Value Theorem can be interpreted as saying that there is a number at which the instantaneous rate of change is equal to the average rate of change over an interval. □

The main significance of the Mean Value Theorem is that it enables us to obtain information about a function from information about its derivative. The next example provides an instance of this principle.

V EXAMPLE 5 Suppose that $f(0) = -3$ and $f'(x) \leqslant 5$ for all values of x. How large can $f(2)$ possibly be?

SOLUTION We are given that f is differentiable (and therefore continuous) everywhere. In particular, we can apply the Mean Value Theorem on the interval $[0, 2]$. There exists a number c such that

$$f(2) - f(0) = f'(c)(2 - 0)$$

so

$$f(2) = f(0) + 2f'(c) = -3 + 2f'(c)$$

We are given that $f'(x) \leqslant 5$ for all x, so in particular we know that $f'(c) \leqslant 5$. Multiplying both sides of this inequality by 2, we have $2f'(c) \leqslant 10$, so

$$f(2) = -3 + 2f'(c) \leqslant -3 + 10 = 7$$

The largest possible value for $f(2)$ is 7. □

The Mean Value Theorem can be used to establish some of the basic facts of differential calculus. One of these basic facts is the following theorem. Others will be found in the following sections.

> **5 THEOREM** If $f'(x) = 0$ for all x in an interval (a, b), then f is constant on (a, b).

PROOF Let x_1 and x_2 be any two numbers in (a, b) with $x_1 < x_2$. Since f is differentiable on (a, b), it must be differentiable on (x_1, x_2) and continuous on $[x_1, x_2]$. By applying the Mean Value Theorem to f on the interval $[x_1, x_2]$, we get a number c such that $x_1 < c < x_2$ and

$$f(x_2) - f(x_1) = f'(c)(x_2 - x_1) \tag{6}$$

Since $f'(x) = 0$ for all x, we have $f'(c) = 0$, and so Equation 6 becomes

$$f(x_2) - f(x_1) = 0 \qquad \text{or} \qquad f(x_2) = f(x_1)$$

Therefore f has the same value at *any* two numbers x_1 and x_2 in (a, b). This means that f is constant on (a, b). □

> **7 COROLLARY** If $f'(x) = g'(x)$ for all x in an interval (a, b), then $f - g$ is constant on (a, b); that is, $f(x) = g(x) + c$ where c is a constant.

PROOF Let $F(x) = f(x) - g(x)$. Then

$$F'(x) = f'(x) - g'(x) = 0$$

for all x in (a, b). Thus, by Theorem 5, F is constant; that is, $f - g$ is constant. □

NOTE Care must be taken in applying Theorem 5. Let

$$f(x) = \frac{x}{|x|} = \begin{cases} 1 & \text{if } x > 0 \\ -1 & \text{if } x < 0 \end{cases}$$

The domain of f is $D = \{x \mid x \neq 0\}$ and $f'(x) = 0$ for all x in D. But f is obviously not a constant function. This does not contradict Theorem 5 because D is not an interval. Notice that f is constant on the interval $(0, \infty)$ and also on the interval $(-\infty, 0)$.

EXAMPLE 6 Prove the identity $\tan^{-1}x + \cot^{-1}x = \pi/2$.

SOLUTION Although calculus isn't needed to prove this identity, the proof using calculus is quitc simplc. If $f(x) = \tan^{-1}x + \cot^{-1}x$, then

$$f'(x) = \frac{1}{1 + x^2} - \frac{1}{1 + x^2} = 0$$

for all values of x. Therefore $f(x) = C$, a constant. To determine the value of C, we put $x = 1$ [because we can evaluate $f(1)$ exactly]. Then

$$C = f(1) = \tan^{-1}1 + \cot^{-1}1 = \frac{\pi}{4} + \frac{\pi}{4} = \frac{\pi}{2}$$

Thus $\tan^{-1}x + \cot^{-1}x = \pi/2$. □

4.2 EXERCISES

1–4 Verify that the function satisfies the three hypotheses of Rolle's Theorem on the given interval. Then find all numbers c that satisfy the conclusion of Rolle's Theorem.

1. $f(x) = 5 - 12x + 3x^2$, $[1, 3]$

2. $f(x) = x^3 - x^2 - 6x + 2$, $[0, 3]$

3. $f(x) = \sqrt{x} - \frac{1}{3}x$, $[0, 9]$

4. $f(x) = \cos 2x$, $[\pi/8, 7\pi/8]$

5. Let $f(x) = 1 - x^{2/3}$. Show that $f(-1) = f(1)$ but there is no number c in $(-1, 1)$ such that $f'(c) = 0$. Why does this not contradict Rolle's Theorem?

6. Let $f(x) = \tan x$. Show that $f(0) = f(\pi)$ but there is no number c in $(0, \pi)$ such that $f'(c) = 0$. Why does this not contradict Rolle's Theorem?

7. Use the graph of f to estimate the values of c that satisfy the conclusion of the Mean Value Theorem for the interval $[0, 8]$.

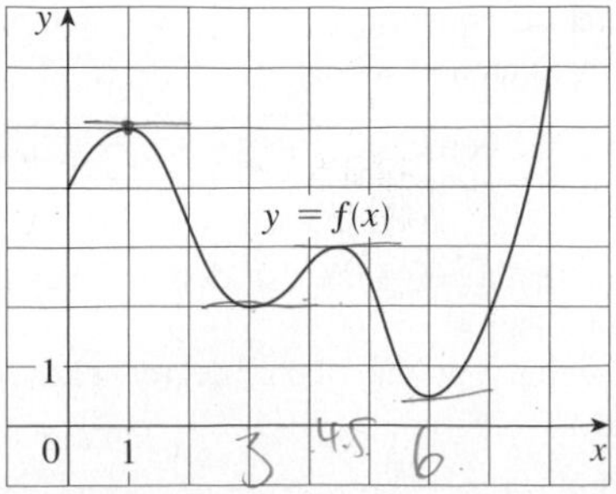

8. Use the graph of f given in Exercise 7 to estimate the values of c that satisfy the conclusion of the Mean Value Theorem for the interval $[1, 7]$.

9. (a) Graph the function $f(x) = x + 4/x$ in the viewing rectangle $[0, 10]$ by $[0, 10]$.
(b) Graph the secant line that passes through the points $(1, 5)$ and $(8, 8.5)$ on the same screen with f.
(c) Find the number c that satisfies the conclusion of the Mean Value Theorem for this function f and the interval $[1, 8]$. Then graph the tangent line at the point $(c, f(c))$ and notice that it is parallel to the secant line.

10. (a) In the viewing rectangle $[-3, 3]$ by $[-5, 5]$, graph the function $f(x) = x^3 - 2x$ and its secant line through the points $(-2, -4)$ and $(2, 4)$. Use the graph to estimate the x-coordinates of the points where the tangent line is parallel to the secant line.
(b) Find the exact values of the numbers c that satisfy the conclusion of the Mean Value Theorem for the interval $[-2, 2]$ and compare with your answers to part (a).

11–14 Verify that the function satisfies the hypotheses of the Mean Value Theorem on the given interval. Then find all numbers c that satisfy the conclusion of the Mean Value Theorem.

11. $f(x) = 3x^2 + 2x + 5, \quad [-1, 1]$

12. $f(x) = x^3 + x - 1, \quad [0, 2]$

13. $f(x) = e^{-2x}, \quad [0, 3]$

14. $f(x) = \dfrac{x}{x + 2}, \quad [1, 4]$

---

15. Let $f(x) = (x - 3)^{-2}$. Show that there is no value of c in $(1, 4)$ such that $f(4) - f(1) = f'(c)(4 - 1)$. Why does this not contradict the Mean Value Theorem?

16. Let $f(x) = 2 - |2x - 1|$. Show that there is no value of c such that $f(3) - f(0) = f'(c)(3 - 0)$. Why does this not contradict the Mean Value Theorem?

17. Show that the equation $1 + 2x + x^3 + 4x^5 = 0$ has exactly one real root.

18. Show that the equation $2x - 1 - \sin x = 0$ has exactly one real root.

19. Show that the equation $x^3 - 15x + c = 0$ has at most one root in the interval $[-2, 2]$.

20. Show that the equation $x^4 + 4x + c = 0$ has at most two real roots.

21. (a) Show that a polynomial of degree 3 has at most three real roots.
(b) Show that a polynomial of degree n has at most n real roots.

22. (a) Suppose that f is differentiable on $\mathbb{R}$ and has two roots. Show that f' has at least one root.
(b) Suppose f is twice differentiable on $\mathbb{R}$ and has three roots. Show that f'' has at least one real root.
(c) Can you generalize parts (a) and (b)?

23. If $f(1) = 10$ and $f'(x) \geqslant 2$ for $1 \leqslant x \leqslant 4$, how small can $f(4)$ possibly be?

24. Suppose that $3 \leqslant f'(x) \leqslant 5$ for all values of x. Show that $18 \leqslant f(8) - f(2) \leqslant 30$.

25. Does there exist a function f such that $f(0) = -1$, $f(2) = 4$, and $f'(x) \leqslant 2$ for all x?

26. Suppose that f and g are continuous on $[a, b]$ and differentiable on (a, b). Suppose also that $f(a) = g(a)$ and $f'(x) < g'(x)$ for $a < x < b$. Prove that $f(b) < g(b)$. [*Hint*: Apply the Mean Value Theorem to the function $h = f - g$.]

27. Show that $\sqrt{1 + x} < 1 + \frac{1}{2}x$ if $x > 0$.

28. Suppose f is an odd function and is differentiable everywhere. Prove that for every positive number b, there exists a number c in $(-b, b)$ such that $f'(c) = f(b)/b$.

29. Use the Mean Value Theorem to prove the inequality

$$|\sin a - \sin b| \leqslant |a - b| \qquad \text{for all } a \text{ and } b$$

30. If $f'(x) = c$ (c a constant) for all x, use Corollary 7 to show that $f(x) = cx + d$ for some constant d.

31. Let $f(x) = 1/x$ and

$$g(x) = \begin{cases} \dfrac{1}{x} & \text{if } x > 0 \\[2mm] 1 + \dfrac{1}{x} & \text{if } x < 0 \end{cases}$$

Show that $f'(x) = g'(x)$ for all x in their domains. Can we conclude from Corollary 7 that $f - g$ is constant?

32. Use the method of Example 6 to prove the identity

$$2 \sin^{-1}x = \cos^{-1}(1 - 2x^2) \qquad x \geqslant 0$$

33. Prove the identity

$$\arcsin \frac{x - 1}{x + 1} = 2 \arctan \sqrt{x} - \frac{\pi}{2}$$

34. At 2:00 PM a car's speedometer reads 30 mi/h. At 2:10 PM it reads 50 mi/h. Show that at some time between 2:00 and 2:10 the acceleration is exactly 120 mi/h^2.

35. Two runners start a race at the same time and finish in a tie. Prove that at some time during the race they have the same speed. [*Hint:* Consider $f(t) = g(t) - h(t)$, where g and h are the position functions of the two runners.]

36. A number a is called a **fixed point** of a function f if $f(a) = a$. Prove that if $f'(x) \neq 1$ for all real numbers x, then f has at most one fixed point.

4.3 HOW DERIVATIVES AFFECT THE SHAPE OF A GRAPH

Many of the applications of calculus depend on our ability to deduce facts about a function f from information concerning its derivatives. Because $f'(x)$ represents the slope of the curve $y = f(x)$ at the point $(x, f(x))$, it tells us the direction in which the curve proceeds at each point. So it is reasonable to expect that information about $f'(x)$ will provide us with information about $f(x)$.

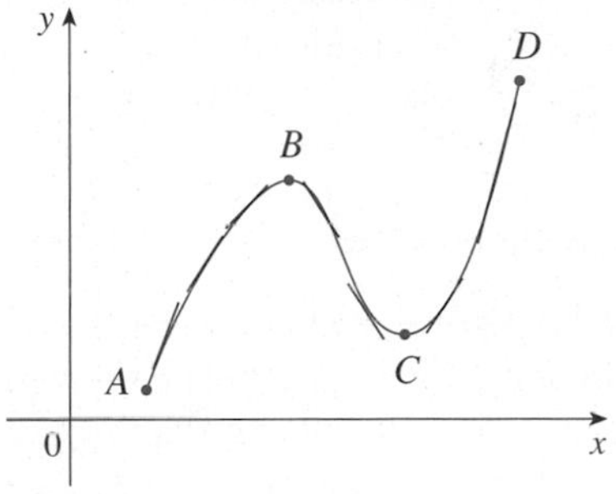

FIGURE 1

WHAT DOES f' SAY ABOUT f?

To see how the derivative of f can tell us where a function is increasing or decreasing, look at Figure 1. (Increasing functions and decreasing functions were defined in Section 1.1.) Between A and B and between C and D, the tangent lines have positive slope and so $f'(x) > 0$. Between B and C, the tangent lines have negative slope and so $f'(x) < 0$. Thus it appears that f increases when $f'(x)$ is positive and decreases when $f'(x)$ is negative. To prove that this is always the case, we use the Mean Value Theorem.

■ Let's abbreviate the name of this test to the I/D Test.

INCREASING/DECREASING TEST

(a) If $f'(x) > 0$ on an interval, then f is increasing on that interval.

(b) If $f'(x) < 0$ on an interval, then f is decreasing on that interval.

PROOF

(a) Let x_1 and x_2 be any two numbers in the interval with $x_1 < x_2$. According to the definition of an increasing function (page 20) we have to show that $f(x_1) < f(x_2)$.

Because we are given that $f'(x) > 0$, we know that f is differentiable on $[x_1, x_2]$. So, by the Mean Value Theorem there is a number c between x_1 and x_2 such that

$$f(x_2) - f(x_1) = f'(c)(x_2 - x_1) \tag{1}$$

Now $f'(c) > 0$ by assumption and $x_2 - x_1 > 0$ because $x_1 < x_2$. Thus the right side of Equation 1 is positive, and so

$$f(x_2) - f(x_1) > 0 \qquad \text{or} \qquad f(x_1) < f(x_2)$$

This shows that f is increasing.

Part (b) is proved similarly. □

V EXAMPLE 1 Find where the function $f(x) = 3x^4 - 4x^3 - 12x^2 + 5$ is increasing and where it is decreasing.

SOLUTION $$f'(x) = 12x^3 - 12x^2 - 24x = 12x(x - 2)(x + 1)$$

To use the I/D Test we have to know where $f'(x) > 0$ and where $f'(x) < 0$. This depends on the signs of the three factors of $f'(x)$, namely, $12x$, $x - 2$, and $x + 1$. We divide the real line into intervals whose endpoints are the critical numbers -1, 0, and 2 and arrange our work in a chart. A plus sign indicates that the given expression is positive, and a minus sign indicates that it is negative. The last column of the chart gives the conclusion based on the I/D Test. For instance, $f'(x) < 0$ for $0 < x < 2$, so f is decreasing on $(0, 2)$. (It would also be true to say that f is decreasing on the closed interval $[0, 2]$.)

Interval	$12x$	$x - 2$	$x + 1$	$f'(x)$	f
$x < -1$	$-$	$-$	$-$	$-$	decreasing on $(-\infty, -1)$
$-1 < x < 0$	$-$	$-$	$+$	$+$	increasing on $(-1, 0)$
$0 < x < 2$	$+$	$-$	$+$	$-$	decreasing on $(0, 2)$
$x > 2$	$+$	$+$	$+$	$+$	increasing on $(2, \infty)$

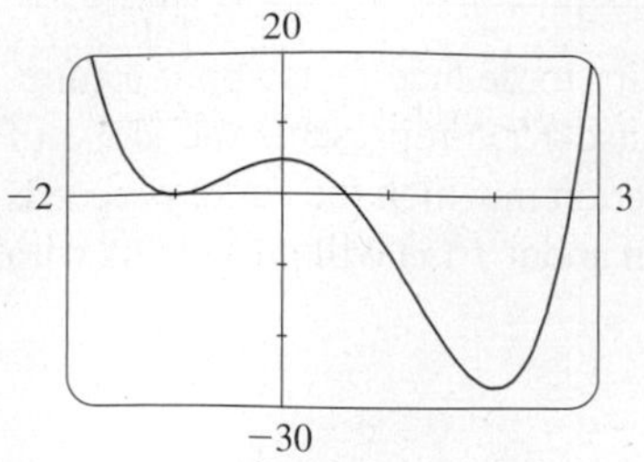

FIGURE 2

The graph of f shown in Figure 2 confirms the information in the chart. □

Recall from Section 4.1 that if f has a local maximum or minimum at c, then c must be a critical number of f (by Fermat's Theorem), but not every critical number gives rise to a maximum or a minimum. We therefore need a test that will tell us whether or not f has a local maximum or minimum at a critical number.

You can see from Figure 2 that $f(0) = 5$ is a local maximum value of f because f increases on $(-1, 0)$ and decreases on $(0, 2)$. Or, in terms of derivatives, $f'(x) > 0$ for $-1 < x < 0$ and $f'(x) < 0$ for $0 < x < 2$. In other words, the sign of $f'(x)$ changes from positive to negative at 0. This observation is the basis of the following test.

THE FIRST DERIVATIVE TEST Suppose that c is a critical number of a continuous function f.

(a) If f' changes from positive to negative at c, then f has a local maximum at c.

(b) If f' changes from negative to positive at c, then f has a local minimum at c.

(c) If f' does not change sign at c (for example, if f' is positive on both sides of c or negative on both sides), then f has no local maximum or minimum at c.

The First Derivative Test is a consequence of the I/D Test. In part (a), for instance, since the sign of $f'(x)$ changes from positive to negative at c, f is increasing to the left of c and decreasing to the right of c. It follows that f has a local maximum at c.

It is easy to remember the First Derivative Test by visualizing diagrams such as those in Figure 3.

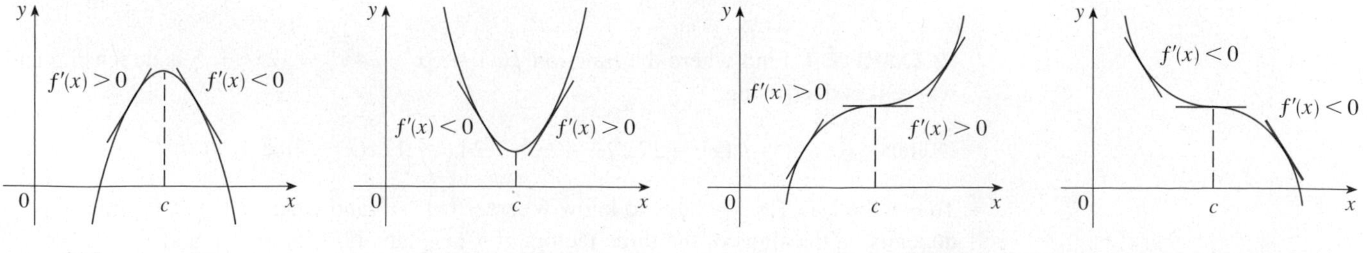

(a) Local maximum (b) Local minimum (c) No maximum or minimum (d) No maximum or minimum

FIGURE 3

V EXAMPLE 2 Find the local minimum and maximum values of the function f in Example 1.

SOLUTION From the chart in the solution to Example 1 we see that $f'(x)$ changes from negative to positive at -1, so $f(-1) = 0$ is a local minimum value by the First Derivative Test. Similarly, f' changes from negative to positive at 2, so $f(2) = -27$ is also a local minimum value. As previously noted, $f(0) = 5$ is a local maximum value because $f'(x)$ changes from positive to negative at 0. □

EXAMPLE 3 Find the local maximum and minimum values of the function

$$g(x) = x + 2\sin x \qquad 0 \leq x \leq 2\pi$$

SOLUTION To find the critical numbers of g, we differentiate:

$$g'(x) = 1 + 2\cos x$$

So $g'(x) = 0$ when $\cos x = -\frac{1}{2}$. The solutions of this equation are $2\pi/3$ and $4\pi/3$. Because g is differentiable everywhere, the only critical numbers are $2\pi/3$ and $4\pi/3$ and so we analyze g in the following table.

■ The + signs in the table come from the fact that $g'(x) > 0$ when $\cos x > -\frac{1}{2}$. From the graph of $y = \cos x$, this is true in the indicated intervals.

Interval	$g'(x) = 1 + 2\cos x$	g
$0 < x < 2\pi/3$	+	increasing on $(0, 2\pi/3)$
$2\pi/3 < x < 4\pi/3$	−	decreasing on $(2\pi/3, 4\pi/3)$
$4\pi/3 < x < 2\pi$	+	increasing on $(4\pi/3, 2\pi)$

Because $g'(x)$ changes from positive to negative at $2\pi/3$, the First Derivative Test tells us that there is a local maximum at $2\pi/3$ and the local maximum value is

$$g(2\pi/3) = \frac{2\pi}{3} + 2\sin\frac{2\pi}{3} = \frac{2\pi}{3} + 2\left(\frac{\sqrt{3}}{2}\right) = \frac{2\pi}{3} + \sqrt{3} \approx 3.83$$

Likewise, $g'(x)$ changes from negative to positive at $4\pi/3$ and so

$$g(4\pi/3) = \frac{4\pi}{3} + 2\sin\frac{4\pi}{3} = \frac{4\pi}{3} + 2\left(-\frac{\sqrt{3}}{2}\right) = \frac{4\pi}{3} - \sqrt{3} \approx 2.46$$

is a local minimum value. The graph of g in Figure 4 supports our conclusion.

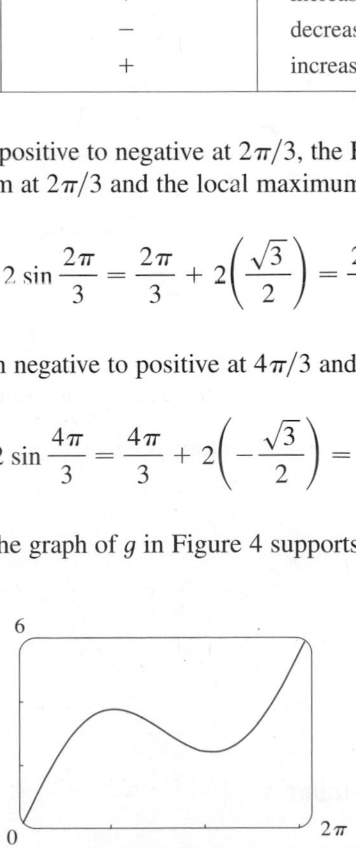

FIGURE 4
$y = x + 2\sin x$

□

WHAT DOES f'' SAY ABOUT f?

Figure 5 shows the graphs of two increasing functions on (a, b). Both graphs join point A to point B but they look different because they bend in different directions. How can we distinguish between these two types of behavior? In Figure 6 tangents to these curves have been drawn at several points. In (a) the curve lies above the tangents and f is called *concave upward* on (a, b). In (b) the curve lies below the tangents and g is called *concave downward* on (a, b).

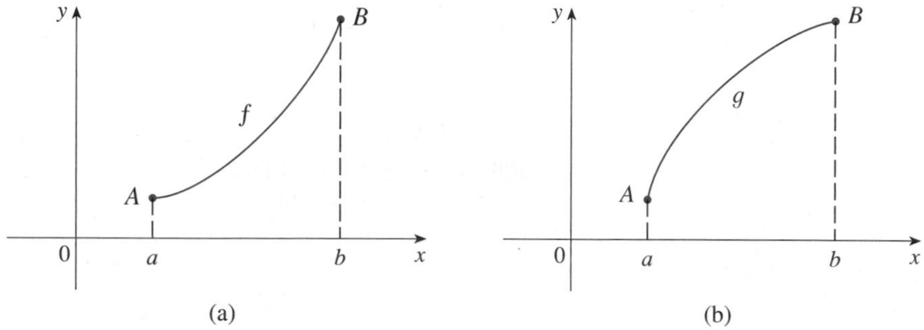

FIGURE 5

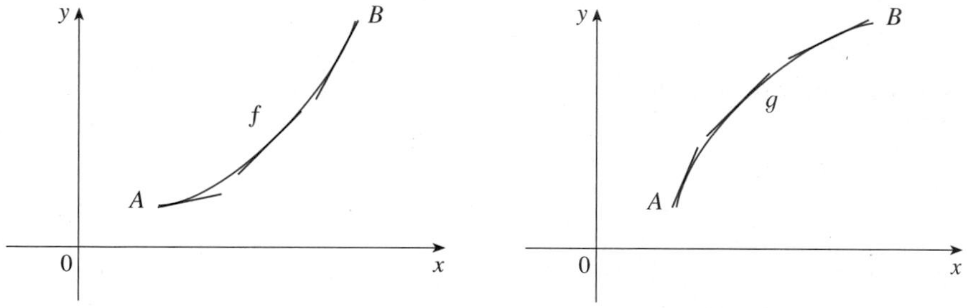

FIGURE 6 (a) Concave upward (b) Concave downward

DEFINITION If the graph of f lies above all of its tangents on an interval I, then it is called **concave upward** on I. If the graph of f lies below all of its tangents on I, it is called **concave downward** on I.

Figure 7 shows the graph of a function that is concave upward (abbreviated CU) on the intervals (b, c), (d, e), and (e, p) and concave downward (CD) on the intervals (a, b), (c, d), and (p, q).

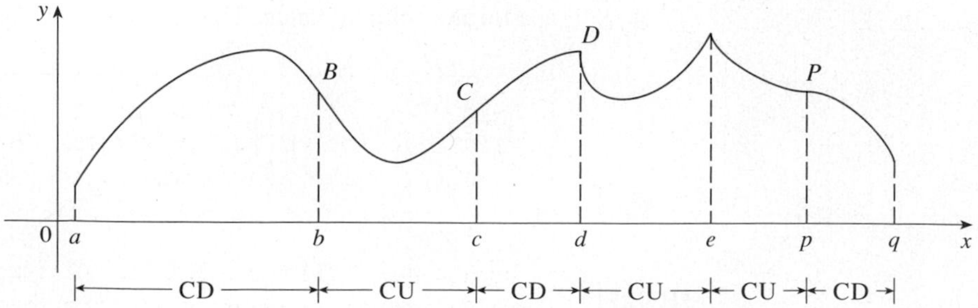

FIGURE 7

Let's see how the second derivative helps determine the intervals of concavity. Looking at Figure 6(a), you can see that, going from left to right, the slope of the tangent increases.

This means that the derivative f' is an increasing function and therefore its derivative f'' is positive. Likewise, in Figure 6(b) the slope of the tangent decreases from left to right, so f' decreases and therefore f'' is negative. This reasoning can be reversed and suggests that the following theorem is true. A proof is given in Appendix F with the help of the Mean Value Theorem.

CONCAVITY TEST

(a) If $f''(x) > 0$ for all x in I, then the graph of f is concave upward on I.

(b) If $f''(x) < 0$ for all x in I, then the graph of f is concave downward on I.

EXAMPLE 4 Figure 8 shows a population graph for Cyprian honeybees raised in an apiary. How does the rate of population increase change over time? When is this rate highest? Over what intervals is P concave upward or concave downward?

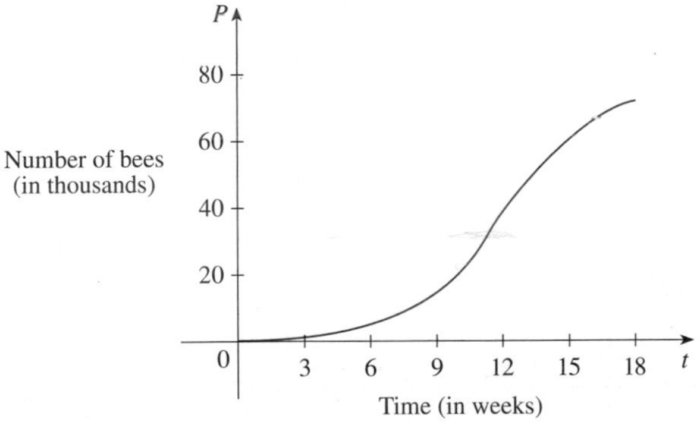

FIGURE 8

SOLUTION By looking at the slope of the curve as t increases, we see that the rate of increase of the population is initially very small, then gets larger until it reaches a maximum at about $t = 12$ weeks, and decreases as the population begins to level off. As the population approaches its maximum value of about 75,000 (called the *carrying capacity*), the rate of increase, $P'(t)$, approaches 0. The curve appears to be concave upward on (0, 12) and concave downward on (12, 18). □

In Example 4, the population curve changed from concave upward to concave downward at approximately the point (12, 38,000). This point is called an *inflection point* of the curve. The significance of this point is that the rate of population increase has its maximum value there. In general, an inflection point is a point where a curve changes its direction of concavity.

DEFINITION A point P on a curve $y = f(x)$ is called an **inflection point** if f is continuous there and the curve changes from concave upward to concave downward or from concave downward to concave upward at P.

For instance, in Figure 7, B, C, D, and P are the points of inflection. Notice that if a curve has a tangent at a point of inflection, then the curve crosses its tangent there.

In view of the Concavity Test, there is a point of inflection at any point where the second derivative changes sign.

EXAMPLE 5 Sketch a possible graph of a function f that satisfies the following conditions:

(i) $f'(x) > 0$ on $(-\infty, 1)$, $f'(x) < 0$ on $(1, \infty)$

(ii) $f''(x) > 0$ on $(-\infty, -2)$ and $(2, \infty)$, $f''(x) < 0$ on $(-2, 2)$

(iii) $\lim_{x \to -\infty} f(x) = -2$, $\lim_{x \to \infty} f(x) = 0$

SOLUTION Condition (i) tells us that f is increasing on $(-\infty, 1)$ and decreasing on $(1, \infty)$. Condition (ii) says that f is concave upward on $(-\infty, -2)$ and $(2, \infty)$, and concave downward on $(-2, 2)$. From condition (iii) we know that the graph of f has two horizontal asymptotes: $y = -2$ and $y = 0$.

We first draw the horizontal asymptote $y = -2$ as a dashed line (see Figure 9). We then draw the graph of f approaching this asymptote at the far left, increasing to its maximum point at $x = 1$ and decreasing toward the x-axis at the far right. We also make sure that the graph has inflection points when $x = -2$ and 2. Notice that we made the curve bend upward for $x < -2$ and $x > 2$, and bend downward when x is between -2 and 2. □

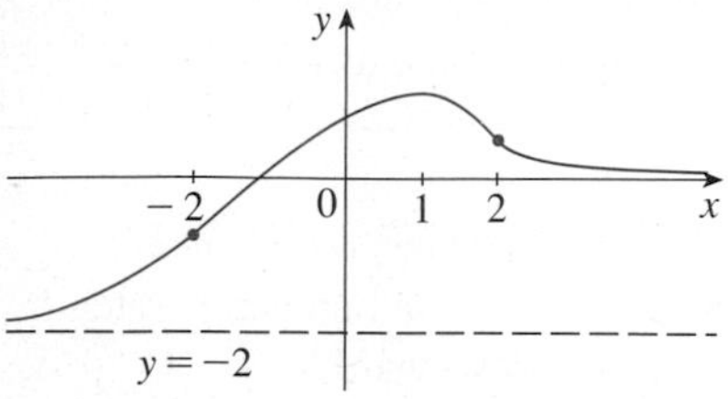

FIGURE 9

Another application of the second derivative is the following test for maximum and minimum values. It is a consequence of the Concavity Test.

> **THE SECOND DERIVATIVE TEST** Suppose f'' is continuous near c.
> (a) If $f'(c) = 0$ and $f''(c) > 0$, then f has a local minimum at c.
> (b) If $f'(c) = 0$ and $f''(c) < 0$, then f has a local maximum at c.

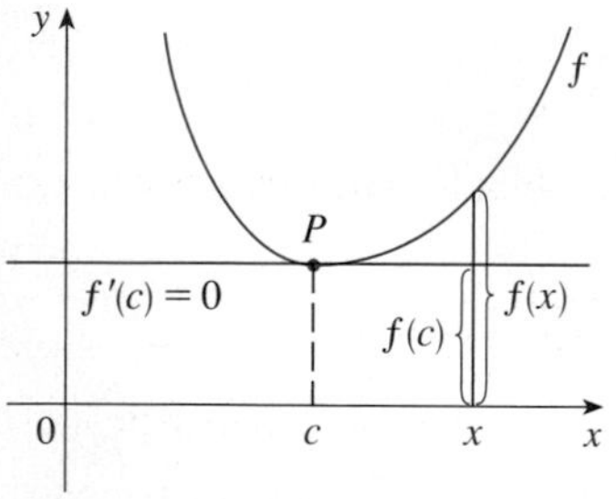

FIGURE 10
$f''(c) > 0$, f is concave upward

For instance, part (a) is true because $f''(x) > 0$ near c and so f is concave upward near c. This means that the graph of f lies *above* its horizontal tangent at c and so f has a local minimum at c. (See Figure 10.)

EXAMPLE 6 Discuss the curve $y = x^4 - 4x^3$ with respect to concavity, points of inflection, and local maxima and minima. Use this information to sketch the curve.

SOLUTION If $f(x) = x^4 - 4x^3$, then

$$f'(x) = 4x^3 - 12x^2 = 4x^2(x - 3)$$

$$f''(x) = 12x^2 - 24x = 12x(x - 2)$$

To find the critical numbers we set $f'(x) = 0$ and obtain $x = 0$ and $x = 3$. To use the Second Derivative Test we evaluate f'' at these critical numbers:

$$f''(0) = 0 \qquad f''(3) = 36 > 0$$

Since $f'(3) = 0$ and $f''(3) > 0$, $f(3) = -27$ is a local minimum. Since $f''(0) = 0$, the Second Derivative Test gives no information about the critical number 0. But since $f'(x) < 0$ for $x < 0$ and also for $0 < x < 3$, the First Derivative Test tells us that f does not have a local maximum or minimum at 0. [In fact, the expression for $f'(x)$ shows that f decreases to the left of 3 and increases to the right of 3.]

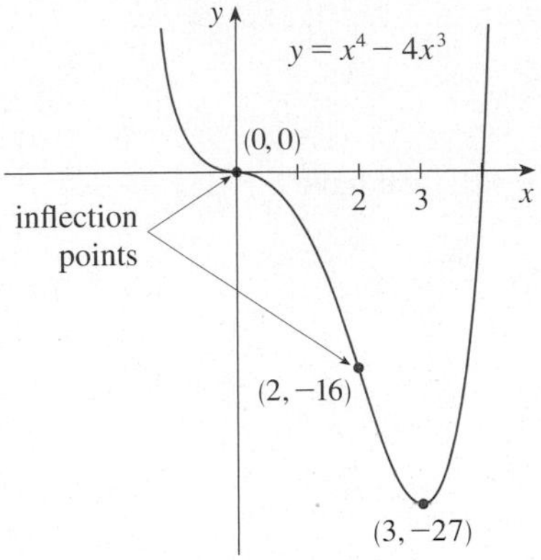

FIGURE 11

Since $f''(x) = 0$ when $x = 0$ or 2, we divide the real line into intervals with these numbers as endpoints and complete the following chart.

Interval	$f''(x) = 12x(x - 2)$	Concavity
$(-\infty, 0)$	+	upward
$(0, 2)$	−	downward
$(2, \infty)$	+	upward

The point (0, 0) is an inflection point since the curve changes from concave upward to concave downward there. Also, $(2, -16)$ is an inflection point since the curve changes from concave downward to concave upward there.

Using the local minimum, the intervals of concavity, and the inflection points, we sketch the curve in Figure 11. □

NOTE The Second Derivative Test is inconclusive when $f''(c) = 0$. In other words, at such a point there might be a maximum, there might be a minimum, or there might be neither (as in Example 6). This test also fails when $f''(c)$ does not exist. In such cases the First Derivative Test must be used. In fact, even when both tests apply, the First Derivative Test is often the easier one to use.

EXAMPLE 7 Sketch the graph of the function $f(x) = x^{2/3}(6 - x)^{1/3}$.

■ Try reproducing the graph in Figure 12 with a graphing calculator or computer. Some machines produce the complete graph, some produce only the portion to the right of the y-axis, and some produce only the portion between $x = 0$ and $x = 6$. For an explanation and cure, see Example 7 in Section 1.4. An equivalent expression that gives the correct graph is

$$y = (x^2)^{1/3} \cdot \frac{6 - x}{|6 - x|} |6 - x|^{1/3}$$

SOLUTION You can use the differentiation rules to check that the first two derivatives are

$$f'(x) = \frac{4 - x}{x^{1/3}(6 - x)^{2/3}} \qquad f''(x) = \frac{-8}{x^{4/3}(6 - x)^{5/3}}$$

Since $f'(x) = 0$ when $x = 4$ and $f'(x)$ does not exist when $x = 0$ or $x = 6$, the critical numbers are 0, 4, and 6.

Interval	$4 - x$	$x^{1/3}$	$(6 - x)^{2/3}$	$f'(x)$	f
$x < 0$	+	−	+	−	decreasing on $(-\infty, 0)$
$0 < x < 4$	+	+	+	+	increasing on $(0, 4)$
$4 < x < 6$	−	+	+	−	decreasing on $(4, 6)$
$x > 6$	−	+	+	−	decreasing on $(6, \infty)$

To find the local extreme values we use the First Derivative Test. Since f' changes from negative to positive at 0, $f(0) = 0$ is a local minimum. Since f' changes from positive to negative at 4, $f(4) = 2^{5/3}$ is a local maximum. The sign of f' does not change at 6, so there is no minimum or maximum there. (The Second Derivative Test could be used at 4, but not at 0 or 6 since f'' does not exist at either of these numbers.)

Looking at the expression for $f''(x)$ and noting that $x^{4/3} \geqslant 0$ for all x, we have $f''(x) < 0$ for $x < 0$ and for $0 < x < 6$ and $f''(x) > 0$ for $x > 6$. So f is concave downward on $(-\infty, 0)$ and $(0, 6)$ and concave upward on $(6, \infty)$, and the only inflection point is (6, 0). The graph is sketched in Figure 12. Note that the curve has vertical tangents at (0, 0) and (6, 0) because $|f'(x)| \to \infty$ as $x \to 0$ and as $x \to 6$. □

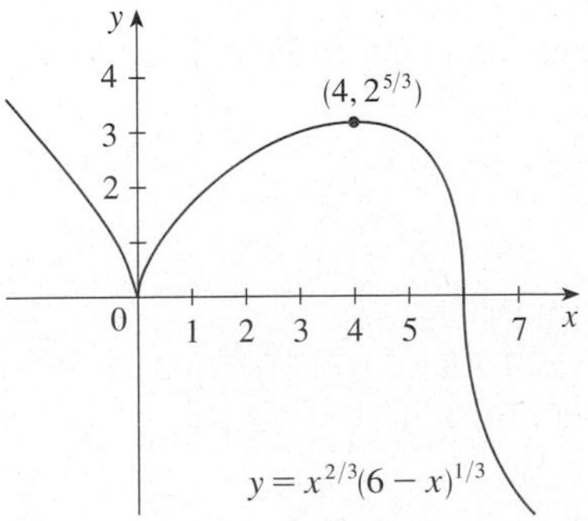

FIGURE 12

EXAMPLE 8 Use the first and second derivatives of $f(x) = e^{1/x}$, together with asymptotes, to sketch its graph.

SOLUTION Notice that the domain of f is $\{x \mid x \neq 0\}$, so we check for vertical asymptotes by computing the left and right limits as $x \to 0$. As $x \to 0^+$, we know that $t = 1/x \to \infty$, so

$$\lim_{x \to 0^+} e^{1/x} = \lim_{t \to \infty} e^t = \infty$$

and this shows that $x = 0$ is a vertical asymptote. As $x \to 0^-$, we have $t = 1/x \to -\infty$, so

$$\lim_{x \to 0^-} e^{1/x} = \lim_{t \to -\infty} e^t = 0$$

TEC In Module 4.3 you can practice using graphical information about f' to determine the shape of the graph of f.

As $x \to \pm\infty$, we have $1/x \to 0$ and so

$$\lim_{x \to \pm\infty} e^{1/x} = e^0 = 1$$

This shows that $y = 1$ is a horizontal asymptote.

Now let's compute the derivative. The Chain Rule gives

$$f'(x) = -\frac{e^{1/x}}{x^2}$$

Since $e^{1/x} > 0$ and $x^2 > 0$ for all $x \neq 0$, we have $f'(x) < 0$ for all $x \neq 0$. Thus f is decreasing on $(-\infty, 0)$ and on $(0, \infty)$. There is no critical number, so the function has no maximum or minimum. The second derivative is

$$f''(x) = -\frac{x^2 e^{1/x}(-1/x^2) - e^{1/x}(2x)}{x^4} = \frac{e^{1/x}(2x + 1)}{x^4}$$

Since $e^{1/x} > 0$ and $x^4 > 0$, we have $f''(x) > 0$ when $x > -\frac{1}{2}$ $(x \neq 0)$ and $f''(x) < 0$ when $x < -\frac{1}{2}$. So the curve is concave downward on $\left(-\infty, -\frac{1}{2}\right)$ and concave upward on $\left(-\frac{1}{2}, 0\right)$ and on $(0, \infty)$. The inflection point is $\left(-\frac{1}{2}, e^{-2}\right)$.

To sketch the graph of f we first draw the horizontal asymptote $y = 1$ (as a dashed line), together with the parts of the curve near the asymptotes in a preliminary sketch [Figure 13(a)]. These parts reflect the information concerning limits and the fact that f is decreasing on both $(-\infty, 0)$ and $(0, \infty)$. Notice that we have indicated that $f(x) \to 0$ as $x \to 0^-$ even though $f(0)$ does not exist. In Figure 13(b) we finish the sketch by incorporating the information concerning concavity and the inflection point. In Figure 13(c) we check our work with a graphing device.

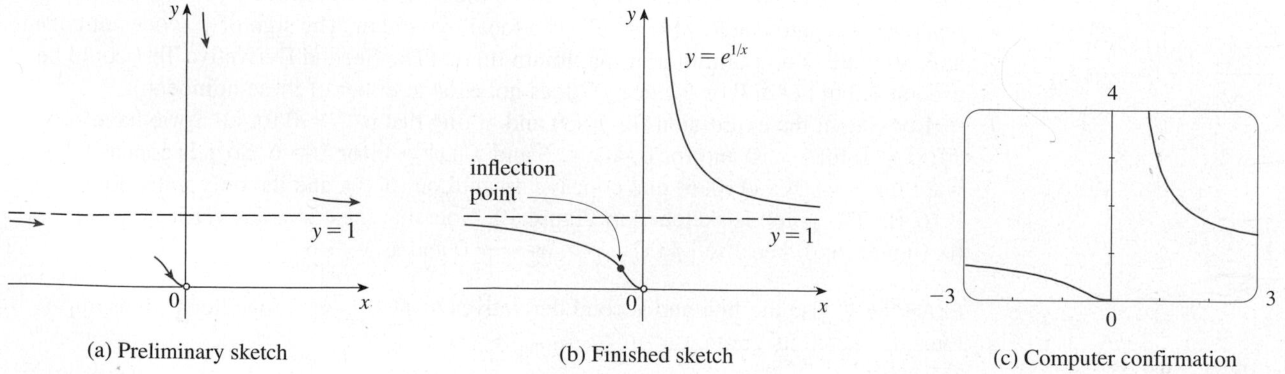

FIGURE 13

□

4.3 EXERCISES

1–2 Use the given graph of f to find the following.
(a) The open intervals on which f is increasing.
(b) The open intervals on which f is decreasing.
(c) The open intervals on which f is concave upward.
(d) The open intervals on which f is concave downward.
(e) The coordinates of the points of inflection.

1.

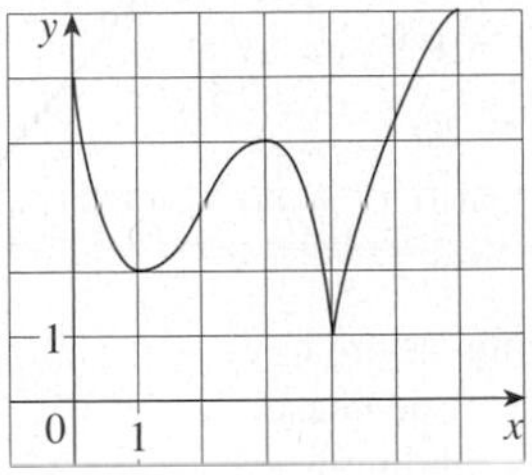

2.

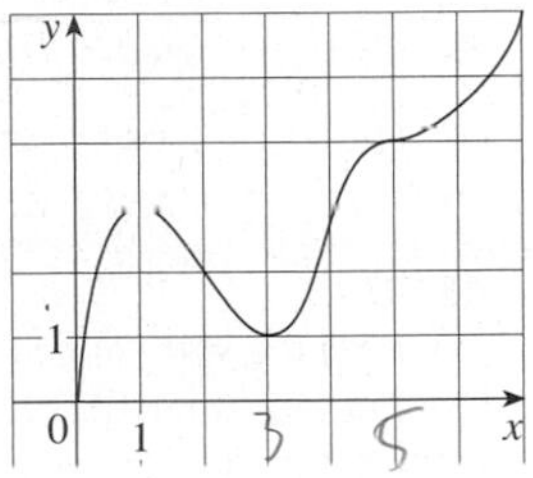

3. Suppose you are given a formula for a function f.
(a) How do you determine where f is increasing or decreasing?
(b) How do you determine where the graph of f is concave upward or concave downward?
(c) How do you locate inflection points?

4. (a) State the First Derivative Test.
(b) State the Second Derivative Test. Under what circumstances is it inconclusive? What do you do if it fails?

5–6 The graph of the *derivative* f' of a function f is shown.
(a) On what intervals is f increasing or decreasing?
(b) At what values of x does f have a local maximum or minimum?

5.

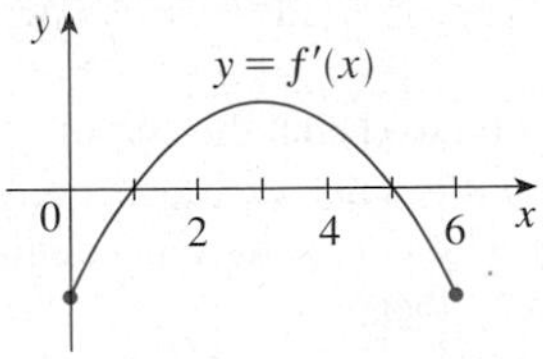

6.

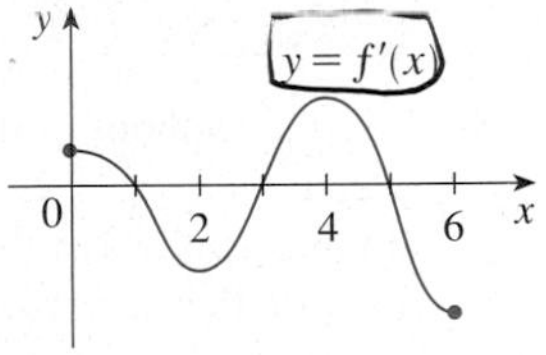

7. The graph of the second derivative f'' of a function f is shown. State the x-coordinates of the inflection points of f. Give reasons for your answers.

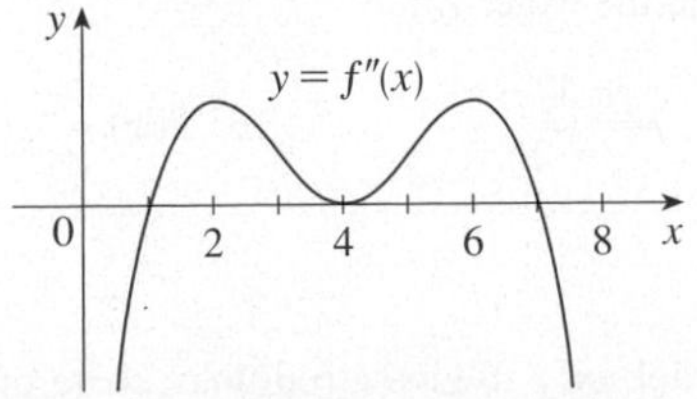

8. The graph of the first derivative f' of a function f is shown.
(a) On what intervals is f increasing? Explain.
(b) At what values of x does f have a local maximum or minimum? Explain.
(c) On what intervals is f concave upward or concave downward? Explain.
(d) What are the x-coordinates of the inflection points of f? Why?

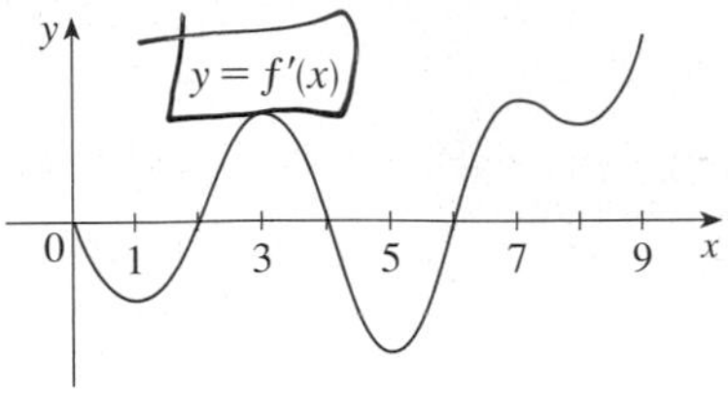

9–18
(a) Find the intervals on which f is increasing or decreasing.
(b) Find the local maximum and minimum values of f.
(c) Find the intervals of concavity and the inflection points.

9. $f(x) = 2x^3 + 3x^2 - 36x$

10. $f(x) = 4x^3 + 3x^2 - 6x + 1$

11. $f(x) = x^4 - 2x^2 + 3$

12. $f(x) = \dfrac{x^2}{x^2 + 3}$

13. $f(x) = \sin x + \cos x, \quad 0 \leqslant x \leqslant 2\pi$

14. $f(x) = \cos^2 x - 2\sin x, \quad 0 \leqslant x \leqslant 2\pi$

15. $f(x) = e^{2x} + e^{-x}$

16. $f(x) = x^2 \ln x$

17. $f(x) = (\ln x)/\sqrt{x}$

18. $f(x) = \sqrt{x}\, e^{-x}$

19–21 Find the local maximum and minimum values of f using both the First and Second Derivative Tests. Which method do you prefer?

19. $f(x) = x^5 - 5x + 3$

20. $f(x) = \dfrac{x}{x^2 + 4}$

21. $f(x) = x + \sqrt{1 - x}$

22. (a) Find the critical numbers of $f(x) = x^4(x - 1)^3$.
(b) What does the Second Derivative Test tell you about the behavior of f at these critical numbers?
(c) What does the First Derivative Test tell you?

23. Suppose f'' is continuous on $(-\infty, \infty)$.
(a) If $f'(2) = 0$ and $f''(2) = -5$, what can you say about f?
(b) If $f'(6) = 0$ and $f''(6) = 0$, what can you say about f?

24–29 Sketch the graph of a function that satisfies all of the given conditions.

24. $f'(x) > 0$ for all $x \neq 1$, vertical asymptote $x = 1$,
$f''(x) > 0$ if $x < 1$ or $x > 3$, $f''(x) < 0$ if $1 < x < 3$

25. $f'(0) = f'(2) = f'(4) = 0$,
$f'(x) > 0$ if $x < 0$ or $2 < x < 4$,
$f'(x) < 0$ if $0 < x < 2$ or $x > 4$,
$f''(x) > 0$ if $1 < x < 3$, $f''(x) < 0$ if $x < 1$ or $x > 3$

26. $f'(1) = f'(-1) = 0$, $f'(x) < 0$ if $|x| < 1$,
$f'(x) > 0$ if $1 < |x| < 2$, $f'(x) = -1$ if $|x| > 2$,
$f''(x) < 0$ if $-2 < x < 0$, inflection point $(0, 1)$

27. $f'(x) > 0$ if $|x| < 2$, $f'(x) < 0$ if $|x| > 2$,
$f'(-2) = 0$, $\lim_{x \to 2} |f'(x)| = \infty$, $f''(x) > 0$ if $x \neq 2$

28. $f'(x) > 0$ if $|x| < 2$, $f'(x) < 0$ if $|x| > 2$,
$f'(2) = 0$, $\lim_{x \to \infty} f(x) = 1$, $f(-x) = -f(x)$,
$f''(x) < 0$ if $0 < x < 3$, $f''(x) > 0$ if $x > 3$

29. $f'(x) < 0$ and $f''(x) < 0$ for all x

---

30. Suppose $f(3) = 2$, $f'(3) = \frac{1}{2}$, and $f'(x) > 0$ and $f''(x) < 0$ for all x.
(a) Sketch a possible graph for f.
(b) How many solutions does the equation $f(x) = 0$ have? Why?
(c) Is it possible that $f'(2) = \frac{1}{3}$? Why?

31–32 The graph of the derivative f' of a continuous function f is shown.
(a) On what intervals is f increasing or decreasing?
(b) At what values of x does f have a local maximum or minimum?
(c) On what intervals is f concave upward or downward?
(d) State the x-coordinate(s) of the point(s) of inflection.
(e) Assuming that $f(0) = 0$, sketch a graph of f.

31.

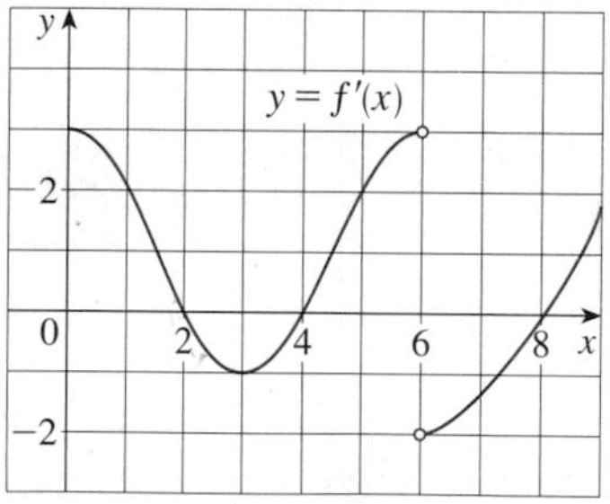

32.

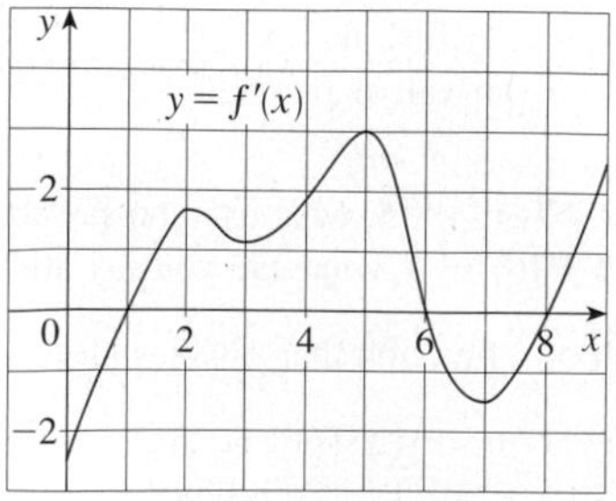

33–44
(a) Find the intervals of increase or decrease.
(b) Find the local maximum and minimum values.
(c) Find the intervals of concavity and the inflection points.
(d) Use the information from parts (a)–(c) to sketch the graph. Check your work with a graphing device if you have one.

33. $f(x) = 2x^3 - 3x^2 - 12x$

34. $f(x) = 2 + 3x - x^3$

35. $f(x) = 2 + 2x^2 - x^4$

36. $g(x) = 200 + 8x^3 + x^4$

37. $h(x) = (x + 1)^5 - 5x - 2$

38. $h(x) = x^5 - 2x^3 + x$

39. $A(x) = x\sqrt{x + 3}$

40. $B(x) = 3x^{2/3} - x$

41. $C(x) = x^{1/3}(x + 4)$

42. $f(x) = \ln(x^4 + 27)$

43. $f(\theta) = 2 \cos \theta + \cos^2\theta$, $0 \leq \theta \leq 2\pi$

44. $f(t) = t + \cos t$, $-2\pi \leq t \leq 2\pi$

45–52
(a) Find the vertical and horizontal asymptotes.
(b) Find the intervals of increase or decrease.
(c) Find the local maximum and minimum values.
(d) Find the intervals of concavity and the inflection points.
(e) Use the information from parts (a)–(d) to sketch the graph of f.

45. $f(x) = \dfrac{x^2}{x^2 - 1}$

46. $f(x) = \dfrac{x^2}{(x - 2)^2}$

47. $f(x) = \sqrt{x^2 + 1} - x$

48. $f(x) = x \tan x$, $-\pi/2 < x < \pi/2$

49. $f(x) = \ln(1 - \ln x)$

50. $f(x) = \dfrac{e^x}{1 + e^x}$

51. $f(x) = e^{-1/(x+1)}$

52. $f(x) = e^{\arctan x}$

---

53. Suppose the derivative of a function f is $f'(x) = (x + 1)^2(x - 3)^5(x - 6)^4$. On what interval is f increasing?

54. Use the methods of this section to sketch the curve $y = x^3 - 3a^2x + 2a^3$, where a is a positive constant. What do the members of this family of curves have in common? How do they differ from each other?

55–56
(a) Use a graph of f to estimate the maximum and minimum values. Then find the exact values.
(b) Estimate the value of x at which f increases most rapidly. Then find the exact value.

55. $f(x) = \dfrac{x + 1}{\sqrt{x^2 + 1}}$

56. $f(x) = x^2 e^{-x}$

57–58
(a) Use a graph of f to give a rough estimate of the intervals of concavity and the coordinates of the points of inflection.
(b) Use a graph of f'' to give better estimates.

57. $f(x) = \cos x + \frac{1}{2} \cos 2x$, $0 \leq x \leq 2\pi$

58. $f(x) = x^3(x - 2)^4$

CAS **59–60** Estimate the intervals of concavity to one decimal place by using a computer algebra system to compute and graph f''.

59. $f(x) = \dfrac{x^4 + x^3 + 1}{\sqrt{x^2 + x + 1}}$ **60.** $f(x) = \dfrac{x^2 \tan^{-1} x}{1 + x^3}$

61. A graph of a population of yeast cells in a new laboratory culture as a function of time is shown.
(a) Describe how the rate of population increase varies.
(b) When is this rate highest?
(c) On what intervals is the population function concave upward or downward?
(d) Estimate the coordinates of the inflection point.

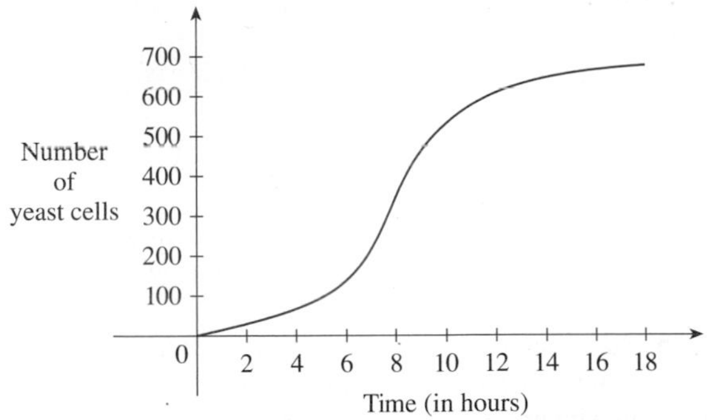

62. Let $f(t)$ be the temperature at time t where you live and suppose that at time $t = 3$ you feel uncomfortably hot. How do you feel about the given data in each case?
(a) $f'(3) = 2,\ f''(3) = 4$ (b) $f'(3) = 2,\ f''(3) = -4$
(c) $f'(3) = -2,\ f''(3) = 4$ (d) $f'(3) = -2,\ f''(3) = -4$

63. Let $K(t)$ be a measure of the knowledge you gain by studying for a test for t hours. Which do you think is larger, $K(8) - K(7)$ or $K(3) - K(2)$? Is the graph of K concave upward or concave downward? Why?

64. Coffee is being poured into the mug shown in the figure at a constant rate (measured in volume per unit time). Sketch a rough graph of the depth of the coffee in the mug as a function of time. Account for the shape of the graph in terms of concavity. What is the significance of the inflection point?

65. A *drug response curve* describes the level of medication in the bloodstream after a drug is administered. A surge function $S(t) = At^p e^{-kt}$ is often used to model the response curve, reflecting an initial surge in the drug level and then a more gradual decline. If, for a particular drug, $A = 0.01$, $p = 4$, $k = 0.07$, and t is measured in minutes, estimate the times corresponding to the inflection points and explain their significance. If you have a graphing device, use it to graph the drug response curve.

66. The family of bell-shaped curves

$$y = \frac{1}{\sigma\sqrt{2\pi}} e^{-(x-\mu)^2/(2\sigma^2)}$$

occurs in probability and statistics, where it is called the *normal density function.* The constant μ is called the *mean* and the positive constant σ is called the *standard deviation.* For simplicity, let's scale the function so as to remove the factor $1/(\sigma\sqrt{2\pi})$ and let's analyze the special case where $\mu = 0$. So we study the function

$$f(x) = e^{-x^2/(2\sigma^2)}$$

(a) Find the asymptote, maximum value, and inflection points of f.
(b) What role does σ play in the shape of the curve?
(c) Illustrate by graphing four members of this family on the same screen.

67. Find a cubic function $f(x) = ax^3 + bx^2 + cx + d$ that has a local maximum value of 3 at -2 and a local minimum value of 0 at 1.

68. For what values of the numbers a and b does the function

$$f(x) = axe^{bx^2}$$

have the maximum value $f(2) = 1$?

69. Show that the curve $y = (1 + x)/(1 + x^2)$ has three points of inflection and they all lie on one straight line.

70. Show that the curves $y = e^{-x}$ and $y = -e^{-x}$ touch the curve $y = e^{-x}\sin x$ at its inflection points.

71. Suppose f is differentiable on an interval I and $f'(x) > 0$ for all numbers x in I except for a single number c. Prove that f is increasing on the entire interval I.

72–74 Assume that all of the functions are twice differentiable and the second derivatives are never 0.

72. (a) If f and g are concave upward on I, show that $f + g$ is concave upward on I.
(b) If f is positive and concave upward on I, show that the function $g(x) = [f(x)]^2$ is concave upward on I.

73. (a) If f and g are positive, increasing, concave upward functions on I, show that the product function fg is concave upward on I.
(b) Show that part (a) remains true if f and g are both decreasing.

(c) Suppose f is increasing and g is decreasing. Show, by giving three examples, that fg may be concave upward, concave downward, or linear. Why doesn't the argument in parts (a) and (b) work in this case?

74. Suppose f and g are both concave upward on $(-\infty, \infty)$. Under what condition on f will the composite function $h(x) = f(g(x))$ be concave upward?

75. Show that $\tan x > x$ for $0 < x < \pi/2$. [*Hint:* Show that $f(x) = \tan x - x$ is increasing on $(0, \pi/2)$.]

76. (a) Show that $e^x \geqslant 1 + x$ for $x \geqslant 0$.
(b) Deduce that $e^x \geqslant 1 + x + \frac{1}{2}x^2$ for $x \geqslant 0$.
(c) Use mathematical induction to prove that for $x \geqslant 0$ and any positive integer n,

$$e^x \geqslant 1 + x + \frac{x^2}{2!} + \cdots + \frac{x^n}{n!}$$

77. Show that a cubic function (a third-degree polynomial) always has exactly one point of inflection. If its graph has three x-intercepts x_1, x_2, and x_3, show that the x-coordinate of the inflection point is $(x_1 + x_2 + x_3)/3$.

78. For what values of c does the polynomial $P(x) = x^4 + cx^3 + x^2$ have two inflection points? One inflection point? None? Illustrate by graphing P for several values of c. How does the graph change as c decreases?

79. Prove that if $(c, f(c))$ is a point of inflection of the graph of f and f'' exists in an open interval that contains c, then $f''(c) = 0$. [*Hint:* Apply the First Derivative Test and Fermat's Theorem to the function $g = f'$.]

80. Show that if $f(x) = x^4$, then $f''(0) = 0$, but $(0, 0)$ is not an inflection point of the graph of f.

81. Show that the function $g(x) = x|x|$ has an inflection point at $(0, 0)$ but $g''(0)$ does not exist.

82. Suppose that f''' is continuous and $f'(c) = f''(c) = 0$, but $f'''(c) > 0$. Does f have a local maximum or minimum at c? Does f have a point of inflection at c?

83. The three cases in the First Derivative Test cover the situations one commonly encounters but do not exhaust all possibilities. Consider the functions f, g, and h whose values at 0 are all 0 and, for $x \neq 0$,

$$f(x) = x^4 \sin\frac{1}{x} \qquad g(x) = x^4\left(2 + \sin\frac{1}{x}\right)$$

$$h(x) = x^4\left(-2 + \sin\frac{1}{x}\right)$$

(a) Show that 0 is a critical number of all three functions but their derivatives change sign infinitely often on both sides of 0.
(b) Show that f has neither a local maximum nor a local minimum at 0, g has a local minimum, and h has a local maximum.

4.4 INDETERMINATE FORMS AND L'HOSPITAL'S RULE

Suppose we are trying to analyze the behavior of the function

$$F(x) = \frac{\ln x}{x - 1}$$

Although F is not defined when $x = 1$, we need to know how F behaves *near* 1. In particular, we would like to know the value of the limit

1 $$\lim_{x \to 1} \frac{\ln x}{x - 1}$$

In computing this limit we can't apply Law 5 of limits (the limit of a quotient is the quotient of the limits, see Section 2.3) because the limit of the denominator is 0. In fact, although the limit in (1) exists, its value is not obvious because both numerator and denominator approach 0 and $\frac{0}{0}$ is not defined.

In general, if we have a limit of the form

$$\lim_{x \to a} \frac{f(x)}{g(x)}$$

where both $f(x) \to 0$ and $g(x) \to 0$ as $x \to a$, then this limit may or may not exist and is called an **indeterminate form of type $\frac{0}{0}$**. We met some limits of this type in Chapter 2. For rational functions, we can cancel common factors:

$$\lim_{x \to 1} \frac{x^2 - x}{x^2 - 1} = \lim_{x \to 1} \frac{x(x - 1)}{(x + 1)(x - 1)} = \lim_{x \to 1} \frac{x}{x + 1} = \frac{1}{2}$$

We used a geometric argument to show that

$$\lim_{x \to 0} \frac{\sin x}{x} = 1$$

But these methods do not work for limits such as (1), so in this section we introduce a systematic method, known as *l'Hospital's Rule,* for the evaluation of indeterminate forms.

Another situation in which a limit is not obvious occurs when we look for a horizontal asymptote of F and need to evaluate the limit

2 $$\lim_{x \to \infty} \frac{\ln x}{x - 1}$$

It isn't obvious how to evaluate this limit because both numerator and denominator become large as $x \to \infty$. There is a struggle between numerator and denominator. If the numerator wins, the limit will be ∞; if the denominator wins, the answer will be 0. Or there may be some compromise, in which case the answer may be some finite positive number.

In general, if we have a limit of the form

$$\lim_{x \to a} \frac{f(x)}{g(x)}$$

where both $f(x) \to \infty$ (or $-\infty$) and $g(x) \to \infty$ (or $-\infty$), then the limit may or may not exist and is called an **indeterminate form of type ∞/∞**. We saw in Section 2.6 that this type of limit can be evaluated for certain functions, including rational functions, by dividing numerator and denominator by the highest power of x that occurs in the denominator. For instance,

$$\lim_{x \to \infty} \frac{x^2 - 1}{2x^2 + 1} = \lim_{x \to \infty} \frac{1 - \dfrac{1}{x^2}}{2 + \dfrac{1}{x^2}} = \frac{1 - 0}{2 + 0} = \frac{1}{2}$$

This method does not work for limits such as (2), but l'Hospital's Rule also applies to this type of indeterminate form.

L'HOSPITAL

L'Hospital's Rule is named after a French nobleman, the Marquis de l'Hospital (1661–1704), but was discovered by a Swiss mathematician, John Bernoulli (1667–1748). You might sometimes see l'Hospital spelled as l'Hôpital, but he spelled his own name l'Hospital, as was common in the 17th century. See Exercise 77 for the example that the Marquis used to illustrate his rule. See the project on page 307 for further historical details.

L'HOSPITAL'S RULE Suppose f and g are differentiable and $g'(x) \neq 0$ on an open interval I that contains a (except possibly at a). Suppose that

$$\lim_{x \to a} f(x) = 0 \quad \text{and} \quad \lim_{x \to a} g(x) = 0$$

or that

$$\lim_{x \to a} f(x) = \pm\infty \quad \text{and} \quad \lim_{x \to a} g(x) = \pm\infty$$

(In other words, we have an indeterminate form of type $\frac{0}{0}$ or ∞/∞.) Then

$$\lim_{x \to a} \frac{f(x)}{g(x)} = \lim_{x \to a} \frac{f'(x)}{g'(x)}$$

if the limit on the right side exists (or is ∞ or $-\infty$).

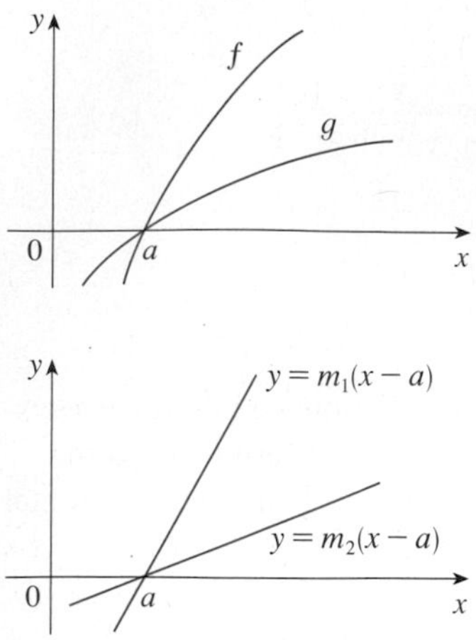

FIGURE 1

■ Figure 1 suggests visually why l'Hospital's Rule might be true. The first graph shows two differentiable functions f and g, each of which approaches 0 as $x \to a$. If we were to zoom in toward the point $(a, 0)$, the graphs would start to look almost linear. But if the functions actually *were* linear, as in the second graph, then their ratio would be

$$\frac{m_1(x-a)}{m_2(x-a)} = \frac{m_1}{m_2}$$

which is the ratio of their derivatives. This suggests that

$$\lim_{x\to a}\frac{f(x)}{g(x)} = \lim_{x\to a}\frac{f'(x)}{g'(x)}$$

NOTE 1 L'Hospital's Rule says that the limit of a quotient of functions is equal to the limit of the quotient of their derivatives, provided that the given conditions are satisfied. It is especially important to verify the conditions regarding the limits of f and g before using l'Hospital's Rule.

NOTE 2 L'Hospital's Rule is also valid for one-sided limits and for limits at infinity or negative infinity; that is, "$x \to a$" can be replaced by any of the symbols $x \to a^+$, $x \to a^-$, $x \to \infty$, or $x \to -\infty$.

NOTE 3 For the special case in which $f(a) = g(a) = 0$, f' and g' are continuous, and $g'(a) \neq 0$, it is easy to see why l'Hospital's Rule is true. In fact, using the alternative form of the definition of a derivative, we have

$$\lim_{x\to a}\frac{f'(x)}{g'(x)} = \frac{f'(a)}{g'(a)} = \frac{\displaystyle\lim_{x\to a}\frac{f(x)-f(a)}{x-a}}{\displaystyle\lim_{x\to a}\frac{g(x)-g(a)}{x-a}} = \lim_{x\to a}\frac{\dfrac{f(x)-f(a)}{x-a}}{\dfrac{g(x)-g(a)}{x-a}}$$

$$= \lim_{x\to a}\frac{f(x)-f(a)}{g(x)-g(a)} = \lim_{x\to a}\frac{f(x)}{g(x)}$$

It is more difficult to prove the general version of l'Hospital's Rule. See Appendix F.

EXAMPLE 1 Find $\displaystyle\lim_{x\to 1}\frac{\ln x}{x-1}$.

SOLUTION Since

$$\lim_{x\to 1}\ln x = \ln 1 = 0 \qquad \text{and} \qquad \lim_{x\to 1}(x-1) = 0$$

we can apply l'Hospital's Rule:

$$\lim_{x\to 1}\frac{\ln x}{x-1} = \lim_{x\to 1}\frac{\dfrac{d}{dx}(\ln x)}{\dfrac{d}{dx}(x-1)} = \lim_{x\to 1}\frac{1/x}{1} = \lim_{x\to 1}\frac{1}{x} = 1$$ □

⊘ Notice that when using l'Hospital's Rule we differentiate the numerator and denominator *separately*. We do *not* use the Quotient Rule.

EXAMPLE 2 Calculate $\displaystyle\lim_{x\to\infty}\frac{e^x}{x^2}$.

SOLUTION We have $\lim_{x\to\infty} e^x = \infty$ and $\lim_{x\to\infty} x^2 = \infty$, so l'Hospital's Rule gives

$$\lim_{x\to\infty}\frac{e^x}{x^2} = \lim_{x\to\infty}\frac{\dfrac{d}{dx}(e^x)}{\dfrac{d}{dx}(x^2)} = \lim_{x\to\infty}\frac{e^x}{2x}$$

Since $e^x \to \infty$ and $2x \to \infty$ as $x \to \infty$, the limit on the right side is also indeterminate, but a second application of l'Hospital's Rule gives

$$\lim_{x\to\infty}\frac{e^x}{x^2} = \lim_{x\to\infty}\frac{e^x}{2x} = \lim_{x\to\infty}\frac{e^x}{2} = \infty$$ □

■ The graph of the function of Example 2 is shown in Figure 2. We have noticed previously that exponential functions grow far more rapidly than power functions, so the result of Example 2 is not unexpected. See also Exercise 69.

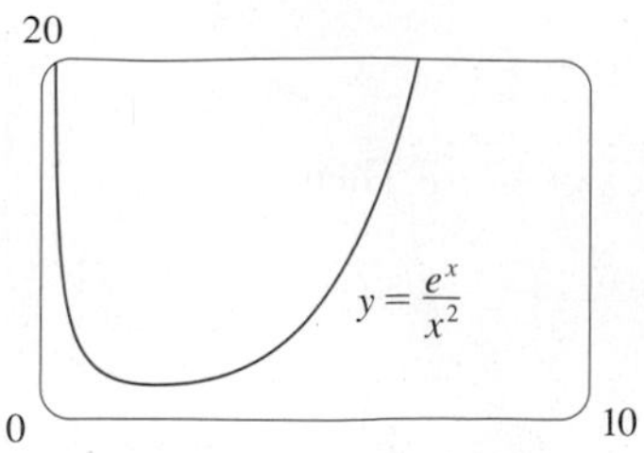

FIGURE 2

■ The graph of the function of Example 3 is shown in Figure 3. We have discussed previously the slow growth of logarithms, so it isn't surprising that this ratio approaches 0 as $x \to \infty$. See also Exercise 70.

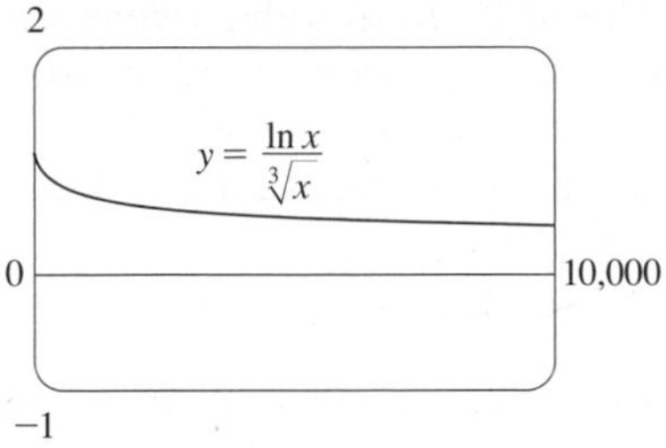

FIGURE 3

V EXAMPLE 3 Calculate $\lim_{x \to \infty} \frac{\ln x}{\sqrt[3]{x}}$.

SOLUTION Since $\ln x \to \infty$ and $\sqrt[3]{x} \to \infty$ as $x \to \infty$, l'Hospital's Rule applies:

$$\lim_{x \to \infty} \frac{\ln x}{\sqrt[3]{x}} = \lim_{x \to \infty} \frac{1/x}{\frac{1}{3}x^{-2/3}}$$

Notice that the limit on the right side is now indeterminate of type $\frac{0}{0}$. But instead of applying l'Hospital's Rule a second time as we did in Example 2, we simplify the expression and see that a second application is unnecessary:

$$\lim_{x \to \infty} \frac{\ln x}{\sqrt[3]{x}} = \lim_{x \to \infty} \frac{1/x}{\frac{1}{3}x^{-2/3}} = \lim_{x \to \infty} \frac{3}{\sqrt[3]{x}} = 0 \qquad \square$$

EXAMPLE 4 Find $\lim_{x \to 0} \frac{\tan x - x}{x^3}$. (See Exercise 38 in Section 2.2.)

■ The graph in Figure 4 gives visual confirmation of the result of Example 4. If we were to zoom in too far, however, we would get an inaccurate graph because $\tan x$ is close to x when x is small. See Exercise 38(d) in Section 2.2.

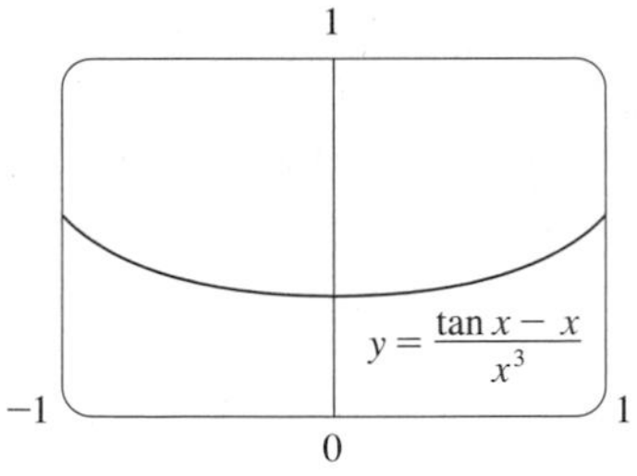

FIGURE 4

SOLUTION Noting that both $\tan x - x \to 0$ and $x^3 \to 0$ as $x \to 0$, we use l'Hospital's Rule:

$$\lim_{x \to 0} \frac{\tan x - x}{x^3} = \lim_{x \to 0} \frac{\sec^2 x - 1}{3x^2}$$

Since the limit on the right side is still indeterminate of type $\frac{0}{0}$, we apply l'Hospital's Rule again:

$$\lim_{x \to 0} \frac{\sec^2 x - 1}{3x^2} = \lim_{x \to 0} \frac{2\sec^2 x \tan x}{6x}$$

Because $\lim_{x \to 0} \sec^2 x = 1$, we simplify the calculation by writing

$$\lim_{x \to 0} \frac{2\sec^2 x \tan x}{6x} = \frac{1}{3} \lim_{x \to 0} \sec^2 x \lim_{x \to 0} \frac{\tan x}{x} = \frac{1}{3} \lim_{x \to 0} \frac{\tan x}{x}$$

We can evaluate this last limit either by using l'Hospital's Rule a third time or by writing $\tan x$ as $(\sin x)/(\cos x)$ and making use of our knowledge of trigonometric limits. Putting together all the steps, we get

$$\lim_{x \to 0} \frac{\tan x - x}{x^3} = \lim_{x \to 0} \frac{\sec^2 x - 1}{3x^2} = \lim_{x \to 0} \frac{2\sec^2 x \tan x}{6x}$$

$$= \frac{1}{3} \lim_{x \to 0} \frac{\tan x}{x} = \frac{1}{3} \lim_{x \to 0} \frac{\sec^2 x}{1} = \frac{1}{3} \qquad \square$$

EXAMPLE 5 Find $\lim_{x \to \pi^-} \frac{\sin x}{1 - \cos x}$.

SOLUTION If we blindly attempted to use l'Hospital's Rule, we would get

⊘

$$\lim_{x \to \pi^-} \frac{\sin x}{1 - \cos x} = \lim_{x \to \pi^-} \frac{\cos x}{\sin x} = -\infty$$

This is **wrong!** Although the numerator $\sin x \to 0$ as $x \to \pi^-$, notice that the denominator $(1 - \cos x)$ does not approach 0, so l'Hospital's Rule can't be applied here.

The required limit is, in fact, easy to find because the function is continuous at π and the denominator is nonzero there:

$$\lim_{x\to\pi^-} \frac{\sin x}{1-\cos x} = \frac{\sin \pi}{1-\cos \pi} = \frac{0}{1-(-1)} = 0 \qquad \square$$

Example 5 shows what can go wrong if you use l'Hospital's Rule without thinking. Other limits *can* be found using l'Hospital's Rule but are more easily found by other methods. (See Examples 3 and 5 in Section 2.3, Example 3 in Section 2.6, and the discussion at the beginning of this section.) So when evaluating any limit, you should consider other methods before using l'Hospital's Rule.

INDETERMINATE PRODUCTS

If $\lim_{x\to a} f(x) = 0$ and $\lim_{x\to a} g(x) = \infty$ (or $-\infty$), then it isn't clear what the value of $\lim_{x\to a} f(x)g(x)$, if any, will be. There is a struggle between f and g. If f wins, the answer will be 0; if g wins, the answer will be ∞ (or $-\infty$). Or there may be a compromise where the answer is a finite nonzero number. This kind of limit is called an **indeterminate form of type $0 \cdot \infty$**. We can deal with it by writing the product fg as a quotient:

$$fg = \frac{f}{1/g} \qquad \text{or} \qquad fg = \frac{g}{1/f}$$

This converts the given limit into an indeterminate form of type $\frac{0}{0}$ or ∞/∞ so that we can use l'Hospital's Rule.

■ Figure 5 shows the graph of the function in Example 6. Notice that the function is undefined at $x = 0$; the graph approaches the origin but never quite reaches it.

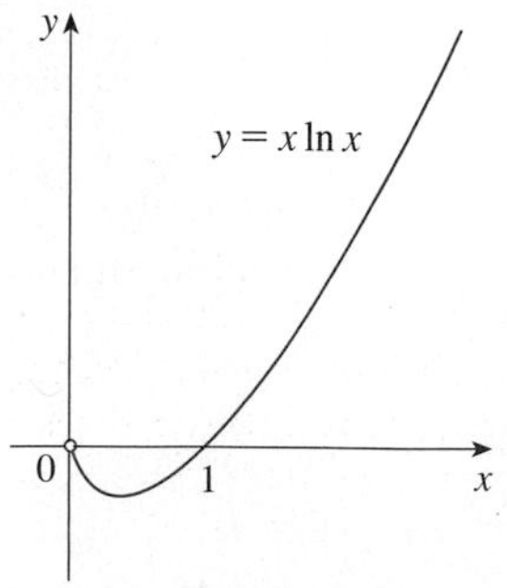

FIGURE 5

V EXAMPLE 6 Evaluate $\lim_{x\to 0^+} x \ln x$.

SOLUTION The given limit is indeterminate because, as $x \to 0^+$, the first factor (x) approaches 0 while the second factor ($\ln x$) approaches $-\infty$. Writing $x = 1/(1/x)$, we have $1/x \to \infty$ as $x \to 0^+$, so l'Hospital's Rule gives

$$\lim_{x\to 0^+} x \ln x = \lim_{x\to 0^+} \frac{\ln x}{1/x} = \lim_{x\to 0^+} \frac{1/x}{-1/x^2} = \lim_{x\to 0^+} (-x) = 0 \qquad \square$$

NOTE In solving Example 6 another possible option would have been to write

$$\lim_{x\to 0^+} x \ln x = \lim_{x\to 0^+} \frac{x}{1/\ln x}$$

This gives an indeterminate form of the type 0/0, but if we apply l'Hospital's Rule we get a more complicated expression than the one we started with. In general, when we rewrite an indeterminate product, we try to choose the option that leads to the simpler limit.

INDETERMINATE DIFFERENCES

If $\lim_{x\to a} f(x) = \infty$ and $\lim_{x\to a} g(x) = \infty$, then the limit

$$\lim_{x\to a} [f(x) - g(x)]$$

is called an **indeterminate form of type $\infty - \infty$**. Again there is a contest between f and g. Will the answer be ∞ (f wins) or will it be $-\infty$ (g wins) or will they compromise on a finite number? To find out, we try to convert the difference into a quotient (for instance, by using a common denominator, or rationalization, or factoring out a common factor) so that we have an indeterminate form of type $\frac{0}{0}$ or ∞/∞.

EXAMPLE 7 Compute $\lim_{x \to (\pi/2)^-} (\sec x - \tan x)$.

SOLUTION First notice that $\sec x \to \infty$ and $\tan x \to \infty$ as $x \to (\pi/2)^-$, so the limit is indeterminate. Here we use a common denominator:

$$\lim_{x \to (\pi/2)^-} (\sec x - \tan x) = \lim_{x \to (\pi/2)^-} \left(\frac{1}{\cos x} - \frac{\sin x}{\cos x} \right)$$

$$= \lim_{x \to (\pi/2)^-} \frac{1 - \sin x}{\cos x} = \lim_{x \to (\pi/2)^-} \frac{-\cos x}{-\sin x} = 0$$

Note that the use of l'Hospital's Rule is justified because $1 - \sin x \to 0$ and $\cos x \to 0$ as $x \to (\pi/2)^-$. □

INDETERMINATE POWERS

Several indeterminate forms arise from the limit

$$\lim_{x \to a} [f(x)]^{g(x)}$$

1. $\lim_{x \to a} f(x) = 0$ and $\lim_{x \to a} g(x) = 0$ type 0^0
2. $\lim_{x \to a} f(x) = \infty$ and $\lim_{x \to a} g(x) = 0$ type ∞^0
3. $\lim_{x \to a} f(x) = 1$ and $\lim_{x \to a} g(x) = \pm\infty$ type 1^∞

Each of these three cases can be treated either by taking the natural logarithm:

$$\text{let} \quad y = [f(x)]^{g(x)}, \quad \text{then} \quad \ln y = g(x) \ln f(x)$$

or by writing the function as an exponential:

$$[f(x)]^{g(x)} = e^{g(x) \ln f(x)}$$

(Recall that both of these methods were used in differentiating such functions.) In either method we are led to the indeterminate product $g(x) \ln f(x)$, which is of type $0 \cdot \infty$.

EXAMPLE 8 Calculate $\lim_{x \to 0^+} (1 + \sin 4x)^{\cot x}$.

SOLUTION First notice that as $x \to 0^+$, we have $1 + \sin 4x \to 1$ and $\cot x \to \infty$, so the given limit is indeterminate. Let

$$y = (1 + \sin 4x)^{\cot x}$$

Then $$\ln y = \ln[(1 + \sin 4x)^{\cot x}] = \cot x \ln(1 + \sin 4x)$$

so l'Hospital's Rule gives

$$\lim_{x \to 0^+} \ln y = \lim_{x \to 0^+} \frac{\ln(1 + \sin 4x)}{\tan x} = \lim_{x \to 0^+} \frac{\dfrac{4 \cos 4x}{1 + \sin 4x}}{\sec^2 x} = 4$$

So far we have computed the limit of $\ln y$, but what we want is the limit of y. To find this we use the fact that $y = e^{\ln y}$:

$$\lim_{x \to 0^+} (1 + \sin 4x)^{\cot x} = \lim_{x \to 0^+} y = \lim_{x \to 0^+} e^{\ln y} = e^4$$ □

■ The graph of the function $y = x^x$, $x > 0$, is shown in Figure 6. Notice that although 0^0 is not defined, the values of the function approach 1 as $x \to 0^+$. This confirms the result of Example 9.

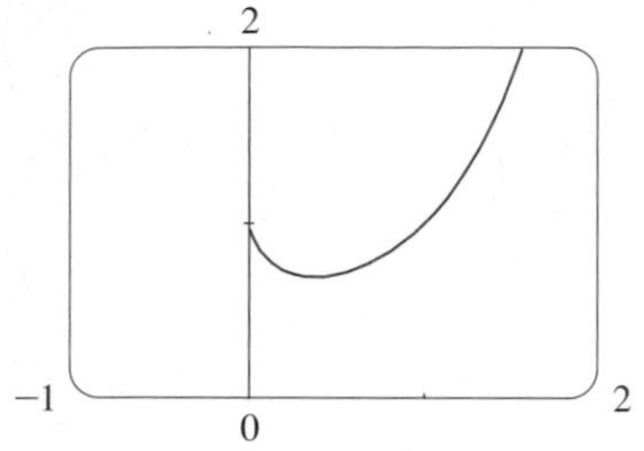

FIGURE 6

EXAMPLE 9 Find $\lim_{x \to 0^+} x^x$.

SOLUTION Notice that this limit is indeterminate since $0^x = 0$ for any $x > 0$ but $x^0 = 1$ for any $x \neq 0$. We could proceed as in Example 8 or by writing the function as an exponential:

$$x^x = (e^{\ln x})^x = e^{x \ln x}$$

In Example 6 we used l'Hospital's Rule to show that

$$\lim_{x \to 0^+} x \ln x = 0$$

Therefore

$$\lim_{x \to 0^+} x^x = \lim_{x \to 0^+} e^{x \ln x} = e^0 = 1$$ □

4.4 EXERCISES

1–4 Given that

$$\lim_{x \to a} f(x) = 0 \qquad \lim_{x \to a} g(x) = 0 \qquad \lim_{x \to a} h(x) = 1$$

$$\lim_{x \to a} p(x) = \infty \qquad \lim_{x \to a} q(x) = \infty$$

which of the following limits are indeterminate forms? For those that are not an indeterminate form, evaluate the limit where possible.

1. (a) $\lim_{x \to a} \frac{f(x)}{g(x)}$ (b) $\lim_{x \to a} \frac{f(x)}{p(x)}$

(c) $\lim_{x \to a} \frac{h(x)}{p(x)}$ (d) $\lim_{x \to a} \frac{p(x)}{f(x)}$

(e) $\lim_{x \to a} \frac{p(x)}{q(x)}$

2. (a) $\lim_{x \to a} [f(x)p(x)]$ (b) $\lim_{x \to a} [h(x)p(x)]$

(c) $\lim_{x \to a} [p(x)q(x)]$

3. (a) $\lim_{x \to a} [f(x) - p(x)]$ (b) $\lim_{x \to a} [p(x) - q(x)]$

(c) $\lim_{x \to a} [p(x) + q(x)]$

4. (a) $\lim_{x \to a} [f(x)]^{g(x)}$ (b) $\lim_{x \to a} [f(x)]^{p(x)}$ (c) $\lim_{x \to a} [h(x)]^{p(x)}$

(d) $\lim_{x \to a} [p(x)]^{f(x)}$ (e) $\lim_{x \to a} [p(x)]^{q(x)}$ (f) $\lim_{x \to a} \sqrt[q(x)]{p(x)}$

5–64 Find the limit. Use l'Hospital's Rule where appropriate. If there is a more elementary method, consider using it. If l'Hospital's Rule doesn't apply, explain why.

5. $\lim_{x \to 1} \frac{x^2 - 1}{x^2 - x}$

6. $\lim_{x \to 2} \frac{x^2 + x - 6}{x - 2}$

7. $\lim_{x \to 1} \frac{x^9 - 1}{x^5 - 1}$

8. $\lim_{x \to 1} \frac{x^a - 1}{x^b - 1}$

9. $\lim_{x \to (\pi/2)^+} \frac{\cos x}{1 - \sin x}$

10. $\lim_{x \to 0} \frac{\sin 4x}{\tan 5x}$

11. $\lim_{t \to 0} \frac{e^t - 1}{t^3}$

12. $\lim_{t \to 0} \frac{e^{3t} - 1}{t}$

13. $\lim_{x \to 0} \frac{\tan px}{\tan qx}$

14. $\lim_{\theta \to \pi/2} \frac{1 - \sin \theta}{\csc \theta}$

15. $\lim_{x \to \infty} \frac{\ln x}{\sqrt{x}}$

16. $\lim_{x \to \infty} \frac{x + x^2}{1 - 2x^2}$

17. $\lim_{x \to 0^+} \frac{\ln x}{x}$

18. $\lim_{x \to \infty} \frac{\ln \ln x}{x}$

19. $\lim_{x \to \infty} \frac{e^x}{x^3}$

20. $\lim_{x \to 1} \frac{\ln x}{\sin \pi x}$

21. $\lim_{x \to 0} \frac{e^x - 1 - x}{x^2}$

22. $\lim_{x \to 0} \frac{e^x - 1 - x - \frac{1}{2}x^2}{x^3}$

23. $\lim_{x \to 0} \dfrac{\tanh x}{\tan x}$

24. $\lim_{x \to 0} \dfrac{x - \sin x}{x - \tan x}$

25. $\lim_{t \to 0} \dfrac{5^t - 3^t}{t}$

26. $\lim_{x \to 0} \dfrac{\sin x - x}{x^3}$

27. $\lim_{x \to 0} \dfrac{\sin^{-1} x}{x}$

28. $\lim_{x \to \infty} \dfrac{(\ln x)^2}{x}$

29. $\lim_{x \to 0} \dfrac{1 - \cos x}{x^2}$

30. $\lim_{x \to 0} \dfrac{\cos mx - \cos nx}{x^2}$

31. $\lim_{x \to 0} \dfrac{x + \sin x}{x + \cos x}$

32. $\lim_{x \to 0} \dfrac{x}{\tan^{-1}(4x)}$

33. $\lim_{x \to 1} \dfrac{1 - x + \ln x}{1 + \cos \pi x}$

34. $\lim_{x \to \infty} \dfrac{\sqrt{x^2 + 2}}{\sqrt{2x^2 + 1}}$

35. $\lim_{x \to 1} \dfrac{x^a - ax + a - 1}{(x - 1)^2}$

36. $\lim_{x \to 0} \dfrac{e^x - e^{-x} - 2x}{x - \sin x}$

37. $\lim_{x \to 0} \dfrac{\cos x - 1 + \frac{1}{2}x^2}{x^4}$

38. $\lim_{x \to a^+} \dfrac{\cos x \ln(x - a)}{\ln(e^x - e^a)}$

39. $\lim_{x \to \infty} x \sin(\pi/x)$

40. $\lim_{x \to -\infty} x^2 e^x$

41. $\lim_{x \to 0} \cot 2x \sin 6x$

42. $\lim_{x \to 0^+} \sin x \ln x$

43. $\lim_{x \to \infty} x^3 e^{-x^2}$

44. $\lim_{x \to \pi/4} (1 - \tan x) \sec x$

45. $\lim_{x \to 1^+} \ln x \tan(\pi x/2)$

46. $\lim_{x \to \infty} x \tan(1/x)$

47. $\lim_{x \to 1} \left(\dfrac{x}{x - 1} - \dfrac{1}{\ln x} \right)$

48. $\lim_{x \to 0} (\csc x - \cot x)$

49. $\lim_{x \to \infty} \left(\sqrt{x^2 + x} - x \right)$

50. $\lim_{x \to 0} \left(\cot x - \dfrac{1}{x} \right)$

51. $\lim_{x \to \infty} (x - \ln x)$

52. $\lim_{x \to \infty} (xe^{1/x} - x)$

53. $\lim_{x \to 0^+} x^{x^2}$

54. $\lim_{x \to 0^+} (\tan 2x)^x$

55. $\lim_{x \to 0} (1 - 2x)^{1/x}$

56. $\lim_{x \to \infty} \left(1 + \dfrac{a}{x} \right)^{bx}$

57. $\lim_{x \to \infty} \left(1 + \dfrac{3}{x} + \dfrac{5}{x^2} \right)^x$

58. $\lim_{x \to \infty} x^{(\ln 2)/(1 + \ln x)}$

59. $\lim_{x \to \infty} x^{1/x}$

60. $\lim_{x \to \infty} (e^x + x)^{1/x}$

61. $\lim_{x \to 0^+} (4x + 1)^{\cot x}$

62. $\lim_{x \to 1} (2 - x)^{\tan(\pi x/2)}$

63. $\lim_{x \to 0^+} (\cos x)^{1/x^2}$

64. $\lim_{x \to \infty} \left(\dfrac{2x - 3}{2x + 5} \right)^{2x+1}$

65–66 Use a graph to estimate the value of the limit. Then use l'Hospital's Rule to find the exact value.

65. $\lim_{x \to \infty} \left(1 + \dfrac{2}{x} \right)^x$

66. $\lim_{x \to 0} \dfrac{5^x - 4^x}{3^x - 2^x}$

67–68 Illustrate l'Hospital's Rule by graphing both $f(x)/g(x)$ and $f'(x)/g'(x)$ near $x = 0$ to see that these ratios have the same limit as $x \to 0$. Also calculate the exact value of the limit.

67. $f(x) = e^x - 1, \quad g(x) = x^3 + 4x$

68. $f(x) = 2x \sin x, \quad g(x) = \sec x - 1$

69. Prove that

$$\lim_{x \to \infty} \frac{e^x}{x^n} = \infty$$

for any positive integer n. This shows that the exponential function approaches infinity faster than any power of x.

70. Prove that

$$\lim_{x \to \infty} \frac{\ln x}{x^p} = 0$$

for any number $p > 0$. This shows that the logarithmic function approaches ∞ more slowly than any power of x.

71. What happens if you try to use l'Hospital's Rule to evaluate

$$\lim_{x \to \infty} \frac{x}{\sqrt{x^2 + 1}}$$

Evaluate the limit using another method.

72. If an object with mass m is dropped from rest, one model for its speed v after t seconds, taking air resistance into account, is

$$v = \frac{mg}{c}(1 - e^{-ct/m})$$

where g is the acceleration due to gravity and c is a positive constant. (In Chapter 9 we will be able to deduce this equation from the assumption that the air resistance is proportional to the speed of the object; c is the proportionality constant.)

(a) Calculate $\lim_{t \to \infty} v$. What is the meaning of this limit?

(b) For fixed t, use l'Hospital's Rule to calculate $\lim_{c \to 0^+} v$. What can you conclude about the velocity of a falling object in a vacuum?

73. If an initial amount A_0 of money is invested at an interest rate r compounded n times a year, the value of the investment after t years is

$$A = A_0\left(1 + \frac{r}{n}\right)^{nt}$$

If we let $n \to \infty$, we refer to the *continuous compounding* of interest. Use l'Hospital's Rule to show that if interest is compounded continuously, then the amount after t years is

$$A = A_0 e^{rt}$$

74. If a metal ball with mass m is projected in water and the force of resistance is proportional to the square of the velocity, then the distance the ball travels in time t is

$$s(t) = \frac{m}{c} \ln \cosh \sqrt{\frac{gc}{mt}}$$

where c is a positive constant. Find $\lim_{c \to 0^+} s(t)$.

75. If an electrostatic field E acts on a liquid or a gaseous polar dielectric, the net dipole moment P per unit volume is

$$P(E) = \frac{e^E + e^{-E}}{e^E - e^{-E}} - \frac{1}{E}$$

Show that $\lim_{E \to 0^+} P(E) = 0$.

76. A metal cable has radius r and is covered by insulation, so that the distance from the center of the cable to the exterior of the insulation is R. The velocity v of an electrical impulse in the cable is

$$v = -c\left(\frac{r}{R}\right)^2 \ln\left(\frac{r}{R}\right)$$

where c is a positive constant. Find the following limits and interpret your answers.

(a) $\lim_{R \to r^+} v$ (b) $\lim_{r \to 0^+} v$

77. The first appearance in print of l'Hospital's Rule was in the book *Analyse des Infiniment Petits* published by the Marquis de l'Hospital in 1696. This was the first calculus *textbook* ever published and the example that the Marquis used in that book to illustrate his rule was to find the limit of the function

$$y = \frac{\sqrt{2a^3x - x^4} - a\sqrt[3]{aax}}{a - \sqrt[4]{ax^3}}$$

as x approaches a, where $a > 0$. (At that time it was common to write aa instead of a^2.) Solve this problem.

78. The figure shows a sector of a circle with central angle θ. Let $A(\theta)$ be the area of the segment between the chord PR and the arc PR. Let $B(\theta)$ be the area of the triangle PQR. Find $\lim_{\theta \to 0^+} A(\theta)/B(\theta)$.

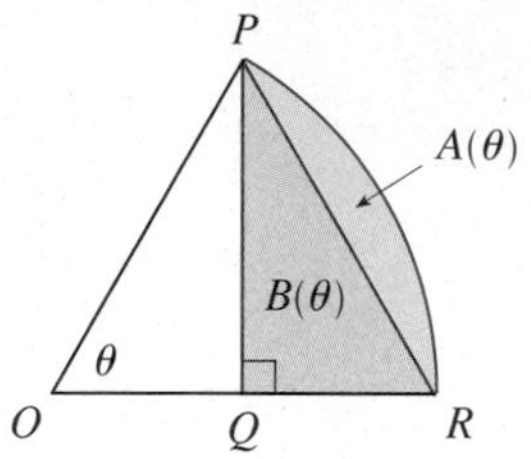

79. If f' is continuous, $f(2) = 0$, and $f'(2) = 7$, evaluate

$$\lim_{x \to 0} \frac{f(2 + 3x) + f(2 + 5x)}{x}$$

80. For what values of a and b is the following equation true?

$$\lim_{x \to 0} \left(\frac{\sin 2x}{x^3} + a + \frac{b}{x^2}\right) = 0$$

81. If f' is continuous, use l'Hospital's Rule to show that

$$\lim_{h \to 0} \frac{f(x + h) - f(x - h)}{2h} = f'(x)$$

Explain the meaning of this equation with the aid of a diagram.

82. If f'' is continuous, show that

$$\lim_{h \to 0} \frac{f(x + h) - 2f(x) + f(x - h)}{h^2} = f''(x)$$

83. Let

$$f(x) = \begin{cases} e^{-1/x^2} & \text{if } x \neq 0 \\ 0 & \text{if } x = 0 \end{cases}$$

(a) Use the definition of derivative to compute $f'(0)$.

(b) Show that f has derivatives of all orders that are defined on $\mathbb{R}$. [*Hint:* First show by induction that there is a polynomial $p_n(x)$ and a nonnegative integer k_n such that $f^{(n)}(x) = p_n(x)f(x)/x^{k_n}$ for $x \neq 0$.]

84. Let

$$f(x) = \begin{cases} |x|^x & \text{if } x \neq 0 \\ 1 & \text{if } x = 0 \end{cases}$$

(a) Show that f is continuous at 0.

(b) Investigate graphically whether f is differentiable at 0 by zooming in several times toward the point (0, 1) on the graph of f.

(c) Show that f is not differentiable at 0. How can you reconcile this fact with the appearance of the graphs in part (b)?

WRITING PROJECT

THE ORIGINS OF L'HOSPITAL'S RULE

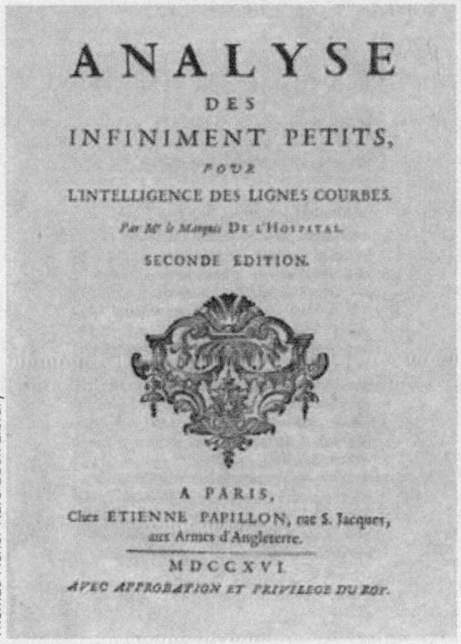

www.stewartcalculus.com
The Internet is another source of information for this project. Click on *History of Mathematics* for a list of reliable websites.

L'Hospital's Rule was first published in 1696 in the Marquis de l'Hospital's calculus textbook *Analyse des Infiniment Petits,* but the rule was discovered in 1694 by the Swiss mathematician John (Johann) Bernoulli. The explanation is that these two mathematicians had entered into a curious business arrangement whereby the Marquis de l'Hospital bought the rights to Bernoulli's mathematical discoveries. The details, including a translation of l'Hospital's letter to Bernoulli proposing the arrangement, can be found in the book by Eves [1].

Write a report on the historical and mathematical origins of l'Hospital's Rule. Start by providing brief biographical details of both men (the dictionary edited by Gillispie [2] is a good source) and outline the business deal between them. Then give l'Hospital's statement of his rule, which is found in Struik's sourcebook [4] and more briefly in the book of Katz [3]. Notice that l'Hospital and Bernoulli formulated the rule geometrically and gave the answer in terms of differentials. Compare their statement with the version of l'Hospital's Rule given in Section 4.4 and show that the two statements are essentially the same.

1. Howard Eves, *In Mathematical Circles (Volume 2: Quadrants III and IV)* (Boston: Prindle, Weber and Schmidt, 1969), pp. 20–22.
2. C. C. Gillispie, ed., *Dictionary of Scientific Biography* (New York: Scribner's, 1974). See the article on Johann Bernoulli by E. A. Fellmann and J. O. Fleckenstein in Volume II and the article on the Marquis de l'Hospital by Abraham Robinson in Volume VIII.
3. Victor Katz, *A History of Mathematics: An Introduction* (New York: HarperCollins, 1993), p. 484.
4. D. J. Struik, ed., *A Sourcebook in Mathematics, 1200–1800* (Princeton, NJ: Princeton University Press, 1969), pp. 315–316.

4.5 SUMMARY OF CURVE SKETCHING

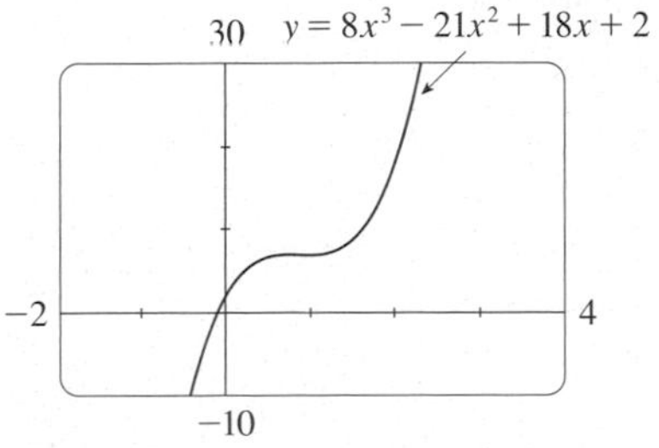

FIGURE 1

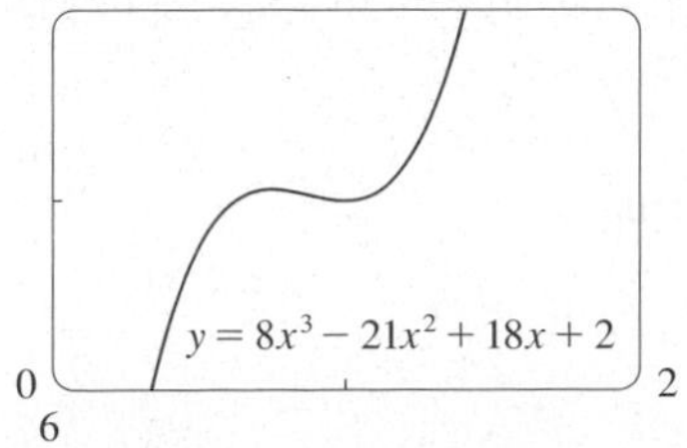

FIGURE 2

So far we have been concerned with some particular aspects of curve sketching: domain, range, and symmetry in Chapter 1; limits, continuity, and asymptotes in Chapter 2; derivatives and tangents in Chapters 2 and 3; and extreme values, intervals of increase and decrease, concavity, points of inflection, and l'Hospital's Rule in this chapter. It is now time to put all of this information together to sketch graphs that reveal the important features of functions.

You might ask: Why don't we just use a graphing calculator or computer to graph a curve? Why do we need to use calculus?

It's true that modern technology is capable of producing very accurate graphs. But even the best graphing devices have to be used intelligently. We saw in Section 1.4 that it is extremely important to choose an appropriate viewing rectangle to avoid getting a misleading graph. (See especially Examples 1, 3, 4, and 5 in that section.) The use of calculus enables us to discover the most interesting aspects of graphs and in many cases to calculate maximum and minimum points and inflection points *exactly* instead of approximately.

For instance, Figure 1 shows the graph of $f(x) = 8x^3 - 21x^2 + 18x + 2$. At first glance it seems reasonable: It has the same shape as cubic curves like $y = x^3$, and it appears to have no maximum or minimum point. But if you compute the derivative, you will see that there is a maximum when $x = 0.75$ and a minimum when $x = 1$. Indeed, if we zoom in to this portion of the graph, we see that behavior exhibited in Figure 2. Without calculus, we could easily have overlooked it.

In the next section we will graph functions by using the interaction between calculus and graphing devices. In this section we draw graphs by first considering the following information. We don't assume that you have a graphing device, but if you do have one you should use it as a check on your work.

GUIDELINES FOR SKETCHING A CURVE

The following checklist is intended as a guide to sketching a curve $y = f(x)$ by hand. Not every item is relevant to every function. (For instance, a given curve might not have an asymptote or possess symmetry.) But the guidelines provide all the information you need to make a sketch that displays the most important aspects of the function.

A. Domain It's often useful to start by determining the domain D of f, that is, the set of values of x for which $f(x)$ is defined.

B. Intercepts The y-intercept is $f(0)$ and this tells us where the curve intersects the y-axis. To find the x-intercepts, we set $y = 0$ and solve for x. (You can omit this step if the equation is difficult to solve.)

C. Symmetry

(i) If $f(-x) = f(x)$ for all x in D, that is, the equation of the curve is unchanged when x is replaced by $-x$, then f is an **even function** and the curve is symmetric about the y-axis. This means that our work is cut in half. If we know what the curve looks like for $x \geqslant 0$, then we need only reflect about the y-axis to obtain the complete curve [see Figure 3(a)]. Here are some examples: $y = x^2$, $y = x^4$, $y = |x|$, and $y = \cos x$.

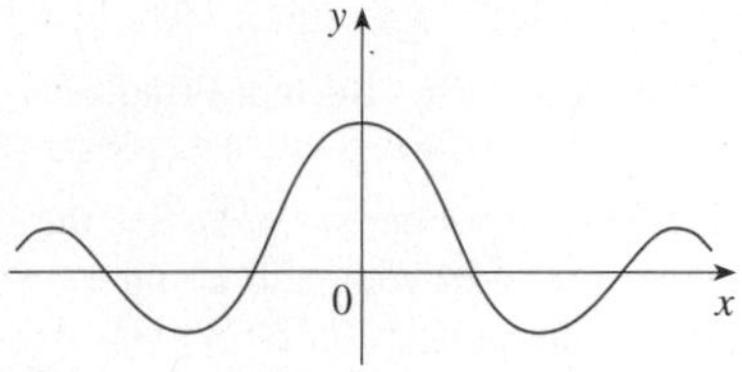

(a) Even function: reflectional symmetry

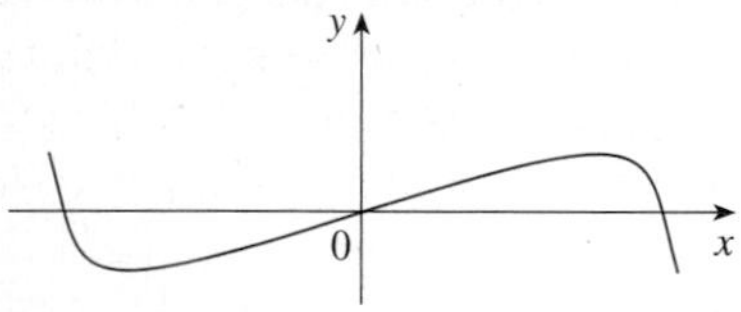

(b) Odd function: rotational symmetry

FIGURE 3

(ii) If $f(-x) = -f(x)$ for all x in D, then f is an **odd function** and the curve is symmetric about the origin. Again we can obtain the complete curve if we know what it looks like for $x \geqslant 0$. [Rotate 180° about the origin; see Figure 3(b).] Some simple examples of odd functions are $y = x$, $y = x^3$, $y = x^5$, and $y = \sin x$.

(iii) If $f(x + p) = f(x)$ for all x in D, where p is a positive constant, then f is called a **periodic function** and the smallest such number p is called the **period.** For instance, $y = \sin x$ has period 2π and $y = \tan x$ has period π. If we know what the graph looks like in an interval of length p, then we can use translation to sketch the entire graph (see Figure 4).

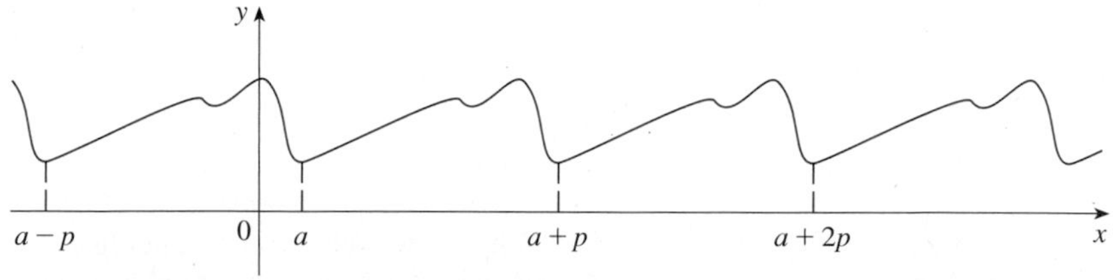

FIGURE 4
Periodic function: translational symmetry

D. Asymptotes

(i) *Horizontal Asymptotes.* Recall from Section 2.6 that if either $\lim_{x\to\infty} f(x) = L$ or $\lim_{x\to-\infty} f(x) = L$, then the line $y = L$ is a horizontal asymptote of the curve $y = f(x)$. If it turns out that $\lim_{x\to\infty} f(x) = \infty$ (or $-\infty$), then we do not have an asymptote to the right, but that is still useful information for sketching the curve.

(ii) *Vertical Asymptotes.* Recall from Section 2.2 that the line $x = a$ is a vertical asymptote if at least one of the following statements is true:

1
$$\lim_{x\to a^+} f(x) = \infty \qquad \lim_{x\to a^-} f(x) = \infty$$

$$\lim_{x\to a^+} f(x) = -\infty \qquad \lim_{x\to a^-} f(x) = -\infty$$

(For rational functions you can locate the vertical asymptotes by equating the denominator to 0 after canceling any common factors. But for other functions this method does not apply.) Furthermore, in sketching the curve it is very useful to know exactly which of the statements in (1) is true. If $f(a)$ is not defined but a is an endpoint of the domain of f, then you should compute $\lim_{x \to a^-} f(x)$ or $\lim_{x \to a^+} f(x)$, whether or not this limit is infinite.

(iii) *Slant Asymptotes.* These are discussed at the end of this section.

E. Intervals of Increase or Decrease Use the I/D Test. Compute $f'(x)$ and find the intervals on which $f'(x)$ is positive (f is increasing) and the intervals on which $f'(x)$ is negative (f is decreasing).

F. Local Maximum and Minimum Values Find the critical numbers of f [the numbers c where $f'(c) = 0$ or $f'(c)$ does not exist]. Then use the First Derivative Test. If f' changes from positive to negative at a critical number c, then $f(c)$ is a local maximum. If f' changes from negative to positive at c, then $f(c)$ is a local minimum. Although it is usually preferable to use the First Derivative Test, you can use the Second Derivative Test if $f'(c) = 0$ and $f''(c) \neq 0$. Then $f''(c) > 0$ implies that $f(c)$ is a local minimum, whereas $f''(c) < 0$ implies that $f(c)$ is a local maximum.

G. Concavity and Points of Inflection Compute $f''(x)$ and use the Concavity Test. The curve is concave upward where $f''(x) > 0$ and concave downward where $f''(x) < 0$. Inflection points occur where the direction of concavity changes.

H. Sketch the Curve Using the information in items A–G, draw the graph. Sketch the asymptotes as dashed lines. Plot the intercepts, maximum and minimum points, and inflection points. Then make the curve pass through these points, rising and falling according to E, with concavity according to G, and approaching the asymptotes. If additional accuracy is desired near any point, you can compute the value of the derivative there. The tangent indicates the direction in which the curve proceeds.

V EXAMPLE 1 Use the guidelines to sketch the curve $y = \dfrac{2x^2}{x^2 - 1}$.

A. The domain is

$$\{x \mid x^2 - 1 \neq 0\} = \{x \mid x \neq \pm 1\} = (-\infty, -1) \cup (-1, 1) \cup (1, \infty)$$

B. The x- and y-intercepts are both 0.

C. Since $f(-x) = f(x)$, the function f is even. The curve is symmetric about the y-axis.

D.
$$\lim_{x \to \pm\infty} \frac{2x^2}{x^2 - 1} = \lim_{x \to \pm\infty} \frac{2}{1 - 1/x^2} = 2$$

Therefore the line $y = 2$ is a horizontal asymptote.

Since the denominator is 0 when $x = \pm 1$, we compute the following limits:

$$\lim_{x \to 1^+} \frac{2x^2}{x^2 - 1} = \infty \qquad \lim_{x \to 1^-} \frac{2x^2}{x^2 - 1} = -\infty$$

$$\lim_{x \to -1^+} \frac{2x^2}{x^2 - 1} = -\infty \qquad \lim_{x \to -1^-} \frac{2x^2}{x^2 - 1} = \infty$$

Therefore the lines $x = 1$ and $x = -1$ are vertical asymptotes. This information about limits and asymptotes enables us to draw the preliminary sketch in Figure 5, showing the parts of the curve near the asymptotes.

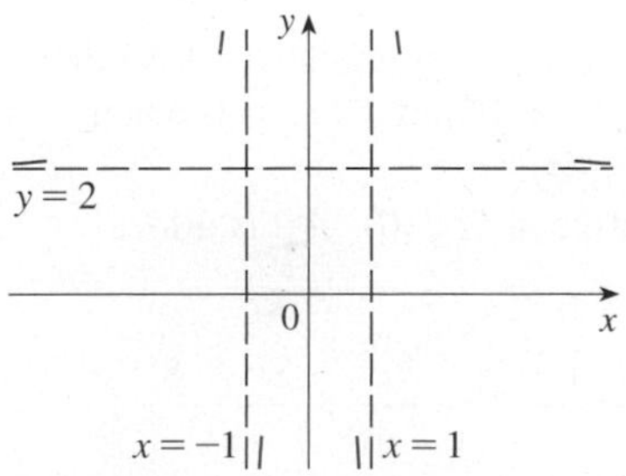

FIGURE 5
Preliminary sketch

■ We have shown the curve approaching its horizontal asymptote from above in Figure 5. This is confirmed by the intervals of increase and decrease.

E. $$f'(x) = \frac{4x(x^2-1) - 2x^2 \cdot 2x}{(x^2-1)^2} = \frac{-4x}{(x^2-1)^2}$$

Since $f'(x) > 0$ when $x < 0$ ($x \neq -1$) and $f'(x) < 0$ when $x > 0$ ($x \neq 1$), f is increasing on $(-\infty, -1)$ and $(-1, 0)$ and decreasing on $(0, 1)$ and $(1, \infty)$.

F. The only critical number is $x = 0$. Since f' changes from positive to negative at 0, $f(0) = 0$ is a local maximum by the First Derivative Test.

G. $$f''(x) = \frac{-4(x^2-1)^2 + 4x \cdot 2(x^2-1)2x}{(x^2-1)^4} = \frac{12x^2+4}{(x^2-1)^3}$$

Since $12x^2 + 4 > 0$ for all x, we have

$$f''(x) > 0 \iff x^2 - 1 > 0 \iff |x| > 1$$

and $f''(x) < 0 \iff |x| < 1$. Thus the curve is concave upward on the intervals $(-\infty, -1)$ and $(1, \infty)$ and concave downward on $(-1, 1)$. It has no point of inflection since 1 and -1 are not in the domain of f.

H. Using the information in E–G, we finish the sketch in Figure 6. □

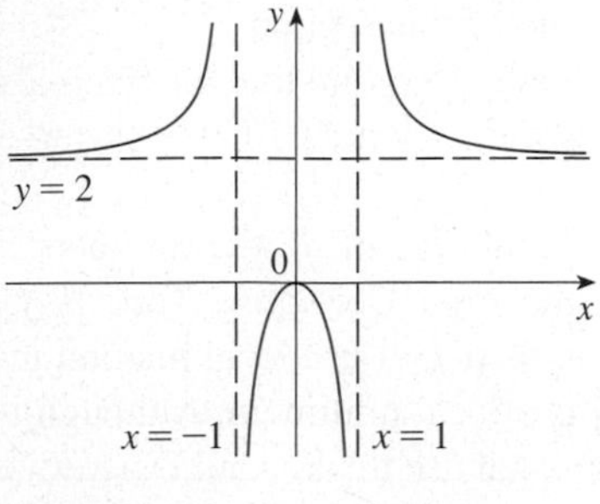

FIGURE 6
Finished sketch of $y = \dfrac{2x^2}{x^2-1}$

EXAMPLE 2 Sketch the graph of $f(x) = \dfrac{x^2}{\sqrt{x+1}}$.

A. Domain $= \{x \mid x + 1 > 0\} = \{x \mid x > -1\} = (-1, \infty)$

B. The x- and y-intercepts are both 0.

C. Symmetry: None

D. Since

$$\lim_{x\to\infty} \frac{x^2}{\sqrt{x+1}} = \infty$$

there is no horizontal asymptote. Since $\sqrt{x+1} \to 0$ as $x \to -1^+$ and $f(x)$ is always positive, we have

$$\lim_{x\to -1^+} \frac{x^2}{\sqrt{x+1}} = \infty$$

and so the line $x = -1$ is a vertical asymptote.

E. $$f'(x) = \frac{2x\sqrt{x+1} - x^2 \cdot 1/(2\sqrt{x+1})}{x+1} = \frac{x(3x+4)}{2(x+1)^{3/2}}$$

We see that $f'(x) = 0$ when $x = 0$ (notice that $-\frac{4}{3}$ is not in the domain of f), so the only critical number is 0. Since $f'(x) < 0$ when $-1 < x < 0$ and $f'(x) > 0$ when $x > 0$, f is decreasing on $(-1, 0)$ and increasing on $(0, \infty)$.

F. Since $f'(0) = 0$ and f' changes from negative to positive at 0, $f(0) = 0$ is a local (and absolute) minimum by the First Derivative Test.

G. $$f''(x) = \frac{2(x+1)^{3/2}(6x+4) - (3x^2+4x)3(x+1)^{1/2}}{4(x+1)^3} = \frac{3x^2+8x+8}{4(x+1)^{5/2}}$$

Note that the denominator is always positive. The numerator is the quadratic $3x^2 + 8x + 8$, which is always positive because its discriminant is $b^2 - 4ac = -32$, which is negative, and the coefficient of x^2 is positive. Thus $f''(x) > 0$ for all x in the domain of f, which means that f is concave upward on $(-1, \infty)$ and there is no point of inflection.

H. The curve is sketched in Figure 7. □

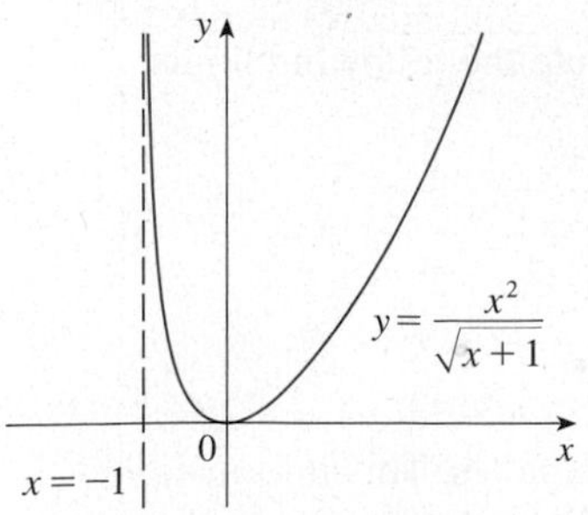

FIGURE 7

V EXAMPLE 3 Sketch the graph of $f(x) = xe^x$.

A. The domain is $\mathbb{R}$.

B. The x- and y-intercepts are both 0.

C. Symmetry: None

D. Because both x and e^x become large as $x \to \infty$, we have $\lim_{x\to\infty} xe^x = \infty$. As $x \to -\infty$, however, $e^x \to 0$ and so we have an indeterminate product that requires the use of l'Hospital's Rule:

$$\lim_{x\to-\infty} xe^x = \lim_{x\to-\infty} \frac{x}{e^{-x}} = \lim_{x\to-\infty} \frac{1}{-e^{-x}} = \lim_{x\to-\infty} (-e^x) = 0$$

Thus the x-axis is a horizontal asymptote.

E.
$$f'(x) = xe^x + e^x = (x + 1)e^x$$

Since e^x is always positive, we see that $f'(x) > 0$ when $x + 1 > 0$, and $f'(x) < 0$ when $x + 1 < 0$. So f is increasing on $(-1, \infty)$ and decreasing on $(-\infty, -1)$.

F. Because $f'(-1) = 0$ and f' changes from negative to positive at $x = -1$, $f(-1) = -e^{-1}$ is a local (and absolute) minimum.

G.
$$f''(x) = (x + 1)e^x + e^x = (x + 2)e^x$$

Since $f''(x) > 0$ if $x > -2$ and $f''(x) < 0$ if $x < -2$, f is concave upward on $(-2, \infty)$ and concave downward on $(-\infty, -2)$. The inflection point is $(-2, -2e^{-2})$.

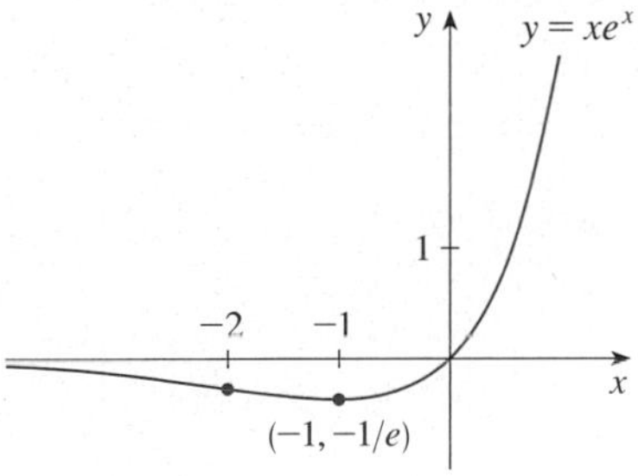

FIGURE 8

H. We use this information to sketch the curve in Figure 8. □

EXAMPLE 4 Sketch the graph of $f(x) = \dfrac{\cos x}{2 + \sin x}$.

A. The domain is $\mathbb{R}$.

B. The y-intercept is $f(0) = \frac{1}{2}$. The x-intercepts occur when $\cos x = 0$, that is, $x = (2n + 1)\pi/2$, where n is an integer.

C. f is neither even nor odd, but $f(x + 2\pi) = f(x)$ for all x and so f is periodic and has period 2π. Thus, in what follows, we need to consider only $0 \leqslant x \leqslant 2\pi$ and then extend the curve by translation in part H.

D. Asymptotes: None

E.
$$f'(x) = \frac{(2 + \sin x)(-\sin x) - \cos x\,(\cos x)}{(2 + \sin x)^2} = -\frac{2 \sin x + 1}{(2 + \sin x)^2}$$

Thus $f'(x) > 0$ when $2 \sin x + 1 < 0 \iff \sin x < -\frac{1}{2} \iff 7\pi/6 < x < 11\pi/6$. So f is increasing on $(7\pi/6, 11\pi/6)$ and decreasing on $(0, 7\pi/6)$ and $(11\pi/6, 2\pi)$.

F. From part E and the First Derivative Test, we see that the local minimum value is $f(7\pi/6) = -1/\sqrt{3}$ and the local maximum value is $f(11\pi/6) = 1/\sqrt{3}$.

G. If we use the Quotient Rule again and simplify, we get

$$f''(x) = -\frac{2 \cos x\,(1 - \sin x)}{(2 + \sin x)^3}$$

Because $(2 + \sin x)^3 > 0$ and $1 - \sin x \geqslant 0$ for all x, we know that $f''(x) > 0$ when $\cos x < 0$, that is, $\pi/2 < x < 3\pi/2$. So f is concave upward on $(\pi/2, 3\pi/2)$ and concave downward on $(0, \pi/2)$ and $(3\pi/2, 2\pi)$. The inflection points are $(\pi/2, 0)$ and $(3\pi/2, 0)$.

H. The graph of the function restricted to $0 \leqslant x \leqslant 2\pi$ is shown in Figure 9. Then we extend it, using periodicity, to the complete graph in Figure 10.

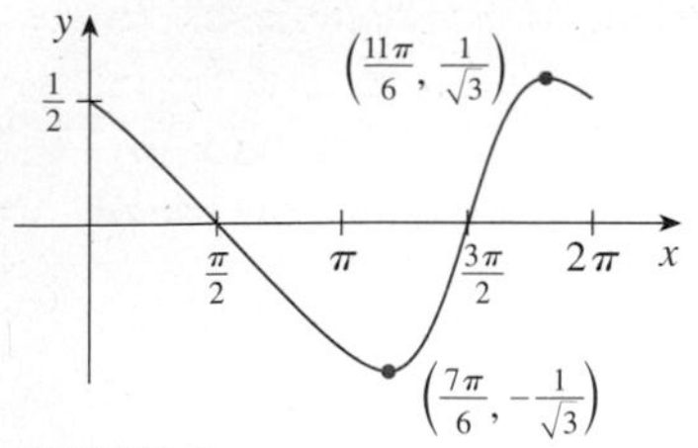

FIGURE 9

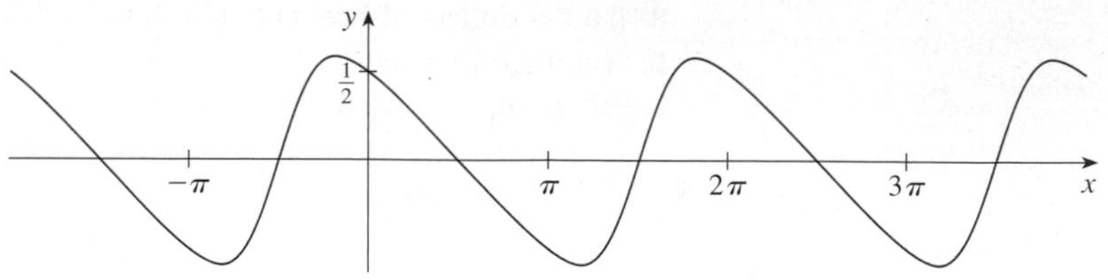

FIGURE 10

□

EXAMPLE 5 Sketch the graph of $y = \ln(4 - x^2)$.

A. The domain is

$$\{x \mid 4 - x^2 > 0\} = \{x \mid x^2 < 4\} = \{x \mid |x| < 2\} = (-2, 2)$$

B. The y-intercept is $f(0) = \ln 4$. To find the x-intercept we set

$$y = \ln(4 - x^2) = 0$$

We know that $\ln 1 = 0$, so we have $4 - x^2 = 1 \Rightarrow x^2 = 3$ and therefore the x-intercepts are $\pm\sqrt{3}$.

C. Since $f(-x) = f(x)$, f is even and the curve is symmetric about the y-axis.

D. We look for vertical asymptotes at the endpoints of the domain. Since $4 - x^2 \to 0^+$ as $x \to 2^-$ and also as $x \to -2^+$, we have

$$\lim_{x \to 2^-} \ln(4 - x^2) = -\infty \qquad \lim_{x \to -2^+} \ln(4 - x^2) = -\infty$$

Thus the lines $x = 2$ and $x = -2$ are vertical asymptotes.

E.
$$f'(x) = \frac{-2x}{4 - x^2}$$

Since $f'(x) > 0$ when $-2 < x < 0$ and $f'(x) < 0$ when $0 < x < 2$, f is increasing on $(-2, 0)$ and decreasing on $(0, 2)$.

F. The only critical number is $x = 0$. Since f' changes from positive to negative at 0, $f(0) = \ln 4$ is a local maximum by the First Derivative Test.

G.
$$f''(x) = \frac{(4 - x^2)(-2) + 2x(-2x)}{(4 - x^2)^2} = \frac{-8 - 2x^2}{(4 - x^2)^2}$$

Since $f''(x) < 0$ for all x, the curve is concave downward on $(-2, 2)$ and has no inflection point.

H. Using this information, we sketch the curve in Figure 11. □

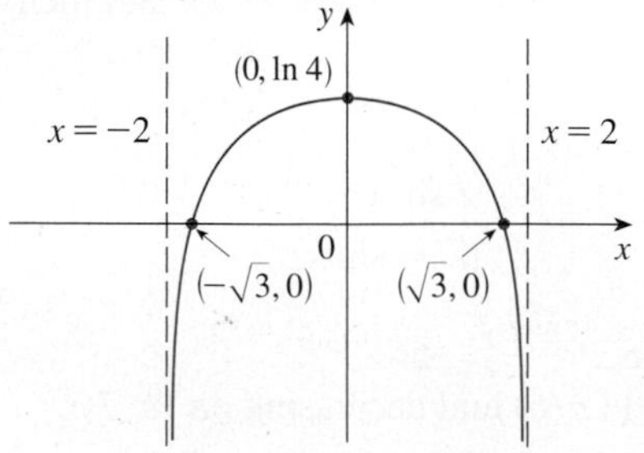

FIGURE 11
$y = \ln(4 - x^2)$

SLANT ASYMPTOTES

Some curves have asymptotes that are *oblique,* that is, neither horizontal nor vertical. If

$$\lim_{x \to \infty} [f(x) - (mx + b)] = 0$$

then the line $y = mx + b$ is called a **slant asymptote** because the vertical distance

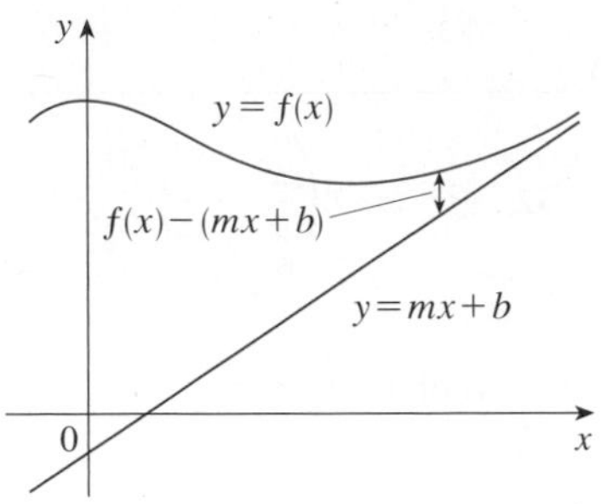

FIGURE 12

between the curve $y = f(x)$ and the line $y = mx + b$ approaches 0, as in Figure 12. (A similar situation exists if we let $x \to -\infty$.) For rational functions, slant asymptotes occur when the degree of the numerator is one more than the degree of the denominator. In such a case the equation of the slant asymptote can be found by long division as in the following example.

V EXAMPLE 6 Sketch the graph of $f(x) = \dfrac{x^3}{x^2 + 1}$.

A. The domain is $\mathbb{R} = (-\infty, \infty)$.
B. The x- and y-intercepts are both 0.
C. Since $f(-x) = -f(x)$, f is odd and its graph is symmetric about the origin.
D. Since $x^2 + 1$ is never 0, there is no vertical asymptote. Since $f(x) \to \infty$ as $x \to \infty$ and $f(x) \to -\infty$ as $x \to -\infty$, there is no horizontal asymptote. But long division gives

$$f(x) = \frac{x^3}{x^2 + 1} = x - \frac{x}{x^2 + 1}$$

$$f(x) - x = -\frac{x}{x^2 + 1} = -\frac{\frac{1}{x}}{1 + \frac{1}{x^2}} \to 0 \quad \text{as} \quad x \to \pm\infty$$

So the line $y = x$ is a slant asymptote.

E.
$$f'(x) = \frac{3x^2(x^2 + 1) - x^3 \cdot 2x}{(x^2 + 1)^2} = \frac{x^2(x^2 + 3)}{(x^2 + 1)^2}$$

Since $f'(x) > 0$ for all x (except 0), f is increasing on $(-\infty, \infty)$.
F. Although $f'(0) = 0$, f' does not change sign at 0, so there is no local maximum or minimum.

G.
$$f''(x) = \frac{(4x^3 + 6x)(x^2 + 1)^2 - (x^4 + 3x^2) \cdot 2(x^2 + 1)2x}{(x^2 + 1)^4} = \frac{2x(3 - x^2)}{(x^2 + 1)^3}$$

Since $f''(x) = 0$ when $x = 0$ or $x = \pm\sqrt{3}$, we set up the following chart:

Interval	x	$3 - x^2$	$(x^2 + 1)^3$	$f''(x)$	f
$x < -\sqrt{3}$	$-$	$-$	$+$	$+$	CU on $(-\infty, -\sqrt{3})$
$-\sqrt{3} < x < 0$	$-$	$+$	$+$	$-$	CD on $(-\sqrt{3}, 0)$
$0 < x < \sqrt{3}$	$+$	$+$	$+$	$+$	CU on $(0, \sqrt{3})$
$x > \sqrt{3}$	$+$	$-$	$+$	$-$	CD on $(\sqrt{3}, \infty)$

The points of inflection are $(-\sqrt{3}, -\frac{3}{4}\sqrt{3})$, $(0, 0)$, and $(\sqrt{3}, \frac{3}{4}\sqrt{3})$.
H. The graph of f is sketched in Figure 13. □

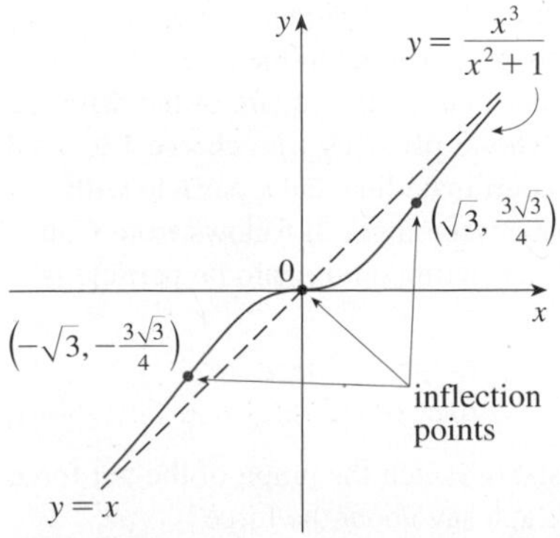

FIGURE 13

4.5 EXERCISES

1–52 Use the guidelines of this section to sketch the curve.

1. $y = x^3 + x$

2. $y = x^3 + 6x^2 + 9x$

3. $y = 2 - 15x + 9x^2 - x^3$

4. $y = 8x^2 - x^4$

5. $y = x^4 + 4x^3$

6. $y = x(x + 2)^3$

7. $y = 2x^5 - 5x^2 + 1$

8. $y = (4 - x^2)^5$

9. $y = \dfrac{x}{x - 1}$

10. $y = \dfrac{x^2 - 4}{x^2 - 2x}$

11. $y = \dfrac{1}{x^2 - 9}$

12. $y = \dfrac{x}{x^2 - 9}$

13. $y = \dfrac{x}{x^2 + 9}$

14. $y = \dfrac{x^2}{x^2 + 9}$

15. $y = \dfrac{x - 1}{x^2}$

16. $y = 1 + \dfrac{1}{x} + \dfrac{1}{x^2}$

17. $y = \dfrac{x^2}{x^2 + 3}$

18. $y = \dfrac{x}{x^3 - 1}$

19. $y = x\sqrt{5 - x}$

20. $y = 2\sqrt{x} - x$

21. $y = \sqrt{x^2 + x - 2}$

22. $y = \sqrt{x^2 + x} - x$

23. $y = \dfrac{x}{\sqrt{x^2 + 1}}$

24. $y = x\sqrt{2 - x^2}$

25. $y = \dfrac{\sqrt{1 - x^2}}{x}$

26. $y = \dfrac{x}{\sqrt{x^2 - 1}}$

27. $y = x - 3x^{1/3}$

28. $y = x^{5/3} - 5x^{2/3}$

29. $y = \sqrt[3]{x^2 - 1}$

30. $y = \sqrt[3]{x^3 + 1}$

31. $y = 3 \sin x - \sin^3 x$

32. $y = x + \cos x$

33. $y = x \tan x, \quad -\pi/2 < x < \pi/2$

34. $y = 2x - \tan x, \quad -\pi/2 < x < \pi/2$

35. $y = \frac{1}{2}x - \sin x, \quad 0 < x < 3\pi$

36. $y = \sec x + \tan x, \quad 0 < x < \pi/2$

37. $y = \dfrac{\sin x}{1 + \cos x}$

38. $y = \dfrac{\sin x}{2 + \cos x}$

39. $y = e^{\sin x}$

40. $y = e^{-x} \sin x, \quad 0 \leqslant x \leqslant 2\pi$

41. $y = 1/(1 + e^{-x})$

42. $y = e^{2x} - e^x$

43. $y = x - \ln x$

44. $y = e^x/x$

45. $y = (1 + e^x)^{-2}$

46. $y = \ln(x^2 - 3x + 2)$

47. $y = \ln(\sin x)$

48. $y = \dfrac{\ln x}{x^2}$

49. $y = xe^{-x^2}$

50. $y = (x^2 - 3)e^{-x}$

51. $y = e^{3x} + e^{-2x}$

52. $y = \tan^{-1}\left(\dfrac{x - 1}{x + 1}\right)$

---

53. In the theory of relativity, the mass of a particle is

$$m = \frac{m_0}{\sqrt{1 - v^2/c^2}}$$

where m_0 is the rest mass of the particle, m is the mass when the particle moves with speed v relative to the observer, and c is the speed of light. Sketch the graph of m as a function of v.

54. In the theory of relativity, the energy of a particle is

$$E = \sqrt{m_0^2 c^4 + h^2 c^2/\lambda^2}$$

where m_0 is the rest mass of the particle, λ is its wave length, and h is Planck's constant. Sketch the graph of E as a function of λ. What does the graph say about the energy?

55. The figure shows a beam of length L embedded in concrete walls. If a constant load W is distributed evenly along its length, the beam takes the shape of the deflection curve

$$y = -\frac{W}{24EI}x^4 + \frac{WL}{12EI}x^3 - \frac{WL^2}{24EI}x^2$$

where E and I are positive constants. (E is Young's modulus of elasticity and I is the moment of inertia of a cross-section of the beam.) Sketch the graph of the deflection curve.

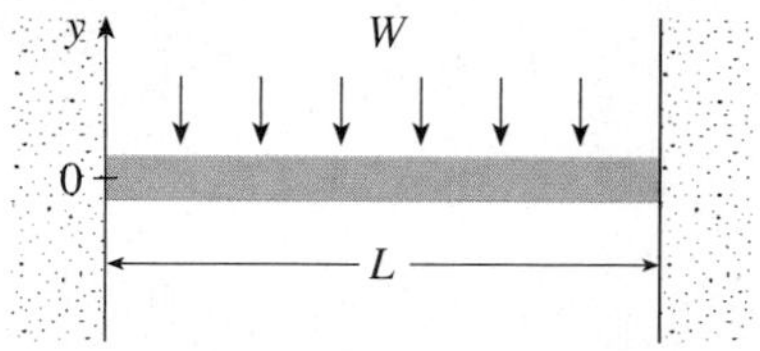

56. Coulomb's Law states that the force of attraction between two charged particles is directly proportional to the product of the charges and inversely proportional to the square of the distance between them. The figure shows particles with charge 1 located at positions 0 and 2 on a coordinate line and a particle with charge -1 at a position x between them. It follows from Coulomb's Law that the net force acting on the middle particle is

$$F(x) = -\frac{k}{x^2} + \frac{k}{(x - 2)^2} \qquad 0 < x < 2$$

where k is a positive constant. Sketch the graph of the net force function. What does the graph say about the force?

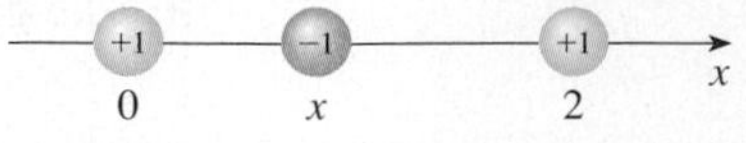

57–60 Find an equation of the slant asymptote. Do not sketch the curve.

57. $y = \dfrac{x^2 + 1}{x + 1}$

58. $y = \dfrac{2x^3 + x^2 + x + 3}{x^2 + 2x}$

59. $y = \dfrac{4x^3 - 2x^2 + 5}{2x^2 + x - 3}$

60. $y = \dfrac{5x^4 + x^2 + x}{x^3 - x^2 + 2}$

61–66 Use the guidelines of this section to sketch the curve. In guideline D find an equation of the slant asymptote.

61. $y = \dfrac{-2x^2 + 5x - 1}{2x - 1}$

62. $y = \dfrac{x^2 + 12}{x - 2}$

63. $xy = x^2 + 4$

64. $y = e^x - x$

65. $y = \dfrac{2x^3 + x^2 + 1}{x^2 + 1}$

66. $y = \dfrac{(x + 1)^3}{(x - 1)^2}$

67. Show that the curve $y = x - \tan^{-1}x$ has two slant asymptotes: $y = x + \pi/2$ and $y = x - \pi/2$. Use this fact to help sketch the curve.

68. Show that the curve $y = \sqrt{x^2 + 4x}$ has two slant asymptotes: $y = x + 2$ and $y = -x - 2$. Use this fact to help sketch the curve.

69. Show that the lines $y = (b/a)x$ and $y = -(b/a)x$ are slant asymptotes of the hyperbola $(x^2/a^2) - (y^2/b^2) = 1$.

70. Let $f(x) = (x^3 + 1)/x$. Show that

$$\lim_{x \to \pm\infty} [f(x) - x^2] = 0$$

This shows that the graph of f approaches the graph of $y = x^2$, and we say that the curve $y = f(x)$ is *asymptotic* to the parabola $y = x^2$. Use this fact to help sketch the graph of f.

71. Discuss the asymptotic behavior of $f(x) = (x^4 + 1)/x$ in the same manner as in Exercise 70. Then use your results to help sketch the graph of f.

72. Use the asymptotic behavior of $f(x) = \cos x + 1/x^2$ to sketch its graph without going through the curve-sketching procedure of this section.

4.6 GRAPHING WITH CALCULUS *AND* CALCULATORS

■ If you have not already read Section 1.4, you should do so now. In particular, it explains how to avoid some of the pitfalls of graphing devices by choosing appropriate viewing rectangles.

The method we used to sketch curves in the preceding section was a culmination of much of our study of differential calculus. The graph was the final object that we produced. In this section our point of view is completely different. Here we *start* with a graph produced by a graphing calculator or computer and then we refine it. We use calculus to make sure that we reveal all the important aspects of the curve. And with the use of graphing devices we can tackle curves that would be far too complicated to consider without technology. The theme is the *interaction* between calculus and calculators.

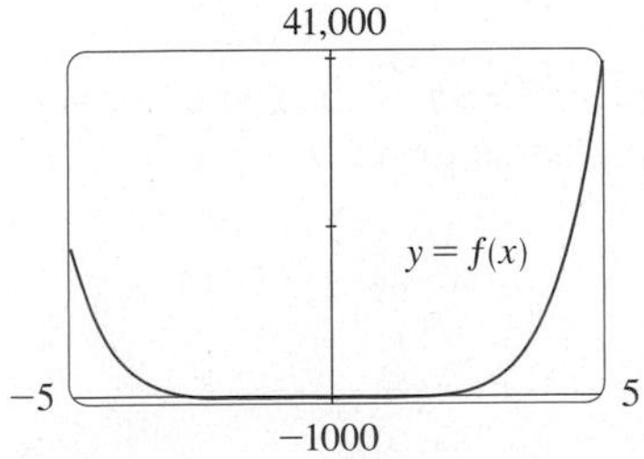

FIGURE 1

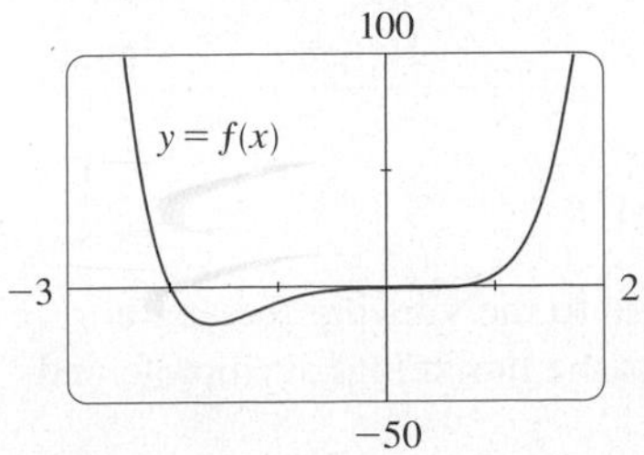

FIGURE 2

EXAMPLE 1 Graph the polynomial $f(x) = 2x^6 + 3x^5 + 3x^3 - 2x^2$. Use the graphs of f' and f'' to estimate all maximum and minimum points and intervals of concavity.

SOLUTION If we specify a domain but not a range, many graphing devices will deduce a suitable range from the values computed. Figure 1 shows the plot from one such device if we specify that $-5 \leqslant x \leqslant 5$. Although this viewing rectangle is useful for showing that the asymptotic behavior (or end behavior) is the same as for $y = 2x^6$, it is obviously hiding some finer detail. So we change to the viewing rectangle $[-3, 2]$ by $[-50, 100]$ shown in Figure 2.

From this graph it appears that there is an absolute minimum value of about -15.33 when $x \approx -1.62$ (by using the cursor) and f is decreasing on $(-\infty, -1.62)$ and increasing on $(-1.62, \infty)$. Also, there appears to be a horizontal tangent at the origin and inflection points when $x = 0$ and when x is somewhere between -2 and -1.

Now let's try to confirm these impressions using calculus. We differentiate and get

$$f'(x) = 12x^5 + 15x^4 + 9x^2 - 4x$$

$$f''(x) = 60x^4 + 60x^3 + 18x - 4$$

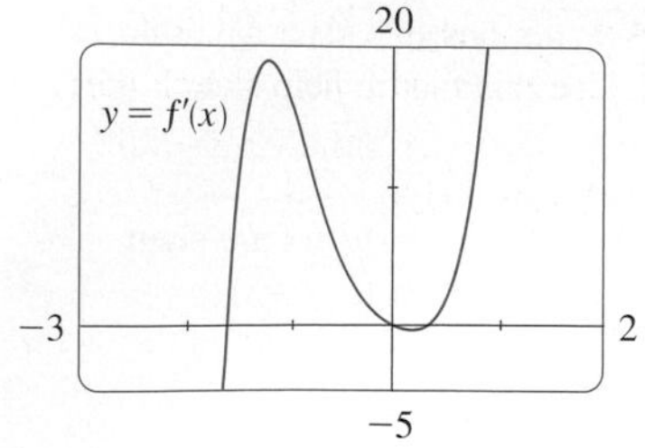

FIGURE 3

When we graph f' in Figure 3 we see that $f'(x)$ changes from negative to positive when $x \approx -1.62$; this confirms (by the First Derivative Test) the minimum value that we found earlier. But, perhaps to our surprise, we also notice that $f'(x)$ changes from positive to negative when $x = 0$ and from negative to positive when $x \approx 0.35$. This means that f has a local maximum at 0 and a local minimum when $x \approx 0.35$, but these were hidden in Figure 2. Indeed, if we now zoom in toward the origin in Figure 4, we see what we missed before: a local maximum value of 0 when $x = 0$ and a local minimum value of about -0.1 when $x \approx 0.35$.

What about concavity and inflection points? From Figures 2 and 4 there appear to be inflection points when x is a little to the left of -1 and when x is a little to the right of 0. But it's difficult to determine inflection points from the graph of f, so we graph the second derivative f'' in Figure 5. We see that f'' changes from positive to negative when $x \approx -1.23$ and from negative to positive when $x \approx 0.19$. So, correct to two decimal places, f is concave upward on $(-\infty, -1.23)$ and $(0.19, \infty)$ and concave downward on $(-1.23, 0.19)$. The inflection points are $(-1.23, -10.18)$ and $(0.19, -0.05)$.

We have discovered that no single graph reveals all the important features of this polynomial. But Figures 2 and 4, when taken together, do provide an accurate picture. □

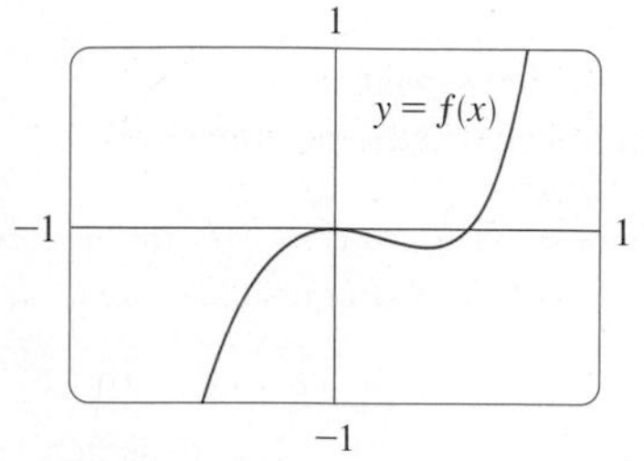

FIGURE 4

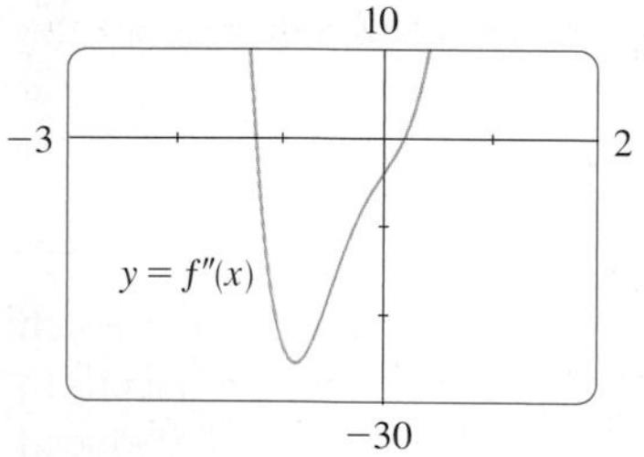

FIGURE 5

V EXAMPLE 2 Draw the graph of the function

$$f(x) = \frac{x^2 + 7x + 3}{x^2}$$

in a viewing rectangle that contains all the important features of the function. Estimate the maximum and minimum values and the intervals of concavity. Then use calculus to find these quantities exactly.

SOLUTION Figure 6, produced by a computer with automatic scaling, is a disaster. Some graphing calculators use $[-10, 10]$ by $[-10, 10]$ as the default viewing rectangle, so let's try it. We get the graph shown in Figure 7; it's a major improvement.

The y-axis appears to be a vertical asymptote and indeed it is because

$$\lim_{x \to 0} \frac{x^2 + 7x + 3}{x^2} = \infty$$

Figure 7 also allows us to estimate the x-intercepts: about -0.5 and -6.5. The exact values are obtained by using the quadratic formula to solve the equation $x^2 + 7x + 3 = 0$; we get $x = (-7 \pm \sqrt{37})/2$.

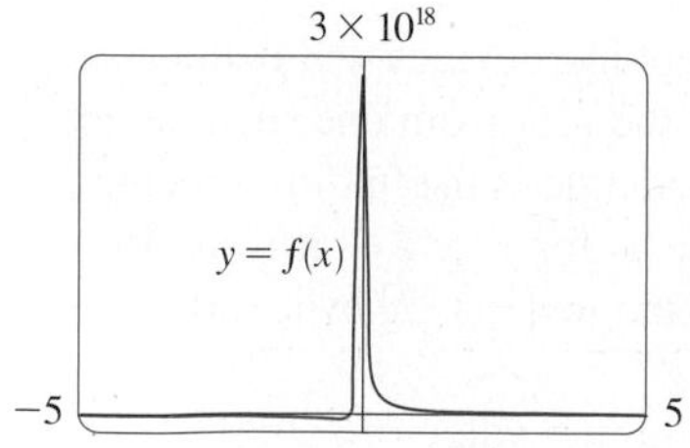

FIGURE 6

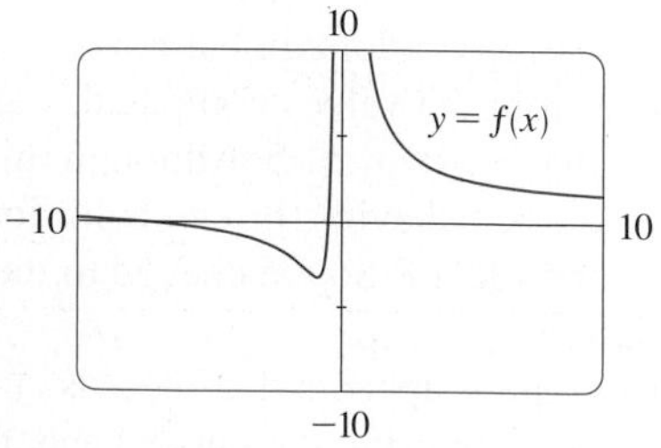

FIGURE 7

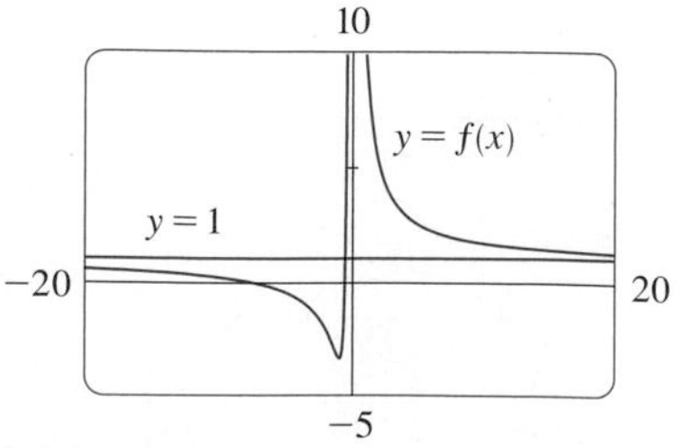

FIGURE 8

To get a better look at horizontal asymptotes, we change to the viewing rectangle $[-20, 20]$ by $[-5, 10]$ in Figure 8. It appears that $y = 1$ is the horizontal asymptote and this is easily confirmed:

$$\lim_{x \to \pm\infty} \frac{x^2 + 7x + 3}{x^2} = \lim_{x \to \pm\infty} \left(1 + \frac{7}{x} + \frac{3}{x^2}\right) = 1$$

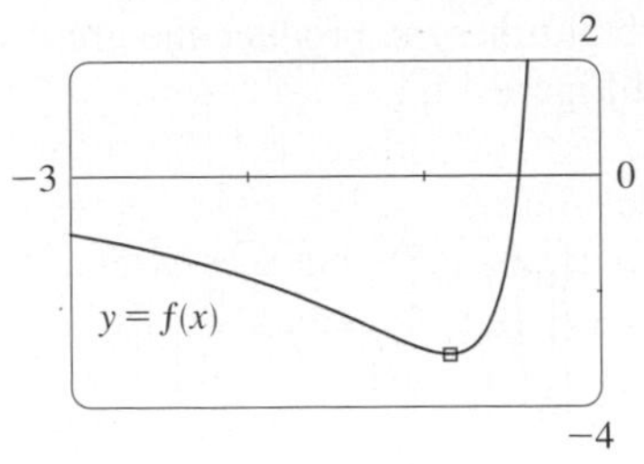

FIGURE 9

To estimate the minimum value we zoom in to the viewing rectangle $[-3, 0]$ by $[-4, 2]$ in Figure 9. The cursor indicates that the absolute minimum value is about -3.1 when $x \approx -0.9$, and we see that the function decreases on $(-\infty, -0.9)$ and $(0, \infty)$ and increases on $(-0.9, 0)$. The exact values are obtained by differentiating:

$$f'(x) = -\frac{7}{x^2} - \frac{6}{x^3} = -\frac{7x+6}{x^3}$$

This shows that $f'(x) > 0$ when $-\frac{6}{7} < x < 0$ and $f'(x) < 0$ when $x < -\frac{6}{7}$ and when $x > 0$. The exact minimum value is $f\left(-\frac{6}{7}\right) = -\frac{37}{12} \approx -3.08$.

Figure 9 also shows that an inflection point occurs somewhere between $x = -1$ and $x = -2$. We could estimate it much more accurately using the graph of the second derivative, but in this case it's just as easy to find exact values. Since

$$f''(x) = \frac{14}{x^3} + \frac{18}{x^4} = \frac{2(7x+9)}{x^4}$$

we see that $f''(x) > 0$ when $x > -\frac{9}{7}$ $(x \neq 0)$. So f is concave upward on $\left(-\frac{9}{7}, 0\right)$ and $(0, \infty)$ and concave downward on $\left(-\infty, -\frac{9}{7}\right)$. The inflection point is $\left(-\frac{9}{7}, -\frac{71}{27}\right)$.

The analysis using the first two derivatives shows that Figures 7 and 8 display all the major aspects of the curve. □

V EXAMPLE 3 Graph the function $f(x) = \dfrac{x^2(x+1)^3}{(x-2)^2(x-4)^4}$.

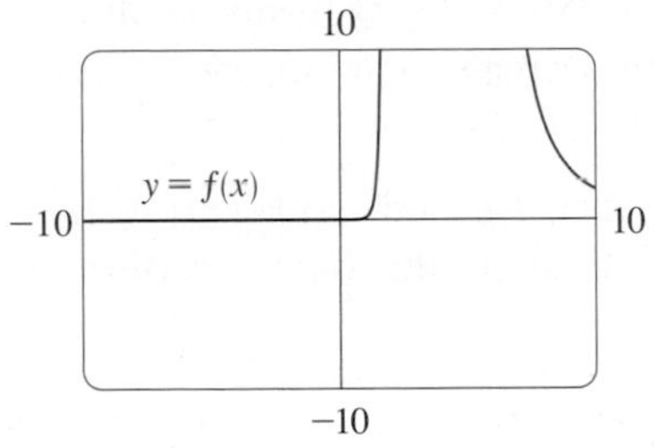

FIGURE 10

SOLUTION Drawing on our experience with a rational function in Example 2, let's start by graphing f in the viewing rectangle $[-10, 10]$ by $[-10, 10]$. From Figure 10 we have the feeling that we are going to have to zoom in to see some finer detail and also zoom out to see the larger picture. But, as a guide to intelligent zooming, let's first take a close look at the expression for $f(x)$. Because of the factors $(x-2)^2$ and $(x-4)^4$ in the denominator, we expect $x = 2$ and $x = 4$ to be the vertical asymptotes. Indeed

$$\lim_{x\to 2} \frac{x^2(x+1)^3}{(x-2)^2(x-4)^4} = \infty \qquad \text{and} \qquad \lim_{x\to 4} \frac{x^2(x+1)^3}{(x-2)^2(x-4)^4} = \infty$$

To find the horizontal asymptotes, we divide numerator and denominator by x^6:

$$\frac{x^2(x+1)^3}{(x-2)^2(x-4)^4} = \frac{\dfrac{x^2}{x^3} \cdot \dfrac{(x+1)^3}{x^3}}{\dfrac{(x-2)^2}{x^2} \cdot \dfrac{(x-4)^4}{x^4}} = \frac{\dfrac{1}{x}\left(1+\dfrac{1}{x}\right)^3}{\left(1-\dfrac{2}{x}\right)^2\left(1-\dfrac{4}{x}\right)^4}$$

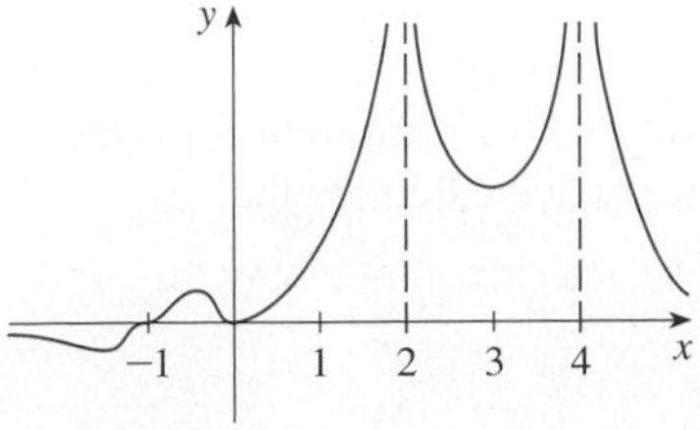

FIGURE 11

This shows that $f(x) \to 0$ as $x \to \pm\infty$, so the x-axis is a horizontal asymptote.

It is also very useful to consider the behavior of the graph near the x-intercepts using an analysis like that in Example 11 in Section 2.6. Since x^2 is positive, $f(x)$ does not change sign at 0 and so its graph doesn't cross the x-axis at 0. But, because of the factor $(x+1)^3$, the graph does cross the x-axis at -1 and has a horizontal tangent there. Putting all this information together, but without using derivatives, we see that the curve has to look something like the one in Figure 11.

Now that we know what to look for, we zoom in (several times) to produce the graphs in Figures 12 and 13 and zoom out (several times) to get Figure 14.

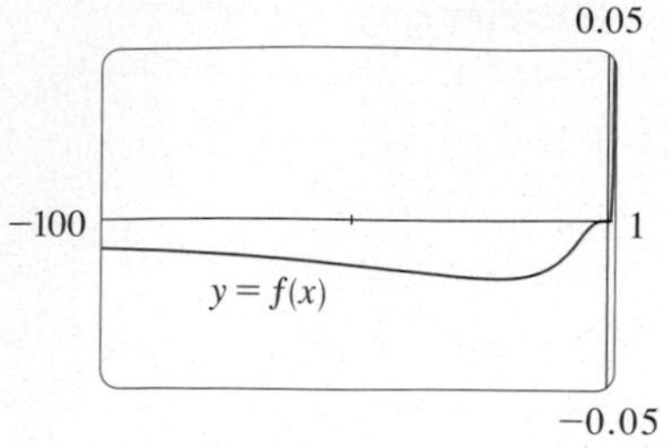

FIGURE 12

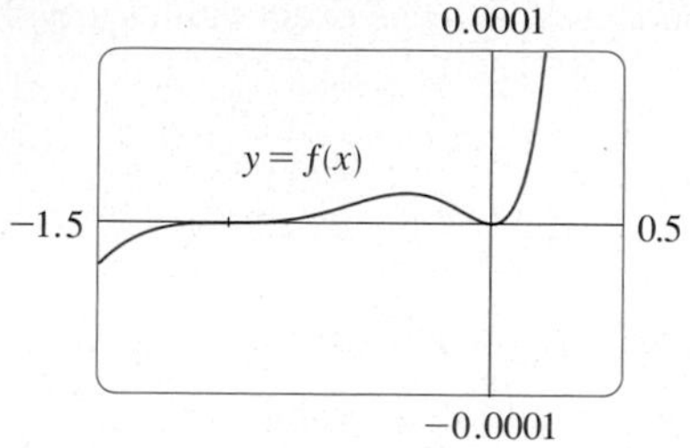

FIGURE 13

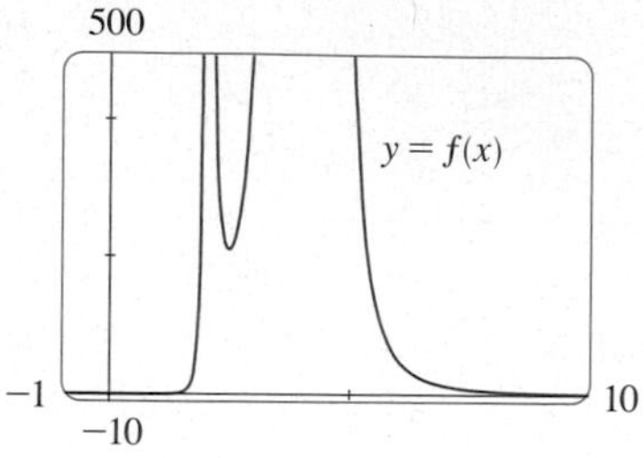

FIGURE 14

We can read from these graphs that the absolute minimum is about -0.02 and occurs when $x \approx -20$. There is also a local maximum ≈ 0.00002 when $x \approx -0.3$ and a local minimum ≈ 211 when $x \approx 2.5$. These graphs also show three inflection points near -35, -5, and -1 and two between -1 and 0. To estimate the inflection points closely we would need to graph f'', but to compute f'' by hand is an unreasonable chore. If you have a computer algebra system, then it's easy to do (see Exercise 15).

We have seen that, for this particular function, *three* graphs (Figures 12, 13, and 14) are necessary to convey all the useful information. The only way to display all these features of the function on a single graph is to draw it by hand. Despite the exaggerations and distortions, Figure 11 does manage to summarize the essential nature of the function. □

■ The family of functions

$$f(x) = \sin(x + \sin cx)$$

where c is a constant, occurs in applications to frequency modulation (FM) synthesis. A sine wave is modulated by a wave with a different frequency ($\sin cx$). The case where $c = 2$ is studied in Example 4. Exercise 25 explores another special case.

EXAMPLE 4 Graph the function $f(x) = \sin(x + \sin 2x)$. For $0 \leqslant x \leqslant \pi$, estimate all maximum and minimum values, intervals of increase and decrease, and inflection points correct to one decimal place.

SOLUTION We first note that f is periodic with period 2π. Also, f is odd and $|f(x)| \leqslant 1$ for all x. So the choice of a viewing rectangle is not a problem for this function: We start with $[0, \pi]$ by $[-1.1, 1.1]$. (See Figure 15.)

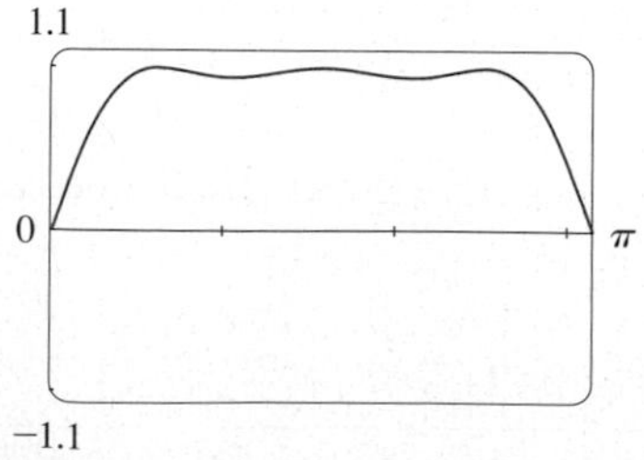

FIGURE 15

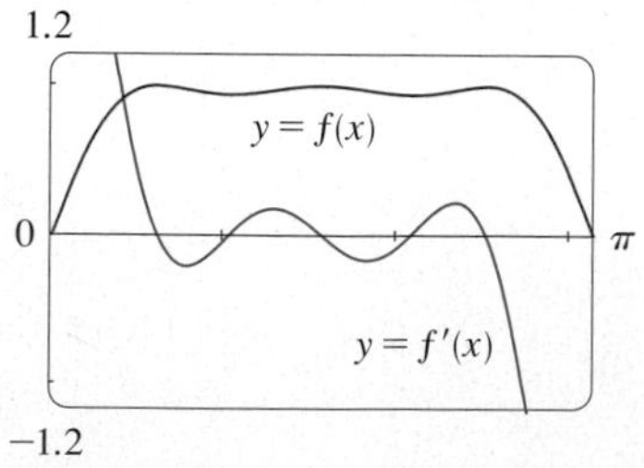

FIGURE 16

It appears that there are three local maximum values and two local minimum values in that window. To confirm this and locate them more accurately, we calculate that

$$f'(x) = \cos(x + \sin 2x) \cdot (1 + 2\cos 2x)$$

and graph both f and f' in Figure 16.

Using zoom-in and the First Derivative Test, we find the following values to one decimal place.

Intervals of increase:	$(0, 0.6),\ (1.0, 1.6),\ (2.1, 2.5)$
Intervals of decrease:	$(0.6, 1.0),\ (1.6, 2.1),\ (2.5, \pi)$
Local maximum values:	$f(0.6) \approx 1,\ f(1.6) \approx 1,\ f(2.5) \approx 1$
Local minimum values:	$f(1.0) \approx 0.94,\ f(2.1) \approx 0.94$

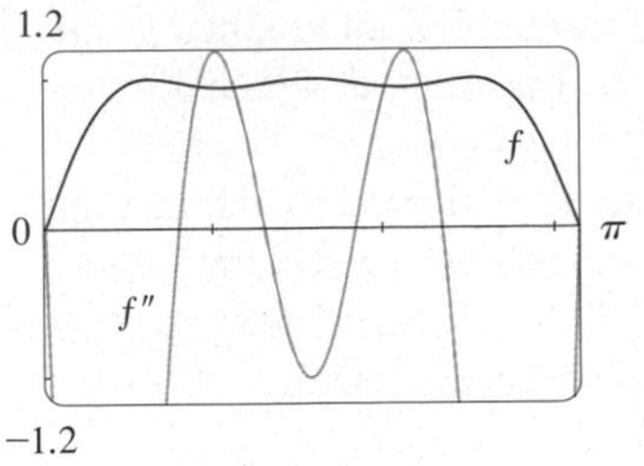

FIGURE 17

The second derivative is

$$f''(x) = -(1 + 2\cos 2x)^2 \sin(x + \sin 2x) - 4 \sin 2x \cos(x + \sin 2x)$$

Graphing both f and f'' in Figure 17, we obtain the following approximate values:

Concave upward on:	$(0.8, 1.3),\ (1.8, 2.3)$
Concave downward on:	$(0, 0.8),\ (1.3, 1.8),\ (2.3, \pi)$
Inflection points:	$(0, 0),\ (0.8, 0.97),\ (1.3, 0.97),\ (1.8, 0.97),\ (2.3, 0.97)$

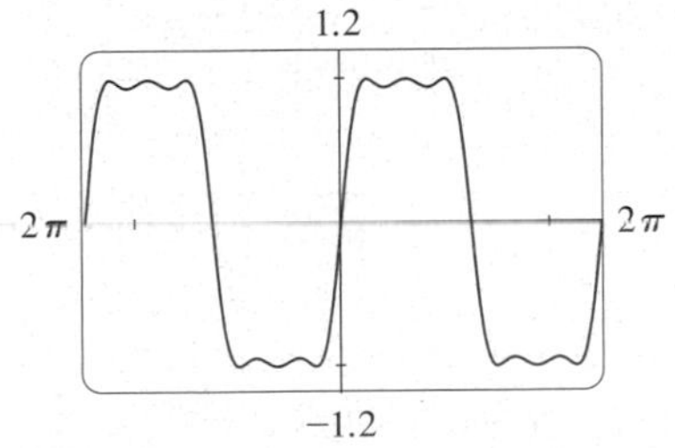

FIGURE 18

Having checked that Figure 15 does indeed represent f accurately for $0 \leqslant x \leqslant \pi$, we can state that the extended graph in Figure 18 represents f accurately for $-2\pi \leqslant x \leqslant 2\pi$. □

Our final example is concerned with *families* of functions. As discussed in Section 1.4, this means that the functions in the family are related to each other by a formula that contains one or more arbitrary constants. Each value of the constant gives rise to a member of the family and the idea is to see how the graph of the function changes as the constant changes.

EXAMPLE 5 How does the graph of $f(x) = 1/(x^2 + 2x + c)$ vary as c varies?

SOLUTION The graphs in Figures 19 and 20 (the special cases $c = 2$ and $c = -2$) show two very different-looking curves. Before drawing any more graphs, let's see what members of this family have in common. Since

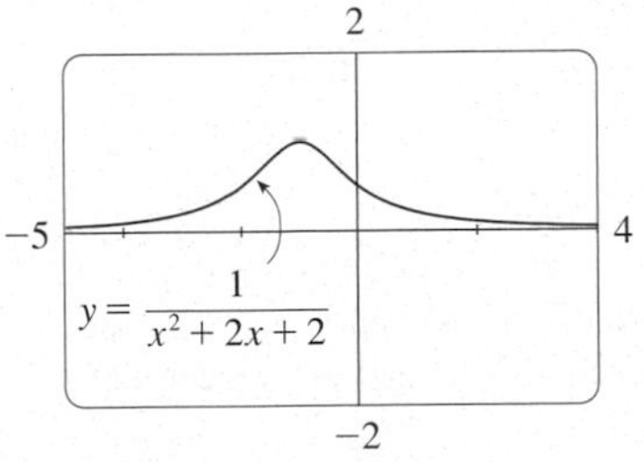

FIGURE 19
$c = 2$

$$\lim_{x \to \pm\infty} \frac{1}{x^2 + 2x + c} = 0$$

for any value of c, they all have the x-axis as a horizontal asymptote. A vertical asymptote will occur when $x^2 + 2x + c = 0$. Solving this quadratic equation, we get $x = -1 \pm \sqrt{1 - c}$. When $c > 1$, there is no vertical asymptote (as in Figure 19). When $c = 1$, the graph has a single vertical asymptote $x = -1$ because

$$\lim_{x \to -1} \frac{1}{x^2 + 2x + 1} = \lim_{x \to -1} \frac{1}{(x + 1)^2} = \infty$$

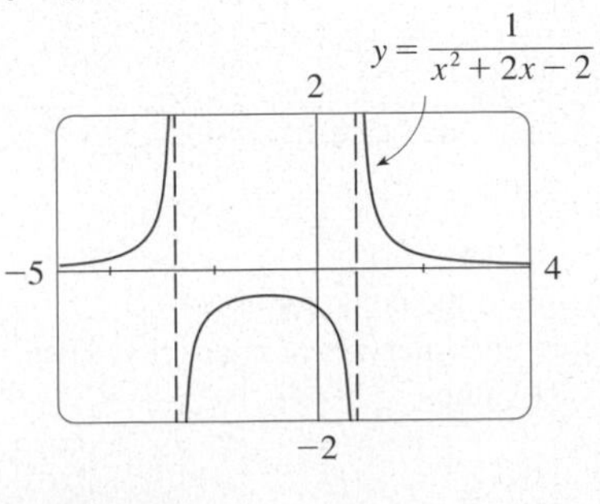

FIGURE 20
$c = -2$

When $c < 1$, there are two vertical asymptotes: $x = -1 \pm \sqrt{1 - c}$ (as in Figure 20).

Now we compute the derivative:

$$f'(x) = -\frac{2x + 2}{(x^2 + 2x + c)^2}$$

This shows that $f'(x) = 0$ when $x = -1$ (if $c \neq 1$), $f'(x) > 0$ when $x < -1$, and

$f'(x) < 0$ when $x > -1$. For $c \geqslant 1$, this means that f increases on $(-\infty, -1)$ and decreases on $(-1, \infty)$. For $c > 1$, there is an absolute maximum value $f(-1) = 1/(c - 1)$. For $c < 1$, $f(-1) = 1/(c - 1)$ is a local maximum value and the intervals of increase and decrease are interrupted at the vertical asymptotes.

Figure 21 is a "slide show" displaying five members of the family, all graphed in the viewing rectangle $[-5, 4]$ by $[-2, 2]$. As predicted, $c = 1$ is the value at which a transition takes place from two vertical asymptotes to one, and then to none. As c increases from 1, we see that the maximum point becomes lower; this is explained by the fact that $1/(c - 1) \to 0$ as $c \to \infty$. As c decreases from 1, the vertical asymptotes become more widely separated because the distance between them is $2\sqrt{1 - c}$, which becomes large as $c \to -\infty$. Again, the maximum point approaches the x-axis because $1/(c - 1) \to 0$ as $c \to -\infty$.

TEC See an animation of Figure 21 in Visual 4.6.

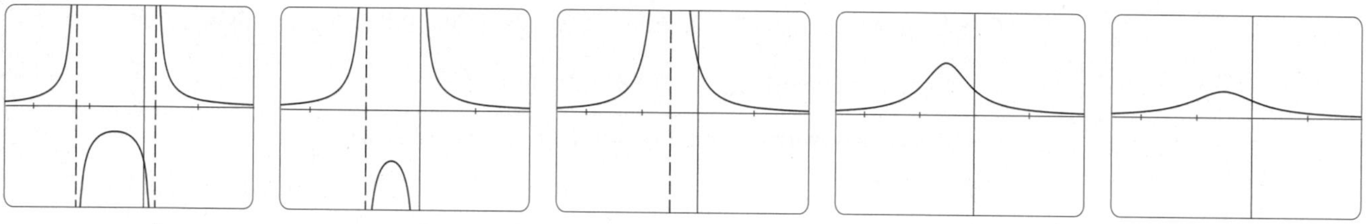

FIGURE 21 The family of functions $f(x) = 1/(x^2 + 2x + c)$

There is clearly no inflection point when $c \leqslant 1$. For $c > 1$ we calculate that

$$f''(x) = \frac{2(3x^2 + 6x + 4 - c)}{(x^2 + 2x + c)^3}$$

and deduce that inflection points occur when $x = -1 \pm \sqrt{3(c - 1)}/3$. So the inflection points become more spread out as c increases and this seems plausible from the last two parts of Figure 21. □

4.6 EXERCISES

1–8 Produce graphs of f that reveal all the important aspects of the curve. In particular, you should use graphs of f' and f'' to estimate the intervals of increase and decrease, extreme values, intervals of concavity, and inflection points.

1. $f(x) = 4x^4 - 32x^3 + 89x^2 - 95x + 29$

2. $f(x) = x^6 - 15x^5 + 75x^4 - 125x^3 - x$

3. $f(x) = x^6 - 10x^5 - 400x^4 + 2500x^3$

4. $f(x) = \dfrac{x^2 - 1}{40x^3 + x + 1}$

5. $f(x) = \dfrac{x}{x^3 - x^2 - 4x + 1}$

6. $f(x) = \tan x + 5 \cos x$

7. $f(x) = x^2 - 4x + 7 \cos x, \quad -4 \leqslant x \leqslant 4$

8. $f(x) = \dfrac{e^x}{x^2 - 9}$

9–10 Produce graphs of f that reveal all the important aspects of the curve. Estimate the intervals of increase and decrease and intervals of concavity, and use calculus to find these intervals exactly.

9. $f(x) = 1 + \dfrac{1}{x} + \dfrac{8}{x^2} + \dfrac{1}{x^3}$

10. $f(x) = \dfrac{1}{x^8} - \dfrac{2 \times 10^8}{x^4}$

11–12
(a) Graph the function.
(b) Use l'Hospital's Rule to explain the behavior as $x \to 0$.
(c) Estimate the minimum value and intervals of concavity. Then use calculus to find the exact values.

11. $f(x) = x^2 \ln x$

12. $f(x) = xe^{1/x}$

13–14 Sketch the graph by hand using asymptotes and intercepts, but not derivatives. Then use your sketch as a guide to producing graphs (with a graphing device) that display the major features of the curve. Use these graphs to estimate the maximum and minimum values.

13. $f(x) = \dfrac{(x+4)(x-3)^2}{x^4(x-1)}$

14. $f(x) = \dfrac{(2x+3)^2(x-2)^5}{x^3(x-5)^2}$

CAS **15.** If f is the function considered in Example 3, use a computer algebra system to calculate f' and then graph it to confirm that all the maximum and minimum values are as given in the example. Calculate f'' and use it to estimate the intervals of concavity and inflection points.

CAS **16.** If f is the function of Exercise 14, find f' and f'' and use their graphs to estimate the intervals of increase and decrease and concavity of f.

CAS **17–22** Use a computer algebra system to graph f and to find f' and f''. Use graphs of these derivatives to estimate the intervals of increase and decrease, extreme values, intervals of concavity, and inflection points of f.

17. $f(x) = \dfrac{\sqrt{x}}{x^2+x+1}$

18. $f(x) = \dfrac{x^{2/3}}{1+x+x^4}$

19. $f(x) = \sqrt{x + 5\sin x}, \quad x \leqslant 20$

20. $f(x) = (x^2 - 1)e^{\arctan x}$

21. $f(x) = \dfrac{1 - e^{1/x}}{1 + e^{1/x}}$

22. $f(x) = \dfrac{1}{1 + e^{\tan x}}$

CAS **23–24**

(a) Graph the function.

(b) Explain the shape of the graph by computing the limit as $x \to 0^+$ or as $x \to \infty$.

(c) Estimate the maximum and minimum values and then use calculus to find the exact values.

(d) Use a graph of f'' to estimate the x-coordinates of the inflection points.

23. $f(x) = x^{1/x}$

24. $f(x) = (\sin x)^{\sin x}$

25. In Example 4 we considered a member of the family of functions $f(x) = \sin(x + \sin cx)$ that occur in FM synthesis. Here we investigate the function with $c = 3$. Start by graphing f in the viewing rectangle $[0, \pi]$ by $[-1.2, 1.2]$. How many local maximum points do you see? The graph has more than are visible to the naked eye. To discover the hidden maximum and minimum points you will need to examine the graph of f' very carefully. In fact, it helps to look at the graph of f'' at the same time. Find all the maximum and minimum values and inflection points. Then graph f in the viewing rectangle $[-2\pi, 2\pi]$ by $[-1.2, 1.2]$ and comment on symmetry.

26–33 Describe how the graph of f varies as c varies. Graph several members of the family to illustrate the trends that you discover. In particular, you should investigate how maximum and minimum points and inflection points move when c changes. You should also identify any transitional values of c at which the basic shape of the curve changes.

26. $f(x) = x^3 + cx$

27. $f(x) = x^4 + cx^2$

28. $f(x) = x\sqrt{c^2 - x^2}$

29. $f(x) = e^{-c/x^2}$

30. $f(x) = \ln(x^2 + c)$

31. $f(x) = \dfrac{cx}{1 + c^2x^2}$

32. $f(x) = \dfrac{1}{(1 - x^2)^2 + cx^2}$

33. $f(x) = cx + \sin x$

34. The family of functions $f(t) = C(e^{-at} - e^{-bt})$, where a, b, and C are positive numbers and $b > a$, has been used to model the concentration of a drug injected into the bloodstream at time $t = 0$. Graph several members of this family. What do they have in common? For fixed values of C and a, discover graphically what happens as b increases. Then use calculus to prove what you have discovered.

35. Investigate the family of curves given by $f(x) = xe^{-cx}$, where c is a real number. Start by computing the limits as $x \to \pm\infty$. Identify any transitional values of c where the basic shape changes. What happens to the maximum or minimum points and inflection points as c changes? Illustrate by graphing several members of the family.

36. Investigate the family of curves given by the equation $f(x) = x^4 + cx^2 + x$. Start by determining the transitional value of c at which the number of inflection points changes. Then graph several members of the family to see what shapes are possible. There is another transitional value of c at which the number of critical numbers changes. Try to discover it graphically. Then prove what you have discovered.

37. (a) Investigate the family of polynomials given by the equation $f(x) = cx^4 - 2x^2 + 1$. For what values of c does the curve have minimum points?

(b) Show that the minimum and maximum points of every curve in the family lie on the parabola $y = 1 - x^2$. Illustrate by graphing this parabola and several members of the family.

38. (a) Investigate the family of polynomials given by the equation $f(x) = 2x^3 + cx^2 + 2x$. For what values of c does the curve have maximum and minimum points?

(b) Show that the minimum and maximum points of every curve in the family lie on the curve $y = x - x^3$. Illustrate by graphing this curve and several members of the family.

4.7 OPTIMIZATION PROBLEMS

The methods we have learned in this chapter for finding extreme values have practical applications in many areas of life. A businessperson wants to minimize costs and maximize profits. A traveler wants to minimize transportation time. Fermat's Principle in optics states that light follows the path that takes the least time. In this section and the next we solve such problems as maximizing areas, volumes, and profits and minimizing distances, times, and costs.

In solving such practical problems the greatest challenge is often to convert the word problem into a mathematical optimization problem by setting up the function that is to be maximized or minimized. Let's recall the problem-solving principles discussed on page 76 and adapt them to this situation:

STEPS IN SOLVING OPTIMIZATION PROBLEMS

1. **Understand the Problem** The first step is to read the problem carefully until it is clearly understood. Ask yourself: What is the unknown? What are the given quantities? What are the given conditions?
2. **Draw a Diagram** In most problems it is useful to draw a diagram and identify the given and required quantities on the diagram.
3. **Introduce Notation** Assign a symbol to the quantity that is to be maximized or minimized (let's call it Q for now). Also select symbols $(a, b, c, \ldots, x, y)$ for other unknown quantities and label the diagram with these symbols. It may help to use initials as suggestive symbols—for example, A for area, h for height, t for time.
4. Express Q in terms of some of the other symbols from Step 3.
5. If Q has been expressed as a function of more than one variable in Step 4, use the given information to find relationships (in the form of equations) among these variables. Then use these equations to eliminate all but one of the variables in the expression for Q. Thus Q will be expressed as a function of *one* variable x, say, $Q = f(x)$. Write the domain of this function.
6. Use the methods of Sections 4.1 and 4.3 to find the *absolute* maximum or minimum value of f. In particular, if the domain of f is a closed interval, then the Closed Interval Method in Section 4.1 can be used.

EXAMPLE 1 A farmer has 2400 ft of fencing and wants to fence off a rectangular field that borders a straight river. He needs no fence along the river. What are the dimensions of the field that has the largest area?

- Understand the problem
- Analogy: Try special cases
- Draw diagrams

SOLUTION In order to get a feeling for what is happening in this problem, let's experiment with some special cases. Figure 1 (not to scale) shows three possible ways of laying out the 2400 ft of fencing.

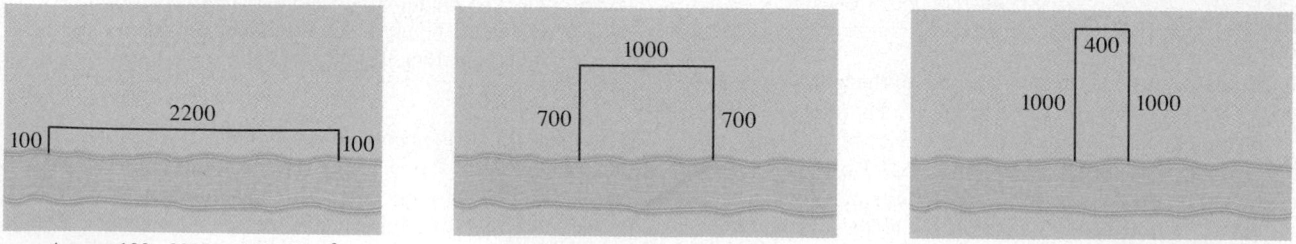

Area $= 100 \cdot 2200 = 220{,}000\ \text{ft}^2$ Area $= 700 \cdot 1000 = 700{,}000\ \text{ft}^2$ Area $= 1000 \cdot 400 = 400{,}000\ \text{ft}^2$

FIGURE 1

We see that when we try shallow, wide fields or deep, narrow fields, we get relatively small areas. It seems plausible that there is some intermediate configuration that produces the largest area.

■ Introduce notation

Figure 2 illustrates the general case. We wish to maximize the area A of the rectangle. Let x and y be the depth and width of the rectangle (in feet). Then we express A in terms of x and y:

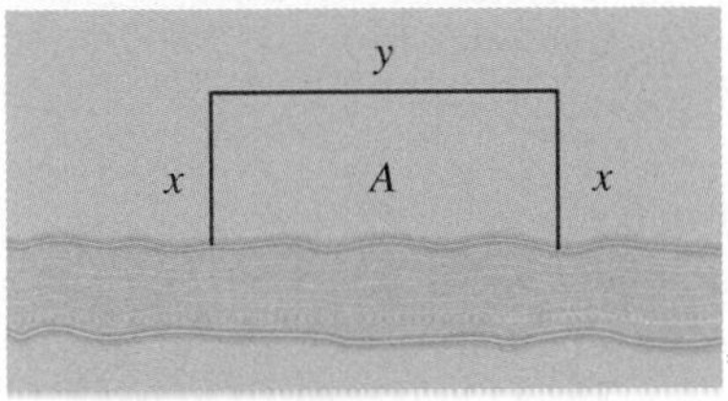

FIGURE 2

$$A = xy$$

We want to express A as a function of just one variable, so we eliminate y by expressing it in terms of x. To do this we use the given information that the total length of the fencing is 2400 ft. Thus

$$2x + y = 2400$$

From this equation we have $y = 2400 - 2x$, which gives

$$A = x(2400 - 2x) = 2400x - 2x^2$$

Note that $x \geqslant 0$ and $x \leqslant 1200$ (otherwise $A < 0$). So the function that we wish to maximize is

$$A(x) = 2400x - 2x^2 \qquad 0 \leqslant x \leqslant 1200$$

The derivative is $A'(x) = 2400 - 4x$, so to find the critical numbers we solve the equation

$$2400 - 4x = 0$$

which gives $x = 600$. The maximum value of A must occur either at this critical number or at an endpoint of the interval. Since $A(0) = 0$, $A(600) = 720{,}000$, and $A(1200) = 0$, the Closed Interval Method gives the maximum value as $A(600) = 720{,}000$.

[Alternatively, we could have observed that $A''(x) = -4 < 0$ for all x, so A is always concave downward and the local maximum at $x = 600$ must be an absolute maximum.]

Thus the rectangular field should be 600 ft deep and 1200 ft wide. □

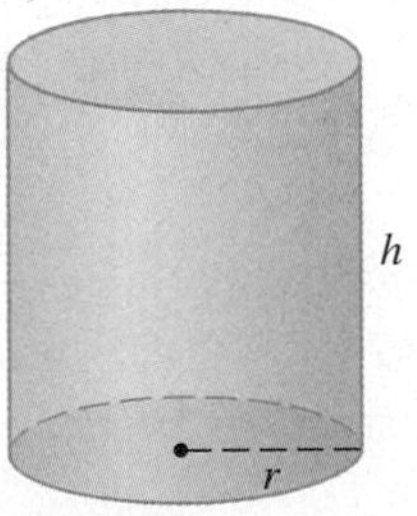

FIGURE 3

V EXAMPLE 2 A cylindrical can is to be made to hold 1 L of oil. Find the dimensions that will minimize the cost of the metal to manufacture the can.

SOLUTION Draw the diagram as in Figure 3, where r is the radius and h the height (both in centimeters). In order to minimize the cost of the metal, we minimize the total surface area of the cylinder (top, bottom, and sides). From Figure 4 we see that the sides are made from a rectangular sheet with dimensions $2\pi r$ and h. So the surface area is

$$A = 2\pi r^2 + 2\pi rh$$

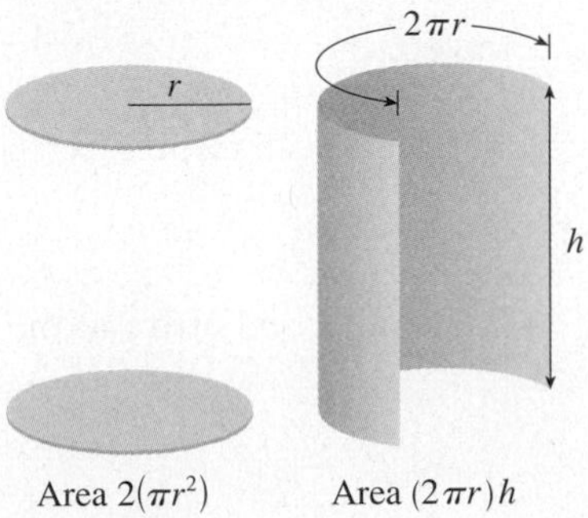

FIGURE 4

To eliminate h we use the fact that the volume is given as 1 L, which we take to be 1000 cm^3. Thus

$$\pi r^2 h = 1000$$

which gives $h = 1000/(\pi r^2)$. Substitution of this into the expression for A gives

$$A = 2\pi r^2 + 2\pi r\left(\frac{1000}{\pi r^2}\right) = 2\pi r^2 + \frac{2000}{r}$$

Therefore the function that we want to minimize is

$$A(r) = 2\pi r^2 + \frac{2000}{r} \qquad r > 0$$

To find the critical numbers, we differentiate:

$$A'(r) = 4\pi r - \frac{2000}{r^2} = \frac{4(\pi r^3 - 500)}{r^2}$$

■ In the Applied Project on page 333 we investigate the most economical shape for a can by taking into account other manufacturing costs.

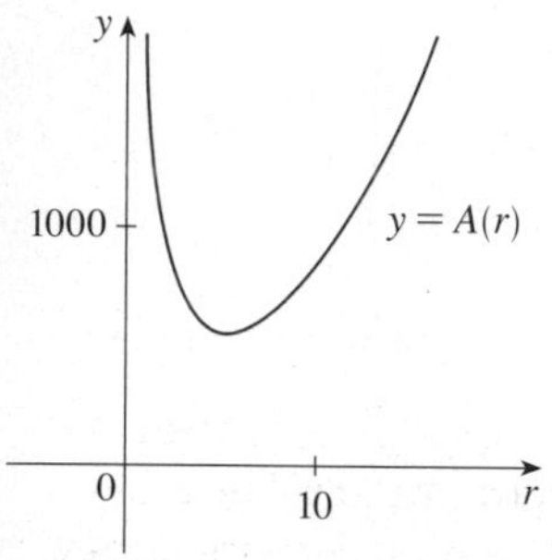

FIGURE 5

Then $A'(r) = 0$ when $\pi r^3 = 500$, so the only critical number is $r = \sqrt[3]{500/\pi}$.

Since the domain of A is $(0, \infty)$, we can't use the argument of Example 1 concerning endpoints. But we can observe that $A'(r) < 0$ for $r < \sqrt[3]{500/\pi}$ and $A'(r) > 0$ for $r > \sqrt[3]{500/\pi}$, so A is decreasing for *all* r to the left of the critical number and increasing for *all* r to the right. Thus $r = \sqrt[3]{500/\pi}$ must give rise to an *absolute* minimum.

[Alternatively, we could argue that $A(r) \to \infty$ as $r \to 0^+$ and $A(r) \to \infty$ as $r \to \infty$, so there must be a minimum value of $A(r)$, which must occur at the critical number. See Figure 5.]

The value of h corresponding to $r = \sqrt[3]{500/\pi}$ is

$$h = \frac{1000}{\pi r^2} = \frac{1000}{\pi(500/\pi)^{2/3}} = 2\sqrt[3]{\frac{500}{\pi}} = 2r$$

Thus, to minimize the cost of the can, the radius should be $\sqrt[3]{500/\pi}$ cm and the height should be equal to twice the radius, namely, the diameter. □

NOTE 1 The argument used in Example 2 to justify the absolute minimum is a variant of the First Derivative Test (which applies only to *local* maximum or minimum values) and is stated here for future reference.

TEC Module 4.7 takes you through six additional optimization problems, including animations of the physical situations.

FIRST DERIVATIVE TEST FOR ABSOLUTE EXTREME VALUES Suppose that c is a critical number of a continuous function f defined on an interval.

(a) If $f'(x) > 0$ for all $x < c$ and $f'(x) < 0$ for all $x > c$, then $f(c)$ is the absolute maximum value of f.

(b) If $f'(x) < 0$ for all $x < c$ and $f'(x) > 0$ for all $x > c$, then $f(c)$ is the absolute minimum value of f.

NOTE 2 An alternative method for solving optimization problems is to use implicit differentiation. Let's look at Example 2 again to illustrate the method. We work with the same equations

$$A = 2\pi r^2 + 2\pi rh \qquad \pi r^2 h = 1000$$

but instead of eliminating h, we differentiate both equations implicitly with respect to r:

$$A' = 4\pi r + 2\pi h + 2\pi rh' \qquad 2\pi rh + \pi r^2 h' = 0$$

The minimum occurs at a critical number, so we set $A' = 0$, simplify, and arrive at the equations

$$2r + h + rh' = 0 \qquad 2h + rh' = 0$$

and subtraction gives $2r - h = 0$, or $h = 2r$.

V EXAMPLE 3 Find the point on the parabola $y^2 = 2x$ that is closest to the point $(1, 4)$.

SOLUTION The distance between the point $(1, 4)$ and the point (x, y) is

$$d = \sqrt{(x-1)^2 + (y-4)^2}$$

(See Figure 6.) But if (x, y) lies on the parabola, then $x = \frac{1}{2}y^2$, so the expression for d becomes

$$d = \sqrt{\left(\tfrac{1}{2}y^2 - 1\right)^2 + (y-4)^2}$$

(Alternatively, we could have substituted $y = \sqrt{2x}$ to get d in terms of x alone.) Instead of minimizing d, we minimize its square:

$$d^2 = f(y) = \left(\tfrac{1}{2}y^2 - 1\right)^2 + (y-4)^2$$

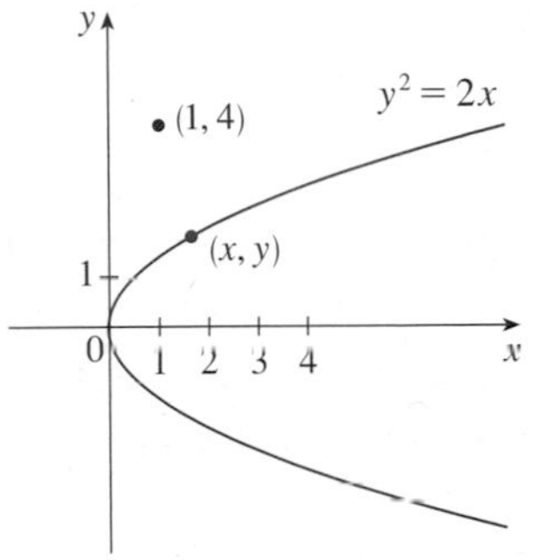

FIGURE 6

(You should convince yourself that the minimum of d occurs at the same point as the minimum of d^2, but d^2 is easier to work with.) Differentiating, we obtain

$$f'(y) = 2\left(\tfrac{1}{2}y^2 - 1\right)y + 2(y-4) = y^3 - 8$$

so $f'(y) = 0$ when $y = 2$. Observe that $f'(y) < 0$ when $y < 2$ and $f'(y) > 0$ when $y > 2$, so by the First Derivative Test for Absolute Extreme Values, the absolute minimum occurs when $y = 2$. (Or we could simply say that because of the geometric nature of the problem, it's obvious that there is a closest point but not a farthest point.) The corresponding value of x is $x = \frac{1}{2}y^2 = 2$. Thus the point on $y^2 = 2x$ closest to $(1, 4)$ is $(2, 2)$. □

EXAMPLE 4 A man launches his boat from point A on a bank of a straight river, 3 km wide, and wants to reach point B, 8 km downstream on the opposite bank, as quickly as possible (see Figure 7). He could row his boat directly across the river to point C and then run to B, or he could row directly to B, or he could row to some point D between C and B and then run to B. If he can row 6 km/h and run 8 km/h, where should he land to reach B as soon as possible? (We assume that the speed of the water is negligible compared with the speed at which the man rows.)

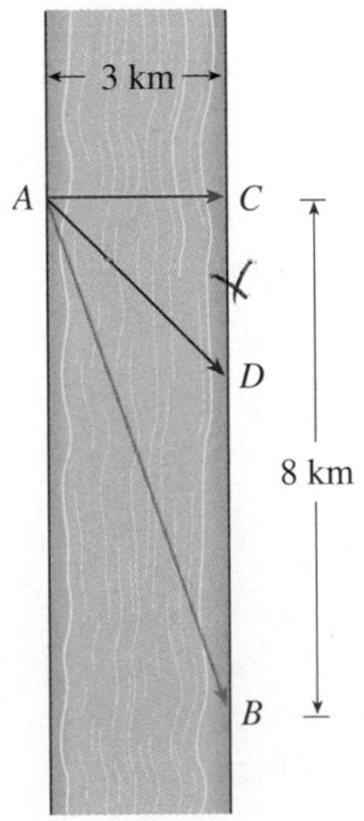

FIGURE 7

SOLUTION If we let x be the distance from C to D, then the running distance is $|DB| = 8 - x$ and the Pythagorean Theorem gives the rowing distance as $|AD| = \sqrt{x^2 + 9}$. We use the equation

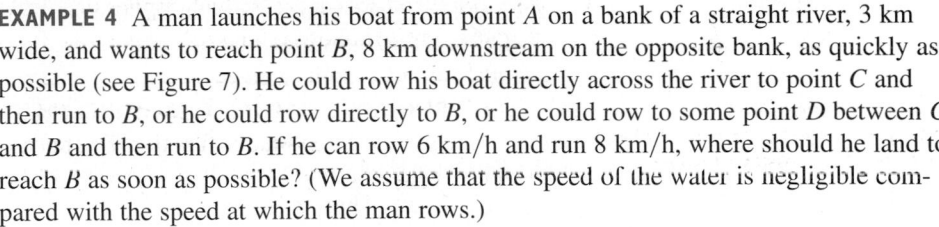

$$\text{time} = \frac{\text{distance}}{\text{rate}}$$

Then the rowing time is $\sqrt{x^2 + 9}/6$ and the running time is $(8 - x)/8$, so the total time T as a function of x is

$$T(x) = \frac{\sqrt{x^2 + 9}}{6} + \frac{8 - x}{8}$$

The domain of this function T is $[0, 8]$. Notice that if $x = 0$, he rows to C and if $x = 8$, he rows directly to B. The derivative of T is

$$T'(x) = \frac{x}{6\sqrt{x^2 + 9}} - \frac{1}{8}$$

Thus, using the fact that $x \geq 0$, we have

$$T'(x) = 0 \iff \frac{x}{6\sqrt{x^2+9}} = \frac{1}{8} \iff 4x = 3\sqrt{x^2+9}$$

$$\iff 16x^2 = 9(x^2+9) \iff 7x^2 = 81$$

$$\iff x = \frac{9}{\sqrt{7}}$$

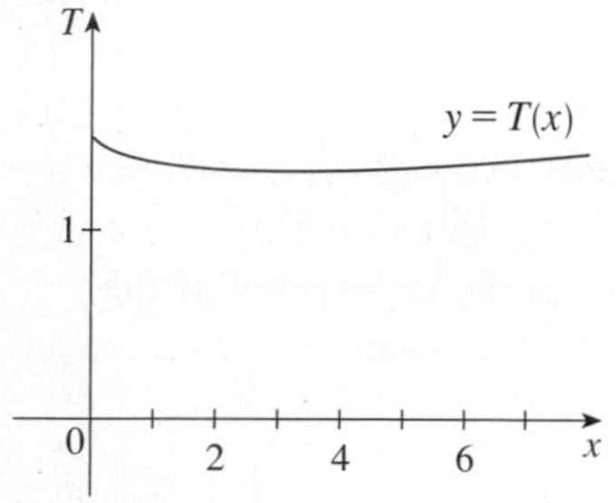

FIGURE 8

The only critical number is $x = 9/\sqrt{7}$. To see whether the minimum occurs at this critical number or at an endpoint of the domain $[0, 8]$, we evaluate T at all three points:

$$T(0) = 1.5 \qquad T\left(\frac{9}{\sqrt{7}}\right) = 1 + \frac{\sqrt{7}}{8} \approx 1.33 \qquad T(8) = \frac{\sqrt{73}}{6} \approx 1.42$$

Since the smallest of these values of T occurs when $x = 9/\sqrt{7}$, the absolute minimum value of T must occur there. Figure 8 illustrates this calculation by showing the graph of T.

Thus the man should land the boat at a point $9/\sqrt{7}$ km ($\approx$3.4 km) downstream from his starting point. □

V EXAMPLE 5 Find the area of the largest rectangle that can be inscribed in a semicircle of radius r.

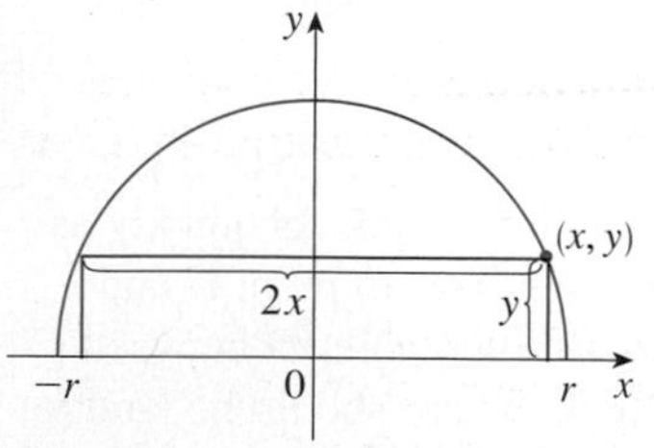

FIGURE 9

SOLUTION 1 Let's take the semicircle to be the upper half of the circle $x^2 + y^2 = r^2$ with center the origin. Then the word *inscribed* means that the rectangle has two vertices on the semicircle and two vertices on the x-axis as shown in Figure 9.

Let (x, y) be the vertex that lies in the first quadrant. Then the rectangle has sides of lengths $2x$ and y, so its area is

$$A = 2xy$$

To eliminate y we use the fact that (x, y) lies on the circle $x^2 + y^2 = r^2$ and so $y = \sqrt{r^2 - x^2}$. Thus

$$A = 2x\sqrt{r^2 - x^2}$$

The domain of this function is $0 \leq x \leq r$. Its derivative is

$$A' = 2\sqrt{r^2 - x^2} - \frac{2x^2}{\sqrt{r^2 - x^2}} = \frac{2(r^2 - 2x^2)}{\sqrt{r^2 - x^2}}$$

which is 0 when $2x^2 = r^2$, that is, $x = r/\sqrt{2}$ (since $x \geq 0$). This value of x gives a maximum value of A since $A(0) = 0$ and $A(r) = 0$. Therefore the area of the largest inscribed rectangle is

$$A\left(\frac{r}{\sqrt{2}}\right) = 2\frac{r}{\sqrt{2}}\sqrt{r^2 - \frac{r^2}{2}} = r^2$$

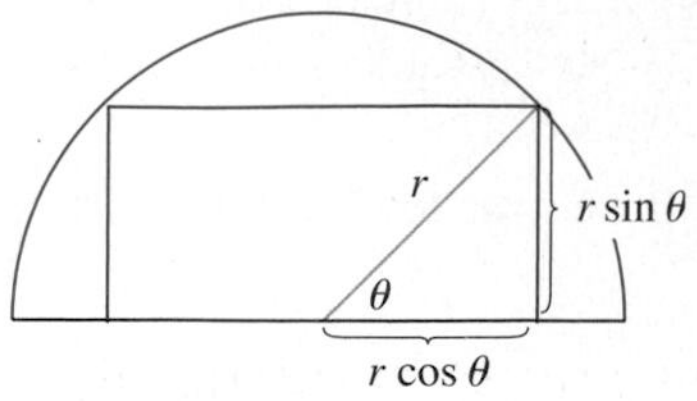

FIGURE 10

SOLUTION 2 A simpler solution is possible if we think of using an angle as a variable. Let θ be the angle shown in Figure 10. Then the area of the rectangle is

$$A(\theta) = (2r\cos\theta)(r\sin\theta) = r^2(2\sin\theta\cos\theta) = r^2\sin 2\theta$$

We know that $\sin 2\theta$ has a maximum value of 1 and it occurs when $2\theta = \pi/2$. So $A(\theta)$ has a maximum value of r^2 and it occurs when $\theta = \pi/4$.

Notice that this trigonometric solution doesn't involve differentiation. In fact, we didn't need to use calculus at all. □

APPLICATIONS TO BUSINESS AND ECONOMICS

In Section 3.7 we introduced the idea of marginal cost. Recall that if $C(x)$, the **cost function**, is the cost of producing x units of a certain product, then the **marginal cost** is the rate of change of C with respect to x. In other words, the marginal cost function is the derivative, $C'(x)$, of the cost function.

Now let's consider marketing. Let $p(x)$ be the price per unit that the company can charge if it sells x units. Then p is called the **demand function** (or **price function**) and we would expect it to be a decreasing function of x. If x units are sold and the price per unit is $p(x)$, then the total revenue is

$$R(x) = xp(x)$$

and R is called the **revenue function**. The derivative R' of the revenue function is called the **marginal revenue function** and is the rate of change of revenue with respect to the number of units sold.

If x units are sold, then the total profit is

$$P(x) = R(x) - C(x)$$

and P is called the **profit function**. The **marginal profit function** is P', the derivative of the profit function. In Exercises 53–58 you are asked to use the marginal cost, revenue, and profit functions to minimize costs and maximize revenues and profits.

V EXAMPLE 6 A store has been selling 200 DVD burners a week at \$350 each. A market survey indicates that for each \$10 rebate offered to buyers, the number of units sold will increase by 20 a week. Find the demand function and the revenue function. How large a rebate should the store offer to maximize its revenue?

SOLUTION If x is the number of DVD burners sold per week, then the weekly increase in sales is $x - 200$. For each increase of 20 units sold, the price is decreased by \$10. So for each additional unit sold, the decrease in price will be $\frac{1}{20} \times 10$ and the demand function is

$$p(x) = 350 - \tfrac{10}{20}(x - 200) = 450 - \tfrac{1}{2}x$$

The revenue function is

$$R(x) = xp(x) = 450x - \tfrac{1}{2}x^2$$

Since $R'(x) = 450 - x$, we see that $R'(x) = 0$ when $x = 450$. This value of x gives an absolute maximum by the First Derivative Test (or simply by observing that the graph of R is a parabola that opens downward). The corresponding price is

$$p(450) = 450 - \tfrac{1}{2}(450) = 225$$

and the rebate is $350 - 225 = 125$. Therefore, to maximize revenue, the store should offer a rebate of \$125. □

4.7 EXERCISES

1. Consider the following problem: Find two numbers whose sum is 23 and whose product is a maximum.

(a) Make a table of values, like the following one, so that the sum of the numbers in the first two columns is always 23. On the basis of the evidence in your table, estimate the answer to the problem.

First number	Second number	Product
1	22	22
2	21	42
3	20	60
⋮	⋮	⋮

(b) Use calculus to solve the problem and compare with your answer to part (a).

2. Find two numbers whose difference is 100 and whose product is a minimum.

3. Find two positive numbers whose product is 100 and whose sum is a minimum.

4. Find a positive number such that the sum of the number and its reciprocal is as small as possible.

5. Find the dimensions of a rectangle with perimeter 100 m whose area is as large as possible.

6. Find the dimensions of a rectangle with area 1000 m^2 whose perimeter is as small as possible.

7. A model used for the yield Y of an agricultural crop as a function of the nitrogen level N in the soil (measured in appropriate units) is

$$Y = \frac{kN}{1 + N^2}$$

where k is a positive constant. What nitrogen level gives the best yield?

8. The rate (in mg carbon/m^3/h) at which photosynthesis takes place for a species of phytoplankton is modeled by the function

$$P = \frac{100I}{I^2 + I + 4}$$

where I is the light intensity (measured in thousands of foot-candles). For what light intensity is P a maximum?

9. Consider the following problem: A farmer with 750 ft of fencing wants to enclose a rectangular area and then divide it into four pens with fencing parallel to one side of the rectangle. What is the largest possible total area of the four pens?

(a) Draw several diagrams illustrating the situation, some with shallow, wide pens and some with deep, narrow pens. Find the total areas of these configurations. Does it appear that there is a maximum area? If so, estimate it.

(b) Draw a diagram illustrating the general situation. Introduce notation and label the diagram with your symbols.

(c) Write an expression for the total area.

(d) Use the given information to write an equation that relates the variables.

(e) Use part (d) to write the total area as a function of one variable.

(f) Finish solving the problem and compare the answer with your estimate in part (a).

10. Consider the following problem: A box with an open top is to be constructed from a square piece of cardboard, 3 ft wide, by cutting out a square from each of the four corners and bending up the sides. Find the largest volume that such a box can have.

(a) Draw several diagrams to illustrate the situation, some short boxes with large bases and some tall boxes with small bases. Find the volumes of several such boxes. Does it appear that there is a maximum volume? If so, estimate it.

(b) Draw a diagram illustrating the general situation. Introduce notation and label the diagram with your symbols.

(c) Write an expression for the volume.

(d) Use the given information to write an equation that relates the variables.

(e) Use part (d) to write the volume as a function of one variable.

(f) Finish solving the problem and compare the answer with your estimate in part (a).

11. A farmer wants to fence an area of 1.5 million square feet in a rectangular field and then divide it in half with a fence parallel to one of the sides of the rectangle. How can he do this so as to minimize the cost of the fence?

12. A box with a square base and open top must have a volume of 32,000 cm^3. Find the dimensions of the box that minimize the amount of material used.

13. If 1200 cm^2 of material is available to make a box with a square base and an open top, find the largest possible volume of the box.

14. A rectangular storage container with an open top is to have a volume of 10 m^3. The length of its base is twice the width. Material for the base costs $10 per square meter. Material for the sides costs $6 per square meter. Find the cost of materials for the cheapest such container.

15. Do Exercise 14 assuming the container has a lid that is made from the same material as the sides.

16. (a) Show that of all the rectangles with a given area, the one with smallest perimeter is a square.

(b) Show that of all the rectangles with a given perimeter, the one with greatest area is a square.

17. Find the point on the line $y = 4x + 7$ that is closest to the origin.

18. Find the point on the line $6x + y = 9$ that is closest to the point $(-3, 1)$.

19. Find the points on the ellipse $4x^2 + y^2 = 4$ that are farthest away from the point $(1, 0)$.

20. Find, correct to two decimal places, the coordinates of the point on the curve $y = \tan x$ that is closest to the point $(1, 1)$.

21. Find the dimensions of the rectangle of largest area that can be inscribed in a circle of radius r.

22. Find the area of the largest rectangle that can be inscribed in the ellipse $x^2/a^2 + y^2/b^2 = 1$.

23. Find the dimensions of the rectangle of largest area that can be inscribed in an equilateral triangle of side L if one side of the rectangle lies on the base of the triangle.

24. Find the dimensions of the rectangle of largest area that has its base on the x-axis and its other two vertices above the x-axis and lying on the parabola $y = 8 - x^2$.

25. Find the dimensions of the isosceles triangle of largest area that can be inscribed in a circle of radius r.

26. Find the area of the largest rectangle that can be inscribed in a right triangle with legs of lengths 3 cm and 4 cm if two sides of the rectangle lie along the legs.

27. A right circular cylinder is inscribed in a sphere of radius r. Find the largest possible volume of such a cylinder.

28. A right circular cylinder is inscribed in a cone with height h and base radius r. Find the largest possible volume of such a cylinder.

29. A right circular cylinder is inscribed in a sphere of radius r. Find the largest possible surface area of such a cylinder.

30. A Norman window has the shape of a rectangle surmounted by a semicircle. (Thus the diameter of the semicircle is equal to the width of the rectangle. See Exercise 56 on page 23.) If the perimeter of the window is 30 ft, find the dimensions of the window so that the greatest possible amount of light is admitted.

31. The top and bottom margins of a poster are each 6 cm and the side margins are each 4 cm. If the area of printed material on the poster is fixed at 384 cm^2, find the dimensions of the poster with the smallest area.

32. A poster is to have an area of 180 in^2 with 1-inch margins at the bottom and sides and a 2-inch margin at the top. What dimensions will give the largest printed area?

33. A piece of wire 10 m long is cut into two pieces. One piece is bent into a square and the other is bent into an equilateral triangle. How should the wire be cut so that the total area enclosed is (a) a maximum? (b) A minimum?

34. Answer Exercise 33 if one piece is bent into a square and the other into a circle.

35. A cylindrical can without a top is made to contain V cm^3 of liquid. Find the dimensions that will minimize the cost of the metal to make the can.

36. A fence 8 ft tall runs parallel to a tall building at a distance of 4 ft from the building. What is the length of the shortest ladder that will reach from the ground over the fence to the wall of the building?

37. A cone-shaped drinking cup is made from a circular piece of paper of radius R by cutting out a sector and joining the edges CA and CB. Find the maximum capacity of such a cup.

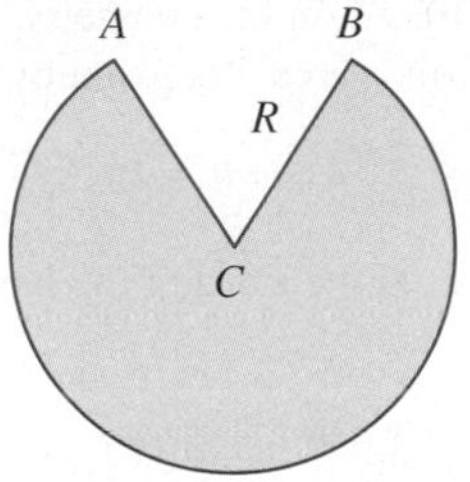

38. A cone-shaped paper drinking cup is to be made to hold 27 cm^3 of water. Find the height and radius of the cup that will use the smallest amount of paper.

39. A cone with height h is inscribed in a larger cone with height H so that its vertex is at the center of the base of the larger cone. Show that the inner cone has maximum volume when $h = \frac{1}{3}H$.

40. An object with weight W is dragged along a horizontal plane by a force acting along a rope attached to the object. If the rope makes an angle θ with a plane, then the magnitude of the force is

$$F = \frac{\mu W}{\mu \sin \theta + \cos \theta}$$

where μ is a constant called the coefficient of friction. For what value of θ is F smallest?

41. If a resistor of R ohms is connected across a battery of E volts with internal resistance r ohms, then the power (in watts) in the external resistor is

$$P = \frac{E^2 R}{(R + r)^2}$$

If E and r are fixed but R varies, what is the maximum value of the power?

42. For a fish swimming at a speed v relative to the water, the energy expenditure per unit time is proportional to v^3. It is believed that migrating fish try to minimize the total energy required to swim a fixed distance. If the fish are swimming against a current u $(u < v)$, then the time required to swim a distance L is $L/(v - u)$ and the total energy E required to swim the distance is given by

$$E(v) = av^3 \cdot \frac{L}{v - u}$$

where a is the proportionality constant.

(a) Determine the value of v that minimizes E.

(b) Sketch the graph of E.

Note: This result has been verified experiment[illegible] migrating fish swim against a current at a speed [illegible] than the current speed.

43. In a beehive, each cell is a regular hexagonal prism, open at one end with a trihedral angle at the other end as in the figure. It is believed that bees form their cells in such a way as to minimize the surface area for a given volume, thus using the least amount of wax in cell construction. Examination of these cells has shown that the measure of the apex angle θ is amazingly consistent. Based on the geometry of the cell, it can be shown that the surface area S is given by

$$S = 6sh - \tfrac{3}{2}s^2 \cot\theta + \left(3s^2\sqrt{3}/2\right)\csc\theta$$

where s, the length of the sides of the hexagon, and h, the height, are constants.

(a) Calculate $dS/d\theta$.

(b) What angle should the bees prefer?

(c) Determine the minimum surface area of the cell (in terms of s and h).

Note: Actual measurements of the angle θ in beehives have been made, and the measures of these angles seldom differ from the calculated value by more than 2°.

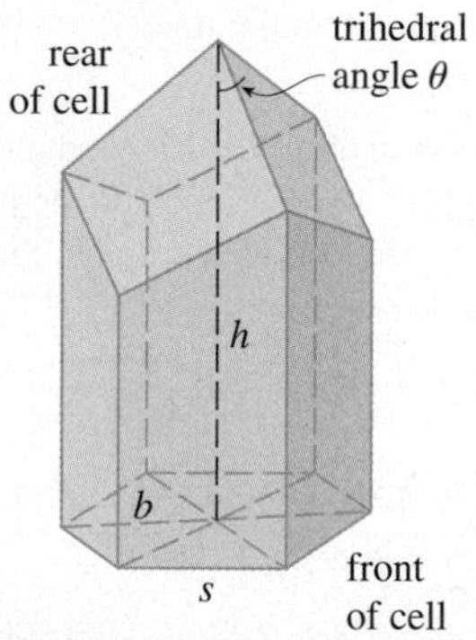

44. A boat leaves a dock at 2:00 PM and travels due south at a speed of 20 km/h. Another boat has been heading due east at 15 km/h and reaches the same dock at 3:00 PM. At what time were the two boats closest together?

45. Solve the problem in Example 4 if the river is 5 km wide and point B is only 5 km downstream from A.

46. A woman at a point A on the shore of a circular lake with radius 2 mi wants to arrive at the point C diametrically opposite A on the other side of the lake in the shortest possible time. She can walk at the rate of 4 mi/h and row a boat at 2 mi/h. How should she proceed?

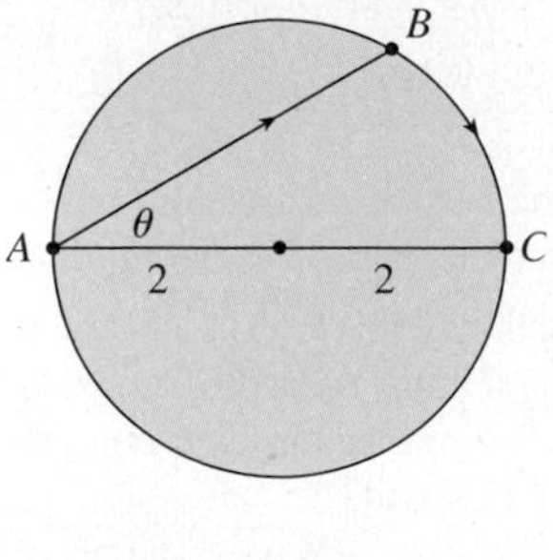

47. An oil refinery is located on the north bank of a straight river that is 2 km wide. A pipeline is to be constructed from the refinery to storage tanks located on the south bank of the river 6 km east of the refinery. The cost of laying pipe is \$400,000/km over land to a point P on the north bank and \$800,000/km under the river to the tanks. To minimize the cost of the pipeline, where should P be located?

48. Suppose the refinery in Exercise 47 is located 1 km north of the river. Where should P be located?

49. The illumination of an object by a light source is directly proportional to the strength of the source and inversely proportional to the square of the distance from the source. If two light sources, one three times as strong as the other, are placed 10 ft apart, where should an object be placed on the line between the sources so as to receive the least illumination?

50. Find an equation of the line through the point (3, 5) that cuts off the least area from the first quadrant.

51. Let a and b be positive numbers. Find the length of the shortest line segment that is cut off by the first quadrant and passes through the point (a, b).

52. At which points on the curve $y = 1 + 40x^3 - 3x^5$ does the tangent line have the largest slope?

53. (a) If $C(x)$ is the cost of producing x units of a commodity, then the **average cost** per unit is $c(x) = C(x)/x$. Show that if the average cost is a minimum, then the marginal cost equals the average cost.

(b) If $C(x) = 16{,}000 + 200x + 4x^{3/2}$, in dollars, find (i) the cost, average cost, and marginal cost at a production level of 1000 units; (ii) the production level that will minimize the average cost; and (iii) the minimum average cost.

54. (a) Show that if the profit $P(x)$ is a maximum, then the marginal revenue equals the marginal cost.

(b) If $C(x) = 16{,}000 + 500x - 1.6x^2 + 0.004x^3$ is the cost function and $p(x) = 1700 - 7x$ is the demand function, find the production level that will maximize profit.

55. A baseball team plays in a stadium that holds 55,000 spectators. With ticket prices at \$10, the average attendance had been 27,000. When ticket prices were lowered to \$8, the average attendance rose to 33,000.

(a) Find the demand function, assuming that it is linear.

(b) How should ticket prices be set to maximize revenue?

56. During the summer months Terry makes and sells necklaces on the beach. Last summer he sold the necklaces for \$10 each and his sales averaged 20 per day. When he increased the price by \$1, he found that the average decreased by two sales per day.

(a) Find the demand function, assuming that it is linear.

(b) If the material for each necklace costs Terry \$6, what should the selling price be to maximize his profit?

57. A manufacturer has been selling 1000 television sets a week at \$450 each. A market survey indicates that for each \$10 rebate offered to the buyer, the number of sets sold will increase by 100 per week.
(a) Find the demand function.
(b) How large a rebate should the company offer the buyer in order to maximize its revenue?
(c) If its weekly cost function is $C(x) = 68{,}000 + 150x$, how should the manufacturer set the size of the rebate in order to maximize its profit?

58. The manager of a 100-unit apartment complex knows from experience that all units will be occupied if the rent is \$800 per month. A market survey suggests that, on average, one additional unit will remain vacant for each \$10 increase in rent. What rent should the manager charge to maximize revenue?

59. Show that of all the isosceles triangles with a given perimeter, the one with the greatest area is equilateral.

CAS **60.** The frame for a kite is to be made from six pieces of wood. The four exterior pieces have been cut with the lengths indicated in the figure. To maximize the area of the kite, how long should the diagonal pieces be?

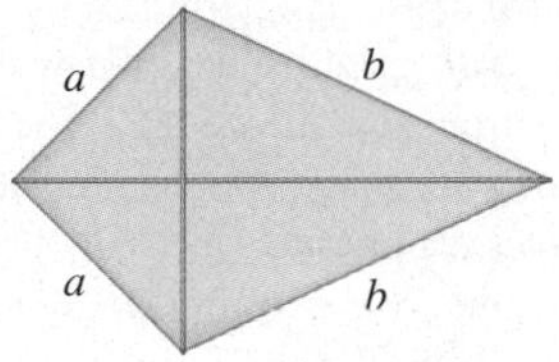

61. A point P needs to be located somewhere on the line AD so that the total length L of cables linking P to the points A, B, and C is minimized (see the figure). Express L as a function of $x = |AP|$ and use the graphs of L and dL/dx to estimate the minimum value.

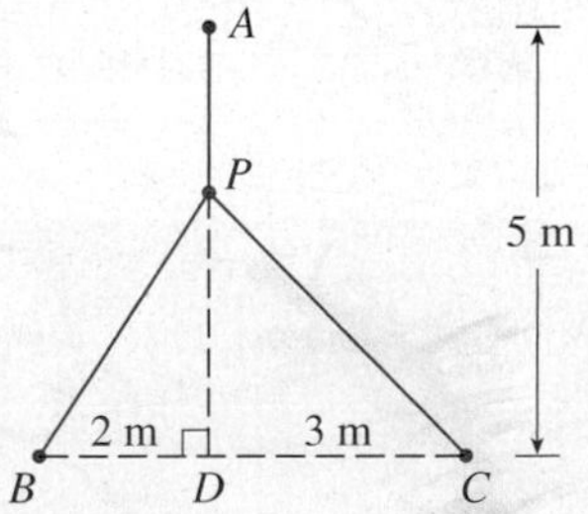

62. The graph shows the fuel consumption c of a car (measured in gallons per hour) as a function of the speed v of the car. At very low speeds the engine runs inefficiently, so initially c decreases as the speed increases. But at high speeds the fuel consumption increases. You can see that $c(v)$ is minimized for this car when $v \approx 30$ mi/h. However, for fuel efficiency, what must be minimized is not the consumption in gallons per hour but rather the fuel consumption in gallons *per mile*. Let's call this consumption G. Using the graph, estimate the speed at which G has its minimum value.

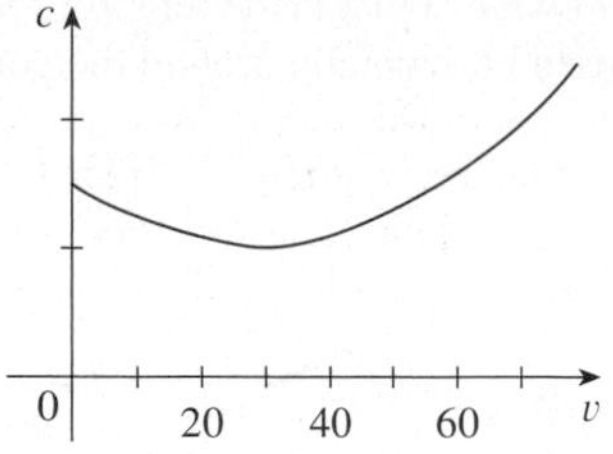

63. Let v_1 be the velocity of light in air and v_2 the velocity of light in water. According to Fermat's Principle, a ray of light will travel from a point A in the air to a point B in the water by a path ACB that minimizes the time taken. Show that

$$\frac{\sin \theta_1}{\sin \theta_2} = \frac{v_1}{v_2}$$

where θ_1 (the angle of incidence) and θ_2 (the angle of refraction) are as shown. This equation is known as Snell's Law.

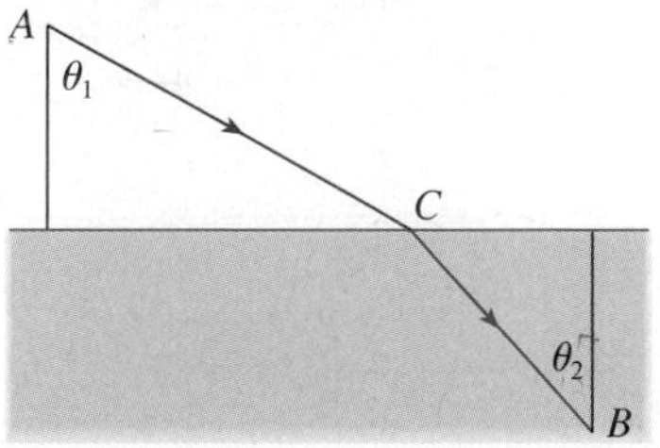

64. Two vertical poles PQ and ST are secured by a rope PRS going from the top of the first pole to a point R on the ground between the poles and then to the top of the second pole as in the figure. Show that the shortest length of such a rope occurs when $\theta_1 = \theta_2$.

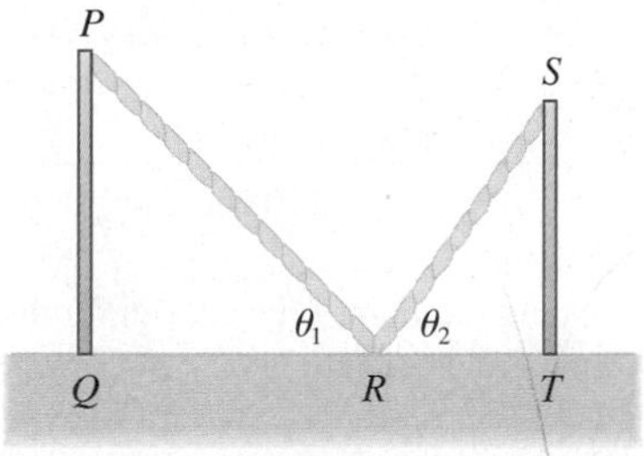

65. The upper right-hand corner of a piece of paper, 12 in. by 8 in., as in the figure, is folded over to the bottom edge. How would you fold it so as to minimize the length of the fold? In other words, how would you choose x to minimize y?

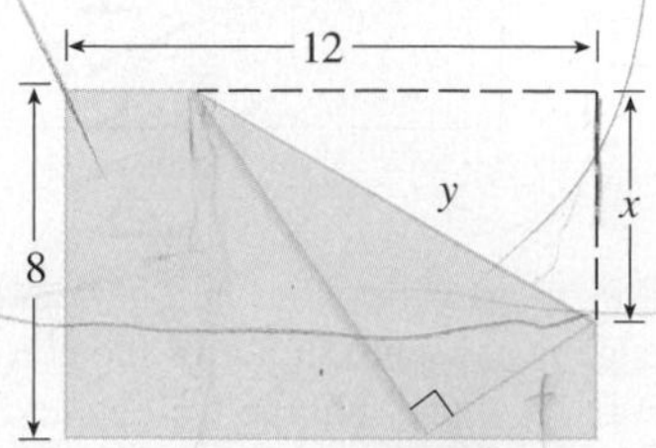

66. A steel pipe is being carried down a hallway 9 ft wide. At the end of the hall there is a right-angled turn into a narrower hallway 6 ft wide. What is the length of the longest pipe that can be carried horizontally around the corner?

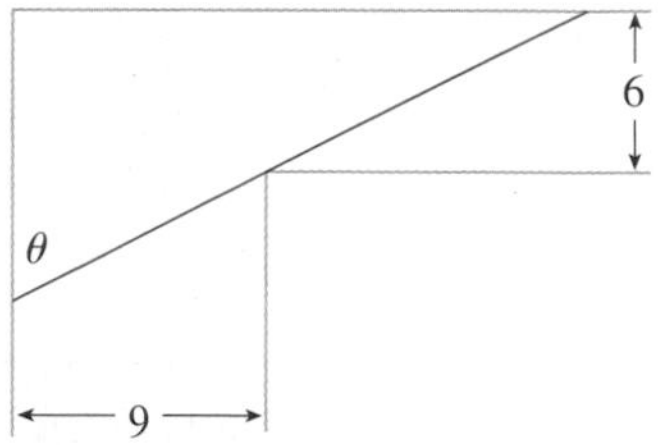

67. An observer stands at a point P, one unit away from a track. Two runners start at the point S in the figure and run along the track. One runner runs three times as fast as the other. Find the maximum value of the observer's angle of sight θ between the runners. [*Hint:* Maximize $\tan \theta$.]

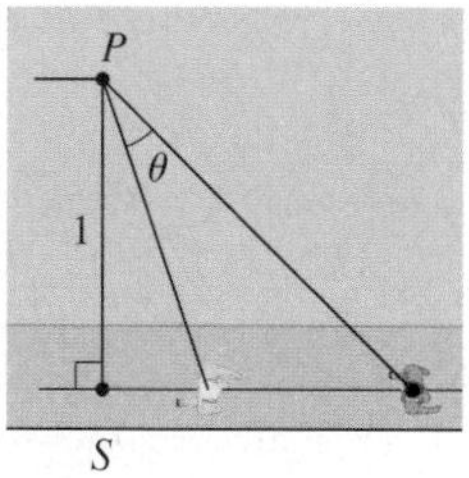

68. A rain gutter is to be constructed from a metal sheet of width 30 cm by bending up one-third of the sheet on each side through an angle θ. How should θ be chosen so that the gutter will carry the maximum amount of water?

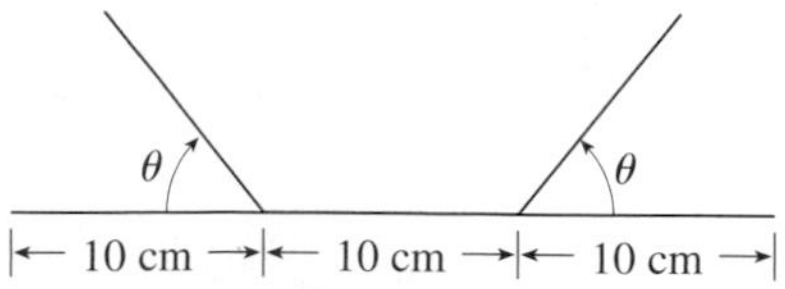

69. Where should the point P be chosen on the line segment AB so as to maximize the angle θ?

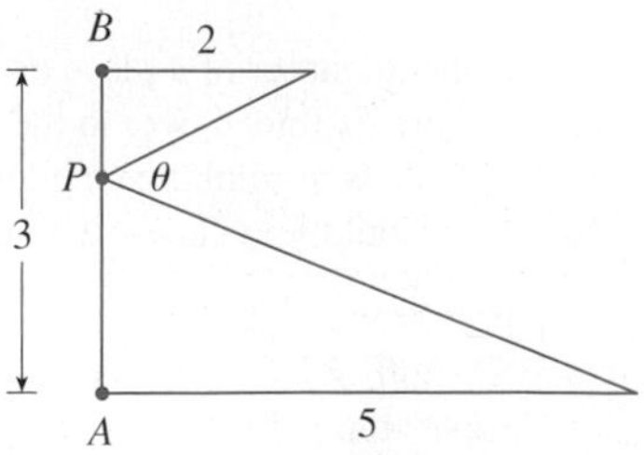

70. A painting in an art gallery has height h and is hung so that its lower edge is a distance d above the eye of an observer (as in the figure). How far from the wall should the observer stand to get the best view? (In other words, where should the observer stand so as to maximize the angle θ subtended at his eye by the painting?)

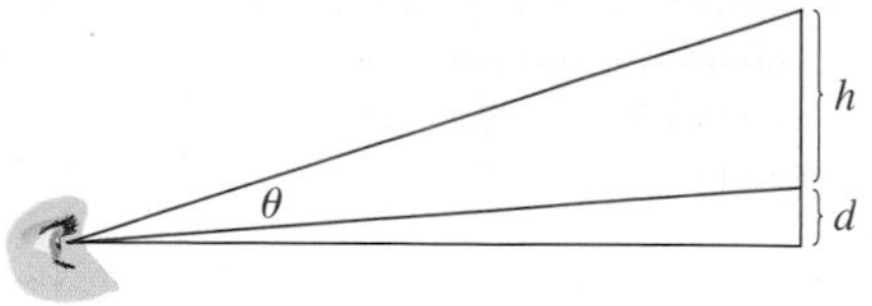

71. Find the maximum area of a rectangle that can be circumscribed about a given rectangle with length L and width W. [*Hint:* Express the area as a function of an angle θ.]

72. The blood vascular system consists of blood vessels (arteries, arterioles, capillaries, and veins) that convey blood from the heart to the organs and back to the heart. This system should work so as to minimize the energy expended by the heart in pumping the blood. In particular, this energy is reduced when the resistance of the blood is lowered. One of Poiseuille's Laws gives the resistance R of the blood as

$$R = C\frac{L}{r^4}$$

where L is the length of the blood vessel, r is the radius, and C is a positive constant determined by the viscosity of the blood. (Poiseuille established this law experimentally, but it also follows from Equation 8.4.2.) The figure shows a main blood vessel with radius r_1 branching at an angle θ into a smaller vessel with radius r_2

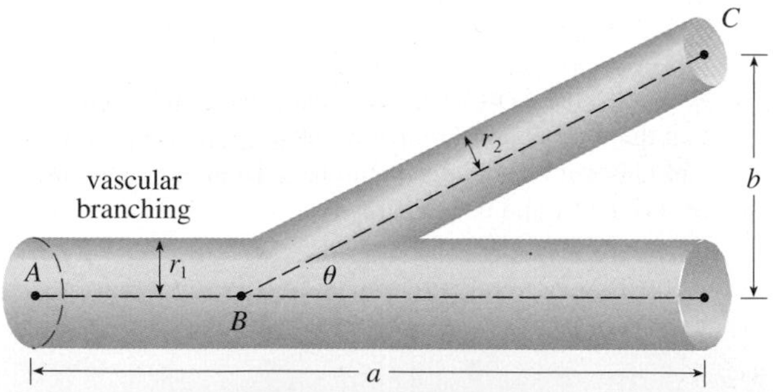

© Manfred Cage / Peter Arnold

(a) Use Poiseuille's Law to show that the total resistance of the blood along the path ABC is

$$R = C\left(\frac{a - b\cot\theta}{r_1^4} + \frac{b\csc\theta}{r_2^4}\right)$$

where a and b are the distances shown in the figure.

(b) Prove that this resistance is minimized when

$$\cos\theta = \frac{r_2^4}{r_1^4}$$

(c) Find the optimal branching angle (correct to the nearest degree) when the radius of the smaller blood vessel is two-thirds the radius of the larger vessel.

73. Ornithologists have determined that some species of birds tend to avoid flights over large bodies of water during daylight hours. It is believed that more energy is required to fly over water than land because air generally rises over land and falls over water during the day. A bird with these tendencies is released from an island that is 5 km from the nearest point B on a straight shoreline, flies to a point C on the shoreline, and then flies along the shoreline to its nesting area D. Assume that the bird instinctively chooses a path that will minimize its energy expenditure. Points B and D are 13 km apart.

(a) In general, if it takes 1.4 times as much energy to fly over water as land, to what point C should the bird fly in order to minimize the total energy expended in returning to its nesting area?

(b) Let W and L denote the energy (in joules) per kilometer flown over water and land, respectively. What would a large value of the ratio W/L mean in terms of the bird's flight? What would a small value mean? Determine the ratio W/L corresponding to the minimum expenditure of energy.

(c) What should the value of W/L be in order for the bird to fly directly to its nesting area D? What should the value of W/L be for the bird to fly to B and then along the shore to D?

(d) If the ornithologists observe that birds of a certain species reach the shore at a point 4 km from B, how many times more energy does it take a bird to fly over water than land?

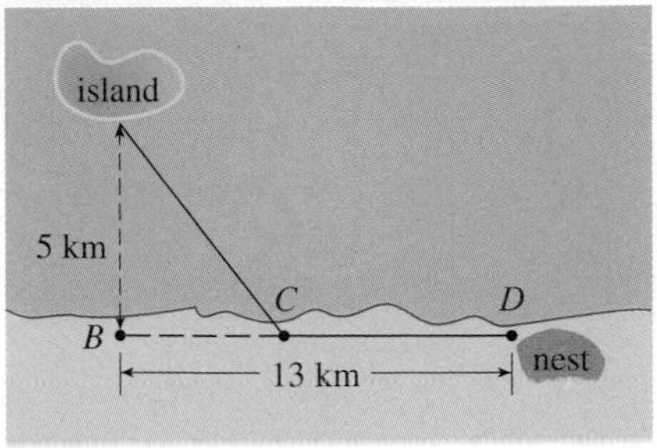

74. Two light sources of identical strength are placed 10 m apart. An object is to be placed at a point P on a line ℓ parallel to the line joining the light sources and at a distance d meters from it (see the figure). We want to locate P on ℓ so that the intensity of illumination is minimized. We need to use the fact that the intensity of illumination for a single source is directly proportional to the strength of the source and inversely proportional to the square of the distance from the source.

(a) Find an expression for the intensity $I(x)$ at the point P.

(b) If $d = 5$ m, use graphs of $I(x)$ and $I'(x)$ to show that the intensity is minimized when $x = 5$ m, that is, when P is at the midpoint of ℓ.

(c) If $d = 10$ m, show that the intensity (perhaps surprisingly) is *not* minimized at the midpoint.

(d) Somewhere between $d = 5$ m and $d = 10$ m there is a transitional value of d at which the point of minimal illumination abruptly changes. Estimate this value of d by graphical methods. Then find the exact value of d.

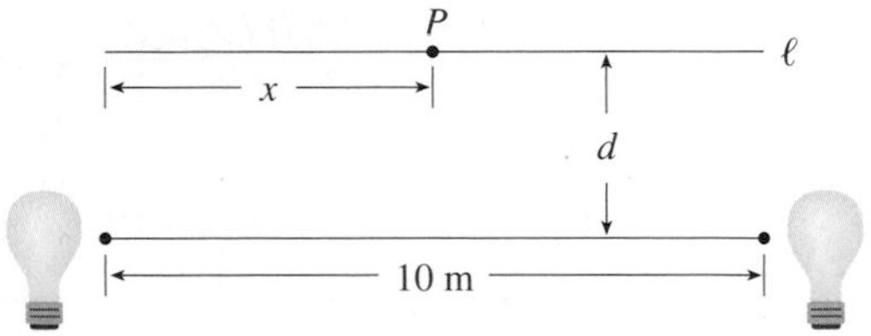

APPLIED PROJECT THE SHAPE OF A CAN

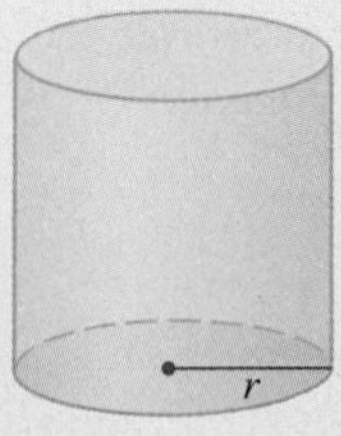

In this project we investigate the most economical shape for a can. We first interpret this to mean that the volume V of a cylindrical can is given and we need to find the height h and radius r that minimize the cost of the metal to make the can (see the figure). If we disregard any waste metal in the manufacturing process, then the problem is to minimize the surface area of the cylinder. We solved this problem in Example 2 in Section 4.7 and we found that $h = 2r$; that is, the height should be the same as the diameter. But if you go to your cupboard or your supermarket with a ruler, you will discover that the height is usually greater than the diameter and the ratio h/r varies from 2 up to about 3.8. Let's see if we can explain this phenomenon.

1. The material for the cans is cut from sheets of metal. The cylindrical sides are formed by bending rectangles; these rectangles are cut from the sheet with little or no waste. But if the

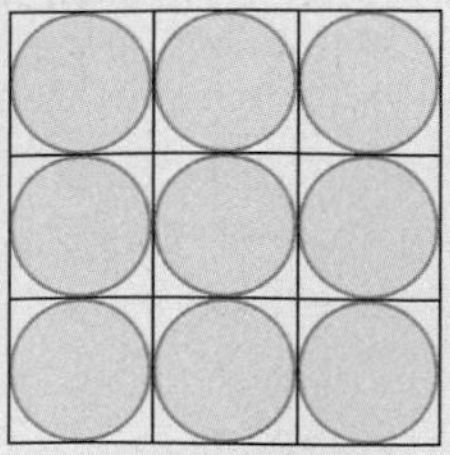

Discs cut from squares

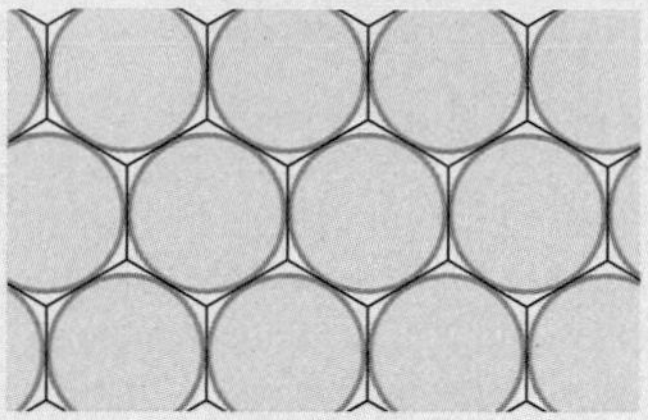

Discs cut from hexagons

top and bottom discs are cut from squares of side $2r$ (as in the figure), this leaves considerable waste metal, which may be recycled but has little or no value to the can makers. If this is the case, show that the amount of metal used is minimized when

$$\frac{h}{r} = \frac{8}{\pi} \approx 2.55$$

2. A more efficient packing of the discs is obtained by dividing the metal sheet into hexagons and cutting the circular lids and bases from the hexagons (see the figure). Show that if this strategy is adopted, then

$$\frac{h}{r} = \frac{4\sqrt{3}}{\pi} \approx 2.21$$

3. The values of h/r that we found in Problems 1 and 2 are a little closer to the ones that actually occur on supermarket shelves, but they still don't account for everything. If we look more closely at some real cans, we see that the lid and the base are formed from discs with radius larger than r that are bent over the ends of the can. If we allow for this we would increase h/r. More significantly, in addition to the cost of the metal we need to incorporate the manufacturing of the can into the cost. Let's assume that most of the expense is incurred in joining the sides to the rims of the cans. If we cut the discs from hexagons as in Problem 2, then the total cost is proportional to

$$4\sqrt{3}\,r^2 + 2\pi rh + k(4\pi r + h)$$

where k is the reciprocal of the length that can be joined for the cost of one unit area of metal. Show that this expression is minimized when

$$\frac{\sqrt[3]{V}}{k} = \sqrt[3]{\frac{\pi h}{r}} \cdot \frac{2\pi - h/r}{\pi h/r - 4\sqrt{3}}$$

4. Plot $\sqrt[3]{V}/k$ as a function of $x = h/r$ and use your graph to argue that when a can is large or joining is cheap, we should make h/r approximately 2.21 (as in Problem 2). But when the can is small or joining is costly, h/r should be substantially larger.

5. Our analysis shows that large cans should be almost square but small cans should be tall and thin. Take a look at the relative shapes of the cans in a supermarket. Is our conclusion usually true in practice? Are there exceptions? Can you suggest reasons why small cans are not always tall and thin?

4.8 NEWTON'S METHOD

Suppose that a car dealer offers to sell you a car for \$18,000 or for payments of \$375 per month for five years. You would like to know what monthly interest rate the dealer is, in effect, charging you. To find the answer, you have to solve the equation

$$48x(1 + x)^{60} - (1 + x)^{60} + 1 = 0 \qquad \boxed{1}$$

(The details are explained in Exercise 41.) How would you solve such an equation?

For a quadratic equation $ax^2 + bx + c = 0$ there is a well-known formula for the roots. For third- and fourth-degree equations there are also formulas for the roots, but they are

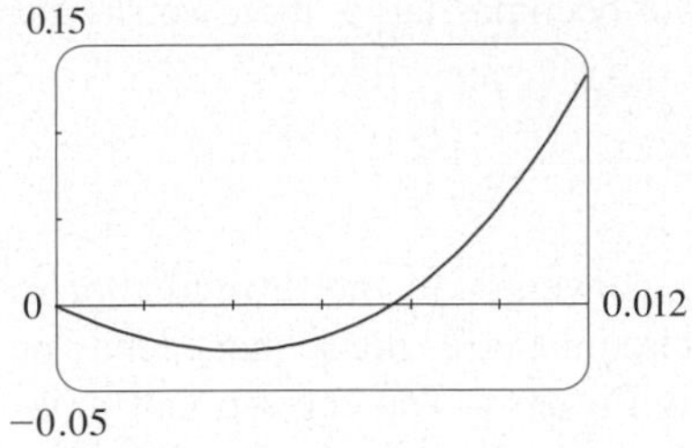

FIGURE 1

■ Try to solve Equation 1 using the numerical rootfinder on your calculator or computer. Some machines are not able to solve it. Others are successful but require you to specify a starting point for the search.

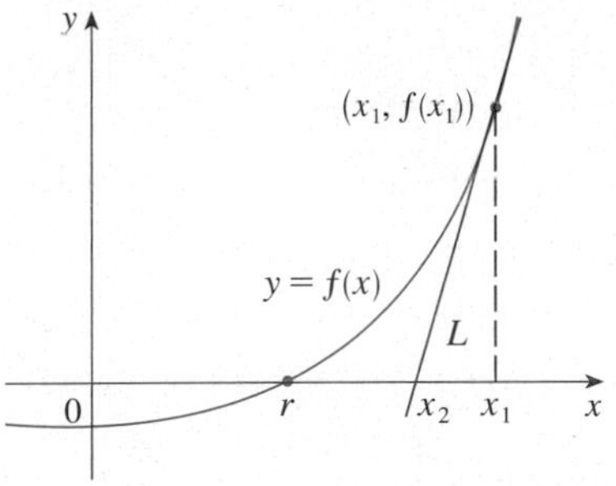

FIGURE 2

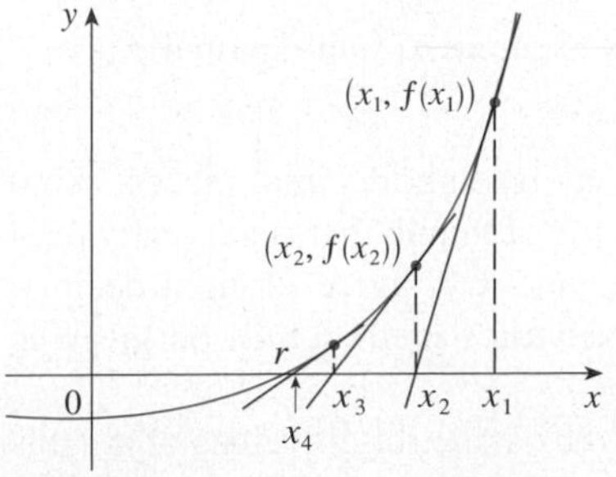

FIGURE 3

extremely complicated. If f is a polynomial of degree 5 or higher, there is no such formula (see the note on page 210). Likewise, there is no formula that will enable us to find the exact roots of a transcendental equation such as $\cos x = x$.

We can find an *approximate* solution to Equation 1 by plotting the left side of the equation. Using a graphing device, and after experimenting with viewing rectangles, we produce the graph in Figure 1.

We see that in addition to the solution $x = 0$, which doesn't interest us, there is a solution between 0.007 and 0.008. Zooming in shows that the root is approximately 0.0076. If we need more accuracy we could zoom in repeatedly, but that becomes tiresome. A faster alternative is to use a numerical rootfinder on a calculator or computer algebra system. If we do so, we find that the root, correct to nine decimal places, is 0.007628603.

How do those numerical rootfinders work? They use a variety of methods, but most of them make some use of **Newton's method**, also called the **Newton-Raphson method**. We will explain how this method works, partly to show what happens inside a calculator or computer, and partly as an application of the idea of linear approximation.

The geometry behind Newton's method is shown in Figure 2, where the root that we are trying to find is labeled r. We start with a first approximation x_1, which is obtained by guessing, or from a rough sketch of the graph of f, or from a computer-generated graph of f. Consider the tangent line L to the curve $y = f(x)$ at the point $(x_1, f(x_1))$ and look at the x-intercept of L, labeled x_2. The idea behind Newton's method is that the tangent line is close to the curve and so its x-intercept, x_2, is close to the x-intercept of the curve (namely, the root r that we are seeking). Because the tangent is a line, we can easily find its x-intercept.

To find a formula for x_2 in terms of x_1 we use the fact that the slope of L is $f'(x_1)$, so its equation is

$$y - f(x_1) = f'(x_1)(x - x_1)$$

Since the x-intercept of L is x_2, we set $y = 0$ and obtain

$$0 - f(x_1) = f'(x_1)(x_2 - x_1)$$

If $f'(x_1) \neq 0$, we can solve this equation for x_2:

$$x_2 = x_1 - \frac{f(x_1)}{f'(x_1)}$$

We use x_2 as a second approximation to r.

Next we repeat this procedure with x_1 replaced by x_2, using the tangent line at $(x_2, f(x_2))$. This gives a third approximation:

$$x_3 = x_2 - \frac{f(x_2)}{f'(x_2)}$$

If we keep repeating this process, we obtain a sequence of approximations $x_1, x_2, x_3, x_4, \ldots$ as shown in Figure 3. In general, if the nth approximation is x_n and $f'(x_n) \neq 0$, then the next approximation is given by

2
$$x_{n+1} = x_n - \frac{f(x_n)}{f'(x_n)}$$

■ Sequences were briefly introduced in *A Preview of Calculus* on page 6. A more thorough discussion starts in Section 11.1.

If the numbers x_n become closer and closer to r as n becomes large, then we say that the sequence *converges* to r and we write

$$\lim_{n \to \infty} x_n = r$$

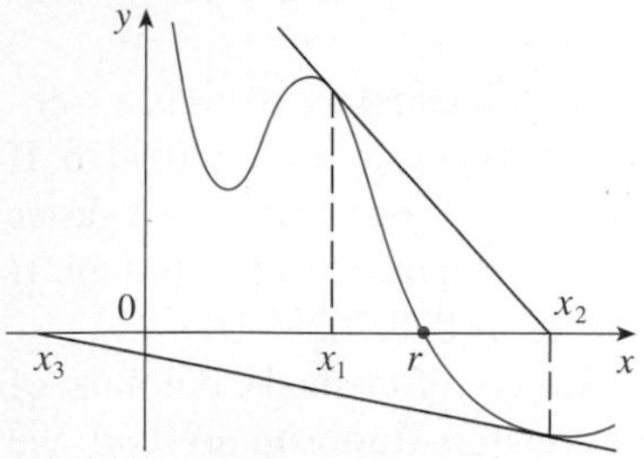

FIGURE 4

Although the sequence of successive approximations converges to the desired root for functions of the type illustrated in Figure 3, in certain circumstances the sequence may not converge. For example, consider the situation shown in Figure 4. You can see that x_2 is a worse approximation than x_1. This is likely to be the case when $f'(x_1)$ is close to 0. It might even happen that an approximation (such as x_3 in Figure 4) falls outside the domain of f. Then Newton's method fails and a better initial approximation x_1 should be chosen. See Exercises 31–34 for specific examples in which Newton's method works very slowly or does not work at all.

EXAMPLE 1 Starting with $x_1 = 2$, find the third approximation x_3 to the root of the equation $x^3 - 2x - 5 = 0$.

SOLUTION We apply Newton's method with

$$f(x) = x^3 - 2x - 5 \qquad \text{and} \qquad f'(x) = 3x^2 - 2$$

TEC In Module 4.8 you can investigate how Newton's Method works for several functions and what happens when you change x_1.

Newton himself used this equation to illustrate his method and he chose $x_1 = 2$ after some experimentation because $f(1) = -6$, $f(2) = -1$, and $f(3) = 16$. Equation 2 becomes

$$x_{n+1} = x_n - \frac{x_n^3 - 2x_n - 5}{3x_n^2 - 2}$$

With $n = 1$ we have

$$x_2 = x_1 - \frac{x_1^3 - 2x_1 - 5}{3x_1^2 - 2}$$

$$= 2 - \frac{2^3 - 2(2) - 5}{3(2)^2 - 2} = 2.1$$

■ Figure 5 shows the geometry behind the first step in Newton's method in Example 1. Since $f'(2) = 10$, the tangent line to $y = x^3 - 2x - 5$ at $(2, -1)$ has equation $y = 10x - 21$ so its x-intercept is $x_2 = 2.1$.

Then with $n = 2$ we obtain

$$x_3 = x_2 - \frac{x_2^3 - 2x_2 - 5}{3x_2^2 - 2}$$

$$= 2.1 - \frac{(2.1)^3 - 2(2.1) - 5}{3(2.1)^2 - 2} \approx 2.0946$$

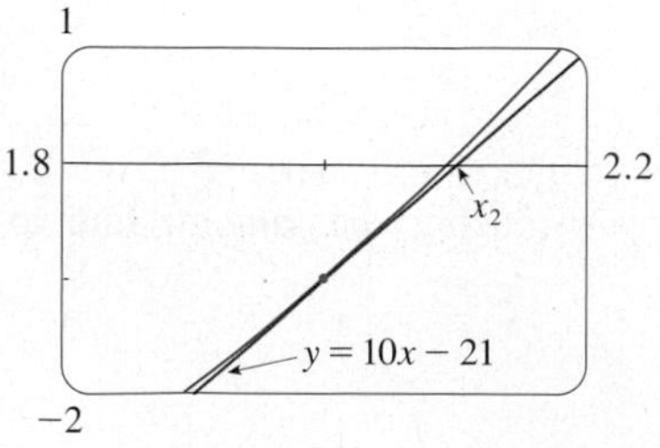

FIGURE 5

It turns out that this third approximation $x_3 \approx 2.0946$ is accurate to four decimal places. □

Suppose that we want to achieve a given accuracy, say to eight decimal places, using Newton's method. How do we know when to stop? The rule of thumb that is generally used is that we can stop when successive approximations x_n and x_{n+1} agree to eight decimal places. (A precise statement concerning accuracy in Newton's method will be given in Exercise 39 in Section 11.11.)

Notice that the procedure in going from n to $n + 1$ is the same for all values of n. (It is called an *iterative* process.) This means that Newton's method is particularly convenient for use with a programmable calculator or a computer.

V EXAMPLE 2 Use Newton's method to find $\sqrt[6]{2}$ correct to eight decimal places.

SOLUTION First we observe that finding $\sqrt[6]{2}$ is equivalent to finding the positive root of the equation

$$x^6 - 2 = 0$$

so we take $f(x) = x^6 - 2$. Then $f'(x) = 6x^5$ and Formula 2 (Newton's method) becomes

$$x_{n+1} = x_n - \frac{x_n^6 - 2}{6x_n^5}$$

If we choose $x_1 = 1$ as the initial approximation, then we obtain

$$x_2 \approx 1.16666667$$
$$x_3 \approx 1.12644368$$
$$x_4 \approx 1.12249707$$
$$x_5 \approx 1.12246205$$
$$x_6 \approx 1.12246205$$

Since x_5 and x_6 agree to eight decimal places, we conclude that

$$\sqrt[6]{2} \approx 1.12246205$$

to eight decimal places. □

V EXAMPLE 3 Find, correct to six decimal places, the root of the equation $\cos x = x$.

SOLUTION We first rewrite the equation in standard form:

$$\cos x - x = 0$$

Therefore we let $f(x) = \cos x - x$. Then $f'(x) = -\sin x - 1$, so Formula 2 becomes

$$x_{n+1} = x_n - \frac{\cos x_n - x_n}{-\sin x_n - 1} = x_n + \frac{\cos x_n - x_n}{\sin x_n + 1}$$

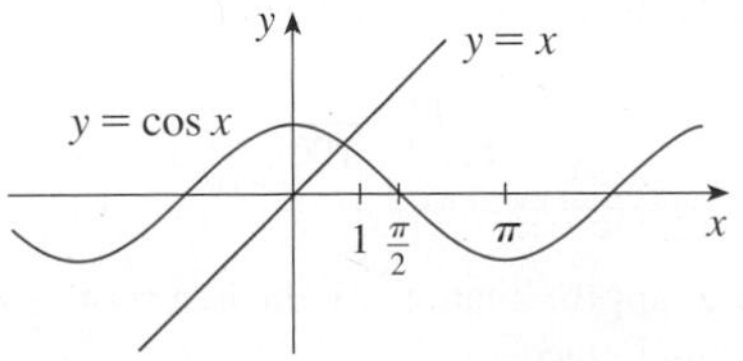

FIGURE 6

In order to guess a suitable value for x_1 we sketch the graphs of $y = \cos x$ and $y = x$ in Figure 6. It appears that they intersect at a point whose x-coordinate is somewhat less than 1, so let's take $x_1 = 1$ as a convenient first approximation. Then, remembering to put our calculator in radian mode, we get

$$x_2 \approx 0.75036387$$
$$x_3 \approx 0.73911289$$
$$x_4 \approx 0.73908513$$
$$x_5 \approx 0.73908513$$

Since x_4 and x_5 agree to six decimal places (eight, in fact), we conclude that the root of the equation, correct to six decimal places, is 0.739085. □

Instead of using the rough sketch in Figure 6 to get a starting approximation for Newton's method in Example 3, we could have used the more accurate graph that a calcu-

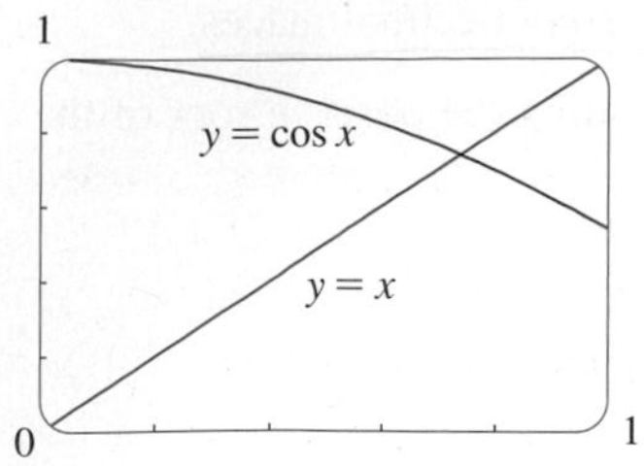

FIGURE 7

lator or computer provides. Figure 7 suggests that we use $x_1 = 0.75$ as the initial approximation. Then Newton's method gives

$$x_2 \approx 0.73911114 \qquad x_3 \approx 0.73908513 \qquad x_4 \approx 0.73908513$$

and so we obtain the same answer as before, but with one fewer step.

You might wonder why we bother at all with Newton's method if a graphing device is available. Isn't it easier to zoom in repeatedly and find the roots as we did in Section 1.4? If only one or two decimal places of accuracy are required, then indeed Newton's method is inappropriate and a graphing device suffices. But if six or eight decimal places are required, then repeated zooming becomes tiresome. It is usually faster and more efficient to use a computer and Newton's method in tandem—the graphing device to get started and Newton's method to finish.

4.8 EXERCISES

1. The figure shows the graph of a function f. Suppose that Newton's method is used to approximate the root r of the equation $f(x) = 0$ with initial approximation $x_1 = 1$.
(a) Draw the tangent lines that are used to find x_2 and x_3, and estimate the numerical values of x_2 and x_3.
(b) Would $x_1 = 5$ be a better first approximation? Explain.

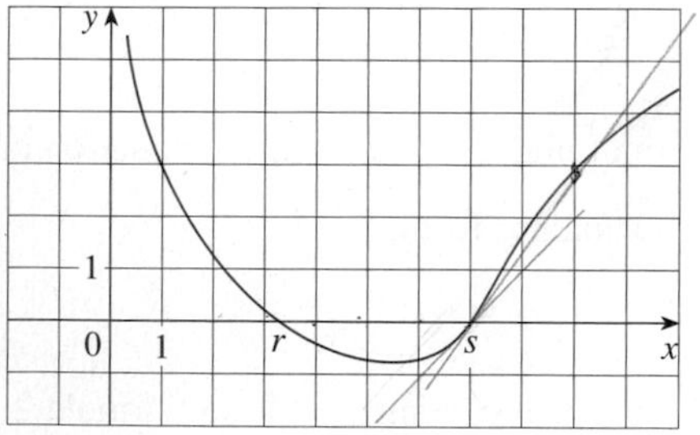

2. Follow the instructions for Exercise 1(a) but use $x_1 = 9$ as the starting approximation for finding the root s.

3. Suppose the line $y = 5x - 4$ is tangent to the curve $y = f(x)$ when $x = 3$. If Newton's method is used to locate a root of the equation $f(x) = 0$ and the initial approximation is $x_1 = 3$, find the second approximation x_2.

4. For each initial approximation, determine graphically what happens if Newton's method is used for the function whose graph is shown.
(a) $x_1 = 0$ (b) $x_1 = 1$ (c) $x_1 = 3$
(d) $x_1 = 4$ (e) $x_1 = 5$

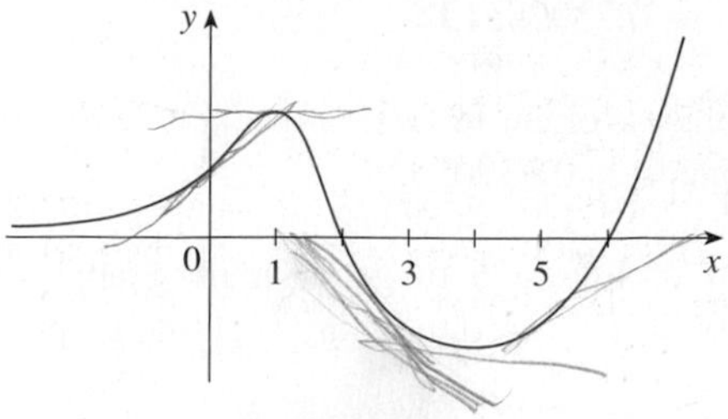

5–8 Use Newton's method with the specified initial approximation x_1 to find x_3, the third approximation to the root of the given equation. (Give your answer to four decimal places.)

5. $x^3 + 2x - 4 = 0, \quad x_1 = 1$

6. $\frac{1}{3}x^3 + \frac{1}{2}x^2 + 3 = 0, \quad x_1 = -3$

7. $x^5 - x - 1 = 0, \quad x_1 = 1$

8. $x^5 + 2 = 0, \quad x_1 = -1$

9. Use Newton's method with initial approximation $x_1 = -1$ to find x_2, the second approximation to the root of the equation $x^3 + x + 3 = 0$. Explain how the method works by first graphing the function and its tangent line at $(-1, 1)$.

10. Use Newton's method with initial approximation $x_1 = 1$ to find x_2, the second approximation to the root of the equation $x^4 - x - 1 = 0$. Explain how the method works by first graphing the function and its tangent line at $(1, -1)$.

11–12 Use Newton's method to approximate the given number correct to eight decimal places.

11. $\sqrt[5]{20}$ **12.** $\sqrt[100]{100}$

13–16 Use Newton's method to approximate the indicated root of the equation correct to six decimal places.

13. The root of $x^4 - 2x^3 + 5x^2 - 6 = 0$ in the interval $[1, 2]$

14. The root of $2.2x^5 - 4.4x^3 + 1.3x^2 - 0.9x - 4.0 = 0$ in the interval $[-2, -1]$

15. The positive root of $\sin x = x^2$

16. The positive root of $2 \cos x = x^4$

17–22 Use Newton's method to find all roots of the equation correct to six decimal places.

17. $x^4 = 1 + x$ **18.** $e^x = 3 - 2x$

19. $(x - 2)^2 = \ln x$

20. $\dfrac{1}{x} = 1 + x^3$

21. $\cos x = \sqrt{x}$

22. $\tan x = \sqrt{1 - x^2}$

23–28 Use Newton's method to find all the roots of the equation correct to eight decimal places. Start by drawing a graph to find initial approximations.

23. $x^6 - x^5 - 6x^4 - x^2 + x + 10 = 0$

24. $x^2(4 - x^2) = \dfrac{4}{x^2 + 1}$

25. $x^2\sqrt{2 - x - x^2} = 1$

26. $3\sin(x^2) = 2x$

27. $4e^{-x^2}\sin x = x^2 - x + 1$

28. $e^{\arctan x} = \sqrt{x^3 + 1}$

29. (a) Apply Newton's method to the equation $x^2 - a = 0$ to derive the following square-root algorithm (used by the ancient Babylonians to compute $\sqrt{a}$):

$$x_{n+1} = \frac{1}{2}\left(x_n + \frac{a}{x_n}\right)$$

(b) Use part (a) to compute $\sqrt{1000}$ correct to six decimal places.

30. (a) Apply Newton's method to the equation $1/x - a = 0$ to derive the following reciprocal algorithm:

$$x_{n+1} = 2x_n - ax_n^2$$

(This algorithm enables a computer to find reciprocals without actually dividing.)

(b) Use part (a) to compute $1/1.6984$ correct to six decimal places.

31. Explain why Newton's method doesn't work for finding the root of the equation $x^3 - 3x + 6 = 0$ if the initial approximation is chosen to be $x_1 = 1$.

32. (a) Use Newton's method with $x_1 = 1$ to find the root of the equation $x^3 - x = 1$ correct to six decimal places.

(b) Solve the equation in part (a) using $x_1 = 0.6$ as the initial approximation.

(c) Solve the equation in part (a) using $x_1 = 0.57$. (You definitely need a programmable calculator for this part.)

(d) Graph $f(x) = x^3 - x - 1$ and its tangent lines at $x_1 = 1$, 0.6, and 0.57 to explain why Newton's method is so sensitive to the value of the initial approximation.

33. Explain why Newton's method fails when applied to the equation $\sqrt[3]{x} = 0$ with any initial approximation $x_1 \neq 0$. Illustrate your explanation with a sketch.

34. If

$$f(x) = \begin{cases} \sqrt{x} & \text{if } x \geq 0 \\ -\sqrt{-x} & \text{if } x < 0 \end{cases}$$

then the root of the equation $f(x) = 0$ is $x = 0$. Explain why Newton's method fails to find the root no matter which initial approximation $x_1 \neq 0$ is used. Illustrate your explanation with a sketch.

35. (a) Use Newton's method to find the critical numbers of the function $f(x) = x^6 - x^4 + 3x^3 - 2x$ correct to six decimal places.

(b) Find the absolute minimum value of f correct to four decimal places.

36. Use Newton's method to find the absolute maximum value of the function $f(x) = x\cos x$, $0 \leq x \leq \pi$, correct to six decimal places.

37. Use Newton's method to find the coordinates of the inflection point of the curve $y = e^{\cos x}$, $0 \leq x \leq \pi$, correct to six decimal places.

38. Of the infinitely many lines that are tangent to the curve $y = -\sin x$ and pass through the origin, there is one that has the largest slope. Use Newton's method to find the slope of that line correct to six decimal places.

39. Use Newton's method to find the coordinates, correct to six decimal places, of the point on the parabola $y = (x - 1)^2$ that is closest to the origin.

40. In the figure, the length of the chord AB is 4 cm and the length of the arc AB is 5 cm. Find the central angle θ, in radians, correct to four decimal places. Then give the answer to the nearest degree.

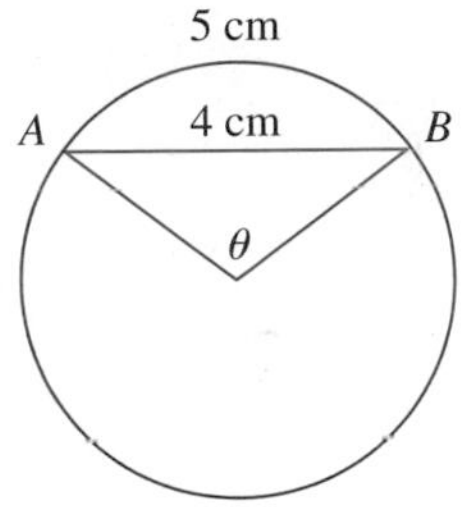

41. A car dealer sells a new car for \$18,000. He also offers to sell the same car for payments of \$375 per month for five years. What monthly interest rate is this dealer charging?

To solve this problem you will need to use the formula for the present value A of an annuity consisting of n equal payments of size R with interest rate i per time period:

$$A = \frac{R}{i}[1 - (1 + i)^{-n}]$$

Replacing i by x, show that

$$48x(1 + x)^{60} - (1 + x)^{60} + 1 = 0$$

Use Newton's method to solve this equation.

42. The figure shows the sun located at the origin and the earth at the point (1, 0). (The unit here is the distance between the centers of the earth and the sun, called an *astronomical unit:* 1 AU $\approx 1.496 \times 10^8$ km.) There are five locations L_1, L_2, L_3, L_4, and L_5 in this plane of rotation of the earth about the sun where a satellite remains motionless with respect to the earth because the forces acting on the satellite (including the gravitational attractions of the earth and the sun) balance each other. These locations are called *libration points.* (A solar research satellite has been placed at one of these libration points.) If m_1 is the mass of the sun, m_2 is the mass of the earth, and $r = m_2/(m_1 + m_2)$, it turns out that the x-coordinate of L_1 is the unique root of the fifth-degree equation

$$p(x) = x^5 - (2 + r)x^4 + (1 + 2r)x^3 - (1 - r)x^2 + 2(1 - r)x + r - 1 = 0$$

and the x-coordinate of L_2 is the root of the equation

$$p(x) - 2rx^2 = 0$$

Using the value $r \approx 3.04042 \times 10^{-6}$, find the locations of the libration points (a) L_1 and (b) L_2.

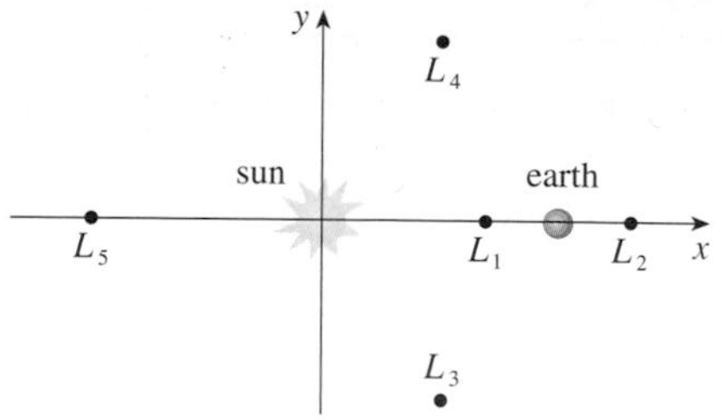

4.9 ANTIDERIVATIVES

A physicist who knows the velocity of a particle might wish to know its position at a given time. An engineer who can measure the variable rate at which water is leaking from a tank wants to know the amount leaked over a certain time period. A biologist who knows the rate at which a bacteria population is increasing might want to deduce what the size of the population will be at some future time. In each case, the problem is to find a function F whose derivative is a known function f. If such a function F exists, it is called an *antiderivative* of f.

> **DEFINITION** A function F is called an **antiderivative** of f on an interval I if $F'(x) = f(x)$ for all x in I.

For instance, let $f(x) = x^2$. It isn't difficult to discover an antiderivative of f if we keep the Power Rule in mind. In fact, if $F(x) = \frac{1}{3}x^3$, then $F'(x) = x^2 = f(x)$. But the function $G(x) = \frac{1}{3}x^3 + 100$ also satisfies $G'(x) = x^2$. Therefore both F and G are antiderivatives of f. Indeed, any function of the form $H(x) = \frac{1}{3}x^3 + C$, where C is a constant, is an antiderivative of f. The question arises: Are there any others?

To answer this question, recall that in Section 4.2 we used the Mean Value Theorem to prove that if two functions have identical derivatives on an interval, then they must differ by a constant (Corollary 4.2.7). Thus if F and G are any two antiderivatives of f, then

$$F'(x) = f(x) = G'(x)$$

so $G(x) - F(x) = C$, where C is a constant. We can write this as $G(x) = F(x) + C$, so we have the following result.

> **1 THEOREM** If F is an antiderivative of f on an interval I, then the most general antiderivative of f on I is
>
> $$F(x) + C$$
>
> where C is an arbitrary constant.

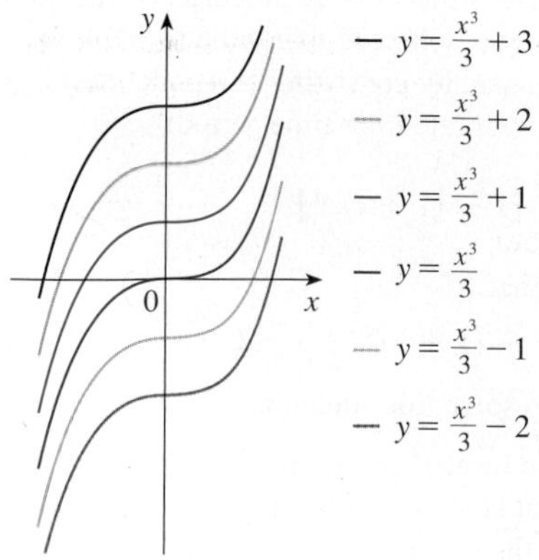

FIGURE 1
Members of the family of antiderivatives of $f(x) = x^2$

Going back to the function $f(x) = x^2$, we see that the general antiderivative of f is $\frac{1}{3}x^3 + C$. By assigning specific values to the constant C, we obtain a family of functions whose graphs are vertical translates of one another (see Figure 1). This makes sense because each curve must have the same slope at any given value of x.

EXAMPLE 1 Find the most general antiderivative of each of the following functions.
(a) $f(x) = \sin x$ (b) $f(x) = 1/x$ (c) $f(x) = x^n$, $n \neq -1$

SOLUTION
(a) If $F(x) = -\cos x$, then $F'(x) = \sin x$, so an antiderivative of $\sin x$ is $-\cos x$. By Theorem 1, the most general antiderivative is $G(x) = -\cos x + C$.

(b) Recall from Section 3.6 that

$$\frac{d}{dx}(\ln x) = \frac{1}{x}$$

So on the interval $(0, \infty)$ the general antiderivative of $1/x$ is $\ln x + C$. We also learned that

$$\frac{d}{dx}(\ln|x|) = \frac{1}{x}$$

for all $x \neq 0$. Theorem 1 then tells us that the general antiderivative of $f(x) = 1/x$ is $\ln|x| + C$ on any interval that doesn't contain 0. In particular, this is true on each of the intervals $(-\infty, 0)$ and $(0, \infty)$. So the general antiderivative of f is

$$F(x) = \begin{cases} \ln x + C_1 & \text{if } x > 0 \\ \ln(-x) + C_2 & \text{if } x < 0 \end{cases}$$

(c) We use the Power Rule to discover an antiderivative of x^n. In fact, if $n \neq -1$, then

$$\frac{d}{dx}\left(\frac{x^{n+1}}{n+1}\right) = \frac{(n+1)x^n}{n+1} = x^n$$

Thus the general antiderivative of $f(x) = x^n$ is

$$F(x) = \frac{x^{n+1}}{n+1} + C$$

This is valid for $n \geqslant 0$ since then $f(x) = x^n$ is defined on an interval. If n is negative (but $n \neq -1$), it is valid on any interval that doesn't contain 0. □

As in Example 1, every differentiation formula, when read from right to left, gives rise to an antidifferentiation formula. In Table 2 we list some particular antiderivatives. Each formula in the table is true because the derivative of the function in the right column appears in the left column. In particular, the first formula says that the antiderivative of a constant times a function is the constant times the antiderivative of the function. The second formula says that the antiderivative of a sum is the sum of the antiderivatives. (We use the notation $F' = f$, $G' = g$.)

2 TABLE OF ANTIDIFFERENTIATION FORMULAS

■ To obtain the most general antiderivative from the particular ones in Table 2, we have to add a constant (or constants), as in Example 1.

Function	Particular antiderivative	Function	Particular antiderivative
$cf(x)$	$cF(x)$	$\sin x$	$-\cos x$
$f(x) + g(x)$	$F(x) + G(x)$	$\sec^2 x$	$\tan x$
$x^n \ (n \neq -1)$	$\dfrac{x^{n+1}}{n+1}$	$\sec x \tan x$	$\sec x$
$1/x$	$\ln\|x\|$	$\dfrac{1}{\sqrt{1-x^2}}$	$\sin^{-1}x$
e^x	e^x	$\dfrac{1}{1+x^2}$	$\tan^{-1}x$
$\cos x$	$\sin x$		

EXAMPLE 2 Find all functions g such that

$$g'(x) = 4\sin x + \frac{2x^5 - \sqrt{x}}{x}$$

SOLUTION We first rewrite the given function as follows:

$$g'(x) = 4\sin x + \frac{2x^5}{x} - \frac{\sqrt{x}}{x} = 4\sin x + 2x^4 - \frac{1}{\sqrt{x}}$$

Thus we want to find an antiderivative of

$$g'(x) = 4\sin x + 2x^4 - x^{-1/2}$$

Using the formulas in Table 2 together with Theorem 1, we obtain

$$g(x) = 4(-\cos x) + 2\frac{x^5}{5} - \frac{x^{1/2}}{\frac{1}{2}} + C$$

$$= -4\cos x + \tfrac{2}{5}x^5 - 2\sqrt{x} + C$$ □

In applications of calculus it is very common to have a situation as in Example 2, where it is required to find a function, given knowledge about its derivatives. An equation that involves the derivatives of a function is called a **differential equation**. These will be studied in some detail in Chapter 9, but for the present we can solve some elementary differential equations. The general solution of a differential equation involves an arbitrary constant (or constants) as in Example 2. However, there may be some extra conditions given that will determine the constants and therefore uniquely specify the solution.

■ Figure 2 shows the graphs of the function f' in Example 3 and its antiderivative f. Notice that $f'(x) > 0$, so f is always increasing. Also notice that when f' has a maximum or minimum, f appears to have an inflection point. So the graph serves as a check on our calculation.

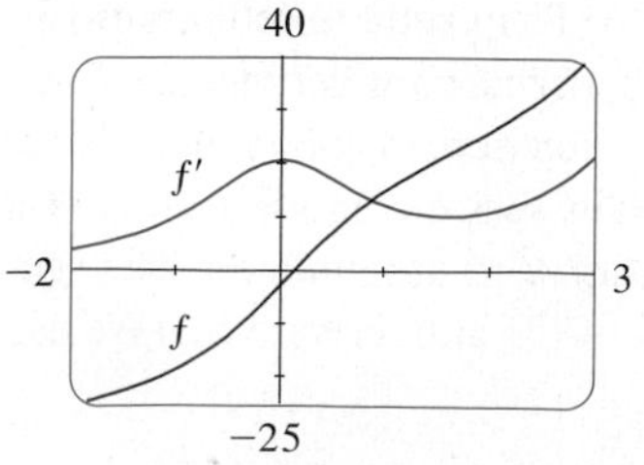

FIGURE 2

EXAMPLE 3 Find f if $f'(x) = e^x + 20(1 + x^2)^{-1}$ and $f(0) = -2$.

SOLUTION The general antiderivative of

$$f'(x) = e^x + \frac{20}{1 + x^2}$$

is

$$f(x) = e^x + 20\tan^{-1}x + C$$

To determine C we use the fact that $f(0) = -2$:

$$f(0) = e^0 + 20\tan^{-1}0 + C = -2$$

Thus we have $C = -2 - 1 = -3$, so the particular solution is

$$f(x) = e^x + 20\tan^{-1}x - 3$$ □

V EXAMPLE 4 Find f if $f''(x) = 12x^2 + 6x - 4$, $f(0) = 4$, and $f(1) = 1$.

SOLUTION The general antiderivative of $f''(x) = 12x^2 + 6x - 4$ is

$$f'(x) = 12\frac{x^3}{3} + 6\frac{x^2}{2} - 4x + C = 4x^3 + 3x^2 - 4x + C$$

Using the antidifferentiation rules once more, we find that

$$f(x) = 4\frac{x^4}{4} + 3\frac{x^3}{3} - 4\frac{x^2}{2} + Cx + D = x^4 + x^3 - 2x^2 + Cx + D$$

To determine C and D we use the given conditions that $f(0) = 4$ and $f(1) = 1$. Since $f(0) = 0 + D = 4$, we have $D = 4$. Since

$$f(1) = 1 + 1 - 2 + C + 4 = 1$$

we have $C = -3$. Therefore the required function is

$$f(x) = x^4 + x^3 - 2x^2 - 3x + 4$$ □

If we are given the graph of a function f, it seems reasonable that we should be able to sketch the graph of an antiderivative F. Suppose, for instance, that we are given that $F(0) = 1$. Then we have a place to start, the point $(0, 1)$, and the direction in which we move our pencil is given at each stage by the derivative $F'(x) = f(x)$. In the next example we use the principles of this chapter to show how to graph F even when we don't have a formula for f. This would be the case, for instance, when $f(x)$ is determined by experimental data.

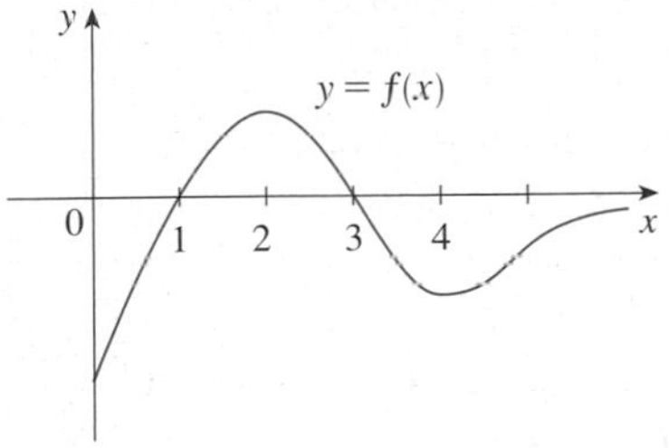

FIGURE 3

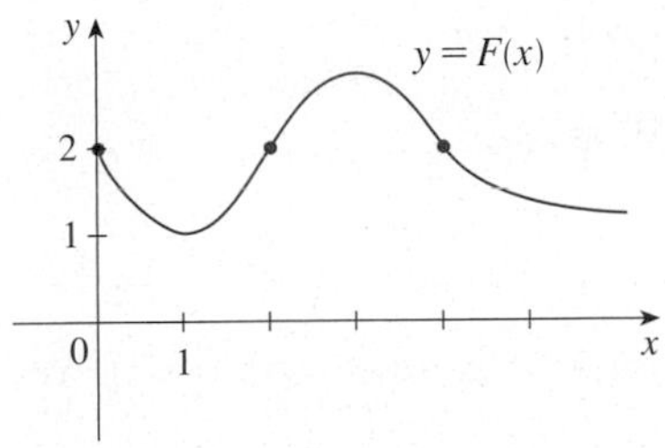

FIGURE 4

V EXAMPLE 5 The graph of a function f is given in Figure 3. Make a rough sketch of an antiderivative F, given that $F(0) = 2$.

SOLUTION We are guided by the fact that the slope of $y = F(x)$ is $f(x)$. We start at the point $(0, 2)$ and draw F as an initially decreasing function since $f(x)$ is negative when $0 < x < 1$. Notice that $f(1) = f(3) = 0$, so F has horizontal tangents when $x = 1$ and $x = 3$. For $1 < x < 3$, $f(x)$ is positive and so F is increasing. We see that F has a local minimum when $x = 1$ and a local maximum when $x = 3$. For $x > 3$, $f(x)$ is negative and so F is decreasing on $(3, \infty)$. Since $f(x) \to 0$ as $x \to \infty$, the graph of F becomes flatter as $x \to \infty$. Also notice that $F''(x) = f'(x)$ changes from positive to negative at $x = 2$ and from negative to positive at $x = 4$, so F has inflection points when $x = 2$ and $x = 4$. We use this information to sketch the graph of the antiderivative in Figure 4. □

RECTILINEAR MOTION

Antidifferentiation is particularly useful in analyzing the motion of an object moving in a straight line. Recall that if the object has position function $s = f(t)$, then the velocity function is $v(t) = s'(t)$. This means that the position function is an antiderivative of the velocity function. Likewise, the acceleration function is $a(t) = v'(t)$, so the velocity function is an antiderivative of the acceleration. If the acceleration and the initial values $s(0)$ and $v(0)$ are known, then the position function can be found by antidifferentiating twice.

V EXAMPLE 6 A particle moves in a straight line and has acceleration given by $a(t) = 6t + 4$. Its initial velocity is $v(0) = -6$ cm/s and its initial displacement is $s(0) = 9$ cm. Find its position function $s(t)$.

SOLUTION Since $v'(t) = a(t) = 6t + 4$, antidifferentiation gives

$$v(t) = 6\frac{t^2}{2} + 4t + C = 3t^2 + 4t + C$$

Note that $v(0) = C$. But we are given that $v(0) = -6$, so $C = -6$ and

$$v(t) = 3t^2 + 4t - 6$$

Since $v(t) = s'(t)$, s is the antiderivative of v:

$$s(t) = 3\frac{t^3}{3} + 4\frac{t^2}{2} - 6t + D = t^3 + 2t^2 - 6t + D$$

This gives $s(0) = D$. We are given that $s(0) = 9$, so $D = 9$ and the required position function is

$$s(t) = t^3 + 2t^2 - 6t + 9$$ □

An object near the surface of the earth is subject to a gravitational force that produces a downward acceleration denoted by g. For motion close to the ground we may assume that g is constant, its value being about 9.8 m/s² (or 32 ft/s²).

EXAMPLE 7 A ball is thrown upward with a speed of 48 ft/s from the edge of a cliff 432 ft above the ground. Find its height above the ground t seconds later. When does it reach its maximum height? When does it hit the ground?

SOLUTION The motion is vertical and we choose the positive direction to be upward. At time t the distance above the ground is $s(t)$ and the velocity $v(t)$ is decreasing. Therefore the acceleration must be negative and we have

$$a(t) = \frac{dv}{dt} = -32$$

Taking antiderivatives, we have

$$v(t) = -32t + C$$

To determine C we use the given information that $v(0) = 48$. This gives $48 = 0 + C$, so

$$v(t) = -32t + 48$$

The maximum height is reached when $v(t) = 0$, that is, after 1.5 s. Since $s'(t) = v(t)$, we antidifferentiate again and obtain

$$s(t) = -16t^2 + 48t + D$$

Using the fact that $s(0) = 432$, we have $432 = 0 + D$ and so

$$s(t) = -16t^2 + 48t + 432$$

■ Figure 5 shows the position function of the ball in Example 7. The graph corroborates the conclusions we reached: The ball reaches its maximum height after 1.5 s and hits the ground after 6.9 s.

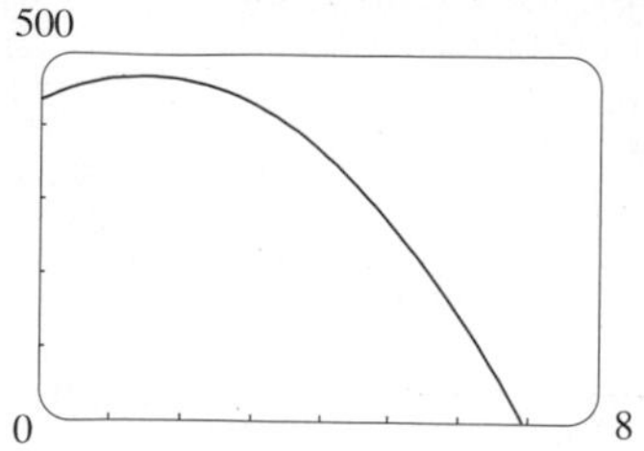

FIGURE 5

The expression for $s(t)$ is valid until the ball hits the ground. This happens when $s(t) = 0$, that is, when

$$-16t^2 + 48t + 432 = 0$$

or, equivalently, $$t^2 - 3t - 27 = 0$$

Using the quadratic formula to solve this equation, we get

$$t = \frac{3 \pm 3\sqrt{13}}{2}$$

We reject the solution with the minus sign since it gives a negative value for t. Therefore the ball hits the ground after $3(1 + \sqrt{13})/2 \approx 6.9$ s. □

4.9 EXERCISES

1–20 Find the most general antiderivative of the function. (Check your answer by differentiation.)

1. $f(x) = x - 3$

2. $f(x) = \frac{1}{2}x^2 - 2x + 6$

3. $f(x) = \frac{1}{2} + \frac{3}{4}x^2 - \frac{4}{5}x^3$

4. $f(x) = 8x^9 - 3x^6 + 12x^3$

5. $f(x) = (x + 1)(2x - 1)$

6. $f(x) = x(2 - x)^2$

7. $f(x) = 5x^{1/4} - 7x^{3/4}$

8. $f(x) = 2x + 3x^{1.7}$

9. $f(x) = 6\sqrt{x} - \sqrt[6]{x}$

10. $f(x) = \sqrt[4]{x^3} + \sqrt[3]{x^4}$

11. $f(x) = \dfrac{10}{x^9}$

12. $g(x) = \dfrac{5 - 4x^3 + 2x^6}{x^6}$

13. $f(u) = \dfrac{u^4 + 3\sqrt{u}}{u^2}$

14. $f(x) = 3e^x + 7\sec^2 x$

15. $g(\theta) = \cos\theta - 5\sin\theta$

16. $f(t) = \sin t + 2\sinh t$

17. $f(x) = 5e^x - 3\cosh x$

18. $f(x) = 2\sqrt{x} + 6\cos x$

19. $f(x) = \dfrac{x^5 - x^3 + 2x}{x^4}$

20. $f(x) = \dfrac{2 + x^2}{1 + x^2}$

21–22 Find the antiderivative F of f that satisfies the given condition. Check your answer by comparing the graphs of f and F.

21. $f(x) = 5x^4 - 2x^5, \quad F(0) = 4$

22. $f(x) = 4 - 3(1 + x^2)^{-1}, \quad F(1) = 0$

23–46 Find f.

23. $f''(x) = 6x + 12x^2$

24. $f''(x) = 2 + x^3 + x^6$

25. $f''(x) = \frac{2}{3}x^{2/3}$

26. $f''(x) = 6x + \sin x$

27. $f'''(t) = e^t$

28. $f'''(t) = t - \sqrt{t}$

29. $f'(x) = 1 - 6x, \quad f(0) = 8$

30. $f'(x) = 8x^3 + 12x + 3, \quad f(1) = 6$

31. $f'(x) = \sqrt{x}\,(6 + 5x), \quad f(1) = 10$

32. $f'(x) = 2x - 3/x^4, \quad x > 0, \quad f(1) = 3$

33. $f'(t) = 2\cos t + \sec^2 t, \quad -\pi/2 < t < \pi/2, \quad f(\pi/3) = 4$

34. $f'(x) = (x^2 - 1)/x, \quad f(1) = \frac{1}{2}, \quad f(-1) = 0$

35. $f'(x) = x^{-1/3}, \quad f(1) = 1, \quad f(-1) = -1$

36. $f'(x) = 4/\sqrt{1 - x^2}, \quad f(\frac{1}{2}) = 1$

37. $f''(x) = 24x^2 + 2x + 10, \quad f(1) = 5, \quad f'(1) = -3$

38. $f''(x) = 4 - 6x - 40x^3, \quad f(0) = 2, \quad f'(0) = 1$

39. $f''(\theta) = \sin\theta + \cos\theta, \quad f(0) = 3, \quad f'(0) = 4$

40. $f''(t) = 3/\sqrt{t}, \quad f(4) = 20, \quad f'(4) = 7$

41. $f''(x) = 2 - 12x, \quad f(0) = 9, \quad f(2) = 15$

42. $f''(x) = 20x^3 + 12x^2 + 4, \quad f(0) = 8, \quad f(1) = 5$

43. $f''(x) = 2 + \cos x, \quad f(0) = -1, \quad f(\pi/2) = 0$

44. $f''(t) = 2e^t + 3\sin t, \quad f(0) = 0, \quad f(\pi) = 0$

45. $f''(x) = x^{-2}, \quad x > 0, \quad f(1) = 0, \quad f(2) = 0$

46. $f'''(x) = \cos x, \quad f(0) = 1, \quad f'(0) = 2, \quad f''(0) = 3$

47. Given that the graph of f passes through the point $(1, 6)$ and that the slope of its tangent line at $(x, f(x))$ is $2x + 1$, find $f(2)$.

48. Find a function f such that $f'(x) = x^3$ and the line $x + y = 0$ is tangent to the graph of f.

49–50 The graph of a function f is shown. Which graph is an antiderivative of f and why?

49.

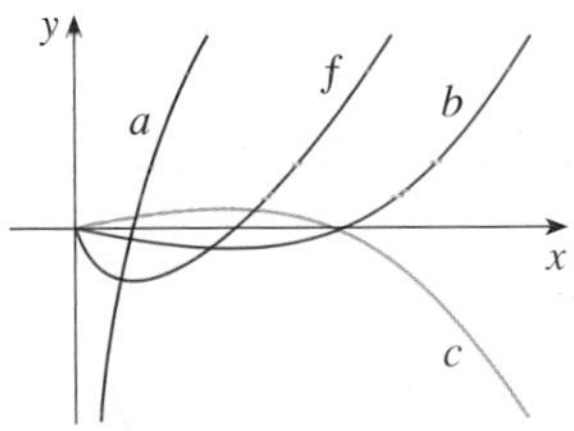

50.

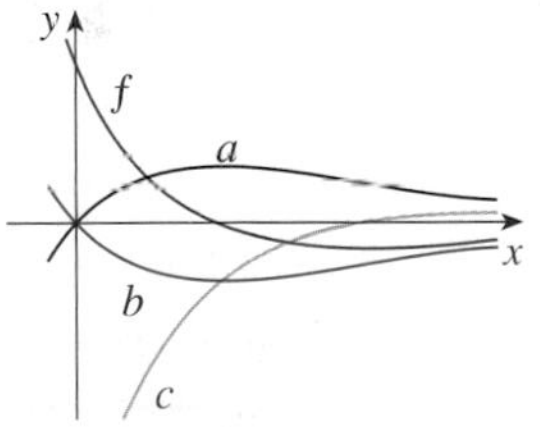

51. The graph of a function is shown in the figure. Make a rough sketch of an antiderivative F, given that $F(0) = 1$.

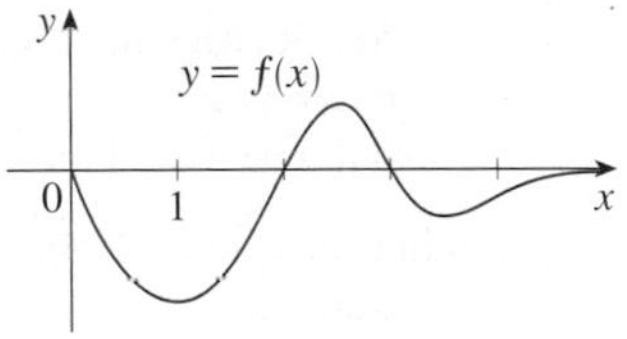

52. The graph of the velocity function of a particle is shown in the figure. Sketch the graph of the position function.

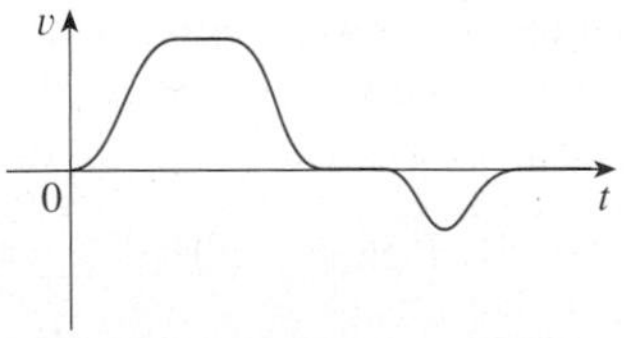

53. The graph of f' is shown in the figure. Sketch the graph of f if f is continuous and $f(0) = -1$.

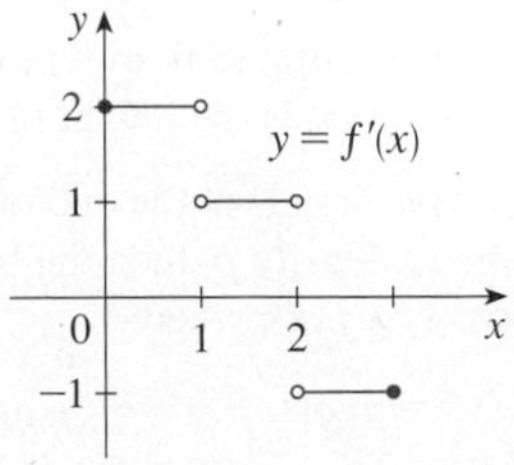

54. (a) Use a graphing device to graph $f(x) = 2x - 3\sqrt{x}$.
(b) Starting with the graph in part (a), sketch a rough graph of the antiderivative F that satisfies $F(0) = 1$.
(c) Use the rules of this section to find an expression for $F(x)$.
(d) Graph F using the expression in part (c). Compare with your sketch in part (b).

55–56 Draw a graph of f and use it to make a rough sketch of the antiderivative that passes through the origin.

55. $f(x) = \dfrac{\sin x}{1 + x^2}, \quad -2\pi \leqslant x \leqslant 2\pi$

56. $f(x) = \sqrt{x^4 - 2x^2 + 2} - 1, \quad -1.5 \leqslant x \leqslant 1.5$

57–62 A particle is moving with the given data. Find the position of the particle.

57. $v(t) = \sin t - \cos t, \quad s(0) = 0$

58. $v(t) = 1.5\sqrt{t}, \quad s(4) = 10$

59. $a(t) = t - 2, \quad s(0) = 1, \quad v(0) = 3$

60. $a(t) = \cos t + \sin t, \quad s(0) = 0, \quad v(0) = 5$

61. $a(t) = 10 \sin t + 3 \cos t, \quad s(0) = 0, \quad s(2\pi) = 12$

62. $a(t) = t^2 - 4t + 6, \quad s(0) = 0, \quad s(1) = 20$

---

63. A stone is dropped from the upper observation deck (the Space Deck) of the CN Tower, 450 m above the ground.
(a) Find the distance of the stone above ground level at time t.
(b) How long does it take the stone to reach the ground?
(c) With what velocity does it strike the ground?
(d) If the stone is thrown downward with a speed of 5 m/s, how long does it take to reach the ground?

64. Show that for motion in a straight line with constant acceleration a, initial velocity v_0, and initial displacement s_0, the displacement after time t is

$$s = \tfrac{1}{2}at^2 + v_0 t + s_0$$

65. An object is projected upward with initial velocity v_0 meters per second from a point s_0 meters above the ground. Show that

$$[v(t)]^2 = v_0^2 - 19.6[s(t) - s_0]$$

66. Two balls are thrown upward from the edge of the cliff in Example 7. The first is thrown with a speed of 48 ft/s and the other is thrown a second later with a speed of 24 ft/s. Do the balls ever pass each other?

67. A stone was dropped off a cliff and hit the ground with a speed of 120 ft/s. What is the height of the cliff?

68. If a diver of mass m stands at the end of a diving board with length L and linear density ρ, then the board takes on the shape of a curve $y = f(x)$, where

$$EIy'' = mg(L - x) + \tfrac{1}{2}\rho g(L - x)^2$$

E and I are positive constants that depend on the material of the board and g (< 0) is the acceleration due to gravity.
(a) Find an expression for the shape of the curve.
(b) Use $f(L)$ to estimate the distance below the horizontal at the end of the board.

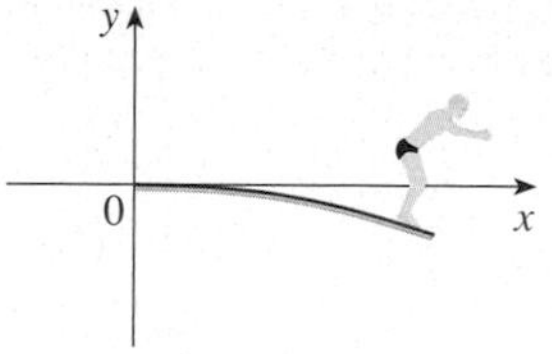

69. A company estimates that the marginal cost (in dollars per item) of producing x items is $1.92 - 0.002x$. If the cost of producing one item is \$562, find the cost of producing 100 items.

70. The linear density of a rod of length 1 m is given by $\rho(x) = 1/\sqrt{x}$, in grams per centimeter, where x is measured in centimeters from one end of the rod. Find the mass of the rod.

71. Since raindrops grow as they fall, their surface area increases and therefore the resistance to their falling increases. A raindrop has an initial downward velocity of 10 m/s and its downward acceleration is

$$a = \begin{cases} 9 - 0.9t & \text{if } 0 \leqslant t \leqslant 10 \\ 0 & \text{if } t > 10 \end{cases}$$

If the raindrop is initially 500 m above the ground, how long does it take to fall?

72. A car is traveling at 50 mi/h when the brakes are fully applied, producing a constant deceleration of 22 ft/s^2. What is the distance traveled before the car comes to a stop?

73. What constant acceleration is required to increase the speed of a car from 30 mi/h to 50 mi/h in 5 s?

74. A car braked with a constant deceleration of 16 ft/s^2, producing skid marks measuring 200 ft before coming to a stop. How fast was the car traveling when the brakes were first applied?

75. A car is traveling at 100 km/h when the driver sees an accident 80 m ahead and slams on the brakes. What constant deceleration is required to stop the car in time to avoid a pileup?

76. A model rocket is fired vertically upward from rest. Its acceleration for the first three seconds is $a(t) = 60t$, at which time the fuel is exhausted and it becomes a freely "falling" body. Fourteen seconds later, the rocket's parachute opens, and the (downward) velocity slows linearly to -18 ft/s in 5 s. The rocket then "floats" to the ground at that rate.
(a) Determine the position function s and the velocity function v (for all times t). Sketch the graphs of s and v.

(b) At what time does the rocket reach its maximum height, and what is that height?

(c) At what time does the rocket land?

77. A high-speed bullet train accelerates and decelerates at the rate of 4 ft/s^2. Its maximum cruising speed is 90 mi/h.

(a) What is the maximum distance the train can travel if it accelerates from rest until it reaches its cruising speed and then runs at that speed for 15 minutes?

(b) Suppose that the train starts from rest and must come to a complete stop in 15 minutes. What is the maximum distance it can travel under these conditions?

(c) Find the minimum time that the train takes to travel between two consecutive stations that are 45 miles apart.

(d) The trip from one station to the next takes 37.5 minutes. How far apart are the stations?

4 REVIEW

CONCEPT CHECK

1. Explain the difference between an absolute maximum and a local maximum. Illustrate with a sketch.

2. (a) What does the Extreme Value Theorem say?
(b) Explain how the Closed Interval Method works.

3. (a) State Fermat's Theorem.
(b) Define a critical number of f.

4. (a) State Rolle's Theorem.
(b) State the Mean Value Theorem and give a geometric interpretation.

5. (a) State the Increasing/Decreasing Test.
(b) What does it mean to say that f is concave upward on an interval I?
(c) State the Concavity Test.
(d) What are inflection points? How do you find them?

6. (a) State the First Derivative Test.
(b) State the Second Derivative Test.
(c) What are the relative advantages and disadvantages of these tests?

7. (a) What does l'Hospital's Rule say?
(b) How can you use l'Hospital's Rule if you have a product $f(x)g(x)$ where $f(x) \to 0$ and $g(x) \to \infty$ as $x \to a$?
(c) How can you use l'Hospital's Rule if you have a difference $f(x) - g(x)$ where $f(x) \to \infty$ and $g(x) \to \infty$ as $x \to a$?
(d) How can you use l'Hospital's Rule if you have a power $[f(x)]^{g(x)}$ where $f(x) \to 0$ and $g(x) \to 0$ as $x \to a$?

8. If you have a graphing calculator or computer, why do you need calculus to graph a function?

9. (a) Given an initial approximation x_1 to a root of the equation $f(x) = 0$, explain geometrically, with a diagram, how the second approximation x_2 in Newton's method is obtained.
(b) Write an expression for x_2 in terms of x_1, $f(x_1)$, and $f'(x_1)$.
(c) Write an expression for x_{n+1} in terms of x_n, $f(x_n)$, and $f'(x_n)$.
(d) Under what circumstances is Newton's method likely to fail or to work very slowly?

10. (a) What is an antiderivative of a function f?
(b) Suppose F_1 and F_2 are both antiderivatives of f on an interval I. How are F_1 and F_2 related?

TRUE-FALSE QUIZ

Determine whether the statement is true or false. If it is true, explain why. If it is false, explain why or give an example that disproves the statement.

1. If $f'(c) = 0$, then f has a local maximum or minimum at c.

2. If f has an absolute minimum value at c, then $f'(c) = 0$.

3. If f is continuous on (a, b), then f attains an absolute maximum value $f(c)$ and an absolute minimum value $f(d)$ at some numbers c and d in (a, b).

4. If f is differentiable and $f(-1) = f(1)$, then there is a number c such that $|c| < 1$ and $f'(c) = 0$.

5. If $f'(x) < 0$ for $1 < x < 6$, then f is decreasing on $(1, 6)$.

6. If $f''(2) = 0$, then $(2, f(2))$ is an inflection point of the curve $y = f(x)$.

7. If $f'(x) = g'(x)$ for $0 < x < 1$, then $f(x) = g(x)$ for $0 < x < 1$.

8. There exists a function f such that $f(1) = -2$, $f(3) = 0$, and $f'(x) > 1$ for all x.

9. There exists a function f such that $f(x) > 0$, $f'(x) < 0$, and $f''(x) > 0$ for all x.

10. There exists a function f such that $f(x) < 0$, $f'(x) < 0$, and $f''(x) > 0$ for all x.

11. If f and g are increasing on an interval I, then $f + g$ is increasing on I.

12. If f and g are increasing on an interval I, then $f - g$ is increasing on I.

13. If f and g are increasing on an interval I, then fg is increasing on I.

14. If f and g are positive increasing functions on an interval I, then fg is increasing on I.

15. If f is increasing and $f(x) > 0$ on I, then $g(x) = 1/f(x)$ is decreasing on I.

16. If f is even, then f' is even.

17. If f is periodic, then f' is periodic.

18. The most general antiderivative of $f(x) = x^{-2}$ is

$$F(x) = -\frac{1}{x} + C$$

19. If $f'(x)$ exists and is nonzero for all x, then $f(1) \neq f(0)$.

20. $\lim_{x \to 0} \frac{x}{e^x} = 1$

EXERCISES

1–6 Find the local and absolute extreme values of the function on the given interval.

1. $f(x) = x^3 - 6x^2 + 9x + 1, \quad [2, 4]$

2. $f(x) = x\sqrt{1 - x}, \quad [-1, 1]$

3. $f(x) = \frac{3x - 4}{x^2 + 1}, \quad [-2, 2]$

4. $f(x) = (x^2 + 2x)^3, \quad [-2, 1]$

5. $f(x) = x + \sin 2x, \quad [0, \pi]$

6. $f(x) = (\ln x)/x^2, \quad [1, 3]$

7–14 Evaluate the limit.

7. $\lim_{x \to 0} \frac{\tan \pi x}{\ln(1 + x)}$

8. $\lim_{x \to 0} \frac{1 - \cos x}{x^2 + x}$

9. $\lim_{x \to 0} \frac{e^{4x} - 1 - 4x}{x^2}$

10. $\lim_{x \to \infty} \frac{e^{4x} - 1 - 4x}{x^2}$

11. $\lim_{x \to \infty} x^3 e^{-x}$

12. $\lim_{x \to 0^+} x^2 \ln x$

13. $\lim_{x \to 1^+} \left(\frac{x}{x - 1} - \frac{1}{\ln x} \right)$

14. $\lim_{x \to (\pi/2)^-} (\tan x)^{\cos x}$

15–17 Sketch the graph of a function that satisfies the given conditions:

15. $f(0) = 0, \quad f'(-2) = f'(1) = f'(9) = 0,$
$\lim_{x \to \infty} f(x) = 0, \quad \lim_{x \to 6} f(x) = -\infty,$
$f'(x) < 0$ on $(-\infty, -2)$, $(1, 6)$, and $(9, \infty)$,
$f'(x) > 0$ on $(-2, 1)$ and $(6, 9)$,
$f''(x) > 0$ on $(-\infty, 0)$ and $(12, \infty)$,
$f''(x) < 0$ on $(0, 6)$ and $(6, 12)$

16. $f(0) = 0, \quad f$ is continuous and even,
$f'(x) = 2x$ if $0 < x < 1, \quad f'(x) = -1$ if $1 < x < 3,$
$f'(x) = 1$ if $x > 3$

17. f is odd, $\quad f'(x) < 0$ for $0 < x < 2$,
$f'(x) > 0$ for $x > 2, \quad f''(x) > 0$ for $0 < x < 3$,
$f''(x) < 0$ for $x > 3, \quad \lim_{x \to \infty} f(x) = -2$

18. The figure shows the graph of the *derivative* f' of a function f.
(a) On what intervals is f increasing or decreasing?
(b) For what values of x does f have a local maximum or minimum?
(c) Sketch the graph of f''.
(d) Sketch a possible graph of f.

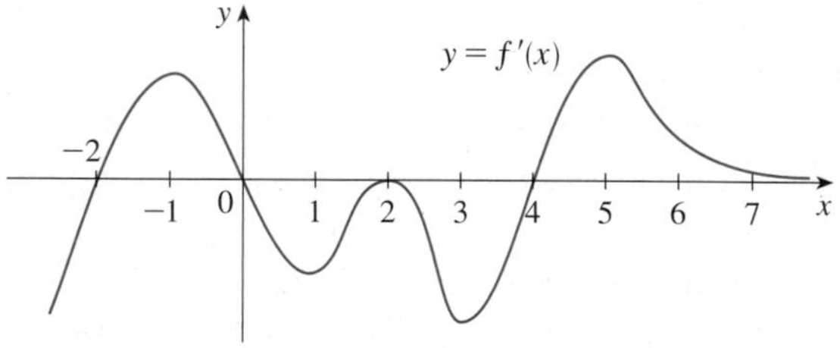

19–34 Use the guidelines of Section 4.5 to sketch the curve.

19. $y = 2 - 2x - x^3$

20. $y = x^3 - 6x^2 - 15x + 4$

21. $y = x^4 - 3x^3 + 3x^2 - x$

22. $y = \frac{1}{1 - x^2}$

23. $y = \frac{1}{x(x - 3)^2}$

24. $y = \frac{1}{x^2} - \frac{1}{(x - 2)^2}$

25. $y = x^2/(x + 8)$

26. $y = \sqrt{1 - x} + \sqrt{1 + x}$

27. $y = x\sqrt{2 + x}$

28. $y = \sqrt[3]{x^2 + 1}$

29. $y = \sin^2 x - 2 \cos x$

30. $y = 4x - \tan x, \quad -\pi/2 < x < \pi/2$

31. $y = \sin^{-1}(1/x)$

32. $y = e^{2x - x^2}$

33. $y = xe^{-2x}$

34. $y = x + \ln(x^2 + 1)$

35–38 Produce graphs of f that reveal all the important aspects of the curve. Use graphs of f' and f'' to estimate the intervals of increase and decrease, extreme values, intervals of concavity, and inflection points. In Exercise 35 use calculus to find these quantities exactly.

35. $f(x) = \frac{x^2 - 1}{x^3}$

36. $f(x) = \frac{x^3 - x}{x^2 + x + 3}$

37. $f(x) = 3x^6 - 5x^5 + x^4 - 5x^3 - 2x^2 + 2$

38. $f(x) = x^2 + 6.5 \sin x, \quad -5 \leqslant x \leqslant 5$

39. Graph $f(x) = e^{-1/x^2}$ in a viewing rectangle that shows all the main aspects of this function. Estimate the inflection points. Then use calculus to find them exactly.

CAS **40.** (a) Graph the function $f(x) = 1/(1 + e^{1/x})$.
(b) Explain the shape of the graph by computing the limits of $f(x)$ as x approaches ∞, $-\infty$, 0^+, and 0^-.
(c) Use the graph of f to estimate the coordinates of the inflection points.
(d) Use your CAS to compute and graph f''.
(e) Use the graph in part (d) to estimate the inflection points more accurately.

CAS **41–42** Use the graphs of f, f', and f'' to estimate the x-coordinates of the maximum and minimum points and inflection points of f.

41. $f(x) = \dfrac{\cos^2 x}{\sqrt{x^2 + x + 1}}, \quad -\pi \leqslant x \leqslant \pi$

42. $f(x) = e^{-0.1x} \ln(x^2 - 1)$

43. Investigate the family of functions $f(x) = \ln(\sin x + C)$. What features do the members of this family have in common? How do they differ? For which values of C is f continuous on $(-\infty, \infty)$? For which values of C does f have no graph at all? What happens as $C \to \infty$?

44. Investigate the family of functions $f(x) = cxe^{-cx^2}$. What happens to the maximum and minimum points and the inflection points as c changes? Illustrate your conclusions by graphing several members of the family.

45. Show that the equation $3x + 2\cos x + 5 = 0$ has exactly one real root.

46. Suppose that f is continuous on $[0, 4]$, $f(0) = 1$, and $2 \leqslant f'(x) \leqslant 5$ for all x in $(0, 4)$. Show that $9 \leqslant f(4) \leqslant 21$.

47. By applying the Mean Value Theorem to the function $f(x) = x^{1/5}$ on the interval $[32, 33]$, show that

$$2 < \sqrt[5]{33} < 2.0125$$

48. For what values of the constants a and b is $(1, 6)$ a point of inflection of the curve $y = x^3 + ax^2 + bx + 1$?

49. Let $g(x) = f(x^2)$, where f is twice differentiable for all x, $f'(x) > 0$ for all $x \neq 0$, and f is concave downward on $(-\infty, 0)$ and concave upward on $(0, \infty)$.
(a) At what numbers does g have an extreme value?
(b) Discuss the concavity of g.

50. Find two positive integers such that the sum of the first number and four times the second number is 1000 and the product of the numbers is as large as possible.

51. Show that the shortest distance from the point (x_1, y_1) to the straight line $Ax + By + C = 0$ is

$$\frac{|Ax_1 + By_1 + C|}{\sqrt{A^2 + B^2}}$$

52. Find the point on the hyperbola $xy = 8$ that is closest to the point $(3, 0)$.

53. Find the smallest possible area of an isosceles triangle that is circumscribed about a circle of radius r.

54. Find the volume of the largest circular cone that can be inscribed in a sphere of radius r.

55. In ΔABC, D lies on AB, $CD \perp AB$, $|AD| = |BD| = 4$ cm, and $|CD| = 5$ cm. Where should a point P be chosen on CD so that the sum $|PA| + |PB| + |PC|$ is a minimum?

56. Solve Exercise 55 when $|CD| = 2$ cm.

57. The velocity of a wave of length L in deep water is

$$v = K\sqrt{\frac{L}{C} + \frac{C}{L}}$$

where K and C are known positive constants. What is the length of the wave that gives the minimum velocity?

58. A metal storage tank with volume V is to be constructed in the shape of a right circular cylinder surmounted by a hemisphere. What dimensions will require the least amount of metal?

59. A hockey team plays in an arena with a seating capacity of 15,000 spectators. With the ticket price set at \$12, average attendance at a game has been 11,000. A market survey indicates that for each dollar the ticket price is lowered, average attendance will increase by 1000. How should the owners of the team set the ticket price to maximize their revenue from ticket sales?

60. A manufacturer determines that the cost of making x units of a commodity is $C(x) = 1800 + 25x - 0.2x^2 + 0.001x^3$ and the demand function is $p(x) = 48.2 - 0.03x$.
(a) Graph the cost and revenue functions and use the graphs to estimate the production level for maximum profit.
(b) Use calculus to find the production level for maximum profit.
(c) Estimate the production level that minimizes the average cost.

61. Use Newton's method to find the root of the equation $x^5 - x^4 + 3x^2 - 3x - 2 = 0$ in the interval $[1, 2]$ correct to six decimal places.

62. Use Newton's method to find all roots of the equation $\sin x = x^2 - 3x + 1$ correct to six decimal places.

63. Use Newton's method to find the absolute maximum value of the function $f(t) = \cos t + t - t^2$ correct to eight decimal places.

64. Use the guidelines in Section 4.5 to sketch the curve $y = x \sin x$, $0 \leqslant x \leqslant 2\pi$. Use Newton's method when necessary.

65–72 Find f.

65. $f'(x) = \cos x - (1 - x^2)^{-1/2}$

66. $f'(x) = 2e^x + \sec x \tan x$

67. $f'(x) = \sqrt{x^3} + \sqrt[3]{x^2}$

68. $f'(x) = \sinh x + 2\cosh x$, $f(0) = 2$

69. $f'(t) = 2t - 3\sin t$, $f(0) = 5$

70. $f'(u) = \dfrac{u^2 + \sqrt{u}}{u}$, $f(1) = 3$

71. $f''(x) = 1 - 6x + 48x^2$, $f(0) = 1$, $f'(0) = 2$

72. $f''(x) = 2x^3 + 3x^2 - 4x + 5$, $f(0) = 2$, $f(1) = 0$

73–74 A particle is moving with the given data. Find the position of the particle.

73. $v(t) = 2t - 1/(1 + t^2)$, $s(0) = 1$

74. $a(t) = \sin t + 3\cos t$, $s(0) = 0$, $v(0) = 2$

75. (a) If $f(x) = 0.1e^x + \sin x$, $-4 \leqslant x \leqslant 4$, use a graph of f to sketch a rough graph of the antiderivative F of f that satisfies $F(0) = 0$.
(b) Find an expression for $F(x)$.
(c) Graph F using the expression in part (b). Compare with your sketch in part (a).

76. Investigate the family of curves given by

$$f(x) = x^4 + x^3 + cx^2$$

In particular you should determine the transitional value of c at which the number of critical numbers changes and the transitional value at which the number of inflection points changes. Illustrate the various possible shapes with graphs.

77. A canister is dropped from a helicopter 500 m above the ground. Its parachute does not open, but the canister has been designed to withstand an impact velocity of 100 m/s. Will it burst?

78. In an automobile race along a straight road, car A passed car B twice. Prove that at some time during the race their accelerations were equal. State the assumptions that you make.

79. A rectangular beam will be cut from a cylindrical log of radius 10 inches.
(a) Show that the beam of maximal cross-sectional area is a square.
(b) Four rectangular planks will be cut from the four sections of the log that remain after cutting the square beam. Determine the dimensions of the planks that will have maximal cross-sectional area.
(c) Suppose that the strength of a rectangular beam is proportional to the product of its width and the square of its depth. Find the dimensions of the strongest beam that can be cut from the cylindrical log.

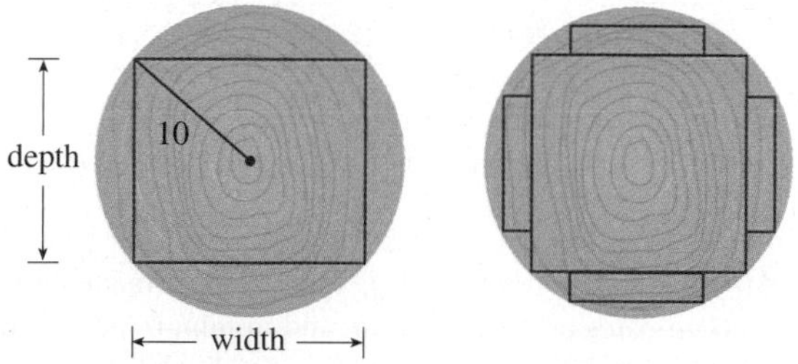

80. If a projectile is fired with an initial velocity v at an angle of inclination θ from the horizontal, then its trajectory, neglecting air resistance, is the parabola

$$y = (\tan\theta)x - \frac{g}{2v^2\cos^2\theta}x^2 \qquad 0 \leqslant \theta \leqslant \frac{\pi}{2}$$

(a) Suppose the projectile is fired from the base of a plane that is inclined at an angle α, $\alpha > 0$, from the horizontal, as shown in the figure. Show that the range of the projectile, measured up the slope, is given by

$$R(\theta) = \frac{2v^2\cos\theta\,\sin(\theta - \alpha)}{g\cos^2\alpha}$$

(b) Determine θ so that R is a maximum.
(c) Suppose the plane is at an angle α *below* the horizontal. Determine the range R in this case, and determine the angle at which the projectile should be fired to maximize R.

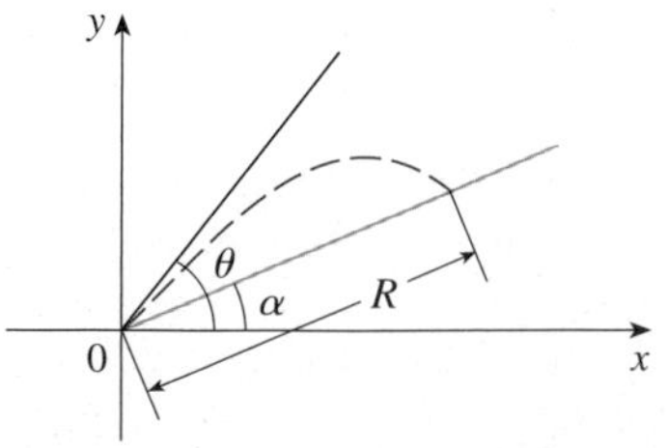

81. Show that, for $x > 0$,

$$\frac{x}{1 + x^2} < \tan^{-1}x < x$$

82. Sketch the graph of a function f such that $f'(x) < 0$ for all x, $f''(x) > 0$ for $|x| > 1$, $f''(x) < 0$ for $|x| < 1$, and $\lim_{x\to\pm\infty}[f(x) + x] = 0$.

PROBLEMS PLUS

One of the most important principles of problem solving is *analogy* (see page 76). If you are having trouble getting started on a problem, it is sometimes helpful to start by solving a similar, but simpler, problem. The following example illustrates the principle. Cover up the solution and try solving it yourself first.

EXAMPLE 1 If x, y, and z are positive numbers, prove that

$$\frac{(x^2+1)(y^2+1)(z^2+1)}{xyz} \geqslant 8$$

SOLUTION It may be difficult to get started on this problem. (Some students have tackled it by multiplying out the numerator, but that just creates a mess.) Let's try to think of a similar, simpler problem. When several variables are involved, it's often helpful to think of an analogous problem with fewer variables. In the present case we can reduce the number of variables from three to one and prove the analogous inequality

$$\boxed{1} \qquad \frac{x^2+1}{x} \geqslant 2 \qquad \text{for } x > 0$$

In fact, if we are able to prove (1), then the desired inequality follows because

$$\frac{(x^2+1)(y^2+1)(z^2+1)}{xyz} = \left(\frac{x^2+1}{x}\right)\left(\frac{y^2+1}{y}\right)\left(\frac{z^2+1}{z}\right) \geqslant 2 \cdot 2 \cdot 2 = 8$$

The key to proving (1) is to recognize that it is a disguised version of a minimum problem. If we let

$$f(x) = \frac{x^2+1}{x} = x + \frac{1}{x} \qquad x > 0$$

then $f'(x) = 1 - (1/x^2)$, so $f'(x) = 0$ when $x = 1$. Also, $f'(x) < 0$ for $0 < x < 1$ and $f'(x) > 0$ for $x > 1$. Therefore the absolute minimum value of f is $f(1) = 2$. This means that

$$\frac{x^2+1}{x} \geqslant 2 \qquad \text{for all positive values of } x$$

and, as previously mentioned, the given inequality follows by multiplication.

The inequality in (1) could also be proved without calculus. In fact, if $x > 0$, we have

$$\frac{x^2+1}{x} \geqslant 2 \iff x^2 + 1 \geqslant 2x \iff x^2 - 2x + 1 \geqslant 0$$

$$\iff (x-1)^2 \geqslant 0$$

Because the last inequality is obviously true, the first one is true too. □

Look Back
What have we learned from the solution to this example?

- To solve a problem involving several variables, it might help to solve a similar problem with just one variable.
- When trying to prove an inequality, it might help to think of it as a maximum or minimum problem.

PROBLEMS PLUS

PROBLEMS

1. If a rectangle has its base on the x-axis and two vertices on the curve $y = e^{-x^2}$, show that the rectangle has the largest possible area when the two vertices are at the points of inflection of the curve.

2. Show that $|\sin x - \cos x| \leqslant \sqrt{2}$ for all x.

3. Show that, for all positive values of x and y,

$$\frac{e^{x+y}}{xy} \geqslant e^2$$

4. Show that $x^2y^2(4 - x^2)(4 - y^2) \leqslant 16$ for all numbers x and y such that $|x| \leqslant 2$ and $|y| \leqslant 2$.

5. If a, b, c, and d are constants such that

$$\lim_{x \to 0} \frac{ax^2 + \sin bx + \sin cx + \sin dx}{3x^2 + 5x^4 + 7x^6} = 8$$

find the value of the sum $a + b + c + d$.

6. Find the point on the parabola $y = 1 - x^2$ at which the tangent line cuts from the first quadrant the triangle with the smallest area.

7. Find the highest and lowest points on the curve $x^2 + xy + y^2 = 12$.

8. Sketch the set of all points (x, y) such that $|x + y| \leqslant e^x$.

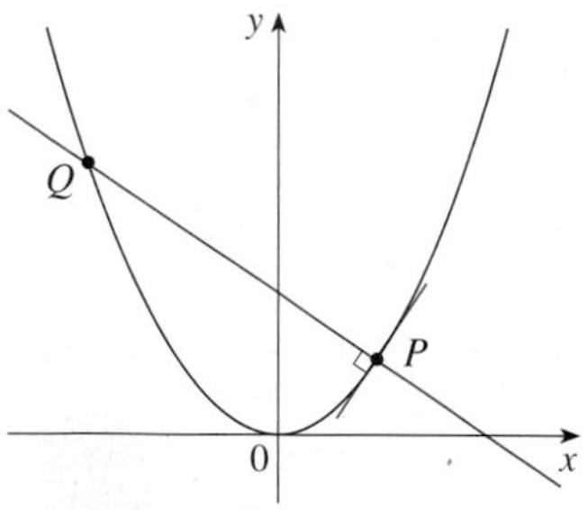

FIGURE FOR PROBLEM 9

9. If $P(a, a^2)$ is any point on the parabola $y = x^2$, except for the origin, let Q be the point where the normal line intersects the parabola again. Show that the line segment PQ has the shortest possible length when $a = 1/\sqrt{2}$.

10. For what values of c does the curve $y = cx^3 + e^x$ have inflection points?

11. Determine the values of the number a for which the function f has no critical number:

$$f(x) = (a^2 + a - 6)\cos 2x + (a - 2)x + \cos 1$$

12. Sketch the region in the plane consisting of all points (x, y) such that

$$2xy \leqslant |x - y| \leqslant x^2 + y^2$$

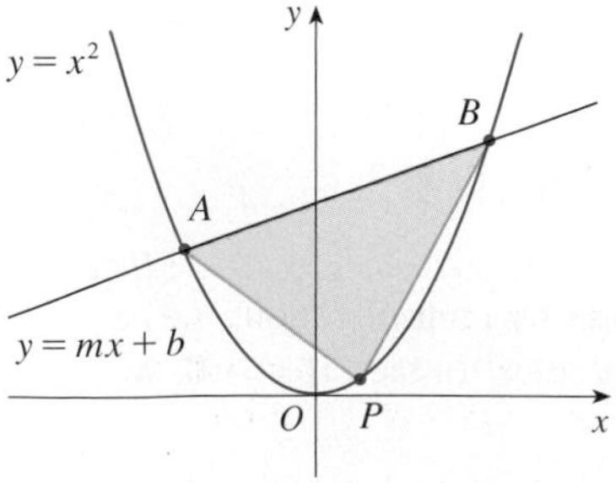

FIGURE FOR PROBLEM 13

13. The line $y = mx + b$ intersects the parabola $y = x^2$ in points A and B. (See the figure.) Find the point P on the arc AOB of the parabola that maximizes the area of the triangle PAB.

14. $ABCD$ is a square piece of paper with sides of length 1 m. A quarter-circle is drawn from B to D with center A. The piece of paper is folded along EF, with E on AB and F on AD, so that A falls on the quarter-circle. Determine the maximum and minimum areas that the triangle AEF can have.

15. For which positive numbers a does the curve $y = a^x$ intersect the line $y = x$?

16. For what value of a is the following equation true?

$$\lim_{x \to \infty} \left(\frac{x + a}{x - a} \right)^x = e$$

17. Let $f(x) = a_1 \sin x + a_2 \sin 2x + \cdots + a_n \sin nx$, where $a_1, a_2, \ldots, a_n$ are real numbers and n is a positive integer. If it is given that $|f(x)| \leqslant |\sin x|$ for all x, show that

$$|a_1 + 2a_2 + \cdots + na_n| \leqslant 1$$

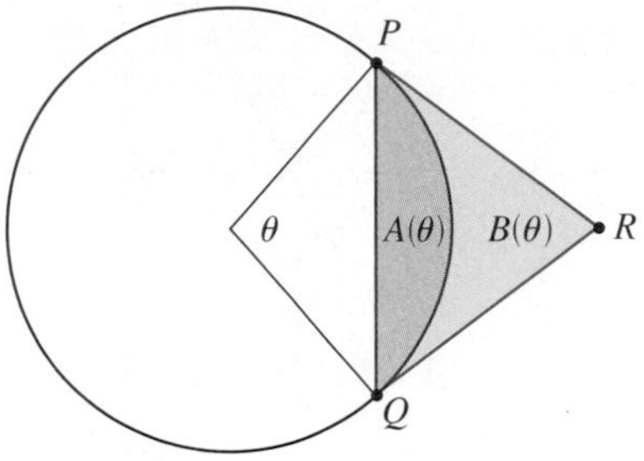

FIGURE FOR PROBLEM 18

18. An arc PQ of a circle subtends a central angle θ as in the figure. Let $A(\theta)$ be the area between the chord PQ and the arc PQ. Let $B(\theta)$ be the area between the tangent lines PR, QR, and the arc. Find

$$\lim_{\theta \to 0^+} \frac{A(\theta)}{B(\theta)}$$

19. The speeds of sound c_1 in an upper layer and c_2 in a lower layer of rock and the thickness h of the upper layer can be determined by seismic exploration if the speed of sound in the lower layer is greater than the speed in the upper layer. A dynamite charge is detonated at a point P and the transmitted signals are recorded at a point Q, which is a distance D from P. The first signal to arrive at Q travels along the surface and takes T_1 seconds. The next signal travels from P to a point R, from R to S in the lower layer, and then to Q, taking T_2 seconds. The third signal is reflected off the lower layer at the midpoint O of RS and takes T_3 seconds to reach Q.

(a) Express T_1, T_2, and T_3 in terms of D, h, c_1, c_2, and θ.

(b) Show that T_2 is a minimum when $\sin\theta = c_1/c_2$.

(c) Suppose that $D = 1$ km, $T_1 = 0.26$ s, $T_2 = 0.32$ s, and $T_3 = 0.34$ s. Find c_1, c_2, and h.

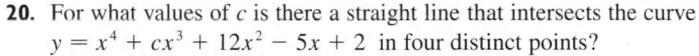

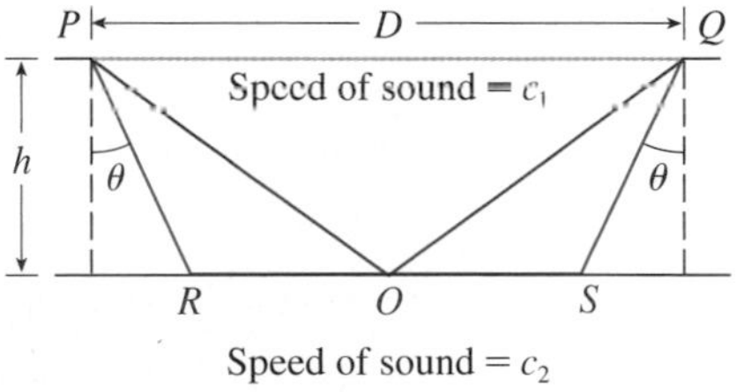

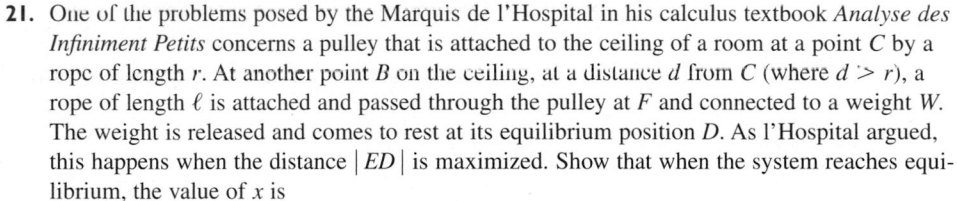

Note: Geophysicists use this technique when studying the structure of the earth's crust, whether searching for oil or examining fault lines.

20. For what values of c is there a straight line that intersects the curve $y = x^4 + cx^3 + 12x^2 - 5x + 2$ in four distinct points?

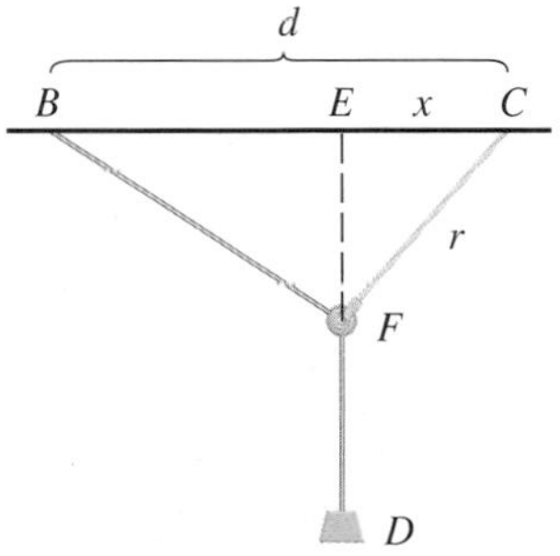

FIGURE FOR PROBLEM 21

21. One of the problems posed by the Marquis de l'Hospital in his calculus textbook *Analyse des Infiniment Petits* concerns a pulley that is attached to the ceiling of a room at a point C by a rope of length r. At another point B on the ceiling, at a distance d from C (where $d > r$), a rope of length ℓ is attached and passed through the pulley at F and connected to a weight W. The weight is released and comes to rest at its equilibrium position D. As l'Hospital argued, this happens when the distance $|ED|$ is maximized. Show that when the system reaches equilibrium, the value of x is

$$\frac{r}{4d}\left(r + \sqrt{r^2 + 8d^2}\right)$$

Notice that this expression is independent of both W and ℓ.

22. Given a sphere with radius r, find the height of a pyramid of minimum volume whose base is a square and whose base and triangular faces are all tangent to the sphere. What if the base of the pyramid is a regular n-gon? (A regular n-gon is a polygon with n equal sides and angles.) (Use the fact that the volume of a pyramid is $\frac{1}{3}Ah$, where A is the area of the base.)

23. Assume that a snowball melts so that its volume decreases at a rate proportional to its surface area. If it takes three hours for the snowball to decrease to half its original volume, how much longer will it take for the snowball to melt completely?

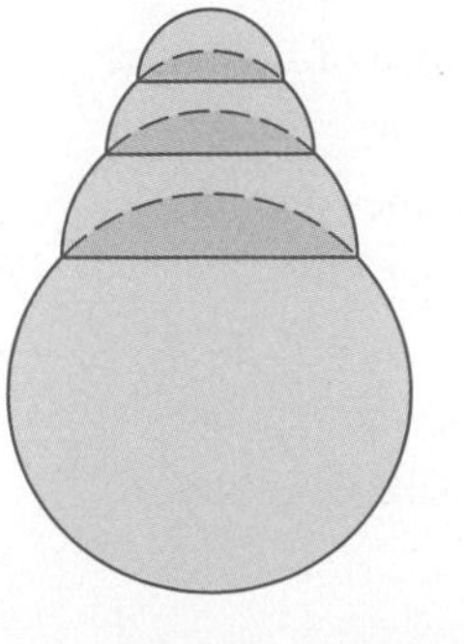

FIGURE FOR PROBLEM 24

24. A hemispherical bubble is placed on a spherical bubble of radius 1. A smaller hemispherical bubble is then placed on the first one. This process is continued until n chambers, including the sphere, are formed. (The figure shows the case $n = 4$.) Use mathematical induction to prove that the maximum height of any bubble tower with n chambers is $1 + \sqrt{n}$.

5
INTEGRALS

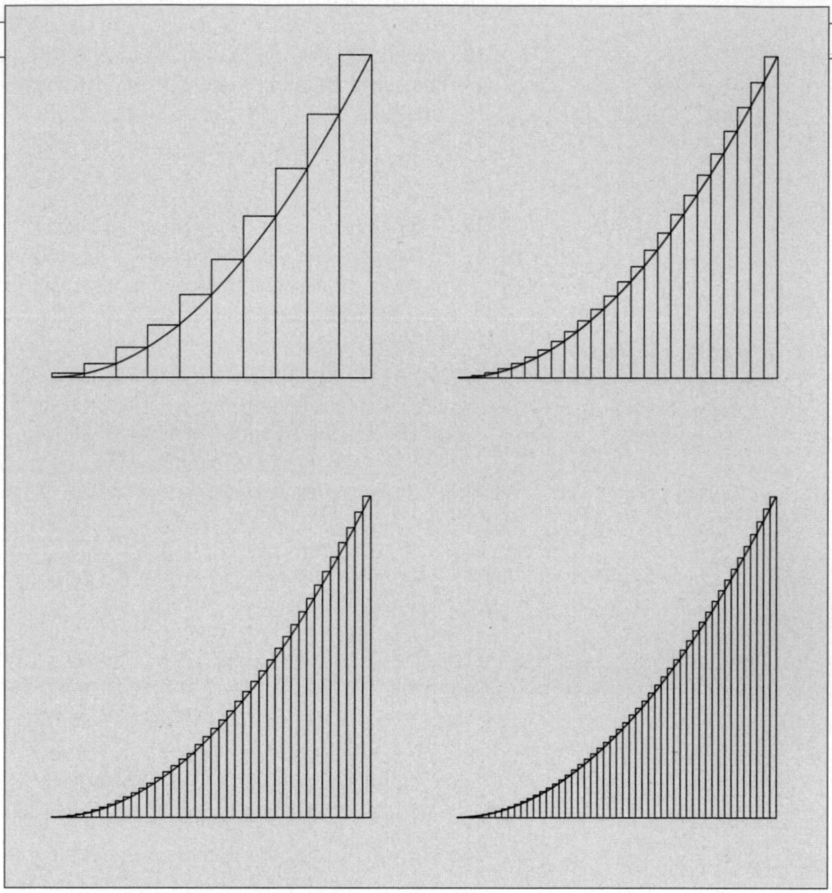

To compute an area we approximate a region by rectangles and let the number of rectangles become large. The precise area is the limit of these sums of areas of rectangles.

In Chapter 2 we used the tangent and velocity problems to introduce the derivative, which is the central idea in differential calculus. In much the same way, this chapter starts with the area and distance problems and uses them to formulate the idea of a definite integral, which is the basic concept of integral calculus. We will see in Chapters 6 and 8 how to use the integral to solve problems concerning volumes, lengths of curves, population predictions, cardiac output, forces on a dam, work, consumer surplus, and baseball, among many others.

There is a connection between integral calculus and differential calculus. The Fundamental Theorem of Calculus relates the integral to the derivative, and we will see in this chapter that it greatly simplifies the solution of many problems.

5.1 AREAS AND DISTANCES

■ Now is a good time to read (or reread) *A Preview of Calculus* (see page 2). It discusses the unifying ideas of calculus and helps put in perspective where we have been and where we are going.

In this section we discover that in trying to find the area under a curve or the distance traveled by a car, we end up with the same special type of limit.

THE AREA PROBLEM

We begin by attempting to solve the *area problem:* Find the area of the region S that lies under the curve $y = f(x)$ from a to b. This means that S, illustrated in Figure 1, is bounded by the graph of a continuous function f [where $f(x) \geqslant 0$], the vertical lines $x = a$ and $x = b$, and the x-axis.

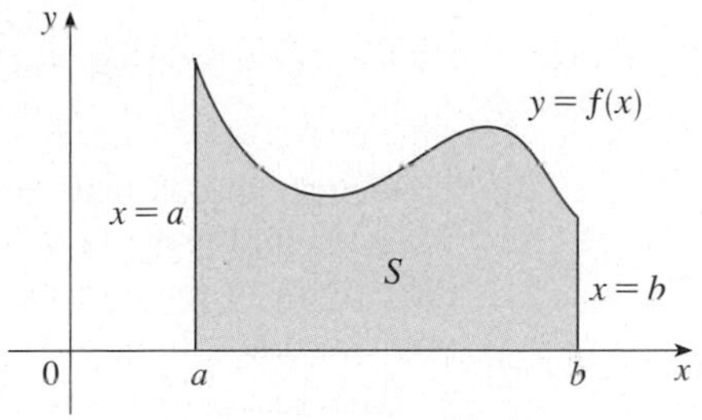

FIGURE 1
$S = \{(x, y) \mid a \leqslant x \leqslant b, 0 \leqslant y \leqslant f(x)\}$

In trying to solve the area problem we have to ask ourselves: What is the meaning of the word *area*? This question is easy to answer for regions with straight sides. For a rectangle, the area is defined as the product of the length and the width. The area of a triangle is half the base times the height. The area of a polygon is found by dividing it into triangles (as in Figure 2) and adding the areas of the triangles.

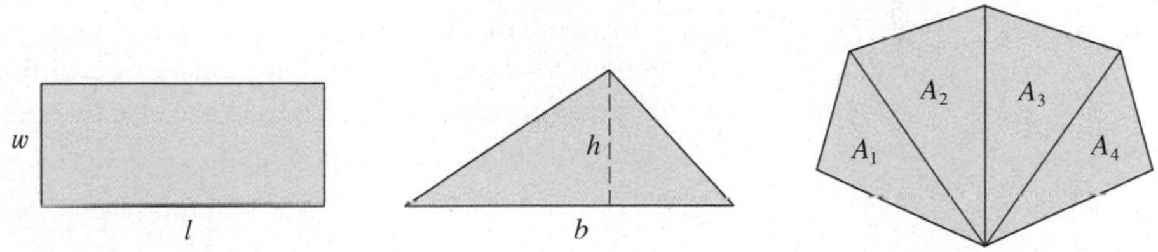

FIGURE 2 $A = lw$ $A = \frac{1}{2}bh$ $A = A_1 + A_2 + A_3 + A_4$

However, it isn't so easy to find the area of a region with curved sides. We all have an intuitive idea of what the area of a region is. But part of the area problem is to make this intuitive idea precise by giving an exact definition of area.

Recall that in defining a tangent we first approximated the slope of the tangent line by slopes of secant lines and then we took the limit of these approximations. We pursue a similar idea for areas. We first approximate the region S by rectangles and then we take the limit of the areas of these rectangles as we increase the number of rectangles. The following example illustrates the procedure.

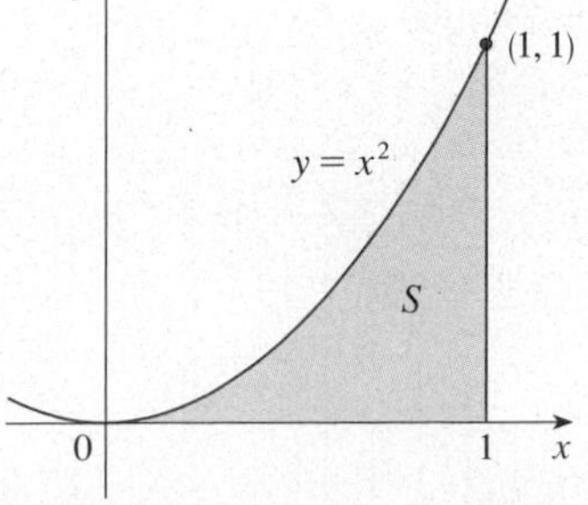

FIGURE 3

V EXAMPLE 1 Use rectangles to estimate the area under the parabola $y = x^2$ from 0 to 1 (the parabolic region S illustrated in Figure 3).

SOLUTION We first notice that the area of S must be somewhere between 0 and 1 because S is contained in a square with side length 1, but we can certainly do better than that.

Suppose we divide S into four strips S_1, S_2, S_3, and S_4 by drawing the vertical lines $x = \frac{1}{4}$, $x = \frac{1}{2}$, and $x = \frac{3}{4}$ as in Figure 4(a).

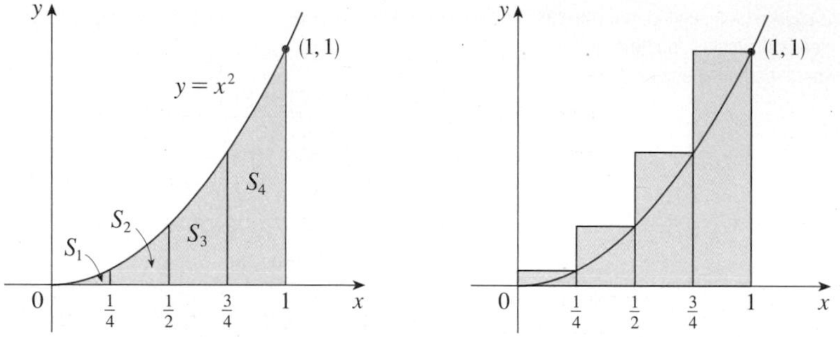

FIGURE 4

We can approximate each strip by a rectangle whose base is the same as the strip and whose height is the same as the right edge of the strip [see Figure 4(b)]. In other words, the heights of these rectangles are the values of the function $f(x) = x^2$ at the right endpoints of the subintervals $\left[0, \frac{1}{4}\right]$, $\left[\frac{1}{4}, \frac{1}{2}\right]$, $\left[\frac{1}{2}, \frac{3}{4}\right]$, and $\left[\frac{3}{4}, 1\right]$.

Each rectangle has width $\frac{1}{4}$ and the heights are $\left(\frac{1}{4}\right)^2$, $\left(\frac{1}{2}\right)^2$, $\left(\frac{3}{4}\right)^2$, and 1^2. If we let R_4 be the sum of the areas of these approximating rectangles, we get

$$R_4 = \tfrac{1}{4} \cdot \left(\tfrac{1}{4}\right)^2 + \tfrac{1}{4} \cdot \left(\tfrac{1}{2}\right)^2 + \tfrac{1}{4} \cdot \left(\tfrac{3}{4}\right)^2 + \tfrac{1}{4} \cdot 1^2 = \tfrac{15}{32} = 0.46875$$

From Figure 4(b) we see that the area A of S is less than R_4, so

$$A < 0.46875$$

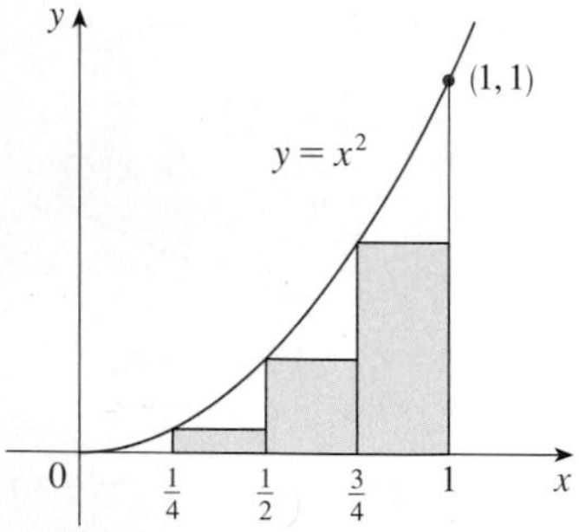

FIGURE 5

Instead of using the rectangles in Figure 4(b) we could use the smaller rectangles in Figure 5 whose heights are the values of f at the left endpoints of the subintervals. (The leftmost rectangle has collapsed because its height is 0.) The sum of the areas of these approximating rectangles is

$$L_4 = \tfrac{1}{4} \cdot 0^2 + \tfrac{1}{4} \cdot \left(\tfrac{1}{4}\right)^2 + \tfrac{1}{4} \cdot \left(\tfrac{1}{2}\right)^2 + \tfrac{1}{4} \cdot \left(\tfrac{3}{4}\right)^2 = \tfrac{7}{32} = 0.21875$$

We see that the area of S is larger than L_4, so we have lower and upper estimates for A:

$$0.21875 < A < 0.46875$$

We can repeat this procedure with a larger number of strips. Figure 6 shows what happens when we divide the region S into eight strips of equal width.

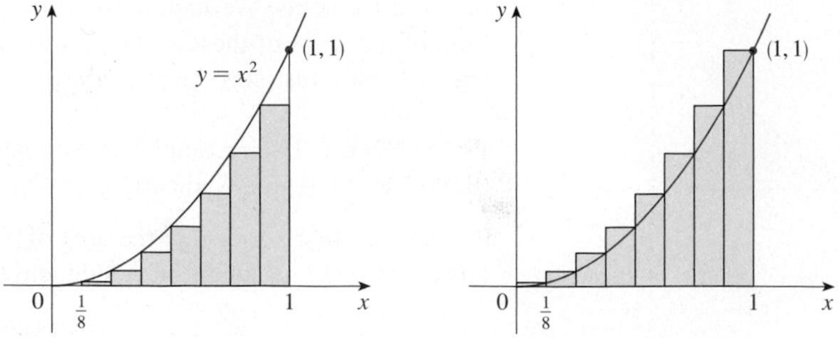

FIGURE 6
Approximating S with eight rectangles

(a) Using left endpoints (b) Using right endpoints

By computing the sum of the areas of the smaller rectangles (L_8) and the sum of the areas of the larger rectangles (R_8), we obtain better lower and upper estimates for A:

$$0.2734375 < A < 0.3984375$$

n	L_n	R_n
10	0.2850000	0.3850000
20	0.3087500	0.3587500
30	0.3168519	0.3501852
50	0.3234000	0.3434000
100	0.3283500	0.3383500
1000	0.3328335	0.3338335

So one possible answer to the question is to say that the true area of S lies somewhere between 0.2734375 and 0.3984375.

We could obtain better estimates by increasing the number of strips. The table at the left shows the results of similar calculations (with a computer) using n rectangles whose heights are found with left endpoints (L_n) or right endpoints (R_n). In particular, we see by using 50 strips that the area lies between 0.3234 and 0.3434. With 1000 strips we narrow it down even more: A lies between 0.3328335 and 0.3338335. A good estimate is obtained by averaging these numbers: $A \approx 0.3333335$. □

From the values in the table in Example 1, it looks as if R_n is approaching $\frac{1}{3}$ as n increases. We confirm this in the next example.

V EXAMPLE 2 For the region S in Example 1, show that the sum of the areas of the upper approximating rectangles approaches $\frac{1}{3}$, that is,

$$\lim_{n\to\infty} R_n = \tfrac{1}{3}$$

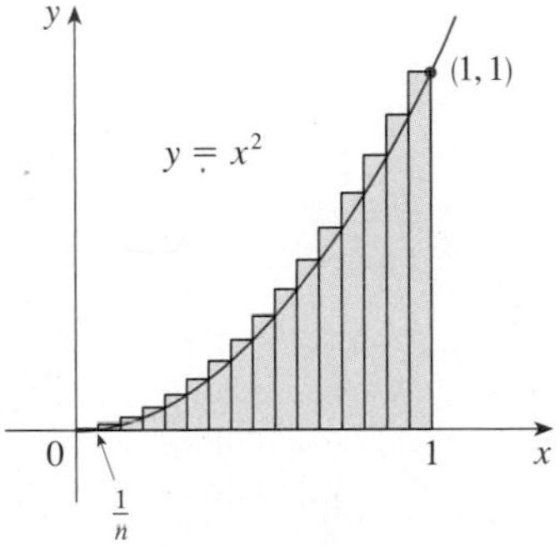

FIGURE 7

SOLUTION R_n is the sum of the areas of the n rectangles in Figure 7. Each rectangle has width $1/n$ and the heights are the values of the function $f(x) = x^2$ at the points $1/n, 2/n, 3/n, \ldots, n/n$; that is, the heights are $(1/n)^2, (2/n)^2, (3/n)^2, \ldots, (n/n)^2$. Thus

$$R_n = \frac{1}{n}\left(\frac{1}{n}\right)^2 + \frac{1}{n}\left(\frac{2}{n}\right)^2 + \frac{1}{n}\left(\frac{3}{n}\right)^2 + \cdots + \frac{1}{n}\left(\frac{n}{n}\right)^2$$

$$= \frac{1}{n} \cdot \frac{1}{n^2}(1^2 + 2^2 + 3^2 + \cdots + n^2)$$

$$= \frac{1}{n^3}(1^2 + 2^2 + 3^2 + \cdots + n^2)$$

Here we need the formula for the sum of the squares of the first n positive integers:

1 $$1^2 + 2^2 + 3^2 + \cdots + n^2 = \frac{n(n+1)(2n+1)}{6}$$

Perhaps you have seen this formula before. It is proved in Example 5 in Appendix E.

Putting Formula 1 into our expression for R_n, we get

$$R_n = \frac{1}{n^3} \cdot \frac{n(n+1)(2n+1)}{6} = \frac{(n+1)(2n+1)}{6n^2}$$

Thus we have

$$\lim_{n\to\infty} R_n = \lim_{n\to\infty} \frac{(n+1)(2n+1)}{6n^2} = \lim_{n\to\infty} \frac{1}{6}\left(\frac{n+1}{n}\right)\left(\frac{2n+1}{n}\right)$$

$$= \lim_{n\to\infty} \frac{1}{6}\left(1 + \frac{1}{n}\right)\left(2 + \frac{1}{n}\right) = \frac{1}{6} \cdot 1 \cdot 2 = \frac{1}{3}$$ □

■ Here we are computing the limit of the sequence $\{R_n\}$. Sequences were discussed in *A Preview of Calculus* and will be studied in detail in Chapter 11. Their limits are calculated in the same way as limits at infinity (Section 2.6). In particular, we know that

$$\lim_{n\to\infty} \frac{1}{n} = 0$$

It can be shown that the lower approximating sums also approach $\frac{1}{3}$, that is,

$$\lim_{n\to\infty} L_n = \frac{1}{3}$$

From Figures 8 and 9 it appears that, as n increases, both L_n and R_n become better and better approximations to the area of S. Therefore, we *define* the area A to be the limit of the sums of the areas of the approximating rectangles, that is,

$$A = \lim_{n\to\infty} R_n = \lim_{n\to\infty} L_n = \frac{1}{3}$$

TEC In Visual 5.1 you can create pictures like those in Figures 8 and 9 for other values of n.

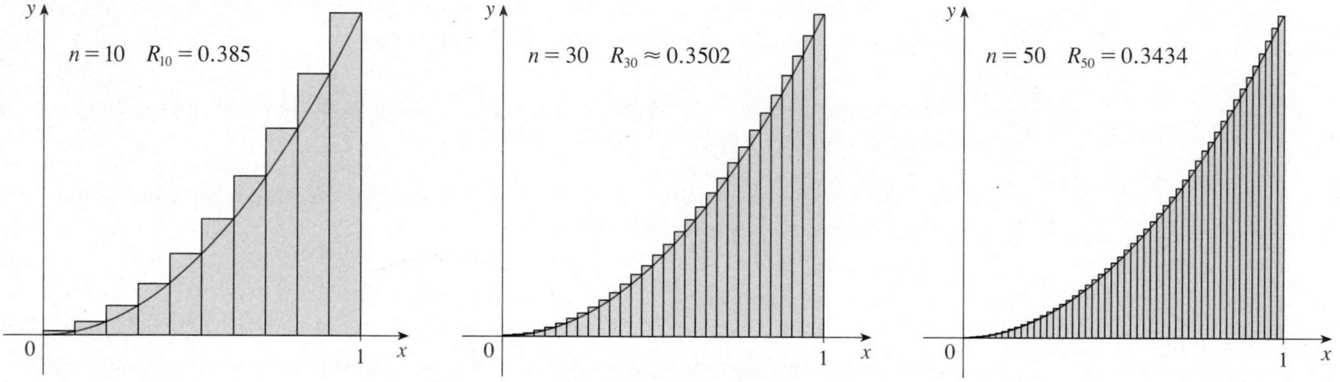

FIGURE 8

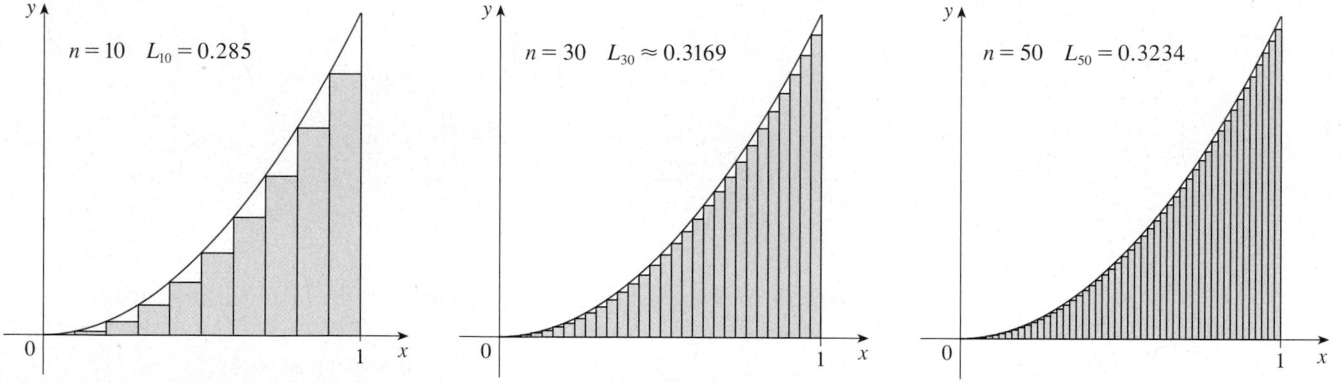

FIGURE 9
The area is the number that is smaller than all upper sums and larger than all lower sums

Let's apply the idea of Examples 1 and 2 to the more general region S of Figure 1. We start by subdividing S into n strips $S_1, S_2, \ldots, S_n$ of equal width as in Figure 10.

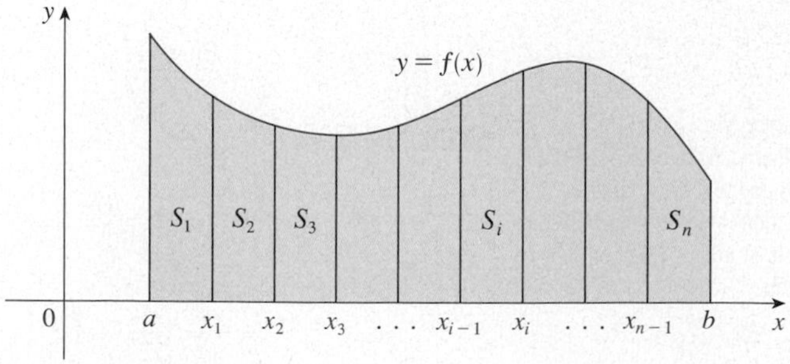

FIGURE 10

The width of the interval $[a, b]$ is $b - a$, so the width of each of the n strips is

$$\Delta x = \frac{b - a}{n}$$

These strips divide the interval $[a, b]$ into n subintervals

$$[x_0, x_1], \quad [x_1, x_2], \quad [x_2, x_3], \quad \ldots, \quad [x_{n-1}, x_n]$$

where $x_0 = a$ and $x_n = b$. The right endpoints of the subintervals are

$$x_1 = a + \Delta x,$$
$$x_2 = a + 2\,\Delta x,$$
$$x_3 = a + 3\,\Delta x,$$
$$\vdots$$

Let's approximate the ith strip S_i by a rectangle with width Δx and height $f(x_i)$, which is the value of f at the right endpoint (see Figure 11). Then the area of the ith rectangle is $f(x_i)\,\Delta x$. What we think of intuitively as the area of S is approximated by the sum of the areas of these rectangles, which is

$$R_n = f(x_1)\,\Delta x + f(x_2)\,\Delta x + \cdots + f(x_n)\,\Delta x$$

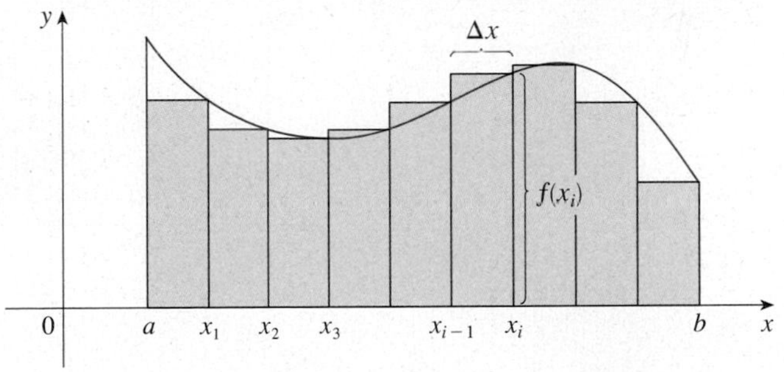

FIGURE 11

Figure 12 shows this approximation for $n = 2, 4, 8$, and 12. Notice that this approximation appears to become better and better as the number of strips increases, that is, as $n \to \infty$. Therefore we define the area A of the region S in the following way.

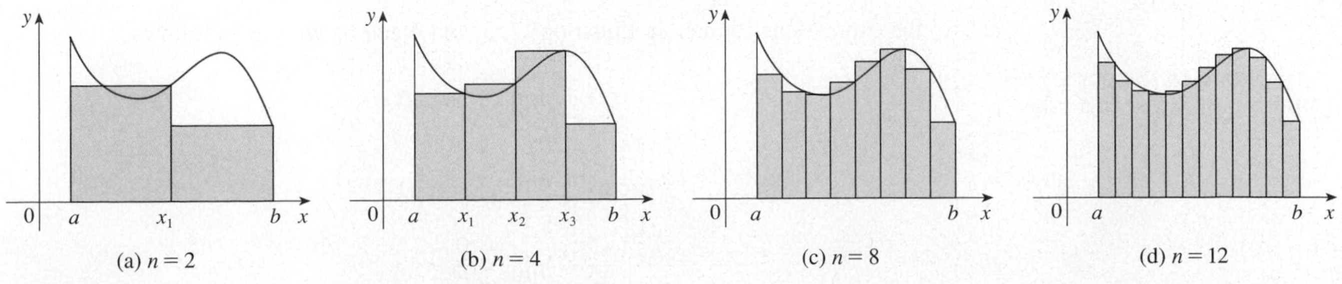

FIGURE 12

2 **DEFINITION** The **area** A of the region S that lies under the graph of the continuous function f is the limit of the sum of the areas of approximating rectangles:

$$A = \lim_{n\to\infty} R_n = \lim_{n\to\infty} [f(x_1)\,\Delta x + f(x_2)\,\Delta x + \cdots + f(x_n)\,\Delta x]$$

It can be proved that the limit in Definition 2 always exists, since we are assuming that f is continuous. It can also be shown that we get the same value if we use left endpoints:

3 $$A = \lim_{n\to\infty} L_n = \lim_{n\to\infty} [f(x_0)\,\Delta x + f(x_1)\,\Delta x + \cdots + f(x_{n-1})\,\Delta x]$$

In fact, instead of using left endpoints or right endpoints, we could take the height of the ith rectangle to be the value of f at *any* number x_i^* in the ith subinterval $[x_{i-1}, x_i]$. We call the numbers $x_1^*, x_2^*, \ldots, x_n^*$ the **sample points**. Figure 13 shows approximating rectangles when the sample points are not chosen to be endpoints. So a more general expression for the area of S is

4 $$A = \lim_{n\to\infty} [f(x_1^*)\,\Delta x + f(x_2^*)\,\Delta x + \cdots + f(x_n^*)\,\Delta x]$$

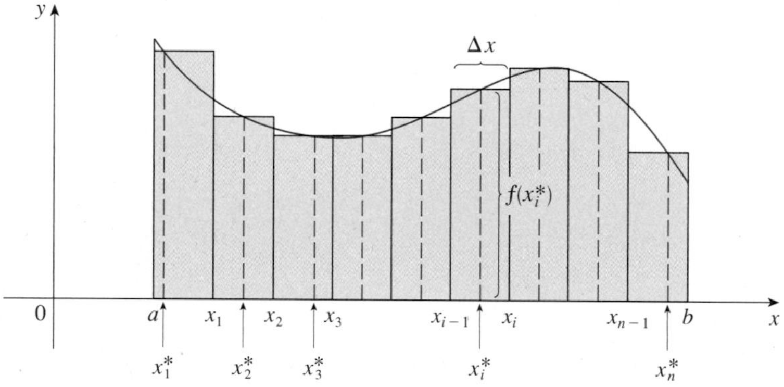

FIGURE 13

This tells us to end with $i = n$.
This tells us to add. $\longrightarrow \sum_{i=m}^{n} f(x_i)\,\Delta x$
This tells us to start with $i = m$.

■ If you need practice with sigma notation, look at the examples and try some of the exercises in Appendix E.

We often use **sigma notation** to write sums with many terms more compactly. For instance,

$$\sum_{i=1}^{n} f(x_i)\,\Delta x = f(x_1)\,\Delta x + f(x_2)\,\Delta x + \cdots + f(x_n)\,\Delta x$$

So the expressions for area in Equations 2, 3, and 4 can be written as follows:

$$A = \lim_{n\to\infty} \sum_{i=1}^{n} f(x_i)\,\Delta x$$

$$A = \lim_{n\to\infty} \sum_{i=1}^{n} f(x_{i-1})\,\Delta x$$

$$A = \lim_{n\to\infty} \sum_{i=1}^{n} f(x_i^*)\,\Delta x$$

We can also rewrite Formula 1 in the following way:

$$\sum_{i=1}^{n} i^2 = \frac{n(n+1)(2n+1)}{6}$$

EXAMPLE 3 Let A be the area of the region that lies under the graph of $f(x) = e^{-x}$ between $x = 0$ and $x = 2$.
(a) Using right endpoints, find an expression for A as a limit. Do not evaluate the limit.
(b) Estimate the area by taking the sample points to be midpoints and using four subintervals and then ten subintervals.

SOLUTION
(a) Since $a = 0$ and $b = 2$, the width of a subinterval is

$$\Delta x = \frac{2-0}{n} = \frac{2}{n}$$

So $x_1 = 2/n$, $x_2 = 4/n$, $x_3 = 6/n$, $x_i = 2i/n$, and $x_n = 2n/n$. The sum of the areas of the approximating rectangles is

$$\begin{aligned} R_n &= f(x_1)\,\Delta x + f(x_2)\,\Delta x + \cdots + f(x_n)\,\Delta x \\ &= e^{-x_1}\,\Delta x + e^{-x_2}\,\Delta x + \cdots + e^{-x_n}\,\Delta x \\ &= e^{-2/n}\left(\frac{2}{n}\right) + e^{-4/n}\left(\frac{2}{n}\right) + \cdots + e^{-2n/n}\left(\frac{2}{n}\right) \end{aligned}$$

According to Definition 2, the area is

$$A = \lim_{n\to\infty} R_n = \lim_{n\to\infty} \frac{2}{n}\left(e^{-2/n} + e^{-4/n} + e^{-6/n} + \cdots + e^{-2n/n}\right)$$

Using sigma notation we could write

$$A = \lim_{n\to\infty} \frac{2}{n} \sum_{i=1}^{n} e^{-2i/n}$$

It is difficult to evaluate this limit directly by hand, but with the aid of a computer algebra system it isn't hard (see Exercise 24). In Section 5.3 we will be able to find A more easily using a different method.
(b) With $n = 4$ the subintervals of equal width $\Delta x = 0.5$ are $[0, 0.5]$, $[0.5, 1]$, $[1, 1.5]$, and $[1.5, 2]$. The midpoints of these subintervals are $x_1^* = 0.25$, $x_2^* = 0.75$, $x_3^* = 1.25$, and $x_4^* = 1.75$, and the sum of the areas of the four approximating rectangles (see Figure 14) is

$$\begin{aligned} M_4 &= \sum_{i=1}^{4} f(x_i^*)\,\Delta x \\ &= f(0.25)\,\Delta x + f(0.75)\,\Delta x + f(1.25)\,\Delta x + f(1.75)\,\Delta x \\ &= e^{-0.25}(0.5) + e^{-0.75}(0.5) + e^{-1.25}(0.5) + e^{-1.75}(0.5) \\ &= \tfrac{1}{2}\left(e^{-0.25} + e^{-0.75} + e^{-1.25} + e^{-1.75}\right) \approx 0.8557 \end{aligned}$$

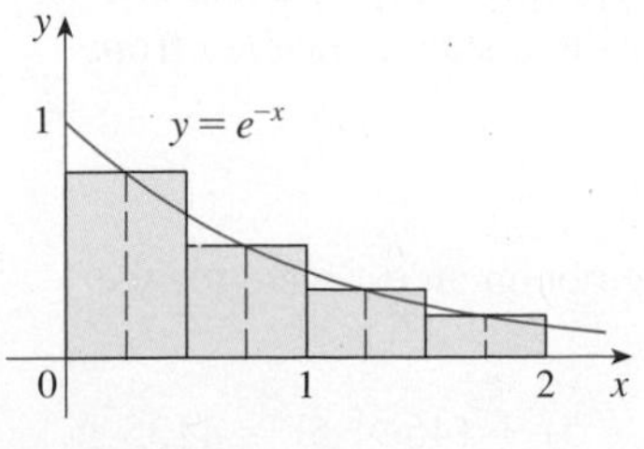

FIGURE 14

So an estimate for the area is

$$A \approx 0.8557$$

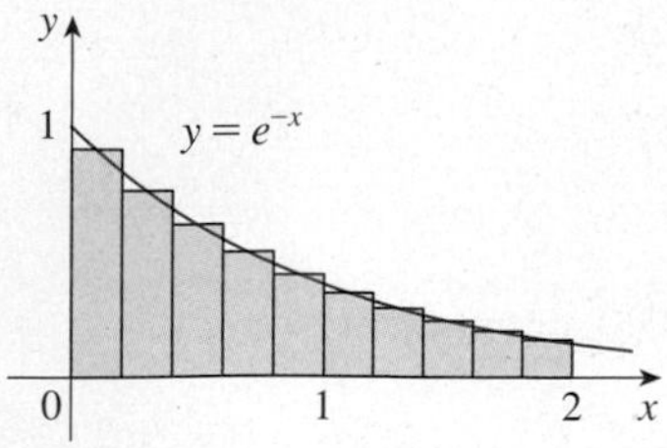

FIGURE 15

With $n = 10$ the subintervals are $[0, 0.2], [0.2, 0.4], \ldots, [1.8, 2]$ and the midpoints are $x_1^* = 0.1, x_2^* = 0.3, x_3^* = 0.5, \ldots, x_{10}^* = 1.9$. Thus

$$A \approx M_{10} = f(0.1)\,\Delta x + f(0.3)\,\Delta x + f(0.5)\,\Delta x + \cdots + f(1.9)\,\Delta x$$
$$= 0.2(e^{-0.1} + e^{-0.3} + e^{-0.5} + \cdots + e^{-1.9}) \approx 0.8632$$

From Figure 15 it appears that this estimate is better than the estimate with $n = 4$. □

THE DISTANCE PROBLEM

Now let's consider the *distance problem:* Find the distance traveled by an object during a certain time period if the velocity of the object is known at all times. (In a sense this is the inverse problem of the velocity problem that we discussed in Section 2.1.) If the velocity remains constant, then the distance problem is easy to solve by means of the formula

$$\text{distance} = \text{velocity} \times \text{time}$$

But if the velocity varies, it's not so easy to find the distance traveled. We investigate the problem in the following example.

V EXAMPLE 4 Suppose the odometer on our car is broken and we want to estimate the distance driven over a 30-second time interval. We take speedometer readings every five seconds and record them in the following table:

Time (s)	0	5	10	15	20	25	30
Velocity (mi/h)	17	21	24	29	32	31	28

In order to have the time and the velocity in consistent units, let's convert the velocity readings to feet per second (1 mi/h = 5280/3600 ft/s):

Time (s)	0	5	10	15	20	25	30
Velocity (ft/s)	25	31	35	43	47	46	41

During the first five seconds the velocity doesn't change very much, so we can estimate the distance traveled during that time by assuming that the velocity is constant. If we take the velocity during that time interval to be the initial velocity (25 ft/s), then we obtain the approximate distance traveled during the first five seconds:

$$25 \text{ ft/s} \times 5 \text{ s} = 125 \text{ ft}$$

Similarly, during the second time interval the velocity is approximately constant and we take it to be the velocity when $t = 5$ s. So our estimate for the distance traveled from $t = 5$ s to $t = 10$ s is

$$31 \text{ ft/s} \times 5 \text{ s} = 155 \text{ ft}$$

If we add similar estimates for the other time intervals, we obtain an estimate for the total distance traveled:

$$(25 \times 5) + (31 \times 5) + (35 \times 5) + (43 \times 5) + (47 \times 5) + (46 \times 5) = 1135 \text{ ft}$$

We could just as well have used the velocity at the *end* of each time period instead of the velocity at the beginning as our assumed constant velocity. Then our estimate becomes

$$(31 \times 5) + (35 \times 5) + (43 \times 5) + (47 \times 5) + (46 \times 5) + (41 \times 5) = 1215 \text{ ft}$$

If we had wanted a more accurate estimate, we could have taken velocity readings every two seconds, or even every second. □

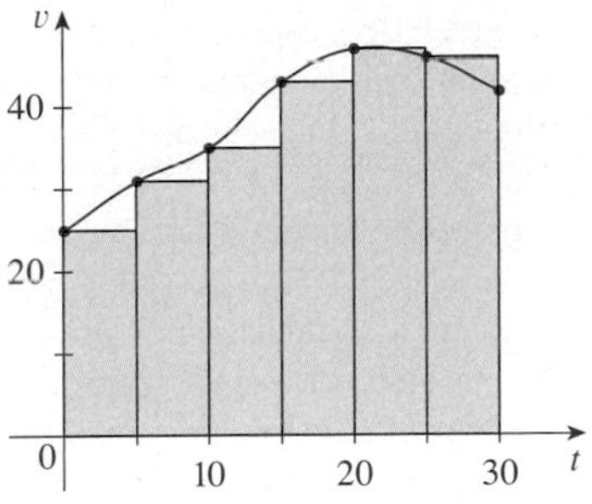

FIGURE 16

Perhaps the calculations in Example 4 remind you of the sums we used earlier to estimate areas. The similarity is explained when we sketch a graph of the velocity function of the car in Figure 16 and draw rectangles whose heights are the initial velocities for each time interval. The area of the first rectangle is $25 \times 5 = 125$, which is also our estimate for the distance traveled in the first five seconds. In fact, the area of each rectangle can be interpreted as a distance because the height represents velocity and the width represents time. The sum of the areas of the rectangles in Figure 16 is $L_6 = 1135$, which is our initial estimate for the total distance traveled.

In general, suppose an object moves with velocity $v = f(t)$, where $a \leqslant t \leqslant b$ and $f(t) \geqslant 0$ (so the object always moves in the positive direction). We take velocity readings at times $t_0 (= a), t_1, t_2, \ldots, t_n (= b)$ so that the velocity is approximately constant on each subinterval. If these times are equally spaced, then the time between consecutive readings is $\Delta t = (b - a)/n$. During the first time interval the velocity is approximately $f(t_0)$ and so the distance traveled is approximately $f(t_0)\,\Delta t$. Similarly, the distance traveled during the second time interval is about $f(t_1)\,\Delta t$ and the total distance traveled during the time interval $[a, b]$ is approximately

$$f(t_0)\,\Delta t + f(t_1)\,\Delta t + \cdots + f(t_{n-1})\,\Delta t = \sum_{i=1}^{n} f(t_{i-1})\,\Delta t$$

If we use the velocity at right endpoints instead of left endpoints, our estimate for the total distance becomes

$$f(t_1)\,\Delta t + f(t_2)\,\Delta t + \cdots + f(t_n)\,\Delta t = \sum_{i=1}^{n} f(t_i)\,\Delta t$$

The more frequently we measure the velocity, the more accurate our estimates become, so it seems plausible that the *exact* distance d traveled is the *limit* of such expressions:

5
$$d = \lim_{n \to \infty} \sum_{i=1}^{n} f(t_{i-1})\,\Delta t = \lim_{n \to \infty} \sum_{i=1}^{n} f(t_i)\,\Delta t$$

We will see in Section 5.4 that this is indeed true.

Because Equation 5 has the same form as our expressions for area in Equations 2 and 3, it follows that the distance traveled is equal to the area under the graph of the velocity function. In Chapters 6 and 8 we will see that other quantities of interest in the natural and social sciences—such as the work done by a variable force or the cardiac output of the heart—can also be interpreted as the area under a curve. So when we compute areas in this chapter, bear in mind that they can be interpreted in a variety of practical ways.

5.1 EXERCISES

1. (a) By reading values from the given graph of f, use five rectangles to find a lower estimate and an upper estimate for the area under the given graph of f from $x = 0$ to $x = 10$. In each case sketch the rectangles that you use.
(b) Find new estimates using ten rectangles in each case.

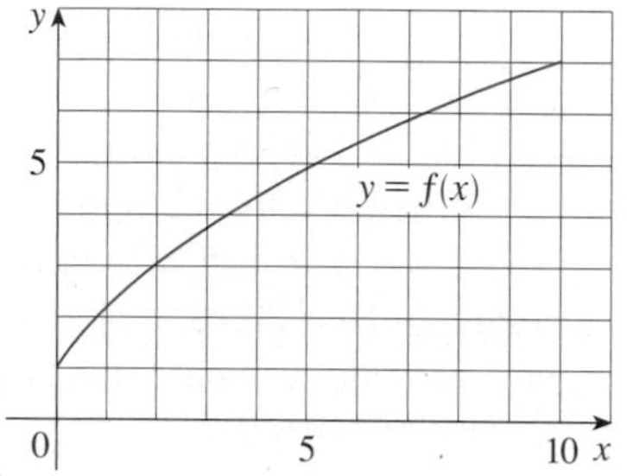

2. (a) Use six rectangles to find estimates of each type for the area under the given graph of f from $x = 0$ to $x = 12$.
(i) L_6 (sample points are left endpoints)
(ii) R_6 (sample points are right endpoints)
(iii) M_6 (sample points are midpoints)
(b) Is L_6 an underestimate or overestimate of the true area?
(c) Is R_6 an underestimate or overestimate of the true area?
(d) Which of the numbers L_6, R_6, or M_6 gives the best estimate? Explain.

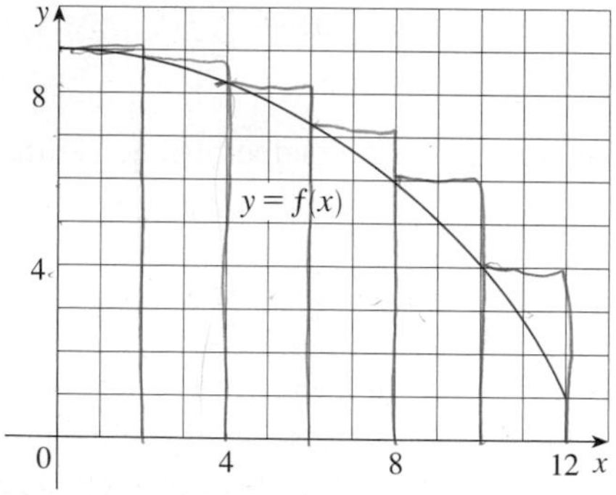

3. (a) Estimate the area under the graph of $f(x) = \cos x$ from $x = 0$ to $x = \pi/2$ using four approximating rectangles and right endpoints. Sketch the graph and the rectangles. Is your estimate an underestimate or an overestimate?
(b) Repeat part (a) using left endpoints.

4. (a) Estimate the area under the graph of $f(x) = \sqrt{x}$ from $x = 0$ to $x = 4$ using four approximating rectangles and right endpoints. Sketch the graph and the rectangles. Is your estimate an underestimate or an overestimate?
(b) Repeat part (a) using left endpoints.

5. (a) Estimate the area under the graph of $f(x) = 1 + x^2$ from $x = -1$ to $x = 2$ using three rectangles and right endpoints. Then improve your estimate by using six rectangles. Sketch the curve and the approximating rectangles.
(b) Repeat part (a) using left endpoints.
(c) Repeat part (a) using midpoints.
(d) From your sketches in parts (a)–(c), which appears to be the best estimate?

6. (a) Graph the function $f(x) = e^{-x^2}$, $-2 \leqslant x \leqslant 2$.
(b) Estimate the area under the graph of f using four approximating rectangles and taking the sample points to be (i) right endpoints and (ii) midpoints. In each case sketch the curve and the rectangles.
(c) Improve your estimates in part (b) by using 8 rectangles.

7–8 With a programmable calculator (or a computer), it is possible to evaluate the expressions for the sums of areas of approximating rectangles, even for large values of n, using looping. (On a TI use the Is> command or a For-EndFor loop, on a Casio use Isz, on an HP or in BASIC use a FOR-NEXT loop.) Compute the sum of the areas of approximating rectangles using equal subintervals and right endpoints for $n = 10, 30, 50$, and 100. Then guess the value of the exact area.

7. The region under $y = x^4$ from 0 to 1

8. The region under $y = \cos x$ from 0 to $\pi/2$

CAS **9.** Some computer algebra systems have commands that will draw approximating rectangles and evaluate the sums of their areas, at least if x_i^* is a left or right endpoint. (For instance, in Maple use `leftbox`, `rightbox`, `leftsum`, and `rightsum`.)
(a) If $f(x) = 1/(x^2 + 1)$, $0 \leqslant x \leqslant 1$, find the left and right sums for $n = 10, 30$, and 50.
(b) Illustrate by graphing the rectangles in part (a).
(c) Show that the exact area under f lies between 0.780 and 0.791.

CAS **10.** (a) If $f(x) = \ln x$, $1 \leqslant x \leqslant 4$, use the commands discussed in Exercise 9 to find the left and right sums for $n = 10$, 30, and 50.
(b) Illustrate by graphing the rectangles in part (a).
(c) Show that the exact area under f lies between 2.50 and 2.59.

11. The speed of a runner increased steadily during the first three seconds of a race. Her speed at half-second intervals is given in the table. Find lower and upper estimates for the distance that she traveled during these three seconds.

t (s)	0	0.5	1.0	1.5	2.0	2.5	3.0
v (ft/s)	0	6.2	10.8	14.9	18.1	19.4	20.2

12. Speedometer readings for a motorcycle at 12-second intervals are given in the table.
(a) Estimate the distance traveled by the motorcycle during this time period using the velocities at the beginning of the time intervals.
(b) Give another estimate using the velocities at the end of the time periods.
(c) Are your estimates in parts (a) and (b) upper and lower estimates? Explain.

t (s)	0	12	24	36	48	60
v (ft/s)	30	28	25	22	24	27

13. Oil leaked from a tank at a rate of $r(t)$ liters per hour. The rate decreased as time passed and values of the rate at two-hour time intervals are shown in the table. Find lower and upper estimates for the total amount of oil that leaked out.

t (h)	0	2	4	6	8	10
$r(t)$ (L/h)	8.7	7.6	6.8	6.2	5.7	5.3

14. When we estimate distances from velocity data, it is sometimes necessary to use times $t_0, t_1, t_2, t_3, \ldots$ that are not equally spaced. We can still estimate distances using the time periods $\Delta t_i = t_i - t_{i-1}$. For example, on May 7, 1992, the space shuttle *Endeavour* was launched on mission STS-49, the purpose of which was to install a new perigee kick motor in an Intelsat communications satellite. The table, provided by NASA, gives the velocity data for the shuttle between liftoff and the jettisoning of the solid rocket boosters. Use these data to estimate the height above the earth's surface of the *Endeavour,* 62 seconds after liftoff.

Event	Time (s)	Velocity (ft/s)
Launch	0	0
Begin roll maneuver	10	185
End roll maneuver	15	319
Throttle to 89%	20	447
Throttle to 67%	32	742
Throttle to 104%	59	1325
Maximum dynamic pressure	62	1445
Solid rocket booster separation	125	4151

15. The velocity graph of a braking car is shown. Use it to estimate the distance traveled by the car while the brakes are applied.

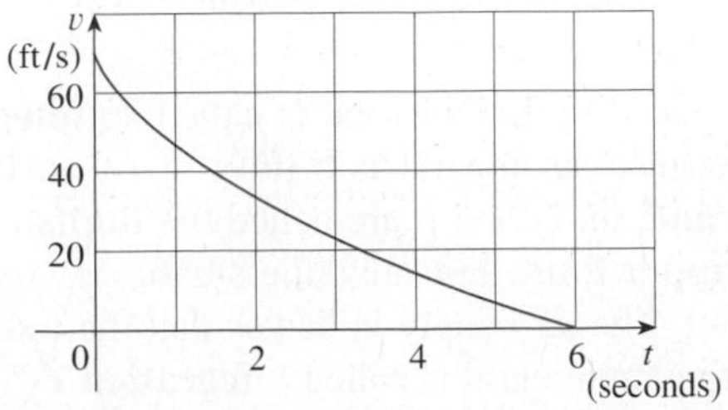

16. The velocity graph of a car accelerating from rest to a speed of 120 km/h over a period of 30 seconds is shown. Estimate the distance traveled during this period.

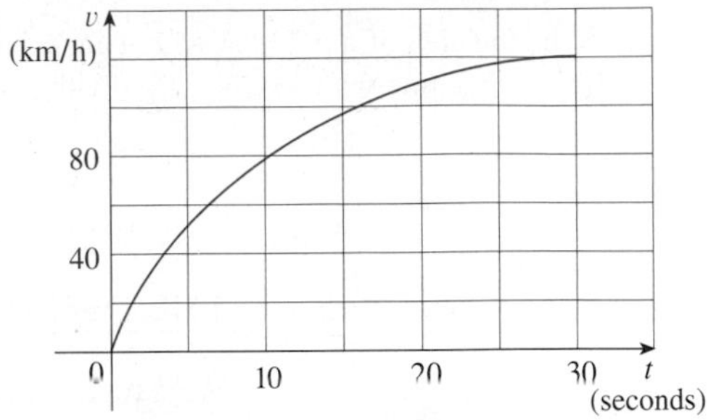

17–19 Use Definition 2 to find an expression for the area under the graph of f as a limit. Do not evaluate the limit.

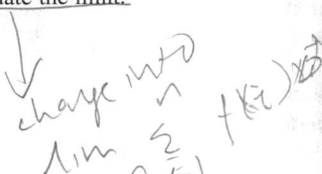

17. $f(x) = \sqrt[4]{x}, \quad 1 \leqslant x \leqslant 16$

18. $f(x) = \dfrac{\ln x}{x}, \quad 3 \leqslant x \leqslant 10$

19. $f(x) = x \cos x, \quad 0 \leqslant x \leqslant \pi/2$

20–21 Determine a region whose area is equal to the given limit. Do not evaluate the limit.

20. $\displaystyle \lim_{n\to\infty} \sum_{i=1}^{n} \frac{2}{n}\left(5 + \frac{2i}{n}\right)^{10}$

21. $\displaystyle \lim_{n\to\infty} \sum_{i=1}^{n} \frac{\pi}{4n} \tan \frac{i\pi}{4n}$

22. (a) Use Definition 2 to find an expression for the area under the curve $y = x^3$ from 0 to 1 as a limit.
(b) The following formula for the sum of the cubes of the first n integers is proved in Appendix E. Use it to evaluate the limit in part (a).

$$1^3 + 2^3 + 3^3 + \cdots + n^3 = \left[\frac{n(n+1)}{2}\right]^2$$

CAS **23.** (a) Express the area under the curve $y = x^5$ from 0 to 2 as a limit.
(b) Use a computer algebra system to find the sum in your expression from part (a).
(c) Evaluate the limit in part (a).

CAS **24.** Find the exact area of the region under the graph of $y = e^{-x}$ from 0 to 2 by using a computer algebra system to evaluate the sum and then the limit in Example 3(a). Compare your answer with the estimate obtained in Example 3(b).

CAS **25.** Find the exact area under the cosine curve $y = \cos x$ from $x = 0$ to $x = b$, where $0 \leqslant b \leqslant \pi/2$. (Use a computer algebra system both to evaluate the sum and compute the limit.) In particular, what is the area if $b = \pi/2$?

26. (a) Let A_n be the area of a polygon with n equal sides inscribed in a circle with radius r. By dividing the polygon into n congruent triangles with central angle $2\pi/n$, show that

$$A_n = \tfrac{1}{2}nr^2 \sin\left(\frac{2\pi}{n}\right)$$

(b) Show that $\lim_{n\to\infty} A_n = \pi r^2$. [*Hint:* Use Equation 3.3.2.]

5.2 THE DEFINITE INTEGRAL

We saw in Section 5.1 that a limit of the form

$$\boxed{1} \qquad \lim_{n\to\infty} \sum_{i=1}^{n} f(x_i^*)\,\Delta x = \lim_{n\to\infty} [f(x_1^*)\,\Delta x + f(x_2^*)\,\Delta x + \cdots + f(x_n^*)\,\Delta x]$$

arises when we compute an area. We also saw that it arises when we try to find the distance traveled by an object. It turns out that this same type of limit occurs in a wide variety of situations even when f is not necessarily a positive function. In Chapters 6 and 8 we will see that limits of the form (1) also arise in finding lengths of curves, volumes of solids, centers of mass, force due to water pressure, and work, as well as other quantities. We therefore give this type of limit a special name and notation.

> **2 DEFINITION OF A DEFINITE INTEGRAL** If f is a function defined for $a \leqslant x \leqslant b$, we divide the interval $[a, b]$ into n subintervals of equal width $\Delta x = (b - a)/n$. We let $x_0\ (= a), x_1, x_2, \ldots, x_n\ (= b)$ be the endpoints of these subintervals and we let $x_1^*, x_2^*, \ldots, x_n^*$ be any **sample points** in these subintervals, so x_i^* lies in the ith subinterval $[x_{i-1}, x_i]$. Then the **definite integral of f from a to b** is
>
> $$\int_a^b f(x)\,dx = \lim_{n\to\infty} \sum_{i=1}^{n} f(x_i^*)\,\Delta x$$
>
> provided that this limit exists. If it does exist, we say that f is **integrable** on $[a, b]$.

The precise meaning of the limit that defines the integral is as follows:

For every number $\varepsilon > 0$ there is an integer N such that

$$\left| \int_a^b f(x)\,dx - \sum_{i=1}^{n} f(x_i^*)\,\Delta x \right| < \varepsilon$$

for every integer $n > N$ and for every choice of x_i^* in $[x_{i-1}, x_i]$.

NOTE 1 The symbol $\int$ was introduced by Leibniz and is called an **integral sign**. It is an elongated S and was chosen because an integral is a limit of sums. In the notation $\int_a^b f(x)\,dx$, $f(x)$ is called the **integrand** and a and b are called the **limits of integration**; a is the **lower limit** and b is the **upper limit**. For now, the symbol dx has no meaning by itself; $\int_a^b f(x)\,dx$ is all one symbol. The dx simply indicates that the independent variable is x. The procedure of calculating an integral is called **integration**.

NOTE 2 The definite integral $\int_a^b f(x)\,dx$ is a number; it does not depend on x. In fact, we could use any letter in place of x without changing the value of the integral:

$$\int_a^b f(x)\,dx = \int_a^b f(t)\,dt = \int_a^b f(r)\,dr$$

NOTE 3 The sum

$$\sum_{i=1}^{n} f(x_i^*)\,\Delta x$$

that occurs in Definition 2 is called a **Riemann sum** after the German mathematician Bernhard Riemann (1826–1866). So Definition 2 says that the definite integral of an integrable function can be approximated to within any desired degree of accuracy by a Riemann sum.

RIEMANN

Bernhard Riemann received his Ph.D. under the direction of the legendary Gauss at the University of Göttingen and remained there to teach. Gauss, who was not in the habit of praising other mathematicians, spoke of Riemann's "creative, active, truly mathematical mind and gloriously fertile originality." The definition (2) of an integral that we use is due to Riemann. He also made major contributions to the theory of functions of a complex variable, mathematical physics, number theory, and the foundations of geometry. Riemann's broad concept of space and geometry turned out to be the right setting, 50 years later, for Einstein's general relativity theory. Riemann's health was poor throughout his life, and he died of tuberculosis at the age of 39.

We know that if f happens to be positive, then the Riemann sum can be interpreted as a sum of areas of approximating rectangles (see Figure 1). By comparing Definition 2 with the definition of area in Section 5.1, we see that the definite integral $\int_a^b f(x)\,dx$ can be interpreted as the area under the curve $y = f(x)$ from a to b. (See Figure 2.)

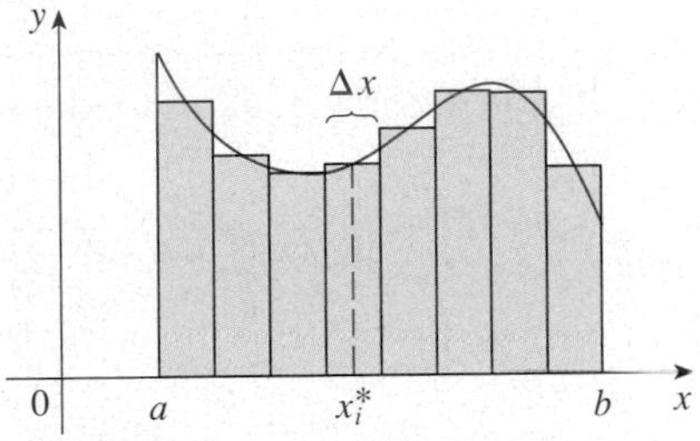

FIGURE 1
If $f(x) \geq 0$, the Riemann sum $\Sigma f(x_i^*)\,\Delta x$ is the sum of areas of rectangles.

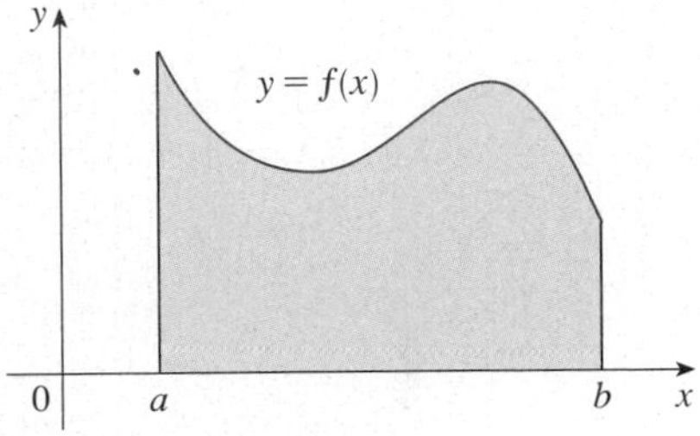

FIGURE 2
If $f(x) \geq 0$, the integral $\int_a^b f(x)\,dx$ is the area under the curve $y = f(x)$ from a to b.

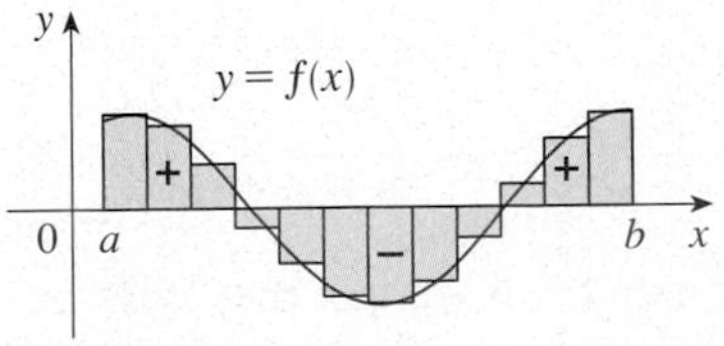

FIGURE 3
$\Sigma f(x_i^*)\,\Delta x$ is an approximation to the net area

If f takes on both positive and negative values, as in Figure 3, then the Riemann sum is the sum of the areas of the rectangles that lie above the x-axis and the *negatives* of the areas of the rectangles that lie below the x-axis (the areas of the gold rectangles *minus* the areas of the blue rectangles). When we take the limit of such Riemann sums, we get the situation illustrated in Figure 4. A definite integral can be interpreted as a **net area**, that is, a difference of areas:

$$\int_a^b f(x)\,dx = A_1 - A_2$$

where A_1 is the area of the region above the x-axis and below the graph of f, and A_2 is the area of the region below the x-axis and above the graph of f.

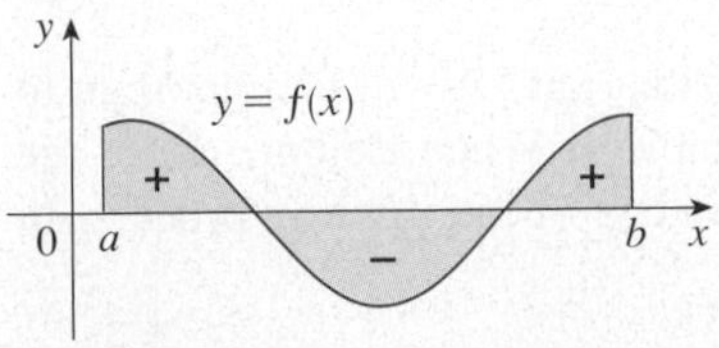

FIGURE 4
$\int_a^b f(x)\,dx$ is the net area

NOTE 4 Although we have defined $\int_a^b f(x)\,dx$ by dividing $[a, b]$ into subintervals of equal width, there are situations in which it is advantageous to work with subintervals of unequal width. For instance, in Exercise 14 in Section 5.1 NASA provided velocity data at times that were not equally spaced, but we were still able to estimate the distance traveled. And there are methods for numerical integration that take advantage of unequal subintervals.

If the subinterval widths are $\Delta x_1, \Delta x_2, \dots, \Delta x_n$, we have to ensure that all these widths approach 0 in the limiting process. This happens if the largest width, $\max \Delta x_i$, approaches 0. So in this case the definition of a definite integral becomes

$$\int_a^b f(x)\,dx = \lim_{\max \Delta x_i \to 0} \sum_{i=1}^{n} f(x_i^*)\,\Delta x_i$$

NOTE 5 We have defined the definite integral for an inegrable function, but not all functions are integrable (see Exercises 67–68). The following theorem shows that the most commonly occurring functions are in fact integrable. It is proved in more advanced courses.

> **3 THEOREM** If f is continuous on $[a, b]$, or if f has only a finite number of jump discontinuities, then f is integrable on $[a, b]$; that is, the definite integral $\int_a^b f(x)\,dx$ exists.

If f is integrable on $[a, b]$, then the limit in Definition 2 exists and gives the same value no matter how we choose the sample points x_i^*. To simplify the calculation of the integral we often take the sample points to be right endpoints. Then $x_i^* = x_i$ and the definition of an integral simplifies as follows.

> **4 THEOREM** If f is integrable on $[a, b]$, then
>
> $$\int_a^b f(x)\,dx = \lim_{n \to \infty} \sum_{i=1}^{n} f(x_i)\,\Delta x$$
>
> where $\qquad \Delta x = \dfrac{b - a}{n} \qquad$ and $\qquad x_i = a + i\,\Delta x$

EXAMPLE 1 Express

$$\lim_{n \to \infty} \sum_{i=1}^{n} (x_i^3 + x_i \sin x_i)\,\Delta x$$

as an integral on the interval $[0, \pi]$.

SOLUTION Comparing the given limit with the limit in Theorem 4, we see that they will be identical if we choose $f(x) = x^3 + x \sin x$. We are given that $a = 0$ and $b = \pi$. Therefore, by Theorem 4, we have

$$\lim_{n \to \infty} \sum_{i=1}^{n} (x_i^3 + x_i \sin x_i)\,\Delta x = \int_0^\pi (x^3 + x \sin x)\,dx \qquad \square$$

Later, when we apply the definite integral to physical situations, it will be important to recognize limits of sums as integrals, as we did in Example 1. When Leibniz chose the notation for an integral, he chose the ingredients as reminders of the limiting process. In general, when we write

$$\lim_{n \to \infty} \sum_{i=1}^{n} f(x_i^*)\,\Delta x = \int_a^b f(x)\,dx$$

we replace $\lim \Sigma$ by $\int$, x_i^* by x, and Δx by dx.

EVALUATING INTEGRALS

When we use a limit to evaluate a definite integral, we need to know how to work with sums. The following three equations give formulas for sums of powers of positive integers. Equation 5 may be familiar to you from a course in algebra. Equations 6 and 7 were discussed in Section 5.1 and are proved in Appendix E.

5 $$\sum_{i=1}^{n} i = \frac{n(n+1)}{2}$$

6 $$\sum_{i=1}^{n} i^2 = \frac{n(n+1)(2n+1)}{6}$$

7 $$\sum_{i=1}^{n} i^3 = \left[\frac{n(n+1)}{2}\right]^2$$

The remaining formulas are simple rules for working with sigma notation:

8 $$\sum_{i=1}^{n} c = nc$$

9 $$\sum_{i=1}^{n} ca_i = c\sum_{i=1}^{n} a_i$$

10 $$\sum_{i=1}^{n} (a_i + b_i) = \sum_{i=1}^{n} a_i + \sum_{i=1}^{n} b_i$$

11 $$\sum_{i=1}^{n} (a_i - b_i) = \sum_{i=1}^{n} a_i - \sum_{i=1}^{n} b_i$$

■ Formulas 8–11 are proved by writing out each side in expanded form. The left side of Equation 9 is

$$ca_1 + ca_2 + \cdots + ca_n$$

The right side is

$$c(a_1 + a_2 + \cdots + a_n)$$

These are equal by the distributive property. The other formulas are discussed in Appendix E.

EXAMPLE 2

(a) Evaluate the Riemann sum for $f(x) = x^3 - 6x$ taking the sample points to be right endpoints and $a = 0$, $b = 3$, and $n = 6$.

(b) Evaluate $\int_0^3 (x^3 - 6x)\,dx$.

SOLUTION

(a) With $n = 6$ the interval width is

$$\Delta x = \frac{b-a}{n} = \frac{3-0}{6} = \frac{1}{2}$$

and the right endpoints are $x_1 = 0.5$, $x_2 = 1.0$, $x_3 = 1.5$, $x_4 = 2.0$, $x_5 = 2.5$, and $x_6 = 3.0$. So the Riemann sum is

$$\begin{aligned} R_6 &= \sum_{i=1}^{6} f(x_i)\,\Delta x \\ &= f(0.5)\,\Delta x + f(1.0)\,\Delta x + f(1.5)\,\Delta x + f(2.0)\,\Delta x + f(2.5)\,\Delta x + f(3.0)\,\Delta x \\ &= \tfrac{1}{2}(-2.875 - 5 - 5.625 - 4 + 0.625 + 9) \\ &= -3.9375 \end{aligned}$$

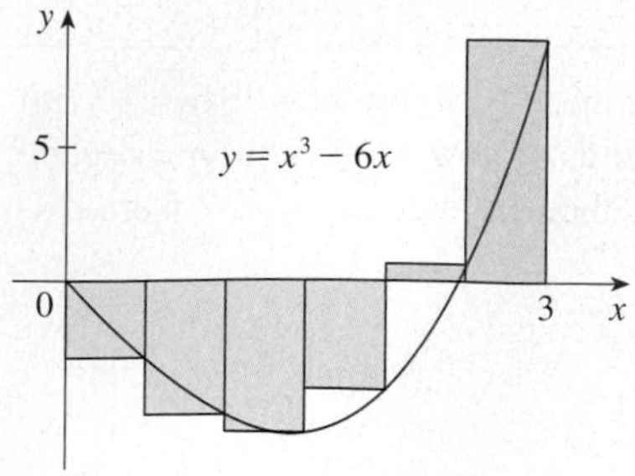

FIGURE 5

Notice that f is not a positive function and so the Riemann sum does not represent a sum of areas of rectangles. But it does represent the sum of the areas of the gold rectangles (above the x-axis) minus the sum of the areas of the blue rectangles (below the x-axis) in Figure 5.

(b) With n subintervals we have

$$\Delta x = \frac{b - a}{n} = \frac{3}{n}$$

Thus $x_0 = 0$, $x_1 = 3/n$, $x_2 = 6/n$, $x_3 = 9/n$, and, in general, $x_i = 3i/n$. Since we are using right endpoints, we can use Theorem 4:

■ In the sum, n is a constant (unlike i), so we can move $3/n$ in front of the Σ sign.

$$\int_0^3 (x^3 - 6x)\,dx = \lim_{n\to\infty} \sum_{i=1}^{n} f(x_i)\,\Delta x = \lim_{n\to\infty} \sum_{i=1}^{n} f\left(\frac{3i}{n}\right)\frac{3}{n}$$

$$= \lim_{n\to\infty} \frac{3}{n} \sum_{i=1}^{n} \left[\left(\frac{3i}{n}\right)^3 - 6\left(\frac{3i}{n}\right)\right] \qquad \text{(Equation 9 with } c = 3/n\text{)}$$

$$= \lim_{n\to\infty} \frac{3}{n} \sum_{i=1}^{n} \left[\frac{27}{n^3} i^3 - \frac{18}{n} i\right]$$

$$= \lim_{n\to\infty} \left[\frac{81}{n^4} \sum_{i=1}^{n} i^3 - \frac{54}{n^2} \sum_{i=1}^{n} i\right] \qquad \text{(Equations 11 and 9)}$$

$$= \lim_{n\to\infty} \left\{\frac{81}{n^4}\left[\frac{n(n+1)}{2}\right]^2 - \frac{54}{n^2}\frac{n(n+1)}{2}\right\} \qquad \text{(Equations 7 and 5)}$$

$$= \lim_{n\to\infty} \left[\frac{81}{4}\left(1 + \frac{1}{n}\right)^2 - 27\left(1 + \frac{1}{n}\right)\right]$$

$$= \frac{81}{4} - 27 = -\frac{27}{4} = -6.75$$

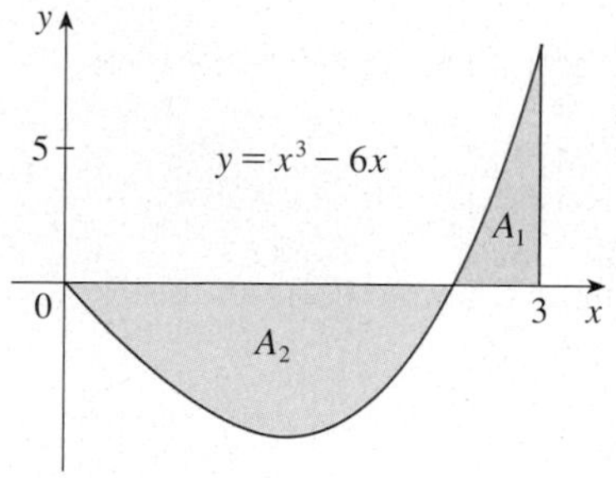

FIGURE 6
$\int_0^3 (x^3 - 6x)\,dx = A_1 - A_2 = -6.75$

This integral can't be interpreted as an area because f takes on both positive and negative values. But it can be interpreted as the difference of areas $A_1 - A_2$, where A_1 and A_2 are shown in Figure 6.

Figure 7 illustrates the calculation by showing the positive and negative terms in the right Riemann sum R_n for $n = 40$. The values in the table show the Riemann sums approaching the exact value of the integral, -6.75, as $n \to \infty$.

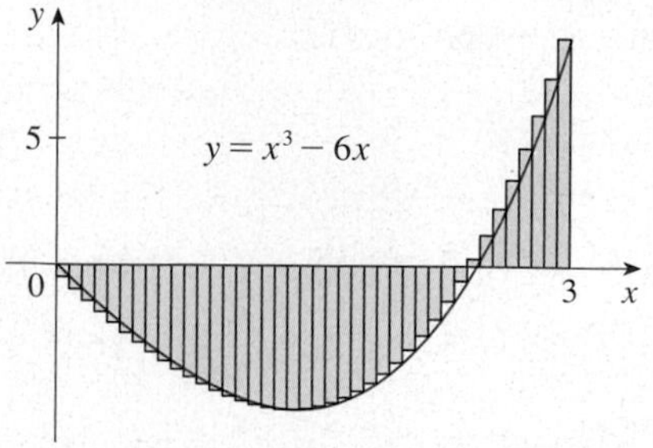

FIGURE 7
$R_{40} \approx -6.3998$

n	R_n
40	−6.3998
100	−6.6130
500	−6.7229
1000	−6.7365
5000	−6.7473

□

A much simpler method for evaluating the integral in Example 2 will be given in Section 5.3.

■ Because $f(x) = e^x$ is positive, the integral in Example 3 represents the area shown in Figure 8.

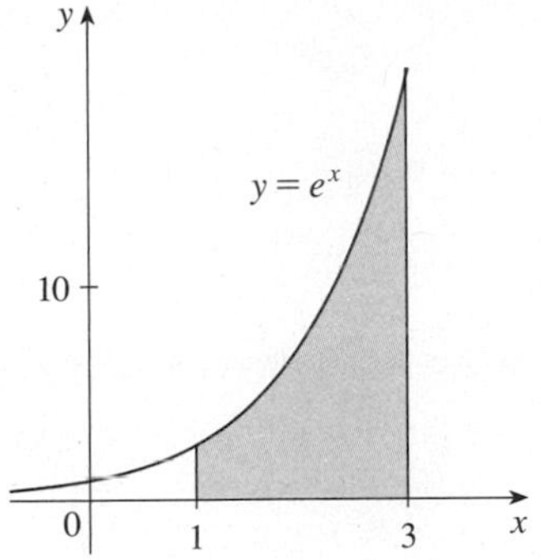

FIGURE 8

EXAMPLE 3

(a) Set up an expression for $\int_1^3 e^x\,dx$ as a limit of sums.

(b) Use a computer algebra system to evaluate the expression.

SOLUTION

(a) Here we have $f(x) = e^x$, $a = 1$, $b = 3$, and

$$\Delta x = \frac{b-a}{n} = \frac{2}{n}$$

So $x_0 = 1$, $x_1 = 1 + 2/n$, $x_2 = 1 + 4/n$, $x_3 = 1 + 6/n$, and

$$x_i = 1 + \frac{2i}{n}$$

From Theorem 4, we get

$$\begin{aligned}\int_1^3 e^x\,dx &= \lim_{n\to\infty}\sum_{i=1}^{n} f(x_i)\,\Delta x\\ &= \lim_{n\to\infty}\sum_{i=1}^{n} f\left(1 + \frac{2i}{n}\right)\frac{2}{n}\\ &= \lim_{n\to\infty}\frac{2}{n}\sum_{i=1}^{n} e^{1+2i/n}\end{aligned}$$

(b) If we ask a computer algebra system to evaluate the sum and simplify, we obtain

■ A computer algebra system is able to find an explicit expression for this sum because it is a geometric series. The limit could be found using l'Hospital's Rule.

$$\sum_{i=1}^{n} e^{1+2i/n} = \frac{e^{(3n+2)/n} - e^{(n+2)/n}}{e^{2/n} - 1}$$

Now we ask the computer algebra system to evaluate the limit:

$$\int_1^3 e^x\,dx = \lim_{n\to\infty}\frac{2}{n}\cdot\frac{e^{(3n+2)/n} - e^{(n+2)/n}}{e^{2/n} - 1} = e^3 - e$$

We will learn a much easier method for the evaluation of integrals in the next section. □

V EXAMPLE 4 Evaluate the following integrals by interpreting each in terms of areas.

(a) $\int_0^1 \sqrt{1 - x^2}\,dx$ (b) $\int_0^3 (x - 1)\,dx$

SOLUTION

(a) Since $f(x) = \sqrt{1 - x^2} \geqslant 0$, we can interpret this integral as the area under the curve $y = \sqrt{1 - x^2}$ from 0 to 1. But, since $y^2 = 1 - x^2$, we get $x^2 + y^2 = 1$, which shows that the graph of f is the quarter-circle with radius 1 in Figure 9. Therefore

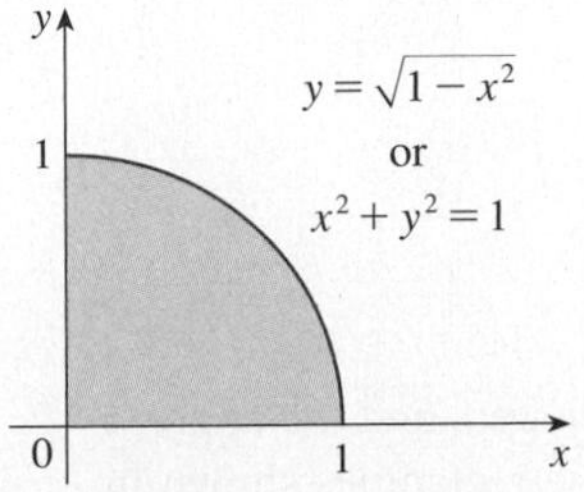

FIGURE 9

$$\int_0^1 \sqrt{1 - x^2}\,dx = \tfrac{1}{4}\pi(1)^2 = \frac{\pi}{4}$$

(In Section 7.3 we will be able to *prove* that the area of a circle of radius r is πr^2.)

(b) The graph of $y = x - 1$ is the line with slope 1 shown in Figure 10. We compute the integral as the difference of the areas of the two triangles:

$$\int_0^3 (x - 1)\,dx = A_1 - A_2 = \tfrac{1}{2}(2 \cdot 2) - \tfrac{1}{2}(1 \cdot 1) = 1.5$$

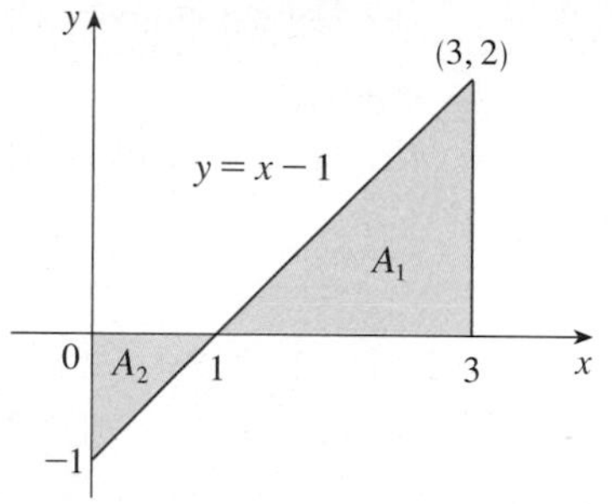

FIGURE 10

□

THE MIDPOINT RULE

We often choose the sample point x_i^* to be the right endpoint of the ith subinterval because it is convenient for computing the limit. But if the purpose is to find an *approximation* to an integral, it is usually better to choose x_i^* to be the midpoint of the interval, which we denote by $\overline{x}_i$. Any Riemann sum is an approximation to an integral, but if we use midpoints we get the following approximation.

TEC Module 5.2/7.7 shows how the Midpoint Rule estimates improve as n increases.

MIDPOINT RULE

$$\int_a^b f(x)\,dx \approx \sum_{i=1}^{n} f(\overline{x}_i)\,\Delta x = \Delta x\,[f(\overline{x}_1) + \cdots + f(\overline{x}_n)]$$

where $\qquad \Delta x = \dfrac{b - a}{n}$

and $\qquad \overline{x}_i = \tfrac{1}{2}(x_{i-1} + x_i) = \text{midpoint of } [x_{i-1}, x_i]$

EXAMPLE 5 Use the Midpoint Rule with $n = 5$ to approximate $\displaystyle\int_1^2 \frac{1}{x}\,dx$.

SOLUTION The endpoints of the five subintervals are 1, 1.2, 1.4, 1.6, 1.8, and 2.0, so the midpoints are 1.1, 1.3, 1.5, 1.7, and 1.9. The width of the subintervals is $\Delta x = (2 - 1)/5 = \frac{1}{5}$, so the Midpoint Rule gives

$$\int_1^2 \frac{1}{x}\,dx \approx \Delta x\,[f(1.1) + f(1.3) + f(1.5) + f(1.7) + f(1.9)]$$

$$= \frac{1}{5}\left(\frac{1}{1.1} + \frac{1}{1.3} + \frac{1}{1.5} + \frac{1}{1.7} + \frac{1}{1.9}\right)$$

$$\approx 0.691908$$

Since $f(x) = 1/x > 0$ for $1 \leqslant x \leqslant 2$, the integral represents an area, and the approximation given by the Midpoint Rule is the sum of the areas of the rectangles shown in Figure 11. □

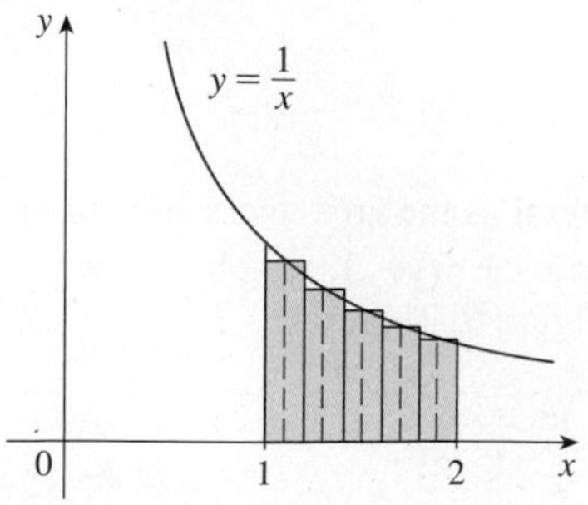

FIGURE 11

At the moment we don't know how accurate the approximation in Example 5 is, but in Section 7.7 we will learn a method for estimating the error involved in using the Midpoint Rule. At that time we will discuss other methods for approximating definite integrals.

If we apply the Midpoint Rule to the integral in Example 2, we get the picture in Figure 12. The approximation $M_{40} \approx -6.7563$ is much closer to the true value -6.75 than the right endpoint approximation, $R_{40} \approx -6.3998$, shown in Figure 7.

TEC In Visual 5.2 you can compare left, right, and midpoint approximations to the integral in Example 2 for different values of n.

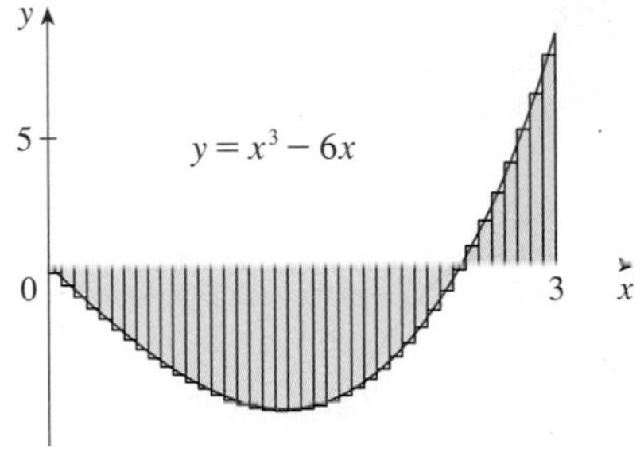

FIGURE 12
$M_{40} \approx -6.7563$

PROPERTIES OF THE DEFINITE INTEGRAL

When we defined the definite integral $\int_a^b f(x)\,dx$, we implicitly assumed that $a < b$. But the definition as a limit of Riemann sums makes sense even if $a > b$. Notice that if we reverse a and b, then Δx changes from $(b - a)/n$ to $(a - b)/n$. Therefore

$$\int_b^a f(x)\,dx = -\int_a^b f(x)\,dx$$

If $a = b$, then $\Delta x = 0$ and so

$$\int_a^a f(x)\,dx = 0$$

We now develop some basic properties of integrals that will help us to evaluate integrals in a simple manner. We assume that f and g are continuous functions.

PROPERTIES OF THE INTEGRAL

1. $\int_a^b c\,dx = c(b - a)$, where c is any constant

2. $\int_a^b [f(x) + g(x)]\,dx = \int_a^b f(x)\,dx + \int_a^b g(x)\,dx$

3. $\int_a^b cf(x)\,dx = c\int_a^b f(x)\,dx$, where c is any constant

4. $\int_a^b [f(x) - g(x)]\,dx = \int_a^b f(x)\,dx - \int_a^b g(x)\,dx$

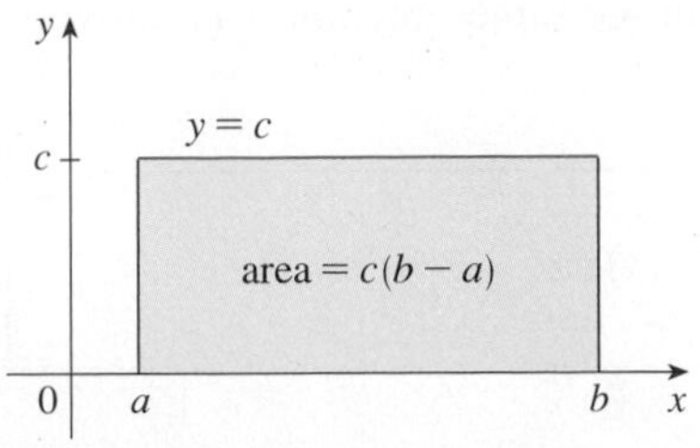

FIGURE 13
$\int_a^b c\,dx = c(b - a)$

Property 1 says that the integral of a constant function $f(x) = c$ is the constant times the length of the interval. If $c > 0$ and $a < b$, this is to be expected because $c(b - a)$ is the area of the shaded rectangle in Figure 13.

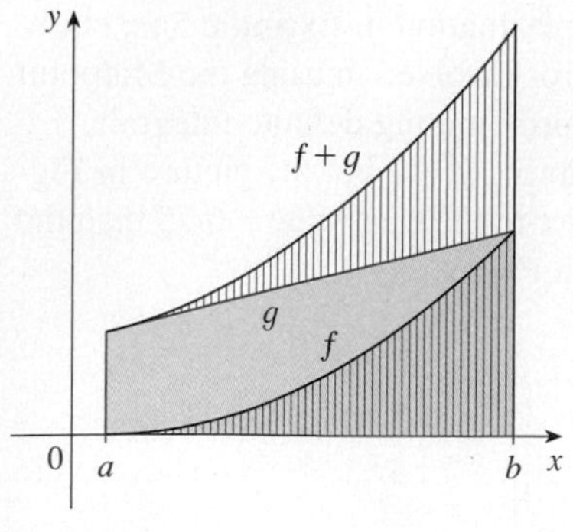

FIGURE 14

$\int_a^b [f(x)+g(x)]\,dx = \int_a^b f(x)\,dx + \int_a^b g(x)\,dx$

Property 2 says that the integral of a sum is the sum of the integrals. For positive functions it says that the area under $f + g$ is the area under f plus the area under g. Figure 14 helps us understand why this is true: In view of how graphical addition works, the corresponding vertical line segments have equal height.

In general, Property 2 follows from Theorem 4 and the fact that the limit of a sum is the sum of the limits:

$$\begin{aligned}\int_a^b [f(x) + g(x)]\,dx &= \lim_{n\to\infty} \sum_{i=1}^{n} [f(x_i) + g(x_i)]\,\Delta x \\ &= \lim_{n\to\infty} \left[\sum_{i=1}^{n} f(x_i)\,\Delta x + \sum_{i=1}^{n} g(x_i)\,\Delta x \right] \\ &= \lim_{n\to\infty} \sum_{i=1}^{n} f(x_i)\,\Delta x + \lim_{n\to\infty} \sum_{i=1}^{n} g(x_i)\,\Delta x \\ &= \int_a^b f(x)\,dx + \int_a^b g(x)\,dx\end{aligned}$$

■ Property 3 seems intuitively reasonable because we know that multiplying a function by a positive number c stretches or shrinks its graph vertically by a factor of c. So it stretches or shrinks each approximating rectangle by a factor c and therefore it has the effect of multiplying the area by c.

Property 3 can be proved in a similar manner and says that the integral of a constant times a function is the constant times the integral of the function. In other words, a constant (but *only* a constant) can be taken in front of an integral sign. Property 4 is proved by writing $f - g = f + (-g)$ and using Properties 2 and 3 with $c = -1$.

EXAMPLE 6 Use the properties of integrals to evaluate $\int_0^1 (4 + 3x^2)\,dx$.

SOLUTION Using Properties 2 and 3 of integrals, we have

$$\int_0^1 (4 + 3x^2)\,dx = \int_0^1 4\,dx + \int_0^1 3x^2\,dx = \int_0^1 4\,dx + 3\int_0^1 x^2\,dx$$

We know from Property 1 that

$$\int_0^1 4\,dx = 4(1 - 0) = 4$$

and we found in Example 2 in Section 5.1 that $\int_0^1 x^2\,dx = \frac{1}{3}$. So

$$\begin{aligned}\int_0^1 (4 + 3x^2)\,dx &= \int_0^1 4\,dx + 3\int_0^1 x^2\,dx \\ &= 4 + 3\cdot\tfrac{1}{3} = 5\end{aligned}$$

□

The next property tells us how to combine integrals of the same function over adjacent intervals:

5. $$\int_a^c f(x)\,dx + \int_c^b f(x)\,dx = \int_a^b f(x)\,dx$$

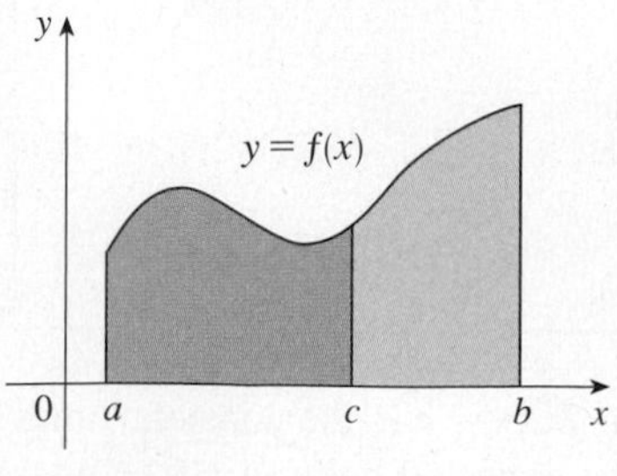

FIGURE 15

This is not easy to prove in general, but for the case where $f(x) \geq 0$ and $a < c < b$ Property 5 can be seen from the geometric interpretation in Figure 15: The area under $y = f(x)$ from a to c plus the area from c to b is equal to the total area from a to b.

V EXAMPLE 7 If it is known that $\int_0^{10} f(x)\,dx = 17$ and $\int_0^8 f(x)\,dx = 12$, find $\int_8^{10} f(x)\,dx$.

SOLUTION By Property 5, we have

$$\int_0^8 f(x)\,dx + \int_8^{10} f(x)\,dx = \int_0^{10} f(x)\,dx$$

so

$$\int_8^{10} f(x)\,dx = \int_0^{10} f(x)\,dx - \int_0^8 f(x)\,dx = 17 - 12 = 5$$

□

Properties 1–5 are true whether $a < b$, $a = b$, or $a > b$. The following properties, in which we compare sizes of functions and sizes of integrals, are true only if $a \leqslant b$.

COMPARISON PROPERTIES OF THE INTEGRAL

6. If $f(x) \geqslant 0$ for $a \leqslant x \leqslant b$, then $\int_a^b f(x)\,dx \geqslant 0$.

7. If $f(x) \geqslant g(x)$ for $a \leqslant x \leqslant b$, then $\int_a^b f(x)\,dx \geqslant \int_a^b g(x)\,dx$.

8. If $m \leqslant f(x) \leqslant M$ for $a \leqslant x \leqslant b$, then

$$m(b-a) \leqslant \int_a^b f(x)\,dx \leqslant M(b-a)$$

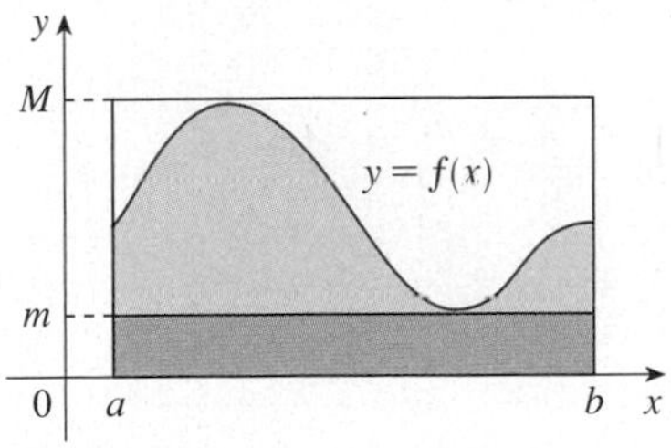

FIGURE 16

If $f(x) \geqslant 0$, then $\int_a^b f(x)\,dx$ represents the area under the graph of f, so the geometric interpretation of Property 6 is simply that areas are positive. But the property can be proved from the definition of an integral (Exercise 64). Property 7 says that a bigger function has a bigger integral. It follows from Properties 6 and 4 because $f - g \geqslant 0$.

Property 8 is illustrated by Figure 16 for the case where $f(x) \geqslant 0$. If f is continuous we could take m and M to be the absolute minimum and maximum values of f on the interval $[a, b]$. In this case Property 8 says that the area under the graph of f is greater than the area of the rectangle with height m and less than the area of the rectangle with height M.

PROOF OF PROPERTY 8 Since $m \leqslant f(x) \leqslant M$, Property 7 gives

$$\int_a^b m\,dx \leqslant \int_a^b f(x)\,dx \leqslant \int_a^b M\,dx$$

Using Property 1 to evaluate the integrals on the left and right sides, we obtain

$$m(b-a) \leqslant \int_a^b f(x)\,dx \leqslant M(b-a)$$

□

Property 8 is useful when all we want is a rough estimate of the size of an integral without going to the bother of using the Midpoint Rule.

EXAMPLE 8 Use Property 8 to estimate $\displaystyle\int_0^1 e^{-x^2}\,dx$.

SOLUTION Because $f(x) = e^{-x^2}$ is a decreasing function on $[0, 1]$, its absolute maximum value is $M = f(0) = 1$ and its absolute minimum value is $m = f(1) = e^{-1}$. Thus, by

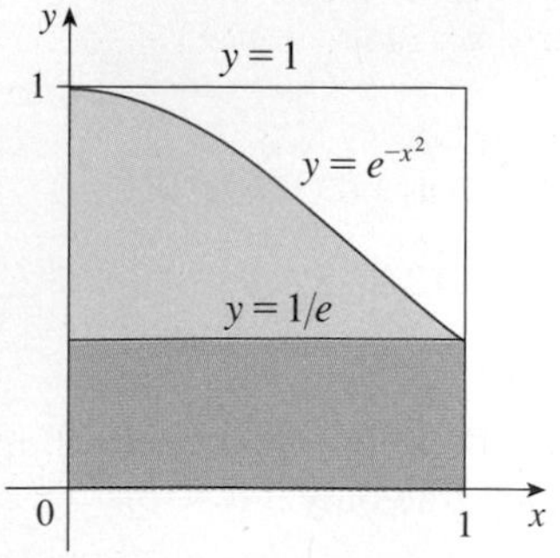

FIGURE 17

Property 8,

$$e^{-1}(1 - 0) \leq \int_0^1 e^{-x^2}\,dx \leq 1(1 - 0)$$

or

$$e^{-1} \leq \int_0^1 e^{-x^2}\,dx \leq 1$$

Since $e^{-1} \approx 0.3679$, we can write

$$0.367 \leq \int_0^1 e^{-x^2}\,dx \leq 1$$ □

The result of Example 8 is illustrated in Figure 17. The integral is greater than the area of the lower rectangle and less than the area of the square.

5.2 EXERCISES

1. Evaluate the Riemann sum for $f(x) = 3 - \frac{1}{2}x$, $2 \leq x \leq 14$, with six subintervals, taking the sample points to be left endpoints. Explain, with the aid of a diagram, what the Riemann sum represents.

2. If $f(x) = x^2 - 2x$, $0 \leq x \leq 3$, evaluate the Riemann sum with $n = 6$, taking the sample points to be right endpoints. What does the Riemann sum represent? Illustrate with a diagram.

3. If $f(x) = e^x - 2$, $0 \leq x \leq 2$, find the Riemann sum with $n = 4$ correct to six decimal places, taking the sample points to be midpoints. What does the Riemann sum represent? Illustrate with a diagram.

4. (a) Find the Riemann sum for $f(x) = \sin x$, $0 \leq x \leq 3\pi/2$, with six terms, taking the sample points to be right endpoints. (Give your answer correct to six decimal places.) Explain what the Riemann sum represents with the aid of a sketch.

(b) Repeat part (a) with midpoints as sample points.

5. The graph of a function f is given. Estimate $\int_0^8 f(x)\,dx$ using four subintervals with (a) right endpoints, (b) left endpoints, and (c) midpoints.

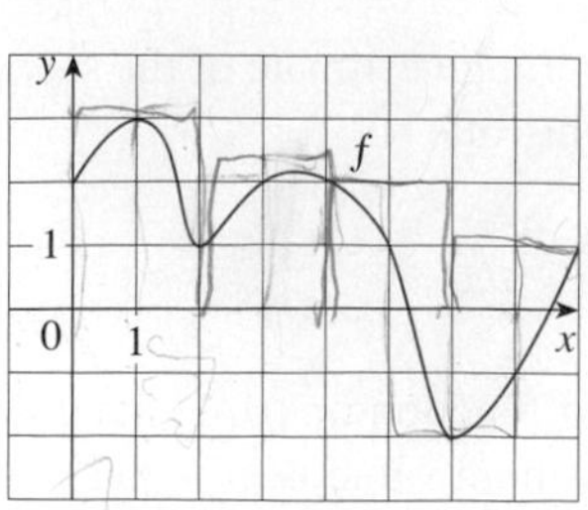

6. The graph of g is shown. Estimate $\int_{-3}^{3} g(x)\,dx$ with six subintervals using (a) right endpoints, (b) left endpoints, and (c) midpoints.

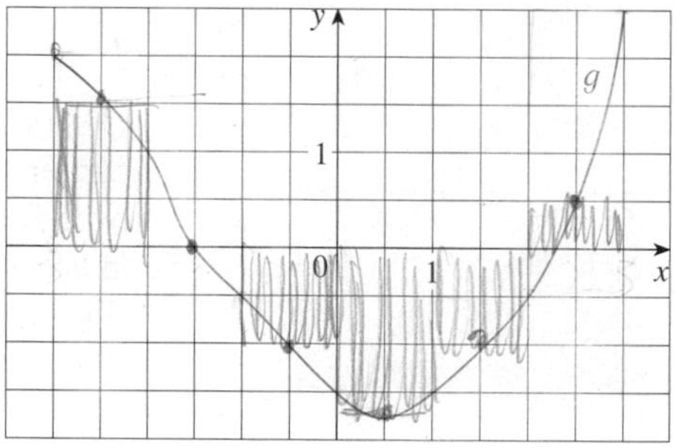

7. A table of values of an increasing function f is shown. Use the table to find lower and upper estimates for $\int_0^{25} f(x)\,dx$.

x	0	5	10	15	20	25
$f(x)$	−42	−37	−25	−6	15	36

8. The table gives the values of a function obtained from an experiment. Use them to estimate $\int_3^9 f(x)\,dx$ using three equal subintervals with (a) right endpoints, (b) left endpoints, and (c) midpoints. If the function is known to be an increasing function, can you say whether your estimates are less than or greater than the exact value of the integral?

x	3	4	5	6	7	8	9
$f(x)$	−3.4	−2.1	−0.6	0.3	0.9	1.4	1.8

9–12 Use the Midpoint Rule with the given value of n to approximate the integral. Round the answer to four decimal places.

9. $\int_2^{10} \sqrt{x^3 + 1}\, dx$, $n = 4$ **10.** $\int_0^{\pi/2} \cos^4 x\, dx$, $n = 4$

11. $\int_0^1 \sin(x^2)\, dx$, $n = 5$ **12.** $\int_1^5 x^2 e^{-x}\, dx$, $n = 4$

CAS **13.** If you have a CAS that evaluates midpoint approximations and graphs the corresponding rectangles (use `middlesum` and `middlebox` commands in Maple), check the answer to Exercise 11 and illustrate with a graph. Then repeat with $n = 10$ and $n = 20$.

14. With a programmable calculator or computer (see the instructions for Exercise 7 in Section 5.1), compute the left and right Riemann sums for the function $f(x) = \sin(x^2)$ on the interval $[0, 1]$ with $n = 100$. Explain why these estimates show that

$$0.306 < \int_0^1 \sin(x^2)\, dx < 0.315$$

Deduce that the approximation using the Midpoint Rule with $n = 5$ in Exercise 11 is accurate to two decimal places.

15. Use a calculator or computer to make a table of values of right Riemann sums R_n for the integral $\int_0^{\pi} \sin x\, dx$ with $n = 5$, 10, 50, and 100. What value do these numbers appear to be approaching?

16. Use a calculator or computer to make a table of values of left and right Riemann sums L_n and R_n for the integral $\int_0^2 e^{-x^2}\, dx$ with $n = 5$, 10, 50, and 100. Between what two numbers must the value of the integral lie? Can you make a similar statement for the integral $\int_{-1}^2 e^{-x^2}\, dx$? Explain.

17–20 Express the limit as a definite integral on the given interval.

17. $\lim_{n\to\infty} \sum_{i=1}^{n} x_i \ln(1 + x_i^2)\, \Delta x$, $[2, 6]$

18. $\lim_{n\to\infty} \sum_{i=1}^{n} \frac{\cos x_i}{x_i}\, \Delta x$, $[\pi, 2\pi]$

19. $\lim_{n\to\infty} \sum_{i=1}^{n} \sqrt{2x_i^* + (x_i^*)^2}\, \Delta x$, $[1, 8]$

20. $\lim_{n\to\infty} \sum_{i=1}^{n} [4 - 3(x_i^*)^2 + 6(x_i^*)^5]\, \Delta x$, $[0, 2]$

21–25 Use the form of the definition of the integral given in Theorem 4 to evaluate the integral.

21. $\int_{-1}^5 (1 + 3x)\, dx$ **22.** $\int_1^4 (x^2 + 2x - 5)\, dx$

23. $\int_0^2 (2 - x^2)\, dx$ **24.** $\int_0^5 (1 + 2x^3)\, dx$

25. $\int_1^2 x^3\, dx$

---

26. (a) Find an approximation to the integral $\int_0^4 (x^2 - 3x)\, dx$ using a Riemann sum with right endpoints and $n = 8$.
(b) Draw a diagram like Figure 3 to illustrate the approximation in part (a).
(c) Use Theorem 4 to evaluate $\int_0^4 (x^2 - 3x)\, dx$.
(d) Interpret the integral in part (c) as a difference of areas and illustrate with a diagram like Figure 4.

27. Prove that $\int_a^b x\, dx = \frac{b^2 - a^2}{2}$.

28. Prove that $\int_a^b x^2\, dx = \frac{b^3 - a^3}{3}$.

29–30 Express the integral as a limit of Riemann sums. Do not evaluate the limit.

29. $\int_2^6 \frac{x}{1 + x^5}\, dx$ **30.** $\int_1^{10} (x - 4 \ln x)\, dx$

CAS **31–32** Express the integral as a limit of sums. Then evaluate, using a computer algebra system to find both the sum and the limit.

31. $\int_0^{\pi} \sin 5x\, dx$ **32.** $\int_2^{10} x^6\, dx$

---

33. The graph of f is shown. Evaluate each integral by interpreting it in terms of areas.

(a) $\int_0^2 f(x)\, dx$ (b) $\int_0^5 f(x)\, dx$

(c) $\int_5^7 f(x)\, dx$ (d) $\int_0^9 f(x)\, dx$

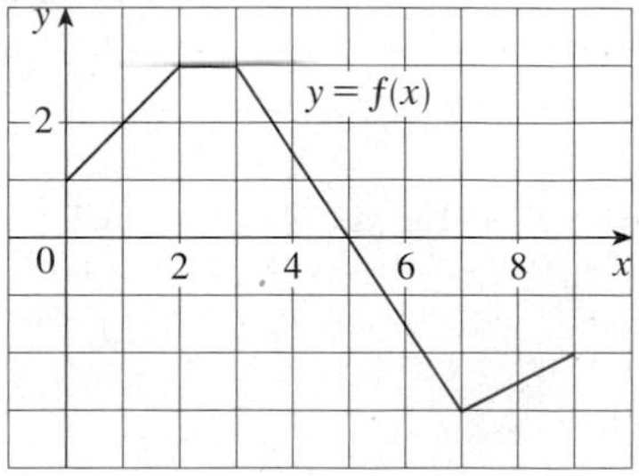

34. The graph of g consists of two straight lines and a semicircle. Use it to evaluate each integral.

(a) $\int_0^2 g(x)\, dx$ (b) $\int_2^6 g(x)\, dx$ (c) $\int_0^7 g(x)\, dx$

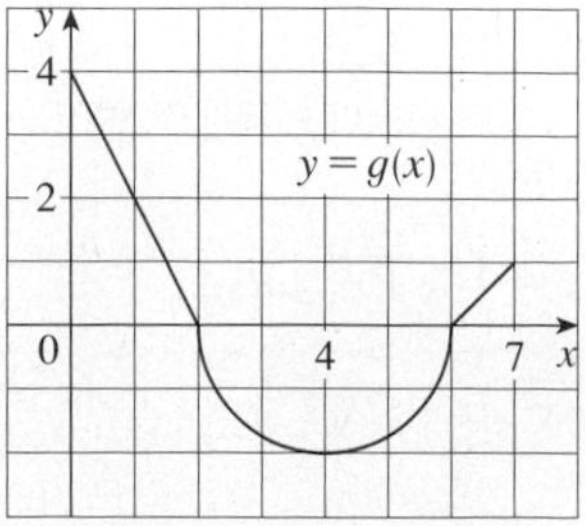

35–40 Evaluate the integral by interpreting it in terms of areas.

35. $\int_0^3 (\frac{1}{2}x - 1)\,dx$

36. $\int_{-2}^{2} \sqrt{4 - x^2}\,dx$

37. $\int_{-3}^{0} (1 + \sqrt{9 - x^2})\,dx$

38. $\int_{-1}^{3} (3 - 2x)\,dx$

39. $\int_{-1}^{2} |x|\,dx$

40. $\int_0^{10} |x - 5|\,dx$

41. Evaluate $\int_\pi^\pi \sin^2 x \cos^4 x\,dx$.

42. Given that $\int_0^1 3x\sqrt{x^2 + 4}\,dx = 5\sqrt{5} - 8$, what is $\int_1^0 3u\sqrt{u^2 + 4}\,du$?

43. In Example 2 in Section 5.1 we showed that $\int_0^1 x^2\,dx = \frac{1}{3}$. Use this fact and the properties of integrals to evaluate $\int_0^1 (5 - 6x^2)\,dx$.

44. Use the properties of integrals and the result of Example 3 to evaluate $\int_1^3 (2e^x - 1)\,dx$.

45. Use the result of Example 3 to evaluate $\int_1^3 e^{x+2}\,dx$.

46. Use the result of Exercise 27 and the fact that $\int_0^{\pi/2} \cos x\,dx = 1$ (from Exercise 25 in Section 5.1), together with the properties of integrals, to evaluate $\int_0^{\pi/2} (2\cos x - 5x)\,dx$.

47. Write as a single integral in the form $\int_a^b f(x)\,dx$:

$$\int_{-2}^{2} f(x)\,dx + \int_2^5 f(x)\,dx - \int_{-2}^{-1} f(x)\,dx$$

48. If $\int_1^5 f(x)\,dx = 12$ and $\int_4^5 f(x)\,dx = 3.6$, find $\int_1^4 f(x)\,dx$.

49. If $\int_0^9 f(x)\,dx = 37$ and $\int_0^9 g(x)\,dx = 16$, find $\int_0^9 [2f(x) + 3g(x)]\,dx$.

50. Find $\int_0^5 f(x)\,dx$ if

$$f(x) = \begin{cases} 3 & \text{for } x < 3 \\ x & \text{for } x \geqslant 3 \end{cases}$$

51. Suppose f has absolute minimum value m and absolute maximum value M. Between what two values must $\int_0^2 f(x)\,dx$ lie? Which property of integrals allows you to make your conclusion?

52–54 Use the properties of integrals to verify the inequality without evaluating the integrals.

52. $\int_0^1 \sqrt{1 + x^2}\,dx \leqslant \int_0^1 \sqrt{1 + x}\,dx$

53. $2 \leqslant \int_{-1}^{1} \sqrt{1 + x^2}\,dx \leqslant 2\sqrt{2}$

54. $\dfrac{\sqrt{2}\,\pi}{24} \leqslant \int_{\pi/6}^{\pi/4} \cos x\,dx \leqslant \dfrac{\sqrt{3}\,\pi}{24}$

55–60 Use Property 8 to estimate the value of the integral.

55. $\int_1^4 \sqrt{x}\,dx$

56. $\int_0^2 \dfrac{1}{1 + x^2}\,dx$

57. $\int_{\pi/4}^{\pi/3} \tan x\,dx$

58. $\int_0^2 (x^3 - 3x + 3)\,dx$

59. $\int_0^2 xe^{-x}\,dx$

60. $\int_\pi^{2\pi} (x - 2\sin x)\,dx$

61–62 Use properties of integrals, together with Exercises 27 and 28, to prove the inequality.

61. $\int_1^3 \sqrt{x^4 + 1}\,dx \geqslant \dfrac{26}{3}$

62. $\int_0^{\pi/2} x\sin x\,dx \leqslant \dfrac{\pi^2}{8}$

63. Prove Property 3 of integrals.

64. Prove Property 6 of integrals.

65. If f is continuous on $[a, b]$, show that

$$\left| \int_a^b f(x)\,dx \right| \leqslant \int_a^b |f(x)|\,dx$$

[*Hint:* $-|f(x)| \leqslant f(x) \leqslant |f(x)|$.]

66. Use the result of Exercise 65 to show that

$$\left| \int_0^{2\pi} f(x)\sin 2x\,dx \right| \leqslant \int_0^{2\pi} |f(x)|\,dx$$

67. Let $f(x) = 0$ if x is any rational number and $f(x) = 1$ if x is any irrational number. Show that f is not integrable on $[0, 1]$.

68. Let $f(0) = 0$ and $f(x) = 1/x$ if $0 < x \leqslant 1$. Show that f is not integrable on $[0, 1]$. [*Hint:* Show that the first term in the Riemann sum, $f(x_i^*)\,\Delta x$, can be made arbitrarily large.]

69–70 Express the limit as a definite integral.

69. $\lim_{n\to\infty} \sum_{i=1}^{n} \dfrac{i^4}{n^5}$ [*Hint:* Consider $f(x) = x^4$.]

70. $\lim_{n\to\infty} \dfrac{1}{n} \sum_{i=1}^{n} \dfrac{1}{1 + (i/n)^2}$

71. Find $\int_1^2 x^{-2}\,dx$. *Hint:* Choose x_i^* to be the geometric mean of x_{i-1} and x_i (that is, $x_i^* = \sqrt{x_{i-1}x_i}$) and use the identity

$$\frac{1}{m(m+1)} = \frac{1}{m} - \frac{1}{m+1}$$

DISCOVERY PROJECT

AREA FUNCTIONS

1. (a) Draw the line $y = 2t + 1$ and use geometry to find the area under this line, above the t-axis, and between the vertical lines $t = 1$ and $t = 3$.
(b) If $x > 1$, let $A(x)$ be the area of the region that lies under the line $y = 2t + 1$ between $t = 1$ and $t = x$. Sketch this region and use geometry to find an expression for $A(x)$.
(c) Differentiate the area function $A(x)$. What do you notice?

2. (a) If $x \geqslant -1$, let

$$A(x) = \int_{-1}^{x} (1 + t^2)\, dt$$

$A(x)$ represents the area of a region. Sketch that region.
(b) Use the result of Exercise 28 in Section 5.2 to find an expression for $A(x)$.
(c) Find $A'(x)$. What do you notice?
(d) If $x \geqslant -1$ and h is a small positive number, then $A(x + h) - A(x)$ represents the area of a region. Describe and sketch the region.
(e) Draw a rectangle that approximates the region in part (d). By comparing the areas of these two regions, show that

$$\frac{A(x + h) - A(x)}{h} \approx 1 + x^2$$

(f) Use part (e) to give an intuitive explanation for the result of part (c).

3. (a) Draw the graph of the function $f(x) = \cos(x^2)$ in the viewing rectangle $[0, 2]$ by $[-1.25, 1.25]$.
(b) If we define a new function g by

$$g(x) = \int_{0}^{x} \cos(t^2)\, dt$$

then $g(x)$ is the area under the graph of f from 0 to x [until $f(x)$ becomes negative, at which point $g(x)$ becomes a difference of areas]. Use part (a) to determine the value of x at which $g(x)$ starts to decrease. [Unlike the integral in Problem 2, it is impossible to evaluate the integral defining g to obtain an explicit expression for $g(x)$.]
(c) Use the integration command on your calculator or computer to estimate $g(0.2)$, $g(0.4)$, $g(0.6), \ldots, g(1.8)$, $g(2)$. Then use these values to sketch a graph of g.
(d) Use your graph of g from part (c) to sketch the graph of g' using the interpretation of $g'(x)$ as the slope of a tangent line. How does the graph of g' compare with the graph of f?

4. Suppose f is a continuous function on the interval $[a, b]$ and we define a new function g by the equation

$$g(x) = \int_{a}^{x} f(t)\, dt$$

Based on your results in Problems 1–3, conjecture an expression for $g'(x)$.

5.3 THE FUNDAMENTAL THEOREM OF CALCULUS

The Fundamental Theorem of Calculus is appropriately named because it establishes a connection between the two branches of calculus: differential calculus and integral calculus. Differential calculus arose from the tangent problem, whereas integral calculus arose from a seemingly unrelated problem, the area problem. Newton's mentor at Cambridge,

Isaac Barrow (1630–1677), discovered that these two problems are actually closely related. In fact, he realized that differentiation and integration are inverse processes. The Fundamental Theorem of Calculus gives the precise inverse relationship between the derivative and the integral. It was Newton and Leibniz who exploited this relationship and used it to develop calculus into a systematic mathematical method. In particular, they saw that the Fundamental Theorem enabled them to compute areas and integrals very easily without having to compute them as limits of sums as we did in Sections 5.1 and 5.2.

The first part of the Fundamental Theorem deals with functions defined by an equation of the form

$$\boxed{1} \qquad g(x) = \int_a^x f(t)\,dt$$

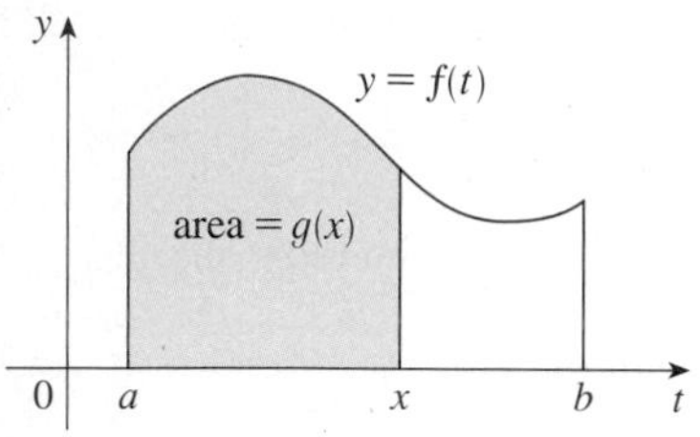

FIGURE 1

where f is a continuous function on $[a, b]$ and x varies between a and b. Observe that g depends only on x, which appears as the variable upper limit in the integral. If x is a fixed number, then the integral $\int_a^x f(t)\,dt$ is a definite number. If we then let x vary, the number $\int_a^x f(t)\,dt$ also varies and defines a function of x denoted by $g(x)$.

If f happens to be a positive function, then $g(x)$ can be interpreted as the area under the graph of f from a to x, where x can vary from a to b. (Think of g as the "area so far" function; see Figure 1.)

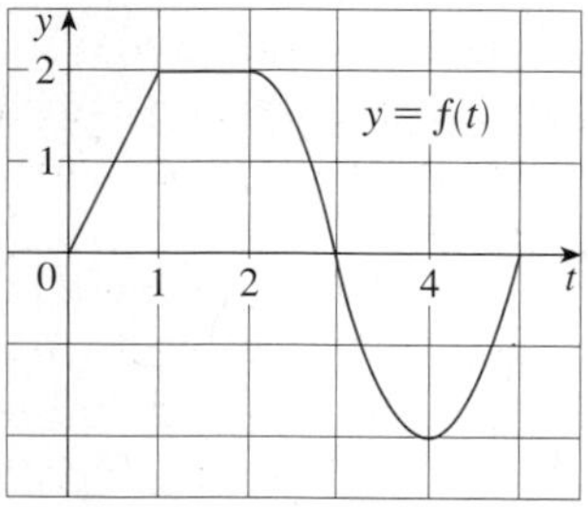

FIGURE 2

V EXAMPLE 1 If f is the function whose graph is shown in Figure 2 and $g(x) = \int_0^x f(t)\,dt$, find the values of $g(0)$, $g(1)$, $g(2)$, $g(3)$, $g(4)$, and $g(5)$. Then sketch a rough graph of g.

SOLUTION First we notice that $g(0) = \int_0^0 f(t)\,dt = 0$. From Figure 3 we see that $g(1)$ is the area of a triangle:

$$g(1) = \int_0^1 f(t)\,dt = \tfrac{1}{2}(1 \cdot 2) = 1$$

To find $g(2)$ we add to $g(1)$ the area of a rectangle:

$$g(2) = \int_0^2 f(t)\,dt = \int_0^1 f(t)\,dt + \int_1^2 f(t)\,dt = 1 + (1 \cdot 2) = 3$$

We estimate that the area under f from 2 to 3 is about 1.3, so

$$g(3) = g(2) + \int_2^3 f(t)\,dt \approx 3 + 1.3 = 4.3$$

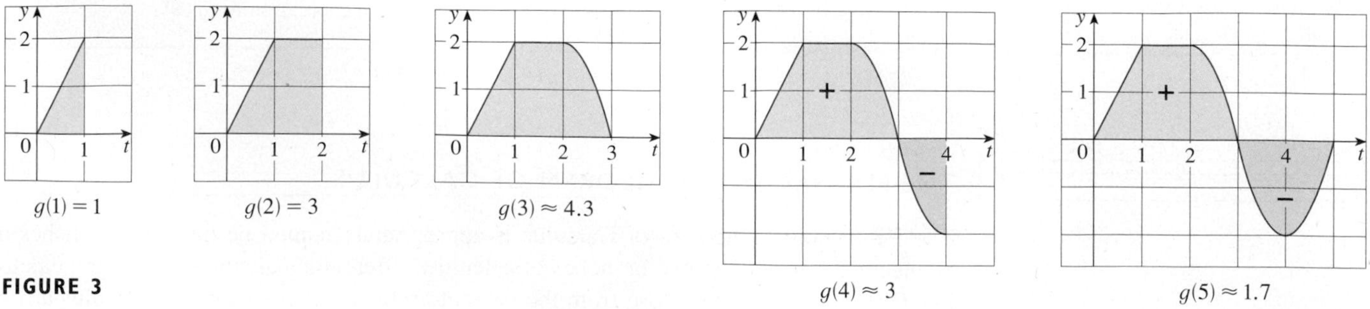

FIGURE 3

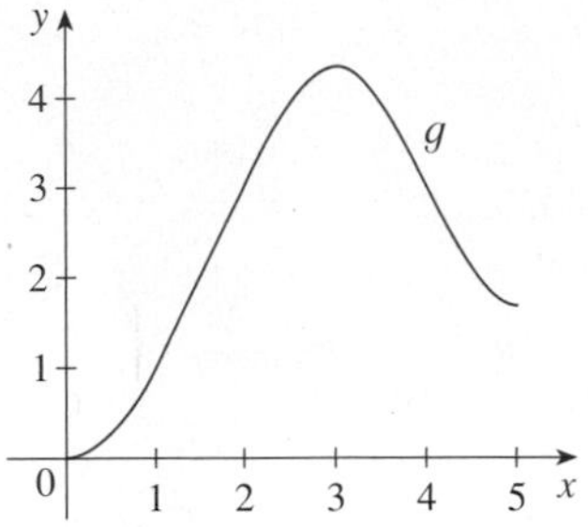

FIGURE 4
$g(x) = \int_a^x f(t)\,dt$

For $t > 3$, $f(t)$ is negative and so we start subtracting areas:

$$g(4) = g(3) + \int_3^4 f(t)\,dt \approx 4.3 + (-1.3) = 3.0$$

$$g(5) = g(4) + \int_4^5 f(t)\,dt \approx 3 + (-1.3) = 1.7$$

We use these values to sketch the graph of g in Figure 4. Notice that, because $f(t)$ is positive for $t < 3$, we keep adding area for $t < 3$ and so g is increasing up to $x = 3$, where it attains a maximum value. For $x > 3$, g decreases because $f(t)$ is negative. □

If we take $f(t) = t$ and $a = 0$, then, using Exercise 27 in Section 5.2, we have

$$g(x) = \int_0^x t\,dt = \frac{x^2}{2}$$

Notice that $g'(x) = x$, that is, $g' = f$. In other words, if g is defined as the integral of f by Equation 1, then g turns out to be an antiderivative of f, at least in this case. And if we sketch the derivative of the function g shown in Figure 4 by estimating slopes of tangents, we get a graph like that of f in Figure 2. So we suspect that $g' = f$ in Example 1 too.

To see why this might be generally true we consider any continuous function f with $f(x) \geqslant 0$. Then $g(x) = \int_a^x f(t)\,dt$ can be interpreted as the area under the graph of f from a to x, as in Figure 1.

In order to compute $g'(x)$ from the definition of derivative we first observe that, for $h > 0$, $g(x + h) - g(x)$ is obtained by subtracting areas, so it is the area under the graph of f from x to $x + h$ (the gold area in Figure 5). For small h you can see from the figure that this area is approximately equal to the area of the rectangle with height $f(x)$ and width h:

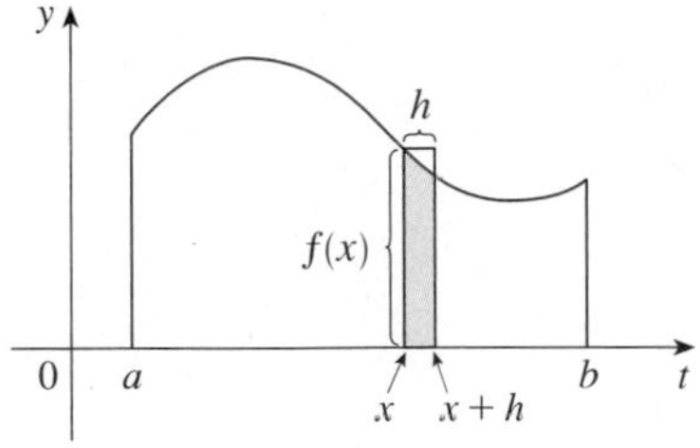

FIGURE 5

$$g(x + h) - g(x) \approx hf(x)$$

so

$$\frac{g(x + h) - g(x)}{h} \approx f(x)$$

Intuitively, we therefore expect that

$$g'(x) = \lim_{h \to 0} \frac{g(x + h) - g(x)}{h} = f(x)$$

The fact that this is true, even when f is not necessarily positive, is the first part of the Fundamental Theorem of Calculus.

■ We abbreviate the name of this theorem as FTC1. In words, it says that the derivative of a definite integral with respect to its upper limit is the integrand evaluated at the upper limit.

THE FUNDAMENTAL THEOREM OF CALCULUS, PART I If f is continuous on $[a, b]$, then the function g defined by

$$g(x) = \int_a^x f(t)\,dt \qquad a \leqslant x \leqslant b$$

is continuous on $[a, b]$ and differentiable on (a, b), and $g'(x) = f(x)$.

PROOF If x and $x + h$ are in (a, b), then

$$g(x+h) - g(x) = \int_a^{x+h} f(t)\,dt - \int_a^x f(t)\,dt$$

$$= \left(\int_a^x f(t)\,dt + \int_x^{x+h} f(t)\,dt\right) - \int_a^x f(t)\,dt \qquad \text{(by Property 5)}$$

$$= \int_x^{x+h} f(t)\,dt$$

and so, for $h \neq 0$,

2 $$\frac{g(x+h) - g(x)}{h} = \frac{1}{h}\int_x^{x+h} f(t)\,dt$$

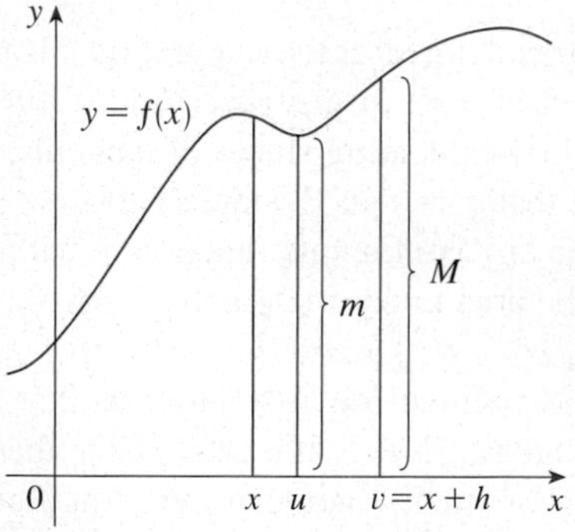

FIGURE 6

For now let us assume that $h > 0$. Since f is continuous on $[x, x + h]$, the Extreme Value Theorem says that there are numbers u and v in $[x, x + h]$ such that $f(u) = m$ and $f(v) = M$, where m and M are the absolute minimum and maximum values of f on $[x, x + h]$. (See Figure 6.)

By Property 8 of integrals, we have

$$mh \leqslant \int_x^{x+h} f(t)\,dt \leqslant Mh$$

that is,

$$f(u)h \leqslant \int_x^{x+h} f(t)\,dt \leqslant f(v)h$$

Since $h > 0$, we can divide this inequality by h:

$$f(u) \leqslant \frac{1}{h}\int_x^{x+h} f(t)\,dt \leqslant f(v)$$

Now we use Equation 2 to replace the middle part of this inequality:

3 $$f(u) \leqslant \frac{g(x+h) - g(x)}{h} \leqslant f(v)$$

TEC Module 5.3 provides visual evidence for FTC1.

Inequality 3 can be proved in a similar manner for the case $h < 0$. (See Exercise 67.)

Now we let $h \to 0$. Then $u \to x$ and $v \to x$, since u and v lie between x and $x + h$. Therefore

$$\lim_{h \to 0} f(u) = \lim_{u \to x} f(u) = f(x)$$

and

$$\lim_{h \to 0} f(v) = \lim_{v \to x} f(v) = f(x)$$

because f is continuous at x. We conclude, from (3) and the Squeeze Theorem, that

$$\boxed{4} \qquad g'(x) = \lim_{h \to 0} \frac{g(x+h) - g(x)}{h} = f(x)$$

If $x = a$ or b, then Equation 4 can be interpreted as a one-sided limit. Then Theorem 2.8.4 (modified for one-sided limits) shows that g is continuous on $[a, b]$. □

Using Leibniz notation for derivatives, we can write FTC1 as

$$\boxed{5} \qquad \frac{d}{dx} \int_a^x f(t)\,dt = f(x)$$

when f is continuous. Roughly speaking, Equation 5 says that if we first integrate f and then differentiate the result, we get back to the original function f.

V EXAMPLE 2 Find the derivative of the function $g(x) = \int_0^x \sqrt{1 + t^2}\,dt$.

SOLUTION Since $f(t) = \sqrt{1 + t^2}$ is continuous, Part 1 of the Fundamental Theorem of Calculus gives

$$g'(x) = \sqrt{1 + x^2}$$ □

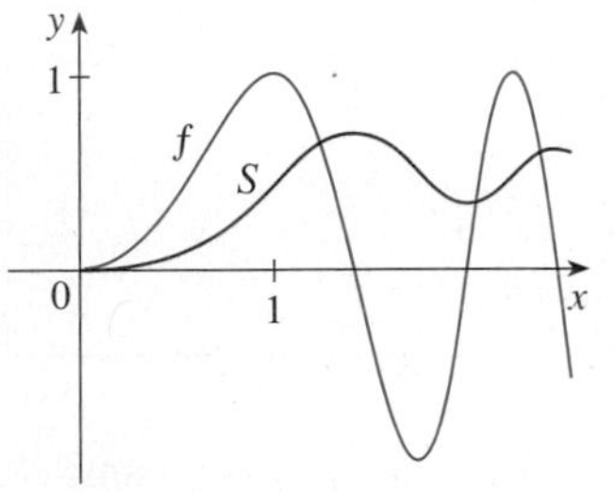

FIGURE 7
$f(x) = \sin(\pi x^2/2)$
$S(x) = \int_0^x \sin(\pi t^2/2)\,dt$

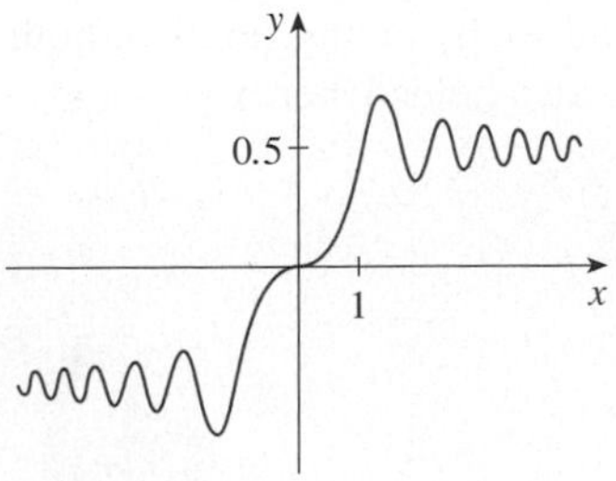

FIGURE 8
The Fresnel function
$S(x) = \int_0^x \sin(\pi t^2/2)\,dt$

EXAMPLE 3 Although a formula of the form $g(x) = \int_a^x f(t)\,dt$ may seem like a strange way of defining a function, books on physics, chemistry, and statistics are full of such functions. For instance, the **Fresnel function**

$$S(x) = \int_0^x \sin(\pi t^2/2)\,dt$$

is named after the French physicist Augustin Fresnel (1788–1827), who is famous for his works in optics. This function first appeared in Fresnel's theory of the diffraction of light waves, but more recently it has been applied to the design of highways.

Part 1 of the Fundamental Theorem tells us how to differentiate the Fresnel function:

$$S'(x) = \sin(\pi x^2/2)$$

This means that we can apply all the methods of differential calculus to analyze S (see Exercise 61).

Figure 7 shows the graphs of $f(x) = \sin(\pi x^2/2)$ and the Fresnel function $S(x) = \int_0^x f(t)\,dt$. A computer was used to graph S by computing the value of this integral for many values of x. It does indeed look as if $S(x)$ is the area under the graph of f from 0 to x [until $x \approx 1.4$, when $S(x)$ becomes a difference of areas]. Figure 8 shows a larger part of the graph of S.

If we now start with the graph of S in Figure 7 and think about what its derivative should look like, it seems reasonable that $S'(x) = f(x)$. [For instance, S is increasing when $f(x) > 0$ and decreasing when $f(x) < 0$.] So this gives a visual confirmation of Part 1 of the Fundamental Theorem of Calculus. □

EXAMPLE 4 Find $\dfrac{d}{dx}\displaystyle\int_1^{x^4} \sec t\,dt$.

SOLUTION Here we have to be careful to use the Chain Rule in conjunction with FTC1. Let $u = x^4$. Then

$$\begin{aligned}\frac{d}{dx}\int_1^{x^4} \sec t\,dt &= \frac{d}{dx}\int_1^{u} \sec t\,dt \\ &= \frac{d}{du}\left(\int_1^{u} \sec t\,dt\right)\frac{du}{dx} \qquad \text{(by the Chain Rule)} \\ &= \sec u\,\frac{du}{dx} \qquad \text{(by FTC1)} \\ &= \sec(x^4)\cdot 4x^3\end{aligned}$$

□

In Section 5.2 we computed integrals from the definition as a limit of Riemann sums and we saw that this procedure is sometimes long and difficult. The second part of the Fundamental Theorem of Calculus, which follows easily from the first part, provides us with a much simpler method for the evaluation of integrals.

■ We abbreviate this theorem as FTC2.

THE FUNDAMENTAL THEOREM OF CALCULUS, PART 2 If f is continuous on $[a, b]$, then

$$\int_a^b f(x)\,dx = F(b) - F(a)$$

where F is any antiderivative of f, that is, a function such that $F' = f$.

PROOF Let $g(x) = \int_a^x f(t)\,dt$. We know from Part 1 that $g'(x) = f(x)$; that is, g is an antiderivative of f. If F is any other antiderivative of f on $[a, b]$, then we know from Corollary 4.2.7 that F and g differ by a constant:

6 $$F(x) = g(x) + C$$

for $a < x < b$. But both F and g are continuous on $[a, b]$ and so, by taking limits of both sides of Equation 6 (as $x \to a^+$ and $x \to b^-$), we see that it also holds when $x = a$ and $x = b$.

If we put $x = a$ in the formula for $g(x)$, we get

$$g(a) = \int_a^a f(t)\,dt = 0$$

So, using Equation 6 with $x = b$ and $x = a$, we have

$$\begin{aligned}F(b) - F(a) &= [g(b) + C] - [g(a) + C] \\ &= g(b) - g(a) = g(b) \\ &= \int_a^b f(t)\,dt\end{aligned}$$

□

Part 2 of the Fundamental Theorem states that if we know an antiderivative F of f, then we can evaluate $\int_a^b f(x)\,dx$ simply by subtracting the values of F at the endpoints of the interval $[a, b]$. It's very surprising that $\int_a^b f(x)\,dx$, which was defined by a complicated procedure involving all of the values of $f(x)$ for $a \leqslant x \leqslant b$, can be found by knowing the values of $F(x)$ at only two points, a and b.

Although the theorem may be surprising at first glance, it becomes plausible if we interpret it in physical terms. If $v(t)$ is the velocity of an object and $s(t)$ is its position at time t, then $v(t) = s'(t)$, so s is an antiderivative of v. In Section 5.1 we considered an object that always moves in the positive direction and made the guess that the area under the velocity curve is equal to the distance traveled. In symbols:

$$\int_a^b v(t)\,dt = s(b) - s(a)$$

That is exactly what FTC2 says in this context.

V EXAMPLE 5 Evaluate the integral $\int_1^3 e^x\,dx$.

SOLUTION The function $f(x) = e^x$ is continuous everywhere and we know that an antiderivative is $F(x) = e^x$, so Part 2 of the Fundamental Theorem gives

■ Compare the calculation in Example 5 with the much harder one in Example 3 in Section 5.2.

$$\int_1^3 e^x\,dx = F(3) - F(1) = e^3 - e$$

Notice that FTC2 says we can use *any* antiderivative F of f. So we may as well use the simplest one, namely $F(x) = e^x$, instead of $e^x + 7$ or $e^x + C$. □

We often use the notation

$$F(x)\Big]_a^b = F(b) - F(a)$$

So the equation of FTC2 can be written as

$$\int_a^b f(x)\,dx = F(x)\Big]_a^b \qquad \text{where} \qquad F' = f$$

Other common notations are $F(x)\big|_a^b$ and $[F(x)]_a^b$.

EXAMPLE 6 Find the area under the parabola $y = x^2$ from 0 to 1.

SOLUTION An antiderivative of $f(x) = x^2$ is $F(x) = \frac{1}{3}x^3$. The required area A is found using Part 2 of the Fundamental Theorem:

■ In applying the Fundamental Theorem we use a particular antiderivative F of f. It is not necessary to use the most general antiderivative.

$$A = \int_0^1 x^2\,dx = \frac{x^3}{3}\Big]_0^1 = \frac{1^3}{3} - \frac{0^3}{3} = \frac{1}{3}$$ □

If you compare the calculation in Example 6 with the one in Example 2 in Section 5.1, you will see that the Fundamental Theorem gives a *much* shorter method.

EXAMPLE 7 Evaluate $\int_3^6 \frac{dx}{x}$.

SOLUTION The given integral is an abbreviation for

$$\int_3^6 \frac{1}{x}\,dx$$

An antiderivative of $f(x) = 1/x$ is $F(x) = \ln|x|$ and, because $3 \leqslant x \leqslant 6$, we can write $F(x) = \ln x$. So

$$\int_3^6 \frac{1}{x}\,dx = \ln x\Big]_3^6 = \ln 6 - \ln 3$$

$$= \ln \frac{6}{3} = \ln 2$$

□

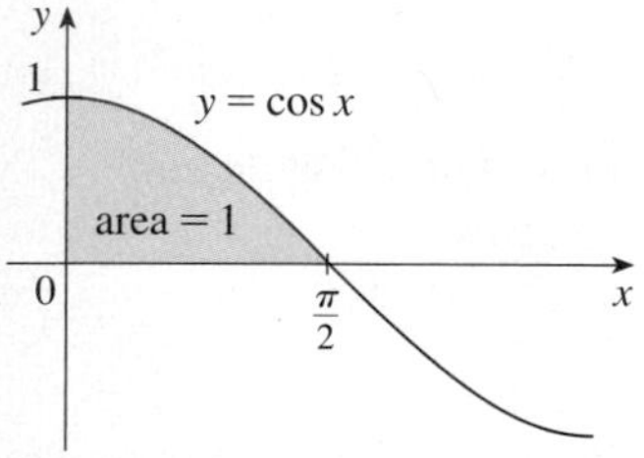

FIGURE 9

EXAMPLE 8 Find the area under the cosine curve from 0 to b, where $0 \leqslant b \leqslant \pi/2$.

SOLUTION Since an antiderivative of $f(x) = \cos x$ is $F(x) = \sin x$, we have

$$A = \int_0^b \cos x\,dx = \sin x\Big]_0^b = \sin b - \sin 0 = \sin b$$

In particular, taking $b = \pi/2$, we have proved that the area under the cosine curve from 0 to $\pi/2$ is $\sin(\pi/2) = 1$. (See Figure 9.) □

When the French mathematician Gilles de Roberval first found the area under the sine and cosine curves in 1635, this was a very challenging problem that required a great deal of ingenuity. If we didn't have the benefit of the Fundamental Theorem, we would have to compute a difficult limit of sums using obscure trigonometric identities (or a computer algebra system as in Exercise 25 in Section 5.1). It was even more difficult for Roberval because the apparatus of limits had not been invented in 1635. But in the 1660s and 1670s, when the Fundamental Theorem was discovered by Barrow and exploited by Newton and Leibniz, such problems became very easy, as you can see from Example 8.

EXAMPLE 9 What is wrong with the following calculation?

⊘

$$\int_{-1}^3 \frac{1}{x^2}\,dx = \frac{x^{-1}}{-1}\bigg]_{-1}^3 = -\frac{1}{3} - 1 = -\frac{4}{3}$$

SOLUTION To start, we notice that this calculation must be wrong because the answer is negative but $f(x) = 1/x^2 \geqslant 0$ and Property 6 of integrals says that $\int_a^b f(x)\,dx \geqslant 0$ when $f \geqslant 0$. The Fundamental Theorem of Calculus applies to continuous functions. It can't be applied here because $f(x) = 1/x^2$ is not continuous on $[-1, 3]$. In fact, f has an infinite discontinuity at $x = 0$, so

$$\int_{-1}^3 \frac{1}{x^2}\,dx \quad \text{does not exist}$$

□

DIFFERENTIATION AND INTEGRATION AS INVERSE PROCESSES

We end this section by bringing together the two parts of the Fundamental Theorem.

THE FUNDAMENTAL THEOREM OF CALCULUS Suppose f is continuous on $[a, b]$.

1. If $g(x) = \int_a^x f(t)\,dt$, then $g'(x) = f(x)$.

2. $\int_a^b f(x)\,dx = F(b) - F(a)$, where F is any antiderivative of f, that is, $F' = f$.

We noted that Part 1 can be rewritten as

$$\frac{d}{dx}\int_a^x f(t)\,dt = f(x)$$

which says that if f is integrated and then the result is differentiated, we arrive back at the original function f. Since $F'(x) = f(x)$, Part 2 can be rewritten as

$$\int_a^b F'(x)\,dx = F(b) - F(a)$$

This version says that if we take a function F, first differentiate it, and then integrate the result, we arrive back at the original function F, but in the form $F(b) - F(a)$. Taken together, the two parts of the Fundamental Theorem of Calculus say that differentiation and integration are inverse processes. Each undoes what the other does.

The Fundamental Theorem of Calculus is unquestionably the most important theorem in calculus and, indeed, it ranks as one of the great accomplishments of the human mind. Before it was discovered, from the time of Eudoxus and Archimedes to the time of Galileo and Fermat, problems of finding areas, volumes, and lengths of curves were so difficult that only a genius could meet the challenge. But now, armed with the systematic method that Newton and Leibniz fashioned out of the Fundamental Theorem, we will see in the chapters to come that these challenging problems are accessible to all of us.

5.3 EXERCISES

1. Explain exactly what is meant by the statement that "differentiation and integration are inverse processes."

2. Let $g(x) = \int_0^x f(t)\,dt$, where f is the function whose graph is shown.

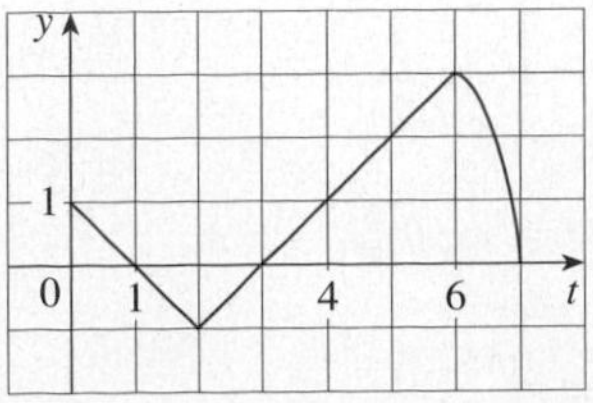

(a) Evaluate $g(x)$ for $x = 0, 1, 2, 3, 4, 5$, and 6.

(b) Estimate $g(7)$.

(c) Where does g have a maximum value? Where does it have a minimum value?

(d) Sketch a rough graph of g.

3. Let $g(x) = \int_0^x f(t)\,dt$, where f is the function whose graph is shown.

(a) Evaluate $g(0)$, $g(1)$, $g(2)$, $g(3)$, and $g(6)$.

(b) On what interval is g increasing?

(c) Where does g have a maximum value?
(d) Sketch a rough graph of g.

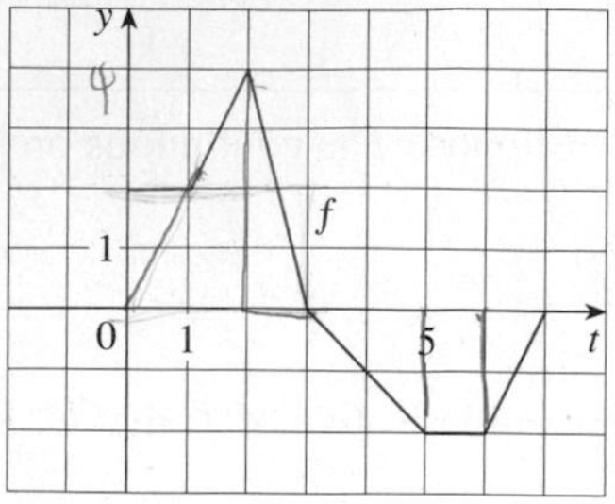

4. Let $g(x) = \int_{-3}^{x} f(t)\,dt$, where f is the function whose graph is shown.
(a) Evaluate $g(-3)$ and $g(3)$.
(b) Estimate $g(-2)$, $g(-1)$, and $g(0)$.
(c) On what interval is g increasing?
(d) Where does g have a maximum value?
(e) Sketch a rough graph of g.
(f) Use the graph in part (e) to sketch the graph of $g'(x)$. Compare with the graph of f.

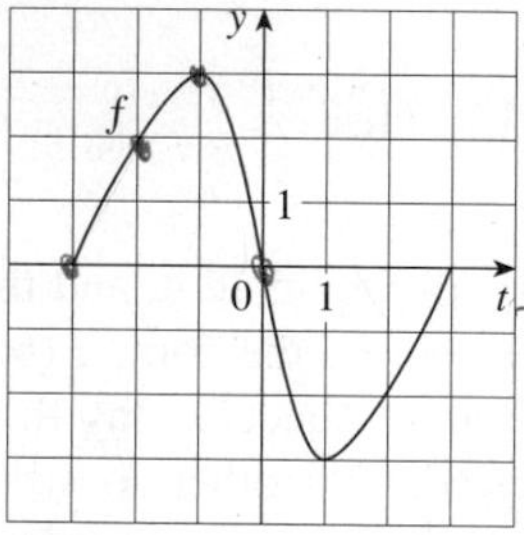

5–6 Sketch the area represented by $g(x)$. Then find $g'(x)$ in two ways: (a) by using Part 1 of the Fundamental Theorem and (b) by evaluating the integral using Part 2 and then differentiating.

5. $g(x) = \int_{1}^{x} t^2\,dt$

6. $g(x) = \int_{0}^{x} (1 + \sqrt{t})\,dt$

7–18 Use Part 1 of the Fundamental Theorem of Calculus to find the derivative of the function.

7. $g(x) = \int_{1}^{x} \frac{1}{t^3 + 1}\,dt$

8. $g(x) = \int_{3}^{x} e^{t^2 - t}\,dt$

9. $g(y) = \int_{2}^{y} t^2 \sin t\,dt$

10. $g(r) = \int_{0}^{r} \sqrt{x^2 + 4}\,dx$

11. $F(x) = \int_{x}^{\pi} \sqrt{1 + \sec t}\,dt$

$\left[\textit{Hint: } \int_{x}^{\pi} \sqrt{1 + \sec t}\,dt = -\int_{\pi}^{x} \sqrt{1 + \sec t}\,dt\right]$

12. $G(x) = \int_{x}^{1} \cos \sqrt{t}\,dt$

13. $h(x) = \int_{2}^{1/x} \arctan t\,dt$

14. $h(x) = \int_{0}^{x^2} \sqrt{1 + r^3}\,dr$

15. $y = \int_{0}^{\tan x} \sqrt{t + \sqrt{t}}\,dt$

16. $y = \int_{1}^{\cos x} (1 + v^2)^{10}\,dv$

17. $y = \int_{1-3x}^{1} \frac{u^3}{1 + u^2}\,du$

18. $y = \int_{e^x}^{0} \sin^3 t\,dt$

19–42 Evaluate the integral.

19. $\int_{-1}^{2} (x^3 - 2x)\,dx$

20. $\int_{-2}^{5} 6\,dx$

21. $\int_{1}^{4} (5 - 2t + 3t^2)\,dt$

22. $\int_{0}^{1} \left(1 + \frac{1}{2}u^4 - \frac{2}{5}u^9\right)du$

23. $\int_{0}^{1} x^{4/5}\,dx$

24. $\int_{1}^{8} \sqrt[3]{x}\,dx$

25. $\int_{1}^{2} \frac{3}{t^4}\,dt$

26. $\int_{\pi}^{2\pi} \cos\theta\,d\theta$

27. $\int_{0}^{2} x(2 + x^5)\,dx$

28. $\int_{0}^{1} (3 + x\sqrt{x})\,dx$

29. $\int_{1}^{9} \frac{x - 1}{\sqrt{x}}\,dx$

30. $\int_{0}^{2} (y - 1)(2y + 1)\,dy$

31. $\int_{0}^{\pi/4} \sec^2 t\,dt$

32. $\int_{0}^{\pi/4} \sec\theta \tan\theta\,d\theta$

33. $\int_{1}^{2} (1 + 2y)^2\,dy$

34. $\int_{0}^{1} \cosh t\,dt$

35. $\int_{1}^{9} \frac{1}{2x}\,dx$

36. $\int_{0}^{1} 10^x\,dx$

37. $\int_{1/2}^{\sqrt{3}/2} \frac{6}{\sqrt{1 - t^2}}\,dt$

38. $\int_{0}^{1} \frac{4}{t^2 + 1}\,dt$

39. $\int_{-1}^{1} e^{u+1}\,du$

40. $\int_{1}^{2} \frac{4 + u^2}{u^3}\,du$

41. $\int_{0}^{\pi} f(x)\,dx$ where $f(x) = \begin{cases} \sin x & \text{if } 0 \leqslant x < \pi/2 \\ \cos x & \text{if } \pi/2 \leqslant x \leqslant \pi \end{cases}$

42. $\int_{-2}^{2} f(x)\,dx$ where $f(x) = \begin{cases} 2 & \text{if } -2 \leqslant x \leqslant 0 \\ 4 - x^2 & \text{if } 0 < x \leqslant 2 \end{cases}$

43–46 What is wrong with the equation?

43. $\int_{-2}^{1} x^{-4}\,dx = \left.\frac{x^{-3}}{-3}\right]_{-2}^{1} = -\frac{3}{8}$

44. $\int_{-1}^{2} \frac{4}{x^3}\,dx = -\left.\frac{2}{x^2}\right]_{-1}^{2} = \frac{3}{2}$

45. $\int_{\pi/3}^{\pi} \sec\theta \tan\theta\,d\theta = \sec\theta\Big]_{\pi/3}^{\pi} = -3$

46. $\int_{0}^{\pi} \sec^2 x\,dx = \tan x\Big]_{0}^{\pi} = 0$

47–50 Use a graph to give a rough estimate of the area of the region that lies beneath the given curve. Then find the exact area.

47. $y = \sqrt[3]{x}, \quad 0 \leqslant x \leqslant 27$

48. $y = x^{-4}, \quad 1 \leqslant x \leqslant 6$

49. $y = \sin x, \quad 0 \leqslant x \leqslant \pi$

50. $y = \sec^2 x, \quad 0 \leqslant x \leqslant \pi/3$

51–52 Evaluate the integral and interpret it as a difference of areas. Illustrate with a sketch.

51. $\int_{-1}^{2} x^3\,dx$

52. $\int_{\pi/4}^{5\pi/2} \sin x\,dx$

53–56 Find the derivative of the function.

53. $g(x) = \int_{2x}^{3x} \frac{u^2 - 1}{u^2 + 1}\,du$

$\left[\textit{Hint: } \int_{2x}^{3x} f(u)\,du = \int_{2x}^{0} f(u)\,du + \int_{0}^{3x} f(u)\,du\right]$

54. $g(x) = \int_{\tan x}^{x^2} \frac{1}{\sqrt{2 + t^4}}\,dt$

55. $y = \int_{\sqrt{x}}^{x^3} \sqrt{t}\,\sin t\,dt$

56. $y = \int_{\cos x}^{5x} \cos(u^2)\,du$

---

57. If $F(x) = \int_1^x f(t)\,dt$, where $f(t) = \int_1^{t^2} \frac{\sqrt{1 + u^4}}{u}\,du$, find $F''(2)$.

58. Find the interval on which the curve $y = \int_0^x \frac{1}{1 + t + t^2}\,dt$ is concave upward.

59. If $f(1) = 12$, f' is continuous, and $\int_1^4 f'(x)\,dx = 17$, what is the value of $f(4)$?

60. The **error function**

$$\text{erf}(x) = \frac{2}{\sqrt{\pi}} \int_0^x e^{-t^2}\,dt$$

is used in probability, statistics, and engineering.

(a) Show that $\int_a^b e^{-t^2}\,dt = \frac{1}{2}\sqrt{\pi}\,[\text{erf}(b) - \text{erf}(a)]$.

(b) Show that the function $y = e^{x^2}\text{erf}(x)$ satisfies the differential equation $y' = 2xy + 2/\sqrt{\pi}$.

61. The Fresnel function S was defined in Example 3 and graphed in Figures 7 and 8.

(a) At what values of x does this function have local maximum values?

(b) On what intervals is the function concave upward?

CAS (c) Use a graph to solve the following equation correct to two decimal places:

$$\int_0^x \sin(\pi t^2/2)\,dt = 0.2$$

CAS **62.** The **sine integral function**

$$\text{Si}(x) = \int_0^x \frac{\sin t}{t}\,dt$$

is important in electrical engineering. [The integrand $f(t) = (\sin t)/t$ is not defined when $t = 0$, but we know that its limit is 1 when $t \to 0$. So we define $f(0) = 1$ and this makes f a continuous function everywhere.]

(a) Draw the graph of Si.

(b) At what values of x does this function have local maximum values?

(c) Find the coordinates of the first inflection point to the right of the origin.

(d) Does this function have horizontal asymptotes?

(e) Solve the following equation correct to one decimal place:

$$\int_0^x \frac{\sin t}{t}\,dt = 1$$

63–64 Let $g(x) = \int_0^x f(t)\,dt$, where f is the function whose graph is shown.

(a) At what values of x do the local maximum and minimum values of g occur?

(b) Where does g attain its absolute maximum value?

(c) On what intervals is g concave downward?

(d) Sketch the graph of g.

63.

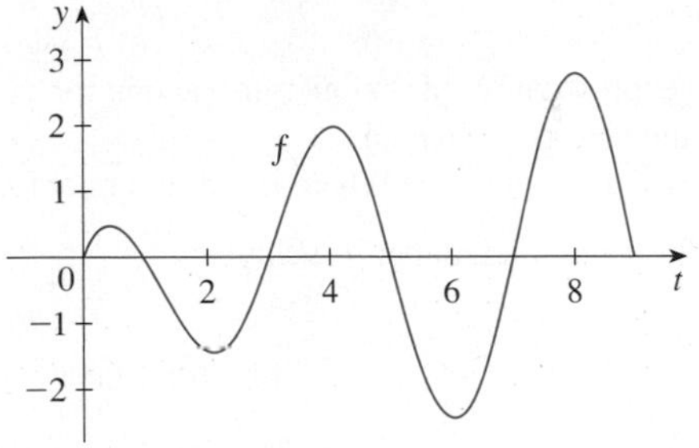

64.

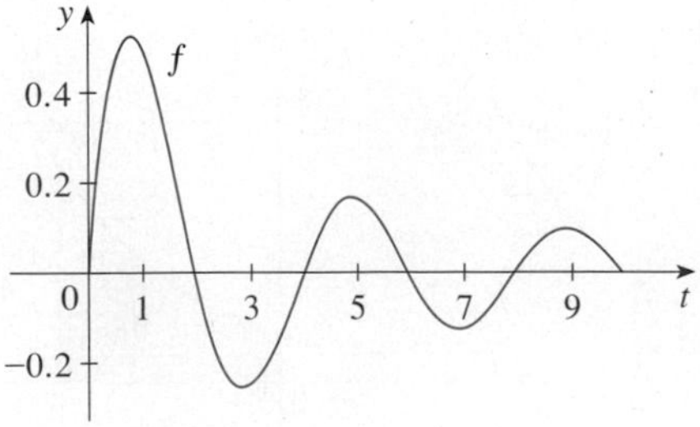

65–66 Evaluate the limit by first recognizing the sum as a Riemann sum for a function defined on [0, 1].

65. $\lim_{n \to \infty} \sum_{i=1}^{n} \frac{i^3}{n^4}$

66. $\lim_{n \to \infty} \frac{1}{n}\left(\sqrt{\frac{1}{n}} + \sqrt{\frac{2}{n}} + \sqrt{\frac{3}{n}} + \cdots + \sqrt{\frac{n}{n}}\right)$

67. Justify (3) for the case $h < 0$.

68. If f is continuous and g and h are differentiable functions, find a formula for

$$\frac{d}{dx}\int_{g(x)}^{h(x)} f(t)\,dt$$

69. (a) Show that $1 \leqslant \sqrt{1 + x^3} \leqslant 1 + x^3$ for $x \geqslant 0$.
(b) Show that $1 \leqslant \int_0^1 \sqrt{1 + x^3}\,dx \leqslant 1.25$.

70. (a) Show that $\cos(x^2) \geqslant \cos x$ for $0 \leqslant x \leqslant 1$.
(b) Deduce that $\int_0^{\pi/6} \cos(x^2)\,dx \geqslant \frac{1}{2}$.

71. Show that

$$0 \leqslant \int_5^{10} \frac{x^2}{x^4 + x^2 + 1}\,dx \leqslant 0.1$$

by comparing the integrand to a simpler function.

72. Let

$$f(x) = \begin{cases} 0 & \text{if } x < 0 \\ x & \text{if } 0 \leqslant x \leqslant 1 \\ 2 - x & \text{if } 1 < x \leqslant 2 \\ 0 & \text{if } x > 2 \end{cases}$$

and $$g(x) = \int_0^x f(t)\,dt$$

(a) Find an expression for $g(x)$ similar to the one for $f(x)$.
(b) Sketch the graphs of f and g.
(c) Where is f differentiable? Where is g differentiable?

73. Find a function f and a number a such that

$$6 + \int_a^x \frac{f(t)}{t^2}\,dt = 2\sqrt{x} \qquad \text{for all } x > 0$$

74. The area labeled B is three times the area labeled A. Express b in terms of a.

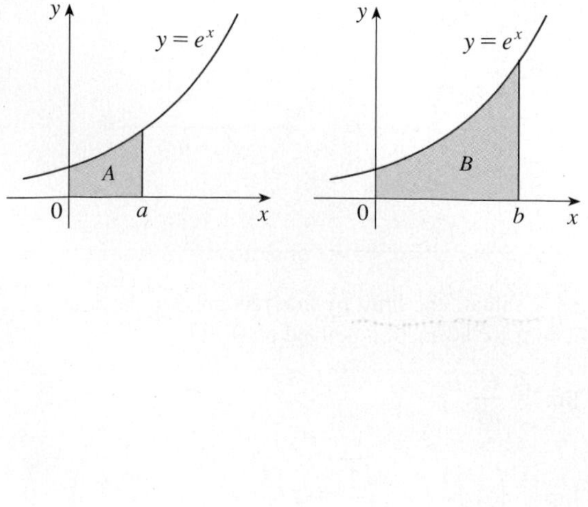

75. A manufacturing company owns a major piece of equipment that depreciates at the (continuous) rate $f = f(t)$, where t is the time measured in months since its last overhaul. Because a fixed cost A is incurred each time the machine is overhauled, the company wants to determine the optimal time T (in months) between overhauls.
(a) Explain why $\int_0^t f(s)\,ds$ represents the loss in value of the machine over the period of time t since the last overhaul.
(b) Let $C = C(t)$ be given by

$$C(t) = \frac{1}{t}\left[A + \int_0^t f(s)\,ds\right]$$

What does C represent and why would the company want to minimize C?
(c) Show that C has a minimum value at the numbers $t = T$ where $C(T) = f(T)$.

76. A high-tech company purchases a new computing system whose initial value is V. The system will depreciate at the rate $f = f(t)$ and will accumulate maintenance costs at the rate $g = g(t)$, where t is the time measured in months. The company wants to determine the optimal time to replace the system.
(a) Let

$$C(t) = \frac{1}{t}\int_0^t [f(s) + g(s)]\,ds$$

Show that the critical numbers of C occur at the numbers t where $C(t) = f(t) + g(t)$.
(b) Suppose that

$$f(t) = \begin{cases} \dfrac{V}{15} - \dfrac{V}{450}t & \text{if } 0 < t \leqslant 30 \\ 0 & \text{if } t > 30 \end{cases}$$

and $$g(t) = \frac{Vt^2}{12{,}900} \qquad t > 0$$

Determine the length of time T for the total depreciation $D(t) = \int_0^t f(s)\,ds$ to equal the initial value V.
(c) Determine the absolute minimum of C on $(0, T]$.
(d) Sketch the graphs of C and $f + g$ in the same coordinate system, and verify the result in part (a) in this case.

5.4 INDEFINITE INTEGRALS AND THE NET CHANGE THEOREM

We saw in Section 5.3 that the second part of the Fundamental Theorem of Calculus provides a very powerful method for evaluating the definite integral of a function, assuming that we can find an antiderivative of the function. In this section we introduce a notation for antiderivatives, review the formulas for antiderivatives, and use them to evaluate definite integrals. We also reformulate FTC2 in a way that makes it easier to apply to science and engineering problems.

INDEFINITE INTEGRALS

Both parts of the Fundamental Theorem establish connections between antiderivatives and definite integrals. Part 1 says that if f is continuous, then $\int_a^x f(t)\,dt$ is an antiderivative of f. Part 2 says that $\int_a^b f(x)\,dx$ can be found by evaluating $F(b) - F(a)$, where F is an antiderivative of f.

We need a convenient notation for antiderivatives that makes them easy to work with. Because of the relation given by the Fundamental Theorem between antiderivatives and integrals, the notation $\int f(x)\,dx$ is traditionally used for an antiderivative of f and is called an **indefinite integral**. Thus

$$\int f(x)\,dx = F(x) \qquad \text{means} \qquad F'(x) = f(x)$$

For example, we can write

$$\int x^2\,dx = \frac{x^3}{3} + C \qquad \text{because} \qquad \frac{d}{dx}\left(\frac{x^3}{3} + C\right) = x^2$$

So we can regard an indefinite integral as representing an entire *family* of functions (one antiderivative for each value of the constant C).

⊘ You should distinguish carefully between definite and indefinite integrals. A definite integral $\int_a^b f(x)\,dx$ is a *number*, whereas an indefinite integral $\int f(x)\,dx$ is a *function* (or family of functions). The connection between them is given by Part 2 of the Fundamental Theorem. If f is continuous on $[a, b]$, then

$$\int_a^b f(x)\,dx = \int f(x)\,dx\Big]_a^b$$

The effectiveness of the Fundamental Theorem depends on having a supply of antiderivatives of functions. We therefore restate the Table of Antidifferentiation Formulas from Section 4.9, together with a few others, in the notation of indefinite integrals. Any formula can be verified by differentiating the function on the right side and obtaining the integrand. For instance

$$\int \sec^2 x\,dx = \tan x + C \qquad \text{because} \qquad \frac{d}{dx}(\tan x + C) = \sec^2 x$$

1 TABLE OF INDEFINITE INTEGRALS

$$\int cf(x)\,dx = c\int f(x)\,dx \qquad \int [f(x) + g(x)]\,dx = \int f(x)\,dx + \int g(x)\,dx$$

$$\int k\,dx = kx + C$$

$$\int x^n\,dx = \frac{x^{n+1}}{n+1} + C \quad (n \neq -1) \qquad \int \frac{1}{x}\,dx = \ln|x| + C$$

$$\int e^x\,dx = e^x + C \qquad \int a^x\,dx = \frac{a^x}{\ln a} + C$$

$$\int \sin x\,dx = -\cos x + C \qquad \int \cos x\,dx = \sin x + C$$

$$\int \sec^2 x\,dx = \tan x + C \qquad \int \csc^2 x\,dx = -\cot x + C$$

$$\int \sec x \tan x\,dx = \sec x + C \qquad \int \csc x \cot x\,dx = -\csc x + C$$

$$\int \frac{1}{x^2+1}\,dx = \tan^{-1}x + C \qquad \int \frac{1}{\sqrt{1-x^2}}\,dx = \sin^{-1}x + C$$

$$\int \sinh x\,dx = \cosh x + C \qquad \int \cosh x\,dx = \sinh x + C$$

Recall from Theorem 4.9.1 that the most general antiderivative *on a given interval* is obtained by adding a constant to a particular antiderivative. **We adopt the convention that when a formula for a general indefinite integral is given, it is valid only on an interval.** Thus we write

$$\int \frac{1}{x^2}\,dx = -\frac{1}{x} + C$$

with the understanding that it is valid on the interval $(0, \infty)$ or on the interval $(-\infty, 0)$. This is true despite the fact that the general antiderivative of the function $f(x) = 1/x^2$, $x \neq 0$, is

$$F(x) = \begin{cases} -\dfrac{1}{x} + C_1 & \text{if } x < 0 \\[2mm] -\dfrac{1}{x} + C_2 & \text{if } x > 0 \end{cases}$$

■ The indefinite integral in Example 1 is graphed in Figure 1 for several values of C. The value of C is the y-intercept.

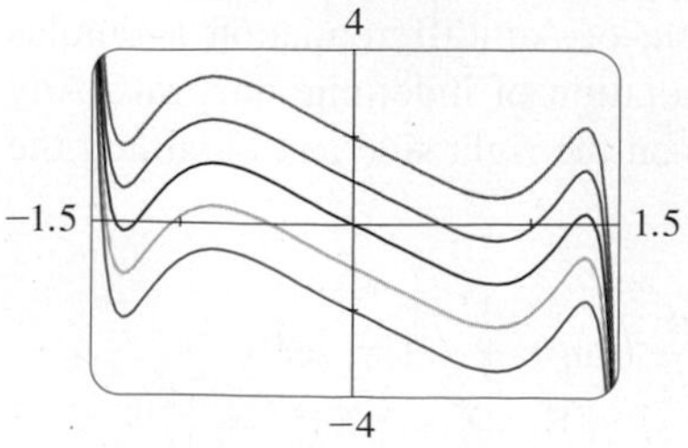

FIGURE 1

EXAMPLE 1 Find the general indefinite integral

$$\int (10x^4 - 2\sec^2 x)\,dx$$

SOLUTION Using our convention and Table 1, we have

$$\int (10x^4 - 2\sec^2 x)\,dx = 10\int x^4\,dx - 2\int \sec^2 x\,dx$$

$$= 10\frac{x^5}{5} - 2\tan x + C = 2x^5 - 2\tan x + C$$

You should check this answer by differentiating it. □

EXAMPLE 2 Evaluate $\displaystyle\int \frac{\cos\theta}{\sin^2\theta}\,d\theta$.

SOLUTION This indefinite integral isn't immediately apparent in Table 1, so we use trigonometric identities to rewrite the function before integrating:

$$\int \frac{\cos\theta}{\sin^2\theta}\,d\theta = \int \left(\frac{1}{\sin\theta}\right)\left(\frac{\cos\theta}{\sin\theta}\right) d\theta$$
$$= \int \csc\theta\cot\theta\,d\theta = -\csc\theta + C$$ □

EXAMPLE 3 Evaluate $\displaystyle\int_0^3 (x^3 - 6x)\,dx$.

SOLUTION Using FTC2 and Table 1, we have

$$\int_0^3 (x^3 - 6x)\,dx = \frac{x^4}{4} - 6\frac{x^2}{2}\bigg]_0^3$$
$$= \left(\tfrac{1}{4}\cdot 3^4 - 3\cdot 3^2\right) - \left(\tfrac{1}{4}\cdot 0^4 - 3\cdot 0^2\right)$$
$$= \tfrac{81}{4} - 27 - 0 + 0 = -6.75$$

Compare this calculation with Example 2(b) in Section 5.2. □

■ Figure 2 shows the graph of the integrand in Example 4. We know from Section 5.2 that the value of the integral can be interpreted as the sum of the areas labeled with a plus sign minus the area labeled with a minus sign.

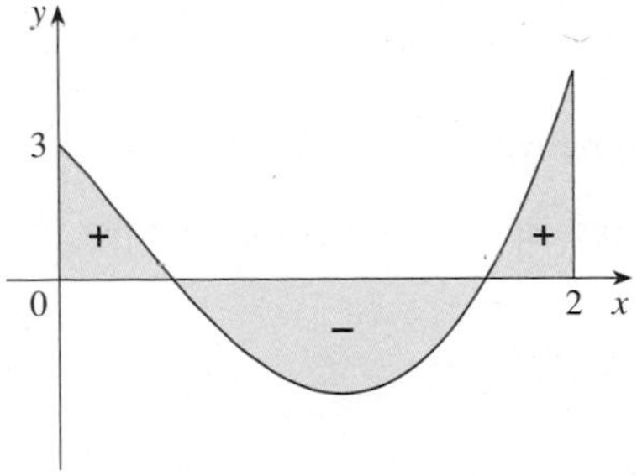

FIGURE 2

EXAMPLE 4 Find $\displaystyle\int_0^2 \left(2x^3 - 6x + \frac{3}{x^2+1}\right) dx$ and interpret the result in terms of areas.

SOLUTION The Fundamental Theorem gives

$$\int_0^2 \left(2x^3 - 6x + \frac{3}{x^2+1}\right) dx = 2\frac{x^4}{4} - 6\frac{x^2}{2} + 3\tan^{-1}x\bigg]_0^2$$
$$= \tfrac{1}{2}x^4 - 3x^2 + 3\tan^{-1}x\Big]_0^2$$
$$= \tfrac{1}{2}(2^4) - 3(2^2) + 3\tan^{-1}2 - 0$$
$$= -4 + 3\tan^{-1}2$$

This is the exact value of the integral. If a decimal approximation is desired, we can use a calculator to approximate $\tan^{-1} 2$. Doing so, we get

$$\int_0^2 \left(2x^3 - 6x + \frac{3}{x^2+1}\right) dx \approx -0.67855$$ □

EXAMPLE 5 Evaluate $\displaystyle\int_1^9 \frac{2t^2 + t^2\sqrt{t} - 1}{t^2}\,dt$.

SOLUTION First we need to write the integrand in a simpler form by carrying out the division:

$$\int_1^9 \frac{2t^2 + t^2\sqrt{t} - 1}{t^2}\,dt = \int_1^9 (2 + t^{1/2} - t^{-2})\,dt$$
$$= 2t + \frac{t^{3/2}}{\frac{3}{2}} - \frac{t^{-1}}{-1}\bigg]_1^9 = 2t + \tfrac{2}{3}t^{3/2} + \frac{1}{t}\bigg]_1^9$$
$$= \left(2\cdot 9 + \tfrac{2}{3}\cdot 9^{3/2} + \tfrac{1}{9}\right) - \left(2\cdot 1 + \tfrac{2}{3}\cdot 1^{3/2} + \tfrac{1}{1}\right)$$
$$= 18 + 18 + \tfrac{1}{9} - 2 - \tfrac{2}{3} - 1 = 32\tfrac{4}{9}$$ □

APPLICATIONS

Part 2 of the Fundamental Theorem says that if f is continuous on $[a, b]$, then

$$\int_a^b f(x)\,dx = F(b) - F(a)$$

where F is any antiderivative of f. This means that $F' = f$, so the equation can be rewritten as

$$\int_a^b F'(x)\,dx = F(b) - F(a)$$

We know that $F'(x)$ represents the rate of change of $y = F(x)$ with respect to x and $F(b) - F(a)$ is the change in y when x changes from a to b. [Note that y could, for instance, increase, then decrease, then increase again. Although y might change in both directions, $F(b) - F(a)$ represents the *net* change in y.] So we can reformulate FTC2 in words as follows.

THE NET CHANGE THEOREM The integral of a rate of change is the net change:

$$\int_a^b F'(x)\,dx = F(b) - F(a)$$

This principle can be applied to all of the rates of change in the natural and social sciences that we discussed in Section 3.7. Here are a few instances of this idea:

- If $V(t)$ is the volume of water in a reservoir at time t, then its derivative $V'(t)$ is the rate at which water flows into the reservoir at time t. So

$$\int_{t_1}^{t_2} V'(t)\,dt = V(t_2) - V(t_1)$$

  is the change in the amount of water in the reservoir between time t_1 and time t_2.

- If $[\mathrm{C}](t)$ is the concentration of the product of a chemical reaction at time t, then the rate of reaction is the derivative $d[\mathrm{C}]/dt$. So

$$\int_{t_1}^{t_2} \frac{d[\mathrm{C}]}{dt}\,dt = [\mathrm{C}](t_2) - [\mathrm{C}](t_1)$$

  is the change in the concentration of C from time t_1 to time t_2.

- If the mass of a rod measured from the left end to a point x is $m(x)$, then the linear density is $\rho(x) = m'(x)$. So

$$\int_a^b \rho(x)\,dx = m(b) - m(a)$$

  is the mass of the segment of the rod that lies between $x = a$ and $x = b$.

- If the rate of growth of a population is dn/dt, then

$$\int_{t_1}^{t_2} \frac{dn}{dt}\,dt = n(t_2) - n(t_1)$$

  is the net change in population during the time period from t_1 to t_2. (The population increases when births happen and decreases when deaths occur. The net change takes into account both births and deaths.)

- If $C(x)$ is the cost of producing x units of a commodity, then the marginal cost is the derivative $C'(x)$. So

$$\int_{x_1}^{x_2} C'(x)\,dx = C(x_2) - C(x_1)$$

is the increase in cost when production is increased from x_1 units to x_2 units.

- If an object moves along a straight line with position function $s(t)$, then its velocity is $v(t) = s'(t)$, so

2 $$\int_{t_1}^{t_2} v(t)\,dt = s(t_2) - s(t_1)$$

is the net change of position, or *displacement,* of the particle during the time period from t_1 to t_2. In Section 5.1 we guessed that this was true for the case where the object moves in the positive direction, but now we have proved that it is always true.

- If we want to calculate the distance the object travels during that time interval, we have to consider the intervals when $v(t) \geqslant 0$ (the particle moves to the right) and also the intervals when $v(t) \leqslant 0$ (the particle moves to the left). In both cases the distance is computed by integrating $|v(t)|$, the speed. Therefore

3 $$\int_{t_1}^{t_2} |v(t)|\,dt = \text{total distance traveled}$$

Figure 3 shows how both displacement and distance traveled can be interpreted in terms of areas under a velocity curve.

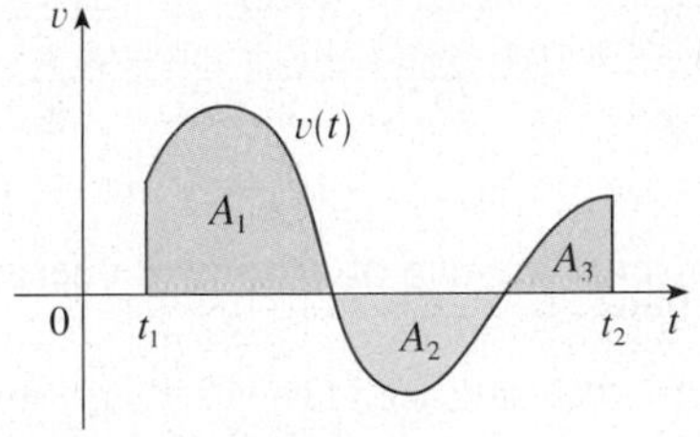

$$\text{displacement} = \int_{t_1}^{t_2} v(t)\,dt = A_1 - A_2 + A_3$$

$$\text{distance} = \int_{t_1}^{t_2} |v(t)|\,dt = A_1 + A_2 + A_3$$

FIGURE 3

- The acceleration of the object is $a(t) = v'(t)$, so

$$\int_{t_1}^{t_2} a(t)\,dt = v(t_2) - v(t_1)$$

is the change in velocity from time t_1 to time t_2.

V EXAMPLE 6 A particle moves along a line so that its velocity at time t is $v(t) = t^2 - t - 6$ (measured in meters per second).
(a) Find the displacement of the particle during the time period $1 \leqslant t \leqslant 4$.
(b) Find the distance traveled during this time period.

SOLUTION
(a) By Equation 2, the displacement is

$$s(4) - s(1) = \int_1^4 v(t)\,dt = \int_1^4 (t^2 - t - 6)\,dt$$

$$= \left[\frac{t^3}{3} - \frac{t^2}{2} - 6t\right]_1^4 = -\frac{9}{2}$$

This means that the particle moved 4.5 m toward the left.

(b) Note that $v(t) = t^2 - t - 6 = (t - 3)(t + 2)$ and so $v(t) \leq 0$ on the interval $[1, 3]$ and $v(t) \geq 0$ on $[3, 4]$. Thus, from Equation 3, the distance traveled is

■ To integrate the absolute value of $v(t)$, we use Property 5 of integrals from Section 5.2 to split the integral into two parts, one where $v(t) \leq 0$ and one where $v(t) \geq 0$.

$$\begin{aligned}\int_1^4 |v(t)|\,dt &= \int_1^3 [-v(t)]\,dt + \int_3^4 v(t)\,dt \\ &= \int_1^3 (-t^2 + t + 6)\,dt + \int_3^4 (t^2 - t - 6)\,dt \\ &= \left[-\frac{t^3}{3} + \frac{t^2}{2} + 6t\right]_1^3 + \left[\frac{t^3}{3} - \frac{t^2}{2} - 6t\right]_3^4 \\ &= \frac{61}{6} \approx 10.17 \text{ m}\end{aligned}$$

□

EXAMPLE 7 Figure 4 shows the power consumption in the city of San Francisco for a day in September (P is measured in megawatts; t is measured in hours starting at midnight). Estimate the energy used on that day.

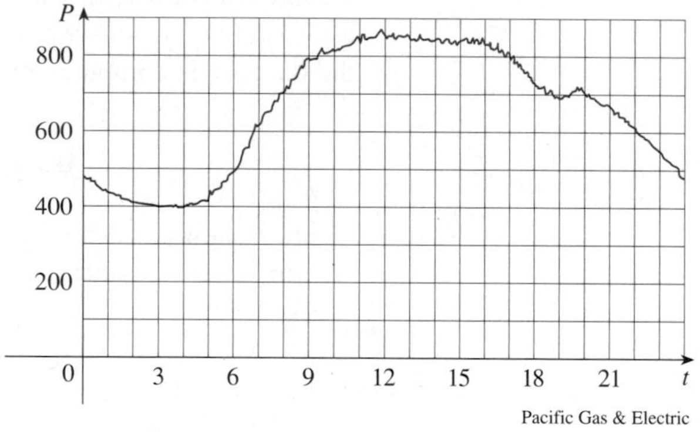

FIGURE 4

SOLUTION Power is the rate of change of energy: $P(t) = E'(t)$. So, by the Net Change Theorem,

$$\int_0^{24} P(t)\,dt = \int_0^{24} E'(t)\,dt = E(24) - E(0)$$

is the total amount of energy used that day. We approximate the value of the integral using the Midpoint Rule with 12 subintervals and $\Delta t = 2$:

$$\begin{aligned}\int_0^{24} P(t)\,dt &\approx [P(1) + P(3) + P(5) + \cdots + P(21) + P(23)]\,\Delta t \\ &\approx (440 + 400 + 420 + 620 + 790 + 840 + 850 \\ &\qquad + 840 + 810 + 690 + 670 + 550)(2) \\ &= 15{,}840\end{aligned}$$

The energy used was approximately 15,840 megawatt-hours. □

■ A note on units

How did we know what units to use for energy in Example 7? The integral $\int_0^{24} P(t)\,dt$ is defined as the limit of sums of terms of the form $P(t_i^*)\,\Delta t$. Now $P(t_i^*)$ is measured in megawatts and Δt is measured in hours, so their product is measured in megawatt-hours. The same is true of the limit. In general, the unit of measurement for $\int_a^b f(x)\,dx$ is the product of the unit for $f(x)$ and the unit for x.

5.4 EXERCISES

1–4 Verify by differentiation that the formula is correct.

1. $\displaystyle\int \frac{x}{\sqrt{x^2+1}}\,dx = \sqrt{x^2+1} + C$

2. $\displaystyle\int x\cos x\,dx = x\sin x + \cos x + C$

3. $\displaystyle\int \cos^3 x\,dx = \sin x - \tfrac{1}{3}\sin^3 x + C$

4. $\displaystyle\int \frac{x}{\sqrt{a+bx}}\,dx = \frac{2}{3b^2}(bx-2a)\sqrt{a+bx} + C$

5–18 Find the general indefinite integral.

5. $\displaystyle\int (x^2 + x^{-2})\,dx$

6. $\displaystyle\int \left(\sqrt{x^3} + \sqrt[3]{x^2}\right) dx$

7. $\displaystyle\int \left(x^4 - \tfrac{1}{2}x^3 + \tfrac{1}{4}x - 2\right) dx$

8. $\displaystyle\int (y^3 + 1.8y^2 - 2.4y)\,dy$

9. $\displaystyle\int (1-t)(2+t^2)\,dt$

10. $\displaystyle\int v(v^2+2)^2\,dv$

11. $\displaystyle\int \frac{x^3 - 2\sqrt{x}}{x}\,dx$

12. $\displaystyle\int \left(x^2 + 1 + \frac{1}{x^2+1}\right) dx$

13. $\displaystyle\int (\sin x + \sinh x)\,dx$

14. $\displaystyle\int (\csc^2 t - 2e^t)\,dt$

15. $\displaystyle\int (\theta - \csc\theta\cot\theta)\,d\theta$

16. $\displaystyle\int \sec t(\sec t + \tan t)\,dt$

17. $\displaystyle\int (1 + \tan^2\alpha)\,d\alpha$

18. $\displaystyle\int \frac{\sin 2x}{\sin x}\,dx$

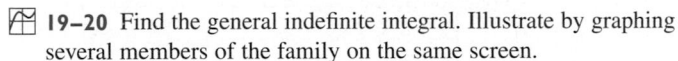

19–20 Find the general indefinite integral. Illustrate by graphing several members of the family on the same screen.

19. $\displaystyle\int \left(\cos x + \tfrac{1}{2}x\right) dx$

20. $\displaystyle\int (e^x - 2x^2)\,dx$

21–44 Evaluate the integral.

21. $\displaystyle\int_0^2 (6x^2 - 4x + 5)\,dx$

22. $\displaystyle\int_1^3 (1 + 2x - 4x^3)\,dx$

23. $\displaystyle\int_{-1}^0 (2x - e^x)\,dx$

24. $\displaystyle\int_{-2}^0 (u^5 - u^3 + u^2)\,du$

25. $\displaystyle\int_{-2}^2 (3u+1)^2\,du$

26. $\displaystyle\int_0^4 (2v+5)(3v-1)\,dv$

27. $\displaystyle\int_1^4 \sqrt{t}\,(1+t)\,dt$

28. $\displaystyle\int_0^9 \sqrt{2t}\,dt$

29. $\displaystyle\int_{-2}^{-1} \left(4y^3 + \frac{2}{y^3}\right) dy$

30. $\displaystyle\int_1^2 \frac{y + 5y^7}{y^3}\,dy$

31. $\displaystyle\int_0^1 x\left(\sqrt[3]{x} + \sqrt[4]{x}\right) dx$

32. $\displaystyle\int_0^5 (2e^x + 4\cos x)\,dx$

33. $\displaystyle\int_1^4 \sqrt{\frac{5}{x}}\,dx$

34. $\displaystyle\int_1^9 \frac{3x-2}{\sqrt{x}}\,dx$

35. $\displaystyle\int_0^{\pi} (4\sin\theta - 3\cos\theta)\,d\theta$

36. $\displaystyle\int_{\pi/4}^{\pi/3} \sec\theta\tan\theta\,d\theta$

37. $\displaystyle\int_0^{\pi/4} \frac{1+\cos^2\theta}{\cos^2\theta}\,d\theta$

38. $\displaystyle\int_0^{\pi/3} \frac{\sin\theta + \sin\theta\tan^2\theta}{\sec^2\theta}\,d\theta$

39. $\displaystyle\int_1^{64} \frac{1+\sqrt[3]{x}}{\sqrt{x}}\,dx$

40. $\displaystyle\int_{-10}^{10} \frac{2e^x}{\sinh x + \cosh x}\,dx$

41. $\displaystyle\int_0^{1/\sqrt{3}} \frac{t^2-1}{t^4-1}\,dt$

42. $\displaystyle\int_1^2 \frac{(x-1)^3}{x^2}\,dx$

43. $\displaystyle\int_{-1}^2 (x - 2|x|)\,dx$

44. $\displaystyle\int_0^{3\pi/2} |\sin x|\,dx$

45. Use a graph to estimate the x-intercepts of the curve $y = x + x^2 - x^4$. Then use this information to estimate the area of the region that lies under the curve and above the x-axis.

46. Repeat Exercise 45 for the curve $y = 2x + 3x^4 - 2x^6$.

47. The area of the region that lies to the right of the y-axis and to the left of the parabola $x = 2y - y^2$ (the shaded region in the figure) is given by the integral $\int_0^2 (2y - y^2)\,dy$. (Turn your head clockwise and think of the region as lying below the curve $x = 2y - y^2$ from $y = 0$ to $y = 2$.) Find the area of the region.

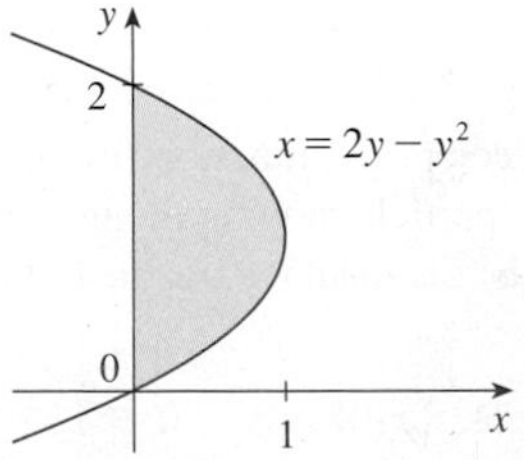

48. The boundaries of the shaded region are the y-axis, the line $y = 1$, and the curve $y = \sqrt[4]{x}$. Find the area of this region by writing x as a function of y and integrating with respect to y (as in Exercise 47).

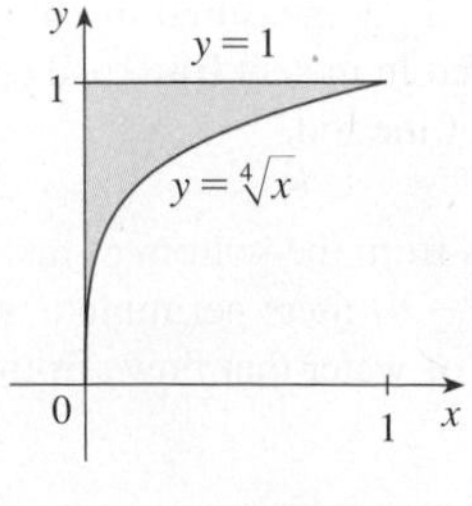

49. If $w'(t)$ is the rate of growth of a child in pounds per year, what does $\int_5^{10} w'(t)\,dt$ represent?

50. The current in a wire is defined as the derivative of the charge: $I(t) = Q'(t)$. (See Example 3 in Section 3.7.) What does $\int_a^b I(t)\,dt$ represent?

51. If oil leaks from a tank at a rate of $r(t)$ gallons per minute at time t, what does $\int_0^{120} r(t)\,dt$ represent?

52. A honeybee population starts with 100 bees and increases at a rate of $n'(t)$ bees per week. What does $100 + \int_0^{15} n'(t)\,dt$ represent?

53. In Section 4.7 we defined the marginal revenue function $R'(x)$ as the derivative of the revenue function $R(x)$, where x is the number of units sold. What does $\int_{1000}^{5000} R'(x)\,dx$ represent?

54. If $f(x)$ is the slope of a trail at a distance of x miles from the start of the trail, what does $\int_3^5 f(x)\,dx$ represent?

55. If x is measured in meters and $f(x)$ is measured in newtons, what are the units for $\int_0^{100} f(x)\,dx$?

56. If the units for x are feet and the units for $a(x)$ are pounds per foot, what are the units for da/dx? What units does $\int_2^8 a(x)\,dx$ have?

57–58 The velocity function (in meters per second) is given for a particle moving along a line. Find (a) the displacement and (b) the distance traveled by the particle during the given time interval.

57. $v(t) = 3t - 5, \quad 0 \leqslant t \leqslant 3$

58. $v(t) = t^2 - 2t - 8, \quad 1 \leqslant t \leqslant 6$

59–60 The acceleration function (in m/s^2) and the initial velocity are given for a particle moving along a line. Find (a) the velocity at time t and (b) the distance traveled during the given time interval.

59. $a(t) = t + 4, \quad v(0) = 5, \quad 0 \leqslant t \leqslant 10$

60. $a(t) = 2t + 3, \quad v(0) = -4, \quad 0 \leqslant t \leqslant 3$

61. The linear density of a rod of length 4 m is given by $\rho(x) = 9 + 2\sqrt{x}$ measured in kilograms per meter, where x is measured in meters from one end of the rod. Find the total mass of the rod.

62. Water flows from the bottom of a storage tank at a rate of $r(t) = 200 - 4t$ liters per minute, where $0 \leqslant t \leqslant 50$. Find the amount of water that flows from the tank during the first 10 minutes.

63. The velocity of a car was read from its speedometer at 10-second intervals and recorded in the table. Use the Midpoint Rule to estimate the distance traveled by the car.

t (s)	v (mi/h)	t (s)	v (mi/h)
0	0	60	56
10	38	70	53
20	52	80	50
30	58	90	47
40	55	100	45
50	51		

64. Suppose that a volcano is erupting and readings of the rate $r(t)$ at which solid materials are spewed into the atmosphere are given in the table. The time t is measured in seconds and the units for $r(t)$ are tonnes (metric tons) per second.

t	0	1	2	3	4	5	6
$r(t)$	2	10	24	36	46	54	60

(a) Give upper and lower estimates for the total quantity $Q(6)$ of erupted materials after 6 seconds.
(b) Use the Midpoint Rule to estimate $Q(6)$.

65. The marginal cost of manufacturing x yards of a certain fabric is $C'(x) = 3 - 0.01x + 0.000006x^2$ (in dollars per yard). Find the increase in cost if the production level is raised from 2000 yards to 4000 yards.

66. Water flows into and out of a storage tank. A graph of the rate of change $r(t)$ of the volume of water in the tank, in liters per day, is shown. If the amount of water in the tank at time $t = 0$ is 25,000 L, use the Midpoint Rule to estimate the amount of water four days later.

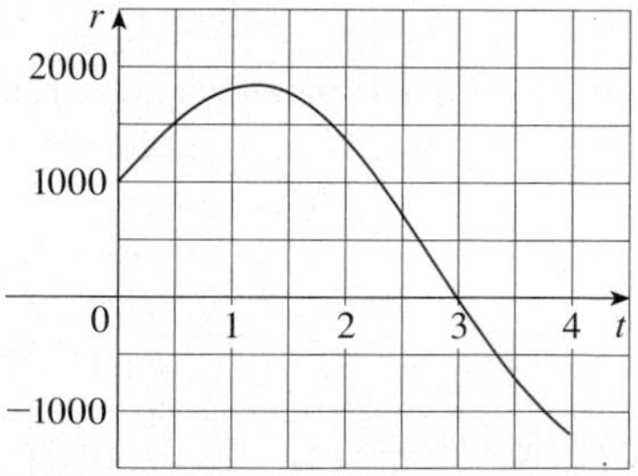

67. Economists use a cumulative distribution called a *Lorenz curve* to describe the distribution of income between households in a given country. Typically, a Lorenz curve is defined on $[0, 1]$ with endpoints $(0, 0)$ and $(1, 1)$, and is continuous, increasing, and concave upward. The points on this curve are determined by ranking all households by income and then computing the percentage of households whose income is less than or equal to a given percentage of the total income of the country. For example, the point $(a/100, b/100)$ is on the Lorenz curve if the bottom $a\%$ of the households receive less than or equal to $b\%$ of the total income. *Absolute equality* of income distribution would occur if the bottom $a\%$ of the households receive $a\%$ of the income, in which case the Lorenz curve would be the line $y = x$. The area between the Lorenz curve and the line $y = x$ measures how much the income distribution differs from absolute equality. The *coefficient of inequality* is the ratio of the area between the Lorenz curve and the line $y = x$ to the area under $y = x$.

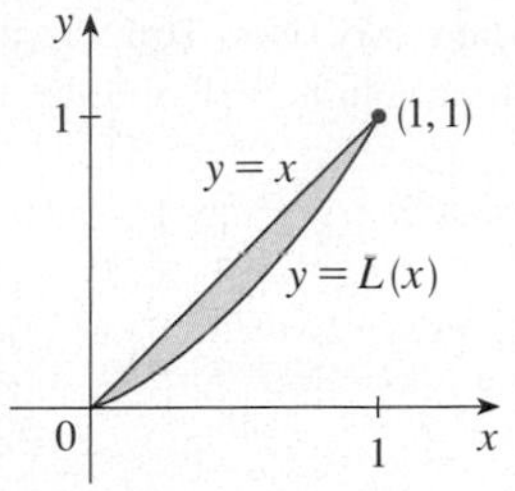

(a) Show that the coefficient of inequality is twice the area between the Lorenz curve and the line $y = x$, that is, show that

$$\text{coefficient of inequality} = 2\int_0^1 [x - L(x)]\,dx$$

(b) The income distribution for a certain country is represented by the Lorenz curve defined by the equation

$$L(x) = \tfrac{5}{12}x^2 + \tfrac{7}{12}x$$

What is the percentage of total income received by the bottom 50% of the households? Find the coefficient of inequality.

68. On May 7, 1992, the space shuttle *Endeavour* was launched on mission STS-49, the purpose of which was to install a new perigee kick motor in an Intelsat communications satellite. The table gives the velocity data for the shuttle between liftoff and the jettisoning of the solid rocket boosters.

Event	Time (s)	Velocity (ft/s)
Launch	0	0
Begin roll maneuver	10	185
End roll maneuver	15	319
Throttle to 89%	20	447
Throttle to 67%	32	742
Throttle to 104%	59	1325
Maximum dynamic pressure	62	1445
Solid rocket booster separation	125	4151

(a) Use a graphing calculator or computer to model these data by a third-degree polynomial.

(b) Use the model in part (a) to estimate the height reached by the *Endeavour*, 125 seconds after liftoff.

WRITING PROJECT

NEWTON, LEIBNIZ, AND THE INVENTION OF CALCULUS

We sometimes read that the inventors of calculus were Sir Isaac Newton (1642–1727) and Gottfried Wilhelm Leibniz (1646–1716). But we know that the basic ideas behind integration were investigated 2500 years ago by ancient Greeks such as Eudoxus and Archimedes, and methods for finding tangents were pioneered by Pierre Fermat (1601–1665), Isaac Barrow (1630–1677), and others. Barrow—who taught at Cambridge and was a major influence on Newton—was the first to understand the inverse relationship between differentiation and integration. What Newton and Leibniz did was to use this relationship, in the form of the Fundamental Theorem of Calculus, in order to develop calculus into a systematic mathematical discipline. It is in this sense that Newton and Leibniz are credited with the invention of calculus.

Read about the contributions of these men in one or more of the given references and write a report on one of the following three topics. You can include biographical details, but the main thrust of your report should be a description, in some detail, of their methods and notations. In particular, you should consult one of the sourcebooks, which give excerpts from the original publications of Newton and Leibniz, translated from Latin to English.

- The Role of Newton in the Development of Calculus
- The Role of Leibniz in the Development of Calculus
- The Controversy between the Followers of Newton and Leibniz over Priority in the Invention of Calculus

References

1. Carl Boyer and Uta Merzbach, *A History of Mathematics* (New York: Wiley, 1987), Chapter 19.

2. Carl Boyer, *The History of the Calculus and Its Conceptual Development* (New York: Dover, 1959), Chapter V.
3. C. H. Edwards, *The Historical Development of the Calculus* (New York: Springer-Verlag, 1979), Chapters 8 and 9.
4. Howard Eves, *An Introduction to the History of Mathematics,* 6th ed. (New York: Saunders, 1990), Chapter 11.
5. C. C. Gillispie, ed., *Dictionary of Scientific Biography* (New York: Scribner's, 1974). See the article on Leibniz by Joseph Hofmann in Volume VIII and the article on Newton by I. B. Cohen in Volume X.
6. Victor Katz, *A History of Mathematics: An Introduction* (New York: HarperCollins, 1993), Chapter 12.
7. Morris Kline, *Mathematical Thought from Ancient to Modern Times* (New York: Oxford University Press, 1972), Chapter 17.

Sourcebooks

1. John Fauvel and Jeremy Gray, eds., *The History of Mathematics: A Reader* (London: MacMillan Press, 1987), Chapters 12 and 13.
2. D. E. Smith, ed., *A Sourcebook in Mathematics* (New York: Dover, 1959), Chapter V.
3. D. J. Struik, ed., *A Sourcebook in Mathematics, 1200–1800* (Princeton, N.J.: Princeton University Press, 1969), Chapter V.

5.5 THE SUBSTITUTION RULE

Because of the Fundamental Theorem, it's important to be able to find antiderivatives. But our antidifferentiation formulas don't tell us how to evaluate integrals such as

$$\boxed{1} \qquad \int 2x\sqrt{1 + x^2}\, dx$$

To find this integral we use the problem-solving strategy of *introducing something extra.* Here the "something extra" is a new variable; we change from the variable x to a new variable u. Suppose that we let u be the quantity under the root sign in (1), $u = 1 + x^2$. Then the differential of u is $du = 2x\,dx$. Notice that if the dx in the notation for an integral were to be interpreted as a differential, then the differential $2x\,dx$ would occur in (1) and so, formally, without justifying our calculation, we could write

■ Differentials were defined in Section 3.10. If $u = f(x)$, then

$$du = f'(x)\,dx$$

$$\boxed{2} \qquad \int 2x\sqrt{1 + x^2}\, dx = \int \sqrt{1 + x^2}\, 2x\, dx = \int \sqrt{u}\, du$$

$$= \tfrac{2}{3}u^{3/2} + C = \tfrac{2}{3}(x^2 + 1)^{3/2} + C$$

But now we can check that we have the correct answer by using the Chain Rule to differentiate the final function of Equation 2:

$$\frac{d}{dx}\left[\tfrac{2}{3}(x^2 + 1)^{3/2} + C\right] = \tfrac{2}{3} \cdot \tfrac{3}{2}(x^2 + 1)^{1/2} \cdot 2x = 2x\sqrt{x^2 + 1}$$

In general, this method works whenever we have an integral that we can write in the form $\int f(g(x))g'(x)\,dx$. Observe that if $F' = f$, then

$$\boxed{3} \qquad \int F'(g(x))g'(x)\,dx = F(g(x)) + C$$

because, by the Chain Rule,

$$\frac{d}{dx}[F(g(x))] = F'(g(x))g'(x)$$

If we make the "change of variable" or "substitution" $u = g(x)$, then from Equation 3 we have

$$\int F'(g(x))g'(x)\,dx = F(g(x)) + C = F(u) + C = \int F'(u)\,du$$

or, writing $F' = f$, we get

$$\int f(g(x))g'(x)\,dx = \int f(u)\,du$$

Thus we have proved the following rule.

4 THE SUBSTITUTION RULE If $u = g(x)$ is a differentiable function whose range is an interval I and f is continuous on I, then

$$\int f(g(x))g'(x)\,dx = \int f(u)\,du$$

Notice that the Substitution Rule for integration was proved using the Chain Rule for differentiation. Notice also that if $u = g(x)$, then $du = g'(x)\,dx$, so a way to remember the Substitution Rule is to think of dx and du in (4) as differentials.

Thus the Substitution Rule says: **It is permissible to operate with *dx* and *du* after integral signs as if they were differentials.**

EXAMPLE 1 Find $\int x^3 \cos(x^4 + 2)\,dx$.

SOLUTION We make the substitution $u = x^4 + 2$ because its differential is $du = 4x^3\,dx$, which, apart from the constant factor 4, occurs in the integral. Thus, using $x^3\,dx = du/4$ and the Substitution Rule, we have

$$\begin{aligned}\int x^3 \cos(x^4 + 2)\,dx &= \int \cos u \cdot \tfrac{1}{4}\,du = \tfrac{1}{4}\int \cos u\,du \\ &= \tfrac{1}{4}\sin u + C \\ &= \tfrac{1}{4}\sin(x^4 + 2) + C\end{aligned}$$

■ Check the answer by differentiating it.

Notice that at the final stage we had to return to the original variable x. □

The idea behind the Substitution Rule is to replace a relatively complicated integral by a simpler integral. This is accomplished by changing from the original variable x to a new variable u that is a function of x. Thus, in Example 1, we replaced the integral $\int x^3 \cos(x^4 + 2)\,dx$ by the simpler integral $\frac{1}{4}\int \cos u\,du$.

The main challenge in using the Substitution Rule is to think of an appropriate substitution. You should try to choose u to be some function in the integrand whose differential also occurs (except for a constant factor). This was the case in Example 1. If that is not possible, try choosing u to be some complicated part of the integrand (perhaps the inner function in a composite function). Finding the right substitution is a bit of an art. It's not unusual to guess wrong; if your first guess doesn't work, try another substitution.

EXAMPLE 2 Evaluate $\int \sqrt{2x+1}\,dx$.

SOLUTION 1 Let $u = 2x + 1$. Then $du = 2\,dx$, so $dx = du/2$. Thus the Substitution Rule gives

$$\int \sqrt{2x+1}\,dx = \int \sqrt{u}\,\frac{du}{2} = \tfrac{1}{2}\int u^{1/2}\,du$$
$$= \frac{1}{2}\cdot\frac{u^{3/2}}{3/2} + C = \tfrac{1}{3}u^{3/2} + C$$
$$= \tfrac{1}{3}(2x+1)^{3/2} + C$$

SOLUTION 2 Another possible substitution is $u = \sqrt{2x+1}$. Then

$$du = \frac{dx}{\sqrt{2x+1}} \qquad \text{so} \qquad dx = \sqrt{2x+1}\,du = u\,du$$

(Or observe that $u^2 = 2x + 1$, so $2u\,du = 2\,dx$.) Therefore

$$\int \sqrt{2x+1}\,dx = \int u\cdot u\,du = \int u^2\,du$$
$$= \frac{u^3}{3} + C = \tfrac{1}{3}(2x+1)^{3/2} + C \qquad \square$$

EXAMPLE 3 Find $\int \frac{x}{\sqrt{1-4x^2}}\,dx$.

SOLUTION Let $u = 1 - 4x^2$. Then $du = -8x\,dx$, so $x\,dx = -\frac{1}{8}\,du$ and

$$\int \frac{x}{\sqrt{1-4x^2}}\,dx = -\tfrac{1}{8}\int \frac{1}{\sqrt{u}}\,du = -\tfrac{1}{8}\int u^{-1/2}\,du$$
$$= -\tfrac{1}{8}(2\sqrt{u}) + C = -\tfrac{1}{4}\sqrt{1-4x^2} + C \qquad \square$$

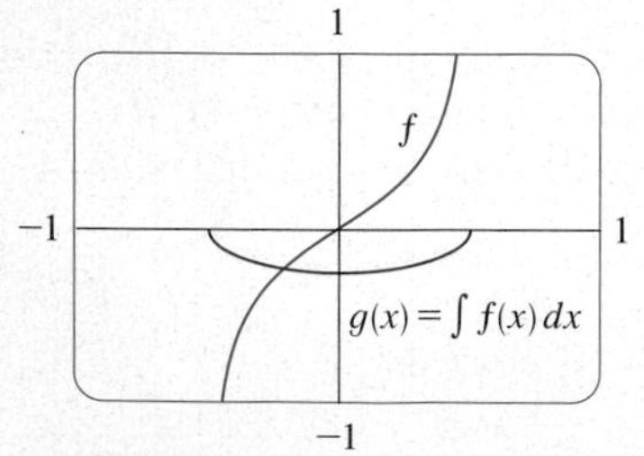

FIGURE 1

$f(x) = \dfrac{x}{\sqrt{1-4x^2}}$

$g(x) = \int f(x)\,dx = -\frac{1}{4}\sqrt{1-4x^2}$

The answer to Example 3 could be checked by differentiation, but instead let's check it with a graph. In Figure 1 we have used a computer to graph both the integrand $f(x) = x/\sqrt{1-4x^2}$ and its indefinite integral $g(x) = -\frac{1}{4}\sqrt{1-4x^2}$ (we take the case $C = 0$). Notice that $g(x)$ decreases when $f(x)$ is negative, increases when $f(x)$ is positive, and has its minimum value when $f(x) = 0$. So it seems reasonable, from the graphical evidence, that g is an antiderivative of f.

EXAMPLE 4 Calculate $\int e^{5x}\,dx$.

SOLUTION If we let $u = 5x$, then $du = 5\,dx$, so $dx = \frac{1}{5}\,du$. Therefore

$$\int e^{5x}\,dx = \tfrac{1}{5}\int e^u\,du = \tfrac{1}{5}e^u + C = \tfrac{1}{5}e^{5x} + C \qquad \square$$

EXAMPLE 5 Find $\int \sqrt{1 + x^2}\, x^5\, dx$.

SOLUTION An appropriate substitution becomes more obvious if we factor x^5 as $x^4 \cdot x$. Let $u = 1 + x^2$. Then $du = 2x\, dx$, so $x\, dx = du/2$. Also $x^2 = u - 1$, so $x^4 = (u - 1)^2$:

$$\begin{aligned}\int \sqrt{1 + x^2}\, x^5\, dx &= \int \sqrt{1 + x^2}\, x^4 \cdot x\, dx \\ &= \int \sqrt{u}\, (u - 1)^2 \frac{du}{2} = \tfrac{1}{2} \int \sqrt{u}\, (u^2 - 2u + 1)\, du \\ &= \tfrac{1}{2} \int (u^{5/2} - 2u^{3/2} + u^{1/2})\, du \\ &= \tfrac{1}{2}\left(\tfrac{2}{7}u^{7/2} - 2 \cdot \tfrac{2}{5}u^{5/2} + \tfrac{2}{3}u^{3/2}\right) + C \\ &= \tfrac{1}{7}(1 + x^2)^{7/2} - \tfrac{2}{5}(1 + x^2)^{5/2} + \tfrac{1}{3}(1 + x^2)^{3/2} + C\end{aligned}$$ □

V EXAMPLE 6 Calculate $\int \tan x\, dx$.

SOLUTION First we write tangent in terms of sine and cosine:

$$\int \tan x\, dx = \int \frac{\sin x}{\cos x}\, dx$$

This suggests that we should substitute $u = \cos x$, since then $du = -\sin x\, dx$ and so $\sin x\, dx = -du$:

$$\begin{aligned}\int \tan x\, dx &= \int \frac{\sin x}{\cos x}\, dx = -\int \frac{du}{u} \\ &= -\ln|u| + C = -\ln|\cos x| + C\end{aligned}$$ □

Since $-\ln|\cos x| = \ln(|\cos x|^{-1}) = \ln(1/|\cos x|) = \ln|\sec x|$, the result of Example 6 can also be written as

5 $$\int \tan x\, dx = \ln|\sec x| + C$$

DEFINITE INTEGRALS

When evaluating a *definite* integral by substitution, two methods are possible. One method is to evaluate the indefinite integral first and then use the Fundamental Theorem. For instance, using the result of Example 2, we have

$$\begin{aligned}\int_0^4 \sqrt{2x + 1}\, dx &= \int \sqrt{2x + 1}\, dx\Big]_0^4 = \tfrac{1}{3}(2x + 1)^{3/2}\Big]_0^4 \\ &= \tfrac{1}{3}(9)^{3/2} - \tfrac{1}{3}(1)^{3/2} = \tfrac{1}{3}(27 - 1) = \tfrac{26}{3}\end{aligned}$$

Another method, which is usually preferable, is to change the limits of integration when the variable is changed.

■ This rule says that when using a substitution in a definite integral, we must put everything in terms of the new variable u, not only x and dx but also the limits of integration. The new limits of integration are the values of u that correspond to $x = a$ and $x = b$.

6 THE SUBSTITUTION RULE FOR DEFINITE INTEGRALS If g' is continuous on $[a, b]$ and f is continuous on the range of $u = g(x)$, then

$$\int_a^b f(g(x))g'(x)\,dx = \int_{g(a)}^{g(b)} f(u)\,du$$

PROOF Let F be an antiderivative of f. Then, by (3), $F(g(x))$ is an antiderivative of $f(g(x))g'(x)$, so by Part 2 of the Fundamental Theorem, we have

$$\int_a^b f(g(x))g'(x)\,dx = F(g(x))\Big]_a^b = F(g(b)) - F(g(a))$$

But, applying FTC2 a second time, we also have

$$\int_{g(a)}^{g(b)} f(u)\,du = F(u)\Big]_{g(a)}^{g(b)} = F(g(b)) - F(g(a))$$ □

EXAMPLE 7 Evaluate $\int_0^4 \sqrt{2x+1}\,dx$ using (6).

SOLUTION Using the substitution from Solution 1 of Example 2, we have $u = 2x + 1$ and $dx = du/2$. To find the new limits of integration we note that

$$\text{when } x = 0,\ u = 2(0) + 1 = 1 \qquad \text{and} \qquad \text{when } x = 4,\ u = 2(4) + 1 = 9$$

Therefore
$$\int_0^4 \sqrt{2x+1}\,dx = \int_1^9 \tfrac{1}{2}\sqrt{u}\,du = \tfrac{1}{2}\cdot\tfrac{2}{3}u^{3/2}\Big]_1^9$$
$$= \tfrac{1}{3}(9^{3/2} - 1^{3/2}) = \tfrac{26}{3}$$

■ The geometric interpretation of Example 7 is shown in Figure 2. The substitution $u = 2x + 1$ stretches the interval $[0, 4]$ by a factor of 2 and translates it to the right by 1 unit. The Substitution Rule shows that the two areas are equal.

Observe that when using (6) we do not return to the variable x after integrating. We simply evaluate the expression in u between the appropriate values of u. □

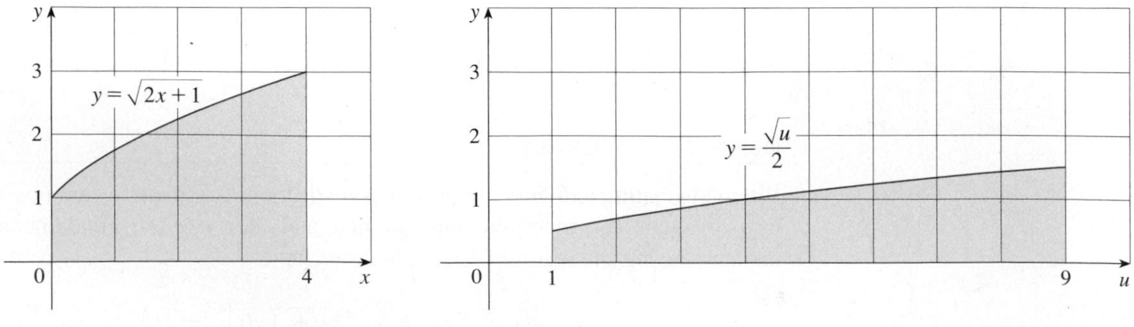

FIGURE 2

■ The integral given in Example 8 is an abbreviation for

$$\int_1^2 \frac{1}{(3-5x)^2}\,dx$$

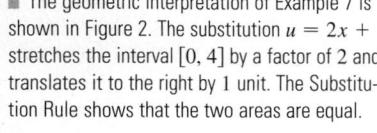

EXAMPLE 8 Evaluate $\displaystyle\int_1^2 \frac{dx}{(3-5x)^2}$.

SOLUTION Let $u = 3 - 5x$. Then $du = -5\,dx$, so $dx = -du/5$. When $x = 1$, $u = -2$ and when $x = 2$, $u = -7$. Thus

$$\int_1^2 \frac{dx}{(3-5x)^2} = -\frac{1}{5}\int_{-2}^{-7} \frac{du}{u^2}$$

$$= -\frac{1}{5}\left[-\frac{1}{u}\right]_{-2}^{-7} = \frac{1}{5u}\Bigg]_{-2}^{-7}$$

$$= \frac{1}{5}\left(-\frac{1}{7} + \frac{1}{2}\right) = \frac{1}{14}$$ □

V EXAMPLE 9 Calculate $\int_1^e \frac{\ln x}{x}\,dx$.

SOLUTION We let $u = \ln x$ because its differential $du = dx/x$ occurs in the integral. When $x = 1$, $u = \ln 1 = 0$; when $x = e$, $u = \ln e = 1$. Thus

$$\int_1^e \frac{\ln x}{x}\,dx = \int_0^1 u\,du = \frac{u^2}{2}\Bigg]_0^1 = \frac{1}{2}$$ □

■ Since the function $f(x) = (\ln x)/x$ in Example 9 is positive for $x > 1$, the integral represents the area of the shaded region in Figure 3.

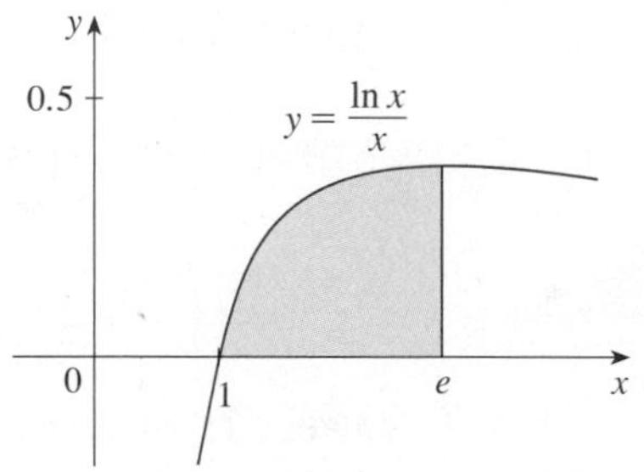

FIGURE 3

SYMMETRY

The next theorem uses the Substitution Rule for Definite Integrals (6) to simplify the calculation of integrals of functions that possess symmetry properties.

> **7 INTEGRALS OF SYMMETRIC FUNCTIONS** Suppose f is continuous on $[-a, a]$.
>
> (a) If f is even $[f(-x) = f(x)]$, then $\int_{-a}^{a} f(x)\,dx = 2\int_0^a f(x)\,dx$.
>
> (b) If f is odd $[f(-x) = -f(x)]$, then $\int_{-a}^{a} f(x)\,dx = 0$.

PROOF We split the integral in two:

$$\boxed{8} \qquad \int_{-a}^{a} f(x)\,dx = \int_{-a}^{0} f(x)\,dx + \int_0^a f(x)\,dx = -\int_0^{-a} f(x)\,dx + \int_0^a f(x)\,dx$$

In the first integral on the far right side we make the substitution $u = -x$. Then $du = -dx$ and when $x = -a$, $u = a$. Therefore

$$-\int_0^{-a} f(x)\,dx = -\int_0^a f(-u)(-du) = \int_0^a f(-u)\,du$$

and so Equation 8 becomes

$$\boxed{9} \qquad \int_{-a}^{a} f(x)\,dx = \int_{0}^{a} f(-u)\,du + \int_{0}^{a} f(x)\,dx$$

(a) If f is even, then $f(-u) = f(u)$ so Equation 9 gives

$$\int_{-a}^{a} f(x)\,dx = \int_{0}^{a} f(u)\,du + \int_{0}^{a} f(x)\,dx = 2\int_{0}^{a} f(x)\,dx$$

(b) If f is odd, then $f(-u) = -f(u)$ and so Equation 9 gives

$$\int_{-a}^{a} f(x)\,dx = -\int_{0}^{a} f(u)\,du + \int_{0}^{a} f(x)\,dx = 0$$ □

Theorem 7 is illustrated by Figure 4. For the case where f is positive and even, part (a) says that the area under $y = f(x)$ from $-a$ to a is twice the area from 0 to a because of symmetry. Recall that an integral $\int_a^b f(x)\,dx$ can be expressed as the area above the x-axis and below $y = f(x)$ minus the area below the axis and above the curve. Thus part (b) says the integral is 0 because the areas cancel.

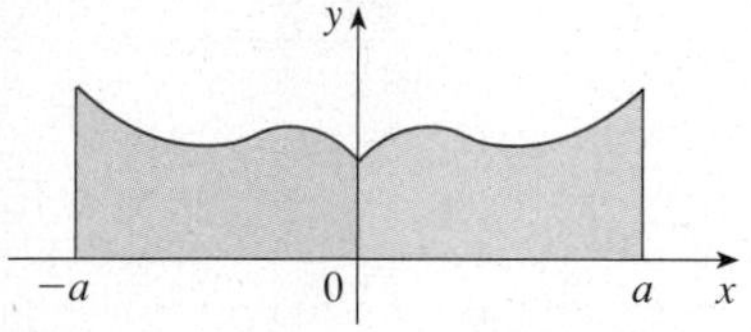

(a) f even, $\int_{-a}^{a} f(x)\,dx = 2\int_{0}^{a} f(x)\,dx$

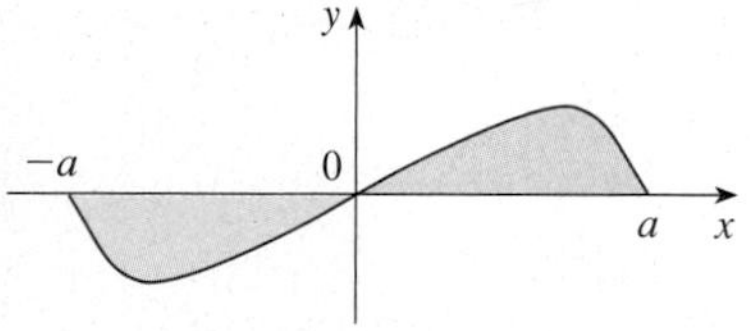

(b) f odd, $\int_{-a}^{a} f(x)\,dx = 0$

FIGURE 4

EXAMPLE 10 Since $f(x) = x^6 + 1$ satisfies $f(-x) = f(x)$, it is even and so

$$\int_{-2}^{2} (x^6 + 1)\,dx = 2\int_{0}^{2} (x^6 + 1)\,dx$$

$$= 2\left[\tfrac{1}{7}x^7 + x\right]_0^2 = 2\left(\tfrac{128}{7} + 2\right) = \tfrac{284}{7}$$ □

EXAMPLE 11 Since $f(x) = (\tan x)/(1 + x^2 + x^4)$ satisfies $f(-x) = -f(x)$, it is odd and so

$$\int_{-1}^{1} \frac{\tan x}{1 + x^2 + x^4}\,dx = 0$$ □

5.5 EXERCISES

1–6 Evaluate the integral by making the given substitution.

1. $\int e^{-x}\,dx, \quad u = -x$

2. $\int x^3(2 + x^4)^5\,dx, \quad u = 2 + x^4$

3. $\int x^2\sqrt{x^3 + 1}\,dx, \quad u = x^3 + 1$

4. $\int \frac{dt}{(1 - 6t)^4}, \quad u = 1 - 6t$

5. $\int \cos^3\theta \sin\theta\,d\theta, \quad u = \cos\theta$

6. $\int \frac{\sec^2(1/x)}{x^2}\,dx, \quad u = 1/x$

7–46 Evaluate the indefinite integral.

7. $\int x\sin(x^2)\,dx$

8. $\int x^2(x^3 + 5)^9\,dx$

9. $\int (3x - 2)^{20}\,dx$

10. $\int (3t + 2)^{2.4}\,dt$

11. $\int (x + 1)\sqrt{2x + x^2}\,dx$

12. $\int \frac{x}{(x^2 + 1)^2}\,dx$

13. $\int \frac{dx}{5 - 3x}$

14. $\int e^x \sin(e^x)\,dx$

15. $\int \sin \pi t\,dt$

16. $\int \frac{x}{x^2 + 1}\,dx$

17. $\int \frac{a + bx^2}{\sqrt{3ax + bx^3}}\,dx$

18. $\int \sec 2\theta \tan 2\theta\,d\theta$

19. $\displaystyle\int \frac{(\ln x)^2}{x}\,dx$

20. $\displaystyle\int \frac{dx}{ax+b} \quad (a \neq 0)$

21. $\displaystyle\int \frac{\cos\sqrt{t}}{\sqrt{t}}\,dt$

22. $\displaystyle\int \sqrt{x}\,\sin(1+x^{3/2})\,dx$

23. $\displaystyle\int \cos\theta\,\sin^6\theta\,d\theta$

24. $\displaystyle\int (1+\tan\theta)^5\sec^2\theta\,d\theta$

25. $\displaystyle\int e^x\sqrt{1+e^x}\,dx$

26. $\displaystyle\int e^{\cos t}\sin t\,dt$

27. $\displaystyle\int \frac{z^2}{\sqrt[3]{1+z^3}}\,dz$

28. $\displaystyle\int \frac{\tan^{-1}x}{1+x^2}\,dx$

29. $\displaystyle\int e^{\tan x}\sec^2 x\,dx$

30. $\displaystyle\int \frac{\sin(\ln x)}{x}\,dx$

31. $\displaystyle\int \frac{\cos x}{\sin^2 x}\,dx$

32. $\displaystyle\int \frac{e^x}{e^x+1}\,dx$

33. $\displaystyle\int \sqrt{\cot x}\,\csc^2 x\,dx$

34. $\displaystyle\int \frac{\cos(\pi/x)}{x^2}\,dx$

35. $\displaystyle\int \frac{\sin 2x}{1+\cos^2 x}\,dx$

36. $\displaystyle\int \frac{\sin x}{1+\cos^2 x}\,dx$

37. $\displaystyle\int \cot x\,dx$

38. $\displaystyle\int \frac{dt}{\cos^2 t\sqrt{1+\tan t}}$

39. $\displaystyle\int \sec^3 x\tan x\,dx$

40. $\displaystyle\int \sin t\,\sec^2(\cos t)\,dt$

41. $\displaystyle\int \frac{dx}{\sqrt{1-x^2}\,\sin^{-1}x}$

42. $\displaystyle\int \frac{x}{1+x^4}\,dx$

43. $\displaystyle\int \frac{1+x}{1+x^2}\,dx$

44. $\displaystyle\int \frac{x^2}{\sqrt{1-x}}\,dx$

45. $\displaystyle\int \frac{x}{\sqrt[4]{x+2}}\,dx$

46. $\displaystyle\int x^3\sqrt{x^2+1}\,dx$

47–50 Evaluate the indefinite integral. Illustrate and check that your answer is reasonable by graphing both the function and its antiderivative (take $C = 0$).

47. $\displaystyle\int x(x^2-1)^3\,dx$

48. $\displaystyle\int \frac{\sin\sqrt{x}}{\sqrt{x}}\,dx$

49. $\displaystyle\int \sin^3 x\cos x\,dx$

50. $\displaystyle\int \tan^2\theta\,\sec^2\theta\,d\theta$

51–70 Evaluate the definite integral.

51. $\displaystyle\int_0^2 (x-1)^{25}\,dx$

52. $\displaystyle\int_0^7 \sqrt{4+3x}\,dx$

53. $\displaystyle\int_0^1 x^2(1+2x^3)^5\,dx$

54. $\displaystyle\int_0^{\sqrt{\pi}} x\cos(x^2)\,dx$

55. $\displaystyle\int_0^{\pi} \sec^2(t/4)\,dt$

56. $\displaystyle\int_{1/6}^{1/2} \csc\pi t\cot\pi t\,dt$

57. $\displaystyle\int_{-\pi/6}^{\pi/6} \tan^3\theta\,d\theta$

58. $\displaystyle\int_0^1 xe^{-x^2}\,dx$

59. $\displaystyle\int_1^2 \frac{e^{1/x}}{x^2}\,dx$

60. $\displaystyle\int_{-\pi/2}^{\pi/2} \frac{x^2\sin x}{1+x^6}\,dx$

61. $\displaystyle\int_0^{13} \frac{dx}{\sqrt[3]{(1+2x)^2}}$

62. $\displaystyle\int_0^{\pi/2} \cos x\,\sin(\sin x)\,dx$

63. $\displaystyle\int_0^a x\sqrt{x^2+a^2}\,dx \quad (a>0)$

64. $\displaystyle\int_0^a x\sqrt{a^2-x^2}\,dx$

65. $\displaystyle\int_1^2 x\sqrt{x-1}\,dx$

66. $\displaystyle\int_0^4 \frac{x}{\sqrt{1+2x}}\,dx$

67. $\displaystyle\int_e^{e^4} \frac{dx}{x\sqrt{\ln x}}$

68. $\displaystyle\int_0^{1/2} \frac{\sin^{-1}x}{\sqrt{1-x^2}}\,dx$

69. $\displaystyle\int_0^1 \frac{e^z+1}{e^z+z}\,dz$

70. $\displaystyle\int_0^{T/2} \sin(2\pi t/T-\alpha)\,dt$

71–72 Use a graph to give a rough estimate of the area of the region that lies under the given curve. Then find the exact area.

71. $y=\sqrt{2x+1}, \quad 0 \leqslant x \leqslant 1$

72. $y = 2\sin x - \sin 2x, \quad 0 \leqslant x \leqslant \pi$

73. Evaluate $\int_{-2}^{2}(x+3)\sqrt{4-x^2}\,dx$ by writing it as a sum of two integrals and interpreting one of those integrals in terms of an area.

74. Evaluate $\int_0^1 x\sqrt{1-x^4}\,dx$ by making a substitution and interpreting the resulting integral in terms of an area.

75. Which of the following areas are equal? Why?

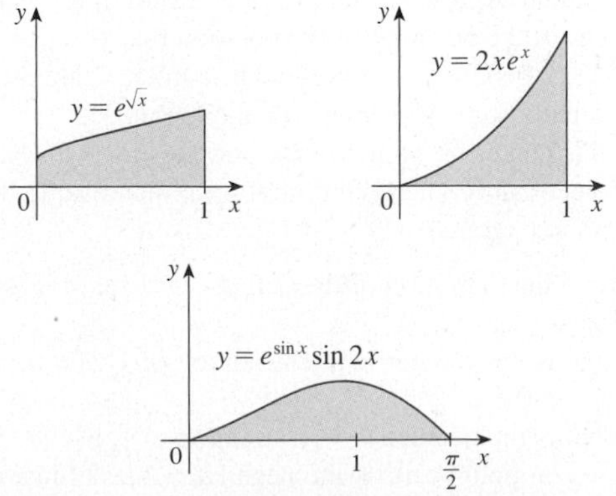

76. A model for the basal metabolism rate, in kcal/h, of a young man is $R(t) = 85 - 0.18\cos(\pi t/12)$, where t is the time in hours measured from 5:00 AM. What is the total basal metabolism of this man, $\int_0^{24} R(t)\,dt$, over a 24-hour time period?

77. An oil storage tank ruptures at time $t = 0$ and oil leaks from the tank at a rate of $r(t) = 100e^{-0.01t}$ liters per minute. How much oil leaks out during the first hour?

78. A bacteria population starts with 400 bacteria and grows at a rate of $r(t) = (450.268)e^{1.12567t}$ bacteria per hour. How many bacteria will there be after three hours?

79. Breathing is cyclic and a full respiratory cycle from the beginning of inhalation to the end of exhalation takes about 5 s. The maximum rate of air flow into the lungs is about 0.5 L/s. This explains, in part, why the function $f(t) = \frac{1}{2}\sin(2\pi t/5)$ has often been used to model the rate of air flow into the lungs. Use this model to find the volume of inhaled air in the lungs at time t.

80. Alabama Instruments Company has set up a production line to manufacture a new calculator. The rate of production of these calculators after t weeks is

$$\frac{dx}{dt} = 5000\left(1 - \frac{100}{(t+10)^2}\right) \text{ calculators/week}$$

(Notice that production approaches 5000 per week as time goes on, but the initial production is lower because of the workers' unfamiliarity with the new techniques.) Find the number of calculators produced from the beginning of the third week to the end of the fourth week.

81. If f is continuous and $\int_0^4 f(x)\,dx = 10$, find $\int_0^2 f(2x)\,dx$.

82. If f is continuous and $\int_0^9 f(x)\,dx = 4$, find $\int_0^3 xf(x^2)\,dx$.

83. If f is continuous on $\mathbb{R}$, prove that

$$\int_a^b f(-x)\,dx = \int_{-b}^{-a} f(x)\,dx$$

For the case where $f(x) \geqslant 0$ and $0 < a < b$, draw a diagram to interpret this equation geometrically as an equality of areas.

84. If f is continuous on $\mathbb{R}$, prove that

$$\int_a^b f(x+c)\,dx = \int_{a+c}^{b+c} f(x)\,dx$$

For the case where $f(x) \geqslant 0$, draw a diagram to interpret this equation geometrically as an equality of areas.

85. If a and b are positive numbers, show that

$$\int_0^1 x^a(1-x)^b\,dx = \int_0^1 x^b(1-x)^a\,dx$$

86. If f is continuous on $[0, \pi]$, use the substitution $u = \pi - x$ to show that

$$\int_0^\pi xf(\sin x)\,dx = \frac{\pi}{2}\int_0^\pi f(\sin x)\,dx$$

87. Use Exercise 86 to evaluate the integral

$$\int_0^\pi \frac{x\sin x}{1+\cos^2 x}\,dx$$

88. (a) If f is continuous, prove that

$$\int_0^{\pi/2} f(\cos x)\,dx = \int_0^{\pi/2} f(\sin x)\,dx$$

(b) Use part (a) to evaluate $\int_0^{\pi/2} \cos^2 x\,dx$ and $\int_0^{\pi/2} \sin^2 x\,dx$.

5 REVIEW

CONCEPT CHECK

1. (a) Write an expression for a Riemann sum of a function f. Explain the meaning of the notation that you use.
   (b) If $f(x) \geqslant 0$, what is the geometric interpretation of a Riemann sum? Illustrate with a diagram.
   (c) If $f(x)$ takes on both positive and negative values, what is the geometric interpretation of a Riemann sum? Illustrate with a diagram.

2. (a) Write the definition of the definite integral of a function from a to b.
   (b) What is the geometric interpretation of $\int_a^b f(x)\,dx$ if $f(x) \geqslant 0$?
   (c) What is the geometric interpretation of $\int_a^b f(x)\,dx$ if $f(x)$ takes on both positive and negative values? Illustrate with a diagram.

3. State both parts of the Fundamental Theorem of Calculus.

4. (a) State the Net Change Theorem.
   (b) If $r(t)$ is the rate at which water flows into a reservoir, what does $\int_{t_1}^{t_2} r(t)\,dt$ represent?

5. Suppose a particle moves back and forth along a straight line with velocity $v(t)$, measured in feet per second, and acceleration $a(t)$.
   (a) What is the meaning of $\int_{60}^{120} v(t)\,dt$?
   (b) What is the meaning of $\int_{60}^{120} |v(t)|\,dt$?
   (c) What is the meaning of $\int_{60}^{120} a(t)\,dt$?

6. (a) Explain the meaning of the indefinite integral $\int f(x)\,dx$.
   (b) What is the connection between the definite integral $\int_a^b f(x)\,dx$ and the indefinite integral $\int f(x)\,dx$?

7. Explain exactly what is meant by the statement that "differentiation and integration are inverse processes."

8. State the Substitution Rule. In practice, how do you use it?

TRUE-FALSE QUIZ

Determine whether the statement is true or false. If it is true, explain why. If it is false, explain why or give an example that disproves the statement.

1. If f and g are continuous on $[a, b]$, then

$$\int_a^b [f(x) + g(x)]\, dx = \int_a^b f(x)\, dx + \int_a^b g(x)\, dx$$

2. If f and g are continuous on $[a, b]$, then

$$\int_a^b [f(x)g(x)]\, dx = \left(\int_a^b f(x)\, dx\right)\left(\int_a^b g(x)\, dx\right)$$

3. If f is continuous on $[a, b]$, then

$$\int_a^b 5f(x)\, dx = 5\int_a^b f(x)\, dx$$

4. If f is continuous on $[a, b]$, then

$$\int_a^b xf(x)\, dx = x\int_a^b f(x)\, dx$$

5. If f is continuous on $[a, b]$ and $f(x) \geqslant 0$, then

$$\int_a^b \sqrt{f(x)}\, dx = \sqrt{\int_a^b f(x)\, dx}$$

6. If f' is continuous on $[1, 3]$, then $\int_1^3 f'(v)\, dv = f(3) - f(1)$.

7. If f and g are continuous and $f(x) \geqslant g(x)$ for $a \leqslant x \leqslant b$, then

$$\int_a^b f(x)\, dx \geqslant \int_a^b g(x)\, dx$$

8. If f and g are differentiable and $f(x) \geqslant g(x)$ for $a < x < b$, then $f'(x) \geqslant g'(x)$ for $a < x < b$.

9. $\displaystyle\int_{-1}^1 \left(x^5 - 6x^9 + \frac{\sin x}{(1 + x^4)^2}\right) dx = 0$

10. $\displaystyle\int_{-5}^5 (ax^2 + bx + c)\, dx = 2\int_0^5 (ax^2 + c)\, dx$

11. $\displaystyle\int_{-2}^1 \frac{1}{x^4}\, dx = -\frac{3}{8}$

12. $\int_0^2 (x - x^3)\, dx$ represents the area under the curve $y = x - x^3$ from 0 to 2.

13. All continuous functions have derivatives.

14. All continuous functions have antiderivatives.

15. If f is continuous on $[a, b]$, then

$$\frac{d}{dx}\left(\int_a^b f(x)\, dx\right) = f(x)$$

EXERCISES

1. Use the given graph of f to find the Riemann sum with six subintervals. Take the sample points to be (a) left endpoints and (b) midpoints. In each case draw a diagram and explain what the Riemann sum represents.

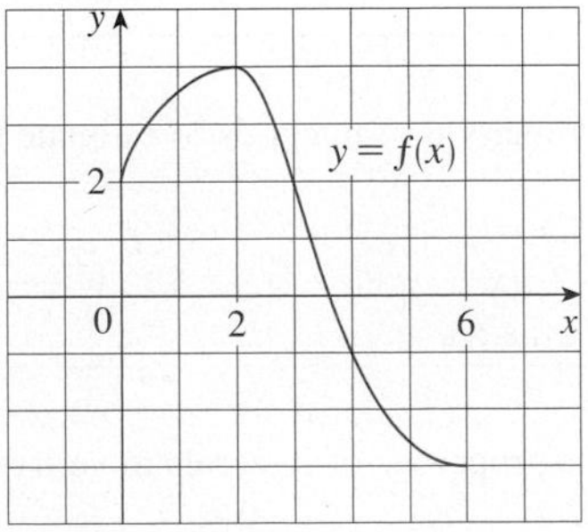

2. (a) Evaluate the Riemann sum for

$$f(x) = x^2 - x \qquad 0 \leqslant x \leqslant 2$$

with four subintervals, taking the sample points to be right endpoints. Explain, with the aid of a diagram, what the Riemann sum represents.

(b) Use the definition of a definite integral (with right endpoints) to calculate the value of the integral

$$\int_0^2 (x^2 - x)\, dx$$

(c) Use the Fundamental Theorem to check your answer to part (b).

(d) Draw a diagram to explain the geometric meaning of the integral in part (b).

3. Evaluate

$$\int_0^1 \left(x + \sqrt{1 - x^2}\right) dx$$

by interpreting it in terms of areas.

4. Express

$$\lim_{n\to\infty} \sum_{i=1}^{n} \sin x_i\, \Delta x$$

as a definite integral on the interval $[0, \pi]$ and then evaluate the integral.

5. If $\int_0^6 f(x)\,dx = 10$ and $\int_0^4 f(x)\,dx = 7$, find $\int_4^6 f(x)\,dx$.

CAS **6.** (a) Write $\int_1^5 (x + 2x^5)\,dx$ as a limit of Riemann sums, taking the sample points to be right endpoints. Use a computer algebra system to evaluate the sum and to compute the limit.
(b) Use the Fundamental Theorem to check your answer to part (a).

7. The following figure shows the graphs of f, f', and $\int_0^x f(t)\,dt$. Identify each graph, and explain your choices.

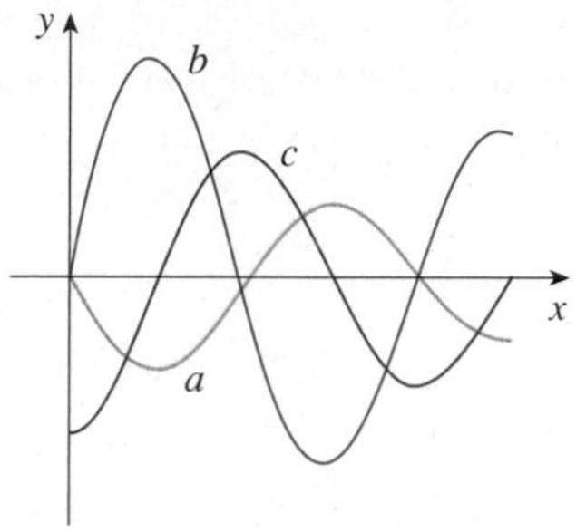

8. Evaluate:

(a) $\int_0^1 \frac{d}{dx}\left(e^{\arctan x}\right) dx$ (b) $\frac{d}{dx}\int_0^1 e^{\arctan x}\,dx$

(c) $\frac{d}{dx}\int_0^x e^{\arctan t}\,dt$

9–38 Evaluate the integral, if it exists.

9. $\int_1^2 (8x^3 + 3x^2)\,dx$

10. $\int_0^T (x^4 - 8x + 7)\,dx$

11. $\int_0^1 (1 - x^9)\,dx$

12. $\int_0^1 (1 - x)^9\,dx$

13. $\int_1^9 \frac{\sqrt{u} - 2u^2}{u}\,du$

14. $\int_0^1 (\sqrt[4]{u} + 1)^2\,du$

15. $\int_0^1 y(y^2 + 1)^5\,dy$

16. $\int_0^2 y^2\sqrt{1 + y^3}\,dy$

17. $\int_1^5 \frac{dt}{(t - 4)^2}$

18. $\int_0^1 \sin(3\pi t)\,dt$

19. $\int_0^1 v^2 \cos(v^3)\,dv$

20. $\int_{-1}^1 \frac{\sin x}{1 + x^2}\,dx$

21. $\int_{-\pi/4}^{\pi/4} \frac{t^4 \tan t}{2 + \cos t}\,dt$

22. $\int_0^1 \frac{e^x}{1 + e^{2x}}\,dx$

23. $\int \left(\frac{1 - x}{x}\right)^2 dx$

24. $\int_1^{10} \frac{x}{x^2 - 4}\,dx$

25. $\int \frac{x + 2}{\sqrt{x^2 + 4x}}\,dx$

26. $\int \frac{\csc^2 x}{1 + \cot x}\,dx$

27. $\int \sin \pi t \cos \pi t\,dt$

28. $\int \sin x \cos(\cos x)\,dx$

29. $\int \frac{e^{\sqrt{x}}}{\sqrt{x}}\,dx$

30. $\int \frac{\cos(\ln x)}{x}\,dx$

31. $\int \tan x \ln(\cos x)\,dx$

32. $\int \frac{x}{\sqrt{1 - x^4}}\,dx$

33. $\int \frac{x^3}{1 + x^4}\,dx$

34. $\int \sinh(1 + 4x)\,dx$

35. $\int \frac{\sec\theta \tan\theta}{1 + \sec\theta}\,d\theta$

36. $\int_0^{\pi/4} (1 + \tan t)^3 \sec^2 t\,dt$

37. $\int_0^3 |x^2 - 4|\,dx$

38. $\int_0^4 |\sqrt{x} - 1|\,dx$

39–40 Evaluate the indefinite integral. Illustrate and check that your answer is reasonable by graphing both the function and its antiderivative (take $C = 0$).

39. $\int \frac{\cos x}{\sqrt{1 + \sin x}}\,dx$

40. $\int \frac{x^3}{\sqrt{x^2 + 1}}\,dx$

41. Use a graph to give a rough estimate of the area of the region that lies under the curve $y = x\sqrt{x}$, $0 \leq x \leq 4$. Then find the exact area.

42. Graph the function $f(x) = \cos^2 x \sin^3 x$ and use the graph to guess the value of the integral $\int_0^{2\pi} f(x)\,dx$. Then evaluate the integral to confirm your guess.

43–48 Find the derivative of the function.

43. $F(x) = \int_0^x \frac{t^2}{1 + t^3}\,dt$

44. $F(x) = \int_x^1 \sqrt{t + \sin t}\,dt$

45. $g(x) = \int_0^{x^4} \cos(t^2)\,dt$

46. $g(x) = \int_1^{\sin x} \frac{1 - t^2}{1 + t^4}\,dt$

47. $y = \int_{\sqrt{x}}^x \frac{e^t}{t}\,dt$

48. $y = \int_{2x}^{3x+1} \sin(t^4)\,dt$

49–50 Use Property 8 of integrals to estimate the value of the integral.

49. $\int_1^3 \sqrt{x^2 + 3}\,dx$

50. $\int_3^5 \frac{1}{x + 1}\,dx$

51–54 Use the properties of integrals to verify the inequality.

51. $\int_0^1 x^2 \cos x\,dx \leq \frac{1}{3}$

52. $\int_{\pi/4}^{\pi/2} \frac{\sin x}{x}\,dx \leq \frac{\sqrt{2}}{2}$

53. $\int_0^1 e^x \cos x\,dx \leq e - 1$

54. $\int_0^1 x \sin^{-1} x\,dx \leq \pi/4$

55. Use the Midpoint Rule with $n = 6$ to approximate $\int_0^3 \sin(x^3)\,dx$.

56. A particle moves along a line with velocity function $v(t) = t^2 - t$, where v is measured in meters per second. Find (a) the displacement and (b) the distance traveled by the particle during the time interval $[0, 5]$.

57. Let $r(t)$ be the rate at which the world's oil is consumed, where t is measured in years starting at $t = 0$ on January 1, 2000, and $r(t)$ is measured in barrels per year. What does $\int_0^8 r(t)\,dt$ represent?

58. A radar gun was used to record the speed of a runner at the times given in the table. Use the Midpoint Rule to estimate the distance the runner covered during those 5 seconds.

t (s)	v (m/s)	t (s)	v (m/s)
0	0	3.0	10.51
0.5	4.67	3.5	10.67
1.0	7.34	4.0	10.76
1.5	8.86	4.5	10.81
2.0	9.73	5.0	10.81
2.5	10.22		

59. A population of honeybees increased at a rate of $r(t)$ bees per week, where the graph of r is as shown. Use the Midpoint Rule with six subintervals to estimate the increase in the bee population during the first 24 weeks.

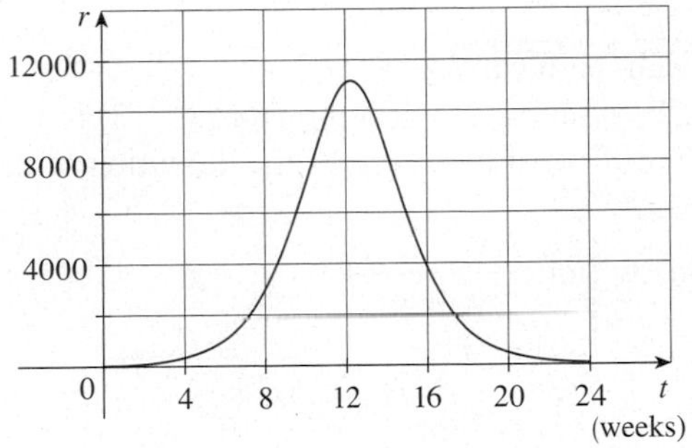

60. Let

$$f(x) = \begin{cases} -x - 1 & \text{if } -3 \leqslant x \leqslant 0 \\ -\sqrt{1 - x^2} & \text{if } 0 \leqslant x \leqslant 1 \end{cases}$$

Evaluate $\int_{-3}^{1} f(x)\,dx$ by interpreting the integral as a difference of areas.

61. If f is continuous and $\int_0^2 f(x)\,dx = 6$, evaluate $\int_0^{\pi/2} f(2\sin\theta)\cos\theta\,d\theta$.

62. The Fresnel function $S(x) = \int_0^x \sin\left(\frac{1}{2}\pi t^2\right) dt$ was introduced in Section 5.3. Fresnel also used the function

$$C(x) = \int_0^x \cos\left(\tfrac{1}{2}\pi t^2\right) dt$$

in his theory of the diffraction of light waves.

(a) On what intervals is C increasing?

(b) On what intervals is C concave upward?

CAS (c) Use a graph to solve the following equation correct to two decimal places:

$$\int_0^x \cos\left(\tfrac{1}{2}\pi t^2\right) dt = 0.7$$

CAS (d) Plot the graphs of C and S on the same screen. How are these graphs related?

63. Estimate the value of the number c such that the area under the curve $y = \sinh cx$ between $x = 0$ and $x = 1$ is equal to 1.

64. Suppose that the temperature in a long, thin rod placed along the x-axis is initially $C/(2a)$ if $|x| \leqslant a$ and 0 if $|x| > a$. It can be shown that if the heat diffusivity of the rod is k, then the temperature of the rod at the point x at time t is

$$T(x, t) = \frac{C}{a\sqrt{4\pi kt}} \int_0^a e^{-(x-u)^2/(4kt)}\,du$$

To find the temperature distribution that results from an initial hot spot concentrated at the origin, we need to compute

$$\lim_{a \to 0} T(x, t)$$

Use l'Hospital's Rule to find this limit.

65. If f is a continuous function such that

$$\int_0^x f(t)\,dt = xe^{2x} + \int_0^x e^{-t} f(t)\,dt$$

for all x, find an explicit formula for $f(x)$.

66. Suppose h is a function such that $h(1) = -2$, $h'(1) = 2$, $h''(1) = 3$, $h(2) = 6$, $h'(2) = 5$, $h''(2) = 13$, and h'' is continuous everywhere. Evaluate $\int_1^2 h''(u)\,du$.

67. If f' is continuous on $[a, b]$, show that

$$2\int_a^b f(x)f'(x)\,dx = [f(b)]^2 - [f(a)]^2$$

68. Find $\displaystyle\lim_{h \to 0} \frac{1}{h} \int_2^{2+h} \sqrt{1 + t^3}\,dt$.

69. If f is continuous on $[0, 1]$, prove that

$$\int_0^1 f(x)\,dx = \int_0^1 f(1 - x)\,dx$$

70. Evaluate

$$\lim_{n \to \infty} \frac{1}{n}\left[\left(\frac{1}{n}\right)^9 + \left(\frac{2}{n}\right)^9 + \left(\frac{3}{n}\right)^9 + \cdots + \left(\frac{n}{n}\right)^9\right]$$

71. Suppose f is continuous, $f(0) = 0$, $f(1) = 1$, $f'(x) > 0$, and $\int_0^1 f(x)\,dx = \frac{1}{3}$. Find the value of the integral $\int_0^1 f^{-1}(y)\,dy$.

PROBLEMS PLUS

Before you look at the solution of the following example, cover it up and first try to solve the problem yourself.

EXAMPLE 1 Evaluate $\displaystyle\lim_{x\to 3}\left(\frac{x}{x-3}\int_3^x \frac{\sin t}{t}\,dt\right)$.

SOLUTION Let's start by having a preliminary look at the ingredients of the function. What happens to the first factor, $x/(x-3)$, when x approaches 3? The numerator approaches 3 and the denominator approaches 0, so we have

$$\frac{x}{x-3} \to \infty \quad \text{as} \quad x \to 3^+ \qquad \text{and} \qquad \frac{x}{x-3} \to -\infty \quad \text{as} \quad x \to 3^-$$

The second factor approaches $\int_3^3 (\sin t)/t\,dt$, which is 0. It's not clear what happens to the function as a whole. (One factor is becoming large while the other is becoming small.) So how do we proceed?

■ The principles of problem solving are discussed on page 76.

One of the principles of problem solving is *recognizing something familiar.* Is there a part of the function that reminds us of something we've seen before? Well, the integral

$$\int_3^x \frac{\sin t}{t}\,dt$$

has x as its upper limit of integration and that type of integral occurs in Part 1 of the Fundamental Theorem of Calculus:

$$\frac{d}{dx}\int_a^x f(t)\,dt = f(x)$$

This suggests that differentiation might be involved.

Once we start thinking about differentiation, the denominator $(x-3)$ reminds us of something else that should be familiar: One of the forms of the definition of the derivative in Chapter 2 is

$$F'(a) = \lim_{x\to a}\frac{F(x)-F(a)}{x-a}$$

and with $a = 3$ this becomes

$$F'(3) = \lim_{x\to 3}\frac{F(x)-F(3)}{x-3}$$

So what is the function F in our situation? Notice that if we define

$$F(x) = \int_3^x \frac{\sin t}{t}\,dt$$

then $F(3) = 0$. What about the factor x in the numerator? That's just a red herring, so let's factor it out and put together the calculation:

■ Another approach is to use l'Hospital's Rule.

$$\begin{aligned}
\lim_{x\to 3}\left(\frac{x}{x-3}\int_3^x \frac{\sin t}{t}\,dt\right) &= \lim_{x\to 3} x \cdot \lim_{x\to 3}\frac{\displaystyle\int_3^x \frac{\sin t}{t}\,dt}{x-3} \\
&= 3\lim_{x\to 3}\frac{F(x)-F(3)}{x-3} \\
&= 3F'(3) = 3\,\frac{\sin 3}{3} \qquad \text{(FTC1)} \\
&= \sin 3
\end{aligned}$$

□

PROBLEMS

1. If $x \sin \pi x = \int_0^{x^2} f(t)\,dt$, where f is a continuous function, find $f(4)$.

2. Find the minimum value of the area of the region under the curve $y = x + 1/x$ from $x = a$ to $x = a + 1.5$, for all $a > 0$.

3. If f is a differentiable function such that $f(x)$ is never 0 and $\int_0^x f(t)\,dt = [f(x)]^2$ for all x, find f.

4. (a) Graph several members of the family of functions $f(x) = (2cx - x^2)/c^3$ for $c > 0$ and look at the regions enclosed by these curves and the x-axis. Make a conjecture about how the areas of these regions are related.
(b) Prove your conjecture in part (a).
(c) Take another look at the graphs in part (a) and use them to sketch the curve traced out by the vertices (highest points) of the family of functions. Can you guess what kind of curve this is?
(d) Find an equation of the curve you sketched in part (c).

5. If $f(x) = \int_0^{g(x)} \frac{1}{\sqrt{1 + t^3}}\,dt$, where $g(x) = \int_0^{\cos x} [1 + \sin(t^2)]\,dt$, find $f'(\pi/2)$.

6. If $f(x) = \int_0^x x^2 \sin(t^2)\,dt$, find $f'(x)$.

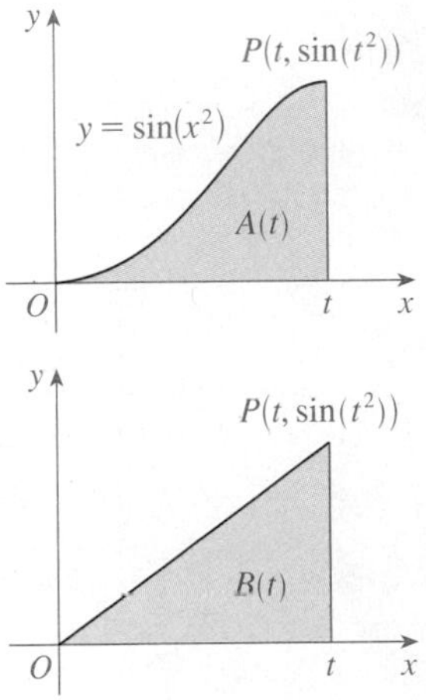

FIGURE FOR PROBLEM 8

7. Evaluate $\lim_{x \to 0} \frac{1}{x} \int_0^x (1 - \tan 2t)^{1/t}\,dt$.

8. The figure shows two regions in the first quadrant: $A(t)$ is the area under the curve $y = \sin(x^2)$ from 0 to t, and $B(t)$ is the area of the triangle with vertices O, P, and $(t, 0)$. Find $\lim_{t \to 0^+} A(t)/B(t)$.

9. Find the interval $[a, b]$ for which the value of the integral $\int_a^b (2 + x - x^2)\,dx$ is a maximum.

10. Use an integral to estimate the sum $\sum_{i=1}^{10000} \sqrt{i}$.

11. (a) Evaluate $\int_0^n [\![x]\!]\,dx$, where n is a positive integer.
(b) Evaluate $\int_a^b [\![x]\!]\,dx$, where a and b are real numbers with $0 \leqslant a < b$.

12. Find $\frac{d^2}{dx^2} \int_0^x \left(\int_1^{\sin t} \sqrt{1 + u^4}\,du \right) dt$.

13. Suppose the coefficients of the cubic polynomial $P(x) = a + bx + cx^2 + dx^3$ satisfy the equation

$$a + \frac{b}{2} + \frac{c}{3} + \frac{d}{4} = 0$$

Show that the equation $P(x) = 0$ has a root between 0 and 1. Can you generalize this result for an nth-degree polynomial?

14. A circular disk of radius r is used in an evaporator and is rotated in a vertical plane. If it is to be partially submerged in the liquid so as to maximize the exposed wetted area of the disk, show that the center of the disk should be positioned at a height $r/\sqrt{1 + \pi^2}$ above the surface of the liquid.

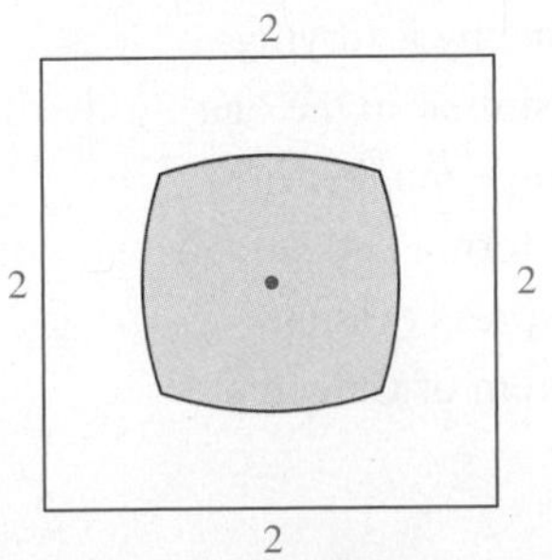

FIGURE FOR PROBLEM 16

15. Prove that if f is continuous, then $\int_0^x f(u)(x - u)\,du = \int_0^x \left(\int_0^u f(t)\,dt \right) du$.

16. The figure shows a region consisting of all points inside a square that are closer to the center than to the sides of the square. Find the area of the region.

17. Evaluate $\lim_{n \to \infty} \left(\frac{1}{\sqrt{n}\sqrt{n+1}} + \frac{1}{\sqrt{n}\sqrt{n+2}} + \cdots + \frac{1}{\sqrt{n}\sqrt{n+n}} \right)$.

18. For any number c, we let $f_c(x)$ be the smaller of the two numbers $(x - c)^2$ and $(x - c - 2)^2$. Then we define $g(c) = \int_0^1 f_c(x)\,dx$. Find the maximum and minimum values of $g(c)$ if $-2 \leqslant c \leqslant 2$.

6

APPLICATIONS OF INTEGRATION

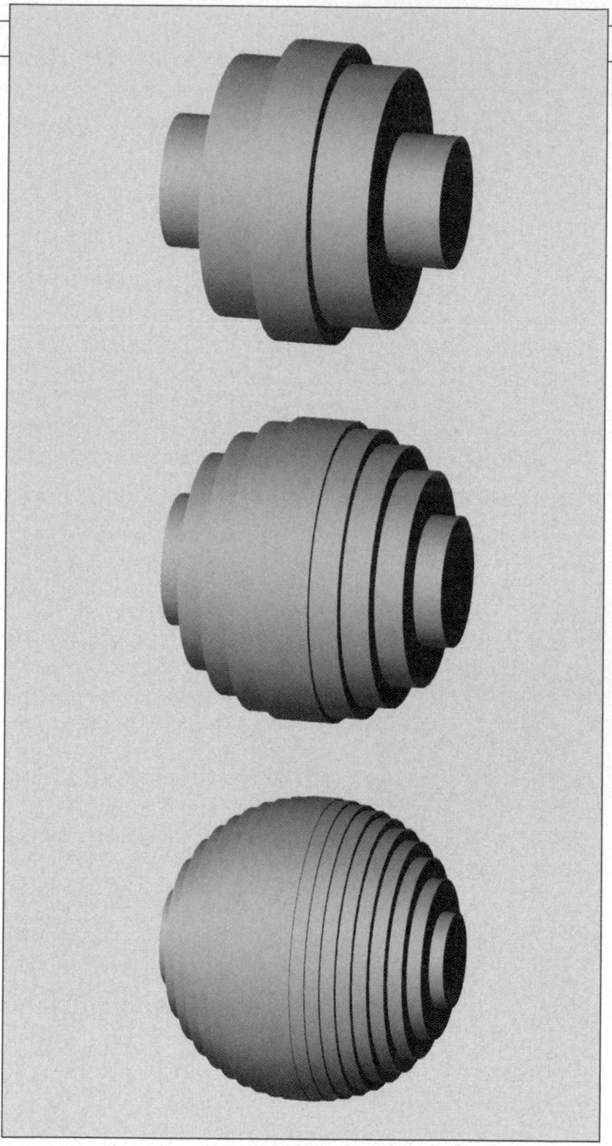

The volume of a sphere is the limit of sums of volumes of approximating cylinders.

In this chapter we explore some of the applications of the definite integral by using it to compute areas between curves, volumes of solids, and the work done by a varying force. The common theme is the following general method, which is similar to the one we used to find areas under curves: We break up a quantity Q into a large number of small parts. We next approximate each small part by a quantity of the form $f(x_i^*)\,\Delta x$ and thus approximate Q by a Riemann sum. Then we take the limit and express Q as an integral. Finally we evaluate the integral using the Fundamental Theorem of Calculus or the Midpoint Rule.

6.1 AREAS BETWEEN CURVES

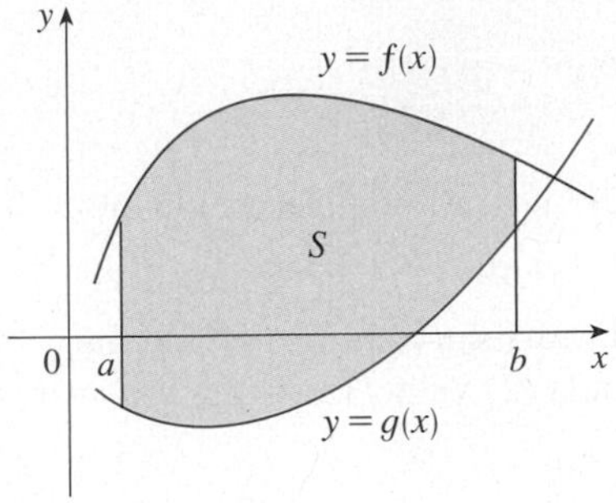

FIGURE 1
$S = \{(x, y) \mid a \leq x \leq b, g(x) \leq y \leq f(x)\}$

In Chapter 5 we defined and calculated areas of regions that lie under the graphs of functions. Here we use integrals to find areas of regions that lie between the graphs of two functions.

Consider the region S that lies between two curves $y = f(x)$ and $y = g(x)$ and between the vertical lines $x = a$ and $x = b$, where f and g are continuous functions and $f(x) \geq g(x)$ for all x in $[a, b]$. (See Figure 1.)

Just as we did for areas under curves in Section 5.1, we divide S into n strips of equal width and then we approximate the ith strip by a rectangle with base Δx and height $f(x_i^*) - g(x_i^*)$. (See Figure 2. If we like, we could take all of the sample points to be right endpoints, in which case $x_i^* = x_i$.) The Riemann sum

$$\sum_{i=1}^{n} [f(x_i^*) - g(x_i^*)]\,\Delta x$$

is therefore an approximation to what we intuitively think of as the area of S.

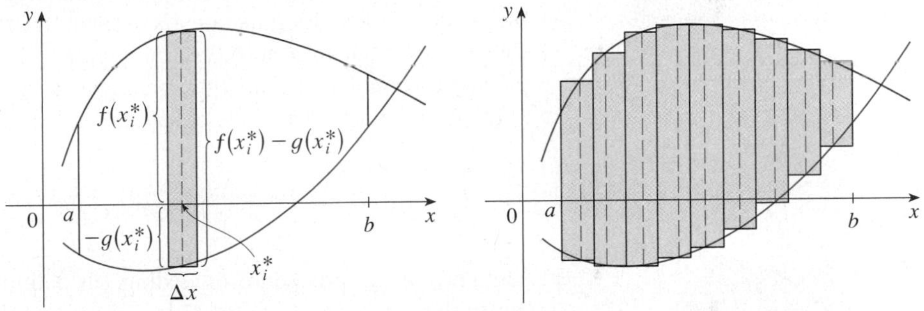

FIGURE 2 (a) Typical rectangle (b) Approximating rectangles

This approximation appears to become better and better as $n \to \infty$. Therefore we define the **area** A of the region S as the limiting value of the sum of the areas of these approximating rectangles.

1
$$A = \lim_{n \to \infty} \sum_{i=1}^{n} [f(x_i^*) - g(x_i^*)]\,\Delta x$$

We recognize the limit in (1) as the definite integral of $f - g$. Therefore we have the following formula for area.

2 The area A of the region bounded by the curves $y = f(x)$, $y = g(x)$, and the lines $x = a$, $x = b$, where f and g are continuous and $f(x) \geq g(x)$ for all x in $[a, b]$, is

$$A = \int_a^b [f(x) - g(x)]\,dx$$

Notice that in the special case where $g(x) = 0$, S is the region under the graph of f and our general definition of area (1) reduces to our previous definition (Definition 2 in Section 5.1).

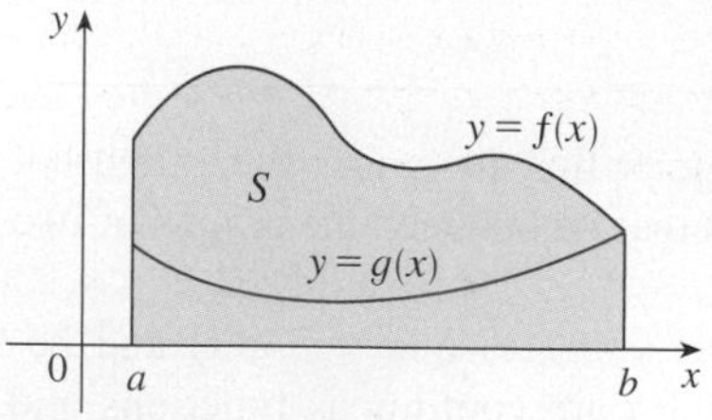

FIGURE 3

$A = \int_a^b f(x)\,dx - \int_a^b g(x)\,dx$

In the case where both f and g are positive, you can see from Figure 3 why (2) is true:

$$A = [\text{area under } y = f(x)] - [\text{area under } y = g(x)]$$
$$= \int_a^b f(x)\,dx - \int_a^b g(x)\,dx = \int_a^b [f(x) - g(x)]\,dx$$

EXAMPLE 1 Find the area of the region bounded above by $y = e^x$, bounded below by $y = x$, and bounded on the sides by $x = 0$ and $x = 1$.

SOLUTION The region is shown in Figure 4. The upper boundary curve is $y = e^x$ and the lower boundary curve is $y = x$. So we use the area formula (2) with $f(x) = e^x$, $g(x) = x$, $a = 0$, and $b = 1$:

$$A = \int_0^1 (e^x - x)\,dx = e^x - \tfrac{1}{2}x^2\Big]_0^1$$
$$= e - \tfrac{1}{2} - 1 = e - 1.5$$ □

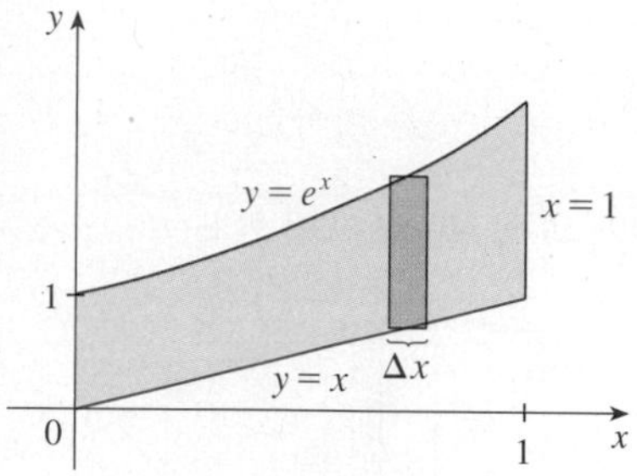

FIGURE 4

In Figure 4 we drew a typical approximating rectangle with width Δx as a reminder of the procedure by which the area is defined in (1). In general, when we set up an integral for an area, it's helpful to sketch the region to identify the top curve y_T, the bottom curve y_B, and a typical approximating rectangle as in Figure 5. Then the area of a typical rectangle is $(y_T - y_B)\,\Delta x$ and the equation

$$A = \lim_{n\to\infty} \sum_{i=1}^{n} (y_T - y_B)\,\Delta x = \int_a^b (y_T - y_B)\,dx$$

summarizes the procedure of adding (in a limiting sense) the areas of all the typical rectangles.

Notice that in Figure 5 the left-hand boundary reduces to a point, whereas in Figure 3 the right-hand boundary reduces to a point. In the next example both of the side boundaries reduce to a point, so the first step is to find a and b.

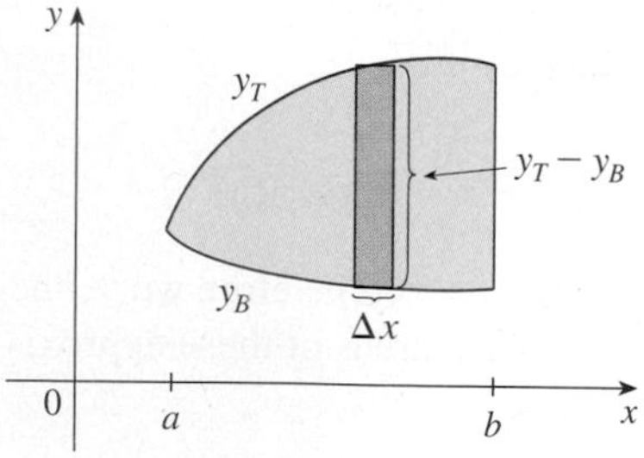

FIGURE 5

V EXAMPLE 2 Find the area of the region enclosed by the parabolas $y = x^2$ and $y = 2x - x^2$.

SOLUTION We first find the points of intersection of the parabolas by solving their equations simultaneously. This gives $x^2 = 2x - x^2$, or $2x^2 - 2x = 0$. Thus $2x(x - 1) = 0$, so $x = 0$ or 1. The points of intersection are $(0, 0)$ and $(1, 1)$.

We see from Figure 6 that the top and bottom boundaries are

$$y_T = 2x - x^2 \quad \text{and} \quad y_B = x^2$$

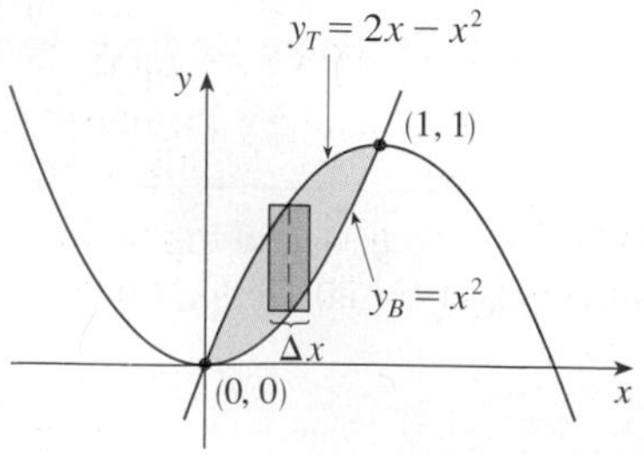

FIGURE 6

The area of a typical rectangle is

$$(y_T - y_B)\,\Delta x = (2x - x^2 - x^2)\,\Delta x$$

and the region lies between $x = 0$ and $x = 1$. So the total area is

$$A = \int_0^1 (2x - 2x^2)\,dx = 2\int_0^1 (x - x^2)\,dx$$
$$= 2\left[\frac{x^2}{2} - \frac{x^3}{3}\right]_0^1 = 2\left(\frac{1}{2} - \frac{1}{3}\right) = \frac{1}{3}$$ □

Sometimes it's difficult, or even impossible, to find the points of intersection of two curves exactly. As shown in the following example, we can use a graphing calculator or computer to find approximate values for the intersection points and then proceed as before.

EXAMPLE 3 Find the approximate area of the region bounded by the curves $y = x/\sqrt{x^2 + 1}$ and $y = x^4 - x$.

SOLUTION If we were to try to find the exact intersection points, we would have to solve the equation

$$\frac{x}{\sqrt{x^2 + 1}} = x^4 - x$$

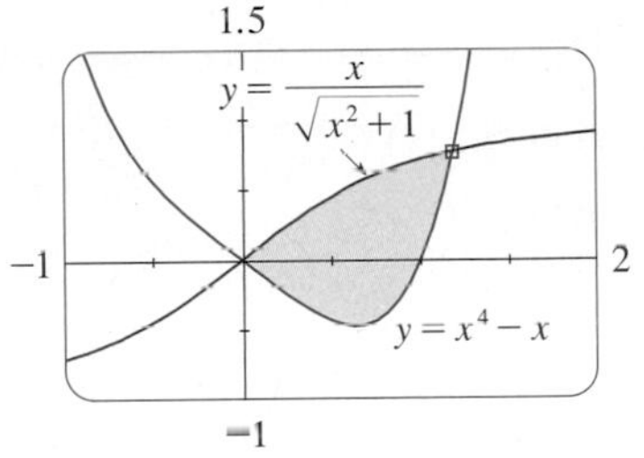

FIGURE 7

This looks like a very difficult equation to solve exactly (in fact, it's impossible), so instead we use a graphing device to draw the graphs of the two curves in Figure 7. One intersection point is the origin. We zoom in toward the other point of intersection and find that $x \approx 1.18$. (If greater accuracy is required, we could use Newton's method or a rootfinder, if available on our graphing device.) Thus an approximation to the area between the curves is

$$A \approx \int_0^{1.18} \left[\frac{x}{\sqrt{x^2 + 1}} - (x^4 - x) \right] dx$$

To integrate the first term we use the subsitution $u = x^2 + 1$. Then $du = 2x\,dx$, and when $x = 1.18$, we have $u \approx 2.39$. So

$$\begin{aligned} A &\approx \tfrac{1}{2} \int_1^{2.39} \frac{du}{\sqrt{u}} - \int_0^{1.18} (x^4 - x)\,dx \\ &= \sqrt{u}\,\Big]_1^{2.39} - \left[\frac{x^5}{5} - \frac{x^2}{2} \right]_0^{1.18} \\ &= \sqrt{2.39} - 1 - \frac{(1.18)^5}{5} + \frac{(1.18)^2}{2} \\ &\approx 0.785 \end{aligned}$$

□

EXAMPLE 4 Figure 8 shows velocity curves for two cars, A and B, that start side by side and move along the same road. What does the area between the curves represent? Use the Midpoint Rule to estimate it.

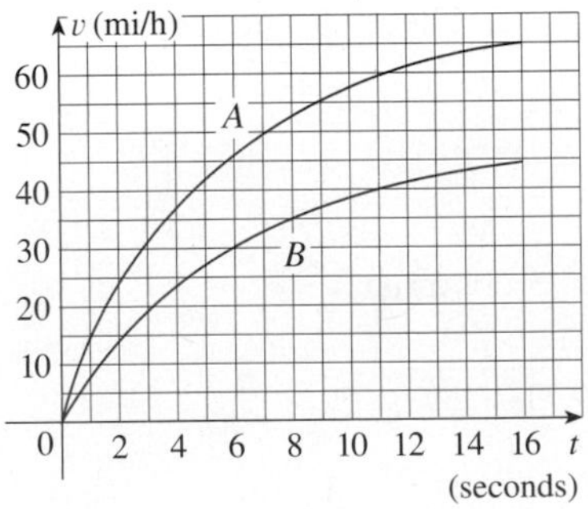

FIGURE 8

SOLUTION We know from Section 5.4 that the area under the velocity curve A represents the distance traveled by car A during the first 16 seconds. Similarly, the area under curve B is the distance traveled by car B during that time period. So the area between these curves, which is the difference of the areas under the curves, is the distance between the cars after 16 seconds. We read the velocities from the graph and convert them to feet per second ($1 \text{ mi/h} = \frac{5280}{3600} \text{ ft/s}$).

t	0	2	4	6	8	10	12	14	16
v_A	0	34	54	67	76	84	89	92	95
v_B	0	21	34	44	51	56	60	63	65
$v_A - v_B$	0	13	20	23	25	28	29	29	30

We use the Midpoint Rule with $n = 4$ intervals, so that $\Delta t = 4$. The midpoints of the intervals are $\bar{t}_1 = 2$, $\bar{t}_2 = 6$, $\bar{t}_3 = 10$, and $\bar{t}_4 = 14$. We estimate the distance between the cars after 16 seconds as follows:

$$\int_0^{16} (v_A - v_B)\,dt \approx \Delta t\,[13 + 23 + 28 + 29]$$
$$= 4(93) = 372 \text{ ft}$$

□

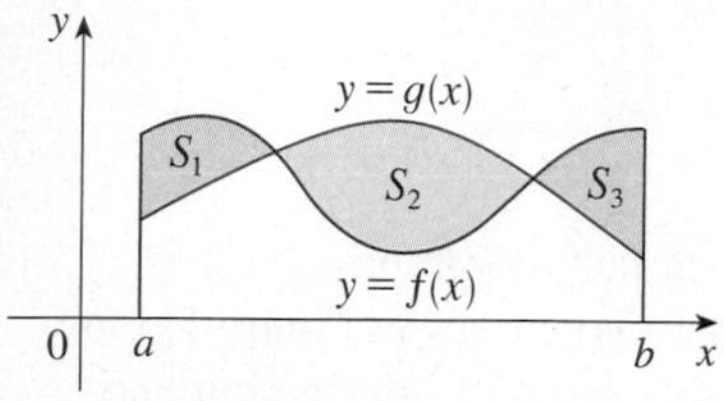

FIGURE 9

If we are asked to find the area between the curves $y = f(x)$ and $y = g(x)$ where $f(x) \geqslant g(x)$ for some values of x but $g(x) \geqslant f(x)$ for other values of x, then we split the given region S into several regions $S_1, S_2, \dots$ with areas $A_1, A_2, \dots$ as shown in Figure 9. We then define the area of the region S to be the sum of the areas of the smaller regions $S_1, S_2, \dots$, that is, $A = A_1 + A_2 + \cdots$. Since

$$|f(x) - g(x)| = \begin{cases} f(x) - g(x) & \text{when } f(x) \geqslant g(x) \\ g(x) - f(x) & \text{when } g(x) \geqslant f(x) \end{cases}$$

we have the following expression for A.

3 The area between the curves $y = f(x)$ and $y = g(x)$ and between $x = a$ and $x = b$ is

$$A = \int_a^b |f(x) - g(x)|\,dx$$

When evaluating the integral in (3), however, we must still split it into integrals corresponding to $A_1, A_2, \dots$.

V EXAMPLE 5 Find the area of the region bounded by the curves $y = \sin x$, $y = \cos x$, $x = 0$, and $x = \pi/2$.

SOLUTION The points of intersection occur when $\sin x = \cos x$, that is, when $x = \pi/4$ (since $0 \leqslant x \leqslant \pi/2$). The region is sketched in Figure 10. Observe that $\cos x \geqslant \sin x$ when $0 \leqslant x \leqslant \pi/4$ but $\sin x \geqslant \cos x$ when $\pi/4 \leqslant x \leqslant \pi/2$. Therefore the required area is

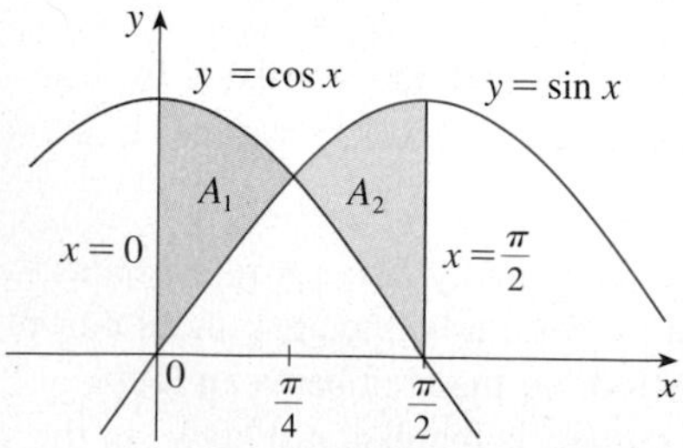

FIGURE 10

$$\begin{aligned} A &= \int_0^{\pi/2} |\cos x - \sin x|\,dx = A_1 + A_2 \\ &= \int_0^{\pi/4} (\cos x - \sin x)\,dx + \int_{\pi/4}^{\pi/2} (\sin x - \cos x)\,dx \\ &= \Big[\sin x + \cos x\Big]_0^{\pi/4} + \Big[-\cos x - \sin x\Big]_{\pi/4}^{\pi/2} \\ &= \left(\frac{1}{\sqrt{2}} + \frac{1}{\sqrt{2}} - 0 - 1\right) + \left(-0 - 1 + \frac{1}{\sqrt{2}} + \frac{1}{\sqrt{2}}\right) \\ &= 2\sqrt{2} - 2 \end{aligned}$$

In this particular example we could have saved some work by noticing that the region is symmetric about $x = \pi/4$ and so

$$A = 2A_1 = 2\int_0^{\pi/4} (\cos x - \sin x)\,dx$$

□

Some regions are best treated by regarding x as a function of y. If a region is bounded by curves with equations $x = f(y)$, $x = g(y)$, $y = c$, and $y = d$, where f and g are continuous and $f(y) \geqslant g(y)$ for $c \leqslant y \leqslant d$ (see Figure 11), then its area is

$$A = \int_c^d [f(y) - g(y)]\,dy$$

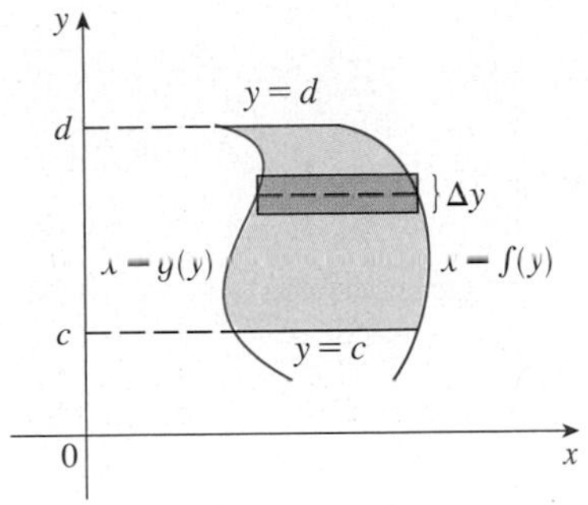

FIGURE 11

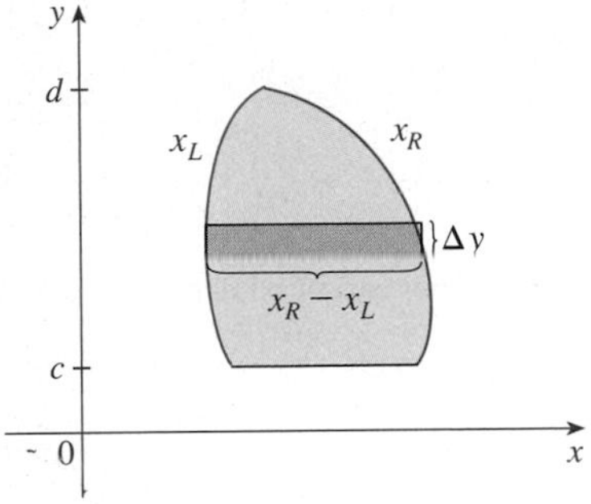

FIGURE 12

If we write x_R for the right boundary and x_L for the left boundary, then, as Figure 12 illustrates, we have

$$A = \int_c^d (x_R - x_L)\,dy$$

Here a typical approximating rectangle has dimensions $x_R - x_L$ and Δy.

EXAMPLE 6 Find the area enclosed by the line $y = x - 1$ and the parabola $y^2 = 2x + 6$.

SOLUTION By solving the two equations we find that the points of intersection are $(-1, -2)$ and $(5, 4)$. We solve the equation of the parabola for x and notice from Figure 13 that the left and right boundary curves are

$$x_L = \tfrac{1}{2}y^2 - 3 \qquad x_R = y + 1$$

We must integrate between the appropriate y-values, $y = -2$ and $y = 4$. Thus

$$\begin{aligned} A &= \int_{-2}^{4} (x_R - x_L)\,dy \\ &= \int_{-2}^{4} \left[(y + 1) - \left(\tfrac{1}{2}y^2 - 3\right)\right] dy \\ &= \int_{-2}^{4} \left(-\tfrac{1}{2}y^2 + y + 4\right) dy \\ &= -\frac{1}{2}\left(\frac{y^3}{3}\right) + \frac{y^2}{2} + 4y \Bigg]_{-2}^{4} \\ &= -\tfrac{1}{6}(64) + 8 + 16 - \left(\tfrac{4}{3} + 2 - 8\right) = 18 \end{aligned}$$

□

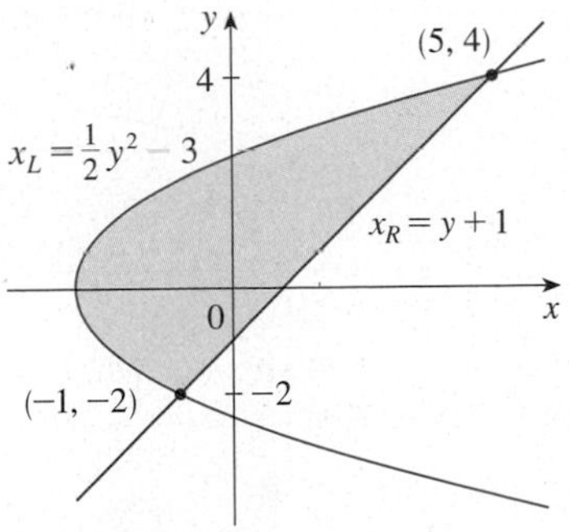

FIGURE 13

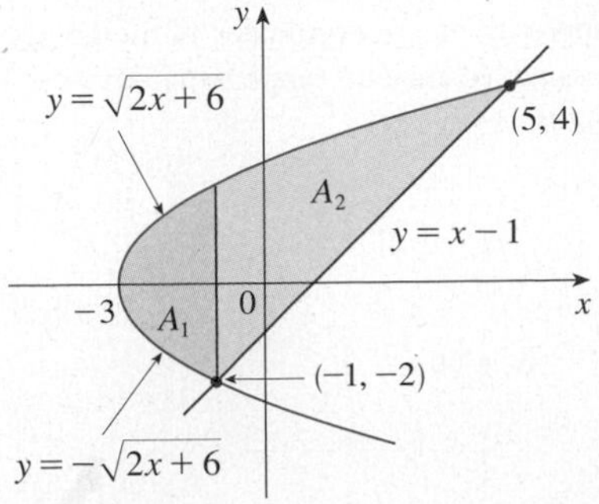

FIGURE 14

We could have found the area in Example 6 by integrating with respect to x instead of y, but the calculation is much more involved. It would have meant splitting the region in two and computing the areas labeled A_1 and A_2 in Figure 14. The method we used in Example 6 is *much* easier.

6.1 EXERCISES

1–4 Find the area of the shaded region.

1.

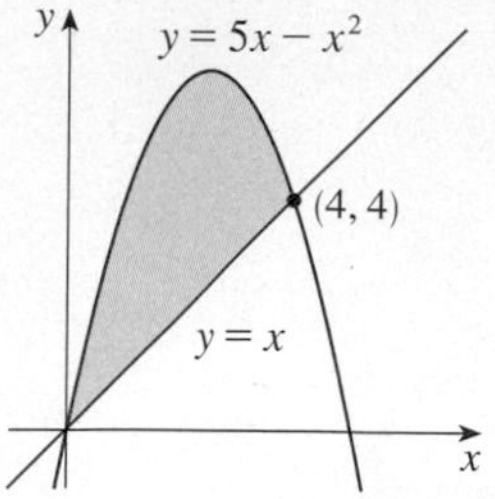

2.

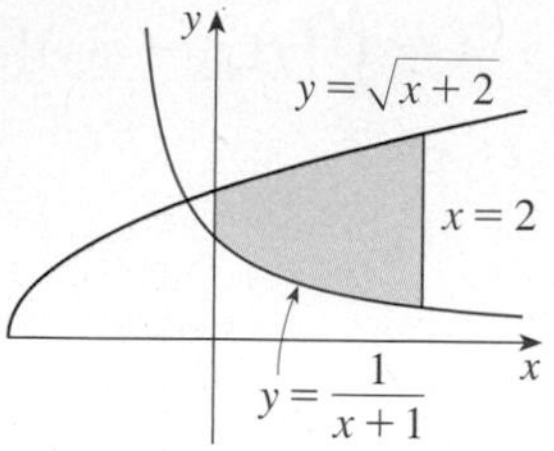

3.

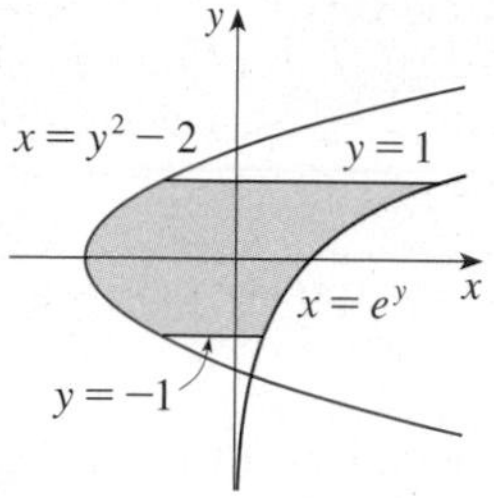

4.

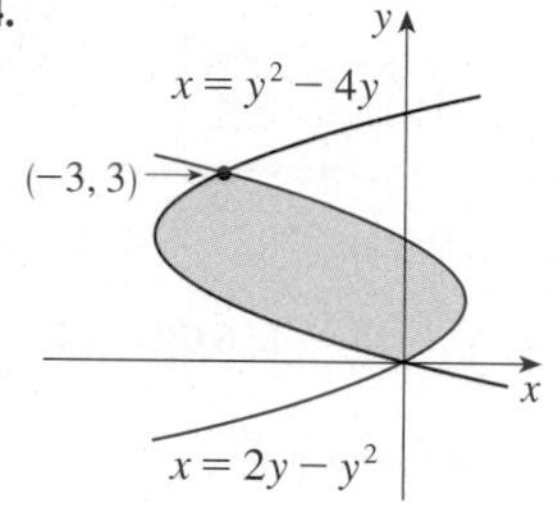

5–28 Sketch the region enclosed by the given curves. Decide whether to integrate with respect to x or y. ~~Draw a typical approximating rectangle and label its height and width.~~ Then find the area of the region.

5. $y = x + 1, \quad y = 9 - x^2, \quad x = -1, \quad x = 2$

6. $y = \sin x, \quad y = e^x, \quad x = 0, \quad x = \pi/2$

7. $y = x, \quad y = x^2$

8. $y = x^2 - 2x, \quad y = x + 4$

9. $y = 1/x, \quad y = 1/x^2, \quad x = 2$

10. $y = 1 + \sqrt{x}, \quad y = (3 + x)/3$

11. $y = x^2, \quad y^2 = x$

12. $y = x^2, \quad y = 4x - x^2$

13. $y = 12 - x^2, \quad y = x^2 - 6$

14. $y = \cos x, \quad y = 2 - \cos x, \quad 0 \leqslant x \leqslant 2\pi$

15. $y = \tan x, \quad y = 2\sin x, \quad -\pi/3 \leqslant x \leqslant \pi/3$

16. $y = x^3 - x, \quad y = 3x$

17. $y = \sqrt{x}, \quad y = \frac{1}{2}x, \quad x = 9$

18. $y = 8 - x^2, \quad y = x^2, \quad x = -3, \quad x = 3$

19. $x = 2y^2, \quad x = 4 + y^2$

20. $4x + y^2 = 12, \quad x = y$

21. $x = 1 - y^2, \quad x = y^2 - 1$

22. $y = \sin(\pi x/2), \quad y = x$

23. $y = \cos x, \quad y = \sin 2x, \quad x = 0, \quad x = \pi/2$

24. $y = \cos x, \quad y = 1 - \cos x, \quad 0 \leqslant x \leqslant \pi$

25. $y = x^2, \quad y = 2/(x^2 + 1)$

26. $y = |x|, \quad y = x^2 - 2$

27. $y = 1/x, \quad y = x, \quad y = \frac{1}{4}x, \quad x > 0$

28. $y = 3x^2, \quad y = 8x^2, \quad 4x + y = 4, \quad x \geqslant 0$

29–30 Use calculus to find the area of the triangle with the given vertices.

29. (0, 0), (2, 1), (−1, 6)

30. (0, 5), (2, −2), (5, 1)

31–32 Evaluate the integral and interpret it as the area of a region. Sketch the region.

31. $\displaystyle\int_0^{\pi/2} |\sin x - \cos 2x|\, dx$

32. $\displaystyle\int_0^{4} |\sqrt{x+2} - x|\, dx$

33–34 Use the Midpoint Rule with $n = 4$ to approximate the area of the region bounded by the given curves.

33. $y = \sin^2(\pi x/4), \quad y = \cos^2(\pi x/4), \quad 0 \leqslant x \leqslant 1$

34. $y = \sqrt[3]{16 - x^3}, \quad y = x, \quad x = 0$

35–38 Use a graph to find approximate x-coordinates of the points of intersection of the given curves. Then find (approximately) the area of the region bounded by the curves.

35. $y = x\sin(x^2), \quad y = x^4$

36. $y = e^x, \quad y = 2 - x^2$

37. $y = 3x^2 - 2x, \quad y = x^3 - 3x + 4$

38. $y = x\cos x, \quad y = x^{10}$

CAS **39.** Use a computer algebra system to find the exact area enclosed by the curves $y = x^5 - 6x^3 + 4x$ and $y = x$.

40. Sketch the region in the xy-plane defined by the inequalities $x - 2y^2 \geqslant 0$, $1 - x - |y| \geqslant 0$ and find its area.

41. Racing cars driven by Chris and Kelly are side by side at the start of a race. The table shows the velocities of each car (in miles per hour) during the first ten seconds of the race. Use the Midpoint Rule to estimate how much farther Kelly travels than Chris does during the first ten seconds.

t	v_C	v_K	t	v_C	v_K
0	0	0	6	69	80
1	20	22	7	75	86
2	32	37	8	81	93
3	46	52	9	86	98
4	54	61	10	90	102
5	62	71			

42. The widths (in meters) of a kidney-shaped swimming pool were measured at 2-meter intervals as indicated in the figure. Use the Midpoint Rule to estimate the area of the pool.

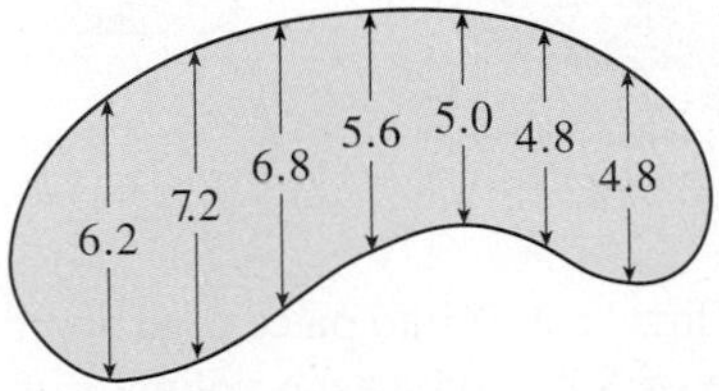

43. A cross-section of an airplane wing is shown. Measurements of the thickness of the wing, in centimeters, at 20-centimeter intervals are 5.8, 20.3, 26.7, 29.0, 27.6, 27.3, 23.8, 20.5, 15.1, 8.7, and 2.8. Use the Midpoint Rule to estimate the area of the wing's cross-section.

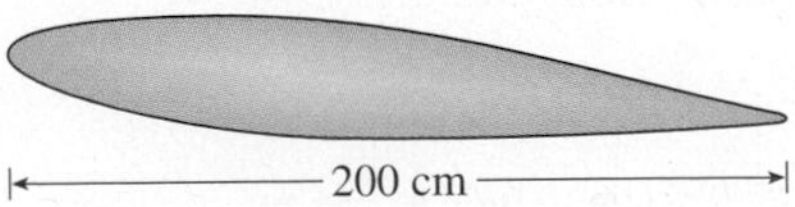

44. If the birth rate of a population is $b(t) = 2200e^{0.024t}$ people per year and the death rate is $d(t) = 1460e^{0.018t}$ people per year, find the area between these curves for $0 \leqslant t \leqslant 10$. What does this area represent?

45. Two cars, A and B, start side by side and accelerate from rest. The figure shows the graphs of their velocity functions.

(a) Which car is ahead after one minute? Explain.

(b) What is the meaning of the area of the shaded region?

(c) Which car is ahead after two minutes? Explain.

(d) Estimate the time at which the cars are again side by side.

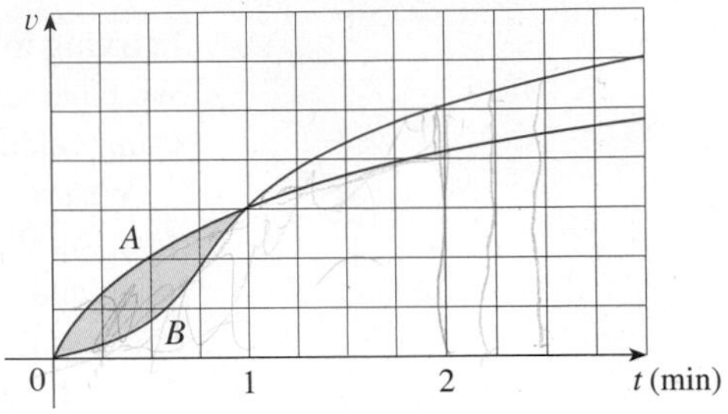

46. The figure shows graphs of the marginal revenue function R' and the marginal cost function C' for a manufacturer. [Recall from Section 4.7 that $R(x)$ and $C(x)$ represent the revenue and cost when x units are manufactured. Assume that R and C are measured in thousands of dollars.] What is the meaning of the area of the shaded region? Use the Midpoint Rule to estimate the value of this quantity.

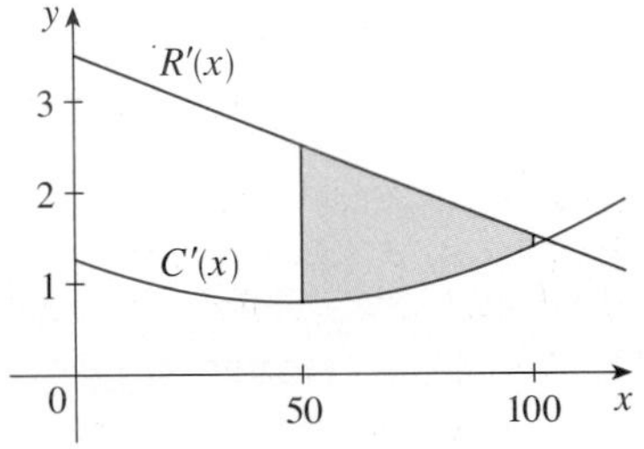

47. The curve with equation $y^2 = x^2(x + 3)$ is called **Tschirnhausen's cubic**. If you graph this curve you will see that part of the curve forms a loop. Find the area enclosed by the loop.

48. Find the area of the region bounded by the parabola $y = x^2$, the tangent line to this parabola at $(1, 1)$, and the x-axis.

49. Find the number b such that the line $y = b$ divides the region bounded by the curves $y = x^2$ and $y = 4$ into two regions with equal area.

50. (a) Find the number a such that the line $x = a$ bisects the area under the curve $y = 1/x^2$, $1 \leqslant x \leqslant 4$.

(b) Find the number b such that the line $y = b$ bisects the area in part (a).

51. Find the values of c such that the area of the region bounded by the parabolas $y = x^2 - c^2$ and $y = c^2 - x^2$ is 576.

52. Suppose that $0 < c < \pi/2$. For what value of c is the area of the region enclosed by the curves $y = \cos x$, $y = \cos(x - c)$, and $x = 0$ equal to the area of the region enclosed by the curves $y = \cos(x - c)$, $x = \pi$, and $y = 0$?

53. For what values of m do the line $y = mx$ and the curve $y = x/(x^2 + 1)$ enclose a region? Find the area of the region.

6.2 VOLUMES

In trying to find the volume of a solid we face the same type of problem as in finding areas. We have an intuitive idea of what volume means, but we must make this idea precise by using calculus to give an exact definition of volume.

We start with a simple type of solid called a **cylinder** (or, more precisely, a *right cylinder*). As illustrated in Figure 1(a), a cylinder is bounded by a plane region B_1, called the **base**, and a congruent region B_2 in a parallel plane. The cylinder consists of all points on line segments that are perpendicular to the base and join B_1 to B_2. If the area of the base is A and the height of the cylinder (the distance from B_1 to B_2) is h, then the volume V of the cylinder is defined as

$$V = Ah$$

In particular, if the base is a circle with radius r, then the cylinder is a circular cylinder with volume $V = \pi r^2 h$ [see Figure 1(b)], and if the base is a rectangle with length l and width w, then the cylinder is a rectangular box (also called a *rectangular parallelepiped*) with volume $V = lwh$ [see Figure 1(c)].

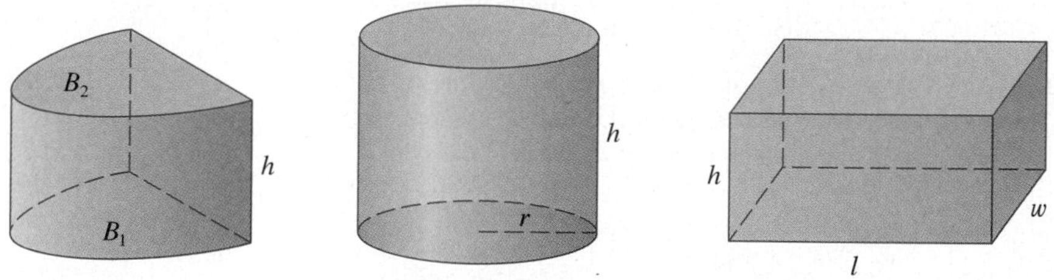

FIGURE 1

(a) Cylinder $V = Ah$ (b) Circular cylinder $V = \pi r^2 h$ (c) Rectangular box $V = lwh$

For a solid S that isn't a cylinder we first "cut" S into pieces and approximate each piece by a cylinder. We estimate the volume of S by adding the volumes of the cylinders. We arrive at the exact volume of S through a limiting process in which the number of pieces becomes large.

We start by intersecting S with a plane and obtaining a plane region that is called a **cross-section** of S. Let $A(x)$ be the area of the cross-section of S in a plane P_x perpendicular to the x-axis and passing through the point x, where $a \leqslant x \leqslant b$. (See Figure 2. Think of slicing S with a knife through x and computing the area of this slice.) The cross-sectional area $A(x)$ will vary as x increases from a to b.

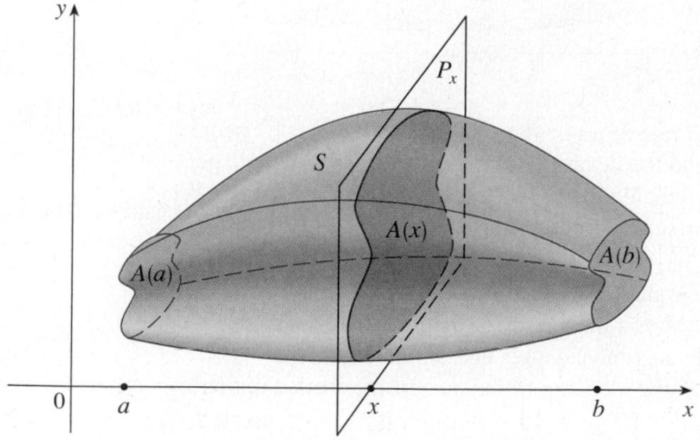

FIGURE 2

Let's divide S into n "slabs" of equal width Δx by using the planes $P_{x_1}, P_{x_2}, \ldots$ to slice the solid. (Think of slicing a loaf of bread.) If we choose sample points x_i^* in $[x_{i-1}, x_i]$, we can approximate the ith slab S_i (the part of S that lies between the planes $P_{x_{i-1}}$ and P_{x_i}) by a cylinder with base area $A(x_i^*)$ and "height" Δx. (See Figure 3.)

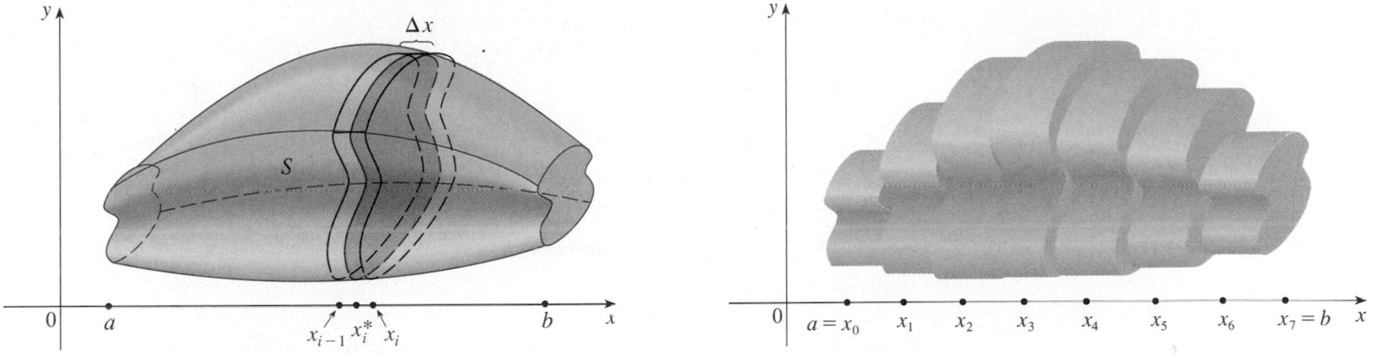

FIGURE 3

The volume of this cylinder is $A(x_i^*)\,\Delta x$, so an approximation to our intuitive conception of the volume of the ith slab S_i is

$$V(S_i) \approx A(x_i^*)\,\Delta x$$

Adding the volumes of these slabs, we get an approximation to the total volume (that is, what we think of intuitively as the volume):

$$V \approx \sum_{i=1}^{n} A(x_i^*)\,\Delta x$$

This approximation appears to become better and better as $n \to \infty$. (Think of the slices as becoming thinner and thinner.) Therefore, we *define* the volume as the limit of these sums as $n \to \infty$. But we recognize the limit of Riemann sums as a definite integral and so we have the following definition.

■ It can be proved that this definition is independent of how S is situated with respect to the x-axis. In other words, no matter how we slice S with parallel planes, we always get the same answer for V.

DEFINITION OF VOLUME Let S be a solid that lies between $x = a$ and $x = b$. If the cross-sectional area of S in the plane P_x, through x and perpendicular to the x-axis, is $A(x)$, where A is a continuous function, then the **volume** of S is

$$V = \lim_{n\to\infty} \sum_{i=1}^{n} A(x_i^*)\,\Delta x = \int_a^b A(x)\,dx$$

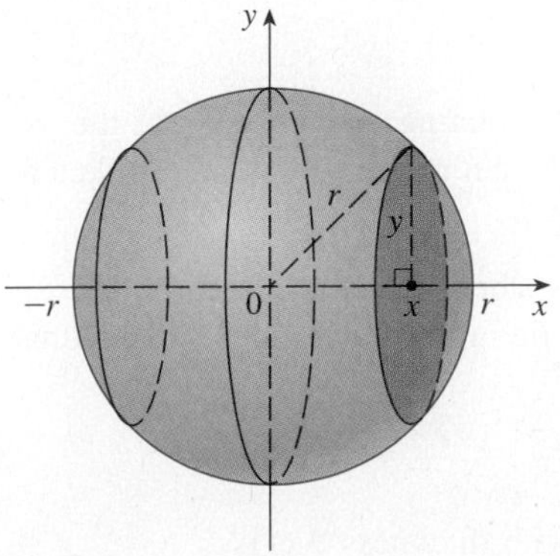

FIGURE 4

When we use the volume formula $V = \int_a^b A(x)\,dx$, it is important to remember that $A(x)$ is the area of a moving cross-section obtained by slicing through x perpendicular to the x-axis.

Notice that, for a cylinder, the cross-sectional area is constant: $A(x) = A$ for all x. So our definition of volume gives $V = \int_a^b A\,dx = A(b - a)$; this agrees with the formula $V = Ah$.

EXAMPLE 1 Show that the volume of a sphere of radius r is $V = \frac{4}{3}\pi r^3$.

SOLUTION If we place the sphere so that its center is at the origin (see Figure 4), then the plane P_x intersects the sphere in a circle whose radius (from the Pythagorean Theorem) is $y = \sqrt{r^2 - x^2}$. So the cross-sectional area is

$$A(x) = \pi y^2 = \pi(r^2 - x^2)$$

Using the definition of volume with $a = -r$ and $b = r$, we have

$$V = \int_{-r}^{r} A(x)\,dx = \int_{-r}^{r} \pi(r^2 - x^2)\,dx$$

$$= 2\pi \int_{0}^{r} (r^2 - x^2)\,dx \qquad \text{(The integrand is even.)}$$

$$= 2\pi\left[r^2x - \frac{x^3}{3}\right]_0^r = 2\pi\left(r^3 - \frac{r^3}{3}\right)$$

$$= \tfrac{4}{3}\pi r^3$$

□

Figure 5 illustrates the definition of volume when the solid is a sphere with radius $r = 1$. From the result of Example 1, we know that the volume of the sphere is $\frac{4}{3}\pi \approx 4.18879$. Here the slabs are circular cylinders, or *disks*, and the three parts of Figure 5 show the geometric interpretations of the Riemann sums

$$\sum_{i=1}^{n} A(\bar{x}_i)\,\Delta x = \sum_{i=1}^{n} \pi(1^2 - \bar{x}_i^2)\,\Delta x$$

TEC Visual 6.2A shows an animation of Figure 5.

when $n = 5$, 10, and 20 if we choose the sample points x_i^* to be the midpoints $\bar{x}_i$. Notice that as we increase the number of approximating cylinders, the corresponding Riemann sums become closer to the true volume.

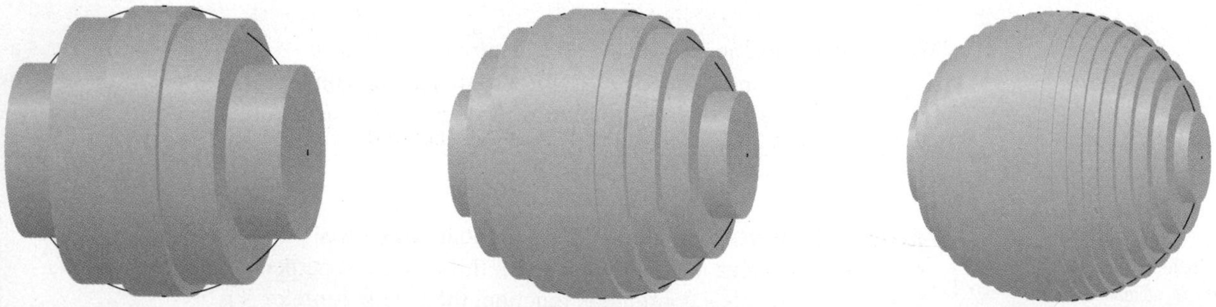

(a) Using 5 disks, $V \approx 4.2726$ (b) Using 10 disks, $V \approx 4.2097$ (c) Using 20 disks, $V \approx 4.1940$

FIGURE 5 Approximating the volume of a sphere with radius 1

V EXAMPLE 2 Find the volume of the solid obtained by rotating about the x-axis the region under the curve $y = \sqrt{x}$ from 0 to 1. Illustrate the definition of volume by sketching a typical approximating cylinder.

SOLUTION The region is shown in Figure 6(a). If we rotate about the x-axis, we get the solid shown in Figure 6(b). When we slice through the point x, we get a disk with radius $\sqrt{x}$. The area of this cross-section is

$$A(x) = \pi\left(\sqrt{x}\right)^2 = \pi x$$

and the volume of the approximating cylinder (a disk with thickness Δx) is

$$A(x)\,\Delta x = \pi x\,\Delta x$$

■ Did we get a reasonable answer in Example 2? As a check on our work, let's replace the given region by a square with base [0, 1] and height 1. If we rotate this square, we get a cylinder with radius 1, height 1, and volume $\pi \cdot 1^2 \cdot 1 = \pi$. We computed that the given solid has half this volume. That seems about right.

The solid lies between $x = 0$ and $x = 1$, so its volume is

$$V = \int_0^1 A(x)\,dx = \int_0^1 \pi x\,dx = \pi \frac{x^2}{2}\Big]_0^1 = \frac{\pi}{2}$$

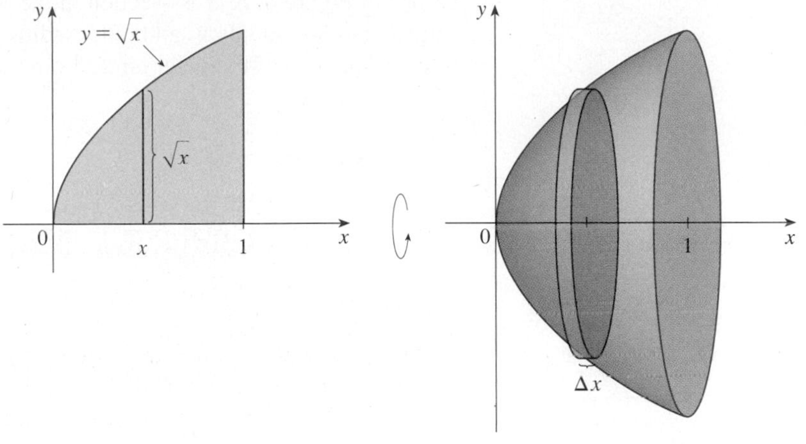

FIGURE 6 (a) (b)

□

V EXAMPLE 3 Find the volume of the solid obtained by rotating the region bounded by $y = x^3$, $y = 8$, and $x = 0$ about the y-axis.

SOLUTION The region is shown in Figure 7(a) and the resulting solid is shown in Figure 7(b). Because the region is rotated about the y-axis, it makes sense to slice the solid perpendicular to the y-axis and therefore to integrate with respect to y. If we slice at height y, we get a circular disk with radius x, where $x = \sqrt[3]{y}$. So the area of a cross-section through y is

$$A(y) = \pi x^2 = \pi(\sqrt[3]{y})^2 = \pi y^{2/3}$$

and the volume of the approximating cylinder pictured in Figure 7(b) is

$$A(y)\,\Delta y = \pi y^{2/3}\,\Delta y$$

Since the solid lies between $y = 0$ and $y = 8$, its volume is

$$V = \int_0^8 A(y)\,dy = \int_0^8 \pi y^{2/3}\,dy = \pi\left[\tfrac{3}{5}y^{5/3}\right]_0^8 = \frac{96\pi}{5}$$

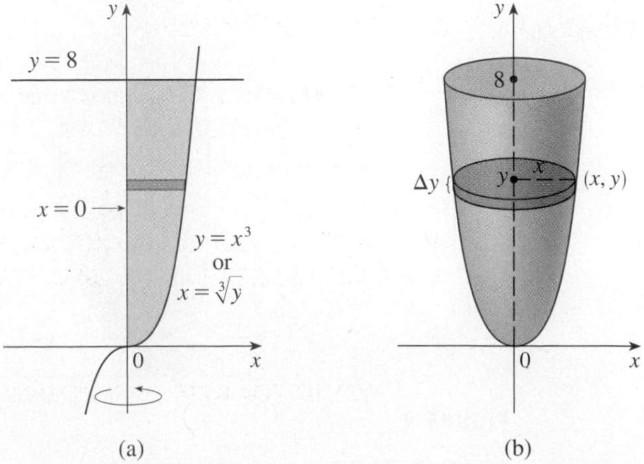

FIGURE 7 (a) (b)

□

EXAMPLE 4 The region $\mathcal{R}$ enclosed by the curves $y = x$ and $y = x^2$ is rotated about the x-axis. Find the volume of the resulting solid.

SOLUTION The curves $y = x$ and $y = x^2$ intersect at the points (0, 0) and (1, 1). The region between them, the solid of rotation, and a cross-section perpendicular to the x-axis are shown in Figure 8. A cross-section in the plane P_x has the shape of a *washer* (an annular ring) with inner radius x^2 and outer radius x, so we find the cross-sectional area by subtracting the area of the inner circle from the area of the outer circle:

$$A(x) = \pi x^2 - \pi(x^2)^2 = \pi(x^2 - x^4)$$

Therefore we have

$$V = \int_0^1 A(x)\,dx = \int_0^1 \pi(x^2 - x^4)\,dx = \pi\left[\frac{x^3}{3} - \frac{x^5}{5}\right]_0^1 = \frac{2\pi}{15}$$

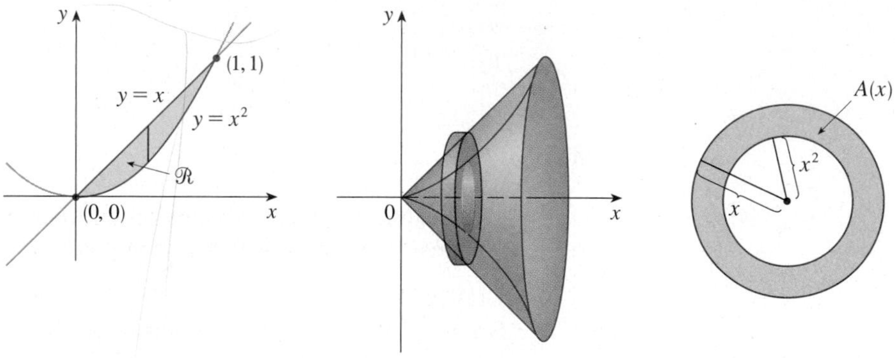

FIGURE 8 (a) (b) (c) □

EXAMPLE 5 Find the volume of the solid obtained by rotating the region in Example 4 about the line $y = 2$.

SOLUTION The solid and a cross-section are shown in Figure 9. Again the cross-section is a washer, but this time the inner radius is $2 - x$ and the outer radius is $2 - x^2$.

TEC Visual 6.2B shows how solids of revolution are formed.

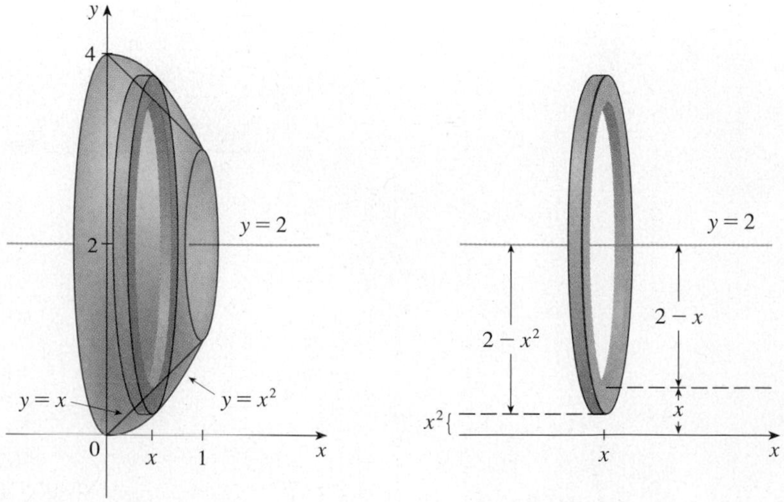

FIGURE 9

The cross-sectional area is

$$A(x) = \pi(2 - x^2)^2 - \pi(2 - x)^2$$

and so the volume of S is

$$\begin{aligned} V &= \int_0^1 A(x)\,dx \\ &= \pi \int_0^1 [(2 - x^2)^2 - (2 - x)^2]\,dx \\ &= \pi \int_0^1 (x^4 - 5x^2 + 4x)\,dx \\ &= \pi \left[\frac{x^5}{5} - 5\frac{x^3}{3} + 4\frac{x^2}{2} \right]_0^1 \\ &= \frac{8\pi}{15} \end{aligned}$$

□

The solids in Examples 1–5 are all called **solids of revolution** because they are obtained by revolving a region about a line. In general, we calculate the volume of a solid of revolution by using the basic defining formula

$$V = \int_a^b A(x)\,dx \qquad \text{or} \qquad V = \int_c^d A(y)\,dy$$

and we find the cross-sectional area $A(x)$ or $A(y)$ in one of the following ways:

- If the cross-section is a disk (as in Examples 1–3), we find the radius of the disk (in terms of x or y) and use

$$A = \pi(\text{radius})^2$$

- If the cross-section is a washer (as in Examples 4 and 5), we find the inner radius r_{in} and outer radius r_{out} from a sketch (as in Figures 8, 9, and 10) and compute the area of the washer by subtracting the area of the inner disk from the area of the outer disk:

$$A = \pi(\text{outer radius})^2 - \pi(\text{inner radius})^2$$

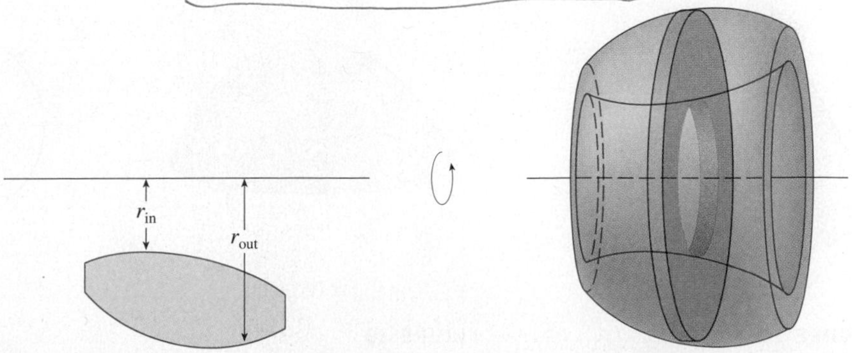

FIGURE 10

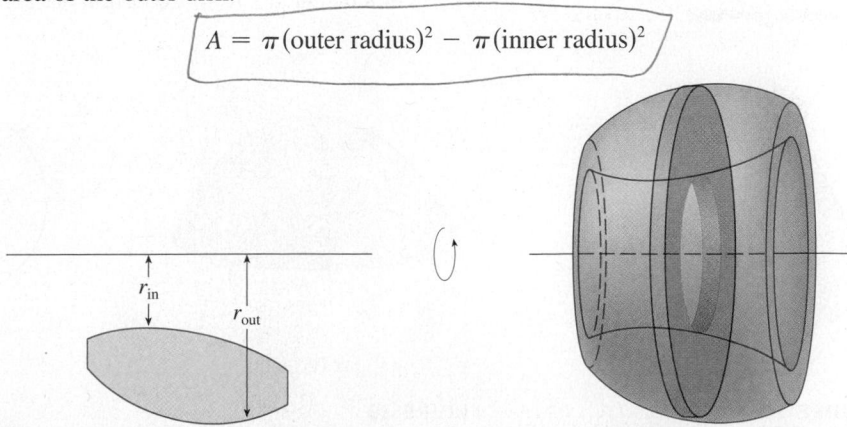

The next example gives a further illustration of the procedure.

EXAMPLE 6 Find the volume of the solid obtained by rotating the region in Example 4 about the line $x = -1$.

SOLUTION Figure 11 shows a horizontal cross-section. It is a washer with inner radius $1 + y$ and outer radius $1 + \sqrt{y}$, so the cross-sectional area is

$$\begin{aligned} A(y) &= \pi(\text{outer radius})^2 - \pi(\text{inner radius})^2 \\ &= \pi\left(1 + \sqrt{y}\right)^2 - \pi(1 + y)^2 \end{aligned}$$

The volume is

$$\begin{aligned} V &= \int_0^1 A(y)\,dy = \pi \int_0^1 \left[\left(1 + \sqrt{y}\right)^2 - (1 + y)^2\right] dy \\ &= \pi \int_0^1 \left(2\sqrt{y} - y - y^2\right) dy = \pi\left[\frac{4y^{3/2}}{3} - \frac{y^2}{2} - \frac{y^3}{3}\right]_0^1 = \frac{\pi}{2} \end{aligned}$$

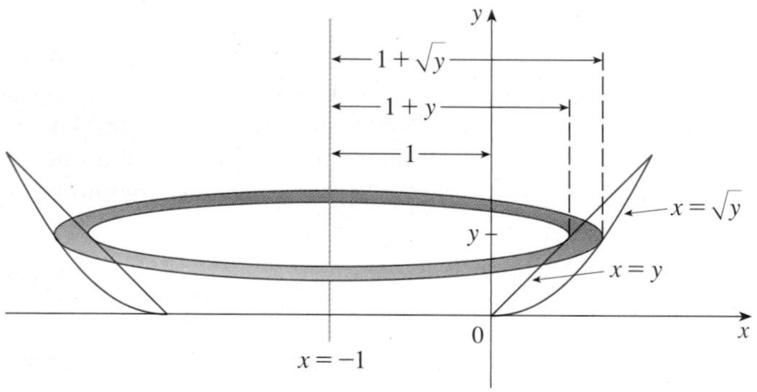

FIGURE 11

□

We now find the volumes of three solids that are *not* solids of revolution.

EXAMPLE 7 Figure 12 shows a solid with a circular base of radius 1. Parallel cross-sections perpendicular to the base are equilateral triangles. Find the volume of the solid.

SOLUTION Let's take the circle to be $x^2 + y^2 = 1$. The solid, its base, and a typical cross-section at a distance x from the origin are shown in Figure 13.

TEC Visual 6.2C shows how the solid in Figure 12 is generated.

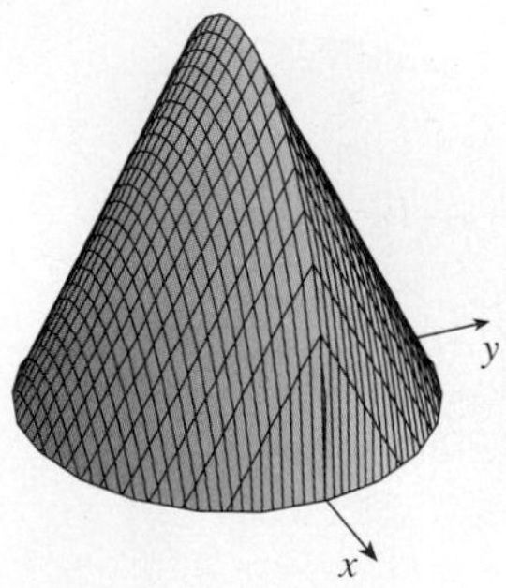

FIGURE 12
Computer-generated picture of the solid in Example 7

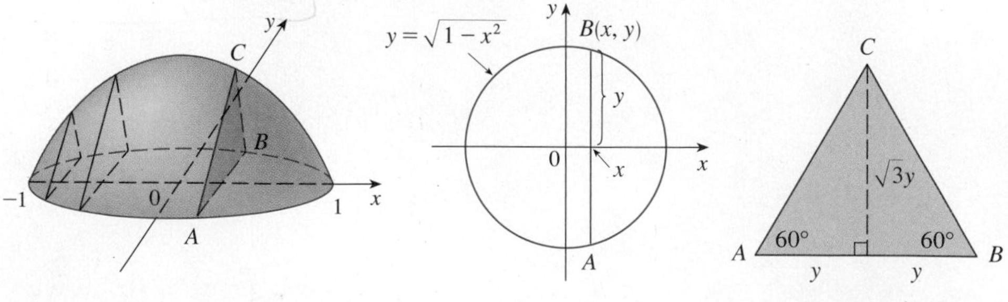

(a) The solid (b) Its base (c) A cross-section

FIGURE 13

Since B lies on the circle, we have $y = \sqrt{1 - x^2}$ and so the base of the triangle ABC is $|AB| = 2\sqrt{1 - x^2}$. Since the triangle is equilateral, we see from Figure 13(c) that its height is $\sqrt{3}\, y = \sqrt{3}\sqrt{1 - x^2}$. The cross-sectional area is therefore

$$A(x) = \tfrac{1}{2} \cdot 2\sqrt{1 - x^2} \cdot \sqrt{3}\sqrt{1 - x^2} = \sqrt{3}\,(1 - x^2)$$

and the volume of the solid is

$$V = \int_{-1}^{1} A(x)\,dx = \int_{-1}^{1} \sqrt{3}\,(1 - x^2)\,dx$$

$$= 2\int_{0}^{1} \sqrt{3}\,(1 - x^2)\,dx = 2\sqrt{3}\left[x - \frac{x^3}{3}\right]_0^1 = \frac{4\sqrt{3}}{3} \qquad \square$$

V EXAMPLE 8 Find the volume of a pyramid whose base is a square with side L and whose height is h.

SOLUTION We place the origin O at the vertex of the pyramid and the x-axis along its central axis as in Figure 14. Any plane P_x that passes through x and is perpendicular to the x-axis intersects the pyramid in a square with side of length s, say. We can express s in terms of x by observing from the similar triangles in Figure 15 that

$$\frac{x}{h} = \frac{s/2}{L/2} = \frac{s}{L}$$

and so $s = Lx/h$. [Another method is to observe that the line OP has slope $L/(2h)$ and so its equation is $y = Lx/(2h)$.] Thus the cross-sectional area is

$$A(x) = s^2 = \frac{L^2}{h^2}x^2$$

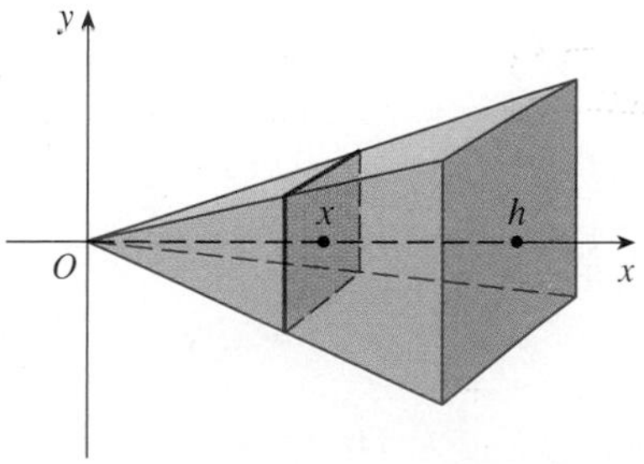

FIGURE 14

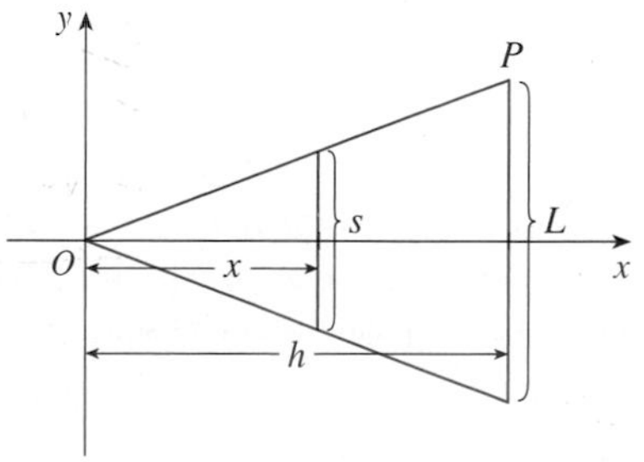

FIGURE 15

The pyramid lies between $x = 0$ and $x = h$, so its volume is

$$V = \int_0^h A(x)\,dx = \int_0^h \frac{L^2}{h^2}x^2\,dx = \frac{L^2}{h^2}\frac{x^3}{3}\bigg]_0^h = \frac{L^2h}{3} \qquad \square$$

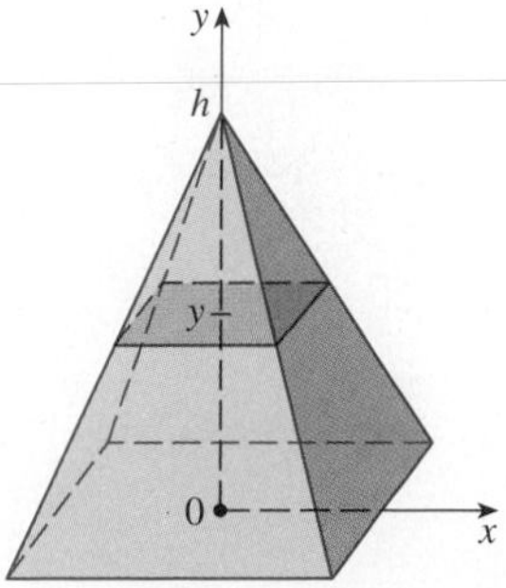

FIGURE 16

NOTE We didn't need to place the vertex of the pyramid at the origin in Example 8. We did so merely to make the equations simple. If, instead, we had placed the center of the base at the origin and the vertex on the positive y-axis, as in Figure 16, you can verify that we would have obtained the integral

$$V = \int_0^h \frac{L^2}{h^2}(h - y)^2\, dy = \frac{L^2 h}{3}$$

EXAMPLE 9 A wedge is cut out of a circular cylinder of radius 4 by two planes. One plane is perpendicular to the axis of the cylinder. The other intersects the first at an angle of 30° along a diameter of the cylinder. Find the volume of the wedge.

SOLUTION If we place the x-axis along the diameter where the planes meet, then the base of the solid is a semicircle with equation $y = \sqrt{16 - x^2}$, $-4 \leqslant x \leqslant 4$. A cross-section perpendicular to the x-axis at a distance x from the origin is a triangle ABC, as shown in Figure 17, whose base is $y = \sqrt{16 - x^2}$ and whose height is $|BC| = y \tan 30° = \sqrt{16 - x^2}/\sqrt{3}$. Thus the cross-sectional area is

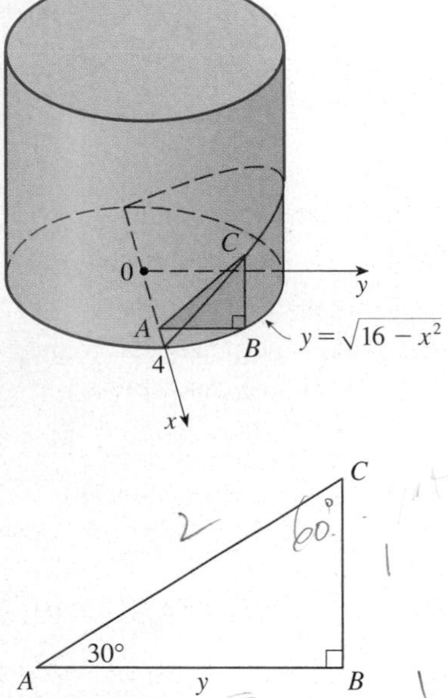

FIGURE 17

$$A(x) = \tfrac{1}{2}\sqrt{16 - x^2} \cdot \frac{1}{\sqrt{3}}\sqrt{16 - x^2} = \frac{16 - x^2}{2\sqrt{3}}$$

and the volume is

$$V = \int_{-4}^{4} A(x)\, dx = \int_{-4}^{4} \frac{16 - x^2}{2\sqrt{3}}\, dx$$

$$= \frac{1}{\sqrt{3}} \int_0^4 (16 - x^2)\, dx = \frac{1}{\sqrt{3}}\left[16x - \frac{x^3}{3}\right]_0^4$$

$$= \frac{128}{3\sqrt{3}}$$

For another method see Exercise 64. □

6.2 EXERCISES

1–18 Find the volume of the solid obtained by rotating the region bounded by the given curves about the specified line. Sketch the region, the solid, and a typical disk or washer.

1. $y = 2 - \frac{1}{2}x$, $y = 0$, $x = 1$, $x = 2$; about the x-axis
2. $y = 1 - x^2$, $y = 0$; about the x-axis
3. $y = 1/x$, $x = 1$, $x = 2$, $y = 0$; about the x-axis
4. $y = \sqrt{25 - x^2}$, $y = 0$, $x = 2$, $x = 4$; about the x-axis
5. $x = 2\sqrt{y}$, $x = 0$, $y = 9$; about the y-axis
6. $y = \ln x$, $y = 1$, $y = 2$, $x = 0$; about the y-axis
7. $y = x^3$, $y = x$, $x \geqslant 0$; about the x-axis
8. $y = \frac{1}{4}x^2$, $y = 5 - x^2$; about the x-axis
9. $y^2 = x$, $x = 2y$; about the y-axis
10. $y = \frac{1}{4}x^2$, $x = 2$, $y = 0$; about the y-axis
11. $y = x$, $y = \sqrt{x}$; about $y = 1$
12. $y = e^{-x}$, $y = 1$, $x = 2$; about $y = 2$
13. $y = 1 + \sec x$, $y = 3$; about $y = 1$
14. $y = 1/x$, $y = 0$, $x = 1$, $x = 3$; about $y = -1$
15. $x = y^2$, $x = 1$; about $x = 1$
16. $y = x$, $y = \sqrt{x}$; about $x = 2$
17. $y = x^2$, $x = y^2$; about $x = -1$
18. $y = x$, $y = 0$, $x = 2$, $x = 4$; about $x = 1$

19–30 Refer to the figure and find the volume generated by rotating the given region about the specified line.

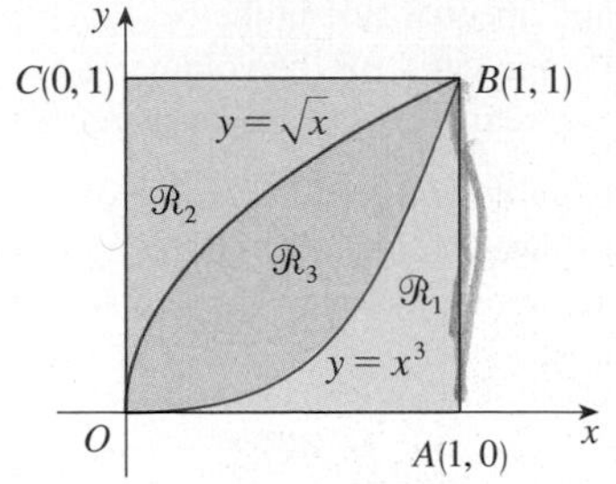

19. $\mathcal{R}_1$ about OA **20.** $\mathcal{R}_1$ about OC

21. $\mathcal{R}_1$ about AB **22.** $\mathcal{R}_1$ about BC

23. $\mathcal{R}_2$ about OA **24.** $\mathcal{R}_2$ about OC

25. $\mathcal{R}_2$ about AB **26.** $\mathcal{R}_2$ about BC

27. $\mathcal{R}_3$ about OA **28.** $\mathcal{R}_3$ about OC

29. $\mathcal{R}_3$ about AB **30.** $\mathcal{R}_3$ about BC

31–36 Set up, but do not evaluate, an integral for the volume of the solid obtained by rotating the region bounded by the given curves about the specified line.

31. $y = \tan^3 x, \; y = 1, \; x = 0$; about $y = 1$

32. $y = (x - 2)^4, \; 8x - y = 16$; about $x = 10$

33. $y = 0, \; y = \sin x, \; 0 \leqslant x \leqslant \pi$; about $y = 1$

34. $y = 0, \; y = \sin x, \; 0 \leqslant x \leqslant \pi$; about $y = -2$

35. $x^2 - y^2 = 1, \; x = 3$; about $x = -2$

36. $y = \cos x, \; y = 2 - \cos x, \; 0 \leqslant x \leqslant 2\pi$; about $y = 4$

37–38 Use a graph to find approximate x-coordinates of the points of intersection of the given curves. Then use your calculator to find (approximately) the volume of the solid obtained by rotating about the x-axis the region bounded by these curves.

37. $y = 2 + x^2 \cos x, \quad y = x^4 + x + 1$

38. $y = 3\sin(x^2), \quad y = e^{x/2} + e^{-2x}$

CAS **39–40** Use a computer algebra system to find the exact volume of the solid obtained by rotating the region bounded by the given curves about the specified line.

39. $y = \sin^2 x, \; y = 0, \; 0 \leqslant x \leqslant \pi$; about $y = -1$

40. $y = x, \; y = xe^{1-x/2}$; about $y = 3$

41–44 Each integral represents the volume of a solid. Describe the solid.

41. $\pi \int_0^{\pi/2} \cos^2 x \, dx$ **42.** $\pi \int_2^5 y \, dy$

43. $\pi \int_0^1 (y^4 - y^8) \, dy$ **44.** $\pi \int_0^{\pi/2} [(1 + \cos x)^2 - 1^2] \, dx$

---

45. A CAT scan produces equally spaced cross-sectional views of a human organ that provide information about the organ otherwise obtained only by surgery. Suppose that a CAT scan of a human liver shows cross-sections spaced 1.5 cm apart. The liver is 15 cm long and the cross-sectional areas, in square centimeters, are 0, 18, 58, 79, 94, 106, 117, 128, 63, 39, and 0. Use the Midpoint Rule to estimate the volume of the liver.

46. A log 10 m long is cut at 1-meter intervals and its cross-sectional areas A (at a distance x from the end of the log) are listed in the table. Use the Midpoint Rule with $n = 5$ to estimate the volume of the log.

x (m)	A (m^2)	x (m)	A (m^2)
0	0.68	6	0.53
1	0.65	7	0.55
2	0.64	8	0.52
3	0.61	9	0.50
4	0.58	10	0.48
5	0.59		

47. (a) If the region shown in the figure is rotated about the x-axis to form a solid, use the Midpoint Rule with $n = 4$ to estimate the volume of the solid.

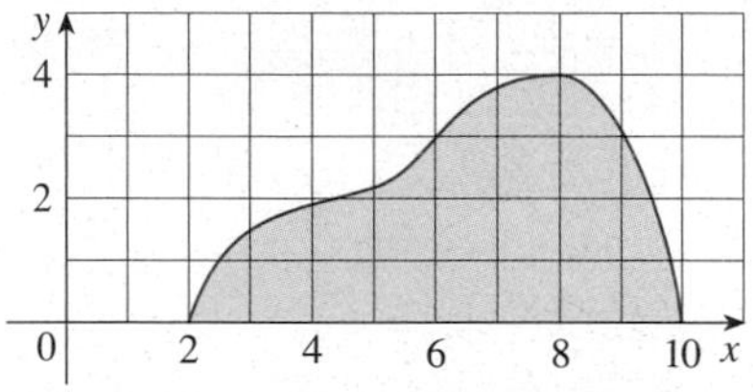

(b) Estimate the volume if the region is rotated about the y-axis. Again use the Midpoint Rule with $n = 4$.

CAS **48.** (a) A model for the shape of a bird's egg is obtained by rotating about the x-axis the region under the graph of

$$f(x) = (ax^3 + bx^2 + cx + d)\sqrt{1 - x^2}$$

Use a CAS to find the volume of such an egg.

(b) For a Red-throated Loon, $a = -0.06$, $b = 0.04$, $c = 0.1$, and $d = 0.54$. Graph f and find the volume of an egg of this species.

49–61 Find the volume of the described solid S.

49. A right circular cone with height h and base radius r

50. A frustum of a right circular cone with height h, lower base radius R, and top radius r

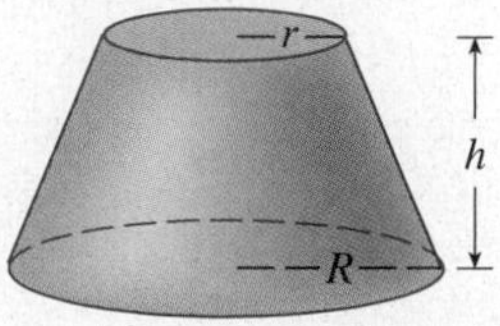

51. A cap of a sphere with radius r and height h

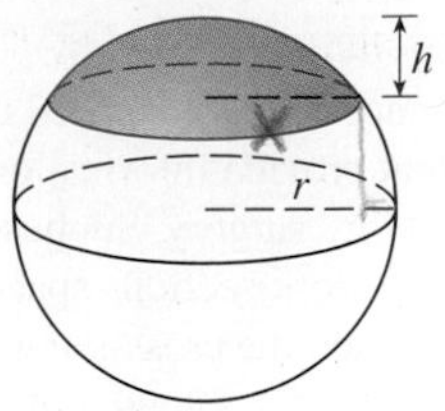

52. A frustum of a pyramid with square base of side b, square top of side a, and height h

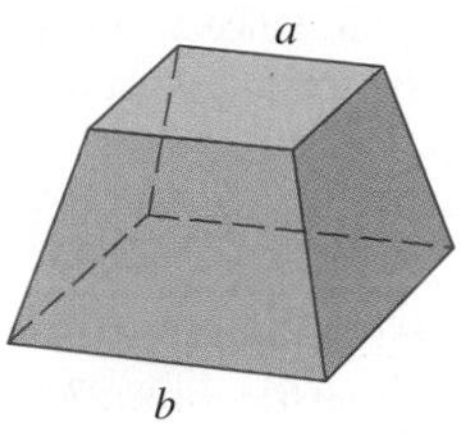

What happens if $a = b$? What happens if $a = 0$?

53. A pyramid with height h and rectangular base with dimensions b and $2b$

54. A pyramid with height h and base an equilateral triangle with side a (a tetrahedron)

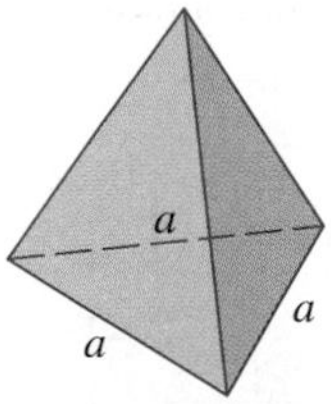

55. A tetrahedron with three mutually perpendicular faces and three mutually perpendicular edges with lengths 3 cm, 4 cm, and 5 cm

56. The base of S is a circular disk with radius r. Parallel cross-sections perpendicular to the base are squares.

57. The base of S is an elliptical region with boundary curve $9x^2 + 4y^2 = 36$. Cross-sections perpendicular to the x-axis are isosceles right triangles with hypotenuse in the base.

58. The base of S is the triangular region with vertices $(0, 0)$, $(1, 0)$, and $(0, 1)$. Cross-sections perpendicular to the y-axis are equilateral triangles.

59. The base of S is the same base as in Exercise 58, but cross-sections perpendicular to the x-axis are squares.

60. The base of S is the region enclosed by the parabola $y = 1 - x^2$ and the x-axis. Cross-sections perpendicular to the y-axis are squares.

61. The base of S is the same base as in Exercise 60, but cross-sections perpendicular to the x-axis are isosceles triangles with height equal to the base.

62. The base of S is a circular disk with radius r. Parallel cross-sections perpendicular to the base are isosceles triangles with height h and unequal side in the base.
(a) Set up an integral for the volume of S.
(b) By interpreting the integral as an area, find the volume of S.

63. (a) Set up an integral for the volume of a solid *torus* (the donut-shaped solid shown in the figure) with radii r and R.
(b) By interpreting the integral as an area, find the volume of the torus.

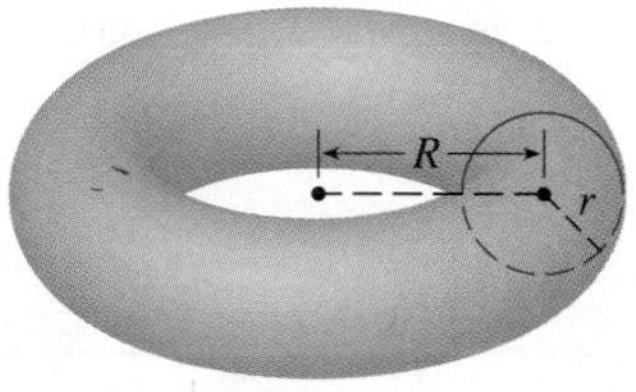

64. Solve Example 9 taking cross-sections to be parallel to the line of intersection of the two planes.

65. (a) Cavalieri's Principle states that if a family of parallel planes gives equal cross-sectional areas for two solids S_1 and S_2, then the volumes of S_1 and S_2 are equal. Prove this principle.
(b) Use Cavalieri's Principle to find the volume of the oblique cylinder shown in the figure.

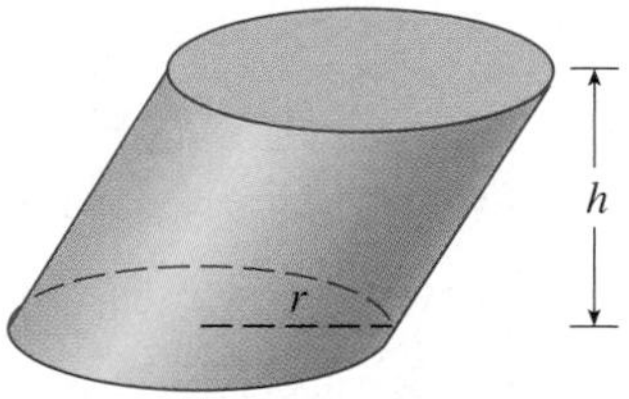

66. Find the volume common to two circular cylinders, each with radius r, if the axes of the cylinders intersect at right angles.

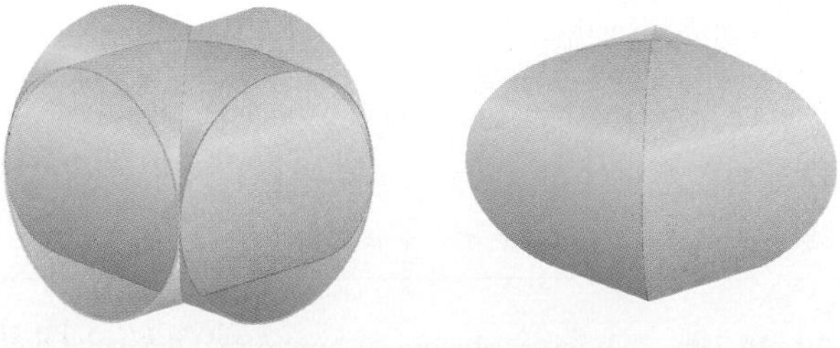

67. Find the volume common to two spheres, each with radius r, if the center of each sphere lies on the surface of the other sphere.

68. A bowl is shaped like a hemisphere with diameter 30 cm. A heavy ball with diameter 10 cm is placed in the bowl and water is poured into the bowl to a depth of h centimeters. Find the volume of water in the bowl.

69. A hole of radius r is bored through the middle of a cylinder of radius $R > r$ at right angles to the axis of the cylinder. Set up, but do not evaluate, an integral for the volume cut out.

70. A hole of radius r is bored through the center of a sphere of radius $R > r$. Find the volume of the remaining portion of the sphere.

71. Some of the pioneers of calculus, such as Kepler and Newton, were inspired by the problem of finding the volumes of wine barrels. (In fact Kepler published a book *Stereometria doliorum* in 1715 devoted to methods for finding the volumes of barrels.) They often approximated the shape of the sides by parabolas.

(a) A barrel with height h and maximum radius R is constructed by rotating about the x-axis the parabola $y = R - cx^2$, $-h/2 \leqslant x \leqslant h/2$, where c is a positive constant. Show that the radius of each end of the barrel is $r = R - d$, where $d = ch^2/4$.

(b) Show that the volume enclosed by the barrel is

$$V = \tfrac{1}{3}\pi h\left(2R^2 + r^2 - \tfrac{2}{5}d^2\right)$$

72. Suppose that a region $\mathcal{R}$ has area A and lies above the x-axis. When $\mathcal{R}$ is rotated about the x-axis, it sweeps out a solid with volume V_1. When $\mathcal{R}$ is rotated about the line $y = -k$ (where k is a positive number), it sweeps out a solid with volume V_2. Express V_2 in terms of V_1, k, and A.

6.3 VOLUMES BY CYLINDRICAL SHELLS

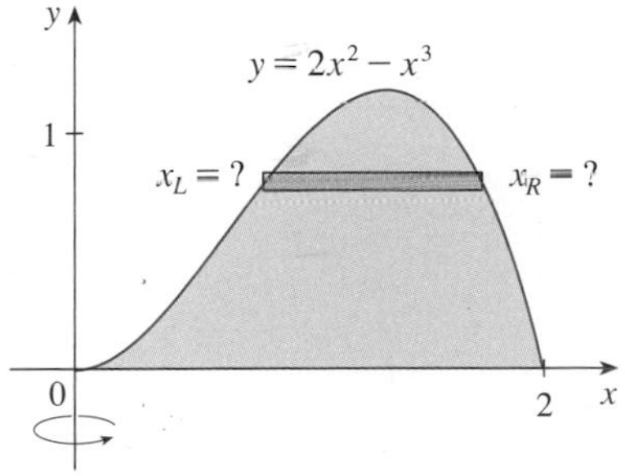

FIGURE 1

Some volume problems are very difficult to handle by the methods of the preceding section. For instance, let's consider the problem of finding the volume of the solid obtained by rotating about the y-axis the region bounded by $y = 2x^2 - x^3$ and $y = 0$. (See Figure 1.) If we slice perpendicular to the y-axis, we get a washer. But to compute the inner radius and the outer radius of the washer, we would have to solve the cubic equation $y = 2x^2 - x^3$ for x in terms of y; that's not easy.

Fortunately, there is a method, called the **method of cylindrical shells**, that is easier to use in such a case. Figure 2 shows a cylindrical shell with inner radius r_1, outer radius r_2, and height h. Its volume V is calculated by subtracting the volume V_1 of the inner cylinder from the volume V_2 of the outer cylinder:

$$V = V_2 - V_1 = \pi r_2^2 h - \pi r_1^2 h = \pi(r_2^2 - r_1^2)h$$

$$= \pi(r_2 + r_1)(r_2 - r_1)h = 2\pi \frac{r_2 + r_1}{2} h(r_2 - r_1)$$

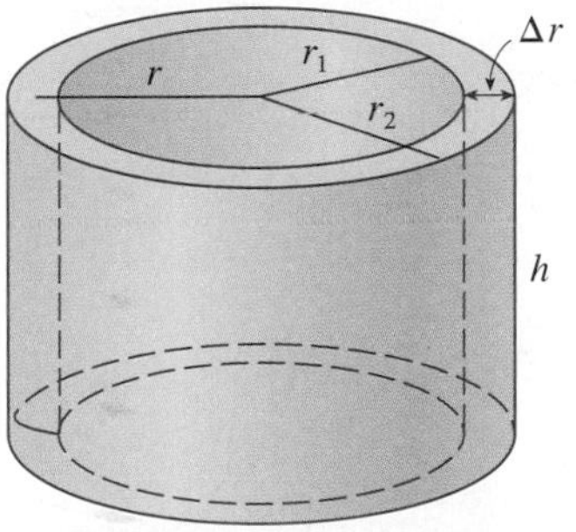

FIGURE 2

If we let $\Delta r = r_2 - r_1$ (the thickness of the shell) and $r = \frac{1}{2}(r_2 + r_1)$ (the average radius of the shell), then this formula for the volume of a cylindrical shell becomes

1 $$V = 2\pi rh\,\Delta r$$

and it can be remembered as

$$V = [\text{circumference}][\text{height}][\text{thickness}]$$

Now let S be the solid obtained by rotating about the y-axis the region bounded by $y = f(x)$ [where $f(x) \geqslant 0$], $y = 0$, $x = a$, and $x = b$, where $b > a \geqslant 0$. (See Figure 3.)

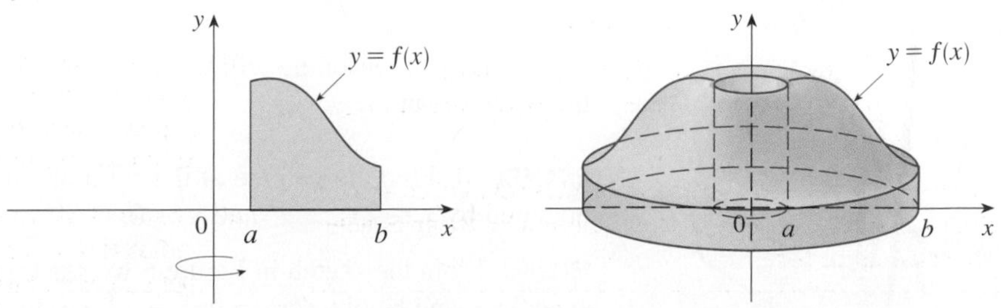

FIGURE 3

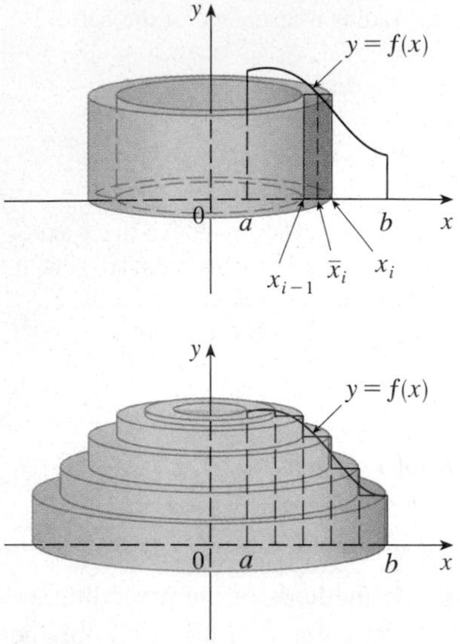

FIGURE 4

We divide the interval $[a, b]$ into n subintervals $[x_{i-1}, x_i]$ of equal width Δx and let $\overline{x}_i$ be the midpoint of the ith subinterval. If the rectangle with base $[x_{i-1}, x_i]$ and height $f(\overline{x}_i)$ is rotated about the y-axis, then the result is a cylindrical shell with average radius $\overline{x}_i$, height $f(\overline{x}_i)$, and thickness Δx (see Figure 4), so by Formula 1 its volume is

$$V_i = (2\pi\overline{x}_i)[f(\overline{x}_i)]\,\Delta x$$

Therefore an approximation to the volume V of S is given by the sum of the volumes of these shells:

$$V \approx \sum_{i=1}^{n} V_i = \sum_{i=1}^{n} 2\pi\overline{x}_i f(\overline{x}_i)\,\Delta x$$

This approximation appears to become better as $n \to \infty$. But, from the definition of an integral, we know that

$$\lim_{n\to\infty} \sum_{i=1}^{n} 2\pi\overline{x}_i f(\overline{x}_i)\,\Delta x = \int_a^b 2\pi x f(x)\,dx$$

Thus the following appears plausible:

2 The volume of the solid in Figure 3, obtained by rotating about the y-axis the region under the curve $y = f(x)$ from a to b, is

$$V = \int_a^b 2\pi x f(x)\,dx \qquad \text{where } 0 \leqslant a < b$$

The argument using cylindrical shells makes Formula 2 seem reasonable, but later we will be able to prove it (see Exercise 67 in Section 7.1).

The best way to remember Formula 2 is to think of a typical shell, cut and flattened as in Figure 5, with radius x, circumference $2\pi x$, height $f(x)$, and thickness Δx or dx:

$$\int_a^b \underbrace{(2\pi x)}_{\text{circumference}} \ \underbrace{[f(x)]}_{\text{height}} \ \underbrace{dx}_{\text{thickness}}$$

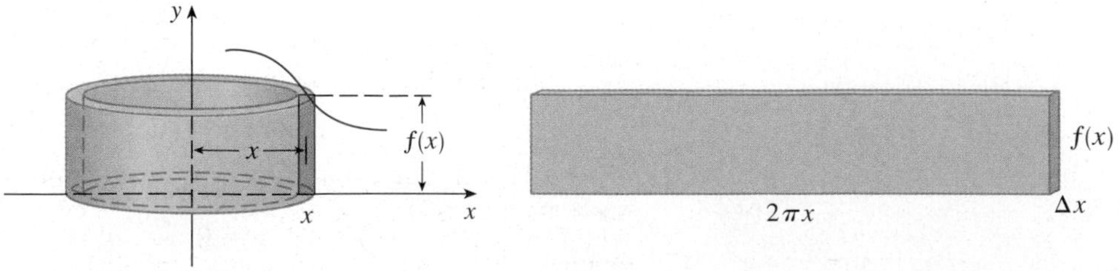

FIGURE 5

This type of reasoning will be helpful in other situations, such as when we rotate about lines other than the y-axis.

EXAMPLE 1 Find the volume of the solid obtained by rotating about the y-axis the region bounded by $y = 2x^2 - x^3$ and $y = 0$.

SOLUTION From the sketch in Figure 6 we see that a typical shell has radius x, circumference $2\pi x$, and height $f(x) = 2x^2 - x^3$. So, by the shell method, the volume is

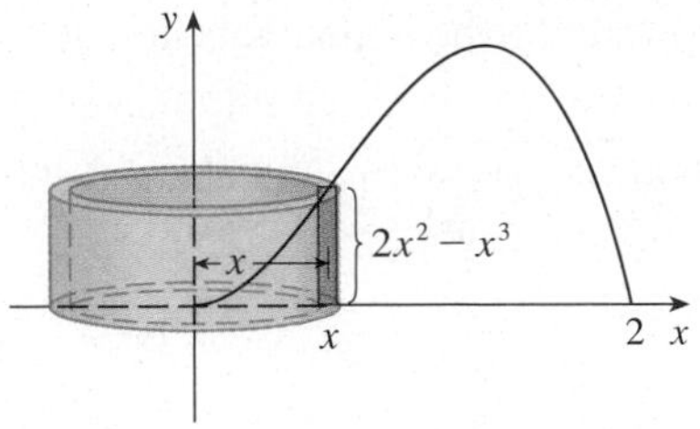

FIGURE 6

$$V = \int_0^2 (2\pi x)(2x^2 - x^3)\,dx = 2\pi \int_0^2 (2x^3 - x^4)\,dx$$

$$= 2\pi\left[\tfrac{1}{2}x^4 - \tfrac{1}{5}x^5\right]_0^2 = 2\pi\left(8 - \tfrac{32}{5}\right) = \tfrac{16}{5}\pi$$

It can be verified that the shell method gives the same answer as slicing. □

■ Figure 7 shows a computer-generated picture of the solid whose volume we computed in Example 1.

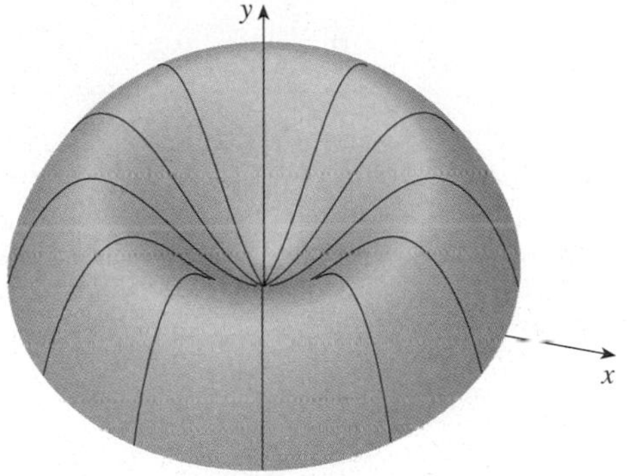

FIGURE 7

NOTE Comparing the solution of Example 1 with the remarks at the beginning of this section, we see that the method of cylindrical shells is much easier than the washer method for this problem. We did not have to find the coordinates of the local maximum and we did not have to solve the equation of the curve for x in terms of y. However, in other examples the methods of the preceding section may be easier.

V EXAMPLE 2 Find the volume of the solid obtained by rotating about the y-axis the region between $y = x$ and $y = x^2$.

SOLUTION The region and a typical shell are shown in Figure 8. We see that the shell has radius x, circumference $2\pi x$, and height $x - x^2$. So the volume is

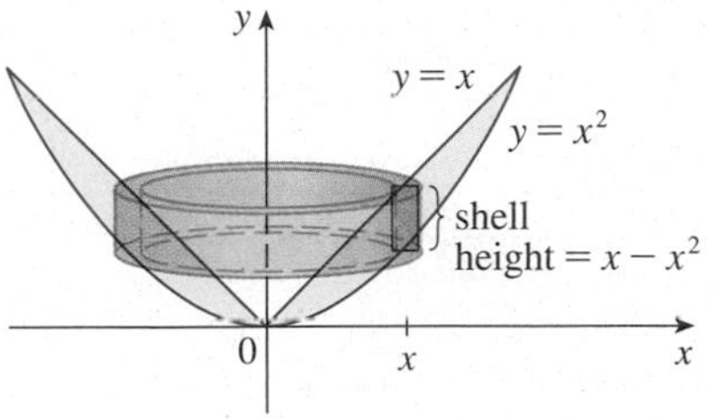

FIGURE 8

$$V = \int_0^1 (2\pi x)(x - x^2)\,dx = 2\pi \int_0^1 (x^2 - x^3)\,dx$$

$$= 2\pi\left[\frac{x^3}{3} - \frac{x^4}{4}\right]_0^1 = \frac{\pi}{6}$$ □

As the following example shows, the shell method works just as well if we rotate about the x-axis. We simply have to draw a diagram to identify the radius and height of a shell.

V EXAMPLE 3 Use cylindrical shells to find the volume of the solid obtained by rotating about the x-axis the region under the curve $y = \sqrt{x}$ from 0 to 1.

SOLUTION This problem was solved using disks in Example 2 in Section 6.2. To use shells we relabel the curve $y = \sqrt{x}$ (in the figure in that example) as $x = y^2$ in Figure 9. For rotation about the x-axis we see that a typical shell has radius y, circumference $2\pi y$, and height $1 - y^2$. So the volume is

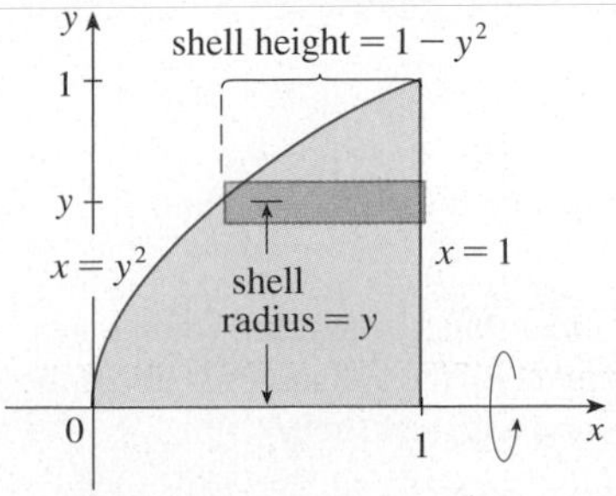

FIGURE 9

$$V = \int_0^1 (2\pi y)(1 - y^2)\,dy = 2\pi \int_0^1 (y - y^3)\,dy = 2\pi\left[\frac{y^2}{2} - \frac{y^4}{4}\right]_0^1 = \frac{\pi}{2}$$

In this problem the disk method was simpler. □

EXAMPLE 4 Find the volume of the solid obtained by rotating the region bounded by $y = x - x^2$ and $y = 0$ about the line $x = 2$.

SOLUTION Figure 10 shows the region and a cylindrical shell formed by rotation about the line $x = 2$. It has radius $2 - x$, circumference $2\pi(2 - x)$, and height $x - x^2$.

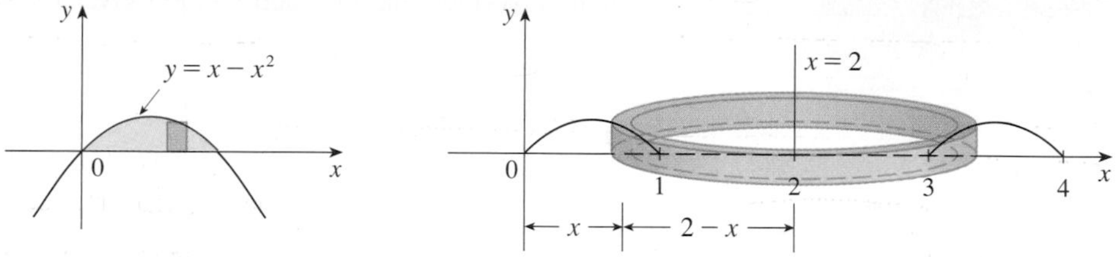

FIGURE 10

The volume of the given solid is

$$V = \int_0^1 2\pi(2 - x)(x - x^2)\,dx = 2\pi \int_0^1 (x^3 - 3x^2 + 2x)\,dx$$

$$= 2\pi \left[\frac{x^4}{4} - x^3 + x^2 \right]_0^1 = \frac{\pi}{2}$$

□

6.3 EXERCISES

1. Let S be the solid obtained by rotating the region shown in the figure about the y-axis. Explain why it is awkward to use slicing to find the volume V of S. Sketch a typical approximating shell. What are its circumference and height? Use shells to find V.

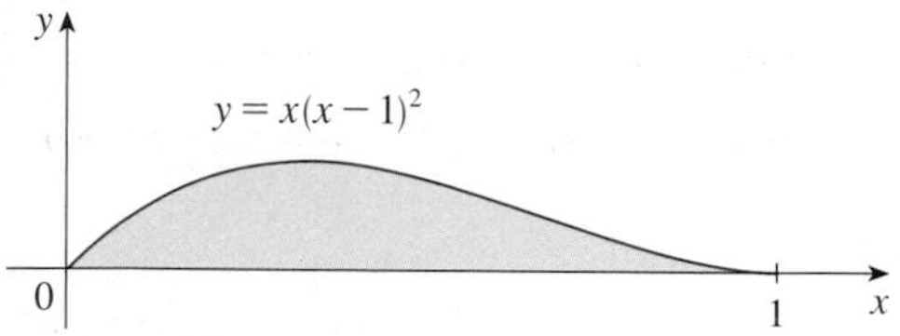

2. Let S be the solid obtained by rotating the region shown in the figure about the y-axis. Sketch a typical cylindrical shell and find its circumference and height. Use shells to find the volume of S. Do you think this method is preferable to slicing? Explain.

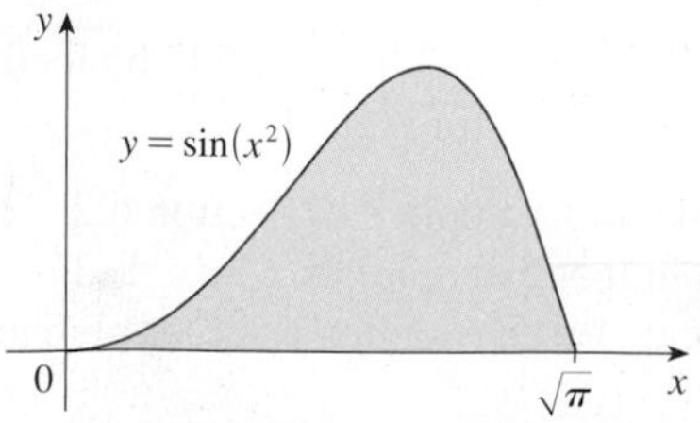

3–7 Use the method of cylindrical shells to find the volume generated by rotating the region bounded by the given curves about the y-axis. Sketch the region and a typical shell.

3. $y = 1/x, \quad y = 0, \quad x = 1, \quad x = 2$

4. $y = x^2, \quad y = 0, \quad x = 1$

5. $y = e^{-x^2}, \quad y = 0, \quad x = 0, \quad x = 1$

6. $y = 3 + 2x - x^2, \quad x + y = 3$

7. $y = 4(x - 2)^2, \quad y = x^2 - 4x + 7$

8. Let V be the volume of the solid obtained by rotating about the y-axis the region bounded by $y = \sqrt{x}$ and $y = x^2$. Find V both by slicing and by cylindrical shells. In both cases draw a diagram to explain your method.

9–14 Use the method of cylindrical shells to find the volume of the solid obtained by rotating the region bounded by the given curves about the x-axis. Sketch the region and a typical shell.

9. $x = 1 + y^2, \quad x = 0, \quad y = 1, \quad y = 2$

10. $x = \sqrt{y}, \quad x = 0, \quad y = 1$

11. $y = x^3, \quad y = 8, \quad x = 0$

12. $x = 4y^2 - y^3, \quad x = 0$

13. $x = 1 + (y - 2)^2, \quad x = 2$

14. $x + y = 3, \quad x = 4 - (y - 1)^2$

15–20 Use the method of cylindrical shells to find the volume generated by rotating the region bounded by the given curves about the specified axis. Sketch the region and a typical shell.

15. $y = x^4, \; y = 0, \; x = 1;$ about $x = 2$

16. $y = \sqrt{x}$, $y = 0$, $x = 1$; about $x = -1$

17. $y = 4x - x^2$, $y = 3$; about $x = 1$

18. $y = x^2$, $y = 2 - x^2$; about $x = 1$

19. $y = x^3$, $y = 0$, $x = 1$; about $y = 1$

20. $y = x^2$, $x = y^2$; about $y = -1$

21–26 Set up, but do not evaluate, an integral for the volume of the solid obtained by rotating the region bounded by the given curves about the specified axis.

21. $y = \ln x$, $y = 0$, $x = 2$; about the y-axis

22. $y = x$, $y = 4x - x^2$; about $x = 7$

23. $y = x^4$, $y = \sin(\pi x/2)$; about $x = -1$

24. $y = 1/(1 + x^2)$, $y = 0$, $x = 0$, $x = 2$; about $x = 2$

25. $x = \sqrt{\sin y}$, $0 \leq y \leq \pi$, $x = 0$; about $y = 4$

26. $x^2 - y^2 = 7$, $x = 4$; about $y = 5$

27. Use the Midpoint Rule with $n = 5$ to estimate the volume obtained by rotating about the y-axis the region under the curve $y = \sqrt{1 + x^3}$, $0 \leq x \leq 1$.

28. If the region shown in the figure is rotated about the y-axis to form a solid, use the Midpoint Rule with $n = 5$ to estimate the volume of the solid.

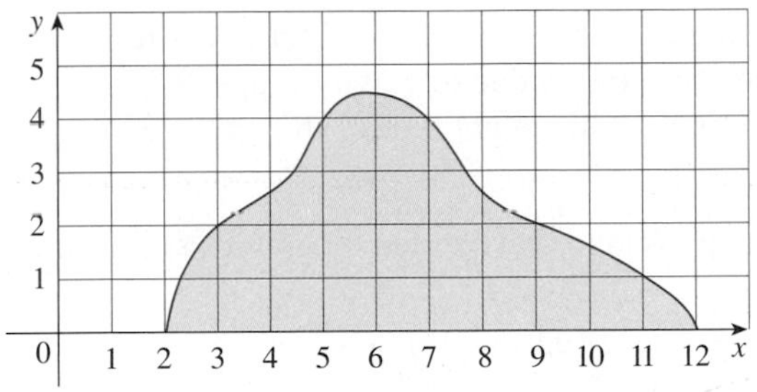

29–32 Each integral represents the volume of a solid. Describe the solid.

29. $\int_0^3 2\pi x^5 \, dx$

30. $2\pi \int_0^2 \frac{y}{1 + y^2} \, dy$

31. $\int_0^1 2\pi(3 - y)(1 - y^2) \, dy$

32. $\int_0^{\pi/4} 2\pi(\pi - x)(\cos x - \sin x) \, dx$

33–34 Use a graph to estimate the x-coordinates of the points of intersection of the given curves. Then use this information and your calculator to estimate the volume of the solid obtained by rotating about the y-axis the region enclosed by these curves.

33. $y = e^x$, $y = \sqrt{x} + 1$

34. $y = x^3 - x + 1$, $y = -x^4 + 4x - 1$

CAS **35–36** Use a computer algebra system to find the exact volume of the solid obtained by rotating the region bounded by the given curves about the specified line.

35. $y = \sin^2 x$, $y = \sin^4 x$, $0 \leq x \leq \pi$; about $x = \pi/2$

36. $y = x^3 \sin x$, $y = 0$, $0 \leq x \leq \pi$; about $x = -1$

37–42 The region bounded by the given curves is rotated about the specified axis. Find the volume of the resulting solid by any method.

37. $y = -x^2 + 6x - 8$, $y = 0$; about the y-axis

38. $y = -x^2 + 6x - 8$, $y = 0$; about the x-axis

39. $y = 5$, $y = x + (4/x)$; about $x = -1$

40. $x = 1 - y^4$, $x = 0$; about $x = 2$

41. $x^2 + (y - 1)^2 = 1$; about the y-axis

42. $x = (y - 3)^2$, $x = 4$; about $y = 1$

43–45 Use cylindrical shells to find the volume of the solid.

43. A sphere of radius r

44. The solid torus of Exercise 63 in Section 6.2

45. A right circular cone with height h and base radius r

46. Suppose you make napkin rings by drilling holes with different diameters through two wooden balls (which also have different diameters). You discover that both napkin rings have the same height h, as shown in the figure.
(a) Guess which ring has more wood in it.
(b) Check your guess: Use cylindrical shells to compute the volume of a napkin ring created by drilling a hole with radius r through the center of a sphere of radius R and express the answer in terms of h.

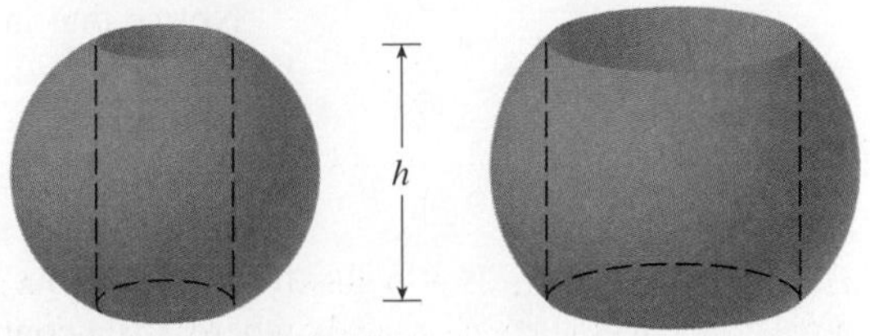

6.4 WORK

The term *work* is used in everyday language to mean the total amount of effort required to perform a task. In physics it has a technical meaning that depends on the idea of a *force*. Intuitively, you can think of a force as describing a push or pull on an object—for example, a horizontal push of a book across a table or the downward pull of the earth's gravity on a ball. In general, if an object moves along a straight line with position function $s(t)$, then the **force** F on the object (in the same direction) is defined by Newton's Second Law of Motion as the product of its mass m and its acceleration:

1
$$F = m\frac{d^2s}{dt^2}$$

In the SI metric system, the mass is measured in kilograms (kg), the displacement in meters (m), the time in seconds (s), and the force in newtons ($\text{N} = \text{kg}\cdot\text{m/s}^2$). Thus a force of 1 N acting on a mass of 1 kg produces an acceleration of 1 m/s^2. In the US Customary system, the fundamental unit is chosen to be the unit of force, which is the pound.

In the case of constant acceleration, the force F is also constant and the work done is defined to be the product of the force F and the distance d that the object moves:

2
$$W = Fd \qquad \text{work} = \text{force} \times \text{distance}$$

If F is measured in newtons and d in meters, then the unit for W is a newton-meter, which is called a joule (J). If F is measured in pounds and d in feet, then the unit for W is a foot-pound (ft-lb), which is about 1.36 J.

V EXAMPLE 1

(a) How much work is done in lifting a 1.2-kg book off the floor to put it on a desk that is 0.7 m high? Use the fact that the acceleration due to gravity is $g = 9.8 \text{ m/s}^2$.
(b) How much work is done in lifting a 20-lb weight 6 ft off the ground?

SOLUTION

(a) The force exerted is equal and opposite to that exerted by gravity, so Equation 1 gives

$$F = mg = (1.2)(9.8) = 11.76 \text{ N}$$

and then Equation 2 gives the work done as

$$W = Fd = (11.76)(0.7) \approx 8.2 \text{ J}$$

(b) Here the force is given as $F = 20$ lb, so the work done is

$$W = Fd = 20 \cdot 6 = 120 \text{ ft-lb}$$

Notice that in part (b), unlike part (a), we did not have to multiply by g because we were given the *weight* (which is a force) and not the mass of the object. □

Equation 2 defines work as long as the force is constant, but what happens if the force is variable? Let's suppose that the object moves along the x-axis in the positive direction, from $x = a$ to $x = b$, and at each point x between a and b a force $f(x)$ acts on the object, where f is a continuous function. We divide the interval $[a, b]$ into n subintervals with endpoints $x_0, x_1, \ldots, x_n$ and equal width Δx. We choose a sample point x_i^* in the ith subinterval $[x_{i-1}, x_i]$. Then the force at that point is $f(x_i^*)$. If n is large, then Δx is small, and since f is continuous, the values of f don't change very much over the interval $[x_{i-1}, x_i]$. In other words, f is almost constant on the interval and so the work W_i that is done in moving the particle from x_{i-1} to x_i is approximately given by Equation 2:

$$W_i \approx f(x_i^*)\,\Delta x$$

Thus we can approximate the total work by

3 $$W \approx \sum_{i=1}^{n} f(x_i^*)\,\Delta x$$

It seems that this approximation becomes better as we make n larger. Therefore we define the **work done in moving the object from a to b** as the limit of this quantity as $n \to \infty$. Since the right side of (3) is a Riemann sum, we recognize its limit as being a definite integral and so

4 $$W = \lim_{n\to\infty} \sum_{i=1}^{n} f(x_i^*)\,\Delta x = \int_a^b f(x)\,dx$$

EXAMPLE 2 When a particle is located a distance x feet from the origin, a force of $x^2 + 2x$ pounds acts on it. How much work is done in moving it from $x = 1$ to $x = 3$?

SOLUTION
$$W = \int_1^3 (x^2 + 2x)\,dx = \frac{x^3}{3} + x^2 \bigg]_1^3 = \frac{50}{3}$$

The work done is $16\frac{2}{3}$ ft-lb. □

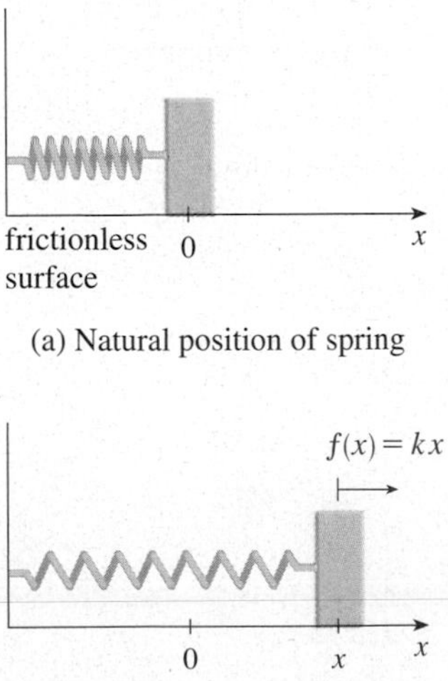

(a) Natural position of spring

(b) Stretched position of spring

FIGURE 1
Hooke's Law

In the next example we use a law from physics: **Hooke's Law** states that the force required to maintain a spring stretched x units beyond its natural length is proportional to x:

$$f(x) = kx$$

where k is a positive constant (called the **spring constant**). Hooke's Law holds provided that x is not too large (see Figure 1).

EXAMPLE 3 A force of 40 N is required to hold a spring that has been stretched from its natural length of 10 cm to a length of 15 cm. How much work is done in stretching the spring from 15 cm to 18 cm?

SOLUTION According to Hooke's Law, the force required to hold the spring stretched x meters beyond its natural length is $f(x) = kx$. When the spring is stretched from 10 cm to 15 cm, the amount stretched is 5 cm = 0.05 m. This means that $f(0.05) = 40$, so

$$0.05k = 40 \qquad k = \frac{40}{0.05} = 800$$

Thus $f(x) = 800x$ and the work done in stretching the spring from 15 cm to 18 cm is

$$W = \int_{0.05}^{0.08} 800x\,dx = 800\,\frac{x^2}{2}\bigg]_{0.05}^{0.08}$$
$$= 400[(0.08)^2 - (0.05)^2] = 1.56 \text{ J}$$ □

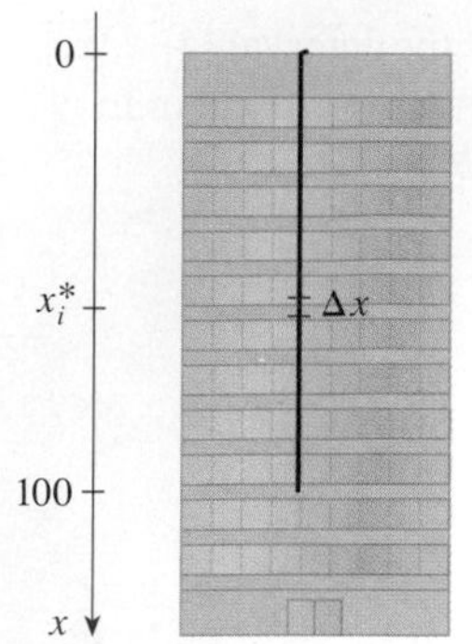

FIGURE 2

■ If we had placed the origin at the bottom of the cable and the x-axis upward, we would have gotten

$$W = \int_0^{100} 2(100 - x)\,dx$$

which gives the same answer.

EXAMPLE 4 A 200-lb cable is 100 ft long and hangs vertically from the top of a tall building. How much work is required to lift the cable to the top of the building?

SOLUTION Here we don't have a formula for the force function, but we can use an argument similar to the one that led to Definition 4.

Let's place the origin at the top of the building and the x-axis pointing downward as in Figure 2. We divide the cable into small parts with length Δx. If x_i^* is a point in the ith such interval, then all points in the interval are lifted by approximately the same amount, namely x_i^*. The cable weighs 2 pounds per foot, so the weight of the ith part is $2\,\Delta x$. Thus the work done on the ith part, in foot-pounds, is

$$\underbrace{(2\,\Delta x)}_{\text{force}}\ \underbrace{x_i^*}_{\text{distance}} = 2x_i^*\,\Delta x$$

We get the total work done by adding all these approximations and letting the number of parts become large (so $\Delta x \to 0$):

$$W = \lim_{n\to\infty} \sum_{i=1}^{n} 2x_i^*\,\Delta x = \int_0^{100} 2x\,dx$$

$$= x^2\Big]_0^{100} = 10{,}000 \text{ ft-lb}$$

□

EXAMPLE 5 A tank has the shape of an inverted circular cone with height 10 m and base radius 4 m. It is filled with water to a height of 8 m. Find the work required to empty the tank by pumping all of the water to the top of the tank. (The density of water is $1000\ \text{kg/m}^3$.)

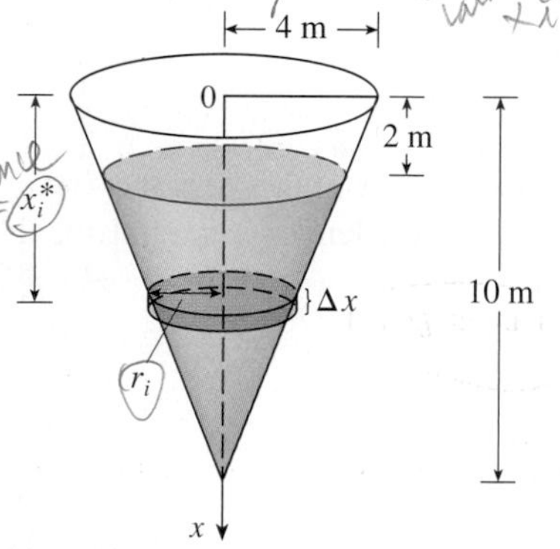

FIGURE 3

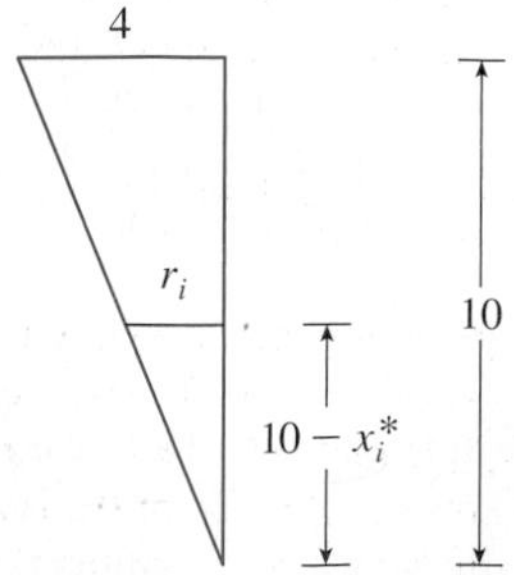

FIGURE 4

SOLUTION Let's measure depths from the top of the tank by introducing a vertical coordinate line as in Figure 3. The water extends from a depth of 2 m to a depth of 10 m and so we divide the interval $[2, 10]$ into n subintervals with endpoints $x_0, x_1, \ldots, x_n$ and choose x_i^* in the ith subinterval. This divides the water into n layers. The ith layer is approximated by a circular cylinder with radius r_i and height Δx. We can compute r_i from similar triangles, using Figure 4, as follows:

$$\frac{r_i}{10 - x_i^*} = \frac{4}{10} \qquad r_i = \tfrac{2}{5}(10 - x_i^*)$$

Thus an approximation to the volume of the ith layer of water is

$$V_i \approx \pi r_i^2\,\Delta x = \frac{4\pi}{25}(10 - x_i^*)^2\,\Delta x$$

and so its mass is

$$m_i = \text{density} \times \text{volume}$$

$$\approx 1000 \cdot \frac{4\pi}{25}(10 - x_i^*)^2\,\Delta x = 160\pi(10 - x_i^*)^2\,\Delta x$$

The force required to raise this layer must overcome the force of gravity and so

$$F_i = m_i g \approx (9.8)160\pi(10 - x_i^*)^2\,\Delta x$$

$$\approx 1570\pi(10 - x_i^*)^2\,\Delta x$$

Each particle in the layer must travel a distance of approximately x_i^*. The work W_i done to raise this layer to the top is approximately the product of the force F_i and the distance x_i^*:

$$W_i \approx F_i x_i^* \approx 1570\pi x_i^*(10 - x_i^*)^2\,\Delta x$$

To find the total work done in emptying the entire tank, we add the contributions of each of the n layers and then take the limit as $n \to \infty$:

$$W = \lim_{n\to\infty} \sum_{i=1}^{n} 1570\pi x_i^*(10 - x_i^*)^2 \,\Delta x = \int_2^{10} 1570\pi x(10 - x)^2 \,dx$$

$$= 1570\pi \int_2^{10} (100x - 20x^2 + x^3)\,dx = 1570\pi \left[50x^2 - \frac{20x^3}{3} + \frac{x^4}{4} \right]_2^{10}$$

$$= 1570\pi\left(\tfrac{2048}{3}\right) \approx 3.4 \times 10^6 \text{ J}$$ □

6.4 EXERCISES

1. How much work is done in lifting a 40-kg sandbag to a height of 1.5 m?

2. Find the work done if a constant force of 100 lb is used to pull a cart a distance of 200 ft.

3. A particle is moved along the x axis by a force that measures $10/(1 + x)^2$ pounds at a point x feet from the origin. Find the work done in moving the particle from the origin to a distance of 9 ft.

4. When a particle is located a distance x meters from the origin, a force of $\cos(\pi x/3)$ newtons acts on it. How much work is done in moving the particle from $x = 1$ to $x = 2$? Interpret your answer by considering the work done from $x = 1$ to $x = 1.5$ and from $x = 1.5$ to $x = 2$.

5. Shown is the graph of a force function (in newtons) that increases to its maximum value and then remains constant. How much work is done by the force in moving an object a distance of 8 m?

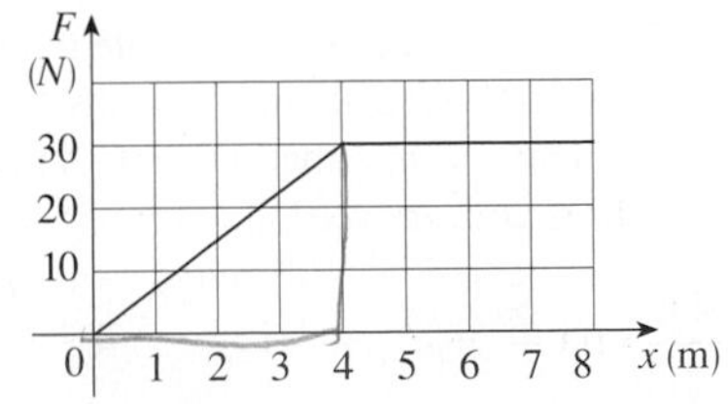

6. The table shows values of a force function $f(x)$, where x is measured in meters and $f(x)$ in newtons. Use the Midpoint Rule to estimate the work done by the force in moving an object from $x = 4$ to $x = 20$.

x	4	6	8	10	12	14	16	18	20
$f(x)$	5	5.8	7.0	8.8	9.6	8.2	6.7	5.2	4.1

7. A force of 10 lb is required to hold a spring stretched 4 in. beyond its natural length. How much work is done in stretching it from its natural length to 6 in. beyond its natural length?

8. A spring has a natural length of 20 cm. If a 25-N force is required to keep it stretched to a length of 30 cm, how much work is required to stretch it from 20 cm to 25 cm?

9. Suppose that 2 J of work is needed to stretch a spring from its natural length of 30 cm to a length of 42 cm.
   (a) How much work is needed to stretch the spring from 35 cm to 40 cm?
   (b) How far beyond its natural length will a force of 30 N keep the spring stretched?

10. If the work required to stretch a spring 1 ft beyond its natural length is 12 ft-lb, how much work is needed to stretch it 9 in. beyond its natural length?

11. A spring has natural length 20 cm. Compare the work W_1 done in stretching the spring from 20 cm to 30 cm with the work W_2 done in stretching it from 30 cm to 40 cm. How are W_2 and W_1 related?

12. If 6 J of work is needed to stretch a spring from 10 cm to 12 cm and another 10 J is needed to stretch it from 12 cm to 14 cm, what is the natural length of the spring?

13–20 Show how to approximate the required work by a Riemann sum. Then express the work as an integral and evaluate it.

13. A heavy rope, 50 ft long, weighs 0.5 lb/ft and hangs over the edge of a building 120 ft high.
   (a) How much work is done in pulling the rope to the top of the building?
   (b) How much work is done in pulling half the rope to the top of the building?

14. A chain lying on the ground is 10 m long and its mass is 80 kg. How much work is required to raise one end of the chain to a height of 6 m?

15. A cable that weighs 2 lb/ft is used to lift 800 lb of coal up a mine shaft 500 ft deep. Find the work done.

16. A bucket that weighs 4 lb and a rope of negligible weight are used to draw water from a well that is 80 ft deep. The bucket is filled with 40 lb of water and is pulled up at a rate of 2 ft/s, but water leaks out of a hole in the bucket at a rate of 0.2 lb/s. Find the work done in pulling the bucket to the top of the well.

17. A leaky 10-kg bucket is lifted from the ground to a height of 12 m at a constant speed with a rope that weighs 0.8 kg/m. Initially the bucket contains 36 kg of water, but the water

leaks at a constant rate and finishes draining just as the bucket reaches the 12 m level. How much work is done?

18. A 10-ft chain weighs 25 lb and hangs from a ceiling. Find the work done in lifting the lower end of the chain to the ceiling so that it's level with the upper end.

19. An aquarium 2 m long, 1 m wide, and 1 m deep is full of water. Find the work needed to pump half of the water out of the aquarium. (Use the fact that the density of water is 1000 kg/m^3.)

20. A circular swimming pool has a diameter of 24 ft, the sides are 5 ft high, and the depth of the water is 4 ft. How much work is required to pump all of the water out over the side? (Use the fact that water weighs 62.5 lb/ft^3.)

21–24 A tank is full of water. Find the work required to pump the water out of the spout. In Exercises 23 and 24 use the fact that water weighs 62.5 lb/ft^3.

21.

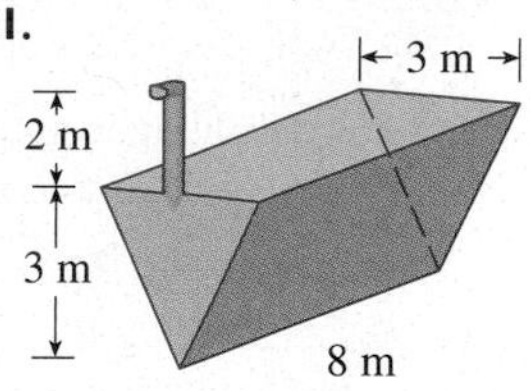

22.

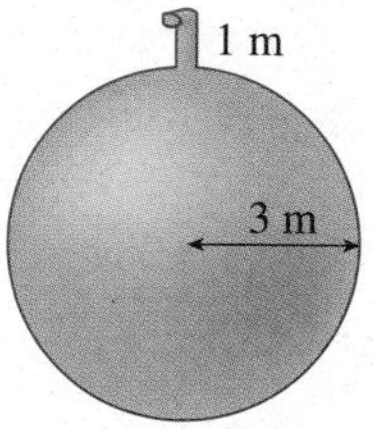

23.

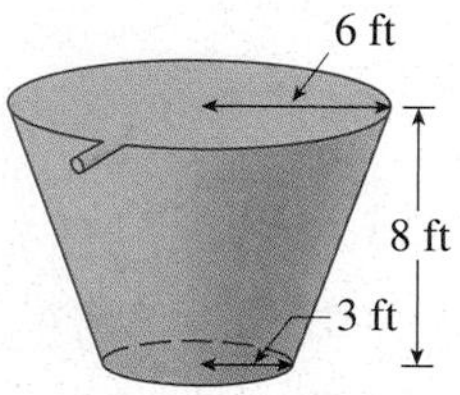

frustum of a cone

24.

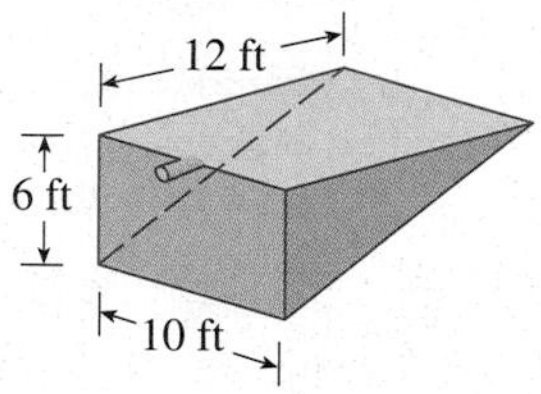

---

25. Suppose that for the tank in Exercise 21 the pump breaks down after 4.7×10^5 J of work has been done. What is the depth of the water remaining in the tank?

26. Solve Exercise 22 if the tank is half full of oil that has a density of 900 kg/m^3.

27. When gas expands in a cylinder with radius r, the pressure at any given time is a function of the volume: $P = P(V)$. The force exerted by the gas on the piston (see the figure) is the product of the pressure and the area: $F = \pi r^2 P$. Show that the work done by the gas when the volume expands from volume V_1 to volume V_2 is

$$W = \int_{V_1}^{V_2} P \, dV$$

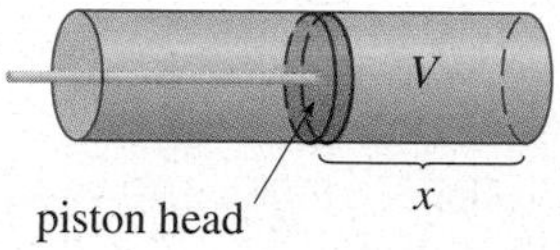

28. In a steam engine the pressure P and volume V of steam satisfy the equation $PV^{1.4} = k$, where k is a constant. (This is true for adiabatic expansion, that is, expansion in which there is no heat transfer between the cylinder and its surroundings.) Use Exercise 27 to calculate the work done by the engine during a cycle when the steam starts at a pressure of 160 lb/in^2 and a volume of 100 in^3 and expands to a volume of 800 in^3.

29. Newton's Law of Gravitation states that two bodies with masses m_1 and m_2 attract each other with a force

$$F = G\frac{m_1 m_2}{r^2}$$

where r is the distance between the bodies and G is the gravitational constant. If one of the bodies is fixed, find the work needed to move the other from $r = a$ to $r = b$.

30. Use Newton's Law of Gravitation to compute the work required to launch a 1000-kg satellite vertically to an orbit 1000 km high. You may assume that the earth's mass is 5.98×10^{24} kg and is concentrated at its center. Take the radius of the earth to be 6.37×10^6 m and $G = 6.67 \times 10^{-11} \text{ N·m}^2/\text{kg}^2$.

6.5 AVERAGE VALUE OF A FUNCTION

It is easy to calculate the average value of finitely many numbers $y_1, y_2, \ldots, y_n$:

$$y_{\text{ave}} = \frac{y_1 + y_2 + \cdots + y_n}{n}$$

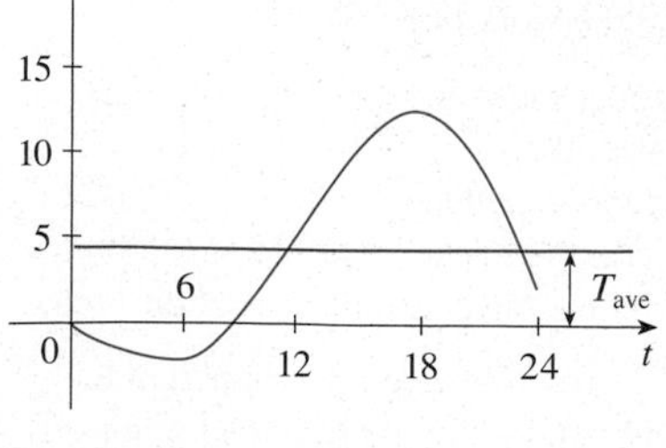

FIGURE 1

But how do we compute the average temperature during a day if infinitely many temperature readings are possible? Figure 1 shows the graph of a temperature function $T(t)$, where t is measured in hours and T in °C, and a guess at the average temperature, T_{ave}.

In general, let's try to compute the average value of a function $y = f(x)$, $a \leqslant x \leqslant b$. We start by dividing the interval $[a, b]$ into n equal subintervals, each with length $\Delta x = (b - a)/n$. Then we choose points $x_1^*, \ldots, x_n^*$ in successive subintervals and calculate the average of the numbers $f(x_1^*), \ldots, f(x_n^*)$:

$$\frac{f(x_1^*) + \cdots + f(x_n^*)}{n}$$

(For example, if f represents a temperature function and $n = 24$, this means that we take temperature readings every hour and then average them.) Since $\Delta x = (b - a)/n$, we can write $n = (b - a)/\Delta x$ and the average value becomes

$$\frac{f(x_1^*) + \cdots + f(x_n^*)}{\dfrac{b-a}{\Delta x}} = \frac{1}{b-a}\left[f(x_1^*)\,\Delta x + \cdots + f(x_n^*)\,\Delta x\right]$$

$$= \frac{1}{b-a}\sum_{i=1}^{n} f(x_i^*)\,\Delta x$$

If we let n increase, we would be computing the average value of a large number of closely spaced values. (For example, we would be averaging temperature readings taken every minute or even every second.) The limiting value is

$$\lim_{n\to\infty} \frac{1}{b-a}\sum_{i=1}^{n} f(x_i^*)\,\Delta x = \frac{1}{b-a}\int_a^b f(x)\,dx$$

by the definition of a definite integral.

Therefore we define the **average value of f** on the interval $[a, b]$ as

■ For a positive function, we can think of this definition as saying

$$\frac{\text{area}}{\text{width}} = \text{average height}$$

$$f_{\text{ave}} = \frac{1}{b-a}\int_a^b f(x)\,dx$$

☑ **EXAMPLE 1** Find the average value of the function $f(x) = 1 + x^2$ on the interval $[-1, 2]$.

SOLUTION With $a = -1$ and $b = 2$ we have

$$f_{\text{ave}} = \frac{1}{b-a}\int_a^b f(x)\,dx = \frac{1}{2-(-1)}\int_{-1}^{2}(1 + x^2)\,dx = \frac{1}{3}\left[x + \frac{x^3}{3}\right]_{-1}^{2} = 2 \qquad \square$$

If $T(t)$ is the temperature at time t, we might wonder if there is a specific time when the temperature is the same as the average temperature. For the temperature function graphed in Figure 1, we see that there are two such times—just before noon and just before midnight. In general, is there a number c at which the value of a function f is exactly equal to the average value of the function, that is, $f(c) = f_{\text{ave}}$? The following theorem says that this is true for continuous functions.

THE MEAN VALUE THEOREM FOR INTEGRALS If f is continuous on $[a, b]$, then there exists a number c in $[a, b]$ such that

$$f(c) = f_{\text{ave}} = \frac{1}{b-a}\int_a^b f(x)\,dx$$

that is, $$\int_a^b f(x)\,dx = f(c)(b - a)$$

The Mean Value Theorem for Integrals is a consequence of the Mean Value Theorem for derivatives and the Fundamental Theorem of Calculus. The proof is outlined in Exercise 23.

The geometric interpretation of the Mean Value Theorem for Integrals is that, for *positive* functions f, there is a number c such that the rectangle with base $[a, b]$ and height $f(c)$ has the same area as the region under the graph of f from a to b. (See Figure 2 and the more picturesque interpretation in the margin note.)

■ You can always chop off the top of a (two-dimensional) mountain at a certain height and use it to fill in the valleys so that the mountaintop becomes completely flat.

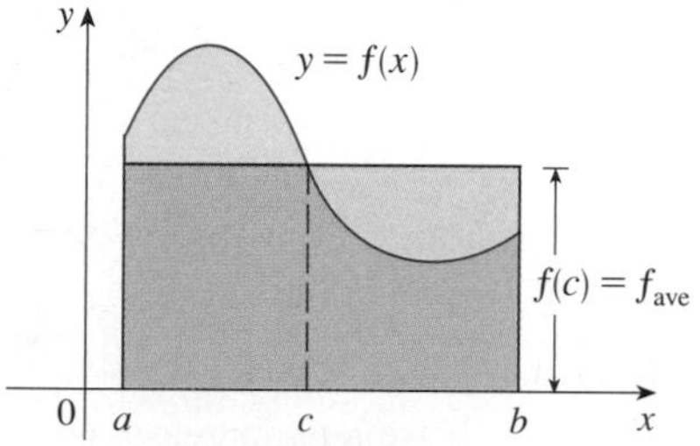

FIGURE 2

EXAMPLE 2 Since $f(x) = 1 + x^2$ is continuous on the interval $[-1, 2]$, the Mean Value Theorem for Integrals says there is a number c in $[-1, 2]$ such that

$$\int_{-1}^{2} (1 + x^2)\,dx = f(c)[2 - (-1)]$$

In this particular case we can find c explicitly. From Example 1 we know that $f_{ave} = 2$, so the value of c satisfies

$$f(c) = f_{ave} = 2$$

Therefore $$1 + c^2 = 2 \qquad \text{so} \qquad c^2 = 1$$

So in this case there happen to be two numbers $c = \pm 1$ in the interval $[-1, 2]$ that work in the Mean Value Theorem for Integrals. □

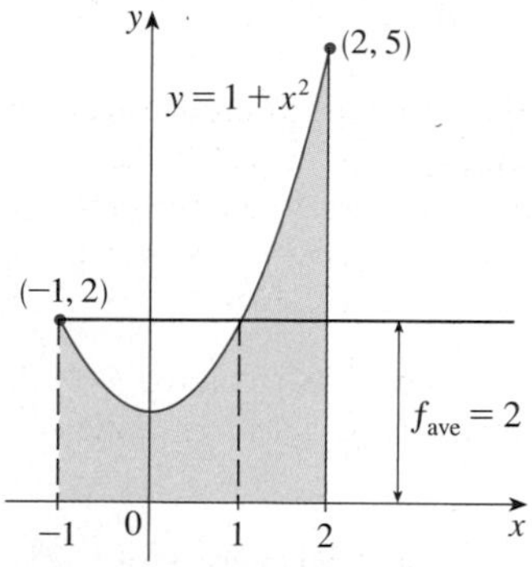

FIGURE 3

Examples 1 and 2 are illustrated by Figure 3.

EXAMPLE 3 Show that the average velocity of a car over a time interval $[t_1, t_2]$ is the same as the average of its velocities during the trip.

SOLUTION If $s(t)$ is the displacement of the car at time t, then, by definition, the average velocity of the car over the interval is

$$\frac{\Delta s}{\Delta t} = \frac{s(t_2) - s(t_1)}{t_2 - t_1}$$

On the other hand, the average value of the velocity function on the interval is

$$\begin{aligned} v_{ave} &= \frac{1}{t_2 - t_1} \int_{t_1}^{t_2} v(t)\,dt = \frac{1}{t_2 - t_1} \int_{t_1}^{t_2} s'(t)\,dt \\ &= \frac{1}{t_2 - t_1} [s(t_2) - s(t_1)] \qquad \text{(by the Net Change Theorem)} \\ &= \frac{s(t_2) - s(t_1)}{t_2 - t_1} = \text{average velocity} \end{aligned}$$

□

6.5 EXERCISES

1–8 Find the average value of the function on the given interval.

1. $f(x) = 4x - x^2$, $[0, 4]$

2. $f(x) = \sin 4x$, $[-\pi, \pi]$

3. $g(x) = \sqrt[3]{x}$, $[1, 8]$

4. $g(x) = x^2\sqrt{1 + x^3}$, $[0, 2]$

5. $f(t) = te^{-t^2}$, $[0, 5]$

6. $f(\theta) = \sec^2(\theta/2)$, $[0, \pi/2]$

7. $h(x) = \cos^4 x \sin x$, $[0, \pi]$

8. $h(u) = (3 - 2u)^{-1}$, $[-1, 1]$

9–12

(a) Find the average value of f on the given interval.

(b) Find c such that $f_{\text{ave}} = f(c)$.

(c) Sketch the graph of f and a rectangle whose area is the same as the area under the graph of f.

9. $f(x) = (x - 3)^2$, $[2, 5]$

10. $f(x) = \sqrt{x}$, $[0, 4]$

11. $f(x) = 2 \sin x - \sin 2x$, $[0, \pi]$

12. $f(x) = 2x/(1 + x^2)^2$, $[0, 2]$

13. If f is continuous and $\int_1^3 f(x)\,dx = 8$, show that f takes on the value 4 at least once on the interval $[1, 3]$.

14. Find the numbers b such that the average value of $f(x) = 2 + 6x - 3x^2$ on the interval $[0, b]$ is equal to 3.

15. The table gives values of a continuous function. Use the Midpoint Rule to estimate the average value of f on $[20, 50]$.

x	20	25	30	35	40	45	50
$f(x)$	42	38	31	29	35	48	60

16. The velocity graph of an accelerating car is shown.

(a) Estimate the average velocity of the car during the first 12 seconds.

(b) At what time was the instantaneous velocity equal to the average velocity?

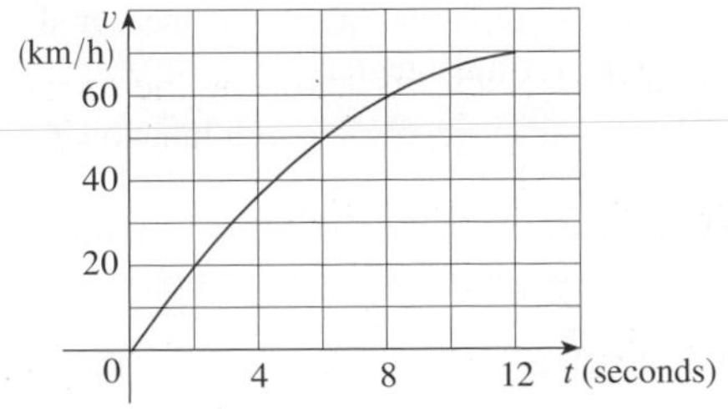

17. In a certain city the temperature (in °F) t hours after 9 AM was modeled by the function

$$T(t) = 50 + 14 \sin \frac{\pi t}{12}$$

Find the average temperature during the period from 9 AM to 9 PM.

18. (a) A cup of coffee has temperature 95°C and takes 30 minutes to cool to 61°C in a room with temperature 20°C. Use Newton's Law of Cooling (Section 3.8) to show that the temperature of the coffee after t minutes is

$$T(t) = 20 + 75e^{-kt}$$

where $k \approx 0.02$.

(b) What is the average temperature of the coffee during the first half hour?

19. The linear density in a rod 8 m long is $12/\sqrt{x + 1}$ kg/m, where x is measured in meters from one end of the rod. Find the average density of the rod.

20. If a freely falling body starts from rest, then its displacement is given by $s = \frac{1}{2}gt^2$. Let the velocity after a time T be v_T. Show that if we compute the average of the velocities with respect to t we get $v_{\text{ave}} = \frac{1}{2}v_T$, but if we compute the average of the velocities with respect to s we get $v_{\text{ave}} = \frac{2}{3}v_T$.

21. Use the result of Exercise 79 in Section 5.5 to compute the average volume of inhaled air in the lungs in one respiratory cycle.

22. The velocity v of blood that flows in a blood vessel with radius R and length l at a distance r from the central axis is

$$v(r) = \frac{P}{4\eta l}(R^2 - r^2)$$

where P is the pressure difference between the ends of the vessel and η is the viscosity of the blood (see Example 7 in Section 3.7). Find the average velocity (with respect to r) over the interval $0 \leqslant r \leqslant R$. Compare the average velocity with the maximum velocity.

23. Prove the Mean Value Theorem for Integrals by applying the Mean Value Theorem for derivatives (see Section 4.2) to the function $F(x) = \int_a^x f(t)\,dt$.

24. If $f_{\text{ave}}[a, b]$ denotes the average value of f on the interval $[a, b]$ and $a < c < b$, show that

$$f_{\text{ave}}[a, b] = \frac{c - a}{b - a} f_{\text{ave}}[a, c] + \frac{b - c}{b - a} f_{\text{ave}}[c, b]$$

APPLIED PROJECT

CAS WHERE TO SIT AT THE MOVIES

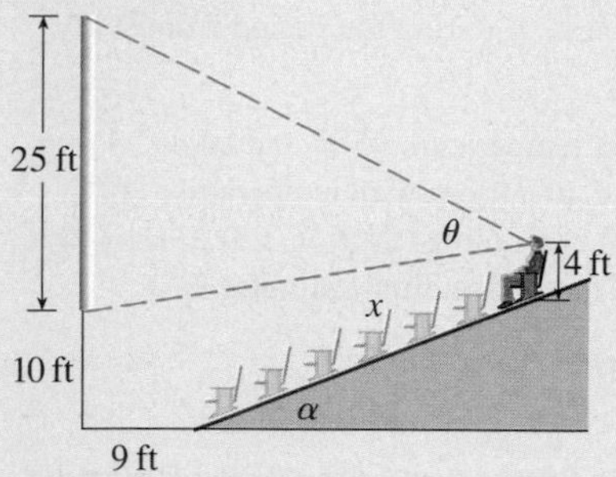

A movie theater has a screen that is positioned 10 ft off the floor and is 25 ft high. The first row of seats is placed 9 ft from the screen and the rows are set 3 ft apart. The floor of the seating area is inclined at an angle of $\alpha = 20°$ above the horizontal and the distance up the incline that you sit is x. The theater has 21 rows of seats, so $0 \leqslant x \leqslant 60$. Suppose you decide that the best place to sit is in the row where the angle θ subtended by the screen at your eyes is a maximum. Let's also suppose that your eyes are 4 ft above the floor, as shown in the figure. (In Exercise 70 in Section 4.7 we looked at a simpler version of this problem, where the floor is horizontal, but this project involves a more complicated situation and requires technology.)

1. Show that

$$\theta = \arccos\left(\frac{a^2 + b^2 - 625}{2ab}\right)$$

where $$a^2 = (9 + x\cos\alpha)^2 + (31 - x\sin\alpha)^2$$

and $$b^2 = (9 + x\cos\alpha)^2 + (x\sin\alpha - 6)^2$$

2. Use a graph of θ as a function of x to estimate the value of x that maximizes θ. In which row should you sit? What is the viewing angle θ in this row?

3. Use your computer algebra system to differentiate θ and find a numerical value for the root of the equation $d\theta/dx = 0$. Does this value confirm your result in Problem 2?

4. Use the graph of θ to estimate the average value of θ on the interval $0 \leqslant x \leqslant 60$. Then use your CAS to compute the average value. Compare with the maximum and minimum values of θ.

6 REVIEW

CONCEPT CHECK

1. (a) Draw two typical curves $y = f(x)$ and $y = g(x)$, where $f(x) \geqslant g(x)$ for $a \leqslant x \leqslant b$. Show how to approximate the area between these curves by a Riemann sum and sketch the corresponding approximating rectangles. Then write an expression for the exact area.
(b) Explain how the situation changes if the curves have equations $x = f(y)$ and $x = g(y)$, where $f(y) \geqslant g(y)$ for $c \leqslant y \leqslant d$.

2. Suppose that Sue runs faster than Kathy throughout a 1500-meter race. What is the physical meaning of the area between their velocity curves for the first minute of the race?

3. (a) Suppose S is a solid with known cross-sectional areas. Explain how to approximate the volume of S by a Riemann sum. Then write an expression for the exact volume.
(b) If S is a solid of revolution, how do you find the cross-sectional areas?

4. (a) What is the volume of a cylindrical shell?
(b) Explain how to use cylindrical shells to find the volume of a solid of revolution.
(c) Why might you want to use the shell method instead of slicing?

5. Suppose that you push a book across a 6-meter-long table by exerting a force $f(x)$ at each point from $x = 0$ to $x = 6$. What does $\int_0^6 f(x)\,dx$ represent? If $f(x)$ is measured in newtons, what are the units for the integral?

6. (a) What is the average value of a function f on an interval $[a, b]$?
(b) What does the Mean Value Theorem for Integrals say? What is its geometric interpretation?

EXERCISES

1–6 Find the area of the region bounded by the given curves.

1. $y = x^2, \quad y = 4x - x^2$

2. $y = 1/x, \quad y = x^2, \quad y = 0, \quad x = e$

3. $y = 1 - 2x^2, \quad y = |x|$

4. $x + y = 0, \quad x = y^2 + 3y$

5. $y = \sin(\pi x/2), \quad y = x^2 - 2x$

6. $y = \sqrt{x}, \quad y = x^2, \quad x = 2$

7–11 Find the volume of the solid obtained by rotating the region bounded by the given curves about the specified axis.

7. $y = 2x, \; y = x^2$; about the x-axis

8. $x = 1 + y^2, \; y = x - 3$; about the y-axis

9. $x = 0, \; x = 9 - y^2$; about $x = -1$

10. $y = x^2 + 1, \; y = 9 - x^2$; about $y = -1$

11. $x^2 - y^2 = a^2, \; x = a + h$ (where $a > 0, h > 0$); about the y-axis

12–14 Set up, but do not evaluate, an integral for the volume of the solid obtained by rotating the region bounded by the given curves about the specified axis.

12. $y = \tan x, \; y = x, \; x = \pi/3$; about the y-axis

13. $y = \cos^2 x, \; |x| \leqslant \pi/2, \; y = \frac{1}{4}$; about $x = \pi/2$

14. $y = \sqrt{x}, \; y = x^2$; about $y = 2$

15. Find the volumes of the solids obtained by rotating the region bounded by the curves $y = x$ and $y = x^2$ about the following lines.
(a) The x-axis (b) The y-axis (c) $y = 2$

16. Let $\mathcal{R}$ be the region in the first quadrant bounded by the curves $y = x^3$ and $y = 2x - x^2$. Calculate the following quantities.
(a) The area of $\mathcal{R}$
(b) The volume obtained by rotating $\mathcal{R}$ about the x-axis
(c) The volume obtained by rotating $\mathcal{R}$ about the y-axis

17. Let $\mathcal{R}$ be the region bounded by the curves $y = \tan(x^2)$, $x = 1$, and $y = 0$. Use the Midpoint Rule with $n = 4$ to estimate the following quantities.
(a) The area of $\mathcal{R}$
(b) The volume obtained by rotating $\mathcal{R}$ about the x-axis

18. Let $\mathcal{R}$ be the region bounded by the curves $y = 1 - x^2$ and $y = x^6 - x + 1$. Estimate the following quantities.
(a) The x-coordinates of the points of intersection of the curves
(b) The area of $\mathcal{R}$
(c) The volume generated when $\mathcal{R}$ is rotated about the x-axis
(d) The volume generated when $\mathcal{R}$ is rotated about the y-axis

19–22 Each integral represents the volume of a solid. Describe the solid.

19. $\int_0^{\pi/2} 2\pi x \cos x \, dx$

20. $\int_0^{\pi/2} 2\pi \cos^2 x \, dx$

21. $\int_0^{\pi} \pi(2 - \sin x)^2 \, dx$

22. $\int_0^{4} 2\pi(6 - y)(4y - y^2) \, dy$

23. The base of a solid is a circular disk with radius 3. Find the volume of the solid if parallel cross-sections perpendicular to the base are isosceles right triangles with hypotenuse lying along the base.

24. The base of a solid is the region bounded by the parabolas $y = x^2$ and $y = 2 - x^2$. Find the volume of the solid if the cross-sections perpendicular to the x-axis are squares with one side lying along the base.

25. The height of a monument is 20 m. A horizontal cross-section at a distance x meters from the top is an equilateral triangle with side $\frac{1}{4}x$ meters. Find the volume of the monument.

26. (a) The base of a solid is a square with vertices located at $(1, 0)$, $(0, 1)$, $(-1, 0)$, and $(0, -1)$. Each cross-section perpendicular to the x-axis is a semicircle. Find the volume of the solid.
(b) Show that by cutting the solid of part (a), we can rearrange it to form a cone. Thus compute its volume more simply.

27. A force of 30 N is required to maintain a spring stretched from its natural length of 12 cm to a length of 15 cm. How much work is done in stretching the spring from 12 cm to 20 cm?

28. A 1600-lb elevator is suspended by a 200-ft cable that weighs 10 lb/ft. How much work is required to raise the elevator from the basement to the third floor, a distance of 30 ft?

29. A tank full of water has the shape of a paraboloid of revolution as shown in the figure; that is, its shape is obtained by rotating a parabola about a vertical axis.
(a) If its height is 4 ft and the radius at the top is 4 ft, find the work required to pump the water out of the tank.
(b) After 4000 ft-lb of work has been done, what is the depth of the water remaining in the tank?

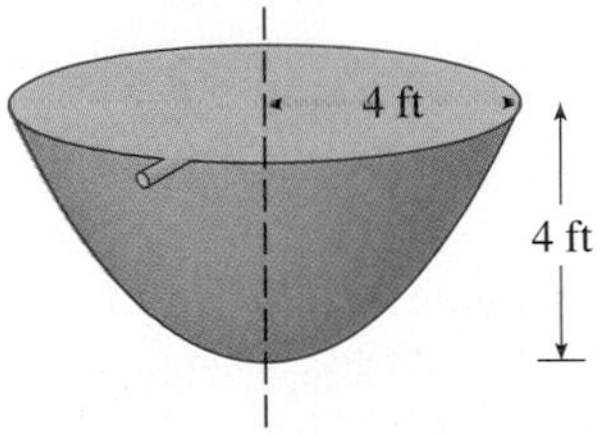

30. Find the average value of the function $f(t) = t \sin(t^2)$ on the interval $[0, 10]$.

31. If f is a continuous function, what is the limit as $h \to 0$ of the average value of f on the interval $[x, x + h]$?

32. Let $\mathcal{R}_1$ be the region bounded by $y = x^2$, $y = 0$, and $x = b$, where $b > 0$. Let $\mathcal{R}_2$ be the region bounded by $y = x^2$, $x = 0$, and $y = b^2$.
(a) Is there a value of b such that $\mathcal{R}_1$ and $\mathcal{R}_2$ have the same area?
(b) Is there a value of b such that $\mathcal{R}_1$ sweeps out the same volume when rotated about the x-axis and the y-axis?
(c) Is there a value of b such that $\mathcal{R}_1$ and $\mathcal{R}_2$ sweep out the same volume when rotated about the x-axis?
(d) Is there a value of b such that $\mathcal{R}_1$ and $\mathcal{R}_2$ sweep out the same volume when rotated about the y-axis?

PROBLEMS PLUS

1. (a) Find a positive continuous function f such that the area under the graph of f from 0 to t is $A(t) = t^3$ for all $t > 0$.

(b) A solid is generated by rotating about the x-axis the region under the curve $y = f(x)$, where f is a positive function and $x \geqslant 0$. The volume generated by the part of the curve from $x = 0$ to $x = b$ is b^2 for all $b > 0$. Find the function f.

2. There is a line through the origin that divides the region bounded by the parabola $y = x - x^2$ and the x-axis into two regions with equal area. What is the slope of that line?

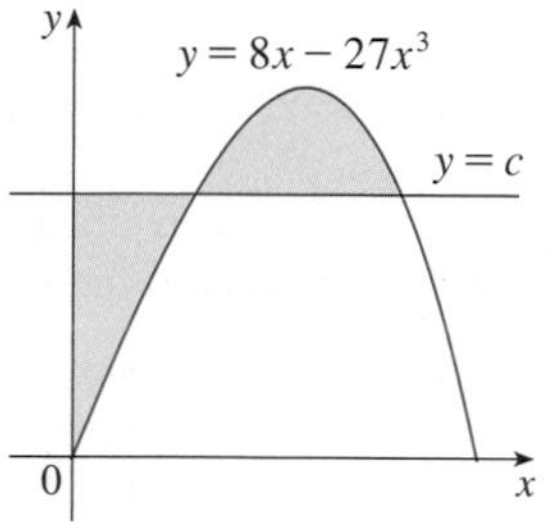

FIGURE FOR PROBLEM 3

3. The figure shows a horizontal line $y = c$ intersecting the curve $y = 8x - 27x^3$. Find the number c such that the areas of the shaded regions are equal.

4. A cylindrical glass of radius r and height L is filled with water and then tilted until the water remaining in the glass exactly covers its base.

(a) Determine a way to "slice" the water into parallel rectangular cross-sections and then *set up* a definite integral for the volume of the water in the glass.

(b) Determine a way to "slice" the water into parallel cross-sections that are trapezoids and then *set up* a definite integral for the volume of the water.

(c) Find the volume of water in the glass by evaluating one of the integrals in part (a) or part (b).

(d) Find the volume of the water in the glass from purely geometric considerations.

(e) Suppose the glass is tilted until the water exactly covers half the base. In what direction can you "slice" the water into triangular cross-sections? Rectangular cross-sections? Cross-sections that are segments of circles? Find the volume of water in the glass.

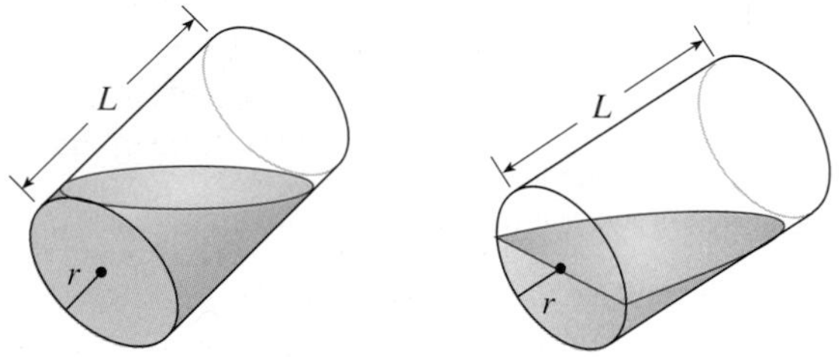

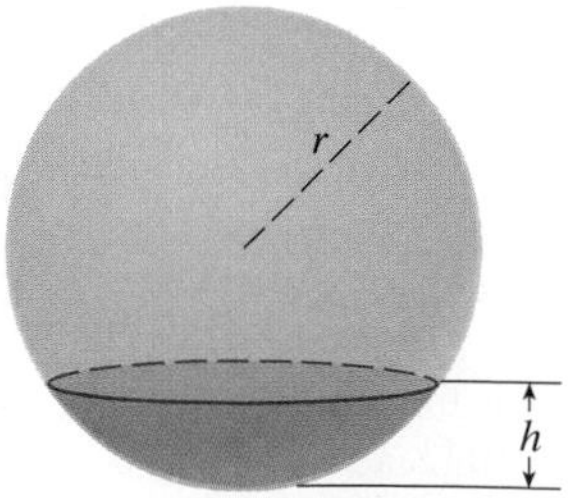

FIGURE FOR PROBLEM 5

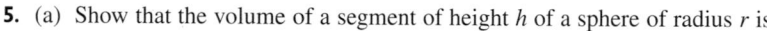

5. (a) Show that the volume of a segment of height h of a sphere of radius r is

$$V = \tfrac{1}{3}\pi h^2(3r - h)$$

(b) Show that if a sphere of radius 1 is sliced by a plane at a distance x from the center in such a way that the volume of one segment is twice the volume of the other, then x is a solution of the equation

$$3x^3 - 9x + 2 = 0$$

where $0 < x < 1$. Use Newton's method to find x accurate to four decimal places.

(c) Using the formula for the volume of a segment of a sphere, it can be shown that the depth x to which a floating sphere of radius r sinks in water is a root of the equation

$$x^3 - 3rx^2 + 4r^3s = 0$$

where s is the specific gravity of the sphere. Suppose a wooden sphere of radius 0.5 m has specific gravity 0.75. Calculate, to four-decimal-place accuracy, the depth to which the sphere will sink.

(d) A hemispherical bowl has radius 5 inches and water is running into the bowl at the rate of $0.2\ \text{in}^3/\text{s}$.

(i) How fast is the water level in the bowl rising at the instant the water is 3 inches deep?

(ii) At a certain instant, the water is 4 inches deep. How long will it take to fill the bowl?

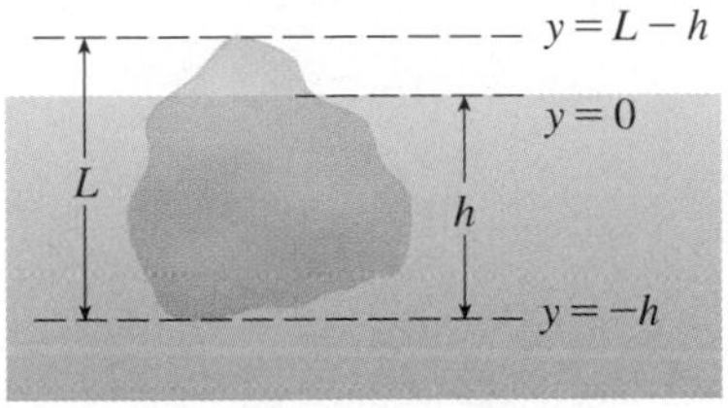

FIGURE FOR PROBLEM 6

6. Archimedes' Principle states that the buoyant force on an object partially or fully submerged in a fluid is equal to the weight of the fluid that the object displaces. Thus, for an object of density ρ_0 floating partly submerged in a fluid of density ρ_f, the buoyant force is given by $F = \rho_f g \int_{-h}^{0} A(y)\,dy$, where g is the acceleration due to gravity and $A(y)$ is the area of a typical cross-section of the object. The weight of the object is given by

$$W = \rho_0 g \int_{-h}^{L-h} A(y)\,dy$$

(a) Show that the percentage of the volume of the object above the surface of the liquid is

$$100\,\frac{\rho_f - \rho_0}{\rho_f}$$

(b) The density of ice is $917\ \text{kg/m}^3$ and the density of seawater is $1030\ \text{kg/m}^3$. What percentage of the volume of an iceberg is above water?

(c) An ice cube floats in a glass filled to the brim with water. Does the water overflow when the ice melts?

(d) A sphere of radius 0.4 m and having negligible weight is floating in a large freshwater lake. How much work is required to completely submerge the sphere? The density of the water is $1000\ \text{kg/m}^3$.

7. Water in an open bowl evaporates at a rate proportional to the area of the surface of the water. (This means that the rate of decrease of the volume is proportional to the area of the surface.) Show that the depth of the water decreases at a constant rate, regardless of the shape of the bowl.

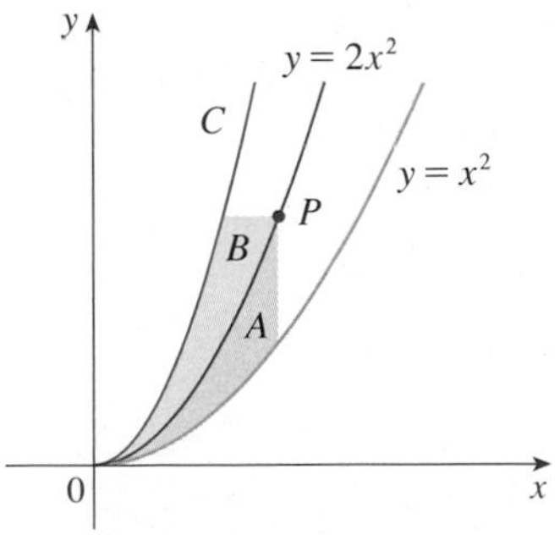

FIGURE FOR PROBLEM 9

8. A sphere of radius 1 overlaps a smaller sphere of radius r in such a way that their intersection is a circle of radius r. (In other words, they intersect in a great circle of the small sphere.) Find r so that the volume inside the small sphere and outside the large sphere is as large as possible.

9. The figure shows a curve C with the property that, for every point P on the middle curve $y = 2x^2$, the areas A and B are equal. Find an equation for C.

10. A paper drinking cup filled with water has the shape of a cone with height h and semivertical angle θ (see the figure). A ball is placed carefully in the cup, thereby displacing some of the water and making it overflow. What is the radius of the ball that causes the greatest volume of water to spill out of the cup?

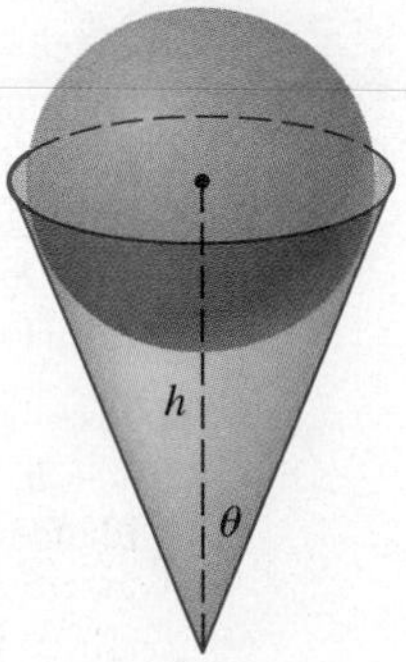

11. A *clepsydra*, or water clock, is a glass container with a small hole in the bottom through which water can flow. The "clock" is calibrated for measuring time by placing markings on the container corresponding to water levels at equally spaced times. Let $x = f(y)$ be continuous on the interval $[0, b]$ and assume that the container is formed by rotating the graph of f about the y-axis. Let V denote the volume of water and h the height of the water level at time t.

(a) Determine V as a function of h.

(b) Show that

$$\frac{dV}{dt} = \pi[f(h)]^2 \frac{dh}{dt}$$

(c) Suppose that A is the area of the hole in the bottom of the container. It follows from Torricelli's Law that the rate of change of the volume of the water is given by

$$\frac{dV}{dt} = kA\sqrt{h}$$

where k is a negative constant. Determine a formula for the function f such that dh/dt is a constant C. What is the advantage in having $dh/dt = C$?

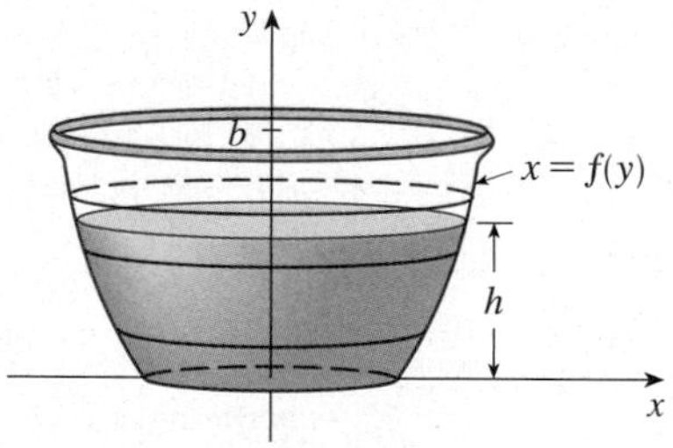

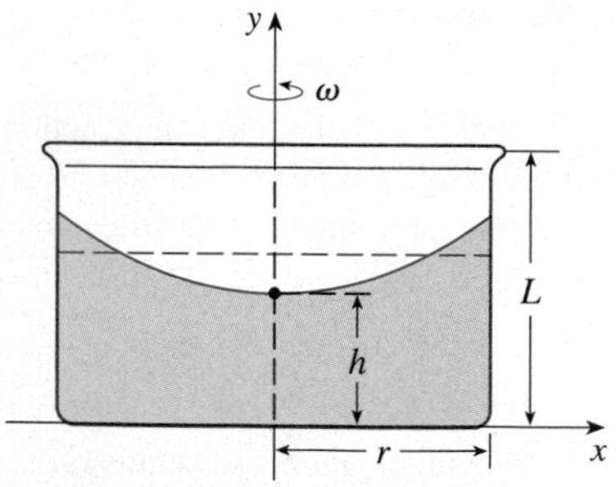

FIGURE FOR PROBLEM 12

12. A cylindrical container of radius r and height L is partially filled with a liquid whose volume is V. If the container is rotated about its axis of symmetry with constant angular speed ω, then the container will induce a rotational motion in the liquid around the same axis. Eventually, the liquid will be rotating at the same angular speed as the container. The surface of the liquid will be convex, as indicated in the figure, because the centrifugal force on the liquid particles increases with the distance from the axis of the container. It can be shown that the surface of the liquid is a paraboloid of revolution generated by rotating the parabola

$$y = h + \frac{\omega^2 x^2}{2g}$$

about the y-axis, where g is the acceleration due to gravity.

(a) Determine h as a function of ω.

(b) At what angular speed will the surface of the liquid touch the bottom? At what speed will it spill over the top?

(c) Suppose the radius of the container is 2 ft, the height is 7 ft, and the container and liquid are rotating at the same constant angular speed. The surface of the liquid is 5 ft below the top of the tank at the central axis and 4 ft below the top of the tank 1 ft out from the central axis.

(i) Determine the angular speed of the container and the volume of the fluid.

(ii) How far below the top of the tank is the liquid at the wall of the container?

13. Suppose the graph of a cubic polynomial intersects the parabola $y = x^2$ when $x = 0$, $x = a$, and $x = b$, where $0 < a < b$. If the two regions between the curves have the same area, how is b related to a?

CAS **14.** Suppose we are planning to make a taco from a round tortilla with diameter 8 inches by bending the tortilla so that it is shaped as if it is partially wrapped around a circular cylinder. We will fill the tortilla to the edge (but no more) with meat, cheese, and other ingredients. Our problem is to decide how to curve the tortilla in order to maximize the volume of food it can hold.

(a) We start by placing a circular cylinder of radius r along a diameter of the tortilla and folding the tortilla around the cylinder. Let x represent the distance from the center of the tortilla to a point P on the diameter (see the figure). Show that the cross-sectional area of the filled taco in the plane through P perpendicular to the axis of the cylinder is

$$A(x) = r\sqrt{16 - x^2} - \tfrac{1}{2}r^2 \sin\left(\frac{2}{r}\sqrt{16 - x^2}\right)$$

and write an expression for the volume of the filled taco.

(b) Determine (approximately) the value of r that maximizes the volume of the taco. (Use a graphical approach with your CAS.)

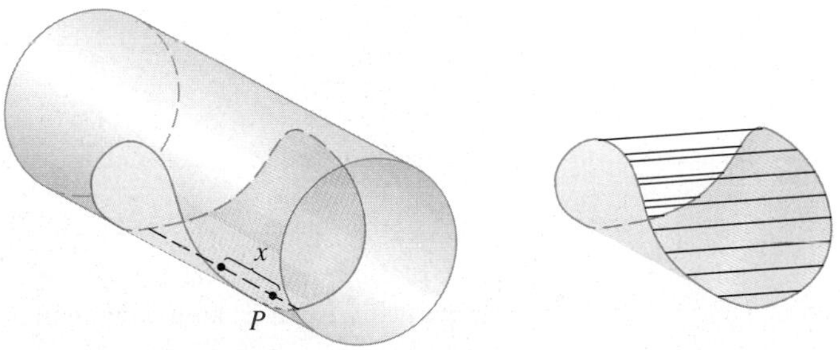

15. If the tangent at a point P on the curve $y = x^3$ intersects the curve again at Q, let A be the area of the region bounded by the curve and the line segment PQ. Let B be the area of the region defined in the same way starting with Q instead of P. What is the relationship between A and B?

7

TECHNIQUES OF INTEGRATION

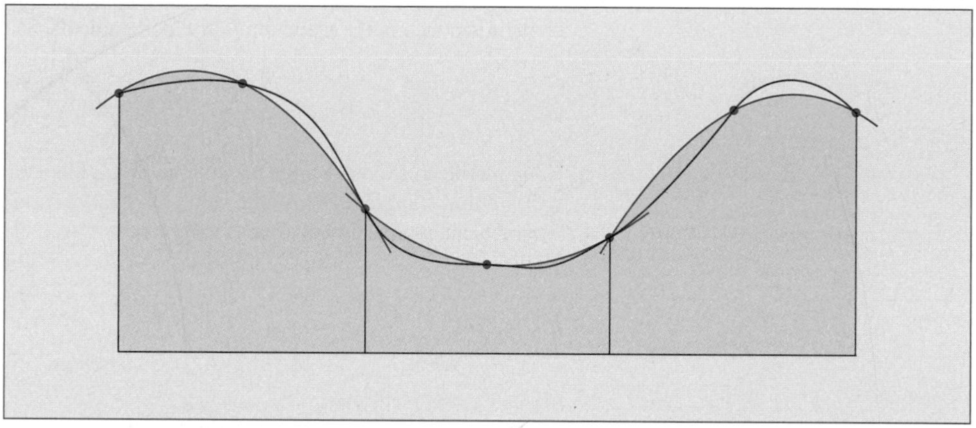

Simpson's Rule estimates integrals by approximating graphs with parabolas.

Because of the Fundamental Theorem of Calculus, we can integrate a function if we know an antiderivative, that is, an indefinite integral. We summarize here the most important integrals that we have learned so far.

$$\int x^n\,dx = \frac{x^{n+1}}{n+1} + C \quad (n \neq -1) \qquad \int \frac{1}{x}\,dx = \ln|x| + C$$

$$\int e^x\,dx = e^x + C \qquad \int a^x\,dx = \frac{a^x}{\ln a} + C$$

$$\int \sin x\,dx = -\cos x + C \qquad \int \cos x\,dx = \sin x + C$$

$$\int \sec^2 x\,dx = \tan x + C \qquad \int \csc^2 x\,dx = -\cot x + C$$

$$\int \sec x \tan x\,dx = \sec x + C \qquad \int \csc x \cot x\,dx = -\csc x + C$$

$$\int \sinh x\,dx = \cosh x + C \qquad \int \cosh x\,dx = \sinh x + C$$

$$\int \tan x\,dx = \ln|\sec x| + C \qquad \int \cot x\,dx = \ln|\sin x| + C$$

$$\int \frac{1}{x^2+a^2}\,dx = \frac{1}{a}\tan^{-1}\left(\frac{x}{a}\right) + C \qquad \int \frac{1}{\sqrt{a^2-x^2}}\,dx = \sin^{-1}\left(\frac{x}{a}\right) + C, \quad a > 0$$

In this chapter we develop techniques for using these basic integration formulas to obtain indefinite integrals of more complicated functions. We learned the most important method of integration, the Substitution Rule, in Section 5.5. The other general technique, integration by parts, is presented in Section 7.1. Then we learn methods that are special to particular classes of functions, such as trigonometric functions and rational functions.

Integration is not as straightforward as differentiation; there are no rules that absolutely guarantee obtaining an indefinite integral of a function. Therefore we discuss a strategy for integration in Section 7.5.

7.1 INTEGRATION BY PARTS

Every differentiation rule has a corresponding integration rule. For instance, the Substitution Rule for integration corresponds to the Chain Rule for differentiation. The rule that corresponds to the Product Rule for differentiation is called the rule for *integration by parts.*

The Product Rule states that if f and g are differentiable functions, then

$$\frac{d}{dx}[f(x)g(x)] = f(x)g'(x) + g(x)f'(x)$$

In the notation for indefinite integrals this equation becomes

$$\int [f(x)g'(x) + g(x)f'(x)]\,dx = f(x)g(x)$$

or

$$\int f(x)g'(x)\,dx + \int g(x)f'(x)\,dx = f(x)g(x)$$

We can rearrange this equation as

1

$$\int f(x)g'(x)\,dx = f(x)g(x) - \int g(x)f'(x)\,dx$$

Formula 1 is called the **formula for integration by parts**. It is perhaps easier to remember in the following notation. Let $u = f(x)$ and $v = g(x)$. Then the differentials are $du = f'(x)\,dx$ and $dv = g'(x)\,dx$, so, by the Substitution Rule, the formula for integration by parts becomes

2

$$\int u\,dv = uv - \int v\,du$$

EXAMPLE 1 Find $\int x \sin x\,dx$.

SOLUTION USING FORMULA 1 Suppose we choose $f(x) = x$ and $g'(x) = \sin x$. Then $f'(x) = 1$ and $g(x) = -\cos x$. (For g we can choose *any* antiderivative of g'.) Thus, using Formula 1, we have

$$\begin{aligned}\int x \sin x\,dx &= f(x)g(x) - \int g(x)f'(x)\,dx \\ &= x(-\cos x) - \int (-\cos x)\,dx \\ &= -x\cos x + \int \cos x\,dx \\ &= -x\cos x + \sin x + C\end{aligned}$$

It's wise to check the answer by differentiating it. If we do so, we get $x \sin x$, as expected.

SOLUTION USING FORMULA 2 Let

■ It is helpful to use the pattern:

$u = \square \quad dv = \square$

$du = \square \quad v = \square$

$$u = x \qquad dv = \sin x\, dx$$

Then $$du = dx \qquad v = -\cos x$$

and so

$$\int x \sin x\, dx = \int \overbrace{x}^{u}\, \overbrace{\sin x\, dx}^{dv} = \overbrace{x}^{u}\, \overbrace{(-\cos x)}^{v} - \int \overbrace{(-\cos x)}^{v}\, \overbrace{dx}^{du}$$

$$= -x \cos x + \int \cos x\, dx$$

$$= -x \cos x + \sin x + C$$ □

NOTE Our aim in using integration by parts is to obtain a simpler integral than the one we started with. Thus in Example 1 we started with $\int x \sin x\, dx$ and expressed it in terms of the simpler integral $\int \cos x\, dx$. If we had instead chosen $u = \sin x$ and $dv = x\, dx$, then $du = \cos x\, dx$ and $v = x^2/2$, so integration by parts gives

$$\int x \sin x\, dx = (\sin x)\frac{x^2}{2} - \frac{1}{2}\int x^2 \cos x\, dx$$

Although this is true, $\int x^2 \cos x\, dx$ is a more difficult integral than the one we started with. In general, when deciding on a choice for u and dv, we usually try to choose $u = f(x)$ to be a function that becomes simpler when differentiated (or at least not more complicated) as long as $dv = g'(x)\, dx$ can be readily integrated to give v.

V EXAMPLE 2 Evaluate $\int \ln x\, dx$.

SOLUTION Here we don't have much choice for u and dv. Let

$$u = \ln x \qquad dv = dx$$

Then $$du = \frac{1}{x}\, dx \qquad v = x$$

Integrating by parts, we get

$$\int \ln x\, dx = x \ln x - \int x \frac{dx}{x}$$

■ It's customary to write $\int 1\, dx$ as $\int dx$.

$$= x \ln x - \int dx$$

■ Check the answer by differentiating it.

$$= x \ln x - x + C$$

Integration by parts is effective in this example because the derivative of the function $f(x) = \ln x$ is simpler than f. □

EXAMPLE 3 Find $\int t^2 e^t\,dt$.

SOLUTION Notice that t^2 becomes simpler when differentiated (whereas e^t is unchanged when differentiated or integrated), so we choose

$$u = t^2 \qquad dv = e^t\,dt$$

Then
$$du = 2t\,dt \qquad v = e^t$$

Integration by parts gives

$$\int t^2 e^t\,dt = t^2 e^t - 2\int t e^t\,dt \tag{3}$$

The integral that we obtained, $\int t e^t\,dt$, is simpler than the original integral but is still not obvious. Therefore, we use integration by parts a second time, this time with $u = t$ and $dv = e^t\,dt$. Then $du = dt$, $v = e^t$, and

$$\int t e^t\,dt = te^t - \int e^t\,dt = te^t - e^t + C$$

L
I
A
T
E

Putting this in Equation 3, we get

$$\begin{aligned}\int t^2 e^t\,dt &= t^2 e^t - 2\int t e^t\,dt \\ &= t^2 e^t - 2(te^t - e^t + C) \\ &= t^2 e^t - 2te^t + 2e^t + C_1 \qquad \text{where } C_1 = -2C\end{aligned}$$ □

■ An easier method, using complex numbers, is given in Exercise 50 in Appendix H.

EXAMPLE 4 Evaluate $\int e^x \sin x\,dx$.

$u = e^x \quad du = e^x\,dx$
$dv = \sin x\,dx \quad v = -\cos x$

SOLUTION Neither e^x nor $\sin x$ becomes simpler when differentiated, but we try choosing $u = e^x$ and $dv = \sin x\,dx$ anyway. Then $du = e^x\,dx$ and $v = -\cos x$, so integration by parts gives

$$\int e^x \sin x\,dx = -e^x \cos x + \int e^x \cos x\,dx \tag{4}$$

The integral that we have obtained, $\int e^x \cos x\,dx$, is no simpler than the original one, but at least it's no more difficult. Having had success in the preceding example integrating by parts twice, we persevere and integrate by parts again. This time we use $u = e^x$ and $dv = \cos x\,dx$. Then $du = e^x\,dx$, $v = \sin x$, and

$$\int e^x \cos x\,dx = e^x \sin x - \int e^x \sin x\,dx \tag{5}$$

At first glance, it appears as if we have accomplished nothing because we have arrived at $\int e^x \sin x\,dx$, which is where we started. However, if we put the expression for $\int e^x \cos x\,dx$ from Equation 5 into Equation 4 we get

$$\int e^x \sin x\,dx = -e^x \cos x + e^x \sin x - \int e^x \sin x\,dx$$

■ Figure 1 illustrates Example 4 by showing the graphs of $f(x) = e^x \sin x$ and $F(x) = \frac{1}{2}e^x(\sin x - \cos x)$. As a visual check on our work, notice that $f(x) = 0$ when F has a maximum or minimum.

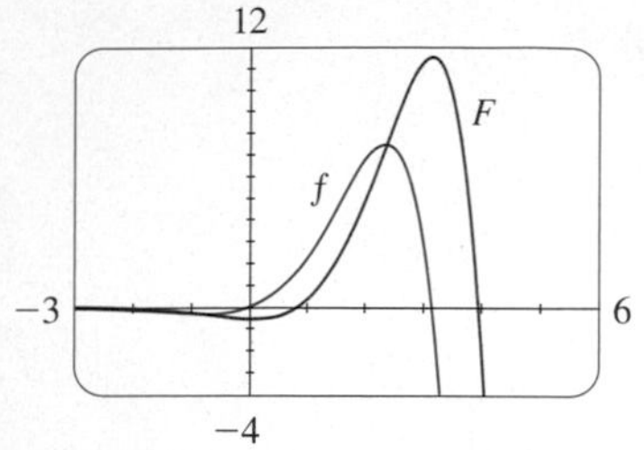

FIGURE 1

This can be regarded as an equation to be solved for the unknown integral. Adding $\int e^x \sin x\,dx$ to both sides, we obtain

$$2\int e^x \sin x\,dx = -e^x \cos x + e^x \sin x$$

Dividing by 2 and adding the constant of integration, we get

$$\int e^x \sin x\,dx = \tfrac{1}{2}e^x(\sin x - \cos x) + C$$ □

If we combine the formula for integration by parts with Part 2 of the Fundamental Theorem of Calculus, we can evaluate definite integrals by parts. Evaluating both sides of Formula 1 between a and b, assuming f' and g' are continuous, and using the Fundamental Theorem, we obtain

6 $$\int_a^b f(x)g'(x)\,dx = f(x)g(x)\Big]_a^b - \int_a^b g(x)f'(x)\,dx$$

EXAMPLE 5 Calculate $\int_0^1 \tan^{-1}x\,dx$.

SOLUTION Let

$$u = \tan^{-1}x \qquad dv = dx$$

Then $$du = \frac{dx}{1+x^2} \qquad v = x$$

So Formula 6 gives

$$\begin{aligned}\int_0^1 \tan^{-1}x\,dx &= x\tan^{-1}x\Big]_0^1 - \int_0^1 \frac{x}{1+x^2}\,dx\\ &= 1\cdot\tan^{-1}1 - 0\cdot\tan^{-1}0 - \int_0^1 \frac{x}{1+x^2}\,dx\\ &= \frac{\pi}{4} - \int_0^1 \frac{x}{1+x^2}\,dx\end{aligned}$$

■ Since $\tan^{-1}x \geqslant 0$ for $x \geqslant 0$, the integral in Example 5 can be interpreted as the area of the region shown in Figure 2.

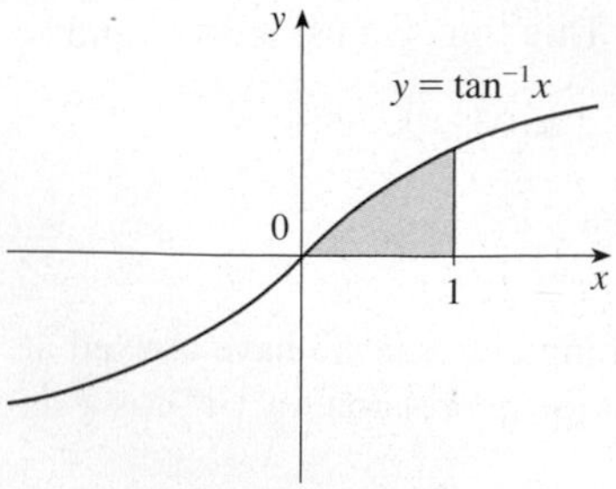

FIGURE 2

To evaluate this integral we use the substitution $t = 1 + x^2$ (since u has another meaning in this example). Then $dt = 2x\,dx$, so $x\,dx = \frac{1}{2}\,dt$. When $x = 0$, $t = 1$; when $x = 1$, $t = 2$; so

$$\begin{aligned}\int_0^1 \frac{x}{1+x^2}\,dx &= \tfrac{1}{2}\int_1^2 \frac{dt}{t} = \tfrac{1}{2}\ln|t|\Big]_1^2\\ &= \tfrac{1}{2}(\ln 2 - \ln 1) = \tfrac{1}{2}\ln 2\end{aligned}$$

Therefore $$\int_0^1 \tan^{-1}x\,dx = \frac{\pi}{4} - \int_0^1 \frac{x}{1+x^2}\,dx = \frac{\pi}{4} - \frac{\ln 2}{2}$$ □

EXAMPLE 6 Prove the reduction formula

■ Equation 7 is called a *reduction formula* because the exponent n has been *reduced* to $n - 1$ and $n - 2$.

$$\boxed{7} \qquad \int \sin^n x\,dx = -\frac{1}{n}\cos x \sin^{n-1}x + \frac{n-1}{n}\int \sin^{n-2}x\,dx$$

where $n \geqslant 2$ is an integer.

SOLUTION Let $\quad u = \sin^{n-1}x \qquad dv = \sin x\,dx$

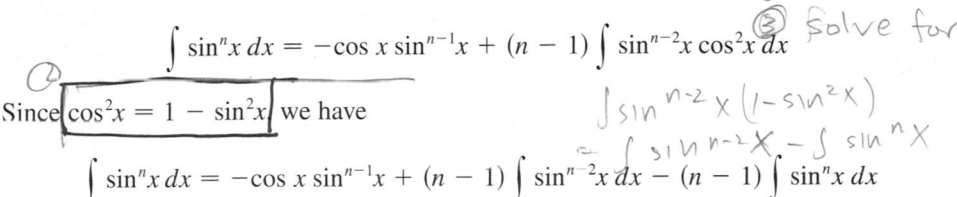

Then $\quad du = (n-1)\sin^{n-2}x\cos x\,dx \qquad v = -\cos x$

so integration by parts gives

$$\int \sin^n x\,dx = -\cos x \sin^{n-1}x + (n-1)\int \sin^{n-2}x\cos^2 x\,dx$$

Since $\cos^2 x = 1 - \sin^2 x$ we have

$$\int \sin^n x\,dx = -\cos x \sin^{n-1}x + (n-1)\int \sin^{n-2}x\,dx - (n-1)\int \sin^n x\,dx$$

As in Example 4, we solve this equation for the desired integral by taking the last term on the right side to the left side. Thus we have

$$n\int \sin^n x\,dx = -\cos x \sin^{n-1}x + (n-1)\int \sin^{n-2}x\,dx$$

or

$$\int \sin^n x\,dx = -\frac{1}{n}\cos x \sin^{n-1}x + \frac{n-1}{n}\int \sin^{n-2}x\,dx \qquad \square$$

The reduction formula (7) is useful because by using it repeatedly we could eventually express $\int \sin^n x\,dx$ in terms of $\int \sin x\,dx$ (if n is odd) or $\int (\sin x)^0\,dx = \int dx$ (if n is even).

7.1 EXERCISES

1–2 Evaluate the integral using integration by parts with the indicated choices of u and dv.

1. $\int x^2 \ln x\,dx;\quad u = \ln x,\ dv = x^2\,dx$

2. $\int \theta \cos\theta\,d\theta;\quad u = \theta,\ dv = \cos\theta\,d\theta$

3–32 Evaluate the integral.

3. $\int x\cos 5x\,dx$

4. $\int xe^{-x}\,dx$

5. $\int re^{r/2}\,dr$

6. $\int t\sin 2t\,dt$

7. $\int x^2\sin \pi x\,dx$

8. $\int x^2\cos mx\,dx$

9. $\int \ln(2x+1)\,dx$

10. $\int \sin^{-1}x\,dx$

11. $\int \arctan 4t\,dt$

12. $\int p^5\ln p\,dp$

13. $\int t\sec^2 2t\,dt$

14. $\int s\,2^s\,ds$

15. $\int (\ln x)^2\,dx$

16. $\int t\sinh mt\,dt$

17. $\int e^{2\theta}\sin 3\theta\,d\theta$

18. $\int e^{-\theta}\cos 2\theta\,d\theta$

19. $\int_0^{\pi} t\sin 3t\,dt$

20. $\int_0^1 (x^2+1)e^{-x}\,dx$

21. $\int_0^1 t\cosh t\,dt$

22. $\int_4^9 \frac{\ln y}{\sqrt{y}}\,dy$

23. $\int_1^2 \frac{\ln x}{x^2}\,dx$

24. $\int_0^{\pi} x^3\cos x\,dx$

25. $\int_0^1 \frac{y}{e^{2y}}\,dy$

26. $\int_1^{\sqrt{3}} \arctan(1/x)\,dx$

27. $\int_0^{1/2} \cos^{-1}x\,dx$

28. $\int_1^2 \frac{(\ln x)^2}{x^3}\,dx$

29. $\int \cos x \ln(\sin x)\,dx$

30. $\int_0^1 \frac{r^3}{\sqrt{4+r^2}}\,dr$

31. $\int_1^2 x^4(\ln x)^2\,dx$

32. $\int_0^t e^s \sin(t-s)\,ds$

33–38 First make a substitution and then use integration by parts to evaluate the integral.

33. $\int \cos\sqrt{x}\,dx$

34. $\int t^3 e^{-t^2}\,dt$

35. $\int_{\sqrt{\pi/2}}^{\sqrt{\pi}} \theta^3 \cos(\theta^2)\,d\theta$

36. $\int_0^{\pi} e^{\cos t} \sin 2t\,dt$

37. $\int x\ln(1+x)\,dx$

38. $\int \sin(\ln x)\,dx$

39–42 Evaluate the indefinite integral. Illustrate, and check that your answer is reasonable, by graphing both the function and its antiderivative (take $C = 0$).

39. $\int (2x+3)e^x\,dx$

40. $\int x^{3/2}\ln x\,dx$

41. $\int x^3\sqrt{1+x^2}\,dx$

42. $\int x^2 \sin 2x\,dx$

43. (a) Use the reduction formula in Example 6 to show that

$$\int \sin^2 x\,dx = \frac{x}{2} - \frac{\sin 2x}{4} + C$$

(b) Use part (a) and the reduction formula to evaluate $\int \sin^4 x\,dx$.

44. (a) Prove the reduction formula

$$\int \cos^n x\,dx = \frac{1}{n}\cos^{n-1}x\sin x + \frac{n-1}{n}\int \cos^{n-2}x\,dx$$

(b) Use part (a) to evaluate $\int \cos^2 x\,dx$.
(c) Use parts (a) and (b) to evaluate $\int \cos^4 x\,dx$.

45. (a) Use the reduction formula in Example 6 to show that

$$\int_0^{\pi/2} \sin^n x\,dx = \frac{n-1}{n}\int_0^{\pi/2} \sin^{n-2}x\,dx$$

where $n \geq 2$ is an integer.

(b) Use part (a) to evaluate $\int_0^{\pi/2} \sin^3 x\,dx$ and $\int_0^{\pi/2} \sin^5 x\,dx$.
(c) Use part (a) to show that, for odd powers of sine,

$$\int_0^{\pi/2} \sin^{2n+1}x\,dx = \frac{2\cdot 4\cdot 6\cdot\cdots\cdot 2n}{3\cdot 5\cdot 7\cdot\cdots\cdot(2n+1)}$$

46. Prove that, for even powers of sine,

$$\int_0^{\pi/2} \sin^{2n}x\,dx = \frac{1\cdot 3\cdot 5\cdot\cdots\cdot(2n-1)}{2\cdot 4\cdot 6\cdot\cdots\cdot 2n}\,\frac{\pi}{2}$$

47–50 Use integration by parts to prove the reduction formula.

47. $\int (\ln x)^n\,dx = x(\ln x)^n - n\int (\ln x)^{n-1}\,dx$

48. $\int x^n e^x\,dx = x^n e^x - n\int x^{n-1}e^x\,dx$

49. $\int \tan^n x\,dx = \frac{\tan^{n-1}x}{n-1} - \int \tan^{n-2}x\,dx \quad (n \neq 1)$

50. $\int \sec^n x\,dx = \frac{\tan x\sec^{n-2}x}{n-1} + \frac{n-2}{n-1}\int \sec^{n-2}x\,dx \quad (n \neq 1)$

51. Use Exercise 47 to find $\int (\ln x)^3\,dx$.

52. Use Exercise 48 to find $\int x^4 e^x\,dx$.

53–54 Find the area of the region bounded by the given curves.

53. $y = xe^{-0.4x}, \quad y = 0, \quad x = 5$

54. $y = 5\ln x, \quad y = x\ln x$

55–56 Use a graph to find approximate x-coordinates of the points of intersection of the given curves. Then find (approximately) the area of the region bounded by the curves.

55. $y = x\sin x, \quad y = (x-2)^2$

56. $y = \arctan 3x, \quad y = \frac{1}{2}x$

57–60 Use the method of cylindrical shells to find the volume generated by rotating the region bounded by the given curves about the specified axis.

57. $y = \cos(\pi x/2),\ y = 0,\ 0 \leq x \leq 1;$ about the y-axis

58. $y = e^x,\ y = e^{-x},\ x = 1;$ about the y-axis

59. $y = e^{-x},\ y = 0,\ x = -1,\ x = 0;$ about $x = 1$

60. $y = e^x,\ x = 0,\ y = \pi;$ about the x-axis

61. Find the average value of $f(x) = x^2 \ln x$ on the interval $[1, 3]$.

62. A rocket accelerates by burning its onboard fuel, so its mass decreases with time. Suppose the initial mass of the rocket at liftoff (including its fuel) is m, the fuel is consumed at rate r, and the exhaust gases are ejected with constant velocity v_e (relative to the rocket). A model for the velocity of the rocket at time t is given by the equation

$$v(t) = -gt - v_e \ln \frac{m - rt}{m}$$

where g is the acceleration due to gravity and t is not too large. If $g = 9.8 \text{ m/s}^2$, $m = 30{,}000$ kg, $r = 160$ kg/s, and $v_e = 3000$ m/s, find the height of the rocket one minute after liftoff.

63. A particle that moves along a straight line has velocity $v(t) = t^2 e^{-t}$ meters per second after t seconds. How far will it travel during the first t seconds?

64. If $f(0) = g(0) = 0$ and f'' and g'' are continuous, show that

$$\int_0^a f(x)g''(x)\,dx = f(a)g'(a) - f'(a)g(a) + \int_0^a f''(x)g(x)\,dx$$

65. Suppose that $f(1) = 2$, $f(4) = 7$, $f'(1) = 5$, $f'(4) = 3$, and f'' is continuous. Find the value of $\int_1^4 xf''(x)\,dx$.

66. (a) Use integration by parts to show that

$$\int f(x)\,dx = xf(x) - \int xf'(x)\,dx$$

(b) If f and g are inverse functions and f' is continuous, prove that

$$\int_a^b f(x)\,dx = bf(b) - af(a) - \int_{f(a)}^{f(b)} g(y)\,dy$$

[*Hint:* Use part (a) and make the substitution $y = f(x)$.]

(c) In the case where f and g are positive functions and $b > a > 0$, draw a diagram to give a geometric interpretation of part (b).

(d) Use part (b) to evaluate $\int_1^e \ln x\,dx$.

67. We arrived at Formula 6.3.2, $V = \int_a^b 2\pi x f(x)\,dx$, by using cylindrical shells, but now we can use integration by parts to prove it using the slicing method of Section 6.2, at least for the case where f is one-to-one and therefore has an inverse function g. Use the figure to show that

$$V = \pi b^2 d - \pi a^2 c - \int_c^d \pi [g(y)]^2\,dy$$

Make the substitution $y = f(x)$ and then use integration by parts on the resulting integral to prove that

$$V = \int_a^b 2\pi x f(x)\,dx$$

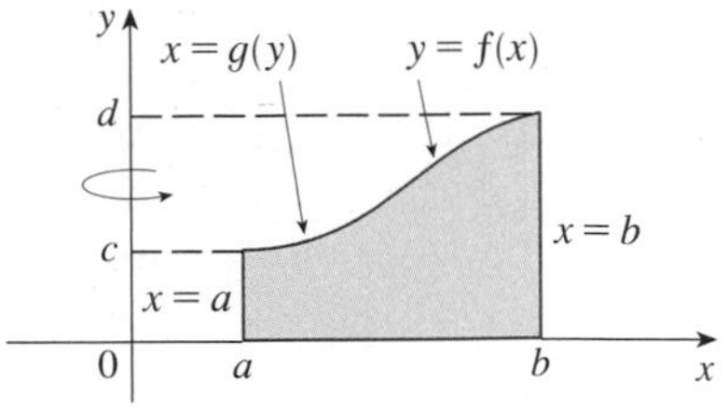

68. Let $I_n = \int_0^{\pi/2} \sin^n x\,dx$.

(a) Show that $I_{2n+2} \leqslant I_{2n+1} \leqslant I_{2n}$.

(b) Use Exercise 46 to show that

$$\frac{I_{2n+2}}{I_{2n}} = \frac{2n+1}{2n+2}$$

(c) Use parts (a) and (b) to show that

$$\frac{2n+1}{2n+2} \leqslant \frac{I_{2n+1}}{I_{2n}} \leqslant 1$$

and deduce that $\lim_{n \to \infty} I_{2n+1}/I_{2n} = 1$.

(d) Use part (c) and Exercises 45 and 46 to show that

$$\lim_{n \to \infty} \frac{2}{1} \cdot \frac{2}{3} \cdot \frac{4}{3} \cdot \frac{4}{5} \cdot \frac{6}{5} \cdot \frac{6}{7} \cdot \cdots \cdot \frac{2n}{2n-1} \cdot \frac{2n}{2n+1} = \frac{\pi}{2}$$

This formula is usually written as an infinite product:

$$\frac{\pi}{2} = \frac{2}{1} \cdot \frac{2}{3} \cdot \frac{4}{3} \cdot \frac{4}{5} \cdot \frac{6}{5} \cdot \frac{6}{7} \cdot \cdots$$

and is called the *Wallis product*.

(e) We construct rectangles as follows. Start with a square of area 1 and attach rectangles of area 1 alternately beside or on top of the previous rectangle (see the figure). Find the limit of the ratios of width to height of these rectangles.

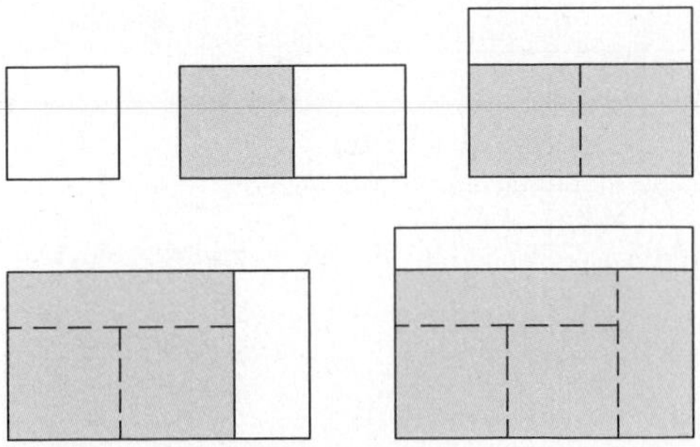

7.2 TRIGONOMETRIC INTEGRALS

In this section we use trigonometric identities to integrate certain combinations of trigonometric functions. We start with powers of sine and cosine.

EXAMPLE 1 Evaluate $\int \cos^3 x\, dx$.

SOLUTION Simply substituting $u = \cos x$ isn't helpful, since then $du = -\sin x\, dx$. In order to integrate powers of cosine, we would need an extra $\sin x$ factor. Similarly, a power of sine would require an extra $\cos x$ factor. Thus here we can separate one cosine factor and convert the remaining $\cos^2 x$ factor to an expression involving sine using the identity $\sin^2 x + \cos^2 x = 1$:

$$\cos^3 x = \cos^2 x \cdot \cos x = (1 - \sin^2 x) \cos x$$

We can then evaluate the integral by substituting $u = \sin x$, so $du = \cos x\, dx$ and

$$\begin{aligned}\int \cos^3 x\, dx &= \int \cos^2 x \cdot \cos x\, dx = \int (1 - \sin^2 x) \cos x\, dx \\ &= \int (1 - u^2)\, du = u - \tfrac{1}{3}u^3 + C \\ &= \sin x - \tfrac{1}{3}\sin^3 x + C\end{aligned}$$

□

In general, we try to write an integrand involving powers of sine and cosine in a form where we have only one sine factor (and the remainder of the expression in terms of cosine) or only one cosine factor (and the remainder of the expression in terms of sine). The identity $\sin^2 x + \cos^2 x = 1$ enables us to convert back and forth between even powers of sine and cosine.

V EXAMPLE 2 Find $\int \sin^5 x \cos^2 x\, dx$.

SOLUTION We could convert $\cos^2 x$ to $1 - \sin^2 x$, but we would be left with an expression in terms of $\sin x$ with no extra $\cos x$ factor. Instead, we separate a single sine factor and rewrite the remaining $\sin^4 x$ factor in terms of $\cos x$:

$$\sin^5 x \cos^2 x = (\sin^2 x)^2 \cos^2 x \sin x = (1 - \cos^2 x)^2 \cos^2 x \sin x$$

Substituting $u = \cos x$, we have $du = -\sin x\, dx$ and so

$$\begin{aligned}\int \sin^5 x \cos^2 x\, dx &= \int (\sin^2 x)^2 \cos^2 x \sin x\, dx \\ &= \int (1 - \cos^2 x)^2 \cos^2 x \sin x\, dx \\ &= \int (1 - u^2)^2 u^2 (-du) = -\int (u^2 - 2u^4 + u^6)\, du \\ &= -\left(\frac{u^3}{3} - 2\frac{u^5}{5} + \frac{u^7}{7}\right) + C \\ &= -\tfrac{1}{3}\cos^3 x + \tfrac{2}{5}\cos^5 x - \tfrac{1}{7}\cos^7 x + C\end{aligned}$$

□

■ Figure 1 shows the graphs of the integrand $\sin^5 x \cos^2 x$ in Example 2 and its indefinite integral (with $C = 0$). Which is which?

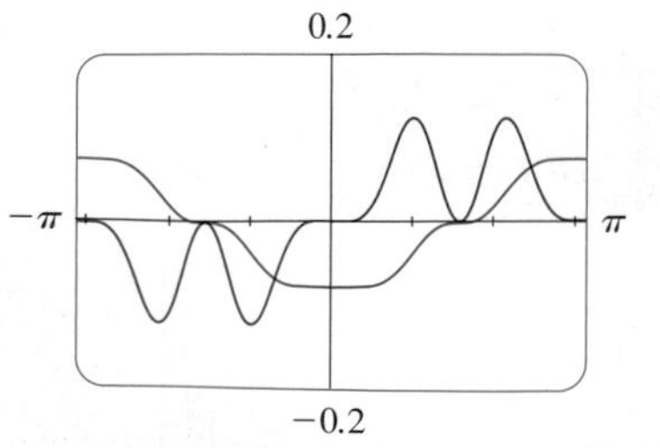

FIGURE 1

In the preceding examples, an odd power of sine or cosine enabled us to separate a single factor and convert the remaining even power. If the integrand contains even powers of both sine and cosine, this strategy fails. In this case, we can take advantage of the following half-angle identities (see Equations 17b and 17a in Appendix D):

$$\sin^2 x = \tfrac{1}{2}(1 - \cos 2x) \qquad \text{and} \qquad \cos^2 x = \tfrac{1}{2}(1 + \cos 2x)$$

■ Example 3 shows that the area of the region shown in Figure 2 is $\pi/2$.

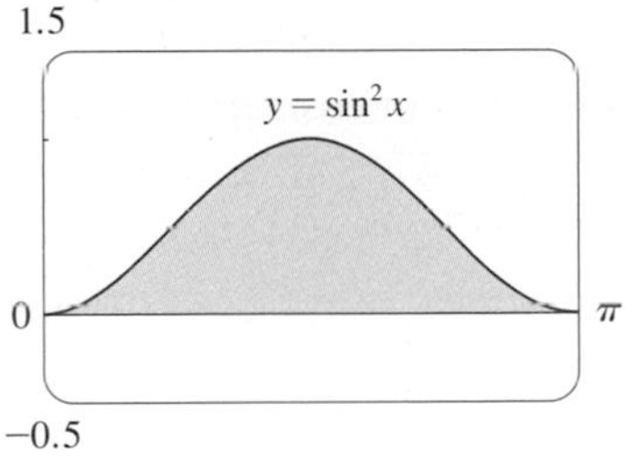

FIGURE 2

EXAMPLE 3 Evaluate $\int_0^\pi \sin^2 x\, dx$.

SOLUTION If we write $\sin^2 x = 1 - \cos^2 x$, the integral is no simpler to evaluate. Using the half-angle formula for $\sin^2 x$, however, we have

$$\int_0^\pi \sin^2 x\, dx = \tfrac{1}{2}\int_0^\pi (1 - \cos 2x)\, dx = \left[\tfrac{1}{2}\left(x - \tfrac{1}{2}\sin 2x\right)\right]_0^\pi$$

$$= \tfrac{1}{2}\left(\pi - \tfrac{1}{2}\sin 2\pi\right) - \tfrac{1}{2}\left(0 - \tfrac{1}{2}\sin 0\right) = \tfrac{1}{2}\pi$$

Notice that we mentally made the substitution $u = 2x$ when integrating $\cos 2x$. Another method for evaluating this integral was given in Exercise 43 in Section 7.1. □

EXAMPLE 4 Find $\int \sin^4 x\, dx$.

SOLUTION We could evaluate this integral using the reduction formula for $\int \sin^n x\, dx$ (Equation 7.1.7) together with Example 3 (as in Exercise 43 in Section 7.1), but a better method is to write $\sin^4 x = (\sin^2 x)^2$ and use a half-angle formula:

$$\int \sin^4 x\, dx = \int (\sin^2 x)^2\, dx$$

$$= \int \left(\frac{1 - \cos 2x}{2}\right)^2 dx$$

$$= \tfrac{1}{4}\int (1 - 2\cos 2x + \cos^2 2x)\, dx$$

Since $\cos^2 2x$ occurs, we must use another half-angle formula

$$\cos^2 2x = \tfrac{1}{2}(1 + \cos 4x)$$

This gives

$$\int \sin^4 x\, dx = \tfrac{1}{4}\int \left[1 - 2\cos 2x + \tfrac{1}{2}(1 + \cos 4x)\right] dx$$

$$= \tfrac{1}{4}\int \left(\tfrac{3}{2} - 2\cos 2x + \tfrac{1}{2}\cos 4x\right) dx$$

$$= \tfrac{1}{4}\left(\tfrac{3}{2}x - \sin 2x + \tfrac{1}{8}\sin 4x\right) + C$$

□

To summarize, we list guidelines to follow when evaluating integrals of the form $\int \sin^m x \cos^n x\, dx$, where $m \geqslant 0$ and $n \geqslant 0$ are integers.

STRATEGY FOR EVALUATING $\int \sin^m x \cos^n x \, dx$

(a) If the power of cosine is odd ($n = 2k + 1$), save one cosine factor and use $\cos^2 x = 1 - \sin^2 x$ to express the remaining factors in terms of sine:

$$\int \sin^m x \cos^{2k+1} x \, dx = \int \sin^m x \, (\cos^2 x)^k \cos x \, dx$$

$$= \int \sin^m x \, (1 - \sin^2 x)^k \cos x \, dx$$

Then substitute $u = \sin x$.

(b) If the power of sine is odd ($m = 2k + 1$), save one sine factor and use $\sin^2 x = 1 - \cos^2 x$ to express the remaining factors in terms of cosine:

$$\int \sin^{2k+1} x \cos^n x \, dx = \int (\sin^2 x)^k \cos^n x \sin x \, dx$$

$$= \int (1 - \cos^2 x)^k \cos^n x \sin x \, dx$$

Then substitute $u = \cos x$. [Note that if the powers of both sine and cosine are odd, either (a) or (b) can be used.]

(c) If the powers of both sine and cosine are even, use the half-angle identities

$$\sin^2 x = \tfrac{1}{2}(1 - \cos 2x) \qquad \cos^2 x = \tfrac{1}{2}(1 + \cos 2x)$$

It is sometimes helpful to use the identity

$$\sin x \cos x = \tfrac{1}{2} \sin 2x$$

We can use a similar strategy to evaluate integrals of the form $\int \tan^m x \sec^n x \, dx$. Since $(d/dx) \tan x = \sec^2 x$, we can separate a $\sec^2 x$ factor and convert the remaining (even) power of secant to an expression involving tangent using the identity $\sec^2 x = 1 + \tan^2 x$. Or, since $(d/dx) \sec x = \sec x \tan x$, we can separate a $\sec x \tan x$ factor and convert the remaining (even) power of tangent to secant.

V EXAMPLE 5 Evaluate $\int \tan^6 x \sec^4 x \, dx$.

SOLUTION If we separate one $\sec^2 x$ factor, we can express the remaining $\sec^2 x$ factor in terms of tangent using the identity $\sec^2 x = 1 + \tan^2 x$. We can then evaluate the integral by substituting $u = \tan x$ so that $du = \sec^2 x \, dx$:

$$\int \tan^6 x \sec^4 x \, dx = \int \tan^6 x \sec^2 x \sec^2 x \, dx$$

$$= \int \tan^6 x \, (1 + \tan^2 x) \sec^2 x \, dx$$

$$= \int u^6 (1 + u^2) \, du = \int (u^6 + u^8) \, du$$

$$= \frac{u^7}{7} + \frac{u^9}{9} + C$$

$$= \tfrac{1}{7} \tan^7 x + \tfrac{1}{9} \tan^9 x + C$$

□

EXAMPLE 6 Find $\int \tan^5\theta \sec^7\theta \, d\theta$.

SOLUTION If we separate a $\sec^2\theta$ factor, as in the preceding example, we are left with a $\sec^5\theta$ factor, which isn't easily converted to tangent. However, if we separate a $\sec\theta \tan\theta$ factor, we can convert the remaining power of tangent to an expression involving only secant using the identity $\tan^2\theta = \sec^2\theta - 1$. We can then evaluate the integral by substituting $u = \sec\theta$, so $du = \sec\theta \tan\theta \, d\theta$:

$$\begin{aligned}
\int \tan^5\theta \sec^7\theta \, d\theta &= \int \tan^4\theta \sec^6\theta \sec\theta \tan\theta \, d\theta \\
&= \int (\sec^2\theta - 1)^2 \sec^6\theta \sec\theta \tan\theta \, d\theta \\
&= \int (u^2 - 1)^2 u^6 \, du \\
&= \int (u^{10} - 2u^8 + u^6) \, du \\
&= \frac{u^{11}}{11} - 2\frac{u^9}{9} + \frac{u^7}{7} + C \\
&= \tfrac{1}{11}\sec^{11}\theta - \tfrac{2}{9}\sec^9\theta + \tfrac{1}{7}\sec^7\theta + C
\end{aligned}$$

□

The preceding examples demonstrate strategies for evaluating integrals of the form $\int \tan^m x \sec^n x \, dx$ for two cases, which we summarize here.

STRATEGY FOR EVALUATING $\int \tan^m x \sec^n x \, dx$

(a) If the power of secant is even ($n = 2k$, $k \geq 2$), save a factor of $\sec^2 x$ and use $\sec^2 x = 1 + \tan^2 x$ to express the remaining factors in terms of $\tan x$:

$$\begin{aligned}
\int \tan^m x \sec^{2k} x \, dx &= \int \tan^m x (\sec^2 x)^{k-1} \sec^2 x \, dx \\
&= \int \tan^m x (1 + \tan^2 x)^{k-1} \sec^2 x \, dx
\end{aligned}$$

Then substitute $u = \tan x$.

(b) If the power of tangent is odd ($m = 2k + 1$), save a factor of $\sec x \tan x$ and use $\tan^2 x = \sec^2 x - 1$ to express the remaining factors in terms of $\sec x$:

$$\begin{aligned}
\int \tan^{2k+1} x \sec^n x \, dx &= \int (\tan^2 x)^k \sec^{n-1} x \sec x \tan x \, dx \\
&= \int (\sec^2 x - 1)^k \sec^{n-1} x \sec x \tan x \, dx
\end{aligned}$$

Then substitute $u = \sec x$.

For other cases, the guidelines are not as clear-cut. We may need to use identities, integration by parts, and occasionally a little ingenuity. We will sometimes need to be able to integrate $\tan x$ by using the formula established in (5.5.5):

$$\int \tan x\,dx = \ln|\sec x| + C$$

We will also need the indefinite integral of secant:

$$\boxed{1} \qquad \int \sec x\,dx = \ln|\sec x + \tan x| + C$$

We could verify Formula 1 by differentiating the right side, or as follows. First we multiply numerator and denominator by $\sec x + \tan x$:

$$\int \sec x\,dx = \int \sec x\,\frac{\sec x + \tan x}{\sec x + \tan x}\,dx$$

$$= \int \frac{\sec^2 x + \sec x \tan x}{\sec x + \tan x}\,dx$$

If we substitute $u = \sec x + \tan x$, then $du = (\sec x \tan x + \sec^2 x)\,dx$, so the integral becomes $\int (1/u)\,du = \ln|u| + C$. Thus we have

$$\int \sec x\,dx = \ln|\sec x + \tan x| + C$$

EXAMPLE 7 Find $\int \tan^3 x\,dx$.

SOLUTION Here only $\tan x$ occurs, so we use $\tan^2 x = \sec^2 x - 1$ to rewrite a $\tan^2 x$ factor in terms of $\sec^2 x$:

$$\int \tan^3 x\,dx = \int \tan x \tan^2 x\,dx = \int \tan x\,(\sec^2 x - 1)\,dx$$

$$= \int \tan x \sec^2 x\,dx - \int \tan x\,dx$$

$$= \frac{\tan^2 x}{2} - \ln|\sec x| + C$$

In the first integral we mentally substituted $u = \tan x$ so that $du = \sec^2 x\,dx$. □

If an even power of tangent appears with an odd power of secant, it is helpful to express the integrand completely in terms of $\sec x$. Powers of $\sec x$ may require integration by parts, as shown in the following example.

EXAMPLE 8 Find $\int \sec^3 x\,dx$.

SOLUTION Here we integrate by parts with

$$u = \sec x \qquad dv = \sec^2 x\,dx$$

$$du = \sec x \tan x\,dx \qquad v = \tan x$$

Then

$$\begin{aligned}\int \sec^3 x\, dx &= \sec x \tan x - \int \sec x \tan^2 x\, dx \\ &= \sec x \tan x - \int \sec x\,(\sec^2 x - 1)\, dx \\ &= \sec x \tan x - \int \sec^3 x\, dx + \int \sec x\, dx\end{aligned}$$

Using Formula 1 and solving for the required integral, we get

$$\int \sec^3 x\, dx = \tfrac{1}{2}\big(\sec x \tan x + \ln|\sec x + \tan x|\big) + C \qquad \square$$

Integrals such as the one in the preceding example may seem very special but they occur frequently in applications of integration, as we will see in Chapter 8. Integrals of the form $\int \cot^m x \csc^n x\, dx$ can be found by similar methods because of the identity $1 + \cot^2 x = \csc^2 x$.

Finally, we can make use of another set of trigonometric identities:

■ These product identities are discussed in Appendix D.

2 To evaluate the integrals (a) $\int \sin mx \cos nx\, dx$, (b) $\int \sin mx \sin nx\, dx$, or (c) $\int \cos mx \cos nx\, dx$, use the corresponding identity:

(a) $\sin A \cos B = \tfrac{1}{2}[\sin(A - B) + \sin(A + B)]$

(b) $\sin A \sin B = \tfrac{1}{2}[\cos(A - B) - \cos(A + B)]$

(c) $\cos A \cos B = \tfrac{1}{2}[\cos(A - B) + \cos(A + B)]$

EXAMPLE 9 Evaluate $\int \sin 4x \cos 5x\, dx$.

SOLUTION This integral could be evaluated using integration by parts, but it's easier to use the identity in Equation 2(a) as follows:

$$\begin{aligned}\int \sin 4x \cos 5x\, dx &= \int \tfrac{1}{2}[\sin(-x) + \sin 9x]\, dx \\ &= \tfrac{1}{2}\int (-\sin x + \sin 9x)\, dx \\ &= \tfrac{1}{2}\big(\cos x - \tfrac{1}{9}\cos 9x\big) + C\end{aligned} \qquad \square$$

7.2 EXERCISES

1–49 Evaluate the integral.

1. $\int \sin^3 x \cos^2 x\, dx$
2. $\int \sin^6 x \cos^3 x\, dx$
3. $\int_{\pi/2}^{3\pi/4} \sin^5 x \cos^3 x\, dx$
4. $\int_0^{\pi/2} \cos^5 x\, dx$
5. $\int \sin^2(\pi x) \cos^5(\pi x)\, dx$
6. $\int \dfrac{\sin^3(\sqrt{x})}{\sqrt{x}}\, dx$
7. $\int_0^{\pi/2} \cos^2\theta\, d\theta$
8. $\int_0^{\pi/2} \sin^2(2\theta)\, d\theta$
9. $\int_0^{\pi} \sin^4(3t)\, dt$
10. $\int_0^{\pi} \cos^6\theta\, d\theta$
11. $\int (1 + \cos\theta)^2\, d\theta$
12. $\int x \cos^2 x\, dx$
13. $\int_0^{\pi/2} \sin^2 x \cos^2 x\, dx$
14. $\int_0^{\pi} \sin^2 t \cos^4 t\, dt$
15. $\int \dfrac{\cos^5\alpha}{\sqrt{\sin\alpha}}\, d\alpha$
16. $\int \cos\theta \cos^5(\sin\theta)\, d\theta$

17. $\int \cos^2 x \tan^3 x\, dx$

18. $\int \cot^5\theta \sin^4\theta\, d\theta$

19. $\int \frac{\cos x + \sin 2x}{\sin x}\, dx$

20. $\int \cos^2 x \sin 2x\, dx$

21. $\int \sec^2 x \tan x\, dx$

22. $\int_0^{\pi/2} \sec^4(t/2)\, dt$

23. $\int \tan^2 x\, dx$

24. $\int (\tan^2 x + \tan^4 x)\, dx$

25. $\int \sec^6 t\, dt$

26. $\int_0^{\pi/4} \sec^4\theta \tan^4\theta\, d\theta$

27. $\int_0^{\pi/3} \tan^5 x \sec^4 x\, dx$

28. $\int \tan^3(2x) \sec^5(2x)\, dx$

29. $\int \tan^3 x \sec x\, dx$

30. $\int_0^{\pi/3} \tan^5 x \sec^6 x\, dx$

31. $\int \tan^5 x\, dx$

32. $\int \tan^6(ay)\, dy$

33. $\int \frac{\tan^3\theta}{\cos^4\theta}\, d\theta$

34. $\int \tan^2 x \sec x\, dx$

35. $\int x \sec x \tan x\, dx$

36. $\int \frac{\sin\phi}{\cos^3\phi}\, d\phi$

37. $\int_{\pi/6}^{\pi/2} \cot^2 x\, dx$

38. $\int_{\pi/4}^{\pi/2} \cot^3 x\, dx$

39. $\int \cot^3\alpha \csc^3\alpha\, d\alpha$

40. $\int \csc^4 x \cot^6 x\, dx$

41. $\int \csc x\, dx$

42. $\int_{\pi/6}^{\pi/3} \csc^3 x\, dx$

43. $\int \sin 8x \cos 5x\, dx$

44. $\int \cos \pi x \cos 4\pi x\, dx$

45. $\int \sin 5\theta \sin\theta\, d\theta$

46. $\int \frac{\cos x + \sin x}{\sin 2x}\, dx$

47. $\int \frac{1 - \tan^2 x}{\sec^2 x}\, dx$

48. $\int \frac{dx}{\cos x - 1}$

49. $\int t \sec^2(t^2) \tan^4(t^2)\, dt$

50. If $\int_0^{\pi/4} \tan^6 x \sec x\, dx = I$, express the value of $\int_0^{\pi/4} \tan^8 x \sec x\, dx$ in terms of I.

51–54 Evaluate the indefinite integral. Illustrate, and check that your answer is reasonable, by graphing both the integrand and its antiderivative (taking $C = 0$).

51. $\int x \sin^2(x^2)\, dx$

52. $\int \sin^3 x \cos^4 x\, dx$

53. $\int \sin 3x \sin 6x\, dx$

54. $\int \sec^4 \frac{x}{2}\, dx$

55. Find the average value of the function $f(x) = \sin^2 x \cos^3 x$ on the interval $[-\pi, \pi]$.

56. Evaluate $\int \sin x \cos x\, dx$ by four methods:
(a) the substitution $u = \cos x$
(b) the substitution $u = \sin x$
(c) the identity $\sin 2x = 2 \sin x \cos x$
(d) integration by parts
Explain the different appearances of the answers.

57–58 Find the area of the region bounded by the given curves.

57. $y = \sin^2 x,\quad y = \cos^2 x,\quad -\pi/4 \leqslant x \leqslant \pi/4$

58. $y = \sin^3 x,\quad y = \cos^3 x,\quad \pi/4 \leqslant x \leqslant 5\pi/4$

59–60 Use a graph of the integrand to guess the value of the integral. Then use the methods of this section to prove that your guess is correct.

59. $\int_0^{2\pi} \cos^3 x\, dx$

60. $\int_0^2 \sin 2\pi x \cos 5\pi x\, dx$

61–64 Find the volume obtained by rotating the region bounded by the given curves about the specified axis.

61. $y = \sin x,\ y = 0,\ \pi/2 \leqslant x \leqslant \pi;$ about the x-axis

62. $y = \sin^2 x,\ y = 0,\ 0 \leqslant x \leqslant \pi;$ about the x-axis

63. $y = \sin x,\ y = \cos x,\ 0 \leqslant x \leqslant \pi/4;$ about $y = 1$

64. $y = \sec x,\ y = \cos x,\ 0 \leqslant x \leqslant \pi/3;$ about $y = -1$

65. A particle moves on a straight line with velocity function $v(t) = \sin \omega t \cos^2 \omega t$. Find its position function $s = f(t)$ if $f(0) = 0$.

66. Household electricity is supplied in the form of alternating current that varies from 155 V to -155 V with a frequency of 60 cycles per second (Hz). The voltage is thus given by the equation

$$E(t) = 155 \sin(120\pi t)$$

where t is the time in seconds. Voltmeters read the RMS (root-mean-square) voltage, which is the square root of the average value of $[E(t)]^2$ over one cycle.
(a) Calculate the RMS voltage of household current.
(b) Many electric stoves require an RMS voltage of 220 V. Find the corresponding amplitude A needed for the voltage $E(t) = A \sin(120\pi t)$.

67–69 Prove the formula, where m and n are positive integers.

67. $\int_{-\pi}^{\pi} \sin mx \cos nx \, dx = 0$

68. $\int_{-\pi}^{\pi} \sin mx \sin nx \, dx = \begin{cases} 0 & \text{if } m \neq n \\ \pi & \text{if } m = n \end{cases}$

69. $\int_{-\pi}^{\pi} \cos mx \cos nx \, dx = \begin{cases} 0 & \text{if } m \neq n \\ \pi & \text{if } m = n \end{cases}$

70. A *finite Fourier series* is given by the sum

$$\begin{aligned} f(x) &= \sum_{n=1}^{N} a_n \sin nx \\ &= a_1 \sin x + a_2 \sin 2x + \cdots + a_N \sin Nx \end{aligned}$$

Show that the mth coefficient a_m is given by the formula

$$a_m = \frac{1}{\pi} \int_{-\pi}^{\pi} f(x) \sin mx \, dx$$

7.3 TRIGONOMETRIC SUBSTITUTION

In finding the area of a circle or an ellipse, an integral of the form $\int \sqrt{a^2 - x^2}\, dx$ arises, where $a > 0$. If it were $\int x\sqrt{a^2 - x^2}\, dx$, the substitution $u = a^2 - x^2$ would be effective but, as it stands, $\int \sqrt{a^2 - x^2}\, dx$ is more difficult. If we change the variable from x to θ by the substitution $x = a \sin \theta$, then the identity $1 - \sin^2\theta = \cos^2\theta$ allows us to get rid of the root sign because

$$\sqrt{a^2 - x^2} = \sqrt{a^2 - a^2 \sin^2\theta} = \sqrt{a^2(1 - \sin^2\theta)} = \sqrt{a^2 \cos^2\theta} = a\,|\cos \theta|$$

Notice the difference between the substitution $u = a^2 - x^2$ (in which the new variable is a function of the old one) and the substitution $x = a \sin \theta$ (the old variable is a function of the new one).

In general we can make a substitution of the form $x = g(t)$ by using the Substitution Rule in reverse. To make our calculations simpler, we assume that g has an inverse function; that is, g is one-to-one. In this case, if we replace u by x and x by t in the Substitution Rule (Equation 5.5.4), we obtain

$$\int f(x)\, dx = \int f(g(t))g'(t)\, dt$$

This kind of substitution is called *inverse substitution*.

We can make the inverse substitution $x = a \sin \theta$ provided that it defines a one-to-one function. This can be accomplished by restricting θ to lie in the interval $[-\pi/2, \pi/2]$.

In the following table we list trigonometric substitutions that are effective for the given radical expressions because of the specified trigonometric identities. In each case the restriction on θ is imposed to ensure that the function that defines the substitution is one-to-one. (These are the same intervals used in Section 1.6 in defining the inverse functions.)

TABLE OF TRIGONOMETRIC SUBSTITUTIONS

Expression	Substitution	Identity
$\sqrt{a^2 - x^2}$	$x = a \sin \theta, \quad -\frac{\pi}{2} \leqslant \theta \leqslant \frac{\pi}{2}$	$1 - \sin^2\theta = \cos^2\theta$
$\sqrt{a^2 + x^2}$	$x = a \tan \theta, \quad -\frac{\pi}{2} < \theta < \frac{\pi}{2}$	$1 + \tan^2\theta = \sec^2\theta$
$\sqrt{x^2 - a^2}$	$x = a \sec \theta, \quad 0 \leqslant \theta < \frac{\pi}{2} \text{ or } \pi \leqslant \theta < \frac{3\pi}{2}$	$\sec^2\theta - 1 = \tan^2\theta$

V EXAMPLE 1 Evaluate $\displaystyle\int \frac{\sqrt{9 - x^2}}{x^2}\,dx$.

SOLUTION Let $x = 3 \sin \theta$, where $-\pi/2 \leqslant \theta \leqslant \pi/2$. Then $dx = 3 \cos \theta \, d\theta$ and

$$\sqrt{9 - x^2} = \sqrt{9 - 9 \sin^2\theta} = \sqrt{9 \cos^2\theta} = 3\,|\cos \theta| = 3 \cos \theta$$

(Note that $\cos \theta \geqslant 0$ because $-\pi/2 \leqslant \theta \leqslant \pi/2$.) Thus the Inverse Substitution Rule gives

$$\begin{aligned}\int \frac{\sqrt{9 - x^2}}{x^2}\,dx &= \int \frac{3 \cos \theta}{9 \sin^2\theta}\, 3 \cos \theta \, d\theta \\ &= \int \frac{\cos^2\theta}{\sin^2\theta}\, d\theta = \int \cot^2\theta \, d\theta \\ &= \int (\csc^2\theta - 1)\, d\theta \\ &= -\cot \theta - \theta + C\end{aligned}$$

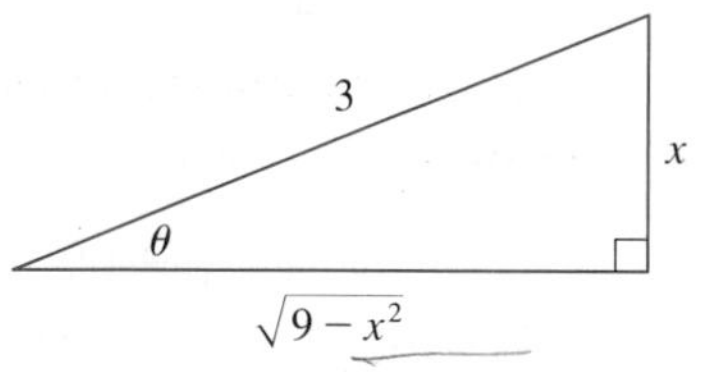

FIGURE 1

$\sin \theta = \dfrac{x}{3}$

Since this is an indefinite integral, we must return to the original variable x. This can be done either by using trigonometric identities to express $\cot \theta$ in terms of $\sin \theta = x/3$ or by drawing a diagram, as in Figure 1, where θ is interpreted as an angle of a right triangle. Since $\sin \theta = x/3$, we label the opposite side and the hypotenuse as having lengths x and 3. Then the Pythagorean Theorem gives the length of the adjacent side as $\sqrt{9 - x^2}$, so we can simply read the value of $\cot \theta$ from the figure:

$$\cot \theta = \frac{\sqrt{9 - x^2}}{x}$$

(Although $\theta > 0$ in the diagram, this expression for $\cot \theta$ is valid even when $\theta < 0$.) Since $\sin \theta = x/3$, we have $\theta = \sin^{-1}(x/3)$ and so

$$\int \frac{\sqrt{9 - x^2}}{x^2}\,dx = -\frac{\sqrt{9 - x^2}}{x} - \sin^{-1}\!\left(\frac{x}{3}\right) + C \qquad \square$$

V EXAMPLE 2 Find the area enclosed by the ellipse

$$\frac{x^2}{a^2} + \frac{y^2}{b^2} = 1$$

SOLUTION Solving the equation of the ellipse for y, we get

$$\frac{y^2}{b^2} = 1 - \frac{x^2}{a^2} = \frac{a^2 - x^2}{a^2} \qquad \text{or} \qquad y = \pm\frac{b}{a}\sqrt{a^2 - x^2}$$

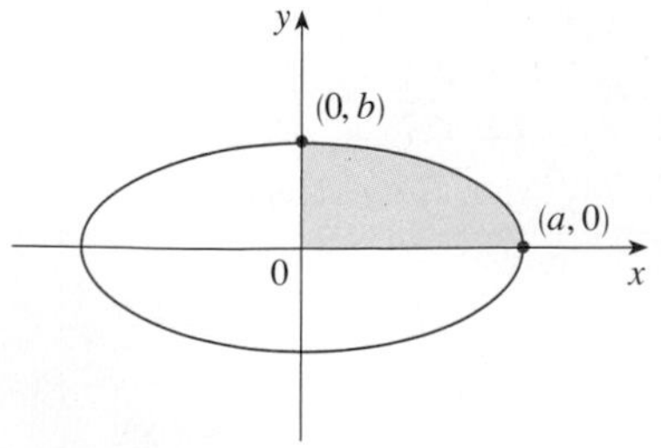

FIGURE 2

$\dfrac{x^2}{a^2} + \dfrac{y^2}{b^2} = 1$

Because the ellipse is symmetric with respect to both axes, the total area A is four times the area in the first quadrant (see Figure 2). The part of the ellipse in the first quadrant is given by the function

$$y = \frac{b}{a}\sqrt{a^2 - x^2} \qquad 0 \leqslant x \leqslant a$$

and so

$$\tfrac{1}{4}A = \int_0^a \frac{b}{a}\sqrt{a^2 - x^2}\,dx$$

To evaluate this integral we substitute $x = a \sin \theta$. Then $dx = a \cos \theta \, d\theta$. To change the limits of integration we note that when $x = 0$, $\sin \theta = 0$, so $\theta = 0$; when $x = a$, $\sin \theta = 1$, so $\theta = \pi/2$. Also

$$\sqrt{a^2 - x^2} = \sqrt{a^2 - a^2 \sin^2\theta} = \sqrt{a^2 \cos^2\theta} = a|\cos \theta| = a \cos \theta$$

since $0 \leq \theta \leq \pi/2$. Therefore

$$\begin{aligned} A &= 4\frac{b}{a}\int_0^a \sqrt{a^2 - x^2}\, dx = 4\frac{b}{a}\int_0^{\pi/2} a\cos\theta \cdot a\cos\theta\, d\theta \\ &= 4ab\int_0^{\pi/2} \cos^2\theta\, d\theta = 4ab\int_0^{\pi/2} \tfrac{1}{2}(1 + \cos 2\theta)\, d\theta \\ &= 2ab\big[\theta + \tfrac{1}{2}\sin 2\theta\big]_0^{\pi/2} = 2ab\left(\frac{\pi}{2} + 0 - 0\right) = \pi ab \end{aligned}$$

We have shown that the area of an ellipse with semiaxes a and b is πab. In particular, taking $a = b = r$, we have proved the famous formula that the area of a circle with radius r is πr^2. □

NOTE Since the integral in Example 2 was a definite integral, we changed the limits of integration and did not have to convert back to the original variable x.

EXAMPLE 3 Find $\displaystyle\int \frac{1}{x^2\sqrt{x^2 + 4}}\, dx$.

SOLUTION Let $x = 2 \tan \theta$, $-\pi/2 < \theta < \pi/2$. Then $dx = 2 \sec^2\theta \, d\theta$ and

$$\sqrt{x^2 + 4} = \sqrt{4(\tan^2\theta + 1)} = \sqrt{4 \sec^2\theta} = 2|\sec \theta| = 2 \sec \theta$$

Thus we have

$$\int \frac{dx}{x^2\sqrt{x^2 + 4}} = \int \frac{2\sec^2\theta\, d\theta}{4\tan^2\theta \cdot 2\sec\theta} = \frac{1}{4}\int \frac{\sec\theta}{\tan^2\theta}\, d\theta$$

To evaluate this trigonometric integral we put everything in terms of $\sin \theta$ and $\cos \theta$:

$$\frac{\sec\theta}{\tan^2\theta} = \frac{1}{\cos\theta} \cdot \frac{\cos^2\theta}{\sin^2\theta} = \frac{\cos\theta}{\sin^2\theta}$$

Therefore, making the substitution $u = \sin \theta$, we have

$$\begin{aligned} \int \frac{dx}{x^2\sqrt{x^2 + 4}} &= \frac{1}{4}\int \frac{\cos\theta}{\sin^2\theta}\, d\theta = \frac{1}{4}\int \frac{du}{u^2} \\ &= \frac{1}{4}\left(-\frac{1}{u}\right) + C = -\frac{1}{4\sin\theta} + C \\ &= -\frac{\csc\theta}{4} + C \end{aligned}$$

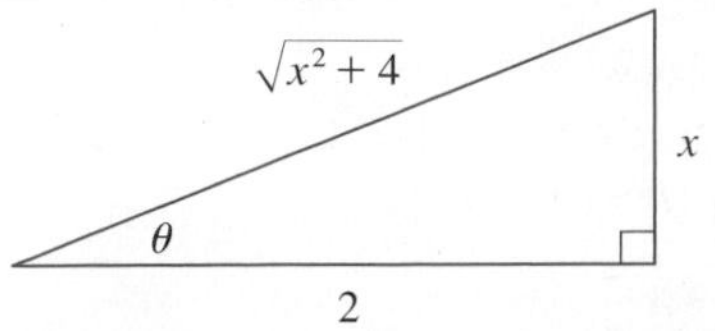

FIGURE 3
$\tan \theta = \dfrac{x}{2}$

We use Figure 3 to determine that $\csc \theta = \sqrt{x^2 + 4}/x$ and so

$$\int \frac{dx}{x^2\sqrt{x^2 + 4}} = -\frac{\sqrt{x^2 + 4}}{4x} + C$$ □

EXAMPLE 4 Find $\int \frac{x}{\sqrt{x^2+4}}\,dx$.

SOLUTION It would be possible to use the trigonometric substitution $x = 2\tan\theta$ here (as in Example 3). But the direct substitution $u = x^2 + 4$ is simpler, because then $du = 2x\,dx$ and

$$\int \frac{x}{\sqrt{x^2+4}}\,dx = \frac{1}{2}\int \frac{du}{\sqrt{u}} = \sqrt{u} + C = \sqrt{x^2+4} + C \qquad \square$$

NOTE Example 4 illustrates the fact that even when trigonometric substitutions are possible, they may not give the easiest solution. You should look for a simpler method first.

EXAMPLE 5 Evaluate $\int \frac{dx}{\sqrt{x^2-a^2}}$, where $a > 0$.

SOLUTION 1 We let $x = a\sec\theta$, where $0 < \theta < \pi/2$ or $\pi < \theta < 3\pi/2$. Then $dx = a\sec\theta\tan\theta\,d\theta$ and

$$\sqrt{x^2-a^2} = \sqrt{a^2(\sec^2\theta - 1)} = \sqrt{a^2\tan^2\theta} = a|\tan\theta| = a\tan\theta$$

Therefore

$$\int \frac{dx}{\sqrt{x^2-a^2}} = \int \frac{a\sec\theta\tan\theta}{a\tan\theta}\,d\theta$$

$$= \int \sec\theta\,d\theta = \ln|\sec\theta + \tan\theta| + C$$

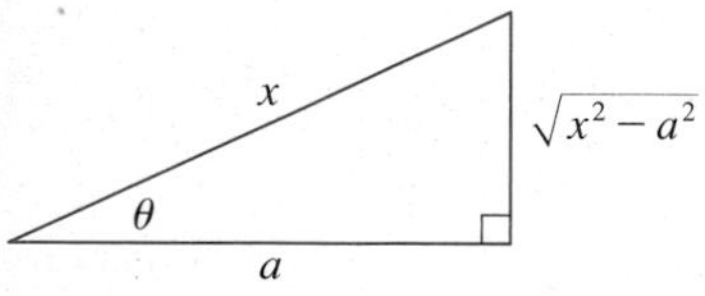

FIGURE 4

$\sec\theta = \frac{x}{a}$

The triangle in Figure 4 gives $\tan\theta = \sqrt{x^2-a^2}/a$, so we have

$$\int \frac{dx}{\sqrt{x^2-a^2}} = \ln\left|\frac{x}{a} + \frac{\sqrt{x^2-a^2}}{a}\right| + C$$

$$= \ln\left|x + \sqrt{x^2-a^2}\right| - \ln a + C$$

Writing $C_1 = C - \ln a$, we have

1 $$\int \frac{dx}{\sqrt{x^2-a^2}} = \ln\left|x + \sqrt{x^2-a^2}\right| + C_1$$

SOLUTION 2 For $x > 0$ the hyperbolic substitution $x = a\cosh t$ can also be used. Using the identity $\cosh^2 y - \sinh^2 y = 1$, we have

$$\sqrt{x^2-a^2} = \sqrt{a^2(\cosh^2 t - 1)} = \sqrt{a^2\sinh^2 t} = a\sinh t$$

Since $dx = a\sinh t\,dt$, we obtain

$$\int \frac{dx}{\sqrt{x^2-a^2}} = \int \frac{a\sinh t\,dt}{a\sinh t} = \int dt = t + C$$

Since $\cosh t = x/a$, we have $t = \cosh^{-1}(x/a)$ and

2 $$\int \frac{dx}{\sqrt{x^2-a^2}} = \cosh^{-1}\left(\frac{x}{a}\right) + C$$

Although Formulas 1 and 2 look quite different, they are actually equivalent by Formula 3.11.4. □

NOTE As Example 5 illustrates, hyperbolic substitutions can be used in place of trigonometric substitutions and sometimes they lead to simpler answers. But we usually use trigonometric substitutions because trigonometric identities are more familiar than hyperbolic identities.

EXAMPLE 6 Find $\displaystyle\int_0^{3\sqrt{3}/2} \frac{x^3}{(4x^2+9)^{3/2}}\,dx$.

SOLUTION First we note that $(4x^2+9)^{3/2} = (\sqrt{4x^2+9}\,)^3$ so trigonometric substitution is appropriate. Although $\sqrt{4x^2+9}$ is not quite one of the expressions in the table of trigonometric substitutions, it becomes one of them if we make the preliminary substitution $u = 2x$. When we combine this with the tangent substitution, we have $x = \frac{3}{2}\tan\theta$, which gives $dx = \frac{3}{2}\sec^2\theta\,d\theta$ and

$$\sqrt{4x^2+9} = \sqrt{9\tan^2\theta + 9} = 3\sec\theta$$

When $x = 0$, $\tan\theta = 0$, so $\theta = 0$; when $x = 3\sqrt{3}/2$, $\tan\theta = \sqrt{3}$, so $\theta = \pi/3$.

$$\begin{aligned}\int_0^{3\sqrt{3}/2} \frac{x^3}{(4x^2+9)^{3/2}}\,dx &= \int_0^{\pi/3} \frac{\frac{27}{8}\tan^3\theta}{27\sec^3\theta}\,\tfrac{3}{2}\sec^2\theta\,d\theta \\ &= \tfrac{3}{16}\int_0^{\pi/3} \frac{\tan^3\theta}{\sec\theta}\,d\theta = \tfrac{3}{16}\int_0^{\pi/3} \frac{\sin^3\theta}{\cos^2\theta}\,d\theta \\ &= \tfrac{3}{16}\int_0^{\pi/3} \frac{1-\cos^2\theta}{\cos^2\theta}\sin\theta\,d\theta\end{aligned}$$

Now we substitute $u = \cos\theta$ so that $du = -\sin\theta\,d\theta$. When $\theta = 0$, $u = 1$; when $\theta = \pi/3$, $u = \frac{1}{2}$. Therefore

$$\begin{aligned}\int_0^{3\sqrt{3}/2} \frac{x^3}{(4x^2+9)^{3/2}}\,dx &= -\tfrac{3}{16}\int_1^{1/2} \frac{1-u^2}{u^2}\,du = \tfrac{3}{16}\int_1^{1/2} (1-u^{-2})\,du \\ &= \tfrac{3}{16}\left[u + \frac{1}{u}\right]_1^{1/2} = \tfrac{3}{16}\left[\left(\tfrac{1}{2}+2\right) - (1+1)\right] = \tfrac{3}{32}\end{aligned}$$ □

EXAMPLE 7 Evaluate $\displaystyle\int \frac{x}{\sqrt{3-2x-x^2}}\,dx$.

SOLUTION We can transform the integrand into a function for which trigonometric substitution is appropriate by first completing the square under the root sign:

$$\begin{aligned}3 - 2x - x^2 &= 3 - (x^2+2x) = 3 + 1 - (x^2+2x+1) \\ &= 4 - (x+1)^2\end{aligned}$$

This suggests that we make the substitution $u = x + 1$. Then $du = dx$ and $x = u - 1$, so

$$\int \frac{x}{\sqrt{3-2x-x^2}}\,dx = \int \frac{u-1}{\sqrt{4-u^2}}\,du$$

■ Figure 5 shows the graphs of the integrand in Example 7 and its indefinite integral (with $C = 0$). Which is which?

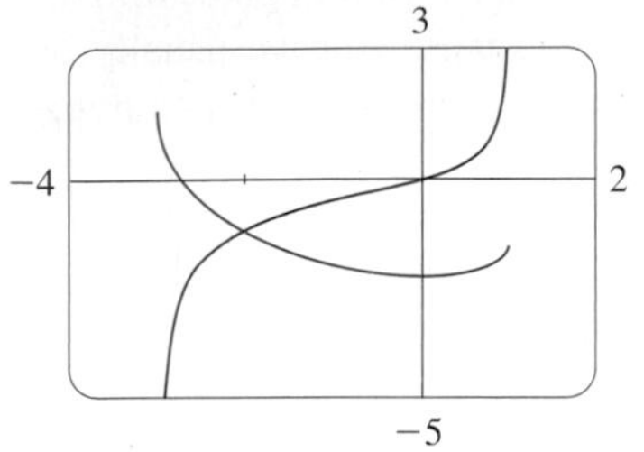

FIGURE 5

We now substitute $u = 2\sin\theta$, giving $du = 2\cos\theta\,d\theta$ and $\sqrt{4 - u^2} = 2\cos\theta$, so

$$\int \frac{x}{\sqrt{3 - 2x - x^2}}\,dx = \int \frac{2\sin\theta - 1}{2\cos\theta}\,2\cos\theta\,d\theta$$

$$= \int (2\sin\theta - 1)\,d\theta$$

$$= -2\cos\theta - \theta + C$$

$$= -\sqrt{4 - u^2} - \sin^{-1}\left(\frac{u}{2}\right) + C$$

$$= -\sqrt{3 - 2x - x^2} - \sin^{-1}\left(\frac{x+1}{2}\right) + C \qquad \square$$

7.3 EXERCISES

1–3 Evaluate the integral using the indicated trigonometric substitution. Sketch and label the associated right triangle.

1. $\displaystyle\int \frac{1}{x^2\sqrt{x^2 - 9}}\,dx;\quad x = 3\sec\theta$

2. $\displaystyle\int x^3\sqrt{9 - x^2}\,dx;\quad x = 3\sin\theta$

3. $\displaystyle\int \frac{x^3}{\sqrt{x^2 + 9}}\,dx;\quad x = 3\tan\theta$

4–30 Evaluate the integral.

4. $\displaystyle\int_0^{2\sqrt{3}} \frac{x^3}{\sqrt{16 - x^2}}\,dx$

5. $\displaystyle\int_{\sqrt{2}}^{2} \frac{1}{t^3\sqrt{t^2 - 1}}\,dt$

6. $\displaystyle\int_1^2 \frac{\sqrt{x^2 - 1}}{x}\,dx$

7. $\displaystyle\int \frac{1}{x^2\sqrt{25 - x^2}}\,dx$

8. $\displaystyle\int \frac{x^3}{\sqrt{x^2 + 100}}\,dx$

9. $\displaystyle\int \frac{dx}{\sqrt{x^2 + 16}}$

10. $\displaystyle\int \frac{t^5}{\sqrt{t^2 + 2}}\,dt$

11. $\displaystyle\int \sqrt{1 - 4x^2}\,dx$

12. $\displaystyle\int_0^1 x\sqrt{x^2 + 4}\,dx$

13. $\displaystyle\int \frac{\sqrt{x^2 - 9}}{x^3}\,dx$

14. $\displaystyle\int \frac{du}{u\sqrt{5 - u^2}}$

15. $\displaystyle\int_0^a x^2\sqrt{a^2 - x^2}\,dx$

16. $\displaystyle\int_{\sqrt{2}/3}^{2/3} \frac{dx}{x^5\sqrt{9x^2 - 1}}$

17. $\displaystyle\int \frac{x}{\sqrt{x^2 - 7}}\,dx$

18. $\displaystyle\int \frac{dx}{[(ax)^2 - b^2]^{3/2}}$

19. $\displaystyle\int \frac{\sqrt{1 + x^2}}{x}\,dx$

20. $\displaystyle\int \frac{t}{\sqrt{25 - t^2}}\,dt$

21. $\displaystyle\int_0^{0.6} \frac{x^2}{\sqrt{9 - 25x^2}}\,dx$

22. $\displaystyle\int_0^1 \sqrt{x^2 + 1}\,dx$

23. $\displaystyle\int \sqrt{5 + 4x - x^2}\,dx$

24. $\displaystyle\int \frac{dt}{\sqrt{t^2 - 6t + 13}}$

25. $\displaystyle\int \frac{x}{\sqrt{x^2 + x + 1}}\,dx$

26. $\displaystyle\int \frac{x^2}{(3 + 4x - 4x^2)^{3/2}}\,dx$

27. $\displaystyle\int \sqrt{x^2 + 2x}\,dx$

28. $\displaystyle\int \frac{x^2 + 1}{(x^2 - 2x + 2)^2}\,dx$

29. $\displaystyle\int x\sqrt{1 - x^4}\,dx$

30. $\displaystyle\int_0^{\pi/2} \frac{\cos t}{\sqrt{1 + \sin^2 t}}\,dt$

31. (a) Use trigonometric substitution to show that

$$\int \frac{dx}{\sqrt{x^2 + a^2}} = \ln\left(x + \sqrt{x^2 + a^2}\right) + C$$

(b) Use the hyperbolic substitution $x = a\sinh t$ to show that

$$\int \frac{dx}{\sqrt{x^2 + a^2}} = \sinh^{-1}\left(\frac{x}{a}\right) + C$$

These formulas are connected by Formula 3.11.3.

32. Evaluate

$$\int \frac{x^2}{(x^2 + a^2)^{3/2}}\,dx$$

(a) by trigonometric substitution.
(b) by the hyperbolic substitution $x = a\sinh t$.

33. Find the average value of $f(x) = \sqrt{x^2 - 1}/x$, $1 \leqslant x \leqslant 7$.

34. Find the area of the region bounded by the hyperbola $9x^2 - 4y^2 = 36$ and the line $x = 3$.

35. Prove the formula $A = \frac{1}{2}r^2\theta$ for the area of a sector of a circle with radius r and central angle θ. [*Hint:* Assume $0 < \theta < \pi/2$ and place the center of the circle at the origin so it has the equation $x^2 + y^2 = r^2$. Then A is the sum of the area of the triangle POQ and the area of the region PQR in the figure.]

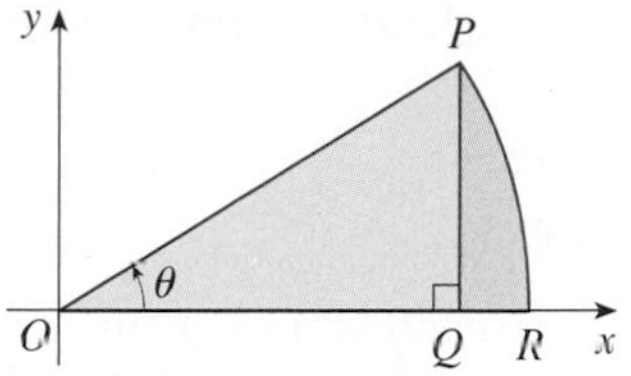

36. Evaluate the integral

$$\int \frac{dx}{x^4\sqrt{x^2 - 2}}$$

Graph the integrand and its indefinite integral on the same screen and check that your answer is reasonable.

37. Use a graph to approximate the roots of the equation $x^2\sqrt{4 - x^2} = 2 - x$. Then approximate the area bounded by the curve $y = x^2\sqrt{4 - x^2}$ and the line $y = 2 - x$.

38. A charged rod of length L produces an electric field at point $P(a, b)$ given by

$$E(P) = \int_{-a}^{L-a} \frac{\lambda b}{4\pi\varepsilon_0(x^2 + b^2)^{3/2}}\, dx$$

where λ is the charge density per unit length on the rod and ε_0 is the free space permittivity (see the figure). Evaluate the integral to determine an expression for the electric field $E(P)$.

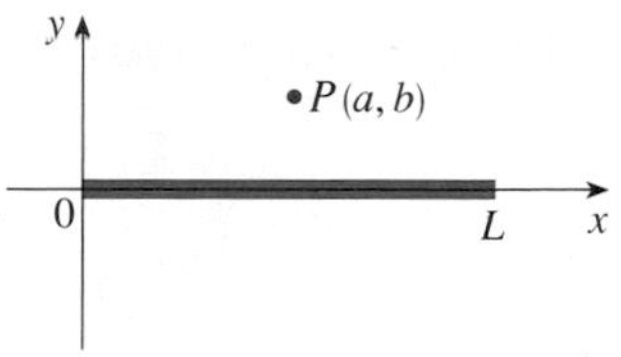

39. (a) Use trigonometric substitution to verify that

$$\int_0^x \sqrt{a^2 - t^2}\, dt = \tfrac{1}{2}a^2 \sin^{-1}(x/a) + \tfrac{1}{2}x\sqrt{a^2 - x^2}$$

(b) Use the figure to give trigonometric interpretations of both terms on the right side of the equation in part (a).

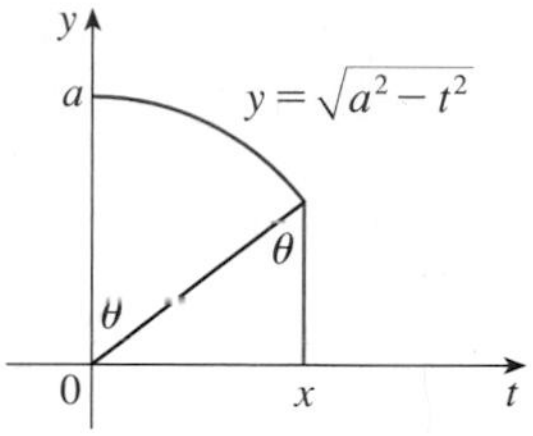

40. The parabola $y = \frac{1}{2}x^2$ divides the disk $x^2 + y^2 \leqslant 8$ into two parts. Find the areas of both parts.

41. Find the area of the crescent shaped region (called a *lune*) bounded by arcs of circles with radii r and R. (See the figure.)

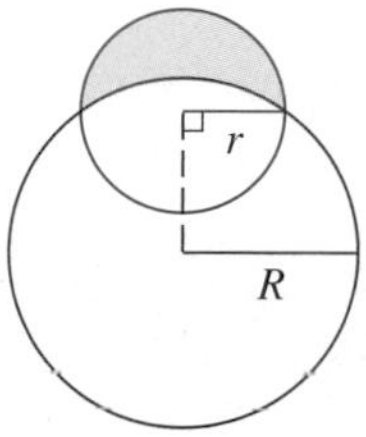

42. A water storage tank has the shape of a cylinder with diameter 10 ft. It is mounted so that the circular cross-sections are vertical. If the depth of the water is 7 ft, what percentage of the total capacity is being used?

43. A torus is generated by rotating the circle $x^2 + (y - R)^2 = r^2$ about the x-axis. Find the volume enclosed by the torus.

7.4 INTEGRATION OF RATIONAL FUNCTIONS BY PARTIAL FRACTIONS

In this section we show how to integrate any rational function (a ratio of polynomials) by expressing it as a sum of simpler fractions, called *partial fractions,* that we already know how to integrate. To illustrate the method, observe that by taking the fractions $2/(x - 1)$ and $1/(x + 2)$ to a common denominator we obtain

$$\frac{2}{x - 1} - \frac{1}{x + 2} = \frac{2(x + 2) - (x - 1)}{(x - 1)(x + 2)} = \frac{x + 5}{x^2 + x - 2}$$

If we now reverse the procedure, we see how to integrate the function on the right side of this equation:

$$\int \frac{x+5}{x^2+x-2}\,dx = \int \left(\frac{2}{x-1} - \frac{1}{x+2} \right) dx$$
$$= 2\ln|x-1| - \ln|x+2| + C$$

To see how the method of partial fractions works in general, let's consider a rational function

$$f(x) = \frac{P(x)}{Q(x)}$$

where P and Q are polynomials. It's possible to express f as a sum of simpler fractions provided that the degree of P is less than the degree of Q. Such a rational function is called *proper.* Recall that if

$$P(x) = a_n x^n + a_{n-1}x^{n-1} + \cdots + a_1 x + a_0$$

where $a_n \neq 0$, then the degree of P is n and we write $\deg(P) = n$.

If f is improper, that is, $\deg(P) \geqslant \deg(Q)$, then we must take the preliminary step of dividing Q into P (by long division) until a remainder $R(x)$ is obtained such that $\deg(R) < \deg(Q)$. The division statement is

$$\boxed{1} \qquad f(x) = \frac{P(x)}{Q(x)} = S(x) + \frac{R(x)}{Q(x)}$$

where S and R are also polynomials.

As the following example illustrates, sometimes this preliminary step is all that is required.

V EXAMPLE 1 Find $\displaystyle\int \frac{x^3+x}{x-1}\,dx$.

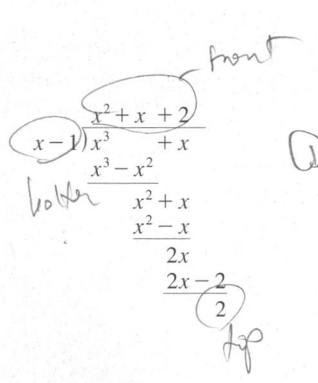

SOLUTION Since the degree of the numerator is greater than the degree of the denominator, we first perform the long division. This enables us to write

$$\int \frac{x^3+x}{x-1}\,dx = \int \left(x^2 + x + 2 + \frac{2}{x-1} \right) dx$$
$$= \frac{x^3}{3} + \frac{x^2}{2} + 2x + 2\ln|x-1| + C$$ □

The next step is to factor the denominator $Q(x)$ as far as possible. It can be shown that any polynomial Q can be factored as a product of linear factors (of the form $ax + b$) and irreducible quadratic factors (of the form $ax^2 + bx + c$, where $b^2 - 4ac < 0$). For instance, if $Q(x) = x^4 - 16$, we could factor it as

$$Q(x) = (x^2-4)(x^2+4) = (x-2)(x+2)(x^2+4)$$

The third step is to express the proper rational function $R(x)/Q(x)$ (from Equation 1) as a sum of **partial fractions** of the form

$$\frac{A}{(ax+b)^i} \qquad \text{or} \qquad \frac{Ax+B}{(ax^2+bx+c)^j}$$

A theorem in algebra guarantees that it is always possible to do this. We explain the details for the four cases that occur.

CASE I ■ The denominator $Q(x)$ is a product of distinct linear factors.

This means that we can write

$$Q(x) = (a_1x + b_1)(a_2x + b_2) \cdots (a_kx + b_k)$$

where no factor is repeated (and no factor is a constant multiple of another). In this case the partial fraction theorem states that there exist constants $A_1, A_2, \ldots, A_k$ such that

$$\boxed{2} \qquad \frac{R(x)}{Q(x)} = \frac{A_1}{a_1x + b_1} + \frac{A_2}{a_2x + b_2} + \cdots + \frac{A_k}{a_kx + b_k}$$

These constants can be determined as in the following example.

EXAMPLE 2 Evaluate $\displaystyle\int \frac{x^2 + 2x - 1}{2x^3 + 3x^2 - 2x}\,dx$.

SOLUTION Since the degree of the numerator is less than the degree of the denominator, we don't need to divide. We factor the denominator as

$$2x^3 + 3x^2 - 2x = x(2x^2 + 3x - 2) = x(2x - 1)(x + 2)$$

Since the denominator has three distinct linear factors, the partial fraction decomposition of the integrand (2) has the form

$$\boxed{3} \qquad \frac{x^2 + 2x - 1}{x(2x - 1)(x + 2)} = \frac{A}{x} + \frac{B}{2x - 1} + \frac{C}{x + 2}$$

■ Another method for finding A, B, and C is given in the note after this example.

To determine the values of A, B, and C, we multiply both sides of this equation by the product of the denominators, $x(2x - 1)(x + 2)$, obtaining

$$\boxed{4} \qquad x^2 + 2x - 1 = A(2x - 1)(x + 2) + Bx(x + 2) + Cx(2x - 1)$$

Expanding the right side of Equation 4 and writing it in the standard form for polynomials, we get

$$\boxed{5} \qquad x^2 + 2x - 1 = (2A + B + 2C)x^2 + (3A + 2B - C)x - 2A$$

The polynomials in Equation 5 are identical, so their coefficients must be equal. The coefficient of x^2 on the right side, $2A + B + 2C$, must equal the coefficient of x^2 on the left side—namely, 1. Likewise, the coefficients of x are equal and the constant terms are equal. This gives the following system of equations for A, B, and C:

$$\begin{aligned} 2A + B + 2C &= 1 \\ 3A + 2B - C &= 2 \\ -2A \qquad\qquad &= -1 \end{aligned}$$

Solving, we get $A = \frac{1}{2}$, $B = \frac{1}{5}$, and $C = -\frac{1}{10}$, and so

■ We could check our work by taking the terms to a common denominator and adding them.

$$\int \frac{x^2 + 2x - 1}{2x^3 + 3x^2 - 2x}\,dx = \int \left(\frac{1}{2}\frac{1}{x} + \frac{1}{5}\frac{1}{2x - 1} - \frac{1}{10}\frac{1}{x + 2} \right) dx$$

$$= \tfrac{1}{2} \ln |x| + \tfrac{1}{10} \ln |2x - 1| - \tfrac{1}{10} \ln |x + 2| + K$$

■ Figure 1 shows the graphs of the integrand in Example 2 and its indefinite integral (with $K = 0$). Which is which?

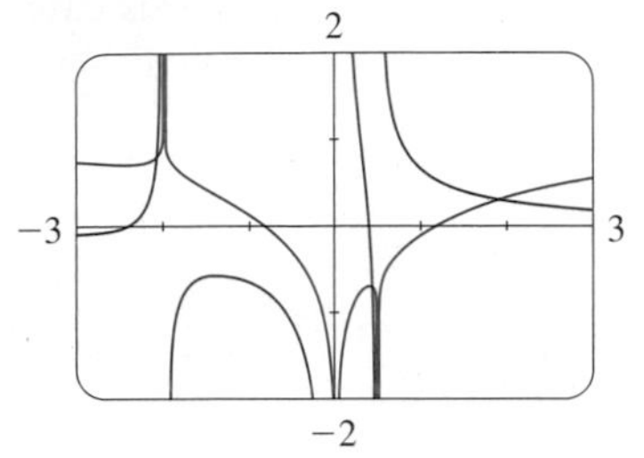

FIGURE 1

In integrating the middle term we have made the mental substitution $u = 2x - 1$, which gives $du = 2\,dx$ and $dx = du/2$. □

NOTE We can use an alternative method to find the coefficients A, B, and C in Example 2. Equation 4 is an identity; it is true for every value of x. Let's choose values of x that simplify the equation. If we put $x = 0$ in Equation 4, then the second and third terms on the right side vanish and the equation then becomes $-2A = -1$, or $A = \frac{1}{2}$. Likewise, $x = \frac{1}{2}$ gives $5B/4 = \frac{1}{4}$ and $x = -2$ gives $10C = -1$, so $B = \frac{1}{5}$ and $C = -\frac{1}{10}$. (You may object that Equation 3 is not valid for $x = 0, \frac{1}{2}$, or -2, so why should Equation 4 be valid for those values? In fact, Equation 4 is true for all values of x, even $x = 0, \frac{1}{2}$, and -2. See Exercise 69 for the reason.)

EXAMPLE 3 Find $\displaystyle\int \frac{dx}{x^2 - a^2}$, where $a \neq 0$.

SOLUTION The method of partial fractions gives

$$\frac{1}{x^2 - a^2} = \frac{1}{(x - a)(x + a)} = \frac{A}{x - a} + \frac{B}{x + a}$$

and therefore

$$A(x + a) + B(x - a) = 1$$

Using the method of the preceding note, we put $x = a$ in this equation and get $A(2a) = 1$, so $A = 1/(2a)$. If we put $x = -a$, we get $B(-2a) = 1$, so $B = -1/(2a)$. Thus

$$\int \frac{dx}{x^2 - a^2} = \frac{1}{2a} \int \left(\frac{1}{x - a} - \frac{1}{x + a} \right) dx$$

$$= \frac{1}{2a} \left(\ln |x - a| - \ln |x + a| \right) + C$$

Since $\ln x - \ln y = \ln(x/y)$, we can write the integral as

6 $$\int \frac{dx}{x^2 - a^2} = \frac{1}{2a} \ln \left| \frac{x - a}{x + a} \right| + C$$

See Exercises 55–56 for ways of using Formula 6. □

CASE II ■ ***Q(x)* is a product of linear factors, some of which are repeated.**

Suppose the first linear factor $(a_1 x + b_1)$ is repeated r times; that is, $(a_1 x + b_1)^r$ occurs in the factorization of $Q(x)$. Then instead of the single term $A_1/(a_1 x + b_1)$ in Equation 2, we would use

$$\boxed{7} \qquad \frac{A_1}{a_1x + b_1} + \frac{A_2}{(a_1x + b_1)^2} + \cdots + \frac{A_r}{(a_1x + b_1)^r}$$

By way of illustration, we could write

$$\frac{x^3 - x + 1}{x^2(x - 1)^3} = \frac{A}{x} + \frac{B}{x^2} + \frac{C}{x - 1} + \frac{D}{(x - 1)^2} + \frac{E}{(x - 1)^3}$$

but we prefer to work out in detail a simpler example.

EXAMPLE 4 Find $\displaystyle\int \frac{x^4 - 2x^2 + 4x + 1}{x^3 - x^2 - x + 1}\,dx$.

SOLUTION The first step is to divide. The result of long division is

$$\frac{x^4 - 2x^2 + 4x + 1}{x^3 - x^2 - x + 1} = x + 1 + \frac{4x}{x^3 - x^2 - x + 1}$$

The second step is to factor the denominator $Q(x) = x^3 - x^2 - x + 1$. Since $Q(1) = 0$, we know that $x - 1$ is a factor and we obtain

$$\begin{aligned} x^3 - x^2 - x + 1 &= (x - 1)(x^2 - 1) = (x - 1)(x - 1)(x + 1) \\ &= (x - 1)^2(x + 1) \end{aligned}$$

Since the linear factor $x - 1$ occurs twice, the partial fraction decomposition is

$$\frac{4x}{(x - 1)^2(x + 1)} = \frac{A}{x - 1} + \frac{B}{(x - 1)^2} + \frac{C}{x + 1}$$

Multiplying by the least common denominator, $(x - 1)^2(x + 1)$, we get

$$\boxed{8} \qquad \begin{aligned} 4x &= A(x - 1)(x + 1) + B(x + 1) + C(x - 1)^2 \\ &= (A + C)x^2 + (B - 2C)x + (-A + B + C) \end{aligned}$$

■ Another method for finding the coefficients:
Put $x = 1$ in (8): $B = 2$.
Put $x = -1$: $C = -1$.
Put $x = 0$: $A = B + C = 1$.

Now we equate coefficients:

$$\begin{aligned} A \qquad + C &= 0 \\ B - 2C &= 4 \\ -A + B + C &= 0 \end{aligned}$$

Solving, we obtain $A = 1$, $B = 2$, and $C = -1$, so

$$\begin{aligned} \int \frac{x^4 - 2x^2 + 4x + 1}{x^3 - x^2 - x + 1}\,dx &= \int \left[x + 1 + \frac{1}{x - 1} + \frac{2}{(x - 1)^2} - \frac{1}{x + 1} \right] dx \\ &= \frac{x^2}{2} + x + \ln|x - 1| - \frac{2}{x - 1} - \ln|x + 1| + K \\ &= \frac{x^2}{2} + x - \frac{2}{x - 1} + \ln\left|\frac{x - 1}{x + 1}\right| + K \end{aligned}$$

□

CASE III ▪ $Q(x)$ contains irreducible quadratic factors, none of which is repeated.
If $Q(x)$ has the factor $ax^2 + bx + c$, where $b^2 - 4ac < 0$, then, in addition to the partial fractions in Equations 2 and 7, the expression for $R(x)/Q(x)$ will have a term of the form

9
$$\frac{Ax + B}{ax^2 + bx + c}$$

where A and B are constants to be determined. For instance, the function given by $f(x) = x/[(x - 2)(x^2 + 1)(x^2 + 4)]$ has a partial fraction decomposition of the form

$$\frac{x}{(x - 2)(x^2 + 1)(x^2 + 4)} = \frac{A}{x - 2} + \frac{Bx + C}{x^2 + 1} + \frac{Dx + E}{x^2 + 4}$$

The term given in (9) can be integrated by completing the square and using the formula

10
$$\int \frac{dx}{x^2 + a^2} = \frac{1}{a} \tan^{-1}\left(\frac{x}{a}\right) + C$$

V EXAMPLE 5 Evaluate $\displaystyle\int \frac{2x^2 - x + 4}{x^3 + 4x}\, dx$.

SOLUTION Since $x^3 + 4x = x(x^2 + 4)$ can't be factored further, we write

$$\frac{2x^2 - x + 4}{x(x^2 + 4)} = \frac{A}{x} + \frac{Bx + C}{x^2 + 4}$$

Multiplying by $x(x^2 + 4)$, we have

$$\begin{aligned} 2x^2 - x + 4 &= A(x^2 + 4) + (Bx + C)x \\ &= (A + B)x^2 + Cx + 4A \end{aligned}$$

Equating coefficients, we obtain

$$A + B = 2 \qquad C = -1 \qquad 4A = 4$$

Thus $A = 1$, $B = 1$, and $C = -1$ and so

$$\int \frac{2x^2 - x + 4}{x^3 + 4x}\, dx = \int \left(\frac{1}{x} + \frac{x - 1}{x^2 + 4}\right) dx$$

In order to integrate the second term we split it into two parts:

$$\int \frac{x - 1}{x^2 + 4}\, dx = \int \frac{x}{x^2 + 4}\, dx - \int \frac{1}{x^2 + 4}\, dx$$

We make the substitution $u = x^2 + 4$ in the first of these integrals so that $du = 2x\, dx$. We evaluate the second integral by means of Formula 10 with $a = 2$:

$$\begin{aligned} \int \frac{2x^2 - x + 4}{x(x^2 + 4)}\, dx &= \int \frac{1}{x}\, dx + \int \frac{x}{x^2 + 4}\, dx - \int \frac{1}{x^2 + 4}\, dx \\ &= \ln|x| + \tfrac{1}{2}\ln(x^2 + 4) - \tfrac{1}{2}\tan^{-1}(x/2) + K \end{aligned}$$

□

EXAMPLE 6 Evaluate $\displaystyle\int \frac{4x^2 - 3x + 2}{4x^2 - 4x + 3}\,dx$.

SOLUTION Since the degree of the numerator is *not less than* the degree of the denominator, we first divide and obtain

$$\frac{4x^2 - 3x + 2}{4x^2 - 4x + 3} = 1 + \frac{x - 1}{4x^2 - 4x + 3}$$

Notice that the quadratic $4x^2 - 4x + 3$ is irreducible because its discriminant is $b^2 - 4ac = -32 < 0$. This means it can't be factored, so we don't need to use the partial fraction technique.

To integrate the given function we complete the square in the denominator:

$$4x^2 - 4x + 3 = (2x - 1)^2 + 2$$

This suggests that we make the substitution $u = 2x - 1$. Then, $du = 2\,dx$ and $x = \frac{1}{2}(u + 1)$, so

$$\begin{aligned}
\int \frac{4x^2 - 3x + 2}{4x^2 - 4x + 3}\,dx &= \int \left(1 + \frac{x - 1}{4x^2 - 4x + 3}\right) dx \\
&= x + \tfrac{1}{2}\int \frac{\frac{1}{2}(u + 1) - 1}{u^2 + 2}\,du = x + \tfrac{1}{4}\int \frac{u - 1}{u^2 + 2}\,du \\
&= x + \tfrac{1}{4}\int \frac{u}{u^2 + 2}\,du - \tfrac{1}{4}\int \frac{1}{u^2 + 2}\,du \\
&= x + \tfrac{1}{8}\ln(u^2 + 2) - \frac{1}{4}\cdot\frac{1}{\sqrt{2}}\tan^{-1}\left(\frac{u}{\sqrt{2}}\right) + C \\
&= x + \tfrac{1}{8}\ln(4x^2 - 4x + 3) - \frac{1}{4\sqrt{2}}\tan^{-1}\left(\frac{2x - 1}{\sqrt{2}}\right) + C
\end{aligned}$$

□

NOTE Example 6 illustrates the general procedure for integrating a partial fraction of the form

$$\frac{Ax + B}{ax^2 + bx + c} \qquad \text{where } b^2 - 4ac < 0$$

We complete the square in the denominator and then make a substitution that brings the integral into the form

$$\int \frac{Cu + D}{u^2 + a^2}\,du = C\int \frac{u}{u^2 + a^2}\,du + D\int \frac{1}{u^2 + a^2}\,du$$

Then the first integral is a logarithm and the second is expressed in terms of $\tan^{-1}$.

CASE IV ▪ $Q(x)$ contains a repeated irreducible quadratic factor.

If $Q(x)$ has the factor $(ax^2 + bx + c)^r$, where $b^2 - 4ac < 0$, then instead of the single partial fraction (9), the sum

$$\boxed{11} \qquad \frac{A_1x + B_1}{ax^2 + bx + c} + \frac{A_2x + B_2}{(ax^2 + bx + c)^2} + \cdots + \frac{A_rx + B_r}{(ax^2 + bx + c)^r}$$

occurs in the partial fraction decomposition of $R(x)/Q(x)$. Each of the terms in (11) can be integrated by first completing the square.

■ It would be extremely tedious to work out by hand the numerical values of the coefficients in Example 7. Most computer algebra systems, however, can find the numerical values very quickly. For instance, the Maple command

convert(f, parfrac, x)

or the Mathematica command

Apart[f]

gives the following values:

$$A = -1, \quad B = \tfrac{1}{8}, \quad C = D = -1,$$
$$E = \tfrac{15}{8}, \quad F = -\tfrac{1}{8}, \quad G = H = \tfrac{3}{4},$$
$$I = -\tfrac{1}{2}, \quad J = \tfrac{1}{2}$$

EXAMPLE 7 Write out the form of the partial fraction decomposition of the function

$$\frac{x^3 + x^2 + 1}{x(x-1)(x^2+x+1)(x^2+1)^3}$$

SOLUTION

$$\frac{x^3 + x^2 + 1}{x(x-1)(x^2+x+1)(x^2+1)^3}$$
$$= \frac{A}{x} + \frac{B}{x-1} + \frac{Cx+D}{x^2+x+1} + \frac{Ex+F}{x^2+1} + \frac{Gx+H}{(x^2+1)^2} + \frac{Ix+J}{(x^2+1)^3} \qquad \square$$

EXAMPLE 8 Evaluate $\displaystyle\int \frac{1 - x + 2x^2 - x^3}{x(x^2+1)^2}\,dx$.

SOLUTION The form of the partial fraction decomposition is

$$\frac{1 - x + 2x^2 - x^3}{x(x^2+1)^2} = \frac{A}{x} + \frac{Bx+C}{x^2+1} + \frac{Dx+E}{(x^2+1)^2}$$

Multiplying by $x(x^2+1)^2$, we have

$$\begin{aligned}
-x^3 + 2x^2 - x + 1 &= A(x^2+1)^2 + (Bx+C)x(x^2+1) + (Dx+E)x \\
&= A(x^4 + 2x^2 + 1) + B(x^4 + x^2) + C(x^3 + x) + Dx^2 + Ex \\
&= (A + B)x^4 + Cx^3 + (2A + B + D)x^2 + (C + E)x + A
\end{aligned}$$

If we equate coefficients, we get the system

$$A + B = 0 \qquad C = -1 \qquad 2A + B + D = 2 \qquad C + E = -1 \qquad A = 1$$

which has the solution $A = 1$, $B = -1$, $C = -1$, $D = 1$, and $E = 0$. Thus

$$\begin{aligned}
\int \frac{1 - x + 2x^2 - x^3}{x(x^2+1)^2}\,dx &= \int \left(\frac{1}{x} - \frac{x+1}{x^2+1} + \frac{x}{(x^2+1)^2} \right) dx \\
&= \int \frac{dx}{x} - \int \frac{x}{x^2+1}\,dx - \int \frac{dx}{x^2+1} + \int \frac{x\,dx}{(x^2+1)^2} \\
&= \ln|x| - \tfrac{1}{2}\ln(x^2+1) - \tan^{-1}x - \frac{1}{2(x^2+1)} + K \qquad \square
\end{aligned}$$

■ In the second and fourth terms we made the mental substitution $u = x^2 + 1$.

We note that sometimes partial fractions can be avoided when integrating a rational function. For instance, although the integral

$$\int \frac{x^2+1}{x(x^2+3)}\,dx$$

could be evaluated by the method of Case III, it's much easier to observe that if $u = x(x^2 + 3) = x^3 + 3x$, then $du = (3x^2 + 3)\,dx$ and so

$$\int \frac{x^2 + 1}{x(x^2 + 3)}\,dx = \tfrac{1}{3}\ln|x^3 + 3x| + C$$

RATIONALIZING SUBSTITUTIONS

Some nonrational functions can be changed into rational functions by means of appropriate substitutions. In particular, when an integrand contains an expression of the form $\sqrt[n]{g(x)}$, then the substitution $u = \sqrt[n]{g(x)}$ may be effective. Other instances appear in the exercises.

EXAMPLE 9 Evaluate $\displaystyle\int \frac{\sqrt{x + 4}}{x}\,dx$.

SOLUTION Let $u = \sqrt{x + 4}$. Then $u^2 = x + 4$, so $x = u^2 - 4$ and $dx = 2u\,du$. Therefore

$$\int \frac{\sqrt{x + 4}}{x}\,dx = \int \frac{u}{u^2 - 4}\,2u\,du = 2\int \frac{u^2}{u^2 - 4}\,du$$

$$= 2\int \left(1 + \frac{4}{u^2 - 4}\right) du$$

We can evaluate this integral either by factoring $u^2 - 4$ as $(u - 2)(u + 2)$ and using partial fractions or by using Formula 6 with $a = 2$:

$$\int \frac{\sqrt{x + 4}}{x}\,dx = 2\int du + 8\int \frac{du}{u^2 - 4}$$

$$= 2u + 8 \cdot \frac{1}{2 \cdot 2}\ln\left|\frac{u - 2}{u + 2}\right| + C$$

$$= 2\sqrt{x + 4} + 2\ln\left|\frac{\sqrt{x + 4} - 2}{\sqrt{x + 4} + 2}\right| + C \qquad \square$$

7.4 EXERCISES

1–6 Write out the form of the partial fraction decomposition of the function (as in Example 7). Do not determine the numerical values of the coefficients.

1. (a) $\dfrac{2x}{(x + 3)(3x + 1)}$ (b) $\dfrac{1}{x^3 + 2x^2 + x}$

2. (a) $\dfrac{x}{x^2 + x - 2}$ (b) $\dfrac{x^2}{x^2 + x + 2}$

3. (a) $\dfrac{x^4 + 1}{x^5 + 4x^3}$ (b) $\dfrac{1}{(x^2 - 9)^2}$

4. (a) $\dfrac{x^3}{x^2 + 4x + 3}$ (b) $\dfrac{2x + 1}{(x + 1)^3(x^2 + 4)^2}$

5. (a) $\dfrac{x^4}{x^4 - 1}$ (b) $\dfrac{t^4 + t^2 + 1}{(t^2 + 1)(t^2 + 4)^2}$

6. (a) $\dfrac{x^4}{(x^3 + x)(x^2 - x + 3)}$ (b) $\dfrac{1}{x^6 - x^3}$

7–38 Evaluate the integral.

7. $\displaystyle\int \frac{x}{x - 6}\,dx$

8. $\displaystyle\int \frac{r^2}{r + 4}\,dr$

9. $\displaystyle\int \frac{x - 9}{(x + 5)(x - 2)}\,dx$

10. $\displaystyle\int \frac{1}{(t + 4)(t - 1)}\,dt$

11. $\displaystyle\int_2^3 \frac{1}{x^2 - 1}\,dx$

12. $\displaystyle\int_0^1 \frac{x - 1}{x^2 + 3x + 2}\,dx$

13. $\displaystyle\int \frac{ax}{x^2 - bx}\,dx$

14. $\displaystyle\int \frac{1}{(x + a)(x + b)}\,dx$

15. $\displaystyle\int_3^4 \frac{x^3 - 2x^2 - 4}{x^3 - 2x^2}\,dx$

16. $\displaystyle\int_0^1 \frac{x^3 - 4x - 10}{x^2 - x - 6}\,dx$

17. $\displaystyle\int_1^2 \frac{4y^2 - 7y - 12}{y(y + 2)(y - 3)}\,dy$

18. $\displaystyle\int \frac{x^2 + 2x - 1}{x^3 - x}\,dx$

19. $\displaystyle\int \frac{1}{(x + 5)^2(x - 1)}\,dx$

20. $\displaystyle\int \frac{x^2 - 5x + 16}{(2x + 1)(x - 2)^2}\,dx$

21. $\displaystyle\int \frac{x^3 + 4}{x^2 + 4}\,dx$

22. $\displaystyle\int \frac{ds}{s^2(s - 1)^2}$

23. $\displaystyle\int \frac{5x^2 + 3x - 2}{x^3 + 2x^2}\,dx$

24. $\displaystyle\int \frac{x^2 - x + 6}{x^3 + 3x}\,dx$

25. $\displaystyle\int \frac{10}{(x - 1)(x^2 + 9)}\,dx$

26. $\displaystyle\int \frac{x^2 + x + 1}{(x^2 + 1)^2}\,dx$

27. $\displaystyle\int \frac{x^3 + x^2 + 2x + 1}{(x^2 + 1)(x^2 + 2)}\,dx$

28. $\displaystyle\int \frac{x^2 - 2x - 1}{(x - 1)^2(x^2 + 1)}\,dx$

29. $\displaystyle\int \frac{x + 4}{x^2 + 2x + 5}\,dx$

30. $\displaystyle\int \frac{3x^2 + x + 4}{x^4 + 3x^2 + 2}\,dx$

31. $\displaystyle\int \frac{1}{x^3 - 1}\,dx$

32. $\displaystyle\int_0^1 \frac{x}{x^2 + 4x + 13}\,dx$

33. $\displaystyle\int_0^1 \frac{x^3 + 2x}{x^4 + 4x^2 + 3}\,dx$

34. $\displaystyle\int \frac{x^3}{x^3 + 1}\,dx$

35. $\displaystyle\int \frac{dx}{x(x^2 + 4)^2}$

36. $\displaystyle\int \frac{x^4 + 3x^2 + 1}{x^5 + 5x^3 + 5x}\,dx$

37. $\displaystyle\int \frac{x^2 - 3x + 7}{(x^2 - 4x + 6)^2}\,dx$

38. $\displaystyle\int \frac{x^3 + 2x^2 + 3x - 2}{(x^2 + 2x + 2)^2}\,dx$

39–50 Make a substitution to express the integrand as a rational function and then evaluate the integral.

39. $\displaystyle\int \frac{1}{x\sqrt{x + 1}}\,dx$

40. $\displaystyle\int \frac{dx}{2\sqrt{x + 3} + x}$

41. $\displaystyle\int_9^{16} \frac{\sqrt{x}}{x - 4}\,dx$

42. $\displaystyle\int_0^1 \frac{1}{1 + \sqrt[3]{x}}\,dx$

43. $\displaystyle\int \frac{x^3}{\sqrt[3]{x^2 + 1}}\,dx$

44. $\displaystyle\int_{1/3}^3 \frac{\sqrt{x}}{x^2 + x}\,dx$

45. $\displaystyle\int \frac{1}{\sqrt{x} - \sqrt[3]{x}}\,dx$ [*Hint:* Substitute $u = \sqrt[6]{x}$.]

46. $\displaystyle\int \frac{\sqrt{1 + \sqrt{x}}}{x}\,dx$

47. $\displaystyle\int \frac{e^{2x}}{e^{2x} + 3e^x + 2}\,dx$

48. $\displaystyle\int \frac{\cos x}{\sin^2 x + \sin x}\,dx$

49. $\displaystyle\int \frac{\sec^2 t}{\tan^2 t + 3\tan t + 2}\,dt$

50. $\displaystyle\int \frac{e^x}{(e^x - 2)(e^{2x} + 1)}\,dx$

51–52 Use integration by parts, together with the techniques of this section, to evaluate the integral.

51. $\displaystyle\int \ln(x^2 - x + 2)\,dx$

52. $\displaystyle\int x\tan^{-1}x\,dx$

53. Use a graph of $f(x) = 1/(x^2 - 2x - 3)$ to decide whether $\int_0^2 f(x)\,dx$ is positive or negative. Use the graph to give a rough estimate of the value of the integral and then use partial fractions to find the exact value.

54. Graph both $y = 1/(x^3 - 2x^2)$ and an antiderivative on the same screen.

55–56 Evaluate the integral by completing the square and using Formula 6.

55. $\displaystyle\int \frac{dx}{x^2 - 2x}$

56. $\displaystyle\int \frac{2x + 1}{4x^2 + 12x - 7}\,dx$

57. The German mathematician Karl Weierstrass (1815–1897) noticed that the substitution $t = \tan(x/2)$ will convert any rational function of $\sin x$ and $\cos x$ into an ordinary rational function of t.

(a) If $t = \tan(x/2)$, $-\pi < x < \pi$, sketch a right triangle or use trigonometric identities to show that

$$\cos\left(\frac{x}{2}\right) = \frac{1}{\sqrt{1 + t^2}} \quad \text{and} \quad \sin\left(\frac{x}{2}\right) = \frac{t}{\sqrt{1 + t^2}}$$

(b) Show that

$$\cos x = \frac{1 - t^2}{1 + t^2} \quad \text{and} \quad \sin x = \frac{2t}{1 + t^2}$$

(c) Show that

$$dx = \frac{2}{1 + t^2}\,dt$$

58–61 Use the substitution in Exercise 57 to transform the integrand into a rational function of t and then evaluate the integral.

58. $\displaystyle\int \frac{dx}{3 - 5\sin x}$

59. $\displaystyle\int \frac{1}{3\sin x - 4\cos x}\,dx$

60. $\displaystyle\int_{\pi/3}^{\pi/2} \frac{1}{1 + \sin x - \cos x}\,dx$

61. $\displaystyle\int_0^{\pi/2} \frac{\sin 2x}{2 + \cos x}\,dx$

62–63 Find the area of the region under the given curve from 1 to 2.

62. $y = \dfrac{1}{x^3 + x}$

63. $y = \dfrac{x^2 + 1}{3x - x^2}$

64. Find the volume of the resulting solid if the region under the curve $y = 1/(x^2 + 3x + 2)$ from $x = 0$ to $x = 1$ is rotated about (a) the x-axis and (b) the y-axis.

65. One method of slowing the growth of an insect population without using pesticides is to introduce into the population a number of sterile males that mate with fertile females but produce no offspring. If P represents the number of female insects in a population, S the number of sterile males introduced each generation, and r the population's natural growth rate, then the female population is related to time t by

$$t = \int \frac{P + S}{P[(r - 1)P - S]}\,dP$$

Suppose an insect population with 10,000 females grows at a rate of $r = 0.10$ and 900 sterile males are added. Evaluate the integral to give an equation relating the female population to time. (Note that the resulting equation can't be solved explicitly for P.)

66. Factor $x^4 + 1$ as a difference of squares by first adding and subtracting the same quantity. Use this factorization to evaluate $\int 1/(x^4 + 1)\,dx$.

CAS **67.** (a) Use a computer algebra system to find the partial fraction decomposition of the function

$$f(x) = \frac{4x^3 - 27x^2 + 5x - 32}{30x^5 - 13x^4 + 50x^3 - 286x^2 - 299x - 70}$$

(b) Use part (a) to find $\int f(x)\,dx$ (by hand) and compare with the result of using the CAS to integrate f directly. Comment on any discrepancy.

CAS **68.** (a) Find the partial fraction decomposition of the function

$$f(x) = \frac{12x^5 - 7x^3 - 13x^2 + 8}{100x^6 - 80x^5 + 116x^4 - 80x^3 + 41x^2 - 20x + 4}$$

(b) Use part (a) to find $\int f(x)\,dx$ and graph f and its indefinite integral on the same screen.

(c) Use the graph of f to discover the main features of the graph of $\int f(x)\,dx$.

69. Suppose that F, G, and Q are polynomials and

$$\frac{F(x)}{Q(x)} = \frac{G(x)}{Q(x)}$$

for all x except when $Q(x) = 0$. Prove that $F(x) = G(x)$ for all x. [*Hint:* Use continuity.]

70. If f is a quadratic function such that $f(0) = 1$ and

$$\int \frac{f(x)}{x^2(x + 1)^3}\,dx$$

is a rational function, find the value of $f'(0)$.

7.5 STRATEGY FOR INTEGRATION

As we have seen, integration is more challenging than differentiation. In finding the derivative of a function it is obvious which differentiation formula we should apply. But it may not be obvious which technique we should use to integrate a given function.

Until now individual techniques have been applied in each section. For instance, we usually used substitution in Exercises 5.5, integration by parts in Exercises 7.1, and partial fractions in Exercises 7.4. But in this section we present a collection of miscellaneous integrals in random order and the main challenge is to recognize which technique or formula to use. No hard and fast rules can be given as to which method applies in a given situation, but we give some advice on strategy that you may find useful.

A prerequisite for strategy selection is a knowledge of the basic integration formulas. In the following table we have collected the integrals from our previous list together with several additional formulas that we have learned in this chapter. Most of them should be memorized. It is useful to know them all, but the ones marked with an asterisk need not be memorized since they are easily derived. Formula 19 can be avoided by using partial fractions, and trigonometric substitutions can be used in place of Formula 20.

TABLE OF INTEGRATION FORMULAS Constants of integration have been omitted.

1. $\int x^n\,dx = \dfrac{x^{n+1}}{n+1} \quad (n \neq -1)$

2. $\int \dfrac{1}{x}\,dx = \ln|x|$

3. $\int e^x\,dx = e^x$

4. $\int a^x\,dx = \dfrac{a^x}{\ln a}$

5. $\int \sin x\,dx = -\cos x$

6. $\int \cos x\,dx = \sin x$

7. $\int \sec^2 x\,dx = \tan x$

8. $\int \csc^2 x\,dx = -\cot x$

9. $\int \sec x \tan x\,dx = \sec x$

10. $\int \csc x \cot x\,dx = -\csc x$

11. $\int \sec x\,dx = \ln|\sec x + \tan x|$

12. $\int \csc x\,dx = \ln|\csc x - \cot x|$

13. $\int \tan x\,dx = \ln|\sec x|$

14. $\int \cot x\,dx = \ln|\sin x|$

15. $\int \sinh x\,dx = \cosh x$

16. $\int \cosh x\,dx = \sinh x$

17. $\int \dfrac{dx}{x^2 + a^2} = \dfrac{1}{a}\tan^{-1}\left(\dfrac{x}{a}\right)$

18. $\int \dfrac{dx}{\sqrt{a^2 - x^2}} = \sin^{-1}\left(\dfrac{x}{a}\right), \quad a > 0$

***19.** $\int \dfrac{dx}{x^2 - a^2} = \dfrac{1}{2a}\ln\left|\dfrac{x-a}{x+a}\right|$

***20.** $\int \dfrac{dx}{\sqrt{x^2 \pm a^2}} = \ln\left|x + \sqrt{x^2 \pm a^2}\right|$

Once you are armed with these basic integration formulas, if you don't immediately see how to attack a given integral, you might try the following four-step strategy.

1. Simplify the Integrand if Possible Sometimes the use of algebraic manipulation or trigonometric identities will simplify the integrand and make the method of integration obvious. Here are some examples:

$$\int \sqrt{x}\,(1 + \sqrt{x})\,dx = \int (\sqrt{x} + x)\,dx$$

$$\int \frac{\tan\theta}{\sec^2\theta}\,d\theta = \int \frac{\sin\theta}{\cos\theta}\cos^2\theta\,d\theta$$

$$= \int \sin\theta\cos\theta\,d\theta = \tfrac{1}{2}\int \sin 2\theta\,d\theta$$

$$\int (\sin x + \cos x)^2\,dx = \int (\sin^2 x + 2\sin x\cos x + \cos^2 x)\,dx$$

$$= \int (1 + 2\sin x\cos x)\,dx$$

2. Look for an Obvious Substitution Try to find some function $u = g(x)$ in the integrand whose differential $du = g'(x)\,dx$ also occurs, apart from a constant factor. For instance, in the integral

$$\int \frac{x}{x^2 - 1}\,dx$$

we notice that if $u = x^2 - 1$, then $du = 2x\,dx$. Therefore we use the substitution $u = x^2 - 1$ instead of the method of partial fractions.

3. Classify the Integrand According to Its Form If Steps 1 and 2 have not led to the solution, then we take a look at the form of the integrand $f(x)$.

(a) *Trigonometric functions.* If $f(x)$ is a product of powers of $\sin x$ and $\cos x$, of $\tan x$ and $\sec x$, or of $\cot x$ and $\csc x$, then we use the substitutions recommended in Section 7.2.

(b) *Rational functions.* If f is a rational function, we use the procedure of Section 7.4 involving partial fractions.

(c) *Integration by parts.* If $f(x)$ is a product of a power of x (or a polynomial) and a transcendental function (such as a trigonometric, exponential, or logarithmic function), then we try integration by parts, choosing u and dv according to the advice given in Section 7.1. If you look at the functions in Exercises 7.1, you will see that most of them are the type just described.

(d) *Radicals.* Particular kinds of substitutions are recommended when certain radicals appear.

  (i) If $\sqrt{\pm x^2 \pm a^2}$ occurs, we use a trigonometric substitution according to the table in Section 7.3.

  (ii) If $\sqrt[n]{ax + b}$ occurs, we use the rationalizing substitution $u = \sqrt[n]{ax + b}$. More generally, this sometimes works for $\sqrt[n]{g(x)}$.

4. Try Again If the first three steps have not produced the answer, remember that there are basically only two methods of integration: substitution and parts.

(a) *Try substitution.* Even if no substitution is obvious (Step 2), some inspiration or ingenuity (or even desperation) may suggest an appropriate substitution.

(b) *Try parts.* Although integration by parts is used most of the time on products of the form described in Step 3(c), it is sometimes effective on single functions. Looking at Section 7.1, we see that it works on $\tan^{-1}x$, $\sin^{-1}x$, and $\ln x$, and these are all inverse functions.

(c) *Manipulate the integrand.* Algebraic manipulations (perhaps rationalizing the denominator or using trigonometric identities) may be useful in transforming the integral into an easier form. These manipulations may be more substantial than in Step 1 and may involve some ingenuity. Here is an example:

$$\int \frac{dx}{1 - \cos x} = \int \frac{1}{1 - \cos x} \cdot \frac{1 + \cos x}{1 + \cos x}\,dx = \int \frac{1 + \cos x}{1 - \cos^2 x}\,dx$$

$$= \int \frac{1 + \cos x}{\sin^2 x}\,dx = \int \left(\csc^2 x + \frac{\cos x}{\sin^2 x}\right) dx$$

(d) *Relate the problem to previous problems.* When you have built up some experience in integration, you may be able to use a method on a given integral that is similar to a method you have already used on a previous integral. Or you may even be able to express the given integral in terms of a previous one. For instance, $\int \tan^2 x \sec x \, dx$ is a challenging integral, but if we make use of the identity $\tan^2 x = \sec^2 x - 1$, we can write

$$\int \tan^2 x \sec x \, dx = \int \sec^3 x \, dx - \int \sec x \, dx$$

and if $\int \sec^3 x \, dx$ has previously been evaluated (see Example 8 in Section 7.2), then that calculation can be used in the present problem.

(e) *Use several methods.* Sometimes two or three methods are required to evaluate an integral. The evaluation could involve several successive substitutions of different types, or it might combine integration by parts with one or more substitutions.

In the following examples we indicate a method of attack but do not fully work out the integral.

EXAMPLE 1 $\displaystyle\int \frac{\tan^3 x}{\cos^3 x} \, dx$

In Step 1 we rewrite the integral:

$$\int \frac{\tan^3 x}{\cos^3 x} \, dx = \int \tan^3 x \sec^3 x \, dx$$

The integral is now of the form $\int \tan^m x \sec^n x \, dx$ with m odd, so we can use the advice in Section 7.2.

Alternatively, if in Step 1 we had written

$$\int \frac{\tan^3 x}{\cos^3 x} \, dx = \int \frac{\sin^3 x}{\cos^3 x} \frac{1}{\cos^3 x} \, dx = \int \frac{\sin^3 x}{\cos^6 x} \, dx$$

then we could have continued as follows with the substitution $u = \cos x$:

$$\int \frac{\sin^3 x}{\cos^6 x} \, dx = \int \frac{1 - \cos^2 x}{\cos^6 x} \sin x \, dx = \int \frac{1 - u^2}{u^6} (-du)$$

$$= \int \frac{u^2 - 1}{u^6} \, du = \int (u^{-4} - u^{-6}) \, du$$ □

V EXAMPLE 2 $\int e^{\sqrt{x}} \, dx$

According to (ii) in Step 3(d), we substitute $u = \sqrt{x}$. Then $x = u^2$, so $dx = 2u \, du$ and

$$\int e^{\sqrt{x}} \, dx = 2 \int u e^u \, du$$

The integrand is now a product of u and the transcendental function e^u so it can be integrated by parts. □

EXAMPLE 3 $\displaystyle\int \frac{x^5 + 1}{x^3 - 3x^2 - 10x}\,dx$

No algebraic simplification or substitution is obvious, so Steps 1 and 2 don't apply here. The integrand is a rational function so we apply the procedure of Section 7.4, remembering that the first step is to divide. □

V EXAMPLE 4 $\displaystyle\int \frac{dx}{x\sqrt{\ln x}}$

Here Step 2 is all that is needed. We substitute $u = \ln x$ because its differential is $du = dx/x$, which occurs in the integral. □

V EXAMPLE 5 $\displaystyle\int \sqrt{\frac{1 - x}{1 + x}}\,dx$

Although the rationalizing substitution

$$u = \sqrt{\frac{1 - x}{1 + x}}$$

works here [(ii) in Step 3(d)], it leads to a very complicated rational function. An easier method is to do some algebraic manipulation [either as Step 1 or as Step 4(c)]. Multiplying numerator and denominator by $\sqrt{1 - x}$, we have

$$\begin{aligned}\int \sqrt{\frac{1 - x}{1 + x}}\,dx &= \int \frac{1 - x}{\sqrt{1 - x^2}}\,dx \\ &= \int \frac{1}{\sqrt{1 - x^2}}\,dx - \int \frac{x}{\sqrt{1 - x^2}}\,dx \\ &= \sin^{-1}x + \sqrt{1 - x^2} + C\end{aligned}$$ □

CAN WE INTEGRATE ALL CONTINUOUS FUNCTIONS?

The question arises: Will our strategy for integration enable us to find the integral of every continuous function? For example, can we use it to evaluate $\int e^{x^2}\,dx$? The answer is No, at least not in terms of the functions that we are familiar with.

The functions that we have been dealing with in this book are called **elementary functions**. These are the polynomials, rational functions, power functions (x^a), exponential functions (a^x), logarithmic functions, trigonometric and inverse trigonometric functions, hyperbolic and inverse hyperbolic functions, and all functions that can be obtained from these by the five operations of addition, subtraction, multiplication, division, and composition. For instance, the function

$$f(x) = \sqrt{\frac{x^2 - 1}{x^3 + 2x - 1}} + \ln(\cosh x) - xe^{\sin 2x}$$

is an elementary function.

If f is an elementary function, then f' is an elementary function but $\int f(x)\,dx$ need not be an elementary function. Consider $f(x) = e^{x^2}$. Since f is continuous, its integral exists, and if we define the function F by

$$F(x) = \int_0^x e^{t^2}\,dt$$

then we know from Part 1 of the Fundamental Theorem of Calculus that

$$F'(x) = e^{x^2}$$

Thus, $f(x) = e^{x^2}$ has an antiderivative F, but it has been proved that F is not an elementary function. This means that no matter how hard we try, we will never succeed in evaluating $\int e^{x^2} dx$ in terms of the functions we know. (In Chapter 11, however, we will see how to express $\int e^{x^2} dx$ as an infinite series.) The same can be said of the following integrals:

$$\int \frac{e^x}{x}\,dx \qquad \int \sin(x^2)\,dx \qquad \int \cos(e^x)\,dx$$

$$\int \sqrt{x^3+1}\,dx \qquad \int \frac{1}{\ln x}\,dx \qquad \int \frac{\sin x}{x}\,dx$$

In fact, the majority of elementary functions don't have elementary antiderivatives. You may be assured, though, that the integrals in the following exercises are all elementary functions.

7.5 EXERCISES

1–80 Evaluate the integral.

1. $\int \cos x\,(1+\sin^2 x)\,dx$
2. $\int \frac{\sin^3 x}{\cos x}\,dx$
3. $\int \frac{\sin x + \sec x}{\tan x}\,dx$
4. $\int \tan^3\theta\,d\theta$
5. $\int_0^2 \frac{2t}{(t-3)^2}\,dt$
6. $\int \frac{x}{\sqrt{3-x^4}}\,dx$
7. $\int_{-1}^1 \frac{e^{\arctan y}}{1+y^2}\,dy$
8. $\int x\csc x\cot x\,dx$
9. $\int_1^3 r^4 \ln r\,dr$
10. $\int_0^4 \frac{x-1}{x^2-4x-5}\,dx$
11. $\int \frac{x-1}{x^2-4x+5}\,dx$
12. $\int \frac{x}{x^4+x^2+1}\,dx$
13. $\int \sin^3\theta\cos^5\theta\,d\theta$
14. $\int \frac{x^3}{\sqrt{1+x^2}}\,dx$
15. $\int \frac{dx}{(1-x^2)^{3/2}}$
16. $\int_0^{\sqrt{2}/2} \frac{x^2}{\sqrt{1-x^2}}\,dx$
17. $\int x\sin^2 x\,dx$
18. $\int \frac{e^{2t}}{1+e^{4t}}\,dt$
19. $\int e^{x+e^x}\,dx$
20. $\int e^2\,dx$
21. $\int \arctan\sqrt{x}\,dx$
22. $\int \frac{\ln x}{x\sqrt{1+(\ln x)^2}}\,dx$
23. $\int_0^1 (1+\sqrt{x})^8\,dx$
24. $\int \ln(x^2-1)\,dx$
25. $\int \frac{3x^2-2}{x^2-2x-8}\,dx$
26. $\int \frac{3x^2-2}{x^3-2x-8}\,dx$
27. $\int \frac{dx}{1+e^x}$
28. $\int \sin\sqrt{at}\,dt$
29. $\int_0^5 \frac{3w-1}{w+2}\,dw$
30. $\int_{-2}^2 |x^2-4x|\,dx$
31. $\int \sqrt{\frac{1+x}{1-x}}\,dx$
32. $\int \frac{\sqrt{2x-1}}{2x+3}\,dx$
33. $\int \sqrt{3-2x-x^2}\,dx$
34. $\int_{\pi/4}^{\pi/2} \frac{1+4\cot x}{4-\cot x}\,dx$
35. $\int_{-1}^1 x^8\sin x\,dx$
36. $\int \sin 4x\cos 3x\,dx$
37. $\int_0^{\pi/4} \cos^2\theta\tan^2\theta\,d\theta$
38. $\int_0^{\pi/4} \tan^5\theta\sec^3\theta\,d\theta$
39. $\int \frac{\sec\theta\tan\theta}{\sec^2\theta-\sec\theta}\,d\theta$
40. $\int \frac{1}{\sqrt{4y^2-4y-3}}\,dy$
41. $\int \theta\tan^2\theta\,d\theta$
42. $\int \frac{\tan^{-1}x}{x^2}\,dx$
43. $\int e^x\sqrt{1+e^x}\,dx$
44. $\int \sqrt{1+e^x}\,dx$
45. $\int x^5 e^{-x^3}\,dx$
46. $\int \frac{1+\sin x}{1-\sin x}\,dx$
47. $\int x^3(x-1)^{-4}\,dx$
48. $\int \frac{x}{x^4-a^4}\,dx$

49. $\displaystyle\int \frac{1}{x\sqrt{4x+1}}\,dx$

50. $\displaystyle\int \frac{1}{x^2\sqrt{4x+1}}\,dx$

51. $\displaystyle\int \frac{1}{x\sqrt{4x^2+1}}\,dx$

52. $\displaystyle\int \frac{dx}{x(x^4+1)}$

53. $\displaystyle\int x^2 \sinh mx\,dx$

54. $\displaystyle\int (x+\sin x)^2\,dx$

55. $\displaystyle\int \frac{dx}{x+x\sqrt{x}}$

56. $\displaystyle\int \frac{dx}{\sqrt{x}+x\sqrt{x}}$

57. $\displaystyle\int x\sqrt[3]{x+c}\,dx$

58. $\displaystyle\int \frac{x\ln x}{\sqrt{x^2-1}}\,dx$

59. $\displaystyle\int \cos x \cos^3(\sin x)\,dx$

60. $\displaystyle\int \frac{dx}{x^2\sqrt{4x^2-1}}$

61. $\displaystyle\int \sqrt{x}\,e^{\sqrt{x}}\,dx$

62. $\displaystyle\int \frac{1}{x+\sqrt[3]{x}}\,dx$

63. $\displaystyle\int \frac{\sin 2x}{1+\cos^4 x}\,dx$

64. $\displaystyle\int_{\pi/4}^{\pi/3} \frac{\ln(\tan x)}{\sin x \cos x}\,dx$

65. $\displaystyle\int \frac{1}{\sqrt{x+1}+\sqrt{x}}\,dx$

66. $\displaystyle\int_2^3 \frac{u^3+1}{u^3-u^2}\,du$

67. $\displaystyle\int_1^{\sqrt{3}} \frac{\sqrt{1+x^2}}{x^2}\,dx$

68. $\displaystyle\int \frac{1}{1+2e^x-e^{-x}}\,dx$

69. $\displaystyle\int \frac{e^{2x}}{1+e^x}\,dx$

70. $\displaystyle\int \frac{\ln(x+1)}{x^2}\,dx$

71. $\displaystyle\int \frac{x+\arcsin x}{\sqrt{1-x^2}}\,dx$

72. $\displaystyle\int \frac{4^x+10^x}{2^x}\,dx$

73. $\displaystyle\int \frac{1}{(x-2)(x^2+4)}\,dx$

74. $\displaystyle\int \frac{dx}{\sqrt{x}(2+\sqrt{x})^4}$

75. $\displaystyle\int \frac{xe^x}{\sqrt{1+e^x}}\,dx$

76. $\displaystyle\int (x^2-bx)\sin 2x\,dx$

77. $\displaystyle\int \frac{\sqrt{x}}{1+x^3}\,dx$

78. $\displaystyle\int \frac{\sec x \cos 2x}{\sin x+\sec x}\,dx$

79. $\displaystyle\int x\sin^2 x\cos x\,dx$

80. $\displaystyle\int \frac{\sin x\cos x}{\sin^4 x+\cos^4 x}\,dx$

81. The functions $y = e^{x^2}$ and $y = x^2 e^{x^2}$ don't have elementary antiderivatives, but $y = (2x^2+1)e^{x^2}$ does. Evaluate $\int (2x^2+1)e^{x^2}\,dx$.

7.6 INTEGRATION USING TABLES AND COMPUTER ALGEBRA SYSTEMS

In this section we describe how to use tables and computer algebra systems to integrate functions that have elementary antiderivatives. You should bear in mind, though, that even the most powerful computer algebra systems can't find explicit formulas for the antiderivatives of functions like e^{x^2} or the other functions described at the end of Section 7.5.

TABLES OF INTEGRALS

Tables of indefinite integrals are very useful when we are confronted by an integral that is difficult to evaluate by hand and we don't have access to a computer algebra system. A relatively brief table of 120 integrals, categorized by form, is provided on the Reference Pages at the back of the book. More extensive tables are available in *CRC Standard Mathematical Tables and Formulae,* 31st ed. by Daniel Zwillinger (Boca Raton, FL: CRC Press, 2002) (709 entries) or in Gradshteyn and Ryzhik's *Table of Integrals, Series, and Products,* 6e (San Diego: Academic Press, 2000), which contains hundreds of pages of integrals. It should be remembered, however, that integrals do not often occur in exactly the form listed in a table. Usually we need to use substitution or algebraic manipulation to transform a given integral into one of the forms in the table.

EXAMPLE 1 The region bounded by the curves $y = \arctan x$, $y = 0$, and $x = 1$ is rotated about the y-axis. Find the volume of the resulting solid.

SOLUTION Using the method of cylindrical shells, we see that the volume is

$$V = \int_0^1 2\pi x \arctan x\,dx$$

■ The Table of Integrals appears on Reference Pages 6–10 at the back of the book.

In the section of the Table of Integrals titled *Inverse Trigonometric Forms* we locate Formula 92:

$$\int u \tan^{-1}u\, du = \frac{u^2+1}{2}\tan^{-1}u - \frac{u}{2} + C$$

Thus the volume is

$$\begin{aligned} V &= 2\pi \int_0^1 x \tan^{-1}x\, dx = 2\pi\left[\frac{x^2+1}{2}\tan^{-1}x - \frac{x}{2}\right]_0^1 \\ &= \pi\left[(x^2+1)\tan^{-1}x - x\right]_0^1 = \pi(2\tan^{-1}1 - 1) \\ &= \pi[2(\pi/4) - 1] = \tfrac{1}{2}\pi^2 - \pi \end{aligned}$$

□

EXAMPLE 2 Use the Table of Integrals to find $\int \frac{x^2}{\sqrt{5-4x^2}}\, dx$.

SOLUTION If we look at the section of the table titled *Forms involving* $\sqrt{a^2-u^2}$, we see that the closest entry is number 34:

$$\int \frac{u^2}{\sqrt{a^2-u^2}}\, du = -\frac{u}{2}\sqrt{a^2-u^2} + \frac{a^2}{2}\sin^{-1}\left(\frac{u}{a}\right) + C$$

This is not exactly what we have, but we will be able to use it if we first make the substitution $u = 2x$:

$$\int \frac{x^2}{\sqrt{5-4x^2}}\, dx = \int \frac{(u/2)^2}{\sqrt{5-u^2}}\frac{du}{2} = \frac{1}{8}\int \frac{u^2}{\sqrt{5-u^2}}\, du$$

Then we use Formula 34 with $a^2 = 5$ (so $a = \sqrt{5}$):

$$\begin{aligned} \int \frac{x^2}{\sqrt{5-4x^2}}\, dx &= \frac{1}{8}\int \frac{u^2}{\sqrt{5-u^2}}\, du = \frac{1}{8}\left(-\frac{u}{2}\sqrt{5-u^2} + \frac{5}{2}\sin^{-1}\frac{u}{\sqrt{5}}\right) + C \\ &= -\frac{x}{8}\sqrt{5-4x^2} + \frac{5}{16}\sin^{-1}\left(\frac{2x}{\sqrt{5}}\right) + C \end{aligned}$$

□

EXAMPLE 3 Use the Table of Integrals to find $\int x^3 \sin x\, dx$.

SOLUTION If we look in the section called *Trigonometric Forms,* we see that none of the entries explicitly includes a u^3 factor. However, we can use the reduction formula in entry 84 with $n = 3$:

$$\int x^3 \sin x\, dx = -x^3\cos x + 3\int x^2 \cos x\, dx$$

85. $\int u^n \cos u\, du = u^n \sin u - n\int u^{n-1}\sin u\, du$

We now need to evaluate $\int x^2 \cos x\, dx$. We can use the reduction formula in entry 85 with $n = 2$, followed by entry 82:

$$\begin{aligned} \int x^2 \cos x\, dx &= x^2 \sin x - 2\int x \sin x\, dx \\ &= x^2\sin x - 2(\sin x - x\cos x) + K \end{aligned}$$

Combining these calculations, we get

$$\int x^3 \sin x\, dx = -x^3 \cos x + 3x^2 \sin x + 6x \cos x - 6 \sin x + C$$

where $C = 3K$. □

V EXAMPLE 4 Use the Table of Integrals to find $\int x\sqrt{x^2 + 2x + 4}\, dx$.

SOLUTION Since the table gives forms involving $\sqrt{a^2 + x^2}$, $\sqrt{a^2 - x^2}$, and $\sqrt{x^2 - a^2}$, but not $\sqrt{ax^2 + bx + c}$, we first complete the square:

$$x^2 + 2x + 4 = (x + 1)^2 + 3$$

If we make the substitution $u = x + 1$ (so $x = u - 1$), the integrand will involve the pattern $\sqrt{a^2 + u^2}$:

$$\int x\sqrt{x^2 + 2x + 4}\, dx = \int (u - 1)\sqrt{u^2 + 3}\, du$$

$$= \int u\sqrt{u^2 + 3}\, du - \int \sqrt{u^2 + 3}\, du$$

The first integral is evaluated using the substitution $t = u^2 + 3$:

$$\int u\sqrt{u^2 + 3}\, du = \tfrac{1}{2}\int \sqrt{t}\, dt = \tfrac{1}{2} \cdot \tfrac{2}{3}t^{3/2} = \tfrac{1}{3}(u^2 + 3)^{3/2}$$

21. $\int \sqrt{a^2 + u^2}\, du = \frac{u}{2}\sqrt{a^2 + u^2} + \frac{a^2}{2}\ln\left(u + \sqrt{a^2 + u^2}\right) + C$

For the second integral we use Formula 21 with $a = \sqrt{3}$:

$$\int \sqrt{u^2 + 3}\, du = \frac{u}{2}\sqrt{u^2 + 3} + \tfrac{3}{2}\ln\left(u + \sqrt{u^2 + 3}\right)$$

Thus

$$\int x\sqrt{x^2 + 2x + 4}\, dx$$

$$= \tfrac{1}{3}(x^2 + 2x + 4)^{3/2} - \frac{x + 1}{2}\sqrt{x^2 + 2x + 4} - \tfrac{3}{2}\ln\left(x + 1 + \sqrt{x^2 + 2x + 4}\right) + C$$

□

COMPUTER ALGEBRA SYSTEMS

We have seen that the use of tables involves matching the form of the given integrand with the forms of the integrands in the tables. Computers are particularly good at matching patterns. And just as we used substitutions in conjunction with tables, a CAS can perform substitutions that transform a given integral into one that occurs in its stored formulas. So it isn't surprising that computer algebra systems excel at integration. That doesn't mean that integration by hand is an obsolete skill. We will see that a hand computation sometimes produces an indefinite integral in a form that is more convenient than a machine answer.

To begin, let's see what happens when we ask a machine to integrate the relatively simple function $y = 1/(3x - 2)$. Using the substitution $u = 3x - 2$, an easy calculation by hand gives

$$\int \frac{1}{3x - 2}\, dx = \tfrac{1}{3}\ln|3x - 2| + C$$

whereas Derive, Mathematica, and Maple all return the answer

$$\tfrac{1}{3}\ln(3x - 2)$$

The first thing to notice is that computer algebra systems omit the constant of integration. In other words, they produce a *particular* antiderivative, not the most general one. Therefore, when making use of a machine integration, we might have to add a constant. Second, the absolute value signs are omitted in the machine answer. That is fine if our problem is concerned only with values of x greater than $\frac{2}{3}$. But if we are interested in other values of x, then we need to insert the absolute value symbol.

In the next example we reconsider the integral of Example 4, but this time we ask a machine for the answer.

EXAMPLE 5 Use a computer algebra system to find $\int x\sqrt{x^2 + 2x + 4}\,dx$.

SOLUTION Maple responds with the answer

$$\tfrac{1}{3}(x^2 + 2x + 4)^{3/2} - \tfrac{1}{4}(2x + 2)\sqrt{x^2 + 2x + 4} - \frac{3}{2}\operatorname{arcsinh}\frac{\sqrt{3}}{3}(1 + x)$$

This looks different from the answer we found in Example 4, but it is equivalent because the third term can be rewritten using the identity

■ This is Equation 3.11.3.

$$\operatorname{arcsinh} x = \ln\left(x + \sqrt{x^2 + 1}\right)$$

Thus

$$\begin{aligned}\operatorname{arcsinh}\frac{\sqrt{3}}{3}(1 + x) &= \ln\left[\frac{\sqrt{3}}{3}(1 + x) + \sqrt{\tfrac{1}{3}(1 + x)^2 + 1}\right] \\ &= \ln\frac{1}{\sqrt{3}}\left[1 + x + \sqrt{(1 + x)^2 + 3}\right] \\ &= \ln\frac{1}{\sqrt{3}} + \ln\left(x + 1 + \sqrt{x^2 + 2x + 4}\right)\end{aligned}$$

The resulting extra term $-\frac{3}{2}\ln\left(1/\sqrt{3}\right)$ can be absorbed into the constant of integration.

Mathematica gives the answer

$$\left(\frac{5}{6} + \frac{x}{6} + \frac{x^2}{3}\right)\sqrt{x^2 + 2x + 4} - \frac{3}{2}\operatorname{arcsinh}\left(\frac{1 + x}{\sqrt{3}}\right)$$

Mathematica combined the first two terms of Example 4 (and the Maple result) into a single term by factoring.

Derive gives the answer

$$\tfrac{1}{6}\sqrt{x^2 + 2x + 4}\,(2x^2 + x + 5) - \tfrac{3}{2}\ln\left(\sqrt{x^2 + 2x + 4} + x + 1\right)$$

The first term is like the first term in the Mathematica answer, and the second term is identical to the last term in Example 4. □

EXAMPLE 6 Use a CAS to evaluate $\int x(x^2 + 5)^8\,dx$.

SOLUTION Maple and Mathematica give the same answer:

$$\tfrac{1}{18}x^{18} + \tfrac{5}{2}x^{16} + 50x^{14} + \tfrac{1750}{3}x^{12} + 4375x^{10} + 21875x^8 + \tfrac{218750}{3}x^6 + 156250x^4 + \tfrac{390625}{2}x^2$$

It's clear that both systems must have expanded $(x^2 + 5)^8$ by the Binomial Theorem and then integrated each term.

If we integrate by hand instead, using the substitution $u = x^2 + 5$, we get

■ Derive and the TI-89/92 also give this answer.

$$\int x(x^2 + 5)^8\, dx = \tfrac{1}{18}(x^2 + 5)^9 + C$$

For most purposes, this is a more convenient form of the answer. □

EXAMPLE 7 Use a CAS to find $\int \sin^5 x \cos^2 x\, dx$.

SOLUTION In Example 2 in Section 7.2 we found that

$$\boxed{1} \qquad \int \sin^5 x \cos^2 x\, dx = -\tfrac{1}{3}\cos^3 x + \tfrac{2}{5}\cos^5 x - \tfrac{1}{7}\cos^7 x + C$$

Derive and Maple report the answer

$$-\tfrac{1}{7}\sin^4 x \cos^3 x - \tfrac{4}{35}\sin^2 x \cos^3 x - \tfrac{8}{105}\cos^3 x$$

whereas Mathematica produces

$$-\tfrac{5}{64}\cos x - \tfrac{1}{192}\cos 3x + \tfrac{3}{320}\cos 5x - \tfrac{1}{448}\cos 7x$$

We suspect that there are trigonometric identities which show these three answers are equivalent. Indeed, if we ask Derive, Maple, and Mathematica to simplify their expressions using trigonometric identities, they ultimately produce the same form of the answer as in Equation 1. □

7.6 EXERCISES

1–4 Use the indicated entry in the Table of Integrals on the Reference Pages to evaluate the integral.

1. $\displaystyle\int \frac{\sqrt{7 - 2x^2}}{x^2}\, dx$; entry 33

2. $\displaystyle\int \frac{3x}{\sqrt{3 - 2x}}\, dx$; entry 55

3. $\displaystyle\int \sec^3(\pi x)\, dx$; entry 71

4. $\displaystyle\int e^{2\theta} \sin 3\theta\, d\theta$; entry 98

5–30 Use the Table of Integrals on Reference Pages 6–10 to evaluate the integral.

5. $\displaystyle\int_0^1 2x \cos^{-1}x\, dx$

6. $\displaystyle\int_2^3 \frac{1}{x^2\sqrt{4x^2 - 7}}\, dx$

7. $\displaystyle\int \tan^3(\pi x)\, dx$

8. $\displaystyle\int \frac{\ln(1 + \sqrt{x})}{\sqrt{x}}\, dx$

9. $\displaystyle\int \frac{dx}{x^2\sqrt{4x^2 + 9}}$

10. $\displaystyle\int \frac{\sqrt{2y^2 - 3}}{y^2}\, dy$

11. $\displaystyle\int_{-1}^0 t^2 e^{-t}\, dt$

12. $\displaystyle\int x^2 \operatorname{csch}(x^3 + 1)\, dx$

13. $\displaystyle\int \frac{\tan^3(1/z)}{z^2}\, dz$

14. $\displaystyle\int \sin^{-1}\sqrt{x}\, dx$

15. $\displaystyle\int e^{2x} \arctan(e^x)\, dx$

16. $\displaystyle\int x \sin(x^2) \cos(3x^2)\, dx$

17. $\displaystyle\int y\sqrt{6 + 4y - 4y^2}\, dy$

18. $\displaystyle\int \frac{dx}{2x^3 - 3x^2}$

19. $\displaystyle\int \sin^2 x \cos x \ln(\sin x)\, dx$

20. $\displaystyle\int \frac{\sin 2\theta}{\sqrt{5 - \sin\theta}}\, d\theta$

21. $\displaystyle\int \frac{e^x}{3 - e^{2x}}\, dx$

22. $\displaystyle\int_0^2 x^3\sqrt{4x^2 - x^4}\, dx$

23. $\displaystyle\int \sec^5 x\, dx$

24. $\displaystyle\int \sin^6 2x\, dx$

25. $\displaystyle\int \frac{\sqrt{4 + (\ln x)^2}}{x}\, dx$

26. $\displaystyle\int_0^1 x^4 e^{-x}\, dx$

27. $\int \sqrt{e^{2x} - 1}\, dx$

28. $\int e^t \sin(\alpha t - 3)\, dt$

29. $\int \frac{x^4\, dx}{\sqrt{x^{10} - 2}}$

30. $\int \frac{\sec^2\theta \tan^2\theta}{\sqrt{9 - \tan^2\theta}}\, d\theta$

31. Find the volume of the solid obtained when the region under the curve $y = x\sqrt{4 - x^2}$, $0 \leqslant x \leqslant 2$, is rotated about the y-axis.

32. The region under the curve $y = \tan^2 x$ from 0 to $\pi/4$ is rotated about the x-axis. Find the volume of the resulting solid.

33. Verify Formula 53 in the Table of Integrals (a) by differentiation and (b) by using the substitution $t = a + bu$.

34. Verify Formula 31 (a) by differentiation and (b) by substituting $u = a \sin\theta$.

CAS **35–42** Use a computer algebra system to evaluate the integral. Compare the answer with the result of using tables. If the answers are not the same, show that they are equivalent.

35. $\int \sec^4 x\, dx$

36. $\int \csc^5 x\, dx$

37. $\int x^2\sqrt{x^2 + 4}\, dx$

38. $\int \frac{dx}{e^x(3e^x + 2)}$

39. $\int x\sqrt{1 + 2x}\, dx$

40. $\int \sin^4 x\, dx$

41. $\int \tan^5 x\, dx$

42. $\int \frac{1}{\sqrt{1 + \sqrt[3]{x}}}\, dx$

CAS **43.** (a) Use the table of integrals to evaluate $F(x) = \int f(x)\, dx$, where

$$f(x) = \frac{1}{x\sqrt{1 - x^2}}$$

What is the domain of f and F?

(b) Use a CAS to evaluate $F(x)$. What is the domain of the function F that the CAS produces? Is there a discrepancy between this domain and the domain of the function F that you found in part (a)?

CAS **44.** Computer algebra systems sometimes need a helping hand from human beings. Try to evaluate

$$\int (1 + \ln x)\sqrt{1 + (x \ln x)^2}\, dx$$

with a computer algebra system. If it doesn't return an answer, make a substitution that changes the integral into one that the CAS *can* evaluate.

CAS **45–48** Use a CAS to find an antiderivative F of f such that $F(0) = 0$. Graph f and F and locate approximately the x-coordinates of the extreme points and inflection points of F.

45. $f(x) = \frac{x^2 - 1}{x^4 + x^2 + 1}$

46. $f(x) = xe^{-x}\sin x, \quad -5 \leqslant x \leqslant 5$

47. $f(x) = \sin^4 x \cos^6 x, \quad 0 \leqslant x \leqslant \pi$

48. $f(x) = \frac{x^3 - x}{x^6 + 1}$

DISCOVERY PROJECT

CAS PATTERNS IN INTEGRALS

In this project a computer algebra system is used to investigate indefinite integrals of families of functions. By observing the patterns that occur in the integrals of several members of the family, you will first guess, and then prove, a general formula for the integral of any member of the family.

1. (a) Use a computer algebra system to evaluate the following integrals.

(i) $\int \frac{1}{(x + 2)(x + 3)}\, dx$

(ii) $\int \frac{1}{(x + 1)(x + 5)}\, dx$

(iii) $\int \frac{1}{(x + 2)(x - 5)}\, dx$

(iv) $\int \frac{1}{(x + 2)^2}\, dx$

(b) Based on the pattern of your responses in part (a), guess the value of the integral

$$\int \frac{1}{(x + a)(x + b)}\, dx$$

if $a \neq b$. What if $a = b$?

(c) Check your guess by asking your CAS to evaluate the integral in part (b). Then prove it using partial fractions.

2. (a) Use a computer algebra system to evaluate the following integrals.

(i) $\int \sin x \cos 2x\, dx$ (ii) $\int \sin 3x \cos 7x\, dx$ (iii) $\int \sin 8x \cos 3x\, dx$

(b) Based on the pattern of your responses in part (a), guess the value of the integral

$$\int \sin ax \cos bx\, dx$$

(c) Check your guess with a CAS. Then prove it using the techniques of Section 7.2. For what values of a and b is it valid?

3. (a) Use a computer algebra system to evaluate the following integrals.

(i) $\int \ln x\, dx$ (ii) $\int x \ln x\, dx$ (iii) $\int x^2 \ln x\, dx$

(iv) $\int x^3 \ln x\, dx$ (v) $\int x^7 \ln x\, dx$

(b) Based on the pattern of your responses in part (a), guess the value of

$$\int x^n \ln x\, dx$$

(c) Use integration by parts to prove the conjecture that you made in part (b). For what values of n is it valid?

4. (a) Use a computer algebra system to evaluate the following integrals.

(i) $\int xe^x\, dx$ (ii) $\int x^2e^x\, dx$ (iii) $\int x^3e^x\, dx$

(iv) $\int x^4e^x\, dx$ (v) $\int x^5e^x\, dx$

(b) Based on the pattern of your responses in part (a), guess the value of $\int x^6e^x\, dx$. Then use your CAS to check your guess.

(c) Based on the patterns in parts (a) and (b), make a conjecture as to the value of the integral

$$\int x^n e^x\, dx$$

when n is a positive integer.

(d) Use mathematical induction to prove the conjecture you made in part (c).

7.7 APPROXIMATE INTEGRATION

There are two situations in which it is impossible to find the exact value of a definite integral.

The first situation arises from the fact that in order to evaluate $\int_a^b f(x)\, dx$ using the Fundamental Theorem of Calculus we need to know an antiderivative of f. Sometimes, however, it is difficult, or even impossible, to find an antiderivative (see Section 7.5). For example, it is impossible to evaluate the following integrals exactly:

$$\int_0^1 e^{x^2}\, dx \qquad \int_{-1}^1 \sqrt{1 + x^3}\, dx$$

The second situation arises when the function is determined from a scientific experiment through instrument readings or collected data. There may be no formula for the function (see Example 5).

In both cases we need to find approximate values of definite integrals. We already know one such method. Recall that the definite integral is defined as a limit of Riemann sums, so any Riemann sum could be used as an approximation to the integral: If we divide $[a, b]$ into n subintervals of equal length $\Delta x = (b - a)/n$, then we have

$$\int_a^b f(x)\,dx \approx \sum_{i=1}^{n} f(x_i^*)\,\Delta x$$

where x_i^* is any point in the ith subinterval $[x_{i-1}, x_i]$. If x_i^* is chosen to be the left endpoint of the interval, then $x_i^* = x_{i-1}$ and we have

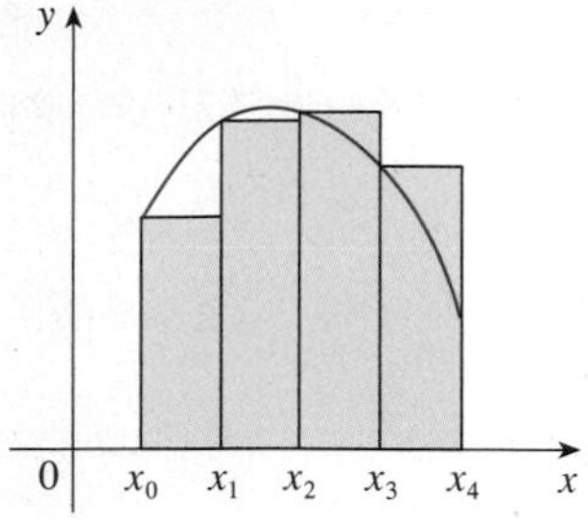

(a) Left endpoint approximation

$$\boxed{1} \qquad \int_a^b f(x)\,dx \approx L_n = \sum_{i=1}^{n} f(x_{i-1})\,\Delta x$$

If $f(x) \geqslant 0$, then the integral represents an area and (1) represents an approximation of this area by the rectangles shown in Figure 1(a). If we choose x_i^* to be the right endpoint, then $x_i^* = x_i$ and we have

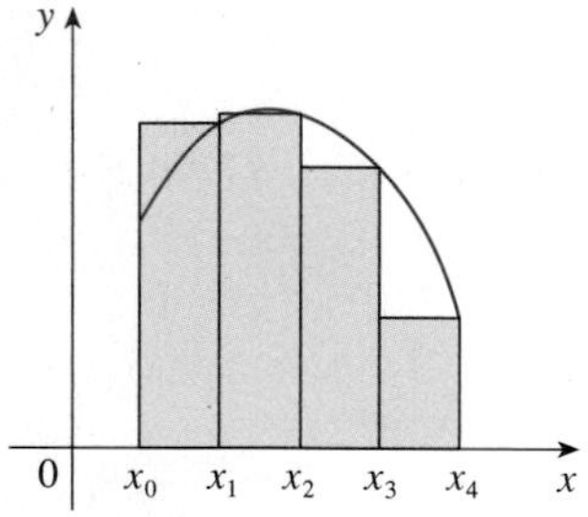

(b) Right endpoint approximation

$$\boxed{2} \qquad \int_a^b f(x)\,dx \approx R_n = \sum_{i=1}^{n} f(x_i)\,\Delta x$$

[See Figure 1(b).] The approximations L_n and R_n defined by Equations 1 and 2 are called the **left endpoint approximation** and **right endpoint approximation**, respectively.

In Section 5.2 we also considered the case where x_i^* is chosen to be the midpoint $\bar{x}_i$ of the subinterval $[x_{i-1}, x_i]$. Figure 1(c) shows the midpoint approximation M_n, which appears to be better than either L_n or R_n.

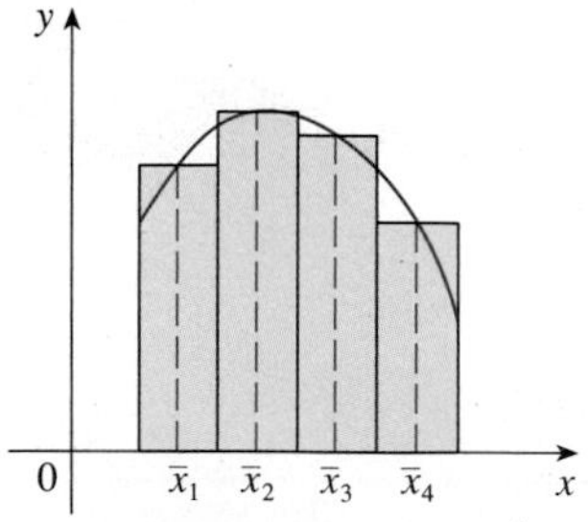

(c) Midpoint approximation

FIGURE 1

MIDPOINT RULE

$$\int_a^b f(x)\,dx \approx M_n = \Delta x\,[f(\bar{x}_1) + f(\bar{x}_2) + \cdots + f(\bar{x}_n)]$$

where $\qquad \Delta x = \dfrac{b - a}{n}$

and $\qquad \bar{x}_i = \tfrac{1}{2}(x_{i-1} + x_i) = \text{midpoint of } [x_{i-1}, x_i]$

Another approximation, called the Trapezoidal Rule, results from averaging the approximations in Equations 1 and 2:

$$\int_a^b f(x)\,dx \approx \frac{1}{2}\left[\sum_{i=1}^{n} f(x_{i-1})\,\Delta x + \sum_{i=1}^{n} f(x_i)\,\Delta x\right] = \frac{\Delta x}{2}\left[\sum_{i=1}^{n} \big(f(x_{i-1}) + f(x_i)\big)\right]$$

$$= \frac{\Delta x}{2}\left[\big(f(x_0) + f(x_1)\big) + \big(f(x_1) + f(x_2)\big) + \cdots + \big(f(x_{n-1}) + f(x_n)\big)\right]$$

$$= \frac{\Delta x}{2}\left[f(x_0) + 2f(x_1) + 2f(x_2) + \cdots + 2f(x_{n-1}) + f(x_n)\right]$$

TRAPEZOIDAL RULE

$$\int_a^b f(x)\,dx \approx T_n = \frac{\Delta x}{2}\,[f(x_0) + 2f(x_1) + 2f(x_2) + \cdots + 2f(x_{n-1}) + f(x_n)]$$

where $\Delta x = (b - a)/n$ and $x_i = a + i\,\Delta x$.

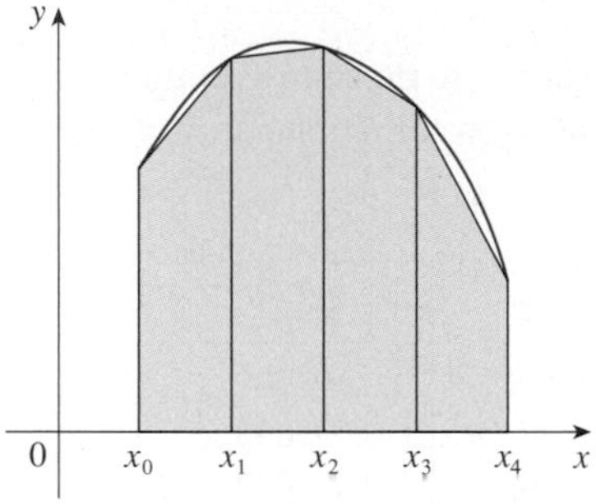

FIGURE 2
Trapezoidal approximation

The reason for the name Trapezoidal Rule can be seen from Figure 2, which illustrates the case $f(x) \geqslant 0$. The area of the trapezoid that lies above the ith subinterval is

$$\Delta x\left(\frac{f(x_{i-1}) + f(x_i)}{2}\right) = \frac{\Delta x}{2}\,[f(x_{i-1}) + f(x_i)]$$

and if we add the areas of all these trapezoids, we get the right side of the Trapezoidal Rule.

EXAMPLE 1 Use (a) the Trapezoidal Rule and (b) the Midpoint Rule with $n = 5$ to approximate the integral $\int_1^2 (1/x)\,dx$.

SOLUTION

(a) With $n = 5$, $a = 1$, and $b = 2$, we have $\Delta x = (2 - 1)/5 = 0.2$, and so the Trapezoidal Rule gives

$$\begin{aligned}\int_1^2 \frac{1}{x}\,dx \approx T_5 &= \frac{0.2}{2}\,[f(1) + 2f(1.2) + 2f(1.4) + 2f(1.6) + 2f(1.8) + f(2)]\\ &= 0.1\left(\frac{1}{1} + \frac{2}{1.2} + \frac{2}{1.4} + \frac{2}{1.6} + \frac{2}{1.8} + \frac{1}{2}\right)\\ &\approx 0.695635\end{aligned}$$

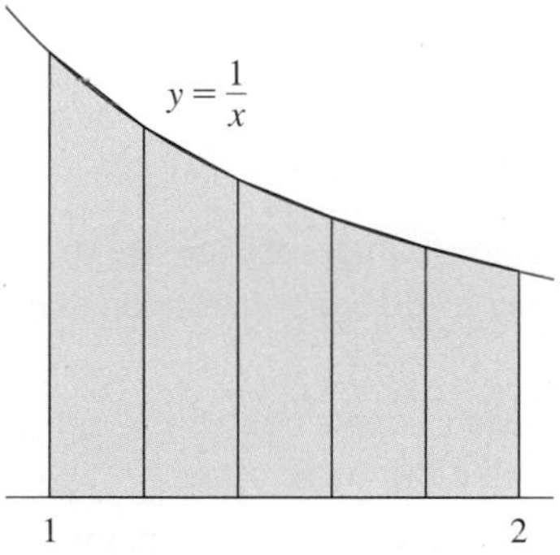

FIGURE 3

This approximation is illustrated in Figure 3.

(b) The midpoints of the five subintervals are 1.1, 1.3, 1.5, 1.7, and 1.9, so the Midpoint Rule gives

$$\begin{aligned}\int_1^2 \frac{1}{x}\,dx &\approx \Delta x\,[f(1.1) + f(1.3) + f(1.5) + f(1.7) + f(1.9)]\\ &= \frac{1}{5}\left(\frac{1}{1.1} + \frac{1}{1.3} + \frac{1}{1.5} + \frac{1}{1.7} + \frac{1}{1.9}\right)\\ &\approx 0.691908\end{aligned}$$

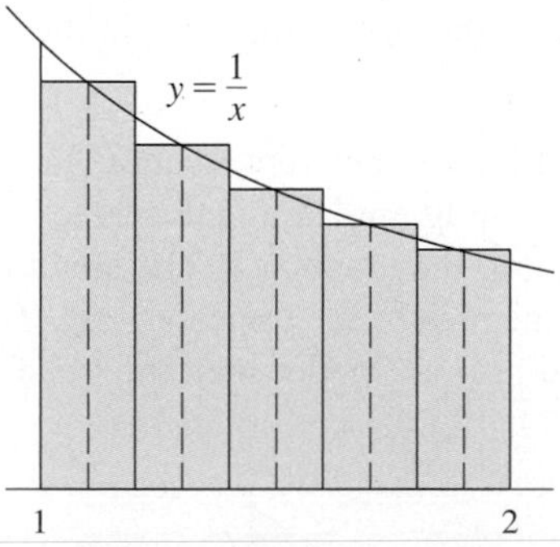

FIGURE 4

This approximation is illustrated in Figure 4. □

In Example 1 we deliberately chose an integral whose value can be computed explicitly so that we can see how accurate the Trapezoidal and Midpoint Rules are. By the Fundamental Theorem of Calculus,

$$\int_1^2 \frac{1}{x}\,dx = \ln x\Big]_1^2 = \ln 2 = 0.693147\ldots$$

$\int_a^b f(x)\,dx = \text{approximation} + \text{error}$

The **error** in using an approximation is defined to be the amount that needs to be added to the approximation to make it exact. From the values in Example 1 we see that the errors in the Trapezoidal and Midpoint Rule approximations for $n = 5$ are

$$E_T \approx -0.002488 \qquad \text{and} \qquad E_M \approx 0.001239$$

In general, we have

$$E_T = \int_a^b f(x)\,dx - T_n \qquad \text{and} \qquad E_M = \int_a^b f(x)\,dx - M_n$$

TEC Module 5.2/7.7 allows you to compare approximation methods.

The following tables show the results of calculations similar to those in Example 1, but for $n = 5$, 10, and 20 and for the left and right endpoint approximations as well as the Trapezoidal and Midpoint Rules.

Approximations to $\int_1^2 \frac{1}{x}\,dx$

n	L_n	R_n	T_n	M_n
5	0.745635	0.645635	0.695635	0.691908
10	0.718771	0.668771	0.693771	0.692835
20	0.705803	0.680803	0.693303	0.693069

Corresponding errors

n	E_L	E_R	E_T	E_M
5	−0.052488	0.047512	−0.002488	0.001239
10	−0.025624	0.024376	−0.000624	0.000312
20	−0.012656	0.012344	−0.000156	0.000078

We can make several observations from these tables:

1. In all of the methods we get more accurate approximations when we increase the value of n. (But very large values of n result in so many arithmetic operations that we have to beware of accumulated round-off error.)
2. The errors in the left and right endpoint approximations are opposite in sign and appear to decrease by a factor of about 2 when we double the value of n.
3. The Trapezoidal and Midpoint Rules are much more accurate than the endpoint approximations.
4. The errors in the Trapezoidal and Midpoint Rules are opposite in sign and appear to decrease by a factor of about 4 when we double the value of n.
5. The size of the error in the Midpoint Rule is about half the size of the error in the Trapezoidal Rule.

■ It turns out that these observations are true in most cases.

Figure 5 shows why we can usually expect the Midpoint Rule to be more accurate than the Trapezoidal Rule. The area of a typical rectangle in the Midpoint Rule is the same as the area of the trapezoid $ABCD$ whose upper side is tangent to the graph at P. The area of this trapezoid is closer to the area under the graph than is the area of the trapezoid $AQRD$ used in the Trapezoidal Rule. [The midpoint error (shaded red) is smaller than the trapezoidal error (shaded blue).]

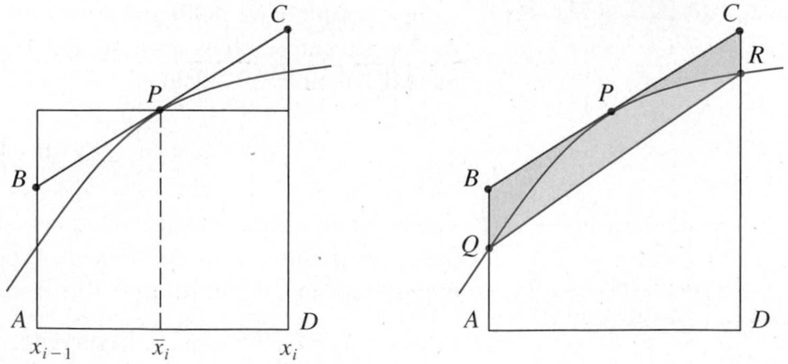

FIGURE 5

These observations are corroborated in the following error estimates, which are proved in books on numerical analysis. Notice that Observation 4 corresponds to the n^2 in each denominator because $(2n)^2 = 4n^2$. The fact that the estimates depend on the size of the second derivative is not surprising if you look at Figure 5, because $f''(x)$ measures how much the graph is curved. [Recall that $f''(x)$ measures how fast the slope of $y = f(x)$ changes.]

3 ERROR BOUNDS Suppose $|f''(x)| \leqslant K$ for $a \leqslant x \leqslant b$. If E_T and E_M are the errors in the Trapezoidal and Midpoint Rules, then

$$|E_T| \leqslant \frac{K(b-a)^3}{12n^2} \qquad \text{and} \qquad |E_M| \leqslant \frac{K(b-a)^3}{24n^2}$$

Let's apply this error estimate to the Trapezoidal Rule approximation in Example 1. If $f(x) = 1/x$, then $f'(x) = -1/x^2$ and $f''(x) = 2/x^3$. Since $1 \leqslant x \leqslant 2$, we have $1/x \leqslant 1$, so

$$|f''(x)| = \left|\frac{2}{x^3}\right| \leqslant \frac{2}{1^3} = 2$$

Therefore, taking $K = 2$, $a = 1$, $b = 2$, and $n = 5$ in the error estimate (3), we see that

■ K can be any number larger than all the values of $|f''(x)|$, but smaller values of K give better error bounds.

$$|E_T| \leqslant \frac{2(2-1)^3}{12(5)^2} = \frac{1}{150} \approx 0.006667$$

Comparing this error estimate of 0.006667 with the actual error of about 0.002488, we see that it can happen that the actual error is substantially less than the upper bound for the error given by (3).

V EXAMPLE 2 How large should we take n in order to guarantee that the Trapezoidal and Midpoint Rule approximations for $\int_1^2 (1/x)\,dx$ are accurate to within 0.0001?

SOLUTION We saw in the preceding calculation that $|f''(x)| \leqslant 2$ for $1 \leqslant x \leqslant 2$, so we can take $K = 2$, $a = 1$, and $b = 2$ in (3). Accuracy to within 0.0001 means that the size of the error should be less than 0.0001. Therefore we choose n so that

$$\frac{2(1)^3}{12n^2} < 0.0001$$

Solving the inequality for n, we get

$$n^2 > \frac{2}{12(0.0001)}$$

or

$$n > \frac{1}{\sqrt{0.0006}} \approx 40.8$$

■ It's quite possible that a lower value for n would suffice, but 41 is the smallest value for which the error bound formula can *guarantee* us accuracy to within 0.0001.

Thus $n = 41$ will ensure the desired accuracy.

For the same accuracy with the Midpoint Rule we choose n so that

$$\frac{2(1)^3}{24n^2} < 0.0001$$

which gives

$$n > \frac{1}{\sqrt{0.0012}} \approx 29$$

□

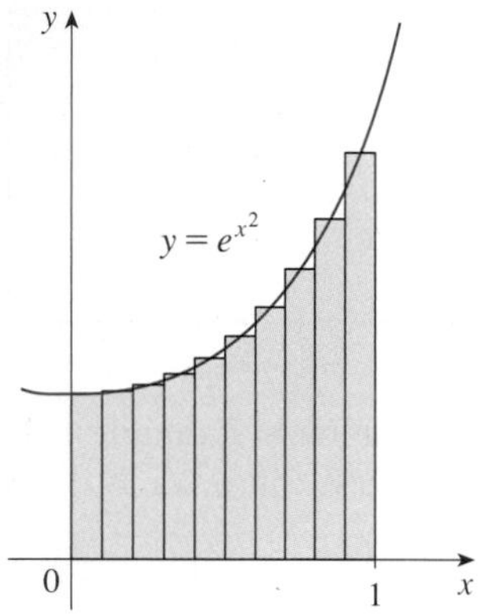

FIGURE 6

EXAMPLE 3

(a) Use the Midpoint Rule with $n = 10$ to approximate the integral $\int_0^1 e^{x^2}\,dx$.
(b) Give an upper bound for the error involved in this approximation.

SOLUTION

(a) Since $a = 0$, $b = 1$, and $n = 10$, the Midpoint Rule gives

$$\begin{aligned}\int_0^1 e^{x^2}\,dx &\approx \Delta x\,[f(0.05) + f(0.15) + \cdots + f(0.85) + f(0.95)]\\ &= 0.1[e^{0.0025} + e^{0.0225} + e^{0.0625} + e^{0.1225} + e^{0.2025} + e^{0.3025}\\ &\qquad + e^{0.4225} + e^{0.5625} + e^{0.7225} + e^{0.9025}]\\ &\approx 1.460393\end{aligned}$$

Figure 6 illustrates this approximation.

(b) Since $f(x) = e^{x^2}$, we have $f'(x) = 2xe^{x^2}$ and $f''(x) = (2 + 4x^2)e^{x^2}$. Also, since $0 \leqslant x \leqslant 1$, we have $x^2 \leqslant 1$ and so

$$0 \leqslant f''(x) = (2 + 4x^2)e^{x^2} \leqslant 6e$$

■ Error estimates give upper bounds for the error. They are theoretical, worst-case scenarios. The actual error in this case turns out to be about 0.0023.

Taking $K = 6e$, $a = 0$, $b = 1$, and $n = 10$ in the error estimate (3), we see that an upper bound for the error is

$$\frac{6e(1)^3}{24(10)^2} = \frac{e}{400} \approx 0.007$$

□

SIMPSON'S RULE

Another rule for approximate integration results from using parabolas instead of straight line segments to approximate a curve. As before, we divide $[a, b]$ into n subintervals of equal length $h = \Delta x = (b - a)/n$, but this time we assume that n is an *even* number. Then on each consecutive pair of intervals we approximate the curve $y = f(x) \geqslant 0$ by a parabola as shown in Figure 7. If $y_i = f(x_i)$, then $P_i(x_i, y_i)$ is the point on the curve lying above x_i. A typical parabola passes through three consecutive points P_i, P_{i+1}, and P_{i+2}.

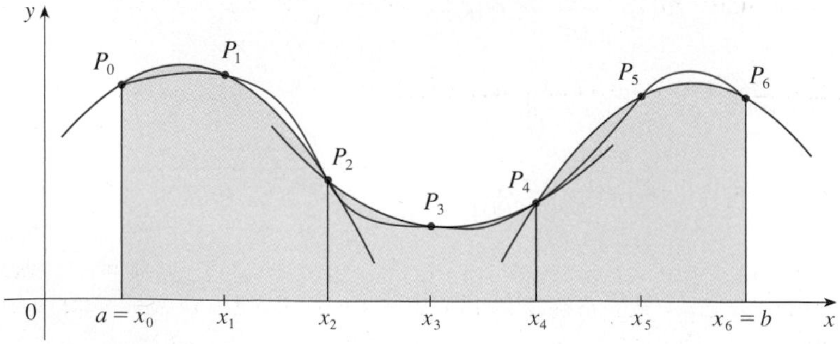

FIGURE 7

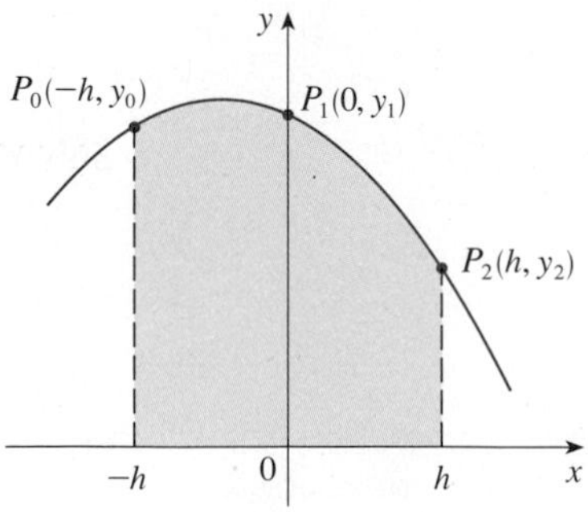

FIGURE 8

To simplify our calculations, we first consider the case where $x_0 = -h$, $x_1 = 0$, and $x_2 = h$. (See Figure 8.) We know that the equation of the parabola through P_0, P_1, and P_2 is of the form $y = Ax^2 + Bx + C$ and so the area under the parabola from $x = -h$ to $x = h$ is

■ Here we have used Theorem 5.5.7. Notice that $Ax^2 + C$ is even and Bx is odd.

$$\begin{aligned}\int_{-h}^{h} (Ax^2 + Bx + C)\,dx &= 2\int_0^h (Ax^2 + C)\,dx \\ &= 2\left[A\frac{x^3}{3} + Cx\right]_0^h \\ &= 2\left(A\frac{h^3}{3} + Ch\right) = \frac{h}{3}(2Ah^2 + 6C)\end{aligned}$$

But, since the parabola passes through $P_0(-h, y_0)$, $P_1(0, y_1)$, and $P_2(h, y_2)$, we have

$$\begin{aligned}y_0 &= A(-h)^2 + B(-h) + C = Ah^2 - Bh + C \\ y_1 &= C \\ y_2 &= Ah^2 + Bh + C\end{aligned}$$

and therefore $$y_0 + 4y_1 + y_2 = 2Ah^2 + 6C$$

Thus we can rewrite the area under the parabola as

$$\frac{h}{3}(y_0 + 4y_1 + y_2)$$

Now, by shifting this parabola horizontally we do not change the area under it. This means that the area under the parabola through P_0, P_1, and P_2 from $x = x_0$ to $x = x_2$ in Figure 7 is still

$$\frac{h}{3}(y_0 + 4y_1 + y_2)$$

Similarly, the area under the parabola through P_2, P_3, and P_4 from $x = x_2$ to $x = x_4$ is

$$\frac{h}{3}(y_2 + 4y_3 + y_4)$$

If we compute the areas under all the parabolas in this manner and add the results, we get

$$\begin{aligned}\int_a^b f(x)\,dx &\approx \frac{h}{3}(y_0 + 4y_1 + y_2) + \frac{h}{3}(y_2 + 4y_3 + y_4) + \cdots + \frac{h}{3}(y_{n-2} + 4y_{n-1} + y_n) \\ &= \frac{h}{3}(y_0 + 4y_1 + 2y_2 + 4y_3 + 2y_4 + \cdots + 2y_{n-2} + 4y_{n-1} + y_n)\end{aligned}$$

Although we have derived this approximation for the case in which $f(x) \geqslant 0$, it is a reasonable approximation for any continuous function f and is called Simpson's Rule after the English mathematician Thomas Simpson (1710–1761). Note the pattern of coefficients: 1, 4, 2, 4, 2, 4, 2, . . . , 4, 2, 4, 1.

SIMPSON

Thomas Simpson was a weaver who taught himself mathematics and went on to become one of the best English mathematicians of the 18th century. What we call Simpson's Rule was actually known to Cavalieri and Gregory in the 17th century, but Simpson popularized it in his best-selling calculus textbook, *A New Treatise of Fluxions*.

SIMPSON'S RULE

$$\int_a^b f(x)\,dx \approx S_n = \frac{\Delta x}{3}[f(x_0) + 4f(x_1) + 2f(x_2) + 4f(x_3) + \cdots + 2f(x_{n-2}) + 4f(x_{n-1}) + f(x_n)]$$

where n is even and $\Delta x = (b - a)/n$.

EXAMPLE 4 Use Simpson's Rule with $n = 10$ to approximate $\int_1^2 (1/x)\,dx$.

SOLUTION Putting $f(x) = 1/x$, $n = 10$, and $\Delta x = 0.1$ in Simpson's Rule, we obtain

$$\begin{aligned}\int_1^2 \frac{1}{x}\,dx &\approx S_{10} \\ &= \frac{\Delta x}{3}[f(1) + 4f(1.1) + 2f(1.2) + 4f(1.3) + \cdots + 2f(1.8) + 4f(1.9) + f(2)] \\ &= \frac{0.1}{3}\left(\frac{1}{1} + \frac{4}{1.1} + \frac{2}{1.2} + \frac{4}{1.3} + \frac{2}{1.4} + \frac{4}{1.5} + \frac{2}{1.6} + \frac{4}{1.7} + \frac{2}{1.8} + \frac{4}{1.9} + \frac{1}{2}\right) \\ &\approx 0.693150\end{aligned}$$

□

Notice that, in Example 4, Simpson's Rule gives us a *much* better approximation ($S_{10} \approx 0.693150$) to the true value of the integral ($\ln 2 \approx 0.693147\ldots$) than does the Trapezoidal Rule ($T_{10} \approx 0.693771$) or the Midpoint Rule ($M_{10} \approx 0.692835$). It turns out (see Exercise 48) that the approximations in Simpson's Rule are weighted averages of those in the Trapezoidal and Midpoint Rules:

$$S_{2n} = \tfrac{1}{3}T_n + \tfrac{2}{3}M_n$$

(Recall that E_T and E_M usually have opposite signs and $|E_M|$ is about half the size of $|E_T|$.)

In many applications of calculus we need to evaluate an integral even if no explicit formula is known for y as a function of x. A function may be given graphically or as a table of values of collected data. If there is evidence that the values are not changing rapidly, then the Trapezoidal Rule or Simpson's Rule can still be used to find an approximate value for $\int_a^b y\,dx$, the integral of y with respect to x.

V EXAMPLE 5 Figure 9 shows data traffic on the link from the United States to SWITCH, the Swiss academic and research network, on February 10, 1998. $D(t)$ is the data throughput, measured in megabits per second (Mb/s). Use Simpson's Rule to estimate the total amount of data transmitted on the link up to noon on that day.

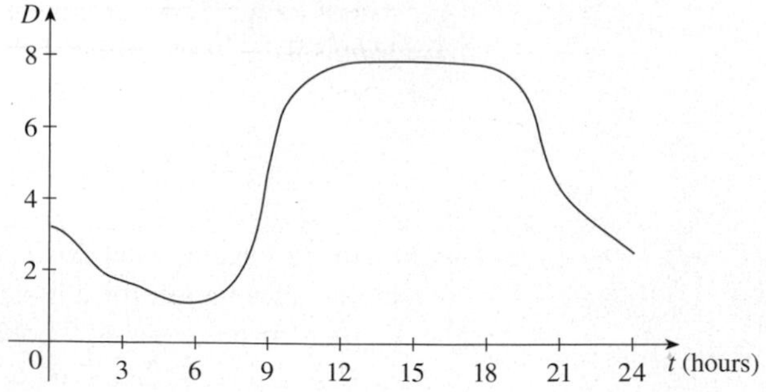

FIGURE 9

SOLUTION Because we want the units to be consistent and $D(t)$ is measured in megabits per second, we convert the units for t from hours to seconds. If we let $A(t)$ be the amount of data (in megabits) transmitted by time t, where t is measured in seconds, then $A'(t) = D(t)$. So, by the Net Change Theorem (see Section 5.4), the total amount of data transmitted by noon (when $t = 12 \times 60^2 = 43{,}200$) is

$$A(43{,}200) = \int_0^{43{,}200} D(t)\, dt$$

We estimate the values of $D(t)$ at hourly intervals from the graph and compile them in the table.

t (hours)	t (seconds)	$D(t)$	t (hours)	t (seconds)	$D(t)$
0	0	3.2	7	25,200	1.3
1	3,600	2.7	8	28,800	2.8
2	7,200	1.9	9	32,400	5.7
3	10,800	1.7	10	36,000	7.1
4	14,400	1.3	11	39,600	7.7
5	18,000	1.0	12	43,200	7.9
6	21,600	1.1			

Then we use Simpson's Rule with $n = 12$ and $\Delta t = 3600$ to estimate the integral:

$$\int_0^{43{,}200} A(t)\, dt \approx \frac{\Delta t}{3}[D(0) + 4D(3600) + 2D(7200) + \cdots + 4D(39{,}600) + D(43{,}200)]$$

$$\approx \frac{3600}{3}[3.2 + 4(2.7) + 2(1.9) + 4(1.7) + 2(1.3) + 4(1.0)$$
$$+ 2(1.1) + 4(1.3) + 2(2.8) + 4(5.7) + 2(7.1) + 4(7.7) + 7.9]$$

$$= 143{,}880$$

Thus the total amount of data transmitted up to noon is about 144,000 megabits, or 144 gigabits. □

n	M_n	S_n
4	0.69121989	0.69315453
8	0.69266055	0.69314765
16	0.69302521	0.69314721

n	E_M	E_S
4	0.00192729	−0.00000735
8	0.00048663	−0.00000047
16	0.00012197	−0.00000003

The table in the margin shows how Simpson's Rule compares with the Midpoint Rule for the integral $\int_1^2 (1/x)\, dx$, whose true value is about 0.69314718. The second table shows how the error E_s in Simpson's Rule decreases by a factor of about 16 when n is doubled. (In Exercises 27 and 28 you are asked to verify this for two additional integrals.) That is consistent with the appearance of n^4 in the denominator of the following error estimate for Simpson's Rule. It is similar to the estimates given in (3) for the Trapezoidal and Midpoint Rules, but it uses the fourth derivative of f.

4 ERROR BOUND FOR SIMPSON'S RULE Suppose that $|f^{(4)}(x)| \leqslant K$ for $a \leqslant x \leqslant b$. If E_S is the error involved in using Simpson's Rule, then

$$|E_S| \leqslant \frac{K(b-a)^5}{180n^4}$$

EXAMPLE 6 How large should we take n in order to guarantee that the Simpson's Rule approximation for $\int_1^2 (1/x)\,dx$ is accurate to within 0.0001?

SOLUTION If $f(x) = 1/x$, then $f^{(4)}(x) = 24/x^5$. Since $x \geq 1$, we have $1/x \leq 1$ and so

$$|f^{(4)}(x)| = \left|\frac{24}{x^5}\right| \leq 24$$

■ Many calculators and computer algebra systems have a built-in algorithm that computes an approximation of a definite integral. Some of these machines use Simpson's Rule; others use more sophisticated techniques such as *adaptive* numerical integration. This means that if a function fluctuates much more on a certain part of the interval than it does elsewhere, then that part gets divided into more subintervals. This strategy reduces the number of calculations required to achieve a prescribed accuracy.

Therefore we can take $K = 24$ in (4). Thus, for an error less than 0.0001, we should choose n so that

$$\frac{24(1)^5}{180n^4} < 0.0001$$

This gives

$$n^4 > \frac{24}{180(0.0001)}$$

or

$$n > \frac{1}{\sqrt[4]{0.00075}} \approx 6.04$$

Therefore $n = 8$ (n must be even) gives the desired accuracy. (Compare this with Example 2, where we obtained $n = 41$ for the Trapezoidal Rule and $n = 29$ for the Midpoint Rule.) □

EXAMPLE 7

(a) Use Simpson's Rule with $n = 10$ to approximate the integral $\int_0^1 e^{x^2}\,dx$.

(b) Estimate the error involved in this approximation.

SOLUTION

(a) If $n = 10$, then $\Delta x = 0.1$ and Simpson's Rule gives

■ Figure 10 illustrates the calculation in Example 7. Notice that the parabolic arcs are so close to the graph of $y = e^{x^2}$ that they are practically indistinguishable from it.

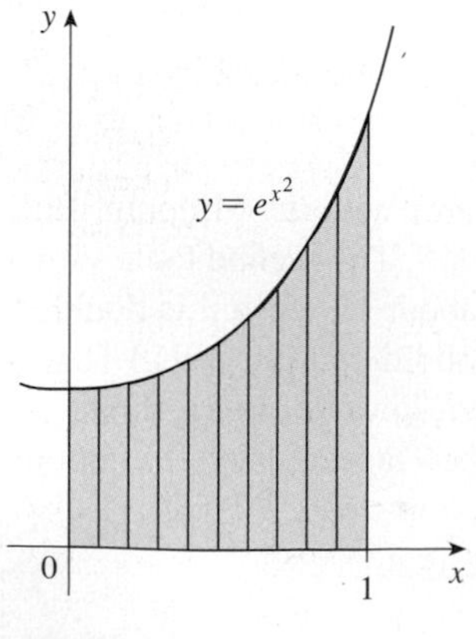

FIGURE 10

$$\int_0^1 e^{x^2}\,dx \approx \frac{\Delta x}{3}[f(0) + 4f(0.1) + 2f(0.2) + \cdots + 2f(0.8) + 4f(0.9) + f(1)]$$

$$= \frac{0.1}{3}[e^0 + 4e^{0.01} + 2e^{0.04} + 4e^{0.09} + 2e^{0.16} + 4e^{0.25} + 2e^{0.36}$$

$$+ 4e^{0.49} + 2e^{0.64} + 4e^{0.81} + e^1]$$

$$\approx 1.462681$$

(b) The fourth derivative of $f(x) = e^{x^2}$ is

$$f^{(4)}(x) = (12 + 48x^2 + 16x^4)e^{x^2}$$

and so, since $0 \leq x \leq 1$, we have

$$0 \leq f^{(4)}(x) \leq (12 + 48 + 16)e^1 = 76e$$

Therefore, putting $K = 76e$, $a = 0$, $b = 1$, and $n = 10$ in (4), we see that the error is at most

$$\frac{76e(1)^5}{180(10)^4} \approx 0.000115$$

(Compare this with Example 3.) Thus, correct to three decimal places, we have

$$\int_0^1 e^{x^2}\,dx \approx 1.463$$

□

7.7 EXERCISES

1. Let $I = \int_0^4 f(x)\,dx$, where f is the function whose graph is shown.
   (a) Use the graph to find L_2, R_2, and M_2.
   (b) Are these underestimates or overestimates of I?
   (c) Use the graph to find T_2. How does it compare with I?
   (d) For any value of n, list the numbers L_n, R_n, M_n, T_n, and I in increasing order.

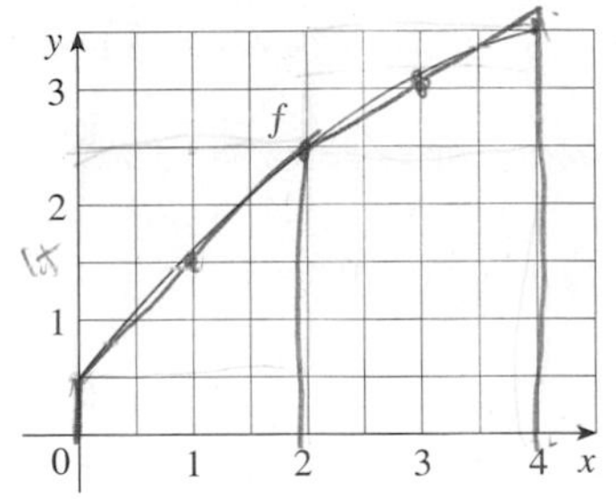

2. The left, right, Trapezoidal, and Midpoint Rule approximations were used to estimate $\int_0^2 f(x)\,dx$, where f is the function whose graph is shown. The estimates were 0.7811, 0.8675, 0.8632, and 0.9540, and the same number of subintervals were used in each case.
   (a) Which rule produced which estimate?
   (b) Between which two approximations does the true value of $\int_0^2 f(x)\,dx$ lie?

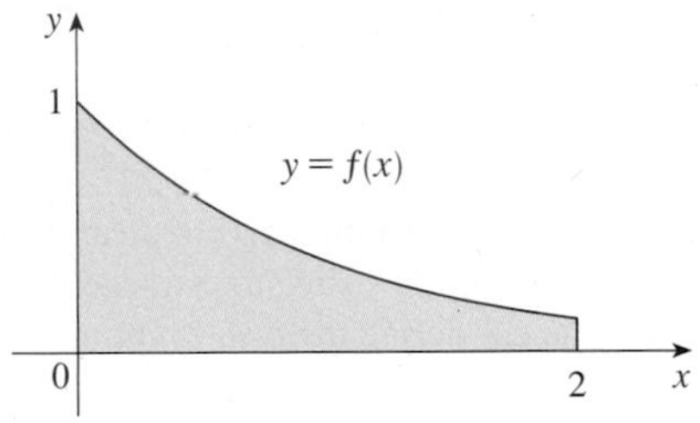

3. Estimate $\int_0^1 \cos(x^2)\,dx$ using (a) the Trapezoidal Rule and (b) the Midpoint Rule, each with $n = 4$. From a graph of the integrand, decide whether your answers are underestimates or overestimates. What can you conclude about the true value of the integral?

4. Draw the graph of $f(x) = \sin\left(\frac{1}{2}x^2\right)$ in the viewing rectangle $[0, 1]$ by $[0, 0.5]$ and let $I = \int_0^1 f(x)\,dx$.
   (a) Use the graph to decide whether L_2, R_2, M_2, and T_2 underestimate or overestimate I.
   (b) For any value of n, list the numbers L_n, R_n, M_n, T_n, and I in increasing order.
   (c) Compute L_5, R_5, M_5, and T_5. From the graph, which do you think gives the best estimate of I?

5–6 Use (a) the Midpoint Rule and (b) Simpson's Rule to approximate the given integral with the specified value of n. (Round your answers to six decimal places.) Compare your results to the actual value to determine the error in each approximation.

5. $\int_0^\pi x^2 \sin x\,dx$, $n = 8$

6. $\int_0^1 e^{-\sqrt{x}}\,dx$, $n = 6$

7–18 Use (a) the Trapezoidal Rule, (b) the Midpoint Rule, and (c) Simpson's Rule to approximate the given integral with the specified value of n. (Round your answers to six decimal places.)

7. $\int_0^2 \sqrt[4]{1 + x^2}\,dx$, $n = 8$

8. $\int_0^{1/2} \sin(x^2)\,dx$, $n = 4$

9. $\int_1^2 \frac{\ln x}{1 + x}\,dx$, $n = 10$

10. $\int_0^3 \frac{dt}{1 + t^2 + t^4}$, $n = 6$

11. $\int_0^{1/2} \sin(e^{t/2})\,dt$, $n = 8$

12. $\int_0^4 \sqrt{1 + \sqrt{x}}\,dx$, $n = 8$

13. $\int_0^4 e^{\sqrt{t}} \sin t\,dt$, $n = 8$

14. $\int_0^1 \sqrt{z}\,e^{-z}\,dz$, $n = 10$

15. $\int_1^5 \frac{\cos x}{x}\,dx$, $n = 8$

16. $\int_4^6 \ln(x^3 + 2)\,dx$, $n = 10$

17. $\int_0^3 \frac{1}{1 + y^5}\,dy$, $n = 6$

18. $\int_0^4 \cos\sqrt{x}\,dx$, $n = 10$

---

19. (a) Find the approximations T_8 and M_8 for the integral $\int_0^1 \cos(x^2)\,dx$.
    (b) Estimate the errors in the approximations of part (a).
    (c) How large do we have to choose n so that the approximations T_n and M_n to the integral in part (a) are accurate to within 0.0001?

20. (a) Find the approximations T_{10} and M_{10} for $\int_1^2 e^{1/x}\,dx$.
    (b) Estimate the errors in the approximations of part (a).
    (c) How large do we have to choose n so that the approximations T_n and M_n to the integral in part (a) are accurate to within 0.0001?

21. (a) Find the approximations T_{10}, M_{10}, and S_{10} for $\int_0^\pi \sin x\,dx$ and the corresponding errors E_T, E_M, and E_S.
    (b) Compare the actual errors in part (a) with the error estimates given by (3) and (4).
    (c) How large do we have to choose n so that the approximations T_n, M_n, and S_n to the integral in part (a) are accurate to within 0.00001?

22. How large should n be to guarantee that the Simpson's Rule approximation to $\int_0^1 e^{x^2}\,dx$ is accurate to within 0.00001?

CAS 23. The trouble with the error estimates is that it is often very difficult to compute four derivatives and obtain a good upper bound K for $|f^{(4)}(x)|$ by hand. But computer algebra systems have no problem computing $f^{(4)}$ and graphing it, so we can easily find a value for K from a machine graph. This exercise deals with approximations to the integral $I = \int_0^{2\pi} f(x)\,dx$, where $f(x) = e^{\cos x}$.

(a) Use a graph to get a good upper bound for $|f''(x)|$.
(b) Use M_{10} to approximate I.
(c) Use part (a) to estimate the error in part (b).
(d) Use the built-in numerical integration capability of your CAS to approximate I.
(e) How does the actual error compare with the error estimate in part (c)?
(f) Use a graph to get a good upper bound for $|f^{(4)}(x)|$.
(g) Use S_{10} to approximate I.
(h) Use part (f) to estimate the error in part (g).
(i) How does the actual error compare with the error estimate in part (h)?
(j) How large should n be to guarantee that the size of the error in using S_n is less than 0.0001?

CAS **24.** Repeat Exercise 23 for the integral $\int_{-1}^{1} \sqrt{4 - x^3}\,dx$.

25–26 Find the approximations L_n, R_n, T_n, and M_n for $n = 5$, 10, and 20. Then compute the corresponding errors E_L, E_R, E_T, and E_M. (Round your answers to six decimal places. You may wish to use the sum command on a computer algebra system.) What observations can you make? In particular, what happens to the errors when n is doubled?

25. $\int_0^1 xe^x\,dx$

26. $\int_1^2 \frac{1}{x^2}\,dx$

27–28 Find the approximations T_n, M_n, and S_n for $n = 6$ and 12. Then compute the corresponding errors E_T, E_M, and E_S. (Round your answers to six decimal places. You may wish to use the sum command on a computer algebra system.) What observations can you make? In particular, what happens to the errors when n is doubled?

27. $\int_0^2 x^4\,dx$

28. $\int_1^4 \frac{1}{\sqrt{x}}\,dx$

29. Estimate the area under the graph in the figure by using (a) the Trapezoidal Rule, (b) the Midpoint Rule, and (c) Simpson's Rule, each with $n = 6$.

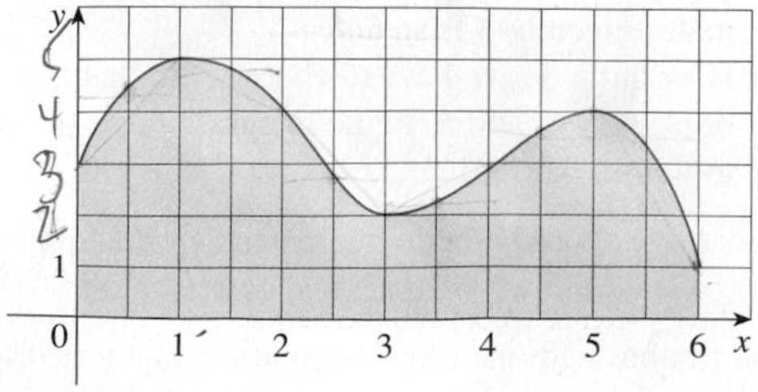

30. The widths (in meters) of a kidney-shaped swimming pool were measured at 2-meter intervals as indicated in the figure. Use Simpson's Rule to estimate the area of the pool.

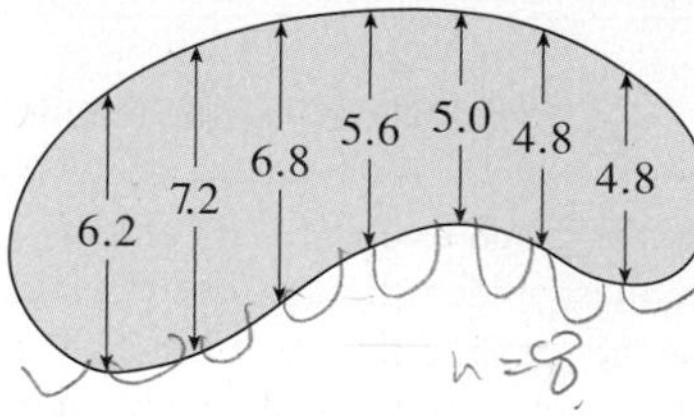

31. (a) Use the Midpoint Rule and the given data to estimate the value of the integral $\int_0^{3.2} f(x)\,dx$.

x	$f(x)$	x	$f(x)$
0.0	6.8	2.0	7.6
0.4	6.5	2.4	8.4
0.8	6.3	2.8	8.8
1.2	6.4	3.2	9.0
1.6	6.9		

(b) If it is known that $-4 \leqslant f''(x) \leqslant 1$ for all x, estimate the error involved in the approximation in part (a).

32. A radar gun was used to record the speed of a runner during the first 5 seconds of a race (see the table). Use Simpson's Rule to estimate the distance the runner covered during those 5 seconds.

t (s)	v (m/s)	t (s)	v (m/s)
0	0	3.0	10.51
0.5	4.67	3.5	10.67
1.0	7.34	4.0	10.76
1.5	8.86	4.5	10.81
2.0	9.73	5.0	10.81
2.5	10.22		

33. The graph of the acceleration $a(t)$ of a car measured in ft/s^2 is shown. Use Simpson's Rule to estimate the increase in the velocity of the car during the 6-second time interval.

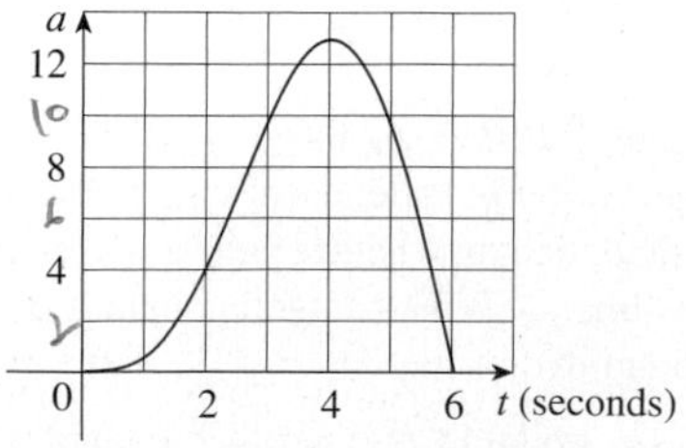

34. Water leaked from a tank at a rate of $r(t)$ liters per hour, where the graph of r is as shown. Use Simpson's Rule to estimate the total amount of water that leaked out during the first 6 hours.

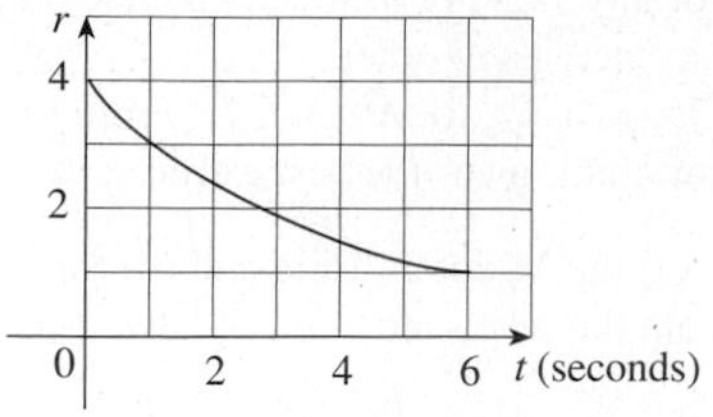

35. The table (supplied by San Diego Gas and Electric) gives the power consumption P in megawatts in San Diego County from midnight to 6:00 AM on December 8, 1999. Use Simpson's Rule to estimate the energy used during that time period. (Use the fact that power is the derivative of energy.)

t	P	t	P
0:00	1814	3:30	1611
0:30	1735	4:00	1621
1:00	1686	4:30	1666
1:30	1646	5:00	1745
2:00	1637	5:30	1886
2:30	1609	6:00	2052
3:00	1604		

36. Shown is the graph of traffic on an Internet service provider's T1 data line from midnight to 8:00 AM. D is the data throughput, measured in megabits per second. Use Simpson's Rule to estimate the total amount of data transmitted during that time period.

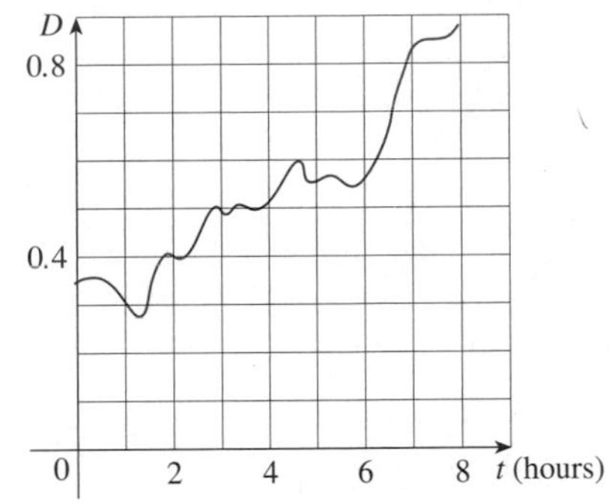

37. If the region shown in the figure is rotated about the y-axis to form a solid, use Simpson's Rule with $n = 8$ to estimate the volume of the solid.

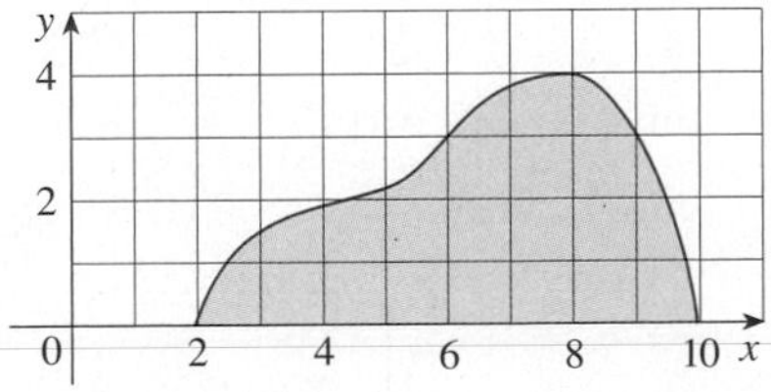

38. The table shows values of a force function $f(x)$, where x is measured in meters and $f(x)$ in newtons. Use Simpson's Rule to estimate the work done by the force in moving an object a distance of 18 m.

x	0	3	6	9	12	15	18
$f(x)$	9.8	9.1	8.5	8.0	7.7	7.5	7.4

39. The region bounded by the curves $y = e^{-1/x}$, $y = 0$, $x = 1$, and $x = 5$ is rotated about the x-axis. Use Simpson's Rule with $n = 8$ to estimate the volume of the resulting solid.

CAS 40. The figure shows a pendulum with length L that makes a maximum angle θ_0 with the vertical. Using Newton's Second Law, it can be shown that the period T (the time for one complete swing) is given by

$$T = 4\sqrt{\frac{L}{g}} \int_0^{\pi/2} \frac{dx}{\sqrt{1 - k^2 \sin^2 x}}$$

where $k = \sin(\frac{1}{2}\theta_0)$ and g is the acceleration due to gravity. If $L = 1$ m and $\theta_0 = 42°$, use Simpson's Rule with $n = 10$ to find the period.

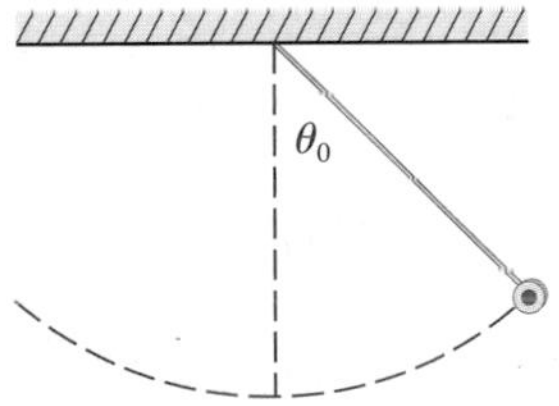

41. The intensity of light with wavelength λ traveling through a diffraction grating with N slits at an angle θ is given by $I(\theta) = N^2 \sin^2 k/k^2$, where $k = (\pi N d \sin\theta)/\lambda$ and d is the distance between adjacent slits. A helium-neon laser with wavelength $\lambda = 632.8 \times 10^{-9}$ m is emitting a narrow band of light, given by $-10^{-6} < \theta < 10^{-6}$, through a grating with 10,000 slits spaced 10^{-4} m apart. Use the Midpoint Rule with $n = 10$ to estimate the total light intensity $\int_{-10^{-6}}^{10^{-6}} I(\theta)\, d\theta$ emerging from the grating.

42. Use the Trapezoidal Rule with $n = 10$ to approximate $\int_0^{20} \cos(\pi x)\, dx$. Compare your result to the actual value. Can you explain the discrepancy?

43. Sketch the graph of a continuous function on $[0, 2]$ for which the Trapezoidal Rule with $n = 2$ is more accurate than the Midpoint Rule.

44. Sketch the graph of a continuous function on $[0, 2]$ for which the right endpoint approximation with $n = 2$ is more accurate than Simpson's Rule.

45. If f is a positive function and $f''(x) < 0$ for $a \leqslant x \leqslant b$, show that

$$T_n < \int_a^b f(x)\, dx < M_n$$

46. Show that if f is a polynomial of degree 3 or lower, then Simpson's Rule gives the exact value of $\int_a^b f(x)\, dx$.

47. Show that $\frac{1}{2}(T_n + M_n) = T_{2n}$.

48. Show that $\frac{1}{3}T_n + \frac{2}{3}M_n = S_{2n}$.

7.8 IMPROPER INTEGRALS

In defining a definite integral $\int_a^b f(x)\,dx$ we dealt with a function f defined on a finite interval $[a, b]$ and we assumed that f does not have an infinite discontinuity (see Section 5.2). In this section we extend the concept of a definite integral to the case where the interval is infinite and also to the case where f has an infinite discontinuity in $[a, b]$. In either case the integral is called an *improper* integral. One of the most important applications of this idea, probability distributions, will be studied in Section 8.5.

TYPE 1: INFINITE INTERVALS

Consider the infinite region S that lies under the curve $y = 1/x^2$, above the x-axis, and to the right of the line $x = 1$. You might think that, since S is infinite in extent, its area must be infinite, but let's take a closer look. The area of the part of S that lies to the left of the line $x = t$ (shaded in Figure 1) is

$$A(t) = \int_1^t \frac{1}{x^2}\,dx = -\frac{1}{x}\Big]_1^t = 1 - \frac{1}{t}$$

Notice that $A(t) < 1$ no matter how large t is chosen.

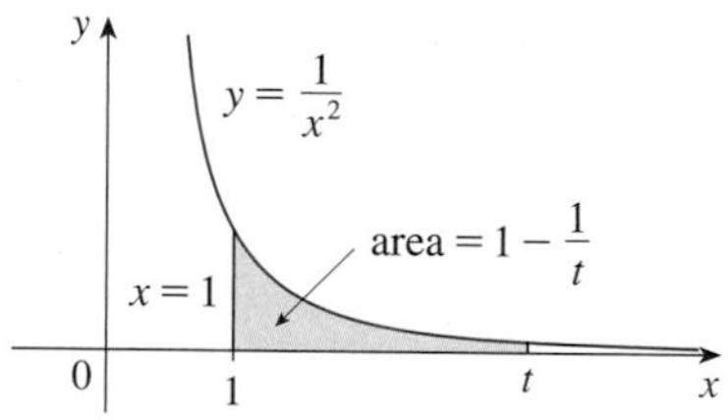

FIGURE 1

We also observe that

$$\lim_{t \to \infty} A(t) = \lim_{t \to \infty} \left(1 - \frac{1}{t}\right) = 1$$

The area of the shaded region approaches 1 as $t \to \infty$ (see Figure 2), so we say that the area of the infinite region S is equal to 1 and we write

$$\int_1^\infty \frac{1}{x^2}\,dx = \lim_{t \to \infty} \int_1^t \frac{1}{x^2}\,dx = 1$$

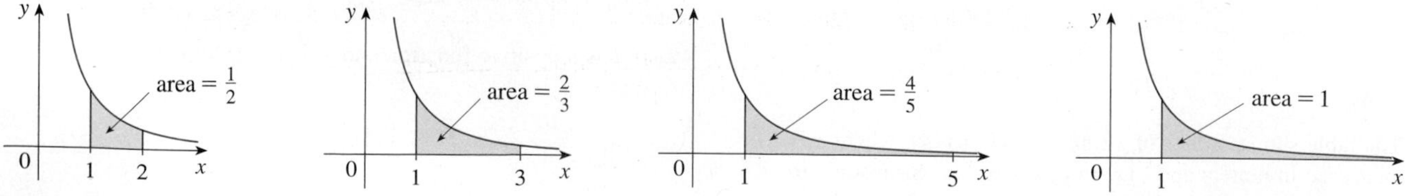

FIGURE 2

Using this example as a guide, we define the integral of f (not necessarily a positive function) over an infinite interval as the limit of integrals over finite intervals.

1 DEFINITION OF AN IMPROPER INTEGRAL OF TYPE 1

(a) If $\int_a^t f(x)\,dx$ exists for every number $t \geqslant a$, then

$$\int_a^\infty f(x)\,dx = \lim_{t\to\infty} \int_a^t f(x)\,dx$$

provided this limit exists (as a finite number).

(b) If $\int_t^b f(x)\,dx$ exists for every number $t \leqslant b$, then

$$\int_{-\infty}^b f(x)\,dx = \lim_{t\to-\infty} \int_t^b f(x)\,dx$$

provided this limit exists (as a finite number).

The improper integrals $\int_a^\infty f(x)\,dx$ and $\int_{-\infty}^b f(x)\,dx$ are called **convergent** if the corresponding limit exists and **divergent** if the limit does not exist.

(c) If both $\int_a^\infty f(x)\,dx$ and $\int_{-\infty}^a f(x)\,dx$ are convergent, then we define

$$\int_{-\infty}^\infty f(x)\,dx = \int_{-\infty}^a f(x)\,dx + \int_a^\infty f(x)\,dx$$

In part (c) any real number a can be used (see Exercise 74).

Any of the improper integrals in Definition 1 can be interpreted as an area provided that f is a positive function. For instance, in case (a) if $f(x) \geqslant 0$ and the integral $\int_a^\infty f(x)\,dx$ is convergent, then we define the area of the region $S = \{(x, y) \mid x \geqslant a, 0 \leqslant y \leqslant f(x)\}$ in Figure 3 to be

$$A(S) = \int_a^\infty f(x)\,dx$$

This is appropriate because $\int_a^\infty f(x)\,dx$ is the limit as $t \to \infty$ of the area under the graph of f from a to t.

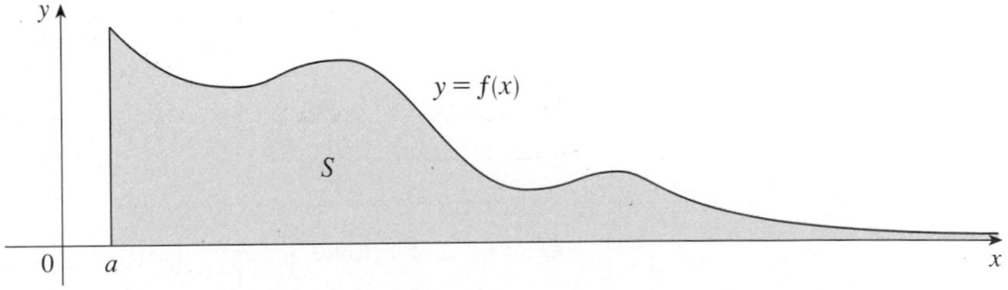

FIGURE 3

EXAMPLE 1 Determine whether the integral $\int_1^\infty (1/x)\,dx$ is convergent or divergent.

SOLUTION According to part (a) of Definition 1, we have

$$\int_1^\infty \frac{1}{x}\,dx = \lim_{t\to\infty} \int_1^t \frac{1}{x}\,dx = \lim_{t\to\infty} \ln|x|\Big]_1^t$$

$$= \lim_{t\to\infty} (\ln t - \ln 1) = \lim_{t\to\infty} \ln t = \infty$$

The limit does not exist as a finite number and so the improper integral $\int_1^\infty (1/x)\,dx$ is divergent. □

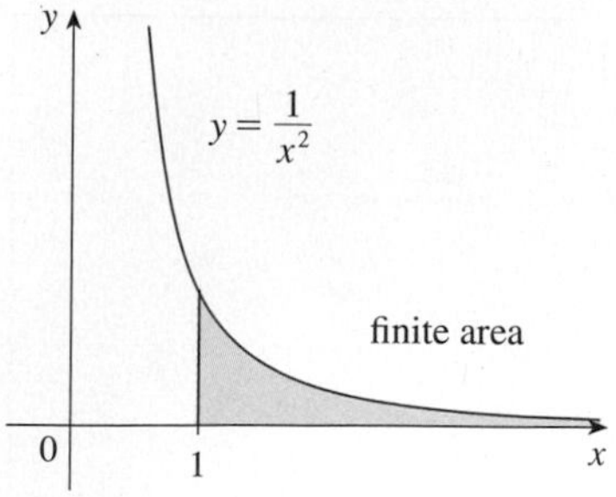

FIGURE 4

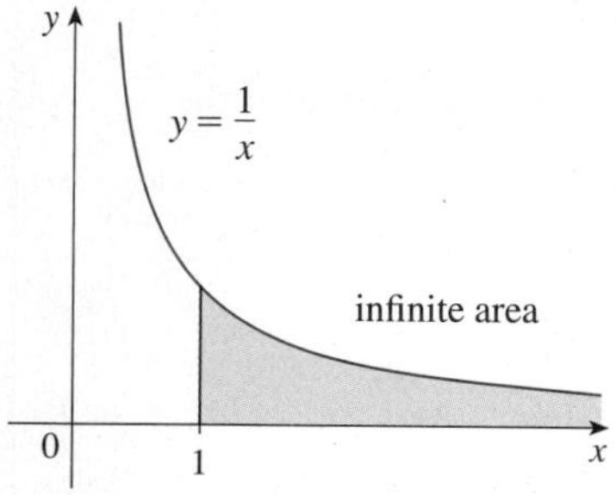

FIGURE 5

TEC In Module 7.8 you can investigate visually and numerically whether several improper integrals are convergent or divergent.

Let's compare the result of Example 1 with the example given at the beginning of this section:

$$\int_1^\infty \frac{1}{x^2}\,dx \text{ converges} \qquad \int_1^\infty \frac{1}{x}\,dx \text{ diverges}$$

Geometrically, this says that although the curves $y = 1/x^2$ and $y = 1/x$ look very similar for $x > 0$, the region under $y = 1/x^2$ to the right of $x = 1$ (the shaded region in Figure 4) has finite area whereas the corresponding region under $y = 1/x$ (in Figure 5) has infinite area. Note that both $1/x^2$ and $1/x$ approach 0 as $x \to \infty$ but $1/x^2$ approaches 0 faster than $1/x$. The values of $1/x$ don't decrease fast enough for its integral to have a finite value.

EXAMPLE 2 Evaluate $\int_{-\infty}^0 xe^x\,dx$.

SOLUTION Using part (b) of Definition 1, we have

$$\int_{-\infty}^0 xe^x\,dx = \lim_{t\to-\infty} \int_t^0 xe^x\,dx$$

We integrate by parts with $u = x$, $dv = e^x\,dx$ so that $du = dx$, $v = e^x$:

$$\int_t^0 xe^x\,dx = xe^x\Big]_t^0 - \int_t^0 e^x\,dx$$

$$= -te^t - 1 + e^t$$

We know that $e^t \to 0$ as $t \to -\infty$, and by l'Hospital's Rule we have

$$\lim_{t\to-\infty} te^t = \lim_{t\to-\infty} \frac{t}{e^{-t}} = \lim_{t\to-\infty} \frac{1}{-e^{-t}}$$

$$= \lim_{t\to-\infty} (-e^t) = 0$$

Therefore

$$\int_{-\infty}^0 xe^x\,dx = \lim_{t\to-\infty} (-te^t - 1 + e^t)$$

$$= -0 - 1 + 0 = -1$$

□

EXAMPLE 3 Evaluate $\int_{-\infty}^\infty \frac{1}{1+x^2}\,dx$.

SOLUTION It's convenient to choose $a = 0$ in Definition 1(c):

$$\int_{-\infty}^\infty \frac{1}{1+x^2}\,dx = \int_{-\infty}^0 \frac{1}{1+x^2}\,dx + \int_0^\infty \frac{1}{1+x^2}\,dx$$

We must now evaluate the integrals on the right side separately:

$$\int_0^\infty \frac{1}{1+x^2}\,dx = \lim_{t\to\infty} \int_0^t \frac{dx}{1+x^2} = \lim_{t\to\infty} \tan^{-1}x\Big]_0^t$$

$$= \lim_{t\to\infty} (\tan^{-1}t - \tan^{-1}0) = \lim_{t\to\infty} \tan^{-1}t = \frac{\pi}{2}$$

$$\int_{-\infty}^{0} \frac{1}{1+x^2}\,dx = \lim_{t\to-\infty}\int_t^0 \frac{dx}{1+x^2} = \lim_{t\to-\infty} \tan^{-1}x\Big]_t^0$$

$$= \lim_{t\to-\infty}(\tan^{-1}0 - \tan^{-1}t)$$

$$= 0 - \left(-\frac{\pi}{2}\right) = \frac{\pi}{2}$$

Since both of these integrals are convergent, the given integral is convergent and

$$\int_{-\infty}^{\infty} \frac{1}{1+x^2}\,dx = \frac{\pi}{2} + \frac{\pi}{2} = \pi$$

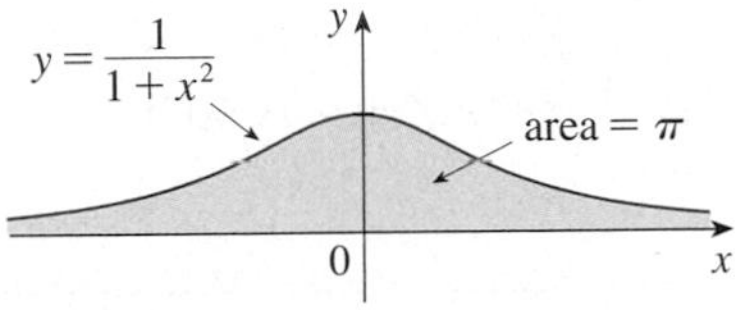

FIGURE 6

Since $1/(1+x^2) > 0$, the given improper integral can be interpreted as the area of the infinite region that lies under the curve $y = 1/(1+x^2)$ and above the x-axis (see Figure 6). □

EXAMPLE 4 For what values of p is the integral

$$\int_1^{\infty} \frac{1}{x^p}\,dx$$

convergent?

SOLUTION We know from Example 1 that if $p = 1$, then the integral is divergent, so let's assume that $p \neq 1$. Then

$$\int_1^{\infty} \frac{1}{x^p}\,dx = \lim_{t\to\infty}\int_1^t x^{-p}\,dx$$

$$= \lim_{t\to\infty} \frac{x^{-p+1}}{-p+1}\Big]_{x=1}^{x=t}$$

$$= \lim_{t\to\infty} \frac{1}{1-p}\left[\frac{1}{t^{p-1}} - 1\right]$$

If $p > 1$, then $p - 1 > 0$, so as $t \to \infty$, $t^{p-1} \to \infty$ and $1/t^{p-1} \to 0$. Therefore

$$\int_1^{\infty} \frac{1}{x^p}\,dx = \frac{1}{p-1} \qquad \text{if } p > 1$$

and so the integral converges. But if $p < 1$, then $p - 1 < 0$ and so

$$\frac{1}{t^{p-1}} = t^{1-p} \to \infty \qquad \text{as } t \to \infty$$

and the integral diverges. □

We summarize the result of Example 4 for future reference:

> **2** $\displaystyle\int_1^{\infty} \frac{1}{x^p}\,dx$ is convergent if $p > 1$ and divergent if $p \leq 1$.

TYPE 2: DISCONTINUOUS INTEGRANDS

Suppose that f is a positive continuous function defined on a finite interval $[a, b)$ but has a vertical asymptote at b. Let S be the unbounded region under the graph of f and above the x-axis between a and b. (For Type 1 integrals, the regions extended indefinitely in a

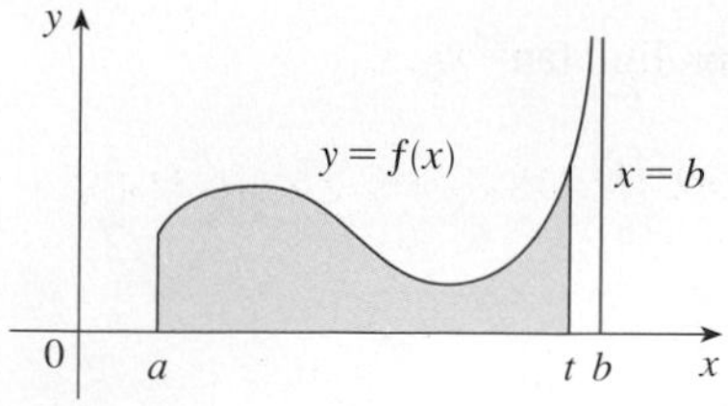

FIGURE 7

horizontal direction. Here the region is infinite in a vertical direction.) The area of the part of S between a and t (the shaded region in Figure 7) is

$$A(t) = \int_a^t f(x)\, dx$$

If it happens that $A(t)$ approaches a definite number A as $t \to b^-$, then we say that the area of the region S is A and we write

$$\int_a^b f(x)\, dx = \lim_{t \to b^-} \int_a^t f(x)\, dx$$

We use this equation to define an improper integral of Type 2 even when f is not a positive function, no matter what type of discontinuity f has at b.

■ Parts (b) and (c) of Definition 3 are illustrated in Figures 8 and 9 for the case where $f(x) \geq 0$ and f has vertical asymptotes at a and c, respectively.

3 DEFINITION OF AN IMPROPER INTEGRAL OF TYPE 2

(a) If f is continuous on $[a, b)$ and is discontinuous at b, then

$$\int_a^b f(x)\, dx = \lim_{t \to b^-} \int_a^t f(x)\, dx$$

if this limit exists (as a finite number).

(b) If f is continuous on $(a, b]$ and is discontinuous at a, then

$$\int_a^b f(x)\, dx = \lim_{t \to a^+} \int_t^b f(x)\, dx$$

if this limit exists (as a finite number).

The improper integral $\int_a^b f(x)\, dx$ is called **convergent** if the corresponding limit exists and **divergent** if the limit does not exist.

(c) If f has a discontinuity at c, where $a < c < b$, and both $\int_a^c f(x)\, dx$ and $\int_c^b f(x)\, dx$ are convergent, then we define

$$\int_a^b f(x)\, dx = \int_a^c f(x)\, dx + \int_c^b f(x)\, dx$$

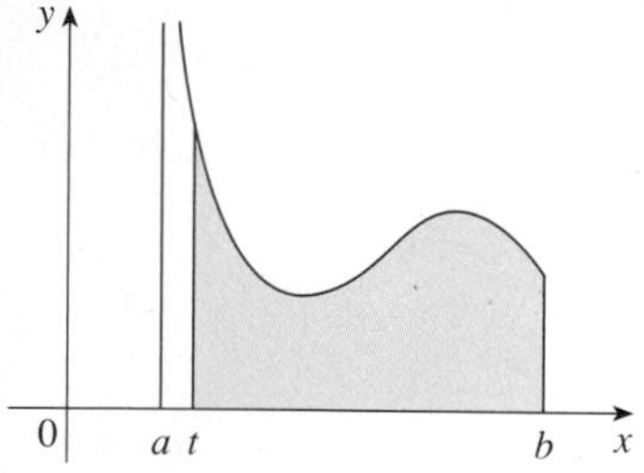

FIGURE 8

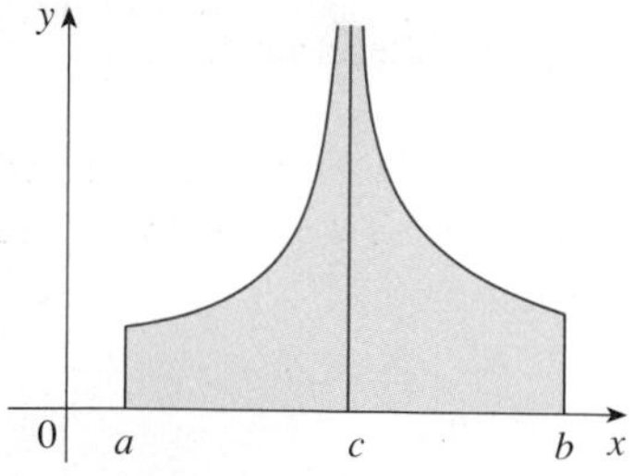

FIGURE 9

EXAMPLE 5 Find $\displaystyle\int_2^5 \frac{1}{\sqrt{x-2}}\, dx$.

SOLUTION We note first that the given integral is improper because $f(x) = 1/\sqrt{x-2}$ has the vertical asymptote $x = 2$. Since the infinite discontinuity occurs at the left endpoint of $[2, 5]$, we use part (b) of Definition 3:

$$\begin{aligned}\int_2^5 \frac{dx}{\sqrt{x-2}} &= \lim_{t \to 2^+} \int_t^5 \frac{dx}{\sqrt{x-2}}\\ &= \lim_{t \to 2^+} 2\sqrt{x-2}\,\Big]_t^5\\ &= \lim_{t \to 2^+} 2\left(\sqrt{3} - \sqrt{t-2}\right)\\ &= 2\sqrt{3}\end{aligned}$$

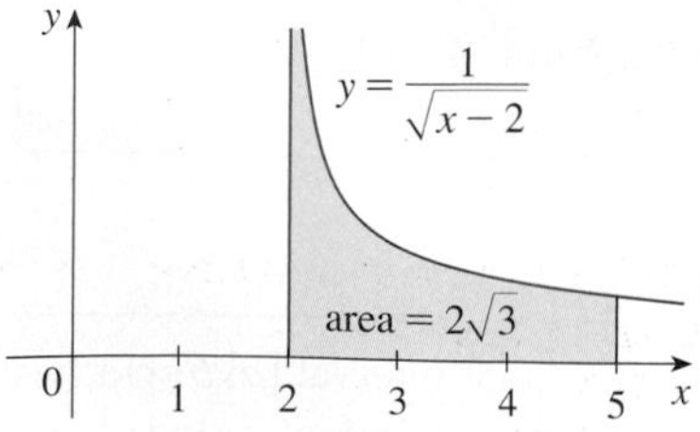

FIGURE 10

Thus the given improper integral is convergent and, since the integrand is positive, we can interpret the value of the integral as the area of the shaded region in Figure 10. □

EXAMPLE 6 Determine whether $\int_0^{\pi/2} \sec x\,dx$ converges or diverges.

SOLUTION Note that the given integral is improper because $\lim_{x\to(\pi/2)^-} \sec x = \infty$. Using part (a) of Definition 3 and Formula 14 from the Table of Integrals, we have

$$\int_0^{\pi/2} \sec x\,dx = \lim_{t\to(\pi/2)^-} \int_0^t \sec x\,dx = \lim_{t\to(\pi/2)^-} \ln|\sec x + \tan x|\Big]_0^t$$

$$= \lim_{t\to(\pi/2)^-} [\ln(\sec t + \tan t) - \ln 1] = \infty$$

because $\sec t \to \infty$ and $\tan t \to \infty$ as $t \to (\pi/2)^-$. Thus the given improper integral is divergent. □

EXAMPLE 7 Evaluate $\displaystyle\int_0^3 \frac{dx}{x-1}$ if possible.

SOLUTION Observe that the line $x = 1$ is a vertical asymptote of the integrand. Since it occurs in the middle of the interval $[0, 3]$, we must use part (c) of Definition 3 with $c = 1$:

$$\int_0^3 \frac{dx}{x-1} = \int_0^1 \frac{dx}{x-1} + \int_1^3 \frac{dx}{x-1}$$

where

$$\int_0^1 \frac{dx}{x-1} = \lim_{t\to1^-} \int_0^t \frac{dx}{x-1} = \lim_{t\to1^-} \ln|x-1|\Big]_0^t$$

$$= \lim_{t\to1^-} \left(\ln|t-1| - \ln|-1|\right)$$

$$= \lim_{t\to1^-} \ln(1-t) = -\infty$$

because $1 - t \to 0^+$ as $t \to 1^-$. Thus $\int_0^1 dx/(x-1)$ is divergent. This implies that $\int_0^3 dx/(x-1)$ is divergent. [We do not need to evaluate $\int_1^3 dx/(x-1)$.] □

WARNING If we had not noticed the asymptote $x = 1$ in Example 7 and had instead confused the integral with an ordinary integral, then we might have made the following erroneous calculation:

$$\int_0^3 \frac{dx}{x-1} = \ln|x-1|\Big]_0^3 = \ln 2 - \ln 1 = \ln 2$$

This is wrong because the integral is improper and must be calculated in terms of limits.

From now on, whenever you meet the symbol $\int_a^b f(x)\,dx$ you must decide, by looking at the function f on $[a, b]$, whether it is an ordinary definite integral or an improper integral.

EXAMPLE 8 Evaluate $\int_0^1 \ln x\,dx$.

SOLUTION We know that the function $f(x) = \ln x$ has a vertical asymptote at 0 since $\lim_{x\to0^+} \ln x = -\infty$. Thus the given integral is improper and we have

$$\int_0^1 \ln x\,dx = \lim_{t\to0^+} \int_t^1 \ln x\,dx$$

Now we integrate by parts with $u = \ln x$, $dv = dx$, $du = dx/x$, and $v = x$:

$$\begin{aligned}\int_t^1 \ln x\,dx &= x \ln x\Big]_t^1 - \int_t^1 dx \\ &= 1 \ln 1 - t \ln t - (1 - t) \\ &= -t \ln t - 1 + t\end{aligned}$$

To find the limit of the first term we use l'Hospital's Rule:

$$\lim_{t\to 0^+} t \ln t = \lim_{t\to 0^+} \frac{\ln t}{1/t} = \lim_{t\to 0^+} \frac{1/t}{-1/t^2} = \lim_{t\to 0^+} (-t) = 0$$

Therefore $$\int_0^1 \ln x\,dx = \lim_{t\to 0^+} (-t \ln t - 1 + t) = -0 - 1 + 0 = -1$$

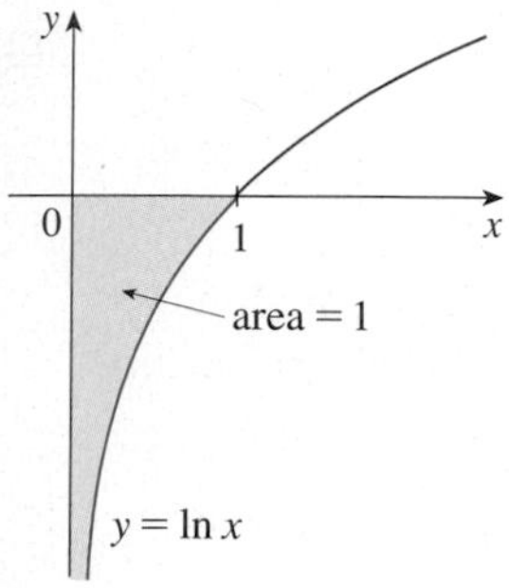

FIGURE 11

Figure 11 shows the geometric interpretation of this result. The area of the shaded region above $y = \ln x$ and below the x-axis is 1. □

A COMPARISON TEST FOR IMPROPER INTEGRALS

Sometimes it is impossible to find the exact value of an improper integral and yet it is important to know whether it is convergent or divergent. In such cases the following theorem is useful. Although we state it for Type 1 integrals, a similar theorem is true for Type 2 integrals.

> **COMPARISON THEOREM** Suppose that f and g are continuous functions with $f(x) \geq g(x) \geq 0$ for $x \geq a$.
>
> (a) If $\int_a^\infty f(x)\,dx$ is convergent, then $\int_a^\infty g(x)\,dx$ is convergent.
>
> (b) If $\int_a^\infty g(x)\,dx$ is divergent, then $\int_a^\infty f(x)\,dx$ is divergent.

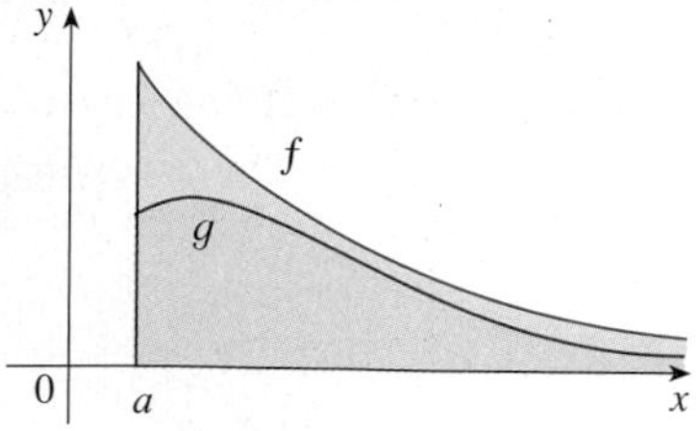

FIGURE 12

We omit the proof of the Comparison Theorem, but Figure 12 makes it seem plausible. If the area under the top curve $y = f(x)$ is finite, then so is the area under the bottom curve $y = g(x)$. And if the area under $y = g(x)$ is infinite, then so is the area under $y = f(x)$. [Note that the reverse is not necessarily true: If $\int_a^\infty g(x)\,dx$ is convergent, $\int_a^\infty f(x)\,dx$ may or may not be convergent, and if $\int_a^\infty f(x)\,dx$ is divergent, $\int_a^\infty g(x)\,dx$ may or may not be divergent.]

EXAMPLE 9 Show that $\int_0^\infty e^{-x^2} dx$ is convergent.

SOLUTION We can't evaluate the integral directly because the antiderivative of e^{-x^2} is not an elementary function (as explained in Section 7.5). We write

$$\int_0^\infty e^{-x^2} dx = \int_0^1 e^{-x^2} dx + \int_1^\infty e^{-x^2} dx$$

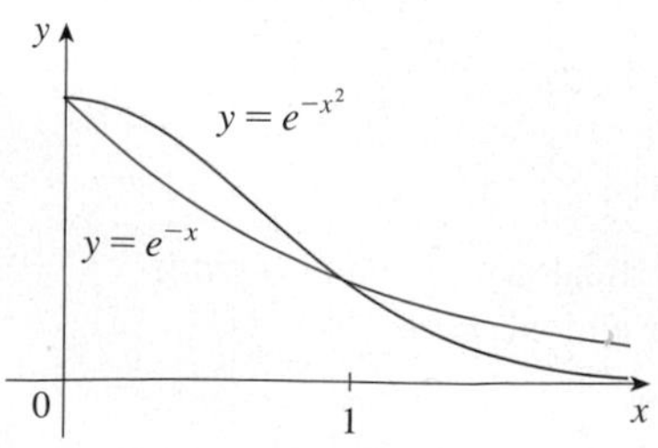

FIGURE 13

and observe that the first integral on the right-hand side is just an ordinary definite integral. In the second integral we use the fact that for $x \geq 1$ we have $x^2 \geq x$, so $-x^2 \leq -x$ and therefore $e^{-x^2} \leq e^{-x}$. (See Figure 13.) The integral of e^{-x} is easy to evaluate:

$$\int_1^\infty e^{-x} dx = \lim_{t\to\infty} \int_1^t e^{-x} dx = \lim_{t\to\infty} (e^{-1} - e^{-t}) = e^{-1}$$

Thus, taking $f(x) = e^{-x}$ and $g(x) = e^{-x^2}$ in the Comparison Theorem, we see that $\int_1^\infty e^{-x^2}\,dx$ is convergent. It follows that $\int_0^\infty e^{-x^2}\,dx$ is convergent. □

TABLE 1

t	$\int_0^t e^{-x^2}\,dx$
1	0.7468241328
2	0.8820813908
3	0.8862073483
4	0.8862269118
5	0.8862269255
6	0.8862269255

In Example 9 we showed that $\int_0^\infty e^{-x^2}\,dx$ is convergent without computing its value. In Exercise 70 we indicate how to show that its value is approximately 0.8862. In probability theory it is important to know the exact value of this improper integral, as we will see in Section 8.5; using the methods of multivariable calculus it can be shown that the exact value is $\sqrt{\pi}/2$. Table 1 illustrates the definition of an improper integral by showing how the (computer-generated) values of $\int_0^t e^{-x^2}\,dx$ approach $\sqrt{\pi}/2$ as t becomes large. In fact, these values converge quite quickly because $e^{-x^2} \to 0$ very rapidly as $x \to \infty$.

TABLE 2

t	$\int_1^t [(1 + e^{-x})/x]\,dx$
2	0.8636306042
5	1.8276735512
10	2.5219648704
100	4.8245541204
1000	7.1271392134
10000	9.4297243064

EXAMPLE 10 The integral $\displaystyle\int_1^\infty \frac{1 + e^{-x}}{x}\,dx$ is divergent by the Comparison Theorem because

$$\frac{1 + e^{-x}}{x} > \frac{1}{x}$$

and $\int_1^\infty (1/x)\,dx$ is divergent by Example 1 [or by (2) with $p = 1$]. □

Table 2 illustrates the divergence of the integral in Example 10. It appears that the values are not approaching any fixed number.

7.8 EXERCISES

1. Explain why each of the following integrals is improper.
   (a) $\displaystyle\int_1^\infty x^4 e^{-x^4}\,dx$ (b) $\displaystyle\int_0^{\pi/2} \sec x\,dx$
   (c) $\displaystyle\int_0^2 \frac{x}{x^2 - 5x + 6}\,dx$ (d) $\displaystyle\int_{-\infty}^0 \frac{1}{x^2 + 5}\,dx$

2. Which of the following integrals are improper? Why?
   (a) $\displaystyle\int_1^2 \frac{1}{2x - 1}\,dx$ (b) $\displaystyle\int_0^1 \frac{1}{2x - 1}\,dx$
   (c) $\displaystyle\int_{-\infty}^\infty \frac{\sin x}{1 + x^2}\,dx$ (d) $\displaystyle\int_1^2 \ln(x - 1)\,dx$

3. Find the area under the curve $y = 1/x^3$ from $x = 1$ to $x = t$ and evaluate it for $t = 10$, 100, and 1000. Then find the total area under this curve for $x \geq 1$.

4. (a) Graph the functions $f(x) = 1/x^{1.1}$ and $g(x) = 1/x^{0.9}$ in the viewing rectangles $[0, 10]$ by $[0, 1]$ and $[0, 100]$ by $[0, 1]$.
   (b) Find the areas under the graphs of f and g from $x = 1$ to $x = t$ and evaluate for $t = 10$, 100, 10^4, 10^6, 10^{10}, and 10^{20}.
   (c) Find the total area under each curve for $x \geq 1$, if it exists.

5–40 Determine whether each integral is convergent or divergent. Evaluate those that are convergent.

5. $\displaystyle\int_1^\infty \frac{1}{(3x + 1)^2}\,dx$

6. $\displaystyle\int_{-\infty}^0 \frac{1}{2x - 5}\,dx$

7. $\displaystyle\int_{-\infty}^{-1} \frac{1}{\sqrt{2 - w}}\,dw$

8. $\displaystyle\int_0^\infty \frac{x}{(x^2 + 2)^2}\,dx$

9. $\displaystyle\int_4^\infty e^{-y/2}\,dy$

10. $\displaystyle\int_{-\infty}^{-1} e^{-2t}\,dt$

11. $\displaystyle\int_{-\infty}^\infty \frac{x}{1 + x^2}\,dx$

12. $\displaystyle\int_{-\infty}^\infty (2 - v^4)\,dv$

13. $\displaystyle\int_{-\infty}^\infty xe^{-x^2}\,dx$

14. $\displaystyle\int_1^\infty \frac{e^{-\sqrt{x}}}{\sqrt{x}}\,dx$

15. $\displaystyle\int_{2\pi}^\infty \sin\theta\,d\theta$

16. $\displaystyle\int_{-\infty}^\infty \cos \pi t\,dt$

17. $\displaystyle\int_1^\infty \frac{x + 1}{x^2 + 2x}\,dx$

18. $\displaystyle\int_0^\infty \frac{dz}{z^2 + 3z + 2}$

19. $\displaystyle\int_0^\infty se^{-5s}\,ds$

20. $\displaystyle\int_{-\infty}^6 re^{r/3}\,dr$

21. $\displaystyle\int_1^\infty \frac{\ln x}{x}\,dx$

22. $\displaystyle\int_{-\infty}^\infty x^3 e^{-x^4}\,dx$

23. $\displaystyle\int_{-\infty}^\infty \frac{x^2}{9 + x^6}\,dx$

24. $\displaystyle\int_0^\infty \frac{e^x}{e^{2x} + 3}\,dx$

25. $\displaystyle\int_e^\infty \frac{1}{x(\ln x)^3}\,dx$

26. $\displaystyle\int_0^\infty \frac{x \arctan x}{(1 + x^2)^2}\,dx$

27. $\displaystyle\int_0^1 \frac{3}{x^5}\,dx$

28. $\displaystyle\int_2^3 \frac{1}{\sqrt{3 - x}}\,dx$

29. $\displaystyle\int_{-2}^{14} \frac{dx}{\sqrt[4]{x+2}}$

30. $\displaystyle\int_{6}^{8} \frac{4}{(x-6)^3}\,dx$

31. $\displaystyle\int_{-2}^{3} \frac{1}{x^4}\,dx$

32. $\displaystyle\int_{0}^{1} \frac{dx}{\sqrt{1-x^2}}$

33. $\displaystyle\int_{0}^{33} (x-1)^{-1/5}\,dx$

34. $\displaystyle\int_{0}^{1} \frac{1}{4y-1}\,dy$

35. $\displaystyle\int_{0}^{3} \frac{dx}{x^2-6x+5}$

36. $\displaystyle\int_{\pi/2}^{\pi} \csc x\,dx$

37. $\displaystyle\int_{-1}^{0} \frac{e^{1/x}}{x^3}\,dx$

38. $\displaystyle\int_{0}^{1} \frac{e^{1/x}}{x^3}\,dx$

39. $\displaystyle\int_{0}^{2} z^2 \ln z\,dz$

40. $\displaystyle\int_{0}^{1} \frac{\ln x}{\sqrt{x}}\,dx$

41–46 Sketch the region and find its area (if the area is finite).

41. $S = \{(x, y) \mid x \leqslant 1,\ 0 \leqslant y \leqslant e^x\}$

42. $S = \{(x, y) \mid x \geqslant -2,\ 0 \leqslant y \leqslant e^{-x/2}\}$

43. $S = \{(x, y) \mid 0 \leqslant y \leqslant 2/(x^2+9)\}$

44. $S = \{(x, y) \mid x \geqslant 0,\ 0 \leqslant y \leqslant x/(x^2+9)\}$

45. $S = \{(x, y) \mid 0 \leqslant x < \pi/2,\ 0 \leqslant y \leqslant \sec^2 x\}$

46. $S = \left\{(x, y) \mid -2 < x \leqslant 0,\ 0 \leqslant y \leqslant 1/\sqrt{x+2}\right\}$

47. (a) If $g(x) = (\sin^2 x)/x^2$, use your calculator or computer to make a table of approximate values of $\int_1^t g(x)\,dx$ for $t = 2, 5, 10, 100, 1000$, and $10{,}000$. Does it appear that $\int_1^\infty g(x)\,dx$ is convergent?
(b) Use the Comparison Theorem with $f(x) = 1/x^2$ to show that $\int_1^\infty g(x)\,dx$ is convergent.
(c) Illustrate part (b) by graphing f and g on the same screen for $1 \leqslant x \leqslant 10$. Use your graph to explain intuitively why $\int_1^\infty g(x)\,dx$ is convergent.

48. (a) If $g(x) = 1/(\sqrt{x} - 1)$, use your calculator or computer to make a table of approximate values of $\int_2^t g(x)\,dx$ for $t = 5, 10, 100, 1000$, and $10{,}000$. Does it appear that $\int_2^\infty g(x)\,dx$ is convergent or divergent?
(b) Use the Comparison Theorem with $f(x) = 1/\sqrt{x}$ to show that $\int_2^\infty g(x)\,dx$ is divergent.
(c) Illustrate part (b) by graphing f and g on the same screen for $2 \leqslant x \leqslant 20$. Use your graph to explain intuitively why $\int_2^\infty g(x)\,dx$ is divergent.

49–54 Use the Comparison Theorem to determine whether the integral is convergent or divergent.

49. $\displaystyle\int_{0}^{\infty} \frac{x}{x^3+1}\,dx$

50. $\displaystyle\int_{1}^{\infty} \frac{2+e^{-x}}{x}\,dx$

51. $\displaystyle\int_{1}^{\infty} \frac{x+1}{\sqrt{x^4-x}}\,dx$

52. $\displaystyle\int_{0}^{\infty} \frac{\arctan x}{2+e^x}\,dx$

53. $\displaystyle\int_{0}^{1} \frac{\sec^2 x}{x\sqrt{x}}\,dx$

54. $\displaystyle\int_{0}^{\pi} \frac{\sin^2 x}{\sqrt{x}}\,dx$

55. The integral

$$\int_0^\infty \frac{1}{\sqrt{x}\,(1+x)}\,dx$$

is improper for two reasons: The interval $[0, \infty)$ is infinite and the integrand has an infinite discontinuity at 0. Evaluate it by expressing it as a sum of improper integrals of Type 2 and Type 1 as follows:

$$\int_0^\infty \frac{1}{\sqrt{x}\,(1+x)}\,dx = \int_0^1 \frac{1}{\sqrt{x}\,(1+x)}\,dx + \int_1^\infty \frac{1}{\sqrt{x}\,(1+x)}\,dx$$

56. Evaluate

$$\int_2^\infty \frac{1}{x\sqrt{x^2-4}}\,dx$$

by the same method as in Exercise 55.

57–59 Find the values of p for which the integral converges and evaluate the integral for those values of p.

57. $\displaystyle\int_0^1 \frac{1}{x^p}\,dx$

58. $\displaystyle\int_e^\infty \frac{1}{x(\ln x)^p}\,dx$

59. $\displaystyle\int_0^1 x^p \ln x\,dx$

60. (a) Evaluate the integral $\int_0^\infty x^n e^{-x}\,dx$ for $n = 0, 1, 2$, and 3.
(b) Guess the value of $\int_0^\infty x^n e^{-x}\,dx$ when n is an arbitrary positive integer.
(c) Prove your guess using mathematical induction.

61. (a) Show that $\int_{-\infty}^\infty x\,dx$ is divergent.
(b) Show that

$$\lim_{t\to\infty} \int_{-t}^{t} x\,dx = 0$$

This shows that we can't define

$$\int_{-\infty}^\infty f(x)\,dx = \lim_{t\to\infty} \int_{-t}^{t} f(x)\,dx$$

62. The *average speed* of molecules in an ideal gas is

$$\bar{v} = \frac{4}{\sqrt{\pi}}\left(\frac{M}{2RT}\right)^{3/2} \int_0^\infty v^3 e^{-Mv^2/(2RT)}\,dv$$

where M is the molecular weight of the gas, R is the gas constant, T is the gas temperature, and v is the molecular speed. Show that

$$\bar{v} = \sqrt{\frac{8RT}{\pi M}}$$

63. We know from Example 1 that the region $\mathcal{R} = \{(x, y) \mid x \geq 1,\ 0 \leq y \leq 1/x\}$ has infinite area. Show that by rotating $\mathcal{R}$ about the x-axis we obtain a solid with finite volume.

64. Use the information and data in Exercises 29 and 30 of Section 6.4 to find the work required to propel a 1000-kg satellite out of the earth's gravitational field.

65. Find the *escape velocity* v_0 that is needed to propel a rocket of mass m out of the gravitational field of a planet with mass M and radius R. Use Newton's Law of Gravitation (see Exercise 29 in Section 6.4) and the fact that the initial kinetic energy of $\frac{1}{2}mv_0^2$ supplies the needed work.

66. Astronomers use a technique called *stellar stereography* to determine the density of stars in a star cluster from the observed (two-dimensional) density that can be analyzed from a photograph. Suppose that in a spherical cluster of radius R the density of stars depends only on the distance r from the center of the cluster. If the perceived star density is given by $y(s)$, where s is the observed planar distance from the center of the cluster, and $x(r)$ is the actual density, it can be shown that

$$y(s) = \int_s^R \frac{2r}{\sqrt{r^2 - s^2}}\, x(r)\, dr$$

If the actual density of stars in a cluster is $x(r) = \frac{1}{2}(R - r)^2$, find the perceived density $y(s)$.

67. A manufacturer of lightbulbs wants to produce bulbs that last about 700 hours but, of course, some bulbs burn out faster than others. Let $F(t)$ be the fraction of the company's bulbs that burn out before t hours, so $F(t)$ always lies between 0 and 1.

(a) Make a rough sketch of what you think the graph of F might look like.
(b) What is the meaning of the derivative $r(t) = F'(t)$?
(c) What is the value of $\int_0^\infty r(t)\, dt$? Why?

68. As we saw in Section 3.8, a radioactive substance decays exponentially: The mass at time t is $m(t) = m(0)e^{kt}$, where $m(0)$ is the initial mass and k is a negative constant. The *mean life* M of an atom in the substance is

$$M = -k \int_0^\infty te^{kt}\, dt$$

For the radioactive carbon isotope, ^{14}C, used in radiocarbon dating, the value of k is -0.000121. Find the mean life of a ^{14}C atom.

69. Determine how large the number a has to be so that

$$\int_a^\infty \frac{1}{x^2 + 1}\, dx < 0.001$$

70. Estimate the numerical value of $\int_0^\infty e^{-x^2}\, dx$ by writing it as the sum of $\int_0^4 e^{-x^2}\, dx$ and $\int_4^\infty e^{-x^2}\, dx$. Approximate the first integral by using Simpson's Rule with $n = 8$ and show that the second integral is smaller than $\int_4^\infty e^{-4x}\, dx$, which is less than 0.0000001.

71. If $f(t)$ is continuous for $t \geq 0$, the *Laplace transform* of f is the function F defined by

$$F(s) = \int_0^\infty f(t)e^{-st}\, dt$$

and the domain of F is the set consisting of all numbers s for which the integral converges. Find the Laplace transforms of the following functions.

(a) $f(t) = 1$ (b) $f(t) = e^t$ (c) $f(t) = t$

72. Show that if $0 \leq f(t) \leq Me^{at}$ for $t \geq 0$, where M and a are constants, then the Laplace transform $F(s)$ exists for $s > a$.

73. Suppose that $0 \leq f(t) \leq Me^{at}$ and $0 \leq f'(t) \leq Ke^{at}$ for $t \geq 0$, where f' is continuous. If the Laplace transform of $f(t)$ is $F(s)$ and the Laplace transform of $f'(t)$ is $G(s)$, show that

$$G(s) = sF(s) - f(0) \qquad s > a$$

74. If $\int_{-\infty}^\infty f(x)\, dx$ is convergent and a and b are real numbers, show that

$$\int_{-\infty}^a f(x)\, dx + \int_a^\infty f(x)\, dx = \int_{-\infty}^b f(x)\, dx + \int_b^\infty f(x)\, dx$$

75. Show that $\int_0^\infty x^2 e^{-x^2}\, dx = \frac{1}{2}\int_0^\infty e^{-x^2}\, dx$.

76. Show that $\int_0^\infty e^{-x^2}\, dx = \int_0^1 \sqrt{-\ln y}\, dy$ by interpreting the integrals as areas.

77. Find the value of the constant C for which the integral

$$\int_0^\infty \left(\frac{1}{\sqrt{x^2 + 4}} - \frac{C}{x + 2} \right) dx$$

converges. Evaluate the integral for this value of C.

78. Find the value of the constant C for which the integral

$$\int_0^\infty \left(\frac{x}{x^2 + 1} - \frac{C}{3x + 1} \right) dx$$

converges. Evaluate the integral for this value of C.

79. Suppose f is continuous on $[0, \infty)$ and $\lim_{x\to\infty} f(x) = 1$. Is it possible that $\int_0^\infty f(x)\, dx$ is convergent?

80. Show that if $a > -1$ and $b > a + 1$, then the following integral is convergent.

$$\int_0^\infty \frac{x^a}{1 + x^b}\, dx$$

7 REVIEW

CONCEPT CHECK

1. State the rule for integration by parts. In practice, how do you use it?

2. How do you evaluate $\int \sin^m x \cos^n x \, dx$ if m is odd? What if n is odd? What if m and n are both even?

3. If the expression $\sqrt{a^2 - x^2}$ occurs in an integral, what substitution might you try? What if $\sqrt{a^2 + x^2}$ occurs? What if $\sqrt{x^2 - a^2}$ occurs?

4. What is the form of the partial fraction expansion of a rational function $P(x)/Q(x)$ if the degree of P is less than the degree of Q and $Q(x)$ has only distinct linear factors? What if a linear factor is repeated? What if $Q(x)$ has an irreducible quadratic factor (not repeated)? What if the quadratic factor is repeated?

5. State the rules for approximating the definite integral $\int_a^b f(x)\,dx$ with the Midpoint Rule, the Trapezoidal Rule, and Simpson's Rule. Which would you expect to give the best estimate? How do you approximate the error for each rule?

6. Define the following improper integrals.
   (a) $\int_a^\infty f(x)\,dx$ (b) $\int_{-\infty}^b f(x)\,dx$ (c) $\int_{-\infty}^\infty f(x)\,dx$

7. Define the improper integral $\int_a^b f(x)\,dx$ for each of the following cases.
   (a) f has an infinite discontinuity at a.
   (b) f has an infinite discontinuity at b.
   (c) f has an infinite discontinuity at c, where $a < c < b$.

8. State the Comparison Theorem for improper integrals.

TRUE-FALSE QUIZ

Determine whether the statement is true or false. If it is true, explain why. If it is false, explain why or give an example that disproves the statement.

1. $\dfrac{x(x^2+4)}{x^2-4}$ can be put in the form $\dfrac{A}{x+2} + \dfrac{B}{x-2}$.

2. $\dfrac{x^2+4}{x(x^2-4)}$ can be put in the form $\dfrac{A}{x} + \dfrac{B}{x+2} + \dfrac{C}{x-2}$.

3. $\dfrac{x^2+4}{x^2(x-4)}$ can be put in the form $\dfrac{A}{x^2} + \dfrac{B}{x-4}$.

4. $\dfrac{x^2-4}{x(x^2+4)}$ can be put in the form $\dfrac{A}{x} + \dfrac{B}{x^2+4}$.

5. $\displaystyle\int_0^4 \frac{x}{x^2-1}\,dx = \tfrac{1}{2}\ln 15$

6. $\displaystyle\int_1^\infty \frac{1}{x^{\sqrt{2}}}\,dx$ is convergent.

7. If f is continuous, then $\int_{-\infty}^\infty f(x)\,dx = \lim_{t\to\infty} \int_{-t}^t f(x)\,dx$.

8. The Midpoint Rule is always more accurate than the Trapezoidal Rule.

9. (a) Every elementary function has an elementary derivative.
   (b) Every elementary function has an elementary antiderivative.

10. If f is continuous on $[0, \infty)$ and $\int_1^\infty f(x)\,dx$ is convergent, then $\int_0^\infty f(x)\,dx$ is convergent.

11. If f is a continuous, decreasing function on $[1, \infty)$ and $\lim_{x\to\infty} f(x) = 0$, then $\int_1^\infty f(x)\,dx$ is convergent.

12. If $\int_a^\infty f(x)\,dx$ and $\int_a^\infty g(x)\,dx$ are both convergent, then $\int_a^\infty [f(x) + g(x)]\,dx$ is convergent.

13. If $\int_a^\infty f(x)\,dx$ and $\int_a^\infty g(x)\,dx$ are both divergent, then $\int_a^\infty [f(x) + g(x)]\,dx$ is divergent.

14. If $f(x) \leqslant g(x)$ and $\int_0^\infty g(x)\,dx$ diverges, then $\int_0^\infty f(x)\,dx$ also diverges.

EXERCISES

Note: Additional practice in techniques of integration is provided in Exercises 7.5.

1–40 Evaluate the integral.

1. $\displaystyle\int_0^5 \frac{x}{x+10}\,dx$

2. $\displaystyle\int_0^5 ye^{-0.6y}\,dy$

3. $\displaystyle\int_0^{\pi/2} \frac{\cos\theta}{1+\sin\theta}\,d\theta$

4. $\displaystyle\int_1^4 \frac{dt}{(2t+1)^3}$

5. $\displaystyle\int_0^{\pi/2} \sin^3\theta\cos^2\theta\,d\theta$

6. $\displaystyle\int \frac{1}{y^2-4y-12}\,dy$

7. $\displaystyle\int \frac{\sin(\ln t)}{t}\,dt$

8. $\displaystyle\int \frac{dx}{\sqrt{e^x-1}}$

9. $\displaystyle\int_1^4 x^{3/2}\ln x\,dx$

10. $\displaystyle\int_0^1 \frac{\sqrt{\arctan x}}{1+x^2}\,dx$

■ Cover up the solution to the exam yourself first.

■ The principles of problem solving discussed on page 76.

■ The computer graphs in Figure 1 seem plausible that all of the integ example have the same value. The integrand is labeled with the corre value of n.

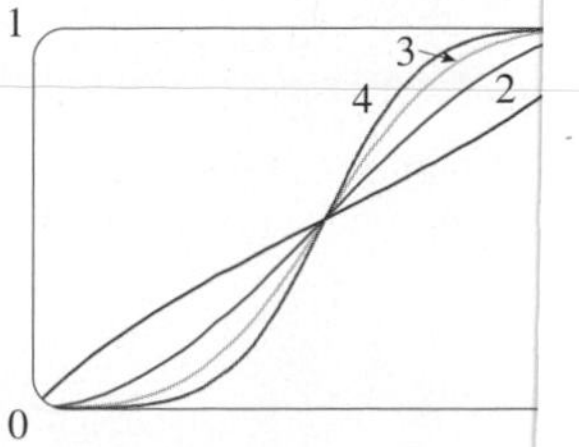

FIGURE 1

11. $\displaystyle\int_1^2 \frac{\sqrt{x^2-1}}{x}\,dx$

12. $\displaystyle\int_{-1}^1 \frac{\sin x}{1+x^2}\,dx$

13. $\displaystyle\int e^{\sqrt[3]{x}}\,dx$

14. $\displaystyle\int \frac{x^2+2}{x+2}\,dx$

15. $\displaystyle\int \frac{x-1}{x^2+2x}\,dx$

16. $\displaystyle\int \frac{\sec^6\theta}{\tan^2\theta}\,d\theta$

17. $\displaystyle\int x \sec x \tan x\,dx$

18. $\displaystyle\int \frac{x^2+8x-3}{x^3+3x^2}\,dx$

19. $\displaystyle\int \frac{x+1}{9x^2+6x+5}\,dx$

20. $\displaystyle\int \tan^5\theta \sec^3\theta\,d\theta$

21. $\displaystyle\int \frac{dx}{\sqrt{x^2-4x}}$

22. $\displaystyle\int te^{\sqrt{t}}\,dt$

23. $\displaystyle\int \frac{dx}{x\sqrt{x^2+1}}$

24. $\displaystyle\int e^x \cos x\,dx$

25. $\displaystyle\int \frac{3x^3-x^2+6x-4}{(x^2+1)(x^2+2)}\,dx$

26. $\displaystyle\int x \sin x \cos x\,dx$

27. $\displaystyle\int_0^{\pi/2} \cos^3 x \sin 2x\,dx$

28. $\displaystyle\int \frac{\sqrt[3]{x}+1}{\sqrt[3]{x}-1}\,dx$

29. $\displaystyle\int_{-1}^1 x^5 \sec x\,dx$

30. $\displaystyle\int \frac{dx}{e^x\sqrt{1-e^{-2x}}}$

31. $\displaystyle\int_0^{\ln 10} \frac{e^x\sqrt{e^x-1}}{e^x+8}\,dx$

32. $\displaystyle\int_0^{\pi/4} \frac{x\sin x}{\cos^3 x}\,dx$

33. $\displaystyle\int \frac{x^2}{(4-x^2)^{3/2}}\,dx$

34. $\displaystyle\int (\arcsin x)^2\,dx$

35. $\displaystyle\int \frac{1}{\sqrt{x+x^{3/2}}}\,dx$

36. $\displaystyle\int \frac{1-\tan\theta}{1+\tan\theta}\,d\theta$

37. $\displaystyle\int (\cos x+\sin x)^2 \cos 2x\,dx$

38. $\displaystyle\int \frac{x^2}{(x+2)^3}\,dx$

39. $\displaystyle\int_0^{1/2} \frac{xe^{2x}}{(1+2x)^2}\,dx$

40. $\displaystyle\int_{\pi/4}^{\pi/3} \frac{\sqrt{\tan\theta}}{\sin 2\theta}\,d\theta$

41–50 Evaluate the integral or show that it is divergent.

41. $\displaystyle\int_1^\infty \frac{1}{(2x+1)^3}\,dx$

42. $\displaystyle\int_1^\infty \frac{\ln x}{x^4}\,dx$

43. $\displaystyle\int_2^\infty \frac{dx}{x\ln x}$

44. $\displaystyle\int_2^6 \frac{y}{\sqrt{y-2}}\,dy$

45. $\displaystyle\int_0^4 \frac{\ln x}{\sqrt{x}}\,dx$

46. $\displaystyle\int_0^1 \frac{1}{2-3x}\,dx$

47. $\displaystyle\int_0^1 \frac{x-1}{\sqrt{x}}\,dx$

48. $\displaystyle\int_{-1}^1 \frac{dx}{x^2-2x}$

49. $\displaystyle\int_{-\infty}^\infty \frac{dx}{4x^2+4x+5}$

50. $\displaystyle\int_1^\infty \frac{\tan^{-1}x}{x^2}\,dx$

51–52 Evaluate the indefinite integral. Illustrate and check that your answer is reasonable by graphing both the function and its antiderivative (take $C = 0$).

51. $\displaystyle\int \ln(x^2+2x+2)\,dx$

52. $\displaystyle\int \frac{x^3}{\sqrt{x^2+1}}\,dx$

53. Graph the function $f(x) = \cos^2 x \sin^3 x$ and use the graph to guess the value of the integral $\int_0^{2\pi} f(x)\,dx$. Then evaluate the integral to confirm your guess.

CAS **54.** (a) How would you evaluate $\int x^5 e^{-2x}\,dx$ by hand? (Don't actually carry out the integration.)
(b) How would you evaluate $\int x^5 e^{-2x}\,dx$ using tables? (Don't actually do it.)
(c) Use a CAS to evaluate $\int x^5 e^{-2x}\,dx$.
(d) Graph the integrand and the indefinite integral on the same screen.

55–58 Use the Table of Integrals on the Reference Pages to evaluate the integral.

55. $\displaystyle\int \sqrt{4x^2-4x-3}\,dx$

56. $\displaystyle\int \csc^5 t\,dt$

57. $\displaystyle\int \cos x\sqrt{4+\sin^2 x}\,dx$

58. $\displaystyle\int \frac{\cot x}{\sqrt{1+2\sin x}}\,dx$

59. Verify Formula 33 in the Table of Integrals (a) by differentiation and (b) by using a trigonometric substitution.

60. Verify Formula 62 in the Table of Integrals.

61. Is it possible to find a number n such that $\int_0^\infty x^n\,dx$ is convergent?

62. For what values of a is $\int_0^\infty e^{ax}\cos x\,dx$ convergent? Evaluate the integral for those values of a.

63–64 Use (a) the Trapezoidal Rule, (b) the Midpoint Rule, and (c) Simpson's Rule with $n = 10$ to approximate the given integral. Round your answers to six decimal places.

63. $\displaystyle\int_2^4 \frac{1}{\ln x}\,dx$

64. $\displaystyle\int_1^4 \sqrt{x}\cos x\,dx$

65. Estimate the errors involved in Exercise 63, parts (a) and (b). How large should n be in each case to guarantee an error of less than 0.00001?

66. Use Simpson's Rule with $n = 6$ to estimate the area under the curve $y = e^x/x$ from $x = 1$ to $x = 4$.

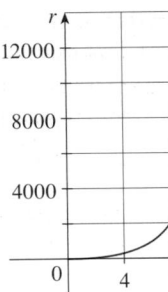

PROBLEMS PLUS

PROBLEMS

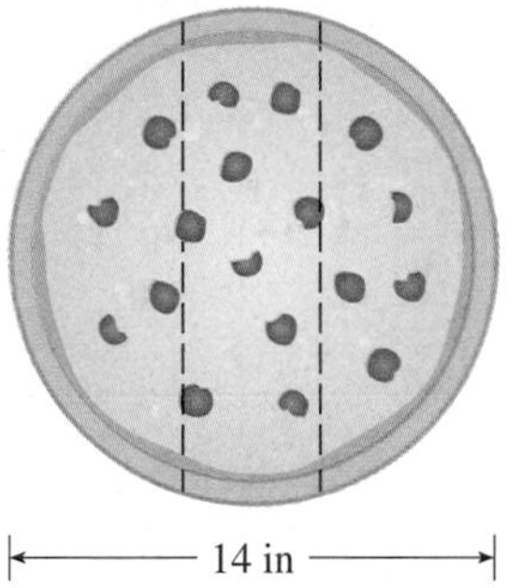

FIGURE FOR PROBLEM 1

1. Three mathematics students have ordered a 14-inch pizza. Instead of slicing it in the traditional way, they decide to slice it by parallel cuts, as shown in the figure. Being mathematics majors, they are able to determine where to slice so that each gets the same amount of pizza. Where are the cuts made?

2. Evaluate $\int \frac{1}{x^7 - x}\, dx$.

The straightforward approach would be to start with partial fractions, but that would be brutal. Try a substitution.

3. Evaluate $\int_0^1 \left(\sqrt[3]{1 - x^7} - \sqrt[7]{1 - x^3}\right) dx$.

4. The centers of two disks with radius 1 are one unit apart. Find the area of the union of the two disks.

5. An ellipse is cut out of a circle with radius a. The major axis of the ellipse coincides with a diameter of the circle and the minor axis has length $2b$. Prove that the area of the remaining part of the circle is the same as the area of an ellipse with semiaxes a and $a - b$.

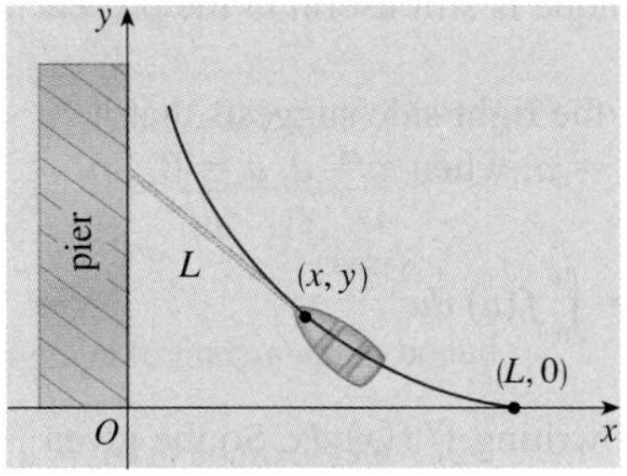

FIGURE FOR PROBLEM 6

6. A man initially standing at the point O walks along a pier pulling a rowboat by a rope of length L. The man keeps the rope straight and taut. The path followed by the boat is a curve called a *tractrix* and it has the property that the rope is always tangent to the curve (see the figure).

(a) Show that if the path followed by the boat is the graph of the function $y = f(x)$, then

$$f'(x) = \frac{dy}{dx} = \frac{-\sqrt{L^2 - x^2}}{x}$$

(b) Determine the function $y = f(x)$.

7. A function f is defined by

$$f(x) = \int_0^{\pi} \cos t \cos(x - t)\, dt \qquad 0 \leqslant x \leqslant 2\pi$$

Find the minimum value of f.

8. If n is a positive integer, prove that

$$\int_0^1 (\ln x)^n\, dx = (-1)^n n!$$

9. Show that

$$\int_0^1 (1 - x^2)^n\, dx = \frac{2^{2n}(n!)^2}{(2n + 1)!}$$

Hint: Start by showing that if I_n denotes the integral, then

$$I_{k+1} = \frac{2k + 2}{2k + 3} I_k$$

10. Suppose that f is a positive function such that f' is continuous.

(a) How is the graph of $y = f(x) \sin nx$ related to the graph of $y = f(x)$? What happens as $n \to \infty$?

(b) Make a guess as to the value of the limit

$$\lim_{n \to \infty} \int_0^1 f(x) \sin nx \, dx$$

based on graphs of the integrand.

(c) Using integration by parts, confirm the guess that you made in part (b). [Use the fact that, since f' is continuous, there is a constant M such that $|f'(x)| \leq M$ for $0 \leq x \leq 1$.]

11. If $0 < a < b$, find $\lim_{t \to 0} \left\{ \int_0^1 [bx + a(1 - x)]^t \, dx \right\}^{1/t}$.

12. Graph $f(x) = \sin(e^x)$ and use the graph to estimate the value of t such that $\int_t^{t+1} f(x)\, dx$ is a maximum. Then find the exact value of t that maximizes this integral.

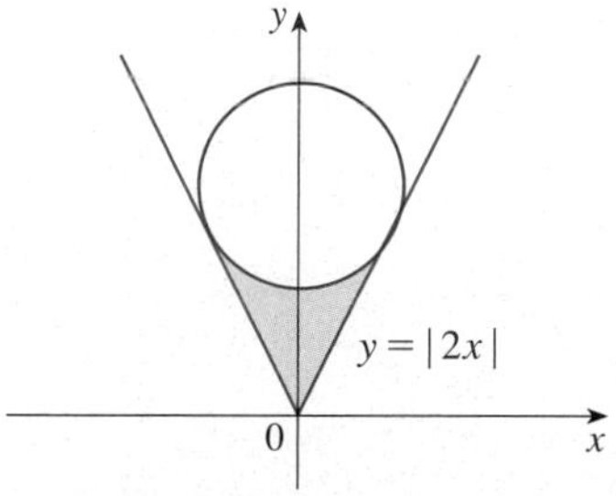

FIGURE FOR PROBLEM 13

13. The circle with radius 1 shown in the figure touches the curve $y = |2x|$ twice. Find the area of the region that lies between the two curves.

14. A rocket is fired straight up, burning fuel at the constant rate of b kilograms per second. Let $v = v(t)$ be the velocity of the rocket at time t and suppose that the velocity u of the exhaust gas is constant. Let $M = M(t)$ be the mass of the rocket at time t and note that M decreases as the fuel burns. If we neglect air resistance, it follows from Newton's Second Law that

$$F = M \frac{dv}{dt} - ub$$

where the force $F = -Mg$. Thus

$$M \frac{dv}{dt} - ub = -Mg \qquad \boxed{1}$$

Let M_1 be the mass of the rocket without fuel, M_2 the initial mass of the fuel, and $M_0 = M_1 + M_2$. Then, until the fuel runs out at time $t = M_2 b$, the mass is $M = M_0 - bt$.

(a) Substitute $M = M_0 - bt$ into Equation 1 and solve the resulting equation for v. Use the initial condition $v(0) = 0$ to evaluate the constant.

(b) Determine the velocity of the rocket at time $t = M_2/b$. This is called the *burnout velocity*.

(c) Determine the height of the rocket $y = y(t)$ at the burnout time.

(d) Find the height of the rocket at any time t.

15. Use integration by parts to show that, for all $x > 0$,

$$0 < \int_0^\infty \frac{\sin t}{\ln(1 + x + t)}\, dt < \frac{2}{\ln(1 + x)}$$

16. Suppose $f(1) = f'(1) = 0$, f'' is continuous on $[0, 1]$ and $|f''(x)| \leq 3$ for all x. Show that

$$\left| \int_0^1 f(x)\, dx \right| \leq \frac{1}{2}$$

8

FURTHER APPLICATIONS OF INTEGRATION

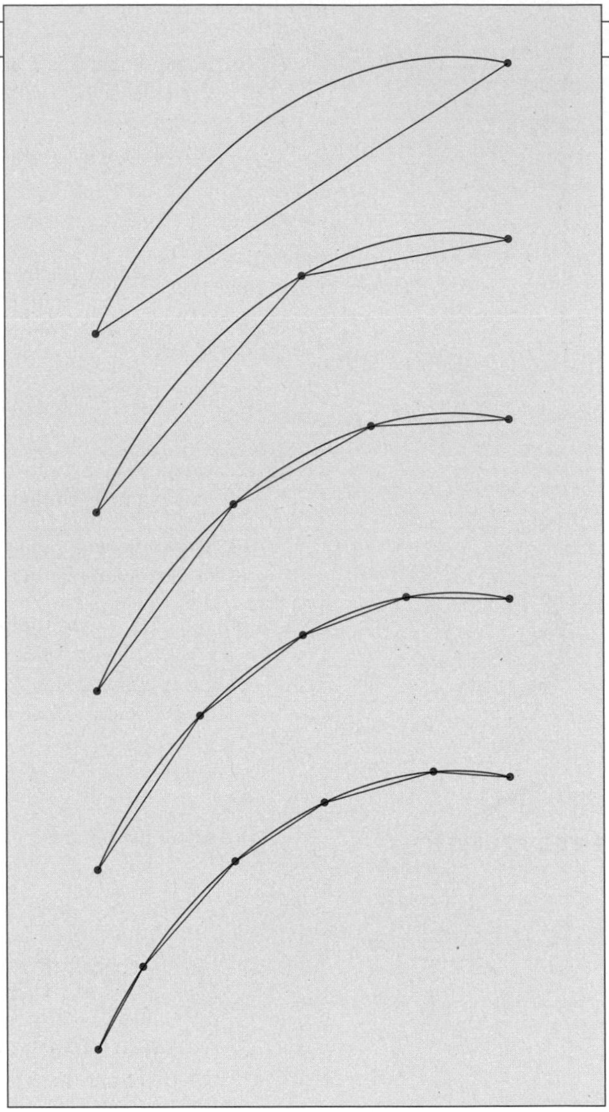

The length of a curve is the limit of lengths of inscribed polygons.

We looked at some applications of integrals in Chapter 6: areas, volumes, work, and average values. Here we explore some of the many other geometric applications of integration—the length of a curve, the area of a surface—as well as quantities of interest in physics, engineering, biology, economics, and statistics. For instance, we will investigate the center of gravity of a plate, the force exerted by water pressure on a dam, the flow of blood from the human heart, and the average time spent on hold during a customer support telephone call.

8.1 ARC LENGTH

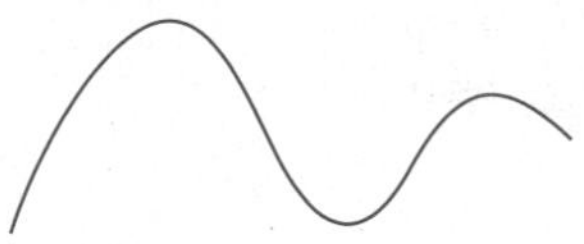

FIGURE 1

TEC Visual 8.1 shows an animation of Figure 2.

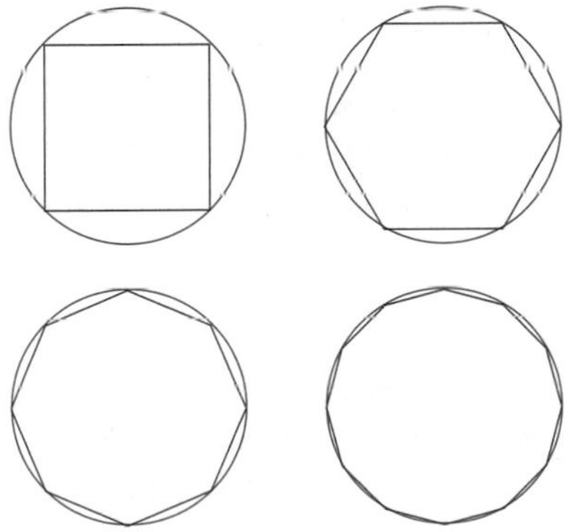

FIGURE 2

What do we mean by the length of a curve? We might think of fitting a piece of string to the curve in Figure 1 and then measuring the string against a ruler. But that might be difficult to do with much accuracy if we have a complicated curve. We need a precise definition for the length of an arc of a curve, in the same spirit as the definitions we developed for the concepts of area and volume.

If the curve is a polygon, we can easily find its length; we just add the lengths of the line segments that form the polygon. (We can use the distance formula to find the distance between the endpoints of each segment.) We are going to define the length of a general curve by first approximating it by a polygon and then taking a limit as the number of segments of the polygon is increased. This process is familiar for the case of a circle, where the circumference is the limit of lengths of inscribed polygons (see Figure 2).

Now suppose that a curve C is defined by the equation $y = f(x)$, where f is continuous and $a \leqslant x \leqslant b$. We obtain a polygonal approximation to C by dividing the interval $[a, b]$ into n subintervals with endpoints $x_0, x_1, \ldots, x_n$ and equal width Δx. If $y_i = f(x_i)$, then the point $P_i(x_i, y_i)$ lies on C and the polygon with vertices $P_0, P_1, \ldots, P_n$, illustrated in Figure 3, is an approximation to C.

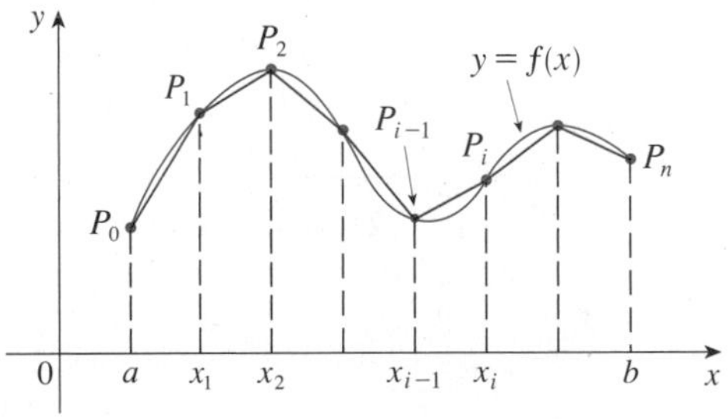

FIGURE 3

The length L of C is approximately the length of this polygon and the approximation gets better as we let n increase. (See Figure 4, where the arc of the curve between P_{i-1} and P_i has been magnified and approximations with successively smaller values of Δx are shown.) Therefore we define the **length** L of the curve C with equation $y = f(x)$, $a \leqslant x \leqslant b$, as the limit of the lengths of these inscribed polygons (if the limit exists):

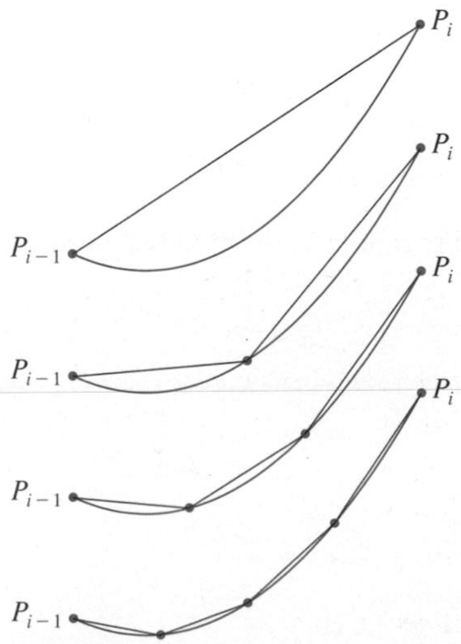

FIGURE 4

1 $$L = \lim_{n \to \infty} \sum_{i=1}^{n} |P_{i-1}P_i|$$

Notice that the procedure for defining arc length is very similar to the procedure we used for defining area and volume: We divided the curve into a large number of small parts. We then found the approximate lengths of the small parts and added them. Finally, we took the limit as $n \to \infty$.

The definition of arc length given by Equation 1 is not very convenient for computational purposes, but we can derive an integral formula for L in the case where f has a continuous derivative. [Such a function f is called **smooth** because a small change in x produces a small change in $f'(x)$.]

If we let $\Delta y_i = y_i - y_{i-1}$, then

$$|P_{i-1}P_i| = \sqrt{(x_i - x_{i-1})^2 + (y_i - y_{i-1})^2} = \sqrt{(\Delta x)^2 + (\Delta y_i)^2}$$

By applying the Mean Value Theorem to f on the interval $[x_{i-1}, x_i]$, we find that there is a number x_i^* between x_{i-1} and x_i such that

$$f(x_i) - f(x_{i-1}) = f'(x_i^*)(x_i - x_{i-1})$$

that is,

$$\Delta y_i = f'(x_i^*)\,\Delta x$$

Thus we have

$$|P_{i-1}P_i| = \sqrt{(\Delta x)^2 + (\Delta y_i)^2} = \sqrt{(\Delta x)^2 + [f'(x_i^*)\,\Delta x]^2}$$

$$= \sqrt{1 + [f'(x_i^*)]^2}\,\sqrt{(\Delta x)^2} = \sqrt{1 + [f'(x_i^*)]^2}\,\Delta x \qquad \text{(since } \Delta x > 0\text{)}$$

Therefore, by Definition 1,

$$L = \lim_{n \to \infty} \sum_{i=1}^{n} |P_{i-1}P_i| = \lim_{n \to \infty} \sum_{i=1}^{n} \sqrt{1 + [f'(x_i^*)]^2}\,\Delta x$$

We recognize this expression as being equal to

$$\int_a^b \sqrt{1 + [f'(x)]^2}\,dx$$

by the definition of a definite integral. This integral exists because the function $g(x) = \sqrt{1 + [f'(x)]^2}$ is continuous. Thus we have proved the following theorem:

2 THE ARC LENGTH FORMULA If f' is continuous on $[a, b]$, then the length of the curve $y = f(x)$, $a \leqslant x \leqslant b$, is

$$L = \int_a^b \sqrt{1 + [f'(x)]^2}\,dx$$

If we use Leibniz notation for derivatives, we can write the arc length formula as follows:

3

$$L = \int_a^b \sqrt{1 + \left(\frac{dy}{dx}\right)^2}\,dx$$

EXAMPLE 1 Find the length of the arc of the semicubical parabola $y^2 = x^3$ between the points $(1, 1)$ and $(4, 8)$. (See Figure 5.)

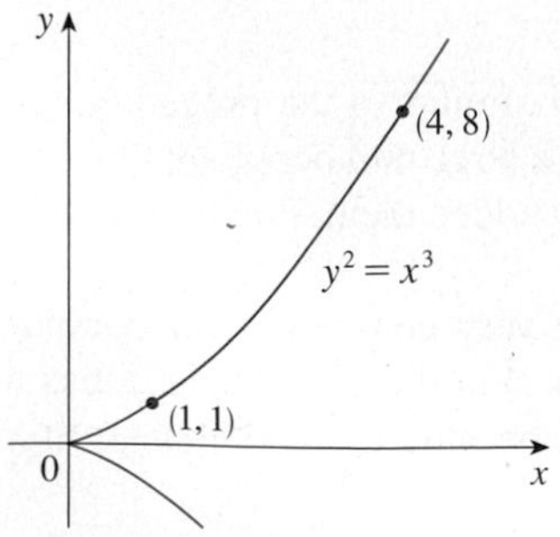

FIGURE 5

SOLUTION For the top half of the curve we have

$$y = x^{3/2} \qquad \frac{dy}{dx} = \tfrac{3}{2}x^{1/2}$$

and so the arc length formula gives

$$L = \int_1^4 \sqrt{1 + \left(\frac{dy}{dx}\right)^2}\,dx = \int_1^4 \sqrt{1 + \tfrac{9}{4}x}\,dx$$

If we substitute $u = 1 + \frac{9}{4}x$, then $du = \frac{9}{4}\,dx$. When $x = 1$, $u = \frac{13}{4}$; when $x = 4$, $u = 10$.

■ As a check on our answer to Example 1, notice from Figure 5 that the arc length ought to be slightly larger than the distance from (1, 1) to (4, 8), which is

$$\sqrt{58} \approx 7.615773$$

According to our calculation in Example 1, we have

$$L = \tfrac{1}{27}\left(80\sqrt{10} - 13\sqrt{13}\right) \approx 7.633705$$

Sure enough, this is a bit greater than the length of the line segment.

Therefore

$$L = \tfrac{4}{9}\int_{13/4}^{10} \sqrt{u}\, du = \tfrac{4}{9} \cdot \tfrac{2}{3} u^{3/2}\Big]_{13/4}^{10}$$

$$= \tfrac{8}{27}\left[10^{3/2} - \left(\tfrac{13}{4}\right)^{3/2}\right] = \tfrac{1}{27}\left(80\sqrt{10} - 13\sqrt{13}\right) \qquad \square$$

If a curve has the equation $x = g(y)$, $c \leqslant y \leqslant d$, and $g'(y)$ is continuous, then by interchanging the roles of x and y in Formula 2 or Equation 3, we obtain the following formula for its length:

4
$$L = \int_c^d \sqrt{1 + [g'(y)]^2}\, dy = \int_c^d \sqrt{1 + \left(\frac{dx}{dy}\right)^2}\, dy$$

V EXAMPLE 2 Find the length of the arc of the parabola $y^2 = x$ from (0, 0) to (1, 1).

SOLUTION Since $x = y^2$, we have $dx/dy = 2y$, and Formula 4 gives

$$L = \int_0^1 \sqrt{1 + \left(\frac{dx}{dy}\right)^2}\, dy = \int_0^1 \sqrt{1 + 4y^2}\, dy$$

We make the trigonometric substitution $y = \frac{1}{2}\tan\theta$, which gives $dy = \frac{1}{2}\sec^2\theta\, d\theta$ and $\sqrt{1 + 4y^2} = \sqrt{1 + \tan^2\theta} = \sec\theta$. When $y = 0$, $\tan\theta = 0$, so $\theta = 0$; when $y = 1$, $\tan\theta = 2$, so $\theta = \tan^{-1}2 = \alpha$, say. Thus

$$L = \int_0^\alpha \sec\theta \cdot \tfrac{1}{2}\sec^2\theta\, d\theta = \tfrac{1}{2}\int_0^\alpha \sec^3\theta\, d\theta$$

$$= \tfrac{1}{2} \cdot \tfrac{1}{2}\Big[\sec\theta\tan\theta + \ln|\sec\theta + \tan\theta|\Big]_0^\alpha \qquad \text{(from Example 8 in Section 7.2)}$$

$$= \tfrac{1}{4}\left(\sec\alpha\tan\alpha + \ln|\sec\alpha + \tan\alpha|\right)$$

(We could have used Formula 21 in the Table of Integrals.) Since $\tan\alpha = 2$, we have $\sec^2\alpha = 1 + \tan^2\alpha = 5$, so $\sec\alpha = \sqrt{5}$ and

$$L = \frac{\sqrt{5}}{2} + \frac{\ln\left(\sqrt{5} + 2\right)}{4} \qquad \square$$

■ Figure 6 shows the arc of the parabola whose length is computed in Example 2, together with polygonal approximations having $n = 1$ and $n = 2$ line segments, respectively. For $n = 1$ the approximate length is $L_1 = \sqrt{2}$, the diagonal of a square. The table shows the approximations L_n that we get by dividing [0, 1] into n equal subintervals. Notice that each time we double the number of sides of the polygon, we get closer to the exact length, which is

$$L = \frac{\sqrt{5}}{2} + \frac{\ln\left(\sqrt{5} + 2\right)}{4} \approx 1.478943$$

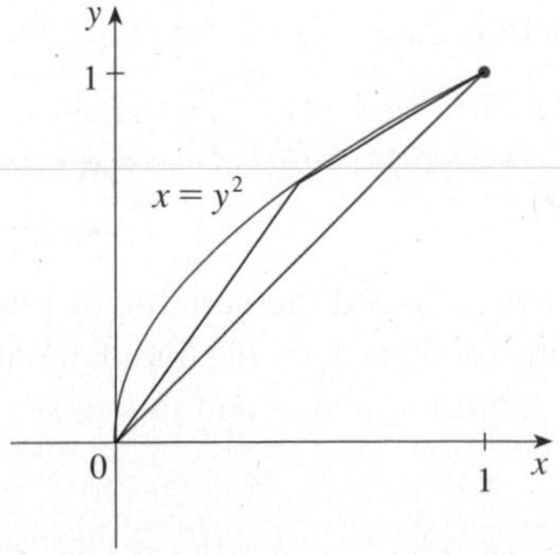

n	L_n
1	1.414
2	1.445
4	1.464
8	1.472
16	1.476
32	1.478
64	1.479

FIGURE 6

Because of the presence of the square root sign in Formulas 2 and 4, the calculation of an arc length often leads to an integral that is very difficult or even impossible to evaluate explicitly. Thus we sometimes have to be content with finding an approximation to the length of a curve, as in the following example.

V EXAMPLE 3
(a) Set up an integral for the length of the arc of the hyperbola $xy = 1$ from the point $(1, 1)$ to the point $\left(2, \frac{1}{2}\right)$.
(b) Use Simpson's Rule with $n = 10$ to estimate the arc length.

SOLUTION
(a) We have

$$y = \frac{1}{x} \qquad \frac{dy}{dx} = -\frac{1}{x^2}$$

and so the arc length is

$$L = \int_1^2 \sqrt{1 + \left(\frac{dy}{dx}\right)^2}\, dx = \int_1^2 \sqrt{1 + \frac{1}{x^4}}\, dx = \int_1^2 \frac{\sqrt{x^4 + 1}}{x^2}\, dx$$

(b) Using Simpson's Rule (see Section 7.7) with $a = 1$, $b = 2$, $n = 10$, $\Delta x = 0.1$, and $f(x) = \sqrt{1 + 1/x^4}$, we have

$$L = \int_1^2 \sqrt{1 + \frac{1}{x^4}}\, dx$$

$$\approx \frac{\Delta x}{3}[f(1) + 4f(1.1) + 2f(1.2) + 4f(1.3) + \cdots + 2f(1.8) + 4f(1.9) + f(2)]$$

$$\approx 1.1321$$ □

■ Checking the value of the definite integral with a more accurate approximation produced by a computer algebra system, we see that the approximation using Simpson's Rule is accurate to four decimal places.

THE ARC LENGTH FUNCTION

We will find it useful to have a function that measures the arc length of a curve from a particular starting point to any other point on the curve. Thus if a smooth curve C has the equation $y = f(x)$, $a \leqslant x \leqslant b$, let $s(x)$ be the distance along C from the initial point $P_0(a, f(a))$ to the point $Q(x, f(x))$. Then s is a function, called the **arc length function**, and, by Formula 2,

5 $$s(x) = \int_a^x \sqrt{1 + [f'(t)]^2}\, dt$$

(We have replaced the variable of integration by t so that x does not have two meanings.) We can use Part 1 of the Fundamental Theorem of Calculus to differentiate Equation 5 (since the integrand is continuous):

6 $$\frac{ds}{dx} = \sqrt{1 + [f'(x)]^2} = \sqrt{1 + \left(\frac{dy}{dx}\right)^2}$$

Equation 6 shows that the rate of change of s with respect to x is always at least 1 and is equal to 1 when $f'(x)$, the slope of the curve, is 0. The differential of arc length is

7 $$ds = \sqrt{1 + \left(\frac{dy}{dx}\right)^2}\, dx$$

and this equation is sometimes written in the symmetric form

8 $$(ds)^2 = (dx)^2 + (dy)^2$$

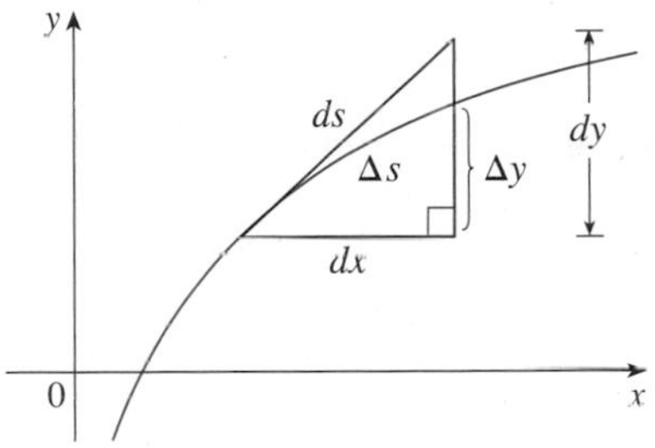

FIGURE 7

The geometric interpretation of Equation 8 is shown in Figure 7. It can be used as a mnemonic device for remembering both of the Formulas 3 and 4. If we write $L = \int ds$, then from Equation 8 either we can solve to get (7), which gives (3), or we can solve to get

$$ds = \sqrt{1 + \left(\frac{dx}{dy}\right)^2}\, dy$$

which gives (4).

V EXAMPLE 4 Find the arc length function for the curve $y = x^2 - \frac{1}{8}\ln x$ taking $P_0(1, 1)$ as the starting point.

SOLUTION If $f(x) = x^2 - \frac{1}{8}\ln x$, then

$$f'(x) = 2x - \frac{1}{8x}$$

$$1 + [f'(x)]^2 = 1 + \left(2x - \frac{1}{8x}\right)^2 = 1 + 4x^2 - \frac{1}{2} + \frac{1}{64x^2}$$

$$= 4x^2 + \frac{1}{2} + \frac{1}{64x^2} = \left(2x + \frac{1}{8x}\right)^2$$

$$\sqrt{1 + [f'(x)]^2} = 2x + \frac{1}{8x}$$

Thus the arc length function is given by

$$s(x) = \int_1^x \sqrt{1 + [f'(t)]^2}\, dt$$

$$= \int_1^x \left(2t - \frac{1}{8t}\right) dt = t^2 + \tfrac{1}{8}\ln t\Big]_1^x$$

$$= x^2 + \tfrac{1}{8}\ln x - 1$$

For instance, the arc length along the curve from (1, 1) to $(3, f(3))$ is

$$s(3) = 3^2 + \tfrac{1}{8}\ln 3 - 1 = 8 + \frac{\ln 3}{8} \approx 8.1373$$ □

■ Figure 8 shows the interpretation of the arc length function in Example 4. Figure 9 shows the graph of this arc length function. Why is $s(x)$ negative when x is less than 1?

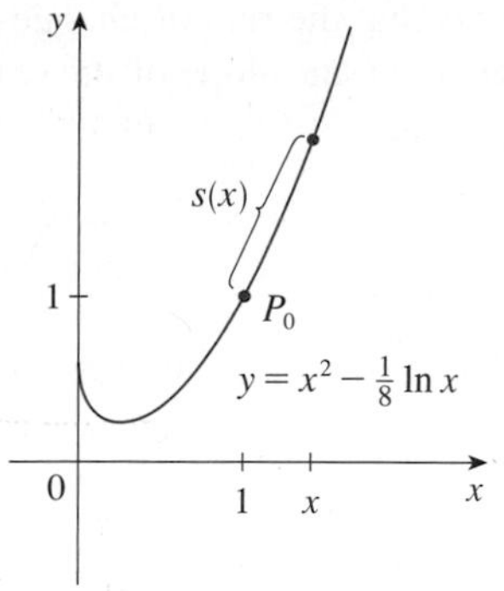

FIGURE 8

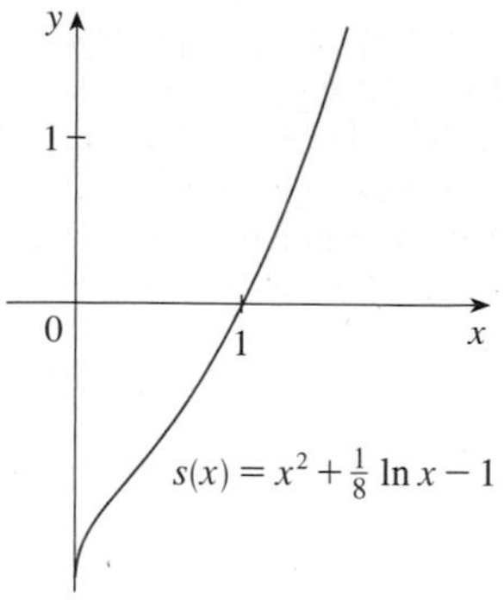

FIGURE 9

8.1 EXERCISES

1. Use the arc length formula (3) to find the length of the curve $y = 2x - 5$, $-1 \leqslant x \leqslant 3$. Check your answer by noting that the curve is a line segment and calculating its length by the distance formula.

2. Use the arc length formula to find the length of the curve $y = \sqrt{2 - x^2}$, $0 \leqslant x \leqslant 1$. Check your answer by noting that the curve is part of a circle.

3–6 Set up, but do not evaluate, an integral for the length of the curve.

3. $y = \cos x, \quad 0 \leqslant x \leqslant 2\pi$

4. $y = xe^{-x^2}, \quad 0 \leqslant x \leqslant 1$

5. $x = y + y^3, \quad 1 \leqslant y \leqslant 4$

6. $\dfrac{x^2}{a^2} + \dfrac{y^2}{b^2} = 1$

7–18 Find the length of the curve.

7. $y = 1 + 6x^{3/2}, \quad 0 \leqslant x \leqslant 1$

8. $y^2 = 4(x + 4)^3, \quad 0 \leqslant x \leqslant 2, \quad y > 0$

9. $y = \dfrac{x^5}{6} + \dfrac{1}{10x^3}, \quad 1 \leqslant x \leqslant 2$

10. $x = \dfrac{y^4}{8} + \dfrac{1}{4y^2}, \quad 1 \leqslant y \leqslant 2$

11. $x = \frac{1}{3}\sqrt{y}\,(y - 3), \quad 1 \leqslant y \leqslant 9$

12. $y = \ln(\cos x), \quad 0 \leqslant x \leqslant \pi/3$

13. $y = \ln(\sec x), \quad 0 \leqslant x \leqslant \pi/4$

14. $y = 3 + \frac{1}{2}\cosh 2x, \quad 0 \leqslant x \leqslant 1$

15. $y = \ln(1 - x^2), \quad 0 \leqslant x \leqslant \frac{1}{2}$

16. $y = \sqrt{x - x^2} + \sin^{-1}(\sqrt{x})$

17. $y = e^x, \quad 0 \leqslant x \leqslant 1$

18. $y = \ln\left(\dfrac{e^x + 1}{e^x - 1}\right), \quad a \leqslant x \leqslant b, \quad a > 0$

19–20 Find the length of the arc of the curve from point P to point Q.

19. $y = \frac{1}{2}x^2, \quad P(-1, \frac{1}{2}), \quad Q(1, \frac{1}{2})$

20. $x^2 = (y - 4)^3, \quad P(1, 5), \quad Q(8, 8)$

21–22 Graph the curve and visually estimate its length. Then find its exact length.

21. $y = \frac{2}{3}(x^2 - 1)^{3/2}, \quad 1 \leqslant x \leqslant 3$

22. $y = \dfrac{x^3}{6} + \dfrac{1}{2x}, \quad \frac{1}{2} \leqslant x \leqslant 1$

23–26 Use Simpson's Rule with $n = 10$ to estimate the arc length of the curve. Compare your answer with the value of the integral produced by your calculator.

23. $y = xe^{-x}, \quad 0 \leqslant x \leqslant 5$

24. $x = y + \sqrt{y}, \quad 1 \leqslant y \leqslant 2$

25. $y = \sec x, \quad 0 \leqslant x \leqslant \pi/3$

26. $y = x \ln x, \quad 1 \leqslant x \leqslant 3$

27. (a) Graph the curve $y = x\sqrt[3]{4 - x}$, $0 \leqslant x \leqslant 4$.
(b) Compute the lengths of inscribed polygons with $n = 1$, 2, and 4 sides. (Divide the interval into equal subintervals.) Illustrate by sketching these polygons (as in Figure 6).
(c) Set up an integral for the length of the curve.
(d) Use your calculator to find the length of the curve to four decimal places. Compare with the approximations in part (b).

28. Repeat Exercise 27 for the curve

$$y = x + \sin x \qquad 0 \leqslant x \leqslant 2\pi$$

CAS **29.** Use either a computer algebra system or a table of integrals to find the *exact* length of the arc of the curve $y = \ln x$ that lies between the points (1, 0) and (2, ln 2).

CAS **30.** Use either a computer algebra system or a table of integrals to find the *exact* length of the arc of the curve $y = x^{4/3}$ that lies between the points (0, 0) and (1, 1). If your CAS has trouble evaluating the integral, make a substitution that changes the integral into one that the CAS can evaluate.

31. Sketch the curve with equation $x^{2/3} + y^{2/3} = 1$ and use symmetry to find its length.

32. (a) Sketch the curve $y^3 = x^2$.
(b) Use Formulas 3 and 4 to set up two integrals for the arc length from (0, 0) to (1, 1). Observe that one of these is an improper integral and evaluate both of them.
(c) Find the length of the arc of this curve from $(-1, 1)$ to (8, 4).

33. Find the arc length function for the curve $y = 2x^{3/2}$ with starting point $P_0(1, 2)$.

34. (a) Graph the curve $y = \frac{1}{3}x^3 + 1/(4x)$, $x > 0$.
(b) Find the arc length function for this curve with starting point $P_0(1, \frac{7}{12})$.
(c) Graph the arc length function.

35. Find the arc length function for the curve $y = \sin^{-1}x + \sqrt{1 - x^2}$ with starting point (0, 1).

36. A steady wind blows a kite due west. The kite's height above ground from horizontal position $x = 0$ to $x = 80$ ft is given by $y = 150 - \frac{1}{40}(x - 50)^2$. Find the distance traveled by the kite.

37. A hawk flying at 15 m/s at an altitude of 180 m accidentally drops its prey. The parabolic trajectory of the falling prey is described by the equation

$$y = 180 - \frac{x^2}{45}$$

until it hits the ground, where y is its height above the ground and x is the horizontal distance traveled in meters. Calculate the distance traveled by the prey from the time it is dropped until the time it hits the ground. Express your answer correct to the nearest tenth of a meter.

38. The Gateway Arch in St. Louis (see the photo on page 256) was constructed using the equation

$$y = 211.49 - 20.96 \cosh 0.03291765x$$

for the central curve of the arch, where x and y are measured in meters and $|x| \leqslant 91.20$. Set up an integral for the length of the arch and use your calculator to estimate the length correct to the nearest meter.

39. A manufacturer of corrugated metal roofing wants to produce panels that are 28 in. wide and 2 in. thick by processing flat sheets of metal as shown in the figure. The profile of the roofing takes the shape of a sine wave. Verify that the sine curve has equation $y = \sin(\pi x/7)$ and find the width w of a flat metal sheet that is needed to make a 28-inch panel. (Use your calculator to evaluate the integral correct to four significant digits.)

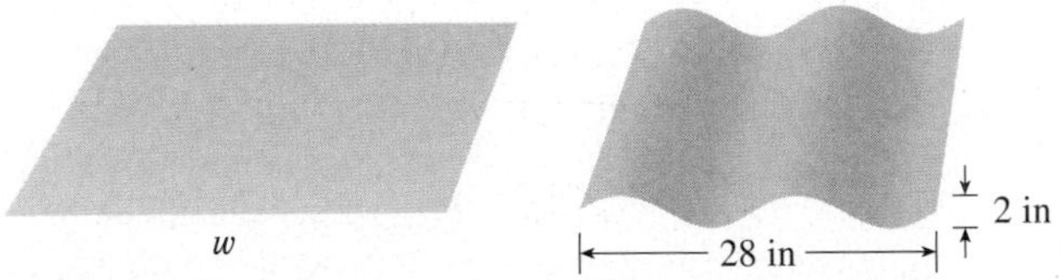

40. (a) The figure shows a telephone wire hanging between two poles at $x = -b$ and $x = b$. It takes the shape of a catenary with equation $y = c + a\cosh(x/a)$. Find the length of the wire.
(b) Suppose two telephone poles are 50 ft apart and the length of the wire between the poles is 51 ft. If the lowest point of the wire must be 20 ft above the ground, how high up on each pole should the wire be attached?

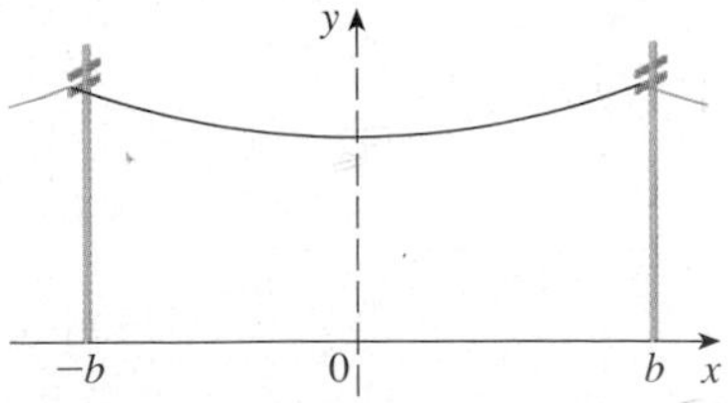

41. Find the length of the curve

$$y = \int_1^x \sqrt{t^3 - 1}\, dt \qquad 1 \leqslant x \leqslant 4$$

42. The curves with equations $x^n + y^n = 1$, $n = 4, 6, 8, \ldots$, are called **fat circles**. Graph the curves with $n = 2$, 4, 6, 8, and 10 to see why. Set up an integral for the length L_{2k} of the fat circle with $n = 2k$. Without attempting to evaluate this integral, state the value of $\lim_{k\to\infty} L_{2k}$.

DISCOVERY PROJECT — ARC LENGTH CONTEST

The curves shown are all examples of graphs of continuous functions f that have the following properties.

1. $f(0) = 0$ and $f(1) = 0$
2. $f(x) \geqslant 0$ for $0 \leqslant x \leqslant 1$
3. The area under the graph of f from 0 to 1 is equal to 1.

The lengths L of these curves, however, are different.

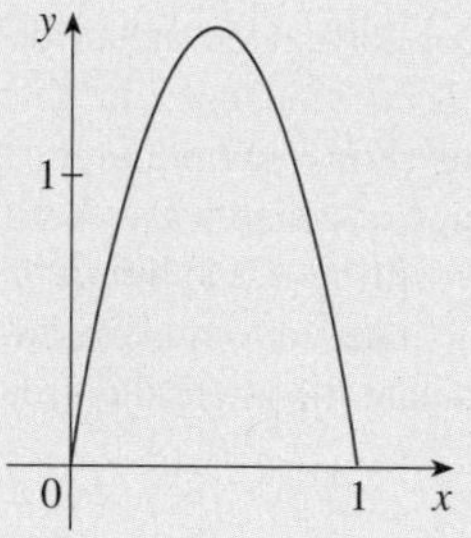

$L \approx 3.249$

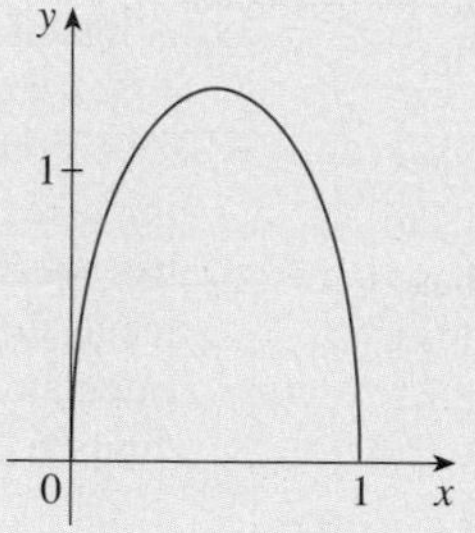

$L \approx 2.919$

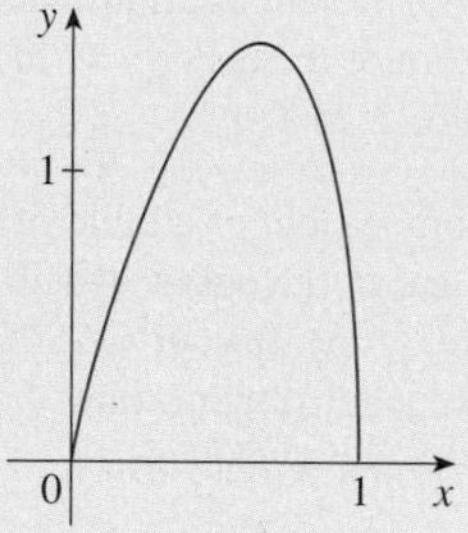

$L \approx 3.152$

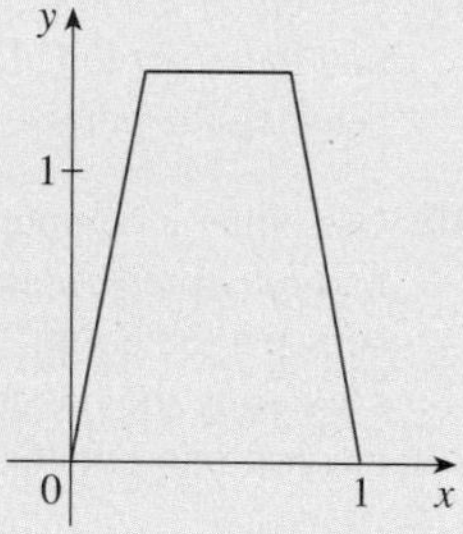

$L \approx 3.213$

Try to discover formulas for two functions that satisfy the given conditions 1, 2, and 3. (Your graphs might be similar to the ones shown or could look quite different.) Then calculate the arc length of each graph. The winning entry will be the one with the smallest arc length.

8.2 AREA OF A SURFACE OF REVOLUTION

A surface of revolution is formed when a curve is rotated about a line. Such a surface is the lateral boundary of a solid of revolution of the type discussed in Sections 6.2 and 6.3.

We want to define the area of a surface of revolution in such a way that it corresponds to our intuition. If the surface area is A, we can imagine that painting the surface would require the same amount of paint as does a flat region with area A.

Let's start with some simple surfaces. The lateral surface area of a circular cylinder with radius r and height h is taken to be $A = 2\pi rh$ because we can imagine cutting the cylinder and unrolling it (as in Figure 1) to obtain a rectangle with dimensions $2\pi r$ and h.

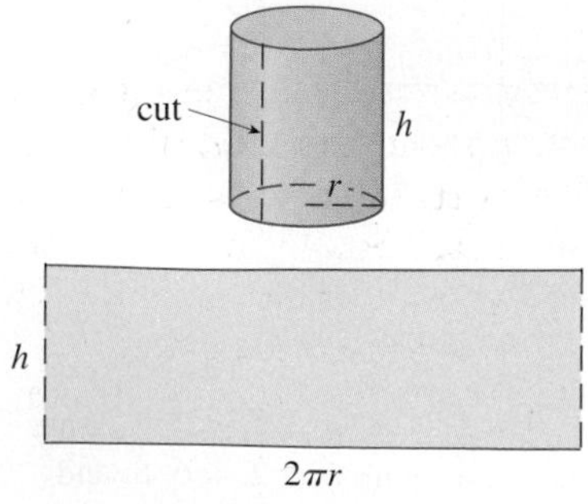

FIGURE 1

Likewise, we can take a circular cone with base radius r and slant height l, cut it along the dashed line in Figure 2, and flatten it to form a sector of a circle with radius l and central angle $\theta = 2\pi r/l$. We know that, in general, the area of a sector of a circle with radius l and angle θ is $\frac{1}{2}l^2\theta$ (see Exercise 35 in Section 7.3) and so in this case the area is

$$A = \tfrac{1}{2}l^2\theta = \tfrac{1}{2}l^2\left(\frac{2\pi r}{l}\right) = \pi rl$$

Therefore we define the lateral surface area of a cone to be $A = \pi rl$.

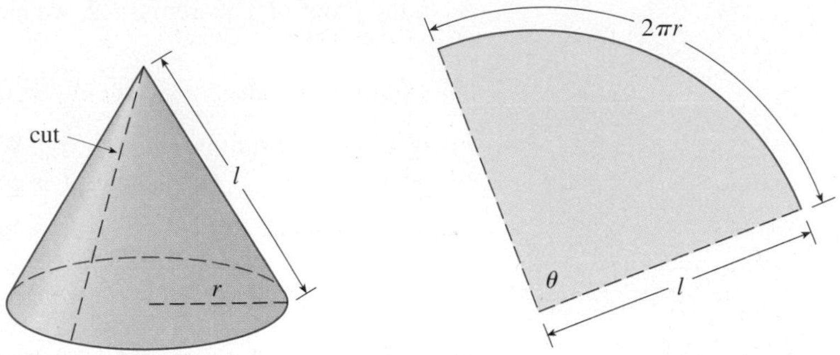

FIGURE 2

What about more complicated surfaces of revolution? If we follow the strategy we used with arc length, we can approximate the original curve by a polygon. When this polygon is rotated about an axis, it creates a simpler surface whose surface area approximates the actual surface area. By taking a limit, we can determine the exact surface area.

The approximating surface, then, consists of a number of *bands,* each formed by rotating a line segment about an axis. To find the surface area, each of these bands can be considered a portion of a circular cone, as shown in Figure 3. The area of the band (or frustum of a cone) with slant height l and upper and lower radii r_1 and r_2 is found by subtracting the areas of two cones:

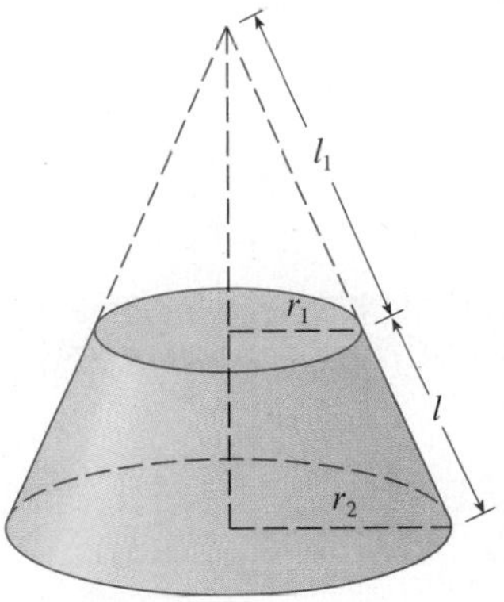

FIGURE 3

$$\boxed{1} \qquad A = \pi r_2(l_1 + l) - \pi r_1 l_1 = \pi[(r_2 - r_1)l_1 + r_2 l]$$

From similar triangles we have

$$\frac{l_1}{r_1} = \frac{l_1 + l}{r_2}$$

which gives

$$r_2 l_1 = r_1 l_1 + r_1 l \qquad \text{or} \qquad (r_2 - r_1)l_1 = r_1 l$$

Putting this in Equation 1, we get

$$A = \pi(r_1 l + r_2 l)$$

or

$$\boxed{2} \qquad \boxed{A = 2\pi r l}$$

where $r = \frac{1}{2}(r_1 + r_2)$ is the average radius of the band.

Now we apply this formula to our strategy. Consider the surface shown in Figure 4, which is obtained by rotating the curve $y = f(x)$, $a \leqslant x \leqslant b$, about the x-axis, where f is positive and has a continuous derivative. In order to define its surface area, we divide the interval $[a, b]$ into n subintervals with endpoints $x_0, x_1, \ldots, x_n$ and equal width Δx, as we did in determining arc length. If $y_i = f(x_i)$, then the point $P_i(x_i, y_i)$ lies on the curve. The part of the surface between x_{i-1} and x_i is approximated by taking the line segment $P_{i-1}P_i$ and rotating it about the x-axis. The result is a band with slant height $l = |P_{i-1}P_i|$ and average radius $r = \frac{1}{2}(y_{i-1} + y_i)$ so, by Formula 2, its surface area is

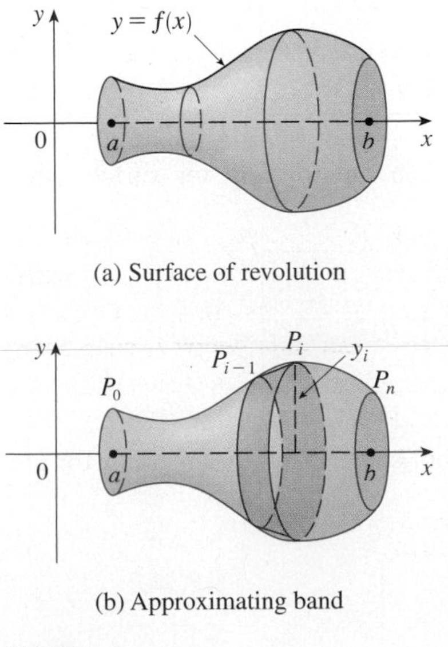

FIGURE 4

$$2\pi \frac{y_{i-1} + y_i}{2} |P_{i-1}P_i|$$

As in the proof of Theorem 8.1.2, we have

$$|P_{i-1}P_i| = \sqrt{1 + [f'(x_i^*)]^2}\,\Delta x$$

where x_i^* is some number in $[x_{i-1}, x_i]$. When Δx is small, we have $y_i = f(x_i) \approx f(x_i^*)$ and also $y_{i-1} = f(x_{i-1}) \approx f(x_i^*)$, since f is continuous. Therefore

$$2\pi \frac{y_{i-1} + y_i}{2} |P_{i-1}P_i| \approx 2\pi f(x_i^*) \sqrt{1 + [f'(x_i^*)]^2}\,\Delta x$$

and so an approximation to what we think of as the area of the complete surface of revolution is

3
$$\sum_{i=1}^{n} 2\pi f(x_i^*) \sqrt{1 + [f'(x_i^*)]^2}\,\Delta x$$

This approximation appears to become better as $n \to \infty$ and, recognizing (3) as a Riemann sum for the function $g(x) = 2\pi f(x)\sqrt{1 + [f'(x)]^2}$, we have

$$\lim_{n\to\infty} \sum_{i=1}^{n} 2\pi f(x_i^*) \sqrt{1 + [f'(x_i^*)]^2}\,\Delta x = \int_a^b 2\pi f(x) \sqrt{1 + [f'(x)]^2}\,dx$$

Therefore, in the case where f is positive and has a continuous derivative, we define the **surface area** of the surface obtained by rotating the curve $y = f(x)$, $a \leq x \leq b$, about the x-axis as

4
$$S = \int_a^b 2\pi f(x) \sqrt{1 + [f'(x)]^2}\,dx$$

With the Leibniz notation for derivatives, this formula becomes

5
$$S = \int_a^b 2\pi y \sqrt{1 + \left(\frac{dy}{dx}\right)^2}\,dx$$

If the curve is described as $x = g(y)$, $c \leq y \leq d$, then the formula for surface area becomes

6
$$S = \int_c^d 2\pi y \sqrt{1 + \left(\frac{dx}{dy}\right)^2}\,dy$$

and both Formulas 5 and 6 can be summarized symbolically, using the notation for arc length given in Section 8.1, as

7
$$S = \int 2\pi y\,ds$$

For rotation about the y-axis, the surface area formula becomes

8
$$S = \int 2\pi x \, ds$$

where, as before, we can use either

$$ds = \sqrt{1 + \left(\frac{dy}{dx}\right)^2} \, dx \qquad \text{or} \qquad ds = \sqrt{1 + \left(\frac{dx}{dy}\right)^2} \, dy$$

These formulas can be remembered by thinking of $2\pi y$ or $2\pi x$ as the circumference of a circle traced out by the point (x, y) on the curve as it is rotated about the x-axis or y-axis, respectively (see Figure 5).

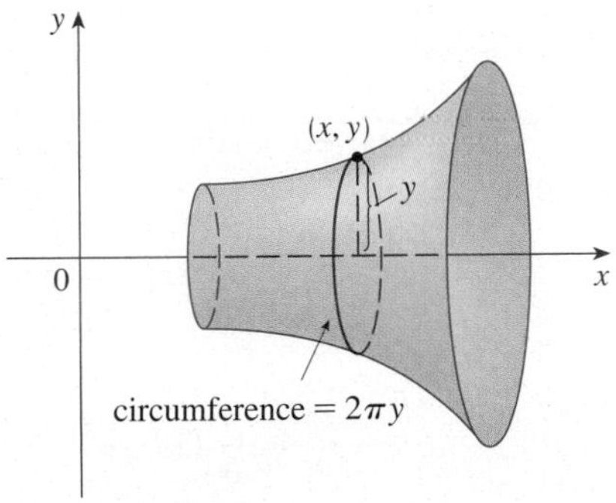

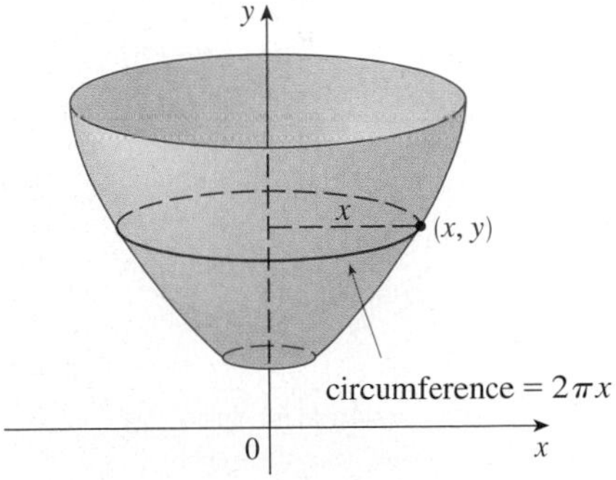

FIGURE 5 (a) Rotation about x-axis: $S = \int 2\pi y \, ds$ (b) Rotation about y-axis: $S = \int 2\pi x \, ds$

EXAMPLE 1 The curve $y = \sqrt{4 - x^2}$, $-1 \leqslant x \leqslant 1$, is an arc of the circle $x^2 + y^2 = 4$. Find the area of the surface obtained by rotating this arc about the x-axis. (The surface is a portion of a sphere of radius 2. See Figure 6.)

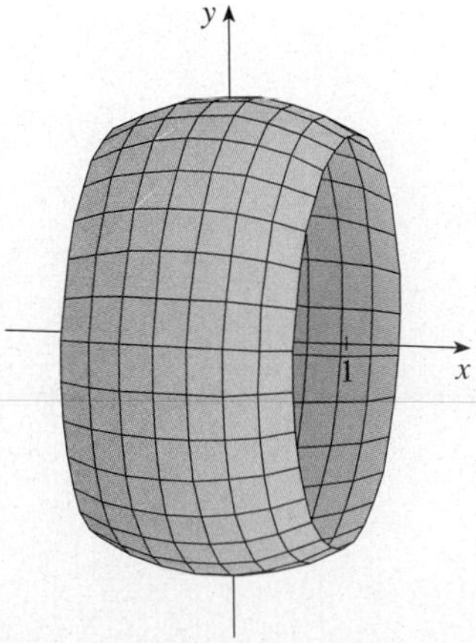

FIGURE 6

■ Figure 6 shows the portion of the sphere whose surface area is computed in Example 1.

SOLUTION We have

$$\frac{dy}{dx} = \tfrac{1}{2}(4 - x^2)^{-1/2}(-2x) = \frac{-x}{\sqrt{4 - x^2}}$$

and so, by Formula 5, the surface area is

$$S = \int_{-1}^{1} 2\pi y \sqrt{1 + \left(\frac{dy}{dx}\right)^2} \, dx$$

$$= 2\pi \int_{-1}^{1} \sqrt{4 - x^2} \sqrt{1 + \frac{x^2}{4 - x^2}} \, dx$$

$$= 2\pi \int_{-1}^{1} \sqrt{4 - x^2} \frac{2}{\sqrt{4 - x^2}} \, dx$$

$$= 4\pi \int_{-1}^{1} 1 \, dx = 4\pi(2) = 8\pi$$

□

■ Figure 7 shows the surface of revolution whose area is computed in Example 2.

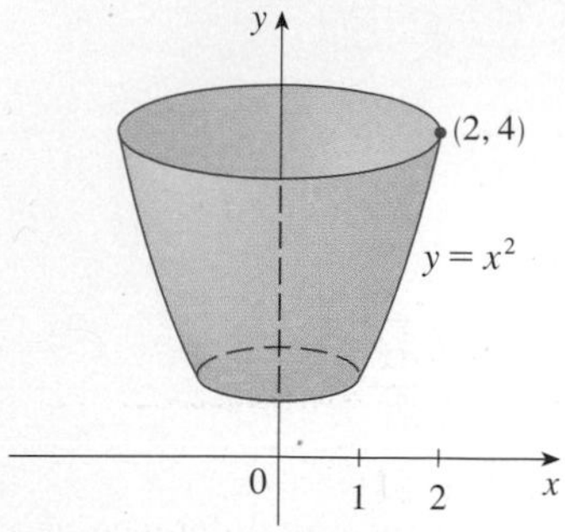

FIGURE 7

EXAMPLE 2 The arc of the parabola $y = x^2$ from $(1, 1)$ to $(2, 4)$ is rotated about the y-axis. Find the area of the resulting surface.

SOLUTION 1 Using

$$y = x^2 \quad \text{and} \quad \frac{dy}{dx} = 2x$$

we have, from Formula 8,

$$\begin{aligned} S &= \int 2\pi x \, ds \\ &= \int_1^2 2\pi x \sqrt{1 + \left(\frac{dy}{dx}\right)^2} \, dx \\ &= 2\pi \int_1^2 x \sqrt{1 + 4x^2} \, dx \end{aligned}$$

Substituting $u = 1 + 4x^2$, we have $du = 8x \, dx$. Remembering to change the limits of integration, we have

$$\begin{aligned} S &= \frac{\pi}{4} \int_5^{17} \sqrt{u} \, du = \frac{\pi}{4} \left[\tfrac{2}{3} u^{3/2}\right]_5^{17} \\ &= \frac{\pi}{6} \left(17\sqrt{17} - 5\sqrt{5}\right) \end{aligned}$$

■ As a check on our answer to Example 2, notice from Figure 7 that the surface area should be close to that of a circular cylinder with the same height and radius halfway between the upper and lower radius of the surface: $2\pi(1.5)(3) \approx 28.27$. We computed that the surface area was

$$\frac{\pi}{6}\left(17\sqrt{17} - 5\sqrt{5}\right) \approx 30.85$$

which seems reasonable. Alternatively, the surface area should be slightly larger than the area of a frustum of a cone with the same top and bottom edges. From Equation 2, this is $2\pi(1.5)(\sqrt{10}) \approx 29.80$.

SOLUTION 2 Using

$$x = \sqrt{y} \quad \text{and} \quad \frac{dx}{dy} = \frac{1}{2\sqrt{y}}$$

we have

$$\begin{aligned} S &= \int 2\pi x \, ds = \int_1^4 2\pi x \sqrt{1 + \left(\frac{dx}{dy}\right)^2} \, dy \\ &= 2\pi \int_1^4 \sqrt{y} \sqrt{1 + \frac{1}{4y}} \, dy = \pi \int_1^4 \sqrt{4y + 1} \, dy \\ &= \frac{\pi}{4} \int_5^{17} \sqrt{u} \, du \qquad \text{(where } u = 1 + 4y\text{)} \\ &= \frac{\pi}{6}\left(17\sqrt{17} - 5\sqrt{5}\right) \qquad \text{(as in Solution 1)} \end{aligned}$$

□

EXAMPLE 3 Find the area of the surface generated by rotating the curve $y = e^x$, $0 \leq x \leq 1$, about the x-axis.

■ Another method: Use Formula 6 with $x = \ln y$.

SOLUTION Using Formula 5 with

$$y = e^x \quad \text{and} \quad \frac{dy}{dx} = e^x$$

we have

$$S = \int_0^1 2\pi y \sqrt{1 + \left(\frac{dy}{dx}\right)^2}\, dx = 2\pi \int_0^1 e^x \sqrt{1 + e^{2x}}\, dx$$

$$= 2\pi \int_1^e \sqrt{1 + u^2}\, du \qquad \text{(where } u = e^x\text{)}$$

$$= 2\pi \int_{\pi/4}^{\alpha} \sec^3\theta\, d\theta \qquad \text{(where } u = \tan\theta \text{ and } \alpha = \tan^{-1}e\text{)}$$

■ Or use Formula 21 in the Table of Integrals.

$$= 2\pi \cdot \tfrac{1}{2}\Big[\sec\theta\tan\theta + \ln|\sec\theta + \tan\theta|\Big]_{\pi/4}^{\alpha} \qquad \text{(by Example 8 in Section 7.2)}$$

$$= \pi\Big[\sec\alpha\tan\alpha + \ln(\sec\alpha + \tan\alpha) - \sqrt{2} - \ln(\sqrt{2} + 1)\Big]$$

Since $\tan\alpha = e$, we have $\sec^2\alpha = 1 + \tan^2\alpha = 1 + e^2$ and

$$S = \pi\Big[e\sqrt{1 + e^2} + \ln\big(e + \sqrt{1 + e^2}\big) - \sqrt{2} - \ln(\sqrt{2} + 1)\Big]$$ □

8.2 EXERCISES

1–4 Set up, but do not evaluate, an integral for the area of the surface obtained by rotating the curve about (a) the x-axis and (b) the y-axis.

1. $y = x^4,\quad 0 \leq x \leq 1$

2. $y = xe^{-x},\quad 1 \leq x \leq 3$

3. $y = \tan^{-1}x,\quad 0 \leq x \leq 1$

4. $x = \sqrt{y - y^2}$

5–12 Find the area of the surface obtained by rotating the curve about the x-axis.

5. $y = x^3,\quad 0 \leq x \leq 2$

6. $9x = y^2 + 18,\quad 2 \leq x \leq 6$

7. $y = \sqrt{1 + 4x},\quad 1 \leq x \leq 5$

8. $y = c + a\cosh(x/a),\quad 0 \leq x \leq a$

9. $y = \sin\pi x,\quad 0 \leq x \leq 1$

10. $y = \dfrac{x^3}{6} + \dfrac{1}{2x},\quad \tfrac{1}{2} \leq x \leq 1$

11. $x = \tfrac{1}{3}(y^2 + 2)^{3/2},\quad 1 \leq y \leq 2$

12. $x = 1 + 2y^2,\quad 1 \leq y \leq 2$

13–16 The given curve is rotated about the y-axis. Find the area of the resulting surface.

13. $y = \sqrt[3]{x},\quad 1 \leq y \leq 2$

14. $y = 1 - x^2,\quad 0 \leq x \leq 1$

15. $x = \sqrt{a^2 - y^2},\quad 0 \leq y \leq a/2$

16. $y = \tfrac{1}{4}x^2 - \tfrac{1}{2}\ln x,\quad 1 \leq x \leq 2$

17–20 Use Simpson's Rule with $n = 10$ to approximate the area of the surface obtained by rotating the curve about the x-axis. Compare your answer with the value of the integral produced by your calculator.

17. $y = \ln x,\quad 1 \leq x \leq 3$

18. $y = x + \sqrt{x},\quad 1 \leq x \leq 2$

19. $y = \sec x,\quad 0 \leq x \leq \pi/3$

20. $y = e^{-x^2},\quad 0 \leq x \leq 1$

CAS **21–22** Use either a CAS or a table of integrals to find the exact area of the surface obtained by rotating the given curve about the x-axis.

21. $y = 1/x,\quad 1 \leq x \leq 2$

22. $y = \sqrt{x^2 + 1},\quad 0 \leq x \leq 3$

CAS **23–24** Use a CAS to find the exact area of the surface obtained by rotating the curve about the y-axis. If your CAS has trouble evaluating the integral, express the surface area as an integral in the other variable.

23. $y = x^3,\quad 0 \leq y \leq 1$

24. $y = \ln(x + 1),\quad 0 \leq x \leq 1$

25. If the region $\mathcal{R} = \{(x, y) \mid x \geq 1,\ 0 \leq y \leq 1/x\}$ is rotated about the x-axis, the volume of the resulting solid is finite (see Exercise 63 in Section 7.8). Show that the surface area is infinite. (The surface is shown in the figure and is known as **Gabriel's horn**.)

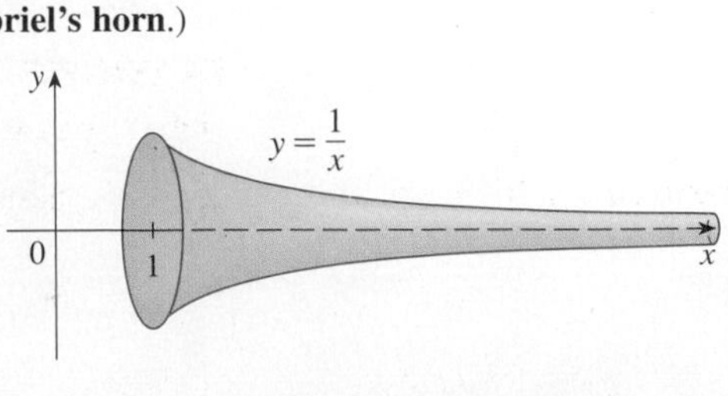

26. If the infinite curve $y = e^{-x}$, $x \geqslant 0$, is rotated about the x-axis, find the area of the resulting surface.

27. (a) If $a > 0$, find the area of the surface generated by rotating the loop of the curve $3ay^2 = x(a - x)^2$ about the x-axis.
(b) Find the surface area if the loop is rotated about the y-axis.

28. A group of engineers is building a parabolic satellite dish whose shape will be formed by rotating the curve $y = ax^2$ about the y-axis. If the dish is to have a 10-ft diameter and a maximum depth of 2 ft, find the value of a and the surface area of the dish.

29. (a) The ellipse

$$\frac{x^2}{a^2} + \frac{y^2}{b^2} = 1 \qquad a > b$$

is rotated about the x-axis to form a surface called an *ellipsoid, or prolate spheroid.* Find the surface area of this ellipsoid.
(b) If the ellipse in part (a) is rotated about its minor axis (the y-axis), the resulting ellipsoid is called an *oblate spheroid.* Find the surface area of this ellipsoid.

30. Find the surface area of the torus in Exercise 63 in Section 6.2.

31. If the curve $y = f(x)$, $a \leqslant x \leqslant b$, is rotated about the horizontal line $y = c$, where $f(x) \leqslant c$, find a formula for the area of the resulting surface.

CAS 32. Use the result of Exercise 31 to set up an integral to find the area of the surface generated by rotating the curve $y = \sqrt{x}$, $0 \leqslant x \leqslant 4$, about the line $y = 4$. Then use a CAS to evaluate the integral.

33. Find the area of the surface obtained by rotating the circle $x^2 + y^2 = r^2$ about the line $y = r$.

34. Show that the surface area of a zone of a sphere that lies between two parallel planes is $S = \pi dh$, where d is the diameter of the sphere and h is the distance between the planes. (Notice that S depends only on the distance between the planes and not on their location, provided that both planes intersect the sphere.)

35. Formula 4 is valid only when $f(x) \geqslant 0$. Show that when $f(x)$ is not necessarily positive, the formula for surface area becomes

$$S = \int_a^b 2\pi |f(x)| \sqrt{1 + [f'(x)]^2}\, dx$$

36. Let L be the length of the curve $y = f(x)$, $a \leqslant x \leqslant b$, where f is positive and has a continuous derivative. Let S_f be the surface area generated by rotating the curve about the x-axis. If c is a positive constant, define $g(x) = f(x) + c$ and let S_g be the corresponding surface area generated by the curve $y = g(x)$, $a \leqslant x \leqslant b$. Express S_g in terms of S_f and L.

DISCOVERY PROJECT — ROTATING ON A SLANT

We know how to find the volume of a solid of revolution obtained by rotating a region about a horizontal or vertical line (see Section 6.2). We also know how to find the surface area of a surface of revolution if we rotate a curve about a horizontal or vertical line (see Section 8.2). But what if we rotate about a slanted line, that is, a line that is neither horizontal nor vertical? In this project you are asked to discover formulas for the volume of a solid of revolution and for the area of a surface of revolution when the axis of rotation is a slanted line.

Let C be the arc of the curve $y = f(x)$ between the points $P(p, f(p))$ and $Q(q, f(q))$ and let $\mathcal{R}$ be the region bounded by C, by the line $y = mx + b$ (which lies entirely below C), and by the perpendiculars to the line from P and Q.

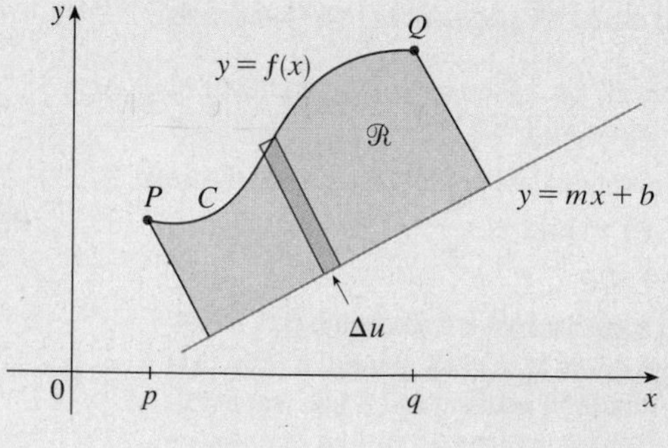

1. Show that the area of $\mathcal{R}$ is

$$\frac{1}{1+m^2}\int_p^q [f(x) - mx - b][1 + mf'(x)]\,dx$$

[*Hint:* This formula can be verified by subtracting areas, but it will be helpful throughout the project to derive it by first approximating the area using rectangles perpendicular to the line, as shown in the figure. Use the figure to help express Δu in terms of Δx.]

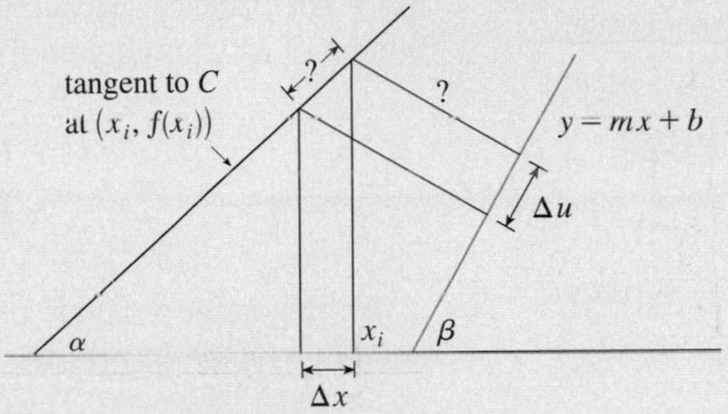

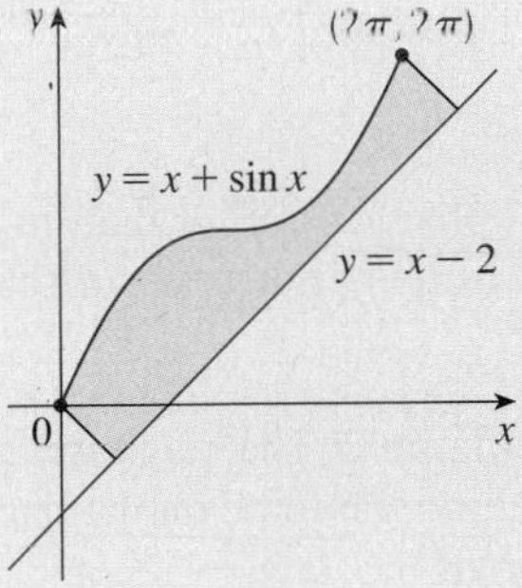

2. Find the area of the region shown in the figure at the left.

3. Find a formula similar to the one in Problem 1 for the volume of the solid obtained by rotating $\mathcal{R}$ about the line $y = mx + b$.

4. Find the volume of the solid obtained by rotating the region of Problem 2 about the line $y = x - 2$.

5. Find a formula for the area of the surface obtained by rotating C about the line $y = mx + b$.

CAS **6.** Use a computer algebra system to find the exact area of the surface obtained by rotating the curve $y = \sqrt{x}$, $0 \leqslant x \leqslant 4$, about the line $y = \frac{1}{2}x$. Then approximate your result to three decimal places.

8.3 APPLICATIONS TO PHYSICS AND ENGINEERING

Among the many applications of integral calculus to physics and engineering, we consider two here: force due to water pressure and centers of mass. As with our previous applications to geometry (areas, volumes, and lengths) and to work, our strategy is to break up the physical quantity into a large number of small parts, approximate each small part, add the results, take the limit, and then evaluate the resulting integral.

HYDROSTATIC FORCE AND PRESSURE

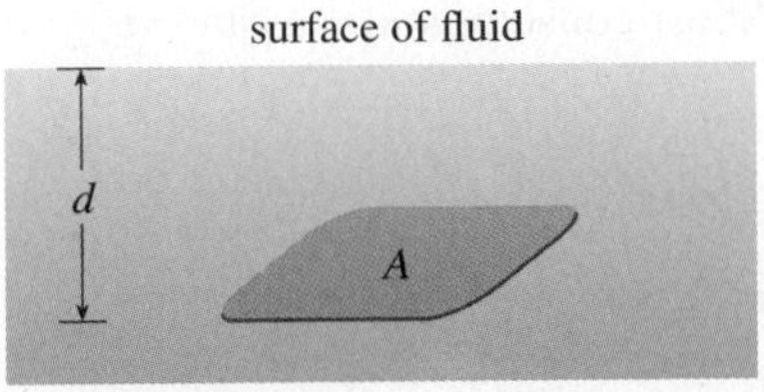

FIGURE 1

Deep-sea divers realize that water pressure increases as they dive deeper. This is because the weight of the water above them increases.

In general, suppose that a thin horizontal plate with area A square meters is submerged in a fluid of density ρ kilograms per cubic meter at a depth d meters below the surface of the fluid as in Figure 1. The fluid directly above the plate has volume $V = Ad$, so its mass is $m = \rho V = \rho Ad$. The force exerted by the fluid on the plate is therefore

$$F = mg = \rho gAd$$

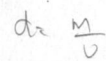

where g is the acceleration due to gravity. The **pressure** P on the plate is defined to be the force per unit area:

$$P = \frac{F}{A} = \rho g d$$

▪ When using US Customary units, we write $P = \rho g d = \delta d$, where $\delta = \rho g$ is the *weight density* (as opposed to ρ, which is the *mass density*). For instance, the weight density of water is $\delta = 62.5 \text{ lb/ft}^3$.

$\delta = 62.5 \text{ lb/ft}^3 = \rho \cdot g$ is the weight density of water

The SI unit for measuring pressure is newtons per square meter, which is called a pascal (abbreviation: $1 \text{ N/m}^2 = 1 \text{ Pa}$). Since this is a small unit, the kilopascal (kPa) is often used. For instance, because the density of water is $\rho = 1000 \text{ kg/m}^3$, the pressure at the bottom of a swimming pool 2 m deep is

$$\begin{aligned} P = \rho g d &= 1000 \text{ kg/m}^3 \times 9.8 \text{ m/s}^2 \times 2 \text{ m} \\ &= 19{,}600 \text{ Pa} = 19.6 \text{ kPa} \end{aligned}$$

An important principle of fluid pressure is the experimentally verified fact that *at any point in a liquid the pressure is the same in all directions.* (A diver feels the same pressure on nose and both ears.) Thus the pressure in *any* direction at a depth d in a fluid with mass density ρ is given by

$$\boxed{1} \qquad P = \rho g d = \delta d$$

This helps us determine the hydrostatic force against a *vertical* plate or wall or dam in a fluid. This is not a straightforward problem because the pressure is not constant but increases as the depth increases.

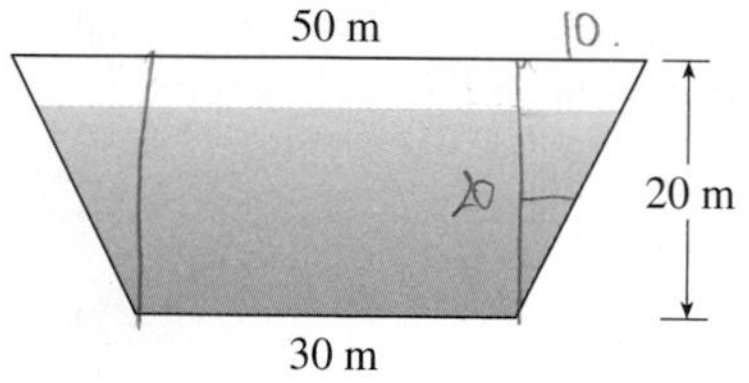

FIGURE 2

V EXAMPLE 1 A dam has the shape of the trapezoid shown in Figure 2. The height is 20 m, and the width is 50 m at the top and 30 m at the bottom. Find the force on the dam due to hydrostatic pressure if the water level is 4 m from the top of the dam.

SOLUTION We choose a vertical x-axis with origin at the surface of the water as in Figure 3(a). The depth of the water is 16 m, so we divide the interval $[0, 16]$ into subintervals of equal length with endpoints x_i and we choose $x_i^* \in [x_{i-1}, x_i]$. The ith horizontal strip of the dam is approximated by a rectangle with height Δx and width w_i, where, from similar triangles in Figure 3(b),

$$\frac{a}{16 - x_i^*} = \frac{10}{20} \qquad \text{or} \qquad a = \frac{16 - x_i^*}{2} = 8 - \frac{x_i^*}{2}$$

and so

$$w_i = 2(15 + a) = 2(15 + 8 - \tfrac{1}{2}x_i^*) = 46 - x_i^*$$

If A_i is the area of the ith strip, then

$$A_i \approx w_i \Delta x = (46 - x_i^*)\,\Delta x$$

If Δx is small, then the pressure P_i on the ith strip is almost constant and we can use Equation 1 to write

$$P_i \approx 1000 g x_i^*$$

The hydrostatic force F_i acting on the ith strip is the product of the pressure and the area:

$$F_i = P_i A_i \approx 1000 g x_i^* (46 - x_i^*)\,\Delta x$$

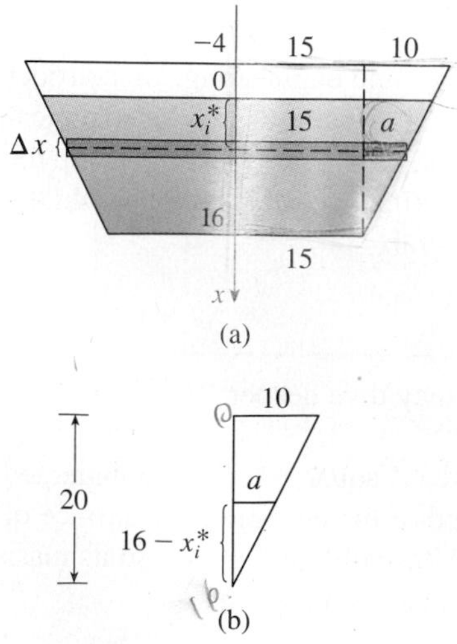

FIGURE 3

Adding these forces and taking the limit as $n \to \infty$, we obtain the total hydrostatic force on the dam:

$$
\begin{aligned}
F &= \lim_{n\to\infty} \sum_{i=1}^{n} 1000gx_i^*(46 - x_i^*)\,\Delta x \\
&= \int_0^{16} 1000gx(46 - x)\,dx \\
&= 1000(9.8)\int_0^{16} (46x - x^2)\,dx \\
&= 9800\left[23x^2 - \frac{x^3}{3}\right]_0^{16} \\
&\approx 4.43 \times 10^7\ \text{N}
\end{aligned}
$$

□

EXAMPLE 2 Find the hydrostatic force on one end of a cylindrical drum with radius 3 ft if the drum is submerged in water 10 ft deep.

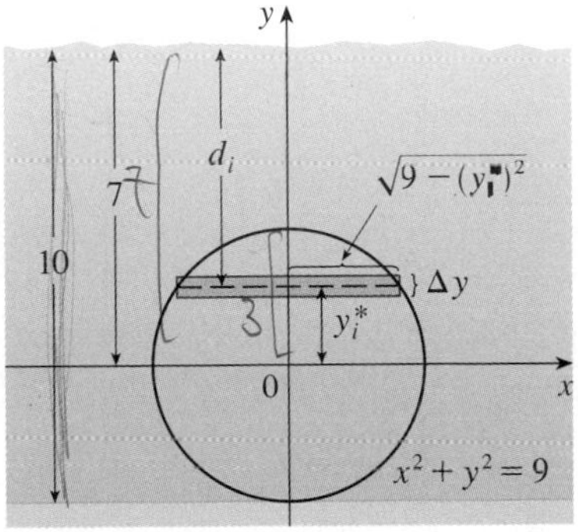

FIGURE 4

SOLUTION In this example it is convenient to choose the axes as in Figure 4 so that the origin is placed at the center of the drum. Then the circle has a simple equation, $x^2 + y^2 = 9$. As in Example 1 we divide the circular region into horizontal strips of equal width. From the equation of the circle, we see that the length of the ith strip is $2\sqrt{9 - (y_i^*)^2}$ and so its area is

$$A_i = 2\sqrt{9 - (y_i^*)^2}\,\Delta y$$

The pressure on this strip is approximately

$$\delta d_i = 62.5(7 - y_i^*)$$

and so the force on the strip is approximately

$$\delta d_i A_i = 62.5(7 - y_i^*)2\sqrt{9 - (y_i^*)^2}\,\Delta y$$

The total force is obtained by adding the forces on all the strips and taking the limit:

$$
\begin{aligned}
F &= \lim_{n\to\infty} \sum_{i=1}^{n} 62.5(7 - y_i^*)2\sqrt{9 - (y_i^*)^2}\,\Delta y \\
&= 125\int_{-3}^{3} (7 - y)\sqrt{9 - y^2}\,dy \\
&= 125 \cdot 7\int_{-3}^{3} \sqrt{9 - y^2}\,dy - 125\int_{-3}^{3} y\sqrt{9 - y^2}\,dy
\end{aligned}
$$

The second integral is 0 because the integrand is an odd function (see Theorem 5.5.7). The first integral can be evaluated using the trigonometric substitution $y = 3\sin\theta$, but it's simpler to observe that it is the area of a semicircular disk with radius 3. Thus

$$
\begin{aligned}
F &= 875\int_{-3}^{3} \sqrt{9 - y^2}\,dy = 875 \cdot \tfrac{1}{2}\pi(3)^2 \\
&= \frac{7875\pi}{2} \approx 12{,}370\ \text{lb}
\end{aligned}
$$

□

MOMENTS AND CENTERS OF MASS

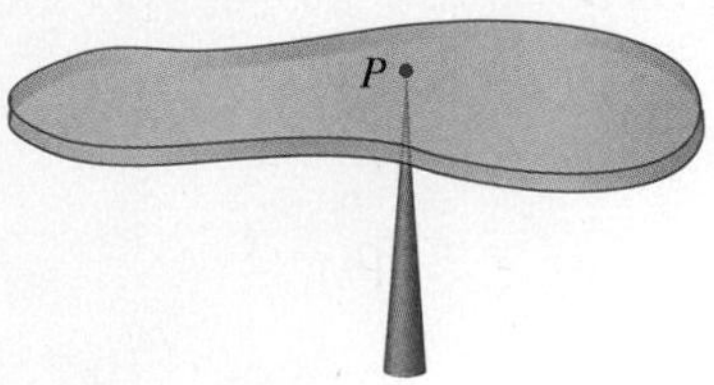

FIGURE 5

Our main objective here is to find the point P on which a thin plate of any given shape balances horizontally as in Figure 5. This point is called the **center of mass** (or center of gravity) of the plate.

We first consider the simpler situation illustrated in Figure 6, where two masses m_1 and m_2 are attached to a rod of negligible mass on opposite sides of a fulcrum and at distances d_1 and d_2 from the fulcrum. The rod will balance if

$$m_1 d_1 = m_2 d_2 \tag{2}$$

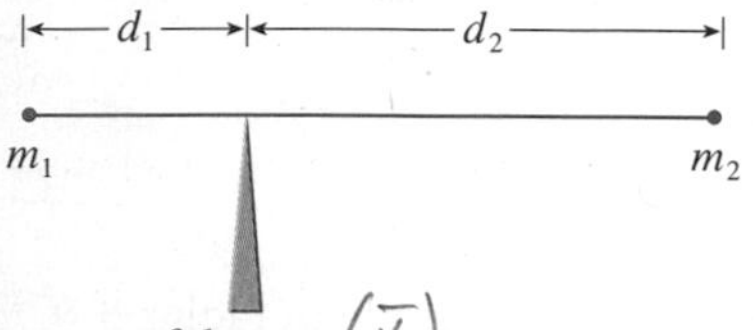

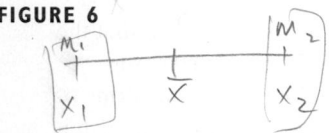

FIGURE 6

This is an experimental fact discovered by Archimedes and called the Law of the Lever. (Think of a lighter person balancing a heavier one on a seesaw by sitting farther away from the center.)

Now suppose that the rod lies along the x-axis with m_1 at x_1 and m_2 at x_2 and the center of mass at $\bar{x}$. If we compare Figures 6 and 7, we see that $d_1 = \bar{x} - x_1$ and $d_2 = x_2 - \bar{x}$ and so Equation 2 gives

$$m_1(\bar{x} - x_1) = m_2(x_2 - \bar{x})$$

$$m_1\bar{x} + m_2\bar{x} = m_1 x_1 + m_2 x_2$$

$$\bar{x} = \frac{m_1 x_1 + m_2 x_2}{m_1 + m_2} \tag{3}$$

The numbers $m_1 x_1$ and $m_2 x_2$ are called the **moments** of the masses m_1 and m_2 (with respect to the origin), and Equation 3 says that the center of mass $\bar{x}$ is obtained by adding the moments of the masses and dividing by the total mass $m = m_1 + m_2$.

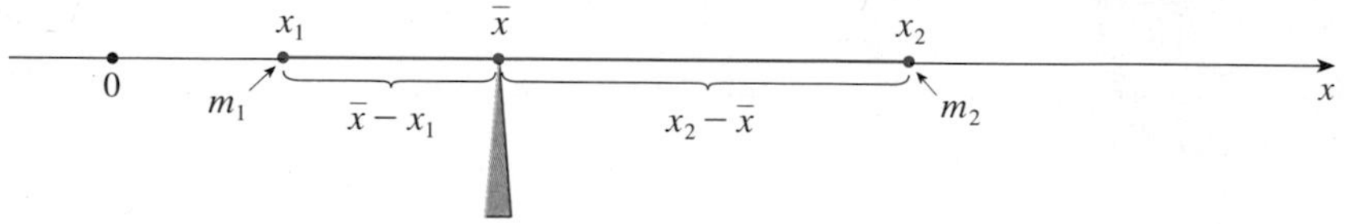

FIGURE 7

In general, if we have a system of n particles with masses $m_1, m_2, \ldots, m_n$ located at the points $x_1, x_2, \ldots, x_n$ on the x-axis, it can be shown similarly that the center of mass of the system is located at

$$\bar{x} = \frac{\sum_{i=1}^{n} m_i x_i}{\sum_{i=1}^{n} m_i} = \frac{\sum_{i=1}^{n} m_i x_i}{m} \tag{4}$$

where $m = \sum m_i$ is the total mass of the system, and the sum of the individual moments

$$M = \sum_{i=1}^{n} m_i x_i$$

is called the **moment of the system about the origin.** Then Equation 4 could be rewritten as $m\bar{x} = M$, which says that if the total mass were considered as being concentrated at the center of mass $\bar{x}$, then its moment would be the same as the moment of the system.

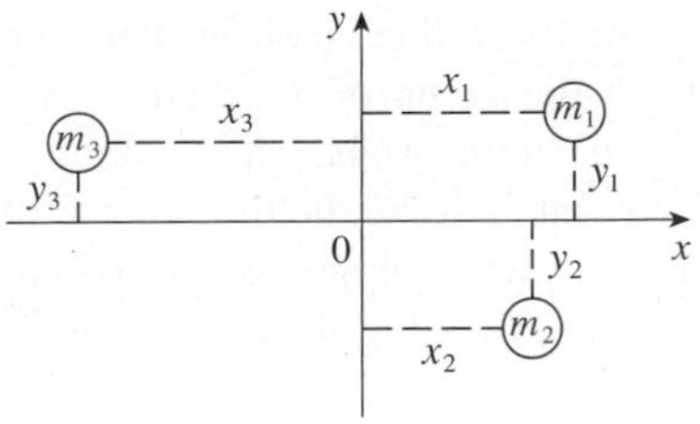

FIGURE 8

Now we consider a system of n particles with masses $m_1, m_2, \ldots, m_n$ located at the points $(x_1, y_1), (x_2, y_2), \ldots, (x_n, y_n)$ in the xy-plane as shown in Figure 8. By analogy with the one-dimensional case, we define the **moment of the system about the y-axis** to be

5 $$M_y = \sum_{i=1}^{n} m_i x_i$$

and the **moment of the system about the x-axis** as

6 $$M_x = \sum_{i=1}^{n} m_i y_i$$

Then M_y measures the tendency of the system to rotate about the y-axis and M_x measures the tendency to rotate about the x-axis.

As in the one-dimensional case, the coordinates $(\bar{x}, \bar{y})$ of the center of mass are given in terms of the moments by the formulas

7 $$\bar{x} = \frac{M_y}{m} \qquad \bar{y} = \frac{M_x}{m}$$

where $m = \sum m_i$ is the total mass. Since $m\bar{x} = M_y$ and $m\bar{y} = M_x$, the center of mass $(\bar{x}, \bar{y})$ is the point where a single particle of mass m would have the same moments as the system.

EXAMPLE 3 Find the moments and center of mass of the system of objects that have masses 3, 4, and 8 at the points $(-1, 1)$, $(2, -1)$, and $(3, 2)$, respectively.

SOLUTION We use Equations 5 and 6 to compute the moments:

$$M_y = 3(-1) + 4(2) + 8(3) = 29$$

$$M_x = 3(1) + 4(-1) + 8(2) = 15$$

Since $m = 3 + 4 + 8 = 15$, we use Equations 7 to obtain

$$\bar{x} = \frac{M_y}{m} = \frac{29}{15} \qquad \bar{y} = \frac{M_x}{m} = \frac{15}{15} = 1$$

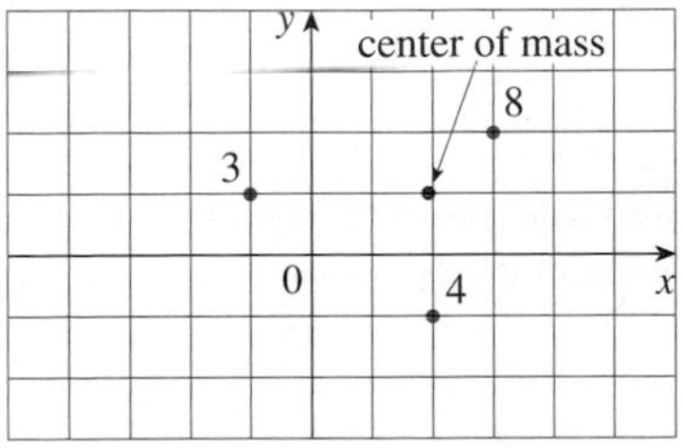

FIGURE 9

Thus the center of mass is $\left(1\frac{14}{15}, 1\right)$. (See Figure 9.) □

Next we consider a flat plate (called a *lamina*) with uniform density ρ that occupies a region $\mathcal{R}$ of the plane. We wish to locate the center of mass of the plate, which is called the **centroid** of $\mathcal{R}$. In doing so we use the following physical principles: The **symmetry principle** says that if $\mathcal{R}$ is symmetric about a line l, then the centroid of $\mathcal{R}$ lies on l. (If $\mathcal{R}$ is reflected about l, then $\mathcal{R}$ remains the same so its centroid remains fixed. But the only fixed points lie on l.) Thus the centroid of a rectangle is its center. Moments should be defined so that if the entire mass of a region is concentrated at the center of mass, then its moments remain unchanged. Also, the moment of the union of two nonoverlapping regions should be the sum of the moments of the individual regions.

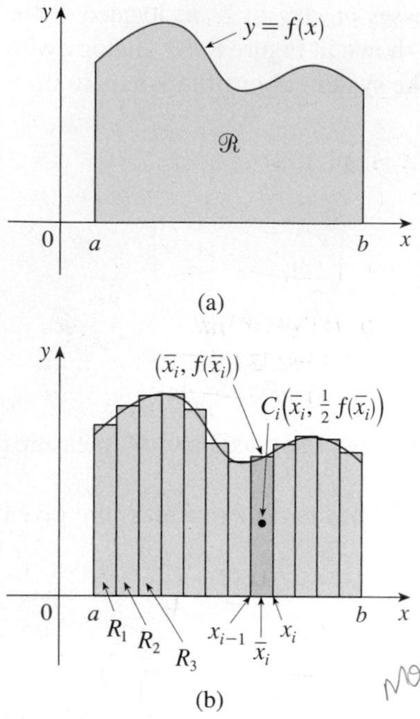

FIGURE 10

Suppose that the region $\mathcal{R}$ is of the type shown in Figure 10(a); that is, $\mathcal{R}$ lies between the lines $x = a$ and $x = b$, above the x-axis, and beneath the graph of f, where f is a continuous function. We divide the interval $[a, b]$ into n subintervals with endpoints $x_0, x_1, \ldots, x_n$ and equal width Δx. We choose the sample point x_i^* to be the midpoint $\bar{x}_i$ of the ith subinterval, that is, $\bar{x}_i = (x_{i-1} + x_i)/2$. This determines the polygonal approximation to $\mathcal{R}$ shown in Figure 10(b). The centroid of the ith approximating rectangle R_i is its center $C_i\left(\bar{x}_i, \tfrac{1}{2} f(\bar{x}_i)\right)$. Its area is $f(\bar{x}_i)\,\Delta x$, so its mass is

$$\rho f(\bar{x}_i)\,\Delta x$$

The moment of R_i about the y-axis is the product of its mass and the distance from C_i to the y-axis, which is $\bar{x}_i$. Thus

$$M_y(R_i) = [\rho f(\bar{x}_i)\,\Delta x]\,\bar{x}_i = \rho \bar{x}_i f(\bar{x}_i)\,\Delta x$$

Adding these moments, we obtain the moment of the polygonal approximation to $\mathcal{R}$, and then by taking the limit as $n \to \infty$ we obtain the moment of $\mathcal{R}$ itself about the y-axis:

$$M_y = \lim_{n\to\infty} \sum_{i=1}^{n} \rho \bar{x}_i f(\bar{x}_i)\,\Delta x = \rho \int_a^b x f(x)\,dx$$

In a similar fashion we compute the moment of R_i about the x-axis as the product of its mass and the distance from C_i to the x-axis:

$$M_x(R_i) = [\rho f(\bar{x}_i)\,\Delta x]\,\tfrac{1}{2} f(\bar{x}_i) = \rho \cdot \tfrac{1}{2}[f(\bar{x}_i)]^2\,\Delta x$$

Again we add these moments and take the limit to obtain the moment of $\mathcal{R}$ about the x-axis:

$$M_x = \lim_{n\to\infty} \sum_{i=1}^{n} \rho \cdot \tfrac{1}{2}[f(\bar{x}_i)]^2\,\Delta x = \rho \int_a^b \tfrac{1}{2}[f(x)]^2\,dx$$

Just as for systems of particles, the center of mass of the plate is defined so that $m\bar{x} = M_y$ and $m\bar{y} = M_x$. But the mass of the plate is the product of its density and its area:

$$m = \rho A = \rho \int_a^b f(x)\,dx$$

and so

$$\bar{x} = \frac{M_y}{m} = \frac{\rho \int_a^b x f(x)\,dx}{\rho \int_a^b f(x)\,dx} = \frac{\int_a^b x f(x)\,dx}{\int_a^b f(x)\,dx}$$

$$\bar{y} = \frac{M_x}{m} = \frac{\rho \int_a^b \tfrac{1}{2}[f(x)]^2\,dx}{\rho \int_a^b f(x)\,dx} = \frac{\int_a^b \tfrac{1}{2}[f(x)]^2\,dx}{\int_a^b f(x)\,dx}$$

Notice the cancellation of the ρ's. The location of the center of mass is independent of the density.

In summary, the center of mass of the plate (or the centroid of $\mathcal{R}$) is located at the point $(\bar{x}, \bar{y})$, where

$$\boxed{8} \qquad \bar{x} = \frac{1}{A}\int_a^b x f(x)\,dx \qquad \bar{y} = \frac{1}{A}\int_a^b \tfrac{1}{2}[f(x)]^2\,dx$$

EXAMPLE 4 Find the center of mass of a semicircular plate of radius r.

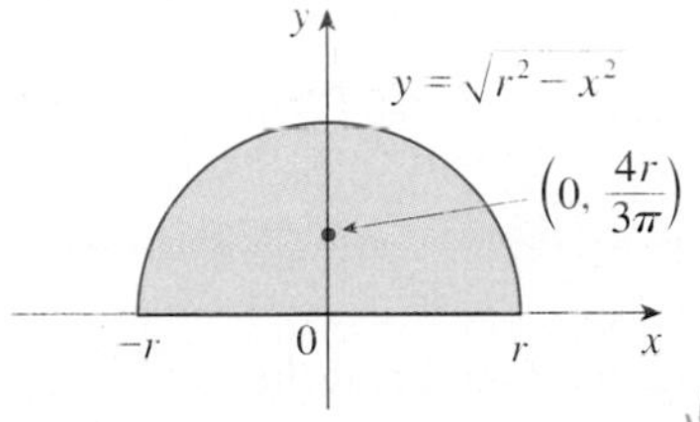

FIGURE 11

SOLUTION In order to use (8) we place the semicircle as in Figure 11 so that $f(x) = \sqrt{r^2 - x^2}$ and $a = -r$, $b = r$. Here there is no need to use the formula to calculate $\bar{x}$ because, by the symmetry principle, the center of mass must lie on the y-axis, so $\bar{x} = 0$. The area of the semicircle is $A = \frac{1}{2}\pi r^2$, so

$$\begin{aligned}
\bar{y} &= \frac{1}{A}\int_{-r}^{r} \tfrac{1}{2}[f(x)]^2\,dx \\
&= \frac{1}{\frac{1}{2}\pi r^2} \cdot \tfrac{1}{2}\int_{-r}^{r} \left(\sqrt{r^2 - x^2}\right)^2 dx \\
&= \frac{2}{\pi r^2}\int_0^r (r^2 - x^2)\,dx = \frac{2}{\pi r^2}\left[r^2 x - \frac{x^3}{3}\right]_0^r \\
&= \frac{2}{\pi r^2}\,\frac{2r^3}{3} = \frac{4r}{3\pi}
\end{aligned}$$

The center of mass is located at the point $(0, 4r/(3\pi))$. □

EXAMPLE 5 Find the centroid of the region bounded by the curves $y = \cos x$, $y = 0$, $x = 0$, and $x = \pi/2$.

SOLUTION The area of the region is

$$A = \int_0^{\pi/2} \cos x\,dx = \sin x\Big]_0^{\pi/2} = 1$$

so Formulas 8 give

$$\begin{aligned}
\bar{x} &= \frac{1}{A}\int_0^{\pi/2} x f(x)\,dx = \int_0^{\pi/2} x \cos x\,dx \\
&= x \sin x\Big]_0^{\pi/2} - \int_0^{\pi/2} \sin x\,dx \qquad \text{(by integration by parts)} \\
&= \frac{\pi}{2} - 1
\end{aligned}$$

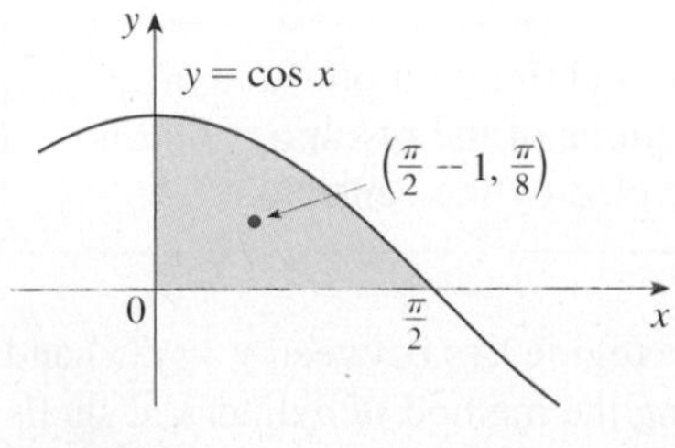

FIGURE 12

$$\begin{aligned}
\bar{y} &= \frac{1}{A}\int_0^{\pi/2} \tfrac{1}{2}[f(x)]^2\,dx = \tfrac{1}{2}\int_0^{\pi/2} \cos^2 x\,dx \\
&= \tfrac{1}{4}\int_0^{\pi/2} (1 + \cos 2x)\,dx = \tfrac{1}{4}\left[x + \tfrac{1}{2}\sin 2x\right]_0^{\pi/2} \\
&= \frac{\pi}{8}
\end{aligned}$$

The centroid is $\left(\frac{1}{2}\pi - 1, \frac{1}{8}\pi\right)$ and is shown in Figure 12. □

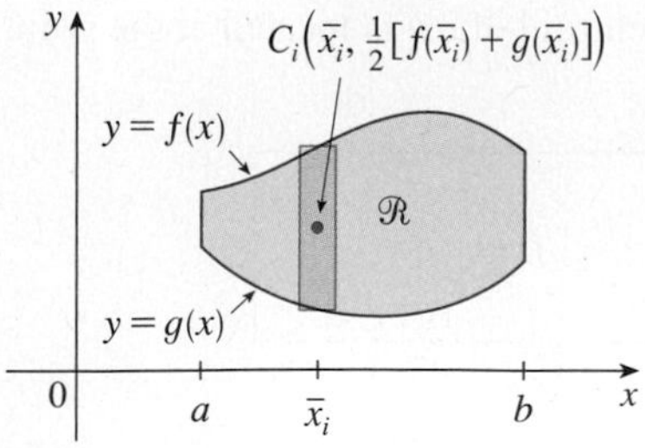

FIGURE 13

If the region ℛ lies between two curves $y = f(x)$ and $y = g(x)$, where $f(x) \geqslant g(x)$, as illustrated in Figure 13, then the same sort of argument that led to Formulas 8 can be used to show that the centroid of ℛ is $(\bar{x}, \bar{y})$, where

9

Centroid between 2 curves

$$\bar{x} = \frac{1}{A} \int_a^b x[f(x) - g(x)]\, dx$$

$$\bar{y} = \frac{1}{A} \int_a^b \tfrac{1}{2}\{[f(x)]^2 - [g(x)]^2\}\, dx$$

(See Exercise 47.)

EXAMPLE 6 Find the centroid of the region bounded by the line $y = x$ and the parabola $y = x^2$.

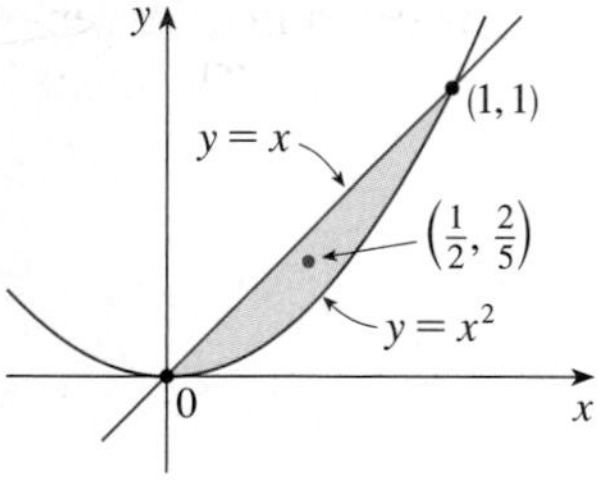

FIGURE 14

SOLUTION The region is sketched in Figure 14. We take $f(x) = x$, $g(x) = x^2$, $a = 0$, and $b = 1$ in Formulas 9. First we note that the area of the region is

$$A = \int_0^1 (x - x^2)\, dx = \frac{x^2}{2} - \frac{x^3}{3}\Bigg]_0^1 = \frac{1}{6}$$

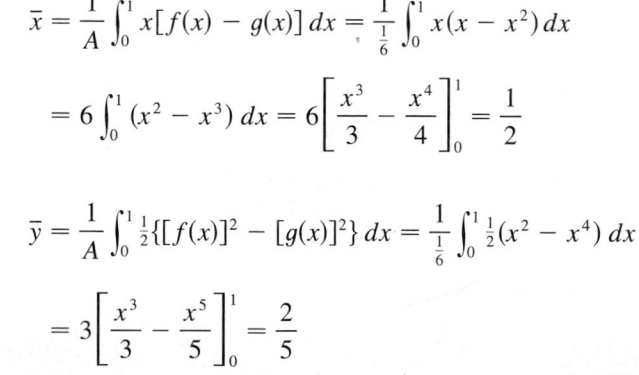

Therefore

$$\bar{x} = \frac{1}{A} \int_0^1 x[f(x) - g(x)]\, dx = \frac{1}{\frac{1}{6}} \int_0^1 x(x - x^2)\, dx$$

$$= 6 \int_0^1 (x^2 - x^3)\, dx = 6\left[\frac{x^3}{3} - \frac{x^4}{4}\right]_0^1 = \frac{1}{2}$$

$$\bar{y} = \frac{1}{A} \int_0^1 \tfrac{1}{2}\{[f(x)]^2 - [g(x)]^2\}\, dx = \frac{1}{\frac{1}{6}} \int_0^1 \tfrac{1}{2}(x^2 - x^4)\, dx$$

$$= 3\left[\frac{x^3}{3} - \frac{x^5}{5}\right]_0^1 = \frac{2}{5}$$

The centroid is $\left(\frac{1}{2}, \frac{2}{5}\right)$. □

We end this section by showing a surprising connection between centroids and volumes of revolution.

■ This theorem is named after the Greek mathematician Pappus of Alexandria, who lived in the fourth century AD.

THEOREM OF PAPPUS Let ℛ be a plane region that lies entirely on one side of a line l in the plane. If ℛ is rotated about l, then the volume of the resulting solid is the product of the area A of ℛ and the distance d traveled by the centroid of ℛ.

PROOF We give the proof for the special case in which the region lies between $y = f(x)$ and $y = g(x)$ as in Figure 13 and the line l is the y-axis. Using the method of cylindrical shells (see Section 6.3), we have

$$\begin{aligned} V &= \int_a^b 2\pi x[f(x) - g(x)]\,dx \\ &= 2\pi \int_a^b x[f(x) - g(x)]\,dx \\ &= 2\pi(\bar{x}A) \qquad \text{(by Formulas 9)} \\ &= (2\pi\bar{x})A = Ad \end{aligned}$$

where $d = 2\pi\bar{x}$ is the distance traveled by the centroid during one rotation about the y-axis. □

EXAMPLE 7 A torus is formed by rotating a circle of radius r about a line in the plane of the circle that is a distance $R\ (> r)$ from the center of the circle. Find the volume of the torus.

SOLUTION The circle has area $A = \pi r^2$. By the symmetry principle, its centroid is its center and so the distance traveled by the centroid during a rotation is $d = 2\pi R$. Therefore, by the Theorem of Pappus, the volume of the torus is

$$V = Ad = (2\pi R)(\pi r^2) = 2\pi^2 r^2 R$$ □

The method of Example 7 should be compared with the method of Exercise 63 in Section 6.2.

8.3 EXERCISES

1. An aquarium 5 ft long, 2 ft wide, and 3 ft deep is full of water. Find (a) the hydrostatic pressure on the bottom of the aquarium, (b) the hydrostatic force on the bottom, and (c) the hydrostatic force on one end of the aquarium.

2. A tank is 8 m long, 4 m wide, 2 m high, and contains kerosene with density 820 kg/m^3 to a depth of 1.5 m. Find (a) the hydrostatic pressure on the bottom of the tank, (b) the hydrostatic force on the bottom, and (c) the hydrostatic force on one end of the tank.

3–11 A vertical plate is submerged (or partially submerged) in water and has the indicated shape. Explain how to approximate the hydrostatic force against one side of the plate by a Riemann sum. Then express the force as an integral and evaluate it.

3.

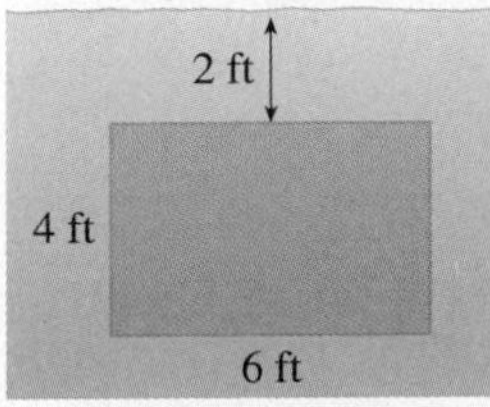

4.

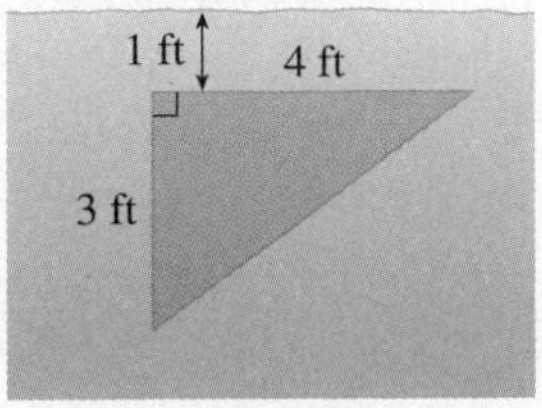

5.

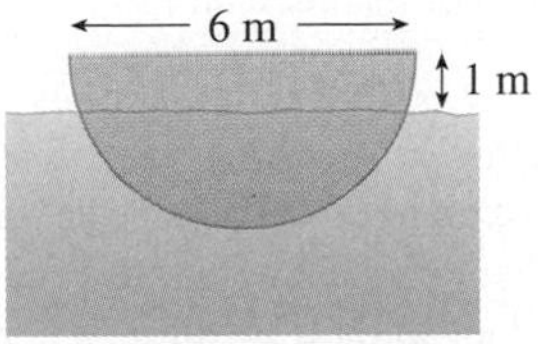

6.

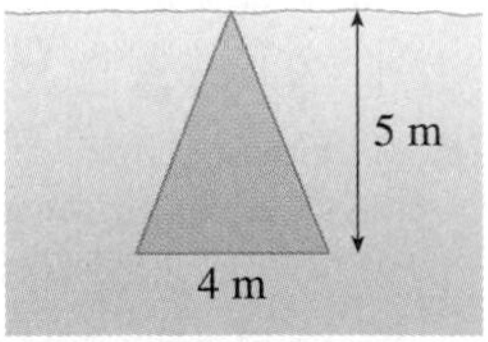

7.

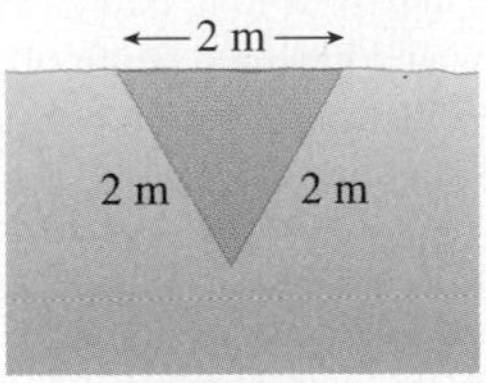

8.

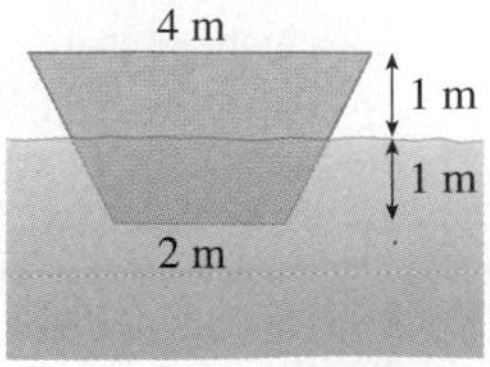

9.

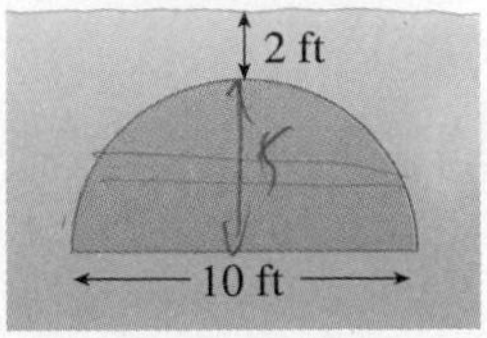

10.

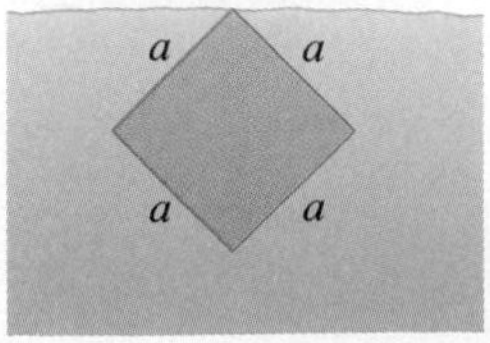

11.

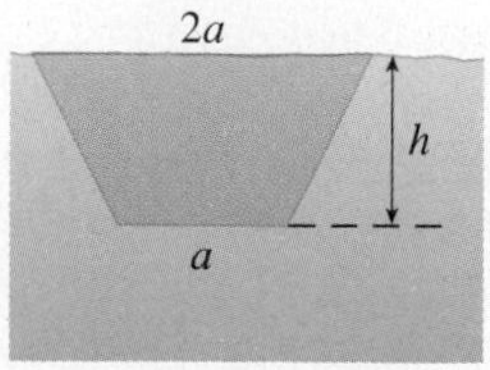

12. A large tank is designed with ends in the shape of the region between the curves $y = \frac{1}{2}x^2$ and $y = 12$, measured in feet. Find the hydrostatic force on one end of the tank if it is filled to a depth of 8 ft with gasoline. (Assume the gasoline's density is 42.0 lb/ft^3.)

13. A trough is filled with a liquid of density 840 kg/m^3. The ends of the trough are equilateral triangles with sides 8 m long and vertex at the bottom. Find the hydrostatic force on one end of the trough.

14. A vertical dam has a semicircular gate as shown in the figure. Find the hydrostatic force against the gate.

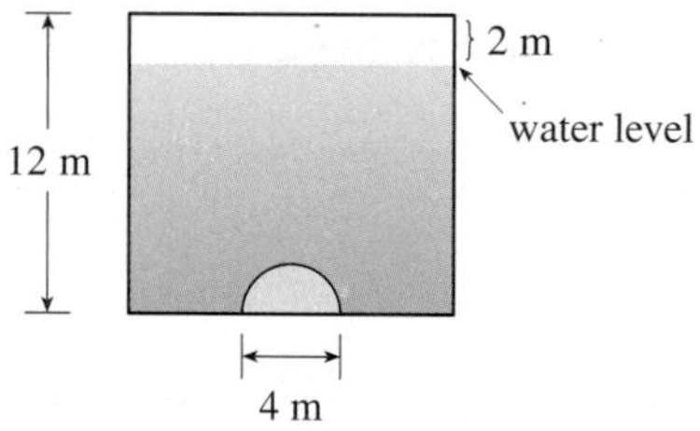

15. A cube with 20-cm-long sides is sitting on the bottom of an aquarium in which the water is one meter deep. Estimate the hydrostatic force on (a) the top of the cube and (b) one of the sides of the cube.

16. A dam is inclined at an angle of 30° from the vertical and has the shape of an isosceles trapezoid 100 ft wide at the top and 50 ft wide at the bottom and with a slant height of 70 ft. Find the hydrostatic force on the dam when it is full of water.

17. A swimming pool is 20 ft wide and 40 ft long and its bottom is an inclined plane, the shallow end having a depth of 3 ft and the deep end, 9 ft. If the pool is full of water, estimate the hydrostatic force on (a) the shallow end, (b) the deep end, (c) one of the sides, and (d) the bottom of the pool.

18. Suppose that a plate is immersed vertically in a fluid with density ρ and the width of the plate is $w(x)$ at a depth of x meters beneath the surface of the fluid. If the top of the plate is at depth a and the bottom is at depth b, show that the hydrostatic force on one side of the plate is

$$F = \int_a^b \rho g x w(x)\, dx$$

19. A vertical, irregularly shaped plate is submerged in water. The table shows measurements of its width, taken at the indicated depths. Use Simpson's Rule to estimate the force of the water against the plate.

Depth (m)	2.0	2.5	3.0	3.5	4.0	4.5	5.0
Plate width (m)	0	0.8	1.7	2.4	2.9	3.3	3.6

20. (a) Use the formula of Exercise 18 to show that

$$F = (\rho g \bar{x})A$$

where $\bar{x}$ is the x-coordinate of the centroid of the plate and A is its area. This equation shows that the hydrostatic force against a vertical plane region is the same as if the region were horizontal at the depth of the centroid of the region.

(b) Use the result of part (a) to give another solution to Exercise 10.

21–22 Point-masses m_i are located on the x-axis as shown. Find the moment M of the system about the origin and the center of mass $\bar{x}$.

21.

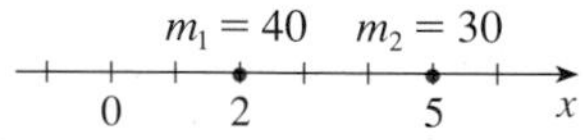

22.

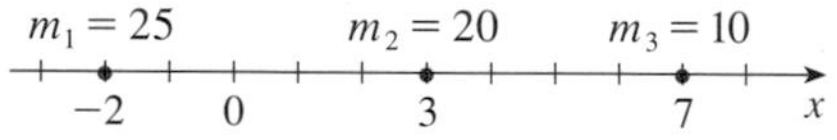

23–24 The masses m_i are located at the points P_i. Find the moments M_x and M_y and the center of mass of the system.

23. $m_1 = 6,\ m_2 = 5,\ m_3 = 10;$
$P_1(1, 5),\ P_2(3, -2),\ P_3(-2, -1)$

24. $m_1 = 6,\ m_2 = 5,\ m_3 = 1,\ m_4 = 4;$
$P_1(1, -2),\ P_2(3, 4),\ P_3(-3, -7),\ P_4(6, -1)$

25–28 Sketch the region bounded by the curves, and visually estimate the location of the centroid. Then find the exact coordinates of the centroid.

25. $y = 4 - x^2,\quad y = 0$

26. $3x + 2y = 6,\quad y = 0,\quad x = 0$

27. $y = e^x,\quad y = 0,\quad x = 0,\quad x = 1$

28. $y = 1/x,\quad y = 0,\quad x = 1,\quad x = 2$

29–33 Find the centroid of the region bounded by the given curves.

29. $y = x^2, \quad x = y^2$

30. $y = x + 2, \quad y = x^2$

31. $y = \sin x, \quad y = \cos x, \quad x = 0, \quad x = \pi/4$

32. $y = x^3, \quad x + y = 2, \quad y = 0$

33. $x = 5 - y^2, \quad x = 0$

34–35 Calculate the moments M_x and M_y and the center of mass of a lamina with the given density and shape.

34. $\rho = 3$

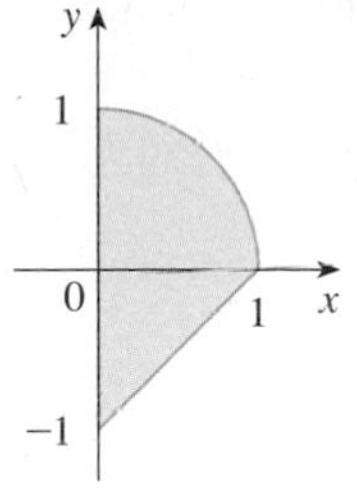

35. $\rho = 10$

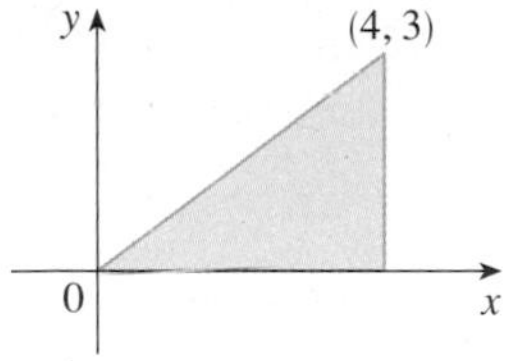

36. Use Simpson's Rule to estimate the centroid of the region shown.

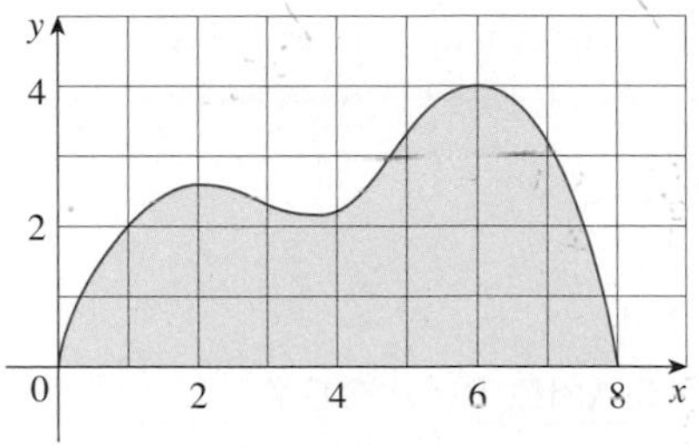

37. Find the centroid of the region bounded by the curves $y = 2^x$ and $y = x^2$, $0 \leqslant x \leqslant 2$, to three decimal places. Sketch the region and plot the centroid to see if your answer is reasonable.

38. Use a graph to find approximate x-coordinates of the points of intersection of the curves $y = x + \ln x$ and $y = x^3 - x$. Then find (approximately) the centroid of the region bounded by these curves.

39. Prove that the centroid of any triangle is located at the point of intersection of the medians. [*Hints:* Place the axes so that the vertices are $(a, 0)$, $(0, b)$, and $(c, 0)$. Recall that a median is a line segment from a vertex to the midpoint of the opposite side. Recall also that the medians intersect at a point two-thirds of the way from each vertex (along the median) to the opposite side.]

40–41 Find the centroid of the region shown, not by integration, but by locating the centroids of the rectangles and triangles (from Exercise 39) and using additivity of moments.

40.

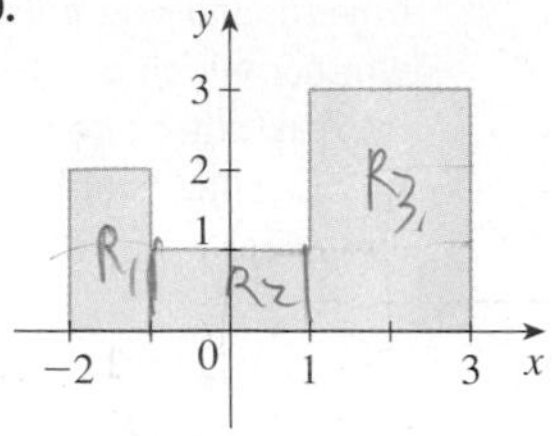

41.

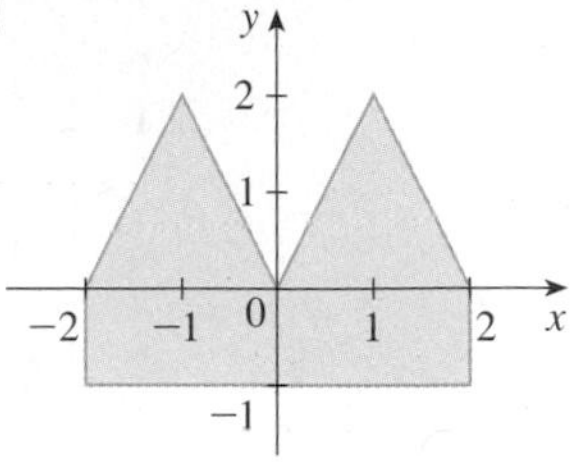

42. A rectangle R with sides a and b is divided into two parts R_1 and R_2 by an arc of a parabola that has its vertex at one corner of R and passes through the opposite corner. Find the centroids of both R_1 and R_2.

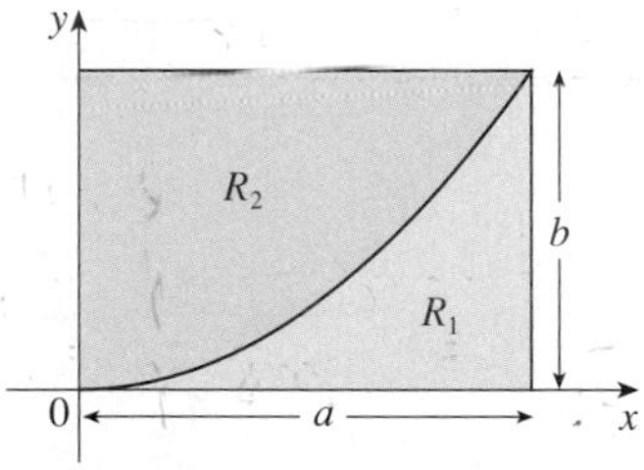

43. If $\bar{x}$ is the x-coordinate of the centroid of the region that lies under the graph of a continuous function f, where $a \leqslant x \leqslant b$, show that

$$\int_a^b (cx + d) f(x)\, dx = (c\bar{x} + d) \int_a^b f(x)\, dx$$

44–46 Use the Theorem of Pappus to find the volume of the given solid.

44. A sphere of radius r (Use Example 4.)

45. A cone with height h and base radius r

46. The solid obtained by rotating the triangle with vertices $(2, 3)$, $(2, 5)$, and $(5, 4)$ about the x-axis

47. Prove Formulas 9.

48. Let $\mathcal{R}$ be the region that lies between the curves $y = x^m$ and $y = x^n$, $0 \leqslant x \leqslant 1$, where m and n are integers with $0 \leqslant n < m$.

(a) Sketch the region $\mathcal{R}$.

(b) Find the coordinates of the centroid of $\mathcal{R}$.

(c) Try to find values of m and n such that the centroid lies *outside* $\mathcal{R}$.

DISCOVERY PROJECT

COMPLEMENTARY COFFEE CUPS

Suppose you have a choice of two coffee cups of the type shown, one that bends outward and one inward, and you notice that they have the same height and their shapes fit together snugly. You wonder which cup holds more coffee. Of course you could fill one cup with water and pour it into the other one but, being a calculus student, you decide on a more mathematical approach. Ignoring the handles, you observe that both cups are surfaces of revolution, so you can think of the coffee as a volume of revolution.

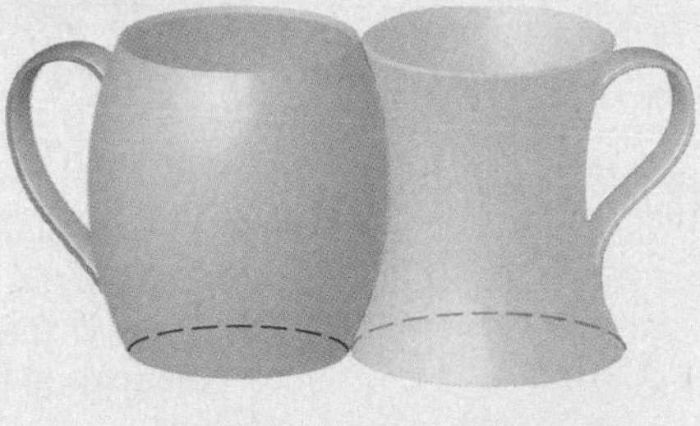

Cup A Cup B

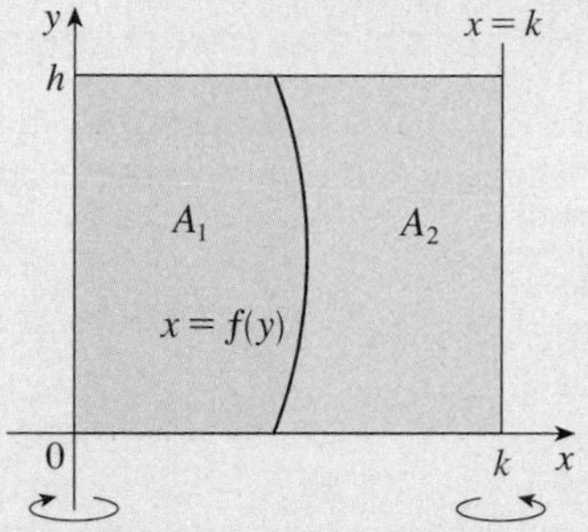

1. Suppose the cups have height h, cup A is formed by rotating the curve $x = f(y)$ about the y-axis, and cup B is formed by rotating the same curve about the line $x = k$. Find the value of k such that the two cups hold the same amount of coffee.

2. What does your result from Problem 1 say about the areas A_1 and A_2 shown in the figure?

3. Use Pappus's Theorem to explain your result in Problems 1 and 2.

4. Based on your own measurements and observations, suggest a value for h and an equation for $x = f(y)$ and calculate the amount of coffee that each cup holds.

8.4 APPLICATIONS TO ECONOMICS AND BIOLOGY

In this section we consider some applications of integration to economics (consumer surplus) and biology (blood flow, cardiac output). Others are described in the exercises.

CONSUMER SURPLUS

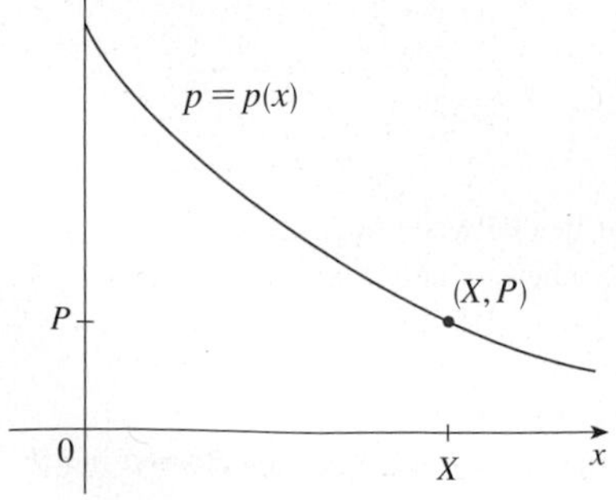

FIGURE 1
A typical demand curve

Recall from Section 4.7 that the demand function $p(x)$ is the price that a company has to charge in order to sell x units of a commodity. Usually, selling larger quantities requires lowering prices, so the demand function is a decreasing function. The graph of a typical demand function, called a **demand curve**, is shown in Figure 1. If X is the amount of the commodity that is currently available, then $P = p(X)$ is the current selling price.

We divide the interval $[0, X]$ into n subintervals, each of length $\Delta x = X/n$, and let $x_i^* = x_i$ be the right endpoint of the ith subinterval, as in Figure 2. If, after the first x_{i-1} units were sold, a total of only x_i units had been available and the price per unit had been set at $p(x_i)$ dollars, then the additional Δx units could have been sold (but no more). The consumers who would have paid $p(x_i)$ dollars placed a high value on the product; they would have paid what it was worth to them. So, in paying only P dollars they have saved an amount of

$$(\text{savings per unit})(\text{number of units}) = [p(x_i) - P]\,\Delta x$$

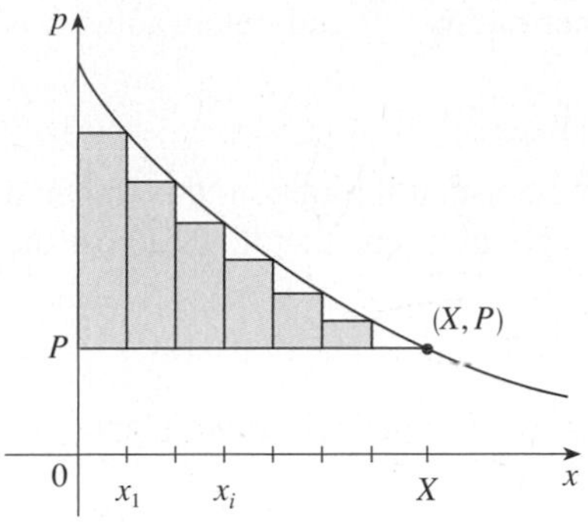

FIGURE 2

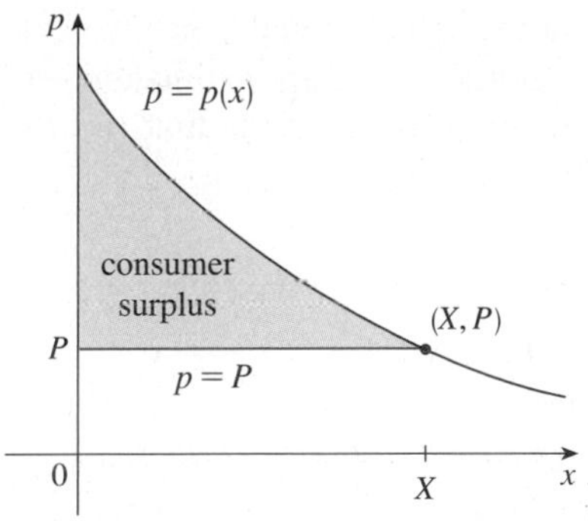

FIGURE 3

Considering similar groups of willing consumers for each of the subintervals and adding the savings, we get the total savings:

$$\sum_{i=1}^{n} [p(x_i) - P]\,\Delta x$$

(This sum corresponds to the area enclosed by the rectangles in Figure 2.) If we let $n \to \infty$, this Riemann sum approaches the integral

$$\boxed{1} \qquad \int_0^X [p(x) - P]\,dx$$

which economists call the **consumer surplus** for the commodity.

The consumer surplus represents the amount of money saved by consumers in purchasing the commodity at price P, corresponding to an amount demanded of X. Figure 3 shows the interpretation of the consumer surplus as the area under the demand curve and above the line $p = P$.

V EXAMPLE 1 The demand for a product, in dollars, is

$$p = 1200 - 0.2x - 0.0001x^2$$

Find the consumer surplus when the sales level is 500.

SOLUTION Since the number of products sold is $X = 500$, the corresponding price is

$$P = 1200 - (0.2)(500) - (0.0001)(500)^2 = 1075$$

Therefore, from Definition 1, the consumer surplus is

$$\begin{aligned}\int_0^{500} [p(x) - P]\,dx &= \int_0^{500} (1200 - 0.2x - 0.0001x^2 - 1075)\,dx \\ &= \int_0^{500} (125 - 0.2x - 0.0001x^2)\,dx \\ &= 125x - 0.1x^2 - (0.0001)\left(\frac{x^3}{3}\right)\Bigg]_0^{500} \\ &= (125)(500) - (0.1)(500)^2 - \frac{(0.0001)(500)^3}{3} \\ &= \$33{,}333.33\end{aligned}$$

□

BLOOD FLOW

In Example 7 in Section 3.7 we discussed the law of laminar flow:

$$v(r) = \frac{P}{4\eta l}(R^2 - r^2)$$

which gives the velocity v of blood that flows along a blood vessel with radius R and length l at a distance r from the central axis, where P is the pressure difference between the ends of the vessel and η is the viscosity of the blood. Now, in order to compute the rate of blood flow, or *flux* (volume per unit time), we consider smaller, equally spaced radii $r_1, r_2, \ldots$.

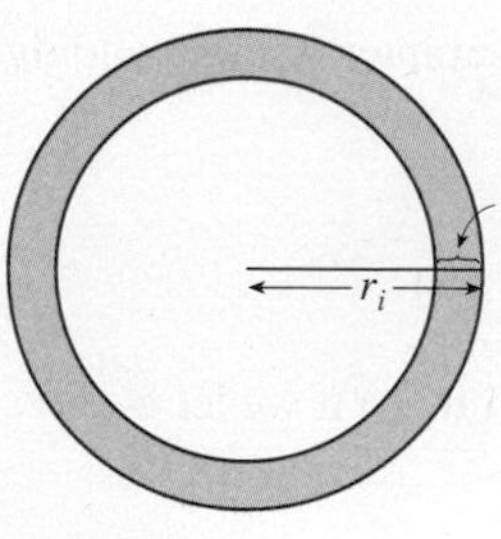

FIGURE 4

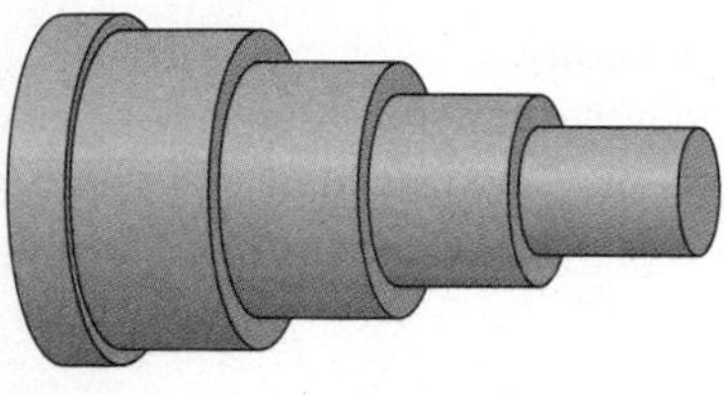

FIGURE 5

The approximate area of the ring (or washer) with inner radius r_{i-1} and outer radius r_i is

$$2\pi r_i \,\Delta r \qquad \text{where} \quad \Delta r = r_i - r_{i-1}$$

(See Figure 4.) If Δr is small, then the velocity is almost constant throughout this ring and can be approximated by $v(r_i)$. Thus the volume of blood per unit time that flows across the ring is approximately

$$(2\pi r_i \,\Delta r)\, v(r_i) = 2\pi r_i\, v(r_i)\,\Delta r$$

and the total volume of blood that flows across a cross-section per unit time is approximately

$$\sum_{i=1}^{n} 2\pi r_i\, v(r_i)\,\Delta r$$

This approximation is illustrated in Figure 5. Notice that the velocity (and hence the volume per unit time) increases toward the center of the blood vessel. The approximation gets better as n increases. When we take the limit we get the exact value of the **flux** (or *discharge*), which is the volume of blood that passes a cross-section per unit time:

$$\begin{aligned} F &= \lim_{n\to\infty} \sum_{i=1}^{n} 2\pi r_i\, v(r_i)\,\Delta r = \int_0^R 2\pi r\, v(r)\,dr \\ &= \int_0^R 2\pi r \frac{P}{4\eta l}(R^2 - r^2)\,dr \\ &= \frac{\pi P}{2\eta l}\int_0^R (R^2 r - r^3)\,dr = \frac{\pi P}{2\eta l}\left[R^2\frac{r^2}{2} - \frac{r^4}{4}\right]_{r=0}^{r=R} \\ &= \frac{\pi P}{2\eta l}\left[\frac{R^4}{2} - \frac{R^4}{4}\right] = \frac{\pi P R^4}{8\eta l} \end{aligned}$$

The resulting equation

2
$$F = \frac{\pi P R^4}{8\eta l}$$

is called **Poiseuille's Law**; it shows that the flux is proportional to the fourth power of the radius of the blood vessel.

CARDIAC OUTPUT

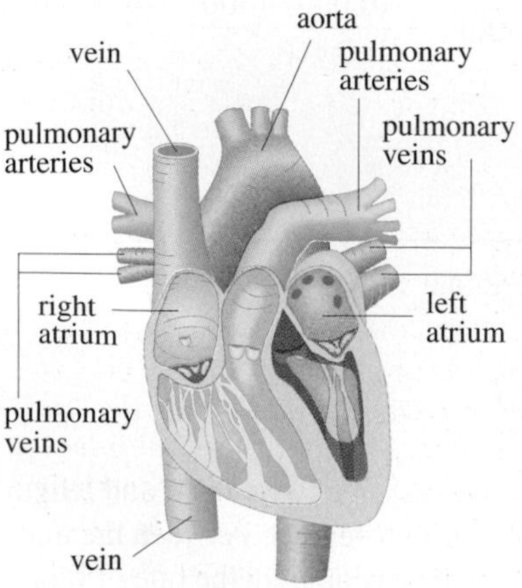

FIGURE 6

Figure 6 shows the human cardiovascular system. Blood returns from the body through the veins, enters the right atrium of the heart, and is pumped to the lungs through the pulmonary arteries for oxygenation. It then flows back into the left atrium through the pulmonary veins and then out to the rest of the body through the aorta. The **cardiac output** of the heart is the volume of blood pumped by the heart per unit time, that is, the rate of flow into the aorta.

The *dye dilution method* is used to measure the cardiac output. Dye is injected into the right atrium and flows through the heart into the aorta. A probe inserted into the aorta measures the concentration of the dye leaving the heart at equally spaced times over a time interval $[0, T]$ until the dye has cleared. Let $c(t)$ be the concentration of the dye at time t. If we divide $[0, T]$ into subintervals of equal length Δt, then the amount of dye that flows past the measuring point during the subinterval from $t = t_{i-1}$ to $t = t_i$ is approximately

$$(\text{concentration})(\text{volume}) = c(t_i)(F\,\Delta t)$$

where F is the rate of flow that we are trying to determine. Thus the total amount of dye is approximately

$$\sum_{i=1}^{n} c(t_i)F\,\Delta t = F\sum_{i=1}^{n} c(t_i)\,\Delta t$$

and, letting $n \to \infty$, we find that the amount of dye is

$$A = F\int_0^T c(t)\,dt$$

Thus the cardiac output is given by

$$\boxed{3} \qquad F = \frac{A}{\int_0^T c(t)\,dt}$$

where the amount of dye A is known and the integral can be approximated from the concentration readings.

t	$c(t)$	t	$c(t)$
0	0	6	6.1
1	0.4	7	4.0
2	2.8	8	2.3
3	6.5	9	1.1
4	9.8	10	0
5	8.9		

EXAMPLE 2 A 5-mg bolus of dye is injected into a right atrium. The concentration of the dye (in milligrams per liter) is measured in the aorta at one-second intervals as shown in the chart. Estimate the cardiac output.

SOLUTION Here $A = 5$, $\Delta t = 1$, and $T = 10$. We use Simpson's Rule to approximate the integral of the concentration:

$$\int_0^{10} c(t)\,dt \approx \tfrac{1}{3}[0 + 4(0.4) + 2(2.8) + 4(6.5) + 2(9.8) + 4(8.9)$$
$$+ 2(6.1) + 4(4.0) + 2(2.3) + 4(1.1) + 0]$$
$$\approx 41.87$$

Thus Formula 3 gives the cardiac output to be

$$F = \frac{A}{\int_0^{10} c(t)\,dt} \approx \frac{5}{41.87} \approx 0.12 \text{ L/s} = 7.2 \text{ L/min}$$ □

8.4 EXERCISES

1. The marginal cost function $C'(x)$ was defined to be the derivative of the cost function. (See Sections 3.7 and 4.7.) If the marginal cost of maufacturing x meters of a fabric is $C'(x) = 5 - 0.008x + 0.000009x^2$ (measured in dollars per meter) and the fixed start-up cost is $C(0) = \$20{,}000$, use the Net Change Theorem to find the cost of producing the first 2000 units.

2. The marginal revenue from the sale of x units of a product is $12 - 0.0004x$. If the revenue from the sale of the first 1000 units is \$12,400, find the revenue from the sale of the first 5000 units.

3. The marginal cost of producing x units of a certain product is $74 + 1.1x - 0.002x^2 + 0.00004x^3$ (in dollars per unit). Find the increase in cost if the production level is raised from 1200 units to 1600 units.

4. The demand function for a certain commodity is $p = 20 - 0.05x$. Find the consumer surplus when the sales level is 300. Illustrate by drawing the demand curve and identifying the consumer surplus as an area.

5. A demand curve is given by $p = 450/(x + 8)$. Find the consumer surplus when the selling price is \$10.

6. The **supply function** $p_S(x)$ for a commodity gives the relation between the selling price and the number of units that manufacturers will produce at that price. For a higher price, manufacturers will produce more units, so p_S is an increasing function of x. Let X be the amount of the commodity currently produced and let $P = p_S(X)$ be the current price. Some producers would be willing to make and sell the commodity for a lower selling price and are therefore receiving more than their minimal price. The excess is called the **producer surplus**. An

argument similar to that for consumer surplus shows that the surplus is given by the integral

$$\int_0^X [P - p_S(x)]\,dx$$

Calculate the producer surplus for the supply function $p_S(x) = 3 + 0.01x^2$ at the sales level $X = 10$. Illustrate by drawing the supply curve and identifying the producer surplus as an area.

7. If a supply curve is modeled by the equation $p = 200 + 0.2x^{3/2}$, find the producer surplus when the selling price is \$400.

8. For a given commodity and pure competition, the number of units produced and the price per unit are determined as the coordinates of the point of intersection of the supply and demand curves. Given the demand curve $p = 50 - \frac{1}{20}x$ and the supply curve $p = 20 + \frac{1}{10}x$, find the consumer surplus and the producer surplus. Illustrate by sketching the supply and demand curves and identifying the surpluses as areas.

9. A company modeled the demand curve for its product (in dollars) by the equation

$$p = \frac{800{,}000e^{-x/5000}}{x + 20{,}000}$$

Use a graph to estimate the sales level when the selling price is \$16. Then find (approximately) the consumer surplus for this sales level.

10. A movie theater has been charging \$7.50 per person and selling about 400 tickets on a typical weeknight. After surveying their customers, the theater estimates that for every 50 cents that they lower the price, the number of moviegoers will increase by 35 per night. Find the demand function and calculate the consumer surplus when the tickets are priced at \$6.00.

11. If the amount of capital that a company has at time t is $f(t)$, then the derivative, $f'(t)$, is called the *net investment flow*. Suppose that the net investment flow is $\sqrt{t}$ million dollars per year (where t is measured in years). Find the increase in capital (the *capital formation*) from the fourth year to the eighth year.

12. If revenue flows into a company at a rate of $f(t) = 9000\sqrt{1 + 2t}$, where t is measured in years and $f(t)$ is measured in dollars per year, find the total revenue obtained in the first four years.

13. *Pareto's Law of Income* states that the number of people with incomes between $x = a$ and $x = b$ is $N = \int_a^b Ax^{-k}\,dx$, where A and k are constants with $A > 0$ and $k > 1$. The average income of these people is

$$\bar{x} = \frac{1}{N}\int_a^b Ax^{1-k}\,dx$$

Calculate $\bar{x}$.

14. A hot, wet summer is causing a mosquito population explosion in a lake resort area. The number of mosquitos is increasing at an estimated rate of $2200 + 10e^{0.8t}$ per week (where t is measured in weeks). By how much does the mosquito population increase between the fifth and ninth weeks of summer?

15. Use Poiseuille's Law to calculate the rate of flow in a small human artery where we can take $\eta = 0.027$, $R = 0.008$ cm, $l = 2$ cm, and $P = 4000$ dynes/cm^2.

16. High blood pressure results from constriction of the arteries. To maintain a normal flow rate (flux), the heart has to pump harder, thus increasing the blood pressure. Use Poiseuille's Law to show that if R_0 and P_0 are normal values of the radius and pressure in an artery and the constricted values are R and P, then for the flux to remain constant, P and R are related by the equation

$$\frac{P}{P_0} = \left(\frac{R_0}{R}\right)^4$$

Deduce that if the radius of an artery is reduced to three-fourths of its former value, then the pressure is more than tripled.

17. The dye dilution method is used to measure cardiac output with 6 mg of dye. The dye concentrations, in mg/L, are modeled by $c(t) = 20te^{-0.6t}$, $0 \leqslant t \leqslant 10$, where t is measured in seconds. Find the cardiac output.

18. After an 8-mg injection of dye, the readings of dye concentration, in mg/L, at two-second intervals are as shown in the table. Use Simpson's Rule to estimate the cardiac output.

t	$c(t)$	t	$c(t)$
0	0	12	3.9
2	2.4	14	2.3
4	5.1	16	1.6
6	7.8	18	0.7
8	7.6	20	0
10	5.4		

19. The graph of the concentration function $c(t)$ is shown after a 7-mg injection of dye into a heart. Use Simpson's Rule to estimate the cardiac output.

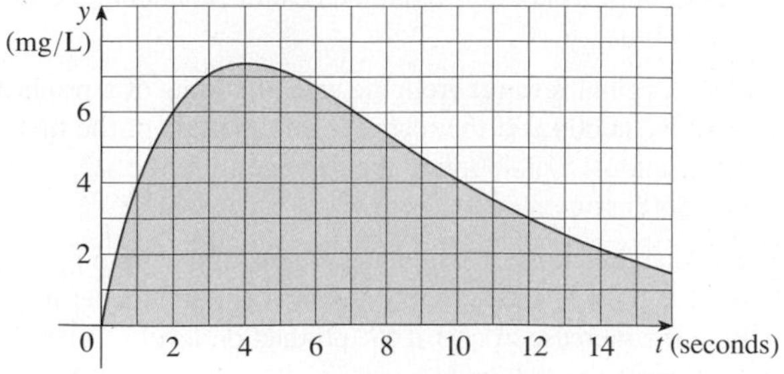

8.5 PROBABILITY

Calculus plays a role in the analysis of random behavior. Suppose we consider the cholesterol level of a person chosen at random from a certain age group, or the height of an adult female chosen at random, or the lifetime of a randomly chosen battery of a certain type. Such quantities are called **continuous random variables** because their values actually range over an interval of real numbers, although they might be measured or recorded only to the nearest integer. We might want to know the probability that a blood cholesterol level is greater than 250, or the probability that the height of an adult female is between 60 and 70 inches, or the probability that the battery we are buying lasts between 100 and 200 hours. If X represents the lifetime of that type of battery, we denote this last probability as follows:

$$P(100 \leqslant X \leqslant 200)$$

According to the frequency interpretation of probability, this number is the long-run proportion of all batteries of the specified type whose lifetimes are between 100 and 200 hours. Since it represents a proportion, the probability naturally falls between 0 and 1.

Every continuous random variable X has a **probability density function** f. This means that the probability that X lies between a and b is found by integrating f from a to b:

1 $$P(a \leqslant X \leqslant b) = \int_a^b f(x)\,dx$$

For example, Figure 1 shows the graph of a model for the probability density function f for a random variable X defined to be the height in inches of an adult female in the United States (according to data from the National Health Survey). The probability that the height of a woman chosen at random from this population is between 60 and 70 inches is equal to the area under the graph of f from 60 to 70.

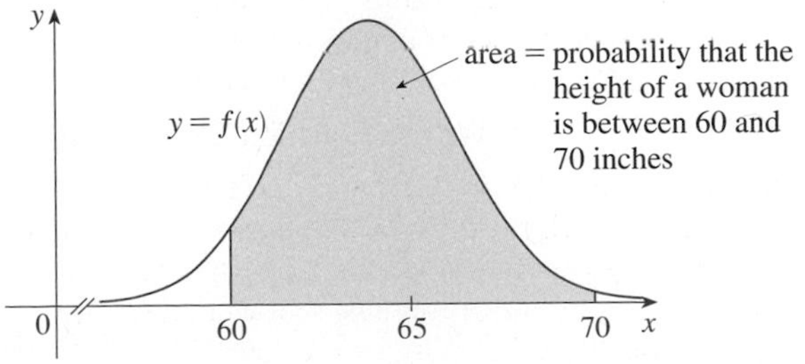

FIGURE 1
Probability density function for the height of an adult female

In general, the probability density function f of a random variable X satisfies the condition $f(x) \geqslant 0$ for all x. Because probabilities are measured on a scale from 0 to 1, it follows that

2 $$\int_{-\infty}^{\infty} f(x)\,dx = 1$$

EXAMPLE 1 Let $f(x) = 0.006x(10 - x)$ for $0 \leqslant x \leqslant 10$ and $f(x) = 0$ for all other values of x.

(a) Verify that f is a probability density function.

(b) Find $P(4 \leqslant X \leqslant 8)$.

SOLUTION
(a) For $0 \leqslant x \leqslant 10$ we have $0.006x(10 - x) \geqslant 0$, so $f(x) \geqslant 0$ for all x. We also need to check that Equation 2 is satisfied:

$$\int_{-\infty}^{\infty} f(x)\,dx = \int_0^{10} 0.006x(10 - x)\,dx = 0.006 \int_0^{10} (10x - x^2)\,dx$$

$$= 0.006\left[5x^2 - \tfrac{1}{3}x^3\right]_0^{10} = 0.006\left(500 - \tfrac{1000}{3}\right) = 1$$

Therefore f is a probability density function.

(b) The probability that X lies between 4 and 8 is

$$P(4 \leqslant X \leqslant 8) = \int_4^8 f(x)\,dx = 0.006 \int_4^8 (10x - x^2)\,dx$$

$$= 0.006\left[5x^2 - \tfrac{1}{3}x^3\right]_4^8 = 0.544$$

□

V EXAMPLE 2 Phenomena such as waiting times and equipment failure times are commonly modeled by exponentially decreasing probability density functions. Find the exact form of such a function.

SOLUTION Think of the random variable as being the time you wait on hold before an agent of a company you're telephoning answers your call. So instead of x, let's use t to represent time, in minutes. If f is the probability density function and you call at time $t = 0$, then, from Definition 1, $\int_0^2 f(t)\,dt$ represents the probability that an agent answers within the first two minutes and $\int_4^5 f(t)\,dt$ is the probability that your call is answered during the fifth minute.

It's clear that $f(t) = 0$ for $t < 0$ (the agent can't answer before you place the call). For $t > 0$ we are told to use an exponentially decreasing function, that is, a function of the form $f(t) = Ae^{-ct}$, where A and c are positive constants. Thus

$$f(t) = \begin{cases} 0 & \text{if } t < 0 \\ Ae^{-ct} & \text{if } t \geqslant 0 \end{cases}$$

We use Equation 2 to determine the value of A:

$$1 = \int_{-\infty}^{\infty} f(t)\,dt = \int_{-\infty}^{0} f(t)\,dt + \int_0^{\infty} f(t)\,dt$$

$$= \int_0^{\infty} Ae^{-ct}\,dt = \lim_{x \to \infty} \int_0^x Ae^{-ct}\,dt$$

$$= \lim_{x \to \infty} \left[-\frac{A}{c}e^{-ct}\right]_0^x = \lim_{x \to \infty} \frac{A}{c}(1 - e^{-cx})$$

$$= \frac{A}{c}$$

Therefore $A/c = 1$ and so $A = c$. Thus every exponential density function has the form

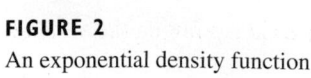

$$f(t) = \begin{cases} 0 & \text{if } t < 0 \\ ce^{-ct} & \text{if } t \geqslant 0 \end{cases}$$

A typical graph is shown in Figure 2.

□

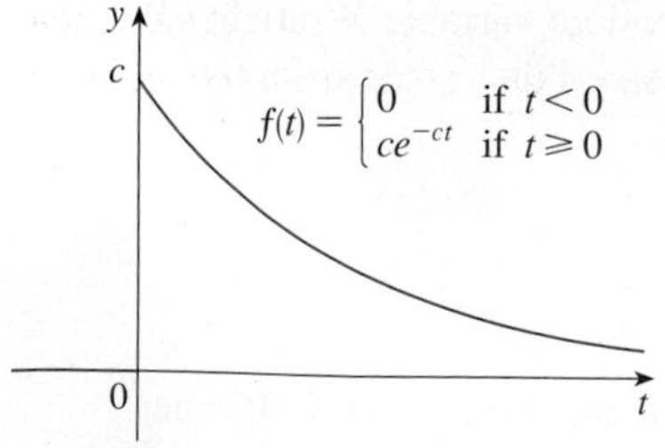

FIGURE 2
An exponential density function

AVERAGE VALUES

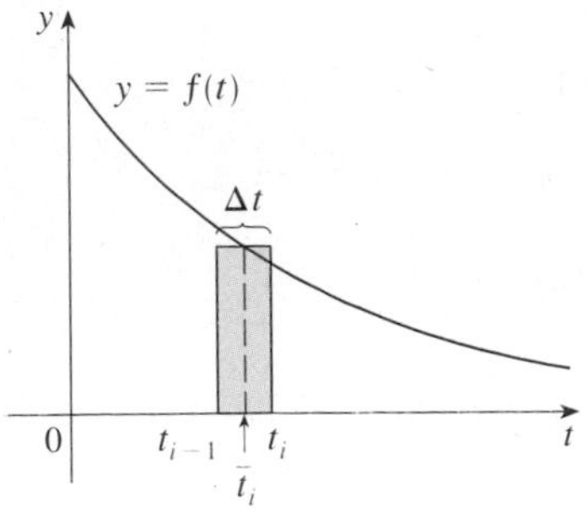

FIGURE 3

Suppose you're waiting for a company to answer your phone call and you wonder how long, on average, you can expect to wait. Let $f(t)$ be the corresponding density function, where t is measured in minutes, and think of a sample of N people who have called this company. Most likely, none of them had to wait more than an hour, so let's restrict our attention to the interval $0 \leqslant t \leqslant 60$. Let's divide that interval into n intervals of length Δt and endpoints $0, t_1, t_2, \ldots t_{60}$. (Think of Δt as lasting a minute, or half a minute, or 10 seconds, or even a second.) The probability that somebody's call gets answered during the time period from t_{i-1} to t_i is the area under the curve $y = f(t)$ from t_{i-1} to t_i, which is approximately equal to $f(\bar{t}_i)\,\Delta t$. (This is the area of the approximating rectangle in Figure 3, where $\bar{t}_i$ is the midpoint of the interval.)

Since the long-run proportion of calls that get answered in the time period from t_{i-1} to t_i is $f(\bar{t}_i)\,\Delta t$, we expect that, out of our sample of N callers, the number whose call was answered in that time period is approximately $Nf(\bar{t}_i)\,\Delta t$ and the time that each waited is about $\bar{t}_i$. Therefore the total time they waited is the product of these numbers: approximately $\bar{t}_i[Nf(\bar{t}_i)\,\Delta t]$. Adding over all such intervals, we get the approximate total of everybody's waiting times:

$$\sum_{i-1}^{n} N\bar{t}_i f(\bar{t}_i)\,\Delta t$$

If we now divide by the number of callers N, we get the approximate *average* waiting time:

$$\sum_{i=1}^{n} \bar{t}_i f(\bar{t}_i)\,\Delta t$$

We recognize this as a Riemann sum for the function $t\,f(t)$. As the time interval shrinks (that is, $\Delta t \to 0$ and $n \to \infty$), this Riemann sum approaches the integral

$$\int_0^{60} t\,f(t)\,dt$$

This integral is called the *mean waiting time*.

■ It is traditional to denote the mean by the Greek letter μ (mu).

In general, the **mean** of any probability density function f is defined to be

$$\mu = \int_{-\infty}^{\infty} x\,f(x)\,dx$$

The mean can be interpreted as the long-run average value of the random variable X. It can also be interpreted as a measure of centrality of the probability density function.

The expression for the mean resembles an integral we have seen before. If $\mathcal{R}$ is the region that lies under the graph of f, we know from Formula 8.3.8 that the x-coordinate of the centroid of $\mathcal{R}$ is

$$\bar{x} = \frac{\int_{-\infty}^{\infty} x\,f(x)\,dx}{\int_{-\infty}^{\infty} f(x)\,dx} = \int_{-\infty}^{\infty} x\,f(x)\,dx = \mu$$

because of Equation 2. So a thin plate in the shape of $\mathcal{R}$ balances at a point on the vertical line $x = \mu$. (See Figure 4.)

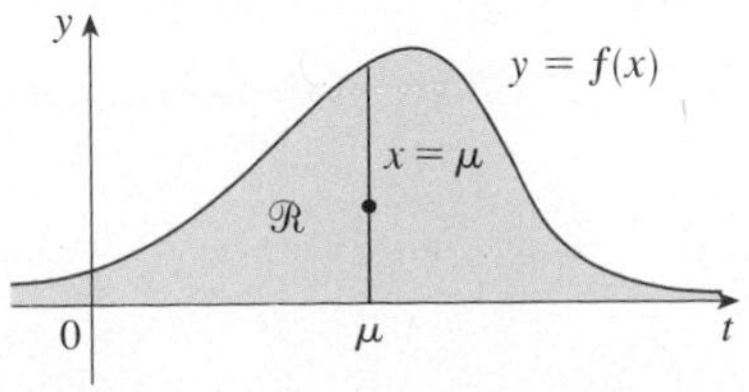

FIGURE 4
$\mathcal{R}$ balances at a point on the line $x = \mu$

EXAMPLE 3 Find the mean of the exponential distribution of Example 2:

$$f(t) = \begin{cases} 0 & \text{if } t < 0 \\ ce^{-ct} & \text{if } t \geqslant 0 \end{cases}$$

SOLUTION According to the definition of a mean, we have

$$\mu = \int_{-\infty}^{\infty} t f(t)\, dt = \int_{0}^{\infty} tce^{-ct}\, dt$$

To evaluate this integral we use integration by parts, with $u = t$ and $dv = ce^{-ct}\, dt$:

$$\int_0^\infty tce^{-ct}\, dt = \lim_{x\to\infty} \int_0^x tce^{-ct}\, dt = \lim_{x\to\infty} \left(-te^{-ct}\Big]_0^x + \int_0^x e^{-ct}\, dt \right)$$

$$= \lim_{x\to\infty} \left(-xe^{-cx} + \frac{1}{c} - \frac{e^{-cx}}{c} \right) = \frac{1}{c}$$

■ The limit of the first term is 0 by l'Hospital's Rule.

The mean is $\mu = 1/c$, so we can rewrite the probability density function as

$$f(t) = \begin{cases} 0 & \text{if } t < 0 \\ \mu^{-1}e^{-t/\mu} & \text{if } t \geqslant 0 \end{cases}$$ □

EXAMPLE 4 Suppose the average waiting time for a customer's call to be answered by a company representative is five minutes.

(a) Find the probability that a call is answered during the first minute.

(b) Find the probability that a customer waits more than five minutes to be answered.

SOLUTION

(a) We are given that the mean of the exponential distribution is $\mu = 5$ min and so, from the result of Example 3, we know that the probability density function is

$$f(t) = \begin{cases} 0 & \text{if } t < 0 \\ 0.2e^{-t/5} & \text{if } t \geqslant 0 \end{cases}$$

Thus the probability that a call is answered during the first minute is

$$P(0 \leqslant T \leqslant 1) = \int_0^1 f(t)\, dt$$

$$= \int_0^1 0.2e^{-t/5}\, dt = 0.2(-5)e^{-t/5}\Big]_0^1$$

$$= 1 - e^{-1/5} \approx 0.1813$$

So about 18% of customers' calls are answered during the first minute.

(b) The probability that a customer waits more than five minutes is

$$P(T > 5) = \int_5^\infty f(t)\, dt = \int_5^\infty 0.2e^{-t/5}\, dt$$

$$= \lim_{x\to\infty} \int_5^x 0.2e^{-t/5}\, dt = \lim_{x\to\infty} (e^{-1} - e^{-x/5})$$

$$= \frac{1}{e} \approx 0.368$$

About 37% of customers wait more than five minutes before their calls are answered. □

Notice the result of Example 4(b): Even though the mean waiting time is 5 minutes, only 37% of callers wait more than 5 minutes. The reason is that some callers have to wait much longer (maybe 10 or 15 minutes), and this brings up the average.

Another measure of centrality of a probability density function is the *median*. That is a number m such that half the callers have a waiting time less than m and the other callers have a waiting time longer than m. In general, the **median** of a probability density function is the number m such that

$$\int_m^\infty f(x)\,dx = \tfrac{1}{2}$$

This means that half the area under the graph of f lies to the right of m. In Exercise 9 you are asked to show that the median waiting time for the company described in Example 4 is approximately 3.5 minutes.

NORMAL DISTRIBUTIONS

Many important random phenomena—such as test scores on aptitude tests, heights and weights of individuals from a homogeneous population, annual rainfall in a given location—are modeled by a **normal distribution**. This means that the probability density function of the random variable X is a member of the family of functions

$$\boxed{3} \qquad f(x) = \frac{1}{\sigma\sqrt{2\pi}}\, e^{-(x-\mu)^2/(2\sigma^2)}$$

■ The standard deviation is denoted by the lowercase Greek letter σ (sigma).

You can verify that the mean for this function is μ. The positive constant σ is called the **standard deviation**; it measures how spread out the values of X are. From the bell-shaped graphs of members of the family in Figure 5, we see that for small values of σ the values of X are clustered about the mean, whereas for larger values of σ the values of X are more spread out. Statisticians have methods for using sets of data to estimate μ and σ.

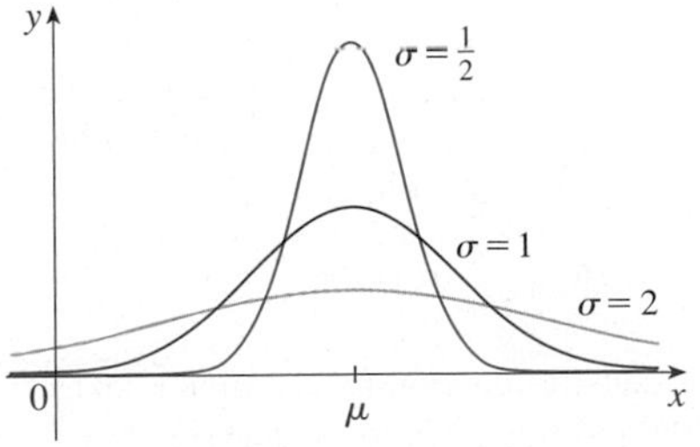

FIGURE 5
Normal distributions

The factor $1/(\sigma\sqrt{2\pi})$ is needed to make f a probability density function. In fact, it can be verified using the methods of multivariable calculus that

$$\int_{-\infty}^{\infty} \frac{1}{\sigma\sqrt{2\pi}}\, e^{-(x-\mu)^2/(2\sigma^2)}\,dx = 1$$

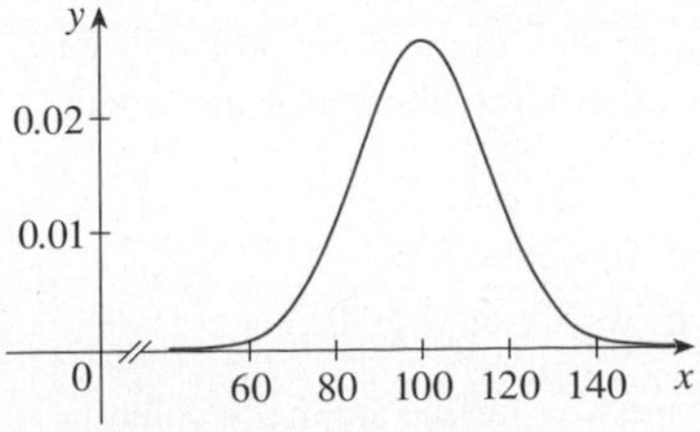

FIGURE 6
Distribution of IQ scores

V EXAMPLE 5 Intelligence Quotient (IQ) scores are distributed normally with mean 100 and standard deviation 15. (Figure 6 shows the corresponding probability density function.)

(a) What percentage of the population has an IQ score between 85 and 115?

(b) What percentage of the population has an IQ above 140?

SOLUTION

(a) Since IQ scores are normally distributed, we use the probability density function given by Equation 3 with $\mu = 100$ and $\sigma = 15$:

$$P(85 \leqslant X \leqslant 115) = \int_{85}^{115} \frac{1}{15\sqrt{2\pi}} e^{-(x-100)^2/(2\cdot 15^2)} \, dx$$

Recall from Section 7.5 that the function $y = e^{-x^2}$ doesn't have an elementary antiderivative, so we can't evaluate the integral exactly. But we can use the numerical integration capability of a calculator or computer (or the Midpoint Rule or Simpson's Rule) to estimate the integral. Doing so, we find that

$$P(85 \leqslant X \leqslant 115) \approx 0.68$$

So about 68% of the population has an IQ between 85 and 115, that is, within one standard deviation of the mean.

(b) The probability that the IQ score of a person chosen at random is more than 140 is

$$P(X > 140) = \int_{140}^{\infty} \frac{1}{15\sqrt{2\pi}} e^{-(x-100)^2/450} \, dx$$

To avoid the improper integral we could approximate it by the integral from 140 to 200. (It's quite safe to say that people with an IQ over 200 are extremely rare.) Then

$$P(X > 140) \approx \int_{140}^{200} \frac{1}{15\sqrt{2\pi}} e^{-(x-100)^2/450} \, dx \approx 0.0038$$

Therefore about 0.4% of the population has an IQ over 140. □

8.5 EXERCISES

1. Let $f(x)$ be the probability density function for the lifetime of a manufacturer's highest quality car tire, where x is measured in miles. Explain the meaning of each integral.

   (a) $\int_{30,000}^{40,000} f(x)\, dx$ (b) $\int_{25,000}^{\infty} f(x)\, dx$

2. Let $f(t)$ be the probability density function for the time it takes you to drive to school in the morning, where t is measured in minutes. Express the following probabilities as integrals.
   (a) The probability that you drive to school in less than 15 minutes
   (b) The probability that it takes you more than half an hour to get to school

3. Let $f(x) = \frac{3}{64} x\sqrt{16 - x^2}$ for $0 \leqslant x \leqslant 4$ and $f(x) = 0$ for all other values of x.
   (a) Verify that f is a probability density function.
   (b) Find $P(X < 2)$.

4. Let $f(x) = xe^{-x}$ if $x \geqslant 0$ and $f(x) = 0$ if $x < 0$.
   (a) Verify that f is a probability density function.
   (b) Find $P(1 \leqslant X \leqslant 2)$.

5. Let $f(x) = c/(1 + x^2)$.
   (a) For what value of c is f a probability density function?
   (b) For that value of c, find $P(-1 < X < 1)$.

6. Let $f(x) = kx^2(1 - x)$ if $0 \leqslant x \leqslant 1$ and $f(x) = 0$ if $x < 0$ or $x > 1$.
   (a) For what value of k is f a probability density function?
   (b) For that value of k, find $P(X \geqslant \frac{1}{2})$.
   (c) Find the mean.

7. A spinner from a board game randomly indicates a real number between 0 and 10. The spinner is fair in the sense that it indicates a number in a given interval with the same probability as it indicates a number in any other interval of the same length.
   (a) Explain why the function

   $$f(x) = \begin{cases} 0.1 & \text{if } 0 \leqslant x \leqslant 10 \\ 0 & \text{if } x < 0 \text{ or } x > 10 \end{cases}$$

   is a probability density function for the spinner's values.
   (b) What does your intuition tell you about the value of the mean? Check your guess by evaluating an integral.

8. (a) Explain why the function whose graph is shown is a probability density function.
(b) Use the graph to find the following probabilities:
(i) $P(X < 3)$ (ii) $P(3 \leq X \leq 8)$
(c) Calculate the mean.

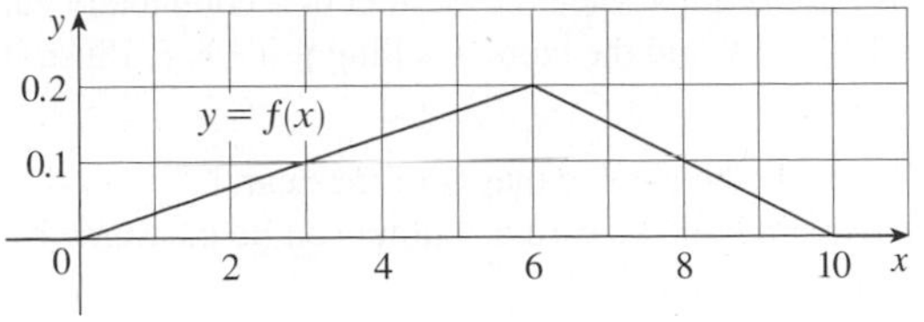

9. Show that the median waiting time for a phone call to the company described in Example 4 is about 3.5 minutes.

10. (a) A type of lightbulb is labeled as having an average lifetime of 1000 hours. It's reasonable to model the probability of failure of these bulbs by an exponential density function with mean $\mu = 1000$. Use this model to find the probability that a bulb
(i) fails within the first 200 hours,
(ii) burns for more than 800 hours.
(b) What is the median lifetime of these lightbulbs?

11. The manager of a fast-food restaurant determines that the average time that her customers wait for service is 2.5 minutes.
(a) Find the probability that a customer has to wait more than 4 minutes.
(b) Find the probability that a customer is served within the first 2 minutes.
(c) The manager wants to advertise that anybody who isn't served within a certain number of minutes gets a free hamburger. But she doesn't want to give away free hamburgers to more than 2% of her customers. What should the advertisement say?

12. According to the National Health Survey, the heights of adult males in the United States are normally distributed with mean 69.0 inches and standard deviation 2.8 inches.
(a) What is the probability that an adult male chosen at random is between 65 inches and 73 inches tall?
(b) What percentage of the adult male population is more than 6 feet tall?

13. The "Garbage Project" at the University of Arizona reports that the amount of paper discarded by households per week is normally distributed with mean 9.4 lb and standard deviation 4.2 lb. What percentage of households throw out at least 10 lb of paper a week?

14. Boxes are labeled as containing 500 g of cereal. The machine filling the boxes produces weights that are normally distributed with standard deviation 12 g.
(a) If the target weight is 500 g, what is the probability that the machine produces a box with less than 480 g of cereal?
(b) Suppose a law states that no more than 5% of a manufacturer's cereal boxes can contain less than the stated weight of 500 g. At what target weight should the manufacturer set its filling machine?

15. The speeds of vehicles on a highway with speed limit 100 km/h are normally distributed with mean 112 km/h and standard deviation 8 km/h.
(a) What is the probability that a randomly chosen vehicle is traveling at a legal speed?
(b) If police are instructed to ticket motorists driving 125 km/h or more, what percentage of motorists are targeted?

16. Show that the probability density function for a normally distributed random variable has inflection points at $x = \mu \pm \sigma$.

17. For any normal distribution, find the probability that the random variable lies within two standard deviations of the mean.

18. The standard deviation for a random variable with probability density function f and mean μ is defined by

$$\sigma = \left[\int_{-\infty}^{\infty} (x - \mu)^2 f(x)\, dx \right]^{1/2}$$

Find the standard deviation for an exponential density function with mean μ.

19. The hydrogen atom is composed of one proton in the nucleus and one electron, which moves about the nucleus. In the quantum theory of atomic structure, it is assumed that the electron does not move in a well-defined orbit. Instead, it occupies a state known as an *orbital,* which may be thought of as a "cloud" of negative charge surrounding the nucleus. At the state of lowest energy, called the *ground state,* or *1s-orbital,* the shape of this cloud is assumed to be a sphere centered at the nucleus. This sphere is described in terms of the probability density function

$$p(r) = \frac{4}{a_0^3} r^2 e^{-2r/a_0} \qquad r \geq 0$$

where a_0 is the *Bohr radius* ($a_0 \approx 5.59 \times 10^{-11}$ m). The integral

$$P(r) = \int_0^r \frac{4}{a_0^3} s^2 e^{-2s/a_0}\, ds$$

gives the probability that the electron will be found within the sphere of radius r meters centered at the nucleus.
(a) Verify that $p(r)$ is a probability density function.
(b) Find $\lim_{r \to \infty} p(r)$. For what value of r does $p(r)$ have its maximum value?
(c) Graph the density function.
(d) Find the probability that the electron will be within the sphere of radius $4a_0$ centered at the nucleus.
(e) Calculate the mean distance of the electron from the nucleus in the ground state of the hydrogen atom.

8 REVIEW

CONCEPT CHECK

1. (a) How is the length of a curve defined?
(b) Write an expression for the length of a smooth curve given by $y = f(x)$, $a \leqslant x \leqslant b$.
(c) What if x is given as a function of y?

2. (a) Write an expression for the surface area of the surface obtained by rotating the curve $y = f(x)$, $a \leqslant x \leqslant b$, about the x-axis.
(b) What if x is given as a function of y?
(c) What if the curve is rotated about the y-axis?

3. Describe how we can find the hydrostatic force against a vertical wall submersed in a fluid.

4. (a) What is the physical significance of the center of mass of a thin plate?
(b) If the plate lies between $y = f(x)$ and $y = 0$, where $a \leqslant x \leqslant b$, write expressions for the coordinates of the center of mass.

5. What does the Theorem of Pappus say?

6. Given a demand function $p(x)$, explain what is meant by the consumer surplus when the amount of a commodity currently available is X and the current selling price is P. Illustrate with a sketch.

7. (a) What is the cardiac output of the heart?
(b) Explain how the cardiac output can be measured by the dye dilution method.

8. What is a probability density function? What properties does such a function have?

9. Suppose $f(x)$ is the probability density function for the weight of a female college student, where x is measured in pounds.
(a) What is the meaning of the integral $\int_0^{130} f(x)\,dx$?
(b) Write an expression for the mean of this density function.
(c) How can we find the median of this density function?

10. What is a normal distribution? What is the significance of the standard deviation?

EXERCISES

1–2 Find the length of the curve.

1. $y = \frac{1}{6}(x^2 + 4)^{3/2}$, $0 \leqslant x \leqslant 3$

2. $y = 2\ln\left(\sin \frac{1}{2}x\right)$, $\pi/3 \leqslant x \leqslant \pi$

3. (a) Find the length of the curve

$$y = \frac{x^4}{16} + \frac{1}{2x^2} \qquad 1 \leqslant x \leqslant 2$$

(b) Find the area of the surface obtained by rotating the curve in part (a) about the y-axis.

4. (a) The curve $y = x^2$, $0 \leqslant x \leqslant 1$, is rotated about the y-axis. Find the area of the resulting surface.
(b) Find the area of the surface obtained by rotating the curve in part (a) about the x-axis.

5. Use Simpson's Rule with $n = 6$ to estimate the length of the curve $y = e^{-x^2}$, $0 \leqslant x \leqslant 3$.

6. Use Simpson's Rule with $n = 6$ to estimate the area of the surface obtained by rotating the curve in Exercise 5 about the x-axis.

7. Find the length of the curve

$$y = \int_1^x \sqrt{\sqrt{t} - 1}\,dt \qquad 1 \leqslant x \leqslant 16$$

8. Find the area of the surface obtained by rotating the curve in Exercise 7 about the y-axis.

9. A gate in an irrigation canal is constructed in the form of a trapezoid 3 ft wide at the bottom, 5 ft wide at the top, and 2 ft high. It is placed vertically in the canal so that the water just covers the gate. Find the hydrostatic force on one side of the gate.

10. A trough is filled with water and its vertical ends have the shape of the parabolic region in the figure. Find the hydrostatic force on one end of the trough.

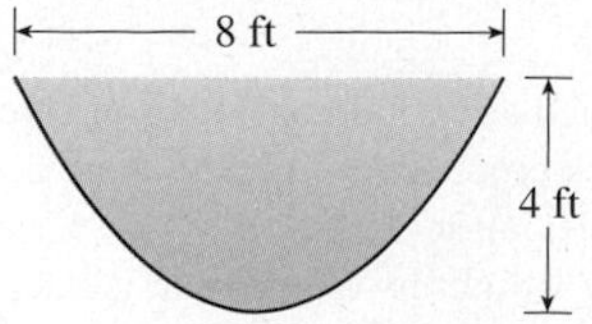

11–12 Find the centroid of the region bounded by the given curves.

11. $y = \frac{1}{2}x, \quad y = \sqrt{x}$

12. $y = \sin x, \quad y = 0, \quad x = \pi/4, \quad x = 3\pi/4$

13–14 Find the centroid of the region shown

13.

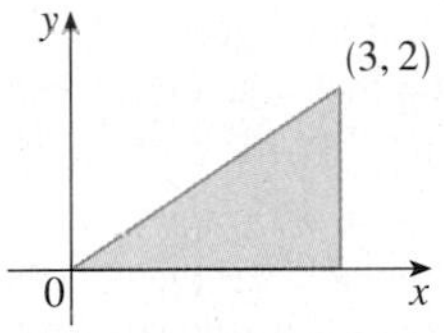

14.

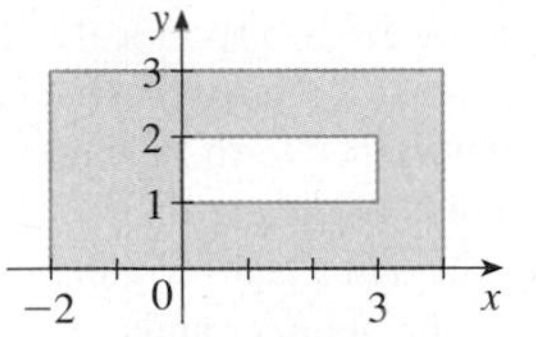

15. Find the volume obtained when the circle of radius 1 with center (1, 0) is rotated about the y-axis.

16. Use the Theorem of Pappus and the fact that the volume of a sphere of radius r is $\frac{4}{3}\pi r^3$ to find the centroid of the semicircular region bounded by the curve $y = \sqrt{r^2 - x^2}$ and the x-axis.

17. The demand function for a commodity is given by

$$p = 2000 - 0.1x - 0.01x^2$$

Find the consumer surplus when the sales level is 100.

18. After a 6-mg injection of dye into a heart, the readings of dye concentration at two-second intervals are as shown in the table. Use Simpson's Rule to estimate the cardiac output.

t	$c(t)$	t	$c(t)$
0	0	14	4.7
2	1.9	16	3.3
4	3.3	18	2.1
6	5.1	20	1.1
8	7.6	22	0.5
10	7.1	24	0
12	5.8		

19. (a) Explain why the function

$$f(x) = \begin{cases} \dfrac{\pi}{20} \sin\left(\dfrac{\pi x}{10}\right) & \text{if } 0 \leqslant x \leqslant 10 \\ 0 & \text{if } x < 0 \text{ or } x > 10 \end{cases}$$

is a probability density function.

(b) Find $P(X < 4)$.

(c) Calculate the mean. Is the value what you would expect?

20. Lengths of human pregnancies are normally distributed with mean 268 days and standard deviation 15 days. What percentage of pregnancies last between 250 days and 280 days?

21. The length of time spent waiting in line at a certain bank is modeled by an exponential density function with mean 8 minutes.

(a) What is the probability that a customer is served in the first 3 minutes?

(b) What is the probability that a customer has to wait more than 10 minutes?

(c) What is the median waiting time?

PROBLEMS PLUS

1. Find the area of the region $S = \{(x, y) \mid x \geq 0,\ y \leq 1,\ x^2 + y^2 \leq 4y\}$.

2. Find the centroid of the region enclosed by the loop of the curve $y^2 = x^3 - x^4$.

3. If a sphere of radius r is sliced by a plane whose distance from the center of the sphere is d, then the sphere is divided into two pieces called segments of one base. The corresponding surfaces are called *spherical zones of one base.*
   (a) Determine the surface areas of the two spherical zones indicated in the figure.
   (b) Determine the approximate area of the Arctic Ocean by assuming that it is approximately circular in shape, with center at the North Pole and "circumference" at 75° north latitude. Use $r = 3960$ mi for the radius of the earth.
   (c) A sphere of radius r is inscribed in a right circular cylinder of radius r. Two planes perpendicular to the central axis of the cylinder and a distance h apart cut off a *spherical zone of two bases* on the sphere. Show that the surface area of the spherical zone equals the surface area of the region that the two planes cut off on the cylinder.
   (d) The *Torrid Zone* is the region on the surface of the earth that is between the Tropic of Cancer (23.45° north latitude) and the Tropic of Capricorn (23.45° south latitude). What is the area of the Torrid Zone?

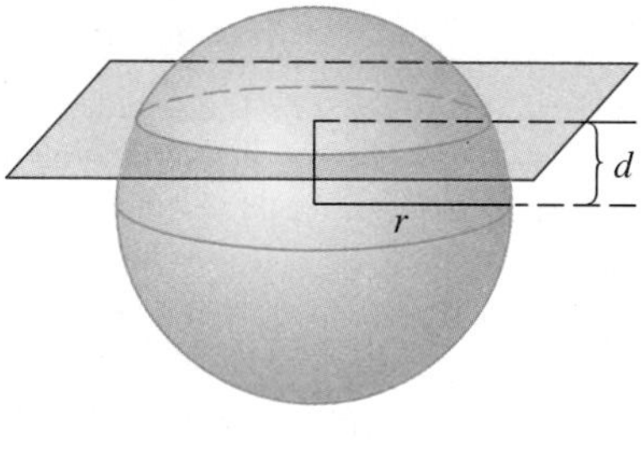

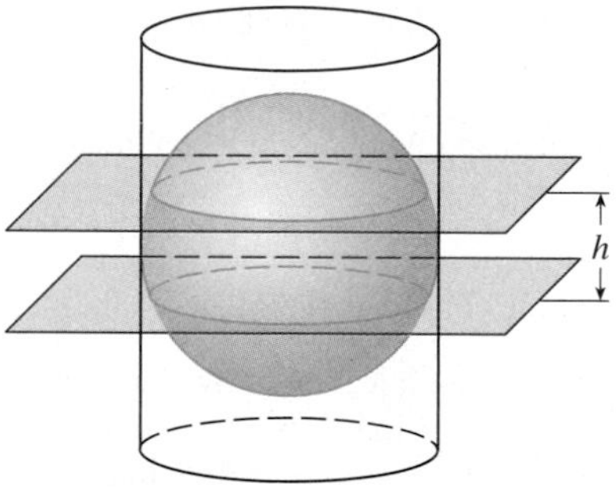

4. (a) Show that an observer at height H above the north pole of a sphere of radius r can see a part of the sphere that has area

$$\frac{2\pi r^2 H}{r + H}$$

   (b) Two spheres with radii r and R are placed so that the distance between their centers is d, where $d > r + R$. Where should a light be placed on the line joining the centers of the spheres in order to illuminate the largest total surface?

5. Suppose that the density of seawater, $\rho = \rho(z)$, varies with the depth z below the surface.
   (a) Show that the hydrostatic pressure is governed by the differential equation

$$\frac{dP}{dz} = \rho(z)g$$

   where g is the acceleration due to gravity. Let P_0 and ρ_0 be the pressure and density at $z = 0$. Express the pressure at depth z as an integral.
   (b) Suppose the density of seawater at depth z is given by $\rho = \rho_0 e^{z/H}$, where H is a positive constant. Find the total force, expressed as an integral, exerted on a vertical circular porthole of radius r whose center is located at a distance $L > r$ below the surface.

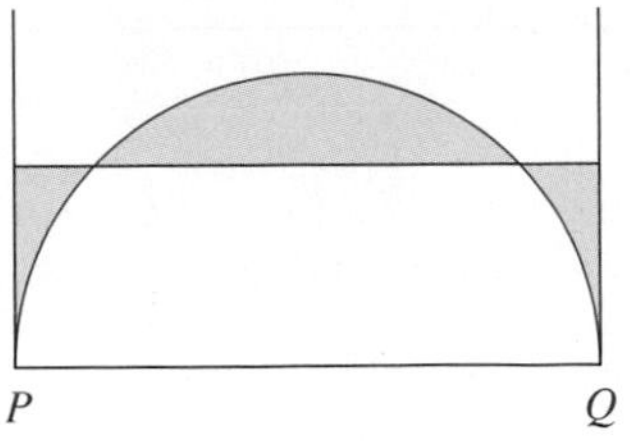

FIGURE FOR PROBLEM 6

6. The figure shows a semicircle with radius 1, horizontal diameter PQ, and tangent lines at P and Q. At what height above the diameter should the horizontal line be placed so as to minimize the shaded area?

7. Let P be a pyramid with a square base of side $2b$ and suppose that S is a sphere with its center on the base of P and S is tangent to all eight edges of P. Find the height of P. Then find the volume of the intersection of S and P.

8. Consider a flat metal plate to be placed vertically under water with its top 2 m below the surface of the water. Determine a shape for the plate so that if the plate is divided into any number of horizontal strips of equal height, the hydrostatic force on each strip is the same.

9. A uniform disk with radius 1 m is to be cut by a line so that the center of mass of the smaller piece lies halfway along a radius. How close to the center of the disk should the cut be made? (Express your answer correct to two decimal places.)

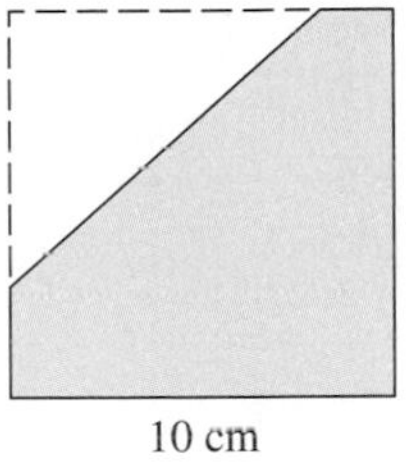

FIGURE FOR PROBLEM 10

10. A triangle with area 30 cm^2 is cut from a corner of a square with side 10 cm, as shown in the figure. If the centroid of the remaining region is 4 cm from the right side of the square, how far is it from the bottom of the square?

11. In a famous 18th-century problem, known as *Buffon's needle problem*, a needle of length h is dropped onto a flat surface (for example, a table) on which parallel lines L units apart, $L \geqslant h$, have been drawn. The problem is to determine the probability that the needle will come to rest intersecting one of the lines. Assume that the lines run east-west, parallel to the x-axis in a rectangular coordinate system (as in the figure). Let y be the distance from the "southern" end of the needle to the nearest line to the north. (If the needle's southern end lies on a line, let $y = 0$. If the needle happens to lie east-west, let the "western" end be the "southern" end.) Let θ be the angle that the needle makes with a ray extending eastward from the "southern" end. Then $0 \leqslant y \leqslant L$ and $0 \leqslant \theta \leqslant \pi$. Note that the needle intersects one of the lines only when $y < h \sin \theta$. The total set of possibilities for the needle can be identified with the rectangular region $0 \leqslant y \leqslant L$, $0 \leqslant \theta \leqslant \pi$, and the proportion of times that the needle intersects a line is the ratio

$$\frac{\text{area under } y = h \sin \theta}{\text{area of rectangle}}$$

This ratio is the probability that the needle intersects a line. Find the probability that the needle will intersect a line if $h = L$. What if $h = \frac{1}{2}L$?

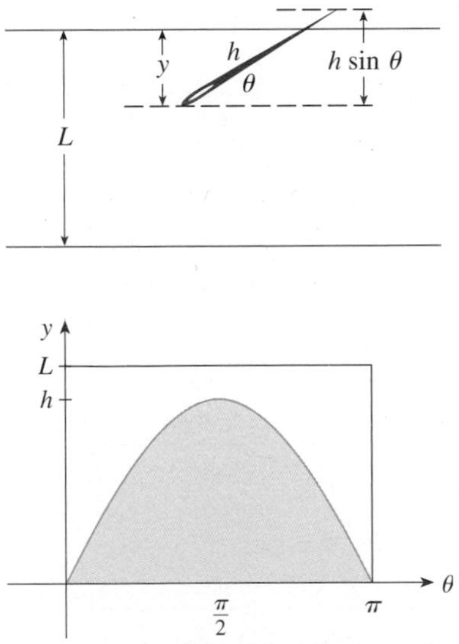

FIGURE FOR PROBLEM 11

12. If the needle in Problem 11 has length $h > L$, it's possible for the needle to intersect more than one line.

(a) If $L = 4$, find the probability that a needle of length 7 will intersect at least one line. [*Hint:* Proceed as in Problem 11. Define y as before; then the total set of possibilities for the needle can be identified with the same rectangular region $0 \leqslant y \leqslant L$, $0 \leqslant \theta \leqslant \pi$. What portion of the rectangle corresponds to the needle intersecting a line?]

(b) If $L = 4$, find the probability that a needle of length 7 will intersect *two* lines.

(c) If $2L < h \leqslant 3L$, find a general formula for the probability that the needle intersects three lines.

9

DIFFERENTIAL EQUATIONS

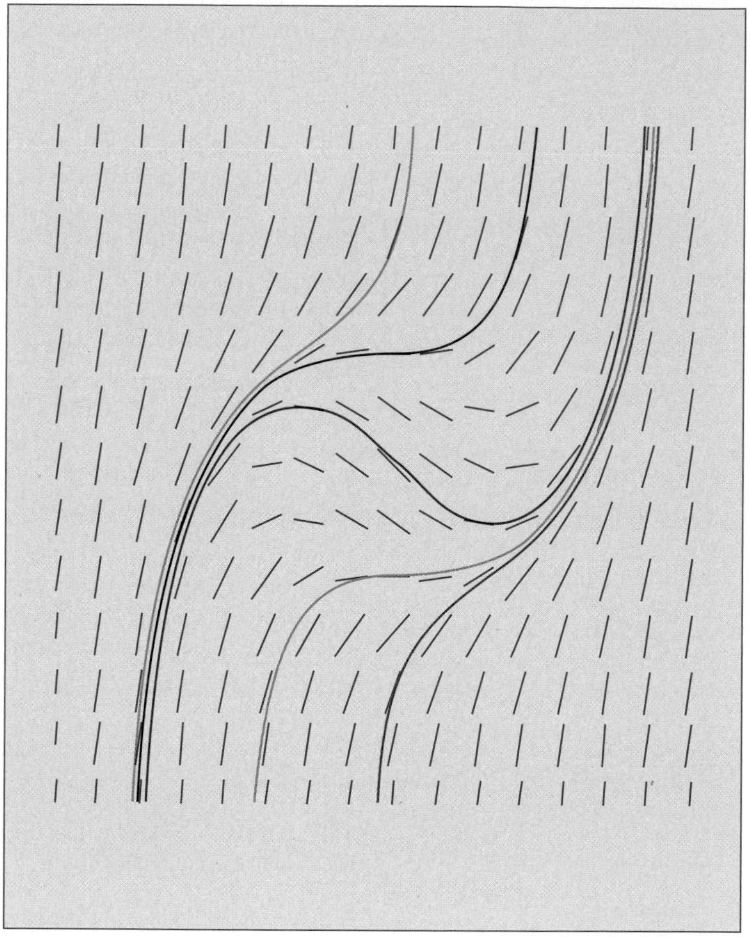

Direction fields enable us to sketch solutions of differential equations without an explicit formula.

Perhaps the most important of all the applications of calculus is to differential equations. When physical scientists or social scientists use calculus, more often than not it is to analyze a differential equation that has arisen in the process of modeling some phenomenon that they are studying. Although it is often impossible to find an explicit formula for the solution of a differential equation, we will see that graphical and numerical approaches provide the needed information.

9.1 MODELING WITH DIFFERENTIAL EQUATIONS

■ Now is a good time to read (or reread) the discussion of mathematical modeling on page 24.

In describing the process of modeling in Section 1.2, we talked about formulating a mathematical model of a real-world problem either through intuitive reasoning about the phenomenon or from a physical law based on evidence from experiments. The mathematical model often takes the form of a *differential equation*, that is, an equation that contains an unknown function and some of its derivatives. This is not surprising because in a real-world problem we often notice that changes occur and we want to predict future behavior on the basis of how current values change. Let's begin by examining several examples of how differential equations arise when we model physical phenomena.

MODELS OF POPULATION GROWTH

One model for the growth of a population is based on the assumption that the population grows at a rate proportional to the size of the population. That is a reasonable assumption for a population of bacteria or animals under ideal conditions (unlimited environment, adequate nutrition, absence of predators, immunity from disease).

Let's identify and name the variables in this model:

t = time (the independent variable)

P = the number of individuals in the population (the dependent variable)

The rate of growth of the population is the derivative dP/dt. So our assumption that the rate of growth of the population is proportional to the population size is written as the equation

$$\frac{dP}{dt} = kP \quad \boxed{1}$$

where k is the proportionality constant. Equation 1 is our first model for population growth; it is a differential equation because it contains an unknown function P and its derivative dP/dt.

Having formulated a model, let's look at its consequences. If we rule out a population of 0, then $P(t) > 0$ for all t. So, if $k > 0$, then Equation 1 shows that $P'(t) > 0$ for all t. This means that the population is always increasing. In fact, as $P(t)$ increases, Equation 1 shows that dP/dt becomes larger. In other words, the growth rate increases as the population increases.

Equation 1 asks us to find a function whose derivative is a constant multiple of itself. We know from Chapter 3 that exponential functions have that property. In fact, if we let $P(t) = Ce^{kt}$, then

$$P'(t) = C(ke^{kt}) = k(Ce^{kt}) = kP(t)$$

Thus any exponential function of the form $P(t) = Ce^{kt}$ is a solution of Equation 1. In Section 9.4 we will see that there is no other solution.

Allowing C to vary through all the real numbers, we get the *family* of solutions $P(t) = Ce^{kt}$ whose graphs are shown in Figure 1. But populations have only positive values and so we are interested only in the solutions with $C > 0$. And we are probably con-

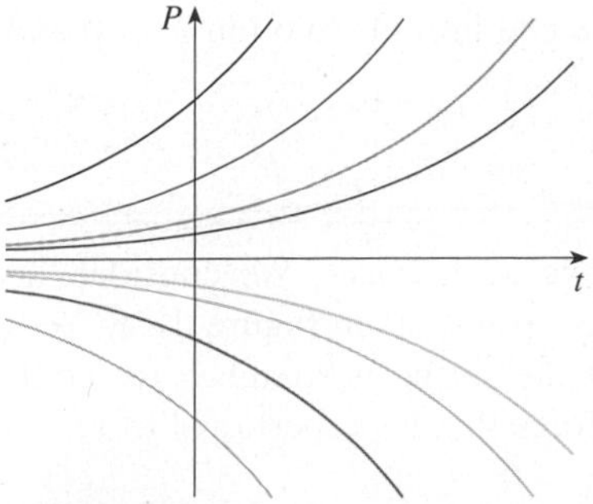

FIGURE 1
The family of solutions of $dP/dt = kP$

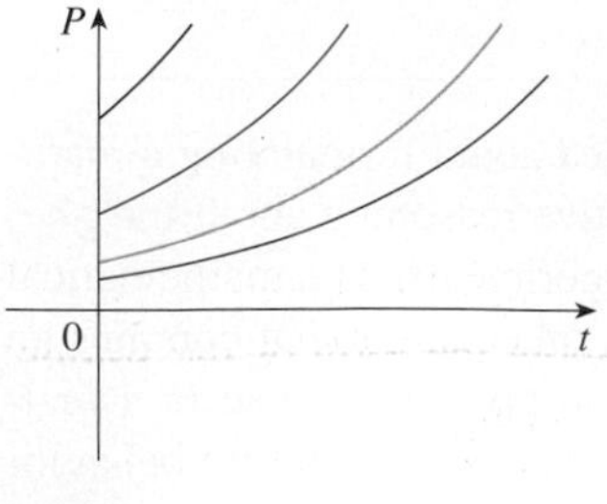

FIGURE 2
The family of solutions $P(t) = Ce^{kt}$ with $C > 0$ and $t \geqslant 0$

cerned only with values of t greater than the initial time $t = 0$. Figure 2 shows the physically meaningful solutions. Putting $t = 0$, we get $P(0) = Ce^{k(0)} = C$, so the constant C turns out to be the initial population, $P(0)$.

Equation 1 is appropriate for modeling population growth under ideal conditions, but we have to recognize that a more realistic model must reflect the fact that a given environment has limited resources. Many populations start by increasing in an exponential manner, but the population levels off when it approaches its *carrying capacity* K (or decreases toward K if it ever exceeds K). For a model to take into account both trends, we make two assumptions:

- $\dfrac{dP}{dt} \approx kP$ if P is small (Initially, the growth rate is proportional to P.)
- $\dfrac{dP}{dt} < 0$ if $P > K$ (P decreases if it ever exceeds K.)

A simple expression that incorporates both assumptions is given by the equation

$$\boxed{2} \qquad \frac{dP}{dt} = kP\left(1 - \frac{P}{K}\right)$$

Notice that if P is small compared with K, then P/K is close to 0 and so $dP/dt \approx kP$. If $P > K$, then $1 - P/K$ is negative and so $dP/dt < 0$.

Equation 2 is called the *logistic differential equation* and was proposed by the Dutch mathematical biologist Pierre-François Verhulst in the 1840s as a model for world population growth. We will develop techniques that enable us to find explicit solutions of the logistic equation in Section 9.4, but for now we can deduce qualitative characteristics of the solutions directly from Equation 2. We first observe that the constant functions $P(t) = 0$ and $P(t) = K$ are solutions because, in either case, one of the factors on the right side of Equation 2 is zero. (This certainly makes physical sense: If the population is ever either 0 or at the carrying capacity, it stays that way.) These two constant solutions are called *equilibrium solutions*.

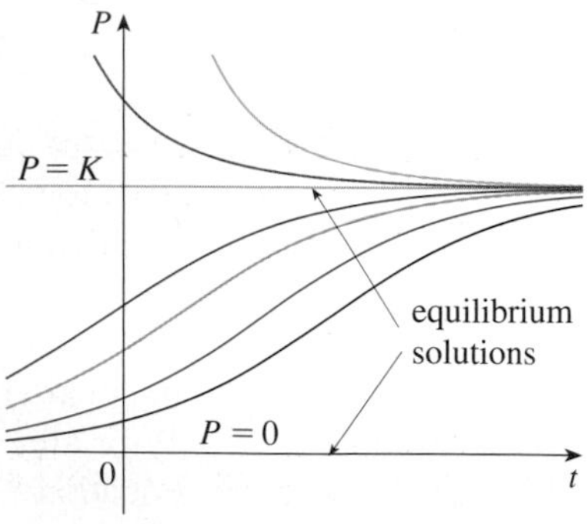

FIGURE 3
Solutions of the logistic equation

If the initial population $P(0)$ lies between 0 and K, then the right side of Equation 2 is positive, so $dP/dt > 0$ and the population increases. But if the population exceeds the carrying capacity ($P > K$), then $1 - P/K$ is negative, so $dP/dt < 0$ and the population decreases. Notice that, in either case, if the population approaches the carrying capacity ($P \to K$), then $dP/dt \to 0$, which means the population levels off. So we expect that the solutions of the logistic differential equation have graphs that look something like the ones in Figure 3. Notice that the graphs move away from the equilibrium solution $P = 0$ and move toward the equilibrium solution $P = K$.

A MODEL FOR THE MOTION OF A SPRING

Let's now look at an example of a model from the physical sciences. We consider the motion of an object with mass m at the end of a vertical spring (as in Figure 4). In Section 6.4 we discussed Hooke's Law, which says that if the spring is stretched (or compressed) x units from its natural length, then it exerts a force that is proportional to x:

$$\text{restoring force} = -kx$$

where k is a positive constant (called the *spring constant*). If we ignore any external resisting forces (due to air resistance or friction) then, by Newton's Second Law (force equals

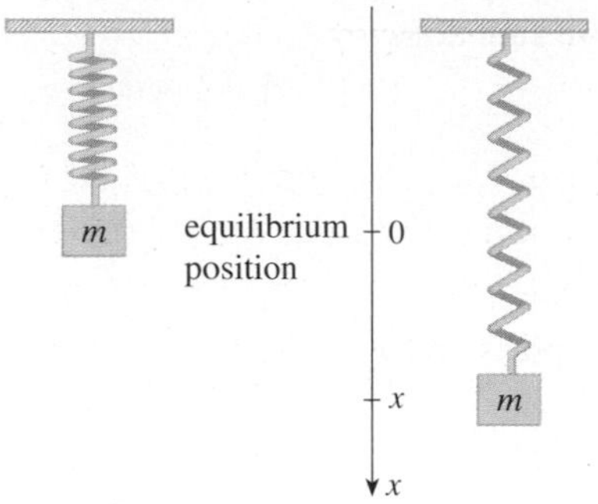

FIGURE 4

mass times acceleration), we have

$$m\frac{d^2x}{dt^2} = -kx \tag{3}$$

This is an example of what is called a *second-order differential equation* because it involves second derivatives. Let's see what we can guess about the form of the solution directly from the equation. We can rewrite Equation 3 in the form

$$\frac{d^2x}{dt^2} = -\frac{k}{m}x$$

which says that the second derivative of x is proportional to x but has the opposite sign. We know two functions with this property, the sine and cosine functions. In fact, it turns out that all solutions of Equation 3 can be written as combinations of certain sine and cosine functions (see Exercise 4). This is not surprising; we expect the spring to oscillate about its equilibrium position and so it is natural to think that trigonometric functions are involved.

GENERAL DIFFERENTIAL EQUATIONS

In general, a **differential equation** is an equation that contains an unknown function and one or more of its derivatives. The **order** of a differential equation is the order of the highest derivative that occurs in the equation. Thus, Equations 1 and 2 are first-order equations and Equation 3 is a second-order equation. In all three of those equations the independent variable is called t and represents time, but in general the independent variable doesn't have to represent time. For example, when we consider the differential equation

$$y' = xy \tag{4}$$

it is understood that y is an unknown function of x.

A function f is called a **solution** of a differential equation if the equation is satisfied when $y = f(x)$ and its derivatives are substituted into the equation. Thus f is a solution of Equation 4 if

$$f'(x) = xf(x)$$

for all values of x in some interval.

When we are asked to *solve* a differential equation we are expected to find all possible solutions of the equation. We have already solved some particularly simple differential equations, namely, those of the form

$$y' = f(x)$$

For instance, we know that the general solution of the differential equation

$$y' = x^3$$

is given by

$$y = \frac{x^4}{4} + C$$

where C is an arbitrary constant.

But, in general, solving a differential equation is not an easy matter. There is no systematic technique that enables us to solve all differential equations. In Section 9.2, however, we will see how to draw rough graphs of solutions even when we have no explicit formula. We will also learn how to find numerical approximations to solutions.

V EXAMPLE 1 Show that every member of the family of functions

$$y = \frac{1 + ce^t}{1 - ce^t}$$

is a solution of the differential equation $y' = \frac{1}{2}(y^2 - 1)$.

SOLUTION We use the Quotient Rule to differentiate the expression for y:

$$y' = \frac{(1 - ce^t)(ce^t) - (1 + ce^t)(-ce^t)}{(1 - ce^t)^2}$$

$$= \frac{ce^t - c^2e^{2t} + ce^t + c^2e^{2t}}{(1 - ce^t)^2} = \frac{2ce^t}{(1 - ce^t)^2}$$

The right side of the differential equation becomes

$$\frac{1}{2}(y^2 - 1) = \frac{1}{2}\left[\left(\frac{1 + ce^t}{1 - ce^t}\right)^2 - 1\right] = \frac{1}{2}\left[\frac{(1 + ce^t)^2 - (1 - ce^t)^2}{(1 - ce^t)^2}\right]$$

$$= \frac{1}{2}\frac{4ce^t}{(1 - ce^t)^2} = \frac{2ce^t}{(1 - ce^t)^2}$$

Therefore, for every value of c, the given function is a solution of the differential equation. □

■ Figure 5 shows graphs of seven members of the family in Example 1. The differential equation shows that if $y \approx \pm 1$, then $y' \approx 0$. That is borne out by the flatness of the graphs near $y = 1$ and $y = -1$.

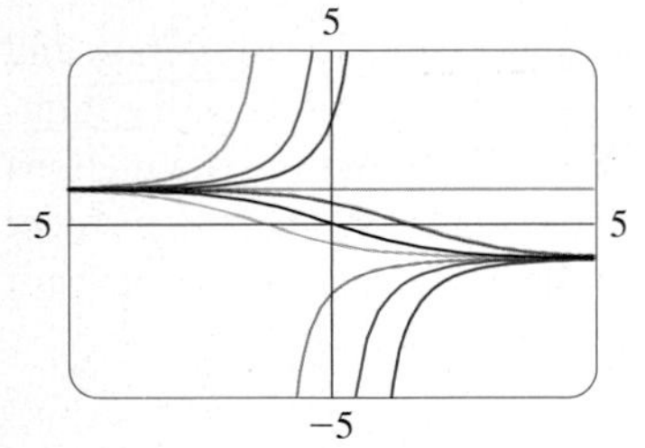

FIGURE 5

When applying differential equations, we are usually not as interested in finding a family of solutions (the *general solution*) as we are in finding a solution that satisfies some additional requirement. In many physical problems we need to find the particular solution that satisfies a condition of the form $y(t_0) = y_0$. This is called an **initial condition**, and the problem of finding a solution of the differential equation that satisfies the initial condition is called an **initial-value problem**.

Geometrically, when we impose an initial condition, we look at the family of solution curves and pick the one that passes through the point (t_0, y_0). Physically, this corresponds to measuring the state of a system at time t_0 and using the solution of the initial-value problem to predict the future behavior of the system.

V EXAMPLE 2 Find a solution of the differential equation $y' = \frac{1}{2}(y^2 - 1)$ that satisfies the initial condition $y(0) = 2$.

SOLUTION Substituting the values $t = 0$ and $y = 2$ into the formula

$$y = \frac{1 + ce^t}{1 - ce^t}$$

from Example 1, we get

$$2 = \frac{1 + ce^0}{1 - ce^0} = \frac{1 + c}{1 - c}$$

Solving this equation for c, we get $2 - 2c = 1 + c$, which gives $c = \frac{1}{3}$. So the solution of the initial-value problem is

$$y = \frac{1 + \frac{1}{3}e^t}{1 - \frac{1}{3}e^t} = \frac{3 + e^t}{3 - e^t}$$ □

9.1 EXERCISES

1. Show that $y = x - x^{-1}$ is a solution of the differential equation $xy' + y = 2x$.

2. Verify that $y = \sin x \cos x - \cos x$ is a solution of the initial-value problem

$$y' + (\tan x)y = \cos^2 x \qquad y(0) = -1$$

on the interval $-\pi/2 < x < \pi/2$.

3. (a) For what values of r does the function $y = e^{rx}$ satisfy the differential equation $2y'' + y' - y = 0$?
(b) If r_1 and r_2 are the values of r that you found in part (a), show that every member of the family of functions $y = ae^{r_1 x} + be^{r_2 x}$ is also a solution.

4. (a) For what values of k does the function $y = \cos kt$ satisfy the differential equation $4y'' = -25y$?
(b) For those values of k, verify that every member of the family of functions $y = A \sin kt + B \cos kt$ is also a solution.

5. Which of the following functions are solutions of the differential equation $y'' + y = \sin x$?
(a) $y = \sin x$ (b) $y = \cos x$
(c) $y = \frac{1}{2}x \sin x$ (d) $y = -\frac{1}{2}x \cos x$

6. (a) Show that every member of the family of functions $y = (\ln x + C)/x$ is a solution of the differential equation $x^2y' + xy = 1$.
(b) Illustrate part (a) by graphing several members of the family of solutions on a common screen.
(c) Find a solution of the differential equation that satisfies the initial condition $y(1) = 2$.
(d) Find a solution of the differential equation that satisfies the initial condition $y(2) = 1$.

7. (a) What can you say about a solution of the equation $y' = -y^2$ just by looking at the differential equation?
(b) Verify that all members of the family $y = 1/(x + C)$ are solutions of the equation in part (a).
(c) Can you think of a solution of the differential equation $y' = -y^2$ that is not a member of the family in part (b)?
(d) Find a solution of the initial-value problem

$$y' = -y^2 \qquad y(0) = 0.5$$

8. (a) What can you say about the graph of a solution of the equation $y' = xy^3$ when x is close to 0? What if x is large?
(b) Verify that all members of the family $y = (c - x^2)^{-1/2}$ are solutions of the differential equation $y' = xy^3$.
(c) Graph several members of the family of solutions on a common screen. Do the graphs confirm what you predicted in part (a)?
(d) Find a solution of the initial-value problem

$$y' = xy^3 \qquad y(0) = 2$$

9. A population is modeled by the differential equation

$$\frac{dP}{dt} = 1.2P\left(1 - \frac{P}{4200}\right)$$

(a) For what values of P is the population increasing?
(b) For what values of P is the population decreasing?
(c) What are the equilibrium solutions?

10. A function $y(t)$ satisfies the differential equation

$$\frac{dy}{dt} = y^4 - 6y^3 + 5y^2$$

(a) What are the constant solutions of the equation?
(b) For what values of y is y increasing?
(c) For what values of y is y decreasing?

11. Explain why the functions with the given graphs *can't* be solutions of the differential equation

$$\frac{dy}{dt} = e^t(y - 1)^2$$

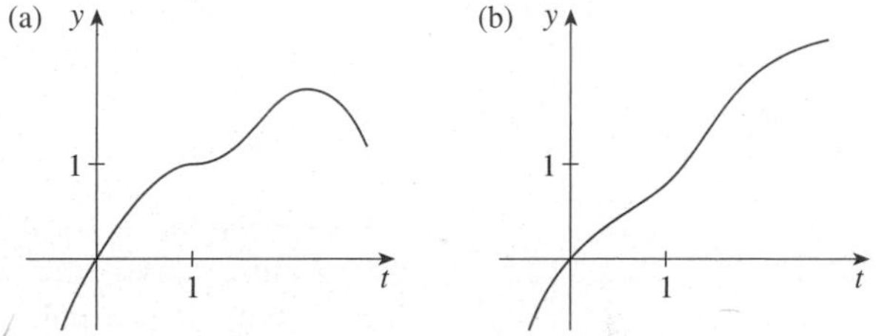

12. The function with the given graph is a solution of one of the following differential equations. Decide which is the correct equation and justify your answer.

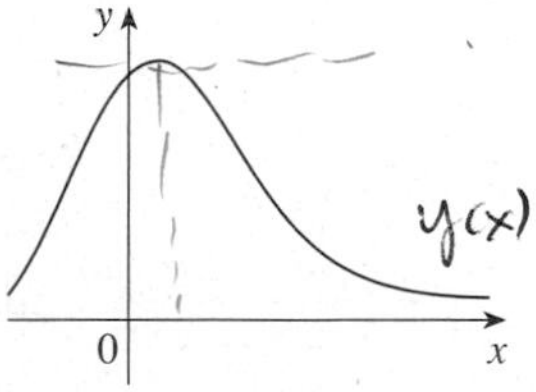

A. $y' = 1 + xy$ **B.** $y' = -2xy$ **C.** $y' = 1 - 2xy$

13. Psychologists interested in learning theory study **learning curves**. A learning curve is the graph of a function $P(t)$, the performance of someone learning a skill as a function of the training time t. The derivative dP/dt represents the rate at which performance improves.
(a) When do you think P increases most rapidly? What happens to dP/dt as t increases? Explain.
(b) If M is the maximum level of performance of which the learner is capable, explain why the differential equation

$$\frac{dP}{dt} = k(M - P) \qquad k \text{ a positive constant}$$

is a reasonable model for learning.

(c) Make a rough sketch of a possible solution of this differential equation.

14. Suppose you have just poured a cup of freshly brewed coffee with temperature 95°C in a room where the temperature is 20°C.
(a) When do you think the coffee cools most quickly? What happens to the rate of cooling as time goes by? Explain.
(b) Newton's Law of Cooling states that the rate of cooling of an object is proportional to the temperature difference between the object and its surroundings, provided that this difference is not too large. Write a differential equation that expresses Newton's Law of Cooling for this particular situation. What is the initial condition? In view of your answer to part (a), do you think this differential equation is an appropriate model for cooling?
(c) Make a rough sketch of the graph of the solution of the initial-value problem in part (b).

9.2 DIRECTION FIELDS AND EULER'S METHOD

Unfortunately, it's impossible to solve most differential equations in the sense of obtaining an explicit formula for the solution. In this section we show that, despite the absence of an explicit solution, we can still learn a lot about the solution through a graphical approach (direction fields) or a numerical approach (Euler's method).

DIRECTION FIELDS

Suppose we are asked to sketch the graph of the solution of the initial-value problem

$$y' = x + y \qquad y(0) = 1$$

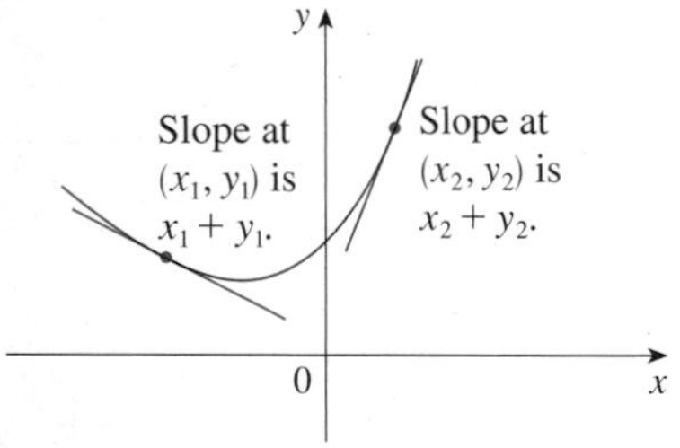

FIGURE 1
A solution of $y' = x + y$

We don't know a formula for the solution, so how can we possibly sketch its graph? Let's think about what the differential equation means. The equation $y' = x + y$ tells us that the slope at any point (x, y) on the graph (called the *solution curve*) is equal to the sum of the x- and y-coordinates of the point (see Figure 1). In particular, because the curve passes through the point (0, 1), its slope there must be $0 + 1 = 1$. So a small portion of the solution curve near the point (0, 1) looks like a short line segment through (0, 1) with slope 1. (See Figure 2.)

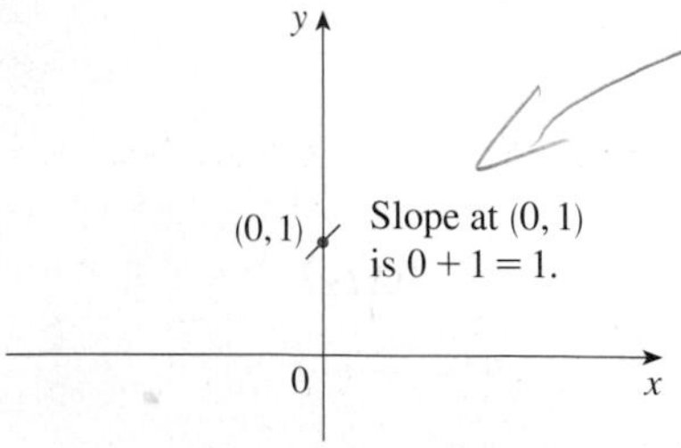

FIGURE 2
Beginning of the solution curve through (0, 1)

As a guide to sketching the rest of the curve, let's draw short line segments at a number of points (x, y) with slope $x + y$. The result is called a *direction field* and is shown in Figure 3. For instance, the line segment at the point (1, 2) has slope $1 + 2 = 3$. The direction field allows us to visualize the general shape of the solution curves by indicating the direction in which the curves proceed at each point.

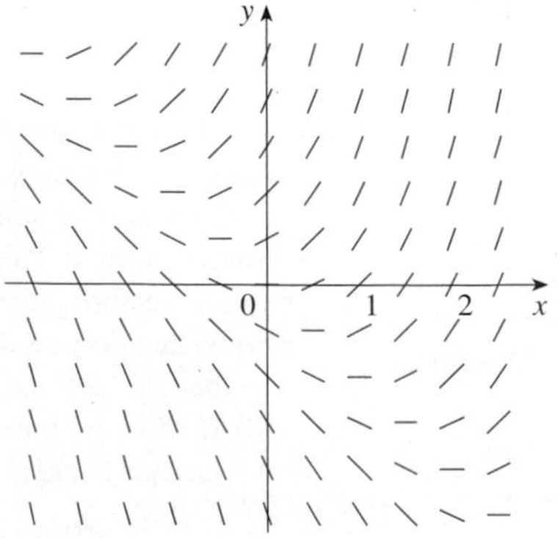

FIGURE 3
Direction field for $y' = x + y$

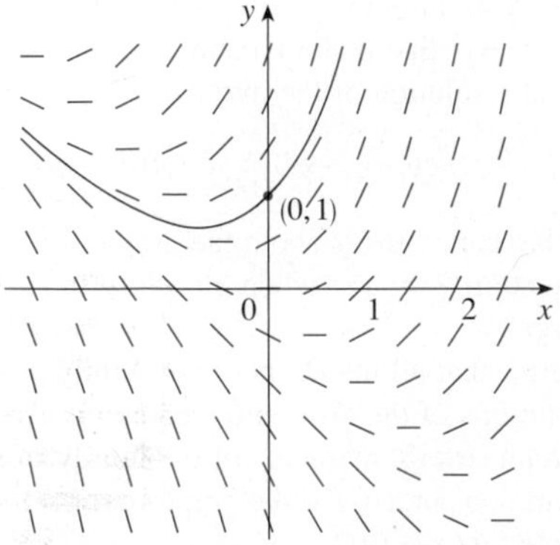

FIGURE 4
The solution curve through (0, 1)

Now we can sketch the solution curve through the point (0, 1) by following the direction field as in Figure 4. Notice that we have drawn the curve so that it is parallel to nearby line segments.

In general, suppose we have a first-order differential equation of the form

$$y' = F(x, y)$$

where $F(x, y)$ is some expression in x and y. The differential equation says that the slope of a solution curve at a point (x, y) on the curve is $F(x, y)$. If we draw short line segments with slope $F(x, y)$ at several points (x, y), the result is called a **direction field** (or **slope field**). These line segments indicate the direction in which a solution curve is heading, so the direction field helps us visualize the general shape of these curves.

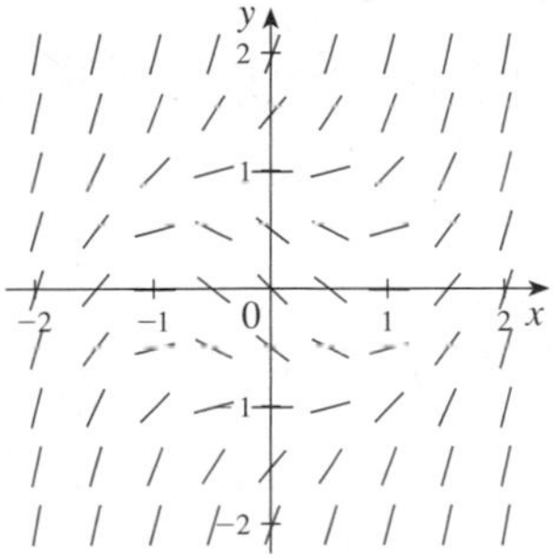

FIGURE 5

V EXAMPLE 1

(a) Sketch the direction field for the differential equation $y' = x^2 + y^2 - 1$.
(b) Use part (a) to sketch the solution curve that passes through the origin.

SOLUTION

(a) We start by computing the slope at several points in the following chart:

x	−2	−1	0	1	2	−2	−1	0	1	2	...
y	0	0	0	0	0	1	1	1	1	1	...
$y' = x^2 + y^2 - 1$	3	0	−1	0	3	4	1	0	1	4	...

Now we draw short line segments with these slopes at these points. The result is the direction field shown in Figure 5.

(b) We start at the origin and move to the right in the direction of the line segment (which has slope −1). We continue to draw the solution curve so that it moves parallel to the nearby line segments. The resulting solution curve is shown in Figure 6. Returning to the origin, we draw the solution curve to the left as well. □

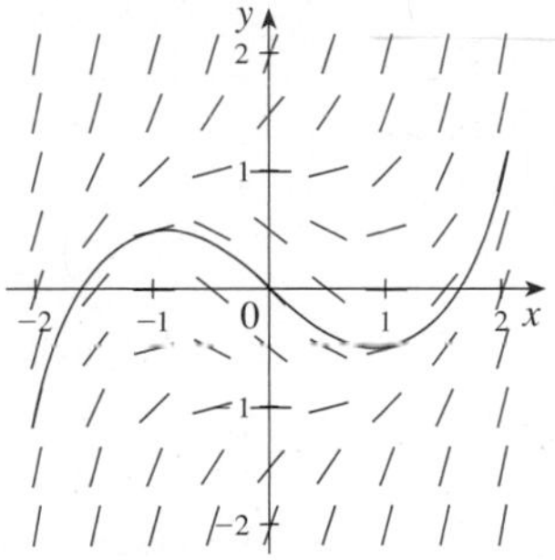

FIGURE 6

The more line segments we draw in a direction field, the clearer the picture becomes. Of course, it's tedious to compute slopes and draw line segments for a huge number of points by hand, but computers are well suited for this task. Figure 7 shows a more detailed, computer-drawn direction field for the differential equation in Example 1. It enables us to draw, with reasonable accuracy, the solution curves shown in Figure 8 with y-intercepts −2, −1, 0, 1, and 2.

TEC Module 9.2A shows direction fields and solution curves for a variety of differential equations.

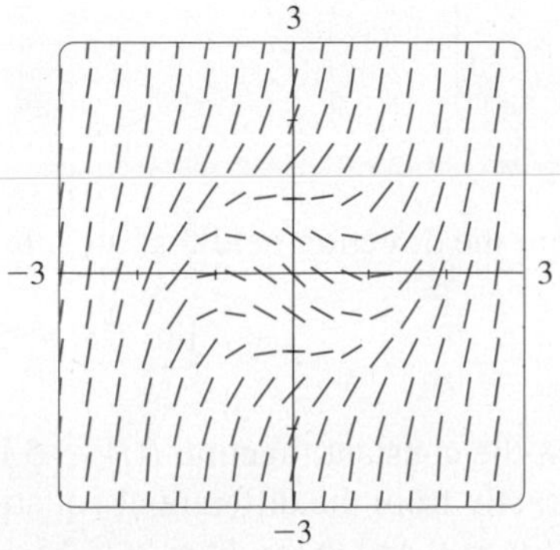

FIGURE 7

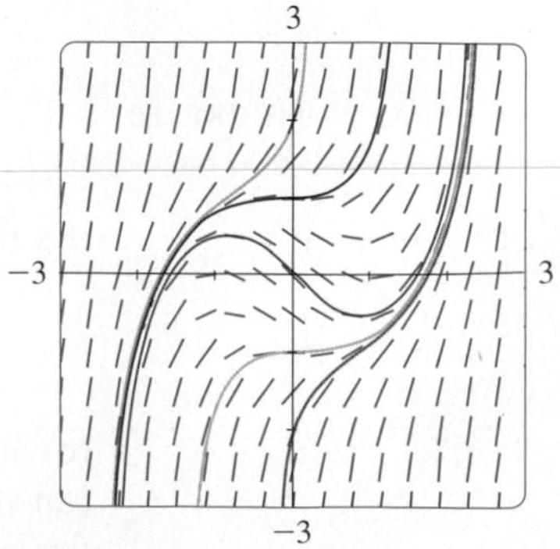

FIGURE 8

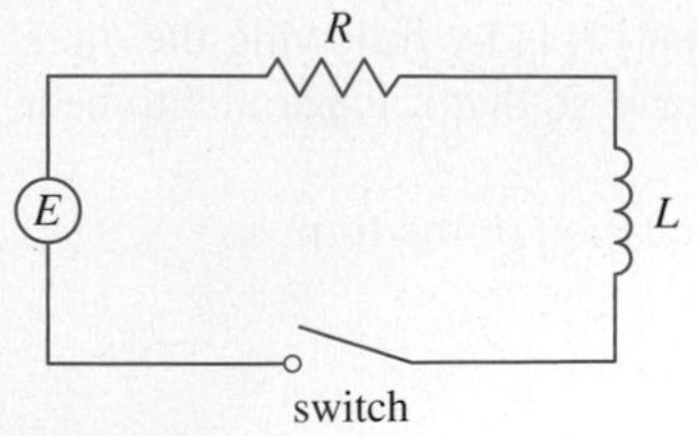

FIGURE 9

Now let's see how direction fields give insight into physical situations. The simple electric circuit shown in Figure 9 contains an electromotive force (usually a battery or generator) that produces a voltage of $E(t)$ volts (V) and a current of $I(t)$ amperes (A) at time t. The circuit also contains a resistor with a resistance of R ohms (Ω) and an inductor with an inductance of L henries (H).

Ohm's Law gives the drop in voltage due to the resistor as RI. The voltage drop due to the inductor is $L(dI/dt)$. One of Kirchhoff's laws says that the sum of the voltage drops is equal to the supplied voltage $E(t)$. Thus we have

$$L\frac{dI}{dt} + RI = E(t) \tag{1}$$

which is a first-order differential equation that models the current I at time t.

V EXAMPLE 2 Suppose that in the simple circuit of Figure 9 the resistance is 12 Ω, the inductance is 4 H, and a battery gives a constant voltage of 60 V.

(a) Draw a direction field for Equation 1 with these values.
(b) What can you say about the limiting value of the current?
(c) Identify any equilibrium solutions.
(d) If the switch is closed when $t = 0$ so the current starts with $I(0) = 0$, use the direction field to sketch the solution curve.

SOLUTION

(a) If we put $L = 4$, $R = 12$, and $E(t) = 60$ in Equation 1, we get

$$4\frac{dI}{dt} + 12I = 60 \qquad \text{or} \qquad \frac{dI}{dt} = 15 - 3I$$

The direction field for this differential equation is shown in Figure 10.

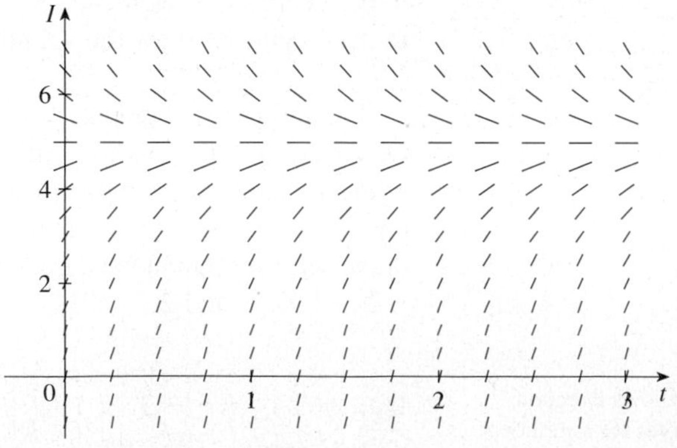

FIGURE 10

(b) It appears from the direction field that all solutions approach the value 5 A, that is,

$$\lim_{t\to\infty} I(t) = 5$$

(c) It appears that the constant function $I(t) = 5$ is an equilibrium solution. Indeed, we can verify this directly from the differential equation $dI/dt = 15 - 3I$. If $I(t) = 5$, then the left side is $dI/dt = 0$ and the right side is $15 - 3(5) = 0$.

(d) We use the direction field to sketch the solution curve that passes through (0, 0), as shown in red in Figure 11.

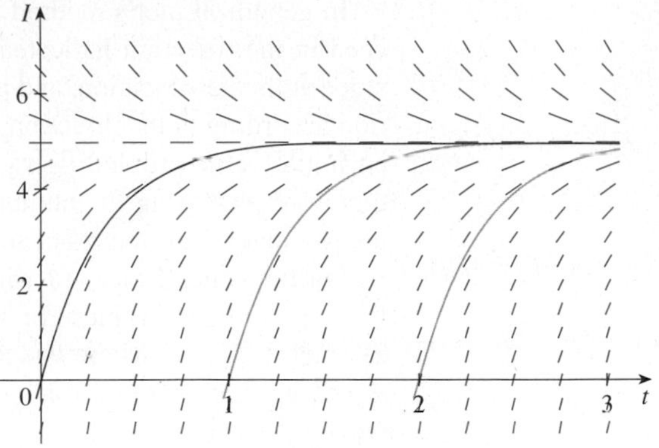

FIGURE 11

□

Notice from Figure 10 that the line segments along any horizontal line are parallel. That is because the independent variable t does not occur on the right side of the equation $I' = 15 - 3I$. In general, a differential equation of the form

$$y' = f(y)$$

in which the independent variable is missing from the right side, is called **autonomous**. For such an equation, the slopes corresponding to two different points with the same y-coordinate must be equal. This means that if we know one solution to an autonomous differential equation, then we can obtain infinitely many others just by shifting the graph of the known solution to the right or left. In Figure 11 we have shown the solutions that result from shifting the solution curve of Example 2 one and two time units (namely, seconds) to the right. They correspond to closing the switch when $t = 1$ or $t = 2$.

EULER'S METHOD

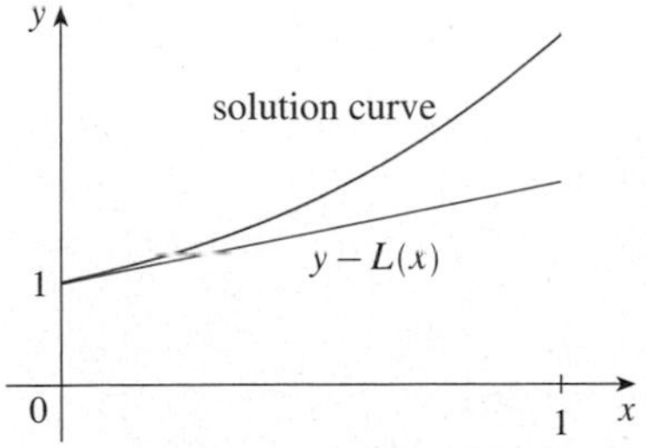

FIGURE 12
First Euler approximation

The basic idea behind direction fields can be used to find numerical approximations to solutions of differential equations. We illustrate the method on the initial-value problem that we used to introduce direction fields:

$$y' = x + y \qquad y(0) = 1$$

The differential equation tells us that $y'(0) = 0 + 1 = 1$, so the solution curve has slope 1 at the point (0, 1). As a first approximation to the solution we could use the linear approximation $L(x) = x + 1$. In other words, we could use the tangent line at (0, 1) as a rough approximation to the solution curve (see Figure 12).

Euler's idea was to improve on this approximation by proceeding only a short distance along this tangent line and then making a midcourse correction by changing direction as indicated by the direction field. Figure 13 shows what happens if we start out along the tangent line but stop when $x = 0.5$. (This horizontal distance traveled is called the *step size.*) Since $L(0.5) = 1.5$, we have $y(0.5) \approx 1.5$ and we take (0.5, 1.5) as the starting point for a new line segment. The differential equation tells us that $y'(0.5) = 0.5 + 1.5 = 2$, so we use the linear function

$$y = 1.5 + 2(x - 0.5) = 2x + 0.5$$

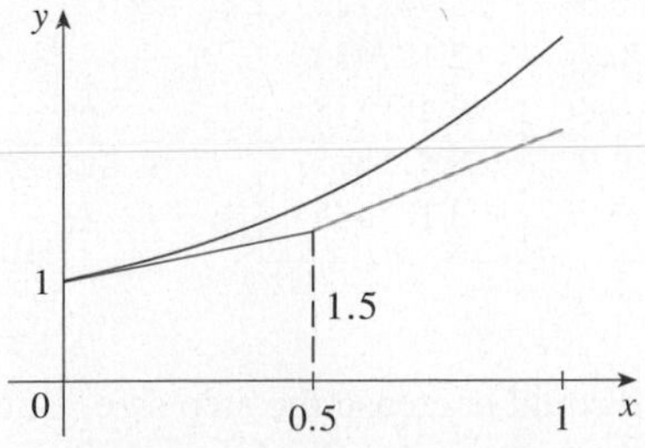

FIGURE 13
Euler approximation with step size 0.5

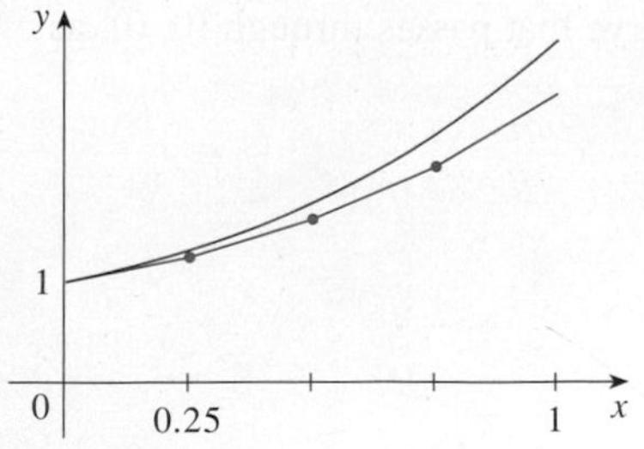

FIGURE 14
Euler approximation with step size 0.25

as an approximation to the solution for $x > 0.5$ (the orange segment in Figure 13). If we decrease the step size from 0.5 to 0.25, we get the better Euler approximation shown in Figure 14.

In general, Euler's method says to start at the point given by the initial value and proceed in the direction indicated by the direction field. Stop after a short time, look at the slope at the new location, and proceed in that direction. Keep stopping and changing direction according to the direction field. Euler's method does not produce the exact solution to an initial-value problem—it gives approximations. But by decreasing the step size (and therefore increasing the number of midcourse corrections), we obtain successively better approximations to the exact solution. (Compare Figures 12, 13, and 14.)

For the general first-order initial-value problem $y' = F(x, y)$, $y(x_0) = y_0$, our aim is to find approximate values for the solution at equally spaced numbers x_0, $x_1 = x_0 + h$, $x_2 = x_1 + h, \ldots$, where h is the step size. The differential equation tells us that the slope at (x_0, y_0) is $y' = F(x_0, y_0)$, so Figure 15 shows that the approximate value of the solution when $x = x_1$ is

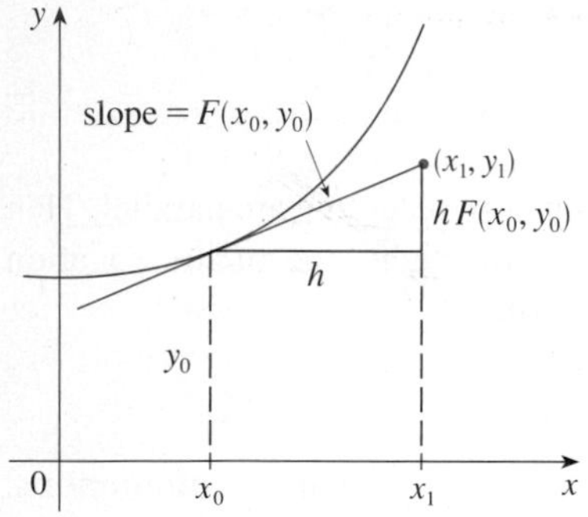

FIGURE 15

$$y_1 = y_0 + hF(x_0, y_0)$$

Similarly,

$$y_2 = y_1 + hF(x_1, y_1)$$

In general,

$$\boxed{y_n = y_{n-1} + hF(x_{n-1}, y_{n-1})}$$

EXAMPLE 3 Use Euler's method with step size 0.1 to construct a table of approximate values for the solution of the initial-value problem

$$y' = x + y \qquad y(0) = 1$$

SOLUTION We are given that $h = 0.1$, $x_0 = 0$, $y_0 = 1$, and $F(x, y) = x + y$. So we have

$$y_1 = y_0 + hF(x_0, y_0) = 1 + 0.1(0 + 1) = 1.1$$

$$y_2 = y_1 + hF(x_1, y_1) = 1.1 + 0.1(0.1 + 1.1) = 1.22$$

$$y_3 = y_2 + hF(x_2, y_2) = 1.22 + 0.1(0.2 + 1.22) = 1.362$$

This means that if $y(x)$ is the exact solution, then $y(0.3) \approx 1.362$.

Proceeding with similar calculations, we get the values in the table:

TEC Module 9.2B shows how Euler's method works numerically and visually for a variety of differential equations and step sizes.

n	x_n	y_n	n	x_n	y_n
1	0.1	1.100000	6	0.6	1.943122
2	0.2	1.220000	7	0.7	2.197434
3	0.3	1.362000	8	0.8	2.487178
4	0.4	1.528200	9	0.9	2.815895
5	0.5	1.721020	10	1.0	3.187485

□

For a more accurate table of values in Example 3 we could decrease the step size. But for a large number of small steps the amount of computation is considerable and so we need to program a calculator or computer to carry out these calculations. The following table shows the results of applying Euler's method with decreasing step size to the initial-value problem of Example 3.

Step size	Euler estimate of $y(0.5)$	Euler estimate of $y(1)$
0.500	1.500000	2.500000
0.250	1.625000	2.882813
0.100	1.721020	3.187485
0.050	1.757789	3.306595
0.020	1.781212	3.383176
0.010	1.789264	3.409628
0.005	1.793337	3.423034
0.001	1.796619	3.433848

■ Computer software packages that produce numerical approximations to solutions of differential equations use methods that are refinements of Euler's method. Although Euler's method is simple and not as accurate, it is the basic idea on which the more accurate methods are based.

Notice that the Euler estimates in the table seem to be approaching limits, namely, the true values of $y(0.5)$ and $y(1)$. Figure 16 shows graphs of the Euler approximations with step sizes 0.5, 0.25, 0.1, 0.05, 0.02, 0.01, and 0.005. They are approaching the exact solution curve as the step size h approaches 0.

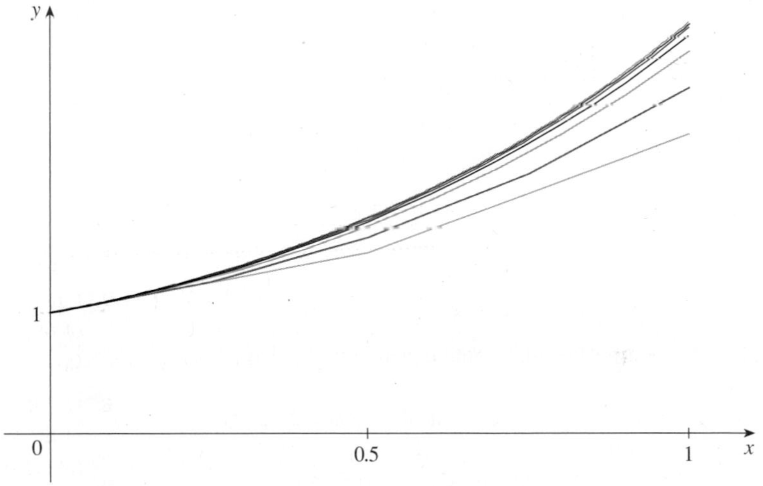

FIGURE 16
Euler approximations approaching the exact solution

V EXAMPLE 4 In Example 2 we discussed a simple electric circuit with resistance 12 Ω, inductance 4 H, and a battery with voltage 60 V. If the switch is closed when $t = 0$, we modeled the current I at time t by the initial-value problem

$$\frac{dI}{dt} = 15 - 3I \qquad I(0) = 0$$

Estimate the current in the circuit half a second after the switch is closed.

SOLUTION We use Euler's method with $F(t, I) = 15 - 3I$, $t_0 = 0$, $I_0 = 0$, and step size $h = 0.1$ second:

$$I_1 = 0 + 0.1(15 - 3 \cdot 0) = 1.5$$

$$I_2 = 1.5 + 0.1(15 - 3 \cdot 1.5) = 2.55$$

$$I_3 = 2.55 + 0.1(15 - 3 \cdot 2.55) = 3.285$$

$$I_4 = 3.285 + 0.1(15 - 3 \cdot 3.285) = 3.7995$$

$$I_5 = 3.7995 + 0.1(15 - 3 \cdot 3.7995) = 4.15965$$

So the current after 0.5 seconds is

$$I(0.5) \approx 4.16 \text{ A}$$

□

9.2 EXERCISES

1. A direction field for the differential equation $y' = y(1 - \frac{1}{4}y^2)$ is shown.

(a) Sketch the graphs of the solutions that satisfy the given initial conditions.

(i) $y(0) = 1$ (ii) $y(0) = -1$

(iii) $y(0) = -3$ (iv) $y(0) = 3$

(b) Find all the equilibrium solutions.

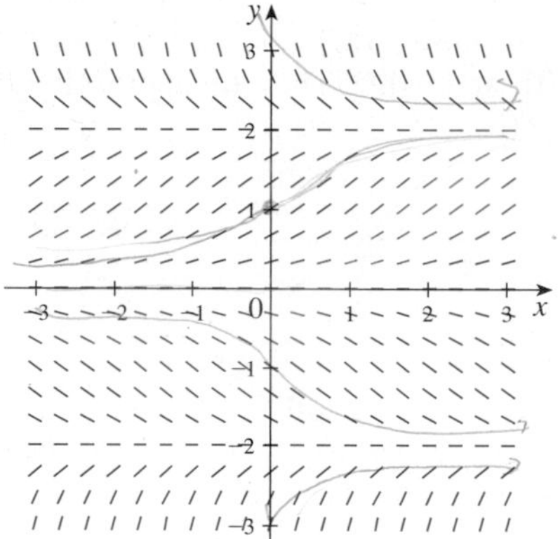

2. A direction field for the differential equation $y' = x \sin y$ is shown.

(a) Sketch the graphs of the solutions that satisfy the given initial conditions.

(i) $y(0) = 1$ (ii) $y(0) = 2$ (iii) $y(0) = \pi$

(iv) $y(0) = 4$ (v) $y(0) = 5$

(b) Find all the equilibrium solutions.

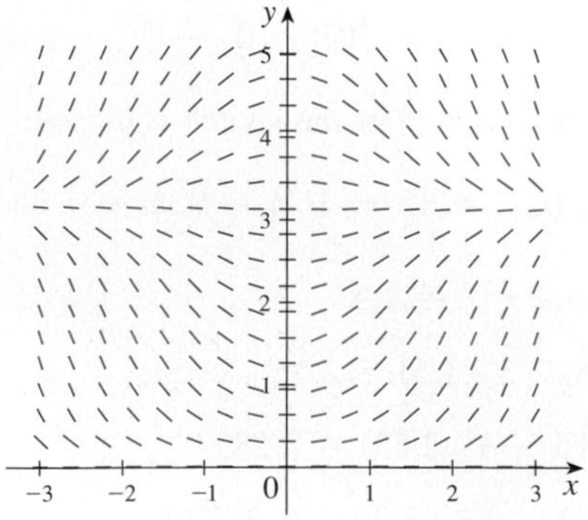

3–6 Match the differential equation with its direction field (labeled I–IV). Give reasons for your answer.

3. $y' = 2 - y$

4. $y' = x(2 - y)$

5. $y' = x + y - 1$

6. $y' = \sin x \sin y$

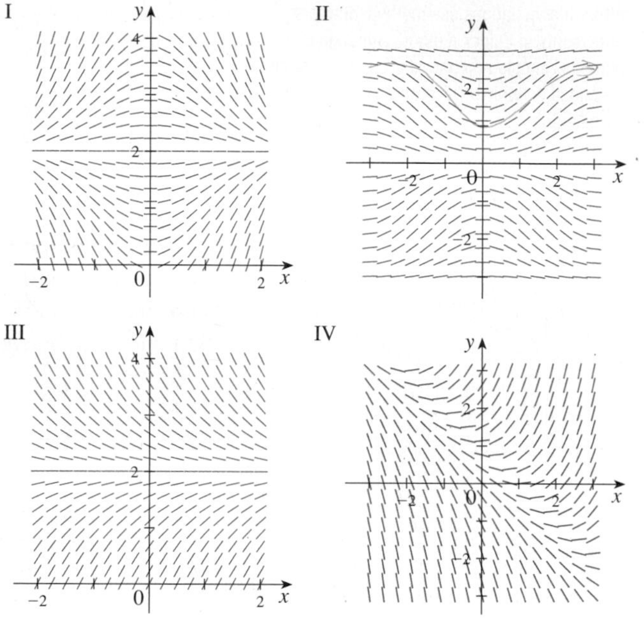

7. Use the direction field labeled II (above) to sketch the graphs of the solutions that satisfy the given initial conditions.

(a) $y(0) = 1$ (b) $y(0) = 2$ (c) $y(0) = -1$

8. Use the direction field labeled IV (above) to sketch the graphs of the solutions that satisfy the given initial conditions.

(a) $y(0) = -1$ (b) $y(0) = 0$ (c) $y(0) = 1$

9–10 Sketch a direction field for the differential equation. Then use it to sketch three solution curves.

9. $y' = 1 + y$

10. $y' = x^2 - y^2$

11–14 Sketch the direction field of the differential equation. Then use it to sketch a solution curve that passes through the given point.

11. $y' = y - 2x$, $(1, 0)$

12. $y' = 1 - xy$, $(0, 0)$

13. $y' = y + xy$, $(0, 1)$

14. $y' = x - xy$, $(1, 0)$

CAS **15–16** Use a computer algebra system to draw a direction field for the given differential equation. Get a printout and sketch on it the solution curve that passes through $(0, 1)$. Then use the CAS to draw the solution curve and compare it with your sketch.

15. $y' = x^2 \sin y$

16. $y' = x(y^2 - 4)$

CAS **17.** Use a computer algebra system to draw a direction field for the differential equation $y' = y^3 - 4y$. Get a printout and sketch on it solutions that satisfy the initial condition $y(0) = c$ for various values of c. For what values of c does $\lim_{t\to\infty} y(t)$ exist? What are the possible values for this limit?

18. Make a rough sketch of a direction field for the autonomous differential equation $y' = f(y)$, where the graph of f is as shown. How does the limiting behavior of solutions depend on the value of $y(0)$?

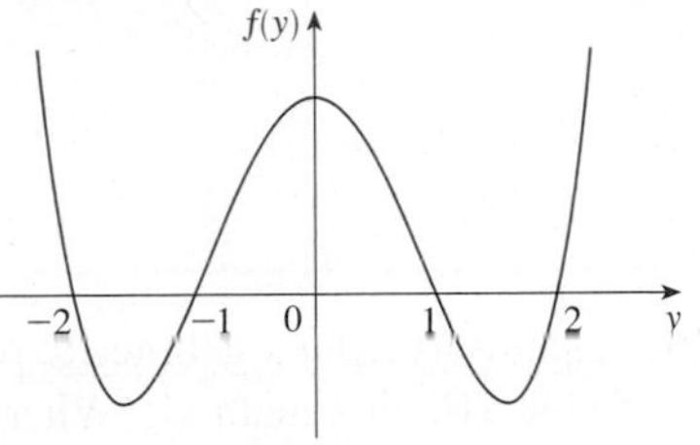

19. (a) Use Euler's method with each of the following step sizes to estimate the value of $y(0.4)$, where y is the solution of the initial-value problem $y' = y$, $y(0) = 1$.
(i) $h = 0.4$ (ii) $h = 0.2$ (iii) $h = 0.1$
(b) We know that the exact solution of the initial-value problem in part (a) is $y = e^x$. Draw, as accurately as you can, the graph of $y = e^x$, $0 \leqslant x \leqslant 0.4$, together with the Euler approximations using the step sizes in part (a). (Your sketches should resemble Figures 12, 13, and 14.) Use your sketches to decide whether your estimates in part (a) are underestimates or overestimates.
(c) The error in Euler's method is the difference between the exact value and the approximate value. Find the errors made in part (a) in using Euler's method to estimate the true value of $y(0.4)$, namely $e^{0.4}$. What happens to the error each time the step size is halved?

20. A direction field for a differential equation is shown. Draw, with a ruler, the graphs of the Euler approximations to the solution curve that passes through the origin. Use step sizes $h = 1$ and $h = 0.5$. Will the Euler estimates be under-estimates or overestimates? Explain.

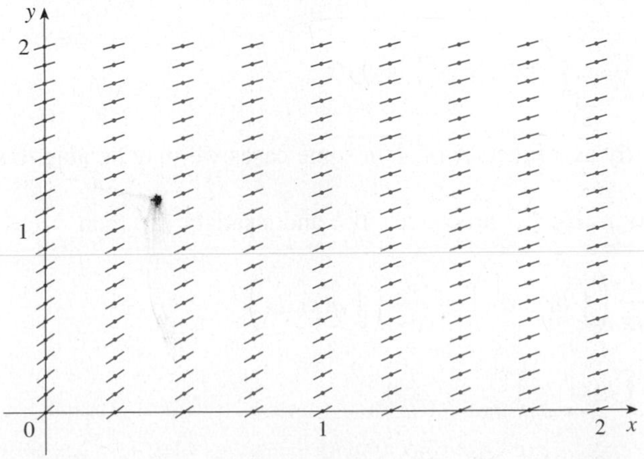

21. Use Euler's method with step size 0.5 to compute the approximate y-values y_1, y_2, y_3, and y_4 of the solution of the initial-value problem $y' = y - 2x$, $y(1) = 0$.

22. Use Euler's method with step size 0.2 to estimate $y(1)$, where $y(x)$ is the solution of the initial-value problem $y' = 1 - xy$, $y(0) = 0$.

23. Use Euler's method with step size 0.1 to estimate $y(0.5)$, where $y(x)$ is the solution of the initial-value problem $y' = y + xy$, $y(0) = 1$.

24. (a) Use Euler's method with step size 0.2 to estimate $y(1.4)$, where $y(x)$ is the solution of the initial-value problem $y' = x - xy$, $y(1) = 0$.
(b) Repeat part (a) with step size 0.1.

25. (a) Program a calculator or computer to use Euler's method to compute $y(1)$, where $y(x)$ is the solution of the initial value problem

$$\frac{dy}{dx} + 3x^2 y = 6x^2 \qquad y(0) = 3$$

(i) $h = 1$ (ii) $h = 0.1$
(iii) $h = 0.01$ (iv) $h = 0.001$
(b) Verify that $y = 2 + e^{-x^3}$ is the exact solution of the differential equation.
(c) Find the errors in using Euler's method to compute $y(1)$ with the step sizes in part (a). What happens to the error when the step size is divided by 10?

CAS **26.** (a) Program your computer algebra system, using Euler's method with step size 0.01, to calculate $y(2)$, where y is the solution of the initial-value problem

$$y' = x^3 - y^3 \qquad y(0) = 1$$

(b) Check your work by using the CAS to draw the solution curve.

27. The figure shows a circuit containing an electromotive force, a capacitor with a capacitance of C farads (F), and a resistor with a resistance of R ohms (Ω). The voltage drop across the capacitor is Q/C, where Q is the charge (in coulombs), so in this case Kirchhoff's Law gives

$$RI + \frac{Q}{C} = E(t)$$

But $I = dQ/dt$, so we have

$$R\frac{dQ}{dt} + \frac{1}{C}Q = E(t)$$

Suppose the resistance is 5 Ω, the capacitance is 0.05 F, and a battery gives a constant voltage of 60 V.
(a) Draw a direction field for this differential equation.
(b) What is the limiting value of the charge?

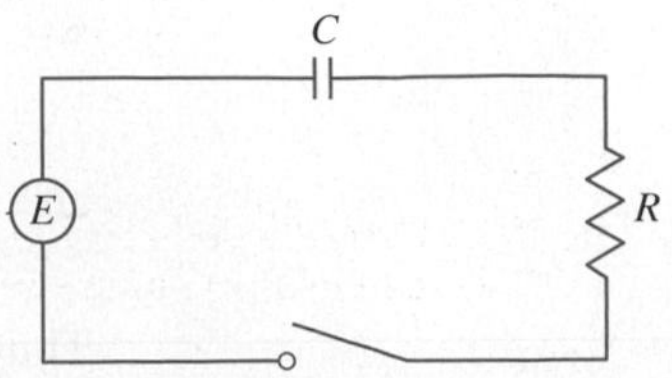

(c) Is there an equilibrium solution?
(d) If the initial charge is $Q(0) = 0$ C, use the direction field to sketch the solution curve.
(e) If the initial charge is $Q(0) = 0$ C, use Euler's method with step size 0.1 to estimate the charge after half a second.

28. In Exercise 14 in Section 9.1 we considered a 95°C cup of coffee in a 20°C room. Suppose it is known that the coffee cools at a rate of 1°C per minute when its temperature is 70°C.
(a) What does the differential equation become in this case?
(b) Sketch a direction field and use it to sketch the solution curve for the initial-value problem. What is the limiting value of the temperature?
(c) Use Euler's method with step size $h = 2$ minutes to estimate the temperature of the coffee after 10 minutes.

9.3 SEPARABLE EQUATIONS

We have looked at first-order differential equations from a geometric point of view (direction fields) and from a numerical point of view (Euler's method). What about the symbolic point of view? It would be nice to have an explicit formula for a solution of a differential equation. Unfortunately, that is not always possible. But in this section we examine a certain type of differential equation that *can* be solved explicitly.

A **separable equation** is a first-order differential equation in which the expression for dy/dx can be factored as a function of x times a function of y. In other words, it can be written in the form

$$\frac{dy}{dx} = g(x)f(y)$$

The name *separable* comes from the fact that the expression on the right side can be "separated" into a function of x and a function of y. Equivalently, if $f(y) \neq 0$, we could write

1 $$\frac{dy}{dx} = \frac{g(x)}{h(y)}$$

where $h(y) = 1/f(y)$. To solve this equation we rewrite it in the differential form

$$h(y)\,dy = g(x)\,dx$$

■ The technique for solving separable differential equations was first used by James Bernoulli (in 1690) in solving a problem about pendulums and by Leibniz (in a letter to Huygens in 1691). John Bernoulli explained the general method in a paper published in 1694.

so that all y's are on one side of the equation and all x's are on the other side. Then we integrate both sides of the equation:

2 $$\int h(y)\,dy = \int g(x)\,dx$$

Equation 2 defines y implicitly as a function of x. In some cases we may be able to solve for y in terms of x.

We use the Chain Rule to justify this procedure: If h and g satisfy (2), then

$$\frac{d}{dx}\left(\int h(y)\,dy\right) = \frac{d}{dx}\left(\int g(x)\,dx\right)$$

so $$\frac{d}{dy}\left(\int h(y)\,dy\right)\frac{dy}{dx} = g(x)$$

and $$h(y)\frac{dy}{dx} = g(x)$$

Thus Equation 1 is satisfied.

EXAMPLE 1

(a) Solve the differential equation $\dfrac{dy}{dx} = \dfrac{x^2}{y^2}$.

(b) Find the solution of this equation that satisfies the initial condition $y(0) = 2$.

SOLUTION

(a) We write the equation in terms of differentials and integrate both sides:

$$y^2 dy = x^2 dx$$

$$\int y^2 dy = \int x^2 dx$$

$$\tfrac{1}{3}y^3 = \tfrac{1}{3}x^3 + C$$

■ Figure 1 shows graphs of several members of the family of solutions of the differential equation in Example 1. The solution of the initial-value problem in part (b) is shown in red.

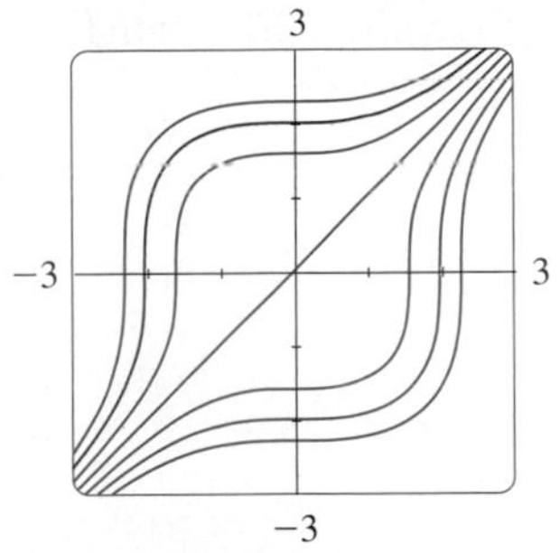

FIGURE 1

where C is an arbitrary constant. (We could have used a constant C_1 on the left side and another constant C_2 on the right side. But then we could combine these constants by writing $C = C_2 - C_1$.)

Solving for y, we get

$$y = \sqrt[3]{x^3 + 3C}$$

We could leave the solution like this or we could write it in the form

$$y = \sqrt[3]{x^3 + K}$$

where $K = 3C$. (Since C is an arbitrary constant, so is K.)

(b) If we put $x = 0$ in the general solution in part (a), we get $y(0) = \sqrt[3]{K}$. To satisfy the initial condition $y(0) = 2$, we must have $\sqrt[3]{K} = 2$ and so $K = 8$.

Thus the solution of the initial-value problem is

$$y = \sqrt[3]{x^3 + 8}$$ □

■ Some computer algebra systems can plot curves defined by implicit equations. Figure 2 shows the graphs of several members of the family of solutions of the differential equation in Example 2. As we look at the curves from left to right, the values of C are 3, 2, 1, 0, -1, -2, and -3.

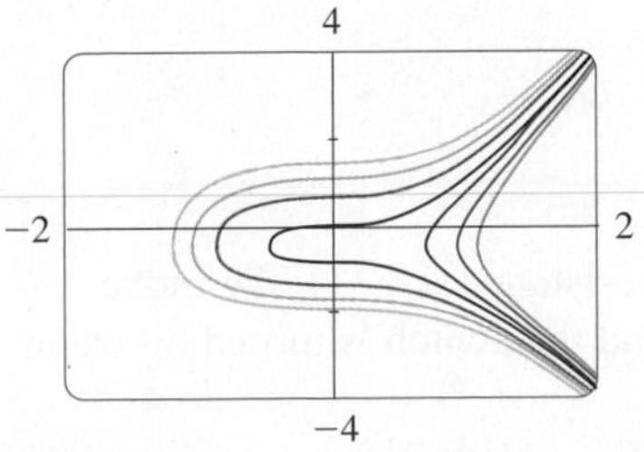

FIGURE 2

V EXAMPLE 2 Solve the differential equation $\dfrac{dy}{dx} = \dfrac{6x^2}{2y + \cos y}$.

SOLUTION Writing the equation in differential form and integrating both sides, we have

$$(2y + \cos y)\,dy = 6x^2 dx$$

$$\int (2y + \cos y)\,dy = \int 6x^2 dx$$

$$\boxed{3} \qquad y^2 + \sin y = 2x^3 + C$$

where C is a constant. Equation 3 gives the general solution implicitly. In this case it's impossible to solve the equation to express y explicitly as a function of x. □

EXAMPLE 3 Solve the equation $y' = x^2 y$.

SOLUTION First we rewrite the equation using Leibniz notation:

$$\frac{dy}{dx} = x^2 y$$

■ If a solution y is a function that satisfies $y(x) \neq 0$ for some x, it follows from a uniqueness theorem for solutions of differential equations that $y(x) \neq 0$ for all x.

If $y \neq 0$, we can rewrite it in differential notation and integrate:

$$\frac{dy}{y} = x^2\,dx \qquad y \neq 0$$

$$\int \frac{dy}{y} = \int x^2\,dx$$

$$\ln|y| = \frac{x^3}{3} + C$$

This equation defines y implicitly as a function of x. But in this case we can solve explicitly for y as follows:

$$|y| = e^{\ln|y|} = e^{(x^3/3)+C} = e^C e^{x^3/3}$$

so

$$y = \pm e^C e^{x^3/3}$$

We can easily verify that the function $y = 0$ is also a solution of the given differential equation. So we can write the general solution in the form

$$y = Ae^{x^3/3}$$

where A is an arbitrary constant ($A = e^C$, or $A = -e^C$, or $A = 0$). □

■ Figure 3 shows a direction field for the differential equation in Example 3. Compare it with Figure 4, in which we use the equation $y = Ae^{x^3/3}$ to graph solutions for several values of A. If you use the direction field to sketch solution curves with y-intercepts 5, 2, 1, -1, and -2, they will resemble the curves in Figure 4.

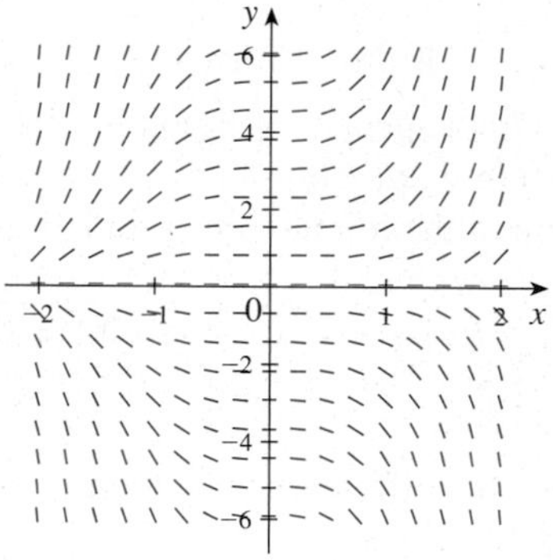

FIGURE 3

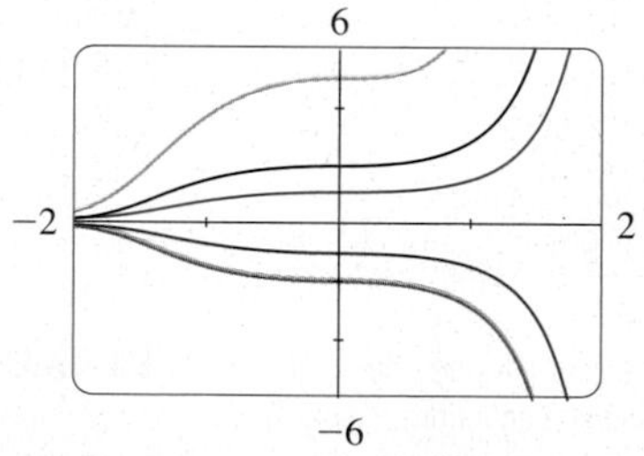

FIGURE 4

EXAMPLE 4 In Section 9.2 we modeled the current $I(t)$ in the electric circuit shown in Figure 5 by the differential equation

$$L\frac{dI}{dt} + RI = E(t)$$

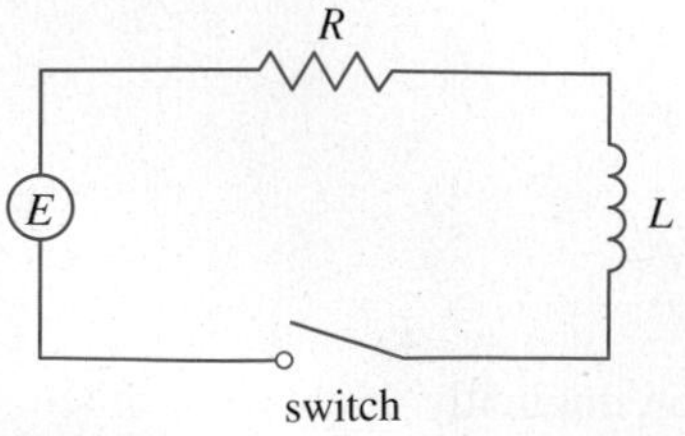

FIGURE 5

Find an expression for the current in a circuit where the resistance is 12 Ω, the inductance is 4 H, a battery gives a constant voltage of 60 V, and the switch is turned on when $t = 0$. What is the limiting value of the current?

SOLUTION With $L = 4$, $R = 12$, and $E(t) = 60$, the equation becomes

$$4\frac{dI}{dt} + 12I = 60 \qquad \text{or} \qquad \frac{dI}{dt} = 15 - 3I$$

and the initial-value problem is

$$\frac{dI}{dt} = 15 - 3I \qquad I(0) = 0$$

We recognize this equation as being separable, and we solve it as follows:

$$\int \frac{dI}{15 - 3I} = \int dt \qquad (15 - 3I \neq 0)$$

$$-\tfrac{1}{3} \ln |15 - 3I| = t + C$$

$$|15 - 3I| = e^{-3(t+C)}$$

$$15 - 3I = \pm e^{-3C} e^{-3t} = Ae^{-3t}$$

$$I = 5 - \tfrac{1}{3}Ae^{-3t}$$

■ Figure 6 shows how the solution in Example 4 (the current) approaches its limiting value. Comparison with Figure 11 in Section 9.2 shows that we were able to draw a fairly accurate solution curve from the direction field.

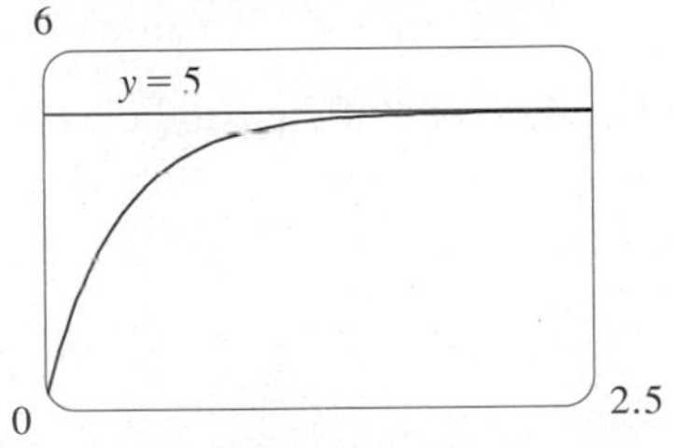

FIGURE 6

Since $I(0) = 0$, we have $5 - \frac{1}{3}A = 0$, so $A = 15$ and the solution is

$$I(t) = 5 - 5e^{-3t}$$

The limiting current, in amperes, is

$$\lim_{t \to \infty} I(t) = \lim_{t \to \infty} (5 - 5e^{-3t}) = 5 - 5 \lim_{t \to \infty} e^{-3t} = 5 - 0 = 5 \qquad \square$$

ORTHOGONAL TRAJECTORIES

An **orthogonal trajectory** of a family of curves is a curve that intersects each curve of the family orthogonally, that is, at right angles (see Figure 7). For instance, each member of the family $y = mx$ of straight lines through the origin is an orthogonal trajectory of the family $x^2 + y^2 = r^2$ of concentric circles with center the origin (see Figure 8). We say that the two families are orthogonal trajectories of each other.

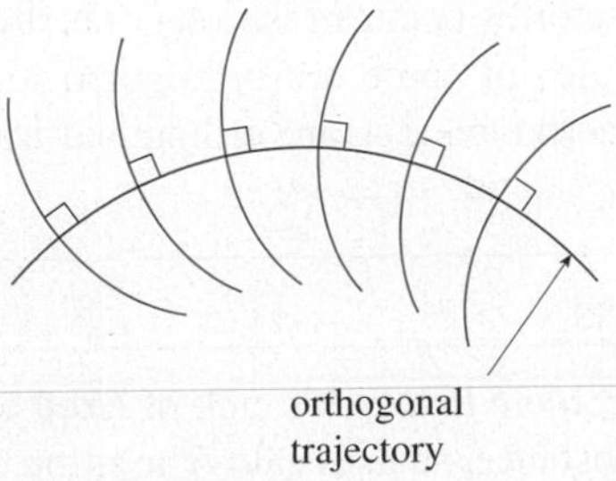

FIGURE 7

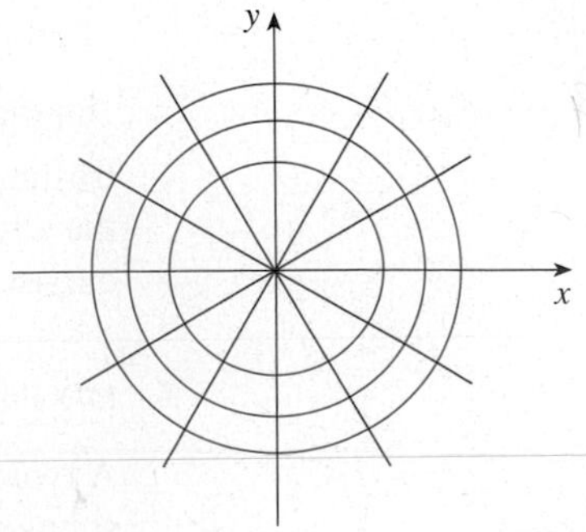

FIGURE 8

EXAMPLE 5 Find the orthogonal trajectories of the family of curves $x = ky^2$, where k is an arbitrary constant.

SOLUTION The curves $x = ky^2$ form a family of parabolas whose axis of symmetry is the x-axis. The first step is to find a single differential equation that is satisfied by all members of the family. If we differentiate $x = ky^2$, we get

$$1 = 2ky \frac{dy}{dx} \qquad \text{or} \qquad \frac{dy}{dx} = \frac{1}{2ky}$$

This differential equation depends on k, but we need an equation that is valid for all values of k simultaneously. To eliminate k we note that, from the equation of the given general parabola $x = ky^2$, we have $k = x/y^2$ and so the differential equation can be written as

$$\frac{dy}{dx} = \frac{1}{2ky} = \frac{1}{2\dfrac{x}{y^2}y}$$

or

$$\frac{dy}{dx} = \frac{y}{2x}$$

This means that the slope of the tangent line at any point (x, y) on one of the parabolas is $y' = y/(2x)$. On an orthogonal trajectory the slope of the tangent line must be the negative reciprocal of this slope. Therefore the orthogonal trajectories must satisfy the differential equation

$$\frac{dy}{dx} = -\frac{2x}{y}$$

This differential equation is separable, and we solve it as follows:

$$\int y\,dy = -\int 2x\,dx$$

$$\frac{y^2}{2} = -x^2 + C$$

4 $$x^2 + \frac{y^2}{2} = C$$

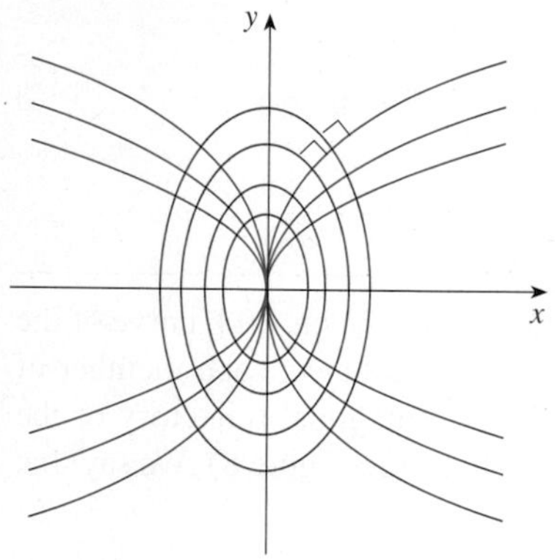

FIGURE 9

where C is an arbitrary positive constant. Thus the orthogonal trajectories are the family of ellipses given by Equation 4 and sketched in Figure 9. □

Orthogonal trajectories occur in various branches of physics. For example, in an electrostatic field the lines of force are orthogonal to the lines of constant potential. Also, the streamlines in aerodynamics are orthogonal trajectories of the velocity-equipotential curves.

MIXING PROBLEMS

A typical mixing problem involves a tank of fixed capacity filled with a thoroughly mixed solution of some substance, such as salt. A solution of a given concentration enters the tank at a fixed rate and the mixture, thoroughly stirred, leaves at a fixed rate, which may differ from the entering rate. If $y(t)$ denotes the amount of substance in the tank at time t, then $y'(t)$ is the rate at which the substance is being added minus the rate at which it is being removed. The mathematical description of this situation often leads to a first-order separable differential equation. We can use the same type of reasoning to model a variety of phenomena: chemical reactions, discharge of pollutants into a lake, injection of a drug into the bloodstream.

EXAMPLE 6 A tank contains 20 kg of salt dissolved in 5000 L of water. Brine that contains 0.03 kg of salt per liter of water enters the tank at a rate of 25 L/min. The solution is kept thoroughly mixed and drains from the tank at the same rate. How much salt remains in the tank after half an hour?

SOLUTION Let $y(t)$ be the amount of salt (in kilograms) after t minutes. We are given that $y(0) = 20$ and we want to find $y(30)$. We do this by finding a differential equation satisfied by $y(t)$. Note that dy/dt is the rate of change of the amount of salt, so

① find rate in and rate out

$$\boxed{5} \qquad \frac{dy}{dt} = (\text{rate in}) - (\text{rate out})$$

② $\frac{dy}{dt}$ = rate in − rate out

③ solve separate eqn

where (rate in) is the rate at which salt enters the tank and (rate out) is the rate at which salt leaves the tank. We have

$$\text{rate in} = \left(0.03\,\frac{\text{kg}}{\text{L}}\right)\left(25\,\frac{\text{L}}{\text{min}}\right) = 0.75\,\frac{\text{kg}}{\text{min}}$$

The tank always contains 5000 L of liquid, so the concentration at time t is $y(t)/5000$ (measured in kilograms per liter). Since the brine flows out at a rate of 25 L/min, we have

$$\text{rate out} = \left(\frac{y(t)}{5000}\,\frac{\text{kg}}{\text{L}}\right)\left(25\,\frac{\text{L}}{\text{min}}\right) = \frac{y(t)}{200}\,\frac{\text{kg}}{\text{min}}$$

Thus, from Equation 5, we get

$$\frac{dy}{dt} = 0.75 - \frac{y(t)}{200} = \frac{150 - y(t)}{200}$$

Solving this separable differential equation, we obtain

$$\int \frac{dy}{150 - y} = \int \frac{dt}{200}$$

$$-\ln|150 - y| = \frac{t}{200} + C$$

■ Figure 10 shows the graph of the function $y(t)$ of Example 6. Notice that, as time goes by, the amount of salt approaches 150 kg.

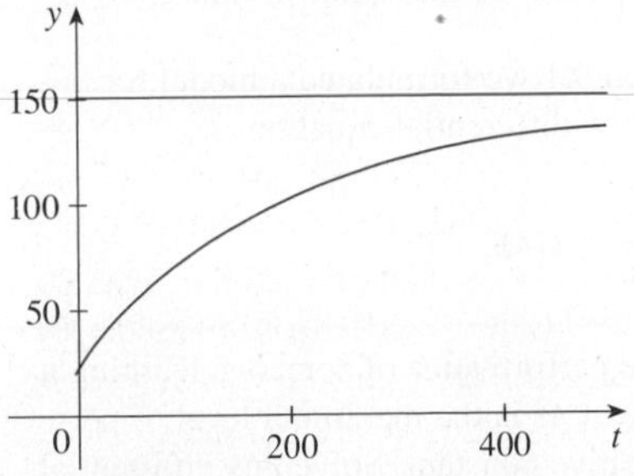

FIGURE 10

Since $y(0) = 20$, we have $-\ln 130 = C$, so

$$-\ln|150 - y| = \frac{t}{200} - \ln 130$$

Therefore $$|150 - y| = 130e^{-t/200}$$

Since $y(t)$ is continuous and $y(0) = 20$ and the right side is never 0, we deduce that $150 - y(t)$ is always positive. Thus $|150 - y| = 150 - y$ and so

$$y(t) = 150 - 130e^{-t/200}$$

The amount of salt after 30 min is

$$y(30) = 150 - 130e^{-30/200} \approx 38.1 \text{ kg}$$

□

9.3 EXERCISES

1–10 Solve the differential equation.

1. $\dfrac{dy}{dx} = \dfrac{y}{x}$

2. $\dfrac{dy}{dx} = \dfrac{\sqrt{x}}{e^y}$

3. $(x^2 + 1)y' = xy$

4. $y' = y^2 \sin x$

5. $(1 + \tan y)y' = x^2 + 1$

6. $\dfrac{du}{dr} = \dfrac{1 + \sqrt{r}}{1 + \sqrt{u}}$

7. $\dfrac{dy}{dt} = \dfrac{te^t}{y\sqrt{1 + y^2}}$

8. $\dfrac{dy}{d\theta} = \dfrac{e^y \sin^2\theta}{y \sec \theta}$

9. $\dfrac{du}{dt} = 2 + 2u + t + tu$

10. $\dfrac{dz}{dt} + e^{t+z} = 0$

11–18 Find the solution of the differential equation that satisfies the given initial condition.

11. $\dfrac{dy}{dx} = \dfrac{x}{y}, \quad y(0) = -3$

12. $\dfrac{dy}{dx} = \dfrac{y \cos x}{1 + y^2}, \quad y(0) = 1$

13. $x \cos x = (2y + e^{3y})y', \quad y(0) = 0$

14. $\dfrac{dP}{dt} = \sqrt{Pt}, \quad P(1) = 2$

15. $\dfrac{du}{dt} = \dfrac{2t + \sec^2 t}{2u}, \quad u(0) = -5$

16. $xy' + y = y^2, \quad y(1) = -1$

17. $y' \tan x = a + y, \; y(\pi/3) = a, \quad 0 < x < \pi/2$

18. $\dfrac{dL}{dt} = kL^2 \ln t, \; L(1) = -1$

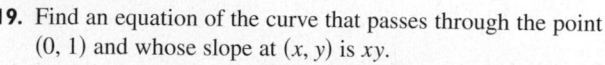

19. Find an equation of the curve that passes through the point $(0, 1)$ and whose slope at (x, y) is xy.

20. Find the function f such that $f'(x) = f(x)(1 - f(x))$ and $f(0) = \frac{1}{2}$.

21. Solve the differential equation $y' = x + y$ by making the change of variable $u = x + y$.

22. Solve the differential equation $xy' = y + xe^{y/x}$ by making the change of variable $v = y/x$.

23. (a) Solve the differential equation $y' = 2x\sqrt{1 - y^2}$.
    (b) Solve the initial-value problem $y' = 2x\sqrt{1 - y^2}$, $y(0) = 0$, and graph the solution.
    (c) Does the initial-value problem $y' = 2x\sqrt{1 - y^2}$, $y(0) = 2$, have a solution? Explain.

24. Solve the equation $e^{-y}y' + \cos x = 0$ and graph several members of the family of solutions. How does the solution curve change as the constant C varies?

CAS 25. Solve the initial-value problem $y' = (\sin x)/\sin y$, $y(0) = \pi/2$, and graph the solution (if your CAS does implicit plots).

CAS 26. Solve the equation $y' = x\sqrt{x^2 + 1}/(ye^y)$ and graph several members of the family of solutions (if your CAS does implicit plots). How does the solution curve change as the constant C varies?

CAS **27–28**
(a) Use a computer algebra system to draw a direction field for the differential equation. Get a printout and use it to sketch some solution curves without solving the differential equation.
(b) Solve the differential equation.
(c) Use the CAS to draw several members of the family of solutions obtained in part (b). Compare with the curves from part (a).

27. $y' = 1/y$

28. $y' = x^2/y$

29–32 Find the orthogonal trajectories of the family of curves. Use a graphing device to draw several members of each family on a common screen.

29. $x^2 + 2y^2 = k^2$

30. $y^2 = kx^3$

31. $y = \dfrac{k}{x}$

32. $y = \dfrac{x}{1 + kx}$

33. Solve the initial-value problem in Exercise 27 in Section 9.2 to find an expression for the charge at time t. Find the limiting value of the charge.

34. In Exercise 28 in Section 9.2 we discussed a differential equation that models the temperature of a 95°C cup of coffee in a 20°C room. Solve the differential equation to find an expression for the temperature of the coffee at time t.

35. In Exercise 13 in Section 9.1 we formulated a model for learning in the form of the differential equation

$$\frac{dP}{dt} = k(M - P)$$

where $P(t)$ measures the performance of someone learning a skill after a training time t, M is the maximum level of performance, and k is a positive constant. Solve this differential equation to find an expression for $P(t)$. What is the limit of this expression?

36. In an elementary chemical reaction, single molecules of two reactants A and B form a molecule of the product C: $A + B \rightarrow C$. The law of mass action states that the rate of reaction is proportional to the product of the concentrations of A and B:

$$\frac{d[C]}{dt} = k[A][B]$$

(See Example 4 in Section 3.7.) Thus, if the initial concentrations are $[A] = a$ moles/L and $[B] = b$ moles/L and we write $x = [C]$, then we have

$$\frac{dx}{dt} = k(a - x)(b - x)$$

CAS (a) Assuming that $a \neq b$, find x as a function of t. Use the fact that the initial concentration of C is 0.

(b) Find $x(t)$ assuming that $a = b$. How does this expression for $x(t)$ simplify if it is known that $[C] = \frac{1}{2}a$ after 20 seconds?

37. In contrast to the situation of Exercise 36, experiments show that the reaction $H_2 + Br_2 \rightarrow 2HBr$ satisfies the rate law

$$\frac{d[HBr]}{dt} = k[H_2][Br_2]^{1/2}$$

and so for this reaction the differential equation becomes

$$\frac{dx}{dt} = k(a - x)(b - x)^{1/2}$$

where $x = [HBr]$ and a and b are the initial concentrations of hydrogen and bromine.

(a) Find x as a function of t in the case where $a = b$. Use the fact that $x(0) = 0$.

(b) If $a > b$, find t as a function of x. [*Hint:* In performing the integration, make the substitution $u = \sqrt{b - x}$.]

38. A sphere with radius 1 m has temperature 15°C. It lies inside a concentric sphere with radius 2 m and temperature 25°C. The temperature $T(r)$ at a distance r from the common center of the spheres satisfies the differential equation

$$\frac{d^2T}{dr^2} + \frac{2}{r}\frac{dT}{dr} = 0$$

If we let $S = dT/dr$, then S satisfies a first-order differential equation. Solve it to find an expression for the temperature $T(r)$ between the spheres.

39. A glucose solution is administered intravenously into the bloodstream at a constant rate r. As the glucose is added, it is converted into other substances and removed from the bloodstream at a rate that is proportional to the concentration at that time. Thus a model for the concentration $C = C(t)$ of the glucose solution in the bloodstream is

$$\frac{dC}{dt} = r - kC$$

where k is a positive constant.

(a) Suppose that the concentration at time $t = 0$ is C_0. Determine the concentration at any time t by solving the differential equation.

(b) Assuming that $C_0 < r/k$, find $\lim_{t\to\infty} C(t)$ and interpret your answer.

40. A certain small country has \$10 billion in paper currency in circulation, and each day \$50 million comes into the country's banks. The government decides to introduce new currency by having the banks replace old bills with new ones whenever old currency comes into the banks. Let $x = x(t)$ denote the amount of new currency in circulation at time t, with $x(0) = 0$.

(a) Formulate a mathematical model in the form of an initial-value problem that represents the "flow" of the new currency into circulation.

(b) Solve the initial-value problem found in part (a).

(c) How long will it take for the new bills to account for 90% of the currency in circulation?

41. A tank contains 1000 L of brine with 15 kg of dissolved salt. Pure water enters the tank at a rate of 10 L/min. The solution is kept thoroughly mixed and drains from the tank at the same rate. How much salt is in the tank (a) after t minutes and (b) after 20 minutes?

42. The air in a room with volume 180 m^3 contains 0.15% carbon dioxide initially. Fresher air with only 0.05% carbon dioxide flows into the room at a rate of 2 m^3/min and the mixed air flows out at the same rate. Find the percentage of carbon dioxide in the room as a function of time. What happens in the long run?

43. A vat with 500 gallons of beer contains 4% alcohol (by volume). Beer with 6% alcohol is pumped into the vat at a rate of 5 gal/min and the mixture is pumped out at the same rate. What is the percentage of alcohol after an hour?

44. A tank contains 1000 L of pure water. Brine that contains 0.05 kg of salt per liter of water enters the tank at a rate of 5 L/min. Brine that contains 0.04 kg of salt per liter of water enters the tank at a rate of 10 L/min. The solution is kept thoroughly mixed and drains from the tank at a rate of 15 L/min. How much salt is in the tank (a) after t minutes and (b) after one hour?

45. When a raindrop falls, it increases in size and so its mass at time t is a function of t, $m(t)$. The rate of growth of the mass is $km(t)$ for some positive constant k. When we apply Newton's Law of Motion to the raindrop, we get $(mv)' = gm$, where v is the velocity of the raindrop (directed downward) and g is the acceleration due to gravity. The *terminal velocity* of the raindrop is $\lim_{t\to\infty} v(t)$. Find an expression for the terminal velocity in terms of g and k.

46. An object of mass m is moving horizontally through a medium which resists the motion with a force that is a function of the velocity; that is,

$$m\frac{d^2s}{dt^2} = m\frac{dv}{dt} = f(v)$$

where $v = v(t)$ and $s = s(t)$ represent the velocity and position of the object at time t, respectively. For example, think of a boat moving through the water.

(a) Suppose that the resisting force is proportional to the velocity, that is, $f(v) = -kv$, k a positive constant. (This model is appropriate for small values of v.) Let $v(0) = v_0$ and $s(0) = s_0$ be the initial values of v and s. Determine v and s at any time t. What is the total distance that the object travels from time $t = 0$?

(b) For larger values of v a better model is obtained by supposing that the resisting force is proportional to the square of the velocity, that is, $f(v) = -kv^2$, $k > 0$. (This model was first proposed by Newton.) Let v_0 and s_0 be the initial values of v and s. Determine v and s at any time t. What is the total distance that the object travels in this case?

47. Let $A(t)$ be the area of a tissue culture at time t and let M be the final area of the tissue when growth is complete. Most cell divisions occur on the periphery of the tissue and the number of cells on the periphery is proportional to $\sqrt{A(t)}$. So a reasonable model for the growth of tissue is obtained by assuming that the rate of growth of the area is jointly proportional to $\sqrt{A(t)}$ and $M - A(t)$.

(a) Formulate a differential equation and use it to show that the tissue grows fastest when $A(t) = \frac{1}{3}M$.

CAS (b) Solve the differential equation to find an expression for $A(t)$. Use a computer algebra system to perform the integration.

48. According to Newton's Law of Universal Gravitation, the gravitational force on an object of mass m that has been projected vertically upward from the earth's surface is

$$F = \frac{mgR^2}{(x + R)^2}$$

where $x = x(t)$ is the object's distance above the surface at time t, R is the earth's radius, and g is the acceleration due to gravity. Also, by Newton's Second Law, $F = ma = m(dv/dt)$ and so

$$m\frac{dv}{dt} = -\frac{mgR^2}{(x + R)^2}$$

(a) Suppose a rocket is fired vertically upward with an initial velocity v_0. Let h be the maximum height above the surface reached by the object. Show that

$$v_0 = \sqrt{\frac{2gRh}{R + h}}$$

[*Hint:* By the Chain Rule, $m(dv/dt) = mv(dv/dx)$.]

(b) Calculate $v_e = \lim_{h\to\infty} v_0$. This limit is called the *escape velocity* for the earth.

(c) Use $R = 3960$ mi and $g = 32\ \text{ft/s}^2$ to calculate v_e in feet per second and in miles per second.

APPLIED PROJECT

HOW FAST DOES A TANK DRAIN?

If water (or other liquid) drains from a tank, we expect that the flow will be greatest at first (when the water depth is greatest) and will gradually decrease as the water level decreases. But we need a more precise mathematical description of how the flow decreases in order to answer the kinds of questions that engineers ask: How long does it take for a tank to drain completely? How much water should a tank hold in order to guarantee a certain minimum water pressure for a sprinkler system?

Let $h(t)$ and $V(t)$ be the height and volume of water in a tank at time t. If water drains through a hole with area a at the bottom of the tank, then Torricelli's Law says that

$$\boxed{1} \qquad \frac{dV}{dt} = -a\sqrt{2gh}$$

where g is the acceleration due to gravity. So the rate at which water flows from the tank is proportional to the square root of the water height.

1. (a) Suppose the tank is cylindrical with height 6 ft and radius 2 ft and the hole is circular with radius 1 inch. If we take $g = 32\ \text{ft/s}^2$, show that h satisfies the differential equation

$$\frac{dh}{dt} = -\frac{1}{72}\sqrt{h}$$

(b) Solve this equation to find the height of the water at time t, assuming the tank is full at time $t = 0$.

(c) How long will it take for the water to drain completely?

2. Because of the rotation and viscosity of the liquid, the theoretical model given by Equation 1 isn't quite accurate. Instead, the model

$$\boxed{2} \qquad \frac{dh}{dt} = k\sqrt{h}$$

is often used and the constant k (which depends on the physical properties of the liquid) is determined from data concerning the draining of the tank.

(a) Suppose that a hole is drilled in the side of a cylindrical bottle and the height h of the water (above the hole) decreases from 10 cm to 3 cm in 68 seconds. Use Equation 2 to find an expression for $h(t)$. Evaluate $h(t)$ for $t = 10, 20, 30, 40, 50, 60$.

■ This part of the project is best done as a classroom demonstration or as a group project with three students in each group: a time-keeper to call out seconds, a bottle keeper to estimate the height every 10 seconds, and a record keeper to record these values.

(b) Drill a 4-mm hole near the bottom of the cylindrical part of a two-liter plastic soft-drink bottle. Attach a strip of masking tape marked in centimeters from 0 to 10, with 0 corresponding to the top of the hole. With one finger over the hole, fill the bottle with water to the 10-cm mark. Then take your finger off the hole and record the values of $h(t)$ for $t = 10, 20, 30, 40, 50, 60$ seconds. (You will probably find that it takes 68 seconds for the level to decrease to $h = 3$ cm.) Compare your data with the values of $h(t)$ from part (a). How well did the model predict the actual values?

3. In many parts of the world, the water for sprinkler systems in large hotels and hospitals is supplied by gravity from cylindrical tanks on or near the roofs of the buildings. Suppose such a tank has radius 10 ft and the diameter of the outlet is 2.5 inches. An engineer has to guarantee that the water pressure will be at least 2160 lb/ft^2 for a period of 10 minutes. (When a fire happens, the electrical system might fail and it could take up to 10 minutes for the emergency generator and fire pump to be activated.) What height should the engineer specify for the tank in order to make such a guarantee? (Use the fact that the water pressure at a depth of d feet is $P = 62.5d$. See Section 8.3.)

4. Not all water tanks are shaped like cylinders. Suppose a tank has cross-sectional area $A(h)$ at height h. Then the volume of water up to height h is $V = \int_0^h A(u)\,du$ and so the Fundamental Theorem of Calculus gives $dV/dh = A(h)$. It follows that

$$\frac{dV}{dt} = \frac{dV}{dh}\frac{dh}{dt} = A(h)\frac{dh}{dt}$$

and so Torricelli's Law becomes

$$A(h)\frac{dh}{dt} = -a\sqrt{2gh}$$

(a) Suppose the tank has the shape of a sphere with radius 2 m and is initially half full of water. If the radius of the circular hole is 1 cm and we take $g = 10$ m/s^2, show that h satisfies the differential equation

$$(4h - h^2)\frac{dh}{dt} = -0.0001\sqrt{20h}$$

(b) How long will it take for the water to drain completely?

APPLIED PROJECT

WHICH IS FASTER, GOING UP OR COMING DOWN?

Suppose you throw a ball into the air. Do you think it takes longer to reach its maximum height or to fall back to earth from its maximum height? We will solve the problem in this project but, before getting started, think about that situation and make a guess based on your physical intuition.

1. A ball with mass m is projected vertically upward from the earth's surface with a positive initial velocity v_0. We assume the forces acting on the ball are the force of gravity and a retarding force of air resistance with direction opposite to the direction of motion and with magnitude $p|v(t)|$, where p is a positive constant and $v(t)$ is the velocity of the ball at time t. In both the ascent and the descent, the total force acting on the ball is $-pv - mg$. [During ascent, $v(t)$ is positive and the resistance acts downward; during descent, $v(t)$ is negative and the resistance acts upward.] So, by Newton's Second Law, the equation of motion is

$$mv' = -pv - mg$$

Solve this differential equation to show that the velocity is

$$v(t) = \left(v_0 + \frac{mg}{p}\right)e^{-pt/m} - \frac{mg}{p}$$

2. Show that the height of the ball, until it hits the ground, is

$$y(t) = \left(v_0 + \frac{mg}{p}\right)\frac{m}{p}(1 - e^{-pt/m}) - \frac{mgt}{p}$$

3. Let t_1 be the time that the ball takes to reach its maximum height. Show that

$$t_1 = \frac{m}{p}\ln\left(\frac{mg + pv_0}{mg}\right)$$

Find this time for a ball with mass 1 kg and initial velocity 20 m/s. Assume the air resistance is $\frac{1}{10}$ of the speed.

4. Let t_2 be the time at which the ball falls back to earth. For the particular ball in Problem 3, estimate t_2 by using a graph of the height function $y(t)$. Which is faster, going up or coming down?

5. In general, it's not easy to find t_2 because it's impossible to solve the equation $y(t) = 0$ explicitly. We can, however, use an indirect method to determine whether ascent or descent is faster; we determine whether $y(2t_1)$ is positive or negative. Show that

$$y(2t_1) = \frac{m^2 g}{p^2}\left(x - \frac{1}{x} - 2\ln x\right)$$

where $x = e^{pt_1/m}$. Then show that $x > 1$ and the function

$$f(x) = x - \frac{1}{x} - 2\ln x$$

is increasing for $x > 1$. Use this result to decide whether $y(2t_1)$ is positive or negative. What can you conclude? Is ascent or descent faster?

■ In modeling force due to air resistance, various functions have been used, depending on the physical characteristics and speed of the ball. Here we use a linear model, $-pv$, but a quadratic model ($-pv^2$ on the way up and pv^2 on the way down) is another possibility for higher speeds (see Exercise 46 in Section 9.3). For a golf ball, experiments have shown that a good model is $-pv^{1.3}$ going up and $p|v|^{1.3}$ coming down. But no matter which force function $-f(v)$ is used [where $f(v) > 0$ for $v > 0$ and $f(v) < 0$ for $v < 0$], the answer to the question remains the same. See F. Brauer, "What Goes Up Must Come Down, Eventually," *Amer. Math. Monthly* 108 (2001), pp. 437–440.

9.4 MODELS FOR POPULATION GROWTH

In this section we investigate differential equations that are used to model population growth: the law of natural growth, the logistic equation, and several others.

THE LAW OF NATURAL GROWTH

One of the models for population growth that we considered in Section 9.1 was based on the assumption that the population grows at a rate proportional to the size of the population:

$$\frac{dP}{dt} = kP$$

Is that a reasonable assumption? Suppose we have a population (of bacteria, for instance) with size $P = 1000$ and at a certain time it is growing at a rate of $P' = 300$ bacteria per hour. Now let's take another 1000 bacteria of the same type and put them with the first population. Each half of the new population was growing at a rate of 300 bacteria per hour. We would expect the total population of 2000 to increase at a rate of 600 bacteria per hour initially (provided there's enough room and nutrition). So if we double the size, we double the growth rate. In general, it seems reasonable that the growth rate should be proportional to the size.

In general, if $P(t)$ is the value of a quantity y at time t and if the rate of change of P with respect to t is proportional to its size $P(t)$ at any time, then

1
$$\frac{dP}{dt} = kP$$

where k is a constant. Equation 1 is sometimes called the **law of natural growth**. If k is positive, then the population increases; if k is negative, it decreases.

Because Equation 1 is a separable differential equation, we can solve it by the methods of Section 9.3:

$$\int \frac{dP}{P} = \int k\,dt$$

$$\ln|P| = kt + C$$

$$|P| = e^{kt+C} = e^C e^{kt}$$

$$P = Ae^{kt}$$

where A $(= \pm e^C$ or $0)$ is an arbitrary constant. To see the significance of the constant A, we observe that

$$P(0) = Ae^{k \cdot 0} = A$$

Therefore A is the initial value of the function.

■ Examples and exercises on the use of (2) are given in Section 3.8.

2 The solution of the initial-value problem

$$\frac{dP}{dt} = kP \qquad P(0) = P_0$$

is

$$P(t) = P_0 e^{kt}$$

Another way of writing Equation 1 is

$$\frac{1}{P}\frac{dP}{dt} = k$$

which says that the **relative growth rate** (the growth rate divided by the population size) is constant. Then (2) says that a population with constant relative growth rate must grow exponentially.

We can account for emigration (or "harvesting") from a population by modifying Equation 1: If the rate of emigration is a constant m, then the rate of change of the population is modeled by the differential equation

$$\boxed{3} \qquad \frac{dP}{dt} = kP - m$$

See Exercise 13 for the solution and consequences of Equation 3.

THE LOGISTIC MODEL

As we discussed in Section 9.1, a population often increases exponentially in its early stages but levels off eventually and approaches its carrying capacity because of limited resources. If $P(t)$ is the size of the population at time t, we assume that

$$\frac{dP}{dt} \approx kP \qquad \text{if } P \text{ is small}$$

This says that the growth rate is initially close to being proportional to size. In other words, the relative growth rate is almost constant when the population is small. But we also want to reflect the fact that the relative growth rate decreases as the population P increases and becomes negative if P ever exceeds its **carrying capacity** K, the maximum population that the environment is capable of sustaining in the long run. The simplest expression for the relative growth rate that incorporates these assumptions is

$$\frac{1}{P}\frac{dP}{dt} = k\left(1 - \frac{P}{K}\right)$$

Multiplying by P, we obtain the model for population growth known as the **logistic differential equation**:

$$\boxed{4} \qquad \boxed{\frac{dP}{dt} = kP\left(1 - \frac{P}{K}\right)}$$

Notice from Equation 4 that if P is small compared with K, then P/K is close to 0 and so $dP/dt \approx kP$. However, if $P \to K$ (the population approaches its carrying capacity), then $P/K \to 1$, so $dP/dt \to 0$. We can deduce information about whether solutions increase or decrease directly from Equation 4. If the population P lies between 0 and K, then the right side of the equation is positive, so $dP/dt > 0$ and the population increases. But if the population exceeds the carrying capacity ($P > K$), then $1 - P/K$ is negative, so $dP/dt < 0$ and the population decreases.

Let's start our more detailed analysis of the logistic differential equation by looking at a direction field.

EXAMPLE 1 Draw a direction field for the logistic equation with $k = 0.08$ and carrying capacity $K = 1000$. What can you deduce about the solutions?

SOLUTION In this case the logistic differential equation is

$$\frac{dP}{dt} = 0.08P\left(1 - \frac{P}{1000}\right)$$

A direction field for this equation is shown in Figure 1. We show only the first quadrant because negative populations aren't meaningful and we are interested only in what happens after $t = 0$.

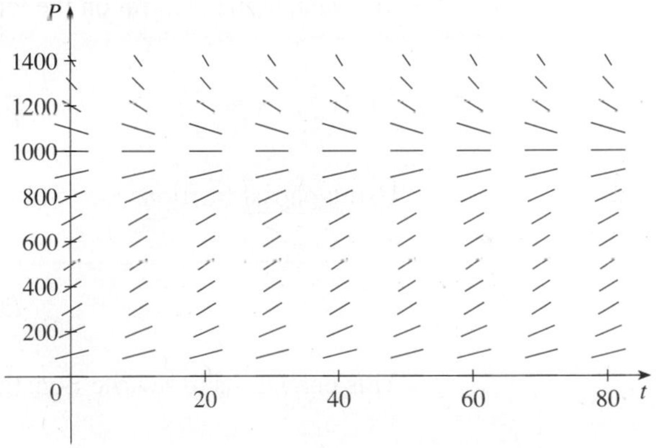

FIGURE 1
Direction field for the logistic equation in Example 1

The logistic equation is autonomous (dP/dt depends only on P, not on t), so the slopes are the same along any horizontal line. As expected, the slopes are positive for $0 < P < 1000$ and negative for $P > 1000$.

The slopes are small when P is close to 0 or 1000 (the carrying capacity). Notice that the solutions move away from the equilibrium solution $P = 0$ and move toward the equilibrium solution $P = 1000$.

In Figure 2 we use the direction field to sketch solution curves with initial populations $P(0) = 100$, $P(0) = 400$, and $P(0) = 1300$. Notice that solution curves that start below $P = 1000$ are increasing and those that start above $P = 1000$ are decreasing. The slopes are greatest when $P \approx 500$ and therefore the solution curves that start below $P = 1000$ have inflection points when $P \approx 500$. In fact we can prove that all solution curves that start below $P = 500$ have an inflection point when P is exactly 500. (See Exercise 9.)

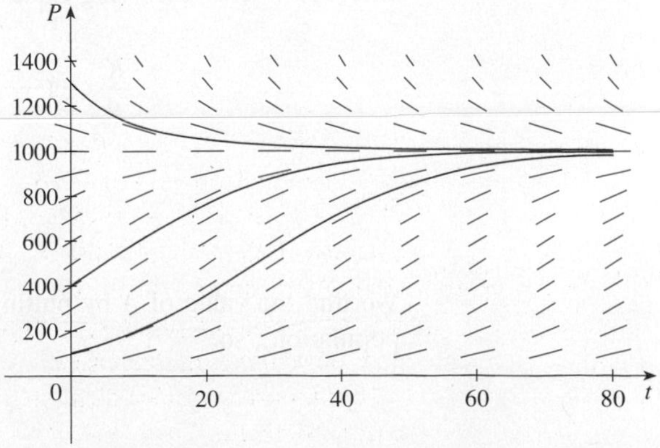

FIGURE 2
Solution curves for the logistic equation in Example 1

□

The logistic equation (4) is separable and so we can solve it explicitly using the method of Section 9.3. Since

$$\frac{dP}{dt} = kP\left(1 - \frac{P}{K}\right)$$

we have

5 $$\int \frac{dP}{P(1 - P/K)} = \int k\,dt$$

To evaluate the integral on the left side, we write

$$\frac{1}{P(1 - P/K)} = \frac{K}{P(K - P)}$$

Using partial fractions (see Section 7.4), we get

$$\frac{K}{P(K - P)} = \frac{1}{P} + \frac{1}{K - P}$$

This enables us to rewrite Equation 5:

$$\int \left(\frac{1}{P} + \frac{1}{K - P}\right) dP = \int k\,dt$$

$$\ln|P| - \ln|K - P| = kt + C$$

$$\ln\left|\frac{K - P}{P}\right| = -kt - C$$

$$\left|\frac{K - P}{P}\right| = e^{-kt-C} = e^{-C}e^{-kt}$$

6 $$\frac{K - P}{P} = Ae^{-kt}$$

where $A = \pm e^{-C}$. Solving Equation 6 for P, we get

$$\frac{K}{P} - 1 = Ae^{-kt} \qquad \Rightarrow \qquad \frac{P}{K} = \frac{1}{1 + Ae^{-kt}}$$

so $$P = \frac{K}{1 + Ae^{-kt}}$$

We find the value of A by putting $t = 0$ in Equation 6. If $t = 0$, then $P = P_0$ (the initial population), so

$$\frac{K - P_0}{P_0} = Ae^0 = A$$

Thus the solution to the logistic equation is

$$\boxed{7} \qquad P(t) = \frac{K}{1 + Ae^{-kt}} \qquad \text{where } A = \frac{K - P_0}{P_0}$$

Using the expression for $P(t)$ in Equation 7, we see that

$$\lim_{t \to \infty} P(t) = K$$

which is to be expected.

EXAMPLE 2 Write the solution of the initial-value problem

$$\frac{dP}{dt} = 0.08P\left(1 - \frac{P}{1000}\right) \qquad P(0) = 100$$

and use it to find the population sizes $P(40)$ and $P(80)$. At what time does the population reach 900?

SOLUTION The differential equation is a logistic equation with $k = 0.08$, carrying capacity $K = 1000$, and initial population $P_0 = 100$. So Equation 7 gives the population at time t as

$$P(t) = \frac{1000}{1 + Ae^{-0.08t}} \qquad \text{where } A = \frac{1000 - 100}{100} = 9$$

Thus

$$P(t) = \frac{1000}{1 + 9e^{-0.08t}}$$

So the population sizes when $t = 40$ and 80 are

$$P(40) = \frac{1000}{1 + 9e^{-3.2}} \approx 731.6 \qquad P(80) = \frac{1000}{1 + 9e^{-6.4}} \approx 985.3$$

The population reaches 900 when

$$\frac{1000}{1 + 9e^{-0.08t}} = 900$$

Solving this equation for t, we get

$$1 + 9e^{-0.08t} = \tfrac{10}{9}$$

$$e^{-0.08t} = \tfrac{1}{81}$$

$$-0.08t = \ln \tfrac{1}{81} = -\ln 81$$

$$t = \frac{\ln 81}{0.08} \approx 54.9$$

So the population reaches 900 when t is approximately 55. As a check on our work, we graph the population curve in Figure 3 and observe where it intersects the line $P = 900$. The cursor indicates that $t \approx 55$. □

■ Compare the solution curve in Figure 3 with the lowest solution curve we drew from the direction field in Figure 2.

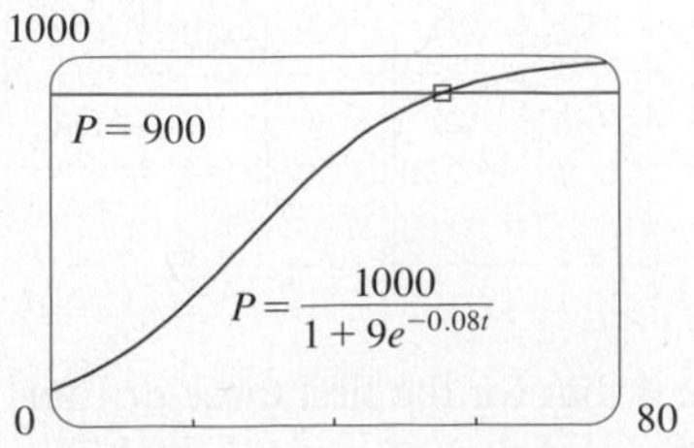

FIGURE 3

COMPARISON OF THE NATURAL GROWTH AND LOGISTIC MODELS

In the 1930s the biologist G. F. Gause conducted an experiment with the protozoan *Paramecium* and used a logistic equation to model his data. The table gives his daily count of the population of protozoa. He estimated the initial relative growth rate to be 0.7944 and the carrying capacity to be 64.

t (days)	0	1	2	3	4	5	6	7	8	9	10	11	12	13	14	15	16
P (observed)	2	3	22	16	39	52	54	47	50	76	69	51	57	70	53	59	57

V EXAMPLE 3 Find the exponential and logistic models for Gause's data. Compare the predicted values with the observed values and comment on the fit.

SOLUTION Given the relative growth rate $k = 0.7944$ and the initial population $P_0 = 2$, the exponential model is

$$P(t) = P_0 e^{kt} = 2e^{0.7944t}$$

Gause used the same value of k for his logistic model. [This is reasonable because $P_0 = 2$ is small compared with the carrying capacity ($K = 64$). The equation

$$\left.\frac{1}{P_0}\frac{dP}{dt}\right|_{t=0} = k\left(1 - \frac{2}{64}\right) \approx k$$

shows that the value of k for the logistic model is very close to the value for the exponential model.]

Then the solution of the logistic equation in Equation 7 gives

$$P(t) = \frac{K}{1 + Ae^{-kt}} = \frac{64}{1 + Ae^{-0.7944t}}$$

where

$$A = \frac{K - P_0}{P_0} = \frac{64 - 2}{2} = 31$$

So

$$P(t) = \frac{64}{1 + 31e^{-0.7944t}}$$

We use these equations to calculate the predicted values (rounded to the nearest integer) and compare them in the following table.

t (days)	0	1	2	3	4	5	6	7	8	9	10	11	12	13	14	15	16
P (observed)	2	3	22	16	39	52	54	47	50	76	69	51	57	70	53	59	57
P (logistic model)	2	4	9	17	28	40	51	57	61	62	63	64	64	64	64	64	64
P (exponential model)	2	4	10	22	48	106	. . .										

We notice from the table and from the graph in Figure 4 that for the first three or four days the exponential model gives results comparable to those of the more sophisticated logistic model. For $t \geqslant 5$, however, the exponential model is hopelessly inaccurate, but the logistic model fits the observations reasonably well.

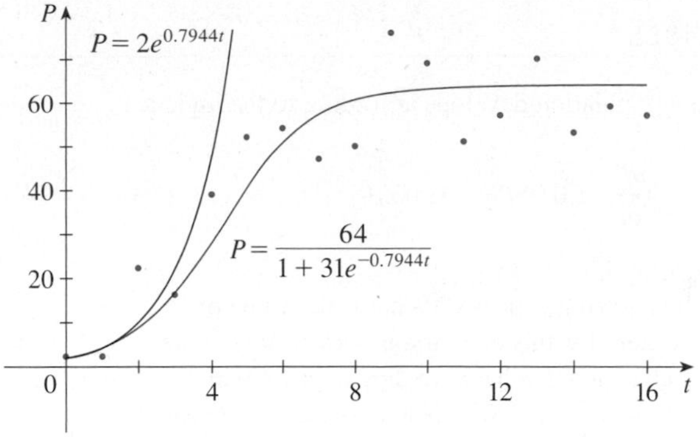

FIGURE 4
The exponential and logistic models for the *Paramecium* data

□

Many countries that formerly experienced exponential growth are now finding that their rates of population growth are declining and the logistic model provides a better model. The table in the margin shows midyear values of $B(t)$, the population of Belgium, in thousands, at time t, from 1980 to 2000. Figure 5 shows these data points together with a shifted logistic function obtained from a calculator with the ability to fit a logistic function to these points by regression. We see that the logistic model provides a very good fit.

t	$B(t)$	t	$B(t)$
1980	9,847	1992	10,036
1982	9,856	1994	10,109
1984	9,855	1996	10,152
1986	9,862	1998	10,175
1988	9,884	2000	10,186
1990	9,962		

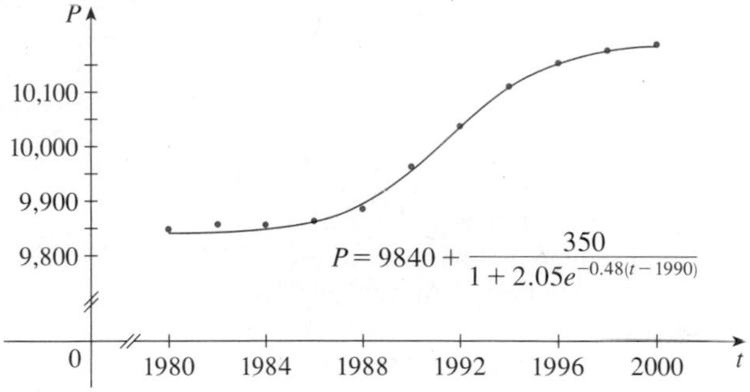

FIGURE 5
Logistic model for the population of Belgium

OTHER MODELS FOR POPULATION GROWTH

The Law of Natural Growth and the logistic differential equation are not the only equations that have been proposed to model population growth. In Exercise 18 we look at the Gompertz growth function and in Exercises 19 and 20 we investigate seasonal-growth models.

Two of the other models are modifications of the logistic model. The differential equation

$$\frac{dP}{dt} = kP\left(1 - \frac{P}{K}\right) - c$$

has been used to model populations that are subject to harvesting of one sort or another. (Think of a population of fish being caught at a constant rate.) This equation is explored in Exercises 15 and 16.

For some species there is a minimum population level m below which the species tends to become extinct. (Adults may not be able to find suitable mates.) Such populations have been modeled by the differential equation

$$\frac{dP}{dt} = kP\left(1 - \frac{P}{K}\right)\left(1 - \frac{m}{P}\right)$$

where the extra factor, $1 - m/P$, takes into account the consequences of a sparse population (see Exercise 17).

9.4 EXERCISES

1. Suppose that a population develops according to the logistic equation

$$\frac{dP}{dt} = 0.05P - 0.0005P^2$$

where t is measured in weeks.

(a) What is the carrying capacity? What is the value of k?

(b) A direction field for this equation is shown. Where are the slopes close to 0? Where are they largest? Which solutions are increasing? Which solutions are decreasing?

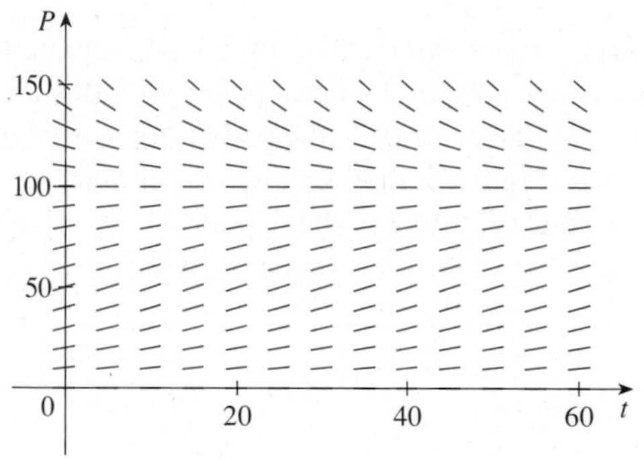

(c) Use the direction field to sketch solutions for initial populations of 20, 40, 60, 80, 120, and 140. What do these solutions have in common? How do they differ? Which solutions have inflection points? At what population levels do they occur?

(d) What are the equilibrium solutions? How are the other solutions related to these solutions?

2. Suppose that a population grows according to a logistic model with carrying capacity 6000 and $k = 0.0015$ per year.

(a) Write the logistic differential equation for these data.

(b) Draw a direction field (either by hand or with a computer algebra system). What does it tell you about the solution curves?

(c) Use the direction field to sketch the solution curves for initial populations of 1000, 2000, 4000, and 8000. What can you say about the concavity of these curves? What is the significance of the inflection points?

(d) Program a calculator or computer to use Euler's method with step size $h = 1$ to estimate the population after 50 years if the initial population is 1000.

(e) If the initial population is 1000, write a formula for the population after t years. Use it to find the population after 50 years and compare with your estimate in part (d).

(f) Graph the solution in part (e) and compare with the solution curve you sketched in part (c).

3. The Pacific halibut fishery has been modeled by the differential equation

$$\frac{dy}{dt} = ky\left(1 - \frac{y}{K}\right)$$

where $y(t)$ is the biomass (the total mass of the members of the population) in kilograms at time t (measured in years), the carrying capacity is estimated to be $K = 8 \times 10^7$ kg, and $k = 0.71$ per year.

(a) If $y(0) = 2 \times 10^7$ kg, find the biomass a year later.

(b) How long will it take for the biomass to reach 4×10^7 kg?

4. The table gives the number of yeast cells in a new laboratory culture.

Time (hours)	Yeast cells	Time (hours)	Yeast cells
0	18	10	509
2	39	12	597
4	80	14	640
6	171	16	664
8	336	18	672

(a) Plot the data and use the plot to estimate the carrying capacity for the yeast population.

(b) Use the data to estimate the initial relative growth rate.

(c) Find both an exponential model and a logistic model for these data.

(d) Compare the predicted values with the observed values, both in a table and with graphs. Comment on how well your models fit the data.

(e) Use your logistic model to estimate the number of yeast cells after 7 hours.

5. The population of the world was about 5.3 billion in 1990. Birth rates in the 1990s ranged from 35 to 40 million per year and death rates ranged from 15 to 20 million per year. Let's assume that the carrying capacity for world population is 100 billion.

(a) Write the logistic differential equation for these data. (Because the initial population is small compared to the carrying capacity, you can take k to be an estimate of the initial relative growth rate.)

(b) Use the logistic model to estimate the world population in the year 2000 and compare with the actual population of 6.1 billion.

(c) Use the logistic model to predict the world population in the years 2100 and 2500.

(d) What are your predictions if the carrying capacity is 50 billion?

6. (a) Make a guess as to the carrying capacity for the US population. Use it and the fact that the population was 250 million in 1990 to formulate a logistic model for the US population.

(b) Determine the value of k in your model by using the fact that the population in 2000 was 275 million.

(c) Use your model to predict the US population in the years 2100 and 2200.

(d) Use your model to predict the year in which the US population will exceed 350 million.

7. One model for the spread of a rumor is that the rate of spread is proportional to the product of the fraction y of the population who have heard the rumor and the fraction who have not heard the rumor.
(a) Write a differential equation that is satisfied by y.
(b) Solve the differential equation. → using the soln
(c) A small town has 1000 inhabitants. At 8 AM, 80 people have heard a rumor. By noon half the town has heard it. At what time will 90% of the population have heard the rumor?

8. Biologists stocked a lake with 400 fish and estimated the carrying capacity (the maximal population for the fish of that species in that lake) to be 10,000. The number of fish tripled in the first year.
(a) Assuming that the size of the fish population satisfies the logistic equation, find an expression for the size of the population after t years.
(b) How long will it take for the population to increase to 5000?

9. (a) Show that if P satisfies the logistic equation (4), then

$$\frac{d^2P}{dt^2} = k^2P\left(1 - \frac{P}{K}\right)\left(1 - \frac{2P}{K}\right)$$

(b) Deduce that a population grows fastest when it reaches half its carrying capacity.

10. For a fixed value of K (say $K = 10$), the family of logistic functions given by Equation 7 depends on the initial value P_0 and the proportionality constant k. Graph several members of this family. How does the graph change when P_0 varies? How does it change when k varies?

11. The table gives the midyear population of Japan, in thousands, from 1960 to 2005.

Year	Population	Year	Population
1960	94,092	1985	120,754
1965	98,883	1990	123,537
1970	104,345	1995	125,341
1975	111,573	2000	126,700
1980	116,807	2005	127,417

Use a graphing calculator to fit both an exponential function and a logistic function to these data. Graph the data points and both functions, and comment on the accuracy of the models. [*Hint:* Subtract 94,000 from each of the population figures. Then, after obtaining a model from your calculator, add 94,000 to get your final model. It might be helpful to choose $t = 0$ to correspond to 1960 or 1980.]

12. The table gives the midyear population of Spain, in thousands, from 1955 to 2000.

Year	Population	Year	Population
1955	29,319	1980	37,488
1960	30,641	1985	38,535
1965	32,085	1990	39,351
1970	33,876	1995	39,750
1975	35,564	2000	40,016

Use a graphing calculator to fit both an exponential function and a logistic function to these data. Graph the data points and both functions, and comment on the accuracy of the models. [*Hint:* Subtract 29,000 from each of the population figures. Then, after obtaining a model from your calculator, add 29,000 to get your final model. It might be helpful to choose $t = 0$ to correspond to 1955 or 1975.]

13. Consider a population $P = P(t)$ with constant relative birth and death rates α and β, respectively, and a constant emigration rate m, where α, β, and m are positive constants. Assume that $\alpha > \beta$. Then the rate of change of the population at time t is modeled by the differential equation

$$\frac{dP}{dt} = kP - m \qquad \text{where } k = \alpha - \beta$$

(a) Find the solution of this equation that satisfies the initial condition $P(0) = P_0$.
(b) What condition on m will lead to an exponential expansion of the population?
(c) What condition on m will result in a constant population? A population decline?
(d) In 1847, the population of Ireland was about 8 million and the difference between the relative birth and death rates was 1.6% of the population. Because of the potato famine in the 1840s and 1850s, about 210,000 inhabitants per year emigrated from Ireland. Was the population expanding or declining at that time?

14. Let c be a positive number. A differential equation of the form

$$\frac{dy}{dt} = ky^{1+c}$$

where k is a positive constant, is called a *doomsday equation* because the exponent in the expression ky^{1+c} is larger than the exponent 1 for natural growth.
(a) Determine the solution that satisfies the initial condition $y(0) = y_0$.
(b) Show that there is a finite time $t = T$ (doomsday) such that $\lim_{t \to T^-} y(t) = \infty$.
(c) An especially prolific breed of rabbits has the growth term $ky^{1.01}$. If 2 such rabbits breed initially and the warren has 16 rabbits after three months, then when is doomsday?

15. Let's modify the logistic differential equation of Example 1 as follows:

$$\frac{dP}{dt} = 0.08P\left(1 - \frac{P}{1000}\right) - 15$$

(a) Suppose $P(t)$ represents a fish population at time t, where t is measured in weeks. Explain the meaning of the term -15.
(b) Draw a direction field for this differential equation.
(c) What are the equilibrium solutions?
(d) Use the direction field to sketch several solution curves. Describe what happens to the fish population for various initial populations.
CAS (e) Solve this differential equation explicitly, either by using partial fractions or with a computer algebra system. Use the initial populations 200 and 300. Graph the solutions and compare with your sketches in part (d).

CAS **16.** Consider the differential equation

$$\frac{dP}{dt} = 0.08P\left(1 - \frac{P}{1000}\right) - c$$

as a model for a fish population, where t is measured in weeks and c is a constant.
(a) Use a CAS to draw direction fields for various values of c.
(b) From your direction fields in part (a), determine the values of c for which there is at least one equilibrium solution. For what values of c does the fish population always die out?
(c) Use the differential equation to prove what you discovered graphically in part (b).
(d) What would you recommend for a limit to the weekly catch of this fish population?

17. There is considerable evidence to support the theory that for some species there is a minimum population m such that the species will become extinct if the size of the population falls below m. This condition can be incorporated into the logistic equation by introducing the factor $(1 - m/P)$. Thus the modified logistic model is given by the differential equation

$$\frac{dP}{dt} = kP\left(1 - \frac{P}{K}\right)\left(1 - \frac{m}{P}\right)$$

(a) Use the differential equation to show that any solution is increasing if $m < P < K$ and decreasing if $0 < P < m$.
(b) For the case where $k = 0.08$, $K = 1000$, and $m = 200$, draw a direction field and use it to sketch several solution curves. Describe what happens to the population for various initial populations. What are the equilibrium solutions?
(c) Solve the differential equation explicitly, either by using partial fractions or with a computer algebra system. Use the initial population P_0.
(d) Use the solution in part (c) to show that if $P_0 < m$, then the species will become extinct. [*Hint:* Show that the numerator in your expression for $P(t)$ is 0 for some value of t.]

18. Another model for a growth function for a limited population is given by the **Gompertz function**, which is a solution of the differential equation

$$\frac{dP}{dt} = c\ln\left(\frac{K}{P}\right)P$$

where c is a constant and K is the carrying capacity.
(a) Solve this differential equation.
(b) Compute $\lim_{t\to\infty} P(t)$.
(c) Graph the Gompertz growth function for $K = 1000$, $P_0 = 100$, and $c = 0.05$, and compare it with the logistic function in Example 2. What are the similarities? What are the differences?
(d) We know from Exercise 9 that the logistic function grows fastest when $P = K/2$. Use the Gompertz differential equation to show that the Gompertz function grows fastest when $P = K/e$.

19. In a **seasonal-growth model**, a periodic function of time is introduced to account for seasonal variations in the rate of growth. Such variations could, for example, be caused by seasonal changes in the availability of food.
(a) Find the solution of the seasonal-growth model

$$\frac{dP}{dt} = kP\cos(rt - \phi) \qquad P(0) = P_0$$

where k, r, and ϕ are positive constants.
(b) By graphing the solution for several values of k, r, and ϕ, explain how the values of k, r, and ϕ affect the solution. What can you say about $\lim_{t\to\infty} P(t)$?

20. Suppose we alter the differential equation in Exercise 19 as follows:

$$\frac{dP}{dt} = kP\cos^2(rt - \phi) \qquad P(0) = P_0$$

(a) Solve this differential equation with the help of a table of integrals or a CAS.
(b) Graph the solution for several values of k, r, and ϕ. How do the values of k, r, and ϕ affect the solution? What can you say about $\lim_{t\to\infty} P(t)$ in this case?

21. Graphs of logistic functions (Figures 2 and 3) look suspiciously similar to the graph of the hyperbolic tangent function (Figure 3 in Section 3.11). Explain the similarity by showing that the logistic function given by Equation 7 can be written as

$$P(t) = \tfrac{1}{2}K\left[1 + \tanh\left(\tfrac{1}{2}k(t - c)\right)\right]$$

where $c = (\ln A)/k$. Thus the logistic function is really just a shifted hyperbolic tangent.

APPLIED PROJECT

CALCULUS AND BASEBALL

In this project we explore three of the many applications of calculus to baseball. The physical interactions of the game, especially the collision of ball and bat, are quite complex and their models are discussed in detail in a book by Robert Adair, *The Physics of Baseball,* 3d ed. (New York: HarperPerennial, 2002).

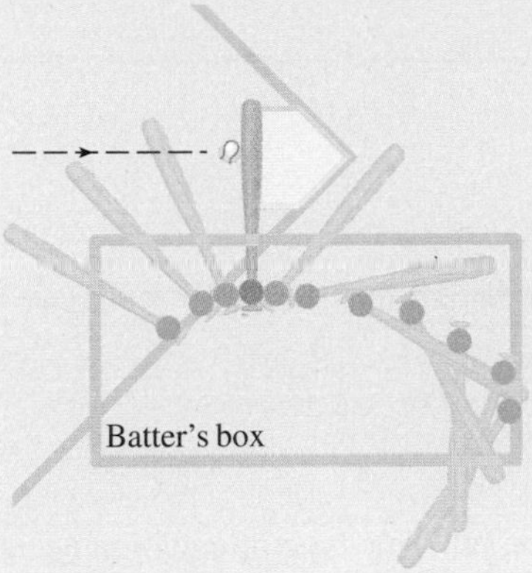

An overhead view of the position of a baseball bat, shown every fiftieth of a second during a typical swing. (Adapted from *The Physics of Baseball*)

1. It may surprise you to learn that the collision of baseball and bat lasts only about a thousandth of a second. Here we calculate the average force on the bat during this collision by first computing the change in the ball's momentum.

The *momentum* p of an object is the product of its mass m and its velocity v, that is, $p = mv$. Suppose an object, moving along a straight line, is acted on by a force $F = F(t)$ that is a continuous function of time.

(a) Show that the change in momentum over a time interval $[t_0, t_1]$ is equal to the integral of F from t_0 to t_1; that is, show that

$$p(t_1) - p(t_0) = \int_{t_0}^{t_1} F(t)\,dt$$

This integral is called the *impulse* of the force over the time interval.

(b) A pitcher throws a 90-mi/h fastball to a batter, who hits a line drive directly back to the pitcher. The ball is in contact with the bat for 0.001 s and leaves the bat with velocity 110 mi/h. A baseball weighs 5 oz and, in US Customary units, its mass is measured in slugs: $m = w/g$ where $g = 32$ ft/s^2.

(i) Find the change in the ball's momentum.

(ii) Find the average force on the bat.

2. In this problem we calculate the work required for a pitcher to throw a 90-mi/h fastball by first considering kinetic energy.

The *kinetic energy* K of an object of mass m and velocity v is given by $K = \frac{1}{2}mv^2$. Suppose an object of mass m, moving in a straight line, is acted on by a force $F = F(s)$ that depends on its position s. According to Newton's Second Law

$$F(s) = ma = m\frac{dv}{dt}$$

where a and v denote the acceleration and velocity of the object.

(a) Show that the work done in moving the object from a position s_0 to a position s_1 is equal to the change in the object's kinetic energy; that is, show that

$$W = \int_{s_0}^{s_1} F(s)\,ds = \tfrac{1}{2}mv_1^2 - \tfrac{1}{2}mv_0^2$$

where $v_0 = v(s_0)$ and $v_1 = v(s_1)$ are the velocities of the object at the positions s_0 and s_1. *Hint:* By the Chain Rule,

$$m\frac{dv}{dt} = m\frac{dv}{ds}\frac{ds}{dt} = mv\frac{dv}{ds}$$

(b) How many foot-pounds of work does it take to throw a baseball at a speed of 90 mi/h?

3. (a) An outfielder fields a baseball 280 ft away from home plate and throws it directly to the catcher with an initial velocity of 100 ft/s. Assume that the velocity $v(t)$ of the ball after t seconds satisfies the differential equation $dv/dt = -\frac{1}{10}v$ because of air resistance. How long does it take for the ball to reach home plate? (Ignore any vertical motion of the ball.)

(b) The manager of the team wonders whether the ball will reach home plate sooner if it is relayed by an infielder. The shortstop can position himself directly between the outfielder and home plate, catch the ball thrown by the outfielder, turn, and throw the ball to the catcher with an initial velocity of 105 ft/s. The manager clocks the relay time of the shortstop (catching, turning, throwing) at half a second. How far from home plate should the shortstop position himself to minimize the total time for the ball to reach home plate? Should the manager encourage a direct throw or a relayed throw? What if the shortstop can throw at 115 ft/s?

(c) For what throwing velocity of the shortstop does a relayed throw take the same time as a direct throw?

9.5 LINEAR EQUATIONS

A first-order **linear** differential equation is one that can be put into the form

$$\boxed{1} \qquad \frac{dy}{dx} + P(x)y = Q(x)$$

where P and Q are continuous functions on a given interval. This type of equation occurs frequently in various sciences, as we will see.

An example of a linear equation is $xy' + y = 2x$ because, for $x \neq 0$, it can be written in the form

$$\boxed{2} \qquad y' + \frac{1}{x}y = 2$$

Notice that this differential equation is not separable because it's impossible to factor the expression for y' as a function of x times a function of y. But we can still solve the equation by noticing, by the Product Rule, that

$$xy' + y = (xy)'$$

and so we can rewrite the equation as

$$(xy)' = 2x$$

If we now integrate both sides of this equation, we get

$$xy = x^2 + C \qquad \text{or} \qquad y = x + \frac{C}{x}$$

If we had been given the differential equation in the form of Equation 2, we would have had to take the preliminary step of multiplying each side of the equation by x.

It turns out that every first-order linear differential equation can be solved in a similar fashion by multiplying both sides of Equation 1 by a suitable function $I(x)$ called an *integrating factor.* We try to find I so that the left side of Equation 1, when multiplied by $I(x)$, becomes the derivative of the product $I(x)y$:

$$\boxed{3} \qquad I(x)(y' + P(x)y) = (I(x)y)'$$

If we can find such a function I, then Equation 1 becomes

$$(I(x)y)' = I(x)\,Q(x)$$

Integrating both sides, we would have

$$I(x)y = \int I(x)\,Q(x)\,dx + C$$

so the solution would be

$$\boxed{4} \qquad y(x) = \frac{1}{I(x)}\left[\int I(x)\,Q(x)\,dx + C\right]$$

To find such an I, we expand Equation 3 and cancel terms:

$$I(x)y' + I(x)\,P(x)y = (I(x)y)' = I'(x)y + I(x)y'$$

$$I(x)\,P(x) = I'(x)$$

This is a separable differential equation for I, which we solve as follows:

$$\int \frac{dI}{I} = \int P(x)\,dx$$

$$\ln|I| = \int P(x)\,dx$$

$$I = Ae^{\int P(x)\,dx}$$

where $A = \pm e^C$. We are looking for a particular integrating factor, not the most general one, so we take $A = 1$ and use

$$\boxed{5} \qquad I(x) = e^{\int P(x)\,dx}$$

Thus a formula for the general solution to Equation 1 is provided by Equation 4, where I is given by Equation 5. Instead of memorizing this formula, however, we just remember the form of the integrating factor.

> To solve the linear differential equation $y' + P(x)y = Q(x)$, multiply both sides by the **integrating factor** $I(x) = e^{\int P(x)\,dx}$ and integrate both sides.

V EXAMPLE 1 Solve the differential equation $\dfrac{dy}{dx} + 3x^2y = 6x^2$.

SOLUTION The given equation is linear since it has the form of Equation 1 with $P(x) = 3x^2$ and $Q(x) = 6x^2$. An integrating factor is

$$I(x) = e^{\int 3x^2\,dx} = e^{x^3}$$

Multiplying both sides of the differential equation by e^{x^3}, we get

$$e^{x^3}\frac{dy}{dx} + 3x^2e^{x^3}y = 6x^2e^{x^3}$$

or

$$\frac{d}{dx}(e^{x^3}y) = 6x^2e^{x^3}$$

■ Figure 1 shows the graphs of several members of the family of solutions in Example 1. Notice that they all approach 2 as $x \to \infty$.

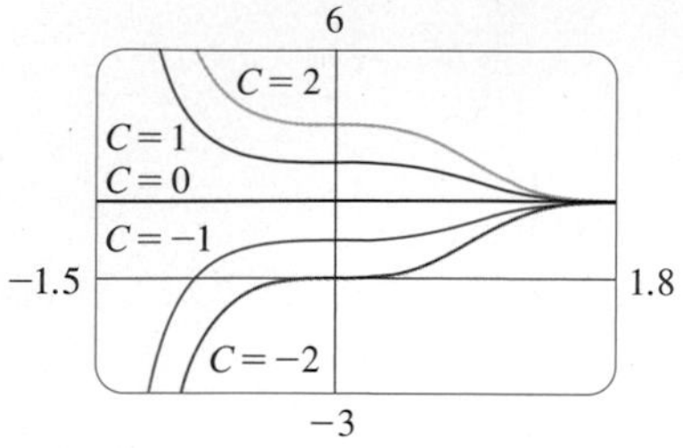

FIGURE 1

Integrating both sides, we have

$$e^{x^3}y = \int 6x^2e^{x^3}\,dx = 2e^{x^3} + C$$

$$y = 2 + Ce^{-x^3}$$ □

EXAMPLE 2 Find the solution of the initial-value problem

$$x^2y' + xy = 1 \qquad x > 0 \qquad y(1) = 2$$

SOLUTION We must first divide both sides by the coefficient of y' to put the differential equation into standard form:

6 $$y' + \frac{1}{x}y = \frac{1}{x^2} \qquad x > 0$$

The integrating factor is

$$I(x) = e^{\int (1/x)\,dx} = e^{\ln x} = x$$

Multiplication of Equation 6 by x gives

$$xy' + y = \frac{1}{x} \qquad \text{or} \qquad (xy)' = \frac{1}{x}$$

Then $$xy = \int \frac{1}{x}\,dx = \ln x + C$$

and so $$y = \frac{\ln x + C}{x}$$

Since $y(1) = 2$, we have

$$2 = \frac{\ln 1 + C}{1} = C$$

Therefore the solution to the initial-value problem is

$$y = \frac{\ln x + 2}{x}$$ □

■ The solution of the initial-value problem in Example 2 is shown in Figure 2.

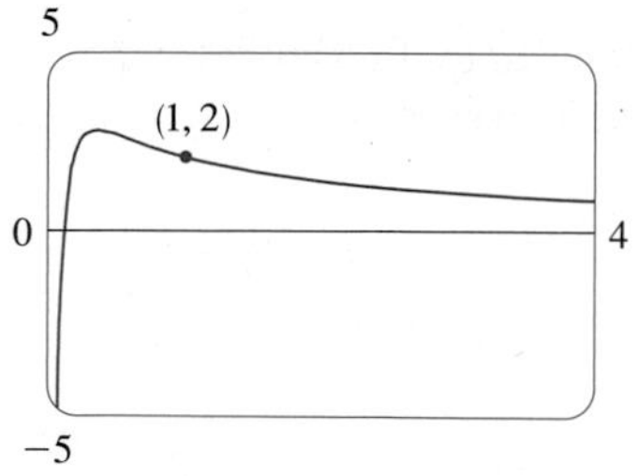

FIGURE 2

EXAMPLE 3 Solve $y' + 2xy = 1$.

SOLUTION The given equation is in the standard form for a linear equation. Multiplying by the integrating factor

$$e^{\int 2x\,dx} = e^{x^2}$$

we get $$e^{x^2}y' + 2xe^{x^2}y = e^{x^2}$$

or $$(e^{x^2}y)' = e^{x^2}$$

Therefore $$e^{x^2}y = \int e^{x^2}\,dx + C$$

■ Even though the solutions of the differential equation in Example 3 are expressed in terms of an integral, they can still be graphed by a computer algebra system (Figure 3).

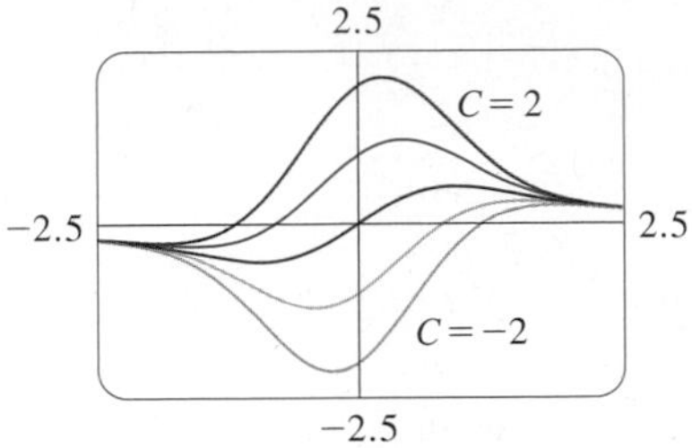

FIGURE 3

Recall from Section 7.5 that $\int e^{x^2}\,dx$ can't be expressed in terms of elementary functions. Nonetheless, it's a perfectly good function and we can leave the answer as

$$y = e^{-x^2}\int e^{x^2}\,dx + Ce^{-x^2}$$

Another way of writing the solution is

$$y = e^{-x^2}\int_0^x e^{t^2}\,dt + Ce^{-x^2}$$

(Any number can be chosen for the lower limit of integration.) □

APPLICATION TO ELECTRIC CIRCUITS

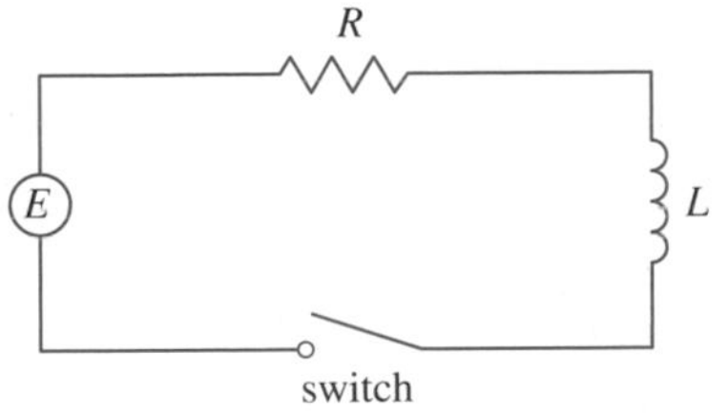

FIGURE 4

In Section 9.2 we considered the simple electric circuit shown in Figure 4: An electromotive force (usually a battery or generator) produces a voltage of $E(t)$ volts (V) and a current of $I(t)$ amperes (A) at time t. The circuit also contains a resistor with a resistance of R ohms (Ω) and an inductor with an inductance of L henries (H).

Ohm's Law gives the drop in voltage due to the resistor as RI. The voltage drop due to the inductor is $L(dI/dt)$. One of Kirchhoff's laws says that the sum of the voltage drops is equal to the supplied voltage $E(t)$. Thus we have

$$\boxed{7} \qquad L\frac{dI}{dt} + RI = E(t)$$

which is a first-order linear differential equation. The solution gives the current I at time t.

V EXAMPLE 4 Suppose that in the simple circuit of Figure 4 the resistance is 12 Ω and the inductance is 4 H. If a battery gives a constant voltage of 60 V and the switch is closed when $t = 0$ so the current starts with $I(0) = 0$, find (a) $I(t)$, (b) the current after 1 s, and (c) the limiting value of the current.

■ The differential equation in Example 4 is both linear and separable, so an alternative method is to solve it as a separable equation (Example 4 in Section 9.3). If we replace the battery by a generator, however, we get an equation that is linear but not separable (Example 5).

SOLUTION

(a) If we put $L = 4$, $R = 12$, and $E(t) = 60$ in Equation 7, we obtain the initial-value problem

$$4\frac{dI}{dt} + 12I = 60 \qquad I(0) = 0$$

or

$$\frac{dI}{dt} + 3I = 15 \qquad I(0) = 0$$

Multiplying by the integrating factor $e^{\int 3\,dt} = e^{3t}$, we get

$$e^{3t}\frac{dI}{dt} + 3e^{3t}I = 15e^{3t}$$

$$\frac{d}{dt}(e^{3t}I) = 15e^{3t}$$

$$e^{3t}I = \int 15e^{3t}\,dt = 5e^{3t} + C$$

$$I(t) = 5 + Ce^{-3t}$$

■ Figure 5 shows how the current in Example 4 approaches its limiting value.

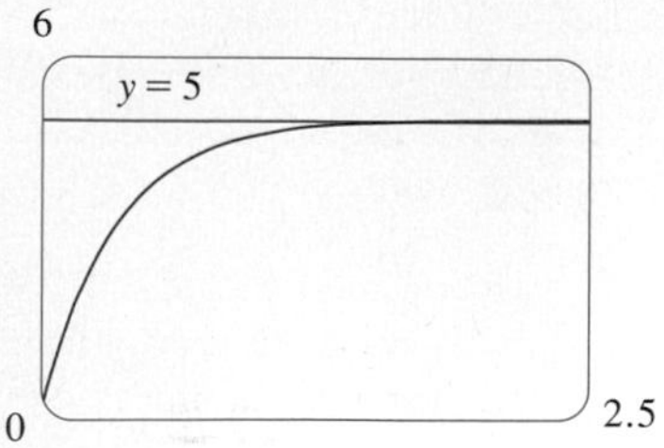

FIGURE 5

Since $I(0) = 0$, we have $5 + C = 0$, so $C = -5$ and

$$I(t) = 5(1 - e^{-3t})$$

(b) After 1 second the current is

$$I(1) = 5(1 - e^{-3}) \approx 4.75 \text{ A}$$

(c) The limiting value of the current is given by

$$\lim_{t\to\infty} I(t) = \lim_{t\to\infty} 5(1 - e^{-3t}) = 5 - 5\lim_{t\to\infty} e^{-3t} = 5 - 0 = 5$$ □

EXAMPLE 5 Suppose that the resistance and inductance remain as in Example 4 but, instead of the battery, we use a generator that produces a variable voltage of $E(t) = 60 \sin 30t$ volts. Find $I(t)$.

SOLUTION This time the differential equation becomes

$$4\frac{dI}{dt} + 12I = 60 \sin 30t \qquad \text{or} \qquad \frac{dI}{dt} + 3I = 15 \sin 30t$$

The same integrating factor e^{3t} gives

$$\frac{d}{dt}(e^{3t}I) = e^{3t}\frac{dI}{dt} + 3e^{3t}I = 15e^{3t} \sin 30t$$

■ Figure 6 shows the graph of the current when the battery is replaced by a generator.

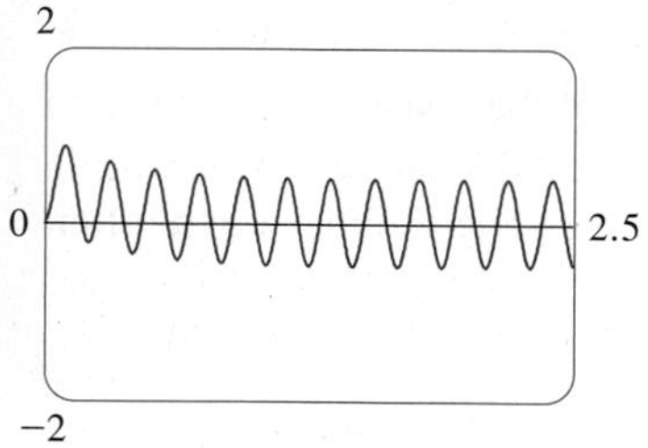

FIGURE 6

Using Formula 98 in the Table of Integrals, we have

$$e^{3t}I = \int 15e^{3t} \sin 30t\, dt = 15\frac{e^{3t}}{909}(3 \sin 30t - 30 \cos 30t) + C$$

$$I = \tfrac{5}{101}(\sin 30t - 10 \cos 30t) + Ce^{-3t}$$

Since $I(0) = 0$, we get

$$-\tfrac{50}{101} + C = 0$$

so

$$I(t) = \tfrac{5}{101}(\sin 30t - 10 \cos 30t) + \tfrac{50}{101}e^{-3t}$$ □

9.5 EXERCISES

1–4 Determine whether the differential equation is linear.

1. $y' + \cos x = y$

2. $y' + \cos y = \tan x$

3. $yy' + xy = x^2$

4. $xy + \sqrt{x} = e^x y'$

5–14 Solve the differential equation.

5. $y' + 2y = 2e^x$

6. $y' = x + 5y$

7. $xy' - 2y = x^2$

8. $x^2y' + 2xy = \cos^2 x$

9. $xy' + y = \sqrt{x}$

10. $y' + y = \sin(e^x)$

11. $\sin x \dfrac{dy}{dx} + (\cos x)y = \sin(x^2)$

12. $x\dfrac{dy}{dx} - 4y = x^4e^x$

13. $(1 + t)\dfrac{du}{dt} + u = 1 + t, \quad t > 0$

14. $t \ln t \dfrac{dr}{dt} + r = te^t$

15–20 Solve the initial-value problem.

15. $y' = x + y, \quad y(0) = 2$

16. $t\dfrac{dy}{dt} + 2y = t^3, \quad t > 0, \quad y(1) = 0$

17. $\dfrac{dv}{dt} - 2tv = 3t^2e^{t^2}, \quad v(0) = 5$

18. $2xy' + y = 6x, \quad x > 0, \quad y(4) = 20$

19. $xy' = y + x^2 \sin x, \quad y(\pi) = 0$

20. $(x^2 + 1)\dfrac{dy}{dx} + 3x(y - 1) = 0, \quad y(0) = 2$

21–22 Solve the differential equation and use a graphing calculator or computer to graph several members of the family of solutions. How does the solution curve change as C varies?

21. $xy' + 2y = e^x$

22. $y' + (\cos x)y = \cos x$

---

23. A **Bernoulli differential equation** (named after James Bernoulli) is of the form

$$\frac{dy}{dx} + P(x)y = Q(x)y^n$$

Observe that, if $n = 0$ or 1, the Bernoulli equation is linear. For other values of n, show that the substitution $u = y^{1-n}$ transforms the Bernoulli equation into the linear equation

$$\frac{du}{dx} + (1 - n)P(x)u = (1 - n)Q(x)$$

24–25 Use the method of Exercise 23 to solve the differential equation.

24. $xy' + y = -xy^2$

25. $y' + \dfrac{2}{x}y = \dfrac{y^3}{x^2}$

---

26. Solve the second-order equation $xy'' + 2y' = 12x^2$ by making the substitution $u = y'$.

27. In the circuit shown in Figure 4, a battery supplies a constant voltage of 40 V, the inductance is 2 H, the resistance is 10 Ω, and $I(0) = 0$.
(a) Find $I(t)$.
(b) Find the current after 0.1 s.

28. In the circuit shown in Figure 4, a generator supplies a voltage of $E(t) = 40 \sin 60t$ volts, the inductance is 1 H, the resistance is 20 Ω, and $I(0) = 1$ A.
(a) Find $I(t)$.
(b) Find the current after 0.1 s.
(c) Use a graphing device to draw the graph of the current function.

29. The figure shows a circuit containing an electromotive force, a capacitor with a capacitance of C farads (F), and a resistor with a resistance of R ohms (Ω). The voltage drop across the capacitor is Q/C, where Q is the charge (in coulombs), so in this case Kirchhoff's Law gives

$$RI + \frac{Q}{C} = E(t)$$

But $I = dQ/dt$ (see Example 3 in Section 3.7), so we have

$$R\frac{dQ}{dt} + \frac{1}{C}Q = E(t)$$

Suppose the resistance is 5 Ω, the capacitance is 0.05 F, a battery gives a constant voltage of 60 V, and the initial charge is $Q(0) = 0$ C. Find the charge and the current at time t.

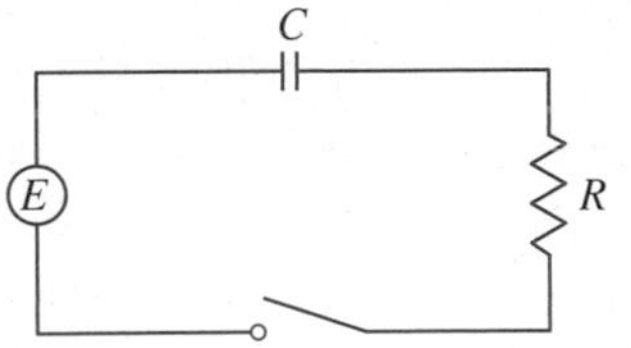

30. In the circuit of Exercise 29, $R = 2$ Ω, $C = 0.01$ F, $Q(0) = 0$, and $E(t) = 10 \sin 60t$. Find the charge and the current at time t.

31. Let $P(t)$ be the performance level of someone learning a skill as a function of the training time t. The graph of P is called a *learning curve*. In Exercise 13 in Section 9.1 we proposed the differential equation

$$\frac{dP}{dt} = k[M - P(t)]$$

as a reasonable model for learning, where k is a positive constant. Solve it as a linear differential equation and use your solution to graph the learning curve.

32. Two new workers were hired for an assembly line. Jim processed 25 units during the first hour and 45 units during the second hour. Mark processed 35 units during the first hour and 50 units the second hour. Using the model of Exercise 31 and assuming that $P(0) = 0$, estimate the maximum number of units per hour that each worker is capable of processing.

33. In Section 9.3 we looked at mixing problems in which the volume of fluid remained constant and saw that such problems give rise to separable equations. (See Example 6 in that section.) If the rates of flow into and out of the system are different, then the volume is not constant and the resulting differential equation is linear but not separable.

A tank contains 100 L of water. A solution with a salt concentration of 0.4 kg/L is added at a rate of 5 L/min. The solution is kept mixed and is drained from the tank at a rate of 3 L/min. If $y(t)$ is the amount of salt (in kilograms) after t minutes, show that y satisfies the differential equation

$$\frac{dy}{dt} = 2 - \frac{3y}{100 + 2t}$$

Solve this equation and find the concentration after 20 minutes.

34. A tank with a capacity of 400 L is full of a mixture of water and chlorine with a concentration of 0.05 g of chlorine per liter. In order to reduce the concentration of chlorine, fresh water is pumped into the tank at a rate of 4 L/s. The mixture is kept stirred and is pumped out at a rate of 10 L/s. Find the amount of chlorine in the tank as a function of time.

35. An object with mass m is dropped from rest and we assume that the air resistance is proportional to the speed of the object. If $s(t)$ is the distance dropped after t seconds, then the speed is $v = s'(t)$ and the acceleration is $a = v'(t)$. If g is the acceleration due to gravity, then the downward force on the object is $mg - cv$, where c is a positive constant, and Newton's Second Law gives

$$m\frac{dv}{dt} = mg - cv$$

(a) Solve this as a linear equation to show that

$$v = \frac{mg}{c}(1 - e^{-ct/m})$$

(b) What is the limiting velocity?
(c) Find the distance the object has fallen after t seconds.

36. If we ignore air resistance, we can conclude that heavier objects fall no faster than lighter objects. But if we take air resistance into account, our conclusion changes. Use the expression for the velocity of a falling object in Exercise 35(a) to find dv/dm and show that heavier objects *do* fall faster than lighter ones.

9.6 PREDATOR-PREY SYSTEMS

We have looked at a variety of models for the growth of a single species that lives alone in an environment. In this section we consider more realistic models that take into account the interaction of two species in the same habitat. We will see that these models take the form of a pair of linked differential equations.

We first consider the situation in which one species, called the *prey,* has an ample food supply and the second species, called the *predator,* feeds on the prey. Examples of prey and predators include rabbits and wolves in an isolated forest, food fish and sharks, aphids and ladybugs, and bacteria and amoebas. Our model will have two dependent variables and both are functions of time. We let $R(t)$ be the number of prey (using R for rabbits) and $W(t)$ be the number of predators (with W for wolves) at time t.

In the absence of predators, the ample food supply would support exponential growth of the prey, that is,

$$\frac{dR}{dt} = kR \qquad \text{where } k \text{ is a positive constant}$$

In the absence of prey, we assume that the predator population would decline at a rate proportional to itself, that is,

$$\frac{dW}{dt} = -rW \qquad \text{where } r \text{ is a positive constant}$$

With both species present, however, we assume that the principal cause of death among the prey is being eaten by a predator, and the birth and survival rates of the predators depend on their available food supply, namely, the prey. We also assume that the two species encounter each other at a rate that is proportional to both populations and is therefore proportional to the product RW. (The more there are of either population, the more encounters there are likely to be.) A system of two differential equations that incorporates these assumptions is as follows:

W represents the predator.
R represents the prey.

$$\boxed{1} \qquad \frac{dR}{dt} = kR - aRW \qquad \frac{dW}{dt} = -rW + bRW$$

where k, r, a, and b are positive constants. Notice that the term $-aRW$ decreases the natural growth rate of the prey and the term bRW increases the natural growth rate of the predators.

■ The Lotka-Volterra equations were proposed as a model to explain the variations in the shark and food-fish populations in the Adriatic Sea by the Italian mathematician Vito Volterra (1860–1940).

The equations in (1) are known as the **predator-prey equations**, or the **Lotka-Volterra equations**. A **solution** of this system of equations is a pair of functions $R(t)$ and $W(t)$ that describe the populations of prey and predator as functions of time. Because the system is coupled (R and W occur in both equations), we can't solve one equation and then the other; we have to solve them simultaneously. Unfortunately, it is usually impossible to find explicit formulas for R and W as functions of t. We can, however, use graphical methods to analyze the equations.

V EXAMPLE 1 Suppose that populations of rabbits and wolves are described by the Lotka-Volterra equations (1) with $k = 0.08$, $a = 0.001$, $r = 0.02$, and $b = 0.00002$. The time t is measured in months.

(a) Find the constant solutions (called the **equilibrium solutions**) and interpret the answer.

(b) Use the system of differential equations to find an expression for dW/dR.

(c) Draw a direction field for the resulting differential equation in the RW-plane. Then use that direction field to sketch some solution curves.

(d) Suppose that, at some point in time, there are 1000 rabbits and 40 wolves. Draw the corresponding solution curve and use it to describe the changes in both population levels.

(e) Use part (d) to make sketches of R and W as functions of t.

SOLUTION

(a) With the given values of k, a, r, and b, the Lotka-Volterra equations become

$$\frac{dR}{dt} = 0.08R - 0.001RW$$

$$\frac{dW}{dt} = -0.02W + 0.00002RW$$

Both R and W will be constant if both derivatives are 0, that is,

$$R' = R(0.08 - 0.001W) = 0$$

$$W' = W(-0.02 + 0.00002R) = 0$$

One solution is given by $R = 0$ and $W = 0$. (This makes sense: If there are no rabbits or wolves, the populations are certainly not going to increase.) The other constant solution is

$$W = \frac{0.08}{0.001} = 80 \qquad R = \frac{0.02}{0.00002} = 1000$$

So the equilibrium populations consist of 80 wolves and 1000 rabbits. This means that 1000 rabbits are just enough to support a constant wolf population of 80. There are neither too many wolves (which would result in fewer rabbits) nor too few wolves (which would result in more rabbits).

(b) We use the Chain Rule to eliminate t:

$$\frac{dW}{dt} = \frac{dW}{dR}\frac{dR}{dt}$$

so

$$\frac{dW}{dR} = \frac{\dfrac{dW}{dt}}{\dfrac{dR}{dt}} = \frac{-0.02W + 0.00002RW}{0.08R - 0.001RW}$$

(c) If we think of W as a function of R, we have the differential equation

$$\frac{dW}{dR} = \frac{-0.02W + 0.00002RW}{0.08R - 0.001RW}$$

We draw the direction field for this differential equation in Figure 1 and we use it to sketch several solution curves in Figure 2. If we move along a solution curve, we observe how the relationship between R and W changes as time passes. Notice that the curves appear to be closed in the sense that if we travel along a curve, we always return to the same point. Notice also that the point (1000, 80) is inside all the solution curves. That point is called an *equilibrium point* because it corresponds to the equilibrium solution $R = 1000$, $W = 80$.

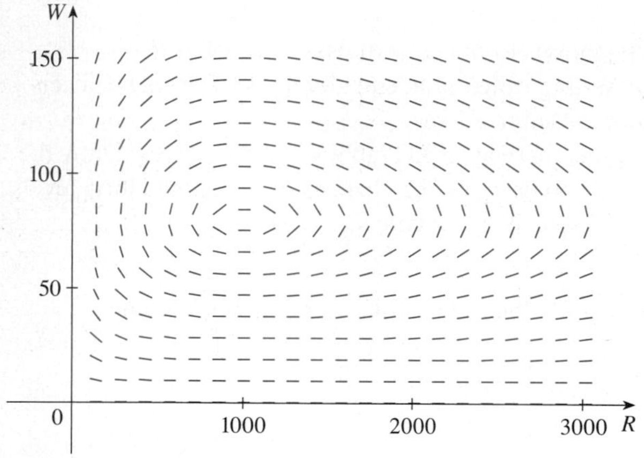

FIGURE 1 Direction field for the predator-prey system

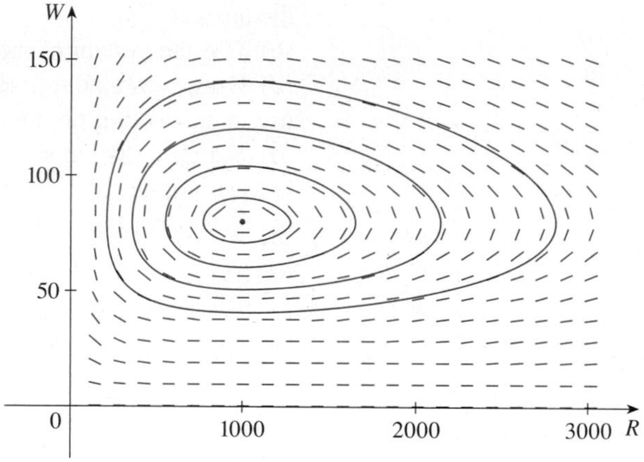

FIGURE 2 Phase portrait of the system

When we represent solutions of a system of differential equations as in Figure 2, we refer to the RW-plane as the **phase plane**, and we call the solution curves **phase trajectories**. So a phase trajectory is a path traced out by solutions (R, W) as time goes by. A **phase portrait** consists of equilibrium points and typical phase trajectories, as shown in Figure 2.

(d) Starting with 1000 rabbits and 40 wolves corresponds to drawing the solution curve through the point $P_0(1000, 40)$. Figure 3 shows this phase trajectory with the direction field removed. Starting at the point P_0 at time $t = 0$ and letting t increase, do we move clockwise or counterclockwise around the phase trajectory? If we put $R = 1000$ and

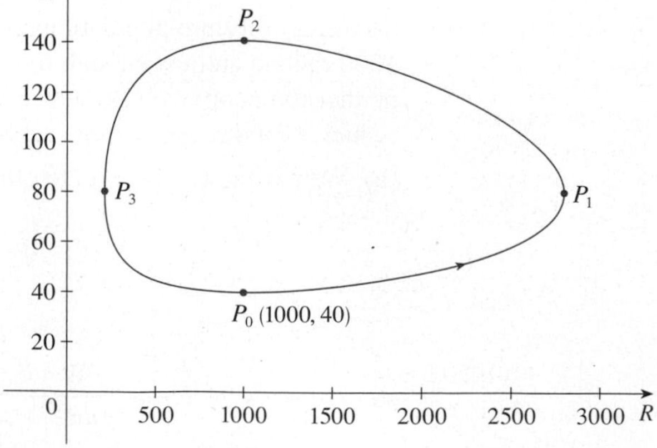

FIGURE 3
Phase trajectory through (1000, 40)

$W = 40$ in the first differential equation, we get

$$\frac{dR}{dt} = 0.08(1000) - 0.001(1000)(40) = 80 - 40 = 40$$

Since $dR/dt > 0$, we conclude that R is increasing at P_0 and so we move counterclockwise around the phase trajectory.

We see that at P_0 there aren't enough wolves to maintain a balance between the populations, so the rabbit population increases. That results in more wolves and eventually there are so many wolves that the rabbits have a hard time avoiding them. So the number of rabbits begins to decline (at P_1, where we estimate that R reaches its maximum population of about 2800). This means that at some later time the wolf population starts to fall (at P_2, where $R = 1000$ and $W \approx 140$). But this benefits the rabbits, so their population later starts to increase (at P_3, where $W = 80$ and $R \approx 210$). As a consequence, the wolf population eventually starts to increase as well. This happens when the populations return to their initial values of $R = 1000$ and $W = 40$, and the entire cycle begins again.

(e) From the description in part (d) of how the rabbit and wolf populations rise and fall, we can sketch the graphs of $R(t)$ and $W(t)$. Suppose the points P_1, P_2, and P_3 in Figure 3 are reached at times t_1, t_2, and t_3. Then we can sketch graphs of R and W as in Figure 4.

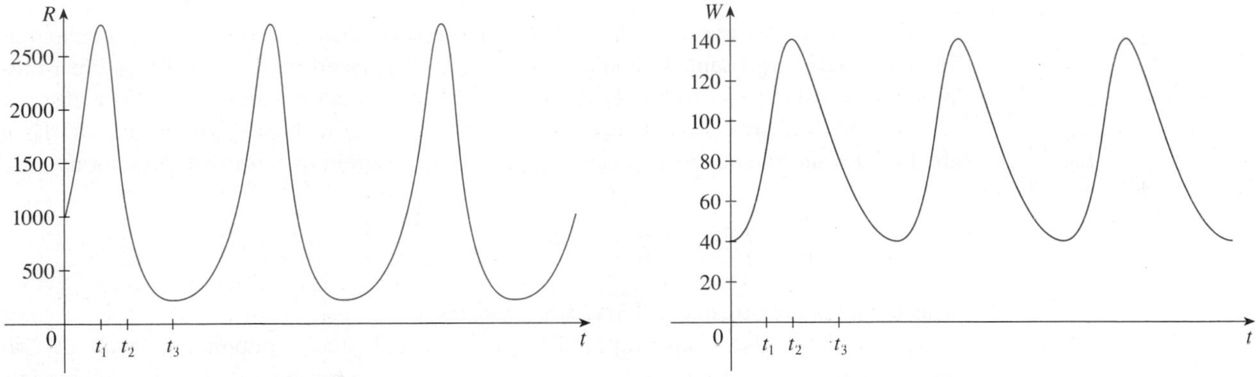

FIGURE 4
Graphs of the rabbit and wolf populations as functions of time

To make the graphs easier to compare, we draw the graphs on the same axes but with different scales for R and W, as in Figure 5. Notice that the rabbits reach their maximum populations about a quarter of a cycle before the wolves.

TEC In Module 9.6 you can change the coefficients in the Lotka-Volterra equations and observe the resulting changes in the phase trajectory and graphs of the rabbit and wolf populations.

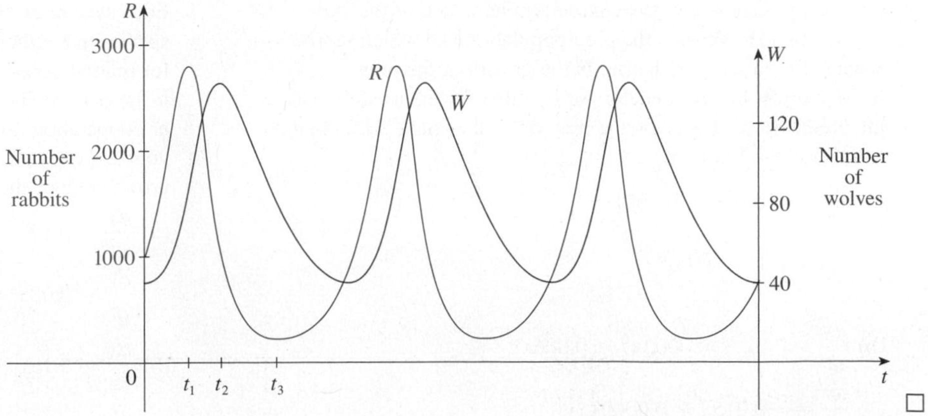

FIGURE 5
Comparison of the rabbit and wolf populations

An important part of the modeling process, as we discussed in Section 1.2, is to interpret our mathematical conclusions as real-world predictions and to test the predictions against real data. The Hudson's Bay Company, which started trading in animal furs in Canada in 1670, has kept records that date back to the 1840s. Figure 6 shows graphs of the number of pelts of the snowshoe hare and its predator, the Canada lynx, traded by the company over a 90-year period. You can see that the coupled oscillations in the hare and lynx populations predicted by the Lotka-Volterra model do actually occur and the period of these cycles is roughly 10 years.

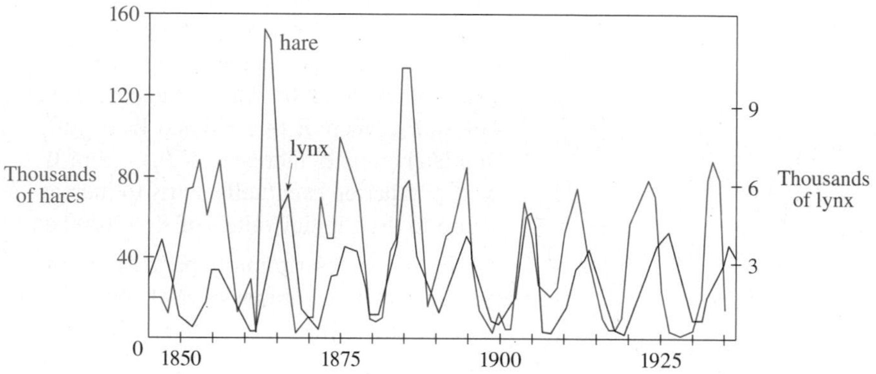

FIGURE 6
Relative abundance of hare and lynx from Hudson's Bay Company records

Although the relatively simple Lotka-Volterra model has had some success in explaining and predicting coupled populations, more sophisticated models have also been proposed. One way to modify the Lotka-Volterra equations is to assume that, in the absence of predators, the prey grow according to a logistic model with carrying capacity K. Then the Lotka-Volterra equations (1) are replaced by the system of differential equations

$$\frac{dR}{dt} = kR\left(1 - \frac{R}{K}\right) - aRW \qquad \frac{dW}{dt} = -rW + bRW$$

This model is investigated in Exercises 9 and 10.

Models have also been proposed to describe and predict population levels of two species that compete for the same resources or cooperate for mutual benefit. Such models are explored in Exercise 2.

9.6 EXERCISES

1. For each predator-prey system, determine which of the variables, x or y, represents the prey population and which represents the predator population. Is the growth of the prey restricted just by the predators or by other factors as well? Do the predators feed only on the prey or do they have additional food sources? Explain.

(a) $\dfrac{dx}{dt} = -0.05x + 0.0001xy$

$\dfrac{dy}{dt} = 0.1y - 0.005xy$

(b) $\dfrac{dx}{dt} = 0.2x - 0.0002x^2 - 0.006xy$

$\dfrac{dy}{dt} = -0.015y + 0.00008xy$

2. Each system of differential equations is a model for two species that either compete for the same resources or cooperate for mutual benefit (flowering plants and insect pollinators, for instance). Decide whether each system describes competition or cooperation and explain why it is a reasonable model. (Ask yourself what effect an increase in one species has on the growth rate of the other.)

(a) $\dfrac{dx}{dt} = 0.12x - 0.0006x^2 + 0.00001xy$

$\dfrac{dy}{dt} = 0.08x + 0.00004xy$

(b) $\dfrac{dx}{dt} = 0.15x - 0.0002x^2 - 0.0006xy$

$\dfrac{dy}{dt} = 0.2y - 0.00008y^2 - 0.0002xy$

3–4 A phase trajectory is shown for populations of rabbits (R) and foxes (F).

(a) Describe how each population changes as time goes by.

(b) Use your description to make a rough sketch of the graphs of R and F as functions of time.

3.

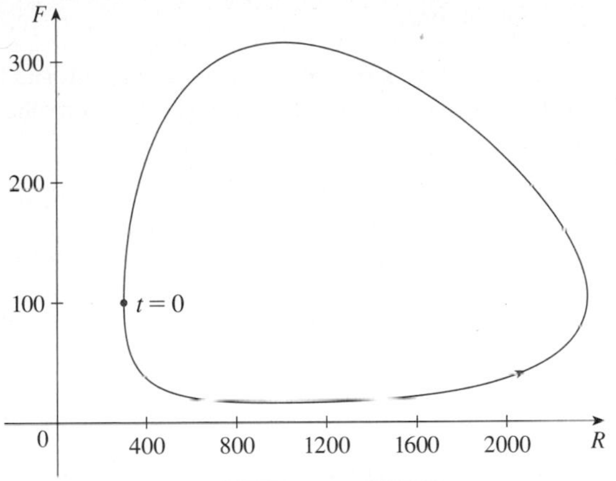

4.

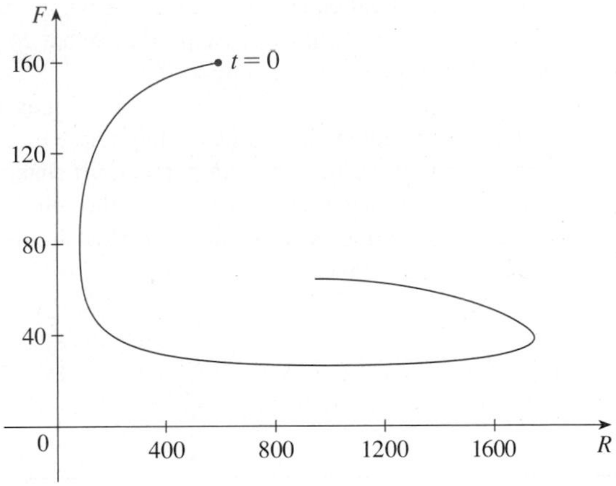

5–6 Graphs of populations of two species are shown. Use them to sketch the corresponding phase trajectory.

5.

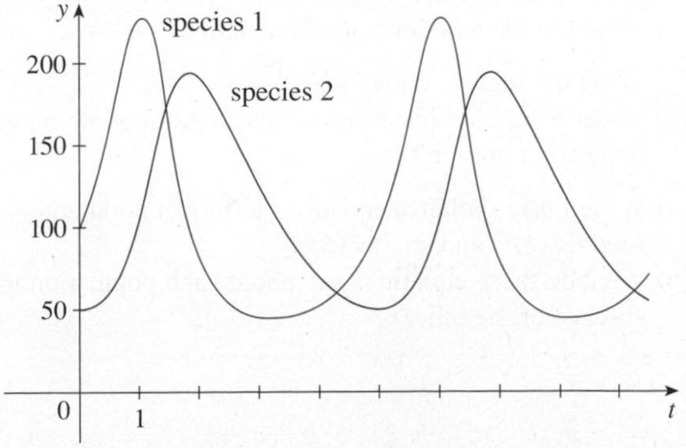

6.

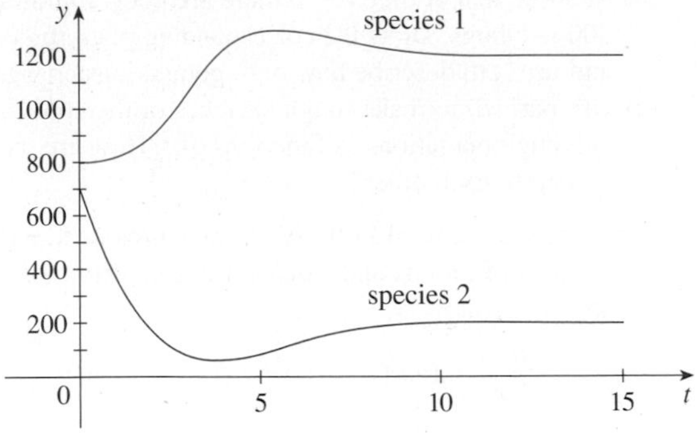

7. In Example 1(b) we showed that the rabbit and wolf populations satisfy the differential equation

$$\frac{dW}{dR} = \frac{-0.02W + 0.00002RW}{0.08R - 0.001RW}$$

By solving this separable differential equation, show that

$$\frac{R^{0.02}W^{0.08}}{e^{0.00002R}e^{0.001W}} = C$$

where C is a constant.

It is impossible to solve this equation for W as an explicit function of R (or vice versa). If you have a computer algebra system that graphs implicitly defined curves, use this equation and your CAS to draw the solution curve that passes through the point (1000, 40) and compare with Figure 3.

8. Populations of aphids and ladybugs are modeled by the equations

$$\frac{dA}{dt} = 2A - 0.01AL$$

$$\frac{dL}{dt} = -0.5L + 0.0001AL$$

(a) Find the equilibrium solutions and explain their significance.

(b) Find an expression for dL/dA.

(c) The direction field for the differential equation in part (b) is shown. Use it to sketch a phase portrait. What do the phase trajectories have in common?

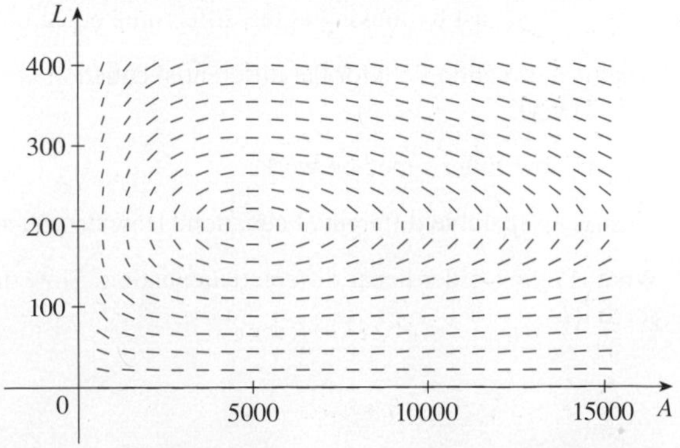

(d) Suppose that at time $t = 0$ there are 1000 aphids and 200 ladybugs. Draw the corresponding phase trajectory and use it to describe how both populations change.
(e) Use part (d) to make rough sketches of the aphid and ladybug populations as functions of t. How are the graphs related to each other?

9. In Example 1 we used Lotka-Volterra equations to model populations of rabbits and wolves. Let's modify those equations as follows:

$$\frac{dR}{dt} = 0.08R(1 - 0.0002R) - 0.001RW$$

$$\frac{dW}{dt} = -0.02W + 0.00002RW$$

(a) According to these equations, what happens to the rabbit population in the absence of wolves?

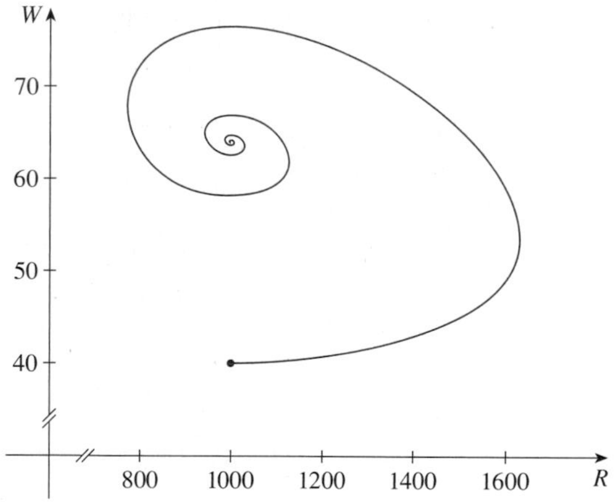

(b) Find all the equilibrium solutions and explain their significance.
(c) The figure shows the phase trajectory that starts at the point (1000, 40). Describe what eventually happens to the rabbit and wolf populations.
(d) Sketch graphs of the rabbit and wolf populations as functions of time.

CAS **10.** In Exercise 8 we modeled populations of aphids and ladybugs with a Lotka-Volterra system. Suppose we modify those equations as follows:

$$\frac{dA}{dt} = 2A(1 - 0.0001A) - 0.01AL$$

$$\frac{dL}{dt} = -0.5L + 0.0001AL$$

(a) In the absence of ladybugs, what does the model predict about the aphids?
(b) Find the equilibrium solutions.
(c) Find an expression for dL/dA.
(d) Use a computer algebra system to draw a direction field for the differential equation in part (c). Then use the direction field to sketch a phase portrait. What do the phase trajectories have in common?
(e) Suppose that at time $t = 0$ there are 1000 aphids and 200 ladybugs. Draw the corresponding phase trajectory and use it to describe how both populations change.
(f) Use part (e) to make rough sketches of the aphid and ladybug populations as functions of t. How are the graphs related to each other?

9 REVIEW

CONCEPT CHECK

1. (a) What is a differential equation?
(b) What is the order of a differential equation?
(c) What is an initial condition?

2. What can you say about the solutions of the equation $y' = x^2 + y^2$ just by looking at the differential equation?

3. What is a direction field for the differential equation $y' = F(x, y)$?

4. Explain how Euler's method works.

5. What is a separable differential equation? How do you solve it?

6. What is a first-order linear differential equation? How do you solve it?

7. (a) Write a differential equation that expresses the law of natural growth. What does it say in terms of relative growth rate?
(b) Under what circumstances is this an appropriate model for population growth?
(c) What are the solutions of this equation?

8. (a) Write the logistic equation.
(b) Under what circumstances is this an appropriate model for population growth?

9. (a) Write Lotka-Volterra equations to model populations of food fish (F) and sharks (S).
(b) What do these equations say about each population in the absence of the other?

TRUE-FALSE QUIZ

Determine whether the statement is true or false. If it is true, explain why. If it is false, explain why or give an example that disproves the statement.

1. All solutions of the differential equation $y' = -1 - y^4$ are decreasing functions.
2. The function $f(x) = (\ln x)/x$ is a solution of the differential equation $x^2y' + xy = 1$.
3. The equation $y' = x + y$ is separable.
4. The equation $y' = 3y - 2x + 6xy - 1$ is separable.
5. The equation $e^x y' = y$ is linear.
6. The equation $y' + xy = e^y$ is linear.
7. If y is the solution of the initial-value problem

$$\frac{dy}{dt} = 2y\left(1 - \frac{y}{5}\right) \qquad y(0) = 1$$

then $\lim_{t\to\infty} y = 5$.

EXERCISES

1. (a) A direction field for the differential equation $y' = y(y - 2)(y - 4)$ is shown. Sketch the graphs of the solutions that satisfy the given initial conditions.
   (i) $y(0) = -0.3$ (ii) $y(0) = 1$
   (iii) $y(0) = 3$ (iv) $y(0) = 4.3$
   (b) If the initial condition is $y(0) = c$, for what values of c is $\lim_{t\to\infty} y(t)$ finite? What are the equilibrium solutions?

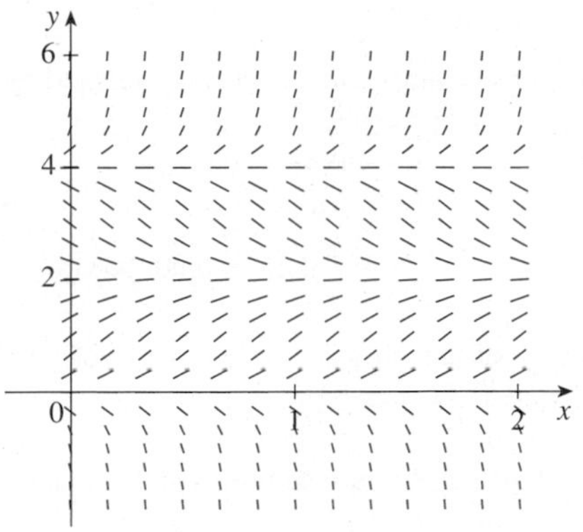

2. (a) Sketch a direction field for the differential equation $y' = x/y$. Then use it to sketch the four solutions that satisfy the initial conditions $y(0) = 1$, $y(0) = -1$, $y(2) = 1$, and $y(-2) = 1$.
   (b) Check your work in part (a) by solving the differential equation explicitly. What type of curve is each solution curve?

3. (a) A direction field for the differential equation $y' = x^2 - y^2$ is shown. Sketch the solution of the initial-value problem

$$y' = x^2 - y^2 \qquad y(0) = 1$$

Use your graph to estimate the value of $y(0.3)$.

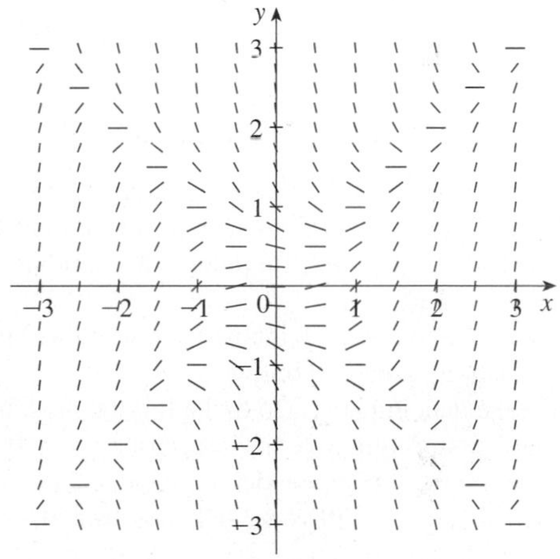

   (b) Use Euler's method with step size 0.1 to estimate $y(0.3)$ where $y(x)$ is the solution of the initial-value problem in part (a). Compare with your estimate from part (a).
   (c) On what lines are the centers of the horizontal line segments of the direction field in part (a) located? What happens when a solution curve crosses these lines?

4. (a) Use Euler's method with step size 0.2 to estimate $y(0.4)$, where $y(x)$ is the solution of the initial-value problem

$$y' = 2xy^2 \qquad y(0) = 1$$

   (b) Repeat part (a) with step size 0.1.
   (c) Find the exact solution of the differential equation and compare the value at 0.4 with the approximations in parts (a) and (b).

5–8 Solve the differential equation.

5. $y' = xe^{-\sin x} - y\cos x$
6. $\dfrac{dx}{dt} = 1 - t + x - tx$
7. $2ye^{y^2}y' = 2x + 3\sqrt{x}$
8. $x^2y' - y = 2x^3e^{-1/x}$

9–11 Solve the initial-value problem.

9. $\dfrac{dr}{dt} + 2tr = r, \quad r(0) = 5$

10. $(1 + \cos x)y' = (1 + e^{-y})\sin x, \quad y(0) = 0$

11. $xy' - y = x \ln x, \quad y(1) = 2$

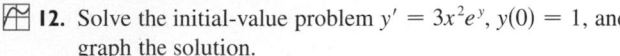

12. Solve the initial-value problem $y' = 3x^2e^y$, $y(0) = 1$, and graph the solution.

13–14 Find the orthogonal trajectories of the family of curves.

13. $y = ke^x$

14. $y = e^{kx}$

---

15. (a) Write the solution of the initial-value problem

$$\frac{dP}{dt} = 0.1P\left(1 - \frac{P}{2000}\right) \qquad P(0) = 100$$

and use it to find the population when $t = 20$.

(b) When does the population reach 1200?

16. (a) The population of the world was 5.28 billion in 1990 and 6.07 billion in 2000. Find an exponential model for these data and use the model to predict the world population in the year 2020.

(b) According to the model in part (a), when will the world population exceed 10 billion?

(c) Use the data in part (a) to find a logistic model for the population. Assume a carrying capacity of 100 billion. Then use the logistic model to predict the population in 2020. Compare with your prediction from the exponential model.

(d) According to the logistic model, when will the world population exceed 10 billion? Compare with your prediction in part (b).

17. The von Bertalanffy growth model is used to predict the length $L(t)$ of a fish over a period of time. If L_∞ is the largest length for a species, then the hypothesis is that the rate of growth in length is proportional to $L_\infty - L$, the length yet to be achieved.

(a) Formulate and solve a differential equation to find an expression for $L(t)$.

(b) For the North Sea haddock it has been determined that $L_\infty = 53$ cm, $L(0) = 10$ cm, and the constant of proportionality is 0.2. What does the expression for $L(t)$ become with these data?

18. A tank contains 100 L of pure water. Brine that contains 0.1 kg of salt per liter enters the tank at a rate of 10 L/min. The solution is kept thoroughly mixed and drains from the tank at the same rate. How much salt is in the tank after 6 minutes?

19. One model for the spread of an epidemic is that the rate of spread is jointly proportional to the number of infected people and the number of uninfected people. In an isolated town of 5000 inhabitants, 160 people have a disease at the beginning of the week and 1200 have it at the end of the week. How long does it take for 80% of the population to become infected?

20. The Brentano-Stevens Law in psychology models the way that a subject reacts to a stimulus. It states that if R represents the reaction to an amount S of stimulus, then the relative rates of increase are proportional:

$$\frac{1}{R}\frac{dR}{dt} = \frac{k}{S}\frac{dS}{dt}$$

where k is a positive constant. Find R as a function of S.

21. The transport of a substance across a capillary wall in lung physiology has been modeled by the differential equation

$$\frac{dh}{dt} = -\frac{R}{V}\left(\frac{h}{k + h}\right)$$

where h is the hormone concentration in the bloodstream, t is time, R is the maximum transport rate, V is the volume of the capillary, and k is a positive constant that measures the affinity between the hormones and the enzymes that assist the process. Solve this differential equation to find a relationship between h and t.

22. Populations of birds and insects are modeled by the equations

$$\frac{dx}{dt} = 0.4x - 0.002xy$$

$$\frac{dy}{dt} = -0.2y + 0.000008xy$$

(a) Which of the variables, x or y, represents the bird population and which represents the insect population? Explain.

(b) Find the equilibrium solutions and explain their significance.

(c) Find an expression for dy/dx.

(d) The direction field for the differential equation in part (c) is shown. Use it to sketch the phase trajectory corre-

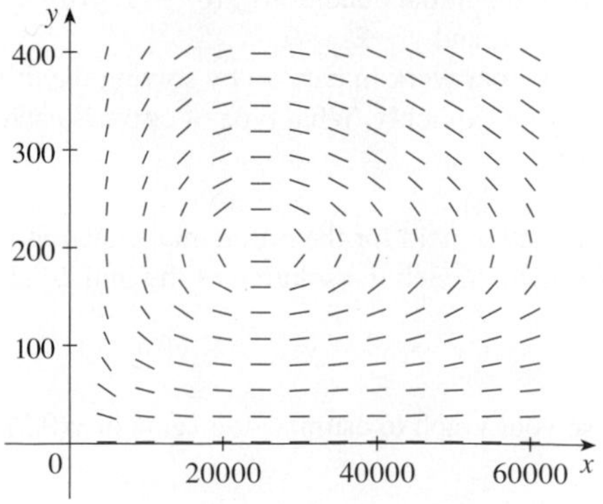

sponding to initial populations of 100 birds and 40,000 insects. Then use the phase trajectory to describe how both populations change.

(e) Use part (d) to make rough sketches of the bird and insect populations as functions of time. How are these graphs related to each other?

23. Suppose the model of Exercise 22 is replaced by the equations

$$\frac{dx}{dt} = 0.4x(1 - 0.000005x) - 0.002xy$$

$$\frac{dy}{dt} = -0.2y + 0.000008xy$$

(a) According to these equations, what happens to the insect population in the absence of birds?

(b) Find the equilibrium solutions and explain their significance.

(c) The figure shows the phase trajectory that starts with 100 birds and 40,000 insects. Describe what eventually happens to the bird and insect populations.

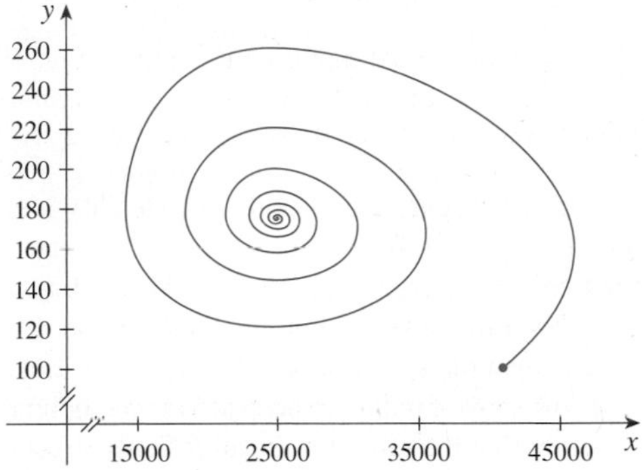

(d) Sketch graphs of the bird and insect populations as functions of time.

24. Barbara weighs 60 kg and is on a diet of 1600 calories per day, of which 850 are used automatically by basal metabolism. She spends about 15 cal/kg/day times her weight doing exercise. If 1 kg of fat contains 10,000 cal and we assume that the storage of calories in the form of fat is 100% efficient, formulate a differential equation and solve it to find her weight as a function of time. Does her weight ultimately approach an equilibrium weight?

25. When a flexible cable of uniform density is suspended between two fixed points and hangs of its own weight, the shape $y = f(x)$ of the cable must satisfy a differential equation of the form

$$\frac{d^2y}{dx^2} = k\sqrt{1 + \left(\frac{dy}{dx}\right)^2}$$

where k is a positive constant. Consider the cable shown in the figure.

(a) Let $z = dy/dx$ in the differential equation. Solve the resulting first-order differential equation (in z), and then integrate to find y.

(b) Determine the length of the cable.

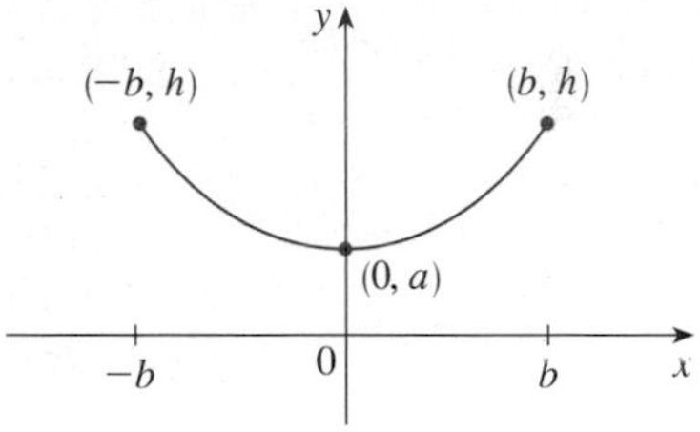

PROBLEMS PLUS

1. Find all functions f such that f' is continuous and

$$[f(x)]^2 = 100 + \int_0^x \{[f(t)]^2 + [f'(t)]^2\}\,dt \qquad \text{for all real } x$$

2. A student forgot the Product Rule for differentiation and made the mistake of thinking that $(fg)' = f'g'$. However, he was lucky and got the correct answer. The function f that he used was $f(x) = e^{x^2}$ and the domain of his problem was the interval $(\frac{1}{2}, \infty)$. What was the function g?

3. Let f be a function with the property that $f(0) = 1, f'(0) = 1$, and $f(a + b) = f(a)f(b)$ for all real numbers a and b. Show that $f'(x) = f(x)$ for all x and deduce that $f(x) = e^x$.

4. Find all functions f that satisfy the equation

$$\left(\int f(x)\,dx\right)\left(\int \frac{1}{f(x)}\,dx\right) = -1$$

5. Find the curve $y = f(x)$ such that $f(x) \geqslant 0, f(0) = 0, f(1) = 1$, and the area under the graph of f from 0 to x is proportional to the $(n + 1)$st power of $f(x)$.

6. A *subtangent* is a portion of the x-axis that lies directly beneath the segment of a tangent line from the point of contact to the x-axis. Find the curves that pass through the point $(c, 1)$ and whose subtangents all have length c.

7. A peach pie is removed from the oven at 5:00 PM. At that time it is piping hot, 100°C. At 5:10 PM its temperature is 80°C; at 5:20 PM it is 65°C. What is the temperature of the room?

8. Snow began to fall during the morning of February 2 and continued steadily into the afternoon. At noon a snowplow began removing snow from a road at a constant rate. The plow traveled 6 km from noon to 1 PM but only 3 km from 1 PM to 2 PM. When did the snow begin to fall? [*Hints:* To get started, let t be the time measured in hours after noon; let $x(t)$ be the distance traveled by the plow at time t; then the speed of the plow is dx/dt. Let b be the number of hours before noon that it began to snow. Find an expression for the height of the snow at time t. Then use the given information that the rate of removal R (in m^3/h) is constant.]

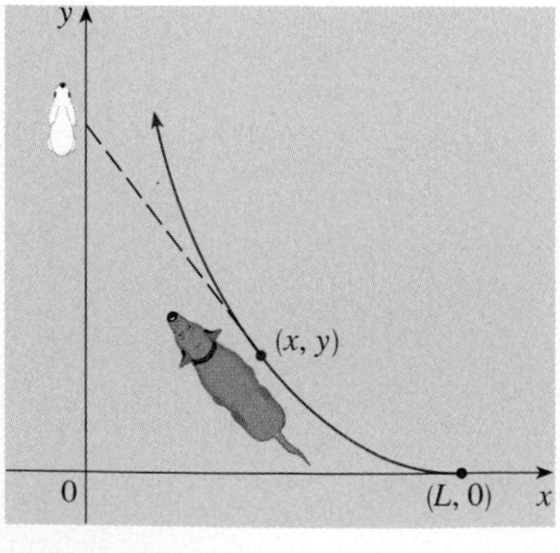

FIGURE FOR PROBLEM 9

9. A dog sees a rabbit running in a straight line across an open field and gives chase. In a rectangular coordinate system (as shown in the figure), assume:

(i) The rabbit is at the origin and the dog is at the point $(L, 0)$ at the instant the dog first sees the rabbit.

(ii) The rabbit runs up the y-axis and the dog always runs straight for the rabbit.

(iii) The dog runs at the same speed as the rabbit.

(a) Show that the dog's path is the graph of the function $y = f(x)$, where y satisfies the differential equation

$$x\frac{d^2y}{dx^2} = \sqrt{1 + \left(\frac{dy}{dx}\right)^2}$$

(b) Determine the solution of the equation in part (a) that satisfies the initial conditions $y = y' = 0$ when $x = L$. [*Hint:* Let $z = dy/dx$ in the differential equation and solve the resulting first-order equation to find z; then integrate z to find y.]

(c) Does the dog ever catch the rabbit?

10. (a) Suppose that the dog in Problem 9 runs twice as fast as the rabbit. Find a differential equation for the path of the dog. Then solve it to find the point where the dog catches the rabbit.

(b) Suppose the dog runs half as fast as the rabbit. How close does the dog get to the rabbit? What are their positions when they are closest?

11. A planning engineer for a new alum plant must present some estimates to his company regarding the capacity of a silo designed to contain bauxite ore until it is processed into alum. The ore resembles pink talcum powder and is poured from a conveyor at the top of the silo. The silo is a cylinder 100 ft high with a radius of 200 ft. The conveyor carries $60{,}000\pi \text{ ft}^3/\text{h}$ and the ore maintains a conical shape whose radius is 1.5 times its height.

(a) If, at a certain time t, the pile is 60 ft high, how long will it take for the pile to reach the top of the silo?

(b) Management wants to know how much room will be left in the floor area of the silo when the pile is 60 ft high. How fast is the floor area of the pile growing at that height?

(c) Suppose a loader starts removing the ore at the rate of $20{,}000\pi \text{ ft}^3/\text{h}$ when the height of the pile reaches 90 ft. Suppose, also, that the pile continues to maintain its shape. How long will it take for the pile to reach the top of the silo under these conditions?

12. Find the curve that passes through the point $(3, 2)$ and has the property that if the tangent line is drawn at any point P on the curve, then the part of the tangent line that lies in the first quadrant is bisected at P.

13. Recall that the normal line to a curve at a point P on the curve is the line that passes through P and is perpendicular to the tangent line at P. Find the curve that passes through the point $(3, 2)$ and has the property that if the normal line is drawn at any point on the curve, then the y-intercept of the normal line is always 6.

14. Find all curves with the property that if the normal line is drawn at any point P on the curve, then the part of the normal line between P and the x-axis is bisected by the y-axis.

11

INFINITE SEQUENCES AND SERIES

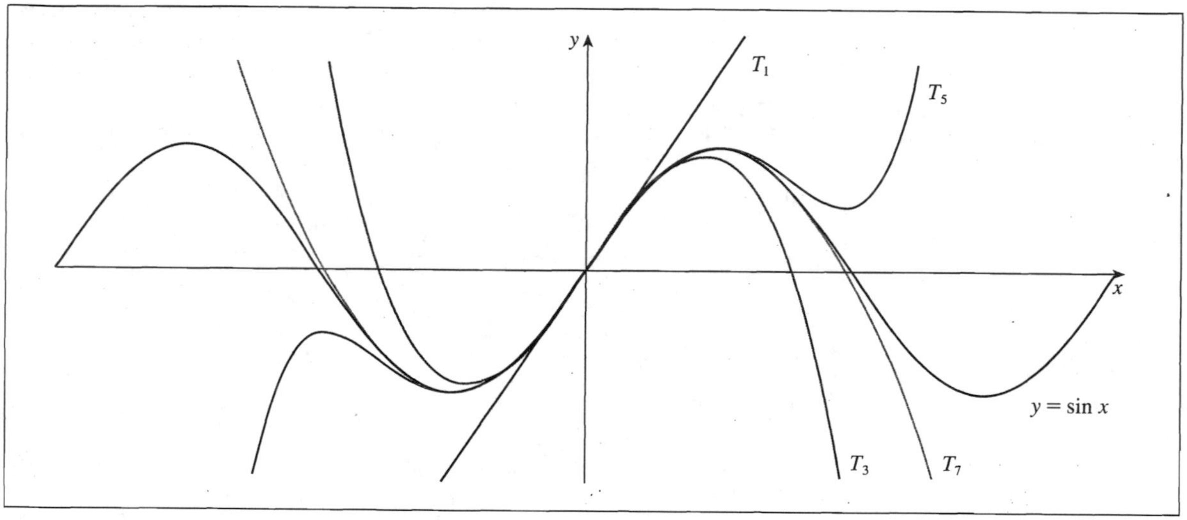

The partial sums T_n of a Taylor series provide better and better approximations to a function as n increases.

Infinite sequences and series were introduced briefly in *A Preview of Calculus* in connection with Zeno's paradoxes and the decimal representation of numbers. Their importance in calculus stems from Newton's idea of representing functions as sums of infinite series. For instance, in finding areas he often integrated a function by first expressing it as a series and then integrating each term of the series. We will pursue his idea in Section 11.10 in order to integrate such functions as e^{-x^2}. (Recall that we have previously been unable to do this.) Many of the functions that arise in mathematical physics and chemistry, such as Bessel functions, are defined as sums of series, so it is important to be familiar with the basic concepts of convergence of infinite sequences and series.

Physicists also use series in another way, as we will see in Section 11.11. In studying fields as diverse as optics, special relativity, and electromagnetism, they analyze phenomena by replacing a function with the first few terms in the series that represents it.

11.1 SEQUENCES

A **sequence** can be thought of as a list of numbers written in a definite order:

$$a_1, a_2, a_3, a_4, \ldots, a_n, \ldots$$

The number a_1 is called the *first term*, a_2 is the *second term*, and in general a_n is the *nth term*. We will deal exclusively with infinite sequences and so each term a_n will have a successor a_{n+1}.

Notice that for every positive integer n there is a corresponding number a_n and so a sequence can be defined as a function whose domain is the set of positive integers. But we usually write a_n instead of the function notation $f(n)$ for the value of the function at the number n.

NOTATION The sequence $\{a_1, a_2, a_3, \ldots\}$ is also denoted by

$$\{a_n\} \qquad \text{or} \qquad \{a_n\}_{n=1}^{\infty}$$

EXAMPLE 1 Some sequences can be defined by giving a formula for the nth term. In the following examples we give three descriptions of the sequence: one by using the preceding notation, another by using the defining formula, and a third by writing out the terms of the sequence. Notice that n doesn't have to start at 1.

(a) $\left\{\dfrac{n}{n+1}\right\}_{n=1}^{\infty}$ $\quad a_n = \dfrac{n}{n+1}$ $\quad \left\{\dfrac{1}{2}, \dfrac{2}{3}, \dfrac{3}{4}, \dfrac{4}{5}, \ldots, \dfrac{n}{n+1}, \ldots\right\}$

(b) $\left\{\dfrac{(-1)^n(n+1)}{3^n}\right\}$ $\quad a_n = \dfrac{(-1)^n(n+1)}{3^n}$ $\quad \left\{-\dfrac{2}{3}, \dfrac{3}{9}, -\dfrac{4}{27}, \dfrac{5}{81}, \ldots, \dfrac{(-1)^n(n+1)}{3^n}, \ldots\right\}$

(c) $\left\{\sqrt{n-3}\right\}_{n=3}^{\infty}$ $\quad a_n = \sqrt{n-3},\ n \geqslant 3$ $\quad \left\{0, 1, \sqrt{2}, \sqrt{3}, \ldots, \sqrt{n-3}, \ldots\right\}$

(d) $\left\{\cos\dfrac{n\pi}{6}\right\}_{n=0}^{\infty}$ $\quad a_n = \cos\dfrac{n\pi}{6},\ n \geqslant 0$ $\quad \left\{1, \dfrac{\sqrt{3}}{2}, \dfrac{1}{2}, 0, \ldots, \cos\dfrac{n\pi}{6}, \ldots\right\}$ □

V EXAMPLE 2 Find a formula for the general term a_n of the sequence

$$\left\{\frac{3}{5}, -\frac{4}{25}, \frac{5}{125}, -\frac{6}{625}, \frac{7}{3125}, \ldots\right\}$$

assuming that the pattern of the first few terms continues.

SOLUTION We are given that

$$a_1 = \frac{3}{5} \qquad a_2 = -\frac{4}{25} \qquad a_3 = \frac{5}{125} \qquad a_4 = -\frac{6}{625} \qquad a_5 = \frac{7}{3125}$$

Notice that the numerators of these fractions start with 3 and increase by 1 whenever we go to the next term. The second term has numerator 4, the third term has numerator 5; in general, the nth term will have numerator $n + 2$. The denominators are the powers of 5, so a_n has denominator 5^n. The signs of the terms are alternately positive and negative, so we need to multiply by a power of -1. In Example 1(b) the factor $(-1)^n$ meant we started with a negative term. Here we want to start with a positive term and so we use $(-1)^{n-1}$ or $(-1)^{n+1}$. Therefore

$$a_n = (-1)^{n-1} \frac{n+2}{5^n}$$ □

EXAMPLE 3 Here are some sequences that don't have a simple defining equation.
(a) The sequence $\{p_n\}$, where p_n is the population of the world as of January 1 in the year n.
(b) If we let a_n be the digit in the nth decimal place of the number e, then $\{a_n\}$ is a well-defined sequence whose first few terms are

$$\{7, 1, 8, 2, 8, 1, 8, 2, 8, 4, 5, \ldots\}$$

(c) The **Fibonacci sequence** $\{f_n\}$ is defined recursively by the conditions

$$f_1 = 1 \qquad f_2 = 1 \qquad f_n = f_{n-1} + f_{n-2} \qquad n \geqslant 3$$

Each term is the sum of the two preceding terms. The first few terms are

$$\{1, 1, 2, 3, 5, 8, 13, 21, \ldots\}$$

This sequence arose when the 13th-century Italian mathematician known as Fibonacci solved a problem concerning the breeding of rabbits (see Exercise 71). □

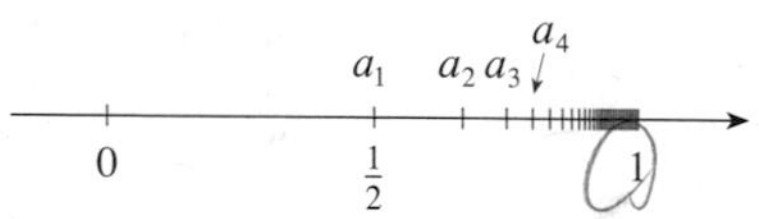

FIGURE 1

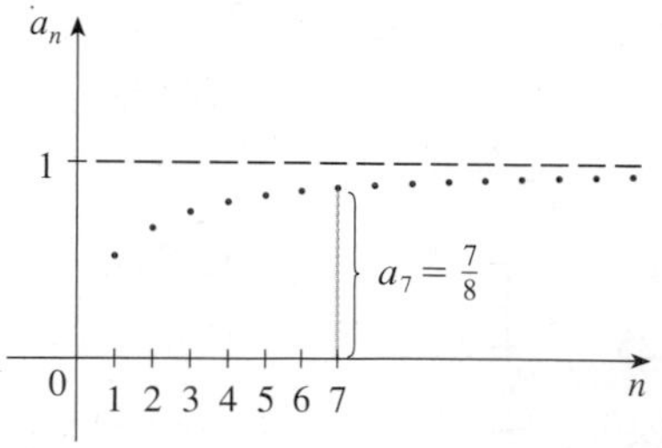

FIGURE 2

A sequence such as the one in Example 1(a), $a_n = n/(n+1)$, can be pictured either by plotting its terms on a number line as in Figure 1 or by plotting its graph as in Figure 2. Note that, since a sequence is a function whose domain is the set of positive integers, its graph consists of isolated points with coordinates

$$(1, a_1) \qquad (2, a_2) \qquad (3, a_3) \qquad \ldots \qquad (n, a_n) \qquad \ldots$$

From Figure 1 or 2 it appears that the terms of the sequence $a_n = n/(n+1)$ are approaching 1 as n becomes large. In fact, the difference

$$1 - \frac{n}{n+1} = \frac{1}{n+1}$$

can be made as small as we like by taking n sufficiently large. We indicate this by writing

$$\lim_{n \to \infty} \frac{n}{n+1} = 1$$

In general, the notation

$$\lim_{n \to \infty} a_n = L$$

means that the terms of the sequence $\{a_n\}$ approach L as n becomes large. Notice that the following definition of the limit of a sequence is very similar to the definition of a limit of a function at infinity given in Section 2.6.

1 DEFINITION A sequence $\{a_n\}$ has the **limit** L and we write

$$\lim_{n \to \infty} a_n = L \qquad \text{or} \qquad a_n \to L \text{ as } n \to \infty$$

if we can make the terms a_n as close to L as we like by taking n sufficiently large. If $\lim_{n\to\infty} a_n$ exists, we say the sequence **converges** (or is **convergent**). Otherwise, we say the sequence **diverges** (or is **divergent**).

Figure 3 illustrates Definition 1 by showing the graphs of two sequences that have the limit L.

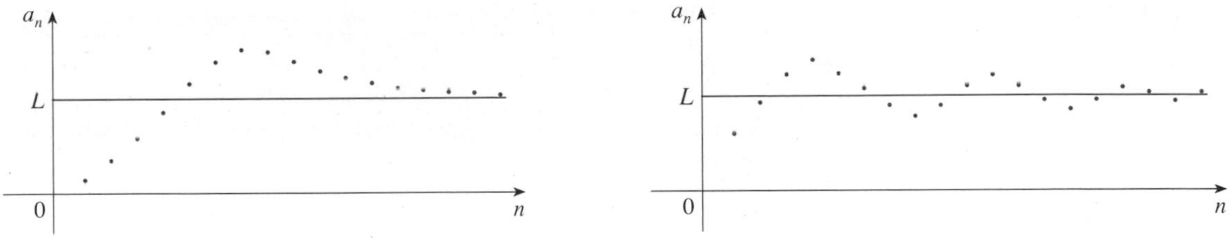

FIGURE 3
Graphs of two sequences with $\lim_{n\to\infty} a_n = L$

A more precise version of Definition 1 is as follows.

■ Compare this definition with Definition 2.6.7.

2 DEFINITION A sequence $\{a_n\}$ has the **limit** L and we write

$$\lim_{n \to \infty} a_n = L \qquad \text{or} \qquad a_n \to L \text{ as } n \to \infty$$

if for every $\varepsilon > 0$ there is a corresponding integer N such that

$$\text{if} \quad n > N \quad \text{then} \quad |a_n - L| < \varepsilon$$

Definition 2 is illustrated by Figure 4, in which the terms $a_1, a_2, a_3, \ldots$ are plotted on a number line. No matter how small an interval $(L - \varepsilon, L + \varepsilon)$ is chosen, there exists an N such that all terms of the sequence from a_{N+1} onward must lie in that interval.

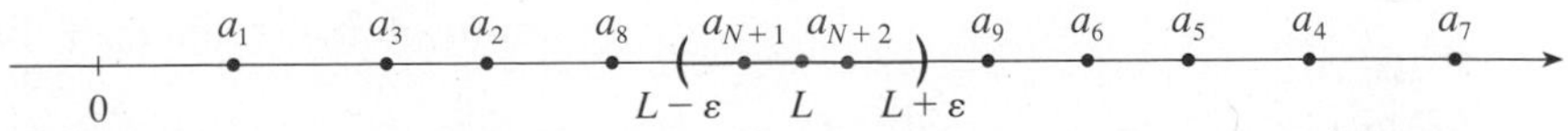

FIGURE 4

Another illustration of Definition 2 is given in Figure 5. The points on the graph of $\{a_n\}$ must lie between the horizontal lines $y = L + \varepsilon$ and $y = L - \varepsilon$ if $n > N$. This picture must be valid no matter how small ε is chosen, but usually a smaller ε requires a larger N.

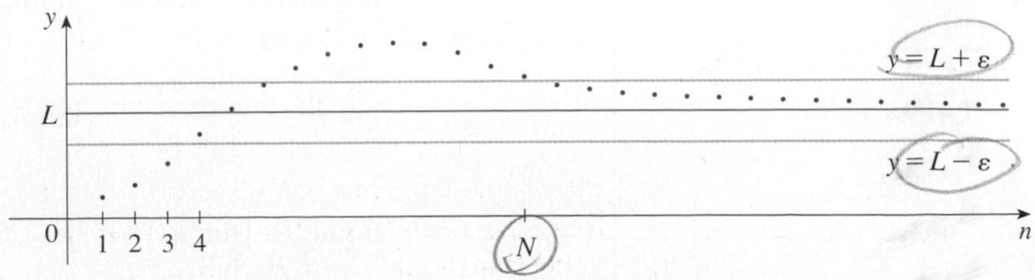

FIGURE 5

If you compare Definition 2 with Definition 2.6.7, you will see that the only difference between $\lim_{n\to\infty} a_n = L$ and $\lim_{x\to\infty} f(x) = L$ is that n is required to be an integer. Thus we have the following theorem, which is illustrated by Figure 6.

3 THEOREM If $\lim_{x\to\infty} f(x) = L$ and $f(n) = a_n$ when n is an integer, then $\lim_{n\to\infty} a_n = L$.

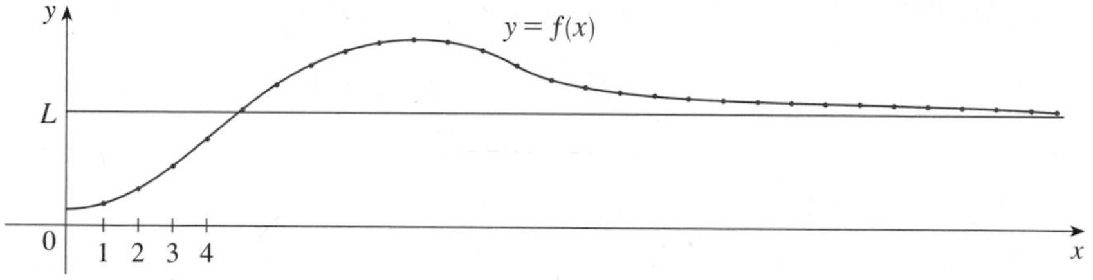

FIGURE 6

In particular, since we know that $\lim_{x\to\infty} (1/x^r) = 0$ when $r > 0$ (Theorem 2.6.5), we have

4
$$\lim_{n\to\infty} \frac{1}{n^r} = 0 \qquad \text{if } r > 0$$

If a_n becomes large as n becomes large, we use the notation $\lim_{n\to\infty} a_n = \infty$. The following precise definition is similar to Definition 2.6.9.

5 DEFINITION $\lim_{n\to\infty} a_n = \infty$ means that for every positive number M there is an integer N such that

$$\text{if} \quad n > N \quad \text{then} \quad a_n > M$$

If $\lim_{n\to\infty} a_n = \infty$, then the sequence $\{a_n\}$ is divergent but in a special way. We say that $\{a_n\}$ diverges to ∞.

The Limit Laws given in Section 2.3 also hold for the limits of sequences and their proofs are similar.

LIMIT LAWS FOR SEQUENCES

If $\{a_n\}$ and $\{b_n\}$ are convergent sequences and c is a constant, then

$$\lim_{n\to\infty} (a_n + b_n) = \lim_{n\to\infty} a_n + \lim_{n\to\infty} b_n$$

$$\lim_{n\to\infty} (a_n - b_n) = \lim_{n\to\infty} a_n - \lim_{n\to\infty} b_n$$

$$\lim_{n\to\infty} ca_n = c \lim_{n\to\infty} a_n \qquad\qquad \lim_{n\to\infty} c = c$$

$$\lim_{n\to\infty} (a_n b_n) = \lim_{n\to\infty} a_n \cdot \lim_{n\to\infty} b_n$$

$$\lim_{n\to\infty} \frac{a_n}{b_n} = \frac{\lim_{n\to\infty} a_n}{\lim_{n\to\infty} b_n} \quad \text{if } \lim_{n\to\infty} b_n \neq 0$$

$$\lim_{n\to\infty} a_n^p = \left[\lim_{n\to\infty} a_n\right]^p \text{ if } p > 0 \text{ and } a_n > 0$$

The Squeeze Theorem can also be adapted for sequences as follows (see Figure 7).

SQUEEZE THEOREM FOR SEQUENCES

If $a_n \leqslant b_n \leqslant c_n$ for $n \geqslant n_0$ and $\lim_{n\to\infty} a_n = \lim_{n\to\infty} c_n = L$, then $\lim_{n\to\infty} b_n = L$.

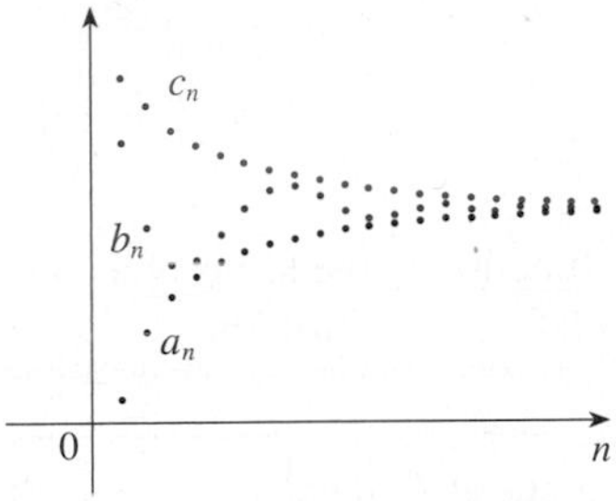

FIGURE 7
The sequence $\{b_n\}$ is squeezed between the sequences $\{a_n\}$ and $\{c_n\}$.

Another useful fact about limits of sequences is given by the following theorem, whose proof is left as Exercise 75.

6 THEOREM If $\lim_{n\to\infty} |a_n| = 0$, then $\lim_{n\to\infty} a_n = 0$.

EXAMPLE 4 Find $\lim_{n\to\infty} \dfrac{n}{n+1}$.

SOLUTION The method is similar to the one we used in Section 2.6: Divide numerator and denominator by the highest power of n and then use the Limit Laws.

$$\lim_{n\to\infty} \frac{n}{n+1} = \lim_{n\to\infty} \frac{1}{1+\dfrac{1}{n}} = \frac{\lim_{n\to\infty} 1}{\lim_{n\to\infty} 1 + \lim_{n\to\infty} \dfrac{1}{n}}$$

$$= \frac{1}{1+0} = 1$$

■ This shows that the guess we made earlier from Figures 1 and 2 was correct.

Here we used Equation 4 with $r = 1$. □

EXAMPLE 5 Calculate $\lim_{n\to\infty} \dfrac{\ln n}{n}$.

SOLUTION Notice that both numerator and denominator approach infinity as $n \to \infty$. We can't apply l'Hospital's Rule directly because it applies not to sequences but to functions of a real variable. However, we can apply l'Hospital's Rule to the related function $f(x) = (\ln x)/x$ and obtain

$$\lim_{x\to\infty} \frac{\ln x}{x} = \lim_{x\to\infty} \frac{1/x}{1} = 0$$

Therefore, by Theorem 3, we have

$$\lim_{n\to\infty} \frac{\ln n}{n} = 0$$ □

EXAMPLE 6 Determine whether the sequence $a_n = (-1)^n$ is convergent or divergent.

SOLUTION If we write out the terms of the sequence, we obtain

$$\{-1, 1, -1, 1, -1, 1, -1, \ldots\}$$

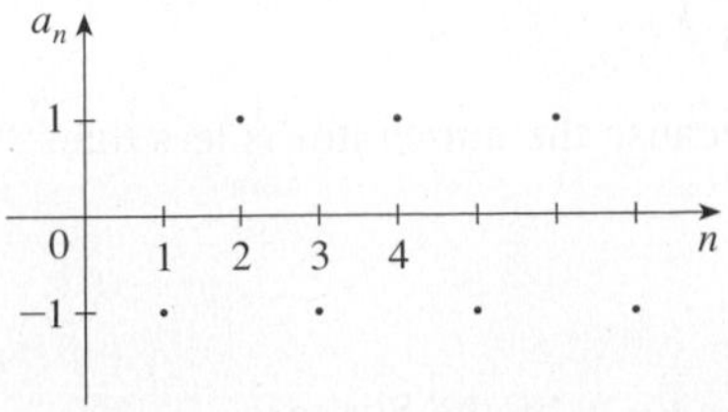

FIGURE 8

The graph of this sequence is shown in Figure 8. Since the terms oscillate between 1 and -1 infinitely often, a_n does not approach any number. Thus $\lim_{n\to\infty} (-1)^n$ does not exist; that is, the sequence $\{(-1)^n\}$ is divergent. □

■ The graph of the sequence in Example 7 is shown in Figure 9 and supports our answer.

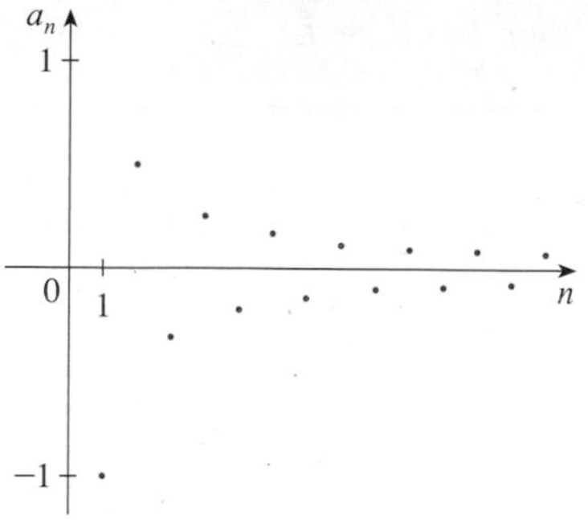

FIGURE 9

EXAMPLE 7 Evaluate $\lim_{n\to\infty} \frac{(-1)^n}{n}$ if it exists.

SOLUTION

$$\lim_{n\to\infty} \left| \frac{(-1)^n}{n} \right| = \lim_{n\to\infty} \frac{1}{n} = 0$$

Therefore, by Theorem 6,

$$\lim_{n\to\infty} \frac{(-1)^n}{n} = 0$$

□

The following theorem says that if we apply a continuous function to the terms of a convergent sequence, the result is also convergent. The proof is left as Exercise 76.

> **7 THEOREM** If $\lim_{n\to\infty} a_n = L$ and the function f is continuous at L, then
>
> $$\lim_{n\to\infty} f(a_n) = f(L)$$

EXAMPLE 8 Find $\lim_{n\to\infty} \sin(\pi/n)$.

SOLUTION Because the sine function is continuous at 0, Theorem 7 enables us to write

$$\lim_{n\to\infty} \sin(\pi/n) = \sin\left(\lim_{n\to\infty} (\pi/n)\right) = \sin 0 = 0$$

□

V EXAMPLE 9 Discuss the convergence of the sequence $a_n = n!/n^n$, where $n! = 1 \cdot 2 \cdot 3 \cdot \cdots \cdot n$.

■ CREATING GRAPHS OF SEQUENCES

Some computer algebra systems have special commands that enable us to create sequences and graph them directly. With most graphing calculators, however, sequences can be graphed by using parametric equations. For instance, the sequence in Example 9 can be graphed by entering the parametric equations

$$x = t \qquad y = t!/t^t$$

and graphing in dot mode, starting with $t = 1$ and setting the t-step equal to 1. The result is shown in Figure 10.

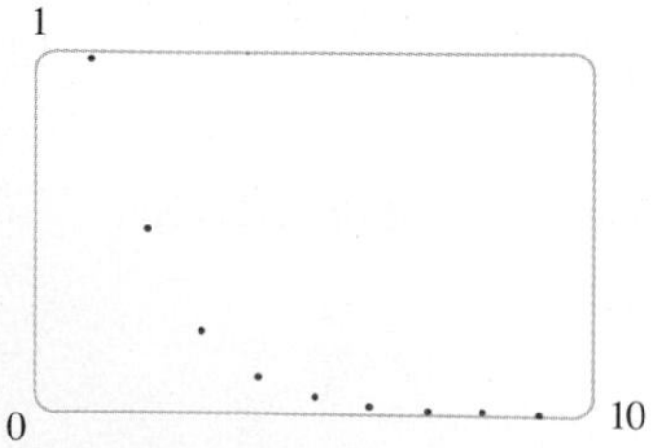

FIGURE 10

SOLUTION Both numerator and denominator approach infinity as $n \to \infty$, but here we have no corresponding function for use with l'Hospital's Rule ($x!$ is not defined when x is not an integer). Let's write out a few terms to get a feeling for what happens to a_n as n gets large:

$$a_1 = 1 \qquad a_2 = \frac{1 \cdot 2}{2 \cdot 2} \qquad a_3 = \frac{1 \cdot 2 \cdot 3}{3 \cdot 3 \cdot 3}$$

8 $$a_n = \frac{1 \cdot 2 \cdot 3 \cdot \cdots \cdot n}{n \cdot n \cdot n \cdot \cdots \cdot n}$$

It appears from these expressions and the graph in Figure 10 that the terms are decreasing and perhaps approach 0. To confirm this, observe from Equation 8 that

$$a_n = \frac{1}{n}\left(\frac{2 \cdot 3 \cdot \cdots \cdot n}{n \cdot n \cdot \cdots \cdot n}\right)$$

Notice that the expression in parentheses is at most 1 because the numerator is less than (or equal to) the denominator. So

$$0 < a_n \leqslant \frac{1}{n}$$

We know that $1/n \to 0$ as $n \to \infty$. Therefore $a_n \to 0$ as $n \to \infty$ by the Squeeze Theorem.

□

EXAMPLE 10 For what values of r is the sequence $\{r^n\}$ convergent?

SOLUTION We know from Section 2.6 and the graphs of the exponential functions in Section 1.5 that $\lim_{x\to\infty} a^x = \infty$ for $a > 1$ and $\lim_{x\to\infty} a^x = 0$ for $0 < a < 1$. Therefore, putting $a = r$ and using Theorem 3, we have

$$\lim_{n\to\infty} r^n = \begin{cases} \infty & \text{if } r > 1 \\ 0 & \text{if } 0 < r < 1 \end{cases}$$

It is obvious that

$$\lim_{n\to\infty} 1^n = 1 \qquad \text{and} \qquad \lim_{n\to\infty} 0^n = 0$$

If $-1 < r < 0$, then $0 < |r| < 1$, so

$$\lim_{n\to\infty} |r^n| = \lim_{n\to\infty} |r|^n = 0$$

and therefore $\lim_{n\to\infty} r^n = 0$ by Theorem 6. If $r \leqslant -1$, then $\{r^n\}$ diverges as in Example 6. Figure 11 shows the graphs for various values of r. (The case $r = -1$ is shown in Figure 8.)

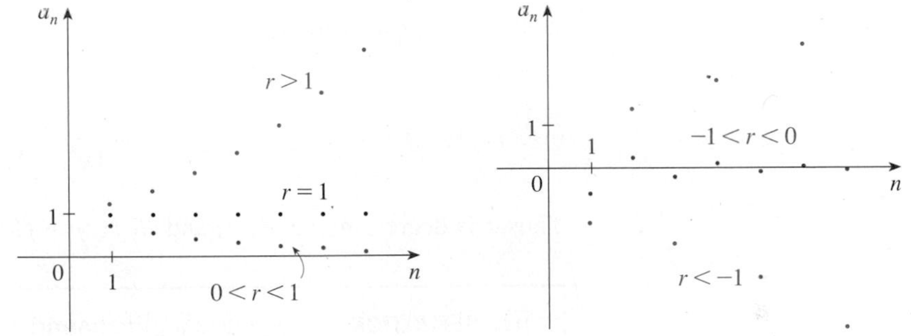

FIGURE 11
The sequence $a_n = r^n$

□

The results of Example 10 are summarized for future use as follows.

9 The sequence $\{r^n\}$ is convergent if $-1 < r \leqslant 1$ and divergent for all other values of r.

$$\lim_{n\to\infty} r^n = \begin{cases} 0 & \text{if } -1 < r < 1 \\ 1 & \text{if } r = 1 \end{cases}$$

10 DEFINITION A sequence $\{a_n\}$ is called **increasing** if $a_n < a_{n+1}$ for all $n \geqslant 1$, that is, $a_1 < a_2 < a_3 < \cdots$. It is called **decreasing** if $a_n > a_{n+1}$ for all $n \geqslant 1$. It is called **monotonic** if it is either increasing or decreasing.

EXAMPLE 11 The sequence $\left\{\dfrac{3}{n+5}\right\}$ is decreasing because

■ The right side is smaller because it has a larger denominator.

$$\frac{3}{n+5} > \frac{3}{(n+1)+5} = \frac{3}{n+6}$$

and so $a_n > a_{n+1}$ for all $n \geqslant 1$. □

EXAMPLE 12 Show that the sequence $a_n = \dfrac{n}{n^2 + 1}$ is decreasing.

SOLUTION 1 We must show that $a_{n+1} < a_n$, that is,

$$\frac{n+1}{(n+1)^2+1} < \frac{n}{n^2+1}$$

This inequality is equivalent to the one we get by cross-multiplication:

$$\begin{aligned} \frac{n+1}{(n+1)^2+1} < \frac{n}{n^2+1} \quad &\iff \quad (n+1)(n^2+1) < n[(n+1)^2+1] \\ &\iff \quad n^3 + n^2 + n + 1 < n^3 + 2n^2 + 2n \\ &\iff \quad 1 < n^2 + n \end{aligned}$$

Since $n \geqslant 1$, we know that the inequality $n^2 + n > 1$ is true. Therefore $a_{n+1} < a_n$ and so $\{a_n\}$ is decreasing.

SOLUTION 2 Consider the function $f(x) = \dfrac{x}{x^2 + 1}$:

$$f'(x) = \frac{x^2 + 1 - 2x^2}{(x^2+1)^2} = \frac{1 - x^2}{(x^2+1)^2} < 0 \qquad \text{whenever } x^2 > 1$$

Thus f is decreasing on $(1, \infty)$ and so $f(n) > f(n + 1)$. Therefore $\{a_n\}$ is decreasing. □

11 DEFINITION A sequence $\{a_n\}$ is **bounded above** if there is a number M such that

$$a_n \leqslant M \qquad \text{for all } n \geqslant 1$$

It is **bounded below** if there is a number m such that

$$m \leqslant a_n \qquad \text{for all } n \geqslant 1$$

If it is bounded above and below, then $\{a_n\}$ is a **bounded sequence**.

For instance, the sequence $a_n = n$ is bounded below $(a_n > 0)$ but not above. The sequence $a_n = n/(n + 1)$ is bounded because $0 < a_n < 1$ for all n.

We know that not every bounded sequence is convergent [for instance, the sequence $a_n = (-1)^n$ satisfies $-1 \leqslant a_n \leqslant 1$ but is divergent from Example 6] and not every monotonic sequence is convergent $(a_n = n \to \infty)$. But if a sequence is both bounded *and* monotonic, then it must be convergent. This fact is proved as Theorem 12, but intuitively you can understand why it is true by looking at Figure 12. If $\{a_n\}$ is increasing and $a_n \leqslant M$ for all n, then the terms are forced to crowd together and approach some number L.

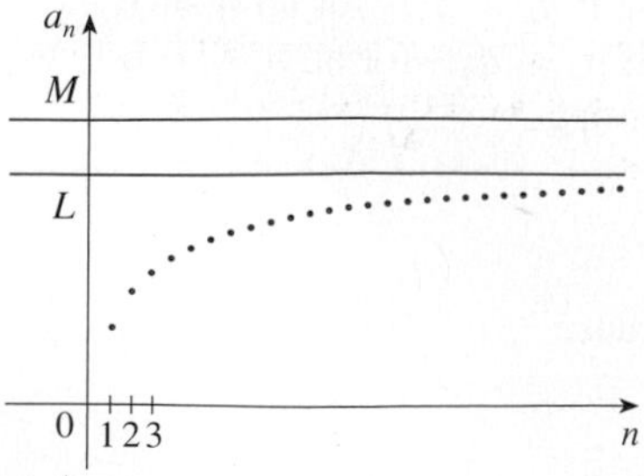

FIGURE 12

The proof of Theorem 12 is based on the **Completeness Axiom** for the set $\mathbb{R}$ of real numbers, which says that if S is a nonempty set of real numbers that has an upper bound M ($x \leqslant M$ for all x in S), then S has a **least upper bound** b. (This means that b is an upper bound for S, but if M is any other upper bound, then $b \leqslant M$.) The Completeness Axiom is an expression of the fact that there is no gap or hole in the real number line.

[12] MONOTONIC SEQUENCE THEOREM Every bounded, monotonic sequence is convergent.

PROOF Suppose $\{a_n\}$ is an increasing sequence. Since $\{a_n\}$ is bounded, the set $S = \{a_n \mid n \geq 1\}$ has an upper bound. By the Completeness Axiom it has a least upper bound L. Given $\varepsilon > 0$, $L - \varepsilon$ is *not* an upper bound for S (since L is the *least* upper bound). Therefore

$$a_N > L - \varepsilon \qquad \text{for some integer } N$$

But the sequence is increasing so $a_n \geq a_N$ for every $n > N$. Thus if $n > N$, we have

$$a_n > L - \varepsilon$$

so

$$0 \leq L - a_n < \varepsilon$$

since $a_n \leq L$. Thus

$$|L - a_n| < \varepsilon \qquad \text{whenever } n > N$$

so $\lim_{n\to\infty} a_n = L$.

A similar proof (using the greatest lower bound) works if $\{a_n\}$ is decreasing. □

The proof of Theorem 12 shows that a sequence that is increasing and bounded above is convergent. (Likewise, a decreasing sequence that is bounded below is convergent.) This fact is used many times in dealing with infinite series.

EXAMPLE 13 Investigate the sequence $\{a_n\}$ defined by the recurrence relation

$$a_1 = 2 \qquad a_{n+1} = \tfrac{1}{2}(a_n + 6) \qquad \text{for } n = 1, 2, 3, \ldots$$

SOLUTION We begin by computing the first several terms:

$$a_1 = 2 \qquad a_2 = \tfrac{1}{2}(2 + 6) = 4 \qquad a_3 = \tfrac{1}{2}(4 + 6) = 5$$

$$a_4 = \tfrac{1}{2}(5 + 6) = 5.5 \qquad a_5 = 5.75 \qquad a_6 = 5.875$$

$$a_7 = 5.9375 \qquad a_8 = 5.96875 \qquad a_9 = 5.984375$$

■ Mathematical induction is often used in dealing with recursive sequences. See page 77 for a discussion of the Principle of Mathematical Induction.

These initial terms suggest that the sequence is increasing and the terms are approaching 6. To confirm that the sequence is increasing, we use mathematical induction to show that $a_{n+1} > a_n$ for all $n \geq 1$. This is true for $n = 1$ because $a_2 = 4 > a_1$. If we assume that it is true for $n = k$, then we have

$$a_{k+1} > a_k$$

so

$$a_{k+1} + 6 > a_k + 6$$

and

$$\tfrac{1}{2}(a_{k+1} + 6) > \tfrac{1}{2}(a_k + 6)$$

Thus

$$a_{k+2} > a_{k+1}$$

We have deduced that $a_{n+1} > a_n$ is true for $n = k + 1$. Therefore the inequality is true for all n by induction.

Next we verify that $\{a_n\}$ is bounded by showing that $a_n < 6$ for all n. (Since the sequence is increasing, we already know that it has a lower bound: $a_n \geqslant a_1 = 2$ for all n.) We know that $a_1 < 6$, so the assertion is true for $n = 1$. Suppose it is true for $n = k$. Then

$$a_k < 6$$

so

$$a_k + 6 < 12$$

$$\tfrac{1}{2}(a_k + 6) < \tfrac{1}{2}(12) = 6$$

Thus

$$a_{k+1} < 6$$

This shows, by mathematical induction, that $a_n < 6$ for all n.

Since the sequence $\{a_n\}$ is increasing and bounded, Theorem 12 guarantees that it has a limit. The theorem doesn't tell us what the value of the limit is. But now that we know $L = \lim_{n\to\infty} a_n$ exists, we can use the recurrence relation to write

$$\lim_{n\to\infty} a_{n+1} = \lim_{n\to\infty} \tfrac{1}{2}(a_n + 6) = \tfrac{1}{2}\left(\lim_{n\to\infty} a_n + 6\right) = \tfrac{1}{2}(L + 6)$$

■ A proof of this fact is requested in Exercise 58.

Since $a_n \to L$, it follows that $a_{n+1} \to L$, too (as $n \to \infty$, $n + 1 \to \infty$ too). So we have

$$L = \tfrac{1}{2}(L + 6)$$

Solving this equation for L, we get $L = 6$, as predicted. □

11.1 EXERCISES

1. (a) What is a sequence?
(b) What does it mean to say that $\lim_{n\to\infty} a_n = 8$?
(c) What does it mean to say that $\lim_{n\to\infty} a_n = \infty$?

2. (a) What is a convergent sequence? Give two examples.
(b) What is a divergent sequence? Give two examples.

3–8 List the first five terms of the sequence.

3. $a_n = 1 - (0.2)^n$

4. $a_n = \dfrac{n + 1}{3n - 1}$

5. $a_n = \dfrac{3(-1)^n}{n!}$

6. $\{2 \cdot 4 \cdot 6 \cdot \cdots \cdot (2n)\}$

7. $a_1 = 3, \quad a_{n+1} = 2a_n - 1$

8. $a_1 = 4, \quad a_{n+1} = \dfrac{a_n}{a_n - 1}$

9–14 Find a formula for the general term a_n of the sequence, assuming that the pattern of the first few terms continues.

9. $\left\{1, \frac{1}{3}, \frac{1}{5}, \frac{1}{7}, \frac{1}{9}, \ldots\right\}$

10. $\left\{1, \frac{1}{3}, \frac{1}{9}, \frac{1}{27}, \frac{1}{81}, \ldots\right\}$

11. $\{2, 7, 12, 17, \ldots\}$

12. $\left\{-\frac{1}{4}, \frac{2}{9}, -\frac{3}{16}, \frac{4}{25}, \ldots\right\}$

13. $\left\{1, -\frac{2}{3}, \frac{4}{9}, -\frac{8}{27}, \ldots\right\}$

14. $\{5, 1, 5, 1, 5, 1, \ldots\}$

15. List the first six terms of the sequence defined by

$$a_n = \frac{n}{2n + 1}$$

Does the sequence appear to have a limit? If so, find it.

16. List the first nine terms of the sequence $\{\cos(n\pi/3)\}$. Does this sequence appear to have a limit? If so, find it. If not, explain why.

17–46 Determine whether the sequence converges or diverges. If it converges, find the limit.

17. $a_n = 1 - (0.2)^n$

18. $a_n = \dfrac{n^3}{n^3 + 1}$

19. $a_n = \dfrac{3 + 5n^2}{n + n^2}$

20. $a_n = \dfrac{n^3}{n + 1}$

21. $a_n = e^{1/n}$

22. $a_n = \dfrac{3^{n+2}}{5^n}$

23. $a_n = \tan\left(\dfrac{2n\pi}{1 + 8n}\right)$

24. $a_n = \sqrt{\dfrac{n + 1}{9n + 1}}$

25. $a_n = \dfrac{(-1)^{n-1}n}{n^2 + 1}$

26. $a_n = \dfrac{(-1)^n n^3}{n^3 + 2n^2 + 1}$

27. $a_n = \cos(n/2)$

28. $a_n = \cos(2/n)$

29. $\left\{\dfrac{(2n - 1)!}{(2n + 1)!}\right\}$

30. $\{\arctan 2n\}$

31. $\left\{\dfrac{e^n + e^{-n}}{e^{2n} - 1}\right\}$

32. $\left\{\dfrac{\ln n}{\ln 2n}\right\}$

33. $\{n^2 e^{-n}\}$

34. $\{n \cos n\pi\}$

35. $a_n = \dfrac{\cos^2 n}{2^n}$

36. $a_n = \ln(n + 1) - \ln n$

37. $a_n = n \sin(1/n)$

38. $a_n = \sqrt[n]{2^{1+3n}}$

39. $a_n = \left(1 + \dfrac{2}{n}\right)^n$

40. $a_n = \dfrac{\sin 2n}{1 + \sqrt{n}}$

41. $a_n = \ln(2n^2 + 1) - \ln(n^2 + 1)$

42. $a_n = \dfrac{(\ln n)^2}{n}$

43. $\{0, 1, 0, 0, 1, 0, 0, 0, 1, \dots\}$

44. $\left\{\frac{1}{1}, \frac{1}{3}, \frac{1}{2}, \frac{1}{4}, \frac{1}{3}, \frac{1}{5}, \frac{1}{4}, \frac{1}{6}, \dots\right\}$

45. $a_n = \dfrac{n!}{2^n}$

46. $a_n = \dfrac{(-3)^n}{n!}$

47–53 Use a graph of the sequence to decide whether the sequence is convergent or divergent. If the sequence is convergent, guess the value of the limit from the graph and then prove your guess. (See the margin note on page 680 for advice on graphing sequences.)

47. $a_n = 1 + (-2/e)^n$

48. $a_n = \sqrt{n}\sin(\pi/\sqrt{n})$

49. $a_n = \sqrt{\dfrac{3 + 2n^2}{8n^2 + n}}$

50. $a_n = \sqrt[n]{3^n + 5^n}$

51. $a_n = \dfrac{n^2 \cos n}{1 + n^2}$

52. $a_n = \dfrac{1 \cdot 3 \cdot 5 \cdot \cdots \cdot (2n - 1)}{n!}$

53. $a_n = \dfrac{1 \cdot 3 \cdot 5 \cdot \cdots \cdot (2n - 1)}{(2n)^n}$

54. (a) Determine whether the sequence defined as follows is convergent or divergent:

$$a_1 = 1 \qquad a_{n+1} = 4 - a_n \qquad \text{for } n \geqslant 1$$

(b) What happens if the first term is $a_1 = 2$?

55. If \$1000 is invested at 6% interest, compounded annually, then after n years the investment is worth $a_n = 1000(1.06)^n$ dollars.
(a) Find the first five terms of the sequence $\{a_n\}$.
(b) Is the sequence convergent or divergent? Explain.

56. Find the first 40 terms of the sequence defined by

$$a_{n+1} = \begin{cases} \frac{1}{2}a_n & \text{if } a_n \text{ is an even number} \\ 3a_n + 1 & \text{if } a_n \text{ is an odd number} \end{cases}$$

and $a_1 = 11$. Do the same if $a_1 = 25$. Make a conjecture about this type of sequence.

57. For what values of r is the sequence $\{nr^n\}$ convergent?

58. (a) If $\{a_n\}$ is convergent, show that

$$\lim_{n\to\infty} a_{n+1} = \lim_{n\to\infty} a_n$$

(b) A sequence $\{a_n\}$ is defined by $a_1 = 1$ and $a_{n+1} = 1/(1 + a_n)$ for $n \geqslant 1$. Assuming that $\{a_n\}$ is convergent, find its limit.

59. Suppose you know that $\{a_n\}$ is a decreasing sequence and all its terms lie between the numbers 5 and 8. Explain why the sequence has a limit. What can you say about the value of the limit?

60–66 Determine whether the sequence is increasing, decreasing, or not monotonic. Is the sequence bounded?

60. $a_n = (-2)^{n+1}$

61. $a_n = \dfrac{1}{2n + 3}$

62. $a_n = \dfrac{2n - 3}{3n + 4}$

63. $a_n = n(-1)^n$

64. $a_n = ne^{-n}$

65. $a_n = \dfrac{n}{n^2 + 1}$

66. $a_n = n + \dfrac{1}{n}$

67. Find the limit of the sequence

$$\left\{\sqrt{2}, \sqrt{2\sqrt{2}}, \sqrt{2\sqrt{2\sqrt{2}}}, \dots\right\}$$

68. A sequence $\{a_n\}$ is given by $a_1 = \sqrt{2}$, $a_{n+1} = \sqrt{2 + a_n}$.
(a) By induction or otherwise, show that $\{a_n\}$ is increasing and bounded above by 3. Apply the Monotonic Sequence Theorem to show that $\lim_{n\to\infty} a_n$ exists.
(b) Find $\lim_{n\to\infty} a_n$.

69. Show that the sequence defined by

$$a_1 = 1 \qquad a_{n+1} = 3 - \frac{1}{a_n}$$

is increasing and $a_n < 3$ for all n. Deduce that $\{a_n\}$ is convergent and find its limit.

70. Show that the sequence defined by

$$a_1 = 2 \qquad a_{n+1} = \frac{1}{3 - a_n}$$

satisfies $0 < a_n \leqslant 2$ and is decreasing. Deduce that the sequence is convergent and find its limit.

71. (a) Fibonacci posed the following problem: Suppose that rabbits live forever and that every month each pair produces a new pair which becomes productive at age 2 months. If we start with one newborn pair, how many pairs of rabbits will we have in the nth month? Show that the answer is f_n, where $\{f_n\}$ is the Fibonacci sequence defined in Example 3(c).
(b) Let $a_n = f_{n+1}/f_n$ and show that $a_{n-1} = 1 + 1/a_{n-2}$. Assuming that $\{a_n\}$ is convergent, find its limit.

72. (a) Let $a_1 = a$, $a_2 = f(a)$, $a_3 = f(a_2) = f(f(a))$, . . . , $a_{n+1} = f(a_n)$, where f is a continuous function. If $\lim_{n\to\infty} a_n = L$, show that $f(L) = L$.
(b) Illustrate part (a) by taking $f(x) = \cos x$, $a = 1$, and estimating the value of L to five decimal places.

73. (a) Use a graph to guess the value of the limit

$$\lim_{n\to\infty} \frac{n^5}{n!}$$

(b) Use a graph of the sequence in part (a) to find the smallest values of N that correspond to $\varepsilon = 0.1$ and $\varepsilon = 0.001$ in Definition 2.

74. Use Definition 2 directly to prove that $\lim_{n\to\infty} r^n = 0$ when $|r| < 1$.

75. Prove Theorem 6.
[*Hint:* Use either Definition 2 or the Squeeze Theorem.]

76. Prove Theorem 7.

77. Prove that if $\lim_{n\to\infty} a_n = 0$ and $\{b_n\}$ is bounded, then $\lim_{n\to\infty} (a_n b_n) = 0$.

78. Let $a_n = \left(1 + \dfrac{1}{n}\right)^n$.
(a) Show that if $0 \leqslant a < b$, then

$$\frac{b^{n+1} - a^{n+1}}{b - a} < (n + 1)b^n$$

(b) Deduce that $b^n[(n + 1)a - nb] < a^{n+1}$.
(c) Use $a = 1 + 1/(n + 1)$ and $b = 1 + 1/n$ in part (b) to show that $\{a_n\}$ is increasing.
(d) Use $a = 1$ and $b = 1 + 1/(2n)$ in part (b) to show that $a_{2n} < 4$.
(e) Use parts (c) and (d) to show that $a_n < 4$ for all n.
(f) Use Theorem 12 to show that $\lim_{n\to\infty} (1 + 1/n)^n$ exists. (The limit is e. See Equation 3.6.6.)

79. Let a and b be positive numbers with $a > b$. Let a_1 be their arithmetic mean and b_1 their geometric mean:

$$a_1 = \frac{a + b}{2} \qquad b_1 = \sqrt{ab}$$

Repeat this process so that, in general,

$$a_{n+1} = \frac{a_n + b_n}{2} \qquad b_{n+1} = \sqrt{a_n b_n}$$

(a) Use mathematical induction to show that

$$a_n > a_{n+1} > b_{n+1} > b_n$$

(b) Deduce that both $\{a_n\}$ and $\{b_n\}$ are convergent.
(c) Show that $\lim_{n\to\infty} a_n = \lim_{n\to\infty} b_n$. Gauss called the common value of these limits the **arithmetic-geometric mean** of the numbers a and b.

80. (a) Show that if $\lim_{n\to\infty} a_{2n} = L$ and $\lim_{n\to\infty} a_{2n+1} = L$, then $\{a_n\}$ is convergent and $\lim_{n\to\infty} a_n = L$.
(b) If $a_1 = 1$ and

$$a_{n+1} = 1 + \frac{1}{1 + a_n}$$

find the first eight terms of the sequence $\{a_n\}$. Then use part (a) to show that $\lim_{n\to\infty} a_n = \sqrt{2}$. This gives the **continued fraction expansion**

$$\sqrt{2} = 1 + \cfrac{1}{2 + \cfrac{1}{2 + \cdots}}$$

81. The size of an undisturbed fish population has been modeled by the formula

$$p_{n+1} = \frac{bp_n}{a + p_n}$$

where p_n is the fish population after n years and a and b are positive constants that depend on the species and its environment. Suppose that the population in year 0 is $p_0 > 0$.
(a) Show that if $\{p_n\}$ is convergent, then the only possible values for its limit are 0 and $b - a$.
(b) Show that $p_{n+1} < (b/a)p_n$.
(c) Use part (b) to show that if $a > b$, then $\lim_{n\to\infty} p_n = 0$; in other words, the population dies out.
(d) Now assume that $a < b$. Show that if $p_0 < b - a$, then $\{p_n\}$ is increasing and $0 < p_n < b - a$. Show also that if $p_0 > b - a$, then $\{p_n\}$ is decreasing and $p_n > b - a$. Deduce that if $a < b$, then $\lim_{n\to\infty} p_n = b - a$.

LABORATORY PROJECT

CAS LOGISTIC SEQUENCES

A sequence that arises in ecology as a model for population growth is defined by the **logistic difference equation**

$$p_{n+1} = kp_n(1 - p_n)$$

where p_n measures the size of the population of the nth generation of a single species. To keep the numbers manageable, p_n is a fraction of the maximal size of the population, so $0 \leqslant p_n \leqslant 1$. Notice that the form of this equation is similar to the logistic differential equation in Section 9.4. The discrete model—with sequences instead of continuous functions—is preferable for modeling insect populations, where mating and death occur in a periodic fashion.

An ecologist is interested in predicting the size of the population as time goes on, and asks these questions: Will it stabilize at a limiting value? Will it change in a cyclical fashion? Or will it exhibit random behavior?

Write a program to compute the first n terms of this sequence starting with an initial population p_0, where $0 < p_0 < 1$. Use this program to do the following.

1. Calculate 20 or 30 terms of the sequence for $p_0 = \frac{1}{2}$ and for two values of k such that $1 < k < 3$. Graph the sequences. Do they appear to converge? Repeat for a different value of p_0 between 0 and 1. Does the limit depend on the choice of p_0? Does it depend on the choice of k?

2. Calculate terms of the sequence for a value of k between 3 and 3.4 and plot them. What do you notice about the behavior of the terms?

3. Experiment with values of k between 3.4 and 3.5. What happens to the terms?

4. For values of k between 3.6 and 4, compute and plot at least 100 terms and comment on the behavior of the sequence. What happens if you change p_0 by 0.001? This type of behavior is called *chaotic* and is exhibited by insect populations under certain conditions.

11.2 SERIES

If we try to add the terms of an infinite sequence $\{a_n\}_{n=1}^{\infty}$ we get an expression of the form

1 $$a_1 + a_2 + a_3 + \cdots + a_n + \cdots$$

which is called an **infinite series** (or just a **series**) and is denoted, for short, by the symbol

$$\sum_{n=1}^{\infty} a_n \quad \text{or} \quad \sum a_n$$

But does it make sense to talk about the sum of infinitely many terms?

It would be impossible to find a finite sum for the series

$$1 + 2 + 3 + 4 + 5 + \cdots + n + \cdots$$

because if we start adding the terms we get the cumulative sums 1, 3, 6, 10, 15, 21, . . . and, after the nth term, we get $n(n + 1)/2$, which becomes very large as n increases.

However, if we start to add the terms of the series

$$\frac{1}{2} + \frac{1}{4} + \frac{1}{8} + \frac{1}{16} + \frac{1}{32} + \frac{1}{64} + \cdots + \frac{1}{2^n} + \cdots$$

n	Sum of first n terms
1	0.50000000
2	0.75000000
3	0.87500000
4	0.93750000
5	0.96875000
6	0.98437500
7	0.99218750
10	0.99902344
15	0.99996948
20	0.99999905
25	0.99999997

we get $\frac{1}{2}, \frac{3}{4}, \frac{7}{8}, \frac{15}{16}, \frac{31}{32}, \frac{63}{64}, \ldots, 1 - 1/2^n, \ldots$. The table shows that as we add more and more terms, these *partial sums* become closer and closer to 1. (See also Figure 11 in *A Preview of Calculus,* page 7.) In fact, by adding sufficiently many terms of the series we can make the partial sums as close as we like to 1. So it seems reasonable to say that the sum of this infinite series is 1 and to write

$$\sum_{n=1}^{\infty} \frac{1}{2^n} = \frac{1}{2} + \frac{1}{4} + \frac{1}{8} + \frac{1}{16} + \cdots + \frac{1}{2^n} + \cdots = 1$$

We use a similar idea to determine whether or not a general series (1) has a sum. We consider the **partial sums**

$$s_1 = a_1$$

$$s_2 = a_1 + a_2$$

$$s_3 = a_1 + a_2 + a_3$$

$$s_4 = a_1 + a_2 + a_3 + a_4$$

and, in general,

$$s_n = a_1 + a_2 + a_3 + \cdots + a_n = \sum_{i=1}^{n} a_i$$

These partial sums form a new sequence $\{s_n\}$, which may or may not have a limit. If $\lim_{n\to\infty} s_n = s$ exists (as a finite number), then, as in the preceding example, we call it the sum of the infinite series $\Sigma\, a_n$.

2 DEFINITION Given a series $\Sigma_{n=1}^{\infty} a_n = a_1 + a_2 + a_3 + \cdots$, let s_n denote its nth partial sum:

$$s_n = \sum_{i=1}^{n} a_i = a_1 + a_2 + \cdots + a_n$$

If the sequence $\{s_n\}$ is convergent and $\lim_{n\to\infty} s_n = s$ exists as a real number, then the series $\Sigma\, a_n$ is called **convergent** and we write

$$a_1 + a_2 + \cdots + a_n + \cdots = s \qquad \text{or} \qquad \sum_{n=1}^{\infty} a_n = s$$

The number s is called the **sum** of the series. Otherwise, the series is called **divergent**.

Thus the sum of a series is the limit of the sequence of partial sums. So when we write $\Sigma_{n=1}^{\infty} a_n = s$, we mean that by adding sufficiently many terms of the series we can get as close as we like to the number s. Notice that

$$\sum_{n=1}^{\infty} a_n = \lim_{n\to\infty} \sum_{i=1}^{n} a_i$$

■ Compare with the improper integral

$$\int_1^{\infty} f(x)\,dx = \lim_{t\to\infty} \int_1^t f(x)\,dx$$

To find this integral, we integrate from 1 to t and then let $t \to \infty$. For a series, we sum from 1 to n and then let $n \to \infty$.

EXAMPLE 1 An important example of an infinite series is the **geometric series**

$$a + ar + ar^2 + ar^3 + \cdots + ar^{n-1} + \cdots = \sum_{n=1}^{\infty} ar^{n-1} \qquad a \neq 0$$

Each term is obtained from the preceding one by multiplying it by the **common ratio** r. (We have already considered the special case where $a = \frac{1}{2}$ and $r = \frac{1}{2}$ on page 687.)

If $r = 1$, then $s_n = a + a + \cdots + a = na \to \pm\infty$. Since $\lim_{n\to\infty} s_n$ doesn't exist, the geometric series diverges in this case.

If $r \neq 1$, we have

■ Figure 1 provides a geometric demonstration of the result in Example 1. If the triangles are constructed as shown and s is the sum of the series, then, by similar triangles,

$$\frac{s}{a} = \frac{a}{a - ar} \quad \text{so} \quad s = \frac{a}{1 - r}$$

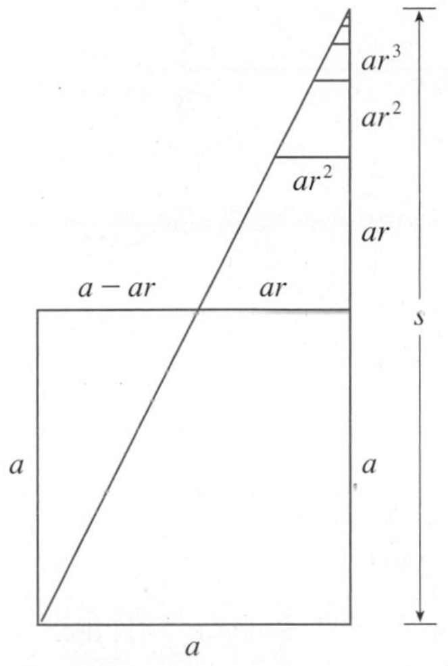

FIGURE 1

$$s_n = a + ar + ar^2 + \cdots + ar^{n-1}$$

and
$$rs_n = \qquad ar + ar^2 + \cdots + ar^{n-1} + ar^n$$

Subtracting these equations, we get

$$s_n - rs_n = a - ar^n$$

3
$$s_n = \frac{a(1 - r^n)}{1 - r}$$

If $-1 < r < 1$, we know from (11.1.9) that $r^n \to 0$ as $n \to \infty$, so

$$\lim_{n\to\infty} s_n = \lim_{n\to\infty} \frac{a(1 - r^n)}{1 - r} = \frac{a}{1 - r} - \frac{a}{1 - r} \lim_{n\to\infty} r^n = \frac{a}{1 - r}$$

Thus when $|r| < 1$ the geometric series is convergent and its sum is $a/(1 - r)$.

If $r \leqslant -1$ or $r > 1$, the sequence $\{r^n\}$ is divergent by (11.1.9) and so, by Equation 3, $\lim_{n\to\infty} s_n$ does not exist. Therefore the geometric series diverges in those cases. □

We summarize the results of Example 1 as follows.

4 The geometric series

$$\sum_{n=1}^{\infty} ar^{n-1} = a + ar + ar^2 + \cdots$$

is convergent if $|r| < 1$ and its sum is

$$\sum_{n=1}^{\infty} ar^{n-1} = \frac{a}{1 - r} \qquad |r| < 1$$

If $|r| \geqslant 1$, the geometric series is divergent.

■ In words: The sum of a convergent geometric series is

$$\frac{\text{first term}}{1 - \text{common ratio}}$$

V EXAMPLE 2 Find the sum of the geometric series

$$5 - \tfrac{10}{3} + \tfrac{20}{9} - \tfrac{40}{27} + \cdots$$

SOLUTION The first term is $a = 5$ and the common ratio is $r = -\frac{2}{3}$. Since $|r| = \frac{2}{3} < 1$, the series is convergent by (4) and its sum is

$$5 - \frac{10}{3} + \frac{20}{9} - \frac{40}{27} + \cdots = \frac{5}{1 - \left(-\frac{2}{3}\right)} = \frac{5}{\frac{5}{3}} = 3 \qquad \square$$

■ What do we really mean when we say that the sum of the series in Example 2 is 3? Of course, we can't literally add an infinite number of terms, one by one. But, according to Definition 2, the total sum is the limit of the sequence of partial sums. So, by taking the sum of sufficiently many terms, we can get as close as we like to the number 3. The table shows the first ten partial sums s_n and the graph in Figure 2 shows how the sequence of partial sums approaches 3.

n	s_n
1	5.000000
2	1.666667
3	3.888889
4	2.407407
5	3.395062
6	2.736626
7	3.175583
8	2.882945
9	3.078037
10	2.947975

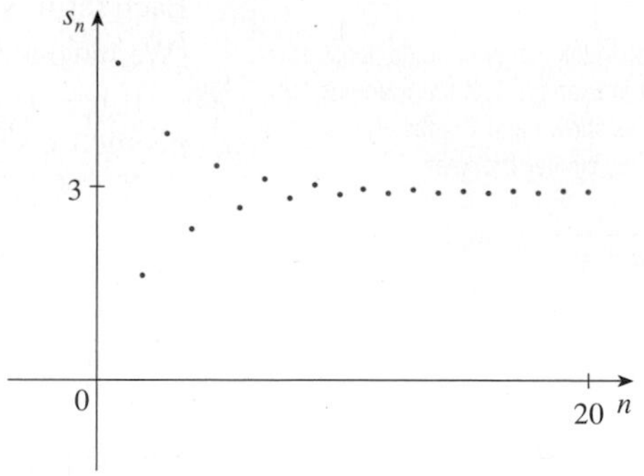

FIGURE 2

EXAMPLE 3 Is the series $\sum_{n=1}^{\infty} 2^{2n}3^{1-n}$ convergent or divergent?

SOLUTION Let's rewrite the nth term of the series in the form ar^{n-1}:

■ Another way to identify a and r is to write out the first few terms:

$$4 + \tfrac{16}{3} + \tfrac{64}{9} + \cdots$$

$$\sum_{n=1}^{\infty} 2^{2n}3^{1-n} = \sum_{n=1}^{\infty} (2^2)^n 3^{-(n-1)} = \sum_{n=1}^{\infty} \frac{4^n}{3^{n-1}} = \sum_{n=1}^{\infty} 4\left(\tfrac{4}{3}\right)^{n-1}$$

We recognize this series as a geometric series with $a = 4$ and $r = \frac{4}{3}$. Since $r > 1$, the series diverges by (4). □

V EXAMPLE 4 Write the number $2.3\overline{17} = 2.3171717\ldots$ as a ratio of integers.

SOLUTION

$$2.3171717\ldots = 2.3 + \frac{17}{10^3} + \frac{17}{10^5} + \frac{17}{10^7} + \cdots$$

After the first term we have a geometric series with $a = 17/10^3$ and $r = 1/10^2$. Therefore

$$2.3\overline{17} = 2.3 + \frac{\dfrac{17}{10^3}}{1 - \dfrac{1}{10^2}} = 2.3 + \frac{\dfrac{17}{1000}}{\dfrac{99}{100}}$$

$$= \frac{23}{10} + \frac{17}{990} = \frac{1147}{495}$$ □

EXAMPLE 5 Find the sum of the series $\sum_{n=0}^{\infty} x^n$, where $|x| < 1$.

SOLUTION Notice that this series starts with $n = 0$ and so the first term is $x^0 = 1$. (With series, we adopt the convention that $x^0 = 1$ even when $x = 0$.) Thus

TEC Module 11.2 explores a series that depends on an angle θ in a triangle and enables you to see how rapidly the series converges when θ varies.

$$\sum_{n=0}^{\infty} x^n = 1 + x + x^2 + x^3 + x^4 + \cdots$$

This is a geometric series with $a = 1$ and $r = x$. Since $|r| = |x| < 1$, it converges and (4) gives

$$\boxed{5} \qquad \sum_{n=0}^{\infty} x^n = \frac{1}{1 - x}$$ □

EXAMPLE 6 Show that the series $\sum_{n=1}^{\infty} \frac{1}{n(n+1)}$ is convergent, and find its sum.

SOLUTION This is not a geometric series, so we go back to the definition of a convergent series and compute the partial sums.

$$s_n = \sum_{i=1}^{n} \frac{1}{i(i+1)} = \frac{1}{1 \cdot 2} + \frac{1}{2 \cdot 3} + \frac{1}{3 \cdot 4} + \cdots + \frac{1}{n(n+1)}$$

We can simplify this expression if we use the partial fraction decomposition

$$\frac{1}{i(i+1)} = \frac{1}{i} - \frac{1}{i+1}$$

(see Section 7.4). Thus we have

$$\begin{aligned} s_n &= \sum_{i=1}^{n} \frac{1}{i(i+1)} = \sum_{i=1}^{n} \left(\frac{1}{i} - \frac{1}{i+1} \right) \\ &= \left(1 - \frac{1}{2}\right) + \left(\frac{1}{2} - \frac{1}{3}\right) + \left(\frac{1}{3} - \frac{1}{4}\right) + \cdots + \left(\frac{1}{n} - \frac{1}{n+1}\right) \\ &= 1 - \frac{1}{n+1} \end{aligned}$$

■ Notice that the terms cancel in pairs. This is an example of a telescoping sum: Because of all the cancellations, the sum collapses (like a pirate's collapsing telescope) into just two terms.

and so

$$\lim_{n \to \infty} s_n = \lim_{n \to \infty} \left(1 - \frac{1}{n+1} \right) = 1 - 0 = 1$$

Therefore the given series is convergent and

$$\sum_{n=1}^{\infty} \frac{1}{n(n+1)} = 1$$ □

■ Figure 3 illustrates Example 6 by showing the graphs of the sequence of terms $a_n = 1/[n(n+1)]$ and the sequence $\{s_n\}$ of partial sums. Notice that $a_n \to 0$ and $s_n \to 1$. See Exercises 62 and 63 for two geometric interpretations of Example 6.

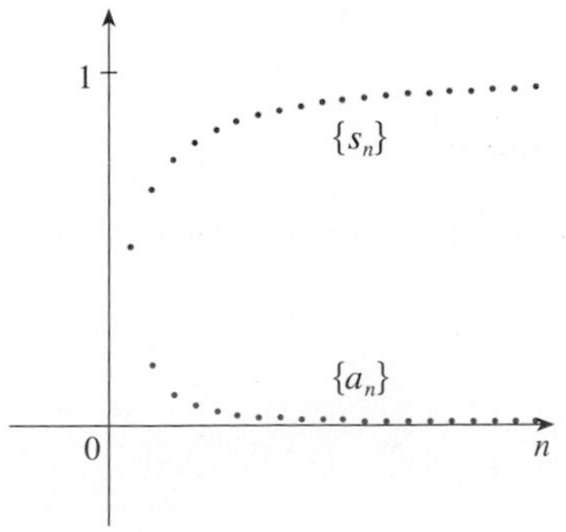

FIGURE 3

V EXAMPLE 7 Show that the **harmonic series**

$$\sum_{n=1}^{\infty} \frac{1}{n} = 1 + \frac{1}{2} + \frac{1}{3} + \frac{1}{4} + \cdots$$

is divergent.

SOLUTION For this particular series it's convenient to consider the partial sums $s_2, s_4, s_8, s_{16}, s_{32}, \ldots$ and show that they become large.

$$s_1 = 1$$

$$s_2 = 1 + \tfrac{1}{2}$$

$$s_4 = 1 + \tfrac{1}{2} + \left(\tfrac{1}{3} + \tfrac{1}{4}\right) > 1 + \tfrac{1}{2} + \left(\tfrac{1}{4} + \tfrac{1}{4}\right) = 1 + \tfrac{2}{2}$$

$$\begin{aligned} s_8 &= 1 + \tfrac{1}{2} + \left(\tfrac{1}{3} + \tfrac{1}{4}\right) + \left(\tfrac{1}{5} + \tfrac{1}{6} + \tfrac{1}{7} + \tfrac{1}{8}\right) \\ &> 1 + \tfrac{1}{2} + \left(\tfrac{1}{4} + \tfrac{1}{4}\right) + \left(\tfrac{1}{8} + \tfrac{1}{8} + \tfrac{1}{8} + \tfrac{1}{8}\right) \\ &= 1 + \tfrac{1}{2} + \tfrac{1}{2} + \tfrac{1}{2} = 1 + \tfrac{3}{2} \end{aligned}$$

$$\begin{aligned} s_{16} &= 1 + \tfrac{1}{2} + \left(\tfrac{1}{3} + \tfrac{1}{4}\right) + \left(\tfrac{1}{5} + \cdots + \tfrac{1}{8}\right) + \left(\tfrac{1}{9} + \cdots + \tfrac{1}{16}\right) \\ &> 1 + \tfrac{1}{2} + \left(\tfrac{1}{4} + \tfrac{1}{4}\right) + \left(\tfrac{1}{8} + \cdots + \tfrac{1}{8}\right) + \left(\tfrac{1}{16} + \cdots + \tfrac{1}{16}\right) \\ &= 1 + \tfrac{1}{2} + \tfrac{1}{2} + \tfrac{1}{2} + \tfrac{1}{2} = 1 + \tfrac{4}{2} \end{aligned}$$

Similarly, $s_{32} > 1 + \frac{5}{2}$, $s_{64} > 1 + \frac{6}{2}$, and in general

$$s_{2^n} > 1 + \frac{n}{2}$$

■ The method used in Example 7 for showing that the harmonic series diverges is due to the French scholar Nicole Oresme (1323–1382).

This shows that $s_{2^n} \to \infty$ as $n \to \infty$ and so $\{s_n\}$ is divergent. Therefore the harmonic series diverges. □

6 THEOREM If the series $\sum_{n=1}^{\infty} a_n$ is convergent, then $\lim_{n\to\infty} a_n = 0$.

PROOF Let $s_n = a_1 + a_2 + \cdots + a_n$. Then $a_n = s_n - s_{n-1}$. Since $\Sigma\, a_n$ is convergent, the sequence $\{s_n\}$ is convergent. Let $\lim_{n\to\infty} s_n = s$. Since $n - 1 \to \infty$ as $n \to \infty$, we also have $\lim_{n\to\infty} s_{n-1} = s$. Therefore

$$\begin{aligned}\lim_{n\to\infty} a_n &= \lim_{n\to\infty} (s_n - s_{n-1}) = \lim_{n\to\infty} s_n - \lim_{n\to\infty} s_{n-1} \\ &= s - s = 0\end{aligned}$$

□

NOTE 1 With any *series* $\Sigma\, a_n$ we associate two *sequences:* the sequence $\{s_n\}$ of its partial sums and the sequence $\{a_n\}$ of its terms. If $\Sigma\, a_n$ is convergent, then the limit of the sequence $\{s_n\}$ is s (the sum of the series) and, as Theorem 6 asserts, the limit of the sequence $\{a_n\}$ is 0.

⊘ NOTE 2 The converse of Theorem 6 is not true in general. If $\lim_{n\to\infty} a_n = 0$, we cannot conclude that $\Sigma\, a_n$ is convergent. Observe that for the harmonic series $\Sigma\, 1/n$ we have $a_n = 1/n \to 0$ as $n \to \infty$, but we showed in Example 7 that $\Sigma\, 1/n$ is divergent.

7 THE TEST FOR DIVERGENCE If $\lim_{n\to\infty} a_n$ does not exist or if $\lim_{n\to\infty} a_n \neq 0$, then the series $\sum_{n=1}^{\infty} a_n$ is divergent.

The Test for Divergence follows from Theorem 6 because, if the series is not divergent, then it is convergent, and so $\lim_{n\to\infty} a_n = 0$.

EXAMPLE 8 Show that the series $\sum_{n=1}^{\infty} \frac{n^2}{5n^2 + 4}$ diverges.

SOLUTION

$$\lim_{n\to\infty} a_n = \lim_{n\to\infty} \frac{n^2}{5n^2 + 4} = \lim_{n\to\infty} \frac{1}{5 + 4/n^2} = \frac{1}{5} \neq 0$$

So the series diverges by the Test for Divergence. □

NOTE 3 If we find that $\lim_{n\to\infty} a_n \neq 0$, we know that $\Sigma\, a_n$ is divergent. If we find that $\lim_{n\to\infty} a_n = 0$, we know *nothing* about the convergence or divergence of $\Sigma\, a_n$. Remember the warning in Note 2: If $\lim_{n\to\infty} a_n = 0$, the series $\Sigma\, a_n$ might converge or it might diverge.

8 THEOREM If Σa_n and Σb_n are convergent series, then so are the series Σca_n (where c is a constant), $\Sigma (a_n + b_n)$, and $\Sigma (a_n - b_n)$, and

(i) $\displaystyle\sum_{n=1}^{\infty} ca_n = c\sum_{n=1}^{\infty} a_n$ (ii) $\displaystyle\sum_{n=1}^{\infty} (a_n + b_n) = \sum_{n=1}^{\infty} a_n + \sum_{n=1}^{\infty} b_n$

(iii) $\displaystyle\sum_{n=1}^{\infty} (a_n - b_n) = \sum_{n=1}^{\infty} a_n - \sum_{n=1}^{\infty} b_n$

These properties of convergent series follow from the corresponding Limit Laws for Sequences in Section 11.1. For instance, here is how part (ii) of Theorem 8 is proved:

Let

$$s_n = \sum_{i=1}^{n} a_i \qquad s = \sum_{n=1}^{\infty} a_n \qquad t_n = \sum_{i=1}^{n} b_i \qquad t = \sum_{n=1}^{\infty} b_n$$

The nth partial sum for the series $\Sigma (a_n + b_n)$ is

$$u_n = \sum_{i=1}^{n} (a_i + b_i)$$

and, using Equation 5.2.10, we have

$$\begin{aligned}\lim_{n\to\infty} u_n = \lim_{n\to\infty} \sum_{i=1}^{n} (a_i + b_i) &= \lim_{n\to\infty} \left(\sum_{i=1}^{n} a_i + \sum_{i=1}^{n} b_i \right) \\ &= \lim_{n\to\infty} \sum_{i=1}^{n} a_i + \lim_{n\to\infty} \sum_{i=1}^{n} b_i = \lim_{n\to\infty} s_n + \lim_{n\to\infty} t_n = s + t\end{aligned}$$

Therefore $\Sigma (a_n + b_n)$ is convergent and its sum is

$$\sum_{n=1}^{\infty} (a_n + b_n) = s + t = \sum_{n=1}^{\infty} a_n + \sum_{n=1}^{\infty} b_n$$ □

EXAMPLE 9 Find the sum of the series $\displaystyle\sum_{n=1}^{\infty} \left(\frac{3}{n(n+1)} + \frac{1}{2^n} \right)$.

SOLUTION The series $\Sigma 1/2^n$ is a geometric series with $a = \frac{1}{2}$ and $r = \frac{1}{2}$, so

$$\sum_{n=1}^{\infty} \frac{1}{2^n} = \frac{\frac{1}{2}}{1 - \frac{1}{2}} = 1$$

In Example 6 we found that

$$\sum_{n=1}^{\infty} \frac{1}{n(n+1)} = 1$$

So, by Theorem 8, the given series is convergent and

$$\sum_{n=1}^{\infty} \left(\frac{3}{n(n+1)} + \frac{1}{2^n} \right) = 3\sum_{n=1}^{\infty} \frac{1}{n(n+1)} + \sum_{n=1}^{\infty} \frac{1}{2^n} = 3 \cdot 1 + 1 = 4$$ □

NOTE 4 A finite number of terms doesn't affect the convergence or divergence of a series. For instance, suppose that we were able to show that the series

$$\sum_{n=4}^{\infty} \frac{n}{n^3 + 1}$$

is convergent. Since

$$\sum_{n=1}^{\infty} \frac{n}{n^3+1} = \frac{1}{2} + \frac{2}{9} + \frac{3}{28} + \sum_{n=4}^{\infty} \frac{n}{n^3+1}$$

it follows that the entire series $\sum_{n=1}^{\infty} n/(n^3+1)$ is convergent. Similarly, if it is known that the series $\sum_{n=N+1}^{\infty} a_n$ converges, then the full series

$$\sum_{n=1}^{\infty} a_n = \sum_{n=1}^{N} a_n + \sum_{n=N+1}^{\infty} a_n$$

is also convergent.

11.2 EXERCISES

1. (a) What is the difference between a sequence and a series?
(b) What is a convergent series? What is a divergent series?

2. Explain what it means to say that $\sum_{n=1}^{\infty} a_n = 5$.

3–8 Find at least 10 partial sums of the series. Graph both the sequence of terms and the sequence of partial sums on the same screen. Does it appear that the series is convergent or divergent? If it is convergent, find the sum. If it is divergent, explain why.

3. $\displaystyle\sum_{n=1}^{\infty} \frac{12}{(-5)^n}$

4. $\displaystyle\sum_{n=1}^{\infty} \frac{2n^2-1}{n^2+1}$

5. $\displaystyle\sum_{n=1}^{\infty} \tan n$

6. $\displaystyle\sum_{n=1}^{\infty} (0.6)^{n-1}$

7. $\displaystyle\sum_{n=1}^{\infty} \left(\frac{1}{\sqrt{n}} - \frac{1}{\sqrt{n+1}}\right)$

8. $\displaystyle\sum_{n=2}^{\infty} \frac{1}{n(n+2)}$

9. Let $a_n = \dfrac{2n}{3n+1}$.
(a) Determine whether $\{a_n\}$ is convergent.
(b) Determine whether $\sum_{n=1}^{\infty} a_n$ is convergent.

10. (a) Explain the difference between

$$\sum_{i=1}^{n} a_i \quad \text{and} \quad \sum_{j=1}^{n} a_j$$

(b) Explain the difference between

$$\sum_{i=1}^{n} a_i \quad \text{and} \quad \sum_{i=1}^{n} a_j$$

11–20 Determine whether the geometric series is convergent or divergent. If it is convergent, find its sum.

11. $3 + 2 + \frac{4}{3} + \frac{8}{9} + \cdots$

12. $\frac{1}{8} - \frac{1}{4} + \frac{1}{2} - 1 + \cdots$

13. $3 - 4 + \frac{16}{3} - \frac{64}{9} + \cdots$

14. $1 + 0.4 + 0.16 + 0.064 + \cdots$

15. $\displaystyle\sum_{n=1}^{\infty} 6(0.9)^{n-1}$

16. $\displaystyle\sum_{n=1}^{\infty} \frac{10^n}{(-9)^{n-1}}$

17. $\displaystyle\sum_{n=1}^{\infty} \frac{(-3)^{n-1}}{4^n}$

18. $\displaystyle\sum_{n=0}^{\infty} \frac{1}{(\sqrt{2})^n}$

19. $\displaystyle\sum_{n=0}^{\infty} \frac{\pi^n}{3^{n+1}}$

20. $\displaystyle\sum_{n=1}^{\infty} \frac{e^n}{3^{n-1}}$

21–34 Determine whether the series is convergent or divergent. If it is convergent, find its sum.

21. $\displaystyle\sum_{n=1}^{\infty} \frac{1}{2n}$

22. $\displaystyle\sum_{n=1}^{\infty} \frac{n+1}{2n-3}$

23. $\displaystyle\sum_{k=2}^{\infty} \frac{k^2}{k^2-1}$

24. $\displaystyle\sum_{k=1}^{\infty} \frac{k(k+2)}{(k+3)^2}$

25. $\displaystyle\sum_{n=1}^{\infty} \frac{1+2^n}{3^n}$

26. $\displaystyle\sum_{n=1}^{\infty} \frac{1+3^n}{2^n}$

27. $\displaystyle\sum_{n=1}^{\infty} \sqrt[n]{2}$

28. $\displaystyle\sum_{n=1}^{\infty} [(0.8)^{n-1} - (0.3)^n]$

29. $\displaystyle\sum_{n=1}^{\infty} \ln\left(\frac{n^2+1}{2n^2+1}\right)$

30. $\displaystyle\sum_{k=1}^{\infty} (\cos 1)^k$

31. $\displaystyle\sum_{n=1}^{\infty} \arctan n$

32. $\displaystyle\sum_{n=1}^{\infty} \left(\frac{3}{5^n} + \frac{2}{n}\right)$

33. $\displaystyle\sum_{n=1}^{\infty} \left(\frac{1}{e^n} + \frac{1}{n(n+1)}\right)$

34. $\displaystyle\sum_{n=1}^{\infty} \frac{e^n}{n^2}$

35–40 Determine whether the series is convergent or divergent by expressing s_n as a telescoping sum (as in Example 6). If it is convergent, find its sum.

35. $\displaystyle\sum_{n=2}^{\infty} \frac{2}{n^2-1}$

36. $\displaystyle\sum_{n=1}^{\infty} \frac{2}{n^2+4n+3}$

37. $\displaystyle\sum_{n=1}^{\infty} \frac{3}{n(n+3)}$

38. $\displaystyle\sum_{n=1}^{\infty} \ln\frac{n}{n+1}$

39. $\sum_{n=1}^{\infty} \left(e^{1/n} - e^{1/(n+1)}\right)$

40. $\sum_{n=1}^{\infty} \left(\cos\frac{1}{n^2} - \cos\frac{1}{(n+1)^2}\right)$

41–46 Express the number as a ratio of integers.

41. $0.\overline{2} = 0.2222\ldots$

42. $0.\overline{73} = 0.73737373\ldots$

43. $3.\overline{417} = 3.417417417\ldots$

44. $6.2\overline{54} = 6.2545454\ldots$

45. $1.53\overline{42}$

46. $7.\overline{12345}$

47–51 Find the values of x for which the series converges. Find the sum of the series for those values of x.

47. $\sum_{n=1}^{\infty} \frac{x^n}{3^n}$

48. $\sum_{n=1}^{\infty} (x-4)^n$

49. $\sum_{n=0}^{\infty} 4^n x^n$

50. $\sum_{n=0}^{\infty} \frac{(x+3)^n}{2^n}$

51. $\sum_{n=0}^{\infty} \frac{\cos^n x}{2^n}$

52. We have seen that the harmonic series is a divergent series whose terms approach 0. Show that

$$\sum_{n=1}^{\infty} \ln\left(1 + \frac{1}{n}\right)$$

is another series with this property.

CAS **53–54** Use the partial fraction command on your CAS to find a convenient expression for the partial sum, and then use this expression to find the sum of the series. Check your answer by using the CAS to sum the series directly.

53. $\sum_{n=1}^{\infty} \frac{3n^2 + 3n + 1}{(n^2+n)^3}$

54. $\sum_{n=2}^{\infty} \frac{1}{n^3 - n}$

55. If the nth partial sum of a series $\sum_{n=1}^{\infty} a_n$ is

$$s_n = \frac{n-1}{n+1}$$

find a_n and $\sum_{n=1}^{\infty} a_n$.

56. If the nth partial sum of a series $\sum_{n=1}^{\infty} a_n$ is $s_n = 3 - n2^{-n}$, find a_n and $\sum_{n=1}^{\infty} a_n$.

57. When money is spent on goods and services, those who receive the money also spend some of it. The people receiving some of the twice-spent money will spend some of that, and so on. Economists call this chain reaction the *multiplier effect*. In a hypothetical isolated community, the local government begins the process by spending D dollars. Suppose that each recipient of spent money spends $100c\%$ and saves $100s\%$ of the money that he or she receives. The values c and s are called the *marginal propensity to consume* and the *marginal propensity to save* and, of course, $c + s = 1$.

(a) Let S_n be the total spending that has been generated after n transactions. Find an equation for S_n.

(b) Show that $\lim_{n\to\infty} S_n = kD$, where $k = 1/s$. The number k is called the *multiplier*. What is the multiplier if the marginal propensity to consume is 80%?

Note: The federal government uses this principle to justify deficit spending. Banks use this principle to justify lending a large percentage of the money that they receive in deposits.

58. A certain ball has the property that each time it falls from a height h onto a hard, level surface, it rebounds to a height rh, where $0 < r < 1$. Suppose that the ball is dropped from an initial height of H meters.

(a) Assuming that the ball continues to bounce indefinitely, find the total distance that it travels. (Use the fact that the ball falls $\frac{1}{2}gt^2$ meters in t seconds.)

(b) Calculate the total time that the ball travels.

(c) Suppose that each time the ball strikes the surface with velocity v it rebounds with velocity $-kv$, where $0 < k < 1$. How long will it take for the ball to come to rest?

59. Find the value of c if

$$\sum_{n=2}^{\infty} (1+c)^{-n} = 2$$

60. Find the value of c such that

$$\sum_{n=0}^{\infty} e^{nc} = 10$$

61. In Example 7 we showed that the harmonic series is divergent. Here we outline another method, making use of the fact that $e^x > 1 + x$ for any $x > 0$. (See Exercise 4.3.76.)

If s_n is the nth partial sum of the harmonic series, show that $e^{s_n} > n + 1$. Why does this imply that the harmonic series is divergent?

62. Graph the curves $y = x^n$, $0 \leqslant x \leqslant 1$, for $n = 0, 1, 2, 3, 4, \ldots$ on a common screen. By finding the areas between successive curves, give a geometric demonstration of the fact, shown in Example 6, that

$$\sum_{n=1}^{\infty} \frac{1}{n(n+1)} = 1$$

63. The figure shows two circles C and D of radius 1 that touch at P. T is a common tangent line; C_1 is the circle that touches C, D, and T; C_2 is the circle that touches C, D, and C_1; C_3 is the circle that touches C, D, and C_2. This procedure can be continued indefinitely and produces an infinite sequence of circles $\{C_n\}$. Find an expression for the diameter of C_n and thus provide another geometric demonstration of Example 6.

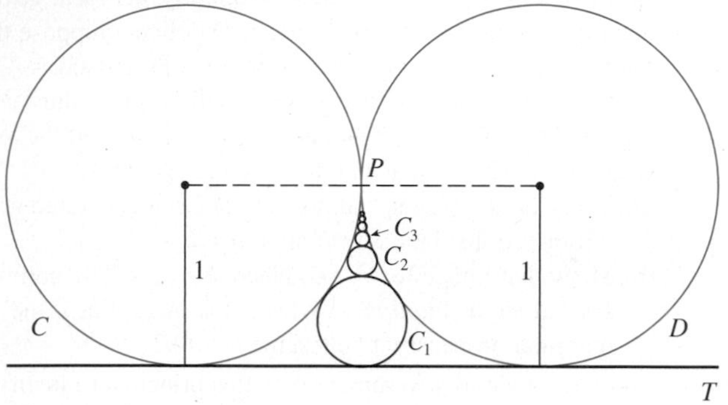

64. A right triangle ABC is given with $\angle A = \theta$ and $|AC| = b$. CD is drawn perpendicular to AB, DE is drawn perpendicular to BC, $EF \perp AB$, and this process is continued indefinitely, as shown in the figure. Find the total length of all the perpendiculars

$$|CD| + |DE| + |EF| + |FG| + \cdots$$

in terms of b and θ.

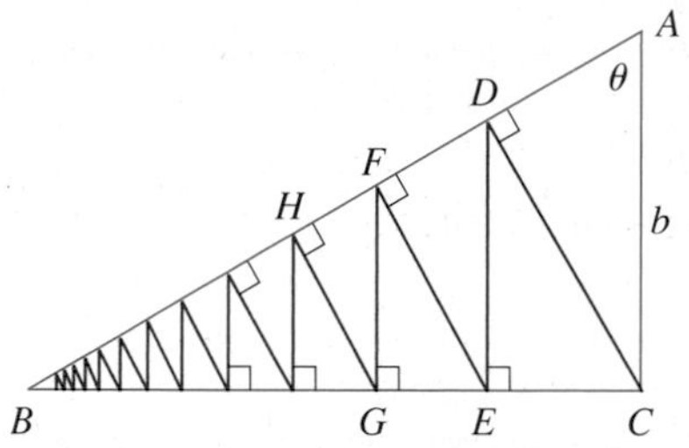

65. What is wrong with the following calculation?

$$\begin{aligned} 0 &= 0 + 0 + 0 + \cdots \\ &= (1 - 1) + (1 - 1) + (1 - 1) + \cdots \\ &= 1 - 1 + 1 - 1 + 1 - 1 + \cdots \\ &= 1 + (-1 + 1) + (-1 + 1) + (-1 + 1) + \cdots \\ &= 1 + 0 + 0 + 0 + \cdots = 1 \end{aligned}$$

(Guido Ubaldus thought that this proved the existence of God because "something has been created out of nothing.")

66. Suppose that $\sum_{n=1}^{\infty} a_n$ ($a_n \neq 0$) is known to be a convergent series. Prove that $\sum_{n=1}^{\infty} 1/a_n$ is a divergent series.

67. Prove part (i) of Theorem 8.

68. If $\Sigma\, a_n$ is divergent and $c \neq 0$, show that $\Sigma\, ca_n$ is divergent.

69. If $\Sigma\, a_n$ is convergent and $\Sigma\, b_n$ is divergent, show that the series $\Sigma\, (a_n + b_n)$ is divergent. [*Hint:* Argue by contradiction.]

70. If $\Sigma\, a_n$ and $\Sigma\, b_n$ are both divergent, is $\Sigma\, (a_n + b_n)$ necessarily divergent?

71. Suppose that a series $\Sigma\, a_n$ has positive terms and its partial sums s_n satisfy the inequality $s_n \leqslant 1000$ for all n. Explain why $\Sigma\, a_n$ must be convergent.

72. The Fibonacci sequence was defined in Section 11.1 by the equations

$$f_1 = 1, \quad f_2 = 1, \quad f_n = f_{n-1} + f_{n-2} \qquad n \geqslant 3$$

Show that each of the following statements is true.

(a) $\dfrac{1}{f_{n-1} f_{n+1}} = \dfrac{1}{f_{n-1} f_n} - \dfrac{1}{f_n f_{n+1}}$

(b) $\displaystyle\sum_{n=2}^{\infty} \frac{1}{f_{n-1} f_{n+1}} = 1$

(c) $\displaystyle\sum_{n=2}^{\infty} \frac{f_n}{f_{n-1} f_{n+1}} = 2$

73. The **Cantor set**, named after the German mathematician Georg Cantor (1845–1918), is constructed as follows. We start with the closed interval [0, 1] and remove the open interval $\left(\frac{1}{3}, \frac{2}{3}\right)$. That leaves the two intervals $\left[0, \frac{1}{3}\right]$ and $\left[\frac{2}{3}, 1\right]$ and we remove the open middle third of each. Four intervals remain and again we remove the open middle third of each of them. We continue this procedure indefinitely, at each step removing the open middle third of every interval that remains from the preceding step. The Cantor set consists of the numbers that remain in [0, 1] after all those intervals have been removed.

(a) Show that the total length of all the intervals that are removed is 1. Despite that, the Cantor set contains infinitely many numbers. Give examples of some numbers in the Cantor set.

(b) The **Sierpinski carpet** is a two-dimensional counterpart of the Cantor set. It is constructed by removing the center one-ninth of a square of side 1, then removing the centers of the eight smaller remaining squares, and so on. (The figure shows the first three steps of the construction.) Show that the sum of the areas of the removed squares is 1. This implies that the Sierpinski carpet has area 0.

74. (a) A sequence $\{a_n\}$ is defined recursively by the equation $a_n = \frac{1}{2}(a_{n-1} + a_{n-2})$ for $n \geqslant 3$, where a_1 and a_2 can be any real numbers. Experiment with various values of a_1 and a_2 and use your calculator to guess the limit of the sequence.
(b) Find $\lim_{n\to\infty} a_n$ in terms of a_1 and a_2 by expressing $a_{n+1} - a_n$ in terms of $a_2 - a_1$ and summing a series.

75. Consider the series

$$\sum_{n=1}^{\infty} \frac{n}{(n+1)!}$$

(a) Find the partial sums s_1, s_2, s_3, and s_4. Do you recognize the denominators? Use the pattern to guess a formula for s_n.
(b) Use mathematical induction to prove your guess.
(c) Show that the given infinite series is convergent, and find its sum.

76. In the figure there are infinitely many circles approaching the vertices of an equilateral triangle, each circle touching other circles and sides of the triangle. If the triangle has sides of length 1, find the total area occupied by the circles.

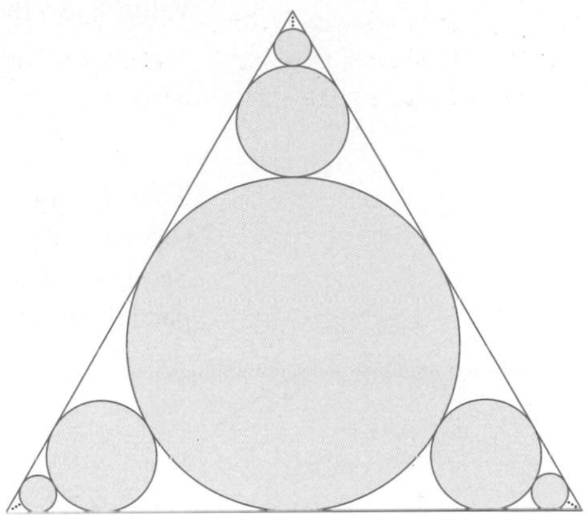

11.3 THE INTEGRAL TEST AND ESTIMATES OF SUMS

In general, it is difficult to find the exact sum of a series. We were able to accomplish this for geometric series and the series $\Sigma\, 1/[n(n+1)]$ because in each of those cases we could find a simple formula for the nth partial sum s_n. But usually it isn't easy to compute $\lim_{n\to\infty} s_n$. Therefore, in the next few sections, we develop several tests that enable us to determine whether a series is convergent or divergent without explicitly finding its sum. (In some cases, however, our methods will enable us to find good estimates of the sum.) Our first test involves improper integrals.

We begin by investigating the series whose terms are the reciprocals of the squares of the positive integers:

$$\sum_{n=1}^{\infty} \frac{1}{n^2} = \frac{1}{1^2} + \frac{1}{2^2} + \frac{1}{3^2} + \frac{1}{4^2} + \frac{1}{5^2} + \cdots$$

n	$s_n = \sum_{i=1}^{n} \frac{1}{i^2}$
5	1.4636
10	1.5498
50	1.6251
100	1.6350
500	1.6429
1000	1.6439
5000	1.6447

There's no simple formula for the sum s_n of the first n terms, but the computer-generated table of values given in the margin suggests that the partial sums are approaching a number near 1.64 as $n \to \infty$ and so it looks as if the series is convergent.

We can confirm this impression with a geometric argument. Figure 1 shows the curve $y = 1/x^2$ and rectangles that lie below the curve. The base of each rectangle is an interval of length 1; the height is equal to the value of the function $y = 1/x^2$ at the right endpoint of the interval. So the sum of the areas of the rectangles is

$$\frac{1}{1^2} + \frac{1}{2^2} + \frac{1}{3^2} + \frac{1}{4^2} + \frac{1}{5^2} + \cdots = \sum_{n=1}^{\infty} \frac{1}{n^2}$$

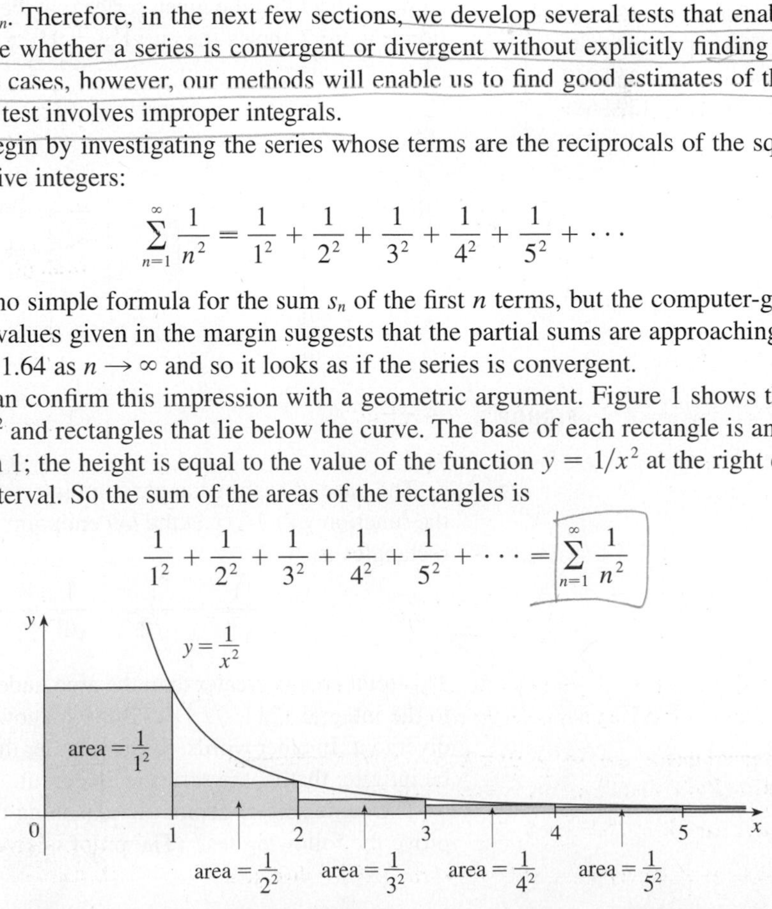

FIGURE 1

If we exclude the first rectangle, the total area of the remaining rectangles is smaller than the area under the curve $y = 1/x^2$ for $x \geqslant 1$, which is the value of the integral $\int_1^\infty (1/x^2)\,dx$. In Section 7.8 we discovered that this improper integral is convergent and has value 1. So the picture shows that all the partial sums are less than

$$\frac{1}{1^2} + \int_1^\infty \frac{1}{x^2}\,dx = 2$$

Thus the partial sums are bounded. We also know that the partial sums are increasing (because all the terms are positive). Therefore the partial sums converge (by the Monotonic Sequence Theorem) and so the series is convergent. The sum of the series (the limit of the partial sums) is also less than 2:

$$\sum_{n=1}^{\infty} \frac{1}{n^2} = \frac{1}{1^2} + \frac{1}{2^2} + \frac{1}{3^2} + \frac{1}{4^2} + \cdots < 2$$

[The exact sum of this series was found by the Swiss mathematician Leonhard Euler (1707–1783) to be $\pi^2/6$, but the proof of this fact is quite difficult. (See Problem 6 in the Problems Plus following Chapter 15.)]

Now let's look at the series

$$\sum_{n=1}^{\infty} \frac{1}{\sqrt{n}} = \frac{1}{\sqrt{1}} + \frac{1}{\sqrt{2}} + \frac{1}{\sqrt{3}} + \frac{1}{\sqrt{4}} + \frac{1}{\sqrt{5}} + \cdots$$

n	$s_n = \sum_{i=1}^{n} \frac{1}{\sqrt{i}}$
5	3.2317
10	5.0210
50	12.7524
100	18.5896
500	43.2834
1000	61.8010
5000	139.9681

The table of values of s_n suggests that the partial sums aren't approaching a finite number, so we suspect that the given series may be divergent. Again we use a picture for confirmation. Figure 2 shows the curve $y = 1/\sqrt{x}$, but this time we use rectangles whose tops lie *above* the curve.

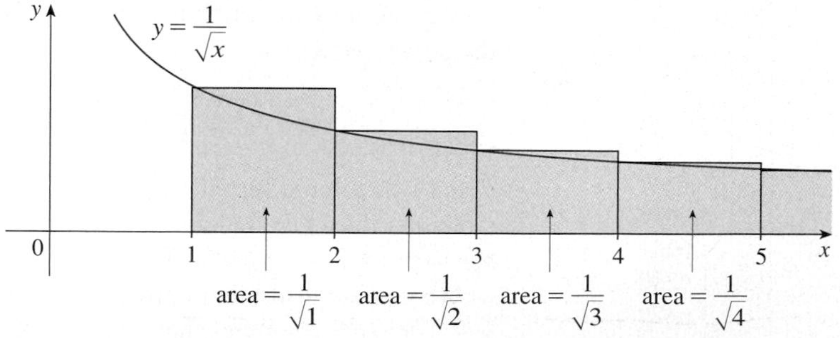

FIGURE 2

The base of each rectangle is an interval of length 1. The height is equal to the value of the function $y = 1/\sqrt{x}$ at the *left* endpoint of the interval. So the sum of the areas of all the rectangles is

$$\frac{1}{\sqrt{1}} + \frac{1}{\sqrt{2}} + \frac{1}{\sqrt{3}} + \frac{1}{\sqrt{4}} + \frac{1}{\sqrt{5}} + \cdots = \sum_{n=1}^{\infty} \frac{1}{\sqrt{n}}$$

This total area is greater than the area under the curve $y = 1/\sqrt{x}$ for $x \geqslant 1$, which is equal to the integral $\int_1^\infty (1/\sqrt{x})\,dx$. But we know from Section 7.8 that this improper integral is divergent. In other words, the area under the curve is infinite. So the sum of the series must be infinite; that is, the series is divergent.

The same sort of geometric reasoning that we used for these two series can be used to prove the following test. (The proof is given at the end of this section.)

THE INTEGRAL TEST Suppose f is a continuous, positive, decreasing function on $[1, \infty)$ and let $a_n = f(n)$. Then the series $\sum_{n=1}^{\infty} a_n$ is convergent if and only if the improper integral $\int_1^{\infty} f(x)\,dx$ is convergent. In other words:

(i) If $\int_1^{\infty} f(x)\,dx$ is convergent, then $\sum_{n=1}^{\infty} a_n$ is convergent.

(ii) If $\int_1^{\infty} f(x)\,dx$ is divergent, then $\sum_{n=1}^{\infty} a_n$ is divergent.

positive
continuous
decreasing

NOTE When we use the Integral Test, it is not necessary to start the series or the integral at $n = 1$. For instance, in testing the series

$$\sum_{n=4}^{\infty} \frac{1}{(n-3)^2} \qquad \text{we use} \qquad \int_4^{\infty} \frac{1}{(x-3)^2}\,dx$$

Also, it is not necessary that f be always decreasing. What is important is that f be *ultimately* decreasing, that is, decreasing for x larger than some number N. Then $\sum_{n=N}^{\infty} a_n$ is convergent, so $\sum_{n=1}^{\infty} a_n$ is convergent by Note 4 of Section 11.2.

EXAMPLE 1 Test the series $\displaystyle\sum_{n=1}^{\infty} \frac{1}{n^2+1}$ for convergence or divergence.

SOLUTION The function $f(x) = 1/(x^2 + 1)$ is continuous, positive, and decreasing on $[1, \infty)$ so we use the Integral Test:

$$\int_1^{\infty} \frac{1}{x^2+1}\,dx = \lim_{t\to\infty} \int_1^t \frac{1}{x^2+1}\,dx = \lim_{t\to\infty} \tan^{-1}x\Big]_1^t$$

$$= \lim_{t\to\infty}\left(\tan^{-1}t - \frac{\pi}{4}\right) = \frac{\pi}{2} - \frac{\pi}{4} = \frac{\pi}{4}$$

Thus $\int_1^{\infty} 1/(x^2+1)\,dx$ is a convergent integral and so, by the Integral Test, the series $\sum 1/(n^2+1)$ is convergent. □

V EXAMPLE 2 For what values of p is the series $\displaystyle\sum_{n=1}^{\infty} \frac{1}{n^p}$ convergent?

■ In order to use the Integral Test we need to be able to evaluate $\int_1^{\infty} f(x)\,dx$ and therefore we have to be able to find an antiderivative of f. Frequently this is difficult or impossible, so we need other tests for convergence too.

SOLUTION If $p < 0$, then $\lim_{n\to\infty}(1/n^p) = \infty$. If $p = 0$, then $\lim_{n\to\infty}(1/n^p) = 1$. In either case, $\lim_{n\to\infty}(1/n^p) \neq 0$, so the given series diverges by the Test for Divergence (11.2.7).

If $p > 0$, then the function $f(x) = 1/x^p$ is clearly continuous, positive, and decreasing on $[1, \infty)$. We found in Chapter 7 [see (7.8.2)] that

$$\int_1^{\infty} \frac{1}{x^p}\,dx \text{ converges if } p > 1 \text{ and diverges if } p \leqslant 1$$

It follows from the Integral Test that the series $\sum 1/n^p$ converges if $p > 1$ and diverges if $0 < p \leqslant 1$. (For $p = 1$, this series is the harmonic series discussed in Example 7 in Section 11.2.) □

The series in Example 2 is called the **p-series**. It is important in the rest of this chapter, so we summarize the results of Example 2 for future reference as follows.

1 The p-series $\sum_{n=1}^{\infty} \frac{1}{n^p}$ is convergent if $p > 1$ and divergent if $p \leq 1$.

EXAMPLE 3
(a) The series

$$\sum_{n=1}^{\infty} \frac{1}{n^3} = \frac{1}{1^3} + \frac{1}{2^3} + \frac{1}{3^3} + \frac{1}{4^3} + \cdots$$

is convergent because it is a p-series with $p = 3 > 1$.
(b) The series

$$\sum_{n=1}^{\infty} \frac{1}{n^{1/3}} = \sum_{n=1}^{\infty} \frac{1}{\sqrt[3]{n}} = 1 + \frac{1}{\sqrt[3]{2}} + \frac{1}{\sqrt[3]{3}} + \frac{1}{\sqrt[3]{4}} + \cdots$$

is divergent because it is a p-series with $p = \frac{1}{3} < 1$. □

NOTE We should *not* infer from the Integral Test that the sum of the series is equal to the value of the integral. In fact,

$$\sum_{n=1}^{\infty} \frac{1}{n^2} = \frac{\pi^2}{6} \qquad \text{whereas} \qquad \int_1^{\infty} \frac{1}{x^2}\,dx = 1$$

Therefore, in general,

$$\sum_{n=1}^{\infty} a_n \neq \int_1^{\infty} f(x)\,dx$$

V EXAMPLE 4 Determine whether the series $\sum_{n=1}^{\infty} \frac{\ln n}{n}$ converges or diverges.

SOLUTION The function $f(x) = (\ln x)/x$ is positive and continuous for $x > 1$ because the logarithm function is continuous. But it is not obvious whether or not f is decreasing, so we compute its derivative:

$$f'(x) = \frac{(1/x)x - \ln x}{x^2} = \frac{1 - \ln x}{x^2}$$

Thus $f'(x) < 0$ when $\ln x > 1$, that is, $x > e$. It follows that f is decreasing when $x > e$ and so we can apply the Integral Test:

$$\int_1^{\infty} \frac{\ln x}{x}\,dx = \lim_{t \to \infty} \int_1^t \frac{\ln x}{x}\,dx = \lim_{t \to \infty} \left. \frac{(\ln x)^2}{2} \right]_1^t = \lim_{t \to \infty} \frac{(\ln t)^2}{2} = \infty$$

Since this improper integral is divergent, the series $\Sigma\,(\ln n)/n$ is also divergent by the Integral Test. □

ESTIMATING THE SUM OF A SERIES

Suppose we have been able to use the Integral Test to show that a series $\Sigma\, a_n$ is convergent and we now want to find an approximation to the sum s of the series. Of course, any partial sum s_n is an approximation to s because $\lim_{n\to\infty} s_n = s$. But how good is such an approximation? To find out, we need to estimate the size of the **remainder**

$$R_n = s - s_n = a_{n+1} + a_{n+2} + a_{n+3} + \cdots$$

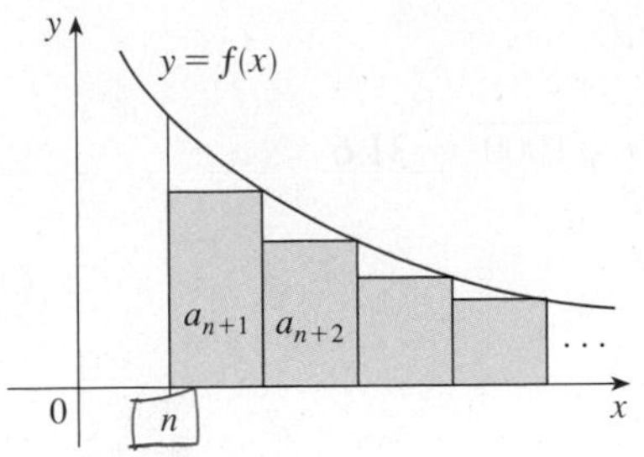

FIGURE 3

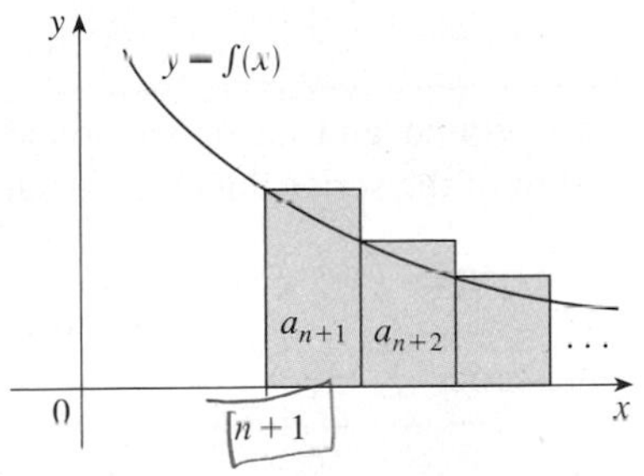

FIGURE 4

The remainder R_n is the error made when s_n, the sum of the first n terms, is used as an approximation to the total sum.

We use the same notation and ideas as in the Integral Test, assuming that f is decreasing on $[n, \infty)$. Comparing the areas of the rectangles with the area under $y = f(x)$ for $x > n$ in Figure 3, we see that

$$R_n = a_{n+1} + a_{n+2} + \cdots \leqslant \int_n^\infty f(x)\,dx$$

Similarly, we see from Figure 4 that

$$R_n = a_{n+1} + a_{n+2} + \cdots \geqslant \int_{n+1}^\infty f(x)\,dx$$

So we have proved the following error estimate.

> **2 REMAINDER ESTIMATE FOR THE INTEGRAL TEST** Suppose $f(k) = a_k$, where f is a continuous, positive, decreasing function for $x \geqslant n$ and $\Sigma\, a_n$ is convergent. If $R_n = s - s_n$, then
>
> $$\int_{n+1}^\infty f(x)\,dx \leqslant R_n \leqslant \int_n^\infty f(x)\,dx$$

V EXAMPLE 5

(a) Approximate the sum of the series $\Sigma\, 1/n^3$ by using the sum of the first 10 terms. Estimate the error involved in this approximation.

(b) How many terms are required to ensure that the sum is accurate to within 0.0005?

SOLUTION In both parts (a) and (b) we need to know $\int_n^\infty f(x)\,dx$. With $f(x) = 1/x^3$, which satisfies the conditions of the Integral Test, we have

$$\int_n^\infty \frac{1}{x^3}\,dx = \lim_{t\to\infty}\left[-\frac{1}{2x^2}\right]_n^t = \lim_{t\to\infty}\left(-\frac{1}{2t^2} + \frac{1}{2n^2}\right) = \frac{1}{2n^2}$$

(a)

$$\sum_{n=1}^\infty \frac{1}{n^3} \approx s_{10} = \frac{1}{1^3} + \frac{1}{2^3} + \frac{1}{3^3} + \cdots + \frac{1}{10^3} \approx 1.1975$$

According to the remainder estimate in (2), we have

$$R_{10} \leqslant \int_{10}^\infty \frac{1}{x^3}\,dx = \frac{1}{2(10)^2} = \frac{1}{200}$$

So the size of the error is at most 0.005.

(b) Accuracy to within 0.0005 means that we have to find a value of n such that $R_n \leqslant 0.0005$. Since

$$R_n \leqslant \int_n^\infty \frac{1}{x^3}\,dx = \frac{1}{2n^2}$$

we want

$$\frac{1}{2n^2} < 0.0005$$

Solving this inequality, we get

$$n^2 > \frac{1}{0.001} = 1000 \qquad \text{or} \qquad n > \sqrt{1000} \approx 31.6$$

We need 32 terms to ensure accuracy to within 0.0005. □

If we add s_n to each side of the inequalities in (2), we get

$$\boxed{3} \qquad s_n + \int_{n+1}^{\infty} f(x)\,dx \leqslant s \leqslant s_n + \int_{n}^{\infty} f(x)\,dx$$

because $s_n + R_n = s$. The inequalities in (3) give a lower bound and an upper bound for s. They provide a more accurate approximation to the sum of the series than the partial sum s_n does.

EXAMPLE 6 Use (3) with $n = 10$ to estimate the sum of the series $\sum_{n=1}^{\infty} \frac{1}{n^3}$.

SOLUTION The inequalities in (3) become

$$s_{10} + \int_{11}^{\infty} \frac{1}{x^3}\,dx \leqslant s \leqslant s_{10} + \int_{10}^{\infty} \frac{1}{x^3}\,dx$$

From Example 5 we know that

$$\int_{n}^{\infty} \frac{1}{x^3}\,dx = \frac{1}{2n^2}$$

so

$$s_{10} + \frac{1}{2(11)^2} \leqslant s \leqslant s_{10} + \frac{1}{2(10)^2}$$

Using $s_{10} \approx 1.197532$, we get

$$1.201664 \leqslant s \leqslant 1.202532$$

If we approximate s by the midpoint of this interval, then the error is at most half the length of the interval. So

$$\sum_{n=1}^{\infty} \frac{1}{n^3} \approx 1.2021 \qquad \text{with error} < 0.0005$$ □

If we compare Example 6 with Example 5, we see that the improved estimate in (3) can be much better than the estimate $s \approx s_n$. To make the error smaller than 0.0005 we had to use 32 terms in Example 5 but only 10 terms in Example 6.

PROOF OF THE INTEGRAL TEST

We have already seen the basic idea behind the proof of the Integral Test in Figures 1 and 2 for the series $\Sigma\ 1/n^2$ and $\Sigma\ 1/\sqrt{n}$. For the general series $\Sigma\ a_n$, look at Figures 5 and 6. The area of the first shaded rectangle in Figure 5 is the value of f at the right endpoint of $[1, 2]$,

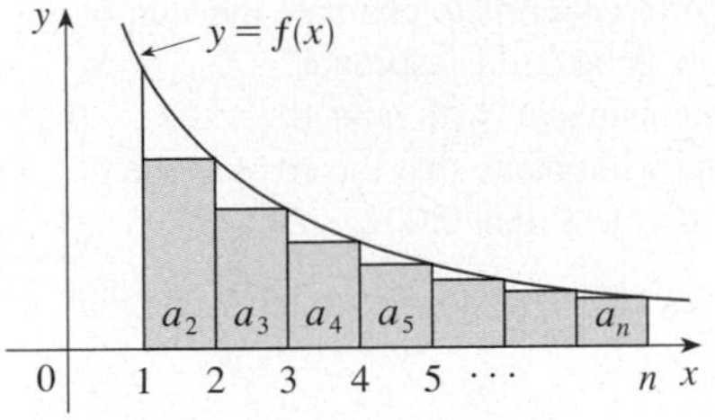

FIGURE 5

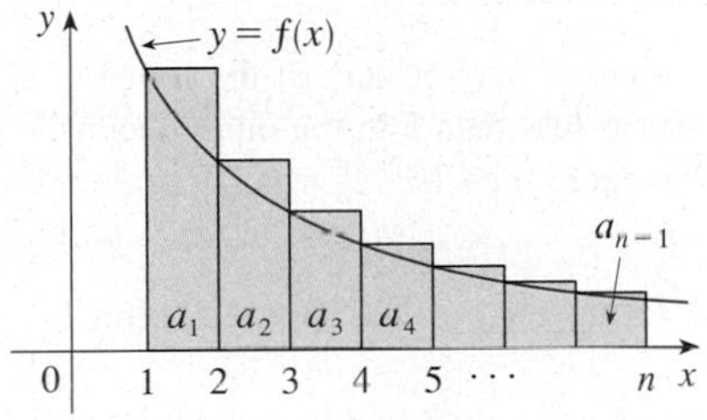

FIGURE 6

that is, $f(2) = a_2$. So, comparing the areas of the shaded rectangles with the area under $y = f(x)$ from 1 to n, we see that

$$\boxed{4} \qquad a_2 + a_3 + \cdots + a_n \leqslant \int_1^n f(x)\,dx$$

(Notice that this inequality depends on the fact that f is decreasing.) Likewise, Figure 6 shows that

$$\boxed{5} \qquad \int_1^n f(x)\,dx \leqslant a_1 + a_2 + \cdots + a_{n-1}$$

(i) If $\int_1^\infty f(x)\,dx$ is convergent, then (4) gives

$$\sum_{i=2}^{n} a_i \leqslant \int_1^n f(x)\,dx \leqslant \int_1^\infty f(x)\,dx$$

since $f(x) \geqslant 0$. Therefore

$$s_n = a_1 + \sum_{i=2}^{n} a_i \leqslant a_1 + \int_1^\infty f(x)\,dx = M, \text{ say}$$

Since $s_n \leqslant M$ for all n, the sequence $\{s_n\}$ is bounded above. Also

$$s_{n+1} = s_n + a_{n+1} \geqslant s_n$$

since $a_{n+1} = f(n+1) \geqslant 0$. Thus $\{s_n\}$ is an increasing bounded sequence and so it is convergent by the Monotonic Sequence Theorem (11.1.12). This means that $\Sigma\, a_n$ is convergent.

(ii) If $\int_1^\infty f(x)\,dx$ is divergent, then $\int_1^n f(x)\,dx \to \infty$ as $n \to \infty$ because $f(x) \geqslant 0$. But (5) gives

$$\int_1^n f(x)\,dx \leqslant \sum_{i=1}^{n-1} a_i = s_{n-1}$$

and so $s_{n-1} \to \infty$. This implies that $s_n \to \infty$ and so $\Sigma\, a_n$ diverges. □

11.3 EXERCISES

1. Draw a picture to show that

$$\sum_{n=2}^{\infty} \frac{1}{n^{1.3}} < \int_1^\infty \frac{1}{x^{1.3}}\,dx$$

What can you conclude about the series?

2. Suppose f is a continuous positive decreasing function for $x \geqslant 1$ and $a_n = f(n)$. By drawing a picture, rank the following three quantities in increasing order:

$$\int_1^6 f(x)\,dx \qquad \sum_{i=1}^{5} a_i \qquad \sum_{i=2}^{6} a_i$$

3–8 Use the Integral Test to determine whether the series is convergent or divergent.

3. $\displaystyle\sum_{n=1}^{\infty} \frac{1}{\sqrt[5]{n}}$

4. $\displaystyle\sum_{n=1}^{\infty} \frac{1}{n^5}$

5. $\displaystyle\sum_{n=1}^{\infty} \frac{1}{(2n+1)^3}$

6. $\displaystyle\sum_{n=1}^{\infty} \frac{1}{\sqrt{n+4}}$

7. $\displaystyle\sum_{n=1}^{\infty} ne^{-n}$

8. $\displaystyle\sum_{n=1}^{\infty} \frac{n+2}{n+1}$

9–26 Determine whether the series is convergent or divergent.

9. $\sum_{n=1}^{\infty} \frac{2}{n^{0.85}}$

10. $\sum_{n=1}^{\infty} (n^{-1.4} + 3n^{-1.2})$

11. $1 + \frac{1}{8} + \frac{1}{27} + \frac{1}{64} + \frac{1}{125} + \cdots$

12. $1 + \frac{1}{2\sqrt{2}} + \frac{1}{3\sqrt{3}} + \frac{1}{4\sqrt{4}} + \frac{1}{5\sqrt{5}} + \cdots$

13. $1 + \frac{1}{3} + \frac{1}{5} + \frac{1}{7} + \frac{1}{9} + \cdots$

14. $\frac{1}{5} + \frac{1}{8} + \frac{1}{11} + \frac{1}{14} + \frac{1}{17} + \cdots$

15. $\sum_{n=1}^{\infty} \frac{5 - 2\sqrt{n}}{n^3}$

16. $\sum_{n=1}^{\infty} \frac{n^2}{n^3 + 1}$

17. $\sum_{n=1}^{\infty} \frac{1}{n^2 + 4}$

18. $\sum_{n=1}^{\infty} \frac{3n + 2}{n(n + 1)}$

19. $\sum_{n=1}^{\infty} \frac{\ln n}{n^3}$

20. $\sum_{n=1}^{\infty} \frac{1}{n^2 - 4n + 5}$

21. $\sum_{n=2}^{\infty} \frac{1}{n \ln n}$

22. $\sum_{n=2}^{\infty} \frac{1}{n(\ln n)^2}$

23. $\sum_{n=1}^{\infty} \frac{e^{1/n}}{n^2}$

24. $\sum_{n=3}^{\infty} \frac{n^2}{e^n}$

25. $\sum_{n=1}^{\infty} \frac{1}{n^3 + n}$

26. $\sum_{n=1}^{\infty} \frac{n}{n^4 + 1}$

27–30 Find the values of p for which the series is convergent.

27. $\sum_{n=2}^{\infty} \frac{1}{n(\ln n)^p}$

28. $\sum_{n=3}^{\infty} \frac{1}{n \ln n\, [\ln(\ln n)]^p}$

29. $\sum_{n=1}^{\infty} n(1 + n^2)^p$

30. $\sum_{n=1}^{\infty} \frac{\ln n}{n^p}$

31. The Riemann zeta-function ζ is defined by

$$\zeta(x) = \sum_{n=1}^{\infty} \frac{1}{n^x}$$

and is used in number theory to study the distribution of prime numbers. What is the domain of ζ?

32. (a) Find the partial sum s_{10} of the series $\sum_{n=1}^{\infty} 1/n^4$. Estimate the error in using s_{10} as an approximation to the sum of the series.
(b) Use (3) with $n = 10$ to give an improved estimate of the sum.
(c) Find a value of n so that s_n is within 0.00001 of the sum.

33. (a) Use the sum of the first 10 terms to estimate the sum of the series $\sum_{n=1}^{\infty} 1/n^2$. How good is this estimate?
(b) Improve this estimate using (3) with $n = 10$.
(c) Find a value of n that will ensure that the error in the approximation $s \approx s_n$ is less than 0.001.

34. Find the sum of the series $\sum_{n=1}^{\infty} 1/n^5$ correct to three decimal places.

35. Estimate $\sum_{n=1}^{\infty} (2n + 1)^{-6}$ correct to five decimal places.

36. How many terms of the series $\sum_{n=2}^{\infty} 1/[n(\ln n)^2]$ would you need to add to find its sum to within 0.01?

37. Show that if we want to approximate the sum of the series $\sum_{n=1}^{\infty} n^{-1.001}$ so that the error is less than 5 in the ninth decimal place, then we need to add more than $10^{11,301}$ terms!

CAS **38.** (a) Show that the series $\sum_{n=1}^{\infty} (\ln n)^2/n^2$ is convergent.
(b) Find an upper bound for the error in the approximation $s \approx s_n$.
(c) What is the smallest value of n such that this upper bound is less than 0.05?
(d) Find s_n for this value of n.

39. (a) Use (4) to show that if s_n is the nth partial sum of the harmonic series, then

$$s_n \leqslant 1 + \ln n$$

(b) The harmonic series diverges, but very slowly. Use part (a) to show that the sum of the first million terms is less than 15 and the sum of the first billion terms is less than 22.

40. Use the following steps to show that the sequence

$$t_n = 1 + \frac{1}{2} + \frac{1}{3} + \cdots + \frac{1}{n} - \ln n$$

has a limit. (The value of the limit is denoted by γ and is called Euler's constant.)
(a) Draw a picture like Figure 6 with $f(x) = 1/x$ and interpret t_n as an area [or use (5)] to show that $t_n > 0$ for all n.
(b) Interpret

$$t_n - t_{n+1} = [\ln(n + 1) - \ln n] - \frac{1}{n + 1}$$

as a difference of areas to show that $t_n - t_{n+1} > 0$. Therefore, $\{t_n\}$ is a decreasing sequence.
(c) Use the Monotonic Sequence Theorem to show that $\{t_n\}$ is convergent.

41. Find all positive values of b for which the series $\sum_{n=1}^{\infty} b^{\ln n}$ converges.

42. Find all values of c for which the following series converges.

$$\sum_{n=1}^{\infty} \left(\frac{c}{n} - \frac{1}{n + 1} \right)$$

11.4 THE COMPARISON TESTS

In the comparison tests the idea is to compare a given series with a series that is known to be convergent or divergent. For instance, the series

$$\boxed{1} \qquad \sum_{n=1}^{\infty} \frac{1}{2^n + 1}$$

reminds us of the series $\sum_{n=1}^{\infty} 1/2^n$, which is a geometric series with $a = \frac{1}{2}$ and $r = \frac{1}{2}$ and is therefore convergent. Because the series (1) is so similar to a convergent series, we have the feeling that it too must be convergent. Indeed, it is. The inequality

$$\frac{1}{2^n + 1} < \frac{1}{2^n}$$

shows that our given series (1) has smaller terms than those of the geometric series and therefore all its partial sums are also smaller than 1 (the sum of the geometric series). This means that its partial sums form a bounded increasing sequence, which is convergent. It also follows that the sum of the series is less than the sum of the geometric series:

$$\sum_{n=1}^{\infty} \frac{1}{2^n + 1} < 1$$

Similar reasoning can be used to prove the following test, which applies only to series whose terms are positive. The first part says that if we have a series whose terms are *smaller* than those of a known *convergent* series, then our series is also convergent. The second part says that if we start with a series whose terms are *larger* than those of a known *divergent* series, then it too is divergent.

THE COMPARISON TEST Suppose that $\sum a_n$ and $\sum b_n$ are series with positive terms.

(i) If $\sum b_n$ is convergent and $a_n \leqslant b_n$ for all n, then $\sum a_n$ is also convergent.

(ii) If $\sum b_n$ is divergent and $a_n \geqslant b_n$ for all n, then $\sum a_n$ is also divergent.

■ It is important to keep in mind the distinction between a sequence and a series. A sequence is a list of numbers, whereas a series is a sum. With every series $\sum a_n$ there are associated two sequences: the sequence $\{a_n\}$ of terms and the sequence $\{s_n\}$ of partial sums.

PROOF

(i) Let $\qquad s_n = \sum_{i=1}^{n} a_i \qquad t_n = \sum_{i=1}^{n} b_i \qquad t = \sum_{n=1}^{\infty} b_n$

Since both series have positive terms, the sequences $\{s_n\}$ and $\{t_n\}$ are increasing $(s_{n+1} = s_n + a_{n+1} \geqslant s_n)$. Also $t_n \to t$, so $t_n \leqslant t$ for all n. Since $a_i \leqslant b_i$, we have $s_n \leqslant t_n$. Thus $s_n \leqslant t$ for all n. This means that $\{s_n\}$ is increasing and bounded above and therefore converges by the Monotonic Sequence Theorem. Thus $\sum a_n$ converges.

(ii) If $\sum b_n$ is divergent, then $t_n \to \infty$ (since $\{t_n\}$ is increasing). But $a_i \geqslant b_i$ so $s_n \geqslant t_n$. Thus $s_n \to \infty$. Therefore $\sum a_n$ diverges. □

In using the Comparison Test we must, of course, have some known series $\sum b_n$ for the purpose of comparison. Most of the time we use one of these series:

Standard Series for Use with the Comparison Test

- A p-series [$\sum 1/n^p$ converges if $p > 1$ and diverges if $p \leqslant 1$; see (11.3.1)]
- A geometric series [$\sum ar^{n-1}$ converges if $|r| < 1$ and diverges if $|r| \geqslant 1$; see (11.2.4)]

EXAMPLE 1 Determine whether the series $\sum_{n=1}^{\infty} \frac{5}{2n^2 + 4n + 3}$ converges or diverges.

SOLUTION For large n the dominant term in the denominator is $2n^2$ so we compare the given series with the series $\Sigma\ 5/(2n^2)$. Observe that

$$\frac{5}{2n^2 + 4n + 3} < \frac{5}{2n^2}$$

because the left side has a bigger denominator. (In the notation of the Comparison Test, a_n is the left side and b_n is the right side.) We know that

$$\sum_{n=1}^{\infty} \frac{5}{2n^2} = \frac{5}{2} \sum_{n=1}^{\infty} \frac{1}{n^2}$$

is convergent because it's a constant times a p-series with $p = 2 > 1$. Therefore

$$\sum_{n=1}^{\infty} \frac{5}{2n^2 + 4n + 3}$$

is convergent by part (i) of the Comparison Test. □

NOTE 1 Although the condition $a_n \leqslant b_n$ or $a_n \geqslant b_n$ in the Comparison Test is given for all n, we need verify only that it holds for $n \geqslant N$, where N is some fixed integer, because the convergence of a series is not affected by a finite number of terms. This is illustrated in the next example.

EXAMPLE 2 Test the series $\sum_{n=1}^{\infty} \frac{\ln n}{n}$ for convergence or divergence.

SOLUTION This series was tested (using the Integral Test) in Example 4 in Section 11.3, but it is also possible to test it by comparing it with the harmonic series. Observe that $\ln n > 1$ for $n \geqslant 3$ and so

$$\frac{\ln n}{n} > \frac{1}{n} \qquad n \geqslant 3$$

We know that $\Sigma\ 1/n$ is divergent (p-series with $p = 1$). Thus the given series is divergent by the Comparison Test. □

NOTE 2 The terms of the series being tested must be smaller than those of a convergent series or larger than those of a divergent series. If the terms are larger than the terms of a convergent series or smaller than those of a divergent series, then the Comparison Test doesn't apply. Consider, for instance, the series

$$\sum_{n=1}^{\infty} \frac{1}{2^n - 1}$$

The inequality

$$\frac{1}{2^n - 1} > \frac{1}{2^n}$$

is useless as far as the Comparison Test is concerned because $\Sigma\ b_n = \Sigma\ (\frac{1}{2})^n$ is convergent and $a_n > b_n$. Nonetheless, we have the feeling that $\Sigma\ 1/(2^n - 1)$ ought to be convergent because it is very similar to the convergent geometric series $\Sigma\ (\frac{1}{2})^n$. In such cases the following test can be used.

■ Exercises 40 and 41 deal with the cases $c = 0$ and $c = \infty$.

THE LIMIT COMPARISON TEST Suppose that $\Sigma\, a_n$ and $\Sigma\, b_n$ are series with positive terms. If

$$\lim_{n\to\infty} \frac{a_n}{b_n} = c$$

where c is a finite number and $c > 0$, then either both series converge or both diverge.

PROOF Let m and M be positive numbers such that $m < c < M$. Because a_n/b_n is close to c for large n, there is an integer N such that

$$m < \frac{a_n}{b_n} < M \qquad \text{when } n > N$$

and so

$$mb_n < a_n < Mb_n \qquad \text{when } n > N$$

If $\Sigma\, b_n$ converges, so does $\Sigma\, Mb_n$. Thus $\Sigma\, a_n$ converges by part (i) of the Comparison Test. If $\Sigma\, b_n$ diverges, so does $\Sigma\, mb_n$ and part (ii) of the Comparison Test shows that $\Sigma\, a_n$ diverges. □

EXAMPLE 3 Test the series $\displaystyle\sum_{n=1}^{\infty} \frac{1}{2^n - 1}$ for convergence or divergence.

SOLUTION We use the Limit Comparison Test with

$$a_n = \frac{1}{2^n - 1} \qquad b_n = \frac{1}{2^n}$$

and obtain

$$\lim_{n\to\infty} \frac{a_n}{b_n} = \lim_{n\to\infty} \frac{1/(2^n - 1)}{1/2^n} = \lim_{n\to\infty} \frac{2^n}{2^n - 1} = \lim_{n\to\infty} \frac{1}{1 - 1/2^n} = 1 > 0$$

Since this limit exists and $\Sigma\, 1/2^n$ is a convergent geometric series, the given series converges by the Limit Comparison Test. □

EXAMPLE 4 Determine whether the series $\displaystyle\sum_{n=1}^{\infty} \frac{2n^2 + 3n}{\sqrt{5 + n^5}}$ converges or diverges.

SOLUTION The dominant part of the numerator is $2n^2$ and the dominant part of the denominator is $\sqrt{n^5} = n^{5/2}$. This suggests taking

$$a_n = \frac{2n^2 + 3n}{\sqrt{5 + n^5}} \qquad b_n = \frac{2n^2}{n^{5/2}} = \frac{2}{n^{1/2}}$$

$$\lim_{n\to\infty} \frac{a_n}{b_n} = \lim_{n\to\infty} \frac{2n^2 + 3n}{\sqrt{5 + n^5}} \cdot \frac{n^{1/2}}{2} = \lim_{n\to\infty} \frac{2n^{5/2} + 3n^{3/2}}{2\sqrt{5 + n^5}}$$

$$= \lim_{n\to\infty} \frac{2 + \dfrac{3}{n}}{2\sqrt{\dfrac{5}{n^5} + 1}} = \frac{2 + 0}{2\sqrt{0 + 1}} = 1$$

Since $\Sigma\, b_n = 2\,\Sigma\, 1/n^{1/2}$ is divergent (p-series with $p = \frac{1}{2} < 1$), the given series diverges by the Limit Comparison Test. □

Notice that in testing many series we find a suitable comparison series $\Sigma\, b_n$ by keeping only the highest powers in the numerator and denominator.

ESTIMATING SUMS

If we have used the Comparison Test to show that a series $\Sigma\, a_n$ converges by comparison with a series $\Sigma\, b_n$, then we may be able to estimate the sum $\Sigma\, a_n$ by comparing remainders. As in Section 11.3, we consider the remainder

$$R_n = s - s_n = a_{n+1} + a_{n+2} + \cdots$$

For the comparison series $\Sigma\, b_n$ we consider the corresponding remainder

$$T_n = t - t_n = b_{n+1} + b_{n+2} + \cdots$$

Since $a_n \leqslant b_n$ for all n, we have $R_n \leqslant T_n$. If $\Sigma\, b_n$ is a p-series, we can estimate its remainder T_n as in Section 11.3. If $\Sigma\, b_n$ is a geometric series, then T_n is the sum of a geometric series and we can sum it exactly (see Exercises 35 and 36). In either case we know that R_n is smaller than T_n.

V EXAMPLE 5 Use the sum of the first 100 terms to approximate the sum of the series $\Sigma\, 1/(n^3 + 1)$. Estimate the error involved in this approximation.

SOLUTION Since

$$\frac{1}{n^3 + 1} < \frac{1}{n^3}$$

the given series is convergent by the Comparison Test. The remainder T_n for the comparison series $\Sigma\, 1/n^3$ was estimated in Example 5 in Section 11.3 using the Remainder Estimate for the Integral Test. There we found that

$$T_n \leqslant \int_n^\infty \frac{1}{x^3}\, dx = \frac{1}{2n^2}$$

Therefore the remainder R_n for the given series satisfies

$$R_n \leqslant T_n \leqslant \frac{1}{2n^2}$$

With $n = 100$ we have

$$R_{100} \leqslant \frac{1}{2(100)^2} = 0.00005$$

Using a programmable calculator or a computer, we find that

$$\sum_{n=1}^{\infty} \frac{1}{n^3 + 1} \approx \sum_{n=1}^{100} \frac{1}{n^3 + 1} \approx 0.6864538$$

with error less than 0.00005. □

11.4 EXERCISES

1. Suppose Σa_n and Σb_n are series with positive terms and Σb_n is known to be convergent.
   (a) If $a_n > b_n$ for all n, what can you say about Σa_n? Why?
   (b) If $a_n < b_n$ for all n, what can you say about Σa_n? Why?

2. Suppose Σa_n and Σb_n are series with positive terms and Σb_n is known to be divergent.
   (a) If $a_n > b_n$ for all n, what can you say about Σa_n? Why?
   (b) If $a_n < b_n$ for all n, what can you say about Σa_n? Why?

3–32 Determine whether the series converges or diverges.

3. $\displaystyle\sum_{n=1}^{\infty} \frac{n}{2n^3 + 1}$

4. $\displaystyle\sum_{n=2}^{\infty} \frac{n^3}{n^4 - 1}$

5. $\displaystyle\sum_{n=1}^{\infty} \frac{n + 1}{n\sqrt{n}}$

6. $\displaystyle\sum_{n=1}^{\infty} \frac{n - 1}{n^2\sqrt{n}}$

7. $\displaystyle\sum_{n=1}^{\infty} \frac{9^n}{3 + 10^n}$

8. $\displaystyle\sum_{n=1}^{\infty} \frac{4 + 3^n}{2^n}$

9. $\displaystyle\sum_{n=1}^{\infty} \frac{\cos^2 n}{n^2 + 1}$

10. $\displaystyle\sum_{n=1}^{\infty} \frac{n^2 - 1}{3n^4 + 1}$

11. $\displaystyle\sum_{n=1}^{\infty} \frac{n - 1}{n4^n}$

12. $\displaystyle\sum_{n=0}^{\infty} \frac{1 + \sin n}{10^n}$

13. $\displaystyle\sum_{n=1}^{\infty} \frac{\arctan n}{n^{1.2}}$

14. $\displaystyle\sum_{n=2}^{\infty} \frac{\sqrt{n}}{n - 1}$

15. $\displaystyle\sum_{n=1}^{\infty} \frac{2 + (-1)^n}{n\sqrt{n}}$

16. $\displaystyle\sum_{n=1}^{\infty} \frac{1}{\sqrt{n^3 + 1}}$

17. $\displaystyle\sum_{n=1}^{\infty} \frac{1}{\sqrt{n^2 + 1}}$

18. $\displaystyle\sum_{n=1}^{\infty} \frac{1}{2n + 3}$

19. $\displaystyle\sum_{n=1}^{\infty} \frac{1 + 4^n}{1 + 3^n}$

20. $\displaystyle\sum_{n=1}^{\infty} \frac{n + 4^n}{n + 6^n}$

21. $\displaystyle\sum_{n=1}^{\infty} \frac{\sqrt{n + 2}}{2n^2 + n + 1}$

22. $\displaystyle\sum_{n=3}^{\infty} \frac{n + 2}{(n + 1)^3}$

23. $\displaystyle\sum_{n=1}^{\infty} \frac{5 + 2n}{(1 + n^2)^2}$

24. $\displaystyle\sum_{n=1}^{\infty} \frac{n^2 - 5n}{n^3 + n + 1}$

25. $\displaystyle\sum_{n=1}^{\infty} \frac{1 + n + n^2}{\sqrt{1 + n^2 + n^6}}$

26. $\displaystyle\sum_{n=1}^{\infty} \frac{n + 5}{\sqrt[3]{n^7 + n^2}}$

27. $\displaystyle\sum_{n=1}^{\infty} \left(1 + \frac{1}{n}\right)^2 e^{-n}$

28. $\displaystyle\sum_{n=1}^{\infty} \frac{e^{1/n}}{n}$

29. $\displaystyle\sum_{n=1}^{\infty} \frac{1}{n!}$

30. $\displaystyle\sum_{n=1}^{\infty} \frac{n!}{n^n}$

31. $\displaystyle\sum_{n=1}^{\infty} \sin\left(\frac{1}{n}\right)$

32. $\displaystyle\sum_{n=1}^{\infty} \frac{1}{n^{1+1/n}}$

33–36 Use the sum of the first 10 terms to approximate the sum of the series. Estimate the error.

33. $\displaystyle\sum_{n=1}^{\infty} \frac{1}{\sqrt{n^4 + 1}}$

34. $\displaystyle\sum_{n=1}^{\infty} \frac{\sin^2 n}{n^3}$

35. $\displaystyle\sum_{n=1}^{\infty} \frac{1}{1 + 2^n}$

36. $\displaystyle\sum_{n=1}^{\infty} \frac{n}{(n + 1)3^n}$

37. The meaning of the decimal representation of a number $0.d_1d_2d_3\ldots$ (where the digit d_i is one of the numbers 0, 1, 2, . . . , 9) is that

$$0.d_1d_2d_3d_4\ldots = \frac{d_1}{10} + \frac{d_2}{10^2} + \frac{d_3}{10^3} + \frac{d_4}{10^4} + \cdots$$

Show that this series always converges.

38. For what values of p does the series $\Sigma_{n=2}^{\infty} 1/(n^p \ln n)$ converge?

39. Prove that if $a_n \geqslant 0$ and Σa_n converges, then Σa_n^2 also converges.

40. (a) Suppose that Σa_n and Σb_n are series with positive terms and Σb_n is convergent. Prove that if

$$\lim_{n\to\infty} \frac{a_n}{b_n} = 0$$

then Σa_n is also convergent.
   (b) Use part (a) to show that the series converges.
   (i) $\displaystyle\sum_{n=1}^{\infty} \frac{\ln n}{n^3}$ (ii) $\displaystyle\sum_{n=1}^{\infty} \frac{\ln n}{\sqrt{n}\, e^n}$

41. (a) Suppose that Σa_n and Σb_n are series with positive terms and Σb_n is divergent. Prove that if

$$\lim_{n\to\infty} \frac{a_n}{b_n} = \infty$$

then Σa_n is also divergent.
   (b) Use part (a) to show that the series diverges.
   (i) $\displaystyle\sum_{n=2}^{\infty} \frac{1}{\ln n}$ (ii) $\displaystyle\sum_{n=1}^{\infty} \frac{\ln n}{n}$

42. Give an example of a pair of series Σa_n and Σb_n with positive terms where $\lim_{n\to\infty} (a_n/b_n) = 0$ and Σb_n diverges, but Σa_n converges. (Compare with Exercise 40.)

43. Show that if $a_n > 0$ and $\lim_{n\to\infty} na_n \neq 0$, then Σa_n is divergent.

44. Show that if $a_n > 0$ and Σa_n is convergent, then $\Sigma \ln(1 + a_n)$ is convergent.

45. If Σa_n is a convergent series with positive terms, is it true that $\Sigma \sin(a_n)$ is also convergent?

46. If Σa_n and Σb_n are both convergent series with positive terms, is it true that $\Sigma a_n b_n$ is also convergent?

11.5 ALTERNATING SERIES

The convergence tests that we have looked at so far apply only to series with positive terms. In this section and the next we learn how to deal with series whose terms are not necessarily positive. Of particular importance are *alternating series,* whose terms alternate in sign.

An **alternating series** is a series whose terms are alternately positive and negative. Here are two examples:

$$1 - \frac{1}{2} + \frac{1}{3} - \frac{1}{4} + \frac{1}{5} - \frac{1}{6} + \cdots = \sum_{n=1}^{\infty} \frac{(-1)^{n-1}}{n}$$

$$-\frac{1}{2} + \frac{2}{3} - \frac{3}{4} + \frac{4}{5} - \frac{5}{6} + \frac{6}{7} - \cdots = \sum_{n=1}^{\infty} (-1)^n \frac{n}{n+1}$$

We see from these examples that the nth term of an alternating series is of the form

$$a_n = (-1)^{n-1}b_n \qquad \text{or} \qquad a_n = (-1)^n b_n$$

where b_n is a positive number. (In fact, $b_n = |a_n|$.)

The following test says that if the terms of an alternating series decrease toward 0 in absolute value, then the series converges.

THE ALTERNATING SERIES TEST If the alternating series

$$\sum_{n=1}^{\infty} (-1)^{n-1}b_n = b_1 - b_2 + b_3 - b_4 + b_5 - b_6 + \cdots \qquad b_n > 0$$

satisfies

$$\text{(i)} \quad b_{n+1} \leqslant b_n \qquad \text{for all } n$$

$$\text{(ii)} \quad \lim_{n\to\infty} b_n = 0$$

then the series is convergent.

Before giving the proof let's look at Figure 1, which gives a picture of the idea behind the proof. We first plot $s_1 = b_1$ on a number line. To find s_2 we subtract b_2, so s_2 is to the left of s_1. Then to find s_3 we add b_3, so s_3 is to the right of s_2. But, since $b_3 < b_2$, s_3 is to the left of s_1. Continuing in this manner, we see that the partial sums oscillate back and forth. Since $b_n \to 0$, the successive steps are becoming smaller and smaller. The even partial sums $s_2, s_4, s_6, \ldots$ are increasing and the odd partial sums $s_1, s_3, s_5, \ldots$ are decreasing. Thus it seems plausible that both are converging to some number s, which is the sum of the series. Therefore we consider the even and odd partial sums separately in the following proof.

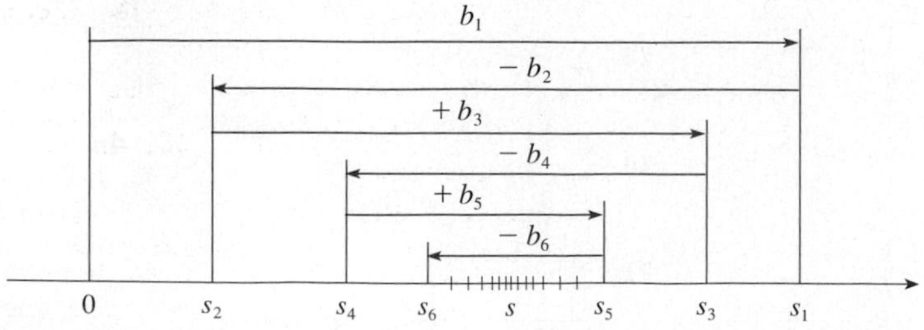

FIGURE 1

PROOF OF THE ALTERNATING SERIES TEST We first consider the even partial sums:

$$s_2 = b_1 - b_2 \geqslant 0 \qquad \text{since } b_2 \leqslant b_1$$

$$s_4 = s_2 + (b_3 - b_4) \geqslant s_2 \qquad \text{since } b_4 \leqslant b_3$$

In general $$s_{2n} = s_{2n-2} + (b_{2n-1} - b_{2n}) \geqslant s_{2n-2} \qquad \text{since } b_{2n} \leqslant b_{2n-1}$$

Thus $$0 \leqslant s_2 \leqslant s_4 \leqslant s_6 \leqslant \cdots \leqslant s_{2n} \leqslant \cdots$$

But we can also write

$$s_{2n} = b_1 - (b_2 - b_3) - (b_4 - b_5) - \cdots - (b_{2n-2} - b_{2n-1}) - b_{2n}$$

Every term in brackets is positive, so $s_{2n} \leqslant b_1$ for all n. Therefore the sequence $\{s_{2n}\}$ of even partial sums is increasing and bounded above. It is therefore convergent by the Monotonic Sequence Theorem. Let's call its limit s, that is,

$$\lim_{n\to\infty} s_{2n} = s$$

Now we compute the limit of the odd partial sums:

$$\begin{aligned} \lim_{n\to\infty} s_{2n+1} &= \lim_{n\to\infty} (s_{2n} + b_{2n+1}) \\ &= \lim_{n\to\infty} s_{2n} + \lim_{n\to\infty} b_{2n+1} \\ &= s + 0 \qquad \text{[by condition (ii)]} \\ &= s \end{aligned}$$

Since both the even and odd partial sums converge to s, we have $\lim_{n\to\infty} s_n = s$ [see Exercise 80(a) in Section 11.1] and so the series is convergent. □

■ Figure 2 illustrates Example 1 by showing the graphs of the terms $a_n = (-1)^{n-1}/n$ and the partial sums s_n. Notice how the values of s_n zigzag across the limiting value, which appears to be about 0.7. In fact, it can be proved that the exact sum of the series is $\ln 2 \approx 0.693$ (see Exercise 36).

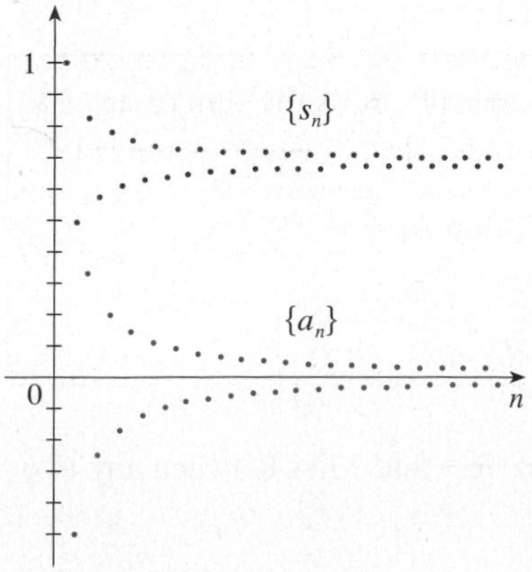

FIGURE 2

V EXAMPLE 1 The alternating harmonic series

$$1 - \frac{1}{2} + \frac{1}{3} - \frac{1}{4} + \cdots = \sum_{n=1}^{\infty} \frac{(-1)^{n-1}}{n}$$

satisfies

$$\text{(i) } b_{n+1} < b_n \qquad \text{because} \qquad \frac{1}{n+1} < \frac{1}{n}$$

$$\text{(ii) } \lim_{n\to\infty} b_n = \lim_{n\to\infty} \frac{1}{n} = 0$$

so the series is convergent by the Alternating Series Test. □

V EXAMPLE 2 The series $\displaystyle\sum_{n=1}^{\infty} \frac{(-1)^n 3n}{4n-1}$ is alternating but

$$\lim_{n\to\infty} b_n = \lim_{n\to\infty} \frac{3n}{4n-1} = \lim_{n\to\infty} \frac{3}{4 - \dfrac{1}{n}} = \frac{3}{4}$$

so condition (ii) is not satisfied. Instead, we look at the limit of the nth term of the series:

$$\lim_{n\to\infty} a_n = \lim_{n\to\infty} \frac{(-1)^n 3n}{4n - 1}$$

This limit does not exist, so the series diverges by the Test for Divergence. □

EXAMPLE 3 Test the series $\displaystyle\sum_{n=1}^{\infty} (-1)^{n+1} \frac{n^2}{n^3 + 1}$ for convergence or divergence.

SOLUTION The given series is alternating so we try to verify conditions (i) and (ii) of the Alternating Series Test.

Unlike the situation in Example 1, it is not obvious that the sequence given by $b_n = n^2/(n^3 + 1)$ is decreasing. However, if we consider the related function $f(x) = x^2/(x^3 + 1)$, we find that

$$f'(x) = \frac{x(2 - x^3)}{(x^3 + 1)^2}$$

Since we are considering only positive x, we see that $f'(x) < 0$ if $2 - x^3 < 0$, that is, $x > \sqrt[3]{2}$. Thus f is decreasing on the interval $(\sqrt[3]{2}, \infty)$. This means that $f(n + 1) < f(n)$ and therefore $b_{n+1} < b_n$ when $n \geqslant 2$. (The inequality $b_2 < b_1$ can be verified directly but all that really matters is that the sequence $\{b_n\}$ is eventually decreasing.)

■ Instead of verifying condition (i) of the Alternating Series Test by computing a derivative, we could verify that $b_{n+1} < b_n$ directly by using the technique of Solution 1 of Example 12 in Section 11.1.

Condition (ii) is readily verified:

$$\lim_{n\to\infty} b_n = \lim_{n\to\infty} \frac{n^2}{n^3 + 1} = \lim_{n\to\infty} \frac{\dfrac{1}{n}}{1 + \dfrac{1}{n^3}} = 0$$

Thus the given series is convergent by the Alternating Series Test. □

ESTIMATING SUMS

A partial sum s_n of any convergent series can be used as an approximation to the total sum s, but this is not of much use unless we can estimate the accuracy of the approximation. The error involved in using $s \approx s_n$ is the remainder $R_n = s - s_n$. The next theorem says that for series that satisfy the conditions of the Alternating Series Test, the size of the error is smaller than b_{n+1}, which is the absolute value of the first neglected term.

■ You can see geometrically why the Alternating Series Estimation Theorem is true by looking at Figure 1 (on page 710). Notice that $s - s_4 < b_5$, $|s - s_5| < b_6$, and so on. Notice also that s lies between any two consecutive partial sums.

ALTERNATING SERIES ESTIMATION THEOREM If $s = \Sigma\, (-1)^{n-1} b_n$ is the sum of an alternating series that satisfies

(i) $0 \leqslant b_{n+1} \leqslant b_n$ and (ii) $\displaystyle\lim_{n\to\infty} b_n = 0$

then

$$|R_n| = |s - s_n| \leqslant b_{n+1}$$

PROOF We know from the proof of the Alternating Series Test that s lies between any two consecutive partial sums s_n and s_{n+1}. It follows that

$$|s - s_n| \leqslant |s_{n+1} - s_n| = b_{n+1}$$ □

EXAMPLE 4 Find the sum of the series $\sum_{n=0}^{\infty} \frac{(-1)^n}{n!}$ correct to three decimal places. (By definition, $0! = 1$.)

SOLUTION We first observe that the series is convergent by the Alternating Series Test because

$$\text{(i)} \quad \frac{1}{(n+1)!} = \frac{1}{n!\,(n+1)} < \frac{1}{n!}$$

$$\text{(ii)} \quad 0 < \frac{1}{n!} < \frac{1}{n} \to 0 \quad \text{so} \quad \frac{1}{n!} \to 0 \text{ as } n \to \infty$$

To get a feel for how many terms we need to use in our approximation, let's write out the first few terms of the series:

$$s = \frac{1}{0!} - \frac{1}{1!} + \frac{1}{2!} - \frac{1}{3!} + \frac{1}{4!} - \frac{1}{5!} + \frac{1}{6!} - \frac{1}{7!} + \cdots$$

$$= 1 - 1 + \tfrac{1}{2} - \tfrac{1}{6} + \tfrac{1}{24} - \tfrac{1}{120} + \tfrac{1}{720} - \tfrac{1}{5040} + \cdots$$

Notice that $\quad b_7 = \frac{1}{5040} < \frac{1}{5000} = 0.0002$

and $\quad s_6 = 1 - 1 + \frac{1}{2} - \frac{1}{6} + \frac{1}{24} - \frac{1}{120} + \frac{1}{720} \approx 0.368056$

By the Alternating Series Estimation Theorem we know that

$$|s - s_6| \leqslant b_7 < 0.0002$$

This error of less than 0.0002 does not affect the third decimal place, so we have $s \approx 0.368$ correct to three decimal places. □

■ In Section 11.10 we will prove that $e^x = \sum_{n=0}^{\infty} x^n/n!$ for all x, so what we have obtained in Example 4 is actually an approximation to the number e^{-1}.

NOTE The rule that the error (in using s_n to approximate s) is smaller than the first neglected term is, in general, valid only for alternating series that satisfy the conditions of the Alternating Series Estimation Theorem. The rule does not apply to other types of series.

11.5 EXERCISES

1. (a) What is an alternating series?
   (b) Under what conditions does an alternating series converge?
   (c) If these conditions are satisfied, what can you say about the remainder after n terms?

2–20 Test the series for convergence or divergence.

2. $-\frac{1}{3} + \frac{2}{4} - \frac{3}{5} + \frac{4}{6} - \frac{5}{7} + \cdots$

3. $\frac{4}{7} - \frac{4}{8} + \frac{4}{9} - \frac{4}{10} + \frac{4}{11} - \cdots$

4. $\frac{1}{\sqrt{2}} - \frac{1}{\sqrt{3}} + \frac{1}{\sqrt{4}} - \frac{1}{\sqrt{5}} + \frac{1}{\sqrt{6}} - \cdots$

5. $\sum_{n=1}^{\infty} \frac{(-1)^{n-1}}{2n+1}$

6. $\sum_{n=1}^{\infty} \frac{(-1)^{n-1}}{\ln(n+4)}$

7. $\sum_{n=1}^{\infty} (-1)^n \frac{3n-1}{2n+1}$

8. $\sum_{n=1}^{\infty} (-1)^n \frac{n}{\sqrt{n^3+2}}$

9. $\sum_{n=1}^{\infty} (-1)^n \frac{n}{10^n}$

10. $\sum_{n=1}^{\infty} (-1)^n \frac{\sqrt{n}}{1+2\sqrt{n}}$

11. $\sum_{n=1}^{\infty} (-1)^{n+1} \frac{n^2}{n^3+4}$

12. $\sum_{n=1}^{\infty} (-1)^{n-1} \frac{e^{1/n}}{n}$

13. $\sum_{n=2}^{\infty} (-1)^n \frac{n}{\ln n}$

14. $\sum_{n=1}^{\infty} (-1)^{n-1} \frac{\ln n}{n}$

15. $\sum_{n=1}^{\infty} \frac{\cos n\pi}{n^{3/4}}$

16. $\sum_{n=1}^{\infty} \frac{\sin(n\pi/2)}{n!}$

17. $\sum_{n=1}^{\infty} (-1)^n \sin\left(\frac{\pi}{n}\right)$

18. $\sum_{n=1}^{\infty} (-1)^n \cos\left(\frac{\pi}{n}\right)$

19. $\sum_{n=1}^{\infty} (-1)^n \frac{n^n}{n!}$

20. $\sum_{n=1}^{\infty} \left(-\frac{n}{5}\right)^n$

21–22 Calculate the first 10 partial sums of the series and graph both the sequence of terms and the sequence of partial sums on the same screen. Estimate the error in using the 10th partial sum to approximate the total sum.

21. $\sum_{n=1}^{\infty} \frac{(-1)^{n-1}}{n^{3/2}}$ **22.** $\sum_{n=1}^{\infty} \frac{(-1)^{n-1}}{n^3}$

23–26 Show that the series is convergent. How many terms of the series do we need to add in order to find the sum to the indicated accuracy?

23. $\sum_{n=1}^{\infty} \frac{(-1)^{n+1}}{n^6}$ $(|\text{error}| < 0.00005)$

24. $\sum_{n=1}^{\infty} \frac{(-1)^{n}}{n\,5^n}$ $(|\text{error}| < 0.0001)$

25. $\sum_{n=0}^{\infty} \frac{(-1)^{n}}{10^n n!}$ $(|\text{error}| < 0.000005)$

26. $\sum_{n=1}^{\infty} (-1)^{n-1} n e^{-n}$ $(|\text{error}| < 0.01)$

27–30 Approximate the sum of the series correct to four decimal places.

27. $\sum_{n=1}^{\infty} \frac{(-1)^{n+1}}{n^5}$ **28.** $\sum_{n=1}^{\infty} \frac{(-1)^{n} n}{8^n}$

29. $\sum_{n=1}^{\infty} \frac{(-1)^{n-1} n^2}{10^n}$ **30.** $\sum_{n=1}^{\infty} \frac{(-1)^{n}}{3^n n!}$

31. Is the 50th partial sum s_{50} of the alternating series $\sum_{n=1}^{\infty} (-1)^{n-1}/n$ an overestimate or an underestimate of the total sum? Explain.

32–34 For what values of p is each series convergent?

32. $\sum_{n=1}^{\infty} \frac{(-1)^{n-1}}{n^p}$

33. $\sum_{n=1}^{\infty} \frac{(-1)^{n}}{n + p}$ **34.** $\sum_{n=2}^{\infty} (-1)^{n-1} \frac{(\ln n)^p}{n}$

35. Show that the series $\sum (-1)^{n-1} b_n$, where $b_n = 1/n$ if n is odd and $b_n = 1/n^2$ if n is even, is divergent. Why does the Alternating Series Test not apply?

36. Use the following steps to show that

$$\sum_{n=1}^{\infty} \frac{(-1)^{n-1}}{n} = \ln 2$$

Let h_n and s_n be the partial sums of the harmonic and alternating harmonic series.

(a) Show that $s_{2n} = h_{2n} - h_n$.

(b) From Exercise 40 in Section 11.3 we have

$$h_n - \ln n \to \gamma \quad \text{as } n \to \infty$$

and therefore

$$h_{2n} - \ln(2n) \to \gamma \quad \text{as } n \to \infty$$

Use these facts together with part (a) to show that $s_{2n} \to \ln 2$ as $n \to \infty$.

11.6 ABSOLUTE CONVERGENCE AND THE RATIO AND ROOT TESTS

Given any series $\sum a_n$, we can consider the corresponding series

$$\sum_{n=1}^{\infty} |a_n| = |a_1| + |a_2| + |a_3| + \cdots$$

whose terms are the absolute values of the terms of the original series.

■ We have convergence tests for series with positive terms and for alternating series. But what if the signs of the terms switch back and forth irregularly? We will see in Example 3 that the idea of absolute convergence sometimes helps in such cases.

1 DEFINITION A series $\sum a_n$ is called **absolutely convergent** if the series of absolute values $\sum |a_n|$ is convergent.

Notice that if $\sum a_n$ is a series with positive terms, then $|a_n| = a_n$ and so absolute convergence is the same as convergence in this case.

EXAMPLE 1 The series

$$\sum_{n=1}^{\infty} \frac{(-1)^{n-1}}{n^2} = 1 - \frac{1}{2^2} + \frac{1}{3^2} - \frac{1}{4^2} + \cdots$$

is absolutely convergent because

$$\sum_{n=1}^{\infty} \left| \frac{(-1)^{n-1}}{n^2} \right| = \sum_{n=1}^{\infty} \frac{1}{n^2} = 1 + \frac{1}{2^2} + \frac{1}{3^2} + \frac{1}{4^2} + \cdots$$

is a convergent p-series ($p = 2$). □

EXAMPLE 2 We know that the alternating harmonic series

$$\sum_{n=1}^{\infty} \frac{(-1)^{n-1}}{n} = 1 - \frac{1}{2} + \frac{1}{3} - \frac{1}{4} + \cdots$$

is convergent (see Example 1 in Section 11.5), but it is not absolutely convergent because the corresponding series of absolute values is

$$\sum_{n=1}^{\infty} \left| \frac{(-1)^{n-1}}{n} \right| = \sum_{n=1}^{\infty} \frac{1}{n} = 1 + \frac{1}{2} + \frac{1}{3} + \frac{1}{4} + \cdots$$

which is the harmonic series (p-series with $p = 1$) and is therefore divergent. □

2 DEFINITION A series $\sum a_n$ is called **conditionally convergent** if it is convergent but not absolutely convergent.

Example 2 shows that the alternating harmonic series is conditionally convergent. Thus it is possible for a series to be convergent but not absolutely convergent. However, the next theorem shows that absolute convergence implies convergence.

3 THEOREM If a series $\sum a_n$ is absolutely convergent, then it is convergent.

PROOF Observe that the inequality

$$0 \leqslant a_n + |a_n| \leqslant 2|a_n|$$

is true because $|a_n|$ is either a_n or $-a_n$. If $\sum a_n$ is absolutely convergent, then $\sum |a_n|$ is convergent, so $\sum 2|a_n|$ is convergent. Therefore, by the Comparison Test, $\sum (a_n + |a_n|)$ is convergent. Then

$$\sum a_n = \sum (a_n + |a_n|) - \sum |a_n|$$

is the difference of two convergent series and is therefore convergent. □

■ Figure 1 shows the graphs of the terms a_n and partial sums s_n of the series in Example 3. Notice that the series is not alternating but has positive and negative terms.

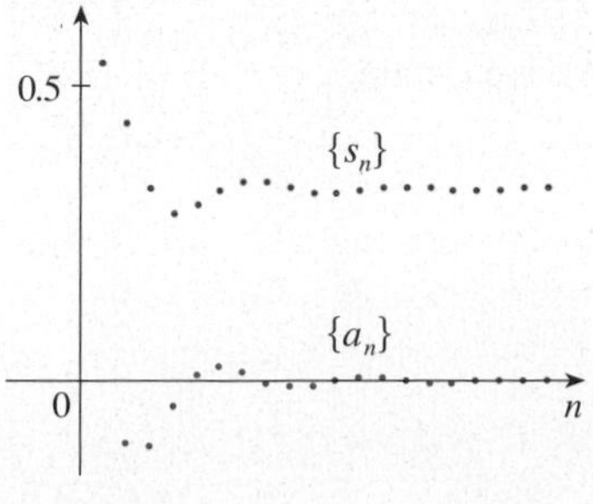

FIGURE 1

EXAMPLE 3 Determine whether the series

$$\sum_{n=1}^{\infty} \frac{\cos n}{n^2} = \frac{\cos 1}{1^2} + \frac{\cos 2}{2^2} + \frac{\cos 3}{3^2} + \cdots$$

is convergent or divergent.

SOLUTION This series has both positive and negative terms, but it is not alternating. (The first term is positive, the next three are negative, and the following three are positive: The signs change irregularly.) We can apply the Comparison Test to the series of absolute values

$$\sum_{n=1}^{\infty} \left| \frac{\cos n}{n^2} \right| = \sum_{n=1}^{\infty} \frac{|\cos n|}{n^2}$$

Since $|\cos n| \leqslant 1$ for all n, we have

$$\frac{|\cos n|}{n^2} \leqslant \frac{1}{n^2}$$

We know that $\Sigma\, 1/n^2$ is convergent (p-series with $p = 2$) and therefore $\Sigma\, |\cos n|/n^2$ is convergent by the Comparison Test. Thus the given series $\Sigma\, (\cos n)/n^2$ is absolutely convergent and therefore convergent by Theorem 3. □

The following test is very useful in determining whether a given series is absolutely convergent.

THE RATIO TEST

(i) If $\lim_{n\to\infty} \left|\frac{a_{n+1}}{a_n}\right| = L < 1$, then the series $\sum_{n=1}^{\infty} a_n$ is absolutely convergent (and therefore convergent).

(ii) If $\lim_{n\to\infty} \left|\frac{a_{n+1}}{a_n}\right| = L > 1$ or $\lim_{n\to\infty} \left|\frac{a_{n+1}}{a_n}\right| = \infty$, then the series $\sum_{n=1}^{\infty} a_n$ is divergent.

(iii) If $\lim_{n\to\infty} \left|\frac{a_{n+1}}{a_n}\right| = 1$, the Ratio Test is inconclusive; that is, no conclusion can be drawn about the convergence or divergence of $\Sigma\, a_n$.

PROOF

(i) The idea is to compare the given series with a convergent geometric series. Since $L < 1$, we can choose a number r such that $L < r < 1$. Since

$$\lim_{n\to\infty} \left|\frac{a_{n+1}}{a_n}\right| = L \qquad \text{and} \qquad L < r$$

the ratio $|a_{n+1}/a_n|$ will eventually be less than r; that is, there exists an integer N such that

$$\left|\frac{a_{n+1}}{a_n}\right| < r \qquad \text{whenever } n \geqslant N$$

or, equivalently,

4 $$|a_{n+1}| < |a_n|\, r \qquad \text{whenever } n \geqslant N$$

Putting n successively equal to $N, N + 1, N + 2, \ldots$ in (4), we obtain

$$|a_{N+1}| < |a_N|\, r$$

$$|a_{N+2}| < |a_{N+1}|\, r < |a_N|\, r^2$$

$$|a_{N+3}| < |a_{N+2}|\, r < |a_N|\, r^3$$

and, in general,

5 $$|a_{N+k}| < |a_N|\, r^k \qquad \text{for all } k \geqslant 1$$

Now the series

$$\sum_{k=1}^{\infty} |a_N| r^k = |a_N| r + |a_N| r^2 + |a_N| r^3 + \cdots$$

is convergent because it is a geometric series with $0 < r < 1$. So the inequality (5), together with the Comparison Test, shows that the series

$$\sum_{n=N+1}^{\infty} |a_n| = \sum_{k=1}^{\infty} |a_{N+k}| = |a_{N+1}| + |a_{N+2}| + |a_{N+3}| + \cdots$$

is also convergent. It follows that the series $\sum_{n=1}^{\infty} |a_n|$ is convergent. (Recall that a finite number of terms doesn't affect convergence.) Therefore $\sum a_n$ is absolutely convergent.

(ii) If $|a_{n+1}/a_n| \to L > 1$ or $|a_{n+1}/a_n| \to \infty$, then the ratio $|a_{n+1}/a_n|$ will eventually be greater than 1; that is, there exists an integer N such that

$$\left|\frac{a_{n+1}}{a_n}\right| > 1 \qquad \text{whenever } n \geqslant N$$

This means that $|a_{n+1}| > |a_n|$ whenever $n \geqslant N$ and so

$$\lim_{n \to \infty} a_n \neq 0$$

Therefore $\sum a_n$ diverges by the Test for Divergence. □

NOTE Part (iii) of the Ratio Test says that if $\lim_{n\to\infty} |a_{n+1}/a_n| = 1$, the test gives no information. For instance, for the convergent series $\sum 1/n^2$ we have

$$\left|\frac{a_{n+1}}{a_n}\right| = \frac{\dfrac{1}{(n+1)^2}}{\dfrac{1}{n^2}} = \frac{n^2}{(n+1)^2} = \frac{1}{\left(1 + \dfrac{1}{n}\right)^2} \to 1 \qquad \text{as } n \to \infty$$

whereas for the divergent series $\sum 1/n$ we have

$$\left|\frac{a_{n+1}}{a_n}\right| = \frac{\dfrac{1}{n+1}}{\dfrac{1}{n}} = \frac{n}{n+1} = \frac{1}{1 + \dfrac{1}{n}} \to 1 \qquad \text{as } n \to \infty$$

Therefore, if $\lim_{n\to\infty} |a_{n+1}/a_n| = 1$, the series $\sum a_n$ might converge or it might diverge. In this case the Ratio Test fails and we must use some other test.

EXAMPLE 4 Test the series $\displaystyle\sum_{n=1}^{\infty} (-1)^n \frac{n^3}{3^n}$ for absolute convergence.

SOLUTION We use the Ratio Test with $a_n = (-1)^n n^3/3^n$:

$$\left|\frac{a_{n+1}}{a_n}\right| = \left|\frac{\dfrac{(-1)^{n+1}(n+1)^3}{3^{n+1}}}{\dfrac{(-1)^n n^3}{3^n}}\right| = \frac{(n+1)^3}{3^{n+1}} \cdot \frac{3^n}{n^3}$$

$$= \frac{1}{3}\left(\frac{n+1}{n}\right)^3 = \frac{1}{3}\left(1 + \frac{1}{n}\right)^3 \to \frac{1}{3} < 1$$

■ ESTIMATING SUMS

In the last three sections we used various methods for estimating the sum of a series—the method depended on which test was used to prove convergence. What about series for which the Ratio Test works? There are two possibilities: If the series happens to be an alternating series, as in Example 4, then it is best to use the methods of Section 11.5. If the terms are all positive, then use the special methods explained in Exercise 34.

Thus, by the Ratio Test, the given series is absolutely convergent and therefore convergent. □

EXAMPLE 5 Test the convergence of the series $\sum_{n=1}^{\infty} \frac{n^n}{n!}$.

SOLUTION Since the terms $a_n = n^n/n!$ are positive, we don't need the absolute value signs.

$$\frac{a_{n+1}}{a_n} = \frac{(n+1)^{n+1}}{(n+1)!} \cdot \frac{n!}{n^n} = \frac{(n+1)(n+1)^n}{(n+1)n!} \cdot \frac{n!}{n^n}$$

$$= \left(\frac{n+1}{n}\right)^n = \left(1 + \frac{1}{n}\right)^n \to e \quad \text{as } n \to \infty$$

(See Equation 3.6.6.) Since $e > 1$, the given series is divergent by the Ratio Test. □

NOTE Although the Ratio Test works in Example 5, an easier method is to use the Test for Divergence. Since

$$a_n = \frac{n^n}{n!} = \frac{n \cdot n \cdot n \cdot \cdots \cdot n}{1 \cdot 2 \cdot 3 \cdot \cdots \cdot n} \geqslant n$$

it follows that a_n does not approach 0 as $n \to \infty$. Therefore the given series is divergent by the Test for Divergence.

The following test is convenient to apply when nth powers occur. Its proof is similar to the proof of the Ratio Test and is left as Exercise 37.

THE ROOT TEST

(i) If $\lim_{n\to\infty} \sqrt[n]{|a_n|} = L < 1$, then the series $\sum_{n=1}^{\infty} a_n$ is absolutely convergent (and therefore convergent).

(ii) If $\lim_{n\to\infty} \sqrt[n]{|a_n|} = L > 1$ or $\lim_{n\to\infty} \sqrt[n]{|a_n|} = \infty$, then the series $\sum_{n=1}^{\infty} a_n$ is divergent.

(iii) If $\lim_{n\to\infty} \sqrt[n]{|a_n|} = 1$, the Root Test is inconclusive.

If $\lim_{n\to\infty} \sqrt[n]{|a_n|} = 1$, then part (iii) of the Root Test says that the test gives no information. The series $\Sigma\, a_n$ could converge or diverge. (If $L = 1$ in the Ratio Test, don't try the Root Test because L will again be 1. And if $L = 1$ in the Root Test, don't try the Ratio Test because it will fail too.)

EXAMPLE 6 Test the convergence of the series $\sum_{n=1}^{\infty} \left(\frac{2n+3}{3n+2}\right)^n$.

SOLUTION

$$a_n = \left(\frac{2n+3}{3n+2}\right)^n$$

$$\sqrt[n]{|a_n|} = \frac{2n+3}{3n+2} = \frac{2 + \frac{3}{n}}{3 + \frac{2}{n}} \to \frac{2}{3} < 1$$

Thus the given series converges by the Root Test. □

REARRANGEMENTS

The question of whether a given convergent series is absolutely convergent or conditionally convergent has a bearing on the question of whether infinite sums behave like finite sums.

If we rearrange the order of the terms in a finite sum, then of course the value of the sum remains unchanged. But this is not always the case for an infinite series. By a **rearrangement** of an infinite series $\Sigma\, a_n$ we mean a series obtained by simply changing the order of the terms. For instance, a rearrangement of $\Sigma\, a_n$ could start as follows:

$$a_1 + a_2 + a_5 + a_3 + a_4 + a_{15} + a_6 + a_7 + a_{20} + \cdots$$

It turns out that

if $\Sigma\, a_n$ is an absolutely convergent series with sum s, then any rearrangement of $\Sigma\, a_n$ has the same sum s.

However, any conditionally convergent series can be rearranged to give a different sum. To illustrate this fact let's consider the alternating harmonic series

$$1 - \tfrac{1}{2} + \tfrac{1}{3} - \tfrac{1}{4} + \tfrac{1}{5} - \tfrac{1}{6} + \tfrac{1}{7} - \tfrac{1}{8} + \cdots = \ln 2 \qquad \boxed{6}$$

(See Exercise 36 in Section 11.5.) If we multiply this series by $\frac{1}{2}$, we get

$$\tfrac{1}{2} - \tfrac{1}{4} + \tfrac{1}{6} - \tfrac{1}{8} + \cdots = \tfrac{1}{2}\ln 2$$

Inserting zeros between the terms of this series, we have

■ Adding these zeros does not affect the sum of the series; each term in the sequence of partial sums is repeated, but the limit is the same.

$$0 + \tfrac{1}{2} + 0 - \tfrac{1}{4} + 0 + \tfrac{1}{6} + 0 - \tfrac{1}{8} + \cdots = \tfrac{1}{2}\ln 2 \qquad \boxed{7}$$

Now we add the series in Equations 6 and 7 using Theorem 11.2.8:

$$1 + \tfrac{1}{3} - \tfrac{1}{2} + \tfrac{1}{5} + \tfrac{1}{7} - \tfrac{1}{4} + \cdots = \tfrac{3}{2}\ln 2 \qquad \boxed{8}$$

Notice that the series in (8) contains the same terms as in (6), but rearranged so that one negative term occurs after each pair of positive terms. The sums of these series, however, are different. In fact, Riemann proved that

if $\Sigma\, a_n$ is a conditionally convergent series and r is any real number whatsoever, then there is a rearrangement of $\Sigma\, a_n$ that has a sum equal to r.

A proof of this fact is outlined in Exercise 40.

11.6 EXERCISES

1. What can you say about the series $\Sigma\, a_n$ in each of the following cases?

(a) $\displaystyle\lim_{n\to\infty}\left|\frac{a_{n+1}}{a_n}\right| = 8$ (b) $\displaystyle\lim_{n\to\infty}\left|\frac{a_{n+1}}{a_n}\right| = 0.8$

(c) $\displaystyle\lim_{n\to\infty}\left|\frac{a_{n+1}}{a_n}\right| = 1$

2–28 Determine whether the series is absolutely convergent, conditionally convergent, or divergent.

2. $\displaystyle\sum_{n=1}^{\infty}\frac{n^2}{2^n}$

3. $\displaystyle\sum_{n=0}^{\infty}\frac{(-10)^n}{n!}$

4. $\displaystyle\sum_{n=1}^{\infty}(-1)^{n-1}\frac{2^n}{n^4}$

5. $\displaystyle\sum_{n=1}^{\infty}\frac{(-1)^{n+1}}{\sqrt[4]{n}}$

6. $\displaystyle\sum_{n=1}^{\infty}\frac{(-1)^n}{n^4}$

7. $\displaystyle\sum_{k=1}^{\infty}k\left(\tfrac{2}{3}\right)^k$

8. $\displaystyle\sum_{n=1}^{\infty}\frac{n!}{100^n}$

9. $\displaystyle\sum_{n=1}^{\infty}(-1)^n\frac{(1.1)^n}{n^4}$

10. $\displaystyle\sum_{n=1}^{\infty}(-1)^n\frac{n}{\sqrt{n^3+2}}$

11. $\displaystyle\sum_{n=1}^{\infty}\frac{(-1)^n e^{1/n}}{n^3}$

12. $\displaystyle\sum_{n=1}^{\infty}\frac{\sin 4n}{4^n}$

13. $\displaystyle\sum_{n=1}^{\infty}\frac{10^n}{(n+1)4^{2n+1}}$

14. $\displaystyle\sum_{n=1}^{\infty}(-1)^{n+1}\frac{n^2 2^n}{n!}$

15. $\displaystyle\sum_{n=1}^{\infty} \frac{(-1)^n \arctan n}{n^2}$

16. $\displaystyle\sum_{n=1}^{\infty} \frac{3 - \cos n}{n^{2/3} - 2}$

17. $\displaystyle\sum_{n=2}^{\infty} \frac{(-1)^n}{\ln n}$

18. $\displaystyle\sum_{n=1}^{\infty} \frac{n!}{n^n}$

19. $\displaystyle\sum_{n=1}^{\infty} \frac{\cos(n\pi/3)}{n!}$

20. $\displaystyle\sum_{n=1}^{\infty} \frac{(-2)^n}{n^n}$

21. $\displaystyle\sum_{n=1}^{\infty} \left(\frac{n^2 + 1}{2n^2 + 1}\right)^n$

22. $\displaystyle\sum_{n=2}^{\infty} \left(\frac{-2n}{n + 1}\right)^{5n}$

23. $\displaystyle\sum_{n=1}^{\infty} \left(1 + \frac{1}{n}\right)^{n^2}$

24. $\displaystyle\sum_{n=2}^{\infty} \frac{n}{(\ln n)^n}$

25. $\displaystyle 1 - \frac{1 \cdot 3}{3!} + \frac{1 \cdot 3 \cdot 5}{5!} - \frac{1 \cdot 3 \cdot 5 \cdot 7}{7!} + \cdots + (-1)^{n-1}\frac{1 \cdot 3 \cdot 5 \cdot \cdots \cdot (2n - 1)}{(2n - 1)!} + \cdots$

26. $\displaystyle \frac{2}{5} + \frac{2 \cdot 6}{5 \cdot 8} + \frac{2 \cdot 6 \cdot 10}{5 \cdot 8 \cdot 11} + \frac{2 \cdot 6 \cdot 10 \cdot 14}{5 \cdot 8 \cdot 11 \cdot 14} + \cdots$

27. $\displaystyle\sum_{n=1}^{\infty} \frac{2 \cdot 4 \cdot 6 \cdot \cdots \cdot (2n)}{n!}$

28. $\displaystyle\sum_{n=1}^{\infty} (-1)^n \frac{2^n n!}{5 \cdot 8 \cdot 11 \cdot \cdots \cdot (3n + 2)}$

29. The terms of a series are defined recursively by the equations

$$a_1 = 2 \qquad a_{n+1} = \frac{5n + 1}{4n + 3} a_n$$

Determine whether $\Sigma\, a_n$ converges or diverges.

30. A series $\Sigma\, a_n$ is defined by the equations

$$a_1 = 1 \qquad a_{n+1} = \frac{2 + \cos n}{\sqrt{n}} a_n$$

Determine whether $\Sigma\, a_n$ converges or diverges.

31. For which of the following series is the Ratio Test inconclusive (that is, it fails to give a definite answer)?

(a) $\displaystyle\sum_{n=1}^{\infty} \frac{1}{n^3}$ (b) $\displaystyle\sum_{n=1}^{\infty} \frac{n}{2^n}$

(c) $\displaystyle\sum_{n=1}^{\infty} \frac{(-3)^{n-1}}{\sqrt{n}}$ (d) $\displaystyle\sum_{n=1}^{\infty} \frac{\sqrt{n}}{1 + n^2}$

32. For which positive integers k is the following series convergent?

$$\sum_{n=1}^{\infty} \frac{(n!)^2}{(kn)!}$$

33. (a) Show that $\Sigma_{n=0}^{\infty} x^n/n!$ converges for all x.
(b) Deduce that $\lim_{n\to\infty} x^n/n! = 0$ for all x.

34. Let $\Sigma\, a_n$ be a series with positive terms and let $r_n = a_{n+1}/a_n$. Suppose that $\lim_{n\to\infty} r_n = L < 1$, so $\Sigma\, a_n$ converges by the Ratio Test. As usual, we let R_n be the remainder after n terms, that is,

$$R_n = a_{n+1} + a_{n+2} + a_{n+3} + \cdots$$

(a) If $\{r_n\}$ is a decreasing sequence and $r_{n+1} < 1$, show, by summing a geometric series, that

$$R_n \leqslant \frac{a_{n+1}}{1 - r_{n+1}}$$

(b) If $\{r_n\}$ is an increasing sequence, show that

$$R_n \leqslant \frac{a_{n+1}}{1 - L}$$

35. (a) Find the partial sum s_5 of the series $\Sigma_{n=1}^{\infty} 1/n2^n$. Use Exercise 34 to estimate the error in using s_5 as an approximation to the sum of the series.
(b) Find a value of n so that s_n is within 0.00005 of the sum. Use this value of n to approximate the sum of the series.

36. Use the sum of the first 10 terms to approximate the sum of the series

$$\sum_{n=1}^{\infty} \frac{n}{2^n}$$

Use Exercise 34 to estimate the error.

37. Prove the Root Test. [*Hint for part (i):* Take any number r such that $L < r < 1$ and use the fact that there is an integer N such that $\sqrt[n]{|a_n|} < r$ whenever $n \geqslant N$.]

38. Around 1910, the Indian mathematician Srinivasa Ramanujan discovered the formula

$$\frac{1}{\pi} = \frac{2\sqrt{2}}{9801} \sum_{n=0}^{\infty} \frac{(4n)!(1103 + 26390n)}{(n!)^4 396^{4n}}$$

William Gosper used this series in 1985 to compute the first 17 million digits of π.
(a) Verify that the series is convergent.
(b) How many correct decimal places of π do you get if you use just the first term of the series? What if you use two terms?

39. Given any series $\Sigma\, a_n$, we define a series $\Sigma\, a_n^+$ whose terms are all the positive terms of $\Sigma\, a_n$ and a series $\Sigma\, a_n^-$ whose terms are all the negative terms of $\Sigma\, a_n$. To be specific, we let

$$a_n^+ = \frac{a_n + |a_n|}{2} \qquad a_n^- = \frac{a_n - |a_n|}{2}$$

Notice that if $a_n > 0$, then $a_n^+ = a_n$ and $a_n^- = 0$, whereas if $a_n < 0$, then $a_n^- = a_n$ and $a_n^+ = 0$.
(a) If $\Sigma\, a_n$ is absolutely convergent, show that both of the series $\Sigma\, a_n^+$ and $\Sigma\, a_n^-$ are convergent.
(b) If $\Sigma\, a_n$ is conditionally convergent, show that both of the series $\Sigma\, a_n^+$ and $\Sigma\, a_n^-$ are divergent.

40. Prove that if $\Sigma\, a_n$ is a conditionally convergent series and r is any real number, then there is a rearrangement of $\Sigma\, a_n$ whose sum is r. [*Hints:* Use the notation of Exercise 39. Take just enough positive terms a_n^+ so that their sum is greater than r. Then add just enough negative terms a_n^- so that the cumulative sum is less than r. Continue in this manner and use Theorem 11.2.6.]

11.7 STRATEGY FOR TESTING SERIES

We now have several ways of testing a series for convergence or divergence; the problem is to decide which test to use on which series. In this respect, testing series is similar to integrating functions. Again there are no hard and fast rules about which test to apply to a given series, but you may find the following advice of some use.

It is not wise to apply a list of the tests in a specific order until one finally works. That would be a waste of time and effort. Instead, as with integration, the main strategy is to classify the series according to its *form*.

1. If the series is of the form $\Sigma\, 1/n^p$, it is a p-series, which we know to be convergent if $p > 1$ and divergent if $p \leq 1$.

2. If the series has the form $\Sigma\, ar^{n-1}$ or $\Sigma\, ar^n$, it is a geometric series, which converges if $|r| < 1$ and diverges if $|r| \geq 1$. Some preliminary algebraic manipulation may be required to bring the series into this form.

3. If the series has a form that is similar to a p-series or a geometric series, then one of the comparison tests should be considered. In particular, if a_n is a rational function or an algebraic function of n (involving roots of polynomials), then the series should be compared with a p-series. Notice that most of the series in Exercises 11.4 have this form. (The value of p should be chosen as in Section 11.4 by keeping only the highest powers of n in the numerator and denominator.) The comparison tests apply only to series with positive terms, but if $\Sigma\, a_n$ has some negative terms, then we can apply the Comparison Test to $\Sigma\, |a_n|$ and test for absolute convergence.

4. If you can see at a glance that $\lim_{n\to\infty} a_n \neq 0$, then the Test for Divergence should be used.

5. If the series is of the form $\Sigma\,(-1)^{n-1}b_n$ or $\Sigma\,(-1)^n b_n$, then the Alternating Series Test is an obvious possibility.

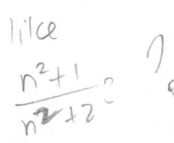

6. Series that involve factorials or other products (including a constant raised to the nth power) are often conveniently tested using the Ratio Test. Bear in mind that $|a_{n+1}/a_n| \to 1$ as $n \to \infty$ for all p-series and therefore all rational or algebraic functions of n. Thus the Ratio Test should not be used for such series.

7. If a_n is of the form $(b_n)^n$, then the Root Test may be useful.

8. If $a_n = f(n)$, where $\int_1^\infty f(x)\,dx$ is easily evaluated, then the Integral Test is effective (assuming the hypotheses of this test are satisfied).

In the following examples we don't work out all the details but simply indicate which tests should be used.

EXAMPLE 1 $\displaystyle\sum_{n=1}^{\infty} \frac{n-1}{2n+1}$

Since $a_n \to \frac{1}{2} \neq 0$ as $n \to \infty$, we should use the Test for Divergence. □

EXAMPLE 2 $\displaystyle\sum_{n=1}^{\infty} \frac{\sqrt{n^3+1}}{3n^3+4n^2+2}$

Since a_n is an algebraic function of n, we compare the given series with a p-series. The comparison series for the Limit Comparison Test is $\Sigma\, b_n$, where

$$b_n = \frac{\sqrt{n^3}}{3n^3} = \frac{n^{3/2}}{3n^3} = \frac{1}{3n^{3/2}}$$ □

EXAMPLE 3 $\displaystyle\sum_{n=1}^{\infty} ne^{-n^2}$

Since the integral $\int_1^\infty xe^{-x^2}\,dx$ is easily evaluated, we use the Integral Test. The Ratio Test also works. □

EXAMPLE 4 $\displaystyle\sum_{n=1}^{\infty} (-1)^n \frac{n^3}{n^4+1}$

Since the series is alternating, we use the Alternating Series Test. □

EXAMPLE 5 $\displaystyle\sum_{k=1}^{\infty} \frac{2^k}{k!}$

Since the series involves $k!$, we use the Ratio Test. □

EXAMPLE 6 $\displaystyle\sum_{n=1}^{\infty} \frac{1}{2+3^n}$

Since the series is closely related to the geometric series $\Sigma\, 1/3^n$, we use the Comparison Test. □

11.7 EXERCISES

1–38 Test the series for convergence or divergence.

1. $\displaystyle\sum_{n=1}^{\infty} \frac{1}{n+3^n}$
2. $\displaystyle\sum_{n=1}^{\infty} \frac{(2n+1)^n}{n^{2n}}$
3. $\displaystyle\sum_{n=1}^{\infty} (-1)^n \frac{n}{n+2}$
4. $\displaystyle\sum_{n=1}^{\infty} (-1)^n \frac{n}{n^2+2}$
5. $\displaystyle\sum_{n=1}^{\infty} \frac{n^2 2^{n-1}}{(-5)^n}$
6. $\displaystyle\sum_{n=1}^{\infty} \frac{1}{2n+1}$
7. $\displaystyle\sum_{n=2}^{\infty} \frac{1}{n\sqrt{\ln n}}$
8. $\displaystyle\sum_{k=1}^{\infty} \frac{2^k k!}{(k+2)!}$
9. $\displaystyle\sum_{k=1}^{\infty} k^2 e^{-k}$
10. $\displaystyle\sum_{n=1}^{\infty} n^2 e^{-n^3}$
11. $\displaystyle\sum_{n=2}^{\infty} \frac{(-1)^{n+1}}{n\ln n}$
12. $\displaystyle\sum_{n=1}^{\infty} \sin n$
13. $\displaystyle\sum_{n=1}^{\infty} \frac{3^n n^2}{n!}$
14. $\displaystyle\sum_{n=1}^{\infty} \frac{\sin 2n}{1+2^n}$
15. $\displaystyle\sum_{n=0}^{\infty} \frac{n!}{2\cdot 5\cdot 8\cdot \cdots \cdot (3n+2)}$
16. $\displaystyle\sum_{n=1}^{\infty} \frac{n^2+1}{n^3+1}$
17. $\displaystyle\sum_{n=1}^{\infty} (-1)^n 2^{1/n}$
18. $\displaystyle\sum_{n=2}^{\infty} \frac{(-1)^{n-1}}{\sqrt{n}-1}$
19. $\displaystyle\sum_{n=1}^{\infty} (-1)^n \frac{\ln n}{\sqrt{n}}$
20. $\displaystyle\sum_{k=1}^{\infty} \frac{k+5}{5^k}$
21. $\displaystyle\sum_{n=1}^{\infty} \frac{(-2)^{2n}}{n^n}$
22. $\displaystyle\sum_{n=1}^{\infty} \frac{\sqrt{n^2-1}}{n^3+2n^2+5}$
23. $\displaystyle\sum_{n=1}^{\infty} \tan(1/n)$
24. $\displaystyle\sum_{n=1}^{\infty} n\sin(1/n)$
25. $\displaystyle\sum_{n=1}^{\infty} \frac{n!}{e^{n^2}}$
26. $\displaystyle\sum_{n=1}^{\infty} \frac{n^2+1}{5^n}$
27. $\displaystyle\sum_{k=1}^{\infty} \frac{k\ln k}{(k+1)^3}$
28. $\displaystyle\sum_{n=1}^{\infty} \frac{e^{1/n}}{n^2}$
29. $\displaystyle\sum_{n=1}^{\infty} \frac{(-1)^n}{\cosh n}$
30. $\displaystyle\sum_{j=1}^{\infty} (-1)^j \frac{\sqrt{j}}{j+5}$
31. $\displaystyle\sum_{k=1}^{\infty} \frac{5^k}{3^k+4^k}$
32. $\displaystyle\sum_{n=1}^{\infty} \frac{(n!)^n}{n^{4n}}$
33. $\displaystyle\sum_{n=1}^{\infty} \frac{\sin(1/n)}{\sqrt{n}}$
34. $\displaystyle\sum_{n=1}^{\infty} \frac{1}{n+n\cos^2 n}$
35. $\displaystyle\sum_{n=1}^{\infty} \left(\frac{n}{n+1}\right)^{n^2}$
36. $\displaystyle\sum_{n=2}^{\infty} \frac{1}{(\ln n)^{\ln n}}$
37. $\displaystyle\sum_{n=1}^{\infty} (\sqrt[n]{2}-1)^n$
38. $\displaystyle\sum_{n=1}^{\infty} (\sqrt[n]{2}-1)$

11.8 POWER SERIES

A **power series** is a series of the form

1
$$\sum_{n=0}^{\infty} c_n x^n = c_0 + c_1 x + c_2 x^2 + c_3 x^3 + \cdots$$

where x is a variable and the c_n's are constants called the **coefficients** of the series. For each fixed x, the series (1) is a series of constants that we can test for convergence or divergence. A power series may converge for some values of x and diverge for other values of x. The sum of the series is a function

$$f(x) = c_0 + c_1 x + c_2 x^2 + \cdots + c_n x^n + \cdots$$

whose domain is the set of all x for which the series converges. Notice that f resembles a polynomial. The only difference is that f has infinitely many terms.

For instance, if we take $c_n = 1$ for all n, the power series becomes the geometric series

$$\sum_{n=0}^{\infty} x^n = 1 + x + x^2 + \cdots + x^n + \cdots$$

which converges when $-1 < x < 1$ and diverges when $|x| \geq 1$ (see Equation 11.2.5).

More generally, a series of the form

power series:

2
$$\sum_{n=0}^{\infty} c_n(x-a)^n = c_0 + c_1(x-a) + c_2(x-a)^2 + \cdots$$

is called a **power series in $(x - a)$** or a **power series centered at a** or a **power series about a**. Notice that in writing out the term corresponding to $n = 0$ in Equations 1 and 2 we have adopted the convention that $(x-a)^0 = 1$ even when $x = a$. Notice also that when $x = a$, all of the terms are 0 for $n \geq 1$ and so the power series (2) always converges when $x = a$.

■ **TRIGONOMETRIC SERIES**

A power series is a series in which each term is a power function. A **trigonometric series**

$$\sum_{n=0}^{\infty} (a_n \cos nx + b_n \sin nx)$$

is a series whose terms are trigonometric functions. This type of series is discussed on the website

www.stewartcalculus.com

Click on **Additional Topics** and then on **Fourier Series**.

EXAMPLE 1 For what values of x is the series $\sum_{n=0}^{\infty} n!x^n$ convergent?

SOLUTION We use the Ratio Test. If we let a_n, as usual, denote the nth term of the series, then $a_n = n!x^n$. If $x \neq 0$, we have

$$\lim_{n\to\infty} \left| \frac{a_{n+1}}{a_n} \right| = \lim_{n\to\infty} \left| \frac{(n+1)!x^{n+1}}{n!x^n} \right| = \lim_{n\to\infty} (n+1)|x| = \infty$$

■ Notice that

$$(n+1)! = (n+1)n(n-1) \cdot \cdots \cdot 3 \cdot 2 \cdot 1$$
$$= (n+1)n!$$

By the Ratio Test, the series diverges when $x \neq 0$. Thus the given series converges only when $x = 0$. □

EXAMPLE 2 For what values of x does the series $\sum_{n=1}^{\infty} \frac{(x-3)^n}{n}$ converge?

need to do endpts

SOLUTION Let $a_n = (x-3)^n/n$. Then

$$\left| \frac{a_{n+1}}{a_n} \right| = \left| \frac{(x-3)^{n+1}}{n+1} \cdot \frac{n}{(x-3)^n} \right|$$

$$= \frac{1}{1 + \frac{1}{n}} |x - 3| \to |x - 3| \qquad \text{as } n \to \infty$$

By the Ratio Test, the given series is absolutely convergent, and therefore convergent, when $|x - 3| < 1$ and divergent when $|x - 3| > 1$. Now

$$|x - 3| < 1 \iff -1 < x - 3 < 1 \iff 2 < x < 4$$

so the series converges when $2 < x < 4$ and diverges when $x < 2$ or $x > 4$.

The Ratio Test gives no information when $|x - 3| = 1$ so we must consider $x = 2$ and $x = 4$ separately. If we put $x = 4$ in the series, it becomes $\Sigma\, 1/n$, the harmonic series, which is divergent. If $x = 2$, the series is $\Sigma\, (-1)^n/n$, which converges by the Alternating Series Test. Thus the given power series converges for $2 \leqslant x < 4$. □

National Film Board of Canada

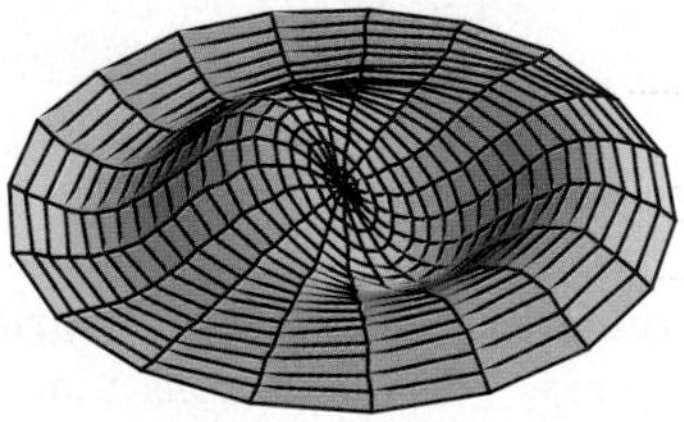

■ Notice how closely the computer-generated model (which involves Bessel functions and cosine functions) matches the photograph of a vibrating rubber membrane.

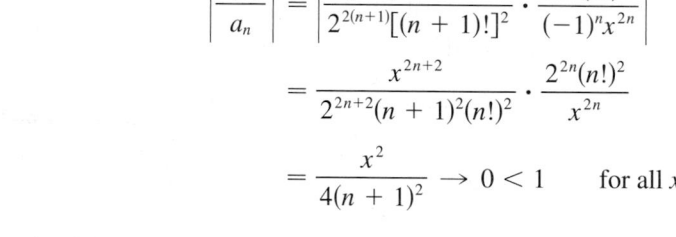

We will see that the main use of a power series is that it provides a way to represent some of the most important functions that arise in mathematics, physics, and chemistry. In particular, the sum of the power series in the next example is called a **Bessel function**, after the German astronomer Friedrich Bessel (1784–1846), and the function given in Exercise 35 is another example of a Bessel function. In fact, these functions first arose when Bessel solved Kepler's equation for describing planetary motion. Since that time, these functions have been applied in many different physical situations, including the temperature distribution in a circular plate and the shape of a vibrating drumhead.

EXAMPLE 3 Find the domain of the Bessel function of order 0 defined by

$$J_0(x) = \sum_{n=0}^{\infty} \frac{(-1)^n x^{2n}}{2^{2n}(n!)^2}$$

SOLUTION Let $a_n = (-1)^n x^{2n}/[2^{2n}(n!)^2]$. Then

$$\left|\frac{a_{n+1}}{a_n}\right| = \left|\frac{(-1)^{n+1}x^{2(n+1)}}{2^{2(n+1)}[(n+1)!]^2} \cdot \frac{2^{2n}(n!)^2}{(-1)^n x^{2n}}\right|$$

$$= \frac{x^{2n+2}}{2^{2n+2}(n+1)^2(n!)^2} \cdot \frac{2^{2n}(n!)^2}{x^{2n}}$$

$$= \frac{x^2}{4(n+1)^2} \to 0 < 1 \qquad \text{for all } x$$

Thus, by the Ratio Test, the given series converges for all values of x. In other words, the domain of the Bessel function J_0 is $(-\infty, \infty) = \mathbb{R}$. □

Recall that the sum of a series is equal to the limit of the sequence of partial sums. So when we define the Bessel function in Example 3 as the sum of a series we mean that, for every real number x,

$$J_0(x) = \lim_{n\to\infty} s_n(x) \qquad \text{where} \qquad s_n(x) = \sum_{i=0}^{n} \frac{(-1)^i x^{2i}}{2^{2i}(i!)^2}$$

The first few partial sums are

$$s_0(x) = 1 \qquad s_1(x) = 1 - \frac{x^2}{4} \qquad s_2(x) = 1 - \frac{x^2}{4} + \frac{x^4}{64}$$

$$s_3(x) = 1 - \frac{x^2}{4} + \frac{x^4}{64} - \frac{x^6}{2304} \qquad s_4(x) = 1 - \frac{x^2}{4} + \frac{x^4}{64} - \frac{x^6}{2304} + \frac{x^8}{147{,}456}$$

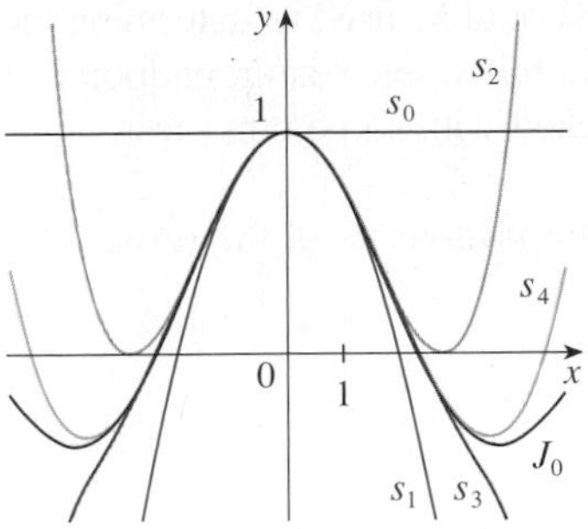

FIGURE 1
Partial sums of the Bessel function J_0

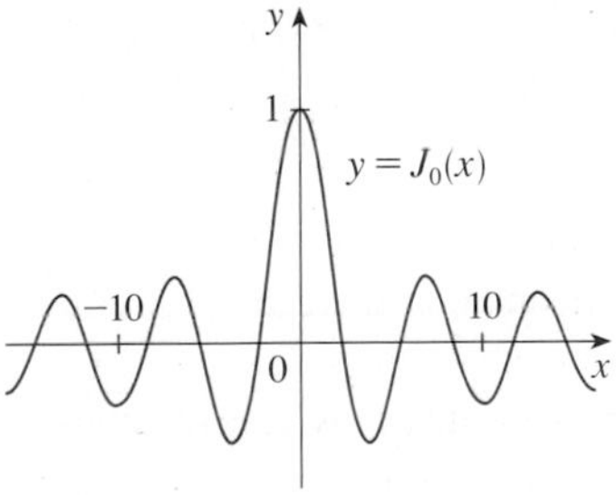

FIGURE 2

Figure 1 shows the graphs of these partial sums, which are polynomials. They are all approximations to the function J_0, but notice that the approximations become better when more terms are included. Figure 2 shows a more complete graph of the Bessel function.

For the power series that we have looked at so far, the set of values of x for which the series is convergent has always turned out to be an interval [a finite interval for the geometric series and the series in Example 2, the infinite interval $(-\infty, \infty)$ in Example 3, and a collapsed interval $[0, 0] = \{0\}$ in Example 1]. The following theorem, proved in Appendix F, says that this is true in general.

3 THEOREM For a given power series $\sum_{n=0}^{\infty} c_n(x-a)^n$, there are only three possibilities:

(i) The series converges only when $x = a$.

(ii) The series converges for all x.

(iii) There is a positive number R such that the series converges if $|x - a| < R$ and diverges if $|x - a| > R$.

The number R in case (iii) is called the **radius of convergence** of the power series. By convention, the radius of convergence is $R = 0$ in case (i) and $R = \infty$ in case (ii). The **interval of convergence** of a power series is the interval that consists of all values of x for which the series converges. In case (i) the interval consists of just a single point a. In case (ii) the interval is $(-\infty, \infty)$. In case (iii) note that the inequality $|x - a| < R$ can be rewritten as $a - R < x < a + R$. When x is an *endpoint* of the interval, that is, $x = a \pm R$, anything can happen—the series might converge at one or both endpoints or it might diverge at both endpoints. Thus in case (iii) there are four possibilities for the interval of convergence:

$$(a - R, a + R) \qquad (a - R, a + R] \qquad [a - R, a + R) \qquad [a - R, a + R]$$

The situation is illustrated in Figure 3.

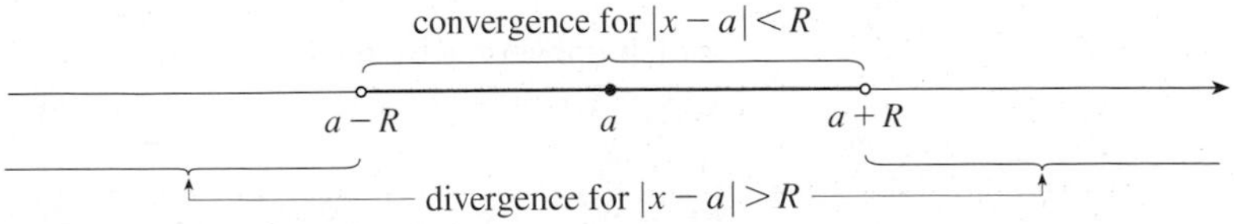

FIGURE 3

We summarize here the radius and interval of convergence for each of the examples already considered in this section.

	Series	Radius of convergence	Interval of convergence
Geometric series	$\sum_{n=0}^{\infty} x^n$	$R = 1$	$(-1, 1)$
Example 1	$\sum_{n=0}^{\infty} n!\, x^n$	$R = 0$	$\{0\}$
Example 2	$\sum_{n=1}^{\infty} \frac{(x-3)^n}{n}$	$R = 1$	$[2, 4)$
Example 3	$\sum_{n=0}^{\infty} \frac{(-1)^n x^{2n}}{2^{2n}(n!)^2}$	$R = \infty$	$(-\infty, \infty)$

In general, the Ratio Test (or sometimes the Root Test) should be used to determine the radius of convergence R. The Ratio and Root Tests always fail when x is an endpoint of the interval of convergence, so the endpoints must be checked with some other test.

EXAMPLE 4 Find the radius of convergence and interval of convergence of the series

$$\sum_{n=0}^{\infty} \frac{(-3)^n x^n}{\sqrt{n+1}}$$

SOLUTION Let $a_n = (-3)^n x^n / \sqrt{n+1}$. Then

$$\left|\frac{a_{n+1}}{a_n}\right| = \left|\frac{(-3)^{n+1} x^{n+1}}{\sqrt{n+2}} \cdot \frac{\sqrt{n+1}}{(-3)^n x^n}\right| = \left|-3x\sqrt{\frac{n+1}{n+2}}\right|$$

$$= 3\sqrt{\frac{1+(1/n)}{1+(2/n)}}\,|x| \to 3|x| \qquad \text{as } n \to \infty$$

By the Ratio Test, the given series converges if $3|x| < 1$ and diverges if $3|x| > 1$. Thus it converges if $|x| < \frac{1}{3}$ and diverges if $|x| > \frac{1}{3}$. This means that the radius of convergence is $R = \frac{1}{3}$.

We know the series converges in the interval $\left(-\frac{1}{3}, \frac{1}{3}\right)$, but we must now test for convergence at the endpoints of this interval. If $x = -\frac{1}{3}$, the series becomes

$$\sum_{n=0}^{\infty} \frac{(-3)^n\left(-\frac{1}{3}\right)^n}{\sqrt{n+1}} = \sum_{n=0}^{\infty} \frac{1}{\sqrt{n+1}} = \frac{1}{\sqrt{1}} + \frac{1}{\sqrt{2}} + \frac{1}{\sqrt{3}} + \frac{1}{\sqrt{4}} + \cdots$$

which diverges. (Use the Integral Test or simply observe that it is a p-series with $p = \frac{1}{2} < 1$.) If $x = \frac{1}{3}$, the series is

$$\sum_{n=0}^{\infty} \frac{(-3)^n\left(\frac{1}{3}\right)^n}{\sqrt{n+1}} = \sum_{n=0}^{\infty} \frac{(-1)^n}{\sqrt{n+1}}$$

which converges by the Alternating Series Test. Therefore the given power series converges when $-\frac{1}{3} < x \leqslant \frac{1}{3}$, so the interval of convergence is $\left(-\frac{1}{3}, \frac{1}{3}\right]$. □

V EXAMPLE 5 Find the radius of convergence and interval of convergence of the series

$$\sum_{n=0}^{\infty} \frac{n(x+2)^n}{3^{n+1}}$$

SOLUTION If $a_n = n(x+2)^n / 3^{n+1}$, then

$$\left|\frac{a_{n+1}}{a_n}\right| = \left|\frac{(n+1)(x+2)^{n+1}}{3^{n+2}} \cdot \frac{3^{n+1}}{n(x+2)^n}\right|$$

$$= \left(1 + \frac{1}{n}\right)\frac{|x+2|}{3} \to \frac{|x+2|}{3} \qquad \text{as } n \to \infty$$

Using the Ratio Test, we see that the series converges if $|x+2|/3 < 1$ and it diverges if $|x+2|/3 > 1$. So it converges if $|x+2| < 3$ and diverges if $|x+2| > 3$. Thus the radius of convergence is $R = 3$.

The inequality $|x + 2| < 3$ can be written as $-5 < x < 1$, so we test the series at the endpoints -5 and 1. When $x = -5$, the series is

$$\sum_{n=0}^{\infty} \frac{n(-3)^n}{3^{n+1}} = \frac{1}{3} \sum_{n=0}^{\infty} (-1)^n n$$

which diverges by the Test for Divergence [$(-1)^n n$ doesn't converge to 0]. When $x = 1$, the series is

$$\sum_{n=0}^{\infty} \frac{n(3)^n}{3^{n+1}} = \frac{1}{3} \sum_{n=0}^{\infty} n$$

which also diverges by the Test for Divergence. Thus the series converges only when $-5 < x < 1$, so the interval of convergence is $(-5, 1)$. □

11.8 EXERCISES

1. What is a power series?

2. (a) What is the radius of convergence of a power series? How do you find it?
(b) What is the interval of convergence of a power series? How do you find it?

3–28 Find the radius of convergence and interval of convergence of the series.

3. $\sum_{n=1}^{\infty} \frac{x^n}{\sqrt{n}}$

4. $\sum_{n=0}^{\infty} \frac{(-1)^n x^n}{n+1}$

5. $\sum_{n=1}^{\infty} \frac{(-1)^{n-1} x^n}{n^3}$

6. $\sum_{n=1}^{\infty} \sqrt{n}\, x^n$

7. $\sum_{n=0}^{\infty} \frac{x^n}{n!}$

8. $\sum_{n=1}^{\infty} n^n x^n$

9. $\sum_{n=1}^{\infty} (-1)^n \frac{n^2 x^n}{2^n}$

10. $\sum_{n=1}^{\infty} \frac{10^n x^n}{n^3}$

11. $\sum_{n=1}^{\infty} \frac{(-2)^n x^n}{\sqrt[4]{n}}$

12. $\sum_{n=1}^{\infty} \frac{x^n}{5^n n^5}$

13. $\sum_{n=2}^{\infty} (-1)^n \frac{x^n}{4^n \ln n}$

14. $\sum_{n=0}^{\infty} (-1)^n \frac{x^{2n}}{(2n)!}$

15. $\sum_{n=0}^{\infty} \frac{(x-2)^n}{n^2+1}$

16. $\sum_{n=0}^{\infty} (-1)^n \frac{(x-3)^n}{2n+1}$

17. $\sum_{n=1}^{\infty} \frac{3^n (x+4)^n}{\sqrt{n}}$

18. $\sum_{n=1}^{\infty} \frac{n}{4^n} (x+1)^n$

19. $\sum_{n=1}^{\infty} \frac{(x-2)^n}{n^n}$

20. $\sum_{n=1}^{\infty} \frac{(3x-2)^n}{n\, 3^n}$

21. $\sum_{n=1}^{\infty} \frac{n}{b^n} (x-a)^n, \quad b > 0$

22. $\sum_{n=1}^{\infty} \frac{n(x-4)^n}{n^3+1}$

23. $\sum_{n=1}^{\infty} n!(2x-1)^n$

24. $\sum_{n=1}^{\infty} \frac{n^2 x^n}{2 \cdot 4 \cdot 6 \cdot \cdots \cdot (2n)}$

25. $\sum_{n=1}^{\infty} \frac{(4x+1)^n}{n^2}$

26. $\sum_{n=2}^{\infty} \frac{x^{2n}}{n(\ln n)^2}$

27. $\sum_{n=1}^{\infty} \frac{x^n}{1 \cdot 3 \cdot 5 \cdot \cdots \cdot (2n-1)}$

28. $\sum_{n=1}^{\infty} \frac{n! x^n}{1 \cdot 3 \cdot 5 \cdot \cdots \cdot (2n-1)}$

29. If $\sum_{n=0}^{\infty} c_n 4^n$ is convergent, does it follow that the following series are convergent?

(a) $\sum_{n=0}^{\infty} c_n(-2)^n$ (b) $\sum_{n=0}^{\infty} c_n(-4)^n$

30. Suppose that $\sum_{n=0}^{\infty} c_n x^n$ converges when $x = -4$ and diverges when $x = 6$. What can be said about the convergence or divergence of the following series?

(a) $\sum_{n=0}^{\infty} c_n$ (b) $\sum_{n=0}^{\infty} c_n 8^n$

(c) $\sum_{n=0}^{\infty} c_n(-3)^n$ (d) $\sum_{n=0}^{\infty} (-1)^n c_n 9^n$

31. If k is a positive integer, find the radius of convergence of the series

$$\sum_{n=0}^{\infty} \frac{(n!)^k}{(kn)!} x^n$$

32. Let p and q be real numbers with $p < q$. Find a power series whose interval of convergence is
(a) (p, q) (b) $(p, q]$
(c) $[p, q)$ (d) $[p, q]$

33. Is it possible to find a power series whose interval of convergence is $[0, \infty)$? Explain.

34. Graph the first several partial sums $s_n(x)$ of the series $\Sigma_{n=0}^{\infty} x^n$, together with the sum function $f(x) = 1/(1 - x)$, on a common screen. On what interval do these partial sums appear to be converging to $f(x)$?

35. The function J_1 defined by

$$J_1(x) = \sum_{n=0}^{\infty} \frac{(-1)^n x^{2n+1}}{n!(n+1)!2^{2n+1}}$$

is called the *Bessel function of order 1*.
(a) Find its domain.
(b) Graph the first several partial sums on a common screen.
CAS (c) If your CAS has built-in Bessel functions, graph J_1 on the same screen as the partial sums in part (b) and observe how the partial sums approximate J_1.

36. The function A defined by

$$A(x) = 1 + \frac{x^3}{2 \cdot 3} + \frac{x^6}{2 \cdot 3 \cdot 5 \cdot 6} + \frac{x^9}{2 \cdot 3 \cdot 5 \cdot 6 \cdot 8 \cdot 9} + \cdots$$

is called the *Airy function* after the English mathematician and astronomer Sir George Airy (1801–1892).
(a) Find the domain of the Airy function.
(b) Graph the first several partial sums on a common screen.
CAS (c) If your CAS has built-in Airy functions, graph A on the same screen as the partial sums in part (b) and observe how the partial sums approximate A.

37. A function f is defined by

$$f(x) = 1 + 2x + x^2 + 2x^3 + x^4 + \cdots$$

that is, its coefficients are $c_{2n} = 1$ and $c_{2n+1} = 2$ for all $n \geqslant 0$. Find the interval of convergence of the series and find an explicit formula for $f(x)$.

38. If $f(x) = \Sigma_{n=0}^{\infty} c_n x^n$, where $c_{n+4} = c_n$ for all $n \geqslant 0$, find the interval of convergence of the series and a formula for $f(x)$.

39. Show that if $\lim_{n\to\infty} \sqrt[n]{|c_n|} = c$, where $c \neq 0$, then the radius of convergence of the power series $\Sigma c_n x^n$ is $R = 1/c$.

40. Suppose that the power series $\Sigma c_n(x - a)^n$ satisfies $c_n \neq 0$ for all n. Show that if $\lim_{n\to\infty} |c_n/c_{n+1}|$ exists, then it is equal to the radius of convergence of the power series.

41. Suppose the series $\Sigma c_n x^n$ has radius of convergence 2 and the series $\Sigma d_n x^n$ has radius of convergence 3. What is the radius of convergence of the series $\Sigma (c_n + d_n)x^n$?

42. Suppose that the radius of convergence of the power series $\Sigma c_n x^n$ is R. What is the radius of convergence of the power series $\Sigma c_n x^{2n}$?

11.9 REPRESENTATIONS OF FUNCTIONS AS POWER SERIES

In this section we learn how to represent certain types of functions as sums of power series by manipulating geometric series or by differentiating or integrating such a series. You might wonder why we would ever want to express a known function as a sum of infinitely many terms. We will see later that this strategy is useful for integrating functions that don't have elementary antiderivatives, for solving differential equations, and for approximating functions by polynomials. (Scientists do this to simplify the expressions they deal with; computer scientists do this to represent functions on calculators and computers.)

We start with an equation that we have seen before:

$$\boxed{1} \qquad \frac{1}{1-x} = 1 + x + x^2 + x^3 + \cdots = \sum_{n=0}^{\infty} x^n \qquad |x| < 1$$

We first encountered this equation in Example 5 in Section 11.2, where we obtained it by observing that it is a geometric series with $a = 1$ and $r = x$. But here our point of view is different. We now regard Equation 1 as expressing the function $f(x) = 1/(1 - x)$ as a sum of a power series.

■ A geometric illustration of Equation 1 is shown in Figure 1. Because the sum of a series is the limit of the sequence of partial sums, we have

$$\frac{1}{1-x} = \lim_{n\to\infty} s_n(x)$$

where

$$s_n(x) = 1 + x + x^2 + \cdots + x^n$$

is the nth partial sum. Notice that as n increases, $s_n(x)$ becomes a better approximation to $f(x)$ for $-1 < x < 1$.

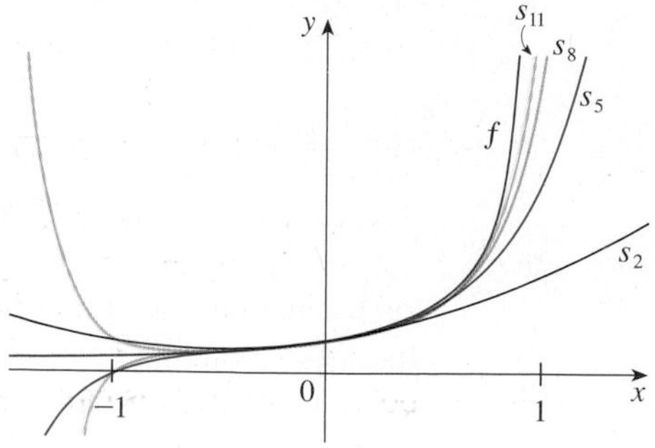

FIGURE 1
$f(x) = \dfrac{1}{1-x}$ and some partial sums

■ When a power series is asked for in this section, it is assumed that the series is centered at 0, unless otherwise specified.

EXAMPLE 1 Express $1/(1+x^2)$ as the sum of a power series and find the interval of convergence.

SOLUTION Replacing x by $-x^2$ in Equation 1, we have

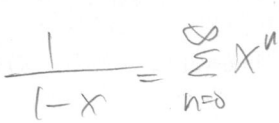

$$\frac{1}{1+x^2} = \frac{1}{1-(-x^2)} = \sum_{n=0}^{\infty} (-x^2)^n$$

$$= \sum_{n=0}^{\infty} (-1)^n x^{2n} = 1 - x^2 + x^4 - x^6 + x^8 - \cdots$$

Because this is a geometric series, it converges when $|-x^2| < 1$, that is, $x^2 < 1$, or $|x| < 1$. Therefore the interval of convergence is $(-1, 1)$. (Of course, we could have determined the radius of convergence by applying the Ratio Test, but that much work is unnecessary here.) □

EXAMPLE 2 Find a power series representation for $1/(x+2)$.

SOLUTION In order to put this function in the form of the left side of Equation 1 we first factor a 2 from the denominator:

$$\frac{1}{2+x} = \frac{1}{2\left(1+\dfrac{x}{2}\right)} = \frac{1}{2\left[1-\left(-\dfrac{x}{2}\right)\right]}$$

$$= \frac{1}{2}\sum_{n=0}^{\infty}\left(-\frac{x}{2}\right)^n = \sum_{n=0}^{\infty}\frac{(-1)^n}{2^{n+1}}x^n$$

This series converges when $|-x/2| < 1$, that is, $|x| < 2$. So the interval of convergence is $(-2, 2)$. □

EXAMPLE 3 Find a power series representation of $x^3/(x+2)$.

SOLUTION Since this function is just x^3 times the function in Example 2, all we have to do is to multiply that series by x^3:

■ It's legitimate to move x^3 across the sigma sign because it doesn't depend on n. [Use Theorem 11.2.8(i) with $c = x^3$.]

$$\frac{x^3}{x+2} = x^3 \cdot \frac{1}{x+2} = x^3 \sum_{n=0}^{\infty}\frac{(-1)^n}{2^{n+1}}x^n = \sum_{n=0}^{\infty}\frac{(-1)^n}{2^{n+1}}x^{n+3}$$

$$= \tfrac{1}{2}x^3 - \tfrac{1}{4}x^4 + \tfrac{1}{8}x^5 - \tfrac{1}{16}x^6 + \cdots$$

Another way of writing this series is as follows:

$$\frac{x^3}{x+2} = \sum_{n=3}^{\infty}\frac{(-1)^{n-1}}{2^{n-2}}x^n$$

As in Example 2, the interval of convergence is $(-2, 2)$. □

DIFFERENTIATION AND INTEGRATION OF POWER SERIES

The sum of a power series is a function $f(x) = \sum_{n=0}^{\infty} c_n(x-a)^n$ whose domain is the interval of convergence of the series. We would like to be able to differentiate and integrate such functions, and the following theorem (which we won't prove) says that we can do so by differentiating or integrating each individual term in the series, just as we would for a polynomial. This is called **term-by-term differentiation and integration**.

2 THEOREM If the power series $\Sigma\, c_n(x-a)^n$ has radius of convergence $R > 0$, then the function f defined by

$$f(x) = c_0 + c_1(x-a) + c_2(x-a)^2 + \cdots = \sum_{n=0}^{\infty} c_n(x-a)^n$$

is differentiable (and therefore continuous) on the interval $(a - R, a + R)$ and

(i) $f'(x) = c_1 + 2c_2(x-a) + 3c_3(x-a)^2 + \cdots = \displaystyle\sum_{n=1}^{\infty} nc_n(x-a)^{n-1}$

(ii) $\displaystyle\int f(x)\,dx = C + c_0(x-a) + c_1\frac{(x-a)^2}{2} + c_2\frac{(x-a)^3}{3} + \cdots$

$$= C + \sum_{n=0}^{\infty} c_n \frac{(x-a)^{n+1}}{n+1}$$

The radii of convergence of the power series in Equations (i) and (ii) are both R.

■ In part (ii), $\int c_0\,dx = c_0x + C_1$ is written as $c_0(x-a) + C$, where $C = C_1 + ac_0$, so all the terms of the series have the same form.

NOTE 1 Equations (i) and (ii) in Theorem 2 can be rewritten in the form

(iii) $\displaystyle\frac{d}{dx}\left[\sum_{n=0}^{\infty} c_n(x-a)^n\right] = \sum_{n=0}^{\infty} \frac{d}{dx}\left[c_n(x-a)^n\right]$

(iv) $\displaystyle\int\left[\sum_{n=0}^{\infty} c_n(x-a)^n\right]dx = \sum_{n=0}^{\infty} \int c_n(x-a)^n\,dx$

We know that, for finite sums, the derivative of a sum is the sum of the derivatives and the integral of a sum is the sum of the integrals. Equations (iii) and (iv) assert that the same is true for infinite sums, provided we are dealing with *power series.* (For other types of series of functions the situation is not as simple; see Exercise 36.)

NOTE 2 Although Theorem 2 says that the radius of convergence remains the same when a power series is differentiated or integrated, this does not mean that the *interval* of convergence remains the same. It may happen that the original series converges at an endpoint, whereas the differentiated series diverges there. (See Exercise 37.)

NOTE 3 The idea of differentiating a power series term by term is the basis for a powerful method for solving differential equations. We will discuss this method in Chapter 17.

EXAMPLE 4 In Example 3 in Section 11.8 we saw that the Bessel function

$$J_0(x) = \sum_{n=0}^{\infty} \frac{(-1)^n x^{2n}}{2^{2n}(n!)^2}$$

is defined for all x. Thus, by Theorem 2, J_0 is differentiable for all x and its derivative is found by term-by-term differentiation as follows:

$$J_0'(x) = \sum_{n=0}^{\infty} \frac{d}{dx}\frac{(-1)^n x^{2n}}{2^{2n}(n!)^2} = \sum_{n=1}^{\infty} \frac{(-1)^n\, 2nx^{2n-1}}{2^{2n}(n!)^2}$$

Note: only X terms change.

□

EXAMPLE 5 Express $1/(1-x)^2$ as a power series by differentiating Equation 1. What is the radius of convergence?

SOLUTION Differentiating each side of the equation

$$\frac{1}{1-x} = 1 + x + x^2 + x^3 + \cdots = \sum_{n=0}^{\infty} x^n$$

we get

$$\frac{1}{(1-x)^2} = 1 + 2x + 3x^2 + \cdots = \sum_{n=1}^{\infty} nx^{n-1}$$

If we wish, we can replace n by $n+1$ and write the answer as

$$\frac{1}{(1-x)^2} = \sum_{n=0}^{\infty} (n+1)x^n$$

According to Theorem 2, the radius of convergence of the differentiated series is the same as the radius of convergence of the original series, namely, $R = 1$. □

EXAMPLE 6 Find a power series representation for $\ln(1-x)$ and its radius of convergence.

SOLUTION We notice that, except for a factor of -1, the derivative of this function is $1/(1-x)$. So we integrate both sides of Equation 1:

$$-\ln(1-x) = \int \frac{1}{1-x}\,dx = \int (1 + x + x^2 + \cdots)\,dx$$

$$= x + \frac{x^2}{2} + \frac{x^3}{3} + \cdots + C = \sum_{n=0}^{\infty} \frac{x^{n+1}}{n+1} + C = \sum_{n=1}^{\infty} \frac{x^n}{n} + C \qquad |x| < 1$$

To determine the value of C we put $x = 0$ in this equation and obtain $-\ln(1-0) = C$. Thus $C = 0$ and

$$\ln(1-x) = -x - \frac{x^2}{2} - \frac{x^3}{3} - \cdots = -\sum_{n=1}^{\infty} \frac{x^n}{n} \qquad |x| < 1$$

The radius of convergence is the same as for the original series: $R = 1$. □

Notice what happens if we put $x = \frac{1}{2}$ in the result of Example 6. Since $\ln \frac{1}{2} = -\ln 2$, we see that

$$\ln 2 = \frac{1}{2} + \frac{1}{8} + \frac{1}{24} + \frac{1}{64} + \cdots = \sum_{n=1}^{\infty} \frac{1}{n\,2^n}$$

EXAMPLE 7 Find a power series representation for $f(x) = \tan^{-1}x$.

SOLUTION We observe that $f'(x) = 1/(1+x^2)$ and find the required series by integrating the power series for $1/(1+x^2)$ found in Example 1.

$$\tan^{-1}x = \int \frac{1}{1+x^2}\,dx = \int (1 - x^2 + x^4 - x^6 + \cdots)\,dx$$

$$= C + x - \frac{x^3}{3} + \frac{x^5}{5} - \frac{x^7}{7} + \cdots$$

To find C we put $x = 0$ and obtain $C = \tan^{-1}0 = 0$. Therefore

$$\tan^{-1}x = x - \frac{x^3}{3} + \frac{x^5}{5} - \frac{x^7}{7} + \cdots = \sum_{n=0}^{\infty} (-1)^n \frac{x^{2n+1}}{2n+1}$$

Since the radius of convergence of the series for $1/(1 + x^2)$ is 1, the radius of convergence of this series for $\tan^{-1}x$ is also 1. □

■ The power series for $\tan^{-1}x$ obtained in Example 7 is called *Gregory's series* after the Scottish mathematician James Gregory (1638–1675), who had anticipated some of Newton's discoveries. We have shown that Gregory's series is valid when $-1 < x < 1$, but it turns out (although it isn't easy to prove) that it is also valid when $x = \pm 1$. Notice that when $x = 1$ the series becomes

$$\frac{\pi}{4} = 1 - \frac{1}{3} + \frac{1}{5} - \frac{1}{7} + \cdots$$

This beautiful result is known as the Leibniz formula for π.

EXAMPLE 8

(a) Evaluate $\int [1/(1 + x^7)]\,dx$ as a power series.

(b) Use part (a) to approximate $\int_0^{0.5} [1/(1 + x^7)]\,dx$ correct to within 10^{-7}.

SOLUTION

(a) The first step is to express the integrand, $1/(1 + x^7)$, as the sum of a power series. As in Example 1, we start with Equation 1 and replace x by $-x^7$:

$$\frac{1}{1 + x^7} = \frac{1}{1 - (-x^7)} = \sum_{n=0}^{\infty} (-x^7)^n$$

$$= \sum_{n=0}^{\infty} (-1)^n x^{7n} = 1 - x^7 + x^{14} - \cdots$$

Now we integrate term by term:

$$\int \frac{1}{1 + x^7}\,dx = \int \sum_{n=0}^{\infty} (-1)^n x^{7n}\,dx = C + \sum_{n=0}^{\infty} (-1)^n \frac{x^{7n+1}}{7n+1}$$

$$= C + x - \frac{x^8}{8} + \frac{x^{15}}{15} - \frac{x^{22}}{22} + \cdots$$

■ This example demonstrates one way in which power series representations are useful. Integrating $1/(1 + x^7)$ by hand is incredibly difficult. Different computer algebra systems return different forms of the answer, but they are all extremely complicated. (If you have a CAS, try it yourself.) The infinite series answer that we obtain in Example 8(a) is actually much easier to deal with than the finite answer provided by a CAS.

This series converges for $|-x^7| < 1$, that is, for $|x| < 1$.

(b) In applying the Fundamental Theorem of Calculus, it doesn't matter which antiderivative we use, so let's use the antiderivative from part (a) with $C = 0$:

$$\int_0^{0.5} \frac{1}{1 + x^7}\,dx = \left[x - \frac{x^8}{8} + \frac{x^{15}}{15} - \frac{x^{22}}{22} + \cdots \right]_0^{1/2}$$

$$= \frac{1}{2} - \frac{1}{8 \cdot 2^8} + \frac{1}{15 \cdot 2^{15}} - \frac{1}{22 \cdot 2^{22}} + \cdots + \frac{(-1)^n}{(7n+1)2^{7n+1}} + \cdots$$

This infinite series is the exact value of the definite integral, but since it is an alternating series, we can approximate the sum using the Alternating Series Estimation Theorem. If we stop adding after the term with $n = 3$, the error is smaller than the term with $n = 4$:

$$\frac{1}{29 \cdot 2^{29}} \approx 6.4 \times 10^{-11}$$

So we have

$$\int_0^{0.5} \frac{1}{1 + x^7}\,dx \approx \frac{1}{2} - \frac{1}{8 \cdot 2^8} + \frac{1}{15 \cdot 2^{15}} - \frac{1}{22 \cdot 2^{22}} \approx 0.49951374$$ □

11.9 EXERCISES

1. If the radius of convergence of the power series $\sum_{n=0}^{\infty} c_n x^n$ is 10, what is the radius of convergence of the series $\sum_{n=1}^{\infty} n c_n x^{n-1}$? Why?

2. Suppose you know that the series $\sum_{n=0}^{\infty} b_n x^n$ converges for $|x| < 2$. What can you say about the following series? Why?

$$\sum_{n=0}^{\infty} \frac{b_n}{n+1} x^{n+1}$$

3–10 Find a power series representation for the function and determine the interval of convergence.

3. $f(x) = \dfrac{1}{1+x}$

4. $f(x) = \dfrac{3}{1-x^4}$

5. $f(x) = \dfrac{2}{3-x}$

6. $f(x) = \dfrac{1}{x+10}$

7. $f(x) = \dfrac{x}{9+x^2}$

8. $f(x) = \dfrac{x}{2x^2+1}$

9. $f(x) = \dfrac{1+x}{1-x}$

10. $f(x) = \dfrac{x^2}{a^3-x^3}$

11–12 Express the function as the sum of a power series by first using partial fractions. Find the interval of convergence.

11. $f(x) = \dfrac{3}{x^2-x-2}$

12. $f(x) = \dfrac{x+2}{2x^2-x-1}$

13. (a) Use differentiation to find a power series representation for

$$f(x) = \frac{1}{(1+x)^2}$$

What is the radius of convergence?

(b) Use part (a) to find a power series for

$$f(x) = \frac{1}{(1+x)^3}$$

(c) Use part (b) to find a power series for

$$f(x) = \frac{x^2}{(1+x)^3}$$

14. (a) Find a power series representation for $f(x) = \ln(1+x)$. What is the radius of convergence?

(b) Use part (a) to find a power series for $f(x) = x\ln(1+x)$.

(c) Use part (a) to find a power series for $f(x) = \ln(x^2+1)$.

15–18 Find a power series representation for the function and determine the radius of convergence.

15. $f(x) = \ln(5-x)$

16. $f(x) = \dfrac{x^2}{(1-2x)^2}$

17. $f(x) = \dfrac{x^3}{(x-2)^2}$

18. $f(x) = \arctan(x/3)$

19–22 Find a power series representation for f, and graph f and several partial sums $s_n(x)$ on the same screen. What happens as n increases?

19. $f(x) = \dfrac{x}{x^2+16}$

20. $f(x) = \ln(x^2+4)$

21. $f(x) = \ln\left(\dfrac{1+x}{1-x}\right)$

22. $f(x) = \tan^{-1}(2x)$

23–26 Evaluate the indefinite integral as a power series. What is the radius of convergence?

23. $\displaystyle\int \frac{t}{1-t^8}\,dt$

24. $\displaystyle\int \frac{\ln(1-t)}{t}\,dt$

25. $\displaystyle\int \frac{x-\tan^{-1}x}{x^3}\,dx$

26. $\displaystyle\int \tan^{-1}(x^2)\,dx$

27–30 Use a power series to approximate the definite integral to six decimal places.

27. $\displaystyle\int_0^{0.2} \frac{1}{1+x^5}\,dx$

28. $\displaystyle\int_0^{0.4} \ln(1+x^4)\,dx$

29. $\displaystyle\int_0^{0.1} x\arctan(3x)\,dx$

30. $\displaystyle\int_0^{0.3} \frac{x^2}{1+x^4}\,dx$

31. Use the result of Example 6 to compute ln 1.1 correct to five decimal places.

32. Show that the function

$$f(x) = \sum_{n=0}^{\infty} \frac{(-1)^n x^{2n}}{(2n)!}$$

is a solution of the differential equation

$$f''(x) + f(x) = 0$$

33. (a) Show that J_0 (the Bessel function of order 0 given in Example 4) satisfies the differential equation

$$x^2 J_0''(x) + xJ_0'(x) + x^2 J_0(x) = 0$$

(b) Evaluate $\int_0^1 J_0(x)\,dx$ correct to three decimal places.

34. The Bessel function of order 1 is defined by

$$J_1(x) = \sum_{n=0}^{\infty} \frac{(-1)^n x^{2n+1}}{n!(n+1)!2^{2n+1}}$$

(a) Show that J_1 satisfies the differential equation

$$x^2 J_1''(x) + xJ_1'(x) + (x^2 - 1)J_1(x) = 0$$

(b) Show that $J_0'(x) = -J_1(x)$.

35. (a) Show that the function

$$f(x) = \sum_{n=0}^{\infty} \frac{x^n}{n!}$$

is a solution of the differential equation

$$f'(x) = f(x)$$

(b) Show that $f(x) = e^x$.

36. Let $f_n(x) = (\sin nx)/n^2$. Show that the series $\Sigma f_n(x)$ converges for all values of x but the series of derivatives $\Sigma f_n'(x)$ diverges when $x = 2n\pi$, n an integer. For what values of x does the series $\Sigma f_n''(x)$ converge?

37. Let

$$f(x) = \sum_{n=1}^{\infty} \frac{x^n}{n^2}$$

Find the intervals of convergence for f, f', and f''.

38. (a) Starting with the geometric series $\Sigma_{n=0}^{\infty} x^n$, find the sum of the series

$$\sum_{n=1}^{\infty} nx^{n-1} \qquad |x| < 1$$

(b) Find the sum of each of the following series.

(i) $\sum_{n=1}^{\infty} nx^n$, $|x| < 1$ (ii) $\sum_{n=1}^{\infty} \frac{n}{2^n}$

(c) Find the sum of each of the following series.

(i) $\sum_{n=2}^{\infty} n(n-1)x^n$, $|x| < 1$

(ii) $\sum_{n=2}^{\infty} \frac{n^2 - n}{2^n}$ (iii) $\sum_{n=1}^{\infty} \frac{n^2}{2^n}$

39. Use the power series for $\tan^{-1}x$ to prove the following expression for π as the sum of an infinite series:

$$\pi = 2\sqrt{3} \sum_{n=0}^{\infty} \frac{(-1)^n}{(2n+1)3^n}$$

40. (a) By completing the square, show that

$$\int_0^{1/2} \frac{dx}{x^2 - x + 1} = \frac{\pi}{3\sqrt{3}}$$

(b) By factoring $x^3 + 1$ as a sum of cubes, rewrite the integral in part (a). Then express $1/(x^3 + 1)$ as the sum of a power series and use it to prove the following formula for π:

$$\pi = \frac{3\sqrt{3}}{4} \sum_{n=0}^{\infty} \frac{(-1)^n}{8^n} \left(\frac{2}{3n+1} + \frac{1}{3n+2} \right)$$

11.10 TAYLOR AND MACLAURIN SERIES

In the preceding section we were able to find power series representations for a certain restricted class of functions. Here we investigate more general problems: Which functions have power series representations? How can we find such representations?

We start by supposing that f is any function that can be represented by a power series

1 $$f(x) = c_0 + c_1(x-a) + c_2(x-a)^2 + c_3(x-a)^3 + c_4(x-a)^4 + \cdots \qquad |x - a| < R$$

Let's try to determine what the coefficients c_n must be in terms of f. To begin, notice that if we put $x = a$ in Equation 1, then all terms after the first one are 0 and we get

$$f(a) = c_0$$

By Theorem 11.9.2, we can differentiate the series in Equation 1 term by term:

2 $$f'(x) = c_1 + 2c_2(x-a) + 3c_3(x-a)^2 + 4c_4(x-a)^3 + \cdots \qquad |x - a| < R$$

and substitution of $x = a$ in Equation 2 gives

$$f'(a) = c_1$$

Now we differentiate both sides of Equation 2 and obtain

$$\boxed{3} \quad f''(x) = 2c_2 + 2 \cdot 3c_3(x - a) + 3 \cdot 4c_4(x - a)^2 + \cdots \qquad |x - a| < R$$

Again we put $x = a$ in Equation 3. The result is

$$f''(a) = 2c_2$$

Let's apply the procedure one more time. Differentiation of the series in Equation 3 gives

$$\boxed{4} \quad f'''(x) = 2 \cdot 3c_3 + 2 \cdot 3 \cdot 4c_4(x - a) + 3 \cdot 4 \cdot 5c_5(x - a)^2 + \cdots \qquad |x - a| < R$$

and substitution of $x = a$ in Equation 4 gives

$$f'''(a) = 2 \cdot 3c_3 = 3!c_3$$

By now you can see the pattern. If we continue to differentiate and substitute $x = a$, we obtain

$$f^{(n)}(a) = 2 \cdot 3 \cdot 4 \cdot \cdots \cdot nc_n = n!c_n$$

Solving this equation for the nth coefficient c_n, we get

$$c_n = \frac{f^{(n)}(a)}{n!}$$

This formula remains valid even for $n = 0$ if we adopt the conventions that $0! = 1$ and $f^{(0)} = f$. Thus we have proved the following theorem.

$\boxed{5}$ **THEOREM** If f has a power series representation (expansion) at a, that is, if

$$f(x) = \sum_{n=0}^{\infty} c_n(x - a)^n \qquad |x - a| < R$$

then its coefficients are given by the formula

$$c_n = \frac{f^{(n)}(a)}{n!}$$

Substituting this formula for c_n back into the series, we see that *if* f has a power series expansion at a, then it must be of the following form.

$$\boxed{6} \quad f(x) = \sum_{n=0}^{\infty} \frac{f^{(n)}(a)}{n!}(x - a)^n$$

$$= f(a) + \frac{f'(a)}{1!}(x - a) + \frac{f''(a)}{2!}(x - a)^2 + \frac{f'''(a)}{3!}(x - a)^3 + \cdots$$

TAYLOR AND MACLAURIN

The Taylor series is named after the English mathematician Brook Taylor (1685–1731) and the Maclaurin series is named in honor of the Scottish mathematician Colin Maclaurin (1698–1746) despite the fact that the Maclaurin series is really just a special case of the Taylor series. But the idea of representing particular functions as sums of power series goes back to Newton, and the general Taylor series was known to the Scottish mathematician James Gregory in 1668 and to the Swiss mathematician John Bernoulli in the 1690s. Taylor was apparently unaware of the work of Gregory and Bernoulli when he published his discoveries on series in 1715 in his book *Methodus incrementorum directa et inversa*. Maclaurin series are named after Colin Maclaurin because he popularized them in his calculus textbook *Treatise of Fluxions* published in 1742.

The series in Equation 6 is called the **Taylor series of the function f at a** (or **about a** or **centered at a**). For the special case $a = 0$ the Taylor series becomes

$$\boxed{7} \qquad f(x) = \sum_{n=0}^{\infty} \frac{f^{(n)}(0)}{n!} x^n = f(0) + \frac{f'(0)}{1!} x + \frac{f''(0)}{2!} x^2 + \cdots$$

This case arises frequently enough that it is given the special name **Maclaurin series**.

NOTE We have shown that *if* f can be represented as a power series about a, then f is equal to the sum of its Taylor series. But there exist functions that are not equal to the sum of their Taylor series. An example of such a function is given in Exercise 70.

EXAMPLE 1 Find the Maclaurin series of the function $f(x) = e^x$ and its radius of convergence.

SOLUTION If $f(x) = e^x$, then $f^{(n)}(x) = e^x$, so $f^{(n)}(0) = e^0 = 1$ for all n. Therefore the Taylor series for f at 0 (that is, the Maclaurin series) is

$$\sum_{n=0}^{\infty} \frac{f^{(n)}(0)}{n!} x^n = \sum_{n=0}^{\infty} \frac{x^n}{n!} = 1 + \frac{x}{1!} + \frac{x^2}{2!} + \frac{x^3}{3!} + \cdots$$

To find the radius of convergence we let $a_n = x^n/n!$. Then

$$\left| \frac{a_{n+1}}{a_n} \right| = \left| \frac{x^{n+1}}{(n+1)!} \cdot \frac{n!}{x^n} \right| = \frac{|x|}{n+1} \to 0 < 1$$

so, by the Ratio Test, the series converges for all x and the radius of convergence is $R = \infty$. □

The conclusion we can draw from Theorem 5 and Example 1 is that *if* e^x has a power series expansion at 0, then

$$e^x = \sum_{n=0}^{\infty} \frac{x^n}{n!}$$

So how can we determine whether e^x *does* have a power series representation?

Let's investigate the more general question: Under what circumstances is a function equal to the sum of its Taylor series? In other words, if f has derivatives of all orders, when is it true that

$$f(x) = \sum_{n=0}^{\infty} \frac{f^{(n)}(a)}{n!} (x - a)^n$$

As with any convergent series, this means that $f(x)$ is the limit of the sequence of partial sums. In the case of the Taylor series, the partial sums are

Partial sum:

$$T_n(x) = \sum_{i=0}^{n} \frac{f^{(i)}(a)}{i!} (x - a)^i$$

$$= f(a) + \frac{f'(a)}{1!}(x - a) + \frac{f''(a)}{2!}(x - a)^2 + \cdots + \frac{f^{(n)}(a)}{n!}(x - a)^n$$

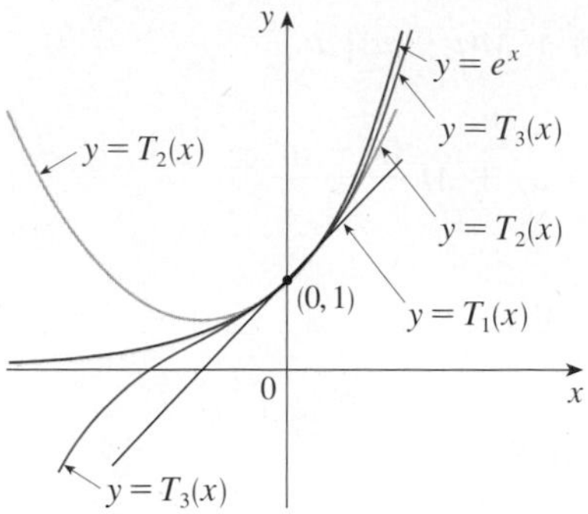

FIGURE 1

■ As n increases, $T_n(x)$ appears to approach e^x in Figure 1. This suggests that e^x is equal to the sum of its Taylor series.

Notice that T_n is a polynomial of degree n called the ***n*th-degree Taylor polynomial of *f* at *a***. For instance, for the exponential function $f(x) = e^x$, the result of Example 1 shows that the Taylor polynomials at 0 (or Maclaurin polynomials) with $n = 1, 2$, and 3 are

$$T_1(x) = 1 + x \qquad T_2(x) = 1 + x + \frac{x^2}{2!} \qquad T_3(x) = 1 + x + \frac{x^2}{2!} + \frac{x^3}{3!}$$

The graphs of the exponential function and these three Taylor polynomials are drawn in Figure 1.

In general, $f(x)$ is the sum of its Taylor series if

$$f(x) = \lim_{n \to \infty} T_n(x)$$

If we let

$$R_n(x) = f(x) - T_n(x) \qquad \text{so that} \qquad f(x) = T_n(x) + R_n(x)$$

then $R_n(x)$ is called the **remainder** of the Taylor series. If we can somehow show that $\lim_{n \to \infty} R_n(x) = 0$, then it follows that

$$\lim_{n \to \infty} T_n(x) = \lim_{n \to \infty} [f(x) - R_n(x)] = f(x) - \lim_{n \to \infty} R_n(x) = f(x)$$

We have therefore proved the following.

8 THEOREM If $f(x) = T_n(x) + R_n(x)$, where T_n is the nth-degree Taylor polynomial of f at a and

$$\lim_{n \to \infty} R_n(x) = 0$$

for $|x - a| < R$, then f is equal to the sum of its Taylor series on the interval $|x - a| < R$.

In trying to show that $\lim_{n \to \infty} R_n(x) = 0$ for a specific function f, we usually use the following fact.

9 TAYLOR'S INEQUALITY If $|f^{(n+1)}(x)| \leq M$ for $|x - a| \leq d$, then the remainder $R_n(x)$ of the Taylor series satisfies the inequality

$$|R_n(x)| \leq \frac{M}{(n+1)!} |x - a|^{n+1} \qquad \text{for } |x - a| \leq d$$

To see why this is true for $n = 1$, we assume that $|f''(x)| \leq M$. In particular, we have $f''(x) \leq M$, so for $a \leq x \leq a + d$ we have

$$\int_a^x f''(t)\,dt \leq \int_a^x M\,dt$$

An antiderivative of f'' is f', so by Part 2 of the Fundamental Theorem of Calculus, we have

$$f'(x) - f'(a) \leq M(x - a) \qquad \text{or} \qquad f'(x) \leq f'(a) + M(x - a)$$

■ As alternatives to Taylor's Inequality, we have the following formulas for the remainder term. If $f^{(n+1)}$ is continuous on an interval I and $x \in I$, then

$$R_n(x) = \frac{1}{n!}\int_a^x (x-t)^n f^{(n+1)}(t)\,dt$$

This is called the *integral form of the remainder term.* Another formula, called *Lagrange's form of the remainder term*, states that there is a number z between x and a such that

$$R_n(x) = \frac{f^{(n+1)}(z)}{(n+1)!}(x-a)^{n+1}$$

This version is an extension of the Mean Value Theorem (which is the case $n = 0$).

Proofs of these formulas, together with discussions of how to use them to solve the examples of Sections 11.10 and 11.11, are given on the website

www.stewartcalculus.com

Click on *Additional Topics* and then on *Formulas for the Remainder Term in Taylor series.*

Thus
$$\int_a^x f'(t)\,dt \leqslant \int_a^x [f'(a) + M(t-a)]\,dt$$

$$f(x) - f(a) \leqslant f'(a)(x-a) + M\frac{(x-a)^2}{2}$$

$$f(x) - f(a) - f'(a)(x-a) \leqslant \frac{M}{2}(x-a)^2$$

But $R_1(x) = f(x) - T_1(x) = f(x) - f(a) - f'(a)(x-a)$. So

$$R_1(x) \leqslant \frac{M}{2}(x-a)^2$$

A similar argument, using $f''(x) \geqslant -M$, shows that

$$R_1(x) \geqslant -\frac{M}{2}(x-a)^2$$

So
$$|R_1(x)| \leqslant \frac{M}{2}|x-a|^2$$

Although we have assumed that $x > a$, similar calculations show that this inequality is also true for $x < a$.

This proves Taylor's Inequality for the case where $n = 1$. The result for any n is proved in a similar way by integrating $n + 1$ times. (See Exercise 69 for the case $n = 2$.)

NOTE In Section 11.11 we will explore the use of Taylor's Inequality in approximating functions. Our immediate use of it is in conjunction with Theorem 8.

In applying Theorems 8 and 9 it is often helpful to make use of the following fact.

10 $$\lim_{n\to\infty} \frac{x^n}{n!} = 0 \qquad \text{for every real number } x$$

This is true because we know from Example 1 that the series $\Sigma\, x^n/n!$ converges for all x and so its nth term approaches 0.

EXAMPLE 2 Prove that e^x is equal to the sum of its Maclaurin series.

SOLUTION If $f(x) = e^x$, then $f^{(n+1)}(x) = e^x$ for all n. If d is any positive number and $|x| \leqslant d$, then $|f^{(n+1)}(x)| = e^x \leqslant e^d$. So Taylor's Inequality, with $a = 0$ and $M = e^d$, says that

$$|R_n(x)| \leqslant \frac{e^d}{(n+1)!}|x|^{n+1} \qquad \text{for } |x| \leqslant d$$

Notice that the same constant $M = e^d$ works for every value of n. But, from Equation 10, we have

$$\lim_{n\to\infty} \frac{e^d}{(n+1)!}|x|^{n+1} = e^d \lim_{n\to\infty} \frac{|x|^{n+1}}{(n+1)!} = 0$$

It follows from the Squeeze Theorem that $\lim_{n\to\infty} |R_n(x)| = 0$ and therefore $\lim_{n\to\infty} R_n(x) = 0$ for all values of x. By Theorem 8, e^x is equal to the sum of its Maclaurin series, that is,

11 $$e^x = \sum_{n=0}^{\infty} \frac{x^n}{n!} \qquad \text{for all } x$$ □

■ In 1748 Leonard Euler used Equation 12 to find the value of e correct to 23 digits. In 2003 Shigeru Kondo, again using the series in (12), computed e to more than 50 billion decimal places. The special techniques employed to speed up the computation are explained on the web page

numbers.computation.free.fr

In particular, if we put $x = 1$ in Equation 11, we obtain the following expression for the number e as a sum of an infinite series:

12 $$e = \sum_{n=0}^{\infty} \frac{1}{n!} = 1 + \frac{1}{1!} + \frac{1}{2!} + \frac{1}{3!} + \cdots$$

EXAMPLE 3 Find the Taylor series for $f(x) = e^x$ at $a = 2$.

SOLUTION We have $f^{(n)}(2) = e^2$ and so, putting $a = 2$ in the definition of a Taylor series (6), we get

$$\sum_{n=0}^{\infty} \frac{f^{(n)}(2)}{n!}(x-2)^n = \sum_{n=0}^{\infty} \frac{e^2}{n!}(x-2)^n$$

Again it can be verified, as in Example 1, that the radius of convergence is $R = \infty$. As in Example 2 we can verify that $\lim_{n\to\infty} R_n(x) = 0$, so

13 $$e^x = \sum_{n=0}^{\infty} \frac{e^2}{n!}(x-2)^n \qquad \text{for all } x$$ □

We have two power series expansions for e^x, the Maclaurin series in Equation 11 and the Taylor series in Equation 13. The first is better if we are interested in values of x near 0 and the second is better if x is near 2.

EXAMPLE 4 Find the Maclaurin series for $\sin x$ and prove that it represents $\sin x$ for all x.

SOLUTION We arrange our computation in two columns as follows:

$$\begin{aligned} f(x) &= \sin x & f(0) &= 0 \\ f'(x) &= \cos x & f'(0) &= 1 \\ f''(x) &= -\sin x & f''(0) &= 0 \\ f'''(x) &= -\cos x & f'''(0) &= -1 \\ f^{(4)}(x) &= \sin x & f^{(4)}(0) &= 0 \end{aligned}$$

Since the derivatives repeat in a cycle of four, we can write the Maclaurin series as follows:

$$f(0) + \frac{f'(0)}{1!}x + \frac{f''(0)}{2!}x^2 + \frac{f'''(0)}{3!}x^3 + \cdots$$

$$= x - \frac{x^3}{3!} + \frac{x^5}{5!} - \frac{x^7}{7!} + \cdots = \sum_{n=0}^{\infty} (-1)^n \frac{x^{2n+1}}{(2n+1)!}$$

Since $f^{(n+1)}(x)$ is $\pm\sin x$ or $\pm\cos x$, we know that $|f^{(n+1)}(x)| \leqslant 1$ for all x. So we can take $M = 1$ in Taylor's Inequality:

14 $$|R_n(x)| \leqslant \frac{M}{(n+1)!}|x^{n+1}| = \frac{|x|^{n+1}}{(n+1)!}$$

By Equation 10 the right side of this inequality approaches 0 as $n \to \infty$, so $|R_n(x)| \to 0$ by the Squeeze Theorem. It follows that $R_n(x) \to 0$ as $n \to \infty$, so $\sin x$ is equal to the sum of its Maclaurin series by Theorem 8. □

■ Figure 2 shows the graph of $\sin x$ together with its Taylor (or Maclaurin) polynomials

$$T_1(x) = x$$

$$T_3(x) = x - \frac{x^3}{3!}$$

$$T_5(x) = x - \frac{x^3}{3!} + \frac{x^5}{5!}$$

Notice that, as n increases, $T_n(x)$ becomes a better approximation to $\sin x$.

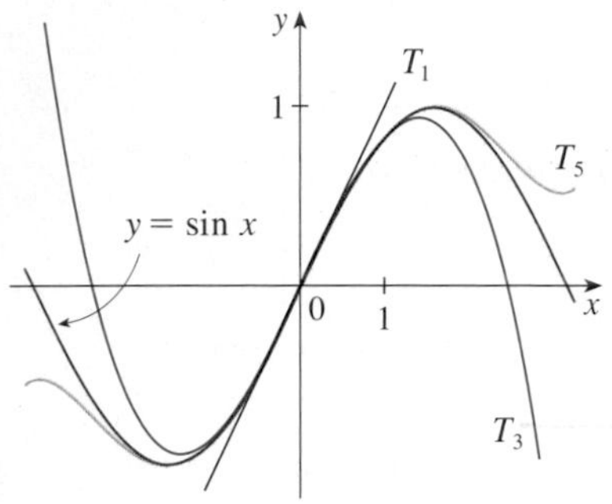

FIGURE 2

We state the result of Example 4 for future reference.

15 $$\sin x = x - \frac{x^3}{3!} + \frac{x^5}{5!} - \frac{x^7}{7!} + \cdots = \sum_{n=0}^{\infty} (-1)^n \frac{x^{2n+1}}{(2n+1)!} \quad \text{for all } x$$

EXAMPLE 5 Find the Maclaurin series for $\cos x$.

SOLUTION We could proceed directly as in Example 4 but it's easier to differentiate the Maclaurin series for $\sin x$ given by Equation 15:

$$\cos x = \frac{d}{dx}(\sin x) = \frac{d}{dx}\left(x - \frac{x^3}{3!} + \frac{x^5}{5!} - \frac{x^7}{7!} + \cdots\right)$$

$$= 1 - \frac{3x^2}{3!} + \frac{5x^4}{5!} - \frac{7x^6}{7!} + \cdots = 1 - \frac{x^2}{2!} + \frac{x^4}{4!} - \frac{x^6}{6!} + \cdots$$

Since the Maclaurin series for $\sin x$ converges for all x, Theorem 2 in Section 11.9 tells us that the differentiated series for $\cos x$ also converges for all x. Thus

■ The Maclaurin series for e^x, $\sin x$, and $\cos x$ that we found in Examples 2, 4, and 5 were discovered, using different methods, by Newton. These equations are remarkable because they say we know everything about each of these functions if we know all its derivatives at the single number 0.

16 $$\cos x = 1 - \frac{x^2}{2!} + \frac{x^4}{4!} - \frac{x^6}{6!} + \cdots = \sum_{n=0}^{\infty} (-1)^n \frac{x^{2n}}{(2n)!} \quad \text{for all } x$$

□

EXAMPLE 6 Find the Maclaurin series for the function $f(x) = x\cos x$.

SOLUTION Instead of computing derivatives and substituting in Equation 7, it's easier to multiply the series for $\cos x$ (Equation 16) by x:

$$x\cos x = x\sum_{n=0}^{\infty} (-1)^n \frac{x^{2n}}{(2n)!} = \sum_{n=0}^{\infty} (-1)^n \frac{x^{2n+1}}{(2n)!}$$ □

EXAMPLE 7 Represent $f(x) = \sin x$ as the sum of its Taylor series centered at $\pi/3$.

SOLUTION Arranging our work in columns, we have

$$\begin{aligned} f(x) &= \sin x & f\left(\frac{\pi}{3}\right) &= \frac{\sqrt{3}}{2} \\ f'(x) &= \cos x & f'\left(\frac{\pi}{3}\right) &= \frac{1}{2} \\ f''(x) &= -\sin x & f''\left(\frac{\pi}{3}\right) &= -\frac{\sqrt{3}}{2} \\ f'''(x) &= -\cos x & f'''\left(\frac{\pi}{3}\right) &= -\frac{1}{2} \end{aligned}$$

■ We have obtained two different series representations for $\sin x$, the Maclaurin series in Example 4 and the Taylor series in Example 7. It is best to use the Maclaurin series for values of x near 0 and the Taylor series for x near $\pi/3$. Notice that the third Taylor polynomial T_3 in Figure 3 is a good approximation to $\sin x$ near $\pi/3$ but not as good near 0. Compare it with the third Maclaurin polynomial T_3 in Figure 2, where the opposite is true.

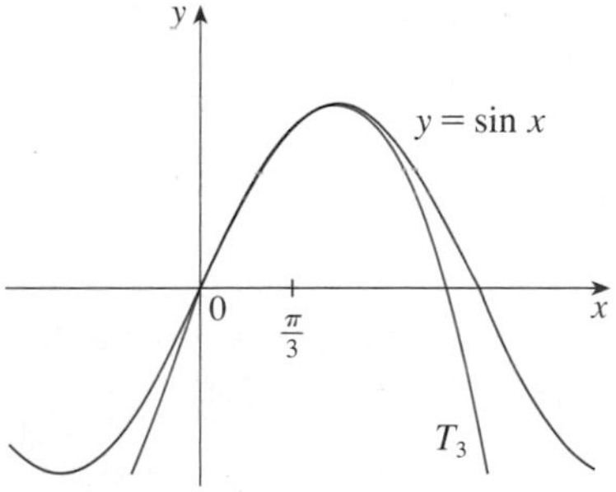

FIGURE 3

and this pattern repeats indefinitely. Therefore the Taylor series at $\pi/3$ is

$$f\left(\frac{\pi}{3}\right) + \frac{f'\left(\frac{\pi}{3}\right)}{1!}\left(x - \frac{\pi}{3}\right) + \frac{f''\left(\frac{\pi}{3}\right)}{2!}\left(x - \frac{\pi}{3}\right)^2 + \frac{f'''\left(\frac{\pi}{3}\right)}{3!}\left(x - \frac{\pi}{3}\right)^3 + \cdots$$

$$= \frac{\sqrt{3}}{2} + \frac{1}{2 \cdot 1!}\left(x - \frac{\pi}{3}\right) - \frac{\sqrt{3}}{2 \cdot 2!}\left(x - \frac{\pi}{3}\right)^2 - \frac{1}{2 \cdot 3!}\left(x - \frac{\pi}{3}\right)^3 + \cdots$$

The proof that this series represents $\sin x$ for all x is very similar to that in Example 4. [Just replace x by $x - \pi/3$ in (14).] We can write the series in sigma notation if we separate the terms that contain $\sqrt{3}$:

$$\sin x = \sum_{n=0}^{\infty} \frac{(-1)^n \sqrt{3}}{2(2n)!}\left(x - \frac{\pi}{3}\right)^{2n} + \sum_{n=0}^{\infty} \frac{(-1)^n}{2(2n+1)!}\left(x - \frac{\pi}{3}\right)^{2n+1}$$ □

The power series that we obtained by indirect methods in Examples 5 and 6 and in Section 11.9 are indeed the Taylor or Maclaurin series of the given functions because Theorem 5 asserts that, no matter how a power series representation $f(x) = \Sigma\, c_n(x - a)^n$ is obtained, it is always true that $c_n = f^{(n)}(a)/n!$. In other words, the coefficients are uniquely determined.

EXAMPLE 8 Find the Maclaurin series for $f(x) = (1 + x)^k$, where k is any real number.

SOLUTION Arranging our work in columns, we have

$$\begin{aligned} f(x) &= (1 + x)^k & f(0) &= 1 \\ f'(x) &= k(1 + x)^{k-1} & f'(0) &= k \\ f''(x) &= k(k - 1)(1 + x)^{k-2} & f''(0) &= k(k - 1) \\ f'''(x) &= k(k - 1)(k - 2)(1 + x)^{k-3} & f'''(0) &= k(k - 1)(k - 2) \\ &\vdots & &\vdots \\ f^{(n)}(x) &= k(k - 1) \cdots (k - n + 1)(1 + x)^{k-n} & f^{(n)}(0) &= k(k - 1) \cdots (k - n + 1) \end{aligned}$$

Therefore the Maclaurin series of $f(x) = (1 + x)^k$ is

$$\sum_{n=0}^{\infty} \frac{f^{(n)}(0)}{n!} x^n = \sum_{n=0}^{\infty} \frac{k(k - 1) \cdots (k - n + 1)}{n!} x^n$$

This series is called the **binomial series**. If its nth term is a_n, then

$$\left|\frac{a_{n+1}}{a_n}\right| = \left|\frac{k(k-1)\cdots(k-n+1)(k-n)x^{n+1}}{(n+1)!} \cdot \frac{n!}{k(k-1)\cdots(k-n+1)x^n}\right|$$

$$= \frac{|k-n|}{n+1}|x| = \frac{\left|1-\dfrac{k}{n}\right|}{1+\dfrac{1}{n}}|x| \to |x| \quad \text{as } n \to \infty$$

Thus, by the Ratio Test, the binomial series converges if $|x| < 1$ and diverges if $|x| > 1$. □

The traditional notation for the coefficients in the binomial series is

$$\binom{k}{n} = \frac{k(k-1)(k-2)\cdots(k-n+1)}{n!}$$

and these numbers are called the **binomial coefficients**.

The following theorem states that $(1+x)^k$ is equal to the sum of its Maclaurin series. It is possible to prove this by showing that the remainder term $R_n(x)$ approaches 0, but that turns out to be quite difficult. The proof outlined in Exercise 71 is much easier.

17 THE BINOMIAL SERIES If k is any real number and $|x| < 1$, then

$$(1+x)^k = \sum_{n=0}^{\infty} \binom{k}{n} x^n = 1 + kx + \frac{k(k-1)}{2!}x^2 + \frac{k(k-1)(k-2)}{3!}x^3 + \cdots$$

Although the binomial series always converges when $|x| < 1$, the question of whether or not it converges at the endpoints, ± 1, depends on the value of k. It turns out that the series converges at 1 if $-1 < k \leqslant 0$ and at both endpoints if $k \geqslant 0$. Notice that if k is a positive integer and $n > k$, then the expression for $\binom{k}{n}$ contains a factor $(k-k)$, so $\binom{k}{n} = 0$ for $n > k$. This means that the series terminates and reduces to the ordinary Binomial Theorem when k is a positive integer. (See Reference Page 1.)

EXAMPLE 9 Find the Maclaurin series for the function $f(x) = \dfrac{1}{\sqrt{4-x}}$ and its radius of convergence.

SOLUTION We write $f(x)$ in a form where we can use the binomial series:

$$\frac{1}{\sqrt{4-x}} = \frac{1}{\sqrt{4\left(1-\dfrac{x}{4}\right)}} = \frac{1}{2\sqrt{1-\dfrac{x}{4}}} = \frac{1}{2}\left(1-\frac{x}{4}\right)^{-1/2}$$

Using the binomial series with $k = -\frac{1}{2}$ and with x replaced by $-x/4$, we have

$$\frac{1}{\sqrt{4-x}} = \frac{1}{2}\left(1 - \frac{x}{4}\right)^{-1/2} = \frac{1}{2}\sum_{n=0}^{\infty}\binom{-\frac{1}{2}}{n}\left(-\frac{x}{4}\right)^n$$

$$= \frac{1}{2}\left[1 + \left(-\frac{1}{2}\right)\left(-\frac{x}{4}\right) + \frac{(-\frac{1}{2})(-\frac{3}{2})}{2!}\left(-\frac{x}{4}\right)^2 + \frac{(-\frac{1}{2})(-\frac{3}{2})(-\frac{5}{2})}{3!}\left(-\frac{x}{4}\right)^3 + \cdots + \frac{(-\frac{1}{2})(-\frac{3}{2})(-\frac{5}{2})\cdots(-\frac{1}{2}-n+1)}{n!}\left(-\frac{x}{4}\right)^n + \cdots\right]$$

this is the Maclaurin Series.

$$= \frac{1}{2}\left[1 + \frac{1}{8}x + \frac{1\cdot 3}{2!8^2}x^2 + \frac{1\cdot 3\cdot 5}{3!8^3}x^3 + \cdots + \frac{1\cdot 3\cdot 5\cdot\cdots\cdot(2n-1)}{n!8^n}x^n + \cdots\right]$$

We know from (17) that this series converges when $|-x/4| < 1$, that is, $|x| < 4$, so the radius of convergence is $R = 4$. □

We collect in the following table, for future reference, some important Maclaurin series that we have derived in this section and the preceding one.

TABLE 1
Important Maclaurin Series and Their Radii of Convergence

$\frac{1}{1-x} = \sum_{n=0}^{\infty} x^n = 1 + x + x^2 + x^3 + \cdots$	$R = 1$
$e^x = \sum_{n=0}^{\infty}\frac{x^n}{n!} = 1 + \frac{x}{1!} + \frac{x^2}{2!} + \frac{x^3}{3!} + \cdots$	$R = \infty$
$\sin x = \sum_{n=0}^{\infty}(-1)^n\frac{x^{2n+1}}{(2n+1)!} = x - \frac{x^3}{3!} + \frac{x^5}{5!} - \frac{x^7}{7!} + \cdots$	$R = \infty$
$\cos x = \sum_{n=0}^{\infty}(-1)^n\frac{x^{2n}}{(2n)!} = 1 - \frac{x^2}{2!} + \frac{x^4}{4!} - \frac{x^6}{6!} + \cdots$	$R = \infty$
$\tan^{-1}x = \sum_{n=0}^{\infty}(-1)^n\frac{x^{2n+1}}{2n+1} = x - \frac{x^3}{3} + \frac{x^5}{5} - \frac{x^7}{7} + \cdots$	$R = 1$
$(1+x)^k = \sum_{n=0}^{\infty}\binom{k}{n}x^n = 1 + kx + \frac{k(k-1)}{2!}x^2 + \frac{k(k-1)(k-2)}{3!}x^3 + \cdots$	$R = 1$

TEC Module 11.10/11.11 enables you to see how successive Taylor polynomials approach the original function.

One reason that Taylor series are important is that they enable us to integrate functions that we couldn't previously handle. In fact, in the introduction to this chapter we mentioned that Newton often integrated functions by first expressing them as power series and then integrating the series term by term. The function $f(x) = e^{-x^2}$ can't be integrated by techniques discussed so far because its antiderivative is not an elementary function (see Section 7.5). In the following example we use Newton's idea to integrate this function.

V EXAMPLE 10

(a) Evaluate $\int e^{-x^2}\,dx$ as an infinite series.

(b) Evaluate $\int_0^1 e^{-x^2}\,dx$ correct to within an error of 0.001.

SOLUTION

(a) First we find the Maclaurin series for $f(x) = e^{-x^2}$. Although it's possible to use the direct method, let's find it simply by replacing x with $-x^2$ in the series for e^x given in Table 1. Thus, for all values of x,

$$e^{-x^2} = \sum_{n=0}^{\infty} \frac{(-x^2)^n}{n!} = \sum_{n=0}^{\infty} (-1)^n \frac{x^{2n}}{n!} = 1 - \frac{x^2}{1!} + \frac{x^4}{2!} - \frac{x^6}{3!} + \cdots$$

Now we integrate term by term:

$$\int e^{-x^2}\,dx = \int \left(1 - \frac{x^2}{1!} + \frac{x^4}{2!} - \frac{x^6}{3!} + \cdots + (-1)^n \frac{x^{2n}}{n!} + \cdots\right) dx$$

$$= C + x - \frac{x^3}{3 \cdot 1!} + \frac{x^5}{5 \cdot 2!} - \frac{x^7}{7 \cdot 3!} + \cdots + (-1)^n \frac{x^{2n+1}}{(2n+1)n!} + \cdots$$

This series converges for all x because the original series for e^{-x^2} converges for all x.

(b) The Fundamental Theorem of Calculus gives

$$\int_0^1 e^{-x^2}\,dx = \left[x - \frac{x^3}{3 \cdot 1!} + \frac{x^5}{5 \cdot 2!} - \frac{x^7}{7 \cdot 3!} + \frac{x^9}{9 \cdot 4!} - \cdots\right]_0^1$$

$$= 1 - \tfrac{1}{3} + \tfrac{1}{10} - \tfrac{1}{42} + \tfrac{1}{216} - \cdots$$

$$\approx 1 - \tfrac{1}{3} + \tfrac{1}{10} - \tfrac{1}{42} + \tfrac{1}{216} \approx 0.7475$$

■ We can take $C = 0$ in the antiderivative in part (a).

The Alternating Series Estimation Theorem shows that the error involved in this approximation is less than

$$\frac{1}{11 \cdot 5!} = \frac{1}{1320} < 0.001$$

□

Another use of Taylor series is illustrated in the next example. The limit could be found with l'Hospital's Rule, but instead we use a series.

EXAMPLE 11 Evaluate $\displaystyle\lim_{x \to 0} \frac{e^x - 1 - x}{x^2}$.

SOLUTION Using the Maclaurin series for e^x, we have

$$\lim_{x \to 0} \frac{e^x - 1 - x}{x^2} = \lim_{x \to 0} \frac{\left(1 + \frac{x}{1!} + \frac{x^2}{2!} + \frac{x^3}{3!} + \cdots\right) - 1 - x}{x^2}$$

$$= \lim_{x \to 0} \frac{\frac{x^2}{2!} + \frac{x^3}{3!} + \frac{x^4}{4!} + \cdots}{x^2}$$

$$= \lim_{x \to 0} \left(\frac{1}{2} + \frac{x}{3!} + \frac{x^2}{4!} + \frac{x^3}{5!} + \cdots\right) = \frac{1}{2}$$

■ Some computer algebra systems compute limits in this way.

because power series are continuous functions.

□

MULTIPLICATION AND DIVISION OF POWER SERIES

If power series are added or subtracted, they behave like polynomials (Theorem 11.2.8 shows this). In fact, as the following example illustrates, they can also be multiplied and divided like polynomials. We find only the first few terms because the calculations for the later terms become tedious and the initial terms are the most important ones.

EXAMPLE 12 Find the first three nonzero terms in the Maclaurin series for (a) $e^x \sin x$ and (b) $\tan x$.

SOLUTION

(a) Using the Maclaurin series for e^x and $\sin x$ in Table 1, we have

$$e^x \sin x = \left(1 + \frac{x}{1!} + \frac{x^2}{2!} + \frac{x^3}{3!} + \cdots\right)\left(x - \frac{x^3}{3!} + \cdots\right)$$

We multiply these expressions, collecting like terms just as for polynomials:

$$\begin{array}{rl}
 & 1 + x + \tfrac{1}{2}x^2 + \tfrac{1}{6}x^3 + \cdots \\
\times & x \qquad\qquad - \tfrac{1}{6}x^3 + \cdots \\
\hline
 & x + x^2 + \tfrac{1}{2}x^3 + \tfrac{1}{6}x^4 + \cdots \\
+ & \qquad - \tfrac{1}{6}x^3 - \tfrac{1}{6}x^4 - \cdots \\
\hline
 & x + x^2 + \tfrac{1}{3}x^3 + \cdots
\end{array}$$

Thus $$e^x \sin x = x + x^2 + \tfrac{1}{3}x^3 + \cdots$$

(b) Using the Maclaurin series in Table 1, we have

$$\tan x = \frac{\sin x}{\cos x} = \frac{x - \dfrac{x^3}{3!} + \dfrac{x^5}{5!} - \cdots}{1 - \dfrac{x^2}{2!} + \dfrac{x^4}{4!} - \cdots}$$

We use a procedure like long division:

$$\begin{array}{r|l}
 & x + \tfrac{1}{3}x^3 + \tfrac{2}{15}x^5 + \cdots \\
\hline
1 - \tfrac{1}{2}x^2 + \tfrac{1}{24}x^4 - \cdots & x - \tfrac{1}{6}x^3 + \tfrac{1}{120}x^5 - \cdots \\
 & x - \tfrac{1}{2}x^3 + \tfrac{1}{24}x^5 - \cdots \\
\cline{2-2}
 & \tfrac{1}{3}x^3 - \tfrac{1}{30}x^5 + \cdots \\
 & \tfrac{1}{3}x^3 - \tfrac{1}{6}x^5 + \cdots \\
\cline{2-2}
 & \tfrac{2}{15}x^5 + \cdots
\end{array}$$

Thus $$\tan x = x + \tfrac{1}{3}x^3 + \tfrac{2}{15}x^5 + \cdots$$ □

Although we have not attempted to justify the formal manipulations used in Example 12, they are legitimate. There is a theorem which states that if both $f(x) = \Sigma\, c_n x^n$ and $g(x) = \Sigma\, b_n x^n$ converge for $|x| < R$ and the series are multiplied as if they were polynomials, then the resulting series also converges for $|x| < R$ and represents $f(x)g(x)$. For division we require $b_0 \neq 0$; the resulting series converges for sufficiently small $|x|$.

11.10 EXERCISES

1. If $f(x) = \sum_{n=0}^{\infty} b_n(x-5)^n$ for all x, write a formula for b_8.

2. The graph of f is shown.

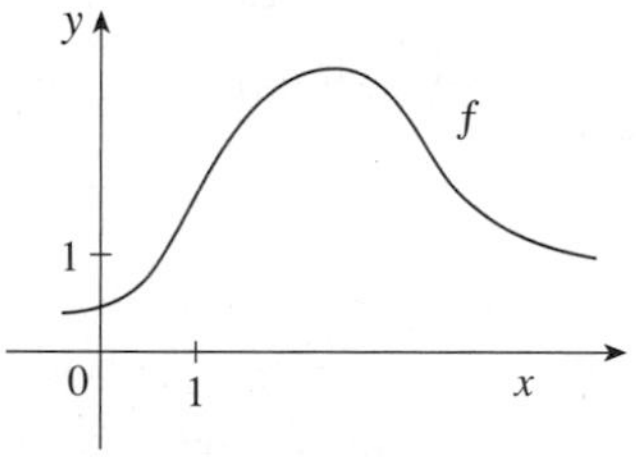

(a) Explain why the series

$$1.6 - 0.8(x-1) + 0.4(x-1)^2 - 0.1(x-1)^3 + \cdots$$

is *not* the Taylor series of f centered at 1.

(b) Explain why the series

$$2.8 + 0.5(x-2) + 1.5(x-2)^2 - 0.1(x-2)^3 + \cdots$$

is *not* the Taylor series of f centered at 2.

3. If $f^{(n)}(0) = (n+1)!$ for $n = 0, 1, 2, \ldots$, find the Maclaurin series for f and its radius of convergence.

4. Find the Taylor series for f centered at 4 if

$$f^{(n)}(4) = \frac{(-1)^n n!}{3^n(n+1)}$$

What is the radius of convergence of the Taylor series?

5–12 Find the Maclaurin series for $f(x)$ using the definition of a Maclaurin series. [Assume that f has a power series expansion. Do not show that $R_n(x) \to 0$.] Also find the associated radius of convergence.

5. $f(x) = (1-x)^{-2}$ **6.** $f(x) = \ln(1+x)$

7. $f(x) = \sin \pi x$ **8.** $f(x) = \cos 3x$

9. $f(x) = e^{5x}$ **10.** $f(x) = xe^x$

11. $f(x) = \sinh x$ **12.** $f(x) = \cosh x$

13–20 Find the Taylor series for $f(x)$ centered at the given value of a. [Assume that f has a power series expansion. Do not show that $R_n(x) \to 0$.]

13. $f(x) = x^4 - 3x^2 + 1, \quad a = 1$

14. $f(x) = x - x^3, \quad a = -2$

15. $f(x) = e^x, \quad a = 3$ **16.** $f(x) = 1/x, \quad a = -3$

17. $f(x) = \cos x, \quad a = \pi$ **18.** $f(x) = \sin x, \quad a = \pi/2$

19. $f(x) = 1/\sqrt{x}, \quad a = 9$ **20.** $f(x) = x^{-2}, \quad a = 1$

21. Prove that the series obtained in Exercise 7 represents $\sin \pi x$ for all x.

22. Prove that the series obtained in Exercise 18 represents $\sin x$ for all x.

23. Prove that the series obtained in Exercise 11 represents $\sinh x$ for all x.

24. Prove that the series obtained in Exercise 12 represents $\cosh x$ for all x.

25–28 Use the binomial series to expand the function as a power series. State the radius of convergence.

25. $\sqrt{1+x}$ **26.** $\dfrac{1}{(1+x)^4}$

27. $\dfrac{1}{(2+x)^3}$ **28.** $(1-x)^{2/3}$

29–38 Use a Maclaurin series in Table 1 to obtain the Maclaurin series for the given function.

29. $f(x) = \sin \pi x$ **30.** $f(x) = \cos(\pi x/2)$

31. $f(x) = e^x + e^{2x}$ **32.** $f(x) = e^x + 2e^{-x}$

33. $f(x) = x\cos\left(\tfrac{1}{2}x^2\right)$ **34.** $f(x) = x^2 \tan^{-1}(x^3)$

35. $f(x) = \dfrac{x}{\sqrt{4+x^2}}$ **36.** $f(x) = \dfrac{x^2}{\sqrt{2+x}}$

37. $f(x) = \sin^2 x \quad \left[\textit{Hint: } \text{Use } \sin^2 x = \tfrac{1}{2}(1-\cos 2x).\right]$

38. $f(x) = \begin{cases} \dfrac{x - \sin x}{x^3} & \text{if } x \neq 0 \\ \tfrac{1}{6} & \text{if } x = 0 \end{cases}$

39–42 Find the Maclaurin series of f (by any method) and its radius of convergence. Graph f and its first few Taylor polynomials on the same screen. What do you notice about the relationship between these polynomials and f?

39. $f(x) = \cos(x^2)$ **40.** $f(x) = e^{-x^2} + \cos x$

41. $f(x) = xe^{-x}$ **42.** $f(x) = \ln(1+x^2)$

43. Use the Maclaurin series for e^x to calculate $e^{-0.2}$ correct to five decimal places.

44. Use the Maclaurin series for $\sin x$ to compute $\sin 3°$ correct to five decimal places.

45. (a) Use the binomial series to expand $1/\sqrt{1 - x^2}$.
(b) Use part (a) to find the Maclaurin series for $\sin^{-1}x$.

46. (a) Expand $1/\sqrt[4]{1 + x}$ as a power series.
(b) Use part (a) to estimate $1/\sqrt[4]{1.1}$ correct to three decimal places.

47–50 Evaluate the indefinite integral as an infinite series.

47. $\int x \cos(x^3)\, dx$

48. $\int \dfrac{e^x - 1}{x}\, dx$

49. $\int \dfrac{\cos x - 1}{x}\, dx$

50. $\int \arctan(x^2)\, dx$

51–54 Use series to approximate the definite integral to within the indicated accuracy.

51. $\int_0^1 x \cos(x^3)\, dx$ (three decimal places)

52. $\int_0^{0.2} [\tan^{-1}(x^3) + \sin(x^3)]\, dx$ (five decimal places)

53. $\int_0^{0.4} \sqrt{1 + x^4}\, dx$ $(|\text{error}| < 5 \times 10^{-6})$

54. $\int_0^{0.5} x^2 e^{-x^2}\, dx$ $(|\text{error}| < 0.001)$

55–57 Use series to evaluate the limit.

55. $\lim_{x \to 0} \dfrac{x - \tan^{-1}x}{x^3}$

56. $\lim_{x \to 0} \dfrac{1 - \cos x}{1 + x - e^x}$

57. $\lim_{x \to 0} \dfrac{\sin x - x + \frac{1}{6}x^3}{x^5}$

58. Use the series in Example 12(b) to evaluate

$$\lim_{x \to 0} \frac{\tan x - x}{x^3}$$

We found this limit in Example 4 in Section 4.4 using l'Hospital's Rule three times. Which method do you prefer?

59–62 Use multiplication or division of power series to find the first three nonzero terms in the Maclaurin series for the function.

59. $y = e^{-x^2} \cos x$

60. $y = \sec x$

61. $y = \dfrac{x}{\sin x}$

62. $y = e^x \ln(1 - x)$

63–68 Find the sum of the series.

63. $\sum_{n=0}^{\infty} (-1)^n \dfrac{x^{4n}}{n!}$

64. $\sum_{n=0}^{\infty} \dfrac{(-1)^n \pi^{2n}}{6^{2n}(2n)!}$

65. $\sum_{n=0}^{\infty} \dfrac{(-1)^n \pi^{2n+1}}{4^{2n+1}(2n + 1)!}$

66. $\sum_{n=0}^{\infty} \dfrac{3^n}{5^n n!}$

67. $3 + \dfrac{9}{2!} + \dfrac{27}{3!} + \dfrac{81}{4!} + \cdots$

68. $1 - \ln 2 + \dfrac{(\ln 2)^2}{2!} - \dfrac{(\ln 2)^3}{3!} + \cdots$

69. Prove Taylor's Inequality for $n = 2$, that is, prove that if $|f'''(x)| \leqslant M$ for $|x - a| \leqslant d$, then

$$|R_2(x)| \leqslant \frac{M}{6}|x - a|^3 \qquad \text{for } |x - a| \leqslant d$$

70. (a) Show that the function defined by

$$f(x) = \begin{cases} e^{-1/x^2} & \text{if } x \neq 0 \\ 0 & \text{if } x = 0 \end{cases}$$

is not equal to its Maclaurin series.
(b) Graph the function in part (a) and comment on its behavior near the origin.

71. Use the following steps to prove (17).
(a) Let $g(x) = \sum_{n=0}^{\infty} \binom{k}{n} x^n$. Differentiate this series to show that

$$g'(x) = \frac{kg(x)}{1 + x} \qquad -1 < x < 1$$

(b) Let $h(x) = (1 + x)^{-k} g(x)$ and show that $h'(x) = 0$.
(c) Deduce that $g(x) = (1 + x)^k$.

72. In Exercise 53 in Section 10.2 it was shown that the length of the ellipse $x = a \sin\theta$, $y = b \cos\theta$, where $a > b > 0$, is

$$L = 4a \int_0^{\pi/2} \sqrt{1 - e^2 \sin^2\theta}\, d\theta$$

where $e = \sqrt{a^2 - b^2}/a$ is the eccentricity of the ellipse. Expand the integrand as a binomial series and use the result of Exercise 46 in Section 7.1 to express L as a series in powers of the eccentricity up to the term in e^6.

LABORATORY PROJECT

CAS AN ELUSIVE LIMIT

This project deals with the function

$$f(x) = \frac{\sin(\tan x) - \tan(\sin x)}{\arcsin(\arctan x) - \arctan(\arcsin x)}$$

1. Use your computer algebra system to evaluate $f(x)$ for $x = 1, 0.1, 0.01, 0.001$, and 0.0001. Does it appear that f has a limit as $x \to 0$?

2. Use the CAS to graph f near $x = 0$. Does it appear that f has a limit as $x \to 0$?

3. Try to evaluate $\lim_{x \to 0} f(x)$ with l'Hospital's Rule, using the CAS to find derivatives of the numerator and denominator. What do you discover? How many applications of l'Hospital's Rule are required?

4. Evaluate $\lim_{x \to 0} f(x)$ by using the CAS to find sufficiently many terms in the Taylor series of the numerator and denominator. (Use the command `taylor` in Maple or `Series` in Mathematica.)

5. Use the limit command on your CAS to find $\lim_{x \to 0} f(x)$ directly. (Most computer algebra systems use the method of Problem 4 to compute limits.)

6. In view of the answers to Problems 4 and 5, how do you explain the results of Problems 1 and 2?

WRITING PROJECT

HOW NEWTON DISCOVERED THE BINOMIAL SERIES

The Binomial Theorem, which gives the expansion of $(a + b)^k$, was known to Chinese mathematicians many centuries before the time of Newton for the case where the exponent k is a positive integer. In 1665, when he was 22, Newton was the first to discover the infinite series expansion of $(a + b)^k$ when k is a fractional exponent (positive or negative). He didn't publish his discovery, but he stated it and gave examples of how to use it in a letter (now called the *epistola prior*) dated June 13, 1676, that he sent to Henry Oldenburg, secretary of the Royal Society of London, to transmit to Leibniz. When Leibniz replied, he asked how Newton had discovered the binomial series. Newton wrote a second letter, the *epistola posterior* of October 24, 1676, in which he explained in great detail how he arrived at his discovery by a very indirect route. He was investigating the areas under the curves $y = (1 - x^2)^{n/2}$ from 0 to x for $n = 0, 1, 2, 3, 4, \ldots$. These are easy to calculate if n is even. By observing patterns and interpolating, Newton was able to guess the answers for odd values of n. Then he realized he could get the same answers by expressing $(1 - x^2)^{n/2}$ as an infinite series.

Write a report on Newton's discovery of the binomial series. Start by giving the statement of the binomial series in Newton's notation (see the *epistola prior* on page 285 of [4] or page 402 of [2]). Explain why Newton's version is equivalent to Theorem 17 on page 742. Then read Newton's *epistola posterior* (page 287 in [4] or page 404 in [2]) and explain the patterns that Newton discovered in the areas under the curves $y = (1 - x^2)^{n/2}$. Show how he was able to guess the areas under the remaining curves and how he verified his answers. Finally, explain how these discoveries led to the binomial series. The books by Edwards [1] and Katz [3] contain commentaries on Newton's letters.

1. C. H. Edwards, *The Historical Development of the Calculus* (New York: Springer-Verlag, 1979), pp. 178–187.

2. John Fauvel and Jeremy Gray, eds., *The History of Mathematics: A Reader* (London: MacMillan Press, 1987).

3. Victor Katz, *A History of Mathematics: An Introduction* (New York: HarperCollins, 1993), pp. 463–466.

4. D. J. Struik, ed., *A Sourcebook in Mathematics, 1200–1800* (Princeton, NJ: Princeton University Press, 1969).

11.11 APPLICATIONS OF TAYLOR POLYNOMIALS

In this section we explore two types of applications of Taylor polynomials. First we look at how they are used to approximate functions—computer scientists like them because polynomials are the simplest of functions. Then we investigate how physicists and engineers use them in such fields as relativity, optics, blackbody radiation, electric dipoles, the velocity of water waves, and building highways across a desert.

APPROXIMATING FUNCTIONS BY POLYNOMIALS

Suppose that $f(x)$ is equal to the sum of its Taylor series at a:

$$f(x) = \sum_{n=0}^{\infty} \frac{f^{(n)}(a)}{n!}(x - a)^n$$

In Section 11.10 we introduced the notation $T_n(x)$ for the nth partial sum of this series and called it the nth-degree Taylor polynomial of f at a. Thus

$$T_n(x) = \sum_{i=0}^{n} \frac{f^{(i)}(a)}{i!}(x - a)^i$$

$$= f(a) + \frac{f'(a)}{1!}(x - a) + \frac{f''(a)}{2!}(x - a)^2 + \cdots + \frac{f^{(n)}(a)}{n!}(x - a)^n$$

Since f is the sum of its Taylor series, we know that $T_n(x) \to f(x)$ as $n \to \infty$ and so T_n can be used as an approximation to f: $f(x) \approx T_n(x)$.

Notice that the first-degree Taylor polynomial

$$T_1(x) = f(a) + f'(a)(x - a)$$

is the same as the linearization of f at a that we discussed in Section 3.10. Notice also that T_1 and its derivative have the same values at a that f and f' have. In general, it can be shown that the derivatives of T_n at a agree with those of f up to and including derivatives of order n (see Exercise 38).

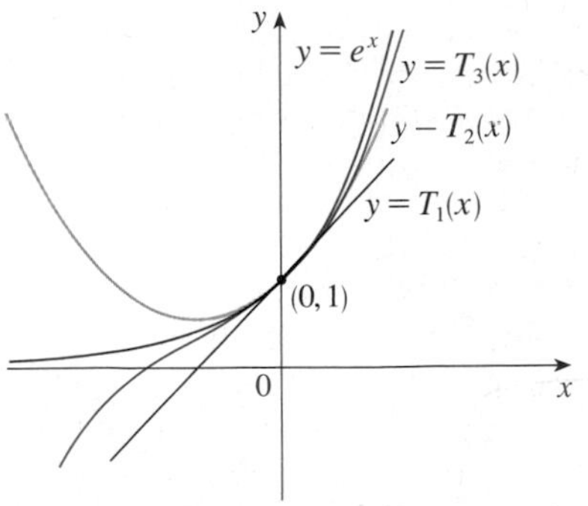

FIGURE 1

To illustrate these ideas let's take another look at the graphs of $y = e^x$ and its first few Taylor polynomials, as shown in Figure 1. The graph of T_1 is the tangent line to $y = e^x$ at (0, 1); this tangent line is the best linear approximation to e^x near (0, 1). The graph of T_2 is the parabola $y = 1 + x + x^2/2$, and the graph of T_3 is the cubic curve $y = 1 + x + x^2/2 + x^3/6$, which is a closer fit to the exponential curve $y = e^x$ than T_2. The next Taylor polynomial T_4 would be an even better approximation, and so on.

	$x = 0.2$	$x = 3.0$
$T_2(x)$	1.220000	8.500000
$T_4(x)$	1.221400	16.375000
$T_6(x)$	1.221403	19.412500
$T_8(x)$	1.221403	20.009152
$T_{10}(x)$	1.221403	20.079665
e^x	1.221403	20.085537

The values in the table give a numerical demonstration of the convergence of the Taylor polynomials $T_n(x)$ to the function $y = e^x$. We see that when $x = 0.2$ the convergence is very rapid, but when $x = 3$ it is somewhat slower. In fact, the farther x is from 0, the more slowly $T_n(x)$ converges to e^x.

When using a Taylor polynomial T_n to approximate a function f, we have to ask the questions: How good an approximation is it? How large should we take n to be in order to achieve a desired accuracy? To answer these questions we need to look at the absolute value of the remainder:

$$|R_n(x)| = |f(x) - T_n(x)|$$

There are three possible methods for estimating the size of the error:

1. If a graphing device is available, we can use it to graph $|R_n(x)|$ and thereby estimate the error.
2. If the series happens to be an alternating series, we can use the Alternating Series Estimation Theorem.
3. In all cases we can use Taylor's Inequality (Theorem 11.10.9), which says that if $|f^{(n+1)}(x)| \leqslant M$, then

$$|R_n(x)| \leqslant \frac{M}{(n+1)!}|x-a|^{n+1}$$

EXAMPLE 1

(a) Approximate the function $f(x) = \sqrt[3]{x}$ by a Taylor polynomial of degree 2 at $a = 8$.
(b) How accurate is this approximation when $7 \leqslant x \leqslant 9$?

SOLUTION

(a)

$$\begin{aligned} f(x) &= \sqrt[3]{x} = x^{1/3} & f(8) &= 2 \\ f'(x) &= \tfrac{1}{3}x^{-2/3} & f'(8) &= \tfrac{1}{12} \\ f''(x) &= -\tfrac{2}{9}x^{-5/3} & f''(8) &= -\tfrac{1}{144} \\ f'''(x) &= \tfrac{10}{27}x^{-8/3} & & \end{aligned}$$

Thus the second-degree Taylor polynomial is

$$\begin{aligned} T_2(x) &= f(8) + \frac{f'(8)}{1!}(x-8) + \frac{f''(8)}{2!}(x-8)^2 \\ &= 2 + \tfrac{1}{12}(x-8) - \tfrac{1}{288}(x-8)^2 \end{aligned}$$

The desired approximation is

$$\sqrt[3]{x} \approx T_2(x) = 2 + \tfrac{1}{12}(x-8) - \tfrac{1}{288}(x-8)^2$$

(b) The Taylor series is not alternating when $x < 8$, so we can't use the Alternating Series Estimation Theorem in this example. But we can use Taylor's Inequality with $n = 2$ and $a = 8$:

$$|R_2(x)| \leqslant \frac{M}{3!}|x-8|^3$$

where $|f'''(x)| \leqslant M$. Because $x \geqslant 7$, we have $x^{8/3} \geqslant 7^{8/3}$ and so

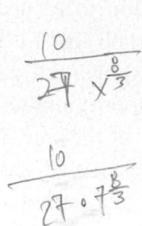

$$f'''(x) = \frac{10}{27} \cdot \frac{1}{x^{8/3}} \leqslant \frac{10}{27} \cdot \frac{1}{7^{8/3}} < 0.0021$$

Therefore we can take $M = 0.0021$. Also $7 \leqslant x \leqslant 9$, so $-1 \leqslant x - 8 \leqslant 1$ and $|x-8| \leqslant 1$. Then Taylor's Inequality gives

$$|R_2(x)| \leqslant \frac{0.0021}{3!} \cdot 1^3 = \frac{0.0021}{6} < 0.0004$$

Thus, if $7 \leqslant x \leqslant 9$, the approximation in part (a) is accurate to within 0.0004. □

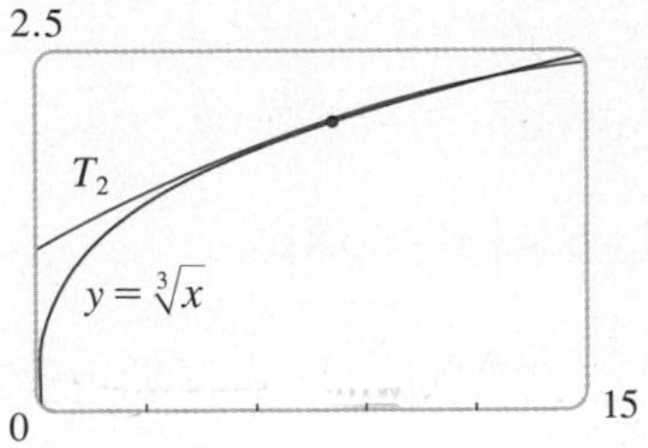

FIGURE 2

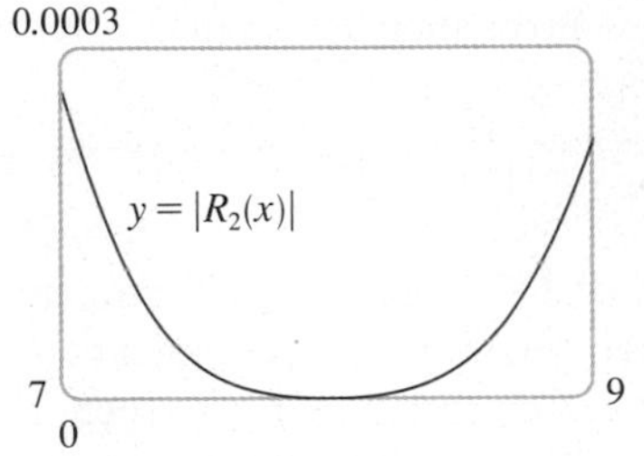

FIGURE 3

Let's use a graphing device to check the calculation in Example 1. Figure 2 shows that the graphs of $y = \sqrt[3]{x}$ and $y = T_2(x)$ are very close to each other when x is near 8. Figure 3 shows the graph of $|R_2(x)|$ computed from the expression

$$|R_2(x)| = |\sqrt[3]{x} - T_2(x)|$$

We see from the graph that

$$|R_2(x)| < 0.0003$$

when $7 \leqslant x \leqslant 9$. Thus the error estimate from graphical methods is slightly better than the error estimate from Taylor's Inequality in this case.

V EXAMPLE 2

(a) What is the maximum error possible in using the approximation

$$\sin x \approx x - \frac{x^3}{3!} + \frac{x^5}{5!}$$

when $-0.3 \leqslant x \leqslant 0.3$? Use this approximation to find $\sin 12°$ correct to six decimal places.

(b) For what values of x is this approximation accurate to within 0.00005?

SOLUTION

(a) Notice that the Maclaurin series

$$\sin x = x - \frac{x^3}{3!} + \frac{x^5}{5!} - \frac{x^7}{7!} + \cdots$$

is alternating for all nonzero values of x, and the successive terms decrease in size because $|x| < 1$, so we can use the Alternating Series Estimation Theorem. The error in approximating $\sin x$ by the first three terms of its Maclaurin series is at most

$$\left|\frac{x^7}{7!}\right| = \frac{|x|^7}{5040}$$

If $-0.3 \leqslant x \leqslant 0.3$, then $|x| \leqslant 0.3$, so the error is smaller than

$$\frac{(0.3)^7}{5040} \approx 4.3 \times 10^{-8}$$

To find $\sin 12°$ we first convert to radian measure.

$$\sin 12° = \sin\left(\frac{12\pi}{180}\right) = \sin\left(\frac{\pi}{15}\right)$$

$$\approx \frac{\pi}{15} - \left(\frac{\pi}{15}\right)^3\frac{1}{3!} + \left(\frac{\pi}{15}\right)^5\frac{1}{5!} \approx 0.20791169$$

Thus, correct to six decimal places, $\sin 12° \approx 0.207912$.

(b) The error will be smaller than 0.00005 if

$$\frac{|x|^7}{5040} < 0.00005$$

Solving this inequality for x, we get

$$|x|^7 < 0.252 \qquad \text{or} \qquad |x| < (0.252)^{1/7} \approx 0.821$$

So the given approximation is accurate to within 0.00005 when $|x| < 0.82$. □

TEC Module 11.10/11.11 graphically shows the remainders in Taylor polynomial approximations.

What if we use Taylor's Inequality to solve Example 2? Since $f^{(7)}(x) = -\cos x$, we have $|f^{(7)}(x)| \leq 1$ and so

$$|R_6(x)| \leq \frac{1}{7!}|x|^7$$

So we get the same estimates as with the Alternating Series Estimation Theorem.

What about graphical methods? Figure 4 shows the graph of

$$|R_6(x)| = \left|\sin x - \left(x - \tfrac{1}{6}x^3 + \tfrac{1}{120}x^5\right)\right|$$

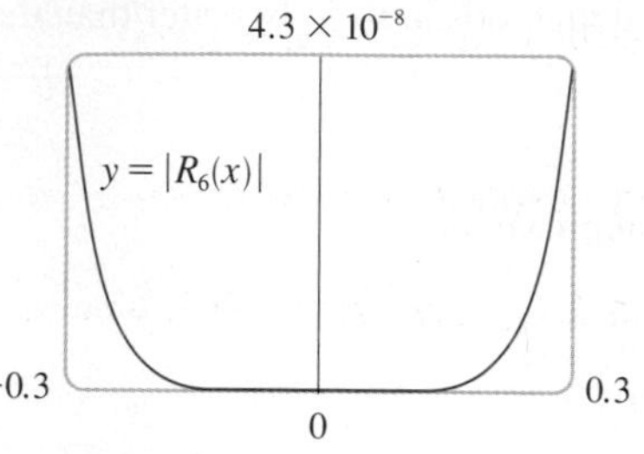

FIGURE 4

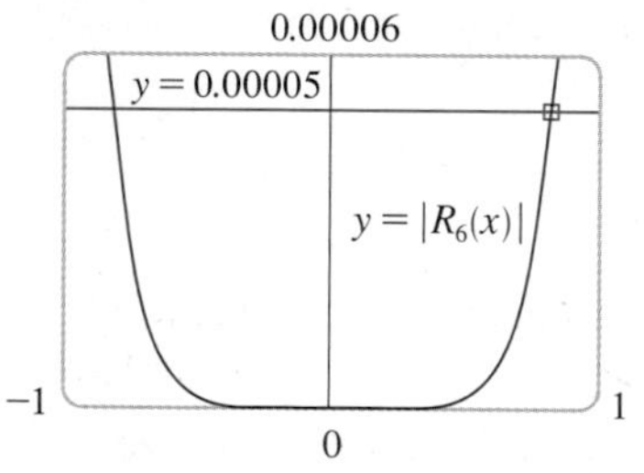

FIGURE 5

and we see from it that $|R_6(x)| < 4.3 \times 10^{-8}$ when $|x| \leq 0.3$. This is the same estimate that we obtained in Example 2. For part (b) we want $|R_6(x)| < 0.00005$, so we graph both $y = |R_6(x)|$ and $y = 0.00005$ in Figure 5. By placing the cursor on the right intersection point we find that the inequality is satisfied when $|x| < 0.82$. Again this is the same estimate that we obtained in the solution to Example 2.

If we had been asked to approximate $\sin 72°$ instead of $\sin 12°$ in Example 2, it would have been wise to use the Taylor polynomials at $a = \pi/3$ (instead of $a = 0$) because they are better approximations to $\sin x$ for values of x close to $\pi/3$. Notice that 72° is close to 60° (or $\pi/3$ radians) and the derivatives of $\sin x$ are easy to compute at $\pi/3$.

Figure 6 shows the graphs of the Maclaurin polynomial approximations

$$T_1(x) = x \qquad\qquad T_3(x) = x - \frac{x^3}{3!}$$

$$T_5(x) = x - \frac{x^3}{3!} + \frac{x^5}{5!} \qquad\qquad T_7(x) = x - \frac{x^3}{3!} + \frac{x^5}{5!} - \frac{x^7}{7!}$$

to the sine curve. You can see that as n increases, $T_n(x)$ is a good approximation to $\sin x$ on a larger and larger interval.

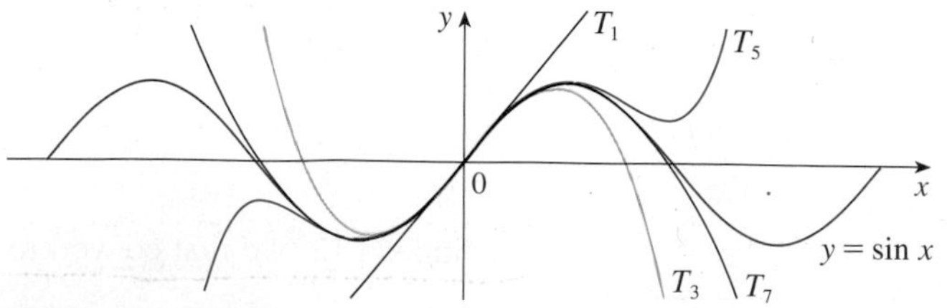

FIGURE 6

One use of the type of calculation done in Examples 1 and 2 occurs in calculators and computers. For instance, when you press the sin or e^x key on your calculator, or when a computer programmer uses a subroutine for a trigonometric or exponential or Bessel function, in many machines a polynomial approximation is calculated. The polynomial is often a Taylor polynomial that has been modified so that the error is spread more evenly throughout an interval.

APPLICATIONS TO PHYSICS

Taylor polynomials are also used frequently in physics. In order to gain insight into an equation, a physicist often simplifies a function by considering only the first two or three terms in its Taylor series. In other words, the physicist uses a Taylor polynomial as an approximation to the function. Taylor's Inequality can then be used to gauge the accuracy of the approximation. The following example shows one way in which this idea is used in special relativity.

EXAMPLE 3 In Einstein's theory of special relativity the mass of an object moving with velocity v is

$$m = \frac{m_0}{\sqrt{1 - v^2/c^2}}$$

where m_0 is the mass of the object when at rest and c is the speed of light. The kinetic energy of the object is the difference between its total energy and its energy at rest:

$$K = mc^2 - m_0c^2$$

(a) Show that when v is very small compared with c, this expression for K agrees with classical Newtonian physics: $K = \frac{1}{2}m_0v^2$.
(b) Use Taylor's Inequality to estimate the difference in these expressions for K when $|v| \leqslant 100$ m/s.

SOLUTION

(a) Using the expressions given for K and m, we get

$$K = mc^2 - m_0c^2 = \frac{m_0c^2}{\sqrt{1 - v^2/c^2}} - m_0c^2$$
$$= m_0c^2\left[\left(1 - \frac{v^2}{c^2}\right)^{-1/2} - 1\right]$$

■ The upper curve in Figure 7 is the graph of the expression for the kinetic energy K of an object with velocity v in special relativity. The lower curve shows the function used for K in classical Newtonian physics. When v is much smaller than the speed of light, the curves are practically identical.

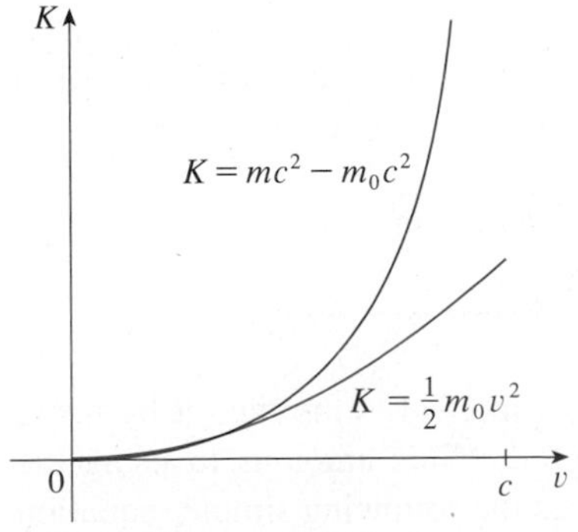

FIGURE 7

With $x = -v^2/c^2$, the Maclaurin series for $(1 + x)^{-1/2}$ is most easily computed as a binomial series with $k = -\frac{1}{2}$. (Notice that $|x| < 1$ because $v < c$.) Therefore we have

$$(1 + x)^{-1/2} = 1 - \tfrac{1}{2}x + \frac{(-\frac{1}{2})(-\frac{3}{2})}{2!}x^2 + \frac{(-\frac{1}{2})(-\frac{3}{2})(-\frac{5}{2})}{3!}x^3 + \cdots$$
$$= 1 - \tfrac{1}{2}x + \tfrac{3}{8}x^2 - \tfrac{5}{16}x^3 + \cdots$$

and

$$K = m_0c^2\left[\left(1 + \frac{1}{2}\frac{v^2}{c^2} + \frac{3}{8}\frac{v^4}{c^4} + \frac{5}{16}\frac{v^6}{c^6} + \cdots\right) - 1\right]$$
$$= m_0c^2\left(\frac{1}{2}\frac{v^2}{c^2} + \frac{3}{8}\frac{v^4}{c^4} + \frac{5}{16}\frac{v^6}{c^6} + \cdots\right)$$

If v is much smaller than c, then all terms after the first are very small when compared with the first term. If we omit them, we get

$$K \approx m_0c^2\left(\frac{1}{2}\frac{v^2}{c^2}\right) = \tfrac{1}{2}m_0v^2$$

(b) If $x = -v^2/c^2$, $f(x) = m_0c^2[(1 + x)^{-1/2} - 1]$, and M is a number such that $|f''(x)| \leqslant M$, then we can use Taylor's Inequality to write

$$|R_1(x)| \leqslant \frac{M}{2!}x^2$$

We have $f''(x) = \frac{3}{4}m_0c^2(1 + x)^{-5/2}$ and we are given that $|v| \leqslant 100$ m/s, so

$$|f''(x)| = \frac{3m_0c^2}{4(1 - v^2/c^2)^{5/2}} \leqslant \frac{3m_0c^2}{4(1 - 100^2/c^2)^{5/2}} \quad (= M)$$

Thus, with $c = 3 \times 10^8$ m/s,

$$|R_1(x)| \leqslant \frac{1}{2} \cdot \frac{3m_0c^2}{4(1 - 100^2/c^2)^{5/2}} \cdot \frac{100^4}{c^4} < (4.17 \times 10^{-10})m_0$$

So when $|v| \leqslant 100$ m/s, the magnitude of the error in using the Newtonian expression for kinetic energy is at most $(4.2 \times 10^{-10})m_0$. □

Another application to physics occurs in optics. Figure 8 is adapted from *Optics,* 4th ed., by Eugene Hecht (San Francisco: Addison-Wesley, 2002), page 153. It depicts a wave from the point source S meeting a spherical interface of radius R centered at C. The ray SA is refracted toward P.

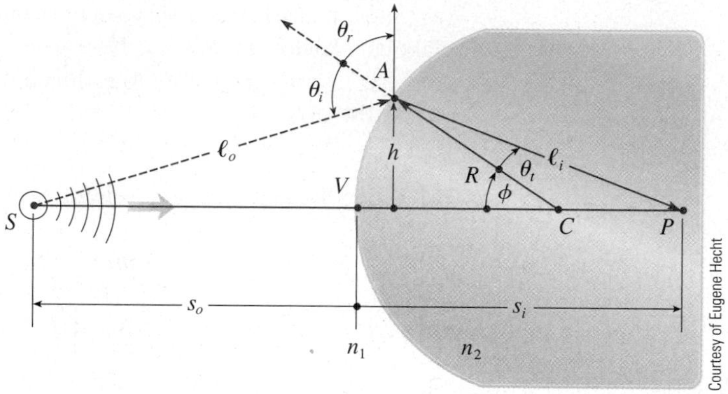

FIGURE 8
Refraction at a spherical interface

Using Fermat's principle that light travels so as to minimize the time taken, Hecht derives the equation

1 $$\frac{n_1}{\ell_o} + \frac{n_2}{\ell_i} = \frac{1}{R}\left(\frac{n_2 s_i}{\ell_i} - \frac{n_1 s_o}{\ell_o}\right)$$

where n_1 and n_2 are indexes of refraction and ℓ_o, ℓ_i, s_o, and s_i are the distances indicated in Figure 8. By the Law of Cosines, applied to triangles ACS and ACP, we have

2 $$\ell_o = \sqrt{R^2 + (s_o + R)^2 - 2R(s_o + R)\cos\phi}$$

$$\ell_i = \sqrt{R^2 + (s_i - R)^2 + 2R(s_i - R)\cos\phi}$$

■ Here we use the identity

$$\cos(\pi - \phi) = -\cos\phi$$

Because Equation 1 is cumbersome to work with, Gauss, in 1841, simplified it by using the linear approximation $\cos\phi \approx 1$ for small values of ϕ. (This amounts to using the Taylor polynomial of degree 1.) Then Equation 1 becomes the following simpler equation [as you are asked to show in Exercise 34(a)]:

3 $$\frac{n_1}{s_o} + \frac{n_2}{s_i} = \frac{n_2 - n_1}{R}$$

The resulting optical theory is known as *Gaussian optics,* or *first-order optics,* and has become the basic theoretical tool used to design lenses.

A more accurate theory is obtained by approximating $\cos\phi$ by its Taylor polynomial of degree 3 (which is the same as the Taylor polynomial of degree 2). This takes into account rays for which ϕ is not so small, that is, rays that strike the surface at greater distances h above the axis. In Exercise 34(b) you are asked to use this approximation to derive the more accurate equation

$$\boxed{4} \quad \frac{n_1}{s_o} + \frac{n_2}{s_i} = \frac{n_2 - n_1}{R} + h^2\left[\frac{n_1}{2s_o}\left(\frac{1}{s_o} + \frac{1}{R}\right)^2 + \frac{n_2}{2s_i}\left(\frac{1}{R} - \frac{1}{s_i}\right)^2\right]$$

The resulting optical theory is known as *third-order optics*.

Other applications of Taylor polynomials to physics and engineering are explored in Exercises 32, 33, 35, 36, and 37 and in the Applied Project on page 757.

11.11 EXERCISES

1. (a) Find the Taylor polynomials up to degree 6 for $f(x) = \cos x$ centered at $a = 0$. Graph f and these polynomials on a common screen.
(b) Evaluate f and these polynomials at $x = \pi/4$, $\pi/2$, and π.
(c) Comment on how the Taylor polynomials converge to $f(x)$.

2. (a) Find the Taylor polynomials up to degree 3 for $f(x) = 1/x$ centered at $a = 1$. Graph f and these polynomials on a common screen.
(b) Evaluate f and these polynomials at $x = 0.9$ and 1.3.
(c) Comment on how the Taylor polynomials converge to $f(x)$.

3–10 Find the Taylor polynomial $T_3(x)$ for the function f at the number a. Graph f and T_3 on the same screen.

3. $f(x) = 1/x, \quad a = 2$

4. $f(x) = x + e^{-x}, \quad a = 0$

5. $f(x) = \cos x, \quad a = \pi/2$

6. $f(x) = e^{-x} \sin x, \quad a = 0$

7. $f(x) = \arcsin x, \quad a = 0$

8. $f(x) = \dfrac{\ln x}{x}, \quad a = 1$

9. $f(x) = xe^{-2x}, \quad a = 0$

10. $f(x) = \tan^{-1} x, \quad a = 1$

CAS **11–12** Use a computer algebra system to find the Taylor polynomials T_n centered at a for $n = 2, 3, 4, 5$. Then graph these polynomials and f on the same screen.

11. $f(x) = \cot x, \quad a = \pi/4$

12. $f(x) = \sqrt[3]{1 + x^2}, \quad a = 0$

13–22
(a) Approximate f by a Taylor polynomial with degree n at the number a.
(b) Use Taylor's Inequality to estimate the accuracy of the approximation $f(x) \approx T_n(x)$ when x lies in the given interval.
(c) Check your result in part (b) by graphing $|R_n(x)|$.

13. $f(x) = \sqrt{x}, \quad a = 4, \quad n = 2, \quad 4 \leqslant x \leqslant 4.2$

14. $f(x) = x^{-2}, \quad a = 1, \quad n = 2, \quad 0.9 \leqslant x \leqslant 1.1$

15. $f(x) = x^{2/3}, \quad a = 1, \quad n = 3, \quad 0.8 \leqslant x \leqslant 1.2$

16. $f(x) = \sin x, \quad a = \pi/6, \quad n = 4, \quad 0 \leqslant x \leqslant \pi/3$

17. $f(x) = \sec x, \quad a = 0, \quad n = 2, \quad -0.2 \leqslant x \leqslant 0.2$

18. $f(x) = \ln(1 + 2x), \quad a = 1, \quad n = 3, \quad 0.5 \leqslant x \leqslant 1.5$

19. $f(x) = e^{x^2}, \quad a = 0, \quad n = 3, \quad 0 \leqslant x \leqslant 0.1$

20. $f(x) = x \ln x, \quad a = 1, \quad n = 3, \quad 0.5 \leqslant x \leqslant 1.5$

21. $f(x) = x \sin x, \quad a = 0, \quad n = 4, \quad -1 \leqslant x \leqslant 1$

22. $f(x) = \sinh 2x, \quad a = 0, \quad n = 5, \quad -1 \leqslant x \leqslant 1$

23. Use the information from Exercise 5 to estimate cos 80° correct to five decimal places.

24. Use the information from Exercise 16 to estimate sin 38° correct to five decimal places.

25. Use Taylor's Inequality to determine the number of terms of the Maclaurin series for e^x that should be used to estimate $e^{0.1}$ to within 0.00001.

26. How many terms of the Maclaurin series for $\ln(1 + x)$ do you need to use to estimate ln 1.4 to within 0.001?

27–29 Use the Alternating Series Estimation Theorem or Taylor's Inequality to estimate the range of values of x for which the given approximation is accurate to within the stated error. Check your answer graphically.

27. $\sin x \approx x - \dfrac{x^3}{6} \quad (|\text{error}| < 0.01)$

28. $\cos x \approx 1 - \dfrac{x^2}{2} + \dfrac{x^4}{24} \quad (|\text{error}| < 0.005)$

29. $\arctan x \approx x - \dfrac{x^3}{3} + \dfrac{x^5}{5} \quad (|\text{error}| < 0.05)$

30. Suppose you know that

$$f^{(n)}(4) = \frac{(-1)^n n!}{3^n(n+1)}$$

and the Taylor series of f centered at 4 converges to $f(x)$ for all x in the interval of convergence. Show that the fifth-degree Taylor polynomial approximates $f(5)$ with error less than 0.0002.

31. A car is moving with speed 20 m/s and acceleration 2 m/s^2 at a given instant. Using a second-degree Taylor polynomial, estimate how far the car moves in the next second. Would it be reasonable to use this polynomial to estimate the distance traveled during the next minute?

32. The resistivity ρ of a conducting wire is the reciprocal of the conductivity and is measured in units of ohm-meters (Ω-m). The resistivity of a given metal depends on the temperature according to the equation

$$\rho(t) = \rho_{20} e^{\alpha(t-20)}$$

where t is the temperature in °C. There are tables that list the values of α (called the temperature coefficient) and ρ_{20} (the resistivity at 20°C) for various metals. Except at very low temperatures, the resistivity varies almost linearly with temperature and so it is common to approximate the expression for $\rho(t)$ by its first- or second-degree Taylor polynomial at $t = 20$.

(a) Find expressions for these linear and quadratic approximations.

(b) For copper, the tables give $\alpha = 0.0039$/°C and $\rho_{20} = 1.7 \times 10^{-8}$ Ω-m. Graph the resistivity of copper and the linear and quadratic approximations for $-250°\text{C} \leqslant t \leqslant 1000°\text{C}$.

(c) For what values of t does the linear approximation agree with the exponential expression to within one percent?

33. An electric dipole consists of two electric charges of equal magnitude and opposite sign. If the charges are q and $-q$ and are located at a distance d from each other, then the electric field E at the point P in the figure is

$$E = \frac{q}{D^2} - \frac{q}{(D+d)^2}$$

By expanding this expression for E as a series in powers of d/D, show that E is approximately proportional to $1/D^3$ when P is far away from the dipole.

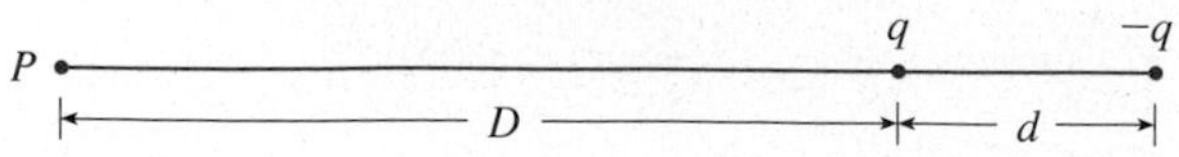

34. (a) Derive Equation 3 for Gaussian optics from Equation 1 by approximating $\cos \phi$ in Equation 2 by its first-degree Taylor polynomial.

(b) Show that if $\cos \phi$ is replaced by its third-degree Taylor polynomial in Equation 2, then Equation 1 becomes Equation 4 for third-order optics. [*Hint:* Use the first two terms in the binomial series for ℓ_o^{-1} and ℓ_i^{-1}. Also, use $\phi \approx \sin \phi$.]

35. If a water wave with length L moves with velocity v across a body of water with depth d, as in the figure, then

$$v^2 = \frac{gL}{2\pi} \tanh \frac{2\pi d}{L}$$

(a) If the water is deep, show that $v \approx \sqrt{gL/(2\pi)}$.

(b) If the water is shallow, use the Maclaurin series for tanh to show that $v \approx \sqrt{gd}$. (Thus in shallow water the velocity of a wave tends to be independent of the length of the wave.)

(c) Use the Alternating Series Estimation Theorem to show that if $L > 10d$, then the estimate $v^2 \approx gd$ is accurate to within $0.014gL$.

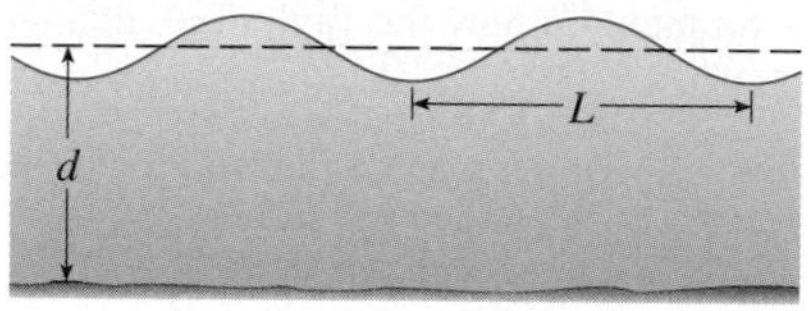

36. The period of a pendulum with length L that makes a maximum angle θ_0 with the vertical is

$$T = 4\sqrt{\frac{L}{g}} \int_0^{\pi/2} \frac{dx}{\sqrt{1 - k^2 \sin^2 x}}$$

where $k = \sin(\frac{1}{2}\theta_0)$ and g is the acceleration due to gravity. (In Exercise 40 in Section 7.7 we approximated this integral using Simpson's Rule.)

(a) Expand the integrand as a binomial series and use the result of Exercise 46 in Section 7.1 to show that

$$T = 2\pi\sqrt{\frac{L}{g}}\left[1 + \frac{1^2}{2^2}k^2 + \frac{1^2 3^2}{2^2 4^2}k^4 + \frac{1^2 3^2 5^2}{2^2 4^2 6^2}k^6 + \cdots\right]$$

If θ_0 is not too large, the approximation $T \approx 2\pi\sqrt{L/g}$, obtained by using only the first term in the series, is often used. A better approximation is obtained by using two terms:

$$T \approx 2\pi\sqrt{\frac{L}{g}}\left(1 + \tfrac{1}{4}k^2\right)$$

(b) Notice that all the terms in the series after the first one have coefficients that are at most $\frac{1}{4}$. Use this fact to compare this series with a geometric series and show that

$$2\pi\sqrt{\frac{L}{g}}\left(1 + \tfrac{1}{4}k^2\right) \leqslant T \leqslant 2\pi\sqrt{\frac{L}{g}}\,\frac{4 - 3k^2}{4 - 4k^2}$$

(c) Use the inequalities in part (b) to estimate the period of a pendulum with $L = 1$ meter and $\theta_0 = 10°$. How does it compare with the estimate $T \approx 2\pi\sqrt{L/g}$? What if $\theta_0 = 42°$?

37. If a surveyor measures differences in elevation when making plans for a highway across a desert, corrections must be made for the curvature of the earth.

(a) If R is the radius of the earth and L is the length of the highway, show that the correction is

$$C = R \sec(L/R) - R$$

(b) Use a Taylor polynomial to show that

$$C \approx \frac{L^2}{2R} + \frac{5L^4}{24R^3}$$

(c) Compare the corrections given by the formulas in parts (a) and (b) for a highway that is 100 km long. (Take the radius of the earth to be 6370 km.)

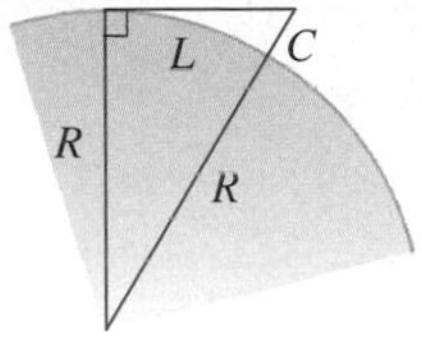

38. Show that T_n and f have the same derivatives at a up to order n.

39. In Section 4.8 we considered Newton's method for approximating a root r of the equation $f(x) = 0$, and from an initial approximation x_1 we obtained successive approximations $x_2, x_3, \ldots$, where

$$x_{n+1} = x_n - \frac{f(x_n)}{f'(x_n)}$$

Use Taylor's Inequality with $n = 1$, $a = x_n$, and $x = r$ to show that if $f''(x)$ exists on an interval I containing r, x_n, and x_{n+1}, and $|f''(x)| \leqslant M$, $|f'(x)| \geqslant K$ for all $x \in I$, then

$$|x_{n+1} - r| \leqslant \frac{M}{2K}|x_n - r|^2$$

[This means that if x_n is accurate to d decimal places, then x_{n+1} is accurate to about $2d$ decimal places. More precisely, if the error at stage n is at most 10^{-m}, then the error at stage $n + 1$ is at most $(M/2K)10^{-2m}$.]

APPLIED PROJECT

RADIATION FROM THE STARS

© Luke Dodd, Photo Researchers, Inc.

Any object emits radiation when heated. A *blackbody* is a system that absorbs all the radiation that falls on it. For instance, a matte black surface or a large cavity with a small hole in its wall (like a blastfurnace) is a blackbody and emits blackbody radiation. Even the radiation from the sun is close to being blackbody radiation.

Proposed in the late 19th century, the Rayleigh-Jeans Law expresses the energy density of blackbody radiation of wavelength λ as

$$f(\lambda) = \frac{8\pi kT}{\lambda^4}$$

where λ is measured in meters, T is the temperature in kelvins (K), and k is Boltzmann's constant. The Rayleigh-Jeans Law agrees with experimental measurements for long wavelengths but disagrees drastically for short wavelengths. [The law predicts that $f(\lambda) \to \infty$ as $\lambda \to 0^+$ but experiments have shown that $f(\lambda) \to 0$.] This fact is known as the *ultraviolet catastrophe*.

In 1900 Max Planck found a better model (known now as Planck's Law) for blackbody radiation:

$$f(\lambda) = \frac{8\pi hc\lambda^{-5}}{e^{hc/(\lambda kT)} - 1}$$

where λ is measured in meters, T is the temperature (in kelvins), and

$$h = \text{Planck's constant} = 6.6262 \times 10^{-34} \text{ J·s}$$

$$c = \text{speed of light} = 2.997925 \times 10^8 \text{ m/s}$$

$$k = \text{Boltzmann's constant} = 1.3807 \times 10^{-23} \text{ J/K}$$

1. Use l'Hospital's Rule to show that

$$\lim_{\lambda \to 0^+} f(\lambda) = 0 \qquad \text{and} \qquad \lim_{\lambda \to \infty} f(\lambda) = 0$$

for Planck's Law. So this law models blackbody radiation better than the Rayleigh-Jeans Law for short wavelengths.

2. Use a Taylor polynomial to show that, for large wavelengths, Planck's Law gives approximately the same values as the Rayleigh-Jeans Law.

3. Graph f as given by both laws on the same screen and comment on the similarities and differences. Use $T = 5700$ K (the temperature of the sun). (You may want to change from meters to the more convenient unit of micrometers: $1\ \mu m = 10^{-6}$ m.)

4. Use your graph in Problem 3 to estimate the value of λ for which $f(\lambda)$ is a maximum under Planck's Law.

5. Investigate how the graph of f changes as T varies. (Use Planck's Law.) In particular, graph f for the stars Betelgeuse ($T = 3400$ K), Procyon ($T = 6400$ K), and Sirius ($T = 9200$ K) as well as the sun. How does the total radiation emitted (the area under the curve) vary with T? Use the graph to comment on why Sirius is known as a blue star and Betelgeuse as a red star.

11 REVIEW

CONCEPT CHECK

1. (a) What is a convergent sequence?
   (b) What is a convergent series?
   (c) What does $\lim_{n\to\infty} a_n = 3$ mean?
   (d) What does $\sum_{n=1}^{\infty} a_n = 3$ mean?

2. (a) What is a bounded sequence?
   (b) What is a monotonic sequence?
   (c) What can you say about a bounded monotonic sequence?

3. (a) What is a geometric series? Under what circumstances is it convergent? What is its sum?
   (b) What is a p-series? Under what circumstances is it convergent?

4. Suppose $\sum a_n = 3$ and s_n is the nth partial sum of the series. What is $\lim_{n\to\infty} a_n$? What is $\lim_{n\to\infty} s_n$?

5. State the following.
   (a) The Test for Divergence
   (b) The Integral Test
   (c) The Comparison Test
   (d) The Limit Comparison Test
   (e) The Alternating Series Test
   (f) The Ratio Test
   (g) The Root Test

6. (a) What is an absolutely convergent series?
   (b) What can you say about such a series?
   (c) What is a conditionally convergent series?

7. (a) If a series is convergent by the Integral Test, how do you estimate its sum?
   (b) If a series is convergent by the Comparison Test, how do you estimate its sum?
   (c) If a series is convergent by the Alternating Series Test, how do you estimate its sum?

8. (a) Write the general form of a power series.
   (b) What is the radius of convergence of a power series?
   (c) What is the interval of convergence of a power series?

9. Suppose $f(x)$ is the sum of a power series with radius of convergence R.
   (a) How do you differentiate f? What is the radius of convergence of the series for f'?
   (b) How do you integrate f? What is the radius of convergence of the series for $\int f(x)\,dx$?

10. (a) Write an expression for the nth-degree Taylor polynomial of f centered at a.
    (b) Write an expression for the Taylor series of f centered at a.
    (c) Write an expression for the Maclaurin series of f.
    (d) How do you show that $f(x)$ is equal to the sum of its Taylor series?
    (e) State Taylor's Inequality.

11. Write the Maclaurin series and the interval of convergence for each of the following functions.
    (a) $1/(1 - x)$ (b) e^x (c) $\sin x$
    (d) $\cos x$ (e) $\tan^{-1}x$

12. Write the binomial series expansion of $(1 + x)^k$. What is the radius of convergence of this series?

TRUE-FALSE QUIZ

Determine whether the statement is true or false. If it is true, explain why. If it is false, explain why or give an example that disproves the statement.

1. If $\lim_{n \to \infty} a_n = 0$, then $\Sigma\, a_n$ is convergent.
2. The series $\Sigma_{n=1}^{\infty} n^{-\sin 1}$ is convergent.
3. If $\lim_{n \to \infty} a_n = L$, then $\lim_{n \to \infty} a_{2n+1} = L$.
4. If $\Sigma\, c_n 6^n$ is convergent, then $\Sigma\, c_n(-2)^n$ is convergent.
5. If $\Sigma\, c_n 6^n$ is convergent, then $\Sigma\, c_n(-6)^n$ is convergent.
6. If $\Sigma\, c_n x^n$ diverges when $x = 6$, then it diverges when $x = 10$.
7. The Ratio Test can be used to determine whether $\Sigma\, 1/n^3$ converges.
8. The Ratio Test can be used to determine whether $\Sigma\, 1/n!$ converges.
9. If $0 \leqslant a_n \leqslant b_n$ and $\Sigma\, b_n$ diverges, then $\Sigma\, a_n$ diverges.
10. $\displaystyle\sum_{n=0}^{\infty} \frac{(-1)^n}{n!} = \frac{1}{e}$
11. If $-1 < \alpha < 1$, then $\lim_{n \to \infty} \alpha^n = 0$.
12. If $\Sigma\, a_n$ is divergent, then $\Sigma\, |a_n|$ is divergent.
13. If $f(x) = 2x - x^2 + \frac{1}{3}x^3 - \cdots$ converges for all x, then $f'''(0) = 2$.
14. If $\{a_n\}$ and $\{b_n\}$ are divergent, then $\{a_n + b_n\}$ is divergent.
15. If $\{a_n\}$ and $\{b_n\}$ are divergent, then $\{a_n b_n\}$ is divergent.
16. If $\{a_n\}$ is decreasing and $a_n > 0$ for all n, then $\{a_n\}$ is convergent.
17. If $a_n > 0$ and $\Sigma\, a_n$ converges, then $\Sigma\, (-1)^n a_n$ converges.
18. If $a_n > 0$ and $\lim_{n \to \infty} (a_{n+1}/a_n) < 1$, then $\lim_{n \to \infty} a_n = 0$.
19. $0.99999\ldots = 1$
20. If $\displaystyle\sum_{n=1}^{\infty} a_n = A$ and $\displaystyle\sum_{n=1}^{\infty} b_n = B$, then $\displaystyle\sum_{n=1}^{\infty} a_n b_n = AB$.

EXERCISES

1–8 Determine whether the sequence is convergent or divergent. If it is convergent, find its limit.

1. $a_n = \dfrac{2 + n^3}{1 + 2n^3}$
2. $a_n = \dfrac{9^{n+1}}{10^n}$
3. $a_n = \dfrac{n^3}{1 + n^2}$
4. $a_n = \cos(n\pi/2)$
5. $a_n = \dfrac{n \sin n}{n^2 + 1}$
6. $a_n = \dfrac{\ln n}{\sqrt{n}}$
7. $\{(1 + 3/n)^{4n}\}$
8. $\{(-10)^n/n!\}$

9. A sequence is defined recursively by the equations $a_1 = 1$, $a_{n+1} = \frac{1}{3}(a_n + 4)$. Show that $\{a_n\}$ is increasing and $a_n < 2$ for all n. Deduce that $\{a_n\}$ is convergent and find its limit.

10. Show that $\lim_{n \to \infty} n^4 e^{-n} = 0$ and use a graph to find the smallest value of N that corresponds to $\varepsilon = 0.1$ in the precise definition of a limit.

11–22 Determine whether the series is convergent or divergent.

11. $\displaystyle\sum_{n=1}^{\infty} \frac{n}{n^3 + 1}$
12. $\displaystyle\sum_{n=1}^{\infty} \frac{n^2 + 1}{n^3 + 1}$
13. $\displaystyle\sum_{n=1}^{\infty} \frac{n^3}{5^n}$
14. $\displaystyle\sum_{n=1}^{\infty} \frac{(-1)^n}{\sqrt{n + 1}}$
15. $\displaystyle\sum_{n=2}^{\infty} \frac{1}{n\sqrt{\ln n}}$
16. $\displaystyle\sum_{n=1}^{\infty} \ln\left(\frac{n}{3n + 1}\right)$
17. $\displaystyle\sum_{n=1}^{\infty} \frac{\cos 3n}{1 + (1.2)^n}$
18. $\displaystyle\sum_{n=1}^{\infty} \frac{n^{2n}}{(1 + 2n^2)^n}$
19. $\displaystyle\sum_{n=1}^{\infty} \frac{1 \cdot 3 \cdot 5 \cdot \cdots \cdot (2n - 1)}{5^n n!}$
20. $\displaystyle\sum_{n=1}^{\infty} \frac{(-5)^{2n}}{n^2 9^n}$
21. $\displaystyle\sum_{n=1}^{\infty} (-1)^{n-1} \frac{\sqrt{n}}{n + 1}$
22. $\displaystyle\sum_{n=1}^{\infty} \frac{\sqrt{n + 1} - \sqrt{n - 1}}{n}$

23–26 Determine whether the series is conditionally convergent, absolutely convergent, or divergent.

23. $\displaystyle\sum_{n=1}^{\infty} (-1)^{n-1} n^{-1/3}$
24. $\displaystyle\sum_{n=1}^{\infty} (-1)^{n-1} n^{-3}$
25. $\displaystyle\sum_{n=1}^{\infty} \frac{(-1)^n (n + 1) 3^n}{2^{2n+1}}$
26. $\displaystyle\sum_{n=2}^{\infty} \frac{(-1)^n \sqrt{n}}{\ln n}$

27–31 Find the sum of the series.

27. $\displaystyle\sum_{n=1}^{\infty} \frac{(-3)^{n-1}}{2^{3n}}$
28. $\displaystyle\sum_{n=1}^{\infty} \frac{1}{n(n + 3)}$
29. $\displaystyle\sum_{n=1}^{\infty} [\tan^{-1}(n + 1) - \tan^{-1} n]$
30. $\displaystyle\sum_{n=0}^{\infty} \frac{(-1)^n \pi^n}{3^{2n}(2n)!}$

31. $1 - e + \dfrac{e^2}{2!} - \dfrac{e^3}{3!} + \dfrac{e^4}{4!} - \cdots$

---

32. Express the repeating decimal 4.17326326326 . . . as a fraction.

33. Show that $\cosh x \geqslant 1 + \frac{1}{2}x^2$ for all x.

34. For what values of x does the series $\Sigma_{n=1}^{\infty} (\ln x)^n$ converge?

35. Find the sum of the series $\displaystyle\sum_{n=1}^{\infty} \frac{(-1)^{n+1}}{n^5}$ correct to four decimal places.

36. (a) Find the partial sum s_5 of the series $\Sigma_{n=1}^{\infty} 1/n^6$ and estimate the error in using it as an approximation to the sum of the series.
(b) Find the sum of this series correct to five decimal places.

37. Use the sum of the first eight terms to approximate the sum of the series $\Sigma_{n=1}^{\infty} (2 + 5^n)^{-1}$. Estimate the error involved in this approximation.

38. (a) Show that the series $\displaystyle\sum_{n=1}^{\infty} \frac{n^n}{(2n)!}$ is convergent.
(b) Deduce that $\displaystyle\lim_{n \to \infty} \frac{n^n}{(2n)!} = 0$.

39. Prove that if the series $\Sigma_{n=1}^{\infty} a_n$ is absolutely convergent, then the series

$$\sum_{n=1}^{\infty} \left(\frac{n+1}{n}\right) a_n$$

is also absolutely convergent.

40–43 Find the radius of convergence and interval of convergence of the series.

40. $\displaystyle\sum_{n=1}^{\infty} (-1)^n \frac{x^n}{n^2 5^n}$

41. $\displaystyle\sum_{n=1}^{\infty} \frac{(x+2)^n}{n\,4^n}$

42. $\displaystyle\sum_{n=1}^{\infty} \frac{2^n(x-2)^n}{(n+2)!}$

43. $\displaystyle\sum_{n=0}^{\infty} \frac{2^n(x-3)^n}{\sqrt{n+3}}$

---

44. Find the radius of convergence of the series

$$\sum_{n=1}^{\infty} \frac{(2n)!}{(n!)^2} x^n$$

45. Find the Taylor series of $f(x) = \sin x$ at $a = \pi/6$.

46. Find the Taylor series of $f(x) = \cos x$ at $a = \pi/3$.

47–54 Find the Maclaurin series for f and its radius of convergence. You may use either the direct method (definition of a Maclaurin series) or known series such as geometric series, binomial series, or the Maclaurin series for e^x, $\sin x$, and $\tan^{-1}x$.

47. $f(x) = \dfrac{x^2}{1 + x}$

48. $f(x) = \tan^{-1}(x^2)$

49. $f(x) = \ln(1 - x)$

50. $f(x) = xe^{2x}$

51. $f(x) = \sin(x^4)$

52. $f(x) = 10^x$

53. $f(x) = 1/\sqrt[4]{16 - x}$

54. $f(x) = (1 - 3x)^{-5}$

---

55. Evaluate $\displaystyle\int \frac{e^x}{x}\,dx$ as an infinite series.

56. Use series to approximate $\int_0^1 \sqrt{1 + x^4}\,dx$ correct to two decimal places.

57–58
(a) Approximate f by a Taylor polynomial with degree n at the number a.
(b) Graph f and T_n on a common screen.
(c) Use Taylor's Inequality to estimate the accuracy of the approximation $f(x) \approx T_n(x)$ when x lies in the given interval.
(d) Check your result in part (c) by graphing $|R_n(x)|$.

57. $f(x) = \sqrt{x}$, $a = 1$, $n = 3$, $0.9 \leqslant x \leqslant 1.1$

58. $f(x) = \sec x$, $a = 0$, $n = 2$, $0 \leqslant x \leqslant \pi/6$

---

59. Use series to evaluate the following limit.

$$\lim_{x \to 0} \frac{\sin x - x}{x^3}$$

60. The force due to gravity on an object with mass m at a height h above the surface of the earth is

$$F = \frac{mgR^2}{(R + h)^2}$$

where R is the radius of the earth and g is the acceleration due to gravity.
(a) Express F as a series in powers of h/R.
(b) Observe that if we approximate F by the first term in the series, we get the expression $F \approx mg$ that is usually used when h is much smaller than R. Use the Alternating Series Estimation Theorem to estimate the range of values of h for which the approximation $F \approx mg$ is accurate to within one percent. (Use $R = 6400$ km.)

61. Suppose that $f(x) = \Sigma_{n=0}^{\infty} c_n x^n$ for all x.
(a) If f is an odd function, show that

$$c_0 = c_2 = c_4 = \cdots = 0$$

(b) If f is an even function, show that

$$c_1 = c_3 = c_5 = \cdots = 0$$

62. If $f(x) = e^{x^2}$, show that $f^{(2n)}(0) = \dfrac{(2n)!}{n!}$.

f of (2n) at 0

PROBLEMS PLUS

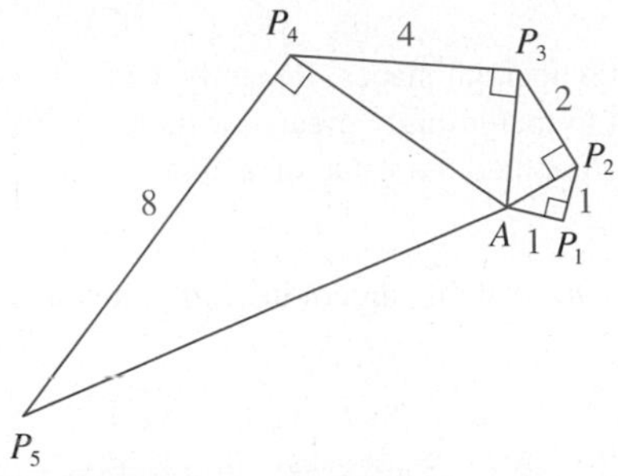

FIGURE FOR PROBLEM 4

1. If $f(x) = \sin(x^3)$, find $f^{(15)}(0)$.

2. A function f is defined by

$$f(x) = \lim_{n \to \infty} \frac{x^{2n} - 1}{x^{2n} + 1}$$

Where is f continuous?

3. (a) Show that $\tan \frac{1}{2}x = \cot \frac{1}{2}x - 2 \cot x$.

(b) Find the sum of the series

$$\sum_{n=1}^{\infty} \frac{1}{2^n} \tan \frac{x}{2^n}$$

4. Let $\{P_n\}$ be a sequence of points determined as in the figure. Thus $|AP_1| = 1$, $|P_nP_{n+1}| = 2^{n-1}$, and angle AP_nP_{n+1} is a right angle. Find $\lim_{n\to\infty} \angle P_nAP_{n+1}$.

5. To construct the **snowflake curve**, start with an equilateral triangle with sides of length 1. Step 1 in the construction is to divide each side into three equal parts, construct an equilateral triangle on the middle part, and then delete the middle part (see the figure). Step 2 is to repeat step 1 for each side of the resulting polygon. This process is repeated at each succeeding step. The snowflake curve is the curve that results from repeating this process indefinitely.

(a) Let s_n, l_n, and p_n represent the number of sides, the length of a side, and the total length of the nth approximating curve (the curve obtained after step n of the construction), respectively. Find formulas for s_n, l_n, and p_n.

(b) Show that $p_n \to \infty$ as $n \to \infty$.

(c) Sum an infinite series to find the area enclosed by the snowflake curve.

Note: Parts (b) and (c) show that the snowflake curve is infinitely long but encloses only a finite area.

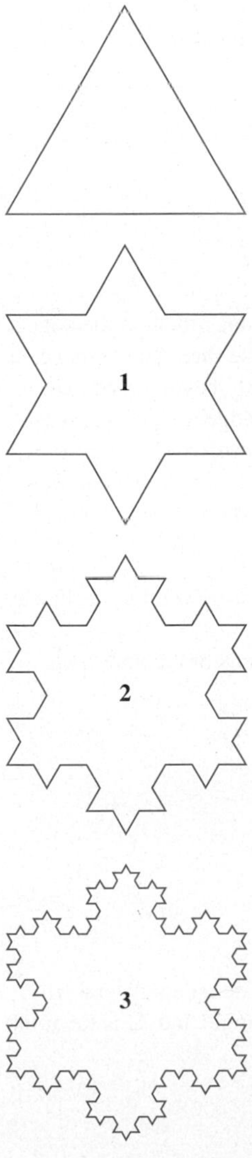

FIGURE FOR PROBLEM 5

6. Find the sum of the series

$$1 + \frac{1}{2} + \frac{1}{3} + \frac{1}{4} + \frac{1}{6} + \frac{1}{8} + \frac{1}{9} + \frac{1}{12} + \cdots$$

where the terms are the reciprocals of the positive integers whose only prime factors are 2s and 3s.

7. (a) Show that for $xy \neq -1$,

$$\arctan x - \arctan y = \arctan \frac{x - y}{1 + xy}$$

if the left side lies between $-\pi/2$ and $\pi/2$.

(b) Show that

$$\arctan \tfrac{120}{119} - \arctan \tfrac{1}{239} = \frac{\pi}{4}$$

(c) Deduce the following formula of John Machin (1680–1751):

$$4 \arctan \tfrac{1}{5} - \arctan \tfrac{1}{239} = \frac{\pi}{4}$$

(d) Use the Maclaurin series for arctan to show that

$$0.197395560 < \arctan \tfrac{1}{5} < 0.197395562$$

(e) Show that

$$0.004184075 < \arctan \tfrac{1}{239} < 0.004184077$$

(f) Deduce that, correct to seven decimal places,

$$\pi \approx 3.1415927$$

Machin used this method in 1706 to find π correct to 100 decimal places. Recently, with the aid of computers, the value of π has been computed to increasingly greater accuracy. Yasumada Kanada of the University of Tokyo recently computed the value of π to a trillion decimal places!

8. (a) Prove a formula similar to the one in Problem 7(a) but involving arccot instead of arctan.
(b) Find the sum of the series

$$\sum_{n=0}^{\infty} \operatorname{arccot}(n^2 + n + 1)$$

9. Find the interval of convergence of $\Sigma_{n=1}^{\infty} n^3 x^n$ and find its sum.

10. If $a_0 + a_1 + a_2 + \cdots + a_k = 0$, show that

$$\lim_{n\to\infty} \left(a_0\sqrt{n} + a_1\sqrt{n+1} + a_2\sqrt{n+2} + \cdots + a_k\sqrt{n+k}\right) = 0$$

If you don't see how to prove this, try the problem-solving strategy of *using analogy* (see page 76). Try the special cases $k = 1$ and $k = 2$ first. If you can see how to prove the assertion for these cases, then you will probably see how to prove it in general.

11. Find the sum of the series $\displaystyle\sum_{n=2}^{\infty} \ln\left(1 - \frac{1}{n^2}\right)$.

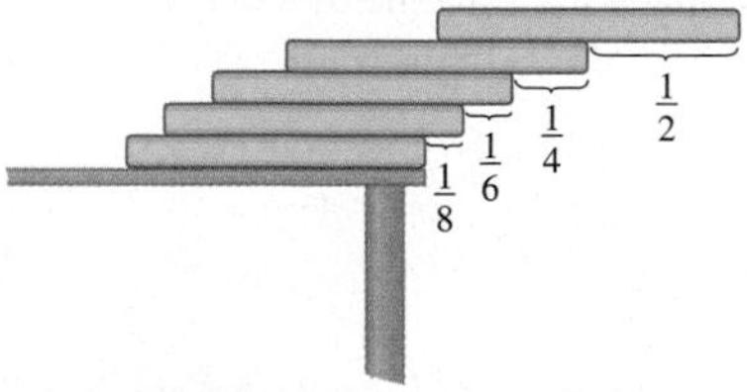

FIGURE FOR PROBLEM 12

12. Suppose you have a large supply of books, all the same size, and you stack them at the edge of a table, with each book extending farther beyond the edge of the table than the one beneath it. Show that it is possible to do this so that the top book extends entirely beyond the table. In fact, show that the top book can extend any distance at all beyond the edge of the table if the stack is high enough. Use the following method of stacking: The top book extends half its length beyond the second book. The second book extends a quarter of its length beyond the third. The third extends one-sixth of its length beyond the fourth, and so on. (Try it yourself with a deck of cards.) Consider centers of mass.

13. If the curve $y = e^{-x/10} \sin x$, $x \geq 0$, is rotated about the x-axis, the resulting solid looks like an infinite decreasing string of beads.
(a) Find the exact volume of the nth bead. (Use either a table of integrals or a computer algebra system.)
(b) Find the total volume of the beads.

14. If $p > 1$, evaluate the expression

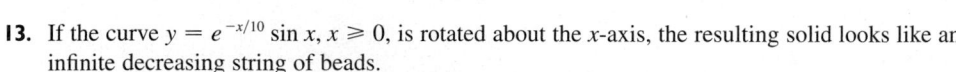

$$\frac{1 + \dfrac{1}{2^p} + \dfrac{1}{3^p} + \dfrac{1}{4^p} + \cdots}{1 - \dfrac{1}{2^p} + \dfrac{1}{3^p} - \dfrac{1}{4^p} + \cdots}$$

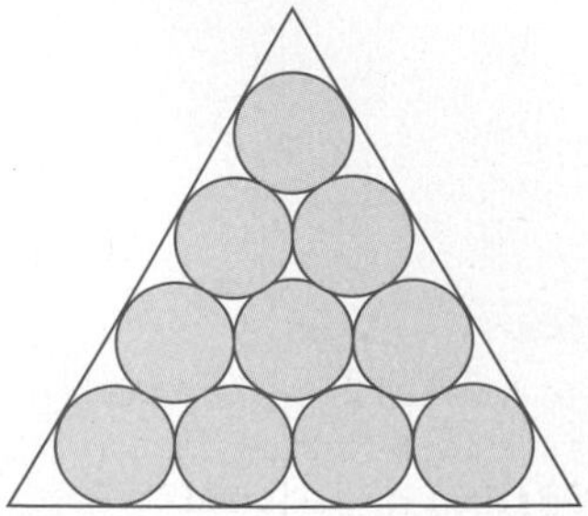

FIGURE FOR PROBLEM 15

15. Suppose that circles of equal diameter are packed tightly in n rows inside an equilateral triangle. (The figure illustrates the case $n = 4$.) If A is the area of the triangle and A_n is the total area occupied by the n rows of circles, show that

$$\lim_{n\to\infty} \frac{A_n}{A} = \frac{\pi}{2\sqrt{3}}$$

16. A sequence $\{a_n\}$ is defined recursively by the equations

$$a_0 = a_1 = 1 \qquad n(n-1)a_n = (n-1)(n-2)a_{n-1} - (n-3)a_{n-2}$$

Find the sum of the series $\sum_{n=0}^{\infty} a_n$.

17. Taking the value of x^x at 0 to be 1 and integrating a series term by term, show that

$$\int_0^1 x^x\,dx = \sum_{n=1}^{\infty} \frac{(-1)^{n-1}}{n^n}$$

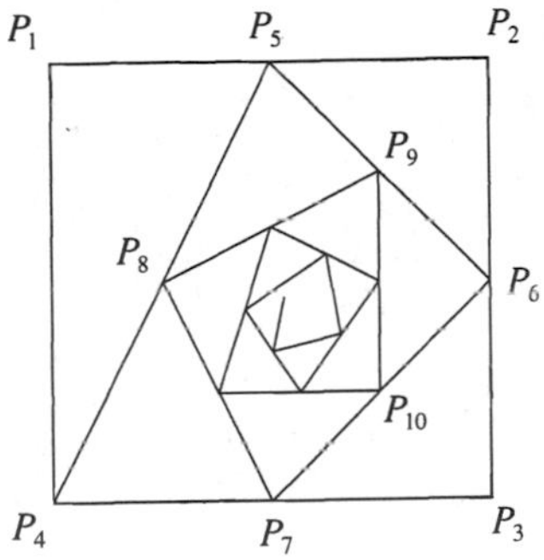

FIGURE FOR PROBLEM 18

18. Starting with the vertices $P_1(0, 1)$, $P_2(1, 1)$, $P_3(1, 0)$, $P_4(0, 0)$ of a square, we construct further points as shown in the figure: P_5 is the midpoint of P_1P_2, P_6 is the midpoint of P_2P_3, P_7 is the midpoint of P_3P_4, and so on. The polygonal spiral path $P_1P_2P_3P_4P_5P_6P_7\ldots$ approaches a point P inside the square.
   (a) If the coordinates of P_n are (x_n, y_n), show that $\frac{1}{2}x_n + x_{n+1} + x_{n+2} + x_{n+3} = 2$ and find a similar equation for the y-coordinates.
   (b) Find the coordinates of P.

19. If $f(x) = \sum_{m=0}^{\infty} c_m x^m$ has positive radius of convergence and $e^{f(x)} = \sum_{n=0}^{\infty} d_n x^n$, show that

$$nd_n = \sum_{i=1}^{n} ic_i d_{n-i} \qquad n \geqslant 1$$

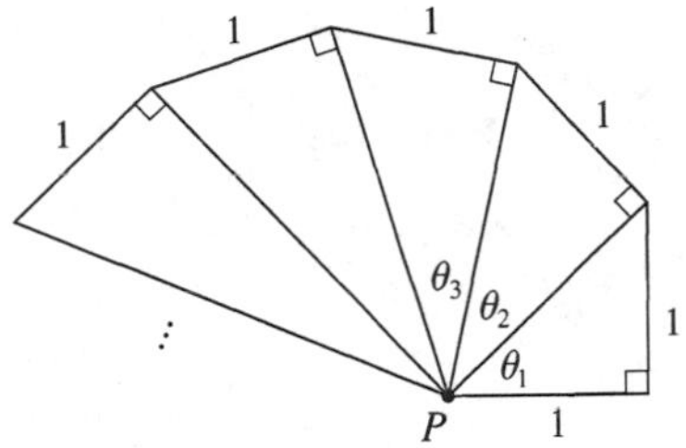

FIGURE FOR PROBLEM 20

20. Right-angled triangles are constructed as in the figure. Each triangle has height 1 and its base is the hypotenuse of the preceding triangle. Show that this sequence of triangles makes indefinitely many turns around P by showing that $\sum \theta_n$ is a divergent series.

21. Consider the series whose terms are the reciprocals of the positive integers that can be written in base 10 notation without using the digit 0. Show that this series is convergent and the sum is less than 90.

22. (a) Show that the Maclaurin series of the function

$$f(x) = \frac{x}{1 - x - x^2} \qquad \text{is} \qquad \sum_{n=1}^{\infty} f_n x^n$$

   where f_n is the nth Fibonacci number, that is, $f_1 = 1$, $f_2 = 1$, and $f_n = f_{n-1} + f_{n-2}$ for $n \geqslant 3$. [*Hint:* Write $x/(1 - x - x^2) = c_0 + c_1x + c_2x^2 + \cdots$ and multiply both sides of this equation by $1 - x - x^2$.]
   (b) By writing $f(x)$ as a sum of partial fractions and thereby obtaining the Maclaurin series in a different way, find an explicit formula for the nth Fibonacci number.

23. Let

$$u = 1 + \frac{x^3}{3!} + \frac{x^6}{6!} + \frac{x^9}{9!} + \cdots$$

$$v = x + \frac{x^4}{4!} + \frac{x^7}{7!} + \frac{x^{10}}{10!} + \cdots$$

$$w = \frac{x^2}{2!} + \frac{x^5}{5!} + \frac{x^8}{8!} + \cdots$$

Show that $u^3 + v^3 + w^3 - 3uvw = 1$.

24. Prove that if $n > 1$, the nth partial sum of the harmonic series is not an integer.
Hint: Let 2^k be the largest power of 2 that is less than or equal to n and let M be the product of all odd integers that are less than or equal to n. Suppose that $s_n = m$, an integer. Then $M2^k s_n = M2^k m$. The right side of this equation is even. Prove that the left side is odd by showing that each of its terms is an even integer, except for the last one.

17

SECOND-ORDER DIFFERENTIAL EQUATIONS

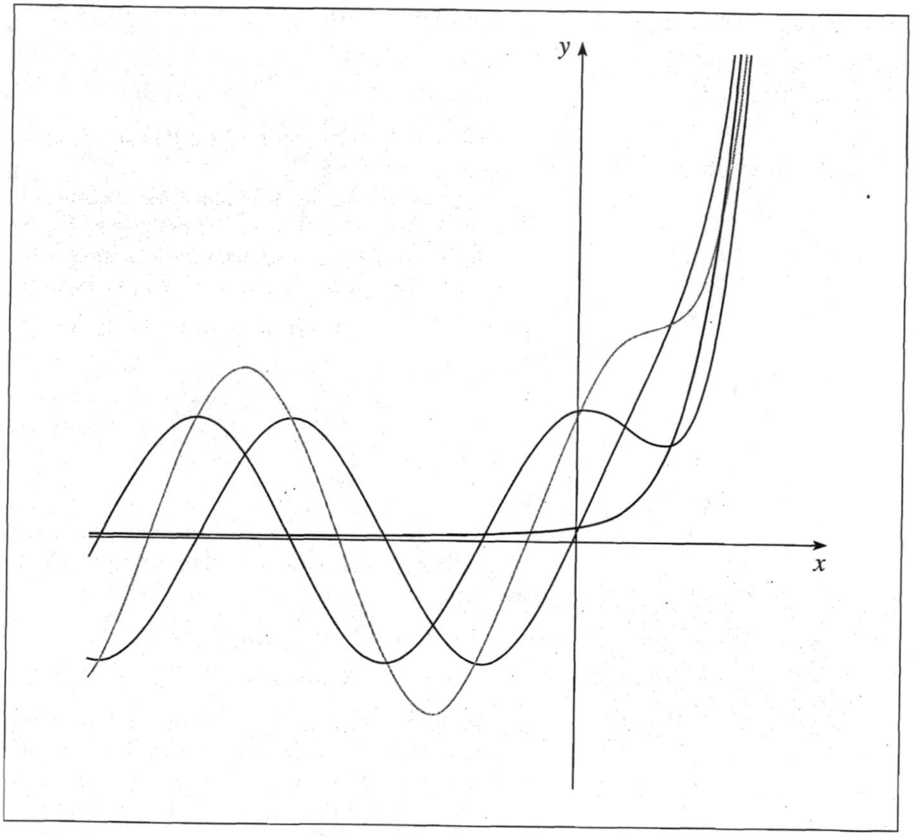

Most of the solutions of the differential equation $y'' + 4y = e^{3x}$ resemble sine functions when x is negative but they all look like exponential functions when x is large.

The basic ideas of differential equations were explained in Chapter 9; there we concentrated on first-order equations. In this chapter we study second-order linear differential equations and learn how they can be applied to solve problems concerning the vibrations of springs and the analysis of electric circuits. We will also see how infinite series can be used to solve differential equations.

17.1 SECOND-ORDER LINEAR EQUATIONS

A **second-order linear differential equation** has the form

$$\boxed{1} \qquad P(x)\frac{d^2y}{dx^2} + Q(x)\frac{dy}{dx} + R(x)y = G(x)$$

where P, Q, R, and G are continuous functions. We saw in Section 9.1 that equations of this type arise in the study of the motion of a spring. In Section 17.3 we will further pursue this application as well as the application to electric circuits.

In this section we study the case where $G(x) = 0$, for all x, in Equation 1. Such equations are called **homogeneous** linear equations. Thus the form of a second-order linear homogeneous differential equation is

$$\boxed{2} \qquad P(x)\frac{d^2y}{dx^2} + Q(x)\frac{dy}{dx} + R(x)y = 0$$

If $G(x) \neq 0$ for some x, Equation 1 is **nonhomogeneous** and is discussed in Section 17.2.

Two basic facts enable us to solve homogeneous linear equations. The first of these says that if we know two solutions y_1 and y_2 of such an equation, then the **linear combination** $y = c_1y_1 + c_2y_2$ is also a solution.

> **3 THEOREM** If $y_1(x)$ and $y_2(x)$ are both solutions of the linear homogeneous equation (2) and c_1 and c_2 are any constants, then the function
>
> $$y(x) = c_1y_1(x) + c_2y_2(x)$$
>
> is also a solution of Equation 2.

PROOF Since y_1 and y_2 are solutions of Equation 2, we have

$$P(x)y_1'' + Q(x)y_1' + R(x)y_1 = 0$$

and

$$P(x)y_2'' + Q(x)y_2' + R(x)y_2 = 0$$

Therefore, using the basic rules for differentiation, we have

$$\begin{aligned}
P(x)y'' &+ Q(x)y' + R(x)y \\
&= P(x)(c_1y_1 + c_2y_2)'' + Q(x)(c_1y_1 + c_2y_2)' + R(x)(c_1y_1 + c_2y_2) \\
&= P(x)(c_1y_1'' + c_2y_2'') + Q(x)(c_1y_1' + c_2y_2') + R(x)(c_1y_1 + c_2y_2) \\
&= c_1[P(x)y_1'' + Q(x)y_1' + R(x)y_1] + c_2[P(x)y_2'' + Q(x)y_2' + R(x)y_2] \\
&= c_1(0) + c_2(0) = 0
\end{aligned}$$

Thus $y = c_1y_1 + c_2y_2$ is a solution of Equation 2. □

The other fact we need is given by the following theorem, which is proved in more advanced courses. It says that the general solution is a linear combination of two **linearly independent** solutions y_1 and y_2. This means that neither y_1 nor y_2 is a constant multiple of the other. For instance, the functions $f(x) = x^2$ and $g(x) = 5x^2$ are linearly dependent, but $f(x) = e^x$ and $g(x) = xe^x$ are linearly independent.

4 THEOREM If y_1 and y_2 are linearly independent solutions of Equation 2, and $P(x)$ is never 0, then the general solution is given by

$$y(x) = c_1 y_1(x) + c_2 y_2(x)$$

where c_1 and c_2 are arbitrary constants.

Theorem 4 is very useful because it says that if we know *two* particular linearly independent solutions, then we know *every* solution.

In general, it is not easy to discover particular solutions to a second-order linear equation. But it is always possible to do so if the coefficient functions P, Q, and R are constant functions, that is, if the differential equation has the form

$$ay'' + by' + cy = 0 \tag{5}$$

where a, b, and c are constants and $a \neq 0$.

It's not hard to think of some likely candidates for particular solutions of Equation 5 if we state the equation verbally. We are looking for a function y such that a constant times its second derivative y'' plus another constant times y' plus a third constant times y is equal to 0. We know that the exponential function $y = e^{rx}$ (where r is a constant) has the property that its derivative is a constant multiple of itself: $y' = re^{rx}$. Furthermore, $y'' = r^2e^{rx}$. If we substitute these expressions into Equation 5, we see that $y = e^{rx}$ is a solution if

$$ar^2e^{rx} + bre^{rx} + ce^{rx} = 0$$

or

$$(ar^2 + br + c)e^{rx} = 0$$

But e^{rx} is never 0. Thus $y = e^{rx}$ is a solution of Equation 5 if r is a root of the equation

$$ar^2 + br + c = 0 \tag{6}$$

Equation 6 is called the **auxiliary equation** (or **characteristic equation**) of the differential equation $ay'' + by' + cy = 0$. Notice that it is an algebraic equation that is obtained from the differential equation by replacing y'' by r^2, y' by r, and y by 1.

Sometimes the roots r_1 and r_2 of the auxiliary equation can be found by factoring. In other cases they are found by using the quadratic formula:

$$r_1 = \frac{-b + \sqrt{b^2 - 4ac}}{2a} \qquad r_2 = \frac{-b - \sqrt{b^2 - 4ac}}{2a} \tag{7}$$

We distinguish three cases according to the sign of the discriminant $b^2 - 4ac$.

■ **CASE I $b^2 - 4ac > 0$**

In this case the roots r_1 and r_2 of the auxiliary equation are real and distinct, so $y_1 = e^{r_1 x}$ and $y_2 = e^{r_2 x}$ are two linearly independent solutions of Equation 5. (Note that $e^{r_2 x}$ is not a constant multiple of $e^{r_1 x}$.) Therefore, by Theorem 4, we have the following fact.

8 If the roots r_1 and r_2 of the auxiliary equation $ar^2 + br + c = 0$ are real and unequal, then the general solution of $ay'' + by' + cy = 0$ is

$$y = c_1 e^{r_1 x} + c_2 e^{r_2 x}$$

■ In Figure 1 the graphs of the basic solutions $f(x) = e^{2x}$ and $g(x) = e^{-3x}$ of the differential equation in Example 1 are shown in blue and red, respectively. Some of the other solutions, linear combinations of f and g, are shown in black.

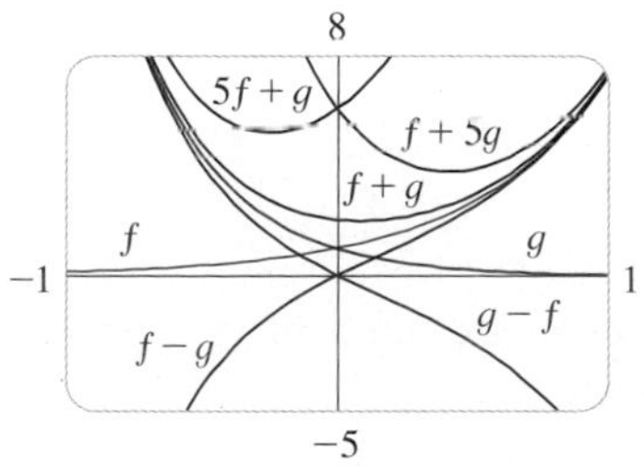

FIGURE 1

EXAMPLE 1 Solve the equation $y'' + y' - 6y = 0$.

SOLUTION The auxiliary equation is

$$r^2 + r - 6 = (r - 2)(r + 3) = 0$$

whose roots are $r = 2, -3$. Therefore, by (8), the general solution of the given differential equation is

$$y = c_1 e^{2x} + c_2 e^{-3x}$$

We could verify that this is indeed a solution by differentiating and substituting into the differential equation. □

EXAMPLE 2 Solve $3\dfrac{d^2y}{dx^2} + \dfrac{dy}{dx} - y = 0$.

SOLUTION To solve the auxiliary equation $3r^2 + r - 1 = 0$, we use the quadratic formula:

$$r = \frac{-1 \pm \sqrt{13}}{6}$$

Since the roots are real and distinct, the general solution is

$$y = c_1 e^{(-1+\sqrt{13})x/6} + c_2 e^{(-1-\sqrt{13})x/6}$$ □

■ **CASE II $b^2 - 4ac = 0$**

In this case $r_1 = r_2$; that is, the roots of the auxiliary equation are real and equal. Let's denote by r the common value of r_1 and r_2. Then, from Equations 7, we have

9 $$r = -\frac{b}{2a} \qquad \text{so} \quad 2ar + b = 0$$

We know that $y_1 = e^{rx}$ is one solution of Equation 5. We now verify that $y_2 = xe^{rx}$ is also a solution:

$$\begin{aligned} ay_2'' + by_2' + cy_2 &= a(2re^{rx} + r^2xe^{rx}) + b(e^{rx} + rxe^{rx}) + cxe^{rx} \\ &= (2ar + b)e^{rx} + (ar^2 + br + c)xe^{rx} \\ &= 0(e^{rx}) + 0(xe^{rx}) = 0 \end{aligned}$$

The first term is 0 by Equations 9; the second term is 0 because r is a root of the auxiliary equation. Since $y_1 = e^{rx}$ and $y_2 = xe^{rx}$ are linearly independent solutions, Theorem 4 provides us with the general solution.

> **10** If the auxiliary equation $ar^2 + br + c = 0$ has only one real root r, then the general solution of $ay'' + by' + cy = 0$ is
>
> $$y = c_1 e^{rx} + c_2 x e^{rx}$$

■ Figure 2 shows the basic solutions $f(x) = e^{-3x/2}$ and $g(x) = xe^{-3x/2}$ in Example 3 and some other members of the family of solutions. Notice that all of them approach 0 as $x \to \infty$.

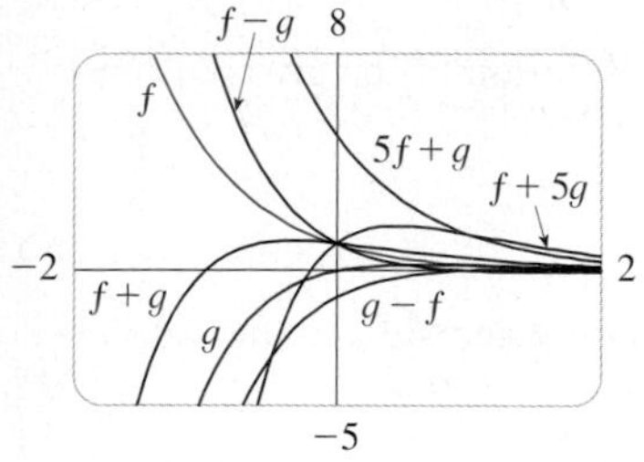

FIGURE 2

EXAMPLE 3 Solve the equation $4y'' + 12y' + 9y = 0$.

SOLUTION The auxiliary equation $4r^2 + 12r + 9 = 0$ can be factored as

$$(2r + 3)^2 = 0$$

so the only root is $r = -\frac{3}{2}$. By (10), the general solution is

$$y = c_1 e^{-3x/2} + c_2 x e^{-3x/2}$$ □

■ CASE III $b^2 - 4ac < 0$

In this case the roots r_1 and r_2 of the auxiliary equation are complex numbers. (See Appendix H for information about complex numbers.) We can write

$$r_1 = \alpha + i\beta \qquad r_2 = \alpha - i\beta$$

where α and β are real numbers. [In fact, $\alpha = -b/(2a)$, $\beta = \sqrt{4ac - b^2}/(2a)$.] Then, using Euler's equation

$$e^{i\theta} = \cos\theta + i\sin\theta$$

from Appendix H, we write the solution of the differential equation as

$$\begin{aligned} y &= C_1 e^{r_1 x} + C_2 e^{r_2 x} = C_1 e^{(\alpha + i\beta)x} + C_2 e^{(\alpha - i\beta)x} \\ &= C_1 e^{\alpha x}(\cos\beta x + i\sin\beta x) + C_2 e^{\alpha x}(\cos\beta x - i\sin\beta x) \\ &= e^{\alpha x}[(C_1 + C_2)\cos\beta x + i(C_1 - C_2)\sin\beta x] \\ &= e^{\alpha x}(c_1\cos\beta x + c_2\sin\beta x) \end{aligned}$$

where $c_1 = C_1 + C_2$, $c_2 = i(C_1 - C_2)$. This gives all solutions (real or complex) of the differential equation. The solutions are real when the constants c_1 and c_2 are real. We summarize the discussion as follows.

> **11** If the roots of the auxiliary equation $ar^2 + br + c = 0$ are the complex numbers $r_1 = \alpha + i\beta$, $r_2 = \alpha - i\beta$, then the general solution of $ay'' + by' + cy = 0$ is
>
> $$y = e^{\alpha x}(c_1\cos\beta x + c_2\sin\beta x)$$

■ Figure 3 shows the graphs of the solutions in Example 4, $f(x) = e^{3x} \cos 2x$ and $g(x) = e^{3x} \sin 2x$, together with some linear combinations. All solutions approach 0 as $x \to -\infty$.

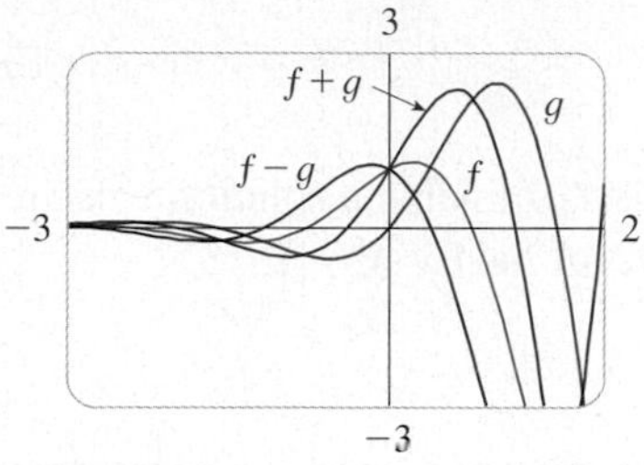

FIGURE 3

V EXAMPLE 4 Solve the equation $y'' - 6y' + 13y = 0$.

SOLUTION The auxiliary equation is $r^2 - 6r + 13 = 0$. By the quadratic formula, the roots are

$$r = \frac{6 \pm \sqrt{36 - 52}}{2} = \frac{6 \pm \sqrt{-16}}{2} = 3 \pm 2i$$

By (11), the general solution of the differential equation is

$$y = e^{3x}(c_1 \cos 2x + c_2 \sin 2x)$$ □

INITIAL-VALUE AND BOUNDARY-VALUE PROBLEMS

An **initial-value problem** for the second-order Equation 1 or 2 consists of finding a solution y of the differential equation that also satisfies initial conditions of the form

$$y(x_0) = y_0 \qquad y'(x_0) = y_1$$

where y_0 and y_1 are given constants. If P, Q, R, and G are continuous on an interval and $P(x) \neq 0$ there, then a theorem found in more advanced books guarantees the existence and uniqueness of a solution to this initial-value problem. Examples 5 and 6 illustrate the technique for solving such a problem.

EXAMPLE 5 Solve the initial-value problem

$$y'' + y' - 6y = 0 \qquad y(0) = 1 \qquad y'(0) = 0$$

SOLUTION From Example 1 we know that the general solution of the differential equation is

$$y(x) = c_1 e^{2x} + c_2 e^{-3x}$$

Differentiating this solution, we get

$$y'(x) = 2c_1 e^{2x} - 3c_2 e^{-3x}$$

■ Figure 4 shows the graph of the solution of the initial-value problem in Example 5. Compare with Figure 1.

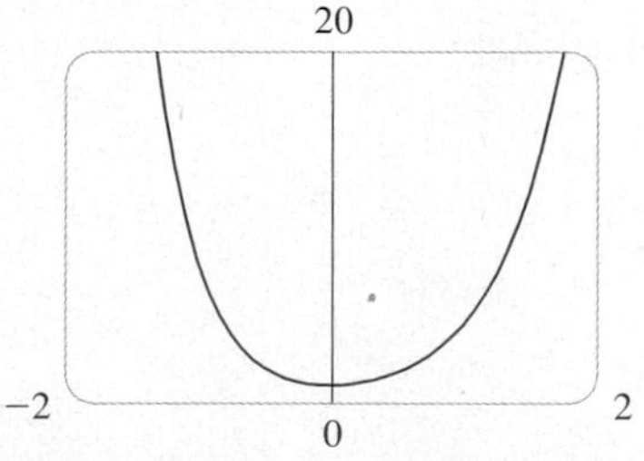

FIGURE 4

To satisfy the initial conditions we require that

12 $$y(0) = c_1 + c_2 = 1$$

13 $$y'(0) = 2c_1 - 3c_2 = 0$$

From (13), we have $c_2 = \frac{2}{3}c_1$ and so (12) gives

$$c_1 + \tfrac{2}{3}c_1 = 1 \qquad c_1 = \tfrac{3}{5} \qquad c_2 = \tfrac{2}{5}$$

Thus the required solution of the initial-value problem is

$$y = \tfrac{3}{5}e^{2x} + \tfrac{2}{5}e^{-3x}$$ □

EXAMPLE 6 Solve the initial-value problem

$$y'' + y = 0 \qquad y(0) = 2 \qquad y'(0) = 3$$

SOLUTION The auxiliary equation is $r^2 + 1 = 0$, or $r^2 = -1$, whose roots are $\pm i$. Thus $\alpha = 0$, $\beta = 1$, and since $e^{0x} = 1$, the general solution is

$$y(x) = c_1 \cos x + c_2 \sin x$$

Since $$y'(x) = -c_1 \sin x + c_2 \cos x$$

■ The solution to Example 6 is graphed in Figure 5. It appears to be a shifted sine curve and, indeed, you can verify that another way of writing the solution is

$$y = \sqrt{13}\sin(x + \phi) \quad \text{where } \tan\phi = \tfrac{2}{3}$$

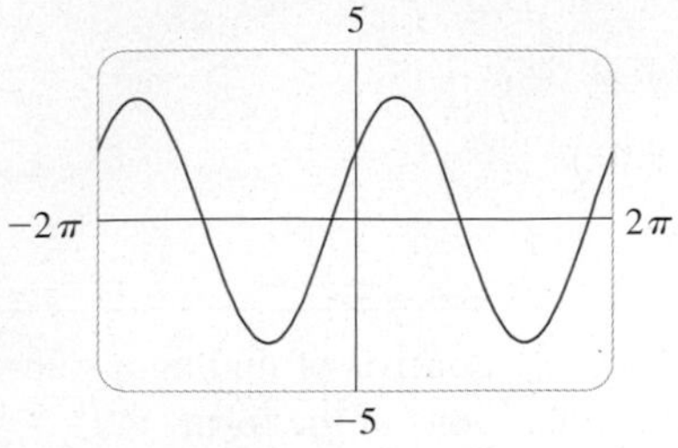

FIGURE 5

the initial conditions become

$$y(0) = c_1 = 2 \qquad y'(0) = c_2 = 3$$

Therefore the solution of the initial-value problem is

$$y(x) = 2\cos x + 3\sin x$$ □

A **boundary-value problem** for Equation 1 or 2 consists of finding a solution y of the differential equation that also satisfies boundary conditions of the form

$$y(x_0) = y_0 \qquad y(x_1) = y_1$$

In contrast with the situation for initial-value problems, a boundary-value problem does not always have a solution. The method is illustrated in Example 7.

V EXAMPLE 7 Solve the boundary-value problem

$$y'' + 2y' + y = 0 \qquad y(0) = 1 \qquad y(1) = 3$$

SOLUTION The auxiliary equation is

$$r^2 + 2r + 1 = 0 \qquad \text{or} \qquad (r + 1)^2 = 0$$

whose only root is $r = -1$. Therefore the general solution is

$$y(x) = c_1e^{-x} + c_2xe^{-x}$$

The boundary conditions are satisfied if

$$y(0) = c_1 = 1$$

$$y(1) = c_1e^{-1} + c_2e^{-1} = 3$$

■ Figure 6 shows the graph of the solution of the boundary-value problem in Example 7.

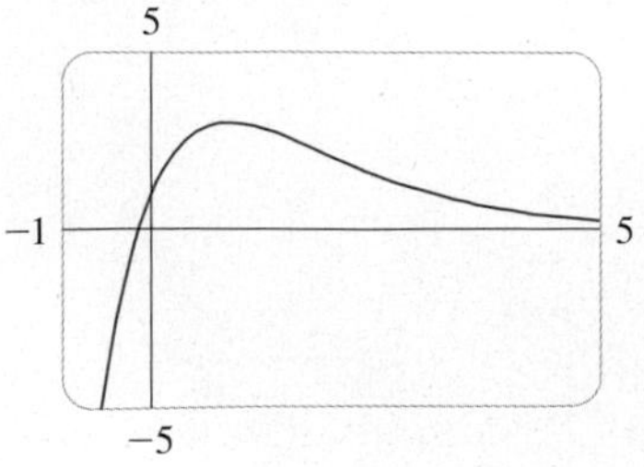

FIGURE 6

The first condition gives $c_1 = 1$, so the second condition becomes

$$e^{-1} + c_2e^{-1} = 3$$

Solving this equation for c_2 by first multiplying through by e, we get

$$1 + c_2 = 3e \qquad \text{so} \qquad c_2 = 3e - 1$$

Thus the solution of the boundary-value problem is

$$y = e^{-x} + (3e - 1)xe^{-x}$$ □

SUMMARY: SOLUTIONS OF $ay'' + by' + c = 0$

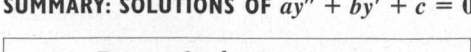

Roots of $ar^2 + br + c = 0$	General solution
r_1, r_2 real and distinct	$y = c_1e^{r_1x} + c_2e^{r_2x}$
$r_1 = r_2 = r$	$y = c_1e^{rx} + c_2xe^{rx}$
r_1, r_2 complex: $\alpha \pm i\beta$	$y = e^{\alpha x}(c_1\cos\beta x + c_2\sin\beta x)$

17.1 EXERCISES

1–13 Solve the differential equation.

1. $y'' - y' - 6y = 0$
2. $y'' + 4y' + 4y = 0$
3. $y'' + 16y = 0$
4. $y'' - 8y' + 12y = 0$
5. $9y'' - 12y' + 4y = 0$
6. $25y'' + 9y = 0$
7. $y' = 2y''$
8. $y'' - 4y' + y = 0$
9. $y'' - 4y' + 13y = 0$
10. $y'' + 3y' = 0$
11. $2\dfrac{d^2y}{dt^2} + 2\dfrac{dy}{dt} - y = 0$
12. $8\dfrac{d^2y}{dt^2} + 12\dfrac{dy}{dt} + 5y = 0$
13. $100\dfrac{d^2P}{dt^2} + 200\dfrac{dP}{dt} + 101P = 0$

14–16 Graph the two basic solutions of the differential equation and several other solutions. What features do the solutions have in common?

14. $\dfrac{d^2y}{dx^2} + 4\dfrac{dy}{dx} + 20y = 0$
15. $5\dfrac{d^2y}{dx^2} - 2\dfrac{dy}{dx} - 3y = 0$
16. $9\dfrac{d^2y}{dx^2} + 6\dfrac{dy}{dx} + y = 0$

17–24 Solve the initial-value problem.

17. $2y'' + 5y' + 3y = 0, \quad y(0) = 3, \quad y'(0) = -4$
18. $y'' + 3y = 0, \quad y(0) = 1, \quad y'(0) = 3$
19. $4y'' - 4y' + y = 0, \quad y(0) = 1, \quad y'(0) = -1.5$
20. $2y'' + 5y' - 3y = 0, \quad y(0) = 1, \quad y'(0) = 4$
21. $y'' + 16y = 0, \quad y(\pi/4) = -3, \quad y'(\pi/4) = 4$
22. $y'' - 2y' + 5y = 0, \quad y(\pi) = 0, \quad y'(\pi) = 2$
23. $y'' + 2y' + 2y = 0, \quad y(0) = 2, \quad y'(0) = 1$
24. $y'' + 12y' + 36y = 0, \quad y(1) = 0, \quad y'(1) = 1$

25–32 Solve the boundary-value problem, if possible.

25. $4y'' + y = 0, \quad y(0) = 3, \quad y(\pi) = -4$
26. $y'' + 2y' = 0, \quad y(0) = 1, \quad y(1) = 2$
27. $y'' - 3y' + 2y = 0, \quad y(0) = 1, \quad y(3) = 0$
28. $y'' + 100y = 0, \quad y(0) = 2, \quad y(\pi) = 5$
29. $y'' - 6y' + 25y = 0, \quad y(0) = 1, \quad y(\pi) = 2$
30. $y'' - 6y' + 9y = 0, \quad y(0) = 1, \quad y(1) = 0$
31. $y'' + 4y' + 13y = 0, \quad y(0) = 2, \quad y(\pi/2) = 1$
32. $9y'' - 18y' + 10y = 0, \quad y(0) = 0, \quad y(\pi) = 1$

33. Let L be a nonzero real number.
    (a) Show that the boundary-value problem $y'' + \lambda y = 0$, $y(0) = 0$, $y(L) = 0$ has only the trivial solution $y = 0$ for the cases $\lambda = 0$ and $\lambda < 0$.
    (b) For the case $\lambda > 0$, find the values of λ for which this problem has a nontrivial solution and give the corresponding solution.
34. If a, b, and c are all positive constants and $y(x)$ is a solution of the differential equation $ay'' + by' + cy = 0$, show that $\lim_{x\to\infty} y(x) = 0$.

17.2 NONHOMOGENEOUS LINEAR EQUATIONS

In this section we learn how to solve second-order nonhomogeneous linear differential equations with constant coefficients, that is, equations of the form

$$ay'' + by' + cy = G(x) \tag{1}$$

where a, b, and c are constants and G is a continuous function. The related homogeneous equation

$$ay'' + by' + cy = 0 \tag{2}$$

is called the **complementary equation** and plays an important role in the solution of the original nonhomogeneous equation (1).

3 THEOREM The general solution of the nonhomogeneous differential equation (1) can be written as

$$y(x) = y_p(x) + y_c(x)$$

where y_p is a particular solution of Equation 1 and y_c is the general solution of the complementary Equation 2.

PROOF All we have to do is verify that if y is any solution of Equation 1, then $y - y_p$ is a solution of the complementary Equation 2. Indeed

$$\begin{aligned} a(y - y_p)'' + b(y - y_p)' + c(y - y_p) &= ay'' - ay_p'' + by' - by_p' + cy - cy_p \\ &= (ay'' + by' + cy) - (ay_p'' + by_p' + cy_p) \\ &= g(x) - g(x) = 0 \end{aligned}$$

□

We know from Section 17.1 how to solve the complementary equation. (Recall that the solution is $y_c = c_1y_1 + c_2y_2$, where y_1 and y_2 are linearly independent solutions of Equation 2.) Therefore Theorem 3 says that we know the general solution of the nonhomogeneous equation as soon as we know a particular solution y_p. There are two methods for finding a particular solution: The method of undetermined coefficients is straightforward but works only for a restricted class of functions G. The method of variation of parameters works for every function G but is usually more difficult to apply in practice.

THE METHOD OF UNDETERMINED COEFFICIENTS

We first illustrate the method of undetermined coefficients for the equation

$$ay'' + by' + cy = G(x)$$

where $G(x)$ is a polynomial. It is reasonable to guess that there is a particular solution y_p that is a polynomial of the same degree as G because if y is a polynomial, then $ay'' + by' + cy$ is also a polynomial. We therefore substitute $y_p(x) =$ a polynomial (of the same degree as G) into the differential equation and determine the coefficients.

V EXAMPLE 1 Solve the equation $y'' + y' - 2y = x^2$.

SOLUTION The auxiliary equation of $y'' + y' - 2y = 0$ is

$$r^2 + r - 2 = (r - 1)(r + 2) = 0$$

with roots $r = 1, -2$. So the solution of the complementary equation is

$$y_c = c_1e^x + c_2e^{-2x}$$

Since $G(x) = x^2$ is a polynomial of degree 2, we seek a particular solution of the form

$$y_p(x) = Ax^2 + Bx + C$$

Then $y_p' = 2Ax + B$ and $y_p'' = 2A$ so, substituting into the given differential equation, we have

$$(2A) + (2Ax + B) - 2(Ax^2 + Bx + C) = x^2$$

or $$-2Ax^2 + (2A - 2B)x + (2A + B - 2C) = x^2$$

Polynomials are equal when their coefficients are equal. Thus

$$-2A = 1 \qquad 2A - 2B = 0 \qquad 2A + B - 2C = 0$$

The solution of this system of equations is

$$A = -\tfrac{1}{2} \qquad B = -\tfrac{1}{2} \qquad C = -\tfrac{3}{4}$$

A particular solution is therefore

$$y_p(x) = -\tfrac{1}{2}x^2 - \tfrac{1}{2}x - \tfrac{3}{4}$$

and, by Theorem 3, the general solution is

$$y = y_c + y_p = c_1 e^x + c_2 e^{-2x} - \tfrac{1}{2}x^2 - \tfrac{1}{2}x - \tfrac{3}{4}$$ □

■ Figure 1 shows four solutions of the differential equation in Example 1 in terms of the particular solution y_p and the functions $f(x) = e^x$ and $g(x) = e^{-2x}$.

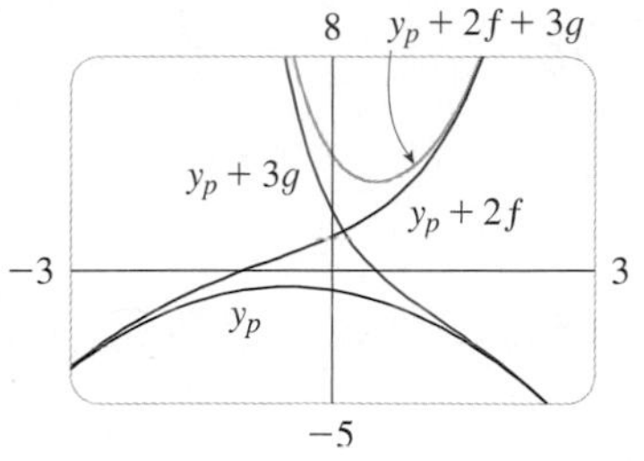

FIGURE 1

If $G(x)$ (the right side of Equation 1) is of the form Ce^{kx}, where C and k are constants, then we take as a trial solution a function of the same form, $y_p(x) = Ae^{kx}$, because the derivatives of e^{kx} are constant multiples of e^{kx}.

EXAMPLE 2 Solve $y'' + 4y = e^{3x}$.

SOLUTION The auxiliary equation is $r^2 + 4 = 0$ with roots $\pm 2i$, so the solution of the complementary equation is

$$y_c(x) = c_1 \cos 2x + c_2 \sin 2x$$

For a particular solution we try $y_p(x) = Ae^{3x}$. Then $y_p' = 3Ae^{3x}$ and $y_p'' = 9Ae^{3x}$. Substituting into the differential equation, we have

$$9Ae^{3x} + 4(Ae^{3x}) = e^{3x}$$

so $13Ae^{3x} = e^{3x}$ and $A = \frac{1}{13}$. Thus a particular solution is

$$y_p(x) = \tfrac{1}{13}e^{3x}$$

and the general solution is

$$y(x) = c_1 \cos 2x + c_2 \sin 2x + \tfrac{1}{13}e^{3x}$$ □

■ Figure 2 shows solutions of the differential equation in Example 2 in terms of y_p and the functions $f(x) = \cos 2x$ and $g(x) = \sin 2x$. Notice that all solutions approach ∞ as $x \to \infty$ and all solutions (except y_p) resemble sine functions when x is negative.

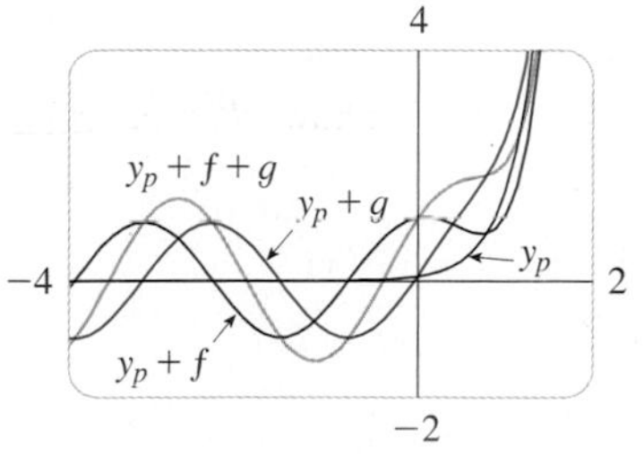

FIGURE 2

If $G(x)$ is either $C \cos kx$ or $C \sin kx$, then, because of the rules for differentiating the sine and cosine functions, we take as a trial particular solution a function of the form

$$y_p(x) = A \cos kx + B \sin kx$$

V EXAMPLE 3 Solve $y'' + y' - 2y = \sin x$.

SOLUTION We try a particular solution

$$y_p(x) = A \cos x + B \sin x$$

Then $$y_p' = -A \sin x + B \cos x \qquad y_p'' = -A \cos x - B \sin x$$

so substitution in the differential equation gives

$$(-A\cos x - B\sin x) + (-A\sin x + B\cos x) - 2(A\cos x + B\sin x) = \sin x$$

or

$$(-3A + B)\cos x + (-A - 3B)\sin x = \sin x$$

This is true if

$$-3A + B = 0 \qquad \text{and} \qquad -A - 3B = 1$$

The solution of this system is

$$A = -\tfrac{1}{10} \qquad B = -\tfrac{3}{10}$$

so a particular solution is

$$y_p(x) = -\tfrac{1}{10}\cos x - \tfrac{3}{10}\sin x$$

In Example 1 we determined that the solution of the complementary equation is $y_c = c_1e^x + c_2e^{-2x}$. Thus the general solution of the given equation is

$$y(x) = c_1e^x + c_2e^{-2x} - \tfrac{1}{10}(\cos x + 3\sin x)$$ □

If $G(x)$ is a product of functions of the preceding types, then we take the trial solution to be a product of functions of the same type. For instance, in solving the differential equation

$$y'' + 2y' + 4y = x\cos 3x$$

we would try

$$y_p(x) = (Ax + B)\cos 3x + (Cx + D)\sin 3x$$

If $G(x)$ is a sum of functions of these types, we use the easily verified *principle of superposition*, which says that if y_{p_1} and y_{p_2} are solutions of

$$ay'' + by' + cy = G_1(x) \qquad ay'' + by' + cy = G_2(x)$$

respectively, then $y_{p_1} + y_{p_2}$ is a solution of

$$ay'' + by' + cy = G_1(x) + G_2(x)$$

EXAMPLE 4 Solve $y'' - 4y = xe^x + \cos 2x$.

SOLUTION The auxiliary equation is $r^2 - 4 = 0$ with roots ± 2, so the solution of the complementary equation is $y_c(x) = c_1e^{2x} + c_2e^{-2x}$. For the equation $y'' - 4y = xe^x$ we try

$$y_{p_1}(x) = (Ax + B)e^x$$

Then $y_{p_1}' = (Ax + A + B)e^x$, $y_{p_1}'' = (Ax + 2A + B)e^x$, so substitution in the equation gives

$$(Ax + 2A + B)e^x - 4(Ax + B)e^x = xe^x$$

or

$$(-3Ax + 2A - 3B)e^x = xe^x$$

Thus $-3A = 1$ and $2A - 3B = 0$, so $A = -\frac{1}{3}$, $B = -\frac{2}{9}$, and

$$y_{p_1}(x) = \left(-\tfrac{1}{3}x - \tfrac{2}{9}\right)e^x$$

For the equation $y'' - 4y = \cos 2x$, we try

$$y_{p_2}(x) = C\cos 2x + D\sin 2x$$

■ In Figure 3 we show the particular solution $y_p = y_{p_1} + y_{p_2}$ of the differential equation in Example 4. The other solutions are given in terms of $f(x) = e^{2x}$ and $g(x) = e^{-2x}$.

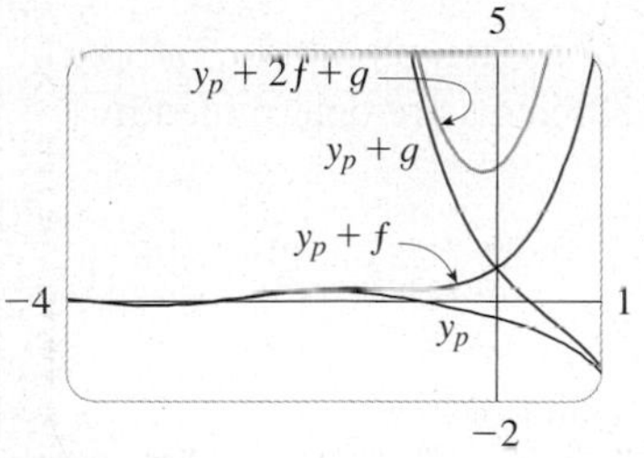

FIGURE 3

Substitution gives

$$-4C\cos 2x - 4D\sin 2x - 4(C\cos 2x + D\sin 2x) = \cos 2x$$

or

$$-8C\cos 2x - 8D\sin 2x = \cos 2x$$

Therefore $-8C = 1$, $-8D = 0$, and

$$y_{p_2}(x) = -\tfrac{1}{8}\cos 2x$$

By the superposition principle, the general solution is

$$y = y_c + y_{p_1} + y_{p_2} = c_1e^{2x} + c_2e^{-2x} - \left(\tfrac{1}{3}x + \tfrac{2}{9}\right)e^x - \tfrac{1}{8}\cos 2x$$ □

Finally we note that the recommended trial solution y_p sometimes turns out to be a solution of the complementary equation and therefore can't be a solution of the nonhomogeneous equation. In such cases we multiply the recommended trial solution by x (or by x^2 if necessary) so that no term in $y_p(x)$ is a solution of the complementary equation.

EXAMPLE 5 Solve $y'' + y = \sin x$.

SOLUTION The auxiliary equation is $r^2 + 1 = 0$ with roots $\pm i$, so the solution of the complementary equation is

$$y_c(x) = c_1\cos x + c_2\sin x$$

Ordinarily, we would use the trial solution

$$y_p(x) = A\cos x + B\sin x$$

but we observe that it is a solution of the complementary equation, so instead we try

$$y_p(x) = Ax\cos x + Bx\sin x$$

Then

$$y_p'(x) = A\cos x - Ax\sin x + B\sin x + Bx\cos x$$

$$y_p''(x) = -2A\sin x - Ax\cos x + 2B\cos x - Bx\sin x$$

Substitution in the differential equation gives

$$y_p'' + y_p = -2A\sin x + 2B\cos x = \sin x$$

■ The graphs of four solutions of the differential equation in Example 5 are shown in Figure 4.

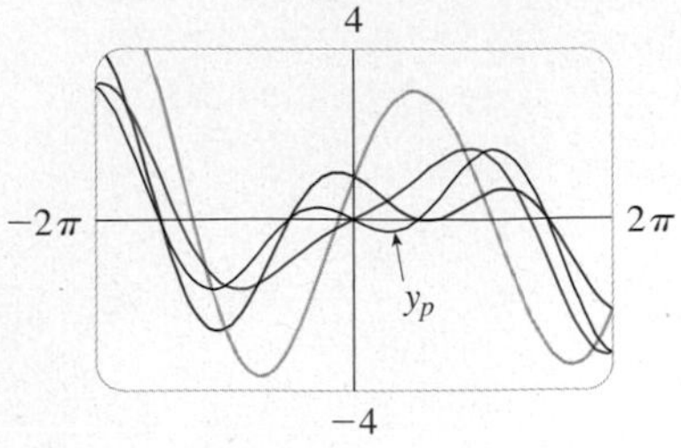

FIGURE 4

so $A = -\frac{1}{2}$, $B = 0$, and

$$y_p(x) = -\tfrac{1}{2}x\cos x$$

The general solution is

$$y(x) = c_1\cos x + c_2\sin x - \tfrac{1}{2}x\cos x$$ □

We summarize the method of undetermined coefficients as follows:

SUMMARY OF THE METHOD OF UNDETERMINED COEFFICIENTS

1. If $G(x) = e^{kx}P(x)$, where P is a polynomial of degree n, then try $y_p(x) = e^{kx}Q(x)$, where $Q(x)$ is an nth-degree polynomial (whose coefficients are determined by substituting in the differential equation).

2. If $G(x) = e^{kx}P(x)\cos mx$ or $G(x) = e^{kx}P(x)\sin mx$, where P is an nth-degree polynomial, then try

$$y_p(x) = e^{kx}Q(x)\cos mx + e^{kx}R(x)\sin mx$$

where Q and R are nth-degree polynomials.

Modification: If any term of y_p is a solution of the complementary equation, multiply y_p by x (or by x^2 if necessary).

EXAMPLE 6 Determine the form of the trial solution for the differential equation $y'' - 4y' + 13y = e^{2x}\cos 3x$.

SOLUTION Here $G(x)$ has the form of part 2 of the summary, where $k = 2$, $m = 3$, and $P(x) = 1$. So, at first glance, the form of the trial solution would be

$$y_p(x) = e^{2x}(A\cos 3x + B\sin 3x)$$

But the auxiliary equation is $r^2 - 4r + 13 = 0$, with roots $r = 2 \pm 3i$, so the solution of the complementary equation is

$$y_c(x) = e^{2x}(c_1\cos 3x + c_2\sin 3x)$$

This means that we have to multiply the suggested trial solution by x. So, instead, we use

$$y_p(x) = xe^{2x}(A\cos 3x + B\sin 3x)$$ □

2

THE METHOD OF VARIATION OF PARAMETERS

Suppose we have already solved the homogeneous equation $ay'' + by' + cy = 0$ and written the solution as

4 $$y(x) = c_1y_1(x) + c_2y_2(x)$$

where y_1 and y_2 are linearly independent solutions. Let's replace the constants (or parameters) c_1 and c_2 in Equation 4 by arbitrary functions $u_1(x)$ and $u_2(x)$. We look for a particular solution of the nonhomogeneous equation $ay'' + by' + cy = G(x)$ of the form

$$\boxed{5} \qquad y_p(x) = u_1(x)\,y_1(x) + u_2(x)\,y_2(x)$$

(This method is called **variation of parameters** because we have varied the parameters c_1 and c_2 to make them functions.) Differentiating Equation 5, we get

$$\boxed{6} \qquad y_p' = (u_1'y_1 + u_2'y_2) + (u_1y_1' + u_2y_2')$$

Since u_1 and u_2 are arbitrary functions, we can impose two conditions on them. One condition is that y_p is a solution of the differential equation; we can choose the other condition so as to simplify our calculations. In view of the expression in Equation 6, let's impose the condition that

$$\boxed{7} \qquad u_1'y_1 + u_2'y_2 = 0$$

Then $$y_p'' = u_1'y_1' + u_2'y_2' + u_1y_1'' + u_2y_2''$$

Substituting in the differential equation, we get

$$a(u_1'y_1' + u_2'y_2' + u_1y_1'' + u_2y_2'') + b(u_1y_1' + u_2y_2') + c(u_1y_1 + u_2y_2) = G$$

or

$$\boxed{8} \qquad u_1(ay_1'' + by_1' + cy_1) + u_2(ay_2'' + by_2' + cy_2) + a(u_1'y_1' + u_2'y_2') = G$$

But y_1 and y_2 are solutions of the complementary equation, so

$$ay_1'' + by_1' + cy_1 = 0 \qquad \text{and} \qquad ay_2'' + by_2' + cy_2 = 0$$

and Equation 8 simplifies to

$$\boxed{9} \qquad a(u_1'y_1' + u_2'y_2') = G$$

Equations 7 and 9 form a system of two equations in the unknown functions u_1' and u_2'. After solving this system we may be able to integrate to find u_1 and u_2 and then the particular solution is given by Equation 5.

EXAMPLE 7 Solve the equation $y'' + y = \tan x$, $0 < x < \pi/2$.

SOLUTION The auxiliary equation is $r^2 + 1 = 0$ with roots $\pm i$, so the solution of $y'' + y = 0$ is $c_1 \sin x + c_2 \cos x$. Using variation of parameters, we seek a solution of the form

$$y_p(x) = u_1(x) \sin x + u_2(x) \cos x$$

Then $$y_p' = (u_1' \sin x + u_2' \cos x) + (u_1 \cos x - u_2 \sin x)$$

Set

$$\boxed{10} \qquad u_1' \sin x + u_2' \cos x = 0$$

Then $$y_p'' = u_1' \cos x - u_2' \sin x - u_1 \sin x - u_2 \cos x$$

For y_p to be a solution we must have

$$\boxed{11} \qquad y_p'' + y_p = u_1' \cos x - u_2' \sin x = \tan x$$

Solving Equations 10 and 11, we get

$$u_1'(\sin^2 x + \cos^2 x) = \cos x \tan x$$

$$u_1' = \sin x \qquad u_1(x) = -\cos x$$

(We seek a particular solution, so we don't need a constant of integration here.) Then, from Equation 10, we obtain

$$u_2' = -\frac{\sin x}{\cos x} u_1' = -\frac{\sin^2 x}{\cos x} = \frac{\cos^2 x - 1}{\cos x} = \cos x - \sec x$$

So $$u_2(x) = \sin x - \ln(\sec x + \tan x)$$

■ Figure 5 shows four solutions of the differential equation in Example 7.

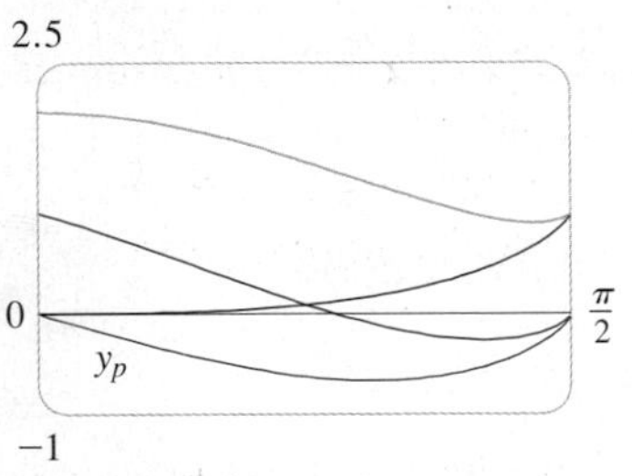

FIGURE 5

(Note that $\sec x + \tan x > 0$ for $0 < x < \pi/2$.) Therefore

$$y_p(x) = -\cos x \sin x + [\sin x - \ln(\sec x + \tan x)] \cos x$$
$$= -\cos x \ln(\sec x + \tan x)$$

and the general solution is

$$y(x) = c_1 \sin x + c_2 \cos x - \cos x \ln(\sec x + \tan x)$$ □

17.2 EXERCISES

1–10 Solve the differential equation or initial-value problem using the method of undetermined coefficients.

1. $y'' + 3y' + 2y = x^2$

2. $y'' + 9y = e^{3x}$

3. $y'' - 2y' = \sin 4x$

4. $y'' + 6y' + 9y = 1 + x$

5. $y'' - 4y' + 5y = e^{-x}$

6. $y'' + 2y' + y = xe^{-x}$

7. $y'' + y = e^x + x^3, \quad y(0) = 2, \quad y'(0) = 0$

8. $y'' - 4y = e^x \cos x, \quad y(0) = 1, \quad y'(0) = 2$

9. $y'' - y' = xe^x, \quad y(0) = 2, \quad y'(0) = 1$

10. $y'' + y' - 2y = x + \sin 2x, \quad y(0) = 1, \quad y'(0) = 0$

11–12 Graph the particular solution and several other solutions. What characteristics do these solutions have in common?

11. $y'' + 3y' + 2y = \cos x$ **12.** $y'' + 4y = e^{-x}$

13–18 Write a trial solution for the method of undetermined coefficients. Do not determine the coefficients.

13. $y'' + 9y = e^{2x} + x^2 \sin x$

14. $y'' + 9y' = xe^{-x} \cos \pi x$

15. $y'' + 9y' = 1 + xe^{9x}$

16. $y'' + 3y' - 4y = (x^3 + x)e^x$

17. $y'' + 2y' + 10y = x^2 e^{-x} \cos 3x$

18. $y'' + 4y = e^{3x} + x \sin 2x$

19–22 Solve the differential equation using (a) undetermined coefficients and (b) variation of parameters.

19. $4y'' + y = \cos x$

20. $y'' - 2y' - 3y = x + 2$

21. $y'' - 2y' + y = e^{2x}$

22. $y'' - y' = e^x$

23–28 Solve the differential equation using the method of variation of parameters.

23. $y'' + y = \sec^2 x,\ 0 < x < \pi/2$

24. $y'' + y = \sec^3 x,\ 0 < x < \pi/2$

25. $y'' - 3y' + 2y = \dfrac{1}{1 + e^{-x}}$

26. $y'' + 3y' + 2y = \sin(e^x)$

27. $y'' - 2y' + y = \dfrac{e^x}{1 + x^2}$

28. $y'' + 4y' + 4y = \dfrac{e^{-2x}}{x^3}$

17.3 APPLICATIONS OF SECOND-ORDER DIFFERENTIAL EQUATIONS

Second-order linear differential equations have a variety of applications in science and engineering. In this section we explore two of them: the vibration of springs and electric circuits.

VIBRATING SPRINGS

We consider the motion of an object with mass m at the end of a spring that is either vertical (as in Figure 1) or horizontal on a level surface (as in Figure 2).

In Section 6.4 we discussed Hooke's Law, which says that if the spring is stretched (or compressed) x units from its natural length, then it exerts a force that is proportional to x:

$$\text{restoring force} = -kx$$

where k is a positive constant (called the **spring constant**). If we ignore any external resisting forces (due to air resistance or friction) then, by Newton's Second Law (force equals mass times acceleration), we have

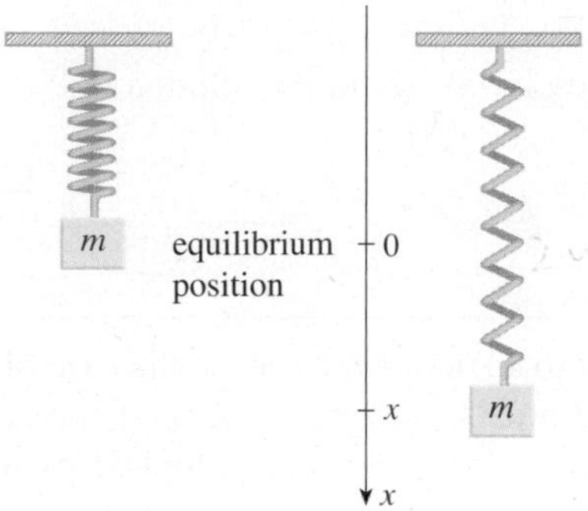

FIGURE 1

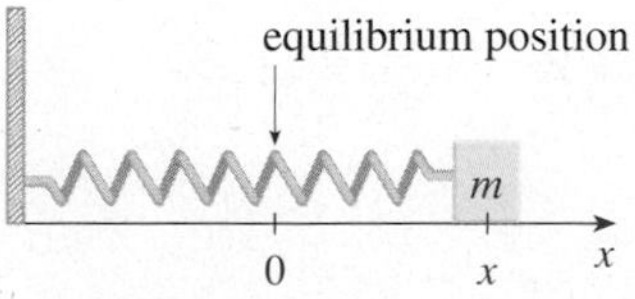

FIGURE 2

$$\boxed{1} \qquad m\frac{d^2x}{dt^2} = -kx \qquad \text{or} \qquad m\frac{d^2x}{dt^2} + kx = 0$$

This is a second-order linear differential equation. Its auxiliary equation is $mr^2 + k = 0$ with roots $r = \pm\omega i$, where $\omega = \sqrt{k/m}$. Thus the general solution is

$$x(t) = c_1 \cos \omega t + c_2 \sin \omega t$$

which can also be written as

$$x(t) = A\cos(\omega t + \delta)$$

where

$$\omega = \sqrt{k/m} \quad \text{(frequency)}$$

$$A = \sqrt{c_1^2 + c_2^2} \quad \text{(amplitude)}$$

$$\cos \delta = \frac{c_1}{A} \qquad \sin \delta = -\frac{c_2}{A} \qquad (\delta \text{ is the phase angle})$$

(See Exercise 17.) This type of motion is called **simple harmonic motion**.

EXAMPLE 1 A spring with a mass of 2 kg has natural length 0.5 m. A force of 25.6 N is required to maintain it stretched to a length of 0.7 m. If the spring is stretched to a length of 0.7 m and then released with initial velocity 0, find the position of the mass at any time *t*.

SOLUTION From Hooke's Law, the force required to stretch the spring is

$$k(0.2) = 25.6$$

so $k = 25.6/0.2 = 128$. Using this value of the spring constant k, together with $m = 2$ in Equation 1, we have

$$2\frac{d^2x}{dt^2} + 128x = 0$$

As in the earlier general discussion, the solution of this equation is

2 $$x(t) = c_1 \cos 8t + c_2 \sin 8t$$

We are given the initial condition that $x(0) = 0.2$. But, from Equation 2, $x(0) = c_1$. Therefore $c_1 = 0.2$. Differentiating Equation 2, we get

$$x'(t) = -8c_1 \sin 8t + 8c_2 \cos 8t$$

Since the initial velocity is given as $x'(0) = 0$, we have $c_2 = 0$ and so the solution is

$$x(t) = \tfrac{1}{5}\cos 8t$$ □

FIGURE 3

Schwinn Cycling and Fitness

DAMPED VIBRATIONS

We next consider the motion of a spring that is subject to a frictional force (in the case of the horizontal spring of Figure 2) or a damping force (in the case where a vertical spring moves through a fluid as in Figure 3). An example is the damping force supplied by a shock absorber in a car or a bicycle.

We assume that the damping force is proportional to the velocity of the mass and acts in the direction opposite to the motion. (This has been confirmed, at least approximately, by some physical experiments.) Thus

$$\text{damping force} = -c\frac{dx}{dt}$$

where c is a positive constant, called the **damping constant**. Thus, in this case, Newton's Second Law gives

$$m\frac{d^2x}{dt^2} = \text{restoring force} + \text{damping force} = -kx - c\frac{dx}{dt}$$

or

3 $$m\frac{d^2x}{dt^2} + c\frac{dx}{dt} + kx = 0$$

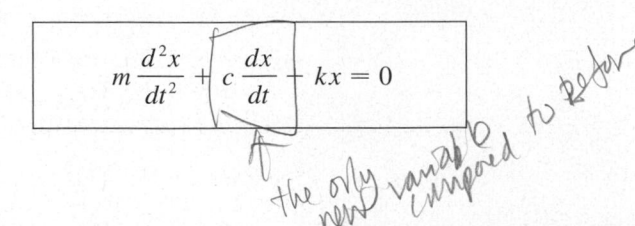

the only new variable compared to before

Equation 3 is a second-order linear differential equation and its auxiliary equation is $mr^2 + cr + k = 0$. The roots are

$$\boxed{4} \qquad r_1 = \frac{-c + \sqrt{c^2 - 4mk}}{2m} \qquad r_2 = \frac{-c - \sqrt{c^2 - 4mk}}{2m}$$

According to Section 17.1 we need to discuss three cases.

■ CASE I $c^2 - 4mk > 0$ (overdamping)
In this case r_1 and r_2 are distinct real roots and

$$x = c_1 e^{r_1 t} + c_2 e^{r_2 t}$$

Since c, m, and k are all positive, we have $\sqrt{c^2 - 4mk} < c$, so the roots r_1 and r_2 given by Equations 4 must both be negative. This shows that $x \to 0$ as $t \to \infty$. Typical graphs of x as a function of t are shown in Figure 4. Notice that oscillations do not occur. (It's possible for the mass to pass through the equilibrium position once, but only once.) This is because $c^2 > 4mk$ means that there is a strong damping force (high-viscosity oil or grease) compared with a weak spring or small mass.

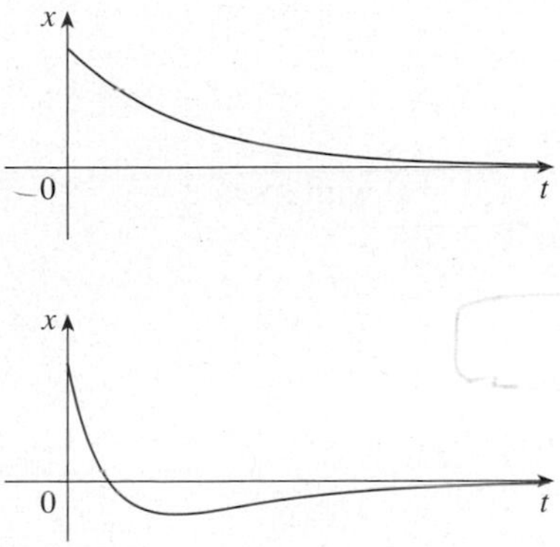

FIGURE 4
Overdamping

■ CASE II $c^2 - 4mk = 0$ (critical damping)
This case corresponds to equal roots

$$r_1 = r_2 = -\frac{c}{2m}$$

and the solution is given by

$$x = (c_1 + c_2 t)e^{-(c/2m)t}$$

It is similar to Case I, and typical graphs resemble those in Figure 4 (see Exercise 12), but the damping is just sufficient to suppress vibrations. Any decrease in the viscosity of the fluid leads to the vibrations of the following case.

■ CASE III $c^2 - 4mk < 0$ (underdamping)
Here the roots are complex:

$$\left.\begin{matrix} r_1 \\ r_2 \end{matrix}\right\} = -\frac{c}{2m} \pm \omega i$$

where

$$\omega = \frac{\sqrt{4mk - c^2}}{2m}$$

The solution is given by

$$x = e^{-(c/2m)t}(c_1 \cos \omega t + c_2 \sin \omega t)$$

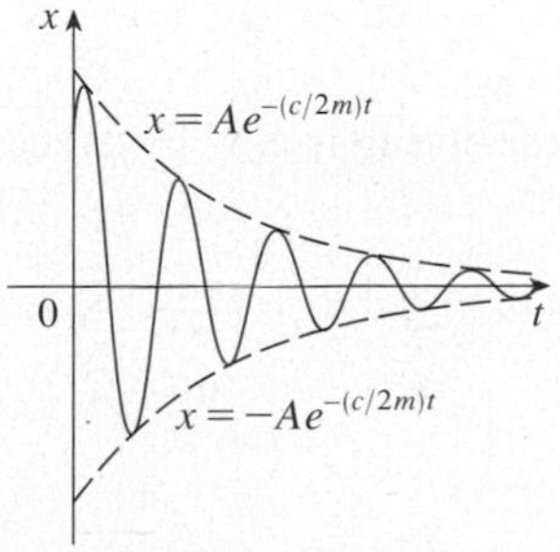

FIGURE 5
Underdamping

We see that there are oscillations that are damped by the factor $e^{-(c/2m)t}$. Since $c > 0$ and $m > 0$, we have $-(c/2m) < 0$ so $e^{-(c/2m)t} \to 0$ as $t \to \infty$. This implies that $x \to 0$ as $t \to \infty$; that is, the motion decays to 0 as time increases. A typical graph is shown in Figure 5.

EXAMPLE 2 Suppose that the spring of Example 1 is immersed in a fluid with damping constant $c = 40$. Find the position of the mass at any time t if it starts from the equilibrium position and is given a push to start it with an initial velocity of 0.6 m/s.

SOLUTION From Example 1, the mass is $m = 2$ and the spring constant is $k = 128$, so the differential equation (3) becomes

$$2\frac{d^2x}{dt^2} + 40\frac{dx}{dt} + 128x = 0$$

or

$$\frac{d^2x}{dt^2} + 20\frac{dx}{dt} + 64x = 0$$

The auxiliary equation is $r^2 + 20r + 64 = (r + 4)(r + 16) = 0$ with roots -4 and -16, so the motion is overdamped and the solution is

$$x(t) = c_1e^{-4t} + c_2e^{-16t}$$

■ Figure 6 shows the graph of the position function for the overdamped motion in Example 2.

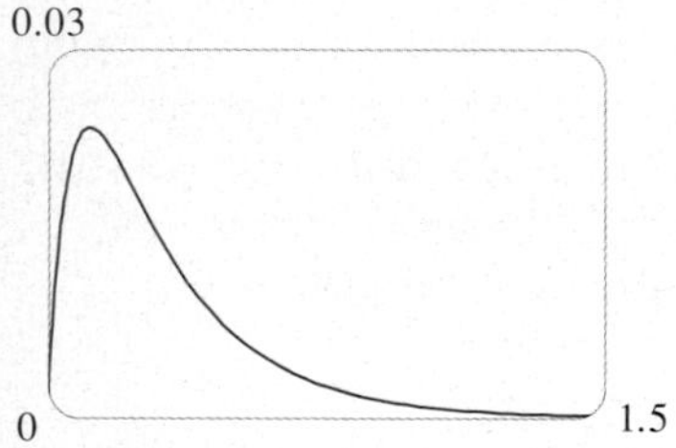

FIGURE 6

We are given that $x(0) = 0$, so $c_1 + c_2 = 0$. Differentiating, we get

$$x'(t) = -4c_1e^{-4t} - 16c_2e^{-16t}$$

so

$$x'(0) = -4c_1 - 16c_2 = 0.6$$

Since $c_2 = -c_1$, this gives $12c_1 = 0.6$ or $c_1 = 0.05$. Therefore

$$x = 0.05(e^{-4t} - e^{-16t})$$ □

FORCED VIBRATIONS

Suppose that, in addition to the restoring force and the damping force, the motion of the spring is affected by an external force $F(t)$. Then Newton's Second Law gives

$$m\frac{d^2x}{dt^2} = \text{restoring force} + \text{damping force} + \text{external force}$$

$$= -kx - c\frac{dx}{dt} + F(t)$$

Thus, instead of the homogeneous equation (3), the motion of the spring is now governed by the following nonhomogeneous differential equation:

5 $$m\frac{d^2x}{dt^2} + c\frac{dx}{dt} + kx = F(t)$$

The motion of the spring can be determined by the methods of Section 17.2.

A commonly occurring type of external force is a periodic force function

$$F(t) = F_0 \cos \omega_0 t \qquad \text{where} \quad \omega_0 \neq \omega = \sqrt{k/m}$$

In this case, and in the absence of a damping force ($c = 0$), you are asked in Exercise 9 to use the method of undetermined coefficients to show that

6 $$x(t) = c_1 \cos \omega t + c_2 \sin \omega t + \frac{F_0}{m(\omega^2 - \omega_0^2)} \cos \omega_0 t$$

If $\omega_0 = \omega$, then the applied frequency reinforces the natural frequency and the result is vibrations of large amplitude. This is the phenomenon of **resonance** (see Exercise 10).

ELECTRIC CIRCUITS

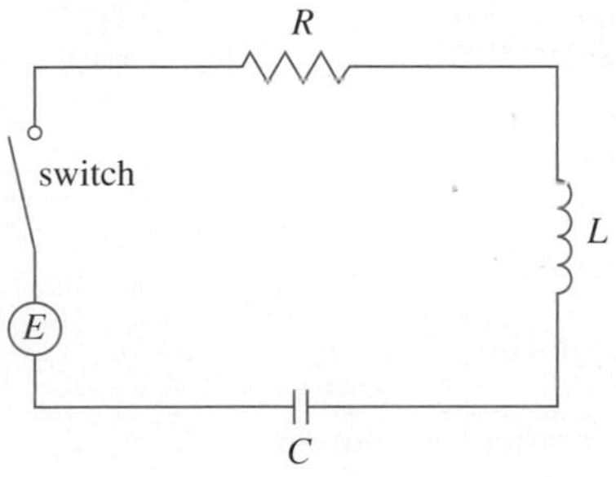

FIGURE 7

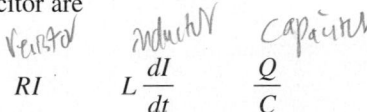

In Sections 9.3 and 9.5 we were able to use first-order separable and linear equations to analyze electric circuits that contain a resistor and inductor (see Figure 5 on page 582 or Figure 4 on page 605) or a resistor and capacitor (see Exercise 29 on page 607). Now that we know how to solve second-order linear equations, we are in a position to analyze the circuit shown in Figure 7. It contains an electromotive force E (supplied by a battery or generator), a resistor R, an inductor L, and a capacitor C, in series. If the charge on the capacitor at time t is $Q = Q(t)$, then the current is the rate of change of Q with respect to t: $I = dQ/dt$. As in Section 9.5, it is known from physics that the voltage drops across the resistor, inductor, and capacitor are

$$RI \qquad L\frac{dI}{dt} \qquad \frac{Q}{C}$$

respectively. Kirchhoff's voltage law says that the sum of these voltage drops is equal to the supplied voltage:

$$L\frac{dI}{dt} + RI + \frac{Q}{C} = E(t)$$

Since $I = dQ/dt$, this equation becomes

7 $$L\frac{d^2Q}{dt^2} + R\frac{dQ}{dt} + \frac{1}{C}Q = E(t)$$

which is a second-order linear differential equation with constant coefficients. If the charge Q_0 and the current I_0 are known at time 0, then we have the initial conditions

$$Q(0) = Q_0 \qquad Q'(0) = I(0) = I_0$$

and the initial-value problem can be solved by the methods of Section 17.2.

A differential equation for the current can be obtained by differentiating Equation 7 with respect to t and remembering that $I = dQ/dt$:

$$L\frac{d^2I}{dt^2} + R\frac{dI}{dt} + \frac{1}{C}I = E'(t)$$

EXAMPLE 3 Find the charge and current at time t in the circuit of Figure 7 if $R = 40\ \Omega$, $L = 1$ H, $C = 16 \times 10^{-4}$ F, $E(t) = 100\cos 10t$, and the initial charge and current are both 0.

SOLUTION With the given values of L, R, C, and $E(t)$, Equation 7 becomes

8 $$\frac{d^2Q}{dt^2} + 40\frac{dQ}{dt} + 625Q = 100\cos 10t$$

The auxiliary equation is $r^2 + 40r + 625 = 0$ with roots

$$r = \frac{-40 \pm \sqrt{-900}}{2} = -20 \pm 15i$$

so the solution of the complementary equation is

$$Q_c(t) = e^{-20t}(c_1\cos 15t + c_2\sin 15t)$$

For the method of undetermined coefficients we try the particular solution

$$Q_p(t) = A\cos 10t + B\sin 10t$$

Then $$Q_p'(t) = -10A\sin 10t + 10B\cos 10t$$

$$Q_p''(t) = -100A\cos 10t - 100B\sin 10t$$

Substituting into Equation 8, we have

$$(-100A\cos 10t - 100B\sin 10t) + 40(-10A\sin 10t + 10B\cos 10t) + 625(A\cos 10t + B\sin 10t) = 100\cos 10t$$

or $$(525A + 400B)\cos 10t + (-400A + 525B)\sin 10t = 100\cos 10t$$

Equating coefficients, we have

$$\begin{aligned} 525A + 400B &= 100 \\ -400A + 525B &= 0 \end{aligned} \qquad \text{or} \qquad \begin{aligned} 21A + 16B &= 4 \\ -16A + 21B &= 0 \end{aligned}$$

The solution of this system is $A = \frac{84}{697}$ and $B = \frac{64}{697}$, so a particular solution is

$$Q_p(t) = \tfrac{1}{697}(84\cos 10t + 64\sin 10t)$$

and the general solution is

$$\begin{aligned} Q(t) &= Q_c(t) + Q_p(t) \\ &= e^{-20t}(c_1\cos 15t + c_2\sin 15t) + \tfrac{4}{697}(21\cos 10t + 16\sin 10t) \end{aligned}$$

Imposing the initial condition $Q(0) = 0$, we get

$$Q(0) = c_1 + \tfrac{84}{697} = 0 \qquad c_1 = -\tfrac{84}{697}$$

To impose the other initial condition, we first differentiate to find the current:

$$I = \frac{dQ}{dt} = e^{-20t}[(-20c_1 + 15c_2)\cos 15t + (-15c_1 - 20c_2)\sin 15t]$$
$$+ \tfrac{40}{697}(-21 \sin 10t + 16 \cos 10t)$$

$$I(0) = -20c_1 + 15c_2 + \tfrac{640}{697} = 0 \qquad c_2 = -\tfrac{464}{2091}$$

Thus the formula for the charge is

$$Q(t) = \frac{4}{697}\left[\frac{e^{-20t}}{3}(-63 \cos 15t - 116 \sin 15t) + (21 \cos 10t + 16 \sin 10t)\right]$$

and the expression for the current is

$$I(t) = \tfrac{1}{2091}[e^{-20t}(-1920 \cos 15t + 13{,}060 \sin 15t) + 120(-21 \sin 10t + 16 \cos 10t)]$$

□

NOTE 1 In Example 3 the solution for $Q(t)$ consists of two parts. Since $e^{-20t} \to 0$ as $t \to \infty$ and both $\cos 15t$ and $\sin 15t$ are bounded functions,

$$Q_c(t) = \tfrac{4}{2091}e^{-20t}(-63 \cos 15t - 116 \sin 15t) \to 0 \qquad \text{as } t \to \infty$$

So, for large values of t,

$$Q(t) \approx Q_p(t) = \tfrac{4}{697}(21 \cos 10t + 16 \sin 10t)$$

and, for this reason, $Q_p(t)$ is called the **steady state solution**. Figure 8 shows how the graph of the steady state solution compares with the graph of Q in this case.

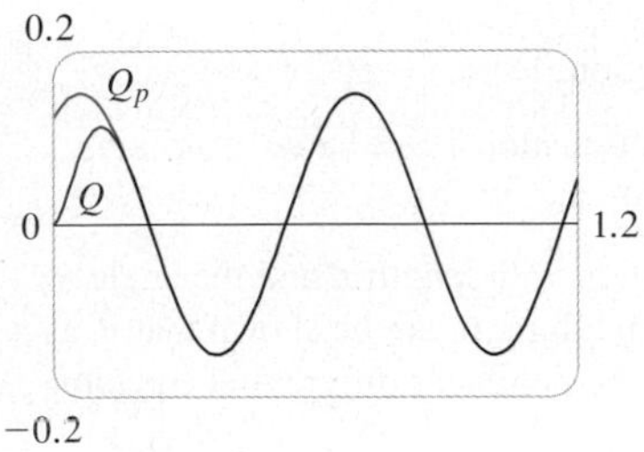

FIGURE 8

NOTE 2 Comparing Equations 5 and 7, we see that mathematically they are identical. This suggests the analogies given in the following chart between physical situations that, at first glance, are very different.

5 $m\dfrac{d^2x}{dt^2} + c\dfrac{dx}{dt} + kx = F(t)$

7 $L\dfrac{d^2Q}{dt^2} + R\dfrac{dQ}{dt} + \dfrac{1}{C}Q = E(t)$

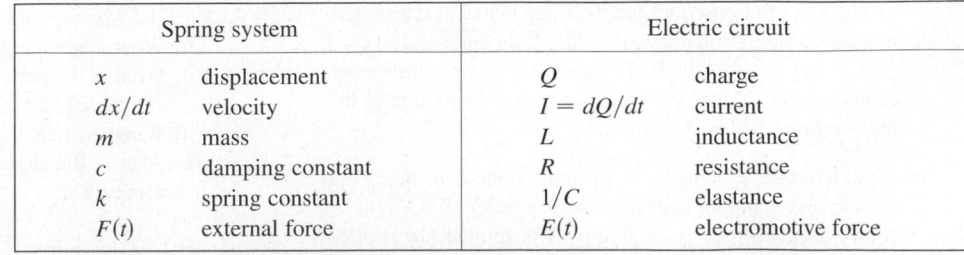

Spring system		Electric circuit	
x	displacement	Q	charge
dx/dt	velocity	$I = dQ/dt$	current
m	mass	L	inductance
c	damping constant	R	resistance
k	spring constant	$1/C$	elastance
$F(t)$	external force	$E(t)$	electromotive force

We can also transfer other ideas from one situation to the other. For instance, the steady state solution discussed in Note 1 makes sense in the spring system. And the phenomenon of resonance in the spring system can be usefully carried over to electric circuits as electrical resonance.

17.3 EXERCISES

1. A spring has natural length 0.75 m and a 5-kg mass. A force of 25 N is needed to keep the spring stretched to a length of 1 m. If the spring is stretched to a length of 1.1 m and then released with velocity 0, find the position of the mass after t seconds.

2. A spring with an 8-kg mass is kept stretched 0.4 m beyond its natural length by a force of 32 N. The spring starts at its equilibrium position and is given an initial velocity of 1 m/s. Find the position of the mass at any time t.

3. A spring with a mass of 2 kg has damping constant 14, and a force of 6 N is required to keep the spring stretched 0.5 m beyond its natural length. The spring is stretched 1 m beyond its natural length and then released with zero velocity. Find the position of the mass at any time t.

4. A force of 13 N is needed to keep a spring with a 2-kg mass stretched 0.25 m beyond its natural length. The damping constant of the spring is $c = 8$.
(a) If the mass starts at the equilibrium position with a velocity of 0.5 m/s, find its position at time t.
(b) Graph the position function of the mass.

5. For the spring in Exercise 3, find the mass that would produce critical damping.

6. For the spring in Exercise 4, find the damping constant that would produce critical damping.

7. A spring has a mass of 1 kg and its spring constant is $k = 100$. The spring is released at a point 0.1 m above its equilibrium position. Graph the position function for the following values of the damping constant c: 10, 15, 20, 25, 30. What type of damping occurs in each case?

8. A spring has a mass of 1 kg and its damping constant is $c = 10$. The spring starts from its equilibrium position with a velocity of 1 m/s. Graph the position function for the following values of the spring constant k: 10, 20, 25, 30, 40. What type of damping occurs in each case?

9. Suppose a spring has mass m and spring constant k and let $\omega = \sqrt{k/m}$. Suppose that the damping constant is so small that the damping force is negligible. If an external force $F(t) = F_0 \cos \omega_0 t$ is applied, where $\omega_0 \neq \omega$, use the method of undetermined coefficients to show that the motion of the mass is described by Equation 6.

10. As in Exercise 9, consider a spring with mass m, spring constant k, and damping constant $c = 0$, and let $\omega = \sqrt{k/m}$. If an external force $F(t) = F_0 \cos \omega t$ is applied (the applied frequency equals the natural frequency), use the method of undetermined coefficients to show that the motion of the mass is given by $x(t) = c_1 \cos \omega t + c_2 \sin \omega t + (F_0/(2m\omega))t \sin \omega t$.

11. Show that if $\omega_0 \neq \omega$, but ω/ω_0 is a rational number, then the motion described by Equation 6 is periodic.

12. Consider a spring subject to a frictional or damping force.
(a) In the critically damped case, the motion is given by $x = c_1 e^{rt} + c_2 t e^{rt}$. Show that the graph of x crosses the t-axis whenever c_1 and c_2 have opposite signs.
(b) In the overdamped case, the motion is given by $x = c_1 e^{r_1 t} + c_2 e^{r_2 t}$, where $r_1 > r_2$. Determine a condition on the relative magnitudes of c_1 and c_2 under which the graph of x crosses the t-axis at a positive value of t.

13. A series circuit consists of a resistor with $R = 20\ \Omega$, an inductor with $L = 1$ H, a capacitor with $C = 0.002$ F, and a 12-V battery. If the initial charge and current are both 0, find the charge and current at time t.

14. A series circuit contains a resistor with $R = 24\ \Omega$, an inductor with $L = 2$ H, a capacitor with $C = 0.005$ F, and a 12-V battery. The initial charge is $Q = 0.001$ C and the initial current is 0.
(a) Find the charge and current at time t.
(b) Graph the charge and current functions.

15. The battery in Exercise 13 is replaced by a generator producing a voltage of $E(t) = 12 \sin 10t$. Find the charge at time t.

16. The battery in Exercise 14 is replaced by a generator producing a voltage of $E(t) = 12 \sin 10t$.
(a) Find the charge at time t.
(b) Graph the charge function.

17. Verify that the solution to Equation 1 can be written in the form $x(t) = A \cos(\omega t + \delta)$.

18. The figure shows a pendulum with length L and the angle θ from the vertical to the pendulum. It can be shown that θ, as a function of time, satisfies the nonlinear differential equation

$$\frac{d^2\theta}{dt^2} + \frac{g}{L} \sin \theta = 0$$

where g is the acceleration due to gravity. For small values of θ we can use the linear approximation $\sin \theta \approx \theta$ and then the differential equation becomes linear.
(a) Find the equation of motion of a pendulum with length 1 m if θ is initially 0.2 rad and the initial angular velocity is $d\theta/dt = 1$ rad/s.
(b) What is the maximum angle from the vertical?
(c) What is the period of the pendulum (that is, the time to complete one back-and-forth swing)?
(d) When will the pendulum first be vertical?
(e) What is the angular velocity when the pendulum is vertical?

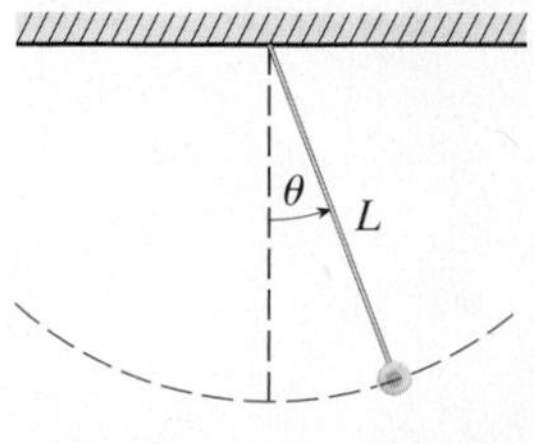

17.4 SERIES SOLUTIONS

Many differential equations can't be solved explicitly in terms of finite combinations of simple familiar functions. This is true even for a simple-looking equation like

$$y'' - 2xy' + y = 0 \tag{1}$$

But it is important to be able to solve equations such as Equation 1 because they arise from physical problems and, in particular, in connection with the Schrödinger equation in quantum mechanics. In such a case we use the method of power series; that is, we look for a solution of the form

$$y = f(x) = \sum_{n=0}^{\infty} c_n x^n = c_0 + c_1 x + c_2 x^2 + c_3 x^3 + \cdots$$

The method is to substitute this expression into the differential equation and determine the values of the coefficients $c_0, c_1, c_2, \ldots$. This technique resembles the method of undetermined coefficients discussed in Section 17.2.

Before using power series to solve Equation 1, we illustrate the method on the simpler equation $y'' + y = 0$ in Example 1. It's true that we already know how to solve this equation by the techniques of Section 17.1, but it's easier to understand the power series method when it is applied to this simpler equation.

V EXAMPLE 1 Use power series to solve the equation $y'' + y = 0$.

SOLUTION We assume there is a solution of the form

$$y = c_0 + c_1 x + c_2 x^2 + c_3 x^3 + \cdots = \sum_{n=0}^{\infty} c_n x^n \tag{2}$$

We can differentiate power series term by term, so

$$y' = c_1 + 2c_2 x + 3c_3 x^2 + \cdots = \sum_{n=1}^{\infty} n c_n x^{n-1}$$

$$y'' = 2c_2 + 2 \cdot 3c_3 x + \cdots = \sum_{n=2}^{\infty} n(n-1) c_n x^{n-2} \tag{3}$$

In order to compare the expressions for y and y'' more easily, we rewrite y'' as follows:

■ By writing out the first few terms of (4), you can see that it is the same as (3). To obtain (4), we replaced n by $n + 2$ and began the summation at 0 instead of 2.

$$y'' = \sum_{n=0}^{\infty} (n+2)(n+1) c_{n+2} x^n \tag{4}$$

Substituting the expressions in Equations 2 and 4 into the differential equation, we obtain

$$\sum_{n=0}^{\infty} (n+2)(n+1) c_{n+2} x^n + \sum_{n=0}^{\infty} c_n x^n = 0$$

or

$$\sum_{n=0}^{\infty} [(n+2)(n+1) c_{n+2} + c_n] x^n = 0 \tag{5}$$

If two power series are equal, then the corresponding coefficients must be equal. Therefore the coefficients of x^n in Equation 5 must be 0:

$$(n+2)(n+1)c_{n+2} + c_n = 0$$

6
$$c_{n+2} = -\frac{c_n}{(n+1)(n+2)} \qquad n = 0, 1, 2, 3, \dots$$

Equation 6 is called a *recursion relation*. If c_0 and c_1 are known, this equation allows us to determine the remaining coefficients recursively by putting $n = 0, 1, 2, 3, \dots$ in succession.

Put $n = 0$: $\qquad c_2 = -\dfrac{c_0}{1 \cdot 2}$

Put $n = 1$: $\qquad c_3 = -\dfrac{c_1}{2 \cdot 3}$

Put $n = 2$: $\qquad c_4 = -\dfrac{c_2}{3 \cdot 4} = \dfrac{c_0}{1 \cdot 2 \cdot 3 \cdot 4} = \dfrac{c_0}{4!}$

Put $n = 3$: $\qquad c_5 = -\dfrac{c_3}{4 \cdot 5} = \dfrac{c_1}{2 \cdot 3 \cdot 4 \cdot 5} = \dfrac{c_1}{5!}$

Put $n = 4$: $\qquad c_6 = -\dfrac{c_4}{5 \cdot 6} = -\dfrac{c_0}{4!\,5 \cdot 6} = -\dfrac{c_0}{6!}$

Put $n = 5$: $\qquad c_7 = -\dfrac{c_5}{6 \cdot 7} = -\dfrac{c_1}{5!\,6 \cdot 7} = -\dfrac{c_1}{7!}$

By now we see the pattern:

For the even coefficients, $c_{2n} = (-1)^n \dfrac{c_0}{(2n)!}$

For the odd coefficients, $c_{2n+1} = (-1)^n \dfrac{c_1}{(2n+1)!}$

Putting these values back into Equation 2, we write the solution as

$$\begin{aligned} y &= c_0 + c_1 x + c_2 x^2 + c_3 x^3 + c_4 x^4 + c_5 x^5 + \cdots \\ &= c_0\left(1 - \frac{x^2}{2!} + \frac{x^4}{4!} - \frac{x^6}{6!} + \cdots + (-1)^n \frac{x^{2n}}{(2n)!} + \cdots\right) \\ &\quad + c_1\left(x - \frac{x^3}{3!} + \frac{x^5}{5!} - \frac{x^7}{7!} + \cdots + (-1)^n \frac{x^{2n+1}}{(2n+1)!} + \cdots\right) \\ &= c_0 \sum_{n=0}^{\infty} (-1)^n \frac{x^{2n}}{(2n)!} + c_1 \sum_{n=0}^{\infty} (-1)^n \frac{x^{2n+1}}{(2n+1)!} \end{aligned}$$

Notice that there are two arbitrary constants, c_0 and c_1. □

NOTE 1 We recognize the series obtained in Example 1 as being the Maclaurin series for $\cos x$ and $\sin x$. (See Equations 11.10.16 and 11.10.15.) Therefore we could write the solution as

$$y(x) = c_0 \cos x + c_1 \sin x$$

But we are not usually able to express power series solutions of differential equations in terms of known functions.

EXAMPLE 2 Solve $y'' - 2xy' + y = 0$.

SOLUTION We assume there is a solution of the form

$$y = \sum_{n=0}^{\infty} c_n x^n$$

Then

$$y' = \sum_{n=1}^{\infty} nc_n x^{n-1}$$

and

$$y'' = \sum_{n=2}^{\infty} n(n-1)c_n x^{n-2} = \sum_{n=0}^{\infty} (n+2)(n+1)c_{n+2} x^n$$

as in Example 1. Substituting in the differential equation, we get

$$\sum_{n=0}^{\infty} (n+2)(n+1)c_{n+2} x^n - 2x \sum_{n=1}^{\infty} nc_n x^{n-1} + \sum_{n=0}^{\infty} c_n x^n = 0$$

$$\sum_{n=0}^{\infty} (n+2)(n+1)c_{n+2} x^n - \sum_{n=1}^{\infty} 2nc_n x^n + \sum_{n=0}^{\infty} c_n x^n = 0$$

$$\sum_{n=1}^{\infty} 2nc_n x^n = \sum_{n=0}^{\infty} 2nc_n x^n$$

$$\sum_{n=0}^{\infty} [(n+2)(n+1)c_{n+2} - (2n-1)c_n]x^n = 0$$

This equation is true if the coefficient of x^n is 0:

$$(n+2)(n+1)c_{n+2} - (2n-1)c_n = 0$$

7 $$c_{n+2} = \frac{2n-1}{(n+1)(n+2)} c_n \qquad n = 0, 1, 2, 3, \ldots$$

We solve this recursion relation by putting $n = 0, 1, 2, 3, \ldots$ successively in Equation 7:

Put $n = 0$: $$c_2 = \frac{-1}{1 \cdot 2} c_0$$

Put $n = 1$: $$c_3 = \frac{1}{2 \cdot 3} c_1$$

Put $n = 2$: $$c_4 = \frac{3}{3 \cdot 4} c_2 = -\frac{3}{1 \cdot 2 \cdot 3 \cdot 4} c_0 = -\frac{3}{4!} c_0$$

Put $n = 3$: $$c_5 = \frac{5}{4 \cdot 5} c_3 = \frac{1 \cdot 5}{2 \cdot 3 \cdot 4 \cdot 5} c_1 = \frac{1 \cdot 5}{5!} c_1$$

$$\text{Put } n = 4: \qquad c_6 = \frac{7}{5 \cdot 6} c_4 = -\frac{3 \cdot 7}{4!\, 5 \cdot 6} c_0 = -\frac{3 \cdot 7}{6!} c_0$$

$$\text{Put } n = 5: \qquad c_7 = \frac{9}{6 \cdot 7} c_5 = \frac{1 \cdot 5 \cdot 9}{5!\, 6 \cdot 7} c_1 = \frac{1 \cdot 5 \cdot 9}{7!} c_1$$

$$\text{Put } n = 6: \qquad c_8 = \frac{11}{7 \cdot 8} c_6 = -\frac{3 \cdot 7 \cdot 11}{8!} c_0$$

$$\text{Put } n = 7: \qquad c_9 = \frac{13}{8 \cdot 9} c_7 = \frac{1 \cdot 5 \cdot 9 \cdot 13}{9!} c_1$$

In general, the even coefficients are given by

4 find eqns for odd & even terms after writing out several terms

$$c_{2n} = -\frac{3 \cdot 7 \cdot 11 \cdot \cdots \cdot (4n - 5)}{(2n)!} c_0$$

and the odd coefficients are given by

$$c_{2n+1} = \frac{1 \cdot 5 \cdot 9 \cdot \cdots \cdot (4n - 3)}{(2n + 1)!} c_1$$

5 write the soln in terms of sums of power series

The solution is

$$\begin{aligned} y &= c_0 + c_1 x + c_2 x^2 + c_3 x^3 + c_4 x^4 + \cdots \\ &= c_0 \left(1 - \frac{1}{2!} x^2 - \frac{3}{4!} x^4 - \frac{3 \cdot 7}{6!} x^6 - \frac{3 \cdot 7 \cdot 11}{8!} x^8 - \cdots \right) \\ &\quad + c_1 \left(x + \frac{1}{3!} x^3 + \frac{1 \cdot 5}{5!} x^5 + \frac{1 \cdot 5 \cdot 9}{7!} x^7 + \frac{1 \cdot 5 \cdot 9 \cdot 13}{9!} x^9 + \cdots \right) \end{aligned}$$

or

8

$$\begin{aligned} y &= c_0 \left(1 - \frac{1}{2!} x^2 - \sum_{n=2}^{\infty} \frac{3 \cdot 7 \cdot \cdots \cdot (4n - 5)}{(2n)!} x^{2n} \right) \\ &\quad + c_1 \left(x + \sum_{n=1}^{\infty} \frac{1 \cdot 5 \cdot 9 \cdot \cdots \cdot (4n - 3)}{(2n + 1)!} x^{2n+1} \right) \end{aligned}$$

□

NOTE 2 In Example 2 we had to assume that the differential equation had a series solution. But now we could verify directly that the function given by Equation 8 is indeed a solution.

NOTE 3 Unlike the situation of Example 1, the power series that arise in the solution of Example 2 do not define elementary functions. The functions

$$y_1(x) = 1 - \frac{1}{2!} x^2 - \sum_{n=2}^{\infty} \frac{3 \cdot 7 \cdot \cdots \cdot (4n - 5)}{(2n)!} x^{2n}$$

and

$$y_2(x) = x + \sum_{n=1}^{\infty} \frac{1 \cdot 5 \cdot 9 \cdot \cdots \cdot (4n - 3)}{(2n + 1)!} x^{2n+1}$$

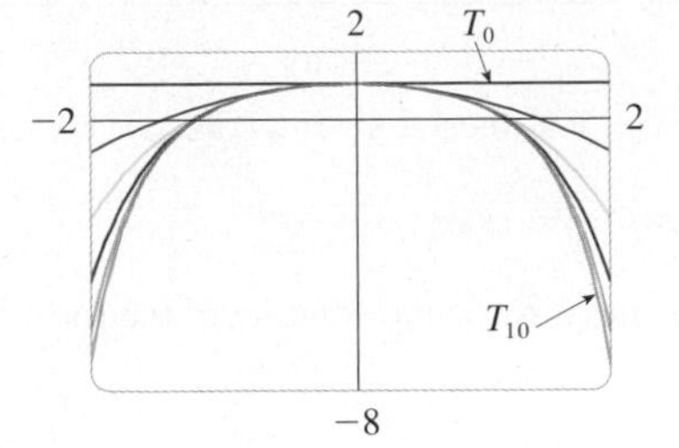

FIGURE 1

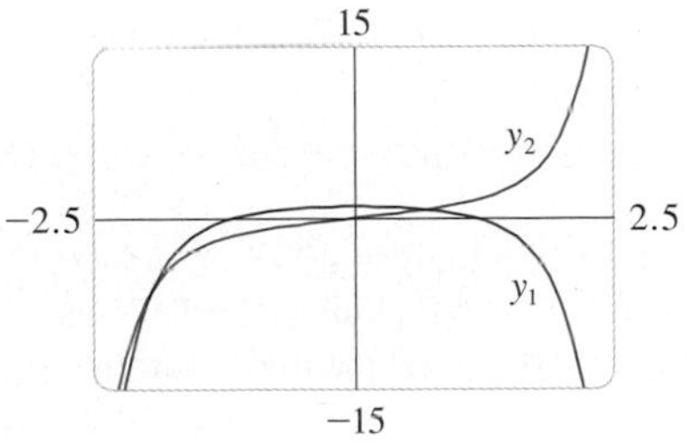

FIGURE 2

are perfectly good functions but they can't be expressed in terms of familiar functions. We can use these power series expressions for y_1 and y_2 to compute approximate values of the functions and even to graph them. Figure 1 shows the first few partial sums $T_0, T_2, T_4, \ldots$ (Taylor polynomials) for $y_1(x)$, and we see how they converge to y_1. In this way we can graph both y_1 and y_2 in Figure 2.

NOTE 4 If we were asked to solve the initial-value problem

$$y'' - 2xy' + y = 0 \qquad y(0) = 0 \qquad y'(0) = 1$$

we would observe from Theorem 11.10.5 that

$$c_0 = y(0) = 0 \qquad c_1 = y'(0) = 1$$

This would simplify the calculations in Example 2, since all of the even coefficients would be 0. The solution to the initial-value problem is

$$y(x) = x + \sum_{n=1}^{\infty} \frac{1 \cdot 5 \cdot 9 \cdot \cdots \cdot (4n-3)}{(2n+1)!} x^{2n+1}$$

17.4 EXERCISES

1–11 Use power series to solve the differential equation.

1. $y' - y = 0$

2. $y' = xy$

3. $y' - x^2 y$

4. $(x - 3)y' + 2y = 0$

5. $y'' + xy' + y = 0$

6. $y'' = y$

7. $(x - 1)y'' + y' = 0$

8. $y'' = xy$

9. $y'' - xy' - y = 0, \quad y(0) = 1, \quad y'(0) = 0$

10. $y'' + x^2 y = 0, \quad y(0) = 1, \quad y'(0) = 0$

11. $y'' + x^2 y' + xy = 0, \quad y(0) = 0, \quad y'(0) = 1$

12. The solution of the initial value problem

$$x^2 y'' + xy' + x^2 y = 0 \qquad y(0) = 1 \qquad y'(0) = 0$$

is called a Bessel function of order 0.

(a) Solve the initial-value problem to find a power series expansion for the Bessel function.

(b) Graph several Taylor polynomials until you reach one that looks like a good approximation to the Bessel function on the interval $[-5, 5]$.

17 REVIEW

CONCEPT CHECK

1. (a) Write the general form of a second-order homogeneous linear differential equation with constant coefficients.

(b) Write the auxiliary equation.

(c) How do you use the roots of the auxiliary equation to solve the differential equation? Write the form of the solution for each of the three cases that can occur.

2. (a) What is an initial-value problem for a second-order differential equation?

(b) What is a boundary-value problem for such an equation?

3. (a) Write the general form of a second-order nonhomogeneous linear differential equation with constant coefficients.

(b) What is the complementary equation? How does it help solve the original differential equation?

(c) Explain how the method of undetermined coefficients works.

(d) Explain how the method of variation of parameters works.

4. Discuss two applications of second-order linear differential equations.

5. How do you use power series to solve a differential equation?

TRUE-FALSE QUIZ

Determine whether the statement is true or false. If it is true, explain why. If it is false, explain why or give an example that disproves the statement.

1. If y_1 and y_2 are solutions of $y'' + y = 0$, then $y_1 + y_2$ is also a solution of the equation.
2. If y_1 and y_2 are solutions of $y'' + 6y' + 5y = x$, then $c_1 y_1 + c_2 y_2$ is also a solution of the equation.
3. The general solution of $y'' - y = 0$ can be written as
$$y = c_1 \cosh x + c_2 \sinh x$$
4. The equation $y'' - y = e^x$ has a particular solution of the form
$$y_p = Ae^x$$

EXERCISES

1–10 Solve the differential equation.

1. $y'' - 2y' - 15y = 0$
2. $y'' + 4y' + 13y = 0$
3. $y'' + 3y = 0$
4. $4y'' + 4y' + y = 0$
5. $\dfrac{d^2y}{dx^2} - 4\dfrac{dy}{dx} + 5y = e^{2x}$
6. $\dfrac{d^2y}{dx^2} + \dfrac{dy}{dx} - 2y = x^2$
7. $\dfrac{d^2y}{dx^2} - 2\dfrac{dy}{dx} + y = x \cos x$
8. $\dfrac{d^2y}{dx^2} + 4y = \sin 2x$
9. $\dfrac{d^2y}{dx^2} - \dfrac{dy}{dx} - 6y = 1 + e^{-2x}$
10. $\dfrac{d^2y}{dx^2} + y = \csc x, \quad 0 < x < \pi/2$

11–14 Solve the initial-value problem.

11. $y'' + 6y' = 0, \quad y(1) = 3, \quad y'(1) = 12$
12. $y'' - 6y' + 25y = 0, \quad y(0) = 2, \quad y'(0) = 1$
13. $y'' - 5y' + 4y = 0, \quad y(0) = 0, \quad y'(0) = 1$
14. $9y'' + y = 3x + e^{-x}, \quad y(0) = 1, \quad y'(0) = 2$

15. Use power series to solve the initial-value problem
$$y'' + xy' + y = 0 \qquad y(0) = 0 \qquad y'(0) = 1$$

16. Use power series to solve the equation
$$y'' - xy' - 2y = 0$$

17. A series circuit contains a resistor with $R = 40\ \Omega$, an inductor with $L = 2$ H, a capacitor with $C = 0.0025$ F, and a 12-V battery. The initial charge is $Q = 0.01$ C and the initial current is 0. Find the charge at time t.

18. A spring with a mass of 2 kg has damping constant 16, and a force of 12.8 N keeps the spring stretched 0.2 m beyond its natural length. Find the position of the mass at time t if it starts at the equilibrium position with a velocity of 2.4 m/s.

19. Assume that the earth is a solid sphere of uniform density with mass M and radius $R = 3960$ mi. For a particle of mass m within the earth at a distance r from the earth's center, the gravitational force attracting the particle to the center is
$$F_r = \frac{-GM_r m}{r^2}$$
where G is the gravitational constant and M_r is the mass of the earth within the sphere of radius r.
    (a) Show that $F_r = \dfrac{-GMm}{R^3} r$.
    (b) Suppose a hole is drilled through the earth along a diameter. Show that if a particle of mass m is dropped from rest at the surface, into the hole, then the distance $y = y(t)$ of the particle from the center of the earth at time t is given by
$$y''(t) = -k^2 y(t)$$
    where $k^2 = GM/R^3 = g/R$.
    (c) Conclude from part (b) that the particle undergoes simple harmonic motion. Find the period T.
    (d) With what speed does the particle pass through the center of the earth?

APPENDIXES

A Numbers, Inequalities, and Absolute Values

B Coordinate Geometry and Lines

C Graphs of Second-Degree Equations

D Trigonometry

E Sigma Notation

F Proofs of Theorems

G The Logarithm Defined as an Integral

H Complex Numbers

I Answers to Odd-Numbered Exercises

A NUMBERS, INEQUALITIES, AND ABSOLUTE VALUES

Calculus is based on the real number system. We start with the **integers**:

$$\ldots, \quad -3, \quad -2, \quad -1, \quad 0, \quad 1, \quad 2, \quad 3, \quad 4, \quad \ldots$$

Then we construct the **rational numbers**, which are ratios of integers. Thus any rational number r can be expressed as

$$r = \frac{m}{n} \qquad \text{where } m \text{ and } n \text{ are integers and } n \neq 0$$

Examples are

$$\tfrac{1}{2} \qquad -\tfrac{3}{7} \qquad 46 = \tfrac{46}{1} \qquad 0.17 = \tfrac{17}{100}$$

(Recall that division by 0 is always ruled out, so expressions like $\frac{3}{0}$ and $\frac{0}{0}$ are undefined.) Some real numbers, such as $\sqrt{2}$, can't be expressed as a ratio of integers and are therefore called **irrational numbers**. It can be shown, with varying degrees of difficulty, that the following are also irrational numbers:

$$\sqrt{3} \qquad \sqrt{5} \qquad \sqrt[3]{2} \qquad \pi \qquad \sin 1^\circ \qquad \log_{10} 2$$

The set of all real numbers is usually denoted by the symbol $\mathbb{R}$. When we use the word *number* without qualification, we mean "real number."

Every number has a decimal representation. If the number is rational, then the corresponding decimal is repeating. For example,

$$\tfrac{1}{2} = 0.5000\ldots = 0.5\overline{0} \qquad \tfrac{2}{3} = 0.66666\ldots = 0.\overline{6}$$

$$\tfrac{157}{495} = 0.317171717\ldots = 0.3\overline{17} \qquad \tfrac{9}{7} = 1.285714285714\ldots = 1.\overline{285714}$$

(The bar indicates that the sequence of digits repeats forever.) On the other hand, if the number is irrational, the decimal is nonrepeating:

$$\sqrt{2} = 1.414213562373095\ldots \qquad \pi = 3.141592653589793\ldots$$

If we stop the decimal expansion of any number at a certain place, we get an approximation to the number. For instance, we can write

$$\pi \approx 3.14159265$$

where the symbol $\approx$ is read "is approximately equal to." The more decimal places we retain, the better the approximation we get.

The real numbers can be represented by points on a line as in Figure 1. The positive direction (to the right) is indicated by an arrow. We choose an arbitrary reference point O, called the **origin**, which corresponds to the real number 0. Given any convenient unit of measurement, each positive number x is represented by the point on the line a distance of x units to the right of the origin, and each negative number $-x$ is represented by the point x units to the left of the origin. Thus every real number is represented by a point on the line, and every point P on the line corresponds to exactly one real number. The number associated with the point P is called the **coordinate** of P and the line is then called a **coor-**

dinate line, or a **real number line**, or simply a **real line**. Often we identify the point with its coordinate and think of a number as being a point on the real line.

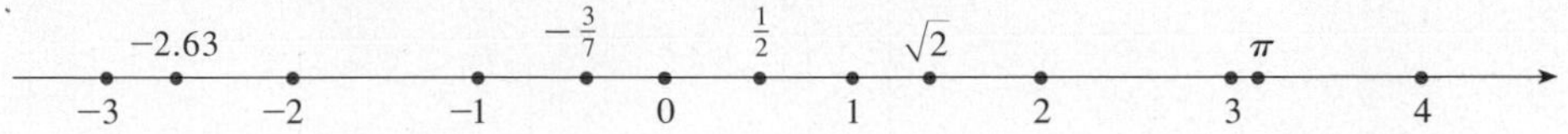

FIGURE 1

The real numbers are ordered. We say *a is less than b* and write $a < b$ if $b - a$ is a positive number. Geometrically this means that a lies to the left of b on the number line. (Equivalently, we say *b is greater than a* and write $b > a$.) The symbol $a \leqslant b$ (or $b \geqslant a$) means that either $a < b$ or $a = b$ and is read "a is less than or equal to b." For instance, the following are true inequalities:

$$7 < 7.4 < 7.5 \qquad -3 > -\pi \qquad \sqrt{2} < 2 \qquad \sqrt{2} \leqslant 2 \qquad 2 \leqslant 2$$

In what follows we need to use *set notation*. A **set** is a collection of objects, and these objects are called the **elements** of the set. If S is a set, the notation $a \in S$ means that a is an element of S, and $a \notin S$ means that a is not an element of S. For example, if Z represents the set of integers, then $-3 \in Z$ but $\pi \notin Z$. If S and T are sets, then their **union** $S \cup T$ is the set consisting of all elements that are in S or T (or in both S and T). The **intersection** of S and T is the set $S \cap T$ consisting of all elements that are in both S and T. In other words, $S \cap T$ is the common part of S and T. The empty set, denoted by $\varnothing$, is the set that contains no element.

Some sets can be described by listing their elements between braces. For instance, the set A consisting of all positive integers less than 7 can be written as

$$A = \{1, 2, 3, 4, 5, 6\}$$

We could also write A in *set-builder notation* as

$$A = \{x \mid x \text{ is an integer and } 0 < x < 7\}$$

which is read "A is the set of x such that x is an integer and $0 < x < 7$."

INTERVALS

Certain sets of real numbers, called **intervals,** occur frequently in calculus and correspond geometrically to line segments. For example, if $a < b$, the **open interval** from a to b consists of all numbers between a and b and is denoted by the symbol (a, b). Using set-builder notation, we can write

$$(a, b) = \{x \mid a < x < b\}$$

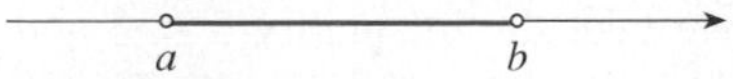

FIGURE 2
Open interval (a, b)

Notice that the endpoints of the interval—namely, a and b—are excluded. This is indicated by the round brackets ( ) and by the open dots in Figure 2. The **closed interval** from a to b is the set

$$[a, b] = \{x \mid a \leqslant x \leqslant b\}$$

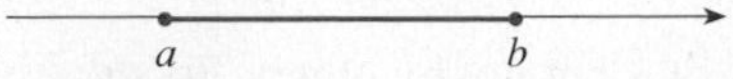

FIGURE 3
Closed interval $[a, b]$

Here the endpoints of the interval are included. This is indicated by the square brackets [ ] and by the solid dots in Figure 3. It is also possible to include only one endpoint in an interval, as shown in Table 1.

1 TABLE OF INTERVALS

Notation	Set description	Picture
(a, b)	$\{x \mid a < x < b\}$	
$[a, b]$	$\{x \mid a \leqslant x \leqslant b\}$	
$[a, b)$	$\{x \mid a \leqslant x < b\}$	
$(a, b]$	$\{x \mid a < x \leqslant b\}$	
(a, ∞)	$\{x \mid x > a\}$	
$[a, \infty)$	$\{x \mid x \geqslant a\}$	
$(-\infty, b)$	$\{x \mid x < b\}$	
$(-\infty, b]$	$\{x \mid x \leqslant b\}$	
$(-\infty, \infty)$	$\mathbb{R}$ (set of all real numbers)	

■ Table 1 lists the nine possible types of intervals. When these intervals are discussed, it is always assumed that $a < b$.

We also need to consider infinite intervals such as

$$(a, \infty) = \{x \mid x > a\}$$

This does not mean that ∞ ("infinity") is a number. The notation (a, ∞) stands for the set of all numbers that are greater than a, so the symbol ∞ simply indicates that the interval extends indefinitely far in the positive direction.

INEQUALITIES

When working with inequalities, note the following rules.

2 RULES FOR INEQUALITIES

1. If $a < b$, then $a + c < b + c$.
2. If $a < b$ and $c < d$, then $a + c < b + d$.
3. If $a < b$ and $c > 0$, then $ac < bc$.
4. If $a < b$ and $c < 0$, then $ac > bc$.
5. If $0 < a < b$, then $1/a > 1/b$.

Rule 1 says that we can add any number to both sides of an inequality, and Rule 2 says that two inequalities can be added. However, we have to be careful with multiplication. Rule 3 says that we can multiply both sides of an inequality by a *positive* number, but Rule 4 says that if we multiply both sides of an inequality by a negative number, then we reverse the direction of the inequality. For example, if we take the inequality $3 < 5$ and multiply by 2, we get $6 < 10$, but if we multiply by -2, we get $-6 > -10$. Finally, Rule 5 says that if we take reciprocals, then we reverse the direction of an inequality (provided the numbers are positive).

EXAMPLE 1 Solve the inequality $1 + x < 7x + 5$.

SOLUTION The given inequality is satisfied by some values of x but not by others. To *solve* an inequality means to determine the set of numbers x for which the inequality is true. This is called the *solution set*.

First we subtract 1 from each side of the inequality (using Rule 1 with $c = -1$):

$$x < 7x + 4$$

Then we subtract $7x$ from both sides (Rule 1 with $c = -7x$):

$$-6x < 4$$

Now we divide both sides by -6 (Rule 4 with $c = -\frac{1}{6}$):

$$x > -\tfrac{4}{6} = -\tfrac{2}{3}$$

These steps can all be reversed, so the solution set consists of all numbers greater than $-\frac{2}{3}$. In other words, the solution of the inequality is the interval $\left(-\frac{2}{3}, \infty\right)$. □

EXAMPLE 2 Solve the inequalities $4 \leqslant 3x - 2 < 13$.

SOLUTION Here the solution set consists of all values of x that satisfy both inequalities. Using the rules given in (2), we see that the following inequalities are equivalent:

$$4 \leqslant 3x - 2 < 13$$

$$6 \leqslant 3x < 15 \qquad \text{(add 2)}$$

$$2 \leqslant x < 5 \qquad \text{(divide by 3)}$$

Therefore the solution set is $[2, 5)$. □

EXAMPLE 3 Solve the inequality $x^2 - 5x + 6 \leqslant 0$.

SOLUTION First we factor the left side:

$$(x - 2)(x - 3) \leqslant 0$$

We know that the corresponding equation $(x - 2)(x - 3) = 0$ has the solutions 2 and 3. The numbers 2 and 3 divide the real line into three intervals:

$$(-\infty, 2) \qquad (2, 3) \qquad (3, \infty)$$

On each of these intervals we determine the signs of the factors. For instance,

$$x \in (-\infty, 2) \quad \Rightarrow \quad x < 2 \quad \Rightarrow \quad x - 2 < 0$$

Then we record these signs in the following chart:

Interval	$x - 2$	$x - 3$	$(x - 2)(x - 3)$
$x < 2$	$-$	$-$	$+$
$2 < x < 3$	$+$	$-$	$-$
$x > 3$	$+$	$+$	$+$

■ A visual method for solving Example 3 is to use a graphing device to graph the parabola $y = x^2 - 5x + 6$ (as in Figure 4) and observe that the curve lies on or below the x-axis when $2 \leqslant x \leqslant 3$.

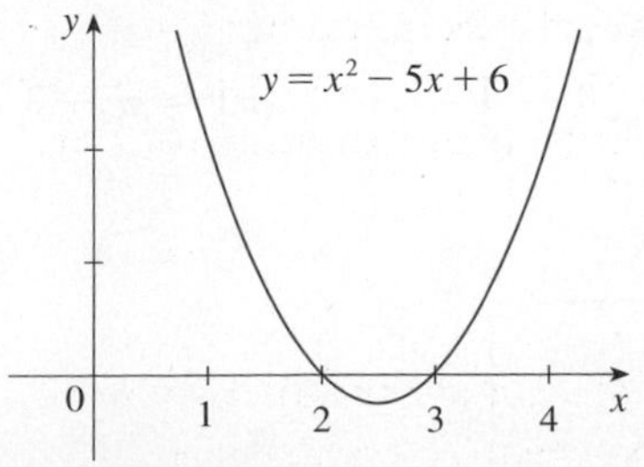

FIGURE 4

Another method for obtaining the information in the chart is to use *test values*. For instance, if we use the test value $x = 1$ for the interval $(-\infty, 2)$, then substitution in $x^2 - 5x + 6$ gives

$$1^2 - 5(1) + 6 = 2$$

The polynomial $x^2 - 5x + 6$ doesn't change sign inside any of the three intervals, so we conclude that it is positive on $(-\infty, 2)$.

Then we read from the chart that $(x - 2)(x - 3)$ is negative when $2 < x < 3$. Thus the solution of the inequality $(x - 2)(x - 3) \leqslant 0$ is

$$\{x \mid 2 \leqslant x \leqslant 3\} = [2, 3]$$

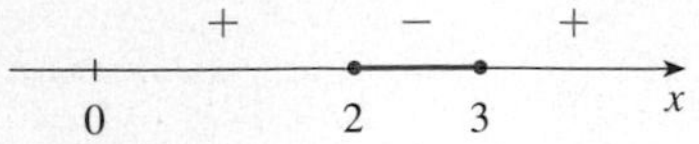

FIGURE 5

Notice that we have included the endpoints 2 and 3 because we are looking for values of x such that the product is either negative or zero. The solution is illustrated in Figure 5. □

EXAMPLE 4 Solve $x^3 + 3x^2 > 4x$.

SOLUTION First we take all nonzero terms to one side of the inequality sign and factor the resulting expression:

$$x^3 + 3x^2 - 4x > 0 \qquad \text{or} \qquad x(x - 1)(x + 4) > 0$$

As in Example 3 we solve the corresponding equation $x(x - 1)(x + 4) = 0$ and use the solutions $x = -4$, $x = 0$, and $x = 1$ to divide the real line into four intervals $(-\infty, -4)$, $(-4, 0)$, $(0, 1)$, and $(1, \infty)$. On each interval the product keeps a constant sign as shown in the following chart:

Interval	x	$x - 1$	$x + 4$	$x(x - 1)(x + 4)$
$x < -4$	−	−	−	−
$-4 < x < 0$	−	−	+	+
$0 < x < 1$	+	−	+	−
$x > 1$	+	+	+	+

Then we read from the chart that the solution set is

$$\{x \mid -4 < x < 0 \text{ or } x > 1\} = (-4, 0) \cup (1, \infty)$$

FIGURE 6

The solution is illustrated in Figure 6. □

ABSOLUTE VALUE

The **absolute value** of a number a, denoted by $|a|$, is the distance from a to 0 on the real number line. Distances are always positive or 0, so we have

$$|a| \geqslant 0 \qquad \text{for every number } a$$

For example,

$$|3| = 3 \qquad |-3| = 3 \qquad |0| = 0 \qquad |\sqrt{2} - 1| = \sqrt{2} - 1 \qquad |3 - \pi| = \pi - 3$$

In general, we have

■ Remember that if a is negative, then $-a$ is positive.

3

$$|a| = a \qquad \text{if } a \geqslant 0$$
$$|a| = -a \qquad \text{if } a < 0$$

EXAMPLE 5 Express $|3x - 2|$ without using the absolute-value symbol.

SOLUTION

$$|3x - 2| = \begin{cases} 3x - 2 & \text{if } 3x - 2 \geqslant 0 \\ -(3x - 2) & \text{if } 3x - 2 < 0 \end{cases}$$

$$= \begin{cases} 3x - 2 & \text{if } x \geqslant \frac{2}{3} \\ 2 - 3x & \text{if } x < \frac{2}{3} \end{cases}$$ □

Recall that the symbol $\sqrt{\ }$ means "the positive square root of." Thus $\sqrt{r} = s$ means $s^2 = r$ and $s \geqslant 0$. Therefore, the equation $\sqrt{a^2} = a$ is not always true. It is true only when $a \geqslant 0$. If $a < 0$, then $-a > 0$, so we have $\sqrt{a^2} = -a$. In view of (3), we then have the equation

4

$$\sqrt{a^2} = |a|$$

which is true for all values of a.

Hints for the proofs of the following properties are given in the exercises.

5 **PROPERTIES OF ABSOLUTE VALUES** Suppose a and b are any real numbers and n is an integer. Then

1. $|ab| = |a||b|$ **2.** $\left|\dfrac{a}{b}\right| = \dfrac{|a|}{|b|} \quad (b \neq 0)$ **3.** $|a^n| = |a|^n$

For solving equations or inequalities involving absolute values, it's often very helpful to use the following statements.

6 Suppose $a > 0$. Then

4. $|x| = a$ if and only if $x = \pm a$

5. $|x| < a$ if and only if $-a < x < a$

6. $|x| > a$ if and only if $x > a$ or $x < -a$

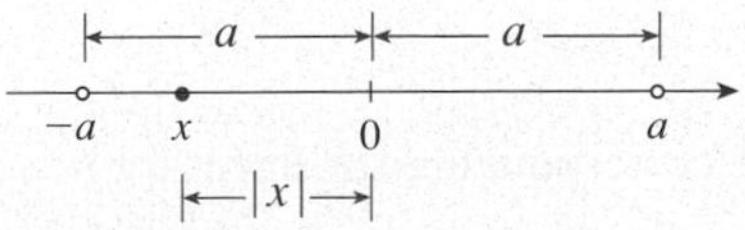

FIGURE 7

For instance, the inequality $|x| < a$ says that the distance from x to the origin is less than a, and you can see from Figure 7 that this is true if and only if x lies between $-a$ and a.

If a and b are any real numbers, then the distance between a and b is the absolute value of the difference, namely, $|a - b|$, which is also equal to $|b - a|$. (See Figure 8.)

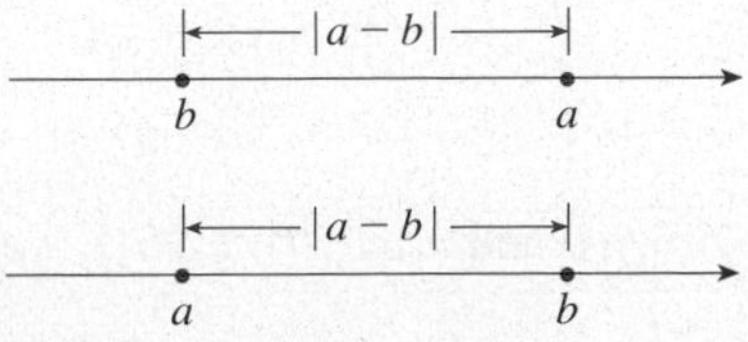

FIGURE 8
Length of a line segment = $|a - b|$

EXAMPLE 6 Solve $|2x - 5| = 3$.

SOLUTION By Property 4 of (6), $|2x - 5| = 3$ is equivalent to

$$2x - 5 = 3 \qquad \text{or} \qquad 2x - 5 = -3$$

So $2x = 8$ or $2x = 2$. Thus $x = 4$ or $x = 1$. □

EXAMPLE 7 Solve $|x - 5| < 2$.

SOLUTION 1 By Property 5 of (6), $|x - 5| < 2$ is equivalent to

$$-2 < x - 5 < 2$$

Therefore, adding 5 to each side, we have

$$3 < x < 7$$

and the solution set is the open interval $(3, 7)$.

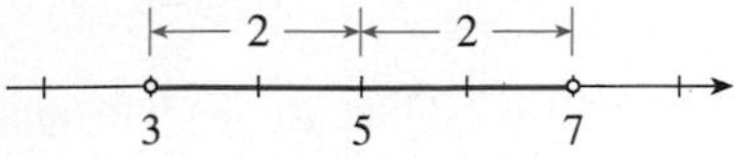

FIGURE 9

SOLUTION 2 Geometrically the solution set consists of all numbers x whose distance from 5 is less than 2. From Figure 9 we see that this is the interval $(3, 7)$. □

EXAMPLE 8 Solve $|3x + 2| \geqslant 4$.

SOLUTION By Properties 4 and 6 of (6), $|3x + 2| \geqslant 4$ is equivalent to

$$3x + 2 \geqslant 4 \qquad \text{or} \qquad 3x + 2 \leqslant -4$$

In the first case $3x \geqslant 2$, which gives $x \geqslant \frac{2}{3}$. In the second case $3x \leqslant -6$, which gives $x \leqslant -2$. So the solution set is

$$\{x \mid x \leqslant -2 \text{ or } x \geqslant \tfrac{2}{3}\} = (-\infty, -2] \cup [\tfrac{2}{3}, \infty)$$ □

Another important property of absolute value, called the Triangle Inequality, is used frequently not only in calculus but throughout mathematics in general.

7 **THE TRIANGLE INEQUALITY** If a and b are any real numbers, then

$$|a + b| \leqslant |a| + |b|$$

Observe that if the numbers a and b are both positive or both negative, then the two sides in the Triangle Inequality are actually equal. But if a and b have opposite signs, the left side involves a subtraction and the right side does not. This makes the Triangle Inequality seem reasonable, but we can prove it as follows.

Notice that

$$-|a| \leqslant a \leqslant |a|$$

is always true because a equals either $|a|$ or $-|a|$. The corresponding statement for b is

$$-|b| \leqslant b \leqslant |b|$$

Adding these inequalities, we get

$$-(|a| + |b|) \leqslant a + b \leqslant |a| + |b|$$

If we now apply Properties 4 and 5 (with x replaced by $a + b$ and a by $|a| + |b|$), we obtain

$$|a + b| \leqslant |a| + |b|$$

which is what we wanted to show.

EXAMPLE 9 If $|x - 4| < 0.1$ and $|y - 7| < 0.2$, use the Triangle Inequality to estimate $|(x + y) - 11|$.

SOLUTION In order to use the given information, we use the Triangle Inequality with $a = x - 4$ and $b = y - 7$:

$$\begin{aligned} |(x + y) - 11| &= |(x - 4) + (y - 7)| \\ &\leqslant |x - 4| + |y - 7| \\ &< 0.1 + 0.2 = 0.3 \end{aligned}$$

Thus $$|(x + y) - 11| < 0.3$$ □

A EXERCISES

1–12 Rewrite the expression without using the absolute value symbol.

1. $|5 - 23|$

2. $|5| - |-23|$

3. $|-\pi|$

4. $|\pi - 2|$

5. $|\sqrt{5} - 5|$

6. $||-2| - |-3||$

7. $|x - 2|$ if $x < 2$

8. $|x - 2|$ if $x > 2$

9. $|x + 1|$

10. $|2x - 1|$

11. $|x^2 + 1|$

12. $|1 - 2x^2|$

13–38 Solve the inequality in terms of intervals and illustrate the solution set on the real number line.

13. $2x + 7 > 3$

14. $3x - 11 < 4$

15. $1 - x \leqslant 2$

16. $4 - 3x \geqslant 6$

17. $2x + 1 < 5x - 8$

18. $1 + 5x > 5 - 3x$

19. $-1 < 2x - 5 < 7$

20. $1 < 3x + 4 \leqslant 16$

21. $0 \leqslant 1 - x < 1$

22. $-5 \leqslant 3 - 2x \leqslant 9$

23. $4x < 2x + 1 \leqslant 3x + 2$

24. $2x - 3 < x + 4 < 3x - 2$

25. $(x - 1)(x - 2) > 0$

26. $(2x + 3)(x - 1) \geqslant 0$

27. $2x^2 + x \leqslant 1$

28. $x^2 < 2x + 8$

29. $x^2 + x + 1 > 0$

30. $x^2 + x > 1$

31. $x^2 < 3$

32. $x^2 \geqslant 5$

33. $x^3 - x^2 \leqslant 0$

34. $(x + 1)(x - 2)(x + 3) \geqslant 0$

35. $x^3 > x$

36. $x^3 + 3x < 4x^2$

37. $\dfrac{1}{x} < 4$

38. $-3 < \dfrac{1}{x} \leqslant 1$

39. The relationship between the Celsius and Fahrenheit temperature scales is given by $C = \frac{5}{9}(F - 32)$, where C is the temperature in degrees Celsius and F is the temperature in degrees Fahrenheit. What interval on the Celsius scale corresponds to the temperature range $50 \leqslant F \leqslant 95$?

40. Use the relationship between C and F given in Exercise 39 to find the interval on the Fahrenheit scale corresponding to the temperature range $20 \leqslant C \leqslant 30$.

41. As dry air moves upward, it expands and in so doing cools at a rate of about 1°C for each 100-m rise, up to about 12 km.
(a) If the ground temperature is 20°C, write a formula for the temperature at height h.
(b) What range of temperature can be expected if a plane takes off and reaches a maximum height of 5 km?

42. If a ball is thrown upward from the top of a building 128 ft high with an initial velocity of 16 ft/s, then the height h above the ground t seconds later will be

$$h = 128 + 16t - 16t^2$$

During what time interval will the ball be at least 32 ft above the ground?

43–46 Solve the equation for x.

43. $|2x| = 3$

44. $|3x + 5| = 1$

45. $|x + 3| = |2x + 1|$

46. $\left|\dfrac{2x - 1}{x + 1}\right| = 3$

47–56 Solve the inequality.

47. $|x| < 3$

48. $|x| \geqslant 3$

49. $|x - 4| < 1$

50. $|x - 6| < 0.1$

51. $|x + 5| \geqslant 2$

52. $|x + 1| \geqslant 3$

53. $|2x - 3| \leqslant 0.4$

54. $|5x - 2| < 6$

55. $1 \leqslant |x| \leqslant 4$

56. $0 < |x - 5| < \frac{1}{2}$

57–58 Solve for x, assuming a, b, and c are positive constants.

57. $a(bx - c) \geqslant bc$

58. $a \leqslant bx + c < 2a$

59–60 Solve for x, assuming a, b, and c are negative constants.

59. $ax + b < c$

60. $\dfrac{ax + b}{c} \leqslant b$

61. Suppose that $|x - 2| < 0.01$ and $|y - 3| < 0.04$. Use the Triangle Inequality to show that $|(x + y) - 5| < 0.05$.

62. Show that if $|x + 3| < \frac{1}{2}$, then $|4x + 13| < 3$.

63. Show that if $a < b$, then $a < \dfrac{a + b}{2} < b$.

64. Use Rule 3 to prove Rule 5 of (2).

65. Prove that $|ab| = |a||b|$. [*Hint:* Use Equation 4.]

66. Prove that $\left|\dfrac{a}{b}\right| = \dfrac{|a|}{|b|}$.

67. Show that if $0 < a < b$, then $a^2 < b^2$.

68. Prove that $|x - y| \geqslant |x| - |y|$. [*Hint:* Use the Triangle Inequality with $a = x - y$ and $b = y$.]

69. Show that the sum, difference, and product of rational numbers are rational numbers.

70. (a) Is the sum of two irrational numbers always an irrational number?
(b) Is the product of two irrational numbers always an irrational number?

B COORDINATE GEOMETRY AND LINES

Just as the points on a line can be identified with real numbers by assigning them coordinates, as described in Appendix A, so the points in a plane can be identified with ordered pairs of real numbers. We start by drawing two perpendicular coordinate lines that intersect at the origin O on each line. Usually one line is horizontal with positive direction to the right and is called the x-axis; the other line is vertical with positive direction upward and is called the y-axis.

Any point P in the plane can be located by a unique ordered pair of numbers as follows. Draw lines through P perpendicular to the x- and y-axes. These lines intersect the axes in points with coordinates a and b as shown in Figure 1. Then the point P is assigned the ordered pair (a, b). The first number a is called the ***x*-coordinate** of P; the second number b is called the ***y*-coordinate** of P. We say that P is the point with coordinates (a, b), and we denote the point by the symbol $P(a, b)$. Several points are labeled with their coordinates in Figure 2.

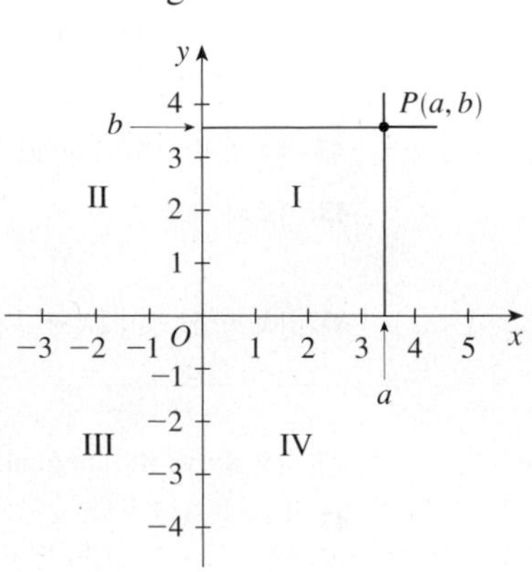

FIGURE 1

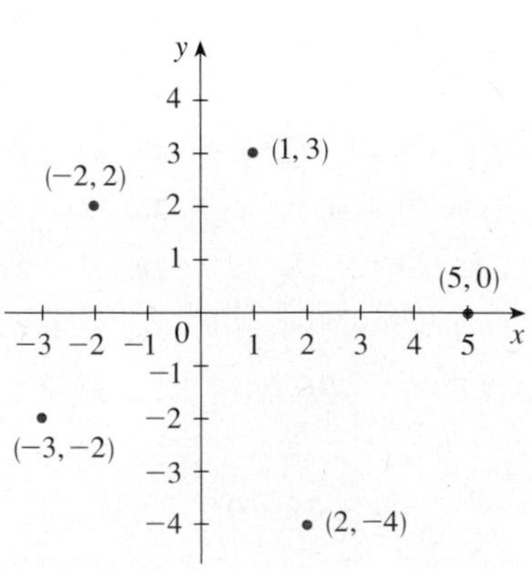

FIGURE 2

By reversing the preceding process we can start with an ordered pair (a, b) and arrive at the corresponding point P. Often we identify the point P with the ordered pair (a, b) and refer to "the point (a, b)." [Although the notation used for an open interval (a, b) is the same as the notation used for a point (a, b), you will be able to tell from the context which meaning is intended.]

This coordinate system is called the **rectangular coordinate system** or the **Cartesian coordinate system** in honor of the French mathematician René Descartes (1596–1650), even though another Frenchman, Pierre Fermat (1601–1665), invented the principles of analytic geometry at about the same time as Descartes. The plane supplied with this coordinate system is called the **coordinate plane** or the **Cartesian plane** and is denoted by $\mathbb{R}^2$.

The x- and y-axes are called the **coordinate axes** and divide the Cartesian plane into four quadrants, which are labeled I, II, III, and IV in Figure 1. Notice that the first quadrant consists of those points whose x- and y-coordinates are both positive.

EXAMPLE 1 Describe and sketch the regions given by the following sets.

(a) $\{(x, y) \mid x \geqslant 0\}$ (b) $\{(x, y) \mid y = 1\}$ (c) $\{(x, y) \mid |y| < 1\}$

SOLUTION

(a) The points whose x-coordinates are 0 or positive lie on the y-axis or to the right of it as indicated by the shaded region in Figure 3(a).

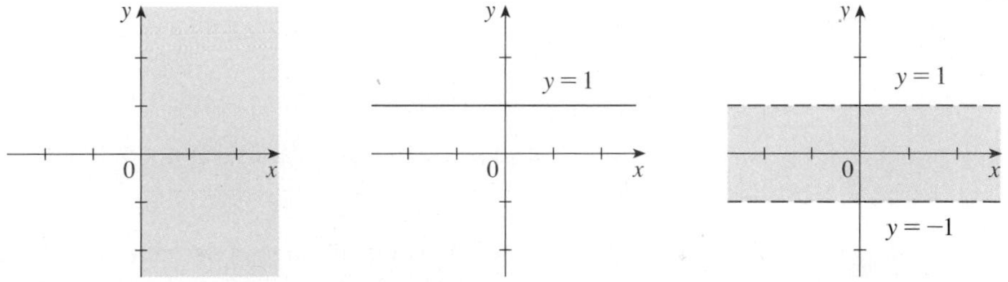

FIGURE 3 (a) $x \geqslant 0$ (b) $y - 1$ (c) $|y| < 1$

(b) The set of all points with y-coordinate 1 is a horizontal line one unit above the x-axis [see Figure 3(b)].

(c) Recall from Appendix A that

$$|y| < 1 \qquad \text{if and only if} \qquad -1 < y < 1$$

The given region consists of those points in the plane whose y-coordinates lie between -1 and 1. Thus the region consists of all points that lie between (but not on) the horizontal lines $y = 1$ and $y = -1$. [These lines are shown as dashed lines in Figure 3(c) to indicate that the points on these lines don't lie in the set.] □

Recall from Appendix A that the distance between points a and b on a number line is $|a - b| = |b - a|$. Thus the distance between points $P_1(x_1, y_1)$ and $P_3(x_2, y_1)$ on a horizontal line must be $|x_2 - x_1|$ and the distance between $P_2(x_2, y_2)$ and $P_3(x_2, y_1)$ on a vertical line must be $|y_2 - y_1|$. (See Figure 4.)

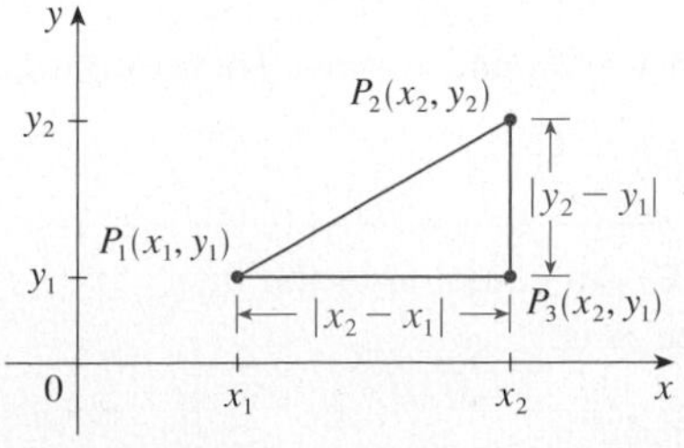

FIGURE 4

To find the distance $|P_1P_2|$ between any two points $P_1(x_1, y_1)$ and $P_2(x_2, y_2)$, we note that triangle $P_1P_2P_3$ in Figure 4 is a right triangle, and so by the Pythagorean Theorem we have

$$|P_1P_2| = \sqrt{|P_1P_3|^2 + |P_2P_3|^2} = \sqrt{|x_2 - x_1|^2 + |y_2 - y_1|^2}$$

$$= \sqrt{(x_2 - x_1)^2 + (y_2 - y_1)^2}$$

1 DISTANCE FORMULA The distance between the points $P_1(x_1, y_1)$ and $P_2(x_2, y_2)$ is

$$|P_1P_2| = \sqrt{(x_2 - x_1)^2 + (y_2 - y_1)^2}$$

EXAMPLE 2 The distance between $(1, -2)$ and $(5, 3)$ is

$$\sqrt{(5-1)^2 + [3-(-2)]^2} = \sqrt{4^2 + 5^2} = \sqrt{41}$$ □

LINES

We want to find an equation of a given line L; such an equation is satisfied by the coordinates of the points on L and by no other point. To find the equation of L we use its *slope*, which is a measure of the steepness of the line.

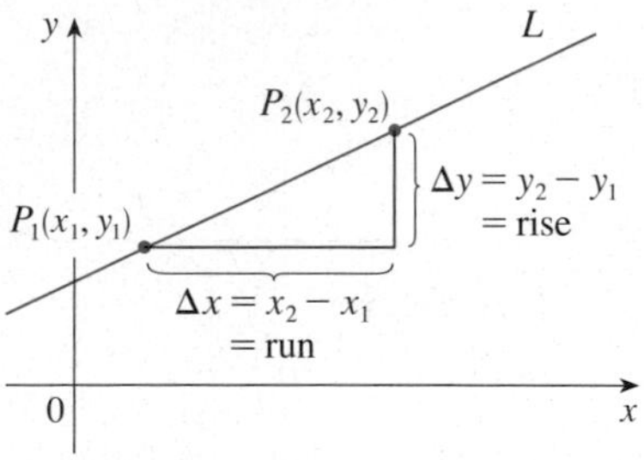

FIGURE 5

2 DEFINITION The **slope** of a nonvertical line that passes through the points $P_1(x_1, y_1)$ and $P_2(x_2, y_2)$ is

$$m = \frac{\Delta y}{\Delta x} = \frac{y_2 - y_1}{x_2 - x_1}$$

The slope of a vertical line is not defined.

Thus the slope of a line is the ratio of the change in y, Δy, to the change in x, Δx. (See Figure 5.) The slope is therefore the rate of change of y with respect to x. The fact that the line is straight means that the rate of change is constant.

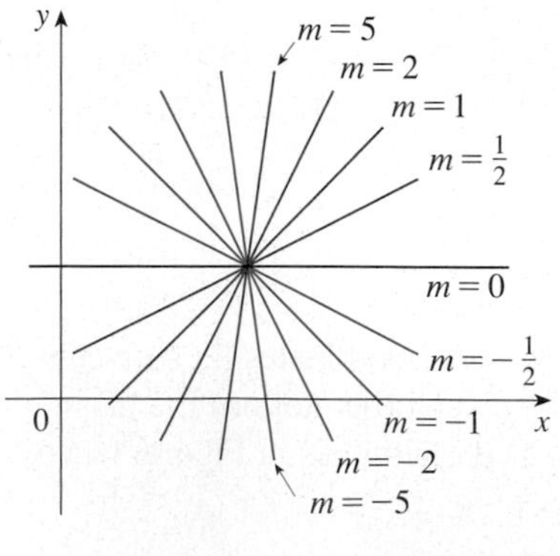

FIGURE 6

Figure 6 shows several lines labeled with their slopes. Notice that lines with positive slope slant upward to the right, whereas lines with negative slope slant downward to the right. Notice also that the steepest lines are the ones for which the absolute value of the slope is largest, and a horizontal line has slope 0.

Now let's find an equation of the line that passes through a given point $P_1(x_1, y_1)$ and has slope m. A point $P(x, y)$ with $x \neq x_1$ lies on this line if and only if the slope of the line through P_1 and P is equal to m; that is,

$$\frac{y - y_1}{x - x_1} = m$$

This equation can be rewritten in the form

$$y - y_1 = m(x - x_1)$$

and we observe that this equation is also satisfied when $x = x_1$ and $y = y_1$. Therefore it is an equation of the given line.

3 POINT-SLOPE FORM OF THE EQUATION OF A LINE An equation of the line passing through the point $P_1(x_1, y_1)$ and having slope m is

$$y - y_1 = m(x - x_1)$$

EXAMPLE 3 Find an equation of the line through $(1, -7)$ with slope $-\frac{1}{2}$.

SOLUTION Using (3) with $m = -\frac{1}{2}$, $x_1 = 1$, and $y_1 = -7$, we obtain an equation of the line as

$$y + 7 = -\tfrac{1}{2}(x - 1)$$

which we can rewrite as

$$2y + 14 = -x + 1 \qquad \text{or} \qquad x + 2y + 13 = 0$$ □

EXAMPLE 4 Find an equation of the line through the points $(-1, 2)$ and $(3, -4)$.

SOLUTION By Definition 2 the slope of the line is

$$m = \frac{-4 - 2}{3 - (-1)} = -\frac{3}{2}$$

Using the point-slope form with $x_1 = -1$ and $y_1 = 2$, we obtain

$$y - 2 = -\tfrac{3}{2}(x + 1)$$

which simplifies to $$3x + 2y = 1$$ □

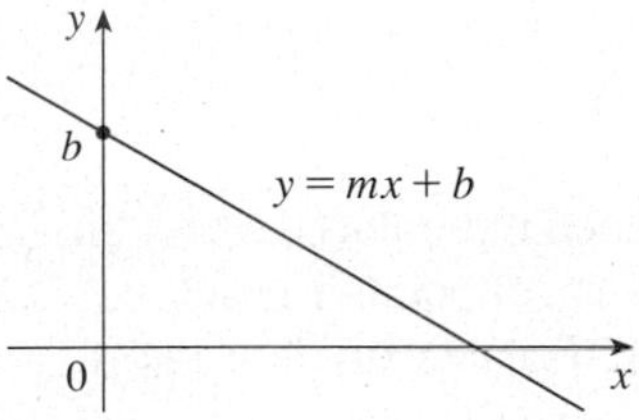

FIGURE 7

Suppose a nonvertical line has slope m and y-intercept b. (See Figure 7.) This means it intersects the y-axis at the point $(0, b)$, so the point-slope form of the equation of the line, with $x_1 = 0$ and $y_1 = b$, becomes

$$y - b = m(x - 0)$$

This simplifies as follows.

> **4 SLOPE-INTERCEPT FORM OF THE EQUATION OF A LINE** An equation of the line with slope m and y-intercept b is
>
> $$y = mx + b$$

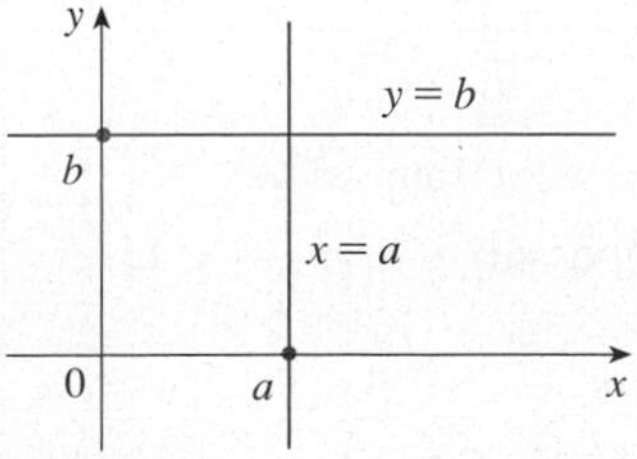

FIGURE 8

In particular, if a line is horizontal, its slope is $m = 0$, so its equation is $y = b$, where b is the y-intercept (see Figure 8). A vertical line does not have a slope, but we can write its equation as $x = a$, where a is the x-intercept, because the x-coordinate of every point on the line is a.

Observe that the equation of every line can be written in the form

5 $$Ax + By + C = 0$$

because a vertical line has the equation $x = a$ or $x - a = 0$ ($A = 1$, $B = 0$, $C = -a$) and a nonvertical line has the equation $y = mx + b$ or $-mx + y - b = 0$ ($A = -m$, $B = 1$, $C = -b$). Conversely, if we start with a general first-degree equation, that is, an equation of the form (5), where A, B, and C are constants and A and B are not both 0, then we can show that it is the equation of a line. If $B = 0$, the equation becomes $Ax + C = 0$ or $x = -C/A$, which represents a vertical line with x-intercept $-C/A$. If $B \neq 0$, the equation can be rewritten by solving for y:

$$y = -\frac{A}{B}x - \frac{C}{B}$$

and we recognize this as being the slope-intercept form of the equation of a line ($m = -A/B$, $b = -C/B$). Therefore an equation of the form (5) is called a **linear equation** or the **general equation of a line**. For brevity, we often refer to "the line $Ax + By + C = 0$" instead of "the line whose equation is $Ax + By + C = 0$."

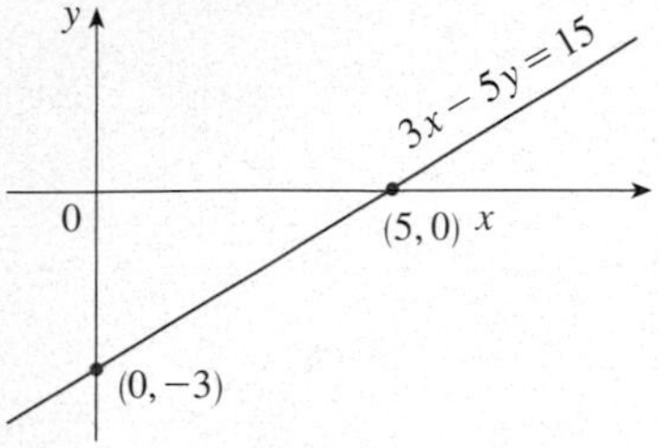

FIGURE 9

EXAMPLE 5 Sketch the graph of the equation $3x - 5y = 15$.

SOLUTION Since the equation is linear, its graph is a line. To draw the graph, we can simply find two points on the line. It's easiest to find the intercepts. Substituting $y = 0$ (the equation of the x-axis) in the given equation, we get $3x = 15$, so $x = 5$ is the x-intercept. Substituting $x = 0$ in the equation, we see that the y-intercept is -3. This allows us to sketch the graph as in Figure 9. □

EXAMPLE 6 Graph the inequality $x + 2y > 5$.

SOLUTION We are asked to sketch the graph of the set $\{(x, y) \mid x + 2y > 5\}$ and we do so by solving the inequality for y:

$$x + 2y > 5$$

$$2y > -x + 5$$

$$y > -\tfrac{1}{2}x + \tfrac{5}{2}$$

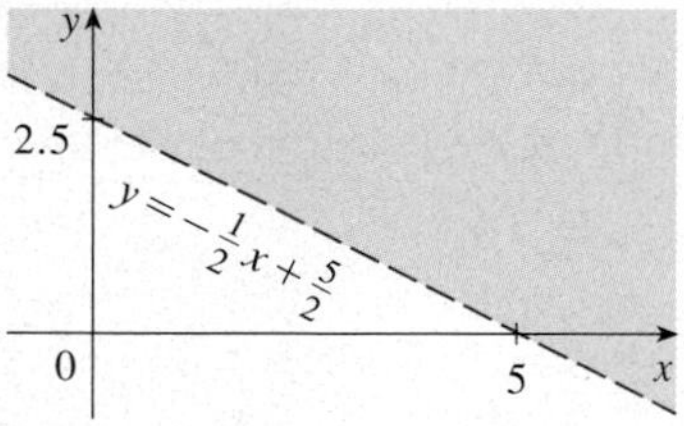

FIGURE 10

Compare this inequality with the equation $y = -\frac{1}{2}x + \frac{5}{2}$, which represents a line with slope $-\frac{1}{2}$ and y-intercept $\frac{5}{2}$. We see that the given graph consists of points whose y-coordinates are *larger* than those on the line $y = -\frac{1}{2}x + \frac{5}{2}$. Thus the graph is the region that lies *above* the line, as illustrated in Figure 10. □

PARALLEL AND PERPENDICULAR LINES

Slopes can be used to show that lines are parallel or perpendicular. The following facts are proved, for instance, in *Precalculus: Mathematics for Calculus, Fifth Edition* by Stewart, Redlin, and Watson (Thomson Brooks/Cole, Belmont, CA, 2006).

6 PARALLEL AND PERPENDICULAR LINES

1. Two nonvertical lines are parallel if and only if they have the same slope.
2. Two lines with slopes m_1 and m_2 are perpendicular if and only if $m_1 m_2 = -1$; that is, their slopes are negative reciprocals:

$$m_2 = -\frac{1}{m_1}$$

EXAMPLE 7 Find an equation of the line through the point (5, 2) that is parallel to the line $4x + 6y + 5 = 0$.

SOLUTION The given line can be written in the form

$$y = -\tfrac{2}{3}x - \tfrac{5}{6}$$

which is in slope-intercept form with $m = -\frac{2}{3}$. Parallel lines have the same slope, so the required line has slope $-\frac{2}{3}$ and its equation in point-slope form is

$$y - 2 = -\tfrac{2}{3}(x - 5)$$

We can write this equation as $2x + 3y = 16$. □

EXAMPLE 8 Show that the lines $2x + 3y = 1$ and $6x - 4y - 1 = 0$ are perpendicular.

SOLUTION The equations can be written as

$$y = -\tfrac{2}{3}x + \tfrac{1}{3} \qquad \text{and} \qquad y = \tfrac{3}{2}x - \tfrac{1}{4}$$

from which we see that the slopes are

$$m_1 = -\tfrac{2}{3} \qquad \text{and} \qquad m_2 = \tfrac{3}{2}$$

Since $m_1m_2 = -1$, the lines are perpendicular. □

B EXERCISES

1–6 Find the distance between the points.

1. $(1, 1)$, $(4, 5)$

2. $(1, -3)$, $(5, 7)$

3. $(6, -2)$, $(-1, 3)$

4. $(1, -6)$, $(-1, -3)$

5. $(2, 5)$, $(4, -7)$

6. (a, b), (b, a)

7–10 Find the slope of the line through P and Q.

7. $P(1, 5)$, $Q(4, 11)$

8. $P(-1, 6)$, $Q(4, -3)$

9. $P(-3, 3)$, $Q(-1, -6)$

10. $P(-1, -4)$, $Q(6, 0)$

11. Show that the triangle with vertices $A(0, 2)$, $B(-3, -1)$, and $C(-4, 3)$ is isosceles.

12. (a) Show that the triangle with vertices $A(6, -7)$, $B(11, -3)$, and $C(2, -2)$ is a right triangle using the converse of the Pythagorean Theorem.
(b) Use slopes to show that ABC is a right triangle.
(c) Find the area of the triangle.

13. Show that the points $(-2, 9)$, $(4, 6)$, $(1, 0)$, and $(-5, 3)$ are the vertices of a square.

14. (a) Show that the points $A(-1, 3)$, $B(3, 11)$, and $C(5, 15)$ are collinear (lie on the same line) by showing that $|AB| + |BC| = |AC|$.
(b) Use slopes to show that A, B, and C are collinear.

15. Show that $A(1, 1)$, $B(7, 4)$, $C(5, 10)$, and $D(-1, 7)$ are vertices of a parallelogram.

16. Show that $A(1, 1)$, $B(11, 3)$, $C(10, 8)$, and $D(0, 6)$ are vertices of a rectangle.

17–20 Sketch the graph of the equation.

17. $x = 3$

18. $y = -2$

19. $xy = 0$

20. $|y| = 1$

21–36 Find an equation of the line that satisfies the given conditions.

21. Through $(2, -3)$, slope 6

22. Through $(-1, 4)$, slope -3

23. Through $(1, 7)$, slope $\frac{2}{3}$

24. Through $(-3, -5)$, slope $-\frac{7}{2}$

25. Through $(2, 1)$ and $(1, 6)$

26. Through $(-1, -2)$ and $(4, 3)$

27. Slope 3, y-intercept -2

28. Slope $\frac{2}{5}$, y-intercept 4

29. x-intercept 1, y-intercept -3

30. x-intercept -8, y-intercept 6

31. Through $(4, 5)$, parallel to the x-axis

32. Through $(4, 5)$, parallel to the y-axis

33. Through $(1, -6)$, parallel to the line $x + 2y = 6$

34. y-intercept 6, parallel to the line $2x + 3y + 4 = 0$

35. Through $(-1, -2)$, perpendicular to the line $2x + 5y + 8 = 0$

36. Through $(\frac{1}{2}, -\frac{2}{3})$, perpendicular to the line $4x - 8y = 1$

37–42 Find the slope and y-intercept of the line and draw its graph.

37. $x + 3y = 0$

38. $2x - 5y = 0$

39. $y = -2$

40. $2x - 3y + 6 = 0$

41. $3x - 4y = 12$

42. $4x + 5y = 10$

43–52 Sketch the region in the xy-plane.

43. $\{(x, y) \mid x < 0\}$

44. $\{(x, y) \mid y > 0\}$

45. $\{(x, y) \mid xy < 0\}$

46. $\{(x, y) \mid x \geqslant 1 \text{ and } y < 3\}$

47. $\{(x, y) \mid |x| \leqslant 2\}$

48. $\{(x, y) \mid |x| < 3 \text{ and } |y| < 2\}$

49. $\{(x, y) \mid 0 \leqslant y \leqslant 4 \text{ and } x \leqslant 2\}$

50. $\{(x, y) \mid y > 2x - 1\}$

51. $\{(x, y) \mid 1 + x \leqslant y \leqslant 1 - 2x\}$

52. $\{(x, y) \mid -x \leqslant y < \frac{1}{2}(x + 3)\}$

53. Find a point on the y-axis that is equidistant from $(5, -5)$ and $(1, 1)$.

54. Show that the midpoint of the line segment from $P_1(x_1, y_1)$ to $P_2(x_2, y_2)$ is

$$\left(\frac{x_1 + x_2}{2}, \frac{y_1 + y_2}{2} \right)$$

55. Find the midpoint of the line segment joining the given points.
(a) $(1, 3)$ and $(7, 15)$ (b) $(-1, 6)$ and $(8, -12)$

56. Find the lengths of the medians of the triangle with vertices $A(1, 0)$, $B(3, 6)$, and $C(8, 2)$. (A median is a line segment from a vertex to the midpoint of the opposite side.)

57. Show that the lines $2x - y = 4$ and $6x - 2y = 10$ are not parallel and find their point of intersection.

58. Show that the lines $3x - 5y + 19 = 0$ and $10x + 6y - 50 = 0$ are perpendicular and find their point of intersection.

59. Find an equation of the perpendicular bisector of the line segment joining the points $A(1, 4)$ and $B(7, -2)$.

60. (a) Find equations for the sides of the triangle with vertices $P(1, 0)$, $Q(3, 4)$, and $R(-1, 6)$.
(b) Find equations for the medians of this triangle. Where do they intersect?

61. (a) Show that if the x- and y-intercepts of a line are nonzero numbers a and b, then the equation of the line can be put in the form

$$\frac{x}{a} + \frac{y}{b} = 1$$

This equation is called the **two-intercept form** of an equation of a line.
(b) Use part (a) to find an equation of the line whose x-intercept is 6 and whose y-intercept is -8.

62. A car leaves Detroit at 2:00 PM, traveling at a constant speed west along I-96. It passes Ann Arbor, 40 mi from Detroit, at 2:50 PM.
(a) Express the distance traveled in terms of the time elapsed.
(b) Draw the graph of the equation in part (a).
(c) What is the slope of this line? What does it represent?

C GRAPHS OF SECOND-DEGREE EQUATIONS

In Appendix B we saw that a first-degree, or linear, equation $Ax + By + C = 0$ represents a line. In this section we discuss second-degree equations such as

$$x^2 + y^2 = 1 \qquad y = x^2 + 1 \qquad \frac{x^2}{9} + \frac{y^2}{4} = 1 \qquad x^2 - y^2 = 1$$

which represent a circle, a parabola, an ellipse, and a hyperbola, respectively.

The graph of such an equation in x and y is the set of all points (x, y) that satisfy the equation; it gives a visual representation of the equation. Conversely, given a curve in the xy-plane, we may have to find an equation that represents it, that is, an equation satisfied by the coordinates of the points on the curve and by no other point. This is the other half of the basic principle of analytic geometry as formulated by Descartes and Fermat. The idea is that if a geometric curve can be represented by an algebraic equation, then the rules of algebra can be used to analyze the geometric problem.

CIRCLES

As an example of this type of problem, let's find an equation of the circle with radius r and center (h, k). By definition, the circle is the set of all points $P(x, y)$ whose distance from

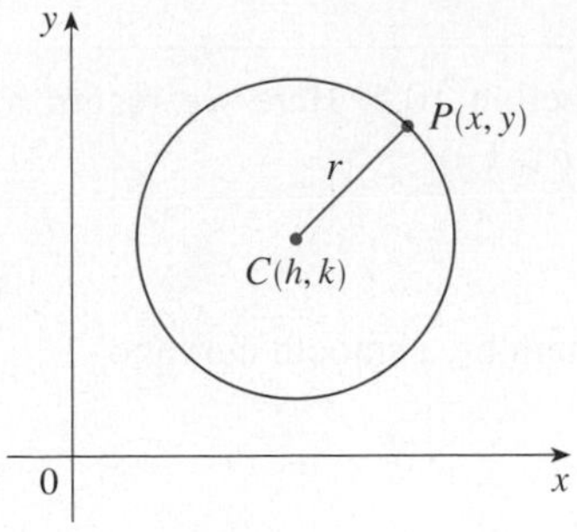

FIGURE 1

the center $C(h, k)$ is r. (See Figure 1.) Thus P is on the circle if and only if $|PC| = r$. From the distance formula, we have

$$\sqrt{(x - h)^2 + (y - k)^2} = r$$

or equivalently, squaring both sides, we get

$$(x - h)^2 + (y - k)^2 = r^2$$

This is the desired equation.

> **1 EQUATION OF A CIRCLE** An equation of the circle with center (h, k) and radius r is
>
> $$(x - h)^2 + (y - k)^2 = r^2$$
>
> In particular, if the center is the origin $(0, 0)$, the equation is
>
> $$x^2 + y^2 = r^2$$

EXAMPLE 1 Find an equation of the circle with radius 3 and center $(2, -5)$.

SOLUTION From Equation 1 with $r = 3$, $h = 2$, and $k = -5$, we obtain

$$(x - 2)^2 + (y + 5)^2 = 9$$ □

EXAMPLE 2 Sketch the graph of the equation $x^2 + y^2 + 2x - 6y + 7 = 0$ by first showing that it represents a circle and then finding its center and radius.

SOLUTION We first group the x-terms and y-terms as follows:

$$(x^2 + 2x) + (y^2 - 6y) = -7$$

Then we complete the square within each grouping, adding the appropriate constants to both sides of the equation:

$$(x^2 + 2x + 1) + (y^2 - 6y + 9) = -7 + 1 + 9$$

or

$$(x + 1)^2 + (y - 3)^2 = 3$$

Comparing this equation with the standard equation of a circle (1), we see that $h = -1$, $k = 3$, and $r = \sqrt{3}$, so the given equation represents a circle with center $(-1, 3)$ and radius $\sqrt{3}$. It is sketched in Figure 2.

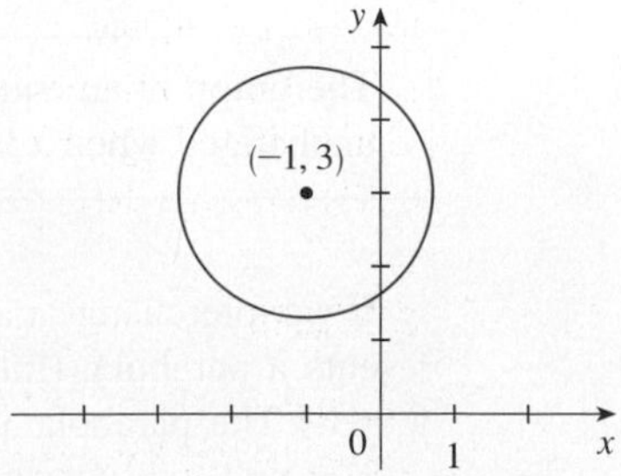

FIGURE 2
$x^2 + y^2 + 2x - 6y + 7 = 0$

□

PARABOLAS

The geometric properties of parabolas are reviewed in Section 10.5. Here we regard a parabola as a graph of an equation of the form $y = ax^2 + bx + c$.

EXAMPLE 3 Draw the graph of the parabola $y = x^2$.

SOLUTION We set up a table of values, plot points, and join them by a smooth curve to obtain the graph in Figure 3.

x	$y = x^2$
0	0
$\pm\frac{1}{2}$	$\frac{1}{4}$
± 1	1
± 2	4
± 3	9

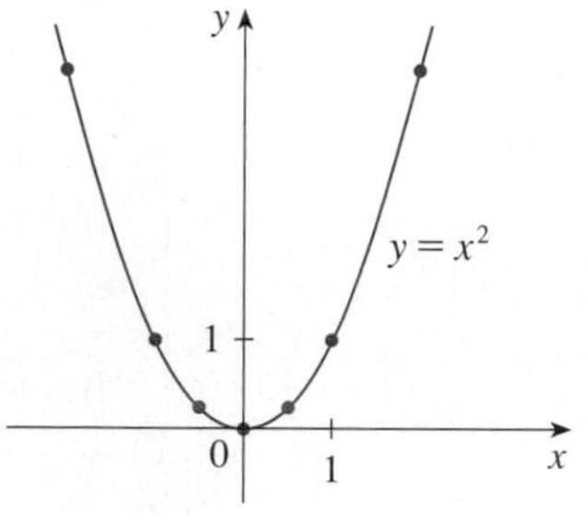

FIGURE 3

□

Figure 4 shows the graphs of several parabolas with equations of the form $y = ax^2$ for various values of the number a. In each case the *vertex*, the point where the parabola changes direction, is the origin. We see that the parabola $y = ax^2$ opens upward if $a > 0$ and downward if $a < 0$ (as in Figure 5).

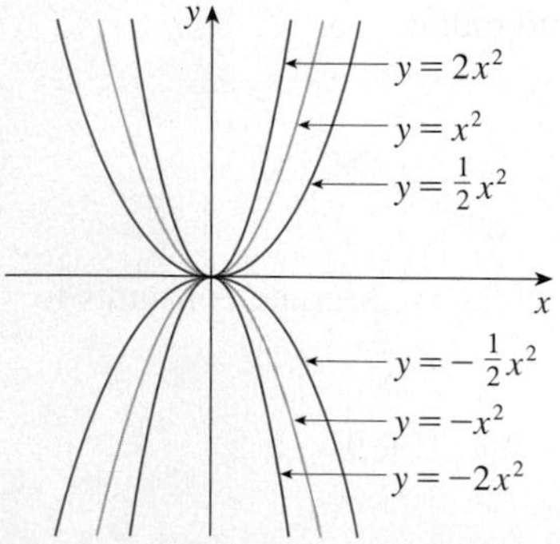

FIGURE 4

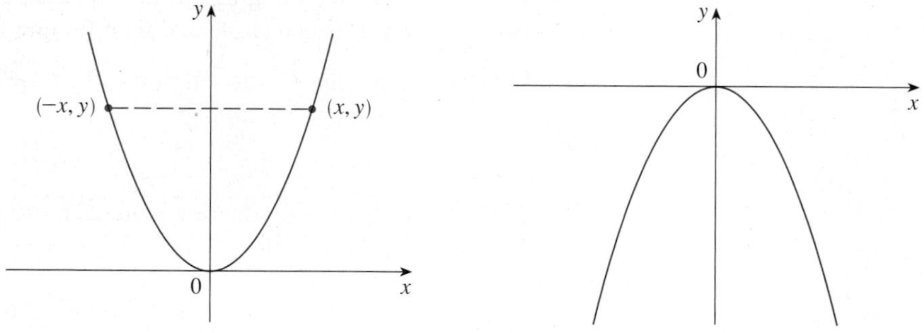

(a) $y = ax^2$, $a > 0$

(b) $y = ax^2$, $a < 0$

FIGURE 5

Notice that if (x, y) satisfies $y = ax^2$, then so does $(-x, y)$. This corresponds to the geometric fact that if the right half of the graph is reflected about the y-axis, then the left half of the graph is obtained. We say that the graph is **symmetric with respect to the y-axis**.

> The graph of an equation is symmetric with respect to the y-axis if the equation is unchanged when x is replaced by $-x$.

If we interchange x and y in the equation $y = ax^2$, the result is $x = ay^2$, which also represents a parabola. (Interchanging x and y amounts to reflecting about the diagonal line $y = x$.) The parabola $x = ay^2$ opens to the right if $a > 0$ and to the left if $a < 0$. (See

Figure 6.) This time the parabola is symmetric with respect to the x-axis because if (x, y) satisfies $x = ay^2$, then so does $(x, -y)$.

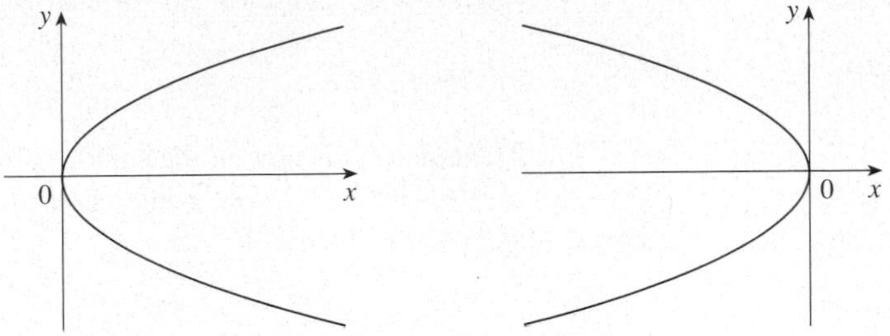

FIGURE 6

(a) $x = ay^2,\ a > 0$

(b) $x = ay^2,\ a < 0$

> The graph of an equation is symmetric with respect to the x-axis if the equation is unchanged when y is replaced by $-y$.

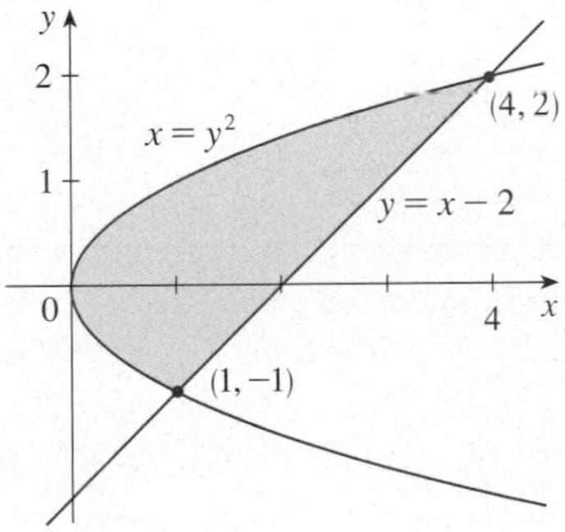

FIGURE 7

EXAMPLE 4 Sketch the region bounded by the parabola $x = y^2$ and the line $y = x - 2$.

SOLUTION First we find the points of intersection by solving the two equations. Substituting $x = y + 2$ into the equation $x = y^2$, we get $y + 2 = y^2$, which gives

$$0 = y^2 - y - 2 = (y - 2)(y + 1)$$

so $y = 2$ or -1. Thus the points of intersection are $(4, 2)$ and $(1, -1)$, and we draw the line $y = x - 2$ passing through these points. We then sketch the parabola $x = y^2$ by referring to Figure 6(a) and having the parabola pass through $(4, 2)$ and $(1, -1)$. The region bounded by $x = y^2$ and $y = x - 2$ means the finite region whose boundaries are these curves. It is sketched in Figure 7. □

ELLIPSES

The curve with equation

2
$$\frac{x^2}{a^2} + \frac{y^2}{b^2} = 1$$

where a and b are positive numbers, is called an **ellipse** in standard position. (Geometric properties of ellipses are discussed in Section 10.5.) Observe that Equation 2 is unchanged if x is replaced by $-x$ or y is replaced by $-y$, so the ellipse is symmetric with respect to both axes. As a further aid to sketching the ellipse, we find its intercepts.

> The ***x*-intercepts** of a graph are the x-coordinates of the points where the graph intersects the x-axis. They are found by setting $y = 0$ in the equation of the graph.
>
> The ***y*-intercepts** are the y-coordinates of the points where the graph intersects the y-axis. They are found by setting $x = 0$ in its equation.

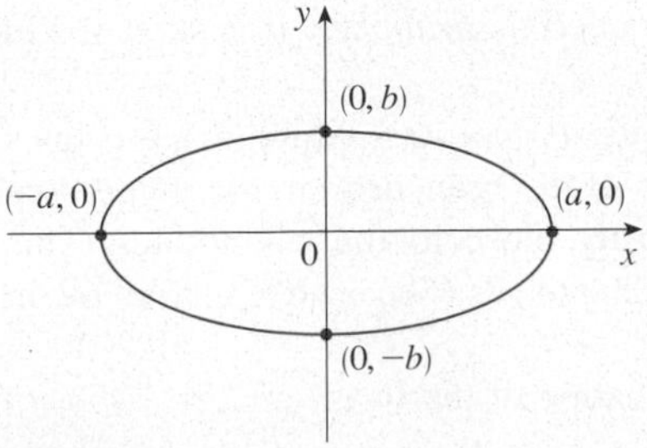

FIGURE 8
$\frac{x^2}{a^2} + \frac{y^2}{b^2} = 1$

If we set $y = 0$ in Equation 2, we get $x^2 = a^2$ and so the x-intercepts are $\pm a$. Setting $x = 0$, we get $y^2 = b^2$, so the y-intercepts are $\pm b$. Using this information, together with symmetry, we sketch the ellipse in Figure 8. If $a = b$, the ellipse is a circle with radius a.

EXAMPLE 5 Sketch the graph of $9x^2 + 16y^2 = 144$.

SOLUTION We divide both sides of the equation by 144:

$$\frac{x^2}{16} + \frac{y^2}{9} = 1$$

The equation is now in the standard form for an ellipse (2), so we have $a^2 = 16$, $b^2 = 9$, $a = 4$, and $b = 3$. The x-intercepts are ± 4; the y-intercepts are ± 3. The graph is sketched in Figure 9.

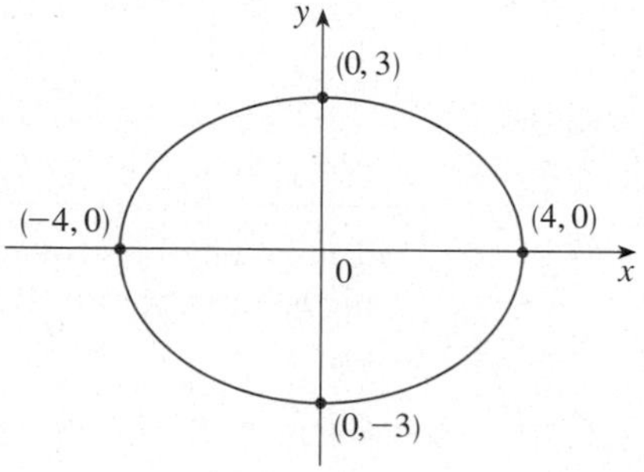

FIGURE 9
$9x^2 + 16y^2 = 144$

□

HYPERBOLAS

The curve with equation

3 $$\frac{x^2}{a^2} - \frac{y^2}{b^2} = 1$$

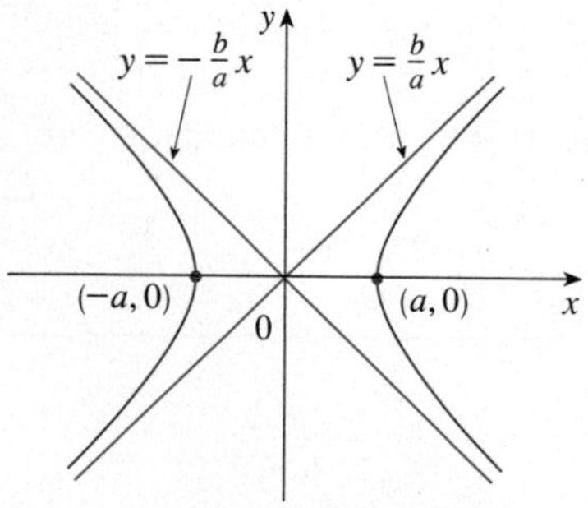

FIGURE 10
The hyperbola $\frac{x^2}{a^2} - \frac{y^2}{b^2} = 1$

is called a **hyperbola** in standard position. Again, Equation 3 is unchanged when x is replaced by $-x$ or y is replaced by $-y$, so the hyperbola is symmetric with respect to both axes. To find the x-intercepts we set $y = 0$ and obtain $x^2 = a^2$ and $x = \pm a$. However, if we put $x = 0$ in Equation 3, we get $y^2 = -b^2$, which is impossible, so there is no y-intercept. In fact, from Equation 3 we obtain

$$\frac{x^2}{a^2} = 1 + \frac{y^2}{b^2} \geqslant 1$$

which shows that $x^2 \geqslant a^2$ and so $|x| = \sqrt{x^2} \geqslant a$. Therefore we have $x \geqslant a$ or $x \leqslant -a$. This means that the hyperbola consists of two parts, called its *branches.* It is sketched in Figure 10.

In drawing a hyperbola it is useful to draw first its *asymptotes,* which are the lines $y = (b/a)x$ and $y = -(b/a)x$ shown in Figure 10. Both branches of the hyperbola approach the asymptotes; that is, they come arbitrarily close to the asymptotes. This involves the idea of a limit, which is discussed in Chapter 2. (See also Exercise 69 in Section 4.5.)

By interchanging the roles of x and y we get an equation of the form

$$\frac{y^2}{a^2} - \frac{x^2}{b^2} = 1$$

which also represents a hyperbola and is sketched in Figure 11.

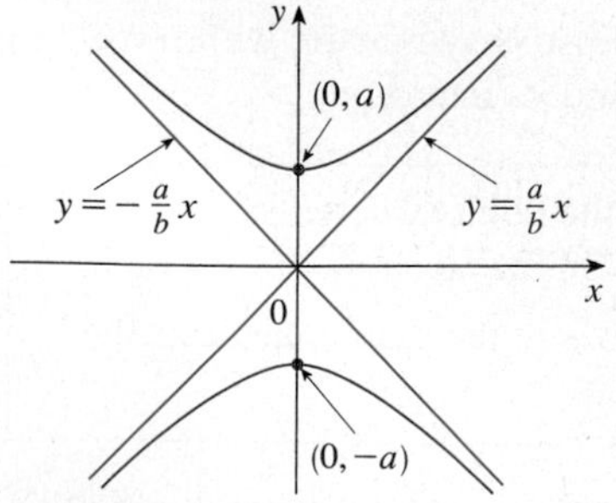

FIGURE 11
The hyperbola $\frac{y^2}{a^2} - \frac{x^2}{b^2} = 1$

EXAMPLE 6 Sketch the curve $9x^2 - 4y^2 = 36$.

SOLUTION Dividing both sides by 36, we obtain

$$\frac{x^2}{4} - \frac{y^2}{9} = 1$$

which is the standard form of the equation of a hyperbola (Equation 3). Since $a^2 = 4$, the x-intercepts are ± 2. Since $b^2 = 9$, we have $b = 3$ and the asymptotes are $y = \pm(\frac{3}{2})x$. The hyperbola is sketched in Figure 12.

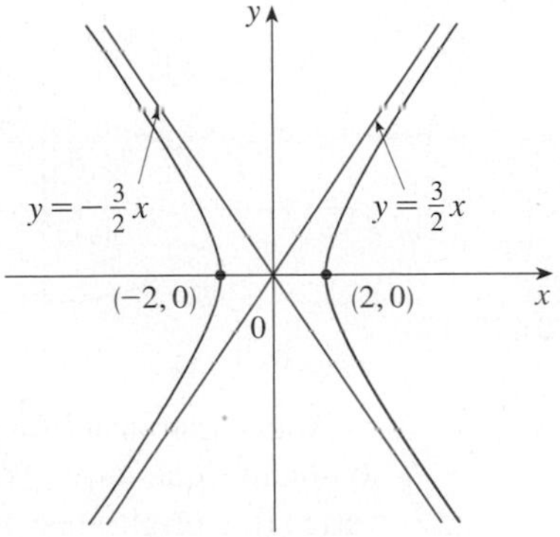

FIGURE 12
The hyperbola $9x^2 - 4y^2 = 36$

□

If $b = a$, a hyperbola has the equation $x^2 - y^2 = a^2$ (or $y^2 - x^2 = a^2$) and is called an *equilateral hyperbola* [see Figure 13(a)]. Its asymptotes are $y = \pm x$, which are perpendicular. If an equilateral hyperbola is rotated by 45°, the asymptotes become the x- and y-axes, and it can be shown that the new equation of the hyperbola is $xy = k$, where k is a constant [see Figure 13(b)].

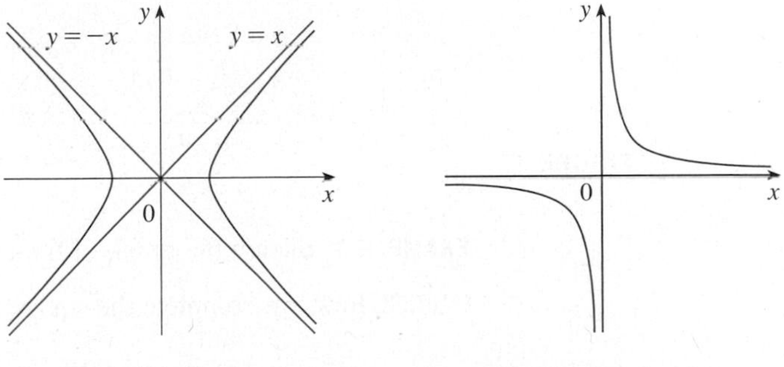

FIGURE 13
Equilateral hyperbolas

(a) $x^2 - y^2 = a^2$

(b) $xy = k \ (k > 0)$

SHIFTED CONICS

Recall that an equation of the circle with center the origin and radius r is $x^2 + y^2 = r^2$, but if the center is the point (h, k), then the equation of the circle becomes

$$(x - h)^2 + (y - k)^2 = r^2$$

Similarly, if we take the ellipse with equation

4
$$\frac{x^2}{a^2} + \frac{y^2}{b^2} = 1$$

and translate it (shift it) so that its center is the point (h, k), then its equation becomes

5 $$\frac{(x-h)^2}{a^2} + \frac{(y-k)^2}{b^2} = 1$$

(See Figure 14.)

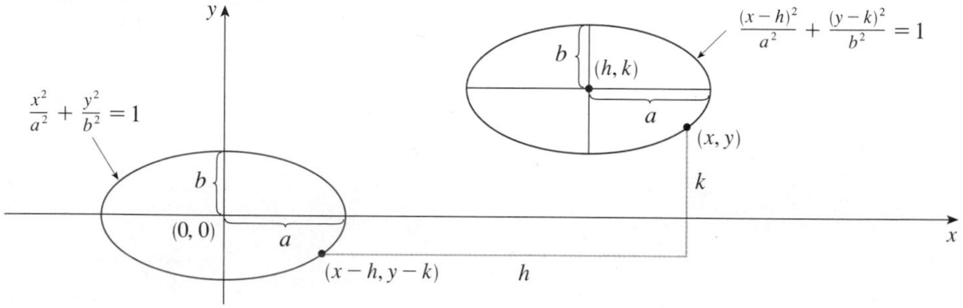

FIGURE 14

Notice that in shifting the ellipse, we replaced x by $x - h$ and y by $y - k$ in Equation 4 to obtain Equation 5. We use the same procedure to shift the parabola $y = ax^2$ so that its vertex (the origin) becomes the point (h, k) as in Figure 15. Replacing x by $x - h$ and y by $y - k$, we see that the new equation is

$$y - k = a(x-h)^2 \qquad \text{or} \qquad y = a(x-h)^2 + k$$

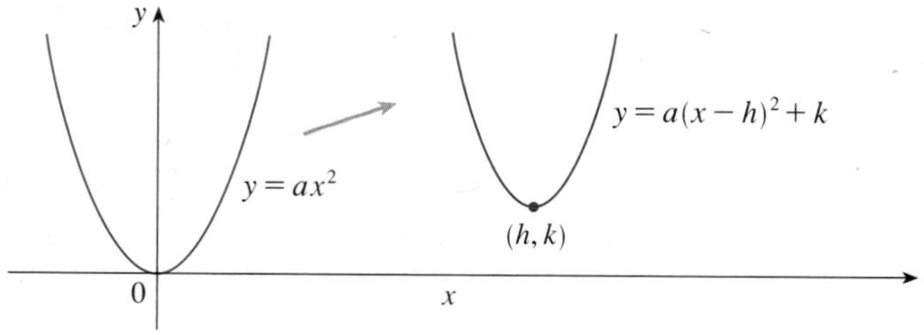

FIGURE 15

EXAMPLE 7 Sketch the graph of the equation $y = 2x^2 - 4x + 1$.

SOLUTION First we complete the square:

$$y = 2(x^2 - 2x) + 1 = 2(x-1)^2 - 1$$

In this form we see that the equation represents the parabola obtained by shifting $y = 2x^2$ so that its vertex is at the point $(1, -1)$. The graph is sketched in Figure 16.

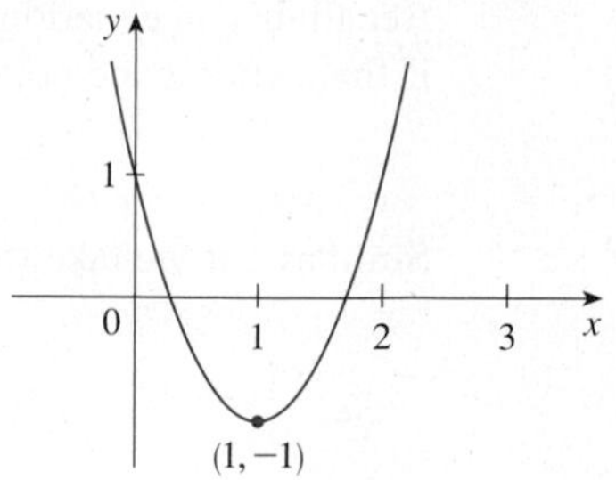

FIGURE 16
$y = 2x^2 - 4x + 1$

□

EXAMPLE 8 Sketch the curve $x = 1 - y^2$.

SOLUTION This time we start with the parabola $x = -y^2$ (as in Figure 6 with $a = -1$) and shift one unit to the right to get the graph of $x = 1 - y^2$. (See Figure 17.)

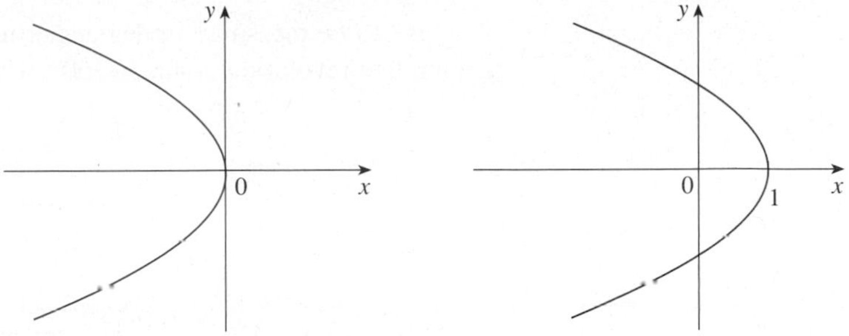

FIGURE 17 (a) $x = -y^2$ (b) $x = 1 - y^2$ □

C EXERCISES

1–4 Find an equation of a circle that satisfies the given conditions.

1. Center $(3, -1)$, radius 5

2. Center $(-2, -8)$, radius 10

3. Center at the origin, passes through $(4, 7)$

4. Center $(-1, 5)$, passes through $(-4, -6)$

5–9 Show that the equation represents a circle and find the center and radius.

5. $x^2 + y^2 - 4x + 10y + 13 = 0$

6. $x^2 + y^2 + 6y + 2 = 0$

7. $x^2 + y^2 + x = 0$

8. $16x^2 + 16y^2 + 8x + 32y + 1 = 0$

9. $2x^2 + 2y^2 - x + y = 1$

10. Under what condition on the coefficients a, b, and c does the equation $x^2 + y^2 + ax + by + c = 0$ represent a circle? When that condition is satisfied, find the center and radius of the circle.

11–32 Identify the type of curve and sketch the graph. Do not plot points. Just use the standard graphs given in Figures 5, 6, 8, 10, and 11 and shift if necessary.

11. $y = -x^2$

12. $y^2 - x^2 = 1$

13. $x^2 + 4y^2 = 16$

14. $x = -2y^2$

15. $16x^2 - 25y^2 = 400$

16. $25x^2 + 4y^2 = 100$

17. $4x^2 + y^2 = 1$

18. $y = x^2 + 2$

19. $x = y^2 - 1$

20. $9x^2 - 25y^2 = 225$

21. $9y^2 - x^2 = 9$

22. $2x^2 + 5y^2 = 10$

23. $xy = 4$

24. $y = x^2 + 2x$

25. $9(x - 1)^2 + 4(y - 2)^2 = 36$

26. $16x^2 + 9y^2 - 36y = 108$

27. $y = x^2 - 6x + 13$

28. $x^2 - y^2 - 4x + 3 = 0$

29. $x = 4 - y^2$

30. $y^2 - 2x + 6y + 5 = 0$

31. $x^2 + 4y^2 - 6x + 5 = 0$

32. $4x^2 + 9y^2 - 16x + 54y + 61 = 0$

33–34 Sketch the region bounded by the curves.

33. $y = 3x$, $y = x^2$

34. $y = 4 - x^2$, $x - 2y = 2$

35. Find an equation of the parabola with vertex $(1, -1)$ that passes through the points $(-1, 3)$ and $(3, 3)$.

36. Find an equation of the ellipse with center at the origin that passes through the points $\left(1, -10\sqrt{2}/3\right)$ and $\left(-2, 5\sqrt{5}/3\right)$.

37–40 Sketch the graph of the set.

37. $\{(x, y) \mid x^2 + y^2 \leqslant 1\}$

38. $\{(x, y) \mid x^2 + y^2 > 4\}$

39. $\{(x, y) \mid y \geqslant x^2 - 1\}$

40. $\{(x, y) \mid x^2 + 4y^2 \leqslant 4\}$

D TRIGONOMETRY

ANGLES

Angles can be measured in degrees or in radians (abbreviated as rad). The angle given by a complete revolution contains 360°, which is the same as 2π rad. Therefore

$$\boxed{1} \qquad \pi \text{ rad} = 180°$$

and

$$\boxed{2} \qquad 1 \text{ rad} = \left(\frac{180}{\pi}\right)^{\circ} \approx 57.3° \qquad 1° = \frac{\pi}{180} \text{ rad} \approx 0.017 \text{ rad}$$

EXAMPLE 1

(a) Find the radian measure of 60°. (b) Express $5\pi/4$ rad in degrees.

SOLUTION

(a) From Equation 1 or 2 we see that to convert from degrees to radians we multiply by $\pi/180$. Therefore

$$60° = 60\left(\frac{\pi}{180}\right) = \frac{\pi}{3} \text{ rad}$$

(b) To convert from radians to degrees we multiply by $180/\pi$. Thus

$$\frac{5\pi}{4} \text{ rad} = \frac{5\pi}{4}\left(\frac{180}{\pi}\right) = 225°$$

□

In calculus we use radians to measure angles except when otherwise indicated. The following table gives the correspondence between degree and radian measures of some common angles.

Degrees	0°	30°	45°	60°	90°	120°	135°	150°	180°	270°	360°
Radians	0	$\frac{\pi}{6}$	$\frac{\pi}{4}$	$\frac{\pi}{3}$	$\frac{\pi}{2}$	$\frac{2\pi}{3}$	$\frac{3\pi}{4}$	$\frac{5\pi}{6}$	π	$\frac{3\pi}{2}$	2π

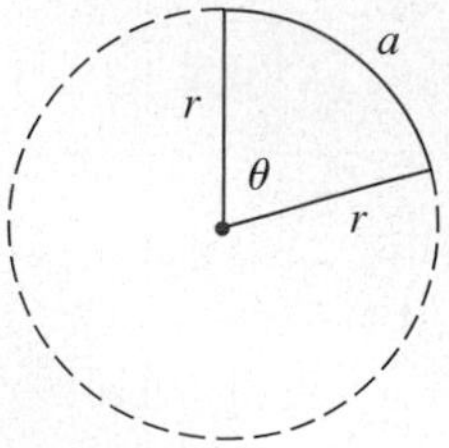

FIGURE 1

Figure 1 shows a sector of a circle with central angle θ and radius r subtending an arc with length a. Since the length of the arc is proportional to the size of the angle, and since the entire circle has circumference $2\pi r$ and central angle 2π, we have

$$\frac{\theta}{2\pi} = \frac{a}{2\pi r}$$

Solving this equation for θ and for a, we obtain

$$\boxed{3} \qquad \theta = \frac{a}{r} \qquad\qquad a = r\theta$$

Remember that Equations 3 are valid only when θ is measured in radians.

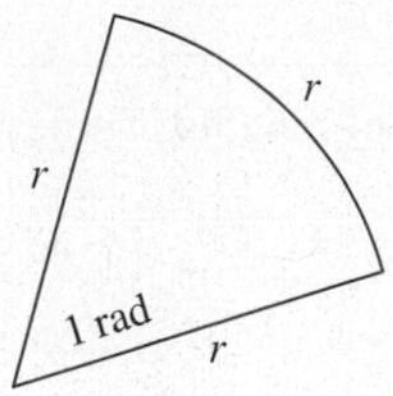

FIGURE 2

In particular, putting $a = r$ in Equation 3, we see that an angle of 1 rad is the angle subtended at the center of a circle by an arc equal in length to the radius of the circle (see Figure 2).

EXAMPLE 2
(a) If the radius of a circle is 5 cm, what angle is subtended by an arc of 6 cm?
(b) If a circle has radius 3 cm, what is the length of an arc subtended by a central angle of $3\pi/8$ rad?

SOLUTION
(a) Using Equation 3 with $a = 6$ and $r = 5$, we see that the angle is

$$\theta = \tfrac{6}{5} = 1.2 \text{ rad}$$

(b) With $r = 3$ cm and $\theta = 3\pi/8$ rad, the arc length is

$$a = r\theta = 3\left(\frac{3\pi}{8}\right) = \frac{9\pi}{8} \text{ cm} \qquad \square$$

The **standard position** of an angle occurs when we place its vertex at the origin of a coordinate system and its initial side on the positive x-axis as in Figure 3. A **positive** angle is obtained by rotating the initial side counterclockwise until it coincides with the terminal side. Likewise, **negative** angles are obtained by clockwise rotation as in Figure 4.

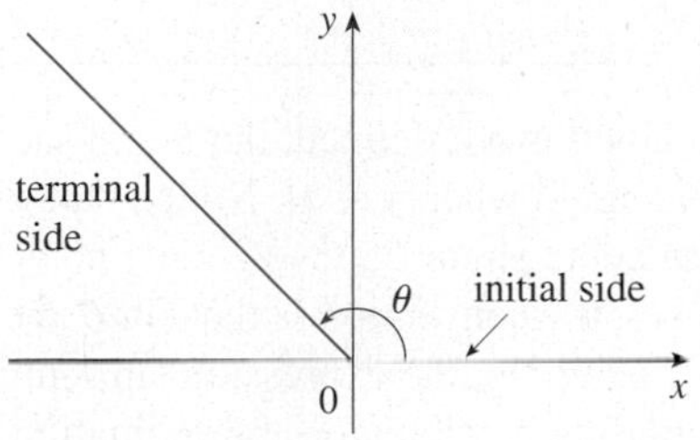

FIGURE 3 $\theta \geqslant 0$

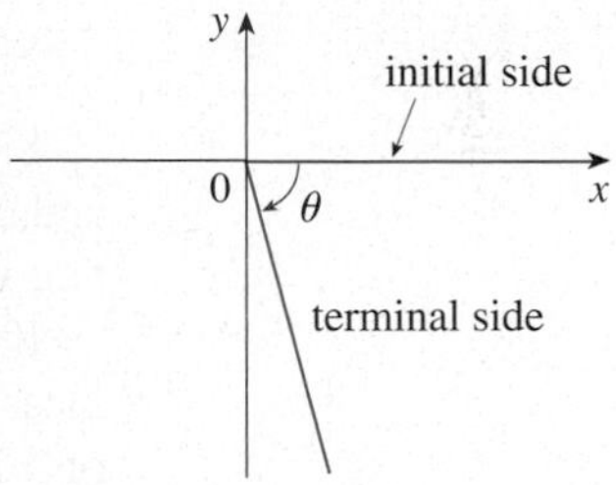

FIGURE 4 $\theta < 0$

Figure 5 shows several examples of angles in standard position. Notice that different angles can have the same terminal side. For instance, the angles $3\pi/4$, $-5\pi/4$, and $11\pi/4$ have the same initial and terminal sides because

$$\frac{3\pi}{4} - 2\pi = -\frac{5\pi}{4} \qquad \frac{3\pi}{4} + 2\pi = \frac{11\pi}{4}$$

and 2π rad represents a complete revolution.

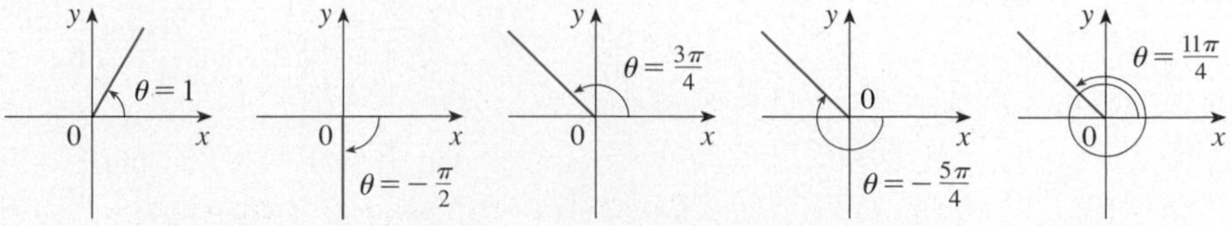

FIGURE 5
Angles in standard position

THE TRIGONOMETRIC FUNCTIONS

For an acute angle θ the six trigonometric functions are defined as ratios of lengths of sides of a right triangle as follows (see Figure 6).

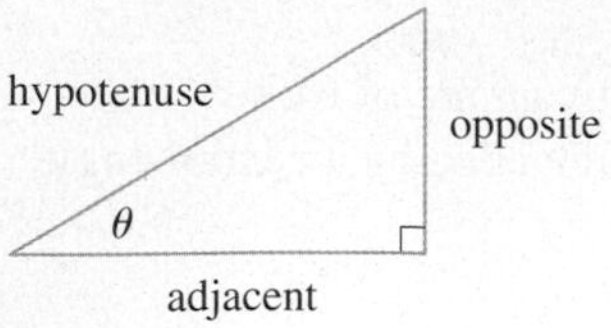

FIGURE 6

4

$$\sin\theta = \frac{\text{opp}}{\text{hyp}} \qquad \csc\theta = \frac{\text{hyp}}{\text{opp}}$$

$$\cos\theta = \frac{\text{adj}}{\text{hyp}} \qquad \sec\theta = \frac{\text{hyp}}{\text{adj}}$$

$$\tan\theta = \frac{\text{opp}}{\text{adj}} \qquad \cot\theta = \frac{\text{adj}}{\text{opp}}$$

This definition doesn't apply to obtuse or negative angles, so for a general angle θ in standard position we let $P(x, y)$ be any point on the terminal side of θ and we let r be the distance $|OP|$ as in Figure 7. Then we define

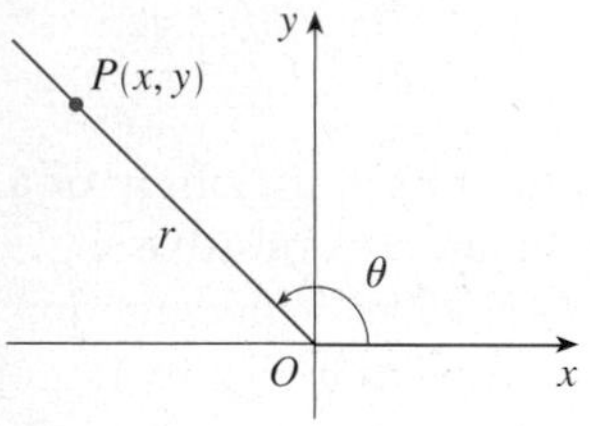

FIGURE 7

5

$$\sin\theta = \frac{y}{r} \qquad \csc\theta = \frac{r}{y}$$

$$\cos\theta = \frac{x}{r} \qquad \sec\theta = \frac{r}{x}$$

$$\tan\theta = \frac{y}{x} \qquad \cot\theta = \frac{x}{y}$$

Since division by 0 is not defined, $\tan\theta$ and $\sec\theta$ are undefined when $x = 0$ and $\csc\theta$ and $\cot\theta$ are undefined when $y = 0$. Notice that the definitions in (4) and (5) are consistent when θ is an acute angle.

If θ is a number, the convention is that $\sin\theta$ means the sine of the angle whose *radian* measure is θ. For example, the expression $\sin 3$ implies that we are dealing with an angle of 3 rad. When finding a calculator approximation to this number, we must remember to set our calculator in radian mode, and then we obtain

$$\sin 3 \approx 0.14112$$

If we want to know the sine of the angle 3° we would write sin 3° and, with our calculator in degree mode, we find that

$$\sin 3^\circ \approx 0.05234$$

The exact trigonometric ratios for certain angles can be read from the triangles in Figure 8. For instance,

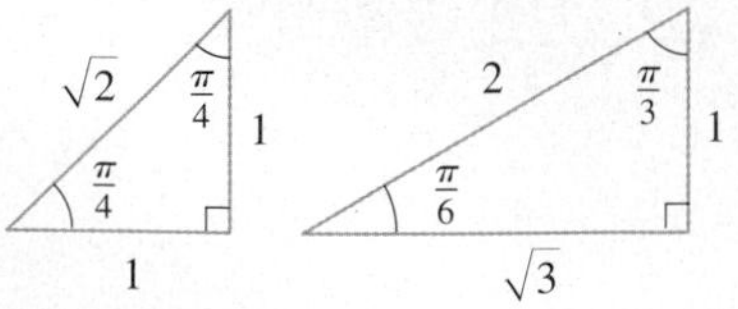

FIGURE 8

$$\sin\frac{\pi}{4} = \frac{1}{\sqrt{2}} \qquad \sin\frac{\pi}{6} = \frac{1}{2} \qquad \sin\frac{\pi}{3} = \frac{\sqrt{3}}{2}$$

$$\cos\frac{\pi}{4} = \frac{1}{\sqrt{2}} \qquad \cos\frac{\pi}{6} = \frac{\sqrt{3}}{2} \qquad \cos\frac{\pi}{3} = \frac{1}{2}$$

$$\tan\frac{\pi}{4} = 1 \qquad \tan\frac{\pi}{6} = \frac{1}{\sqrt{3}} \qquad \tan\frac{\pi}{3} = \sqrt{3}$$

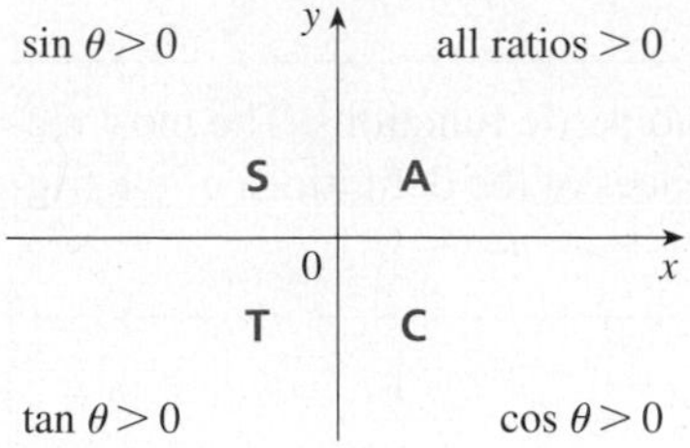

FIGURE 9

The signs of the trigonometric functions for angles in each of the four quadrants can be remembered by means of the rule "All Students Take Calculus" shown in Figure 9.

EXAMPLE 3 Find the exact trigonometric ratios for $\theta = 2\pi/3$.

SOLUTION From Figure 10 we see that a point on the terminal line for $\theta = 2\pi/3$ is $P(-1, \sqrt{3})$. Therefore, taking

$$x = -1 \qquad y = \sqrt{3} \qquad r = 2$$

in the definitions of the trigonometric ratios, we have

$$\sin\frac{2\pi}{3} = \frac{\sqrt{3}}{2} \qquad \cos\frac{2\pi}{3} = -\frac{1}{2} \qquad \tan\frac{2\pi}{3} = -\sqrt{3}$$

$$\csc\frac{2\pi}{3} = \frac{2}{\sqrt{3}} \qquad \sec\frac{2\pi}{3} = -2 \qquad \cot\frac{2\pi}{3} = -\frac{1}{\sqrt{3}}$$ □

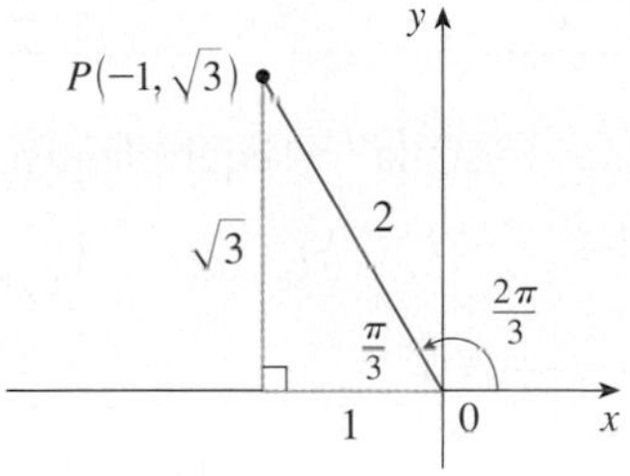

FIGURE 10

The following table gives some values of $\sin\theta$ and $\cos\theta$ found by the method of Example 3.

θ	0	$\frac{\pi}{6}$	$\frac{\pi}{4}$	$\frac{\pi}{3}$	$\frac{\pi}{2}$	$\frac{2\pi}{3}$	$\frac{3\pi}{4}$	$\frac{5\pi}{6}$	π	$\frac{3\pi}{2}$	2π
$\sin\theta$	0	$\frac{1}{2}$	$\frac{1}{\sqrt{2}}$	$\frac{\sqrt{3}}{2}$	1	$\frac{\sqrt{3}}{2}$	$\frac{1}{\sqrt{2}}$	$\frac{1}{2}$	0	-1	0
$\cos\theta$	1	$\frac{\sqrt{3}}{2}$	$\frac{1}{\sqrt{2}}$	$\frac{1}{2}$	0	$-\frac{1}{2}$	$-\frac{1}{\sqrt{2}}$	$-\frac{\sqrt{3}}{2}$	-1	0	1

EXAMPLE 4 If $\cos\theta = \frac{2}{5}$ and $0 < \theta < \pi/2$, find the other five trigonometric functions of θ.

SOLUTION Since $\cos\theta = \frac{2}{5}$, we can label the hypotenuse as having length 5 and the adjacent side as having length 2 in Figure 11. If the opposite side has length x, then the Pythagorean Theorem gives $x^2 + 4 = 25$ and so $x^2 = 21$, $x = \sqrt{21}$. We can now use the diagram to write the other five trigonometric functions:

$$\sin\theta = \frac{\sqrt{21}}{5} \qquad \tan\theta = \frac{\sqrt{21}}{2}$$

$$\csc\theta = \frac{5}{\sqrt{21}} \qquad \sec\theta = \frac{5}{2} \qquad \cot\theta = \frac{2}{\sqrt{21}}$$ □

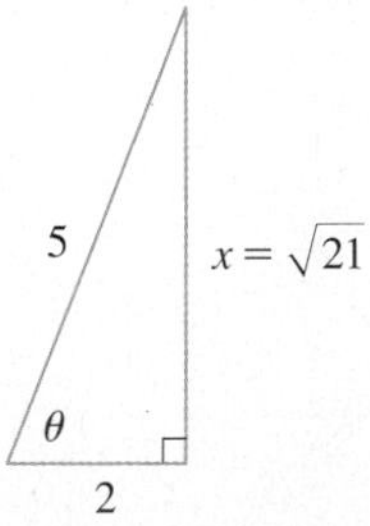

FIGURE 11

EXAMPLE 5 Use a calculator to approximate the value of x in Figure 12.

SOLUTION From the diagram we see that

$$\tan 40^\circ = \frac{16}{x}$$

Therefore $$x = \frac{16}{\tan 40^\circ} \approx 19.07$$ □

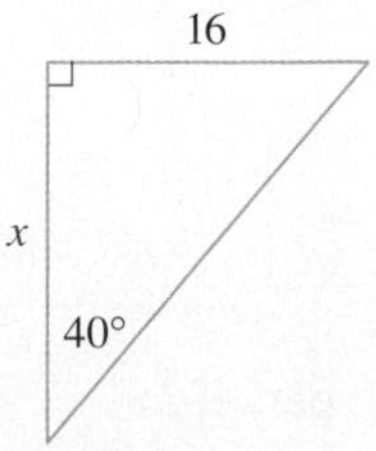

FIGURE 12

TRIGONOMETRIC IDENTITIES

A trigonometric identity is a relationship among the trigonometric functions. The most elementary are the following, which are immediate consequences of the definitions of the trigonometric functions.

6
$$\csc\theta = \frac{1}{\sin\theta} \qquad \sec\theta = \frac{1}{\cos\theta} \qquad \cot\theta = \frac{1}{\tan\theta}$$
$$\tan\theta = \frac{\sin\theta}{\cos\theta} \qquad \cot\theta = \frac{\cos\theta}{\sin\theta}$$

For the next identity we refer back to Figure 7. The distance formula (or, equivalently, the Pythagorean Theorem) tells us that $x^2 + y^2 = r^2$. Therefore

$$\sin^2\theta + \cos^2\theta = \frac{y^2}{r^2} + \frac{x^2}{r^2} = \frac{x^2 + y^2}{r^2} = \frac{r^2}{r^2} = 1$$

We have therefore proved one of the most useful of all trigonometric identities:

7
$$\sin^2\theta + \cos^2\theta = 1$$

If we now divide both sides of Equation 7 by $\cos^2\theta$ and use Equations 6, we get

8
$$\tan^2\theta + 1 = \sec^2\theta$$

Similarly, if we divide both sides of Equation 7 by $\sin^2\theta$, we get

9
$$1 + \cot^2\theta = \csc^2\theta$$

The identities

10a
$$\sin(-\theta) = -\sin\theta$$

10b
$$\cos(-\theta) = \cos\theta$$

■ Odd functions and even functions are discussed in Section 1.1.

show that sin is an odd function and cos is an even function. They are easily proved by drawing a diagram showing θ and $-\theta$ in standard position (see Exercise 39).

Since the angles θ and $\theta + 2\pi$ have the same terminal side, we have

11
$$\sin(\theta + 2\pi) = \sin\theta \qquad \cos(\theta + 2\pi) = \cos\theta$$

These identities show that the sine and cosine functions are periodic with period 2π.

The remaining trigonometric identities are all consequences of two basic identities called the **addition formulas**:

$$\text{12a} \quad \sin(x + y) = \sin x \cos y + \cos x \sin y$$

$$\text{12b} \quad \cos(x + y) = \cos x \cos y - \sin x \sin y$$

The proofs of these addition formulas are outlined in Exercises 85, 86, and 87.

By substituting $-y$ for y in Equations 12a and 12b and using Equations 10a and 10b, we obtain the following **subtraction formulas**:

$$\text{13a} \quad \sin(x - y) = \sin x \cos y - \cos x \sin y$$

$$\text{13b} \quad \cos(x - y) = \cos x \cos y + \sin x \sin y$$

Then, by dividing the formulas in Equations 12 or Equations 13, we obtain the corresponding formulas for $\tan(x \pm y)$:

$$\text{14a} \quad \tan(x + y) = \frac{\tan x + \tan y}{1 - \tan x \tan y}$$

$$\text{14b} \quad \tan(x - y) = \frac{\tan x - \tan y}{1 + \tan x \tan y}$$

If we put $y = x$ in the addition formulas (12), we get the **double-angle formulas**:

$$\text{15a} \quad \sin 2x = 2 \sin x \cos x$$

$$\text{15b} \quad \cos 2x = \cos^2 x - \sin^2 x$$

Then, by using the identity $\sin^2 x + \cos^2 x = 1$, we obtain the following alternate forms of the double-angle formulas for $\cos 2x$:

$$\text{16a} \quad \cos 2x = 2 \cos^2 x - 1$$

$$\text{16b} \quad \cos 2x = 1 - 2 \sin^2 x$$

If we now solve these equations for $\cos^2 x$ and $\sin^2 x$, we get the following **half-angle formulas**, which are useful in integral calculus:

$$\text{17a} \quad \cos^2 x = \frac{1 + \cos 2x}{2}$$

$$\text{17b} \quad \sin^2 x = \frac{1 - \cos 2x}{2}$$

Finally, we state the **product formulas**, which can be deduced from Equations 12 and 13:

18a $$\sin x \cos y = \tfrac{1}{2}[\sin(x + y) + \sin(x - y)]$$

18b $$\cos x \cos y = \tfrac{1}{2}[\cos(x + y) + \cos(x - y)]$$

18c $$\sin x \sin y = \tfrac{1}{2}[\cos(x - y) - \cos(x + y)]$$

There are many other trigonometric identities, but those we have stated are the ones used most often in calculus. If you forget any of them, remember that they can all be deduced from Equations 12a and 12b.

EXAMPLE 6 Find all values of x in the interval $[0, 2\pi]$ such that $\sin x = \sin 2x$.

SOLUTION Using the double-angle formula (15a), we rewrite the given equation as

$$\sin x = 2 \sin x \cos x \qquad \text{or} \qquad \sin x (1 - 2 \cos x) = 0$$

Therefore, there are two possibilities:

$$\sin x = 0 \qquad \text{or} \qquad 1 - 2 \cos x = 0$$

$$x = 0, \pi, 2\pi \qquad\qquad \cos x = \tfrac{1}{2}$$

$$x = \frac{\pi}{3}, \frac{5\pi}{3}$$

The given equation has five solutions: 0, $\pi/3$, π, $5\pi/3$, and 2π. □

GRAPHS OF THE TRIGONOMETRIC FUNCTIONS

The graph of the function $f(x) = \sin x$, shown in Figure 13(a), is obtained by plotting points for $0 \leqslant x \leqslant 2\pi$ and then using the periodic nature of the function (from Equation 11) to complete the graph. Notice that the zeros of the sine function occur at the

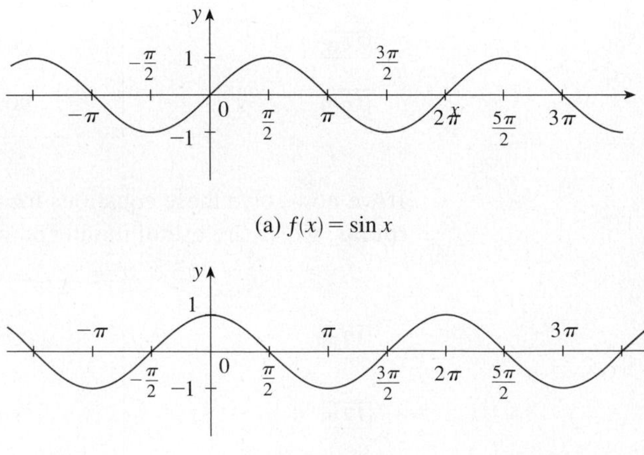

(a) $f(x) = \sin x$

(b) $g(x) = \cos x$

FIGURE 13

integer multiples of π, that is,

$$\sin x = 0 \qquad \text{whenever } x = n\pi, \quad n \text{ an integer}$$

Because of the identity

$$\cos x = \sin\left(x + \frac{\pi}{2}\right)$$

(which can be verified using Equation 12a), the graph of cosine is obtained by shifting the graph of sine by an amount $\pi/2$ to the left [see Figure 13(b)]. Note that for both the sine and cosine functions the domain is $(-\infty, \infty)$ and the range is the closed interval $[-1, 1]$. Thus, for all values of x, we have

$$-1 \leqslant \sin x \leqslant 1 \qquad -1 \leqslant \cos x \leqslant 1$$

The graphs of the remaining four trigonometric functions are shown in Figure 14 and their domains are indicated there. Notice that tangent and cotangent have range $(-\infty, \infty)$, whereas cosecant and secant have range $(-\infty, -1] \cup [1, \infty)$. All four functions are periodic: tangent and cotangent have period π, whereas cosecant and secant have period 2π.

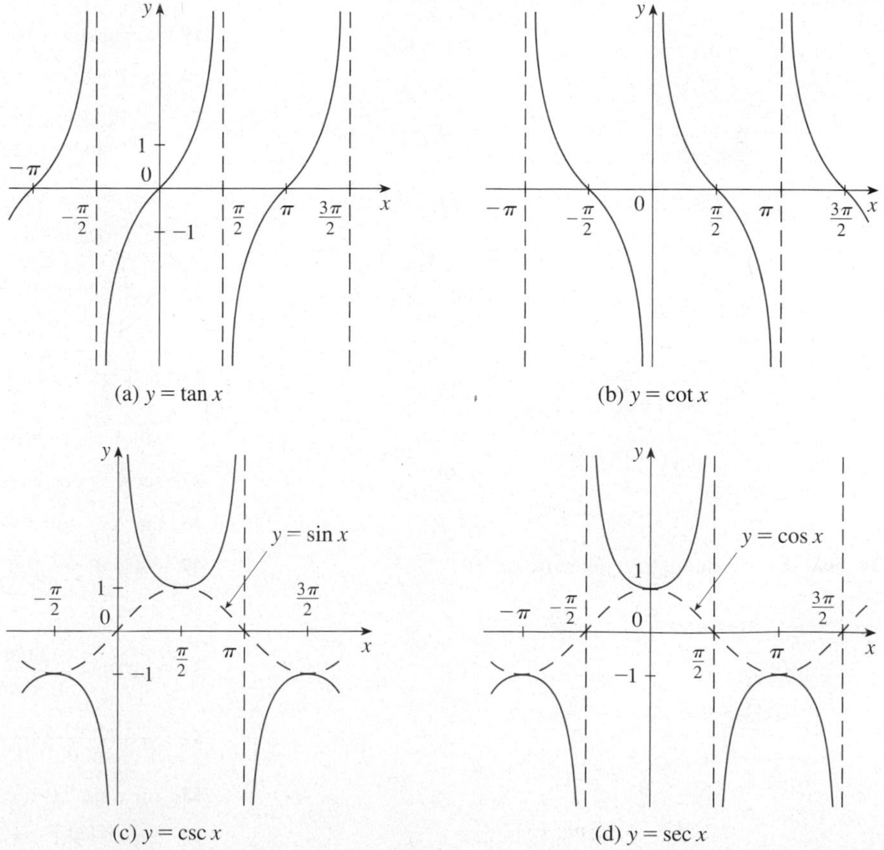

FIGURE 14

D EXERCISES

1–6 Convert from degrees to radians.

1. $210°$ **2.** $300°$ **3.** $9°$

4. $-315°$ **5.** $900°$ **6.** $36°$

7–12 Convert from radians to degrees.

7. 4π **8.** $-\frac{7\pi}{2}$ **9.** $\frac{5\pi}{12}$

10. $\frac{8\pi}{3}$ **11.** $-\frac{3\pi}{8}$ **12.** 5

13. Find the length of a circular arc subtended by an angle of $\pi/12$ rad if the radius of the circle is 36 cm.

14. If a circle has radius 10 cm, find the length of the arc subtended by a central angle of $72°$.

15. A circle has radius 1.5 m. What angle is subtended at the center of the circle by an arc 1 m long?

16. Find the radius of a circular sector with angle $3\pi/4$ and arc length 6 cm.

17–22 Draw, in standard position, the angle whose measure is given.

17. $315°$ **18.** $-150°$ **19.** $-\frac{3\pi}{4}$ rad

20. $\frac{7\pi}{3}$ rad **21.** 2 rad **22.** -3 rad

23–28 Find the exact trigonometric ratios for the angle whose radian measure is given.

23. $\frac{3\pi}{4}$ **24.** $\frac{4\pi}{3}$ **25.** $\frac{9\pi}{2}$

26. -5π **27.** $\frac{5\pi}{6}$ **28.** $\frac{11\pi}{4}$

29–34 Find the remaining trigonometric ratios.

29. $\sin\theta = \frac{3}{5}, \quad 0 < \theta < \frac{\pi}{2}$

30. $\tan\alpha = 2, \quad 0 < \alpha < \frac{\pi}{2}$

31. $\sec\phi = -1.5, \quad \frac{\pi}{2} < \phi < \pi$

32. $\cos x = -\frac{1}{3}, \quad \pi < x < \frac{3\pi}{2}$

33. $\cot\beta = 3, \quad \pi < \beta < 2\pi$

34. $\csc\theta = -\frac{4}{3}, \quad \frac{3\pi}{2} < \theta < 2\pi$

35–38 Find, correct to five decimal places, the length of the side labeled x.

35.

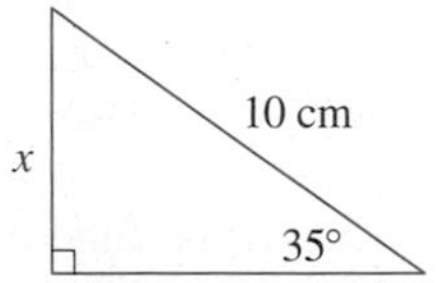

36.

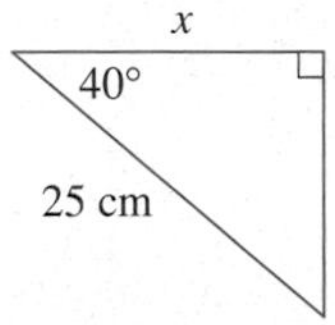

37.

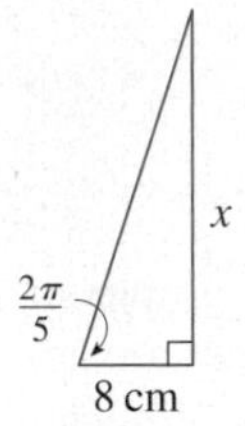

38.

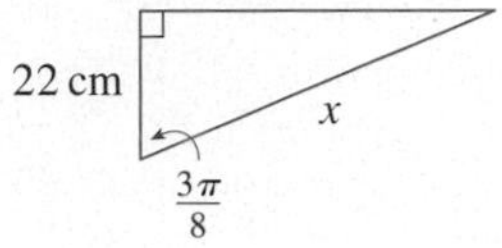

39–41 Prove each equation.

39. (a) Equation 10a (b) Equation 10b

40. (a) Equation 14a (b) Equation 14b

41. (a) Equation 18a (b) Equation 18b
(c) Equation 18c

42–58 Prove the identity.

42. $\cos\left(\frac{\pi}{2} - x\right) = \sin x$

43. $\sin\left(\frac{\pi}{2} + x\right) = \cos x$

44. $\sin(\pi - x) = \sin x$

45. $\sin\theta\cot\theta = \cos\theta$

46. $(\sin x + \cos x)^2 = 1 + \sin 2x$

47. $\sec y - \cos y = \tan y \sin y$

48. $\tan^2\alpha - \sin^2\alpha = \tan^2\alpha\sin^2\alpha$

49. $\cot^2\theta + \sec^2\theta = \tan^2\theta + \csc^2\theta$

50. $2\csc 2t = \sec t \csc t$

51. $\tan 2\theta = \frac{2\tan\theta}{1 - \tan^2\theta}$

52. $\frac{1}{1 - \sin\theta} + \frac{1}{1 + \sin\theta} = 2\sec^2\theta$

53. $\sin x \sin 2x + \cos x \cos 2x = \cos x$

54. $\sin^2 x - \sin^2 y = \sin(x + y)\sin(x - y)$

55. $\frac{\sin\phi}{1 - \cos\phi} = \csc\phi + \cot\phi$

56. $\tan x + \tan y = \dfrac{\sin(x + y)}{\cos x \cos y}$

57. $\sin 3\theta + \sin \theta = 2 \sin 2\theta \cos \theta$

58. $\cos 3\theta = 4 \cos^3\theta - 3 \cos \theta$

59–64 If $\sin x = \frac{1}{3}$ and $\sec y = \frac{5}{4}$, where x and y lie between 0 and $\pi/2$, evaluate the expression.

59. $\sin(x + y)$

60. $\cos(x + y)$

61. $\cos(x - y)$

62. $\sin(x - y)$

63. $\sin 2y$

64. $\cos 2y$

65–72 Find all values of x in the interval $[0, 2\pi]$ that satisfy the equation.

65. $2 \cos x - 1 = 0$

66. $3 \cot^2 x = 1$

67. $2 \sin^2 x = 1$

68. $|\tan x| = 1$

69. $\sin 2x = \cos x$

70. $2 \cos x + \sin 2x = 0$

71. $\sin x = \tan x$

72. $2 + \cos 2x = 3 \cos x$

73–76 Find all values of x in the interval $[0, 2\pi]$ that satisfy the inequality.

73. $\sin x \leqslant \frac{1}{2}$

74. $2 \cos x + 1 > 0$

75. $-1 < \tan x < 1$

76. $\sin x > \cos x$

77–82 Graph the function by starting with the graphs in Figures 13 and 14 and applying the transformations of Section 1.3 where appropriate.

77. $y = \cos\left(x - \dfrac{\pi}{3}\right)$

78. $y = \tan 2x$

79. $y = \dfrac{1}{3} \tan\left(x - \dfrac{\pi}{2}\right)$

80. $y = 1 + \sec x$

81. $y = |\sin x|$

82. $y = 2 + \sin\left(x + \dfrac{\pi}{4}\right)$

---

83. Prove the **Law of Cosines**: If a triangle has sides with lengths a, b, and c, and θ is the angle between the sides with lengths a and b, then

$$c^2 = a^2 + b^2 - 2ab \cos \theta$$

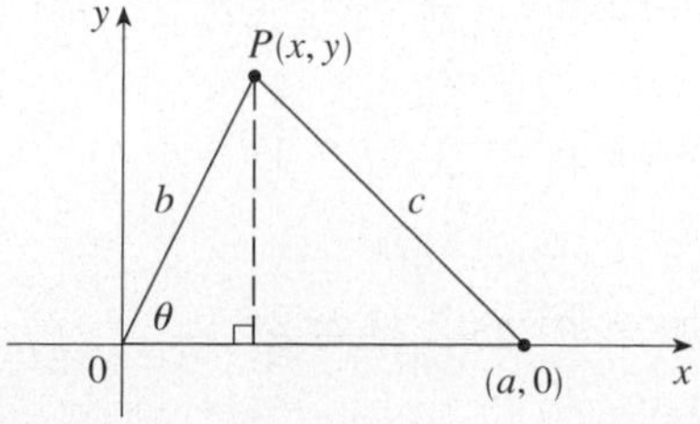

[*Hint:* Introduce a coordinate system so that θ is in standard position as in the figure. Express x and y in terms of θ and then use the distance formula to compute c.]

84. In order to find the distance $|AB|$ across a small inlet, a point C is located as in the figure and the following measurements were recorded:

$$\angle C = 103° \qquad |AC| = 820 \text{ m} \qquad |BC| = 910 \text{ m}$$

Use the Law of Cosines from Exercise 83 to find the required distance.

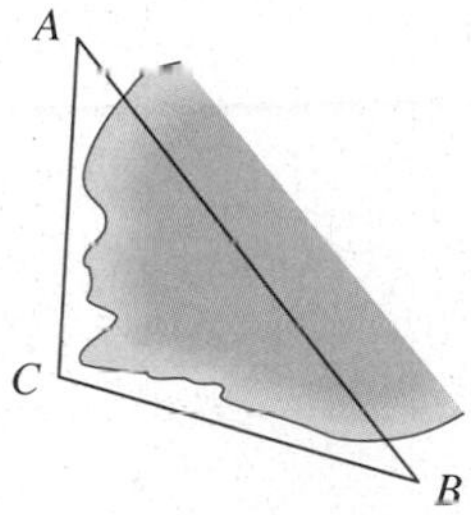

85. Use the figure to prove the subtraction formula

$$\cos(\alpha - \beta) = \cos \alpha \cos \beta + \sin \alpha \sin \beta$$

[*Hint:* Compute c^2 in two ways (using the Law of Cosines from Exercise 83 and also using the distance formula) and compare the two expressions.]

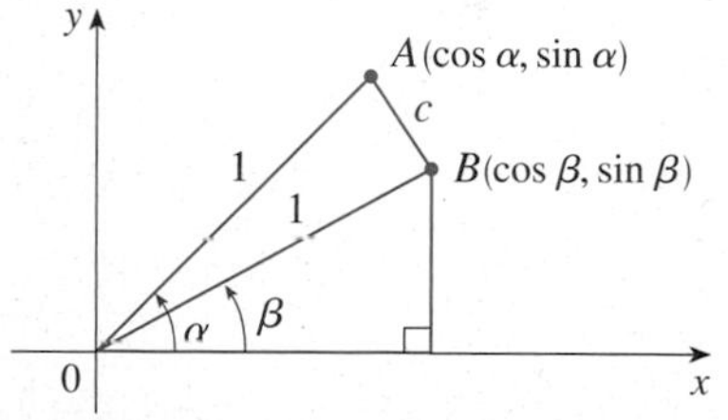

86. Use the formula in Exercise 85 to prove the addition formula for cosine (12b).

87. Use the addition formula for cosine and the identities

$$\cos\left(\frac{\pi}{2} - \theta\right) = \sin \theta \qquad \sin\left(\frac{\pi}{2} - \theta\right) = \cos \theta$$

to prove the subtraction formula for the sine function.

88. Show that the area of a triangle with sides of lengths a and b and with included angle θ is

$$A = \tfrac{1}{2}ab \sin \theta$$

89. Find the area of triangle ABC, correct to five decimal places, if

$$|AB| = 10 \text{ cm} \qquad |BC| = 3 \text{ cm} \qquad \angle ABC = 107°$$

E SIGMA NOTATION

A convenient way of writing sums uses the Greek letter Σ (capital sigma, corresponding to our letter S) and is called **sigma notation**.

This tells us to end with $i = n$.
This tells us to add.
This tells us to start with $i = m$.

$$\sum_{i=m}^{n} a_i$$

1 DEFINITION If $a_m, a_{m+1}, \ldots, a_n$ are real numbers and m and n are integers such that $m \leqslant n$, then

$$\sum_{i=m}^{n} a_i = a_m + a_{m+1} + a_{m+2} + \cdots + a_{n-1} + a_n$$

With function notation, Definition 1 can be written as

$$\sum_{i=m}^{n} f(i) = f(m) + f(m+1) + f(m+2) + \cdots + f(n-1) + f(n)$$

Thus the symbol $\Sigma_{i=m}^{n}$ indicates a summation in which the letter i (called the **index of summation**) takes on consecutive integer values beginning with m and ending with n, that is, $m, m+1, \ldots, n$. Other letters can also be used as the index of summation.

EXAMPLE 1

(a) $\displaystyle\sum_{i=1}^{4} i^2 = 1^2 + 2^2 + 3^2 + 4^2 = 30$

(b) $\displaystyle\sum_{i=3}^{n} i = 3 + 4 + 5 + \cdots + (n-1) + n$

(c) $\displaystyle\sum_{j=0}^{5} 2^j = 2^0 + 2^1 + 2^2 + 2^3 + 2^4 + 2^5 = 63$

(d) $\displaystyle\sum_{k=1}^{n} \frac{1}{k} = 1 + \frac{1}{2} + \frac{1}{3} + \cdots + \frac{1}{n}$

(e) $\displaystyle\sum_{i=1}^{3} \frac{i-1}{i^2+3} = \frac{1-1}{1^2+3} + \frac{2-1}{2^2+3} + \frac{3-1}{3^2+3} = 0 + \frac{1}{7} + \frac{1}{6} = \frac{13}{42}$

(f) $\displaystyle\sum_{i=1}^{4} 2 = 2 + 2 + 2 + 2 = 8$ □

EXAMPLE 2 Write the sum $2^3 + 3^3 + \cdots + n^3$ in sigma notation.

SOLUTION There is no unique way of writing a sum in sigma notation. We could write

$$2^3 + 3^3 + \cdots + n^3 = \sum_{i=2}^{n} i^3$$

or

$$2^3 + 3^3 + \cdots + n^3 = \sum_{j=1}^{n-1} (j+1)^3$$

or

$$2^3 + 3^3 + \cdots + n^3 = \sum_{k=0}^{n-2} (k+2)^3$$ □

The following theorem gives three simple rules for working with sigma notation.

2 THEOREM If c is any constant (that is, it does not depend on i), then

(a) $\sum_{i=m}^{n} ca_i = c \sum_{i=m}^{n} a_i$ (b) $\sum_{i=m}^{n} (a_i + b_i) = \sum_{i=m}^{n} a_i + \sum_{i=m}^{n} b_i$

(c) $\sum_{i=m}^{n} (a_i - b_i) = \sum_{i=m}^{n} a_i - \sum_{i=m}^{n} b_i$

PROOF To see why these rules are true, all we have to do is write both sides in expanded form. Rule (a) is just the distributive property of real numbers:

$$ca_m + ca_{m+1} + \cdots + ca_n = c(a_m + a_{m+1} + \cdots + a_n)$$

Rule (b) follows from the associative and commutative properties:

$$\begin{aligned}(a_m + b_m) &+ (a_{m+1} + b_{m+1}) + \cdots + (a_n + b_n)\\ &= (a_m + a_{m+1} + \cdots + a_n) + (b_m + b_{m+1} + \cdots + b_n)\end{aligned}$$

Rule (c) is proved similarly. □

EXAMPLE 3 Find $\sum_{i=1}^{n} 1$.

SOLUTION

$$\sum_{i=1}^{n} 1 = \underbrace{1 + 1 + \cdots + 1}_{n \text{ terms}} = n$$

□

EXAMPLE 4 Prove the formula for the sum of the first n positive integers:

$$\sum_{i=1}^{n} i = 1 + 2 + 3 + \cdots + n = \frac{n(n + 1)}{2}$$

SOLUTION This formula can be proved by mathematical induction (see page 77) or by the following method used by the German mathematician Karl Friedrich Gauss (1777–1855) when he was ten years old.

Write the sum S twice, once in the usual order and once in reverse order:

$$\begin{aligned}S &= 1 + 2 + 3 + \cdots + (n - 1) + n\\ S &= n + (n - 1) + (n - 2) + \cdots + 2 + 1\end{aligned}$$

Adding all columns vertically, we get

$$2S = (n + 1) + (n + 1) + (n + 1) + \cdots + (n + 1) + (n + 1)$$

On the right side there are n terms, each of which is $n + 1$, so

$$2S = n(n + 1) \qquad \text{or} \qquad S = \frac{n(n + 1)}{2}$$

□

EXAMPLE 5 Prove the formula for the sum of the squares of the first n positive integers:

$$\sum_{i=1}^{n} i^2 = 1^2 + 2^2 + 3^2 + \cdots + n^2 = \frac{n(n + 1)(2n + 1)}{6}$$

SOLUTION 1 Let S be the desired sum. We start with the *telescoping sum* (or collapsing sum):

Most terms cancel in pairs.

$$\sum_{i=1}^{n} [(1+i)^3 - i^3] = (2^3 - 1^3) + (3^3 - 2^3) + (4^3 - 3^3) + \cdots + [(n+1)^3 - n^3]$$
$$= (n+1)^3 - 1^3 = n^3 + 3n^2 + 3n$$

On the other hand, using Theorem 2 and Examples 3 and 4, we have

$$\sum_{i=1}^{n} [(1+i)^3 - i^3] = \sum_{i=1}^{n} [3i^2 + 3i + 1] = 3\sum_{i=1}^{n} i^2 + 3\sum_{i=1}^{n} i + \sum_{i=1}^{n} 1$$
$$= 3S + 3\,\frac{n(n+1)}{2} + n = 3S + \tfrac{3}{2}n^2 + \tfrac{5}{2}n$$

Thus we have

$$n^3 + 3n^2 + 3n = 3S + \tfrac{3}{2}n^2 + \tfrac{5}{2}n$$

Solving this equation for S, we obtain

$$3S = n^3 + \tfrac{3}{2}n^2 + \tfrac{1}{2}n$$

or

$$S = \frac{2n^3 + 3n^2 + n}{6} = \frac{n(n+1)(2n+1)}{6}$$

■ **PRINCIPLE OF MATHEMATICAL INDUCTION**

Let S_n be a statement involving the positive integer n. Suppose that

1. S_1 is true.
2. If S_k is true, then S_{k+1} is true.

Then S_n is true for all positive integers n.

■ See pages 77 and 80 for a more thorough discussion of mathematical induction.

SOLUTION 2 Let S_n be the given formula.

1. S_1 is true because $\qquad 1^2 = \dfrac{1(1+1)(2\cdot 1+1)}{6}$

2. Assume that S_k is true; that is,

$$1^2 + 2^2 + 3^2 + \cdots + k^2 = \frac{k(k+1)(2k+1)}{6}$$

Then

$$1^2 + 2^2 + 3^2 + \cdots + (k+1)^2 = (1^2 + 2^2 + 3^2 + \cdots + k^2) + (k+1)^2$$
$$= \frac{k(k+1)(2k+1)}{6} + (k+1)^2$$
$$= (k+1)\,\frac{k(2k+1) + 6(k+1)}{6}$$
$$= (k+1)\,\frac{2k^2 + 7k + 6}{6}$$
$$= \frac{(k+1)(k+2)(2k+3)}{6}$$
$$= \frac{(k+1)[(k+1)+1][2(k+1)+1]}{6}$$

So S_{k+1} is true.

By the Principle of Mathematical Induction, S_n is true for all n. □

We list the results of Examples 3, 4, and 5 together with a similar result for cubes (see Exercises 37–40) as Theorem 3. These formulas are needed for finding areas and evaluating integrals in Chapter 5.

3 THEOREM Let c be a constant and n a positive integer. Then

(a) $\sum_{i=1}^{n} 1 = n$

(b) $\sum_{i=1}^{n} c = nc$

(c) $\sum_{i=1}^{n} i = \frac{n(n+1)}{2}$

(d) $\sum_{i=1}^{n} i^2 = \frac{n(n+1)(2n+1)}{6}$

(e) $\sum_{i=1}^{n} i^3 = \left[\frac{n(n+1)}{2}\right]^2$

EXAMPLE 6 Evaluate $\sum_{i=1}^{n} i(4i^2 - 3)$.

SOLUTION Using Theorems 2 and 3, we have

$$\begin{aligned}\sum_{i=1}^{n} i(4i^2 - 3) &= \sum_{i=1}^{n} (4i^3 - 3i) = 4\sum_{i=1}^{n} i^3 - 3\sum_{i=1}^{n} i \\ &= 4\left[\frac{n(n+1)}{2}\right]^2 - 3\,\frac{n(n+1)}{2} \\ &= \frac{n(n+1)[2n(n+1) - 3]}{2} \\ &= \frac{n(n+1)(2n^2 + 2n - 3)}{2}\end{aligned}$$

□

■ The type of calculation in Example 7 arises in Chapter 5 when we compute areas.

EXAMPLE 7 Find $\lim_{n\to\infty} \sum_{i=1}^{n} \frac{3}{n}\left[\left(\frac{i}{n}\right)^2 + 1\right]$.

SOLUTION

$$\begin{aligned}\lim_{n\to\infty} \sum_{i=1}^{n} \frac{3}{n}\left[\left(\frac{i}{n}\right)^2 + 1\right] &= \lim_{n\to\infty} \sum_{i=1}^{n} \left[\frac{3}{n^3} i^2 + \frac{3}{n}\right] \\ &= \lim_{n\to\infty} \left[\frac{3}{n^3}\sum_{i=1}^{n} i^2 + \frac{3}{n}\sum_{i=1}^{n} 1\right] \\ &= \lim_{n\to\infty} \left[\frac{3}{n^3}\,\frac{n(n+1)(2n+1)}{6} + \frac{3}{n}\cdot n\right] \\ &= \lim_{n\to\infty} \left[\frac{1}{2}\cdot\frac{n}{n}\cdot\left(\frac{n+1}{n}\right)\left(\frac{2n+1}{n}\right) + 3\right] \\ &= \lim_{n\to\infty} \left[\frac{1}{2}\cdot 1\left(1 + \frac{1}{n}\right)\left(2 + \frac{1}{n}\right) + 3\right] \\ &= \tfrac{1}{2}\cdot 1 \cdot 1 \cdot 2 + 3 = 4\end{aligned}$$

□

E EXERCISES

1–10 Write the sum in expanded form.

1. $\sum_{i=1}^{5} \sqrt{i}$

2. $\sum_{i=1}^{6} \frac{1}{i+1}$

3. $\sum_{i=4}^{6} 3^i$

4. $\sum_{i=4}^{6} i^3$

5. $\sum_{k=0}^{4} \frac{2k-1}{2k+1}$

6. $\sum_{k=5}^{8} x^k$

7. $\sum_{i=1}^{n} i^{10}$

8. $\sum_{j=n}^{n+3} j^2$

9. $\sum_{j=0}^{n-1} (-1)^j$

10. $\sum_{i=1}^{n} f(x_i)\,\Delta x_i$

11–20 Write the sum in sigma notation.

11. $1 + 2 + 3 + 4 + \cdots + 10$

12. $\sqrt{3} + \sqrt{4} + \sqrt{5} + \sqrt{6} + \sqrt{7}$

13. $\frac{1}{2} + \frac{2}{3} + \frac{3}{4} + \frac{4}{5} + \cdots + \frac{19}{20}$

14. $\frac{3}{7} + \frac{4}{8} + \frac{5}{9} + \frac{6}{10} + \cdots + \frac{23}{27}$

15. $2 + 4 + 6 + 8 + \cdots + 2n$

16. $1 + 3 + 5 + 7 + \cdots + (2n - 1)$

17. $1 + 2 + 4 + 8 + 16 + 32$

18. $\frac{1}{1} + \frac{1}{4} + \frac{1}{9} + \frac{1}{16} + \frac{1}{25} + \frac{1}{36}$

19. $x + x^2 + x^3 + \cdots + x^n$

20. $1 - x + x^2 - x^3 + \cdots + (-1)^n x^n$

21–35 Find the value of the sum.

21. $\sum_{i=4}^{8} (3i - 2)$

22. $\sum_{i=3}^{6} i(i + 2)$

23. $\sum_{j=1}^{6} 3^{j+1}$

24. $\sum_{k=0}^{8} \cos k\pi$

25. $\sum_{n=1}^{20} (-1)^n$

26. $\sum_{i=1}^{100} 4$

27. $\sum_{i=0}^{4} (2^i + i^2)$

28. $\sum_{i=-2}^{4} 2^{3-i}$

29. $\sum_{i=1}^{n} 2i$

30. $\sum_{i=1}^{n} (2 - 5i)$

31. $\sum_{i=1}^{n} (i^2 + 3i + 4)$

32. $\sum_{i=1}^{n} (3 + 2i)^2$

33. $\sum_{i=1}^{n} (i + 1)(i + 2)$

34. $\sum_{i=1}^{n} i(i + 1)(i + 2)$

35. $\sum_{i=1}^{n} (i^3 - i - 2)$

36. Find the number n such that $\sum_{i=1}^{n} i = 78$.

37. Prove formula (b) of Theorem 3.

38. Prove formula (e) of Theorem 3 using mathematical induction.

39. Prove formula (e) of Theorem 3 using a method similar to that of Example 5, Solution 1 [start with $(1 + i)^4 - i^4$].

40. Prove formula (e) of Theorem 3 using the following method published by Abu Bekr Mohammed ibn Alhusain Alkarchi in about AD 1010. The figure shows a square $ABCD$ in which sides AB and AD have been divided into segments of lengths 1, 2, 3, . . . , n. Thus the side of the square has length $n(n + 1)/2$ so the area is $[n(n + 1)/2]^2$. But the area is also the sum of the areas of the n "gnomons" $G_1, G_2, \ldots, G_n$ shown in the figure. Show that the area of G_i is i^3 and conclude that formula (e) is true.

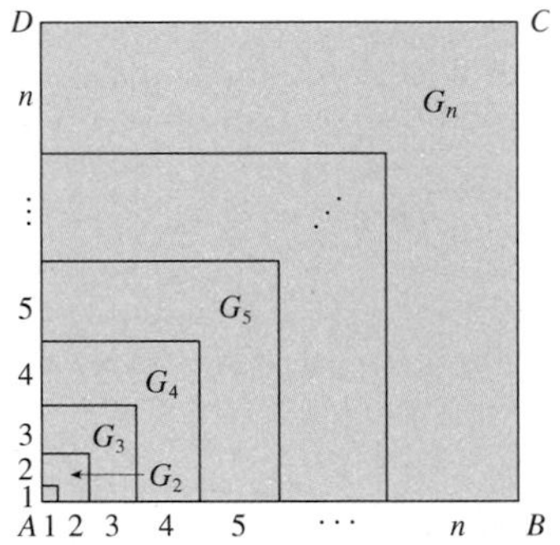

41. Evaluate each telescoping sum.

(a) $\sum_{i=1}^{n} [i^4 - (i - 1)^4]$ (b) $\sum_{i=1}^{100} (5^i - 5^{i-1})$

(c) $\sum_{i=3}^{99} \left(\frac{1}{i} - \frac{1}{i+1}\right)$ (d) $\sum_{i=1}^{n} (a_i - a_{i-1})$

42. Prove the generalized triangle inequality:

$$\left|\sum_{i=1}^{n} a_i\right| \leqslant \sum_{i=1}^{n} |a_i|$$

43–46 Find the limit.

43. $\lim_{n\to\infty} \sum_{i=1}^{n} \frac{1}{n}\left(\frac{i}{n}\right)^2$

44. $\lim_{n\to\infty} \sum_{i=1}^{n} \frac{1}{n}\left[\left(\frac{i}{n}\right)^3 + 1\right]$

45. $\lim_{n\to\infty} \sum_{i=1}^{n} \frac{2}{n}\left[\left(\frac{2i}{n}\right)^3 + 5\left(\frac{2i}{n}\right)\right]$

46. $\displaystyle\lim_{n\to\infty}\sum_{i=1}^{n}\frac{3}{n}\left[\left(1+\frac{3i}{n}\right)^3-2\left(1+\frac{3i}{n}\right)\right]$

47. Prove the formula for the sum of a finite geometric series with first term a and common ratio $r \neq 1$:

$$\sum_{i=1}^{n} ar^{i-1} = a + ar + ar^2 + \cdots + ar^{n-1} = \frac{a(r^n - 1)}{r - 1}$$

48. Evaluate $\displaystyle\sum_{i=1}^{n}\frac{3}{2^{i-1}}$.

49. Evaluate $\displaystyle\sum_{i=1}^{n}(2i + 2^i)$.

50. Evaluate $\displaystyle\sum_{i=1}^{m}\left[\sum_{j=1}^{n}(i + j)\right]$.

F PROOFS OF THEOREMS

In this appendix we present proofs of several theorems that are stated in the main body of the text. The sections in which they occur are indicated in the margin.

SECTION 2.3

LIMIT LAWS Suppose that c is a constant and the limits

$$\lim_{x\to a} f(x) = L \qquad \text{and} \qquad \lim_{x\to a} g(x) = M$$

exist. Then

1. $\displaystyle\lim_{x\to a}[f(x) + g(x)] = L + M$

2. $\displaystyle\lim_{x\to a}[f(x) - g(x)] = L - M$

3. $\displaystyle\lim_{x\to a}[cf(x)] = cL$

4. $\displaystyle\lim_{x\to a}[f(x)g(x)] = LM$

5. $\displaystyle\lim_{x\to a}\frac{f(x)}{g(x)} = \frac{L}{M} \quad \text{if } M \neq 0$

PROOF OF LAW 4 Let $\varepsilon > 0$ be given. We want to find $\delta > 0$ such that

$$\text{if} \quad 0 < |x - a| < \delta \quad \text{then} \quad |f(x)g(x) - LM| < \varepsilon$$

In order to get terms that contain $|f(x) - L|$ and $|g(x) - M|$, we add and subtract $Lg(x)$ as follows:

$$\begin{aligned}
|f(x)g(x) - LM| &= |f(x)g(x) - Lg(x) + Lg(x) - LM| \\
&= |[f(x) - L]g(x) + L[g(x) - M]| \\
&\leqslant |[f(x) - L]g(x)| + |L[g(x) - M]| \qquad \text{(Triangle Inequality)} \\
&= |f(x) - L||g(x)| + |L||g(x) - M|
\end{aligned}$$

We want to make each of these terms less than $\varepsilon/2$.

Since $\lim_{x\to a} g(x) = M$, there is a number $\delta_1 > 0$ such that

$$\text{if} \quad 0 < |x - a| < \delta_1 \quad \text{then} \quad |g(x) - M| < \frac{\varepsilon}{2(1 + |L|)}$$

Also, there is a number $\delta_2 > 0$ such that if $0 < |x - a| < \delta_2$, then

$$|g(x) - M| < 1$$

and therefore

$$|g(x)| = |g(x) - M + M| \leqslant |g(x) - M| + |M| < 1 + |M|$$

Since $\lim_{x \to a} f(x) = L$, there is a number $\delta_3 > 0$ such that

$$\text{if} \quad 0 < |x - a| < \delta_3 \quad \text{then} \quad |f(x) - L| < \frac{\varepsilon}{2(1 + |M|)}$$

Let $\delta = \min\{\delta_1, \delta_2, \delta_3\}$. If $0 < |x - a| < \delta$, then we have $0 < |x - a| < \delta_1$, $0 < |x - a| < \delta_2$, and $0 < |x - a| < \delta_3$, so we can combine the inequalities to obtain

$$\begin{aligned} |f(x)g(x) - LM| &\leqslant |f(x) - L||g(x)| + |L||g(x) - M| \\ &< \frac{\varepsilon}{2(1 + |M|)}(1 + |M|) + |L| \frac{\varepsilon}{2(1 + |L|)} \\ &< \frac{\varepsilon}{2} + \frac{\varepsilon}{2} = \varepsilon \end{aligned}$$

This shows that $\lim_{x \to a} f(x)g(x) = LM$. □

PROOF OF LAW 3 If we take $g(x) = c$ in Law 4, we get

$$\begin{aligned} \lim_{x \to a} [cf(x)] &= \lim_{x \to a} [g(x)f(x)] = \lim_{x \to a} g(x) \cdot \lim_{x \to a} f(x) \\ &= \lim_{x \to a} c \cdot \lim_{x \to a} f(x) \\ &= c \lim_{x \to a} f(x) \qquad \text{(by Law 7)} \end{aligned}$$ □

PROOF OF LAW 2 Using Law 1 and Law 3 with $c = -1$, we have

$$\begin{aligned} \lim_{x \to a} [f(x) - g(x)] &= \lim_{x \to a} [f(x) + (-1)g(x)] = \lim_{x \to a} f(x) + \lim_{x \to a} (-1)g(x) \\ &= \lim_{x \to a} f(x) + (-1) \lim_{x \to a} g(x) = \lim_{x \to a} f(x) - \lim_{x \to a} g(x) \end{aligned}$$ □

PROOF OF LAW 5 First let us show that

$$\lim_{x \to a} \frac{1}{g(x)} = \frac{1}{M}$$

To do this we must show that, given $\varepsilon > 0$, there exists $\delta > 0$ such that

$$\text{if} \quad 0 < |x - a| < \delta \quad \text{then} \quad \left| \frac{1}{g(x)} - \frac{1}{M} \right| < \varepsilon$$

Observe that
$$\left| \frac{1}{g(x)} - \frac{1}{M} \right| = \frac{|M - g(x)|}{|Mg(x)|}$$

We know that we can make the numerator small. But we also need to know that the denominator is not small when x is near a. Since $\lim_{x \to a} g(x) = M$, there is a number $\delta_1 > 0$ such that, whenever $0 < |x - a| < \delta_1$, we have

$$|g(x) - M| < \frac{|M|}{2}$$

and therefore
$$\begin{aligned} |M| = |M - g(x) + g(x)| &\leqslant |M - g(x)| + |g(x)| \\ &< \frac{|M|}{2} + |g(x)| \end{aligned}$$

This shows that

$$\text{if} \quad 0 < |x - a| < \delta_1 \quad \text{then} \quad |g(x)| > \frac{|M|}{2}$$

and so, for these values of x,

$$\frac{1}{|Mg(x)|} = \frac{1}{|M||g(x)|} < \frac{1}{|M|} \cdot \frac{2}{|M|} = \frac{2}{M^2}$$

Also, there exists $\delta_2 > 0$ such that

$$\text{if} \quad 0 < |x - a| < \delta_2 \quad \text{then} \quad |g(x) - M| < \frac{M^2}{2}\,\varepsilon$$

Let $\delta = \min\{\delta_1, \delta_2\}$. Then, for $0 < |x - a| < \delta$, we have

$$\left|\frac{1}{g(x)} - \frac{1}{M}\right| = \frac{|M - g(x)|}{|Mg(x)|} < \frac{2}{M^2}\,\frac{M^2}{2}\,\varepsilon = \varepsilon$$

It follows that $\lim_{x\to a} 1/g(x) = 1/M$. Finally, using Law 4, we obtain

$$\lim_{x\to a} \frac{f(x)}{g(x)} = \lim_{x\to a} f(x)\left(\frac{1}{g(x)}\right) = \lim_{x\to a} f(x) \lim_{x\to a} \frac{1}{g(x)} = L \cdot \frac{1}{M} = \frac{L}{M}$$ □

2 THEOREM If $f(x) \leqslant g(x)$ for all x in an open interval that contains a (except possibly at a) and

$$\lim_{x\to a} f(x) = L \quad \text{and} \quad \lim_{x\to a} g(x) = M$$

then $L \leqslant M$.

PROOF We use the method of proof by contradiction. Suppose, if possible, that $L > M$. Law 2 of limits says that

$$\lim_{x\to a} [g(x) - f(x)] = M - L$$

Therefore, for any $\varepsilon > 0$, there exists $\delta > 0$ such that

$$\text{if} \quad 0 < |x - a| < \delta \quad \text{then} \quad |[g(x) - f(x)] - (M - L)| < \varepsilon$$

In particular, taking $\varepsilon = L - M$ (noting that $L - M > 0$ by hypothesis), we have a number $\delta > 0$ such that

$$\text{if} \quad 0 < |x - a| < \delta \quad \text{then} \quad |[g(x) - f(x)] - (M - L)| < L - M$$

Since $a \leqslant |a|$ for any number a, we have

$$\text{if} \quad 0 < |x - a| < \delta \quad \text{then} \quad [g(x) - f(x)] - (M - L) < L - M$$

which simplifies to

$$\text{if} \quad 0 < |x - a| < \delta \quad \text{then} \quad g(x) < f(x)$$

But this contradicts $f(x) \leqslant g(x)$. Thus the inequality $L > M$ must be false. Therefore $L \leqslant M$. □

3 THE SQUEEZE THEOREM If $f(x) \leqslant g(x) \leqslant h(x)$ for all x in an open interval that contains a (except possibly at a) and

$$\lim_{x \to a} f(x) = \lim_{x \to a} h(x) = L$$

then
$$\lim_{x \to a} g(x) = L$$

PROOF Let $\varepsilon > 0$ be given. Since $\lim_{x \to a} f(x) = L$, there is a number $\delta_1 > 0$ such that

$$\text{if} \quad 0 < |x - a| < \delta_1 \quad \text{then} \quad |f(x) - L| < \varepsilon$$

that is,
$$\text{if} \quad 0 < |x - a| < \delta_1 \quad \text{then} \quad L - \varepsilon < f(x) < L + \varepsilon$$

Since $\lim_{x \to a} h(x) = L$, there is a number $\delta_2 > 0$ such that

$$\text{if} \quad 0 < |x - a| < \delta_2 \quad \text{then} \quad |h(x) - L| < \varepsilon$$

that is,
$$\text{if} \quad 0 < |x - a| < \delta_2 \quad \text{then} \quad L - \varepsilon < h(x) < L + \varepsilon$$

Let $\delta = \min\{\delta_1, \delta_2\}$. If $0 < |x - a| < \delta$, then $0 < |x - a| < \delta_1$ and $0 < |x - a| < \delta_2$, so

$$L - \varepsilon < f(x) \leqslant g(x) \leqslant h(x) < L + \varepsilon$$

In particular,
$$L - \varepsilon < g(x) < L + \varepsilon$$

and so $|g(x) - L| < \varepsilon$. Therefore $\lim_{x \to a} g(x) = L$. □

SECTION 2.5

THEOREM If f is a one-to-one continuous function defined on an interval (a, b), then its inverse function f^{-1} is also continuous.

PROOF First we show that if f is both one-to-one and continuous on (a, b), then it must be either increasing or decreasing on (a, b). If it were neither increasing nor decreasing, then there would exist numbers x_1, x_2, and x_3 in (a, b) with $x_1 < x_2 < x_3$ such that $f(x_2)$ does not lie between $f(x_1)$ and $f(x_3)$. There are two possibilities: either (1) $f(x_3)$ lies between $f(x_1)$ and $f(x_2)$ or (2) $f(x_1)$ lies between $f(x_2)$ and $f(x_3)$. (Draw a picture.) In case (1) we apply the Intermediate Value Theorem to the continuous function f to get a number c between x_1 and x_2 such that $f(c) = f(x_3)$. In case (2) the Intermediate Value Theorem gives a number c between x_2 and x_3 such that $f(c) = f(x_1)$. In either case we have contradicted the fact that f is one-to-one.

Let us assume, for the sake of definiteness, that f is increasing on (a, b). We take any number y_0 in the domain of f^{-1} and we let $f^{-1}(y_0) = x_0$; that is, x_0 is the number in (a, b) such that $f(x_0) = y_0$. To show that f^{-1} is continuous at y_0 we take any $\varepsilon > 0$ such that the interval $(x_0 - \varepsilon, x_0 + \varepsilon)$ is contained in the interval (a, b). Since f is increasing, it maps the numbers in the interval $(x_0 - \varepsilon, x_0 + \varepsilon)$ onto the numbers in the interval $(f(x_0 - \varepsilon), f(x_0 + \varepsilon))$ and f^{-1} reverses the correspondence. If we let δ denote the smaller of the numbers $\delta_1 = y_0 - f(x_0 - \varepsilon)$ and $\delta_2 = f(x_0 + \varepsilon) - y_0$, then the interval $(y_0 - \delta, y_0 + \delta)$ is contained in the interval $(f(x_0 - \varepsilon), f(x_0 + \varepsilon))$ and so is mapped into the interval $(x_0 - \varepsilon, x_0 + \varepsilon)$ by f^{-1}. (See the arrow diagram in Figure 1.) We have therefore found a number $\delta > 0$ such that

$$\text{if} \quad |y - y_0| < \delta \quad \text{then} \quad |f^{-1}(y) - f^{-1}(y_0)| < \varepsilon$$

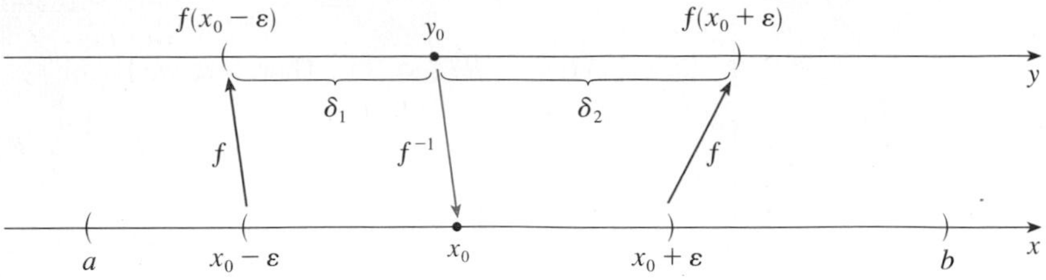

FIGURE 1

This shows that $\lim_{y \to y_0} f^{-1}(y) = f^{-1}(y_0)$ and so f^{-1} is continuous at any number y_0 in its domain. □

> **8 THEOREM** If f is continuous at b and $\lim_{x \to a} g(x) = b$, then
>
> $$\lim_{x \to a} f(g(x)) = f(b)$$

PROOF Let $\varepsilon > 0$ be given. We want to find a number $\delta > 0$ such that

$$\text{if} \quad 0 < |x - a| < \delta \quad \text{then} \quad |f(g(x)) - f(b)| < \varepsilon$$

Since f is continuous at b, we have

$$\lim_{y \to b} f(y) = f(b)$$

and so there exists $\delta_1 > 0$ such that

$$\text{if} \quad 0 < |y - b| < \delta_1 \quad \text{then} \quad |f(y) - f(b)| < \varepsilon$$

Since $\lim_{x \to a} g(x) = b$, there exists $\delta > 0$ such that

$$\text{if} \quad 0 < |x - a| < \delta \quad \text{then} \quad |g(x) - b| < \delta_1$$

Combining these two statements, we see that whenever $0 < |x - a| < \delta$ we have $|g(x) - b| < \delta_1$, which implies that $|f(g(x)) - f(b)| < \varepsilon$. Therefore we have proved that $\lim_{x \to a} f(g(x)) = f(b)$. □

SECTION 3.3

The proof of the following result was promised when we proved that $\displaystyle\lim_{\theta \to 0} \frac{\sin \theta}{\theta} = 1$.

> **THEOREM** If $0 < \theta < \pi/2$, then $\theta \leqslant \tan \theta$.

PROOF Figure 2 shows a sector of a circle with center O, central angle θ, and radius 1. Then

$$|AD| = |OA| \tan \theta = \tan \theta$$

We approximate the arc AB by an inscribed polygon consisting of n equal line segments

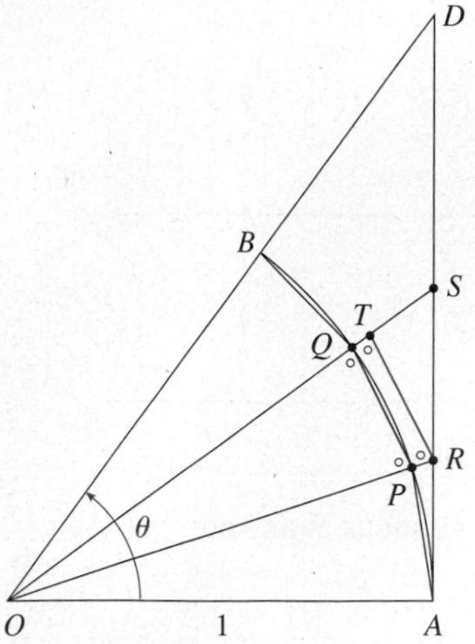

FIGURE 2

and we look at a typical segment PQ. We extend the lines OP and OQ to meet AD in the points R and S. Then we draw $RT \parallel PQ$ as in Figure 2. Observe that

$$\angle RTO = \angle PQO < 90^\circ$$

and so $\angle RTS > 90^\circ$. Therefore we have

$$|PQ| < |RT| < |RS|$$

If we add n such inequalities, we get

$$L_n < |AD| = \tan\theta$$

where L_n is the length of the inscribed polygon. Thus, by Theorem 2.3.2, we have

$$\lim_{n\to\infty} L_n \leqslant \tan\theta$$

But the arc length is defined in Equation 8.1.1 as the limit of the lengths of inscribed polygons, so

$$\theta = \lim_{n\to\infty} L_n \leqslant \tan\theta$$

□

SECTION 4.3

CONCAVITY TEST

(a) If $f''(x) > 0$ for all x in I, then the graph of f is concave upward on I.

(b) If $f''(x) < 0$ for all x in I, then the graph of f is concave downward on I.

PROOF OF (a) Let a be any number in I. We must show that the curve $y = f(x)$ lies above the tangent line at the point $(a, f(a))$. The equation of this tangent is

$$y = f(a) + f'(a)(x - a)$$

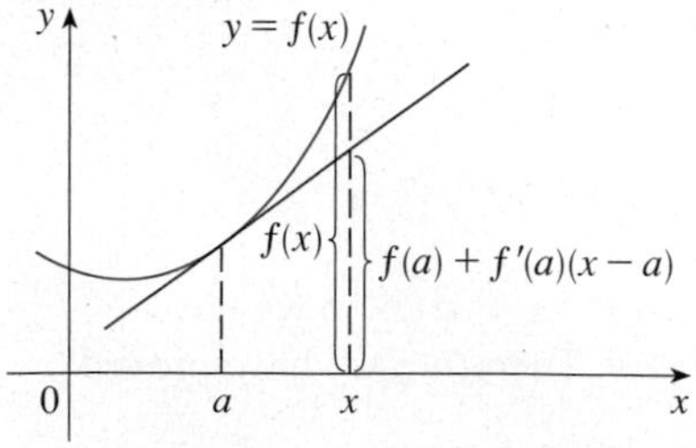

FIGURE 3

So we must show that

$$f(x) > f(a) + f'(a)(x - a)$$

whenever $x \in I$ $(x \neq a)$. (See Figure 3.)

First let us take the case where $x > a$. Applying the Mean Value Theorem to f on the interval $[a, x]$, we get a number c, with $a < c < x$, such that

$$\boxed{1} \qquad f(x) - f(a) = f'(c)(x - a)$$

Since $f'' > 0$ on I, we know from the Increasing/Decreasing Test that f' is increasing on I. Thus, since $a < c$, we have

$$f'(a) < f'(c)$$

and so, multiplying this inequality by the positive number $x - a$, we get

$$\boxed{2} \qquad f'(a)(x - a) < f'(c)(x - a)$$

Now we add $f(a)$ to both sides of this inequality:

$$f(a) + f'(a)(x - a) < f(a) + f'(c)(x - a)$$

But from Equation 1 we have $f(x) = f(a) + f'(c)(x - a)$. So this inequality becomes

$$\boxed{3} \qquad f(x) > f(a) + f'(a)(x - a)$$

which is what we wanted to prove.

For the case where $x < a$ we have $f'(c) < f'(a)$, but multiplication by the negative number $x - a$ reverses the inequality, so we get (2) and (3) as before. □

SECTION 4.4

In order to give the promised proof of l'Hospital's Rule, we first need a generalization of the Mean Value Theorem. The following theorem is named after another French mathematician, Augustin-Louis Cauchy (1789–1857).

■ See the biographical sketch of Cauchy on page 113.

1 CAUCHY'S MEAN VALUE THEOREM Suppose that the functions f and g are continuous on $[a, b]$ and differentiable on (a, b), and $g'(x) \neq 0$ for all x in (a, b). Then there is a number c in (a, b) such that

$$\frac{f'(c)}{g'(c)} = \frac{f(b) - f(a)}{g(b) - g(a)}$$

Notice that if we take the special case in which $g(x) = x$, then $g'(c) = 1$ and Theorem 1 is just the ordinary Mean Value Theorem. Furthermore, Theorem 1 can be proved in a similar manner. You can verify that all we have to do is change the function h given by Equation 4.2.4 to the function

$$h(x) = f(x) - f(a) - \frac{f(b) - f(a)}{g(b) - g(a)}[g(x) - g(a)]$$

and apply Rolle's Theorem as before.

L'HOSPITAL'S RULE Suppose f and g are differentiable and $g'(x) \neq 0$ on an open interval I that contains a (except possibly at a). Suppose that

$$\lim_{x \to a} f(x) = 0 \qquad \text{and} \qquad \lim_{x \to a} g(x) = 0$$

or that

$$\lim_{x \to a} f(x) = \pm\infty \qquad \text{and} \qquad \lim_{x \to a} g(x) = \pm\infty$$

(In other words, we have an indeterminate form of type $\frac{0}{0}$ or ∞/∞.) Then

$$\lim_{x \to a} \frac{f(x)}{g(x)} = \lim_{x \to a} \frac{f'(x)}{g'(x)}$$

if the limit on the right side exists (or is ∞ or $-\infty$).

PROOF OF L'HOSPITAL'S RULE We are assuming that $\lim_{x \to a} f(x) = 0$ and $\lim_{x \to a} g(x) = 0$. Let

$$L = \lim_{x \to a} \frac{f'(x)}{g'(x)}$$

We must show that $\lim_{x \to a} f(x)/g(x) = L$. Define

$$F(x) = \begin{cases} f(x) & \text{if } x \neq a \\ 0 & \text{if } x = a \end{cases} \qquad G(x) = \begin{cases} g(x) & \text{if } x \neq a \\ 0 & \text{if } x = a \end{cases}$$

Then F is continuous on I since f is continuous on $\{x \in I \mid x \neq a\}$ and

$$\lim_{x \to a} F(x) = \lim_{x \to a} f(x) = 0 = F(a)$$

Likewise, G is continuous on I. Let $x \in I$ and $x > a$. Then F and G are continuous on $[a, x]$ and differentiable on (a, x) and $G' \neq 0$ there (since $F' = f'$ and $G' = g'$). Therefore, by Cauchy's Mean Value Theorem, there is a number y such that $a < y < x$ and

$$\frac{F'(y)}{G'(y)} = \frac{F(x) - F(a)}{G(x) - G(a)} = \frac{F(x)}{G(x)}$$

Here we have used the fact that, by definition, $F(a) = 0$ and $G(a) = 0$. Now, if we let $x \to a^+$, then $y \to a^+$ (since $a < y < x$), so

$$\lim_{x \to a^+} \frac{f(x)}{g(x)} = \lim_{x \to a^+} \frac{F(x)}{G(x)} = \lim_{y \to a^+} \frac{F'(y)}{G'(y)} = \lim_{y \to a^+} \frac{f'(y)}{g'(y)} = L$$

A similar argument shows that the left-hand limit is also L. Therefore

$$\lim_{x \to a} \frac{f(x)}{g(x)} = L$$

This proves l'Hospital's Rule for the case where a is finite.

If a is infinite, we let $t = 1/x$. Then $t \to 0^+$ as $x \to \infty$, so we have

$$\begin{aligned} \lim_{x \to \infty} \frac{f(x)}{g(x)} &= \lim_{t \to 0^+} \frac{f(1/t)}{g(1/t)} \\ &= \lim_{t \to 0^+} \frac{f'(1/t)(-1/t^2)}{g'(1/t)(-1/t^2)} \qquad \text{(by l'Hospital's Rule for finite } a\text{)} \\ &= \lim_{t \to 0^+} \frac{f'(1/t)}{g'(1/t)} = \lim_{x \to \infty} \frac{f'(x)}{g'(x)} \end{aligned}$$

□

SECTION 11.8

In order to prove Theorem 11.8.3, we first need the following results.

THEOREM

1. If a power series $\Sigma\, c_n x^n$ converges when $x = b$ (where $b \neq 0$), then it converges whenever $|x| < |b|$.

2. If a power series $\Sigma\, c_n x^n$ diverges when $x = d$ (where $d \neq 0$), then it diverges whenever $|x| > |d|$.

PROOF OF 1 Suppose that $\Sigma\, c_n b^n$ converges. Then, by Theorem 11.2.6, we have $\lim_{n\to\infty} c_n b^n = 0$. According to Definition 11.1.2 with $\varepsilon = 1$, there is a positive integer N such that $|c_n b^n| < 1$ whenever $n \geqslant N$. Thus, for $n \geqslant N$, we have

$$|c_n x^n| = \left|\frac{c_n b^n x^n}{b^n}\right| = |c_n b^n| \left|\frac{x}{b}\right|^n < \left|\frac{x}{b}\right|^n$$

If $|x| < |b|$, then $|x/b| < 1$, so $\Sigma\, |x/b|^n$ is a convergent geometric series. Therefore, by the Comparison Test, the series $\Sigma_{n=N}^{\infty} |c_n x^n|$ is convergent. Thus the series $\Sigma\, c_n x^n$ is absolutely convergent and therefore convergent. □

PROOF OF 2 Suppose that $\Sigma\, c_n d^n$ diverges. If x is any number such that $|x| > |d|$, then $\Sigma\, c_n x^n$ cannot converge because, by part 1, the convergence of $\Sigma\, c_n x^n$ would imply the convergence of $\Sigma\, c_n d^n$. Therefore $\Sigma\, c_n x^n$ diverges whenever $|x| > |d|$. □

THEOREM For a power series $\Sigma\, c_n x^n$ there are only three possibilities:

1. The series converges only when $x = 0$.
2. The series converges for all x.
3. There is a positive number R such that the series converges if $|x| < R$ and diverges if $|x| > R$.

PROOF Suppose that neither case 1 nor case 2 is true. Then there are nonzero numbers b and d such that $\Sigma\, c_n x^n$ converges for $x = b$ and diverges for $x = d$. Therefore the set $S = \{x \mid \Sigma\, c_n x^n \text{ converges}\}$ is not empty. By the preceding theorem, the series diverges if $|x| > |d|$, so $|x| \leqslant |d|$ for all $x \in S$. This says that $|d|$ is an upper bound for the set S. Thus, by the Completeness Axiom (see Section 11.1), S has a least upper bound R. If $|x| > R$, then $x \notin S$, so $\Sigma\, c_n x^n$ diverges. If $|x| < R$, then $|x|$ is not an upper bound for S and so there exists $b \in S$ such that $b > |x|$. Since $b \in S$, $\Sigma\, c_n b^n$ converges, so by the preceding theorem $\Sigma\, c_n x^n$ converges. □

3 THEOREM For a power series $\Sigma\, c_n(x - a)^n$ there are only three possibilities:

1. The series converges only when $x = a$.
2. The series converges for all x.
3. There is a positive number R such that the series converges if $|x - a| < R$ and diverges if $|x - a| > R$.

PROOF If we make the change of variable $u = x - a$, then the power series becomes $\Sigma\, c_n u^n$ and we can apply the preceding theorem to this series. In case 3 we have convergence for $|u| < R$ and divergence for $|u| > R$. Thus we have convergence for $|x - a| < R$ and divergence for $|x - a| > R$. □

SECTION 14.3

CLAIRAUT'S THEOREM Suppose f is defined on a disk D that contains the point (a, b). If the functions f_{xy} and f_{yx} are both continuous on D, then $f_{xy}(a, b) = f_{yx}(a, b)$.

PROOF For small values of h, $h \neq 0$, consider the difference

$$\Delta(h) = [f(a + h, b + h) - f(a + h, b)] - [f(a, b + h) - f(a, b)]$$

Notice that if we let $g(x) = f(x, b + h) - f(x, b)$, then

$$\Delta(h) = g(a + h) - g(a)$$

By the Mean Value Theorem, there is a number c between a and $a + h$ such that

$$g(a + h) - g(a) = g'(c)h = h[f_x(c, b + h) - f_x(c, b)]$$

Applying the Mean Value Theorem again, this time to f_x, we get a number d between b and $b + h$ such that

$$f_x(c, b + h) - f_x(c, b) = f_{xy}(c, d)h$$

Combining these equations, we obtain

$$\Delta(h) = h^2 f_{xy}(c, d)$$

If $h \to 0$, then $(c, d) \to (a, b)$, so the continuity of f_{xy} at (a, b) gives

$$\lim_{h \to 0} \frac{\Delta(h)}{h^2} = \lim_{(c, d) \to (a, b)} f_{xy}(c, d) = f_{xy}(a, b)$$

Similarly, by writing

$$\Delta(h) = [f(a + h, b + h) - f(a, b + h)] - [f(a + h, b) - f(a, b)]$$

and using the Mean Value Theorem twice and the continuity of f_{yx} at (a, b), we obtain

$$\lim_{h \to 0} \frac{\Delta(h)}{h^2} = f_{yx}(a, b)$$

It follows that $f_{xy}(a, b) = f_{yx}(a, b)$. □

SECTION 14.4

8 THEOREM If the partial derivatives f_x and f_y exist near (a, b) and are continuous at (a, b), then f is differentiable at (a, b).

PROOF Let

$$\Delta z = f(a + \Delta x, b + \Delta y) - f(a, b)$$

According to (14.4.7), to prove that f is differentiable at (a, b) we have to show that we can write Δz in the form

$$\Delta z = f_x(a, b)\,\Delta x + f_y(a, b)\,\Delta y + \varepsilon_1\,\Delta x + \varepsilon_2\,\Delta y$$

where ε_1 and $\varepsilon_2 \to 0$ as $(\Delta x, \Delta y) \to (0, 0)$.

Referring to Figure 4, we write

$$\boxed{1} \quad \Delta z = [f(a + \Delta x, b + \Delta y) - f(a, b + \Delta y)] + [f(a, b + \Delta y) - f(a, b)]$$

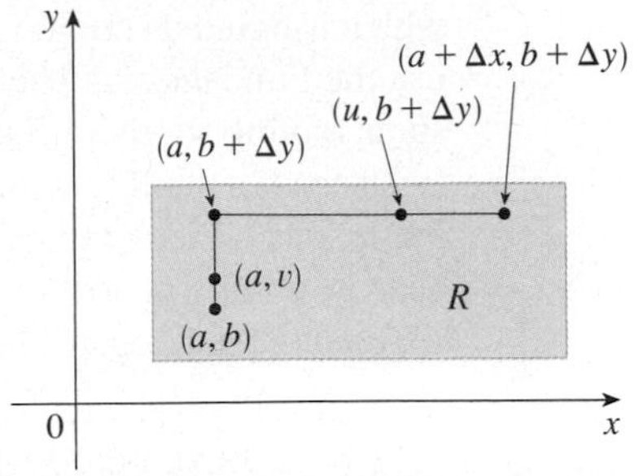

FIGURE 4

Observe that the function of a single variable

$$g(x) = f(x, b + \Delta y)$$

is defined on the interval $[a, a + \Delta x]$ and $g'(x) = f_x(x, b + \Delta y)$. If we apply the Mean Value Theorem to g, we get

$$g(a + \Delta x) - g(a) = g'(u)\,\Delta x$$

where u is some number between a and $a + \Delta x$. In terms of f, this equation becomes

$$f(a + \Delta x, b + \Delta y) - f(a, b + \Delta y) = f_x(u, b + \Delta y)\,\Delta x$$

This gives us an expression for the first part of the right side of Equation 1. For the second part we let $h(y) = f(a, y)$. Then h is a function of a single variable defined on the interval $[b, b + \Delta y]$ and $h'(y) = f_y(a, y)$. A second application of the Mean Value Theorem then gives

$$h(b + \Delta y) - h(b) = h'(v)\,\Delta y$$

where v is some number between b and $b + \Delta y$. In terms of f, this becomes

$$f(a, b + \Delta y) - f(a, b) = f_y(a, v)\,\Delta y$$

We now substitute these expressions into Equation 1 and obtain

$$\begin{aligned} \Delta z &= f_x(u, b + \Delta y)\,\Delta x + f_y(a, v)\,\Delta y \\ &= f_x(a, b)\,\Delta x + [f_x(u, b + \Delta y) - f_x(a, b)]\,\Delta x + f_y(a, b)\,\Delta y \\ &\qquad + [f_y(a, v) - f_y(a, b)]\,\Delta y \\ &= f_x(a, b)\,\Delta x + f_y(a, b)\,\Delta y + \varepsilon_1\,\Delta x + \varepsilon_2\,\Delta y \end{aligned}$$

where
$$\varepsilon_1 = f_x(u, b + \Delta y) - f_x(a, b)$$
$$\varepsilon_2 = f_y(a, v) - f_y(a, b)$$

Since $(u, b + \Delta y) \to (a, b)$ and $(a, v) \to (a, b)$ as $(\Delta x, \Delta y) \to (0, 0)$ and since f_x and f_y are continuous at (a, b), we see that $\varepsilon_1 \to 0$ and $\varepsilon_2 \to 0$ as $(\Delta x, \Delta y) \to (0, 0)$.

Therefore f is differentiable at (a, b). □

G THE LOGARITHM DEFINED AS AN INTEGRAL

Our treatment of exponential and logarithmic functions until now has relied on our intuition, which is based on numerical and visual evidence. (See Sections 1.5, 1.6, and 3.1.) Here we use the Fundamental Theorem of Calculus to give an alternative treatment that provides a surer footing for these functions.

Instead of starting with a^x and defining $\log_a x$ as its inverse, this time we start by defining $\ln x$ as an integral and then define the exponential function as its inverse. You should bear in mind that we do not use any of our previous definitions and results concerning exponential and logarithmic functions.

THE NATURAL LOGARITHM

We first define $\ln x$ as an integral.

1 DEFINITION The **natural logarithmic function** is the function defined by

$$\ln x = \int_1^x \frac{1}{t}\,dt \qquad x > 0$$

The existence of this function depends on the fact that the integral of a continuous function always exists. If $x > 1$, then $\ln x$ can be interpreted geometrically as the area under the hyperbola $y = 1/t$ from $t = 1$ to $t = x$. (See Figure 1.) For $x = 1$, we have

$$\ln 1 = \int_1^1 \frac{1}{t}\,dt = 0$$

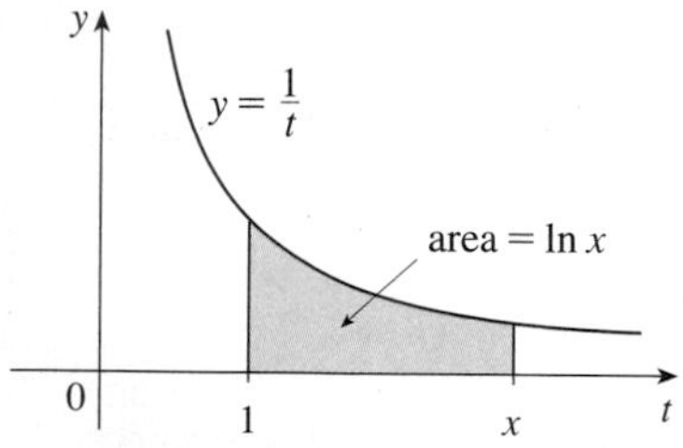

FIGURE 1

For $0 < x < 1$,
$$\ln x = \int_1^x \frac{1}{t}\,dt = -\int_x^1 \frac{1}{t}\,dt < 0$$

and so $\ln x$ is the negative of the area shown in Figure 2.

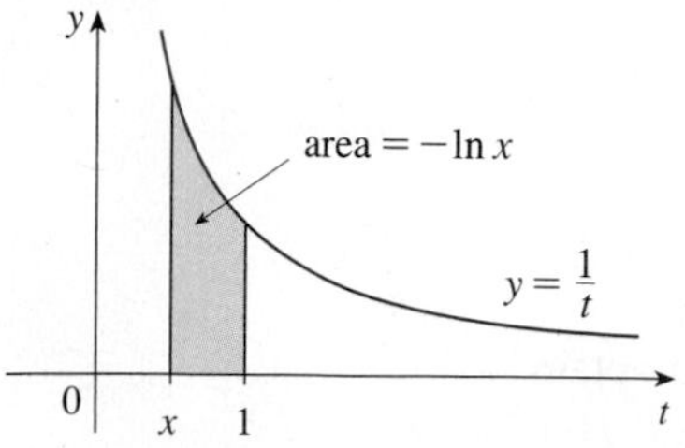

FIGURE 2

V EXAMPLE 1
(a) By comparing areas, show that $\frac{1}{2} < \ln 2 < \frac{3}{4}$.
(b) Use the Midpoint Rule with $n = 10$ to estimate the value of $\ln 2$.

SOLUTION

(a) We can interpret $\ln 2$ as the area under the curve $y = 1/t$ from 1 to 2. From Figure 3 we see that this area is larger than the area of rectangle $BCDE$ and smaller than the area of trapezoid $ABCD$. Thus we have

$$\tfrac{1}{2} \cdot 1 < \ln 2 < 1 \cdot \tfrac{1}{2}\left(1 + \tfrac{1}{2}\right)$$
$$\tfrac{1}{2} < \ln 2 < \tfrac{3}{4}$$

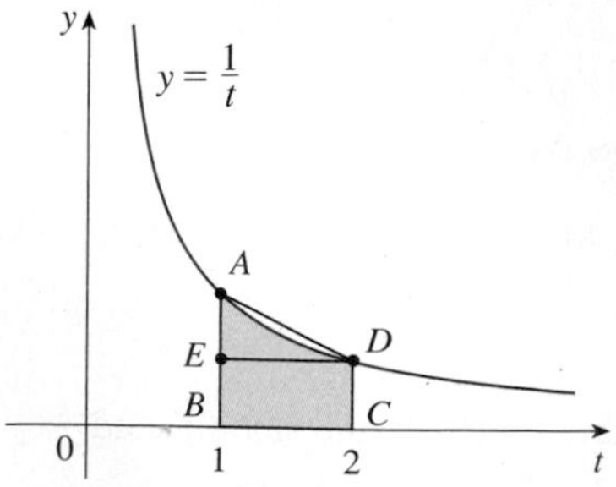

FIGURE 3

(b) If we use the Midpoint Rule with $f(t) = 1/t$, $n = 10$, and $\Delta t = 0.1$, we get

$$\ln 2 = \int_1^2 \frac{1}{t}\,dt \approx (0.1)[f(1.05) + f(1.15) + \cdots + f(1.95)]$$

$$= (0.1)\left(\frac{1}{1.05} + \frac{1}{1.15} + \cdots + \frac{1}{1.95}\right) \approx 0.693 \qquad \square$$

Notice that the integral that defines $\ln x$ is exactly the type of integral discussed in Part 1 of the Fundamental Theorem of Calculus (see Section 5.3). In fact, using that theorem, we have

$$\frac{d}{dx}\int_1^x \frac{1}{t}\,dt = \frac{1}{x}$$

and so

2
$$\frac{d}{dx}(\ln x) = \frac{1}{x}$$

We now use this differentiation rule to prove the following properties of the logarithm function.

3 LAWS OF LOGARITHMS If x and y are positive numbers and r is a rational number, then

1. $\ln(xy) = \ln x + \ln y$ **2.** $\ln\left(\dfrac{x}{y}\right) = \ln x - \ln y$ **3.** $\ln(x^r) = r\ln x$

PROOF

1. Let $f(x) = \ln(ax)$, where a is a positive constant. Then, using Equation 2 and the Chain Rule, we have

$$f'(x) = \frac{1}{ax}\frac{d}{dx}(ax) = \frac{1}{ax}\cdot a = \frac{1}{x}$$

Therefore $f(x)$ and $\ln x$ have the same derivative and so they must differ by a constant:

$$\ln(ax) = \ln x + C$$

Putting $x = 1$ in this equation, we get $\ln a = \ln 1 + C = 0 + C = C$. Thus

$$\ln(ax) = \ln x + \ln a$$

If we now replace the constant a by any number y, we have

$$\ln(xy) = \ln x + \ln y$$

2. Using Law 1 with $x = 1/y$, we have

$$\ln\frac{1}{y} + \ln y = \ln\left(\frac{1}{y}\cdot y\right) = \ln 1 = 0$$

and so
$$\ln\frac{1}{y} = -\ln y$$

Using Law 1 again, we have

$$\ln\left(\frac{x}{y}\right) = \ln\left(x\cdot\frac{1}{y}\right) = \ln x + \ln\frac{1}{y} = \ln x - \ln y$$

The proof of Law 3 is left as an exercise. □

In order to graph $y = \ln x$, we first determine its limits:

4 (a) $\lim_{x \to \infty} \ln x = \infty$ (b) $\lim_{x \to 0^+} \ln x = -\infty$

PROOF

(a) Using Law 3 with $x = 2$ and $r = n$ (where n is any positive integer), we have $\ln(2^n) = n \ln 2$. Now $\ln 2 > 0$, so this shows that $\ln(2^n) \to \infty$ as $n \to \infty$. But $\ln x$ is an increasing function since its derivative $1/x > 0$. Therefore $\ln x \to \infty$ as $x \to \infty$.

(b) If we let $t = 1/x$, then $t \to \infty$ as $x \to 0^+$. Thus, using (a), we have

$$\lim_{x \to 0^+} \ln x = \lim_{t \to \infty} \ln\left(\frac{1}{t}\right) = \lim_{t \to \infty} (-\ln t) = -\infty \qquad \square$$

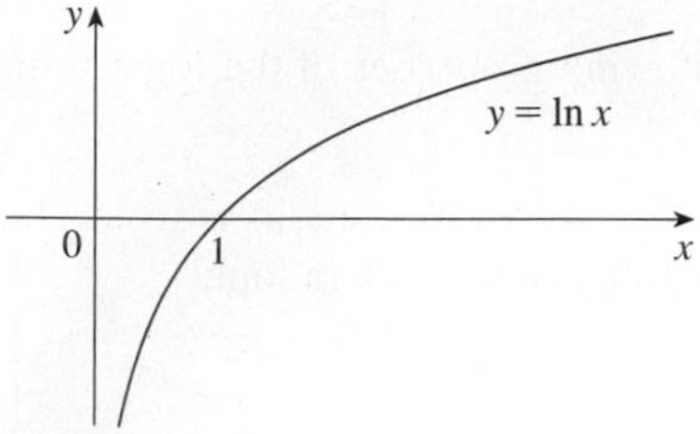

FIGURE 4

If $y = \ln x$, $x > 0$, then

$$\frac{dy}{dx} = \frac{1}{x} > 0 \qquad \text{and} \qquad \frac{d^2y}{dx^2} = -\frac{1}{x^2} < 0$$

which shows that $\ln x$ is increasing and concave downward on $(0, \infty)$. Putting this information together with (4), we draw the graph of $y = \ln x$ in Figure 4.

Since $\ln 1 = 0$ and $\ln x$ is an increasing continuous function that takes on arbitrarily large values, the Intermediate Value Theorem shows that there is a number where $\ln x$ takes on the value 1. (See Figure 5.) This important number is denoted by e.

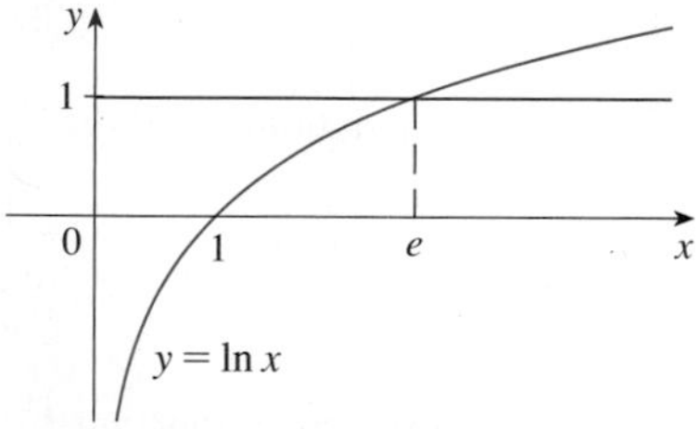

FIGURE 5

5 DEFINITION e is the number such that $\ln e = 1$.

We will show (in Theorem 19) that this definition is consistent with our previous definition of e.

THE NATURAL EXPONENTIAL FUNCTION

Since ln is an increasing function, it is one-to-one and therefore has an inverse function, which we denote by exp. Thus, according to the definition of an inverse function,

$f^{-1}(x) = y \iff f(y) = x$

6 $\exp(x) = y \iff \ln y = x$

and the cancellation equations are

$f^{-1}(f(x)) = x$
$f(f^{-1}(x)) = x$

7 $\exp(\ln x) = x$ and $\ln(\exp x) = x$

In particular, we have

$$\exp(0) = 1 \quad \text{since} \quad \ln 1 = 0$$

$$\exp(1) = e \quad \text{since} \quad \ln e = 1$$

We obtain the graph of $y = \exp x$ by reflecting the graph of $y = \ln x$ about the line $y = x$.

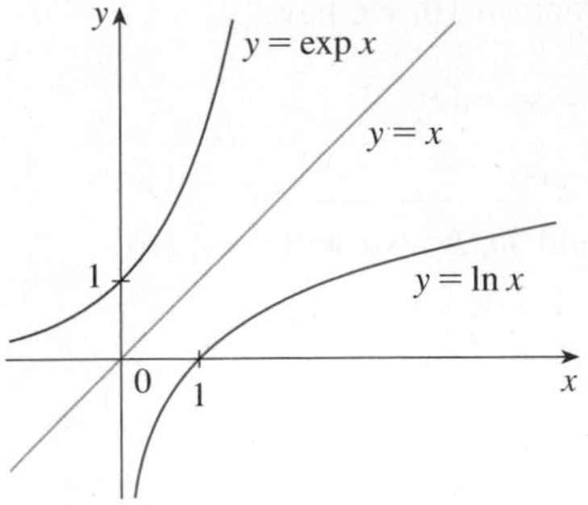

FIGURE 6

(See Figure 6.) The domain of exp is the range of ln, that is, $(-\infty, \infty)$; the range of exp is the domain of ln, that is, $(0, \infty)$.

If r is any rational number, then the third law of logarithms gives

$$\ln(e^r) = r \ln e = r$$

Therefore, by (6), $$\exp(r) = e^r$$

Thus $\exp(x) = e^x$ whenever x is a rational number. This leads us to define e^x, even for irrational values of x, by the equation

$$\boxed{e^x = \exp(x)}$$

In other words, for the reasons given, we define e^x to be the inverse of the function $\ln x$. In this notation (6) becomes

8 $$\boxed{e^x = y \iff \ln y = x}$$

and the cancellation equations (7) become

9 $$\boxed{e^{\ln x} = x \qquad x > 0}$$

10 $$\boxed{\ln(e^x) = x \qquad \text{for all } x}$$

The natural exponential function $f(x) = e^x$ is one of the most frequently occurring functions in calculus and its applications, so it is important to be familiar with its graph (Figure 7) and its properties (which follow from the fact that it is the inverse of the natural logarithmic function).

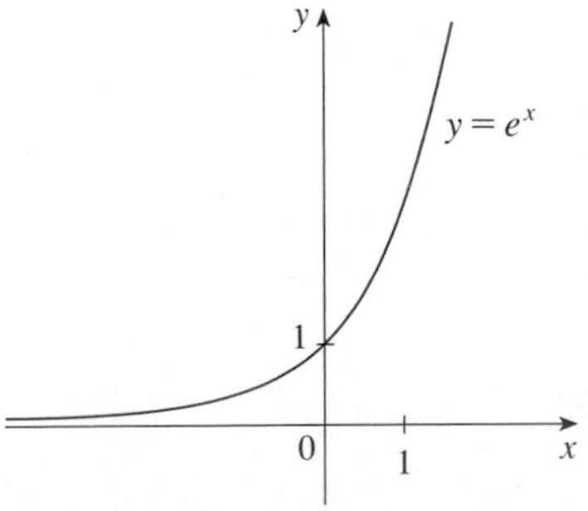

FIGURE 7
The natural exponential function

PROPERTIES OF THE EXPONENTIAL FUNCTION The exponential function $f(x) = e^x$ is an increasing continuous function with domain $\mathbb{R}$ and range $(0, \infty)$. Thus $e^x > 0$ for all x. Also

$$\lim_{x \to -\infty} e^x = 0 \qquad \lim_{x \to \infty} e^x = \infty$$

So the x-axis is a horizontal asymptote of $f(x) = e^x$.

We now verify that f has the other properties expected of an exponential function.

11 **LAWS OF EXPONENTS** If x and y are real numbers and r is rational, then

1. $e^{x+y} = e^x e^y$ **2.** $e^{x-y} = \dfrac{e^x}{e^y}$ **3.** $(e^x)^r = e^{rx}$

PROOF OF LAW 1 Using the first law of logarithms and Equation 10, we have

$$\ln(e^x e^y) = \ln(e^x) + \ln(e^y) = x + y = \ln(e^{x+y})$$

Since ln is a one-to-one function, it follows that $e^x e^y = e^{x+y}$.

Laws 2 and 3 are proved similarly (see Exercises 6 and 7). As we will soon see, Law 3 actually holds when r is any real number. □

We now prove the differentiation formula for e^x.

12

$$\frac{d}{dx}(e^x) = e^x$$

PROOF The function $y = e^x$ is differentiable because it is the inverse function of $y = \ln x$, which we know is differentiable with nonzero derivative. To find its derivative, we use the inverse function method. Let $y = e^x$. Then $\ln y = x$ and, differentiating this latter equation implicitly with respect to x, we get

$$\frac{1}{y}\frac{dy}{dx} = 1$$

$$\frac{dy}{dx} = y = e^x$$

□

GENERAL EXPONENTIAL FUNCTIONS

If $a > 0$ and r is any rational number, then by (9) and (11),

$$a^r = (e^{\ln a})^r = e^{r \ln a}$$

Therefore, even for irrational numbers x, we *define*

13

$$a^x = e^{x \ln a}$$

Thus, for instance,

$$2^{\sqrt{3}} = e^{\sqrt{3}\ln 2} \approx e^{1.20} \approx 3.32$$

The function $f(x) = a^x$ is called the **exponential function with base *a***. Notice that a^x is positive for all x because e^x is positive for all x.

Definition 13 allows us to extend one of the laws of logarithms. We already know that $\ln(a^r) = r \ln a$ when r is rational. But if we now let r be *any* real number we have, from Definition 13,

$$\ln a^r = \ln(e^{r \ln a}) = r \ln a$$

Thus

14 $\ln a^r = r \ln a$ for any real number r

The general laws of exponents follow from Definition 13 together with the laws of exponents for e^x.

15 LAWS OF EXPONENTS If x and y are real numbers and $a, b > 0$, then

1. $a^{x+y} = a^x a^y$ **2.** $a^{x-y} = a^x/a^y$ **3.** $(a^x)^y = a^{xy}$ **4.** $(ab)^x = a^x b^x$

PROOF

1. Using Definition 13 and the laws of exponents for e^x, we have

$$\begin{aligned} a^{x+y} &= e^{(x+y)\ln a} = e^{x \ln a + y \ln a} \\ &= e^{x \ln a} e^{y \ln a} = a^x a^y \end{aligned}$$

3. Using Equation 14 we obtain

$$(a^x)^y = e^{y \ln(a^x)} = e^{yx \ln a} = e^{xy \ln a} = a^{xy}$$

The remaining proofs are left as exercises. □

The differentiation formula for exponential functions is also a consequence of Definition 13:

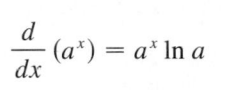

16
$$\frac{d}{dx}(a^x) = a^x \ln a$$

PROOF

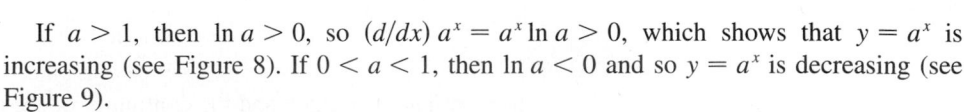

$$\frac{d}{dx}(a^x) = \frac{d}{dx}(e^{x \ln a}) = e^{x \ln a}\frac{d}{dx}(x \ln a) = a^x \ln a$$ □

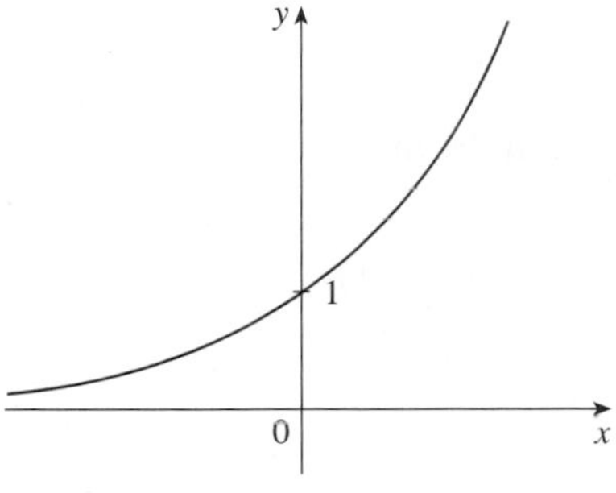

$\lim_{x\to-\infty} a^x = 0, \ \lim_{x\to\infty} a^x = \infty$

FIGURE 8 $y = a^x, \ a > 1$

If $a > 1$, then $\ln a > 0$, so $(d/dx)\, a^x = a^x \ln a > 0$, which shows that $y = a^x$ is increasing (see Figure 8). If $0 < a < 1$, then $\ln a < 0$ and so $y = a^x$ is decreasing (see Figure 9).

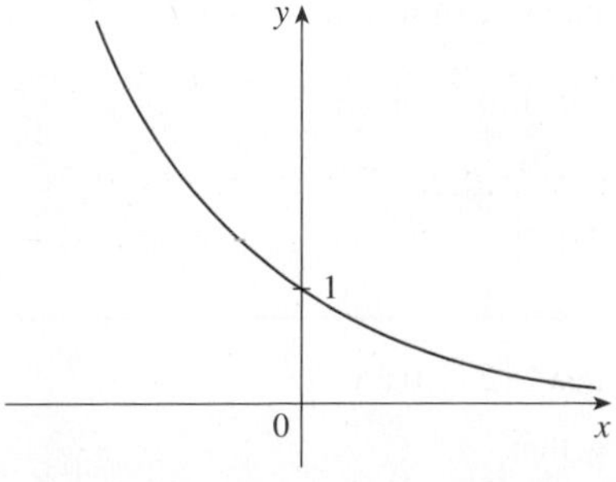

$\lim_{x\to-\infty} a^x = \infty, \ \lim_{x\to\infty} a^x = 0$

FIGURE 9 $y = a^x, \ 0 < a < 1$

GENERAL LOGARITHMIC FUNCTIONS

If $a > 0$ and $a \neq 1$, then $f(x) = a^x$ is a one-to-one function. Its inverse function is called the **logarithmic function with base *a*** and is denoted by $\log_a$. Thus

17
$$\log_a x = y \iff a^y = x$$

In particular, we see that

$$\log_e x = \ln x$$

The laws of logarithms are similar to those for the natural logarithm and can be deduced from the laws of exponents (see Exercise 10).

To differentiate $y = \log_a x$, we write the equation as $a^y = x$. From Equation 14 we have $y \ln a = \ln x$, so

$$\log_a x = y = \frac{\ln x}{\ln a}$$

Since $\ln a$ is a constant, we can differentiate as follows:

$$\frac{d}{dx}(\log_a x) = \frac{d}{dx}\frac{\ln x}{\ln a} = \frac{1}{\ln a}\frac{d}{dx}(\ln x) = \frac{1}{x \ln a}$$

18

$$\frac{d}{dx}(\log_a x) = \frac{1}{x \ln a}$$

THE NUMBER e EXPRESSED AS A LIMIT

In this section we defined e as the number such that $\ln e = 1$. The next theorem shows that this is the same as the number e defined in Section 3.1 (see Equation 3.6.5).

19

$$e = \lim_{x \to 0} (1 + x)^{1/x}$$

PROOF Let $f(x) = \ln x$. Then $f'(x) = 1/x$, so $f'(1) = 1$. But, by the definition of derivative,

$$f'(1) = \lim_{h \to 0} \frac{f(1 + h) - f(1)}{h} = \lim_{x \to 0} \frac{f(1 + x) - f(1)}{x}$$

$$= \lim_{x \to 0} \frac{\ln(1 + x) - \ln 1}{x} = \lim_{x \to 0} \frac{1}{x} \ln(1 + x) = \lim_{x \to 0} \ln(1 + x)^{1/x}$$

Because $f'(1) = 1$, we have

$$\lim_{x \to 0} \ln(1 + x)^{1/x} = 1$$

Then, by Theorem 2.5.8 and the continuity of the exponential function, we have

$$e = e^1 = e^{\lim_{x \to 0} \ln(1+x)^{1/x}} = \lim_{x \to 0} e^{\ln(1+x)^{1/x}} = \lim_{x \to 0} (1 + x)^{1/x}$$ □

G EXERCISES

1. (a) By comparing areas, show that

$$\tfrac{1}{3} < \ln 1.5 < \tfrac{5}{12}$$

(b) Use the Midpoint Rule with $n = 10$ to estimate $\ln 1.5$.

2. Refer to Example 1.

(a) Find the equation of the tangent line to the curve $y = 1/t$ that is parallel to the secant line AD.

(b) Use part (a) to show that $\ln 2 > 0.66$.

3. By comparing areas, show that

$$\frac{1}{2} + \frac{1}{3} + \cdots + \frac{1}{n} < \ln n < 1 + \frac{1}{2} + \frac{1}{3} + \cdots + \frac{1}{n - 1}$$

4. (a) By comparing areas, show that $\ln 2 < 1 < \ln 3$.

(b) Deduce that $2 < e < 3$.

5. Prove the third law of logarithms. [*Hint:* Start by showing that both sides of the equation have the same derivative.]

6. Prove the second law of exponents for e^x [see (11)].

7. Prove the third law of exponents for e^x [see (11)].

8. Prove the second law of exponents [see (15)].

9. Prove the fourth law of exponents [see (15)].

10. Deduce the following laws of logarithms from (15):
 (a) $\log_a(xy) = \log_a x + \log_a y$
 (b) $\log_a(x/y) = \log_a x - \log_a y$
 (c) $\log_a(x^y) = y \log_a x$

H COMPLEX NUMBERS

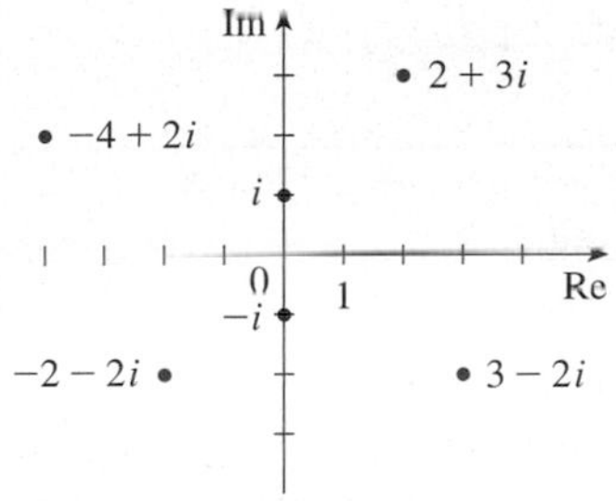

FIGURE 1
Complex numbers as points in the Argand plane

A **complex number** can be represented by an expression of the form $a + bi$, where a and b are real numbers and i is a symbol with the property that $i^2 = -1$. The complex number $a + bi$ can also be represented by the ordered pair (a, b) and plotted as a point in a plane (called the Argand plane) as in Figure 1. Thus the complex number $i = 0 + 1 \cdot i$ is identified with the point $(0, 1)$.

The **real part** of the complex number $a + bi$ is the real number a and the **imaginary part** is the real number b. Thus the real part of $4 - 3i$ is 4 and the imaginary part is -3. Two complex numbers $a + bi$ and $c + di$ are **equal** if $a = c$ and $b = d$; that is, their real parts are equal and their imaginary parts are equal. In the Argand plane the horizontal axis is called the real axis and the vertical axis is called the imaginary axis.

The sum and difference of two complex numbers are defined by adding or subtracting their real parts and their imaginary parts:

$$(a + bi) + (c + di) = (a + c) + (b + d)i$$

$$(a + bi) - (c + di) = (a - c) + (b - d)i$$

For instance,

$$(1 - i) + (4 + 7i) = (1 + 4) + (-1 + 7)i = 5 + 6i$$

The product of complex numbers is defined so that the usual commutative and distributive laws hold:

$$\begin{aligned}(a + bi)(c + di) &= a(c + di) + (bi)(c + di)\\ &= ac + adi + bci + bdi^2\end{aligned}$$

Since $i^2 = -1$, this becomes

$$(a + bi)(c + di) = (ac - bd) + (ad + bc)i$$

EXAMPLE 1

$$\begin{aligned}(-1 + 3i)(2 - 5i) &= (-1)(2 - 5i) + 3i(2 - 5i)\\ &= -2 + 5i + 6i - 15(-1) = 13 + 11i\end{aligned}$$ □

Division of complex numbers is much like rationalizing the denominator of a rational expression. For the complex number $z = a + bi$, we define its **complex conjugate** to be $\bar{z} = a - bi$. To find the quotient of two complex numbers we multiply numerator and denominator by the complex conjugate of the denominator.

EXAMPLE 2 Express the number $\dfrac{-1 + 3i}{2 + 5i}$ in the form $a + bi$.

SOLUTION We multiply numerator and denominator by the complex conjugate of $2 + 5i$, namely $2 - 5i$, and we take advantage of the result of Example 1:

$$\frac{-1 + 3i}{2 + 5i} = \frac{-1 + 3i}{2 + 5i} \cdot \frac{2 - 5i}{2 - 5i} = \frac{13 + 11i}{2^2 + 5^2} = \frac{13}{29} + \frac{11}{29} i \qquad \square$$

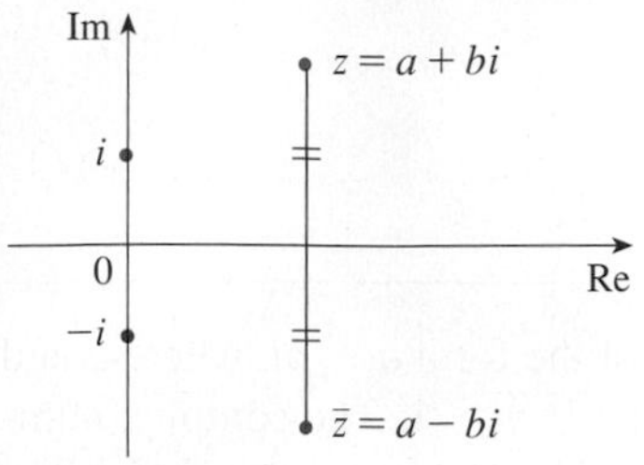

FIGURE 2

The geometric interpretation of the complex conjugate is shown in Figure 2: $\bar{z}$ is the reflection of z in the real axis. We list some of the properties of the complex conjugate in the following box. The proofs follow from the definition and are requested in Exercise 18.

PROPERTIES OF CONJUGATES

$$\overline{z + w} = \bar{z} + \bar{w} \qquad \overline{zw} = \bar{z}\,\bar{w} \qquad \overline{z^n} = \bar{z}^n$$

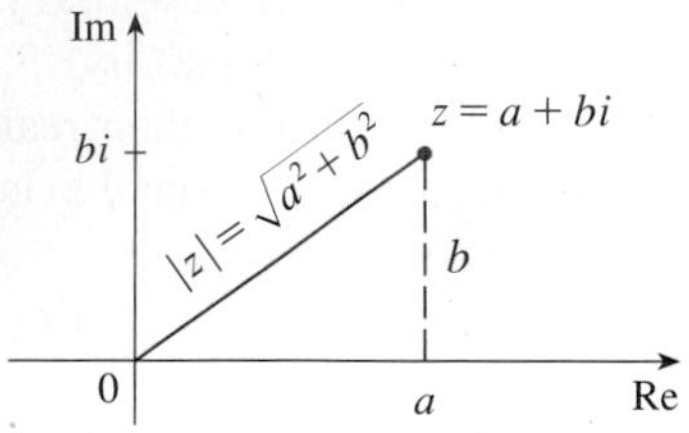

FIGURE 3

The **modulus**, or **absolute value**, $|z|$ of a complex number $z = a + bi$ is its distance from the origin. From Figure 3 we see that if $z = a + bi$, then

$$|z| = \sqrt{a^2 + b^2}$$

Notice that

$$z\bar{z} = (a + bi)(a - bi) = a^2 + abi - abi - b^2i^2 = a^2 + b^2$$

and so

$$z\bar{z} = |z|^2$$

This explains why the division procedure in Example 2 works in general:

$$\frac{z}{w} = \frac{z\bar{w}}{w\bar{w}} = \frac{z\bar{w}}{|w|^2}$$

Since $i^2 = -1$, we can think of i as a square root of -1. But notice that we also have $(-i)^2 = i^2 = -1$ and so $-i$ is also a square root of -1. We say that i is the **principal square root** of -1 and write $\sqrt{-1} = i$. In general, if c is any positive number, we write

$$\sqrt{-c} = \sqrt{c}\, i$$

With this convention, the usual derivation and formula for the roots of the quadratic equation $ax^2 + bx + c = 0$ are valid even when $b^2 - 4ac < 0$:

$$x = \frac{-b \pm \sqrt{b^2 - 4ac}}{2a}$$

EXAMPLE 3 Find the roots of the equation $x^2 + x + 1 = 0$.

SOLUTION Using the quadratic formula, we have

$$x = \frac{-1 \pm \sqrt{1^2 - 4 \cdot 1}}{2} = \frac{-1 \pm \sqrt{-3}}{2} = \frac{-1 \pm \sqrt{3}\, i}{2} \qquad \square$$

We observe that the solutions of the equation in Example 3 are complex conjugates of each other. In general, the solutions of any quadratic equation $ax^2 + bx + c = 0$ with real coefficients a, b, and c are always complex conjugates. (If z is real, $\bar{z} = z$, so z is its own conjugate.)

We have seen that if we allow complex numbers as solutions, then every quadratic equation has a solution. More generally, it is true that every polynomial equation

$$a_n x^n + a_{n-1} x^{n-1} + \cdots + a_1 x + a_0 = 0$$

of degree at least one has a solution among the complex numbers. This fact is known as the Fundamental Theorem of Algebra and was proved by Gauss.

POLAR FORM

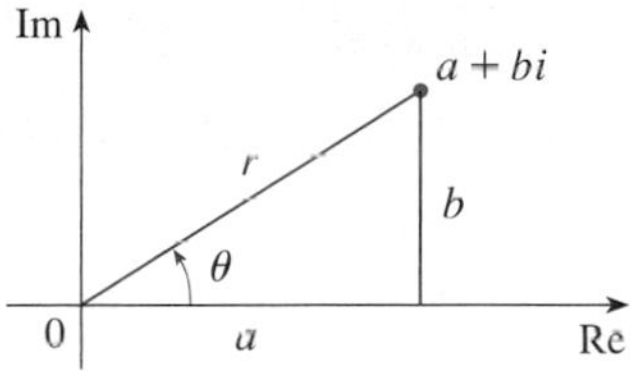

FIGURE 4

We know that any complex number $z = a + bi$ can be considered as a point (a, b) and that any such point can be represented by polar coordinates (r, θ) with $r \geqslant 0$. In fact,

$$a = r\cos\theta \qquad b = r\sin\theta$$

as in Figure 4. Therefore we have

$$z = a + bi = (r\cos\theta) + (r\sin\theta)i$$

Thus we can write any complex number z in the form

$$\boxed{z = r(\cos\theta + i\sin\theta)}$$

where $\qquad r = |z| = \sqrt{a^2 + b^2} \qquad$ and $\qquad \tan\theta = \dfrac{b}{a}$

The angle θ is called the **argument** of z and we write $\theta = \arg(z)$. Note that $\arg(z)$ is not unique; any two arguments of z differ by an integer multiple of 2π.

EXAMPLE 4 Write the following numbers in polar form.

(a) $z = 1 + i$ (b) $w = \sqrt{3} - i$

SOLUTION

(a) We have $r = |z| = \sqrt{1^2 + 1^2} = \sqrt{2}$ and $\tan\theta = 1$, so we can take $\theta = \pi/4$. Therefore the polar form is

$$z = \sqrt{2}\left(\cos\frac{\pi}{4} + i\sin\frac{\pi}{4}\right)$$

(b) Here we have $r = |w| = \sqrt{3 + 1} = 2$ and $\tan\theta = -1/\sqrt{3}$. Since w lies in the fourth quadrant, we take $\theta = -\pi/6$ and

$$w = 2\left[\cos\left(-\frac{\pi}{6}\right) + i\sin\left(-\frac{\pi}{6}\right)\right]$$

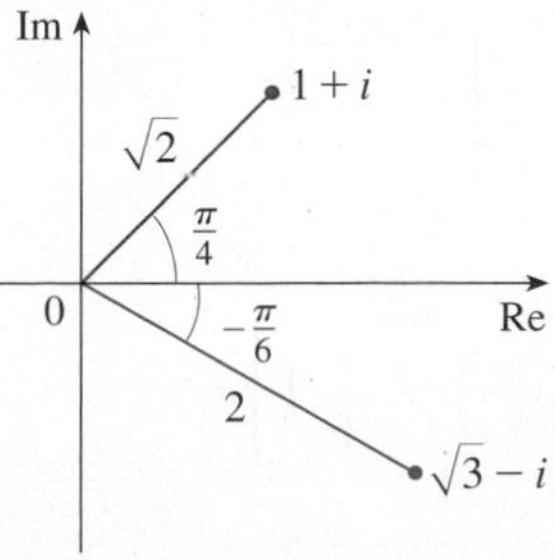

FIGURE 5

The numbers z and w are shown in Figure 5. □

The polar form of complex numbers gives insight into multiplication and division. Let

$$z_1 = r_1(\cos\theta_1 + i\sin\theta_1) \qquad z_2 = r_2(\cos\theta_2 + i\sin\theta_2)$$

be two complex numbers written in polar form. Then

$$\begin{aligned} z_1z_2 &= r_1r_2(\cos\theta_1 + i\sin\theta_1)(\cos\theta_2 + i\sin\theta_2) \\ &= r_1r_2[(\cos\theta_1\cos\theta_2 - \sin\theta_1\sin\theta_2) + i(\sin\theta_1\cos\theta_2 + \cos\theta_1\sin\theta_2)] \end{aligned}$$

Therefore, using the addition formulas for cosine and sine, we have

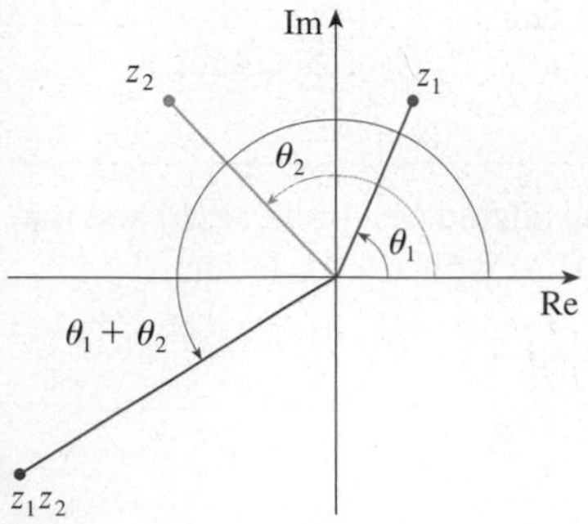

FIGURE 6

1
$$z_1z_2 = r_1r_2[\cos(\theta_1 + \theta_2) + i\sin(\theta_1 + \theta_2)]$$

This formula says that *to multiply two complex numbers we multiply the moduli and add the arguments.* (See Figure 6.)

A similar argument using the subtraction formulas for sine and cosine shows that *to divide two complex numbers we divide the moduli and subtract the arguments.*

$$\frac{z_1}{z_2} = \frac{r_1}{r_2}[\cos(\theta_1 - \theta_2) + i\sin(\theta_1 - \theta_2)] \qquad z_2 \neq 0$$

In particular, taking $z_1 = 1$ and $z_2 = z$ (and therefore $\theta_1 = 0$ and $\theta_2 = \theta$), we have the following, which is illustrated in Figure 7.

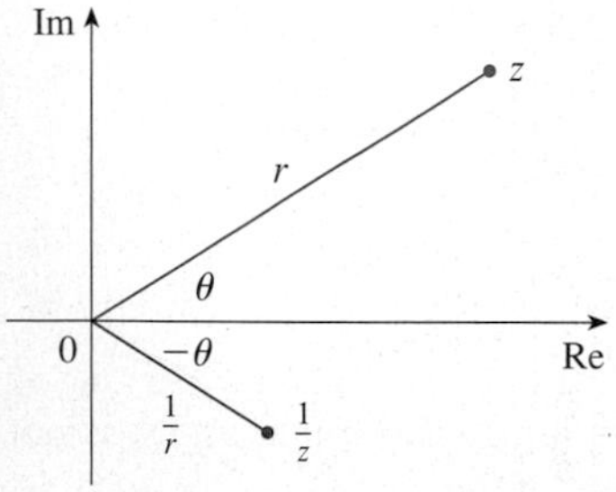

FIGURE 7

$$\text{If} \quad z = r(\cos\theta + i\sin\theta), \quad \text{then} \quad \frac{1}{z} = \frac{1}{r}(\cos\theta - i\sin\theta).$$

EXAMPLE 5 Find the product of the complex numbers $1 + i$ and $\sqrt{3} - i$ in polar form.

SOLUTION From Example 4 we have

$$1 + i = \sqrt{2}\left(\cos\frac{\pi}{4} + i\sin\frac{\pi}{4}\right)$$

and

$$\sqrt{3} - i = 2\left[\cos\left(-\frac{\pi}{6}\right) + i\sin\left(-\frac{\pi}{6}\right)\right]$$

So, by Equation 1,

$$\begin{aligned} (1 + i)(\sqrt{3} - i) &= 2\sqrt{2}\left[\cos\left(\frac{\pi}{4} - \frac{\pi}{6}\right) + i\sin\left(\frac{\pi}{4} - \frac{\pi}{6}\right)\right] \\ &= 2\sqrt{2}\left(\cos\frac{\pi}{12} + i\sin\frac{\pi}{12}\right) \end{aligned}$$

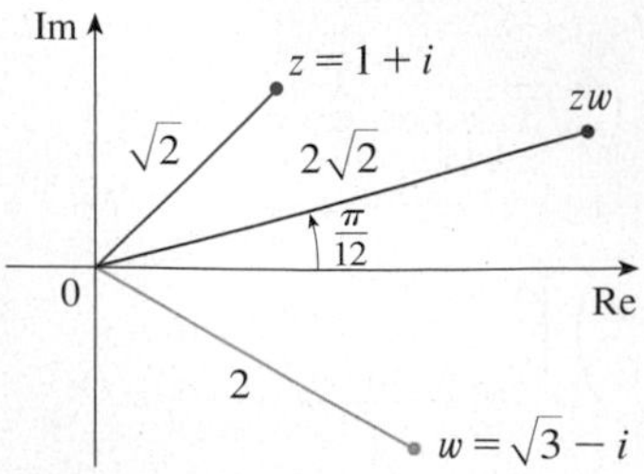

FIGURE 8

This is illustrated in Figure 8. □

Repeated use of Formula 1 shows how to compute powers of a complex number. If

$$z = r(\cos\theta + i\sin\theta)$$

then $$z^2 = r^2(\cos 2\theta + i\sin 2\theta)$$

and $$z^3 = zz^2 = r^3(\cos 3\theta + i\sin 3\theta)$$

In general, we obtain the following result, which is named after the French mathematician Abraham De Moivre (1667–1754).

2 DE MOIVRE'S THEOREM If $z = r(\cos\theta + i\sin\theta)$ and n is a positive integer, then

$$z^n = [r(\cos\theta + i\sin\theta)]^n = r^n(\cos n\theta + i\sin n\theta)$$

This says that *to take the nth power of a complex number we take the nth power of the modulus and multiply the argument by n.*

EXAMPLE 6 Find $(\frac{1}{2} + \frac{1}{2}i)^{10}$.

SOLUTION Since $\frac{1}{2} + \frac{1}{2}i = \frac{1}{2}(1 + i)$, it follows from Example 4(a) that $\frac{1}{2} + \frac{1}{2}i$ has the polar form

$$\frac{1}{2} + \frac{1}{2}i = \frac{\sqrt{2}}{2}\left(\cos\frac{\pi}{4} + i\sin\frac{\pi}{4}\right)$$

So by De Moivre's Theorem,

$$\begin{aligned}\left(\frac{1}{2} + \frac{1}{2}i\right)^{10} &= \left(\frac{\sqrt{2}}{2}\right)^{10}\left(\cos\frac{10\pi}{4} + i\sin\frac{10\pi}{4}\right)\\ &= \frac{2^5}{2^{10}}\left(\cos\frac{5\pi}{2} + i\sin\frac{5\pi}{2}\right) = \frac{1}{32}i\end{aligned}$$ □

De Moivre's Theorem can also be used to find the nth roots of complex numbers. An nth root of the complex number z is a complex number w such that

$$w^n = z$$

Writing these two numbers in trigonometric form as

$$w = s(\cos\phi + i\sin\phi) \qquad \text{and} \qquad z = r(\cos\theta + i\sin\theta)$$

and using De Moivre's Theorem, we get

$$s^n(\cos n\phi + i\sin n\phi) = r(\cos\theta + i\sin\theta)$$

The equality of these two complex numbers shows that

$$s^n = r \qquad \text{or} \qquad s = r^{1/n}$$

and $$\cos n\phi = \cos\theta \qquad \text{and} \qquad \sin n\phi = \sin\theta$$

From the fact that sine and cosine have period 2π it follows that

$$n\phi = \theta + 2k\pi \qquad \text{or} \qquad \phi = \frac{\theta + 2k\pi}{n}$$

Thus
$$w = r^{1/n}\left[\cos\left(\frac{\theta + 2k\pi}{n}\right) + i\sin\left(\frac{\theta + 2k\pi}{n}\right)\right]$$

Since this expression gives a different value of w for $k = 0, 1, 2, \ldots, n - 1$, we have the following.

3 ROOTS OF A COMPLEX NUMBER Let $z = r(\cos\theta + i\sin\theta)$ and let n be a positive integer. Then z has the n distinct nth roots

$$w_k = r^{1/n}\left[\cos\left(\frac{\theta + 2k\pi}{n}\right) + i\sin\left(\frac{\theta + 2k\pi}{n}\right)\right]$$

where $k = 0, 1, 2, \ldots, n - 1$.

Notice that each of the nth roots of z has modulus $|w_k| = r^{1/n}$. Thus all the nth roots of z lie on the circle of radius $r^{1/n}$ in the complex plane. Also, since the argument of each successive nth root exceeds the argument of the previous root by $2\pi/n$, we see that the nth roots of z are equally spaced on this circle.

EXAMPLE 7 Find the six sixth roots of $z = -8$ and graph these roots in the complex plane.

SOLUTION In trigonometric form, $z = 8(\cos\pi + i\sin\pi)$. Applying Equation 3 with $n = 6$, we get

$$w_k = 8^{1/6}\left(\cos\frac{\pi + 2k\pi}{6} + i\sin\frac{\pi + 2k\pi}{6}\right)$$

We get the six sixth roots of -8 by taking $k = 0, 1, 2, 3, 4, 5$ in this formula:

$$w_0 = 8^{1/6}\left(\cos\frac{\pi}{6} + i\sin\frac{\pi}{6}\right) = \sqrt{2}\left(\frac{\sqrt{3}}{2} + \frac{1}{2}i\right)$$

$$w_1 = 8^{1/6}\left(\cos\frac{\pi}{2} + i\sin\frac{\pi}{2}\right) = \sqrt{2}\,i$$

$$w_2 = 8^{1/6}\left(\cos\frac{5\pi}{6} + i\sin\frac{5\pi}{6}\right) = \sqrt{2}\left(-\frac{\sqrt{3}}{2} + \frac{1}{2}i\right)$$

$$w_3 = 8^{1/6}\left(\cos\frac{7\pi}{6} + i\sin\frac{7\pi}{6}\right) = \sqrt{2}\left(-\frac{\sqrt{3}}{2} - \frac{1}{2}i\right)$$

$$w_4 = 8^{1/6}\left(\cos\frac{3\pi}{2} + i\sin\frac{3\pi}{2}\right) = -\sqrt{2}\,i$$

$$w_5 = 8^{1/6}\left(\cos\frac{11\pi}{6} + i\sin\frac{11\pi}{6}\right) = \sqrt{2}\left(\frac{\sqrt{3}}{2} - \frac{1}{2}i\right)$$

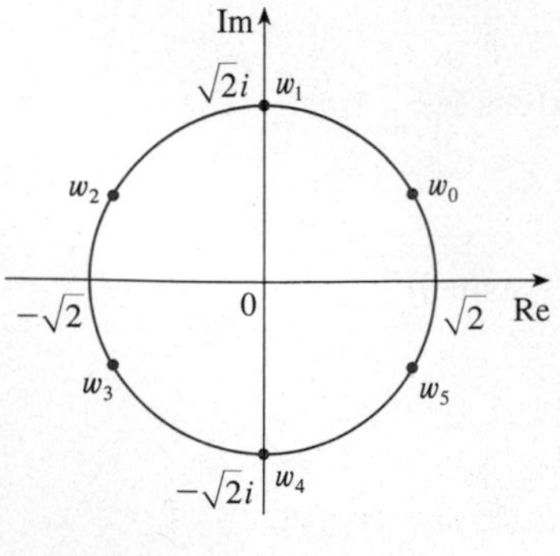

FIGURE 9
The six sixth roots of $z = -8$

All these points lie on the circle of radius $\sqrt{2}$ as shown in Figure 9. □

COMPLEX EXPONENTIALS

We also need to give a meaning to the expression e^z when $z = x + iy$ is a complex number. The theory of infinite series as developed in Chapter 11 can be extended to the case where the terms are complex numbers. Using the Taylor series for e^x (11.10.11) as our guide, we define

4 $$e^z = \sum_{n=0}^{\infty} \frac{z^n}{n!} = 1 + z + \frac{z^2}{2!} + \frac{z^3}{3!} + \cdots$$

and it turns out that this complex exponential function has the same properties as the real exponential function. In particular, it is true that

5 $$e^{z_1+z_2} = e^{z_1}e^{z_2}$$

If we put $z = iy$, where y is a real number, in Equation 4, and use the facts that

$$i^2 = -1, \quad i^3 = i^2 i = -i, \quad i^4 = 1, \quad i^5 = i, \quad \ldots$$

we get
$$\begin{aligned} e^{iy} &= 1 + iy + \frac{(iy)^2}{2!} + \frac{(iy)^3}{3!} + \frac{(iy)^4}{4!} + \frac{(iy)^5}{5!} + \cdots \\ &= 1 + iy - \frac{y^2}{2!} - i\frac{y^3}{3!} + \frac{y^4}{4!} + i\frac{y^5}{5!} + \cdots \\ &= \left(1 - \frac{y^2}{2!} + \frac{y^4}{4!} - \frac{y^6}{6!} + \cdots\right) + i\left(y - \frac{y^3}{3!} + \frac{y^5}{5!} - \cdots\right) \\ &= \cos y + i \sin y \end{aligned}$$

Here we have used the Taylor series for $\cos y$ and $\sin y$ (Equations 11.10.16 and 11.10.15). The result is a famous formula called **Euler's formula**:

6 $$e^{iy} = \cos y + i \sin y$$

Combining Euler's formula with Equation 5, we get

7 $$e^{x+iy} = e^x e^{iy} = e^x(\cos y + i \sin y)$$

EXAMPLE 8 Evaluate: (a) $e^{i\pi}$ (b) $e^{-1+i\pi/2}$

SOLUTION

■ We could write the result of Example 8(a) as

$$e^{i\pi} + 1 = 0$$

This equation relates the five most famous numbers in all of mathematics: 0, 1, e, i, and π.

(a) From Euler's equation (6) we have

$$e^{i\pi} = \cos \pi + i \sin \pi = -1 + i(0) = -1$$

(b) Using Equation 7 we get

$$e^{-1+i\pi/2} = e^{-1}\left(\cos \frac{\pi}{2} + i \sin \frac{\pi}{2}\right) = \frac{1}{e}[0 + i(1)] = \frac{i}{e}$$ □

Finally, we note that Euler's equation provides us with an easier method of proving De Moivre's Theorem:

$$[r(\cos \theta + i \sin \theta)]^n = (re^{i\theta})^n = r^n e^{in\theta} = r^n(\cos n\theta + i \sin n\theta)$$

H EXERCISES

1–14 Evaluate the expression and write your answer in the form $a + bi$.

1. $(5 - 6i) + (3 + 2i)$

2. $\left(4 - \frac{1}{2}i\right) - \left(9 + \frac{5}{2}i\right)$

3. $(2 + 5i)(4 - i)$

4. $(1 - 2i)(8 - 3i)$

5. $\overline{12 + 7i}$

6. $\overline{2i\left(\frac{1}{2} - i\right)}$

7. $\dfrac{1 + 4i}{3 + 2i}$

8. $\dfrac{3 + 2i}{1 - 4i}$

9. $\dfrac{1}{1 + i}$

10. $\dfrac{3}{4 - 3i}$

11. i^3

12. i^{100}

13. $\sqrt{-25}$

14. $\sqrt{-3}\sqrt{-12}$

15–17 Find the complex conjugate and the modulus of the number.

15. $12 - 5i$

16. $-1 + 2\sqrt{2}\,i$

17. $-4i$

18. Prove the following properties of complex numbers.
(a) $\overline{z + w} = \overline{z} + \overline{w}$ (b) $\overline{zw} = \overline{z}\,\overline{w}$
(c) $\overline{z^n} = \overline{z}^n$, where n is a positive integer
[*Hint:* Write $z = a + bi$, $w = c + di$.]

19–24 Find all solutions of the equation.

19. $4x^2 + 9 = 0$

20. $x^4 = 1$

21. $x^2 + 2x + 5 = 0$

22. $2x^2 - 2x + 1 = 0$

23. $z^2 + z + 2 = 0$

24. $z^2 + \frac{1}{2}z + \frac{1}{4} = 0$

25–28 Write the number in polar form with argument between 0 and 2π.

25. $-3 + 3i$

26. $1 - \sqrt{3}\,i$

27. $3 + 4i$

28. $8i$

29–32 Find polar forms for zw, z/w, and $1/z$ by first putting z and w into polar form.

29. $z = \sqrt{3} + i$, $w = 1 + \sqrt{3}\,i$

30. $z = 4\sqrt{3} - 4i$, $w = 8i$

31. $z = 2\sqrt{3} - 2i$, $w = -1 + i$

32. $z = 4(\sqrt{3} + i)$, $w = -3 - 3i$

33–36 Find the indicated power using De Moivre's Theorem.

33. $(1 + i)^{20}$

34. $(1 - \sqrt{3}\,i)^5$

35. $(2\sqrt{3} + 2i)^5$

36. $(1 - i)^8$

37–40 Find the indicated roots. Sketch the roots in the complex plane.

37. The eighth roots of 1

38. The fifth roots of 32

39. The cube roots of i

40. The cube roots of $1 + i$

41–46 Write the number in the form $a + bi$.

41. $e^{i\pi/2}$

42. $e^{2\pi i}$

43. $e^{i\pi/3}$

44. $e^{-i\pi}$

45. $e^{2+i\pi}$

46. $e^{\pi+i}$

47. Use De Moivre's Theorem with $n = 3$ to express $\cos 3\theta$ and $\sin 3\theta$ in terms of $\cos\theta$ and $\sin\theta$.

48. Use Euler's formula to prove the following formulas for $\cos x$ and $\sin x$:

$$\cos x = \frac{e^{ix} + e^{-ix}}{2} \qquad \sin x = \frac{e^{ix} - e^{-ix}}{2i}$$

49. If $u(x) = f(x) + ig(x)$ is a complex-valued function of a real variable x and the real and imaginary parts $f(x)$ and $g(x)$ are differentiable functions of x, then the derivative of u is defined to be $u'(x) = f'(x) + ig'(x)$. Use this together with Equation 7 to prove that if $F(x) = e^{rx}$, then $F'(x) = re^{rx}$ when $r = a + bi$ is a complex number.

50. (a) If u is a complex-valued function of a real variable, its indefinite integral $\int u(x)\,dx$ is an antiderivative of u. Evaluate

$$\int e^{(1+i)x}\,dx$$

(b) By considering the real and imaginary parts of the integral in part (a), evaluate the real integrals

$$\int e^x \cos x\,dx \qquad \text{and} \qquad \int e^x \sin x\,dx$$

(c) Compare with the method used in Example 4 in Section 7.1.

I ANSWERS TO ODD-NUMBERED EXERCISES

CHAPTER 1

EXERCISES 1.1 ■ PAGE 20

1. (a) -2 (b) 2.8 (c) $-3, 1$ (d) $-2.5, 0.3$
(e) $[-3, 3], [-2, 3]$ (f) $[-1, 3]$

3. $[-85, 115]$ **5.** No

7. Yes, $[-3, 2], [-3, -2) \cup [-1, 3]$

9. Diet, exercise, or illness

11.

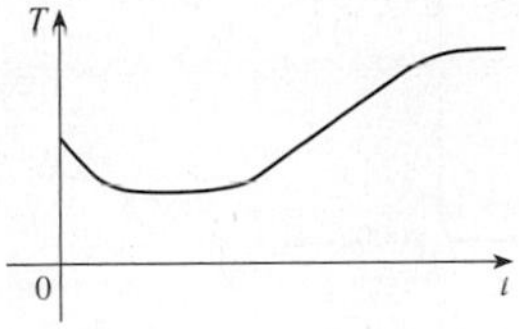

13.

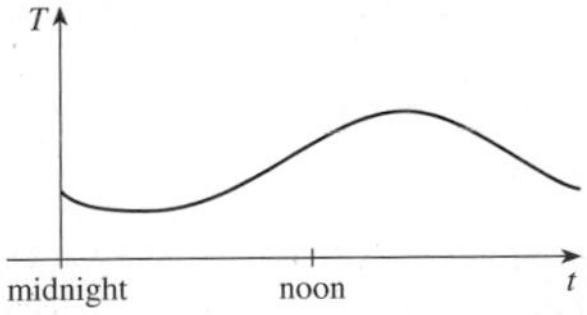

15.

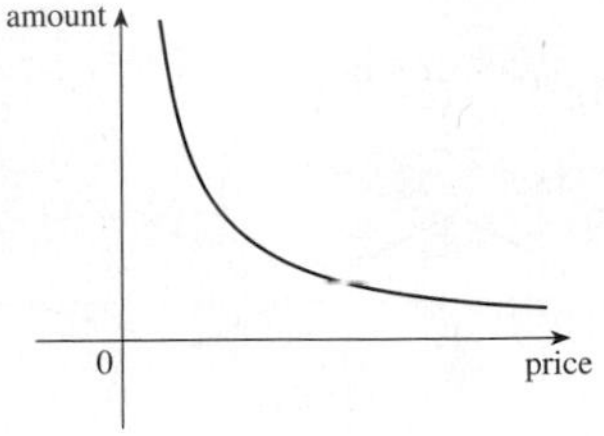

17.

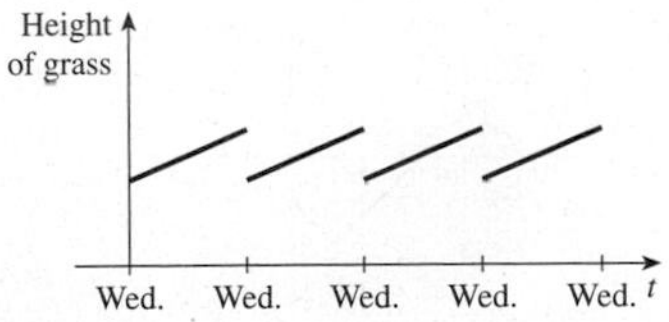

19. (a)

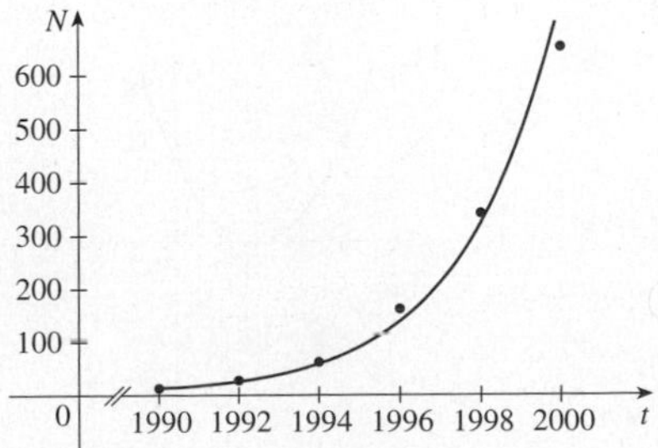

(b) In millions: 92; 485

21. 12, 16, $3a^2 - a + 2$, $3a^2 + a + 2$, $3a^2 + 5a + 4$, $6a^2 - 2a + 4$, $12a^2 - 2a + 2$, $3a^4 - a^2 + 2$, $9a^4 - 6a^3 + 13a^2 - 4a + 4$, $3a^2 + 6ah + 3h^2 - a - h + 2$

23. $-3 - h$ **25.** $-1/(ax)$

27. $\{x \mid x \neq \frac{1}{3}\} = (-\infty, \frac{1}{3}) \cup (\frac{1}{3}, \infty)$

29. $[0, \infty)$ **31.** $(-\infty, 0) \cup (5, \infty)$

33. $(-\infty, \infty)$

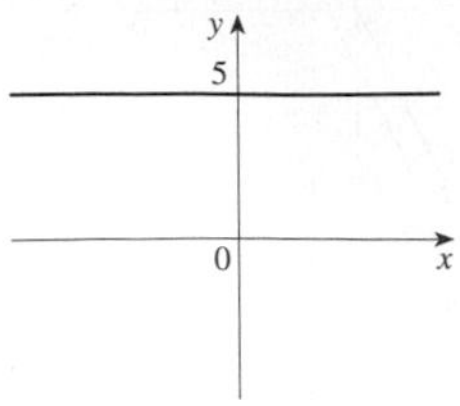

35. $(-\infty, \infty)$

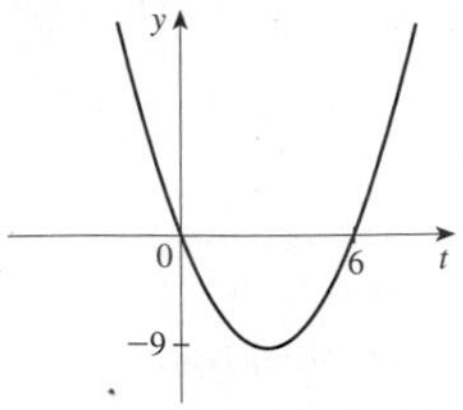

37. $[5, \infty)$

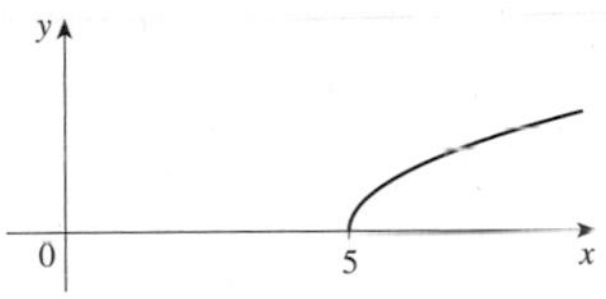

39. $(-\infty, 0) \cup (0, \infty)$

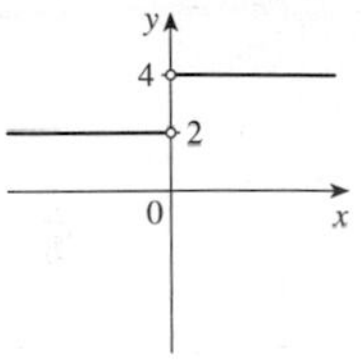

41. $(-\infty, \infty)$

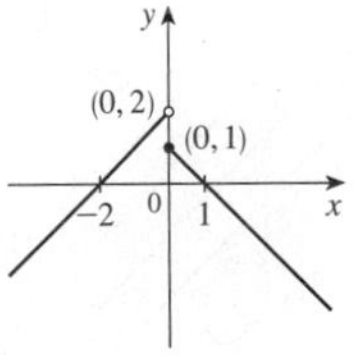

43. $(-\infty, \infty)$

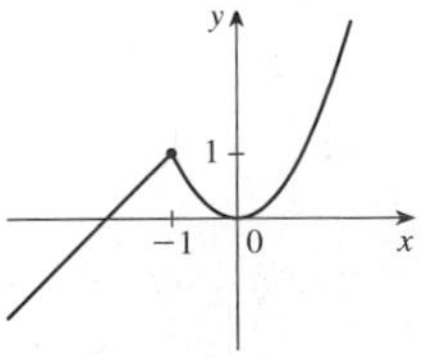

45. $f(x) = \frac{5}{2}x - \frac{11}{2}, 1 \leqslant x \leqslant 5$ **47.** $f(x) = 1 - \sqrt{-x}$

49. $f(x) = \begin{cases} -x + 3 & \text{if } 0 \leqslant x \leqslant 3 \\ 2x - 6 & \text{if } 3 < x \leqslant 5 \end{cases}$

51. $A(L) = 10L - L^2, 0 < L < 10$

53. $A(x) = \sqrt{3}x^2/4, x > 0$ **55.** $S(x) = x^2 + (8/x), x > 0$

57. $V(x) = 4x^3 - 64x^2 + 240x, 0 < x < 6$

59. (a)

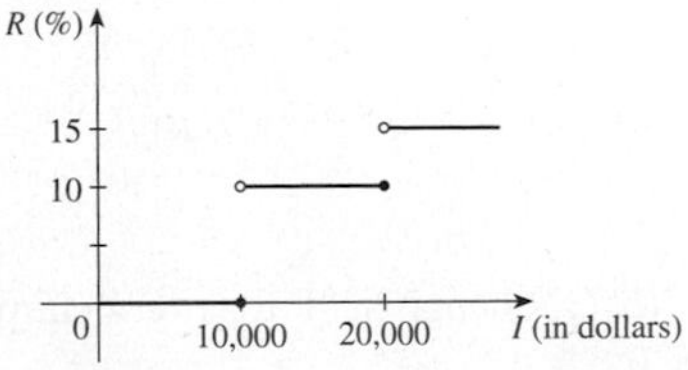

(b) \$400, \$1900

(c)

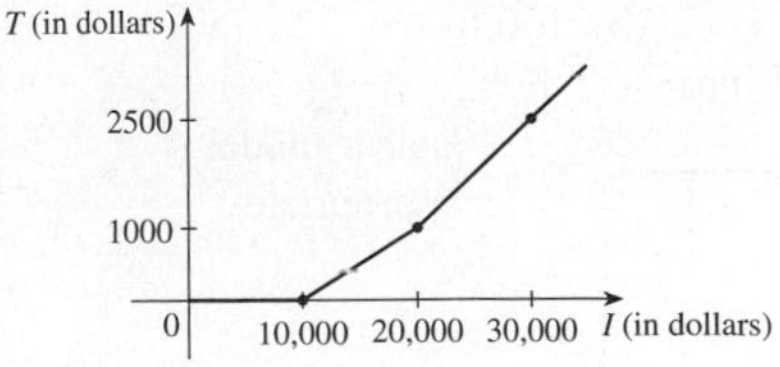

61. f is odd, g is even

63. (a) $(-5, 3)$ (b) $(-5, -3)$

65. Odd **67.** Neither **69.** Even

EXERCISES 1.2 ■ PAGE 34

1. (a) Root (b) Algebraic (c) Polynomial (degree 9)
(d) Rational (e) Trigonometric (f) Logarithmic

3. (a) h (b) f (c) g

5. (a) $y = 2x + b$, where b is the y-intercept.

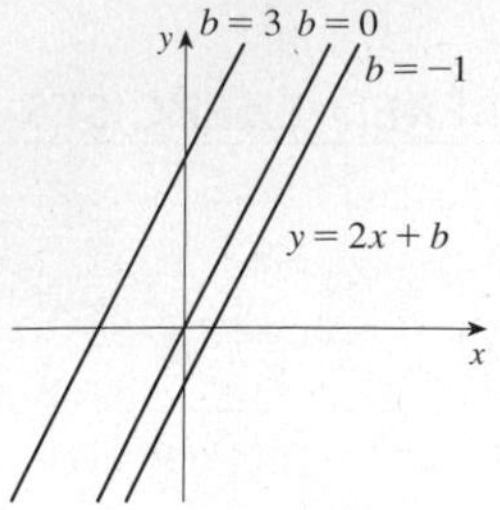

(b) $y = mx + 1 - 2m$, where m is the slope. See graph at right.
(c) $y = 2x - 3$

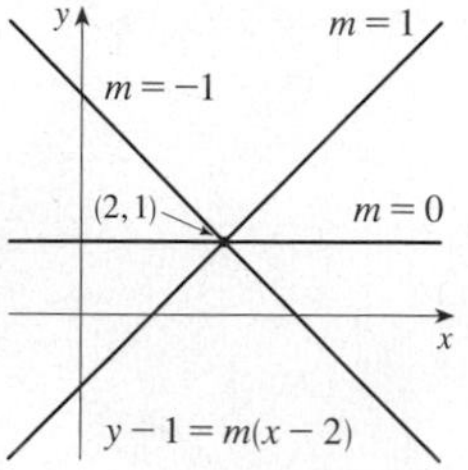

7. Their graphs have slope -1.

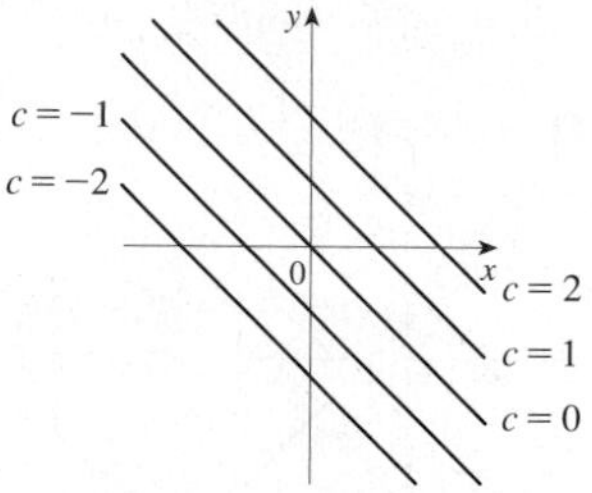

9. $f(x) = -3x(x + 1)(x - 2)$

11. (a) 8.34, change in mg for every 1 year change (b) 8.34 mg

13. (a)

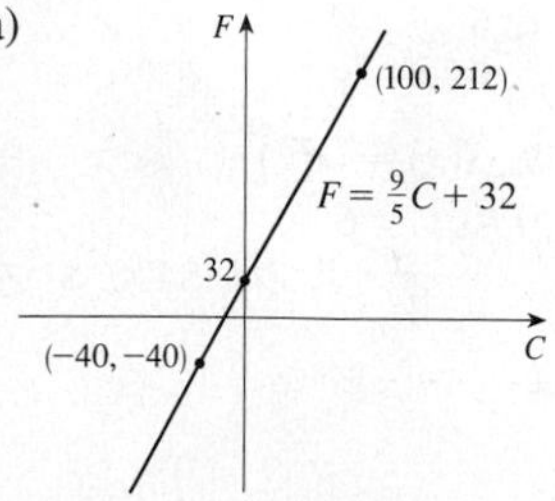

(b) $\frac{9}{5}$, change in °F for every 1°C change; 32, Fahrenheit temperature corresponding to 0°C

15. (a) $T = \frac{1}{6}N + \frac{307}{6}$ (b) $\frac{1}{6}$, change in °F for every chirp per minute change (c) 76°F

17. (a) $P = 0.434d + 15$ (b) 196 ft

19. (a) Cosine (b) Linear

21. (a)

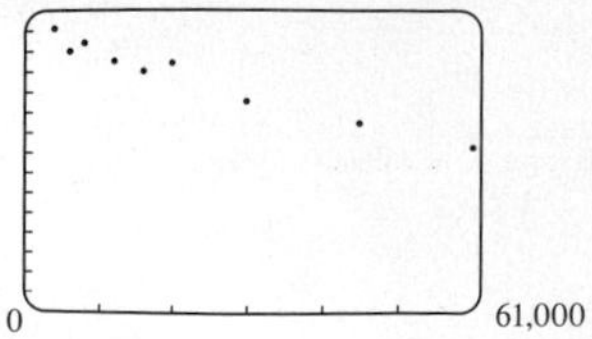

Linear model is appropriate

(b) $y = -0.000105x + 14.521$

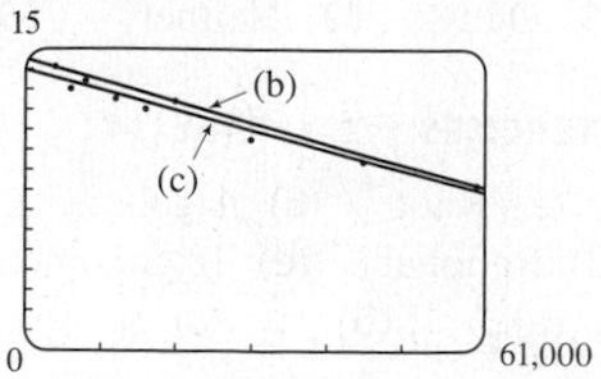

(c) $y = -0.00009979x + 13.951$ [See graph in (b).]
(d) About 11.5 per 100 population (e) About 6% (f) No

23. (a)

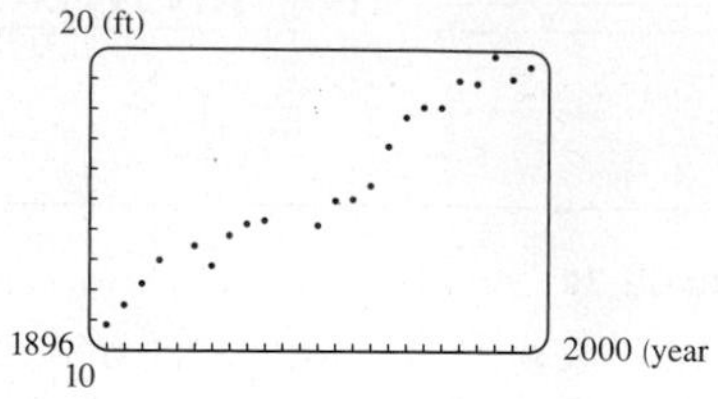

Linear model is appropriate
(b) $y = 0.08912x - 158.24$ (c) 20 ft (d) No

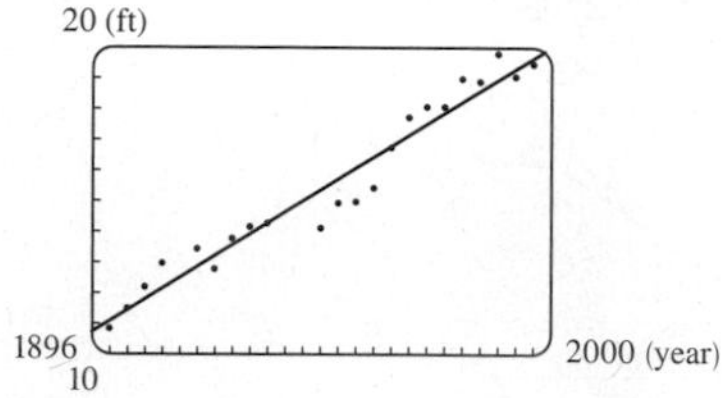

25. $y \approx 0.0012937x^3 - 7.06142x^2 + 12{,}823x - 7{,}743{,}770$; 1914 million

EXERCISES 1.3 ▪ PAGE 43

1. (a) $y = f(x) + 3$ (b) $y = f(x) - 3$ (c) $y = f(x - 3)$
(d) $y = f(x + 3)$ (e) $y = -f(x)$ (f) $y = f(-x)$
(g) $y = 3f(x)$ (h) $y = \frac{1}{3}f(x)$

3. (a) 3 (b) 1 (c) 4 (d) 5 (e) 2

5. (a)

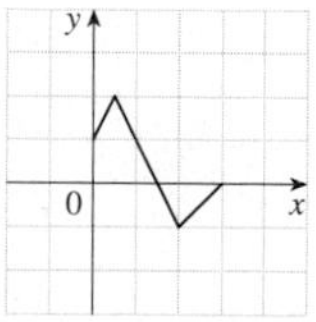

(b)

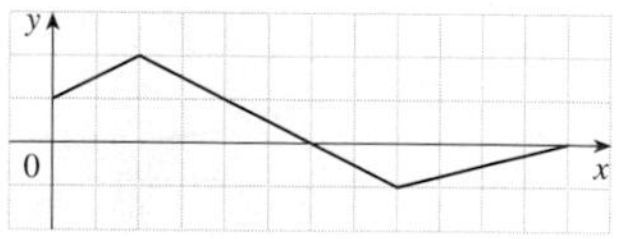

(c)

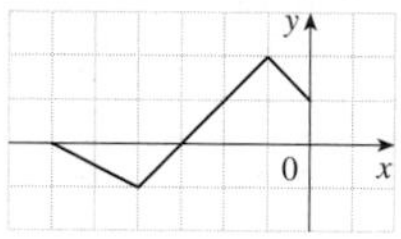

(d)

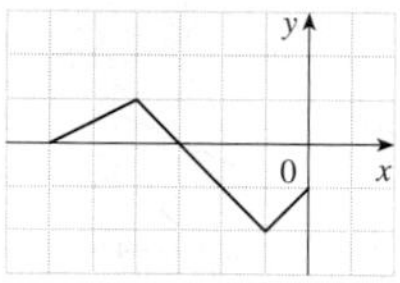

7. $y = -\sqrt{-x^2 - 5x - 4} - 1$

9.

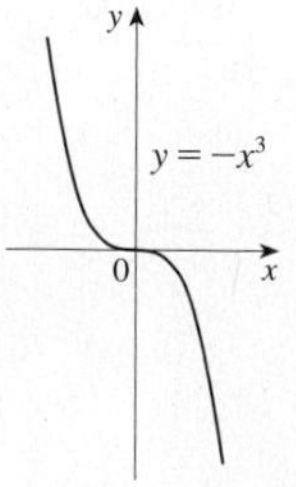

11.

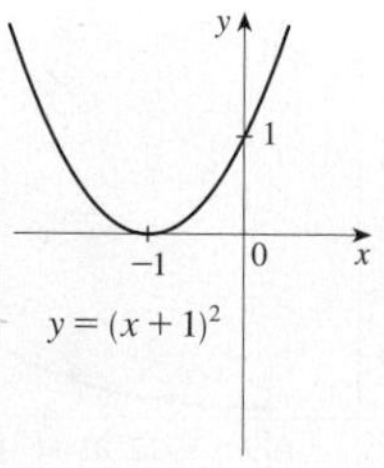

13.

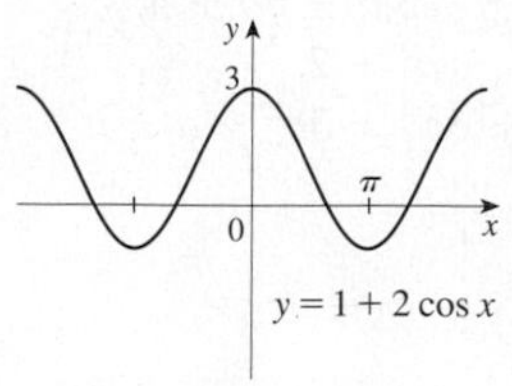

15.

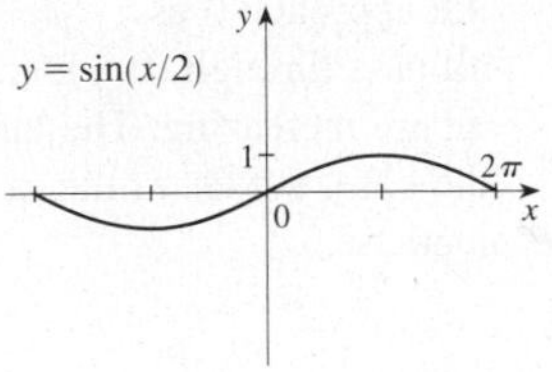

17.

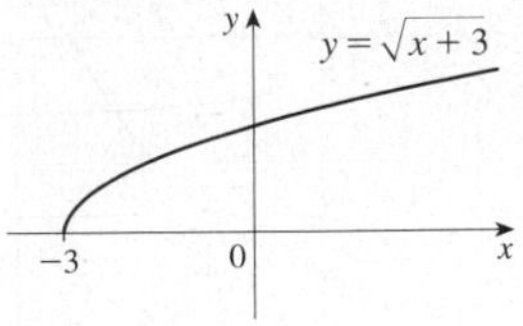

19.

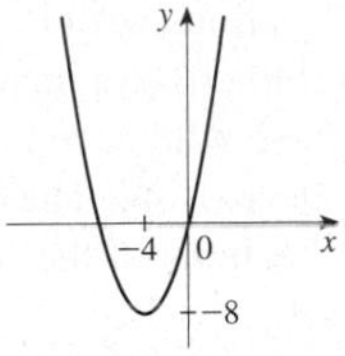

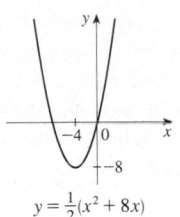

$y = \frac{1}{2}(x^2 + 8x)$

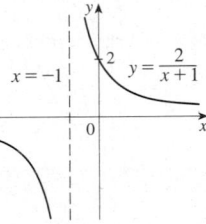

21.

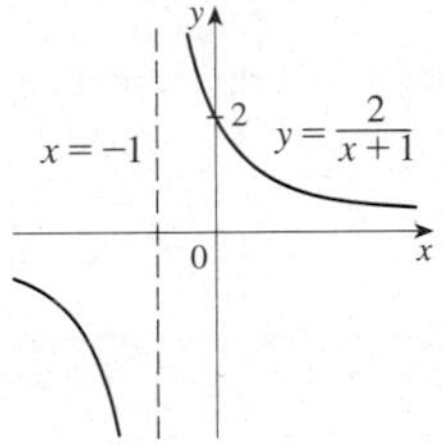

23.

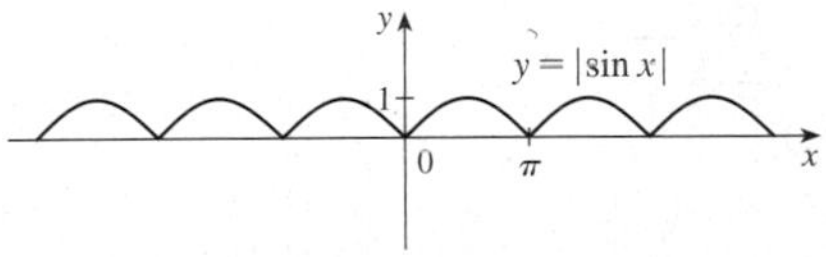

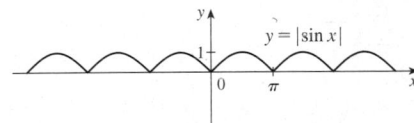

25. $L(t) = 12 + 2\sin\left[\dfrac{2\pi}{365}(t - 80)\right]$

27. (a) The portion of the graph of $y = f(x)$ to the right of the y-axis is reflected about the y-axis.

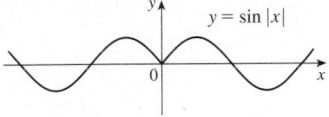

(b)

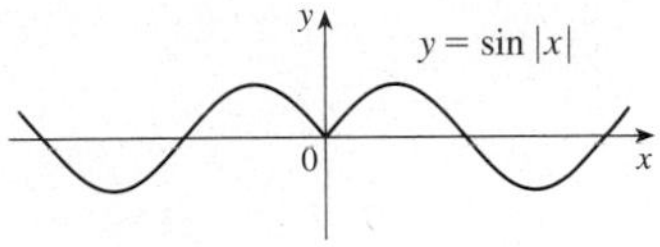

(c)

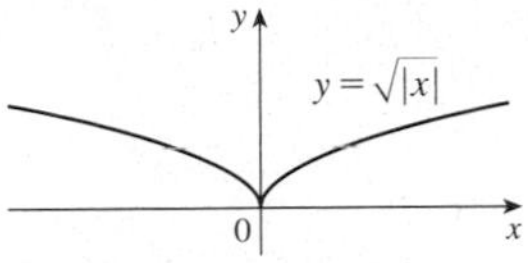

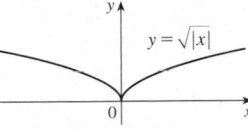

29. $(f + g)(x) = x^3 + 5x^2 - 1,\ (-\infty, \infty)$
$(f - g)(x) = x^3 - x^2 + 1,\ (-\infty, \infty)$
$(fg)(x) = 3x^5 + 6x^4 - x^3 - 2x^2,\ (-\infty, \infty)$
$(f/g)(x) = (x^3 + 2x^2)/(3x^2 - 1),\ \{x \mid x \neq \pm 1/\sqrt{3}\}$

31. (a) $(f \circ g)(x) = 4x^2 + 4x,\ (-\infty, \infty)$
(b) $(g \circ f)(x) = 2x^2 - 1,\ (-\infty, \infty)$
(c) $(f \circ f)(x) = x^4 - 2x^2,\ (-\infty, \infty)$
(d) $(g \circ g)(x) = 4x + 3,\ (-\infty, \infty)$

33. (a) $(f \circ g)(x) = 1 - 3\cos x,\ (-\infty, \infty)$
(b) $(g \circ f)(x) = \cos(1 - 3x),\ (-\infty, \infty)$
(c) $(f \circ f)(x) = 9x - 2,\ (-\infty, \infty)$
(d) $(g \circ g)(x) = \cos(\cos x),\ (-\infty, \infty)$

35. (a) $(f \circ g)(x) = (2x^2 + 6x + 5)/[(x + 2)(x + 1)]$, $\{x \mid x \neq -2, -1\}$
(b) $(g \circ f)(x) = (x^2 + x + 1)/(x + 1)^2,\ \{x \mid x \neq -1, 0\}$
(c) $(f \circ f)(x) = (x^4 + 3x^2 + 1)/[x(x^2 + 1)],\ \{x \mid x \neq 0\}$
(d) $(g \circ g)(x) = (2x + 3)/(3x + 5),\ \{x \mid x \neq -2, -\frac{5}{3}\}$

37. $(f \circ g \circ h)(x) = 2x - 1$

39. $(f \circ g \circ h)(x) = \sqrt{x^6 + 4x^3 + 1}$

41. $g(x) = x^2 + 1, f(x) = x^{10}$

43. $g(x) = \sqrt[3]{x},\ f(x) = x/(1 + x)$

45. $g(t) = \cos t,\ f(t) = \sqrt{t}$

47. $h(x) = x^2,\ g(x) = 3^x,\ f(x) = 1 - x$

49. $h(x) = \sqrt{x},\ g(x) = \sec x,\ f(x) = x^4$

51. (a) 4 (b) 3 (c) 0 (d) Does not exist; $f(6) = 6$ is not in the domain of g. (e) 4 (f) −2

53. (a) $r(t) = 60t$ (b) $(A \circ r)(t) = 3600\pi t^2$; the area of the circle as a function of time

55. (a) $s = \sqrt{d^2 + 36}$ (b) $d = 30t$
(c) $s = \sqrt{900t^2 + 36}$; the distance between the lighthouse and the ship as a function of the time elapsed since noon

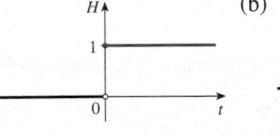

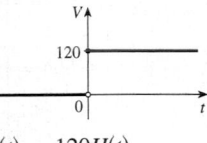

57. (a)

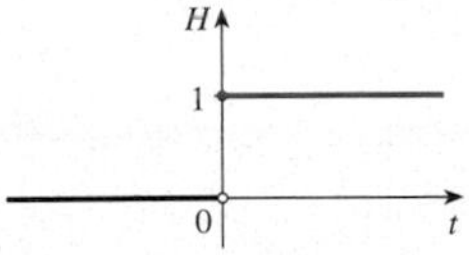

(b)

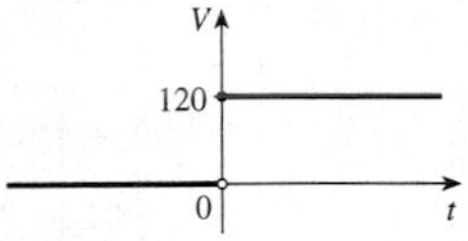

$V(t) = 120H(t)$

(c)

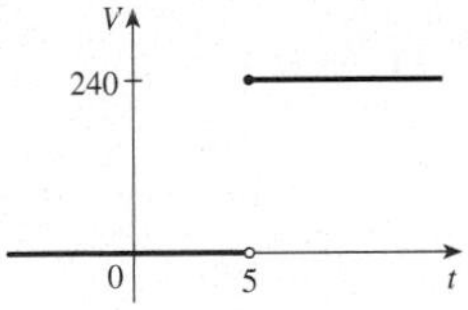

$V(t) = 240H(t - 5)$

59. Yes; $m_1 m_2$

61. (a) $f(x) = x^2 + 6$ (b) $g(x) = x^2 + x - 1$

63. (a) Even; even (b) Odd; even

65. Yes

EXERCISES 1.4 ▪ PAGE 51

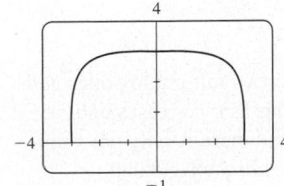

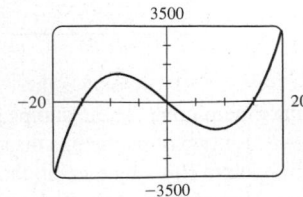

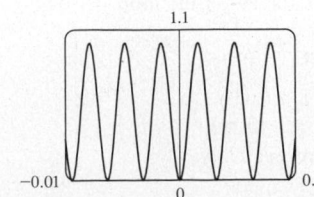

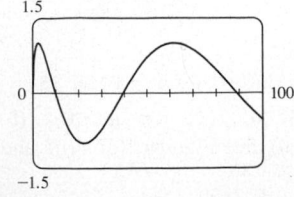

1. (c) **3.**

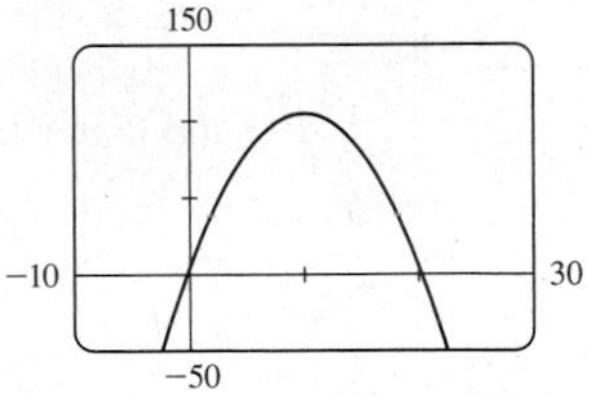

5.

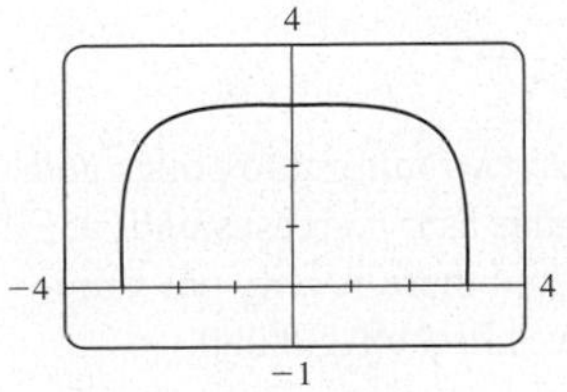

7.

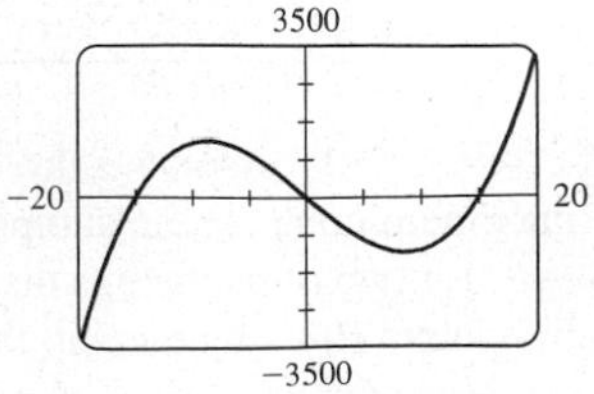

9.

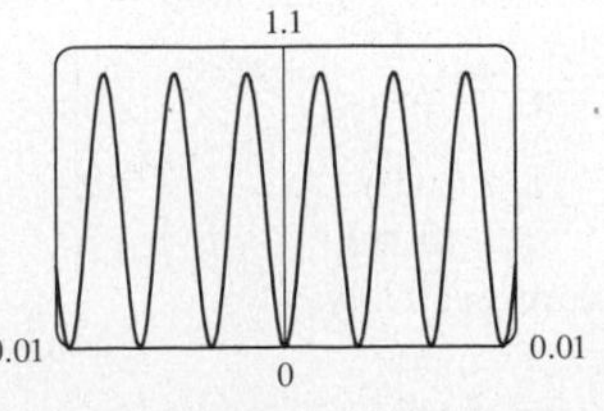

11.

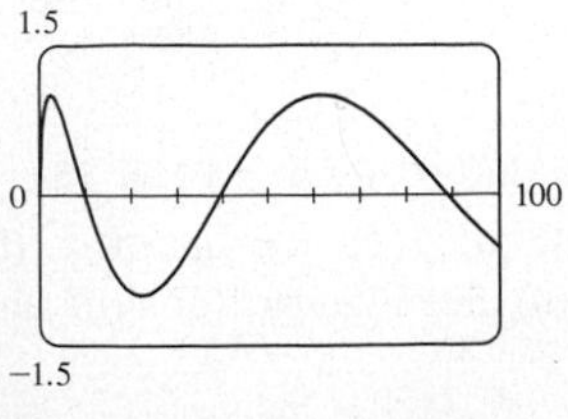

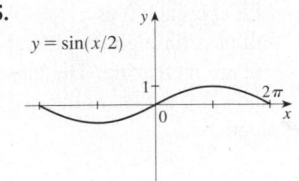

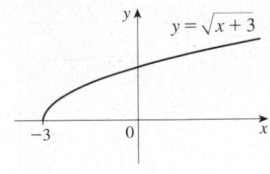

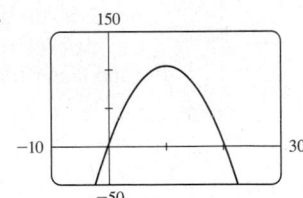

13.

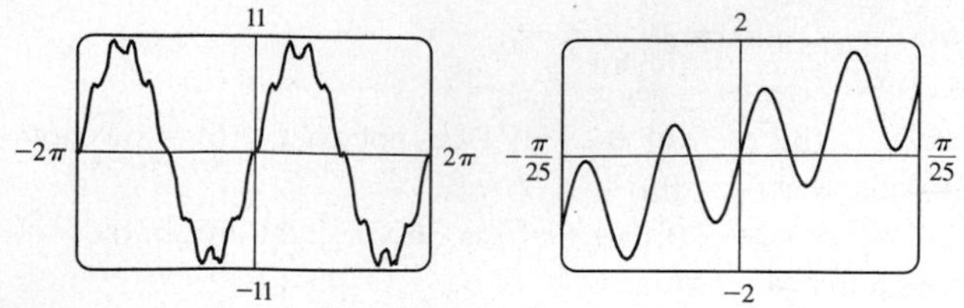

15.

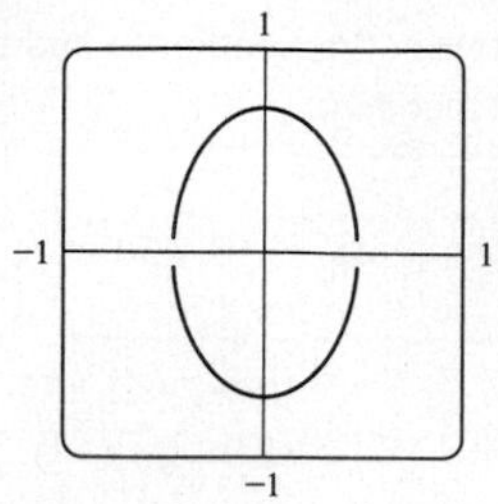

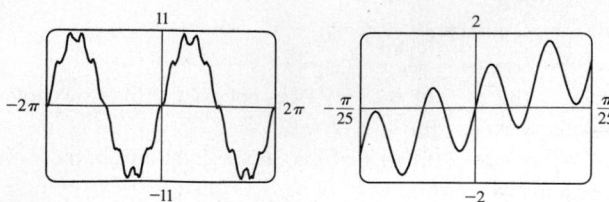

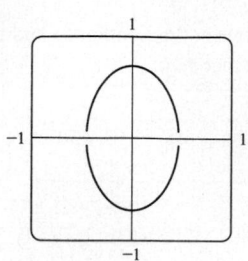

17. No **19.** 9.05 **21.** 0, 0.88 **23.** g

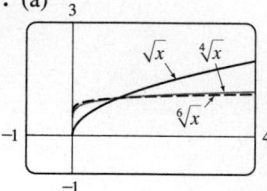

25. $-0.85 < x < 0.85$

27. (a)

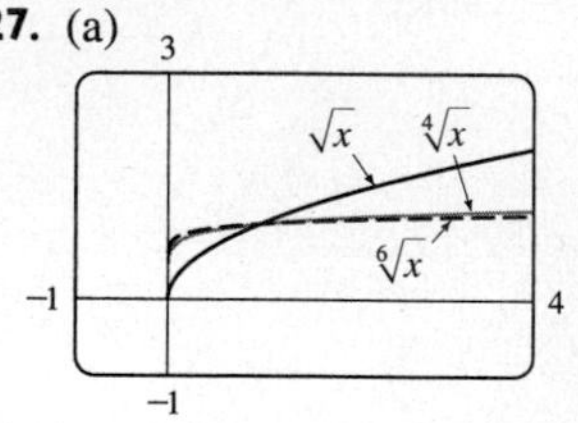

(b)

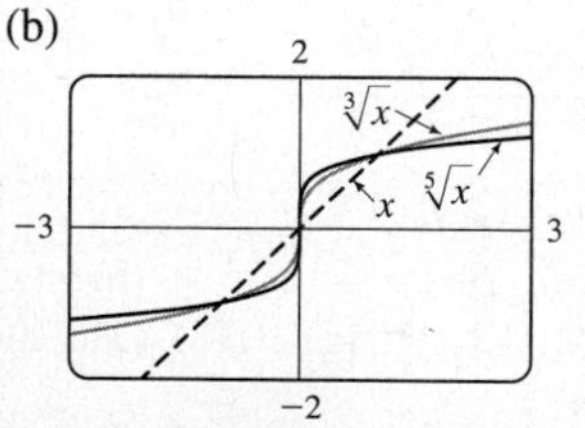

(c)

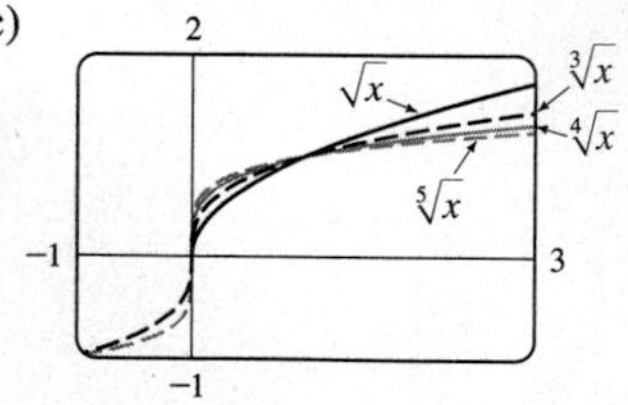

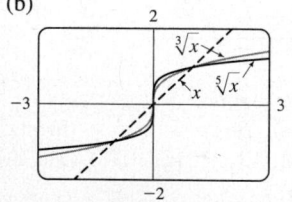

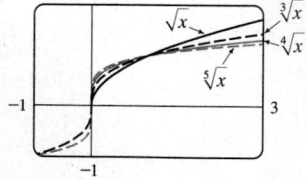

(d) Graphs of even roots are similar to $\sqrt{x}$, graphs of odd roots are similar to $\sqrt[3]{x}$. As n increases, the graph of $y = \sqrt[n]{x}$ becomes steeper near 0 and flatter for $x > 1$.

29.

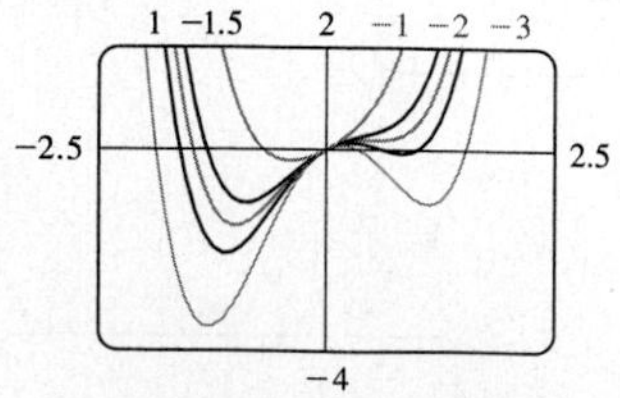

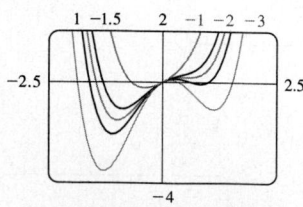

If $c < -1.5$, the graph has three humps: two minimum points and a maximum point. These humps get flatter as c increases until at $c = -1.5$ two of the humps disappear and there is only one minimum point. This single hump then moves to the right and approaches the origin as c increases.

31. The hump gets larger and moves to the right.

33. If $c < 0$, the loop is to the right of the origin; if $c > 0$, the loop is to the left. The closer c is to 0, the larger the loop.

EXERCISES 1.5 ▪ PAGE 58

1. (a) $f(x) = a^x, a > 0$ (b) $\mathbb{R}$ (c) $(0, \infty)$
(d) See Figures 4(c), 4(b), and 4(a), respectively.

3.

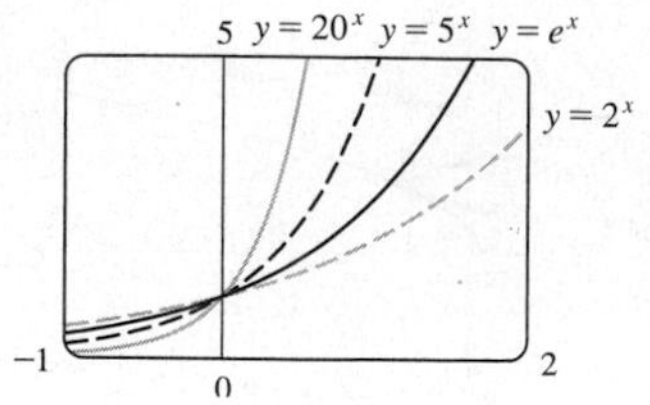

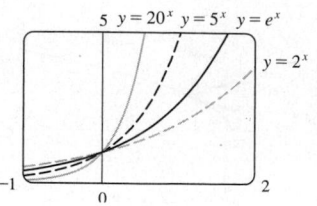

All approach 0 as $x \to -\infty$, all pass through (0, 1), and all are increasing. The larger the base, the faster the rate of increase.

5.

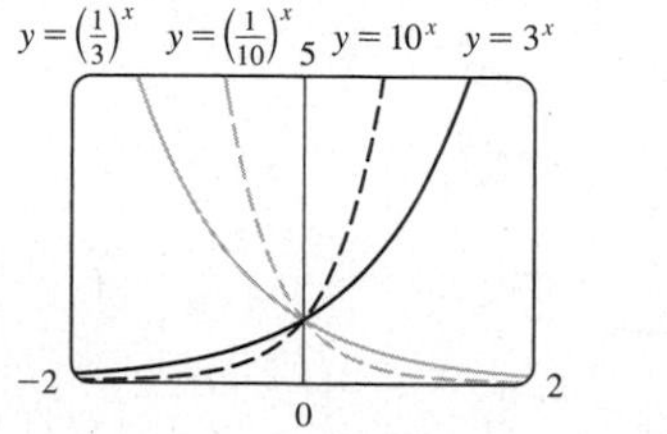

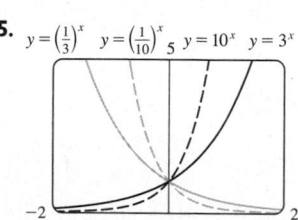

The functions with base greater than 1 are increasing and those with base less than 1 are decreasing. The latter are reflections of the former about the y-axis.

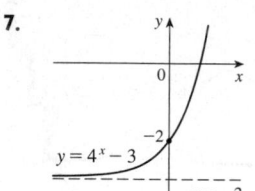

7.

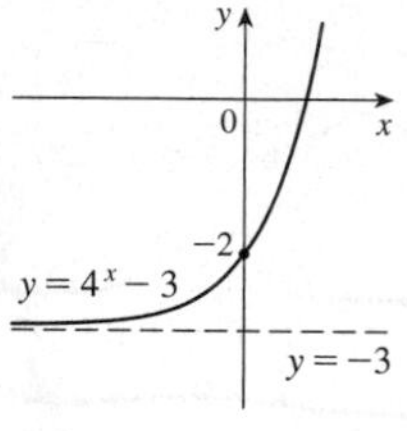

9.

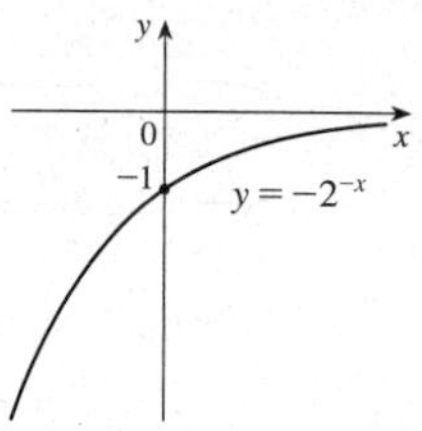

11.

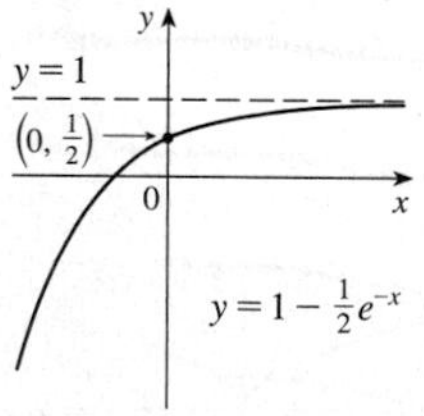

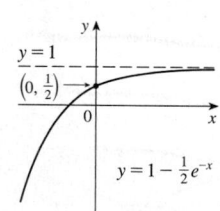

13. (a) $y = e^x - 2$ (b) $y = e^{x-2}$ (c) $y = -e^x$
(d) $y = e^{-x}$ (e) $y = -e^{-x}$

15. (a) $(-\infty, \infty)$ (b) $(-\infty, 0) \cup (0, \infty)$

17. $f(x) = 3 \cdot 2^x$ **23.** At $x \approx 35.8$

25. (a) 3200 (b) $100 \cdot 2^{t/3}$ (c) 10,159
(d)

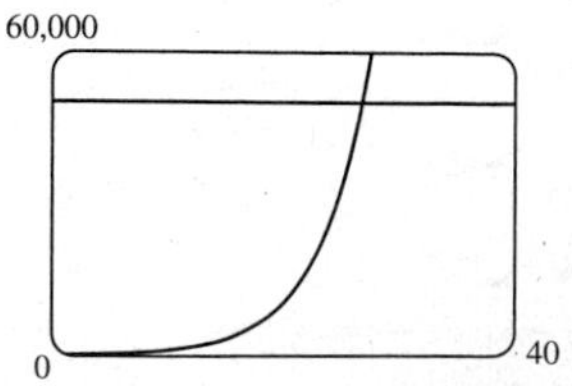

$t \approx 26.9$ h

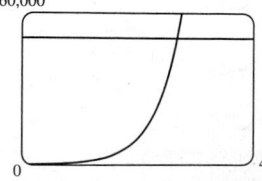

27. $y = ab^t$, where $a \approx 3.154832569 \times 10^{-12}$ and $b \approx 1.017764706$; 5498 million; 7417 million

EXERCISES 1.6 ▪ PAGE 70

1. (a) See Definition 1.
(b) It must pass the Horizontal Line Test.

3. No **5.** Yes **7.** No **9.** No **11.** Yes

13. No **15.** 2 **17.** 0

19. $F = \frac{9}{5}C + 32$; the Fahrenheit temperature as a function of the Celsius temperature; $[-273.15, \infty)$

21. $f^{-1}(x) = -\frac{1}{3}x^2 + \frac{10}{3}, \ x \geqslant 0$ **23.** $f^{-1}(x) = \sqrt[3]{\ln x}$

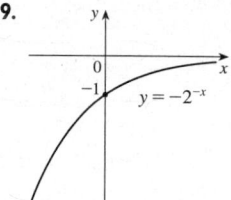

25. $y = e^x - 3$

27. $f^{-1}(x) = \sqrt[4]{x - 1}$

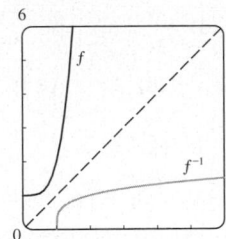

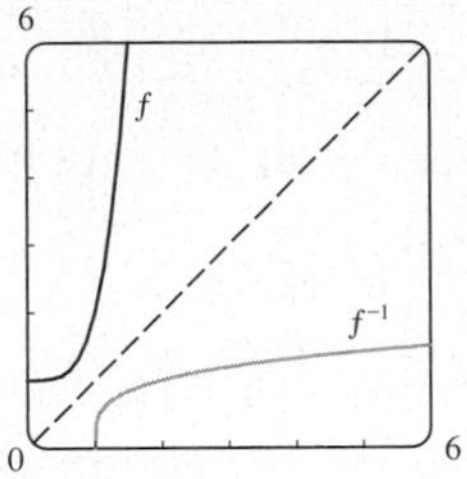

29.

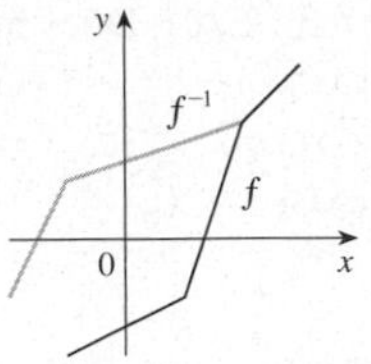

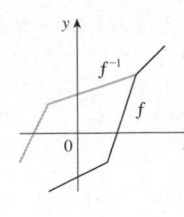

31. (a) It's defined as the inverse of the exponential function with base a, that is, $\log_a x = y \iff a^y = x$.
(b) $(0, \infty)$ (c) $\mathbb{R}$ (d) See Figure 11.

33. (a) 3 (b) -3 **35.** (a) 3 (b) -2 **37.** $\ln 1215$

39. $\ln \dfrac{(1 + x^2)\sqrt{x}}{\sin x}$

41.

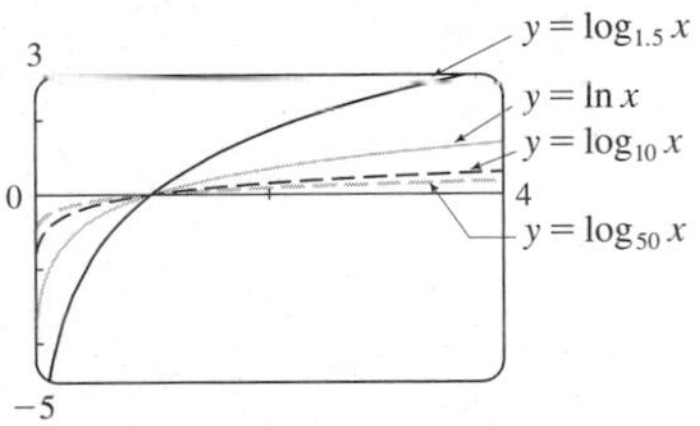

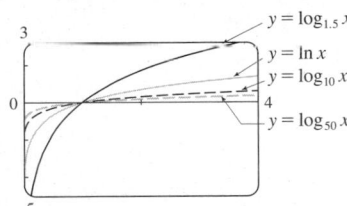

All graphs approach $-\infty$ as $x \to 0^+$, all pass through (1, 0), and all are increasing. The larger the base, the slower the rate of increase.

43. About 1,084,588 mi

45. (a)

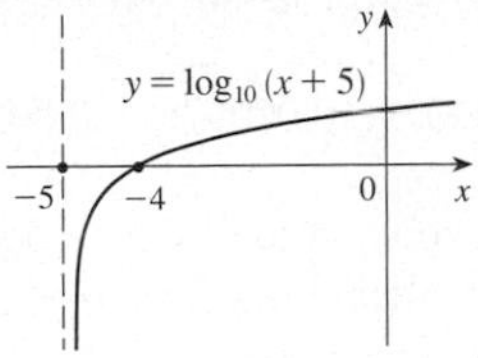

(b)

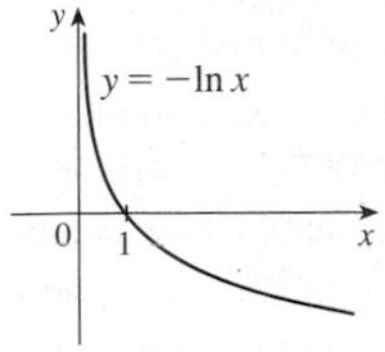

47. (a) $\sqrt{e}$ (b) $-\ln 5$

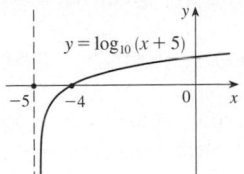

49. (a) $5 + \log_2 3$ or $5 + (\ln 3)/\ln 2$ (b) $\frac{1}{2}(1 + \sqrt{1 + 4e})$

51. (a) $x < \ln 10$ (b) $x > 1/e$

53. (a) $\left(-\infty, \frac{1}{2}\ln 3\right]$ (b) $f^{-1}(x) = \frac{1}{2}\ln(3 - x^2)$, $\left[0, \sqrt{3}\right)$

55.

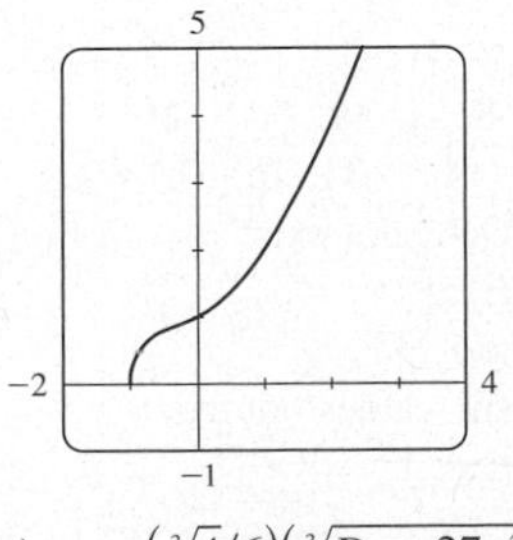

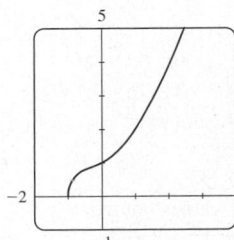

The graph passes the Horizontal Line Test.

$f^{-1}(x) = -\left(\sqrt[3]{4}/6\right)\left(\sqrt[3]{D - 27x^2 + 20} - \sqrt[3]{D + 27x^2 - 20} + \sqrt[3]{2}\right)$, where $D = 3\sqrt{3}\sqrt{27x^4 - 40x^2 + 16}$; two of the expressions are complex.

57. (a) $f^{-1}(n) = (3/\ln 2)\ln(n/100)$; the time elapsed when there are n bacteria (b) After about 26.9 hours

59. (a) $\pi/3$ (b) π **61.** (a) $\pi/4$ (b) $\pi/4$

63. (a) 10 (b) $\pi/3$ **67.** $x/\sqrt{1 + x^2}$

69.

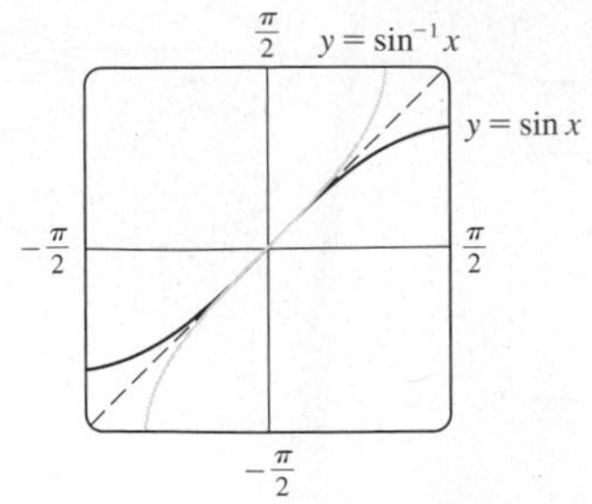

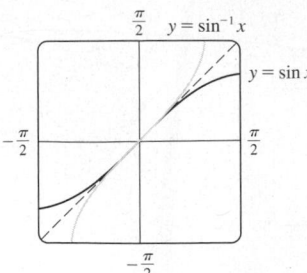

The second graph is the reflection of the first graph about the line $y = x$.

71. (a) $\left[-\frac{2}{3}, 0\right]$ (b) $[-\pi/2, \pi/2]$

73. (a) $g^{-1}(x) = f^{-1}(x) - c$ (b) $h^{-1}(x) = (1/c)f^{-1}(x)$

CHAPTER 1 REVIEW ▪ PAGE 73

True-False Quiz

1. False **3.** False **5.** True **7.** False **9.** True
11. False **13.** False

Exercises

1. (a) 2.7 (b) 2.3, 5.6 (c) $[-6, 6]$ (d) $[-4, 4]$
(e) $[-4, 4]$ (f) No; it fails the Horizontal Line Test.
(g) Odd; its graph is symmetric about the origin.

3. $2a + h - 2$ **5.** $\left(-\infty, \frac{1}{3}\right) \cup \left(\frac{1}{3}, \infty\right)$, $(-\infty, 0) \cup (0, \infty)$

7. $(-6, \infty)$, $\mathbb{R}$

9. (a) Shift the graph 8 units upward.
(b) Shift the graph 8 units to the left.
(c) Stretch the graph vertically by a factor of 2, then shift it 1 unit upward.
(d) Shift the graph 2 units to the right and 2 units downward.
(e) Reflect the graph about the x-axis.
(f) Reflect the graph about the line $y = x$ (assuming f is one-to-one).

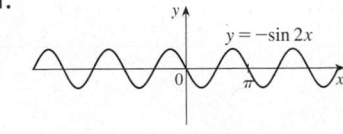

11.

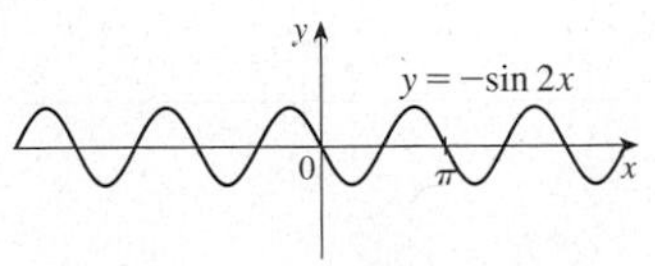

13.

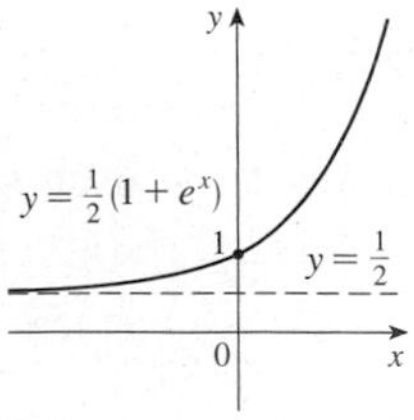

15.

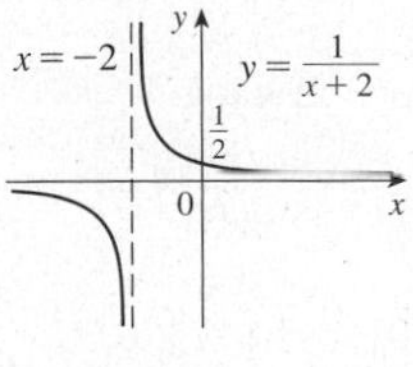

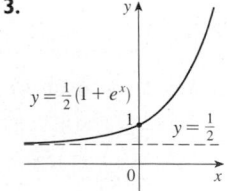

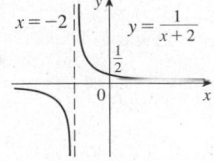

17. (a) Neither (b) Odd (c) Even (d) Neither

19. (a) $(f \circ g)(x) = \ln(x^2 - 9)$, $(-\infty, -3) \cup (3, \infty)$
(b) $(g \circ f)(x) = (\ln x)^2 - 9$, $(0, \infty)$
(c) $(f \circ f)(x) = \ln \ln x$, $(1, \infty)$
(d) $(g \circ g)(x) = (x^2 - 9)^2 - 9$, $(-\infty, \infty)$

21. $y = 0.2493x - 423.4818$; about 77.6 years

23. 1 **25.** (a) 9 (b) 2 (c) $1/\sqrt{3}$ (d) $\frac{3}{5}$

27. (a) ≈ 4.4 years

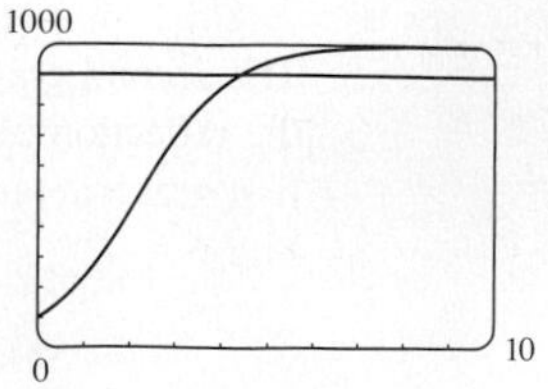

(b) $t = -\ln\left(\dfrac{1000 - P}{9P}\right)$; the time required for the population to reach a given number P.
(c) $\ln 81 \approx 4.4$ years

PRINCIPLES OF PROBLEM SOLVING ■ PAGE 81

1. $a = 4\sqrt{h^2 - 16}/h$, where a is the length of the altitude and h is the length of the hypotenuse

3. $-\frac{7}{3}, 9$

5.

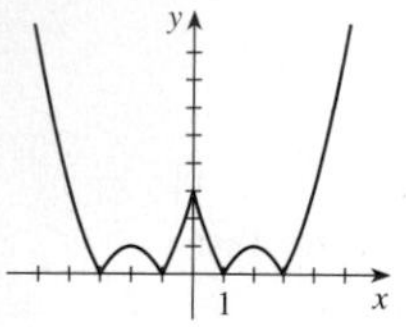

7.

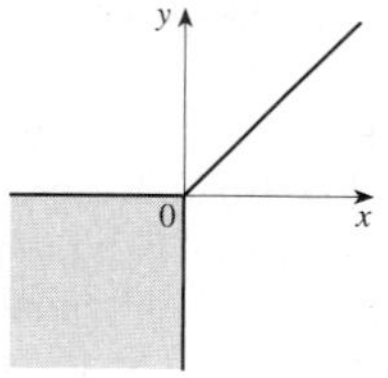

9.

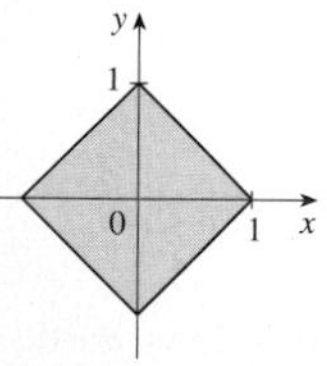

11. 5 **13.** $x \in [-1, 1 - \sqrt{3}) \cup (1 + \sqrt{3}, 3]$

15. 40 mi/h **19.** $f_n(x) = x^{2^{n+1}}$

CHAPTER 2

EXERCISES 2.1 ■ PAGE 87

1. (a) $-44.4, -38.8, -27.8, -22.2, -16.\overline{6}$
(b) -33.3 (c) $-33\frac{1}{3}$

3. (a) (i) 0.333333 (ii) 0.263158 (iii) 0.251256
(iv) 0.250125 (v) 0.2 (vi) 0.238095 (vii) 0.248756
(viii) 0.249875 (b) $\frac{1}{4}$ (c) $y = \frac{1}{4}x + \frac{1}{4}$

5. (a) (i) -32 ft/s (ii) -25.6 ft/s (iii) -24.8 ft/s
(iv) -24.16 ft/s (b) -24 ft/s

7. (a) (i) 4.65 m/s (ii) 5.6 m/s (iii) 7.55 m/s
(iv) 7 m/s (b) 6.3 m/s

9. (a) 0, 1.7321, -1.0847, -2.7433, 4.3301, -2.8173, 0, -2.1651, -2.6061, -5, 3.4202; no (c) -31.4

EXERCISES 2.2 ■ PAGE 96

1. Yes

3. (a) $\lim_{x \to -3} f(x) = \infty$ means that the values of $f(x)$ can be made arbitrarily large (as large as we please) by taking x sufficiently close to -3 (but not equal to -3).
(b) $\lim_{x \to 4^+} f(x) = -\infty$ means that the values of $f(x)$ can be made arbitrarily large negative by taking x sufficiently close to 4 through values larger than 4.

5. (a) 2 (b) 3 (c) Does not exist (d) 4
(e) Does not exist

7. (a) -1 (b) -2 (c) Does not exist (d) 2 (e) 0
(f) Does not exist (g) 1 (h) 3

9. (a) $-\infty$ (b) ∞ (c) ∞ (d) $-\infty$ (e) ∞
(f) $x = -7, x = -3, x = 0, x = 6$

11. (a) 1 (b) 0 (c) Does not exist

13.

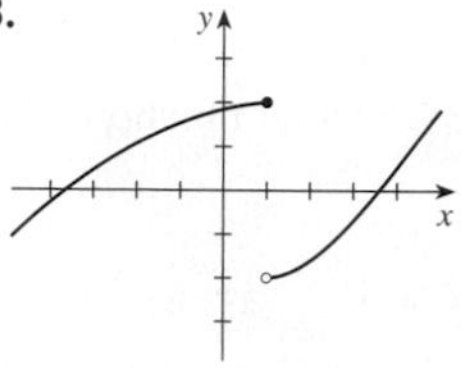

15.

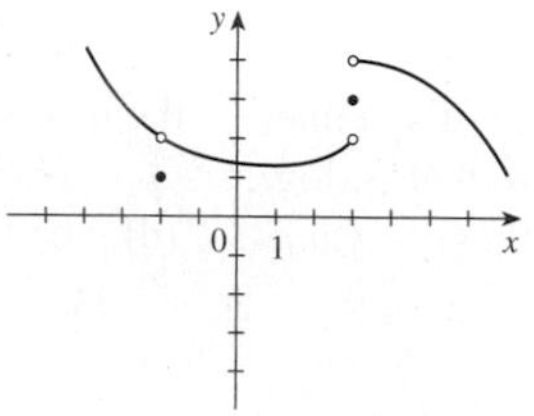

17. $\frac{2}{3}$ **19.** $\frac{1}{2}$ **21.** $\frac{1}{4}$ **23.** $\frac{3}{5}$ **25.** $-\infty$

27. ∞ **29.** $-\infty$ **31.** $-\infty$ **33.** $-\infty$; ∞

35. (a) 2.71828 (b)

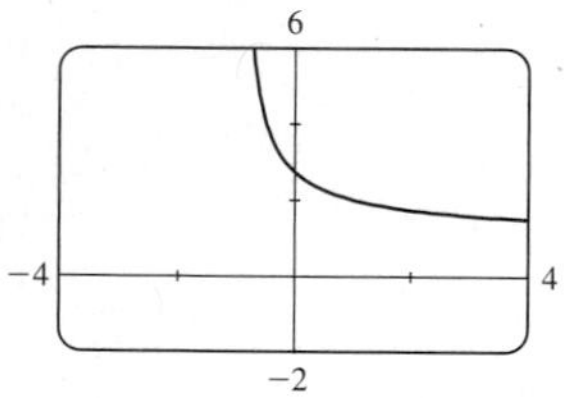

37. (a) 0.998000, 0.638259, 0.358484, 0.158680, 0.038851, 0.008928, 0.001465; 0
(b) 0.000572, -0.000614, -0.000907, -0.000978, -0.000993, -0.001000; -0.001

39. No matter how many times we zoom in toward the origin, the graph appears to consist of almost-vertical lines. This indicates more and more frequent oscillations as $x \to 0$.

41. $x \approx \pm 0.90, \pm 2.24$; $x = \pm\sin^{-1}(\pi/4), \pm(\pi - \sin^{-1}(\pi/4))$

EXERCISES 2.3 ■ PAGE 106

1. (a) -6 (b) -8 (c) 2 (d) -6
(e) Does not exist (f) 0

3. 59 **5.** 390 **7.** $\frac{1}{8}$ **9.** 0 **11.** 5

13. Does not exist **15.** $\frac{6}{5}$ **17.** 8 **19.** $\frac{1}{12}$ **21.** 6

23. $\frac{1}{6}$ **25.** $-\frac{1}{16}$ **27.** $\frac{1}{128}$ **29.** $-\frac{1}{2}$ **31.** (a), (b) $\frac{2}{3}$

35. 7 **39.** 6 **41.** -4 **43.** Does not exist

45. (a)

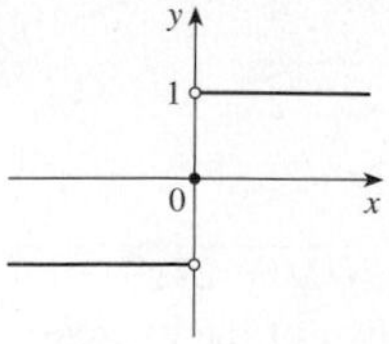

(b) (i) 1
(ii) -1
(iii) Does not exist
(iv) 1

47. (a) (i) 2 (ii) -2 (b) No (c)

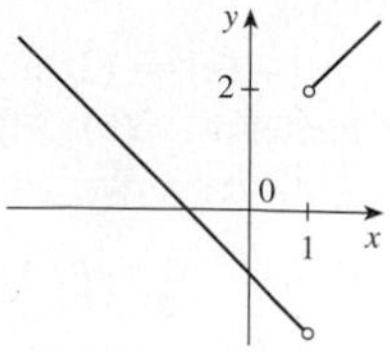

49. (a) (i) -2 (ii) Does not exist (iii) -3
(b) (i) $n-1$ (ii) n (c) a is not an integer.
55. 8 **61.** 15; -1

EXERCISES 2.4 ▪ PAGE 117

1. $\frac{4}{7}$ (or any smaller positive number)
3. 1.44 (or any smaller positive number)
5. 0.0906 (or any smaller positive number)
7. 0.11, 0.012 (or smaller positive numbers)
9. (a) 0.031 (b) 0.010
11. (a) $\sqrt{1000/\pi}$ cm (b) Within approximately 0.0445 cm
(c) Radius; area; $\sqrt{1000/\pi}$; 1000; 5; ≈ 0.0445
13. (a) 0.025 (b) 0.0025
35. (a) 0.093 (b) $\delta = (B^{2/3} - 12)/(6B^{1/3}) - 1$, where $B = 216 + 108\varepsilon + 12\sqrt{336 + 324\varepsilon + 81\varepsilon^2}$
41. Within 0.1

EXERCISES 2.5 ▪ PAGE 128

1. $\lim_{x\to 4} f(x) = f(4)$
3. (a) $f(-4)$ is not defined and $\lim_{x\to a} f(x)$ [for $a = -2$, 2, and 4] does not exist
(b) -4, neither; -2, left; 2, right; 4, right
5.

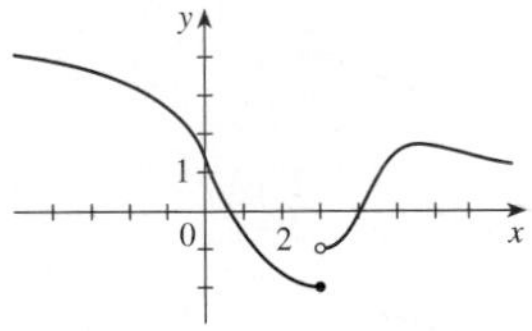

7. (a) (b) Discontinuous at $t = 1, 2, 3, 4$

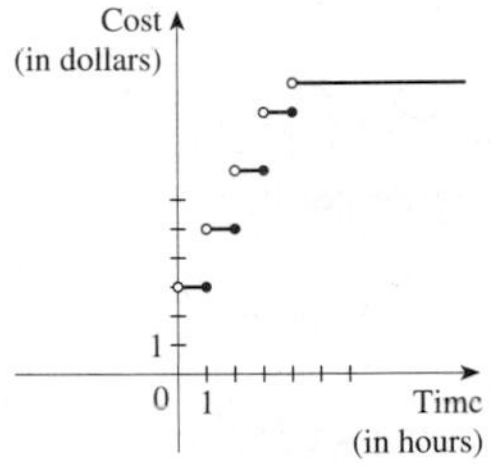

9. 6
15. $f(2)$ is not defined.

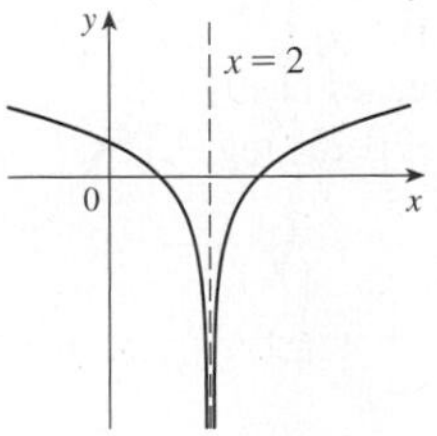

17. $\lim_{x\to 0} f(x)$ does not exist.

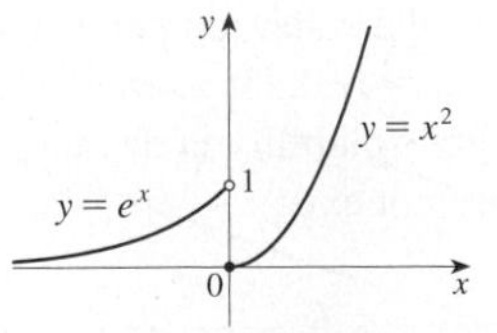

19. $\lim_{x\to 0} f(x) \neq f(0)$

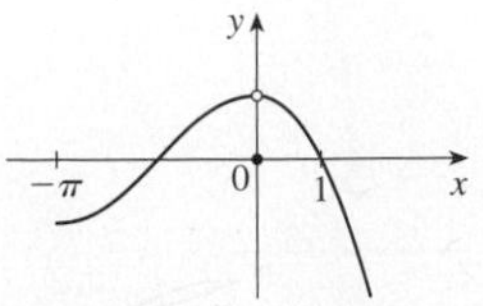

21. $\{x \mid x \neq -3, -2\}$
23. $\left[\frac{1}{2}, \infty\right)$ **25.** $(-\infty, \infty)$ **27.** $(-\infty, -1) \cup (1, \infty)$
29. $x = 0$

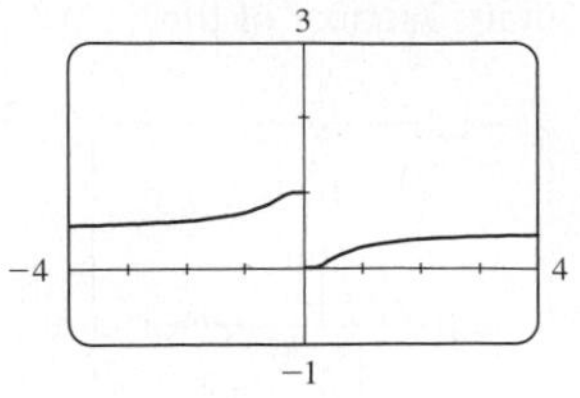

31. $\frac{7}{3}$ **33.** 1
37. 0, left

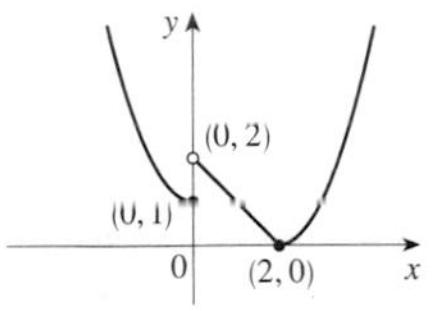

39. 0, right; 1, left

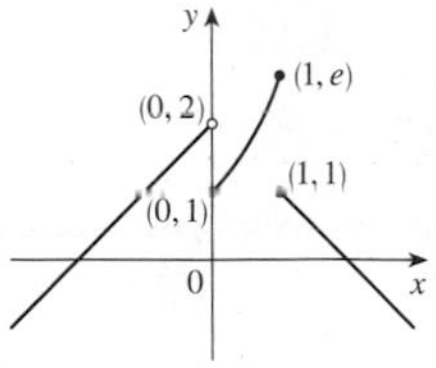

41. $\frac{2}{3}$ **43.** (a) $g(x) = x^3 + x^2 + x + 1$ (b) $g(x) = x^2 + x$
51. (b) (0.86, 0.87) **53.** (b) 70.347
59. None **61.** Yes

EXERCISES 2.6 ▪ PAGE 140

1. (a) As x becomes large, $f(x)$ approaches 5.
(b) As x becomes large negative, $f(x)$ approaches 3.
3. (a) ∞ (b) ∞ (c) $-\infty$ (d) 1 (e) 2
(f) $x = -1$, $x = 2$, $y = 1$, $y = 2$
5.

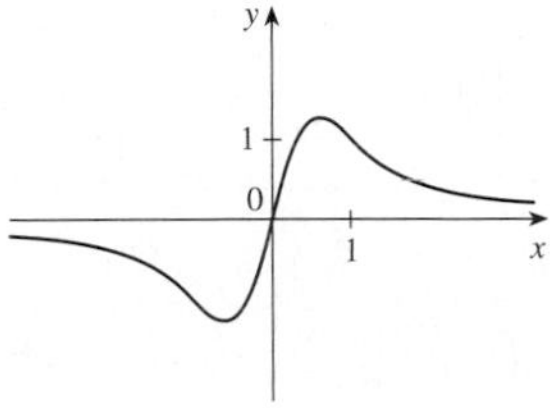

7.

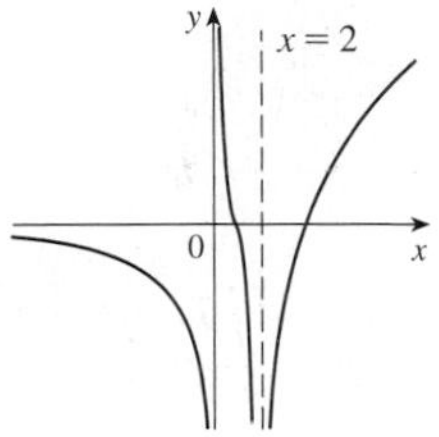

9.

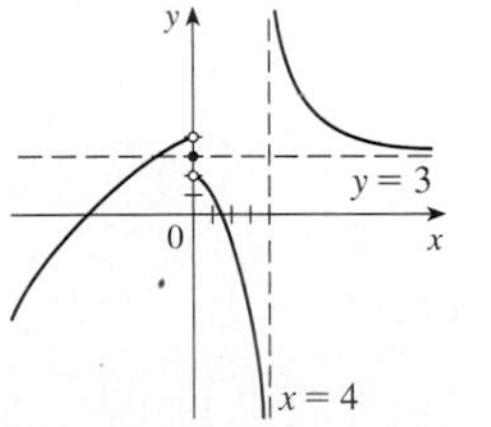

11. 0 **13.** $\frac{3}{2}$ **15.** 0 **17.** $-\frac{1}{2}$ **19.** $\frac{1}{2}$ **21.** 2
23. 3 **25.** $\frac{1}{6}$ **27.** $\frac{1}{2}(a-b)$ **29.** ∞ **31.** $-\infty$
33. $-\frac{1}{2}$ **35.** 0 **37.** (a), (b) $-\frac{1}{2}$ **39.** $y = 2$; $x = 2$
41. $y = 2$; $x = -2$, $x = 1$ **43.** $x = 5$ **45.** $y = 3$
47. $f(x) = \dfrac{2-x}{x^2(x-3)}$
49. $-\infty$, $-\infty$

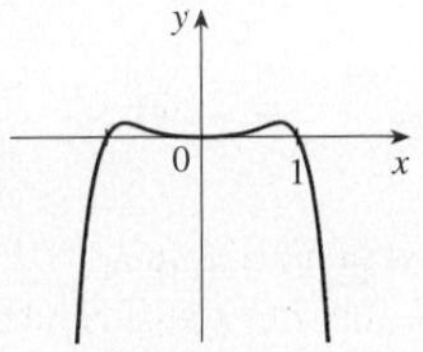

51. $-\infty$, ∞

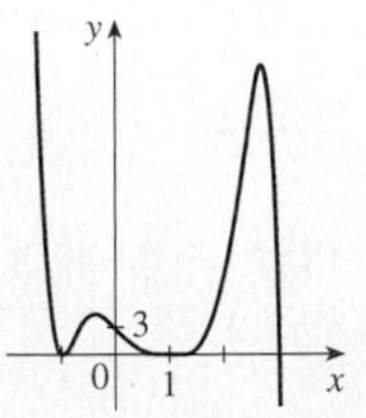

53. (a) 0 (b) An infinite number of times

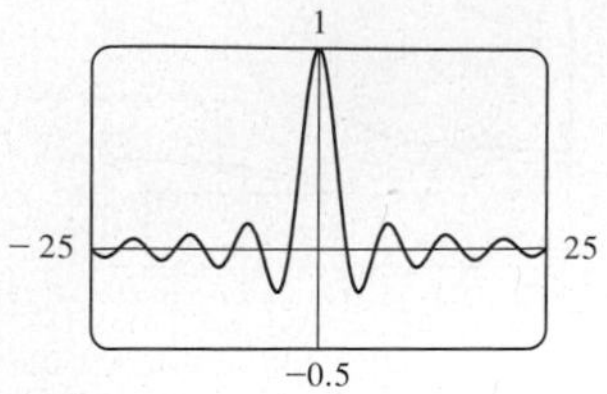

55. (a) 0 (b) $\pm\infty$ **57.** 5

59. (a) v^* (b) ≈ 0.47 s

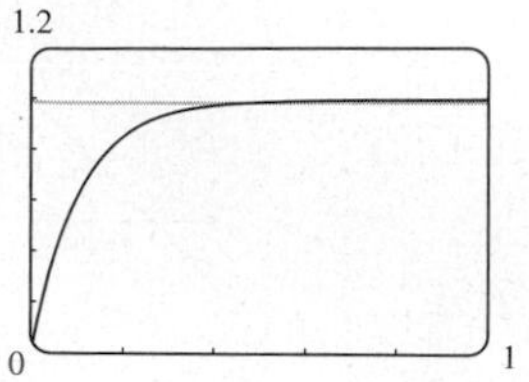

61. $N \geqslant 15$ **63.** $N \leqslant -6, N \leqslant -22$ **65.** (a) $x > 100$

EXERCISES 2.7 ▪ PAGE 150

1. (a) $\dfrac{f(x) - f(3)}{x - 3}$ (b) $\displaystyle\lim_{x \to 3} \frac{f(x) - f(3)}{x - 3}$

3. (a) 2 (b) $y = 2x + 1$ (c)

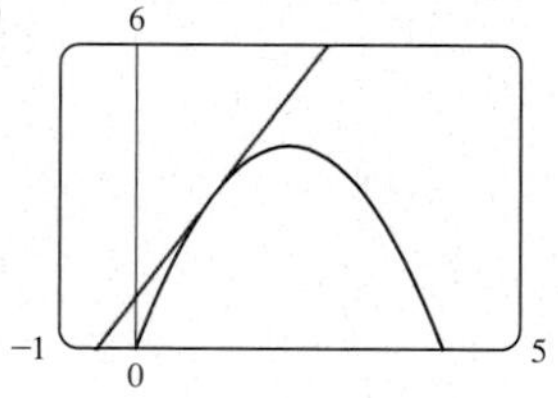

5. $y = -x + 5$ **7.** $y = \frac{1}{2}x + \frac{1}{2}$

9. (a) $8a - 6a^2$ (b) $y = 2x + 3, y = -8x + 19$

(c)

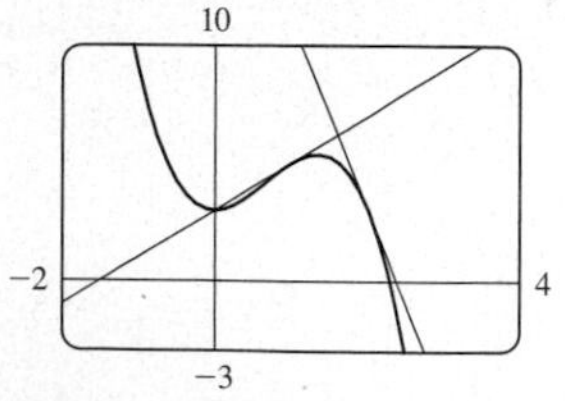

11. (a) Right: $0 < t < 1$ and $4 < t < 6$; left: $2 < t < 3$; standing still: $1 < t < 2$ and $3 < t < 4$

(b)

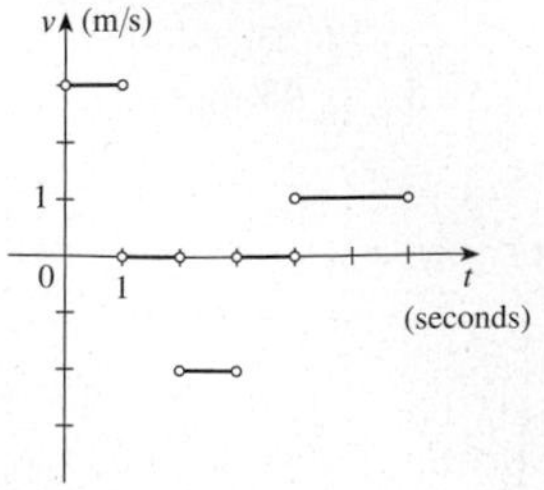

13. -24 ft/s **15.** $-2/a^3$ m/s; -2 m/s; $-\frac{1}{4}$ m/s; $-\frac{2}{27}$ m/s

17. $g'(0), 0, g'(4), g'(2), g'(-2)$

19.

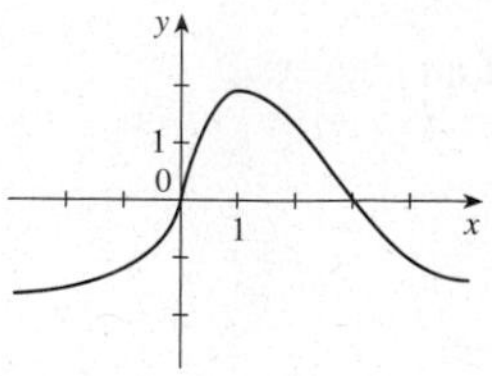

21. 7; $y = 7x - 12$

23. (a) $-\frac{3}{5}$; $y = -\frac{3}{5}x + \frac{16}{5}$ (b)

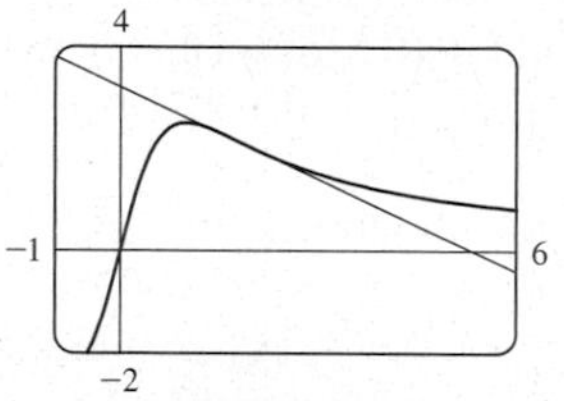

25. $-2 + 8a$ **27.** $\dfrac{5}{(a + 3)^2}$ **29.** $\dfrac{-1}{2(a + 2)^{3/2}}$

31. $f(x) = x^{10}, a = 1$ or $f(x) = (1 + x)^{10}, a = 0$

33. $f(x) = 2^x, a = 5$

35. $f(x) = \cos x, a = \pi$ or $f(x) = \cos(\pi + x), a = 0$

37. 1 m/s; 1 m/s

39. Greater (in magnitude)

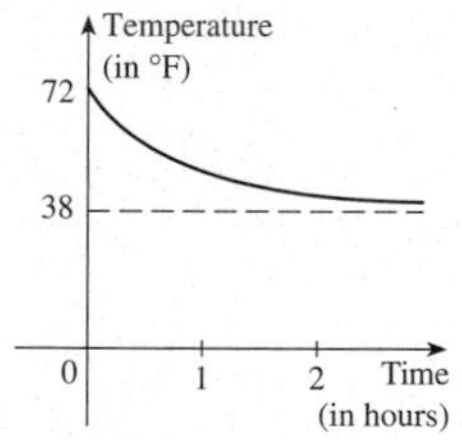

41. (a) (i) 11 percent/year (ii) 13 percent/year (iii) 16 percent/year
(b) 14.5 percent/year (c) 15 percent/year

43. (a) (i) \$20.25/unit (ii) \$20.05/unit (b) \$20/unit

45. (a) The rate at which the cost is changing per ounce of gold produced; dollars per ounce
(b) When the 800th ounce of gold is produced, the cost of production is \$17/oz.
(c) Decrease in the short term; increase in the long term

47. The rate at which the temperature is changing at 10:00 AM; 4°F/h

49. (a) The rate at which the oxygen solubility changes with respect to the water temperature; (mg/L)/°C
(b) $S'(16) \approx -0.25$; as the temperature increases past 16°C, the oxygen solubility is decreasing at a rate of 0.25 (mg/L)/°C.

51. Does not exist

EXERCISES 2.8 ▪ PAGE 162

1. (a) 1.5
(b) 1
(c) 0
(d) −4
(e) 0
(f) 1
(g) 1.5

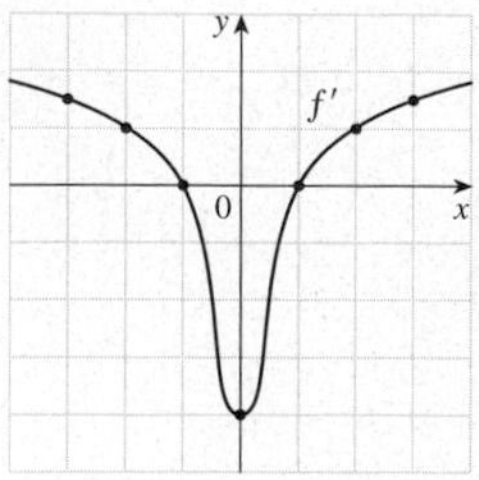

3. (a) II (b) IV (c) I (d) III

5.
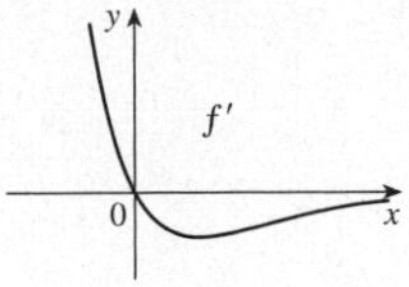

7.
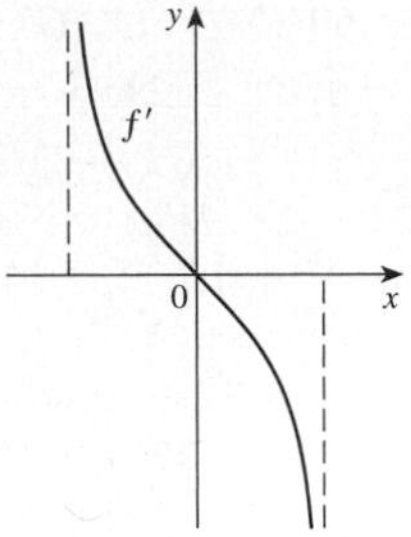

9.
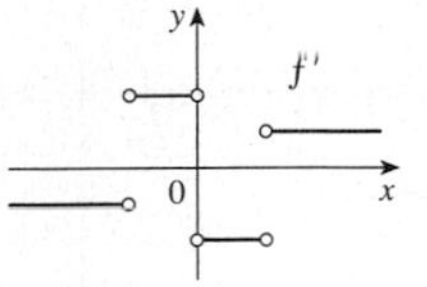

11.
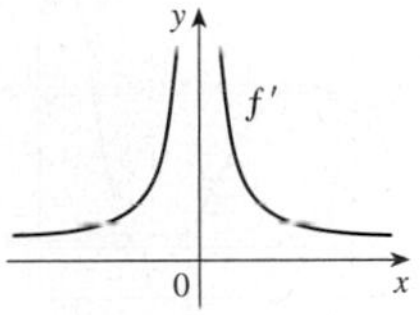

13.
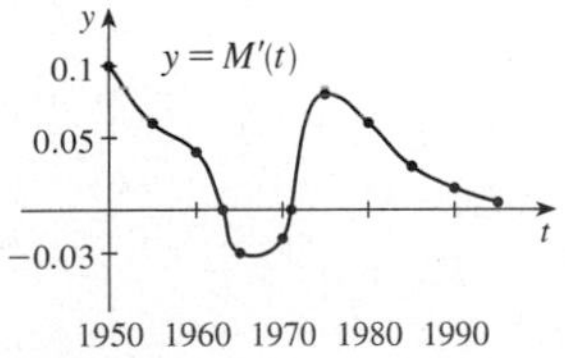

1963 to 1971

15.
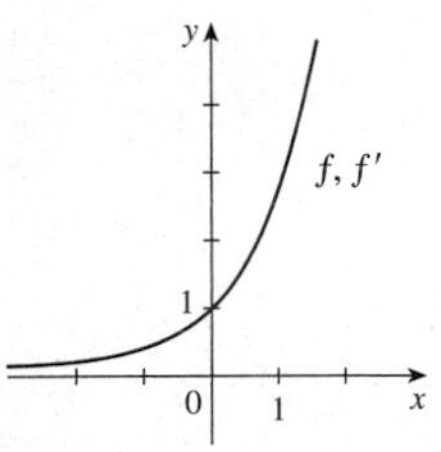

$f'(x) = e^x$

17. (a) 0, 1, 2, 4 (b) $-1, -2, -4$ (c) $f'(x) = 2x$

19. $f'(x) = \frac{1}{2}$, $\mathbb{R}$, $\mathbb{R}$ **21.** $f'(t) = 5 - 18t$, $\mathbb{R}$, $\mathbb{R}$

23. $f'(x) = 3x^2 - 3$, $\mathbb{R}$, $\mathbb{R}$

25. $g'(x) = 1/\sqrt{1 + 2x}$, $[-\frac{1}{2}, \infty)$, $(-\frac{1}{2}, \infty)$

27. $G'(t) = \dfrac{4}{(t+1)^2}$, $(-\infty, -1) \cup (-1, \infty)$, $(-\infty, -1) \cup (-1, \infty)$

29. $f'(x) = 4x^3$, $\mathbb{R}$, $\mathbb{R}$ **31.** (a) $f'(x) = 4x^3 + 2$

33. (a) The rate at which the unemployment rate is changing, in percent unemployed per year

(b)

t	$U'(t)$	t	$U'(t)$
1993	−0.80	1998	−0.35
1994	−0.65	1999	−0.25
1995	−0.35	2000	0.25
1996	−0.35	2001	0.90
1997	−0.45	2002	1.10

35. −4 (corner); 0 (discontinuity)

37. −1 (vertical tangent); 4 (corner)

39.
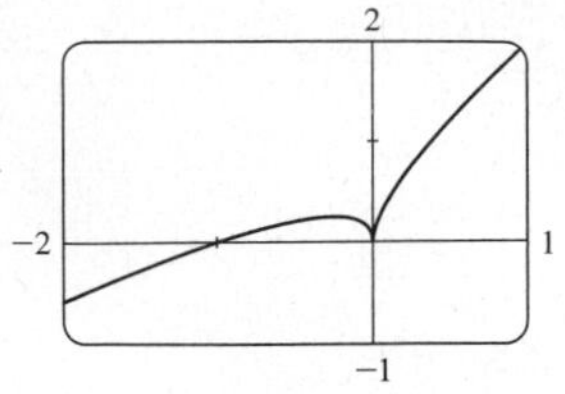

Differentiable at −1; not differentiable at 0

41. $a = f, b = f', c = f''$

43. a = acceleration, b = velocity, c = position

45.
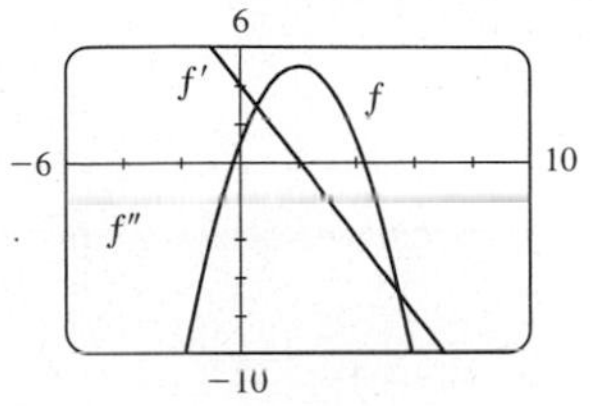

$f'(x) = 4 - 2x$,
$f''(x) = -2$

47.
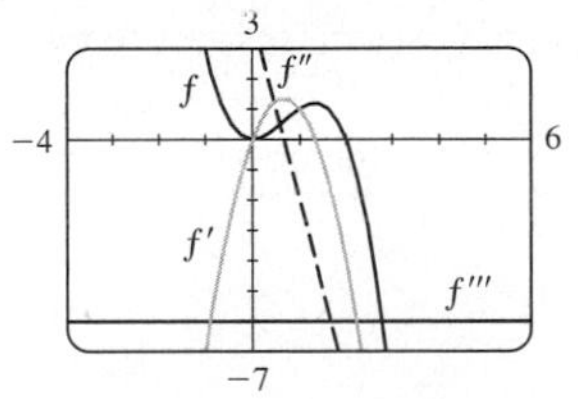

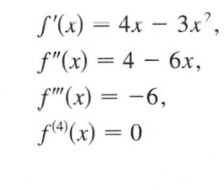

$f'(x) = 4x - 3x^2$,
$f''(x) = 4 - 6x$,
$f'''(x) = -6$,
$f^{(4)}(x) = 0$

49. (a) $\frac{1}{3}a^{-2/3}$

51. $f'(x) = \begin{cases} -1 & \text{if } x < 6 \\ 1 & \text{if } x > 6 \end{cases}$

or $f'(x) = \dfrac{x - 6}{|x - 6|}$

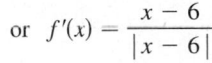

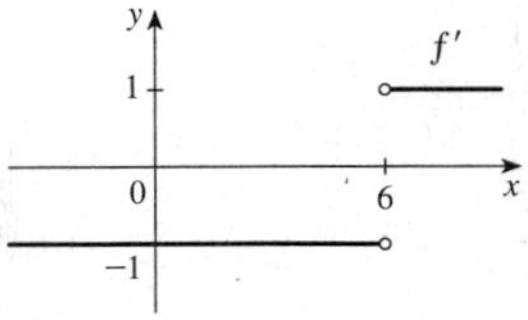

53. (a)
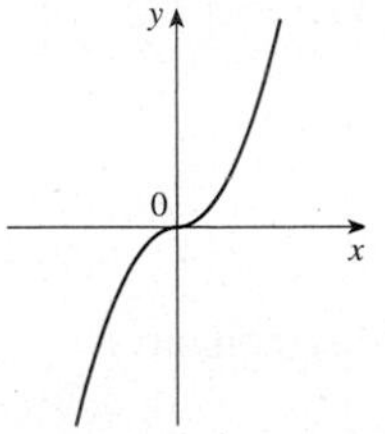

(b) All x
(c) $f'(x) = 2|x|$

57. 63°

CHAPTER 2 REVIEW ▪ PAGE 166

True-False Quiz

1. False **3.** True **5.** False **7.** True **9.** True
11. False **13.** True **15.** True **17.** False **19.** False

Exercises

1. (a) (i) 3 (ii) 0 (iii) Does not exist (iv) 2
(v) ∞ (vi) −∞ (vii) 4 (viii) −1
(b) $y = 4, y = -1$ (c) $x = 0, x = 2$ (d) −3, 0, 2, 4

3. 1 **5.** $\frac{3}{2}$ **7.** 3 **9.** ∞ **11.** $\frac{4}{7}$ **13.** $\frac{1}{2}$

15. $-\infty$ **17.** 2 **19.** $\pi/2$ **21.** $x = 0, y = 0$ **23.** 1

29. (a) (i) 3 (ii) 0 (iii) Does not exist (iv) 0 (v) 0 (vi) 0

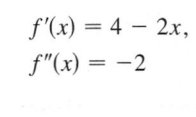

(b) At 0 and 3 (c)

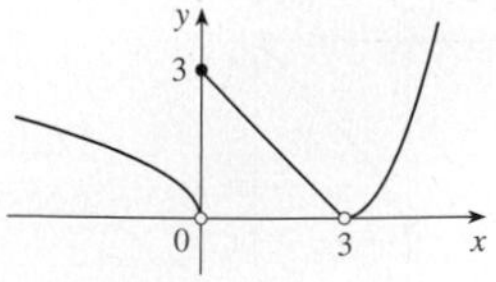

31. $\mathbb{R}$ **35.** (a) -8 (b) $y = -8x + 17$
37. (a) (i) 3 m/s (ii) 2.75 m/s (iii) 2.625 m/s
(iv) 2.525 m/s (b) 2.5 m/s
39. (a) 10 (b) $y = 10x - 16$
(c)

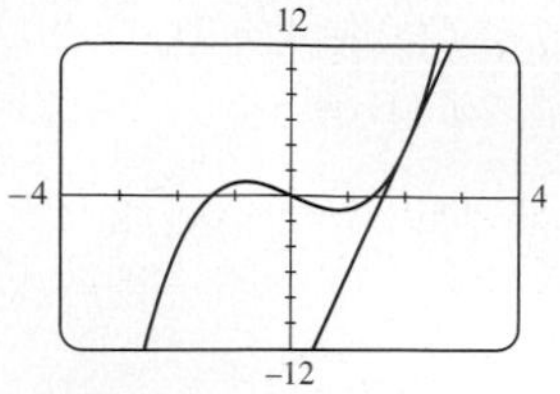

41. (a) The rate at which the cost changes with respect to the interest rate; dollars/(percent per year)
(b) As the interest rate increases past 10%, the cost is increasing at a rate of \$1200/(percent per year).
(c) Always positive
43.

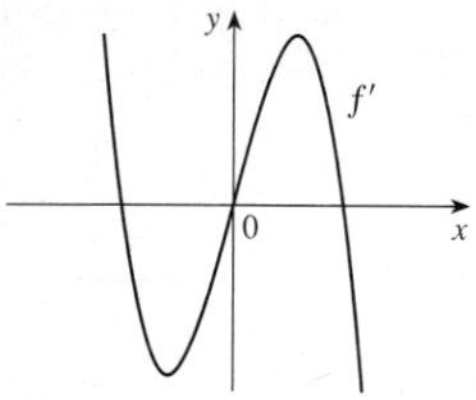

45. (a) $f'(x) = -\frac{5}{2}(3 - 5x)^{-1/2}$ (b) $\left(-\infty, \frac{3}{5}\right], \left(-\infty, \frac{3}{5}\right)$
(c)

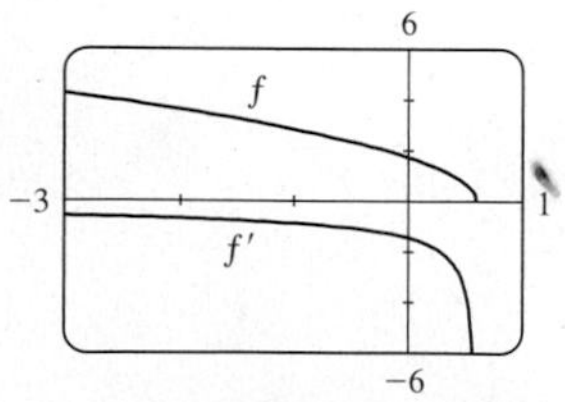

47. -4 (discontinuity), -1 (corner), 2 (discontinuity), 5 (vertical tangent)
49. The rate at which the total value of US currency in circulation is changing in billions of dollars per year; \$22.2 billion/year
51. 0

PROBLEMS PLUS ▪ PAGE 170

1. $\frac{2}{3}$ **3.** -4 **5.** 1 **7.** $a = \frac{1}{2} \pm \frac{1}{2}\sqrt{5}$
9. $\frac{3}{4}$ **11.** (b) Yes (c) Yes; no
13. (a) 0 (b) 1 (c) $f'(x) = x^2 + 1$

CHAPTER 3

EXERCISES 3.1 ▪ PAGE 180

1. (a) See Definition of the Number e (page 179).
(b) 0.99, 1.03; $2.7 < e < 2.8$
3. $f'(x) = 0$ **5.** $f'(t) = -\frac{2}{3}$ **7.** $f'(x) = 3x^2 - 4$
9. $f'(t) = t^3$ **11.** $y' = -\frac{2}{5}x^{-7/5}$ **13.** $V'(r) = 4\pi r^2$
15. $A'(s) = 60/s^6$ **17.** $G'(x) = 1/(2\sqrt{x}) - 2e^x$
19. $F'(x) = \frac{5}{32}x^4$ **21.** $y' = 2ax + b$
23. $y' = \frac{3}{2}\sqrt{x} + (2/\sqrt{x}) - 3/(2x\sqrt{x})$
25. $y' = 0$ **27.** $H'(x) = 3x^2 + 3 - 3x^{-2} - 3x^{-4}$
29. $u' = \frac{1}{5}t^{-4/5} + 10t^{3/2}$ **31.** $z' = -10A/y^{11} + Be^y$
33. $y = \frac{1}{4}x + \frac{3}{4}$
35. Tangent: $y = 2x + 2$; normal: $y = -\frac{1}{2}x + 2$
37. $y = 3x - 1$ **39.** $e^x - 5$ **41.** $45x^{14} - 15x^2$
43. (a) (c) $4x^3 - 9x^2 - 12x + 7$

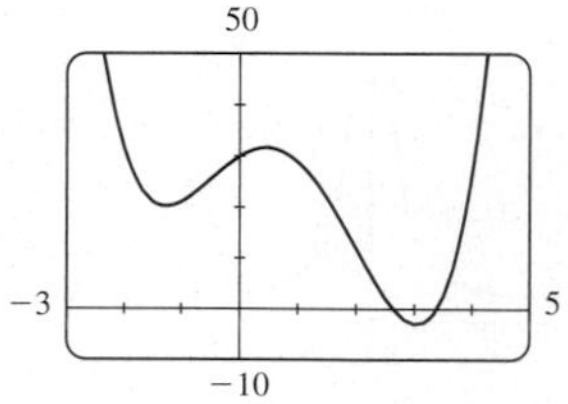

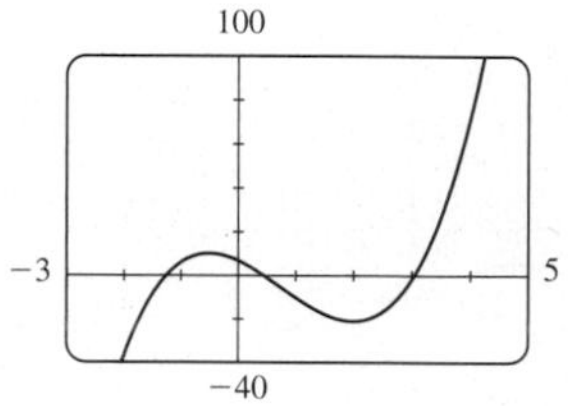

45. $f'(x) = 4x^3 - 9x^2 + 16, f''(x) = 12x^2 - 18x$
47. $f'(x) = 2 - \frac{15}{4}x^{-1/4}, f''(x) = \frac{15}{16}x^{-5/4}$
49. (a) $v(t) = 3t^2 - 3, a(t) = 6t$ (b) 12 m/s^2
(c) $a(1) = 6 \text{ m/s}^2$ **51.** $(-2, 21), (1, -6)$
55. $y = 12x - 15, y = 12x + 17$ **57.** $y = \frac{1}{3}x - \frac{1}{3}$
59. $(\pm 2, 4)$ **63.** $P(x) = x^2 - x + 3$
65. $y = \frac{3}{16}x^3 - \frac{9}{4}x + 3$
67. No

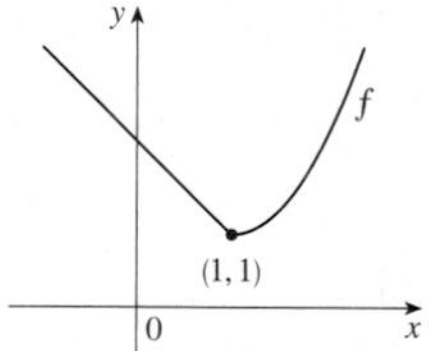

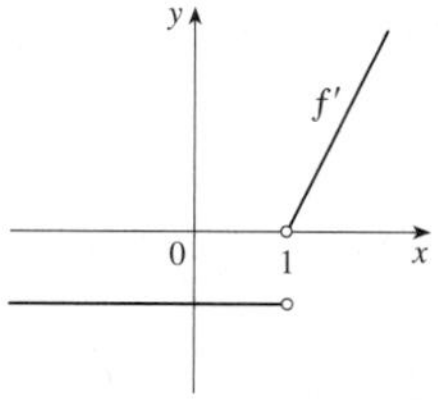

69. (a) Not differentiable at 3 or -3

$$f'(x) = \begin{cases} 2x & \text{if } |x| > 3 \\ -2x & \text{if } |x| < 3 \end{cases}$$

(b)

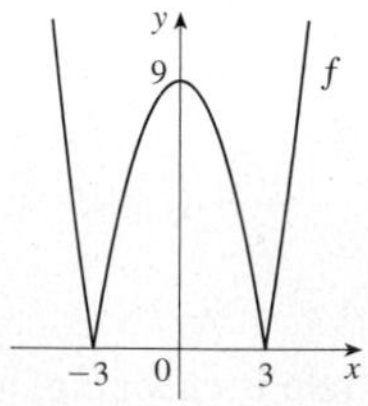

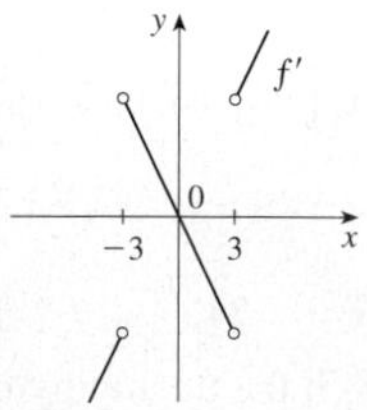

71. $y = 2x^2 - x$ **73.** $a = -\frac{1}{2}, b = 2$ **75.** $m = 4, b = -4$
77. 1000 **79.** 3; 1

EXERCISES 3.2 ▪ PAGE 187

1. $y' = 5x^4 + 3x^2 + 2x$
3. $f'(x) = e^x(x^3 + 3x^2 + 2x + 2)$
5. $y' = (x - 2)e^x/x^3$ **7.** $g'(x) = 5/(2x + 1)^2$
9. $V'(x) = 14x^6 - 4x^3 - 6$
11. $F'(y) = 5 + 14/y^2 + 9/y^4$
13. $y' = \dfrac{x^2(3 - x^2)}{(1 - x^2)^2}$ **15.** $y' = \dfrac{2t(-t^4 - 4t^2 + 7)}{(t^4 - 3t^2 + 1)^2}$

17. $y' = (r^2 - 2)e^r$ **19.** $y' = 2v - 1/\sqrt{v}$

21. $f'(t) = \dfrac{4 + t^{1/2}}{(2 + \sqrt{t})^2}$ **23.** $f'(x) = -ACe^x/(B + Ce^x)^2$

25. $f'(x) = 2cx/(x^2 + c)^2$

27. $(x^4 + 4x^3)e^x$; $(x^4 + 8x^3 + 12x^2)e^x$

29. $\dfrac{2x^2 + 2x}{(1 + 2x)^2}$; $\dfrac{2}{(1 + 2x)^3}$

31. $y = \frac{1}{2}x + \frac{1}{2}$ **33.** $y = 2x$; $y = -\frac{1}{2}x$

35. (a) $y = \frac{1}{2}x + 1$ (b)

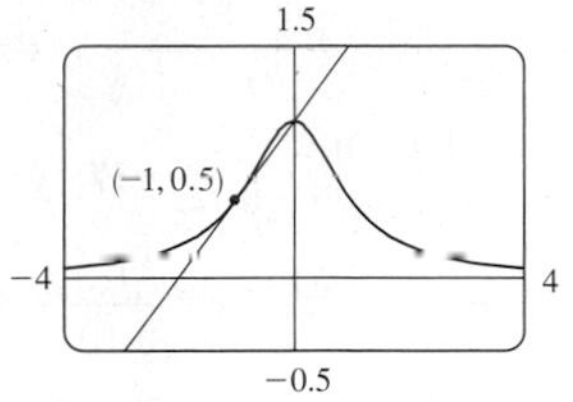

37. (a) $e^x(x - 3)/x^4$ **39.** xe^x, $(x + 1)e^x$

41. $\frac{1}{4}$ **43.** (a) -16 (b) $-\frac{20}{9}$ (c) 20

45. 7 **47.** (a) 0 (b) $-\frac{2}{3}$

49. (a) $y' = xg'(x) + g(x)$ (b) $y' = [g(x) - xg'(x)]/[g(x)]^2$
(c) $y' = [xg'(x) - g(x)]/x^2$

51. Two, $(-2 \pm \sqrt{3}, (1 \mp \sqrt{3})/2)$

53. \$1.627 billion/year **55.** (c) $3e^{3x}$

57. $f'(x) = (x^2 + 2x)e^x$, $f''(x) = (x^2 + 4x + 2)e^x$,
$f'''(x) = (x^2 + 6x + 6)e^x$, $f^{(4)}(x) = (x^2 + 8x + 12)e^x$,
$f^{(5)}(x) = (x^2 + 10x + 20)e^x$; $f^{(n)}(x) = [x^2 + 2nx + n(n - 1)]e^x$

EXERCISES 3.3 ■ PAGE 195

1. $f'(x) = 6x + 2\sin x$ **3.** $f'(x) = \cos x - \frac{1}{2}\csc^2 x$

5. $g'(t) = 3t^2 \cos t - t^3 \sin t$

7. $h'(\theta) = -\csc\theta \cot\theta + e^\theta(\cot\theta - \csc^2\theta)$

9. $y' = \dfrac{2 - \tan x + x\sec^2 x}{(2 - \tan x)^2}$ **11.** $f'(\theta) = \dfrac{\sec\theta\tan\theta}{(1 + \sec\theta)^2}$

13. $y' = (x\cos x - 2\sin x)/x^3$

15. $f'(x) = e^x \csc x(-x\cot x + x + 1)$

21. $y = 2\sqrt{3}x - \frac{2}{3}\sqrt{3}\pi + 2$ **23.** $y = x + 1$

25. (a) $y = 2x$ (b)

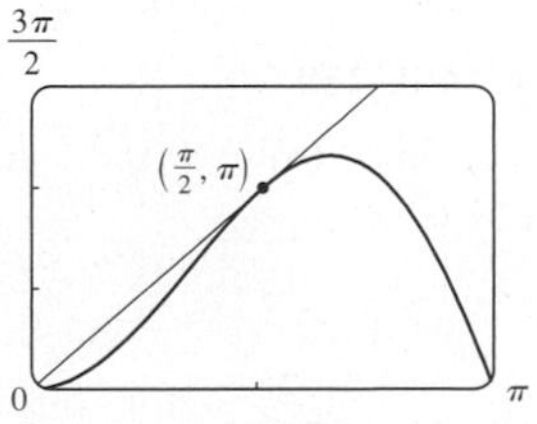

27. (a) $\sec x \tan x - 1$

29. $\theta\cos\theta + \sin\theta$; $2\cos\theta - \theta\sin\theta$

31. (a) $f'(x) = (1 + \tan x)/\sec x$ (b) $f'(x) = \cos x + \sin x$

33. $(2n + 1)\pi \pm \frac{1}{3}\pi$, n an integer

35. (a) $v(t) = 8\cos t$, $a(t) = -8\sin t$
(b) $4\sqrt{3}, -4, -4\sqrt{3}$; to the left

37. 5 ft/rad **39.** 3 **41.** 3 **43.** sin 1

45. $\frac{1}{2}$ **47.** $-\sqrt{2}$

49. (a) $\sec^2 x = 1/\cos^2 x$ (b) $\sec x \tan x = (\sin x)/\cos^2 x$
(c) $\cos x - \sin x = (\cot x - 1)/\csc x$

51. 1

EXERCISES 3.4 ■ PAGE 203

1. $4\cos 4x$ **3.** $-20x(1 - x^2)^9$ **5.** $e^{\sqrt{x}}/(2\sqrt{x})$

7. $F'(x) = 10x(x^4 + 3x^2 - 2)^4(2x^2 + 3)$

9. $F'(x) = \dfrac{2 + 3x^2}{4(1 + 2x + x^3)^{3/4}}$ **11.** $g'(t) = -\dfrac{12t^3}{(t^4 + 1)^4}$

13. $y' = -3x^2 \sin(a^3 + x^3)$ **15.** $y' = e^{-kx}(-kx + 1)$

17. $g'(x) = 4(1 + 4x)^4(3 + x - x^2)^7(17 + 9x - 21x^2)$

19. $y' = 8(2x - 5)^3(8x^2 - 5)^{-4}(-4x^2 + 30x - 5)$

21. $y' = \dfrac{-12x(x^2 + 1)^2}{(x^2 - 1)^4}$ **23.** $y' = (\cos x - x\sin x)e^{x\cos x}$

25. $F'(z) = 1/[(z - 1)^{1/2}(z + 1)^{3/2}]$

27. $y' = (r^2 + 1)^{-3/2}$ **29.** $y' = 2\cos(\tan 2x)\sec^2(2x)$

31. $y' = 2^{\sin \pi x}(\pi \ln 2)\cos \pi x$ **33.** $y' = 4\sec^2 x \tan x$

35. $y' = \dfrac{4e^{2x}}{(1 + e^{2x})^2}\sin\dfrac{1 - e^{2x}}{1 + e^{2x}}$

37. $y' = -2\cos\theta\cot(\sin\theta)\csc^2(\sin\theta)$

39. $f'(t) = \sec^2(e^t)e^t + e^{\tan t}\sec^2 t$

41. $f'(t) = 4\sin(e^{\sin^2 t})\cos(e^{\sin^2 t})\,e^{\sin^2 t}\sin t\cos t$

43. $g'(x) = 2r^2 p(\ln a)(2ra^{rx} + n)^{p-1}a^{rx}$

45. $y' = \dfrac{-\pi\cos(\tan \pi x)\sec^2(\pi x)\sin\sqrt{\sin(\tan \pi x)}}{2\sqrt{\sin(\tan \pi x)}}$

47. $h'(x) = x/\sqrt{x^2 + 1}$, $h''(x) = 1/(x^2 + 1)^{3/2}$

49. $e^{\alpha x}(\beta\cos\beta x + \alpha\sin\beta x)$; $e^{\alpha x}[(\alpha^2 - \beta^2)\sin\beta x + 2\alpha\beta\cos\beta x]$

51. $y = 20x + 1$ **53.** $y = -x + \pi$

55. (a) $y = \frac{1}{2}x + 1$ (b)

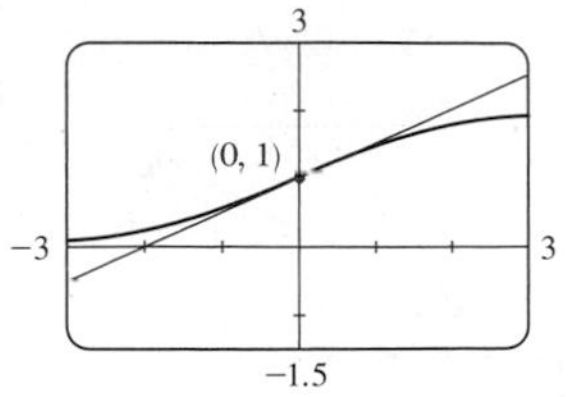

57. (a) $f'(x) = (2 - 2x^2)/\sqrt{2 - x^2}$

59. $((\pi/2) + 2n\pi, 3)$, $((3\pi/2) + 2n\pi, -1)$, n an integer

61. 24 **63.** (a) 30 (b) 36

65. (a) $\frac{3}{4}$ (b) Does not exist (c) -2

67. (a) $F'(x) = e^x f'(e^x)$ (b) $G'(x) = e^{f(x)}f'(x)$

69. 120 **71.** 96 **75.** $-2^{50}\cos 2x$

77. $v(t) = \frac{5}{2}\pi\cos(10\pi t)$ cm/s

79. (a) $\dfrac{dB}{dt} = \dfrac{7\pi}{54}\cos\dfrac{2\pi t}{5.4}$ (b) 0.16

81. $v(t) = 2e^{-1.5t}(2\pi\cos 2\pi t - 1.5\sin 2\pi t)$

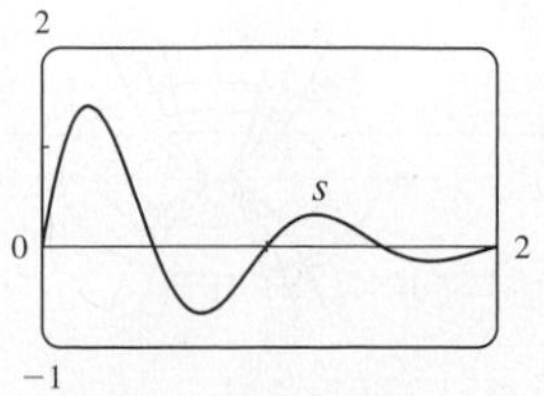

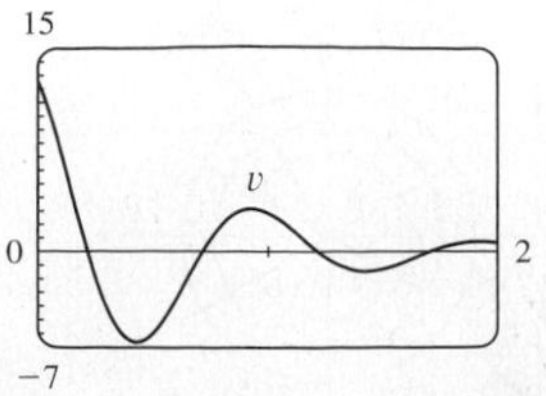

83. dv/dt is the rate of change of velocity with respect to time; dv/ds is the rate of change of velocity with respect to displacement
85. (a) $y = ab^t$ where $a \approx 100.01244$ and $b \approx 0.000045146$
(b) -670.63 μA
87. (b) The factored form
91. (b) $-n \cos^{n-1}x \sin[(n+1)x]$

EXERCISES 3.5 ▪ PAGE 213

1. (a) $y' = -(y + 2 + 6x)/x$
(b) $y = (4/x) - 2 - 3x$, $y' = -(4/x^2) - 3$
3. (a) $y' = -y^2/x^2$ (b) $y = x/(x-1)$, $y' = -1/(x-1)^2$
5. $y' = -x^2/y^2$
7. $y' = \dfrac{2x + y}{2y - x}$ **9.** $y' = \dfrac{3y^2 - 5x^4 - 4x^3y}{x^4 + 3y^2 - 6xy}$
11. $y' = \dfrac{-2xy^2 - \sin y}{2x^2y + x\cos y}$ **13.** $y' = \tan x \tan y$
15. $y' = \dfrac{y(y - e^{x/y})}{y^2 - xe^{x/y}}$ **17.** $y' = \dfrac{4xy\sqrt{xy} - y}{x - 2x^2\sqrt{xy}}$
19. $y' = \dfrac{e^y \sin x + y\cos(xy)}{e^y \cos x - x\cos(xy)}$ **21.** $-\frac{16}{13}$
23. $x' = \dfrac{-2x^4y + x^3 - 6xy^2}{4x^3y^2 - 3x^2y + 2y^3}$ **25.** $y = -x + 2$
27. $y = x + \frac{1}{2}$ **29.** $y = -\frac{9}{13}x + \frac{40}{13}$
31. (a) $y = \frac{9}{2}x - \frac{5}{2}$ (b)

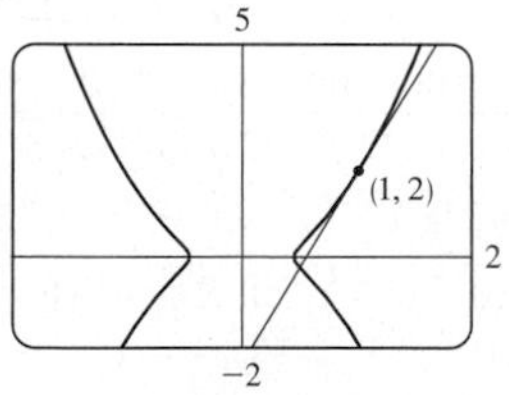

33. $-81/y^3$ **35.** $-2x/y^5$
37. (a)

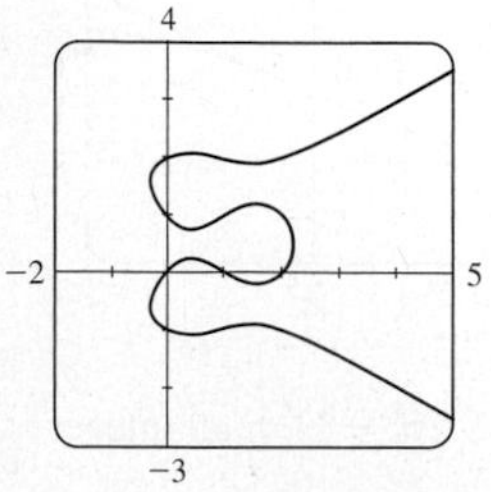

Eight; $x \approx 0.42, 1.58$

(b) $y = -x + 1$, $y = \frac{1}{3}x + 2$ (c) $1 \mp \frac{1}{3}\sqrt{3}$
39. $(\pm\frac{5}{4}\sqrt{3}, \pm\frac{5}{4})$ **41.** $(x_0x/a^2) - (y_0y/b^2) = 1$
45. $y' = \dfrac{1}{2\sqrt{x}(1 + x)}$ **47.** $y' = \dfrac{1}{\sqrt{-x^2 - x}}$
49. $G'(x) = -1 - \dfrac{x \arccos x}{\sqrt{1 - x^2}}$ **51.** $h'(t) = 0$
53. $y' = -2e^{2x}/\sqrt{1 - e^{4x}}$ **55.** $1 - \dfrac{x \arcsin x}{\sqrt{1 - x^2}}$
59.

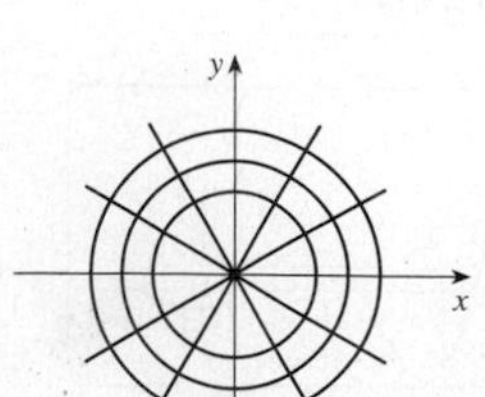

61.

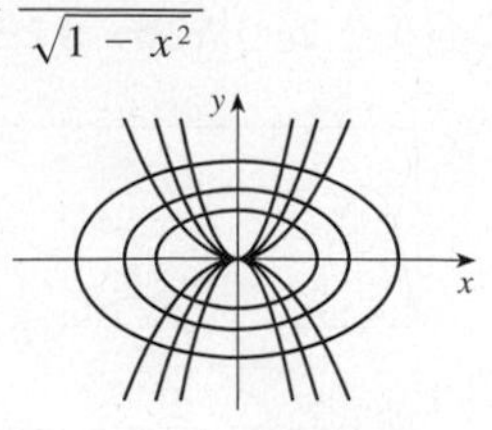

63. $(\pm\sqrt{3}, 0)$ **65.** $(-1, -1), (1, 1)$ **67.** (b) $\frac{3}{2}$ **69.** 2

EXERCISES 3.6 ▪ PAGE 220

1. The differentiation formula is simplest.
3. $f'(x) = \dfrac{\cos(\ln x)}{x}$ **5.** $f'(x) = \dfrac{3}{(3x - 1)\ln 2}$
7. $f'(x) = \dfrac{1}{5x\sqrt[5]{(\ln x)^4}}$ **9.** $f'(x) = \dfrac{\sin x}{x} + \cos x \ln(5x)$
11. $F'(t) = \dfrac{6}{2t + 1} - \dfrac{12}{3t - 1}$ **13.** $g'(x) = \dfrac{2x^2 - 1}{x(x^2 - 1)}$
15. $f'(u) = \dfrac{1 + \ln 2}{u[1 + \ln(2u)]^2}$ **17.** $y' = \dfrac{10x + 1}{5x^2 + x - 2}$
19. $y' = \dfrac{-x}{1 + x}$ **21.** $y' = \dfrac{1}{\ln 10} + \log_{10} x$
23. $y' = x + 2x\ln(2x)$; $y'' = 3 + 2\ln(2x)$
25. $y' = \dfrac{1}{\sqrt{1 + x^2}}$; $y'' = \dfrac{-x}{(1 + x^2)^{3/2}}$
27. $f'(x) = \dfrac{2x - 1 - (x - 1)\ln(x - 1)}{(x - 1)[1 - \ln(x - 1)]^2}$;
$(1, 1 + e) \cup (1 + e, \infty)$
29. $f'(x) = \dfrac{2(x - 1)}{x(x - 2)}$; $(-\infty, 0) \cup (2, \infty)$
31. 1 **33.** $y = 3x - 2$ **35.** $\cos x + 1/x$
37. $y' = (2x + 1)^5(x^4 - 3)^6\left(\dfrac{10}{2x + 1} + \dfrac{24x^3}{x^4 - 3}\right)$
39. $y' = \dfrac{\sin^2 x \tan^4 x}{(x^2 + 1)^2}\left(2\cot x + \dfrac{4\sec^2 x}{\tan x} - \dfrac{4x}{x^2 + 1}\right)$
41. $y' = x^x(1 + \ln x)$
43. $y' = x^{\sin x}\left(\dfrac{\sin x}{x} + \cos x \ln x\right)$
45. $y' = (\cos x)^x(-x\tan x + \ln\cos x)$
47. $y' = (\tan x)^{1/x}\left(\dfrac{\sec^2 x}{x\tan x} - \dfrac{\ln\tan x}{x^2}\right)$
49. $y' = \dfrac{2x}{x^2 + y^2 - 2y}$ **51.** $f^{(n)}(x) = \dfrac{(-1)^{n-1}(n - 1)!}{(x - 1)^n}$

EXERCISES 3.7 ▪ PAGE 230

1. (a) $3t^2 - 24t + 36$ (b) -9 ft/s (c) $t = 2, 6$
(d) $0 \leqslant t < 2$, $t > 6$
(e) 96 ft
(f)

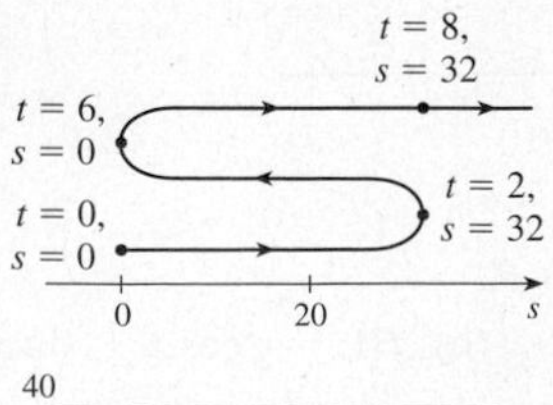

(g) $6t - 24$; -6 m/s²
(h)

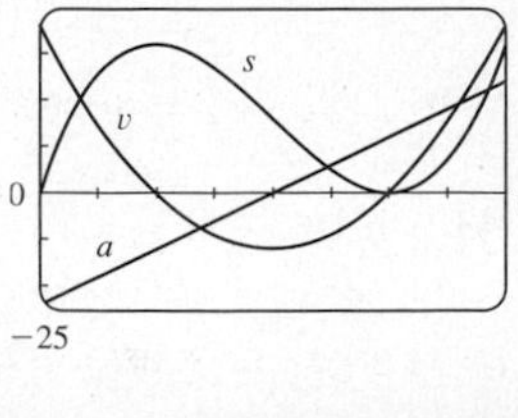

(i) Speeding up when $2 < t < 4$ or $t > 6$; slowing down when $0 \leqslant t < 2$ or $4 < t < 6$

3. (a) $-\dfrac{\pi}{4}\sin\left(\dfrac{\pi t}{4}\right)$ (b) $-\frac{1}{8}\pi\sqrt{2}$ ft/s (c) $t = 0, 4, 8$
(d) $4 < t < 8$ (e) 4 ft
(f)

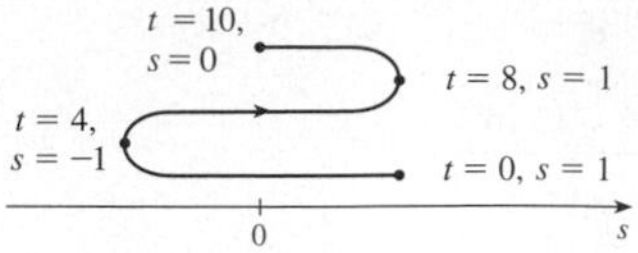

(g) $-\frac{1}{16}\pi^2\cos(\pi t/4)$; $\frac{1}{32}\pi^2\sqrt{2}$ ft/s^2
(h)

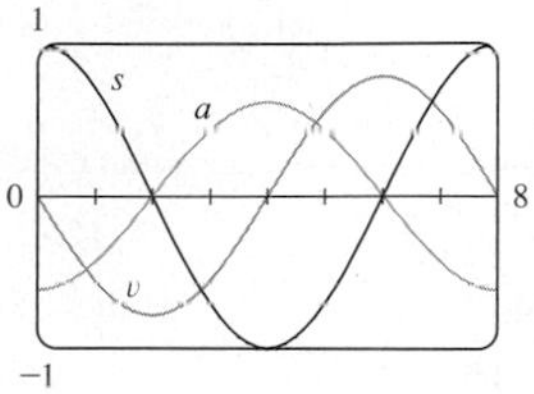

(i) Speeding up when $0 < t < 2$, $4 < t < 6$, $8 < t < 10$; slowing down when $2 < t < 4$, $6 < t < 8$
5. (a) Speeding up when $0 < t < 1$ or $2 < t < 3$; slowing down when $1 < t < 2$
(b) Speeding up when $1 < t < 2$ or $3 < t < 4$; slowing down when $0 < t < 1$ or $2 < t < 3$
7. (a) $t = 4$ s
(b) $t = 1.5$ s; the velocity has an absolute minimum.
9. (a) 5.02 m/s (b) $\sqrt{17}$ m/s
11. (a) 30 mm^2/mm; the rate at which the area is increasing with respect to side length as x reaches 15 mm
(b) $\Delta A \approx 2x\,\Delta x$
13. (a) (i) 5π (ii) 4.5π (iii) 4.1π
(b) 4π (c) $\Delta A \approx 2\pi r\,\Delta r$
15. (a) 8π ft^2/ft (b) 16π ft^2/ft (c) 24π ft^2/ft
The rate increases as the radius increases.
17. (a) 6 kg/m (b) 12 kg/m (c) 18 kg/m
At the right end; at the left end
19. (a) 4.75 A (b) 5 A; $t = \frac{2}{3}$ s
21. (a) $dV/dP = -C/P^2$ (b) At the beginning
23. $400(3^t)\ln 3$; ≈ 6850 bacteria/h
25. (a) 16 million/year; 78.5 million/year
(b) $P(t) = at^3 + bt^2 + ct + d$, where $a \approx 0.00129371$, $b \approx -7.061422$, $c \approx 12{,}822.979$, $d \approx -7{,}743{,}770$
(c) $P'(t) = 3at^2 + 2bt + c$
(d) 14.48 million/year; 75.29 million/year (smaller)
(e) 81.62 million/year
27. (a) 0.926 cm/s; 0.694 cm/s; 0
(b) 0; -92.6 (cm/s)/cm; -185.2 (cm/s)/cm
(c) At the center; at the edge
29. (a) $C'(x) = 12 - 0.2x + 0.0015x^2$
(b) \$32/yard; the cost of producing the 201st yard
(c) \$32.20
31. (a) $[xp'(x) - p(x)]/x^2$; the average productivity increases as new workers are added.
33. -0.2436 K/min
35. (a) 0 and 0 (b) $C = 0$
(c) (0, 0), (500, 50); it is possible for the species to coexist.

EXERCISES 3.8 ■ PAGE 239

1. About 235
3. (a) $100(4.2)^t$ (b) ≈ 7409 (c) $\approx 10{,}632$ bacteria/h
(d) $(\ln 100)/(\ln 4.2) \approx 3.2$ h
5. (a) 1508 million, 1871 million (b) 2161 million
(c) 3972 million; wars in the first half of century, increased life expectancy in second half
7. (a) $Ce^{-0.0005t}$ (b) $-2000\ln 0.9 \approx 211$ s
9. (a) $100 \times 2^{-t/30}$ mg (b) ≈ 9.92 mg (c) ≈ 199.3 years
11. ≈ 2500 years **13.** (a) ≈ 137°F (b) ≈ 116 min
15. (a) $13.\overline{3}$°C (b) ≈ 67.74 min
17. (a) ≈ 64.5 kPa (b) ≈ 39.9 kPa
19. (a) (i) \$3828.84 (ii) \$3840.25 (iii) \$3850.08
(iv) \$3851.61 (v) \$3852.01 (vi) \$3852.08
(b) $dA/dt = 0.05A$, $A(0) = 3000$

EXERCISES 3.9 ■ PAGE 245

1. $dV/dt = 3x^2\,dx/dt$ **3.** 48 cm^2/s **5.** $3/(25\pi)$ m/min
7. 70 **9.** $\pm\frac{46}{13}$
11. (a) The plane's altitude is 1 mi and its speed is 500 mi/h.
(b) The rate at which the distance from the plane to the station is increasing when the plane is 2 mi from the station
(c)

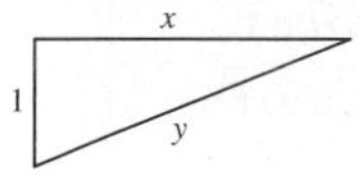

(d) $y^2 = x^2 + 1$
(e) $250\sqrt{3}$ mi/h
13. (a) The height of the pole (15 ft), the height of the man (6 ft), and the speed of the man (5 ft/s)
(b) The rate at which the tip of the man's shadow is moving when he is 40 ft from the pole
(c)

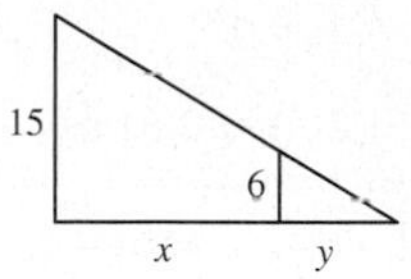

(d) $\dfrac{15}{6} = \dfrac{x + y}{y}$ (e) $\frac{25}{3}$ ft/s
15. 65 mi/h **17.** $837/\sqrt{8674} \approx 8.99$ ft/s
19. -1.6 cm/min **21.** $\frac{720}{13} \approx 55.4$ km/h
23. $(10{,}000 + 800{,}000\pi/9) \approx 2.89 \times 10^5$ cm^3/min
25. $\frac{10}{3}$ cm/min **27.** $6/(5\pi) \approx 0.38$ ft/min **29.** 0.3 m^2/s
31. 80 cm^3/min **33.** $\frac{107}{810} \approx 0.132$ Ω/s **35.** 0.396 m/min
37. (a) 360 ft/s (b) 0.096 rad/s **39.** $\frac{10}{9}\pi$ km/min
41. $1650/\sqrt{31} \approx 296$ km/h **43.** $\frac{7}{4}\sqrt{15} \approx 6.78$ m/s

EXERCISES 3.10 ■ PAGE 252

1. $L(x) = -10x - 6$ **3.** $L(x) = -x + \pi/2$
5. $\sqrt{1 - x} \approx 1 - \frac{1}{2}x$;
$\sqrt{0.9} \approx 0.95$,
$\sqrt{0.99} \approx 0.995$

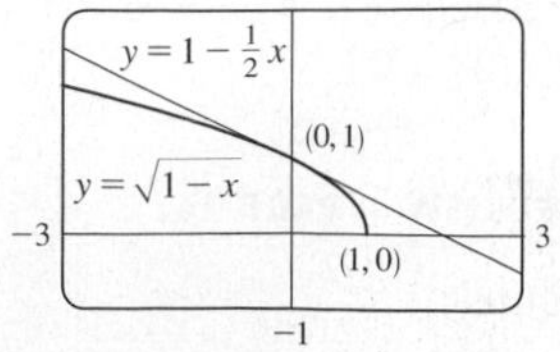

7. $-1.204 < x < 0.706$ **9.** $-0.045 < x < 0.055$
11. (a) $dy = 2x(x\cos 2x + \sin 2x)\,dx$ (b) $dy = \dfrac{t}{1 + t^2}\,dt$

13. (a) $dy = -\dfrac{2}{(u-1)^2}\,du$ (b) $dy = -\dfrac{6r^2}{(1+r^3)^3}\,dr$

15. (a) $dy = \frac{1}{10}e^{x/10}\,dx$ (b) 0.01; 0.0101

17. (a) $dy = \sec^2 x\,dx$ (b) -0.2

19. $\Delta y = 0.64$, $dy = 0.8$

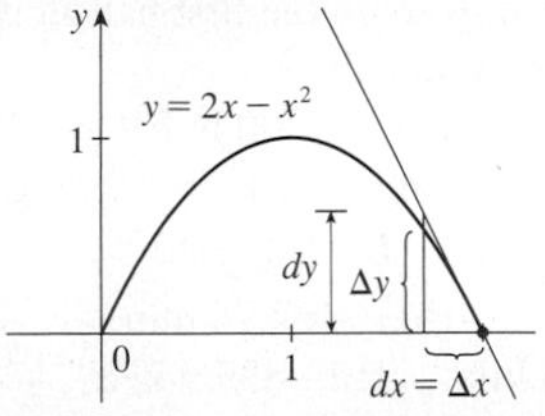

21. $\Delta y = -0.1$, $dy = -0.125$

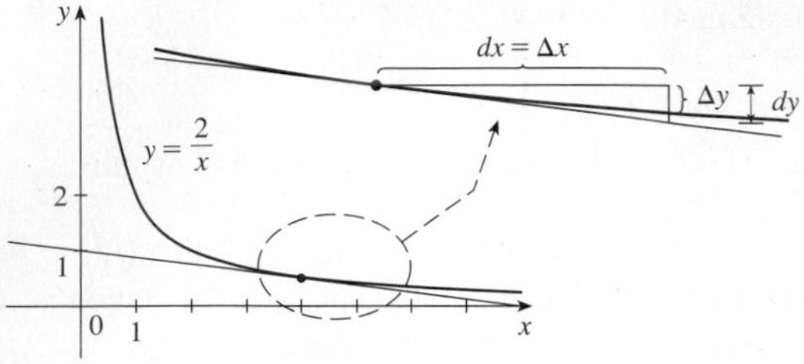

23. 32.08 **25.** 4.02 **27.** $1 - \pi/90 \approx 0.965$

33. (a) 270 cm^3, 0.01, 1% (b) 36 cm^2, $0.00\overline{6}$, $0.\overline{6}$%

35. (a) $84/\pi \approx 27$ cm^2; $\frac{1}{84} \approx 0.012$

(b) $1764/\pi^2 \approx 179$ cm^3; $\frac{1}{56} \approx 0.018$

37. (a) $2\pi rh\,\Delta r$ (b) $\pi(\Delta r)^2 h$

43. (a) 4.8, 5.2 (b) Too large

EXERCISES 3.11 ▪ PAGE 259

1. (a) 0 (b) 1 **3.** (a) $\frac{3}{4}$ (b) $\frac{1}{2}(e^2 - e^{-2}) \approx 3.62686$

5. (a) 1 (b) 0

21. $\operatorname{sech} x = \frac{3}{5}$, $\sinh x = \frac{4}{3}$, $\operatorname{csch} x = \frac{3}{4}$, $\tanh x = \frac{4}{5}$, $\coth x = \frac{5}{4}$

23. (a) 1 (b) -1 (c) ∞ (d) $-\infty$ (e) 0 (f) 1
(g) ∞ (h) $-\infty$ (i) 0

31. $f'(x) = x\cosh x$ **33.** $h'(x) = \tanh x$

35. $y' = 3e^{\cosh 3x}\sinh 3x$ **37.** $f'(t) = -2e^t\operatorname{sech}^2(e^t)\tanh(e^t)$

39. $y' = \dfrac{\operatorname{sech}^2 x}{1 + \tanh^2 x}$ **41.** $G'(x) = \dfrac{-2\sinh x}{(1+\cosh x)^2}$

43. $y' = \dfrac{1}{2\sqrt{x}(1-x)}$ **45.** $y' = \sinh^{-1}(x/3)$

47. $y' = \dfrac{-1}{x\sqrt{x^2+1}}$

51. (a) 0.3572 (b) 70.34°

53. (b) $y = 2\sinh 3x - 4\cosh 3x$

55. $\left(\ln(1+\sqrt{2}), \sqrt{2}\right)$

CHAPTER 3 REVIEW ▪ PAGE 261

True-False Quiz

1. True **3.** True **5.** False **7.** False **9.** True
11. True

Exercises

1. $6x(x^4 - 3x^2 + 5)^2(2x^2 - 3)$ **3.** $\dfrac{1}{2\sqrt{x}} - \dfrac{4}{3\sqrt[3]{x^7}}$

5. $\dfrac{2(2x^2+1)}{\sqrt{x^2+1}}$ **7.** $2\cos 2\theta\, e^{\sin 2\theta}$

9. $\dfrac{t^2+1}{(1-t^2)^2}$ **11.** $\dfrac{\cos\sqrt{x} - \sqrt{x}\sin\sqrt{x}}{2\sqrt{x}}$

13. $-\dfrac{e^{1/x}(1+2x)}{x^4}$ **15.** $\dfrac{1 - y^4 - 2xy}{4xy^3 + x^2 - 3}$

17. $\dfrac{2\sec 2\theta(\tan 2\theta - 1)}{(1+\tan 2\theta)^2}$ **19.** $(1+c^2)e^{cx}\sin x$

21. $3^{x\ln x}(\ln 3)(1 + \ln x)$ **23.** $-(x-1)^{-2}$

25. $\dfrac{2x - y\cos(xy)}{x\cos(xy) + 1}$ **27.** $\dfrac{2}{(1+2x)\ln 5}$

29. $\cot x - \sin x\cos x$ **31.** $\dfrac{4x}{1+16x^2} + \tan^{-1}(4x)$

33. $5\sec 5x$ **35.** $-6x\csc^2(3x^2+5)$

37. $\cos\left(\tan\sqrt{1+x^3}\right)\left(\sec^2\sqrt{1+x^3}\right)\dfrac{3x^2}{2\sqrt{1+x^3}}$

39. $2\cos\theta\tan(\sin\theta)\sec^2(\sin\theta)$

41. $\dfrac{(x-2)^4(3x^2 - 55x - 52)}{2\sqrt{x+1}(x+3)^8}$ **43.** $2x^2\cosh(x^2) + \sinh(x^2)$

45. $3\tanh 3x$ **47.** $\dfrac{\cosh x}{\sqrt{\sinh^2 x - 1}}$

49. $\dfrac{-3\sin\left(e^{\sqrt{\tan 3x}}\right)e^{\sqrt{\tan 3x}}\sec^2(3x)}{2\sqrt{\tan 3x}}$ **51.** $-\frac{4}{27}$ **53.** $-5x^4/y^{11}$

57. $y = 2\sqrt{3}x + 1 - \pi\sqrt{3}/3$ **59.** $y = 2x + 1$

61. $y = -x + 2$; $y = x + 2$

63. (a) $\dfrac{10 - 3x}{2\sqrt{5-x}}$ (b) $y = \frac{7}{4}x + \frac{1}{4}$, $y = -x + 8$

(c)

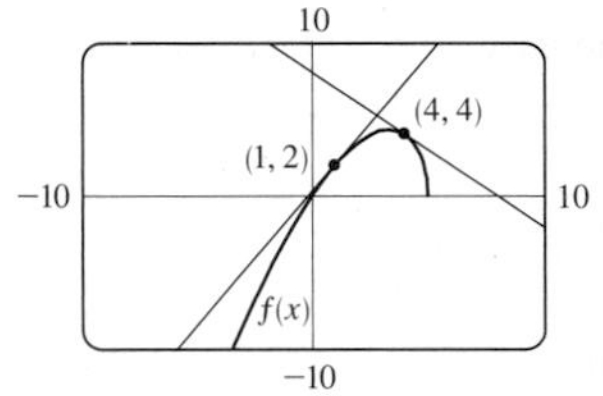

65. $(\pi/4, \sqrt{2})$, $(5\pi/4, -\sqrt{2})$ **69.** (a) 2 (b) 44

71. $2xg(x) + x^2 g'(x)$ **73.** $2g(x)g'(x)$ **75.** $g'(e^x)e^x$

77. $g'(x)/g(x)$ **79.** $\dfrac{f'(x)[g(x)]^2 + g'(x)[f(x)]^2}{[f(x)+g(x)]^2}$

81. $f'(g(\sin 4x))g'(\sin 4x)(\cos 4x)(4)$ **83.** $(-3, 0)$

85. $y = -\frac{2}{3}x^2 + \frac{14}{3}x$

87. $v(t) = -Ae^{-ct}[c\cos(\omega t + \delta) + \omega\sin(\omega t + \delta)]$,
$a(t) = Ae^{-ct}[(c^2 - \omega^2)\cos(\omega t + \delta) + 2c\omega\sin(\omega t + \delta)]$

89. (a) $v(t) = 3t^2 - 12$; $a(t) = 6t$ (b) $t > 2$; $0 \leqslant t < 2$
(c) 23 (d)

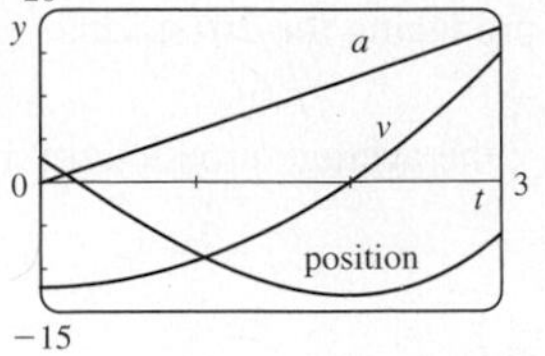

(e) $t > 2$; $0 < t < 2$

91. 4 kg/m **93.** (a) $200(3.24)^t$ (b) $\approx 22{,}040$
(c) $\approx 25{,}910$ bacteria/h (d) $(\ln 50)/(\ln 3.24) \approx 3.33$ h
95. (a) $C_0 e^{-kt}$ (b) ≈ 100 h **97.** $\frac{4}{3}$ cm²/min
99. 13 ft/s **101.** 400 ft/h
103. (a) $L(x) = 1 + x$; $\sqrt[3]{1 + 3x} \approx 1 + x$; $\sqrt[3]{1.03} \approx 1.01$
(b) $-0.23 < x < 0.40$
105. $12 + \frac{3}{2}\pi \approx 16.7$ cm² **107.** $\frac{1}{32}$ **109.** $\frac{1}{4}$ **111.** $\frac{1}{8}x^2$

PROBLEMS PLUS ▪ PAGE 266

1. $\left(\pm\frac{1}{2}\sqrt{3}, \frac{1}{4}\right)$ **9.** $\left(0, \frac{5}{4}\right)$
11. (a) $4\pi\sqrt{3}/\sqrt{11}$ rad/s (b) $40\left(\cos\theta + \sqrt{8 + \cos^2\theta}\right)$ cm
(c) $-480\pi \sin\theta \left(1 + \cos\theta/\sqrt{8 + \cos^2\theta}\right)$ cm/s
15. $x_T \in (3, \infty)$, $y_T \in (2, \infty)$, $x_N \in \left(0, \frac{5}{3}\right)$, $y_N \in \left(-\frac{5}{2}, 0\right)$
17. (b) (i) 53° (or 127°) (ii) 63° (or 117°)
19. R approaches the midpoint of the radius AO.
21. $-\sin a$ **23.** $2\sqrt{e}$ **27.** $(1, -2), (-1, 0)$
29. $\sqrt{29}/58$ **31.** $2 + \frac{375}{128}\pi \approx 11.204$ cm³/min

CHAPTER 4

EXERCISES 4.1 ▪ PAGE 277

Abbreviations: abs., absolute; loc., local; max., maximum; min., minimum

1. Absolute minimum: smallest function value on the entire domain of the function; local minimum at c: smallest function value when x is near c
3. Abs. max. at s, abs. min. at r, loc. max. at c, loc. min. at b and r
5. Abs. max. $f(4) = 5$, loc. max. $f(4) = 5$ and $f(6) = 4$, loc. min. $f(2) = 2$ and $f(5) = 3$

7.
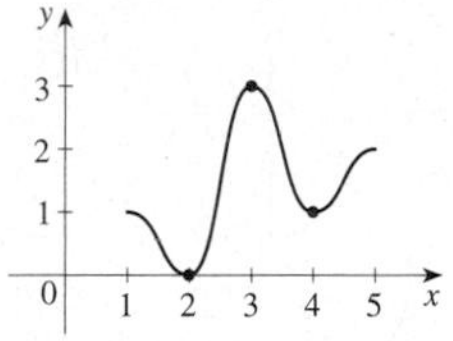

9.
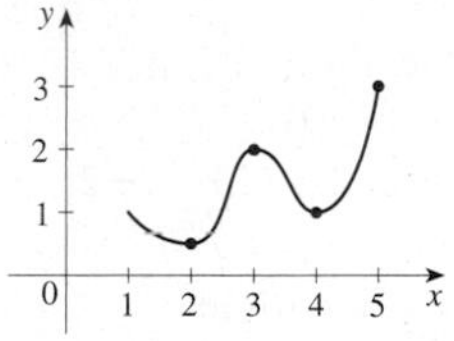

11. (a)
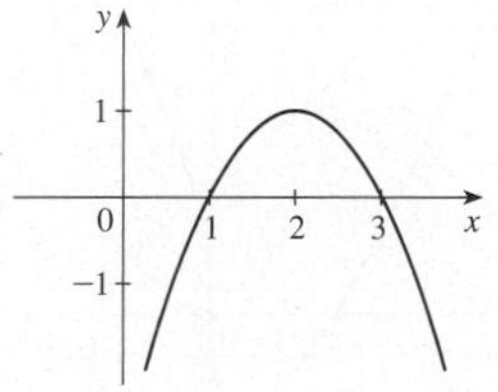

(b)
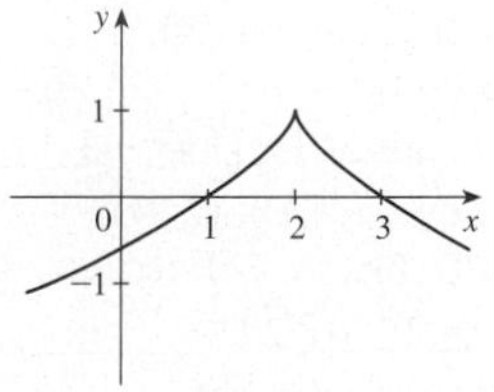

(c)
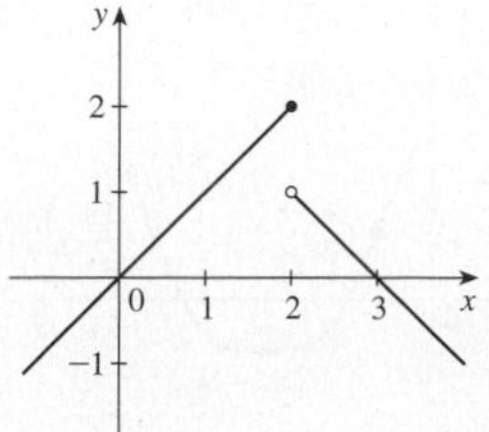

13. (a)
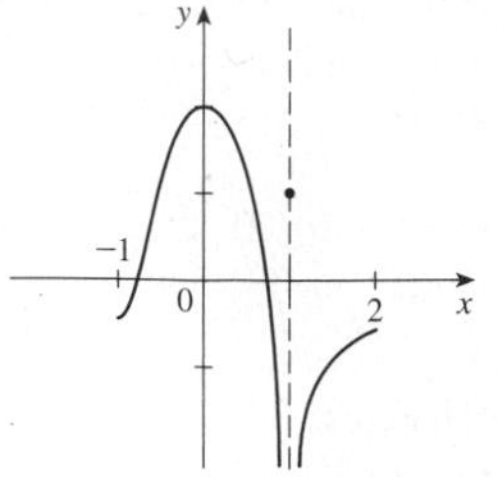

(b)
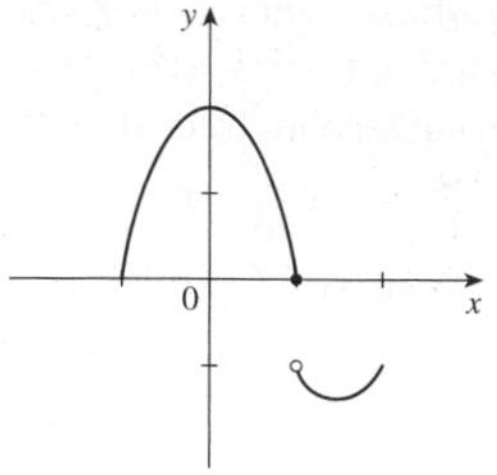

15. Abs. max. $f(1) = 5$ **17.** None
19. Abs. min. $f(0) = 0$
21. Abs. max. $f(-3) = 9$, abs. and loc. min. $f(0) = 0$
23. Abs. max. $f(2) = \ln 2$
25. Abs. max. $f(0) = 1$ **27.** Abs. max. $f(3) = 2$
29. $-\frac{2}{5}$ **31.** $-4, 2$ **33.** $0, \frac{1}{2}\left(-1 \pm \sqrt{5}\right)$ **35.** $0, 2$
37. $0, \frac{4}{9}$ **39.** $0, \frac{8}{7}, 4$ **41.** $n\pi$ (n an integer) **43.** $0, \frac{2}{3}$
45. 10 **47.** $f(0) = 5$, $f(2) = -7$
49. $f(-1) = 8$, $f(2) = -19$
51. $f(3) = 66$, $f(\pm 1) = 2$ **53.** $f(1) = \frac{1}{2}$, $f(0) = 0$
55. $f\left(\sqrt{2}\right) = 2$, $f(-1) = -\sqrt{3}$
57. $f(\pi/6) = \frac{3}{2}\sqrt{3}$, $f(\pi/2) = 0$
59. $f(2) = 2/\sqrt{e}$, $f(-1) = -1/\sqrt[8]{e}$
61. $f(1) = \ln 3$, $f\left(-\frac{1}{2}\right) = \ln\frac{3}{4}$
63. $f\left(\dfrac{a}{a + b}\right) = \dfrac{a^a b^b}{(a + b)^{a+b}}$
65. (a) 2.19, 1.81 (b) $\frac{6}{25}\sqrt{\frac{3}{5}} + 2$, $-\frac{6}{25}\sqrt{\frac{3}{5}} + 2$
67. (a) 0.32, 0.00 (b) $\frac{3}{16}\sqrt{3}, 0$ **69.** ≈ 3.9665°C
71. Cheapest, $t \approx 0.855$ (June 1994); most expensive, $t \approx 4.618$ (March 1998)
73. (a) $r = \frac{2}{3}r_0$ (b) $v = \frac{4}{27}kr_0^3$
(c)
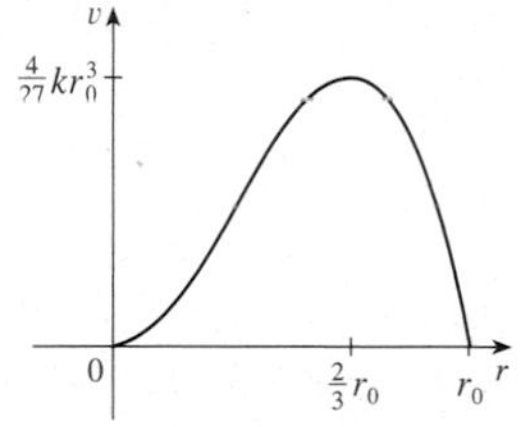

EXERCISES 4.2 ▪ PAGE 285

1. 2 **3.** $\frac{9}{4}$ **5.** f is not differentiable on $(-1, 1)$
7. 0.8, 3.2, 4.4, 6.1
9. (a), (b)
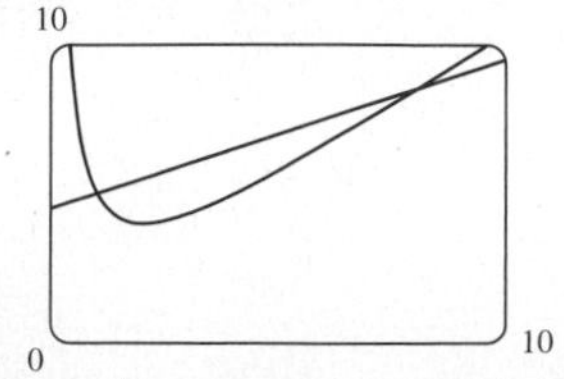

(c) $2\sqrt{2}$
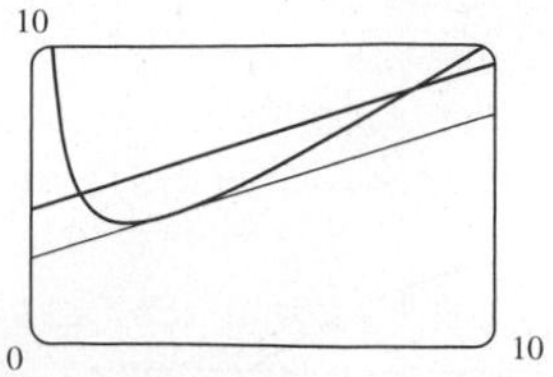

11. 0 **13.** $-\frac{1}{2}\ln\left[\frac{1}{6}\left(1 - e^{-6}\right)\right]$ **15.** f is not continous at 3
23. 16 **25.** No **31.** No

EXERCISES 4.3 ▪ PAGE 295

Abbreviations: inc., increasing; dec., decreasing; CD, concave downward; CU, concave upward; HA, horizontal asymptote; VA, vertical asymptote; IP, inflection point(s)

1. (a) $(1, 3), (4, 6)$ (b) $(0, 1), (3, 4)$ (c) $(0, 2)$
(d) $(2, 4), (4, 6)$ (e) $(2, 3)$

3. (a) I/D Test (b) Concavity Test
(c) Find points at which the concavity changes.

5. (a) Inc. on $(1, 5)$; dec. on $(0, 1)$ and $(5, 6)$
(b) Loc. max. at $x = 5$, loc. min. at $x = 1$

7. $x = 1, 7$

9. (a) Inc. on $(-\infty, 3)$, $(2, \infty)$; dec. on $(-3, 2)$
(b) Loc. max. $f(-3) = 81$; loc. min. $f(2) = -44$
(c) CU on $(-\frac{1}{2}, \infty)$; CD on $(-\infty, -\frac{1}{2})$; IP $(-\frac{1}{2}, \frac{37}{2})$

11. (a) Inc. on $(-1, 0)$, $(1, \infty)$; dec. on $(-\infty, -1)$, $(0, 1)$
(b) Loc. max. $f(0) = 3$; loc. min. $f(\pm 1) = 2$
(c) CU on $(-\infty, -\sqrt{3}/3)$, $(\sqrt{3}/3, \infty)$;
CD on $(-\sqrt{3}/3, \sqrt{3}/3)$; IP $(\pm\sqrt{3}/3, \frac{22}{9})$

13. (a) Inc. on $(0, \pi/4)$, $(5\pi/4, 2\pi)$; dec. on $(\pi/4, 5\pi/4)$
(b) Loc. max. $f(\pi/4) = \sqrt{2}$; loc. min. $f(5\pi/4) = -\sqrt{2}$
(c) CU on $(3\pi/4, 7\pi/4)$; CD on $(0, 3\pi/4)$, $(7\pi/4, 2\pi)$;
IP $(3\pi/4, 0)$, $(7\pi/4, 0)$

15. (a) Inc. on $(-\frac{1}{3}\ln 2, \infty)$; dec. on $(-\infty, -\frac{1}{3}\ln 2)$
(b) Loc. min. $f(-\frac{1}{3}\ln 2) = 2^{-2/3} + 2^{1/3}$ (c) CU on $(-\infty, \infty)$

17. (a) Inc. on $(0, e^2)$; dec. on (e^2, ∞)
(b) Loc. max. $f(e^2) = 2/e$
(c) CU on $(e^{8/3}, \infty)$; CD on $(0, e^{8/3})$; IP $(e^{8/3}, \frac{8}{3}e^{-4/3})$

19. Loc. max. $f(-1) = 7$, loc. min. $f(1) = -1$

21. Loc. max. $f(\frac{3}{4}) = \frac{5}{4}$

23. (a) f has a local maximum at 2.
(b) f has a horizontal tangent at 6.

25.

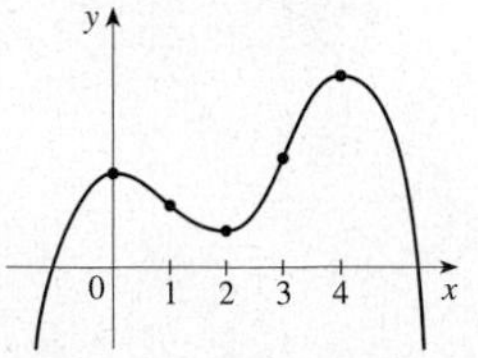

27.

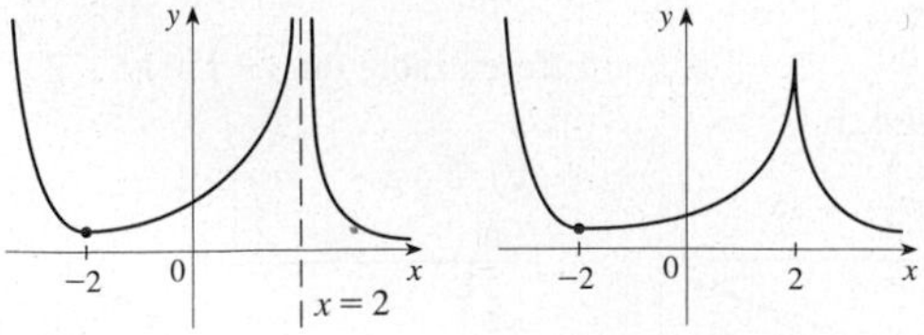

29.

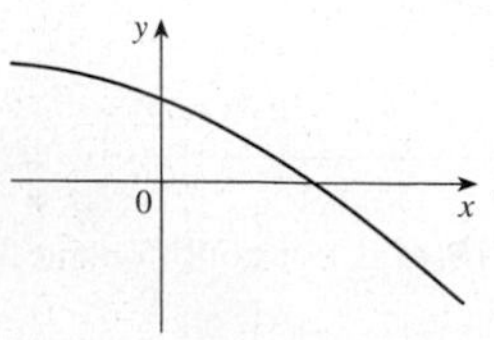

31. (a) Inc. on $(0, 2)$, $(4, 6)$, $(8, \infty)$;
dec. on $(2, 4)$, $(6, 8)$
(b) Loc. max. at $x = 2, 6$;
loc. min. at $x = 4, 8$
(c) CU on $(3, 6)$, $(6, \infty)$;
CD on $(0, 3)$
(d) 3 (e) See graph at right.

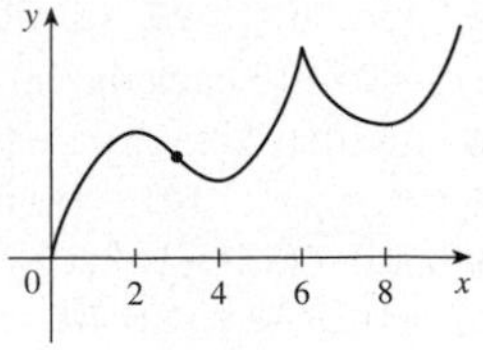

33. (a) Inc. on $(-\infty, -1)$, $(2, \infty)$;
dec. on $(-1, 2)$
(b) Loc. max. $f(-1) = 7$;
loc. min. $f(2) = -20$
(c) CU on $(\frac{1}{2}, \infty)$; CD on $(-\infty, \frac{1}{2})$;
IP $(\frac{1}{2}, -\frac{13}{2})$
(d) See graph at right.

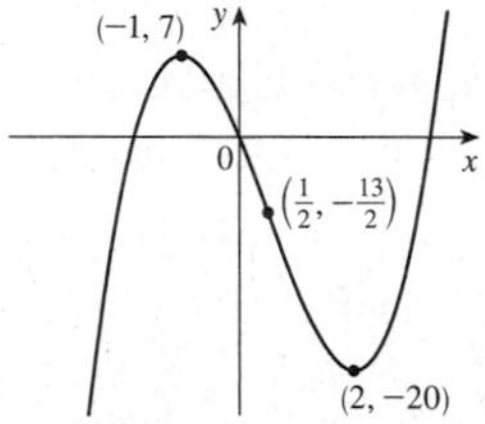

35. (a) Inc. on $(-\infty, -1)$, $(0, 1)$;
dec. on $(-1, 0)$, $(1, \infty)$
(b) Loc. max. $f(-1) = 3$, $f(1) = 3$;
loc. min. $f(0) = 2$
(c) CU on $(-1/\sqrt{3}, 1/\sqrt{3})$;
CD on $(-\infty, -1/\sqrt{3})$, $(1/\sqrt{3}, \infty)$;
IP $(\pm 1/\sqrt{3}, \frac{23}{9})$
(d) See graph at right.

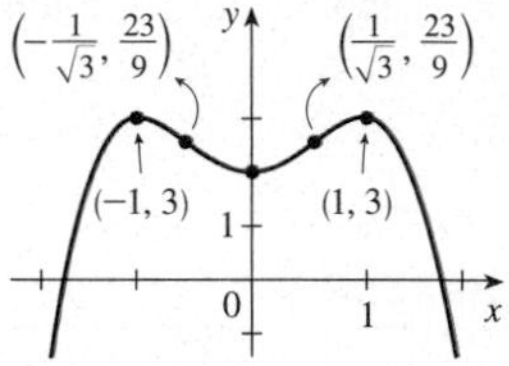

37. (a) Inc. on $(-\infty, -2)$, $(0, \infty)$;
dec. on $(-2, 0)$
(b) Loc. max. $h(-2) = 7$;
loc. min. $h(0) = -1$
(c) CU on $(-1, \infty)$;
CD on $(-\infty, -1)$; IP $(-1, 3)$
(d) See graph at right.

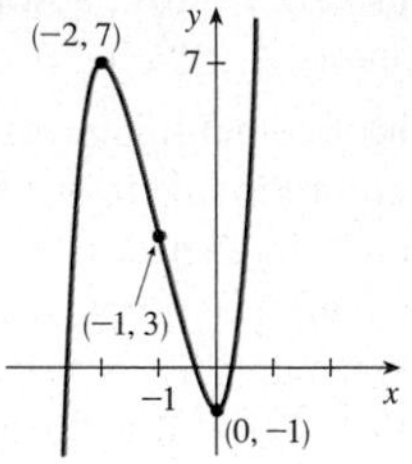

39. (a) Inc. on $(-2, \infty)$;
dec. on $(-3, -2)$
(b) Loc. min. $A(-2) = -2$
(c) CU on $(-3, \infty)$
(d) See graph at right.

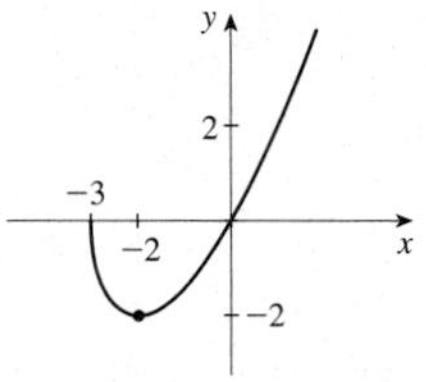

41. (a) Inc. on $(-1, \infty)$;
dec. on $(-\infty, -1)$
(b) Loc. min. $C(-1) = -3$
(c) CU on $(-\infty, 0)$, $(2, \infty)$;
CD on $(0, 2)$;
IPs $(0, 0)$, $(2, 6\sqrt[3]{2})$
(d) See graph at right.

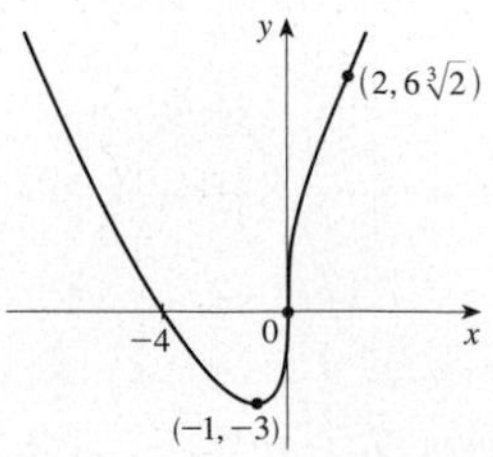

43. (a) Inc. on $(\pi, 2\pi)$;
dec. on $(0, \pi)$
(b) Loc. min. $f(\pi) = -1$
(c) CU on $(\pi/3, 5\pi/3)$;
CD on $(0, \pi/3)$, $(5\pi/3, 2\pi)$;
IP $(\pi/3, \frac{5}{4})$, $(5\pi/3, \frac{5}{4})$
(d) See graph at right.

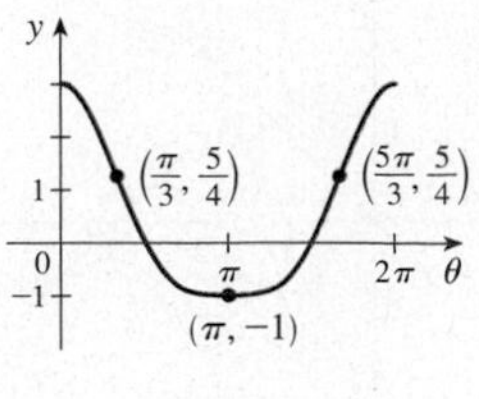

45. (a) HA $y = 1$, VA $x = -1$, $x = 1$
(b) Inc. on $(-\infty, -1)$, $(-1, 0)$;
dec. on $(0, 1)$, $(1, \infty)$
(c) Loc. max. $f(0) = 0$
(d) CU on $(-\infty, -1)$, $(1, \infty)$;
CD on $(-1, 1)$
(e) See graph at right.

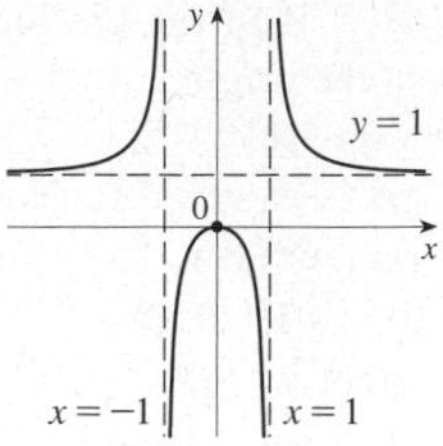

47. (a) HA $y = 0$
(b) Dec. on $(-\infty, \infty)$
(c) None
(d) CU on $(-\infty, \infty)$
(e) See graph at right.

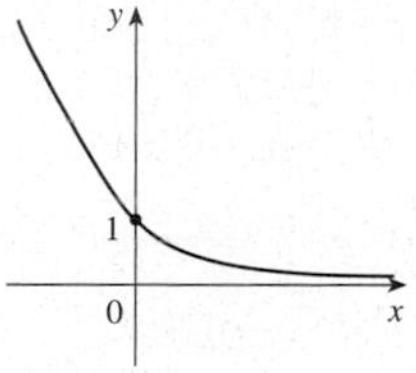

49. (a) VA $x = 0$, $x = e$
(b) Dec. on $(0, e)$
(c) None
(d) CU on $(0, 1)$; CD on $(1, e)$;
IP $(1, 0)$
(e) See graph at right.

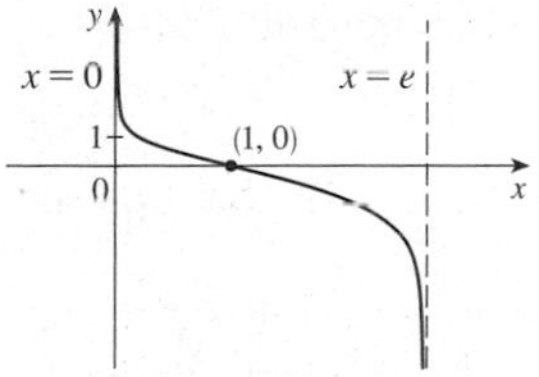

51. (a) HA $y = 1$, VA $x = -1$
(b) Inc. on $(-\infty, -1)$, $(-1, \infty)$
(c) None
(d) CU on $(-\infty, -1)$, $\left(-1, -\frac{1}{2}\right)$;
CD on $\left(-\frac{1}{2}, \infty\right)$; IP $\left(-\frac{1}{2}, 1/e^2\right)$
(e) See graph at right.

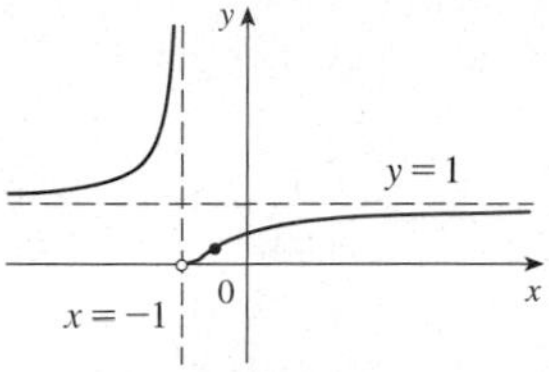

53. $(3, \infty)$
55. (a) Loc. and abs. max. $f(1) = \sqrt{2}$, no min.
(b) $\frac{1}{4}(3 - \sqrt{17})$
57. (b) CU on $(0.94, 2.57)$, $(3.71, 5.35)$;
CD on $(0, 0.94)$, $(2.57, 3.71)$, $(5.35, 2\pi)$;
IP $(0.94, 0.44)$, $(2.57, -0.63)$, $(3.71, -0.63)$, $(5.35, 0.44)$
59. CU on $(-\infty, -0.6)$, $(0.0, \infty)$; CD on $(-0.6, 0.0)$
61. (a) The rate of increase is initially very small, increases to a maximum at $t \approx 8$ h, then decreases toward 0.
(b) When $t = 8$ (c) CU on $(0, 8)$; CD on $(8, 18)$ (d) $(8, 350)$
63. $K(3) - K(2)$; CD
65. 28.57 min, when the rate of increase of drug level in the bloodstream is greatest; 85.71 min, when rate of decrease is greatest
67. $f(x) = \frac{1}{9}(2x^3 + 3x^2 - 12x + 7)$

EXERCISES 4.4 ■ PAGE 304

1. (a) Indeterminate (b) 0 (c) 0
(d) ∞, $-\infty$, or does not exist (e) Indeterminate
3. (a) $-\infty$ (b) Indeterminate (c) ∞
5. 2 **7.** $\frac{9}{5}$ **9.** $-\infty$ **11.** ∞ **13.** p/q
15. 0 **17.** $-\infty$ **19.** ∞ **21.** $\frac{1}{2}$ **23.** 1
25. $\ln \frac{5}{3}$ **27.** 1 **29.** $\frac{1}{2}$ **31.** 0 **33.** $-1/\pi^2$
35. $\frac{1}{2}a(a - 1)$ **37.** $\frac{1}{24}$ **39.** π **41.** 3 **43.** 0
45. $-2/\pi$ **47.** $\frac{1}{2}$ **49.** $\frac{1}{2}$ **51.** ∞ **53.** 1
55. e^{-2} **57.** e^3 **59.** 1 **61.** e^4
63. $1/\sqrt{e}$ **65.** e^2 **67.** $\frac{1}{4}$ **71.** 1 **77.** $\frac{16}{9}a$ **79.** 56
83. (a) 0

EXERCISES 4.5 ■ PAGE 314

1. A. $\mathbb{R}$ B. y-int. 0; x-int. 0
C. About $(0, 0)$ D. None
E. Inc. on $(-\infty, \infty)$ F. None
G. CU on $(0, \infty)$; CD on $(-\infty, 0)$;
IP $(0, 0)$
H. See graph at right.

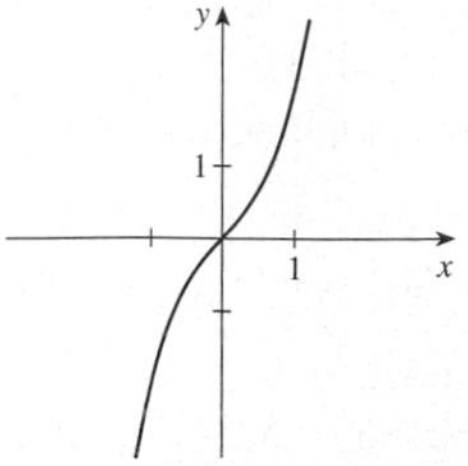

3. A. $\mathbb{R}$ B. y int. 2; x-int. 2, $\frac{1}{2}(7 \pm 3\sqrt{5})$
C. None D. None
E. Inc. on $(1, 5)$;
dec. on $(-\infty, 1)$, $(5, \infty)$
F. Loc. min. $f(1) = -5$;
loc. max. $f(5) = 27$
G. CU on $(-\infty, 3)$;
CD on $(3, \infty)$; IP $(3, 11)$
H. See graph at right.

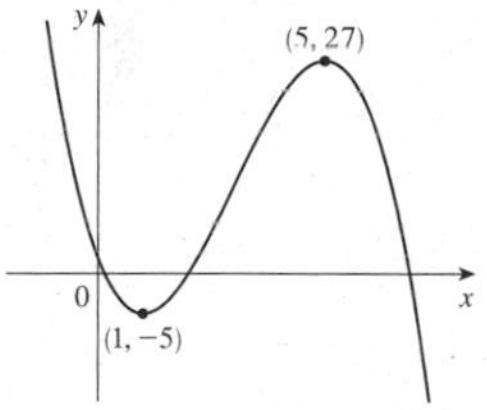

5. A. $\mathbb{R}$ B. y-int. 0; x-int. -4, 0
C. None D. None
E. Inc. on $(-3, \infty)$;
dec. on $(-\infty, -3)$
F. Loc. min. $f(-3) = -27$
G. CU on $(-\infty, -2)$, $(0, \infty)$;
CD on $(-2, 0)$; IP $(0, 0)$, $(-2, -16)$
H. See graph at right.

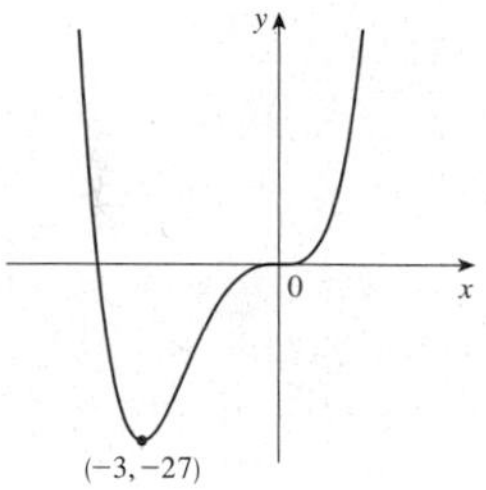

7. A. $\mathbb{R}$ B. y-int. 1
C. None D. None
E. Inc. on $(-\infty, 0)$, $(1, \infty)$;
dec. on $(0, 1)$
F. Loc. max. $f(0) = 1$;
loc. min. $f(1) = -2$
G. CU on $\left(1/\sqrt[3]{4}, \infty\right)$;
CD on $\left(-\infty, 1/\sqrt[3]{4}\right)$;
IP $\left(1/\sqrt[3]{4}, 1 - 9/\left(2\sqrt[3]{16}\right)\right)$
H. See graph at right.

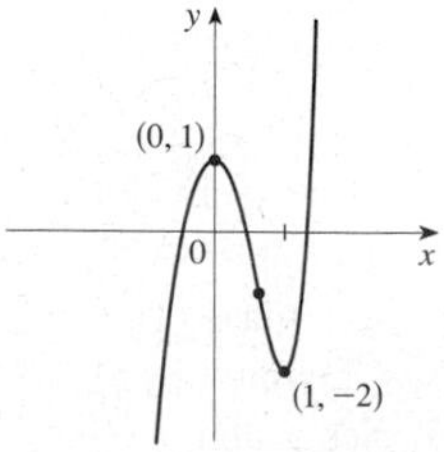

9. A. $\{x \mid x \neq 1\}$ B. y-int. 0; x-int. 0
C. None D. VA $x = 1$, HA $y = 1$
E. Dec. on $(-\infty, 1)$, $(1, \infty)$
F. None
G. CU on $(1, \infty)$; CD on $(-\infty, 1)$
H. See graph at right.

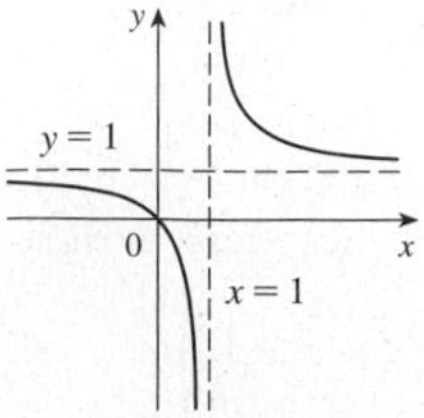

11. A. $\{x \mid x \neq \pm 3\}$ B. y-int. $-\frac{1}{9}$
C. About y-axis D. VA $x = \pm 3$, HA $y = 0$
E. Inc. on $(-\infty, -3)$, $(-3, 0)$;
dec. on $(0, 3)$, $(3, \infty)$
F. Loc. max. $f(0) = -\frac{1}{9}$
G. CU on $(-\infty, -3)$, $(3, \infty)$;
CD on $(-3, 3)$
H. See graph at right.

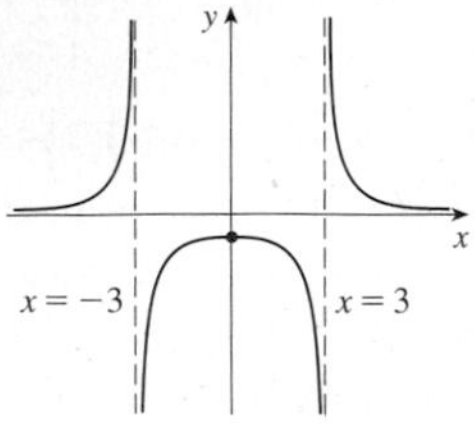

13. A. $\mathbb{R}$ B. y-int. 0; x-int. 0
C. About (0, 0) D. HA $y = 0$
E. Inc. on $(-3, 3)$;
dec. on $(-\infty, -3)$, $(3, \infty)$
F. Loc. min. $f(-3) = -\frac{1}{6}$;
loc. max. $f(3) = \frac{1}{6}$;
G. CU on $(-3\sqrt{3}, 0)$, $(3\sqrt{3}, \infty)$;
CD on $(-\infty, -3\sqrt{3})$, $(0, 3\sqrt{3})$;
IP $(0, 0)$, $(\pm 3\sqrt{3}, \pm\sqrt{3}/12)$
H. See graph at right.

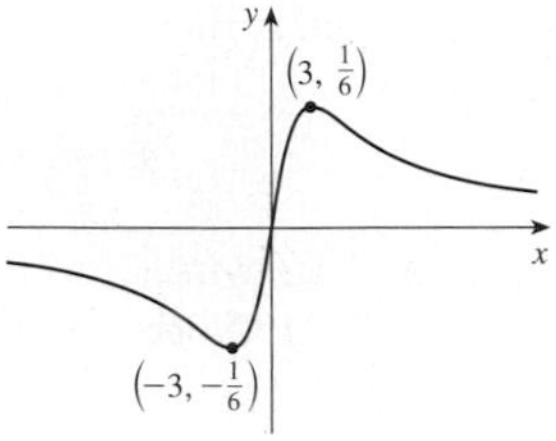

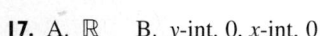

15. A. $(-\infty, 0) \cup (0, \infty)$ B. x-int. 1
C. None D. HA $y = 0$; VA $x = 0$
E. Inc. on $(0, 2)$;
dec. on $(-\infty, 0)$, $(2, \infty)$
F. Loc. max. $f(2) = \frac{1}{4}$
G. CU on $(3, \infty)$;
CD on $(-\infty, 0)$, $(0, 3)$; IP $(3, \frac{2}{9})$
H. See graph at right

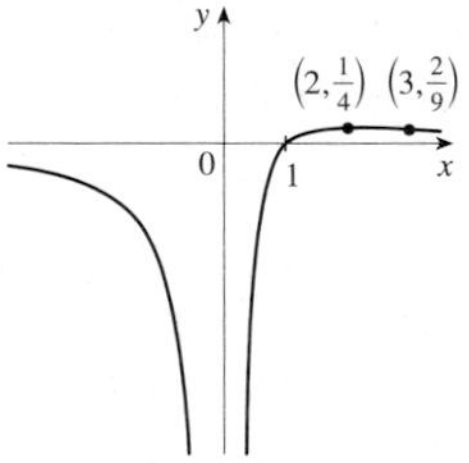

17. A. $\mathbb{R}$ B. y-int. 0, x-int. 0
C. About y-axis D. HA $y = 1$
E. Inc. on $(0, \infty)$; dec. on $(-\infty, 0)$
F. Loc. min. $f(0) = 0$
G. CU on $(-1, 1)$;
CD on $(-\infty, -1)$, $(1, \infty)$; IP $(\pm 1, \frac{1}{4})$
H. See graph at right

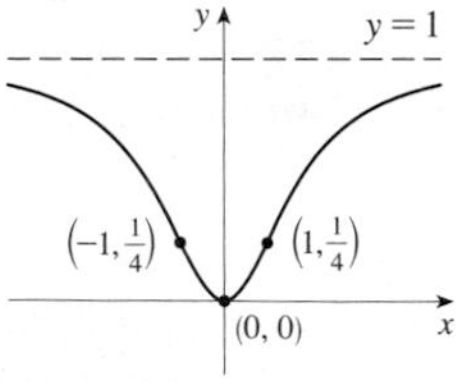

19. A. $(-\infty, 5]$ B. y-int. 0; x-int. 0, 5
C. None D. None
E. Inc. on $(-\infty, \frac{10}{3})$; dec. on $(\frac{10}{3}, 5)$
F. Loc. max. $f(\frac{10}{3}) = \frac{10}{9}\sqrt{15}$
G. CD on $(-\infty, 5)$
H. See graph at right.

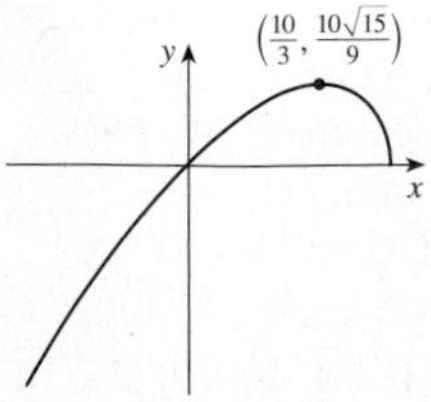

21. A. $(-\infty, -2) \cup (1, \infty)$
B. x-int. -2, 1
C. None D. None
E. Inc. on $(1, \infty)$; dec. on $(-\infty, -2)$
F. None
G. CD on $(-\infty, -2)$, $(1, \infty)$
H. See graph at right.

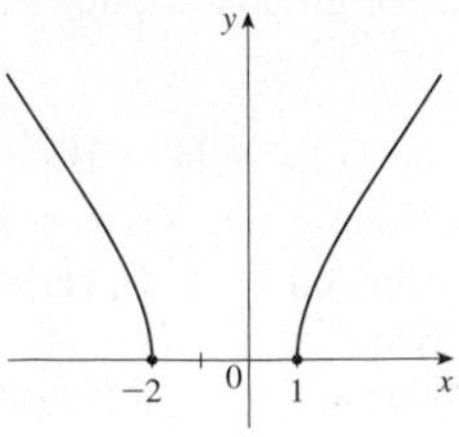

23. A. $\mathbb{R}$ B. y-int. 0; x-int. 0
C. About the origin
D. HA $y = \pm 1$
E. Inc. on $(-\infty, \infty)$ F. None
G. CU on $(-\infty, 0)$;
CD on $(0, \infty)$; IP $(0, 0)$
H. See graph at right.

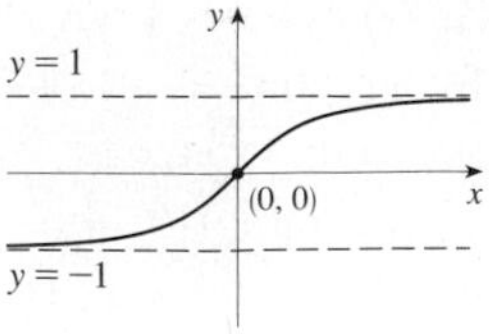

25. A. $\{x \mid |x| \leqslant 1, x \neq 0\} = [-1, 0) \cup (0, 1]$
B. x-int. ± 1 C. About (0, 0)
D. VA $x = 0$
E. Dec. on $(-1, 0)$, $(0, 1)$
F. None
G. CU on $(-1, -\sqrt{2/3})$, $(0, \sqrt{2/3})$;
CD on $(-\sqrt{2/3}, 0)$, $(\sqrt{2/3}, 1)$;
IP $(\pm\sqrt{2/3}, \pm 1/\sqrt{2})$
H. See graph at right.

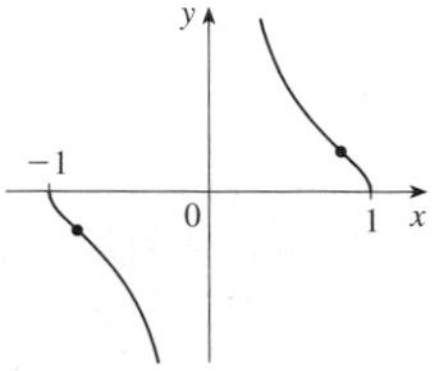

27. A. $\mathbb{R}$ B. y-int. 0; x-int. 0, $\pm 3\sqrt{3}$ C. About the origin
D. None E. Inc. on $(-\infty, -1)$, $(1, \infty)$; dec. on $(-1, 1)$
F. Loc. max. $f(-1) = 2$;
loc. min. $f(1) = -2$
G. CU on $(0, \infty)$; CD on $(-\infty, 0)$;
IP $(0, 0)$
H. See graph at right.

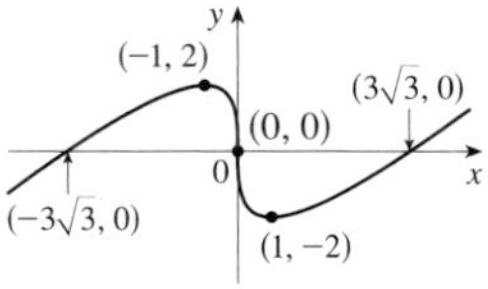

29. A. $\mathbb{R}$ B. y-int. -1; x-int. ± 1
C. About y-axis D. None
E. Inc. on $(0, \infty)$; dec. on $(-\infty, 0)$
F. Loc. min. $f(0) = -1$
G. CU on $(-1, 1)$;
CD on $(-\infty, -1)$, $(1, \infty)$;
IP $(\pm 1, 0)$
H. See graph at right.

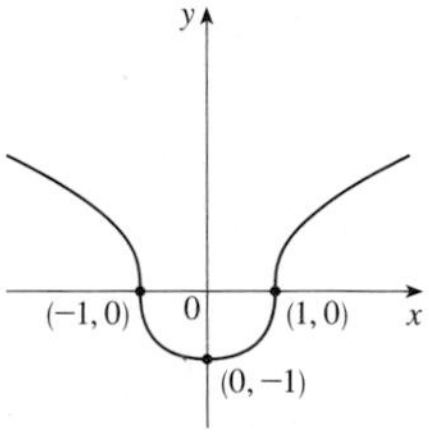

31. A. $\mathbb{R}$ B. y-int. 0; x-int. $n\pi$ (n an integer)
C. About the origin, period 2π D. None
E. Inc. on $(2n\pi - \pi/2, 2n\pi + \pi/2)$;
dec. on $(2n\pi + \pi/2, 2n\pi + 3\pi/2)$
F. Loc. max. $f(2n\pi + \pi/2) = 2$;
loc. min. $f(2n\pi + 3\pi/2) = -2$
G. CU on $((2n - 1)\pi, 2n\pi)$;
CD on $(2n\pi, (2n + 1)\pi)$; IP $(n\pi, 0)$
H. See graph at right.

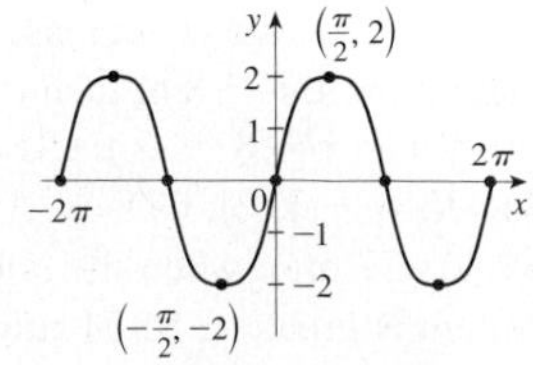

33. A. $(-\pi/2, \pi/2)$ B. y-int. 0; x-int. 0 C. About y-axis
D. VA $x = \pm\pi/2$
E. Inc. on $(0, \pi/2)$;
dec. on $(-\pi/2, 0)$
F. Loc. min. $f(0) = 0$
G. CU on $(-\pi/2, \pi/2)$
H. See graph at right.

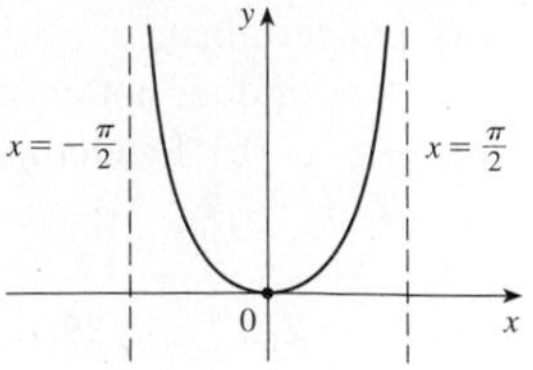

35. A. $(0, 3\pi)$ C. None D. None
E. Inc. on $(\pi/3, 5\pi/3)$, $(7\pi/3, 3\pi)$;
dec. on $(0, \pi/3)$, $(5\pi/3, 7\pi/3)$
F. Loc. min. $f(\pi/3) = (\pi/6) - \frac{1}{2}\sqrt{3}$, $f(7\pi/3) = (7\pi/6) - \frac{1}{2}\sqrt{3}$;
loc. max. $f(5\pi/3) = (5\pi/6) + \frac{1}{2}\sqrt{3}$
G. CU on $(0, \pi)$, $(2\pi, 3\pi)$;
CD on $(\pi, 2\pi)$;
IP $(\pi, \pi/2)$, $(2\pi, \pi)$
H. See graph at right.

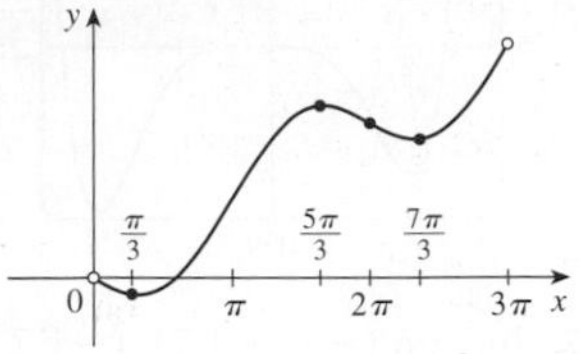

37. A. All reals except $(2n + 1)\pi$ (n an integer)
B. y-int. 0; x-int. $2n\pi$
C. About the origin, period 2π
D. VA $x = (2n + 1)\pi$
E. Inc. on $((2n - 1)\pi, (2n + 1)\pi)$ F. None
G. CU on $(2n\pi, (2n + 1)\pi)$; CD on $((2n - 1)\pi, 2n\pi)$;
IP $(2n\pi, 0)$
H.

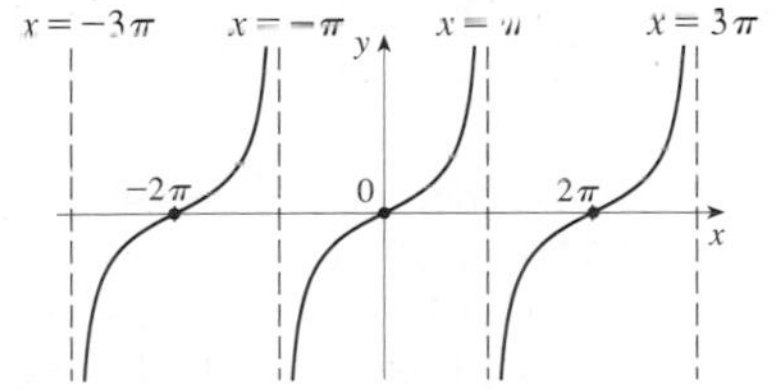

39. A. $\mathbb{R}$ B. y-int. 1 C. Period 2π D. None
Answers for E–G are for the interval $[0, 2\pi]$.
E. Inc. on $(0, \pi/2)$, $(3\pi/2, 2\pi)$; dec. on $(\pi/2, 3\pi/2)$
F. Loc. max. $f(\pi/2) = e$; loc. min. $f(3\pi/2) = e^{-1}$
G. CU on $(0, \alpha)$, $(\beta, 2\pi)$ where $\alpha = \sin^{-1}\left(\frac{1}{2}(-1 + \sqrt{5})\right)$,
$\beta = \pi - \alpha$; CD on (α, β); IP when $x = \alpha, \beta$
H.

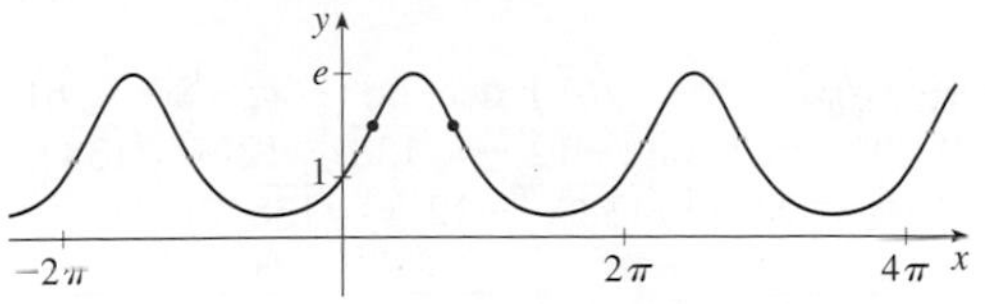

41. A. $\mathbb{R}$ B. y-int. $\frac{1}{2}$ C. None
D. HA $y = 0$, $y = 1$
E. Inc. on $\mathbb{R}$ F. None
G. CU on $(-\infty, 0)$; CD on $(0, \infty)$;
IP $\left(0, \frac{1}{2}\right)$ H. See graph at right.

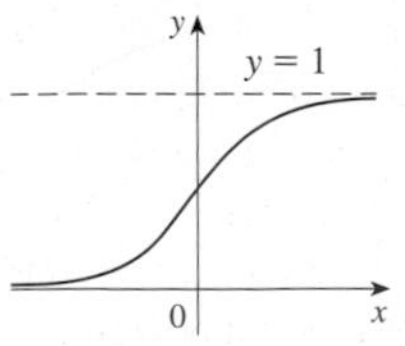

43. A. $(0, \infty)$ B. None
C. None D. VA $x = 0$
E. Inc. on $(1, \infty)$; dec. on $(0, 1)$
F. Loc. min. $f(1) = 1$
G. CU on $(0, \infty)$
H. See graph at right.

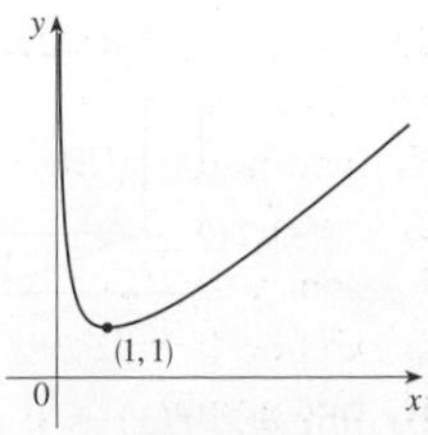

45. A. $\mathbb{R}$ B. y-int. $\frac{1}{4}$ C. None
D. HA $y = 0$, $y = 1$
E. Dec. on $\mathbb{R}$ F. None
G. CU on $\left(\ln \frac{1}{2}, \infty\right)$; CD on $\left(-\infty, \ln \frac{1}{2}\right)$;
IP $\left(\ln \frac{1}{2}, \frac{4}{9}\right)$
H. See graph at right.

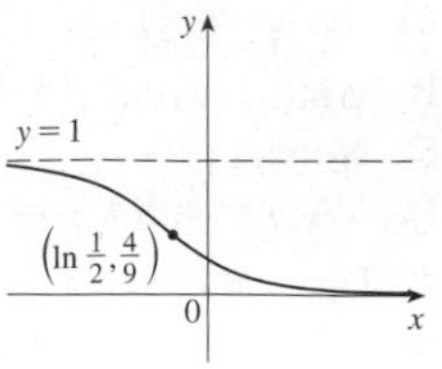

47. A. All x in $(2n\pi, (2n + 1)\pi)$ (n an integer)
B. x-int. $\pi/2 + 2n\pi$ C. Period 2π D. VA $x = n\pi$
E. Inc. on $(2n\pi, \pi/2 + 2n\pi)$; dec. on $(\pi/2 + 2n\pi, (2n + 1)\pi)$
F. Loc. max. $f(\pi/2 + 2n\pi) = 0$ G. CD on $(2n\pi, (2n + 1)\pi)$
H.

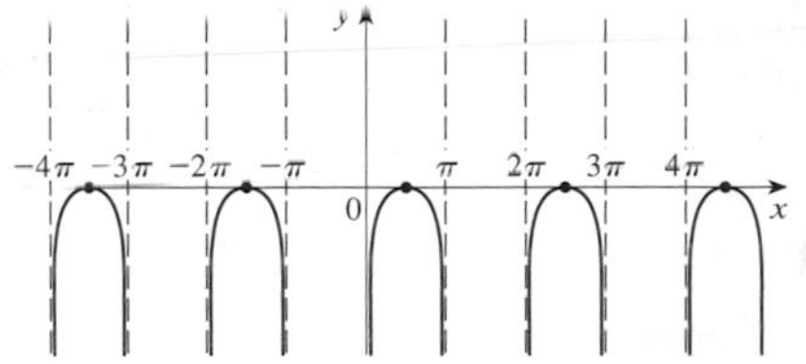

49. A. $\mathbb{R}$ B. y-int. 0; x-int. 0 C. About $(0, 0)$ D. HA $y = 0$
E. Inc. on $\left(-1/\sqrt{2}, 1/\sqrt{2}\right)$; dec. on $\left(-\infty, -1/\sqrt{2}\right)$, $\left(1/\sqrt{2}, \infty\right)$
F. Loc. min. $f\left(-1/\sqrt{2}\right) = -1/\sqrt{2e}$; loc. max. $f\left(1/\sqrt{2}\right) = 1/\sqrt{2e}$
G. CU on $\left(-\sqrt{3/2}, 0\right)$, $\left(\sqrt{3/2}, \infty\right)$; CD on $\left(-\infty, -\sqrt{3/2}\right)$, $\left(0, \sqrt{3/2}\right)$;
IP $\left(\pm\sqrt{3/2}, \pm\sqrt{3/2}e^{-3/2}\right)$, $(0, 0)$
H.

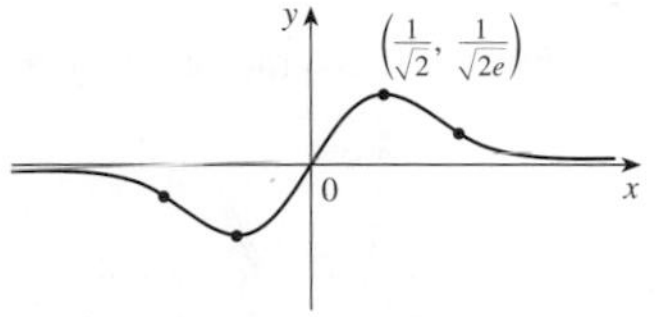

51. A. $\mathbb{R}$ B. y-int. 2
C. None D. None
E. Inc. on $\left(\frac{1}{5} \ln \frac{2}{3}, \infty\right)$; dec. on $\left(-\infty, \frac{1}{5} \ln \frac{2}{3}\right)$
F. Loc. min. $f\left(\frac{1}{5} \ln \frac{2}{3}\right) = \left(\frac{2}{3}\right)^{3/5} + \left(\frac{2}{3}\right)^{-2/5}$
G. CU on $(-\infty, \infty)$
H. See graph at right.

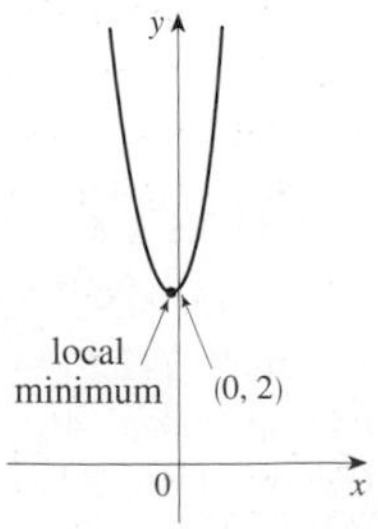

53.

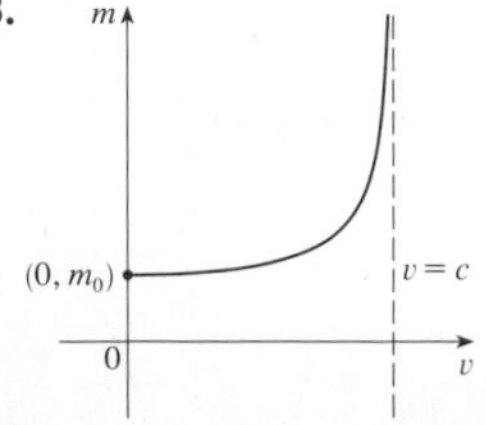

55.

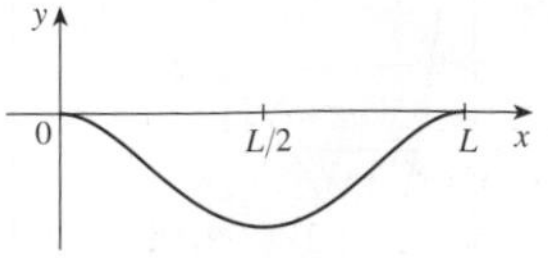

57. $y = x - 1$ **59.** $y = 2x - 2$

61. A. $\left(-\infty, \frac{1}{2}\right) \cup \left(\frac{1}{2}, \infty\right)$
B. y-int. 1; x-int. $\frac{1}{4}\left(5 \pm \sqrt{17}\right)$
C. None
D. VA $x = \frac{1}{2}$; SA $y = -x + 2$
E. Dec. on $\left(-\infty, \frac{1}{2}\right), \left(\frac{1}{2}, \infty\right)$
F. None
G. CU on $\left(\frac{1}{2}, \infty\right)$; CD on $\left(-\infty, \frac{1}{2}\right)$
H. See graph at right

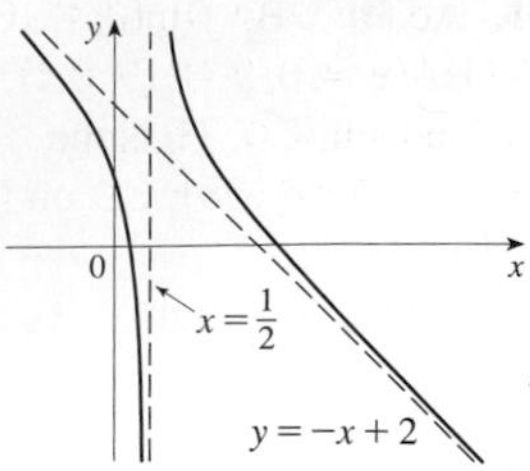

63. A. $\{x \mid x \neq 0\}$ B. None
C. About (0, 0) D. VA $x = 0$; SA $y = x$
E. Inc. on $(-\infty, -2)$, $(2, \infty)$;
dec. on $(-2, 0)$, $(0, 2)$
F. Loc. max. $f(-2) = -4$;
loc. min. $f(2) = 4$
G. CU on $(0, \infty)$; CD on $(-\infty, 0)$
H. See graph at right.

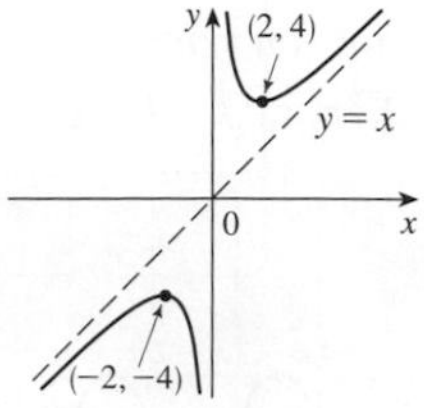

65. A. $\mathbb{R}$ B. y-int. 1; x-int. -1
C. None D. SA $y = 2x + 1$
E. Inc. on $(-\infty, \infty)$ F. None
G. CU on $\left(-\infty, -\sqrt{3}\right)$,
$\left(0, \sqrt{3}\right)$;
CD on $\left(-\sqrt{3}, 0\right), \left(\sqrt{3}, \infty\right)$;
IP $\left(\pm\sqrt{3}, 1 \pm \frac{3}{2}\sqrt{3}\right)$, (0, 1)
H. See graph at right.

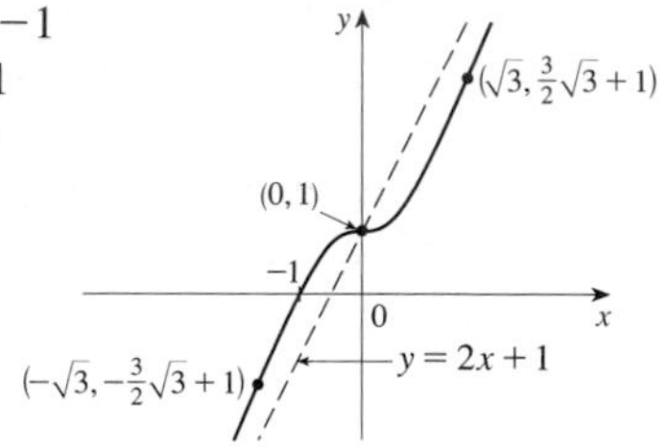

67.

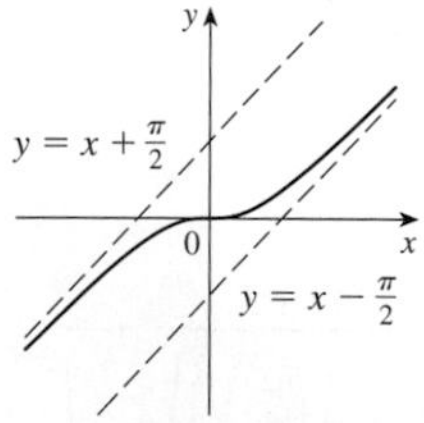

71. VA $x = 0$, asymptotic to $y = x^3$

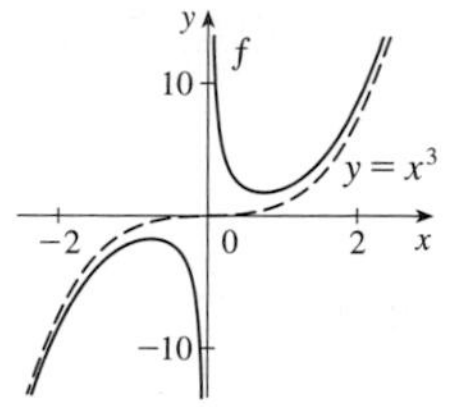

EXERCISES 4.6 ▪ PAGE 320

1. Inc. on (0.92, 2.5), (2.58, ∞); dec. on ($-\infty$, 0.92), (2.5, 2.58); loc. max. $f(2.5) \approx 4$; loc. min. $f(0.92) \approx -5.12$, $f(2.58) \approx 3.998$; CU on ($-\infty$, 1.46), (2.54, ∞); CD on (1.46, 2.54); IP (1.46, -1.40), (2.54, 3.999)

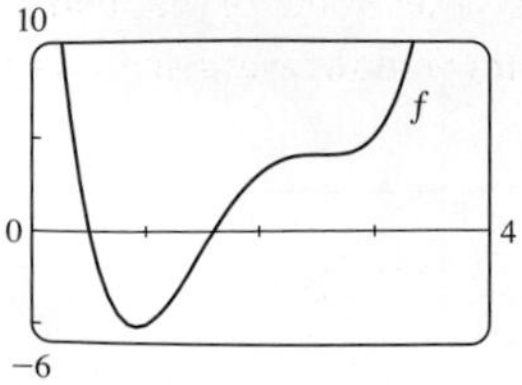

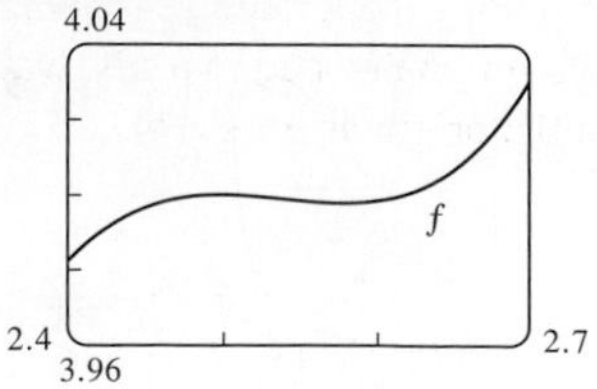

3. Inc. on (-15, 4.40), (18.93, ∞); dec. on ($-\infty$, -15), (4.40, 18.93); loc. max. $f(4.40) \approx 53{,}800$; loc. min. $f(-15) \approx -9{,}700{,}000$, $f(18.93) \approx -12{,}700{,}000$; CU on ($-\infty$, -11.34), (0, 2.92), (15.08, ∞); CD on (-11.34, 0), (2.92, 15.08); IP (0, 0), $\approx$ (-11.34, $-6{,}250{,}000$), (2.92, 31,800), (15.08, $-8{,}150{,}000$)

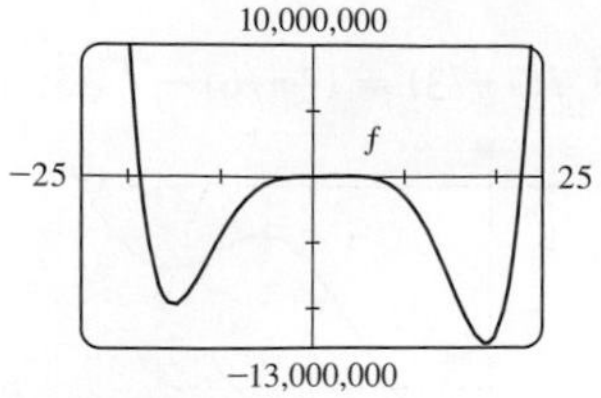

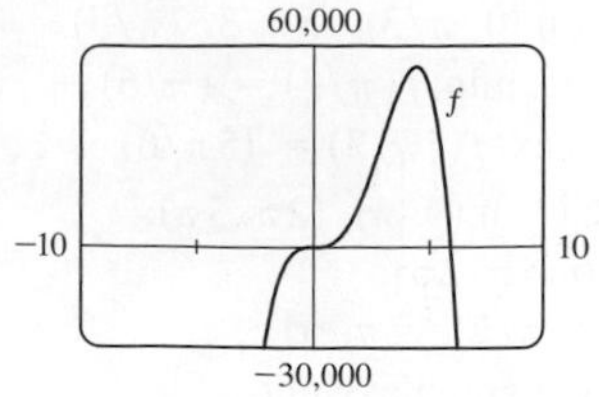

5. Inc. on ($-\infty$, -1.7), (-1.7, 0.24), (0.24, 1); dec. on (1, 2.46), (2.46, ∞); loc. max. $f(1) = -\frac{1}{3}$; CU on ($-\infty$, -1.7), (-0.506, 0.24), (2.46, ∞); CD on (-1.7, -0.506), (0.24, 2.46); IP (-0.506, -0.192)

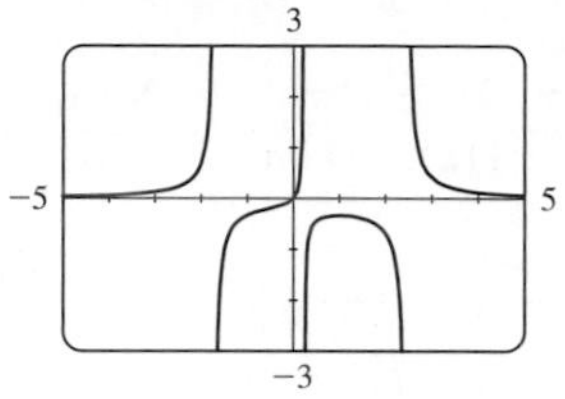

7. Inc. on (-1.49, -1.07), (2.89, 4); dec. on (-4, -1.49), (-1.07, 2.89); loc. max. $f(-1.07) \approx 8.79$; loc. min. $f(-1.49) \approx 8.75$, $f(2.89) \approx -9.99$; CU on (-4, -1.28), (1.28, 4); CD on (-1.28, 1.28); IP (-1.28, 8.77), (1.28, -1.48)

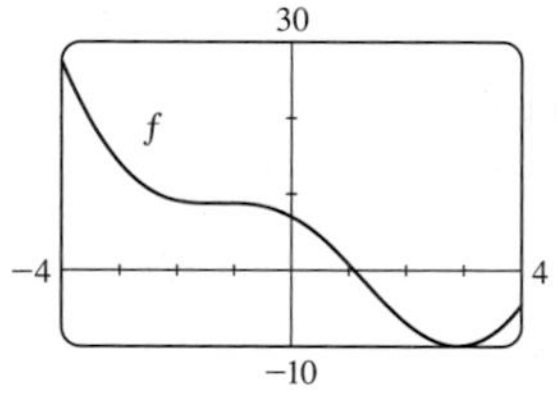

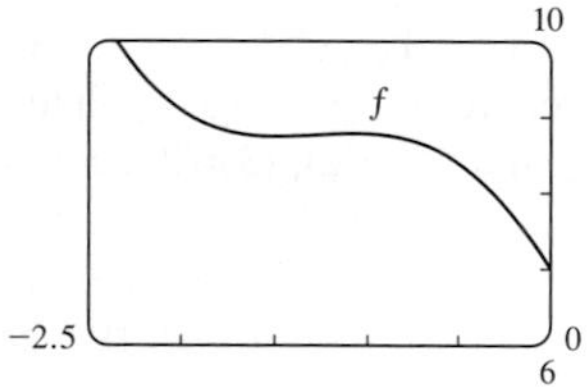

9. Inc. on $\left(-8 - \sqrt{61}, -8 + \sqrt{61}\right)$; dec. on $\left(-\infty, -8 - \sqrt{61}\right)$, $\left(-8 + \sqrt{61}, 0\right)$, $(0, \infty)$; CU on $\left(-12 - \sqrt{138}, -12 + \sqrt{138}\right)$, $(0, \infty)$; CD on $\left(-\infty, -12 - \sqrt{138}\right)$, $\left(-12 + \sqrt{138}, 0\right)$

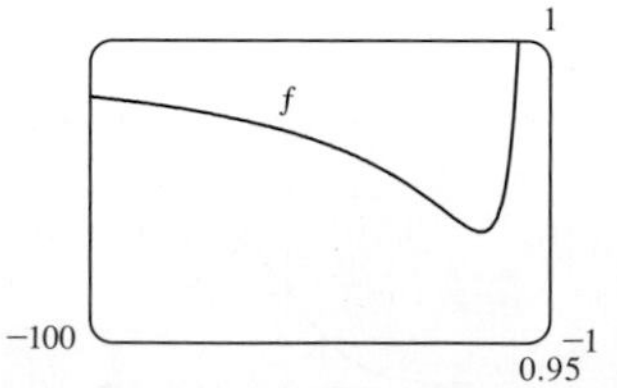

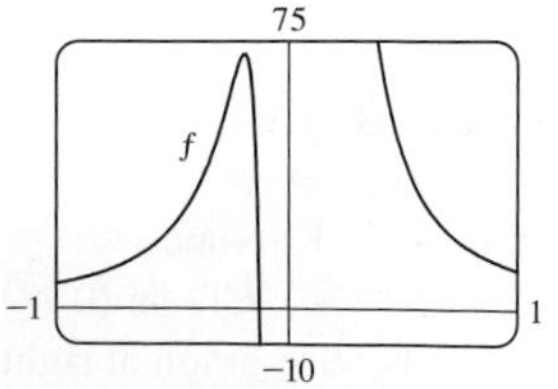

11. (a)

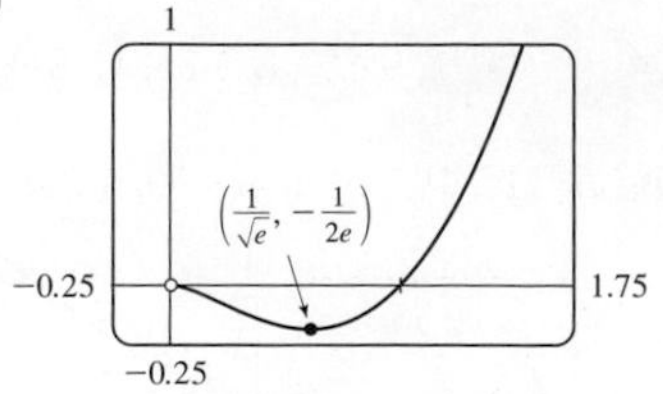

(b) $\lim_{x \to 0^+} f(x) = 0$
(c) Loc. min. $f\left(1/\sqrt{e}\right) = -1/(2e)$; CD on $(0, e^{-3/2})$; CU on $(e^{-3/2}, \infty)$

13. Loc. max. $f(-5.6) \approx 0.018$, $f(0.82) \approx -281.5$, $f(5.2) \approx 0.0145$; loc. min. $f(3) = 0$

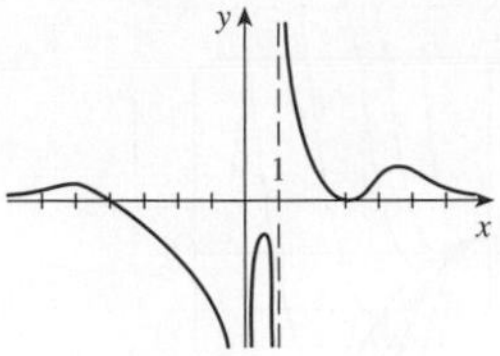

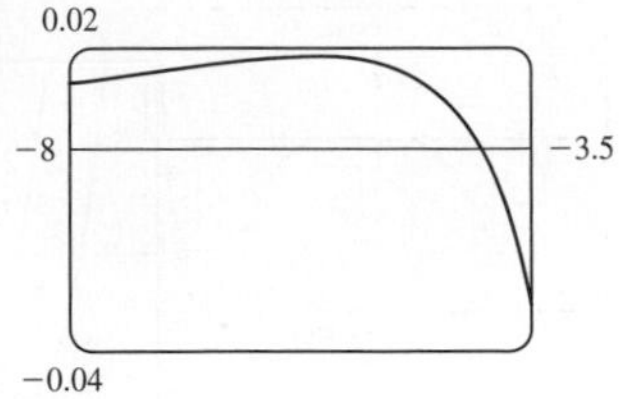

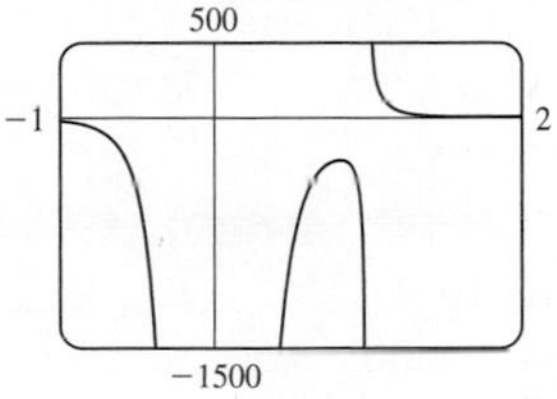

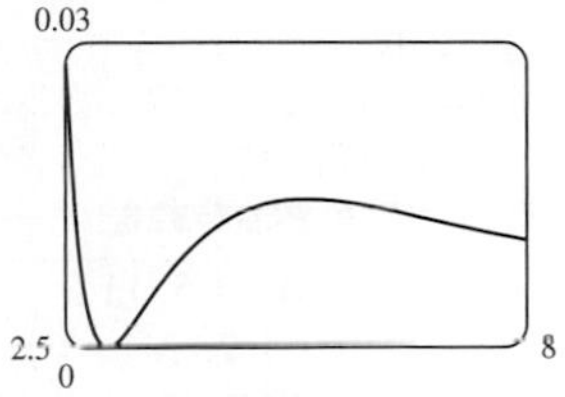

15. $f'(x) = -\dfrac{x(x+1)^2(x^3 + 18x^2 - 44x - 16)}{(x-2)^3(x-4)^5}$

$$f''(x) = 2\frac{(x+1)(x^6 + 36x^5 + 6x^4 - 628x^3 + 684x^2 + 672x + 64)}{(x-2)^4(x-4)^6}$$

CU on $(-35.3, -5.0)$, $(-1, -0.5)$, $(-0.1, 2)$, $(2, 4)$, $(4, \infty)$;
CD on $(-\infty, -35.3)$, $(-5.0, -1)$, $(-0.5, -0.1)$;
IP $(-35.3, -0.015)$, $(-5.0, -0.005)$, $(-1, 0)$, $(-0.5, 0.00001)$, $(-0.1, 0.0000066)$

17. Inc. on $(0, 0.43)$; dec. on $(0.43, \infty)$; loc. max. $f(0.43) \approx 0.41$; CU on $(0.94, \infty)$; CD on $(0, 0.94)$; IP $(0.94, 0.34)$

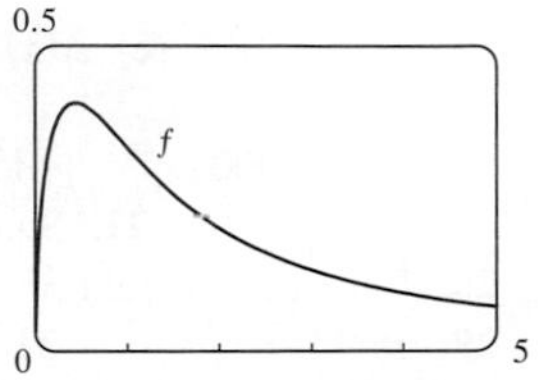

19. Inc. on $(-4.91, -4.51)$, $(0, 1.77)$, $(4.91, 8.06)$, $(10.79, 14.34)$, $(17.08, 20)$;
dec. on $(-4.51, -4.10)$, $(1.77, 4.10)$, $(8.06, 10.79)$, $(14.34, 17.08)$;
loc. max. $f(-4.51) \approx 0.62$, $f(1.77) \approx 2.58$, $f(8.06) \approx 3.60$, $f(14.34) \approx 4.39$;
loc. min. $f(10.79) \approx 2.43$, $f(17.08) \approx 3.49$; CU on $(9.60, 12.25)$, $(15.81, 18.65)$;
CD on $(-4.91, -4.10)$, $(0, 4.10)$, $(4.91, 9.60)$, $(12.25, 15.81)$, $(18.65, 20)$;
IPs at $(9.60, 2.95)$, $(12.25, 3.27)$, $(15.81, 3.91)$, $(18.65, 4.20)$

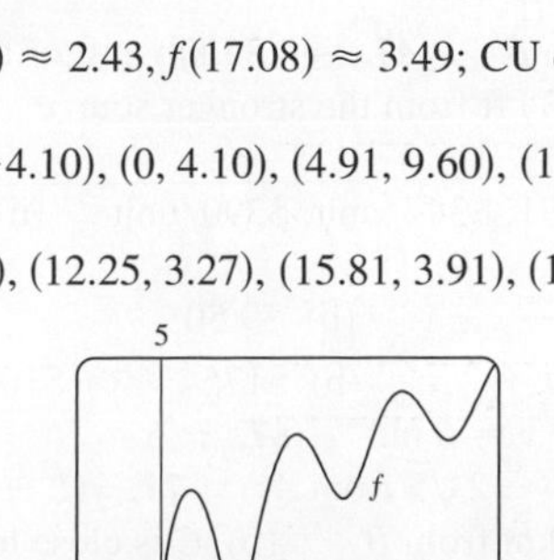

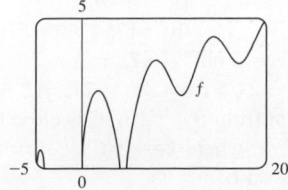

21. Inc. on $(-\infty, 0)$, $(0, \infty)$; CU on $(-\infty, -0.4)$, $(0, 0.4)$; CD on $(-0.4, 0)$, $(0.4, \infty)$; IP $(\mp 0.4, \pm 0.8)$

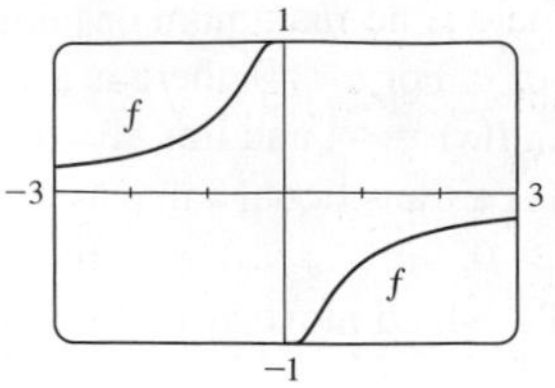

23. (a)

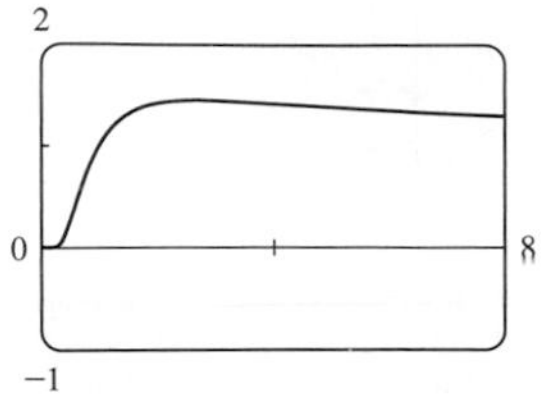

(b) $\lim_{x\to 0^+} x^{1/x} = 0$, $\lim_{x\to\infty} x^{1/x} = 1$
(c) Loc. max. $f(e) = e^{1/e}$ (d) IP at $x \approx 0.58, 4.37$

25. Max. $f(0.59) \approx 1$, $f(0.68) \approx 1$, $f(1.96) \approx 1$; min. $f(0.64) \approx 0.99996$, $f(1.46) \approx 0.49$, $f(2.73) \approx -0.51$; IP $(0.61, 0.99998)$, $(0.66, 0.99998)$, $(1.17, 0.72)$, $(1.75, 0.77)$, $(2.28, 0.34)$

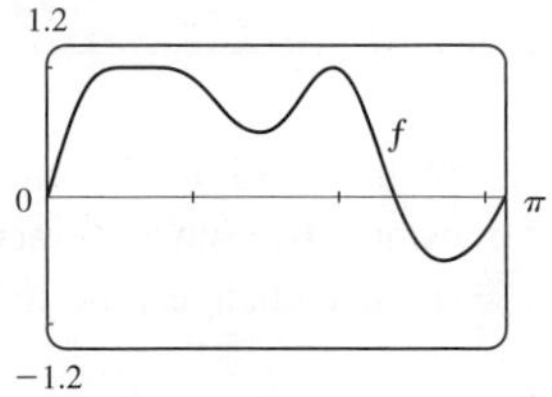

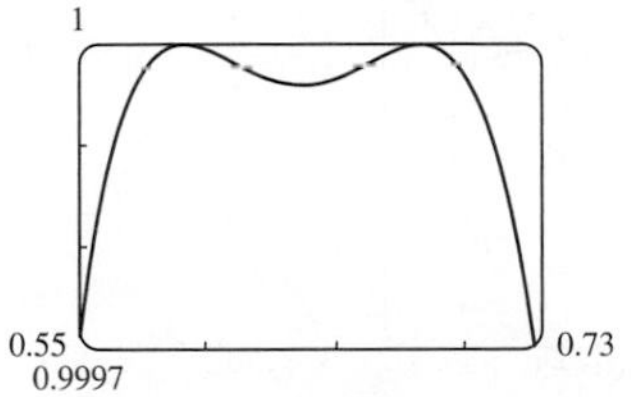

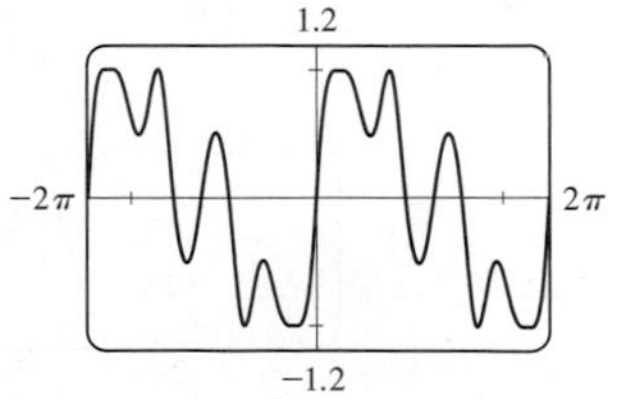

27. For $c \geqslant 0$, there is no IP and only one extreme point, the origin. For $c < 0$, there is a maximum point at the origin, two minimum points, and two IPs, which move downward and away from the origin as $c \to -\infty$.

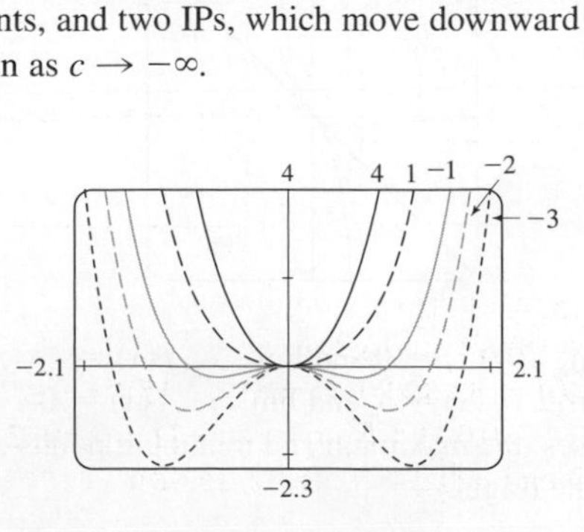

29. There is no maximum or minimum, regardless of the value of c. For $c < 0$, there is a vertical asymptote at $x = 0$, $\lim_{x\to 0} f(x) = \infty$, and $\lim_{x\to\pm\infty} f(x) = 1$.
$c = 0$ is a transitional value at which $f(x) = 1$ for $x \neq 0$.
For $c > 0$, $\lim_{x\to 0} f(x) = 0$, $\lim_{x\to\pm\infty} f(x) = 1$, and there are two IPs, which move away from the y-axis as $c \to \infty$.

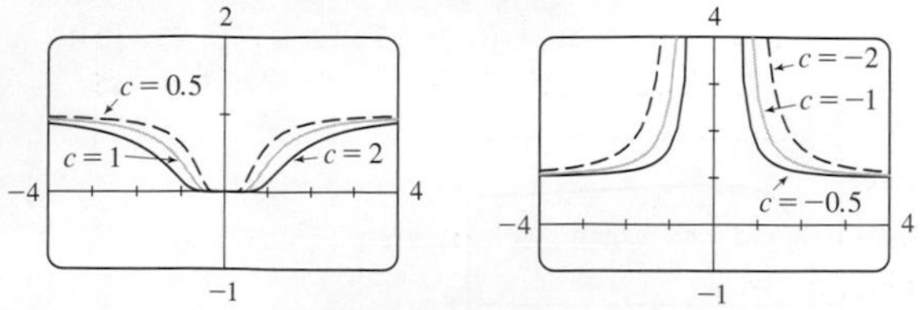

31. For $c > 0$, the maximum and minimum values are always $\pm\frac{1}{2}$, but the extreme points and IPs move closer to the y-axis as c increases. $c = 0$ is a transitional value: when c is replaced by $-c$, the curve is reflected in the x-axis.

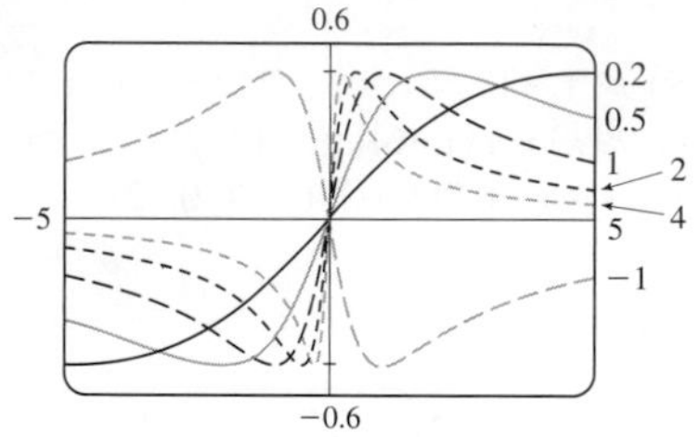

33. For $|c| < 1$, the graph has local maximum and minimum values; for $|c| \geq 1$ it does not. The function increases for $c \geq 1$ and decreases for $c \leq -1$. As c changes, the IPs move vertically but not horizontally.

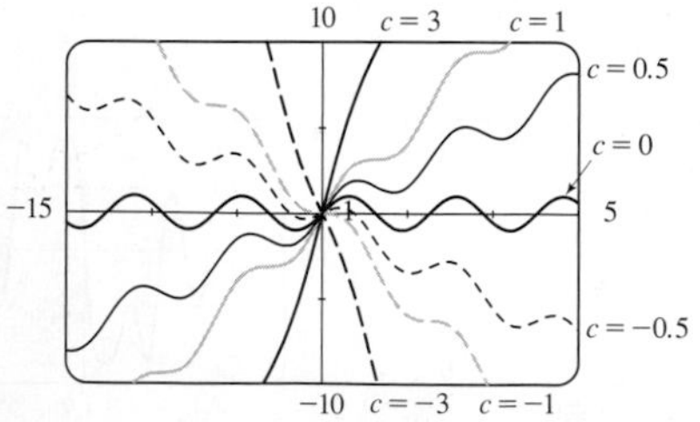

35.

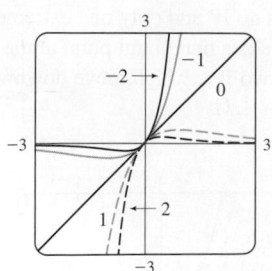

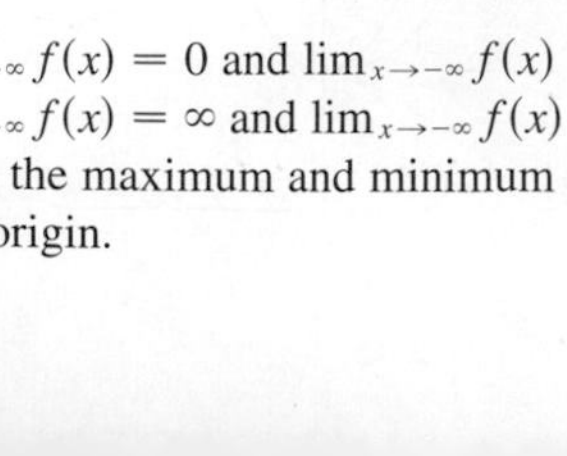

For $c > 0$, $\lim_{x\to\infty} f(x) = 0$ and $\lim_{x\to-\infty} f(x) = -\infty$.
For $c < 0$, $\lim_{x\to\infty} f(x) = \infty$ and $\lim_{x\to-\infty} f(x) = 0$.
As $|c|$ increases, the maximum and minimum points and the IPs get closer to the origin.

37. (a) Positive (b)

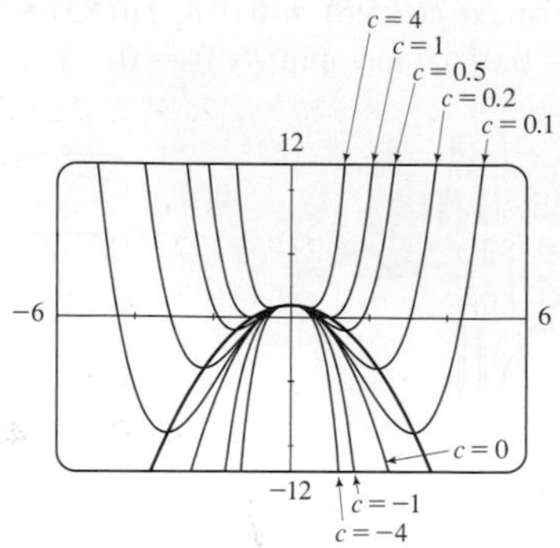

EXERCISES 4.7 ■ PAGE 328

1. (a) 11, 12 (b) 11.5, 11.5 **3.** 10, 10
5. 25 m by 25 m **7.** $N = 1$
9. (a)

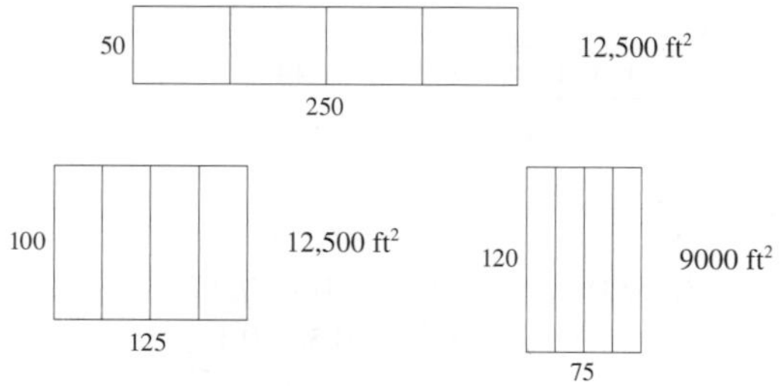

(b)

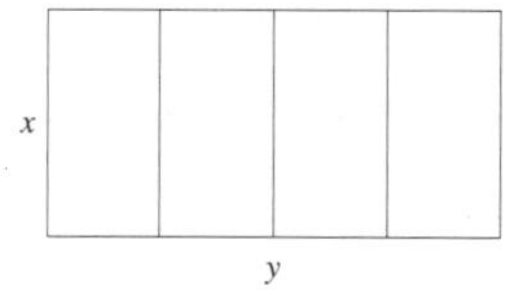

(c) $A = xy$ (d) $5x + 2y = 750$ (e) $A(x) = 375x - \frac{5}{2}x^2$
(f) $14{,}062.5\ \text{ft}^2$
11. 1000 ft by 1500 ft **13.** $4000\ \text{cm}^3$ **15.** \$191.28
17. $\left(-\frac{28}{17}, \frac{7}{17}\right)$ **19.** $\left(-\frac{1}{3}, \pm\frac{4}{3}\sqrt{2}\right)$ **21.** Square, side $\sqrt{2}\,r$
23. $L/2$, $\sqrt{3}\,L/4$ **25.** Base $\sqrt{3}\,r$, height $3r/2$
27. $4\pi r^3/(3\sqrt{3})$ **29.** $\pi r^2(1 + \sqrt{5})$ **31.** 24 cm, 36 cm
33. (a) Use all of the wire for the square
(b) $40\sqrt{3}/(9 + 4\sqrt{3})$ m for the square
35. Height = radius = $\sqrt[3]{V/\pi}$ cm **37.** $V = 2\pi R^3/(9\sqrt{3})$
41. $E^2/(4r)$
43. (a) $\frac{3}{2}S^2 \csc\theta\,(\csc\theta - \sqrt{3}\cot\theta)$ (b) $\cos^{-1}(1/\sqrt{3}) \approx 55°$
(c) $6s\left[h + s/(2\sqrt{2})\right]$
45. Row directly to B **47.** $\approx$4.85 km east of the refinery
49. $10\sqrt[3]{3}/(1 + \sqrt[3]{3})$ ft from the stronger source
51. $(a^{2/3} + b^{2/3})^{3/2}$
53. (b) (i) \$342,491; \$342/unit; \$390/unit (ii) 400
(iii) \$320/unit
55. (a) $p(x) = 19 - \frac{1}{3000}x$ (b) \$9.50
57. (a) $p(x) = 550 - \frac{1}{10}x$ (b) \$175 (c) \$100
61. 9.35 m **65.** $x = 6$ in. **67.** $\pi/6$
69. At a distance $5 - 2\sqrt{5}$ from A **71.** $\frac{1}{2}(L + W)^2$
73. (a) About 5.1 km from B (b) C is close to B; C is close to D; $W/L = \sqrt{25 + x^2}/x$, where $x = |BC|$ (c) $\approx$1.07; no such value (d) $\sqrt{41}/4 \approx 1.6$

EXERCISES 4.8 ■ PAGE 338

1. (a) $x_2 \approx 2.3, x_3 \approx 3$ (b) No **3.** $\frac{4}{5}$ **5.** 1.1797
7. 1.1785 **9.** -1.25 **11.** 1.82056420 **13.** 1.217562
15. 0.876726 **17.** $-0.724492, 1.220744$
19. 1.412391, 3.057104 **21.** 0.641714
23. $-1.93822883, -1.21997997, 1.13929375, 2.98984102$
25. $-1.97806681, -0.82646233$
27. 0.21916368, 1.08422462 **29.** (b) 31.622777
35. (a) $-1.293227, -0.441731, 0.507854$ (b) -2.0212
37. (0.904557, 1.855277) **39.** (0.410245, 0.347810)
41. 0.76286%

EXERCISES 4.9 ■ PAGE 345

1. $F(x) = \frac{1}{2}x^2 - 3x + C$ **3.** $F(x) = \frac{1}{2}x + \frac{1}{4}x^3 - \frac{1}{5}x^4 + C$
5. $F(x) = \frac{2}{3}x^3 + \frac{1}{2}x^2 - x + C$ **7.** $F(x) = 4x^{5/4} - 4x^{7/4} + C$
9. $F(x) = 4x^{3/2} - \frac{6}{7}x^{7/6} + C$
11. $F(x) = \begin{cases} -5/(4x^8) + C_1 & \text{if } x < 0 \\ -5/(4x^8) + C_2 & \text{if } x > 0 \end{cases}$
13. $F(u) = \frac{1}{3}u^3 - 6u^{-1/2} + C$
15. $G(\theta) = \sin\theta + 5\cos\theta + C$
17. $F(x) = 5e^x - 3\sinh x + C$
19. $F(x) = \frac{1}{2}x^2 - \ln|x| - 1/x^2 + C$
21. $F(x) = x^5 - \frac{1}{3}x^6 + 4$ **23.** $x^3 + x^4 + Cx + D$
25. $\frac{3}{20}x^{8/3} + Cx + D$ **27.** $e^t + \frac{1}{2}Ct^2 + Dt + E$
29. $x - 3x^2 + 8$ **31.** $4x^{3/2} + 2x^{5/2} + 4$
33. $2\sin t + \tan t + 4 - 2\sqrt{3}$
35. $\frac{3}{2}x^{2/3} - \frac{1}{2}$ if $x > 0$; $\frac{3}{2}x^{2/3} - \frac{5}{2}$ if $x < 0$
37. $2x^4 + \frac{1}{3}x^3 + 5x^2 - 22x + \frac{59}{3}$
39. $-\sin\theta - \cos\theta + 5\theta + 4$ **41.** $x^2 - 2x^3 + 9x + 9$
43. $x^2 - \cos x - \frac{1}{2}\pi x$ **45.** $-\ln x + (\ln 2)x - \ln 2$
47. 10 **49.** b
51.

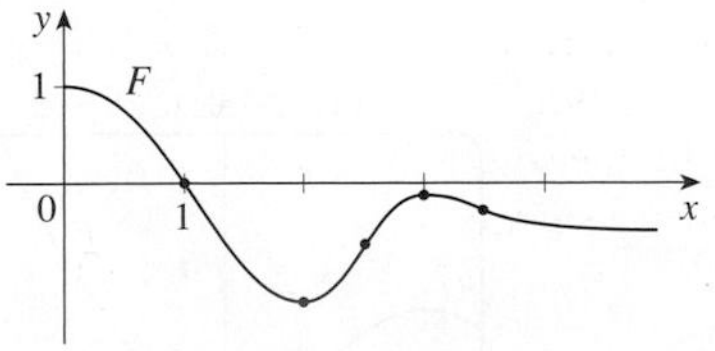

53.

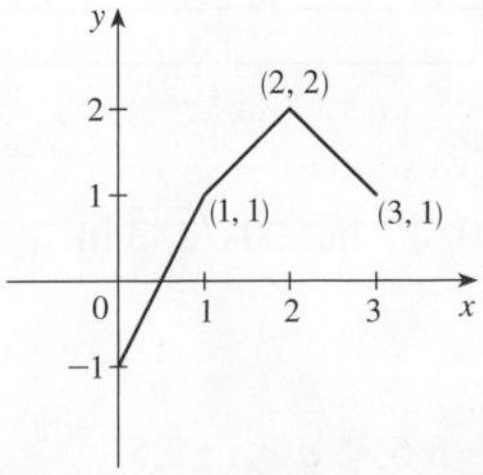

55.

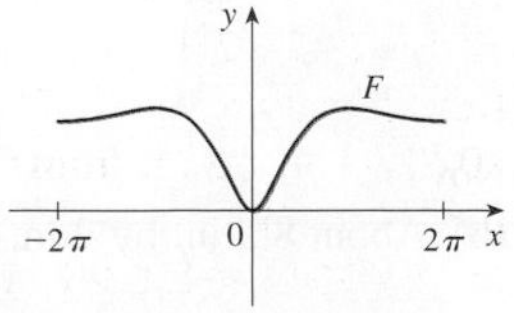

57. $s(t) = 1 - \cos t - \sin t$ **59.** $s(t) = \frac{1}{6}t^3 - t^2 + 3t + 1$
61. $s(t) = -10\sin t - 3\cos t + (6/\pi)t + 3$
63. (a) $s(t) = 450 - 4.9t^2$ (b) $\sqrt{450/4.9} \approx 9.58$ s
(c) $-9.8\sqrt{450/4.9} \approx -93.9$ m/s (d) About 9.09 s
67. 225 ft **69.** \$742.08 **71.** $\frac{130}{11} \approx 11.8$ s
73. $\frac{88}{15} \approx 5.87$ ft/s^2 **75.** 62,500 km/h^2 ≈ 4.82 m/s^2
77. (a) 22.9125 mi (b) 21.675 mi (c) 30 min 33 s
(d) 55.425 mi

CHAPTER 4 REVIEW ■ PAGE 347

True-False Quiz

1. False **3.** False **5.** True **7.** False **9.** True
11. True **13.** False **15.** True **17.** True **19.** True

Exercises

1. Abs. max. $f(4) = 5$, abs. and loc. min. $f(3) = 1$;
loc. min. $f(3) = 1$
3. Abs. max. $f(2) = \frac{2}{5}$, abs. and loc. min. $f\left(-\frac{1}{3}\right) = -\frac{9}{2}$
5. Abs. max. $f(\pi) = \pi$; abs. min. $f(0) = 0$; loc. max.
$f(\pi/3) = (\pi/3) + \frac{1}{2}\sqrt{3}$; loc. min. $f(2\pi/3) = (2\pi/3) - \frac{1}{2}\sqrt{3}$
7. π **9.** 8 **11.** 0 **13.** $\frac{1}{2}$
15.

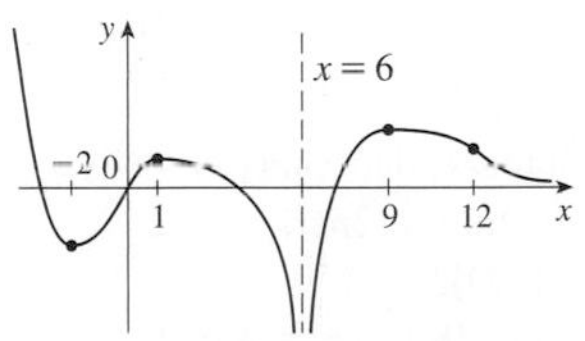

17.

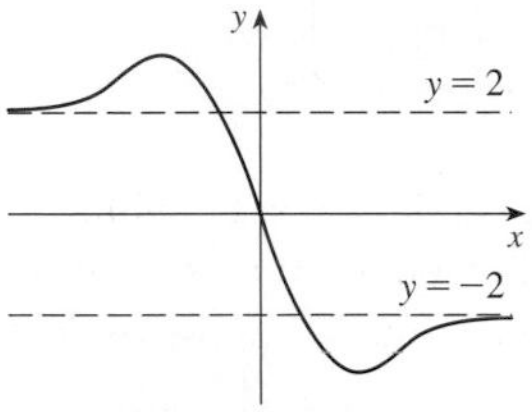

19. A. ℝ B. y-int. 2
C. None D. None
E. Dec. on $(-\infty, \infty)$ F. None
G. CU on $(-\infty, 0)$;
CD on $(0, \infty)$; IP (0, 2)
H. See graph at right.

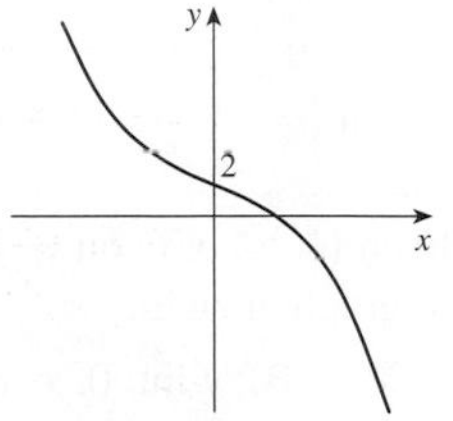

21. A. ℝ B. y-int. 0; x-int. 0, 1
C. None D. None
E. Inc. on $\left(\frac{1}{4}, \infty\right)$, dec. on $\left(-\infty, \frac{1}{4}\right)$
F. Loc. min. $f\left(\frac{1}{4}\right) = -\frac{27}{256}$
G. CU on $\left(-\infty, \frac{1}{2}\right)$, $(1, \infty)$;
CD on $\left(\frac{1}{2}, 1\right)$; IP $\left(\frac{1}{2}, -\frac{1}{16}\right)$, (1, 0)
H. See graph at right.

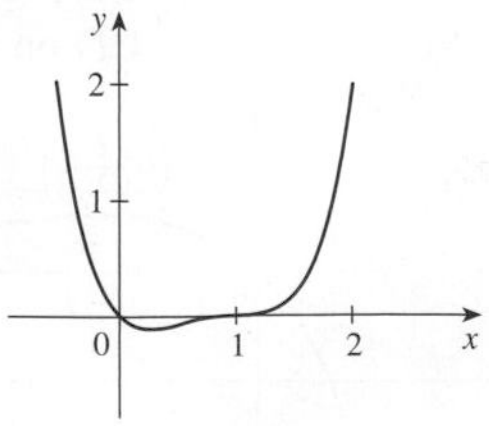

23. A. $\{x \mid x \neq 0, 3\}$
B. None C. None
D. HA $y = 0$; VA $x = 0$, $x = 3$
E. Inc. on (1, 3); dec. on $(-\infty, 0)$,
(0, 1), $(3, \infty)$
F. Loc. min. $f(1) = \frac{1}{4}$
G. CU on (0, 3), $(3, \infty)$; CD on $(-\infty, 0)$
H. See graph at right.

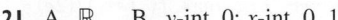

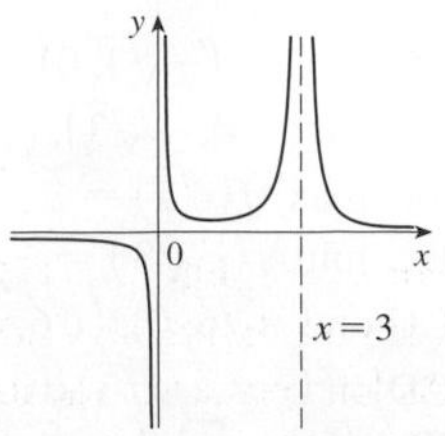

25. A. $\{x \mid x \neq -8\}$
B. y-int. 0, x-int. 0 C. None
D. VA $x = -8$; SA $y = x - 8$
E. Inc. on $(-\infty, -16)$, $(0, \infty)$;
dec. on $(-16, -8)$, $(-8, 0)$
F. Loc. max. $f(-16) = -32$;
loc. min. $f(0) = 0$
G. CU on $(-8, \infty)$; CD on $(-\infty, -8)$
H. See graph at right.

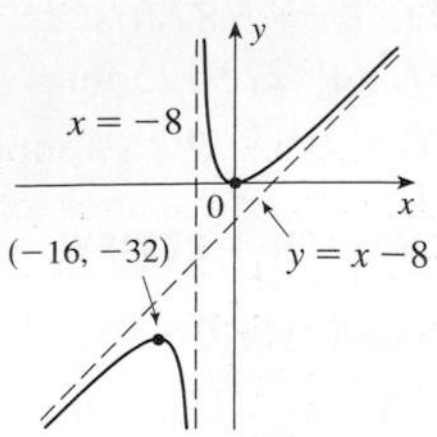

27. A. $[-2, \infty)$
B. y-int. 0; x-int. $-2, 0$
C. None D. None
E. Inc. on $\left(-\frac{4}{3}, \infty\right)$, dec. on $\left(-2, -\frac{4}{3}\right)$
F. Loc. min. $f\left(-\frac{4}{3}\right) = -\frac{4}{9}\sqrt{6}$
G. CU on $(-2, \infty)$
H. See graph at right.

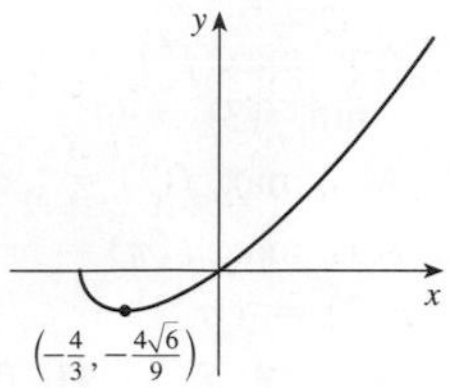

29. A. $\mathbb{R}$ B. y-int. -2
C. About y-axis, period 2π D. None
E. Inc. on $(2n\pi, (2n + 1)\pi)$, n an integer; dec. on $((2n - 1)\pi, 2n\pi)$
F. Loc. max. $f((2n + 1)\pi) = 2$; loc. min. $f(2n\pi) = -2$
G. CU on $(2n\pi - (\pi/3), 2n\pi + (\pi/3))$;
CD on $(2n\pi + (\pi/3), 2n\pi + (5\pi/3))$; IP $\left(2n\pi \pm (\pi/3), -\frac{1}{4}\right)$
H.

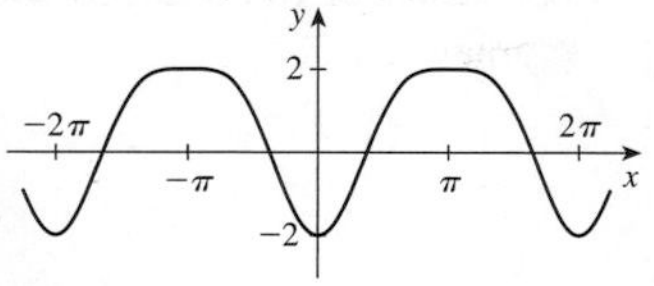

31. A. $\{x \mid |x| \geqslant 1\}$
B. None C. About (0, 0)
D. HA $y = 0$
E. Dec. on $(-\infty, -1)$, $(1, \infty)$
F. None
G. CU on $(1, \infty)$; CD on $(-\infty, -1)$
H. See graph at right.

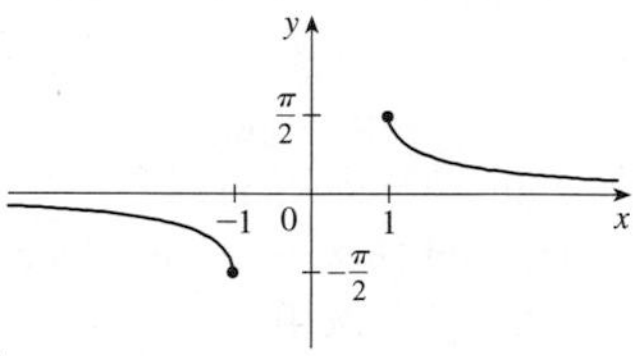

33. A. $\mathbb{R}$ B. y-int. 0, x-int. 0 C. None D. HA $y = 0$
E. Inc. on $\left(-\infty, \frac{1}{2}\right)$, dec. on $\left(\frac{1}{2}, \infty\right)$ F. Loc. max. $f\left(\frac{1}{2}\right) = 1/(2e)$
G. CU on $(1, \infty)$; CD on $(-\infty, 1)$; IP $(1, e^{-2})$
H.

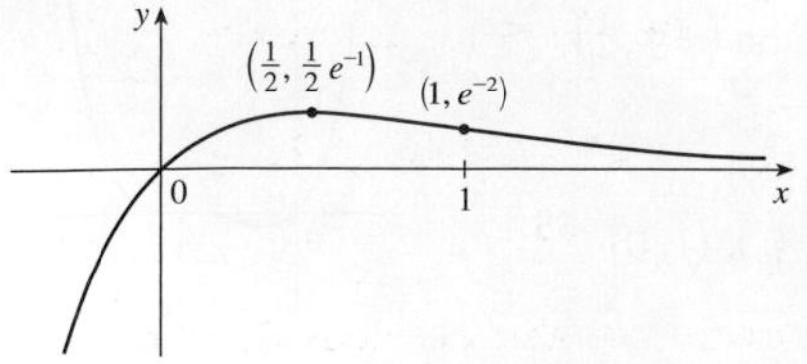

35. Inc. on $(-\sqrt{3}, 0)$, $(0, \sqrt{3})$;
dec. on $(-\infty, -\sqrt{3})$, $(\sqrt{3}, \infty)$;
loc. max. $f(\sqrt{3}) = \frac{2}{9}\sqrt{3}$,
loc. min. $f(-\sqrt{3}) = -\frac{2}{9}\sqrt{3}$;
CU on $(-\sqrt{6}, 0)$, $(\sqrt{6}, \infty)$;
CD on $(-\infty, -\sqrt{6})$, $(0, \sqrt{6})$;
IP $\left(\sqrt{6}, \frac{5}{36}\sqrt{6}\right)$, $\left(-\sqrt{6}, -\frac{5}{36}\sqrt{6}\right)$

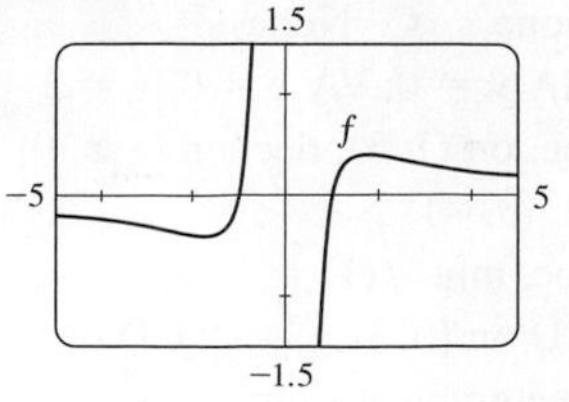

37. Inc. on $(-0.23, 0)$, $(1.62, \infty)$; dec. on $(-\infty, -0.23)$, $(0, 1.62)$;
loc. max. $f(0) = 2$; loc. min. $f(-0.23) \approx 1.96$, $f(1.62) \approx -19.2$;
CU on $(-\infty, -0.12)$, $(1.24, \infty)$;
CD on $(-0.12, 1.24)$; IP $(-0.12, 1.98)$, $(1.24, -12.1)$

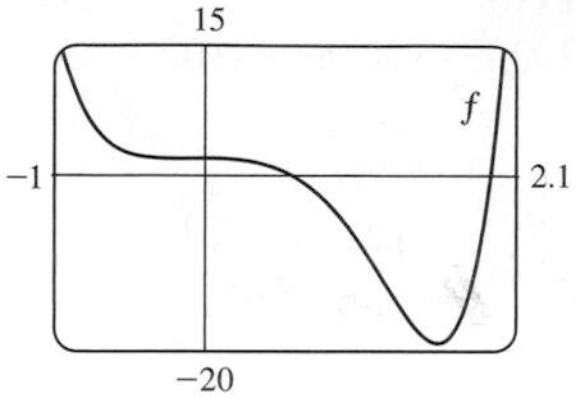

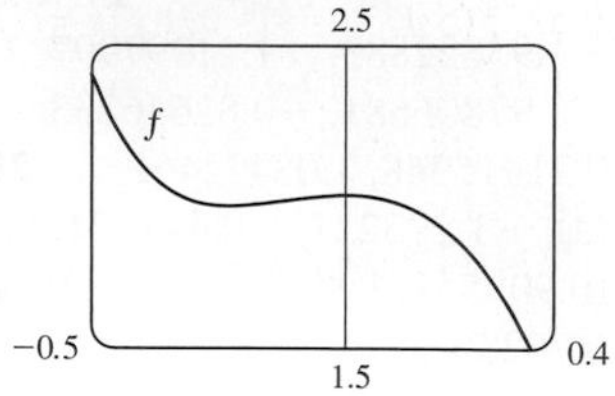

39. $(\pm 0.82, 0.22)$; $\left(\pm\sqrt{2/3}, e^{-3/2}\right)$

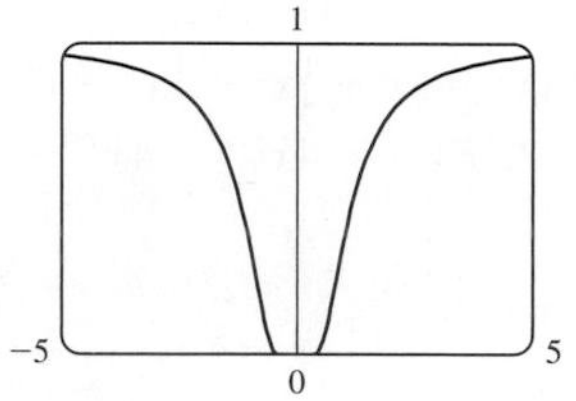

41. $-2.96, -0.18, 3.01$; $-1.57, 1.57$; $-2.16, -0.75, 0.46, 2.21$
43. For $C > -1$, f is periodic with period 2π and has local maxima at $2n\pi + \pi/2$, n an integer. For $C \leqslant -1$, f has no graph. For $-1 < C \leqslant 1$, f has vertical asymptotes. For $C > 1$, f is continuous on $\mathbb{R}$. As C increases, f moves upward and its oscillations become less pronounced.
49. (a) 0 (b) CU on $\mathbb{R}$ **53.** $3\sqrt{3}r^2$
55. $4/\sqrt{3}$ cm from D **57.** $L = C$ **59.** \$11.50
61. 1.297383 **63.** 1.16718557
65. $f(x) = \sin x - \sin^{-1}x + C$
67. $f(x) = \frac{2}{5}x^{5/2} + \frac{3}{5}x^{5/3} + C$
69. $f(t) = t^2 + 3\cos t + 2$
71. $f(x) = \frac{1}{2}x^2 - x^3 + 4x^4 + 2x + 1$
73. $s(t) = t^2 - \tan^{-1}t + 1$
75. (b) $0.1e^x - \cos x + 0.9$ (c)

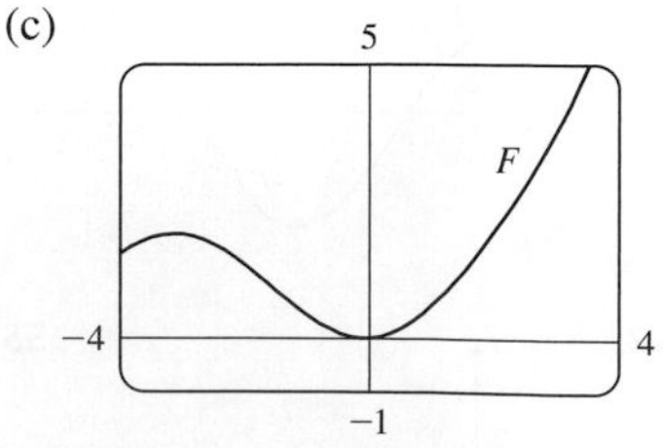

77. No
79. (b) About 8.5 in. by 2 in. (c) $20/\sqrt{3}$ in., $20\sqrt{2/3}$ in.

PROBLEMS PLUS ▪ PAGE 352

5. 24 **7.** $(-2, 4)$, $(2, -4)$ **11.** $-3.5 < a < -2.5$
13. $(m/2, m^2/4)$ **15.** $a \leqslant e^{1/e}$
19. (a) $T_1 = D/c_1$, $T_2 = (2h\sec\theta)/c_1 + (D - 2h\tan\theta)/c_2$, $T_3 = \sqrt{4h^2 + D^2}/c_1$
(c) $c_1 \approx 3.85$ km/s, $c_2 \approx 7.66$ km/s, $h \approx 0.42$ km
23. $3/\left(\sqrt[3]{2} - 1\right) \approx 11\frac{1}{2}$ h

CHAPTER 5

EXERCISES 5.1 ▪ PAGE 364

1. (a) 40, 52 (b) 43.2, 49.2

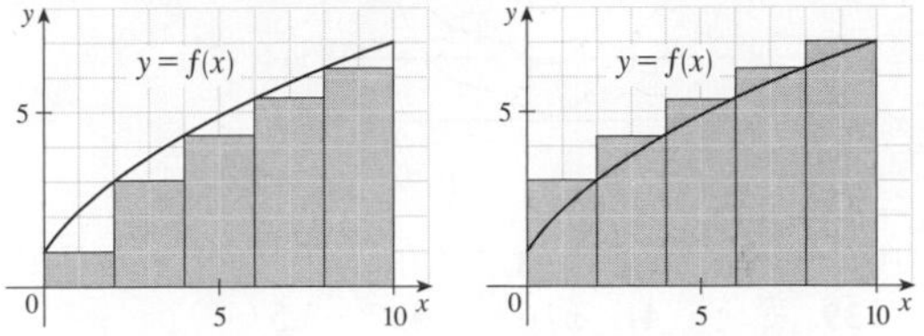

3. (a) 0.7908, underestimate (b) 1.1835, overestimate

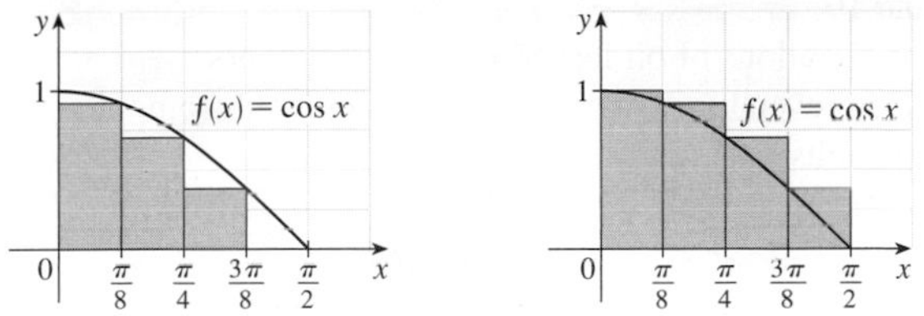

5. (a) 8, 6.875 (b) 5, 5.375

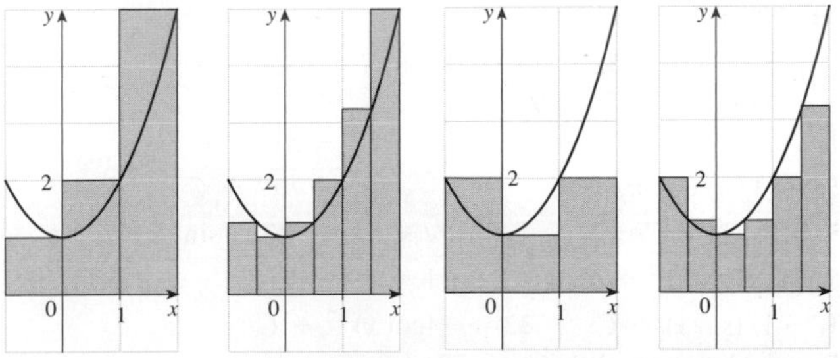

(c) 5.75, 5.9375

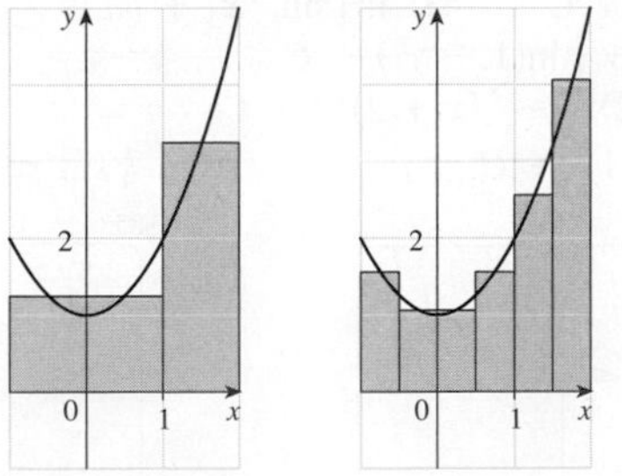

(d) M_6

7. 0.2533, 0.2170, 0.2101, 0.2050; 0.2

9. (a) Left: 0.8100, 0.7937, 0.7904;
right: 0.7600, 0.7770, 0.7804

11. 34.7 ft, 44.8 ft **13.** 63.2 L, 70 L **15.** 155 ft

17. $\lim_{n\to\infty} \sum_{i=1}^{n} \sqrt[4]{1 + 15i/n} \cdot (15/n)$ **19.** $\lim_{n\to\infty} \sum_{i=1}^{n} \left(\frac{i\pi}{2n} \cos \frac{i\pi}{2n}\right) \frac{\pi}{2n}$

21. The region under the graph of $y = \tan x$ from 0 to $\pi/4$

23. (a) $\lim_{n\to\infty} \frac{64}{n^6} \sum_{i=1}^{n} i^5$ (b) $\frac{n^2(n+1)^2(2n^2+2n-1)}{12}$ (c) $\frac{32}{3}$

25. $\sin b$, 1

EXERCISES 5.2 ▪ PAGE 376

1. −6

The Riemann sum represents the sum of the areas of the two rectangles above the x-axis minus the sum of the areas of the three rectangles below the x-axis; that is, the net area of the rectangles with respect to the x-axis.

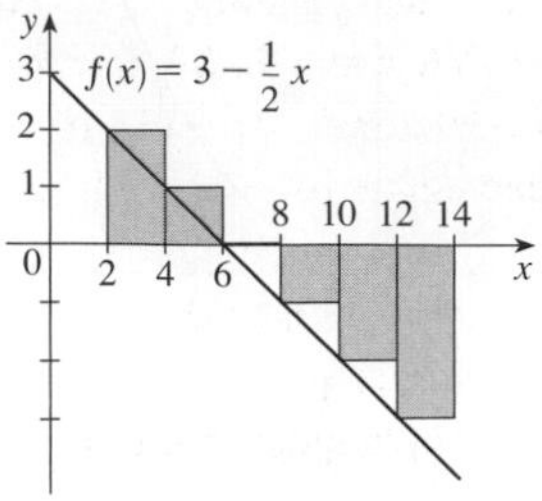

3. 2.322986

The Riemann sum represents the sum of the areas of the three rectangles above the x-axis minus the area of the rectangle below the x-axis.

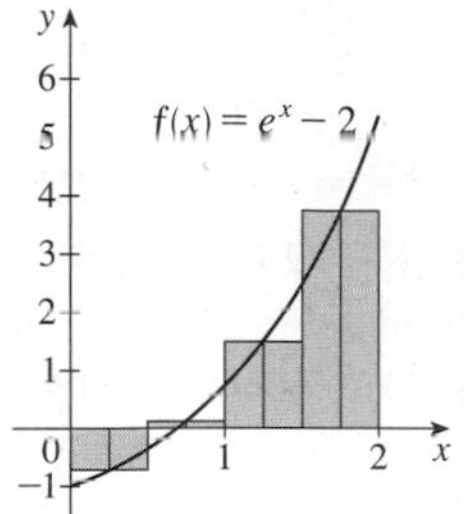

5. (a) 4 (b) 6 (c) 10 **7.** −475, −85 **9.** 124.1644

11. 0.3084 **13.** 0.30843908, 0.30981629, 0.31015563

15.

n	R_n
5	1.933766
10	1.983524
50	1.999342
100	1.999836

The values of R_n appear to be approaching 2.

17. $\int_2^6 x \ln(1 + x^2)\, dx$ **19.** $\int_1^8 \sqrt{2x + x^2}\, dx$ **21.** 42

23. $\frac{4}{3}$ **25.** 3.75 **29.** $\lim_{n\to\infty} \sum_{i=1}^{n} \frac{2 + 4i/n}{1 + (2 + 4i/n)^5} \cdot \frac{4}{n}$

31. $\lim_{n\to\infty} \sum_{i=1}^{n} \left(\sin \frac{5\pi i}{n}\right) \frac{\pi}{n} = \frac{2}{5}$

33. (a) 4 (b) 10 (c) −3 (d) 2 **35.** $-\frac{3}{4}$

37. $3 + \frac{9}{4}\pi$ **39.** 2.5 **41.** 0 **43.** 3 **45.** $e^5 - e^3$

47. $\int_{-1}^{5} f(x)\, dx$ **49.** 122

51. $2m \leqslant \int_0^2 f(x)\, dx < 2M$ by Comparison Property 8

55. $3 \leqslant \int_1^4 \sqrt{x}\, dx \leqslant 6$ **57.** $\frac{\pi}{12} \leqslant \int_{\pi/4}^{\pi/3} \tan x\, dx \leqslant \frac{\pi}{12}\sqrt{3}$

59. $0 \leqslant \int_0^2 xe^{-x}\, dx \leqslant 2/e$ **69.** $\int_0^1 x^4\, dx$ **71.** $\frac{1}{2}$

EXERCISES 5.3 ▪ PAGE 387

1. One process undoes what the other one does. See the Fundamental Theorem of Calculus, page 387.

3. (a) 0, 2, 5, 7, 3
(b) (0, 3)
(c) $x = 3$
(d)

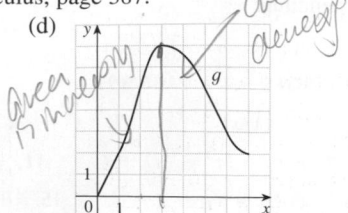

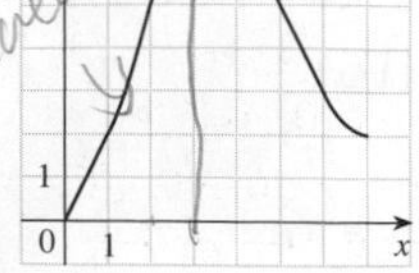

5. 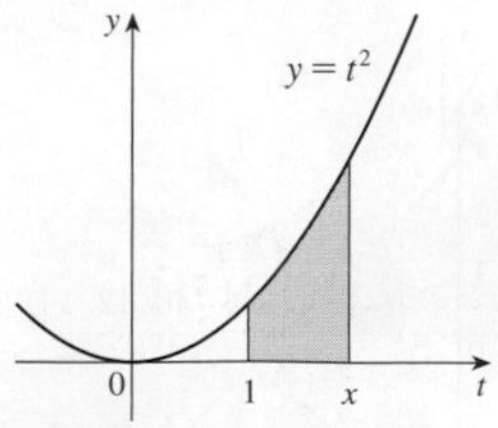

(a), (b) x^2

7. $g'(x) = 1/(x^3 + 1)$

9. $g'(y) = y^2 \sin y$ **11.** $F'(x) = -\sqrt{1 + \sec x}$

13. $h'(x) = -\dfrac{\arctan(1/x)}{x^2}$ **15.** $y' = \sqrt{\tan x + \sqrt{\tan x}}\,\sec^2 x$

17. $y' = \dfrac{3(1 - 3x)^3}{1 + (1 - 3x)^2}$ **19.** $\frac{3}{4}$ **21.** 63

23. $\frac{5}{9}$ **25.** $\frac{7}{8}$ **27.** $\frac{156}{7}$ **29.** $\frac{40}{3}$ **31.** 1 **33.** $\frac{49}{3}$

35. ln 3 **37.** π **39.** $e^2 - 1$ **41.** 0

43. The function $f(x) = x^{-4}$ is not continuous on the interval $[-2, 1]$, so FTC2 cannot be applied.

45. The function $f(\theta) = \sec\theta \tan\theta$ is not continuous on the interval $[\pi/3, \pi]$, so FTC2 cannot be applied.

47. $\frac{243}{4}$ **49.** 2

51. 3.75 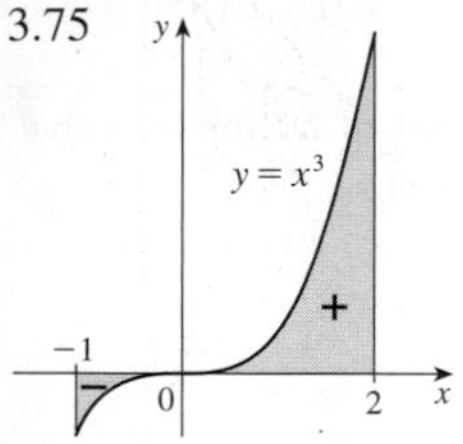

53. $g'(x) = \dfrac{-2(4x^2 - 1)}{4x^2 + 1} + \dfrac{3(9x^2 - 1)}{9x^2 + 1}$

55. $y' = 3x^{7/2}\sin(x^3) - \dfrac{\sin\sqrt{x}}{2\sqrt[4]{x}}$ **57.** $\sqrt{257}$ **59.** 29

61. (a) $-2\sqrt{n}$, $\sqrt{4n - 2}$, n an integer > 0
(b) $(0, 1)$, $(-\sqrt{4n - 1}, -\sqrt{4n - 3})$, and $(\sqrt{4n - 1}, \sqrt{4n + 1})$, n an integer > 0 (c) 0.74

63. (a) Loc. max. at 1 and 5; loc. min. at 3 and 7
(b) $x = 9$
(c) $(\frac{1}{2}, 2)$, $(4, 6)$, $(8, 9)$
(d) See graph at right.

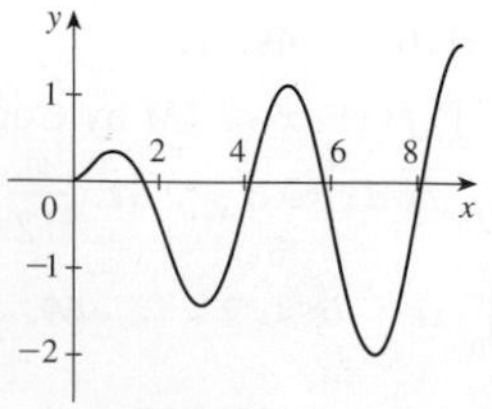

65. $\frac{1}{4}$ **73.** $f(x) = x^{3/2}$, $a = 9$

75. (b) Average expenditure over $[0, t]$; minimize average expenditure

EXERCISES 5.4 ■ PAGE 397

5. $\frac{1}{3}x^3 - (1/x) + C$ **7.** $\frac{1}{5}x^5 - \frac{1}{8}x^4 + \frac{1}{8}x^2 - 2x + C$

9. $2t - t^2 + \frac{1}{3}t^3 - \frac{1}{4}t^4 + C$ **11.** $\frac{1}{3}x^3 - 4\sqrt{x} + C$

13. $-\cos x + \cosh x + C$ **15.** $\frac{1}{2}\theta^2 + \csc\theta + C$

17. $\tan\alpha + C$

19. $\sin x + \frac{1}{4}x^2 + C$

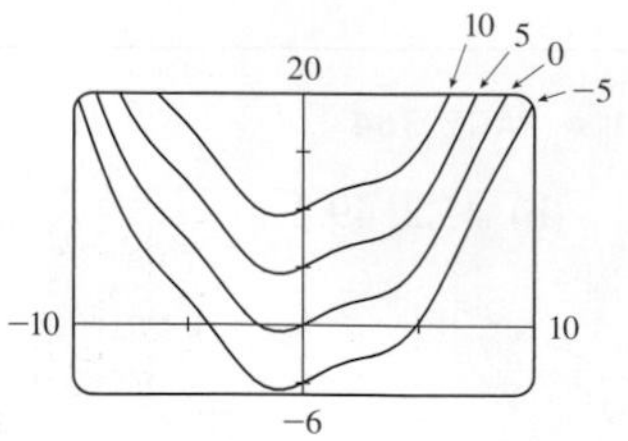

21. 18 **23.** $-2 + 1/e$ **25.** 52

27. $\frac{256}{15}$ **29.** $-\frac{63}{4}$ **31.** $\frac{55}{63}$ **33.** $2\sqrt{5}$ **35.** 8

37. $1 + \pi/4$ **39.** $\frac{256}{5}$ **41.** $\pi/6$ **43.** -3.5

45. 0, 1.32; 0.84 **47.** $\frac{4}{3}$

49. The increase in the child's weight (in pounds) between the ages of 5 and 10

51. Number of gallons of oil leaked in the first 2 hours

53. Increase in revenue when production is increased from 1000 to 5000 units

55. Newton-meters (or joules) **57.** (a) $-\frac{3}{2}$ m (b) $\frac{41}{6}$ m

59. (a) $v(t) = \frac{1}{2}t^2 + 4t + 5$ m/s (b) $416\frac{2}{3}$ m

61. $46\frac{2}{3}$ kg **63.** 1.4 mi **65.** \$58,000

67. (b) At most 40%; $\frac{5}{36}$

EXERCISES 5.5 ■ PAGE 406

1. $-e^{-x} + C$ **3.** $\frac{2}{9}(x^3 + 1)^{3/2} + C$ **5.** $-\frac{1}{4}\cos^4\theta + C$

7. $-\frac{1}{2}\cos(x^2) + C$ **9.** $\frac{1}{63}(3x - 2)^{21} + C$

11. $\frac{1}{3}(2x + x^2)^{3/2} + C$ **13.** $-\frac{1}{3}\ln|5 - 3x| + C$

15. $-(1/\pi)\cos \pi t + C$ **17.** $\frac{2}{3}\sqrt{3ax + bx^3} + C$

19. $\frac{1}{3}(\ln x)^3 + C$ **21.** $2\sin\sqrt{t} + C$ **23.** $\frac{1}{7}\sin^7\theta + C$

25. $\frac{2}{3}(1 + e^x)^{3/2} + C$ **27.** $\frac{1}{2}(1 + z^3)^{2/3} + C$ **29.** $e^{\tan x} + C$

31. $-1/(\sin x) + C$ **33.** $-\frac{2}{3}(\cot x)^{3/2} + C$

35. $-\ln(1 + \cos^2 x) + C$ **37.** $\ln|\sin x| + C$

39. $\frac{1}{3}\sec^3 x + C$ **41.** $\ln|\sin^{-1}x| + C$

43. $\tan^{-1}x + \frac{1}{2}\ln(1 + x^2) + C$

45. $\frac{4}{7}(x + 2)^{7/4} - \frac{8}{3}(x + 2)^{3/4} + C$

47. $\frac{1}{8}(x^2 - 1)^4 + C$ **49.** $\frac{1}{4}\sin^4 x + C$

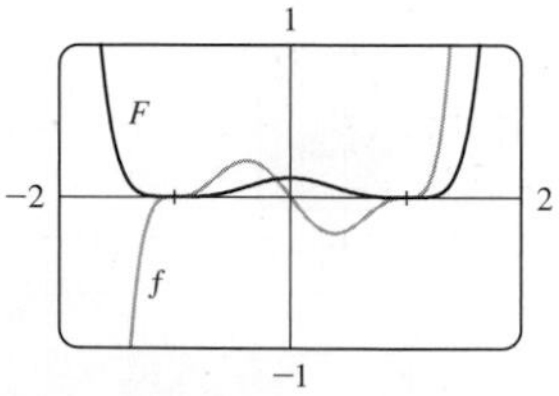

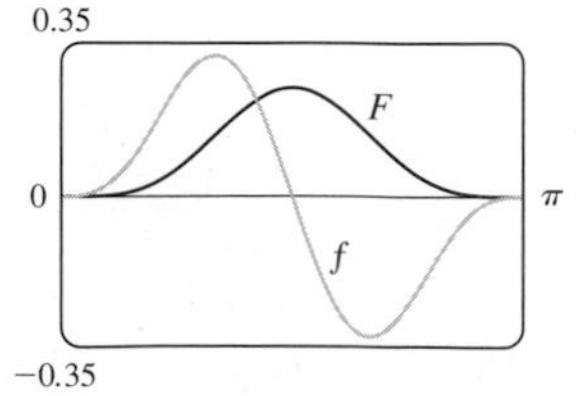

51. 0 **53.** $\frac{182}{9}$ **55.** 4

57. 0 **59.** $e - \sqrt{e}$ **61.** 3 **63.** $\frac{1}{3}(2\sqrt{2} - 1)a^3$

65. $\frac{16}{15}$ **67.** 2 **69.** $\ln(e + 1)$ **71.** $\sqrt{3} - \frac{1}{3}$

73. 6π **75.** All three areas are equal. **77.** ≈ 4512 L

79. $\dfrac{5}{4\pi}\left(1 - \cos\dfrac{2\pi t}{5}\right)$ L **81.** 5 **87.** $\pi^2/4$

CHAPTER 5 REVIEW ■ PAGE 409

True-False Quiz

1. True **3.** True **5.** False **7.** True **9.** True

11. False **13.** False **15.** False

Exercises

1. (a) 8 (b) 5.7

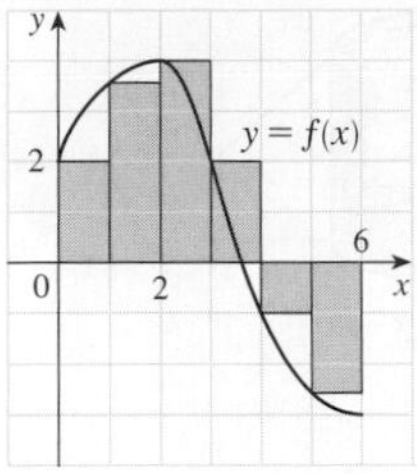

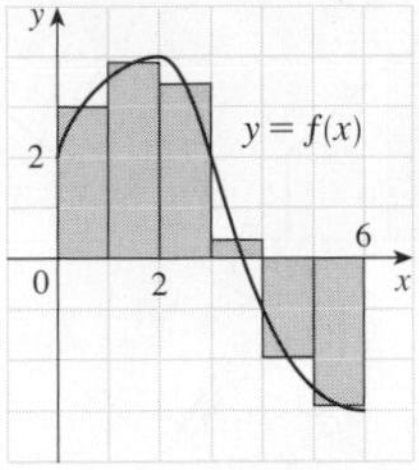

3. $\frac{1}{2} + \pi/4$ **5.** 3 **7.** f is c, f' is b, $\int_0^x f(t)\,dt$ is a
9. 37 **11.** $\frac{9}{10}$ **13.** -76 **15.** $\frac{21}{4}$ **17.** Does not exist
19. $\frac{1}{3}\sin 1$ **21.** 0 **23.** $-(1/x) - 2\ln|x| + x + C$
25. $\sqrt{x^2 + 4x} + C$ **27.** $[1/(2\pi)]\sin^2\pi t + C$
29. $2e^{\sqrt{x}} + C$ **31.** $-\frac{1}{2}[\ln(\cos x)]^2 + C$
33. $\frac{1}{4}\ln(1 + x^4) + C$ **35.** $\ln|1 + \sec\theta| + C$ **37.** $\frac{23}{3}$
39. $2\sqrt{1 + \sin x} + C$ **41.** $\frac{64}{5}$ **43.** $F'(x) = x^2/(1 + x^3)$
45. $g'(x) = 4x^3\cos(x^8)$ **47.** $y' = (2e^x - e^{\sqrt{x}})/(2x)$
49. $4 \leqslant \int_1^3 \sqrt{x^2 + 3}\,dx \leqslant 4\sqrt{3}$ **55.** 0.280981
57. Number of barrels of oil consumed from Jan. 1, 2000, through Jan. 1, 2008
59. 72,400 **61.** 3 **63.** $c \approx 1.62$
65. $f(x) = e^{2x}(1 + 2x)/(1 - e^{-x})$ **71.** $\frac{2}{3}$

PROBLEMS PLUS ▪ PAGE 413

1. $\pi/2$ **3.** $f(x) = \frac{1}{2}x$ **5.** -1 **7.** e^{-2} **9.** $[-1, 2]$
11. (a) $\frac{1}{2}(n - 1)n$ (b) $\frac{1}{2}\llbracket b \rrbracket(2b - \llbracket b \rrbracket - 1) - \frac{1}{2}\llbracket a \rrbracket(2a - \llbracket a \rrbracket - 1)$
17. $2(\sqrt{2} - 1)$

CHAPTER 6

EXERCISES 6.1 ▪ PAGE 420

1. $\frac{32}{3}$ **3.** $e - (1/e) + \frac{10}{3}$ **5.** 19.5 **7.** $\frac{1}{6}$ **9.** $\ln 2 - \frac{1}{2}$
11. $\frac{1}{3}$ **13.** 72 **15.** $2 - 2\ln 2$ **17.** $\frac{59}{12}$ **19.** $\frac{32}{3}$
21. $\frac{8}{3}$ **23.** $\frac{1}{2}$ **25.** $\pi - \frac{2}{3}$ **27.** $\ln 2$ **29.** 6.5
31. $\frac{3}{2}\sqrt{3} - 1$ **33.** 0.6407 **35.** 0, 0.90; 0.04 **37.** 8.38
39. $12\sqrt{6} - 9$ **41.** $117\frac{1}{3}$ ft **43.** 4232 cm^2
45. (a) Car A (b) The distance by which A is ahead of B after 1 minute (c) Car A (d) $t \approx 2.2$ min
47. $\frac{24}{5}\sqrt{3}$ **49.** $4^{2/3}$ **51.** ± 6
53. $0 < m < 1$; $m - \ln m - 1$

EXERCISES 6.2 ▪ PAGE 430

1. $19\pi/12$

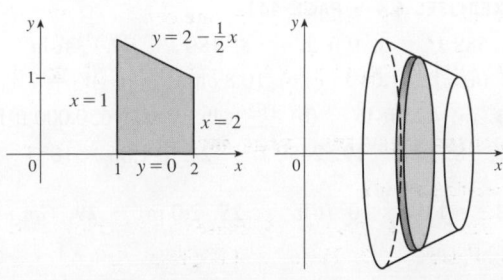

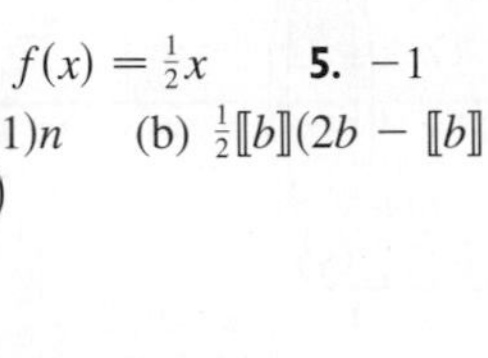

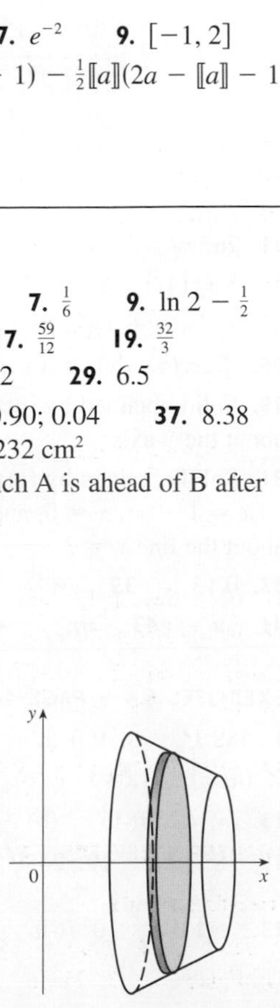

3. $\pi/2$

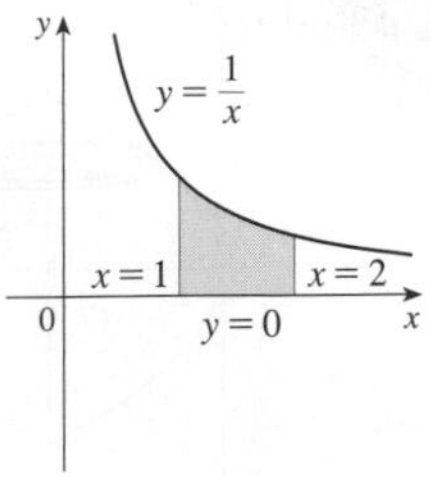

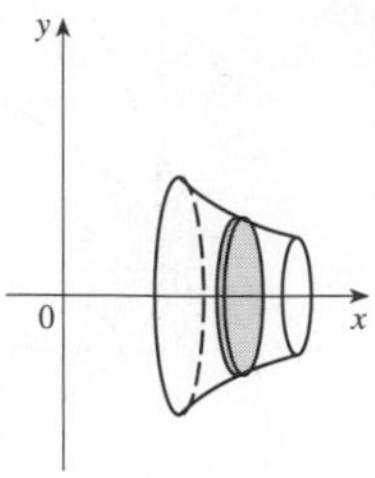

5. 162π

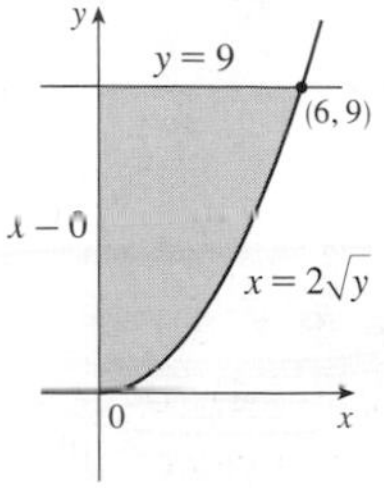

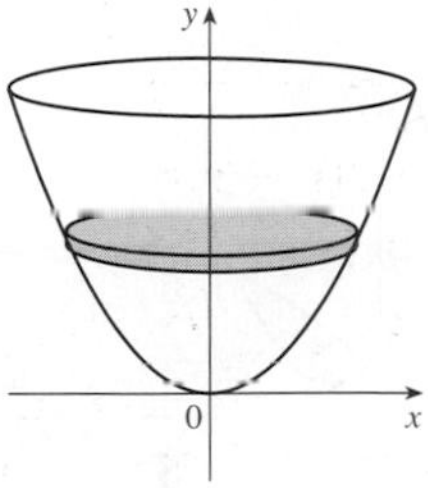

7. $4\pi/21$

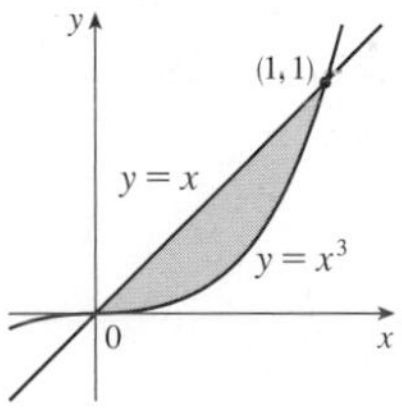

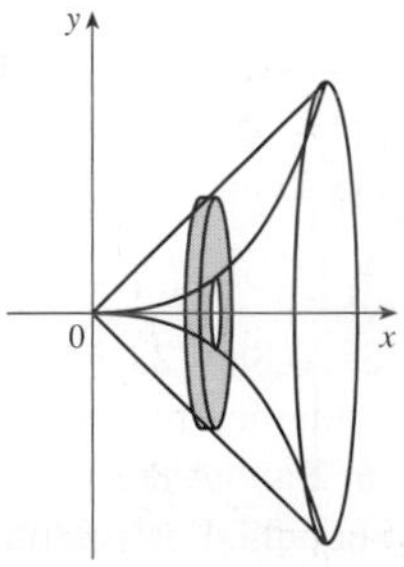

9. $64\pi/15$

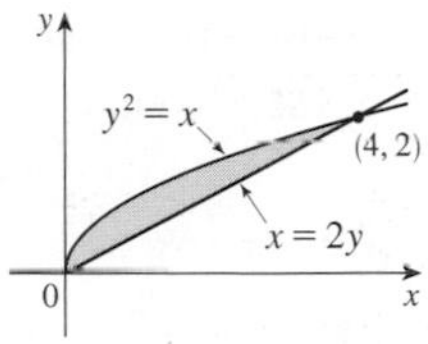

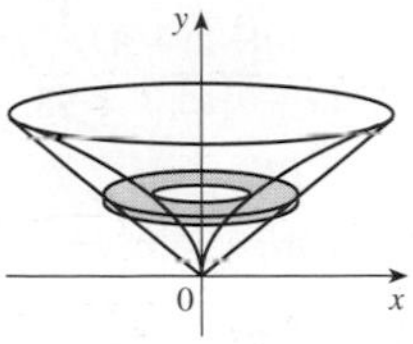

11. $\pi/6$

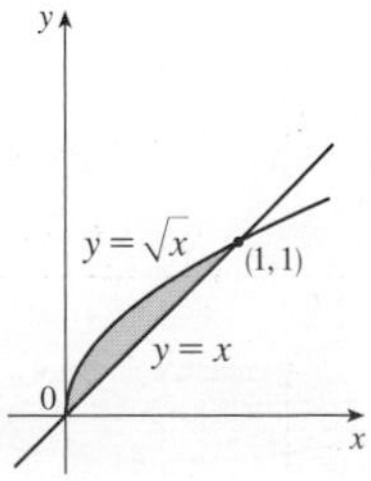

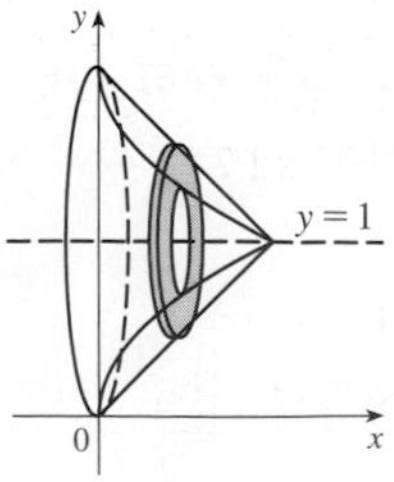

13. $2\pi(\frac{4}{3}\pi - \sqrt{3})$

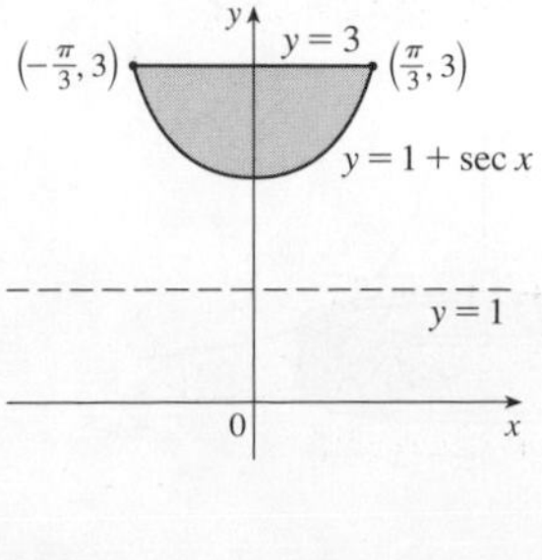

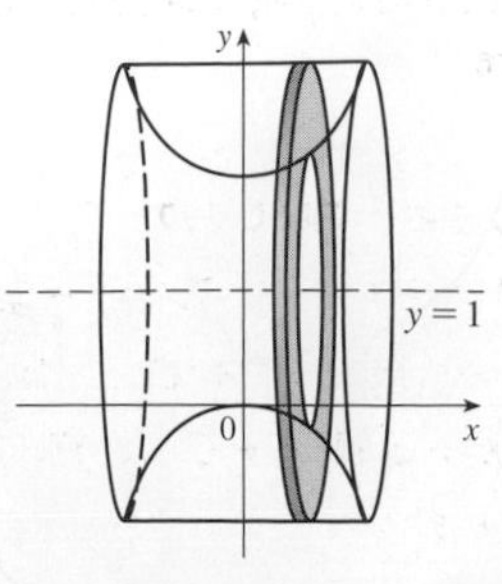

15. $16\pi/15$

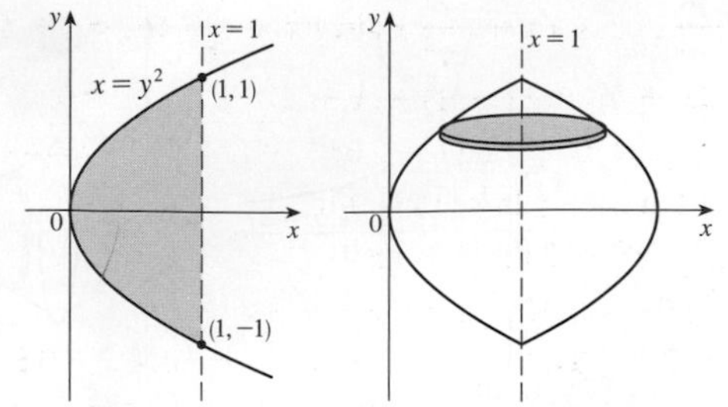

17. $29\pi/30$

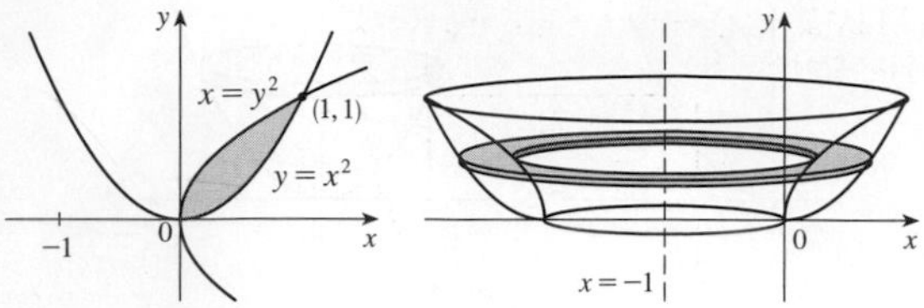

19. $\pi/7$ **21.** $\pi/10$ **23.** $\pi/2$ **25.** $7\pi/15$
27. $5\pi/14$ **29.** $13\pi/30$ **31.** $\pi\int_0^{\pi/4}(1-\tan^3 x)^2\,dx$
33. $\pi\int_0^{\pi}[1^2-(1-\sin x)^2]\,dx$
35. $\pi\int_{-2\sqrt{2}}^{2\sqrt{2}}\left[5^2-\left(\sqrt{1+y^2}+2\right)^2\right]dy$
37. $-1.288, 0.884; 23.780$ **39.** $\frac{11}{8}\pi^2$
41. Solid obtained by rotating the region $0 \leqslant y \leqslant \cos x$, $0 \leqslant x \leqslant \pi/2$ about the x-axis
43. Solid obtained by rotating the region above the x-axis bounded by $x = y^2$ and $x = y^4$ about the y-axis
45. 1110 cm^3 **47.** (a) 196 (b) 838 **49.** $\frac{1}{3}\pi r^2 h$
51. $\pi h^2\left(r-\frac{1}{3}h\right)$ **53.** $\frac{2}{3}b^2 h$ **55.** 10 cm^3 **57.** 24
59. $\frac{1}{3}$ **61.** $\frac{8}{15}$
63. (a) $8\pi R\int_0^r\sqrt{r^2-y^2}\,dy$ (b) $2\pi^2 r^2 R$
65. (b) $\pi r^2 h$ **67.** $\frac{5}{12}\pi r^3$ **69.** $8\int_0^r\sqrt{R^2-y^2}\sqrt{r^2-y^2}\,dy$

EXERCISES 6.3 ■ PAGE 436

1. Circumference $= 2\pi x$, height $= x(x-1)^2$; $\pi/15$

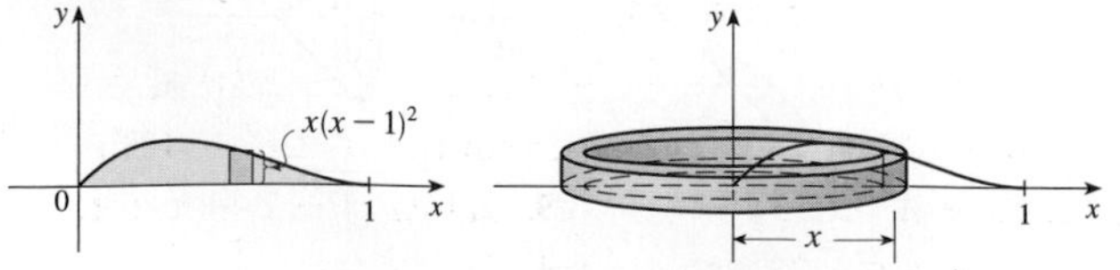

3. 2π

5. $\pi(1-1/e)$

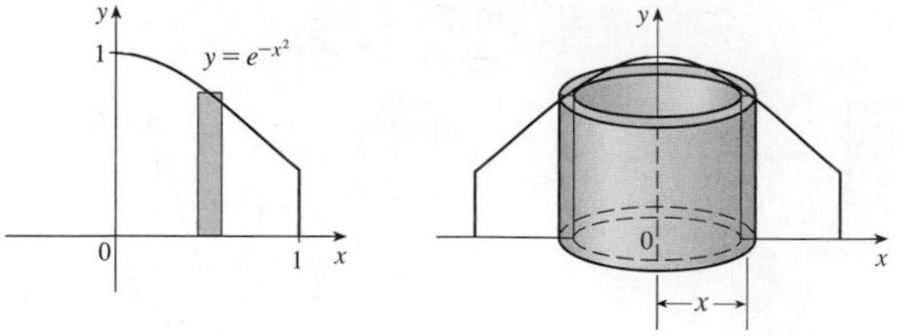

7. 16π

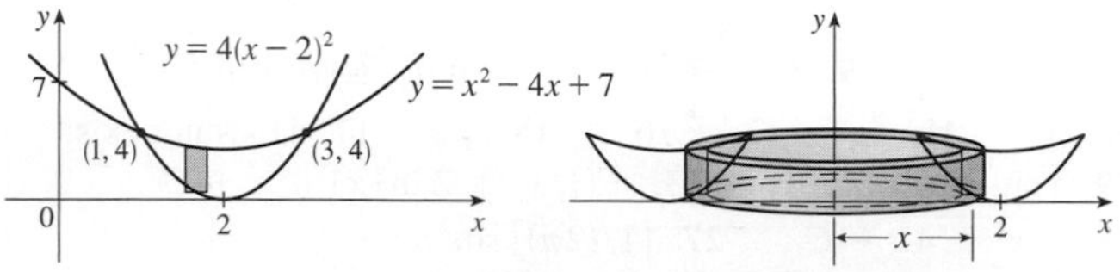

9. $21\pi/2$

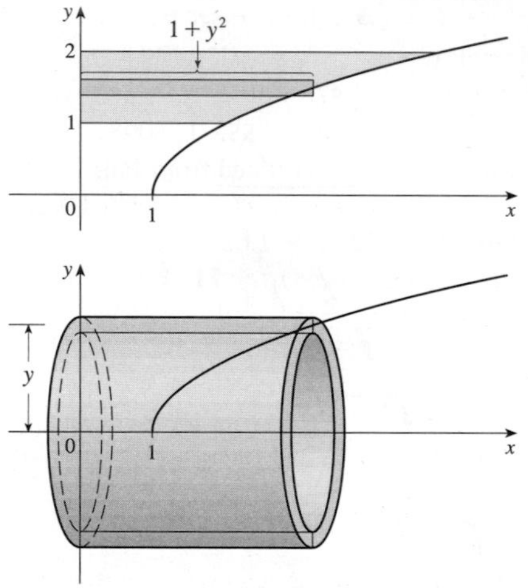

11. $768\pi/7$ **13.** $16\pi/3$ **15.** $7\pi/15$ **17.** $8\pi/3$
19. $5\pi/14$ **21.** $\int_1^2 2\pi x\ln x\,dx$
23. $\int_0^1 2\pi(x+1)[\sin(\pi x/2)-x^4]\,dx$
25. $\int_0^{\pi}2\pi(4-y)\sqrt{\sin y}\,dy$ **27.** 3.68
29. Solid obtained by rotating the region $0 \leqslant y \leqslant x^4$, $0 \leqslant x \leqslant 3$ about the y-axis
31. Solid obtained by rotating the region bounded by (i) $x = 1 - y^2$, $x = 0$, and $y = 0$, or (ii) $x = y^2$, $x = 1$, and $y = 0$ about the line $y = 3$
33. 0.13 **35.** $\frac{1}{32}\pi^3$ **37.** 8π **39.** $2\pi(12 - 4\ln 4)$
41. $\frac{4}{3}\pi$ **43.** $\frac{4}{3}\pi r^3$ **45.** $\frac{1}{3}\pi r^2 h$

EXERCISES 6.4 ■ PAGE 441

1. 588 J **3.** 9 ft-lb **5.** 180 J **7.** $\frac{15}{4}$ ft-lb
9. (a) $\frac{25}{24} \approx 1.04$ J (b) 10.8 cm **11.** $W_2 = 3W_1$
13. (a) 625 ft-lb (b) $\frac{1875}{4}$ ft-lb **15.** 650,000 ft-lb
17. 3857 J **19.** 2450 J **21.** $\approx 1.06 \times 10^6$ J
23. $\approx 1.04 \times 10^5$ ft-lb **25.** 2.0 m **29.** $Gm_1m_2\left(\frac{1}{a}-\frac{1}{b}\right)$

EXERCISES 6.5 ■ PAGE 445

1. $\frac{8}{3}$ **3.** $\frac{45}{28}$ **5.** $\frac{1}{10}(1 - e^{-25})$ **7.** $2/(5\pi)$

9. (a) 1 (b) 2, 4 (c)

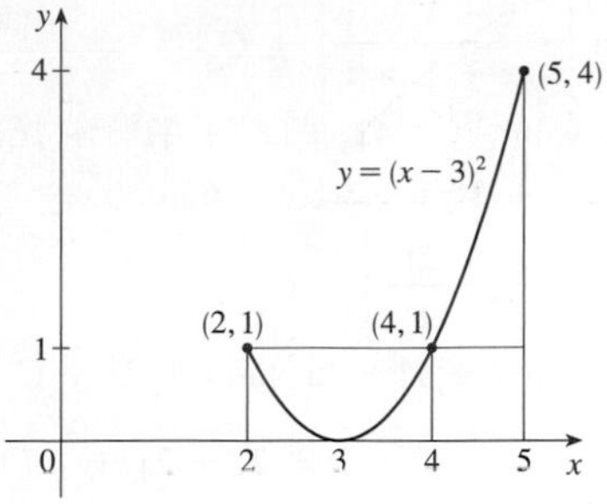

11. (a) $4/\pi$ (b) $\approx 1.24, 2.81$

(c)

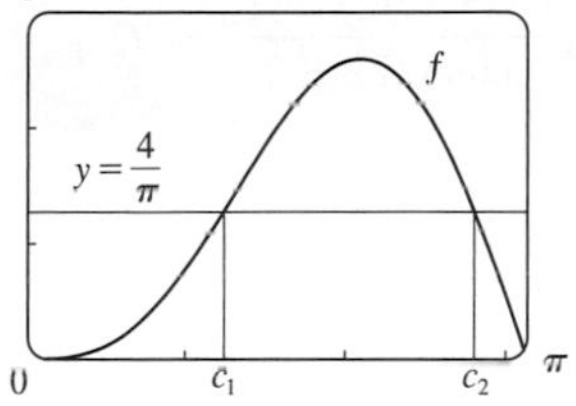

15. $38\frac{1}{3}$ **17.** $(50 + 28/\pi)°\text{F} \approx 59°\text{F}$ **19.** 6 kg/m

21. $5/(4\pi) \approx 0.4$ L

CHAPTER 6 REVIEW ■ PAGE 446

Exercises

1. $\frac{8}{3}$ **3.** $\frac{7}{12}$ **5.** $\frac{4}{3} + 4/\pi$ **7.** $64\pi/15$ **9.** $1656\pi/5$

11. $\frac{4}{3}\pi(2ah + h^2)^{3/2}$ **13.** $\int_{-\pi/3}^{\pi/3} 2\pi(\pi/2 - x)\left(\cos^2 x - \frac{1}{4}\right) dx$

15. (a) $2\pi/15$ (b) $\pi/6$ (c) $8\pi/15$

17. (a) 0.38 (b) 0.87

19. Solid obtained by rotating the region $0 \leqslant y \leqslant \cos x$, $0 \leqslant x \leqslant \pi/2$ about the y-axis

21. Solid obtained by rotating the region $0 \leqslant x \leqslant \pi$, $0 \leqslant y \leqslant 2 - \sin x$ about the x-axis

23. 36 **25.** $\frac{125}{3}\sqrt{3}$ m^3 **27.** 3.2 J

29. (a) $8000\pi/3 \approx 8378$ ft-lb (b) 2.1 ft **31.** $f(x)$

PROBLEMS PLUS ■ PAGE 448

1. (a) $f(t) = 3t^2$ (b) $f(x) = \sqrt{2x/\pi}$ **3.** $\frac{32}{27}$

5. (b) 0.2261 (c) 0.6736 m

(d) (i) $1/(105\pi) \approx 0.003$ in/s (ii) $370\pi/3$ s ≈ 6.5 min

9. $y = \frac{32}{9}x^2$

11. (a) $V = \int_0^h \pi[f(y)]^2\, dy$ (c) $f(y) = \sqrt{kA/(\pi C)}\, y^{1/4}$

Advantage: the markings on the container are equally spaced.

13. $b = 2a$ **15.** $B = 16A$

CHAPTER 7

EXERCISES 7.1 ■ PAGE 457

1. $\frac{1}{3}x^3 \ln x - \frac{1}{9}x^3 + C$ **3.** $\frac{1}{5}x \sin 5x + \frac{1}{25}\cos 5x + C$

5. $2(r - 2)e^{r/2} + C$

7. $-\frac{1}{\pi}x^2 \cos \pi x + \frac{2}{\pi^2} x \sin \pi x + \frac{2}{\pi^3}\cos \pi x + C$

9. $\frac{1}{2}(2x + 1)\ln(2x + 1) - x + C$

11. $t \arctan 4t - \frac{1}{8}\ln(1 + 16t^2) + C$

13. $\frac{1}{2}t \tan 2t - \frac{1}{4}\ln|\sec 2t| + C$

15. $x(\ln x)^2 - 2x\ln x + 2x + C$

17. $\frac{1}{13}e^{2\theta}(2\sin 3\theta - 3\cos 3\theta) + C$

19. $\pi/3$ **21.** $1 - 1/e$ **23.** $\frac{1}{2} - \frac{1}{2}\ln 2$ **25.** $\frac{1}{4} - \frac{3}{4}e^{-2}$

27. $\frac{1}{6}\left(\pi + 6 - 3\sqrt{3}\right)$ **29.** $\sin x\,(\ln \sin x - 1) + C$

31. $\frac{32}{5}(\ln 2)^2 - \frac{64}{25}\ln 2 + \frac{62}{125}$

33. $2\sqrt{x}\sin\sqrt{x} + 2\cos\sqrt{x} + C$ **35.** $-\frac{1}{2} - \pi/4$

37. $\frac{1}{2}(x^2 - 1)\ln(1 + x) - \frac{1}{4}x^2 + \frac{1}{2}x + \frac{3}{4} + C$

39. $(2x + 1)e^x + C$

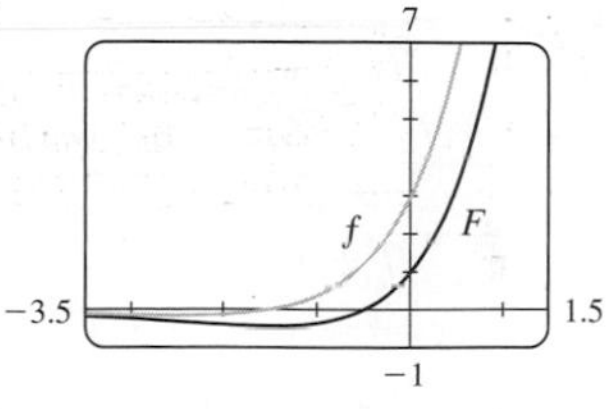

41. $\frac{1}{3}x^2(1 + x^2)^{3/2} - \frac{2}{15}(1 + x^2)^{5/2} + C$

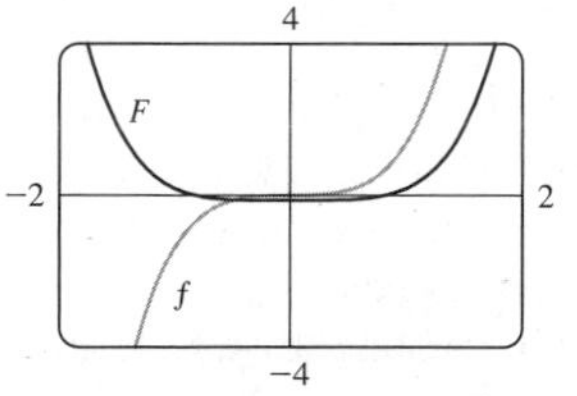

43. (b) $-\frac{1}{4}\cos x \sin^3 x + \frac{3}{8}x - \frac{3}{16}\sin 2x + C$

45. (b) $\frac{2}{3}, \frac{8}{15}$ **51.** $x(\ln x)^3 - 3x(\ln x)^2 + 6x\ln x - 6x + C$

53. $\frac{25}{4} - \frac{75}{4}e^{-2}$ **55.** 1.0475, 2.8731; 2.1828 **57.** $4 - 8/\pi$

59. $2\pi e$ **61.** $\frac{9}{2}\ln 3 - \frac{13}{9}$ **63.** $2 - e^{-t}(t^2 + 2t + 2)$ m

65. 2

EXERCISES 7.2 ■ PAGE 465

1. $\frac{1}{5}\cos^5 x - \frac{1}{3}\cos^3 x + C$ **3.** $-\frac{11}{384}$

5. $\frac{1}{3\pi}\sin^3(\pi x) - \frac{2}{5\pi}\sin^5(\pi x) + \frac{1}{7\pi}\sin^7(\pi x) + C$

7. $\pi/4$ **9.** $3\pi/8$ **11.** $\frac{3}{2}\theta + 2\sin\theta + \frac{1}{4}\sin 2\theta + C$

13. $\pi/16$ **15.** $\frac{2}{45}\sqrt{\sin\alpha}\,(45 - 18\sin^2\alpha + 15\sin^4\alpha) + C$

17. $\frac{1}{2}\cos^2 x - \ln|\cos x| + C$ **19.** $\ln|\sin x| + 2\sin x + C$

21. $\frac{1}{2}\tan^2 x + C$ **23.** $\tan x - x + C$

25. $\frac{1}{5}\tan^5 t + \frac{2}{3}\tan^3 t + \tan t + C$ **27.** $\frac{117}{8}$

29. $\frac{1}{3}\sec^3 x - \sec x + C$

31. $\frac{1}{4}\sec^4 x - \tan^2 x + \ln|\sec x| + C$

33. $\frac{1}{6}\tan^6\theta + \frac{1}{4}\tan^4\theta + C$

35. $x\sec x - \ln|\sec x + \tan x| + C$ **37.** $\sqrt{3} - \frac{1}{3}\pi$

39. $\frac{1}{3}\csc^3\alpha - \frac{1}{5}\csc^5\alpha + C$ **41.** $\ln|\csc x - \cot x| + C$

43. $-\frac{1}{6}\cos 3x - \frac{1}{26}\cos 13x + C$ **45.** $\frac{1}{8}\sin 4\theta - \frac{1}{12}\sin 6\theta + C$

47. $\frac{1}{2}\sin 2x + C$ **49.** $\frac{1}{10}\tan^5(t^2) + C$

51. $\frac{1}{4}x^2 - \frac{1}{4}\sin(x^2)\cos(x^2) + C$ **53.** $\frac{1}{6}\sin 3x - \frac{1}{18}\sin 9x + C$

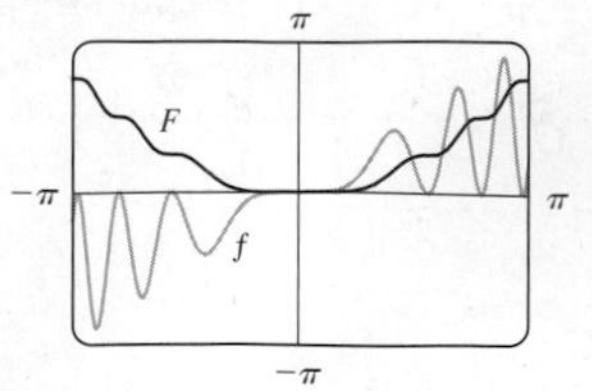

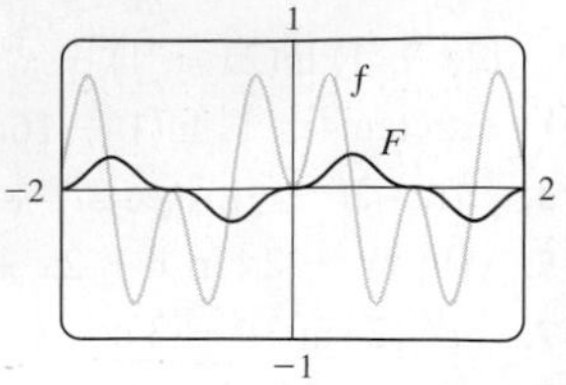

55. 0 **57.** 1 **59.** 0 **61.** $\pi^2/4$ **63.** $\pi(2\sqrt{2} - \frac{5}{2})$

65. $s = (1 - \cos^3\omega t)/(3\omega)$

EXERCISES 7.3 ■ PAGE 472

1. $\sqrt{x^2 - 9}/(9x) + C$ **3.** $\frac{1}{3}(x^2 - 18)\sqrt{x^2 + 9} + C$

5. $\pi/24 + \sqrt{3}/8 - \frac{1}{4}$ **7.** $-\sqrt{25 - x^2}/(25x) + C$

9. $\ln(\sqrt{x^2 + 16} + x) + C$ **11.** $\frac{1}{4}\sin^{-1}(2x) + \frac{1}{2}x\sqrt{1 - 4x^2} + C$

13. $\frac{1}{6}\sec^{-1}(x/3) - \sqrt{x^2 - 9}/(2x^2) + C$

15. $\frac{1}{16}\pi a^4$ **17.** $\sqrt{x^2 - 7} + C$

19. $\ln\left|(\sqrt{1 + x^2} - 1)/x\right| + \sqrt{1 + x^2} + C$ **21.** $\frac{9}{500}\pi$

23. $\frac{9}{2}\sin^{-1}((x - 2)/3) + \frac{1}{2}(x - 2)\sqrt{5 + 4x - x^2} + C$

25. $\sqrt{x^2 + x + 1} - \frac{1}{2}\ln(\sqrt{x^2 + x + 1} + x + \frac{1}{2}) + C$

27. $\frac{1}{2}(x + 1)\sqrt{x^2 + 2x} - \frac{1}{2}\ln\left|x + 1 + \sqrt{x^2 + 2x}\right| + C$

29. $\frac{1}{4}\sin^{-1}(x^2) + \frac{1}{4}x^2\sqrt{1 - x^4} + C$

33. $\frac{1}{6}(\sqrt{48} - \sec^{-1} 7)$ **37.** 0.81, 2; 2.10

41. $r\sqrt{R^2 - r^2} + \pi r^2/2 - R^2\arcsin(r/R)$ **43.** $2\pi^2Rr^2$

EXERCISES 7.4 ■ PAGE 481

1. (a) $\dfrac{A}{x + 3} + \dfrac{B}{3x + 1}$ (b) $\dfrac{A}{x} + \dfrac{B}{x + 1} + \dfrac{C}{(x + 1)^2}$

3. (a) $\dfrac{A}{x} + \dfrac{B}{x^2} + \dfrac{C}{x^3} + \dfrac{Dx + E}{x^2 + 4}$

(b) $\dfrac{A}{x + 3} + \dfrac{B}{(x + 3)^2} + \dfrac{C}{x - 3} + \dfrac{D}{(x - 3)^2}$

5. (a) $1 + \dfrac{A}{x - 1} + \dfrac{B}{x + 1} + \dfrac{Cx + D}{x^2 + 1}$

(b) $\dfrac{At + B}{t^2 + 1} + \dfrac{Ct + D}{t^2 + 4} + \dfrac{Et + F}{(t^2 + 4)^2}$

7. $x + 6\ln|x - 6| + C$

9. $2\ln|x + 5| - \ln|x - 2| + C$ **11.** $\frac{1}{2}\ln\frac{3}{2}$

13. $a\ln|x - b| + C$ **15.** $\frac{7}{6} + \ln\frac{2}{3}$

17. $\frac{27}{5}\ln 2 - \frac{9}{5}\ln 3$ (or $\frac{9}{5}\ln\frac{8}{3}$)

19. $-\dfrac{1}{36}\ln|x + 5| + \dfrac{1}{6}\dfrac{1}{x + 5} + \dfrac{1}{36}\ln|x - 1| + C$

21. $\frac{1}{2}x^2 - 2\ln(x^2 + 4) + 2\tan^{-1}(x/2) + C$

23. $2\ln|x| + (1/x) + 3\ln|x + 2| + C$

25. $\ln|x - 1| - \frac{1}{2}\ln(x^2 + 9) - \frac{1}{3}\tan^{-1}(x/3) + C$

27. $\frac{1}{2}\ln(x^2 + 1) + (1/\sqrt{2})\tan^{-1}(x/\sqrt{2}) + C$

29. $\frac{1}{2}\ln(x^2 + 2x + 5) + \frac{3}{2}\tan^{-1}\left(\dfrac{x + 1}{2}\right) + C$

31. $\frac{1}{3}\ln|x - 1| - \frac{1}{6}\ln(x^2 + x + 1) - \dfrac{1}{\sqrt{3}}\tan^{-1}\dfrac{2x + 1}{\sqrt{3}} + C$

33. $\frac{1}{4}\ln\frac{8}{3}$ **35.** $\frac{1}{16}\ln|x| - \frac{1}{32}\ln(x^2 + 4) + \dfrac{1}{8(x^2 + 4)} + C$

37. $\frac{7}{8}\sqrt{2}\tan^{-1}\left(\dfrac{x - 2}{\sqrt{2}}\right) + \dfrac{3x - 8}{4(x^2 - 4x + 6)} + C$

39. $\ln\left|\dfrac{\sqrt{x + 1} - 1}{\sqrt{x + 1} + 1}\right| + C$

41. $2 + \ln\frac{25}{9}$ **43.** $\frac{3}{10}(x^2 + 1)^{5/3} - \frac{3}{4}(x^2 + 1)^{2/3} + C$

45. $2\sqrt{x} + 3\sqrt[3]{x} + 6\sqrt[6]{x} + 6\ln\left|\sqrt[6]{x} - 1\right| + C$

47. $\ln\left[\dfrac{(e^x + 2)^2}{e^x + 1}\right] + C$

49. $\ln|\tan t + 1| - \ln|\tan t + 2| + C$

51. $(x - \frac{1}{2})\ln(x^2 - x + 2) - 2x + \sqrt{7}\tan^{-1}\left(\dfrac{2x - 1}{\sqrt{7}}\right) + C$

53. $-\frac{1}{2}\ln 3 \approx -0.55$

55. $\frac{1}{2}\ln\left|\dfrac{x - 2}{x}\right| + C$ **59.** $\frac{1}{5}\ln\left|\dfrac{2\tan(x/2) - 1}{\tan(x/2) + 2}\right| + C$

61. $4\ln\frac{2}{3} + 2$ **63.** $-1 + \frac{11}{3}\ln 2$

65. $t = -\ln P - \frac{1}{9}\ln(0.9P + 900) + C$, where $C \approx 10.23$

67. (a) $\dfrac{24{,}110}{4879}\dfrac{1}{5x + 2} - \dfrac{668}{323}\dfrac{1}{2x + 1} - \dfrac{9438}{80{,}155}\dfrac{1}{3x - 7} + \dfrac{1}{260{,}015}\dfrac{22{,}098x + 48{,}935}{x^2 + x + 5}$

(b) $\dfrac{4822}{4879}\ln|5x + 2| - \dfrac{334}{323}\ln|2x + 1| - \dfrac{3146}{80{,}155}\ln|3x - 7| + \dfrac{11{,}049}{260{,}015}\ln(x^2 + x + 5) + \dfrac{75{,}772}{260{,}015\sqrt{19}}\tan^{-1}\dfrac{2x + 1}{\sqrt{19}} + C$

The CAS omits the absolute value signs and the constant of integration.

EXERCISES 7.5 ■ PAGE 488

1. $\sin x + \frac{1}{3}\sin^3 x + C$

3. $\sin x + \ln|\csc x - \cot x| + C$

5. $4 - \ln 9$ **7.** $e^{\pi/4} - e^{-\pi/4}$

9. $\frac{243}{5}\ln 3 - \frac{242}{25}$ **11.** $\frac{1}{2}\ln(x^2 - 4x + 5) + \tan^{-1}(x - 2) + C$

13. $\frac{1}{8}\cos^8\theta - \frac{1}{6}\cos^6\theta + C$ (or $\frac{1}{4}\sin^4\theta - \frac{1}{3}\sin^6\theta + \frac{1}{8}\sin^8\theta + C$)

15. $x/\sqrt{1 - x^2} + C$

17. $\frac{1}{4}x^2 - \frac{1}{2}x\sin x\cos x + \frac{1}{4}\sin^2 x + C$
(or $\frac{1}{4}x^2 - \frac{1}{4}x\sin 2x - \frac{1}{8}\cos 2x + C$)

19. $e^{e^x} + C$ **21.** $(x + 1)\arctan\sqrt{x} - \sqrt{x} + C$

23. $\frac{4097}{45}$ **25.** $3x + \frac{23}{3}\ln|x - 4| - \frac{5}{3}\ln|x + 2| + C$

27. $x - \ln(1 + e^x) + C$ **29.** $15 + 7\ln\frac{2}{7}$

31. $\sin^{-1}x - \sqrt{1 - x^2} + C$

33. $2\sin^{-1}\left(\dfrac{x + 1}{2}\right) + \dfrac{x + 1}{2}\sqrt{3 - 2x - x^2} + C$

35. 0 **37.** $\pi/8 - \frac{1}{4}$ **39.** $\ln|\sec\theta - 1| - \ln|\sec\theta| + C$

41. $\theta\tan\theta - \frac{1}{2}\theta^2 - \ln|\sec\theta| + C$ **43.** $\frac{2}{3}(1 + e^x)^{3/2} + C$

45. $-\frac{1}{3}(x^3 + 1)e^{-x^3} + C$

47. $\ln|x - 1| - 3(x - 1)^{-1} - \frac{3}{2}(x - 1)^{-2} - \frac{1}{3}(x - 1)^{-3} + C$

49. $\ln\left|\dfrac{\sqrt{4x + 1} - 1}{\sqrt{4x + 1} + 1}\right| + C$ **51.** $-\ln\left|\dfrac{\sqrt{4x^2 + 1} + 1}{2x}\right| + C$

53. $\dfrac{1}{m}x^2\cosh(mx) - \dfrac{2}{m^2}x\sinh(mx) + \dfrac{2}{m^3}\cosh(mx) + C$

55. $2 \ln \sqrt{x} - 2 \ln(1 + \sqrt{x}) + C$

57. $\frac{3}{7}(x + c)^{7/3} - \frac{3}{4}c(x + c)^{4/3} + C$

59. $\sin(\sin x) - \frac{1}{3}\sin^3(\sin x) + C$ **61.** $2(x - 2\sqrt{x} + 2)e^{\sqrt{x}} + C$

63. $-\tan^{-1}(\cos^2 x) + C$ **65.** $\frac{2}{3}[(x + 1)^{3/2} - x^{3/2}] + C$

67. $\sqrt{2} - 2/\sqrt{3} + \ln(2 + \sqrt{3}) - \ln(1 + \sqrt{2})$

69. $e^x - \ln(1 + e^x) + C$

71. $-\sqrt{1 - x^2} + \frac{1}{2}(\arcsin x)^2 + C$

73. $\frac{1}{8}\ln|x - 2| - \frac{1}{16}\ln(x^2 + 4) - \frac{1}{8}\tan^{-1}(x/2) + C$

75. $2(x - 2)\sqrt{1 + e^x} + 2 \ln \dfrac{\sqrt{1 + e^x} + 1}{\sqrt{1 + e^x} - 1} + C$

77. $\frac{2}{3}\tan^{-1}(x^{3/2}) + C$

79. $\frac{1}{3}x \sin^3 x + \frac{1}{3}\cos x - \frac{1}{9}\cos^3 x + C$ **81.** $xe^{x^2} + C$

EXERCISES 7.6 ■ PAGE 493

1. $(-1/x)\sqrt{7 - 2x^2} - \sqrt{2}\sin^{-1}(\sqrt{2}x/\sqrt{7}) + C$

3. $\dfrac{1}{2\pi}\sec(\pi x)\tan(\pi x) + \dfrac{1}{2\pi}\ln|\sec(\pi x) + \tan(\pi x)| + C$

5. $\pi/4$ **7.** $\dfrac{1}{2\pi}\tan^2(\pi x) + \dfrac{1}{\pi}\ln|\cos(\pi x)| + C$

9. $-\sqrt{4x^2 + 9}/(9x) + C$ **11.** $e - 2$

13. $-\frac{1}{2}\tan^2(1/z) - \ln|\cos(1/z)| + C$

15. $\frac{1}{2}(e^{2x} + 1)\arctan(e^x) - \frac{1}{2}e^x + C$

17. $\dfrac{2y - 1}{8}\sqrt{6 + 4y - 4y^2} + \frac{7}{8}\sin^{-1}\left(\dfrac{2y - 1}{\sqrt{7}}\right)$
$- \frac{1}{12}(6 + 4y - 4y^2)^{3/2} + C$

19. $\frac{1}{9}\sin^3 x\,[3\ln(\sin x) - 1] + C$

21. $\dfrac{1}{2\sqrt{3}}\ln\left|\dfrac{e^x + \sqrt{3}}{e^x - \sqrt{3}}\right| + C$

23. $\frac{1}{4}\tan x \sec^3 x + \frac{3}{8}\tan x \sec x + \frac{3}{8}\ln|\sec x + \tan x| + C$

25. $\frac{1}{2}(\ln x)\sqrt{4 + (\ln x)^2} + 2\ln\left[\ln x + \sqrt{4 + (\ln x)^2}\right] + C$

27. $\sqrt{e^{2x} - 1} - \cos^{-1}(e^{-x}) + C$

29. $\frac{1}{5}\ln|x^5 + \sqrt{x^{10} - 2}| + C$ **31.** $2\pi^2$

35. $\frac{1}{3}\tan x \sec^2 x + \frac{2}{3}\tan x + C$

37. $\frac{1}{4}x(x^2 + 2)\sqrt{x^2 + 4} - 2\ln(\sqrt{x^2 + 4} + x) + C$

39. $\frac{1}{10}(1 + 2x)^{5/2} - \frac{1}{6}(1 + 2x)^{3/2} + C$

41. $-\ln|\cos x| - \frac{1}{2}\tan^2 x + \frac{1}{4}\tan^4 x + C$

43. (a) $-\ln\left|\dfrac{1 + \sqrt{1 - x^2}}{x}\right| + C$;
both have domain $(-1, 0) \cup (0, 1)$

45. $F(x) = \frac{1}{2}\ln(x^2 - x + 1) - \frac{1}{2}\ln(x^2 + x + 1)$;
max. at -1, min. at 1; IP at -1.7, 0, and 1.7

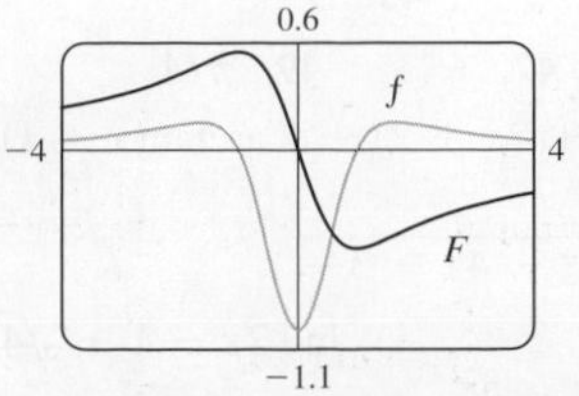

47. $F(x) = -\frac{1}{10}\sin^3 x \cos^7 x - \frac{3}{80}\sin x \cos^7 x + \frac{1}{160}\sin x \cos^5 x$
$+ \frac{1}{128}\sin x \cos^3 x + \frac{3}{256}\sin x \cos x + \frac{3}{256}x$;
max. at π, min. at 0; IP at 0.7, $\pi/2$, and 2.5

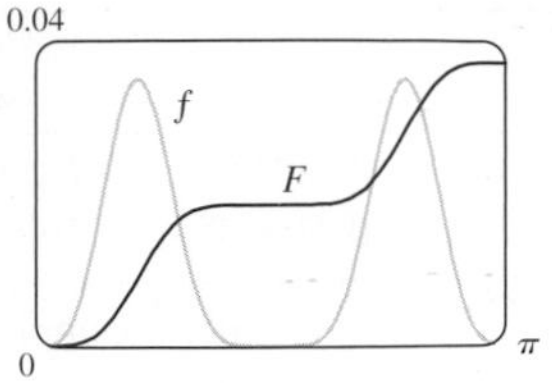

EXERCISES 7.7 ■ PAGE 505

1. (a) $L_2 = 6$, $R_2 = 12$, $M_2 \approx 9.6$
(b) L_2 is an underestimate, R_2 and M_2 are overestimates.
(c) $T_2 = 9 < I$ (d) $L_n < T_n < I < M_n < R_n$

3. (a) $T_4 \approx 0.895759$ (underestimate)
(b) $M_4 \approx 0.908907$ (overestimate)
$T_4 < I < M_4$

5. (a) 5.932957, $E_M \approx -0.063353$
(b) 5.869247, $E_S \approx 0.000357$

7. (a) 2.413790 (b) 2.411453 (c) 2.412232

9. (a) 0.146879 (b) 0.147391 (c) 0.147219

11. (a) 0.451948 (b) 0.451991 (c) 0.451976

13. (a) 4.513618 (b) 4.748256 (c) 4.675111

15. (a) −0.495333 (b) −0.543321 (c) −0.526123

17. (a) 1.064275 (b) 1.067416 (c) 1.074915

19. (a) $T_8 \approx 0.902333$, $M_8 \approx 0.905620$
(b) $|E_T| \leqslant 0.0078$, $|E_M| \leqslant 0.0039$
(c) $n = 71$ for T_n, $n = 50$ for M_n

21. (a) $T_{10} \approx 1.983524$, $E_T \approx 0.016476$;
$M_{10} \approx 2.008248$, $E_M \approx -0.008248$;
$S_{10} \approx 2.000110$, $E_S \approx -0.000110$
(b) $|E_T| \leqslant 0.025839$, $|E_M| \leqslant 0.012919$, $|E_S| \leqslant 0.000170$
(c) $n = 509$ for T_n, $n = 360$ for M_n, $n = 22$ for S_n

23. (a) 2.8 (b) 7.954926518 (c) 0.2894
(d) 7.954926521 (e) The actual error is much smaller.
(f) 10.9 (g) 7.953789422 (h) 0.0593
(i) The actual error is smaller. (j) $n \geqslant 50$

25.

n	L_n	R_n	T_n	M_n
5	0.742943	1.286599	1.014771	0.992621
10	0.867782	1.139610	1.003696	0.998152
20	0.932967	1.068881	1.000924	0.999538

n	E_L	E_R	E_T	E_M
5	0.257057	−0.286599	−0.014771	0.007379
10	0.132218	−0.139610	−0.003696	0.001848
20	0.067033	−0.068881	−0.000924	0.000462

Observations are the same as after Example 1.

27.

n	T_n	M_n	S_n
6	6.695473	6.252572	6.403292
12	6.474023	6.363008	6.400206

n	E_T	E_M	E_S
6	−0.295473	0.147428	−0.003292
12	−0.074023	0.036992	−0.000206

Observations are the same as after Example 1.

29. (a) 19.8 (b) 20.6 (c) $20.5\overline{3}$

31. (a) 23.44 (b) $0.341\overline{3}$ **33.** $37.7\overline{3}$ ft/s

35. 10,177 megawatt-hours **37.** 828 **39.** 6.0 **41.** 59.4

43.

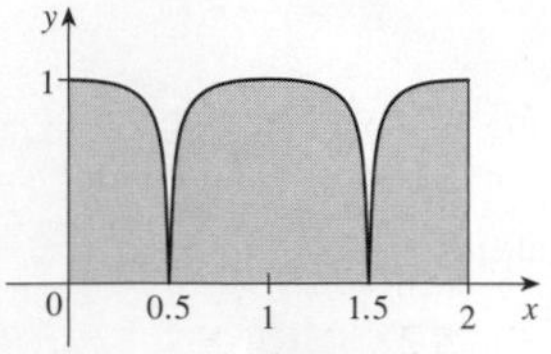

EXERCISES 7.8 ▪ PAGE 515

Abbreviations: C, convergent; D, divergent

1. (a) Infinite interval (b) Infinite discontinuity (c) Infinite discontinuity (d) Infinite interval

3. $\frac{1}{2} - 1/(2t^2)$; 0.495, 0.49995, 0.4999995; 0.5

5. $\frac{1}{12}$ **7.** D **9.** $2e^{-2}$ **11.** D **13.** 0 **15.** D

17. D **19.** $\frac{1}{25}$ **21.** D **23.** $\pi/9$

25. $\frac{1}{2}$ **27.** D **29.** $\frac{32}{3}$ **31.** D **33.** $\frac{75}{4}$

35. D **37.** $-2/e$ **39.** $\frac{8}{3}\ln 2 - \frac{8}{9}$

41. e **43.** $2\pi/3$

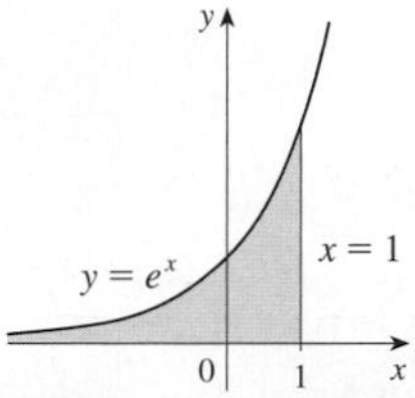

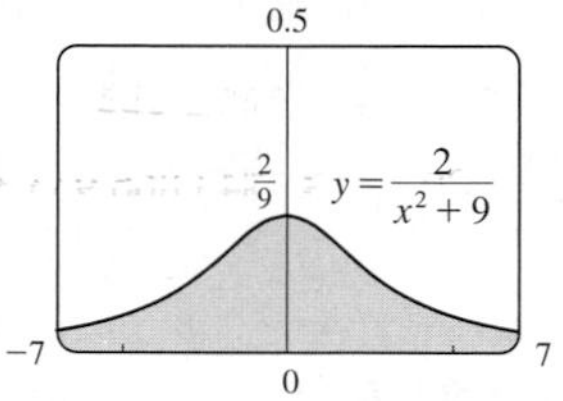

45. Infinite area

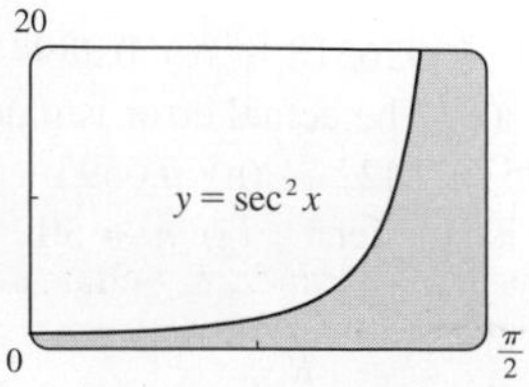

47. (a)

t	$\int_1^t [(\sin^2 x)/x^2]\,dx$
2	0.447453
5	0.577101
10	0.621306
100	0.668479
1,000	0.672957
10,000	0.673407

It appears that the integral is convergent.

(c)

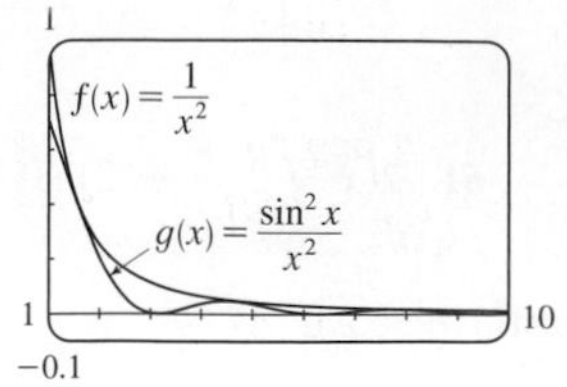

49. C **51.** D **53.** D **55.** π **57.** $p < 1,\ 1/(1-p)$

59. $p > -1,\ -1/(p+1)^2$ **65.** $\sqrt{2GM/R}$

67. (a)

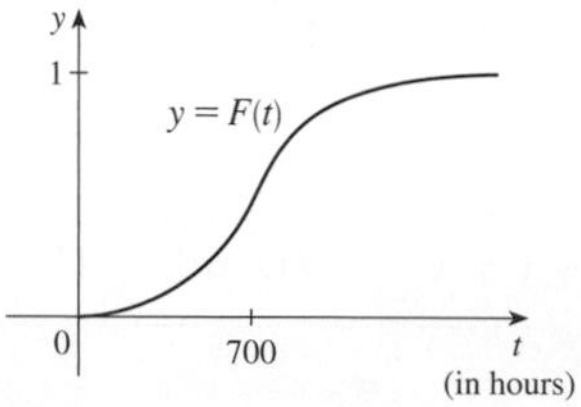

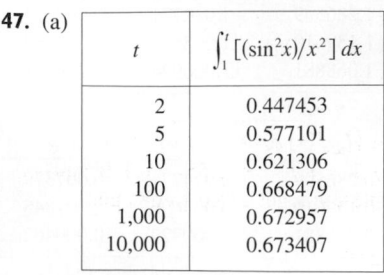

(b) The rate at which the fraction $F(t)$ increases as t increases

(c) 1; all bulbs burn out eventually

69. 1000

71. (a) $F(s) = 1/s,\ s > 0$ (b) $F(s) = 1/(s-1),\ s > 1$ (c) $F(s) = 1/s^2,\ s > 0$

77. $C = 1$; ln 2 **79.** No

CHAPTER 7 REVIEW ▪ PAGE 518

True-False Quiz

1. False **3.** False **5.** False **7.** False

9. (a) True (b) False **11.** False **13.** False

Exercises

1. $5 + 10\ln\frac{2}{3}$ **3.** ln 2 **5.** $\frac{2}{15}$

7. $-\cos(\ln t) + C$ **9.** $\frac{64}{5}\ln 4 - \frac{124}{25}$

11. $\sqrt{3} - \frac{1}{3}\pi$ **13.** $3e^{\sqrt[3]{x}}(\sqrt[3]{x^2} - 2\sqrt[3]{x} + 2) + C$

15. $-\frac{1}{2}\ln|x| + \frac{3}{2}\ln|x+2| + C$

17. $x\sec x - \ln|\sec x + \tan x| + C$

19. $\frac{1}{18}\ln(9x^2 + 6x + 5) + \frac{1}{9}\tan^{-1}\left[\frac{1}{2}(3x+1)\right] + C$

21. $\ln\left|x - 2 + \sqrt{x^2 - 4x}\right| + C$

23. $\ln\left|\dfrac{\sqrt{x^2+1} - 1}{x}\right| + C$

25. $\frac{3}{2}\ln(x^2+1) - 3\tan^{-1}x + \sqrt{2}\tan^{-1}(x/\sqrt{2}) + C$

27. $\frac{2}{5}$ **29.** 0 **31.** $6 - \frac{3}{2}\pi$

33. $\dfrac{x}{\sqrt{4 - x^2}} - \sin^{-1}\left(\dfrac{x}{2}\right) + C$

35. $4\sqrt{1 + \sqrt{x}} + C$ **37.** $\frac{1}{2}\sin 2x - \frac{1}{8}\cos 4x + C$

39. $\frac{1}{8}e - \frac{1}{4}$ **41.** $\frac{1}{36}$ **43.** D

45. $4\ln 4 - 8$ **47.** $-\frac{4}{3}$ **49.** $\pi/4$

51. $(x+1)\ln(x^2 + 2x + 2) + 2\arctan(x+1) - 2x + C$

53. 0

55. $\frac{1}{4}(2x-1)\sqrt{4x^2 - 4x - 3} - \ln\left|2x - 1 + \sqrt{4x^2 - 4x - 3}\right| + C$

57. $\frac{1}{2}\sin x\sqrt{4+\sin^2 x}+2\ln\left(\sin x+\sqrt{4+\sin^2 x}\right)+C$
61. No
63. (a) 1.925444 (b) 1.920915 (c) 1.922470
65. (a) 0.01348, $n \geqslant 368$ (b) 0.00674, $n \geqslant 260$
67. 8.6 mi
69. (a) 3.8 (b) 1.7867, 0.000646 (c) $n \geqslant 30$
71. C **73.** 2 **75.** $\frac{3}{16}\pi^2$

PROBLEMS PLUS ■ PAGE 521

1. About 1.85 inches from the center **3.** 0
7. $f(\pi) = -\pi/2$ **11.** $(b^b a^{-a})^{1/(b-a)}e^{-1}$
13. $2-\sin^{-1}(2/\sqrt{5})$

CHAPTER 8

EXERCISES 8.1 ■ PAGE 530

1. $4\sqrt{5}$ **3.** $\int_0^{2\pi}\sqrt{1+\sin^2 x}\,dx$ **5.** $\int_1^4\sqrt{9y^4+6y^2+2}\,dy$
7. $\frac{2}{243}(82\sqrt{82}-1)$ **9.** $\frac{1261}{240}$ **11.** $\frac{32}{3}$
13. $\ln(\sqrt{2}+1)$ **15.** $\ln 3-\frac{1}{2}$
17. $\sqrt{1+e^2}-\sqrt{2}+\ln\left(\sqrt{1+e^2}-1\right)-1-\ln\left(\sqrt{2}-1\right)$
19. $\sqrt{2}+\ln(1+\sqrt{2})$ **21.** $\frac{46}{3}$ **23.** 5.115840
25. 1.569619
27. (a), (b)

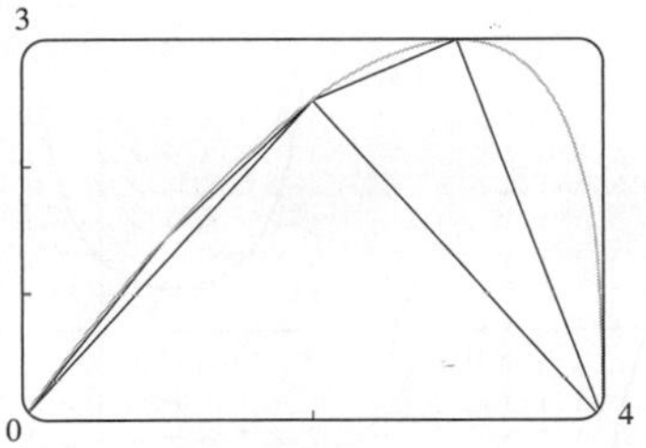

$L_1 = 4$,
$L_2 \approx 6.43$,
$L_4 \approx 7.50$

(c) $\int_0^4\sqrt{1+\left[4(3-x)/(3(4-x)^{2/3})\right]^2}\,dx$ (d) 7.7988
29. $\sqrt{5}-\ln\left(\frac{1}{2}(1+\sqrt{5})\right)-\sqrt{2}+\ln(1+\sqrt{2})$
31. 6

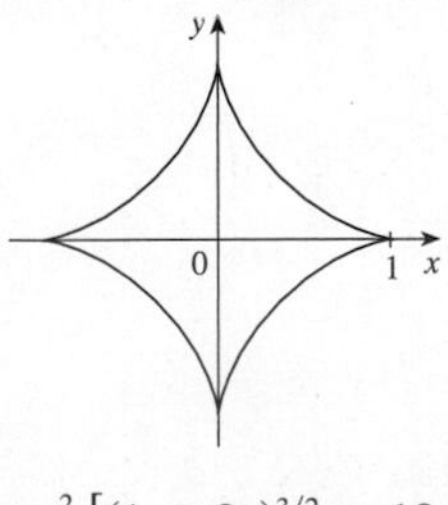

33. $s(x)=\frac{2}{27}\left[(1+9x)^{3/2}-10\sqrt{10}\right]$ **35.** $2\sqrt{2}\left(\sqrt{1+x}-1\right)$
37. 209.1 m **39.** 29.36 in. **41.** 12.4

EXERCISES 8.2 ■ PAGE 537

1. (a) $\int_0^1 2\pi x^4\sqrt{1+16x^6}\,dx$ (b) $\int_0^1 2\pi x\sqrt{1+16x^6}\,dx$

3. (a) $\displaystyle\int_0^1 2\pi\tan^{-1}x\sqrt{1+\frac{1}{(1+x^2)^2}}\,dx$

(b) $\displaystyle\int_0^1 2\pi x\sqrt{1+\frac{1}{(1+x^2)^2}}\,dx$

5. $\frac{1}{27}\pi(145\sqrt{145}-1)$ **7.** $\frac{98}{3}\pi$

9. $2\sqrt{1+\pi^2}+(2/\pi)\ln\left(\pi+\sqrt{1+\pi^2}\right)$ **11.** $\frac{21}{2}\pi$
13. $\frac{1}{27}\pi(145\sqrt{145}-10\sqrt{10})$ **15.** πa^2
17. 9.023754 **19.** 13.527296
21. $\frac{1}{4}\pi\left[4\ln(\sqrt{17}+4)-4\ln(\sqrt{2}+1)-\sqrt{17}+4\sqrt{2}\right]$
23. $\frac{1}{6}\pi\left[\ln(\sqrt{10}+3)+3\sqrt{10}\right]$
27. (a) $\frac{1}{3}\pi a^2$ (b) $\frac{56}{45}\pi\sqrt{3}\,a^2$

29. (a) $\displaystyle 2\pi\left[b^2+\frac{a^2 b\sin^{-1}\left(\sqrt{a^2-b^2}/a\right)}{\sqrt{a^2-b^2}}\right]$

(b) $\displaystyle 2\pi\left[a^2+\frac{ab^2\sin^{-1}\left(\sqrt{b^2-a^2}/b\right)}{\sqrt{b^2-a^2}}\right]$

31. $\int_a^b 2\pi[c-f(x)]\sqrt{1+[f'(x)]^2}\,dx$ **33.** $4\pi^2 r^2$

EXERCISES 8.3 ■ PAGE 547

1. (a) 187.5 lb/ft² (b) 1875 lb (c) 562.5 lb
3. 6000 lb **5.** 6.7×10^4 N **7.** 9.8×10^3 N
9. 1.2×10^4 lb **11.** $\frac{2}{3}\delta ah$ **13.** 5.27×10^5 N
15. (a) 314 N (b) 353 N
17. (a) 5.63×10^3 lb (b) 5.06×10^4 lb
(c) 4.88×10^4 lb (d) 3.03×10^5 lb
19. 2.5×10^5 N **21.** 230; $\frac{23}{7}$ **23.** 10; 1; $\left(\frac{1}{21},\frac{10}{21}\right)$

25. (0, 1.6) **27.** $\left(\dfrac{1}{e-1},\dfrac{e+1}{4}\right)$ **29.** $\left(\frac{9}{20},\frac{9}{20}\right)$

31. $\left(\dfrac{\pi\sqrt{2}-4}{4(\sqrt{2}-1)},\dfrac{1}{4(\sqrt{2}-1)}\right)$ **33.** (2, 0)

35. 60; 160; $\left(\frac{8}{3},1\right)$ **37.** (0.781, 1.330) **41.** $\left(0,\frac{1}{12}\right)$
45. $\frac{1}{3}\pi r^2 h$

EXERCISES 8.4 ■ PAGE 553

1. \$38,000 **3.** \$43,866,933.33 **5.** \$407.25
7. \$12,000 **9.** 3727; \$37,753
11. $\frac{2}{3}(16\sqrt{2}-8)\approx$ \$9.75 million **13.** $\dfrac{(1-k)(b^{2-k}-a^{2-k})}{(2-k)(b^{1-k}-a^{1-k})}$
15. 1.19×10^{-4} cm³/s
17. 6.60 L/min **19.** 5.77 L/min

EXERCISES 8.5 ■ PAGE 560

1. (a) The probability that a randomly chosen tire will have a lifetime between 30,000 and 40,000 miles
(b) The probability that a randomly chosen tire will have a lifetime of at least 25,000 miles
3. (a) $f(x)\geqslant 0$ for all x and $\int_{-\infty}^{\infty}f(x)\,dx=1$
(b) $1-\frac{3}{8}\sqrt{3}\approx 0.35$
5. (a) $1/\pi$ (b) $\frac{1}{2}$
7. (a) $f(x)\geqslant 0$ for all x and $\int_{-\infty}^{\infty}f(x)\,dx=1$ (b) 5
11. (a) $e^{-4/2.5}\approx 0.20$ (b) $1-e^{-2/2.5}\approx 0.55$ (c) If you aren't served within 10 minutes, you get a free hamburger.
13. $\approx$44%
15. (a) 0.0668 (b) $\approx$5.21%
17. $\approx$0.9545

19. (b) 0; a_0 (c)

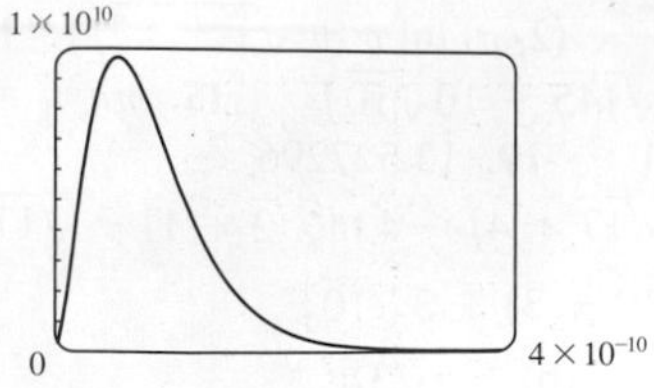

(d) $1 - 41e^{-8} \approx 0.986$ (e) $\frac{3}{2}a_0$

CHAPTER 8 REVIEW ■ PAGE 562

Exercises

1. $\frac{15}{2}$ **3.** (a) $\frac{21}{16}$ (b) $\frac{41}{10}\pi$ **5.** 3.292287 **7.** $\frac{124}{5}$

9. ≈ 458 lb **11.** $\left(\frac{8}{5}, 1\right)$ **13.** $\left(2, \frac{2}{3}\right)$ **15.** $2\pi^2$

17. \$7166.67

19. (a) $f(x) \geqslant 0$ for all x and $\int_{-\infty}^{\infty} f(x)\,dx = 1$
(b) ≈ 0.3455 (c) 5, yes

21. (a) $1 - e^{-3/8} \approx 0.31$ (b) $e^{-5/4} \approx 0.29$
(c) $8 \ln 2 \approx 5.55$ min

PROBLEMS PLUS ■ PAGE 564

1. $\frac{2}{3}\pi - \frac{1}{2}\sqrt{3}$

3. (a) $2\pi r(r \pm d)$ (b) $\approx 3.36 \times 10^6$ mi^2
(d) $\approx 7.84 \times 10^7$ mi^2

5. (a) $P(z) = P_0 + g \int_0^z \rho(x)\,dx$
(b) $(P_0 - \rho_0 gH)(\pi r^2) + \rho_0 gHe^{L/H} \int_{-r}^{r} e^{x/H} \cdot 2\sqrt{r^2 - x^2}\,dx$

7. Height $\sqrt{2}\,b$, volume $\left(\frac{28}{27}\sqrt{6} - 2\right)\pi b^3$ **9.** 0.14 m

11. $2/\pi$, $1/\pi$

CHAPTER 9

EXERCISES 9.1 ■ PAGE 571

3. (a) $\frac{1}{2}$, -1 **5.** (d)

7. (a) It must be either 0 or decreasing
(c) $y = 0$ (d) $y = 1/(x + 2)$

9. (a) $0 < P < 4200$ (b) $P > 4200$
(c) $P = 0$, $P = 4200$

13. (a) At the beginning; stays positive, but decreases
(c)

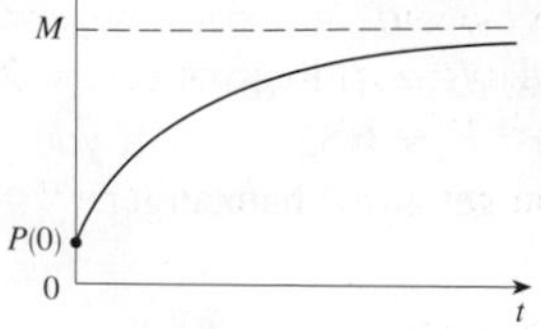

EXERCISES 9.2 ■ PAGE 578

1. (a)

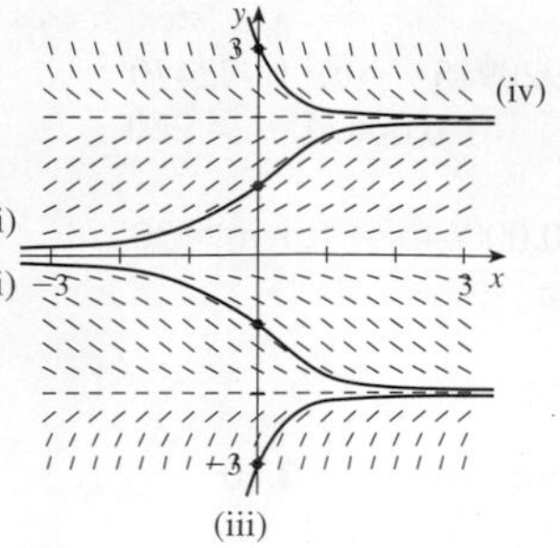

(b) $y = 0$, $y = 2$, $y = -2$

3. III **5.** IV

7.

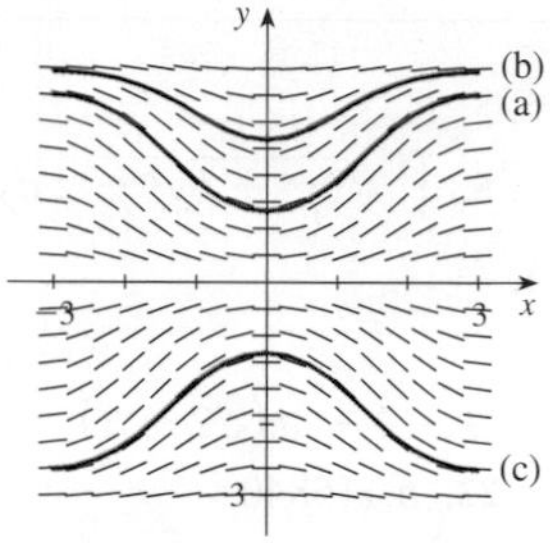

9.

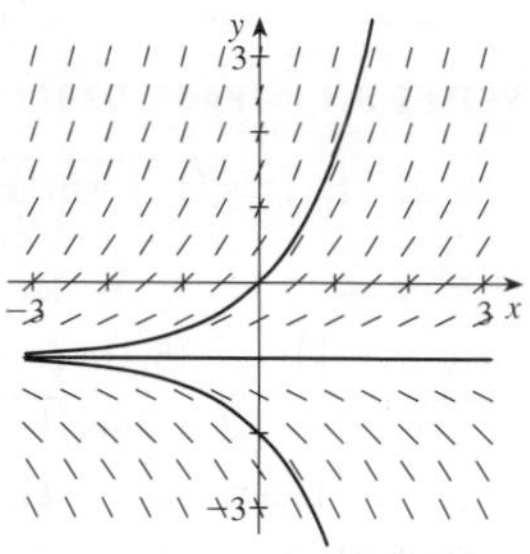

11.

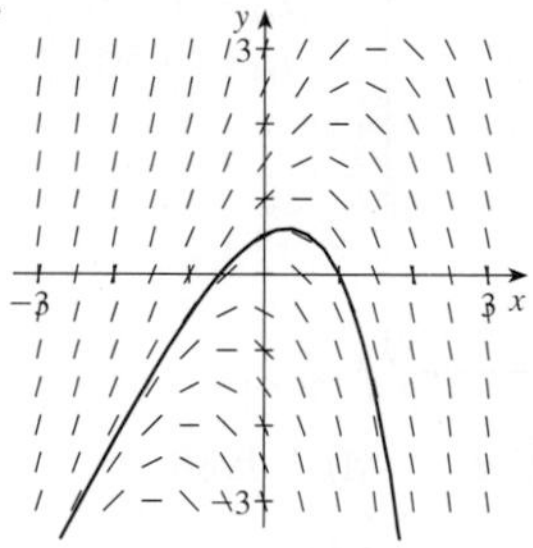

13.

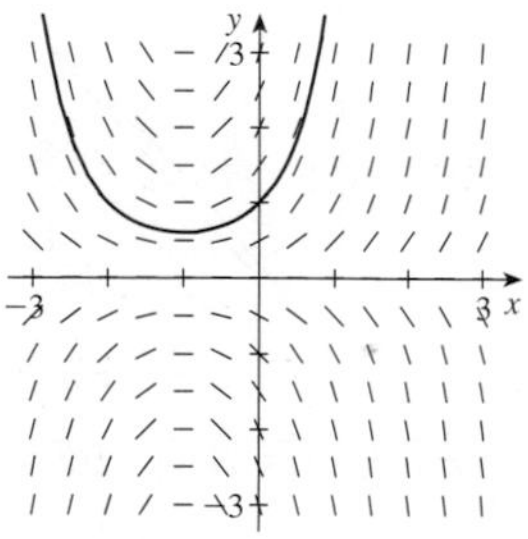

15.

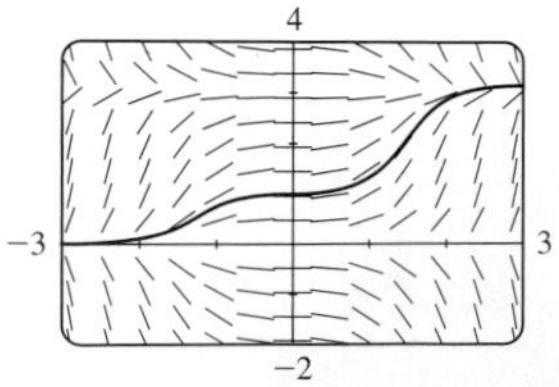

17.

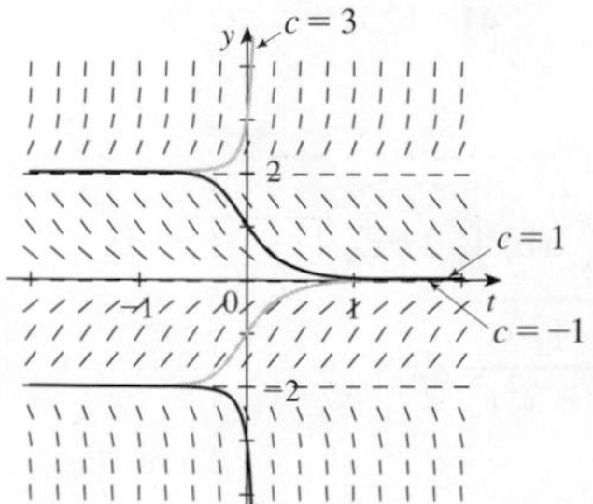

$-2 \leqslant c \leqslant 2$; $-2, 0, 2$

19. (a) (i) 1.4 (ii) 1.44 (iii) 1.4641

(b) Underestimates

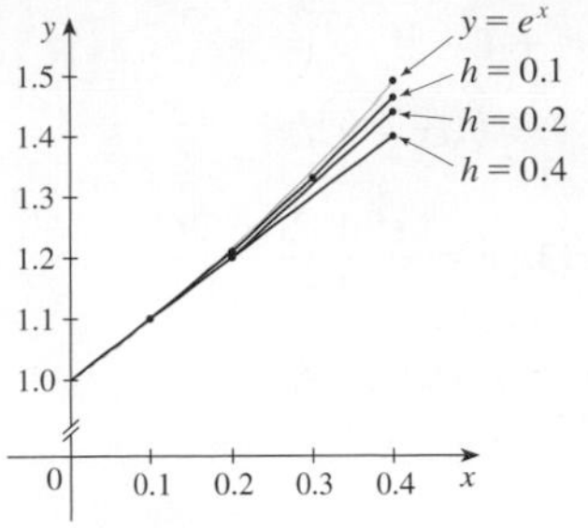

(c) (i) 0.0918 (ii) 0.0518 (iii) 0.0277

It appears that the error is also halved (approximately).

21. $-1, -3, -6.5, -12.25$ **23.** 1.7616

25. (a) (i) 3 (ii) 2.3928 (iii) 2.3701 (iv) 2.3681

(c) (i) -0.6321 (ii) -0.0249 (iii) -0.0022 (iv) -0.0002

It appears that the error is also divided by 10 (approximately).

27. (a), (d) (b) 3

(c) Yes; $Q = 3$

(e) 2.77 C

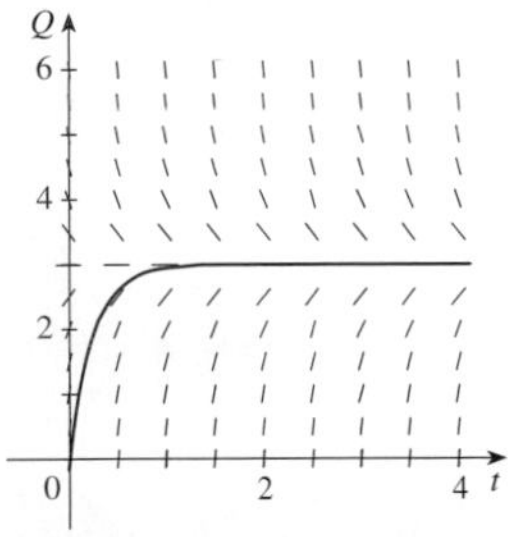

EXERCISES 9.3 ■ PAGE 586

1. $y = Kx$ **3.** $y = K\sqrt{x^2 + 1}$

5. $y + \ln|\sec y| = \frac{1}{3}x^3 + x + C$

7. $y = \pm\sqrt{[3(te^t - e^t + C)]^{2/3} - 1}$ **9.** $u = Ae^{2t + t^2/2} - 1$

11. $y = -\sqrt{x^2 + 9}$ **13.** $\cos x + x \sin x = y^2 + \frac{1}{3}e^{3y} + \frac{2}{3}$

15. $u = -\sqrt{t^2 + \tan t + 25}$ **17.** $y = \dfrac{4a}{\sqrt{3}} \sin x - a$

19. $y = e^{x^2/2}$ **21.** $y = Ke^x - x - 1$

23. (a) $\sin^{-1} y = x^2 + C$

(b) $y = \sin(x^2)$, $-\sqrt{\pi/2} \leqslant x \leqslant \sqrt{\pi/2}$ (c) No

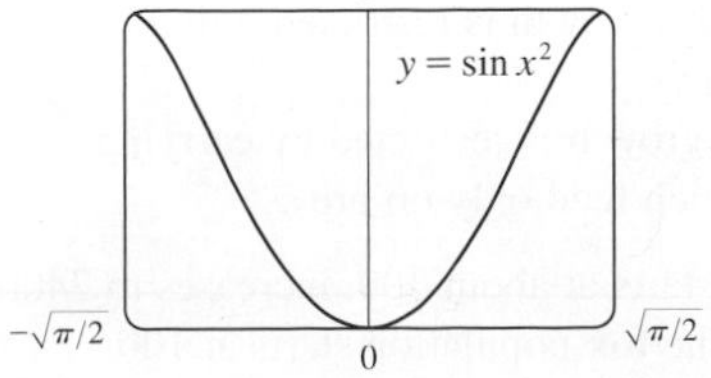

25. $\cos y = \cos x - 1$

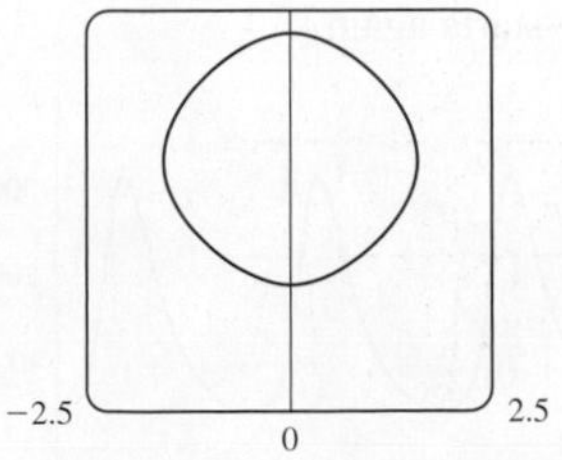

27. (a), (c) (b) $y = \pm\sqrt{2(x + C)}$

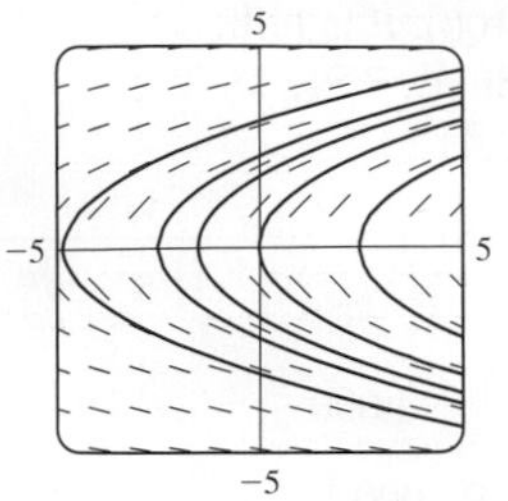

29. $y = Cx^2$ **31.** $x^2 - y^2 = C$

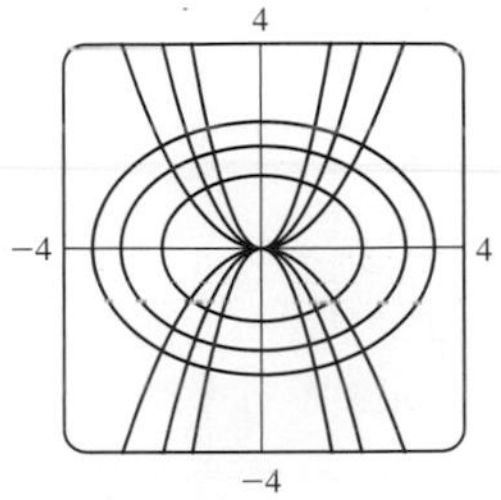

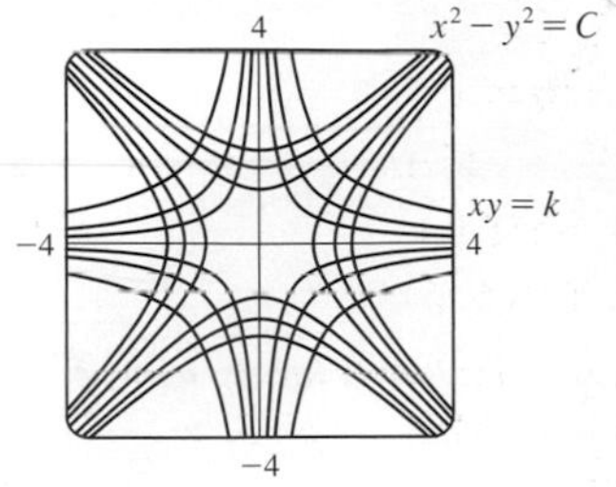

33. $Q(t) = 3 - 3e^{-4t}$; 3 **35.** $P(t) = M - Me^{-kt}$; M

37. (a) $x = a - \dfrac{4}{(kt + 2/\sqrt{a})^2}$

(b) $t = \dfrac{2}{k\sqrt{a - b}}\left(\tan^{-1}\sqrt{\dfrac{b}{a - b}} - \tan^{-1}\sqrt{\dfrac{b - x}{a - b}}\right)$

39. (a) $C(t) = (C_0 - r/k)e^{-kt} + r/k$

(b) r/k; the concentration approaches r/k regardless of the value of C_0

41. (a) $15e^{-t/100}$ kg (b) $15e^{-0.2} \approx 12.3$ kg

43. About 4.9% **45.** g/k

47. (a) $dA/dt = k\sqrt{A}\,(M - A)$ (b) $A(t) = M\left(\dfrac{Ce^{\sqrt{M}kt} - 1}{Ce^{\sqrt{M}kt} + 1}\right)^2$,

where $C = \dfrac{\sqrt{M} + \sqrt{A_0}}{\sqrt{M} - \sqrt{A_0}}$ and $A_0 = A(0)$

EXERCISES 9.4 ■ PAGE 598

1. (a) 100; 0.05 (b) Where P is close to 0 or 100; on the line $P = 50$; $0 < P_0 < 100$; $P_0 > 100$

(c)

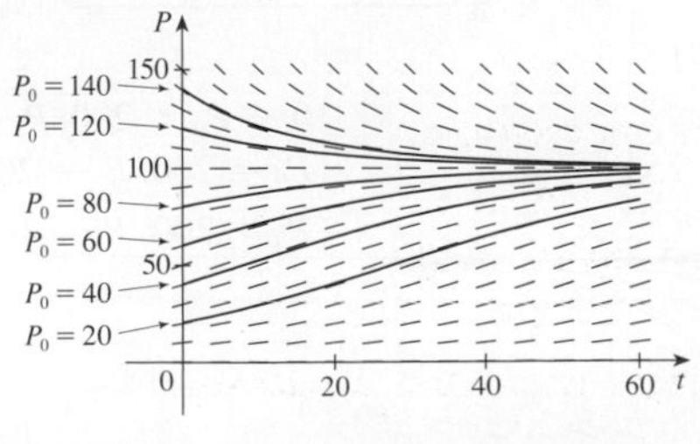

Solutions approach 100; some increase and some decrease, some have an inflection point but others don't; solutions with $P_0 = 20$ and $P_0 = 40$ have inflection points at $P = 50$

(d) $P = 0$, $P = 100$; other solutions move away from $P = 0$ and toward $P = 100$

3. (a) 3.23×10^7 kg (b) ≈ 1.55 years

5. (a) $dP/dt = \frac{1}{265}P(1 - P/100)$, P in billions
(b) 5.49 billion (c) In billions: 7.81, 27.72
(d) In billions: 5.48, 7.61, 22.41

7. (a) $dy/dt = ky(1 - y)$ (b) $y = \dfrac{y_0}{y_0 + (1 - y_0)e^{-kt}}$
(c) 3:36 PM

11. $P_E(t) = 1578.3(1.0933)^t + 94{,}000$;
$P_L(t) = \dfrac{32{,}658.5}{1 + 12.75e^{-0.1706t}} + 94{,}000$

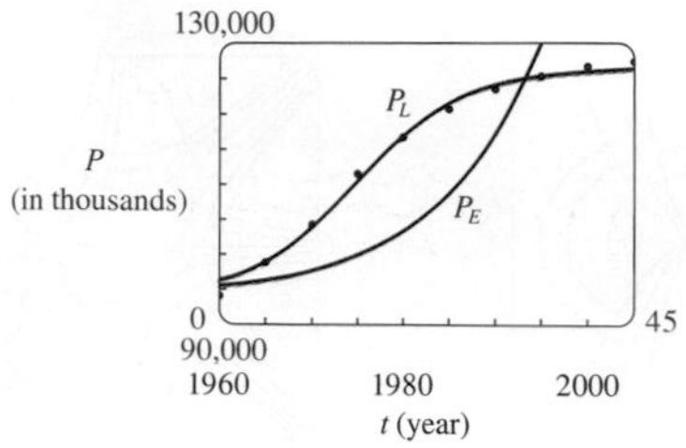

13. (a) $P(t) = \dfrac{m}{k} + \left(P_0 - \dfrac{m}{k}\right)e^{kt}$ (b) $m < kP_0$
(c) $m = kP_0$, $m > kP_0$ (d) Declining

15. (a) Fish are caught at a rate of 15 per week.
(b) See part (d) (c) $P = 250$, $P = 750$
(d)

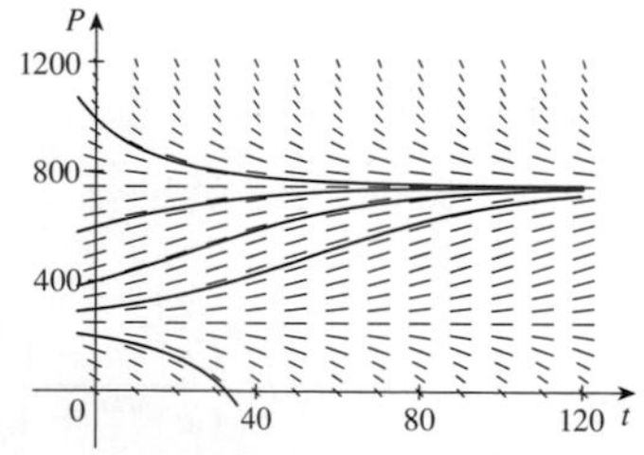

$0 < P_0 < 250$: $P \to 0$;
$P_0 = 250$: $P \to 250$;
$P_0 > 250$: $P \to 750$

(e) $P(t) = \dfrac{250 - 750ke^{t/25}}{1 - ke^{t/25}}$
where $k = \frac{1}{11}, -\frac{1}{9}$

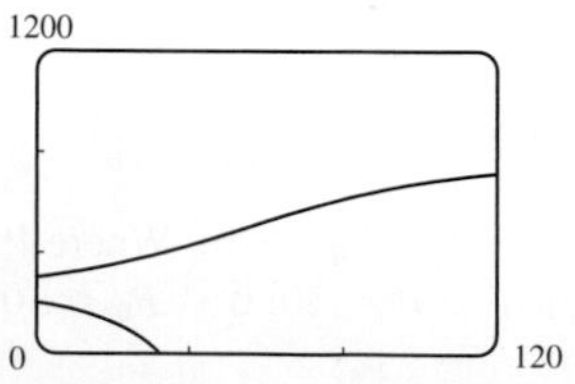

17. (b)

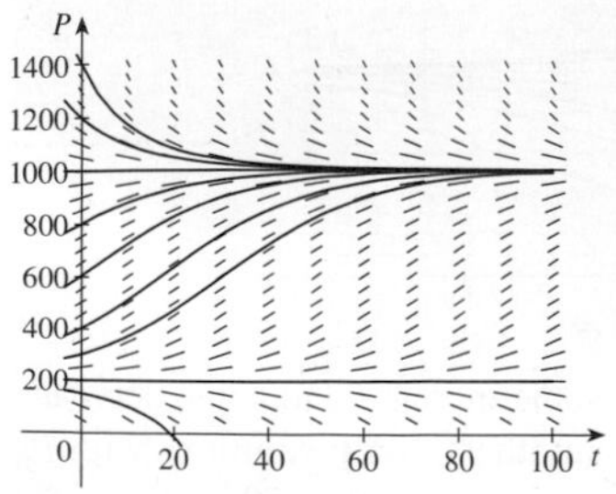

$0 < P_0 < 200$: $P \to 0$;
$P_0 = 200$: $P \to 200$;
$P_0 > 200$: $P \to 1000$

(c) $P(t) = \dfrac{m(K - P_0) + K(P_0 - m)e^{(K-m)(k/K)t}}{K - P_0 + (P_0 - m)e^{(K-m)(k/K)t}}$

19. (a) $P(t) = P_0 e^{(k/r)[\sin(rt - \phi) + \sin\phi]}$ (b) Does not exist

EXERCISES 9.5 ■ PAGE 606

1. Yes **3.** No **5.** $y = \frac{2}{3}e^x + Ce^{-2x}$

7. $y = x^2 \ln|x| + Cx^2$ **9.** $y = \frac{2}{3}\sqrt{x} + C/x$

11. $y = \dfrac{\int \sin(x^2)\,dx + C}{\sin x}$ **13.** $u = \dfrac{t^2 + 2t + 2C}{2(t + 1)}$

15. $y = -x - 1 + 3e^x$ **17.** $v = t^3e^{t^2} + 5e^{t^2}$

19. $y = -x\cos x - x$

21. $y = \dfrac{(x - 1)e^x + C}{x^2}$

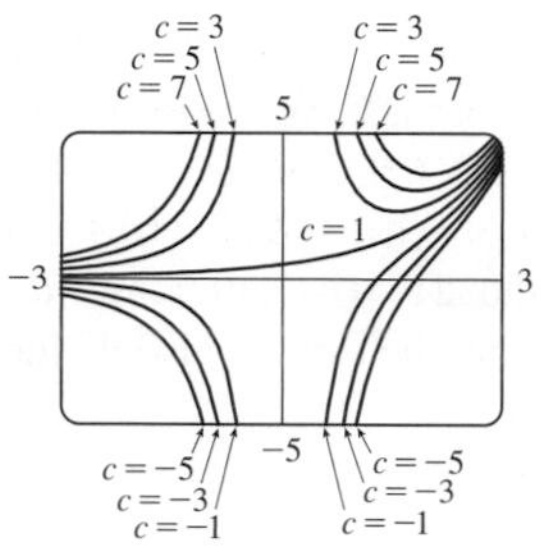

25. $y = \pm\left(Cx^4 + \dfrac{2}{5x}\right)^{-1/2}$

27. (a) $I(t) = 4 - 4e^{-5t}$ (b) $4 - 4e^{-1/2} \approx 1.57$ A

29. $Q(t) = 3(1 - e^{-4t})$, $I(t) = 12e^{-4t}$

31. $P(t) = M + Ce^{-kt}$

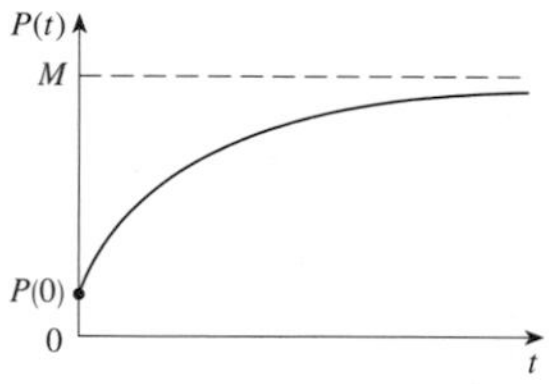

33. $y = \frac{2}{5}(100 + 2t) - 40{,}000(100 + 2t)^{-3/2}$; 0.2275 kg/L

35. (b) mg/c (c) $(mg/c)[t + (m/c)e^{-ct/m}] - m^2g/c^2$

EXERCISES 9.6 ■ PAGE 612

1. (a) x = predators, y = prey; growth is restricted only by predators, which feed only on prey.
(b) x = prey, y = predators; growth is restricted by carrying capacity and by predators, which feed only on prey.

3. (a) The rabbit population starts at about 300, increases to 2400, then decreases back to 300. The fox population starts at 100, decreases to about 20, increases to about 315, decreases to 100, and the cycle starts again.
(b)

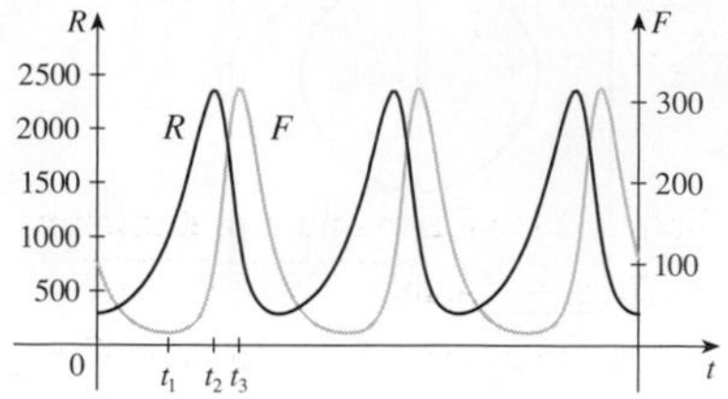

5.

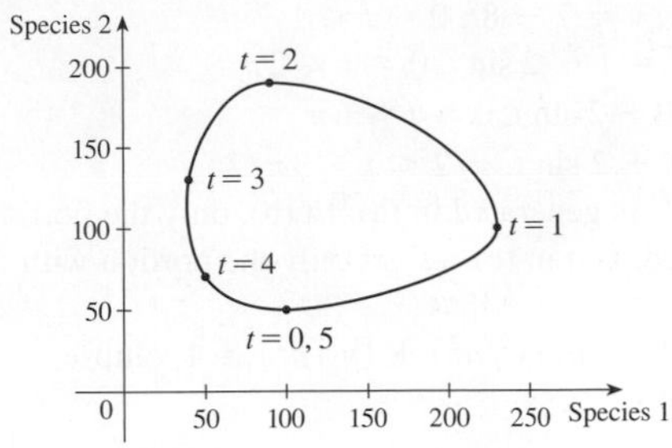

9. (a) Population stabilizes at 5000.
(b) (i) $W = 0, R = 0$: Zero populations
(ii) $W = 0, R = 5000$: In the absence of wolves, the rabbit population is always 5000.
(iii) $W = 64, R = 1000$: Both populations are stable.
(c) The populations stabilize at 1000 rabbits and 64 wolves.
(d)

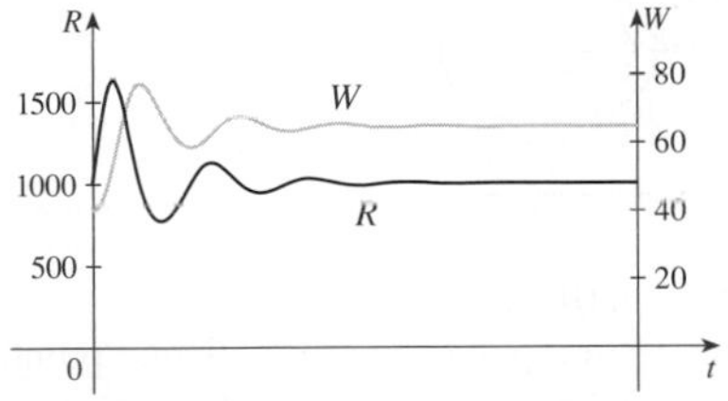

15. (a) $P(t) = \dfrac{2000}{1 + 19e^{-0.1t}}$; ≈ 560 (b) $t = -10 \ln \frac{2}{57} \approx 33.5$

17. (a) $L(t) = L_\infty - [L_\infty - L(0)]e^{-kt}$ (b) $L(t) = 53 - 43e^{-0.2t}$

19. 15 days **21.** $k \ln h + h = (-R/V)t + C$

23. (a) Stabilizes at 200,000
(b) (i) $x = 0, y = 0$: Zero populations
(ii) $x = 200{,}000, y = 0$: In the absence of birds, the insect population is always 200,000.
(iii) $x = 25{,}000, y = 175$: Both populations are stable.
(c) The populations stabilize at 25,000 insects and 175 birds.
(d)

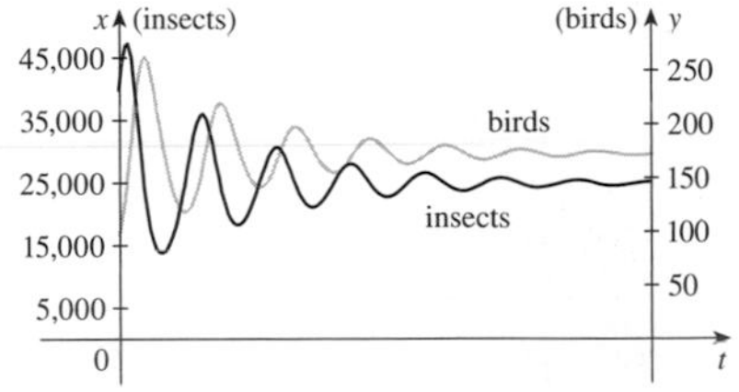

25. (a) $y = (1/k) \cosh kx + a - 1/k$ or
$y = (1/k) \cosh kx - (1/k) \cosh kb + h$ (b) $(2/k) \sinh kb$

CHAPTER 9 REVIEW ■ PAGE 615

True-False Quiz

1. True **3.** False **5.** True **7.** True

Exercises

1. (a)

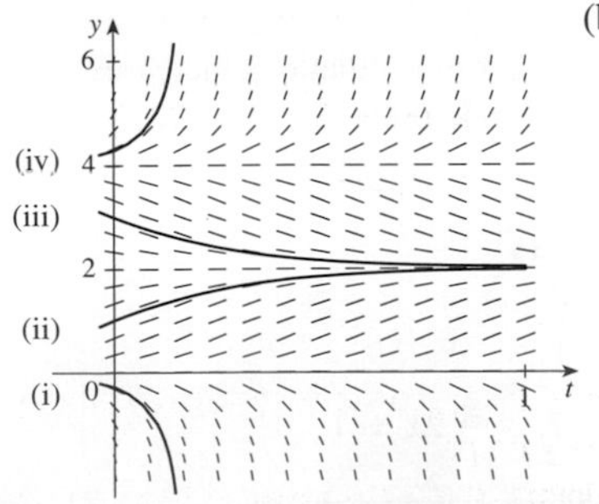

(b) $0 \leqslant c \leqslant 4$;
$y = 0, y = 2, y = 4$

3. (a)

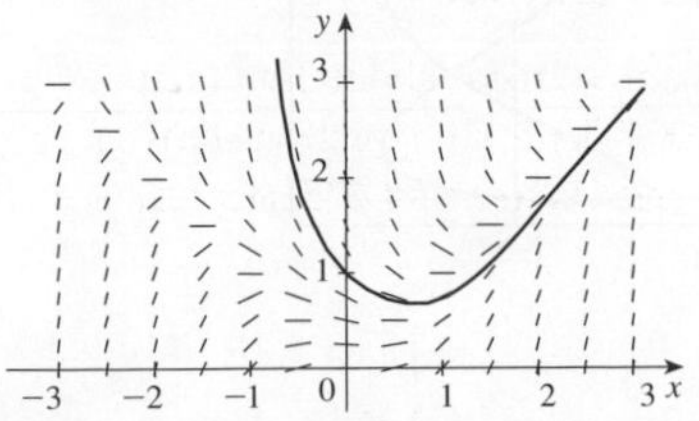

$y(0.3) \approx 0.8$

(b) 0.75676
(c) $y = x$ and $y = -x$; there is a local maximum or minimum

5. $y = (\frac{1}{2}x^2 + C)e^{-\sin x}$ **7.** $y = \pm\sqrt{\ln(x^2 + 2x^{3/2} + C)}$

9. $r(t) = 5e^{t - t^2}$ **11.** $y = \frac{1}{2}x(\ln x)^2 + 2x$ **13.** $x = C - \frac{1}{2}y^2$

PROBLEMS PLUS ■ PAGE 618

1. $f(x) = \pm 10e^x$ **5.** $y = x^{1/n}$ **7.** 20°C

9. (b) $f(x) = \dfrac{x^2 - L^2}{4L} - \frac{1}{2}L \ln\left(\dfrac{x}{L}\right)$ (c) No

11. (a) 9.8 h (b) $31{,}900\pi \approx 100{,}000$ ft^2; 6283 ft^2/h
(c) 5.1 h

13. $x^2 + (y - 6)^2 = 25$

CHAPTER 10

EXERCISES 10.1 ■ PAGE 626

1.

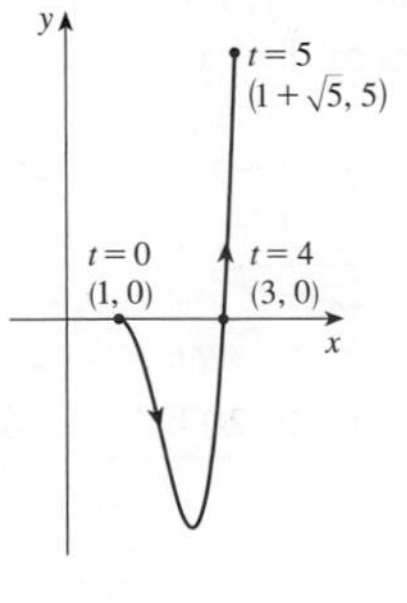

3.

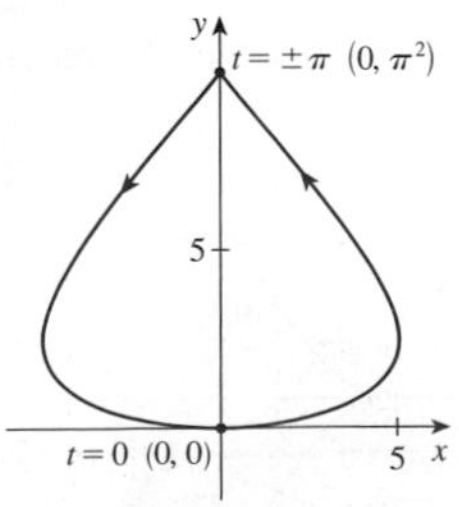

5. (a)

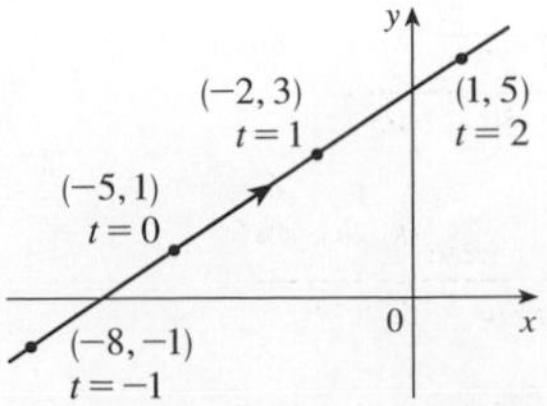

(b) $y = \frac{2}{3}x + \frac{13}{3}$

7. (a)

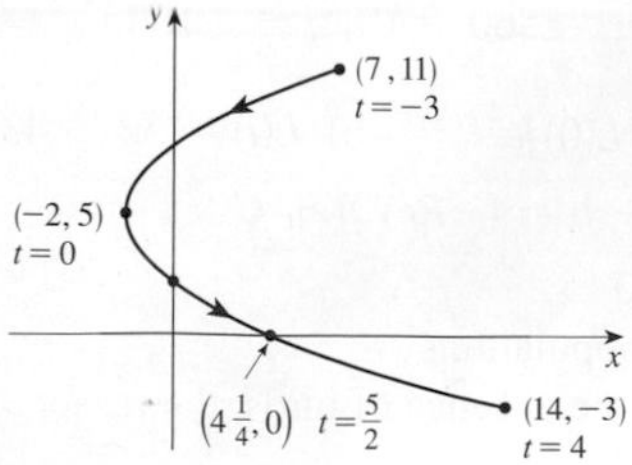

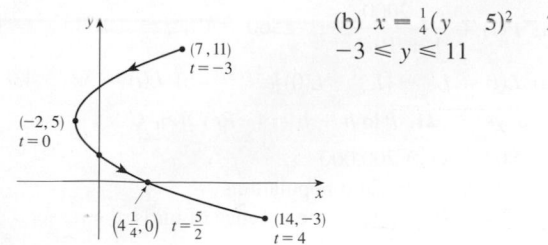

(b) $x = \frac{1}{4}(y \quad 5)^2 \quad 2$, $-3 \leqslant y \leqslant 11$

9. (a)

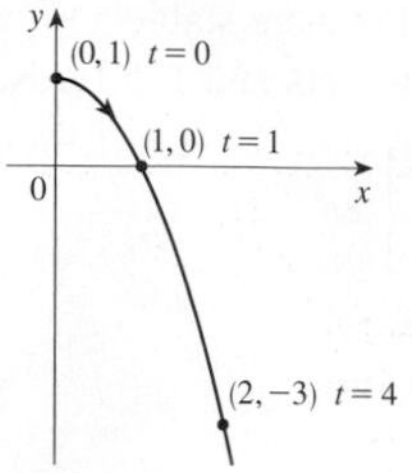

(b) $y = 1 - x^2, x \geqslant 0$

11. (a) $x^2 + y^2 = 1, x \geqslant 0$
(b)

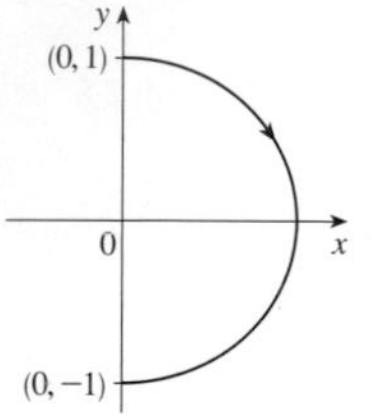

13. (a) $y = 1/x, y > 1$
(b)

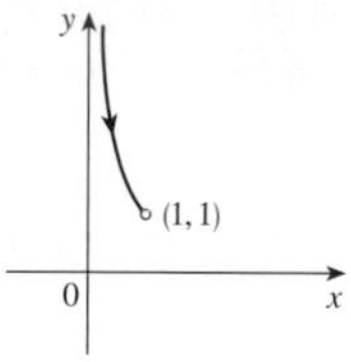

15. (a) $y = \frac{1}{2}\ln x + 1$
(b)

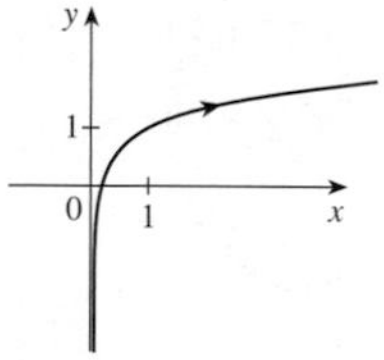

17. (a) $y^2 - x^2 = 1, y \geqslant 1$
(b)

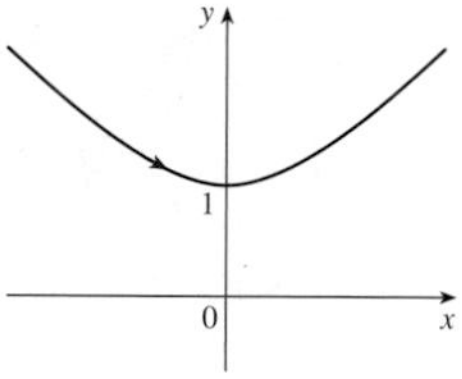

19. Moves counterclockwise along the circle $(x - 3)^2 + (y - 1)^2 = 4$ from (3, 3) to (3, −1)
21. Moves 3 times clockwise around the ellipse $(x^2/25) + (y^2/4) = 1$, starting and ending at (0, −2)
23. It is contained in the rectangle described by $1 \leqslant x \leqslant 4$ and $2 \leqslant y \leqslant 3$.
25.

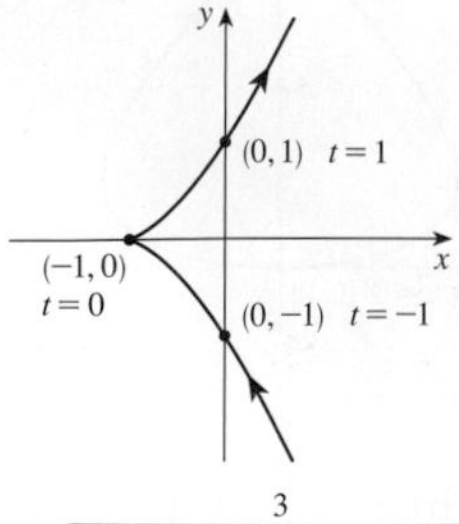

27.

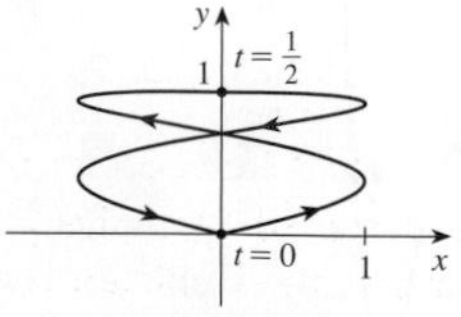

29.

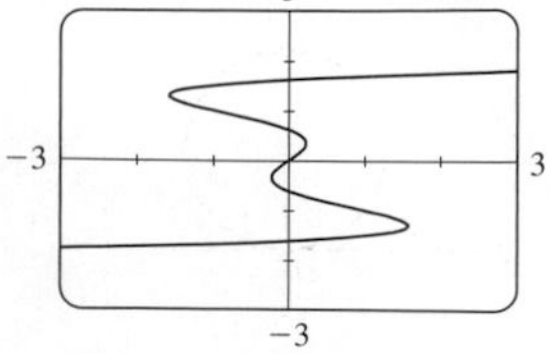

31. (b) $x = -2 + 5t, y = 7 - 8t, 0 \leqslant t \leqslant 1$
33. (a) $x = 2\cos t, y = 1 - 2\sin t, 0 \leqslant t \leqslant 2\pi$
(b) $x = 2\cos t, y = 1 + 2\sin t, 0 \leqslant t \leqslant 6\pi$
(c) $x = 2\cos t, y = 1 + 2\sin t, \pi/2 \leqslant t \leqslant 3\pi/2$
37. The curve $y = x^{2/3}$ is generated in (a). In (b), only the portion with $x \geqslant 0$ is generated, and in (c) we get only the portion with $x > 0$.
41. $x = a\cos\theta, y = b\sin\theta$; $(x^2/a^2) + (y^2/b^2) = 1$, ellipse
43.

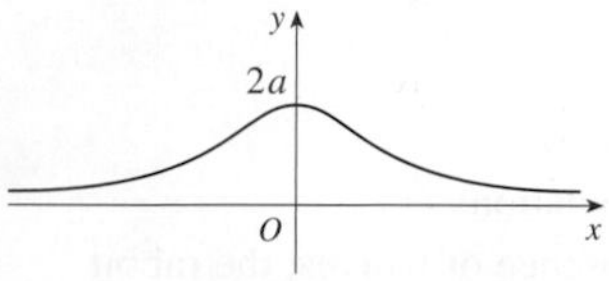

45. (a) Two points of intersection

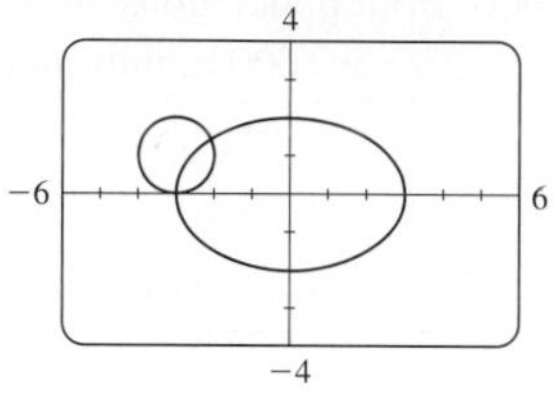

(b) One collision point at (−3, 0) when $t = 3\pi/2$
(c) There are still two intersection points, but no collision point.
47. For $c = 0$, there is a cusp; for $c > 0$, there is a loop whose size increases as c increases.

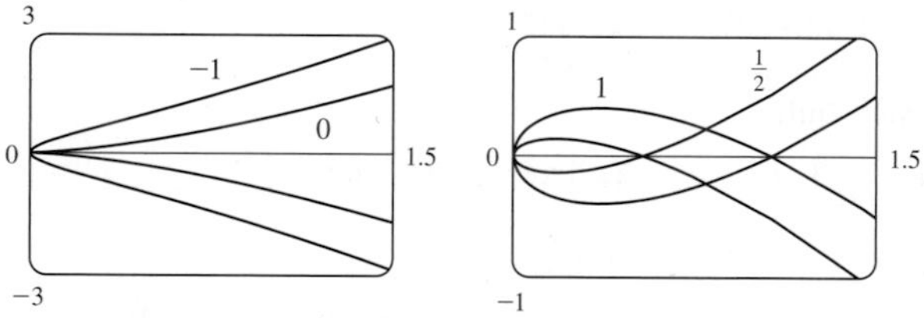

49. As n increases, the number of oscillations increases; a and b determine the width and height.

EXERCISES 10.2 ▪ PAGE 636

1. $\dfrac{2t + 1}{t\cos t + \sin t}$ **3.** $y = -x$
5. $y = -(2/e)x + 3$ **7.** $y = 2x + 1$
9. $y = \frac{1}{6}x$

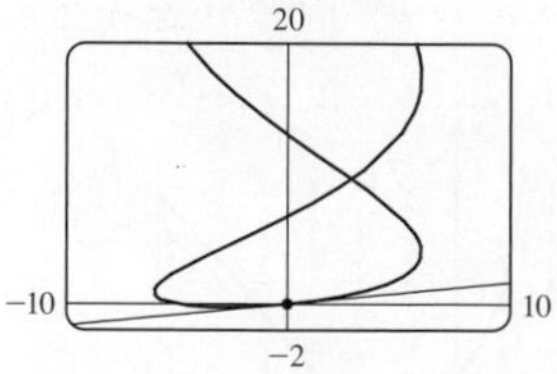

11. $1 + \frac{3}{2}t, 3/(4t), t > 0$
13. $-e^{-t}, e^{-t}/(1 - e^t), t < 0$
15. $-\frac{3}{2}\tan t, -\frac{3}{4}\sec^3 t, \pi/2 < t < 3\pi/2$
17. Horizontal at (6, ±16), vertical at (10, 0)
19. Horizontal at $(\pm\sqrt{2}, \pm 1)$ (four points), vertical at (±2, 0)
21. (0.6, 2); $(5 \cdot 6^{-6/5}, e^{6^{-1/5}})$

23.

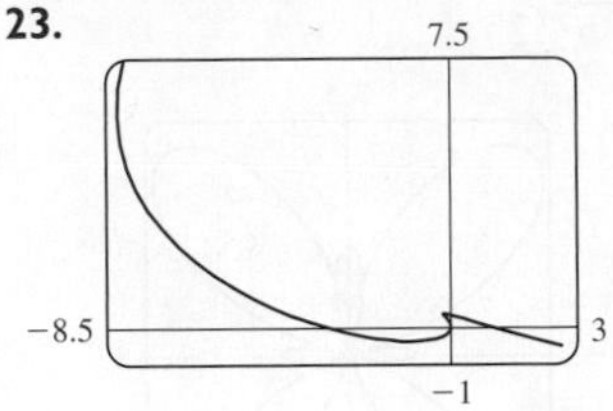

25. $y = x, y = -x$

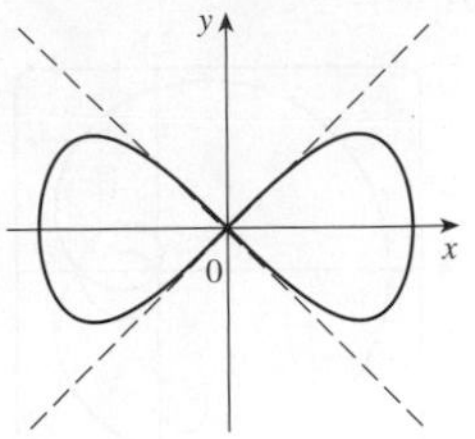

27. (a) $d \sin \theta/(r - d \cos \theta)$ **29.** $\left(\frac{16}{27}, \frac{29}{9}\right), (-2, -4)$

31. πab **33.** $3 - e$ **35.** $2\pi r^2 + \pi d^2$

37. $\int_1^2 \sqrt{1 + 4t^2}\, dt \approx 3.1678$

39. $\int_0^{2\pi} \sqrt{3 - 2\sin t - 2\cos t}\, dt \approx 10.0367$ **41.** $4\sqrt{2} - 2$

43. $-\sqrt{10}/3 + \ln(3 + \sqrt{10}) + \sqrt{2} - \ln(1 + \sqrt{2})$

45. $\sqrt{2}\,(e^{\pi} - 1)$

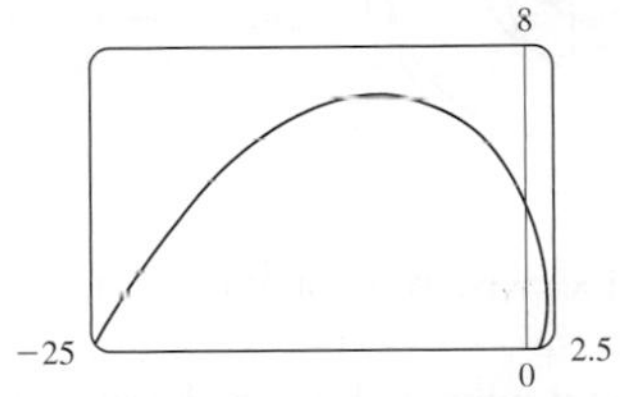

47. $e^3 + 11 - e^{-8}$

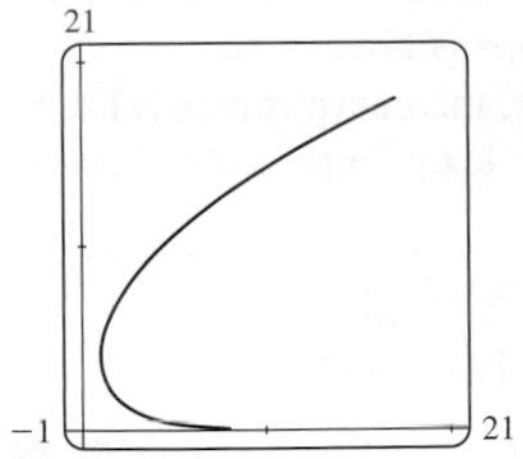

49. 612.3053 **51.** $6\sqrt{2}, \sqrt{2}$

55. (a) $t \in [0, 4\pi]$

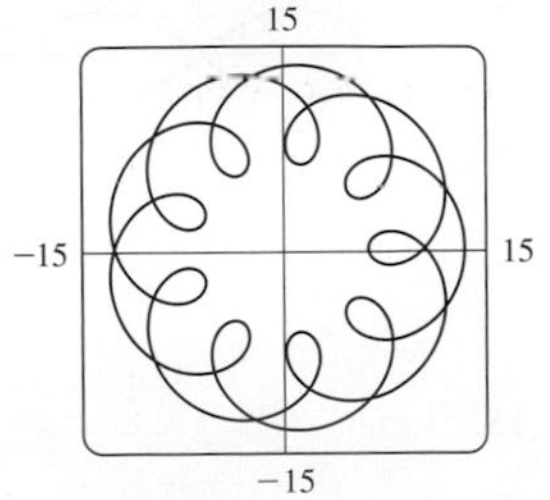

(b) ≈ 294

57. $\int_0^1 2\pi(t^2 + 1)e^t\sqrt{e^{2t}(t + 1)^2(t^2 + 2t + 2)}\, dt \approx 103.5999$

59. $\frac{2}{1215}\pi(247\sqrt{13} + 64)$ **61.** $\frac{6}{5}\pi a^2$ **63.** 59.101

65. $\frac{24}{5}\pi(949\sqrt{26} + 1)$ **71.** $\frac{1}{4}$

EXERCISES 10.3 ■ PAGE 647

1. (a)

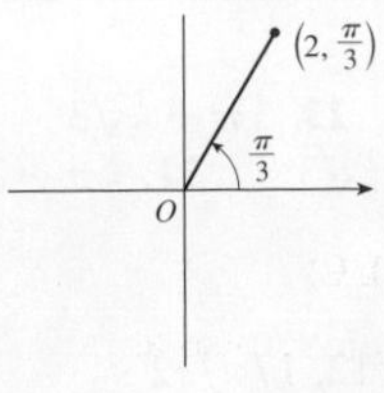

$(2, 7\pi/3), (-2, 4\pi/3)$

(b)

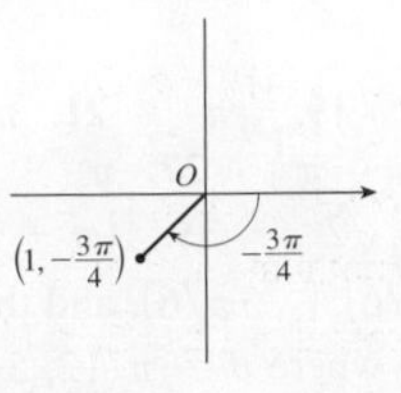

$(1, 5\pi/4), (-1, \pi/4)$

(c)

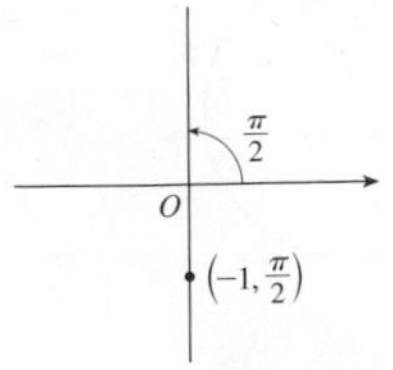

$(1, 3\pi/2), (-1, 5\pi/2)$

3. (a)

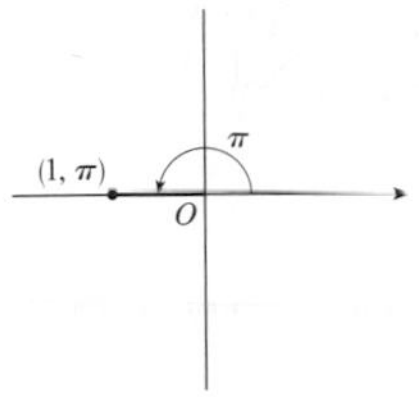

$(-1, 0)$

(b)

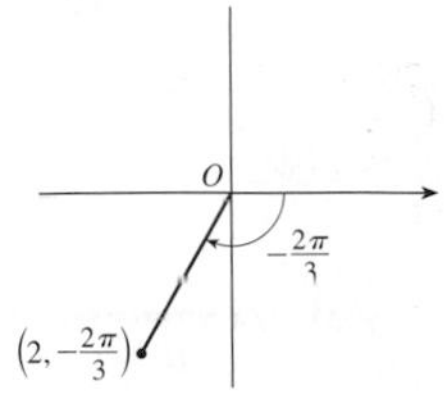

$(-1, -\sqrt{3})$

(c)

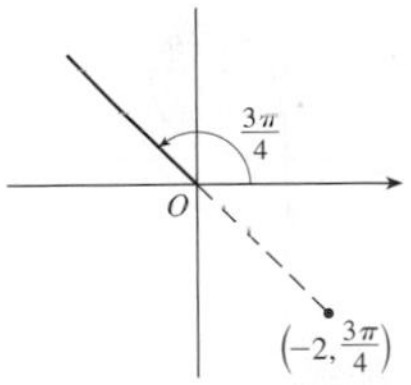

$(\sqrt{2}, -\sqrt{2})$

5. (a) (i) $(2\sqrt{2}, 7\pi/4)$ (ii) $(-2\sqrt{2}, 3\pi/4)$
(b) (i) $(2, 2\pi/3)$ (ii) $(-2, 5\pi/3)$

7.

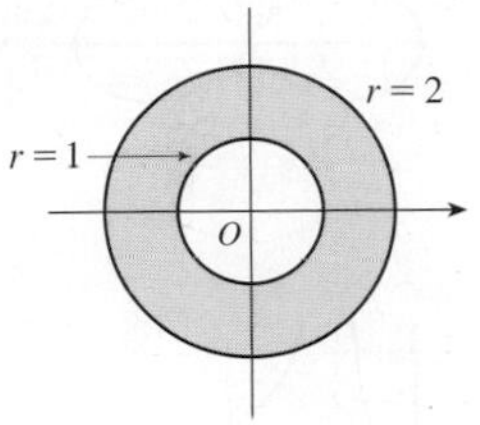

9.

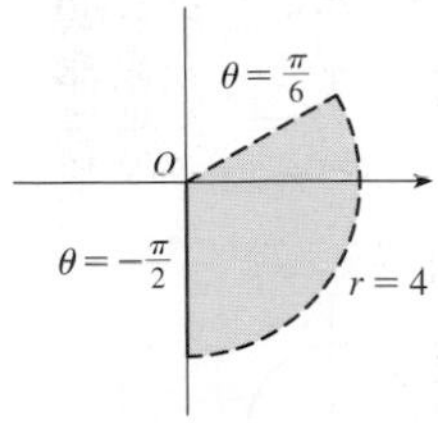

11.

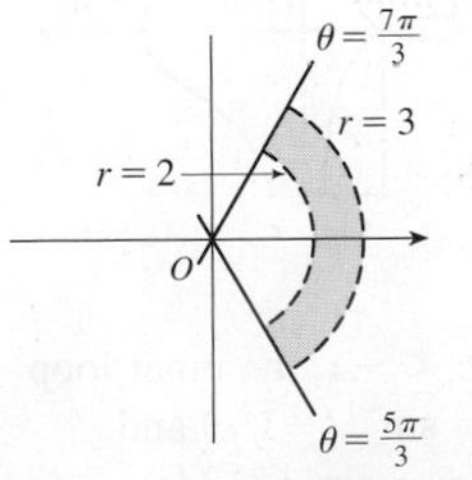

13. $2\sqrt{3}$ **15.** Circle, center O, radius 2

17. Circle, center $\left(0, \frac{3}{2}\right)$, radius $\frac{3}{2}$

19. Horizontal line, 1 unit above the x-axis

21. $r = 3 \sec \theta$ **23.** $r = -\cot \theta \csc \theta$ **25.** $r = 2c \cos \theta$

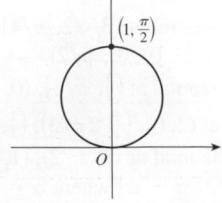

27. (a) $\theta = \pi/6$ (b) $x = 3$

29.

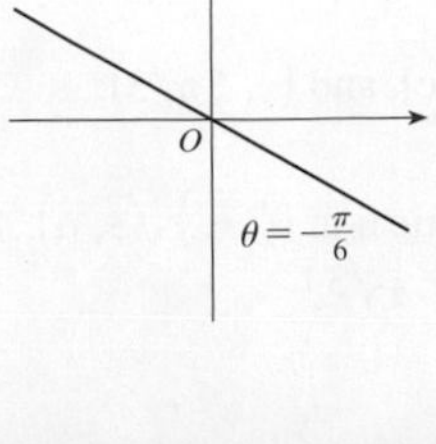

31.

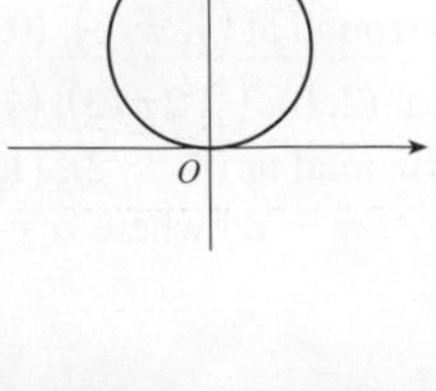

33.

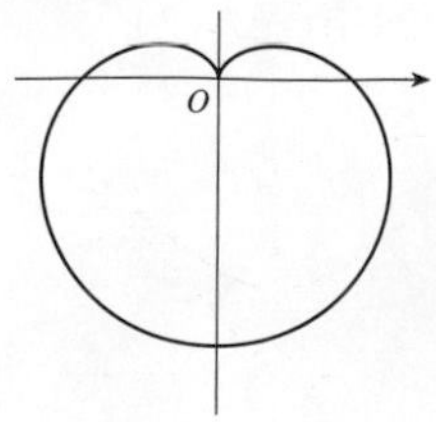

35.

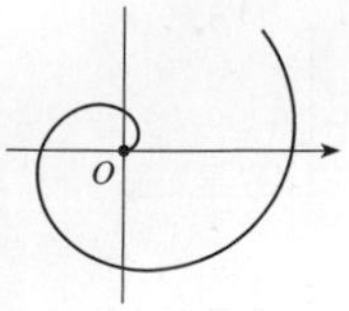

37.

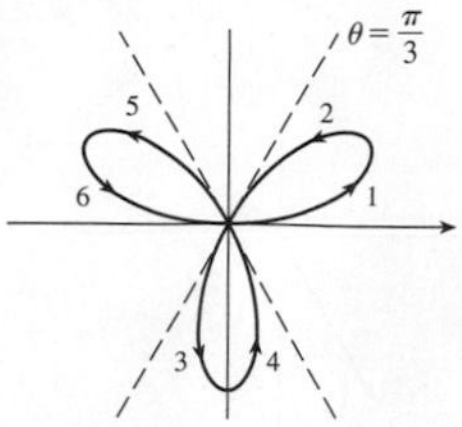

39.

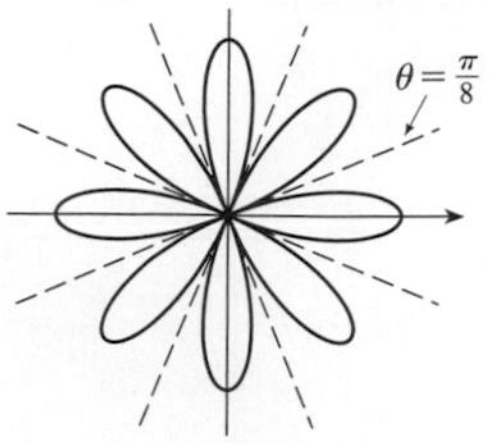

41.

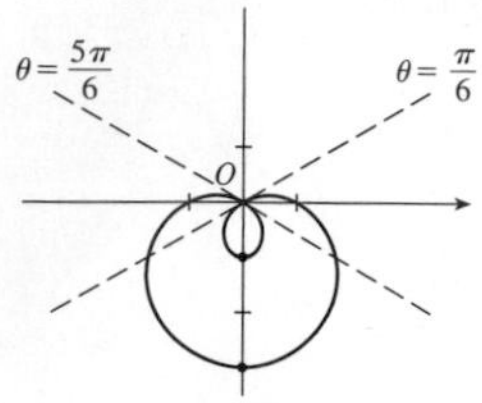

43.

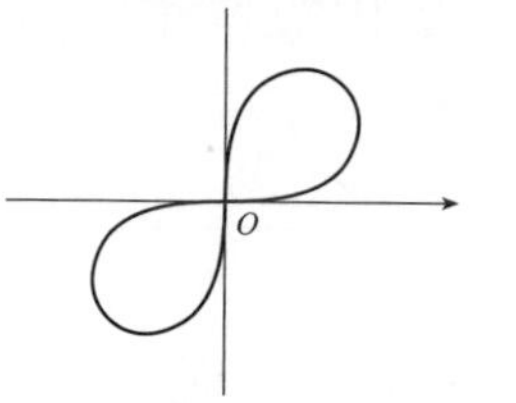

45.

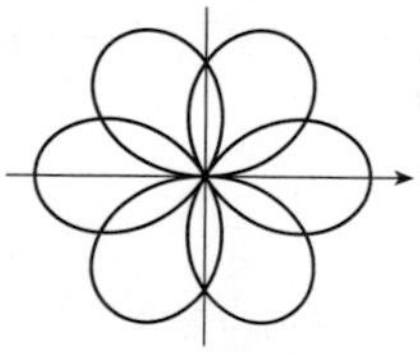

47.

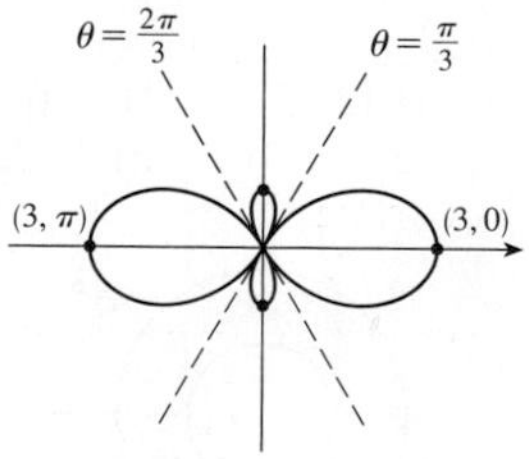

49.

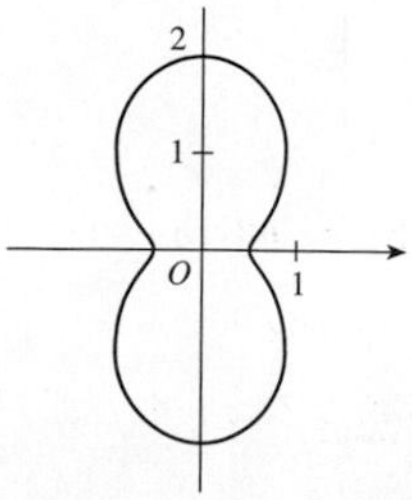

51.

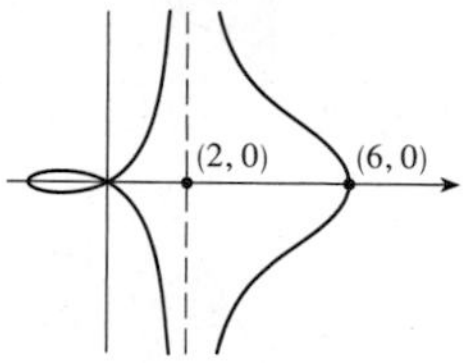

53.

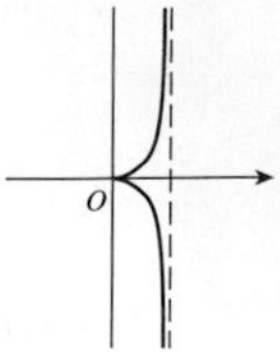

55. (a) For $c < -1$, the inner loop begins at $\theta = \sin^{-1}(-1/c)$ and ends at $\theta = \pi - \sin^{-1}(-1/c)$; for $c > 1$, it begins at $\theta = \pi + \sin^{-1}(1/c)$ and ends at $\theta = 2\pi - \sin^{-1}(1/c)$.

57. $\sqrt{3}$ **59.** $-\pi$ **61.** 1

63. Horizontal at $(3/\sqrt{2}, \pi/4)$, $(-3/\sqrt{2}, 3\pi/4)$; vertical at $(3, 0)$, $(0, \pi/2)$

65. Horizontal at $(\frac{3}{2}, \pi/3)$, $(0, \pi)$ [the pole], and $(\frac{3}{2}, 5\pi/3)$; vertical at $(2, 0)$, $(\frac{1}{2}, 2\pi/3)$, $(\frac{1}{2}, 4\pi/3)$

67. Horizontal at $(3, \pi/2)$, $(1, 3\pi/2)$; vertical at $(\frac{3}{2} + \frac{1}{2}\sqrt{3}, \alpha)$, $(\frac{3}{2} + \frac{1}{2}\sqrt{3}, \pi - \alpha)$ where $\alpha = \sin^{-1}(-\frac{1}{2} + \frac{1}{2}\sqrt{3})$

69. Center $(b/2, a/2)$, radius $\sqrt{a^2 + b^2}/2$

71.

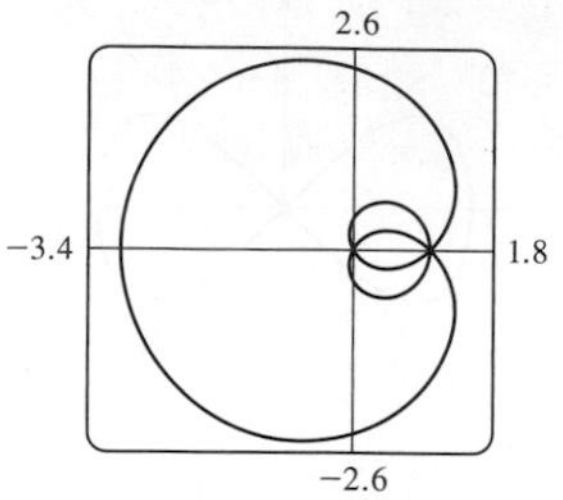

73.

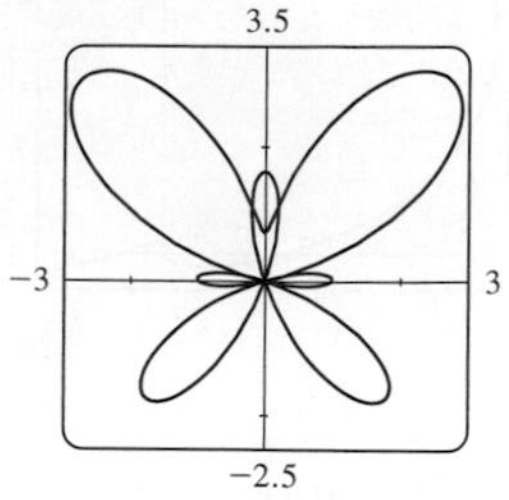

75.

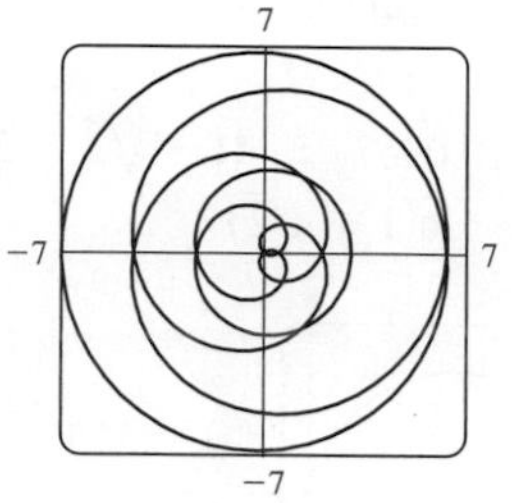

77. By counterclockwise rotation through angle $\pi/6$, $\pi/3$, or α about the origin

79. (a) A rose with n loops if n is odd and $2n$ loops if n is even
(b) Number of loops is always $2n$

81. For $0 < a < 1$, the curve is an oval, which develops a dimple as $a \to 1^-$. When $a > 1$, the curve splits into two parts, one of which has a loop.

EXERCISES 10.4 ■ PAGE 653

1. $\pi^5/10{,}240$ **3.** $\pi/12 + \frac{1}{8}\sqrt{3}$ **5.** π^2 **7.** $\frac{41}{4}\pi$

9. $\frac{9}{4}\pi$

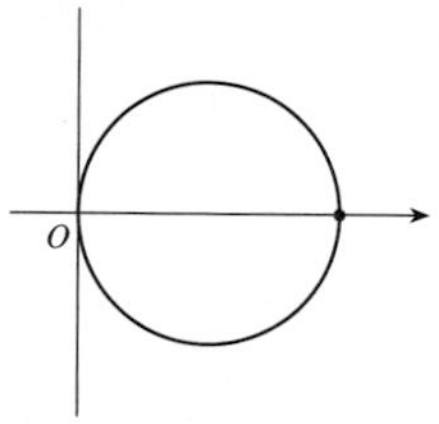

11. 4

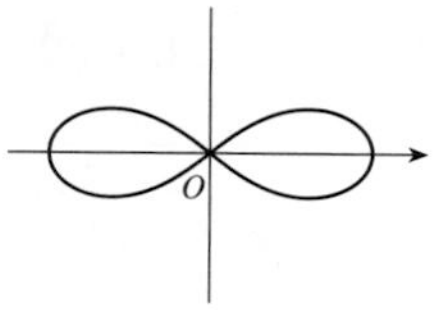

13. π

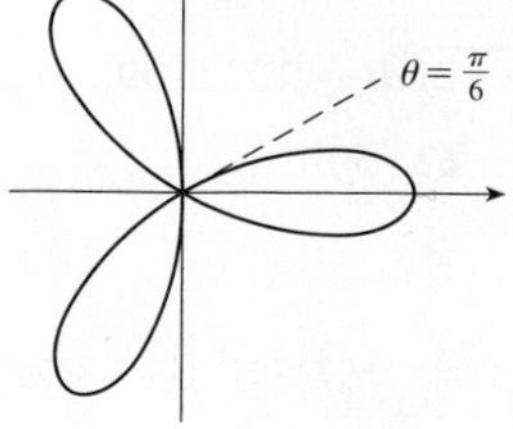

15. 3π

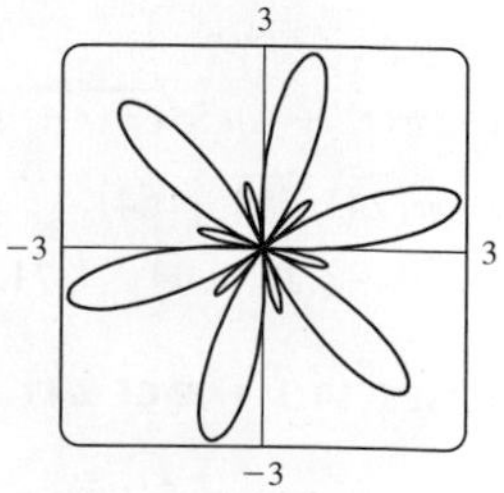

17. $\frac{1}{8}\pi$ **19.** $\frac{9}{20}\pi$ **21.** $\pi - \frac{3}{2}\sqrt{3}$ **23.** $\frac{1}{3}\pi + \frac{1}{2}\sqrt{3}$

25. $4\sqrt{3} - \frac{4}{3}\pi$ **27.** π **29.** $\frac{5}{24}\pi - \frac{1}{4}\sqrt{3}$ **31.** $\frac{1}{2}\pi - 1$

33. $1 - \frac{1}{2}\sqrt{2}$ **35.** $\frac{1}{4}(\pi + 3\sqrt{3})$

37. $(\frac{3}{2}, \pi/6)$, $(\frac{3}{2}, 5\pi/6)$, and the pole

39. $(1, \theta)$ where $\theta = \pi/12, 5\pi/12, 13\pi/12, 17\pi/12$ and $(-1, \theta)$ where $\theta = 7\pi/12, 11\pi/12, 19\pi/12, 23\pi/12$

41. $(\frac{1}{2}\sqrt{3}, \pi/3), (\frac{1}{2}\sqrt{3}, 2\pi/3)$, and the pole

43. Intersection at $\theta \approx 0.89, 2.25$; area ≈ 3.46 **45.** π

47. $\frac{8}{3}[(\pi^2 + 1)^{3/2} - 1]$ **49.** 29.0653 **51.** 9.6884

53. $\frac{16}{3}$ **55.** (b) $2\pi(2 - \sqrt{2})$

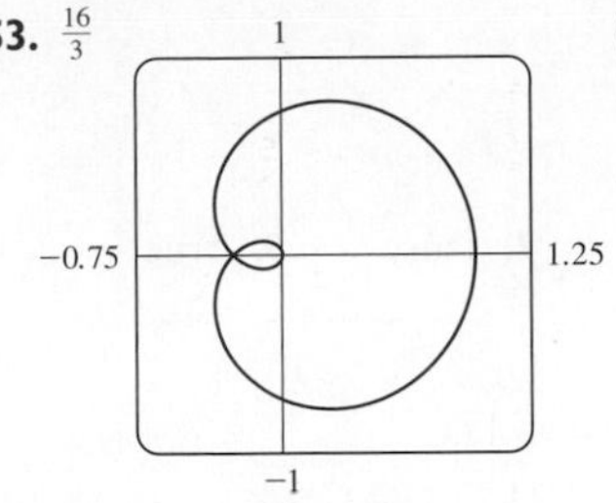

EXERCISES 10.5 ■ PAGE 660

1. $(0, 0), (\frac{1}{8}, 0), x = -\frac{1}{8}$

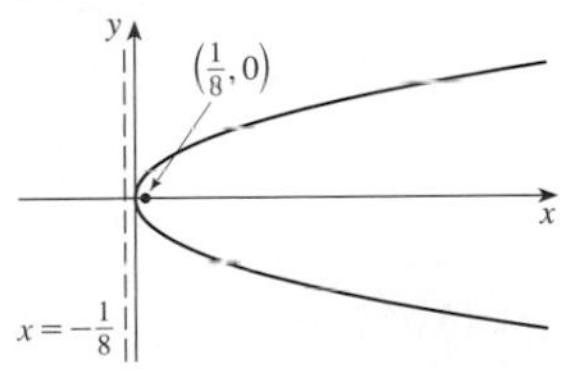

3. $(0, 0), (0, -\frac{1}{16}), y = \frac{1}{16}$

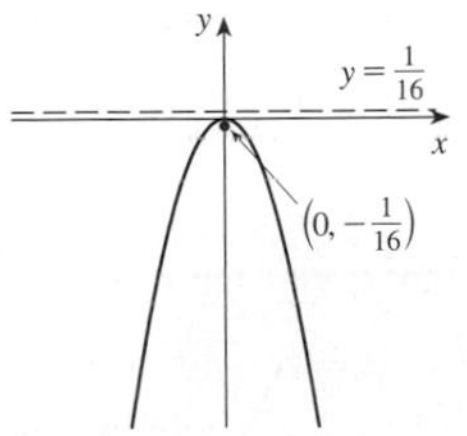

5. $(-2, 3), (-2, 5), y = 1$

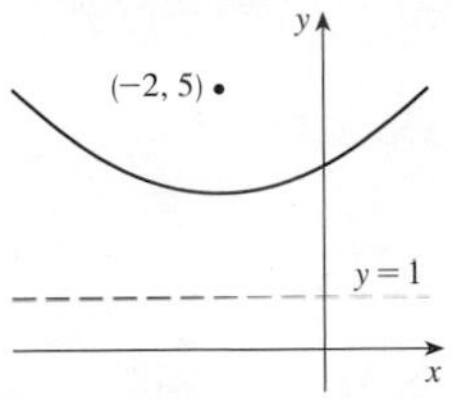

7. $(-2, -1), (-5, -1), x = 1$

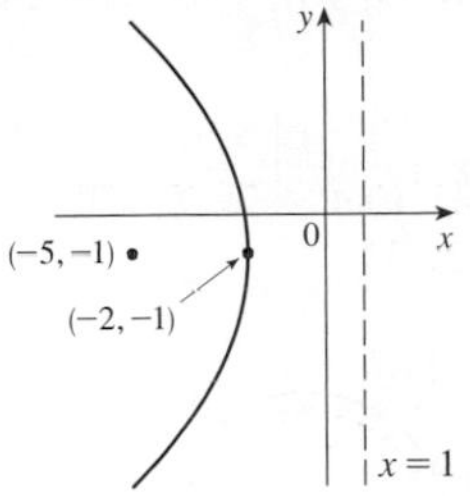

9. $x = -y^2$, focus $(-\frac{1}{4}, 0)$, directrix $x = \frac{1}{4}$

11. $(\pm 3, 0), (\pm 2, 0)$

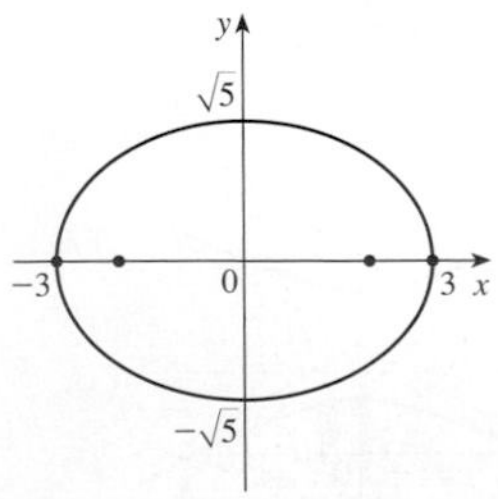

13. $(0, \pm 4), (0, \pm 2\sqrt{3})$

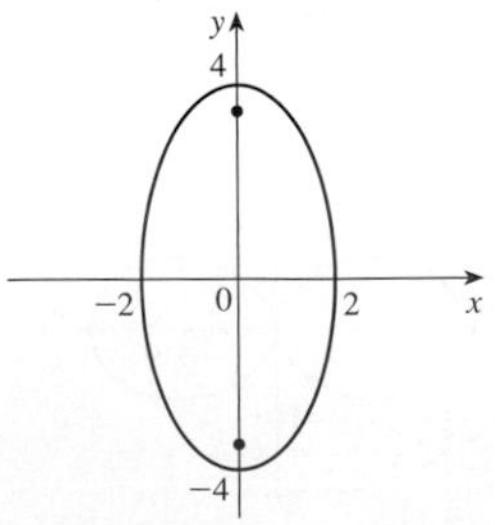

15. $(1, \pm 3), (1, \pm \sqrt{5})$

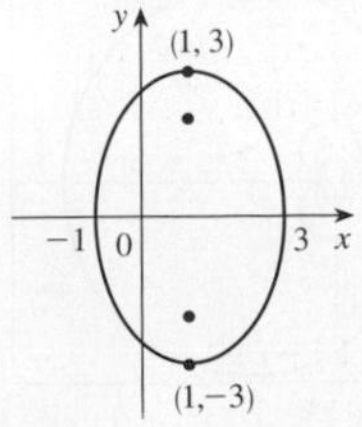

17. $\dfrac{x^2}{4} + \dfrac{y^2}{9} = 1$, foci $(0, \pm\sqrt{5})$

19. $(\pm 12, 0), (\pm 13, 0)$, $y = \pm\frac{5}{12}x$

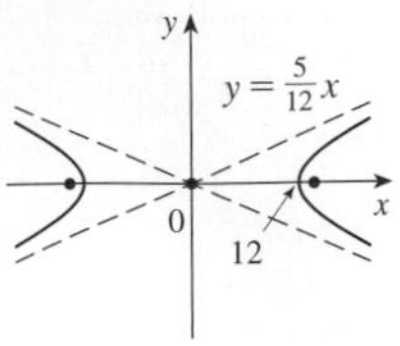

21. $(0, \pm 2), (0, \pm 2\sqrt{2})$, $y = \pm x$

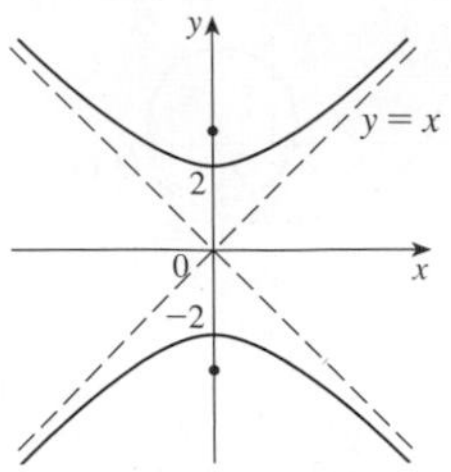

23. $(4, -2), (2, -2)$; $(3 \pm \sqrt{5}, -2)$; $y + 2 = \pm 2(x - 3)$

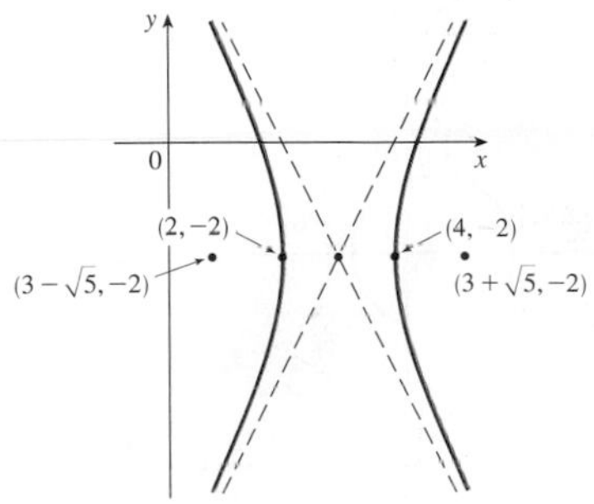

25. Parabola, $(0, -1), (0, -\frac{3}{4})$

27. Ellipse, $(\pm\sqrt{2}, 1), (\pm 1, 1)$

29. Hyperbola, $(0, 1), (0, -3)$; $(0, -1 \pm \sqrt{5})$ **31.** $x^2 = -8y$

33. $y^2 = -12(x + 1)$ **35.** $y - 3 = 2(x - 2)^2$

37. $\dfrac{x^2}{25} + \dfrac{y^2}{21} = 1$ **39.** $\dfrac{x^2}{12} + \dfrac{(y - 4)^2}{16} = 1$

41. $\dfrac{(x + 1)^2}{12} + \dfrac{(y - 4)^2}{16} = 1$ **43.** $\dfrac{x^2}{9} - \dfrac{y^2}{16} = 1$

45. $\dfrac{(y - 1)^2}{25} - \dfrac{(x + 3)^2}{39} = 1$ **47.** $\dfrac{x^2}{9} - \dfrac{y^2}{36} = 1$

49. $\dfrac{x^2}{3{,}763{,}600} + \dfrac{y^2}{3{,}753{,}196} = 1$

51. (a) $\dfrac{121x^2}{1{,}500{,}625} - \dfrac{121y^2}{3{,}339{,}375} = 1$ (b) ≈ 248 mi

55. (a) Ellipse (b) Hyperbola (c) No curve

59. 9.69 **61.** $\dfrac{b^2c}{a} + ab\ln\left(\dfrac{a}{b + c}\right)$ where $c^2 = a^2 + b^2$

EXERCISES 10.6 ■ PAGE 668

1. $r = \dfrac{42}{4 + 7\sin\theta}$ **3.** $r = \dfrac{15}{4 - 3\cos\theta}$

5. $r = \dfrac{8}{1 - \sin\theta}$ **7.** $r = \dfrac{4}{2 + \cos\theta}$

9. (a) 1 (b) Parabola (c) $y = 1$

(d)

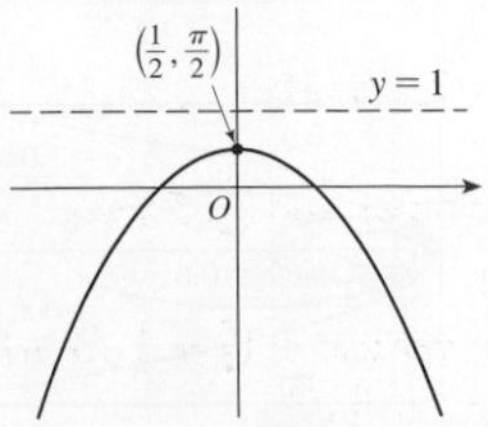

11. (a) $\frac{1}{4}$ (b) Ellipse (c) $y = -12$
(d)

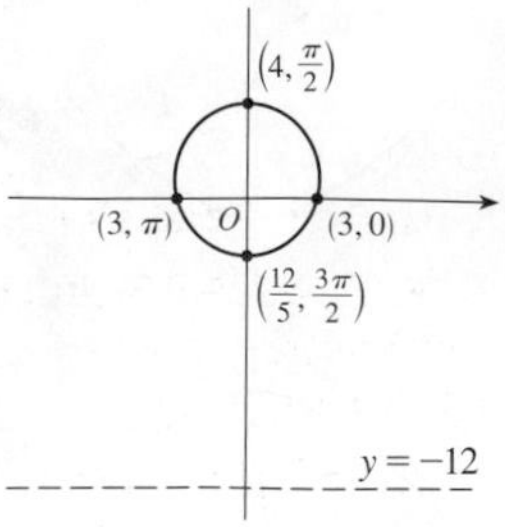

13. (a) $\frac{1}{3}$ (b) Ellipse (c) $x = \frac{9}{2}$
(d)

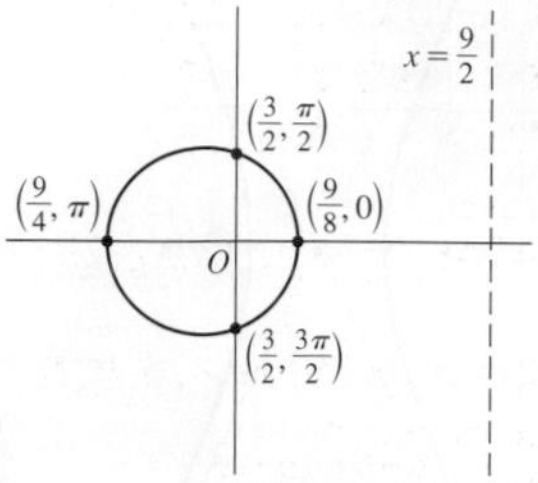

15. (a) 2 (b) Hyperbola (c) $x = -\frac{3}{8}$
(d)

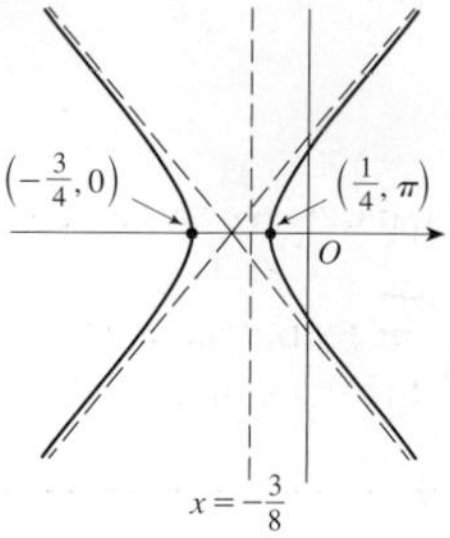

17. (a) $2, y = -\frac{1}{2}$

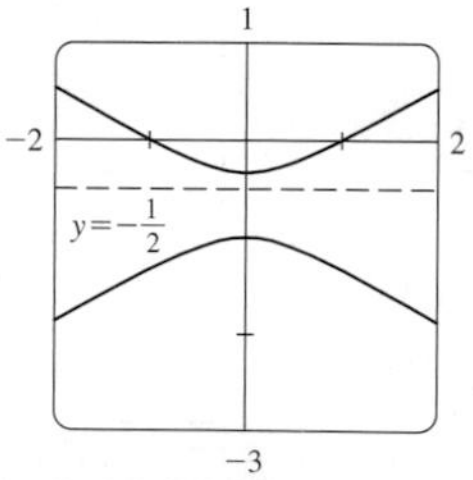

(b) $r = \dfrac{1}{1 - 2\sin(\theta - 3\pi/4)}$

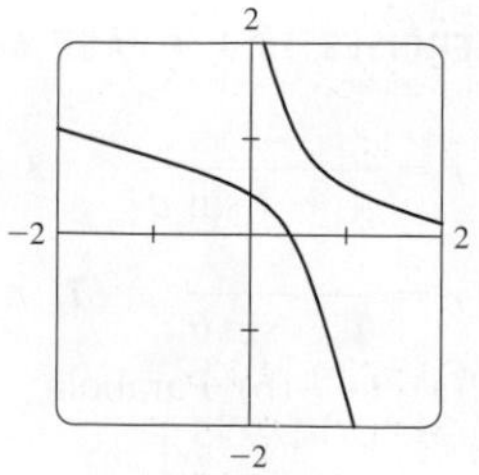

19. The ellipse is nearly circular when e is close to 0 and becomes more elongated as $e \to 1^-$. At $e = 1$, the curve becomes a parabola.

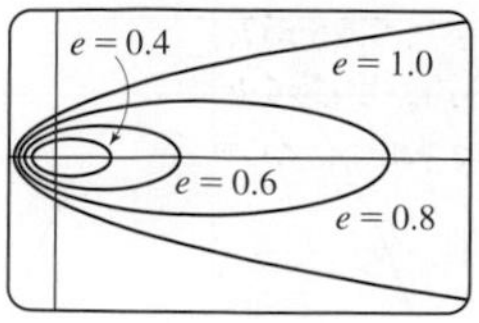

25. $r = \dfrac{2.26 \times 10^8}{1 + 0.093\cos\theta}$

27. 35.64 AU **29.** 7.0×10^7 km **31.** 3.6×10^8 km

CHAPTER 10 REVIEW ■ PAGE 669

True-False Quiz

1. False **3.** False **5.** True **7.** False **9.** True

Exercises

1. $x = y^2 - 8y + 12$

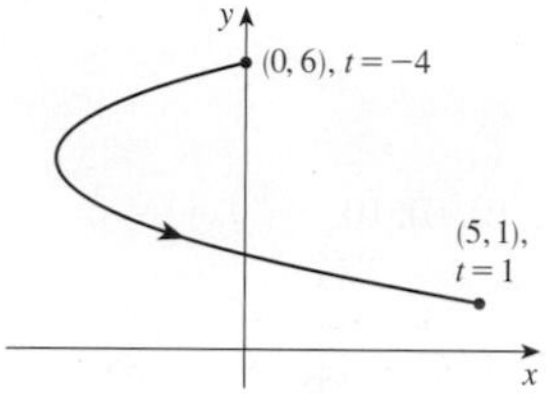

3. $y = 1/x$

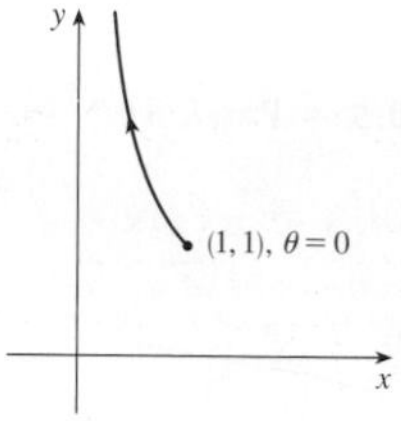

5. $x = t, y = \sqrt{t}$; $x = t^4, y = t^2$;
$x = \tan^2 t, y = \tan t, 0 \leqslant t < \pi/2$

7. (a)

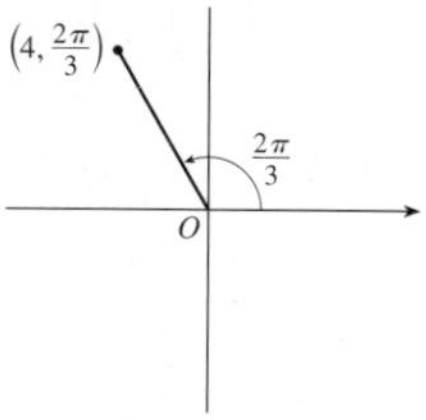

(b) $(3\sqrt{2}, 3\pi/4)$, $(-3\sqrt{2}, 7\pi/4)$

$(-2, 2\sqrt{3})$

9.

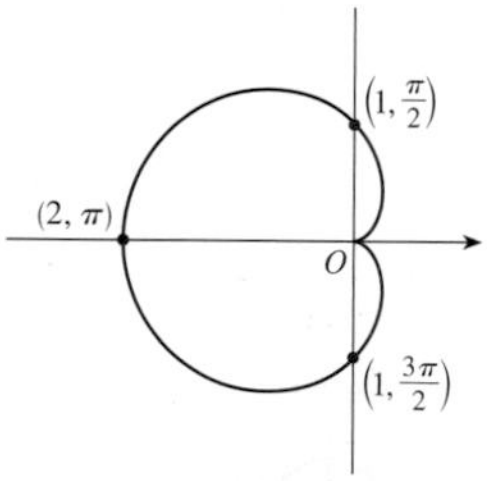

11.

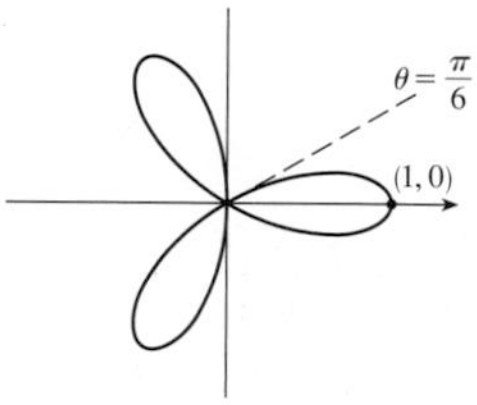

13.

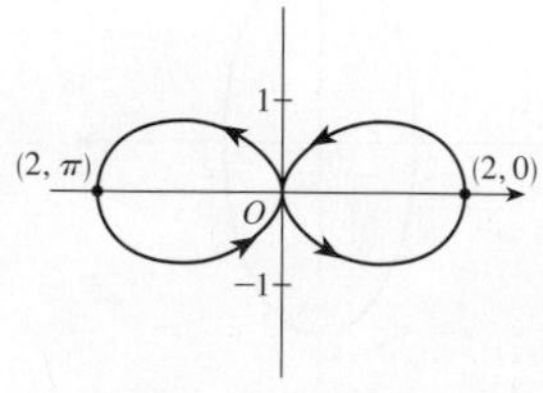

15.

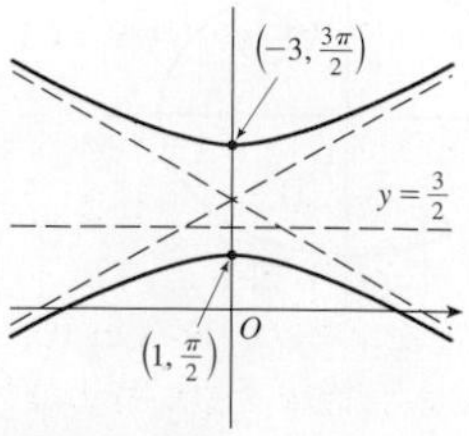

17. $r = \dfrac{2}{\cos\theta + \sin\theta}$

19.

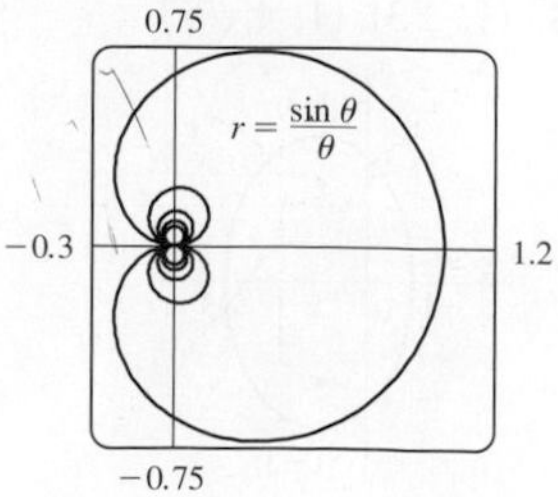

21. 2 **23.** -1

25. $\dfrac{1 + \sin t}{1 + \cos t}, \dfrac{1 + \cos t + \sin t}{(1 + \cos t)^3}$ **27.** $\left(\frac{11}{8}, \frac{3}{4}\right)$

29. Vertical tangent at $\left(\frac{3}{2}a, \pm\frac{1}{2}\sqrt{3}\,a\right), (-3a, 0)$; horizontal tangent at $(a, 0), \left(-\frac{1}{2}a, \pm\frac{3}{2}\sqrt{3}a\right)$

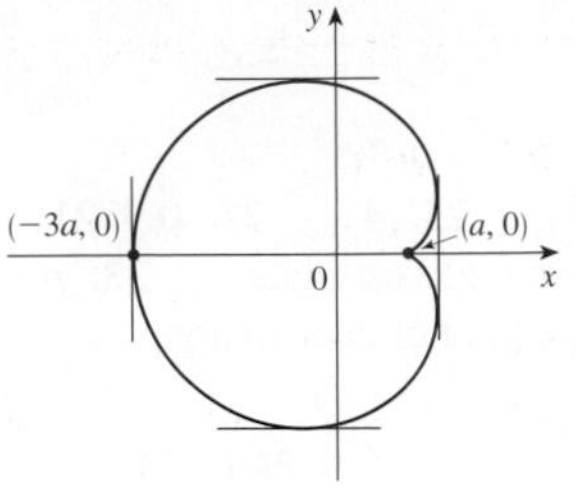

31. 18 **33.** $(2, \pm\pi/3)$ **35.** $\frac{1}{2}(\pi - 1)$

37. $2(5\sqrt{5} - 1)$

39. $\dfrac{2\sqrt{\pi^2 + 1} - \sqrt{4\pi^2 + 1}}{2\pi} + \ln\left(\dfrac{2\pi + \sqrt{4\pi^2 + 1}}{\pi + \sqrt{\pi^2 + 1}}\right)$

41. $471{,}295\pi/1024$

43. All curves have the vertical asymptote $x = 1$. For $c < -1$, the curve bulges to the right. At $c = -1$, the curve is the line $x = 1$. For $-1 < c < 0$, it bulges to the left. At $c = 0$ there is a cusp at $(0, 0)$. For $c > 0$, there is a loop.

45. $(\pm 1, 0), (\pm 3, 0)$ **47.** $\left(-\frac{25}{24}, 3\right), (-1, 3)$

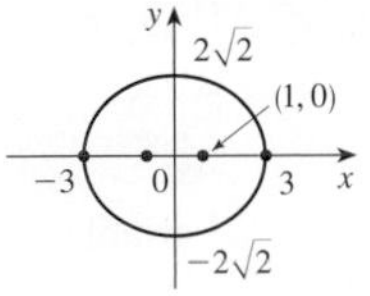

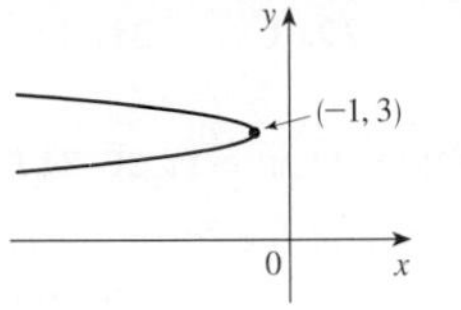

49. $\dfrac{x^2}{25} + \dfrac{y^2}{9} = 1$ **51.** $\dfrac{y^2}{72/5} - \dfrac{x^2}{8/5} = 1$

53. $\dfrac{x^2}{25} + \dfrac{(8y - 399)^2}{160{,}801} = 1$ **55.** $r = \dfrac{4}{3 + \cos\theta}$

57. $x = a(\cot\theta + \sin\theta\cos\theta),\ y = a(1 + \sin^2\theta)$

PROBLEMS PLUS ▪ PAGE 672

1. $\ln(\pi/2)$

3. $\left[-\frac{3}{4}\sqrt{3}, \frac{3}{4}\sqrt{3}\right] \times [-1, 2]$

5. (a) At $(0, 0)$ and $\left(\frac{3}{2}, \frac{3}{2}\right)$

(b) Horizontal tangents at $(0, 0)$ and $\left(\sqrt[3]{2}, \sqrt[3]{4}\right)$; vertical tangents at $(0, 0)$ and $\left(\sqrt[3]{4}, \sqrt[3]{2}\right)$

(d)

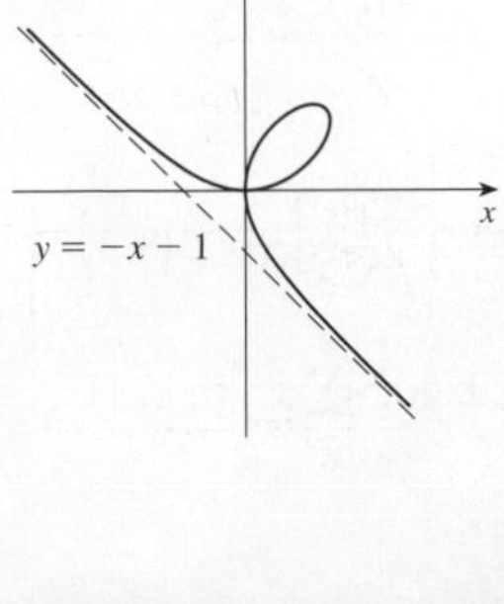

(g) $\frac{3}{2}$

CHAPTER 11

EXERCISES 11.1 ▪ PAGE 684

Abbreviations: C, convergent; D, divergent

1. (a) A sequence is an ordered list of numbers. It can also be defined as a function whose domain is the set of positive integers.
(b) The terms a_n approach 8 as n becomes large.
(c) The terms a_n become large as n becomes large.

3. 0.8, 0.96, 0.992, 0.9984, 0.99968 **5.** $-3, \frac{3}{2}, -\frac{1}{2}, \frac{1}{8}, -\frac{1}{40}$

7. 3, 5, 9, 17, 33 **9.** $a_n = 1/(2n - 1)$ **11.** $a_n = 5n - 3$

13. $a_n = \left(-\frac{2}{3}\right)^{n-1}$ **15.** $\frac{1}{3}, \frac{2}{5}, \frac{3}{7}, \frac{4}{9}, \frac{5}{11}, \frac{6}{13}$; yes; $\frac{1}{2}$

17. 1 **19.** 5 **21.** 1 **23.** 1 **25.** 0 **27.** D

29. 0 **31.** 0 **33.** 0 **35.** 0 **37.** 1 **39.** e^2

41. ln 2 **43.** D **45.** D **47.** 1 **49.** $\frac{1}{2}$

51. D **53.** 0

55. (a) 1060, 1123.60, 1191.02, 1262.48, 1338.23 (b) D

57. $-1 < r < 1$

59. Convergent by the Monotonic Sequence Theorem; $5 \leq L < 8$

61. Decreasing; yes **63.** Not monotonic; no

65. Decreasing; yes **67.** 2 **69.** $\frac{1}{2}(3 + \sqrt{5})$

71. (b) $\frac{1}{2}(1 + \sqrt{5})$ **73.** (a) 0 (b) 9, 11

EXERCISES 11.2 ▪ PAGE 694

1. (a) A sequence is an ordered list of numbers whereas a series is the *sum* of a list of numbers.
(b) A series is convergent if the sequence of partial sums is a convergent sequence. A series is divergent if it is not convergent.

3. $-2.40000, -1.92000, -2.01600, -1.99680, -2.00064, -1.99987, -2.00003, -1.99999, -2.00000, -2.00000$; convergent, sum $= -2$

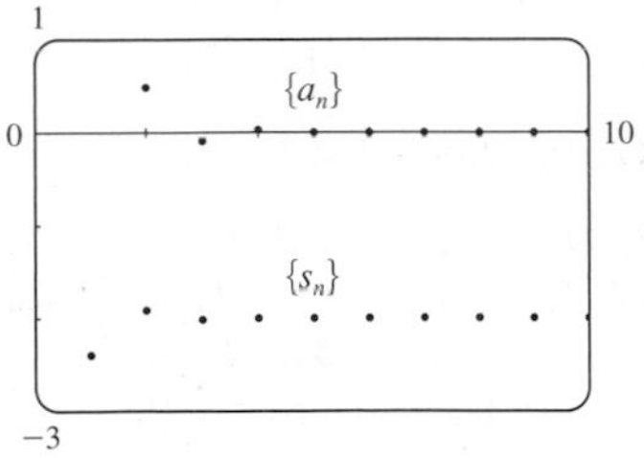

5. $1.55741, -0.62763, -0.77018, 0.38764, -2.99287, -3.28388, -2.41243, -9.21214, -9.66446, -9.01610$; divergent

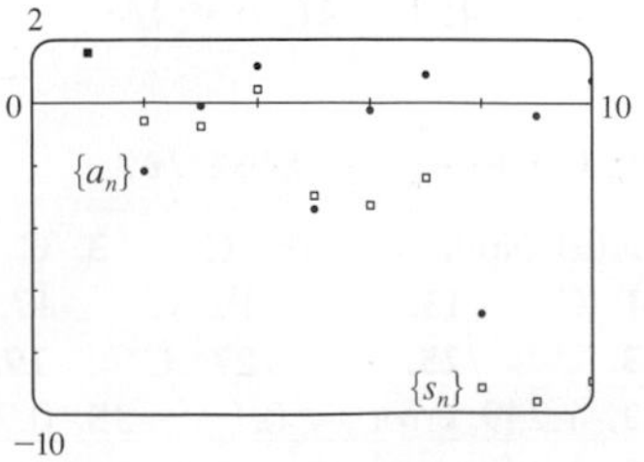

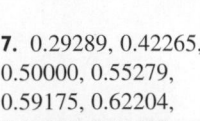

7. 0.29289, 0.42265, 0.50000, 0.55279, 0.59175, 0.62204, 0.64645, 0.66667, 0.68377, 0.69849; convergent, sum $= 1$

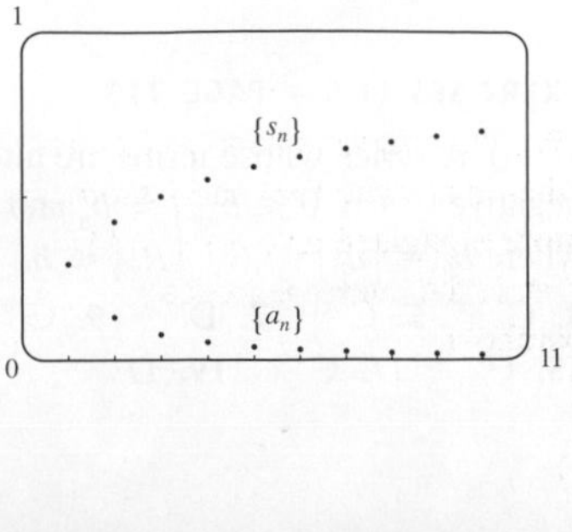

9. (a) C (b) D **11.** 9 **13.** D **15.** 60 **17.** $\frac{1}{7}$
19. D **21.** D **23.** D **25.** $\frac{5}{2}$ **27.** D **29.** D
31. D **33.** $e/(e-1)$ **35.** $\frac{3}{2}$ **37.** $\frac{11}{6}$ **39.** $e-1$
41. $\frac{2}{9}$ **43.** 1138/333 **45.** 5063/3300
47. $-3 < x < 3$; $\dfrac{x}{3-x}$ **49.** $-\frac{1}{4} < x < \frac{1}{4}$; $\dfrac{1}{1-4x}$
51. All x; $\dfrac{2}{2-\cos x}$ **53.** 1
55. $a_1 = 0$, $a_n = \dfrac{2}{n(n+1)}$ for $n > 1$, sum $= 1$
57. (a) $S_n = \dfrac{D(1-c^n)}{1-c}$ (b) 5 **59.** $\frac{1}{2}(\sqrt{3}-1)$
63. $\dfrac{1}{n(n+1)}$ **65.** The series is divergent.
71. $\{s_n\}$ is bounded and increasing.
73. (a) $0, \frac{1}{9}, \frac{2}{9}, \frac{1}{3}, \frac{2}{3}, \frac{7}{9}, \frac{8}{9}, 1$
75. (a) $\frac{1}{2}, \frac{5}{6}, \frac{23}{24}, \frac{119}{120}$; $\dfrac{(n+1)!-1}{(n+1)!}$ (c) 1

EXERCISES 11.3 ▪ PAGE 703

1. C

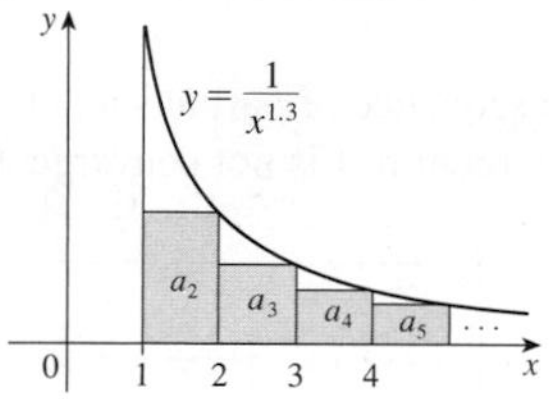

3. D **5.** C **7.** C **9.** D **11.** C **13.** D **15.** C
17. C **19.** C **21.** D **23.** C **25.** C **27.** $p > 1$
29. $p < -1$ **31.** $(1, \infty)$
33. (a) 1.54977, error $\leqslant 0.1$ (b) 1.64522, error $\leqslant 0.005$
(c) $n > 1000$
35. 0.00145 **41.** $b < 1/e$

EXERCISES 11.4 ▪ PAGE 709

1. (a) Nothing (b) C **3.** C **5.** D **7.** C **9.** C
11. C **13.** C **15.** C **17.** D **19.** D **21.** C
23. C **25.** D **27.** C **29.** C **31.** D
33. 1.249, error < 0.1 **35.** 0.76352, error < 0.001
45. Yes

EXERCISES 11.5 ▪ PAGE 713

1. (a) A series whose terms are alternately positive and negative (b) $0 < b_{n+1} \leqslant b_n$ and $\lim_{n\to\infty} b_n = 0$, where $b_n = |a_n|$ (c) $|R_n| \leqslant b_{n+1}$
3. C **5.** C **7.** D **9.** C **11.** C **13.** D
15. C **17.** C **19.** D
21. 1.0000, 0.6464, 0.8389, 0.7139, 0.8033, 0.7353, 0.7893, 0.7451, 0.7821, 0.7505; error < 0.0275

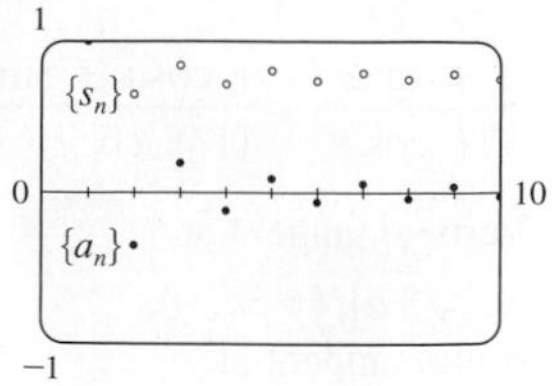

23. 5 **25.** 4 **27.** 0.9721 **29.** 0.0676
31. An underestimate **33.** p is not a negative integer
35. $\{b_n\}$ is not decreasing

EXERCISES 11.6 ▪ PAGE 719

Abbreviations: AC, absolutely convergent; CC, conditionally convergent

1. (a) D (b) C (c) May converge or diverge
3. AC **5.** CC **7.** AC **9.** D **11.** AC **13.** AC
15. AC **17.** CC **19.** AC **21.** AC **23.** D
25. AC **27.** D **29.** D **31.** (a) and (d)
35. (a) $\frac{661}{960} \approx 0.68854$, error < 0.00521
(b) $n \geqslant 11$, 0.693109

EXERCISES 11.7 ▪ PAGE 722

1. C **3.** D **5.** C **7.** D **9.** C **11.** C **13.** C
15. C **17.** D **19.** C **21.** C **23.** D **25.** C
27. C **29.** C **31.** D **33.** C **35.** C **37.** C

EXERCISES 11.8 ▪ PAGE 727

1. A series of the form $\sum_{n=0}^{\infty} c_n(x-a)^n$, where x is a variable and a and the c_n's are constants
3. $1, [-1, 1)$ **5.** $1, [-1, 1]$ **7.** $\infty, (-\infty, \infty)$
9. $2, (-2, 2)$ **11.** $\frac{1}{2}, \left(-\frac{1}{2}, \frac{1}{2}\right]$ **13.** $4, (-4, 4]$
15. $1, [1, 3]$ **17.** $\frac{1}{3}, \left[-\frac{13}{3}, -\frac{11}{3}\right)$ **19.** $\infty, (-\infty, \infty)$
21. $b, (a-b, a+b)$ **23.** $0, \left\{\frac{1}{2}\right\}$ **25.** $\frac{1}{4}, \left[-\frac{1}{2}, 0\right]$
27. $\infty, (-\infty, \infty)$ **29.** (a) Yes (b) No **31.** k^k **33.** No
35. (a) $(-\infty, \infty)$
(b), (c)

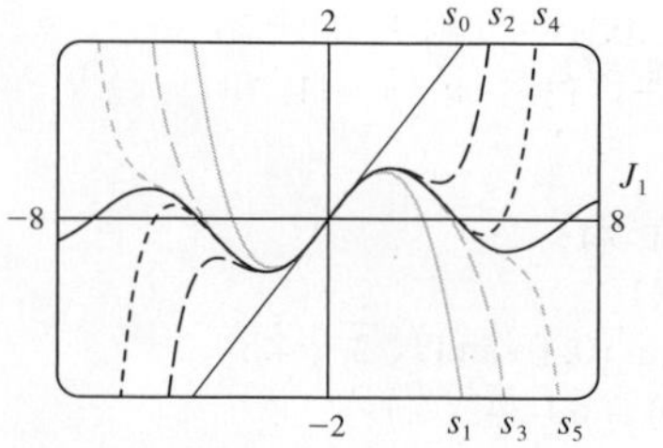

37. $(-1, 1)$, $f(x) = (1+2x)/(1-x^2)$ **41.** 2

EXERCISES 11.9 ▪ PAGE 733

1. 10 **3.** $\displaystyle\sum_{n=0}^{\infty} (-1)^n x^n, (-1, 1)$ **5.** $2\displaystyle\sum_{n=0}^{\infty} \frac{1}{3^{n+1}} x^n, (-3, 3)$
7. $\displaystyle\sum_{n=0}^{\infty} (-1)^n \frac{1}{9^{n+1}} x^{2n+1}, (-3, 3)$ **9.** $1 + 2\displaystyle\sum_{n=1}^{\infty} x^n, (-1, 1)$

11. $\displaystyle\sum_{n=0}^{\infty}\left[(-1)^{n+1}-\frac{1}{2^{n+1}}\right]x^n,\ (-1, 1)$

13. (a) $\displaystyle\sum_{n=0}^{\infty}(-1)^n(n+1)x^n,\ R=1$

(b) $\displaystyle\frac{1}{2}\sum_{n=0}^{\infty}(-1)^n(n+2)(n+1)x^n,\ R=1$

(c) $\displaystyle\frac{1}{2}\sum_{n=2}^{\infty}(-1)^n n(n-1)x^n,\ R=1$

15. $\displaystyle\ln 5-\sum_{n=1}^{\infty}\frac{x^n}{n5^n},\ R=5$ **17.** $\displaystyle\sum_{n=3}^{\infty}\frac{n-2}{2^{n-1}}x^n,\ R=2$

19. $\displaystyle\sum_{n=0}^{\infty}(-1)^n\frac{1}{16^{n+1}}x^{2n+1},\ R=4$

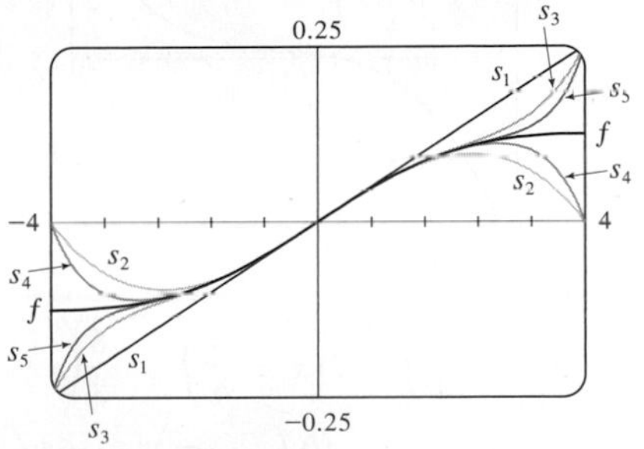

21. $\displaystyle\sum_{n=0}^{\infty}\frac{2x^{2n+1}}{2n+1},\ R=1$

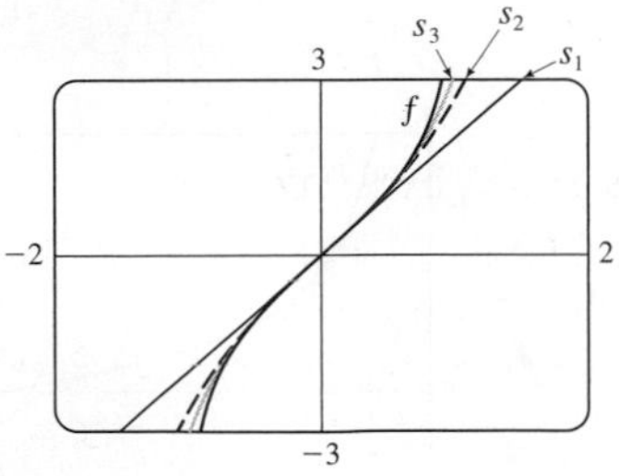

23. $\displaystyle C+\sum_{n=0}^{\infty}\frac{t^{8n+2}}{8n+2},\ R=1$

25. $\displaystyle C+\sum_{n=1}^{\infty}(-1)^{n+1}\frac{x^{2n-1}}{4n^2-1},\ R=1$

27. 0.199989 **29.** 0.000983 **31.** 0.09531

33. (b) 0.920 **37.** $[-1, 1], [-1, 1), (-1, 1)$

EXERCISES 11.10 ■ PAGE 746

1. $b_8 = f^{(8)}(5)/8!$ **3.** $\displaystyle\sum_{n=0}^{\infty}(n+1)x^n,\ R=1$

5. $\displaystyle\sum_{n=0}^{\infty}(n+1)x^n,\ R=1$

7. $\displaystyle\sum_{n=0}^{\infty}(-1)^n\frac{\pi^{2n+1}}{(2n+1)!}x^{2n+1},\ R=\infty$

9. $\displaystyle\sum_{n=0}^{\infty}\frac{5^n}{n!}x^n,\ R=\infty$ **11.** $\displaystyle\sum_{n=0}^{\infty}\frac{x^{2n+1}}{(2n+1)!},\ R=\infty$

13. $-1-2(x-1)+3(x-1)^2+4(x-1)^3+(x-1)^4$, $R=\infty$

15. $\displaystyle\sum_{n=0}^{\infty}\frac{e^3}{n!}(x-3)^n,\ R=\infty$

17. $\displaystyle\sum_{n=0}^{\infty}(-1)^{n+1}\frac{1}{(2n)!}(x-\pi)^{2n},\ R=\infty$

19. $\displaystyle\sum_{n=0}^{\infty}(-1)^n\frac{1\cdot3\cdot5\cdot\cdots\cdot(2n-1)}{2^n\cdot3^{2n+1}\cdot n!}(x-9)^n,\ R=9$

25. $\displaystyle 1+\frac{x}{2}+\sum_{n=2}^{\infty}(-1)^{n-1}\frac{1\cdot3\cdot5\cdot\cdots\cdot(2n-3)}{2^n n!}x^n,\ R=1$

27. $\displaystyle\sum_{n=0}^{\infty}(-1)^n\frac{(n+1)(n+2)}{2^{n+4}}x^n,\ R=2$

29. $\displaystyle\sum_{n=0}^{\infty}(-1)^n\frac{\pi^{2n+1}}{(2n+1)!}x^{2n+1},\ R=\infty$

31. $\displaystyle\sum_{n=0}^{\infty}\frac{2^n+1}{n!}x^n,\ R=\infty$

33. $\displaystyle\sum_{n=0}^{\infty}(-1)^n\frac{1}{2^{2n}(2n)!}x^{4n+1},\ R=\infty$

35. $\displaystyle\tfrac{1}{2}x+\sum_{n=1}^{\infty}(-1)^n\frac{1\cdot3\cdot5\cdot\cdots\cdot(2n-1)}{n!\,2^{3n+1}}x^{2n+1},\ R=2$

37. $\displaystyle\sum_{n=1}^{\infty}(-1)^{n+1}\frac{2^{2n-1}}{(2n)!}x^{2n},\ R=\infty$

39. $\displaystyle\sum_{n=0}^{\infty}(-1)^n\frac{1}{(2n)!}x^{4n},\ R=\infty$

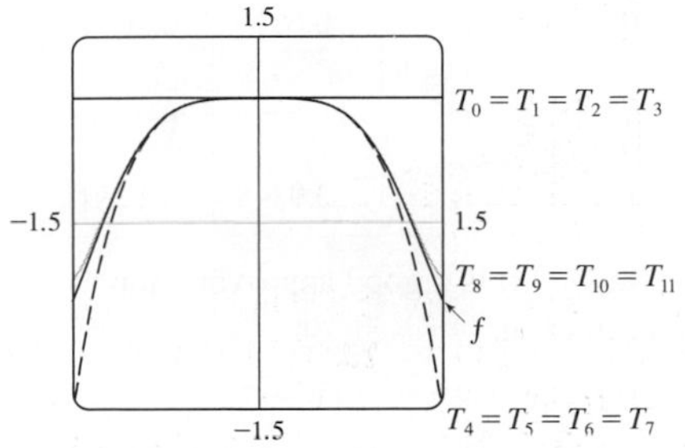

41. $\displaystyle\sum_{n=1}^{\infty}\frac{(-1)^{n-1}}{(n-1)!}x^n,\ R=\infty$

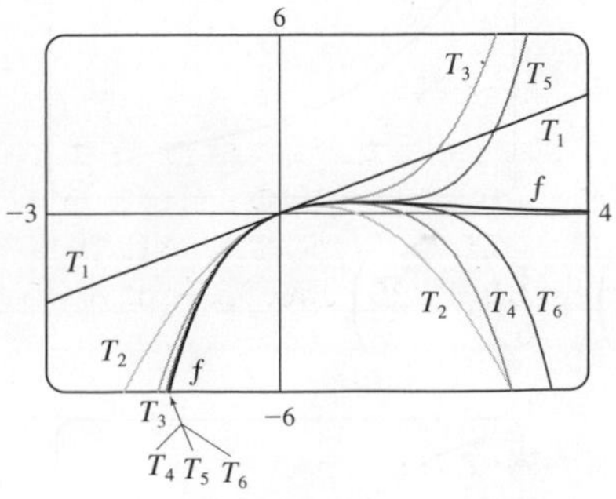

43. 0.81873

45. (a) $\displaystyle 1+\sum_{n=1}^{\infty}\frac{1\cdot3\cdot5\cdot\cdots\cdot(2n-1)}{2^n n!}x^{2n}$

(b) $\displaystyle x+\sum_{n=1}^{\infty}\frac{1\cdot3\cdot5\cdot\cdots\cdot(2n-1)}{(2n+1)2^n n!}x^{2n+1}$

47. $C + \sum_{n=0}^{\infty} (-1)^n \dfrac{x^{6n+2}}{(6n+2)(2n)!}$, $R = \infty$

49. $C + \sum_{n=1}^{\infty} (-1)^n \dfrac{1}{2n\,(2n)!} x^{2n}$, $R = \infty$ **51.** 0.440

53. 0.40102 **55.** $\frac{1}{3}$ **57.** $\frac{1}{120}$ **59.** $1 - \frac{3}{2}x^2 + \frac{25}{24}x^4$

61. $1 + \frac{1}{6}x^2 + \frac{7}{360}x^4$ **63.** e^{-x^4}

65. $1/\sqrt{2}$ **67.** $e^3 - 1$

EXERCISES 11.11 ■ PAGE 755

1. (a) $T_0(x) = 1 = T_1(x)$, $T_2(x) = 1 - \frac{1}{2}x^2 = T_3(x)$,
$T_4(x) = 1 - \frac{1}{2}x^2 + \frac{1}{24}x^4 = T_5(x)$,
$T_6(x) = 1 - \frac{1}{2}x^2 + \frac{1}{24}x^4 - \frac{1}{720}x^6$

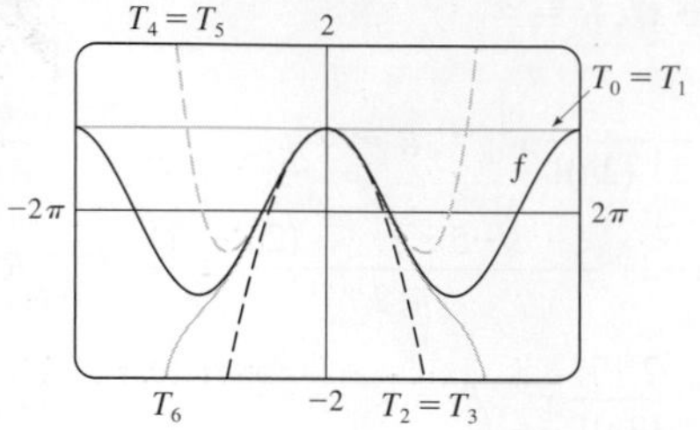

(b)

x	f	$T_0 = T_1$	$T_2 = T_3$	$T_4 = T_5$	T_6
$\frac{\pi}{4}$	0.7071	1	0.6916	0.7074	0.7071
$\frac{\pi}{2}$	0	1	−0.2337	0.0200	−0.0009
π	−1	1	−3.9348	0.1239	−1.2114

(c) As n increases, $T_n(x)$ is a good approximation to $f(x)$ on a larger and larger interval.

3. $\frac{1}{2} - \frac{1}{4}(x-2) + \frac{1}{8}(x-2)^2 - \frac{1}{16}(x-2)^3$

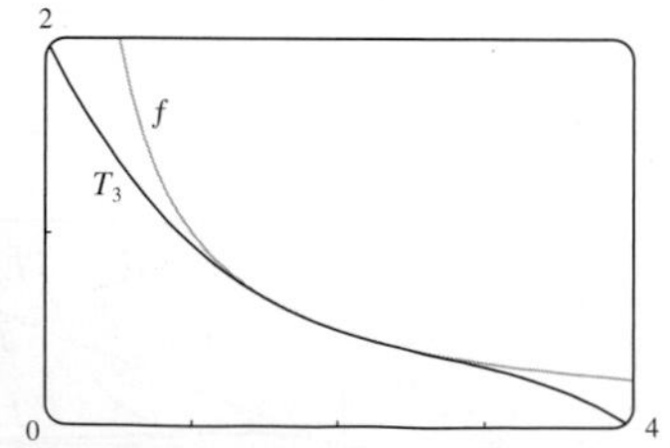

5. $-\left(x - \dfrac{\pi}{2}\right) + \dfrac{1}{6}\left(x - \dfrac{\pi}{2}\right)^3$

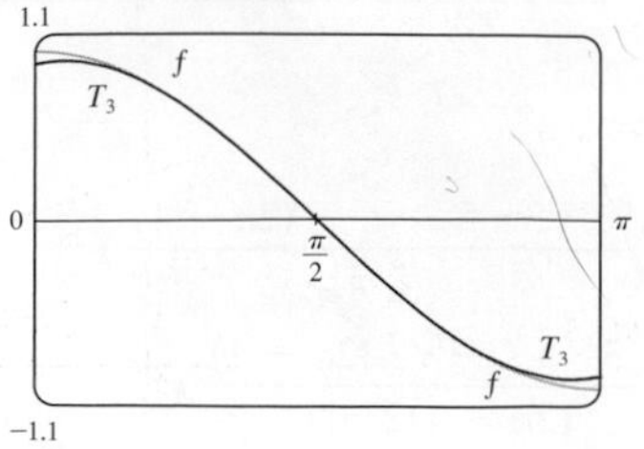

7. $x + \frac{1}{6}x^3$

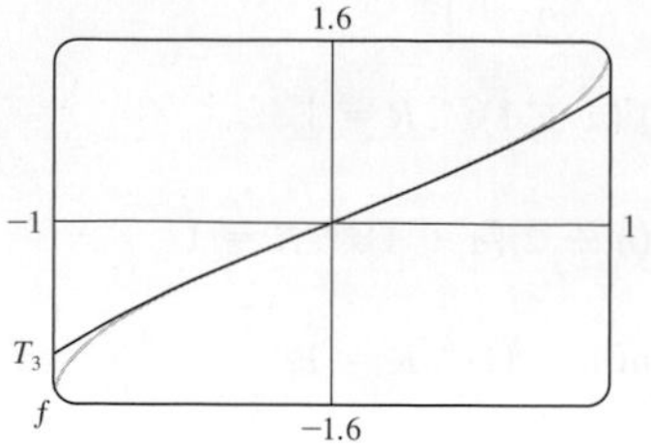

9. $x - 2x^2 + 2x^3$

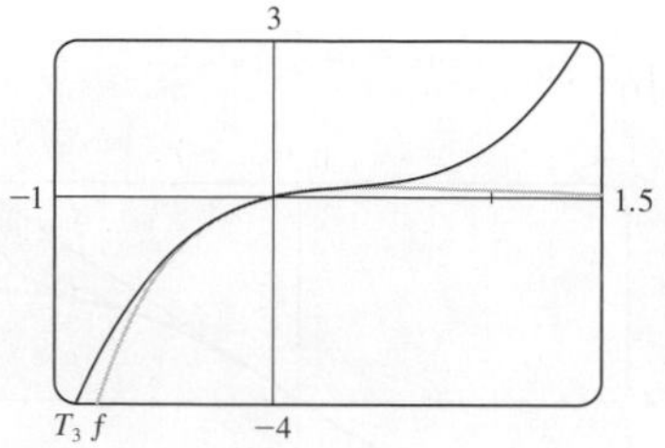

11. $T_5(x) = 1 - 2\left(x - \dfrac{\pi}{4}\right) + 2\left(x - \dfrac{\pi}{4}\right)^2 - \dfrac{8}{3}\left(x - \dfrac{\pi}{4}\right)^3 + \dfrac{10}{3}\left(x - \dfrac{\pi}{4}\right)^4 - \dfrac{64}{15}\left(x - \dfrac{\pi}{4}\right)^5$

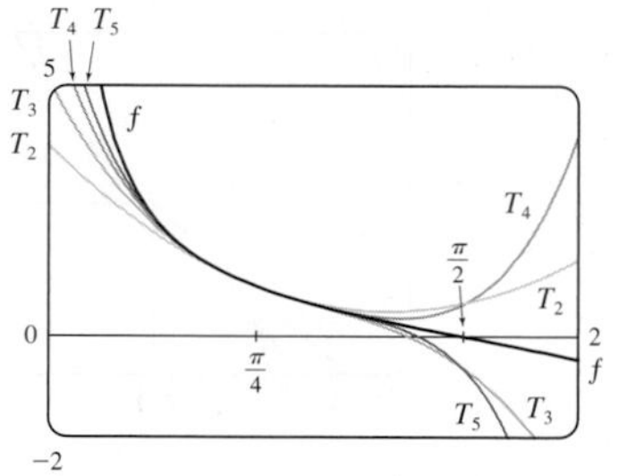

13. (a) $2 + \frac{1}{4}(x-4) - \frac{1}{64}(x-4)^2$ (b) 1.5625×10^{-5}

15. (a) $1 + \frac{2}{3}(x-1) - \frac{1}{9}(x-1)^2 + \frac{4}{81}(x-1)^3$ (b) 0.000097

17. (a) $1 + \frac{1}{2}x^2$ (b) 0.0015 **19.** (a) $1 + x^2$ (b) 0.00006

21. (a) $x^2 - \frac{1}{6}x^4$ (b) 0.042 **23.** 0.17365 **25.** Four

27. $-1.037 < x < 1.037$ **29.** $-0.86 < x < 0.86$

31. 21 m, no **37.** (c) They differ by about 8×10^{-9} km.

CHAPTER 11 REVIEW ■ PAGE 759

True-False Quiz

1. False **3.** True **5.** False **7.** False
9. False **11.** True **13.** True **15.** False
17. True **19.** True

Exercises

1. $\frac{1}{2}$ **3.** D **5.** 0 **7.** e^{12} **9.** 2 **11.** C
13. C **15.** D **17.** C **19.** C **21.** C **23.** CC
25. AC **27.** $\frac{1}{11}$ **29.** $\pi/4$ **31.** e^{-e} **35.** 0.9721

37. 0.18976224, error $< 6.4 \times 10^{-7}$
41. $4, [-6, 2)$ **43.** $0.5, [2.5, 3.5)$

45. $\frac{1}{2}\sum_{n=0}^{\infty}(-1)^n\left[\frac{1}{(2n)!}\left(x-\frac{\pi}{6}\right)^{2n}+\frac{\sqrt{3}}{(2n+1)!}\left(x-\frac{\pi}{6}\right)^{2n+1}\right]$

47. $\sum_{n=0}^{\infty}(-1)^n x^{n+2}, R = 1$ **49.** $-\sum_{n=1}^{\infty}\frac{x^n}{n}, R = 1$

51. $\sum_{n=0}^{\infty}(-1)^n\frac{x^{8n+4}}{(2n+1)!}, R = \infty$

53. $\frac{1}{2}+\sum_{n=1}^{\infty}\frac{1\cdot 5\cdot 9\cdot \cdots \cdot (4n-3)}{n!\,2^{6n+1}}x^n, R = 16$

55. $C + \ln|x| + \sum_{n=1}^{\infty}\frac{x^n}{n \cdot n!}$

57. (a) $1 + \frac{1}{2}(x-1) - \frac{1}{8}(x-1)^2 + \frac{1}{16}(x-1)^3$
(b)

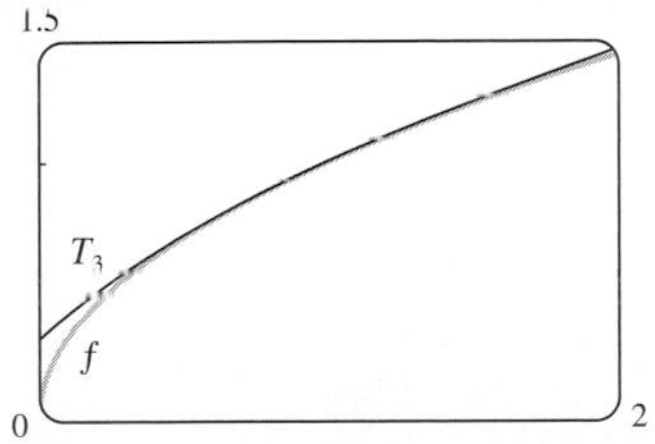

(c) 0.000006

59. $-\frac{1}{6}$

PROBLEMS PLUS ■ PAGE 762

1. $15!/5! = 10{,}897{,}286{,}400$
3. (b) 0 if $x = 0$, $(1/x) - \cot x$ if $x \neq k\pi$, k an integer
5. (a) $s_n = 3 \cdot 4^n$, $l_n = 1/3^n$, $p_n = 4^n/3^{n-1}$ (c) $\frac{2}{5}\sqrt{3}$
9. $(-1, 1)$, $\frac{x^3 + 4x^2 + x}{(1-x)^4}$ **11.** $\ln \frac{1}{2}$
13. (a) $\frac{250}{101}\pi(e^{-(n-1)\pi/5} - e^{-n\pi/5})$ (b) $\frac{250}{101}\pi$

CHAPTER 12

EXERCISES 12.1 ■ PAGE 769

1. $(4, 0, -3)$ **3.** Q; R

5. A vertical plane that intersects the xy-plane in the line $y = 2 - x$, $z = 0$ (see graph at right)

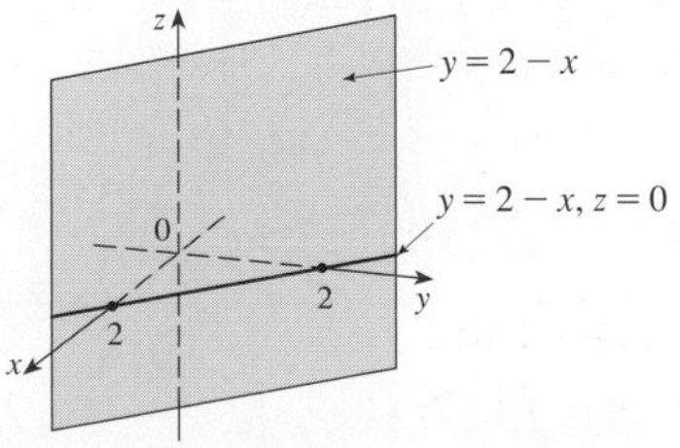

7. $|PQ| = 6$, $|QR| = 2\sqrt{10}$, $|RP| = 6$; isosceles triangle
9. (a) No (b) Yes
11. $(x-1)^2 + (y+4)^4 + (z-3)^2 = 25$;
$(x-1)^2 + (z-3)^2 = 9$, $y = 0$ (a circle)
13. $(x-3)^2 + (y-8)^2 + (z-1)^2 = 30$
15. $(3, -2, 1)$, 5
17. $(2, 0, -6)$, $9/\sqrt{2}$ **19.** (b) $\frac{5}{2}, \frac{1}{2}\sqrt{94}, \frac{1}{2}\sqrt{85}$
21. (a) $(x-2)^2 + (y+3)^2 + (z-6)^2 = 36$
(b) $(x-2)^2 + (y+3)^2 + (z-6)^2 = 4$
(c) $(x-2)^2 + (y+3)^2 + (z-6)^2 = 9$
23. A plane parallel to the xz-plane and 4 units to the left of it
25. A half-space consisting of all points in front of the plane $x = 3$
27. All points on or between the horizontal planes $z = 0$ and $z = 6$
29. All points on or inside a sphere with radius $\sqrt{3}$ and center O
31. All points on or inside a circular cylinder of radius 3 with axis the y-axis
33. $0 < x < 5$ **35.** $r^2 < x^2 + y^2 + z^2 < R^2$
37. (a) $(2, 1, 4)$ (b)

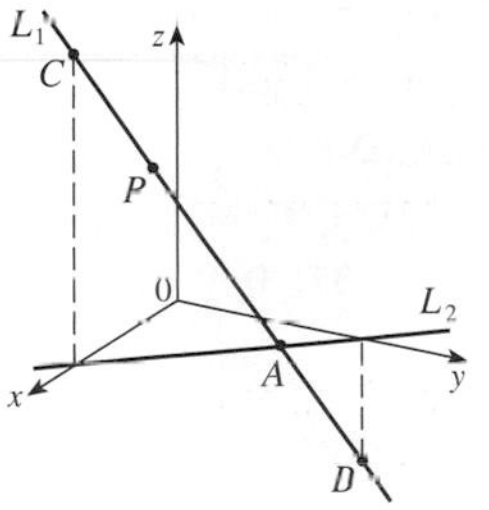

39. $14x - 6y - 10z = 9$, a plane perpendicular to AB

EXERCISES 12.2 ■ PAGE 777

1. (a) Scalar (b) Vector (c) Vector (d) Scalar
3. $\overrightarrow{AB} = \overrightarrow{DC}$, $\overrightarrow{DA} = \overrightarrow{CB}$, $\overrightarrow{DE} = \overrightarrow{EB}$, $\overrightarrow{EA} = \overrightarrow{CE}$
5.

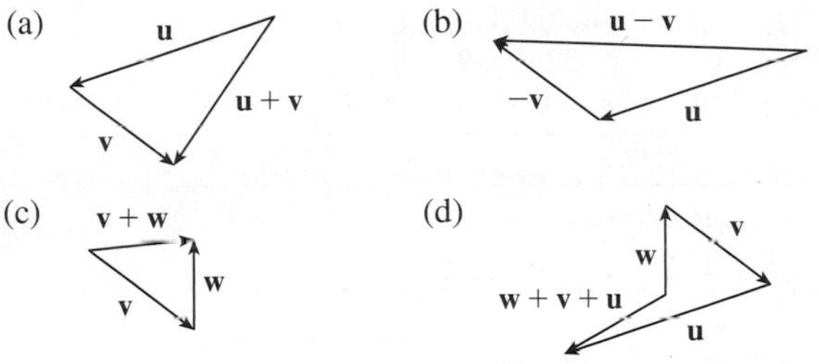

7. $\mathbf{a} = \langle -4, -2 \rangle$

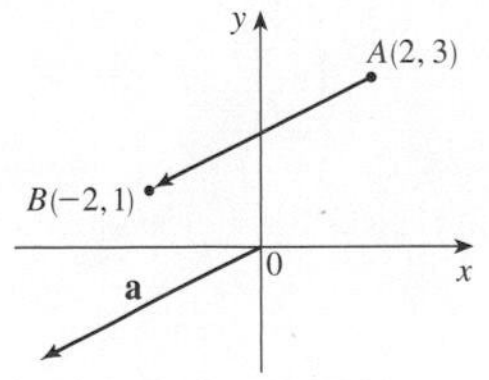

9. $\mathbf{a} = \langle 3, -1 \rangle$

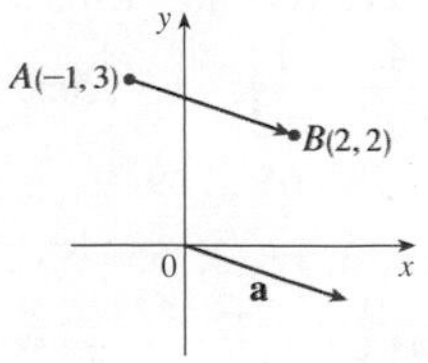

11. $\mathbf{a} = \langle 2, 0, -2 \rangle$

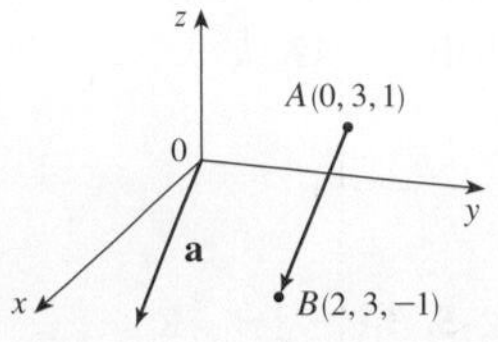

13. $\langle 5, 2 \rangle$

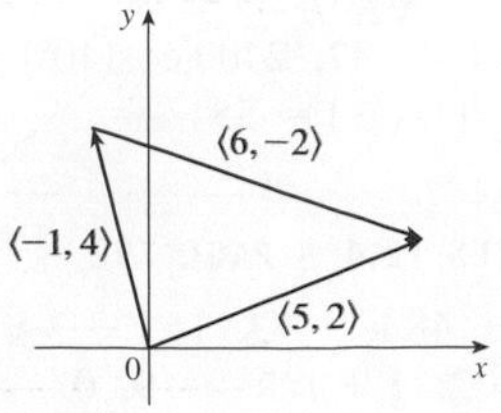

15. $\langle 0, 1, -1 \rangle$

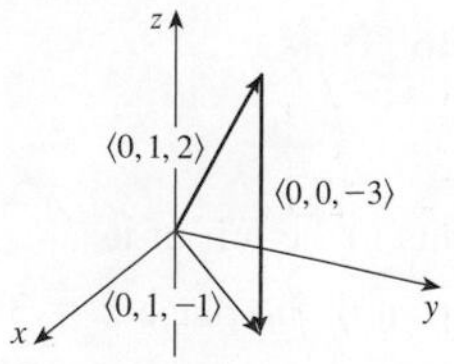

17. $\langle 2, -18 \rangle$, $\langle 1, -42 \rangle$, 13, 10
19. $-\mathbf{i} + \mathbf{j} + 2\mathbf{k}$, $-4\mathbf{i} + \mathbf{j} + 9\mathbf{k}$, $\sqrt{14}$, $\sqrt{82}$
21. $-\dfrac{3}{\sqrt{58}}\mathbf{i} + \dfrac{7}{\sqrt{58}}\mathbf{j}$ **23.** $\frac{8}{9}\mathbf{i} - \frac{1}{9}\mathbf{j} + \frac{4}{9}\mathbf{k}$
25. $\langle 2, 2\sqrt{3} \rangle$ **27.** ≈ 45.96 ft/s, ≈ 38.57 ft/s
29. $100\sqrt{7} \approx 264.6$ N, $\approx 139.1°$
31. $\sqrt{493} \approx 22.2$ mi/h, N8°W
33. $\mathbf{T}_1 \approx -196\,\mathbf{i} + 3.92\,\mathbf{j}$, $\mathbf{T}_2 \approx 196\,\mathbf{i} + 3.92\,\mathbf{j}$
35. $\pm(\mathbf{i} + 4\,\mathbf{j})/\sqrt{17}$ **37.** $\mathbf{0}$
39. (a), (b)

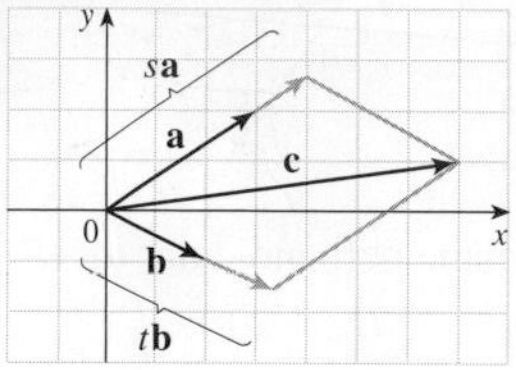

(d) $s = \frac{9}{7}$, $t = \frac{11}{7}$

41. A sphere with radius 1, centered at (x_0, y_0, z_0)

EXERCISES 12.3 ▪ PAGE 784

1. (b), (c), (d) are meaningful
3. 14 **5.** 19 **7.** 32 **9.** −15
11. $\mathbf{u} \cdot \mathbf{v} = \frac{1}{2}$, $\mathbf{u} \cdot \mathbf{w} = -\frac{1}{2}$
15. $\cos^{-1}\left(\dfrac{9 - 4\sqrt{7}}{20}\right) \approx 95°$ **17.** $\cos^{-1}\left(\dfrac{5}{\sqrt{1015}}\right) \approx 81°$
19. $\cos^{-1}\left(\dfrac{-1}{2\sqrt{7}}\right) \approx 101°$ **21.** 45°, 45°, 90°
23. (a) Neither (b) Orthogonal
(c) Orthogonal (d) Parallel
25. Yes **27.** $(\mathbf{i} - \mathbf{j} - \mathbf{k})/\sqrt{3}$ $[\text{or } (-\mathbf{i} + \mathbf{j} + \mathbf{k})/\sqrt{3}]$
29. $\dfrac{3}{5\sqrt{2}}, \dfrac{4}{5\sqrt{2}}, \dfrac{1}{\sqrt{2}}$; 65°, 56°, 45°
31. $\frac{2}{7}, \frac{3}{7}, -\frac{6}{7}$; 73°, 65°, 149°
33. $1/\sqrt{3}, 1/\sqrt{3}, 1/\sqrt{3}$; 55°, 55°, 55°
35. 3, $\langle \frac{9}{5}, -\frac{12}{5} \rangle$ **37.** $\frac{9}{7}$, $\langle \frac{27}{49}, \frac{54}{49}, -\frac{18}{49} \rangle$
39. $1/\sqrt{21}$, $\frac{2}{21}\,\mathbf{i} - \frac{1}{21}\,\mathbf{j} + \frac{4}{21}\,\mathbf{k}$
43. $\langle 0, 0, -2\sqrt{10} \rangle$ or any vector of the form $\langle s, t, 3s - 2\sqrt{10} \rangle$, $s, t \in \mathbb{R}$
45. 144 J **47.** $2400\cos(40°) \approx 1839$ ft-lb **49.** $\frac{13}{5}$
51. $\cos^{-1}(1/\sqrt{3}) \approx 55°$

EXERCISES 12.4 ▪ PAGE 792

1. $16\,\mathbf{i} + 48\,\mathbf{k}$ **3.** $15\,\mathbf{i} - 3\,\mathbf{j} + 3\,\mathbf{k}$ **5.** $\frac{1}{2}\mathbf{i} - \mathbf{j} + \frac{3}{2}\mathbf{k}$
7. $t^4\mathbf{i} - 2t^3\mathbf{j} + t^2\mathbf{k}$ **9.** $\mathbf{0}$ **11.** $\mathbf{i} + \mathbf{j} + \mathbf{k}$
13. (a) Scalar (b) Meaningless (c) Vector
(d) Meaningless (e) Meaningless (f) Scalar
15. 24; into the page **17.** $\langle 5, -3, 1 \rangle$, $\langle -5, 3, -1 \rangle$
19. $\langle -2/\sqrt{6}, -1/\sqrt{6}, 1/\sqrt{6} \rangle$, $\langle 2/\sqrt{6}, 1/\sqrt{6}, -1/\sqrt{6} \rangle$
27. 16 **29.** (a) $\langle 6, 3, 2 \rangle$ (b) $\frac{7}{2}$
31. (a) $\langle 13, -14, 5 \rangle$ (b) $\frac{1}{2}\sqrt{390}$
33. 82 **35.** 3 **39.** $10.8 \sin 80° \approx 10.6$ N·m
41. ≈ 417 N **43.** (b) $\sqrt{97/3}$
49. (a) No (b) No (c) Yes

EXERCISES 12.5 ▪ PAGE 802

1. (a) True (b) False (c) True (d) False (e) False
(f) True (g) False (h) True (i) True (j) False
(k) True
3. $\mathbf{r} = (2\,\mathbf{i} + 2.4\,\mathbf{j} + 3.5\,\mathbf{k}) + t(3\,\mathbf{i} + 2\,\mathbf{j} - \mathbf{k})$;
$x = 2 + 3t$, $y = 2.4 + 2t$, $z = 3.5 - t$
5. $\mathbf{r} = (\mathbf{i} + 6\mathbf{k}) + t(\mathbf{i} + 3\mathbf{j} + \mathbf{k})$;
$x = 1 + t$, $y = 3t$, $z = 6 + t$
7. $x = 1 - 5t$, $y = 3$, $z = 2 - 2t$; $\dfrac{x - 1}{-5} = \dfrac{z - 2}{-2}$, $y = 3$
9. $x = 2 + 2t$, $y = 1 + \frac{1}{2}t$,
$z = -3 - 4t$;
$(x - 2)/2 = 2y - 2 = (z + 3)/(-4)$
11. $x = 1 + t$, $y = -1 + 2t$, $z = 1 + t$;
$x - 1 = (y + 1)/2 = z - 1$
13. Yes
15. (a) $(x - 1)/(-1) = (y + 5)/2 = (z - 6)/(-3)$
(b) $(-1, -1, 0)$, $(-\frac{3}{2}, 0, -\frac{3}{2})$, $(0, -3, 3)$
17. $\mathbf{r}(t) = (2\mathbf{i} - \mathbf{j} + 4\mathbf{k}) + t(2\mathbf{i} + 7\mathbf{j} - 3\mathbf{k})$, $0 \leqslant t \leqslant 1$
19. Parallel **21.** Skew
23. $-2x + y + 5z = 1$ **25.** $x + y - z = -1$
27. $2x - y + 3z = 0$ **29.** $3x - 7z = -9$
31. $x + y + z = 2$ **33.** $-13x + 17y + 7z = -42$
35. $33x + 10y + 4z = 190$ **37.** $x - 2y + 4z = -1$
39.

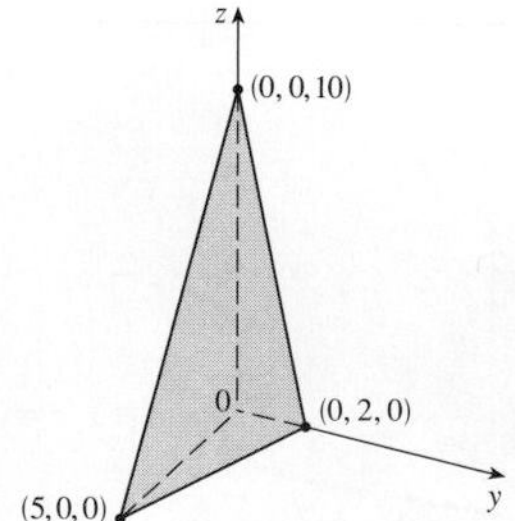

41.

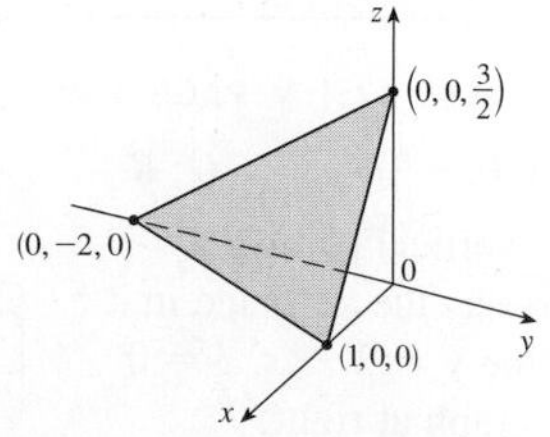

43. (2, 3, 5) **45.** (2, 3, 1) **47.** 1, 0, −1
49. Perpendicular **51.** Neither, $\approx 70.5°$ **53.** Parallel
55. (a) $x = 1$, $y = -t$, $z = t$ (b) $\cos^{-1}\left(\dfrac{5}{3\sqrt{3}}\right) \approx 15.8°$
57. $x = 1$, $y - 2 = -z$
59. $x + 2y + z = 5$ **61.** $(x/a) + (y/b) + (z/c) = 1$

63. $x = 3t$, $y = 1 - t$, $z = 2 - 2t$

65. P_1 and P_3 are parallel, P_2 and P_4 are identical

67. $\sqrt{61/14}$ **69.** $\frac{18}{7}$ **71.** $5/(2\sqrt{14})$ **75.** $1/\sqrt{6}$

EXERCISES 12.6 ▪ PAGE 810

1. (a) Parabola
(b) Parabolic cylinder with rulings parallel to the z-axis
(c) Parabolic cylinder with rulings parallel to the x-axis

3. Elliptic cylinder

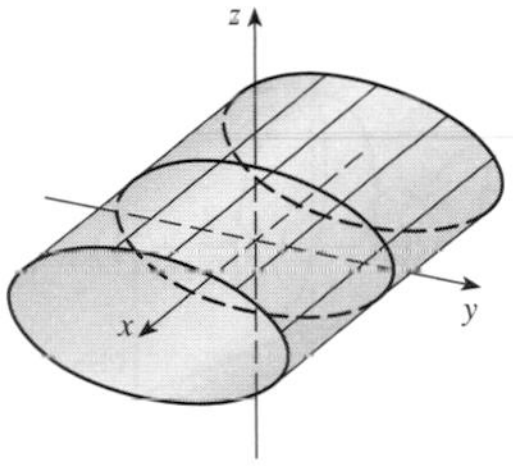

5. Parabolic cylinder

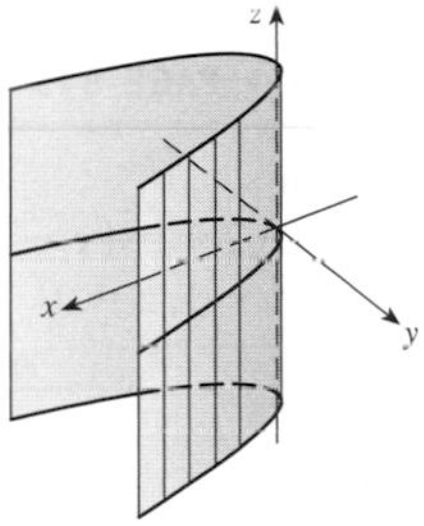

7. Cylindrical surface

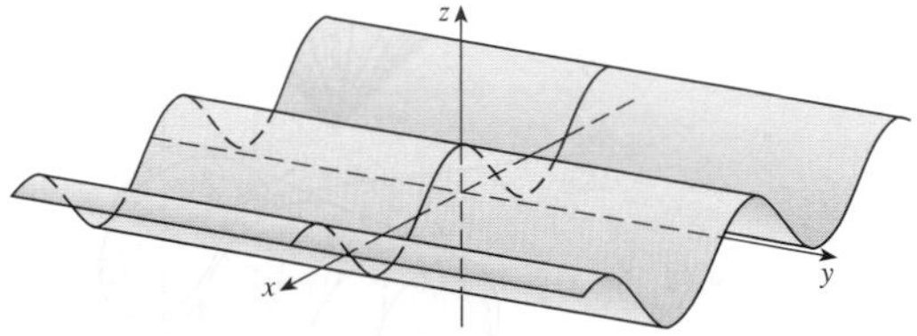

9. (a) $x = k$, $y^2 - z^2 = 1 - k^2$, hyperbola ($k \neq \pm 1$);
$y = k$, $x^2 - z^2 = 1 - k^2$, hyperbola ($k \neq \pm 1$);
$z = k$, $x^2 + z^2 = 1 + k^2$, circle
(b) The hyperboloid is rotated so that it has axis the y-axis
(c) The hyperboloid is shifted one unit in the negative y-direction

11. Elliptic paraboloid with axis the x-axis

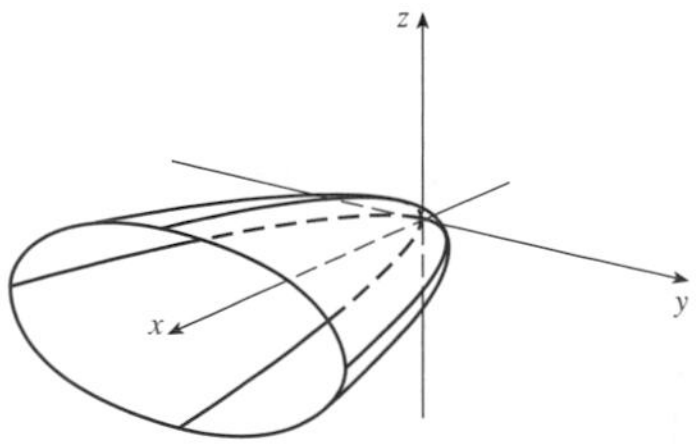

13. Elliptic cone with axis the x-axis

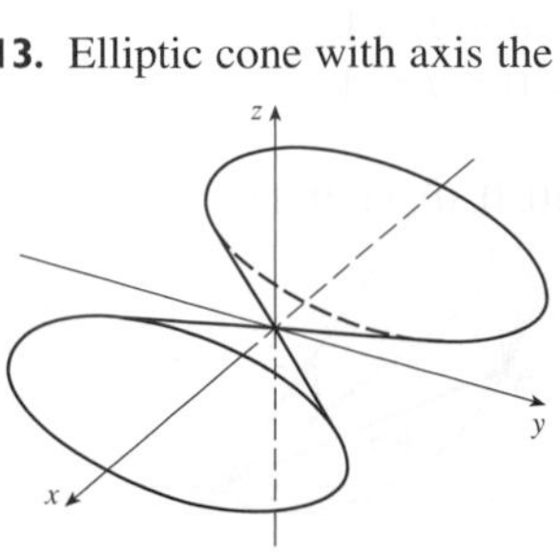

15. Hyperboloid of two sheets

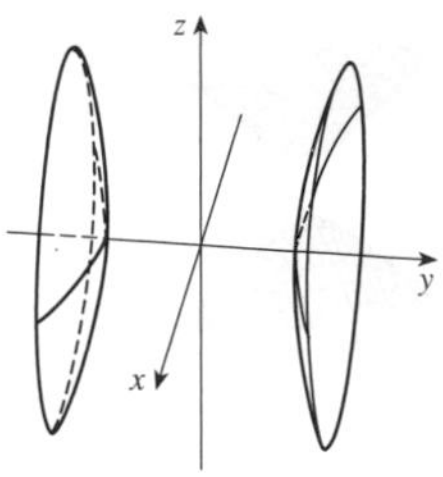

17. Ellipsoid

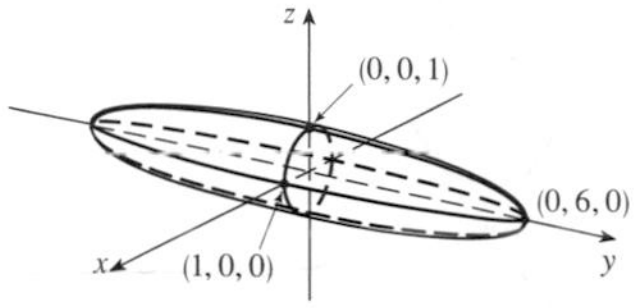

19. Hyperbolic paraboloid

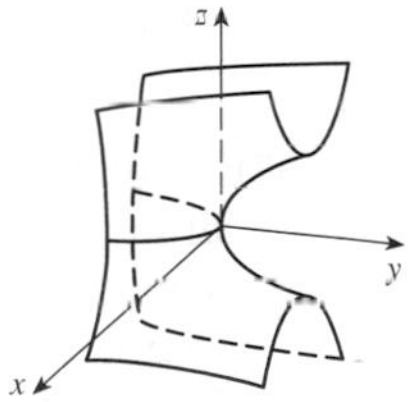

21. VII **23.** II **25.** VI **27.** VIII

29. $-\dfrac{x^2}{9} - \dfrac{y^2}{4} + \dfrac{z^2}{36} = 1$
Hyperboloid of two sheets with axis the z-axis

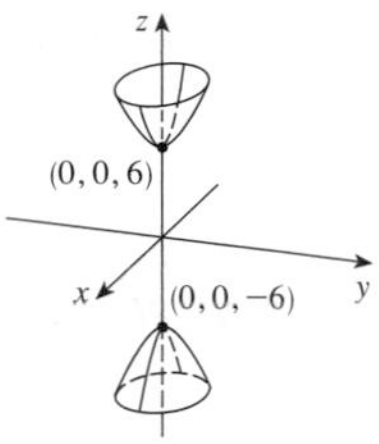

31. $\dfrac{x}{6} = \dfrac{y^2}{3} + \dfrac{z^2}{2}$

Elliptic paraboloid with vertex (0, 0, 0) and axis the x-axis

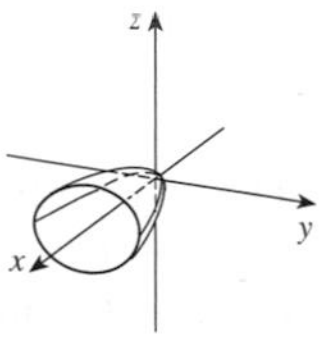

33. $x^2 + \dfrac{(y-2)^2}{4} + (z-3)^2 = 1$

Ellipsoid with center (0, 2, 3)

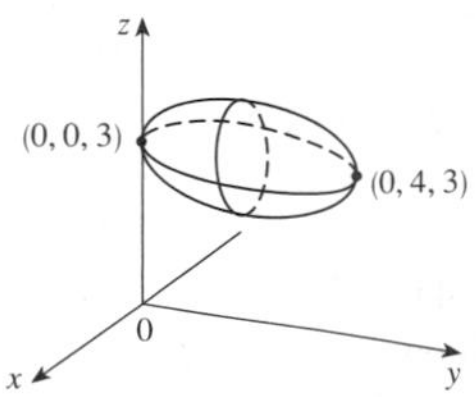

35. $(y + 1)^2 = (x - 2)^2 + (z - 1)^2$
Circular cone with vertex $(2, -1, 1)$ and axis parallel to the y-axis

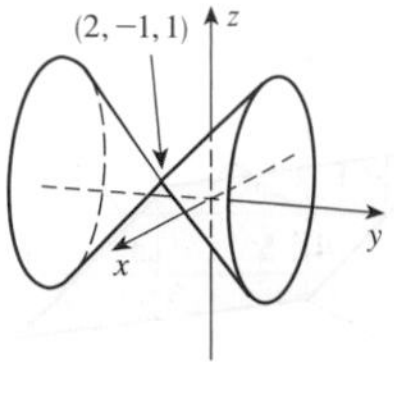

35.

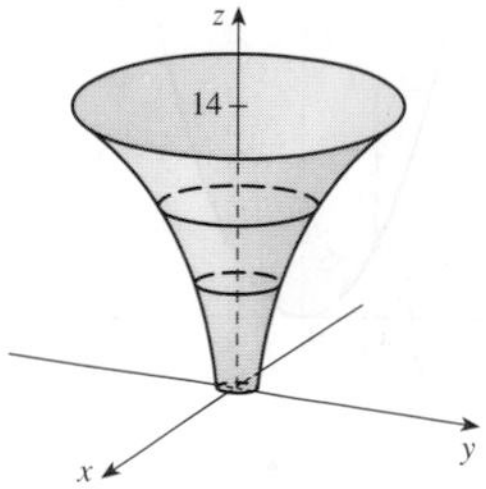

37.

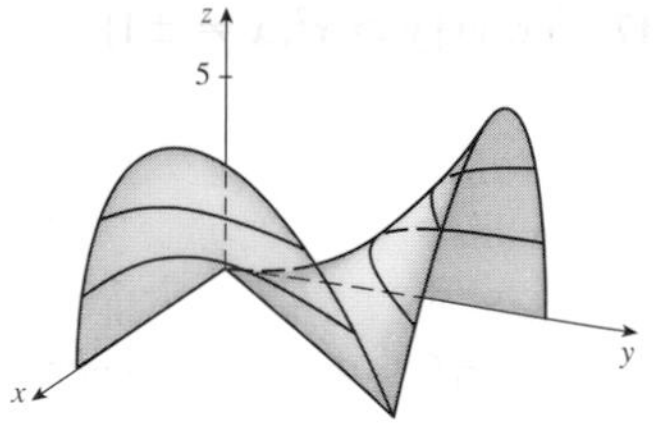

39. $(y - 2x)^2 = k$

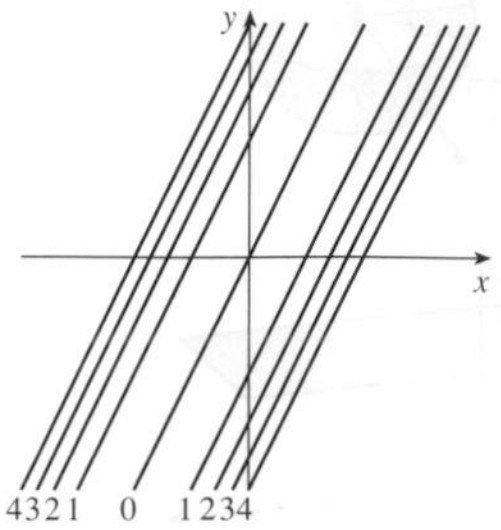

41. $y = \ln x + k$

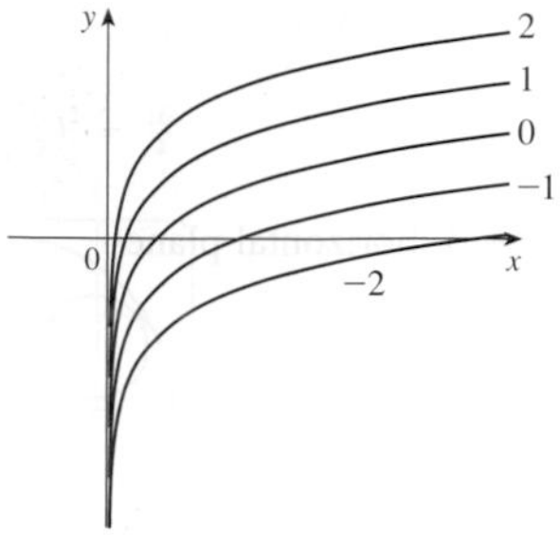

43. $y = ke^{-x}$

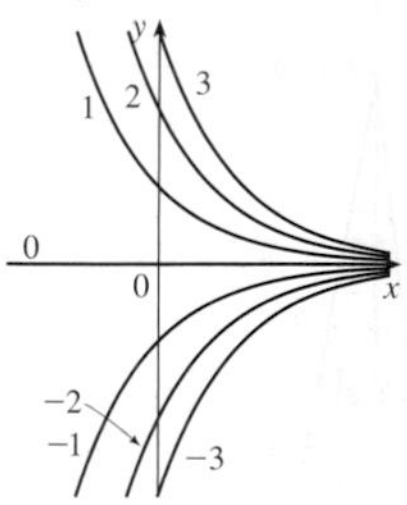

45. $y^2 - x^2 = k^2$

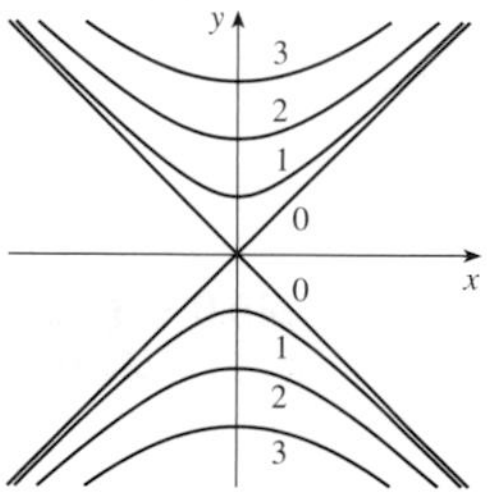

47. $x^2 + 9y^2 = k$

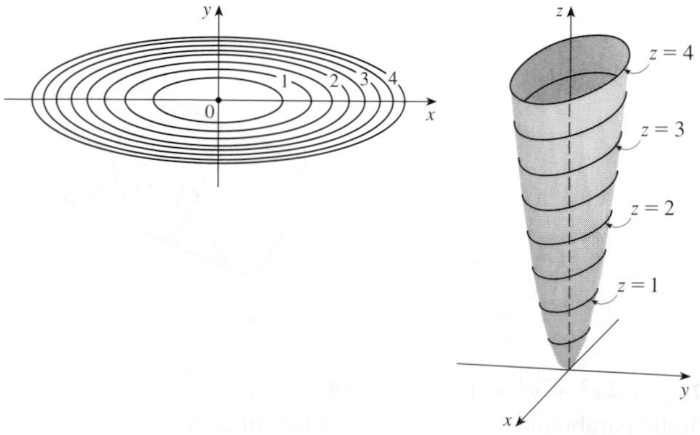

49.

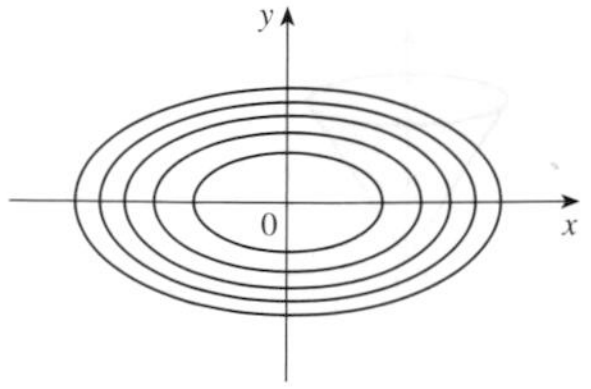

51.

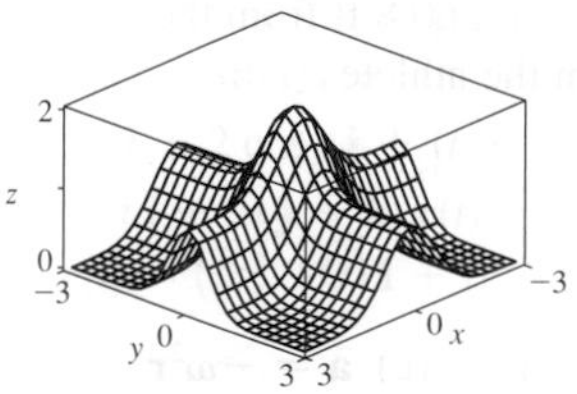

53.

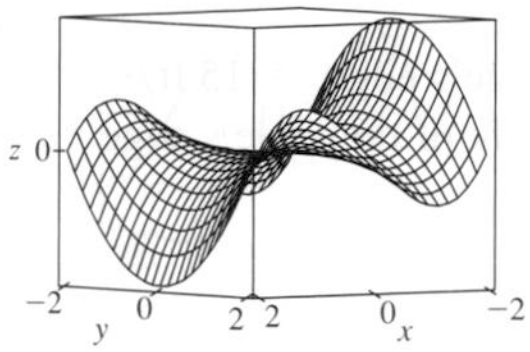

55. (a) C (b) II **57.** (a) F (b) I

59. (a) B (b) VI

61. Family of parallel planes

63. Family of hyperboloids of one or two sheets with axis the y-axis

65. (a) Shift the graph of f upward 2 units

(b) Stretch the graph of f vertically by a factor of 2

(c) Reflect the graph of f about the xy-plane

(d) Reflect the graph of f about the xy-plane and then shift it upward 2 units

67.

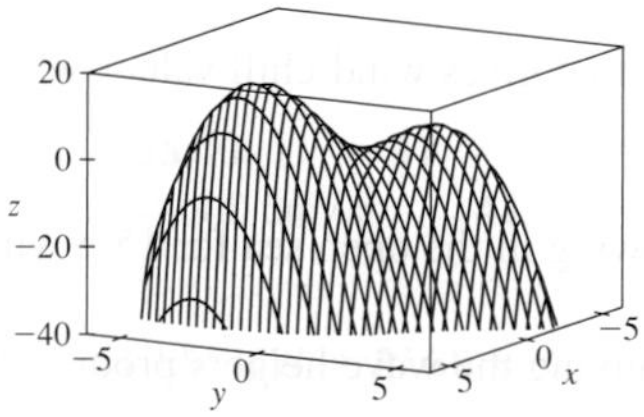

f appears to have a maximum value of about 15. There are two local maximum points but no local minimum point.

69.

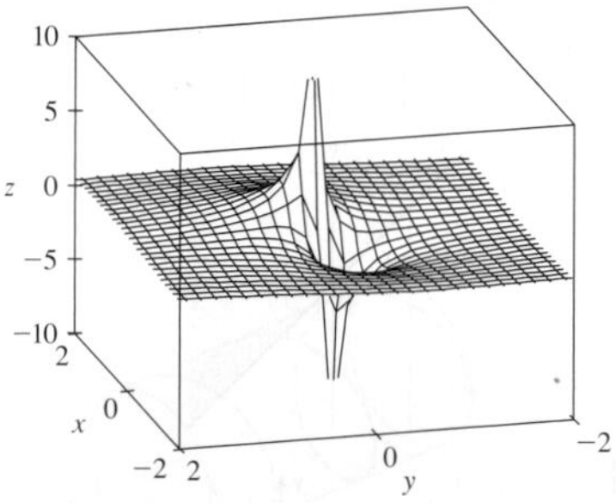

The function values approach 0 as x, y become large; as (x, y) approaches the origin, f approaches $\pm\infty$ or 0, depending on the direction of approach.

71. If $c = 0$, the graph is a cylindrical surface. For $c > 0$, the level curves are ellipses. The graph curves upward as we leave the origin, and the steepness increases as c increases. For $c < 0$, the level curves are hyperbolas. The graph curves upward in the y-direction and downward, approaching the xy-plane, in the x-direction giving a saddle-shaped appearance near $(0, 0, 1)$.

73. $c = -2, 0, 2$ **75.** (b) $y = 0.75x + 0.01$

EXERCISES 14.2 ▪ PAGE 877

1. Nothing; if f is continuous, $f(3, 1) = 6$ **3.** $-\frac{5}{2}$

5. 1 **7.** $\frac{2}{7}$ **9.** Does not exist **11.** Does not exist

13. 0 **15.** Does not exist **17.** 2 **19.** 1

21. Does not exist

23. The graph shows that the function approaches different numbers along different lines.

25. $h(x, y) = (2x + 3y - 6)^2 + \sqrt{2x + 3y - 6}$; $\{(x, y) \mid 2x + 3y \geq 6\}$

27. Along the line $y = x$ **29.** $\{(x, y) \mid y \neq \pm e^{x/2}\}$

31. $\{(x, y) \mid y \geq 0\}$ **33.** $\{(x, y) \mid x^2 + y^2 > 4\}$

35. $\{(x, y, z) \mid y \geq 0, y \neq \sqrt{x^2 + z^2}\}$

37. $\{(x, y) \mid (x, y) \neq (0, 0)\}$ **39.** 0 **41.** -1

43.

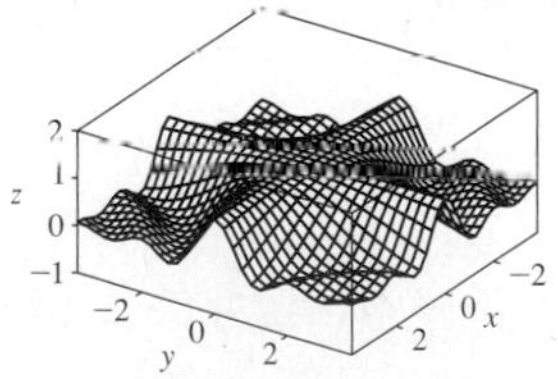

f is continuous on $\mathbb{R}^2$

EXERCISES 14.3 ▪ PAGE 888

1. (a) The rate of change of temperature as longitude varies, with latitude and time fixed; the rate of change as only latitude varies; the rate of change as only time varies.
(b) Positive, negative, positive

3. (a) $f_T(-15, 30) \approx 1.3$; for a temperature of $-15°C$ and wind speed of 30 km/h, the wind-chill index rises by 1.3°C for each degree the temperature increases. $f_v(-15, 30) \approx -0.15$; for a temperature of $-15°C$ and wind speed of 30 km/h, the wind-chill index decreases by 0.15°C for each km/h the wind speed increases.
(b) Positive, negative (c) 0

5. (a) Positive (b) Negative

7. (a) Positive (b) Negative

9. $c = f, b = f_x, a = f_y$

11. $f_x(1, 2) = -8 =$ slope of C_1, $f_y(1, 2) = -4 =$ slope of C_2

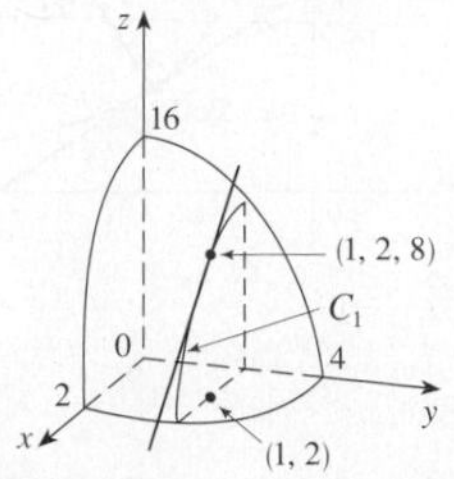

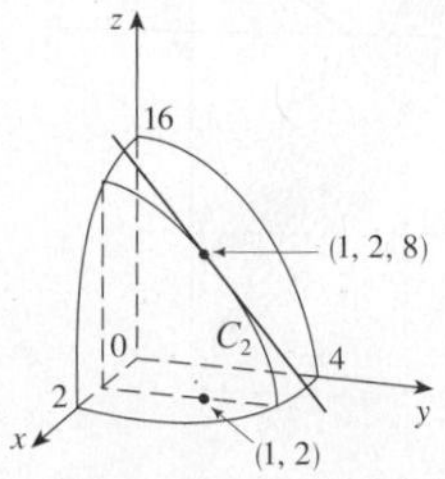

13. $f_x = 2x + 2xy$, $f_y = 2y + x^2$

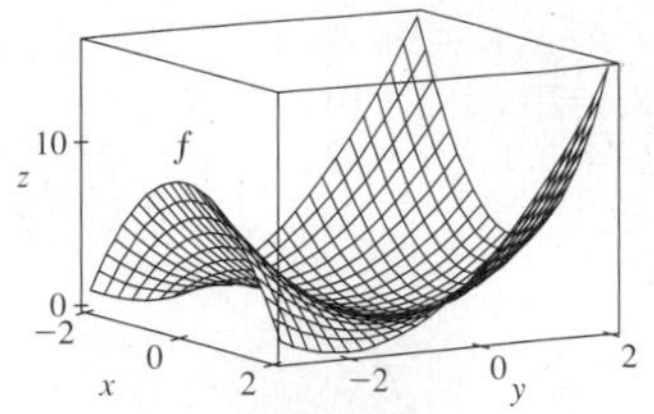

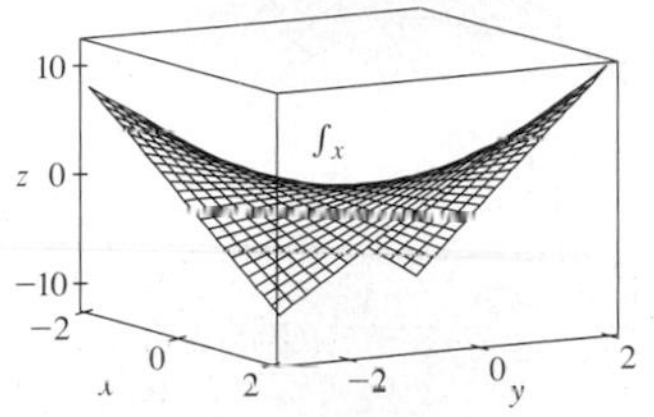

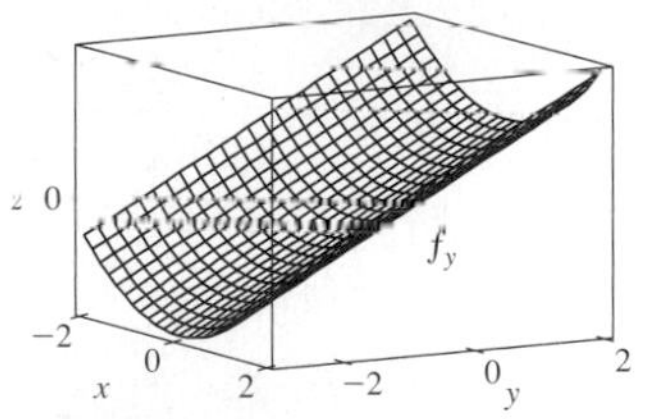

15. $f_x(x, y) = -3y, f_y(x, y) = 5y^4 - 3x$

17. $f_x(x, t) = -\pi e^{-t} \sin \pi x, f_t(x, t) = -e^{-t} \cos \pi x$

19. $\partial z/\partial x = 20(2x + 3y)^9, \partial z/\partial y = 30(2x + 3y)^9$

21. $f_x(x, y) = 2y/(x + y)^2, f_y(x, y) = -2x/(x + y)^2$

23. $\partial w/\partial \alpha = \cos \alpha \cos \beta, \partial w/\partial \beta = -\sin \alpha \sin \beta$

25. $f_r(r, s) = \dfrac{2r^2}{r^2 + s^2} + \ln(r^2 + s^2), f_s(r, s) = \dfrac{2rs}{r^2 + s^2}$

27. $\partial u/\partial t = e^{w/t}(1 - w/t), \partial u/\partial w = e^{w/t}$

29. $f_x = z - 10xy^3z^4, f_y = -15x^2y^2z^4, f_z = x - 20x^2y^3z^3$

31. $\partial w/\partial x = 1/(x + 2y + 3z), \partial w/\partial y = 2/(x + 2y + 3z),$
$\partial w/\partial z = 3/(x + 2y + 3z)$

33. $\partial u/\partial x = y \sin^{-1}(yz), \partial u/\partial y = x \sin^{-1}(yz) + xyz/\sqrt{1 - y^2z^2},$
$\partial u/\partial z = xy^2/\sqrt{1 - y^2z^2}$

35. $f_x = yz^2 \tan(yt), f_y = xyz^2t \sec^2(yt) + xz^2 \tan(yt),$
$f_z = 2xyz \tan(yt), f_t = xy^2z^2 \sec^2(yt)$

37. $\partial u/\partial x_i = x_i/\sqrt{x_1^2 + x_2^2 + \cdots + x_n^2}$

39. $\frac{1}{5}$ **41.** $\frac{1}{4}$

43. $f_x(x, y) = y^2 - 3x^2y, f_y(x, y) = 2xy - x^3$

45. $\dfrac{\partial z}{\partial x} = \dfrac{3yz - 2x}{2z - 3xy}, \dfrac{\partial z}{\partial y} = \dfrac{3xz - 2y}{2z - 3xy}$

47. $\dfrac{\partial z}{\partial x} = \dfrac{1 + y^2z^2}{1 + y + y^2z^2}, \dfrac{\partial z}{\partial y} = \dfrac{-z}{1 + y + y^2z^2}$

49. (a) $f'(x), g'(y)$ (b) $f'(x + y), f'(x + y)$

51. $f_{xx} = 6xy^5 + 24x^2y, f_{xy} = 15x^2y^4 + 8x^3 = f_{yx}, f_{yy} = 20x^3y^3$

53. $w_{uu} = v^2/(u^2 + v^2)^{3/2}, w_{uv} = -uv/(u^2 + v^2)^{3/2} = w_{vu},$
$w_{vv} = u^2/(u^2 + v^2)^{3/2}$

55. $z_{xx} = -2x/(1 + x^2)^2, z_{xy} = 0 = z_{yx}, z_{yy} = -2y/(1 + y^2)^2$

61. $12xy$, $72xy$
63. $24\sin(4x + 3y + 2z)$, $12\sin(4x + 3y + 2z)$
65. $\theta e^{r\theta}(2\sin\theta + \theta\cos\theta + r\theta\sin\theta)$ **67.** $4/(y + 2z)^3$, 0
69. ≈ 12.2, ≈ 16.8, ≈ 23.25 **81.** R^2/R_1^2
87. No **89.** $x = 1 + t$, $y = 2$, $z = 2 - 2t$
93. -2
95. (a)

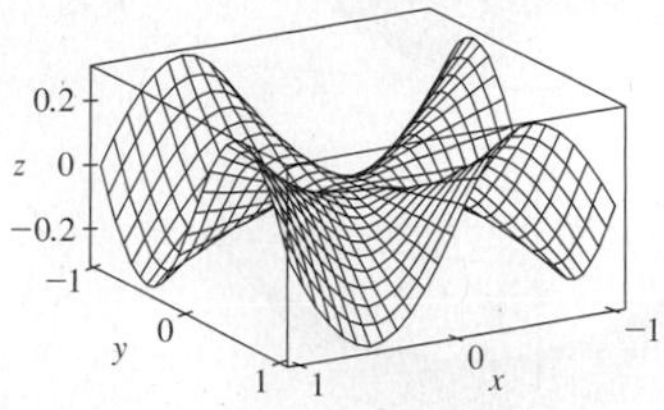

(b) $f_x(x, y) = \dfrac{x^4y + 4x^2y^3 - y^5}{(x^2 + y^2)^2}$, $f_y(x, y) = \dfrac{x^5 - 4x^3y^2 - xy^4}{(x^2 + y^2)^2}$
(c) 0, 0 (e) No, since f_{xy} and f_{yx} are not continuous.

EXERCISES 14.4 ▪ PAGE 899

1. $z = -8x - 2y$
3. $x + y - 2z = 0$
5. $z = y$
7.

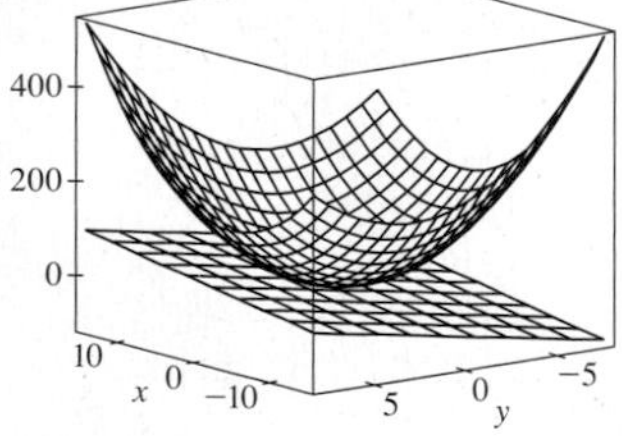

9.

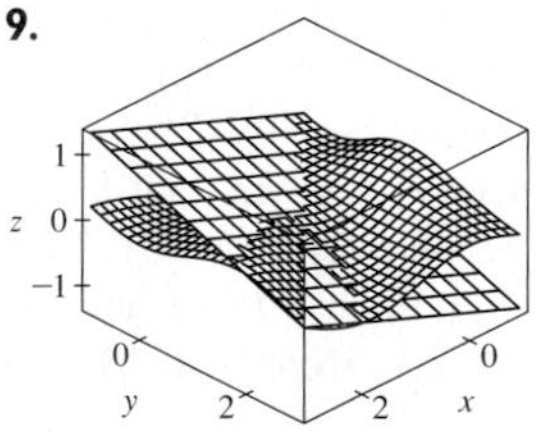

11. $2x + \frac{1}{4}y - 1$ **13.** $\frac{1}{9}x - \frac{2}{9}y + \frac{2}{3}$ **15.** $1 - \pi y$
19. $-\frac{2}{3}x - \frac{7}{3}y + \frac{20}{3}$; $2.8\overline{46}$ **21.** $\frac{3}{7}x + \frac{2}{7}y + \frac{6}{7}z$; 6.9914
23. $4T + H - 329$; 129°F
25. $dz = 3x^2\ln(y^2)\,dx + (2x^3/y)\,dy$
27. $dm = 5p^4q^3\,dp + 3p^5q^2\,dq$
29. $dR = \beta^2\cos\gamma\,d\alpha + 2\alpha\beta\cos\gamma\,d\beta - \alpha\beta^2\sin\gamma\,d\gamma$
31. $\Delta z = 0.9225$, $dz = 0.9$ **33.** 5.4 cm^2 **35.** 16 cm^3
37. 150 **39.** $\frac{1}{17} \approx 0.059\ \Omega$ **41.** 2.3%
43. $\varepsilon_1 = \Delta x$, $\varepsilon_2 = \Delta y$

EXERCISES 14.5 ▪ PAGE 907

1. $(2x + y)\cos t + (2y + x)e^t$
3. $[(x/t) - y\sin t]/\sqrt{1 + x^2 + y^2}$
5. $e^{y/z}[2t - (x/z) - (2xy/z^2)]$
7. $\partial z/\partial s = 2xy^3\cos t + 3x^2y^2\sin t$,
$\partial z/\partial t = -2sxy^3\sin t + 3sx^2y^2\cos t$
9. $\partial z/\partial s = t^2\cos\theta\cos\phi - 2st\sin\theta\sin\phi$,
$\partial z/\partial t = 2st\cos\theta\cos\phi - s^2\sin\theta\sin\phi$

11. $\dfrac{\partial z}{\partial s} = e^r\left(t\cos\theta - \dfrac{s}{\sqrt{s^2 + t^2}}\sin\theta\right)$,
$\dfrac{\partial z}{\partial t} = e^r\left(s\cos\theta - \dfrac{t}{\sqrt{s^2 + t^2}}\sin\theta\right)$

13. 62 **15.** 7, 2
17. $\dfrac{\partial u}{\partial r} = \dfrac{\partial u}{\partial x}\dfrac{\partial x}{\partial r} + \dfrac{\partial u}{\partial y}\dfrac{\partial y}{\partial r}$, $\dfrac{\partial u}{\partial s} = \dfrac{\partial u}{\partial x}\dfrac{\partial x}{\partial s} + \dfrac{\partial u}{\partial y}\dfrac{\partial y}{\partial s}$,
$\dfrac{\partial u}{\partial t} = \dfrac{\partial u}{\partial x}\dfrac{\partial x}{\partial t} + \dfrac{\partial u}{\partial y}\dfrac{\partial y}{\partial t}$
19. $\dfrac{\partial w}{\partial x} = \dfrac{\partial w}{\partial r}\dfrac{\partial r}{\partial x} + \dfrac{\partial w}{\partial s}\dfrac{\partial s}{\partial x} + \dfrac{\partial w}{\partial t}\dfrac{\partial t}{\partial x}$,
$\dfrac{\partial w}{\partial y} = \dfrac{\partial w}{\partial r}\dfrac{\partial r}{\partial y} + \dfrac{\partial w}{\partial s}\dfrac{\partial s}{\partial y} + \dfrac{\partial w}{\partial t}\dfrac{\partial t}{\partial y}$
21. 85, 178, 54 **23.** $\frac{9}{7}, \frac{9}{7}$ **25.** 36, 24, 30
27. $\dfrac{4(xy)^{3/2} - y}{x - 2x^2\sqrt{xy}}$ **29.** $\dfrac{\sin(x - y) + e^y}{\sin(x - y) - xe^y}$
31. $\dfrac{3yz - 2x}{2z - 3xy}, \dfrac{3xz - 2y}{2z - 3xy}$
33. $\dfrac{1 + y^2z^2}{1 + y + y^2z^2}, -\dfrac{z}{1 + y + y^2z^2}$
35. 2°C/s **37.** ≈ -0.33 m/s per minute
39. (a) 6 m^3/s (b) 10 m^2/s (c) 0 m/s
41. ≈ -0.27 L/s **43.** $-1/(12\sqrt{3})$ rad/s
45. (a) $\partial z/\partial r = (\partial z/\partial x)\cos\theta + (\partial z/\partial y)\sin\theta$,
$\partial z/\partial\theta = -(\partial z/\partial x)r\sin\theta + (\partial z/\partial y)r\cos\theta$
51. $4rs\,\partial^2z/\partial x^2 + (4r^2 + 4s^2)\partial^2z/\partial x\,\partial y + 4rs\,\partial^2z/\partial y^2 + 2\,\partial z/\partial y$

EXERCISES 14.6 ▪ PAGE 920

1. ≈ -0.08 mb/km **3.** ≈ 0.778 **5.** $2 + \sqrt{3}/2$
7. (a) $\nabla f(x, y) = \langle 2\cos(2x + 3y), 3\cos(2x + 3y)\rangle$
(b) $\langle 2, 3\rangle$ (c) $\sqrt{3} - \frac{3}{2}$
9. (a) $\langle e^{2yz}, 2xze^{2yz}, 2xye^{2yz}\rangle$ (b) $\langle 1, 12, 0\rangle$ (c) $-\frac{22}{3}$
11. 23/10 **13.** $-8/\sqrt{10}$ **15.** $4/\sqrt{30}$ **17.** $9/(2\sqrt{5})$
19. 2/5 **21.** $4\sqrt{2}$, $\langle -1, 1\rangle$ **23.** 1, $\langle 0, 1\rangle$
25. 1, $\langle 3, 6, -2\rangle$ **27.** (b) $\langle -12, 92\rangle$
29. All points on the line $y = x + 1$
31. (a) $-40/(3\sqrt{3})$
33. (a) $32/\sqrt{3}$ (b) $\langle 38, 6, 12\rangle$ (c) $2\sqrt{406}$ **35.** $\frac{327}{13}$
39. (a) $x + y + z = 11$ (b) $x - 3 = y - 3 = z - 5$
41. (a) $4x - 5y - z = 4$ (b) $\dfrac{x - 2}{4} = \dfrac{y - 1}{-5} = \dfrac{z + 1}{-1}$
43. (a) $x + y - z = 1$ (b) $x - 1 = y = -z$
45.

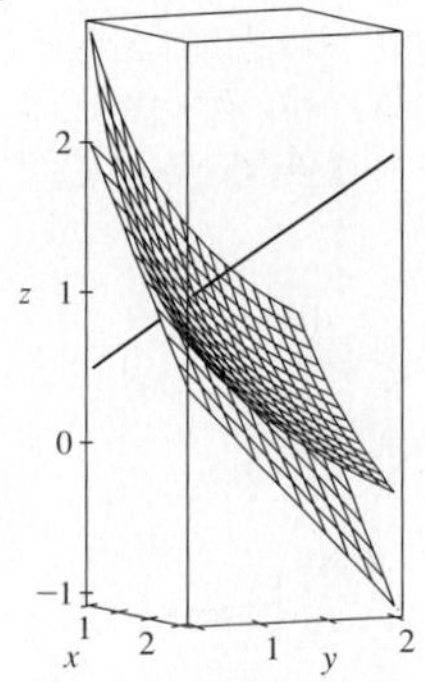

47. $\langle 2, 3\rangle$, $2x + 3y = 12$

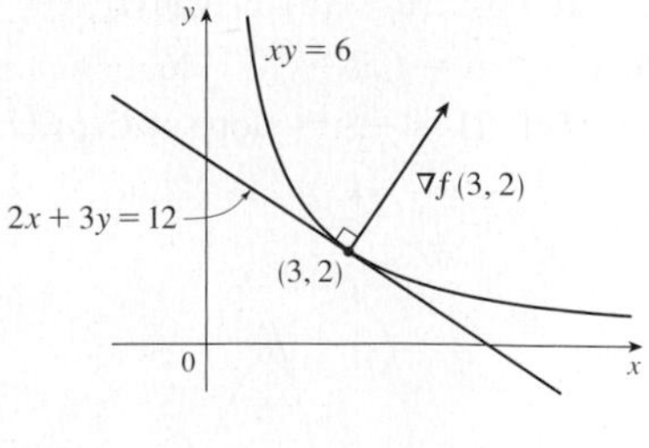

53. No **59.** $x = -1 - 10t$, $y = 1 - 16t$, $z = 2 - 12t$
63. If $\mathbf{u} = \langle a, b\rangle$ and $\mathbf{v} = \langle c, d\rangle$, then $af_x + bf_y$ and $cf_x + df_y$ are known, so we solve linear equations for f_x and f_y.

EXERCISES 14.7 ▪ PAGE 930

1. (a) f has a local minimum at (1, 1).
(b) f has a saddle point at (1, 1).
3. Local minimum at (1, 1), saddle point at (0, 0)
5. Maximum $f(-1, \frac{1}{2}) = 11$
7. Minima $f(1, 1) = 0$, $f(-1, -1) = 0$, saddle point at (0, 0)
9. Saddle points at $(1, -1)$, $(-1, 1)$
11. Minimum $f(2, 1) = -8$, saddle point at (0, 0)
13. None **15.** Minimum $f(0, 0) = 0$, saddle points at $(\pm 1, 0)$
17. Minima $f(0, 1) = f(\pi, -1) = f(2\pi, 1) = -1$,
saddle points at $(\pi/2, 0)$, $(3\pi/2, 0)$
21. Minima $f(1, \pm 1) = 3$, $f(-1, \pm 1) = 3$
23. Maximum $f(\pi/3, \pi/3) = 3\sqrt{3}/2$,
minimum $f(5\pi/3, 5\pi/3) = -3\sqrt{3}/2$, saddle point at (π, π)
25. Minima $f(-1.714, 0) \approx -9.200$, $f(1.402, 0) \approx 0.242$,
saddle point (0.312, 0), lowest point $(-1.714, 0, -9.200)$
27. Maxima $f(-1.267, 0) \approx 1.310$, $f(1.629, \pm 1.063) \approx 8.105$,
saddle points $(-0.259, 0)$, $(1.526, 0)$,
highest points $(1.629, \pm 1.063, 8.105)$
29. Maximum $f(2, 0) = 9$, minimum $f(0, 3) = -14$
31. Maximum $f(\pm 1, 1) = 7$, minimum $f(0, 0) = 4$
33. Maximum $f(3, 0) = 83$, minimum $f(1, 1) = 0$
35. Maximum $f(1, 0) = 2$, minimum $f(-1, 0) = -2$
37.

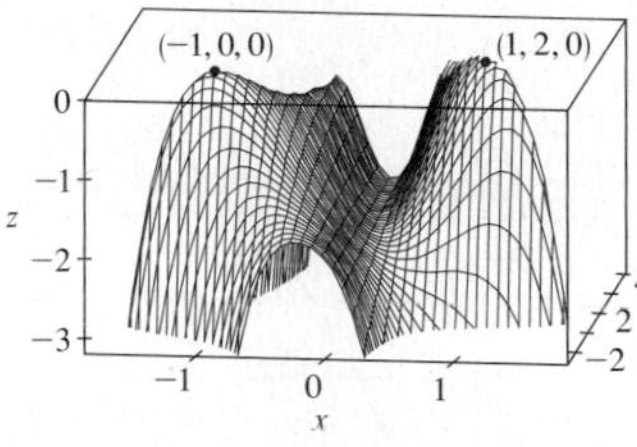

39. $\sqrt{3}$ **41.** $(2, 1, \sqrt{5})$, $(2, 1, -\sqrt{5})$ **43.** $\frac{100}{3}, \frac{100}{3}, \frac{100}{3}$
45. $8r^3/(3\sqrt{3})$
47. $\frac{4}{3}$ **49.** Cube, edge length $c/12$
51. Square base of side 40 cm, height 20 cm **53.** $L^3/(3\sqrt{3})$

EXERCISES 14.8 ▪ PAGE 940

1. $\approx 59, 30$
3. No maximum, minima $f(1, 1) = f(-1, -1) = 2$
5. Maxima $f(\pm 2, 1) = 4$, minima $f(\pm 2, -1) = -4$
7. Maximum $f(1, 3, 5) = 70$, minimum $f(-1, -3, -5) = -70$
9. Maximum $2/\sqrt{3}$, minimum $-2/\sqrt{3}$
11. Maximum $\sqrt{3}$, minimum 1
13. Maximum $f(\frac{1}{2}, \frac{1}{2}, \frac{1}{2}, \frac{1}{2}) = 2$,
minimum $f(-\frac{1}{2}, -\frac{1}{2}, -\frac{1}{2}, -\frac{1}{2}) = -2$
15. Maximum $f(1, \sqrt{2}, -\sqrt{2}) = 1 + 2\sqrt{2}$,
minimum $f(1, -\sqrt{2}, \sqrt{2}) = 1 - 2\sqrt{2}$
17. Maximum $\frac{3}{2}$, minimum $\frac{1}{2}$
19. Maxima $f(\pm 1/\sqrt{2}, \mp 1/(2\sqrt{2})) = e^{1/4}$,
minima $f(\pm 1/\sqrt{2}, \pm 1/(2\sqrt{2})) = e^{-1/4}$
27–37. See Exercises 39–49 in Section 14.7.
39. $L^3/(3\sqrt{3})$
41. Nearest $(\frac{1}{2}, \frac{1}{2}, \frac{1}{2})$, farthest $(-1, -1, 2)$
43. Maximum ≈ 9.7938, minimum ≈ -5.3506
45. (a) c/n (b) When $x_1 = x_2 = \cdots = x_n$

CHAPTER 14 REVIEW ▪ PAGE 944

True-False Quiz

1. True **3.** False **5.** False **7.** True **9.** False
11. True

Exercises

1. $\{(x, y) \mid y > -x - 1\}$

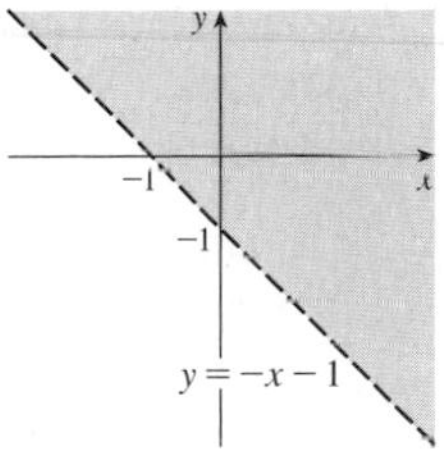

3.

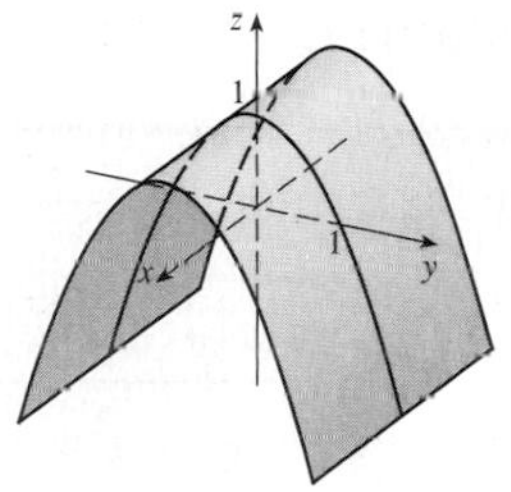

5.

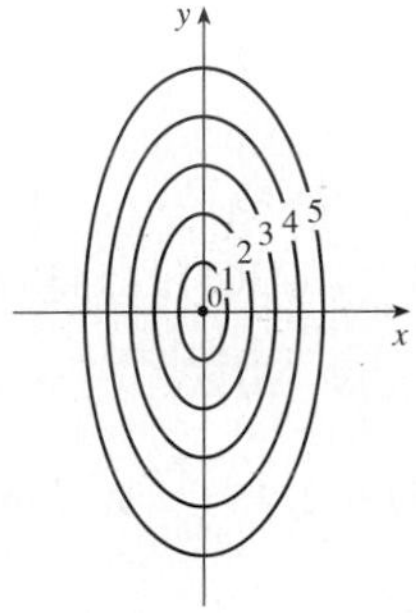

7.

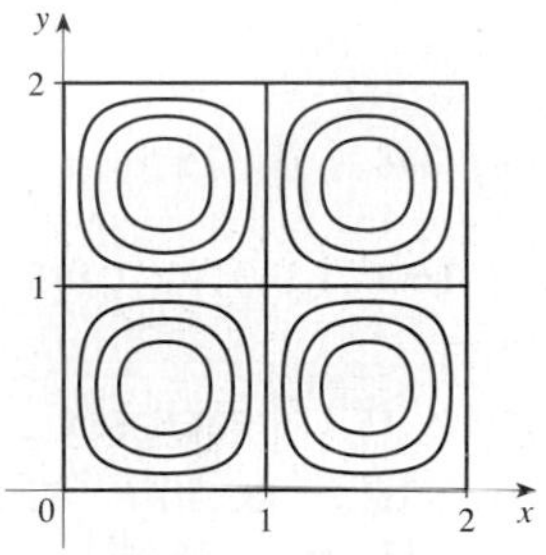

9. $\frac{2}{3}$
11. (a) $\approx 3.5°C/m$, $-3.0°C/m$ (b) $\approx 0.35°C/m$ by Equation 14.6.9 (Definition 14.6.2 gives $\approx 1.1°C/m$.)
(c) -0.25
13. $f_x = 1/\sqrt{2x + y^2}$, $f_y = y/\sqrt{2x + y^2}$
15. $g_u = \tan^{-1} v$, $g_v = u/(1 + v^2)$
17. $T_p = \ln(q + e^r)$, $T_q = p/(q + e^r)$, $T_r = pe^r/(q + e^r)$
19. $f_{xx} = 24x$, $f_{xy} = -2y = f_{yx}$, $f_{yy} = -2x$
21. $f_{xx} = k(k - 1)x^{k-2}y^l z^m$, $f_{xy} = klx^{k-1}y^{l-1}z^m = f_{yx}$,
$f_{xz} = kmx^{k-1}y^l z^{m-1} = f_{zx}$, $f_{yy} = l(l - 1)x^k y^{l-2} z^m$,
$f_{yz} = lmx^k y^{l-1} z^{m-1} = f_{zy}$, $f_{zz} = m(m - 1)x^k y^l z^{m-2}$
25. (a) $z = 8x + 4y + 1$ (b) $\dfrac{x - 1}{8} = \dfrac{y + 2}{4} = 1 - z$
27. (a) $2x - 2y - 3z = 3$ (b) $\dfrac{x - 2}{4} = \dfrac{y + 1}{-4} = \dfrac{z - 1}{-6}$
29. (a) $4x - y - 2z = 6$
(b) $x = 3 + 8t$, $y = 4 - 2t$, $z = 1 - 4t$
31. $(2, \frac{1}{2}, -1)$, $(-2, -\frac{1}{2}, 1)$
33. $60x + \frac{24}{5}y + \frac{32}{5}z - 120$; 38.656
35. $2xy^3(1 + 6p) + 3x^2y^2(pe^p + e^p) + 4z^3(p\cos p + \sin p)$
37. $-47, 108$ **43.** $ze^{x\sqrt{y}}\langle z\sqrt{y}, xz/(2\sqrt{y}), 2\rangle$ **45.** $\frac{43}{5}$
47. $\sqrt{145}/2$, $\langle 4, \frac{9}{2}\rangle$ **49.** $\approx \frac{5}{8}$ knot/mi

51. Minimum $f(-4, 1) = -11$
53. Maximum $f(1, 1) = 1$; saddle points $(0, 0)$, $(0, 3)$, $(3, 0)$
55. Maximum $f(1, 2) = 4$, minimum $f(2, 4) = -64$
57. Maximum $f(-1, 0) = 2$, minima $f(1, \pm 1) = -3$, saddle points $(-1, \pm 1)$, $(1, 0)$
59. Maximum $f(\pm\sqrt{2/3}, 1/\sqrt{3}) = 2/(3\sqrt{3})$, minimum $f(\pm\sqrt{2/3}, -1/\sqrt{3}) = -2/(3\sqrt{3})$
61. Maximum 1, minimum -1
63. $(\pm 3^{-1/4}, 3^{-1/4}\sqrt{2}, \pm 3^{1/4})$, $(\pm 3^{-1/4}, -3^{-1/4}\sqrt{2}, \pm 3^{1/4})$
65. $P(2 - \sqrt{3})$, $P(3 - \sqrt{3})/6$, $P(2\sqrt{3} - 3)/3$

PROBLEMS PLUS ■ PAGE 948

1. $L^2W^2, \frac{1}{4}L^2W^2$ **3.** (a) $x = w/3$, base $= w/3$ (b) Yes
7. $\sqrt{6}/2, 3\sqrt{2}/2$

CHAPTER 15

EXERCISES 15.1 ■ PAGE 958

1. (a) 288 (b) 144
3. (a) $\pi^2/2 \approx 4.935$ (b) 0
5. (a) -6 (b) -3.5
7. $U < V < L$
9. (a) ≈ 248 (b) 15.5
11. 60 **13.** 3
15. 1.141606, 1.143191, 1.143535, 1.143617, 1.143637, 1.143642

EXERCISES 15.2 ■ PAGE 964

1. $500y^3, 3x^2$ **3.** 10 **5.** 2 **7.** 261,632/45 **9.** $\frac{21}{2}\ln 2$
11. 0 **13.** π **15.** $\frac{21}{2}$ **17.** $9 \ln 2$
19. $\frac{1}{2}(\sqrt{3} - 1) - \frac{1}{12}\pi$ **21.** $\frac{1}{2}(e^2 - 3)$
23.

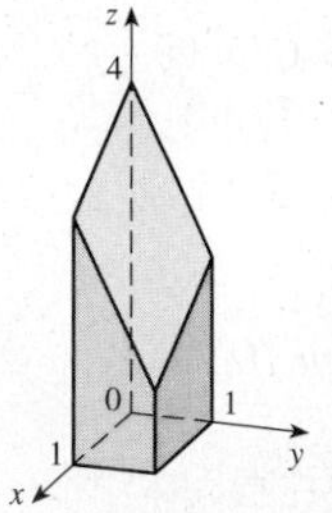

25. 47.5 **27.** $\frac{166}{27}$ **29.** 2 **31.** $\frac{64}{3}$
33. $21e - 57$

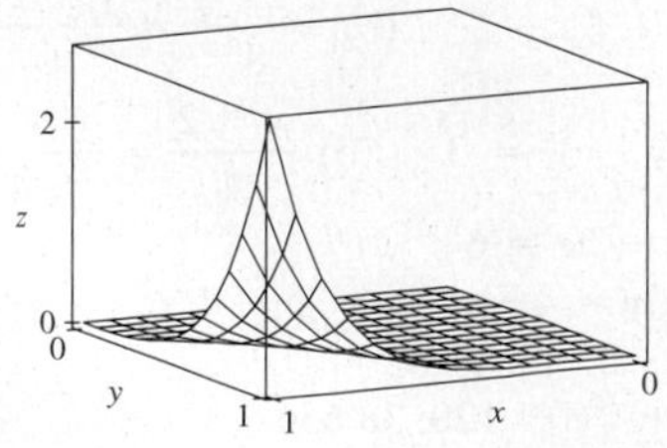

35. $\frac{5}{6}$
37. Fubini's Theorem does not apply. The integrand has an infinite discontinuity at the origin.

EXERCISES 15.3 ■ PAGE 972

1. 32 **3.** $\frac{3}{10}$ **5.** $e - 1$ **7.** $\frac{4}{3}$ **9.** π **11.** $\frac{1}{2}e^{16} - \frac{17}{2}$
13. $\frac{1}{2}(1 - \cos 1)$ **15.** $\frac{147}{20}$ **17.** 0 **19.** $\frac{7}{18}$ **21.** $\frac{31}{8}$
23. 6 **25.** $\frac{128}{15}$ **27.** $\frac{1}{3}$ **29.** 0, 1.213, 0.713 **31.** $\frac{64}{3}$
33.

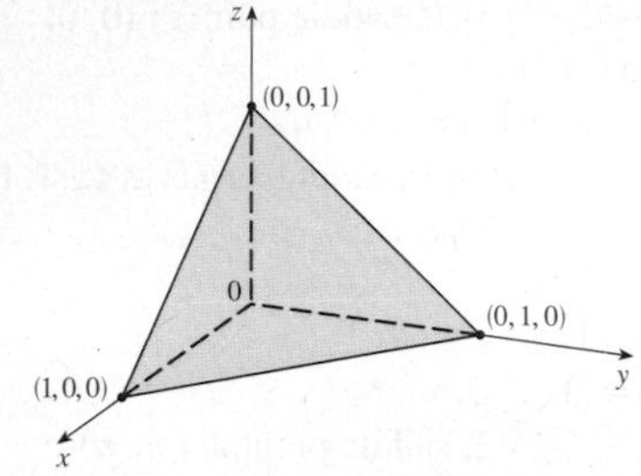

35. 13,984,735,616/14,549,535 **37.** $\pi/2$
39. $\int_0^2 \int_{y^2}^4 f(x, y)\,dx\,dy$ **41.** $\int_{-3}^3 \int_0^{\sqrt{9-x^2}} f(x, y)\,dy\,dx$

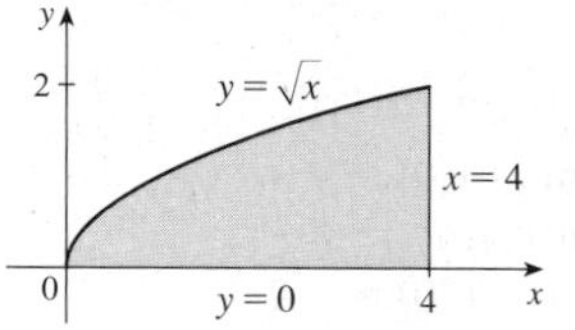

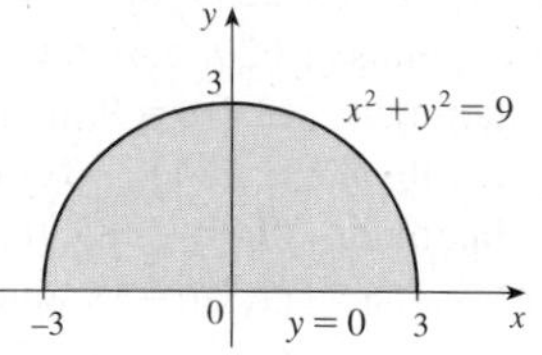

43. $\int_0^{\ln 2} \int_{e^y}^2 f(x, y)\,dx\,dy$

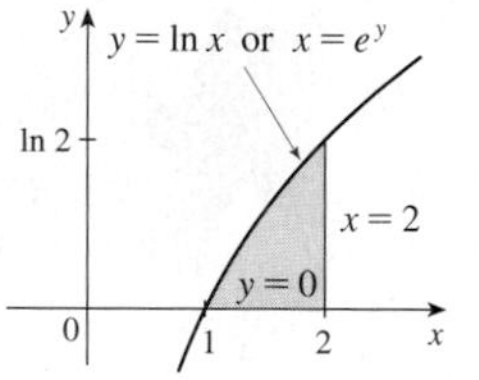

45. $\frac{1}{6}(e^9 - 1)$ **47.** $\frac{1}{3}\ln 9$ **49.** $\frac{1}{3}(2\sqrt{2} - 1)$ **51.** 1
53. $(\pi/16)e^{-1/16} \leqslant \iint_Q e^{-(x^2+y^2)^2}\,dA \leqslant \pi/16$ **55.** $\frac{3}{4}$
59. 8π **61.** $2\pi/3$

EXERCISES 15.4 ■ PAGE 978

1. $\int_0^{3\pi/2} \int_0^4 f(r\cos\theta, r\sin\theta)\,r\,dr\,d\theta$ **3.** $\int_{-1}^1 \int_0^{(x+1)/2} f(x, y)\,dy\,dx$
5. $33\pi/2$

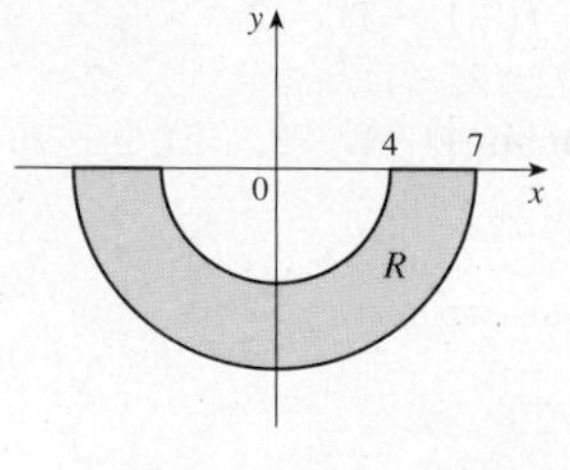

7. 0 **9.** $\frac{1}{2}\pi \sin 9$ **11.** $(\pi/2)(1 - e^{-4})$ **13.** $\frac{3}{64}\pi^2$
15. $\pi/12$ **17.** $\frac{1}{8}(\pi - 2)$ **19.** $\frac{16}{3}\pi$ **21.** $\frac{4}{3}\pi$
23. $\frac{4}{3}\pi a^3$ **25.** $(2\pi/3)[1 - (1/\sqrt{2})]$
27. $(8\pi/3)(64 - 24\sqrt{3})$
29. $\frac{1}{2}\pi(1 - \cos 9)$ **31.** $2\sqrt{2}/3$
33. 1800π ft^3 **35.** $\frac{15}{16}$ **37.** (a) $\sqrt{\pi}/4$ (b) $\sqrt{\pi}/2$

EXERCISES 15.5 ▪ PAGE 988

1. $\frac{64}{3}$ C **3.** $\frac{4}{3}, \left(\frac{4}{3}, 0\right)$ **5.** $6, \left(\frac{3}{4}, \frac{3}{2}\right)$

7. $\frac{1}{4}(e^2 - 1), \left(\dfrac{e^2+1}{2(e^2-1)}, \dfrac{4(e^3-1)}{9(e^2-1)}\right)$

9. $L/4, (L/2, 16/(9\pi))$ **11.** $\left(\frac{3}{8}, 3\pi/16\right)$ **13.** $(0, 45/(14\pi))$

15. $(2a/5, 2a/5)$ if vertex is $(0, 0)$ and sides are along positive axes

17. $\frac{1}{16}(e^4 - 1), \frac{1}{8}(e^2 - 1), \frac{1}{16}(e^4 + 2e^2 - 3)$

19. $7ka^6/180, 7ka^6/180, 7ka^6/90$ if vertex is $(0, 0)$ and sides are along positive axes

21. $m = \pi^2/8, (\bar{x}, \bar{y}) = \left(\dfrac{2\pi}{3} - \dfrac{1}{\pi}, \dfrac{16}{9\pi}\right), I_x = 3\pi^2/64,$
$I_y = \frac{1}{16}(\pi^4 - 3\pi^2), I_0 = \pi^4/16 - 9\pi^2/64$

23. $\rho bh^3/3, \rho b^3h/3; b/\sqrt{3}, h/\sqrt{3}$

25. $\rho a^4\pi/16, \rho a^4\pi/16; a/2, a/2$

27. (a) $\frac{1}{2}$ (b) 0.375 (c) $\frac{5}{48} \approx 0.1042$

29. (b) (i) $e^{-0.2} \approx 0.8187$
(ii) $1 + e^{-1.8} - e^{-0.8} - e^{-1} \approx 0.3481$ (c) 2, 5

31. (a) ≈ 0.500 (b) ≈ 0.632

33. (a) $\iint_D (k/20)\left[20 - \sqrt{(x - x_0)^2 + (y - y_0)^2}\right] dA$, where D is the disk with radius 10 mi centered at the center of the city
(b) $200\pi k/3 \approx 209k, 200\left(\pi/2 - \frac{8}{9}\right)k \approx 136k$, on the edge

EXERCISES 15.6 ▪ PAGE 998

1. $\frac{27}{4}$ **3.** 1 **5.** $\frac{1}{3}(e^3 - 1)$ **7.** $-\frac{1}{3}$ **9.** 4 **11.** $\frac{65}{28}$

13. $8/(3e)$ **15.** $\frac{1}{60}$ **17.** $16\pi/3$ **19.** $\frac{16}{3}$ **21.** 36π

23. (a) $\int_0^1 \int_0^x \int_0^{\sqrt{1-y^2}} dz\,dy\,dx$ (b) $\frac{1}{4}\pi - \frac{1}{3}$

25. 60.533

27.

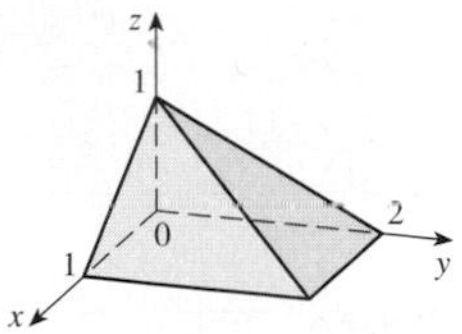

29. $\int_{-2}^{2} \int_0^{4-x^2} \int_{-\sqrt{4-x^2-y}/2}^{\sqrt{4-x^2-y}/2} f(x, y, z)\,dz\,dy\,dx$
$= \int_0^4 \int_{-\sqrt{4-y}}^{\sqrt{4-y}} \int_{-\sqrt{4-x^2-y}/2}^{\sqrt{4-x^2-y}/2} f(x, y, z)\,dz\,dx\,dy$
$= \int_{-1}^{1} \int_0^{4-4z^2} \int_{-\sqrt{4-y-4z^2}}^{\sqrt{4-y-4z^2}} f(x, y, z)\,dx\,dy\,dz$
$= \int_0^4 \int_{-\sqrt{4-y}/2}^{\sqrt{4-y}/2} \int_{-\sqrt{4-y-4z^2}}^{\sqrt{4-y-4z^2}} f(x, y, z)\,dx\,dz\,dy$
$= \int_{-2}^{2} \int_{-\sqrt{4-x^2}/2}^{\sqrt{4-x^2}/2} \int_0^{4-x^2-4z^2} f(x, y, z)\,dy\,dz\,dx$
$= \int_{-1}^{1} \int_{-\sqrt{4-4z^2}}^{\sqrt{4-4z^2}} \int_0^{4-x^2-4z^2} f(x, y, z)\,dy\,dx\,dz$

31. $\int_{-2}^{2} \int_{x^2}^{4} \int_0^{2-y/2} f(x, y, z)\,dz\,dy\,dx$
$= \int_0^4 \int_{-\sqrt{y}}^{\sqrt{y}} \int_0^{2-y/2} f(x, y, z)\,dz\,dx\,dy$
$= \int_0^2 \int_0^{4-2z} \int_{-\sqrt{y}}^{\sqrt{y}} f(x, y, z)\,dx\,dy\,dz$
$= \int_0^4 \int_0^{2-y/2} \int_{-\sqrt{y}}^{\sqrt{y}} f(x, y, z)\,dx\,dz\,dy$
$= \int_{-2}^{2} \int_0^{2-x^2/2} \int_{x^2}^{4-2z} f(x, y, z)\,dy\,dz\,dx$
$= \int_0^2 \int_{-\sqrt{4-2z}}^{\sqrt{4-2z}} \int_{x^2}^{4-2z} f(x, y, z)\,dy\,dx\,dz$

33. $\int_0^1 \int_{\sqrt{x}}^{1} \int_0^{1-y} f(x, y, z)\,dz\,dy\,dx$
$= \int_0^1 \int_0^{y^2} \int_0^{1-y} f(x, y, z)\,dz\,dx\,dy$
$= \int_0^1 \int_0^{1-z} \int_0^{y^2} f(x, y, z)\,dx\,dy\,dz$
$= \int_0^1 \int_0^{1-y} \int_0^{y^2} f(x, y, z)\,dx\,dz\,dy$
$= \int_0^1 \int_0^{1-\sqrt{x}} \int_{\sqrt{x}}^{1-z} f(x, y, z)\,dy\,dz\,dx$
$= \int_0^1 \int_0^{(1-z)^2} \int_{\sqrt{x}}^{1-z} f(x, y, z)\,dy\,dx\,dz$

35. $\int_0^1 \int_y^1 \int_0^y f(x, y, z)\,dz\,dx\,dy = \int_0^1 \int_0^x \int_0^y f(x, y, z)\,dz\,dy\,dx$
$= \int_0^1 \int_z^1 \int_y^1 f(x, y, z)\,dx\,dy\,dz = \int_0^1 \int_0^y \int_y^1 f(x, y, z)\,dx\,dz\,dy$
$= \int_0^1 \int_0^x \int_z^x f(x, y, z)\,dy\,dz\,dx = \int_0^1 \int_z^1 \int_z^x f(x, y, z)\,dy\,dx\,dz$

37. $\frac{79}{30}, \left(\frac{358}{553}, \frac{33}{79}, \frac{571}{553}\right)$ **39.** $a^5, (7a/12, 7a/12, 7a/12)$

41. $I_x = I_y = I_z = \frac{2}{3}kL^5$ **43.** $\frac{1}{2}\pi kha^4$

45. (a) $m = \int_{-3}^{3} \int_{-\sqrt{9-x^2}}^{\sqrt{9-x^2}} \int_1^{5-y} \sqrt{x^2 + y^2}\,dz\,dy\,dx$
(b) $(\bar{x}, \bar{y}, \bar{z})$, where
$\bar{x} = (1/m) \int_{-3}^{3} \int_{-\sqrt{9-x^2}}^{\sqrt{9-x^2}} \int_1^{5-y} x\sqrt{x^2 + y^2}\,dz\,dy\,dx$
$\bar{y} = (1/m) \int_{-3}^{3} \int_{-\sqrt{9-x^2}}^{\sqrt{9-x^2}} \int_1^{5-y} y\sqrt{x^2 + y^2}\,dz\,dy\,dx$
$\bar{z} = (1/m) \int_{-3}^{3} \int_{-\sqrt{9-x^2}}^{\sqrt{9-x^2}} \int_1^{5-y} z\sqrt{x^2 + y^2}\,dz\,dy\,dx$
(c) $\int_{-3}^{3} \int_{-\sqrt{9-x^2}}^{\sqrt{9-x^2}} \int_1^{5-y} (x^2 + y^2)^{3/2}\,dz\,dy\,dx$

47. (a) $\frac{3}{32}\pi + \frac{11}{24}$
(b) $(\bar{x}, \bar{y}, \bar{z}) = \left(\dfrac{28}{9\pi + 44}, \dfrac{30\pi + 128}{45\pi + 220}, \dfrac{45\pi + 208}{135\pi + 660}\right)$
(c) $\frac{1}{240}(68 + 15\pi)$

49. (a) $\frac{1}{8}$ (b) $\frac{1}{64}$ (c) $\frac{1}{5760}$

51. $L^3/8$

53. The region bounded by the ellipsoid $x^2 + 2y^2 + 3z^2 = 1$

EXERCISES 15.7 ▪ PAGE 1004

1. (a)

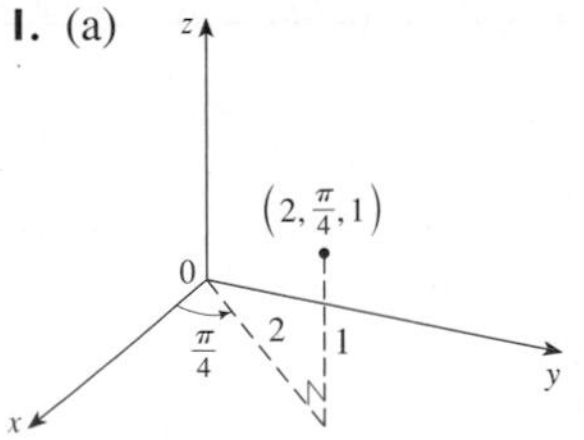

$(\sqrt{2}, \sqrt{2}, 1)$

(b)

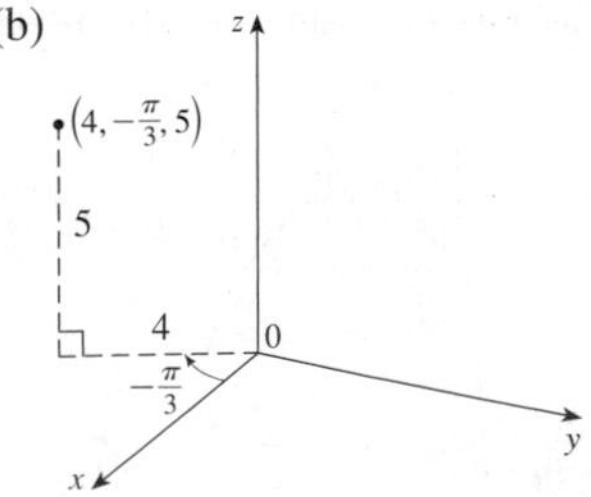

$(2, -2\sqrt{3}, 5)$

3. (a) $(\sqrt{2}, 7\pi/4, 4)$ (b) $(2, 4\pi/3, 2)$

5. Vertical half-plane through the z-axis **7.** Circular paraboloid

9. (a) $z = r^2$ (b) $r = 2\sin\theta$

11.

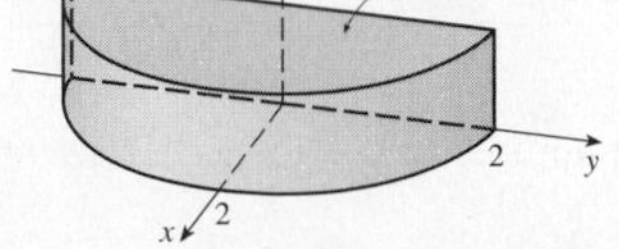

13. Cylindrical coordinates: $6 \leqslant r \leqslant 7, 0 \leqslant \theta \leqslant 2\pi, 0 \leqslant z \leqslant 20$

15.

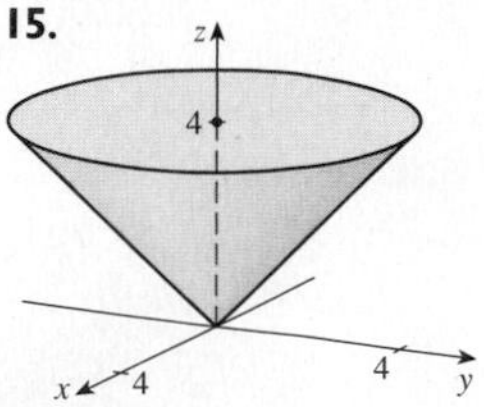

$64\pi/3$

17. 384π **19.** $\pi(e^6 - e - 5)$ **21.** $2\pi/5$

23. (a) 162π (b) $(0, 0, 15)$

25. $\pi K a^2/8, (0, 0, 2a/3)$ **27.** 0

29. (a) $\iiint_C h(P)g(P)\,dV$, where C is the cone

(b) $\approx 3.1 \times 10^{19}$ ft-lb

EXERCISES 15.8 ▪ PAGE 1010

1. (a)

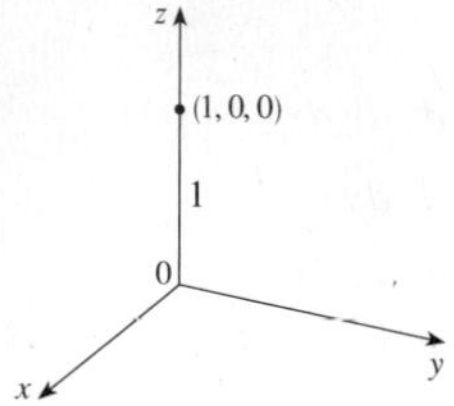

$(0, 0, 1)$

(b)

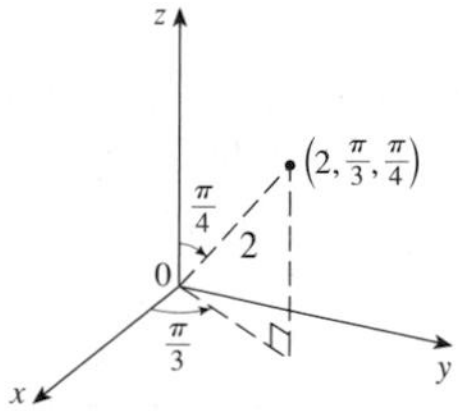

$\left(\frac{1}{2}\sqrt{2}, \frac{1}{2}\sqrt{6}, \sqrt{2}\right)$

3. (a) $(4, \pi/3, \pi/6)$ (b) $\left(\sqrt{2}, 3\pi/2, 3\pi/4\right)$

5. Half-cone

7. Sphere, radius $\frac{1}{2}$, center $\left(0, \frac{1}{2}, 0\right)$

9. (a) $\cos^2\phi = \sin^2\phi$ (b) $\rho^2(\sin^2\phi\cos^2\theta + \cos^2\phi) = 9$

11.

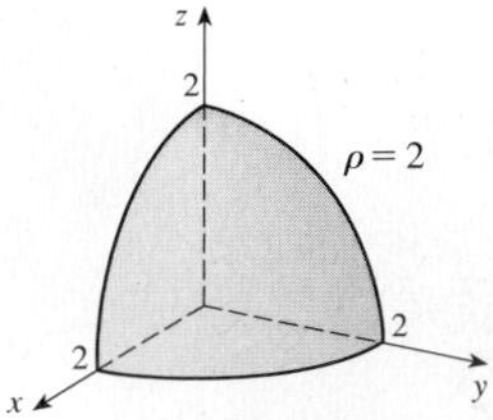

13.

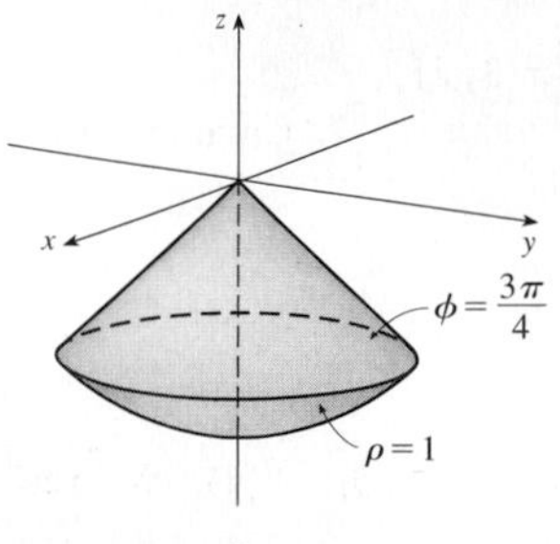

15. $0 \leqslant \phi \leqslant \pi/4, 0 \leqslant \rho \leqslant \cos\phi$

17.

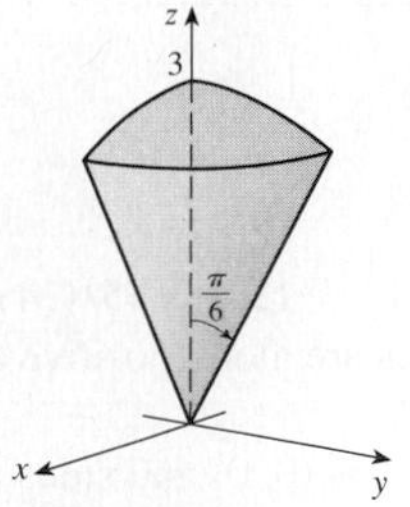

$(9\pi/4)(2 - \sqrt{3})$

19. $\int_0^{\pi/2} \int_0^3 \int_0^2 f(r\cos\theta, r\sin\theta, z)\, r\,dz\,dr\,d\theta$

21. $312{,}500\pi/7$ **23.** $15\pi/16$ **25.** $1562\pi/15$

27. $\left(\sqrt{3} - 1\right)\pi a^3/3$ **29.** (a) 10π (b) $(0, 0, 2.1)$

31. $\left(0, \frac{525}{296}, 0\right)$

33. (a) $\left(0, 0, \frac{3}{8}a\right)$ (b) $4K\pi a^5/15$

35. $(2\pi/3)\left[1 - \left(1/\sqrt{2}\right)\right], \left(0, 0, 3/\left[8\left(2 - \sqrt{2}\right)\right]\right)$

37. $5\pi/6$ **39.** $\left(4\sqrt{2} - 5\right)/15$

41.

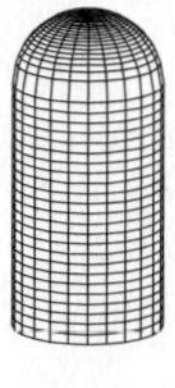

43. $136\pi/99$

EXERCISES 15.9 ▪ PAGE 1020

1. 16 **3.** $\sin^2\theta - \cos^2\theta$ **5.** 0

7. The parallelogram with vertices $(0, 0), (6, 3), (12, 1), (6, -2)$

9. The region bounded by the line $y = 1$, the y-axis, and $y = \sqrt{x}$

11. -3 **13.** 6π **15.** $2\ln 3$

17. (a) $\frac{4}{3}\pi abc$ (b) 1.083×10^{12} km^3

19. $\frac{8}{5}\ln 8$ **21.** $\frac{3}{2}\sin 1$ **23.** $e - e^{-1}$

CHAPTER 15 REVIEW ▪ PAGE 1021

True-False Quiz

1. True **3.** True **5.** True **7.** False

Exercises

1. ≈ 64.0 **3.** $4e^2 - 4e + 3$ **5.** $\frac{1}{2}\sin 1$ **7.** $\frac{2}{3}$

9. $\int_0^{\pi} \int_2^4 f(r\cos\theta, r\sin\theta)\, r\,dr\,d\theta$

11. The region inside the loop of the four-leaved rose $r = \sin 2\theta$ in the first quadrant

13. $\frac{1}{2}\sin 1$ **15.** $\frac{1}{2}e^6 - \frac{7}{2}$ **17.** $\frac{1}{4}\ln 2$ **19.** 8

21. $81\pi/5$ **23.** 40.5 **25.** $\pi/96$ **27.** $\frac{64}{15}$ **29.** 176

31. $\frac{2}{3}$ **33.** $2ma^3/9$

35. (a) $\frac{1}{4}$ (b) $\left(\frac{1}{3}, \frac{8}{15}\right)$

(c) $I_x = \frac{1}{12}, I_y = \frac{1}{24}; \bar{\bar{y}} = 1/\sqrt{3}, \bar{\bar{x}} = 1/\sqrt{6}$

37. $(0, 0, h/4)$

39. 97.2 **41.** 0.0512

43. (a) $\frac{1}{15}$ (b) $\frac{1}{3}$ (c) $\frac{1}{45}$

45. $\int_0^1 \int_0^{1-z} \int_{-\sqrt{y}}^{\sqrt{y}} f(x, y, z)\,dx\,dy\,dz$ **47.** $-\ln 2$ **49.** 0

PROBLEMS PLUS ▪ PAGE 1024

1. 30 **3.** $\frac{1}{2}\sin 1$ **7.** (b) 0.90

CHAPTER 16

EXERCISES 16.1 ■ PAGE 1032

1.

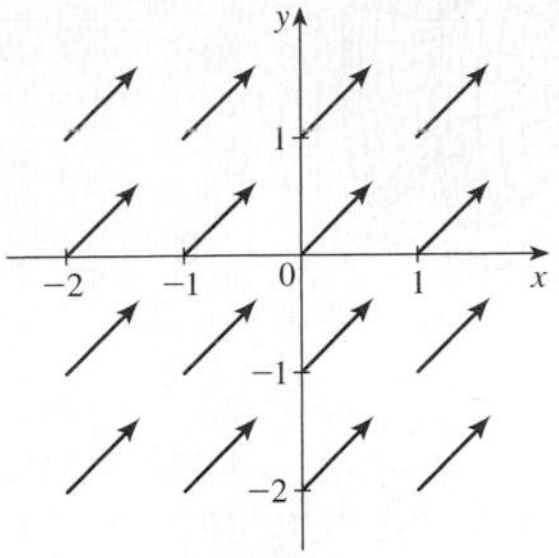

3.

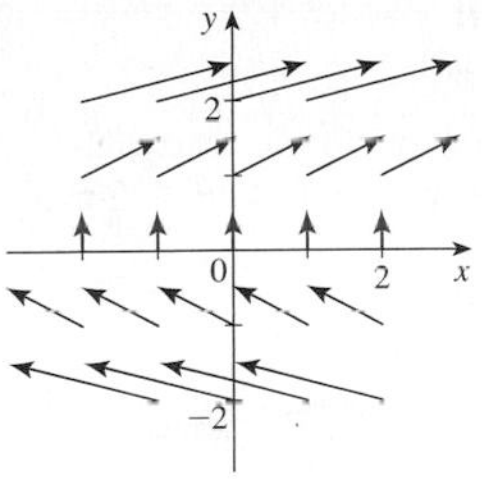

5.

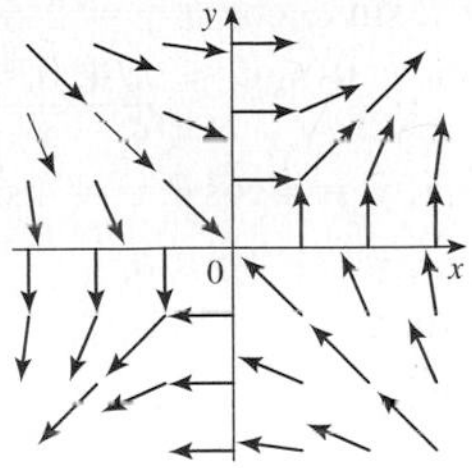

7.

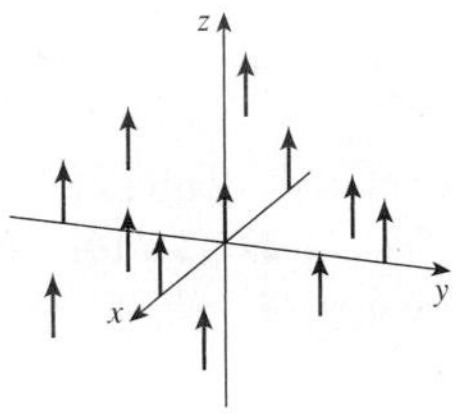

9.

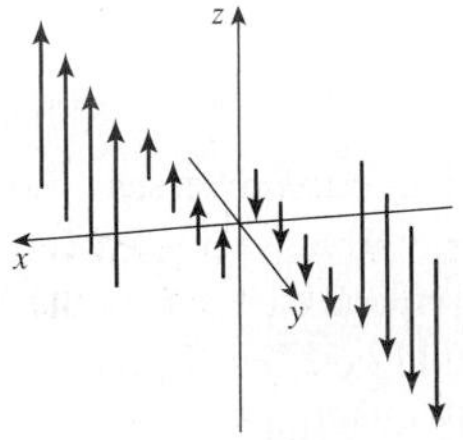

11. II **13.** I **15.** IV **17.** III

19.

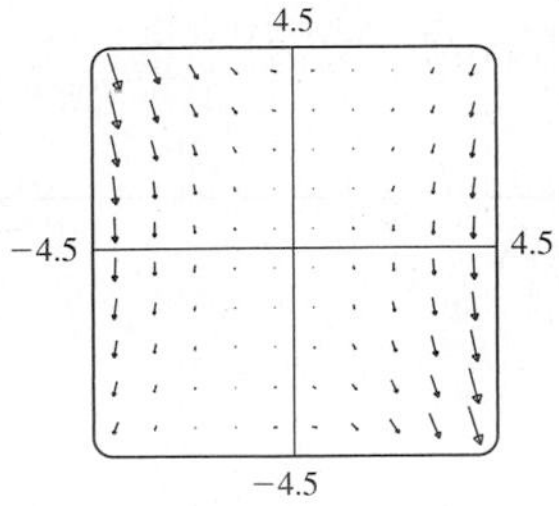

The line $y = 2x$

21. $\nabla f(x, y) = (xy + 1)e^{xy}\,\mathbf{i} + x^2e^{xy}\,\mathbf{j}$

23. $\nabla f(x, y, z) = \dfrac{x}{\sqrt{x^2 + y^2 + z^2}}\,\mathbf{i} + \dfrac{y}{\sqrt{x^2 + y^2 + z^2}}\,\mathbf{j} + \dfrac{z}{\sqrt{x^2 + y^2 + z^2}}\,\mathbf{k}$

25. $\nabla f(x, y) = 2x\,\mathbf{i} - \mathbf{j}$

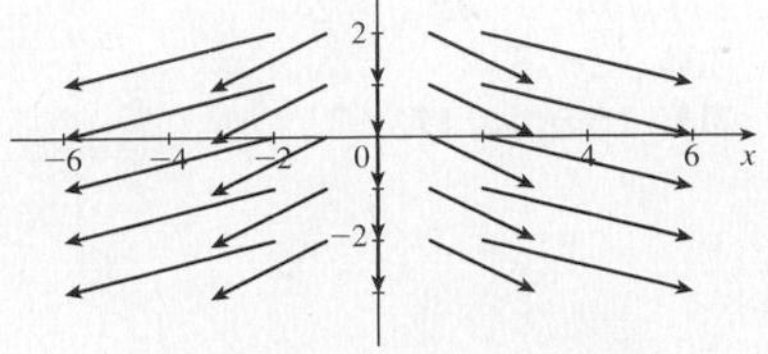

27.

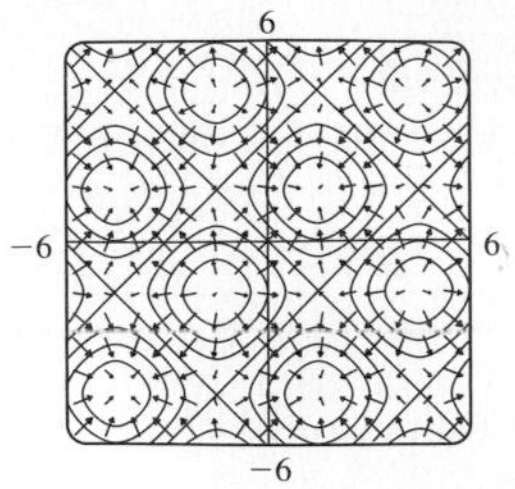

29. III **31.** II **33.** (2.04, 1.03)

35. (a)

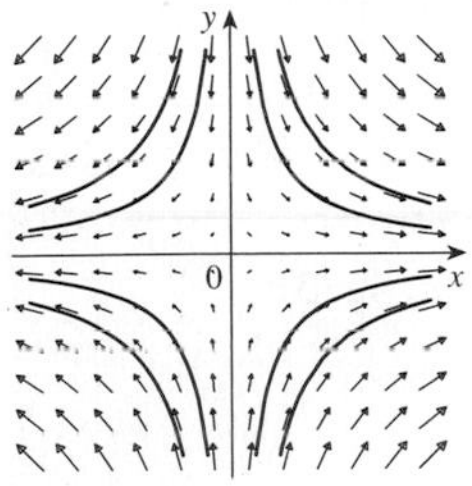

$y = C/x$

(b) $y = 1/x,\ x > 0$

EXERCISES 16.2 ■ PAGE 1043

1. $\frac{1}{54}(145^{3/2} - 1)$ **3.** 1638.4 **5.** $\frac{243}{8}$ **7.** $\frac{17}{3}$ **9.** $\sqrt{5}\,\pi$

11. $\frac{1}{12}\sqrt{14}(e^6 - 1)$ **13.** $\frac{1}{5}$ **15.** $\frac{97}{3}$

17. (a) Positive (b) Negative

19. 45 **21.** $\frac{6}{5} - \cos 1 - \sin 1$ **23.** 1.9633 **25.** 15.0074

27. $3\pi + \frac{2}{3}$

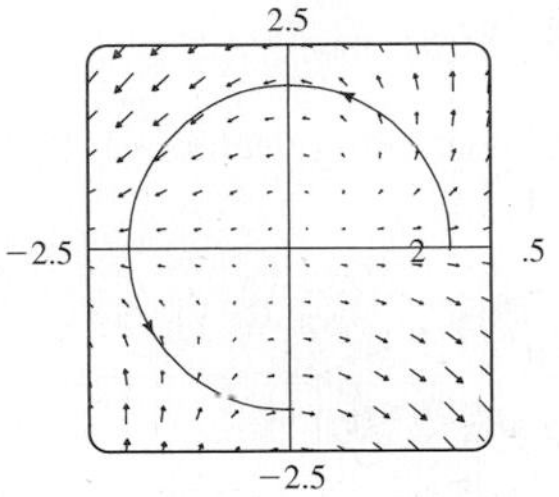

29. (a) $\frac{11}{8} - 1/e$ (b)

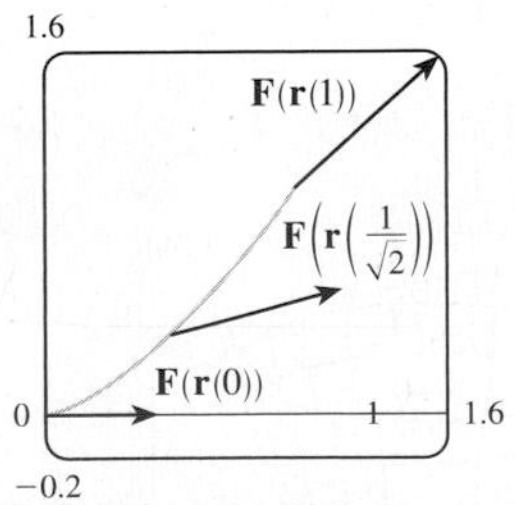

31. $\frac{172{,}704}{5{,}632{,}705}\sqrt{2}(1 - e^{-14\pi})$ **33.** $2\pi k,\ (4/\pi, 0)$

35. (a) $\bar{x} = (1/m)\int_C x\rho(x, y, z)\,ds$,
$\bar{y} = (1/m)\int_C y\rho(x, y, z)\,ds$,
$\bar{z} = (1/m)\int_C z\rho(x, y, z)\,ds$, where $m = \int_C \rho(x, y, z)\,ds$
(b) $(0, 0, 3\pi)$

37. $I_x = k\left(\frac{1}{2}\pi - \frac{4}{3}\right),\ I_y = k\left(\frac{1}{2}\pi - \frac{2}{3}\right)$

39. $2\pi^2$ **41.** 26 **43.** 1.67×10^4 ft-lb **45.** (b) Yes

47. $\approx$22 J

EXERCISES 16.3 ■ PAGE 1053

1. 40 **3.** $f(x, y) = x^2 - 3xy + 2y^2 - 8y + K$
5. $f(x, y) = e^x \sin y + K$ **7.** $f(x, y) = ye^x + x \sin y + K$
9. $f(x, y) = x \ln y + x^2y^3 + K$
11. (b) 16 **13.** (a) $f(x, y) = \frac{1}{2}x^2y^2$ (b) 2
15. (a) $f(x, y, z) = xyz + z^2$ (b) 77
17. (a) $f(x, y, z) = xy^2 \cos z$ (b) 0
19. 2 **21.** 30 **23.** No **25.** Conservative
29. (a) Yes (b) Yes (c) Yes
31. (a) Yes (b) Yes (c) No

EXERCISES 16.4 ■ PAGE 1060

1. 8π **3.** $\frac{2}{3}$ **5.** 12 **7.** $\frac{1}{3}$ **9.** -24π **11.** $\frac{4}{3} - 2\pi$
13. $\frac{625}{2}\pi$ **15.** $-8e + 48e^{-1}$ **17.** $-\frac{1}{12}$
19. 3π **21.** (c) $\frac{9}{2}$ **23.** $(4a/3\pi, 4a/3\pi)$

EXERCISES 16.5 ■ PAGE 1068

1. (a) $-x^2\,\mathbf{i} + 3xy\,\mathbf{j} - xz\,\mathbf{k}$ (b) yz
3. (a) $(x - y)\,\mathbf{i} - y\,\mathbf{j} + \mathbf{k}$ (b) $z - 1/(2\sqrt{z})$
5. (a) $\mathbf{0}$ (b) $2/\sqrt{x^2 + y^2 + z^2}$
7. (a) $\langle 1/y, -1/x, 1/x \rangle$ (b) $1/x + 1/y + 1/z$
9. (a) Negative (b) curl $\mathbf{F} = \mathbf{0}$
11. (a) Zero (b) curl $\mathbf{F}$ points in the negative z-direction
13. $f(x, y, z) = xy^2z^3 + K$ **15.** $f(x, y, z) = x^2y + y^2z + K$
17. Not conservative **19.** No

EXERCISES 16.6 ■ PAGE 1078

1. P: no; Q: yes
3. Plane through (0, 3, 1) containing vectors $\langle 1, 0, 4 \rangle$, $\langle 1, -1, 5 \rangle$
5. Hyperbolic paraboloid
7.

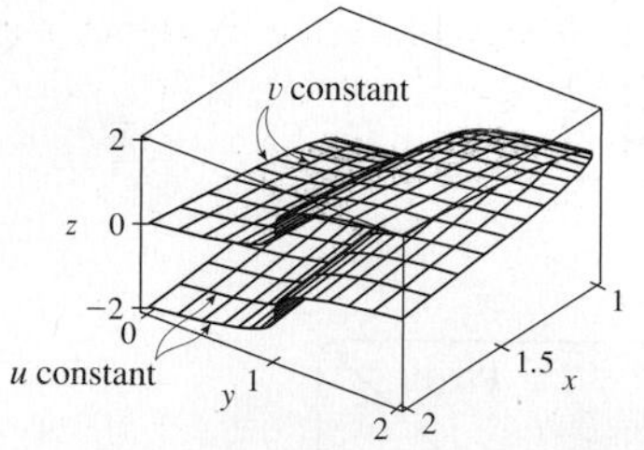

9.

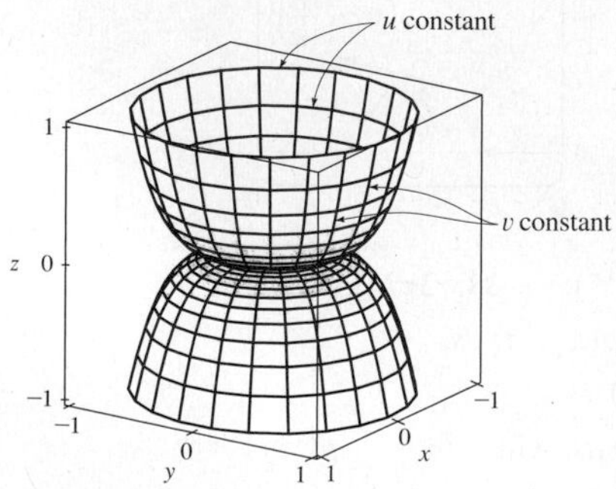

11.

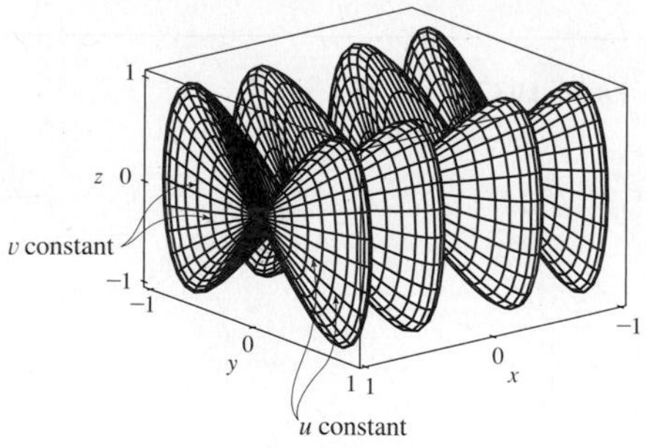

13. IV **15.** II **17.** III
19. $x = 1 + u + v,\ y = 2 + u - v,\ z = -3 - u + v$
21. $x = x,\ z = z,\ y = \sqrt{1 - x^2 + z^2}$
23. $x = 2 \sin \phi \cos \theta,\ y = 2 \sin \phi \sin \theta,$
$z = 2 \cos \phi,\ 0 \le \phi \le \pi/4,\ 0 \le \theta \le 2\pi$
$\left[\text{or } x = x,\ y = y,\ z = \sqrt{4 - x^2 - y^2},\ x^2 + y^2 \le 2\right]$
25. $x = x,\ y = 4 \cos \theta,\ z = 4 \sin \theta,\ 0 \le x \le 5,\ 0 \le \theta \le 2\pi$
29. $x = x,\ y = e^{-x} \cos \theta,$
$z = e^{-x} \sin \theta,\ 0 \le x \le 3,$
$0 \le \theta \le 2\pi$

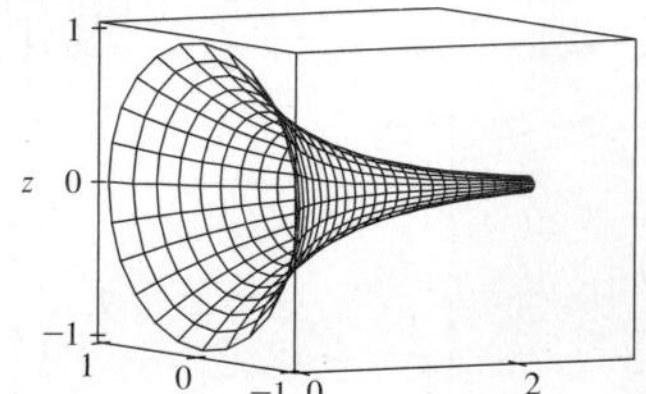

31. (a) Direction reverses (b) Number of coils doubles
33. $3x - y + 3z = 3$ **35.** $-x + 2z = 1$ **37.** $3\sqrt{14}$
39. $\frac{4}{15}(3^{5/2} - 2^{7/2} + 1)$ **41.** $(2\pi/3)(2\sqrt{2} - 1)$
43. $(\pi/6)(17\sqrt{17} - 5\sqrt{5})$
45. $\frac{1}{2}\sqrt{21} + \frac{17}{4}\left[\ln(2 + \sqrt{21}) - \ln\sqrt{17}\right]$ **47.** 4
49. 13.9783
51. (a) 24.2055 (b) 24.2476
53. $\frac{45}{8}\sqrt{14} + \frac{15}{16}\ln\left[(11\sqrt{5} + 3\sqrt{70})/(3\sqrt{5} + \sqrt{70})\right]$
55. (b)

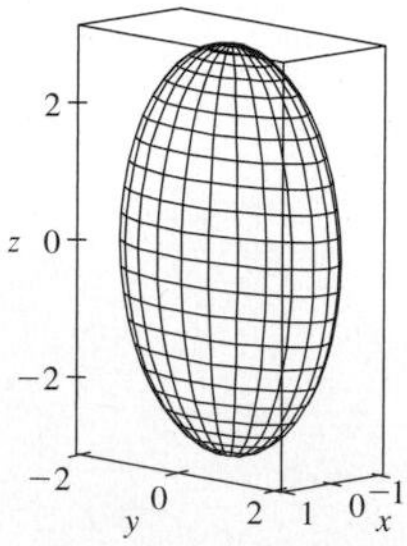

(c) $\int_0^{2\pi}\int_0^{\pi} \sqrt{36 \sin^4 u \cos^2 v + 9 \sin^4 u \sin^2 v + 4 \cos^2 u \sin^2 u}\; du\, dv$
57. 4π **59.** $2a^2(\pi - 2)$

EXERCISES 16.7 ■ PAGE 1091

1. 49.09 **3.** 900π **5.** $171\sqrt{14}$ **7.** $\sqrt{3}/24$
9. $5\sqrt{5}/48 + 1/240$ **11.** $364\sqrt{2}\pi/3$
13. $(\pi/60)(391\sqrt{17} + 1)$ **15.** 16π **17.** 12

19. $\frac{713}{180}$ **21.** $-\frac{1}{6}$ **23.** $-\frac{4}{3}\pi$ **25.** 0 **27.** 48
29. $2\pi + \frac{8}{3}$ **31.** 0.1642 **33.** 3.4895
35. $\iint_S \mathbf{F} \cdot d\mathbf{S} = \iint_D [P(\partial h/\partial x) - Q + R(\partial h/\partial z)]\, dA$, where D = projection of S on xz-plane
37. $(0, 0, a/2)$
39. (a) $I_z = \iint_S (x^2 + y^2)\rho(x, y, z)\, dS$ (b) $4329\sqrt{2}\,\pi/5$
41. 0 kg/s **43.** $\frac{8}{3}\pi a^3 \varepsilon_0$ **45.** 1248π

EXERCISES 16.8 ■ PAGE 1097

3. 0 **5.** 0 **7.** −1 **9.** 80π
11. (a) $81\pi/2$ (b)

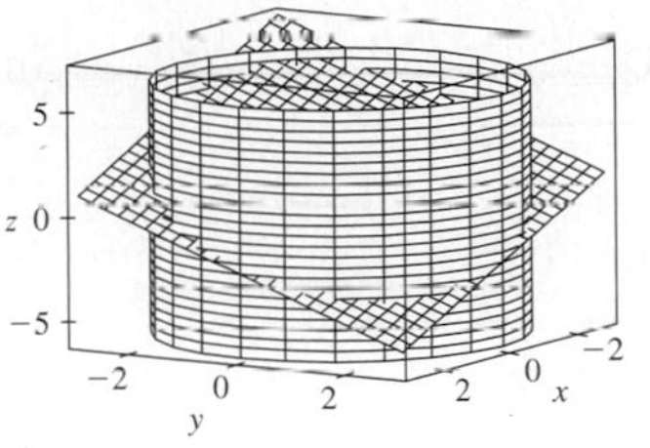

(c) $x = 3\cos t$, $y = 3\sin t$, $z = 1 - 3(\cos t + \sin t)$, $0 \leqslant t \leqslant 2\pi$

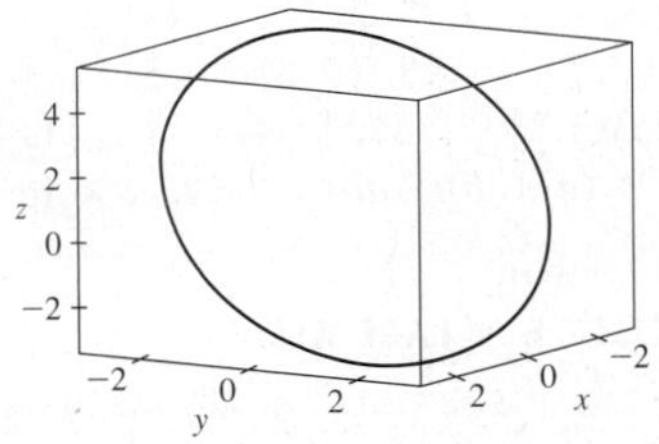

17. 3

EXERCISES 16.9 ■ PAGE 1103

5. 2 **7.** $9\pi/2$
9. 0 **11.** $32\pi/3$ **13.** 0
15. $341\sqrt{2}/60 + \frac{81}{20}\arcsin(\sqrt{3}/3)$ **17.** $13\pi/20$
19. Negative at P_1, positive at P_2
21. div $\mathbf{F} > 0$ in quadrants I, II; div $\mathbf{F} < 0$ in quadrants III, IV

CHAPTER 16 REVIEW ■ PAGE 1106

True-False Quiz

1. False **3.** True **5.** False **7.** True

Exercises

1. (a) Negative (b) Positive **3.** $6\sqrt{10}$ **5.** $\frac{4}{15}$
7. $\frac{110}{3}$ **9.** $\frac{11}{12} - 4/e$ **11.** $f(x, y) = e^y + xe^{xy}$ **13.** 0
17. -8π **25.** $\frac{1}{6}(27 - 5\sqrt{5})$
27. $(\pi/60)(391\sqrt{17} + 1)$ **29.** $-64\pi/3$
33. $-\frac{1}{2}$ **37.** −4 **39.** 21

CHAPTER 17

EXERCISES 17.1 ■ PAGE 1117

1. $y = c_1 e^{3x} + c_2 e^{-2x}$ **3.** $y = c_1 \cos 4x + c_2 \sin 4x$
5. $y = c_1 e^{2x/3} + c_2 x e^{2x/3}$ **7.** $y = c_1 + c_2 e^{x/2}$
9. $y = e^{2x}(c_1 \cos 3x + c_2 \sin 3x)$
11. $y = c_1 e^{(\sqrt{3}-1)t/2} + c_2 e^{-(\sqrt{3}+1)t/2}$
13. $P = e^{-t}\left[c_1 \cos\left(\frac{1}{10}t\right) + c_2 \sin\left(\frac{1}{10}t\right)\right]$
15.

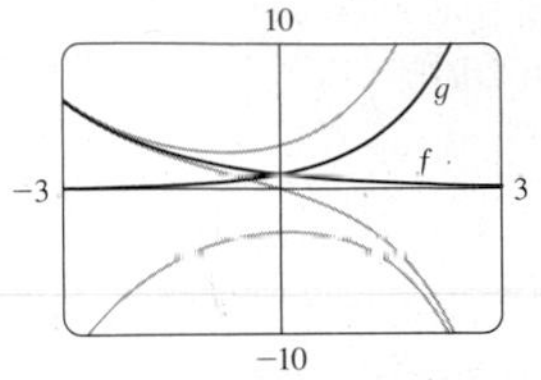

All solutions approach either 0 or $\pm\infty$ as $x \to \pm\infty$.

17. $y = 2e^{-3x/2} + e^{-x}$ **19.** $y = e^{x/2} - 2xe^{x/2}$
21. $y = 3\cos 4x - \sin 4x$ **23.** $y = e^{-x}(2\cos x + 3\sin x)$
25. $y = 3\cos\left(\frac{1}{2}x\right) - 4\sin\left(\frac{1}{2}x\right)$ **27.** $y = \dfrac{e^{x+3}}{e^3 - 1} + \dfrac{e^{2x}}{1 - e^3}$
29. No solution
31. $y = e^{-2x}(2\cos 3x - e^{\pi}\sin 3x)$
33. (b) $\lambda = n^2\pi^2/L^2$, n a positive integer; $y = C\sin(n\pi x/L)$

EXERCISES 17.2 ■ PAGE 1124

1. $y = c_1 e^{-2x} + c_2 e^{-x} + \frac{1}{2}x^2 - \frac{3}{2}x + \frac{7}{4}$
3. $y = c_1 + c_2 e^{2x} + \frac{1}{40}\cos 4x - \frac{1}{20}\sin 4x$
5. $y = e^{2x}(c_1 \cos x + c_2 \sin x) + \frac{1}{10}e^{-x}$
7. $y = \frac{3}{2}\cos x + \frac{11}{2}\sin x + \frac{1}{2}e^x + x^3 - 6x$
9. $y = e^x\left(\frac{1}{2}x^2 - x + 2\right)$
11.

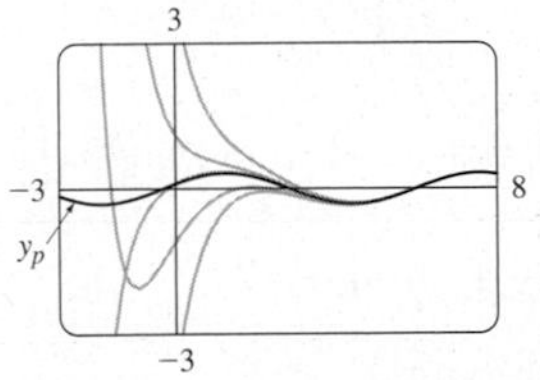

The solutions are all asymptotic to $y_p = \frac{1}{10}\cos x + \frac{3}{10}\sin x$ as $x \to \infty$. Except for y_p, all solutions approach either ∞ or $-\infty$ as $x \to -\infty$.

13. $y_p = Ae^{2x} + (Bx^2 + Cx + D)\cos x + (Ex^2 + Fx + G)\sin x$
15. $y_p = Ax + (Bx + C)e^{9x}$
17. $y_p = xe^{-x}[(Ax^2 + Bx + C)\cos 3x + (Dx^2 + Ex + F)\sin 3x]$
19. $y = c_1\cos\left(\frac{1}{2}x\right) + c_2\sin\left(\frac{1}{2}x\right) - \frac{1}{3}\cos x$
21. $y = c_1 e^x + c_2 x e^x + e^{2x}$
23. $y = c_1 \sin x + c_2 \cos x + \sin x \ln(\sec x + \tan x) - 1$
25. $y = [c_1 + \ln(1 + e^{-x})]e^x + [c_2 - e^{-x} + \ln(1 + e^{-x})]e^{2x}$
27. $y = e^x\left[c_1 + c_2 x - \frac{1}{2}\ln(1 + x^2) + x\tan^{-1}x\right]$

EXERCISES 17.3 ■ PAGE 1132

1. $x = 0.35\cos(2\sqrt{5}\,t)$ **3.** $x = -\frac{1}{5}e^{-6t} + \frac{6}{5}e^{-t}$ **5.** $\frac{49}{12}$ kg

7.

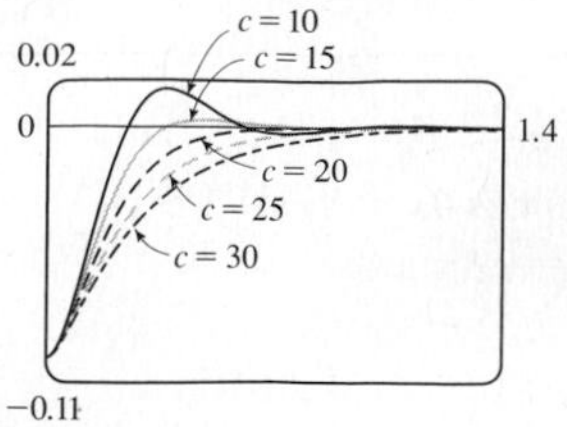

13. $Q(t) = (-e^{-10t}/250)(6\cos 20t + 3\sin 20t) + \frac{3}{125}$, $I(t) = \frac{3}{5}e^{-10t}\sin 20t$
15. $Q(t) = e^{-10t}\left[\frac{3}{250}\cos 20t - \frac{3}{500}\sin 20t\right] - \frac{3}{250}\cos 10t + \frac{3}{125}\sin 10t$

EXERCISES 17.4 ▪ PAGE 1137

1. $c_0 \sum_{n=0}^{\infty} \frac{x^n}{n!} = c_0 e^x$ **3.** $c_0 \sum_{n=0}^{\infty} \frac{x^{3n}}{3^n n!} = c_0 e^{x^3/3}$
5. $c_0 \sum_{n=0}^{\infty} \frac{(-1)^n}{2^n n!} x^{2n} + c_1 \sum_{n=0}^{\infty} \frac{(-2)^n n!}{(2n+1)!} x^{2n+1}$
7. $c_0 + c_1 \sum_{n=1}^{\infty} \frac{x^n}{n} = c_0 - c_1 \ln(1-x)$ for $|x| < 1$
9. $\sum_{n=0}^{\infty} \frac{x^{2n}}{2^n n!} = e^{x^2/2}$
11. $x + \sum_{n=1}^{\infty} \frac{(-1)^n 2^2 5^2 \cdot \cdots \cdot (3n-1)^2}{(3n+1)!} x^{3n+1}$

CHAPTER 17 REVIEW ▪ PAGE 1138

True-False Quiz

1. True **3.** True

Exercises

1. $y = c_1 e^{5x} + c_2 e^{-3x}$ **3.** $y = c_1 \cos(\sqrt{3}x) + c_2 \sin(\sqrt{3}x)$
5. $y = e^{2x}(c_1 \cos x + c_2 \sin x + 1)$
7. $y = c_1 e^x + c_2 x e^x - \frac{1}{2}\cos x - \frac{1}{2}(x+1)\sin x$
9. $y = c_1 e^{3x} + c_2 e^{-2x} - \frac{1}{6} - \frac{1}{5}xe^{-2x}$
11. $y = 5 - 2e^{-6(x-1)}$ **13.** $y = (e^{4x} - e^x)/3$
15. $\sum_{n=0}^{\infty} \frac{(-2)^n n!}{(2n+1)!} x^{2n+1}$
17. $Q(t) = -0.02e^{-10t}(\cos 10t + \sin 10t) + 0.03$
19. (c) $2\pi/k \approx 85$ min (d) $\approx 17{,}600$ mi/h

APPENDIXES

EXERCISES A ▪ PAGE A9

1. 18 **3.** π **5.** $5 - \sqrt{5}$ **7.** $2 - x$

9. $|x+1| = \begin{cases} x+1 & \text{for } x \geq -1 \\ -x-1 & \text{for } x < -1 \end{cases}$ **11.** $x^2 + 1$

13. $(-2, \infty)$

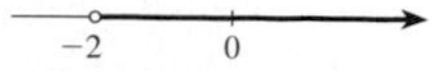

15. $[-1, \infty)$

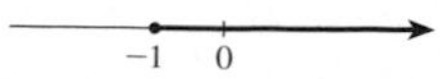

17. $(3, \infty)$

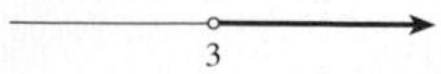

19. $(2, 6)$

21. $(0, 1]$

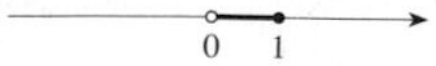

23. $\left[-1, \frac{1}{2}\right)$

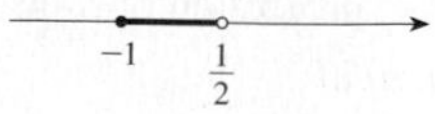

25. $(-\infty, 1) \cup (2, \infty)$

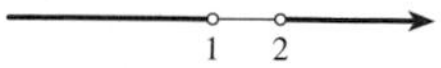

27. $\left[-1, \frac{1}{2}\right]$

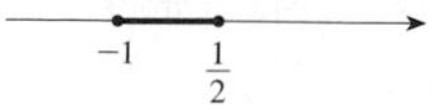

29. $(-\infty, \infty)$

31. $(-\sqrt{3}, \sqrt{3})$

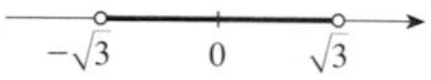

33. $(-\infty, 1]$

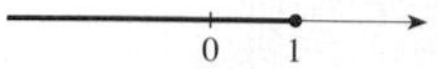

35. $(-1, 0) \cup (1, \infty)$

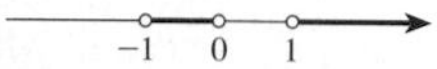

37. $(-\infty, 0) \cup \left(\frac{1}{4}, \infty\right)$

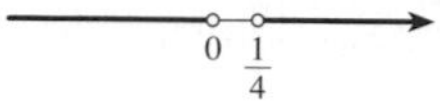

39. $10 \leq C \leq 35$ **41.** (a) $T = 20 - 10h$, $0 \leq h \leq 12$ (b) $-30°\text{C} \leq T \leq 20°\text{C}$ **43.** $\pm\frac{3}{2}$ **45.** $2, -\frac{4}{3}$
47. $(-3, 3)$ **49.** $(3, 5)$ **51.** $(-\infty, -7] \cup [-3, \infty)$
53. $[1.3, 1.7]$ **55.** $[-4, -1] \cup [1, 4]$
57. $x \geq (a+b)c/(ab)$ **59.** $x > (c-b)/a$

EXERCISES B ▪ PAGE A15

1. 5 **3.** $\sqrt{74}$ **5.** $2\sqrt{37}$ **7.** 2 **9.** $-\frac{9}{2}$

17.

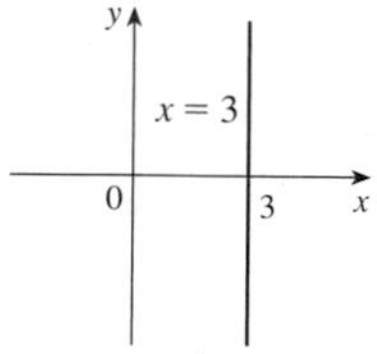

19.

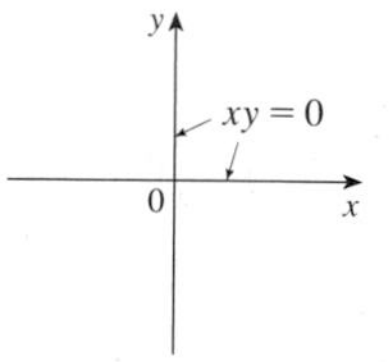

21. $y = 6x - 15$ **23.** $2x - 3y + 19 = 0$
25. $5x + y = 11$ **27.** $y = 3x - 2$ **29.** $y = 3x - 3$
31. $y = 5$ **33.** $x + 2y + 11 = 0$ **35.** $5x - 2y + 1 = 0$
37. $m = -\frac{1}{3}$, $b = 0$ **39.** $m = 0$, $b = -2$ **41.** $m = \frac{3}{4}$, $b = -3$

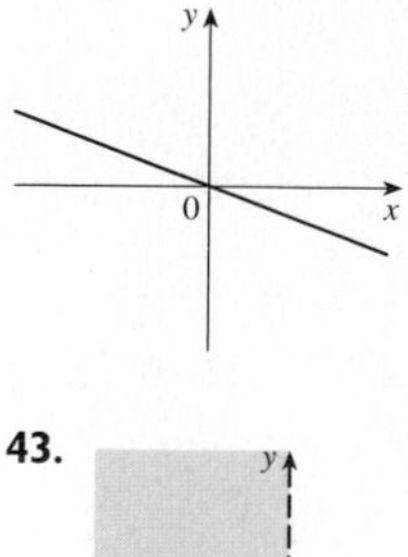

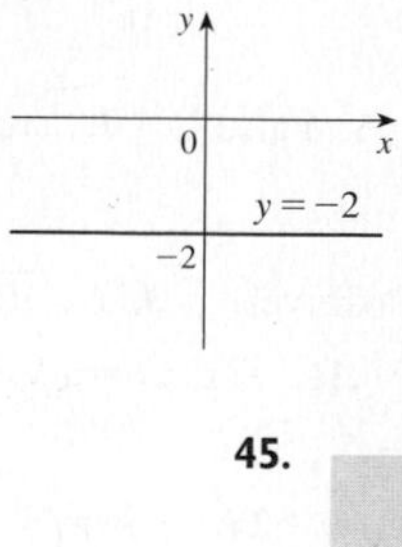

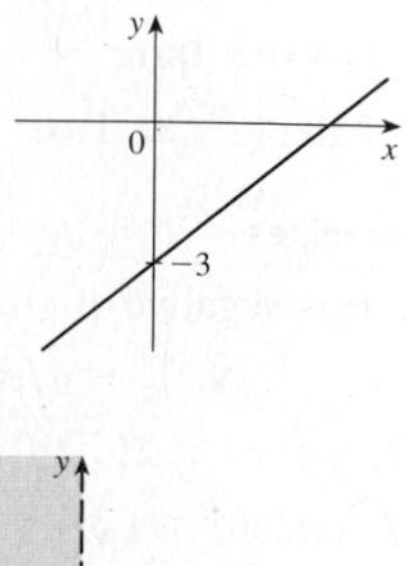

43.

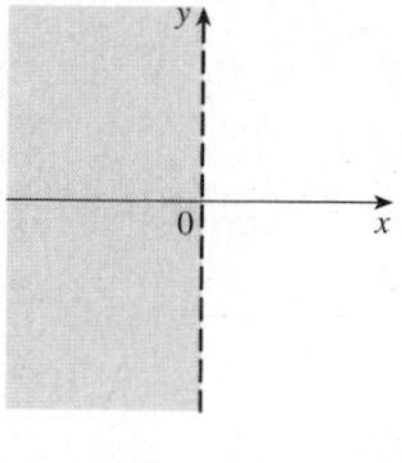

45.

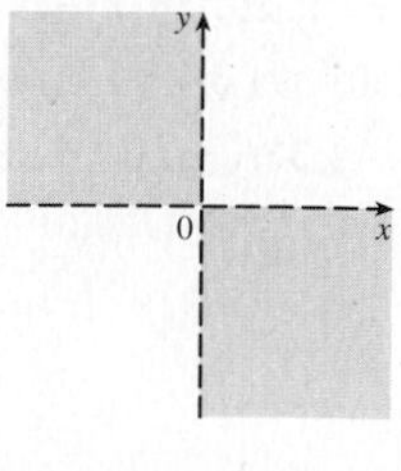

47.

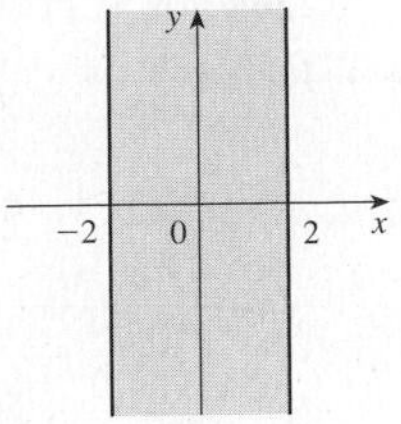

49.

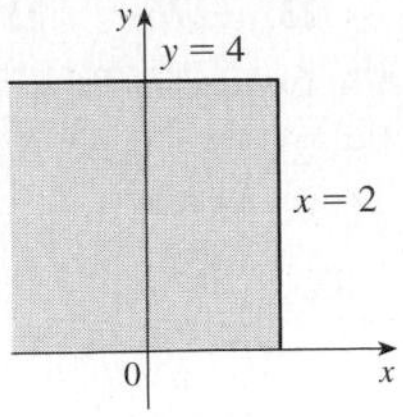

51.

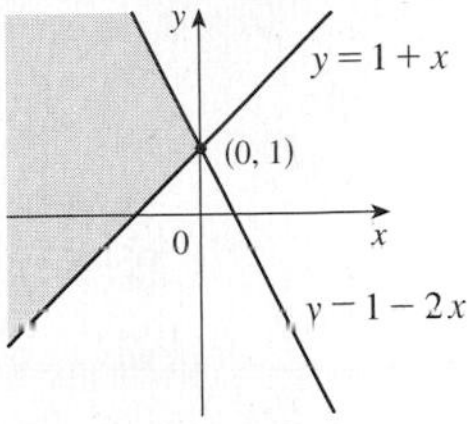

53. $(0, -4)$ **55.** (a) $(4, 9)$ (b) $(3.5, -3)$ **57.** $(1, -2)$

59. $y = x - 3$ **61.** (b) $4x - 3y - 24 = 0$

EXERCISES C ▪ PAGE A23

1. $(x - 3)^2 + (y + 1)^2 = 25$ **3.** $x^2 + y^2 = 65$

5. $(2, -5), 4$ **7.** $\left(-\frac{1}{2}, 0\right), \frac{1}{2}$ **9.** $\left(\frac{1}{4}, -\frac{1}{4}\right), \sqrt{10}/4$

11. Parabola

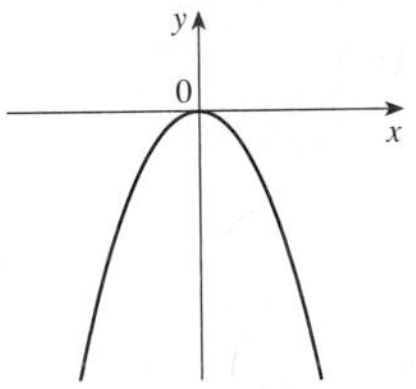

13. Ellipse

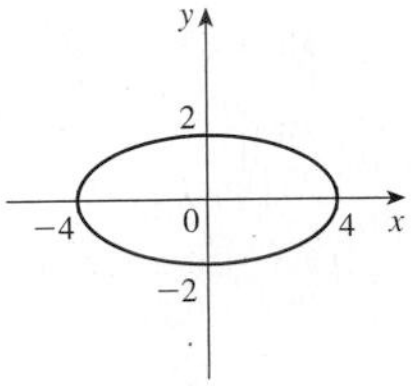

15. Hyperbola

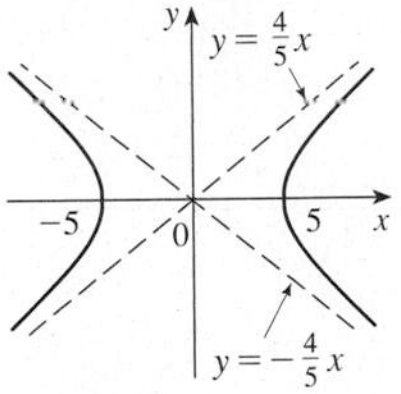

17. Ellipse

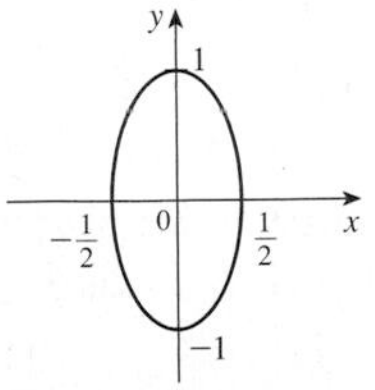

19. Parabola

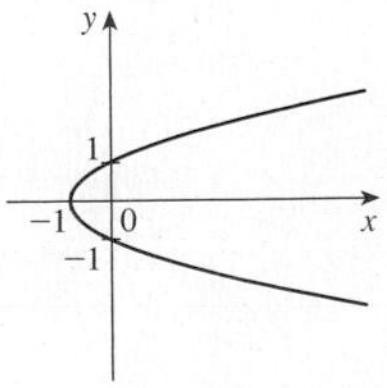

21. Hyperbola

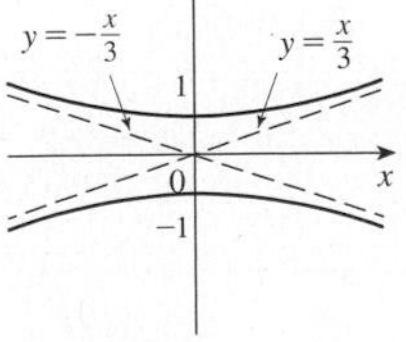

23. Hyperbola

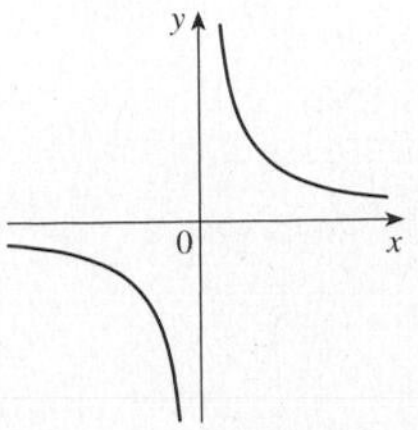

25. Ellipse

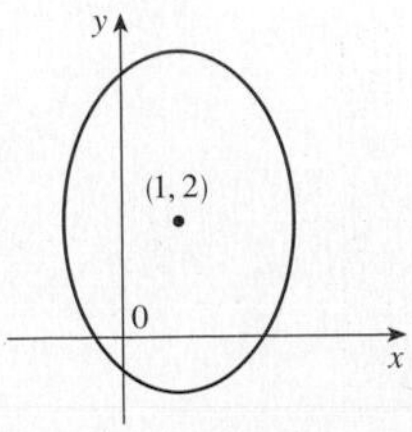

27. Parabola

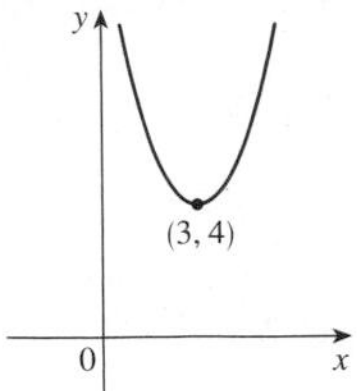

29. Parabola

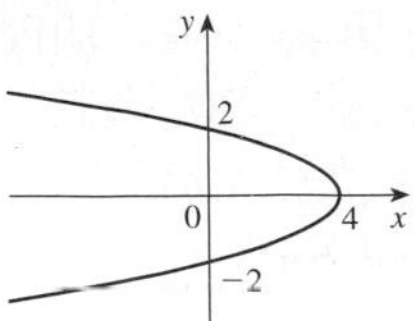

31. Ellipse

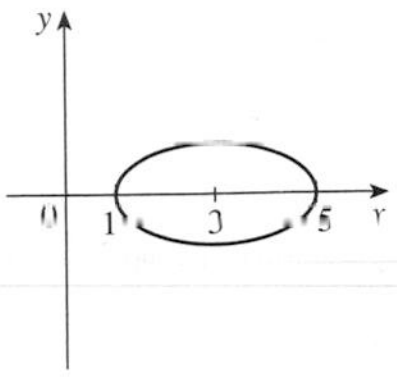

33.

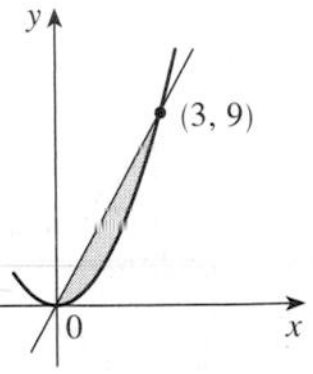

35. $y = x^2 - 2x$

37.

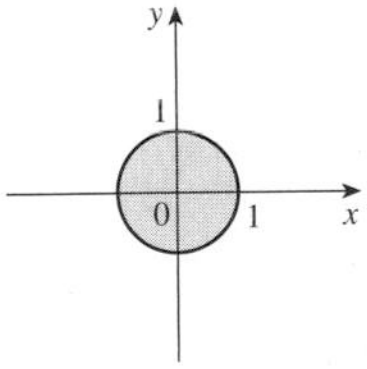

39.

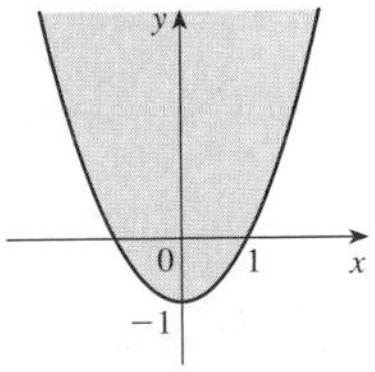

EXERCISES D ▪ PAGE A32

1. $7\pi/6$ **3.** $\pi/20$ **5.** 5π **7.** $720°$ **9.** $75°$

11. $-67.5°$ **13.** 3π cm **15.** $\frac{2}{3}$ rad $= (120/\pi)°$

17.

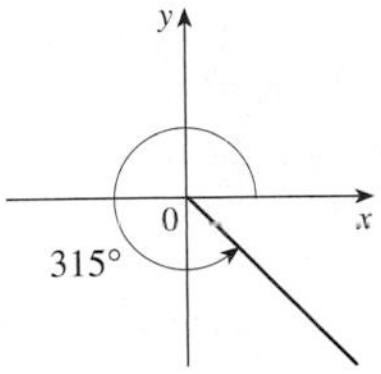

19.

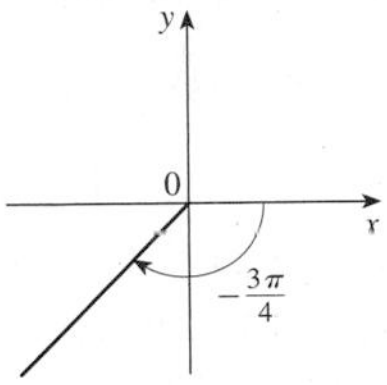

21.

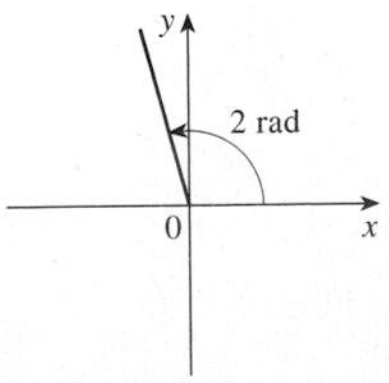

23. $\sin(3\pi/4) = 1/\sqrt{2}$, $\cos(3\pi/4) = -1/\sqrt{2}$, $\tan(3\pi/4) = -1$, $\csc(3\pi/4) = \sqrt{2}$, $\sec(3\pi/4) = -\sqrt{2}$, $\cot(3\pi/4) = -1$

25. $\sin(9\pi/2) = 1$, $\cos(9\pi/2) = 0$, $\csc(9\pi/2) = 1$, $\cot(9\pi/2) = 0$, $\tan(9\pi/2)$ and $\sec(9\pi/2)$ undefined

27. $\sin(5\pi/6) = \frac{1}{2}$, $\cos(5\pi/6) = -\sqrt{3}/2$, $\tan(5\pi/6) = -1/\sqrt{3}$, $\csc(5\pi/6) = 2$, $\sec(5\pi/6) = -2/\sqrt{3}$, $\cot(5\pi/6) = -\sqrt{3}$

29. $\cos\theta = \frac{4}{5}$, $\tan\theta = \frac{3}{4}$, $\csc\theta = \frac{5}{3}$, $\sec\theta = \frac{5}{4}$, $\cot\theta = \frac{4}{3}$

31. $\sin\phi = \sqrt{5}/3$, $\cos\phi = -\frac{2}{3}$, $\tan\phi = -\sqrt{5}/2$, $\csc\phi = 3/\sqrt{5}$, $\cot\phi = -2/\sqrt{5}$

33. $\sin\beta = -1/\sqrt{10}$, $\cos\beta = -3/\sqrt{10}$, $\tan\beta = \frac{1}{3}$, $\csc\beta = -\sqrt{10}$, $\sec\beta = -\sqrt{10}/3$

35. 5.73576 cm **37.** 24.62147 cm **59.** $\frac{1}{15}(4 + 6\sqrt{2})$

61. $\frac{1}{15}(3 + 8\sqrt{2})$ **63.** $\frac{24}{25}$ **65.** $\pi/3, 5\pi/3$

67. $\pi/4, 3\pi/4, 5\pi/4, 7\pi/4$ **69.** $\pi/6, \pi/2, 5\pi/6, 3\pi/2$

71. $0, \pi, 2\pi$ **73.** $0 \leq x \leq \pi/6$ and $5\pi/6 \leq x \leq 2\pi$

75. $0 \leq x < \pi/4$, $3\pi/4 < x < 5\pi/4$, $7\pi/4 < x \leq 2\pi$

77.

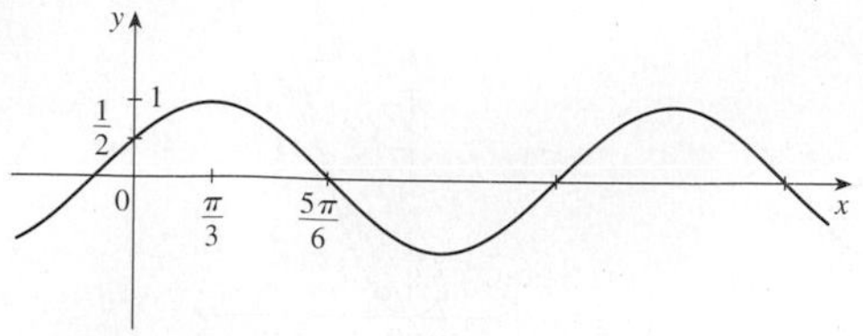

79.

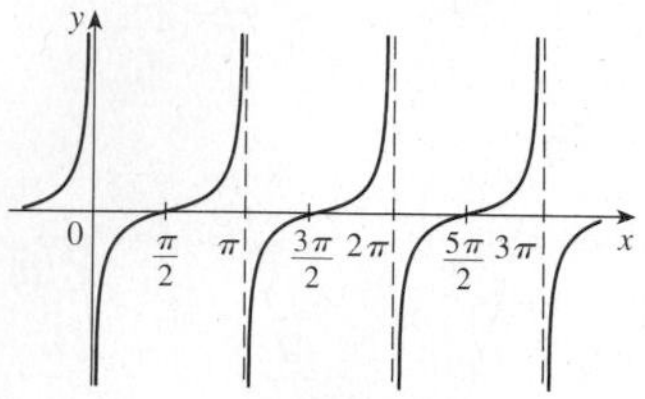

81.

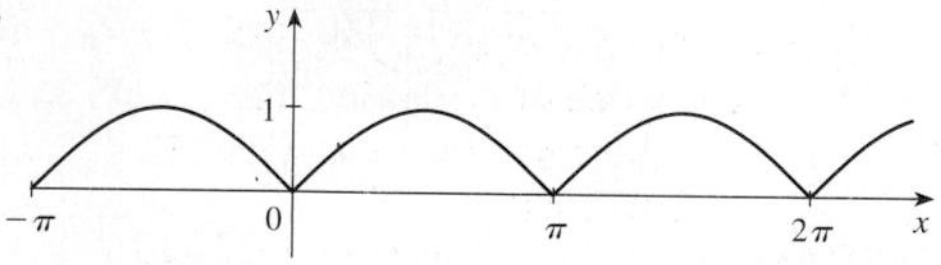

89. 14.34457 cm^2

EXERCISES E ▪ PAGE A38

1. $\sqrt{1} + \sqrt{2} + \sqrt{3} + \sqrt{4} + \sqrt{5}$ **3.** $3^4 + 3^5 + 3^6$

5. $-1 + \frac{1}{3} + \frac{3}{5} + \frac{5}{7} + \frac{7}{9}$ **7.** $1^{10} + 2^{10} + 3^{10} + \cdots + n^{10}$

9. $1 - 1 + 1 - 1 + \cdots + (-1)^{n-1}$ **11.** $\sum_{i=1}^{10} i$

13. $\sum_{i=1}^{19} \frac{i}{i+1}$ **15.** $\sum_{i=1}^{n} 2i$ **17.** $\sum_{i=0}^{5} 2^i$ **19.** $\sum_{i=1}^{n} x^i$

21. 80 **23.** 3276 **25.** 0 **27.** 61 **29.** $n(n + 1)$

31. $n(n^2 + 6n + 17)/3$ **33.** $n(n^2 + 6n + 11)/3$

35. $n(n^3 + 2n^2 - n - 10)/4$

41. (a) n^4 (b) $5^{100} - 1$ (c) $\frac{97}{300}$ (d) $a_n - a_0$

43. $\frac{1}{3}$ **45.** 14 **49.** $2^{n+1} + n^2 + n - 2$

EXERCISES G ▪ PAGE A56

1. (b) 0.405

EXERCISES H ▪ PAGE A64

1. $8 - 4i$ **3.** $13 + 18i$ **5.** $12 - 7i$ **7.** $\frac{11}{13} + \frac{10}{13}i$

9. $\frac{1}{2} - \frac{1}{2}i$ **11.** $-i$ **13.** $5i$ **15.** $12 + 5i, 13$

17. $4i, 4$ **19.** $\pm\frac{3}{2}i$ **21.** $-1 \pm 2i$

23. $-\frac{1}{2} \pm (\sqrt{7}/2)i$ **25.** $3\sqrt{2}\,[\cos(3\pi/4) + i\sin(3\pi/4)]$

27. $5\{\cos[\tan^{-1}(\frac{4}{3})] + i\sin[\tan^{-1}(\frac{4}{3})]\}$

29. $4[\cos(\pi/2) + i\sin(\pi/2)]$, $\cos(-\pi/6) + i\sin(-\pi/6)$, $\frac{1}{2}[\cos(-\pi/6) + i\sin(-\pi/6)]$

31. $4\sqrt{2}\,[\cos(7\pi/12) + i\sin(7\pi/12)]$, $(2\sqrt{2})[\cos(13\pi/12) + i\sin(13\pi/12)]$, $\frac{1}{4}[\cos(\pi/6) + i\sin(\pi/6)]$

33. -1024 **35.** $-512\sqrt{3} + 512i$

37. $\pm 1, \pm i, (1/\sqrt{2})(\pm 1 \pm i)$ **39.** $\pm(\sqrt{3}/2) + \frac{1}{2}i, -i$

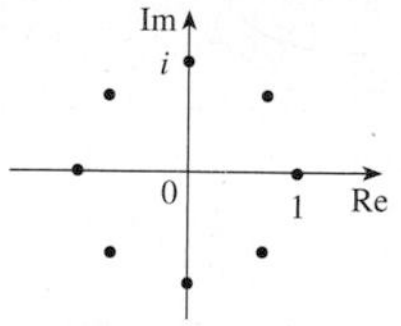

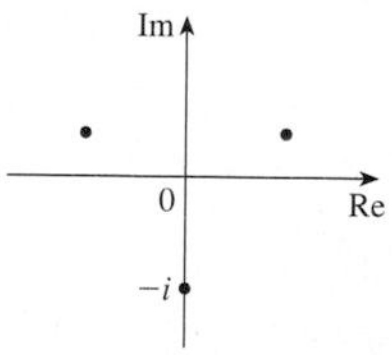

41. i **43.** $\frac{1}{2} + (\sqrt{3}/2)i$ **45.** $-e^2$

47. $\cos 3\theta = \cos^3\theta - 3\cos\theta\sin^2\theta$, $\sin 3\theta = 3\cos^2\theta\sin\theta - \sin^3\theta$

INDEX

RP denotes Reference Page numbers.

Abel, Niels, 210
absolute maximum and minimum, 271, 923, 928
absolute maximum and minimum values, 271, 923, 928
absolute value, 17, A6, A58
absolute value function, 17
absolutely convergent series, 714
acceleration as a rate of change, 161, 221
acceleration of a particle, 839
  components of, 842
  as a vector, 839
Achilles and the tortoise, 6
adaptive numerical integration, 504
addition formulas for sine and cosine, A28, A29
addition of vectors, 770, 773
Airy, Sir George, 728
Airy function, 728
algebraic function, 31
alternating harmonic series, 711
alternating series, 710
Alternating Series Estimation Theorem, 712
Alternating Series Test, 710
analytic geometry, A10
angle, A24
  between curves, 268
  of deviation, 279
  negative, A25
  positive, A25
  standard position, A25
  between vectors, 779
angular momentum, 848
angular speed, 840
antiderivative, 340
antidifferentiation formulas, 341
aphelion, 667
apolune, 661
approach path of an aircraft, 206
approximate integration, 495
approximating cylinder, 424
approximating surface, 533
approximation
  by differentials, 250
  to e, 179
  linear
  by the Midpoint Rule, 496
  by Newton's method, 335
  by an nth-degree Taylor polynomial, 253
  quadratic, 253
  by Riemann sums, 367
  by Simpson's Rule, 500, 502
  tangent line, 247
  by Taylor polynomials, 253, 749
  by Taylor's Inequality, 737
  by the Trapezoidal Rule, 497
approximation, linear, 247, 894, 898
  to a tangent plane, 894
Archimedes' Principle, 449, 1104
arc length, 525
  of a parametric curve, 633
  of a polar curve, 652
  of a space curve, 830, 831
arc length contest, 532
arc length formula, 526
arc length function, 528, 831
area, 3, 355
  of a circle, 469
  under a curve, 355, 360, 366
  between curves, 415, 418
  of an ellipse, 468
  by exhaustion, 3
  by Green's Theorem, 1058
  enclosed by a parametric curve, 632
  in polar coordinates, 639
  of a sector of a circle, 650
  surface, 635, 1075, 1077
  of a surface of a revolution, 532, 538
area, surface, 635, 1075, 1077
area function, 379
Area Problem, 3, 355
argument of a complex number, A59
arithmetic-geometric mean, 686
arrow diagram, 12
astroid, 213, 629
asymptote(s)
  in graphing, 308
  horizontal, 132, 308
  of a hyperbola, 658, A20
  slant, 312
  vertical, 95, 308
asymptotic curve, 315
autonomous differential equation, 575
auxiliary equation, 1112
  complex roots of, 1114
  real roots of, 1113
average cost function, 330
average rate of change, 148, 221
average speed of molecules, 516
average value of a function, 443, 557, 956, 1000
average velocity, 5, 85, 145, 221, 838
axes, coordinate, 765, A11
axes of ellipse, A19
axis of a parabola, 655

bacterial growth, 591, 596
Barrow, Isaac, 4, 153, 380
baseball and calculus, 601
base of a cylinder, 422
base of a logarithm, 63, A53
  change of, 66
basis vectors, 774, 775
Bernoulli, James, 580, 607
Bernoulli, John, 307, 580, 625, 736
Bernoulli differential equation, 607
Bessel, Friedrich, 724
Bessel function, 724, 728
Bézier, Pierre, 639
Bézier curves, 624, 639
binomial coefficients, 742
binomial series, 742, 748
  discovery by Newton, 748
Binomial Theorem, RP1
binormal vector, 834
blackbody radiation, 757
blood flow, 227, 332, 551
boundary curve, 1093
boundary-value problem, 1116
bounded sequence, 682
bounded set, 928

Boyle's Law, 231, 246
Brache, Tycho, 844
brachistochrone problem, 625
branches of hyperbola, 658, A20
Buffon's needle problem, 565
bullet-nose curve, 51, 204

C^1 transformation, 1013
cable (hanging), 255
calculator, graphing, 46, 315. *See also* computer algegra system
calculus, 9
  invention of, 399
cancellation equations
  for inverse functions, 62
  for inverse trigonometric functions, 68
  for logarithms, 64
cans, minimizing manufacturing cost of, 333
Cantor, Georg, 696
Cantor set, 696
capital formation, 554
cardiac output, 552
cardioid, 213, 643
carrying capacity, 233, 568
Cartesian coordinate system, A11
Cartesian plane, A11
Cassini, Giovanni, 649
catenary, 255
CAS. *See* computer algebra system
Cauchy, Augustin-Louis, 113, 961, A45
Cauchy's Mean Value Theorem, A45
Cauchy-Schwarz Inequality, 786
Cavalieri's Principle, 432
center of gravity. *See* center of mass
center of mass, 542
  of a lamina, 981
  of a plate, 545
  of a solid, 996
  of a surface, 1083
  of a wire, 1036
centripetal force, 852
centroid
  of a plane region, 543
  of a solid, 996
Chain Rule, 197, 200
  for several variables, 901, 903, 904
change of base, formula for, 66
change of variables
  in a double integral, 975, 1013, 1016
  in integration, 401
  in a triple integral, 1002, 1007, 1019
characteristic equation, 1112
charge, electric, 224, 980, 996
  in a circuit, 1129
charge density, 980, 996
chemical reaction, 224
circle, area of, 469
circle, equation of, A16
circle of curvature, 835
circular cylinder, 422
circular paraboloid, 810
circulation of a vector field, 1096
cissoid of Diocles, 628, 648
Clairaut, Alexis, 885
Clairaut's Theorem, 885, A3
clipping planes, 804
closed curve, 1048
closed interval, A3
Closed Interval Method, 275
  for a function of two variables, 929
closed set, 928
closed surface, 1086
Cobb, Charles, 856
Cobb-Douglas production function, 857, 887, 940
cochleoid, 670
coefficient(s)
  binomial, 742
  of friction, 196, 278
  of inequality, 399
  of a polynomial, 28
  of a power series, 723
  of static friction, 815
combinations of functions, 41
comets, orbits of, 668
common ratio, 689
comparison properties of the integral, 375
comparison test for improper integrals, 514
Comparison Test for series, 705
Comparison Theorem for integrals, 514
complementary equation, 1117
Completeness Axiom, 682
complex conjugate, A57
complex exponentials, A63
complex number(s), A57
  addition and subtraction of, A57
  argument of, A59
  equality of, A57
  imaginary part of, A57
  modulus of, A58
  multiplication and division of, A57, A60
  polar form, A59
  powers of, A61
  principal square root of, A58
  real part of, A57
  roots of, A62
component function, 817, 1028
components of acceleration, 842
components of a vector, 772, 782
composition of functions, 41, 197
  continuity of, 125, 875
  derivative of, 199
compound interest, 238, 306
compressibility, 225
computer algebra system
  integration with, 491, 732
  pitfalls of using, 91
computer algebra system, graphing with, 46
  for creating a three-dimensional scene, 804
  a curve, 315
  function of two variables, 859
  level curves, 864
  parametric equations, 624
  parametric surface, 1072
  partial derivatives, 885
  polar curve, 646
  sequence, 680
  space curve, 820
  vector field, 1029
concavity, 290
Concavity Test, 291, A44
concentration, 224
conchoid, 626, 648
conditionally convergent series, 715
conductivity (of a substance), 1090
cone, 808
  parametrization of, 1073
conic section, 654, 662
  directrix, 662
  eccentricity, 662
  focus, 662
  polar equation, 664
  shifted, 659, A21
  vertex (vertices), 655
conjugates, properties of, A58
connected region, 1048
conservation of energy, 1052
conservative vector field, 1032, 1053
constant function, 173
Constant Multiple Law of limits, 100
Constant Multiple Rule, 176
constraint, 934, 938
consumer surplus, 550
continued fraction expansion, 686
continuity
  of a function, 119
  of a function of three variables, 876
  of a function of two variables, 874
  on an interval, 121
  from the left, 121
  from the right, 121
  of a vector function, 818
continuous compounding of interest, 238, 306
continuous random variable, 555
contour curves, 860
contour map, 860
convergence
  absolute, 714
  conditional, 715
  of an improper integral, 509, 512
  interval of, 725
  radius of, 725
  of a sequence, 677
  of a series, 688

convergent improper integral, 509, 512
convergent sequence, 677
convergent series, 688
properties of, 693
cooling tower, hyperbolic, 810
coordinate system, A2
cylindrical, 1001
Cartesian, A11
polar, 639
rectangular, A11
spherical, 1005
three-dimensional rectangular, 766
coordinate axes, 765, A11
coordinate planes, 765
coplanar vectors, 791
Coriolis acceleration, 851
Cornu's spiral, 637
cosine function, A26
derivative of, 192
graph of, 32, A31
power series for, 740
cost function, 228, 327
critically damped vibration, 1127
critical number, 274
critical point(s), 923, 933
cross product, 786
direction of, 788
geometric characterization of, 789
magnitude of, 789
properties of, 790
cross-section, 422
of a surface, 804
cubic function, 28
current, 224
curl of a vector field, 1062
curvature, 638, 832
curve(s)
asymptotic, 315
Bézier, 624, 639
boundary, 1093
bullet-nose, 51, 204
cissoid of Diocles, 628, 648
closed, 1048
devil's, 213
dog saddle, 868
epicycloid, 630
equipotential, 868
grid, 1071
helix, 818
length of, 525, 830
level, 854, 860
monkey saddle, 868
orientation of, 1039, 1055
orthogonal, 214
ovals of Cassini, 649
parametric, 621, 818
piecewise-smooth, 1035
polar, 641
serpentine, 188
simple, 1049
smooth, 831
space, 818
strophoid, 653, 671
swallowtail catastrophe, 629
toroidal spiral, 820
trefoil knot, 820
trochoid, 628
twisted cubic, 820
witch of Maria Agnesi, 628
curve fitting, 25
curve-sketching procedure, 308
cusp, 626
cycloid, 624
cylinder, 422
parabolic, 805
parametrization of, 1073
cylindrical coordinate system, 1001
conversion equations for, 1001
triple integrals in, 1002
cylindrical shell, 433

damped vibration, 1126
damping constant, 1126
decay, law of natural, 236
decay, radioactive, 236
decreasing function, 20
decreasing sequence, 681
definite integral, 366, 951
properties of, 373
Substitution Rule for, 404
of a vector function, 827
definite integration
by parts, 453, 456
by substitution, 404
degree of a polynomial, 28
del (∇), 913
delta (Δ) notation, 147, 148
demand curve, 327, 550
demand function, 327, 550
De Moivre, Abraham, A61
De Moivre's Theorem, A61
density
of a lamina, 980
linear, 223
liquid, 540
mass vs. weight, 540
of a solid, 996
dependent variable, 11, 855, 903
derivative(s), 143,154
of a composite function, 199
of a constant function, 173
directional, 910, 911, 914
domain of, 154
of exponential functions, 180, 201, A54, A55
as a function, 154
higher, 160
higher partial, 884
of hyperbolic functions, 256
of an integral, 381
of an inverse function, 215
of inverse trigonometric functions, 211, 213
left-hand, 165
of logarithmic functions, 215, A49, A54
notation, 157
notation for partial, 880
normal, 1069
partial, 879, 880
of a power function, 174
of a power series, 729
of a product, 183, 184
of a quotient, 185, 186
as a rate of change, 148
right-hand, 165
second, 160
second partial, 826
as the slope of a tangent, 147
third, 161
of trigonometric functions, 189, 193
of a vector function, 824
Descartes, René, A11
descent of aircraft, determining start of, 206
determinant, 786
devil's curve, 213
Difference Law of limits, 100
Difference Rule, 177
differentiable function, 157, 895
differential, 250, 896, 898
differential equation, 234, 342, 566, 569
autonomous, 575
Bernoulli, 607
first-order, 569
general solution of, 570
homogeneous, 1111
linear, 602
linearly independent solutions, 1112
logistic, 592, 687
nonhomogeneous, 1111, 1117
order of, 569
partial, 886
second-order, 569, 1111
separable, 580
solution of, 569
differentiation, 157
formulas for, 187, RP5
formulas for vector functions, 826
implicit, 207, 208, 883, 905
logarithmic, 217
partial, 878, 883, 884
of a power series, 729
term by term, 729
of a vector function, 824
differentiation operator, 157

Direct Substitution Property, 102
direction angles, 781
direction cosines, 781
direction field, 572, 573
direction numbers, 795
directional derivative, 910, 911, 914
  maximum value of, 915
  of a temperature function, 910, 911
directrix, 655, 662
discontinuity, 119, 120
discontinuous function, 119
discontinuous integrand, 511
disk method for approximting volume, 424
dispersion, 280
displacement, 145, 395
displacement vector, 770, 783
distance
  between point and line in space, 793
  between point and plane, 793, 800
  between planes, 801
  between points in a plane, A11
  between points in space, 767
  between real numbers, A7
distance formula, A12
  in three dimensions, 767
distance problem, 362
divergence
  of an improper integral, 509, 512
  of an infinite series, 688
  of a sequence, 677
  of a vector field, 1065
Divergence, Test for, 692
Divergence Theorem, 1099
divergent improper integral, 509, 512
divergent sequence, 677
divergent series, 688
division of power series, 745
DNA, helical shape of, 819
dog saddle, 868
domain of a function, 11, 855
Doppler effect, 909
dot product, 779
  in component form, 779
  properties of, 779
double integral, 951, 953
  change of variable in, 1013, 1016
  over general regions, 965, 966
  Midpoint Rule for, 955
  in polar coordinates, 974, 975
  properties of, 958, 970
  over rectangles, 951
double Riemann sum, 954
double-angle formulas, A29
Douglas, Paul, 856
dumpster design, minimizing construction cost, 933
dye dilution method, 552

e (the number), 56, 179, A50
  as a limit, 219
  as a sum of an infinite series, 739
eccentricity, 662
electric charge, 980, 996
electric circuit, 605
  analysis of, 1129
electric field (force per unit charge), 1031
electric force, 1031
electric flux, 1089
elementary function, integrability of, 487
element of a set, A3
ellipse, 213, 656, 662, A19
  area, 491
  directrix, 662
  eccentricity, 662
  foci, 656, 662
  major axis, 657, 667
  polar equation, 664, 667
  reflection property, 658
  rotated, 214
  vertices, 657
ellipsoid, 806, 808
elliptic paraboloid, 806, 808
empirical model, 25
end behavior of a function, 142
endpoint extreme values, 272
energy
  conservation of, 1052
  kinetic, 1052
  potential, 1053
epicycloid, 630
equation(s)
  of a circle, A17
  differential. *See* differential equation
  of an ellipse, 657, 664, A19
  of a graph, A16
  heat conduction, 890
  of a hyperbola, 658, 659, 664, A20
  Laplace's, 886, 1066
  of a line, A12, A13, A14, A16
  of a line in space, 794, 795, 796
  linear, 798, A14
  logistic difference, 687
  logistic differential, 568, 600
  Lotka-Volterra, 609
  *n*th-degree, 210
  of a parabola, 655, 664, A18
  parametric, 621, 795, 818, 1070
  of a plane, 798
  point-slope, 19, A12
  polar, 641, 664
  predator-prey, 609
  second-degree, A16
  slope-intercept, A13
  of a space curve, 818
  of a sphere, 768
  symmetric, 795
  two-intercept form, A16
  vector, 794
  wave, 886
equilateral hyperbola, A21
equilibrium point, 610
equilibrium solution, 568, 609
equipotential curves, 868
equivalent vectors, 770
error
  in approximate integration, 497, 498
  percentage, 251
  relative, 251
  in Taylor approximation, 750
error bounds, 499, 503
error estimate
  for alternating series, 712
  for the Midpoint Rule, 497, 498
  for Simpson's Rule, 503
  for the Trapezoidal Rule, 497, 498
escape velocity, 517
estimate of the sum of a series, 700, 708, 712, 717
Eudoxus, 3
Euler, Leonard, 56, 698, 739
Euler's formula, A63
Euler's Method, 575
even function, 19, 308
expected values, 987
exponential decay, 233
exponential function(s), 33, 52, 180
  with base *a*, A54, A55
  derivative of, 180, 201, A54, A55
  graphs of, 53, 179
  integration of, 371, 385, 402, 743, 744
  limits of, 136, A53
  power series for, 736
  properties of, A53
exponential graph, 53
exponential growth, 233
exponents, laws of, 54, A53, A55
extrapolation, 27
extreme value, 271
Extreme Value Theorem, 272, 928

family
  of functions, 50, 318, 320
  of hypocycloids, 629
  of parametric curves, 625
  of solutions, 568
fat circles, 211, 531
Fermat, Pierre, 4, 153, 273
Fermat's Principle, 331
Fermat's Theorem, 273
Fibonacci, 686
Fibonacci sequence, 676
field
  conservative, 1032
  electric, 1031

force, 1031
gradient, 919, 1031
gravitational, 1031
incompressible, 1066
irrotational, 1064
scalar, 1029
vector, 1027, 1028
velocity, 1027, 1030
First Derivative Test, 288
for Absolute Extreme Values, 324
first octant, 765
first-order linear differential equation, 602
first-order optics, 754
fixed point of a function, 171, 286
flash bulb, current to, 84
flow lines, 1033
fluid flow, 1030, 1066, 1088
flux, 551, 552, 1087, 1089
flux integral, 1087
FM synthesis, 318
focus
of a conic section, 662
of an ellipse, 656, 662
of a hyperbola, 658
of a parabola, 655
folium of Descartes, 208, 672
force, 438
centripetal, 852
constant, 783
exerted by fluid, 539
resultant, 776
torque, 791
force field, 1027, 1031
forced vibrations, 1128
Fourier, Joseph, 230
Fourier series, finite, 467
four-leaved rose, 643
fractions (partial), 473
Frenet-Serret formulas, 838
Fresnel, Augustin, 383
Fresnel function, 383
frustum, 431, 432
Fubini, Guido, 961
Fubini's Theorem, 961, 991
function(s), 11
absolute value, 17
Airy function, 728
algebraic, 31
arc length, 528, 830, 831
area, 379
arrow diagram of, 12
average cost, 330
average value of, 433, 557, 956, 1000
Bessel, 724, 728
Cobb-Douglas production, 857, 887, 940
combinations of, 41
component, 817, 1028
composite, 41, 197, 875
constant, 173
continuity of, 119, 818, 874, 876
cost, 228, 327
cubic, 28
decreasing, 20
demand, 327, 550
derivative of, 146
differentiability of, 157, 895
discontinuous, 119
domain of, 11, 855
elementary, 487
even, 19, 308
exponential, 33, 52
extreme values of, 271
family of, 50, 318, 320
fixed point of, 171, 286
Fresnel, 383
Gompertz, 600
gradient of, 913, 915
graph of, 12, 858
greatest integer, 105
harmonic, 886
Heaviside, 45, 92
homogeneous, 909
hyperbolic, 254
implicit, 207
increasing, 20
integrable, 953
inverse, 59, 61
inverse hyperbolic, 257
inverse trigonometric, 67, 68
joint density, 985, 996
limit of, 88, 109, 871, 876
linear, 24, 858
logarithmic, 34, 63, A50, A53
machine diagram of, 12
marginal cost, 229, 327
marginal profit, 327
marginal revenue, 327
maximum and minimum values of, 271, 922, 923
natural logarithmic, 64
nondifferentiable, 159
of *n* variables, 865
odd, 19, 308
one-to-one, 60
periodic, 308
piecewise defined, 17
polynomial, 28, 874
position, 145
potential, 1032
power, 29
probability density, 555, 985
profit, 327
quadratic, 28
ramp, 45
range of, 11, 855
rational, 31, 874
reciprocal, 31
reflected, 38
representation as a power series, 728
representations of, 12
revenue, 327
root, 30
of several variables, 855, 864
shifted, 37
sine integral, 389
smooth, 525
step, 18
stretched, 38
of three variables, 864
transcendental, 34
transformation of, 37, 38
translation of, 38
trigonometric, 32, A26
of two variables, 855
value of, 11
vector-valued, 817
Fundamental Theorem of Calculus, 381, 384, 387
higher-dimensional versions, 1105
for line integrals, 1046
for vector functions, 828

G (gravitational constant), 231, 442
Gabriel's horn, 537
Galileo, 625, 633
Galois, Evariste, 210
Gause, G. F., 596
Gauss, Karl Friedrich, 1099, A35
Gaussian optics, 754
Gauss's Law, 1090
Gauss's Theorem, 1099
geometric series, 688
geometry of a tetrahedron, 794
Gompertz function, 600
gradient, 913, 915
gradient vector, 913, 915
interpretations of, 919, 920
gradient vector field, 919, 1031
graph(s)
of an equation, A16
of exponential functions, 53, 179
of a function, 12
of a function of two variables, 858
of logarithmic functions, 66
of a parametric curve, 622
of a parametric surface, 1083
polar, 641
of power functions, 30, RP3
of a sequence, 680
of trignometric functions, A30, RP2
graphing calculator, 46, 315, 624, 646
graphing device. *See* computer algebra system
gravitation law, 231, 442
gravitational acceleration, 438

gravitational field, 1031
great circle, 1011
Green, George, 1056, 1098
Green's identities, 1069
Green's Theorem, 1055, 1098
  vector forms, 1066, 1067
greatest integer function, 105
Gregory, James, 732, 736
Gregory's series, 732
grid curve, 1071
ground speed, 778
growth, law of natural, 234, 591
growth rate, 226
  relative, 234, 592

half-angle formulas, A29
half-life, 236
half-space, 864
hare-lynx system, 612
harmonic function, 886
harmonic series, 691
harmonic series, alternating , 711
heat conductivity, 1090
heat conduction equation, 890
heat flow, 1090
heat index, 878
Heaviside, Oliver, 92
Heaviside function, 45, 92
Hecht, Eugene, 250, 253, 754
helix, 818
higher derivatives, 160
higher partial derivatives, 884
homogeneous differential equation, 1111
homogeneous function, 909
Hooke's Law, 439, 1125
horizontal asymptote, 132. 308
horizontal line, equation of, A13
horizontal plane, equation of, 766
Horizontal Line Test, 60
Hubble Space Telescope, 276
Huygens, Christiaan, 625
hydrostatic pressure and force, 539
hydro-turbine optimization, 943
hyperbola, 658, 662, A20
  asymptotes, 658, A20
  branches, 658, A20
  directrix, 662
  eccentricity, 662
  equation, 658, 659, 664, A20
  equilateral, A21
  foci, 658, 662
  polar equation, 664
  reflection property, 662
  vertices, 658
hyperbolic function(s), 254
  derivatives, 256
  inverse, 257
hyperbolic identities, 255
hyperbolic paraboloid, 807, 808
hyperbolic substitution, 470, 471
hyperboloid, 808, 810
hypersphere, 1000
hypocycloid, 629

i (imaginary number), A55
i (standard basis vector), 774
I/D Test, 287
ideal gas law, 233, 891
image of a point, 1013
image of a region, 1013
implicit differentiation, 207, 208, 883, 905
implicit function, 207
Implicit Function Theorem, 906
improper integral, 508
impulse of a force, 601
incompressible velocity field, 1066
increasing function, 20
increasing sequence, 681
Increasing/Decreasing Test, 287
increment, 147, 898
indefinite integrals, 391
  table of, 392
independence of path, 1047
independent random variable, 986
independent variable, 11, 855, 903
indeterminate difference, 302
indeterminate forms of limits, 298
indeterminate power, 303
indeterminate product, 302
index of summation, A34
inequalities, rules for, A4
inertia (moment of), 983, 996, 1045
infinite discontinuity, 120
infinite interval, 508, 509
infinite limit, 94, 116, 136
infinite sequence. *See* sequence
infinite series, *See* Series
inflection point, 291
initial condition, 570
initial point
  of a parametric curve, 622
  of a vector, 770, 1115
initial-value problem, 570
inner product, 779
instantaneous rate of change, 85, 148, 221
instantaneous rate of growth, 226
instantaneous rate of reaction, 225
instantaneous velocity, 86, 145, 221
integer, A2
integrable function, 953
integral(s)
  approximations to, 372
  change of variables in, 400, 1011, 1016, 1019
  comparison properties of, 375
  conversion to cylindrical coordinates, 1002
  conversion to polar coordinates, 975
  conversion to spherical coordinates, 1007
  definite, 366, 827, 951
  derivative of, 381
  double, 951, 953. *See also* double integral
  evaluating, 369
  improper, 508
  indefinite, 391
  iterated, 959, 960
  line, 1034. *See also* line integral
  patterns in, 494
  properties of, 373
  surface, 1081, 1087
  of symmetric functions, 405
  table of, 452, 484, RP6–10
  triple, 990. *See also* triple integral
  units for, 396
Integral Test, 697, 699
integrand, 366
  discontinuous, 511
integration, 366
  approximate, 495
  by computer algebra system, 491
  of exponential functions, 371, 385, 402
  formulas, 452, 484, RP6–10
  indefinite, 391
  limits of, 366
  numerical, 495
  partial, 960
  by partial fractions, 473
  by parts, 45
  of a power series, 729
  by a rationalizing substitution, 481
  reversing order of, 962, 970
  substitution in, 401
  term by term, 729
  of a vector function, 827
intercepts, 308, A19
Intermediate Value Theorem, 126
intermediate variable, 903
interpolation, 27
intersection of planes, 799
intersection of polar graphs, area of, 651
intersection of sets, A3
intersection of three cylinders, 1005
interval, A3
interval of convergence, 725
inverse function(s), 59, 61
inverse transformation, 1013
inverse trigonometric functions, 67, 68
irrational number, A2
irrotational vector field, 1064
isobars, 854, 861
isothermal compressibility, 225
isothermals, 861, 868
iterated integral, 959, 960

j (standard basis vector), 774
Jacobi, Carl, 1015

Jacobian of a transformation, 1015, 1019
jerk, 161
joint density function, 985, 996
joule, 438
jump discontinuity, 120

k (standard basis vector), 774
Kampyle of Eudoxus, 213
Kepler, Johannes, 844, 848
Kepler's Laws, 844, 848
kinetic energy, 1052
Kirchhoff's Laws, 1129
Kondo, Shigeru, 739

Lagrange, Joseph, 282, 283, 935
Lagrange multiplier, 934, 935
lamina, 543, 980
Laplace, Pierre, 886, 1066
Laplace operator, 1066
Laplace's equation, 886, 1066
lattice point, 269
Law of Conservation of Angular Momentum, 848
Law of Conservation of Energy, 1053
law of cosines, A33
law of gravitation, 231, 442
law of laminar flow, 227
learning curve, 571
least squares method, 27, 932
least upper bound, 682
left-hand derivative, 165
left-hand limit, 93, 113
Leibniz, Gottfried Wilhelm, 4, 157, 399, 580, 748
Leibniz notation, 157
lemniscate, 213
length
  of a curve, 525
  of a line segment, A7, A12
  of a parametric curve, 633
  of a polar curve, 652
  of a space curve, 830
  of a vector, 773
level curve(s), 854, 860
  of barometric pressure, 854
  of temperatures, 861
level surface, 865
  tangent plane to, 917
l'Hospital, Marquis de, 299, 307
l'Hospital's Rule, 299, 307
  origins of, 307
libration point, 340
limaçon, 647
Limit Comparison Test, 707
Limit Laws, 99, A39
  for functions of two variables, 873
  for sequences, 678
limit(s), 3, 88
  calculating, 99
  of exponential functions, 136, 137
  of a function, 88, 110
  of a function of three variables, 876
  of a function of two variables, 871
  infinite, 94, 116, 136
  at infinity, 130, 131, 136
  of integration, 366
  left-hand, 93, 113
  of logarithmic functions, 96, A50
  one-sided, 93, 113
  precise definitions, 109, 113, 116, 138, 140
  properties of, 99
  right-hand, 93, 113
  of a sequence, 6, 357, 677
  involving sine and cosine functions, 190, 192
  of a vector function, 817
linear approximation, 247, 894, 898
linear combination, 1111
linear density, 223
linear differential equation, 602, 1111
linear equation, A14
  of a plane, 798
linear function, 24 858
linearity of an integral, 958
linearization, 48, 894
linearly independent solutions, 1112
linear model, 24
linear regression, 27
line(s) in the plane, A12
  equations of, A12, A13, A14
  horizontal, A13
  normal, 175
  parallel, A14
  perpendicular, A14
  secant, 4, 83, 84
  slope of, A12
  tangent, 4, 83, 84, 144
line (in space)
  normal, 918
  parametric equations of, 795
  skew, 797
  symmetric equations of, 795
  tangent, 824
  vector equation of, 794, 795
line integral, 1034
  Fundamental Theorem for, 1046
  for a plane curve, 1034
  with respect to arc length, 1037
  for a space curve, 1039
  of vector fields, 1041, 1042
  work defined as, 1041
liquid force, 539, 540
Lissajous figure, 629
lithotripsy, 658
local maximum and minimum values, 271, 923
logarithm(s), 34, 63
  laws of, 64, A49
  natural, 64, A48
  notation for, 64
logarithmic differentiation, 217
logarithmic function(s), 34, 63
  with base *a*, A53
  derivatives of, 213, A51, A53
  graphs of, 64, 66
  limits of, 96, A50
  properties of, 64, A49
logistic difference equation, 687
logistic differential equation, 568, 592
logistic model, 568
logistic sequence, 687
LORAN system, 661
Lotka-Volterra equations, 609

machine diagram of a function, 12
Maclaurin, Colin, 736
Maclaurin series, 734, 736
  table of, 743
magnitude of a vector, 773
major axis of ellipse, 657
marginal cost function, 229, 327
marginal profit function, 327
marginal productivity, 887
marginal propensity to consume or save, 695
marginal revenue function, 327
mass
  of a lamina, 980
  of a solid, 996
  of a surface, 1083
  of a wire, 1036
mass, center of. *See* center of mass
mathematical induction, principle of, 77, 80, A36
mathematical model, 14, 24
  Cobb-Douglas, for production costs, 857, 887, 940
  for vibration of membrane, 724
maximum and minimum values, 271, 922, 923
mean life of an atom, 517
mean of a probability density function, 557
Mean Value Theorem, 282
  for double integrals, 1023
  for integrals, 443
mean waiting time, 557
median of a probability density function, 559
method of cylindrical shells, 433
method of exhaustion, 3, 102
method of Lagrange multipliers, 934, 935, 938
method of least squares, 27, 932

method of undetermined
coefficients, 1118, 1122
midpoint formula, A16
Midpoint Rule, 372, 496
for double integrals, 955
error in using, 497
for triple integrals, 998
mixing problems, 584
Möbius, August, 1085
Möbius strip, 1079, 1085
modeling
with differential equations, 567
motion of a spring, 568
population growth, 55, 567, 591, 597, 600, 616
vibration of membrane, 724
model(s), mathematical, 24
comparison of natural growth vs. logistic, 596
empirical, 25
exponential, 33
Gompertz function, 600
linear, 24
logarithmic, 34
polynomial, 28
power function, 29
predator-prey, 233, 609
rational function, 31
seasonal-growth, 600
trigonometric, 32, 33
von Bertalanffy, 616
modulus, A58
moment
about an axis, 543, 981
of inertia, 983, 996, 1045
of a lamina, 543, 981
of a mass, 542
about a plane, 996
polar, 983
second, 983
of a solid, 995
of a system of particles, 543
momentum of an object, 601
monkey saddle, 868
monotonic sequence, 681
Monotonic Sequence Theorem, 683
motion in space, 838
motion of a spring, force affecting
restoring, 1125
damping, 1126
resonance, 1129
movie theater seating, 446
multiple integrals. *See* double integral; triple integral
multiplication, scalar, of vectors, 771, 773
multiplication of power series, 745
multiplier (Lagrange), 934, 935, 938
multiplier effect, 695

natural exponential function, 56, A50
derivative of, 180, A52
graph of, 179
power series for, 736
properties of, A51
natural growth law, 234, 591
natural logarithm function, 64, A50
derivative of, 215, A51
limits of, A50
properties of, A51
n-dimensional vector, 774
negative angle, A25
net area, 367
Net Change Theorem, 394
net investment flow, 554
newton (unit of force), 438
Newton, Sir Isaac, 4, 9, 102, 153, 157, 380, 399, 748, 844, 848
Newton's Law of Cooling, 237
Newton's Law of Gravitation, 231, 442, 844, 1030
Newton's method, 334, 335
Newton's Second Law of Motion, 438, 840, 844, 1125
Nicomedes, 626
nondifferentiable function, 159
nonhomogeneous differential equation, 1111, 1117
nonparallel planes, 799
normal component of acceleration, 842
normal derivative, 1069
normal distribution, 559
normal line, 175, 918
normal plane, 835
normal vector, 797, 834
*n*th-degree equation, roots of, 210
*n*th-degree Taylor polynomial, 254, 737
number
complex, A55
integer, A2
irrational, A2
rational, A2
real, A2
numerical integration, 495

octant, 765
odd function, 19, 308
one-sided limits, 93, 113
one-to-one function, 60
one-to-one transformation, 1013
open interval, A3
open region, 1048
optics
first-order, 754
Gaussian, 754
third-order, 755
optimization problems, 271, 322
orbits of planets, 844, 848
order of a differential equation, 569
ordered pair, A10
ordered triple, 765
order of integration, reversed, 962, 970
Oresme, Nicole, 692
orientation of a curve, 1039, 1055
orientation of a surface, 1086
oriented surface, 1085, 1086
origin, A2, A10
orthogonal curves, 214
orthogonal projection, 785
orthogonal surfaces, 922
orthogonal trajectory, 214, 583
orthogonal vectors, 781
osculating circle, 835
osculating plane, 835
Ostrogradsky, Mikhail, 1099
ovals of Cassini, 649
overdamped vibration, 1127

Pappus, Theorem of, 546
Pappus of Alexandria, 546
parabola, 655, 662, A18
axis, 655
directrix, 655, 662
equation, 655, 656
focus, 655, 662
polar equation, 664
reflection property, 268, 269
vertex, 655
parabolic cylinder, 805
paraboloid, 806, 810
paradoxes of Zeno, 6
parallel lines, A14
parallel planes, 799
parallel vectors, 771
parallelepiped, 422
volume of, 791
Parallelogram Law, 771, 786
parameter, 621, 795, 818
parametric curve, 621, 818
arc length of, 633
area under, 632
slope of tangent line to, 630
parametric equations, 621
of a line, 795
of a space curve, 818
of a surface, 1070
of a trajectory, 841
parametric surface, 1070
graph of, 1083
surface area of, 1075, 1076
surface integral over, 1081
tangent plane to, 1974
parametrization of a space curve, 820
smooth, 831
with respect to arc length, 831
paraxial rays, 249

partial derivative(s), 879, 880
of a function of more than three variables, 883
interpretations of, 881
notations for, 880
as rates of change, 880
rules for finding, 880
second, 884
as slopes of tangent lines, 881
partial differential equation, 886
partial fractions, 473
partial integration, 960
partial sum of a series, 688
particle, motion of, 838
parts, integration by, 453
path, 1047
patterns in integrals, 494
pendulum, approximating the period of, 249, 253
percentage error, 251
perihelion, 667
perilune, 661
period, 308
periodic function, 308
perpendicular lines, A14
perpendicular vectors, 781
phase plane, 610
phase portrait, 610
phase trajectory, 610
piecewise defined function, 17
piecewise-smooth curve, 1035
Planck's Law, 757
plane(s), 797
coordinate, 765
equation(s) of, 797, 798
horizontal, 766
normal, 835
osculating, 835
parallel, 799
tangent to a surface, 892, 917, 1074
vertical, 766
plane region of type I, 966
plane region of type II, 967
planetary motion, 844
point of inflection, 291
point(s) in space
coordinates of, 765
distance between, 767
projection of, 766
point-slope equation of a line, 18, A12
Poiseuille, Jean-Louis-Marie, 227
Poiseuille's Laws, 253, 332, 552
polar axis, 639
polar coordinate system, 639
area in, 650
conic sections in, 662
conversion equations for Cartesian coordinates, 640,641
conversion of double integral to, 974, 975
polar curve, 641
arc length of, 652
graph of, 641
symmetry in, 644
tangent line to, 644
polar equation, graph of, 641
polar equation of a conic, 664
polar form of a complex number, A59
polar graph, 641
polar moment of inertia, 983
polar rectangle, 974
polar region, area of, 650
pole, 639
polynomial, 28
polynomial function of two variables, 874
population growth, 591
of bacteria, 226, 591, 596
of insects, 483
models, 567
world, 55, 235
position function, 145
position vector, 773
positive angle, A25
positive orientation
of a boundary curve, 1093
of a closed curve, 1055
of a surface, 1086
potential, 520
potential energy, 1053
potential function, 1032
pound (unit of force), 438
power consumption, approximation of, 396
power function, 29
Power Law of limits, 101
Power Rule, 174, 218
power series, 723
coefficients of, 723
for cosine and sine, 740
differentiation of, 729
division of, 745
for exponenial function, 740
integration of, 729
interval of convergence, 725
multiplication of, 745
radius of convergence, 725
representations of functions as, 728
predator, 608
predator-prey model, 233, 609
pressure exerted by a fluid, 539
prey, 609
prime notation, 146, 177
principal square root of a complex number, A58
principal unit normal vector, 834
principle of mathematical induction, 77, 80, A36
principle of superposition, 1120
probability, 985
probability density function, 555, 985
problem-solving principles, 76
producer surplus, 553
product formulas, A29
Product Law of limits, 100
Product Rule, 183, 184
product
cross, 786. *See also* cross product
dot, 779. *See also* dot product
scalar, 779
scalar triple, 790
triple, 790
profit function, 327
projectile, path of, 629, 841
projection, 766, 782, 783, 785
p-series, 699

quadrant, A11
quadratic approximation, 253, 933
quadratic function, 28
quadric surface(s), 805
cone, 808
cylinder, 805
ellipsoid, 806, 808
hyperboloid, 808, 810
paraboloid, 806, 810
table of graphs, 808
Quotient Law of limits, 100
Quotient Rule, 185, 186

radian measure, 189, A24
radiation from stars, 757
radioactive decay, 235
radiocarbon dating, 240
radius of convergence, 725
radius of gyration, 984
rainbow, formation and location of, 279
rainbow angle, 279
ramp function, 45
range of a function, 11, 855
rate of change
average, 148, 221
derivative as, 148
instantaneous, 86, 148, 221
rate of growth, 226
rate of reaction, 225
rational function, 31, 874
integration of, 473
rational number, A2
rationalizing substitution for integration, 481
Ratio Test, 716
Rayleigh-Jeans Law, 757
real line, A3
real number, A2
rearrangement of a series, 719
reciprocal function, 31
Reciprocal Rule, 189

rectangular coordinate system, A11
conversion to cylindrical coordinates, 1001
conversion to spherical coordinates, 1006
three-dimensional, 766
rectilinear motion, 343
recursion relation, 1134
reduction formula, 457
reflecting a function, 38
reflection property
of an ellipse, 658
of a hyperbola, 662
of a parabola, 268, 269
region
connected, 1048
open, 1048
under a graph, 355, 360
plane, of type I or II, 966, 967
simple plane, 1056
simple solid, 1099
simply-connected, 1049
solid (of type 1, 2, or 3), 991, 993
between two graphs, 415
related rates, 241
relative error, 251
relative growth rate, 234, 592
relative maximum and minimum, 271
remainder estimates
for the Alternating Series, 712
for the Integral Test, 701
remainder of the Taylor series, 737
removable discontinuity, 120
representation(s) of a function, 12
as a power series, 728
resonance, 1129
restoring force, 1125
resultant force, 776
revenue function, 327
reversing order of integration, 962, 970
revolution, solid of, 427
revolution, surface of, 532
Riemann, Georg Bernhard, 367
Riemann sum(s), 367
for multiple integrals, 954, 990
right circular cylinder, 422
right-hand derivative, 165
right-hand limit, 92, 113
right-hand rule, 765, 788
Roberval, Gilles de, 386, 633
rocket science, 941
Rolle, Michel, 280
roller coaster, design of, 182
roller derby, 1012
Rolle's Theorem, 280
root function, 30
Root Test, 718
roots of a complex number, A62
roots of an *n*th-degree equation, 210
ruled surface, 812
ruling of a surface, 804

saddle point, 924
sample point, 360, 952
satellite dish, parabolic, 810
scalar, 771
scalar equation of a plane, 798
scalar field, 1028
scalar multiple of a vector, 771
scalar product, 779
scalar projection, 782, 783
scalar triple product, 790
geometric characterization of, 791
scatter plot, 14
seasonal-growth model, 600
secant function, A26
derivative of, 193
graph of, A31
secant line, 4, 83, 86
secant vector, 824
second derivative, 160
Second Derivative Test, 292
second derivative of a vector function, 826
Second Derivatives Test, 924
second moment of inertia, 983
second partial derivative, 884
second-order differential equation, 569
solutions of, 1111, 1116
sector of a circle, area of, 650
separable differential equation, 580
sequence, 6, 675
bounded, 682
convergent, 677
decreasing, 681
divergent, 677
Fibonacci, 676
graph of, 680
increasing, 681
limit of, 6, 357, 677
monotonic, 681
of partial sums, 688
term of, 675
series, 7, 687
absolutely convergent, 714
alternating, 710
alternating harmonic, 711, 715
binomial, 742, 748
coefficients of, 723
conditionally convergent, 715
convergent, 688
divergent, 688
geometric, 688
Gregory's, 732
harmonic, 691
infinite, 687
Maclaurin, 734, 736
p-, 699
partial sum of, 688
power, 723
rearrangement of, 719
strategy for testing, 721
sum of, 7, 688
Taylor, 734, 736
term of, 687
trigonometric, 723
series solution of a differential equation, 1133
set, bounded or closed, 928
set notation, A3
serpentine, 188
shell method for approximating volume, 433
shift of a function, 37
shifted conics, 659, A21
shock absorber, 1126
Sierpinski carpet, 696
sigma notation, 360, A34
simple curve, 1049
simple harmonic motion, 205
simple plane region, 1056
simple solid region, 1099
simply-connected region, 1049
Simpson, Thomas, 501, 502, 949
Simpson's Rule, 500, 502
error bounds for, 503
sine function, A26
derivative of, 193
graph of, 32, A31
power series for, 740
sine integral function, 389
sink, 1103
skew lines, 797
slant asymptote, 312
slope, A12
slope field, 573
slope-intercept equation of a line, A13
smooth curve, 831
smooth function, 525
smooth parametrization, 831
smooth surface, 1075
Snell's Law, 331
snowflake curve, 761
solid, 422
volume of, 423, 991, 992
solid angle, 1109
solid region, 1099
solid of revolution, 427
rotated on a slant, 538
volume of, 430, 434, 538
solution curve, 572
solution of predator-prey equations, 609
source, 1103
space, three-dimensional, 765
space curve, 818
arc length of, 830, 831
speed of a particle, 148, 839

- sphere
  - equation of, 768
  - flux across, 1088
  - parametrization of, 1072
  - surface area of, 1076
- spherical coordinate system, 1005
  - conversion equations for, 1006
  - triple integrals in, 1006
- spherical wedge, 1007
- spherical zones, 564
- spring constant, 439, 568, 1125
- Squeeze Theorem, 105, A42
  - for sequences, 679
- standard position of an angle, A25
- standard basis vectors, 774, 775
- standard deviation, 559
- static friction, coefficient of, 815
- stationary points, 923
- steady state solution, 1131
- step function, 18
- Stokes, Sir George, 1093, 1098
- Stokes' Theorem, 1092, 1093
- strategy
  - for integration, 483, 484
  - for optimization problems, 322
  - for problem solving, 76
  - for related rates, 243
  - for testing series, 721
  - for trigonometric integrals, 462, 463
- streamlines, 1033
- stretching a function, 38
- strophoid, 653, 671
- Substitution Rule, 400, 401, 404
- subtraction formulas for sine and cosine, A29
- sum
  - of a geometric series, 689
  - of an infinite series, 688
  - of partial fractions, 474
  - Riemann, 367
  - telescoping, 691
  - of vectors, 770, 773
- Sum Law of limits, 100
- Sum Rule, 177
- summation notation, A34
- supply function, 553
- surface(s), 766
  - closed, 1086
  - graph of, 1083
  - level, 865
  - oriented, 1086
  - parametric, 1070
  - positive orientation of, 1086
  - quadric, 805. *See also* quadric surface
  - smooth, 1075
- surface area, 534
  - of a parametric surface, 635, 1075
  - of a sphere, 1076
  - of a surface $z = f(x, y)$, 1077
- surface integral, 1081
  - over a parametric surface, 1081
  - of a vector field, 1087
- surface of revolution, 532
  - surface area of, 534
  - parametric representation of, 1073
- swallowtail catastrophe curve, 629
- symmetric equations of a line, 795
- symmetric functions, integrals of, 405
- symmetry, 19, 308, 405
  - in polar graphs, 644
- symmetry principle, 543

- T and T^{-1} transformations, 1013
- table of differentiation formulas, 187, RP5
- tables of integrals, 484, RP6–10
  - use of, 489
- tangent function, A26
  - derivative, 193
  - graph, 33, A31
- tangent line(s), 143
  - to a curve, 4, 83, 144
  - early methods of finding, 153
  - to a parametric curve, 630
  - to a polar curve, 644
  - to a space curve, 824
- tangent line approximation, 247
- tangent plane
  - to a level surface, 917
  - to a parametric surface, 1074
  - to a surface $F(x, y, z) = k$, 917
  - to a surface $z = f(x, y)$, 892
- tangent plane approximation, 894
- tangent problem, 4, 83, 144
- tangent vector, 824
- tangential component of acceleration, 842
- tautochrone problem, 625
- Taylor, Brook, 736
- Taylor polynomial, 254, 737, 933
  - applications of, 749
- Taylor series, 734, 736
- Taylor's Inequality, 737
- techniques of integration, summary, 484
- telescoping sum, 691
- temperature-humidity index, 866, 878
- term of a sequence, 675
- term of a series, 687
- term-by-term differentiation and integration, 729
- terminal point of a parametric curve, 622
- terminal point of a vector, 770
- terminal velocity, 587
- Test for Divergence, 692
- tests for convergence and divergence of series
  - Alternating Series Test, 710
  - Comparison Test, 705
  - Integral Test, 697, 699
  - Limit Comparison Test, 707
  - Ratio Test, 716
  - Root Test, 718
  - summary of tests, 721
- tetrahedron, 794
- third derivative, 161
- Thomson, Sir William (Lord Kelvin), 1056, 1093, 1098
- three-dimensional coordinate system, 766
- TNB frame, 835
- toroidal spiral, 820
- torque, 791, 848
- Torricelli, Evangelista, 633
- Torricelli's Law, 231
- torsion of a space curve, 838
- torus, 432, 1081
- total differential, 896
- total electric charge, 980, 996
- total fertility rate, 169
- trace of a surface, 804
- trajectory, parametric equations for, 841
- transcendental function, 34
- transfer curve, 851
- transformation, 1013
  - of a function, 37
  - inverse, 1013
  - Jacobian of, 1015, 1019
  - one-to-one, 1013
- translation of a function, 38
- Trapezoidal Rule, 497
  - error in, 497
- tree diagram, 903
- trefoil knot, 820
- Triangle Inequality, A8
  - for vectors, 786
- Triangle Law, 771
- trigonometric functions, 32, A26
  - derivatives of, 189, 193
  - graphs of, A30, A31
  - integrals of, 460
  - inverse, 67, 68
  - limits involving, 190, 192
- trigonometric identities, A28
- trigonometric integrals, 460
  - strategy for evaluating, 462, 463
- trigonometric series, 723
- trigonometric substitutions, 467
  - table of, 467
- triple integral(s), 990
  - applications of, 995
  - in cylindrical coordinates, 1002
  - over a general bounded region, 991
  - Midpoint Rule for, 998
  - in spherical coordinates
- triple product, 790
- triple Riemann sum, 990
- trochoid, 628
- Tschirnhausen cubic, 214, 4

twisted cubic, 820
type I or type II plane region, 966, 967
type 1, 2, or 3 solid region, 991, 993

ultraviolet catastrophe, 757
underdamped vibration, 1127
undetermined coefficients, method of, 1118, 1122
union of sets, A3
unit normal vector, 834
unit tangent vector, 824
unit vector, 775

value of a function, 11
variable
    continuous random, 555
    dependent, 11, 855, 903
    independent, 11, 855, 903
    independent random, 986
    intermediate, 903
variables, change of. *See* change of variables
variation of parameters, method of, 1122, 1123
vascular branching, 332
vector(s), 770
    acceleration as, 839
    addition of, 770, 773
    algebraic, 772
    angle between, 779
    basis, 774, 775
    binormal, 834
    components of, 772
    coplanar, 791
    cross product of, 786
    difference of, 771
    displacement, 770, 783
    dot product, 779, 780
    equivalent, 770
    force, 1030
    geometric representations of, 772
    gradient, 913, 915
    i, **j**, and **k**, 775
    initial point of, 770
    length of, 773
    magnitude of, 773
    multiplication of, 771, 773
    n-dimensional, 774
    negative, 771
    normal, 797, 834
    orthogonal, 781
    parallel, 771
    perpendicular, 781
    position, 773
    principal unit normal, 834
    projection of, 782, 783
    properties of, 774
    scalar multiple of, 771
    standard basis, 775
    subtraction of, 773
    sum of, 770, 773
    tangent, 824
    terminal point of, 770
    three-dimensional, 772
    triple product, 790, 791
    two-dimensional, 772
    unit, 772
    unit normal, 834
    unit tangent, 824
    velocity, 831
    wind velocity, 764
    zero, 770
vector equation
    of a line, 794, 795
    of a line segment, 797
    of a plane, 798
    of a plane curve, 818
vector field, 1027, 1028
    conservative, 1032
    curl of, 1062
    divergence of, 1065
    electric flux of, 1089
    force, 1027, 1031
    flux of, 1087
    gradient, 1031
    gravitational, 1031
    incompressible, 1066
    irrotational, 1064
    line integral of, 1041, 1042
    surface integral of, 1087
    velocity, 1027
vector function, 817
    continuity of, 818
    derivative of, 824
    differentiation formulas for, 826
    integration of, 827
    limit of, 817
    parametric equations of, 818
vector product, 786
    properties of, 790
vector projection, 782, 783
vector triple product, 791
vector-valued function. *See* vector function
velocity, 4, 85, 145, 221
    average, 5, 86, 145, 221
    instantaneous, 86, 145, 221
velocity field, 1030
    air flow, 1027
    ocean currents, 1027
    wind patterns, 1027
velocity gradient, 228
velocity problem, 85, 145
velocity vector, 831
velocity vector field, 1027
Verhulst, Pierre-François, 568
vertex of a parabola, 655
vertical asymptote, 95, 308
vertical line, A13
Vertical Line Test, 16
vertical plane, equation of, 766
vertical tangent line, 159
vertical translation of a graph, 37
vertices
    of an ellipse, 657
    of a hyperbola, 658
vibration of a rubber membrane, 724
vibration of a spring, 1125
vibrations, 1125, 1126, 1128
viewing rectangle, 46
volume, 423
    by double integrals, 951
    by cross-sections, 422
    by cylindrical shells, 433
    by disks, 424, 427
    of a hypersphere, 1000
    of a solid, 422, 953
    of a solid of revolution, 427, 538
    of a solid on a slant, 538
    by triple integrals, 995
    by washers, 426, 427
Volterra, Vito, 609
Von Bertalanffy model, 616

Wallis, John, 4
Wallis product, 459
washer method, 426
wave equation, 886
Weierstrass, Karl, 482
weight (force), 438
wind-chill index, 856
wind patterns in San Francisco Bay area, 1027
witch of Maria Agnesi, 188, 628
work (force), 438, 783
    defined as a line integral, 1041
Wren, Sir Christopher, 635

x-axis, 765, A10
x-coordinate, 765, A10
x-intercept, A19
X-mean, 987
xy-plane, 766
xz-plane, 766

y-axis, 765, A10
y-coordinate, 765, A10
y-intercept, A19
Y-mean, 987
yz-plane, 766

z-axis, 765
z-coordinate, 765
Zeno, 6
Zeno's paradoxes, 6
zero vector, 770

SPECIAL FUNCTIONS

POWER FUNCTIONS $f(x) = x^a$

(i) $f(x) = x^n$, n a positive integer

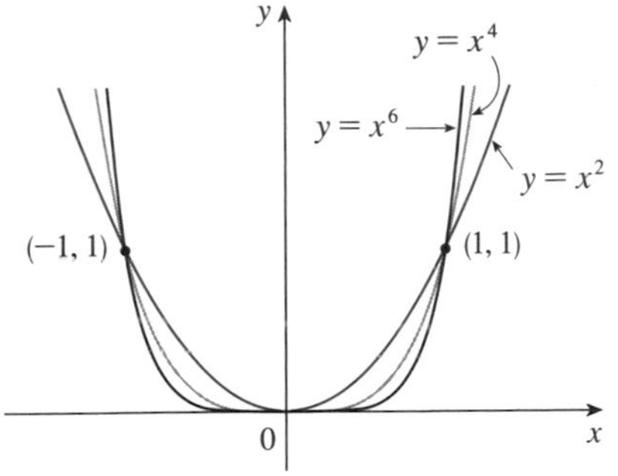

n even

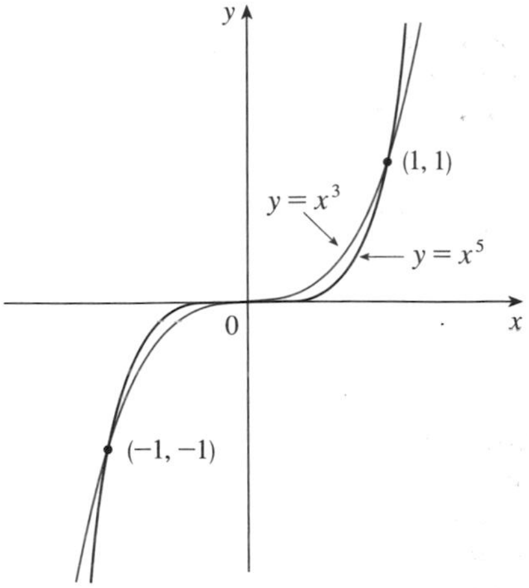

n odd

(ii) $f(x) = x^{1/n} = \sqrt[n]{x}$, n a positive integer

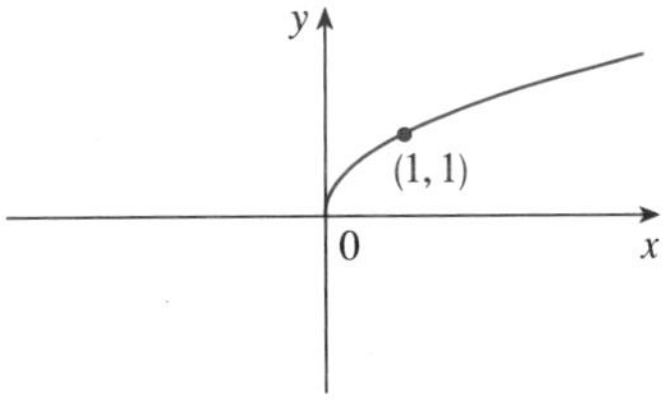

$f(x) = \sqrt{x}$

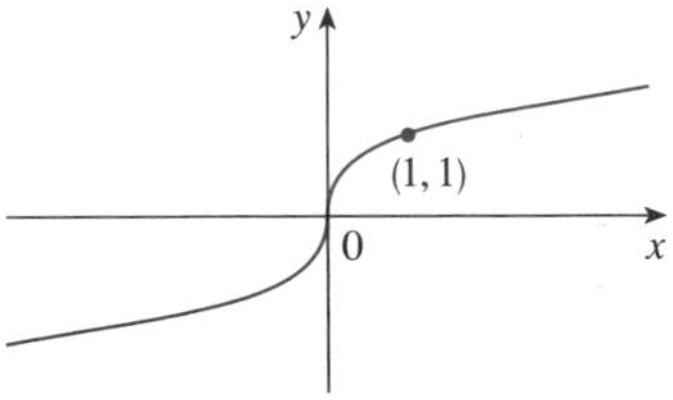

$f(x) = \sqrt[3]{x}$

(iii) $f(x) = x^{-1} = \dfrac{1}{x}$

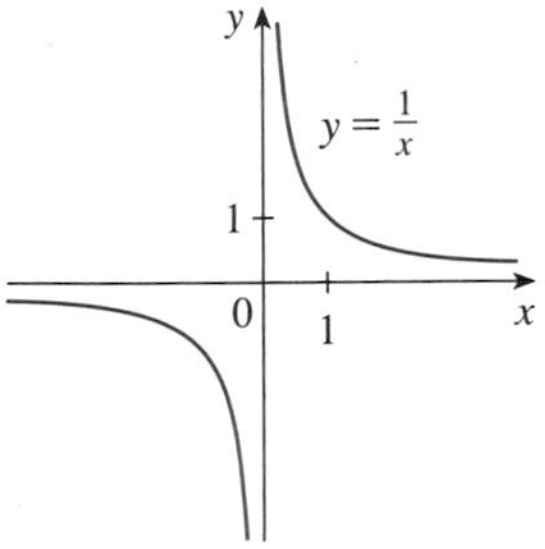

INVERSE TRIGONOMETRIC FUNCTIONS

$\arcsin x = \sin^{-1}x = y \iff \sin y = x$ and $-\dfrac{\pi}{2} \leqslant y \leqslant \dfrac{\pi}{2}$

$\arccos x = \cos^{-1}x = y \iff \cos y = x$ and $0 \leqslant y \leqslant \pi$

$\arctan x = \tan^{-1}x = y \iff \tan y = x$ and $-\dfrac{\pi}{2} < y < \dfrac{\pi}{2}$

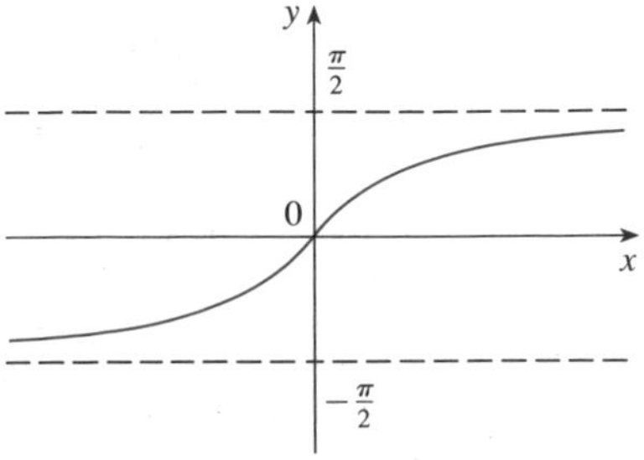

$y = \tan^{-1}x = \arctan x$

$$\lim_{x \to -\infty} \tan^{-1}x = -\frac{\pi}{2}$$

$$\lim_{x \to \infty} \tan^{-1}x = \frac{\pi}{2}$$

SPECIAL FUNCTIONS

EXPONENTIAL AND LOGARITHMIC FUNCTIONS

$\log_a x = y \iff a^y = x$

$\ln x = \log_e x$, where $\ln e = 1$

$\ln x = y \iff e^y = x$

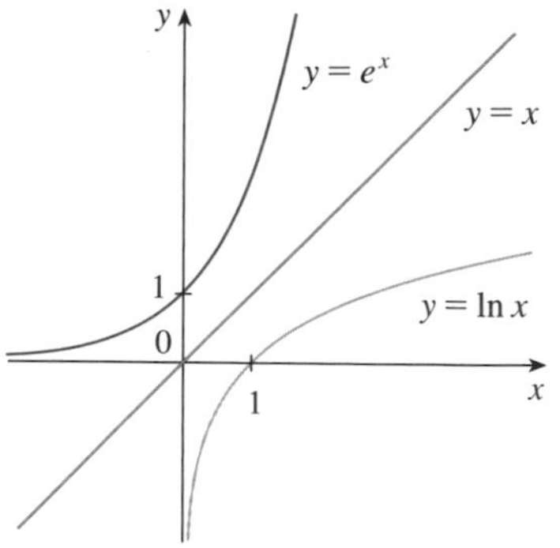

$\lim_{x \to -\infty} e^x = 0$ $\lim_{x \to \infty} e^x = \infty$

$\lim_{x \to 0^+} \ln x = -\infty$ $\lim_{x \to \infty} \ln x = \infty$

Cancellation Equations

$\log_a(a^x) = x$ $a^{\log_a x} = x$

$\ln(e^x) = x$ $e^{\ln x} = x$

Laws of Logarithms

1. $\log_a(xy) = \log_a x + \log_a y$

2. $\log_a\left(\dfrac{x}{y}\right) = \log_a x - \log_a y$

3. $\log_a(x^r) = r \log_a x$

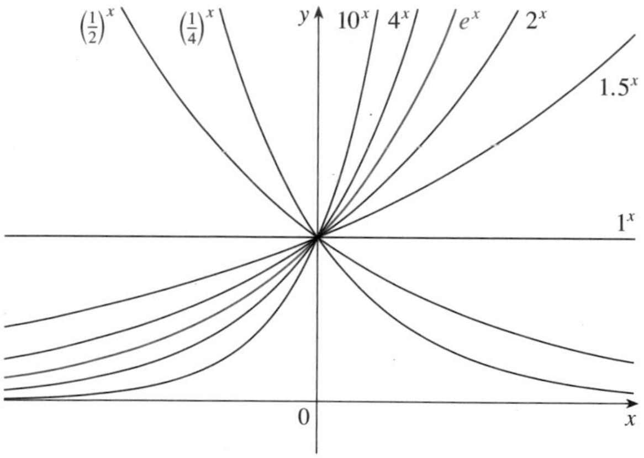

Exponential functions

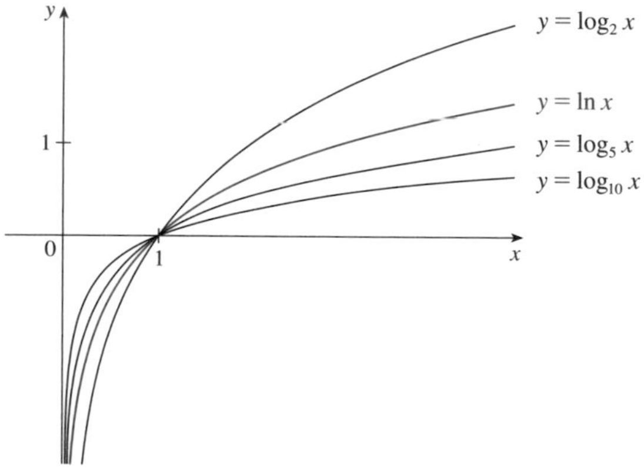

Logarithmic functions

HYPERBOLIC FUNCTIONS

$\sinh x = \dfrac{e^x - e^{-x}}{2}$ $\operatorname{csch} x = \dfrac{1}{\sinh x}$

$\cosh x = \dfrac{e^x + e^{-x}}{2}$ $\operatorname{sech} x = \dfrac{1}{\cosh x}$

$\tanh x = \dfrac{\sinh x}{\cosh x}$ $\coth x = \dfrac{\cosh x}{\sinh x}$

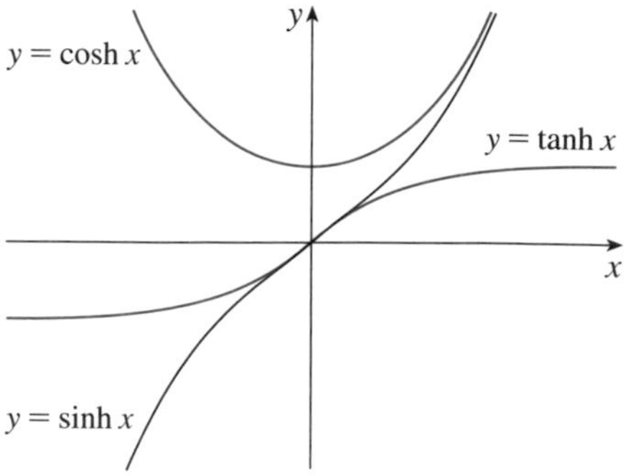

INVERSE HYPERBOLIC FUNCTIONS

$y = \sinh^{-1}x \iff \sinh y = x$ $\sinh^{-1}x = \ln\left(x + \sqrt{x^2 + 1}\right)$

$y = \cosh^{-1}x \iff \cosh y = x$ and $y \geqslant 0$ $\cosh^{-1}x = \ln\left(x + \sqrt{x^2 - 1}\right)$

$y = \tanh^{-1}x \iff \tanh y = x$ $\tanh^{-1}x = \frac{1}{2}\ln\left(\dfrac{1 + x}{1 - x}\right)$

DIFFERENTIATION RULES

GENERAL FORMULAS

1. $\frac{d}{dx}(c) = 0$
2. $\frac{d}{dx}[cf(x)] = cf'(x)$
3. $\frac{d}{dx}[f(x) + g(x)] = f'(x) + g'(x)$
4. $\frac{d}{dx}[f(x) - g(x)] = f'(x) - g'(x)$
5. $\frac{d}{dx}[f(x)g(x)] = f(x)g'(x) + g(x)f'(x)$ (Product Rule)
6. $\frac{d}{dx}\left[\frac{f(x)}{g(x)}\right] = \frac{g(x)f'(x) - f(x)g'(x)}{[g(x)]^2}$ (Quotient Rule)
7. $\frac{d}{dx} f(g(x)) = f'(g(x))g'(x)$ (Chain Rule)
8. $\frac{d}{dx}(x^n) = nx^{n-1}$ (Power Rule)

EXPONENTIAL AND LOGARITHMIC FUNCTIONS

9. $\frac{d}{dx}(e^x) = e^x$
10. $\frac{d}{dx}(a^x) = a^x \ln a$
11. $\frac{d}{dx} \ln|x| = \frac{1}{x}$
12. $\frac{d}{dx}(\log_a x) = \frac{1}{x \ln a}$

TRIGONOMETRIC FUNCTIONS

13. $\frac{d}{dx}(\sin x) = \cos x$
14. $\frac{d}{dx}(\cos x) = -\sin x$
15. $\frac{d}{dx}(\tan x) = \sec^2 x$
16. $\frac{d}{dx}(\csc x) = -\csc x \cot x$
17. $\frac{d}{dx}(\sec x) = \sec x \tan x$
18. $\frac{d}{dx}(\cot x) = -\csc^2 x$

INVERSE TRIGONOMETRIC FUNCTIONS

19. $\frac{d}{dx}(\sin^{-1}x) = \frac{1}{\sqrt{1 - x^2}}$
20. $\frac{d}{dx}(\cos^{-1}x) = -\frac{1}{\sqrt{1 - x^2}}$
21. $\frac{d}{dx}(\tan^{-1}x) = \frac{1}{1 + x^2}$
22. $\frac{d}{dx}(\csc^{-1}x) = -\frac{1}{x\sqrt{x^2 - 1}}$
23. $\frac{d}{dx}(\sec^{-1}x) = \frac{1}{x\sqrt{x^2 - 1}}$
24. $\frac{d}{dx}(\cot^{-1}x) = -\frac{1}{1 + x^2}$

HYPERBOLIC FUNCTIONS

25. $\frac{d}{dx}(\sinh x) = \cosh x$
26. $\frac{d}{dx}(\cosh x) = \sinh x$
27. $\frac{d}{dx}(\tanh x) = \operatorname{sech}^2 x$
28. $\frac{d}{dx}(\operatorname{csch} x) = -\operatorname{csch} x \coth x$
29. $\frac{d}{dx}(\operatorname{sech} x) = -\operatorname{sech} x \tanh x$
30. $\frac{d}{dx}(\coth x) = -\operatorname{csch}^2 x$

INVERSE HYPERBOLIC FUNCTIONS

31. $\frac{d}{dx}(\sinh^{-1}x) = \frac{1}{\sqrt{1 + x^2}}$
32. $\frac{d}{dx}(\cosh^{-1}x) = \frac{1}{\sqrt{x^2 - 1}}$
33. $\frac{d}{dx}(\tanh^{-1}x) = \frac{1}{1 - x^2}$
34. $\frac{d}{dx}(\operatorname{csch}^{-1}x) = -\frac{1}{|x|\sqrt{x^2 + 1}}$
35. $\frac{d}{dx}(\operatorname{sech}^{-1}x) = -\frac{1}{x\sqrt{1 - x^2}}$
36. $\frac{d}{dx}(\coth^{-1}x) = \frac{1}{1 - x^2}$

TABLE OF INTEGRALS

BASIC FORMS

1. $\int u\,dv = uv - \int v\,du$

2. $\int u^n\,du = \frac{u^{n+1}}{n+1} + C, \quad n \neq -1$

3. $\int \frac{du}{u} = \ln|u| + C$

4. $\int e^u\,du = e^u + C$

5. $\int a^u\,du = \frac{a^u}{\ln a} + C$

6. $\int \sin u\,du = -\cos u + C$

7. $\int \cos u\,du = \sin u + C$

8. $\int \sec^2 u\,du = \tan u + C$

9. $\int \csc^2 u\,du = -\cot u + C$

10. $\int \sec u \tan u\,du = \sec u + C$

11. $\int \csc u \cot u\,du = -\csc u + C$

12. $\int \tan u\,du = \ln|\sec u| + C$

13. $\int \cot u\,du = \ln|\sin u| + C$

14. $\int \sec u\,du = \ln|\sec u + \tan u| + C$

15. $\int \csc u\,du = \ln|\csc u - \cot u| + C$

16. $\int \frac{du}{\sqrt{a^2 - u^2}} = \sin^{-1}\frac{u}{a} + C, \quad a > 0$

17. $\int \frac{du}{a^2 + u^2} = \frac{1}{a}\tan^{-1}\frac{u}{a} + C$

18. $\int \frac{du}{u\sqrt{u^2 - a^2}} = \frac{1}{a}\sec^{-1}\frac{u}{a} + C$

19. $\int \frac{du}{a^2 - u^2} = \frac{1}{2a}\ln\left|\frac{u+a}{u-a}\right| + C$

20. $\int \frac{du}{u^2 - a^2} = \frac{1}{2a}\ln\left|\frac{u-a}{u+a}\right| + C$

FORMS INVOLVING $\sqrt{a^2 + u^2}$, $a > 0$

21. $\int \sqrt{a^2 + u^2}\,du = \frac{u}{2}\sqrt{a^2 + u^2} + \frac{a^2}{2}\ln\left(u + \sqrt{a^2 + u^2}\right) + C$

22. $\int u^2\sqrt{a^2 + u^2}\,du = \frac{u}{8}(a^2 + 2u^2)\sqrt{a^2 + u^2} - \frac{a^4}{8}\ln\left(u + \sqrt{a^2 + u^2}\right) + C$

23. $\int \frac{\sqrt{a^2 + u^2}}{u}\,du = \sqrt{a^2 + u^2} - a\ln\left|\frac{a + \sqrt{a^2 + u^2}}{u}\right| + C$

24. $\int \frac{\sqrt{a^2 + u^2}}{u^2}\,du = -\frac{\sqrt{a^2 + u^2}}{u} + \ln\left(u + \sqrt{a^2 + u^2}\right) + C$

25. $\int \frac{du}{\sqrt{a^2 + u^2}} = \ln\left(u + \sqrt{a^2 + u^2}\right) + C$

26. $\int \frac{u^2\,du}{\sqrt{a^2 + u^2}} = \frac{u}{2}\sqrt{a^2 + u^2} - \frac{a^2}{2}\ln\left(u + \sqrt{a^2 + u^2}\right) + C$

27. $\int \frac{du}{u\sqrt{a^2 + u^2}} = -\frac{1}{a}\ln\left|\frac{\sqrt{a^2 + u^2} + a}{u}\right| + C$

28. $\int \frac{du}{u^2\sqrt{a^2 + u^2}} = -\frac{\sqrt{a^2 + u^2}}{a^2 u} + C$

29. $\int \frac{du}{(a^2 + u^2)^{3/2}} = \frac{u}{a^2\sqrt{a^2 + u^2}} + C$

TABLE OF INTEGRALS

FORMS INVOLVING $\sqrt{a^2 - u^2}$, $a > 0$

30. $$\int \sqrt{a^2 - u^2}\,du = \frac{u}{2}\sqrt{a^2 - u^2} + \frac{a^2}{2}\sin^{-1}\frac{u}{a} + C$$

31. $$\int u^2\sqrt{a^2 - u^2}\,du = \frac{u}{8}(2u^2 - a^2)\sqrt{a^2 - u^2} + \frac{a^4}{8}\sin^{-1}\frac{u}{a} + C$$

32. $$\int \frac{\sqrt{a^2 - u^2}}{u}\,du = \sqrt{a^2 - u^2} - a\ln\left|\frac{a + \sqrt{a^2 - u^2}}{u}\right| + C$$

33. $$\int \frac{\sqrt{a^2 - u^2}}{u^2}\,du = -\frac{1}{u}\sqrt{a^2 - u^2} - \sin^{-1}\frac{u}{a} + C$$

34. $$\int \frac{u^2\,du}{\sqrt{a^2 - u^2}} = -\frac{u}{2}\sqrt{a^2 - u^2} + \frac{a^2}{2}\sin^{-1}\frac{u}{a} + C$$

35. $$\int \frac{du}{u\sqrt{a^2 - u^2}} = -\frac{1}{a}\ln\left|\frac{a + \sqrt{a^2 - u^2}}{u}\right| + C$$

36. $$\int \frac{du}{u^2\sqrt{a^2 - u^2}} = -\frac{1}{a^2u}\sqrt{a^2 - u^2} + C$$

37. $$\int (a^2 - u^2)^{3/2}\,du = -\frac{u}{8}(2u^2 - 5a^2)\sqrt{a^2 - u^2} + \frac{3a^4}{8}\sin^{-1}\frac{u}{a} + C$$

38. $$\int \frac{du}{(a^2 - u^2)^{3/2}} = \frac{u}{a^2\sqrt{a^2 - u^2}} + C$$

FORMS INVOLVING $\sqrt{u^2 - a^2}$, $a > 0$

39. $$\int \sqrt{u^2 - a^2}\,du = \frac{u}{2}\sqrt{u^2 - a^2} - \frac{a^2}{2}\ln\left|u + \sqrt{u^2 - a^2}\right| + C$$

40. $$\int u^2\sqrt{u^2 - a^2}\,du = \frac{u}{8}(2u^2 - a^2)\sqrt{u^2 - a^2} - \frac{a^4}{8}\ln\left|u + \sqrt{u^2 - a^2}\right| + C$$

41. $$\int \frac{\sqrt{u^2 - a^2}}{u}\,du = \sqrt{u^2 - a^2} - a\cos^{-1}\frac{a}{|u|} + C$$

42. $$\int \frac{\sqrt{u^2 - a^2}}{u^2}\,du = -\frac{\sqrt{u^2 - a^2}}{u} + \ln\left|u + \sqrt{u^2 - a^2}\right| + C$$

43. $$\int \frac{du}{\sqrt{u^2 - a^2}} = \ln\left|u + \sqrt{u^2 - a^2}\right| + C$$

44. $$\int \frac{u^2\,du}{\sqrt{u^2 - a^2}} = \frac{u}{2}\sqrt{u^2 - a^2} + \frac{a^2}{2}\ln\left|u + \sqrt{u^2 - a^2}\right| + C$$

45. $$\int \frac{du}{u^2\sqrt{u^2 - a^2}} = \frac{\sqrt{u^2 - a^2}}{a^2u} + C$$

46. $$\int \frac{du}{(u^2 - a^2)^{3/2}} = -\frac{u}{a^2\sqrt{u^2 - a^2}} + C$$

TABLE OF INTEGRALS

FORMS INVOLVING $a + bu$

47. $\displaystyle\int \frac{u\,du}{a+bu} = \frac{1}{b^2}\left(a + bu - a\ln|a+bu|\right) + C$

48. $\displaystyle\int \frac{u^2\,du}{a+bu} = \frac{1}{2b^3}\left[(a+bu)^2 - 4a(a+bu) + 2a^2\ln|a+bu|\right] + C$

49. $\displaystyle\int \frac{du}{u(a+bu)} = \frac{1}{a}\ln\left|\frac{u}{a+bu}\right| + C$

50. $\displaystyle\int \frac{du}{u^2(a+bu)} = -\frac{1}{au} + \frac{b}{a^2}\ln\left|\frac{a+bu}{u}\right| + C$

51. $\displaystyle\int \frac{u\,du}{(a+bu)^2} = \frac{a}{b^2(a+bu)} + \frac{1}{b^2}\ln|a+bu| + C$

52. $\displaystyle\int \frac{du}{u(a+bu)^2} = \frac{1}{a(a+bu)} - \frac{1}{a^2}\ln\left|\frac{a+bu}{u}\right| + C$

53. $\displaystyle\int \frac{u^2\,du}{(a+bu)^2} = \frac{1}{b^3}\left(a + bu - \frac{a^2}{a+bu} - 2a\ln|a+bu|\right) + C$

54. $\displaystyle\int u\sqrt{a+bu}\,du = \frac{2}{15b^2}(3bu - 2a)(a+bu)^{3/2} + C$

55. $\displaystyle\int \frac{u\,du}{\sqrt{a+bu}} = \frac{2}{3b^2}(bu - 2a)\sqrt{a+bu} + C$

56. $\displaystyle\int \frac{u^2\,du}{\sqrt{a+bu}} = \frac{2}{15b^3}(8a^2 + 3b^2u^2 - 4abu)\sqrt{a+bu} + C$

57. $\displaystyle\int \frac{du}{u\sqrt{a+bu}} = \frac{1}{\sqrt{a}}\ln\left|\frac{\sqrt{a+bu}-\sqrt{a}}{\sqrt{a+bu}+\sqrt{a}}\right| + C, \quad \text{if } a > 0$

$\displaystyle\qquad = \frac{2}{\sqrt{-a}}\tan^{-1}\sqrt{\frac{a+bu}{-a}} + C, \quad \text{if } a < 0$

58. $\displaystyle\int \frac{\sqrt{a+bu}}{u}\,du = 2\sqrt{a+bu} + a\int \frac{du}{u\sqrt{a+bu}}$

59. $\displaystyle\int \frac{\sqrt{a+bu}}{u^2}\,du = -\frac{\sqrt{a+bu}}{u} + \frac{b}{2}\int \frac{du}{u\sqrt{a+bu}}$

60. $\displaystyle\int u^n\sqrt{a+bu}\,du = \frac{2}{b(2n+3)}\left[u^n(a+bu)^{3/2} - na\int u^{n-1}\sqrt{a+bu}\,du\right]$

61. $\displaystyle\int \frac{u^n\,du}{\sqrt{a+bu}} = \frac{2u^n\sqrt{a+bu}}{b(2n+1)} - \frac{2na}{b(2n+1)}\int \frac{u^{n-1}\,du}{\sqrt{a+bu}}$

62. $\displaystyle\int \frac{du}{u^n\sqrt{a+bu}} = -\frac{\sqrt{a+bu}}{a(n-1)u^{n-1}} - \frac{b(2n-3)}{2a(n-1)}\int \frac{du}{u^{n-1}\sqrt{a+bu}}$

TABLE OF INTEGRALS

TRIGONOMETRIC FORMS

63. $\displaystyle\int \sin^2 u\,du = \tfrac{1}{2}u - \tfrac{1}{4}\sin 2u + C$

64. $\displaystyle\int \cos^2 u\,du = \tfrac{1}{2}u + \tfrac{1}{4}\sin 2u + C$

65. $\displaystyle\int \tan^2 u\,du = \tan u - u + C$

66. $\displaystyle\int \cot^2 u\,du = -\cot u - u + C$

67. $\displaystyle\int \sin^3 u\,du = -\tfrac{1}{3}(2 + \sin^2 u)\cos u + C$

68. $\displaystyle\int \cos^3 u\,du = \tfrac{1}{3}(2 + \cos^2 u)\sin u + C$

69. $\displaystyle\int \tan^3 u\,du = \tfrac{1}{2}\tan^2 u + \ln|\cos u| + C$

70. $\displaystyle\int \cot^3 u\,du = -\tfrac{1}{2}\cot^2 u - \ln|\sin u| + C$

71. $\displaystyle\int \sec^3 u\,du = \tfrac{1}{2}\sec u\tan u + \tfrac{1}{2}\ln|\sec u + \tan u| + C$

72. $\displaystyle\int \csc^3 u\,du = -\tfrac{1}{2}\csc u\cot u + \tfrac{1}{2}\ln|\csc u - \cot u| + C$

73. $\displaystyle\int \sin^n u\,du = -\frac{1}{n}\sin^{n-1}u\cos u + \frac{n-1}{n}\int \sin^{n-2}u\,du$

74. $\displaystyle\int \cos^n u\,du = \frac{1}{n}\cos^{n-1}u\sin u + \frac{n-1}{n}\int \cos^{n-2}u\,du$

75. $\displaystyle\int \tan^n u\,du = \frac{1}{n-1}\tan^{n-1}u - \int \tan^{n-2}u\,du$

76. $\displaystyle\int \cot^n u\,du = \frac{-1}{n-1}\cot^{n-1}u - \int \cot^{n-2}u\,du$

77. $\displaystyle\int \sec^n u\,du = \frac{1}{n-1}\tan u\sec^{n-2}u + \frac{n-2}{n-1}\int \sec^{n-2}u\,du$

78. $\displaystyle\int \csc^n u\,du = \frac{-1}{n-1}\cot u\csc^{n-2}u + \frac{n-2}{n-1}\int \csc^{n-2}u\,du$

79. $\displaystyle\int \sin au\sin bu\,du = \frac{\sin(a-b)u}{2(a-b)} - \frac{\sin(a+b)u}{2(a+b)} + C$

80. $\displaystyle\int \cos au\cos bu\,du = \frac{\sin(a-b)u}{2(a-b)} + \frac{\sin(a+b)u}{2(a+b)} + C$

81. $\displaystyle\int \sin au\cos bu\,du = -\frac{\cos(a-b)u}{2(a-b)} - \frac{\cos(a+b)u}{2(a+b)} + C$

82. $\displaystyle\int u\sin u\,du = \sin u - u\cos u + C$

83. $\displaystyle\int u\cos u\,du = \cos u + u\sin u + C$

84. $\displaystyle\int u^n\sin u\,du = -u^n\cos u + n\int u^{n-1}\cos u\,du$

85. $\displaystyle\int u^n\cos u\,du = u^n\sin u - n\int u^{n-1}\sin u\,du$

86. $\displaystyle\int \sin^n u\cos^m u\,du = -\frac{\sin^{n-1}u\cos^{m+1}u}{n+m} + \frac{n-1}{n+m}\int \sin^{n-2}u\cos^m u\,du$

$\displaystyle\qquad = \frac{\sin^{n+1}u\cos^{m-1}u}{n+m} + \frac{m-1}{n+m}\int \sin^n u\cos^{m-2}u\,du$

INVERSE TRIGONOMETRIC FORMS

87. $\displaystyle\int \sin^{-1}u\,du = u\sin^{-1}u + \sqrt{1-u^2} + C$

88. $\displaystyle\int \cos^{-1}u\,du = u\cos^{-1}u - \sqrt{1-u^2} + C$

89. $\displaystyle\int \tan^{-1}u\,du = u\tan^{-1}u - \tfrac{1}{2}\ln(1+u^2) + C$

90. $\displaystyle\int u\sin^{-1}u\,du = \frac{2u^2-1}{4}\sin^{-1}u + \frac{u\sqrt{1-u^2}}{4} + C$

91. $\displaystyle\int u\cos^{-1}u\,du = \frac{2u^2-1}{4}\cos^{-1}u - \frac{u\sqrt{1-u^2}}{4} + C$

92. $\displaystyle\int u\tan^{-1}u\,du = \frac{u^2+1}{2}\tan^{-1}u - \frac{u}{2} + C$

93. $\displaystyle\int u^n\sin^{-1}u\,du = \frac{1}{n+1}\left[u^{n+1}\sin^{-1}u - \int \frac{u^{n+1}\,du}{\sqrt{1-u^2}}\right], \quad n \neq -1$

94. $\displaystyle\int u^n\cos^{-1}u\,du = \frac{1}{n+1}\left[u^{n+1}\cos^{-1}u + \int \frac{u^{n+1}\,du}{\sqrt{1-u^2}}\right], \quad n \neq -1$

95. $\displaystyle\int u^n\tan^{-1}u\,du = \frac{1}{n+1}\left[u^{n+1}\tan^{-1}u - \int \frac{u^{n+1}\,du}{1+u^2}\right], \quad n \neq -1$

TABLE OF INTEGRALS

EXPONENTIAL AND LOGARITHMIC FORMS

96. $\displaystyle\int ue^{au}\,du = \frac{1}{a^2}(au - 1)e^{au} + C$

97. $\displaystyle\int u^n e^{au}\,du = \frac{1}{a}u^n e^{au} - \frac{n}{a}\int u^{n-1}e^{au}\,du$

98. $\displaystyle\int e^{au}\sin bu\,du = \frac{e^{au}}{a^2 + b^2}(a\sin bu - b\cos bu) + C$

99. $\displaystyle\int e^{au}\cos bu\,du = \frac{e^{au}}{a^2 + b^2}(a\cos bu + b\sin bu) + C$

100. $\displaystyle\int \ln u\,du = u\ln u - u + C$

101. $\displaystyle\int u^n \ln u\,du = \frac{u^{n+1}}{(n + 1)^2}[(n + 1)\ln u - 1] + C$

102. $\displaystyle\int \frac{1}{u\ln u}\,du = \ln|\ln u| + C$

HYPERBOLIC FORMS

103. $\displaystyle\int \sinh u\,du = \cosh u + C$

104. $\displaystyle\int \cosh u\,du = \sinh u + C$

105. $\displaystyle\int \tanh u\,du = \ln\cosh u + C$

106. $\displaystyle\int \coth u\,du = \ln|\sinh u| + C$

107. $\displaystyle\int \operatorname{sech} u\,du = \tan^{-1}|\sinh u| + C$

108. $\displaystyle\int \operatorname{csch} u\,du = \ln\left|\tanh \tfrac{1}{2}u\right| + C$

109. $\displaystyle\int \operatorname{sech}^2 u\,du = \tanh u + C$

110. $\displaystyle\int \operatorname{csch}^2 u\,du = -\coth u + C$

111. $\displaystyle\int \operatorname{sech} u\tanh u\,du = -\operatorname{sech} u + C$

112. $\displaystyle\int \operatorname{csch} u\coth u\,du = -\operatorname{csch} u + C$

FORMS INVOLVING $\sqrt{2au - u^2}$, $a > 0$

113. $\displaystyle\int \sqrt{2au - u^2}\,du = \frac{u - a}{2}\sqrt{2au - u^2} + \frac{a^2}{2}\cos^{-1}\left(\frac{a - u}{a}\right) + C$

114. $\displaystyle\int u\sqrt{2au - u^2}\,du = \frac{2u^2 - au - 3a^2}{6}\sqrt{2au - u^2} + \frac{a^3}{2}\cos^{-1}\left(\frac{a - u}{a}\right) + C$

115. $\displaystyle\int \frac{\sqrt{2au - u^2}}{u}\,du = \sqrt{2au - u^2} + a\cos^{-1}\left(\frac{a - u}{a}\right) + C$

116. $\displaystyle\int \frac{\sqrt{2au - u^2}}{u^2}\,du = -\frac{2\sqrt{2au - u^2}}{u} - \cos^{-1}\left(\frac{a - u}{a}\right) + C$

117. $\displaystyle\int \frac{du}{\sqrt{2au - u^2}} = \cos^{-1}\left(\frac{a - u}{a}\right) + C$

118. $\displaystyle\int \frac{u\,du}{\sqrt{2au - u^2}} = -\sqrt{2au - u^2} + a\cos^{-1}\left(\frac{a - u}{a}\right) + C$

119. $\displaystyle\int \frac{u^2\,du}{\sqrt{2au - u^2}} = -\frac{(u + 3a)}{2}\sqrt{2au - u^2} + \frac{3a^2}{2}\cos^{-1}\left(\frac{a - u}{a}\right) + C$

120. $\displaystyle\int \frac{du}{u\sqrt{2au - u^2}} = -\frac{\sqrt{2au - u^2}}{au} + C$